Practical Handbook of Microbiology

Fourth Edition

Edited by

LORRENCE H. GREEN AND EMANUEL GOLDMAN

CRC Press
Taylor & Francis Group
Boca Raton London New York

CRC Press is an imprint of the
Taylor & Francis Group, an **informa** business

Fourth edition published 2021
by CRC Press
6000 Broken Sound Parkway NW, Suite 300, Boca Raton, FL 33487-2742

and by CRC Press
2 Park Square, Milton Park, Abingdon, Oxon, OX14 4RN

First edition published by CRC Press 1989
Second edition published by CRC Press 2009
Third edition published by CRC Press 2015

CRC Press is an imprint of Taylor & Francis Group, LLC

ISBN: 9780367567637 (hbk)
ISBN: 9780367567644 (pbk)
ISBN: 9781003099277 (ebk)

DOI: 10.1201/9781003099277

Typeset in Times LT Std
by KnowledgeWorks Global Ltd

Contents

Part I   Practical Information and Procedures

Part III Applied Practical Microbiology

Preface

As the Practical Handbook Series has gone from Dr. O'Leary's original first edition published in 1989 to the editions published in 2009 and 2015, we have tried to expand the group of people we are trying to help with our book. Dr. O'Leary's first edition was geared toward people generally trained as microbiologists. In the second edition, we added a second group; people that had been trained in other disciplines, but now found that they were using microorganisms as simply another tool or "chemical reagent." The third edition added yet another group, business people requiring assistance in understanding the microbiology-related companies they were considering supporting.

The fourth edition broadens our audience even further to include people in occupations and trades that might have limited training in microbiology, but who require specific practical information about their field. We hope that adding these chapters will give everyone who reads this book a sense of the importance of microbiology is in our everyday lives.

We salute all our authors, especially Dr. Rita Colwell and Dr. Stephan Lerner who have contributed to all four editions, for sharing their knowledge and expertise and helping to make the *Practical Handbook of Microbiology* an essential resource for the field.

About the Editors

Emanuel Goldman is a professor in the Department of Microbiology, Biochemistry, and Molecular Genetics of the New Jersey Medical School (NJMS), Rutgers Biomedical and Health Sciences (RBHS), a division of Rutgers University, Newark, New Jersey. He graduated with honors from the Bronx High School of Science in 1962, received a BA (cum laude) from Brandeis University in 1966, where he was a chemistry major and music minor, and completed his PhD in biochemistry at the Massachusetts Institute of Technology in 1972. He performed postdoctoral research at Harvard Medical School and at the University of California, Irvine, before joining the faculty of the New Jersey Medical School in 1979, where he rose through the ranks to professor in 1993. Among his awards and honors, Dr. Goldman was a Damon Runyon fellow, a Lievre senior fellow of the California Division, American Cancer Society, and a recipient of a Research Career Development Award from the National Cancer Institute. Among his service activities, he was an officer and organizer of the New York–New Jersey Molecular Biology Club, served as a full member of an American Cancer Society Study Section, and continues to serve on the editorial boards of *Protein Expression and Purification* and *Applied and Environmental Microbiology*. He was also twice elected by his colleagues to serve as the president of his university's chapter of the American Association of University Professors, and he was elected to serve as the president of the Faculty Organization of NJMS. Among several areas of research activity, he has focused on the role of tRNA in the elongation of bacterial protein synthesis, including uncharged tRNA, codon bias, and programmed translational frameshifts. In addition to numerous scientific peer-reviewed publications and publications in the lay press, he has contributed a chapter to Zubay's *Biochemistry* textbook and four chapters to the *Encyclopedia of Life Sciences*. His recently published Comment in *Lancet*, "Exaggerated risk of transmission of COVID-19 by Fomites," has attracted significant international attention.

Lorrence H. Green, PhD, President of Westbury Diagnostics, Inc., earned his PhD in Cell and Molecular Biology from Indiana University, Bloomington, Indiana, in 1978. He followed this with 3 years of recombinant DNA and genetic research at Harvard University. In 1981, he moved into Industry by joining Analytab Products Inc., a major manufacturer of *in vitro* diagnostic test kits. During the next 12 years he helped to invent and manufacture over 40 diagnostic test kits, and rose to become the Director of New Product Development and Product Support.

In 1993, Dr. Green founded Westbury Diagnostics Inc., a microbiology-biotechnology-based contract research and development laboratory also offering consulting services. Mixing his love of business with his love of teaching, Dr. Green has served as an adjunct associate professor of microbiology at the New York College of Osteopathic Medicine, and is currently an adjunct assistant professor of biology at Farmingdale State College and Director of the Fundamentals of the Bioscience Industry Program at Stony Brook University of the State University of New York.

Dr. Green is on the steering committee, and is a former Chairman, of the Microbiology Section of the New York Academy of Sciences. He was also the long-time Treasurer of the NYC Branch of the ASM. From 2001 to 2004 he was a member of the Advisory Committee on Emerging Pathogens and Bioterrorism to the New York City Commissioner of Health. In 2013 he was appointed to the Board of Directors of the Long Island Advancement of Small Business.

His main interests involve using technology in the development of commercial products and in being an entrepreneur who invests in and develops companies. He enjoys providing mentorship and career advice to students at all levels. He has spoken at many career day events, judged many regional science fairs, and has helped dozens of young people with applications to medical school, nursing school, physician's assistant school, and with starting companies.

Most recently, he has become involved in government and is currently the Chairman of the Town of Mamakating Planning Board.

Contributors

Charles Adair
Farmingdale State College
Farmingdale, New York

Elisabeth Adderson
Department of Infectious Diseases
St. Jude Children's Research Hospital
    and Department of Pediatrics
The University of Tennessee Health Science Center
Memphis, Tennessee

Negin Alizadeh Shaygh
Department of Pharmaceutical and Biomedical Sciences
Touro College of Pharmacy
New York, New York

Nicholas Allenby
The Biosphere
Draymans Way
Newcastle Helix
Newcastle upon Tyne, United Kingdom

Sandra B. Andersen
Novo Nordisk Foundation Center for Basic Metabolic Research
University of Copenhagen
Copenhagen, Denmark

Gregory G. Anderson
Department of Biology
Indiana University-Purdue University
Indianapolis, Indiana

Guido Antonelli
Laboratory of Virology
Department of Molecular Medicine
Sapienza University
Rome, Italy

Rachael D. Aubert
Enteric Diseases Laboratory Branch
Centers for Disease Control and Prevention
Atlanta, Georgia

Rita Austin
Department of Medical Laboratory Technology
Farmingdale State College
Farmingdale, New York

Abdu F. Azad
Department of Microbiology and Immunology
School of Medicine
University of Maryland
Baltimore, Maryland

Lourdes G. Bahamonde
Internal Medicine/Gastroenterology/Medical Ethics
Private Practice
Los Angeles, California

Matthew E. Bahamonde
Department of Bioethics and Natural Sciences
College of Arts and Sciences
American Jewish University
Los Angeles, California

Sukhadeo Barbuddhe
Indian Council of Agricultural Research
National Research Centre on Meat
Hyderabad, India

Paramita Basu
Department of Pharmaceutical and
    Biomedical Sciences
Touro College of Pharmacy
New York, New York

Magda Beier-Sexton
Department of Microbiology and Immunology
School of Medicine
University of Maryland
Baltimore, Maryland

Victoria Benvenuto
NYU College of Dentistry
Department of Dental Hygiene and Dental Assisting
New York City, New York

Purnima Bhanot
Department of Microbiology,
    Biochemistry and Molecular Genetics
New Jersey Medical School Rutgers,
    The State University of New Jersey
Newark, New Jersey

Martin J. Blaser
Center for Advanced Biotechnology and Medicine
Rutgers University
New Brunswick, New Jersey

Michael Boadu
Touro College of Pharmacy
New York City, New York

Nagamani Bora
Faculty of Science
Nottingham University
Leicestershire, United Kingdom

Michael Bott
Institute for Bio- and Geosciences
Biotechnology Research Centre Jülich
Jülich, Germany

Tyler S. Brown
Division of Infectious Diseases
Massachusetts General Hospital
Boston, Massachusetts

Joseph Adrian L. Buensalido
Division of Infectious Diseases
University of the Philippines
Manila, Philippines

Hayat Caidi
Enteric Diseases Laboratory Branch
Centers for Disease Control and Prevention
Atlanta, Georgia

Donna L. Catapano
NYU College of Dentistry
Department of Dental Hygiene
    and Dental Assisting
New York, New York

Trinad Chakraborty
German Center for Infection Research
Partner Site Giessen-Marburg-Langen
Justus-Liebig University
Giessen, Germany

Shubham Chakravarty
Department of Biochemistry
Université de Sherbrooke
Quebec, Canada

Stuart Chaskes
Department of Biology
Farmingdale State College
Farmingdale, New York

Violeta Chávez
Department of Pathology
The University of Texas
    Medical School at Houston
Houston, Texas

Seon Young Choi
Maryland Pathogen Research Institute
University of Maryland
College Park, Maryland, and
CosmosID
Rockville, Maryland

Sari Cogneau
BCCM/ITM Collection of Mycobacteria
Antwerp, Belgium

Peter M. Colaninno
Department of Laboratory Sciences
Sunrise Medical Laboratories
A Division of Sonic Healthcare USA
Hicksville, New York

Rebecca E. Colman
Division of Pulmonary, Critical Care &
    Sleep Medicine
University of California San Diego
San Diego, California

Rita R. Colwell
Maryland Pathogen Research Institute
University of Maryland
College Park, Maryland

Stefania De Benedetti
Department of Chemistry and Biochemistry
University of Notre Dame
Notre Dame, Indiana

Carmen E. DeMarco
Division of Infectious Diseases
Providence Hospital
Southfield, Michigan

Jenileima Devi
The Biosphere
Draymans Way
Newcastle Helix
Newcastle upon Tyne, United Kingdom

Swapnil P. Doijad
German Center for Infection Research
Partner site Giessen-Marburg-Langen
Justus-Liebig University
Giessen, Germany

Nina Dombrowski
Department of Marine Microbiology and Biogeochemistry
NIOZ, Royal Netherlands Institute for Sea Research
AB Den Burg, the Netherlands

Timothy P. Driscoll
Department of Biology
West Virginia University
Morgantown, West Virginia

Nellie B. Dumas
Wadsworth Center
New York State Department of Health
Albany, New York

Peter Dürre
Institute of Microbiology and Biotechnology
University of Ulm
Ulm, Germany

Lothar Eggeling
Institute for Bio- and Geosciences
Biotechnology Research Centre Jülich
Jülich, Germany

Laura Eme
CNRS, Université Paris-Saclay
AgroParisTech, Ecologie Systématique Evolution
Orsay, France

Alexander Escasinas
The Biosphere
Draymans Way
Newcastle Helix
Newcastle upon Tyne, United Kingdom

Francesca Falasca
Laboratory of Virology
Department of Molecular Medicine
Sapienza University of Rome
Rome, Italy

Vincent A. Fischetti
Laboratory of Bacterial Pathogenesis
    and Immunology
The Rockefeller University
New York, New York

Jed F. Fisher
Department of Chemistry
    and Biochemistry
University of Notre Dame
Notre Dame, Indiana

Collette Fitzgerald
Enteric Diseases Laboratory Branch
Centers for Disease Control and Prevention
Atlanta, Georgia

Steven L. Foley
Division of Microbiology
National Center for Toxicological Research
U.S. Food and Drug Administration
Jefferson, Arkansas

Joshua Garcia
Marshall B. Ketchum University
Fullerton, California

Joseph J. Gillespie
Department of Microbiology
    and Immunology
School of Medicine
University of Maryland
Baltimore, Maryland

Edmund R. Giugliano (Deceased)
Long Island Jewish Medical Center
New Hyde Park, New York

Emanuel Goldman
Department of Microbiology, Biochemistry
    and Molecular Genetics
New Jersey Medical School
Rutgers, The State University of New Jersey
Newark, New Jersey

Lorrence H. Green
Westbury Diagnostics, Inc.
Farmingdale, New York

Sarah T. Gross
Department of Biology
Farmingdale State College
Farmingdale, New York

Torsten Hain
German Center for Infection Research
Partner site Giessen-Marburg-Langen
Justus-Liebig University
Giessen, Germany

Violet I. Haraszthy
Department of Restorative Dentistry
School of Dental Medicine
State University of New York
University at Buffalo
Buffalo, New York

William E. Herrmann
Retired

Irvin N. Hirshfield
Department of Biological Sciences
St. John's University
Jamaica, New York

Naomi Hoyle
Phagebiotics Research Foundation
Olympia, Washington

Anwar Huq
Maryland Pathogen Research Institute
University of Maryland
College Park, Maryland

J. Michael Janda
Public Health Laboratory at Kern County
San Pablo, California

Ashley M. Joseph
Division of Biology
Kansas State University
Manhattan, Kansas

Lavin Joseph
Enteric Diseases Laboratory Branch
Centers for Disease Control and Prevention
Atlanta, Georgia

Maria Karlsson
Enteric Diseases Laboratory Branch
Centers for Disease Control and Prevention
Atlanta, Georgia

Rebecca K. Kavanagh
Touro College of Pharmacy
New York, New York

Megan L. Kempher
Institute for Environmental Genomics and
    Department of Microbiology and
    Plant Biology
University of Oklahoma
Norman, Oklahoma

Donna J. Kohlerschmidt
Wadsworth Center
New York State Department of Health
Albany, New York

Priyank Kumar
Marshall B. Ketchum University
Fullerton, California

Elizabeth Kutter
Evergreen Bacteriophage Lab
The Evergreen State College
Olympia, Washington

Patrick Kwan
Enteric Diseases Laboratory Branch
Centers for Disease Control and Prevention
Atlanta, Georgia

Mark Laughlin
Enteric Diseases Epidemiology Branch
Centers for Disease Control
    and Prevention
Atlanta, Georgia

Peter S. Lee
Kimberly-Clark
Maumelle, Arkansas

Stephen A. Lerner
Division of Infectious Diseases
School of Medicine
Wayne State University
Detroit, Michigan

Menghan Liu
Sackler Institute of Graduate
    Biomedical Sciences
New York University Langone Health
New York City, New York

Martin J. Loessner
Institute of Food Science and Nutrition
ETH Center
Swiss Federal Institute of Technology
Zürich, Switzerland

Denise L. Lopez
Tulare County Public Health Laboratory
Tulare, California

Geir Åge Løset
Department of Biosciences
University of Oslo and Nextera
Oslo, Norway

Yvonne A. Lue
Accurate Diagnostic Labs
South Plainfield, New Jersey

Tara Mahendrarajah
Department of Marine Microbiology and Biogeochemistry
NIOZ, Royal Netherlands Institute for Sea Research
Den Burg, The Netherlands

Wlodek Mandecki
PharmaSeq, Inc.
Monmouth Junction, New Jersey, and
Department of Microbiology, Biochemistry
    and Molecular Genetics
New Jersey Medical School
Rutgers University
Newark, New Jersey

Barun Mathema
Department of Epidemiology
Columbia University
New York, New York

Meghan A. May
Department of Biomedical Sciences
College of Osteopathic Medicine
University of New England
Biddeford, Maine

Lisa A. Mingle
Wadsworth Center
New York State Department of Health
Albany, New York

Dominique Missiakas
Department of Microbiology
The University of Chicago
Chicago, Illinois

Arindam Mitra
Adamas University
West Bengal, India

Shahriar Mobashery
Department of Chemistry and Biochemistry
University of Notre Dame
Notre Dame, Indiana

Ammara Mushtaq
Division of Infectious Diseases
Icahn School of Medicine at Mount Sinai
New York City, New York

Madge Nanney
Darnell Cookman School of the Medical Arts
Duval County Public Schools
Jacksonville, Florida

Geetha Nattanmai
Wadsworth Center
New York State Department of Health
Albany, New York

Michael C. Nugent
KIROSA Company, Inc.
Baltimore, Maryland

Leonard Osser
Interim CEO Milestone Scientific
Roseland, New Jersey

Charles S. Pavia
Departments of Medical Education and
    Biomedical Sciences
NYIT College of Osteopathic Medicine
Old Westbury, New York

John B. Perkins
The Charles A. Dana Research Institute for Scientists
    Emeriti (RISE)
Drew University
Madison, New Jersey

Janet Pruckler
Enteric Diseases Laboratory Branch
Centers for Disease Control and Prevention
Atlanta, Georgia

Layla Ramos-Hegazy
Department of Biology
Indiana University-Purdue University
Indianapolis, Indiana

Catherine E.D. Rees
School of Biosciences
University of Nottingham
Leicestershire, United Kingdom

Ryan F. Relich
Department of Pathology and Laboratory Medicine
Indiana University School of Medicine
Indianapolis, Indiana

Leen Rigouts
Institute of Tropical Medicine
Antwerp, Belgium

D. Ashley Robinson
Department of Microbiology and Immunology
University of Mississippi Medical Center
Jackson, Mississippi

P. David Rogers
Department of Clinical Pharmacy
The University of Tennessee Health Science Center
Memphis, Tennessee

Ken S. Rosenthal
Northeastern Ohio Medical University
Rootstown, Ohio
Augusta University/University of Georgia Medical Partnership
Athens, Georgia

Bahman Rostama
Department of Biomedical Sciences
University of New England
Biddeford, Maine

Patricia Ryan
Laboratory of Bacterial Pathogenesis and Immunology
The Rockefeller University
New York, New York

Jeffrey M. Rybak
Department of Clinical Pharmacy and Translational Science
University of Tennessee Health Science Center
Memphis, Tennessee

Jason W. Sahl
Pathogen and Microbiome Institute
Northern Arizona University
Flagstaff, Arizona

Inger Sandlie
Department of Biosciences
University of Oslo and Institute of Immunology
Oslo University Hospital
Oslo, Norway

Divya Sarvaiya
Department of Pharmaceutical and Biomedical Sciences
Touro College of Pharmacy
New York City, New York

Michael G. Schmidt
Department of Microbiology and Immunology
Medical University of South Carolina
Charleston, South Carolina

Olaf Schneewind (Deceased)
Department of Microbiology
The University of Chicago
Chicago, Illinois

Stephanie R. Shames
Division of Biology
Kansas State University
Manhattan, Kansas

Tonya Shearin-Patterson
Department of Health Professions
York College
City University of New York
Jamaica, New York

Sanjay K. Shukla
Center for Precision Medicine Research
Marshfield Clinic Research Foundation
Marshfield, Wisconsin

Brad A. Slominski
Kimberly-Clark
Maumelle, Arkansas

Rimsha Sohail
Department of Pharmacy
University of Karachi
Karachi, Pakistan

Scott Sowell
Darnell Cookman School of the Medical Arts
Duval County Public Schools
Jacksonville, Florida

Anja Spang
Department of Marine Microbiology and Biogeochemistry
NIOZ, Royal Netherlands Institute for Sea Research
Den Burg, The Netherlands
Department of Cell and Molecular Biology
Uppsala University
Uppsala, Sweden

Xuanyu Tao
Institute for Environmental Genomics and Department of
    Microbiology and Plant Biology
University of Oklahoma
Norman, Oklahoma

Ombretta Turriziani
Laboratory of Virology
Department of Molecular Medicine
Sapienza University
Rome, Italy

Ernestine M. Vellozzi
Department of Biomedical Sciences
Long Island University
Brookville, New York

Victoria I. Verhoeve
Department of Microbiology and Immunology
University of Maryland School of Medicine
Baltimore, Maryland

Tao Xu
Section on Pathophysiology and Molecular Pharmacology
Joslin Diabetes Center
Boston, Massachusetts

Audrey Wanger
Department of Pathology
The University of Texas Medical School at Houston
Houston, Texas

Alan C. Ward
Department of Biology
Newcastle University
Newcastle upon Tyne, United Kingdom

Volker Winstel
Centre for Experimental and Clinical Infection Research
    and Institute of Medical Microbiology and Hospital
    Epidemiology
Hannover Medical School
Hannover, Germany

Joseph J. Zambon
Department of Periodontics and Endodontics
School of Dental Medicine
State University of New York
University at Buffalo
Buffalo, New York

Daniel R. Zeigler
Bacillus Genetic Stock Center
The Ohio State University
Columbus, Ohio

Aifen Zhou
Institute for Environmental Genomics and Department of
    Microbiology and Plant Biology
University of Oklahoma
Norman, Oklahoma

Jizhong Zhou
Institute for Environmental Genomics and Department of
    Microbiology and Plant Biology
University of Oklahoma
Norman, Oklahoma

Part I

Practical Information and Procedures

1

Sterilization, Disinfection, and Antisepsis

Michael G. Schmidt

CONTENTS

Background

The techniques to control the growth of microorganisms have been shaped by both cultural and scientific advances. From the days of early food preservation—fermentation of milk products and smoking of meats—to extend the shelf life of foods, practical needs have contributed to the development of techniques and tools required for sterilization, disinfection, and antisepsis.

Formal development of clinical medical settings led to an awareness of the causes and effects of disease. This awareness contributed to the development of infection-prevention practices and techniques by Joseph Lister, Oliver Wendell Holmes, Ignaz Semmelweis, and Florence Nightingale, each making formal contributions to the mitigation of disease transmission among patients.

Society continues to shape the infection prevention needs of today. In the last 65 years, we have witnessed the rampant evolution of drug-resistant microbes, a steady increase to the nosocomial-infection rates in healthcare facilities, and the development of exploratory (e.g., endoscopes) and permanently inserted medical devices that have resulted in types of infections never before witnessed. These coupled with a threat from emerging pathogens, as consequence of habitat disruption and destruction, global climate change, modern transportation, and possible bioterrorism, brings together the potential for the introduction of pathogenic microbes into environments not considered in the past.

Our requirements for sterilization, antisepsis, and disinfection thus go beyond the historical needs of research and clinical laboratories and production requirements of commercial facilities. This chapter will review commonly used laboratory techniques, specifically addressing their limitations when used to reduce or eliminate bacteria on materials that were not originally designed to be sterilized or disinfected, and will introduce us to no-touch

solutions for the elimination of microbes to mitigate the risk of transference of these microbes to susceptible hosts.

Methods of Sterilization

Sterilization is a process that results in the complete inactivation of all microbes, including endospores; it can be achieved by mechanical means, heat, chemicals, electromagnetic radiation, or sonic energy (Table 1.1). When using heat, its application may be either moist or dry. Traditionally, moist heat requires increased pressure to facilitate the transfer of energy. This is routinely accomplished with autoclave; dry heat can be applied using an oven (see Table 1.1). However, the dead microbes and their associated cellular debris remain, making even sterilized material a potential hazard.

Disinfection is the process that inactivates most or all microorganisms with the exception of endospores and some acid-fast bacilli. Disinfectants can be subcategorized as high-level disinfectants, which kill all microorganisms with the exception of endospores with an exposure time of less than 45 minutes; intermediate-level disinfectants, destroy the majority of bacteria and viruses but fail to inactivate endospores; and low-level disinfectants, kill the majority of vegetative bacteria, some fungi, and some viruses typically with exposure times of less than 10 minutes (1). Disinfectants are primarily used on inanimate objects and can be sporostatic but are typically not sporicidal (2). Antiseptics, a subset of disinfectants, destroy or inhibit the growth of microorganisms within or present on living tissues and are often referred to as biocides (2).

Steam sterilization and dry heat inactivation of infectious microbes can be monitored using biological indicators or by chemical test strips that turn color upon having met set physical

DOI: 10.1201/9781003099277-2

TABLE 1.1

Routine Methods of Sterilization[a]

Method [Ref.]	Temperature	Pressure	Time	Radiation (Mrad)
Dry heat (119)	150–160°C 302–320°F		>3 hours	
Dry heat (119)	160–170°C 320–338°F		2–3 hours	
Dry heat (119)	170–180°C 338–356°F		1–2 hours	
Moist heat[a]	135°C 275°F	31.5 psig	40 minutes	
Boiling— indirect (119)			1 hour	
Boiling—direct[b]			2 minutes	
Radiation— cobalt 60[c] (119)	Ambient		hours	2–3
Radiation— cesium 137[c] (119)	Ambient		hours	2–3
Electronic accelerators[d] (119, 120)			<1 second	2.5
Ozone (121, 122)		See cited references	See cited references	
Sonic energy (123, 124)	Ambient or in combination with heat or pressure	See review by (124)		

[a] Steam under pressure (i.e., autoclave).
[b] Boiling point of liquids, e.g., cumene (isopropylbenzene) 152°C (306°F).
[c] Dependent on curies in source.
[d] Electrostatic (Van de Graaff), electromagnetic (Linac), direct current, pulsed transformer.

parameters. Usually the spores of species of *Geobacillus* or *Bacillus* are used in either a test strip or suspension, as these organisms are more difficult to kill than most organisms of clinical interest (3). Growth of the spores in liquid media after the cycle of sterilization is complete indicates the load was not successfully processed and should not be considered sterile.

In the days following September 11, 2001, letters containing the spores and cells of *Bacillus anthracis* were mailed to several news media offices, and two U.S. senators resulting in the death of 5 people and the infection of 17 others. This introduced a pathogen into locations where methods of safe handling and eradication of such pathogens were not defined nor anticipated. Further complicating matters was that the microbes contained within the packaging were optimized for airborne dispersal resulting in substantial contamination of the built environment necessitating isolation and/or closure of the buildings.

As a consequence of the Anthrax attacks of 2001, a number of laboratories have conducted experiments on the potential dispersal, detection, and eradication of organisms spread by these unconventional methods. Lemieux and others (4) conducted experiments with simulated building decontamination residue (BDR) contaminated with 10^6 spores of *Geobacillus stearothermophilus* to simulate *Bacillus anthracis* contamination. A single

cycle did not effectively decontaminate the BDR. Only an autoclave cycle of 120 minutes at 31.5 pounds per square inch gauge, relative to atmosphere (psig) [2.17 bar]/275°F (135°C) and 75 minutes at 45 psig [3.10 bar]/292°F (144°C) effectively decontaminated the BDR. Two standard cycles at 40 minutes and 31.5 psig/275°F run in sequence were found even more effective. The authors state that the second cycle's evacuation step probably pulled condensed water out of the pores of the materials, allowing better steam penetration. It was found that both the packing density and material type of the BDR significantly impacted the effectiveness of the decontamination process.

CONVERSION FACTORS FOR COMMON MEASURES USED TO EXPRESS PRESSURES

Pounds per square inch (psi) or precisely, pound-force per square inch (lbf/in^2) is a unit of pressure or stress expressed using the avoirdupois system of measurement resulting from one pound-force of pressure applied to an area of one square inch. Using the International System of Units (SI) 1 PSI is approximately equal to 6,895 N/m^2.

Pounds per square inch absolute (psia) is a term used to delineate a pressure value being reported relative to vacuum rather than ambient atmospheric pressure. Given that 1 atmosphere of pressure at sea level is 14.7 psi, any value need expressed using psia requires the addition of the barometric relative to psia.

Pounds per square inch gauge (psig) the converse to **psia** expresses the pressure value relative to atmospheric pressure. Consequently, gauge pressure takes into consideration the effects of changes to barometric pressure or as a consequence of pressure changes due to altitude differentials.

Other common units used to express pressure are:

1 psi = 0.068046 atmospheres
0.689476 Bar
6,894.76 Pascals
51.7149 Torr
2.03602 inches of Hg
27.7076 inches of water column 6,895 N/m^2

When materials made of unusual substances and/or densities are to be sterilized, the method of sterilization must be monitored and adapted according to the material. Common strategies include alterations of placement and/or lower packing densities within the autoclave. Other common methods of sterilization of problematic materials and substance include use of gases such as ozone or ethylene oxide, radiation, or less commonly, electronic accelerators (see Table 1.1).

Solutions containing heat-labile components require a different approach. Filtration is generally the most accepted and easiest method. The Food and Drug Administration of the United States (FDA) and industry consider 0.22-μm filters to be sterilization grade based on logarithmic reductions of one of the smallest bacteria *Brevundimonas diminuta*, a nonlactose fermenting environmental Gram-negative rod (5). Usually an incubation

TABLE 1.2

Filter Sterilization

Size(μm)	Purpose
0.1	Mycoplasmal removal
0.22	Routine bacterial removal
0.45	Plate counts of water samples
>0.45	Removal of particulates, some bacteria, yeast, and filamentous fungi

period of at least 48 hours is required for colony development of *Br. diminuta* before the colonies are large enough to be viewed for counting. Griffiths and others (6) describe a Tn5 recombination method in which *Br. diminuta* is augmented with the genes for bioluminescence (luxABCDE) and fluorescence (*gfp*). Use of the modified microbe enabled detection within 24 hours of incubation by either bioluminescence or fluorescence. They state that this method may aid in preventing quality control backlogs during filter manufacturing processes.

A 0.22-μm filter will adequately sterilize the majority of heat labile solutions used for molecular biology and microbiology. Labile tissue culture reagents might also require to be certified free from mycoplasma contamination. A 0.1-μm filter is used to remove mycoplasma from tissue-culture solutions (7) (Table 1.2). Today, with the widespread availability of PCR machines, routine testing for *Mycoplasma*, *Acholeplasma*, and *Ureaplasma* via the amplification of sentinel genes has become the established method of choice for highest sensitivity detection of contamination of filtered liquids by these microbes. Detection limits of less than 1 cfu/mL are easily achieved within a few hours. This has transformed filtration-based sterile product release; rather than waiting on results from a 28-day biological culture test, manufacturers can now easily rely on this established methodology taking less than a few hours to complete.

The removal of contaminants from air may be necessary in the case of fermentation, drug manufacturing, or chemical reactions requiring some form of gas. Various types of filters are available for the removal of organisms from air. One such filter is the Aerex 2 (Millipore). It can withstand 200 steam-in-place cycles at 293°F (145°C), it can resist hydraulic pressures of 4.1 bar (60 psig), and has the ability to retain all phages when challenged with the ΦX-174 bacteriophage at 29 nm in diameter at a concentration of between 10^7 and 10^{10} bacteriophages per cartridge (8). In this particular scenario, the operator must assess their needs for how critical the product be free of contaminating materials (i.e., drugs versus topical cosmetics), as well as the ability of the product to withstand pressure, heat, flow-rate, and other parameters of the manufacturing process, and select an appropriate filter.

In some cases, a filter is used to recover bacteria from dilute solutions such as environmental water samples. The organisms of significance whose retention on the filter is most important should determine the filter pore size. The smaller the filter pore size, the slower the sample flow rate and throughput. Standard methods define an acceptable recovery rate for filter-based recovery of bacteria as being 90% of the number of bacteria (colony-forming units, [CFU]) recovered from the same sample by the spread plate method under typically aerobic conditions. In one study performed by Millipore, filtration with a 0.45-μm pore-size filter provided 90% recovery of 12 microorganisms used, ranging in size from *Br. diminuta* to *Candida albicans* (Table 1.2).

It is important to note that all filter sterilization is relative. Filter integrity for sterilization is usually done by a bubble test to confirm the pore size of the manufactured filter (5, 9). The bubble point is based on the fact that liquid is held in the pores of the filter (usually membrane) by surface tension and capillary forces, and the bubble pressure detects the least amount of pressure than can displace the liquid out of the pores of the filter (9). While a 0.22-μm filter will remove most bacteria, it will not remove viruses, mycoplasma, prions, and other small contaminants, including toxins. Every sample type and the level of permissible substances in the filtrate must be assessed on a case-by-case basis. For example, filter-sterilization of water or solutions used in atomic force microscopy might still allow sufficient numbers of small viruses and other particulates to pass through the filter and impede sample interpretation; under such conditions ultrapure water (UPW) may be needed. UPW comprises H_2O, H^+, and OH^- ions in equilibrium with an electrical conductivity of approximately 0.055 μS/cm at 25°C, which may be alternatively expressed by its resistance coefficient of 18.2 MOhm (10).

Ultrasonic Energy

Ultrasonic energy in the range of 20–500 kilohertz (kHz) makes use of physical and chemical phenomena induced by an interaction of the ultrasonic energy with the structures and chemical constituents within the microbial cell. Principally, the sonic energy generates a wave front within the liquid medium. Alternating regions of compression and expansion occur causing pressure changes resulting in cavitation and bubble formation. During the expansion phase within the wave front the expanding bubbles reach a state where they implode resulting in rapid condensation. The condensed molecules collide violently with the subsequent formation of a high-pressure shock wave of approximately 50 MPa resulting in a localized temperature of 5,500°C (11). These rapid changes in localized pressure and extremes of temperature serve to inactivate the microorganisms with temperature being the main contributor to the cell's inactivation (11).

In a study of the use of high-energy ultrasound in the range of 24 kHz over 1 hour where the nominal energy administered was applied at 1,500 W/L or 5,400 kJ/L, the mean killing observed for typical Gram-negative microbes found associated with wastewater, such as total coliforms, fecal coliforms, and/or *Pseudomonas* spp., was 99.5%, 99.2%, and 99.7%, respectively (12). Gram-positive microbes were found to be more recalcitrant to the sonic inactivation using this power density (W/L) and frequency (kHz) of energy, where the mean inactivation observed for *Clostridium perfringens* and fecal streptococci was 66% and 84%, respectively. Sonic inactivation of microbes was found to follow first-order kinetics with respect to the bacterial population and was not affected significantly by the medium in which they were suspended (12). The addition of 5 g/L of TiO_2 particles generally enhanced the destruction of Gram-negative bacteria, yielding a 3–5 log fold reduction in viability, while the addition of the titanium particles to Gram-positive microbes was found to be only mildly more effective. Nevertheless, the application of ultrasonic energy at 4,000 kJ/L in the presence of 5 g/L of TiO_2 particles was able to meet the United States Environmental Protection Agency required limit, $<10^3$ CFU/100 mL total coliforms for water reuse (12).

Other workers have found similar antimicrobial effects when using sonic irradiation to inactivate microbes. Mesophilic aerobes and fungi suspended in orange juice were inactivated within 15 minutes through an application of 240 W of power at a frequency of 500 kHz under ambient temperature conditions (13). The coapplication of heat with and without pressure was found to accelerate the rate with which the population of microbes could be inactivated. For *Escherichia coli* suspended in a phosphate buffer, using a frequency of 20 kHz at a power setting of 100 W using temperature ranges between 40°C and 61°C and 100–500 kPa of pressure was found to be bactericidal between 0.25 minutes and 4 minutes (14).

Disinfectants and Antiseptics

The regulation of disinfectants and antiseptics falls under the jurisdiction of different agencies, depending on where the chemical will be used. In the United States, the Environmental Protection Agency (EPA) regulates disinfectants used on environmental surfaces (e.g., floors, laboratory surfaces, patient bathrooms), and the FDA regulates liquid high-level sterilants used on critical and semicritical patient-care devices.

Labeling terminology can also be confusing, as the FDA uses the same terminology as the United States Centers for Disease Control and Prevention (CDC) (critical [sterilization], semicritical [high and intermediate], and noncritical [low-level disinfection]), while the EPA registers environmental disinfectants based on the claims of the manufacturer at the time of registration (i.e., EPA hospital disinfectant with tuberculocidal claim = CDC

intermediate level disinfectant) (15). In addition, the use of disinfectants varies globally. For example, surface disinfectants containing aldehydes or quaternary ammonium compounds are less often used in American, British, and Italian hospitals than in German ones (16), thus making it difficult to define overall best methods or practices. Whatever method is employed—mechanical, heat, chemical, or radiation—a method of assessment must be in place to test the initial effectiveness of the method on either the organism of the highest level of resistance to the method or a surrogate; and a standard must be included to monitor the effectiveness of the method for its regular use by general staff. While such forms of assessment are common for daily runs from the autoclave, they are used less often when general low-level disinfection occurs on noncritical surfaces. A number of commonly used disinfectants, along with some of the microbes against which they have activity, may be found in Table 1.3.

Disinfectants are often tested against cultures of the following bacteria: *Pseudomonas aeruginosa, Staphylococcus aureus, Salmonella typhimurium, Mycobacterium smegmatis, Prevotella intermedia, Streptococcus mutans, Aggregatibacter actinomycetemcomitans* (formerly *Actinobacillus actinomycetemcomitans), Bacteroides fragilis,* and *Escherichia coli* (17). One common approach is to use the broth dilution method, wherein a standard concentration of the organism is tested against increasing dilutions of the disinfectant. The minimum inhibitory concentration (MIC) is defined as the lowest concentration of the disinfectant that prevented visible growth of the microbe.

While there are many reports of liquid disinfectant activity against liquid/planktonic cultures, biofilms of the aforementioned organisms have survived when the liquid/planktonic culture of the same organism has been killed (17). Therefore, biofilm disinfection must be evaluated separately. Better efficacy in biofilm prevention and removal has been demonstrated by the

TABLE 1.3

Partial Listing of Some Useful Disinfectants

Substance[a]	Concentration	Time	Types of Organisms Killed	Ref.
Inorganic hypochlorite (bleach)	6,000 ppm	5 minutes at 22°C	Adenovirus 8	(125)
	5,000 ppm	10 minutes at 23–27°C	Bacteria, viruses, and fungi, hepatitis A and B, HSV-1 and 2	(1)
	2,000 ppm	pH 7, 5°C, 102 minutes	*Bacillus anthracis* Ames (spores)[b]	(126)
	1,000 ppm	pH 7, 5°C, 68 minutes	*Bacillus anthracis* Sterne (spores)[b]	(127)
		10 minutes at 20 ± 2°C	*Feline calicivirus* (norovirus surrogate)	(128)
Peracetic acid[b]	0.3%		Sporocidal, bactericidal, viricidal, and fungicidal	(2)
Glutaraldehyde	2.65% in hard water (380–420 ppm calcium carbonate)	5 minutes at 22°C	Adenovirus 8	(125)
Ethyl alcohol	70%	5 minutes at 22°C	Disinfectant	(125)
Surfacine	Manufacturer's instructions		VRE	(19)
R-82 (quaternary ammonium compound)	856 ppm (1:256 dilution)	10 minutes at 20°C ± 2°C	*Feline calicivirus* (norovirus surrogate)	(128)
Ozone disinfection of drinking water	0.37 mg/L	Contact-time 5 minutes at 5°C	Noroviruses	(122)
Acidic sodium dodecyl sulfate	1% SDS plus either 0.5% acetic acid and 50 mM glycine (pH 3.7) or 0.2% peracetic acid	30 minutes at 37°C	Prions: specifically PrP^{Sc}, see Ref. for details	(129)

[a] Highest concentration given for use with most resistant organism tested. See cited reference for refined values for each organism.

[b] The authors observed $\log_{10}$ inactivation of populations of spores, the inactivation increasing with contact time, but not sterilization.

use antibiofilm products as compared to detergent disinfectants containing quaternary ammonium compounds (18).

One problem with most disinfectants and antiseptics is their short effective life span. Hospitals and laboratories today are challenged with multiple drug-resistant organisms that may be transmitted from surfaces. These surfaces may be routinely recontaminated by either patient, visitor, or staff activity. A new agent, a combination of antimicrobial silver iodide and a surface-immobilized coating (polyhexamethylene biguanide), Surfacine™, is both immediately active and has residual disinfectant activity (19). Additionally, it can be used on animate or inanimate objects. In tests on Formica™ with vancomycin-resistant *Enterococcus*, Surfacine™ was found to be 100% effective on surface-challenge levels of 100 CFU per square inch for 13 days (19). Rutala and Weber (19) also state that the manufacturer's test data demonstrated inactivation of bacteria, yeast, fungi, and viruses with inactivation time varying by organism. Similarly, Schmidt and colleagues reported on the *in situ* clinical evaluation of a persistently active disinfectant, *Firebird F130*™ (70% ethanol and <1% mixed quaternary ammonium chloride compounds along with proprietary agents designed to maintain activity for 24 hours subsequent its application) (Microban, Huntersville, NC) (20). Here, they learned that this new formulation of two agents known for their ability to disinfect the built clinical environment performed significantly better and for longer than the individual agents by themselves; this particular compound's persistently active material was found superior for all time points when compared to a dilutable quaternary ammonium agent, and it was significantly better for controlling bioburden for two of the three time points for the disinfectant with ethanol and quaternary ammonium as its agent (20).

Inducing a Viable but Nonculturable State upon Exposure to Sublethal Concentrations of Disinfectants and Other Environmental Stressors

Essential to measuring the effectiveness of sterilization and/or exposure to disinfectants is the ability of a microbe to form a colony after exposure to proper growth conditions on routine bacteriological medium on which the microbe would normally grow. Failure of some microbes to form a colony under all circumstances is not always the result of the microbe being killed but rather being in a state referred to as viable but nonculturable (VBNC) (21). This was first appreciated by Xu and colleagues in 1982 in the human pathogen *Vibrio cholerae*, and then during the ensuing period by other workers who investigated how exposure of populations of other microbes to environmental stressors triggered the VBNC condition. Upon entering a VBNC state, the community confers an impression that materials treated have fewer viable bacteria than they actually do. Unfortunately, many of the routine treatment processes used to render materials free or possessing reduced concentrations of bacteria (e.g., preparation of "ready to eat" salads and other produce by subjecting the produce to submersion into a dilute concentration of sodium hypochlorite [3–100 ppm] followed by drying) are in themselves responsible for triggering the VBNC phenotype (22–25). This is especially problematic as presumably treated materials retain substantial concentrations of bacteria capable of causing disease (23, 26–30). In addition to dilute sodium

hypochlorite, many factors have been reported to induce the condition. They include many common household cleaners and inorganic salts (28), nutrient deprivation (31), growth outside the preferred/normal range of temperatures (31–34), alterations to the osmotic pressure exerted by the growth environment (35, 36), oxygen tension (37), commonly used food preservation materials (38), metals (39), exposures to thermosonication (40), and electromagnetic radiation in the visible (41) and ultraviolet spectra (42). Regardless of how the cells achieved a VBNC state, culture-based assessment of a product's sterility may lead to an underestimation in the number of microbes removed from the material and should be accounted for prior to the release of the "sterile" lot.

To address the limitation imposed by the presence of VBNC microbes in products, a nonculture-based system should be used to estimate the concentration of microbes remaining in the treated material. Commonly used methods include those that measure membrane integrity with redox sensitive dyes. As an example, one product marketed under the trade name of Live/Dead®*BacLight*™ assesses whether the microbes have a compromised membrane. Damage to the membrane of bacteria results in their death via a multifactor mechanism. Often commencing with the collapse of the membrane potential, the bacterial cell faces an accumulation of free oxygen radicals secondary to the effects of routine metabolism that are responsible for membrane damage leading to loss of integrity with cytoplasmic leakage and destruction of the nucleic acid within the microbe. Such dead bacterial cells stain red, while those with an intact membrane stain green. The disadvantage of methods relying on membrane integrity to assess life is the need to visualize what fraction of the population is live vs. those dead. Alternatively, total nucleic acid may be recovered from the treated materials whereupon RNA is isolated, subjected to reverse transcription, using a primer targeted against a routinely transcribed messenger RNA, where its presence signals the cell is still alive. This technique draws from the fact that since the average bacterial mRNA has a half-life of 3–5 minutes, those cells continuing to transcribe their genome are considered to be alive. Each of the alternate methods requires time, often specialized equipment, and skilled personnel to accomplish the tests.

Disinfection and Sterilization of Prion-Contaminated Materials

Prions are infectious agents composed solely of protein in a nonnative or misfolded form. Clinically these agents are considered transmissible spongiform encephalopathies (TSEs) and result in a family of rare progressive neurodegenerative disorders that affect both humans and animals. In humans they have been implicated in neurodegenerative diseases ranging from Alzheimer's, Creutzfeldt-Jakob (CJD), variant Creutzfeldt-Jakob (vCJD), Gerstmann-Straussler-Scheinker syndrome, fatal familial insomnia, to Kuru (laughing sickness) (43). This nonliving, but infectious proteinaceous material manifests disease by affecting the structure of the brain or other neural tissue of the central nervous system. In the United States, the incidence of CJD is approximately one case per million population (44–46). Unlike other infectious agents, TSEs do not elicit an immune response (43). To date, they are untreatable and universally fatal following a long incubation period and a noninflammatory pathologic progression of disease within the central nervous system (43, 47). CJD occurs both sporadically, accounting for approximately 85% of the cases, and can also be inherited (15%) (43). However, there have been estimates that approximately 1% of the CJD in the United States may result from a healthcare-associated exposure from contaminated tissues, grafts, or via transmission from contaminated instruments (43).

Prions that cause CJD and other TSEs are recalcitrant to traditional means routinely used for the inactivation of fungi, bacteria, and viruses. The infectious material is introduced to the instrument as a by-product of the medical procedure. Tissues at high risk of containing infectious concentrations of these proteinaceous agents implicated in TSEs include the brain, spinal cord, pituitary gland, and the posterior eye (43). Critical and semicritical devices, as defined by Spaulding (48), must be rendered prion-free prior to their reuse. Presently, there are no antimicrobial products registered with the U.S. EPA having a registration specifically for the inactivation of prions on environmental surfaces, and there is no sterilization processes cleared by the FDA for sterilization of reusable surgical instruments (43). With that caveat, operationally, inactivation of a prion is defined when the infectious concentration has been reduced by $5-6 \log_{10}$ in the number of prion particles associated with the critical/semicritical item (43). Rutala and Weber, writing the guidelines for the disinfection and sterilization of prion contaminated medical instruments for the *Society for Healthcare Epidemiology of America*, summarized data from the literature suggesting that autoclaving at 134°C for at least 18 minutes (prevacuum) or 132°C for 60 minutes (gravity displacement) was effective for the inactivation of prions (43). Further, the addition of chemicals in concert with heat such as sodium hydroxide (0.09N or 0.9N) for 2 hours with subsequent steam sterilization at 121°C for 1 hour was sufficient for the partial or complete loss of infectivity (49–53). The addition of proteolytic enzymes to the disinfection protocol may also facilitate the removal with subsequent inactivation of the infectious material. To enable the removal of the infectious protein, suspect instruments should be subjected to decontamination with subsequent sterilization as soon as possible after coming in contact with suspect tissues. Dilution to extinction remains an adjunctive method to facilitate the mitigation of the risk of exposure to prions. Thus, keeping the instruments wet or damp until they are subjected to decontamination might help mitigate the risk of prion exposure by facilitating removal of the agent from the suspected or contaminated instruments or devices.

Ding and colleagues described the kinetics of ozone inactivation of infectious prion proteins (54). Ozone possesses the property of having the highest oxidation potential among the majority of chemical disinfectant used for the treatment of water (55) and was found to inactivate the prion PrP^{sc} (scrapie 263K) as quantitatively assessed by measuring the templating properties using the protein misfolding cyclic amplification assay (PMCA) (56). Here they learned that pH and temperature, in addition to the concentration of ozone, was able to effectively reduce the concentration of PrP^{sc} by $2\text{-}\log_{10}$, $3\text{-}\log_{10}$, and $4\text{-}\log_{10}$ increments. This led them to conclude in systems using ozone-demand-free water that ozone is effective for prion inactivation (54).

Similarly, hypochlorous acid (HOCL) commonly referred to as bleach was able to inactivate prions and other self-propagating protein amyloid seeds (57). Here, a weekly acidic aqueous formulation of hypochlorous acid, suitable for application directly to skin and mucous membranes or aerosolization to treat entire rooms without harmful effects, was able to eliminate all detectable prion seeding activity of the human Creutzfeldt-Jakob disease, bovine spongiform encephalopathy, cervine chronic wasting disease, sheep scrapie, and hamster scrapie with reductions seen between $\geq 10^3$ and 10^6 fold as measured by real-time quaking-induced conversion assays (RT-QuIC; for more information see refs. (58, 59)). Briefly, this type of assay exploits the self-propagating activity of these proteins or prion seeds by coupling their ability for self-amplification as much as trillionfold with the evolution of a fluorescent signal through the conversion of a recombinant prion protein in the assay medium (rPrP) into detectable amounts of a misfolded, aggregated form of rPrP. The rPrP-aggregates then interact with a specific dye, causing a measurable change in the dye's fluorescence emission spectrum, enabling use of multiwell plates and a fluorescent detection system to reveal the presence of a prion seed with greater sensitivity than animal bioassays (58, 59).

What has been hindering the adoption of a universal method for the inactivation of prions has been the absence of a rapid method, with sufficient sensitivity and specificity, to validate that prion "sterilization" method was indeed effective at rendering the material prion free. To date, alkaline and enzymatic detergents that mineralize proteins have been found to be the most effective at eliminating the infectious potential of prions associated with the instruments (60). The European Community Notified Body (CE) has independently certified some of the protein degrading agents, but recommends that their use remain at the discretion of the user who, in concert with the manufacturer, will assess the scientific qualities of the product (61).

Disinfection of Hands

One of the most important areas for infection control is debulking of the built environment with subsequent disinfection of the hands of their resident microbial flora. Usually, mechanical methods such as scrubbing with a brush are most used, but alternatively, alcohol-based rubs may also be used for hand disinfection. Widmer (62) states in the year 2000, the compliance rate for handwashing was less than 41% and speculated that understaffing may further complicate the problem. The proper methodology for handwashing includes wetting the hands and wrists with water, dispensing a dose of soap with the forearm or elbow, rubbing the hands and wrists with the soap for 10–15 seconds, rinsing the hands, and then drying the hands without rubbing. Additionally, the faucet is turned off while holding the paper towel, so as not to contaminate the hands (62).

Many authors now advocate waterless alcohol rubs. Kampf et al. (63) advocate a 1-minute handwashing at the beginning of the day, a 10-minute drying time, and then unless visibly soiled, a rub-in, alcohol-based hand disinfectant for routine disinfection throughout the remainder of the day. Gupta et al. (64) compared an alcohol-based waterless and an alcohol-based water-aided scrub solution against a brush-based iodine solution under conditions encountered in community hospital operating rooms. The brush-based iodine solution performed best on the first day of the study; but when colony-count reductions were studied between the three options over the course of 5 days, no significant difference was found. The participants found the alcohol-based waterless solution the easiest of the three to use with the lowest complaints of skin irritation. Another reason to advocate non-abrasive hand sanitizing methods is the fact that brush-based systems can damage skin and cause microscopic cuts that can harbor microbes (64).

Cooper et al. (65) found that even small increases in the frequency of effective hand washes were enough to bring endemic nosocomial infective organisms under control in a computer model. Raboud et al. (66), using a Monte Carlo simulation, determined factors that would reduce transmission of nosocomial infection. The factors that were relevant included reducing the nursing patient load from 4.3 (day) and 6.8 (night) to 3.8 (day) and 5.7 (night), increasing the hand-washing rates for visitors, and screening patients for methicillin-resistant *S. aureus* (MRSA) upon admission.

In spite of the publication of guidelines for hand hygiene in healthcare facilities in 2002 (67), the subsequent widespread adoption of hand-hygiene compliance protocols, advances in understanding the mechanics and utility of new products for increased convenience and efficacy, hand hygiene remains a substantial problem for modern healthcare. McGuckin and colleagues reported in 2009 that the three most frequently reported methods for measuring hand hygiene compliance were (1) direct observation, (2) self-reporting by individual healthcare workers (HCWs), and (3) an indirect calculation based on the consumption of products directly tied to hand hygiene compliance (68). Their results showed that hand-hygiene compliance rates at baseline within intensive care units (ICU) were at an extremely low rate of 26%, while the average rate for non-ICU units was found to be only slightly better with a rate of 36% (68). Upon monitoring the usage of the products employed in hand-hygiene compliance protocols in concert with providing feedback to the staff, they witnessed a significant increase in compliance to 37% for the ICUs and 51% for the non-ICUs (ICU, $p = 0.019$; non-ICU, $p < 0.001$) leading them to conclude that in the United States, hand-hygiene compliance can increase with monitoring combined with feedback (68).

In 2018 an alarm within the infection control community sounded loudly enough to be heard by the general public when it was appreciated that some hospital isolates of *Enterococcus faecium* developed a tenfold higher tolerance against alcohol inactivation (69). This observation is especially troubling in that the microbe acquiring tolerance to this vital infection control intervention is also one of the leading causes of healthcare acquired infections. The natural selection of alcohol-tolerant strains of *E. faecium* was not unexpected given the emphasis that the infection control community has placed on the ubiquitous placement of point-of-care disinfection stations in order to facilitate convenient hand-hygiene compliance. While strains of *Enterococcus* have yet to "break-through" becoming tolerant to the concentration of alcohols used in the majority of hand rubs, the genetic modifications accumulating as heritable mutations within genes involved in carbohydrate uptake and metabolism (69) are sufficiently concerning to worry that it is only a matter of time before natural selection succeeds with an emergence of an *Enterococcus* completely tolerant to the lethal effects of exposure to short-chained alcohols used in alcohol-based disinfectant hand hygiene products.

No-Touch Disinfection Technologies

Hospital-associated infections (HAIs) continue to be one of the most common and significant complications associated with hospitalization. There is significant interest in learning how to manage and provide best-practice applications for infection control for hospitals (70–73). Microbes have an intrinsic ability to survive and ultimately colonize common touch surfaces where acquisition and transport from surfaces to humans is common. HCWs have the potential to transfer these microbiological contaminants not only from patient to patient but among themselves and back to surfaces, refreshing or adding to the complexity of the microbial reservoir involved in transmission. There have been many studies looking at the control of contamination of common hospital touch surfaces both from hand to surface contact and vice versa. Investigators have shown that the gloves of nurses frequently collected viable methicillin-resistant *S. aureus* (MRSA) after touching inanimate objects near colonized patients (74). In concert with aggressive hand-hygiene campaigns, recent hygiene guidelines specifically recommend that particular attention be paid to the disinfection of patient-care surfaces, especially surfaces designated "high touch objects" (HTOs) as a target of infection prevention and control (75). The guidelines note that such objects could potentially contribute to secondary transmission by contaminating hands of HCWs or by contacting medical equipment that subsequently contacts patients (1, 70, 76–80). Routine or daily cleaning coupled with cleaning immediately after patient discharge (terminal cleaning) of the surfaces and objects within the room with subsequent application of a hospital grade disinfectant has been an accepted method for controlling and limiting the spread of infectious agents. (81) A concentration of less than 250 aerobic colony-forming units (CFUs) per 100 square centimeters has been proposed as the benchmark, where bacterial levels below this value are considered low risk, while concentrations greater are suggestive of an increased infectious risk to patients (82, 83).

No-touch solutions for the disinfection of at-risk environments within healthcare settings are quickly gaining acceptance as the technologies have been found to be an effective and comprehensive addition to systems-based solutions for infection control. To date, they have not been studied in concert with aggressive hand-hygiene campaigns, appropriate routine and terminal cleaning of patient-care environments, or an active surveillance and isolation protocol for patients entering care who are already colonized with VRE, MRSA, *Clostridioides* (formerly, *Clostridium*) *difficile* (CDI or C-diff) or other multidrug-resistant microbes such as *Klebsiella pneumoniae* carbapenemase (KPC). Since the time of the drafting of the previous edition, studies have yet to be completed to learn the outcome from this call to action, again leading us to wonder whether the antimicrobial effectiveness will have an additive effect or whether it will act synergistically. Only through careful studies will we come to know the answer to this question.

As the name suggests, no-touch technologies do not come in direct contact with colonized, contaminated, or soiled surfaces.

Rather, they distribute their microbiocidal activity through the atmosphere by either delivering a lethal concentration of electromagnetic energy in the ultraviolet spectrum or by the real-time distribution of reactive oxygen species, such as hydrogen peroxide, singlet oxygen, hydroxyl radical, or oxyanions. In general, both systems have been found to effectively reduce the concentration of microbes by at least 4 $\log_{10}$ (84). Both systems have their limitations. Each requires skilled labor to place the equipment and commence the disinfection cycle in the location subjected to disinfection. While studies report on the effectiveness of ultraviolet-C (UV-C) to save healthcare institutions on the costs associated with infections the studies often fail to consider the cost of the intervention (85, 86). In general, the cost associated with UV-C disinfection per room for a typical healthcare institution, at prevailing labor rates for 2019 for the United States, is approximately $9.50 per room. The bulk of the cost, 64%, results from the skilled labor required to operate the equipment (~$15 per hour in 2019 wages), where the estimated time per room subsequent to discharge-based cleaning and routine terminal disinfection is approximately 30 minutes with 30 rooms being serviced per day (assuming two shifts per day). The remaining 36% of the cost results from amortization of the capital cost of the device, including maintenance, service, and supplies (capital equipment ranges $82,000–$124,000 per unit and monthly recurring costs of $3,000–$3,500 for maintenance, service, and supplies [bulbs]).

The disinfection reach of ultraviolet light is subject to the effects of shadowing and "cornering" where the light is unable to reach objects not within the sight lines of the device emitting the disinfecting irradiation. This typically requires that the equipment be placed in the center of the room to ensure uniform coverage. Additionally, rooms must be vacant of patients, staff, and visitors as any associated ultraviolet energy needs to be prevented from leaching into areas occupied by people as UV light energy can cause damage to sight and result in burns to exposed skin. The energy can also shorten the life of equipment in the room as routine exposure to UV light can accelerate the decay of the materials by increasing the brittleness of many of the plastics used in the fabrication of equipment associated with the delivery of healthcare.

Despite the limitations of ultraviolet-assisted disinfection, in one comprehensive study, Anderson and colleagues learned that an automated UV-C emitter was capable of decreasing the bioburden of significant pathogens in hospital rooms (87, 88). In their pragmatic, cluster-randomized, crossover trial at nine hospitals in the southeastern United States, they learned that a contaminated healthcare environment is an important reservoir from which patients can acquire pathogens and, importantly, that enhanced terminal-room disinfection decreases HAI risk. Post hoc analysis of their data showed that after the discharge of a previous occupant with a *C. difficile* infection, there was a decrease in multidrug-resistant microbes (primary outcome for their study) as a result of the inclusion of ultraviolet light to the cleaning regimen (88). Similarly, Rutala and colleagues concluded that an enhanced form of disinfection, here again UV-C energy, can lead to a reduction in both the microbial burden in the built environment with a subsequent decrease in patient colonization and infection by four epidemiologically relevant pathogens (MRSA, VRE, *C. difficile*, and multidrug-resistant *Acinetobacter*) (89).

In 2015, Ghantoji and colleagues assessed the noninferiority of pulsed xenon UV-C energy against the effectiveness of bleach for its ability to reduce the concentration of *C. difficile* resident on high-touch services in rooms housing isolation patients with *C. difficile* infections. Here they similarly learned that for high-touch surfaces in rooms previously occupied by patients with *C. difficile* infections, both methods were essentially equivalent in their ability to decrease environmental contamination by spores of *C. difficile* (90).

The cost of enhanced disinfection via vapor-phase disinfection of a built environment is approximately an order of magnitude greater than the cost of UV-C energy with an estimated cost of approximately $107 per single-patient room, again controlling for the capital investment, labor, and supplies. Like UV-C disinfection, this form of enhanced disinfection of the built environment similarly has limitations in that ventilation/air flow to the room must be controlled/and or limited for the duration of the disinfection cycle. This time can vary depending upon the concentration of peroxide or disinfecting gas used. Nevertheless, the two technologies have been found to be effective for the disinfection of inanimate objects and surfaces. However, neither technology is intended as a substitute for cleaning or for the removal of soil from the resident objects and surfaces within the built patient-care environment. An appropriately trained environmental service team must accomplish cleaning, with subsequent disinfection of the built environment.

Recently, we have begun to witness the incorporation of another no-touch technology. However, unlike UV and vapor-phase oxygen radicals (H_2O_2) that distribute their antimicrobial activity through the atmosphere, this technology requires the microbe come in contact with the material in order to exert its antimicrobial activity. Additionally, once placed, such no-touch systems relying on contact for killing do not require any further user intervention. One such example of this type of no-touch technology is solid antimicrobial copper. The resident microbial burden associated with the built environment is continuously reduced through the strategic placement of solid copper surfaces onto critical high-touch surfaces within the patient-care setting (91). Copper has been used by humans for millennia, first as tools and then as a measure to fight the spread of infectious agents. Metallic copper intrinsically displays a strong antibacterial activity both in aquatic systems (92, 93) and on dry surfaces (94–98). In 2008, the EPA registered five families of copper-containing alloys as antimicrobial, establishing that products manufactured from one of the registered alloys kill 99.9% ($\log_{10}$ 2.0) of bacteria within 2 hours of exposure (99). It is anticipated that the solid antimicrobial copper surfaces will remain microbiocidal for the life of the product (>10 years). A variety of controlled studies have looked at the antimicrobial activity of copper surfaces against specific human pathogens (95, 97, 98, 100–102). Copper has been used to limit the bacterial burden found on commonly touched surfaces and objects in active healthcare environments. In a recent hospital trial, bacterial reductions up to one-third were recorded using copper alloys in place of plastic or aluminum surfaces on light switches, door knobs, and push plates (103). Casey and others (104) observed a median microbial reduction of between 90% and 100% ($\log_{10}$ 1.95 and 2.0) on copper surfaced push plates, faucet handles, and toilet seats, while Schmidt and colleagues demonstrated significantly lower bacterial burdens on six HTOs, averaging an 83% ($\log_{10}$ 1.93) reduction for all of the objects over the course of a 43-month multicenter trial (91).

Current cleaning methods can effectively remove pathogens from surfaces, but studies have shown that more than half of the trial surfaces were not adequately terminally cleaned and may become recontaminated within minutes (105, 106). The rails of hospital beds, as a consequence of coincident interactions with patients, HCWs, and visitors, are one of the most frequently touched items found in the built patient-care environment. Schmidt and colleagues found when they quantitatively assessed the bacterial burden present on bed rails, metallic copper was able to continuously limit the concentration of bacteria resident on frequently touched surfaces before and after routine cleaning (107).

Environmental monitoring of bed frames has shown that beds exceed the suggested bacterial burden more than any other object in patient's rooms (108, 109). The bacterial burden observed was consistently lower on copper surfaced bed rails before and after cleaning. Thus, the copper surfaces were able to augment cleaning by continuously limiting the concentration of aerobic bacteria associated with the rails of beds. This observation was consistently maintained despite the kinetic nature of care present in the environment of the ICU. Low-risk concentrations were associated with 77% of the sampled beds (107). Further, MRSA, VRE, and Gram-negative bacteria were absent from all of the copper samples, arguing that the risk mitigation provided by copper surfaces might be greater than the data suggest (107).

Weber and Rutala (110) in their commentary of the evaluation of no-touch copper conducted by Karpanen and colleagues argued that it was impractical or impossible to coat each of the environmental surfaces with copper (111). However, the data provided by Schmidt and colleagues suggest that the strategic placement of copper in key, high-touch areas offers a novel strategy to limit the bacterial burden on a continuous basis (91). The other no-touch methods for room disinfection rely on discontinuous modalities of application to reduce the environmental bacterial burden (84). Hydrogen peroxide vapor (HPV) is introduced as a gas into a sealed room. Ultraviolet light (UV) achieves its effectiveness through the transient transmission of germicidal radiation within an unoccupied room. Consequently, like the EPA registered disinfectants regularly used to disinfect patient rooms, subsequent to cleaning, both UV and HPV will likely suffer from the same limitations of the rapid restoration of the bacterial burden intrinsic to HTOs. In contrast, copper-alloyed surfaces offer a continuous way to limit and/or control the environmental burden. Hospital and environmental services need not perform additional steps, follow complex treatment algorithms, obtain "buy-in" from other providers, or require additional training or oversight.

In addressing the question of whether or not the strategic placement of copper might ameliorate the rate with which HAIs are acquired, Salgado and colleagues (112) found from the conduct of a multicenter trial that the limited placement of copper as described by Schmidt and colleagues (91) resulted in a significant reduction to the HAI rate and/or MRSA or VRE colonization rate in medical intensive care rooms (ICU). The collective rate for HAI infection or MRSA/VRE colonization was found to be significantly lower 42% (7.1%) in the copper arm of the study when compared against the (12.3%) rate observed in the control rooms ($p = 0.020$). When the data were considered separately for HAI alone, the rate of infection was significantly reduced (58%)

from 8.1% to 3.4% ($p = 0.013$). The investigators also were able to demonstrate that burden and infection were directly linked. In the analysis of the quartile distribution of HAIs stratified by microbial burden measured in the ICU rooms during the patient's stay they learned that there was a significant association between burden and HAI risk ($p = 0.038$), with 89% of HAIs occurring among patients cared for in a room with a burden of more than 500 colony-forming units (CFUs) (112).

The results of the effectiveness of antimicrobial copper surfaces being able to debulk the built environment have been confirmed in a number of subsequent trials (113–115). The first was a trial conducted in a pediatric ICU (115, 116). Here, copper surfaces were found to be equivalently antimicrobial in pediatric settings to activities observed in adult medical ICUs. In a companion trial to the one of Salgado and colleagues (112), copper surfaced bed rails were found to limit and continuously reduce the microbial burden to levels 99% lower than the equivalent control areas, leading the authors to conclude that copper surfaces warranted serious consideration when contemplating the introduction of no-touch disinfection technologies for reducing burden to limit acquisition of HAIs (113). In another trial, set in the medical-surgical suites for a 49-bed rural facility, 20 surfaces and components were outfitted with continuously active antimicrobial copper surfaces for 12 months (114). All of the components fabricated using copper alloys were found to have significantly lower concentrations of bacteria, at or below levels targeted upon completion of terminal cleaning (2.5 cfu/cm^2). Surprisingly vacant, control rooms were found to harbor significant concentrations of bacteria, whereas those fabricated from copper alloys were found to be at or below those concentrations prescribed subsequent to terminal cleaning, leading the authors to conclude that such an intervention using copper alloys can significantly decrease the burden resident on high-touch surfaces implicated in HAIs, and thus warrant inclusion as component for an integrated infection control strategy for rural hospitals (114).

Taken collectively, the data from the no-touch technologies support an argument that any method to minimize the patient's exposure to potentially infectious microbes, whether it be hand hygiene, routine cleaning, or the deployment of no-touch disinfection technologies will likely translate into better patient outcomes. Of the no-touch technologies, the demonstration that antimicrobial activity of copper surfaces was continuous in its effectiveness likely offers the best explanation as to its effectiveness at reducing the rate with which HAIs were acquired.

Hazard Analysis Critical Control Point and Risk Assessment Methods

Food-related industries have used Hazard Analysis Critical Control Point (HACCP) as a method to identify critical control points where lack of compliance or failure to meet standards is where foodborne illness or contamination of a manufactured product may occur.

The HACCP can be used to promote two positive outcomes. First, a well-designed HACCP can be used as a diagnostic tool to find points in a system where failure might occur. For example, if a certain floor disinfectant has a maximum disinfectant activity at 80°F, a critical control point on the HACCP would be that the temperature was measured and remained at that temperature throughout use. The second area of use is in education, where those using the disinfectant could be asked during regular and routine surveys if they were measuring and using the disinfectant at that temperature. Meeting compliance standards by writing elegant in-house documents of required methods, and actually monitoring that those standards are maintained by all staff regardless of station, are two different goals—both of which have equal importance.

The HACCP is a qualitative rather than a quantitative tool. Larson and Aiello (117) list the seven steps of HACCP as follows:

1. Analyze the hazards.
2. Identify the critical control points.
3. Establish preventative measures with critical limits for each control point.
4. Establish procedures to monitor the critical control points.
5. Establish corrective actions to be taken when monitoring shows that a critical limit has not been met.
6. Establish procedures to verify the system is working properly.
7. Establish effective record keeping for documentation.

When more quantitative measures are required, a microbial risk assessment model may be the method of choice. Larson and Aiello (117) use four categories of environmental site-associated contamination risks and need for contamination based on the work of Covello and Merkhoffer (118). These include:

1. Reservoirs (e.g., wet sites: humidifiers, ventilators, sinks) with a high probability (80–100%) of significant contamination and an occasional risk of contamination
2. Reservoir/disseminators (e.g., mops, sponges, cleaning materials) with a medium risk (24–40%) of significant contamination and a constant risk of contamination transfer
3. Hand and food contact surfaces (e.g., chopping boards, kitchen surfaces, cutlery, cooking utensils) with a medium risk (24–40%) of significant contamination and a constant risk of contamination transfer
4. Other sites (e.g., environmental surfaces, curtains, floors) with a low risk (3–40%) of significant contamination and an occasional risk of contamination transfer

Covello and Merkhoffer (118) use six criteria in the evaluation of a risk-assessment-based framework: logical soundness, completeness, accuracy, acceptability, practicality, and effectiveness.

Whether a formal HACCP or risk-assessment or a site-specific method/process-specific model developed by the operator is used, a formal method of assessment of structure, appropriate use, compliance, education, and record-keeping is needed. The methods must be kept current with the practices of the laboratory, hospital, or manufacturing entity, and also with the evolution of resistance in microbes and introduction of new microbes into the environment.

The tables in this chapter represent a selective starting point for choosing a method of sterilization or disinfection. Many of the cited articles contain in-depth information and should be consulted for further details, as there are space limitations herein. New methods are also being developed daily.

REFERENCES

1. Rutala WA, Weber DJ. Uses of inorganic hypochlorite (bleach) in health-care facilities. Clinical Microbiology Reviews. 1997;10(4):597–610.
2. McDonnell G, Russell AD. Antiseptics and disinfectants: activity, action, and resistance. Clinical Microbiology Reviews. 1999;12(1):147–79.
3. Bond WW, Ott BJ, Franke K, McCracken JE. Effective use of liquid chemical germicides on medical devices, instrument design problems. In: Block SS, editor. Disinfection, Sterilization and Preservation. 4th ed. Philadelphia, PA: Lea and Febiger; 1991.
4. Lemieux P, Sieber R, Osborne A, Woodard A. Destruction of spores on building decontamination residue in a commercial autoclave. Applied and Environmental Microbiology. 2006;72(12):7687–93. doi: 10.1128/AEM.02563-05.
5. Carter J. Evaluation of recovery filters for use in bacterial retention testing of sterilizing-grade filters. PDA Journal of Pharmaceutical Science and Technology/PDA. 1996;50(3):147–53.
6. Griffiths MH, Andrew PW, Ball PR, Hall GM. Rapid methods for testing the efficacy of sterilization-grade filter membranes. Applied and Environmental Microbiology. 2000;66(8):3432–7.
7. Stanbridge EJ. Mycoplasma detection-an obligation to scientific accuracy. Israel Journal of Medical Sciences. 1981;17(7):563–8.
8. Lenntech-Millipore. Aerex 2 hydrophobic cartridge filters: high performance gas filters for use in industrial fermentation applications. [Electronic]. Web: EMD-Millipore; 2001 [cited November 12, 2013]. Available from: https://www.lenntech.com/Data-sheets/Aerex-2-Hydrophobic-Cartridge-Filters-L.pdf.
9. Healthcare G. NFF integrity testing united states: GE Healthcare; 2013 [updated and cited October 30, 2013]. Available from: https://www.gelifesciences.com/gehcls_images/GELS/Related%20Content/Files/1314787424814/litdoc28913776_20140611213958.pdf.
10. ASTM-E1153-14. Standard test method for efficacy of sanitizers recommended for inanimate, hard, nonporous non-food contact surfaces. West Conshohocken, PA: ASTM; 2014.
11. Dolatowski ZJ, Stadnik J, Stasiak D. Applications of ultrasound in food technology. Acta Scientiarum Polonorum, Technologia Alimentaria. 2007;6(3):89–99.
12. Drakopoulou S, Terzakis S, Fountoulakis MS, Mantzavinos D, Manios T. Ultrasound-induced inactivation of gram-negative and gram-positive bacteria in secondary treated municipal wastewater. Ultrasonics Sonochemistry. 2009;16(5):629–34. doi: 10.1016/j.ultsonch.2008.11.011.
13. Mason TJ, Paniwnyk L, Lorimer JP. The uses of ultrasound in food technology. Ultrasonics Sonochemistry. 1996;3:s253–60.
14. Lee H, Zhou B, Liang W, Feng H, Martin SE. Inactivation of *Escherichia coli* cells with sonication, manosonication, thermosonication, and manothermosonication: microbial responses and kinetics modeling. Journal of Food Engineering. 2009;93(3):354–64. doi: http://dx.doi.org/10.1016/j.jfoodeng.2009.01.037.

15. Centers for Disease C, Prevention. Appendix A: regulatory framework for disinfectants and sterilants. MMWR Morbidity and Mortality Weekly Report. 2003;52(50): 62–4.

16. Vitali M, Agolini G. Prevention of infection spreading by cleaning and disinfecting: different approaches and difficulties in communicating. American Journal of Infection Control. 2006;34(1):49–50. doi: 10.1016/j.ajic. 2005.05.026.

17. Vieira CD, Farias Lde M, Diniz CG, Alvarez-Leite ME, Camargo ER, Carvalho MA. New methods in the evaluation of chemical disinfectants used in health care services. American Journal of Infection Control. 2005;33(3):162–9. doi: 10.1016/j.ajic.2004.10.007.

18. Marion K, Freney J, James G, Bergeron E, Renaud FN, Costerton JW. Using an efficient biofilm detaching agent: an essential step for the improvement of endoscope reprocessing protocols. The Journal of Hospital Infection. 2006;64(2): 136–42. doi: 10.1016/j.jhin.2006.06.011.

19. Rutala WA, Weber DJ. New disinfection and sterilization methods. Emerging Infectious Diseases. 2001;7(2):348–53. doi: 10.3201/eid0702.700348.

20. Schmidt MG, Fairey SE, Attaway HH. In situ evaluation of a persistent disinfectant provides continuous decontamination within the clinical environment. American Journal of Infection Control. 2019;47(6):732–4. doi: 10.1016/j.ajic.2019.02.013.

21. Xu HS, Roberts N, Singleton FL, Attwell RW, Grimes DJ, Colwell RR. Survival and viability of nonculturable *Escherichia coli* and *Vibrio cholerae* in the estuarine and marine environment. Microbial Ecology. 1982;8(4):313–23. doi: 10.1007/BF02010671.

22. Dinu LD, Bach S. Induction of viable but nonculturable *Escherichia coli* O157:H7 in the phyllosphere of lettuce: a food safety risk factor. Applied and Environmental Microbiology. 2011;77(23):8295–302. doi: 10.1128/AEM.05020-11.

23. Highmore CJ, Warner JC, Rothwell SD, Wilks SA, Keevil CW. Viable-but-nonculturable *Listeria monocytogenes* and *Salmonella enterica* serovar thompson induced by chlorine stress remain infectious. mBio. 2018;9(2). doi: 10.1128/mBio.00540-18.

24. Lin H, Ye C, Chen S, Zhang S, Yu X. Viable but non-culturable *E. coli* induced by low level chlorination have higher persistence to antibiotics than their culturable counterparts. Environmental Pollution. 2017;230:242–9. doi: 10.1016/j.envpol.2017.06.047.

25. Pasquaroli S, Zandri G, Vignaroli C, Vuotto C, Donelli G, Biavasco F. Antibiotic pressure can induce the viable but nonculturable state in *Staphylococcus aureus* growing in biofilms. Journal of Antimicrobial Chemotherapy. 2013;68(8):1812–7. doi: 10.1093/jac/dkt086.

26. Alleron L, Khemiri A, Koubar M, Lacombe C, Coquet L, Cosette P, Jouenne T, Frere J. VBNC Legionella pneumophila cells are still able to produce virulence proteins. Water Research. 2013;47(17):6606–17. doi: 10.1016/j.watres.2013.08.032.

27. Dietersdorfer E, Kirschner A, Schrammel B, Ohradanova-Repic A, Stockinger H, Sommer R, Walochnik J, Cervero-Arago S. Starved viable but non-culturable (VBNC) Legionella strains can infect and replicate in amoebae and human macrophages. Water Research. 2018;141:428–38. doi: 10.1016/j.watres.2018.01.058.

28. Robben C, Fister S, Witte AK, Schoder D, Rossmanith P, Mester P. Induction of the viable but non-culturable state in bacterial pathogens by household cleaners and inorganic salts. Scientific Reports. 2018;8(1):15132. doi: 10.1038/s41598-018-33595-5.

29. Zhang S, Guo L, Yang K, Zhang Y, Ye C, Chen S, Yu X, Huang WE, Cui L. Induction of *Escherichia coli* into a VBNC state by continuous-flow UVC and subsequent changes in metabolic activity at the single-cell level. Frontiers in Microbiology. 2018;9:2243. doi: 10.3389/fmicb.2018.02243.

30. Zhang S, Ye C, Lin H, Lv L, Yu X. UV disinfection induces a VBNC state in *Escherichia coli* and *Pseudomonas aeruginosa*. Environmental Science and Technology. 2015;49(3):1721–8. doi: 10.1021/es505211e.

31. Cook KL, Bolster CH. Survival of *Campylobacter jejuni* and *Escherichia coli* in groundwater during prolonged starvation at low temperatures. Journal of Applied Microbiology. 2007;103(3):573–83. doi: 10.1111/j.1365-2672.2006.03285.x.

32. Besnard V, Federighi M, Declerq E, Jugiau F, Cappelier JM. Environmental and physico-chemical factors induce VBNC state in Listeria monocytogenes. Veterinary Research. 2002;33(4):359–70. doi: 10.1051/vetres:2002022.

33. Maalej S, Denis M, Dukan S. Temperature and growth-phase effects on Aeromonas hydrophila survival in natural seawater microcosms: role of protein synthesis and nucleic acid content on viable but temporarily nonculturable response. Microbiology. 2004;150(Pt 1):181–7. doi: 10.1099/mic.0.26639-0.

34. Wong HC, Wang P. Induction of viable but nonculturable state in Vibrio parahaemolyticus and its susceptibility to environmental stresses. Journal of Applied Microbiology. 2004;96(2):359–66.

35. Asakura H, Kawamoto K, Haishima Y, Igimi S, Yamamoto S, Makino S. Differential expression of the outer membrane protein W (OmpW) stress response in enterohemorrhagic Escherichia coli O157:H7 corresponds to the viable but nonculturable state. Research in Microbiology. 2008;159(9–10): 709–17. doi: 10.1016/j.resmic.2008.08.005.

36. Wong HC, Liu SH. Characterization of the low-salinity stress in Vibrio vulnificus. Journal of Food Protection. 2008;71(2):416–9.

37. Oh E, McMullen L, Jeon B. Impact of oxidative stress defense on bacterial survival and morphological change in Campylobacter jejuni under aerobic conditions. Frontiers in Microbiology. 2015;6:295. doi: 10.3389/fmicb.2015.00295.

38. Ogane H, Sato TA, Shinokawa C, Sawai J. Low-concentration sorbic acid promotes the induction of *Escherichia coli* into a viable but nonculturable state. Biocontrol Science. 2019;24(1):67–71. doi: 10.4265/bio.24.67.

39. Dopp E, Richard J, Dwidjosiswojo Z, Simon A, Wingender J. Influence of the copper-induced viable but non-culturable state on the toxicity of *Pseudomonas aeruginosa* towards human bronchial epithelial cells in vitro. International Journal of Hygiene and Environmental Health. 2017;220(8):1363–9. doi: 10.1016/j.ijheh.2017.09.007.

40. Liao H, Jiang L, Zhang R. Induction of a viable but nonculturable state in *Salmonella typhimurium* by thermosonication and factors affecting resuscitation. FEMS Microbiology Letters. 2018;365(2). doi: 10.1093/femsle/fnx249.

41. Gourmelon M, Cillard J, Pommepuy M. Visible light damage to *Escherichia coli* in seawater: oxidative stress hypothesis. Journal of Applied Bacteriology. 1994;77(1):105–12.

42. Schottroff F, Frohling A, Zunabovic-Pichler M, Krottenthaler A, Schluter O, Jager H. Sublethal injury and viable but non-culturable (VBNC) state in microorganisms during preservation of food and biological materials by non-thermal processes. Frontiers in Microbiology. 2018;9:2773. doi: 10.3389/fmicb.2018.02773.

43. Rutala WA, Weber DJ, Society for healthcare epidemiology of America. Guideline for disinfection and sterilization of prion-contaminated medical instruments. Infection Control and Hospital Epidemiology. 2010;31(2):107–17. doi: 10.1086/650197.

44. Centers for Disease C, Prevention. Surveillance for Creutzfeldt-Jakob disease–United States. MMWR Morbidity and Mortality Weekly Report. 1996;45(31):665–8.

45. Johnson RT, Gibbs CJ, Jr. Creutzfeldt-Jakob disease and related transmissible spongiform encephalopathies. The New England Journal of Medicine. 1998;339(27):1994–2004. doi: 10.1056/NEJM199812313392707.

46. Collins SJ, Lawson VA, Masters CL. Transmissible spongiform encephalopathies. Lancet. 2004;363(9402):51–61. doi: 10.1016/S0140-6736(03)15171-9.

47. Prusiner SB. Prions. Proceedings of the National Academy of Sciences of the United States of America. 1998;95(23):13363–83.

48. Spaulding EH. Chemical disinfection of medical and surgical materials. In: Lawrence C, Block SS, editors. Disinfection, Sterilization, and Preservation. Philadelphia, PA: Lea & Febiger; 1968. pp. 517–31.

49. Yan ZX, Stitz L, Heeg P, Pfaff E, Roth K. Infectivity of prion protein bound to stainless steel wires: a model for testing decontamination procedures for transmissible spongiform encephalopathies. Infection Control and Hospital Epidemiology. 2004;25(4):280–3. doi: 10.1086/502392.

50. Yan ZX, Stitz L, Heeg P, Roth K, Mauz P-S. Low-temperature inactivation of prion-protein on surgical steel surfaces with hydrogen peroxide gas plasma sterilization. Zentralbl Steril. 2008;16:26–34.

51. Jackson GS, McKintosh E, Flechsig E, Prodromidou K, Hirsch P, Linehan J, Brandner S, Clarke AR, Weissmann C, Collinge J. An enzyme-detergent method for effective prion decontamination of surgical steel. The Journal of General Virology. 2005;86(Pt 3):869–78. doi: 10.1099/vir.0.80484-0.

52. Taylor DM, Fernie K, McConnell I. Inactivation of the 22A strain of scrapie agent by autoclaving in sodium hydroxide. Veterinary Microbiology. 1997;58(2-4):87–91.

53. Ernst DR, Race RE. Comparative analysis of scrapie agent inactivation methods. Journal of Virological Methods. 1993;41(2):193–201.

54. Ding N, Neumann NF, Price LM, Braithwaite SL, Balachandran A, Mitchell G, Belosevic M, Gamal El-Din M. Kinetics of ozone inactivation of infectious prion protein. Applied and Environmental Microbiology. 2013;79(8):2721–30. doi: 10.1128/AEM.03698-12.

55. Agency USEP. Alternative disinfectants and oxidants guidance manual. In: Office of Water publication 4607. Washington, DC: United States Environmental Protection Agency; 1999. p. 328.

56. Ding N, Neumann NF, Price LM, Braithwaite SL, Balachandran A, Belosevic M, El-Din MG. Inactivation of template-directed misfolding of infectious prion protein by ozone. Applied and Environmental Microbiology. 2012;78(3):613–20. doi: 10.1128/AEM.06791-11.

57. Hughson AG, Race B, Kraus A, Sangare LR, Robins L, Groveman BR, Saijo E, Phillips K, Contreras L, Dhaliwal V, Manca M, Zanusso G, Terry D, Williams JF, Caughey B. Inactivation of prions and amyloid seeds with hypochlorous acid. PLoS Pathogens. 2016;12(9):e1005914. doi: 10.1371/journal.ppat.1005914.

58. Cheng K, Vendramelli R, Sloan A, Waitt B, Podhorodecki L, Godal D, Knox JD. Endpoint quaking-induced conversion: a sensitive, specific, and high-throughput method for antemortem diagnosis of creutzfeldt-jacob disease. Journal of Clinical Microbiology. 2016;54(7):1751–4. doi: 10.1128/JCM.00542-16.

59. Caughey B, Orru CD, Groveman BR, Hughson AG, Manca M, Raymond LD, Raymond GJ, Race B, Saijo E, Kraus A. Amplified detection of prions and other amyloids by RT-QuIC in diagnostics and the evaluation of therapeutics and disinfectants. Progress in Molecular Biology and Translational Science. 2017;150:375–88. doi: 10.1016/bs.pmbts.2017.06.003.

60. Sutton JM, Dickinson J, Walker JT, Raven ND. Methods to minimize the risks of Creutzfeldt-Jakob disease transmission by surgical procedures: where to set the standard? Clinical Infectious Diseases. 2006;43(6):757–64. doi: 10.1086/507030.

61. Walker JT, Dickinson J, Sutton JM, Marsh PD, Raven ND. Implications for Creutzfeldt-Jakob disease (CJD) in dentistry: a review of current knowledge. Journal of Dental Research. 2008;87(6):511–9.

62. Widmer AF. Replace hand washing with use of a waterless alcohol hand rub?. Clinical Infectious Diseases. 2000;31(1):136–43. doi: 10.1086/313888.

63. Kampf G, Kramer A, Rotter M, Widmer A. Optimizing surgical hand disinfection. Zentralblatt fur Chirurgie. 2006;131(4):322–6. doi: 10.1055/s-2006-947277.

64. Gupta C, Czubatyj AM, Briski LE, Malani AK. Comparison of two alcohol-based surgical scrub solutions with an iodine-based scrub brush for presurgical antiseptic effectiveness in a community hospital. Journal of Hospital Infection. 2007;65(1):65–71. doi: http://dx.doi.org/10.1016/j.jhin.2006.06.026.

65. Cooper BS, Medley GF, Scott GM. Preliminary analysis of the transmission dynamics of nosocomial infections: stochastic and management effects. Journal of Hospital Infection. 1999;43(2):131–47. doi: http://dx.doi.org/10.1053/jhin.1998.0647.

66. Raboud J, Saskin R, Simor A, Loeb M, Green K, Low DE, McGeer A. Modeling transmission of methicillin-resistant *Staphylococcus aureus* among patients admitted to a hospital. Infection Control and Hospital Epidemiology. 2005;26(7):607–15. doi: 10.1086/502589.

67. Boyce JM, Pittet D, Healthcare Infection Control Practices Advisory C, HICPAC/SHEA/APIC/IDSA task force. Guideline for hand hygiene in health-care settings. Recommendations of the healthcare infection control practices advisory committee and the HICPAC/SHEA/APIC/IDSA hand hygiene task force. Society for Healthcare Epidemiology of America/Association for Professionals in Infection Control/Infectious Diseases Society of America. MMWR Recommendations and Reports. 2002;51(RR-16):1–45.

68. McGuckin M, Waterman R, Govednik J. Hand hygiene compliance rates in the United States–a one-year multicenter collaboration using product/volume usage measurement and feedback. American Journal of Medical Quality. 2009;24(3):205–13. doi: 10.1177/1062860609332369.

69. Pidot SJ, Gao W, Buultjens AH, Monk IR, Guerillot R, Carter GP, Lee JYH, Lam MMC, Grayson ML, Ballard SA, Mahony AA, Grabsch EA, Kotsanas D, Korman TM, Coombs GW, Robinson JO, Goncalves da Silva A, Seemann T, Howden BP, Johnson PDR, Stinear TP. Increasing tolerance of hospital *Enterococcus faecium* to handwash alcohols. Science Translational Medicine. 2018;10(452). doi: 10.1126/scitranslmed.aar6115.

70. Bhalla A, Pultz NJ, Gries DM, Ray AJ, Eckstein EC, Aron DC, Donskey CJ. Acquisition of nosocomial pathogens on hands after contact with environmental surfaces near hospitalized patients. Infection Control and Hospital Epidemiology. 2004;25(2):164–7. doi: 10.1086/502369.

71. Boyce JM. Environmental contamination makes an important contribution to hospital infection. The Journal of Hospital Infection. 2007;65(Suppl 2):50–4. doi: 10.1016/S0195-6701(07)60015-2.

72. Boyce JM. Vancomycin-resistant enterococcus. Detection, epidemiology, and control measures. Infectious Disease Clinics of North America. 1997;11(2):367–84.

73. Boyce JM, Pittet D. Guideline for hand hygiene in health-care settings: recommendations of the healthcare infection control practices advisory committee and the HICPAC/SHEA/APIC/IDSA hand hygiene task force. Infection Control and Hospital Epidemiology. 2002;23(12 Suppl):S3–40. doi: 10.1086/503164.

74. Boyce JM, Potter-Bynoe G, Chenevert C, King T. Environmental contamination due to methicillin-resistant Staphylococcus aureus: possible infection control implications. Infection Control and Hospital Epidemiology. 1997;18(9):622–7.

75. Sehulster L, Chinn RY. Guidelines for environmental infection control in health-care facilities. Recommendations of CDC and the Healthcare Infection Control Practices Advisory Committee (HICPAC). MMWR Recommendations and Reports: Morbidity and Mortality Weekly Report Recommendations and reports/Centers for Disease Control. 2003;52(RR-10):1–42.

76. Favero MS, Bond WW. Chemical disinfection of medical and surgical materials. In: Block SS, editor. Disinfection, Sterilization, and Preservation. Philadelphia, PA: Lippincott Williams & Wilkins; 2001. pp. 881–917.

77. Ray AJ, Hoyen CK, Taub TF, Eckstein EC, Donskey CJ. Nosocomial transmission of vancomycin-resistant enterococci from surfaces. JAMA. 2002;287(11):1400–1.

78. Sattar SA, Jacobsen H, Springthorpe VS, Cusack TM, Rubino JR. Chemical disinfection to interrupt transfer of rhinovirus type 14 from environmental surfaces to hands. Applied and Environmental Microbiology. 1993;59(5):1579–85.

79. Sattar SA, Lloyd-Evans N, Springthorpe VS, Nair RC. Institutional outbreaks of rotavirus diarrhoea: potential role of fomites and environmental surfaces as vehicles for virus transmission. The Journal of Hygiene. 1986;96(2):277–89.

80. Ward RL, Bernstein DI, Knowlton DR, Sherwood JR, Young EC, Cusack TM, Rubino JR, Schiff GM. Prevention of surface-to-human transmission of rotaviruses by treatment with disinfectant spray. Journal of Clinical Microbiology. 1991;29(9):1991–6.

81. Rutala WA, Weber DJ, (HICPAC) HICPAC. Guideline for Disinfection and Sterilization in Healthcare Facilities, 2008. Healthcare Infection Control Practices Advisory Committee (HICPAC) [Internet]. 2008. Available from: https://www.cdc.gov/infectioncontrol/pdf/guidelines/disinfection-guidelines-H.pdf.

82. Dancer SJ. How do we assess hospital cleaning? A proposal for microbiological standards for surface hygiene in hospitals. The Journal of Hospital Infection. 2004;56(1):10–5. doi: S0195670103003955 [pii].

83. Malik RE, Cooper RA, Griffith CJ. Use of audit tools to evaluate the efficacy of cleaning systems in hospitals. American Journal of Infection Control. 2003;31(3):181–7.

84. Havill NL, Moore BA, Boyce JM. Comparison of the microbiological efficacy of hydrogen peroxide vapor and ultraviolet light processes for room decontamination. Infection Control and Hospital Epidemiology. 2012;33(5):507–12. doi: 10.1086/665326.

85. Raggi R, Archulet K, Haag CW, Tang W. Clinical, operational, and financial impact of an ultraviolet-C terminal disinfection intervention at a community hospital. American Journal of Infection Control. 2018;46(11):1224–9. doi: 10.1016/j.ajic.2018.05.012.

86. de Kraker MEA, Harbarth S, Dancer SJ. Shining a light on ultraviolet-C disinfection: no golden promises for infection prevention. American Journal of Infection Control. 2018;46(12):1422–3. doi: 10.1016/j.ajic.2018.07.026.

87. Anderson DJ, Gergen MF, Smathers E, Sexton DJ, Chen LF, Weber DJ, Rutala WA. Decontamination of targeted pathogens from patient rooms using an automated ultraviolet-C-emitting device. Infection Control and Hospital Epidemiology. 2013;34(5):466–71. doi: 10.1086/670215.

88. Anderson DJ, Chen LF, Weber DJ, Moehring RW, Lewis SS, Triplett PF, Blocker M, Becerer P, Schwab JC, Knelson LP, Lokhnygina Y, Rutala WA, Kanamori H, Gergen MF, Sexton DJ, Program CDCPE. Enhanced terminal room disinfection and acquisition and infection caused by multidrug-resistant organisms and clostridium difficile (the benefits of enhanced terminal room disinfection study): a cluster-randomised, multicentre, crossover study. Lancet. 2017;389(10071):805–14. doi: 10.1016/S0140-6736(16)31588-4.

89. Rutala WA, Kanamori H, Gergen MF, Knelson LP, Sickbert-Bennett EE, Chen LF, Anderson DJ, Sexton DJ, Weber DJ, the CDCPEP. Enhanced disinfection leads to reduction of microbial contamination and a decrease in patient colonization and infection. Infection Control and Hospital Epidemiology: The Official Journal of the Society of Hospital Epidemiologists of America. 2018;39(9):1118–21. doi: 10.1017/ice.2018.165.

90. Ghantoji SS, Stibich M, Stachowiak J, Cantu S, Adachi JA, Raad, II, Chemaly RF. Non-inferiority of pulsed xenon UV light versus bleach for reducing environmental *Clostridium difficile* contamination on high-touch surfaces in *Clostridium difficile* infection isolation rooms. Journal of Medical Microbiology. 2015;64(Pt 2):191–4. doi: 10.1099/jmm.0.000004-0.

91. Schmidt MG, Attaway HH, Sharpe PA, John J, Jr., Sepkowitz KA, Morgan A, Fairey SE, Singh S, Steed LL, Cantey JR, Freeman KD, Michels HT, Salgado CD. Sustained reduction of microbial burden on common hospital surfaces through introduction of copper. Journal of Clinical Microbiology. 2012;50(7):2217–23. doi: 10.1128/JCM.01032-12.

92. Albright LJ, Wilson EM. Sub-lethal effects of several metallic salts organic compound combinations upon the heterotrophic microflora of a natural water. Water Research. 1974;8:101–5.

93. Jonas RB. Acute copper and cupric ion toxicity in an estuarine microbial community. Applied and Environmental Microbiology. 1989;55(1):43–9.

94. Grass G, Rensing C, Solioz M. Metallic copper as an antimicrobial surface. Applied and Environmental Microbiology. 2011;77(5):1541–7. doi: 10.1128/AEM.02766-10.

95. Noyce JO, Michels H, Keevil CW. Potential use of copper surfaces to reduce survival of epidemic meticillin-resistant Staphylococcus aureus in the healthcare environment. The Journal of Hospital Infection. 2006;63(3):289–97. doi: 10.1016/j.jhin.2005.12.008.

96. Weaver L, Michels HT, Keevil CW. Potential for preventing spread of fungi in air-conditioning systems constructed using copper instead of aluminium. Letters in Applied Microbiology. 2010;50(1):18–23. doi: 10.1111/j.1472-765X.2009.02753.x.

97. Wheeldon LJ, Worthington T, Lambert PA, Hilton AC, Lowden CJ, Elliott TS. Antimicrobial efficacy of copper surfaces against spores and vegetative cells of Clostridium difficile: the germination theory. Journal of Antimicrobial Chemotherapy. 2008;62(3):522–5. doi: 10.1093/jac/dkn219.

98. Wilks SA, Michels H, Keevil CW. The survival of Escherichia coli O157 on a range of metal surfaces. International Journal of Food Microbiology. 2005;105(3):445–54. doi: 10.1016/j.ijfoodmicro.2005.04.021.

99. United States Environmental Protection Agency. EPA registers copper-containing alloy products. http://wwwepagov/opp00001/factsheets/copper-alloy-productshtm. 2008.

100. Quaranta D, Krans T, Espirito Santo C, Elowsky CG, Domaille DW, Chang CJ, Grass G. Mechanisms of contact-mediated killing of yeast cells on dry metallic copper surfaces. Applied Environmental Microbiology. 2010;77(2):416–26. doi: AEM.01704-10 [pii]10.1128/AEM.01704-10.

101. Warnes SL, Green SM, Michels HT, Keevil CW. Biocidal efficacy of copper alloys against pathogenic enterococci involves degradation of genomic and plasmid DNAs. Applied and Environmental Microbiology. 2010;76(16):5390–401. doi: 10.1128/AEM.03050-09.

102. Weaver L, Noyce JO, Michels HT, Keevil CW. Potential action of copper surfaces on meticillin-resistant *Staphylococcus aureus*. Journal of Applied Microbiology. 2010;109(6): 2200–5. doi: 10.1111/j.1365-2672.2010.04852.x.

103. Mikolay A, Huggett S, Tikana L, Grass G, Braun J, Nies DH. Survival of bacteria on metallic copper surfaces in a hospital trial. Applied Microbiology and Biotechnology. 2010;87(5):1875–9. doi: 10.1007/s00253-010-2640-1.

104. Casey AL, Adams D, Karpanen TJ, Lambert PA, Cookson BD, Nightingale P, Miruszenko L, Shillam R, Christian P, Elliott TS. Role of copper in reducing hospital environment contamination. The Journal of Hospital Infection. 2010;74(1):72–7. doi: 10.1016/j.jhin.2009.08.018.

105. Attaway HH, 3rd, Fairey S, Steed LL, Salgado CD, Michels HT, Schmidt MG. Intrinsic bacterial burden associated with intensive care unit hospital beds: effects of disinfection on population recovery and mitigation of potential infection risk. American Journal of Infection Control. 2012;40(10):907–12. doi: 10.1016/j.ajic.2011.11.019.

106. Carling PC, Bartley JM. Evaluating hygienic cleaning in health care settings: what you do not know can harm your patients. American Journal of Infection Control. 2010; 38(5 Suppl 1):S41–50. doi: 10.1016/j.ajic.2010.03.004.

107. Schmidt MG, Attaway HH, Fairey SE, Steed LL, Michels HT, Salgado CD. Copper continuously limits the concentration of bacteria resident on bed rails within the ICU. Infection Control and Hospital Epidemiology. 2013;34(5).

108. Mulvey D, Redding P, Robertson C, Woodall C, Kingsmore P, Bedwell D, Dancer SJ. Finding a benchmark for monitoring hospital cleanliness. The Journal of Hospital Infection. 2011;77(1):25–30. doi: 10.1016/j.jhin.2010.08.006.

109. White LF, Dancer SJ, Robertson C, McDonald J. Are hygiene standards useful in assessing infection risk?. American Journal of Infection Control. 2008;36(5):381–4. doi: 10.1016/j.ajic.2007.10.015.

110. Weber DJ, Rutala WA. Commentary: self-disinfecting surfaces. Infection Control and Hospital Epidemiology. 2012;33(1):10–3.

111. Karpanen TJ, Casey AL, Lambert PA, Cookson BD, Nightingale P, Miruszenko L, Elliott TS. The antimicrobial efficacy of copper alloy furnishing in the clinical environment: a crossover study. Infection Control and Hospital Epidemiology. 2012;33(1):3–9. doi: 10.1086/663644.

112. Salgado CD, Sepkowitz KA, John JF, Cantey JR, Attaway HH, Freeman KD, Sharpe PA, Michels HT, Schmidt MG. Copper surfaces reduce the rate of healthcare-acquired infections in the intensive care unit. Infection Control and Hospital Epidemiology. 2013;34(5):479–86. doi: 10.1086/670207.

113. Schmidt MG, Attaway Iii HH, Fairey SE, Steed LL, Michels HT, Salgado CD. Copper continuously limits the concentration of bacteria resident on bed rails within the intensive care unit. Infection Control and Hospital Epidemiology. 2013;34(5):530–3. doi: 10.1086/670224.

114. Hinsa-Leasure SM, Nartey Q, Vaverka J, Schmidt MG. Copper alloy surfaces sustain terminal cleaning levels in a rural hospital. American Journal of Infection Control. 2016;44(11): e195–e203. doi: 10.1016/j.ajic.2016.06.033.

115. Schmidt MG, von Dessauer B, Benavente C, Benadof D, Cifuentes P, Elgueta A, Duran C, Navarrete MS. Copper surfaces are associated with significantly lower concentrations of bacteria on selected surfaces within a pediatric intensive care unit. American Journal of Infection Control. 2016;44(2): 203–9. doi: 10.1016/j.ajic.2015.09.008.

116. von Dessauer B, Navarrete MS, Benadof D, Benavente C, Schmidt MG. Potential effectiveness of copper surfaces in reducing health care-associated infection rates in a pediatric intensive and intermediate care unit: a nonrandomized controlled trial. American Journal of Infection Control. 2016;44(8):e133–9. doi: 10.1016/j.ajic.2016.03.053.

117. Larson E, Aiello AE. Systematic risk assessment methods for the infection control professional. American Journal of Infection Control. 2006;34(5):323–6. doi: http://dx.doi.org/10.1016/j.ajic.2005.10.009.

118. Covello V, Merkhoffer M. Risk Assessment Methods: Approaches for Assessing Health and Environmental Risks. New York, NY: Plenum Press; 1993.

119. Gaughran ERI, Borick PR. Sterilization, Disinfection, and Antisepsis. In: O'Leary W, editor. Practical Handbook of Microbiology. Boca Raton, FL: CRC Press; 1989. p. 297.

120. Helfinstine SL, Vargas-Aburto C, Uribe RM, Woolverton CJ. Inactivation of bacillus endospores in envelopes by electron beam irradiation. Applied and Environmental Microbiology. 2005;71(11):7029–32. doi: 10.1128/AEM.71.11.7029-7032.2005.

121. Finch GR, Black EK, Labatiuk CW, Gyurek L, Belosevic M. Comparison of Giardia lamblia and Giardia muris cyst inactivation by ozone. Applied and Environmental Microbiology. 1993;59(11):3674–80.

122. Shin GA, Sobsey MD. Reduction of norwalk virus, poliovirus 1, and bacteriophage MS2 by ozone disinfection of water. Applied and Environmental Microbiology. 2003;69(7):3975–8.

123. Piyasena P, Mohareb E, McKellar RC. Inactivation of microbes using ultrasound: a review. International Journal of Food Microbiology. 2003;87(3):207–16.

124. Chemat F, Zill e H, Khan MK. Applications of ultrasound in food technology: processing, preservation and extraction. Ultrasonics Sonochemistry. 2011;18(4):813–35. doi: http://dx.doi.org/10.1016/j.ultsonch.2010.11.023.

125. Rutala WA, Peacock JE, Gergen MF, Sobsey MD, Weber DJ. Efficacy of hospital germicides against adenovirus 8, a common cause of epidemic keratoconjunctivitis in health care facilities. Antimicrobial Agents and Chemotherapy. 2006;50(4):1419–24. doi: 10.1128/AAC.50.4.1419-1424.2006.

126. Rose LJ, Rice EW, Jensen B, Murga R, Peterson A, Donlan RM, Arduino MJ. Chlorine inactivation of bacterial bioterrorism agents. Applied and Environmental Microbiology. 2005;71(1):566–8. doi: 10.1128/AEM.71.1.566-568.2005.

127. Rice EW, Adcock NJ, Sivaganesan M, Rose LJ. Inactivation of spores of *Bacillus anthracis* sterne, bacillus cereus, and bacillus thuringiensis subsp. israelensis by chlorination. Applied and Environmental Microbiology. 2005;71(9):5587–9. doi: 10.1128/AEM.71.9.5587-5589.2005.

128. Jimenez L, Chiang M. Virucidal activity of a quaternary ammonium compound disinfectant against feline calicivirus: a surrogate for norovirus. American Journal of Infection Control. 2006;34(5):269–73. doi: 10.1016/j.ajic.2005.11.009.

129. Peretz D, Supattapone S, Giles K, Vergara J, Freyman Y, Lessard P, Safar JG, Glidden DV, McCulloch C, Nguyen HO, Scott M, Dearmond SJ, Prusiner SB. Inactivation of prions by acidic sodium dodecyl sulfate. Journal of Virology. 2006;80(1):322–31. doi: 10.1128/JVI.80.1.322-331.2006.

2

Quantitation of Microorganisms

Brad A. Slominski and Peter S. Lee

CONTENTS

Quantitative Microbial Enumeration

This chapter provides general information on the various methods used to estimate the number of microorganisms in a given sample. The scope of this chapter is limited to methods used for the enumeration of microbial cells and not the measurement of microbial cellular mass.

Most Probable Number (MPN) Method

General

The most probable number (MPN) method is a microbial estimate method used to enumerate viable cell counts by diluting the microorganisms, followed by growing the diluted microorganisms in replicate liquid medium dilution tubes. An example of a MPN dilution scheme used to prepare the various dilution

DOI: 10.1201/9781003099277-3

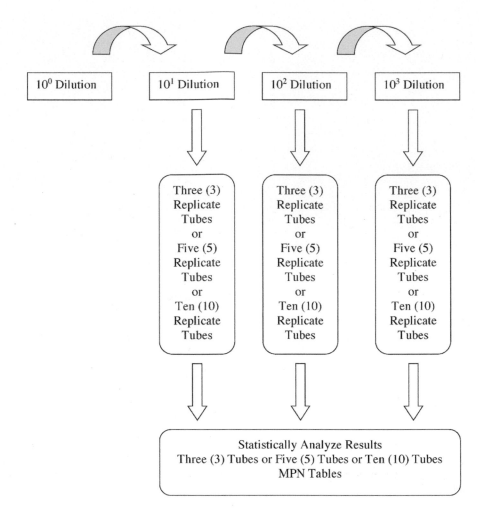

FIGURE 2.1 MPN serial dilution scheme.

test tube replicates is shown in Figure 2.1. After optimal microbial growth incubation, the positive and negative test results are based on the positive (visible turbidity) and negative (clear) replicate dilution tubes. The MPN method can also be referred to as the multiple tube fermentation method if fermentation tubes (Durham tubes) are also used inside the serial dilution tubes to measure gas production. The viable cell counts are then compared to a specific MPN statistical table to interpret the test results observed. There are also several versions of statistical tables used based on the sample matrix tested, test dilutions used, number of replicates per dilution used, and the statistical considerations used.[1–7] Table 2.1 and Table 2.2 provide two examples of MPN statistical tables for three tenfold dilutions with three or five tubes at each dilution, respectively.[8] The confidence limits of the MPN are narrowed when a higher number of replicate dilution tubes is inoculated in the dilution series. A number of assumptions are also considered when using the MPN method of microbial estimation. The microorganisms to be estimated are distributed randomly and evenly separated within the samples tested in the liquid dilution tubes. The microorganisms to be estimated are also separated individually and are not clustered together, nor do they repel each other in the liquid dilution tubes. In addition, the utilization of optimal growth medium, incubation temperature, and incubation period is needed to allow any

single viable cell to grow and become quantifiable in the liquid dilution tubes used. Primary equipment and materials used for this method are serial dilution tubes (may include use of Durham fermentation tubes), dilution replicate tubes, pipettes, specific growth medium/reagents, and incubator with appropriate optimal temperature setting. Sources of error using this method are improper or inadequate preparation of the test samples, serial dilution or dilution factor calculation error, and statistical interpretation error of observed results.

Applications

The MPN method is useful in the estimation of low microorganism counts where particulate matter or turbidity is present in the sample matrix, such as in milk, food, water, and soil. The MPN method is not particularly useful in the enumeration of fungi. It is mainly used in the enumeration of samples with low bacteria concentrations or bacteria that does not grow well on solid medium or use in the enumeration of total aerobic bacterial counts. MPN methods are used in the areas of food,[8–17] pharmaceutical,[18–21] environment,[22–24] water,[25–31] agriculture,[32,33] petroleum,[34–36] and poultry[37,38] applications. Modified or alternate versions of the MPN method have also been developed to further expand the use of the standard MPN method to estimate bacteria by utilizing

TABLE 2.1

Three Tubes Each at 0.1, 0.01, and 0.001 g Inocula and the MPNs per Gram with 95% Confidence Intervals[8]

Positive Tubes			MPN/g	Conf. Limit		Positive Tubes			MPN/g	Conf. Limit	
0.10	0.01	0.001		Low	High	0.10	0.01	0.001		Low	High
0	0	0	<3.0	—	9.5	2	2	0	21	4.5	42
0	0	1	3.0	0.15	9.6	2	2	1	28	8.7	94
0	1	0	3.0	0.15	11	2	2	2	35	8,7	94
0	1	1	6.1	1.2	18	2	3	0	29	8.7	94
0	2	0	6.2	1.2	18	2	3	1	36	8.7	94
0	3	0	9.4	3.6	38	3	0	0	23	4.6	94
1	0	0	3.6	0.17	18	3	0	1	38	8.7	110
1	0	1	7.2	1.3	18	3	0	2	64	17	180
1	0	2	11	3.6	38	3	1	0	43	9	180
1	1	0	7.4	1.3	20	3	1	1	75	17	200
1	1	1	11	3.6	38	3	1	2	120	37	420
1	2	0	11	3.6	42	3	1	3	160	40	420
1	2	1	15	4.5	42	3	2	0	93	18	420
1	3	0	16	4.5	42	3	2	1	150	37	420
2	0	0	9.2	1.4	38	3	2	2	210	40	430
2	0	1	14	3.6	42	3	2	3	290	90	1,000
2	0	2	20	4.5	42	3	3	0	240	42	1,000
2	1	0	15	3.7	42	3	3	1	460	90	2,000
2	1	1	20	4.5	42	3	3	2	1,100	180	4,100
2	1	2	27	8.7	94	3	3	3	>1,000	420	—

solid medium,[39–41] microtiter plates,[42–51] colorimetric reagents or defined substrate medium,[52–57] automated MPN,[58–61] and polymerase chain reaction.[62–66]

Turbidimetric Method

General

The turbidimetric method is used to measure the turbidity in a suitable liquid medium inoculated with a given microorganism.[67–71] Turbidimetric measurement is a function of microorganism cell growth. This method of microbial enumeration is based on the correlation between the turbidity observed and changes in the microbial cell numbers. Standard optical density (O.D.) or turbidimetric curves constructed may be used to estimate the number of microbial cells for the observed turbidity values using a turbidimetric instrument. This is achieved by determining the turbidity of different concentrations of a given species of microorganism in a particular liquid medium using the standard plate count method to determine the number of viable microorganisms per millimeter of sample tested. The turbidimetric calibration standard curve generated is subsequently used as a visual comparison to any given suspension of that microorganism. The use of this method assumes that there is control of the physiological state of the microorganisms. It is based on the species of microorganism and the optimal condition of the turbidimetric instrument used. Primary equipment and materials used for this method are the turbidimetric reading instrument, turbidimetric calibration standards, optically matched tubes or cuvettes, appropriate recording instruments, dilution tubes, pipettes, specific growth medium and reagents, and incubator with appropriate optimal temperature setting. The turbidimetric reading instrument can be, for example, be a spectrophotometer or turbidimeter (wavelength range of 420–615 nm). The typical turbidimetric standard used is the McFarland turbidimetric standard. These standards are used as a reference to adjust the turbidity of microbial suspensions to enable the number of microorganisms to be within a given quantitative range. These standards are made by mixing specified amounts of barium chloride and sulfuric acid to form a barium sulfate precipitate that causes turbidity in the solution mixture. If *Escherichia coli* is used, the 0.5 McFarland standard is equivalent to an approximate concentration of 1.5×10^8 colony forming units (or CFUs) per mL. The 0.5 McFarland standard is a commonly used standard, which is prepared by mixing 0.05 mL of 1% barium chloride dehydrate with 9.95 mL of 1% sulfuric acid. There are other turbidimetric standards that can be prepared from suspensions of latex particles that can extend the expiration use and stability of these suspensions.[72,73] Sources of error for this method include using a noncalibrated instrument, damaged or misaligned optical matched tubes or cuvettes, color or clarity of the liquid suspension medium used, condition of the filter or detector used, and quality of the turbidimetric instrument lamp output. Sources of error for this method can also be microbial related, such as clumping or settling of the microbial suspension, inadequate optimal growth condition used; or analyst related, such as lack of understanding the turbidimetric technique used, sample preparation error, and incorrect use of the turbidimetric standards.[74,75]

TABLE 2.2

Five Tubes Each at 0.1, 0.01, and 0.001 g Inocula and the MPNs per Gram with 95% Confidence Intervals[8]

Positive Tubes			MPN/g	Conf. Limit		Positive Tubes			MPN/g	Conf. Limit	
0.10	**0.01**	**0.001**		**Low**	**High**	**0.10**	**0.01**	**0.001**		**Low**	**High**
0	0	0	<1.8	—	6.8	4	0	2	21	6.8	40
0	0	1	1.8	0.09	6.8	4	0	3	25	9.8	70
0	1	0	1.8	0.09	6.8	4	1	0	17	6	40
0	1	1	3.6	0.7	10	4	1	1	21	6.8	42
0	2	0	3.7	0.7	10	4	1	2	26	9.8	70
0	2	1	5.5	1.8	15	4	1	3	31	10	70
0	3	0	5.6	1.8	15	4	2	0	22	6.8	50
1	0	0	2	0.1	10	4	2	1	26	9.8	70
1	0	1	4	0.7	10	4	2	2	32	10	70
1	0	2	6	1.8	15	4	2	3	38	14	100
1	1	0	4	0.7	12	4	3	0	27	9.9	70
1	1	1	6.1	1.8	15	4	3	1	33	10	70
1	1	2	8.1	3.4	22	4	3	2	39	14	100
1	2	0	6.1	1.8	15	4	4	0	34	14	100
1	2	1	8.2	3.4	22	4	4	1	40	14	100
1	3	0	8.3	3.4	22	4	4	2	47	15	120
1	3	1	10	3.5	22	4	5	0	41	14	100
1	4	0	11	3.5	22	4	5	1	48	15	120
2	0	0	4.5	0.79	15	5	0	0	23	6.8	70
2	0	1	6.8	1.8	15	5	0	1	31	10	70
2	0	2	9.1	3.4	22	5	0	2	43	14	100
2	1	0	6.8	1.8	17	5	0	3	58	22	150
2	1	1	9.2	3.4	22	5	1	0	33	10	100
2	1	2	12	4.1	26	5	1	1	46	14	120
2	2	0	9.3	3.4	22	5	1	2	63	22	150
2	2	1	12	4.1	26	5	1	3	84	34	220
2	2	2	14	5.9	36	5	2	0	49	15	150
2	3	0	12	4.1	26	5	2	1	70	22	170
2	3	1	14	5.9	36	5	2	2	94	34	230
2	4	0	15	5.9	36	5	2	3	120	36	250
3	0	0	7.8	2.1	11	5	2	4	150	58	400
3	0	1	11	3.5	23	5	3	0	79	22	220
3	0	2	13	5.6	35	5	3	1	110	34	250
3	1	0	11	3.5	26	5	3	2	140	52	400
3	1	1	14	5.6	36	5	3	3	180	70	400
3	1	2	17	6	36	5	3	4	210	70	400
3	2	0	14	5.7	36	5	4	0	130	36	400
3	2	1	17	6.8	40	5	4	1	170	58	400
3	2	2	20	6.8	40	5	4	2	220	70	440
3	3	0	17	6.8	40	5	4	3	280	100	710
3	3	1	21	6.8	40	5	4	4	350	100	710
3	3	2	24	9.8	70	5	4	5	430	150	1100
3	4	0	21	6.8	40	5	5	0	240	70	710
3	4	1	24	9.8	70	5	5	1	350	100	1100
3	5	0	25	9.8	70	5	5	2	540	150	1700
4	0	0	13	4.1	35	5	5	3	920	220	2600
4	0	1	17	5.9	36	5	5	4	1600	400	4600
						5	5	5	>1600	700	—

Another recent use of the turbidimetric technique is the use of the turbidimetric method in combination with the use of modern digital photography method.[76]

Applications

Turbidimetric methods are used where a large number of microorganisms is to be enumerated or when the inoculum size of a specific microbial suspension is to be determined. These areas include antimicrobial, preservative, or biocide susceptibility assays of bacteria and fungi,[77–91] growth of microorganisms,[92–96] effects of chemicals on microorganisms,[97,98] and control of biological production.[99]

Plate Count Method

General

The plate count method is based on viable cell counts. The plate count method is performed by diluting the original sample in serial dilution tubes, followed by the plating of aliquots of the prepared serial dilutions into appropriate plate count agar plates by the pour plate or spread plate technique. The pour-plate technique utilizes tempered molten plate count agar poured into the respective plate and mixed with the diluted aliquot sample in the plate, whereas the spread-plate technique utilizes the addition and spreading of the diluted aliquot sample on the surface of the preformed solid plate count agar in the respective plate.[100] An example of a plate count serial dilution scheme is shown in Figure 2.2. These prepared plate count agar plates are then optimally incubated and the colonies observed on these plate count agar plates are then counted as the number of colony forming units (CFUs). The counting of CFUs assumes that every colony is separate and founded by a single viable microbial cell. The total colony counts obtained in CFU from the incubated agar plates and the respective dilution factor used can then be combined to calculate the original number of microorganisms in the sample in CFU per mL. The typical counting ranges are

25–250 CFU or 30–300 CFU per standard plate count agar plate. Additional considerations for counting colony forming units (or CFUs) are counting of plate spreaders, too numerous to count (TNTC) reporting and statistics, rounding and averaging of observed plate counts, limit of detection, and limit of quantification of plate counts.[101–103] There are also optimal condition assumptions for the plate count method as changes to the plate count agar nutrient level or temperature can affect the surface growth of bacteria.[104,105] Primary equipment and materials used for this method are serial dilution tubes (bottles); petri plates or dishes; pipettes; specific growth medium, diluents, and reagents; incubator and water bath with appropriate optimal temperature setting; commercial colony counter (manual or automated); and plate spreader or rod. Total bacteria and fungi can be enumerated separately using the plate count method based on the type of culture medium utilized.[106–111] Specific or selective culture medium can also be used in place of the standard plate count agar media more specific microbial enumeration.[112,113] Sources of error using this method are improper or inadequate preparation of the test samples, serial dilution error, suboptimal incubation conditions, undercounting due to cell aggregation or clumping, and analyst error in the colony counting or calculation of observed results.

Applications

The plate count method is used primarily in the enumeration of samples with high microorganism numbers or microorganisms that do not grow well in liquid media. Plate count methods are used in the areas of food,[106,107,114–117] pharmaceutical,[19–21] environmental including drinking water applications,[30,118–122] and biofilm testing.[123,124] Modified or alternate versions of the plate count methods have also been developed to further enhance the use of the standard plate count method approach to estimate bacteria or fungi by utilizing the roll-tube method,[125–128] drop-plate method,[129–131] spiral plate count method,[132–135] Petrifilm™,[136–151] SimPlate™,[54,152–157] RODAC™ (replicate organism detection and counting plate) for environmental surface sampling,[158–165] dipslide or dipstick paddle method,[166–170] and adhesive sheet method.[171–177]

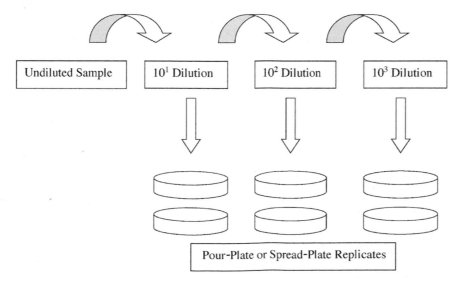

| Undiluted Sample | 10^1 Dilution | 10^2 Dilution | 10^3 Dilution |

Pour-Plate or Spread-Plate Replicates

FIGURE 2.2   Plate count serial dilution scheme.

Membrane Filtration Method

General

The membrane filtration (MF) method is based on viable cell enumeration. The MF method is conducted by filtering a known volume (typical volumes range from 100 mL to 1,000 mL) of liquid sample through a sterile membrane filter (0.22 μm or 0.45 μm pore size). The filter membrane will retain microorganisms during membrane filtration of the liquid sample. After filtration, the membrane filter is placed onto a membrane filter agar medium plate or membrane filter medium pad. These membrane filter plates or pads are then optimally incubated and the colonies observed on these membrane filter plates or pads are counted as the number of colony forming units (CFUs). The total colony counts obtained in CFU from the incubated membrane filter plates and pads and the respective membrane filtered sample volume used can then be combined to calculate the original number of microorganisms in the sample volume used (in CFU per mL). The typical counting ranges are 20–80 CFU per membrane filter used. Primary equipment and materials used for this method are the membrane filtration manifold unit (single or multiple); specific membrane filter with an appropriate membrane pore size and membrane material; filter funnels; pipettes; serial dilution tubes; specific growth medium; sterile buffers or diluents, and reagents, incubator with appropriate optimal temperature setting; vacuum pump; and sterile forceps. Sources of error using this method are improper or inadequate preparation of the test samples, sample dilution error, suboptimal incubation conditions, solid particles and membrane-adsorbed chemicals may interfere with microbial growth, and analyst error in the colony counting or calculation of observed membrane filtration results.

Applications

The membrane filtration method is used to concentrate the number of microorganisms from large volumes of liquid sample with low numbers of microorganisms. This flexible sample volume range allows for the use of larger sample volumes and increases in the sensitivity of microbial enumeration. The membrane filtration method is not particularly useful with turbid or highly particulate samples that may block the membrane pores. Membrane filtration methods are used in the areas of food,[178,179] pharmaceutical,[19–21,180,181] and environmental including drinking and recreational water application.[30,182–190] Modified or alternate versions of the membrane filtration estimate have also been developed to further enhance the use of this method in estimating bacteria by utilizing the defined substrate medium membrane filtration method (selective and/or differential medium)[191–197] and the higher microbial enumeration range hydrophobic grid-membrane filtration (HGMF) method.[198–209]

Direct Count Method

General

The number of microbial cells can be determined by direct microscopic examination of the sample within a demarcated region or field of the microscopic slide and counting the number of microbial cells observed per field under the microscope.

The enumeration of microbial cells in the original sample can be rapidly calculated using an aliquot of known volume used for direct enumeration under the microscope slide (numbers of cells counted per field and number of microscopic fields counted) in number of microbial cells per milliliter or by using a counting standard.[210,211] This direct count method requires a higher number of microorganisms per milliliter for enumeration, and there is the potential for counting both dead and living microbial cells. An extension of the direct count method is the use of fluorescent stains with the epifluorescence microscope or the direct epifluorescence filtration technique (DEFT).[212–219] Primary equipment and materials used for this method are the microscope (light or phase contrast or epifluorescent), microbial counting chambers (Petroff-Hauser, Helber, or Nebauer) or membrane filter, microscope slide support, internal counting standards (latex particles), biological stains or fluorochromes for certain applications, and capillary tubes. Sources of error using this method include the experience of the analysts performing the microbial cell count observed under the microscope, cell aggregation or clumping, inadequate staining, and inadequate microscopic slide sample preservation or preparation.

Applications

Direct count methods are used in the areas of food,[220,221] biocides,[222,223] metal working fluids,[224] and environmental including soil, air, water, or terrestrial ecological applications.[225–232] Modified or alternate versions of the direct count method have also been developed, such as the direct viable count method (DVC),[233–242] fluorescence *in situ* hybridization (FISH) method,[239–242] and combined DVC-FISH method.[243–246]

Particle Counting Method

General

The coulter counting method, or electrical sensing zone (ESZ) method, is a particle count method in which microbial cells suspended in a weak electrolyte pass through an electric field of standard resistance within a small aperture. The measured resistance changes as the microbial cells pass through the aperture. The voltage applied across the aperture creates a sensing zone. If a constant voltage is maintained, a microbial cell passing through will cause a transient decrease in current as the resistance changes. A transient increase in voltage will occur if constant current is maintained. These transient impedance changes in the sensing zone can then be amplified and recorded electronically, thus allowing one to enumerate the number of microbial cells flowing through the aperture opening. The estimation of the total number of microbial cells in a given sample can be correlated with the microbial cell counts obtained for either a metered sample volume or a fixed time period for flow under constant pressure.[247–251]

The flow cytometry method is another particle count method that analyzes the light scatter and fluorescence emitted from individual microbial cells as these cells flow in a single-file manner through an intensely focused laser light source. The sample is injected into a pressurized fluid-delivery system or sheath that forces the suspension of cells into laminar flow and aligned in single file along the central core through the sensing zone. As the cells pass through the sensing zone, a laser light source generates a light-scatter pulse that is collected at small forward angles and right angles to the cell. The collected light pulse from the single cell is converted to an electrical signal (voltage) by either a photodiode or a photomultiplier. The captured data can then be analyzed as two-parameter histograms or by using multiparametric analysis. Fluorescent dyes or stain can be used to further enhance the use of the flow cytometry to discriminate between live or dead microbial cells. A combination of internal calibrating counting beads used per test, the number of events in a region containing microbial cell population and bead population, dilution factor used, and test volume will provide the estimated total number of microbial cells in a given sample.[252–259]

Primary equipment and materials used for the particle count method are commercially available counters (coulter counters or flow cytometry counters), internal liquid bead counting standards (polystyrene beads with or without fluorophores), biological stains or fluorochromes for certain applications, specific reagents (electrolyte solutions or sheathing liquids or preservatives), disposable test tubes, micropipettes plus pipette tips, and vortex mixer. Sources of error using this method include the presence of bubbles or foreign particles while cell counting, which causes high background noise; partial blocking of the counting aperture; a higher number of cells, which may lead to higher probability of the coincident passage of two or more cells through the aperture; and background noise, which will become higher for enumerating low concentrations of cells.

Applications

Particle counting methods are used in the areas of food,[260–265] pharmaceutical or dental,[266–270] and environmental or water treatment applications.[270–281] Flow cytometry methods can also be used for virus enumeration.[282–286] Modified or alternate versions of the standard particle counting methods have also been developed, including the solid state laser cytometry method.[287,288]

Focal Lesion Method

General

The focal lesion method considers an infection of a suitable host by a specific virus to produce a focal lesion response that can be quantitated if it is within an appropriate range of the dilution of the viral sample used. The average number of focal lesions is a linear function of the concentration of virus employed. This method is used to enumerate the number of viruses because each lesion formed is focused at the site of the activity of an individual viral particle. The relationship between the lesion counts obtained and the virus dilution used is also statistically consistent. Primary equipment and materials used depend on the test system employed. Focal lesions can be generated on the skin of whole animals, on the chorioallantoic membrane of chicken embryo, or in cell culture monolayers. Focal lesions can also be produced in agar plate cultures of bacterial viruses and on the leaves of a plant rubbed with a mixture of virus plus abrasive. Sources of error using this method are nonspecific inhibition of focal lesion response by unknown factors, dilution or counting errors, overlapping of focal lesions, size of focal lesions and the numbers countable on the surface area tested, and insufficient viral replication to provide statistical significance.

Applications

The focal lesion method can be used for the enumeration of animal, plant, and bacterial viruses.[289–291]

Quantal Assay Method

General

The quantal assay method uses a statistical method with an all-or-none (cell culture mortality or cytopathic effect or death of experimental animals) response or outcome to enumerate the number of viruses in a given sample. Using a serial dilution approach, the endpoint of activity or infectivity is considered the highest dilution of the virus at which there is 50% or more positive response of the inoculated host. The exact endpoint is determined by interpolation from the cumulative frequencies of positive and negative responses observed at the various dilutions. The endpoint of the quantal titration is the viral dilution that has 50% positive and 50% negative responses. The reciprocal of the dilution yields an estimate of the number of viral units per inoculum volume of the undiluted sample and is expressed in multiples of the 50% endpoint. Primary equipment and materials used for the quantal assay method will depend on the test system employed either as groups of animals, use of chick embryo, or tubes/flasks of cell cultures. Sources of error using this method include dilution or counting errors and insufficient replication to provide statistical significance, and sample titration.

Applications

The quantal assay method is used in infectivity assays in animals or tissue-culture-infectious-dose assay applications in which the animal or cell culture is scored as either infected or not infected. The infectivity titer is proportionately measured by the animal or tissue culture infected. The main use of this method is in the enumeration of culturable waterborne viruses.[292–295]

Plaque Assay Method

General

The plaque assay method utilizes serial dilution of a given viral suspension sample that is inoculated onto a confluent monolayer of a specific host cell culture. Following the adsorption or attachment step, the monolayer is overlaid with either a semisolid or solid medium to restrict the movement of the progeny virus in the vicinity of the infected host cells. After incubation, the infectious plaque plate is enumerated and examined for plaque formation. Each virus unit makes one plaque (an observed localized focus of infected and dead cells) after the cell monolayers are stained with general cellular stain or dye. The infectivity titer of the original viral suspension sample is recorded as plaque-forming units (PFUs) per milliliter. Primary equipment and materials used for this method include serial dilution tubes (bottles), petri plates or dishes, pipettes, specific growth medium, specific cell culture or bacterial host, diluents, reagents, membrane filtration setup, stains/dyes as needed, and incubator and water bath with appropriate optimal temperature setting. Sources of error using this method include improper or inadequate preparation of the test samples, serial dilution error, suboptimal incubation conditions, and analyst error in the plaque counting or calculation of observed results.

Applications

The plaque assay method can be used for the enumeration of animal, plant, and bacterial viruses.[296–310]

Rapid Methods

General

There are several types of rapid microbiological technologies that fall into one or sometimes overlap into two of three general categories: Identification, qualitative (i.e. presence/absence), or quantitative (i.e. countable). Quantitative rapid technologies specifically can be further subdivided into colony forming units (CFU), countable (growth-based), or cellular countable technologies.

Growth-based rapid methods typically require several hours or up to a day or longer to yield final results because these technologies rely on the detection of microcolonies using ATP bioluminescence, cellular fluorescence, or high magnification images. These technologies often have a sensitivity of one CFU. A benefit of these technologies is the ability to continue incubation after detection of growth to yield larger colonies to perform subsequent identification. They also align well with compendial sample preparation methods such as membrane filtration.

Cellular countable-based rapid methods can generate results in seconds to minutes. Due to the speed at which these results can be generated, these methods can be particularly useful when continuous monitoring is desired, such as in active air or purified water systems. One example is the use of Raman spectroscopy to compare microscopic particulate spectral signatures to a library of known microorganisms. Another example is the cleavage of enzymatically fluorescent substrates by metabolically active cells combined with solid phase cytometry, Mie scattering or flow cytometry (see Particle Counting Method section). When samples pass through a flow cell, a laser diode illuminates the sample. Mie scattering sizes any particles present while fluorescence detects biological metabolites like NADH (nicotinamide adenine dinucleotide [NAD] + hydrogen [H]) and riboflavin. These technologies can be sensitive to the single-cell level. An advantage for some cellular-based rapid methods have over growth-based rapid methods is that they can detect microbes that are viable but not culturable using traditional growth media.

Primary equipment and materials used for rapid methods are commercially available including reagents and instruments. Sources of error using these methods vary depending on the instrument being used. Regardless of instrument being used, user training should be a primary consideration.

Applications

Different rapid methods vary in their application, but as a whole they allow for a broad spectrum of detection of bacteria, yeast, and mold in raw material in addition to finished product bioburden testing, environmental monitoring, water analysis, and sterility testing. Some technologies are applicable across multiple types of sample analysis, such as air monitoring or water monitoring. It is common for a single piece of equipment to be dedicated to a certain type of analysis, such as air monitoring or water analysis, even though the underlying technology being used in different pieces of equipment may be essentially the same.[311–323]

Summary

There are a number of test methods used for the quantitative enumeration of microorganisms. A more formal validation approach can also be used to compare new or alternate test methods for the quantitative enumeration of microorganisms.[324–331] This may include rapid test methods for the quantitative enumeration of microorganisms compared to the above standard microbial enumeration test methods.[332–339] Methods for quantitative enumeration of microorganisms will continue to be developed and validated to meet the various industry demands and the associated microbial quantitative enumeration applications, as evidenced by more recent statistical approaches to quantitative microbial estimation or counting methods and the uncertainty associated with microbiological testing.[340–347]

REFERENCES

1. Cochran, W.G., Estimation of bacterial densities by means of the "Most Probable Number," *Biometrics*, 6, 105, 1950.
2. Russek, E. and Colwell, R.R., Computation of most probable numbers, *Appl. Environ. Microbiol.*, 45, 1646, 1983.
3. de Man, J.C., MPN tables, corrected, *J. Appl. Microbiol. Biotechnol.*, 17, 301, 1983.
4. Tillett, H.E. and Coleman, R., Estimated numbers of bacteria in samples from non-homogeneous bodies of water, *J. Appl. Bacteriol.*, 59, 381, 1985.
5. Haas, C.N. and Heller, B., Test of the validity of the Poisson assumption for analysis of MPN results, *Appl. Environ. Microbiol.*, 54, 2996, 1988.
6. Hass, C.N., Estimation of microbial densities from dilution count experiments, *Appl. Environ. Microbiol.*, 55, 1934, 1989.
7. Garthright, W.E. and Blodgett, R.J., FDA's preferred MPN methods for standard, large or unusual tests, with a spreadsheet, *Food Microbiol.*, 20, 439, 2003.
8. U.S. Food and Drug Administration, Most probable number from serial dilutions, *BAM Appendix 2, Bacteriological Analytical Manual Online*, Department of Health and Human Services, 2017.
9. Health Canada, Calculation of the most probable number (MPN), *Official Methods for the Microbiological Analysis of Food, Volume 1, The Compendium of Analytical Methods, Appendix D*, Health Product and Food Branch, 1993.
10. Health Canada, Enumeration of coliforms, faecal coliforms and of *E. coli* in foods using the MPN method, *Official Methods for the Microbiological Analysis of Food, Volume 2, The Compendium of Analytical Methods, MFHPB-19*, Health Product and Food Branch, 2002.

11. Bansal, A. et al., Most-probable-number determination of Salmonella levels in naturally contaminated raw almonds using two sample preparation methods, *J. Food Prot.*, 73, 1986, 2010.

12. Health Canada, Calculation of the most probable number of growth units for the hydrophobic grid membrane filter (HGMF) methods, *Official Methods for the Microbiological Analysis of Food, Volume 1, The Compendium of Analytical Methods, Appendix C*, Health Product and Food Branch, 2014.

13. Gill, A. and Oudit, D., Enumeration of *Escherichia coli* 0157 in outbreak-associated Gouda cheese made with raw milk, *J. Food Prot.*, 78, 1733, 2015.

14. Grevskott, D.H. et al., The species accuracy of the most probable numbers (MPN) European Union reference method for enumeration of *Escherichia coli* in marine bivalves, *J. Microbiol. Methods*, 131, 73, 2016.

15. Chen, Y. et al., Comparative evaluation of direct plating and most probable number for enumeration of low levels of *Listeria monocytogenes* in naturally contaminated ice cream products, *Int. J. Food Microbiol.*, 241, 15, 2017.

16. Walker, D.I. et al., *Escherichia coli* testing and enumeration in live bivalve shell fish, *Food Microbiol.*, 73, 29, 2018.

17. Gill, A. et al., Bacteriological analysis of wheat flour associated with an outbreak of Shiga toxin producing *Escherichia coli* 0121, *Food Microbiol.*, 82, 474, 2019.

18. Sutton, S., The most probable number method and its use in enumeration, qualification, and validation, *J. Validation Technol.*, 16, 35, 2010.

19. JP, 4.05 Microbiological examination of non-sterile products, *Japanese Pharmacopoeia*, 17th ed., 2016.

20. EP, 2.6.12 Microbiological examination of non-sterile products: microbial enumeration tests, *European Pharmacopoeia*, 10th ed., 2019.

21. USP, <61> Microbiological examination of nonsterile products: microbial enumeration tests, USP42-NF37, *The United States Pharmacopeial Convention,* 2019.

22. Munoz, E.F. and Silverman, M.P., Rapid single-step most-probable-number method for enumerating fecal coliforms in effluents from sewage treatment plants, *Appl. Environ. Microbiol.*, 37, 527, 1979.

23. Guerinot, M.L. and Colwell, R.R., Enumeration, isolation, and characterization of N2-fixing bacteria from seawater, *Appl. Environ. Microbiol.*, 50, 350, 1985.

24. Cuba, R.M. et.al., A most-probable number technique for methanotrophic bacteria determination in biological reactors using methane as an electron donor for denitrification, *Environ. Technol.*, 34, 585, 2013.

25. Maul, A., Block, J.C., and El-Shaarawi, A.H., Statistical approach for comparison between methods of bacterial enumeration in drinking water, *J. Microbiol. Meth.*, 4, 67, 1985.

26. Jagals, P. et al., Evaluation of selected membrane filtration and most probable number methods for the enumeration of faecal coliforms, *Escherichia coli* and *Enterococci* in environmental waters, *Quant. Microbiol.*, 2, 1388, 2000.

27. Jenkins, M.B., Endale, D.M., and Fisher, D.S. et al., Most probable number methodology for quantifying dilute concentrations and fluxes of Salmonella in surface waters, *J. Appl. Microbial.*, 104, 1562, 2008.

28. Jenkins, M.B. et al., Most probable number methodology for quantifying dilute concentrations and fluxes of Escherichia coli 0157:H7 in surface waters, *J. Appl. Microbiol.*, 106, 572, 2009.

29. Environment Agency, Methods for the isolation and enumeration of coliform bacteria and Escherichia coli (including *E. coli* O157:H7): C. The enumeration of coliform bacteria and Escherichia coli by a multiple tube most probable number technique using minerals modified glutamate medium incubated at 37°C, in *The Microbiology of Drinking Water-Part 4*, Environment Agency, UK, 2016.

30. Rice, E.W., Baird, R.B., and Eaton, A.D., *Standard Methods for the Examination of Water and Wastewater*, 23rd ed., American Public Health Association, Washington, D.C., 2017.

31. Kirschner, A.K.T. et al., Application of three different methods to determine prevalence, the abundance and the environmental drivers of culturable *Vibrio cholerae* in fresh and brackish bathing waters, *J. Appl. Microbiol.*, 125, 1186, 2018.

32. Dehority, B.A., Tirabasso, P.A., and Grifo Jr., A.P., Most-probable-number procedures for enumerating ruminal bacteria, including the simultaneous estimation of total and cellulolytic numbers in one medium, *Appl. Environ. Microbiol.*, 55, 2789, 1989.

33. Shockey, W.L. and Dehority, B.A., Comparison of two methods for enumeration of anaerobe numbers on forage and evaluation of ethylene oxide treatment for forage sterilization, *Appl. Environ. Microbiol.*, 55, 1766, 1989.

34. Fedorak, P.M., Semple, K.M., and Westlake, D.W.S., A statistical comparison of two culturing methods for enumerating sulfate-reducing bacteria, *J. Microbiol. Meth.*, 7, 19, 1987.

35. Eckford, R.E. and Fedorak, P.M., Applying a most probable number method for enumerating planktonic, dissimilatory, ammonium producing, nitrate reducing bacteria in oil field waters, *Can. J. Microbiol.*, 51, 725, 2005.

36. She, Y.H. et al., Investigation of indigenous microbial enhanced oil recovery in a middle salinity petroleum reservoir, *Adv. Mater. Res.*, 365, 326, 2012.

37. Capita, R. and Alonso-Calleja, C., Comparison of different most-probable-number methods for enumeration of *Listeria* spp. in poultry, *J. Food Prot.*, 66, 65, 2003.

38. Trimble, L.M. et al., Prevalence and concentration of Salmonella and Campylobacter in the processing environment of small-scale pasteurized broiler farms, *Poult. Sci.*, 92, 3060, 2013.

39. Carvalhal, M.L.C., Oliveira, M.S., and Alterthum, F., An economical and time saving alternative to the most-probable-number method for the enumeration of microorganisms, *J. Microbiol. Meth.*, 14, 165, 1991.

40. Pavic, A. et al., A validated miniaturized MPN method based on ISO-6579:2002, for the enumeration of Salmonella from poultry matrices, *J. Appl. Microbiol.*, 109, 25, 2010.

41. Cabrera-Diaz, E. et al., Simultaneous and individual quantitative estimation of Salmonella, Shigella and *Listeria monocytogenes* on inoculated Roma tomatoes (*Lycopersicon esculentum var. Pyriforme*) and Serrano peppers (*Capsicum annum*) using MPN technique, *Food Microbiol.*, 73, 282, 2018.

42. Rowe, R., Todd, R., and Waide, J., Microtechnique for most-probable-number analysis, *Appl. Environ. Microbiol.*, 33, 675, 1977.

43. Gamo, M. and Shoji, T., A method of profiling microbial communities based on a most-probable-number assay that uses BIOLOG plates and multiple sole carbon sources, *Appl. Environ. Microbiol.*, 65, 4419, 1999.

44. Saitoh, S., Iwasaki, K., and Yagi, O., Development of a new most-probable-number method for enumerating methanotrophs, using 48-well microtiter plates, *Microbes Environ*, 17, 191, 2002.

45. Wallenius, K. et al., Simplified MPN method for enumeration of soil naphthalene degraders using gaseous substrate, *Biodegradation*, 23, 47, 2012.

46. Sartory, D.P. et al., Evaluation of an MPN test for rapid enumeration of *Pseudomonas aeruginosa* in hospital waters, *J. Water Health*, 13, 427, 2015.

47. Sartory, D.P. et al., Evaluation of the Pseudomonas/Quanti-Tray MPN test for rapid enumeration of *Pseudomonas aeruginosa*, *Curr. Microbiol.*, 71, 699, 2015.

48. Sartory, D.P. et al., Evaluation of a most probable number method for the enumeration of *Legionella pneumophila* from potable and related water samples, *Lett. Appl. Microbiol.*, 64, 271, 2017.

49. Petrisek, R. and Hall, J., Evaluation of a most probable method for the enumeration of *Legionella pneumophila* from North American potable and nonpotable water samples, *J. Water Health*, 16, 25, 2018.

50. Rech, M.M., Swalla, B.M., and Dobranic, J.K., Evaluation of Legiolert for quantification of *Legionella pneumophila* from non-potable water, *Curr. Microbiol.*, 75, 1282, 2018.

51. Barette, I, Comparison of Legiolert and a conventional culture method for detection of *Legionella pneumophila* from cooling towers in Quebec, *J. AOAC Int.*, 102, 1235, 2019.

52. Edberg, S.C. et al., Enumeration of total coliform and *Escherichia coli* from source water by the defined substrate technology, *Appl. Environ. Microbiol.*, 56, 366, 1990.

53. Harriott, O.T. and Frazer, A.C., Enumeration of acetogens by a colorimetric most-probable-number assay, *Appl. Environ. Microbiol.*, 63, 296, 1997.

54. Jackson, R.W. et al., Multiregional evaluation of the SimPlate heterotrophic plate count method compared to the standard plate count agar pour plate method in water, *Appl. Environ. Microbiol.*, 66, 453, 2000.

55. Kodaka, H., Saito, M., and Matsuoka, H., Evaluation of a new most-probable-number (MPN) dilution plate method for the enumeration of *Escherichia coli* in water samples, *Biocontrol Sci.*, 14, 123, 2009.

56. Boubetra, A. et al., Validation of alternative methods for analysis of drinking water and their application to *Escherichia coli*, *Appl. Environ. Microbiol.*, 77, 3360, 2011.

57. Environment Agency, Methods for the isolation and enumeration of coliform bacteria and Escherichia coli (including *E. coli* O157:H7): D. The enumeration of coliform bacteria and Escherichia coli by a defined substrate most probable number technique Incubated at 37°C, in *The Microbiology of Drinking Water-Part 4*, Environment Agency, UK, 2016.

58. Crowley, E.S. et al., TEMPO TVC for the enumeration of aerobic mesophilic flora in foods: collaborative study, *J. AOAC Int.*, 92, 165, 2009.

59. Crowley, E.S. et al., TEMPO EC for the enumeration of *Escherichia coli* in foods: collaborative study, *J. AOAC Int.*, 93, 576, 2010.

60. Owen, M., Willis, C., and Lamph, D., Evaluation of the TEMPO® most probable number technique for the enumeration of *Enterobacteriaceae* in food and dairy products, *J. Appl. Microbiol.*, 109, 1810, 2010.

61. Lindemann, S. et al., Matrix-specific method validation of an automated most-probable-number system for use in measuring bacteriological quality of grade "A" milk products, *J. Food Prot.*, 79, 1911, 2016.

62. Inoue, D. et al., Comparative evaluation of quantitative polymerase chain reaction methods for routine enumeration of specific bacterial DNA in aquatic samples, *World J. Microbiol. Biotech.*, 21, 1029, 2005.

63. Wright, A.C. et al., Evaluation of postharvest-processed oysters by using PCR-based most–probable-number enumeration of *Vibrio vulnificus* bacteria, *Appl. Microbial.*, 73, 7477, 2007.

64. Henry, R. et al., Environmental monitoring of waterborne Campylobacter: evaluation of the Australian standard and a hybrid extraction-free MPN-PCR method, *Front. Microbiol.*, 6, 74, 2015.

65. Banting, G.S. et al., Evaluation of various Campylobacter-specific quantitative PCR (qPCR) assays for the detection and enumeration of *Campylobacteraceae* in irrigation water and wastewater via a miniaturized most-probable-number-qPCR assay, *Appl. Environ. Microbiol.*, 82, 4743, 2016.

66. Kim, S.A. et al., Development of a rapid method to quantify *Salmonella typhimurium* using a combination of MPN with qPCR and a shortened time incubation, *Food Microbiol.*, 65, 7, 2017.

67. McFarland, J., The nephelometer: an instrument for estimating the number of bacteria in suspensions used for calculating the opsonic index for vaccines, *JAMA*, 49, 1176, 1907.

68. Kurokawa, M., A new method for the turbidimetric measurement of bacterial density, *J. Bacteriol.*, 83, 14, 1962.

69. Fujita, T. and Nunomura, K., New turbidimetric device for measuring cell concentrations in thick microbial suspensions, *Appl. Microbiol.*, 16, 212, 1968.

70. Koch, A.L., Turbidimetric measurements of bacterial cultures in some available commercial instruments, *Anal. Biochem.*, 38, 252, 1970.

71. Li, R.C., Nix, D.E., and Schentag, J.J., New turbidimetric assay for quantitation of viable bacteria densities, *Antimicrob. Agents Chemother.*, 37, 371, 1993.

72. Sadar, M.J., Turbidity science, *Technical Information Series-Booklet No.11*, Hach Company, Loveland, CO, 1998.

73. Sutton, S., Measurement of cell concentration in suspension by optical density, *Pharm.Microbiol. Forum Newslett.*, 12, 3, 2006.

74. Pugh, T.L. and Heller, W., Density of polystyrene and polyvinyl toluene latex particles, *J. Colloid. Sci.*, 12, 173, 1957.

75. Roessler, W.G. and Brewer, C.R., Permanent turbidity standards, *Appl. Microbiol.*, 15, 1114, 1967.

76. Zamora, L.L. and Pérez-Garcia, M.T., Using digital photography to implement the McFarland method, *J. R. Soc. Interface*, 9, 1892, 2012.

77. Kavanaugh, F., Turbidimetric assays: the antibiotic dose-response line, *Appl. Microbiol.*, 16, 777, 1968.

78. Galgiani, J.N. and Stevens, D.A., Turbidimetric studies of growth inhibition of yeasts with three drugs: inquiry into inoculum dependent susceptibility testing, time of onset, of drug effect, and implications for current and new methods, *Antimicrob. Agents Chemother.*, 13, 249, 1978.

79. Hughes, C.E., Bennett, R.L., and Beggs, W.H., Broth dilution testing of *Candida albicans* susceptibility to ketoconazole, *Antimicrob. Agents Chemother.*, 31, 643, 1987.

80. Odds, F.C., Quantitative microculture system with standardize inocula for strain typing, susceptibility testing, and other physiologic measurement with *Candida albicans* and other yeasts, *J. Clin. Microbiol.*, 29, 2735, 1991.

81. Klepser, M.E. et al., Influence of test conditions on antifungal time-kill curve results: proposal for standardized methods, *Antimicrob. Agents Chemother.*, 42, 1207, 1998.

82. Llop, C. et al., Comparison of three methods of determining MICs for filamentous fungi using different end point criteria and incubation periods, *Antimicrob. Agents Chemother.*, 44, 239, 2000.

83. Garg, P. et al., Antimicrobial effect of Chitosan on the growth of lactic acid bacteria strain known to spoil beer, *J. Expt. Microbiol. Immunol.*, 14, 7, 2010.

84. Vieira, D.C., Fiuza, T.F., and Salgado, H.R., Development and validation of a rapid turbidimetric assay to determine the potency of cefuroxime sodium in powder for dissolution for injection, *Pathogens*, 3, 656, 2014.

85. da Silva, L.M. and Salgado, H.R., Rapid turbidimetric assay to potency evaluation of tigecycline in lyophilized powder, *J. Microbiol. Methods*, 110, 49, 2015.

86. Tsai, D.S. et al., Disinfection effects of undoped and silver-doped ceria powders of nanometer crystalline size, *Int. J. Nanomedicine*, 11, 2531, 2016.

87. JP, 4.02 Microbial assay for antibiotics, *Japanese Pharmacopoiea*, 17th ed., 2016.

88. Sardella, D., Gatt, R., and Valdramidis, V.P., Assessing the efficacy of zinc oxide nanoparticles against *Penicillium expansum* by automated turbidimetric analysis, *Mycology*, 9, 43, 2017.

89. Sardella, D., Gatt, R., and Valdramidis, V.P., Turbidimetric assessment of the growth of filamentous fungi and the antifungal activity of zinc oxide nanoparticles, *J. Food Prot.*, 81, 934, 2018.

90. EP, 5.1.3 Efficacy of antimicrobial preservation, *European Pharmacopoeia*, 10th ed., 2019.

91. USP, <51> Antimicrobial effectiveness testing, USP42-NF37, *The United States Pharmacopeial Convention*, 2019.

92. Begot, C. et al., Recommendation for calculating growth parameters by optical density measurements, *J. Microbiol. Meth.*, 25, 225, 1996.

93. Domínguez, M.C., de la Rosa, M., and Borobio, M.V., Application of a spectrophotometric method for the determination of post-antibiotic effect and comparison with viable counts in agar, *J. Antimicrobiol. Chemother.*, 47, 391, 2001.

94. Baty, F., Flandrois, J.P., and Delignette-Muller, M.L., Modeling the lag time of *Listeria monocytogenes* from viable count enumeration and optical density data, *Appl. Microbiol.*, 68, 5816, 2002

95. Meletiadis, J., te Dorsthorst, D.T.A., and Verweij, P.E., Use of turbidimetric growth curves for early determination of antifungal drug resistance of filamentous fungi, *J. Clin. Microbiol.*, 41, 4718, 2003.

96. Maia, M.R. et al., Simple and versatile turbidimetric monitoring of bacterial growth in liquid cultures using a customized 3D print culture tube holder and a miniaturized spectrophotometer: application to facultative and strict anaerobic bacteria, *Front. Microbiolog.*, 7, 1381, 2016.

97. Myllyniemi, A. et al., An automated turbidimetric method for the identification of certain antibiotic groups in incurred kidney samples, *Analyst*, 129, 265, 2004.

98. Pianetti, A. et al., Determination of the variability of *Aeromonas hydrophila* in different types of water by flow cytometry, and comparison with classical methods, *Appl. Environ. Microbiol.*, 71, 7948, 2005.

99. Vaishali, J.P., Tendulkar, S.R., and Chatoo, B.B., Bioprocessing development for the production of an antifungal molecule by *Bacillus licheniformis* BC98, *J. Biosci. Bioeng.*, 98, 231, 2004.

100. Clark, D.S., Comparison of pour and surface plate methods for determination of bacterial counts, *Can. J. Microbiol.*, 13, 1409, 1967.

101. Jennison, M.W. and Wadsworth, G.P., Evaluation of the errors involved in estimating bacterial numbers by the plating method, *J. Bacteriol.*, 39, 389, 1940.

102. Haas, C.H. and Heller, B., Averaging of TNTC counts, *Appl. Environ. Microbiol.*, 54, 2069, 1988.

103. Sutton, S., Counting colonies, *Pharm.Microbiol. Forum Newslett.*, 12, 2, 2006.

104. Salvesen, I. and Vadstein, O., Evaluation of plate count methods for determination of maximum specific growth rate in mixed microbial communities, and its possible application for diversity assessment, *J. Appl. Microbiol.*, 88, 442, 2000.

105. Fujikawa, H. and Morozumi, S., Modeling surface growth of *Escherichia coli* on agar plates, *Appl. Environ. Microbiol.*, 71, 7920, 2005.

106. U.S. Food and Drug Administration, Aerobic plate count, *Bacteriological Analytical Manual Online*, Department of Health and Human Services, 2001, chap. 3.

107. U.S. Food and Drug Administration, Yeast, mold and mycotoxins, *Bacteriological Analytical Manual Online*, Department of Health and Human Services, 2001, chap. 18.

108. Beuchat, L.R., Media for detecting and enumerating yeasts and moulds, *Int. J. Food. Microbiol.*, 17, 145, 1992.

109. Beuchat, L.R., Selective media for detecting and enumerating foodborne yeasts, *Int. J. Food. Microbiol.*, 19, 1, 1993.

110. Beuchat, L.R. et al., Performance of mycological media in enumerating desiccated food spoilage yeast: an interlaboratory study, *Int. J. Food. Microbiol.*, 70, 89, 2001.

111. Beuchat, L.R. and Mann, D.A., Comparison of new and traditional culture-dependent media for enumeration of foodborne yeasts and molds, *J. Food Prot.*, 79, 95, 2016.

112. Nwamaioha, N.O. and Ibrahim, S.A., A selective medium for the enumeration and differentiation of *Lactobacillus delbrueckii ssp. bulgaricus*, *J. Dairy Sci.*, 101, 4953, 2018.

113. EP, 2.6.13 Microbiological examination of non-sterile products (test for specified micro-organisms), *European Pharmacopoeia*, 10th ed., 2019.

114. Reasoner, D.J., Heterotrophic plate count methodology in the United States, *Int. J. Food Microbiol.*, 92, 307, 2004.

115. Gracias, K.S. and McKillip, J.L., A review of conventional detection and enumeration methods for pathogenic bacteria in food, *Can. J. Microbiol.*, 50, 883, 2004.

116. Health Canada, Determination of the aerobic colony count in foods, *Official Methods for the Microbiological Analysis of Food, Volume 2, The Compendium of Analytical Methods, MFHPB-18*, Health Product and Food Branch, 2015.

117. Health Canada, Enumeration of yeasts and moulds in foods, *Official Methods for the Microbiological Analysis of Food, Volume 2, The Compendium of Analytical Methods, MFHPB-22*, Health Product and Food Branch, 2018.

118. Pepper, I.L. et al., Tracking the concentration of heterotrophic plate count bacteria from the source to the consumer's tap, *Int. J. Food Microbiol.*, 92, 289, 2004.

119. Mueller, S.A. et al., Comparison of plate counts, Petrifilm, dipslides, and adenosine triphosphate bioluminescence for monitoring bacteria in cooling-tower water, *Water Environ. Res.*, 81, 401, 2009.

120. Health Canada, *Guidance on the use of heterotrophic plate counts in Canadian drinking water supplies*, Healthy Environments and Consumer Safety Branch, Ottawa, 2012.

121. Allen, M.J. and Edberg, S.C., Health concerns of heterotrophic plate count (HPC) bacteria in dental equipment water line, *Am. J. Dent.*, 29, 137, 2016.

122. Yari, A.R. et al., Assessment of microbial quality of household water output from desalination systems by the heterotrophic plate count method, *J. Water Health*, 16, 930, 2018.

123. Cerca, N. et al., Comparative assessment of antibiotic susceptibility of coagulase-negative *Staphylococci* in biofilm versus planktonic culture as assessed by bacterial enumeration or rapid XTT colorimetry, *J. Antimicrob. Chemother.*, 56, 331, 2005.

124. Ripolles-Avial, C. et al., Quantification of mature *Listeria monocytogenes* biofilm cells formed by an in vitro model: a comparison of different methods, *Int. J. Food Microbiol.*, 289, 209, 2019.

125. Vervaeke, I.J. and Van Nevel, C.J., Comparison of three techniques for the total count of anaerobics from intestinal contents of pigs, *Appl. Microbiol.*, 24, 513, 1972.

126. Moore, W.E.C. and Holdeman, L.V., Special problems associated with the isolation and identification of intestinal bacteria in fecal floral studies, *Am. J. Clin. Nutri.*, 27, 1450, 1974.

127. Grubb, J.A. and Dehority, B.A., Variation in colony counts of total viable anaerobic rumen bacteria as influenced by media and cultural methods, *Appl. Environ. Microbiol.*, 31, 262, 1976.

128. Nagamine, I. et al., A comparison of features and the microbial constitution of the fresh feces of pigs fed diets supplemented with or without dietary microbes, *J. Gen. Appl. Microbiol.*, 44, 375, 1998.

129. Neblett, T.R., Use of droplet plating method and cystine-lactose electrolyte-deficient medium in routine quantitative urine culturing procedure, *J. Clin. Microbiol.*, 4, 296, 1976.

130. Herigstad, B., Hamilton, M., and Heersink, J., How to optimize the drop plate method for enumerating bacteria, *J. Microbiol. Meth.*, 44, 121, 2001.

131. Naghili, H. et al. Validation of drop plate technique for bacterial enumeration by parametric and nonparametric tests, *Vet. Res. Forum*, 4, 179, 2013.

132. Gilchrist, J.E. et al., Spiral plate method for bacterial determination, *Appl. Microbiol.*, 25, 244, 1973.

133. Peeler, J.T. et al., A collaborative study of the spiral plate method for examining milk samples, *J. Food Prot.*, 40, 462, 1977.

134. Zipkes, M.R., Gilchrist, J.E., and Peeler, J.T., Comparison of yeast mold counts by spiral, pour, and streak plate methods, *J. AOAC Int*, 64, 1465, 1981.

135. Alonso-Calleja, C. et al., Evaluation of the spiral plating method for the enumeration of microorganisms throughout the manufacturing and ripening of a raw goat's milk cheese, *J. Food Prot.*, 65, 339, 2002.

136. Knight, L.J. et al., Comparison of the Petrifilm dry rehydratable film and conventional methods for enumeration of yeasts and molds in foods: collaborative study, *J. AOAC Int.*, 80, 806, 1997.

137. Health Canada, Enumeration of coliforms in food products and food ingredients using 3M™ Petrifilm™ coliform count plates, *Official Methods for the Microbiological Analysis of Food, Volume 2, The Compendium of Analytical Methods, MFHPB-35*, Health Product and Food Branch, 2001.

138. Ellis, P. and Meldrum, R., Comparison of the compact dry TC and 3M Petrifilm ACP dry sheet media methods with the spiral plate method for the examination of randomly selected foods for obtaining aerobic colony counts, *J. Food Prot.*, 65, 423, 2002.

139. Health Canada, Enumeration of yeast and mould in food products and food ingredients using 3M™ Petrifilm™ yeast and mold count plates, *Official Methods for the Microbiological Analysis of Food, Volume 2, The Compendium of Analytical Methods, MFHPB-32*, Health Product and Food Branch, 2003.

140. Beloti, V. et al., Evaluation of Petrifilm™ EC and HS for total coliforms and *Escherichia coli* enumeration in water, *Braz. J. Microbiol.*, 34, 310, 2003.

141. Vail, J.H. et al., Enumeration of waterborne *Escherichia coli* with Petrifilm plates: comparison to standard methods, *J. Environ. Qual.*, 32, 368, 2003.

142. Ortolani, M.B. et al., Screening and enumeration of lactic acid bacteria in milk using three different culture media in Petrifilm aerobic count plates and conventional pour plate methodology, *J. Dairy Res.*, 74, 387, 2007.

143. Kudaka, J. et al., Evaluation of the Petrifilm aerobic count plate for enumeration of aerobic marine bacteria from seawater and *Caulerpa lentillifera*, *J. Food. Prot.*, 73, 1529, 2010.

144. Bonesh, D.L., Crowley, E.S., and Bird, P.M., 3M Petrifilm environmental Listeria plate, *J. AOAC Int.*, 96, 225, 2013.

145. Nelson, M.T. et al., Comparison of 3M Petrifilm aerobic count plates to standard methodology for use with AOAC antimicrobial efficacy methods 955.14, 955.15, 964.02, and 964.04 as an alternative enumeration procedure: collaborative study, *J. AOAC Int.*, 96, 717, 2013.

146. Bird, P. et al., Evaluation of the 3M™ Petrifilm™ Salmonella express system for the detection of Salmonella species in selected foods: collaborative study, *J. AOAC Int.*, 97, 1563, 2014.

147. Fritz, B.G. et al., Evaluation of Petrifilm™ aerobic plate counts as an equivalent alternative to drop plating on R2A agar plates in a biofilm disinfectant efficacy test, *Curr. Microbiol.*, 70, 450, 2015.

148. Health Canada, Enumeration of total aerobic bacteria in food products and food ingredients using 3M™ Petrifilm™ aerobic count plates, *Official Methods for the Microbiological Analysis of Food, Volume 2, The Compendium of Analytical Methods, MFHPB-33*, Health Product and Food Branch, 2015.

149. Bird, P. et al., Evaluation of the 3M™ Petrifilm™ rapid yeast and mold count plate for the enumeration of yeast and mold in food: collaborative study, First action 2014.05, *J. AOAC Int.*, 98, 767, 2015.

150. Health Canada, Enumeration of *Escherichia coli* and coliforms in food products and food ingredients using 3M™ Petrifilm™ *E. coli* count plates, *Official Methods for the Microbiological Analysis of Food, Volume 2, The Compendium of Analytical Methods, MFHPB-34*, Health Product and Food Branch, 2016.

151. Bird, P. et al., Evaluation of the 3M™ Petrifilm™ rapid aerobic count plate for the enumeration of aerobic bacteria in food: collaborative study, First action 2015.13, *J. AOAC Int.*, 99, 664, 2016.

152. Townsend, D.E. and Naqui, A., Comparison of SimPlate™ total plate count test with plate count agar method for detection and quantification of bacteria in food, *J. AOAC Int.*, 3, 563, 1998.

153. Feldsine, P.T. et al., Enumeration of total aerobic microorganisms in foods by SimPlate™ total count-color indicator methods and conventional culture methods: collaborative study, *J. AOAC Int.*, 86, 2573, 2003.

154. Ferrati, A.R. et al., A comparison of ready-to-use systems for evaluating the microbiological quality of acidic fruit juices using non-pasteurized orange juices as an experimental model, *Int. Microbiol.*, 8, 49, 2005.

155. Haugue, S.J. et al., Evaluation of the SimPlate™ method for the enumeration of *Escherichia coli* in swab samples from beef and lamb carcasses, *Int. J. Food Microbiol.*, 142, 229, 2010.

156. Porteous, N. et al., A comparison of 2 laboratory methods to test dental unit water line water quality, *Diagn. Microbiol. Infect. Dis.*, 77, 206, 2013.

157. Hague, S.J. et al., Enumeration of *Escherichia coli* in swab samples from pre- and post-chilled pork and lamb carcasses using 3M™ Petrifilm™ select *E. coli* and SimPlate™ coliforms/*E. coli*, *Meat Sci.*, 130, 26, 2017.

158. Bruch, M.K. and Smith, F.W., Improved method for pouring RODAC plates, *Appl. Microbiol.*, 16, 1427, 1968.

159. Hacek, D.M. et al., Comparison of the RODAC imprint method to selective enrichment broth for recovery of vancomycin-resistant Enterococci and drug-resistant *Enterobacteriaceae* from environmental surfaces, *J. Clin. Microbiol.*, 38, 4646, 2000.

160. Collins, S.M. et al., Contamination of the clinical microbiological laboratory with vancomycin-resistant Enterococci and multidrug resistant *Enterobacteriaceae*: implications for hospital and laboratory workers, *J. Clin. Microbiol.*, 39, 3772, 2001.

161. Frabetti, A. et al., Experimental evaluation of the efficacy of sanitation procedure in operating rooms, *Am. J. Infect. Control*, 37, 658, 2009.

162. Health Canada, Environmental sampling for the detection of microorganisms, *Official Methods for the Microbiological Analysis of Food, Volume 3, The Compendium of Analytical Methods, MFLP-41*, Health Product and Food Branch, 2010.

163. Cheng, V.C.C. et al., Hospital outbreak of pulmonary and cutaneous zygomycosis due to contaminated linen items from substandard laundry, *Clin. Infect. Dis.*, 62, 714, 2016.

164. Daneau, G. et al., use of RODAC plates to measure contaminant of *Mycobacterium tubercolosis* in a class IIB biosafety cabinet during routine operations, *Int. J. Mycobacteriol.*, 5, 148, 2016.

165. Okamoto, K. et al., Flocked nylon swabs versus RODAC plates for detection of multidrug-resistant organisms on environmental surfaces in intensive care units, *J. Hosp. Infect.*, 98, 105, 2018.

166. Salo, S. et al., Validation of the microbiological methods Hygicult dipslide, contact plate, and swabbing in surface hygiene control: a Nordic collaborative study, *J. AOAC Intl.*, 83, 1357, 2000.

167. Salo, S. et al., Validation of the Hygicult E dipslides method in surface hygiene control: a Nordic collaborative study, *J. AOAC Int.*, 85, 388, 2002.

168. Scarparo, C. et al., Evaluation of the DipStreak, a new device with an original streaking mechanism for detecting, counting, and presumptive identification of urinary tract pathogens, *J. Clin. Microbiol.*, 40, 2169, 2002.

169. Simpson, A.T. et al., Occupational exposure to metalworking fluid mist and sump fluid contaminants, *Am. Occup. Hyg.*, 47, 17, 2003.

170. Ibfelt, T., Foged, C., and Anderson, L.P., Validation of dipslides as a tool for environmental sampling in a real-life hospital setting, *Eur. J. Clin. Microbiol. Infect. Dis.*, 33, 809, 2014.

171. Yamaguchi, N. et al., Development of an adhesive sheet for direct counting of bacteria on solid surfaces, *J. Microbiol. Meth.*, 53, 405, 2003.

172. Ushiyama, M. and Iwasaki, M., Evaluation of Sanita-kun *E. coli* and coliform sheet medium for enumeration of total coliforms and *E. coli*, *J. AOAC Int.*, 93, 163, 2010.

173. Mizuochi, S. et al., Matrix extension study: validation of the compact dry CF method for enumeration of total coliform bacteria in selected foods, *J. AOAC Int.*, 99, 444, 2016.

174. Mizuochi, S. et al., Matrix extension study: validation of the compact dry EC method for enumeration of *Escherichia coli* and *non-E.coli* coliform bacteria in selected foods, *J. AOAC Int.*, 99, 451, 2016.

175. Mizuochi, S. et al., Matrix extension study: validation of the compact dry TC method for enumeration of total aerobic bacteria in selected foods, *J. AOAC Intl.*, 99, 461, 2016.

176. Teramura, H. et al., MC-Media pad SA (Sanita-kun SA) for the enumeration of *Staphylococcus aureus* in a variety of foods, *J. AOAC Int.*, 101, 456, 2018.

177. Teramura, H. et al., MC-Media pad ACplus ™ for the enumeration of aerobic counts in a variety of foods, *J. AOAC Int.*, 101, 769, 2018.

178. U.S. Food and Drug Administration, Enumeration of *Escherichia coli* and total coliform bacteria, *Bacteriological Analytical Manual Online*, Department of Health and Human Services, Revised 2017, chap. 4.

179. Barre, L. et al., Sensitive enumeration of *Listeria monocytogenes* and other Listeria species in various naturally contaminated matrices using a membrane filtration method, *Food Microbiol.*, 48, 171, 2015.

180. Bauters, T.G. and Nelis, H.J., Comparison of chromogenic and fluorogenic membrane filtration methods for detection of four *Candida* species, *J. Clin. Microbiol.*, 40, 1838, 2002.

181. Jones, D. and Cundell, T. et al., method verification requirements for an advanced imaging system for microbial plate enumeration, *PDA J. Pharm. Sci. Tech.*, 72, 199, 2018.

182. USEPA, Improved enumeration method for recreational water quality indicators: *Enterococci* and *Escherichia coli*, U.S. Environmental Protection Agency Office of Science and Technology, Washington, D.C., 2000.

183. USEPA, Method 1604 Total coliforms and *Escherichia coli* in water by membrane filtration using a simultaneous detection technique (MI medium), U.S. Environmental Protection Agency Office of Water, Washington, D.C., 2002.

184. USEPA, Method 1106.1 Enterococci in water by membrane filtration using a membrane-Enterococcus esculin iron agar (mE-EIA), U.S. Environmental Protection Agency Office of Water, Washington, D.C., 2009.

185. USEPA, Method 1103.1 *Escherichia coli* (*E. coli*) in water by membrane filtration using a membrane-thermotolerant *Escherichia coli* agar (mTEC), U.S. Environmental Protection Agency Office of Water, Washington, D.C., 2010.

186. Environment Agency, Methods for isolation and enumeration of enterococci, in *The Microbiology of Drinking Water-Part 5*, Environment Agency, U.K., 2012.

187. USEPA, Method 1600 Enterococci in water by membrane filtration using membrane-Enterococcus indoxyl-β-D-glucoside agar (mEI), U.S. Environmental Protection Agency Office of Water, Washington, D.C., 2014.

188. USEPA, Method 1603 *Escherichia coli* (E. coli) in water by membrane filtration using modified membrane-thermotolerant *Escherichia coli* agar (modified mTEC), U.S. Environmental Protection Agency Office of Water, Washington, D.C., 2014.

189. Bushon, R.N., Brady, A.M., and Lindsey, B.D., Holding-time and method comparisons for the analysis of fecal-indicator bacteria from groundwater, *Environ. Monit. Assess.*, 187, 672, 2015.

190. Vergine, P. et al., Identification of the faecal indicator *Escherichia coli* in wastewater through the β-D-glucuronidase activity: comparison between two enumeration methods, membrane filtration with TBX agar, and Colilert®-18, *J. Water Health*, 15, 209, 2017.

191. Brenner, K.P. et al., A new medium for the simultaneous detection of total coliforms and *Escherichia coli* in water, *Appl. Environ. Microbiol.*, 59, 3534, 1993.

192. Ciebin, B.W. et al., Comparative evaluation modified m-FC and m-TEC media for membrane filter enumeration of *Escherichia coli* in water, *Appl. Microbiol.*, 61, 3940, 1995.

193. Grant, M.A., A new membrane filtration medium for the simultaneous detection of *Escherichia coli* and total coliforms, *Appl. Environ. Microbiol.*, 63, 3526, 1997.

194. Rompré, A. et al., Detection and enumeration of coliforms in drinking water: current methods and emerging approaches, *J. Microbiol. Meth.*, 49, 31, 2002.

195. Lange, B., Strathmann, M., and Obmer, R., Performance validation of chromogenic coliform agar for the enumeration of *Escherichia coli* and coliform bacteria, *Lett. Appl. Microbiol.*, 57, 547, 2013.

196. Manafi, M. et al., Evaluation of CP chromo select agar for the enumeration of *Clostridium perfringens* from water, *Int. J. Food Microbiol.*, 167, 92, 2013.

197. Watkins, J., Sartory, D.P., and U.K. Standing Committee of Analysts, evaluation of a membrane filtration method for the rapid enumeration of confirmed *Clostridium perfrigens* from water, *Lett. Appl. Microbiol.*, 60, 367, 2015.

198. Sharpe, A.N. and Michaud, G.L., Enumeration of high numbers of bacteria using hydrophobic grid-membrane filters, *Appl. Microbiol.*, 30, 519, 1975.

199. Tsuji, K. and Bussey, D.M., Automation of microbial enumeration: development of a disposable hydrophobic grid-membrane filter unit, *Appl. Environ. Microbiol.*, 52, 857, 1986.

200. Parrington, L.J., Sharpe, A.N., and Peterkin, P.I., Improved aerobic colony count technique for hydrophobic grid-membrane, *Appl. Environ. Microbiol.*, 59, 2784, 1993.

201. Health Canada, Determination of aerobic colony count in foods and environmental samples by the hydrophobic grid-membrane filter (HGMF) method, *Official Methods for the Microbiological Analysis of Food, Volume 3, The Compendium of Analytical Methods, MFLP-56*, Health Protection Branch, 2003.

202. Heikinheimo, A., Linström, M., and Korkeala, H., Enumeration and isolation of cpe-positive *Clostridium perfringens* spores from feces, *J. Clin. Microbiol.*, 42, 3992, 2004.

203. Park, Y.J. and Chen, J., Microbial quality of soft drinks served by the dispensing machines in fast food restaurants and convenience stores in Griffin, Georgia, and surrounding areas, *J. Food Prot.*, 72, 2607, 2009.

204. Tournas, V.H., Evaluation of hydrophobic grid membrane filter for the enumeration of moulds and yeasts in naturally-contaminated foods, *Microbial. Insights*, 2, 31, 2009.

205. Health Canada, Enumeration of *Aeromonas hydrophila* in ice and water by the hydrophobic grid-membrane filter (HGMF) method, *Official Methods for the Microbiological Analysis of Food, Volume 3, The Compendium of Analytical Methods, MFLP-58B*, Health Protection Branch, 2012.

206. Health Canada, Enumeration of Yeast and Mould in foods using the iso-grid membrane filtration, *Official Methods for the Microbiological Analysis of Food, Volume 3, The Compendium of Analytical Methods, MFLP-57*, Health Protection Branch, 2014.

207. Health Canada, Enumeration of *Aeromonas hydrophila* in foods by the hydrophobic grid-membrane filter or direct plating od, *Official Methods for the Microbiological Analysis of Food, Volume 3, The Compendium of Analytical Methods, MFLP-58*, Health Protection Branch, 2014.

208. Health Canada, Enumeration of faecal coliforms in foods by hydrophobic grid-membrane filter (HGMF) method, *Official Methods for the Microbiological Analysis of Food, Volume 3, The Compendium of Analytical Methods, MFLP-55*, Health Protection Branch, 2015.

209. Health Canada, Enumeration of *Pseudomonas aeruginosa* in foods and food Ingredients by the hydrophobic grid-membrane filter (HGMF) method, *Official Methods for the Microbiological Analysis of Food, Volume 3, The Compendium of Analytical Methods, MFLP-61*, Health Protection Branch, 2015.

210. Kirchman, D. et al., Statistical analysis of the direct count method for enumerating bacteria, *Appl. Environ. Microbiol.*, 44, 376, 1982.

211. Humberd, C.M., Short report: enumeration of *Leptospires* using the coulter counter, *Am.J. Trop. Med. Hyg.*, 73, 962, 2005.

212. Hobbie, J.E., Daley, R.J., and Jasper, S., Use of nucleopore filters for counting bacteria by fluorescence microscopy, *Appl. Environ. Microbiol.*, 33, 1225, 1977.

213. Kepner Jr., R.L. and Pratt, J.R., Use of fluorochromes for direct enumeration of total bacteria in environmental samples: past and present, *Microbiol. Rev.*, 58, 603, 1994.

214. Brehm-Stecher, B.F. and Johnson, E.A., Single-cell microbiology: tools, technologies, and applications, *Microbiol. Mol. Biol. Rev.*, 68, 538, 2004.

215. Seo, E.Y., Ahn, T.S., and Zo, Y.G., Agreement, precision, and accuracy of epifluorescence microscopy methods for enumeration of total bacteria numbers, *Appl. Environ. Microbiol.*, 76, 1981, 2010.

216. Cragg, B.A. and Parkes, R.J., Bacteria and archaeal direct counts: a faster method of enumeration, for enrichment cultures and aqueous environmental samples, *J. Microbiol. Methods*, 98, 35, 2014.

217. Campagna, M.C. et al., application of microbiological method direct epifluorescence filter technique/aerobic plate count agar in the identification of irradiated herbs and spices, *Ital. J Food Saf.*, 3,1650, 2014.

218. Guo, R. et al., A rapid and low-cost estimation of bacteria counts in solution using fluorescence spectroscopy, *Anal. Bioanal. Chem.*, 409, 3959, 2017.

219. Robertso, J. et al., Optimization of the protocol for the LIVE/DEAD® BacLight™ bacterial viability kit for rapid determination of bacteria load, *Front. Microbiol.*, 10, 801, 2019.

220. Pettipher, G.L. and Rodrigues, U.M., Rapid enumeration of microorganisms in foods by the direct epifluorescent filter technique, *Appl. Environ. Microbiol.*, 44, 809, 1982.

221. Rapposch, S., Zangeri, P., and Ginzinger, W., Influence of fluorescence of bacteria stained with Acridine Orange on the enumeration of microorganisms in raw milk, *J. Dairy Sci.*, 83, 2753, 2000.

222. Matsunaga, T., Okochi, M., and Nakasono, S., Direct count of bacteria using fluorescent dyes: application to assessment of electrochemical disinfection, *Anal. Chem.*, 67, 4487, 1995.

223. McBain, A.J. et al., Exposure of sink drain microcosms to Triclosan: population dynamics and antimicrobial susceptibility, *Appl. Environ. Microbiol.*, 69, 5433, 2003.

224. Veillette, M. et al., Six months tracking of microbial growth in a metalworking fluid after system cleaning and recharging, *Ann. Occup. Hyg.*, 48, 541, 2004.

225. Bloem, J., Veninga, M., and Shepherd, J., Fully automated determination of soil bacterium numbers, cell volumes, and frequencies of dividing cells by confocal laser scanning microscopy and image analysis, *Appl. Environ. Microbiol.*, 61, 926, 1995.

226. Yu, W. et al., Optimal staining and sample storage time for direct microscopic enumeration of total and active bacteria in soil with two fluorescent dyes, *Appl. Environ. Microbiol.*, 61, 3367, 1995.

227. Bölter, M. et al., Enumeration and biovolume determination of microbial cells — a methodological review and recommendations for applications in ecological research, *Biol. Fertil. Soils*, 36, 249, 2002.

228. Terzieva, S. et al., Comparison of methods for detection and enumeration of airborne microorganisms collected by liquid impingement, *Appl. Environ. Microbiol.*, 62, 2264, 1996.

229. Ramalho, R. et al., Improved methods for the enumeration of heterotrophic bacteria in bottled waters, *J. Microbiol. Meth.*, 44, 97, 2001.

230. Lisle, J.T. et al., Comparison of fluorescence microscopy and solid-phase cytometry methods for counting bacteria in water, *Appl. Environ. Microbiol.*, 70, 5343, 2004.

231. Angenent, L.T. et al., Molecular identification of potential pathogens in water and air of a hospital therapy pool, *PNAS*, 102, 4860, 2005.

232. Helton, R.R., Liu, L., and Wommack, K.E., Assessment of factors influencing direct enumeration of viruses within estuarine sediments, *Appl. Environ. Microbiol.*, 72, 4767, 2006.

233. Kogure, K. et al., Correlation of direct viable count with heterotrophic activity for marine bacteria, *Appl. Environ. Microbiol.*, 53, 2332, 1987.

234. Roszak, D.B. and Colwell, R.R., Metabolic activity of bacterial cells enumerated by direct viable count, *Appl. Environ. Microbiol.*, 53, 2889, 1987.

235. Joux, F. and Lebaron, P., Ecological implication of an improved direct viable count method for aquatic bacteria, *Appl. Environ. Microbiol.*, 63, 3643, 1997.

236. Yokomaku, D., Yamaguchi, N., and Nasu, M., Improved direct viable count procedure for quantitative estimation of bacterial viability in freshwater environments, *Appl. Environ. Microbiol.*, 66, 5544, 2000.

237. Zimmerman, A.M. et al., *Escherichia coli* determination using mTEC agar and fluorescent antibody direct visible counting on coastal recreational water samples, *Lett. Appl. Microbiol.*, 49, 478, 2009.

238. Altuq, G. et al., The application of viable count procedure for measuring viable cells in the various marine environments, *J. Appl. Microbiol.*, 108, 88, 2010.

239. Moter, A. and Göbel, U.B., Fluorescence *in situ* hybridization (FISH) for direct visualization of microorganisms, *J. Microbiol. Meth.*, 41, 85, 2000.

240. Kenzaka, T. et al., Rapid monitoring of *Escherichia coli* in Southeast Asian urban canals by fluorescent-bacteriophage assay, *J. Health Sci.*, 52, 666, 2006.

241. Baudart, J. et al., Rapid quantification of viable Legionella in nuclear cooling tower waters user filter cultivating fluorescent in situ hybridization and solid phase cytometry, *J. Appl. Microbiol.*, 118, 1238, 2015.

242. Lukumbuzya, M. et al., A multicolor fluorescence *in situ* hybridization approach using an extended set of fluorophores to visualize microorganisms, *Front. Microbiol.*, 10, 1383, 2019.

243. Baudart, J. et al., Rapid and sensitive enumeration of viable diluted cells of members of the family *Enterobacteriaceae* in freshwater and drinking water, *Appl. Environ. Microbiol.*, 68, 5057, 2002.

244. Servais, P. et al., Abundance of culturable versus viable *Escherichia coli* in freshwater, *Can. J. Microbiol.*, 55, 905, 2009.

245. García-Hernández, J. et al., A combination of direct viable count and fluorescence *in situ* hybridization for specific enumeration of viable *Lactobacillus delbrueckii subsp. bulgaricus* and *Streptococcus thermophilus*, *Lett. Appl. Microbiol.*, 54, 247, 2012.

246. Rohde, A. et al., Differential detection of pathogenic *Yersinia* spp. by fluorescence *in situ* hybridization, *Food Microbiol.*, 62, 39, 2017.

247. Kubitschek, H.E. and Friske, J.A., Determination of bacterial cell volume with the coulter counter, *J. Bacteriol.*, 168, 1466, 1986.

248. Kogure, K. and Koike, I., Particle counter determination of bacterial biomass in seawater, *Appl. Environ. Microbiol.*, 53, 274, 1987.

249. Shapiro, H.M., Microbial analysis at the single-cell level: tasks and techniques, *J. Microbiol. Meth.*, 42, 3, 2000.

250. Khan, M.M., Pyle, B.H., and Camper, A.K., Specific and rapid enumeration of viable but non-culturable and viable-culturable gram-negative bacteria by using flow cytometry, *Appl. Environ. Microbiol.*, 76, 5088, 2010.

251. Aggrawal, S., Jeon, Y., and Hozalski, R.M., Feasibility of using a particle counter or flow-cytometer for bacterial enumeration in the assimilable organic carbon (AOC) analysis method, *Biodegradation*, 26, 387, 2015.

252. Jepras, R.I. et al., Development of a robust flow cytometric assay for determining numbers of viable bacteria, *Appl. Environ. Microbiol.*, 61, 2969, 1995.

253. Davey, H.M. and Kell, D.B., Flow cytometry and cell sorting of heterogeneous microbial populations: the importance of single-cell analyses, *Microbiol. Rev.*, 60, 641, 1996.

254. Assunção, P. et al., Use of flow cytometry for enumeration of *Mycoplasma mycoides subsp. mycoides* large-colony type in broth medium, *J. Appl. Microbiol.*, 100, 878, 2006.

255. Kerstens, M. et al., Quantification of *Candida albicans* by flow cytometry using TO-PRO®-3 iodide as a single stain viability dye, *J. Microbiol. Methods*, 92, 189, 2013.

256. Kerstens, M. et al., Flow cytometric enumeration of bacteria using TO-PRO®-3 iodide as a single stain viability dye, *J. Lab Autom.*, 19, 555, 2014.

257. Buysschaert, B. et al., Reevaluating multicolor flow cytometry to assess microbial viability, *Appl. Microbiol. Biotechnol.*, 100, 9037, 2016.

258. Ou, F. et al., Absolute bacterial cell enumeration using flow cytometry, *J. Appl. Microbiol.*, 123, 464, 2017.

259. Ou, F. et al., Bead-based flow-cytometric cell counting of live and dead bacteria, *Methods Mol. Biol.*, 1968, 123, 2019.

260. Wang, X. and Slavik, M.F., Rapid detection of Salmonella in chicken washes by immunomagnetic separation and flow cytometry, *J. Food Protect.*, 62, 717, 1999.

261. Gunasekera, T.S., Attfield, P.V., and Veal, D.A., A flow cytometry method for rapid detection and enumeration of total bacteria in milk, *Appl. Environ. Microbiol.*, 66, 1228, 2000.

262. Flint, S., Description and validation of a rapid (1 h) flow cytometry test for enumerating thermophilic bacteria in milk powders, *J. Appl. Microbiol.*, 102, 909, 2007.

263. Wilkes, J.G. et al., Reduction of food matrix interference by a combination of sample preparation and multi-dimensional gating techniques to facilitate rapid, high sensitivity analysis for *Escherichia coli serotype 0157* by flow cytometry, *Food Microbiol.*, 30, 281, 2012.

264. Raymond, Y. and Champagne, C.P., The use of flow cytometry to accurately ascertain total and viable counts of *Lactobacillus rhamnosus* in chocolate, *Food Microbiol.*, 46, 176, 2015.

265. Longin, C. et al., Application of flow cytometry to wine microorganisms, *Food Microbiol.*, 62, 221, 2017.

266. Li, R.C., Lee, S.W., and Lam, J.S., Novel method for assessing post antibiotic effect by using the coulter counter, *Antimicrob. Agents Chemother.*, 40, 1751, 1996.

267. Okada, H. et al., Enumeration of bacterial cell numbers and detection of significant bacteriuria by use of a new flow cytometry-based device, *J. Clin. Microbiol.*, 44, 3596, 2006.

268. Orth, R. et al., An efficient method for enumerating oral spirochetes using flow cytometry, *J. Microbiol. Methods*, 80, 123, 2010.

269. Vollmer, T. et al., Detection of bacterial contamination in platelet concentrates by a sensitive flow cytometric assay (BactiFlow): a multicenter validation study, *Transfusion, Med.* 22, 262, 2012.

270. Fontana, C. et al., Use of low cytometry for rapid and accurate enumeration of live pathogenic Leptospira strains, *J. Microbiol. Methods*, 132, 34, 2017.

271. Lebaron, P., Parthuisot, N., and Catala, P., Comparison of blue nucleic acid dyes for flow cytometric enumeration of bacteria in aquatic systems, *Appl. Environ. Microbiol.*, 64, 1725, 1998.

272. Chen, P. and Li, C., Real-time quantitative PCR with gene probe, fluorochrome and flow cytometry for microorganisms, *J. Environ. Monit.*, 7, 257, 2005.

273. Hammes, F. et al., Flow-cytometric total bacterial cell counts as a descriptive microbiological parameter for drinking water treatment processes, *Water Res.*, 42, 269, 2008.

274. Taguri, T. et al., A rapid detection method using flow cytometry to monitor the risk of Legionella in bath water, *J. Microbiol. Methods*, 86, 25, 2011.

275. Gillespie, S. et al., Assessing microbiological water quality in drinking water distribution systems with disinfectant residual using flow cytometry, *Water Res.*, 65, 224, 2014.

276. Frossard, A., Hammes, F., and Gessner, M.O., Flow cytometric assessment of bacterial abundance in soils, sediments and sludge, *Front. Microbiol.*, 7, 903, 2016.

277. Van Nevel, S. et al., Flow cytometric bacterial cell counts challenge conventional heterotrophic plate counts for routine microbiological drinking water monitoring, *Water Res.*, 113, 191, 2017.

278. Helmi, K. et al., Assessment of flow cytometry for microbial water quality monitoring in cooling tower and oxidizing biocide treatment efficiency, *J. Microbiol. Methods*, 152, 201, 2018.

279. Vignola, M. et al., Flow-cytometric quantification of microbial cell on sand from water biofilters, *Water Res.*, 143, 66, 2018.

280. Cheswick, R. et al., Comparing flow cytometry with culture-based methods for microbial monitoring and as a diagnostic tool for assessing drinking water treatment process, *Environ. Intl.*, 130, 104893, 2019.

281. Fujioka, T. et al., Assessment of online bacterial particle counts for monitoring the performance of reverse osmosis membrane process in potable reuse, *Sci. Total Environ.*, 667, 540, 2019.

282. Marie, D. et al., Enumeration of marine virus in culture and natural samples by flow cytometry, *Appl. Environ. Microbiol.*, 65, 45, 1999.

283. Gates, I.V. et al., Quantitative measurement of varicella-zoster virus infection by semi-automated flow cytometry, *Appl. Environ. Microbiol.*, 75, 2027, 2009.

284. Dawson, E., Rapid, direct quantification of viruses in solution using the VriCyt virus counter, *J. Biomol. Tech.*, 23, S10, 2012.

285. Ma, L. et al., Rapid quantification of bacteria and viruses in influent, settled water, activated sludge and effluent from wastewater treatment plant using flow cytometry, *Water Sci. Technol.*, 68, 1763, 2013.

286. Brown, M.R. et al., Flow cytometric quantification of virus in activated sludge, *Water Res.*, 68, 414, 2015.

287. Broadaway, S.C., Barton, S.A., and Pyle, B.H., Rapid staining and enumeration of small numbers of total bacteria in water by solid-phase laser cytometry, *Appl. Environ. Microbiol.*, 69, 4272, 2003.

288. Delgado-Viscogliosi, P. et al., Rapid method for enumeration of viable *Legionella pneumophila* and other *Legionella* spp. in water, *Appl. Environ. Microbiol.*, 71, 4086, 2005.

289. Marennikova, S.S., Gurvich, E.B., and Shelukhina, E.M., Comparison of the properties of five pox virus strains isolated from monkeys, *Arch. Virol.*, 33, 201, 1971.

290. Ajello, F., Massenti, M.F., and Brancato, P., Foci of degeneration produced by measles virus in cell cultures with antibody free liquid medium, *Med. Microbiol. Immunol.*, 159, 121, 1974.

291. Brick, D.C., Oh, J.O., and Sicher, S.E., Ocular lesions associated with dissemination of type 2 herpes simplex virus from skin infection in newborn rabbits, *Invest. Ophthalmol. Vis. Sci.*, 21, 681, 1981.

292. USEPA, Total culturable virus quantal assay, in *USEPA Manual of Methods for Virology*, U.S. Environmental Protection Agency Office of Science and Technology, Washington, D.C., 2001, chap. 15.

293. Lee, H.K. and Jeong, Y.S., Comparison of total culturable virus assay and multiplex integrated cell culture-PCR for reliability of waterborne virus detection, *Appl. Environ. Microbiol.*, 70, 3632, 2004.

294. Cashdollar, J.L. et al., Development and evaluation of EPA method 1615 for detection of Enterovirus and Norovirus in water, *Appl. Environ. Microbiol.*, 79, 215, 2013.

295. USEPA, Method 1615: measurement of Enterovirus and Norovirus occurrence in water by culture and RT-qPCR, *U.S.* Environmental Protection Agency Office of Research and Development, Cincinnati, *Ohio*, 2014.

296. Cooper, R.D., The plaque assay of animal viruses, *Adv. Virus Res.*, 8, 319, 1961.

297. USEPA, Cell culture procedures for assaying plaque-forming viruses, in *USEPA Manual of Methods for Virology*, U.S. Environmental Protection Agency Office of Science and Technology, Washington, D.C., 1987, chap. 10.

298. Cornax, R. et al., Application of direct plaque assay for detection and enumeration of bacteriophages for *Bacteroides fragilis* from contaminated-water samples, *Appl. Environ. Microbiol.*, 56, 3170, 1990.

299. USEPA, Procedures for detecting coliphages, in *USEPA Manual of Methods for Virology*, U.S. Environmental Protection Agency Office of Science and Technology, Washington, D.C., 2001, chap. 16.

300. USEPA, Method 1602: male-specific (F+) and somatic coliphage in water by single agar layer (SAL) procedure, U.S. Environmental Protection Agency Office of Water, Washington, D.C., 2001.

301. Mocé-Llivina, L., Lucena, F., and Jofre, J., Double-layer plaque assay for quantification of Enteroviruses, *Appl. Environ. Microbiol.*, 70, 2801, 2004.

302. Matrosovich, M. et al., New low-viscosity overall medium for viral plaque assay, *Virol. J.*, 3, 63, 2006.

303. Mazzocco, A. et al., Enumeration of bacteriophages by the direct plating plaque assay, *Methods Mol. Biol.*, 501, 77, 2009.

304. Gonzales-Hernandez, M.B., Braqazzi, C.J., and Wobus, C.E., Plaque assay for murine nonovirus, *J. Vis. Exp.*, 66, 4297, 2012.

305. Shurtleff, A. et al., Standardization of the filovirus plaque assay for use in pre-clinical studies, *Viruses*, 4, 3511, 2012.

306. Smither, C.J. et al., Comparison of the plaque assay and 50% tissue culture infectious dose assay as methods for measuring filovirus infectivity, *J. Virol. Methods*, 193, 565, 2013.

307. Cormier, J. and Janes, M., A double layer plaque assay using spread plate technique for enumeration of bacteriophage MS2, *J. Virol. Methods*, 196, 86, 2014.

308. Kimmitt, P.T. and Redway, K.F., Evaluation of potential for virus dispersal during hand drying: a comparison of the three methods, *J. Appl. Microbiol.*, 120, 478, 2016.

309. McMinn, B.R. et al., Concentration and quantification of somatic and F+ coliphages from recreational waters, *J. Virol. Methods*, 249, 58, 2017.

310. McMinn, BR. et al., Comparison of somatic and F+ coliphage enumeration methods with large volume surface water samples, *J. Virol. Methods*, 261, 63, 2018.

311. Miller, M.J., *Encyclopedia of Rapid Microbiological Methods*, DHI Publishing, LLC, River Grove, IL, 2005.

312. Cundell, A.M., Opportunities for rapid microbial methods, *Eur. Pharm. Rev.*, 1, 62, 2006.

313. Chollet, R. et al., Rapid detection and enumeration of contaminants by ATP bioluminescence using the Milliflex® rapid microbiology detection and enumeration system, *J. Rapid Methods Automation Microbiol.*, 16, 256, 2008.

314. Moldenhauer, J., Overview of rapid microbiological methods, *Principles of Bacterial Detection: Biosensors, Recognition Receptors and Microsystems*, Springer, New York, 2008.

315. Miller, M.J. Rapid microbiological methods, *Microbiology in Pharmaceutical Manufacturing*, Vol. 2, DHI Publishing, River Grove, IL and PDA, Bethesda, MD, 2008.

316. Miller, M.J., Evaluation of a continuous and instantaneous viable and nonviable environmental monitoring technology based on optical spectroscopy, *PDA Annual Meeting-Science Driven Manufacturing: The Application of Emerging Technologies*, Colorado Springs, Colorado, 2008.

317. Miller, M.J., Walsh, M.R., Shrake, J.L., Dukes, R.E., and Hill, D.B., Evaluation of the BioVigilant IMD-A®, a novel optical spectroscopy technology for the continuous and real-time environmental monitoring of viable and nonviable particles. Part II: case studies in environmental monitoring during aseptic filling, intervention assessments and glove integrity testing in manufacturing isolators, *PDA J. Pharm. Sci. Technol.* 63, 258, 2009.

318. Smith, R., Von Tress, M., Tubb, C., and Vanhaecke, E., Evaluation of the ScanRDI(R) as a rapid alternative to the pharmacopoeial sterility test method: comparison of the limits of detection. *PDA J. Pharm. Sci. Technol.*, 64, 356, 2010.

319. Baumstummler, A. et al., Detection of microbial contaminants in mammalian cell cultures using a new fluorescence based staining method, *Lett. Appl. Microbiol.*, 51, 671, 2010.

320. Baumstummler, A. et al., Development of a nondestructive fluorescence-based enzymatic staining of microcolonies for enumerating bacterial contamination in filterable products, *J. Appl. Microbiol.*, 110, 69, 2011.

321. Miller, M.J., Case study of a new growth-based rapid microbiological method (RMM) that detects the presence of specific organisms and provides an estimation of viable cell count. *Am. Pharm. Rev.* 15, 18, 2012.

322. Cundell, A. et al., Novel concept for online water bioburden analysis: key considerations, applications, and business benefits for microbiological risk reduction, *Am. Pharm. Rev.*, 16, 26–31, 2013.

323. Xu, J., Biosensor-based rapid pathogen detection: are we there yet? *Food Safety Mag.*, 32, 2019.

324. Boubetra, A. et al., Validation of alternative methods for the analysis of drinking water and their application to *Escherichia coli*, *Appl. Environ. Microbiol.*, 77, 3360, 2011.

325. PDA, Technical Report #33 (Revised 2013): evaluation, validation and implementation of alternative and rapid microbiological methods, *Parenteral Drug Association Inc.*, Bethesda, Maryland, 2013.

326. Ijzerman-Boon, P.C. and van den Heuval, E.R., Validation of qualitative microbiological test methods, *Pharm. Stat.*, 14, 120, 2015.

327. Murphy, T. et al., Evaluation of PDA Technical Report No.33 statistical testing recommendations for a rapid microbiological method case study, *PDA J. Pharm. Sci. Technol.*, 69, 526, 2015.

328. Limberg, B.J. et al., Performance equivalence and validation of the Soleris automated system for quantitative microbial content testing using pure suspension cultures, *J. AOAC Int.*, 99, 131, 2016.

329. Gordon, O., Goverde, M., Staerk, A., and Roesti, D., Validation of Milliflex® Quantum for Bioburden Testing of Pharmaceutical Products, *PDA J. Pharm. Sci. Technol.*, 71, 206, 2017.

330. EP, 5.1.6 Alternative methods for control of microbiological quality, *European Pharmacopoeia*, 10th ed., 2019.

331. USP, <1223> Validation of alternative microbiological methods, USP42-NF37, *The United States Pharmacopeial Convention*, 2019.

332. de Boer, E. and Beumer, R.R., Methodology for detecting and typing foodborne microorganisms, *Int. J. Food Microbiol.*, 50, 119, 1999.

333. Fung, D.Y.C., Rapid methods and automation in microbiology, *Comp. Rev. Food Sci. Food Safety*, 1, 3, 2002.

334. Hussong, D. and Mello, R., Alternative microbiology methods and pharmaceutical quality control, *Am. Pharm. Rev.*, 9, 62, 2006.

335. Duguid, J. Top ten validation considerations when implementing a rapid mycoplasma test, *Am. Pharm. Rev.* 13, 26, 2010.

336. Miller, M.J., Developing a validation strategy for rapid microbiological methods, *Am. Pharm. Rev.* 13: 28, 2010.

337. Sanders, D.L. et al., Design and validation of a novel quantitative method for the rapid bacterial enumeration using programmed stage movement scanning electron microscopy, *J. Microbiol. Methods*, 91, 544, 2012.

338. Osono, E. et al., Rapid detection of microbes in the dialysis solution by the microcolony fluorescence staining method (Milliflex quantum), *Biocontrol Sci.*, 19, 57, 2014.

339. Miller, M.J., van den Heuvel, E.R., and Roesti, D., The role of statistical analysis in validating rapid microbiological methods, *Eur. Pharm. Rev.*, 21, 46, 2016.

340. Friedman, E.M. et al., In-process microbial testing and statistical properties of a rapid alternative to compendial enumeration methods, *PDA J. Pharm. Sci. Technol.*, 69, 264, 2015.

341. Feldsine, P, Abeyta, C., and Andrews, W.H., AOAC international methods committee guidelines for validation of qualitative and quantitative food microbiological official methods of analysis, *J. AOAC Int.*, 85, 1187, 2002.

342. Clough, H.E. et al., Quantifying uncertainty associated with microbial count data: a Bayesian approach, *Biometrics*, 61, 610, 2005.

343. Lombard, B. et al., Experimental evaluation of different precision criteria applicable to microbiological counting methods, *J. AOAC Int.*, 88, 830, 2005.

344. Corry, J.E. et al., A critical review of the enumeration of food micro-organisms, *Food Microbiol.*, 24, 230, 2007.

345. Jarvis, B., Hedges, A.J., and Corry, J.E., Assessment of measurement uncertainty for quantitative methods of analysis: comparative assessment of the precision (uncertainty) of bacterial colony counts, *Int. J. Food Microbiol.*, 116, 44, 2007.

346. Jarvis, B., Hedges, A.J., and Correy, J.E., The contribution of sampling uncertainty to total measurement uncertainty in the enumeration of microorganisms in foods, *Food Microbiol.*, 30, 362, 2012.

347. Sossé, S.A., Saffaj, T., and Ihssane, B., Validation and measurement uncertainty assessment of a microbiological method using generalized pivotal quantity procedure and Monte-Carlo simulation, *J. AOAC Int.*, 101, 1205, 2018.

3

Culturing and Preserving Microorganisms

Lorrence H. Green

CONTENTS

Introduction

The use of culture as a diagnostic technique is described in Chapter 10. The purpose of this chapter to discuss only the principles of enrichment culture. This technique has been in use for more than 120 years.[1,2] It consists of incubating a sample in a medium that encourages the growth of an organism of interest, while inhibiting the growth of others. In this way, it can assist the technician in isolating pure colonies of microorganisms from mixtures in which the organisms represent only a very small percentage of the overall flora, and in which isolation by streaking may not be practical. In some cases, as in the isolation of salmonella, first growing a sample in an enrichment culture is a necessary first step to increase the relatively small number of organisms usually found in the primary sample.[3] While the use of enrichment broths in food and environmental microbiology has been demonstrated, their use in clinical microbiology has not fully been established.[4] In some cases, the use of enrichment broths has done little more than add unnecessary cost and extra testing.[5]

Designing an Enrichment Medium

Introduction

While the media used to enrich for specific microorganisms might differ greatly, they should all contain an energy source, a carbon source, and a source of major and trace elements. Factors such as the pH, temperature, and oxygen tension should be appropriate to enhance the growth of the microorganism of interest.[1] In addition to providing components that select for specific microorganisms, enrichment media may also contain antibiotics or other compounds that inhibit the growth of microbiological competitors. In the case of autotrophic microorganisms, such as cyanobacteria or algae, enrichment media may lack the carbon sources required by heterotrophic microorganisms.

Historically, enrichment cultures have been developed by selecting chemically defined components, and empirically testing their ability to enrich for the organism of interest. While this might be useful in some settings, recent studies have found that the microbial interactions during the enrichment process can be important. This interaction can be either positive in that some microorganisms will only grow in culture if they can interact with others; or negative in that the presence of other organisms normally found in the same biosphere might inhibit growth.[6–8]

In recent years, molecular methods have been employed in the development of enrichment media. This is especially true in environmental studies in which the biome may have hundreds of different species in a starting sample. Mu et al. performed an initial enrichment culture of marine sediment and obtained isolated colonies. These were then subjected to 16S RNA analysis. By altering the parameters of the enrichment process, they were able to affect the relative ratios of 16S RNA sequences that they observed.

DOI: 10.1201/9781003099277-4

In addition to providing an enrichment method, this technique also allowed them to identify viable but dormant microorganisms.[9]

Energy Source

Most microorganisms are heterotrophs, and as a result, obtain their energy by ingesting organic molecules. When these molecules are broken down, energy contained in the chemical bonds is liberated.[10] Some microorganisms are autotrophs and are capable of obtaining energy either by oxidizing inorganic chemical compounds (chemoautotrophs) or by directly utilizing light (photosynthetic).[10] Chemoautotrophs derive energy from reduced inorganic molecules or ions, which can include H_2, NH_4^+, Fe^{2+}, and S^0.[1,11,12] Photosynthetic microorganisms derive their energy directly from light sources, with bacteria and photosynthetic eukaryotes favoring light in the red portion of the spectrum, while cyanobacteria favor light in the blue portion.[1]

Source of Carbon

Because autotrophic organisms derive their energy from light, this usually requires little more than a source of simple carbon, such as bubbling CO_2 gas into the media or by adding a solid carbonate source. Heterotrophic organisms can obtain their carbon from a wide variety of sources. The most common of these are carbohydrates. Many of these are actually used as the component parts of biochemical tests used to identify microorganisms. These include carbohydrates (i.e., glucose, maltose, lactose, sucrose, etc.), acids found in the TCA cycle (i.e., succinic, oxaloacetic, malic, α-ketogluaric, etc.), amino acids (i.e., alanine, arginine, glycine, serine, etc.), and fatty acids (i.e., lactic, pyruvic, butyric, propionic, etc.).

In the past few years, it has been realized that microorganisms exist that are capable of using toxic compounds as carbon sources. Xu et al. have recently reviewed the use of microorganisms in bioremediation.[13]

Trace and Major Elements

All microorganisms require varying amounts of certain elements. Major elements such as nitrogen, potassium, sodium, magnesium, and calcium can be found as salts. These can include chlorides (NH_4Cl, KCl, $NaCl$), sulfates ($MgSO_4.7H_2O$, Na_2SO_4, $(NH_4)_2SO_4$), carbonates ($MgCO_3$, $CaCO_3$), nitrates ($(CaNO_3)_2.4H_2O$), and phosphates (K_2HPO_4, KH_2PO_4).[1] Other elements may also be required in trace amounts. These can include nickel, iron, zinc, manganese, copper, cobalt, boron, molybdenum, vanadium, strontium, aluminum, rubidium, lithium, iodine, and bromine.[14]

pH

While most organisms prefer to grow in a neutral pH environment, the pH level can be used as an enrichment method. Acidophiles, for instance, can be enriched by lowering the pH in the media.[15]

Physical Components

Although researchers usually tend to focus on ingredients that can be added to a liquid medium, it should be remembered that manipulating the physical environment in which the culture is placed for temperature, oxygen tension, and even pressure, may offer some microorganisms a competitive advantage in reproduction. The levels of these agents will be determined by which organism the technician is trying to isolate.[16] Thermophiles, such as *Thermogladius shocki*, for, instance can be enhanced by growing cultures at temperatures as high as 80–95°C.[17] Selective enrichment can also be obtained by varying the concentrations of media.[18]

Antimicrobial Agents

Enrichment for specific groups or species of microorganisms can not only be achieved by defining conditions that will preferentially allow them to grow, but can also be achieved through the use of agents that will inhibit the growth of competing organisms. In some instances, fast-growing organisms can be inhibited, giving fastidious or slower-growing organisms an advantage in a particular sample. An interesting clinical consequence of the use of antimicrobial agents as a method of enrichment is the observation of resistant organisms in patients who have been hospitalized. Hui et al.[19] have found that previous antibiotic exposure of nosocomial patients using mechanical ventilation can lead to an increase in antibiotic resistance. Antibiotics can also be used for enriching specific resistant strains. It is not known if this technique provides any advantages over the standard methods of testing isolated colonies (i.e., MIC or KB) for antibiotic resistance.[20]

Preservation of Microorganisms

Introduction

Many laboratory test procedures require the use of microorganisms as reagents. Many of these are naturally occurring, but advances in biotechnology have also resulted in the creation of engineered microorganisms. In both instances, to obtain consistent results, these organisms must be preserved in a manner that will allow for their genetic stability and long-term survival. Preservation of microorganisms can be accomplished by a variety of methods. These can include subculturing them, reducing their metabolic rate, or putting them into a state of quasisuspended animation.[21] The chosen method will usually depend on which organism one is trying to preserve. Despite the introduction of newer methodologies, many of the techniques that are used have not changed much since the previous editions of this book. Most still involve the use of drying, lyophilization, or storage in freezing or subfreezing temperatures. A flowchart as well as specific techniques for preservation can be found at www.cabri.org (accessed January 29, 2020).

Methods

Serial Subculture

This is a simple method in which the cultures are periodically passed in liquid or agar media. Some cultures can be stored on agar media in sealed tubes and survive for as long as 10 years.[22] This method has been used extensively for fungi, but usually requires storage under mineral oil.[21] Despite its simplicity and applicability for cultures that cannot survive harsher preservation methods, serial subculture is not a very satisfactory method. In our hands we have had problems with contamination, culture death, and the unintended selection of mutants. This has been particularly difficult in cases in which we were trying to develop new products for the identification of microorganisms. It was imperative to periodically confirm the identity of the organisms being used in the database.

Storage at Low Temperatures

Some species are capable of being stored for long periods at 4–8°C on agar plates or on slants. Sorokulova et al. have found that the use of acacia gum in storage media will allow long-term storage of organisms even at room temperature.[23]

In our lab, we have been able to store cultures at –20°C for extended periods. We have found that this can be accomplished by growing the cultures in liquid media for 24–48 hours and then mixing one part sterile glycerol to three parts culture. While this does not work for all organisms, it allows the use of a standard refrigerator for preserving cultures. Many laboratories store organisms at –70°C to –80°C by first mixing them with 10% glycerol.[24] Cultures can also be stored indefinitely in liquid nitrogen. While the use of liquid nitrogen is a good method for storing microorganisms, for most laboratories, it requires the expense of constantly refilling a specialized thermos. Tedeschi and De Paoli in their review of cryogenic methods of preserving microorganisms point out that in this era of being able to manipulate the microorganism genome, that it is important to pick a cryogenic preservation method that will preserve the genotype and phenotype of the original organism.[25]

Freeze-Drying

This is a widely used method in which a suspension of microorganisms is frozen and then subjected to a vacuum to sublimate the liquid. The resulting dried powder is usually stored in vials sealed in a vacuum. Many factors can affect the stability of the culture. These include the growth media, the age, phase and concentration of the organisms, lyoprotectant added, and the mass of the cotton added to the storage ampule.[21,26–28]

The length of sample viability can vary substantially, but cultures surviving for up to 20 years have been reported.[29] Freeze-drying requires a cryoprotective agent to provide maximal stability. Usually, the organisms are suspended in 10% skim milk.[30]

Storage in Distilled Water

This was cited as a method of preservation of *Pseudomonas* species and fungi in the first edition of this book.[21] Recent studies have found that it is still an effective method of preserving fungi,[31] as well as a wide variety of bacteria (including *Pseudomonas fluorescens*, *Erwinia* spp., *Xanthomonas campestris*, *Salmonella* spp., *Yersinia enterocolitica*, *Escherichia coli* O157:H7, *Listeria monocytogenes*, and *Staphylococcus aureus*).[32] Stationary-phase organisms grown on agar media and then suspended in 10 mL of sterile water were found to be stable when sealed with parafilm membranes and then stored at room temperature in the dark. Even greater stability could be obtained by suspending the organisms in a screw-capped tube with phosphate buffered saline (PBS) at pH 7.2, containing 15.44 μM KH_2PO_4, 1.55 mM NaCl, and 27.09 μM Na_2HPO_4.[24]

Drying

Sterile soil or sand has been used for preserving spore-forming organisms by adding suspensions and then drying at room temperature. In addition, bacterial suspensions have been mixed with melted gelatin and then dried in a desiccator. Both methods have produced samples that are stable for long periods of time.[21] A newer method involves using a microliter quantity of a bacterial suspension that is mixed with a predried activated charcoal cloth-based matrix contained within a resealable system. This can then be stored. Experiments with *Escherichia coli* have found that viability of over a year at 4°C can be obtained.[33]

Recovery and Viability of Preserved Microorganisms

Several factors can affect the viability of microorganisms that have been stored by either freezing or drying. These can include the temperature at which microorganisms are recovered, the type and volume of the media used for recovery, and even how quickly microorganisms are solubilized in a recovery medium.[21] Once reconstituted, microorganisms should be evaluated for cell survival.

Usually, the losses of preservation can be overcome by initially preserving large numbers of microorganisms. In doing this, one must be careful to avoid two problems. The first is that if only a very small fraction of organisms is recovered, this can result in the selection of biochemically distinct strains. The second is that if a culture has inadvertently been contaminated with even a few microorganisms, the preservation technique may lead to a selection for the contaminant.[34]

Drying and freeze-drying have been known to cause changes in several characteristics of preserved microorganisms. These can include colonial appearance and pathogenicity.[21] After revival from a frozen or dried state, many protocols usually advise subculturing the organisms at least two or three times in an attempt to restore any characteristics that may have been lost, and to confirm the genetic identity.[21]

REFERENCES

1. Aaronson, S., Enrichment culture, *CRC Practical Handbook of Microbiology*, O'Leary, W., Ed., CRC Press, Boca Raton, FL, 337, 1989.
2. Prescott, S.C. and Winslow, C-E.A., The relative value of dextrose broth, phenol broth and lactose bile as an enrichment media for the isolation of B.Coli, *Am J Public Hygiene*, 18, 19, 1908
3. Keen, J., Durso, I. and Meehan, L., Isolation of *Salmonella enterica* and Shiga-toxigenic *Escherichia coli* O157 from feces of animals in public contact areas of United States zoological parks, *Appl Environ Microbiol*, 73, 362, 2007.
4. Miles, K.I., Wren, M.W.D. and Benson, S., Is enrichment culture necessary for clinical samples?, *Br J Biomed Sci*, 63, 87, 2006.

5. Samir, S., Shah, S.S., Hines, E.M. and McGowan, K.L., Cerebrospinal fluid enrichment broth cultures rarely contribute to the diagnosis of bacterial meningitis, *Ped Infec Dis J*, 31, 318, 2012.

6. Stewart, EJ., Growing unculturable bacteria, *J Bacteriol*, 194, 4151, 2012.

7. Ottesen, A.R., Gonzalez, A., Bell, R., Arce, C., Rideout, S., Allard, M., Evans, P., Strain, E., Musser, S., Knight, R., Brown, E. and Pettengill, J.B., Co-enriching microflora associated with culture based methods to detect salmonella from tomato phyllosphere, *Plos One*, 8 (9), e73079, 2013

8. Al-Zeyara, S.A., Jarvis, B. and Mackey, B.M., The inhibitory effect of natural microflora of food on growth of Listeria monocytogenes in enrichment broths, *Int J Food Microbiol*, 145, 98, 2011.

9. Mu, D.-S., Liang, Q.-Y., Wang, X.-M., Lu, D.-C., Shi, M.-J., Chen, G.-J. and Du, Z.-J., Metatranscriptomic and comparative genomic insights into resuscitation mechanisms during enrichment culturing, *Microbiome*, 6, 230, 2018.

10. Urry, L.A., Cain, M.L., Wasserman, S.A., Minorsky, P.V., Orr, R.B. and Campbell, N.A., *Biology in Focus*, 3rd ed., Pearson, Hoboken, NJ, 489, 2020.

11. Ohmura, N., Sasaki, K., Matsumoto, N. and Saiki, H., Anaerobic respiration using Fe^{3+}, S^0, and H_2 in the chemolithoautotrophic bacterium, Acidithiobacillus ferrooxidans, *J Bacteriol*, 184, 2081, 2002.

12. Harrold, Z.R., Skidmore, M.L., Hamilton, T.L., Desch, L., Amada, K., van Gelder, W., Glover, K., Roden, E.E. and Boyd, E.S., Aerobic and anaerobic thiosulfate oxidation by a cold-adapted, subglacial chemoautotroph, *Appl Environ Microbiol*, 82, 1486, 2016.

13. Xu, X., Liu, W., Tian, S., Wang, W., Qige, Q., Jiang, P., Gao, X., Li, F., Li, H. and Yu, H., Petroleum hydrocarbon-degrading bacteria for the remediation of oil pollution under aerobic conditions: a perspective analysis, *Front Microbiol*, 9, 1, 2018.

14. Nurfarahin, A.H., Mohamed, M.S. and Phang. L.Y., Culture medium development for microbial-derived surfactants production—an overview, *Molecules*, 23, 1049, 2018.

15. Johnson, D.B. and Hallberg, K.B., The microbiology of acidic mine waters, *Res Microbiol*, 154, 466, 2003.

16. Ohmura, M., Kaji, S., Oomi, G., Miyaki, S. and Kanazawa, S., Effect of pressure on the viability of soil microorganisms, *High Pres Res*, 27, 129, 2007.

17. Meyer-Dombard, D.R., Shock, E. L. and Amend, J.P., Effects of trace element concentrations on culturing thermophiles, *Extremophiles*, 16, 317, 2012.

18. Böllmann, J. and Martienssen, M., Comparison of different media for the detection of denitrifying and nitrate-reducing bacteria in mesotrophic aquatic environments by the most probable number method, *J Microbiol Methods*, 168, 1, 2020.

19. Hui, C., Lin, M.-C., Jao, M.-S., Ne, R., Liu, T.-C. and Wu, R.-G., Previous antibiotic exposure and evolution of antibiotic resistance in mechanically ventilated patients with nosocomial infections, *J Crit Care*, 28, 728, 2013.

20. Frickmann, H., Hahn, A., Schwarz, N.G., Hagen, R.M., Dekker, D., Hinz, R., Micheel, V., Hogan, B., May, J. and Rakotozandrindrainy, R., Influence of broth enrichment as well as storage and transport time on the sensitivity of MRSA Surveillance in the tropics, *Euro J Microbiol Immunol*, 7, 274–277, 2017.

21. Lapage, S., Redway, K. and Rudge, R., Preservation of microorganisms, *CRC Practical Handbook of Microbiology*, O'Leary, W., Ed., CRC Press, Boca Raton, FL, 321, 1989.

22. Antheunisse, J., Preservation of microorganisms, Antonie van Leeuwenhoek, *J Microbiol Serol*, 38, 617, 1972.

23. Sorokulova, I., Watt, J., Olsen, E.,Globa, L., Moore, T., Barbaree, J. and Vodyanoy, V., Natural biopolymer for preservation of microorganisms during sampling and storage, *J Microbiol Methods*, 88, 140, 2012.

24. Cameotra, S.S., Preservation of microorganisms as deposits for patent application, *Biochem Biophys Res Commun*, 353, 849, 2007.

25. Tedeschi, R. and De Paoli, P., Collection and preservation of frozen microorganisms, *Methods Mol Biol*, 675, 313, 2011.

26. Morgan, C.A., Herman, N., White, P.A. and Vesey, G., Preservation of micro-organisms by drying: a review, *J Microbiol Methods*, 66, 183, 2006.

27. Peiren, J., Hellemans A. and De Vos, P., Impact of the freeze-drying process on product appearance, residual moisture content, viability, and batch uniformity of freeze-dried bacterial cultures safeguarded at culture collections, *Appl Microbiol Biotechnol*, 100, 6239, 2016.

28. Yu, C., Reddy, A.P., Simmons, C.W., Simmons, B.A., Singer, S.W. and VanderGheynst, J.S., Preservation of microbial communities enriched on lignocellulose under thermophilic and high-solid conditions, *Biotechnol Biofuels*, 206, 2, 2015.

29. Miyamoto-Shinohara, Y., Sukenobe, J., Imaizumi, T. and Nakahara, T., Survival curves for microbial species stored by freeze-drying, *Cryobiology*, 52, 27, 2006.

30. Crespo, M.J., Abarca, M.L. and Cabañes, F.J., Evaluation of different preservation and storage methods for *Malassezia* spp., *J Clin Microbiol*, 38, 3872, 2000.

31. Richter, D.L., Dixon, T.G. and Smith, J.K., Revival of saprotrophic and mycorrhizal basidiomycete cultures after 30 years in cold storage in sterile water, *Can J Microbiol*, 62, 932, 2016.

32. Liao, C.-H. and Shollenberger, L.M., Survivability and long-term preservation of bacteria in water and in phosphate-buffered saline, *Lett Appl Microbiol*, 37, 45, 2003.

33. Hays, H.C.W., Millner, P.A., Jones, J.K. and Rayner-Brandes, M.H., A novel and convenient self-drying system for bacterial preservation, *J Microbiol Methods*, 63, 29, 2005.

34. Wessman, P., Håkansson, S., Leifer, K. and Rubino S., Formulations for freeze-drying of bacteria and their influence on cell survival, *J Vis Exp*, 78, 2013.

4

Stains for Light Microscopy

Stuart Chaskes and Rita Austin

CONTENTS

DOI: 10.1201/9781003099277-5

The following is a selection of staining methods that may be of assistance to microbiologists.

Gram Stain[1-5]

Background for the Gram Stain

The Gram stain was developed by Christian Gram in 1884 and modified by Hucker in 1921. The Gram stain separates bacteria into two groups: (1) Gram-positive microorganisms that retain the primary dye (crystal violet) and (2) Gram-negative microorganisms that take the color of the counterstain (usually safranin O). These results are due to differences in the structure of the cell wall. Crystal violet is attracted to both Gram-positive and Gram-negative microorganisms. The second step (Gram's iodine, a mordant) stabilizes the crystal violet into the peptidoglycan layer of the cell wall. The peptidoglycan layer is much thicker in Gram-positive bacteria than in Gram-negative bacteria; hence, the crystal violet is more extensively entrapped in the peptidoglycan of Gram-positive bacteria. The third step (alcohol decolorization) dissolves lipids in the outer membrane of Gram-negative bacteria and removes the crystal violet from the peptidoglycan layer. In contrast, the crystal violet is relatively inaccessible in Gram-positive microorganisms and cannot readily be removed by alcohol in Gram-positive microorganisms. After the alcohol step, only the colorless Gram-negative microorganisms can accept the safranin (counterstain). Carbolfuchsin and basic fuchsin are sometimes employed in the counterstain to stain anaerobes and other weakly staining Gram-negative bacteria (including *Legionella* spp., *Campylobacter* spp., and *Brucella* spp.). The *Bacillus-Butyrivibrio-Clostridium* groups can be considered to stain Gram-variable because of changes in their cell wall, which is linked to the growth curve. They typically stain Gram-positive during the lag and exponential phases of growth and then usually stain as a Gram-negative beginning in the stationary growth phase. The S-layer of the cell wall becomes substantially thinner and diffuse as the growth cycle proceeds, which results in more fragile cells that usually stain as a Gram-negative. *Gardnerella vaginalis* has a very thin Gram-positive wall so that it can stain as either a Gram-positive or negative. Mycoplasma lacks a cell wall and stain as a Gram-negative bacterium. Details of the three-step Gram stain by Meszaros and Strenkoski are available.[6]

The three-step method simultaneously decolorizes and counterstains Gram-negative bacteria. Most clinical laboratories are currently using the four-step Gram stain. A modified Brown and Brenn Gram stain can be used to detect Gram-negative and Gram-positive bacteria in tissue.[7] In this method, Gram-positive bacteria stain blue, Gram-negative bacteria stain red, and the background color is yellow.

Standard Gram Stain Procedure

1. Fix the specimen with heat or use 95% methanol for 2 minutes.
2. Apply the primary stain (crystal violet) for 1 minute. Wash with tap water.
3. Apply the mordant (Gram's iodine) for 1 minute. Wash with tap water.
4. Decolorize for 5–15 seconds. Wash with tap water.
5. Counterstain with Safranin for 1 minute. Wash with tap water.

Gram Stain Reagents

1. Primary stain: 2 g crystal violet, 20 mL 95% ethyl alcohol, 0.8 g ammonium oxalate, and 100 mL distilled water
2. Gram's iodine: 2 g potassium iodide, 1 g iodine crystals, and 100 mL distilled water

3. Decolorizer: 50 mL acetone and 50 mL ethanol
4. Counterstain: 4.0 g Safranin, 200 mL 95% ethanol, and 800 mL distilled water

Decolorizer/Counterstain for the Three-Step Gram Stain

Combine 0.40% safranin, 0.30% basic fuchsin, 90% ethanol, and 10% distilled water, and acidify to pH 4.5.

Variations for Gram Stain Reagents

Many variations exist for the Gram stain. The Gram iodine mordant can be stabilized using a polyvinylpyrrolidone–iodine complex. Slowing the decolorizing step is accomplished using only 95% ethanol. Some laboratories prefer to decolorize with isopropanol/acetone (3:1 vol/vol).

Technical Concerns When Gram Staining

The optimum age for staining microorganisms is 18–24 hours. Bacteria that are more than 48 hours old are more likely to have cell wall damage, and this is one reason why Gram-positive bacteria may stain a pink color and appear to be Gram-negative bacteria. Additionally, excessive heat fixing may also damage the cell wall and this renders the bacteria to decolorize excessively. The end result can be that a Gram-positive will stain as Gram-negative bacteria. Occasionally, a smear is uneven in its thickness with the middle portion containing a heavy concentration of bacteria. The end result is that the heavy center portion may not properly decolorize sufficiently. This is the major reason why the bacteria on the edge of a Gram smear are more likely to give true results, and those in the center are often unreliable for reading a Gram stain. Clinical samples that contain bacteria are sometimes damaged by either an inflammatory response or antibiotics. The bacteria may have an atypical morphology and Gram staining can give atypical results. Additional information on the pitfalls and technical trouble spots was discussed by McCelland[3] and Sutton.[4]

Automated Staining Instruments

There are automated Gram-staining instruments available for use in the laboratory that have been found to produce adequate Gram-stain slides as compared to manual staining. Advantages to automated staining were reported to be more uniform, a reduction in hands-on staining time, and conservation of reagent.[8]

Acridine Orange Stain[1,2,14,16]

Background for the Acridine Orange Stain

The acridine orange stain is a sensitive method for detecting low numbers of organisms in cerebrospinal fluid (CSF), blood, buffy coat preparations from neonates, and tissue specimens. The acridine orange stain can be employed in rapidly screening normal sterile specimens (blood and CSF) where low numbers of microorganisms usually exist. Acridine orange is a fluorochrome that binds to nucleic acids. Bacteria and yeasts stain bright orange/red,

and tissue cells stain black to yellowish green. The filter system used on a fluorescent microscope can affect the observed colors. Because red blood cells are not fluorescent, acridine orange stain can be useful for screening blood cultures for the growth of microorganisms. The stain maybe useful in interpreting thick purulent specimens that failed to give clear results with the Gram stain. With experience and training, acridine orange can also be used for detection of malarial parasites in the blood.[7] Acridine orange may also be useful for staining miscellaneous microorganisms such as *Mycoplasma*, *Pneumocystis jiroveci* trophozoites, *Borrelia burgdorferi*, *Acanthamoeba*, *Leishmania*, and *Helicobacter pylori* purulent specimens. Additional information on the acridine orange stain can be found at www.histonet.org.

Acridine Orange Standard Staining Procedure

1. Fix the smear in methanol for 2 minutes.
2. Stain with acridine orange for 1 minutes.
3. Rinse with tap water.

Paraffin Acridine Orange Staining Procedure

1. Deparaffinize and hydrate to distilled water.
2. Stain sections with acridine orange for 30 minutes.
3. Rinse sections in 0.5% acetic acid in 100% alcohol for about one minute.
4. Rinse sections 2 × in 100% alcohol.
5. Rinse sections 2 × in xylene.
6. Mount sections in Fluoromount.

Reagents for Standard Stain

1. 0.2 M sodium acetate buffer (pH 3.75)
2. Add acridine orange to buffer (0.02 g)

Reagents for Paraffin Procedure

1. Add 5 mL acetic acid to 500 mL distilled water.
2. Add 0.05 g acridine orange to the diluted acetic acid.

Macchiavello Stain Modified for *Chlamydia* and *Rickettsia*

Background for the Macchiavello-Gimenez Stain

Chlamydia and *Rickettsia* will stain red, whereas cellular material stains blue. A longer exposure than 1 or 2 seconds with citric acid may stain the organisms blue instead of red. Pinkerton's adaptation of Macchiavello's stain is used to detect *Rickettsia* in tissue samples. *Legionella* can also be stained by this method.

Macchiavello Stain Procedure (Modified) for *Chlamydia*

1. Heat-fix the smear or air-dry.
2. Stain the slide with basic fuchsin for 5 minutes.

3. Wash in tap water and then place the slide in a Coplin jar that contains citric acid. The slide should remain in the citric acid for a few seconds. Decolorization of the *Chlamydia* will occur if the citric acid is left on too long.

4. Wash the slide thoroughly with tap water.

5. Stain the slide for 20–30 seconds with 1% methylene blue.

6. Wash the slide with tap water and air-dry.

Pinkerton's Adaptation of Macchiavello's Stain for *Rickettsia* in Tissues

1. Deparaffinize and hydrate to distilled water.

2. Stain overnight with methylene blue (1%).

3. Decolorize with alcohol (95%).

4. Wash with distilled water.

5. Stain for 30 minutes with basic fuchsin (0.25%).

6. Decolorize 1–2 seconds in citric acid (0.5%).

7. Differentiate quickly with alcohol (100%).

8. Dehydrate with alcohol (95%), followed by 3 × alcohol (100%).

9. Clear in xylene 3 × and mount in Permount.

Reagents for the Macchiavello Stain

1. Basic fuchsin: Dissolve 0.25 g basic fuchsin in 100 mL distilled water.

2. Citric acid: Dissolve 0.5 g citric acid in 100 mL distilled water. Some procedures use 0.5 g in 200 mL distilled water. This reagent must always be fresh.

3. Methylene blue: Dissolve 1.0 g in 100 mL distilled water.

Acid-fast Stains[1,2,8,11]

Background for the Acid-Fast Stain

The cell wall of mycobacteria contains a large amount of lipid that makes it difficult for aqueous-based staining solution to enter the cell. Gram staining is not acceptable because the results are often Gram-variable, beaded Gram-positive rods, or negatively stained images. In contrast, the acid-fast stains contain phenol, which allows basic fuchsin (red) or auramine (fluorescent) to penetrate the cell wall. The primary stain will remain bound to the cell wall mycolic acid residues after acid alcohol is applied. The resistance of the mycobacteria group to acid alcohol decolorization has led to the designation of this group as acid-fast bacilli (AFB). A counterstain such as methylene blue is applied to contrast the mycobacteria (red) from the non-AFB organisms (blue). Detailed information on acid-fast stains can be found at the Centers for Disease Control and Prevention (CDC) *Acid Fast Direct Smear Microscopy Manual*. The CDC also supplies detailed information on the use of fluorochrome staining for the detection of acid-fast mycobacteria.

Ziehl-Neelsen Staining Procedure (Hot Method)

1. Fix the prepared slide with gentle heat.

2. Flood the slide with carbolfuchsin and heat to steaming only once.

3. Leave for 10 minutes.

4. Rinse with distilled water.

5. Decolorize with acid alcohol for 3 minutes.

6. Rinse with distilled water.

7. Counterstain with methylene blue for 1 minute.

8. Rinse with distilled water, drain, and air-dry.

Ziehl-Neelsen Staining Reagents

1. Primary stain: 0.3% carbolfuchsin. Dissolve 50 g phenol in 100 mL ethanol (95%) or methanol (95%). Dissolve 3 g basic fuchsin in the mixture and add distilled water to bring the volume to 1 L.

2. Decolorization solution: Add 30 mL hydrochloric acid to 1 L of 95% denatured alcohol. Cool and mix well before use. Alternate decolorizing reagent (without alcohol): Slowly add 250 mL sulfuric acid (at least 95%) to 750 mL distilled water. Cool and mix well before using.

3. Counterstain: 0.3% methylene blue. Dissolve 3 g methylene blue in 1 L distilled water.

Kinyoun Staining Procedure (Cold Method)

1. Fix and prepare the slide with gentle heat.

2. Stain with Kinyoun carbolfuchsin for 3–5 minutes.

3. Rinse with distilled water.

4. Decolorize with acid alcohol until the red color no longer appears in the washing (about 2 minutes).

5. Rinse with distilled water.

6. Counterstain with methylene blue for 30–60 seconds.

7. Rinse with distilled water, drain, and air-dry.

Kinyoun Staining Reagents

1. Primary stain: Dissolve 40 g basic fuchsin in 200 mL of 95% ethanol. Then add 1000 mL distilled water and 80 g liquefied phenol.

2. Decolorizer: 3% acid. Mix 970 mL of 95% ethanol and 30 mL concentrated hydrochloric acid.

3. Counterstain: 0.3% methylene blue. Dissolve 3 g methylene blue in 1 L distilled water.

Reporting System for Acid-Fast Stains Using Bright Field Microscopy

Table 4.1 follows the CDC recommendation and should be used to report the results of the Kinyoun acid-fast stain.

Truant Fluorochrome Staining Procedure

1. Air-dry the smears and fix on a slide warmer at 70°C for at least 2 hours. A Bunsen burner may be substituted.

TABLE 4.1

Result Reporting of AFB Smear for Bright-Field Microscopy

Read at Total Magnification (1000×)	Report
0 AFB/300 fields	No AFB seen on smear
1–2 AFB/300 fields	Report actual count and suggest a repeat test
1–9 AFB/100 fields	Sparse (1+)
1–9 AFB/10 fields	Few (2+)
1–9 AFB/field	Moderate (3+)
>9 AFB/field	Numerous (4+)

TABLE 4.2

Result Reporting of AFB Smear for Fluorescent Microscopy

Read at Total Magnification (200× to 250×)	Read at Total Magnification (400× to 450×)	Report
0 AFB/30 fields	0 AFB/55 fields	No AFB seen on smear.
1–2 AFB/30 fields	1–2 AFB/70 fields	Report actual count and suggest a repeat test.
1–9 AFB/10 fields	2–18 AFB/50 fields	Sparse (1+)
1–9 AFB/field	4–36 AFB/10 fields	Few (2+)
10–90 AFB/field	4–36 AFB/field	Moderate (3+)
>90 AFB/field	>36 AFB/field	Numerous (4+)

2. Stain with auramine O–rhodamine B solution for 15 minutes.

3. Rinse the slide with distilled water.

4. Decolorize with acid alcohol for approximately 2 minutes.

5. Rinse with distilled water and drain.

6. Counterstain with potassium permanganate for 2 minutes. Applying the counterstain for a longer period may quench the fluorescence of the *Mycobacterium* species. Acridine orange is not recommended as a counterstain by the CDC.

7. Smears are scanned with a 10× objective. It is sometimes necessary to use the 40× objective to confirm the bacterial morphology.

8. Examine at least 30 fields when viewing the slide under a magnification of 200× or 250×. Examine at least 55 fields when viewing the slide under a magnification of 400×. Examine at least 70 fields when viewing the slide under a magnification of 450×.

9. If only one or two AFB are observed in 70 fields, the results are reported as doubtful and should be repeated.

Truant Staining Reagents

1. Auramine O–rhodamine B: Dissolve 0.75 g rhodamine B and 1.5 g auramine O in 75 mL glycerol.

2. Decolorizer: Add 0.5 mL concentrated hydrochloric acid to 100 mL of 70% ethanol.

3. Counterstain: 0.5% potassium permanganate. Dissolve 0.5 g potassium permanganate in 100 mL distilled water.

Reporting System for Acid-Fast Stains Using Fluorescent Microscopy

Table 4.2 follows the CDC recommendation and should be used as a guide for result reporting acid-fast stains using fluorescent microscopy procedures.

Modified Acid-Fast Stain for Detecting Aerobic Actinomycetes (Including *Nocardia*), *Rhodococcus*, *Gordonia*, and *Tsukamurella*

The actinomycetes are a diverse group of Gram-positive rods, often with branching filamentous forms, and are partially acid fast. A modified acid-fast stain using 1% sulfuric acid rather than the usual 3% hydrochloric acid can be used if Gram-positive branching or partially branching organisms are observed.

Modified Acid-fast Stain for Detecting *Cryptosporidium*, *Isospora*, and *Cyclospora*

The modified acid-fast stain is used to identify the oocysts of the coccidian species, which are difficult to detect with the trichrome stain. Fresh or formalin-preserved stools as well as duodenal fluid, bile, and pulmonary samples can be stained.

Modified Acid-Fast Staining Procedure

1. Prepare a thin smear by applying one or two drops of sample to a slide. The specimen is dried via a slide warmer at 60°C and then fixed with 100% methanol for approximately 30 seconds.

2. Apply Kinyoun carbolfuchsin for 1 minute and rinse the slide with distilled water.

3. Decolorize with acid alcohol for about 2 minutes.

4. Counterstain for 2 minutes with malachite green. Rinse with distilled water.

5. Dry (a slide warmer can be used) and apply mounting media and a cover slip.

6. Examine several hundred fields under 40× magnification and confirm morphology under oil immersion.

7. The parasites will stain a pinkish-red color.

Modified Acid-Fast Staining Reagents

1. Kinyoun carbolfuchsin: See earlier procedure.

2. Decolorizer: 10 mL sulfuric acid and 90 mL of 100% ethanol.

3. Counterstain: 3% malachite green. Dissolve 3 g malachite green in 100 mL distilled water.

Mycology Preparations and Stains

Potassium Hydroxide and Lactophenol Cotton Blue Wet Mounts[11–13]

Wet mounts using 10% KOH (potassium hydroxide) will dissolve keratin and cellular material and unmask fungal elements that may be difficult to observe. Lactophenol cotton blue (LPCB) preparations are useful because the phenol in the stain will kill the organisms and the lactic acid preserves fungal structures. Chitin in the fungal cell wall is stained by the cotton blue. The two wet mount preparations (KOH and LPCB) can be combined. Additional variations include using KOH and calcofluor white (CFW) or KOH and dimethylsulfoxide for thicker specimens of skin or nail. An extensive online *Mycology Procedure Manual* (Toronto Medical Laboratories/Mount Sinai Hospital Microbiology Department) is available at http://www.mountsinai. on.ca/education/staff-professionals/microbiology/microbiology-laboratory-manual/MYCOLOGY.doc. Additional staining procedures for the fungi can be found at doctorfungus.org.

KOH Procedure

1. Thin smear scrapings from the margin of a lesion are placed on a slide.
2. Add one or two drops of KOH, place on a cover slip, and allow digestion to occur over 5–30 minutes. Alternately, the slide can be gently heated by passing it through a flame several times. Cool and examine under low power. The fungal cell wall is partially resistant to the effects of KOH. However, the fungi may eventually dissolve in the KOH if left in contact with the reagent for an excessive period.

KOH Reagent

Dissolve 10 g KOH in 80 mL distilled water and then add 20 mL glycerol.

LPCB Procedure

1. Place a thin specimen on the slide and add one drop LPCB. Mix well and place a cover slip over the slide. Nail polish can be used to make a semipermanent slide.
2. Observe under the microscope for fungal elements.

LPCB Reagent

Dissolve 0.5 g cotton blue in 20 mL distilled water and then add 20 mL lactic acid and 20 mL concentrated phenol. Mix after adding 40 mL glycerol.

10% KOH with LPCB Procedure

1. Place a thin specimen on a slide and add one or two drops KOH. Place a cover slip over the preparation and wait for 5–30 minutes at room temperature or gently heat for a few seconds.
2. Add one drop LPCB and add a cover slip; examine under the microscope.

India Ink or Nigrosin Wet Mount

India ink or nigrosin will outline the capsule of *Cryptococcus neoformans*. The stain is far less effective than the latex agglutination procedure when examining CSF. In addition, human red or white blood cells can mimic the appearance of yeast cells. A drop of KOH can be added because human cells are disrupted and yeast cells remain intact. A positive India ink preparation is not always definitive for *C. neoformans* because additional yeasts such as *Rhodotorula* may be encapsulated.

Capsule Detection with India Ink

Mix one drop of the specimen with one drop India ink or one drop nigrosin on a slide. Place a cover slip over the preparation, and let it rest for 10 minutes before examining under the microscope.

Calcofluor White Stain[12–14]

Background for the Calcofluor White Stain

The calcofluor white (CFW) stain can be used for the direct examination of most fungal specimens. This stain binds to cellulose and chitin in the fungal cell wall. CFW is used in conjunction with KOH to enhance the visualization of the fungal cell wall. Positive results are indicated by a bright green to blue fluorescence using a fluorescent microscope. A nonspecific fluorescence from human cellular materials sometimes occurs. A bright yellow-green fluorescence is observed when collagen or elastin is present. The CFW staining technique usually provides better contrast than lactophenol aniline blue stains. The CFW stain can be used for the rapid screening of clinical specimens for fungal elements.[14] Furthermore, a CFW preparation subsequently can be stained with Grocott-Gomori methenamine silver (GMS) and periodic acid-Schiff (PAS) stain without interference.

General CFW Staining Procedure

1. Mix equal volumes of 0.1% CFW and 15% KOH before staining.
2. Place the specimen on the slide and add a few drops of the mixed CFW–KOH solution. Place a cover slip over the material.
3. The slide can be warmed for a few minutes if the material does not clear at 25°C.
4. Observe the specimen under a fluorescence microscope that uses broadband or barrier filters between 300 nm and 412 nm. The maximum absorbance of CFW is at 347 nm.
5. A negative control consists of equal mixtures of CFW–KOH. A positive control consists of a recent *Candida albicans* culture.

CFW Staining Reagents

1. 0.1% CFW (wt/vol) solution. Store the solution in the dark. Gentle heating and/or filtration is necessary to eliminate precipitate formation. CFW is available as a cellufluor solution from Polysciences (Washington, PA) or as fluorescent brightener from Sigma (St. Louis, MO).

2. 15% KOH. Dissolve 15 g KOH in 80 mL distilled water and add 20 mL glycerol. Store both reagents at 25°C.

CFW Stain for *Acanthamoeba* spp., *Pneumocystis Jiroveci, Microsporidium,* and *Cryptosporidium*

This procedure is used only as a quick screening method and not for species identification. The CFW stain is not specific because many objects other than fungi and parasites will fluoresce. Fresh or preserved stool, urine, and other types of specimens can be used in the following procedure (https://www.cdc.gov/dpdx/diagnosticprocedures/index.html). In 2009, Harrington[15] modified the succeeding procedure to detect *Cryptosporidium* oocytes in fecal smears. The modified procedure detected *Cryptosporidium* as well as coinfections with *Microsporidium.*

CFW Staining Procedure for *Acanthamoeba* and other Genera

1. Use approximately 10 µL of preserved or fresh fecal, or urine specimen to prepare a thin smear.
2. Fix the smear in 100% methanol for 30 seconds.
3. Stain with CFW solution for 1 minute. A 0.01% CFW solution in 0.1 M Tris-buffered saline with a final pH of 7.2 constitutes the staining reagent.
4. Rinse with distilled water, air-dry the slide, and mount.
5. Examine under a UV fluorescence microscope using a wavelength at or below 400 nm.
6. The spores of *Microsporidium* will also exhibit a blue-white fluorescent color.

Histopathologic Stain for Fungi: Periodic Acid-Schiff (PAS) Stain[1,2,12,13]

Background for the PAS Stain

Some laboratories prefer to use phase contrast microscopy in place of the periodic acid-Schiff (PAS) stain because similar results are often obtained. In the PAS stain, carbohydrates in the cell wall of the fungi and carbohydrates in human cells are oxidized by periodic acid to form aldehydes, which then react with the fuchsin-sulfurous acid to form the magenta color. Identification of fungal elements in tissue can be enhanced if a counterstain such as light green is used. The background stains green and the yeast cells or hyphae will accumulate the magenta color. The online *Surgical Pathology-Histology Staining Manual* discusses several PAS methods.[16]

PAS Staining Procedure

1. Deparaffinize and hydrate to distilled water.
2. Treat slides with 0.5% periodic acid for 5 minutes, followed by a rinse in distilled water.
3. Stain in Schiff's reagent for 15–30 minutes at room temperature. An alternate method is to microwave on high power for 45–60 seconds. The solution should be a deep magenta color. *Schiff's reagent requires extreme caution because it is a known carcinogen.*
4. Wash in running water (about 5 minutes) to develop the pink color.
5. Counterstain in hematoxylin for 3–6 minutes. Light green can be substituted when fungi are suspected. Go to step 8 if using light green.
6. If using hematoxylin, wash in tap water, followed by a rinse in distilled water.
7. Place in alcohol to dehydrate and apply a cover slip and mount.
8. If using light green, wash in tap water followed by 0.3% ammonia water.
9. Place in 95% alcohol and later in 100% alcohol (2×), followed by clear xylene (2×), and then mount.

Reagents

1. Periodic acid: Dissolve 0.5 g periodic acid in 100 mL distilled water. The reagent is stable for 1 year.
2. Harris' hematoxylin: Dissolve 2.5 g hematoxylin in 25 mL of 100% ethanol. Dissolve 50 g potassium or ammonium alum in 0.5 L heated distilled water. Mix the two solutions without heat. After mixing, the two solutions are boiled very rapidly with stirring (approximate time to reach a boil is 1 minute). After removing from the heat, add 1.25 g mercury oxide (red). Simmer until the stain becomes dark purple in color. Immediately plunge the container into a vessel containing cold water. Add 2–4 mL glacial acetic acid to increase the staining efficiency for the nucleus. The stain must be filtered before use.
3. Light green reagent: Dissolve 0.2 g light green in 100 mL distilled water. Dilute 1:5 with distilled water before use. Determining optimum concentration of light green often requires several attempts.

Grocott-Gomori Methenamine Silver Stain (GMS)[11]

Background for the GMS Stain

Silver stains are useful in detecting fungal elements in tissues. The fungal cell wall contains mucopolysaccharides that are oxidized by the Grocott-Gomori methenamine silver (GMS) stain to release aldehyde groups, which later react with silver nitrate. Silver nitrate is converted to metallic silver, which becomes visible in the tissue. *P. jiroveci* can be detected by several staining techniques, including toluidine blue O, CFW, GMS, and Giemsa. Multiple studies have compared the ease and accuracy in detecting *Pneumocystis* with these stains. A single stain has not emerged as being consistently superior to the others. The procedure described uses the microwave method and employs 2% chromic acid. The conventional method employs 5% chromic acid. The chromic acid solution must be changed if it turns brown. Additional information is available online.[17]

GMA Staining Procedure

1. Deparaffinize and hydrate to distilled water.
2. Oxidize with chromic acid (2%) in a microwave set at high power for 40–50 seconds and wait an additional 5 minutes.
3. Wash in tap water for a few seconds and then wash in distilled water 3×.
4. Rinse the slide in sodium metabisulfate (1%) at room temperature for 1 minute to remove the residual chromic acid.
5. Place the working methenamine silver solution in the microwave (high power) for 60–80 seconds. Agitate the slide in the hot solution. The fungi should stain a light brown color at this stage.
6. Rinse the slide in distilled water 2×.
7. Tone in gold chloride (0.5%) for 1 minute or until gray.
8. Rinse in distilled water 2×.
9. Remove the unreduced silver with sodium thiosulfate (2%) for 2–5 minutes.
10. Rinse in tap water, followed by distilled water.
11. Counterstain with diluted light green (1:5) for 1 minute. The optimal dilution of light green can vary.
12. Rinse in distilled water.
13. Dehydrate, clear, and mount.

Reagents

1. Chromic acid (2%): Dissolve 2.0 g chromium trioxide in 100 mL distilled water.
2. Sodium metabisulfate (1%): Dissolve 1.0 g sodium metabisulfate in 100 mL distilled water.
3. Borax (5%): Dissolve 5.0 g sodium borate in 100 mL distilled water.
4. Stock solution of methenamine silver: Add 5 mL silver nitrate (5%) to 100 mL of methenamine (3%). Store the reagent in the refrigerator in a brown bottle.
5. Working solution of methenamine silver: Add 25 mL distilled water and 2.5 mL borax (5%) to 25 mL methenamine silver stock solution.
6. Gold chloride (0.5%): Dissolve 0.5 g gold chloride in 100 mL distilled water.
7. Light green (0.2%): Dissolve 0.2 g light green in 100 mL distilled water and add 0.2 mL glacial acetic acid.

Trichrome Stain[1–22]

Background for the Trichrome Stain

The Wheatley trichrome stain is the last part of a complete fecal examination, and it usually follows a direct iodine wet mount and/or parasite concentration technique. The Wheatley procedure is a modification of the Gomori tissue stain. The procedure is the definitive method for the identification of protozoan parasites. Small protozoa that have been missed on wet mounts or concentration methods are often detected with the trichrome stain. The trichrome stain detects both trophozoites and cysts and documents a permanent record for each observed parasite. The background material stains green; the cytoplasm of the cysts and trophozoites stain blue green; and chromatoidal bodies (RNA), chromatin material, bacteria, and red blood cells stain red or purplish-red. Larvae or ova of some metazoans also stain red. Thin-shelled ova may collapse when mounting fluid is used. The fecal specimen may be fresh, fixed in polyvinyl alcohol (PVA), Schaudinn, or sodium acetate-acetic acid-formalin (SAF). The CDC has used various components of the trichrome stain to detect *Microsporidium* spores from the fecal component. The complete chromotrope staining procedures are available at the CDC's website (https://www.cdc.gov/dpdx/diagnosticprocedures/index.html).

Trichrome Staining Procedure

1. Place the PVA smears in 70% ethanol/iodine solution for 10 minutes. This step can be eliminated if the fixative does not contain mercuric chloride.
2. Transfer slide to 70% ethanol for 5 minutes. Repeat with a second 70% ethanol treatment for 3–5 minutes.
3. Stain in trichrome for 10 minutes.
4. Rinse quickly in 90% ethanol, acidified, with 1% acetic acid, for 2 or 3 seconds.
5. Rinse quickly using multiple dips in 100% ethanol. Repeat the dipping in a fresh 100% ethanol. Make sure that each slide is destained separately using fresh alcohol. The most common mistake is excessive destaining.
6. Transfer slides into 100% ethanol for 3–5 minutes. Repeat a second time.
7. Transfer slide to xylene for 5 minutes. Repeat a second time.
8. Mount with Permount.
9. A 10× objective can be used to locate a good area of the smear. Then examine the smear under oil immersion and analyze 200–300 fields.

Reagents

1. 70% ethanol with iodine: Add sufficient iodine to 70% ethanol to turn the alcohol a dark tea color (reddish brown). If the ethanol/iodine reagent is too weak, the mercuric chloride will not be extracted from the specimen. The end result will be the formation of a crystalline residue that will hamper the examination of the specimen.
2. D'Antoni's iodine: Dissolve 1 g potassium iodide and 1.5 g powdered iodine crystals in 100 mL distilled water.
3. Trichrome stain: Add 1 mL glacial acetic acid to the dry components of the stain, which are 0.60 g chromotrope 2R, 0.15 g light green SF, 0.15 g fast green, and 0.70 g phosphotungstic acid. The color should be purple. Dissolve all the reagents in 100 mL distilled water.

Iron Hematoxylin Stain[11,18,21–23]

Background for the Iron Hematoxylin Stain

The iron hematoxylin stain is used to identify the trophozoites and cysts of the protozoa group. It is less commonly used than the trichrome stain. The cysts and the trophozoites stain a blue gray to black, and the background material stains blue or pale gray. Helminth eggs and larvae are usually difficult to identify because of excessive stain retention. Yeasts and human cells (red blood cells, neutrophils, and macrophages) are also detected by the stain. The stain can be used with fixatives that include PVA, SAF, or Schaudinn. Good fixation is the key in obtaining a well-stained fecal preparation. The simplest variation of the iron hematoxylin stain is the method of Tompkins and Miller. A modified iron hematoxylin stain that incorporates a carbolfuchsin step will detect some acid-fast parasites (including *Cryptosporidium parvum* and *Isospora belli*).

Iron Hematoxylin Stain Procedure

1. Prepare a thin layer of fecal smear and place it in a fixative. If SAF is used, proceed to step 4. Mayer's albumin can be used to ensure that the specimen will adhere to the slide.
2. Dehydrate the slide in ethanol (70%) for 5 minutes.
3. Transfer slide in iodine ethanol (70%) for 2–5 minutes. The solution should have a strong tea color.
4. Transfer the slide to 50% ethanol for 5 minutes.
5. Wash the slide in a constant stream of tap water for 3 minutes.
6. Transfer the slide into 4% ferric ammonium sulfate mordant for 5 minutes.
7. Wash the slide in a constant stream of tap water for 1 minutes.
8. Stain with hematoxylin (0.5%) for 3–5 minutes.
9. Wash in tap water for about 1 minutes.
10. Destain the slide with 2% phosphotungstic acid for 2 minutes.
11. Wash the slide in a constant stream of tap water.
12. Transfer the slide to 70% ethanol that contains several drops of lithium carbonate (saturated).
13. Transfer the slide to 95% ethanol for 5 minutes.
14. Transfer the slide to 100% ethanol for 5 minutes. Repeat a second time.
15. Transfer slide to xylene for 5 minutes. Repeat a second time.
16. Mount with Permount.

Reagents

1. Seventy percent ethanol/iodine: Add sufficient iodine to 70% ethanol to turn the alcohol a dark tea color (reddish brown).
2. D'Antoni's iodine: Dissolve 1 g potassium iodide and 1.5 g powdered iodine crystals in 100 mL distilled water.
3. Iron hematoxylin stain: Dissolve 10 g hematoxylin in 1000 mL of 100% ethanol. Store the reagent at room temperature.
4. Mordant: Dissolve 10 g ferrous ammonium sulfate and 10 g ferric ammonium sulfate in 990 mL distilled water. Add 10 mL concentrated hydrochloric acid.
5. Working hematoxylin stain: Mix (1:1) the mordant and the hematoxylin stain.
6. Saturated lithium carbonate: Dissolve 1 g lithium carbonate in 100 mL distilled water.

Preparation of Blood Smears for Parasite Examination[3,11,18–20]

Trypanosoma, *Babesia*, *Plasmodium*, and *Leishmania*, as well as most microfilariae, are detected from blood smears. Identification of these parasites is based on the examination of permanent blood films. The blood samples from malaria and *Babesia* patients are best collected toward the end of a paroxysmal episode. Blood samples can be collected randomly from patients with trypanosomiasis. Blood samples should be collected after 10 p.m. on those microfilariae that exhibit nocturnal periodicity (*Wuchereria* and *Brugia*). Blood samples should be collected between 11 a.m. and 1 p.m. if *Loa loa* (diurnal periodicity) is suspected. Thick blood smears are typically used as a screening tool, and thin blood smears are used to observe detailed parasite morphology. Blood parasites are usually identified from the thin films. The Giemsa, Wright, and Wright-Giemsa combination can be used to stain the blood smears. The Giemsa stains the cytoplasm of *Plasmodium* spp. blue, whereas the nuclear material stains red to purple, and Schuffner's dots stain red. The cytoplasm of trypanosomes *Leishmania* and *Babesia* will also stain blue and the nucleus stains red to purple. The sheath of microfilariae often fails to stain but the nuclei will stain blue/purple.

Giemsa Staining Procedure for Thin Films

1. Fix the blood films in 100% methanol for 1 minute.
2. Air-dry the slides.
3. Place the slides into the working Giemsa solution. The working solution is one part Giemsa stock (commercial liquid stain) and 10–50 parts phosphate (pH 7.0–7.2) buffer. Experimentation with several different staining dilutions/times is often required to obtain optimum results.
4. Stain for 10–60 minutes.
5. Briefly wash underwater or in phosphate buffer.
6. Wipe the stain off the bottom of the slide and air-dry.

Reagents

1. Stock Giemsa commercial liquid stain: Dilute 1:10 with buffer for thin smears.
2. Phosphate buffer, pH 7.0: Dissolve 9.3 g Na_2HPO_4 in 1 L distilled water (solution 1), and dissolve 9.2 g NaH_2PO_4 H_2O in 1 L distilled water (solution 2).

Add 900 mL distilled water to a beaker and 61.1 mL solution 1 and 38.9 mL solution 2. This reagent is used to dilute the Giemsa stain.

3. Triton phosphate buffered wash: Add 0.1 mL Triton X-100 to 1000 mL of the pH 7.0 phosphate buffer. This is the wash; tap water can also be used.

The Giemsa stain has been the recommended stain for identification of malarial parasites because it reliably shows the red staining Schuffner's dots in the infected erythrocytes when present.[24]

Variations of the traditional Giemsa and Wright stains, including Wright-Giemsa combination stains, have been commercially developed in response to the urgent nature of staining and reading blood films quickly in the clinical laboratory setting. These variations are based on multiple modifications of the classic formulas and involve repeated dips of a dry thin blood film into each of the three solutions. The first is a methanol fixative solution containing malachite green or fast green. The second solution contains buffer and eosin. The third is a mixture of buffer and thiazine dye. Use of a commercially modified staining procedure can result in a more rapid reporting of blood films because of the shortened staining times that can range from 15 seconds to 10 minutes depending on the actual product used. It is suggested that the staining characteristics of any polymorphonuclear leukocytes in the blood film can help to monitor the quality of these nontraditional staining methods.[25,26] As a good laboratory practice, control slides should be stained as directed.

REFERENCES

1. Atlas, M.A. and Synder, J.W., Reagents, stains, and media: Bacteriology, in: *Manual of Clinical Microbiology*, 10th ed., Murray, P.R., ed., ASM Press, Washington, DC, 2011, Chapter 17.
2. Clarridge, J.E. and Mullins, J.M., Microscopy and staining, in: *Clinical and Pathogenic Microbiology*, 2nd ed., Howard, B.J., ed., Mosby, St. Louis, MO, 1994, Chapter 6.
3. McClelland, R., Gram's stain: The key to microbiology, *MLO Med. Lab. Obs.*, 33(4), 20–22, 2001.
4. Sutton, S., The Gram stain, *Pharmaceut. Microbiol. Forum Newslett.*, 12(2), 4, 2006.
5. Beveridge, T.J. Mechanisms of gram variability in select bacteria, *J. Bacteriol.*, 172(3), 1609–1620, 1990.
6. Mezaros, A., and Strenkoski, L. United States Patent 5,393,661: Three reagent gram staining method and kit, February 28, 1995. http://freepatentsonline.com/5393661.html
7. Surgical Pathology—Histology Staining Manual. Accessed: February 2, 2014. http://library.med.utah.edu/WebPath/HISTHTML/MANUALS/GRAM.PDF
8. Baron, E.J., Mix, S., and Moradi, W., Clinical utility of an automated instrument for gram staining single slides, *J. Clin. Microbiol.*, 48(6), 2014, 2010.
9. Moody, A., Rapid diagnostic tests for malaria parasites, *Clin. Microbiol. Rev.*, 15, 66, 2002.
10. Traunt, J.P., Brett, W.A., and Thomas, W., Jr., Fluorescence microscopy of tubercle bacilli stained with Auramine and Rhodamine, *Henry Ford Hosp. Med. Bull.*, 10, 287, 1962.
11. Woods, G.L. and Walker, H.W., Detection of infectious agents by use of cytological and histological stains, *Clin. Microbiol. Rev.*, 9, 382, 1996.
12. Kern, M.E. and Blevins, S.B., Laboratory procedures for fungal culture and isolation, in: *Medical Mycology*, 2nd ed., F.A. Davis, Philadelphia, PA, 1997, Chapter 2.
13. Larone, D.H., Laboratory procedures; staining methods; media, in: *Medically Important Fungi: A Guide to Identification*, 5th ed., ASM Press, Washington, DC, 2011, Part iv.
14. Mount Sanai Hospital. Mycology Manual, September. 2014. http://www.mountsinai.on.ca/education/staff-professionals/microbiology/microbiology-laboratory-manual/MYCOLOGY.doc
15. Harrington, B.J., The staining of oocysts of cryptosporidium with the fluorescent brighteners Uvitex 2B and Calcoflor White, *Lab. Med.*, 40, 219–223, 2009.
16. Surgical Pathology—Histology Staining Manual. Accessed: February. 2, 2014. http://library.med.utah.edu/WebPath/HISTHTML/MANUALS/PAS.PDF
17. Surgical Pathology—Histology Staining Manual. Accessed: February. 2, 2014. http://library.med.utah.edu/WebPath/HISTHTML/MANUALS/GMS.PDF
18. Heelan, J.S. and Ingersoll, F.W., Processing specimens for recovery of parasites, in: *Essentials of Human Parasitology*, Delmar Thompson Learning, Albany, NY, 2002, Chapter 2.
19. Garcia, L.S., Macroscopic and microscopic examination of fecal specimens, in: *Diagnostic Medical Parasitology*, 5th ed., ASM Press, Washington, DC, 2006, Chapter 27.
20. Leventhal, R. and Cheadle, R., Clinical laboratory procedures, in: *Medical Parasitology: A Self- Instructional Text*, 6th ed., F.A. Davis, Philadelphia, PA, 2012, Chapter 7.
21. Tille, P.M., Parasitology, Part IV, in: *Bailey & Scott's Diagnostic Microbiology*, 13th ed., Mosby Elsevier, St. Louis, MO, 2013, Chapters 47–58.
22. Wheatley, W.B., A rapid staining procedure for intestinal amoebae and flagellates, *Am. J. Clin. Pathol.*, 2, 990, 1951.
23. Tompkins, V.N. and Miller, J.K., Staining intestinal protozoa with iron-hematoxylin-phosphotungstic acid, *Am. J. Clin. Pathol.*, 17, 755, 1947.
24. Centers for Disease Control and Prevention, Malaria surveillance—United States, 2005, *Morbid. Mort. Wkly. Surveill. Summ.*, 56(SS-6), 39, 2007
25. Linscott, A.J. and Sharp, S.E., Reagents, stains and media: Parasitology, in: *Manual of Clinical Microbiology*, 10th ed., Murray, P.R., ed., ASM Press, Washington, DC, 2011, Chapter 131.
26. Garcia, L.S., Section 5, Specific test procedures and algorithms, in: Practical Guide to Diagnostic Parasitology, 2nd ed., ASM Press, Washington, DC, 2009.

5

Identification of Gram-Positive Organisms

Peter M. Colaninno

CONTENTS

Introduction

Human infections caused by Gram-positive organisms have increased over the years, both as nosocomial infections and community acquired infections, so it is imperative that these organisms are identified as expeditiously as possible. However, because of the sheer multitude of Gram-positive organisms implicated in disease, identification can be challenging. Certainly the advent of the matrix-assisted laser desorption/ionization time of flight mass spectrometry (MALDI-TOF MS) has given clinical laboratories the tool to quickly identify organisms, but this technology is expensive and may not be suited for midsize to smaller clinical microbiology laboratories. What follows is a synopsis of identification tests for Gram-positive organisms that can make the process easier.

Gram-Positive Cocci

Staphylococcus

Colony morphology: Colonies of *Staphylococcus* on sheep blood agar present themselves as smooth, yellow, white, or off-white colonies somewhere in the area of 1–2 mm in diameter.[1] Colonies may exhibit β-hemolysis and may show varying degrees of growth. Sometimes the β-hemolysis is not evident after 24 hours of incubation and requires further incubation. Because of the varying degrees of colony size and color that may mimic other Gram-positive organisms, such as streptococci and micrococci, identification of *Staphylococcus* can sometimes be problematic. CHROMagar (CHROMagar Company) and ChromID (bioMerieux) are types of media that can alleviate this problem. They contain chromogenic substrates that yield a certain colony color.[2]

Quick tests: Perhaps the most common quick test employed to help identify colonies of *Staphylococcus* is the catalase test. This simple test can differentiate off-white or gray colonies of *Staphylococcus* from *Streptococcus* and is an invaluable tool. The modified oxidase test is another quick test that can differentiate *Staphylococcus* from *Micrococcus* as is the lysostaphin test (Remel). Differentiation among the staphylococci can be achieved by the coagulase test, testing for both bound and free coagulase. Alternatively, there are a multitude of latex agglutination tests available. Slidex Staph Plus (bioMerieux), Staphy Latex (Tulip Diagnostics),

DOI: 10.1201/9781003099277-6

Staphtex (Hardy Diagnostics), Staphyloslide (Becton-Dickinson), and Staphaurex (Remel) are all latex agglutination tests for the identification of *Staphylococcus aureus*.[3,4]

Conventional methods: *S. aureus* can further be identified by its ability to produce DNase and ferment mannitol. Both DNase agar and mannitol salt agar are readily available. When needed to speciate coagulase negative staphylococci (CNS), there are a number of tests that can be employed. Carbohydrate utilization such as sucrose, xylose, trehalose, fructose, maltose, mannose, and lactose as well as such tests as urease, nitrate reduction, and phosphatase will all aid in the identification of CNS. *Staphylococcus saprophyticus* is a frequent cause of urinary tract infections and can be identified by its resistance to novobiocin. Bacitracin and acid production from glucose are tests that can be employed to differentiate *Staphylococcus* from *Micrococcus*.[5]

Identification strips: There are a number of manufacturers that have developed identification strips containing many of the aforementioned biochemicals. API STAPH (bioMerieux), ID 32 Staph (bioMerieux), BBL Crystal Gram-positive ID (Becton-Dickinson), and RapID Staph Plus (Remel) utilize a number of biochemical tests on their strips and are reliable in identifying most strains of CNS.[6]

Automated methods: The VitekII GPID card (bioMerieux-Vitek), the MicroScan Rapid Pos Combo Panel and Pos ID Panels (Dade/MicroScan) and the Phoenix Automated Microbiology System (Becton Dickinson) are a few of the more common automated identification panels for staphylococci as well as other Gram-positive organisms.[3,7,8]

Molecular methods: Molecular methods have gained widespread popularity in the field of medical microbiology. GenProbe introduced their AccuProbe for *S. aureus* culture confirmation a number of years ago and Roche Molecular Systems has a RT-PCR test for the direct detection of the *mecA* gene for use on their LightCycler.[9,10] Table 5.1 summarizes all of the available tests for *Staphylococcus* species.

TABLE 5.1

Identification Tests for *Staphylococcus*

Catalase test
Coagulase test
Lysostaphin test
Latex agglutination tests
DNase
Mannitol salt
Carbohydrate utilization
Novobiocin disc
Bacitracin disc
API STAPH
ID 32 Staph
RapID Staph Plus
Vitek GPID
MicroScan Pos ID
BD Crystal Gram-positive ID
AccuProbe

TABLE 5.2

Identification Tests for *Micrococcus* and Related Species

Catalase test
Modified oxidase test
Nitrate reduction
Lysostaphin
Furazolidone
Glucose fermentation
Bacitracin disc
API STAPH
RapID Staph
MicroScan

Micrococcus and Related Species

Colony morphology: *Micrococcus* presents as dull, white colonies that may produce a tan, pink, or orange color. They are generally 1–2 mm in diameter and can stick to the surface of the agar plate.[1] A Gram stain will assist in the preliminary identification as micrococci appear as larger Gram-positive cocci arranged in tetrads. Other related species that have a microscopic morphology similar to *Micrococcus* include *Kocuria* species, *Kytococcus* species, and *Arthrobacter agilis*. They all appear as Gram-positive cocci in tetrads and their colony morphology is characterized by pigmentation. *Kocuria* species have been implicated in bacteremia and peritonitis and have a yellow to orange to pink pigment.[11,12] *Kytococcus* species have caused cases of endocarditis and appear as a creamy white to yellow colony.[13] *A. agilis* appears as a red colony and is predominately found in soil.[14]

Quick tests: The catalase test and the modified oxidase test are invaluable tools in helping to differentiate micrococci from staphylococci.

Conventional tests: Glucose fermentation, nitrate reduction, lysostaphin, furazolidone and bacitracin can all be used to separate micrococci from staphylococci.[15]

Identification strips: The API STAPH identification strip (bioMerieux) and RapID Staph Plus (Remel) can be utilized to identify *Micrococcus* species.

Automated methods: MicroScan has 2-hour rapid Gram-positive panels that can identify *Micrococcus* species.

Table 5.2 summarizes the available tests for *Micrococcus* and related species.

Streptococcus, Including *Enterococcus*

Colony morphology: *Streptococcus* colonies are typically smaller on sheep blood agar than *Staphylococcus* colonies and measure about 1 mm. Agar such as Columbia CNA and PEA can enhance the growth of streptococci resulting in larger colony size. Perhaps the one characteristic that can aid in the separation of *Streptococcus* species is hemolysis. ß-hemolysis, α-hemolysis, α-prime hemolysis, and γ-hemolysis are all produced by members of the genus.

Quick tests: The catalase test is very useful in differentiating *Staphylococcus* colonies from *Streptococcus* colonies, especially γ-hemolytic streptococci. After the colony has been tentatively identified as strep, there are several latex agglutination and coagglutination tests that can provide a definitive identification. Kits include Phadebact (MKL Diagnostics), Streptex (Remel),

Streptocard (Becton-Dickinson), Patho-Dx (Remel), and Slidex Strepto Plus (bioMerieux).[3] Two useful tests to help aid in the identification of enterococci are the pyrrolidonyl arylamidase (PYR) and leucine aminopeptidase (LAP) tests. These are available as discs (Remel) and can be inoculated directly with colonies for a rapid identification. The DrySpot Pneumo (Oxoid), PneumoSlide (Becton-Dickinson), and Slidex Pneumo (bioMerieux) are rapid tests for the identification of *Streptococcus pneumoniae.*

Conventional tests: Because there are a multitude of *Streptococcus* species implicated in human disease, there so exist a number of conventional tests to aid in the identification of streptococci. Conventional tests that aid in the identification of ß-hemolytic strep include the bacitracin disc, CAMP test, susceptibility to SXT, hippurate hydrolysis, Vogues-Proskauer (VP), and PYR. Alpha-hemolytic strep can be identified by using the optochin disc, bile solubility, esculin hydrolysis, 6.5% NaCl, arginine hydrolysis, and acid from mannitol. Enterococci can be identified by utilizing bile esculin, 6.5% NaCl, SXT susceptibility, motility, pigment production, and VP.[15]

Identification strips: The API 20 Strep and API Rapid ID 32 Strep (bioMerieux) and RapID STR strips (Remel) provide a number of biochemicals for the identification of streptococci, enterococci, and nutritionally variant strep.[16] The Remel BactiCard Strep is a card that has PYR, LAP, and ESC reactions encompassed on it.

Automated methods: The MicroScan Pos ID panel and the Vitek II GPID card can be utilized for identification of strep.

Molecular methods: The AccuProbe *S. pneumoniae* system is available.

Table 5.3 summarizes the available tests for *Streptococcus* and *Enterococcus* species.

TABLE 5.3

Identification Tests for *Streptococcus* and *Enterococcus*

Hemolysis
Catalase test
Latex agglutination tests
Coagglutination tests
PYR test
LAP test
Bacitracin disc
CAMP test
SXT disc
Hippurate hydrolysis
Voges-Proskauer
Optochin disc
Bile solubility
Esculin hydrolysis
Arginine hydrolysis
Bile esculin
6.5% NaCl
Motility
Pigment production
Carbohydrate utilization
API 20 Strep
API Rapid ID 32 Strep
RapID STR
Vitek II GPID
MicroScan Pos ID
LightCycler

Aerococcus

Colony morphology: Aerococci appear as α-hemolytic colonies that closely resemble enterococci and the viridans streptococci.

Quick tests: Two tests that aid in the identification of *Aerococcus* species are PYR and LAP.

Conventional tests: Laboratory tests utilized for the identification and speciation of *Aerococcus* include bile esculin, 6.5% NaCl, hippurate hydrolysis, and acid production from sorbitol, lactose, maltose, and trehalose.[17]

ID strips: The API 20 Strep and Rapid ID 32 Strep strip contain biochemical that can identify *Aerococcus.*[18]

Automated methods: The Vitek II GPID can be used to identify *Aerococcus.*[15]

Leuconostoc

Colony morphology: *Leuconostoc* are small, α-hemolytic colonies.

Quick tests: The PYR and LAP discs can be utilized.

Conventional tests: Arginine dihydrolase, gas from glucose, resistance to vancomycin, lack of growth at 45°C, and esculin hydrolysis are all useful tests in the identification of *Leuconostoc.*[15,19,20]

ID strips: API 50 CHL.

Automated methods: Vitek II GPID card.

Pediococcus

Colony morphology: Pediococci are small, α-hemolytic colonies that resemble viridans strep.

Quick tests: The PYR and LAP discs can be used.

Conventional tests: Conventional tests include acid production from glucose, arabinose, maltose, and xylose; lack of gas production in glucose; growth at 45°C; and resistance to vancomycin.[15,19]

ID strips: API 50 CHL.

Automated methods: Vitek II GPID card.

Table 5.4 summarizes the available tests for *Aerococcus, Leuconostoc,* and *Pediococcus.*

TABLE 5.4

Identification Tests for *Aerococcus, Leuconostoc,* and *Pediococcus*

PYR test
LAP test
Bile esculin
6.5% NaCl
Hippurate hydrolysis
Arginine hydrolysis
Gas from glucose
Resistance to vancomycin
Growth/lack of growth at 45°C
Carbohydrate utilization
API 20 Strep
API Rapid ID 32 Strep
API 50 CHL
Vitek II GPID

Lactococcus

Colony morphology: Lactococci are small, α-hemolytic colonies that resemble enterococci and viridans strep.

Quick tests: LAP and PYR discs.

Conventional tests: Bile esculin, 6.5% NaCl, arginine dihydrolase, hippurate hydrolysis, and acid from glucose, maltose, lactose, sucrose, mannitol, raffinose, and trehalose are all conventional tests that can be used.[15,19]

ID strips: The API Rapid ID 32 Strep strip contains biochemicals for the identification of *Lactococcus* species.

Automated methods: Vitek II GPID card.

Gemella

Colony morphology: Colonies of *Gemella* closely resemble colonies of viridans streptococci.

Quick tests: The PYR disc can be used to aid in identification.

Conventional tests: Conventional tests include esculin hydrolysis, reduction of nitrite, arginine dihydrolase, and acid production from glucose, mannitol, and maltose.[15,19]

ID strips: The API Rapid ID 32 Strep, RapID STR, and API 20A (bioMerieux) can be used to identify *Gemella*.

Automated methods: The Vitek II GPID card

Table 5.5 summarizes the available tests for *Lactococcus* and *Gemella*.

Miscellaneous Gram-Positive Cocci

There are a number of Gram-positive cocci whose implication in human disease is questionable. *Alloiococcus* has been isolated from tympanocentesis fluid[21]; *Abiotrophia* has caused endocarditis[22]; *Rothia* (*Stomatococcus*) has been isolated from blood cultures[23]; *Globicatella* has been implicated in bacteremia[24]; *Helcococcus* has been isolated from abscesses[25]; *Facklamia*, *Granulicatella*, *Dolosigranulum*, *Ignavigranum*, and *Dolosicoccus* have caused disease in humans, such as endocarditis, keratitis, pneumonia, and synovitis[26–32]; and *Vagococcus* is primarily zoonotic. Biochemical tests are included in Table 5.6.[19,23]

TABLE 5.5

Identification Tests for *Lactococcus* and *Gemella*

LAP test
PYR test
Bile esculin
6.5% NaCl
Arginine hydrolysis
Hippurate hydrolysis
Nitrate reduction
Carbohydrate utilization
API Rapid ID 32 Strep
API 20A
Vitek II GPID

TABLE 5.6

Identification Tests for Miscellaneous Gram-positive Cocci

Hemolysis
Catalase
LAP test
PYR test
Bile Esculin
6.5% NaCl
Motility
Hippurate hydrolysis
Gelatin hydrolysis
Carbohydrate utilization
Arginine hydrolysis
Growth/lack of growth at 45°C
Vancomycin susceptibility
API 20 Strep
API Rapid ID 32 Strep
API Rapid ID 32 Staph

Gram-Positive Bacilli

Corynebacterium

Colony morphology: Colonies of *Corynebacterium* on 5% sheep blood agar appear as tiny, gray-to-white colonies that are generally nonhemolytic. Special media such as cysteine tellurite blood agar and Tinsdale agar should be utilized whenever *Corynebacterium diphtheriae* group are to be isolated.

Quick tests: The Gram stain can be an invaluable tool for the identification of *Corynebacterium* as they can be differentiated from *Bacillus* by their coryneform shape and size and lack of spores. The catalase test is useful in differentiating *Corynebacterium* from *Erysipelothrix* and *Lactobacillus*.

Conventional tests: Useful tests for the identification of *Corynebacterium* include motility, lack of H_2S production in a TSI slant; esculin; nitrate reduction; urease; as well as the carbohydrates glucose, mannitol, maltose, sucrose, xylose, and salicin.[33,34]

ID strips: API Coryne strip (bioMerieux) and RapID CB Plus (Remel) are identification strips that can be used for the identification of *Corynebacterium*.[35,36]

Automated methods: Vitek II GPID card.

Table 5.7 summarizes the available tests for *Corynebacterium*.

TABLE 5.7

Identification Tests for *Corynebacterium*

Catalase test
Motility
Esculin hydrolysis
Nitrate reduction
Lack of H_2S production
Urease
Carbohydrate utilization
API Coryne
RapID CB Plus
Vitek II GPID

TABLE 5.8

Identification Tests for *Bacillus*

Spore stain
Motility
Gas from glucose
Starch hydrolysis 6.5% NaCl
Indole
Nitrate reduction
Vogues-Proskauer
Citrate
Gelatin hydrolysis
Esculin hydrolysis
Growth at 42°C
Carbohydrate fermentation
Susceptibility to penicillin
API 50 CHB
Microgen Bacillus

TABLE 5.9

Identification Tests for *Listeria*

ß-hemolysis
Catalase test
Motility (wet mount)
Growth at 4°C
Esculin hydrolysis
Motility media
Lack of H$_2$S production
CAMP test
Carbohydrate fermentation
API Listeria
Rapidec L. mono
API 20 Strep
API Rapid ID 32 Strep
API Coryne strip
Microgen Listeria
Vitek II GPID

Bacillus

Colony morphology: *Bacillus* species can have many variations in colony morphology. For example, *Bacillus anthracis* appear as nonhemolytic medium-to-large gray, flat colonies with irregular swirls; *Bacillus cereus* appear as large, spreading ß-hemolytic colonies; and *Bacillus subtilis* appear as large, flat colonies that may be pigmented or ß-hemolytic.

Quick tests: The Gram stain and spore stain are useful tools in identification.

Conventional tests: Motility; gas from glucose; starch hydrolysis; growth in nutrient broth with 6.5% NaCl; indole; nitrate reduction; VP; citrate; gelatin hydrolysis; esculin hydrolysis; growth at 42°C, and various carbohydrate fermentations such as glucose, maltose, mannitol, xylose, and salicin. Susceptibility to penicillin can also help in the identification of *B. anthracis*.[37]

ID strips: Microgen Bacillus ID (Microgen Bioproducts) and the API 50 CHB (bioMerieux).

Table 5.8 summarizes the available tests for *Bacillus*.

Listeria

Colony morphology: Colonies appear on sheep blood agar as small, translucent, and gray with a small zone of β-hemolysis.

Quick tests: Catalase and motility on a wet mount are useful.

Conventional tests: Growth at 4°C, esculin hydrolysis, motility, fermentation of glucose, trehalose, and salicin; lack of H$_2$S production; CAMP test.[38] *Listeria monocytogenes*, chromogenic agar (bioMerieux) is available.

ID strips: API 20 Strep, API Listeria strip, Rapidec L. mono (bioMerieux), API Rapid ID 32 Strep, API Coryne strip (bioMerieux), and Microgen Listeria ID (Microgen Bioproducts).[39]

Automated methods: Vitek II GPID card.

Table 5.9 summarizes the available tests for *Listeria*.

Lactobacillus

Colony morphology: *Lactobacillus* can have a variety of colony morphologies ranging from tiny, pinpoint, α-hemolytic colonies resembling viridans strep to small, rough, gray colonies.

Quick tests: Gram stain and catalase tests are two tests that are useful in aiding in the identification.

Conventional tests: Fermentation of glucose, maltose, and sucrose as well as urease and nitrate reduction tests are helpful.[3]

ID strips: API 20A and API 50 CHL (bioMerieux) can be used to identify some strains of *Lactobacillus*.

Automated methods: Vitek II ANC ID card.

Table 5.10 summarizes the available tests for *Lactobacillus*.

Actinomyces, Nocardia, Streptomyces

Colony morphology: Colonies of *Nocardia* can vary but are commonly dry, chalky white in appearance, and can sometimes be ß-hemolytic on sheep blood agar with a late-developing pigment. *Streptomyces* can appear as waxy, glabrous, heaped colonies. *Actinomyces* appear as white colonies with or without ß-hemolysis.

Quick tests: Modified acid-fast stain and catalase tests.

Conventional tests: Casein; tyrosine; xanthine; urease; gelatin hydrolysis; starch hydrolysis; nitrate reduction; lactose, xylose, rhamnose, and arabinose fermentation. The BBL Nocardia ID Quad Plate (Becton-Dickinson) combines casein, tyrosine, xanthine, and starch on one plate.[3,40–42]

ID strips: *Actinomyces*: API Coryne and API 20A (bioMerieux); BBL Crystal rapid Gram-positive (Becton-Dickinson); RapID ANA II (Remel); Rap ID CB Plus (Remel).[43–45]

Nocardia species can be identified with additional tests by the API Coryne strip.

Automated methods: Vitek II ANC ID card.

TABLE 5.10

Identification Tests for *Lactobacillus*

Catalase test
Urease
Nitrate reduction
Carbohydrate fermentation
API 50 CHL
API 20A
Vitek II ANC ID card

TABLE 5.11

Identification Tests for _Actinomyces_, _Nocardia_, and _Streptomyces_

Modified acid-fast stain
Casein
Tyrosine
Xanthine
Urease
Gelatin hydrolysis
Nitrate reduction
Carbohydrate fermentation
API Coryne
BBL Crystal rapid Gram-positive
RapID ANA II
RapID CB Plus
Vitek II ANC ID card

Table 5.11 summarizes the available tests for _Actinomyces_, _Nocardia_, and _Streptomyces_.

Gardnerella

Colony morphology: Special media such as V agar and HBT agar must be used to isolate _Gardnerella vaginalis_. Colonies appear as tiny and β-hemolytic after 48 hours.

Quick tests: Gram stain; catalase; oxidase.

Conventional tests: Hippurate hydrolysis; urease; nitrate reduction; a zone of inhibition with trimethoprim, sulfonamide, and metronidazole; fermentation of glucose, maltose, and sucrose.[46]

ID strips: The API Rapid ID 32 C strip can be used to identify _Gardnerella_.

Automated methods: MicroScan Rapid HNID panel (Dade/MicroScan) and Vitek II NH ID card (bioMerieux).

Table 5.12 summarizes the available tests for _Gardnerella_.

Miscellaneous Gram-Positive Rods

Arcanobacterium, _Erysipelothrix_, _Kurthia_, _Rhodococcus_, _Gordonia_, _Oerskovia_, _Dermatophilus_, _Actinomadura_, _Nocardiopsis_, _Arthrobacter_, _Paenibacillus_, _Cellulomonas_, _Actinobaculum_, _Brevibacterium_, _Cellulosimicrobium_, _Dermabacter_, _Microbacterium_, _Turicella_, and _Tsukamurella_ have all been implicated in human disease.[40,47–61] Identification tests for these organisms appears in Table 5.13.

Anaerobic Gram-Positive Organisms

Clostridium, _Bifidobacterium_, _Peptostreptococcus_, _Finegoldia_, _Peptococcus_, _Micromonas_, _Eggerthella_, _Eubacterium_,

TABLE 5.12

Identification Tests for _Gardnerella_

Catalase test
Oxidase test
Hippurate hydrolysis
Urease
Nitrate reduction
Carbohydrate fermentation
Zone of inhibition with trimethoprim, sulfonamide, and metronidazole
API Rapid ID 32 C
MicroScan Rapid HNID
Vitek II NH ID card

TABLE 5.13

Identification Tests for Miscellaneous Gram-positive Rods

Arcanobacterium, Erysipelothrix, Kurthia
Catalase test
Nitrate reduction
H_2S production
Motility
Urease
CAMP test
Carbohydrate fermentation
API Coryne
API Rapid ID 32 Strep
Vitek II GPID card
Rhodococcus, Gordonia, Oerskovia, Dermatophilus, Actinomadura, Nocardiopsis, Tsukamurella
Pigment production
Modified Acid Fast stain
Urease
Nitrate reduction
Anaerobic growth
Motility
Aerial mycelium
API Coryne
Arthrobacter, Paenibacillus, Cellulomonas, Actinobaculum, Brevibacterium, Cellulosimicrobium, Dermabacter, Microbacterium, Turicella
Microscopic morphologyPigment production
Catalase
Motility
LAP
PYR
Urease
Gelatin hydrolysis
Esculin hydrolysis
Carbohydrate utilization
API Coryne

Pseudoramibacteri, and _Propionibacterium_ are all organisms that have been implicated in human disease.[62–69] Identification tests are listed in Table 5.14.

TABLE 5.14

Identification Tests for Anaerobic Gram-positive Organisms

Clostridium, Bifidobacterium, Eggerthella, Eubacterium, Pseudoramibacter, Propionibacterium
Anaerobic growth
ß-hemolysis
Kanamycin disc (1 mg)
Colistin disc (10 ug)
Vancomycin disc (5 ug)
Indole
Catalase
Lecithinase
Nagler test
Reverse CAMP test
Nitrate reduction
Urease
Arginine dihydrolase
API 20A
API Rapid ID 32A
BBL Crystal ANA ID
Vitek II ANC ID card
Peptostreptococcus, Peptococcus, Finegoldia, Micromonas
Catalase
Indole
API 20A
API Rapid ID 32A
BBL Crystal ANA ID
Vitek II ANC ID card

REFERENCES

1. Kloos, W.E. and Bannerman, T.L. *Staphylococcus* and *Micrococcus*, in *Manual of Clinical Microbiology*, 7th ed., Murray, P.R., Baron, E.J., Pfaller, M.A., Tenover, F.C., and Yolken R.H., Eds., American Society for Microbiology, Washington, DC, 1999, 264.
2. Hedin, G. and Fang, H. Evaluation of two new chromogenic media, CHROMagar MRSA and *S. aureus* ID, for identifying *Staphylococcus aureus* and screening methicillin-resistant S. aureus. *J. Clin. Microbiol.*, 43, 4242, 2005.
3. Forbes, B.A., Sahm, D.F., and Weissfeld, A.S., Eds. *Staphylococcus*, *Micrococcus* and similar organisms, in *Bailey and Scott's Diagnostic Microbiology*, Vol. 11. Mosby, Inc., St. Louis, MO, 2002, 285
4. Personne, P., Bes, M., Lina, G., Vandenesch, F., Brun, Y., and Etienne, J. Comparative performances of six agglutination kits assessed by using typical and atypical strains of Staphylococcus aureus. *J. Clin. Microbiol.*, 35, 1138, 1997.
5. Hansen-Gahrn, B., Heltberg, O., Rosdahl, V., and Sogaard, P. Evaluation of a conventional routine method for identification of clinical isolates of coagulase-negative *Staphylococcus* and *Micrococcus* species. Comparison with API-Staph and API Staph-Ident. *Acta Pathol. Microbiol. Immunol. Scand.*, 95, 283, 1987.
6. Layer, F., Ghebremedhin, B., Moder, K.A., Konig, W., and Konig, B. Comparative study using various methods for identification of *Staphylococcus* species in clinical specimens. *J. Clin. Microbiol.*, 44, 2824, 2006.
7. Bannerman, T.L., Kleeman, K.T., and Kloos, W. Evaluation of the Vitek systems gram-positive identification card for species identification of coagulase-negative *Staphylococci*. *J. Clin. Microbiol.*, 31, 1322, 1993.
8. Salomon, J., Dunk, T., Yu, C., Pollitt, J., and Reuben, J. Rapid automated identification of gram-positive and gram-negative bacteria in the phoenix system. *Abstract presented at 99th General Meeting ASM*, May 1999.
9. Chapin, K., and Musgnug, M. Evaluation of three rapid methods for the direct identification of *Staphylococcus aureus* from positive blood cultures. *J. Clin. Micrbiol.*, 41, 4324, 2003.
10. Shrestha, N.K., Tuohy, M.J., Padmanabhan, R.A., Hall, G.S., and Procop, G.W. Evaluation of the LightCycler Staphylococcus M GRADE kits on positive blood cultures that contained gram-positive cocci in clusters. *J. Clin. Microbiol.*, 43, 6144, 2005.
11. Dunn, R., Bares, S., and David, M. Central venous catheter-related bacteremia caused by *Kocuria kristinae*: Case report and review of the literature. *Ann. Clin. Microbiol. Antimicrob.*, 10, 31, 2011.
12. Dotis, J., Printza, N., Stabouli, S., and Papachristou, F. *Kocuria* species peritonitis: Although rare, we have to care. *Periton Dial Int.*, 35, 26–30, 2015.
13. LeBrun, C., Bouet, J., Gautier, P., Avril, J.L., and Gaillot, O. *Kytococcus schroeteri* endocarditis. *Emerg. Inf. Dis.*, 11, 179, 2005.
14. Koch, C., Schumann, P., and Stackebrandt, E. Reclassification of *Micrococcus agilis* to the genus Arthrobacter as *Arthrobacter agilis* comb. nov. and emendation of the genus *Arthrobacter*. *Int. J. Syst. Bacteriol.*, 45, 837, 1995
15. Bascomb, S., and Manafi, M. Use of enzyme tests in characterization and identification of aerobic and faculatatively anaerobic Gram-positive cocci. *Clin. Microbiol. Rev.*, 11, 318, 1998
16. Sader, H.S., Biedenbach, D., and Jones, R.N. Evaluation of Vitek and API 20S for species identification of enterococci. *Diagn. Microbiol. Infect. Dis.*, 22, 315, 1995.
17. Zhang, Q., Kwoh, C., Attorra, S., and Clarridge III, J.E. *Aerococcus urinae* in urinary tract infections. *J. Clin. Microbiol.*, 38, 1703, 2000.
18. You, M., and Facklam, R. New test system for identification of *Aerococcus*, *Enterococcus* and *Streptococcus* species. *J. Clin. Microbiol.*, 24, 607, 1986
19. Facklam, R., and Elliott, J. Identification, classification and clinical relevance of catalase-negative, Gram-positive cocci, excluding the streptococci and enterococci. *Clin. Microbiol. Rev.*, 8, 479, 1995.
20. Bjorkroth, K.J., Vandamme, P., and Korkeala, H.J. Identification and characterization of *Leuconostoc carnosum*, associated with production and spoilage of vacuum-packaged, sliced, cooked ham. *Appl. Environ. Microbiol.*, 64, 3313, 1998.
21. Harimaya, A., Takeda, R., Hendolin, P., Fujii, N., Ylikoski, J., and Himi, T. High incidence of *Alloiococcus otitidis* in children with otitis media, despite treatment with antibiotic. *J. Clin. Microbiol.*, 44, 946, 2006.
22. Yemisen, M., Koksal, F., Mete, B., Yarimcam, F., Okcun, B., Kucukoglu, S., Samasti, M., Kocazeybek, B., and Ozturk, R. *Abiotrophia defectiva*: A rare cause of infective endocarditis. *Scand. J. Infect. Dis.*, 38, 939, 2006.
23. Clinical Microbiology Proficiency Testing Synopsis. Blood Culture Isolate: *Rothia mucilaginosa (Stomatococcus mucilaginosus)*-Nomenclature Change. M031-4, 2003.
24. Lau, S., Woo, P., Li, N., Teng, J., Leung, K., Ng, K., Que, T., and Yuen, K. *Globicatella* bacteremia identified by 16S ribosomal RNA gene sequencing. *J. Clin. Pathol.*, 59, 303, 2006.
25. Chagla, A., Borczyk, A., Facklam, R., and Lovgren, M. Breast abscess associated with *Helcococcus kunzii*. *J. Clin. Microbiol.*, 36, 2377, 1998.
26. Ananthakrishna, R., Shankarappa, R., Jagadeesan, N., Math, R., Karur, S., and Nanjappa, M. Infective endocarditis: A rare organism in an uncommon setting. *Case Rep. Infect. Dis.*, Article ID 307852, 2012.
27. Gardenier, J., Hranjec, T., Sawyer, R., and Bonatti, H. *Granulicatella adiacens* bacteremia in an elderly trauma patient. *Surg. Infect.*, 12, 251, 2011.
28. Lecuyer, H., Audibert, J., Bobigny, A., Eckert, C., Janiere-Nartey, C., Buu-Hoi, A., Mainardi, J.L., and Podglajen, I. *Dolosigranulum pigrum* causing nosocomial pneumonia and septicemia. *J. Clin. Microbiol.*, 45, 3474, 2007.
29. Hall, G., Gordon, S., Schroeder, S., Smith, K., Anthony, K., and Procop, G. Case of synovitis potentially caused by *Dolosigranulum pigrum*. *J. Clin. Microbiol.*, 39, 1202, 2001.
30. Sampo, M., Ghazouani, O., Cadiou, D., Trichet, E., Hoffart, L., and Drancourt, M. *Dolosigranulum pigrum* keratitis: A three-case series. *BMC Opthamal.*, 13, 31, 2013.
31. Collins, M., Rodriguez, J., Hutson, R., Falsen, E., Sjoden, B., and Facklam, R. *Dolosicoccus paucivorans* gen. nov., sp. Nov., isolated from human blood. *Int. J. Syst. Bacteriol.*, 49, 1439, 1999.
32. Collins, M., Lawson, P., Monasterio, R., Falsen, E., Sjoden, B., and Faclam, R. *Ignavigranum ruoffiae* sp. nov., isolated from human clinical specimens. *Int. J. Syst. Bacteriol.*, 49, 97, 1999.
33. von Graevenitz, A., and Funke, G. An identification scheme for rapidly and aerobically growing Gram-positive rods. *Zentralbl. Bakteriol.*, 284, 246, 1996.
34. Fruh, M., von Graevenitz, A., and Funke, G. Use of second-line biochemical and susceptibility tests for the differential identification of coryneform bacteria. *Clin. Microbiol. Infect.*, 4, 332, 1998.

35. Hudspeth, M.K., Gerardo, S.H., Citron, D.M., and Goldstein, E.J. Evaluation of the RapID CB Plus system for identification of *Corynebacterium* species and other Gram-positive rods. *J. Clin. Microbiol.*, 36, 543, 1998.

36. Gavin, S.E., Leonard, R.B., Briselden, A.M., and Coyle, M.B. Evaluation of the rapid CORYNE identification system for Corynebacterium species and other coryneforms. *J. Clin. Microbiol.*, 30, 1692, 1992.

37. Koneman, E., Allen, S., Janda, W., Schreckenberger, P., and Winn, W. 1997. *Color Atlas and Textbook of Diagnostic Microbiology*, 5th ed., Lippincott Williams & Wilkins, Philadelphia, PA.

38. Kerr, K.G. and Lacey, R.W. Isolation and identification of *Listeria monocytogenes. J. Clin. Pathol.*, 44, 624, 1991.

39. Kerr, K.G., Hawkey, P.M., and Lacey, R.W. Evaluation of the API Coryne system for identification of *Listeria* species. *J. Clin. Microbiol.*, 31, 749, 1993.

40. Funke, G., von Graevenitz, A., Clarridge III, J.E., and Bernard, K.A. Clinical microbiology of coryneform bacteria. *Clin. Microbiol Rev.*, 10, 125, 1997.

41. Kiska, D.L., Hicks, K., and Pettit, D.J. Identification of medically relevant *Nocardia* species with a battery of tests. *J. Clin. Microbiol.*, 40, 1346, 2002.

42. Wauters, G., Avesani, V., Charlier, J., Janssens, M., Vaneechoutte, M., and Delmee, M. Distribution of *Nocardia* species in clinical samples and their routine rapid identification in the laboratory. *J. Clin. Microbiol.*, 43, 2624, 2005.

43. Mossad, S., Tomford, J., Stewert, R., Ratliff, N., and Hall, G. Case report of Streptomyces endocarditis of a prosthetic aortic valve. *J. Clin. Microbiol.*, 33, 3335, 1995.

44. Santala, A., Sarkonen, N., Hall, V., Carlson, P., Somer, K., and Kononen, E. Evaluation of four commercial test systems for identification of actinomyces and some closely related species. *J. Clin. Microbiol.*, 42, 418, 2004.

45. Morrison, J.R. and Tillotson, G.S. Identification of *Actinomyces (Corynebacterium) pyogenes* with the API 20 Strep System. *J. Clin. Microbiol.*, 26, 1865, 1988.

46. Piot, P., Van Dyck, E., Totten, P., and Holmes, K. Identification of *Gardnerella (Haemophilus) vaginalis. J. Clin. Microbiol.*, 15, 19, 1982.

47. Dobinsky, S., Noesselt, T., Rucker, A., Maerker, J., and Mack, D. Three cases of *Arcanobacterium haemolyticum* associated with abscess formation and cellulitis. *Eur. J. Clin. Microbiol Inf. Dis.*, 18, 804, 1999.

48. Reboli, A.C., and Farrer, W.E. *Erysipelothrix rhusiopathiae*: An occupational pathogen. *Clin. Microbiol. Rev.*, 2, 354, 1989.

49. Lejbkowicz, F., Kudinsky, R., Samet, L., Belavsky, L., Barzilai, M., and Predescu, S. Identification of *Nocardiopsis dassonvillei* in a blood sample from a child. *Am. J. Infect. Dis.*, 1, 1, 2005.

50. Rihs, J.D., McNeil, M.M., Brown, J.M., and Yu, V.L. *Oerskovia xanthineolytica* implicated in peritonitis associated with peritoneal dialysis: Case report and review of Oerskovia infections in humans. *J. Clin. Microbiol.*, 28, 1934, 1990.

51. Gil-Sande, E., Brun-Otewro, M., Campo-Cerecedo, F., Esteban, E., Aguilar, L., and Garcia-de-Lomas, J. Etiological misidentification by routine biochemical tests of bacteremia caused by *Gordonia terrea* infection in the course of an episode of acute cholecystitis. *J. Clin. Microbiol.*, 44, 2645, 2006.

52. Woo, P.C., Ngan, A.H., Lau, S.K., and Yuen, K.Y. Tsukamurella conjunctivitis: A novel clinical syndrome. *J. Clin. Microbiol.*, 41, 3368, 2003.

53. Funke, G., Pagano-Niederer, M., Sjoden, B., and Falsen, E. Characteristics of *Arthrobacter cumminsii*, the most frequently encountered *Arthrobacter* species in human clinical specimens. *J. Clin. Microbiol.*, 36, 1539, 1998.

54. Rieg, S., Bauer, T., Peyeri-Hoffmann, G., Held, J., Ritter, W., Wagner, D., Kern, W., and Serr, A. *Paenibacillus larvae* bacteremia in injection drug users. *Emerg. Inf. Dis.*, 13, 487, 2010.

55. Lai, P., Chen, Y., and Lee, S. Infective endocarditis and osteomyelitis caused by *Cellulomonas*: A case report and review of the literature. *Diagn. Microbiol. Infect. Dis.*, 65, 184, 2009.

56. Bank, S., Jensen, A., Hansen, T., Soby, K., and Prag, J. *Actinobaculum schaalii*, a common uropathogen in elderly patients, Denmark. *Emerg. Inf. Dis.*, 16, 76, 2010.

57. Roux, V. and Raoult, D. *Brevibacterium massiliense* sp. nov., isolated from a human ankle discharge. *Int. J. Syst. Evol. Microbiol.*, 59, 1960, 2009.

58. Petkar, H., Li, A., Bunce, N., Duffy, K., Malnick, H., and Shah, J. *Cellulosimicrobium funkei*: First report of infection in a nonimmunocompromised patient and useful phenotypic tests for differentiation from *Cellulosimicrobium cellulans* and *Cellulosimicrobium terreum. J. Clin. Microbiol.*, 49, 1175, 2011.

59. Cercenado, E., Marin, M., Gama, B., Alcala, L., and Santiago, E. Daptomycin-resistant *Dermabacter hominis*: An emerging gram-positive coryneform rod causing human infections. European Society of Clinical Microbiology and Infectious Diseases, Session April 29, 2013.

60. Gneiding, K., Frodl, R., and Funke, G. Identities of *Microbacterium* ssp. encountered in human clinical specimens. *J. Clin. Microbiol.*, 46, 3646, 2008.

61. Gomez-Garces, J., Alhambra, A., Alos, J., Barrera, B., and Garcia, G. Acute and chronic otitis media and *Turicella otitidis*: A controversial association. *Clin. Microbiol. Infect.*, 10, 854, 2004.

62. Brazier, J., Duerden, B., Hall, V., Salmon, J., Hood, J., Brett, M., McLauchlin, J., and George, R. Isolation and identification of *Clostridium* ssp. from infections associated with the injection of drugs: Experiences of a microbiological investigation team. *J. Med. Microbiol.*, 51, 985, 2002

63. Ha, G.Y., Yang, C.H., Kim, H., and Chong, Y. Case of sepsis caused by *Bifidobacterium longum. J. Clin. Microbiol.*, 37, 1227, 1999.

64. Kageyama, A., Benno, Y., and Naskase, T. Phylogenic evidence for the transfer of *Eubacterium lentum* to the genus *Eggerthella* as *Eggerthella lenta* gen. nov., comb. nov. *Inter J. Sys. Bacter.*, 49, 1725, 1999.

65. Bassetti, S., Laifer, G., Goy, G., Fluckiger, U., and Frei, R. Endocarditis caused by *Finegoldia magna* (formally *Peptostreptococcus magnus*): Diagnosis depends on the blood culture system used. *Diag. Microbiol. Inf. Dis.*, 47, 359, 2003.

66. Collins, M., Lawso, P., Williems, A., Cordoba, J., Fernandez-Garayzabal, J., and Garcia, P. The phylogeny of the genus *Clostridium*: Proposal of five new genera and eleven new species combinations. *Int. J. Syst. Bacteriol.*, 44, 812, 1994.

67. Jouseimies-Somer, H., Summanen, P. Citron, D., Baron, E., Wexler, H., and Finegold, S. 2002. *Wadsworth-KTL Anaerobic Bacteriology manual*, 6th ed., Star Publishing, Belmont, CA.

68. Cavallaro, J., Wiggs, L., and Miller, M. Evaluation of the BBL crystal anaerobic identification system. *J. Clin. Microbiol.*, 35, 3186, 1997.

69. Siqueira, J. and Rocas, I. *Pseudoramibacter alactolyticus* in primary endodontic infections. *J. Endod.*, 29, 735, 2003.

6

Identification of Aerobic Gram-Negative Bacteria

Donna J. Kohlerschmidt, Lisa A. Mingle, Nellie B. Dumas, and Geetha Nattanmai

CONTENTS

Introduction

Aerobic Gram-negative bacteria are ubiquitous. Many are found throughout the environment and distributed worldwide. Others are established as normal flora in human and animal mucosa, intestinal tract, and skin. Many of these bacteria are typically harmless, and several are even beneficial. Others account for a large percentage of foodborne illness, and some have been identified as potential weapons of bioterrorism. The identification of these organisms is necessary in many circumstances. The clinical microbiologist identifying the pathogen responsible for the severe, bloody diarrhea of a hospitalized young child; the environmental laboratorian identifying the bacterial contaminant that forced a recall of a commercial product; and the researcher identifying and characterizing the bacteria that are degrading chemical pollutants in a river all find themselves faced with the task of identifying aerobic Gram-negative bacteria. This task can be accomplished using a variety of methods. It often involves a combination of conventional phenotypic tests based on biochemical reactions, commercial kits or systems, and/or molecular analysis.

This chapter addresses the Gram stain and cellular morphology of bacteria, which are key features for preliminary identification of the Gram-negative genera. Descriptions are given of some fundamental phenotypic characteristics including growth on trypticase soy agar (TSA) with 5% sheep blood, pigment production, growth on MacConkey agar, glucose fermentation, and production of oxidase.

In general, aerobic Gram-negative bacteria can be divided into two groups based on their ability to ferment glucose and produce acid in triple sugar iron (TSI) agar or Kligler's iron agar as shown in Figures 6.1 and 6.2. Descriptions of some additional conventional biochemical tests that can be used to further characterize an aerobic Gram-negative bacterium are included in this chapter: motility, acid from carbohydrates, catalase, citrate, decarboxylases, esculin hydrolysis, gelatin liquefaction, indole, methyl red (MR), Voges-Proskauer (VP), urease, and nitrate reduction. However, there are many bacteria that are difficult to identify solely by conventional biochemical tests. Reference laboratories may utilize mass spectrometry or polymerase chain reaction (PCR) assays to assist in the identification of aerobic Gram-negative bacteria.

DOI: 10.1201/9781003099277-7

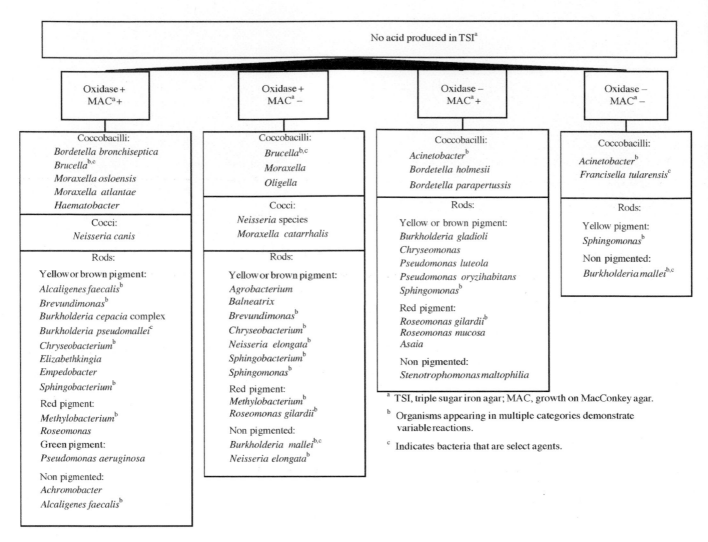

FIGURE 6.1   Gram-negative organisms that grow on TSA with 5% sheep blood and do not ferment glucose. (Data from Refs. [8, 9, 11, 15, 20–24, 27, 28].)

Additionally, 16S rRNA gene sequence analysis can be used to identify unusual members of this group that give problematic results.

The identification algorithms in Figures 6.1 and 6.2 do not include those aerobic Gram-negative bacteria that are incapable of growing on TSA with 5% sheep blood. For example, most *Haemophilus* species will grow only on chocolate agar, *Legionella* species require l-cysteine and will grow on buffered charcoal yeast extract media, and growth of *Bordetella pertussis* requires the supplements contained in Bordet-Gengou or charcoal-horse blood agar. The reader can consult Edelstein and Wadlin,[1] Leber,[2] Ledeboer and Doern,[3] Konig et al.,[4] and Edelstein and Luck[5] for identification algorithms for these organisms.

Laboratory protocols for the identification of unknown aerobic Gram-negative bacteria should also include procedures for the safe handling of potential pathogens. If a select agent (potential bioterrorism threat agent), such as *Francisella tularensis*, *Burkholderia mallei*, *Burkholderia pseudomallei*, *Brucella* species, or *Yersinia pestis*, is suspected at any point during the identification process, culture work on the open lab bench should cease immediately, and a biological safety cabinet should be used if further testing is necessary. An appropriate reference laboratory should be contacted to arrange the shipment of the specimen for complete or confirmatory identification.

Characteristics for Initial Identification

Identification schemes for aerobic Gram-negative organisms that grow on TSA with 5% sheep blood agar begin with performing a Gram stain, observing colony morphology, and assessing the organism's ability to ferment glucose and grow on MacConkey agar. Figures 6.1 and 6.2 show a strategy to divide aerobic Gram-negative organisms based on Gram-stain cellular morphology, TSI reaction, cytochrome oxidase production, pigment production, and growth on MacConkey agar. Detailed algorithms for the preliminary identification of aerobic Gram-negative bacteria have been published by Schreckenberger and Wong[6]; York, Schreckenberger, and Church[7]; Weyant et al.[8]; and Wauters and Vaneechoutte.[9]

Gram Stain

Performing a Gram stain of bacterial growth from isolated colonies should be one of the first steps in identifying an unknown bacterium. Gram-negative organisms have a cell wall composed of a single peptidoglycan layer attached to a lipopolysaccharide-phospholipid bilayer outer membrane with

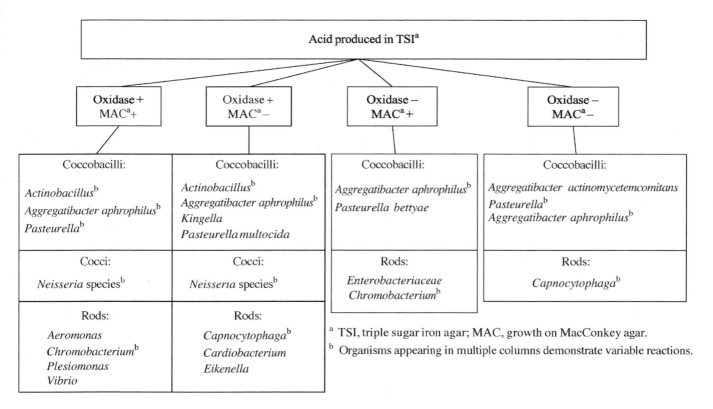

FIGURE 6.2   Gram-negative organisms that grow on TSA with 5% sheep blood and ferment glucose. (Data from Refs. [8, 11, 14, 19, 20, 27, 28, 78].)

protein molecules interspersed in the lipids. Decolorizing with alcohol damages this outer membrane and allows the crystal violet-iodine complex to leak out. The organism then takes up the red color of the safranin counterstain.[10] Once proven to be Gram-negative, bacteria can be further differentiated based on their cellular morphology. They may appear as pronounced rods as seen in most *Enterobacteriaceae* and many *Pseudomonas* species; as Gram-negative cocci, such as most *Neisseria* species and *Moraxella catarrhalis*; or as oval coccobacilli, as seen in *Haemophilus*, *Aggregatibacter*, *Francisella*, and *Acinetobacter* species.

Even an experienced microbiologist might have difficulty distinguishing between true cocci and coccobacilli in some cases. Growth of cultures in the presence of low levels of penicillin can cause the bacteria to reveal their true form. Organisms that normally divide and produce short rods will instead continue to grow and produce filaments when exposed to penicillin. Gram stains made from growth at the edge of the zone of inhibition around a 10μg penicillin disk will show either enlarged cocci if the organism is a true coccus or long filaments if the organism is a Gram-negative coccobacillus or rod.[8]

Colony Morphology

Another key feature for preliminary identification of aerobic Gram-negative organisms is observation of the size, shape, and texture of bacterial colonies growing on agar media. Colonies of most members of the *Enterobacteriaceae* family tend to be large and moist and grow well on TSA with 5% sheep blood after 24 hours of incubation. *Neisseria* species have smaller but distinct colonies after 24 hours. Nonfermentative Gram-negative rods

and fastidious fermenters may require 48 hours or more of incubation before growth is evident. Many Gram-negative organisms produce extracellular enzymes that lyse the red blood cells in the agar. *Escherichia coli*, *Morganella morganii*, *Pseudomonas aeruginosa*, *Aeromonas hydrophila*, and commensal *Neisseria* species can be recognized by the presence of complete clearing of the red blood cells around the colonies (beta-hemolysis) on blood agar. Colonies of *Eikenella corrodens* and *Alcaligenes faecalis* produce only partial lysis of the cells, forming a greenish discoloration (alpha-hemolysis) around the colony.[11] Colonies of *Proteus mirabilis* and *Proteus vulgaris* are distinctive because of their swarming edges that can spread to cover the entire plate. *Photorhabdus* species, usually associated with nematodes but recently reported to cause human infections, are bioluminescent: the colonies emit light when examined under conditions of total darkness.[12,13]

Pigment Production

Gram-negative bacteria can produce several pigments that are helpful in establishing identification. Some pigments are water insoluble and give color to the bacterial colonies. These include carotenoids (yellow-orange), violacein pigments (violet or purple), and phenazines (red, maroon, or yellow). Other pigments, such as pyocyanin and pyoverdine, are water-soluble and diffusible. They confer color to the medium surrounding the colonies as well as to the colonies themselves. Media formulations developed specifically to detect pigment production contain special peptones and an increased concentration of magnesium and sulfate ions.[11] Pigment can also be observed in growth on TSA with 5% sheep blood, brain heart infusion agar, and Loeffler's agar.

Enterobacter species, most strains of *Cronobacter saka-zakii*, *Pantoea agglomerans*, *Leclercia adecarboxylata*, and the nonfermentative Gram-negative rods *Chryseobacterium*, *Elizabethkingia*, *Myroides*, *Flavobacterium*, *Balneatrix*, *Empedobacter*, and *Sphingomonas* species can be distinguished from related organisms by their yellow pigment. *Burkholderia cepacia* complex or *Burkholderia gladioli* can be suspected when a colony of a nonfermentative Gram-negative rod displays a yellow-brown pigment. *Chromobacterium violaceum* produces violet pigment, while colonies of *Roseomonas*, *Methylobacterium*, *Asaia*, *Azospirillum*, and some species of *Serratia* appear pink to red. The fluorescent *Pseudomonas* group is distinguished from other nonfermentative Gram-negative rods by the production of the fluorescent pigment pyoverdine. *P. aeruginosa* isolates may produce both pyoverdine and blue-green pigment pyocyanin.[11,14,15]

Growth on MacConkey Agar

The ability of an organism to grow and to ferment lactose on MacConkey agar offers clues to the organism's identification. MacConkey agar contains crystal violet and bile salts, which are inhibitory to most Gram-positive organisms and certain Gram negatives.[16] *Enterobacteriaceae* will grow on MacConkey agar in 24–48 hours, while fastidious fermenters and nonfermentative Gram-negative organisms may take up to 7 days for growth to appear. Most *Moraxella* species and most species of *Neisseria* can be distinguished from related organisms by their inability to grow on MacConkey agar. MacConkey agar also contains lactose and a neutral red indicator, which can further help characterize the organism. Gram-negative organisms that ferment lactose, such as *Klebsiella pneumoniae* and strains of *E. coli*, will develop into purple colonies as the neutral red indicator changes due to the production of acids. Organisms that do not ferment lactose, such as *Proteus* or *Pseudomonas* species, appear colorless or transparent.

Triple Sugar Iron Agar

Triple sugar iron (TSI) agar can be used to determine whether an organism is capable of fermenting glucose. This medium contains glucose, lactose, and sucrose in a peptone and casein base with a phenol red pH indicator.[11,17,18] An agar slant is inoculated by first stabbing through the center of the medium and then streaking the surface of the slant, using an inoculum of 18–24 hours growth. If the organism ferments glucose, acid will be produced in the agar, lowering the pH and changing the color of the medium from red to yellow. If gas is produced during fermentation, bubbles will form along the stab line, fracturing or displacing the medium. Most species of *Enterobacteriaceae* and *Aeromonas* will produce an acid reaction, often with the production of gas, after 24 hours of incubation at 35°C–37°C. *Pasteurella*, *Eikenella*, and other fastidious fermentative Gram-negative rods will produce a weak acid reaction that may take up to 48 hours to appear, and typically will have no gas formation. *Pseudomonas* species, *Moraxella* species, and other nonfermentative Gram-negative

organisms will show no color change on TSI medium or will produce a pink, alkaline reaction as the bacteria break down the peptones in the agar base.

TSI agar also contains iron salts and sodium thiosulfate to detect the ability of a Gram-negative organism to produce hydrogen sulfide (H_2S). H_2S, if produced, will react with the iron salts to form a black precipitate in the butt of the TSI tube.[11] The ability to produce H_2S is a key test to differentiate *Salmonella*, *Proteus*, *Citrobacter*, and *Edwardsiella tarda* from other members of the *Enterobacteriaceae*. H_2S production helps to separate the genus *Shewanella* from other similar nonfermenters.[9]

Oxidase

The oxidase test determines the ability of an organism to produce cytochrome oxidase enzymes. A phenylenediamine reagent is used as an electron acceptor. In the presence of atmospheric oxygen and the enzyme cytochrome oxidase, the reagent is oxidized to form a deep blue compound, indole phenol. Two different reagents are commercially available: (1) tetramethyl-*p*-phenylenediamine dihydrochloride (Kovac's formulation) and (2) dimethyl-*p*-phenylenediamine dihydrochloride (Gordon and McLeod's reagent). The tetramethyl derivative is the more stable of the two, is less toxic, and is more sensitive in detecting weak results.[8,11] Oxidase can be detected by adding drops of the test reagent directly onto 18–24 hours old growth on a culture plate. Colonies will become blue if cytochrome oxidase is produced. An alternate, more sensitive oxidase test method uses filter paper saturated with reagent. A loopful of organism from an 18–24-hour-old culture is inoculated onto filter paper. Growth should be taken from a medium that contains no glucose. Development of a blue color at the site of inoculation within 10–30 seconds indicates a positive reaction. Care must be taken to avoid using metal loops, which can give a false-positive reaction.[8] All members of the *Enterobacteriaceae* except *Plesiomonas shigelloides* are negative for oxidase. *Aeromonas*, *Vibrio*, *Neisseria*, *Moraxella*, and *Campylobacter* are all positive for oxidase, as are most *Pseudomonas* species. *Acinetobacter* species can be distinguished from similar nonfermentative Gram-negative rods by a negative oxidase reaction.

Additional Conventional Biochemical Tests

Gram-negative organisms can be further characterized phenotypically using a battery of media and substrates. The pattern of reactions is then compared with charts in the literature [8,11,12,19–28] to arrive at a definitive identification. What follows is a brief description of some key tests used for identifying Gram-negative organisms to the species level. More detailed descriptions of these test methods have been published elsewhere.[11,17,18,29]

Acid from Carbohydrates

Many identification schemes for Gram-negative organisms rely on the determination of the organism's ability to produce acid from a variety of carbohydrates. Acid can be produced by either fermentation or oxidation. Fermentative organisms are tested using a 1% solution of carbohydrate in peptone water containing a pH indicator. Different media formulations may incorporate bromocresol purple, phenol red, or Andrade's indicator with acid fuchsin.[8,11,18,29] Selected carbohydrates are inoculated with a suspension of the organism in saline or in broth without glucose. After 1–5 days of incubation at 35°C–37°C, if the organism ferments the carbohydrate, the resulting acid will change the color of the pH indicator. The medium will become yellow if bromocresol purple or phenol red is used as an indicator, or pink if Andrade's indicator is used.

Nonfermentative organisms may produce acid from carbohydrates by oxidation. This can be detected using Hugh and Leifson's oxidation–fermentation basal medium, a semisolid peptone base with a bromothymol blue indicator and 1% concentration of carbohydrates. Two tubes of each carbohydrate are inoculated with the test organism by stabbing an inoculation needle into the center of the semisolid medium. The medium in one tube of each set is overlaid with sterile mineral oil. The second tube is left with the cap loosened. Both tubes are incubated at 35°C–37°C for up to 10 days. If the isolate oxidizes the carbohydrate, acid will be produced in the presence of oxygen, causing a yellow color change in the tube with no oil. If the isolate ferments the carbohydrate, an acid change will occur in the tube overlaid with oil. An organism that is neither oxidative nor fermentative may produce an alkaline (blue) reaction in the open tube if the organism breaks down peptones in the medium base.[8,11,18,29]

Catalase

The enzyme catalase breaks down hydrogen peroxide to form oxygen gas and water. When drops of 3% hydrogen peroxide are added to 18–24 hours of growth of an organism, bubbles will form if the organism is producing the catalase enzyme. Any medium containing whole blood should not be used for catalase testing because red blood cells contain catalase and can cause a false-positive result. Although it contains blood, chocolate agar can be used for the catalase test because the red blood cells have been destroyed during media preparation.[17] The catalase reaction is useful to differentiate *Kingella* species, which are catalase negative, from most other fastidious fermenters and from *Moraxella* species, which are catalase positive. *Neisseria gonorrhoeae* is strongly positive when 30% hydrogen peroxide (superoxol) is used. This characteristic is often included as part of a presumptive identification for *N. gonorrhoeae* from selective media.[19,29]

Citrate Utilization

The ability of a bacterium to use citrate as a sole source of carbon is an important characteristic in the identification of *Enterobacteriaceae*. Simmons citrate medium formulation includes sodium citrate as the sole source of carbon, ammonium phosphate as the sole source of nitrogen, and the pH indicator bromothymol blue. The agar slant is inoculated lightly and incubated at 35°C–37°C for up to four days, keeping the cap loosened because the reaction requires oxygen. If the organism utilizes the citrate, it will grow on the agar slant

and break down the ammonium phosphate to form ammonia. This will cause an alkaline pH change, turning the pH indicator from green to blue.[11,17,18] *E. coli* and *Shigella* species are citrate negative, while many other *Enterobacteriaceae* are citrate positive. Among nonfermentative Gram-negative rods, *A. faecalis* and *Achromobacter xylosoxidans* are consistently positive for citrate utilization, while *Moraxella* species are consistently negative. A positive citrate reaction helps to distinguish *Roseomonas gilardii* from *Methylobacterium* species.

Decarboxylation of Lysine, Arginine, and Ornithine

This test determines whether an organism possesses decarboxylase enzymes specific for the amino acids lysine, arginine, and ornithine. Moeller's decarboxylase medium is commonly used. This medium contains glucose, the pH indicators bromocresol purple and cresol red, and 1% of the l (levo) form of the amino acid to be tested. A control tube containing the decarboxylase base with no amino acid is inoculated in parallel. All tubes are overlaid with oil because decarboxylation requires anaerobic conditions. If the organism possesses the appropriate enzyme, hydrolyzation of the amino acid will occur, forming an amine and producing an alkaline pH change. A positive reaction will result in a purple color change. A negative reaction will retain the color of the base or become acidic (yellow) due to fermentation of glucose in the medium.[11,17,18,29] Fermentative Gram-negative rods require a light inoculum and 24–48 hours of incubation at 35°C–37°C. Nonfermenters, however, may produce a weaker reaction and often need a heavy inoculum and prolonged incubation at 35°C–37°C before a color change occurs. Decreasing the volume of substrate used can also enhance reactions in weakly positive organisms.[11]

Lysine, arginine, and ornithine reactions are key tests that help differentiate among the genera and species of the *Enterobacteriaceae*. For example, many species of *Citrobacter* and *Enterobacter* are positive for arginine and ornithine, while *Klebsiella* species are often only positive for lysine, and *Yersinia* species are typically positive for ornithine. *Leclercia adecarboxylata* is unable to decarboxylate any of the three amino acids, while *Plesiomonas shigelloides* is positive for all three. Among nonfermentative Gram-negative rods, a positive arginine dihydrolase reaction is characteristic of *P. aeruginosa* and the select agents *Burkholderia mallei* and *Burkholderia pseudomallei*. A positive arginine reaction also helps differentiate *Haematobacter* species from *Psychrobacter phenylpyruvicus*.[15] A positive lysine decarboxylase reaction is useful in distinguishing *Stenotrophomonas maltophilia* and members of the *Burkholderia cepacia* complex from most other nonfermenting Gram-negative rods.

Esculin Hydrolysis

Esculin agar formulations consist of 0.1% esculin in a heart infusion basal medium containing ferric citrate. Agar slants are inoculated with isolated growth and incubated at 35°C–37°C for 24 hours, or up to seven days for slowly growing Gram-negative rods. Hydrolysis of esculin forms glucose and esculetin.

The esculetin reacts with ferric ions to form a black compound in the media. Esculin is also a fluorescent compound that appears white-blue under long-wave ultraviolet light at 360 nm. When esculin is hydrolyzed, fluorescence is lost. The development of a black color in the agar or the loss of fluorescence after incubation is interpreted as a positive test. If the agar remains fluorescent or there is no black color formed, the test is interpreted as negative. Gram-negative organisms that produce H_2S may cause a false-positive reaction as H_2S also reacts with ferric ions to produce a black complex. The esculin hydrolysis reaction for these organisms should be interpreted by checking for loss of fluorescence.[11,16,18,29]

Esculin hydrolysis is a key reaction in the identification of *Chryseobacterium* and *Sphingobacterium* species. *Pseudomonas (Chryseomonas) luteola*, *Brevundimonas vesicularis*, *Stenotrophomonas maltophilia*, and *Ochrobactrum anthropi* are all positive for esculin hydrolysis as are some members of the *Burkholderia cepacia* complex and the select agent *Burkholderia pseudomallei*.[11]

Gelatin Liquefaction

The gelatin test detects an organism's ability to produce the proteolytic enzyme gelatinase. Nutrient gelatin is incorporated into a semisolid basal medium. The medium is stabbed with a heavy inoculum of the bacteria to be tested. An uninoculated control tube is included. Both tubes are incubated for up to two weeks. *Enterobacteriaceae* and *Pseudomonas* species should be incubated at 22°C. Nonfermenting Gram-negative organisms are incubated at 30°C. Other organisms are incubated at 35°C.[17] If gelatinase is produced, the inoculated media will become liquefied as the enzymes break down the gelatin present. Since the basal medium is a liquid at incubation temperatures, the tubes must be moved from the incubator to the refrigerator before a final reaction can be assessed. A positive reaction will remain liquid at refrigerator temperature, while a negative reaction will return to a semisolid state.

Alternatively, strips of undeveloped x-ray film coated with gelatin can be used to test for gelatinase production. The strip is placed in a suspension of the organism to be tested. A positive reaction will cause the coating of the strip to be hydrolyzed so that a clear plastic strip remains. The x-ray method requires 4–48 hours of incubation at 35°C.[17,29]

Gelatin is a key reaction in distinguishing *Pseudomonas fluorescens* group (positive) from *Pseudomonas putida* (negative). Among the *Enterobacteriaceae*, *Serratia* species, *Pantoea agglomerans*, and several species of *Proteus* are gelatin positive. A positive gelatin reaction helps to differentiate *Acinetobacter haemolyticus* from other *Acinetobacter* species and distinguishes *Capnocytophaga sputigena* from gelatin-negative *Capnocytophaga* species such as *C. canimorsus*, *C. cynodegmi*, and *C. gingivalis*.[17,29]

Indole Production

The indole production test distinguishes organisms that produce tryptophanase, an enzyme capable of splitting the amino acid tryptophan to produce indole. Aldehyde in the test reagent reacts with indole to form a red-violet compound.[11,29] The test can be performed in a variety of ways. A spot indole test is useful for rapid indole producers such as *E. coli*. A drop of Kovac's reagent, containing 5% *p*-dimethylaminobenzaldehyde, is placed on the colony on an agar surface or on a colony picked up on a swab. The colony will form a red to pink color if positive.

Liquid or semisolid media with a high-tryptone content can be used for a tube test for indole. The medium is inoculated and incubated for 24–48 hours. A few drops of Kovac's reagent are added. If the organism produces tryptophanase, a pink color will develop at the meniscus.[11,17,29] *E. coli* and several other *Enterobacteriaceae* are indole positive and can be tested using this method.

For weakly reacting nonfermenters and fastidious organisms, Ehrlich's reagent, containing 1% *p*-dimethylaminobenzaldehyde, can be used. In this method, 0.5 mL xylene is added to the organism growing in tryptone broth to extract small quantities of indole, if present. The mixture is vortexed and allowed to settle. Six drops of Ehrlich's reagent are then added to the tube. If positive, a pink color will form below the xylene layer.[11,17,29] *Cardiobacterium hominis* and several nonfermentative Gram-negative rods, including *Chryseobacterium indologenes*, will be indole positive by this method.

Motility

Motility is demonstrated macroscopically by making a straight stab of growth into semisolid medium, incubating for 24–48 hours at 35°C–37°C, and then observing whether the organism migrates out from the stab line, causing visible turbidity in the surrounding medium.[17] Various formulations of motility medium are available. *Enterobacteriaceae* can be tested in a 5% agar medium with a tryptose base. Incorporation of 2,3,5-triphenyltetrazolium chloride (TTC) into this medium can help in visualizing the bacterial growth. TTC causes the bacterial cells to develop a red pigment that can aid in the interpretation of the motility reaction. Nonfermentative Gram-negative organisms that fail to grow in 5% agar medium can be tested using 3% agar in a base containing casein peptone and yeast extract. Fastidious organisms may demonstrate motility better in a 1% agar base containing heart infusion and tryptose.[16] *Shigella* species and *Klebsiella* species are always nonmotile. *P. aeruginosa* and *Achromobacter xylosoxidans* are both motile. *Yersinia enterocolitica* is often nonmotile at 35°C–37°C, but motile at 28°C.

Motility may be difficult to demonstrate in semisolid medium if the organism grows slowly. Use of a wet mount made from a suspension of growth in water or broth may enable weak motility to be detected. The mount is observed with the 40× or 100× objective of a light microscope to determine whether the organisms move and change direction.[8]

Methyl Red (MR) and Voges-Proskauer (VP) Tests

The methyl red (MR) test is used to determine an organism's ability to both produce strong acid from glucose and to maintain a low pH after prolonged (48–72 hours) incubation. MR–VP medium formulated by Clark and Lubs is inoculated and incubated for at least 48 hours. Five drops of 0.02% MR reagent are added to a 1 mL aliquot of test broth. Production of a red color in the medium indicates a positive reaction.[11,17,29]

The Voges-Proskauer (VP) test determines an organism's ability to produce acetylmethylcarbinol (acetoin) from glucose metabolism. MR–VP broth is incubated for a minimum of 48 hours, followed by the addition of 0.6 mL of 5% alpha-naphthol and 0.2 mL of 40% potassium hydroxide to a 1 mL aliquot of test broth with gentle mixing. The 40% potassium hydroxide causes acetylmethylcarbinol, if present, to be converted to diacetyl. Alpha-naphthol causes the diacetyl to form a red color.[6,11,17]

The MR and VP tests are used as part of the identification of *Vibrio* and *Aeromonas* species and help to differentiate among the genera and species of the *Enterobacteriaceae*.

For example, *E. coli*, *Citrobacter*, *Salmonella*, and *Shigella* species are positive for MR but negative for VP, while some species of *Enterobacter* are positive for VP but negative for MR. Several species of *Serratia* and *Klebsiella* can be distinguished from other *Enterobacteriaceae* because they are positive for both MR and VP. The tests are usually performed at 35°C–37°C, but *Yersinia enterocolitica* is VP positive at 25°C and variable at 35°C–37°C.

Nitrate Reduction

The ability of an organism to reduce nitrate (NO_3) to nitrite (NO_2) or to nitrogen gas is an important characteristic used to identify and differentiate Gram-negative bacteria. The organism is grown in a peptone or heart infusion broth containing potassium nitrate. An inverted Durham tube is included in the growth medium to detect the reduction of nitrites to nitrogen gas. After 2–5 days of incubation, five drops each of 0.8% sulfanilic acid and 0.5% *N,N*-dimethyl-naphthylamine are added to the test broth. If the organism has reduced the nitrate in the medium to nitrites, the sulfanilic acid will combine with the nitrite to form a diazonium salt. The dimethyl-naphthylamine couples with the diazonium compound to form a red, water-soluble azo dye. If there is no color development and no gas in the Durham tube, zinc dust can be used to determine whether nitrate has been reduced to nongaseous end products. Zinc dust will reduce any nitrate remaining in the medium to nitrite, and a red color will develop. If no red color develops after the addition of zinc, all nitrate in the medium has been reduced, indicating a positive nitrate reduction reaction. The development of a red color after the addition of zinc dust confirms a true negative reaction.[11,17,18,29] All members of the *Enterobacteriaceae*, except for certain biotypes of *Pantoea agglomerans* and certain species of *Serratia* and *Yersinia*, are able to reduce nitrate. *P. aeruginosa*, *Moraxella catarrhalis*, *Neisseria mucosa*, and *Kingella denitrificans* are characterized by a positive nitrate reaction. *A. faecalis* can be distinguished from similar nonfermentative Gram-negative organisms by its inability to reduce nitrate.

Urease

Detection of the ability of an organism to produce the enzyme urease can be accomplished using either broth or an agar medium containing urea and the pH indicator phenol red. Urea medium is inoculated with isolated, 24 hours old colonies and incubated at 35°C–37°C. If the organism produces urease, the urea will be split to produce ammonia and carbon dioxide. This creates an alkaline pH, causing the phenol red indicator to become bright pink.[18,29] Some bacteria, such as *Proteus* species, *Brucella* species, and *Bordetella bronchiseptica*, are capable of producing a strongly positive urease reaction in less than 4 hours. A rapidly positive urease reaction is useful in differentiating the select agent *Brucella* species from other oxidase-positive, slow-growing, aerobic Gram-negative coccobacilli. Other organisms must be incubated 24 hours or longer before a positive urease reaction becomes visible. Among the *Enterobacteriaceae*, *Proteus*, *Providencia*, and *Morganella* species are urease positive, as are some *Citrobacter* and *Yersinia* species. A positive urease reaction helps distinguish *Psychrobacter phenylpyruvicus*, *Haematobacter* species, and *Oligella ureolytica* from related organisms.[15] Urease is also a key test in the identification of *Helicobacter pylori* in gastric and duodenal specimens.

Commercial Identification Systems

There are several commercial systems available for the identification of aerobic Gram-negative organisms. The Vitek® System (bioMerieux, Inc., Hazelwood, MO), Sensititre Automated Reading and Incubation System (Thermofisher Scientific, Waltham, MA), MicroScan Walkaway (Beckman Coulter, Brea, CA), and the Phoenix™ System (BD Biosciences, Sparks, MD) are miniaturized, sealed identification panels containing 30–51 dried biochemical substrates, depending on the particular test kit chosen. The systems incorporate modifications of conventional biochemicals and also test for hydrolysis of chromogenic and fluorogenic substrates and for utilization of single carbon source substrates. A standardized inoculum is used, and the panels are continuously incubated in an automated reader. Reactions are detected by the instrument via various indicator systems. Results are compared with known bacteria in the system's database to arrive at a final identification.[11,30–33]

Semiautomated systems such as MicroScan® panels (Beckman Coulter, Brea, CA) and the BBL Crystal Identification System (Becton Dickinson Microbiology Systems, Sparks, MD) use substrates similar to the fully automated methods. They can be placed in an automated reader for interpretation or read visually by the operator. API® kits (bioMerieux, Inc., Hazelwood, MO), TREK Sensititre plates (Thermofisher Scientific, Waltham, MA), and BD BBL Enterotube II (Becton Dickinson Microbiology Systems, Sparks, MD) also use miniaturized modifications of conventional biochemicals, but the systems are open, allowing reagents to be added by the operator. The wells are read manually, giving the opportunity for visual confirmation of biochemical results. For the semiautomated and manual systems, a 5–10 digit code is generated as the biochemical reactions are interpreted. The code is translated into an identification using a code book or software provided by the manufacturer.[11,32–34]

The Biolog MicroPlate™ system (Biolog, Inc., Hayward, CA) identifies aerobic Gram-negative organisms by testing their ability to utilize 95 separate carbon sources. Reactions are based on the exchange of electrons generated during respiration, leading to a tetrazolium-based color change. Biolog microplates can be read visually, achieving a reaction pattern that can be classified using database software,[11,32,34] or used as part of the fully automated OmniLog ID System (Biolog, Hayward, CA). The OmniLog system allows the user to choose different incubation temperatures to enable maximum growth.[11,32,34]

There are two FDA-cleared identification systems that rely on matrix-assisted laser desorption/ionization time of flight mass spectrometry (MALDI-TOF MS) technology, Bruker Biotyper MALDI-TOF MS platform (Bruker Daltonics Inc., Billerica, MA) and the Vitek MS system (bioMerieux, Inc., Durham, NC). MALDI-TOF MS is a form of mass spectrometry that employs a soft ionization technique. Soft ionization keeps the molecules intact with minimal fragmentation during ion formation, which enables the analysis of proteins.[35] MALDI-TOF MS allows for the identification of microorganisms by analyzing the expression of their intrinsic proteins, particularly the ribosomal proteins that are stable under various culture conditions.[36,37] The proteins present in the sample are separated by mass and charge to yield a spectrum. This spectrum functions as a fingerprint for the particular species that is then compared to a reference database to identify the bacteria (or yeast).[37] This technology provides a fast, reliable, low-cost method for identifying bacterial organisms.

Commercial systems are useful when a large number of isolates must be processed because such systems offer a decreased turnaround time and decreased hands-on time, as compared to conventional biochemical schemes. The biochemical-based commercial platforms perform with high levels of accuracy when identifying metabolically active organisms such as the *Enterobacteriaceae*, *P. aeruginosa*, or *Stenotrophomonas maltophilia*. However, some systems are not designed to detect weak or delayed reactions, which can result in false-negative results. Some fastidious aerobic Gram-negative organisms may not grow well in commercial systems and may require the use of additional testing methods. The MALDI-TOF MS–based platforms are based on ribosomal proteins and often perform as well on inert organisms as on metabolically active organisms. Commercial kits and systems may be limited by the range and number of bacteria included in the particular system's database. Organisms that are not included in the database will not be recognized and may be misidentified.

A number of references provide detailed descriptions and evaluations of biochemical-based commercial systems,[11,32,38] MALDI-TOF MS systems that can be used on bacterial isolates,[39–42] and MALDI-TOF MS identification performed directly from blood cultures.[43–48]

Molecular Testing and Analysis

Molecular analysis is becoming an increasingly important tool for the clinical bacteriologist, especially in the identification of aerobic Gram-negative bacteria. PCR and real-time PCR assays offer specific and rapid identifications that can be the key to the detection of pathogens. This is particularly true for specimen types where classical approaches may be compromised due to a delay in transport or the initiation of antimicrobial therapy prior to specimen collection.

Molecular analysis can quickly provide identification to the species level for certain genera of aerobic Gram-negative

bacteria that cannot be differentiated by commercially available systems, or that would require the use of highly specialized, time-consuming, and labor-intensive protocols. Examples include species determination for organisms of the *Burkholderia cepacia* complex[49] and the genera *Bordetella*,[50] *Brucella*,[51,52] *Achromobacter*,[53] *Yersinia*,[54] and *Campylobacter*.[55] PCR and real-time PCR assays offer specific and rapid identification of foodborne pathogens in clinical specimens as well as in food and other environmental samples.[56,57] Further, many Gram-negative bacteria, such as *Legionella*, *Bordetella*, *Neisseria*, and *Haemophilus*,[58–61] are difficult to cultivate; PCR analyses offer superior sensitivity in identifying these organisms. Additionally, molecular typing and the rapid characterization of antibiotic resistance, pathogenic plasmids, or toxin genes can be achieved. Examples include the typing of *E. coli*,[62] *Salmonella*,[63] and *Neisseria meningitidis*[64] and the detection of genes associated with *Klebsiella pneumoniae* carbapenemase-mediated carbapenem resistance in *K. pneumoniae*[65] and other *Enterobacteriaceae*,[66] the virulence plasmids of *Yersinia pestis*,[54] and the presence of Shiga toxins in *E. coli*.[57] Finally, molecular analyses are utilized for ruling in or out the presence of Gram-negative bacterial select agents in clinical and environmental samples; in such a situation, rapid issuance of test results by a reference laboratory can be crucial in detecting and averting the intentional dissemination of potentially deadly pathogens.[67]

When working with a primary specimen or an unknown aerobic Gram-negative bacterium, the use of syndromic panels, 16S rRNA gene sequence analysis, or other broad-range PCR may be successful.[68] There are a number of commercially available molecular assays that can detect numerous microorganisms simultaneously, including Gram-negative bacteria. There are assays that can detect Gram-negative bacteria directly from clinical specimens such as stool (*E. coli*, *Campylobacter*, *Shigella*, *Yersinia*, *Salmonella*),[69–71] respiratory specimens (*N. meningitidis*, *Haemophilus influenzae*, *Legionella pneumophila*, *Acinetobacter baumannii*),[72–74] and blood or blood cultures (*A. baumannii*, *Enterobacter*, *E. coli*, *H. influenzae*, *Kingella kingae*, *Klebsiella*, *Neisseria*, *Proteus*, *P. aeruginosa*, *Salmonella*, *Serratia marcescens*, *Stenotrophomonas maltophilia*, *Bacteroides fragilis*, *Campylobacter*).[75–78] A number of these assays are FDA-cleared, including the xTAG Gastrointestinal Pathogen Panel (Luminex Corp., Austin, TX), the Prodesse ProGastro SSCS (Hologic Gen-Probe Inc., San Diego, CA), and FilmArray Blood Culture Identification Panel and Respiratory Panel (BioFire Diagnostics, Inc., Salt Lake City, UT), with more to follow in the future. These assays provide more rapid results without the need to wait for bacterial growth. Additionally, broad-range PCR tests are useful when conventional methods provide inconclusive results or if a result is needed rapidly. Broad-range PCR tests are based on the fact that some genes present in all bacterial organisms contain both regions of highly conserved sequence and regions of variable sequence. The highly conserved regions can serve as primer binding sites for PCR, while the variable regions can provide unique identifiers enabling the precise identification of a particular strain or species. These sophisticated PCR applications require adequate laboratory facilities, trained and experienced staff with expertise in the use of molecular databases, and protocols in place to avoid contamination.[79,80] Although such tests are time-consuming and costly, and may only be available at specialized laboratories such as reference laboratories, they can provide an answer when other modes of testing give problematic results or are not feasible.

Acknowledgments

The authors thank Ronald J. Limberger, Kimberlee Musser, Mehdi Shayegani, and Andrea Carpenter, Wadsworth Center, New York State Department of Health, for critical readings of this chapter.

REFERENCES

1. Edelstein, P.H. and Wadlin, J., *Legionella* cultures, in *Clinical Microbiology Procedures Handbook*, 4th edn., Vol. 1, Section 3.11.4, Leber, A.L. (ed.), ASM Press, Washington, DC, 2016.

2. Leber, A.L., *Bordetella* cultures, in *Clinical Microbiology Procedures Handbook*, 4th edn., Vol. 1, Section 3.11.6, Leber, A.L. (ed.), ASM Press, Washington, DC, 2015.

3. Ledeboer, N.A. and Doern, G.V., *Haemophilus*, in *Manual of Clinical Microbiology*, 11th edn., Chap. 36, Jorgensen, H.J. et al. (eds.), ASM Press, Washington, DC, 2015.

4. Konig, C.W., Riffelmann, M. and Coenye, T., *Bordetella* and related genera, in *Manual of Clinical Microbiology*, 11th edn., Chap. 45, Jorgensen, H.J. et al. (eds.), ASM Press, Washington, DC, 2015.

5. Edelstein, P.H. and Luck, C., *Legionella*, in *Manual of Clinical Microbiology*, 11th edn., Chap. 49, Jorgensen, H.J. et al. (eds.), ASM Press, Washington, DC, 2015.

6. Schreckenberger, P.C. and Wong, J.D., Algorithms for identification of aerobic gram-negative bacteria, in *Manual of Clinical Microbiology*, 8th edn., Chap. 23, Murray, P.R. et al. (eds.), ASM Press, Washington, DC, 2003.

7. Church, D.L. and Schreckenberger, P.C., Identification of gram-negative bacteria, in *Clinical Microbiology Procedures Handbook*, 4th edn., Vol. 1, Leber, A.L. (ed.), ASM Press, Washington, DC, 2015.

8. Weyant, R.S., Moss, C.W., Weaver, R.E., Hollis, D.G., Jordan, J.G., Cook, E.C. and Daneshvar, M.I., *Identification of Unusual Pathogenic Gram-Negative Aerobic and Facultatively Anaerobic Bacteria*, 2nd ed., Williams & Wilkins, Baltimore, MD, 1995.

9. Wauters, G. and Vaneechoutte, M., Approaches to the identification of aerobic Gram-negative bacteria, in *Manual of Clinical Microbiology*, 11th edn., Chap. 33, Jorgensen, H.J. et al. (eds.), ASM Press, Washington, DC, 2015.

10. Chan, W.W., Gram stain, in *Clinical Microbiology Procedures Handbook*, 4th edn., Vol. 1, Section 3.2.1, Leber, A.L. (ed.), ASM Press, Washington, DC, 2015.

11. Winn, W.C., Allen, S.D., Janda, W.M., Koneman, E.W., Procop, G.W., Schreckenberger, P.C. and Woods, G.L., *Koneman's Color Atlas and Textbook of Diagnostic Microbiology*, 6th edn., Lippincott Williams & Wilkins, Philadelphia, PA, 2006.

12. Farmer, J.J., *Enterobacteriaceae*: Introduction and identification, in *Manual of Clinical Microbiology*, 8th edn., Chap. 41, Murray, P.R. et al. (eds.), ASM Press, Washington, DC, 2003.

13. Gerrard, J.G., McNevin, S., Alfredson, D., Forgan-Smith, R. and Fraser, N., *Photorhabdus* species: Bioluminescent bacteria as emerging human pathogens? *Emerg. Infect. Dis.*, 9, 251, 2003.

14. Forsythe, S.J., Abbott, S.L. and Pitout, J., *Klebsiella, Enterobacter, Citrobacter, Cronobacter, Serratia, Plesiomonas*, and other *Enterobacteriaceae*, in *Manual of Clinical Microbiology*, 11th edn., Chap. 38, Jorgensen, H.J. et al. (eds.), ASM Press, Washington, DC, 2015.

15. Vaneechoutte, M., Nemec, A., Kampfer, P., Cools, P. and Wauters, G., *Acinetobacter, Chryseobacterium, Moraxella*, and other nonfermentative Gram-negative rods, in *Manual of Clinical Microbiology*, 11th edn., Chap. 44, Jorgensen, H.J. et al. (eds.), ASM Press, Washington, DC, 2015.

16. Atlas, R. and Snyder, J., Reagents, stains, and media: Bacteriology, in *Manual of Clinical Microbiology*, 11th edn., Chap. 19, Jorgensen, H.J. et al. (eds.), ASM Press, Washington, DC, 2015.

17. MacFaddin, J.F., *Biochemical Tests for the Identification of Medical Bacteria*, 3rd edn., Lippincott, Williams & Wilkins Co., Philadelphia, PA, 2000.

18. Forbes, B.A., Sahm, D.F. and Weissfeld, A.S., *Bailey and Scott's Diagnostic Microbiology*, 10th edn., Mosby, St. Louis, MO, 1998.

19. Elias, J., Frosch, M. and Vogel, U., *Neisseria*, in *Manual of Clinical Microbiology*, 11th edn., Chap. 34, Jorgensen, H.J. et al. (eds.), ASM Press, Washington, DC, 2015.

20. Zbinden, R., *Aggregatibacter, Capnocytophaga, Eikenella, Kingella, Pasteurella*, and other fastidious or rarely encountered Gram-negative rods, in *Manual of Clinical Microbiology*, 11th edn., Chap. 35, Jorgensen, H.J. et al. (eds.), ASM Press, Washington, DC, 2015.

21. Hoiby, N., Ciofu, O. and Bjarnsholt, T., *Pseudomonas*, in *Manual of Clinical Microbiology*, 11th edn., Chap. 42, Jorgensen, H.J. et al. (eds.), ASM Press, Washington, DC, 2015.

22. Lipuma, J.J., Currie, B.J., Peacock, S.J. and Vandamme, P.A., *Burkholderia, Stenotrophomonas, Ralstonia, Cupriavidus, Pandoraea, Brevundimonas, Comamonas, Delftia*, and *Acidovorax*, in *Manual of Clinical Microbiology*, 11th edn., Chap. 43, Jorgensen, H.J. et al. (eds.), ASM Press, Washington, DC, 2015.

23. Vaneechoutte, M., Nemec, A., Kampfer, P., Cools, P. and Wauters, G., *Acinetobacter, Chryseobacterium, Moraxella*, and other nonfermentative Gram-negative rods, in *Manual of Clinical Microbiology*, 11th edn., Chap. 44, Jorgensen, H.J. et al. (eds.), ASM Press, Washington, DC, 2015.

24. Chu, M.C. and Weyant, R.S., *Francisella and Brucella*, in Manual of Clinical Microbiology, 8th edn., Chap. 51, Murray, P.R. et al. (eds.), ASM Press, Washington, DC, 2003.

25. Ewing, W.H., *Edwards and Ewing's Identification of Enterobacteriaceae*, 4th edn., Elsevier Scientific, New York, 1986.

26. Janda, J.M. and Abbott, S.L., *The Enterobacteria*, 2nd edn., ASM Press, Washington, DC, 2006.

27. Garrity, G.M., Brenner, D.J., Krieg, N.R. and Staley, J.T., *Bergey's Manual of Systematic Bacteriology*, 2nd edn., Springer, New York, 2005.

28. Holt, J.G., Krieg, N.R., Sneath, P.H.A., Staley, J.T. and Williams, S.T., *Bergey's Manual of Determinative Bacteriology*, 9th edn., Lippincott, Williams & Wilkins, Philadelphia, PA, 2000.

29. Church, D.L., Biochemical tests for the identification of aerobic bacteria, in *Clinical Microbiology Procedures Handbook*, 4th edn., Vol. 1, Section 3.17, Leber, A.L. (ed.), ASM Press, Washington, DC, 2015.

30. Salomon, J., Butterworth, A., Almog, V., Pollitt, J., Williams, W. and Dunk, T., Identification of gram-negative bacteria in the Phoenix™ system, presented at *9th European Congress of Clinical Microbiology and Infectious Diseases (ECCMID)*, Berlin, Germany, 1999.

31. Jorgensen, J.H., New phenotypic methods for the clinical microbiology laboratory, in *Abstract Interscience Conference on Antimicrobial Agents and Chemotherapy*, Vol. 42, Abstract 398, ASM Press, Washington, DC, 2002.

32. O'Hara, C.M., Weinstein, M.P. and Miller, J.M., Manual and automated systems for detection and identification of microorganisms, in *Manual of Clinical Microbiology*, 8th edn., Chap. 14, Murray, P.R. et al. (eds.), ASM Press, Washington, DC, 2003.

33. O'Hara, C.M., Tenover, F.C. and Miller, J.M., Parallel comparison of accuracy of API 20E, Vitek GN, MicroScan Walk/Away Rapid ID and Becton Dickinson Cobas Micro ID-E/NF for identification of members of the family *Enterobacteriaceae* and common gram-negative, nonglucose-fermenting bacilli, *J. Clin. Microbiol.*, 31, 3165, 1993.

34. Truu, J., Talpsep, E., Heinaru, E., Stottmeister, U., Wand, H. and Heinaru, A., Comparison of API 20 NE and Biolog GN identification systems assessed by techniques of multivariate analyses, *J. Microbiol. Methods*, 36, 193, 1999.

35. Karas M. and Hillenkamp F., Laser desorption ionization of proteins with molecular masses exceeding 10,000 daltons, *Anal. Chem.*, 60, 2299, 1988.

36. Fenselau C. and Demirev P.A., Characterization of intact microorganisms by MALDI mass spectrometry, *Mass Spectrom. Rev.*, 20, 157, 2001.

37. Wieser A., Schneider, L., Jung, J. and Schubert, S., MALDI-TOF MS in microbiological diagnostics—Identification of microorganisms and beyond (mini review), *Appl. Microbiol. Biotechnol.*, 93, 965, 2012.

38. Church, D.L., Guidelines for identification of aerobic bacteria, in *Clinical Microbiology Procedures Handbook*, 4th edn., Vol. 1, Section 3.16, Leber, A.L. (ed.), ASM Press, Washington, DC, 2015.

39. Ford, B.A. and Burnham, C.D., Optimization of routine identification of clinically relevant gram-negative bacteria by use of matrix-assisted laser desorption ionization-time of flight mass spectrometry and the Bruker Biotyper, *J. Clin. Microbiol.*, 51, 1412, 2013.

40. Alby, K., Giligan, P.H. and Miller, M.B., Comparison of matrix-assisted laser desorption ionization-time of flight (MALDI-TOF) mass spectrometry platforms for the identification of gram-negative rods from patients with cystic fibrosis. *J. Clin. Microbiol.*, 51, 3852, 2013.

41. Marko, D.C., Saffert, R.T., Cunningham, S.A., Hyman, J., Walsh, J., Arbefeville, S., Howard W. et al., Evaluation of the Bruker Biotyper and Vitek MS matrix-assisted laser desorption ionization-time of flight mass spectrometry systems for identification of nonfermenting gram-negative bacilli isolated from cultures from cystic fibrosis patients, *J. Clin. Microbiol.*, 50, 2034, 2012.

42. Manji, R., Bythrow, M., Branda, J.A., Burnham, C.-A.D., Ferraro, M.J., Garner, O.B., Jennemann, R. et al., Multi-center evaluation of the VITEK MS system for mass spectrometric identification of non-*Enterobacteriaceae* gram-negative bacilli, *Eur. J. Clin. Microbiol. Infect. Dis.*, 33(3), 337–346, 2014.

43. Leli, C., Cenci, E., Cardaccia, A., Moretti, A., D'Alo, F., Pagliochini, R., Barcaccia, M. et al., Rapid identification of bacterial and fungal pathogens from positive blood cultures by MALDI-TOF MS, *Int. J. Med. Microbiol.*, 303, 205, 2013.

44. Gray, T.J., Thomas, L., Olma, T., Iredell, J.R. and Chen, S.C.-A., Rapid identification of gram-negative organisms from blood culture bottles using a modified extraction method and MALDI-TOF mass spectrometry, *Diagn. Microbiol. Infect. Dis.*, 77, 110, 2013.

45. Jamal, W., Saleem, R. and Rotimi, V.O., Rapid identification of pathogens directly from blood culture bottles by Bruker matrix-assisted laser desorption laser ionization-time of flight mass spectrometry versus routine methods. *Diagn. Microbiol. Infect. Dis.*, 76, 404, 2013.

46. Foster, A.G., Rapid identification of microbes in positive blood cultures by use of the Vitek MS matrix- assisted laser desorption ionization-time of flight mass spectrometry system, *J. Clin. Microbiol.*, 51, 3717, 2013.

47. Samaranayake, W.A.M.P., Dempsey, S., Howard-Jones, A.R., Outhred, A.C. and Kesson, A.M., Rapid direct identification of positive paediatric blood cultures by MALDI-TOF MS technology and its clinical impact in the paediatric hospital setting, *BMC Res. Notes*, 13(1), 12, 2020.

48. Azrad, M., Keness, Y., Nitzan, O., Pastukh, N., Tkhawkho, L., Freidus, V. and Peretz A., Cheap and rapid in-house method for direct identification of positive blood cultures by MALDI-TOF MS technology. *BMC Infect. Dis.*, 19(1), 72, 2019.

49. Mahenthiralingam, E., Bischof, J., Byrne, S.K., Radomski, C., Davies, J.E., Av-Gay, Y. and Vandamme, P., DNA-based diagnostic approaches for identification of *Burkholderia cepacia* complex, *Burkholderia vietnamiensis*, *Burkholderia multivorans*, *Burkholderia stabilis*, and *Burkholderia cepacia* genomovars I and III, *J. Clin. Microbiol.*, 38, 3165, 2000.

50. Koidl, C., Bozic, M., Burmeister, A., Hess, M., Marth, E. and Kessler, H.H., Detection and differentiation of *Bordetella* spp. by real-time PCR, *J. Clin. Microbiol.*, 45, 347, 2007.

51. Bricker, B.J., PCR as a diagnostic tool for brucellosis, *Vet. Microbiol.*, 90(1–4), 435, 2002.

52. Noviello, S., Gallo, R., Kelly, M., Limberger, R.J., DeAngelis, K., Cain, L., Wallace, B. and Dumas, N., Laboratory-acquired brucellosis, *Emerg. Infect. Dis.*, 10, 1848, 2004.

53. Rocchetti, T.T., Silbert, S., Gostnell, A., Kubasek, C., Jerris, R., Yong, J. and Widen, R., Rapid detection of four non-fermenting Gram-negative bacteria directly from cystic fibrosis patient's respiratory samples on the BD MAX™ system, *Pract. Lab Med.*, 12, e00102, 2018.

54. Woron, A.M., Nazarian, E.J., Egan, C., McDonough, K.A., Cirino, N.M., Limberger, R.J. and Musser, K.A., Development and evaluation of a 4-target multiplex real-time polymerase chain reaction assay for the detection and characterization of *Yersinia pestis*, *Diagn. Microbiol. Infect. Dis.*, 56, 261, 2006.

55. LaGier, M.J., Joseph, L.A., Passaretti, T.V., Musser, K.A. and Cirino, N.M., A real-time multiplexed PCR assay for rapid detection and differentiation of *Campylobacter jejuni* and *Campylobacter coli*, *Mol. Cell. Probes*, 18, 275, 2004.

56. Harwood, V.J., Gandhi, J.P. and Wright, A.C., Methods for isolation and confirmation of *Vibrio vulnificus* from oysters and environmental sources: A review, *J. Microbiol. Methods*, 59, 301, 2004.

57. Bopp, D.J., Sauders, B.D., Waring, A.L., Ackelsberg, J., Dumas, N., Braun-Howland, E., Dziewulski, D. et al.,

Detection, isolation, and molecular subtyping of *Escherichia coli* O157:H7 and *Campylobacter jejuni* associated with a large waterborne outbreak, *J. Clin. Microbiol.*, 41, 174, 2003.

58. She, R.C., Billetdeaux, E., Phansalkar, A.R. and Petti, C.A., Limited applicability of direct fluorescent- antibody testing for *Bordetella* sp. and *Legionella* sp. specimens for the clinical microbiology laboratory, *J. Clin. Microbiol.*, 45, 2212, 2007.

59. Khanna, M., Fan, J., Pehler-Harrington, K., Waters, C., Douglass, P., Stallock, J., Kehl, S. and Henrickson, K.J., The pneumoplex assays, a multiplex PCR-enzyme hybridization assay that allows simultaneous detection of five organisms, *Mycoplasma pneumoniae*, *Chlamydia* (*Chlamydophila*) *pneumoniae*, *Legionella pneumophila*, *Legionella micdadei*, and *Bordetella pertussis*, and its real-time counterpart, *J. Clin. Microbiol.*, 43, 565, 2005.

60. Hjelmevoll, S.O., Olsen, M.E., Sollid, J.U., Haaheim, H., Unemo, M. and Skogen, V., A fast real-time polymerase chain reaction method for sensitive and specific detection of the *Neisseria gonorrhoeae* porA pseudogene, *J. Mol. Diagn.*, 8, 574, 2006.

61. Poppert, S., Essig, A., Stoehr, B., Steingruber, A., Wirths, B., Juretschko, S., Reischl, U. and Wellinghausen, N., Rapid diagnosis of bacterial meningitis by real-time PCR and fluorescence in situ hybridization, *J. Clin. Microbiol.*, 43, 3390, 2005.

62. Mingle, L.A., Garcia, D.L., Root, T.P., Halse, T.A., Quinlan, T.M., Armstrong, L.R., Chiefari, A.K., Schoonmaker-Bopp, D.J., Dumas, N.B., Limberger, R.J. and Musser, K.A., Enhanced identification and characterization of non-O157 Shiga toxin-producing *Escherichia coli*: a six-year study, *Foodborne Pathog Dis.*, 9(11), 1028–1036, 2012.

63. Kim, S., Frye, J.G., Hu, J., Fedorka-Cray, P.J., Gautom, R. and Boyle, D.S., Multiplex PCR-based method for identification of common clinical serotypes of *Salmonella enterica* subsp. *enterica*, *J. Clin. Microbiol.*, 44, 3608, 2006.

64. Bennett, D.E. and Cafferkey, M.T., Consecutive use of two multiplex PCR-based assays for simultaneous identification and determination of capsular status of nine common *Neisseria meningitidis* serogroups associated with invasive disease, *J. Clin. Microbiol.*, 44, 1127, 2006.

65. Yigit, H., Queenan, A.M., Anderson, G.J., Domenech-Sanchez, A., Biddle, J.W., Steward, C.D., Alberti, S., Bush, K. and Tenover, F.C., Novel carbapenem-hydrolyzing beta-lactamase, KPC-1, from a carbapenem-resistant strain of *Klebsiella pneumoniae*, *Antimicrob. Agents Chemother.*, 45, 1151, 2001.

66. Bratu, S., Brooks, S., Burney, S., Kochar, S., Gupta, J., Landman, D. and Quale, J., Detection and spread of *Escherichia coli* possessing the plasmid-borne carbapenemase KPC-2 in Brooklyn, New York, *Clin. Infect. Dis.*, 44, 972, 2007.

67. Cirino, N.M., Musser, K.A. and Egan, C., Multiplex diagnostic platforms for detection of biothreat agents, *Expert Rev. Mol. Diagn.*, 4, 841, 2004. Review.

68. Zbinden, A., Bottger, E.C., Bosshard, P.P. and Zbinden, R., Evaluation of the colorimetric VITEK 2 card for identification of gram-negative nonfermentative rods: Comparison to 16S rRNA gene sequencing, *J. Clin. Microbiol.*, 45, 2270, 2007.

69. McAuliffe, G.N., Anderson, T.P., Stevens, M., Adams, J., Coleman, R., Mahagamasekera, P., Young, S. et al., Systemic application of multiplex PCR enhances the detection of bacteria, parasites, and viruses in stool samples, *J. Infect.*, 67, 122, 2013.

70. Buchan, B.W., Olson, W.J., Pezewski, M., Marcon, M.J., Novicki, T., Uphoff, T.S., Chandramohan, L., Revell, P. and Ledeboer, N.A., Clinical evaluation of a real-time PCR assay for identification of *Salmonella*, *Shigella*, *Campylobacter* (*Campylobacter jejuni* and *C. coli*), and Shiga toxin-producing *Escherichia coli* isolates in stool specimens, *J. Clin. Microbiol.*, 51, 4001, 2013.

71. Coupland, L.J., McElarney, I., Meader, E., Cowley, K., Alcock, L., Naunton, J. and Gray, J., Simultaneous detection of viral and bacterial enteric pathogens using the Seeplex® Diarrhea ACE detection system, *Epidemiol. Infect.*, 141, 2111, 2013.

72. Benson, R., Tondella, M.L., Bhatnagar, J., Carvalho, M.S., Sampson, J.S., Talkington, D.F., Whitney, A.M. et al., Development and evaluation of a novel multiplex PCR technology for molecular differential detection of bacterial respiratory disease pathogens, *J. Clin. Microbiol.*, 46, 2074, 2008.

73. Nolte, F.S., Molecular diagnostics for detection of bacterial and viral pathogens in community acquired pneumonia, *Clin. Infect. Dis.*, 47, S123, 2008.

74. Pillet, S., Lardeux, M., Dina, J., Grattard, F., Verhoeven, P., Le Goff, J., Vabret, A. and Pozzetto, B., Comparative evaluation of six commercialized multiplex PCR kits for the diagnosis of respiratory infections, *PLOS ONE*, 8, e72174, 2013.

75. Dierkes, C., Ehrenstein, B., Siebig, S., Linde, H., Reischl, U. and Salzberger, B., Clinical impact of a commercially available multiplex PCR system for rapid detection of pathogens in patients with presumed sepsis, *BMC Infect. Dis.*, 9, 126, 2009.

76. Westh, H., Lisby, G., Breysse, F., Boddinghaus, B., Chomarat, M., Gant, V., Goglio, A. et al., Multiplex real-time PCR and blood culture for identification of bloodstream pathogens in patients with suspected sepsis, *Clin. Microbiol. Infect.*, 15, 544, 2009.

77. Laakso, S., Kirveskari, J., Tissari, P. and Maki, M., Evaluation of high-throughput PCR and microarray-based assay in conjunction with automated DNA extraction instruments for diagnosis of sepsis, *PLOS ONE*, 6, e26655, 2011.

78. Ramanan, P., Bryson, A.L., Binnicker, M.J., Pritt, B.S., and Patel, R., Syndromic panel-based testing in clinical microbiology, *Clin Microbiol Rev.*, 31(1), e00024–17, 2017.

79. Maiwald, M., Broad-range PCR for detection and identification of bacteria, in *Molecular Microbiology: Diagnostic Principles and Practice*, Chap. 30, Persing, D.H. et al. (eds.), ASM Press, Washington, DC, 2004.

80. Fredricks, D.N. and Relman, D.A., Application of polymerase chain reaction to the diagnosis of infectious diseases, *Clin. Infect. Dis.*, 29, 475, 1999.

7

Plaque Assay for Bacteriophage

Emanuel Goldman

CONTENTS

History

The plaque assay, an indispensable tool for the study of bacteriophage, was described in the earliest publications on the discovery of these viruses. Felix d'Hérelle, credited as co-discoverer of phage (along with Frederick Twort) described, in effect, the first plaque assay in 1917 [1], when he wrote the following:

> ...if one adds to a culture of *Shiga* as little as a millionfold dilution of a previously lysed culture and if one spreads a droplet of this mixture on an agar slant, then one obtains after incubation a lawn of dysentery bacilli containing a certain number of circular areas of 1 mm diameter where there is no bacterial growth; these points cannot represent but colonies of the antagonistic microbes: a chemical substance cannot concentrate itself over definite points." (Translation from Stent [2].) The "antagonistic microbes" were then named "bacteriophage."

Subsequently, the technique was refined by Gratia [3] and Hershey et al. [4], by pouring onto an agar plate a top layer of molten agar containing a phage-bacteria mixture. The molten agar solidified, fixing the phage particles in the semisolid agar substrate, and as the bacteria grew during incubation to cover the plate as a lawn of bacterial growth, the embedded phage particles killed those bacteria in their vicinity, with the progeny phage infecting other bacteria in the vicinity to produce a zone of clearing, or a "plaque," on the lawn of bacterial growth. This method was described by Delbrück [5] as follows:

> In this technique a few milliliters of melted agar of low concentration, containing the bacteria, are mixed with the sample to be plated. The mixture is poured on the surface of an ordinary nutrient agar plate. The mixture distributes itself uniformly over the plate in a very thin layer and solidifies immediately. We used about 20 mL of 1.3 per cent agar for the lower layer and 1.5 mL of 0.7 per cent agar for the superimposed layer on petri plates of 7 cm diameter. With this technique 60 samples can be plated in about 20 minutes. The plaques are well-developed and can be counted after 4 hours' incubation.

The plaque assay was validated as showing that a single phage infecting a bacterium was sufficient to initiate formation of a plaque, because there was linear proportionality between the plaques observed and the dilution of the phage sample [6]. Electron microscopy later indicated that under optimal conditions, there was a close approximation between physical phage particles and the presumptive phage particles that initiated plaque formation [7].

However, the "efficiency of plating" of a bacteriophage preparation can vary extensively depending on conditions, especially the bacterial host strain. Varying susceptibility of different bacterial strains to different bacteriophage in fact was exploited for many years in the identification of bacteria, in a procedure called "phage typing." Although this procedure is not used as extensively today as in the past, a brief description of the principle will be provided later in this chapter. Current use of phage for identification of bacteria, including phage typing, is discussed in Chapter 8.

Methodology

A suspension of bacteriophage of unknown titer is first subjected to serial dilution, illustrated in Figure 7.1. A small portion (0.1 mL) of the suspension is added to 9.9 mL of dilution fluid, for a 100-fold dilution. A small portion (0.1 mL) of this dilution, in turn, is added to a fresh tube containing 9.9 mL of dilution fluid, for another 100-fold dilution. At this point, the original phage suspension has been diluted by a factor of 10,000. The process is repeated two more times, to a fourth tube where the dilution factor is 10^{-8}. Dilutions, of course, do not always have to be 1:100; other dilutions are perfectly acceptable, as long as the dilution factor is duly noted, since this is needed to obtain the titer of the original stock. In Figure 7.1, there is a fifth dilution tube with the addition of 1–9 mL of dilution fluid, to provide a 1:10 dilution, bringing the overall dilution factor to 10^{-9}. The need for large dilutions is because phage titers can be quite high.

Dilution fluid can be the growth medium for the host bacteria, but since this would be wasteful and expensive, in practice a simpler liquid diluent is used. Sterile saline solution can serve this

DOI: 10.1201/9781003099277-8

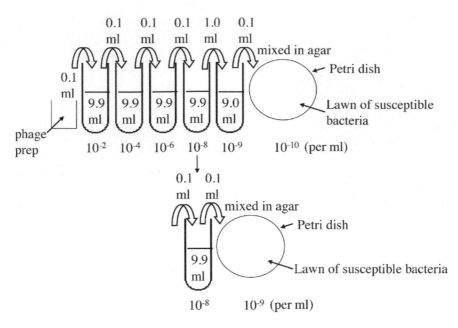

FIGURE 7.1   Serial dilutions.

purpose, or variations of saline with lower salt and a small addition of a growth nutrient, for example, sterile 0.25% (w/v) sodium chloride, 0.1% (w/v) tryptone.

Once the last dilution is made, 0.1 mL is added to tubes containing 2.5 mL soft agar that has been melted in a boiling water bath and equilibrated to ~45°C. Tubes can be kept in a water bath or, more commonly, a dry-block heating unit. The dry block is preferable to avoid unwanted contamination of water-bath water from the outside of the tubes when the contents are poured onto plates. Tubes should be plated fairly soon after addition of the phage dilution, to avoid incubating the phage at the high temperature for an extended period of time. The soft agar contains the nutrients for growth of the bacteria (and can be a minimal media) with 0.65% (w/v) agar. Two drops of a fresh saturated bacterial suspension (about 0.1 mL, ~10^8 bacteria total) are added to the tubes, which are then mixed and poured onto nutrient agar (1–1.5% w/v) petri plates (100 × 15 mm) that have been previously prepared (bottom agar) and are at room temperature. The plate is briefly rocked to facilitate uniform distribution of the top agar across the surface of the bottom agar before the top agar solidifies. If the plates have been refrigerated and are not at room temperature, top agar may solidify too quickly and not distribute uniformly, which will make plaque counting more difficult. The plates are then incubated at 37°C for several hours, usually overnight, to allow the bacteria to make a lawn and for the plaques to develop.

A recent report demonstrated improved visibility and consistency of plaques when a small volume of liquid suspension of the phage dilution was applied to the top agar layer after it had solidified, by a spreading technique with a glass rod (as opposed to mixing the phage in the molten top agar before pouring onto the bottom agar) [8].

Because the titer is unknown at the outset, more than one dilution of the phage stock is plated in this way, so that a reasonable (countable) number of plaques will be obtained at least at one of the dilutions tested. Figure 7.1 illustrates two dilutions that are plated, 10^{-8} and 10^{-9}. If one obtained, for example, 15 plaques

from the 10^{-9} dilution (and ~150 plaques from the 10^{-8} dilution), the titer of the original phage stock would be 1.5×10^{11} phage/mL in the original stock suspension. This number of phage is the reciprocal of the dilution factor and is derived by taking into account that only 0.1 mL of the dilution was added to the plate, which represents an additional dilution factor of 10^{-1} when considered on a per mL basis. Hence, 15 plaques obtained from 0.1 mL of a 10^{-9} dilution = 15×10^{10} phage/mL in the original stock.

An example of actual plaques of a lytic phage grown on *Escherichia coli* is shown in Figure 7.2. The lawn is shown as a black background, although in an actual assay it would likely be tan or brown, and the plaques are shown as clear white circular spots on the lawn. If this plate resulted from plating 0.1 mL of a 10^{-8} dilution (as in Figure 7.1, bottom), then the titer of the stock would be ~1.5×10^{11}, since there are about 150 plaques on this plate. It is important to remember that each plaque represents the progeny phage from a single individual phage that was embedded in the top agar. The plaque as a whole may contain in the order of

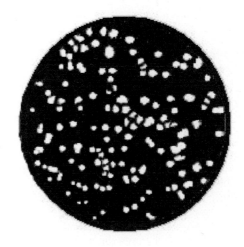

FIGURE 7.2   Bacteriophage plaques.

10^6 phage. A good description of preparation of phage and plaque assays can be found in a laboratory manual by Miller [9].

Different phage exhibit different morphologies in plaque assays. Some plaques are small and clear, with sharply defined borders. Some are large and can be more diffuse. A recent study compared isogenic phage mutants for differences in plaque morphologies and found that differences could be ascribed to specific alleles in the phage [10]. Some phage exhibit turbid plaques, in which there is some bacterial growth within the plaque, as if those bacteria had developed resistance to the phage. This is the hallmark of a temperate (lysogenic) phage, which, at low frequency, integrates its genome into the bacterial host and renders the host immune to reinfection by phage of the same type (see Chapter 48 in Part II). This characteristic appears to make plaque assays more difficult for some temperate phages, for example with *Lactococcus* species. However, improved methodology of plaque assays for temperate lactococcal phages has been reported, with a major enhancement of plaque size and visibility resulting from addition of glycine (between 0.25% and 1.25%) to the medium [11]. New approaches replacing the plaque assay for quantitating phage have been tested, using either of two alternative methods: quantitative PCR (polymerase chain reaction), and NanoSight Limited technology (using laser optical microscopy to visualize nanoparticles) [12]. However, these alternate techniques cannot confirm the viability of phage in the sample.

Application to Phage Typing

The ability of phage to form plaques on a lawn of susceptible bacteria was exploited to develop systematic methods for identifying strains of bacteria by their susceptibility to known stocks of individual phages, a technique known as "phage typing." A chapter of the first edition of this Handbook was devoted to this method [13], which was particularly useful for identifying strains of *Staphylococcus aureus* and *Salmonella typhi*, among others. An update on contemporary use of phage typing is included in Chapter 8.

The method requires a standard set of typing phages that are different for each species to be typed. A culture of the bacteria to be identified is spread over the surface of a nutrient agar plate. Drops from a set of typing phages are placed in a grid pattern on the surface of the plate, which is then incubated to allow the bacteria to form a lawn and for the phage to produce plaques. Following incubation, those phage in the set that can inhibit growth of the bacteria will form plaques. Since the phages were applied in a grid pattern, one can identify which phages the bacterial host is susceptible to, which will then be given a "phage-type" designation.

This principle is illustrated in Figure 7.3. The plate on the left shows a grid pattern of 12 individual phage that are spotted on a surface containing the bacteria to be identified. The plate is incubated to allow the lawn (and phage) to grow. Following incubation, plaques are observed at the locations where phages 2, 5, and 11 were spotted, and not at the locations of the other 9 phages in the set. If the bacteria to be identified had been a strain of

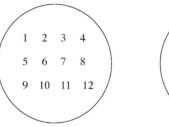
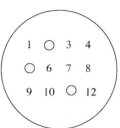

Different phage spotted on lawn of test bacteria

After incubation, bacteria identified as phage-type 2/5/11

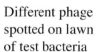

FIGURE 7.3 Phage typing.

Neisseria meningitidis, for example, the strain would thus be identified as *Neisseria meningitidis* phage type 2/5/11.

REFERENCES

1. d'Hérelle, F. Sur un microbe invisible antagoniste des bacilles dysentériques. *C. R. Acad. Sci. Paris*, 165, 373, 1917.
2. Stent, G. S. *Molecular Biology of Bacterial Viruses*, W. H. Freeman and Co., San Francisco, CA, 1963, 5.
3. Gratia, A. Des relations numériques entre bactéries lysogènes et particules de bactériophage. *Ann. Inst. Pasteur*, 57, 652, 1936.
4. Hershey A. D., Kalmanson, G. and Bronfenbrenner, J. Quantitative methods in the study of the phage-antiphage reaction. *J. Immunol.*, 46, 267, 1943.
5. Delbrück, M. The burst size distribution in the growth of bacterial viruses (bacteriophages). *J. Bacteriol.*, 50, 131, 1945.
6. Ellis, E. L. and Delbrück, M. The growth of bacteriophage. *J. Gen. Physiol.*, 22, 365, 1939.
7. Luria, S. E., Williams, R. C. and Backus, R. C. Electron micrographic counts of bacteriophage particles. *J. Bacteriol.*, 61, 179, 1951.
8. Cormier, J. and Janes, M. A double layer plaque assay using spread plate technique for enumeration of bacteriophage MS2. *J. Virol. Methods*, 196, 86, 2014.
9. Miller, J. H. Preparation and plaque assay of a phage stock, *Experiments in Molecular Genetics*, Cold Spring Harbor Laboratory, Cold Spring Harbor, NY, 1972, 37.
10. Gallet, R., Kannoly, S. and Wang, I-N. Effects of bacteriophage traits on plaque formation. *BMC Microbiol.*, 11, 181, 2011.
11. Lillehaug, D. An improved plaque assay for poor plaque-producing temperate lactococcal bacteriophages. *J. Appl. Microbiol.*, 83, 85, 1997.
12. Anderson, B., Rashid, M. H., Carter, C., Pasternack, G., Rajanna, C., Revazishvili, T., Dean, T., Senecal, A. and Sulakvelidze, A. Enumeration of bacteriophage particles: Comparative analysis of the traditional plaque assay and real-time QPCR- and nanosight-based assays. *Bacteriophage*, 1, 86, 2011.
13. Kasatiya, S. S. and Nicolle, P. Phage typing, *Practical Handbook of Microbiology*, 1st ed., O'Leary, W., Ed., CRC Press, Boca Raton, FL, 1989, 279.

8

Phage Identification of Bacteria

Catherine E.D. Rees and Martin J. Loessner

CONTENTS

Introduction

As the name suggests, bacteriophage (literally "bacteria eating" from the Greek) were discovered as lytic agents that destroyed bacterial cells by Edward Twort and Felix D'Herelle (see [1] for a review of the history of their discovery and application). Bacteriophage* are in fact viruses that specifically infect members of the Bacterial kingdom [2]. In the Eukaryotic kingdom, the visible diversity of biological forms makes it unsurprising that viruses have a specific host range, since it is clear that the organisms affected by the different viruses are very different. However, in the Bacterial kingdom, differences between bacterial genera are not as easy to detect and differentiation between members of a species is often only reliably determined at the molecular level. Hence, in the field of bacteriology, the fact that viruses have evolved to specifically infect only certain members of a genus or species seems to be surprising. However, as the molecular recognition events involved in Eukaryotic virus infection are elucidated, it is clear that even subtle difference in cell surface proteins have profound effects on binding and infection of Eukaryotic viruses (see [3] for a review), and this is also true for the bacteria-virus interaction.

Historically, it was in 1926 that D'Herelle first reported the fact that not all members of a species could be infected by the same phage when he was investigating the use of bacteriophage

to treat cholera [4]. He divided isolates of *Vibrio cholerae* into two groups on the basis of phage sensitivity. Until molecular methods of cell discrimination became more widely adopted, differentiation of cell type on the basis of phage sensitivity (or phage typing) was one of the best ways to discriminate between different isolates within a species and much research was put into developing and refining the phage typing sets for a range of different bacteria. Antibody typing (serotyping) is also able to differentiate strains at the subspecies level but, as phage isolation and propagation is easier and cheaper than characterization of polyclonal or monoclonal antibodies, phage typing has often been the first subspecies typing method developed for bacterial pathogens. Where robust and widely adopted phage-typing schemes exist, their use has continued, and phage typing is still used in some settings to subtype isolates of both Gram-negative and Gram-positive bacteria. However, since DNA-based typing methods have become more widely adopted, new phage typing sets are not being developed since generally, WGS analysis is found to be more discriminatory than the use of phage-typing to subspeciate strains, especially when tracking outbreaks of disease [5, 6]. However, so that historical data can be linked to modern methods of whole genome sequence analysis, the genetic differences that underlie phage types are now being derived from genome sequences, and linking phage sensitivity profiles to information gained from WGS continues to inform our understanding of the bacterial cell (e.g., see [7–11]).

In contrast to the use of phage typing, methods for the detection of bacteria using bacteriophage continue to be developed. This has resulted in the emergence of commercial tests, employing

---

* There is an ongoing debate about the plural form: bacteriophage or phages. For a discussion see [2].

DOI: 10.1201/9781003099277-9

a variety of assay formats, for use in both medical and environmental applications. This chapter will provide an overview of these technologies and describes some of the newer findings that may be developed into practical applications in the near future.

Phage Typing

Choice of Phage

Bacteriophage can be divided into two broad groups on the basis of the biological outcome of infection. Virulent, or lytic, phage always lyse the host cell at the end of the replication cycle. Temperate, or lysogenic, phage may cause lysis of the host cell after completing a lytic replication cycle or, alternatively, they may enter a dormant phase of replication where their genome is maintained within the host cell without causing harm to that cell so that the virus is replicated as part of the cell during cell division. Because of this choice of replication pathway, infection with temperate phage will not always result in the formation of plaques (lytic areas of clearing in a lawn of bacteria), and therefore these are often not chosen for use in phage typing schemes. Much work has been put into the isolation of virulent phage that can be evaluated for use in phage typing sets, and these are often found in the natural environment of the target bacterium (for instance, sewage is a good source of bacteria specific for enteric pathogens) [see 12–19]. However, other workers have chosen to try and identify phage variants or mutants (adapted phage) of previously isolated phage that have a new or altered host range. The fact that it was possible to select for such mutants was first shown by Craigie and Yen [20] as early as 1938. These workers identified the fact that phage could be selected from an infection that were now more lytic against that same strain on which they were last propagated and the practice of phage adaptation is still used to improve the discriminatory power of phage sets [18, 21]. More recently, this process has been taken further by the use of genetic engineering to develop recombinant phage with new host ranges that can be rapidly built and selected using high-throughput screening methods [22, 23].

Perhaps more important for science as a whole was the observation that phage host range could be affected by the host strain used to propagate the phage. This led to the discovery of phage restriction barriers (see next section and [24, 25]) and ultimately to restriction enzymes [26–28], which revolutionized the study of cellular molecular biology. In the same vein, more recently, the discovery of the host-encoded CRISPR-Cas phage immunity system has led to the development of an efficient, targeted genome engineering tool that has been widely adopted in many fields [29, 30].

Factors Affecting Phage Infection

There are many factors that influence the ability of a phage to infect a particular cell type. For successful infection, the phage needs to recognize its host by binding to primary surface receptors (see [31]). These are often molecules containing carbohydrates such as lipopolysaccharides [32] (O antigen) in Gram-negative bacteria, teichoic acids [33, 34] (O antigen) in Gram-positive bacteria, or capsular polysaccharides [35] (K antigen). The fact that structural variations in these molecules occur with a relative high frequency, which is used to distinguish cell types on the basis of serology, begins to explain why phage bind to different host cell types with variable efficiencies (see [36, 37]). In addition to carbohydrates, different cell surface proteins can be utilized as receptors such as flagella or pili [38–41], sugar uptake proteins [42], membrane proteins [43, 44], or S-layer proteins [45]. Again, variations in these proteins occur that can be detected using specific antisera (for instance, flagella represent H antigens), and accordingly phage binding occurs with a different affinity depending on the match between phage and host receptor molecules. Another level of variation can occur depending on the pattern of expression of the receptor by the host cell, which may be regulated by environmental conditions. This is particularly true for the expression of flagella and sugar transporters, and therefore the particular culture conditions used can determine whether or not a phage can infect a cell.

Once the phage has penetrated the cell envelope, replication may be prevented by a variety of mechanisms. For instance, if the cell contains a prophage (a dormant lysogenic phage), the regulatory proteins that suppress the lytic replication of the phage may also suppress replication of the newly infecting phage (see [46]).

Cells that possess DNA restriction-modification systems may reduce levels of infection by the destruction of incoming foreign DNA (the restriction barrier [47]) and host cell genes (abortive phage infection genes [48–50]) have been described that specifically block phage replication (for more details see [51]). Most recently, the role of acquired immunity through the CRISPR systems has been recognized as a major driver of phage resistance (see [52]). In effect then, phage typing is an indirect way of investigating variations in cell structure and genetics using the ability of a phage to infect cells (or not) as a biological indicator of the differences that can exist in closely related bacterial cells. Therefore, even though phage typing methods have fallen out of regular usage, the methods used for phage typing are still valuable research tools. In particular, these basic phage typing methods are used to determine the host range of newly isolated phage to be included in phage therapy trials [53].

Phage Typing Methods

Determining Phage Titer and RTD

To determine phage titer, 10-fold serial dilutions of the phage are prepared in a buffer that contains divalent cations (often Mg^{2+}), to facilitate phage binding to the cell surface, and gelatin, to stabilize phage particle structure. Commonly used buffers are Lambda buffer (50 mM Tris-HCl [pH 7.5], 100 mM NaCl, 1 mM MgSO4, 0.2% gelatin) [54] or SM buffer (50 mM Tris-HCl [pH 7.5], 100 mM NaCl, 8 mM $MgSO_4$, 0.01% gelatin) [54]. Alternatively, the broth used for host cell infection can be used. Host cells are freshly grown in broth (usually overnight), diluted into fresh broth, and grown to mid-exponential phase of growth to approximately 1×10^8 cfu mL^{-1}. This broth is often supplemented with Ca^{2+} to promote phage binding. The cells are then spread over an agar plate, allowed to air dry and then 10–20 μl samples [55] of each phage dilution spotted onto the lawn of the bacteria to produce discrete areas of infection (see Figure 8.1). Once the samples have dried, the plates are incubated at the appropriate temperature overnight and the numbers of plaques counted only

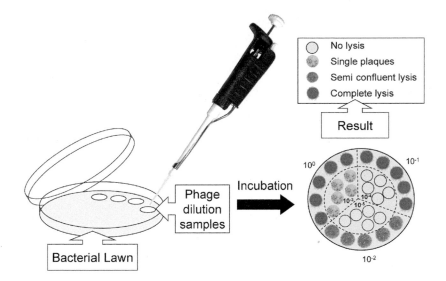

FIGURE 8.1 Phage titration and interpretation of results.

in samples where discrete plaques are seen. This value is then used to either determine the titer of the phage stock, or the routine test dilution (RTD, [56]). The RTD is defined as the highest titer that just fails to produce confluent (complete) lysis of the cells within the sample area. The appearance of semiconfluent lysis indicates that the number of phage present are reaching a point of dilution where numbers are low enough to allow individual infection events to be detected and the growth of plaques from a point of infection can be visualized. This is an important point since the applications of high numbers of phage to a lawn of bacteria can produce confluent lysis even though productive infection does not occur. This phenomenon can be due to the process of "lysis from without" [57]; in essence, large numbers of phage binding to the cell surface damages the cell wall integrity and cells lyse before the phage have replicated. Alternatively, phage preparations may contain defective phage particles or bacteriocins [58] that again can cause complete cell lysis, or the phage suspension itself can contain chemical residues that inhibit lawn growth, giving the appearance of confluent lysis in the area where a sample has been applied. Any of these possibilities may lead to false areas of confluent lysis but are ruled out by dilution of the phage sample to the point where semiconfluent lysis occurs which will not occur unless phage infect their host cells and grow.

Freshly grown host cells are spread over an agar plate to form a bacterial lawn. A 10-fold serial dilution of the phage is prepared and then 10–20 μl samples of each phage dilution spotted onto the lawn and allowed to air dry. After incubation, numbers of plaques are counted only in samples where discrete plaques are formed. The RTD [42] is defined as the highest titer that just fails to produce confluent (complete) lysis of the cells within the sample area.

The phage concentration to produce semiconfluent lysis depends on plaque morphology (phage producing fast growing, large plaques will require fewer numbers to form semiconfluent lysis than phage that produce small discrete plaques) but is generally between 5×10^4 and 2×10^5 pfu mL^{-1} [58] and should be 2- to 10-fold lower than the concentration required to produce confluent lysis. The bacterial strain used to determine the RTD is always that used to propagate the phage so that host adaptation (discussed previously in this chapter) does not affect the titer results.

Determining Efficiency of Plating

Due to the subtle differences in cell surface components and cellular biochemistry that influence phage binding (described in section Factors Affecting Phage Infection), even without host adaptation, the efficiency with which a particular phage will infect different strains of a bacterium will vary. This basically reflects the tightness of the interaction between the phage and host receptors, with stronger binding leading more often to productive infection events and weaker binding leading to fewer, and hence a lower titer. By comparing the titer of a phage preparation when tested against different strains of bacteria, a ratio of productive infection events can be determined. This value is termed the efficiency of plating (or EOP) with a value of 1 representing 100% efficiency (i.e., the titer on the two test strains is the same) and 0.9 representing 90% efficiency or a titer of one log$_{10}$ lower. Since the binding of the phage to the cell can also be influenced by the differential expression of surface proteins and carbohydrates (see section Factors Affecting Phage Infection), and this is influenced by the particular growth state of the cell, some variation in titer of phage on the same strain can occur when the experiment is performed on different days. Therefore, a consistent difference of at least 5- to 10-fold in titer is required before a reliable difference in EOP can be established. This degree of variability in expected titer is another reason for choosing semiconfluent lysis as the RTD when performing phage typing (see next section). If a dilution were chosen that gave in the range of 10–20 discrete plaques, sometimes a result would be positive, and sometimes negative. The use of a higher phage dilution takes this inherent variation into account and allows a more robust typing result to be achieved.

Phage Typing Methods

For phage typing, a panel of phage characterized to have a discriminating host range is chosen and strains are infected with each phage at standard concentration (the RTD). The method is similar to that for determining phage titer, in that a bacterial lawn is prepared and samples of phage are applied to the surface

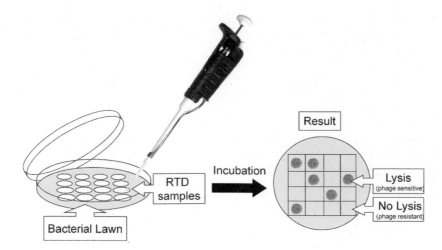

FIGURE 8.2 Phage typing.

(see Figure 8.2). Now samples of different phage from the phage typing set, each diluted to the RTD, are spotted onto the lawn in a grid pattern. In some typing schemes, both the RTD and 100× the RTD are tested. After incubation of the plates, infection is detected by the presence of plaques and patterns of susceptibility to individual phage are determined, leading to the determination of a phage type based on the pattern of sensitivity (see Figure 8.2). The results are not only scored as negative or positive, but are also given a semiquantitative description that in some way reflects the EOP (i.e., at the RTD; if the phage infects with the same efficiency that it infects the host strain, semiconfluent lysis should occur; if confluent lysis is seen on the test strain, then the EOP is greater than 1; if single plaques appear, the EOP is less than 1). Different phage typing schemes use different conventions of scoring; commonly letter descriptors are used for confluent lysis (CL), semiconfluent lysis (SCL), and opaque lysis (OL; this refers to the situation where growth of the lawn occurs within the zone of lysis so that the area is not cleared). These can be combined with < symbols to indicated intermediate degrees of lysis (e.g., <CL or <SCL) and then ± to +++ to indicate increasing numbers of discrete plaques. In typing schemes where more than one phage dilution is used, the scoring is more directly related to the EOP where 5 = ++ reaction in the same dilution as on the propagating strain, 4 = ++ reaction in a dilution 10- to 100-fold more concentrated that that giving ++ on the propagating strain, 3 = ++ in a dilution 10^3- to 10^4-fold more concentrated that that giving ++ on the propagating strain, 2 = in a dilution 10^5 to 10^6 fold more concentrated that that giving ++ on the propagating strain, 1 indicates very weak lysis, and a minus symbol (−) for no reaction. The number 0 is used to record evidence of lysis from without. The typical reaction patterns for a variety of phage can be found in the published literature and Public Health Agencies (e.g., see *Salmonella enterica* [59, 60, 72], *Escherichia coli* O157 [61, 62], *Shigella* [12, 13], *V. cholerae*[63, 64], *Staphylococcus aureus* [58, 65], *Listeria monocytogenes* [66–68], and *Bacillus* spp. [69–71]).

A panel of phage with discriminating host range is each diluted to the RTD. Ten to twenty microliters of each diluted phage are applied to the surface of a bacterial lawn of the test strain in a grid pattern. After incubation of the plates, infection is detected by the presence of plaques and patterns of susceptibility

to individual phage are determined. See text form more details of qualitative scoring schemes used.

The Use of Phage Typing

In the past, these phage typing sets have been successfully used in epidemiological studies to monitor changes in the predominant organism causing disease in a population and the emergence of new dominant clones. For *S. enterica*, the phage typing system described by Callow [72] still provides a useful reference for the differentiation of strains from within the same serovar, although WGS methods have replaced them in epidemiological studies [5, 6]. However, phage typing has been historically used worldwide to investigate a variety sources of bacterial contamination, environmental spread of the organism, and outbreaks of food poisoning involving not only *S. enterica* [73–77] and other members of the *Enterobacteriacae* [77] but also to follow the dissemination of strains *V. cholerae* in India [18, 78] and for studies of the spread of clones *Staphylococcus* in hospital populations [79–80]. The effectiveness of these studies demonstrates the usefulness of this simple microbiological method that can still be a useful tool in settings where access to WGS is not readily available.

Phage-Based Rapid Detection Methods

Traditional culture-based methods of detection exploit the specific growth requirements or cellular biochemistry of particular bacteria to achieve selective growth and detection of specific phenotypic traits through careful formulation of diagnostic media. Further identification can then be achieved using a variety of techniques including antibodies (or phage typing) to reveal more information about variation in cell surface structures or molecular analysis to interrogate the genome structure. However, to carry out this process of identification, single cells present in a sample must be separated from the sample matrix and then grown until discrete colonies are formed. This is inevitably time consuming, and the drive in both the fields of medical microbiology and food microbiology is for rapid, more sensitive diagnostic tests that can provide detection and identification without the need for culture. Using phage to detect the presence of their host cells is an obvious way to address this need. Phage-based tests use the specificity of the interaction between the virus and the host cell as a diagnostic tool, and in this sense is not so different to the use of phage in typing schemes. However, the specificity of the phage-host interaction also means that the target cell can be detected in the presence of competitive flora, thus reducing the need for selective growth of the target bacterium before it is detected. Finally, the fact that the replication of the phage is so much faster than that of the host cell also allows methods to be developed that are more rapid than traditional culture.

Unlike phage selected for typing sets, phage with the widest host range are chosen for rapid detection methods, preferably those that can encompass all members of the genus or species to be detected. There are many examples of well characterized broad host-range phage, but one that has been used more than most to develop these assays is *Salmonella* phage Felix 01, which has been shown to infect more than 95% of *Salmonella* isolates [81]. Similarly, two broad host-range phage have been identified for *Listeria*, A511 [82] that infects 95% of those serotypes most commonly associated with human disease (1/2a and 4b) and P100 [83] that can infect and lyse more than 95% of serotypes 1/2 and 4 (*L. monocytogenes*) and 5 (*Listeria ivanovii*). Both phage have been used to develop rapid methods for the identification of *Listeria* in food samples. Broad host-range phage specific for other bacteria can be readily isolated, often from those environments where the host cells are found naturally [52, 84–86]. Using these broad host-range phage, many different phage-based assay formats have been described, but in this section we will review the three approaches that have been developed into commercial assays. For an in-depth review of the broader use of methods used to engineer bacteriophages and their use in a range of applications see [87–90].

Reporter Phage

In the reporter phage assay, phage are genetically engineered to deliver a reporter gene into the target bacterium. While the specificity of the assay comes from the phage host range, the speed and sensitivity of the assay come from the fact that the reporter genes are inserted into the phage in a way that they are expressed at high level following phage infection, and sensitive instrumentation means that the production of the reporter gene signal can be detected at very low levels.

Reporter phage were first described in 1987 [91] when the principle of reporter phage assays was demonstrated using a Lambda cloning vector containing the bacterial *lux* operon that was used for detection of *E. coli* in milk samples. Even this simple model was able to detect as few as 10 *E. coli* cells within 1 hour using a luminometer to detect the light produced by the infected cells [92]. This group simply created wild type *lux* phage by introducing the reporter genes by random mini Tn*10-luxAB* mutagenesis. These reporter phage were used to detect enteric pathogens on swab samples from abattoir, meat-processing factory surfaces, and animal carcasses [93]. In this study, light levels were compared with viable count, and a good correspondence between the two measurements was found. The phage assay allowed detection of 10^4 cfu g^{-1} or cm^{-2} within 1 hour or <10 cfu g^{-1} or cm^{-2} following a 4-hour enrichment, demonstrating good sensitivity and that a far faster result could be achieved than by conventional culture. A similar level of sensitivity was also reported using a Lambda *lux* phage to detect enteric bacteria without enrichment [94]. These early studies highlight one of the biggest advantages of the phage assays over many modern molecular tests which is that to generate the signal detected, the host cell must be viable to allow the reporter genes encoded by the phage to be expressed. This allows the same live/dead differentiation achieved using conventional culture techniques, which is an important factor when carrying out analysis of processed foods which may contain bacteria that have been inactivated during production of the product; detecting DNA or cell components alone in this case would give a false-positive result.

After these early experiments proved the value of reporter phage, more recently, reporter phage have been created in a more directed manner, by using homologous recombination to introduce the reporter gene into a precise location in the phage genome. For instance, the *luxAB* genes have been introduced into the *Salmonella* phage that infect groups B, D, and some group C serovars using this method [95]. A mixed-phage preparation was used to detect as few as 10 cfu mL^{-1} of *Salmonella* cells with 6 hours of preincubation. However, without preincubation the limit of detection was only 10^8 cfu mL^{-1} in 1–3 hours, indicating that the enrichment phase allowed the number of bacteria to reach a critical threshold that allowed the signal to be detected. The same method was used to introduce *luxAB* genes into the late gene region of the broad host range *Listeria* phage A511 [96]. When this *lux* phage was tested in a range of different food types, detection levels of between 0.1 cells g^{-1} and 10 cells g^{-1} were achieved after 20 hours of selective enrichment (giving a total assay time of 24 hours), depending on the food type tested [97]. This work highlighted the fact that the nature of the matrix that contains the bacteria is an important factor in determining test sensitivity, as clearly the efficiency of phage infection was affected by either the food components or the high levels of competitive microflora present. Thouand et al. [98] also reported that the sensitivity of the P22::*lux* reporter phage assays was affected by the nature of the sample matrix when trying to develop a rapid method to detect the presence of *Salmonella* in samples taken from poultry houses. As the binding to phage to host cell surfaces is essentially a specific protein-ligand interaction, it is not that surprising that complex, organic samples can interfere with the kinetics of binding, and therefore care must be taken before extrapolating proof of concept data using pure laboratory cultures to results that can be obtained when testing more complex samples.

Nevertheless, these early proofs of principle experiments have led to the successful engineering of bacterial bioluminescence genes into a number of different bacteriophage to create reporter phage with specificity for range of different bacterial hosts and different types of application. For instance, the *luxAB* genes have been introduced into the broad host range *Bacillus anthracis*-specific Wβ phage and shown to be able to specifically detect as few as 10 *B. anthrax* spores inoculated into pond water within 8 hours using a simple sampling process without needing to first extract spores from the sample [99, 100]. However, again the effect of the sample on assay sensitivity was noted, and the limit of detection increased to 100 spores when lake water was used that had a higher salt content and higher levels of competitive microflora [100]. The same group also introduced the *luxAB* genes into engineered phage φA1122 used by the CDC as a diagnostic tool for the identification of *Yersinia pestis* [101]. When this was used to test clinical blood samples, it could successfully detect samples containing <10² CFU mL⁻¹ within 5 hours and importantly excluded samples where the infection was due to species typically associated with bacteremia or those relevant to plague diagnosis [102]. A *luxA*B reporter phage that could be used to detect dysentery caused by *Shigella flexneri* has also been described by this group, and although the limit of detection was reduced to ~10³ CFU g⁻¹ when detecting the pathogen in spiked stool samples this is more than sufficient to detect the levels of pathogen normally found in clinical samples in the early stages of infection [103]. Outside the area of human pathogens, *luxAB* reporter phage have also been developed for the detection of economically important plant pathogens such as *Pseudomonas cannabina* pv. *Alisalensis*, which causes bacterial blight of cruciferous vegetables [104], and *Erwinia amylovora*, which is the causative agent of fire blight, with the latter of these being shown to be able to detect the pathogen in natural infected plant samples within 1 hour [105].

Bioluminescence is a particularly good signal when testing biological samples as there is very little background signal generated from the material, producing good signal-to-noise ratios so that sensitive detection can be achieved. The majority of the bioluminescent reporter phage described so far have utilized just the *luxAB* genes that are approximately 2 kbp in size and encode the bacterial luciferase enzyme. This produces light by the conversion of an aldehyde substrate which is converted to carboxylic acid in a reaction that also produces a photon of light. In bacteria, these two genes are part of the larger *lux* operon that also encodes three other genes (*luxCDE*) that produce the aldehyde substrate that drives bioluminescence production (see [106] for more details). However, the full operon is ~6 kbp in size, which is too large to easily introduce into bacteriophage genomes unless some phage genes are deleted [91]. Hence, for these reporter phage assays, the aldehyde is added after phage infection has been allowed to proceed. As the ability to sequence and analyze phage genomes has improved, and more methods for phage engineering have become available (see [89]), full *lux* operon reporter phage have been developed [107–109] in the hope that introducing the full operon would allow fewer cells to be detected and reduce the limit of detection of phage assays. However, more recently, there has been a move toward using the newly discovered eukaryotic enzymes such as NanoLuc, which is a single subunit 19 kDa enzyme derived from sea shrimp. This is 100 times more active than other bioluminescent enzymes and when incorporated into a T7 phage, less than 20 *E. coli* E cells could be detected in a 100 mL sample of drinking water within 5 hours [110]. Similarly, NanoLuc was chosen as the best enzyme to introduce in the *Listeria* phage A511, and in this case, the reporter phage was able to detect one *L. monocytogenes* cell in 25 g of artificially contaminated milk, cold cuts, and lettuce within less than 24 hours and successfully detected *Listeria* spp. in potentially contaminated natural food samples without producing false-positive or false-negative results [111].

In addition to the *lux* genes, a number of other reporter genes have been introduced into phage. Early examples included the green fluorescent protein, favored because its small size (~800 bp) made it easy to introduce into phage genomes when little genetic analysis had been performed. For instance, *gfp* was introduced into the *E. coli* O157:H7-specific phage PP01 to create a fluorescent reporter gene phage that was shown to be able to specifically detect *E. coli* O157:H7 cells in the presence of other competitive *E. coli* cells [112]. The same group also introduced the *gfp* gene into phage T4 that was unable to lyse the *E. coli* host cells at the end of the replication cycle, leading to sustained expression of the *gfp* gene within the infected cells. These highly fluorescent cells could then be detected by automated methods moving toward the production of integrated biosensor assays [113]. Another reporter phage assay for *E. coli* O157:H7 has been developed by inserting the *lacZ* gene into phage T4 [114]. In this case, the specificity of the assay does not come from the phage alone, but it has been combined in a three-step assay comprising enrichment, immunomagnetic separation (IMS), and the use of a colored enzyme substrate in a simple to use sample tube format (Phast Swab [115]). The advantage here is that by using traditional culture and separation methods, they have overcome the requirement to find a phage with a defined specificity. The advantage of the Phast Swab assay format was that the phage were packaged into a unit that does not require dispensing of liquids or opening once the sample has been added to the tube. This type of practical consideration made the assay more accessible to workers in routine Quality Assurance labs. However, despite promising trials, this format did not result in a fully commercial product. Hagens et al. [116] investigated the use of an extremely thermostable glycosidase (CelB) as an alternative to *lacZ* to try and improve the sensitivity of the *Listeria* A511::*luxAB* phage [96]. In this case, the thermostable properties of the reporter protein were used to allow heating of the samples to try and remove background noise from endogenous enzymes, but even though the more sensitive fluorescent substrate methylumbelliferyl-α-D-glucopyranoside (MUG) was used, no increase in sensitivity in real food samples was achieved. However, this group also found that combining reporter phage with selective enrichment of the target bacteria improved the sensitivity of reporter phage assays [117]. In this case, they took advantage of another aspect of phage biology, in this case the high affinity of the carbohydrate-binding domains (CBDs) of the phage lysin enzymes for the unique carbohydrate components of *Listeria* cell wall peptidoglycan that can be utilized as efficient cell capture agents [118].

A similar idea of IMS concentration of cells and reporter phage technology was also used to detect *Salmonella*, this time using phage P22 containing the ice nucleation reporter gene (*ina*) as a reporter [119, 120]. As few as 10 *Salmonella* Dublin cells mL⁻¹ could be detected in the presence of a high level of competitor organisms using the phage alone, but the sensitivity of the assay for food samples was further increased by using it in combination with *Salmonellae*-specific immunomagnetic bead separation [121]. This method was developed into a commercial assay by the Idetek Corporation but is also no longer available.

The most extensively studied reporter phage are those produced for the rapid detection of *Mycobacterium tuberculosis*, since circumventing the need to culture this slow-growing organism reduces detection times from weeks to days. There is also less concern in the area of medical diagnosis about GM, and accordingly, the reporter phage have been extensively evaluated, although no commercial product has been produced. Reporter phage produced for the detection of *Mycobacteria* containing the reporter gene firefly luciferase (*luc* or *Fflux*) [114] have been produced using the lytic phage TM4 [122, 123] and the temperate phage L5 [124], which was surprisingly effective at detecting cells despite the fact that it could form lysogens. However, the most effective *luc* reporter phage (or LRP) has been developed using phage TM4 and this has been extensively modified to optimize reporter gene expression [123]. Using these phage in combination with refinements of the sample preparation, mycobacteria were successfully detected in smear-positive human sputum samples within 24–48 hours [125] and, when compared with other standard microbiological testing methods, was found to perform favorably in the detection of positive clinical samples [123, 126]. Similarly, fluorescent (*gfp*, ZsYellow and *mCherry*) reporter phage based on the broad host range mycobacteriophage TM4 for the detection of *M. tuberculosis* have been developed [127, 128]. Due to the natural fluorescence of many different biological molecules, a different color of fluorescent reporter will be more appropriate for specific sample types. In this case, the *mCherry bombΦ* was shown to be a promising tool for fluorescence microscopy-based methods of diagnosing tuberculosis infections in low income settings [129].

In response to some concerns about the use of GM viruses in routine diagnostic laboratories, nonreplicative forms of reporter phage have become favored since they minimize any possible risk of release of the GM phage. An example of this is the *lux* phage created by random mini Tn*10-luxAB* mutagenesis of phage NV10 [62] that is specific for *E. coli* O157:H7. In this case, the insertion of the *lux* genes into the phage genome created a defective phage that could infect but not replicate [130]. Using this reporter phage, *E. coli* O157:H7 cells were detected within 1 hour of infection. Rather than relying on random mutagensis effects, Kuhn et al. [131] created a genetically "locked" phage that can express the luciferase gene after infecting any wildtype *Salmonella* cells but can only replicate in specially engineered host cells. However, this type of engineering program requires a considerable amount of genetic analysis of the phage to be used and is probably not practicable for a large number of different phage, and is only a concern for nonclinical applications as medical diagnostic laboratories do not have the same regulatory constraints as food or environmental testing laboratories. Another problem facing the food industry is that in addition to speed of identification of contamination problems to prevent release of contamination product, manufacturer also needs to isolate the organism from the product for possible epidemiological trace-back investigations at a later date. Hence, although *lux* reporter phage were launched as a commercial testing service by the Sample 6 company based in the United States, which offered near online testing results for *Listeria* during food production runs [132], the product is no longer available after the company restructured.

Phage Amplification

In contrast to the reporter phage, this technology uses nonengineered phage and so does not suffer from any of the complications surrounding the release of GM organisms. This also makes the development of new assays easier and less expensive, as demonstrated by the fact that one of the earliest publications describes the simultaneous development of tests for *Pseudomonas aeruginosa*, *Salmonella typhimurium*, and *S. aureus* [133]. The other components of the assay are also standard microbiological materials and so there is no need for the purchase of specialist equipment or for highly trained staff to carry out the assay. Although termed the phage amplification assay, as growth of the phage is used to indicate the presence of the target bacterium, the test is effectively a phage protection assay. Samples to be tested for the presence of a target bacterium are mixed with phage and incubated to allow phage infection to occur. A virucide is then added to the sample to destroy any phage that have not infected its host cell. Next, the virucide is neutralized, the infected cells mixed with more bacteria that will support phage replication (helper or sensor cells), and the whole sample is plated out using soft agar. During incubation of the plates, the phage finish their replication cycle, lyse the target cell, and the released phage go on to form plaques by infecting the helper cells present in the lawn. Each plaque represents a phage that was protected from the action of the virucide by infecting its specific target host (see [134]). To develop an assay, all that is required is a phage with an appropriate host range and a virucide that will inactivate the phage while not damaging the host cell. This latter point is key to the assay; the phage must be protected within the cell and the cell must be viable to allow phage replication to be completed [135]. Hence, like the reporter phage assays, phage amplification assays also provide live/dead differentiation.

The test still requires overnight incubation to allow growth of the helper cells to form a lawn and because of this, although this assay method can be used to detect any bacterial cell, it is in the detection of slow growing organisms—such as pathogenic *Mycobacteria*—that the biggest advantage is gained. A phage amplification method for the detection of *M. tuberculosis* in human sputum samples has been developed that only takes 2 days, compared to several weeks for other methods of detection, and was previously sold as the commercial test FAST*PlaqueTB*® [136, 137]. To achieve rapid detection using this assay, the phage chosen was D29 that has a broad host range and can infect both *M. tuberculosis* and also *Mycobacterium smegmatis*, a member of the fast growing group of *Mycobacteria* that can form colonies in 18 hours. *M. smegmatis* is used as the helper cells and lawn develops for plaque visualization after overnight incubation. The limitation of this assay is that, because the broad host range of the phage, it will detect any type of *Mycobacterium*, including nonpathogens. However, when used to test human sputum samples, it is unlikely that high levels of nonpathogenic Mycobacteria will be present and detection of high numbers of Mycobacterial cells indicated that the patient requires treatment. A range of workers have carried out evaluations of this test and generally it performs as well as other culture-based tests [138, 139]. However, the main use of the assay was in clinical settings in developing countries where it is difficult to afford the reagents and machines required for automated culture-based rapid methods, such as MGIT [140].

The broad host range of this phage has made its application in this assay format problematic for samples where both pathogenic and nonpathogenic, environmental Mycobacteria may be present. However, the broad host range of the phage means that it can be used to detect other Mycobacteria if it is used in combination with other tests that improve the specificity of the assay. To this end, the FAST*PlaqueTB*® reagents have been used for the detection of viable *Mycobacterium bovis* or *Mycobacterium avium* subsp. *paratuberculosis* (*Map*) by combining the phage amplification assay with PCR confirmation of cell identity [141]. To do this, DNA is extracted from plaques at the end of the phage assay and then PCR used to detect signature IS elements (IS900 for *Map* and IS1081 for *M. bovis*). This combined Phage Amplification PCR assay was shown to be able to detect viable *Map* in milk samples from infected cattle within 24 hours and has been used to conduct surveys of a range of retail dairy products [142–144]. More recently, methods to detect the presence of mycobacterial pathogens in blood samples have been described and used to detect *Map* in cattle suffering from Johne's disease [145], *M. bovis* in the blood of cattle [146]. This assay has been further refined in to a 6-hour test that has been commercialized as the Actiphage® test by PBD Biotech Ltd. Actiphage is a one tube assay that removes the need for virucide, agar plates and overnight incubation of samples and, in addition to being able to rapidly detect *Map* and *M. bovis* in animal samples [147], has been shown to be able to detect *M. tuberculosis* in the blood of human tuberculosis patients, including those in the early stages of disease [148].

A different approach to improve specificity of the phage amplification assay has been to selectively remove the mycobacterial cells from the sample by immunomagentic separation methods prior to performing the phage assay [149, 150]. However, there have been variable reports of the robustness of this approach, as the efficiency capture by the magnetic beads seems to influence the positivity rate of the test [151, 152].

Another combined assay that uses a variation of the phage amplification assay has been described for the detection of *Salmonella* [153]. Immunomagnetic separation is used to separate the target bacteria the food sample and then the concentrated *Salmonella* cells are then infected with phage. Phage that have not infected a target cell are removed by washing, which also separates the *Salmonella* from the magnetic beads. These infected cells are incubated so that the phage can complete their replication cycle and then they are added to a culture of a phage propagating strain (signal amplifying cells or SACs). The optical density of the culture is monitored and cell lysis indicates that phage, protected within an infected cell, have been carried through the washing steps, indicating the presence of *Salmonella* in the original sample. Using this method, a detection limit of less than 10^4 cells mL^{-1} was achieved in 4–5 hours.

Phage Growth Assays

In this form of assays, the detection event is an increase in the number of phage particles present in a sample indicating that a suitable host strain is present to support phage replication. Typically low numbers of phage are added to the sample which is below a threshold of detection, after a period of incubation, the increase in phage particles in the sample is then detected either by an antibody-based assay or by qPCR. Confusingly, these phage growth assays are also referred to as phage amplification assays, but the two methods are very distinct.

The most successful example of this type of assay is the one that was developed and marketed commercially as MicroPhage as the KeyPath™ MRSA/MSSA Blood Culture Test which both allowed identifying of *S. aureus* in blood samples and also differentiated between MSSA (methicillin-susceptible *S. aureus*) and MRSA (methicillin-resistant *S. aureus*). The test required a short

(5 hours) period of incubation with a mixture, or cocktail, of phage that specifically infect a wide range of *S. aureus* cell types. To determine antibiotic sensitivity, parallel samples were inoculated into tubes with or without the addition of cefoxitin. If after incubation the phage levels in the sample reached the threshold level (1.4×10^8 phage particles per mL), a positive detection result was indicated. If a positive result also occurred in the sample containing the antibiotic, this indicated that the bacterial growth (and therefore phage growth) was not inhibited, and therefore the strain detected was antibiotic resistant (MRSA). In trials, the method had 100% specificity and good levels of sensitivity [154].

Other more complicated assays have been described that rely on specifically detecting only newly synthesized phage molecules. For instance, a method that uses isotopically labeled ^{15}N bacteriophage and matrix-assisted laser desorption/ionization-time-of-flight mass spectrometry (MALDI-TOF MS) has been described. To discriminate between phage added to the assay and those newly synthesized as part of the tests, input phage were isotopically labeled with ^{15}N. After 90 minutes of incubation, MALDI-TOF MS was then used to specifically detect ^{14}N capsid proteins produced in the phage-amplified culture, which indicated the presence of the host bacteria that supported phage growth [155]. This approach overcomes the problem associated with methods that rely on detecting phage growth alone which cannot distinguish the phage used to inoculate the sample from the new progeny phage produced after phage replication has occurred, and causes a high background noise level that reduces test sensitivity.

A similar method of identifying the production of new phage particles combined phage engineering with phage growth assays. In this case, phage capsid proteins were engineered to display a small biotinylation peptide on the surface of the phage protein by fusing it to the major capsid protein. The engineered phage are produced in a specialized strain that lacks the biotin ligase, and therefore input phage are not biotinylated. If new phage are produced in a wildtype strain that naturally encodes this enzyme, the phage produced will have been biotinylated. These phage are then mixed with streptavidin-functionalized Quantum Dots that can then be sensitively detected using flow cytometry. Using this approach and an engineered T7 phage, it was possible to detect less than 10 *E. coli* cells in the presence of a 10^5-fold excess of competitor organisms in as little as 1 hour [156].

An alternative method of measuring phage growth is qPCR, where an increase in the number in phage genomes is monitored after incubation with a sample [157, 158]. Two such assays have been described for the detection of *Brucella abortus* [159] and *Y. pestis* [160] in blood samples and, since both of these organisms require growth at BS level 3, any method that removes the need for culture is a clear benefit. In both cases, the methods were shown to be sensitive and compatible with detection in blood without the need to perform extractions, which inevitably reduces the sensitivity of any DNA-based detection method. Recently, a variant of this method was described, where rather than detecting the increase in bacteriophage genomes, transcription of phage-encoded genes was monitored by reverse transcriptase qPCR [161]. In this case, only a short period of incubation is required because highly transcribed phage genes are targeted, so the number of mRNA molecules increases rapidly after infection, shortening the time required for this to reach a threshold level of detection. In addition, as there is no transcription of phage genes prior to infection, there is no background signal from phage added to the assay which reduces the signal-to-noise ratio, and hence increases sensitivity. The simplicity of this approach is that very little information is required about phage molecular biology before a new assay can be developed, and accordingly in their proof of concept paper, the authors were able to demonstrate successful detection of both *S. aureus* and *B. anthracis* within 5 hours.

Phage Capture Assays

Since phage have evolved to interact very specifically with the surface of their target host cells, proteins that make up phage receptors and enzymes have the same type of specific interaction with bacterial surface molecules as antibodies have with their cognate antigens. Exploiting the specificity of the CBD of phage lysin has already been described, but Galikowska et al. [162] have demonstrated that phage can replace specific antibodies in a standard enzyme immunoassay (EIA) for the detection of *S. enterica* and *E. coli* (see Chapter 10 for figures demonstrating the steps of an EIA assay). In their technique, phage that are specific for the pathogen are adsorbed to the polystyrene surface of microtiter plate wells. The sample is then added, and if the pathogen is present, it binds to the phage. Following this, an antibody raised in rabbits is added, which binds to the pathogen. Detection is achieved by using a biotinylated antirabbit secondary antibody followed by an avidin-alkaline phosphatase conjugate, and the appropriate substrate. This technique offers the advantage of having a reagent that is very stable and - unlike antibodies—the detection agent is self-propagating, thus making it a very inexpensive reagent to produce. The assay can detect as few as 1×10^5 bacteria, which gives it a sensitivity in line with standard EIA assays [162].

In a more simple approach, phage can be linked to carboxylic acid-functionalized magnetic beads and then used to separate and concentrate target bacteria from samples, in the same way, that IMS concentration of samples has already been described. An example of this is the use of the *E. coli* O157-specific phage ECML-117 (ATCC PTA-7950) and ECML-134 (ATCC PTA-7949) to recover *E. coli* O157 from water samples [163]. It was found that the capture efficiency of the coupled assay under extreme conditions was approximately 20% higher than that of antibody-based separation, suggesting that phage binding to host cell is less susceptible to environmental perturbation than are antibody-antigen interactions, which is perhaps not surprising given that phage and host bacterial cells will encounter one another is a range of different environmental conditions and so the host cell binding event must have a certain degree of robustness.

Overview

The history of phage typing has been a little like phage therapy; since its discovery it has gone through different phases of popularity, but even today it continues to be a useful and sensitive tool for rapid discrimination of bacterial cell types. The advances in sequencing technologies means that phage typing *per se* has been superseded by molecular typing methods for routine subspecies identification of strains, but the technique remains a useful and sensitive method of investigating differences in cell structure, and in particular is of interest when identifying which phage are best to include in phage therapy cocktails. Rapid methods of detection have also shown promise and then fallen out of favor, probably because no one phage can provide the Holy Grail of identification—the one test that detects all target cells with 100% specificity and sensitivity. However, the recent trend of combining these with other rapid methods of bacterial detection to improve the speed and specificity of phage-based test shows promise in producing assays that do indeed meet the requirements of the consumer; that is simple, rapid tests with a high degree of discrimination. Phage identification methods have been in use for some 90 years now, and this recent revival of its fortunes in the form of commercial tests that are beginning to emerge suggests that they will continue to be used for some time to come.

REFERENCES

1. Summers, W.C., Bacteriophage research: early history, in *Bacteriophages: Biology and Applications*, Kutter E., and Sulakvelidze A., Eds., CRC Press, Boca Raton, FL, 2005, Chap.2.
2. Ackermann H.-W., Phage or phages, *Bacteriophage Ecol. Group News*, 14, 2002. http://www.mansfield.ohio-state.edu/~sabedon/
3. Dimitrov, D.S., Virus entry: molecular mechanisms and biomedical applications, *Nat. Revs. Microbiol.*, 2, 109, 2004.
4. D'Herelle, F., *The Bacteriophage and Its Behaviour*, The Williams and Wilkins Co, Baltimore, MD, 1926.
5. Ashton, P.M., Nair, S., Peters TM, et al. Identification of *Salmonella* for public health surveillance using whole genome sequencing, *Peer J.*, 2016;4:e1752. Apr 5, 2016. Doi:10.7717/peerj.1752.
6. Chattaway, M.A., Dallman, T.J., Larkin, L., et al., The transformation of reference microbiology methods and surveillance for *Salmonella* with the use of whole genome sequencing in England and Wales, *Front. Public Health*, 7, 317, 2019. Doi: 10.3389/fpubh.2019.00317.
7. Eugster, M.R., Morax, L.S., Hüls, V.J., Huwiler, S.G., Leclercq, A., Lecuit, M., and Loessner, M.J., Bacteriophage predation promotes serovar diversification in *Listeria monocytogenes*, *Molec. Microbiol.*, 97, 33, 2015.
8. Mohammed, M., and Cormican, M., Whole genome sequencing provides possible explanations for the difference in phage susceptibility among two *Salmonella typhimurium* phage types (DT8 and DT30) associated with a single foodborne outbreak, *BMC Res. Notes*, 8, 728, 2015. Doi: 10.1186/s13104-015-1687-6.
9. Duerkop, B.A., Huo, W., Bhardwaj, P., Palmer, K.L., and Hooper, L.V., Molecular basis for lytic bacteriophage resistance in Enterococci, *mBio*, 7 (4), e01304–16, 2016. Doi: 10.1128/mBio.01304-16.
10. Wright, R.C.T., Friman, V-P, Smith, M.C.M., and Brockhurst, M.A. Cross-resistance is modular in bacteria–phage interactions, *PLoS Biol.*, 16(10): e2006057, 2018. https://doi.org/10.1371/journal.pbio.2006057
11. Dunne, M., Hupfeld, M., Klumpp, J., and Loessner, M.J., Molecular basis of bacterial host interactions by Gram-positive targeting bacteriophages, *Viruses*, 10, 397, 2018.

12. Kallings, L.O., Lindberg, A.A., and Sjoberg, L., Phage typing of *Shigella sonnei*, *Arch. Immun. Ther. Exp.*, 16, 280, 1968.

13. Pruneda, R.C., and Farmer, J.J., Bacteriophage typing of *Shigella sonnei*, *J. Clin. Microbiol.*, 5, 66, 1977.

14. Adlakha, S., Sharma, K.B., and Prakash, K., Sewage as a source of phages for typing strains of *Salmonella typhimurium* isolated in India, *Indian J. Med. Res.*, 84, 1, 1986.

15. Baker, P.M., and Farmer, J.J., New bacteriophage typing system for *Yersinia enterocolitica*, *Yersinia kristensenii*, *Yersinia frederiksenii*, and *Yersinia intermedia*: correlation with serotyping, biotyping, and antibiotic susceptibility, *J. Clin. Microbiol.*, 15, 491, 1982.

16. Bhatia, T.R., Phage typing of *Escherichia coli* isolated from chickens, *Can. J. Microbiol.*, 23, 11, 1977.

17. Brown, D.R., Holt, J.G., and Pattee, P.A., Isolation and characterization of *Arthrobacter bacteriophages* and their application to phage typing of soil arthrobacters, *Appl. Environ. Microbiol.*, 35, 185, 1978.

18. Chakrabarti, A.K., Ghosh, A.N., Nair, G.B., Niyogi, S.K., Bhattacharya, S.K., and Sarkar, B.L., Development and evaluation of a phage typing scheme for *Vibrio cholerae* O139, *J. Clin. Microbiol.*, 38, 44, 2000.

19. Gaston, M.A., Isolation and selection of a bacteriophage-typing set for *Enterobacter cloaca*, *J. Med. Microbiol.*, 24, 285, 1987.

20. Craigie, J., and Yen, C.N., The demonstration of types of *B. typhosus* by means of preparations of type II Vi phage, *Can. Public Health J.*, 29, 448, 1938.

21. de Gialluly, C., Loulergue, J., Bruant, G., Mereghetti, L., Massuard, S., van der Mee, N., Audurier, A., and Quentin, R., Identification of new phages to type *Staphylococcus aureus* strains and comparison with a genotypic method, *J. Hosp. Infect.*, 55, 61, 2003.

22. Dunne, M., Rupf, B., Tala, M., Qabrati, X., Ernst, P., Shen, Y., Sumrall, E., Heeb, L., Plückthun, A., Loessner, M.J., and Kilcher, S., Reprogramming bacteriophage host range through structure-guided design of chimeric receptor binding proteins, *Cell Rep.*, 29, 1336–1350.e4, 2019. Doi: 10.1016/j.celrep.2019.09.062.

23. Weynberg, K.D., and Jaschke, P.R., Building better bacteriophage with biofoundries to combat antibiotic-resistant bacteria, *PHAGE: Therapy, Applications, and Research*, 1, 23, 2019. Doi: org/10.1089/phage.2019.0005.

24. Luria, S.E., and Human, M.L., A nonhereditary, host-induced variation of bacteria viruses, *J. Bacteriol.*, 64, 557, 1952.

25. Bertani, G., and Weigle, J.J., Host controlled variation in bacterial viruses, *J. Bacteriol.*, 65, 113, 1953.

26. Linn, S., and Arber, S., A restriction enzyme from *Hemophilus influenzae*; purification and general properties, *Proc. Natl. Acad. Sci. USA*, 59, 1300, 1968.

27. Meselson, M., and Yuan, R., DNA restriction enzyme from *E. coli*, *Nature*, 217, 1110, 1968.

28. Smith, H.O., and Wilcox, K.W., A restriction enzyme from *Hemophilus influenzae*: purification and general properties, *J. Molecl. Biol.*, 51, 379, 1970.

29. Labrie, S., Samson, J., and Moineau, S. Bacteriophage resistance mechanisms. *Nat. Rev. Microbiol.*, 8, 317, 2010. https://doi.org/10.1038/nrmicro2315

30. Garcia-Robledo, J.E., Barrera, M.C., and Tobón, G.J., CRISPR/Cas: from adaptive immune system in prokaryotes to therapeutic weapon against immune-related diseases, *Int. Rev. Immunol.*, 39, 11, 2020. Doi: 10.1080/08830185.2019.1677645.

31. Rakhuba, D.V., Kolomiets, E.I., Dey, E.S., and Novik, G.I., Bacteriophage receptors, mechanisms of phage adsorption and penetration into host cell, *Pol. J. Microbiol.*, 59, 145, 2010.

32. Eriksson, U., Adsorption of phage P22 to *Salmonella typhimurium*, *J. Gen. Virol.*, 34, 207, 1977.

33. Estrela, A.I., Pooley, H.M., de Lencastre, H., and Karamata, D., Genetic and biochemical characterization of *Bacillus subtilis* 168 mutants specifically blocked in the synthesis of the teichoic acid poly(3-O-beta-D-glucopyranosyl-N-acetylgalactosamine 1-phosphate) – *gneA*, a new locus, is associated with UDP-*N*-acetylglucosamine 4-epimerase activity, *J. Gen. Microbiol.*, 137, 943, 1991.

34. Wendlinger, G., Loessner, M.J., and Scherer, S., Bacteriophage receptors on *Listeria monocytogenes* cells are the N-acetylglucosamine and rhamnose substituents of teichoic acids or the peptidoglycan itself, *Microbiol.*, 142, 985, 1996.

35. Hung, C.H., Wu, H.C., and Tseng, Y.H., Mutation in the *Xanthomonas campestris xanA* gene required for synthesis of xanthan and lipopolysaccharide drastically reduces the efficiency of bacteriophage phi L7 adsorption, *Biochem. Biophys. Res. Comm.*, 291, 338, 2002.

36. Baggesen, D.L., Wegener, H.C., and Madsen, M., Correlation of conversion of *Salmonella enterica* serovar enteritidis phage type 1, 4, or 6 to phage type 7 with loss of lipopolysaccharide, *J. Clin. Microbiol.*, 35, 330, 1997.

37. Lawson, A.J., Chart, H., Dassama, M.U., and Threlfall, E.J., Heterogeneity in expression of lipopolysaccharide by strains of *Salmonella enterica* serotype typhimurium definitive phage type 104 and related phage types, *Lett. Appl. Microbiol.*, 34, 428, 2002.

38. Joys, T.M., Correlation between susceptibility to bacteriophage PBS1 and motility in *Bacillus subtilis*, *J. Bacteriol.*, 90, 1575, 1965.

39. Iino, T., and Mitani, M., Flagellar-shape mutants in *Salmonella*, *J. Gen. Microbiol.*, 44, 27, 1966.

40. Merino, S., Camprubí, S., and Tomás, J.M., Isolation and characterization of bacteriophage PM3 from *Aeromonas hydrophila* the bacterial receptor for which is the monopolar flagellum, *FEMS Microbiol. Letts.*, 69, 277, 1990.

41. Romantschuk, M., and Bamford, D.H., Function of pili in bacteriophage phi6 penetration, *J. Gen. Virol.*, 66, 2461, 1985.

42. Schwaartz, M., Phage Lambda receptor (LamB protein) in *Escherichia coli*, *Meths. Enzymol.*, 97, 100, 1983.

43. Heilpern, A.J., and Waldor, M.K., CTX phi, infection of *Vibrio cholerae* requires the *tolQRA* gene products, *J. Bacteriol.*, 182, 1739, 2000.

44. Sun, T.P., and Webster, R.E., Nucleotide sequence of a gene cluster involved in entry of E colicins and single-stranded DNA of infecting filamentous bacteriophages into *Escherichia coli*, *J. Bacteriol.*, 169, 2667, 1987.

45. Callegari, M.L., Sechaud, L., Rousseau, M., Bottazzi, V., and Accolas, J.P., Le recepteur du phage 832-B1 de Lactobacillus helveticus CNRZ 892 est une protéine, in Proc. 23th Dairy Congress, Vol. I., Montreal, 1990.

46. Harvey, D., Harrington, C., Heuzenroeder, M.W., and Murray, C., Lysogenic phage in *Salmonella enterica* serovar Heidelberg (*Salmonella* Heidelberg): implications for organism tracing, *FEMS Microbiol. Letts.*, 103, 291, 1993.

47. Frank, S.A., Polymorphism of bacterial restriction-modification systems – the advantage of diversity, *Evolution*, 48, 1470, 1994.

48. Chopin, M.C., Chopin, A., and Bidnenko, E., Phage abortive infection in lactococci: variations on a theme, *Curr. Opin. Microbiol.*, 8, 473, 2005.

49. Tangney, M., and Fitzgerald, G.F., AbiA, a lactococcal abortive infection mechanism functioning in *Streptococcus thermophilus*, *Appl. Environ. Microbiol.*, 68, 6388, 2002.

50. Tran, L.S.P., Szabo, L., Ponyi, T., Orosz, L., Sik, T., and Holczinger, A., Phage abortive infection of *Bacillus licheniformis* ATCC 9800; identification of the *abi*BL11 gene and localisation and sequencing of its promoter region, *Appl. Microbiol. Biotechnol.*, 52, 845, 1999.

51. Petty, N.K., Evans, T.J., Fineran, P.C., and Salmond, G.P., Biotechnological exploitation of bacteriophage research, *Trends Biotechnol.*, 25, 7, 2007.

52. Hyman, P. Phages for phage therapy: isolation, characterization, and host range breadth. *Pharmaceuticals (Basel).* 12, 35, 2019. Doi: 10.3390/ph12010035.

53. Sambrook, J., Fritsch, E.F., and Maniatis, T., *Molecular Cloning: A Laboratory Manual*, 2nd ed., Cold Spring Harbor Laboratory Press, Cold Spring Harbor, NY, 1989.

54. Miles, A.A., and Misra, S.S., The estimation of the bactericidal power of the blood, *J. Hygiene*, 38, 732, 1938.

55. Adams, M.H., *Bacteriophages*, Interscience, New York, NY, 1959.

56. Heagy, R.C., The effect of 2-4 DNP and phage T2 on *E. coli*, *J. Bacteriol.*, 59, 367, 1950.

57. Kekessy, D.A., and Piguet, J.D., New method for detecting bacteriocin production, *Appl. Microbiol.*, 20, 282, 1970.

58. Blair, J.E., and Williams, R.E.O., Phage typing of *Staphylococci*, *Bull. WHO*, 24, 771, 1961.

59. Anderson, E.S., Ward, L.R., Saxe, M.J., and de Sa, J.D., Bacteriophage-typing designations of *Salmonella typhimurium*, *J. Hyg. (London)*, 78, 297, 1977.

60. Laszlo, V.G., and Csorian, E.S., Subdivision of common *Salmonella* serotypes – phage typing of *S. virchow*, *S. manhattan*, *S. thompson*, *S. oranienburg* and *S. bareilly*, *Acta Microbiol. Hung.* 35, 289, 1988.

61. Ahmed, R., Bopp, C., Borczyk, A., and Kasatiya, S., Phage-typing scheme for *Escherichia coli* O157:H7, *J. Infect. Dis.*, 155, 806, 1987.

62. Khakhria, R., Duck, D., and Lior, H., Extended phage-typing scheme for *Escherichia coli* O157:H7, *Epidemiol. Infect.*, 105, 511, 1990.

63. Basu, S., and Mukerjee, S., Bacteriophage typing of Vibrio ElTor, *Experimenta.*, 24, 299, 1968.

64. Sarkar, B.L., Cholera bacteriophages revisited, *ICMR Bulletin*, 32, (4), ICMR Offset Press, New Delhi, India, 2002. https://main.icmr.nic.in/sites/default/files/icmr_bulletins/buapril02.pdf

65. Richardson, J.F., Rosdahl, V.T., van Leeuwen, W.J., Vickery, A.M., Vindel, A., and Witte, W., Phages for methicillin-resistant *Staphylococcus aureus*: an international trial, *Epidemlol. Infect.*, 122, 227, 1999.

66. Mclauchlin, J., Audurier, A., Frommelt, A., Gerner-Smidt, P., Jacquet, C., Loessner, M.J., van der Mee-Marquet, N., Rocourt, J., Shah, S., and Wilhelms, D., WHO study on subtyping *Listeria monocytogenes*: results of phage-typing, *Int. J. Food Microbiol.*, 32, 289, 1996.

67. Rocourt, J., Taxonomy of the *Listeria* genus and typing of *L. monocytogenes*, *Pathologie Biologie.*, 44, 749, 1996.

68. Sword, C.P., and Pickett, M.J., Isolation and characterization of bacteriophages from *Listeria monocytogenes*, *J. Gen. Microbiol.*, 25, 241, 1961.

69. Abshire, T.G., Brown, J.E., and Ezzell, J.W., Production and validation of the use of gamma phage for identification of *Bacillus anthracis*, *J. Clin. Microbiol.*, 43, 4780, 2005.

70. Brown, E.R., and Cherry, W.B., Specific identification of *Bacillus anthracis* by means of a variant bacteriophage, *J. Infect. Dis.*, 96, 34, 1955.

71. Ackermann, H-W., Azizbekyan, R.R., Bernier, R.L., de Barjac, S., Saindoux, J.R., Valéro, J.R., and Yu, M,X., Phage typing of *Bacillus subtilis* and *B. thuringiensis*, *Res. Microbiol.*, 146, 643, 1995.

72. Callow, B.R., A new phage-typing scheme for *Salmonella typhimurium*, *J. Hyg. (Lond.)*, 57, 346, 1959.

73. Cooke, F.J., Day, M., Wain, J., Ward, L.R., and Threlfall, E.J., Cases of typhoid fever imported into England, Scotland and Wales (2000-2003), *Trans. Roy, Soc. Tropical Med. Hyg.*, 101, 398, 2007.

74. Michel, P. Martin, L.J., Tinga, C.E., and Dore, K., Regional, seasonal, and antimicrobial resistance distributions of *Salmonella typhimurium* in Canada – a multi-provincial study, *Can. J. Pub. Health*, 97, 470, 2006.

75. Matsui, T., Suzuki, S., Takahashi, H., Ohyama, T., Kobayashi, J., Izumiya, H., Watanabe, H., Kasuga, F., Kijima, H., Shibata, K., and Okabe, N., *Salmonella Enteritidis* outbreak associated with a school lunch dessert: cross-contamination and a long incubation period, Japan, 2001, *Epidemiol. Infect.*, 132, 873, 2004.

76. HPA, Salmonella *enteritidis* PT56 in Durham – final report, *Comm. Dis. Rep. CDR Wkly*, 14(5), 2004 [serial online], http://www.hpa.org.uk/cdr/archives/2004/cdr0504.pdf.

77. He, X., and Pan, R., Bacteriophage lytic patterns for identification of *Salmonellae*, *Shigellae*, *Escherichia coli*, *Citrobacter freundii*, and *Enterobactera cloacae*, *J. Clin. Microbiol.*, 30, 590, 1992.

78. Sarkar, B.L., Roy, M.K., Chakrabarti, A.K., and Niyogi, S.K., Distribution of phage type of *Vibrio cholerae* O1 biotype ElTor in Indian scenario (1991-98), *Indian. J. Med Res.*, 109, 204, 1999.

79. Gomes, A.R., Westh, H., and de Lencastre, H., Origins and evolution of methicillin-resistant *Staphylococcus aureus* clonal lineages, *Antimicrob. Agents Chemother.*, 50, 3237, 2006.

80. Salmenlinna, S., Lyytikainen, O., and Vuopio-Varkila, J., Community-acquired methicillin-resistant *Staphylococcus aureus*, Finland, *Emerg. Intect. Dis.*, 8, 602, 2002.

81. Cherry, W.B., Davis, B.R., Edwards, P.R., and Hogan, R.B., A simple procedure for the identification of the genus *Salmonella* by means of a specific bacteriophage, *J. Lab. Clin. Med.*, 44, 51, 1954.

82. Loessner, M.J., and Busse, M., Bacteriophage typing of *Listeria* species, *Appl. Environ. Microbiol.*, 56, 1912, 1990.

83. Carlton, R.M., Noordman, W.H., Biswas, B., de Meester, E.D., and Loessner, M.J., Bacteriophage P100 for control of *Listeria monocytogenes* in foods: genome sequence, bioinformatic analyses, oral toxicity study, and application, *Regul. Tox. Pharmacol.*, 43, 301, 2005.

84. Barman, S., and Majumdar, S., Intracellular replication of choleraphage phi 92, *Intervirol.*, 42, 238, 1999.

85. Jensen, E.C., Schrader, H.S., Rieland, B., Thompson, T.L., Lee, K.W., Nickerson, K.W., and Kokjohn, T.A., Prevalence of broad-host-range lytic bacteriophages of *Sphaerotilus natans*, *Escherichia coli*, and *Pseudomonas aeruginosa*, *Appl. Environ. Microbiol.*, 64, 575, 1998.

86. Sullivan, M.B., Waterbury, J.B., and Chisholm, S.W., Cyanophages infecting the oceanic *Cyanobacterium prochlorococcus*, *Nature*, 424, 1047, 2003.

87. Smartt, A.E., Xu, T., Jegier, P., Carswell, J.J., Blount, S.A., Sayler, G.S., and Ripp, S. Pathogen detection using engineered Bacteriophages. *Anal. Bioanal. Chem.*, 402, 3127, 2012.

88. Bárdy, P., Pantůček, R., Benešík, M., and Doškař, J., Genetically modified bacteriophages in applied microbiology, *J. Appl. Microbiol.*, 121, 618, 2016. Doi: 10. 1111/jam.13207..

89. Pires, D.P., Cleto, S., Sillankorva, S., Azeredo, J., and Lu, T.K., Genetically engineered phages: a review of advances over the last decade. *Microbiol. Mol. Biol. Rev.*, 80, 523, 2016. Doi: 10.1128/MMBR.00069-15.

90. Sagona, A.P., Grigonyte, A.M., MacDonald, P.R., and Jaramillo, A., Genetically modified bacteriophages. *Integr. Biol. (Camb.)*, 8, 465, 2016. Doi: 10.1039/c5ib00267b.

91. Ulitzur, S., and Kuhn, J., Introduction of *lux genes* into bacteria; a new approach for specific determination of bacteria and their antibiotic susceptibility. In: Slomerich, R., Andreesen, R., Kapp, A., Ernst, M., and Woods, W.G., Eds., *Bioluminescence and Chemiluminescence new perspectives*, John Wiley & Sons, New York, NY, 1987, 463.

92. Ulitzur, S., and Kuhn, J., US Patent 4,861,709, Detection and/or identification of microorganisms in a test sample using bioluminescence or other exogenous genetically introduced marker, 1989.

93. Kodikara, C.P., Crew, H.H., and Stewart, G.S.A.B., Near on-line detection of enteric bacteria using *lux* recombinant Bacteriophage, *FEMS Microbiol. Letts.*, 83, 261, 1991.

94. Duzhii, D.E., and Zavilgelskii, G.B., Bacteriophage Lambda: *lux*: design and expression of bioluminescence in *E. coli* cells, *Molec. Gen. Mikrobiol. Virusol.*, 3, 36, 1994.

95. Chen, J., and Griffiths, M.W., *Salmonella* detection in eggs using *lux*+ bacteriophages, *J. Food Protect.*, 59, 908, 1996.

96. Loessner, M.J., Rees, E., Stewart, G.S., and Scherer, S., Construction of luciferase reporter bacteriophage A511::*luxAB* for rapid and sensitive detection of viable *Listeria* cells, *Appl. Environ. Microbiol.*, 62, 1133, 1996.

97. Loessner, M.J., Rudolf, M., and Scherer, S., Evaluation of luciferase reporter bacteriophage A511::*luxAB* for detection of *Listeria monocytogenes* in contaminated foods, *Appl. Environ. Microbiol.*, 63, 2961, 1997.

98. Thouand, G. Vachon, P., Liu., S., Dayre, M., and Griffiths, M.W., Optimization and validation of a simple method using P22::*luxAB* bacteriophage for rapid detection of *Salmonella enterica* serotypes A, B, and D in poultry samples, *J. Food. Prot.*, 7, 380, 2008.

99. Schofield, D.A., and Westwater, C. Phage-mediated bioluminescent detection of *Bacillus anthracis*, *J. Appl. Microbiol.*, 107, 1468, 2009. Doi: 10.1111/j.1365-2672.2009.04332.x.

100. Sharp, N.J., Molineux, I.J., Page, M.A., and Schofield, D.A., Rapid detection of viable *Bacillus anthracis* spores in environmental samples by using engineered reporter phages, *Appl. Environ. Microbiol.* 82, 2380, 2016. Doi: 10.1128/AEM.03772-15.

101. Schofield, D.A., Molineux, I.J., and Westwater, C., Diagnostic bioluminescent phage for detection of *Yersinia pestis*, *J. Clin. Microbiol.*, 47, 3887, 2009. Doi: 10.1128/JCM.01533-09.

102. Vandamm, J.P., Rajanna, C., Sharp, N.J., Molineux, I.J., and Schofield, D.A. Rapid detection and simultaneous antibiotic susceptibility analysis of *Yersinia pestis* directly from clinical specimens by use of reporter phage, *J. Clin. Microbiol.*, 52, 2998, 2014. Doi: 10.1128/JCM.00316-14.

103. Schofield, D.A., Wray, D.J., and Molineux, I.J. Isolation and development of bioluminescent reporter phages for bacterial dysentery, *Eur. J. Clin. Microbiol. Infect. Dis.*, 34, 395, 2015. Doi: org/10.1007/s10096-014-2246-0.

104. Schofield, D.A., Bull, C.T., Rubio, I., Wechter, W. P., Westwater, C., and Molineux, I.J., Development of an engineered bioluminescent reporter phage for detection of bacterial blight of Crucifers, *Appl. Environ. Microbiol.*, 78, 3592, 2012. Doi: 10.1128/AEM.00252-12.

105. Born, Y., Fieseler, L., Thöny, V., Leimer, N., Duffy, B., and Loessner, M.J., Engineering of bacteriophages Y2::*dpoL*1-C and Y2::*luxAB* for efficient control and rapid detection of the fire blight pathogen, *Erwinia amylovora*, *Appl. Environ. Microbiol.*, 83:e00341–17. 2017. Doi: org/10.1128/AEM.00341-17.

106. Klumpp, J., and Loessner, M.J., Detection of bacteria with bioluminescent reporter bacteriophage. In: Thouand, G., and Marks, R., Eds., *Bioluminescence: Fundamentals and Applications in Biotechnology – Vol 1. Advances in Biochemical Engineering/Biotechnology*, vol 144. Springer, Berlin, Heidelberg, 2014.

107. Kim, S., Kim, M., and Ryu, S., Development of an engineered bioluminescent reporter phage for the sensitive detection of viable *Salmonella typhimurium*, *Anal. Chem.*, 86, 5858, 2014. Doi: org/10.1021/ac500645c.

108. Franche, N., Vinaym, M., and Ansaldi, M., Substrate-independent luminescent phage-based biosensor to specifically detect enteric bacteria such as *E. coli.*, *Environ. Sci. Pollut. Res. Int.*, 24, 42, 2017. Doi: 10.1007/s11356-016-6288-y.

109. Kim, J., Kim, M., Kim, S., and Ryu, S., Sensitive detection of viable *Escherichia coli* O157:H7 from foods using a luciferase-reporter phage phiV10*lux*, *Int. J. Food. Microbiol.*, 254, 11, 2017. Doi: 10.1016/j.ijfoodmicro.2017.05.002.

110. Hinkley, T.C. et al. A syringe-based biosensor to rapidly detect low levels of *Escherichia coli* (ECOR13) in drinking water using engineered bacteriophages. *Sensors (Basel)*, 20, 1953, 2020. Doi: 10.3390/s20071953.

111. Meile, S., Sarbach, A., Du, J., Schuppler, M., Saez, C., Loessner, M.J., and Kilcher, S. Engineered reporter phages for rapid bioluminescence-based detection and differentiation of viable *Listeria* cells, *Appl. Environ. Microbiol.*, 86, e00442–20; 2020. Doi: 10.1128/AEM.00442-20.

112. Oda, M., Morita, M., Unno, H., and Tanji, Y., Rapid detection of *Escherichia coli* O157:H7 by using green fluorescent protein-labelled PP01 bacteriophage, *Appl. Environ. Microbiol.*, 70, 527, 2004.

113. Tanji, Y., Furukawa, C., Na, S.H., Hijikata, T., Miyanaga, K., and Unno, H., *Escherichia coli* detection by GFP-labeled lysozyme-inactivated T4 bacteriophage, *J. Biotechnol.*, 114, 11, 2004.

114. Goodridge, L., and Griffiths, M., Reporter bacteriophage assays as a means to detect foodborne pathogenic bacteria, *Food Res. Int.*, 35, 863, 2002.

115. Goodridge, L., Willford, J., and Goodridge, L. D. A Simple Reporter-Phage Assay for Rapid Detection of *E. coli* O157:H7. Presented at 16th Evergreen International Phage Biology Meeting, Washington, DC, August 7–12, 2005.

116. Hagens, S., de Wouters, T., Vollenweider, P., and Loessner, M.J., Reporter bacteriophage A511::*celB* transduces a hyper-thermostable glycosidase from *Pyrococcus furiosus* for rapid and simple detection of viable *Listeria* cells *Bacteriophage*, 1, 143, 2011. Doi: 10.4161/bact.1.3.16710.

117. Kretzer, J.W., Schmelcher, M., and Loessner, M.J. Ultrasensitive and fast diagnostics of viable *Listeria* cells by CBD magnetic separation combined with A511::*luxAB* detection, *Viruses*, 10, 626, 2018.

118. Kretzer, J.W., Lehmann, R., Schmelcher, M., Banz, M., Kim, K.P., Korn, C., and Loessner, M.J., Use of high-affinity cell wall-binding domains of bacteriophage endolysins for immobilization and separation of bacterial cells, *Appl. Environ. Microbiol.*, 73, 1992, 2007.

119. Wolber, P.K., and Green, R.L., Detection of bacteria by transduction of ice nucleation genes, *Trends Biotechnol.*, 8, 276, 1990.

120. Wolber, P.K., and Green, R.L., New rapid method for the detection of *Salmonella* in foods, *Trends Food Sci. Technol.*, 1, 80, 1990.

121. Irwin, P., Gehring, A., Tu, S.I., Brewster, J., Fanelli, J., and Ehrenfeld, E., Minimum detectable level of *Salmonellae* using a binomial-based bacterial ice nucleation detection assay (BIND), *J. AOAC Int.*, 83, 1087, 2000.

122. Jacobs, W.R., Barletta, R.G., Udani, R., Chan, J., Kalkut, G., Sosne, G., Kieser, T., Sarkis, G., Hatfull, G., and Bloom, B., Rapid assessment of drug susceptibilities of *Mycobacterium tuberculosis* by means of luciferase reporter phages, *Science*, 260, 819, 1993.

123. Bardarov, S., Dou, H., Eisenach, K., Banaiee, N., Ya, S., Chan, J., Jacobs, W.R., and Riska, P.F., Detection and drug-susceptibility testing of M-tuberculosis from sputum samples using luciferase reporter phage: comparison with the mycobacteria growth indicator tube (MGIT) system, *Diag. Microbiol. Infect. Dis.*, 45, 53, 2003.

124. Sarkis, G.J., Jacobs, W.R., and Hatfull, G.F., L5 Luciferase reporter mycobacteriophages – A sensitive tool for the detection and assay of live *Mycobacteria*, *Molec. Microbiol.*, 15, 1055, 1995.

125. Riska, P.F., Jacobs, W.R., Bloom, B.R., McKitrick, J., and Chan, J., Specific identification of *Mycobacterium tuberculosis* with the luciferase reporter mycobacteriophage: use of *p*-Nitro-∀-Actylamino-∃-hydroxy Propiophenone, *J. Clin. Microbiol.*, 35, 3225, 1997.

126. Jain, P., Thaler, D.S., Maiga, M. et al. Reporter phage and breath tests: emerging phenotypic assays for diagnosing active tuberculosis, antibiotic resistance, and treatment efficacy, *J. Infect. Dis.* 204 Suppl. 4:S1142, 2011. Doi: 10.1093/infdis/jir454.

127. Piuri, M., Jacobs, W.R. Jr, and Hatfull, G.F., Fluoromycobacteriophages for rapid, specific, and sensitive antibiotic susceptibility testing of *Mycobacterium tuberculosis*. *PLoS ONE*, 4, e4870. 2009. Doi: 10.1371/journal.pone.0004870.

128. Urdániz, E., Rondón, L., Martí, M.A., Hatfull, G.F., and Piuri, M., Rapid whole-cell assay of antitubercular drugs using second-generation fluoromycobacteriophages. *Antimicrob. Agents Chemother.*, 60, 3253, 2016. Doi: 10.1128/AAC.03016-15.

129. Rondón, L., Urdániz, E., Latini, C., et al. Fluoromycobacteriophages can detect viable *Mycobacterium tuberculosis* and determine phenotypic rifampicin resistance in 3-5 days from sputum collection. *Front. Microbiol.*, 9, 1471, 2018. Doi: 10.3389/fmicb.2018.01471.

130. Waddell, T.E., and Poppe, C., Construction of mini-Tn*10luxAB*cam/Ptac-ATS and its use for developing a bacteriophage that transduces bioluminescence to *Escherichia coli* O157:H7, *FEMS Microbiol. Letts.*, 182, 285, 2000.

131. Kuhn, J., Suissa, M., Wyse, J., Cohen, I., Weiser, I., Reznick, S., Lubinsky-Mink, S., Stewart, G., and Ulitzur, S., Detection of bacteria using foreign DNA: the development of a bacteriophage reagent for *Salmonella*, *Int. J. Food Microbiol.*, 74, 229, 2002.

132. Cappillino, M., Shivers, R.P., Brownell, D.R., Jacobson, B., King, J., Kocjan, P., Koeris, M., Tekeian, E., Tempesta, A., Bowers, J., Crowley, E., Bird, P., Benzinger, J., and Fisher, K., Sample6 DETECT/L: an in-plant, in-shift, enrichment-free *Listeria* environmental assay, *J. AOAC Int.*, 98, 436, 2015. Doi: org/10.5740/jaoacint.14-213.

133. Stewart, G.S.A.B., Jassim, S.A., Denyer, S.P., Newby, P., Linley, K., and Dhir, V.K., The specific and sensitive detection of bacterial pathogens within 4 h using bacteriophage amplification, *J. Appl. Microbiol.*, 84, 777, 1998.

134. Mole, R.J., and Maskell, T.W.O.'C., Phage as a diagnostic – the use of phage in TB diagnosis, *J. Chem. Technol. Biotechnol*, 76, 683, 2001.

135. de Siqueira, R.S., Dodd, C.E.R., and Rees, C.E.D., Evaluation of the natural virucidal activity of teas for use in the phage amplification assay, *Int. J. Food Microbiol.*, 111, 259, 2006.

136. McNerney, R., Wilson, S.M., Sidhu, A.M., Harley, V.S., Al Suwaidi, Z., Nye, P.M., Parish, T.. and Stoker, N.G., Inactivation of mycobacteriophage D29 using ferrous ammonium sulphate as a tool for the detection of viable *Mycobacterium smegmatis* and *M. tuberculosis*, *Res. Microbiol.*, 149, 487, 1998.

137. Stewart, G.S.A.B., Jassim, S.A.A., Denyer, S.P., Park, S., Rostas-Mulligan, K., and Rees, C.E.D., Methods for rapid microbial detection. PCT Patent WO 92/02633, 1992.

138. Muzaffar, R. Batool, S., Aziz, F., Naqvi, A., and Rizvi, A., Evaluation of the FASTPlaqueTB assay for direct detection of *Mycobacterium tuberculosis* in sputum specimens, *Int. J. Tubercul. Lung Dis.*, 6, 635, 2002.

139. Dinnes, J., Deeks, J., Kunst, H., Gibson, A., Cummins, E., Waugh, N., Drobniewski, F. and Lalvani, A., A systematic review of rapid diagnostic tests for the detection of tuberculosis infection, *Health Technol. Assess.*, 11(3), 2007.

140. Reisner, B.S., Gatson, A.M., and Woods, G.L., Evaluation of mycobacteria growth indicator tubes for susceptibility testing *of Mycobacterium tuberculosis* to isoniazid and rifampin, *Diag. Microbiol. Infect. Dis.*, 22, 325, 1995.

141. Stanley, E.C., Mole, R.J., Smith, R.J., Glenn, S.M., Barer, M.R., McGowan, M. and Rees, C.E.D., Development of a new, combined rapid method using phage and PCR for detection and identification of viable *Mycobacterium paratuberculosis* bacteria within 48 hours, *Appl. Environ. Microbiol.*, 73, 1851, 2007.

142. Botsaris, G., Slana, I., Liapi, M., Dodd, C., Economides, C., Rees, C., and Pavlik, I. Rapid detection methods for viable *Mycobacterium avium* subspecies *paratuberculosis* in milk and cheese, *Int. J. Food. Microbiol.*, 141, S87–S90, 2010. Doi: org/10.1016/j.ijfoodmicro.2010.03.016.

143. Botsaris,G., Liapi, M., Kakogiannis, C., Dodd, C.E.R., and Rees, C.E.D. Detection of *Mycobacterium avium* subsp. paratuberculosis in bulk tank milk by combined phage-PCR assay: evidence that plaque number is a good predictor of MAP, *Int. J. Food Microbiol.*, 164, 76, 2013.

144. Gerrard, Z.E., Swift, B.M.C., Botsaris, G., et al. Survival of *Mycobacterium avium* subspecies *paratuberculosis* in retail pasteurised milk, *Food Microbiol.*, 74, 57, 2018. Doi: 10.1016/j.fm.2018.03.004.

145. Swift, B.M.C., Denton, E.J., Mahendran, S.A., Huxley, J.N., and Rees, C.E.D., Development of a rapid phage-based method for the detection of viable *Mycobacterium avium* subsp. *paratuberculosis* in blood within 48 hours, *J. Microbiol. Meths.*, 94, 175, 2013.

146. Swift, B.M.C., Convery, T.W., and Rees, C.E.D., Evidence of *Mycobacterium tuberculosis* complex bacteraemia in intra-dermal skin test positive cattle detected using phage-RPA, *Virulence*, 7, 779, 2016. Doi: 10.1080/21505594.2016.1191729.

147. Swift, B.M.C., Meade, N., Sandoval Barron, E., Bennett, M., Perehinec, T., Hughes, V., Stevenson K., and Rees, C.E.D., The development and use of Actiphage® to detect viable mycobacteria from bovine tuberculosis and Johne's disease-infected animals, *Microb. Biotechnol.*, 13, 738, 2020. Doi: org/10.1111/1751-7915.13518.

148. Verma, R., Swift, B.M.C., Handley-Hartill, W., et al., A novel, high-sensitivity, bacteriophage-based assay identifies low-level *Mycobacterium tuberculosis* bacteremia in immunocompetent patients with active and incipient tuberculosis, *Clin. Infect. Dis.*, 70, 933, 2020. Doi: 10.1093/cid/ciz548.

149. Foddai, A., Elliott, C.T., and Grant, I.R., Maximizing capture efficiency and specificity of magnetic separation for *Mycobacterium avium* subsp. *paratuberculosis* cells, *Appl. Environ. Microbiol.*, 76, 7550, 2010. Doi: 10.1128/AEM.01432-10.

150. Foddai, A., Strain, S., Whitlock, R.H., Elliott, C.T., and Grant, I.R., Application of a peptide-mediated magnetic separation-phage assay for detection of viable *Mycobacterium avium* subsp. *paratuberculosis* to bovine bulk tank milk and feces samples, *J. Clin. Microbiol.* 49, 2017, 2011. Doi: 10.1128/JCM.00429-11.

151. Butot, S, Ricchi, M., Sevilla, I.A., et al., Estimation of performance characteristics of analytical methods for *Mycobacterium avium* subsp. paratuberculosis detection in dairy products, *Front Microbiol.*, 10, 509, 2019. Doi: 10.3389/fmicb.2019.00509.

152. Swift. B.M.C., Huxley, J.N., Plain, K.M., et al., Evaluation of the limitations and methods to improve rapid phage-based detection of viable *Mycobacterium avium* subsp. *paratuberculosis* in the blood of experimentally infected cattle, *BMC Vet. Res.*, 12, 115, 2016. Doi: 10.1186/s12917-016-0728-2.

153. Favrin, S.J., Jassim, S.A., and Griffiths, M.W., Development and optimisation of a novel immunomagnetic separation-bacteriophage assay for detection of *Salmonella enterica* serovar Enteritidis in broth, *Appl. Environ. Microbiol.*, 67, 217, 2001.

154. Sullivan, K.V., Turner, N.N., Roundtree, S.S, and McGowan K.L., Rapid detection of methicillin-resistant *Staphylococcus aureus* (MRSA) and methicillin-Susceptible *Staphylococcus aureus* (MSSA) using the KeyPath MRSA/MSSA blood culture test and the BacT/ALERT system in a pediatric population, *Arch. Path. Lab. Med.*, 137, 1103, 2013. https://doi.org/10.5858/arpa.2012-0422-OA.

155. Pierce, C.L, Rees, J.C., Fernández, F.M., and Barr, J.R. Detection of *Staphylococcus aureus* using 15N-labeled bacteriophage amplification coupled with matrix-assisted laser desorption/ionization-time-of-flight mass spectrometry, *Analyt. Chem.*, 83, 2286, 2011.

155. Edgar, R., McKinstry, M., Hwang, J., Oppenheim, A.B., and Fekete, R.A., High-sensitivity bacterial detection using biotin-tagged phage and quantum-dot nanocomplexes, *Proc. Natl. Acad. Sci. USA* 103, 4841, 2006.

157. Kutin, R.K., Alvarez, A., and Jenkins, D.M., Detection of *Ralstonia solanacearum* in natural substrates using phage amplification integrated with real-time PCR assay, *J. Microbiol. Methods*, 76, 241, 2009. Doi: 10.1016/j.mimet.2008.11.008.

158. Luo, J., Jiang, M., Xiong, J., Li, J., Zhang, X., Wei, H., and Yu, J. Exploring a phage-based real-time PCR assay for diagnosing *Acinetobacter baumannii* bloodstream infections with high sensitivity, *Anal. Chim. Acta.*, 1044, 147, 2018. Doi: 10.1016/j.aca.2018.09.038.

159. Sergueev, K.V., Filippov, A.A., and Nikolich, M.P., Highly sensitive bacteriophage-based detection of *Brucella abortus* in mixed culture and spiked blood, *Viruses*, 9, 144, 2017. Doi: 10.3390/v9060144.

160. Sergueev, K.V., He, Y., Borschel, R.H., Nikolich, M.P., and Filippov, A.A., Rapid and sensitive detection of *Yersinia pestis* using amplification of plague diagnostic bacteriophages monitored by real-time PCR, *PLoS ONE*, 5, e11337, 2010.

161. Malagon, F., Estrella, L.A., Stockelman, M.G., Hamilton, T., Teneza-Mora, N., and Biswas, B., Phage-mediated molecular detection (PMMD): a novel rapid method for phage-specific bacterial detection, *Viruses*. 12, 435, 2020. Doi: 10.3390/v12040435.

162. Galikowska, E., Kunikowska, D.,Tokarska-Pietrzak, E., Dziadziuszko, H., Łoś, J.M, Golec, P., Węgrzyn. G., and Łoś, M., Specific detection of *Salmonella enterica* and *Escherichia coli* strains by using ELISA with bacteriophages as recognition agents, *Eur. J. Clin. Microbiol. Infect. Dis.*, 30, 1067, 2011.

163. Wang, Z., Wang, D., Kinchla, A.J. et al., Rapid screening of waterborne pathogens using phage-mediated separation coupled with real-time PCR detection, *Anal. Bioanal. Chem.*, 408, 4169, 2016. https://doi.org/10.1007/s00216-016-9511-2

9

Phage Display and Selection of Protein Ligands

Geir Åge Løset, Wlodek Mandecki, and Inger Sandlie

CONTENTS

Introduction

One of the principal modes of cell-cell communication, as well as control of biochemical processes in cells, relies upon proteins binding to a ligand. The ligand could be another protein, a nucleic acid, or a small molecule. Phage display is a tool that makes possible both presentation of large repertoires of naturally occurring proteins, and the derivation of protein ligands that have never previously been known to exist in nature. The very essence of phage display is that the displayed protein is encoded in the nucleic acid encapsulated by the phage particle, hence creating a physical coupling between genotype and phenotype [1]. Different protein sequences can be displayed by cloning the genes of interest as they occur in nature, or simply modifying the sequence of the DNA through molecular engineering. The simultaneous display of many different molecules on a population of phage then collectively makes what is called a phage library.

The power of phage display is derived from the fact that an extremely large number of displayed ligands in the form of a library can be studied in a single experiment. This is possible due to the fact that the filamentous phage strains such as f1, M13, and fd, of *Escherichia coli* (*E. coli*) are the most prolific phages

in nature, giving rise to titers of up to 10^{13} particles per milliliter culture [2]. Thus, libraries with diversities exceeding 10^{10} unique members have been made. All library members then compete for binding to a defined target, allowing for the identification and characterization of binders with desired specificity and affinity, in a process often termed panning. During panning, the noninteracting library members are removed by washing, and the identity of those interacting with the target is subsequently decoded by virtue of sequencing the DNA encapsulated by the retrieved phage particles. This large combinatorial power, combined with the extreme robustness exhibited by the phage particle, which allows it to withstand the physicochemical challenges during different selection regimens, is the main reason for its success [3, 4].

The most common form of phage display is based on fd and M13, and their two coat proteins, minor protein pIII and major coat protein pVIII. Both have been appended with random peptides or engineered proteins to create phage libraries. In 1985, George Smith was the first to demonstrate that a protein, can be displayed on the filamentous phage particle fused to pIII and selected for in an enrichment experiment from among a large excess of ordinary phage [1]. Five years later, in 1990, he and coworker Jamie K. Scott reported on the construction of the first true phage library, which comprised more than 10 million members, all displaying six amino acid-long random peptides [5]. They panned this library toward two monoclonal antibody targets and identified specific peptide ligands. The significance of this discovery quickly became clear to the scientific community, and the new field of phage display was established with explosive growth following.

Despite its early introduction more than 30 years ago, phage display is still the most important molecular evolution technology available [4, 6, 7], at least in part due to its major contribution to design of new therapeutics [8, 9]. Consequently, a variety of protein scaffolds have been displayed on the phage and served as starting point for a multitude of library designs [3]. Thus, phage display technology has found applications within both basic and applied biological research, as well as in combinatorial chemistry and material sciences [2, 10, 11]. One of the most important events was the report that showed functional display of antibody fragments, which opened the field of antibody discovery by use of this method [12–14]. Today, phage display is an important source of therapeutically and diagnostically useful antibodies, and the complexity of libraries have increased from 10^6 clones early on to now more than 10^{11} clones [15].

The importance of the birth of phage display technology and the achievement of merging it with antibody technology can hardly be underrated with respect to impact it has had on modern life science. Acknowledging this, The Royal Swedish Academy of Sciences awarded the 2018 Nobel prize in Chemistry with one half to Frances H. Arnold and the other half jointly to George P. Smith and Sir Gregory P. Winter exactly for these reasons as given in their statement:

> The 2018 Nobel Laureates in Chemistry have taken control of evolution and used it for purposes that bring the greatest benefit to humankind. Enzymes produced through directed evolution are used to manufacture everything from biofuels to pharmaceuticals. Antibodies evolved using a method called phage display can combat autoimmune diseases and in some cases cure metastatic cancer.

The transition from phage vector systems that rely on the large and genetically unstable phage genome, into the genetically stable low valence phagemid display system greatly improved the performance of phage display [16]. Thus, heterologous capsid fusion proteins may be encoded either in a phage genome or by a phagemid; in the latter case, complementation by a helper phage is needed to support virion production [16]. Thus, a critical component for the effective use of phagemids was the construction of interference-resistant helper phages, such as M13K07, that allowed for high phage titer production and preferential phagemid packaging [17, 18]. Easy manipulation and high transformation frequencies are additional assets that have made the phagemid format the most popular display system for folded domains [13, 14].

In addition, theoretical approaches to biopanning have emerged, guiding investigators in how to implement the most efficient techniques for selecting binders [19–22], as well as evaluating the characteristics of the isolated binders [23–25]. Of particular importance is the emergence of next generation sequencing, a very powerful tool when evaluating panning experiments [26, 27]. Most notable, several excellent books [28–30] and review articles [7, 31–35] that focus on different aspects of phage display have been published.

Filamentous Phage Biology

The filamentous bacteriophages belong to a group of viruses that harbor a circular single-stranded DNA carried in an elongated protein capsid cylinder. The genus is termed *Inoviridae* (UniProtKB taxon identifier 10861) and comprises 22 strains of which seven belong to the group of Enterobacteria phage, namely f1, fd, I2-2, If1, IKe, M13, ZJ/2, Pf1, and Pf3. Based on X-ray diffraction studies, the viruses have been divided into two morphologically distinct classes, and the *E. coli* viruses M13, fd, f1, If1, and IKe belong to class I [36, 37]. Infection depends on the presence of specific pili encoded by conjugative plasmids carried by the host cell. M13, fd, and f1 only infect *E. coli* strains that harbor the conjugative plasmid F, hence the common annotation Ff for F-specific filamentous phage. The nucleotide sequences of M13, fd, and f1 show that these viruses are very closely related (homology ~97%) and, therefore, rather recently evolved from a common ancestor [38].

The Ff class of filamentous phages is by far dominating the phage display field in combination with its natural host *E. coli* (Figure 9.1). Though pIII grafting experiments show that the host range can be expanded in that the phages can become M pili specific, this has yet to be implemented in phage display [39]. A complete list of phage proteins and the genes encoding them is provided in Table 9.1.

M13 Virion

The filamentous phage genome consists of a 6,407 nt single-stranded DNA and encodes 11 proteins. Five structural proteins include a minor coat protein pIII, the major coat protein pVIII, and three other proteins, pVI, pVII, and pIX (Figure 9.2). Proteins pIII and pVI cluster on the adsorption end of the virion, while proteins pVII and pIX cluster on the maturation end. These proteins are present in very few (three to five) copies in the virion, in sharp contrast with pVIII (2,700 copies). The six nonstructural proteins are replication proteins pII and pX, single-stranded DNA binding protein pV, and morphogenic proteins pI, pIV, and pXI (arabic numerals are sometimes used to designate the proteins and genes instead of roman numerals).

Information about the virion structure comes primarily from x-ray diffraction studies and electron micrography. The M13 genome is packaged into a phage capsid 65 Å in diameter and 9,300 Å in length (Figure 9.3). The main building block for the capsid is the 50 amino acid residue-long protein pVIII, oriented at a 20° angle from the particle axis and assembled in a shingle-type array. The assembly of pVIII proteins leaves a cavity of diameter

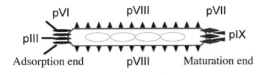

FIGURE 9.1 Schematic of coliphage M13 virion.

TABLE 9.1

F-specific Filamentous Phage Genes/Proteins and Properties

Gene	Protein	Size (aa)	Location	Function	Display Mode/Valence
gI	pI	348	Inner membrane	Assembly	*NA*
	pXI	108	Inner membrane	Assembly	*NA*
gII	pII	409	Cytoplasm	Replication (nickase)	*NA*
	pX	111	Cytoplasm	Replication	*NA*
gIII	pIII	406[a]	Virion tip (end)	Virion component	Yes (N-term)/1-5
gIV	pIV	405[a]	Outer membrane	Assembly (exit channel)	*NA*
gV	pV	87	Cytoplasm	Replication (ssDNA bp)	*NA*
gVI	pVI	112	Virion tip (end)	Virion component	Yes (C-term)/1-5
gVII	pVII	33[b]	Virion tip (start)	Virion component	Yes (N-term)/1-5
gVIII	pVIII	50[a]	Virion filament	Virion component	Yes (N- and C-term)/>50
gIX	pIX	32[b]	Virion tip (start)	Virion component	Yes (N-term)/1-5

[a] The mature protein without signal sequence.
[b] Allows signal sequence independent display [83].

of 25–35 Å, filled with phage DNA throughout the length of the virion. The phage DNA is oriented in the particle, determined by the packaging signal (PS), which forms an imperfect hairpin and is positioned at the maturation end of the phage particle. The PS is necessary and sufficient for incorporation of the DNA into the phage particle. The bulk of the pVIII protein (residues 6–50) forms an alpha helix. Five N-terminal residues are surface-exposed, while residues near the C terminus interact with the genomic DNA. The amino terminus of pVIII has been a subject of extensive engineering in the process of creating phage libraries.

The ends of the particle have different appearance in electron micrographs. The pointed end (also known as the adsorption, proximal, or infectious end) consists of about five copies each of pIII and pVI. The 406-residue pIII consists of three domains, N1, N2, and CT, separated by two glycine-rich linkers (Figure 9.3, Panels B and C). Domains N1 and N2 appear in electron micrographs as knobby structures emanating from the pointed end of the particle into the medium and are key for infectivity. The

crystal structure of a polypeptide comprising both domains has been determined (Figure 9.3, Panel C). The C-terminal 132 residues of the CT domain are proposed to interact with pVIII and form the proximal end of the particle; they are also thought to be buried. Most importantly, they are necessary and sufficient for pIII to be incorporated into the phage particle and for termination of assembly and release of phage from the cell.

The role of the 112 amino acid long protein pVI is not fully understood. It may interact with pVIII at the tip of the particle and is partially buried. The C-terminus of pVI has been used to display heterologous proteins [40], suggesting that it is accessible on the surface of the phage particle. For this reason, pVI appears particularly fit for the display of cDNA libraries [41].

The other phage particle end, also known as the distal or maturation end, appears blunt in electron micrographs. It has approximately five copies each of two small proteins pVII (33 amino acids in length) and pIX (32 amino acids in length). While protein interactions at this particle end are not fully understood,

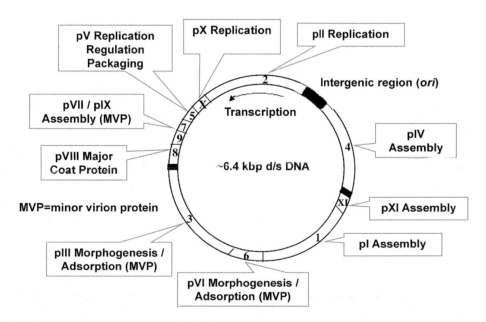

FIGURE 9.2 Phage M13 genome. The replicative form DNA (RF DNA) is depicted. The genes are designated by Arabic numerals, while the proteins they encode by Roman numerals. Proteins X and XI are not encoded by separate genes, but are derived from gene *2* and gene *1*, respectively, as discussed in the text. Modified from a GP Smith PowerPoint slide.

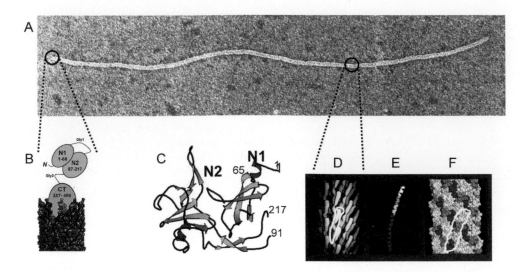

FIGURE 9.3 Panel A: Structure and electron micrograph of a filamentous ssDNA phage. The orientation that the DNA molecule is thought to have within the elongated filamentous particle is indicated by gene numbers; the morphogenetic signal sequence is believed to be in the intergenic (IG) region at the right, possibly bound to gene VII and IX proteins. The diagram also shows how the major coat protein is polymerized in helical array around the DNA. Panel B: Cartoon of the pIII structure, including domains N1, N2, and CT, and glycine-rich linkers Gly1 and Gly2. Numbers indicate amino residues comprising the domains. The pVIII virion body structure (purple) was made by PyMOL (PDB: 1IFJ) and the pIII-CT unit arbitrarily positioned—the real stoichiometry of pIII per phage is about 5. Panel C: Crystal structure of the N1 and N2 domains of pIII [84]. The image was generated using KiNG software (PDB: 1GP3). Numbering is according to the amino acid residues: 1-amino terminus and the site for display of peptides and proteins fused to pIII; 1–65–N1 domain; 91–217-N2 domain. The structure of the 66–90 region, a glycine linker, has not been determined. Panel D: Computer rendering of the electron density map of M13 based on x-ray diffraction studies [85]. The elongated rod-like structures are pVIII subunits forming the phage coat. The highlighted structure corresponds to a single pVIII subunit. Panel E: Model of the alpha-helical domain of pVIII [86] based on x-ray fiber diffraction and solid-state NMR data [87]. Four differently shaded areas are as follows (from top to bottom): (1) the foreign peptide fused to the amino terminal region of pVIII; (2) amino acids 6–11 of pVIII; (3) amino acids 12–19 of the engineered pVIII; and (4) buried residues of pVIII (amino acids 20–50). Panel F: Space-filling model of the phage. About 1% of the phage length is shown. The highlighted structure corresponds to a single pVIII subunit. ([A] Courtesy of Professor R. Webster. Source: GP Smith PowerPoint slide; [D] with permission, from Lee Makowski and the publisher, ©1992 Elsevier; [E] with permission, from Valery Petrenko and the publisher, © 2008 Elsevier; [F] with permission, from Valery Petrenko and the publisher, © 2008 Elsevier.)

it has been shown that the amino termini of both proteins can be used for heterologous protein display [42], which has been exploited for efficient peptide and antibody discovery [43–45].

M13 Life Cycle

In describing the M13 life cycle, we will focus on the aspects of direct significance for experimentation with phage display. More detailed descriptions are available[2].

To infect the bacterium, the bacteriophage uses the tip of the F conjugative pilus of *E. coli* as the receptor. The three bacterial proteins required during phage infection are TolQ, TolR, and TolA, which appear to form a complex anchored in the cytoplasmic membrane; the complex protrudes into the periplasm of *E. coli*. Infection starts with the binding of the N2 domain of the phage pIII protein to the tip of the F pilus, followed by a retraction of the pilus, which brings the adsorption end of the particle to the periplasm. As a result of this binding, the N1 domain of pIII is released from the N2 domain and allowed to interact with TolA. Thus, the tip of the pilus can be considered a receptor, and TolA, a coreceptor. The processes of phage particle disassembly and entrance of the single stranded phage DNA into the cytoplasm that follow are not well understood. Phage pVIII protein is reused in the making of the new phage, and DNA, which is a (+) strand (same orientation as phage mRNA), undergoes replication.

The (+) strand becomes a template for synthesis of the complementary (−) strand. The synthesis by bacterial enzymes and the introduction of negative supercoils by a gyrase produces a covalently closed, supercoiled double-stranded DNA called the parental replicative form DNA (RF DNA). This RF DNA serves a triple role: initially it is a template for transcription of mRNA and for synthesis of additional RF DNA, and after the accumulation of sufficient amounts of phage protein, it serves as a template for synthesis of the (+) strand to be packaged into new phage. Phage protein pII is necessary for nicking/closing of the DNA to be replicated, while the accumulation of pV results in formation of a complex between pV and (+) strands, thereby preventing the conversion of the (+) strand to RF DNA. The RF DNA/pV complex serves as a substrate for phage assembly. An additional protein, pX, created by an in-frame translational start within the p*II* mRNA and thus having a sequence identical to the 111 carboxy terminal residues of pII, helps to regulate the replication of phage DNA.

Phage assembly is a complex process requiring five structural proteins (pIII, pVIII, pVI, pVII, and pIX), three assembly proteins (pI, pIV, and pXI), and at least one bacterial protein, thioredoxin. Protein pXI is created by an in-frame translational start within the p*I* mRNA and thus has a sequence identical to the 108 carboxy terminal residues of pI. At the core of the assembly process is the removal of pV from the pV/RF DNA complex and the association of structural proteins around double-stranded phage DNA in the periplasm, where the inner and outer membranes are in close contact. At the final stages of phage assembly, the phage emerges from the bacterial cell without killing the cell.

Phage Proteins Used for Display

While all five phage coat proteins have been used for display of heterologous peptides or proteins, only two have been widely used, pIII and pVIII. As a result, there are a large amount of data on the design of display experiments and library construction using these proteins.

Protein pIII

Of the two proteins, pIII and pVIII, pIII is by far the most commonly used (alternatively referred to as 3 and 8 type display) (Figure 9.4). Regarding pIII display, the insert location of choice is the amino terminus—that is, the heterologous gene is cloned between the p*III* gene regions that correspond to the signal peptide of pIII and domain N1. Upon secretion of pIII into the periplasmic space of the cell, the signal peptide is removed, and the heterologous peptide ends up as a fusion to the amino terminus of the mature pIII protein.

There are several reasons for this cloning site preference. First, the replication process of the phage seems to tolerate a wide range of insert sizes. This can be rationalized by realizing that insertion places the heterologous peptide at the adsorption end of the phage particle, which creates less steric hindrance during phage maturation; pIII emerges last during the transport of

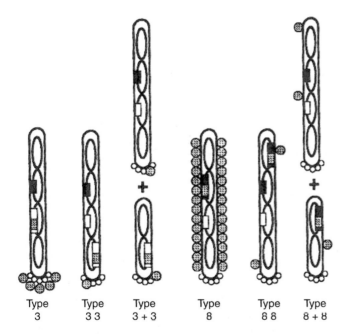

FIGURE 9.4 Types of phage vectors for display experiments. A Type 3 vector has a single recombinant gene *III* (open box) bearing a foreign DNA insert (hatched segment): all five copies of pIII are decorated with the foreign peptide encoded by the insert (hatched circles at the tip). The other phage genes, including *VIII* (black box), are normal. Type 33 vectors provide two genes *III*, one wild-type, the other an insert-bearing recombinant. The virions display a mixture of pIII molecules, only some of which are fused to the foreign peptide. Type 3 + 3 vectors resemble Type 33 in having two genes *III*; in this case, however, the recombinant gene is on a phagemid, a plasmid that contains, in addition to a plasmid replication origin and an antibiotic resistance gene, the filamentous phage "intergenic region." Type 8, 88 and 8 + 8 vectors are the gene-*VIII* counterparts of the gene-*III* vectors. (With permission from GP Smith [*Gene* 128, 1-2] and the publisher [© 1993 Elsevier].)

the particle from the bacterial cell into the medium. Proteins as large as alkaline phosphatase have been expressed as fusions to pIII [46]. On the other hand, many experiments with fusions of short peptides to pIII have also been quite successful.

Second, the structure of the amino-terminal half of pIII, comprising the N1 and N2 domains (Figure 9.3, Panels B and C) is highly conducive to phage display experimentation. Both domains, but particularly the amino terminus, are exposed on the surface of the phage particle. Thus, one can expect that the foreign peptide will be able to adopt its native fold and be exposed to the medium, which would allow for successful selection experiments.

Third, the purpose of many phage experiments is the selection of high-affinity binders to a biomolecule of choice. Under such conditions, a monovalent display is often desired, with expression on pIII being closer to this ideal (five copies) than a clearly polyvalent (up to 2,700 copies) display on pVIII. In real life, the valence of display is difficult to measure; it cannot be excluded that due to unavoidable proteolysis, only a small fraction of phage coat proteins, whether pIII or pVIII, display a foreign peptide. Thus, pIII display might be closer to monovalent than the number of pIII molecules in the virion might indicate. However, it is still most often alluded to as high valence display, to distinguish it from low valence display, which is achieved by phagemid display (discussed later in this chapter).

Alternative display locations in pIII, namely between the N1 and N2 domains and between the N1 and CT domains, have been also explored [47]; however, they have not been widely used.

Protein pVIII

Display experiments on pVIII target the amino terminus of pVIII. That is, the heterologous gene is inserted between the region corresponding to the signal peptide and the mature pVIII protein. Unlike pIII display, there is a significant length limit of six to eight amino acid residues on the size of the peptide displayed; this size limit seems to be connected to the interference of larger peptides with phage packaging and in particular with the transport of the phage particle through the pIV channel. Display of larger peptides fused to pVIII is only possible with hybrid phage display systems (see "Phagemid and Helper Phage," discussed later in the chapter).

Examples of Strategies to Alleviate pIII and pVIII Display Constraints

Even though a large variety of different proteins have been successfully displayed on phage [3], there are still many that are refractory to functional display due to their folding characteristics, a notion that is relevant both to peptide and domain display. Such constraints may be alleviated by manipulating the periplasmic chaperone environment [48, 49], altering the periplasmic targeting propensity [50], route [51, 52], or even by modifying the zwitterionic characteristics of the phage particle [53]. Capsid engineering has also been explored to improve functional display of heterologous fusion proteins. This involves trimming of pIII as in CT display [54], or removing the N1–N2 cysteines of the full length pIII [55]. With the exception of the latter approach, it is notable that all these avenues rely on hybrid display systems (see "Phagemid and Helper Phage," later in this chapter). An additional exception here is some types of pVIII display where modifying the solvent exposed *N*-termini may increase display levels [56].

Nomenclature of Phage Display Experiments

As alluded to above, it is desirable in many phage display experiments to reduce the number of foreign peptides displayed on the phage, relative to the native number of pIII or pVIII proteins (five or 2,700 each, respectively). This is done by introducing into the *E. coli* cell additional copies of the wild-type pIII or pVIII genes; the gene can reside on the same phage vector that harbors the heterologous gene, in which case the system is called 33 or 88, respectively. Alternatively, the gene can be harbored on another vector, and such systems are called 3+3 or 8+8. The single-gene systems are known as type 3 (pIII only) or type 8 (pVIII only). The systems are depicted in Figure 9.4. Because the expression level of the wild-type genes can be controlled, the ratio of the heterologous to wild-type coat protein can be adjusted as needed.

Protein pVII and pIX

The use of pVII and pIX have emerged in recent years [7]. Both peptides and folded-protein domains have been successfully displayed and applied in selection experiments, and in both 7 and 9 as well as 7 + 7 and 9 + 9 versions. Results equal to or better than those obtained with classical pIII and pVIII have been achieved, and both pVII and pIX are now attractive fusion partners (insert PMID reference 27766617). They have some distinct advantages compared to pIII (see "Common difficulties" below).

Phagemid and Helper Phage

A popular implementation of the hybrid system involves the use of a phagemid, which is a cloning vector based on a plasmid (such as pBR322) and containing the replication origin of a filamentous phage as well as either gene *III* or gene *VIII* modified to accept gene inserts for display. The phagemid can be readily propagated in *E. coli* leading to the accumulation of the pIII or pVIII fusion protein. To allow formation of viable phage, a coinfection with so-called M13 helper phage is required. Compared with the wild-type phage, the helper phage has a compromised origin of replication and is unable to be propagated efficiently under normal conditions. However, an *E. coli* cell harboring the phagemid and superinfected with helper phage will produce large amounts of a smaller version of the phage carrying phagemid DNA and displaying the heterologous proteins. Such particles are then used in, for example, affinity-selection experiments.

An often-used version of the M13 helper phage is the interference resistant M13K07 [18] available from New England Biolabs. It is an M13 derivative that carries the mutation Met40Ile in gII, with the origin of replication from P15A and the kanamycin resistance gene from Tn903 both inserted within the M13 origin of replication [18]. Other types of M13 helper phage are R408 and VCSM13 (Stratagene).

One may opt to alternate between different helpers throughout a selection campaign, which is usually based on multiple rounds of target selection and amplification of eluted phages (see discussed later in this chapter). For instance, it can be very useful to alternate between low valence and high valence display on pIII or pIX. This has been made possible by the use of special helper phages that lack functional expression of relevant capsids and thus mimicking the pure 3 and 9 genomic systems [57, 58].

Considerations for Successful Phage-Display Experiments

While many applications of phage display have been reported, most phage-display experiments fall into the following three categories based on the type of protein immobilized on the phage:

1. Preparation and screening of a random peptide library, with the goal of identifying a peptide capable of binding to a given protein under study
2. Cloning of a naturally occurring repertoire of gene segments based on a distinct protein scaffold, such as antibody variable genes, which have been diversified by endogenous mechanisms of the chosen donor, followed by selection, with the goal of finding a variant of the protein possessing a desired property, typically a binding specificity to another molecule
3. Cloning of a protein, followed by randomization and selection, with the goal of finding a variant (mutant) of the protein possessing a desired property, typically a binding specificity to another molecule

Peptide Library

The first question to ask is whether an existing peptide library will be adequate for the particular experiment. Throughout the years, a comprehensive list of random peptide libraries have been reported [31]. In addition, New England Biolabs (www.neb.com) supplies the Ph.D. phage display system and several types of libraries, in which 7 or 12 amino acid random or disulfide-linked peptides are displayed on the phage. The diversity of these libraries is in the range of $1–3 \times 10^9$ different peptides.

There is a focus now on how to optimize the library generation protocols to improve the speed of the process as well as to increase library diversity and functionality. Thus, using current protocols one can manufacture peptide libraries in a swift and streamlined manner that reaches end diversities of more than 10^{10} to 10^{11} different peptides [59–61].

If the decision is made to construct a new peptide library by cloning a random or partially random nucleic acid sequence within the pIII or pVIII gene, the desired length of the peptide needs to be determined as well as the choice of cloning vector, cloning site, and signal sequence. The majority of peptide libraries constructed are of type 3 (Figure 9.4) and the gene insert is placed downstream of the signal sequence in front of gene III, often in a derivative of phage M13; this is a recommended (default) approach. It is more difficult to recommend the peptide length because, while longer peptides provide a higher level of sequence heterogeneity (proportional to the peptide length) and more possibilities for formation of secondary/tertiary protein structures, these longer peptides might be more rapidly degraded proteolytically and the library more difficult to handle. Interpretation of the results might be more difficult as well. However, recent improvements in library generation protocols and the very recent introduction of next generation sequencing as a phage selection screening tool has greatly improved the ability to delineate also highly diverse and complex selection protocols.

Display of Proteins

Many different proteins have been displayed on the phage [3, 31], ranging from enzymes (alkaline phosphatase, trypsin, lysozyme, β-lactamase, glutathione transferase), hormones (human growth hormone, angiotensin), inhibitors (bovine pancreatic trypsin inhibitor, cystatin, tendamistan), toxins (ricin B chain), receptors (α subunit of IgE receptor, T cell receptor, B domain of protein A), ligands (substance P, neurokinin A), and cytokines (IL3, IL-4), two different forms of antibodies (scFv, Fab), and T cell receptors (scTCR and dsTCR). The size of the protein, up to around 60,000 daltons per subunit (e.g., alkaline phosphatase), does not seem to be of major concern (except on pVIII), nor is the existence of a protein in a dimeric form, since a dimeric protein as large as alkaline phosphatase has been successfully displayed on the phage [46]. Similarly, a successful phage display experiment does not depend on whether or not the protein is secreted in its native form, since many intracellular proteins have been displayed on the phage (glutathione S-transferase, cytochrome b_{562}, FK506 binding protein, and others). The typical approach involves use of the 3 + 3 system (with an alternative being the single vector 3 system), with the gene cloned for expression on the amino terminus of pIII.

Antibody Libraries

Special attention needs to be given to the display of antibodies on filamentous phage due to its popularity and wide number of possible applications on the one hand, but common difficulties encountered with such libraries on the other. Antibodies are disulfide-linked heterodimers composed of light and heavy chains, with their total molecular weight often exceeding 150,000 Da. The complexity of the antibody molecule in its native state renders it prohibitively difficult to display on the phage, which is why approaches have been developed to display a truncated single-chain version of the antibody molecule. The single chain Fv (scFv) antibody consists of the variable regions of the heavy and light chains, comprising the full Fv antigen binding site of the parent antibody, connected with a short and flexible peptide linker (Figure 9.5, Panel A). Thus, the antigen-binding region of the antibody is preserved; however, functionalities associated with the remainder of both heavy and light chains, such as

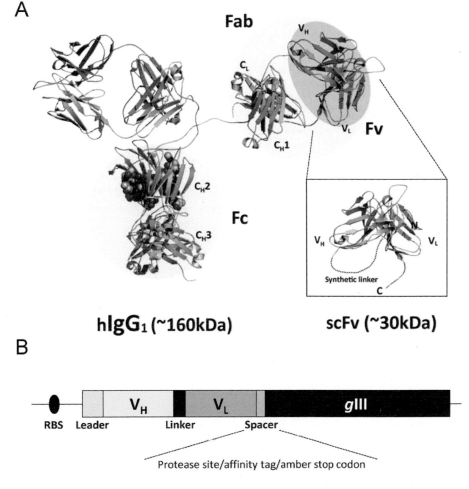

FIGURE 9.5   Antibody structure and recombinant forms used for phage display. Panel A: The single chain Fv (scFv) contains the antigen binding V_H and V_L domains from, for example, immunoglobulin G (IgG) and can be displayed on phage. In the case of the Fab and the diabody, one of the chains is secreted as a pIII fusion and the other is secreted as native nonfusion protein. The figure was prepared using PyMOL based on PDB: 1HZH. Panel B: Schematic illustration of a typical scFv-pIII fusion protein expression cassette used for antibody phage display. The scFv is directly fused to pIII by use of a spacer sequence typically containing tags for detection and purification of the soluble version. The latter can easily be made, for example, by incorporating an amber stop codon in the spacer, which comes into effect dependent on the *E. coli* host used for expression.

binding to various cell receptors and complement proteins, are removed upon the conversion to the single-chain form. Another popular format is to display the full fragment antigen binding (Fab). Throughout the years, several engineered antibody formats have been developed [62], many of which have been displayed [63], significantly contributing to the strong position of phage display as a combinatorial platform [9].

The construction of scFv libraries, which is still the most common type, typically involves the joining of PCR amplified gene fragments corresponding to the F_v regions of the heavy and light chains through a glycine-rich linker, placing the signal sequence in front of the construct and followed by cloning between a suitable ribosome binding site/signal sequence and the 5'-terminal region of pIII (Figure 9.5, Panel B). Often, both mAb-tag, histag, and an amber stop codon are introduced into the construct as well.

Construction of the Phage Library

After a phage vector system is chosen for library construction, a cloning strategy needs to be identified. A typical approach for inserting a peptide-encoding randomized sequence into a vector relies on the cloning of synthetic DNA. An example is presented in Figure 9.6, and the experiment is described in detail by Ravera et al. [64].

The pCANTAB5E phage vector (GE Healthcare) was chosen for cloning, and its two unique restriction sites, *Sfi*I and *Not*I, were selected as the insertion sites for randomized DNA. As a result, the randomized peptide was expressed fused to the pIII protein and had on its amino terminus the FLAG peptide. The random insert sequence was of the type $(NNK)_m$ where N = A, C, G, or T; K = T or G; and m is the number of codons in the random sequence. In the example quoted, m = 40; however, m can vary greatly depending on the experimental design but it typically ranges from 8 to 30 codons. The purpose of the NNK codon design is to avoid two out of three stop codons. The third stop codon (TAG) can be suppressed if the phage is prepared in an *E. coli* suppressor strain, such as TG1, [F' *traD*36 *lacI*q *lac*ZM15 *proA+B+/supE* Δ(*hsdM-mcrB*)5(rk⁻mk⁻ *mcrB*⁻) *thi*, Δ(*lac-proAB*)], leading to incorporation of glutamine into the polypeptide chain in response to TAG codons. Despite this limitation, which effectively can be alleviated by the use of oligos based on defined triblock codon synthesis [65], all amino acid residues could thus be incorporated into the polypeptide due to redundancy in the genetic code. In this example, the random sequence resided on a synthetic oligonucleotide, 145 nt long, which was converted to double-stranded DNA in a PCR reaction, digested with *Sfi*I and *Not*I, ligated into the pCANTAB5E vector, and electroporated into *E. coli*.

There are many methods to clone DNA into phage vectors. Other than the PCR-based cloning outlined previously, approaches often used to clone randomized cDNAs that encode larger proteins include Kunkel mutagenesis [35, 60], a variety of oligonucleotide-based mutagenesis methods, random mutagenesis, combinatorial infection and recombination, and DNA shuffling [30].

Library Diversity

The potential library diversity D goes up very rapidly with the number of randomized positions m, according to the formula $D = 20^m$ (assuming all 20 amino acids can be incorporated). For m = 8, the potential diversity exceeds 2×10^{10}, which is already larger than the number of independent clones in a typical phage library, and if based on NNK randomization, this number increased to above 10^{12} due to the degenerate use of 32 codons. This means that libraries in which more than 8 positions are fully randomized will actually not contain a representation of all possible sequences—a reflection of the enormous diversity of protein sequences. It is reassuring for those who work with phage peptide libraries that binders to many (if not most) proteins are obtainable from such libraries despite a striking contrast between the number of clones in the phage library and the number of allowable sequences. The selection of the binder can be represented as a search for function in a multidimensional protein sequence space [31, 66].

Achieving a library diversity of more than 10^{10} independent clones has traditionally not been easy. While electroporation efficiency can be more than 10^{10} clones per 1 mL of electrocompetent cells and the amount of DNA is typically not an issue, one needs to remember that the electroporation efficiency quoted applies to ideal conditions (i.e., native double-stranded supercoiled DNA), while the *in vitro* constructed DNAs typically transform with significantly lower efficiency. In the example quoted previously [64], the peptide library of 1.5×10^{10} clones was constructed by combining five sublibraries, each of which was obtained from about 100 electroporations. As of today, it is only the improved Kunkel-based protocols that have been able to alleviate this bottle neck, by using the improved preparation of high-density ultracompetent SS320 *E. coli* [60, 61], which now may be purchased from, for example, Lucigen (www.lucigen.com).

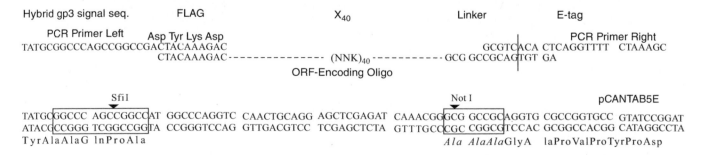

FIGURE 9.6 Cloning of a DNA insert carrying a randomized sequence into the pCANTAB5E phagemid. (See Figure 1, p. 1994 in Ref. [64].)

Selection (Biopanning) Experiment

The principle of phage selection (Figure 9.7) relies on exposing the target proteins immobilized on solid phase to the phage library, allowing peptides carried on phage to bind specifically to the target proteins and be retained on solid phase, and removing of the excess phage by washing. Since this biopanning step will always result in an undesired retention of nonspecifically bound phage, the step is repeated with the eluted and amplified phage. The total number of biopanning steps can vary from two to four or more, depending on the experimental strategy and conditions. The final product of the phage enrichment procedure is a group of phage clones that are considered potential binders to the target protein. Typically, the target-binding properties of the phage clone are confirmed in a phage ELISA assay, and the phage DNA is subjected to DNA sequencing, which identifies peptide sequences capable of binding to the target of interest. The biopanning cycle is schematically represented in Figure 9.8, and a comprehensive selection of biopanning protocols is presented in Refs. [29] and [28].

A great deal of variation is allowed in biopanning procedures ranging from solid phase inorganic materials to selection on live cells [10]. Classically, different solid phase materials have been used for target immobilization, such as polystyrene, nitrocellulose, or streptavidin-coats on beads or plates. The binding

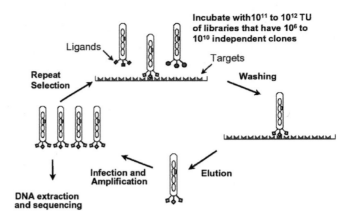

FIGURE 9.8 A schematic for the biopanning cycle.

reaction can take place in solution; in this case, the target protein is biotinylated and the phage-target complex is harvested by binding it to, for example, streptavidin-coated microtiter plates [67]. The biopanning stringency can be varied as well by adjusting the binding/elution buffer composition, time, temperature, etc. A general rule for protein targets known to bind to short linear epitopes (e.g., antibodies recognizing such epitopes) is to use more stringent conditions and fewer biopanning rounds, while target proteins that bind to more complex structures often require an optimization of the biopanning approach.

The analysis of phage clones after biopanning may take several case-specific directions, but traditionally begins with a phage capture ELISA. The prototypic ELISA screen involves immobilizing the target and relevant controls as coat, followed by phage clone capture, where the same phage samples are tested in parallel. Successful phage capture is then detected by anti-phage antibodies directed toward the pVIII, which offers very high sensitivity to the assay as this capsid is found in thousands of copies per phage.

In the next analysis step, several (often as many as 100 or more) ELISA-positive clones are subjected to DNA sequencing over the randomized region of the heterologous gene or cloned insert. The obtained DNA sequences are translated and the amino acid sequence of the peptide or the randomized region is defined. Often, a short linear region of homology can be readily spotted amongst the sequences and a consensus sequence can be derived.

Typical results from a simple biopanning experiment [68] are presented in Figure 9.9. In Panel A, biopanning of a peptide library against an HCV monoclonal antibody (mAb) led to a simple linear sequence of six amino acid residues, RGRRQP, as a consensus. In Panel B, with the target being an anti-HBsAg mAb, the consensus is still a short linear sequence, C*TC, where * is any amino acid residue. However, the presence of two cysteine residues strongly suggests that the binding form of the peptide adopts a looped conformation that is anchored by a disulfide bond formed by two cysteine residues. Nonetheless, the confirmation of such a secondary structure as a binding motif requires additional experimentation.

In view of the phage display complexities that can develop due to the high number of different molecules (over 10^{10}) involved and largely unknown binding properties, it may be of help to seek guidance by theoretical models and simulations of phage

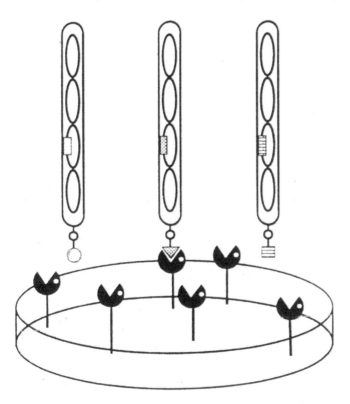

FIGURE 9.7 Phage display principle. Schematic diagram of a display library being "panned" on a petri dish to which an open-jawed receptor has been tethered. Phage displaying peptides that bind the receptor are captured on the dish, while the other phage are washed away; the captured phage are eluted, cloned, and propagated by infecting fresh host cells, and the relevant part of their DNA sequenced to identify the peptide responsible for specific binding. (With permission from GP Smith [*Gene* 128, 1-2] and the publisher [© 1993 Elsevier].)

A

```
                    SLIPRGARTPRQCRSACPPREYSSADHWKS
          AGMWRVPRAENYSAPVRTRPKRQPWAQGSY
       LSADTLRSNSVDHDRVQNEVRSSRERRQPR
       LLEVGRDWVGGNNMWVGRMRERRKNERRQP
                              IWDRRRQRQPGRVENYPVGKPASHQLYILN
                  MKLPVDENKSGRRERRQPTPAGERELIRFD
        YLPYRAGVDKSGGREVGSHMRFFTRERRHA
          AELTGYEDVRRVEKRKGLAGTRERRQPAAY
                    KMLRERRQPSSNLDYEKEVGQFYVVVAKSD
           SYGSGAARVSKLQETGGRRGRRQPSGSYIG
                    FPARSEASGRQRRQPGRDVTHGEAVRVNIL
      IGPRRKWDALASDSGCNPSSHSQRCHRLKP
      PCQNTYRGLMLNEDCRQGRRTRRQPPATTL
      IGQGQQKEALGSRQRFDLRGRRQPVGSGKW
      RSGGVSELRAEGVNRRQARRKEQRRQPPRY
                 RGSYDRRRERRQPRGLR
           QAVVSGERGARPRRQPRTPGVAACARSAGG
      QGYVDAGSISRFGGRGVWRQRRQPLLNGSF
             KIVRLTDAGHRTRRQPRTEEMWKVSTWLLN
      QVLKGGVSKRTMKGRACRQQACPKTVPSVV
      KELHTVEANERLKLREGRLRDRRQPISQWN

   ILLPRRGPRLGVRATRKTSERSQPRGRRQPIPKARRPEGRTWAQPG
```

HBsAg sequence (aa 35-80)

B

MQRKESNPNLG	CVTC	GFRVRQTVGESDGGS	
VNWRGPSATLEGTNSNTGRRGQAVA	CRTC	F	4x
RYSGVVGNAVSEGERLNGLSSS	CVTC	LGWR	
	CRTC	GEVGLMTRPGVRMNA	
GAGQVERLREAKDP	CRTC	GGSRWRGEPFWM	
DERQIQRQEPMVRNSERDAMR	CRTC	AFKEL	2x
TITNSASGLHF	CKTC	WKNSGGPAVAGKQDE	2x
SLDRWPEHLATMGNRLGMTRQ	CKTC	VGSTL	2x
VQWRWNDTEWMR	CKTC	MLSE	
RHPHKRVRQYDGMRGAGGDWS	CKTC	LRPGY	
PKRQISMERWLQVTQGEEVTP	CATC	NPWVA	
L	CKTC	VRSHQERTVKGDQVTGTRICQTC	
WDKRPVVWLRFEESQRLSR	CATC	GVGGVE	
NVNEPGIRQGPAASVGWKVVRLAGI	CKTC	V	
GMKIVVFPKRSVPDVTGSQGAPP	CRTC	TST	
SVLRQAAQFGNFELYVRREGN	CRALTGC	MR	

H166 epitope: C+TC   all
                      | |
("+" is R or K) 121 124

FIGURE 9.9 Examples of biopanning results: Consensus sequences obtained from biopanning against two mAbs, against the HCV core protein (Panel A) and HBV surface antigen (Panel B). ([A] From Figure 5, p. 192 in Ref. [88]; [B] from Figure 1A, p. 1999 in Ref. [68].)

display [19, 20, 22]. Complexities arise because randomized proteins displayed on the phage in the library have vastly differing binding affinities to the target. Depending on the target, the binding clones may or may not be abundant, and depending on the biopanning procedure, particularly its stringency and number of washes, different groups of clones can be retained or lost at varying rates, which can lead to uncertainties in obtaining a desired type of binder. Theoretical analyses as in Ref. [19] provide insight regarding the number of biopanning rounds and washes per round needed while also helping to track the quantitative increase of the number of desired binding clones. Models can help one to understand how the affinity distributions of phage populations change during panning. Though such initial considerations are recommended during planning of a selection campaign, it will nonetheless often be necessary to empirically adjust many of the parameters in the lab. This may range from specific blocking and elution condition to the choice of target immobilization strategy, which have major impact on the number and type of binders retrieved during the actual selection [69, 70].

phagemid to a nonsuppressor strain of *E. coli* can allow for expression of the protein as an entity not fused to pIII. Moreover, the vector may provide the his-tag or an antibody-tag, which is retained on the expressed protein in the nonsuppressor strain; such a tag may facilitate protein purification and subsequent characterization by ELISA.

Second, the gene encoding the protein can be recloned into an expression system. Many expression systems are available, and a good choice to start with is the system typically used to make the native protein in quantity.

Third, for random peptide libraries on phage, peptide synthesis is a very good method to obtain the peptide in quantity (see, e.g., Ref. [71]). Alternatively, the short genes encoding those peptides can be fused to a well-characterized affinity protein, such as maltose binding protein (MBP), and studies of the fusion protein can confirm the properties of the peptide from phage display. Alternatively, the candidate peptides may be indirectly coupled to carriers that allow for further characterization of the complexes by immobilizing biotinylated versions of the peptides to streptavidin [72, 73].

Expression of Identified Protein or Synthesis of Peptide

Obtaining consensus sequences can sometimes provide an important clue as to the significance of a particular region within a protein sequence, particularly when there is a match between the two sequences. But more often than not, further characterization of the binding clones is needed, and toward that goal, the protein displayed on the phage needs to be made in appropriate quantity and then purified and analyzed. Several approaches to this are possible.

First, certain phagemid vectors, such as those presented in Figure 9.5, have an amber stop codon (TAG) between the gene encoding the displayed protein and pIII. A transfer of the

Common Difficulties and the Source of Repertoire Bias

While phage display is a powerful tool to select for a peptide or protein that binds specifically to a ligand, certain problems might arise due to the complexity of the approach. First, the custom library quality is difficult to judge, and in particular, the number of independent clones in the library and the valency of display (the actual number of displayed peptides/proteins). Second, certain clones could be under- or over-represented due to expression inefficiencies or toxicity. Third, procedural difficulties with selection (often, over-selection) may lead to elimination of clones otherwise expected to be identified.

In recent years, the ability to tap into a more representative portion of the enormous diversity found in phage display libraries has been made possible due to the introduction of next generation sequencing. Several studies have illuminated the intrinsic bias introduced during selection, but also identified experimental means to counteract the effects [26, 27, 74–76]. Combined with tools, such as the SAROTUP database (http://i.uestc.edu.cn/sarotup3), researcher today has powerful tools to successfully identify truly target-specific clones by selection [23].

However, in some circumstances, the sequence encoding the displayed peptide cannot be readily identified by sequencing. In our hands, screening a phage display random peptide library for binders to specific targets resulted in some instances with as many as 50% of all selected clones not having an obvious open reading frame, nevertheless the peptide had to have been synthesized or the phage would not have bound to the target ligand [71].

In a series of experiments, we pursued the explanation for this anomaly for one specific clone and its derivatives by fusing the peptide-encoding sequence to a β-galactosidase reporter and expressing it in *E. coli* [77–79]. Ultimately, our experiments indicated that an unexpected translational reinitiation phenomenon was responsible for the anomalous expression [79], however, other unexpected translational events are also possible in such cases [80].

Conclusions

Phage display provides a wealth of approaches to identify protein ligands. In skilled hands, the procedures are quite rapid (a few days), and the results provide a wealth of information about the protein ligands and the target protein itself, as well as their interactions [81]. Despite the emergence in recent years of alternatives to filamentous phage display, such as aptamers, ribosome display, yeast two-hybrid system, bacterial display, other phage systems (T4, T7 and lambda) (see [35] for references), the filamentous phage display nonetheless offers a very reliable set of tools and tested procedures and has been improved over many years by a large number of investigators. Undoubtedly, the merger of phage display library screening by deep sequencing will further fuel this development [82], as may the possibility of alternative display avenues such as offered by pVII and pIX display [7]. Thus, phage display remains an excellent starting point for the identification and activation of protein ligands.

REFERENCES

1. Smith GP: Filamentous fusion phage: Novel expression vectors that display cloned antigens on the virion surface. *Science* 1985, 228(4705):1315–1317.
2. Rakonjac J, Bennett NJ, Spagnuolo J, Gagic D, Russel M: Filamentous bacteriophage: Biology, phage display and nanotechnology applications. *Curr Issues Mol Biol* 2011, 13(2):51–76.
3. Hosse RJ, Rothe A, Power BE: A new generation of protein display scaffolds for molecular recognition. *Protein Sci* 2006, 15(1):14–27.
4. Rothe A, Hosse RJ, Power BE: In vitro display technologies reveal novel biopharmaceutics. *FASEB J* 2006, 20(10):1599–1610.
5. Scott JK, Smith GP: Searching for peptide ligands with an epitope library. *Science* 1990, 249(4967):386–390.
6. Ullman CG, Frigotto L, Cooley RN: In vitro methods for peptide display and their applications. *Brief Funct Genomics* 2011, 10(3):125–134.
7. Løset GÅ, Sandlie I: Next generation phage display by use of pVII and pIX as display scaffolds. *Methods* 2012, 58(1):40–46.
8. Taussig MJ, Stoevesandt O, Borrebaeck CA, Bradbury AR, Cahill D, Cambillau C, de Daruvar A, Dubel S, Eichler J, Frank R *et al*: ProteomeBinders: Planning a European resource of affinity reagents for analysis of the human proteome. *Nature Methods* 2007, 4(1):13–17.
9. Bradbury AR, Sidhu S, Dubel S, McCafferty J: Beyond natural antibodies: The power of in vitro display technologies. *Nat Biotech* 2011, 29(3):245–254.
10. Bratkovic T: Progress in phage display: Evolution of the technique and its applications. *Cell Mol Life Sci* 2010, 67:749–767.
11. Ng S, Jafari MR, Derda R: Bacteriophages and viruses as a support for organic synthesis and combinatorial chemistry. *ACS Chem Biol* 2012, 7(1):123–138.
12. McCafferty J, Griffiths AD, Winter G, Chiswell DJ: Phage antibodies: Filamentous phage displaying antibody variable domains. *Nature* 1990, 348(6301):552–554.
13. Kang AS, Barbas CF, Janda KD, Benkovic SJ, Lerner RA: Linkage of recognition and replication functions by assembling combinatorial antibody Fab libraries along phage surfaces. *PNAS* 1991, 88(10):4363–4366.
14. Barbas CF, 3rd, Kang AS, Lerner RA, Benkovic SJ: Assembly of combinatorial antibody libraries on phage surfaces: The gene III site. *Proc Nat Acad Sci U S A* 1991, 88(18):7978–7982.
15. Ponsel D, Neugebauer J, Ladetzki-Baehs K, Tissot K: High affinity, developability and functional size: The holy grail of combinatorial antibody library generation. *Molecules* 2011, 16(5):3675–3700.
16. Bass S, Greene R, Wells JA: Hormone phage: An enrichment method for variant proteins with altered binding properties. *Proteins* 1990, 8(4):309–314.
17. Levinson A, Silver D, Seed B: Minimal size plasmids containing an M13 origin for production of single-strand transducing particles. *J Mol Appl Genet* 1984, 2(6):507–517.
18. Vieira J, Messing J: Production of single-stranded plasmid DNA. *Methods Enzymol* 1987, 153:3–11.
19. Mandecki W, Chen YC, Grihalde N: A mathematical model for biopanning (affinity selection) using peptide libraries on filamentous phage. *J Theo Biol* 1995, 176(4):523–530.
20. Levitan B: Stochastic modeling and optimization of phage display. *J Mol Biol* 1998, 277(4):893–916.
21. Zhuang G, Katakura Y, Furuta T, Omasa T, Kishimoto M, Suga K: A kinetic model for a biopanning process considering antigen desorption and effective antigen concentration on a solid phase. *J Biosci Bioeng* 2001, 91(5):474–481.
22. Zahnd C, Sarkar CA, Pluckthun A: Computational analysis of off-rate selection experiments to optimize affinity maturation by directed evolution. *Protein Eng Des Sel* 2010, 23(4):175–184.
23. Huang J, Ru B, Li S, Lin H, Guo FB: SAROTUP: Scanner and reporter of target-unrelated peptides. *J Biomed Biotech* 2010, 2010:101932.
24. Kim T, Tyndel MS, Huang H, Sidhu SS, Bader GD, Gfeller D, Kim PM: MUSI: An integrated system for identifying multiple specificity from very large peptide or nucleic acid data sets. *Nucleic Acids Res* 2012, 40(6):e47.

25. Ru B, t Hoen PA, Nie F, Lin H, Guo FB, Huang J: PhD7Faster: Predicting clones propagating faster from the Ph.D.-7 phage display peptide library. *J Bioinform Comput Biol* 2014, 12(1):1450005.

26. Matochko WL, Cory Li S, Tang SKY, Derda R: Prospective identification of parasitic sequences in phage display screens. *Nucleic Acids Res* 2013, 42(3):1784–1979.

27. Matochko WL, Derda R: Error analysis of deep sequencing of phage libraries: Peptides censored in sequencing. *Comput Math Methods Med* 2013, 2013:491612.

28. Clackson T, Lowman HB: *Phage Display: A Practical Approach*: OUP, Oxford; 2004.

29. Barbas CF, Burton DR, Scott JK, Silverman GJ: *Phage Display: A Laboratory Manual*: Cold Spring Harbor Laboratory Press, Cold Spring Harbor, New York; 2004.

30. Sidhu SS, Geyer CR: Phage display, *in Biotechnology and Drug Discovery*: Taylor & Francis, Boca Raton, FL; 2005.

31. Smith GP, Petrenko VA: Phage display. *Chem Rev* 1997, 97(2):391–410.

32. Hoess RH: Protein design and phage display. *Chem Rev* 2001, 101(10):3205–3218.

33. Hoogenboom HR: Overview of antibody phage-display technology and its applications. *Methods Mol Biol* 2002, 178:1–37.

34. Bradbury AR, Marks JD: Antibodies from phage antibody libraries. *J Immunol Methods* 2004, 290(1–2):29–49.

35. Kehoe JW, Kay BK: Filamentous phage display in the new millennium. *Chem Rev* 2005, 105(11):4056–4072.

36. Marvin DA, Pigram WJ, Wiseman RL, Wachtel EJ, Marvin FJ: Filamentous bacterial viruses: XII. Molecular architecture of the class I (fd, If1, IKe) virion. *J Mol Biol* 1974, 88(3):581–598.

37. Marvin DA, Hale RD, Nave C, Helmer-Citterich M: Molecular models and structural comparisons of native and mutant class I filamentous bacteriophages Ff (fd, f1, M13), If1 and IKe. *J Mol Biol* 1994, 235(1):260–286.

38. Hill DF, Petersen GB: Nucleotide sequence of bacteriophage f1 DNA. *J Virol* 1982, 44(1):32–46.

39. Marzari R, Sblattero D, Righi M, Bradbury A: Extending filamentous phage host range by the grafting of a heterologous receptor binding domain. *Gene* 1997, 185(1):27–33.

40. Fransen M, Van Veldhoven PP, Subramani S: Identification of peroxisomal proteins by using M13 phage protein VI phage display: Molecular evidence that mammalian peroxisomes contain a 2,4-dienoyl-CoA reductase. *Biochem J* 1999, 340 (Pt 2):561–568.

41. Hufton SE, Moerkerk PT, Meulemans EV, de Bruine A, Arends JW, Hoogenboom HR: Phage display of cDNA repertoires: The pVI display system and its applications for the selection of immunogenic ligands. *J Immunol Methods* 1999, 231(1–2):39–51.

42. Gao C, Mao S, Chih-Hung L. Lo, Wirsching P, Lerner RA, Janda KD: Making artificial antibodies: A format for phage display of combinatorial heterodimeric arrays. *PNAS* 1999, 96(11):6025–6030.

43. Gao C, Mao S, Ditzel HJ, Farnaes L, Wirsching P, Lerner RA, Janda KD: A cell-penetrating peptide from a novel pVII-pIX phage-displayed random peptide library. *Bioorganic Med Chem* 2002, 10(12):4057–4065.

44. Gao C, Mao S, Kaufmann G, Wirsching P, Lerner RA, Janda KD: A method for the generation of combinatorial antibody libraries using pIX phage display. *PNAS* 2002, 99(20):12612–12616.

45. Shi L, Wheeler JC, Sweet RW, Lu J, Luo J, Tornetta M, Whitaker B, Reddy R, Brittingham R, Borozdina L *et al*: De novo selection of high-affinity antibodies from synthetic fab libraries displayed on phage as pIX fusion proteins. *J Mol Biol* 2010, 397(2):385–396.

46. McCafferty J, Jackson RH, Chiswell DJ: Phage-enzymes: Expression and affinity chromatography of functional alkaline phosphatase on the surface of bacteriophage. *Protein Eng* 1991, 4(8):955–961.

47. Krebber C, Spada S, Desplancq D, Pluckthun A: Co-selection of cognate antibody-antigen pairs by selectively-infective phages. *FEBS Lett* 1995, 377(2):227–231.

48. Bothmann H, Pluckthun A: Selection for a periplasmic factor improving phage display and functional periplasmic expression. *Nat Biotech* 1998, 16(4):376–380.

49. Bothmann H, Pluckthun A: The periplasmic Escherichia coli peptidylprolyl cis, trans-isomerase FkpA. I. Increased functional expression of antibody fragments with and without cis-prolines. *J Bio Chem* 2000, 275(22):17100–17105.

50. Jestin JL, Volioti G, Winter G: Improving the display of proteins on filamentous phage. *Res Microbiol* 2001, 152(2):187–191.

51. Thammawong P, Kasinrerk W, Turner RJ, Tayapiwatana C: Twin-arginine signal peptide attributes effective display of CD147 to filamentous phage. *App Microbiol Biotech* 2006, 69(6):697–703.

52. Steiner D, Forrer P, Stumpp MT, Pluckthun A: Signal sequences directing cotranslational translocation expand the range of proteins amenable to phage display. *Nat Biotech* 2006, 24(7):823–831.

53. Vithayathil R, Hooy RM, Cocco MJ, Weiss GA: The scope of phage display for membrane proteins. *J Mol Biol* 2011, 414(4):499–510.

54. Orum H, Andersen PS, Oster A, Johansen LK, Riise E, Bjornvad M, Svendsen I, Engberg J: Efficient method for constructing comprehensive murine Fab antibody libraries displayed on phage. *Nucleic Acids Res* 1993, 21(19):4491–4498.

55. Heinis C, Rutherford T, Freund S, Winter G: Phage-encoded combinatorial chemical libraries based on bicyclic peptides. *Nat Chem Biol* 2009, 5(7):502–507.

56. Kuzmicheva GA, Jayanna PK, Eroshkin AM, Grishina MA, Pereyaslavskaya ES, Potemkin VA, Petrenko VA: Mutations in fd phage major coat protein modulate affinity of the displayed peptide. *Protein Eng Des Sel* 2009, 22(10):631–639.

57. Soltes G, Hust M, Ng KKY, Bansal A, Field J, Stewart DIH, Dubel S, Cha S, Wiersma EJ: On the influence of vector design on antibody phage display. *J Biotechnol* 2007, 127(4):626–637.

58. Nilssen NR, Frigstad T, Pollmann S, Roos N, Bogen B, Sandlie I, Løset GÅ: DeltaPhage—a novel helper phage for high-valence pIX phagemid display. *Nucleic Acids Res* 2012, 40(16):e120.

59. Scholle MD, Kehoe JW, Kay BK: Efficient construction of a large collection of phage-displayed combinatorial peptide libraries. *Comb Chem High Throughput Screen* 2005, 8(6):545–551.

60. Tonikian R, Zhang Y, Boone C, Sidhu SS: Identifying specificity profiles for peptide recognition modules from phage-displayed peptide libraries. *Nat Protoc* 2007, 2(6):1368–1386.

61. Huang R, Fang P, Kay BK: Improvements to the Kunkel mutagenesis protocol for constructing primary and secondary phage-display libraries. *Methods* 2012, 58(1):10–17.

62. Holliger P, Hudson PJ: Engineered antibody fragments and the rise of single domains. *Nat Biotech* 2005, 23(9):1126–1136.

63. Hoogenboom HR: Selecting and screening recombinant antibody libraries. *Nat Biotech* 2005, 23(9):1105–1116.

64. Ravera MW, Carcamo J, Brissette R, Alam-Moghe A, Dedova O, Cheng W, Hsiao KC, Klebanov D, Shen H, Tang P *et al*: Identification of an allosteric binding site on the transcription factor p53 using a phage-displayed peptide library. *Oncogene* 1998, 16(15):1993–1999.

65. Krumpe L, Schumacher K, McMahon J, Makowski L, Mori T: Trinucleotide cassettes increase diversity of T7 phage-displayed peptide library. *BMC Biotech* 2007, 7(1):65.

66. Mandecki W: A method for construction of long randomized open reading frames and polypeptides. *Protein Eng* 1990, 3(3):221–226.

67. Scholle MD, Collart FR, Kay BK: In vivo biotinylated proteins as targets for phage-display selection experiments. *Protein Expr Purif* 2004, 37(1):243–252.

68. Chen YC, Delbrook K, Dealwis C, Mimms L, Mushahwar IK, Mandecki W: Discontinuous epitopes of hepatitis B surface antigen derived from a filamentous phage peptide library. *Pro Nat Acad Sci U S A* 1996, 93(5):1997–2001.

69. de Bruin R, Spelt K, Mol J, Koes R, Quattrocchio F: Selection of high-affinity phage antibodies from phage display libraries. *Nat Biotech* 1999, 17(4):397–399.

70. Lou J, Marzari R, Verzillo V, Ferrero F, Pak D, Sheng M, Yang C, Sblattero D, Bradbury A: Antibodies in haystacks: How selection strategy influences the outcome of selection from molecular diversity libraries. *J Immunol Methods* 2001, 253(1–2):233–242.

71. Carcamo J, Ravera MW, Brissette R, Dedova O, Beasley JR, Alam-Moghe A, Wan C, Blume A, Mandecki W: Unexpected frameshifts from gene to expressed protein in a phage-displayed peptide library. *Pro Nat Acad Sci U S A* 1998, 95(19):11146–11151.

72. Berntzen G, Brekke OH, Mousavi SA, Andersen JT, Michaelsen TE, Berg T, Sandlie I, Lauvrak V: Characterization of an FcgammaRI-binding peptide selected by phage display. *Protein Eng Des Sel* 2006, 19(3):121–128.

73. Berntzen G, Andersen JT, Ustgard K, Michaelsen TE, Mousavi SA, Qian JD, Kristiansen PE, Lauvrak V, Sandlie I: Identification of a high affinity FcgammaRIIA-binding peptide that distinguishes FcgammaRIIA from FcgammaRIIB and exploits FcgammaRIIA-mediated phagocytosis and degradation. *J Bio Chem* 2009, 284(2):1126–1135.

74. t Hoen PA, Jirka SM, Ten Broeke BR, Schultes EA, Aguilera B, Pang KH, Heemskerk H, Aartsma-Rus A, van Ommen GB, den Dunnen JT: Phage display screening without repetitious selection rounds. *Anal Biochem* 2012, 421(2):622–631.

75. Matochko WL, Chu K, Jin B, Lee SW, Whitesides GM, Derda R: Deep sequencing analysis of phage libraries using Illumina platform. *Methods* 2012, 58(1):47–55.

76. Ngubane NAC, Gresh L, Ioerger TR, Sacchettini JC, Zhang YJ, Rubin EJ, Pym A, Khati M: High-throughput sequencing enhanced phage display identifies peptides that bind mycobacteria. *PLoS ONE* 2013, 8(11):e77844.

77. Goldman E, Korus M, Mandecki W: Efficiencies of translation in three reading frames of unusual non-ORF sequences isolated from phage display. *FASEB J* 2000, 14(3):603–611.

78. Zemsky J, Mandecki W, Goldman E: Genetic analysis of the basis of translation in the -1 frame of an unusual non-ORF sequence isolated from phage display. *Gene Expression* 2002, 10(3):109–114.

79. Song L, Mandecki W, Goldman E: Expression of non-open reading frames isolated from phage display due to translation reinitiation. *FASEB J* 2003, 17(12):1674–1681.

80. Shu P, Dai H, Mandecki W, Goldman E: CCC CGA is a weak translational recoding site in *Escherichia coli*. *Gene* 2004, 343(1):127–132.

81. Ernst A, Gfeller D, Kan Z, Seshagiri S, Kim PM, Bader GD, Sidhu SS: Coevolution of PDZ domain-ligand interactions analyzed by high-throughput phage display and deep sequencing. *Mol Biosyst* 2010, 6(10):1782–1790.

82. Ravn U, Didelot G, Venet S, Ng KT, Gueneau F, Rousseau F, Calloud S, Kosco-Vilbois M, Fischer N: Deep sequencing of phage display libraries to support antibody discovery. *Methods* 2013, 60(1):99–110.

83. Løset GÅ, Roos N, Bogen B, Sandlie I: Expanding the versatility of phage display II: Improved affinity selection of folded domains on protein VII and IX of the filamentous phage. *PLoS ONE* 2011, 6(2):e17433.

84. Lubkowski J, Hennecke F, Pluckthun A, Wlodawer A: The structural basis of phage display elucidated by the crystal structure of the N-terminal domains of g3p. *Nat Struct Biol* 1998, 5(2):140–147.

85. Glucksman MJ, Bhattacharjee S, Makowski L: Three-dimensional structure of a cloning vector. X-ray diffraction studies of filamentous bacteriophage M13 at 7 A resolution. *J Mol Bio* 1992, 226(2):455–470.

86. Petrenko VA: Landscape phage as a molecular recognition interface for detection devices. *Microelectronics J* 2008, 39(2):202–207.

87. Marvin DA, Welsh LC, Symmons MF, Scott WR, Straus SK: Molecular structure of fd (f1, M13) filamentous bacteriophage refined with respect to X-ray fibre diffraction and solid-state NMR data supports specific models of phage assembly at the bacterial membrane. *J Mol Biol* 2006, 355(2):294–309.

88. Grihalde ND, Chen YC, Golden A, Gubbins E, Mandecki W: Epitope mapping of anti-HIV and anti-HCV monoclonal antibodies and characterization of epitope mimics using a filamentous phage peptide library. *Gene* 1995, 166(2):187–195.

10

Diagnostic Medical Microbiology

Lorrence H. Green

CONTENTS

Introduction

Microbiologists have been identifying organisms since Louis Pasteur in the 1800s first realized that, in addition to producing the fine wines of France, microorganisms also caused disease. Since that time, it appears is if the chief goal of diagnostic microbiology, has been to come up with the one test that would separate the helpful organisms from the harmful ones. Unfortunately, it seems that every time someone developed this ultimate test, eventually someone else would discover that there was overlap. The purpose of this chapter is to describe the major methods being used to identify microorganisms in the laboratory during the 19th and 20th centuries. Diagnostic methods in use in the 21st century will be reviewed in Chapter 11.

Culture

This is the oldest technique used in the laboratory today and has a very simple premise. Organisms are grown on media, and then identified using biochemical tests. Despite the advances in the laboratory since this technique was first used by Pasteur and Koch over 145 years ago, it is still the number one method used in clinical laboratories today (personal observation). Many sources and formulations for media exist and a description

DOI: 10.1201/9781003099277-11

of them can be found in the Difco™ & BBL™ *Manual of Microbiological Culture Media*, available at: http://www.bd.com/ds/technicalCenter/misc/difcobblmanual_2nded_preview.pdf (Accessed April 1, 2020).

Selecting the appropriate media seems like a daunting task, but it can be made much easier if one realizes that all media can be categorized into one of three groups: General, selective, or differential. In determining which medium is best-suited for a particular application, the technician should first determine which of the three groups best suits the needs of the research.

Media Types

General media. These have been developed to allow the widest variety of microorganisms to grow. Typically, they are used in applications in which the sample contains a mixture of microorganisms, and the specific one responsible for causing a particular medical or environmental effect is not known. As a result, a medium that will support the growth of all organisms in a sample is needed. One of the most widely used general media is trypticase soy blood agar. These are usually prepared with 5% defibrinated sheep blood in a nutrient agar base.

Selective media. These have been developed to allow only certain microorganisms to grow. Typically, they are used in applications in which the organism of interest comprises only a small percentage of the organisms contained in a sample. They are used when one wants to isolate specific pathogens in samples with an extensive normal flora. An example of a selective medium is Thayer Martin Agar that is used in the isolation of *Neisseria*.

Differential media. These are a subset of selective media that have been developed to not only select organisms, but also help to identify them. Not only will the organisms of interest grow on the plate, but they will also produce characteristically colored colonies. An example of a differential medium is MacConkey agar, which is used to differentiate lactose fermenting *Enterobacteriacae* from nonlactose fermenters. While both types of colonies will grow on the media, only the lactose fermenting organisms will produce purple colonies. Recently, there has been a trend toward the use of differential media for the screening of antibiotic resistant microorganisms.[1] Examples of this are chromID OXA-48, which screens for carbapenem resistant *Enterobacteriaceae*[2] BBL CHROMagar MRSA II[3] and Oxoid Brilliance™ MRSA, which screen for methicillin-resistant *Staphylococcus aureus* (MRSA).[4]

Two other factors that should be considered when choosing a growth medium are whether the medium is defined or undefined. A defined medium is one in which all components are chemically defined. They are rarely encountered in the microbiology laboratory. Most media are undefined. Undefined media are usually prepared from digests of animal or plant proteins, and as such do not have a defined chemical formula. These media are usually tested for growth with a small specific battery of microorganisms, prior to release for sale by the manufacturer. This can sometimes result in lot to lot variations in growth patterns for organisms that are not found on the manufacturer's quality control testing battery of microorganisms. If a researcher is using an organism not being tested by the manufacturer, it is always a good idea to test any new lots of media with their own battery to determine if that lot is suitable.

Colonial Morphology

Different species of organisms will usually produce colonies that have characteristic appearances following incubation on solid media. Some common characteristics used to describe colonies are:

1. Size, measured in mm: Can vary from very small (punctiform, pinpoint) to very large
2. General shape or form of the colony: Round, irregular, filamentous
3. Edge or margin of the colony: Rounded (entire), irregular, lobed
4. Chromogenesis: Some microorganisms may produce pigments as they grow. The pigment can either remain in the colony or diffuse into the surrounding media.
5. Colony texture: Rough, smooth, wrinkled, mucoid butyrous, viscid, brittle
6. Opacity: Transparent, translucent, opaque, iridescent
7. Colony elevation when viewed in cross section: Flat, domed, convex.[5–7]

Gram Stain

After streaking microorganisms on a solid media so that individual colonies are obtained, the first test routinely performed is the Gram stain. A standard method is described in Chapter 4. After staining, the organisms are examined under a microscope and characterized by reaction (+ = purple, − = pink) and by morphology (cocci = spheres, bacilli = rods, spirochete = helix). While these morphologies represent the basic patterns, variations in shape exist.

Biochemical Tests

Once the Gram stain reaction has been ascertained, conventional biochemical testing is then performed. Organisms are suspended, usually in physiological saline, and then inoculated into test tubes containing chemicals and indicators, and incubated. The choice of chemicals will depend upon the results of the Gram stain. Other chapters in this book will describe the chemicals used to identify specific microorganisms. Following incubation, results are read and the identification of the organism is determined. These tests can consist of solutions of simple carbohydrates, which contain a dye that changes color with a change in pH; solutions of chromogenic substrates that change color when they are cleaved; or antibody tests that will specifically react to one organism. Tests can also consist of the extraction and identification of cellular components by chromatography.

Most laboratories no longer routinely perform conventional testing, but rely instead on commercial biochemical test systems. The trend over the past few years has been to use single biochemical tests in instances where a clinician already suspects the identity of the microorganism. While historically these tests used chromogenic or fluorogenic compounds such as the Oxoid™ O.B.I.S. PYR test for *Streptococcus pyogenes*, today, many of these one-step tests are immunologically based and will be discussed later.

TABLE 10.1

List of Manual and Automated Identification Products

Company	Manual Product	Automated Product	Website
Becton Dickinson/BBL	BBL™ Crystal™ Identification Systems		www.bd.com
Becton Dickinson	Enterotube II, Oxi/Ferm II	BD Phoenix™ Automated Microbiology System. BD Viper™, BD MAX™	www.bd.com
Biofire		BioFire®FilmArray®Panels	Biofiredx.com
Biolog	Gen III Biolog's MicroLog® manual microbial ID system	GEN III OmniLog® ID System	www.biolog.com
bioMerieux, Inc	API tests (20E, 20A, Staph)	VITEK®2, VIDAS®3	www.biomerieux-usa.com
MIDI, Inc		The Sherlock® Microbial Identification System (MIDI)	www.midi-inc.com/
BeckmanCoulter	MicroScan® Panels	MicroScan WalkAway plus	www.beckmancoulter.com/
Thermo Scientific	The RapID™ System		www.thermoscientific.com

Note: All websites accessed April 1, 2020.

Most of the time, the clinician will use one of several multitest commercial kits. These kits are all similar in that they contain many biochemical tests packaged together. The tests are inoculated with a suspension of a single microorganism, and incubated. Following incubation, the results of the tests are read and the identity of the organism is determined. For example, the API 20E strip (bioMerieux, Inc.) is used for identifying Gram-negative bacilli. After incubation, positive and negative results are obtained for each chemical test and are used to generate a seven-digit code number. The company's database then uses this number to determine the microorganism. Table 10.1 lists several of commercially available kits, as well as websites for the companies that supply them.

While commercial manual tests are easier than conventional tests, they are still labor intensive and cumbersome in high-volume laboratories. Since the mid-1980s, instruments have been introduced to perform as much of the work as possible. Instruments can perform the inoculation, incubation, reading, and determination of steps automatically. Table 10.1 also lists some of the commercially available systems. Ohara has written an excellent in-depth review of many of these systems.[8]

In addition to bacteria and fungi, viruses can also be identified using culture techniques. This is usually performed by inoculating samples of suspected virus into cultures containing specific cell lines instead of into tubes of specific chemicals. Following incubation, the culture will then be examined for the presence of a distortion of the cell morphology called a cytopathic effect (CPE), which can be diagnostic for certain viruses. Alternatively, as described next in this chapter, the inoculated cells can be stained using specific antiviral immunofluorescent reagents and observed under the microscope. Specific cell cultures can be obtained from sources listed in Chapter 14.

Immunology-Based Assays

Agglutination Assays

Antibodies or antigens, depending upon the test, are attached to a solid support. The support can consist of latex, polymers (such as styrene), red blood cells, bacteria (coagglutination), charcoal, liposomes, or colloidal gold. The sample is added, and, if present, the analyte binds to the corresponding molecule on the support. The support molecules then form a lattice, which is visible as a clump. The tests can be direct or indirect. A direct test is one in which the clinician is attempting to detect a pathogen in the sample. An indirect test is one in which the clinician is looking for the body's response to a pathogen (e.g. an antibody). In a direct test, an antibody is bound to the support. In an indirect test, an antigen (or an antibody) is bound to the support.

Figure 10.1 shows an example of a direct and an indirect test. Tests for identifying common pathogens are available from many commercial suppliers. There are also several suppliers for latex particles for researchers that wish to develop their own tests. Practical information for binding molecules to inert particles can be found at the website of Bangs Laboratories (www.bangslabs.com, accessed April 1, 2020).

Antibody Labeling Techniques

An antibody is labeled with a tag. This tag can be an enzyme, a fluorescent molecule, a radioactive element, or even a colored particle such as latex or gold. As stated previously, tests can be direct or indirect. Figure 10.2 shows an example of an enzyme immunoassays (EIA). This test is carried out on a solid matrix; the most frequently used is the plastic microtiter tray. The analyte is specifically trapped on the matrix by using either an antibody (direct test) or an antigen (indirect test). The matrix is then washed and the labeled antibody is added. Following another washing step, the label is then detected. In EIA or enzyme-linked immunosorbent assays (ELISA), the label is detected by adding a substrate and observing the formation of a color. In fluorescent immunoassays (FIA), the label is detected using a fluorimeter. In radio immunoassays (RIA), the label is detected using a Geiger or scintillation counter.

Enzyme Immunoassay Generations

Diseases such as AIDS and hepatitis C have prompted the need for rapid, easy-to-use tests to identify people that are unknowingly infected before they inadvertently spread the disease. Several generations of tests have been developed over time as technology and information about the diseases improved. In the

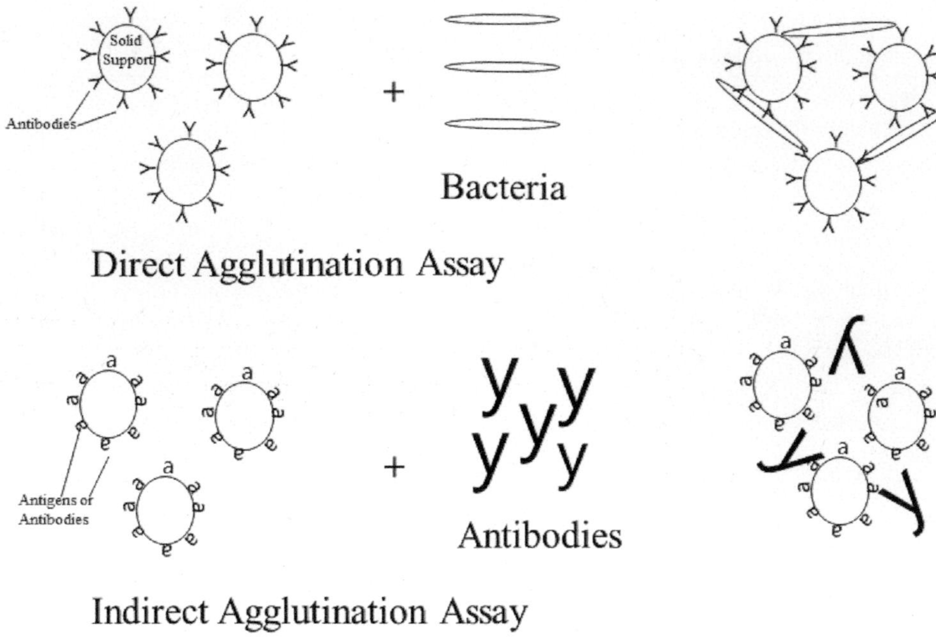

Bacteria

Direct Agglutination Assay

Antibodies

+

Antigens or
Antibodies

Antibodies

+

Indirect Agglutination Assay

FIGURE 10.1   Agglutination assay.

case of AIDS, the first generation of test looked only for IgG molecules generated in response to infection by the HIV-1 virus. AIDS was later found to be caused by both HIV-1 and HIV-2, and as a result, second-generation tests looked for IgG molecules generated in response to both viruses. Since detectable levels of IgG antibodies might not appear for weeks after the initial infection, the third generation of AIDS tests detected both IgG and IgM molecules generated as a result of infection by both HIV-1 and HIV-2. Fourth-generation tests followed, which not only tested for the antibodies, but also had the sensitivity to simultaneously test for the HIV p24 antigen as well.[9,10] Recently a fifth-generation test was introduced, which not only tested for antibodies and the HIV p24 antigen, but reported separate results for both HIV-1 and HIV-2 antibodies and antigens.[11] Fourth-generation tests are

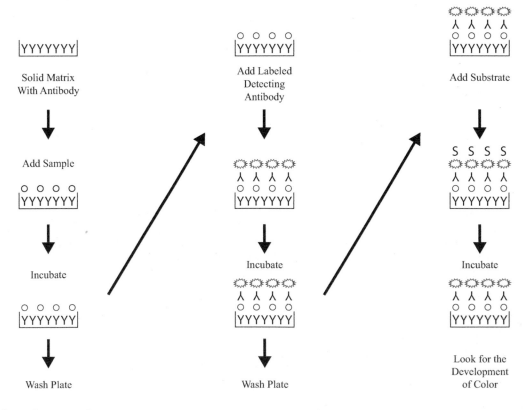

Solid Matrix
With Antibody

Add Sample

Incubate

Wash Plate

Add Labeled
Detecting
Antibody

Incubate

Wash Plate

Add Substrate

Incubate

Look for the
Development
of Color

FIGURE 10.2   Steps of an enzyme immunoassay.

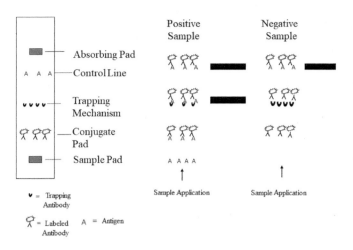

FIGURE 10.3 Lateral flow assay.

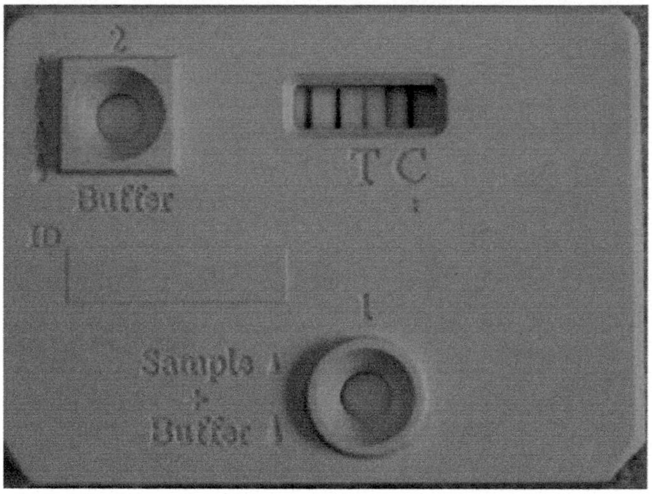

FIGURE 10.4 Dual Path Platform.

also available for hepatitis C virus detection.[12,13] In recent years, labeled antibody assays have been improved with the introduction of lateral flow assays (see Posthuma-Trumpie et al.[14] for a review).

As shown in Figure 10.3, the lateral flow test consists of a strip with several components printed on it. First, there is the conjugate pad, which contains the labeled antibody. There is also a line of trapping antibody and, finally, a control line containing the analyte being assayed. The sample is added and moved through the conjugate pad by capillary action. If the sample contains the analyte, it will bind to the labeled antibody conjugate and pull it forward as it migrates down the strip. When this labeled complex encounters the trapping antibody, if the analyte is present, the complex will bind, forming a colored line. To allow for a positive control, these tests are designed to have an excess of labeled antibody. As the front continues to flow, this excess labeled antibody will bind to the analyte in the control line. If the analyte is present, two lines will be visualized; if the analyte is absent, only the control line will be observed. There are two major ways of visualizing the lines. In one type of test, the antibody is labeled with an enzyme and the line is visualized after a substrate is added. In another type of test, the antibody is labeled with colored markers (such as latex or gold particles) and can be viewed without any additional steps.

A variation of the lateral flow assay is the Dual Path Platform (DPP™). As shown in Figure 10.4, the sample and conjugate have independent flow paths, which allows for greater sensitivity and the ability to identify multiple analytes in a single sample (see www.chembio.com, accessed April 1, 2020, for full description of the technology.)

Antibody Technique Variations

Second antibody. This procedure is used to increase sensitivity, and to minimize the number of labeled antibodies that the lab must maintain. The steps are as described in Figure 10.2, except that the "detecting" antibody is unlabeled. After this antibody has bound to the analyte, it is detected by using a generic labeled second antibody prepared in a different species. For example, individual unlabeled antibodies to separate microorganisms can

be prepared in a rabbit. These rabbit antibodies can be used to bind the pathogen in the plate as in Figure 10.2. If one now adds a labeled antirabbit immunoglobulin antibody prepared in another animal, such as a goat, this will bind to the rabbit antibodies and be detected. Since there are many more sites to bind, this will also increase the sensitivity of the assay.

Another advantage of this technique is that this single reagent can be used to detect any antibodies produced in a rabbit. As a result, a laboratory can produce rabbit antibodies to an entire array of pathogens, but only need produce the single labeled antirabbit immunoglobulin antibody to detect them.

Enhanced EIA. This is used to increase sensitivity. The steps are as described in Figure 10.2, except that the first antibody does not produce a colored end product, it produces a substrate for a second antibody. This second antibody now reacts with an increased amount of substrate and produces the colored end-product.

Biotin avidin. Biotin is a vitamin that can be bound to an antibody and which also has a strong affinity for the glycoprotein avidin. After the antibody has bound to the analyte, it is detected by adding an avidin molecule that has been labeled with an enzyme. Upon adding the substrate, a colored end-product is produced.[15] This variation is used to increase sensitivity and provide a system where multiple analytes can be detected using only a single labeled reagent.

Magnetic beads. Antibodies to analytes of interest are bound to magnetic beads. These are placed into a suspension of the sample. After the antibody has bound to the analyte, the beads can be extracted by use of a magnet and the extracted analyte can then be tested. Some researchers have reported being able to treat the beads so that they can be reused; saving both antibodies and costs.[16] This technique has also been used to isolate specific microorganisms from complex mixtures that can then be tested using antibody assays or PCR.[17,18]

Neutralization. This is a technique that is usually used to detect low levels of antibody in a patient's serum. A known amount of antigen is bound to a solid support. A labeled antibody is then added, and the amount of the end-product is recorded. Following this, the procedure is repeated, but increasing amounts of the patient sample are mixed with the labeled antibody. If the patient sample contains an antibody to the antigen, it will compete with the labeled antibody and reduce (neutralize) the amount of end-product detected. The clinician can calculate the level of antibody in the patient's blood by noting the quantities of patient sample and the decrease in end-product.

Slide tests. Labeled antibodies can also be used to specifically stain pathogens on a slide. One type of very sensitive assay is the direct fluoroimmunoassay. In this procedure, the sample is fixed to a slide and then the slide is incubated with a fluorescently labeled antibody. Following a washing step, the slide is viewed under a fluorescent microscope. If the analyte is present, fluorescence will be observed. This test can be used to detect viruses growing in cell culture.

Many of the companies listed in Table 10.1 also supply antibody-based tests. Suppliers of specific antibodies as well as contracting services can be found at the website www.antibodyresource.com (accessed April 1, 2020).

Nucleic Acid Probes

A labeled DNA probe that is complementary to a specific sequence in the organism of interest is synthesized. This probe is added to the sample and allowed to anneal. Following a washing step, the sample is assayed for the presence of the label. As with antibodies, the label can be a radioactive molecule, an enzyme, biotin/avidin, or a fluorescent molecule. Labeled probes have also been used to identify species on the bases of their ribosomal RNA sequences. Figure 10.5 shows an example of the use of a DNA probe.

A disadvantage of this technique is that many DNA sequences used to identify an organism may only occur once in the genome. As a result, there are usually only limited molecules to detect.

Sample Amplification Methods

Polymerase Chain Reaction

To begin, a specific region of DNA is copied exponentially. The DNA is amplified by supplying primers flanking the region of

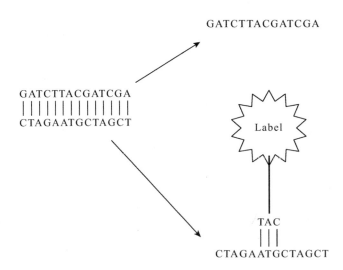

FIGURE 10.5 DNA probe.

interest, nucleotides, and DNA polymerase, and allowing them to undergo many rounds of replication. Usually, the polymerase that is used is the one derived from the thermophilic bacterium *Thermus aquaticus* (Taq polymerase). Since the enzyme is heat-resistant, this allows multiple cycles of the replication to take place without having to constantly add fresh polymerase. Figure 10.6 shows the basic steps. After the first round, there will be four strands. In addition to the original two strands, there will be one strand which will be missing the bases at the 5' end, and one missing them at the 3' end. After the second round of replication, strands that contain only the area of interest will have been produced. In subsequent rounds, these strands will be produced exponentially. Cycle times of 15 minutes are now possible, meaning that in 8 hours, the region of interest can be multiplied millions of times.[19] Once the DNA has been amplified, specific probes can now be used to detect the pathogen. While the steps are fairly standard, different types of DNA may necessitate variations in the temperatures and in the times of incubation. In many instances these parameters must be determined by the investigator. Protocols for performing PCR can be found at http://www.protocol-online.org/prot/Molecular_Biology/PCR/index.html (accessed April 1, 2020).

Real-Time PCR

In recent years, the trend has been to detect the nucleic acid sequence, as it is replicated in real time. Two examples of commercial systems are the Light Cycler® and Taqman®.

Taqman

Similar to conventional PCR, but in addition, a small probe is annealed to the amplicon. This probe has a fluorescent molecule at one end, and a fluorescent quencher at the other. As replication takes place, the Taq polymerase, in addition to adding bases to the primer, also uses its nuclease activity to destroy the probe. When the probe is destroyed, the fluorescent molecule is freed from the quencher, which results in fluorescence that can be measured.

LightCycler

In addition to the primers that are used in the standard PCR protocol, two additional fluorescent probes are also included with LightCycler. Each fluorescent probe hybridizes to the same strand of DNA being copied, but contains a different fluorescent molecule. Both fluorescent probes are incorporated as the amplicon is synthesized. To detect them, light is directed at the sample at a wavelength that excites the first fluorescent molecule. When the first molecule is excited by this wavelength, it fluoresces at a second wavelength. This wavelength will only excite the second molcule if it is on the same strand of the copy. When the second molecule is excited, it fluoresces at a third wavelength. The instrument then detects the third wavelength. Light at the third wavelength will only be detected if the amplicon contains both probes.

Information about Taqman can be found at: www.lifetechnologies.com (accessed April 1, 2020)

Information about LightCycler can be found at: www.roche-applied-science.com (accessed April 1, 2020)

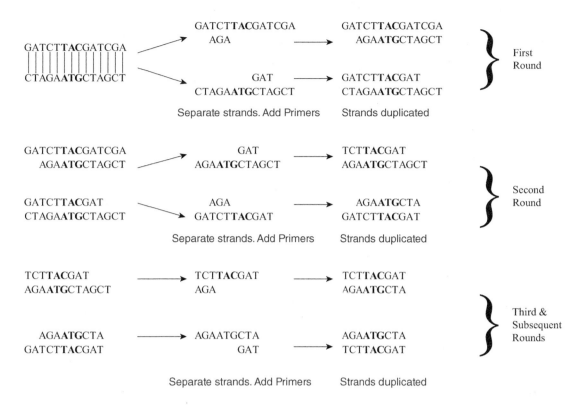

FIGURE 10.6 Polymerase chain reaction.

Variations of PCR

Nested PCR. Amplification occurs in two steps using two different primer pairs. This can increase the sensitivity and specificity of the PCR reaction.[20,21]

Hot start PCR. The PCR mixture, except for one ingredient, is kept at temperatures above the nonspecific binding threshold. The missing component is then added and cycling begins at an elevated temperature. This reduces nonspecific binding.[22]

Multiplex PCR. Several probes are used at one time. This technique can be used to distinguish multiple organisms in a single sample. This helps overcome the expense of having to test a single sample, such as a sputum, with assays for several different microorganisms that might be present.[23]

Random amplified polymorphic DNA. Random primers are used for PCR in a sample. Following replication, the amplified products are separated by size using gel electrophoresis. Patterns obtained from different samples are compared. This can be used to identify similar organism biotypes.[24]

Fluorescent amplified-fragment length polymorphism. Genomic DNA is specifically cut in a limited number of locations, using restriction endonucleases. These fragments are then ligated to double-stranded oligonucleotides. Fluorescene-labeled primers are then added, and several cycles of PCR are run. The labeled fragments are separated on a gel, by size. Different strains of an organism will give different patterns of fluorescent bands, which can be used as a fingerprint.[25]

Ligase Chain Reaction

The steps in ligase chain reaction (LCR) are similar to PCR. Oligonucleotide primers are constructed, but in LCR, the four primers actually flank the gene on both sides of each strand. The sample is heated to denature the DNA and then cooled to allow the probes to anneal to their complementary strand. Ligase is added and fills in the gaps between the two primers, but only if they are aligned. As the reaction proceeds, the newly formed ligated primers, as well as the initial DNA, can now be used as templates. This leads to an exponential replication of the DNA region of interest.[26]

Nanoparticle-assisted PCR. An innovative technique in which nanoparticles are used in the PCR reaction. Their presence reduces the time required to reach the target temperature and thus lowers the time per PCR cycle.[27]

Reverse transcription PCR. In this variation, the enzyme reverse transcriptase is used to make cDNA copies from RNA. This DNA is then amplified using PCR. This technique is useful for identifying RNA viruses, and also for identifying regions of the genome being transcribed.[28]

Other Sample Amplification Methods

Amplicor™. A specific DNA primer is first annealed to a target RNA molecule. An enzyme with both DNA polymerase and reverse transcriptase activity is then used to complete the complementary DNA copy, using this cDNA as a template. PCR is then carried out, using biotinylated primers. The biotinylated amplicon is then bound to a specific probe coated microwell. The amplicon is detected using an avidin-HRP conjugate, which yields a colorimetric result.[29,30] Commercially developed by Roche Molecular Systems.

Reverse hybridization. A series of probes are placed on an inert support. The organism DNA is labeled and hybridized to the support. The label will be visible in areas where there is homology between the organism and the probe and this information is used to identify the organism. Reverse hybridization is especially useful in epidemiological studies in which one is trying to identify organism biotypes. Reverse hybridization and PCR are frequently combined to produce tests to identify species or even strains of microorganisms.[31]

One of the disadvantages of PCR and technologies that employ it is that it requires expensive thermocycling equipment to amplify nucleic acids. The newer techniques described next are all isothermal.

Transcription-mediated amplification (TMA). In TMA, RNA polymerase and reverse transcriptase are used to amplify RNA to produce a DNA copy. A double-stranded DNA molecule is then synthesized and used for transcription to produce new RNA copies of the original sequence. The most recent version of TMA can also amplify DNA. Commercially developed by Hologic (see www.hologic.com, accessed April 9, 2020).

Nucleic acid sequence-based amplification (NASBA). The process uses reverse transcriptase, RNA polymerase, and RNase H to amplify RNA molecules.[32] It is also isothermal. Molecular beacons are now being used to perform real time NASBA.[33] Commercially developed by Biomerieux Organon/Technika.

Strand displacement. This is a technology that involves using a restriction endonuclease to create a nick in a DNA molecule. Polymerases and primers are subsequently used to continuously produce new strands downstream from the site of the nick. As each new strand is started, it replaces the one that had been previously been created.[34]

Other methods of sample nucleic acid amplification include reverse transcription loop-mediated isothermal amplification (RT-LAMP)[35]; rolling circle replication (RCR),[36] and helicase-dependent amplification (HDA).[37]

Signal Amplification Methods

Branched-Chain DNA

Unlike the technologies discussed earlier in this chapter, this technology increases assay sensitivity by amplifying the signal molecules in a reaction. As described in Figure 10.7, a trapping sequence is immobilized on the bottom of a microtiter tray. The sample is added and hybridizes to it. An extender sequence, which is complementary to the sample and to an arm of the amplification molecule, is then added and allowed to hybridize to the sample. An amplification molecule with many branches on it is then added and hybridizes to the other side of the extender sequence. In the next step, enzyme-labeled sequences that are complementary to the chains of the branch are added. Multiple numbers of these will hybridize with the amplification molecule and, in the final step, the label is detected. These assays are frequently used to estimate viral load in HIV and HCV infection.[38]

Hybrid Capture

This technique was originally developed to test for the presence of DNA from papilloma virus in cases of cervical dysplasia.[39] In this technique, RNA probes for the pathogen are added to the sample. If the pathogen is present, an RNA-DNA duplex is formed, which is then added to a microtiter plate with a trapping mechanism that captures RNA-DNA hybrids. Following this, a chemiluminescently labeled antibody specific for RNA-DNA duplexes is added. Following the washing steps, the chemiluminescent reagent is added and the duplex is detected. Commercial systems are available from Quiagen, Inc. (www.qiagen.com, accessed April 1, 2020).

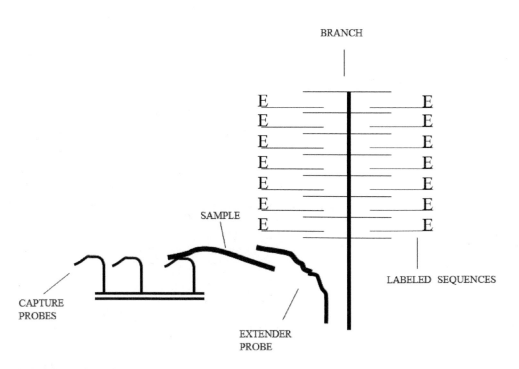

FIGURE 10.7 Branched-chain reaction.

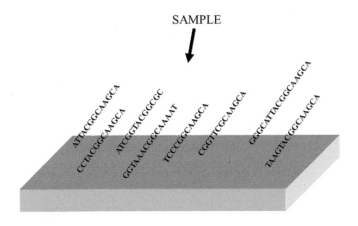

FIGURE 10.8 DNA microarray technology.

DNA Microarray Technology

DNA microarray technology refers to a technique in which hundreds or thousands of oligonucleotides are attached to a solid support (Figure 10.8). The sample DNA or RNA is then hybridized to the support. Following this, sites of binding are detected. Information can be found at http://www.genome.gov/10000533 (accessed April 1, 2020). This technology has been used in detecting genetic variations among a single species of microorganisms[40] and in trying to identify multiple pathogens from a single isolate.[41] Currently labeled nucleic acids are used to identify points of hybridization. However, newer techniques using biosensors may eliminate the need for labels.[42,43]

The advantages and disadvantages of this technology have been reviewed by Lupo et al.[44]

Electrophoresis

This is a method based on separating proteins or nucleic acids from microorganisms based upon their charge (Figure 10.9). A support matrix, usually a gel (acrylamide or agarose) or acetate strip, is prepared and the sample, prepared in an appropriate buffer, is placed in or on the support. When a current is applied, charged molecules in the sample will migrate at a speed relative to their charge. There are variations of this technique that create a chemical environment in which the molecules can also be separated by size or isoelectric point. Following electrophoresis, the support is stained and specific molecules can then be identified based on how far they have migrated. Alternatively, the molecules can be removed from the support by blotting onto an absorbing media and used for further analysis.[45]

Polyacrylamide Gel Electrophoresis (PAGE)

Microorganism proteins are run in the support and stained. Coomassie blue and methyl green are commonly used. Specific protein bands are identified.

Two-Dimensional Gels

The support matrix is subjected to an electric field, and then rotated 90° and run again. This provides much greater resolution.

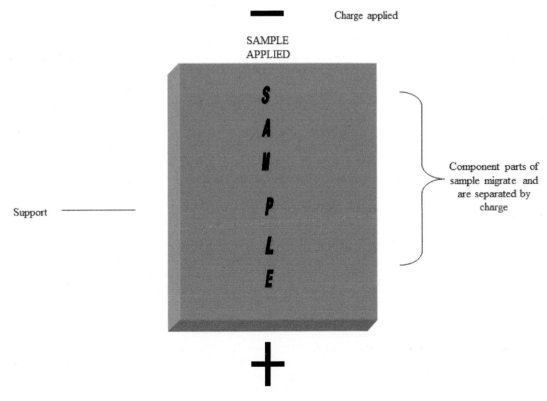

FIGURE 10.9 Electrophoresis.

Pulse Field Gels

The support matrix is alternately exposed to electric fields in varying directions. The advantage of this technique is that it can be used for very large molecules. It is useful in identifying the sources of hospital-acquired infections.

Southern Blots

Microorganism DNA is digested using restriction endonucleases and is then run in the support. Following this, the material in the gel is transferred to either a nitrocellulose or nylon sheet. Staining the sheet with a DNA-specific stain or by hybridizing a labeled specific DNA probe, can then identify bands.

Northern Blots

Similar to Southern blot, except that the sample that is run is RNA. Bands can be identified by staining or by using labeled DNA probes.

Western Blots

Proteins are run and bands are stained using labeled antibodies. This is a confirmatory test in HIV testing.

REFERENCES

1. Tierney, D., Copsey, S.D., Morris, T. and Perry, J.D. A New Chromogenic Medium for Isolation of Bacteroides Fragilis Suitable for Screening for Strains with Antimicrobial Resistance. *Anaerobe*. 39:168, 2016.
2. Girlich, D., Anglade, C., Zambardi, G. and Nordmann, P. Comparative Evaluation of a Novel Chromogenic Medium (chromID OXA-48) for Detection of OXA-48 Producing Enterobacteriaceae. *Diagn Microbiol Infect Dis*. 77:296–300, 2013.
3. Van Vaerenbergh, K., Cartuyvels, R., Coppens, G., Frans, J., Van den Abeele, A. and De Beenhouwer, H. Performance of a New Chromogenic Medium, BBL CHROMagar MRSA II (BD), for Detection of Methicillin-Resistant *Staphylococcus aureus* in Screening Samples. *J. Clin. Microbiol*. 48(4): 1450–1, 2010.
4. Xu, Z., Hou, Y., Peters, B.M., Chen, D., Li, B., Li, L. and Shirtliff, M.E, Chromogenic Media for MRSA Diagnostics. *Mol Biol Rep*. 43:1205, 2016.
5. Smibert, R.M. and Krieg, N.R. Phenotypic Characterization, In P. Gerhardt, R. Murray, W. Wood, and N. Krieg (eds.), *Methods for General and Molecular Bacteriology*. ASM Press, Washington, DC, p. 615. 1994.
6. Sousa, A.M., Machado, I., Nicolau, A. and Pereira, M.O. Improvements on Colony Morphology Identification towards Bacterial Profiling. *J Microbiol Methods*. 95:327–335, 2013.
7. https://bio.libretexts.org/Bookshelves/Ancillary_Materials/Laboratory_Experiments/Microbiology_Labs/Microbiology_Labs_I/08%3A_Bacterial_Colony_Morphology. October 19, 2020.
8. O'Hara, C.M. Manual and Automated Instrumentation for Identification of *Enterobacteriaceae* and Other Aerobic Gram-Negative Bacilli. *Clin Microbiol Rev*. 18:147, 2005.
9. Daskalakis, D. HIV Diagnostic Testing: Evolving Technology and Testing Strategies. *Top Antivir Med*. 19(1):18–22, 2011.
10. Branson, B.M. HIV Diagnostics: Current Recommendations and Opportunities for Improvement. *Infect Dis Clin North Am*. 33:611, 2019.
11. Alexander, T.S. Human Immunodeficiency Virus Diagnostic Testing: 30 Years of Evolution. *Clin Vaccine Immunol*. 23:249, 2016.
12. Ross, R.S., Viazov, S., Salloum, S., Hilgard, P., Gerken, G. and Roggendor, M. Analytical Performance Characteristics and Clinical Utility of a Novel Assay for Total Hepatitis C Virus Core Antigen Quantification. *J Clin Microbiol*. April; 48(4):1161–1168, 2010.
13. Safi, M.A. Hepatitis C: An Overview of Various Laboratory Assays with their Mode of Diagnostic Cooperation. *Clin Lab*. 63:855, 2017.
14. Posthuma-Trumpie, G.A., Korf, J. and van Amerongen, A. Lateral Flow (Immuno)Assay: Its Strengths, Weaknesses, Opportunities and Threats. A Literature Survey. *Anal Bioanal Chem*. 393:569–582, 2009.
15. Sonmez, C., Coplu, N., Gozalan, A., Akin, L. and Esen, B. Comparison of In-House Biotin-Avidin Tetanus IgG Enzyme-Linked-Immunosorbent Assay (ELISA) with Gold Standard in Vivo Mouse Neutralization Test for the Detection of Low Level Antibodies. *J Immunol Methods*. 67(445):67, 2017.
16. Zhao, L., Whiteaker, J.R., Voytovich, U.J., Ivey, R.G. and Paulovich, A.G. Antibody-Coupled Magnetic Beads can be Reused in Immuno-MRM Assays to Reduce Cost and Extend Antibody Supply. *J Proteome Res*. 14:4425, 2015.
17. Schreier, S., Doungchawee, G., Triampo, D., Wangroongsarb, P., Hartskeerl, R.A. and Triampo, W. Development of a Magnetic Bead Fluorescence Microscopy Immunoassay to Detect and Quantify *Leptospira* in Environmental Water Samples. *Acta Trop*. 122(1):119–125, 2012.
18. Taylor, M.J., Ellis, W.A., Montgomery, J.M., Yan, K.-T., McDowell, S.W.J. and Mackie, D.P. Magnetic Immuno Capture PCR Assay (MIPA): Detection of *Leptospira Borgpetersenii* Serovar *Hardjo*. *Vet Microbiol*. 56:135–145, 1997.
19. Lorenz, T.C. Polymerase Chain Reaction: Basic Protocol Plus Troubleshooting and Optimization Strategies. *J Vis Exp*. 63:e3998, doi:10.3791/3998, 2012.
20. Andrade, P., Fioravanti, M., Anjos, E., De Oliveira, C., Albuquerque, D. and Costa, S. Peripheral Blood Leukocytes and Serum Nested Polymerase Chain Reaction are Complementary Methods for Monitoring Active Cytomegalovirus Infection in Transplant Patients. *Can J Infect Dis Med Microbiol*. 24(3):e69–74, 2013.
21. Lin, C., Ying, F., Lai, Y., Li, X., Xue, X., Zhou, T. and Hu, D. Use of Nested PCR for the Detection of Trichomonads in Bronchoalveolar Lavage Fluid. *BMC Infect Dis*. 10:512, 2019.
22. Paul N., Shum J., Le T. Hot Start PCR. *Methods Mol Biol*. 630:301–318, 2010.
23. Garrido, A., Chapela, M.-J., Roman, B., Fajardo, P., Lago, J., Vieites, J.M. and Cabado, A.G. A New Multiplex Real-Time PCR Developed Method for Salmonella spp. and Listeria Monocytogenes Detection in Food and Environmental Samples. *Food Control*. 30:76–85, 2013.
24. Babu, K.N., Rajesh. M.K., Samsudeen, K., Minoo, D., Suraby, E.J., Anupama, K. and Ritto, P. Randomly Amplified Polymorphic DNA (RAPD) and Derived Techniques. *Methods Mol Biol*. 1115:191–209, 2014.

25. Romalde, J.L., Dieguez, A.L., Lasa, A. and Balboa, S. New Vibrio Species Associated to Molluscan Microbiota: A Review. *Front Microbiol.* 4:413, 2014.

26. Zhu, W., Su, X., Gao, X., Dai, Z., and Zou, X. A Label-Free and PCR-Free Electrochemical Assay for Multiplexed MicroRNA Profiles by Ligase Chain Reaction Coupling with Quantum Dots Barcodes. *Biosens Bioelectron.* 53:414–149, 2014.

27. Cui, Y., Wang, Z., Ma, X., Liu, J. and Cui, S. A Sensitive and Specific Nanoparticle-Assisted PCR Assay for Rapid Detection of Porcine Parvovirus. *Lett Appl Microbiol.* 58(2):163–167, 2014.

28. Sansonetti, P., Sali, M., Fabbiani, M., Morandi, M., Martucci, R., Danesh, A., Delogu, G., Bermejo-Martin, J., Sanguinetti, M., Kelvin, D., Cauda, R., Fadda, G. and Rubino, S. Immune Response to Influenza A(H1N1)v in HIV-Infected Patients. *J Infect Dev Ctries.* 15;8(1):101–109, 2014.

29. Schuurs, T.A., Verweij, S.P., Weel, J.F.L., Ouburg, S. and Morré, S.A. Detection of *Chlamydia Trachomatis* and *Neisseria Gonorrhoeae* in an STI Population: Performances of the Presto CT-NG Assay, the Lightmix Kit 480 HT CT/NG and the COBAS Amplicor with Urine Specimens and Urethral/Cervicovaginal Samples. *BMJ Open.* 3(12):e003607, 2013.

30. Wattal, C. and Raveendran, R. Newer Diagnostic Tests and their Application in Pediatric TB. *Indian J Pediatr.* 86:441, 2019.

31. Kim, H.S., Lee, Y., Lee, S., Kim, Y.A. and Sun, Y.K. Recent Trends in Clinically Significant Nontuberculous Mycobacteria Isolates at a Korean General Hospital. *Ann Lab Med.* 34(1): 56–59, 2014.

32. Compton, J. Nucleic Acid sequence Based Amplification. *Nature,* 350:91, 1991.

33. Nadal, A., Coll, A., Cook, N. and Pla, M. A Molecular Beacon-Based Real Time NASBA Assay for Detection of Listeria Monocytogenes in Food Products: Role of Target mRNA Secondary Structure on NASBA Design. *J Microbiol Methods.* 68(3):623–632, 2007.

34. Fakruddin, M., Shahnewaj Bin Mannan, K., Chowdhury, A., Mazumdar, R.M., Hossain, M.N., Islan, S. and Chowdhury, M.D. Nucleic Acid Amplification: Alternative Methods of Polymerase Chain Reaction. *J Pharm Bioallied Sci.* 5(4):245, 2013.

35. Lee, S.H., Ahn, G., Kim, M.S., Jeong, O.C., Lee, J.H., Kwon, H.G., Kim, Y.H. and Ahn, J.Y. 2018. Poly-Adenine-Coupled LAMP Barcoding to Detect Apple Scar Skin Viroid. *ACS Comb Sci.* 20:472–481.

36. Banér, J., Nilsson, M., Mendel-Hartvig, M. and Landegren, U. Signal Amplification of Padlock Probes by Rolling Circle Replication. *Nucleic Acids Res.* 26:5073, 1998.

37. Vincent, M., Xu, Y. and Kong, H. Helicase-Dependent Isothermal DNA Amplification. *EMBO Rep.* 5:795, 2004.

38. Senechal, B. and James, V.L. Ten Years of Extern al Quality Assessment of Human Immunodeficiency Virus Type 1 RNA Quantification. *J Clin Microbiol.* 50(11):3614–3619, 2012.

39. Brown, D.R., Bryan, J.T., Cramer, H. and Fife, K.H. Analysis of Human Papillomavirus Types in Exophytic Condylomata Acuminata by Hybrid Capture and Southern Blot Techniques. *J Clin Microbiol.* 31:2667, 1993.

40. Neverov, A.A., Riddell, M.A., Moss, W.J., Volokhov, D.V., Rota, P.A., Lowe, L.E., Chibo, D., Smit, S.B, Griffin, D.E., Chumakov, K.M. and Chizhikov, V.E. Genotyping of Measles Virus in Clinical Specimens on the Basis of Oligonucleotide Microarray Hybridization Patterns. *J Clin Microbiol.* 44:3752, 2006.

41. Li, Y., Liu, D., Cao, B., Han, W., Liu, Y., Liu, F., Guo, X., Bastin, D.A., Feng, L. and Wang, L. Development of a Serotype-Specific DNA Microarray for Identification of Some Shigella and Pathogenic *Escherichia Coli* Strains. *J Clin Microbiol.* 44:4376, 2006.

42. Ozkumur, E., Ahn, S., Yalcin, A., Lopez, C., Cevik, E., Irani, R.J., DeLisi, C., Chiari, M., and Unlu, M.S. Label-Free Microarray Imaging for Direct Detection of DNA Hybridization and Single-Nucleotide Mismatches. *Biosens Bioelectron.* 25(7):1789–1795, 2010.

43. Aygun, U., Avci, O., Seymour, E., Urey, H., Ünlü, M.S. and Ozkumur, A.Y. Label-Free and High-Throughput Detection of Biomolecular Interactions Using a Flatbed Scanner Biosensor. *ACS Sens.* 2:1424, 2017.

44. Lupo, A., Papp-Wallace, K.M., Sendi, P., Bonomo, R.A. and Endimiani, A. Non-Phenotypic Tests to Detect and Characterize Antibiotic Resistance Mechanisms in Enterobacteriaceae. *Diagn Microbiol Infect Dis.* 77:179, 2013.

45. Lodish, H., Berk, A., Zipursky, L.S., Matsudaira, P., Baltimore, D. and Darnell J. *Molecular Cell Biology* 4th ed., W. H. Freeman, New York, 2000, chap 3, chap 7.

11

Modern Diagnostic Methods in the 21st Century

Lorrence H. Green and Alan C. Ward

CONTENTS

The previous chapter summarized many of the now classic technologies and methods being used to identify microorganisms in the laboratory today. Some of these can trace their lineage as far back as the 19th century. They allowed a very simplistic approach to identifying microorganisms, and more specifically those that caused disease. Koch's postulates, espoused in the 1870s, held that diseases were caused by a single species of microorganism, which could be isolated, grown, and then be used to infect an animal and cause the same disease.[1] With the advent of molecular biology in the 1980s and 1990s, newer methods started finding their way into the clinical laboratories and began challenging that concept. The challenge was brought about because technologies used to identify microorganisms changed from those that used phenotypic markers, such as the presence or absence of an enzyme or specific antigen, to those that used genotypic markers, such as the use of specific nucleic acid sequences.

The advent of 21st-century technologies, such as next-generation sequencing and metagenomics, has challenged that concept even further. The result has been that our understanding of what constitutes a bacterial species, as well as their role in causing disease or affecting the environment, has changed dramatically in just the past few years. The suspected organism identified by Koch's postulates may represent only one of hundreds of genomes that may be present. Disease may not simply be caused by the actions of one organism but may depend upon many complex interdependent interactions in a microbiome. As this book goes to press the world is dealing with a coronavirus pandemic. While the situation is still developing, researchers using modern technologies have found that there are thousands of different genomic combinations associated with the disease.[2] Future research may find that this explains the different outcomes that have been reported.

Matrix-Assisted Laser Desorption Ionization-Time-of-Flight Mass Spectrometry (MALDI-TOF MS)

The application of mass spectrometry to microbial identification has led to the development of instruments that can identify microorganisms directly from a patient sample in a short amount of time. The technique requires 10^4 to 10^6 CFU, so current attention has been focused on samples that are likely to have these numbers, such as positive blood culture bottles and urine samples. The technology also allows for the rapid identification of organisms grown on culture plates.

In the first step of the protocol, the sample is extracted to remove extraneous materials and then is placed on a target plate. Next, a matrix solution composed of a small organic molecule such as 2,5-dihydroxybenzoic acid or ferulic acid is added and allowed to dry.[3] The sample is then exposed to short pulses of high-intensity laser beams. The net effect of this exposure is to ionize the proteins in the sample. Ionized ribosomal proteins appear to provide the best source of information for organism identification. These ionized proteins are then subjected to an electromagnetic field that separates them based on their size and charge. Each microorganism produces its own unique pattern of ionized proteins. When an unknown sample is placed into the test instrument, the pattern that is generated can be compared to a library of patterns produced by known organisms.

Studies have found that the accuracy rate in identifying common organisms using MALDI-TOF was comparable to that obtained using automated instruments based on biochemical tests. The rate was even better for less common isolates.[4]

Since identification using MALDI-TOF does not require that the organism either grow or metabolize many biochemicals, it is much faster, less labor intensive, and less expensive than identification systems based on biochemical testing.[5]

Currently, there are two commercial MALDI-TOF instruments in the clinical microbiology market: the VITEK® MS Biomerieux and the Bruker Microflex, both of which were given FDA approval in 2013.

Next-Generation Sequencing

Perhaps the greatest advances in the microbiology laboratory in the past 10 years have been in the field of DNA sequencing. In 1977, Sanger developed a method of sequencing DNA.[6] The first step of the process used enzymes to cut the sample DNA into fragments. The strands of these fragments were then separated and used as templates for DNA replication using DNA polymerase and nucleotides. Four polymerization chambers were used and a specific dideoxynucleotide analog of either G, C, T, or A was added to each chamber. These analogs inhibited polymerization. Polymerization would proceed until the analog was added. As a result fragments of different lengths were generated in each chamber. These fragments were then run on gels and separated by size. By noting the different fragment lengths generated in each of the chambers, one could determine the sequence of the original fragment. This method worked well for DNA fragments up to as many as 900 bases. By using different restriction endonucleases, overlapping fragments would be generated and the genome sequenced. While very useful, this method was time consuming, required specialized equipment, and required large numbers of DNA strands in the original sample. In addition, the steps of polymerization and detection were two different processes.

Next-generation sequencing (NGS) of bacterial genomes emerged with the whole genome of *Mycoplasma genitalium*, which was sequenced by 454 pyrosequencing in one run in 2005[7] and a human genome[8] relatively soon after. Other techniques soon followed. Among these are Illumina (Solexa) and Ion Torrent:Proton/PGM sequencing. These second-generation methods differed from the Sanger technique in that they combined both the polymerization and detection steps.[9] They do, however, all have the same basic steps. After cutting the DNA into fragments, one strand of the fragment is bound to a support. DNA polymerase and labeled nucleotides are then added to the sample. As a nucleotide is incorporated during polymerization a signal is observed. In this way sequencing is done quickly in real time. The techniques differ in how they are immobilized, and how nucleotide addition is detected.

The demand for high throughput, and low cost per base, has meant that the sample DNA must be cut into short fragments. These are usually 150 bases or fewer.[10] Using fragments of this length means that it is difficult to try and match overlapping fragments. This problem has been partially solved by using huge computing power and innovative software development. Thus, the prediction that sequencing the human genome can be done for $1,000[11,12] will also entail performing a $100,000 analysis. This analysis will be required even in the field of smaller and simpler prokaryotic genomes.[13]

An in-depth tutorial of second-generation sequencing can be found at: https://www.ebi.ac.uk/training/online/course/ebi-next-generation-sequencing-practical-course/what-you-will-learn/what-next-generation-dna- (accessed October 23, 2020).

Sequencing, which can produce long-read, usually single molecules, and often real-time sequencing, has the potential to overcome many of the problems of short-read sequences and add new analytical strategies. Referred to as both long-read and third-generation sequencing, these technologies initially came with the downsides of lower throughput, higher cost, and much higher error rates. Equally, many of the bioinformatic tools for data analysis developed for short-read sequencing were inappropriate for long-read, high-error sequence data.

PacBio, the most mature technology,[14] is based on picoliter-sized wells, each containing a single polymerase. The sequence of a single DNA molecule is detected, in real time through the transparent well bottom (zero-mode waveguide) by fluorescence, and induced as each fluorescent-labeled base is incorporated. Initially, PacBio came as a very large sequencer with read lengths around 1–2 kb, and an error rate about 13–14%.[15] Currently, Wenger et al. achieved ~2 Gb per SMRT cell using an accurate circular consensus (CCS) long-read sequencing protocol, which achieved an average read length >13.5 kb with 99.8% accuracy.[16] CCS in which hairpin adapters are ligated to both ends of double-stranded DNA and the circularized DNA molecule is sequenced multiple times, sacrifices read length for multiple rounds of sequencing across a single fragment, and typically gives ten copies of 2 kb fragments, limited by the processivity of the sequencing DNA polymerase.[17] Wenger et al. noted increasing read lengths >100 kb and exploited this to achieve much longer reads.[16] However, the impact on prokaryotic sequencing is revealed by the increasing number of complete bacterial genomes based on combined PacBio and Illumina sequencing in the databases.

Another long-read technology, nanopore sequencing, is based on a completely different paradigm, both in size, sequencing technology and commercial introduction. In nanopore sequencing, single-strand DNA is driven through a membrane-embedded open pore with a transmembrane voltage; the flow of ions through the pore is measured as ionic current by a pico-current amplifier. The ion current trace (known as a squiggle) is then translated to base sequence.[18]

The initial experience with this technique was somewhat underwhelming. Mikheyev and Tin found 5 kb average read length and about 150 Mb data on the $1,000 MinION from the MinION Access Program, and cited less than 50% read identity, 16× coverage resequencing the 48 kb lambda phage standard DNA.[19] However, over the last 5 years, the development of the MinION has followed the trajectory of third-generation sequencing with dramatic gains. Currently, library preparation from high-quality, high-molecular-weight DNA can be achieved in a few minutes, with transpososome adapter, and barcode attachment. Yields of 20–30 Gb are expected (but not always achieved) from a single-flow cell, at least using the longer ligation-based library preparation. The current record for sequence length of a single read is almost a million bases[20] or longer, 2 Mb,[21] with bases translocated through the pore at speeds of 450 bases/second. Direct sequencing of RNA is underway and direct determination of methylated bases is possible.[22] With sufficient coverage and postanalysis nanopolishing of the signal-level reads in the whole genome assembly of *Escherichia coli*, Loman et al. resequenced the full genome *de novo* with 99.5% identity to the reference.[23]

Metagenomics

The ability to perform rapid nucleic acid sequencing, combined with the fields of genomics and microbiology, has led to the birth of metagenomics. It is both a technique and a field of study.[24] In a very simplified description, total DNA is extracted from all of the organisms in an environment (i.e., Rondon et al. described performing a metagenomics study on a soil sample).[25] The DNA is then processed and cut into fragments using a restriction endonuclease.[26] Fragments are then placed into a plasmid, which as the bacterial artificial chromosome (BAC) might be specially constructed.[27,28]. Following transformation, isolates containing the fragments are cultured and screened for various markers. These might be specific enzymes, or more commonly, the DNA coding for 16S RNA. The fragments are purified and then amplified using PCR, after which they are sequenced. At this point, very intricate software programs are used to establish total organism genomes. This is a daunting task, as samples might have millions of different DNA fragments from thousands of different species.[25]

Metagenomics has allowed scientists to identify microorganisms in a sample that cannot be cultured. Estimates are that in soil samples up to 99% of species may not be culturable.[24] This number is probably equally as high in clinical samples.[29] In addition, it has given scientists the ability to not only study specific genomes, but also see how those genomes might interact with each other in a biome. For a full review of metagenomics see Breitwieser et al.[30]

Metagenomics in the Clinical Laboratory

Traditionally, after a sample arrives in the clinical microbiology laboratory, it is incubated in various media and under different conditions, depending upon the sample source. Following isolation, organisms are identified by looking for phenotypic traits, such as the presence of specific enzymes or other molecular markers. This allows them to then be classified by genus and species. Since these traits are one level removed from DNA, this system is actually classifying microorganisms in a very gross fashion. As a result, organisms are always being renamed and moved from one taxa to another. (See Chapter 17 for a full description)

The introduction of metagenomics into the clinical microbiology laboratory has been a mixed blessing. While it has allowed clinicians to very rapidly identify a pathogenic microorganism in a patient sample without the need to either culture or do traditional biochemical tests, it has also added a great deal of expense, and with routine samples, has not provided much additional information that can be used in treatment. As described previously, clinical microbiology is still very much guided by Koch's postulates. A sample is taken, processed in the standard manner, and a genus and species are determined. This information is then used to treat the patient. This system, which has worked well for more than a century, is based on the premise that the concept of a defined species is valid for bacteria. Metagenomic studies are demonstrating that this may not be true. Assigning a genus and species name to *E. coli* may not be as valid as *Felis catus* is for the cat or *Canus lupus familiaris* is for the dog. In the case of a routine clinical sample, urine for example, the patient is best served by being told that they have an *E. coli* infection and receiving appropriate treatment. Being told that their infection might be caused by multiple subvariants of *E.coli*, some of which might no longer be considered to be *E. coli*, does little to affect treatment, but adds greatly to the cost.[31]

All samples that come into the laboratory are not routine, however. Metagenomic studies are finding that the simple paradigm espoused by Koch does not apply to instances in which many interacting organisms in the biome may ultimately be responsible

TABLE 11.1

List of Metagenomic-Based Products

Company	Product/Service	Website
Aperiomics	Xplore-Patho[SM] test	aperiomics.com
Arc Bio	Arc Bio Galileo™ Pathogen Solution	arcbio.com
MicroGen[DX]	Next-generation sequencing	microgendx.com

for causing pathogenesis. Ruppe et al. examined 24 samples of severe bone and joint infections. Metagenomic sequencing was especially useful in those instances in which the samples were polymicrobial and contained organisms that could not be cultured by conventional means. Furthermore, the information was used to infer the correct antibiotic susceptibility patterns in 94.1% of the monomicrobial samples and in 76.5% of the polymicrobial samples. This is clearly a technology still in its infancy, but one that will probably revolutionize treatment.[32]

Several commercial metagenomics instruments and services for the clinical laboratory are in development and a few are now commercially available (see Table 11.1).

Acknowledgment

The authors thank Dana Johnston, CMP for her assistance in preparing this chapter.

REFERENCES

1. Koch, R. Untersuchungen über Bakterien: V. Die Ätiologie der Milzbrand-Krankheit, begründet auf die Entwicklungsgeschichte des Bacillus anthracis. Cohns Beitrage zur Biologie der Pflanzen, 2, 277, 1876.
2. Bajaj, A. and Purohit, H.J. Understanding SARS-CoV-2: Genetic diversity, transmission and cure in human. *Indian J Microbiol.*, 60, 398–401, 2020. doi.org/10.1007/s12088-020-00869-4.
3 Fitzgerald, M.C., Parr, G.R. and Smith, L. Basic matrices for the matrix-assisted laser desorption ionization mass spectrometry of proteins and oligonucleotides. *Anal Chem.*, 65, 3204–3211, 1993.
4. Saffert, R.T., Cunningham, S.A., Ihde, S.M., et al. Comparison of Bruker Biotyper matrix-assisted laser desorption ionization–time of flight mass spectrometer to BD Phoenix automated microbiology system for identification of Gram-negative bacilli. *J Clin Microbiol.*, 49, 887, 2011.
5. Tan, K.E., Ellis, B.C., Lee, R., Stamper, P.D., Zhang, S.X., and Carroll, K.C. Prospective evaluation of a matrix-assisted laser desorption ionization–time of flight mass spectrometry system in a hospital clinical microbiology laboratory for identification of bacteria and yeasts: A bench-by-bench study for assessing the impact on time to identification and cost-effectiveness. *J Clin Microbiol.*, 50, 3301, 2012.
6. Sanger, F., Nicklen, S., and Coulson, A.R. DNA sequencing with chain-terminating inhibitors. *Proc Natl Acad Sci.*, 74, 5463, 1977.
7. Margulies, M., Egholm, M., Altman, W., et al. Genome sequencing in microfabricated high-density picolitre reactors. *Nature*, 437, 376, 2005.

8. Wheeler, D.A., Srinivasan, M., Egholm, M., Shen, Y., Chen, L., McGuire, A., He, W., Chen, Y.J., Makhijani, V., Roth, G.T., Gomes, X., Tartaro, K., Niazi, F., Turcotte, C.L., Irzyk, G.P., Lupski, J.R., Chinault, C., Song, X.-Z., Liu, Y., Yuan, Y., Nazareth, L., Qin, X., Muzny, D.M., Margulies, M., Weinstock, G.M., Gibbs, R.A., and Rothberg, J.M. The complete genome of an individual by massively parallel DNA sequencing. *Nature*, 452, 872, 2008.

9. Meyerson, M., Gabriel, S., and Getz, G. Advances in understanding cancer genomes through second-generation sequencing. *Nat Rev Genet.*, 11, 685, 2010.

10. Goodwin, S., McPherson, J.D., and McCombie, W.R. Coming of age: Ten years of next-generation sequencing technologies. *Nat Rev Genet.*, 17, 333–351, 2016. Advance online publication, doi: 10.1038/nrg.2016.49.

11. Schloss, J.A. How to get genomes at one ten-thousandth the cost. *Nat. Biotechnol.*, 26, 1113, 2008.

12. Van Dijk, E.L., Auger, H., Jaszczyszyn, Y., and Thermes, C. Ten years of next-generation sequencing technology. *Trends Genet.*, 30, 418, 2014.

13. Mardis, E.R. The $1,000 genome, the $100,000 analysis? *Genome Med.*, 26, 84, 2010.

14. Eid, J., Fehr, A., Gray, J., Luong, K., Lyle, J., Otto, G., Peluso, P., Rank, D., Baybayan, P., Bettman, B., et al. Real-time DNA sequencing from single polymerase molecules. *Science.*, 323, 133, 2009.

15. Carneiro, M.O., Russ, C., Ross, M.G., Gabriel, S.B., Nusbaum, C., and DePristo, M.A. Pacific biosciences sequencing technology for genotyping and variation discovery in human data. *BMC Genom.*, 13, 375, 2012.

16. Wenger, A.M., Peluso, P., Rowell, W.J., et al. Accurate circular consensus long-read sequencing improves variant detection and assembly of a human genome. *Nat Biotechnol.*, 37, 1155, 2019.

17. Travers, K.J., Chin, C.S., Rank, D.R., Eid, J.S., and Turner, S.W. A flexible and efficient template format for circular consensus sequencing and SNP detection. *Nucleic Acids Res.*, 38, e159, 2010.

18. Schreiber, J. and Karplus, K. Segmentation of noisy signals generated by a nanopore. *bioRxiv*, 2015. Advance online publication. doi: 10.1101/014258.

19. Mikheyev, S.A. and Tin, M.M-Y. A first look at the Oxford nanopore MinION sequencer. *Mol Ecol Resour.*, 14, 1097, 2014.

20. Jain, M., Koren, S., Miga, K.H., Quick, J., et al. Nanopore sequencing and assembly of a human genome with ultra-long reads. *Nat Biotechnol.*, 36, 338, 2018.

21. Payne, A., Holmes, N., Rakyan, V., and Loose, M. Whale watching with BulkVis: A graphical viewer for Oxford Nanopore bulk FAST5 files *Bioinformatics*, 35, 2193, 2018.

22. Wescoe, Z.L., Schreiber, J., and Akeson, M. Nanopores discriminate among five C5-cytosine variants in DNA. *J Am Chem Soc.*, 136, 16582, 2014.

23 Loman, N.J., Quick, J., and Simpson, J.T. A complete bacterial genome assembled de novo using only nanopore sequencing data. *Nat Methods.*, 12, 733, 2015.

24. The National Research Council. The New Science of Metagenomics. Revealing the Secrets of Our Microbial Planet. 2007.

25. Rondon, M.R., August, P.R., Bettermann, A.D, et al. Cloning the soil metagenome: A strategy for accessing the genetic and functional diversity of uncultured microorganisms. *Appl Environ Microbiol.*, 66, 2541, 2000.

26. Thomas, T., Gilbert, J., and Meyer, F. Metagenomics—A guide from sampling to data analysis. *Microb Inform Exp.*, 2, 1, 2012.

27. Rondon, M.R., Raffel, S.J., Goodman, R.M., and Handelsman, J. Toward functional genomics in bacteria: Analysis of gene expression in *Escherichia coli* from a bacterial artificial chromosome library of *Bacillus cereus. Proc Natl Acad Sci USA.*, 96, 6451, 1999.

28. Shizuya,H., Birren, B.,Kim,J.J., Mancino,V., Slepak,T., Tachiiri, Y., and Simon, M. Cloning and stable maintenance of 300-kilobase-pair fragments of human DNA in *Escherichia coli* using an F-factor-based vector. *Proc Natl Acad Sci.*, 89, 8794, 1992.

29. Edouard, S., Million, M., Bachar, D., et al. The nasopharyngeal microbiota in patients with viral respiratory tract infections is enriched in bacterial pathogens. *Eur J Clin Microbiol Infect Dis.*, 37, 1725, 2018.

30. Breitwieser, F.P., Lu, J., and Salzberg, S.L. A review of methods and databases for metagenomic classification and assembly. *Brief Bioinform.*, 20, 1125, 2019.

31. Greninger, A. The challenge of diagnostic metagenomics. *Expert Rev Mol Diagn.*, 18, 605, 2018.

32. Ruppé, E., Lazarevic, V., Girard M., et al. Clinical metagenomics of bone and joint infections: A proof of concept study. *Sci Rep.*, 7, 7718, 2017.

12

Antibiotic Susceptibility Testing

Audrey Wanger and Violeta Chávez

CONTENTS

Basic Principles of Antimicrobial Susceptibility Testing (AST)

The purpose of performing antimicrobial susceptibility testing (AST) is to provide *in vitro* data to help ensure that appropriate and adequate antimicrobial therapy is used to optimize treatment outcomes. In addition, the AST data generated daily can be statistically analyzed on an annual basis to generate an antibiogram that reflects the antimicrobial susceptibility and resistance patterns of important pathogens that prevail in a particular hospital. These hospital-specific AST epidemiologic data provide valuable guidance to the clinicians for the appropriate selection of empiric therapy, prior to the availability of culture and susceptibility results that often takes 2–3 days. The purpose of the AST of the culture pathogen is to help clinicians correct and/or modify empiric therapy as soon as the results become available.

Two basic methods of AST are available to laboratories: (1) qualitative and (2) quantitative. Qualitative methods, such as disk diffusion and abbreviated breakpoint dilution systems, are acceptable options for the testing of isolates from "healthy" patients with intact immune defenses and for less serious infections such as uncomplicated urinary tract infections. Both disk diffusion and breakpoint agar or broth dilution systems are considered satisfactory for predicting treatment outcome in these cases (Craig 1993). Quantitative systems that provide a minimal inhibitory concentration (MIC) value are more important in the treatment of serious infections such as endocarditis or osteomyelitis, and for infections in high-risk patient groups such as immunocompromised patients (e.g., transplant patients) and those who are critically ill.

Depending on the size of the workloads, personnel resources available, and issues of convenience, many clinical laboratories prefer to use one or more automated systems for AST regardless of whether results obtained are qualitative or quantitative. These systems handle the majority of the higher volume specimens, such as urines, and are usually supplemented with manual methods, such as disk diffusion and other MIC tests, to handle the limitations of automation as well as provide MIC results for different clinical situations (Jorgensen and Ferraro 1998). Clinical laboratories should continue to develop and improve their AST algorithms to include different types of test systems in order to provide meaningful and accurate data depending on the type of patient, source of specimen, organism species, and anticipated problems in detecting various types of resistance mechanisms.

Disk Diffusion

The principle behind the disk diffusion method, or "Kirby Bauer" test, is the use of a paper disk with a defined amount of antibiotic to generate a dynamically changing gradient of antibiotic concentrations in the agar in the vicinity of the disk (Bauer et al. 1966). The disk is applied to the surface of an agar plate inoculated with the test organism; and while the antibiotic diffuses out of the disk to form the gradient, the test organism starts to divide and grow and progresses toward a critical mass of cells. The so-called inhibition zone edge is formed at the critical time, where a particular concentration of the antibiotic is just able to inhibit the organism before it reaches an overwhelming cell mass or critical mass (Acar and Goldstein 1996). At the zone demarcation point, the density of cells on the growth side is sufficiently large to absorb antibiotic in the immediate vicinity, thus maintaining the concentration at a subinhibitory level and enabling the test organism to grow. The critical times for the demarcation of the inhibition zone edge for most rapidly growing aerobic and facultative anaerobic bacteria vary between 3 and 6 hours, while critical times of fastidious organisms and anaerobic bacteria can vary from 6 to 12 hours or longer.

A density of approximately 10^8 CFU mL^{-1} (CFU = colony-forming unit) of the test organism in the inoculum suspension is used to obtain semiconfluent growth on the agar. The inoculum is prepared by suspending enough well-isolated colonies from an 18- to 24-hour agar plate in broth or physiologic saline (0.85%) to achieve a turbidity matching a 0.5 McFarland Standard. The Prompt™ (BD Diagnostic Systems, Sparks, MD) inoculum preparation system, which consists of a sampling wand and inoculum solution in a plastic bottle, can also be used to optimize workflow because the organisms are maintained at the same density for up to 6 hours. However, this system is not suitable for mucoid strains or fastidious organisms because the amount of cell mass absorbed onto the tip of the sampling wand may be insufficient to give the appropriate inoculum density. Furthermore, the inoculum solution contains Tween, a surfactant that can have potential antimicrobial effects on certain organism groups and can interfere with the antimicrobial activity of some classes of antibiotics. Colony counts of the inoculum suspension should be regularly performed to verify that the inoculum density is correct in terms of CFU mL^{-1}.

Because the disk diffusion is based on the use of a dynamically changing gradient, inoculum density variations can directly affect the critical times and influence the inhibition zone sizes, regardless of the susceptibility of the test organism. A heavy inoculum will shorten the critical time and lead to falsely smaller inhibition zone sizes, resulting in potentially false resistant results, while a light inoculum will cause the reverse effect and generate potentially false susceptible results.

The inoculum suspension should be used within 15 minutes of preparation. This is particularly important for fastidious organisms that lose their viability rapidly. A sterile cotton swab is dipped into the suspension and pressed firmly on the inside of the tube to remove excess liquid. The dried surface of the appropriate agar plate is inoculated by streaking the entire surface and then repeating this twice, rotating the plate 60° each time. This will result in an even distribution of the inoculum. The inoculated plate is then allowed to dry with the lid left ajar for no more than 15 minutes. Once the agar plate is completely dry, the different antibiotic disks are applied either manually or with a dispensing apparatus. In general, no more than 12 disks should be placed on a 150-mm agar plate or five disks on a 90-mm plate. Fewer disks are used when anticipating highly susceptible organisms.

Optimally, disks should be positioned at a distance of 30 mm apart and no closer than 24 mm apart when measured center to center, to minimize the overlap of inhibition zones. Most dispenser devices are self-tamping (disks are tapped or pressed onto the agar surface). If applied manually, the disk must be pressed down to make immediate and complete contact with the agar surface. Once in contact with the agar, the disk cannot be moved because of instantaneous diffusion of antibiotics from the disk to the agar.

Agar plates are incubated in an inverted position (agar side up) under conditions appropriate for the test organism. Plates should not be stacked more than five high to ensure that the plate in the middle reaches incubator temperature within the same time frame as the other plates. After 16–18 hours of incubation at 35°C for rapidly growing aerobic bacteria, or longer where appropriate for fastidious organisms or specific resistance detection conditions, the agar plate is examined to determine if a semiconfluent and even lawn of growth has been obtained before reading the plate. If individual colonies are seen, the inoculum is too light and the test should be repeated because zone sizes may be falsely larger. The same holds true for excessively heavy inoculum, where zone edges may be very hazy and difficult to read and zone sizes may be falsely small.

If the lawn of growth is satisfactory, the zone diameter is read to the nearest millimeter using a ruler or sliding calipers. For Mueller Hinton agar (without blood supplements), zone diameters are read from the back of the plate. For blood-containing agar, zone diameters are read from the surface of the agar. The zone margin, unless otherwise specified, is identified as the area in which no obvious visible growth is seen by the naked eye. Faint growth or microcolonies detectable only with a magnifying glass or by tilting the plate should be ignored in disk diffusion tests.

Reading of zone sizes can be simplified using automated zone readers that use a camera-based image analysis system such as the BIOMIC (Giles Scientific, New York, NY); Aura (Oxoid, Basingstoke, UK); Protozone (Microbiology International, Frederick, MD); or SirScan (i2a., France) (Korgenski and Daly 1998). The agar plate is placed into the zone reader instrument that is connected to a computer system. The system then reads the different zone sizes directly from the agar plate and converts the result to a susceptibility category interpretation using various zone-MIC interpretive guidelines in the software. Although zone sizes are automatically read, the user can intervene and adjust results. The BIOMIC system also converts zone sizes to MIC equivalents using a database of regression analyses based on the assumption of a linear correlation between the zone size and the MIC value. This may not always be the case, however, particularly for fastidious, slow-growing organisms.

The main advantages of disk diffusion testing are the simplicity, lack of need for expensive equipment, and the cost-effective and flexible choice of antibiotics for testing. The main disadvantage of the disk diffusion method is that the information is qualitative, providing only susceptibility category results. It cannot be used to compare the potential efficacy of different agents or to fine-tune the therapy by making dosage adjustments. The other disadvantage is the amount of technologist time required for setting up the test and for manually reading the zone sizes and then consulting the interpretive standards prior to the interpretation of results. Certain antibiotics can be problematic to test with the disk diffusion method due to specific physicochemical properties of the molecule. Glycopeptides such as vancomycin have a high molecular weight (1450) and diffuse very slowly in agar. The slow diffusion results in a poorly resolved concentration gradient around the vancomycin disk, and an inhibition zone size that differs by only a few millimeters between susceptible and resistant strains (CDC 1997; Walsh et al. 2001).

Quantitative Testing (Minimum Inhibitory Concentration)

In critical infections such as endocarditis, meningitis, osteomyelitis, or in the immunocompromised host, accurate quantitative determination of the exact minimum inhibitory concentration (MIC) value of the infecting organism is crucial to guide choices of optimal therapy. In these situations, an MIC value of 0.016 µg mL^{-1} has significantly different therapeutic implications than does an MIC of 1 µg mL^{-1} in influencing the antibiotic choice and dose relative to factors such as body weight, potential drug toxicity, and drug concentrations at the site of infection (Craig 1993).

Broth Macrodilution, Microdilution, or Agar Dilution

Broth macrodilution and agar dilution methods are infrequently performed today. These techniques are predominantly performed in reference or research laboratories because they are not easily automated and require that the laboratory acquire and prepare stock solutions, perform dilutions accurately, and dispense the aliquots either into tubes with broth or agar medium. Quality control of the broth and agar dilution plates also needs to be performed to verify the quality of the reagents before they can be used for testing the clinical isolates. The limited stability of many antibiotic agents in broth and/or agar also necessitates the preparation of fresh reagents, which makes these methods very cumbersome and time consuming. The setup and reading of these methods also require significant experience.

Semiautomated or automated broth microdilution provided by commercial manufacturers is most commonly used in clinical laboratories. The various systems and instruments currently available include Phoenix (Becton Dickinson, Sparks, MD); Vitek and Vitek 2 (bioMerieux, St. Louis, MO); Microscan (Dade Behring, Sacramento, CA); and Sensititre (Trek Diagnostic Systems, Westlake, OH). The advantage of these instrument-based systems includes automation in reading and interpretation. Some of these systems even include automated setup and inoculation. Most systems have both a species identification and AST function. In addition, they include computerized systems for the interpretation of results, and a database "expert" system to help look for inconsistencies between the organism identification and the anticipated susceptibility patterns against multiple antibiotics. Predetermined panels with different types of antibiotics, and to some extent preferred dilution ranges, are selected depending on the requirements of the particular hospital. Most of these systems work well for rapidly growing routine organisms that form the bulk of organisms that are routinely tested (Evangelista et al. 2002).

Limitations may exist, however, that are associated with the resistance mechanisms. These include inducible and/or heteroresistances, testing of mucoid organisms, and slower growing and/or fastidious organisms. The primary underlying reason for these potential limitations is the combination of the use of a relatively small amount of the inoculum of the test organism and the shortened incubation time in the rapid same-day reading systems. The Vitek or Vitek 2 system is promoted as a rapid system with results available in as little as 4–6 hours for identification and slightly longer for susceptibility results. Inducible β-lactamases produced by Gram-negative aerobes such as *Enterobacter*, *Serratia*, and *Citrobacter* species can be difficult to detect, while resistance to cephalosporins may be falsely reported as susceptible in these automated systems. The Sensititre and MicroScan systems both use an "overnight" broth microtiter format. Results are read either automatically or manually after 18–24 hours of incubation. Both the Vitek and the Phoenix can only be read by the instrument. Sensititre can only be used for susceptibility testing and not for organism identification. It can be customized to provide specific panels for individual needs, however, particularly for small volume use.

Etest®

Etest is an agar-based predefined gradient method specifically developed for the determination of exact MIC values and is used to complement other routinely used AST methods. The product was approved by the U.S. Food and Drug Administration (FDA) in 1991 and is currently cleared for clinical testing of Gram-positive and -negative aerobes, anaerobes, pneumococci, streptococci, haemophilus, gonococci, and yeasts (*Candida* species) with a wide variety of antimicrobial agents. Etest is a gradient technique that combines the principles of both the disk diffusion and agar dilution methods. A preformed and predefined gradient of antibiotic concentrations that spans across 15 dilutions of the reference MIC method is immobilized in a dry format onto the surface of a plastic strip. When applied to the surface of an inoculated agar plate, the antibiotic on the plastic strip is instantaneously transferred to the agar in the form of a stable, continuous gradient directly beneath and in the immediate vicinity of the strip. The stability of the gradient is maintained for up to 18–20 hours, depending on the antibiotic. This covers the critical times of a wide range of pathogens, from rapid-growing aerobic bacteria to slow-growing fastidious organisms such as anaerobes and fungi as well as mucoid organisms. The predefined and stable gradient in Etest, unlike the unstable and dynamic gradient around a disk, provides inoculum tolerance. This means that a 100-fold variation in colony-forming units per milliliter (CFU mL^{-1}) has minimal effect on the results of homogeneously susceptible strains (Sanchez and Jones 1992).

A unique feature exclusive to Etest is the ability to use the method in a macro format through the use of a heavier inoculum. This helps optimize the detection of low-level resistance, inducible resistance, and heteroresistance and resistant subpopulations that may occur in varying frequencies. Important examples of heterogeneous resistance are oxacillin resistance in *Staphylococcus aureus* and heterovancomycin intermediate *S. aureus* (hVISA), both of which are well detected by Etest and the latter by the macro method format (Hubert et al. 1999).

After incubation, a parabola-shaped inhibition zone centered alongside the test strip is seen. The MIC is read at the point where the growth/inhibition margin of the organism intersects the edge of the MIC calibrated strip. This can easily be seen with the naked eye for the majority of organisms. Tilting the plate or using a magnifying glass can be helpful to visualize microcolonies, hazes, or other colonies within the ellipse of inhibition when reading the endpoints for bactericidal drugs and for detecting different resistance phenomena such as heteroresistance to oxacillin in *S. aureus*.

More recently, the company Liofilchem® began producing gradient diffusion testing and in 2017 received U.S. FDA approval of their MIC Test Strip (MTS™). These gradient diffusion test strips are similar in principle of methodology to Etest strips. MTS™ are made of special high quality paper impregnated with a predefined concentration gradient of antibiotic, across 15 twofold dilutions like those of a conventional MIC method. When the MTS™ is applied onto an inoculated agar surface, the pre-formed exponential gradient of antimicrobial agent diffuses into the agar for more than an hour. After incubation, a symmetrical inhibition ellipse centered along the strip is formed. The MIC is read directly from the scale in terms of μg/mL at the point where the edge of the inhibition ellipse intersects the strip MTS™ (Liofilchem Inc., Waltham, MA).

AST of Fastidious Organisms

Most fastidious organisms do not grow well enough in routine anti-microbial testing systems and require some type of supplementa-tion—for example, blood. Disk diffusion was initially developed for susceptibility testing of rapidly growing aerobic bacteria, including enterococci, staphylococci, Enterobacteriaceae, and *Pseudomonas aeruginosa*. Modifications have been described by the CLSI (Clinical Laboratory Standards Institute, for-merly NCCLS, United States) for certain fastidious organisms, including *Haemophilus influenzae*, *Neisseria gonorrhoeae*, *Streptococcus pneumoniae*, and *Streptococcus* species (CLSI 2020). Modifications for broth microdilution can also be used for testing fastidious organisms, although panels must be read manually and not by automated instruments. The MicroScan MicroStrep panel (Dade Behring, Sacramento, CA) is a freeze-dried panel containing Mueller Hinton broth with lysed horse blood for testing of *S. pneumoniae* and other *Streptococcus* spe-cies. Similar panels are available on Sensititre (Trek Diagnostics). Vitek 2 (bioMerieux) also has a card specifically designed for susceptibility testing of *Streptococcus* species. Etest, being an agar-based system that can be easily adapted to various growth media, can be used for testing of fastidious organisms, includ-ing *Streptococcus*, *Haemophilus*, and *Neisseria* as well as many other fastidious and very slow-growing organisms.

The CLSI developed a document with guidelines for testing and interpretation of several groups of fastidious organisms such as *Corynebacterium*, *Listeria*, and *Bacillus* species (CLSI 2016b).Other unusual and/or rarely encountered opportunistic pathogens for which no CLSI guidelines are available can also be tested by an MIC method if the system supports good growth of these organisms and only the MIC value is reported without an interpretation. Scientific references can be used to guide the choice of appropriate media, incubation conditions, and antibiot-ics to test for these unusual organisms.

AST of Anaerobes

Antimicrobial resistance among many clinically important spe-cies of anaerobes has increased, which has made empiric therapy choices unpredictable. Few antimicrobials remain where these pathogens are expected to be uniformly susceptible (Hecht 2002, 2006). Although rare, metronidazole resistance has been found in *Bacteroides* species as well as other anaerobes. Equally impor-tant is the need to recognize inherent resistance to metronida-zole in anaerobes such as *Propionibacterium*. In 2004, the CLSI published a document describing agar dilution as the reference method for anaerobic susceptibility testing. Shortly afterward, a broth microdilution method for rapidly growing anaerobes (*Bacteroides* species only) was included in the reference stan-dard; however, the method is only described for a limited number

of antimicrobials and the panels are not commercially available (CLSI 2018a). Anaerobic susceptibility testing is indicated for the management of patients with serious infections who may require long-term therapy and for the guidance of appropriate therapy for organisms that are not predictably susceptible. The CLSI also recommends that clinical laboratories save significant anaerobes and perform AST at the end of the year in a batch format in order to create an anaerobic antibiogram to help guide empiric therapy. The media recommended by the CLSI is Brucella agar or broth supplemented with vitamin K and hemin. After inoculation in agar or broth, results can be read after 24–48 hours of incubation in an anaerobic environment.

AST of Fungi

In 1997, the CLSI published a standard method for antifungal susceptibility testing of yeasts, and later for filamentous fungi (molds) (2002) (CLSI 2016c, 2012). Although no MIC interpre-tive breakpoints exist for molds with any of the currently avail-able antifungal agents, reproducible results have been achieved with the CLSI as well as alternative methods as documented in the literature. The CLSI reference methods for fungi include disk diffusion (yeast only), as well as broth macro and microdilu-tion (CLSI 2017). A commercially available broth microdilution method (Yeast One, Trek Diagnostics, Cleveland, OH), as well as an automated method (Vitek) and Etest, is FDA-cleared for antifungal testing for a limited set of organism and antifungal agent combinations. The recommended medium is RPMI 1640 buffered with MOPS in either a broth or agar format, the latter of which requires supplementation with 2% glucose for Etest. After incubating for 24–48 hours, the MIC results for *Candida* species can be read. Endpoints are determined based on the class of anti-fungal agent with fungicidal agents such as amphotericin being read at 100% inhibition and fungistatic agents such as the azoles read at 80% inhibition; or the first dilution showing a significant decrease in turbidity or growth. MIC breakpoints are published by the CLSI for fluconazole, itraconazole, voriconazole, and flucytosine for some *Candida* species as well as anidulafungin, caspofungin, and micafungin (CLSI 2017; Pfaller et al. 2011).

For *Candida* species with no current breakpoints or for Cryptococcus and mold isolates, epidemiological cutoff values (ECVs) can be used. The ECV cutoff is based on the susceptibil-ity of a wild type population of the isolate. Available values can be found in CLSI M59 (CLSI, 2016a). When testing echinocan-dins, reading of the endpoints is recommended to be performed based on the minimal effective concentration (MEC). The MEC refers to the lowest concentration of an antifungal agent, which results with the growth of small, rounded, compact hyphal forms when compared to the hyphal growth seen in the control well (Pfaller et al. 2010; Pfaller, Boyken, et al. 2011)

AST of Mycobacteria

The reference method for AST of slow-growing mycobacteria, in particular *Mycobacterium tuberculosis*, is the agar proportion method. This technique uses a single critical concentration of the antimycobacterial agent to define susceptibility/resistance. A standardized inoculum of the organisms is added to a Middlebrook 7H11 agar plate. The agent is then added to a single quadrant. Drugs tested include isoniazid, rifampin, ethambutol, and streptomycin. Colonies are counted after 3 weeks of incubation, and the number on the drug-containing quadrant is compared to the control quadrant, which does not contain any drug. Organisms are considered resistant to the drug if the number of colonies on the drug-containing quadrant of resistant colonies is greater than 1% of the control population (CLSI 2018b).

To provide more timely results, clinical laboratories use an automated or semiautomated broth-based system for the detection of growth of *M. tuberculosis* in the presence of a defined critical concentration of the drug in broth as described by the CLSI. Due to the hazards of radiometric testing, nonradiometric automated systems for susceptibility testing of *M. tuberculosis* have been developed and cleared by the FDA. These systems include VersaTREK (ThermoScientific) and Mycobacteria Growth Indicator Tube (MGIT, Becton Dickinson, Sparks, MD) (Woods et al. 2011). Both systems have been FDA-cleared for susceptibility testing of rifampin, isoniazid, ethambutol, pyrazinamide, and streptomycin. The MGIT system consists of a florescent indicator tube, which will emit an orange glow for positive cultures that are inoculated with the prepared specimen and incubated and read daily. The VersaTREK system detects mycobacterial growth by monitoring the rate of oxygen consumption in the culture bottle every 24 minutes automatically or can be read manually.

Susceptibility testing of rapidly growing mycobacteria can be performed by either broth microdilution (CLSI 2018b) or by Etest. Broth dilution can be prepared by the laboratory, or Sensititre™ 96-well plates can be purchased commercially through ThermoFisher Scientific. The laboratory can purchase a plate with antibiotics for susceptibility testing for rapidly growing mycobacteria or other nontuberculous mycobacteria as well.

Susceptibility testing of nontuberculous mycobacteria (NTM) should only be considered for clinically significant isolates, especially those isolated from sterile body fluids. Microbroth dilution is the recommended method for susceptibility testing. In addition, for the *Mycobacterium avium/intracellulare* complex, CLSI and the American Thoracic Society (ATS) recommend susceptibility testing for patients who have been treated with a macrolide and/or those who develop bacteremia while on macrolide prophylaxis or from patients who relapse while on macrolide therapy. Testing is also recommended on initial isolates from blood, tissue, or significant respiratory samples (CLSI) 2018b; Griffith et al. 2007). Additionally, some mycobacteria species have been shown to carry the *erm* gene, which can result in inducible resistance to macrolides. To detect inducible macrolide resistance, the CLSI recommends that a final reading for clarithromycin should be at least after 14 days, unless the isolate has an MIC of ≥8 µg/mL in which case it can be finalized as being resistant (CLSI 2018b).

Susceptibility Interpretation

Information on drug levels, the site of infection, and MIC distributions of biological populations of bacteria and clinical outcome data from clinical trials are used to select MIC cutoffs or breakpoints to predict whether a pathogen with a certain zone-MIC correlate would respond to therapy with standard dosages. These breakpoints are published in the CLSI AST documents and are updated on a yearly basis as needed. The FDA also approves specific MIC breakpoints for antimicrobial agents when the drug is approved for clinical use, and these breakpoints are published in the package insert for the drug. Most clinical laboratories in the United States use the CLSI breakpoints; however, commercially available systems are required to use FDA breakpoints in their interpretive software. For several new agents, only FDA breakpoints may be available; and for some agents, FDA and CLSI breakpoints may be different, causing confusion for laboratorians.

Currently four categories of interpretation are used for reporting of AST results:

1. *Susceptible.* An infection caused by the organism will most probably respond to treatment with a standard dosage of the antibiotic as indicated for use for that infection

2. *Susceptible-dose dependent.* This category defines the likely susceptibility of an organism that is dependent on the dosing regimen used in the patient. Higher exposure in a maximum approved dosage regimen will give the highest probability of adequate coverage. The SDD category is assigned when doses well above those used to calculate the susceptible breakpoint are approved and used clinically (CLSI 2020).

3. *Intermediate.* These results are usually regarded as indicative of nonuseful therapeutic options similar to the resistant category for treatment purposes. It also serves as a buffer zone to help prevent major categorical errors caused by slight changes in the zone sizes due to the influence of technical variables. However, infections at sites where the antibiotic is likely to be highly concentrated, such as β-lactams in the urinary tract, may respond to therapy.

4. *Resistant.* The infection is not likely to respond to therapy.

Agar Screen Methods

Agar-based methods are available for screening of certain resistance mechanisms (Swenson et al. 2007). The agar screen method uses a certain cutoff concentration of the antibiotic that is expected to inhibit the growth of normal susceptible strains and allow resistant phenotypes to grow. The test is performed by inoculating a defined volume of a suspension of the test strain from a pure culture onto the agar, followed by incubation for 18–24 hours. Examples of these agar screen methods include the use of Mueller Hinton agar supplemented with 4% NaCl and 6 µg mL⁻¹ oxacillin to detect oxacillin resistance in *S. aureus* (MRSA screen plate). The agar screen plate is inoculated with a 1-µL loop dipped into a bacterial suspension with a turbidity equivalent to 0.5 McFarland standard. After 24 hours of incubation at 35°C, the appearance of more than one colony is reported as MRSA screen positive and requires the laboratory to perform a confirmatory test to verify the resistance.

Brain heart infusion (BHI) agar supplemented with 6 µg mL⁻¹ of vancomycin is used as a screen for vancomycin resistance in *S. aureus* (VISA) or *Enterococcus* (VRE). The same procedure and interpretation as for the MRSA screen agar plates are used for these tests. Additional screening agars are available for detecting high-level aminoglycoside resistance (HLAR) in enterococci using BHI agar supplemented with 500 µg mL^{-1} of gentamicin and 2000 µg mL^{-1} of streptomycin. A positive HLAR screen result indicates that a synergistic effect will not be seen when the aminoglycoside is combined with either ampicillin, penicillin, or vancomycin. These screen agar plates are commercially available from several manufacturers.

Latex agglutination methods are also available as a direct method for detecting oxacillin resistance in *S. aureus*. A suspension of pure colonies of the test organism is added to latex beads bound to antibody against the PBP2a antigen (marker for MRSA), and the presence of agglutination is interpreted as a positive MRSA result (Cavassini et al. 1999). Due to the expense of this assay and the recent availability of molecular methods for detection of oxacillin resistance, these assays are not often used in clinical laboratories.

Various companies manufacture media that can differentiate various drug-resistant organisms based on a chromogenic medium with the supplementation of the inhibitory antibiotic. For example, the first chromogenic media, CHROMagar™ MRSA, inhibited methicillin-sensitive strains with the addition of methicillin and oxacillin (Merlino et al. 2000). Later, cefoxitin was demonstrated as superior for the selection of MRSA from other staphylococci (Perry et al. 2004). Since then, other companies have marketed chromogenic agar for MRSA, including CHROMID® (bioMérieux), MRSASelect™ (Bio-Rad), Brilliance™ MRSA Agar (Oxoid), and Spectra™ MRSA Medium (Thermo Fisher Scientific).

The first medium for the isolation of vancomycin-resistant enterococci, CHROMID® VRE (bioMérieux), can identify vancomycin resistant enterococci and will also differentiate between *Enterococcus faecium* and *Enterococcus faecalis* (Ledeboer et al. 2007). Other media available include CHROMID VRE (bioMérieux) and Brilliance™ VRE Agar (Oxoid) CHROMagar™ VRE, Spectra™ VRE(Thermo Fisher Scientific), and VRESelect Medium (Bio-Rad).

The same commercial companies listed above have produced media for extended spectrum beta lactamase detection. Typically, these media contain a combination of chromogenic substrates that detect a β-galactosidase or β-glucuronidase. For carbapenemase-producing Enterobacteriaceae, various products can detect isolates that contain genes responsible for the majority of resistance, including IMP, NDM, and VIM metalloenzymes, *Klebsiella pneumoniae* carbapenemase, and OXA-48. These are reviewed by J.D. Perry (2017). The modified Hodge test, MHT, was previously used to test for suspected carbapenemase production in Enterobacteriaceae. The MHT was performed by using a carbapenem disk placed on a lawn of a carbapenem susceptible *Escherichia coli* strain with the clinical isolate streaked in a line away from the disk. If the carbapenem susceptible isolate grew near the streak line and produced a cloverleaf appearance, the isolate was thought to decrease the local concentration of the carbapenem antibiotic (Tamma and Simner 2018). Due to poor specificity and sensitivity, this test was deemed suboptimal and was removed from the 2018 CLSI M100.

Molecular Methods

Molecular-based methods are becoming more common as adjuncts for specific testing needs in the clinical microbiology laboratory (Rasheed et al. 2007). Several molecular tests are now packaged in an easy-to-use kit format where the amplification reaction takes place in a closed environment. This obviates the need for a separate room to perform the testing. Although most of the assays are more expensive than culture methods, the turnaround time of same-day results, theoretically available in several hours, is a major advantage for patient care and can contribute to cost savings to the hospital. Another advantage of these assays is that they do not rely on viable organisms for detection. The most commonly used molecular test for antimicrobial resistance detection is the polymerase chain reaction (PCR) method for the mecA gene to detect oxacillin resistance in *S. aureus* (MRSA). Molecular assays that require a pure culture of the test organism are being replaced by newer assays that directly detect the presence of MRSA in patient specimens. These assays are currently approved by the FDA for use on nasal swabs, skin, and soft tissue, as well as positive blood cultures (Paule et al. 2007; Kelley et al. 2011).

Multiplex assays performed directly on positive blood-culture bottles are now available that can detect the most common organisms found in blood cultures and the most common resistance mechanisms. Results can be read within 2–3 hours, require little hands on time and are closed systems that can be performed in routine microbiology laboratories (Beal et al. 2013). These assays have been shown to shorten the time to administration of appropriate antibiotics to patients when combined with an antibiotic stewardship program. Major companies that are FDA approved include the VERIGENE® (Luminex Corporation), Biofire® Diagnostics FilmArray®, and GenMarkDx ePlex® blood culture identification panels.

Use of molecular testing on colonies can be used as an adjunct to routine susceptibility testing to specify the mechanism of resistance in multidrug resistant Gram-negative bacilli. This can be very useful for treatment selection and also for epidemiology to define the particular resistant mechanisms in individual hospitals. One recently FDA-cleared example is the NG-Test® (Takissian et al. 2019).

Molecular assays have become commonplace; however, antimicrobial resistance is complex and requires critical thinking when comparing genotypic results with phenotypic results. In some instances, the antibiotic susceptibility testing will result in a different phenotypic pattern than what is predicted with molecular testing. The presence of a resistance marker may not predict the failure of the antimicrobial agent and the absence of a genetic marker does not necessarily indicate susceptibility. The CLSI has published guidelines to follow to ensure provision of relevant clinical guidance as these methods evolve (CLSI 2020).

For example, poor growth of the organisms may mask higher MICs, while some resistance markers may be truncated and not cause the expected resistance. Recently, a letter was published regarding an isolate of *E. coli* and *K. pneumoniae* both of which were found to have a truncated carbapenemase that was detected by the molecular testing; however, when susceptibilities were completed, the carbapenem was within a susceptible range (Salimnia et al. 2020).

Another method is matrix-assisted laser desorption-time-of-flight (MALDI-TOF) mass spectroscopy (MS) for rapid AST. A direct-on-target microdroplet growth assay (DOT-MGA) was developed to be used directly from positive blood cultures. In this assay, a bacterial isolate was inoculated into various concentrations of an antibiotic and spotted onto the target for analysis. A score of ≥1.7 was considered a nonsusceptible result, while a score of ≤1.7 was defined as susceptible (Idelevich et al. 2018).

A newer method by ACCELERATE Diagnostics™, the ACCELERATE Pheno™ system uses fluorescent *in situ* hybridization (FISH) probes for organism identification in positive blood cultures and morphokinetic cell analysis for AST. This system evaluates cell morphology and light intensity of a growing clone over time in the presence of each test antibiotic. Three panels exist, one for Gram-positive bacteria, one for Gram-negatives, and one for yeast. Identification is carried out by using a multiplexed FISH, which is a rapid hybridization of monolabeled DNA oligonucleotides that target ribosomal RNA and use a nucleic acid stain for organism detection. The isolate identification is confirmed by comparing the signal from a target probe with a signal from universal bacterial and eukaryotic probes. This technology also allows for polymicrobial identification, AST is carried out by analyzing the ratio of individual cells by recording division and growth patterns compared to a growth control. The instrument will generate a profile that includes growth probability and time and can identify a susceptible versus resistant cell (Marschal et al. 2017). ACCELERATE Pheno™ provides rapid and accurate identification and antimicrobial susceptibility results for organisms routinely found in blood cultures. The system reduces hands on time and enables clinically actionable results to be obtained earlier than with other systems currently available (Pancholi et al. 2018; Charnot-Katsikas et al. 2018).

Quality Control

The CLSI-recommended media for AST is cation-adjusted Mueller Hinton broth for broth methods or the standard Mueller Hinton agar for agar-based methods. The CLSI provides guidelines for the selection of quality control (QC) strains and the frequency with which QC should be performed for both disk diffusion and MIC methods. QC strains chosen for MIC testing should be such that the expected MIC value for the strain falls in the middle of the concentration range being tested for that antibiotic. However, this is seldom the case in MIC test systems that use abbreviated or limited dilution ranges of the drug. QC strains can be purchased in dehydrated form either as Culti-loops® (Remel), pellets, swabs (KWIK-STIK, Hardy Diagnostics), or lyophilized in multiple unit containers (LYFO-DISK, MicroBioLogics, St. Cloud, MN). After processing according to the manufacturer's instructions, organisms should be maintained on an agar slant, stored at 2–8°C, and subcultured weekly for up to 1 month, after which a new QC stock organism should be used. QC organisms can be stored for prolonged periods of time at −70°C in broth (such as brain heart infusion or *Brucella*) or skim milk. Stock organisms should be subcultured at least twice when processed from the freezer storage before being used for quality control testing. Testing of QC strains should be done either on each day of clinical testing for infrequently performed tests or weekly after qualifying an initial 30-day QC validation. The CLSI provides flow diagrams for frequency of QC testing, troubleshooting, and corrective action for out-of-control results.

Antibiotic disks are available from several manufacturers (e.g., Becton Dickinson, Remel, and Hardy Diagnostics). They should be handled and stored according to the manufacturer's instructions. Disks can be stored at 4°C in disk dispensers containing desiccant for day-to-day use for up to 1 week for short-term purposes. Long-term storage of unopened cartridges should be at −20°C. Packages should be allowed to reach room temperature prior to opening to prevent moisture condensing on the outer surface from penetrating into the cartridge and deteriorating the antibiotic.

The CLSI also provides recommendations for detecting unusual AST results and suggests that laboratories modify the list of agents to test based on the prevalence of resistance mechanisms in their institutions. Unusual or questionable results include carbapenem (imipenem, meropenem, ertapenem) resistance in strains belonging to the family Enterobacteriaceae that are susceptible to extended-spectrum cephalosporins (cefotaxime, ceftriaxone, or cefepime). Labile compounds such as imipenem may give unexpected resistant results due to degradation of the compound even with proper storage (CLSI 2020).

REFERENCES

Acar, J. and F. W. Goldstein. 1996. "Disk susceptibility testing." In *Antibiotics in Laboratory Medicine*, edited by V. Lorian, pp. 1–51. Baltimore, MD: Williams and Wilkins.

Bauer, A. W., W. M. Kirby, J. C. Sherris, and M. Turck. 1966. "Antibiotic susceptibility testing by a standardized single disk method." *Am J Clin Pathol* 45 (4):493–6.

Beal, S. G., J. Ciurca, G. Smith, J. John, F. Lee, C. D. Doern, and R. M. Gander. 2013. "Evaluation of the nanosphere verigene gram-positive blood culture assay with the VersaTREK blood culture system and assessment of possible impact on selected patients." *J Clin Microbiol* 51 (12):3988–92. doi: 10.1128/JCM.01889-13.

Cavassini, M., A. Wenger, K. Jaton, D. S. Blanc, and J. Bille. 1999. "Evaluation of MRSA-Screen, a simple anti-PBP 2a slide latex agglutination kit, for rapid detection of methicillin resistance in *Staphylococcus aureus*." *J Clin Microbiol* 37 (5):1591–4.

Charnot-Katsikas, A., V. Tesic, N. Love, B. Hill, C. Bethel, S. Boonlayangoor, and K. G. Beavis. 2018. "Use of the accelerate Pheno system for identification and antimicrobial susceptibility testing of pathogens in positive blood cultures and impact on time to results and workflow." *J Clin Microbiol* 56 (1). doi: 10.1128/JCM.01166-17.

Craig, W. A. 1993. "Qualitative susceptibility tests versus quantitative MIC tests." *Diagn Microbiol Infect Dis* 16 (3):231–6. doi: 10.1016/0732-8893(93)90115-n.

Center for Disease Control (CDC). 1997. "Morbidity and mortality weekly report: MMWR." In *Interim Guidelines for Prevention and Control of Staphylococcal Infection Associated with Reduced Susceptibility to Vancomycin*. Atlanta, GA: Center for Disease Control.

Clinical and Laboratory Standards Institute (CLSI). 2012. "Reference method for broth dilution antifungal susceptibility testing of yeasts; fourth informational supplement." In *CLSI Standard M27-S4*. Wayne, PA: Clinical and Laboratory Standards Institute.

Practical Handbook of Microbiology

Clinical and Laboratory Standards Institute (CLSI). 2016a. "Epidemiological cutoff values for antifungal susceptibility testing." In *CLSI Supplement M59*, 2nd ed. Wayne, PA: Clinical and Laboratory Standards Institute.

Clinical and Laboratory Standards Institute (CLSI). 2016b. "Methods for antimicrobial dilution and disk susceptibility testing of infrequently isolated or fastidious bacteria." In *CLSI Guideline M45*, 3rd ed. Wayne, PA: Clinical and Laboratory Standards Institute.

Clinical and Laboratory Standards Institute (CLSI). 2016c. "Reference method for broth dilution antifungal susceptibility testing of filamentous fungi." In *CLSI Document M57*, 3rd ed. Wayne, PA: Clinical and Laboratory Standards Institute.

Clinical and Laboratory Standards Institute (CLSI). 2017. "Reference method for broth dilution antifungal susceptibility testing of yeasts." In *CLSI Standard M27*, 4th ed. Wayne, PA: Clinical and Laboratory Standards Institute.

Clinical and Laboratory Standards Institute (CLSI). 2018a. "Methods for antimicrobial susceptibility testing of anaerobic bacteria; approved standard." In *CLSI Document M11*, 9th ed. Wayne, PA: Clinical and Laboratory Standards Institute.

Clinical and Laboratory Standards Institute (CLSI). 2018b. "Susceptibility testing of Mycobacteria, Nocardia spp., and other aerobic Actinomycetes." In *CLSI Standard M38*, 3rd ed. Wayne, PA: Clinical and Laboratory Standards Institute.

Clinical and Laboratory Standards Institute (CLSI). 2020. "Performance standards for antimicrobial susceptibility testing," 30th ed. In *CLSI Supplement M100*. Wayne, PA: Clinical and Laboratory Standards Institute.

Evangelista, A. T., A. L. Truant, and P. P. Bourbeau. 2002. "Rapid systems and instruments for antimicrobial susceptibility testing of bacteria." In *Manual of Commercial Methods in Clinical Microbiology*, edited by Allan L. Truant. Philadelphia, PA: ASM.

Griffith, D. E., T. Aksamit, B. A. Brown-Elliott, A. Catanzaro, C. Daley, F. Gordin, S. M. Holland, R. Horsburgh, G. Huitt, M. F. Iademarco, M. Iseman, K. Olivier, S. Ruoss, C. F. von Reyn, R. J. Wallace, Jr., K. Winthrop, ATS Mycobacterial Diseases Subcommittee, American Thoracic Society, and Infectious Disease Society of America. 2007. "An official ATS/IDSA statement: diagnosis, treatment, and prevention of nontuberculous mycobacterial diseases." *Am J Respir Crit Care Med* 175 (4):367–416. doi: 10.1164/rccm.200604-571ST.

Hecht, D. W. 2002. "Evolution of anaerobe susceptibility testing in the United States." *Clin Infect Dis* 35 (Suppl 1):S28–35. doi: 10.1086/341917.

Hecht, D. W. 2006. "Anaerobes: antibiotic resistance, clinical significance, and the role of susceptibility testing." *Anaerobe* 12 (3):115–21. doi: 10.1016/j.anaerobe.2005.10.004.

Hubert, S. K., J. M. Mohammed, S. K. Fridkin, R. P. Gaynes, J. E. McGowan, Jr., and F. C. Tenover. 1999. "Glycopeptide-intermediate *Staphylococcus aureus*: evaluation of a novel screening method and results of a survey of selected U.S. hospitals." *J Clin Microbiol* 37 (11):3590–3.

Idelevich, E. A., L. M. Storck, K. Sparbier, O. Drews, M. Kostrzewa, and K. Becker. 2018. "Rapid direct susceptibility testing from positive blood cultures by the matrix-assisted laser desorption ionization-time of flight mass spectrometry-based direct-on-target microdroplet growth assay." *J Clin Microbiol* 56 (10). doi: 10.1128/JCM.00913-18.

Jorgensen, J. H., and M. J. Ferraro. 1998. "Antimicrobial susceptibility testing: general principles and contemporary practices." *Clin Infect Dis* 26 (4):973–80. doi: 10.1086/513938.

Kelley, P. G., E. A. Grabsch, J. Farrell, S. Xie, J. Montgomery, B. Mayall, and B. P. Howden. 2011. "Evaluation of the Xpert MRSA/SA Blood culture assay for the detection of *Staphylococcus aureus* including strains with reduced vancomycin susceptibility from blood culture specimens." *Diagn Microbiol Infect Dis* 70 (3):404–7. doi: 10.1016/j.diagmicrobio.2011.02.006.

Korgenski, E. K., and J. A. Daly. 1998. "Evaluation of the BIOMIC video reader system for determining interpretive categories of isolates on the basis of disk diffusion susceptibility results." *J Clin Microbiol* 36 (1):302–4.

Ledeboer, N. A., K. Das, M. Eveland, C. Roger-Dalbert, S. Mailler, S. Chatellier, and W. M. Dunne. 2007. "Evaluation of a novel chromogenic agar medium for isolation and differentiation of vancomycin-resistant *Enterococcus faecium* and *Enterococcus faecalis* isolates." *J Clin Microbiol* 45 (5):1556–60. doi: 10.1128/JCM.02116-06.

Marschal, M., J. Bachmaier, I. Autenrieth, P. Oberhettinger, M. Willmann, and S. Peter. 2017. "Evaluation of the accelerate Pheno system for fast identification and antimicrobial susceptibility testing from positive blood cultures in bloodstream infections caused by gram-negative pathogens." *J Clin Microbiol* 55 (7):2116–2126. doi: 10.1128/JCM.00181-17.

Merlino, J., M. Leroi, R. Bradbury, D. Veal, and C. Harbour. 2000. "New chromogenic identification and detection of *Staphylococcus aureus* and methicillin-resistant *S. aureus*." *J Clin Microbiol* 38 (6):2378–80.

Pancholi, P., K. C. Carroll, B. W. Buchan, R. C. Chan, N. Dhiman, B. Ford, P. A. Granato, A. T. Harrington, D. R. Hernandez, R. M. Humphries, M. R. Jindra, N. A. Ledeboer, S. A. Miller, A. B. Mochon, M. A. Morgan, R. Patel, P. C. Schreckenberger, P. D. Stamper, P. J. Simner, N. E. Tucci, C. Zimmerman, and D. M. Wolk. 2018. "Multicenter evaluation of the accelerate Pheno test BC kit for rapid identification and phenotypic antimicrobial susceptibility testing using morphokinetic cellular analysis." *J Clin Microbiol* 56 (4). doi: 10.1128/JCM.01329-17.

Paule, S. M., D. M. Hacek, B. Kufner, K. Truchon, R. B. Thomson, Jr., K. L. Kaul, A. Robicsek, and L. R. Peterson. 2007. "Performance of the BD GeneOhm methicillin-resistant *Staphylococcus aureus* test before and during high-volume clinical use." *J Clin Microbiol* 45 (9):2993–8. doi: 10.1128/JCM.00670-07.

Perry, J. D. 2017. "A decade of development of chromogenic culture media for clinical microbiology in an era of molecular diagnostics." *Clin Microbiol Rev* 30 (2):449–479. doi: 10.1128/CMR.00097-16.

Perry, J. D., A. Davies, L. A. Butterworth, A. L. Hopley, A. Nicholson, and F. K. Gould. 2004. "Development and evaluation of a chromogenic agar medium for methicillin-resistant *Staphylococcus aureus*." *J Clin Microbiol* 42 (10):4519–23. doi: 10.1128/JCM.42.10.4519-4523.2004.

Pfaller, M. A., D. Andes, D. J. Diekema, A. Espinel-Ingroff, D. Sheehan, and CLSI Subcommittee for Antifungal Susceptibility Testing. 2010. "Wild-type MIC distributions, epidemiological cutoff values and species-specific clinical breakpoints for fluconazole and Candida: time for harmonization of CLSI and EUCAST broth microdilution methods." *Drug Resist Updat* 13 (6):180–95. doi: 10.1016/j.drup.2010.09.002.

Pfaller, M. A., L. Boyken, R. J. Hollis, J. Kroeger, S. A. Messer, S. Tendolkar, and D. J. Diekema. 2011. "Wild-type MIC distributions and epidemiological cutoff values for posaconazole and voriconazole and Candida spp. as determined by 24-hour CLSI broth microdilution." *J Clin Microbiol* 49 (2):630–7. doi: 10.1128/JCM.02161-10.

Pfaller, M. A., D. J. Diekema, D. Andes, M. C. Arendrup, S. D. Brown, S. R. Lockhart, M. Motyl, D. S. Perlin, and CLSI Subcommittee for Antifungal Testing. 2011. "Clinical breakpoints for the echinocandins and Candida revisited: integration of molecular, clinical, and microbiological data to arrive at species-specific interpretive criteria." *Drug Resist Updat* 14 (3):164–76. doi: 10.1016/j.drup.2011.01.004.

Rasheed, J. K., F. Cockerill, and F. C. Tenover. 2007. "Detection and Characterization of Antimicrobial Resistance Genes in Pathogenic Bacteria." In *Manual of Clinical Microbiology*, 9th ed. Washington, DC: ASM Press.

Salimnia, H., J. Veltman, P. H. Chandrasekar, J. M. Pogue, R. Mynatt, T. Salimnia, S. H. Marshall, A. M. Hujer, and R. A. Bonomo. 2020. "Carbapenem-susceptible *Klebsiella pneumoniae* and *Escherichia coli* isolates carrying a truncated KPC carbapenemase: a challenge for rapid molecular diagnostics." *J Clin Microbiol* 58 (3). doi: 10.1128/JCM.01627-19.

Sanchez, M. L. and Ronald N. Jones. 1992. "E Test, an antimicrobial susceptibility testing method with broad clinical and epidemiologic application." *Antimicrobic Newsletter*, 8, 1–7.

Swenson, J. M., J. B. Patel, and J. H. Jorgensen. 2007. "Special Phenotypic Methods for Detecting Antibacterial Resistance," pp. 1173–1192. In *Manual of Clinical Microbiology*, 9th ed. Washington, DC: ASM Press.

Takissian, J., R. A. Bonnin, T. Naas, and L. Dortet. 2019. "NG-Test Carba 5 for rapid detection of carbapenemase-producing Enterobacterales from positive blood cultures." *Antimicrob Agents Chemother* 63 (5). doi: 10.1128/AAC.00011-19.

Tamma, P. D., and P. J. Simner. 2018. "Phenotypic detection of carbapenemase-producing organisms from clinical isolates." *J Clin Microbiol* 56 (11). doi: 10.1128/JCM.01140-18.

Walsh, T. R., A. Bolmstrom, A. Qwarnstrom, P. Ho, M. Wootton, R. A. Howe, A. P. MacGowan, and D. Diekema. 2001. "Evaluation of current methods for detection of staphylococci with reduced susceptibility to glycopeptides." *J Clin Microbiol* 39 (7):2439–44. doi: 10.1128/JCM.39.7.2439-2444.2001.

Woods G. L., Shou-Yean Grace Lin, and Edward P. Desmond. 2011. "Susceptibility test methods: Mycobacteria, Nocardia, and other Actinomycetes." In *Manual of Clinical Microbiology*, 11th ed. Washington, DC: ASM Press.

13

Bacterial Cell Breakage or Lysis

Matthew E. Bahamonde

CONTENTS

Introduction

While bacterial cell lysis is a relatively straight-forward process, there are significant variations in its practice. The type of lysis is often dependent upon what subcellular component is to be purified; such as the isolation of DNA versus protein, the scale of the project; miniprep versus industrial-scale, and the type of microbe; Gram-negative versus Gram-positive. Generally, microbial lysis methods can be broken down into three categories based on the relative harshness/intensity of the method: gentle, moderate, or vigorous (Table 13.1).[1] Many of these methods give comparable results and only personal preference plays a role in selection. Others are better suited to specific tasks and should be looked on with preference.

The methods of lysis have been categorized by type of subcellular component to be isolated. The most common components isolated are: nucleic acids, DNA (either genomic or plasmid) and RNA, proteins (either endogenous or recombinant), cell membranes (ghosts), and cell wall (liposaccharide). For both nucleic acids and proteins, there is also discussion of the scale of the project with attention paid to the requirements of large-scale lysis. Differences in lysis between Gram-positive and Gram-negative bacteria as well as Archaea are pointed out where appropriate.

Nucleic Acid Lysis

The two primary types of DNA (genomic and plasmid) extracted from bacteria require differing lysis methods. The extraction of RNA requires additional care due to the fragility of the RNA molecule. For either type, a gentle method is best as to prevent damage to the DNA in the form of cleavage or denaturation or degradation of the RNA.

Plasmid DNA

During the preparation of plasmid DNA, lysis should allow for the plasmids to exit the cell without releasing any of the genomic DNA. This is commonly accomplished via alkaline/sodium dodecyl sulfate (SDS) lysis. During this procedure, the bacteria are exposed to a mild (0.2N) sodium hydroxide (NaOH) solution in the presence of 1% SDS. Exposure to these chemicals is kept relatively short, usually around 5 minutes, and it is often done on ice to further slow the reaction. The reason is to prevent complete loss of the cell membrane due to solubilization by the SDS. The SDS also begins to denature proteins. In the NaOH both proteins and DNA are denatured and RNA is degraded. The genomic DNA is often sheared during the lysis process and, upon denaturation, complementary strands may drift away from each other. The plasmid DNA is also denatured, but because it is relatively small and circular the complementary strands will be joined like links on a chain. When reannealing occurs, the strands will base-pair properly reconstituting the double-stranded DNA plasmids. The genomic DNA strands will not be near their complement and will reanneal into a large mass of intertwined DNA. This reannealing occurs due to neutralization by addition of acidic potassium or sodium acetate. The addition of these salts also causes proteins and SDS to precipitate. This precipitation traps the larger genomic DNA fragments. The preparation is cleared by centrifugation leaving plasmid DNA, some RNA, and a few

DOI: 10.1201/9781003099277-14

TABLE 13.1

Various Lysis Methods Categorized by Their Intensity

Gentle	Moderate	Vigorous
Alkaline lysis	Mortar/Pestle	French press
Triton X-100	Blender	Manton-Gaulin Homogenizer
Phenol		Bead mill
Boiling		Sonication
Proteinase K		
Freeze/Thaw		
Lysozyme/EDTA		

proteins in the supernatant. RNase treatment followed by extraction with organic solvents leaves a solution of nearly pure plasmid DNA, which can be precipitated with alcohol.[2,3] The alkaline lysis method has become the method of choice for the preparation of plasmid DNA and is the method found in the commercial plasmid DNA kits listed in Table 13.2.

A second common method of lysis uses heat (boiling) instead of high pH and a nonionic detergent such as Triton X-100 or Tween-20 instead of SDS. Though these detergents are not as disruptive as SDS, the heat completely denatures both the

TABLE 13.2

List of Commercial Suppliers of Bacterial Nucleic Acid Purification Kits

Supplier[a]	Plasmid DNA[b]	Genomic DNA	RNA
Qiagen Inc. qiagen.com	QIAprep Spin kits QIAprep 96 Plus kits HiSpeed Plasmid kits QIAfilter Plasmid kits QIAGEN Plasmid kits QIAGEN Large-Construct kit EndoFree Plasmid kits DirectPrep 96 kits QuickLyse kits CompactPrep Plasmid kits	DNeasy Blood & Tissue kits QIASymphony DSP kits Blood & Culture DNA kits REPLI-g kits Yeast/Bact. kit Gentra Puregene QIAamp 96 DNA QIAGEN Genomic-tip 20/G	RNeasy Protect Bacteria Mini kits
Promega Corp. promega.com	Wizard® MagneSil® DNA Purification Systems Wizard® MagneSil Tfx™ System PureYield™ Plasmid Systems Wizard® *Plus* DNA Purification Systems Wizard® SV 96 & SV9600 Plasmid DNA Purification Systems	Wizard® Genomic DNA Purification kits Maxwell® RSC Cultured Cells DNA kit	PureYield™ RNA Midiprep System
Bio-Rad Laboratories biorad.com	Aurum™ Plasmid kits Quantum® Prep kits	InstaGene™ Matrix Chelex® 100 Resin	Aurum™ Total RNA 96 kits PureZol™ Isolation Reagent
Sigma-Aldrich sigmaaldrich.com	GenElute™ Plasmid kits GenElute™ HP Plasmid kits GenElute™ 5-Minute Plasmid kits GenElute™ Endotoxin-Free Plasmid kits High Pure Plasmid Isolation kits	GenElute™ Bacterial Genomic kits	GeneElute™ Total RNA Purification kits
Zymo Research zymoresearch.com	ZymoPURE Plasmid Zyppy Plasmid ZR Plasmid ZR BAC	Quick-DNA Fungal/Bacterial ZymoBIOMICS DNA	Direct-zol RNA Quick-RNA Fungal/Bacterial Quick-RNA Fecal/Soil Microbe
Beckman Coulter beckman.com	CosMC Prep	FormaPure XL Total	FormaPure XL Total
Lucigen Corporation lucigen.com	FosmidMAX™ DNA Purification kit	MasterPure™ Gram Positive DNA Purification kit Meta-G-Nome DNA Isolation kit Metagenomic DNA Isolation kit for Water	MasterPure™ Complete RNA Purification kit
New England Biolabs neb.com	Monarch® Plasmid kit	Monarch® Genomic DNA Purification kit	Monarch® Total RNA kit
Biovision Incorporated biovision.com	Express Plasmid kits Plasmid ezFilter kits Plasmid kits	Bacterial Genomic DNA Isolation kit	EasyRNA Bacterial RNA kit
G-Biosciences gbiosciences.com	GET™ Plasmid DNA kit	OmniPrep™ Genomic DNA kit GET™ Genomic DNA kit XIT™ Genomic DNA kit MegaLong™ Genomic DNA kit	GET™ Total RNA kit

TABLE 13.2 *(Continued)*

List of Commercial Suppliers of Bacterial Nucleic Acid Purification Kits

Supplier[a]	Plasmid DNA[b]	Genomic DNA	RNA
As One International, Inc. asone-int.com	Invisorb Plasmid HTS kit Invisorb Spin Plasmid kit	InviMag Bacterial DNA kit InviMag Pathogen kit InviMag Stool kit InviMag Universal kit PSP Spin Stool DNA Plus kit RTP Spin Bacteria DNA kit RTP Spin Pathogen kit	InviMag Universal kit
Takara takarabio.com	NucleoBond® PC Plasmid Purification kits NucleoSpin® Plasmid Purification kits NucleoSpin® Plasmid Easy Pure kit NucleoSpin® Plasmid QuickPure kit NucleoSnap® Plasmid Midi kit NucleoBond® Xtra kits	NucleoSpin® Microbial DNA kit NucleoSpin® Soil kit NucleoSpin® DNA/RNA kit NucleoSpin® DNA/RNA Water kit	NucleoBond® RNA Soil kit NucleoMag® Pathogen kit
ThermoFisher Scientific thermofisher.com	PureLink™ Quick Plasmid Purification kits PureLink™ HiPure Plasmid Purification kits PureLink™ ProQuick 96 Plasmid Purification kits ChargeSwitch® Plasmid kits CloneChecker™ System PureLink™ Plasmid kits S.N.A.P. ™ Plasmid kits ChargeSwitch™ Pro Filter Plasmid kit	PureLink™ Microbiome DNA kit PureLink™ Pro Genomic DNA kits MagMax-96 DNA Multisample	TRIzol® Max™ Bacteria RNA Isolation kit

[a] The suppliers shown are the most common, but are by no means the only suppliers. The kits listed are representative of the supplier's product line and do not constitute their entire catalog.

[b] Plasmid DNA kits are often available for varying yields of DNA; mini up to 20 µg, midi up to 100 µg, maxi up to 500 µg, mega up to 2500 µg, and giga up to 10,000 µg.

proteins and the DNA. Upon cooling, coagulated proteins trap genomic DNA and both precipitate out of solution. The boiling method does contain more RNA as it is not destroyed by the heat as it is by high pH. This method is often augmented by the addition of lysozyme (activity explained later in this chapter).[4,5]

Prior to the lysis steps outlined previously, there is usually the requirement of concentrating the bacterial cells via centrifugation and resuspension in a smaller volume. Typically, the buffered solution used to resuspend contains a chelating agent, typically, ethylenediaminetetraacetic acid (EDTA), and an osmotic regulator such as glucose. EDTA is a chelator of divalent cations and helps to inactivate DNases by removing Mg^{2+}. It also aids in disrupting the outer envelope of Gram-negative bacteria by removing Mg^{2+} from the lipopolysaccharide (LPS) layer. The osmotic regulator is important to prevent the bacterial cells from bursting in the hypotonic solution which would defeat the purpose of the controlled lysis procedures.

Genomic DNA

Unlike a plasmid DNA preparation, a genomic or chromosomal DNA preparation requires the complete degradation of the cell's peptidoglycan layer and cell membrane. Gram-negative bacteria also need their LPS layer disrupted. Digestion with the enzyme lysozyme is required for gram-positive bacteria and is useful, but not required for Gram-negative bacteria. Lysozyme hydrolyzes β(1-4)-linkages between N-acetylmuramic acid and N-acetyl-D-glucosamine residues in peptidoglycans.[6] Other enzymes (Table 13.3) may also prove useful for specific bacterial types that may be resistant to lysozyme. The cells are

then subjected to extended incubation (>15 minutes) in 1% SDS at elevated temperature (37–45°C). This is either followed or accompanied by proteolytic enzyme treatment to further remove proteins. Proteinase K is the enzyme of choice since it actually works better in the presence of SDS. Treatment with DNase-free RNase is also required to remove RNA contamination.

TABLE 13.3

Enzymes Used During Bacterial Lysis

Enzyme	Mode of Action	Sensitive Microbe
Achromopeptidase	A lysyl endopeptidase	Gram-positive bacteria resistant to lysozyme
Labiase	Contains β-N-acetyl-D-glucosaminidase and lysozyme activity	Gram-positive bacteria such as *Lactobacillus*, *Aerococcus*, and *Streptococcus*
Lysostaphin	Cleaves polyglycine cross-links in peptidoglycan	*Staphylococcus* species
Lysozyme	Cleaves between N-acetylmuramic acid and N-acetyl-D-glucosamine residues in peptidoglycans	Gram-positive cells with Gram-negative cells being less susceptible
Mutanolysin	Cleaves N-acetylmuramyl-β(1-4)-N-acetylglucosamine linkages	Lyses *Listeria* and other Gram-positive bacteria such as *Lactobacillus* and *Lactococcus*

Source: Modified from: http://www.sigmaaldrich.com/life-science/metabolomics/enzyme-explorer/learning-center/lysing-enzymes.html#bacterial

As with plasmid preparations chelation of divalent cations such as Mg^{2+} with EDTA is important for lysis and for maintaining the integrity of the chromosomal DNA.[7] Finally, it is critical to treat the lysate gently to prevent shearing of the chromosomal DNA. The purification of genomic DNA, like plasmid DNA, has been reduced to a commercial kit. A list of suppliers is found in Table 13.2.

Mechanical disruption has been found to increase DNA yields when used in conjunction with enzymatic lysis. This is of particular concern when trying to increase the sensitivity of PCR reactions.[8] Some commercial kits rely on mechanical, concussive disruption in the form of bead beating (see Table 13.2), which breaks up bacterial cell wall mechanically by vibrating bacteria with microbeads at high speed.[9] This method can be used with any bacterial strain and is particularly usefully for pathogenic bacteria as there are virtually no unlysed cells, a possibility with enzymatic disruption due to uneven mixing or inactivated enzyme.

RNA

During RNA isolation, the most important issue is stabilization of the RNA. RNases, unlike DNases, are very stable and are found throughout the lab. In addition, because bacterial mRNAs lack a 5' cap and 3' poly(A) tail they are more unstable than their eukaryotic counterparts, with an average half-life of approximately 3 minutes.[10] Traditionally chaotropic (chaos producing) agents, such as guanidinium isothiocyanate or guanidinium hydrochloride, are used to inactivate RNases and preserve RNA. However, there are proprietary reagents such as RNAprotect® by Qiagen Inc. (www.qiagen.com), which can be added to a bacterial culture prior to lysis to protect RNA and reduce artifactual changes in expression that may occur during lysis. In addition to treating the culture, all solutions, glassware, and plasticware must be RNase-free. Solutions except those containing Tris must be treated with 0.1% diethyl pyrocarbonate (DEPC) to inactivate RNases. Autoclaving of solutions after DEPC treatment destroys the DEPC. Tris solutions cannot be treated because Tris reacts with DEPC to inactivate it. Instead, Tris solutions are made using DEPC-treated water. Glassware must be baked at 300°C for at least 4 hours (autoclaving does not inactivate all RNases).[5] Plasticware should be bought already RNase-free. Alternatively, plasticware and equipment may be cleaned with RNase *Zap*™.

The actual lysis can be done either enzymatically or mechanically. Gram-negative bacteria may be lysed with lysozyme alone or in combination with Proteinase K. Gram-positive bacteria should be lysed with both lysozyme and Proteinase K. Certain difficult-to-lyse Gram-positive bacteria may require mechanical disruption with bead beating or a tissue homogenizer in addition to enzymatic lysis. A tissue homogenizer, sold under various brand names, is a rapidly turning rotor that draws liquid suspended cells into a core region. The cells are repeatedly cycled through narrow slits, which mechanically shears them. Mechanical lysis on its own is suitable for a wide range of bacteria and is rapid, but overall, RNA yields are generally lower than when used in conjunction with enzymatic methods.[11] Depending on the purification method, RNase-free DNase may be required to remove any contaminating DNA.

Large-Scale Lysis

Most bacterial lysis for plasmid DNA is done in the laboratory with yields varying in size from 20 µg (miniprep) to 10 mg (gigaprep). The advent of applications such as gene therapy and DNA vaccines will require multigram to kilogram amounts of purified plasmid DNA to be produced in a cost-effective manner. Traditional protocols such as alkaline lysis do not scale up well due to complex mixing requirements that can leave local regions of extremely high pH, which can irreversibly damage plasmids. Use of lysozyme is also problematic due to costs. Large-scale lysis of bacterial filter cakes via fragilization followed by heat has been shown to be viable.[12]

Protein Lysis

Lysis for protein extraction either endogenous or recombinant proteins requires mechanical or chemical (enzymatic and mild detergent) disruption since protocols such as alkaline lysis deliberately destroy proteins. The addition of serine protease inhibitors, such as phenylmethylsulfonyl fluoride (PMSF), or aprotinin are also strongly encouraged to improve yield. Lysis also needs to be carried out either on ice or at 4°C to protect the proteins from denaturing heat.

These methods are primarily for the recovery of exogenous, recombinant proteins, but they are also useful for the recovery of endogenous bacterial proteins.

Mechanical Lysis

Traditionally, bacteria are lysed through liquid homogenization. In this method, cells are sheared by forcing them through a small opening. With bacteria, this is often done with a hand-held tissue homogenizer or the larger French press. A French press consists of a piston that applies high pressure to a sample, forcing it through a tiny hole in the press. Only two passes are required for efficient lysis due to the high pressures used with this process. The equipment is expensive, but the French press is often the method of choice for breaking bacterial cells mechanically. Typical sample volumes range from 40 mL to 250 mL. Sonication or the use of high frequency sound waves is commonly used for samples <100 mL. A vibrating probe is immersed in a liquid cell suspension and energy from the probe creates shock waves that reverberate through the sample. The primary disadvantage of mechanical disruption besides a requirement of expensive specialized equipment is the potential for sample overheating. Either generalized sample heating or localized regions of overheating within a sample can denature proteins, lowering final yield. Lysis must not only be performed at low temperature, but particularly with sonication, the lysis needs to be done in short bursts with enough time in between to allow the sample to cool again.

Because mechanical lysis breaks cells completely all intracellular components, including nucleic acids are released. These can increase viscosity of the lysate and complicate downstream processing. To resolve this, DNase and RNase are added to the sample or Benzonase® (registered trademark of Merck KGaA,

Darmstadt, Germany) alone. Benzonase® is an endonuclease that degrades all forms of DNA and RNA (single-stranded, double-stranded, linear, and circular). One final method, freeze/thaw, can also be used to lyse bacterial cells for the release of recombinant proteins.[13] This technique involves repeated cycles of freezing a cell suspension and then thawing to room temperature or 37°C. Ice crystals that form during the freezing process cause the cells to swell and ultimately burst. The problem is the process can consume a lot of time and is effective only on relatively small samples.

Chemical Lysis

For small-scale lysis, chemical methods are effective and economical. Many companies sell protein lysis kits based on this method (Table 13.4). These kits contain a nonionic detergent-based lysis buffer, lysozyme, and DNase/RNase or Benzonase®. Nonionic detergents are considered more gentle than SDS and less likely to denature proteins.[14] However, most purification schemes will require the extra step of removal of the detergent.

TABLE 13.4

List of Commercial Suppliers of Bacterial Protein Purification Kits

Supplier	Kits
Bio-Rad Laboratories biorad.com	MicroRotofor™ Cell Lysis kit ReadyPrep™ Protein Extraction kit
MilliporeSigma emdmillipore.com	BugBuster® Protein Extraction Reagent BugBuster® + Benzonase® Protein Extraction Reagent BugBuster® + Lysonase Protein Extraction Reagent BugBuster® 10X Protein Extraction Reagent
G-Biosciences gbiosciences.com	PopLysis™ Bacterial Lysis and Extraction Bacterial PE LB™ Bacterial PE LB™ in Phosphate Buffer FOCUS™ Bacterial Proteome
Thermo Fisher Scientific thermofisher.com	B-PER II (2x) Bacterial Protein Extraction Reagent B-PER Complete Bacterial Protein Extraction Reagent B-PER Bacterial Protein Extraction Reagent B-PER (PBS) Bacterial Protein Extraction Reagent
Promega Corp. promega.com	HaloTag® Protein Purification System Magne® HaloTag® Protein Purification Beads HisLink™ Protein Purification System FastBreak™ Cell Lysis Reagent
Qiagen Inc. qiagen.com	Qproteome Bacterial Protein Prep kit Ni-NTA Fast Start kit NoviPure Microbial Protein kit NoviPure Soil Protein kit
Sigma-Aldrich sigmaaldrich.com	CelLytic™ Express CelLytic™ B Plus kit GenElute™ RNA/DNA/Protein Purification kit
Zymo Research zymoresearch.com	His-Spin Protein Miniprep kit Strep-Spin Protein Miniprep kit MBP-Spin Protein Miniprep kit
As One International, Inc. asone-int.com	Purkine GST-tag Protein Purification kit Purkine His-tag Protein Purification kit Purkine MBP-tag Protein Purification kit

The major disadvantages of these methods are their relative expense due primarily to the enzymes involved.

Large-Scale Lysis

In commercial or very large-scale laboratory settings, it is common to require lysis from tens of liters to thousands of liters of cells. As stated earlier, chemical means are not viable at large-scale due to expense, so mechanical lysis is the most common method of large-scale lysis. Of the methods discussed, neither French press nor sonication is amenable to large-scale protein production. The most common methods are bead beating with a mechanical ball mill or Manton-Gaulin Homogenizer. These are effective at lysing large volumes of cells; continuously operating ball mills have the capability of lysing up to 40 L per hour. Due to the tremendous heat generated by the process, ball mills are jacket-cooled, some employing liquid nitrogen to prevent overheating of samples.[15]

Bacterial Ghosts

Bacterial ghosts are empty cells with intact nondenatured envelopes.[16] They have potential uses as drug carriers and vaccine adjuvants.[16–18] They retain all the functional and antigenic determinants found in the living bacteria. Bacterial ghosts are generated through controlled expression of the PhiX174 gene *E* in Gram-negative bacteria. PhiX174 gene *E* encodes a membrane protein that fuses the inner and outer membranes, creating a tunnel through which the cytoplasmic contents of the bacteria are expelled. Except for the tunnel generated by the *E* gene product, the morphology and cell surface structures are unaffected by the lysis. The lumen of the ghost may be filled with vaccine[17] or drugs.[16] In addition to the greater carrying capacity as compared to other delivery systems, ghosts also have the potential to target tissues such as the mucosal surfaces of the gastrointestinal and respiratory tract, and their uptake by phagocytes and M cells.[16]

LPS Lysis

While not technically lysis, the liposaccharide cell envelope of Gram-positive bacteria can be removed and purified via extraction. The most common method uses phenol and hot water.[19] Alternatives include butanol and water as well as detergent.[20]

Archaea

Because the cell walls of Archaea differ from those of eubacteria, enzymatic lysis is not the optimal method of disruption. Archaeal cell walls are composed of the polysaccharide pseudomurein and contain no peptidoglycan.[21] This precludes the use of lysozyme. However, most mechanical methods of lysis, such as the French press or bead beating, are effective and allow archaea to be treated the same as bacteria.

Conclusions

The lysis methods discussed in this chapter are the most popular, but not necessarily the sole methods of lysis. Other methods may be more suitable depending on your cell type, volume, or final application.

REFERENCES

1. Scopes, R. K., *Protein Purification: Principles and Practice*, 3rd ed. Springer-Verlag, New York, 1994.
2. Birnboim, H. C., A rapid alkaline extraction method for the isolation of plasmid DNA, *Methods Enzymol* 100, 243–55, 1983.
3. Birnboim, H. C. and Doly, J., A rapid alkaline extraction procedure for screening recombinant plasmid DNA, *Nucleic Acids Res* 7 (6), 1513–23, 1979.
4. Ausubel, F. M., *Current Protocols in Molecular Biology*, John Wiley & Sons, New York, 2001.
5. Sambrook, J. and Russell, D. W., *Molecular Cloning: A Laboratory Manual*, 3rd ed. Cold Spring Harbor Laboratory Press, Cold Spring Harbor, New York, 2001.
6. Feiner, R. R., Meyer, K., and Steinberg, A., Bacterial lysis by lysozyme, *J Bacteriol* 52 (3), 375–84, 1946.
7. Brown, W. C., Sandine, W. E., and Elliker, P. R., Lysis of lactic acid bacteria by lysozyme and ethylenediaminetetraacetic acid, *J Bacteriol* 83, 697–8, 1962.
8. Zoetendal, E. G., Ben-Amor, K., Akkermans, A. D., Abee, T., and de Vos, W. M., DNA isolation protocols affect the detection limit of PCR approaches of bacteria in samples from the human gastrointestinal tract, *Syst Appl Microbiol* 24 (3), 405–10, 2001.
9. Odumeru, J., Gao, A., Chen, S., Raymond, M., and Mutharia, L., Use of the bead beater for preparation of Mycobacterium paratuberculosis template DNA in milk, *Can J Vet Res* 65 (4), 201–5, 2001.
10. McGarvey, D. J. and Quandt, A., *Stabilization of RNA Prior to Isolation*, QIAGEN GmbH, Hilden, Germany, 2003.
11. QIAGEN. *RNAprotect® Bacteria Reagent Handbook*, 2015.
12. O'Mahony, K., Freitag, R., Hilbrig, F., Muller, P., and Schumacher, I., Proposal for a better integration of bacterial lysis into the production of plasmid DNA at large scale, *J Biotechnol* 119 (2), 118–32, 2005.
13. Johnson, B. H. and Hecht, M. H., Recombinant proteins can be isolated from *E. coli* cells by repeated cycles of freezing and thawing, *Biotechnology* 12 (13), 1357–60, 1994.
14. Godson, G. N. and Sinsheimer, R. L., Lysis of Escherichia coli with a neutral detergent, *Biochim Biophys Acta* 149 (2), 476–88, 1967.
15. Majors, R.E., Sample preparation for large-scale protein purification, *LCGC Europe* 19 (2), 82–92, 2006.
16. Huter, V., Szostak, M. P., Gampfer, J., Prethaler, S., Wanner, G., Gabor, F., and Lubitz, W., Bacterial ghosts as drug carrier and targeting vehicles, *J Control Release* 61 (1–2), 51–63, 1999.
17. Jalava, K., Hensel, A., Szostak, M., Resch, S., and Lubitz, W., Bacterial ghosts as vaccine candidates for veterinary applications, *J Control Release* 85 (1–3), 17–25, 2002.
18. Lubitz, W., Witte, A., Eko, F. O., Kamal, M., Jechlinger, W., Brand, E., Marchart, J., Haidinger, W., Huter, V., Felnerova, D., Stralis-Alves, N., Lechleitner, S., Melzer, H., Szostak, M. P., Resch, S., Mader, H., Kuen, B., Mayr, B., Mayrhofer, P., Geretschlager, R., Haslberger, A., and Hensel, A., Extended recombinant bacterial ghost system, *J Biotechnol* 73 (2–3), 261–73, 1999.
19. Westphal, O. and Jann, K., Bacterial lipopolysaccharides extraction with phenol-water and further applications of the procedure, *Methods Carbohydr Chem* 5, 83–93, 1965.
20. Joiner, K. A., McAdam, K. P., and Kasper, D. L., Lipopolysaccharides from Bacteroides fragilis are mitogenic for spleen cells from endotoxin responder and nonresponder mice, *Infect Immun* 36 (3), 1139–45, 1982.
21. Balch, W. E., Fox, G. E., Magrum, L. J., Woese, C. R., and Wolfe, R. S., Methanogens: reevaluation of a unique biological group, *Microbiol Rev* 43 (2), 260–96, 1979.

14

Major Culture Collections and Sources

Lorrence H. Green

Since the publication of the previous edition of this book, one of the major changes that has taken place in the area of culture collections has been the efforts to standardize how repositories function. To this end, the United States Culture Collection Network was founded in 2012 [1]. The network's mission is to:

> facilitate the safe and responsible utilization of microbial resources for research, education, industry, medicine, and agriculture for the betterment of humankind by providing opportunities for US culture collection workers to engage with each other and with the broader culture collection community

Their main focus has been to ensure standardization of infrastructure items such as access to strains, robust taxonomy, quality control, stock preservation, and collection succession, among others. As methods for identifying microorganisms have come to rely more and more on methods such as DNA sequencing and metagenomics, instead of the traditional enzyme activity tests; the need for absolute consistency in standard strains has become of paramount importance. More information about the network can be found at http://www.usccn.org/Pages/default.aspx (accessed October 23, 2020).

Many sources exist for obtaining microbiological species in pure culture, which can be used as references. Foremost among these is the American Type Culture Collection (ATCC) (Manassas, VA, USA) (www.ATCC.org, accessed October 23, 2020). Several commercial companies such as Microbiologics (www.microbiologics.com, accessed October 23, 2020) and Remel (www.remel.com/About.aspx, accessed October 23, 2020) also supply ATCC microorganisms.

The World Federation for Culture Collections (www.wfcc.info/index.php/collections/display, accessed October 23, 2020) lists 789 culture collections in 77 countries and regions in the world. Among these are:

> Japan Collection of Microorganisms (jcm.brc.riken.jp/en/, accessed October 23, 2020).
>
> HAMBI Culture Collection (Finland) (www.helsinki.fi/en/infrastructures/biodiversity-collections/infrastructures/microbial-domain-biological-resource-centre-hambi# section-63074, accessed October 23, 2020).
>
> Biotech Culture Collection Laboratory (Thailand) (www.biotec.or.th/bcc/, accessed October 23, 2020).
>
> Culture Collection of Switzerland (CCOS) (www.ccos.ch/material/ccos_strains, accessed October 23, 2020).
>
> Microbial Culture Collection (India) (https://mtccindia.res.in/, accessed October 23, 2020).
>
> The Chinese General Culture Collection Center (www.cgmcc.net/english/, accessed October 23, 2020).
>
> Amazon Bacteria Collection (Brazil) (http://cbam.fiocruz.br/, accessed October 23, 2020)
>
> The National Collection of Industrial, Marine and Food Bacteria (Scotland) (www.ncimb.com, accessed October 23, 2020).

Another change since the last edition of this book has been the increasing number of collections that specialize in specific organisms. For instance:

> Coli Genetic Stock Center (https://cgsc.biology.yale.edu/, accessed October 23, 2020): a collection that contains non-pathogenic strains of *Escherichia coli*: predominantly K-12 derivatives, but a few B strains.
>
> UC Davis Viticulture and Enology (https://wineserver.ucdavis.edu/industry-info/enology/culture-collection, accessed October 23, 2020): a collection that specializes in wine yeast and bacteria from around the world.
>
> Culture Collection of Cryophilic Algae (http://cccryo.fraunhofer.de, accessed 3/9/2020): a collection that specializes in extremophiles.

REFERENCE

1. Kyria Boundy-Mills, et al. The United States Culture Collection Network (USCCN): Enhancing Microbial Genomics Research through Living Microbe Culture Collections. *Applied, and Environmental Microbiology* 81, 5671, 2015.

15

Epidemiological Methods in Microbiology

Tyler S. Brown, Barun Mathema, and D. Ashley Robinson

CONTENTS

Introduction

Epidemiological methods in microbiology include laboratory and analytical tools that are used to study the microbial distributions and determinants of infectious disease in host populations. The scope of this chapter is limited to a discussion of the tools that are used to study the molecular and genomic epidemiology of bacterial infectious disease, but these tools have broader application for studying diseases caused by fungi, parasites, and viruses. Natural populations of named bacterial species can accumulate immense levels of genetic variation, and the now widespread use of whole genome sequencing has enabled this variation to be detected at the base-pair level across nearly complete genomes. Analysis of this genetic variation provides the basis for making epidemiological and evolutionary inferences about short-term bacterial transmission, longer-term bacterial evolution, and dispersal, and for identifying the precise bacterial determinants of disease. In this chapter, we discuss the criteria that are used to evaluate different bacterial strain typing tools, the concepts that are used to interpret the genetic variation that is revealed by these tools, and the epidemiological applications for which these tools are deployed.

Overview of Strain Typing

Some Definitions and Assumptions

Although the fundamental concept of bacterial species is still subject to debate, some intraspecies taxonomic units of epidemiological relevance have been defined [1–3]. An isolate has been defined as a group of bacterial cells that represent a pure culture obtained from a single colony grown on a solid medium [4, 5]. Isolates are collected from sources such as clinical specimens, fomites, and environmental samples. A strain has been defined as a group of isolates that are genotypically or phenotypically identical to each other and distinct from other such isolates [4, 5]. A clone has been defined as a group of isolates that are genotypically identical to each other due to common ancestry. It has been emphasized that multiple genotypic markers must be examined in order to classify isolates into a clone [4, 5]. Strain typing tools are the laboratory techniques that are used to assess variation in the phenotypes and genotypes of bacterial isolates. With the increased resolution to distinguish between isolates, which has been made possible in recent years by advances in whole genome sequencing technologies, it has become apparent that substantial

DOI: 10.1201/9781003099277-16

genetic variation exists among groups of isolates previously designated as strains and clones [6–8]. Consequently, in recent years, there has been less focus on developing new strain typing tools and more focus on using sequence data to estimate the degree of genetic relatedness between isolates that is most epidemiologically relevant [9].

Isolates that are part of the same chain of transmission are defined as epidemiologically related. A fundamental assumption that underlies the epidemiological use of strain typing is that epidemiologically related isolates are also genetically closely related isolates [10, 11]. This assumption is based on the reasoning that if isolates share a recent common ancestor, they are also likely to share a common set of biological attributes, including virulence factors that enable bacteria to damage their host and cause disease. It is reasoned that a common set of virulence factors will result in a common disease pathophysiology.

Criteria of Strain Typing Tools

Numerous strain typing tools are available to distinguish between isolates of a bacterial species. However, the selection of a strain typing tool should be tailored to the particular epidemiological problem that is presented [5, 12, 13]. Not all epidemiological problems require the use of high-resolution strain typing tools such as whole genome sequencing. For example, a crowded, inadequately ventilated environment may be linked to an outbreak of pneumococcal pneumonia and an intervention may be devised with little knowledge of the genetic relatedness of the isolates collected from the outbreak; knowledge of the capsular serotype and the antibiotic susceptibility of the isolates may be sufficient to devise an intervention [14]. On the other hand, preventing the spread of an epidemic methicillin-resistant *Staphylococcus aureus* (MRSA) strain in a nosocomial environment may require a genotyping tool in order to distinguish cases from those infected with endemic MRSA strains [15, 16]. For bacterial populations with relatively low genetic diversity, which includes some circulating strains of *Mycobacterium tuberculosis*, the higher resolution provided by whole genome sequencing may be required to make epidemiologically meaningful inferences about clustering or transmission [6, 17].

Strain typing should be viewed as complimentary to an epidemiological investigation [13, 18–21]. Strain typing does not replace the examination of space, time, and other risk factors associated with an infectious disease; rather, the results of strain typing are evaluated in the context of these other factors. Finally, it has been emphasized that strain typing should be done to achieve the public health goal of epidemiology to prevent infectious disease [18]. Strain typing is not exclusively within the domain of epidemiology. The need to distinguish among bacterial isolates also arises in systematics and taxonomy, population genetics, phylogenetics, and other basic endeavors.

Several considerations must be made in order to select an appropriate strain typing tool for a particular epidemiological application [5, 12, 13]. Not all biological variation provides a valid target for strain typing. The validity of a strain typing tool is based on its ability to distinguish between epidemiologically related and unrelated isolates, which defines the sensitivity and specificity of the tool, respectively [10]. A strain typing tool should be validated for a particular epidemiological application, which means that it should be able to confirm established epidemiological findings. No "gold standard" exists that can be used for validation. Validation of a strain typing tool is done empirically by comparison of the tool with previously validated tools and by demonstration that epidemiologically related and unrelated isolates can be distinguished [10]. Lastly, the choice of appropriate tool should be considered along with the sampling strategy for collecting isolates, the statistical framework used for hypothesis testing, and the analytical methods used to estimate relatedness between isolates [22].

Three basic performance criteria, which include discriminatory ability, reproducibility, and typeability, have been used to evaluate strain typing tools [5, 20, 23]. Discriminatory ability is defined as the ability of a strain typing tool to detect differences between isolates. Simpson's index of diversity is widely used to quantify discriminatory ability and it takes into account both the richness and abundance of types in a sample of isolates [24]. Simpson's index estimates the probability that two isolates randomly selected from a sample will have different types. Confidence intervals can be calculated for Simpson's index, which enables statistical comparisons of the discriminatory ability of different strain typing tools applied to the same sample of isolates and statistical comparisons of different samples of isolates (e.g., from cases and controls) characterized with the same typing tools [25]. Reproducibility is defined as the ability of a strain typing tool to yield the same results for the same isolates upon repeated testing. Both within- and between-laboratory reproducibility are evaluated. Typeability is defined as the ability of a strain typing tool to score a positive result for each isolate. Nontypeable isolates produce either null or ambiguous results. All three performance criteria are affected by experimental variation in laboratory techniques and by biological variation in the isolates [5, 20, 23].

Numerous convenience criteria also have been used to evaluate strain typing tools [5]. These criteria include the ease of performance (technical simplicity), the speed of performance (rapidity), the applicability to a broad taxonomic range of species (versatility), and cost. Additional criteria, such as concordance, stability, and ease of interpretation [5, 23], also have been used to evaluate strain typing tools. Epidemiological concordance is analogous to validity, as defined above. On the other hand, strain typing concordance describes the relative agreement of different strain typing tools in classifying isolates as "same" or "different." The Rand and Wallace indices have been used to quantify concordance [26]. Adjusted versions of these indices are available that provide chance-corrected estimates of concordance, and confidence intervals for these indices can be calculated to enable statistical comparisons of concordance [26–28].

Description of Strain Typing Tools

Subgenomic Tools

Classical Strain Typing

Strain typing tools are commonly classified according to whether they assess phenotypes or genotypes. Historically, strain typing tools were based on phenotypes, but these tools have been largely replaced with genetic methods. Phenotyping tools have been used to assess variation in metabolic characteristics (biotyping), susceptibility to lytic bacteriophage (phage typing), susceptibility to antibiotics (resistotyping), immunological reactivity to antibodies (serotyping), and other characteristics. While phenotypic characteristics can be highly correlated with epidemiological relationships, they are generally poor indicators of genetic relationships [29, 30]. The expression of phenotypes can be strongly influenced by environment (e.g., temperature, nutrients, oxygen availability), so a shared phenotype might reflect a shared environmental variable rather than a shared ancestry in a sample of isolates [29, 30].

Several classical strain typing tools, previously in widespread use, are important to describe. Multilocus enzyme electrophoresis (MLEE) is a phenotyping tool that was pivotal to the development of the field of bacterial population genetics [31, 32]. MLEE indirectly assesses variation in gene sequences based on differences in the electrostatic charge of different alleles at expressed enzyme loci [33]. MLEE was the forerunner of a genotyping tool called multilocus sequence typing (MLST) that directly assesses variation in gene sequence from fragments of enzyme loci and other loci that encode products with cellular housekeeping functions [34].

Another category of classical strain typing tools called band-based, or molecular fingerprinting, tools produce data in the form of electrophoretic banding patterns. These tools were the basis for early advancements in molecular epidemiology in the 1980s and 1990s [35, 36]. Among these tools, restriction fragment length polymorphism (RFLP), pulsed-field gel electrophoresis (PFGE), and variable number of tandem repeats (VNTR), remain the most relevant. RFLP refers to variation that is revealed through the digestion of DNA templates with restriction endonucleases [20, 37]. These enzymes recognize short nucleotide sequences (~4–6 bp) called restriction sites with varying levels of specificity depending on the enzyme. If the restriction sites occur within a DNA template, then the enzymes will precisely cut the template at these sites resulting in pieces of a template called restriction fragments. These fragments can be separated by agarose gel electrophoresis. The number and location of the restriction sites can differ among isolates. Banding patterns from RFLP analysis can be made less complex by Southern blotting and hybridization with probes of rRNA genes (ribotyping), insertion sequences (IS typing), and other sequences [20, 37, 38]. PFGE analysis uses restriction endonucleases that recognize relatively few restriction sites in a chromosome, resulting in a small number (~10–30) of large fragments (~10–800 kb) [4, 20]. These large restriction fragments are separated by special electrophoresis equipment that alternates the direction of current. Lastly, VNTR analysis employs PCR amplification of various kinds repetitive sequences including microsatellites (~1–10 bp) and minisatellites (~10–100 bp) [39, 40]. Arrays of these repeats can differ between isolates in length and in sequence, and the length variation can be distinguished by electrophoresis. Due to relatively high discriminatory ability, band-based genotyping tools have been very useful for local outbreak investigations [41]. On the other hand, reproducibility has been low for many of these tools.

Sequence-Based Genotyping

Sequence-based genotyping tools assess variation directly from the nucleotide sequences of either a single locus or multiple loci. These tools provide a fundamental view of genetic variation [34]. The Sanger method of sequencing has been used widely for epidemiological and evolutionary applications. PCR templates are linearly replicated using a reaction mixture that includes fluorescently-labeled dideoxynucleotides and other reagents. Incorporation of dideoxynucleotides interrupts replication, leading to a ladder of amplicons each with their 3' terminus marked by a fluorescent label that can be detected by an automated sequencer. Single locus sequence typing (SLST) has often targeted highly variable sequences, such as VNTR sequences [42]. In contrast, MLST has often targeted relatively conserved sequences, such as the sort of housekeeping loci used for phenotyping by MLEE [43]. The MLST technique utilizes partial nucleotide sequences (~500 bp) from a set of housekeeping loci (~7), which are selected for a given species to be present in all tested isolates, physically distant from each other on the chromosome, and not subject to strong diversifying selection. Unique nucleotide sequences at each locus define alleles, and unique combinations of alleles across loci define sequence types (STs). MLST databases of alleles and STs are publicly accessible via the internet for dozens of microbial species, most of which are bacterial species [44].

Sequence-based genotyping tools achieve high reproducibility. MLST has become one of the most important tools for elucidating bacterial population genetic structure at a coarse scale [45, 46]. Public health networks have used SLST and MLST for international surveillance of bacterial pathogens [44, 47, 48]. Moreover, some SLST schemes have exhibited high enough discriminatory ability to be used for outbreak investigations [49, 50]. The discriminatory ability of VNTR sequences can be greatly improved when characterized by sequencing rather than by electrophoresis [51]. Convenience criteria, such as high time requirement and cost, are the principal weaknesses of sequence-based genotyping tools.

Binary-Based Genotyping

Binary-based genotyping tools assess the presence or absence of specific genetic markers. These tools are generally based on hybridization of labeled genomic DNA to an array of sequences that are fixed onto a surface, such as a membrane or glass slide (microarray or gene-chip) [52]. The reverse of this technique can also be performed, whereby labeled sequences are hybridized to an array of genomic DNA preparations that are fixed onto a surface (library on a slide) [53]. The sequences can consist of parts of a chromosome, such as virulence factors, or they can be more representative of the entire contents of a reference strain's genome using arrayed amplicons or oligonucleotides as is done with comparative genomic hybridization (CGH) [54–57]. CGH studies provided early estimates of gene content variation between strains of a species [52, 54, 55]. Binary-based genotyping tools can exhibit high discriminatory ability, but they can also present problems with reproducibility and they can be time-consuming and costly to use. These tools have most often been applied to studies of the association between virulence factors and infectious disease.

Genotyping of single nucleotide polymorphisms (SNPs) can be viewed as a binary-based genotyping tool. A wide variety of different techniques can be used to distinguish between alleles at polymorphic nucleotide sites in a cost-efficient manner [58]. These techniques make use of allele-specific hybridization, amplification, extension, and ligation reactions, and various characteristics of DNA (e.g., melting temperature) that differ in an allele-specific fashion. Some of these techniques allow the discovery of new SNPs, whereas others are only used to genotype known SNPs. The SNP discovery process needs to be carefully considered in order to reduce biases in subsequent population genetic and phylogenetic analyses [59]. SNPs have been used to provide information about antimicrobial resistance mechanisms [60], and bacterial population genetic structure at the coarse scale of MLST [61, 62] and at a more granular scale [63, 64]. SNP genotyping protocols can be standardized and can provide high reproducibility and discriminatory ability [65].

Genome Sequencing

Second-generation sequencing platforms provide nearly complete, draft-quality genome sequences of bacteria. While these platforms use different underlying technologies that have different strengths and weaknesses [66, 67], they share several characteristics that distinguish them from first-generation Sanger sequencing: (1) libraries of fragmented DNA are produced and platform-specific adapters are ligated to the ends of the fragments; (2) PCR amplification of the library is performed; (3) sequencing is done in parallel, generating millions of relatively short (~50–500 bp) sequence reads. By ligating unique sequences, or barcodes, to each isolate's library, multiple isolates can be sequenced at the same time and the data from each isolate can be separated during downstream analysis. This multiplexing capability decreases the cost per isolate and can make second-generation sequencing competitive in cost with other genotyping tools. Of course, there is a trade-off between the cost per isolate and the number of reads per isolate, which impacts the quality of the resulting genome sequence. Therefore, the goals of the study and the desired quality of the genome sequences need to be carefully considered.

The sequence reads can be assembled into a genome sequence by two basic analysis techniques. With the read-mapping technique, the reads are aligned to a reference sequence, which is often chosen to be a complete genome sequence from a closely related isolate. This technique is useful for discovering SNPs and short insertions and deletions [16, 68, 69]. Several read-mapping quality metrics, such as base depth, quality, and "heterozygosity," should be evaluated [70]. With the *de novo* assembly technique, the reads are aligned to each other to build as complete a genome sequence for an isolate as possible. This technique is useful for assessing gene content [68]. Assembly quality metrics, such as the number and size distribution of contiguous sequences (contigs), should be evaluated [70]. Read mapping also can be used to assess gene content; for example, reads can be mapped onto a customized reference sequence of known virulence factors [71]. Likewise, *de novo* assemblies from multiple isolates can be subsequently aligned to discover SNPs, insertions, and deletions [69]. A large portion (>90%) of a bacterial genome can be assembled with reads from second-generation platforms [66], but repetitive sequences are difficult to sequence and assemble from these platforms.

So-called third- and fourth-generation sequencing platforms, single-molecule real-time (SMRT) sequencing and nanopore-based platforms, respectively, are increasingly used to improve the sequencing of regions where second-generation sequencing is difficult or unreliable [72, 73]. In brief, the core process underlying SMRT sequencing records the sequential addition of individual fluorophore-bound nucleotides to a circular DNA template [72]. This process is parallelized over tens of thousands of specialized reaction cells, which have internal dimensions that are small with respect to the wavelength of light being used. The fluorescence wavelengths measured after each round of a long series of light pulses records the elongating nucleotide sequence in each cell. The DNA template used in SMRT is circular, such that multiple passes over the template are possible during a single sequencing run, allowing for relatively high sequencing depth [72]. In addition, variations in the kinetics of fluorescence emission following each nucleotide addition can be used to detect base-pair modifications such as DNA methylation.

Nanopore-based sequencing uses specialized protein structures (nanopores) that allow for the passage of single nucleotide strands [73]. As each nucleotide of the strand passes through the nanopore, it introduces a distinct disturbance in an electrical field applied across an array of nanopores, which is recorded and used to infer the nucleotide sequence of the strand. This technology is highly compact and portable, allowing for its use in contexts where other sequencers are impractical. Use of nanopore sequencing in infectious disease epidemiology is limited in some situations by this platform's relatively high sequencing error rate [73]. This is most problematic for bacterial species with low genetic diversity where the error rate of nanopore sequencing can potentially exceed the basal mutation rate of the species, but ongoing substantial improvements in this technology and related computational approaches for variant calling from nanopore data may help to address this issue.

Like other sequence-based strain typing tools, genome sequencing should have a high reproducibility. However, it should be recognized that *all* current sequencing platforms have limitations [66, 67, 72, 73]. Standardized or "best practices" protocols are needed for both the sequencing and analysis steps, as with other genotyping tools, to maximize reproducibility. The importance of the reproducibility criterion becomes apparent when considering that a single SNP may be relevant to hypotheses of transmission among outbreak isolates, such as some isolates from the 2011 *Escherichia coli* O104:H4 outbreak in Europe [74]. In contrast, the high discriminatory ability of genome sequencing has been demonstrated repeatedly. Isolates that are indistinguishable by other genotyping tools, even using highly discriminatory markers such as VNTRs, can often be distinguished by genome sequencing [68, 75].

Whole genome sequencing can also resolve between otherwise indistinguishable isolates from the same host and, consequently, has become an important tool for studying within-host evolution of different bacterial species [76–78]. Likewise, whole genome sequencing has become a central tool for identifying the constituents of entire microbial communities isolated from hosts, fomites, and the environment—a field known as metagenomics. Although metagenomic analysis techniques for species-level identification are well-established, additional work is needed to develop and validate tools for strain typing from these complex data [79, 80]. It is difficult to overstate the impact that genome sequencing has on the epidemiology of bacterial infectious disease [81, 82]. Clearly, the challenge has moved from developing high-throughput, high-coverage genotyping tools that satisfy various performance and convenience criteria, to the daunting task of interpreting the data.

Interpretation of Strain Typing Data

Some Considerations

The interpretation of strain typing data can be complicated even if the performance of the selected strain typing tool is well-characterized. In some cases, the fundamental assumption underlying the epidemiological use of strain typing is violated; epidemiologically related isolates will not always be genetically closely related isolates. For example, contamination of food and water with sewage can result in diarrheal outbreaks due to multiple bacterial species [12]. Furthermore, horizontal gene transfer, recombination, and convergent evolution, can spread a common set of virulence factors or drug-resistance mutations to multiple bacterial strains [83]. In other cases, genetically closely related isolates will not always be epidemiologically related in a discernable fashion; this situation can occur with low-incidence infections or those with long latency periods such as tuberculosis. The selected strain typing tool may not have high enough resolution to identify the epidemiologically relevant variation. For example, undetected virulence factors that are imported on mobile genetic elements or point mutations that slightly alter virulence gene expression, may cause isolates that appear to be identical based on subgenomic typing tools to behave with distinct epidemiological patterns [84, 85]. To reinforce the notion that subgenomic typing tools do not detect all of the variation that occurs within a bacterial genome, it has been emphasized that genotypically identical isolates be referred to as "indistinguishable" rather than "identical" [4, 20]. Finally, it is important to remember that infectious disease is a particular outcome of an interaction that is influenced both by bacterial virulence factors and by host risk factors. Strain typing provides information about the bacterial determinants of infectious disease but, especially for bacterial pathogens that sometimes harmlessly colonize hosts and sometimes cause disease (facultative pathogens), the host's risk factors can decisively influence the outcome of the bacteria-host interaction [10].

Strain typing data is more straightforward to interpret when isolates are shown to be either indistinguishable or completely different. The major difficulty in interpretation arises when isolates are shown to be similar with some differences, because the level of similarity that is epidemiologically relevant must be established [5, 11, 20, 86, 87]. Interpretive criteria have been proposed, but these criteria are specific for particular strain typing tools and particular epidemiological applications. For example, PFGE data are interpreted in local outbreak investigations according to the number of banding pattern differences that can arise per genetic event [4, 20]. The loss of a restriction site due to a single point mutation can result in a three-band difference. Thus, to allow for some variation to accumulate in a strain during the course of an outbreak, isolates that differ by two to three

bands are interpreted as probably part of the same outbreak. Sequence-based strain typing tools allow genetic events to be more precisely and comprehensively discerned, but judgement calls regarding the epidemiological relevance of bacterial genetic variation are made with all strain typing tools. In the context of whole genome sequence data, the number of SNPs between isolates is often used to estimate their degree of genetic relatedness [88]. However, the number of SNPs can be influenced by species biology (mutation rate), density with which isolates are sampled from a given transmission network, and technical details related to the sequencing and analysis protocols. In addition, polyclonal infection and within-host heterogeneity can confound inference of epidemiological relationships between hosts. Thus, the importance of properly validating particular strain typing tools for particular epidemiological applications should be underscored. Interpretive criteria are dependent on how rapidly bacteria evolve and the elapsed time since the isolates shared a common ancestor [11, 87]. For most strain typing tools, these two parameters are not known. However, these two parameters can be estimated from sequence data [19].

Insights from Population Genetics and Phylogenetics

Origin and Fate of Genetic Variation

To interpret the results of strain typing tools, the stability of the markers should be taken into account. Stability is defined as the constancy of the markers over time. Stability is best considered within a population genetics framework, which focuses study on the origin and fate of genetic variation [46, 87, 89]. Bacteria are haploid, which means that they contain a single copy of their chromosome(s). Bacteria also reproduce asexually, which means that two identical daughter cells, or clones, are produced from a single parent cell. Mutations refer to the heritable genetic variations that occur within DNA. These variations include nucleotide substitutions, insertions, and deletions. The movements of mobile genetic elements (e.g., transposons, lysogenic phage, integrative conjugative elements) also produce heritable changes in a genome. When the bacterial DNA mismatch-repair machinery undergoes mutations that impair its function, clones can experience a high rate of mutation [90].

Horizontal gene transfer and recombination between clones can occur if the clones have both a mechanism (genetic factors) and opportunity (ecological factors) to exchange genes. Parasexual processes, such as conjugation, transduction, and transformation, enable clones to horizontally transfer genetic material [91]. Recombination between clones can reduce genetic variation if it occurs frequently within a population. In this sense, recombination is a homogenizing force that can speed the replacement of less favorable mutations. Recombination between clones can preserve genetic variation if it occurs frequently between different populations. In this sense, recombination can allow rare alleles to escape extinction. Recombination between clones can also increase genetic variation by combining mutations, thereby creating new alleles. In population genetics parlance, a unique position on a chromosome is called a locus, and a unique combination of mutations at a locus is called an allele. The consequences of recombination between clones are profound for bacterial population genetic structure and infectious disease epidemiology since it causes virulence factors to embark on independent chains of transmission. If recombination occurs frequently enough, clones can become too ephemeral to be epidemiologically relevant taxonomic units [31].

The frequency of alleles in a population can be altered through stochastic processes (genetic drift) and deterministic processes (natural selection) [59]. Some alleles can increase the fitness of a clone in a particular environment, and thus its frequency, while other alleles have no effect on fitness or are deleterious and are slowly replaced. In other words, natural selection can promote (positive selection) or prevent (negative selection) the accumulation of variation at a locus. So-called neutral alleles, which are considered neither advantageous nor deleterious, are thought to vary in frequency primarily under the influence of genetic drift, although many such alleles may be linked to others that are under positive or negative selection. Demographic processes, such as population subdivisions, growth, and bottlenecks, also alter the frequency of alleles in profound ways. Both neutral demography and natural selection can contribute to the evolution of bacteria [59]. These processes combine to cause bacteria to naturally form a clonal population genetic structure [46, 87, 88].

Criteria of Clonality

Two basic criteria have been used to assess the role of recombination in disrupting clonal population genetic structure [92]. These criteria include the repeated isolation of a spatially and temporally stable clone and the nonrandom association of alleles at different loci. It has been emphasized that a clone should occur more frequently than a random assortment of alleles would produce [31]. The repeated isolation of a clone in a bacterial population that experienced frequent recombination would occur only if natural selection favored a unique combination of alleles across loci. Such selection would be unlikely if the study examined genetic variation in multiple loci that encode products of unrelated physiological function. In this case, the simplest explanation is that recombination is rare and that periodic selection is influencing genetic variation in a relatively clonal population [32].

As clones propagate and accumulate variation, clonal lineages are developed. Because the chromosomes within a clonal lineage share a common ancestry, the alleles that occur at one locus will be statistically associated with the alleles that occur at other loci. This nonrandom occurrence of genetic variation is called linkage disequilibrium (LD) [92]. However, processes other than clonal population growth can cause LD. For example, bacterial populations that consist of genetically isolated subpopulations, where recombination occurs frequently within but not between subpopulations, can produce apparent LD if the strain typing data of the subpopulations are analyzed together. In this case, the LD that is detected is due to fixed mutations that distinguish the subpopulations. Such bacterial subpopulations have been called cryptic species [31]. Additionally, the spread of highly fit clones can produce temporary LD even if recombination occurs frequently within the bacterial population. These highly fit clones can achieve broad geographic distributions, but their genetic identity will eventually be eroded by recombination with other clones. This situation describes an epidemic population genetic structure and, interestingly, may have been first recognized by epidemiologists [31, 93]. In summary, bacteria display

a spectrum of population genetic structures, ranging from the extremes of clonal to freely recombining (panmictic) [31, 45]. Even within the boundaries of a named bacterial species, there can be differences in the relative influence of recombination versus mutation on population genetic structure [83, 94].

Evolutionary Epidemiology

Provided that recombination between clones is relatively rare compared to mutation, the analysis of genetic relatedness can be greatly facilitated by phylogenetic clustering techniques. These techniques enable the tree-like (bifurcating) structures of clonal lineages to be reconstructed from strain typing data, primarily sequence data, and they enable the statistical support for these structures to be quantified [95]. On the other hand, when recombination is relatively common compared to mutation, different loci will have different histories and a single phylogenetic tree will not provide an adequate description of the clone's history. In population genetics terminology, if two isolates share the same allele at a given locus, they are considered *identical by state* at that locus, but this shared allele is only considered *identical by descent* if both isolates inherited it from the same shared ancestor. Thus, in populations with high rates of recombination, the presence of sites that are identical by state, but not by descent can confound efforts to infer the genetic relatedness of isolates. In this case, population genetics techniques can be used to assign isolates to "populations," while accounting for recombination between populations (admixture) [59, 62, 96] and identity by descent can estimated directly as a measurement of relatedness between isolate pairs. Both phylogenetic clustering and population genetics assignment techniques provide a means to classify the variation that is detected through strain typing into biological groups that might have some epidemiological relevance.

Much information about the transmission history of a clone can be extracted from a phylogenetic tree. Not only can the rates of evolution and dates of origin of clonal lineages be estimated, patterns of spatial spread can be studied quantitatively [7, 69, 97, 98]. The powerful insights that arise from phylogenetic analysis of genome sequences raise ethical considerations [21], since the precise role of individuals in the chain of transmission might be revealed [16]. Additional insights about the dynamics of bacterial infectious disease, including estimates of some parameters from standard epidemiological models, may be extractable from population genetics analysis [99]. These advances highlight the potential for strain typing to complement epidemiological investigations. Some advantages of more closely integrating epidemiology and evolution have been noted [46, 75, 100].

Epidemiological Applications of Strain Typing

Local Epidemiology

Local epidemiology refers to the investigation of chains of transmission that unfold over limited spatial and temporal scales. Strain typing in this context has been referred to as comparative typing [86]. Applications include the testing of prior epidemiological hypotheses that have identified a source as a possible cause of an outbreak of infectious disease. Strain typing is essentially confirmatory for this application because it is expected that the epidemiologically related isolates will also be genetically closely related; infection control decisions may be made without the strain typing data [12]. Strain typing can also be applied to identify unsuspected sources in a spatial or temporal cluster of infectious disease. With this application, strain typing can point to a need for an epidemiological investigation or it can prevent a superfluous investigation [10, 12]. Finally, strain typing can be applied to distinguish between reinfection of a person with a new strain versus reactivation of a chronic infection caused by a persistently-colonizing strain (e.g., tuberculosis) [101–103].

Strain typing tools with high discriminatory ability are needed for local epidemiology because the putative outbreak isolates need to be distinguished from potentially closely related sporadic isolates that constitute the endemic bacterial population [16]. VNTR analysis, SLST, and genome sequencing, are examples of tools that can be used for local epidemiology [41, 50, 75] and multiple approaches have been developed to characterize transmission patterns from these data. These tools allow closely related isolates to be identified, and their routes of transmission and reservoirs to be studied. This information can contribute to the development of infection control practices within an institution or within a community [12, 41, 86].

Global Epidemiology

Global epidemiology refers to the investigation of chains of transmission that unfold over broad spatial and temporal scales. Strain typing in this context has been referred to as library typing [86]. Applications include monitoring the global circulation and distribution of strains in different host populations. These data allow the emergence or re-emergence of strains to be detected, and they can distinguish between the spread of strains versus the independent spread of virulence factors [11, 104].

Strain typing tools that exhibit high reproducibility and assess relatively stable markers are needed for global epidemiology. In addition, these tools should be high-throughput, should have standardized protocols, and should provide a nomenclature that facilitates between-laboratory communication about the strains [86]. PFGE, MLST, and genome sequencing are examples of strain typing tools that can be used for global epidemiology [44, 47, 105]. These tools allow distantly related isolates to be clustered into clonal lineages, and their patterns of spread and virulence trends to be studied [106, 107]. This information can contribute to the development of public health strategies at national and international levels [87, 102].

Association Mapping

Association mapping refers to the use of strain typing tools to identify the precise bacterial determinants (virulence factors) of infectious disease [108]. Association mapping can be used to identify virulence factors even when it is not clear whether strains differ in their virulence potential [109–111]. The virulence of a strain is often determined by its gene content, with more virulent strains differing from less virulent strains by the presence of conspicuous genetic elements. The loss of genes from commensal ancestors can also lead to increased virulence [112]. Association mapping studies resemble epidemiological case-control studies, whereby isolates that are sampled from persons with infectious disease (cases) are compared for gene content and other sequence variations with isolates that are sampled from persons without infectious disease (controls). Principles of sound epidemiological study design should be considered [12]. Moreover, since bacterial chromosomes can exhibit strong LD, it is necessary to subsequently assess whether the putative virulence factors that are identified have a causal association with infectious disease or are merely genetically linked to the authentic virulence factors [108, 113, 114]. Experimental studies are needed for these subsequent steps.

Association mapping provides an observational approach to virulence factor discovery that is based on discerning bacterial exposures in human populations. In contrast, the experimental approach to virulence factor discovery that is often used in microbiology is based on assigning bacterial exposures in a laboratory microcosm. The observational approach allows multiple bacterial exposures to be simultaneously evaluated in their natural setting, but its results can be confounded by imperfect study design. By comparison, the experimental approach addresses the biological plausibility of causation between bacterial exposures and infectious disease, but its results are often restricted to the simultaneous evaluation of a single bacterial exposure in a model of infectious disease that imperfectly represents the natural setting. Thus, the observational and experimental approaches to virulence factor discovery should be viewed as complimentary to each other.

Concluding Remarks

Recent major advances in the laboratory techniques available for bacterial strain typing, in particular the advances in whole genome sequencing technologies, have led to substantial progress in understanding the bacterial-specific determinants of infectious disease. These technological advances have led to the unraveling of the sometimes cryptic epidemiology of important old threats to human health, while also raising a host of important new questions. Knowledge of the characteristics of different strain typing tools, the population genetic structure of different bacterial species, and the peculiarities of different epidemiological applications still need to be integrated within an overarching conceptual framework. These will remain essential knowledge areas as whole genome sequencing becomes more routinely used to address epidemiological problems.

REFERENCES

1. Dijkshoorn, L., Ursing, B.M., and Ursing, J.B., Strain, clone and species: comments on three basic concepts of bacteriology, *J. Med. Microbiol.*, 49, 397, 2000.
2. Konstantinidis, K.T., Ramette, A., and Tiedje, J.M., The bacterial species definition in the genomic era, *Philos. Trans. R. Soc. Lond. B*, 361, 1929, 2006.
3. Stackebrandt, E. et al., Report of the ad hoc committee for the re-evaluation of the species definition in bacteriology, *Int. J. Syst. Bacteriol.*, 52, 1043, 2002.
4. Tenover, F.C. et al., Interpreting chromosomal DNA restriction patterns produced by pulsed-field gel electrophoresis: criteria for bacterial strain typing, *J. Clin. Microbiol.*, 33, 2233, 1995.

5. van Belkum, A. et al., Guidelines for the validation and application of typing methods for use in bacterial epidemiology, *Clin. Microbiol. Infect. Dis.*, 13, 1, 2007.

6. Brown, T.S. et al., Genomic epidemiology of Lineage 4 *Mycobacterium tuberculosis* subpopulations in New York City and New Jersey, 1999-2009, *BMC Genomics*, 17, 947, 2016.

7. Challagundla, L. et al., Range expansion and the origin of USA300 North American epidemic methicillin-resistant *Staphylococcus aureus*, *mBio*, 9, e02016, 2018.

8. DeLeo, F.R. et al., Molecular dissection of the evolution of carbapenem-resistant multilocus sequence type 258 *Klebsiella pneumoniae*, *Proc. Natl. Acad. Sci. U S A*, 111, 4988, 2014.

9. Nübel, U. et al., From types to trees: reconstructing the spatial spread of *Staphylococcus aureus* based on DNA variation, *Int. J. Med. Microbiol.*, 301, 614, 2011.

10. Riley, L.W., *Molecular Epidemiology of Infectious Diseases: Principles and Practices*, ASM Press, Washington, DC, 2004, chap. 1.

11. Struelens, M., Molecular typing: a key tool for the surveillance and control of nosocomial infection, *Curr. Opin. Infect. Dis.*, 15, 383, 2002.

12. Foxman, B. et al., Choosing an appropriate bacterial typing technique for epidemiologic studies, *Epidemiol. Perspect. Innov.*, 2, 10, 2005.

13. Goering, R.V. et al., From theory to practice: molecular strain typing for the clinical and public health setting, *Euro Surveill.*, 18, 20383, 2013.

14. Hoge, C.W. et al., An epidemic of pneumococcal disease in an overcrowded, inadequately ventilated jail, *N. Engl. J. Med.*, 331, 643, 1994.

15. Coombs, G.W. et al., Controlling a multicenter outbreak involving the New York/Japan methicillin-resistant *Staphylococcus aureus* clone, *Infect. Control Hosp. Epidemiol.*, 28, 845, 2007.

16. Köser, C.U. et al., Rapid whole-genome sequencing for investigation of a neonatal MRSA outbreak, *N. Engl. J. Med.*, 366, 2267, 2012.

17. Walker, T.M. et al., Whole-genome sequencing for prediction of *Mycobacterium tuberculosis* drug susceptibility and resistance: a retrospective cohort study, *Lancet Infect. Dis.*, 15, 1193, 2015.

18. Foxman, B. and Riley, L., Molecular epidemiology: focus on infection, *Am. J. Epidemiol.*, 153, 1135, 2001.

19. Pybus, O.G., Fraser, C. and Rambaut, A., Evolutionary epidemiology: preparing for an age of genomic plenty, *Phil. Trans. R. Soc. Lond. B*, 368, 20120193, 2013.

20. Tenover, F.C., Arbeit, R.D., and Goering, R.V., How to select and interpret molecular strain typing methods for epidemiological studies of bacterial infections: a review for healthcare epidemiologists, *Infect. Control Hosp. Epidemiol.*, 18, 426, 1997.

21. Tong, S.Y.C., Genomic polish for shoe-leather epidemiology, *Nat. Rev. Microbiol.*, 11, 2013.

22. Meehan, C.J. et al., Whole genome sequencing of *Mycobacterium tuberculosis*: current standards and open issues, *Nat. Rev. Microbiol.*, 17, 533, 2019.

23. Maslow, J. and Mulligan, M.E., Epidemiologic typing systems, *Infect. Control Hosp. Epidemiol.*, 17, 595, 1996.

24. Hunter, P.R. and Gaston, M.A., Numerical index of the discriminatory ability of typing systems: an application of Simpson's index of diversity, *J. Clin. Microbiol.*, 26, 2465, 1988.

25. Grundmann, H., Hori, S., and Tanner, G., Determining confidence intervals when measuring genetic diversity and the discriminatory abilities of typing methods for microorganisms, *J. Clin. Microbiol.*, 39, 4190, 2001.

26. Severiano, A. et al., Evaluation of jackknife and bootstrap for defining confidence intervals for pairwise agreement measures. *PLOS One*, 6, e19539, 2011.

27. Severiano, A. et al., Adjusted Wallace coefficient as a measure of congruence between typing methods, *J. Clin. Microbiol.*, 49, 3997, 2011.

28. Smyth, D.S., Wong, A., and Robinson, D.A., Cross-species spread of SCC*mec* IV subtypes in staphylococci, *Infect. Genet. Evol.*, 11, 446, 2011.

29. Tenover, F.C. et al., Comparison of traditional and molecular methods of typing isolates of *Staphylococcus aureus*, *J. Clin. Microbiol.*, 32, 407, 1994.

30. Zaidi, N., Konstantinou, K., and Zervos, M., The role of molecular biology and nucleic acid technology in the study of human infection and epidemiology, *Arch. Pathol. Lab. Med.*, 127, 1098, 2003.

31. Maynard Smith, J. et al., How clonal are bacteria? *Proc. Natl. Acad. Sci. U S A*, 90, 4384, 1993.

32. Selander, R.K. and Levin, B.R., Genetic diversity and structure in *Escherichia coli* populations, *Science*, 210, 545, 1980.

33. Selander, R.K. et al., Methods of multilocus enzyme electrophoresis for bacterial population genetics and systematics, *Appl. Environ. Microbiol.*, 51, 873, 1986.

34. Maiden, M.C. et al., Multilocus sequence typing: a portable approach to the identification of clones within populations of pathogenic microorganisms, *Proc. Natl. Acad. Sci. U S A*, 95, 3140, 1998.

35. Achtman, M., A surfeit of YATMs, *J. Clin. Microbiol.*, 34, 1870, 1996.

36. Witte, W., Strommenger, B., and Werner, G., Diagnostics, typing and taxonomy, in *Gram Positive Pathogens*, 2nd ed., Fishetti, V.A. et al., Eds., ASM Press, Washington, DC, 2006, chap. 31.

37. Olive, D.M. and Bean, P., Principles and applications of methods for DNA-based typing of microbial organisms, *J. Clin. Microbiol.*, 37, 1661, 1999.

38. Bouchet, V., Huot, H. and Goldstein, R., Molecular genetic basis of ribotyping, *Clin. Microbiol. Rev.*, 21, 262, 2008.

39. Lindstedt, B.A., Multiple-locus variable number tandem repeats analysis for genetic fingerprinting of pathogenic bacteria, *Electrophoresis*, 26, 2567, 2005.

40. van Belkum, A. et al., Short-sequence DNA repeats in prokaryotic genomes, *Microbiol. Mol. Biol. Rev.*, 62, 275, 1998.

41. Singh, A. et al., Application of molecular techniques to the study of hospital infection, *Clin. Microbiol. Rev.*, 19, 512, 2006.

42. Shopsin, B. et al., Evaluation of protein A gene polymorphic region DNA sequencing for typing of *Staphylococcus aureus* strains, *J. Clin. Microbiol.*, 37, 3556, 1999.

43. Enright, M.C. et al., Multilocus sequence typing for characterization of methicillin-resistant and methicillin-susceptible clones of *Staphylococcus aureus*, *J. Clin. Microbiol.*, 38, 1008, 2000.

44. Maiden, M.C., Multilocus sequence typing of bacteria, *Annu. Rev. Microbiol.*, 60, 561, 2006.

45. Feil, E.J. et al., Recombination within natural populations of pathogenic bacteria: short-term empirical estimates and long-term phylogenetic consequences, *Proc. Natl. Acad. Sci. U S A*, 98, 182, 2001.

46. Tibayrenc, M., Bridging the gap between molecular epidemiologists and evolutionists, *Trends. Microbiol.*, 13, 575, 2005.

47. Achtman, M. et al., Multilocus sequence typing as a replacement for serotyping in *Salmonella enterica*, *PLoS Pathog.*, 8, e1002776, 2012.

48. Grundmann, H. et al., Geographic distribution of *Staphylococcus aureus* causing invasive infections in Europe: a molecular epidemiological analysis, *PLoS Med.*, 7, e1000215, 2010.

49. Feavers, I.M. et al., Multilocus sequence typing and antigen gene sequencing in the investigation of a meningococcal disease outbreak, *J. Clin. Microbiol.*, 37, 3883, 1999.

50. Mellmann, A. et al., Automated DNA sequence-based early warning system for the detection of methicillin-resistant *Staphylococcus aureus* outbreaks, *PLOS Med.*, 3, e33, 2006.

51. Rakov, A., Ubukata, K., and Robinson, D.A., Population structure of hyperinvasive serotype 12F, clonal complex 218 *Streptococcus pneumoniae* revealed by multilocus *boxB* sequence typing, *Infect. Genet. Evol.*, 11, 1929, 2011.

52. Fitzgerald, J.R. et al., Evolutionary genomics of *Staphylococcus aureus*: insights into the origin of methicillin-resistant strains and the toxic shock syndrome epidemic. *Proc. Natl. Acad. Sci. U S A*, 98, 8821, 2001.

53. Zhang, L. et al., Library on a slide for bacterial comparative genomics, *BMC Microbiol.*, 22, 4, 2004.

54. Dunman, P.M. et al., Uses of *Staphylococcus aureus* GeneChips in genotyping and genetic composition analysis, *J. Clin. Microbiol.*, 42, 4275, 2004.

55. Lindsay, J.A. et al., Microarrays reveal that each of the ten dominant lineages of *Staphylococcus aureus* has a unique combination of surface-associated and regulatory genes, *J. Bacteriol.*, 188, 669, 2006.

56. Saunders, N.A. et al., A virulence-associated gene microarray: a tool for investigation of the evolution and pathogenic potential of *Staphylococcus aureus*, *Microbiol.*, 150, 3763, 2004.

57. van Leeuwen, W., Validation of binary typing for *Staphylococcus aureus* strains, *J. Clin. Microbiol.*, 37, 664, 1999.

58. Kwok, P.Y. and Chen, X., Detection of single nucleotide polymorphisms, *Curr. Issues Mol. Biol.*, 5, 43, 2003.

59. Robinson, D.A., Thomas, J.C., and Hanage, W.P., Population structure of pathogenic bacteria, in *Genetics and Evolution of Infectious Diseases*, Tibayrenc, M., Ed., Elsevier, London, 2011, chap. 3.

60. Zhang, L. et al., Multilocus sequence typing and further genetic characterization of the enigmatic pathogen, *Staphylococcus hominis*, *PLOS One*, 8, e66496, 2013.

61. Stephens, A.J. et al., Methicillin-resistant *Staphylococcus aureus* genotyping using a small set of polymorphisms, *J. Med. Microbiol.*, 55, 43, 2006.

62. Tolo, I. et al., Do *Staphylococcus epidermidis* genetic clusters predict isolation sources? *J. Clin. Microbiol.*, 54, 1711, 2016.

63. Gutacker, M.M. et al., Single-nucleotide polymorphism-based population genetic analysis of *Mycobacterium tuberculosis* strains from 4 geographic sites, *J. Infect. Dis.*, 193, 121, 2006.

64. Nübel, U. et al., Frequent emergence and limited geographic dispersal of methicillin-resistant *Staphylococcus aureus*, *Proc. Natl. Acad. Sci. U S A*, 105, 14130, 2008.

65. Ward, T.J. et al., Multilocus genotyping assays for single nucleotide polymorphism-based subtyping of *Listeria monocytogenes*, *Appl. Environ. Microbiol.*, 74, 7629, 2008.

66. Loman, N.J. et al., Performance comparison of benchtop high-throughput sequencing platforms, *Nat. Biotechnol.*, 30, 434, 2012.

67. Quail, M.A. et al., A tale of three next generation sequencing platforms: comparison of Ion Torrent, Pacific Biosciences and Illumina MiSeq sequencers, *BMC Genomics*, 13, 341, 2012.

68. Harris, S.R. et al., Evolution of MRSA during hospital transmission and intercontinental spread, *Science*, 327, 469, 2010.

69. McAdam, P.R. et al., Molecular tracing of the emergence, adaptation, and transmission of hospital-associated methicillin-resistant *Staphylococcus aureus*, *Proc. Natl. Acad. Sci. U S A*, 109, 9107, 2012.

70. Harris, S.R. et al., Read and assembly metrics inconsequential for clinical utility of whole-genome sequencing in mapping outbreaks, *Nat. Biotechnol.*, 31, 592, 2013.

71. Challagundla, L. et al., Phylogenomic classification and the evolution of clonal complex 5 methicillin-resistant *Staphylococcus aureus* in the Western Hemisphere, *Front. Microbiol.*, 9, 1901, 2018.

72. Koren, S. and Phillippy, A.M., One chromosome, one contig: complete microbial genomes from long-read sequencing and assembly, *Curr. Opin. Microbiol.*, 23, 110, 2015.

73. Tyler, A.D. et al., Evaluation of Oxford Nanopore's MinION sequencing device for microbial whole genome sequencing applications, *Sci. Rep.*, 8, 10931, 2018.

74. Grad, Y.H. et al., Genomic epidemiology of the *Escherichia coli* O104:H4 outbreaks in Europe, 2011, *Proc. Natl. Acad. Sci. U S A*, 109, 3065, 2012.

75. Gardy, J.L. et al., Whole-genome sequencing and social-network analysis of a tuberculosis outbreak, *N. Engl. J. Med.*, 364, 730, 2011.

76. Ley, S.D. et al., Deciphering within-host microevolution of *Mycobacterium tuberculosis* through whole-genome sequencing: the phenotypic impact and way forward, *Microbiol. Mol. Biol. Rev.*, 83, e00062, 2019.

77. Viberg, L.T. et al., Within-host evolution of *Burkholderia pseudomallei* during chronic infection of seven Australasian cystic fibrosis patients, *mBio*, 8, e00356, 2017.

78. Young, B.C. et al., Evolutionary dynamics of *Staphylococcus aureus* during progression from carriage to disease, *Proc. Natl. Acad. Sci. U S A*, 109, 4550, 2012.

79. Brito, I.L. and Alm, E.J., Tracking strains in the microbiome: insights from metagenomics and models, *Front. Microbiol.*, 7, 712, 2016.

80. Segata, N., On the road to strain-resolved comparative metagenomics, *mSystems*, 3, e00190, 2018.

81. Chan, J.Z.M. et al., Genome sequencing in clinical microbiology, *Nat. Biotechnol.*, 30, 1068, 2012.

82. Struelens, M.J. and Brisse, S., From molecular to genomic epidemiology: transforming surveillance and control of infectious diseases, *Euro Surveill.*, 18, 20386, 2013.

83. Robinson, D.A. et al., Evolution and virulence of serogroup 6 pneumococci on a global scale, *J. Bacteriol.*, 184, 6367, 2002.

84. Sandgren, A. et al., Effect of clonal and serotype-specific properties on the invasive capacity of *Streptococcus pneumoniae, J. Infect. Dis.*, 189, 785, 2004.

85. Silva, N.A. et al., Genomic diversity between strains of the same serotype and multilocus sequence type among pneumococcal clinical isolates, *Infect. Immun.*, 74, 3513, 2006.

86. Struelens, M.J., De Gheldre, Y., and Deplano, A., Comparative and library epidemiological typing systems: outbreak investigations versus surveillance systems, *Infect. Control Hosp. Epidemiol.*, 19, 565, 1998.

87. van Belkum, A. et al., Role of genomic typing in taxonomy, evolutionary genetics, and microbial epidemiology, *Clin. Microbiol. Rev.*, 14, 547, 2001.

88. Stimson, J. et al., Beyond the SNP threshold: identifying outbreak clusters using inferred transmissions, *Mol. Biol. Evol.*, 36, 587, 2019.

89. Tibayrenc, M., Toward an integrated genetic epidemiology of parasitic protozoa and other pathogens, *Annu. Rev. Genet.*, 33, 449, 1999.

90. Khil, P.P. et al., Dynamic emergence of mismatch repair deficiency facilitates rapid evolution of ceftazidime-avibactam resistance in *Pseudomonas aeruginosa* acute infection, *mBio*, 10, e01822, 2019.

91. Maynard Smith, J., Dowson, C.G., and Spratt, B.G., Localized sex in bacteria, *Nature*, 349, 29, 1991.

92. Maynard Smith, J., Do bacteria have population genetics? in *Population Genetics of Bacteria*, Baumberg, S., et al., Eds., Cambridge University Press, Cambridge, 1995, chap. 1.

93. Ørskov, F. and Ørskov, I., Summary of a workshop on the clone concept in the epidemiology, taxonomy, and evolution of the Enterobacteriaceae and other bacteria, *J. Infect. Dis.*, 148, 346, 1983.

94. Robinson, D.A. et al., Evolution and global dissemination of macrolide-resistant group A streptococci, *Antimicrob. Agents Chemother.*, 50, 2903, 2006.

95. Didelot, X., Sequence-based analysis of bacterial population structures, in *Bacterial Population Genetics in Infectious Disease*, Robinson, D.A., Falush, D., and Feil, E.J., Eds., John Wiley & Sons, Hoboken, NJ, 2010, chap. 3.

96. Thomas, J.C., Zhang, L., and Robinson, D.A., Differing lifestyles of *Staphylococcus epidermidis* as revealed through Bayesian clustering of multilocus sequence types, *Infect. Genet. Evol.*, 1348, 245, 2013.

97. Brown, T.S. et al., Pre-detection history of extensively drug-resistant tuberculosis in KwaZulu-Natal, South Africa, *Proc. Natl. Acad. Sci. U S A*, 116, 23284, 2019.

98. Smyth, D.S. et al., Population structure of a hybrid clonal group of methicillin-resistant *Staphylococcus aureus*, ST239-MRSA-III, *PLOS One*, 5, e8582, 2010.

99. Dearlove, B. and Wilson, D.J., Coalescent inference for infectious disease: meta-analysis of hepatitis C, *Phil. Trans. R. Soc. B*, 368, 2013.

100. Fittipaldi, N. et al., Integrated whole-genome sequencing and temporospatial analysis of a continuing group A *Streptococcus* epidemic, *Emerg. Microbe Infect.*, 2, e13, 2013.

101. Murray, M. and Alland, D., Methodological problems in the molecular epidemiology of tuberculosis, *Am. J. Epidemiol.*, 155, 565, 2002.

102. Perri, B.R. et al., *Mycobacterium tuberculosis* cluster with developing drug resistance, New York, New York, USA, 2003-2009, *Emerg. Infect. Dis.*, 17, 372, 2011.

103. van Rie, A. et al., Exogenous reinfection as a cause of recurrent tuberculosis after curative treatment, *N. Engl. J. Med.*, 341, 1174, 1999.

104. van Belkum, A., High-throughput epidemiologic typing in clinical microbiology, *Clin. Microbiol. Infect.*, 9, 86, 2003.

105. Swaminathan, B. et al., PulseNet: the molecular subtyping network for foodborne bacterial disease surveillance, United States, *Emerg. Infect. Dis.*, 7, 382, 2001.

106. Enright, M.C. et al., The evolutionary history of methicillin-resistant *Staphylococcus aureus* (MRSA), *Proc. Natl. Acad. Sci. U S A*, 99, 7687, 2002.

107. Robinson, D.A. et al., Re-emergence of early pandemic *Staphylococcus aureus* as a community-acquired methicillin-resistant clone, *Lancet*, 365, 1256, 2005.

108. Falush, D. and Bowden, R., Genome-wide association mapping in bacteria, *Trends. Microbiol.*, 14, 353, 2006.

109. Oleastro, M. et al., Identification of markers for *Helicobacter pylori* strains isolated from children with peptic ulcer disease by suppressive subtractive hybridization, *Infect. Immun.*, 74, 4064, 2006.

110. Xie, J. et al., Molecular epidemiologic identification of *Escherichia coli* genes that are potentially involved in movement of the organism from the intestinal tract to the vagina and bladder, *J. Clin. Microbiol.*, 44, 2434, 2006.

111. Zhang, L. et al., Molecular epidemiologic approaches to urinary tract infection gene discovery in uropathogenic *Escherichia coli, Infect. Immun.*, 68, 2009, 2000.

112. Maurelli, A.T. et al., "Black holes" and bacterial pathogenicity: a large genomic deletion that enhances the virulence of *Shigella* spp. and enteroinvasive *Escherichia coli, Proc. Natl. Acad. Sci. U S A*, 95, 3943, 1998.

113. Johnson, J.R. et al., Experimental mouse lethality of *Escherichia coli* isolates, in relation to accessory traits, phylogenetic group, and ecological source, *J. Infect. Dis.*, 194, 1141, 2006.

114. Sheppard, S.K. et al., Genome-wide association study identifies vitamin B5 biosynthesis as a host specificity factor in *Campylobacter, Proc. Natl. Acad. Sci. U S A*, 110, 11923, 2013.

16

CRISPR

Tao Xu, Megan L. Kempher, Xuanyu Tao, Aifen Zhou, and Jizhong Zhou

CONTENTS

Introduction

History of CRISPR-Cas Biology

The clustered regularly interspaced short palindromic repeat (CRISPR)/CRISPR-associated (Cas) protein system is an adaptive immune system in prokaryotic organisms. It defends against invading bacteriophages and conjugative plasmids.[1] Besides the genes that encode Cas proteins, CRISPR loci harbor short repeats (usually 20–50 bp) separated by nonrepeating spacers (21–72 bp) derived from the invading DNA (Figure 16.1). New spacers are acquired over time and when expressed, guide the Cas effector protein(s) to cleave the invading DNA thus providing immunity against reinfection. Diverse CRISPR-Cas systems have been identified from many bacteria and are now divided into two classes (1 and 2) and further subdivided into six types (I–VI).[2] Class 1 includes type I, III, and IV CRISPR-Cas systems that use multisubunit effector complexes to degrade foreign nucleic acids. Class 2 includes type II, V, and VI and uses a single large Cas protein

for the same purposes. Distinct signature Cas proteins are classified in different types, for example, Cas3 in type I, Cas9 in type II, and Cas10 in type III. Understanding the molecular mechanisms of these CRISPR-Cas systems launched the birth of a diversity of novel, programmable CRISPR-based tools for genome editing,[3–5] transcriptional control,[6] nucleic acid detection,[7,8] as well as DNA cloning and imaging.[9,10] Here we summarize the history of CRISPR-Cas biology including major discoveries that led to our mechanistic understanding of these systems and their subsequent translational applications (Figure 16.2).

The CRISPR sequence was fortuitously found in *Escherichia coli* in 1987;[11] however, it was not until 1993 when it was first characterized by Mojica who described the pattern of palindromic repeated sequences separated by short DNA spacers in salt-tolerant *Haloferax mediterranei*.[12] Soon, various types of CRISPR loci were found in many microbes, such as *Mycobacterium tuberculosis* and *Yersinia pestis*. The discovery of other features, such as the presence of specific CRISPR-associated (*cas*) genes in the immediate vicinity and the identification of the protospacer

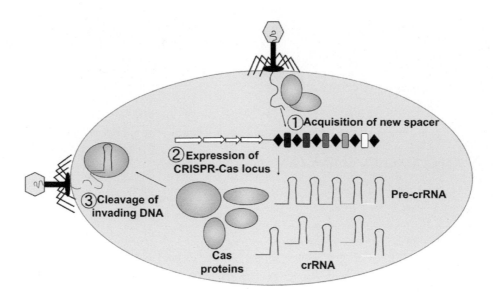

FIGURE 16.1   General mechanism of CRISPR-Cas system in bacterial immunity. (1) A new spacer is added to the CRISPR-Cas locus from an invading phage or conjugal plasmid. (2) Expression of the CRISPR-Cas locus and processing of crRNA. (3) Detection and degradation of foreign genetic material via a crRNA and Cas effector complex.

adjacent motif (PAM), soon followed.[13,14] In 2005, Mojica and other researchers hypothesized that CRISPR was an adaptive immune system based on the sequence of the spacers homologous to bacteriophages, prophages, and plasmids.[15,16] Barrangou et al. proved this hypothesis in 2007 by isolating phage-resistant *Streptococcus thermophilus* that were originally phage-sensitive, where they found the resistant strains had acquired phage-derived sequences in the CRISPR loci and the insertion of multiple spacers correlated with increased phage resistance.[17] They also experimentally identified the role of type II CRISPR *cas9* in phage resistance and *cas7* in generating new spacers and repeats to gain resistance.

The early mechanistic understanding of the CRISPR-Cas system can be ascribed to four major discoveries. First, in 2008 Brouns et al. showed that the CRISPR spacer sequences in

E. coli produced small RNAs, termed CRISPR RNAs (crRNAs) that were programmable and guided type I CRISPR Cas proteins to target essential genes of lambda phage.[18] Second, Marraffini and Sontheimer reconstituted the type III-A CRISPR system of *Staphylococcus epidermidis in vitro* and demonstrated that it acted on DNA instead of RNA.[19] Third, Moineau et al. demonstrated that CRISPR-Cas9 in conjunction with crRNAs created double-stranded breaks in target DNA at precise positions, 3-nt upstream of the PAM,[20] and Cas9 was the only protein of the CRISPR-Cas9 system required for cleavage. Fourth, by using small RNA sequencing of *Streptococcus pyogenes* with a Cas9-containing CRISPR-Cas system, Deltcheva et al. discovered the trans-activating CRISPR RNA (tracrRNA) in 2011,[21] which forms a duplex with crRNA to guide Cas9 to bind a target. These components (Cas nuclease, crRNA, and tracrRNA) are

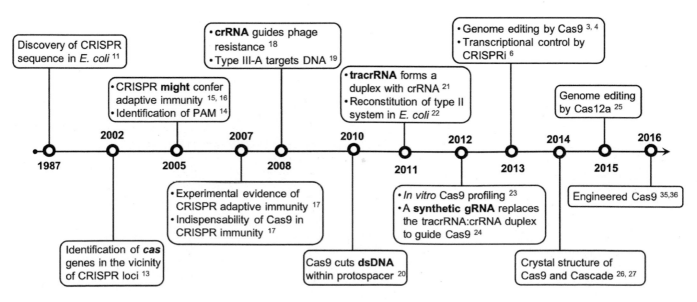

FIGURE 16.2   Major discoveries in the history of CRISPR-Cas biology that lead to CRISPR-Cas technologies.

characterized as major components for immunity in different types of CRISPR-Cas systems.

Further biochemical studies of the CRISPR-Cas components elucidated the detailed molecular mechanisms of action. In 2011, Sapranauskas et al. demonstrated that heterologous expression of a type II CRISPR-Cas system in *E. coli* gained plasmid resistance found in *S. thermophilus*.[22] They biochemically characterized *S. thermophilus* Cas9, including the cleavage site, the requirement for the PAM, and the role of RuvC and HNH domains in cleaving DNA.[23] More importantly, they successfully reprogrammed Cas9 to target a specific site by customizing the 20-nt sequence of the crRNA. At almost the same time, Charpentier and Doudna reported that a single, synthetic guide RNA (gRNA) could replace the tracrRNA:crRNA duplex to guide Cas9 to a target site.[24]

In 2013, multiple research groups repurposed the CRISPR-Cas system for sequence-specific genome editing and transcriptional regulation in both prokaryotes and eukaryotes. Zhang's team first adapted Cas9 for genome editing in eukaryotic cells.[4] Marraffini's group established a two-plasmid based CRISPR-Cas9 system to efficiently counter-select edited genomes in *Streptococcus pneumoniae* and *E. coli* by customizing the crRNA array.[3] Qi et al. employed a catalytically dead Cas9 (dCas9) lacking endonuclease activity to repress expression of targeted genes in *E. coli*, which is also called CRISPR interference (CRISPRi).[6] These exciting reports strongly inspired later studies in many other organisms for a broad range of applied and fundamental research purposes. The continued discovery and dissection of novel CRISPR-Cas systems dramatically expands our understanding of bacterial defense systems and yields novel nucleotide detection and editing tools. For example, the core nuclease of the type V system, Cas12a (also known as Cpf1), presents distinct molecular and biochemical characteristics from Cas9 and has also been applied to edit genomes.[25] Studies on the crystal structure of Cas proteins, such as Cas9,[26,27] Cas12a,[28] Cas module-repeat-associated mysterious proteins (Cmr complex),[29] and Cascade proteins, provides unprecedented insights into how these nucleases work at the molecular level. Extensive studies on Cas9 protein engineering, gRNA modification, optimization of gRNA design and screening tools, have evolved the dominant Cas9 genome editing tool to be a reliable, versatile, and robust system with an exceptional editing efficiency and accuracy.

Cas Nucleases, the Core of CRISPR-Cas Systems

Immunization and immunity are two steps of the CRISPR-Cas mechanism by which prokaryotes become resistant to foreign nucleic acids (Figure 16.1).[30] During the immunization process, exogenous DNA is recognized by a Cas complex and integrated into the CRISPR locus. During the immunity process, the CRISPR locus is transcribed and processed to mature crRNA for interacting with Cas nucleases to form enzymatically active effector complexes. Notably, the effector complexes from different bacteria generally differ in their signature Cas nucleases that are directly responsible for target cleavage and also other accompanying molecular features (e.g., RNA structure, PAM sequence, and PAM position) that the gRNA and target sites must contain for Cas nucleases to work properly. For instance, Cas9 orthologs have distinct preferences for the PAM sequence. Many Cas nucleases have been characterized biochemically, however, here

we will focus on two representatives, Cas9 (type II) and Cas12a (type V), both of which have been well studied and widely used for genome editing. A side-by-side comparison of their mechanistic characteristics is shown in Table 16.1.

Cas9 (formerly known as Csn1 or Csx12) is the major component of type II CRISPR-Cas systems and can cleave double-stranded DNA (dsDNA) in a sequence-specific manner. *S. pyogenes* Cas9 (SpCas9) is the most well-known Cas9 that has been widely used in genome editing. It is a large multidomain protein with two nuclease domains, a RuvC-like nuclease domain near the amino terminus and an HNH nuclease domain in the middle.[24] Its cleavage activity is guided by an incorporated tracrRNA:crRNA duplex or a single-stranded chimeric gRNA. The 20-nt sequence preceding the 83-nt conserved gRNA scaffold can be customized to target different DNA sites. The endonuclease activity of SpCas9 creates blunt dsDNA breaks located 3-bp upstream of the 3'-terminal complementarity region formed between the gRNA recognition sequence and the genomic protospacer. Mutagenesis of either of the catalytic sites in the RuvC or HNH motifs abolishes the ability to create DSBs, leaving only nickase activity. Mutations in these active sites did not alter the affinity of protospacer binding. Importantly, a PAM consensus sequence, a short conserved nucleotide stretch next to the protospacer, is required for Cas9 binding and cleavage; however, Cas9 orthologs may require different PAM sequences, for example, *S. pyogenes* (5'-NGG-3'), *N. meningitidis* (5'-NNNNGATT-3'),[31] *Streptococcus thermophilus* CRISPR1 (5'-NNAGAA-3'),[32] *Treponema denticola* (5'-NAAAAN-3').[33] Therefore, the target site of SpCas9 is a 20-bp protospacer sequence followed by the PAM consensus, NGG. Structure-guided protein engineering of SpCas9 developed an "enhanced specificity" SpCas9 with three amino acids mutated (K848A/K1003A/R1060A),[34] which reduced off-target effects without sacrificing its robust on-target cleavage. Engineered Cas9 variants, such as spCas9-NG and xCas9, dramatically expand our targeting space with relaxed PAMs required for targeting.[35,36]

Some Cas9 enzymes from both subtypes II-A and II-C CRISPR systems are able to recognize and cleave single-stranded RNA (ssRNA) with no requirement of a PAM. *Staphylococcus aureus* Cas9 (SauCas9) and *Campylobacter jejuni* Cas9 (CjeCas9) demonstrated RNA targeting capability when provided with both guide RNAs and divalent metal ions.[37] Apart from the programmable and site-specific characteristics, this RNA-guided RNA cleavage is capable of reducing infection by ssRNA phage *in vivo*. The catalytically inactive SauCas9 enables direct PAM-independent repression of gene expression in *E. coli*, which dramatically expands target accessibility. However, structured or occluded RNA regions will block SauCas9 binding and thereby prevent cleavage or repression activity.

Cas12a is a single RNA-guided DNA nuclease type V CRISPR-Cas system in bacteria such as *Prevotella* and *Francisella*.[25] Cas12a typically contains a RuvC-like endonuclease domain but lacks the second HNH endonuclease domain found in Cas9. The RuvC-like domain of *Francisella novicida* Cas12a (FnCas12a) cleaves both strands of the target DNA, perhaps in a dimeric configuration.[25] Instead of requiring both crRNA and tracrRNA like Cas9, FnCas12a only requires crRNA for sufficient DNA cleavage; a minimum of 18 nt of guide sequence in the crRNA will achieve efficient DNA cleavage *in vitro*. The PAM for FnCas12a,

TABLE 16.1

Model of Representative CRISPR-Cas Systems

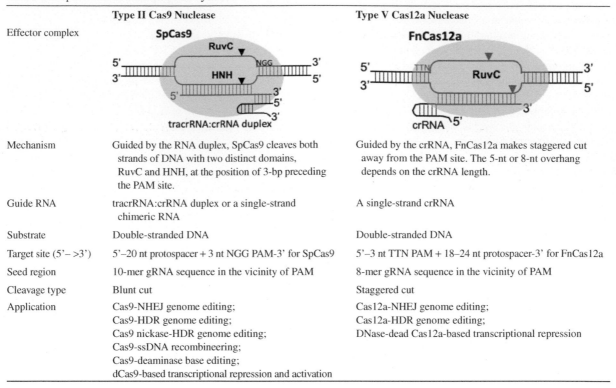

	Type II Cas9 Nuclease	Type V Cas12a Nuclease
Mechanism	Guided by the RNA duplex, SpCas9 cleaves both strands of DNA with two distinct domains, RuvC and HNH, at the position of 3-bp preceding the PAM site.	Guided by the crRNA, FnCas12a makes staggered cut away from the PAM site. The 5-nt or 8-nt overhang depends on the crRNA length.
Guide RNA	tracrRNA:crRNA duplex or a single-strand chimeric RNA	A single-strand crRNA
Substrate	Double-stranded DNA	Double-stranded DNA
Target site (5'– >3')	5'–20 nt protospacer + 3 nt NGG PAM-3' for SpCas9	5'–3 nt TTN PAM + 18–24 nt protospacer-3' for FnCas12a
Seed region	10-mer gRNA sequence in the vicinity of PAM	8-mer gRNA sequence in the vicinity of PAM
Cleavage type	Blunt cut	Staggered cut
Application	Cas9-NHEJ genome editing; Cas9-HDR genome editing; Cas9 nickase-HDR genome editing; Cas9-ssDNA recombineering; Cas9-deaminase base editing; dCas9-based transcriptional repression and activation	Cas12a-NHEJ genome editing; Cas12a-HDR genome editing; DNase-dead Cas12a-based transcriptional repression

which is 5'-TTN-3', is located upstream of the 5' end of the displaced strand of the protospacer. The PAM-proximal seed region that drastically determines target binding is the first 5-nt on the 5' end of the spacer sequence. Two Cas12a enzymes from *Acidaminococcus* and *Lachnospiraceae* have demonstrated efficient genome-editing activity in human cells.[25,38] FnCas12a has been adapted to edit the genome of *Corynebacterium glutamicum*, an anaerobic Gram-positive bacterium for amino acid production.[39] The most efficient *C. glutamicum* crRNA contained a 5'-NYTV-3' PAM and a 21 bp target sequence.

Comparison of Targeted Genetic Engineering Tools

A variety of targeted genetic engineering tools have been developed to modify DNA sequences based on DNA or RNA homology (Table 16.2). TargeTron is based on the group IIA intron Ll.LtrB from *Lactococcus lactis*. As a type of retrotransposon, group II introns achieve site-specific DNA integration through base-pairing of intron RNA with specific DNA sequences assisted by an intron-encoded multifunctional reverse transcriptase. By modifying the intron RNA sequence according to the desired DNA targets, group II intron can

TABLE 16.2

Comparison of Mainstream Targeted Genetic Engineering Tools

Method	Targeting Steps	Factors for Targeting Specificity	Efficiency	Multiplex Targets	Feasible to Study Essential Genes	Targeting Spectrum	Off-Target Effect	Application
CRISPR-Cas	1	guide RNA	high, but varies in organisms	Yes	Yes	everywhere	low	deletion, insertion, point mutation, transcriptional repression, and activation
TargeTron	1 or 2	intron RNA	varies, depends on the insertion sites and species	No	No	limited, biased near the replication origin	high	gene disruption
Double crossover recombination	2	DNA homology	high, but varies in organisms	No	No	everywhere	very low	deletion, insertion, point mutation
Recombineering	1 or 2	DNA homology	low	No	Yes	everywhere	low	deletion, insertion, point mutation
RNA interference	1	RNA homology	varies	Yes	Yes	everywhere	severe	transcriptional repression

be programmed to disrupt a target gene by intron insertion. This technique has been intensively used in microbes with a low DNA repair capacity and transformation efficiency such as solventogenic *Clostridium*;[40,41] however, it might be challenging to find target sites in small genes. Double crossover recombination is based on homologous recombination and requires a suitable counterselectable marker. It involves single crossover integration of the whole plasmid at the target site and then recombination to excise the plasmid backbone to yield a scarless mutant. Recombineering is also a homologous recombination-based method, which is usually assisted by coliphage λ Red systems or the RecET system from Rac prophage. Linear DNA molecules, either double stranded PCR products or single-stranded synthetic oligonucleotides, serve as homologous substrates. Seamless mutants can be generated by two rounds of targeting, recombination, and counterselection.[42] RNA interference is an RNA-dependent gene silencing process to knockdown gene expression. In bacteria, the antisense RNA expressed from a vector will bind to messenger RNA to form an RNA duplex, which will not only promote RNA degradation but also slow down translation to reduce the production of functional proteins. Compared to these mainstream nucleotide-directed targeted genetic engineering tools for genome editing and gene regulation in bacteria, the major advantage of CRISPR-Cas system is the simplicity of the design and the potential of application in multiplexed editing and genome-scale mutagenesis.

CRISPR-Cas Tools for Genome Editing

CRISPR-Cas genome editing tools, irrespective of which Cas nucleases are employed, rely on DNA repair systems to introduce modifications at the target site; involved Cas nucleases either destroy unmutated genomes or create DNA damage to boost DNA repair.[43] Nonhomologous end-joining (NHEJ) and homology-directed repair (HDR) are two major repair systems that are involved with CRISPR-Cas genome editing in both eukaryotes and prokaryotes.[44] Beyond that, irreversible DNA base conversion by cytidine deaminases and introduction of phage recombination machineries have also been implemented during genome editing.[3,45,46] Here we describe four well-established CRISPR-Cas9 editing strategies (Figure 16.3), Cas9-NHEJ, Cas9-HDR, Cas9-deaminase, and Cas9-ssDNA recombineering.

NHEJ-Based Editing

In NHEJ, DSBs are repaired without a homologous template. The Ku protein directly binds and stabilizes the DSB ends, then recruits other NHEJ factors for DNA ligation.[47] NHEJ often generates undefined insertions/deletions (indels) at the target site when the ends of DSBs are remodeled by nucleases and polymerases before DNA ligation (Figure 16.3).[48] NHEJ is mainly found in three bacterial genera, *Bacillus*, *Mycobacteria* and *Pseudomonas*[48,49]; however, the majority of other bacteria lack homologous proteins of Ku and LigD. There are few cases of Cas9-based gene editing in

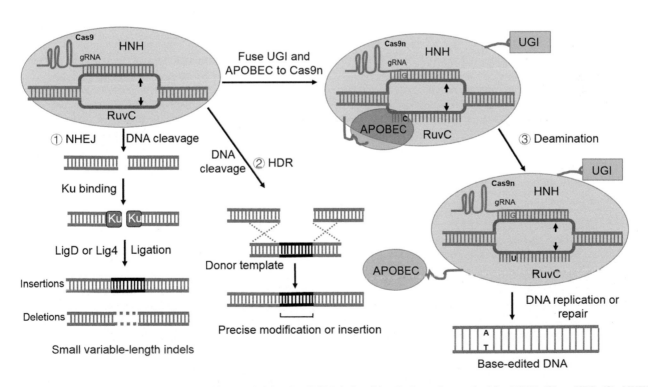

FIGURE 16.3 Cas9-based genome editing. Double-stranded breaks (DSBs) induced by Cas9 can be repaired by NHEJ (①) or HDR (②). NHEJ is Ku-dependent and ends of DSB are repaired without a template and sealed by DNA ligase (Lig4 in Eukarya and LigD in bacteria). As it is error-prone, NHEJ usually introduces small insertion/deletion mutations of various lengths by which the translational frame can be altered. Compared to NHEJ, HDR requires a homologous template sequence to repair the DSB. By recombination, HDR can introduce desired modifications including insertions, deletions, or substitutions. Additionally, Cas9 nickase (Cas9n) can be fused with a cytidine deaminase (APOBEC) and uracil glycosylase inhibitor (UGI) to generate a "base editor" (BE) for direct and irreversible conversion of one target DNA base to another (③). After targeting, the APOBEC will convert cytosine (C) to uracil (U) and then the G:U heteroduplex will be permanently altered to A:T during DNA replication and repair.

prokaryotes using the endogenous NHEJ pathway for DSB repair. For instance, Tong et al. inactivated two genes in *Streptomyces coelicolor* using the Cas9-NHEJ editing strategy.[50] Large chromosomal DNA fragments were deleted efficiently in *E. coli* based on the CRISPR-Cas9 system with a heterologously expressed NHEJ system.[51] The Cas9-NHEJ editing strategy is a powerful tool for bacterial genome reduction and accelerating genome evolution.[51] As NHEJ is a favored DNA repair pathway for DSB repair in most filamentous fungi, particularly in oomycetes,[52,53] the Cas9-NHEJ editing strategy can efficiently introduce indels causing frameshifts or premature stop codons to inactivate gene functions in the absence of homologous templates.

Homologous Recombination-Based Editing

HDR requires an intact copy of the broken chromosomal site to serve as a template for DNA repair.[48,54] It is a faithful and nonmutagenic process[55,56]; its primary function in bacteria is to rescue a stalled or collapsed replication fork. The RecA protein plays a critical role in HDR and is present in all sequenced bacterial species.[55] Cas9 (or its variants) mediated target cleavage is often coupled with HDR for precise genome engineering in bacteria. This Cas9-HDR editing strategy has been successfully applied in many bacteria, including *E. coli*,[57] *Streptococcus*,[3] *Streptomyces*,[58,59] *Tatumella*[57] and *Clostridium*.[5,60] To do so, three components, including *cas9* gene, a target specific gRNA, and a user-designed homologous DNA template, need to be delivered into bacterial cells typically by a single "all-in-one" plasmid or two separate plasmids. Combined with massively parallel oligomer synthesis, Cas9-HDR has enabled genome-wide trackable editing for protein engineering and reconstruction of adaptive laboratory evolution experiments in *E. coli*.[61] While efficient in most bacteria, HDR-based genome engineering is inefficient in most yeasts. Deletion of *ku70* and/or *ku80* genes interestingly promoted HDR events.[52,53] The presence of large homologous donor templates improved Cas9-based editing efficiency in fungi,[62] such as in *Trichoderma reesei*, *Myceliophthora thermophila*, and *Myceliophthora heterothallica*.[63,64] Compared with Cas9-NHEJ editing, Cas9-HDR editing can provide precise editing at intended genomic loci, especially for DNA deletion, insertion, or codon change via substitution.[5,54]

Deaminase-Assisted Base Editing

In addition to insertions, deletions, and substitutions, the CRISPR-Cas9 system has been applied for editing a target base (G:C to A:T conversion), which has been named "base editor" (BE).[46,65] The system consists of two components: (i) a catalytically dead Cas9 (dCas9) or a Cas9 nickase (Cas9n) for targeted DNA binding; (ii) a deaminase tethered to dCas9 for converting cytosine to uracil (Figure 16.3). The original G:C base pair is temporarily converted to a G:U heteroduplex by the deaminase; the G:U will be permanently converted to A:T during DNA replication and repair. This base editing system advances both the scope and efficiency of generating point mutations.[46] Liu et al developed a third generation BE system (BE3) in *E. coli*, which uses Cas9n instead of dCas9 and contains a uracil glycosylase inhibitor at the C terminus of Cas9n. BE3 was able to carry out G:C to A:T conversions precisely and more efficiently than the previous base editing systems.[66] In addition to the G:C to A:T conversion, another new class of base editors, named adenine base editors (ABEs), converts A:T to G:C in *E. coli* and human genomic DNA.[45] The BE3 deaminase is replaced by an evolved tRNA adenosine deaminase in ABE, which is able to deaminate adenine to inosine (I). The inosine can be read as guanosine by the polymerase enzyme and finally the I:T heteroduplex is converted to G:C. Taken together, the application of BE and ABE systems enables precise base editing for all transitions (C↔T and A↔G) in DNA. A major advantage to these base editing systems is that they do not require a donor template, greatly simplifying the entire editing process. They may provide new insights into evolutionary studies of bacteria, especially for studying single nucleotide polymorphisms (SNPs), by improving the construction of this type of mutant.

Cas9-Assisted ssDNA Recombineering

Recombineering (homologous recombination-mediated genetic engineering) is a useful *in vivo* method for genetic engineering, primarily used in *E. coli*.[42,67–70] Based on homologous recombination, it can circumvent the need for the restriction enzyme sites of *in vitro* cloning techniques, and can precisely insert, delete or alter any sequence by homologous recombination of linear DNA including dsDNA from PCR products or ssDNA from synthetic oligonucleotides.[69,71] The λ Red system is now the most commonly used system for recombineering, and contains three phage recombination proteins, Gam, Exo and Beta.[42] Gam prevents the degradation of linear DNA fragments by the RecBCD nuclease in *E. coli*.[72,73] Exo is a 5' to 3' dsDNA exonuclease,[74,75] which generates 3'-overhangs of linear DNA. Beta binds to ssDNA and promotes the annealing of two complementary DNA molecules.[76–78] Using these enzymes, the transformed linear DNA will guide the generation of genetic recombinants.[68,79] As recombineering itself has a low efficiency in promoting homologous recombination, there is little chance of directly selecting the desired mutations without selection pressure.[3,42] Relying on the toxicity of DSBs as a result of inefficient DSB repair in *E. coli*, Wenyan et al. combined Cas9 DNA cleavage with ssDNA recombineering to kill nonmutated cells, thus increasing the targeted editing efficiency.[3] This Cas9-recombineering editing strategy has been applied in other bacteria, such as *Lactobacillus reuteri*,[80] *Corynebacterium glutamicum*[81] and *Pseudomonas putida*.[82]

CRISPR-Cas Tools for Gene Regulation

The ability to precisely control gene regulation is a powerful tool that allows for the interrogation of gene function and aids in improving our understanding of cellular pathways. In addition to gene editing, CRISPR-Cas systems have been repurposed to suppress or activate gene expression. This has presented biologists with an unprecedented ability to reprogram cells, investigate gene function including essential genes, and build new genetic pathways.

Transcriptional Suppression

The first use of CRISPR-Cas tools to suppress gene expression was achieved by engineering a catalytically dead Cas9 (dCas9), named CRISPRi. This was accomplished by altering an amino acid in both the RuvC (D10A) and the HNH (H840A) endonuclease domains

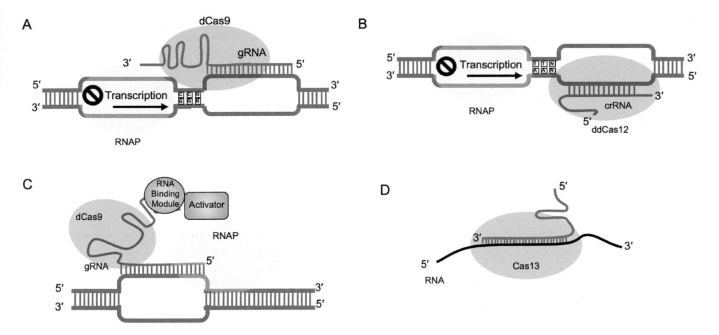

FIGURE 16.4 CRISPR-Cas transcriptional control methods. (A and B) Transcriptional suppression via CRISPRi is achieved by using a catalytically inactive form of either (A) Cas9 or (B) Cas12 to sterically block RNAP from transcribing a gene. (C) Transcriptional activation via CRISPRa is achieved through recruitment of RNAP to the promoter site of a gene. This can be achieved by fusing an activator to either a catalytically inactive form of a Cas protein or to an RNA binding module that interacts with the gRNA. (D) Cas13 directly targets RNA transcripts which can lead to transcriptional suppression of a gene.

(Figure 16.4A).[3,6,23,24,83] Expression of dCas9 along with a gRNA designed to target a gene of interest can precisely and reversibly modify transcription without changing the target sequence. CRISPRi relies on steric hindrance of the RNA polymerase (RNAP) by the catalytically inactive Cas nuclease-gRNA complex (Figure 16.4A and B).[6] The level of silencing can be precisely controlled by moving the target region for the gRNA, altering the length of the gRNA, targeting the same gene with two or more gRNAs, or by using inducible promoters which can be finely tuned to drive the expression of the gRNA and dCas9. Moreover, multiple target genes can be silenced simultaneously just by the expression of additional gRNAs allowing for suppression of an entire operon, pathway, or signaling network.[6] CRISPRi can also be used to interrogate promoter elements and the activity of transcriptional factors by using gRNAs designed to sterically block cognate binding motifs.

Although dCas9 is the most widely used form of CRISPRi, an additional transcriptional suppression system was recently designed using a DNase dead Cas12a (ddCas12a, originally named ddCpf1) from *Acidaminococcus* sp. Cpf1 (Figure 16.4B).[84] Since Cas12a lacks an HNH domain, a single mutation in the RuvC domain (E993A) results in the loss of DNA cleavage. Unlike Cas9, Cas12a possesses its own RNase activity which allows for the processing of the necessary crRNA without tracrRNA or RNase III. This ability offers an advantage over the dCas9 system for multiplexed gene targeting. The dCas9 system requires that each gRNA be expressed separately. However, since ddCpf1 retains its RNase activity, a single CRISPR array can be designed to target multiple genes, greatly simplifying multiplexed gene repression studies.[84]

Transcriptional Activation

The catalytically inactive dCas9 has also been repurposed to activate gene expression, termed CRISPR activation (CRISPRa).

However, compared to CRISPRi, targeted gene activation systems in bacteria have lagged behind. While CRISPRi depends on steric hindrance, CRISPRa is inherently more complicated as it requires transcriptional activators that recruit RNAP to the promoter site of a target gene (Figure 16.4C). While the availability of many broadly applicable gene activators in eukaryotes has led to numerous applications of CRISPRa, a paucity of reports in bacteria has stemmed from the lack of usable, well-characterized transcriptional activation domains. The first report of a successful CRISPRa system in *E. coli* utilized the ω subunit (*rpoZ*) of RNA polymerase, which was fused to dCas9 and with the co-expression of a gRNA was able to achieve induction of target gene expression.[85,86] However, this system has not been widely adopted suggesting that it may not be easily adaptable to other types of promoters or other bacteria.

Another group sought to expand CRISPRa by using modified gRNAs, named scaffold RNAs (scRNA), that were designed to contain an RNA hairpin to recruit RNA binding proteins.[87] These RNA binding proteins were then coupled to potential transcriptional activators and used to screen the efficacy of a variety of activation domains. The most prominent induction was produced using the *E. coli* regulator SoxS and was substantially stronger than the RpoZ-based system. Additionally, simultaneous activation of one gene and suppression of a second gene was achieved by expressing both a scRNA and a gRNA in the same system, which allows for unprecedented control over cellular pathways. Fine-tuning of the system can be achieved by using inducible promoters to control the expression of the scRNA, gRNA, and dCas9.

The initial CRISPRa systems all relied on σ^{70} promoters in *E. coli* limiting their applicability and flexibility. Recently, a CRISPRa system was developed that can achieve high activation (over 1,000-fold when all components were induced) from σ^{54} dependent promoters.[88] The activity of σ^{54} is quite different from σ^{70} and, when bound with RNAP to the promoter, forms a stable closed complex by a way that requires ATP-dependent activation.[89,90] These activators are referred to as bacterial enhancer binding proteins (bEBPs), bind upstream activating domains, and are similar to RNAP II activation in eukaryotes. An engineered activator was constructed by fusing the activation domain of an *E. coli* bEBP PspF (PspFΔHTH) to an RNA-binding peptide λN22plus. The gRNA was designed to contain two BoxB aptamer domains, which recruits λN22plus peptide containing proteins. This system was able to activate expression from several σ54 dependent promoters in *E. coli* and in a nonmodel bacteria *Klebsiella oxytoca* and greatly expands the applicability and flexibility of CRISPRa tools.[88]

Direct RNA Cleavage by Cas Proteins

While the majority of CRISPR-Cas systems are known to target dsDNA, the ability to target RNA has recently been established for Cas9 homologs from *S. pyogenes*, *C. jejuni*, *S. aureus*, and *Neisseria meningitidis*.[37] Initial studies showed that if SpCas9 was supplied with a DNA oligo that contained the PAM sequence (referred to as a PAMmer), it was able to bind and cut ssRNA and, in the absence of a PAMmer, repress translation.[91–94] As mentioned above, SauCas9 was able to protect *E. coli* against an infecting RNA phage.[37] Whether or not this ability exists in nature remains to be seen but was an integral step in confirming the ability of a Cas9 to target RNA *in vivo*. Additionally, SauCas9, along with a gRNA, was able to suppress translation of a GFP reporter sequence chromosomally inserted into *E. coli* as long as the targeted RNA was accessible (unstructured). While it is speculated that repression was caused by occluding the RBS, further investigation will be required to confirm the mechanism of RNA targeting by SauCas9.[37] A programmable, RNA targeted Cas9 greatly expands the application of CRISPR-Cas tools and provides researchers with another level of control over gene expression. Additionally, since SauCas9 can bind ssRNA, this provides the potential to couple this activity to other effector proteins to test their ability to alter translational activity.

Recent efforts to discover novel CRISPR-Cas systems identified a Cas protein (Cas13) that contained an endoRNAse domain suggesting its native activity was to target RNA.[95,96] Cas13 was shown to be able to protect *E. coli* against invading RNA phage infections (Figure 16.4D). However, despite the ability of programmable RNA targeting, Cas13 will indiscriminately cleave surrounding RNA transcripts, which has limited its use in bacteria. This activity does not seem to occur in eukaryotic cells and thus the Cas13 system has been engineered for RNA knockdown, RNA base editing, and live cell imaging.[97]

Other Applications

Nucleotide Acid Detection

Cas nucleases, such as Cas12a and Cas13a, have been used for nucleic acid detection, with the advantages of low cost, high sensitivity, and high specificity. Cas12a and Cas13a are guided by crRNAs; however, Cas12a cleaves dsDNA whereas Cas13a acts on ssRNA. Notably, after target cleavage by these two nucleases, their activated Cas-crRNA complexes present a "collateral"

cleavage activity on any nearby nontargeted sequences. Based on the robust collateral cleavage of Cas12a against nonspecific ssDNA, researchers developed a DNA Endonuclease Targeted CRISPR Trans Reporter method (DETECTR),[8] which couples a Cas12a-based DNA reporter with isothermal preamplification. It has been successfully applied for rapid and accurate qualitative detection of carcinoma-associated human papillomavirus (HPV) types from clinical specimens with 1 attomolar detection limit. Similarly, the specific high-sensitivity enzymatic reporter unlocking method (SHERLOCK),[98] employs the collateral cleavage activity of *Leptotrichia wadei* Cas13a to cleave and activate a quenched fluorescent reporter RNA by which the presence of the target sequence of interest is determined. SHERLOCK also demonstrated a remarkable sensitivity detecting target RNA on an attomolar scale, which is more sensitive than currently reported PCR-based methods that have a detection limit on a femtomolar scale. In addition, it has a strong specificity able to distinguish virus strains and pathogenic bacterial strains with low cross-reactivity. Additionally, the SHERLOCK system was upgraded to simultaneously incorporate three Cas13a orthologs and Cas12a,[7] enabling the detection of three ssRNA targets and one dsDNA target in a single reaction in less than 90 minutes.

In Situ DNA Imaging

Fluorescence in situ hybridization (FISH) with oligonucleotide probes plays an important role in studying the composition and distribution of microorganisms in phylogenetic, ecological, diagnostic, and environmental studies. Cas9-mediated fluorescence in situ hybridization (CASFISH) uses a fluorescently labeled dCas9 protein assembled with various gRNAs to rapidly label genomic loci in fixed cells,[9] even at repetitive DNA elements in a G-rich context. This method does not require global DNA denaturation. The experimental duration for detection takes less than an hour, instead of the hours to days that traditional DNA FISH usually takes. Since the dCas9/gRNA complexes bind target DNA with high affinity and stability, it allows for sequential or simultaneous probing of multiple targets and multicolor labeling of target loci in cells if differently colored dCas9/gRNA complexes are applied. This rapid, robust, less disruptive, and cost-effective technology is a valuable tool for basic research and genetic diagnosis.

DNA Cloning

In light of the programmable sequence-specific cleavage by Cas9, researchers have applied it to perform targeted cloning of large gene clusters that are usually expensive to synthesize or difficult to clone using traditional PCR or restriction enzyme-based methods. The Cas9-Assisted Targeting of CHromosome segments (CATCH) method liberates 150 kb DNA from a bacterial chromosome by in-gel Cas9-based cleavage on both sides of the target DNA, followed by *in vitro* Gibson assembly to ligate with a BAC vector.[10] In contrast, the transformation-associated recombination (TAR)-CRISPR method integrates Cas9-based DNA liberation with intracellular homologous recombination in yeast cells to accomplish DNA ligation.[99] TAR uses homology of up to a few hundred bp for homologous recombination, instead of the ~20 bp of homology required in Gibson assembly. These methods overcome the longstanding challenge of heterologously

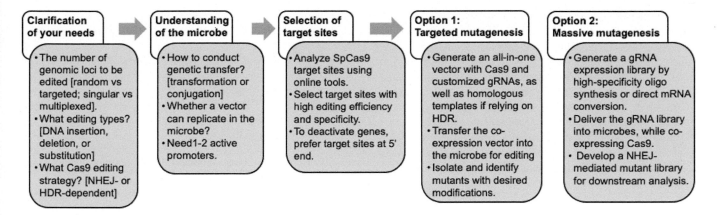

FIGURE 16.5 A guideline for the development of Cas9-based genetic engineering for microbes of interest.

studying large DNA fragments and hopefully promote the engineering of microorganisms to study microbial biosynthetic gene clusters that potentially produce valuable biomolecules, such as pharmaceuticals and biofuels.

Harnessing Cas9 for Your Editing Interest

As the prominent advantages of CRISPR-Cas genome editing tools have been proven in all three domains of life, here we describe a general guideline for users to harness the well-characterized SpCas9 to edit bacterial genomes of interest. We aim to clarify the overall feasibility, preparatory work, and experimental implementation of SpCas9 adaptation (Figure 16.5).

Knowing Your Purpose

The user should specify what DNA sequences or genes will be targeted and by what editing types (e.g. deletion, insertion, and substitution). The SpCas9 genome editing tool provides strong flexibility, allowing for singular and multiplexed editing, as well as the generation of genome wide random mutant libraries. The targeting capability is determined by whether the gRNA is singular or multiplexed in a cell, or if many singular gRNAs are pooled for editing in a large cell population. To achieve a precise modification at the target site, a homologous DNA template that carries the desired modification should be provided along with the SpCas9-gRNA co-expression system in order to guide the endogenous HDR system to fix the DNA break made by SpCas9 cleavage. If no homologous sequence is present, the error-prone NHEJ system will repair the cut and introduce unpredictable indels at the joint site. As unpredictable, these indels may cause a frameshift mutation but this does not guarantee the creation of a loss-of-function mutant. As most bacteria lack functional NHEJ, the toxicity of the SpCas9-induced dsDNA breaks can often lead to failure of successful genome editing.[5]

Preparation of Basic Genetic Toolkit

The Cas9 nuclease and its corresponding gRNA are two essential components that need to be co-expressed in host cells. The user should have prior knowledge of whether the microbe of interest is genetically tractable in terms of genetic transfer (e.g., transduction or transformation) and basic genetic toolkit (e.g., shuttle vectors and functional elements, like promoters, for gene expression). More specifically, the user should determine the antibiotic resistance of the microbe of interest, a suitable vector backbone, a genetic transfer protocol, as well as one or two active promoters to drive the expression of SpCas9 and gRNA.

Selection of Your Target Sites

Cas9 nucleases may require distinct characteristics at the target sites. SpCas9 recognizes a ~20 bp protospacer ending with a 5'-NGG-3' PAM site. Extensive studies have shown that the GC content, secondary structure, and nucleotide preference in the protospacer, as well as the location of mismatches between the target site and gRNA, are all important for editing efficiency and accuracy.[100–103] A few computational programs or online tools, such as CRISPOR (http://crispor.org),[104] CHOPCHOP (https://chopchop.cbu.uib.no),[105] and CRISPR library designer (http://www.e-crisp.org/E-CRISP/index.html),[106] provide user-friendly interfaces to search for qualified target sites in bacterial genes or genomes. Qualified target sites closer to the 5' end of a gene are preferred to avoid the potential for a truncated product that could result from a target site toward the end of a gene.

Plasmid Construction and Mutagenesis

Homologous template-dependent genome editing by Cas9 or Cas9n can circumvent the toxicity of DNA breaks, allowing for predictable, precise, and efficient editing. To do so, the user needs to clone three components into a host-applicable vector, including a Cas9n cassette, a customized gRNA cassette, and a homologous DNA template. The vector backbone with a suitable selection marker gene can be modified to an intermediate plasmid containing just the Cas9n expression cassette. It is worthwhile to determine if the microbial host has a biased codon usage, which may require a codon optimized Cas9 for efficient translation. The intermediate plasmid will be further devised to target a specific gene by incorporating a customized gRNA expression cassette and corresponding homologous donor template. The all-in-one plasmid can be constructed in a single step using Gibson assembly. Multiplexed targeting is also feasible by integrating more than one gRNA and their corresponding donor templates into a single plasmid.

Once the plasmid is transferred into the host cells, antibiotic-resistant recipient cells will express Cas9n and the gRNA to form catalytic effector complexes to create a nick at the target site, which triggers the HDR system to introduce a faithful modification based on the homologous donor template. It is encouraged to check the expression of SpCas9 nickase and gRNAs by real-time PCR. Targeted editing sites are identifiable by PCR amplification of the region harboring the modification and Sanger sequencing of PCR amplicons. Passaging the antibiotic-resistant cells in a selective liquid medium is an efficient way to enrich edited cells for isolation. If the plasmid has a temperature-sensitive replication origin, elevating the growth temperature will cure the plasmid to get plasmid-free cells with desired genetic modifications. Plasmid curing is necessary to iterate editing using the same plasmid backbone.

Massive Mutagenesis

To generate a genome-wide mutant library using Cas9-NHEJ or Cas9n-NHEJ strategies, high-quality gRNAs for all genes should be computationally designed and then chemically synthesized; however, it may be costly to synthesize a large gRNA pool. Alternatively, a method that converts mRNA into a gRNA library for random editing has been established,[107] such that neither chemically synthesized gRNAs nor prior knowledge of target DNA sequences is required. Its major limitation is the potential off-targeting of some resulting gRNAs in the pool. For genome-wide transcriptional repression with the CRISPRi technology, the user should use dCas9, instead of Cas9.

Conclusion and Future Outlook

CRISPR-Cas technologies have revolutionized our ability to edit genomes, finely tune gene expression, and image genomic loci. Major advantages of these tools are their efficiency, cost-effectiveness, simplicity, and applicability to a wide range of cell types. Engineering and directed evolution have already greatly expanded the application of the most widely used Cas9 system by increasing its targeting ability while decreasing off-targeting effects. The diversity of CRISPR-Cas systems continues to grow as new systems are being identified at a rapid pace. As researchers adapt these technologies to fit their specific needs, innovative applications continue to emerge. These tools will continue to transform all areas of biological research for years to come.

REFERENCES

1. Marraffini, L. A. & Sontheimer, E. J. CRISPR interference: RNA-directed adaptive immunity in bacteria and archaea. *Nat. Rev. Genet.* **11**, 181–190 (2010).
2. Mohanraju, P. *et al.* Diverse evolutionary roots and mechanistic variations of the CRISPR-Cas systems. *Science* **353**, aad5147 (2016).
3. Jiang, W., Bikard, D., Cox, D., Zhang, F. & Marraffini, L. A. RNA-guided editing of bacterial genomes using CRISPR-Cas systems. *Nat. Biotechnol.* **31**, 233–239 (2013).
4. Cong, L. *et al.* Multiplex genome engineering using CRISPR/Cas systems. *Science* **339**, 819–823 (2013).
5. Xu, T. *et al.* Efficient genome editing in *clostridium cellulolyticum* via CRISPR-Cas9 nickase. *Appl. Environ. Microbiol.* **81**, 4423–4431 (2015).
6. Qi, L. S. *et al.* Repurposing CRISPR as an RNA-guided platform for sequence-specific control of gene expression. *Cell* **152**, 1173–1183 (2013).
7. Gootenberg, J. S. *et al.* Multiplexed and portable nucleic acid detection platform with Cas13, Cas12a, and Csm6. *Science* **360**, 439–444 (2018).
8. Chen, J. S. *et al.* CRISPR-Cas12a target binding unleashes indiscriminate single-stranded DNase activity. *Science* **360**, 436–439 (2018).
9. Deng, W., Shi, X., Tjian, R., Lionnet, T. & Singer, R. H. CASFISH: CRISPR/Cas9-mediated in situ labeling of genomic loci in fixed cells. *Proc. Natl. Acad. Sci. U. S. A.* **112**, 11870–11875 (2015).
10. Jiang, W. *et al.* Cas9-assisted targeting of CHromosome segments CATCH enables one-step targeted cloning of large gene clusters. *Nat. Commun.* **6**, 8101 (2015).
11. Ishino, Y., Shinagawa, H., Makino, K., Amemura, M. & Nakata, A. Nucleotide sequence of the iap gene, responsible for alkaline phosphatase isozyme conversion in *Escherichia coli*, and identification of the gene product. *J. Bacteriol.* **169**, 5429–5433 (1987).
12. Mojica, F. J., Juez, G. & Rodríguez-Valera, F. Transcription at different salinities of *Haloferax mediterranei* sequences adjacent to partially modified PstI sites. *Mol. Microbiol.* **9**, 613–621 (1993).
13. Jansen, R., van Embden, J. D. A., Gaastra, W. & Schouls, L. M. Identification of genes that are associated with DNA repeats in prokaryotes. *Mol. Microbiol.* **43**, 1565–1575 (2002).
14. Bolotin, A., Quinquis, B., Sorokin, A. & Ehrlich, S. D. Clustered regularly interspaced short palindrome repeats (CRISPRs) have spacers of extrachromosomal origin. *Microbiology* **151**, 2551–2561 (2005).
15. Mojica, F. J. M., Díez-Villaseñor, C., García-Martínez, J. & Soria, E. Intervening sequences of regularly spaced prokaryotic repeats derive from foreign genetic elements. *J. Mol. Evol.* **60**, 174–182 (2005).
16. Pourcel, C., Salvignol, G. & Vergnaud, G. CRISPR elements in *Yersinia pestis* acquire new repeats by preferential uptake of bacteriophage DNA, and provide additional tools for evolutionary studies. *Microbiology* **151**, 653–663 (2005).
17. Barrangou, R. *et al.* CRISPR provides acquired resistance against viruses in prokaryotes. *Science* **315**, 1709–1712 (2007).
18. Brouns, S. J. J. *et al.* Small CRISPR RNAs guide antiviral defense in prokaryotes. *Science* **321**, 960–964 (2008).
19. Marraffini, L. A. & Sontheimer, E. J. CRISPR interference limits horizontal gene transfer in *staphylococci* by targeting DNA. *Science* **322**, 1843–1845 (2008).
20. Garneau, J. E. *et al.* The CRISPR/Cas bacterial immune system cleaves bacteriophage and plasmid DNA. *Nature* **468**, 67–71 (2010).
21. Deltcheva, E. *et al.* CRISPR RNA maturation by trans-encoded small RNA and host factor RNase III. *Nature* **471**, 602 (2011).
22. Sapranauskas, R. *et al.* The *Streptococcus thermophilus* CRISPR/Cas system provides immunity in *Escherichia coli*. *Nucleic Acids Res.* **39**, 9275–9282 (2011).
23. Gasiunas, G., Barrangou, R., Horvath, P. & Siksnys, V. Cas9–crRNA ribonucleoprotein complex mediates specific DNA cleavage for adaptive immunity in bacteria. *Proc. Natl. Acad. Sci. U. S. A.* **109**, E2579–E2586 (2012).

24. Jinek, M. *et al.* A programmable dual-RNA-guided DNA endonuclease in adaptive bacterial immunity. *Science* **337**, 816–821 (2012).

25. Zetsche, B. *et al.* Cpf1 is a single RNA-guided endonuclease of a class 2 CRISPR-Cas system. *Cell* **163**, 759–771 (2015).

26. Jinek, M. *et al.* Structures of Cas9 endonucleases reveal RNA-mediated conformational activation. *Science* **343**, 1247997 (2014).

27. Nishimasu, H. *et al.* Crystal structure of Cas9 in complex with guide RNA and target DNA. *Cell* **156**, 935–949 (2014).

28. Swarts, D. C., van der Oost, J. & Jinek, M. Structural basis for guide RNA processing and seed-dependent DNA targeting by CRISPR-Cas12a. *Mol. Cell* **66**, 221–233.e4 (2017).

29. Taylor, D. W. *et al.* Structural biology. Structures of the CRISPR-Cmr complex reveal mode of RNA target positioning. *Science* **348**, 581–585 (2015).

30. Horvath, P. & Barrangou, R. CRISPR/Cas, the immune system of bacteria and archaea. *Science* **327**, 167–170 (2010).

31. Zhang, Y. *et al.* Processing-independent CRISPR RNAs limit natural transformation in *Neisseria meningitidis*. *Mol. Cell* **50**, 488–503 (2013).

32. Karvelis, T. *et al.* Rapid characterization of CRISPR-Cas9 protospacer adjacent motif sequence elements. *Genome Biol.* **16**, 253 (2015).

33. Esvelt, K. M. *et al.* Orthogonal Cas9 proteins for RNA-guided gene regulation and editing. *Nat. Methods* **10**, 1116–1121 (2013).

34. Slaymaker, I. M. *et al.* Rationally engineered Cas9 nucleases with improved specificity. *Science* **351**, 84–88 (2016).

35. Nishimasu, H. *et al.* Engineered CRISPR-Cas9 nuclease with expanded targeting space. *Science* **361**, 1259–1262 (2018).

36. Lee, J. K. *et al.* Directed evolution of CRISPR-Cas9 to increase its specificity. doi:10.1101/237040

37. Strutt, S. C., Torrez, R. M., Kaya, E., Negrete, O. A. & Doudna, J. A. RNA-dependent RNA targeting by CRISPR-Cas9. *Elife* **7**, (2018).

38. Kim, D. *et al.* Genome-wide analysis reveals specificities of Cpf1 endonucleases in human cells. *Nat. Biotechnol.* **34**, 863–868 (2016).

39. Zhang, J., Yang, F., Yang, Y., Jiang, Y. & Huo, Y.-X. Optimizing a CRISPR-Cpf1-based genome engineering system for *Corynebacterium glutamicum*. *Microb. Cell Fact.* **18**, 60 (2019).

40. Liu, Y.-J., Zhang, J., Cui, G.-Z. & Cui, Q. Current progress of targetron technology: development, improvement and application in metabolic engineering. *Biotechnol. J.* **10**, 855–865 (2015).

41. Wen, Z. *et al.* TargeTron technology applicable in solventogenic clostridia: revisiting 12 Years' advances. *Biotechnol. J.* e1900284 (2019).

42. Sharan, S. K., Thomason, L. C., Kuznetsov, S. G. & Court, D. L. Recombineering: a homologous recombination-based method of genetic engineering. *Nat. Protoc.* **4**, 206–223 (2009).

43. Barrangou, R. & Doudna, J. A. Applications of CRISPR technologies in research and beyond. *Nat. Biotechnol.* **34**, 933–941 (2016).

44. Gaj, T., Gersbach, C. A. & Barbas, C. F., 3rd. ZFN, TALEN, and CRISPR/Cas-based methods for genome engineering. *Trends Biotechnol.* **31**, 397–405 (2013).

45. Gaudelli, N. M. *et al.* Programmable base editing of A•T to G•C in genomic DNA without DNA cleavage. *Nature* **551**, 464–471 (2017).

46. Komor, A. C., Kim, Y. B., Packer, M. S., Zuris, J. A. & Liu, D. R. Programmable editing of a target base in genomic DNA without double-stranded DNA cleavage. *Nature* **533**, 420–424 (2016).

47. Davis, A. J. & Chen, D. J. DNA double strand break repair via non-homologous end-joining. *Transl. Cancer Res.* **2**, 130–143 (2013).

48. Shuman, S. & Glickman, M. S. Bacterial DNA repair by non-homologous end joining. *Nat. Rev. Microbiol.* **5**, 852–861 (2007).

49. Bowater, R. & Doherty, A. J. Making ends meet: repairing breaks in bacterial DNA by non-homologous end-joining. *PLOS Genet.* **2**, e8 (2006).

50. Tong, Y., Charusanti, P., Zhang, L., Weber, T. & Lee, S. Y. CRISPR-Cas9 based engineering of actinomycetal genomes. *ACS Synth. Biol.* **4**, 1020–1029 (2015).

51. Su, T. *et al.* A CRISPR-Cas9 Assisted non-homologous end-joining strategy for one-step engineering of bacterial genome. *Scientific Reports* **6**, (2016).

52. Löbs, A.-K., Schwartz, C. & Wheeldon, I. Genome and metabolic engineering in non-conventional yeasts: current advances and applications. *Synth Syst Biotechnol.* **2**, 198–207 (2017).

53. Schuster, M. & Kahmann, R. CRISPR-Cas9 genome editing approaches in filamentous fungi and oomycetes. *Fungal Genet. Biol.* **130**, 43–53 (2019).

54. Xu, T., Li, Y., Van Nostrand, J. D., He, Z. & Zhou, J. Cas9-based tools for targeted genome editing and transcriptional control. *Appl. Environ. Microbiol.* **80**, 1544–1552 (2014).

55. Lusetti, S. L. & Cox, M. M. The bacterial RecA protein and the recombinational DNA repair of stalled replication forks. *Annu. Rev. Biochem.* **71**, 71–100 (2002).

56. Smith, G. R. Homologous recombination near and far from DNA breaks: alternative roles and contrasting views. *Annu. Rev. Genet.* **35**, 243–274 (2001).

57. Jiang, Y. *et al.* Multigene editing in the *Escherichia coli* genome via the CRISPR-Cas9 system. *Appl. Environ. Microbiol.* **81**, 2506–2514 (2015).

58. Cobb, R. E., Wang, Y. & Zhao, H. High-efficiency multiplex genome editing of *streptomyces* species using an engineered CRISPR/Cas system. *ACS Synthetic Biology* **4**, 723–728 (2015).

59. Huang, H., Zheng, G., Jiang, W., Hu, H. & Lu, Y. One-step high-efficiency CRISPR/Cas9-mediated genome editing in *Streptomyces*. *Acta Biochim. Biophys. Sin.* **47**, 231–243 (2015).

60. Li, Q. *et al.* CRISPR-based genome editing and expression control systems in Clostridium acetobutylicum and *Clostridium beijerinckii*. *Biotechnol. J.* **11**, 961–972 (2016).

61. Garst, A. D. *et al.* Genome-wide mapping of mutations at single-nucleotide resolution for protein, metabolic and genome engineering. *Nat. Biotechnol.* **35**, 48–55 (2017).

62. Shi, T.-Q. *et al.* CRISPR/Cas9-based genome editing of the filamentous fungi: the state of the art. *Appl. Microbiol. Biotechnol.* **101**, 7435–7443 (2017).

63. Liu, R., Chen, L., Jiang, Y., Zhou, Z. & Zou, G. Efficient genome editing in filamentous fungus *Trichoderma reesei* using the CRISPR/Cas9 system. *Cell Discovery* **1**, (2015).

64. Liu, Q. *et al.* Development of a genome-editing CRISPR/Cas9 system in thermophilic fungal *Myceliophthora* species and its application to hyper-cellulase production strain engineering. *Biotechnol. Biofuels* **10**, 1 (2017).

65. Komor, A. C. *et al.* Improved base excision repair inhibition and bacteriophage Mu Gam protein yields C:G-to-T: a base editors with higher efficiency and product purity. *Sci. Adv.* **3**, eaao4774 (2017).

66. Zheng, K. *et al.* Highly efficient base editing in bacteria using a Cas9-cytidine deaminase fusion. *Commun Biol* **1**, 32 (2018).

67. Copeland, N. G., Jenkins, N. A. & Court, D. L. Recombineering: a powerful new tool for mouse functional genomics. *Nat. Rev. Genet.* **2**, 769–779 (2001).

68. Ellis, H. M., Yu, D., DiTizio, T. & Court, D. L. High efficiency mutagenesis, repair, and engineering of chromosomal DNA using single-stranded oligonucleotides. *Proc. Natl. Acad. Sci. U. S. A.* **98**, 6742–6746 (2001).

69. Yu, D. *et al.* An efficient recombination system for chromosome engineering in *Escherichia coli*. *Proc. Natl. Acad. Sci. U. S. A.* **97**, 5978–5983 (2000).

70. Zhang, Y., Buchholz, F., Muyrers, J. P. & Stewart, A. F. A new logic for DNA engineering using recombination in *Escherichia coli*. *Nat. Genet.* **20**, 123–128 (1998).

71. Murphy, K. C., Campellone, K. G. & Poteete, A. R. PCR-mediated gene replacement in *Escherichia coli*. *Gene* **246**, 321–330 (2000).

72. Karu, A. E., Sakaki, Y., Echols, H. & Linn, S. The gamma protein specified by bacteriophage gamma. Structure and inhibitory activity for the recBC enzyme of *Escherichia coli*. *J. Biol. Chem.* **250**, 7377–7387 (1975).

73. Murphy, K. C. Lambda Gam protein inhibits the helicase and chi-stimulated recombination activities of *Escherichia coli* RecBCD enzyme. *J. Bacteriol.* **173**, 5808–5821 (1991).

74. Cassuto, E. & Radding, C. M. Mechanism for the action of λ exonuclease in genetic recombination. *Nat. New Biol.* **229**, 13–16 (1971).

75. Little, J. W. An Exonuclease Induced by Bacteriophage λ: II. Nature of the enzymatic reaction. *J. Biol. Chem.* **242**, 679–686 (1967).

76. Karakousis, G. *et al.* The beta protein of phage λ binds preferentially to an intermediate in DNA renaturation. *J. Mol. Biol.* **276**, 721–731 (1998).

77. Kmiec, E. & Holloman, W. K. Beta protein of bacteriophage lambda promotes renaturation of DNA. *J. Biol. Chem.* **256**, 12636–12639 (1981).

78. Muniyappa, K. & Mythili, E. Phage lambda beta protein, a component of general recombination, is associated with host ribosomal S1 protein. *Biochem. Mol. Biol. Int.* **31**, 1–11 (1993).

79. Cassuto, E., Lash, T., Sriprakash, K. S. & Radding, C. M. Role of exonuclease and protein of phage in genetic recombination, V. recombination of DNA in Vitro. *Proc. Natl. Acad. Sci. U. S. A.* **68**, 1639–1643 (1971).

80. Oh, J.-H. & van Pijkeren, J.-P. CRISPR–Cas9-assisted recombineering in *Lactobacillus reuteri*. *Nucleic Acids Res.* **42**, e131–e131 (2014).

81. Cho, J. S. *et al.* CRISPR/Cas9-coupled recombineering for metabolic engineering of *Corynebacterium glutamicum*. *Metab. Eng.* **42**, 157–167 (2017).

82. Aparicio, T., de Lorenzo, V. & Martínez-García, E. CRISPR/Cas9-based counterselection boosts recombineering efficiency in *Pseudomonas putida*. *Biotechnol. J.* **13**, 1700161 (2018).

83. Mali, P. *et al.* CAS9 transcriptional activators for target specificity screening and paired nickases for cooperative genome engineering. *Nat. Biotechnol.* **31**, 833–838 (2013).

84. Zhang, X. *et al.* Multiplex gene regulation by CRISPR-ddCpf1. *Cell Discov* **3**, 17018 (2017).

85. Hsu, P. D. *et al.* DNA targeting specificity of RNA-guided Cas9 nucleases. *Nat. Biotechnol.* **31**, 827–832 (2013).

86. Bikard, D. *et al.* Programmable repression and activation of bacterial gene expression using an engineered CRISPR-Cas system. *Nucleic Acids Res.* **41**, 7429–7437 (2013).

87. Dong, C., Fontana, J., Patel, A., Carothers, J. M. & Zalatan, J. G. Synthetic CRISPR-Cas gene activators for transcriptional reprogramming in bacteria. *Nat. Commun.* **9**, 2489 (2018).

88. Liu, Y., Wan, X. & Wang, B. Engineered CRISPRa enables programmable eukaryote-like gene activation in bacteria. *Nat. Commun.* **10**, 3693 (2019).

89. Yang, Y. *et al.* Transcription. Structures of the RNA polymerase-σ54 reveal new and conserved regulatory strategies. *Science* **349**, 882–885 (2015).

90. Shingler, V. Signal sensory systems that impact σ54-dependent transcription. *FEMS Microbiol. Rev.* **35**, 425–440 (2011).

91. O'Connell, M. R. *et al.* Programmable RNA recognition and cleavage by CRISPR/Cas9. *Nature* **516**, 263–266 (2014).

92. Nelles, D. A. *et al.* Programmable RNA tracking in live cells with CRISPR/Cas9. *Cell* **165**, 488–496 (2016).

93. Liu, Y. *et al.* Targeting cellular mRNAs translation by CRISPR-Cas9. *Sci. Rep.* **6**, 29652 (2016).

94. Batra, R. *et al.* Elimination of toxic microsatellite repeat expansion RNA by RNA-targeting Cas9. *Cell* **170**, 899–912.e10 (2017).

95. Abudayyeh, O. O. *et al.* RNA targeting with CRISPR-Cas13. *Nature* **550**, 280–284 (2017).

96. Abudayyeh, O. O. *et al.* C2c2 is a single-component programmable RNA-guided RNA-targeting CRISPR effector. *Science* **353**, aaf5573 (2016).

97. Cox, D. B. T. *et al.* RNA editing with CRISPR-Cas13. *Science* **358**, 1019–1027 (2017).

98. Gootenberg, J. S. *et al.* Nucleic acid detection with CRISPR-Cas13a/C2c2. *Science* **356**, 438–442 (2017).

99. Lee, N. C. O., Larionov, V. & Kouprina, N. Highly efficient CRISPR/Cas9-mediated TAR cloning of genes and chromosomal loci from complex genomes in yeast. *Nucleic Acids Res.* **43**, e55 (2015).

100. Lin, Y. *et al.* CRISPR/Cas9 systems have off-target activity with insertions or deletions between target DNA and guide RNA sequences. *Nucleic Acids Res.* **42**, 7473–7485 (2014).

101. Ren, X. *et al.* Enhanced specificity and efficiency of the CRISPR/Cas9 system with optimized sgRNA parameters in Drosophila. *Cell Rep.* **9**, 1151–1162 (2014).

102. Shalem, O. *et al.* Genome-scale CRISPR-Cas9 knockout screening in human cells. *Science* **343**, 84–87 (2014).

103. Wong, N., Liu, W. & Wang, X. WU-CRISPR: characteristics of functional guide RNAs for the CRISPR/Cas9 system. *Genome Biol.* **16**, 218 (2015).

104. Haeussler, M. *et al.* Evaluation of off-target and on-target scoring algorithms and integration into the guide RNA selection tool CRISPOR. *Genome Biol.* **17**, 148 (2016).

105. Labun, K. *et al.* CHOPCHOP v3: expanding the CRISPR web toolbox beyond genome editing. *Nucleic Acids Res.* **47**, W171–W174 (2019).

106. Heigwer, F., Kerr, G. & Boutros, M. E-CRISP: fast CRISPR target site identification. *Nat. Methods* **11**, 122–123 (2014).

107. Arakawa, H. A method to convert mRNA into a gRNA library for CRISPR/Cas9 editing of any organism. *Sci Adv* **2**, e1600699 (2016).

Part II

Survey of Microorganisms

17

Taxonomic Classification of Bacteria

J. Michael Janda

CONTENTS

Introduction

The ability of humans to interact and communicate effectively with one another is dependent upon a language system that places similar objects together in distinct categories that are easily recognizable by the general population. Items such as saws, screwdrivers, and hammers are grouped together under the general term "tools," while meat, bread, and milk are collectively referred to as "foods."

The field of microbiology is no different than those examples. Bacteria recovered from a variety of sources, such as soil, water, aquatic and terrestrial ecosystems, and vertebrates need to be "labeled" such that they can be associated with various functions, including industrial processes, bioremediation, and plant growth. In no area of the microbial world are these labels more important than in the arena of medical and diagnostic microbiology, infectious diseases, and related disciplines. Here, the ability to communicate locally, nationally, or internationally regarding infection, disease outbreak investigations, epidemics, and pandemics is critical in the detection, identification, epidemiologic monitoring, and control of infectious diseases of bacterial etiology. For these subdisciplines, labels must be created that can be effectively communicated worldwide. To achieve these goals, a defined set of rules and guidelines have been developed that cover these labels better known as bacterial taxonomy.

History of Bacterial Taxonomy

The history of bacterial taxonomy began when Antony van Leeuwenhoek through his mastery of developing lenses and construction of more than 500 microscopes in his lifetime, described "tiny animacules" for the first time.[1] Visualization of bacteria, protozoa, and molds from lakes, human saliva, and tooth scrapings by Leeuwenhoek between 1673 and 1674 ushered in a new era.

However, initial attempts at naming and classifying these bacteria did not occur until the late 19th century.[2] Almost 200 years later, the field of bacteriology had advanced significantly, such that by the mid- to late-19th-century scientists had created various media, some biochemical tests, and stains to culture, identify, and describe many bacteria associated with various processes including life-threatening human infections. Examples of cardinal achievements produced during this era include the description/isolation of *Shigella dysenteriae* type 1 by Kiyoshi Shiga, *Salmonella* Typhi by Eberth and Gaffky, and the development of Endo agar to help differentiate typhoid bacilli from *Escherichia coli*.[3–5]

Along with the multitude of scientific discoveries during the "golden era" of bacteriology came a litany of names associated with disease-causing bacteria or those associated with commensal processes. Most of these bacteria were given scientific names (genus and species) although acronyms or scientific jargon were also used such as "Shiga's bacillus." Despite these achievements, several problematic issues arose. Since there was a very limited number of phenotypic tests available at the time, bacteria that were isolated were often poorly characterized. It was also difficult to determine quickly whether two independent researchers were working on identical, similar, or different bacterial species. Various bacteriologists described *Bacillus coli commune* (1885), *Bacillus coli* (1895), and *Bacterium coli* (1895), which were actually synonyms of *E. coli*, while *Micrococcus pneumoniae* (1884) and *Diplococcus pneumoniae* (1886) hold the same taxonomic status in relation to *Streptococcus pneumoniae*. The multiplicity of names for a single bacterial species during the infancy of bacteriology led to uncertainty, taxonomic confusion, and numerous issues regarding isolation and identification.

Today's bacterial taxonomy that is used worldwide to effectively communicate between clinicians, scientists, and investigators is a continuum of three ongoing major processes, namely (1) the attempt to put stability and order to bacterial systematics regarding the naming and description of new species; (2) the evolution from biochemical (phenetic analysis) to molecular and phylogenetic

DOI: 10.1201/9781003099277-19

tests (cladistic analysis) with which to assess the uniqueness or relatedness of one taxon to another, and; (3) the current rules regarding the identification, naming, and description of new taxa.

Bacterial Systematics

By the 1930s, bacterial taxonomy had already become quite complex with more than 2,700 species recognized in the fourth edition of *Bergey's Manual of Systematic Bacteriology* (note: most of these species no longer exist).[2] Numbers of new species and genera continued to be described as more metabolic, phenotypic, and structural data became available.[6] Where bacteria should be classified was highly speculative with many scientists believing they should be grouped with plants— even the seventh edition of *Bergey's* in 1957 still classified them with plants.[6]

During the mid-to-late 1930s, however, a movement began that culminated in a proposal by Stanier and van Niel in 1962 to accept a clear division between bacteria (prokaryotes) and animals and plants (eukaryotes).[6] This separation was also reflected in the 1974 eighth edition of *Bergey's Manual*. Concurrent to these changes was a drive to codify the nomenclature of prokaryotes.[7] The first edition of the *International Code of Nomenclature of Prokaryotes* was published in 1957 with the goal for those who use scientific names of microorganisms "to use correct names and to use them correctly."[8] However, reclassification of bacteria and establishment of *The Code* did not alleviate the myriad problems associated with the existing taxonomy, including duplication of names, lack of type cultures, and poorly characterized or described taxa. To rectify this situation, the Judicial Commission of the International Committee on Systematic Bacteriology authorized a review of all taxa and to only retain those names that were adequately described, and if cultivatable, with a type or neotype strain.[9] That list was published in 1980 as the Approved Lists of Bacterial Names and consisted of 1,792 species and 290 genera.[9] If the species name did not appear on that list it had no standing in nomenclature. Essentially, this 1980 list set the bacterial taxonomy time clock at zero, providing us with a fresh start.

Evolution of Methodologies in the Description of Prokaryotic Species

Figure 17.1 represents a graphic depiction of the transition in methodologies used to describe and propose new prokaryotic species between 1960 and 2010.[2,6,10] Prior to 1960, description and classification of bacterial species was almost totally reliant on cellular morphology, metabolic requirements, and in some instances, pathogenicity.[6] By the early 1960s, phenotypic characterization of bacteria had taken a major step forward with the introduction of numerical taxonomy (NT) into bacterial systematics. NT of bacteria, pioneered by Sokal and Sneath, involves determining the overall similarity of groups to each other based upon the analysis of many weighted or unweighted phenotypic features or characters.[11].

Results were originally coded into a computer to generate matching coefficients such that groups were linked together based upon highest similarity. Matching similarity (S_{sm}) indicating essentially identical groups at the species level required

threshold values around 85% using NT methods.[12] Coupled to the use of NT was the mol% G+C content of bacteria. Strains belonging to the same species should not exhibit more than a 2–3% variation in G+C content.[13,14]

NT had faded from playing a principle role in the identification and description of new taxa by the 1980s and was largely replaced by DNA-DNA hybridization (DDH). DDH started to be introduced in the field of bacterial systematics in the early 1970s and was immediately used to describe many new species of fermentative as well as nonfermentative bacteria of clinical importance. Instead of phenotypic markers, DDH was an overall assessment of the genetic similarity or relatedness between two DNA genomes. The phylogenetic definition for a species using DDH was subsequently determined to be 70% or more relatedness with a ΔT_m (stability of the heteroduplex) of 5°C or less.[15] This method was highly successful and has remained the "gold standard" for many years in determining the uniqueness of a newly proposed taxon and the relatedness of existing species. Traditional DDH, however, suffers from several drawbacks being expensive to perform, technically challenging, labor-intensive and restricted to only a few major research laboratories worldwide. Chemotaxonomy, which was commonly used in describing new species by studying the composition of cell walls, isoprenoid quinones, and bacterial cytochromes, was primarily utilized in this aspect between 1960 and 1980.[6]

In the mid-1990s, an additional molecular platform became available commercially with the introduction of 16S rDNA gene sequencing.[16] The technology allows for the sequencing of the 16S housekeeping genes, which occur in multiple copies within the bacterial chromosome and codes for rRNA. Commercial products contained database matrices that could be compared to variable regions (or full reads) within the 16S gene of test strains sequenced by users. Strains exhibiting <98.7% sequence similarity to established taxa strongly suggested a new species.[12,16] However, the converse statement of ≥98.7% sequence similarity did not necessarily imply species identity to the top match as many studies have indicated that for many groups rRNA sequence similarities are highly conserved with little variation detected among species of a given genus and in some instances a family.[16] Higher sequence similarity values require other techniques, such as DDH, to determine true relatedness.

A novel molecular-phenotypic technique, MALDI-TOF or matrix-assisted laser desorption/ionization-time of flight mass spectrometry had made its way from the research laboratories of the 1980s into diagnostic laboratories by 2010. The technique, based upon the spectral analysis of surface proteins (phenotypic), determines a spectrum for each strain analyzed.[17] The technique has been commercialized with a standard database matrix similar to 16S rDNA sequencing. However, unlike 16S, final results for a possible identification can be rapid (minutes). Like 16S, it has limitations, such as it cannot accurately detect certain groups at the species level. While it is not a standalone technology, such as DDH or 16S, with defined thresholds to determine species relatedness, it has been used in conjunction with sequence analysis (polyphasic approach) to screen for novel species. The term "taxonogenomics" has been coined by Fournier and Drancourt to describe such an application.[18]

The replacement of Sanger sequencing by the advent of high-throughput technologies, often referred to as whole genome

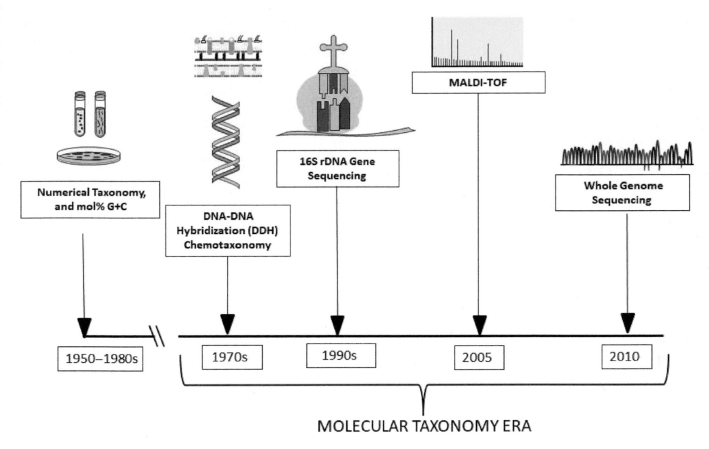

FIGURE 17.1   Molecular taxonomy era.

sequencing (WGS) or next-generation sequencing, has revolutionized our ability to redefine bacterial species and phylogenies on a molecular basis since 2010. Now, the availability of WGS to many laboratories allows for the side-by-side comparison of complete genetic sequences of both genomes. A numeric value, called the average nucleotide identity or ANI, provides for a similarity comparison between two chromosomes. It has been determined that an ANI of 95% is similar to DDH relatedness values of 70% in defining a genomospecies.[6,19] Many other potential applications of WGS are available that are not discussed here. These include routine species identification, outbreak investigations, and epidemiologic disease monitoring.

The Taxonomic Revolution: Domains, Divisions, and Diversifications

The collective impact of the emergence of molecular technologies to delineate bacterial taxonomy and phylogeny has been enormous and cannot be underestimated. It has led to major scientific discoveries of immense importance regarding prokaryotes and their biology, ecology, and disease spectrum. Only a few will be mentioned here.[20] Probably the single greatest achievement in modern microbial systematics was the discovery by Carl Woese of a third "domain" of life. Comparing 16S and 18S sequence catalogues of prokaryotes and eukaryotes revealed to Woese that two types of prokaryotes existed, one called "Bacteria" and the other called "Archaea."[21] These two domains, despite being

prokaryotes, are only distantly related to each other. The third domain is "Eucarya."[22] This discovery by Woese has revolutionized our concept of the tree of life with researchers trying to discover an ancestral form from which some or all of these domains originated.

Molecular methodologies have led to many other discoveries regarding prokaryotes. DDH supported by WGS has determined that all *Shigella* species and *E. coli* constitute a single species at the DNA level.[23,24] Separate names are retained because of the immense clinical and public health importance of *Shigella*. Similarly, *Salmonella* serotypes, previously thought to each represent a different species (at one time more than 2,000 species existed), have been determined to fall within a single species designated *Salmonella enterica*.[25] Thus, such common foodborne pathogens like *S.* Typhimurium, *S.* Enteritidis, and *S.* Newport all belong to a single subspecies (there are five) of *S. enterica*. The outcome of these taxonomic investigations has been a complete revision of *Salmonella* nomenclature.[26]

The availability of less expensive high-throughput WGS has led to the development of new fields of endeavor involving bacterial taxonomy. Metagenomics and culturomics have the ability to characterize and study microbial populations composed of noncultivable and well as culturable bacteria.[27] These technologies are being used to study the NIH-funded Human Microbiome Project (HMP), that is, the resident microbial flora of humans. One aspect of the HMP is to determine the microbial composition of the gastrointestinal tract, what constitutes homeostasis, and how this population changes in disease states.[28] Disease states that such data might have a profound influence on include ulcerative colitis and Crohn's disease.

Bacterial Taxonomy: Nomenclature, Classification, and Rules for the Road

Bacterial taxonomy is composed of two distinct yet related entities, nomenclature and classification. Nomenclature involves the naming and description of new taxa, principally at the species level. It also involves to a lesser extent the correct naming of various groups within taxonomic rank such as genus, family, order, and higher levels. Classification, on the other hand, involves the appropriate placement of related groups or clades next to each other based upon phylogenetic analyses. Taxa, such as species, are placed next to their "nearest neighbors" using one or more bioinformatic methods (such as maximum likelihood) to create trees. Together, nomenclature and classification create dendritic trees or trees of life for the domain, Bacteria.

Bacterial nomenclature is codified, that is, it is governed by an international set of rules and guidelines that are required to be followed for the valid publication of a new taxon such as a bacterial species (only taxonomic rank with defined criteria). Valid publication requires the proposed species to be formally published in the *International Journal of Systematic and Evolutionary Microbiology* (IJSEM) or be published elsewhere, meet all the requirements for valid publication, and subsequently appear in a Validation List in IJSEM. Some of the rules for valid publication include deposition of the nomenclature type strain in at least two recognized culture collections, the proposed name must be clearly indicated, the etymology of the new name must be stated, and the properties of the taxon described.[29,30] A set of minimal standards for the use of genome data regarding the taxonomy of prokaryotes has also been proposed.[20] A complete set of rules can be found in the latest edition of the *International Code of Nomenclature of Prokaryotes*.[8] The ultimate goal or aim of bacterial nomenclature is to maintain stability of names.

In contrast to bacterial nomenclature, there is no official classification system for bacteria. No rules or codes exist. Bacterial classification is currently applied using an operational-based model.[6] This model uses a defined set of criteria with thresholds or limits to identify groups or clades that share phylogenetic similarities.[6] A polyphasic approach to this type of analysis includes phenotypic (biochemical, chemotaxonomic, MALDI-TOF) as well as genetic (G+C, 16S, WGS) data. Unlike bacterial nomenclature, classification is always in a state of flux, constantly evolving, as new techniques and methods generate different approaches, theories, and proposals as to where taxa should be placed or linked on phylogenetic trees. Acceptance of a classification scheme is primarily based upon general acceptance and broad-based use by the medical and scientific communities. A perfect example of this is the species *Pseudomonas maltophilia* (1961), which has been reclassified as *Xanthomonas maltophilia* (1980), and then *Stenotrophomonas maltophilia* (1983) where it resides today. Same taxon (species) but found by phylogenetic investigations not to be a pseudomonad nor a member of the genus *Xanthomonas*.

Bacterial nomenclature and classification are currently intertwined in the proposal, description, and creation of new taxa and taxonomic ranks, although what involves nomenclature versus classification can be confusing to the average microbiologist. As an example, the major taxonomic changes occurring in the genus *Aeromonas* from its proposal in 1943 through 1988 are depicted in Table 17.1. The genus has undergone multiple transformations including reclassification into three separate families (*Pseudomonadaceae* → *Vibrionaceae* → *Aeromonadaceae*) and creation of several new species based initially on phenotypic methods and later using DDH. Today, the genus is composed of more than 30 validly published species (http://www.bacterio. net/aeromonas.html) with several species containing multiple subspecies. The dramatic increase in the number of *Aeromonas* species since 1988 is a testimony to the general availability of a number of techniques besides DDH such as 16S gene sequencing, MALDI-TOF, and the phylogenetic analysis of housekeeping genes including *rpoD, gyrB, dnaJ,* and *recA.*

Bacterial Taxonomy Issues

The complexity of bacterial systematics is not without its issues. These issues fall into two main categories, namely **practical issues** and **theoretical issues**. Practical issues for bacterial taxonomy are those matters that affect everyday scientific communications in the literature to laboratory reports identifying infectious agents. Theoretical issues involve conceptual topics concerning the status of taxonomic ranks, their definition and description, and real and virtual methods of such descriptions encompassing each rank.

Practical Issues

In the practical arena, there are important concerns regarding the proliferation and description of new species. Due to the emergence of WGS, more than 1,000 new bacterial species are proposed each year in major systematic journals.[40] As of 2017, there are more than 15,448 validly published species. To add to this logarithmic explosion of taxa are two simple facts, that the vast majority (>90%) of newly proposed species are described on the basis of a single type strain and that most of these species originate from environmental rather than clinical sources.

Such trends have a profound effect on both the medical and scientific community. Companies producing commercial identification/susceptibility panels will undoubtably be unable to keep up with the increasing number of proposed taxa. This is due to several factors including the gross number of species described, the fact that most new taxa cannot be identified by phenotypic methods, and finally the clinical/medical significance of these single-strain species is unknown. Even updating in-house MALDI-TOF systems will be problematic as the cost of acquiring each type strain from a recognized culture collection could be prohibitive. For international culture collections the increasing number of new taxa may overwhelm their capabilities.

What can be done? The publication of single isolates as new species has been questioned as early as 2001.[2,41] A leading systematic publication, *Antonie van Leeuwenhoek*, has recently changed its editorial policy such that the "one species, one paper" concept will only be considered for publication for studies where the authors can demonstrate novel biology and features of interest such as ecological insights relevant to sources of isolation.[42] A theoretical proposal to establish plural taxonomies for cultured

TABLE 17.1

Taxonomic Evolution of the Genus *Aeromonas* 1943–1988

Date	Family	Genus	Species	Bacterial Taxonomy	Taxonomic Application	Comments	Ref
1943	*Pseudomonadaceae*	*Aeromonas*	*hydrophila*	Nomenclature & Classification	Phenetic	Previously known as: *P. hydrophilus*, (*Aerobacter liquefaciens*)	31
1953	*Pseudomonadaceae*	*Aeromonas*	*salmonicida*	Nomenclature & Classification	Phenetic	*Bacterium salmonicida* Fish pathogen	32
1965	*Vibrionaceae*	*Aeromonas*	All species	Classification	NT	Transfer to new family	33
1976	*Vibrionaceae*	*Aeromonas*	*sobria*	Nomenclature	NT	New species	34
1980	*Vibrionaceae*	*Aeromonas*	*hydrophila punctata salmonicida*	Nomenclature & Classification	Previous data	Approved List	9
1983	*Vibrionaceae*	*Aeromonas*	*media*	Nomenclature	NT	New species	35
1984	*Vibrionaceae*	*Aeromonas*	*caviae*	Nomenclature	Previous data	Revived name, Validation List 15	36
1986	*Aeromonadaceae*	*Aeromonas*	All species	Classification	16S, 5S, rRNA/DNA hybridization	Transfer to a new family	37
1987	*Aeromonadaceae*	*Aeromonas*	*veronii*	Nomenclature	DDH	New species	38
1988	*Aeromonadaceae*	*Aeromonas*	*schubertii*	Nomenclature	DDH	New species	39

Abbreviations: DDH = DNA-DNA hybridization; NT = numerical taxonomy.

and uncultured bacteria could be expanded.[43] Such a concept could have separate tracts for bacterial taxonomy involving taxa of clinical, public health, and medical importance versus environmental species. With a minimum estimate of at least 10 million microbial species on the planet, the vast majority of unnamed taxa described in the future will originate from soil, water, vegetation, invertebrates, and vertebrate species other than humans.

Theoretical Issues

There has always been some controversy regarding what constitutes a useful and workable modern bacterial taxonomy, particularly since the advent of the molecular era. The cornerstone or foundation for all bacterial taxonomy is the definition of a bacterial species, which is the only taxonomic rank that, at present, is clearly defined based on DNA relatedness, 16S rDNA sequence similarities, and ANI. However, what constitutes a bacterial species or the "species concept" in general continues to be a major topic of discussion and contention.[44] The species definition has always been evaluated and re-evaluated as technology and methods change.[45] Many taxonomists believe the present definition of a bacterial species should be completely revised[2,6,46–48] while others suggest that it be entirely rebuilt from scratch.[49] One of the current proposed revisions is the designation of DNA sequence information in lieu of a type culture for uncultured bacteria.[50]

While taxonomic updates are required to address emerging definitions, concepts, and theories on bacterial taxonomy, such changes are not without potential consequences, particularly for the medical community, including clinical microbiologists. Some of these severe implications have been recently addressed in a multiauthored international commentary.[51] Whether a universal approach to bacterial taxonomy can ever be achieved is highly debatable as pure taxonomists and those that use scientific names daily (physicians, nurses, laboratorians) have entirely different perspectives on the topic. Above all, taxonomy for those in clinical

practices needs to be understandable, useful, and workable on a day-to-day basis. Communication is important as we deal with infections, outbreaks, and epidemics from a human perspective.

REFERENCES

1. Jay V. Antony van Leeuwenhoek. *Arch Pathol Lab Med* 126, 658, 2002.
2. Oren A, Garrity GM. Then and now: a systematic review of the systematics of prokaryotes in the last 80 years. *Antonie van Leeuwenhoek* 106, 43, 2014.
3. Hardy SP, Köhler W. Investigating bacillary dysentery: the role of laboratory, technique and people. *Int J Med Microbiol* 296, 171, 2006.
4. Köhler W. Zentralblatt für Bakteriologie—100 years ago the cultural differentiation between typhoid bacilli and *Bacterium coli*: Endo's agar started its triumphal march into the laboratories. *Int J Med Microbiol* 295, 1, 2005.
5. Austrian R. The Gram stain and the etiology of lobar pneumonia, an historical note. *Bacteriol Rev* 24, 261, 1960.
6. Schleifer KH. Classification of bacteria and archaea: past, present and future. *Syst Appl Microbiol* 32, 533, 2009.
7. Weissfeld AS. Bacterial nomenclature: how organisms are named and renamed and renamed. *Clin Microbiol Newslett* 31, 1, 2009.
8. Parker CT, Tindall BJ, Garrity GM (editors). International code of nomenclature of prokaryotes. Prokaryotic code (2008 revision). *Int J Syst Evol Microbiol* 69, S1, 2019.
9. Skerman VBD, McGowan V, Sneath PHA. Approved lists of bacterial names. *Int J Syst Bacteriol* 30, 225, 1980.
10. Moore ERB, et al. Microbial systematics and taxonomy: relevance for a microbial commons. *Res Microbiol* 161, 430, 2010.
11. Sneath PHA. Thirty years of numerical taxonomy. *Syst Biol* 44, 281, 1995.
12. Janda JM. Clinical decisions: how relevant is modern bacterial taxonomy for clinical microbiologists? *Clin Microbiol Newslett* 40, 51, 2018.

13. Thompson CC, et al. Microbial taxonomy in the post-genomic era: rebuilding from scratch? *Arch Microbiol* 197, 359, 2015.

14. Meier-Kolthoff JP, Klenk H-P, Göker M. Taxonomic use of DNA G + C content and DNA-DNA hybridization in the genomic age. *Int J Syst Evol Microbiol* 64, 352, 2014.

15. Wayne LG, et al. Report of the ad hoc committee on the reconciliation of approaches to bacterial systematics. *Int J Syst Bacteriol* 37, 463, 1987.

16. Janda JM, Abbott SL. 16S rRNA gene sequencing for bacterial identification in the diagnostic laboratory: pluses, perils, and pitfalls. *J Clin Microbiol* 45, 2761, 2007.

17. Croxatto A, Prod'hom G, Greub G. Applications of MALDI-TOF mass spectrometry in clinical diagnostic microbiology. *FEMS Microbiol Rev* 36, 380, 2012.

18. Fournier P-E, Drancourt M. New microbes new infections promotes modern prokaryotic taxonomy: a new section "Taxonogenomic: new genomes of microorganisms in humans". *New Microbes New Infect* 7, 48, 2015.

19. Chun J, et al. Proposed minimal standards for the use of genome data for the taxonomy of prokaryotes. *Int J Syst Evol Microbiol* 68, 461, 2018.

20. Kozińska A, Seweryn P, Sitkiewicz I. A crash course in sequencing for a microbiologist. *J Appl Genetics* 60, 103, 2019.

21. Erne L, Doolittle WF. *Archaea. Curr Biol* 25, R845, 2015.

22. Woese CR, Kandler O, Wheelis ML. Towards a natural system of organisms: proposal for the domains Archaea, Bacteria, and Eucarya. *Proc Nat Acad Sci U S A* 87, 4576, 1990.

23. van den Beld MJC, Reubsaet FAG. Differentiation between *Shigella,* enteroinvasive *Escherichia coli* (EIEC) and noninvasive *Escherichia coli. Eur J Clin Microbiol Infect Dis* 31, 899, 2012.

24. Chattaway MA, et al. Identification of *Escherichia coli* and *Shigella* species from whole-genome sequences. *J Clin Microbiol* 55, 616, 2017.

25. Criscuolo A, et al. The speciation and hybridization history of the genus Salmonella. *Microb Genomics* 5(8):e000284, 2019.

26. Brenner FW, et al. *Salmonella* nomenclature. *J Clin Microbiol* 38, 2465, 2000.

27. Greub G. Culturomics: a new approach to study the human microbiome. *Clin Microbiol Infect* 18, 1157, 2012.

28. Barko PC, et al. The gastrointestinal microbiome: a review. *J Vet Intern Med* 32, 9, 2018.

29. Tindall BJ, et al. Valid publication of names of prokaryotes according to the rules of nomenclature: past history and current practice. *Int J Syst Evol Microbiol* 56, 2715, 2006.

30. Parte AC. LPSN—List of prokaryotic names with standing in nomenclature (bacterio.net), 20 years on. *Int J Syst Evol Microbiol* 68, 1825, 2018.

31. Stanier RY. A note on the taxonomy of *Proteus hydrophilus. J Bacteriol* 46, 213, 1943.

32. Griffin PJ, Snieszko SF, Friddle SB. A more comprehensive description of *Bacterium salmonicida. Trans Am Fisheries Soc* 82, 129, 1953.

33. Véron M. The taxonomic position of *Vibrio* and certain comparable bacteria. *C R Acad Sci Hebd Seances Acad Sci D* 261, 5243, 1965.

34. Popoff M, Véron M. A taxonomic study of the *Aeromonas hydrophila-Aeromonas punctata* group. *J Gen Microbiol* 94, 11, 1976.

35. Allen DA, Austin B, Colwell RR. *Aeromonas media,* a new species isolated from river water. *Int J Syst Bacteriol* 33, 599, 1983.

36. Validation List 15. Validation of the publication of new names and new combinations previously effectively published outside the IJSB. *Int J Syst Bacteriol* 34, 355, 1984.

37. Colwell RR, MacDonell MT, De Ley J. Proposal to recognize the family *Aeromonadaceae* fam. nov. *Int J Syst Bacteriol* 36, 473, 1986.

38. Hickman-Brenner FW, et al. *Aeromonas veronii,* a new ornithine decarboxylase-positive species that may cause diarrhea. *J Clin Microbiol* 25, 900, 1987.

39. Hickman-Brenner FW, et al. *Aeromonas schubertii,* a new mannitol-negative species found in human clinical specimens. *J Clin Microbiol* 26, 1561, 1988.

40. Janda JM. Taxonomic update on proposed nomenclature and classification changes for bacteria of medical importance, 2016. *Diagn Microbiol Infect Dis* 88,100, 2017.

41. Christensen H, et al. Is characterization of a single isolate sufficient for valid publication of a new genus or species? Proposal to modify Recommendation 30b of the *Bacteriological Code* (1990 Revision). *Int J Syst Evol Microbiol* 51, 2221, 2001.

42. Sutcliffe IC. Valediction: description of novel prokaryotic taxa published in *Antonie van Leeuwenhoek* – change in editorial policy and a signpost to the future? *Antonie van Leeuwenhoek* 112, 1281, 2019.

43. Rosselló-Móra R, Whitman WB. Dialogue on the nomenclature and classification of prokaryotes. *Syst Appl Microbiol* 42, 5, 2019.

44. Kämpfer P. Systematics of prokaryotes: the state of the art. *Antonie van Leeuwenhoek* 101, 3, 2014.

45. Stackebrandt E, et al. Report of the ad hoc committee for the re-evaluation of the species definition in bacteriology. *Int J Syst Evol Microbiol* 52, 1043, 2002.

46. Krichevsky MI. What is a bacterial species? I will know it when I see it. *Bull BISMiS* 7, 17, 2011.

47. Rosselló-Móra R, Amann R. Past and future species definitions for *Bacteria* and *Archaea. Syst Appl Microbiol* 38, 209, 2015.

48. Stackebrandt E, Smith D. Paradigm shift in species description: the need to move towards a tabular format. *Arch Microbiol* 201, 143, 2019.

49. Thompson CC, et al. Microbial taxonomy in the post-genomic era: rebuilding from scratch? *Arch Microbiol* 197, 359, 2015.

50. Chuvochica M, et al. The importance of designating type material for uncultured taxa. *Syst Appl Microbiol* 42, 15, 2019.

51. Bisgaard M, et al. The use of genomic DNA sequences as type material for valid publication of bacterial species names will have severe implications for clinical microbiology and related disciplines. *Diagn Microbiol Infect Dis* 95, 102, 2019.

18

Bacterial Cell Wall: Morphology and Biochemistry

Stefania De Benedetti, Jed F. Fisher, and Shahriar Mobashery

CONTENTS

Introduction and Summary

The exterior structure of the unicellular microorganism contains and protects its intracellular structure, while regulating solute ingress and egress. This exterior structure is called the *cell envelope*. Within the domain of the archaea the structure of the cell envelope is a proteinaceous surface layer.[1–3] Within the domain of the bacteria, different cell envelope structures are found.[4–7] A key difference is whether the bacterium has a single membrane (is monoderm) or has two membranes (is diderm).[8] A key similarity for most monoderm and diderm bacteria is the presence of a structural *peptidoglycan* polymer that surrounds the entire bacterium.[9–14] This polymer is also called the bacterial *cell wall*. Although the terms cell envelope and cell wall often are used interchangeably, the terms are distinct. The cell envelope is the multilayer barrier that separates and protects the cytoplasm of the bacterium from its external environment. It includes the peptidoglycan cell wall, the membrane(s), the periplasmic space between the two membranes of the diderm bacteria, and the macromolecules that are associated with all of these structures. The location of the cell wall within the cell envelope is a fundamental distinction. The cell wall is the exoskeleton of the Gram-positive bacteria, albeit with extensive integration of a second glycopolymer, the teichoic acids[15–17] as well as capsular polysaccharides.[18] This peptidoglycan cell surface overlays an interstitial space, itself above the single bilayered, cytoplasmic membrane.[19] The cell wall of Gram-negative bacteria is an endoskeleton that is located within the *periplasm*, an interstitial space that separates the cytoplasmic membrane from the outer membrane on the surface of the bacterium. Both of its membranes are bilayered.[20] The dominant structural entity of the surface layer of the Gram-negative bacterium is the lipopolysaccharide.[4,21]

The nonmotile, strictly aerobic, nonfermenter Gram-negative pathogen *Acinetobacter baumannii* is an excellent example, from the dual perspectives of structure and pathogenicity, of the overall organization of the Gram-negative cell envelope.[22] The peptidoglycan cell wall of the *mycobacteria* is likewise an endoskeleton that is located above the cytoplasmic membrane.[23–27] The mycobacterial peptidoglycan is extensively modified on its outward face by a galactofuran polysaccharide outer surface, to which in turn are attached oligoarabinofuran strands that themselves have mycolic acid lipids. In numerous cartoon representations of the mycobacterial cell envelope, these mycolic lipids are represented as the inner leaflet of a second membrane (mycomembrane), with the exterior surface of the mycobacterium cell envelope a structurally heterogeneous (lipid, protein) "capsule."[28] These three envelopes are represented in cartoon form in Figure 18.1A. Bacteria further diversify their cell envelopes. The cell wall is absent in the phylogenically Gram-positive *Mycoplasma* (within the class *Mollicutes*),[11,29,30] and appears in cryptic structural form in the "cell-wall-less" and obligate intracellular pathogens of the class *Chlamydiae*.[31–33] Figure 18.1B shows in cartoon form the proteinaceous Z-ring that initiates septal peptidoglycan synthesis in Gram-negative bacteria, Gram-positive bacteria, and mycobacteria in contrast with the peptidoglycan ring of the cell-wall-less *Chlamydiae*. Other unusual adaptations with respect to the peptidoglycan cell wall are encountered. The symbiotic relationship between the mealybug and its *Moranella* bacterial symbiont derives from the horizontal transfer of genes for peptidoglycan biosynthesis from the genome of the bacterium to the genome of the insect.[34,35]

Peptidoglycan is a synonym for the cell wall. The term peptidoglycan is encountered especially when the molecular structure of the cell wall is emphasized, while a second synonym *murein* (Latin, wall) emphasizes its etymology. In the mature

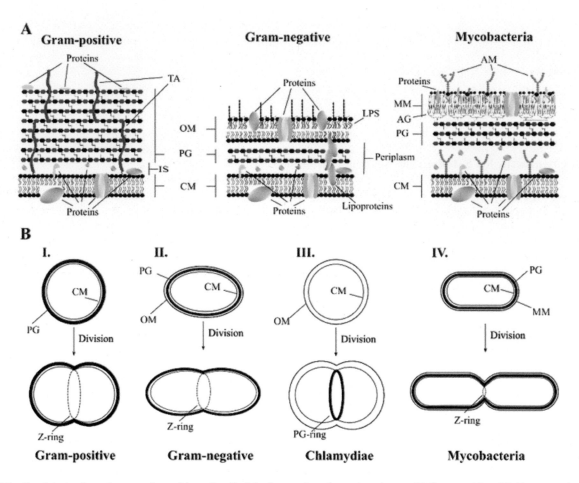

FIGURE 18.1 Panel A. A schematic comparison of the cell wall of the three major eukaryotic pathogens (I.) Gram-positive, (II.) Gram-negative, and (III.) mycobacteria. The notable differences are the presence of one or two membranes, the thickness of the peptidoglycan layer, and the modifications made to the membranes. Panel B. Representation of the peptidoglycan localization in bacteria: in (I.) Gram-positive, (II.) Gram-negative, and (IV.) mycobacteria, peptidoglycan is present throughout the whole cell-cycle (including in cell division). In the Gram-negative *Chlamydiae* (III.) peptidoglycan is synthesized at the septum during cell division, and recruits the division proteins.[31,36-38] (*Abbreviations:* CM: cytoplasmic membrane, PG: peptidoglycan, OM: outer membrane.)

bacterium the peptidoglycan (the cell wall) is a polymer—a single molecule—called the *sacculus*. In this final form it surrounds the three-dimensional bacterium and both imparts, and captures, the shape of the bacterium.[39-44] As its name implies, the molecular structure of the peptidoglycan has both saccharide and peptide components. The bacterium biosynthesizes glycan strands that have, on alternate saccharides, a peptide stem. The structure of the glycan strands is a repeating β-(1,4)-linked disaccharide. This disaccharide, [(*N*-acetylglucosamine)-(*N*-acetylmuramic acid)]$_n$, is often referred to by the acronym "NAG-NAM." This disaccharide is highly conserved among the Gram-positive bacteria, Gram-negative bacteria, and the mycobacteria. Depending on the bacterium, the NAG-NAM glycan is further structurally modified. For example, *O*-acetylation of the saccharide is a common resistance mechanism that protects the cell wall from degradation by the [NAM-NAG]$_n$-cleaving hydrolase lysozyme. *O*-Acetylation also regulates autolysin-dependent transformation of the peptidoglycan during cell division.[45-48] Spore-forming Gram-positive bacteria extensively transform their peptidoglycan during spore formation and germination, notably involving the formation of an intramolecular δ-lactam moiety in the NAG saccharide during spore formation.[49-52] The oligopeptide stem distinguishes the NAM saccharide from the

NAG saccharide. The polymeric cell wall is made by the covalent crosslinking of the stem peptide of one glycan strand, with the stem peptide of an adjacent glycan strand. The result is a polymeric mesh of cross-linked glycan strands surrounding the bacterium. Although Gram-positive bacteria, Gram-negative bacteria, and mycobacteria show common structural features with respect to the oligopeptide structure of the stem, the precise stem structure is characteristic of the particular bacterium. Stem structure is particularly varied among the Gram-positive bacteria (as discussed later). An elegant scientific history of the peptidoglycan, written from the perspective of the numerous methods used for the imaging of this polymer, is available.[53]

The functions of the cell wall—whether exposed to the environment (as is the case for the Gram-positive bacteria) or protected from the environment (as is the case for the Gram-negative bacteria and mycobacteria)—are the preservation of the integrity of the bacterium through both cell growth and cell division, the creation of cell shape, and the creation of a scaffold for the attachment of macromolecular complexes. Examples of these complexes include the porin proteins used for solute ingress and egress, and complexes used for adhesion, environmental sensing, molecular transport and secretion, and propulsion (the pili and the flagella). The peptidoglycan segregates the bacterium from its environment while containing the turgor pressure of the bacterial cytoplasm. This polymeric barrier is not static. Not only does the peptidoglycan undergo continuous remodeling as the bacterium grows, cell division requires the creation of a cell-wall septum that ultimately fractures as the cells divide. At all stages of bacterial growth, structural integrity of the cell wall is paramount. The ability of the cell wall to simultaneously possess structural strength, integrity, and malleability; to contain the turgor pressure and to impart cell shape; to embed numerous complexes; and to accommodate the complex restructuring demanded by cell growth and division is truly remarkable. Moreover, all of these tasks must coordinate with the bacterial cell cycle.[54-61]

As a consequence of the essentiality of the cell wall to bacterial viability, and the relative accessibility of many of the enzymes involved in cell-wall synthesis and remodeling, these enzymes are important targets for antimicrobial agents.[62-66] The search for new targets for new antimicrobials, for structural improvements of existing antimicrobials, and for the identification of new synergistic antimicrobial combinations, are moving forces for research on the cell wall.

Gram-Positive and Gram-Negative Bacteria

In 1884, the Danish bacteriologist Hans Christian Gram developed a staining process that categorized most prokaryotes as either stain-retaining (positive) or stain-negative. Gram-staining involves the gentle drying of a bacterial suspension on the microscopy slide, the addition of a solution of crystal violet (or methylene blue), the rinsing with an iodine solution, the decolorizing with ethanol, and the final addition of safranin as a counterstain. Bacteria that retain the crystal violet-iodine complex, and thus appear purple-brown colored under microscope, are designated as Gram-positive. Those bacteria that do not retain the crystal violet stain appear red-colored and are designated as Gram-negative.[67-69] Numerous bacteria (such as the mycobacteria), however, do not stain decisively. Although it was intuitively clear the staining differentiated between structurally different cell envelopes, it took more than half a century to confirm this hypothesis.[70] Developments in transmission electron microscopy (TEM) in the 1950s gave the first images of bacterial cell envelopes. These images revealed the two membranes of the Gram-negative bacteria. The 5–7 nm-thick (in the Gram-negative bacterium *Escherichia coli*) cell wall is located in the periplasmic space between the inner and the outer membranes.[14] Imaging places the cell wall centrally located within the periplasm and close (if not adjacent) to the inner leaflet of the outer membrane where the peptidoglycan anchors cell envelope-penetrating structures such as the flagellum and secretion systems.[71-74] The exterior surface of Gram-negative bacteria—that is, the outer leaflet of the outer-membrane bilayer—consists primarily of specialized glycolipids (the lipopolysaccharides, LPS) and proteins. The approximately 20 nm-thick cell wall of the Gram-positive bacteria is located above their single membrane.[75-79] This peptidoglycan is extensively modified by the addition of teichoic lipopolysaccharides, capsular polysaccharides, and proteins. A superbly illustrated review presents the current molecular-level understanding of the key structural complexes of the Gram-positive bacterium using *Streptococcus pneumoniae* as the teaching example.[80] The

absence of an outer membrane, and the greater thickness of the Gram-positive cell wall compared to Gram-negative bacteria, result in better retention of the iodine-dye complex and thus preservation of the characteristic color of the stain.[81] Gram-staining facilitates mass-spectral identification of bacteria in blood culture.[82] Efforts toward alternative bacterial "staining" methods include modification of the crystal violet with fluorophores, gold nanoparticles, and magnetic nanoparticles[83]; the development of fluorescent stains that are selective for Gram-negative bacteria[84] or for Gram-positive bacteria[85]; and fluorophores that are incorporated metabolically into the peptidoglycan to facilitate imaging.[86,87] The development of imaging modalities that might have diagnostic value—especially with respect to distinguishing resistance mechanisms—is an active area of research.[88-90] For example, the incorporation of 11C-radiolabeled methionine into the peptidoglycan has promise for PET diagnosis of bacterial infection in whole animals.[91,92]

Chemical Structure of the Peptidoglycan

The nuance of peptidoglycan structure for Gram-positive bacteria is more diverse than that for Gram-negative bacteria, reflecting (for example) the growth circumstance.[93] The glycan strand component of the peptidoglycan for most eubacteria peptidoglycan is a repeating β-1,4-linked disaccharide.[94] The structure of the repeating disaccharide is *N*-acetylglucosamine (abbreviated NAG or GlcNAc) linked to an *N*-acetylmuramic acid (NAM or MurNAc) saccharide (Figure 18.2). The NAG-NAM disaccharide

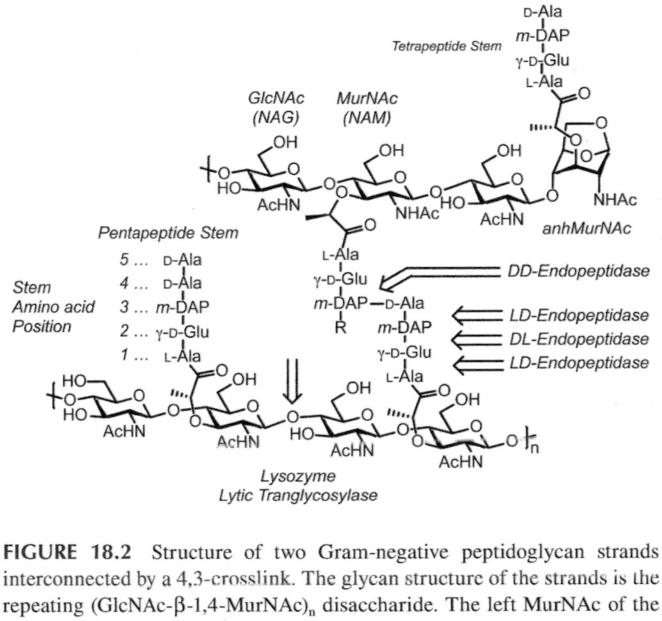

FIGURE 18.2 Structure of two Gram-negative peptidoglycan strands interconnected by a 4,3-crosslink. The glycan structure of the strands is the repeating (GlcNAc-β-1,4-MurNAc)$_n$ disaccharide. The left MurNAc of the lower strand has a full-length pentapeptide stem terminating with the customary D-Ala-D-Ala as the amino acids 4 and 5. The left MurNAc of the upper strand is crosslinked, using the ε-amine of the m-DAP amino acid at position 3, to the D-Ala at position 4 of the right MurNAc of the lower strand. The upper strand shows an anhydroMurNAc (anhMurNAc) terminus created by a lytic glycosylase (the structure of the glycan strand terminus in Gram-positive bacteria is not proven). The arrows indicate the position and generic identity of the enzymes that cleave the peptide and the glycan of the peptidoglycan.

core is identical for most Gram-positive bacteria, Gram-negative bacteria, and mycobacteria. An exception is the Gram-positive methanogenic Archaebacteria, that have a pseudomurein peptidoglycan structure having a β-D-N-acetylglucosamine (NAG) β-1,3-linked to α-L-N-acetyltalosaminuronic acid (NAT) as the repeating disaccharide.[3,29,95] The length of the $(NAG\text{-}NAM)_n$ strands can be analyzed by exhaustive amidase degradation of the peptide stem of the NAM saccharide, followed by chromatography with mass-spectrometry analysis. However, this analysis is uncommon due to the lack of availability of the specific amidases required to degrade the specific peptidoglycan of the bacterium.[96] This analysis has been done for a number of common bacteria. A key basis for this analysis is recognition that the terminating step in glycan strand biosynthesis is the nonhydrolytic, enzyme-catalyzed cleavage of the NAM-NAG disaccharide. The cleavage reaction is an intramolecular transacetalization at the NAM saccharide, cleaving the glycosidic bond, to create an anhydro-muramic (anhNAM) glycan strand terminus.[97,98] This reaction is catalyzed by members of a family of lysozyme-like enzymes, called the lytic transglycosylases.[99,100] Glycan strand lengths, in the few bacteria where this length has been measured, indicate a distribution of lengths. The average glycan strand length for the Gram-negative, rod-shaped bacterium *E. coli* is 54 saccharides corresponding to the $[NAG\text{-}NAM]_{27}$ structure. The average length value for the Gram-negative, curve-shaped bacterium *Vibrio cholerae* is $[NAG\text{-}NAM]_{12}$.[101,102] A concise perspective on the intimate relationship among the bacterial cytoskeleton, peptidoglycan biosynthesis and structure, and cell shape is given in a complementary review.[103] The average glycan length in the PAO1 laboratory strain of *P. aeruginosa* is $[NAG\text{-}NAM]_{60}$ (measured by the terminal anhNAM content of the muropeptides) with 42% cross-linking.[104] The distribution of the length of most glycan strands in the Gram-positive *Staphylococcus aureus* ranges from $[NAG\text{-}NAM]_3$ to $[NAG\text{-}NAM]_{10}$ with a maximum of $[NAG\text{-}NAM]_{26}$ disaccharide units per strand.[105]

Unlike the repeating disaccharide backbone, the amino acids comprising the NAM stem demonstrate variability, especially among the Gram-positive bacteria. The most common peptide stem structure is the (full-length) pentapeptide attached to the D-lactyl moiety of the muramic (NAM) saccharide. The customary pentapeptide sequence is L-Ala (the first amino acid, position-1), γ-D-Glu (amino acid-2), *meso*-diaminopimelate (abbreviated $m\text{-}A_2pm$ or m-DAP, and customary to Gram-negative bacteria and some Gram-positive bacteria such as *B. subtilis*) or L-Lys (customary to Gram-positive bacteria) as amino acid-3, and D-Ala in positions four and five. In many Gram-positive bacteria additional amino acids (usually a mono-, di-, tri-, tetra-, or pentapeptide) are added to the ε-amine of the position-3 L-Lys. These additional amino acids are referred to as the bridging oligopeptide of the peptidoglycan stem. Peptidoglycan polymerization involves the crosslinking of the stem found on one glycan chain to the stem found on a neighboring glycan strand. Cross-linking occurs in either of two ways. The predominant way is DD-4,3-crosslinking (sometimes written as DD-3,4-crosslinking). Figure 18.2 shows the structural schematic for two DD-4,3-crosslinked Gram-negative glycan strands. This crosslink is synthesized by a penicillin-binding protein (PBP) DD-transpeptidase enzyme. As suggested by the historical appellation penicillin-binding protein, PBPs are inactivated by

members of the β-lactam antibiotic family (penicillins, cephalosporins, carbapenems, monobactams, and monosulfactams). This inactivation predisposes the bacterium to the bactericidal effect of the β-lactam. In DD-transpeptidase catalysis, the first step is the transfer of the acyl moiety of the position-4 D-Ala to an active-site serine of the transpeptidase, with loss of the position-5 D-Ala as a leaving group. This first step is the acylation half-reaction. In Gram-negative bacteria, the second step is transfer of this acyl moiety to the ε-amine of the D-stereocenter of the position-3 $m\text{-}A_2pm$ amino acid of a neighboring strand. In Gram-positive bacteria, acyl transfer is made to the ε-amine of the L-Lys, or to the terminal amine of the bridging oligopeptide attached to the ε-amine of the L-Lys. This step is the acyl-transfer, or deacylation, half-reaction. β-Lactam antibiotics are structural surrogates for the -D-Ala-D-Ala dipeptide terminus of the full-length stem. The β-lactam antibiotics readily acylate the active-site serine of the PBPs (with opening of the β-lactam ring), but then cannot transfer the resulting acyl-moiety. The PBP is inactivated as the acyl-enzyme.[106] All bacteria contain multiple PBPs. The pattern for PBP inactivation by β-lactams is absolutely dependent on β-lactam structure. Some PBPs are essential, and some less essential. Accordingly, the selection of a β-lactam with high affinity for an essential PBP has clinical importance. PBP profiles for many of the clinically used β-lactams are available for the Gram-positive bacterium *Streptococcus pneumoniae*,[107] and the Gram-negative bacteria *E. coli* and *Klebsiella pneumoniae*.[108,109]

The second way of crosslinking is LD-3,3-crosslinking. Here, the acyl-donor moiety is the α-L-stereocenter of the position-3 amino acid, with the position-4 D-Ala as the leaving group. The acyl-acceptor used to accomplish crosslinking is (as is also the case in DD-4,3-crosslinking) the terminal amine of the position-3 amino acid. LD-3,3-Crosslinking is catalyzed by cysteine-dependent LD-transpeptidases (Ldts). The Ldt transpeptidases are inactivated by some carbapenem β-lactams.[110–115] The widely used penicillin and cephalosporin β-lactam antibiotics do not inactivate the Ldt transpeptidases. Ldt transpeptidases are found in mycobacteria, Gram-positive bacteria, and Gram-negative bacteria. Indeed, LD-3,3-crosslinking is the dominant crosslinking pattern in stationary-phase mycobacteria.[116,117] In *E. coli* LD-3,3-crosslinking accounts for 2–10% of the peptidoglycan crosslinks in exponential growth, and up to 16% at stationary growth.[118] Microbiological interest in LD-3,3-crosslinking has followed recognition of its abundance in mycobacteria, of its possible role as an emerging mechanism for resistance of the Gram-negative bacteria to clinically used β-lactams,[119,120] and of the importance of LD-carboxypeptidase activity to the creation of the helical shape of *Helicobacter pylori* (and other "bent" bacteria).[44,121,122]

Stem peptide structure is the basis for a semisystematic nomenclature that is used to describe the peptidoglycan.[123,124] In this three-fold alphanumeric nomenclature, a first letter "A" denotes a DD-4,3-crosslink, a second number denotes the type of bridge peptide structure, and the third Greek letter denotes the amino acid at position-3 of the peptide stem. The common crosslink of *E. coli* (a DD-4,3-crosslink, $m\text{-}A_2pm$ at position-3 with no bridging amino acid, and thus a "direct" link) is designated as "A1γ". The common crosslink of *S. aureus* (a DD-4,3-crosslink, L-Lys at position-3 with a pentaglycine crossbridge) has "A3α" designation. Although this nomenclature is cumbersome and imprecise, it is still used. Thus the peptidoglycan of the Gram-positive, rod-shaped, nonmotile, catalase-negative bacterium *Weissella cryptocerci* (isolated from the gut of an insect) is described as an "A4α" type (DD-4,3-crosslink, L-Lys at position-3 with a -Gly-D-Glu dipeptide bridge).[125] Common stem crosslinking patterns are shown in Figure 18.3.

The presence of D-amino acids in the stem (and in some bridge) peptides is widely understood to prevent their degradation by ordinary peptidases and proteases. This objective is further abetted by the D-Glu position-2 amino acid, linked through its γ-carboxyl rather than through its α-carboxyl. The differing amino acids found in the bridge, and the differing oligopeptide lengths of the bridge, are understood to provide the basis for recognition by the PBPs (and other cell-wall enzymes) involved in peptidoglycan synthesis, remodeling, and degradation. The pentaglycine bridge structure that is found in methicillin-resistant *S. aureus* (MRSA) is notable.[127,128] MRSA uses a PBP (termed PBP2a) that is intrinsically resistant to irreversible acylation by the β-lactams as the key basis for its β-lactam resistance.[129] In the presence of β-lactams, PBP2a is synthesized and functions as a transpeptidase where the other PBP transpeptidases of *S. aureus* are impaired. The *fem* factor genes—encoding factors essential for methicillin resistance—facilitate the transpeptidase function of PBP2a.[130,131] The most notable of the Fem proteins are three acyl transferases that assemble the Gly_5 bridge using glycyl-tRNA as the glycyl donor. FemX adds the first Gly, FemA adds the second and third Gly, and FemB adds sequentially the fourth and fifth Gly residues. The pentaglycine bridge is essential to the cell integrity of the MRSA bacterium.[132] The extent of the peptidoglycan crosslinking reaction is variable. In *E. coli* the degree of crosslinking varies from 30% to 60%,[101,133] while the extent of crosslinking for *S. aureus* is higher (74–92%).[105]

The stem peptide is also an attachment point in Gram-positive bacteria for macromolecules, including polysaccharides and proteins. These attached proteins function in cell-wall remodeling, adhesion (adhesins and pili), resistance, and virulence.[1,134–138] The catalysts for protein attachment are the membrane-bound sortases. The best-studied sortase system is that of *S. aureus*.[139] Attachment of the wall teichoic acids to the peptidoglycan contributes significantly to the viability of the Gram-positive bacterium,[15,140,141] and attachment of the pili and adhesins contributes to virulence.[142–146] Sortase inhibition is advantageous to antibacterial chemotherapy.[147–151] Sortases recognize an LPXTG motif in the leader sequence of proteins. As this leader sequence emerges on the periplasmic side from the Sec transport pathway, the threonine acyl moiety is transferred to the active-site cysteine of the sortase as the threonine-glycine bond of the LPXTG motif is cleaved.[152–155] The sortase catalyzes transfer of this acyl moiety to the stem peptide of the key biosynthetic intermediate in peptidoglycan biosynthesis, Lipid II.[138] This modified Lipid II serves as the substrate for transglycosylase-catalyzed synthesis of the glycan strands of the peptidoglycan, which are subsequently cross-linked by transpeptidase catalysis.

Sacculus degradation is a robust method for evaluating peptidoglycan structure. Chemical dissolution of the bacterium (accomplished by boiling the bacterium in SDS detergent, or alternatively with hot brine)[156] recovers the polymeric sacculus in the form of an insoluble shell that preserves the shape of the bacterium—a ghost of the bacterial shape stripped of proteins and membranes. Exhaustive hydrolytic degradation of the sacculus, using lysozymes tolerant of NAG-NAM structural variation, transforms the insoluble sacculus polymer into an array of soluble *muropeptides*.[157] The muropeptides are separated by chromatography[96,156,158–160] or by electrophoresis.[161] Muropeptide characterization is done using mass spectrometry.[162,163] The muropeptide array includes both uncrosslinked NAM-NAG disaccharide derivatives, crosslinked (two NAM-NAG disaccharides, connected through the stem), and small amounts of anhNAM structures arising from lytic transglycosylase termination of glycan strand biosynthesis. Mass spectrometry gives the mass of the muropeptide, including the crosslink position and functional group modifications (such as amidation)[164] to the stem peptide. The presumably distinct muropeptide (peptidoglycan) domains within the sacculus—for example, in a rod-shaped bacterium the different peptidoglycan composition of its poles compared to its sidewalls—are not differentiated. The different methodologies for muropeptide analysis of Gram-negative, Gram-positive, and acid-fast bacteria are reviewed elsewhere.[165] The application of this methodology is exemplified. The peptidoglycan structures of both *Staphylococcus carnosus* and *Deinococcus indicus* reflect metabolic influence.[93,166] Muropeptide analysis identified the role of the *msaABCR* operon in the control of the balance between cell-wall synthesis and cell-wall hydrolysis in *S. aureus*.[167] Significant change in the peptidoglycan structure of *Streptomyces coelicolor* accompanies the transition from mycelial growth to spores, with identification of a role for specific muropeptide structures in spore development.[168] *Enterococcus faecium* releases a peptidoglycan hydrolase (SagA) to the gut for the purpose of cleaving crosslinked muropeptides derived from its peptidoglycan, to give structures that are better recognized by the NOD-dependent innate immunity pathway. As a result the intestinal barrier function is enhanced and host immunity toward competitive pathogens is improved.[169] Muropeptide analysis clarified the subtle structural differences with respect to muropeptide-dependent pathogenicity of *Neisseria* species,[170] guided the identification of a membrane-spanning complex for helical shape formation in *H. pylori*,[122] and demonstrated that the *Campylobacter jejuni* helical to coccoid transition correlated to specific alteration of peptidoglycan structure and diminished the ability of the muropeptides to effect NOD1-dependent and NOD2-dependent activation of innate immunity.[171] The peptidoglycan is different between the body and the stalk of *Caulobacter crescentus*.[172] Muropeptide analysis identified a new peptidoglycan hydrolase (TagX) that is essential for the biogenesis of the type VI secretion system of *A. baumannii*.[173] The peptidoglycan of the spirochete *Borrelia burgdorferi* is atypical, with an L-ornithine-glycine dipeptide replacing the much more common *meso*-diaminopimelic acid of other Gram-negative bacteria at the third position of the stem. The resulting muropeptides are persistent antigens in Lyme arthritis.[174] The use of the sacculus as a substrate allowed characterization of the entire family of the eleven lytic transglycosylases of *Pseudomonas aeruginosa*.[175] Last, in response to outer membrane lipopolysaccharide stress, *E. coli* compensates by activation of an LD-transpeptidase to accomplish 3,3-crosslinking transpeptidation (to replace the more common DD-4,3-crosslinking).[118] The distinction between LD-3,3-crosslinked peptidoglycan and DD-4,3-crosslinked

Peptide Stem Cross-linking Patterns of Representative Bacteria

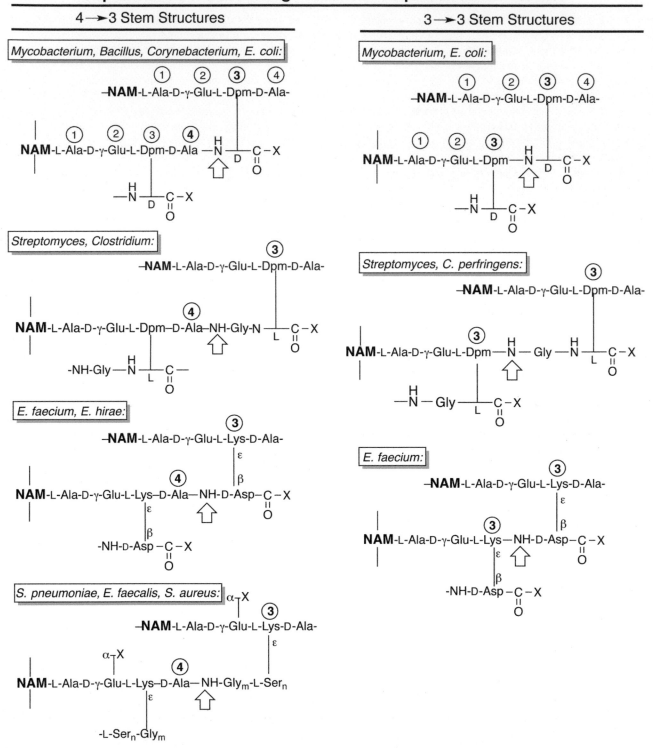

FIGURE 18.3 Stem cross-linking patterns found in different bacteria.[126] The left column exemplifies 4,3-peptidoglycan cross-linking, and the right column 3,3-cross-linking. For each exemplification, the acyl-donor stem of the cross-link is the lower NAM-stem structure, and the acyl-acceptor stem is the upper NAM-stem structure. The nitrogen atom that accepts the acyl moiety, mediated by PBP catalysis, is identified with an arrow. The diamino-containing amino acids found at position 3 of the stems are *meso*-diaminopimelic acid (L,D-Dpm), L,L-aminopimelic acid (L,L-Dpm), or L-lysine. The distal amine of these diamino acid residues is the acyl acceptor in *Mycobacterium*, *Corynebacterium*, *Bacillus* spp., and *E. coli*. Mycobacteria use both 4,3 and 3,3 cross-linking and the extent of 3,3 cross-linking is a resistance mechanism in Gram-positive bacteria. The stem may be substituted by the presence of one or more bridging amino acids. For example, in some *S. aureus* strains the 4,3 stem has a two L-serine and three glycine residue bridge where the terminal glycine serves as the acyl acceptor (right column, bottom entry: $m = 2$, $n = 3$), while in most *S. aureus* strains the stem has a pentaglycine ($m = 0$, $n = 5$) bridge. In *S. pneumoniae* and *S. aureus* the α-carboxylate of the γ-D-glutamate residue at position 2 of the stem is amidated ($X = NH_2$). This same amidation, as well as extensive other modifications to the peptidoglycan, is also found for *E. faecium*.

peptidoglycan is an emerging focus with respect to bacterial resistance to the β-lactam antibiotics.[176,177] A concluding observation is that muropeptide analysis of the *core* peptidoglycans of laboratory and clinical strains of the Gram-negative rod-shaped bacterium *P. aeruginosa* are nearly indistinguishable.[102,178,179] However, comprehensive analysis of a more complete set of 160 muropeptides identified a subset of less abundant muropeptides, that are termed adaptive muropeptides (contrasting planktonic and biofilm growth states). A notable discovery was the presence of NAG de-*N*-acetyl muropeptides (six identified, total relative abundance of 0.5%) among the core planktonic muropeptides that diminished significantly in abundance compared to the biofilm-derived muropeptides.[179] These observations attest to the importance of the preservation of the core peptidoglycan, and the likelihood that specific peptidoglycan structure correlates—in ways yet to be discovered—with bacterial shape, nutrient availability, and growth condition.

Overall muropeptide structure varies as a function of the stage of the bacterium, its growth phase, its shape, the growth medium, the presence of resistance determinants relating to the cell wall, and the presence of antibiotics. The ability of certain D-amino acids (such as D-Cys) to incorporate into the peptidoglycan was a critical methodology in early studies on peptidoglycan growth and to the visualization within the sacculus of new peptidoglycan domains.[180,181] This methodology is still used to analyze the patterns of peptidoglycan growth, as exemplified by analysis of glycine incorporation into the stem peptide (at the fifth position, a position normally occupied by the terminal D-Ala) in growing *C. crescentus*.[182] Muropeptide analysis shows incorporation of other D-amino acids during the stationary phase transition of *Bacillus subtilis* (Gram-positive) and *V. cholerae* (Gram-negative), and the possible correlation with stationary phase cell-wall remodeling and with biofilm dispersal.[183–185] The use of fluorescent D-amino acids has matured into a robust method for the study of peptidoglycan biosynthesis by fluorescent microscopy.[86,186–191] Detailed protocols for the preparation and incorporation of these amino acids in cell culture,[192] and in high-throughput assay for the discovery of inhibitors of peptidoglycan biosynthesis,[193] are available. Alternatively, *N*-acetyl muramic acid derivatives may be used.[194] Fluorescent microscopy has validated the treadmilling by FtsZ filaments as a driving force for peptidoglycan biosynthesis in Gram-negative bacteria,[195,196] the role of penicillin-binding protein 4 (PBP4) in extraseptal peptidoglycan crosslinking in *S. aureus*,[197,198] the evaluation of the spatiotemporal dynamics of peptidoglycan synthesis in both *S. aureus*[199] and the mycobacteria,[200,201] establishing that the biosynthesis of peptidoglycan and the incorporation of wall teichoic acid are directly coordinated in *S. pneumoniae*,[202] and demonstrating the presence of functional peptidoglycan in Chlamydia species.[31] The complementary use of fluorescent antibiotics that inhibit peptidoglycan biosynthesis, or fluorescent indicators of bacterial structure, is a powerful method for the cytological profiling of antibiotic mechanism (differentiating the cell wall, ribosome, DNA, RNA and lipids as the bacterial target).[203–205] Further applications of fluorescence for the study of the peptidoglycan are exemplified by the validation of Lipid II binding as the antibacterial mechanism of ribosomally synthesized post-translationally modified peptides,[206] the study of peptidoglycan remodeling during

B. subtilis sporulation[207] and *S. coelicolor* spore wall synthesis,[208] the selection of β-lactam structures that optimally synergize with vancomycin and daptomycin against *S. aureus*,[209] imaging peptidoglycan biosynthesis in *B. subtilis*,[210] and imaging bacterial infection.[211,212]

The Peptidoglycan-Binding Proteins of the Cell Envelope

The peptidoglycan is attached, by both covalent and noncovalent interactions, to both the inner and outer membranes of the Gram-negative bacterium. Three protein-peptidoglycan interactions dominate: covalent attachment to the Braun Lpp lipoprotein, and the noncovalent attachment of the peptidoglycan to the outer membrane porins and to the Pal (peptidoglycan-associated lipoprotein) protein of the Tol-Pal protein system that structurally bridges the inner and outer membranes. These three systems are important (if not essential) to the viability of the Gram-negative bacterium.

The Braun lipoprotein (Lpp) is the most abundant protein of *E. coli* (1×10^6 copies per cell).[213–216] Lpp is a 58-residue (*E. coli*, mature protein) α-helical peptide, approximately 76 Å in length that assembles as a coiled-coil homotrimer.[217,218] It is encountered primarily as a lipoprotein, covalently attached through the sulfur atom of its N-terminal cysteine to a lipid of the inner leaflet of the outer membrane. A lysine at the other end of the Lpp homotrimer is covalently attached to the L-center of the *meso*-diaminopimelic acid residue at position 3 of the peptide stem of the peptidoglycan. This connection is catalyzed by one of the three (in *E. coli*) L,D-tranpspeptidases of the periplasm.[219] These two covalent linking events stabilize the outer membrane structure.[220,221] Manipulation of the length of the Lpp helix effectively alters the size of the periplasm, and consequently compromises the ability of *E. coli* to sense envelope defects and to respond to stress signals[222] and to assemble its flagellum.[223] Lpp is found to a lesser extent as an integral outer-membrane protein—its "free" form, not bound covalently to peptidoglycan—wherein its C-terminus lysine is exposed on the cell surface (growing *E. coli*, approximately 33% of the Lpp copies).[213,214] The free form of Lpp is involved in the recognition of some cationic antimicrobial peptides[224] but not others.[225] The beneficial, and presumably structural, function of the free Lpp protein in the outer leaflet of the outer membrane is not known.

The OmpA porin is an abundant protein (*E. coli*, 2×10^5 copies per cell) of the outer membrane.[226] Deletion of the OmpA porin in *E. coli* gives a viable, but fragile, phenotype. Although its function as a porin demands that the protein is integral to the outer membrane, OmpA of *E. coli* (and other porins of other bacteria)[227,228] has a peptidoglycan-interacting domain.[229,230] The fragility of the bacterium in the absence of OmpA results from the loss of the cooperative binding by OmpA (in the outer membrane) with the peptidoglycan, concurrent with noncovalent binding of OmpA with the TolR stator protein of the Tol-Pal complex[231] located in the inner membrane.[232] Moreover, a protein-protein interaction between Lpp and OmpA facilitates the noncovalent binding of peptidoglycan by OmpA.[233,234] Structures of several peptidoglycan-OmpA complexes have been determined.[227,235] The inability of OmpA to diffuse in the outer membrane of *E. coli* is not caused by the peptidoglycan binding.[236] Peptidoglycan binding may facilitate recovery of the Omp structure following its reversible deformation in response to mechanical stress.[237]

The third peptidoglycan-associated protein complex in Gram-negative bacteria is the Tol-Pal system. The Pal lipoprotein has an N-terminal lipoprotein domain that localizes the protein to the inner leaflet of the outer membrane, and a C-terminal peptidoglycan-binding domain that is similar in structure to that of OmpA.[238,239] The Tol-Pal system comprises four additional proteins, thus forming a five-protein complex that interconnects the outer membrane, the peptidoglycan, and the inner membrane across the periplasm[240] and that interacts additionally with the outer-membrane porins.[241] The essential functions of the Tol-Pal system are the proton-motive force-dependent maintenance of the cell-envelope integrity (especially with respect to membrane invagination during cell division),[242–246] the transport of macromolecules across the periplasm,[247,248] and the polar localization of proteins and chemoreceptors.[245,249] The Tol-Pal system contributes to the function of the outer membrane as a permeability barrier, whereas Lpp does not.[250] The mechanism of the bacterial colicin toxins is compromise of Tol-Pal function.[251–254] The structure of the *Haemophilus influenzae* Pal-peptidoglycan complex shows Pal recognition of the peptide stem of the peptidoglycan, centered at an uncrosslinked diaminopimelate residue of the stem.[255]

These three systems are the most abundant but emphatically are not all of the important protein-peptidoglycan interactions. Peptidoglycan recognition is important to the structural integration of numerous bacterial machines and systems.[256] The stator proteins of the bacterial flagellum[257–259] have the dual function of ion transport (either H^+ or Na^+) and anchoring the flagellar protein assembly to the peptidoglycan.[260–265] Peptidoglycan binding and ion channel competence are coupled events.[266,267] A functional identity between the *E. coli* Pal and MotB stator peptidoglycan-binding domains was proven by replacement of the MotB domain with that of the Pal domain with retention of function.[268] Excavation of the peptidoglycan to enable insertion of secretion systems uses both hydrolytic excision (amidases)[269–271] and nonhydrolytic excision (lytic transglycosylases)[272–274] of the peptidoglycan. In addition to the OmpA-like peptidoglycan-binding motifs, the LysM carbohydrate-recognition motif is encountered in innumerable peptidoglycan-binding proteins.[275–277]

A compilation of peptidoglycan-binding proteins would be remiss without mention of the lysozymes, the catalysts of peptidoglycan hydrolytic degradation by cleavage of the NAM-NAG glycosidic bond.[278,279] The lysozymes encompass three different protein motifs, isolated from phages, bacteria, plants, and eukaryotes. Their purposes are antibacterial defense, bacterial digestion, bacterial penetration, and bacterial transformation. Peptidoglycan-recognition proteins (PGRPs) are multisubstrate recognition proteins[280] of the innate immune systems.[281] As peptidoglycan-binding proteins they are a primary component of the immune response of insects.[282] The structures of several PGRPs in complex with peptidoglycan fragments are known.[283] PGRP binding to the peptidoglycan results in steric interference with peptidoglycan remodeling

and amidase-dependent peptidoglycan remodeling, culminating in the induction of a reactive-oxygen stress response that is ultimately bactericidal.[284]

Peptidoglycan Biosynthesis

Bacteria display an astonishing breadth of shape and morphological differences as exemplified by the spherical (coccus) shape of the Gram-positive *S. aureus*, the rod shape of the Gram-negative *E. coli*, the spiral shape of the Gram-negative *H. pylori*. Peptidoglycan biosynthesis is fundamentally similar (especially in its early stages) in the Gram-positive bacteria, the Gram-negative bacteria, and the mycobacteria.[10,13,61,285] While the nearly identical molecular structures of the Gram-negative and Gram-positive peptidoglycans are consistent with biosynthetic similarity, the variety of bacterial shapes indicates structural complexity. The basis of this complexity—such as how variation in peptidoglycan structure and interconnectivity controls the shape of the bacterium—is not known. Our imperfect understanding is exemplified by the dimorphic life cycle of the Gram-negative alphaproteobacterium *C. crescentus*. Under conditions of phosphate limitation, it divides asymmetrically to give a motile, swarmer cell and an adherent, stalked cell.[286] The peptidoglycan of the two cells is substantially different: that of the motile cell is dominated by 4,3-DD-transpeptidation, while that of the adherent cell is dominated by 3,3-LD-transpeptidation.[172] Presumably, the two represent different spatial interconnection of otherwise structurally similar glycan strands. Peptidoglycan biosynthesis is customarily studied with respect to the 4,3-DD-transpeptidation in the Gram-positive bacteria *B. subtilis* (rod-shaped), *S. pneumoniae* (coccus), and *S. aureus* (coccus); and in the Gram-negative bacterium *E. coli* (rod-shaped). In rod-shaped bacteria there is presumptive difference between the new peptidoglycan integrated into the old peptidoglycan of the sidewall, and the wholly new peptidoglycan of the septum.[287] The following summary has as its focus the 4,3-DD-transpeptidation event of Gram-negative peptidoglycan biosynthesis. A complementary perspective on peptidoglycan biosynthesis in Gram-positive bacteria is available.[164,288]

The first cytoplasmic step is the Glm-catalyzed conversion of fructose-6-phosphate to UDP-*N*-acetylglucosamine (Figure 18.4). Sequential addition of the stem peptide amino-acid residues is catalyzed by the enzymes of the Mur pathway, to give UDP-MurNAc-pentapeptide.[291–293] Transfer of the MurNAc-pentapeptide-*O*-phosphate to the phosphate group of the membrane-embedded C_{55}-isoprenoid alcohol (undecaprenol) phosphate[294] is catalyzed by the MraY enzyme[295] with the glycolipid *Lipid I* as product.[296,297] The MurNAc-pentapeptide moiety of Lipid I rests as a head group protruding from the inner leaflet of the cytoplasmic membrane.[298] The ensuing reactions of cell-wall biosynthesis, through the release of the new, stem-cross-linked glycan strand in the periplasm, involve membrane-bound intermediates. The MurG enzyme adds *N*-acetylglucosamine to the C-4 position of the NAM (MurNAc) to give the *Lipid II* disaccharide.[299] MurG additionally provides a scaffold for the cytoplasmic organization of the Mur enzymes (as observed in both *S. pneumoniae* and the Gram-negative coccobacillus *Bordetella pertussis*).[300,301]

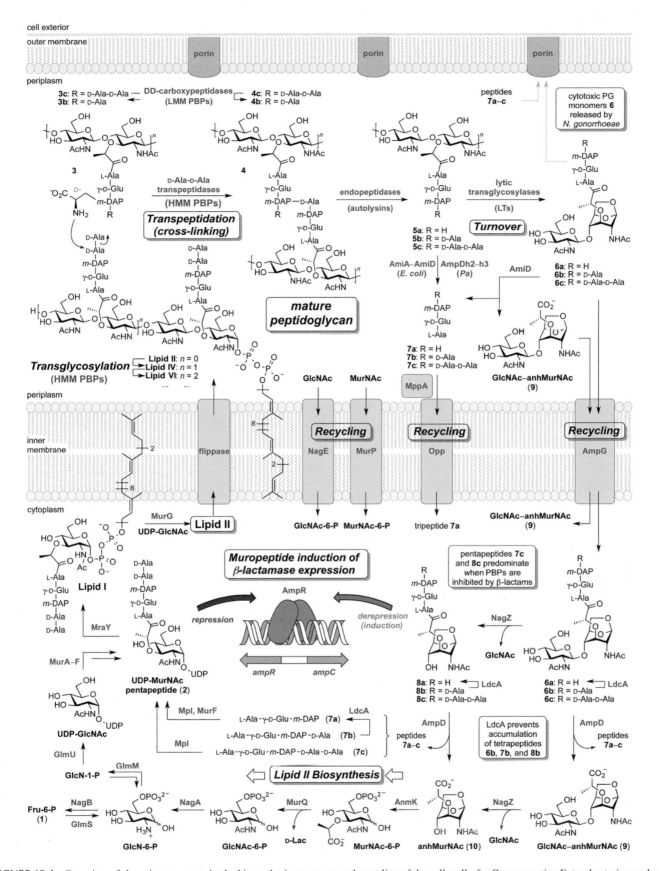

FIGURE 18.4 Overview of the primary events in the biosynthesis, turnover, and recycling of the cell wall of a Gram-negative Enterobacteriaceae bacterium. Lipid II biosynthesis commences with the synthesis of glucosamine-6-phosphate from fructose-6-phosphate (extreme lower left), catalyzed by the GlmS enzyme. GlcNAc-6-P is also a product of peptidoglycan recycling. It is transformed through the Mur pathway, and finally by MraY and MurG, to Lipid II. Lipid II is translocated to the periplasm by the MurJ flippase. The SEDS transglycosylases, the PBP transpeptidases, the PBP bifunctional transglycosylases/transpeptidases, and PBP carboxypeptidases/endopeptidases of the periplasm catalyze the glycan strand assembling, peptide stem crosslinking, and stem structure refinement to give the "mature" peptidoglycan polymer. In many Gram-negative pathogens the inactivation of a low-molecular mass PBP carboxypeptidase by a β-lactam antibiotic alters the pool of recycling muropeptide structures, resulting in the derepression of the gene encoding the AmpC β-lactamase resistance enzyme through the AmpR transcription factor (lower center, in the cytoplasm). In contrast cytoplasmic recycling of muropeptides in *P. aeruginosa* does not use the MurQ etherase pathway as is shown here for *E. coli*.[289,290]

The final event in the cytoplasmic biosynthesis of the peptidoglycan is the membrane-potential requiring translocation of the NAG-NAM disaccharide of Lipid II, as a head group of the glycolipid, from the cytoplasmic side of the membrane to the periplasmic side of the membrane.[302] In the rod-shaped *E. coli* bacterium, peptidoglycan biosynthesis occurs initially in the sidewall, and finally to form a septum. The former (elongasome-dependent synthesis) involves coordination with and insertion into existing peptidoglycan, whereas the latter (divisome-dependent synthesis) does not. Elongasome-dependent synthesis is not connected *intimately* with the cell cycle, whereas divisome-dependent synthesis connects directly and is coordinated by the bacterial cytoskeletal system. Nonetheless, the two systems must coordinate to preserve the cell diameter and for midcell septum localization, and a critical aspect of this coordination occurs when the two intersect in a transition zone of peptidoglycan synthesis from the sidewall to the septum. Each is a multiprotein and multienzyme system centered around penicillin-binding proteins (and are discussed further in the following paragraphs). The PBPs are membrane-bound enzymes, with active sites projecting into the periplasm. They fall into three primary protein classes. Two classes are the high-molecular-mass PBPs (HMM PBPs). Class A HMM PBPs (aPBPs) are bifunctional, with a peptidoglycan glycosyltransferase (PGT) active site close to the outer leaflet of the cytoplasmic membrane and a transpeptidase crosslinking site at the apex of the protein coinciding with the middle of the periplasm. The second class (bPBPs) has comparable mass, but lacks the glycan active site and acts as a monofunctional enzyme through its single transpeptidase active site. The substrate for both the aPBPs and bPBPs is Lipid II. The low-molecular-mass PBPs (cPBPs) are smaller monofunctional enzymes with either (and in some cases, both) DD-carboxypeptidase or DD-endopeptidase activity. The substructure recognized by all PBPs is the -D-Ala-D-Ala terminus of the peptidoglycan stem. All bacteria have at least one PBP of each of the three classes. *E. coli* has three aPBPs (PBP1a, *ponA*; PBP1b, *ponB*; PBP1c, *pbpC*), two bPBPs (PBP2, *pbpA*; PBP3, *ftsI*), and seven cPBPs (PBP4, *dacB*; PBP5, *dacA*; PBP6, *dacC*; PBP6b, *dacD*; PBP7, *pbpG*; PBP4b, *yefW*; AmpH, *ampH*).[303] In *E. coli* both bPBPs are essential, with PBP2 functional in the elongasome and PBP3 functional in the divisome. PBP1a and PBP1b are semiredundant (i.e., the catalytic activity of one can compensate for the loss of activity of the other, under most but not all growth circumstances).[304] Hence, a functional PBP1a or PBP1b is essential. PBP1a is associated primarily, but not exclusively, with PBP2 in the elongasome. PBP1b is associated primarily, but not exclusively, with the divisome.[305] The cPBPs control the degree of cross-linking of the peptidoglycan by hydrolysis of the D-Ala-D-Ala bond of the stem peptides, participate in murein degradation and recycling by hydrolytic cleavage of the peptide cross-bridges, and facilitate (by an unknown mechanism) the bacterial cytoskeleton-dependent organization of cell-wall biosynthesis and ultimately bacterial shape.[306–308] Functional redundancy within the cPBPs ensures both the shape and the integrity of the bacterium.[42] The cPBPs further provide to the bacterium—as a result of their abundance as PBPs and (for some) an ability to hydrolytically turnover β-lactam antibiotics—intrinsic β-lactam resistance.[309,310] Remarkably, a single mutation in the Ω-like loop of the active site of *E. coli* PBP5 abolishes both the intrinsic resistance as well as the ability of this enzyme to maintain the cell shape.[311]

Recognition of two separate systems for peptidoglycan biosynthesis in the rod-shaped bacteria—sidewall biosynthesis by the elongasome proteins and enzymes, and septal biosynthesis by the divisome proteins and enzymes—has been known for some time. Three significant, and complementary, advances characterize the past decade.[189,312] The first is the core (as distinct from the complete) identities of the proteins and enzymes in each system. The second is the recognition that peptidoglycan glycan strand synthesis, in both complexes, involves explicit coordination of a SEDS (shape, elongation, division, sporulation) membrane enzyme with a PBP enzyme.[304,313,314] A further notable advance is recognition that the membrane protein MurJ functions as the translocase to the periplasmic leaflet side of the membrane of the Lipid II molecules (whose biosynthesis is completed at the inner leaflet of the cytoplasmic membrane by MraY as the penultimate catalyst and MurG as the ultimate catalyst) for both complexes.[315–317] A mechanism for the membrane potential-dependent[302] Lipid II translocation is suggested from crystallographic studies of MurJ.[318–321] The third advance is that the dynamics of peptidoglycan biosynthesis at the elongasome and at the divisome are fundamentally different. Each of these three advances are summarized further.

Both the elongasome and the divisome organize under the guidance of protein oligomers that assemble as filaments against the inner leaflet of the cytoplasmic membrane.[322] The actin-like protein component that generates the filament used by the elongasome is the MreB protein.[323,324] MreB polymerizes as an antiparallel double filament that localizes to the sidewall of the rod-shaped bacterium aligned with the curvature of the sidewall.[325–327] MreB is excluded from the pole of the bacterium by its incompatibility with the cardiolipin-rich membrane required by the pole curvature.[328,329] The enzyme catalysts of the elongasome are the binary complex of RodA (the SEDS glycosyltransferase)[330] and PBP2 (the transpeptidase). This binary complex alone is non-catalytic. Recruitment of additional elongasome proteins (notably but not exclusively the membrane protein MreC) effects a conformational change.[331] On the periplasmic side of the membrane, the MreC·RodA·PBP2 complex acquires catalytic competence. With this competence, MreB is recruited to the complex on its cytoplasmic face. This event initiates the recruitment of the remaining proteins of the elongasome.[332] This recruitment presumptively includes the Mur components—the Mur enzymes of *S. pneumoniae* assemble as a multiprotein complex[301]—of Lipid II biosynthesis.[333] Lipid II-dependent peptidoglycan biosynthesis ensues.[334] During elongation growth the MreB filament rotates around the short axis of the cell thus giving direction to the new glycan strands. Movement of the MreB filaments is a response to sidewall peptidoglycan synthesis. β-Lactam inactivation of PBP2, for example, stops the motion.[335,336] A dynamic balance between RodZ and MreB is one basis for the preservation of the rod shape by patch-wise peptidoglycan insertion.[337] Key questions remain unanswered. The complete list of elongasome proteins is not known. Sidewall growth requires integration of new peptidoglycan into old peptidoglycan under a presumptive make-before-break mechanism (so as to preserve cell integrity). A concept for the operation of such a mechanism, incorporating the experimental observation of extensive peptidoglycan release of muropeptides for recycling concurrent with new peptidoglycan synthesis, is discussed.[304,338] The molecular mechanism for sequential synthesis, attachment, and excision of peptidoglycan to achieve net growth is unknown.

The filament that guides the divisome is composed of the tubulin-like protein FtsZ. As the divisome has universal bacterial relevance (encompassing both the Gram-positive coccus and the Gram-negative rod) it is better understood than the elongasome.[339] Peptidoglycan biosynthesis is studied in *E. coli* and in *B. subtilis* with complementary observations.[340,341] Regulation of the divisome in Gram-positive bacteria is, however, different.[342–347] In rod-shaped bacteria the FtsZ filaments assemble circumferentially at precisely the midcell prior to cell division, by a GTP-dependent mechanism. The filament is tethered to the inner leaflet of the cytoplasmic membrane through engagement of anchor proteins (FtsA, ZipA) and is oriented to align with the curvature of the membrane surface by the EzrA regulatory protein.[348] An ensemble of similarly oriented filaments anneal together through lateral interaction to form a ring—the Z-ring—around the diameter of the bacterial membrane at the mid-cell. Recruitment of the remaining 20–30 proteins of the divisome follows. These proteins include the structural and regulatory FtsBLQ sub-complex; the peptidoglycan-synthesizing enzymes FtsW, PBP3, and PBP1b; and the key regulatory protein FtsN. FtsW is the SEDS glycotransferase and PBP3 is the transpeptidase that cross-links the emerging glycan strands from FtsW.[349] The organization of the divisome is suggested to resemble three interconnected layers. The proteins and enzymes of peptidoglycan synthesis (including FtsQ, FtsL, PBP3, and FtsN) comprise as the outer layer; the protoring proteins FtsA, FtsZ, and FtsL comprise the middle ring; and the regulatory proteins ZapB and MatP comprise the inner ring.[350,351] Upon recruitment of the Lipid II–synthesizing MraY/Mur complex, Lipid II is synthesized on the cytoplasmic side of the membrane. Its synthesis recruits to the divisome, as possibly the final step of divisome assembly, MurJ. The Lipid II molecule is translocated to the periplasm, and the assembly of the septal peptidoglycan commences. The FtsZ filaments migrate at the expense of GTP hydrolysis by a treadmilling mechanism, carrying forward the divisome.[195,196] One end of the FtsZ filament shortens by FtsZ monomer dissociation, as the other end grows by FtsZ monomer accretion. The septal peptidoglycan is made by growth in increasingly smaller concentric rings.

These terse summaries of the elongasome and divisome necessarily are oversimplifications and do not address key questions. As septal peptidoglycan synthesis nears completion, FtsZ dissociates[352] and the divisome disassembles. How disassembly reconciles with completion of the intact septum by final peptidoglycan polymerization is not known. Septum synthesis is energy intensive beyond the cost of the chemical synthesis of Lipid II and its polymerization. Whereas sidewall peptidoglycan synthesis is driven by Lipid II polymerization, septal peptidoglycan biosynthesis requires FtsZ-catalyzed GTP hydrolysis. The mechanism for how GTP hydrolysis drives FtsZ-filament treadmilling and divisome motion is not known. Figure 18.5 gives a cartoon

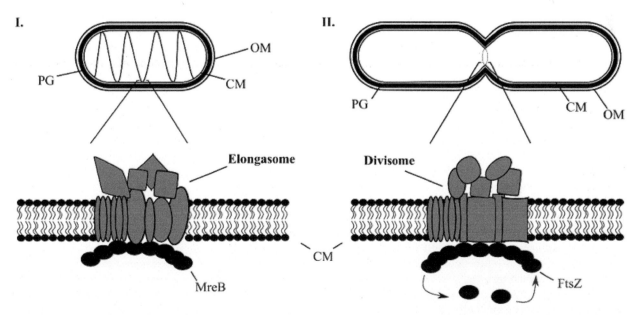

FIGURE 18.5 Schematic representation of the (I.) elongasome and the (II.) divisome machineries of Gram-negative bacteria. Part of both machineries are the cytoplasmic Mur enzymes which synthesize the ultimate peptidoglycan precursor Lipid II; the transmembrane MraY (attachment of the soluble Lipid I to the membrane) and MurJ, the flippase that transfers the Lipid II in the periplasm; and the transpeptidase PBP2 and carboxypeptidase PBP5 in the periplasm. Notable proteins of the elongasome are the cytoplasmic MreB and RodZ, the transmembrane RodA and MreC, and the periplasmic PBP1a. Notable proteins of the divisome are FtsZ (cytoplasm), FtsW (transmembrane), and the bifunctional PBP1b and the transpeptidase PBP3 in the periplasm. The protein stoichiometry of the core of the divisome is suggested by Egan and Vollmer to correspond to two molecules of PBP3, two of PBP1b, two of LpoB, four of FtsN, two of FtsW, two of FtsQ, two of FtsL, two of FtsB, three of FtsK, and eight of FtsA.[338] CM: cytoplasmic membrane, PG: peptidoglycan, OM: outer membrane.

summary of the difference between the translocation motions of the elongasome and the divisome. Moreover, the motion of the divisome must align with the concurrent force, exerted in a different spatial direction, that is necessary to constrict the membrane.[246,353] As complicated as they are individually, conceptualizing the elongasome and divisome as independent entities for peptidoglycan synthesis is wrong. They both function in partnership with respect to the events of the cell cycle; in coordination with the assembly of the lipids, saccharides, and proteins of the entire cell envelope; and in the regulation of the cell shape and size.[56,354,355] Moreover, it should be evident that the elongasome and divisome are not independent of each other. A critical event in peptidoglycan biosynthesis in the rod-shaped bacterium is the decision at midcell to initiate synthesis of a septum. This initiation—the creation of preseptal peptidoglycan as a transition from sidewall to septal peptidoglycan—requires colocalization of both systems.[14,356] Successful Z-ring formation requires sequential division-site localization of the membrane protein RodZ followed by MreB.[357] RodZ has the important task of aligning MreB with the membrane curvature and for the engagement of PBP2 with MreB.[325,337] Preseptal peptidoglycan requires FtsZ, the ZipA anchor protein for FtsZ,[358] and the recruitment by ZipA of either PBP1a or PBP1b.[359] PBP2 and PBP3 are present (or proximal) but both activities are suppressed. While the surface dimension of the preseptal peptidoglycan in *E. coli* is small (not the case for other bacteria)[359] circumstantial evidence indicates that its structural integrity is less than either the sidewall or the septal peptidoglycan. This evidence includes the separation of the pole peptidoglycan during the isolation of the sacculus (loss of the pole peptidoglycan as a result of the rupture of the preseptal peptidoglycan under mechanical stress), and the appearance of "bulges" at the presumptive location of the preseptal peptidoglycan under conditions of antibiotic-induced loss of function of PBPs[360,361] or of lytic transglycosylases.[362,363] The assumption that the bactericidal mechanism of the β-lactam antibiotics was the loss of PBP function has given way to recognition of PBP loss of function as the initiation (and not the culmination) of bactericidal pathway(s). SEDS-dependent glycan synthesis in the absence of PBP crosslinking, or disruption in the clearance and recycling of uncross-linked peptidoglycan, weakens the peptidoglycan polymer and predispose it to catastrophic failure.[364] In Gram-positive bacteria the loss of ordered peptidoglycan synthesis, as a result of cell-wall-targeting antibiotics, disrupts the balance between lipid-anchored teichoic acids and wall-anchored teichoic acids and thus dysregulates autolysin activity as a direct contributing cause to the antibiotic mechanism.[365,366]

This section of peptidoglycan synthesis concludes with consideration of the extraordinary complexity of PBP function as exemplified by the *E. coli* class A bifunctional PBP1b enzyme.[367,368] PBP1b assembles as a homodimer.[369,370] PBP1b is the biosynthetic catalyst of the PBP3-independent, but ZipA-FtsZ-dependent, synthesis of the transitional peptidoglycan between the sidewall and the septum.[358] It locates at this peptidoglycan interface by protein-protein interaction with PBP3, FtsW, FtsBLQ, FtsN, CpoB, MipA, PgpB, and ZipA.[314,359,371,372] PBP1b additionally complexes *in vitro* with the lytic transglycosylase MltA and Slt70.[373–375] Its activity is regulated concurrently by the protein components of the divisome,[371] by the CpoB-Tol system for membrane invagination,[353] and by complexation with the LpoB lipoprotein.[376–379] Upregulation of PBP1b and LpoB compensates for the loss of PBP2 as a result of specific inactivation of PBP2 by the β-lactam mecillinam (amidinocillin).[380] Both the glycosyltransferase and transpeptidase activities of PBP1b contribute (albeit differently) to the multi-generational recovery of the natural rod shape of *E. coli* from spheroplasts.[381–383] Overproduction of inactive PBP1b variants in *E. coli* results in cell lysis.[384] Cell lysis also occurs in a ΔPBP1b *E. coli* strain under conditions of impaired undecaprenyl phosphate recycling.[385] A subinhibitory concentration of a β-lactam that specifically inhibits PBP1b (or genetic deletion of PBP1b) sensitizes *E. coli* to other β-lactams.[386] PBP1b contributes directly to the motility and biofilm-capability of *E. coli*, and to its stationary phase-competitive survival.[387,388] For these reasons, conceptualizing this PBP (and indeed, all PBPs) as single task-specific catalysts is not possible. The breadth of properties of PBP1b emphasizes the extraordinary complexity of these enzymes in peptidoglycan synthesis.

Polymeric Structure of the Peptidoglycan

The spatial relationships and dimensions of the components of the cell envelope have been analyzed by different imaging methods. Cryotransmission electron microscopy of cross-sectioned, vitreously frozen bacteria show the peptidoglycan of the Gram-negative bacterium *E. coli* underneath the outer membrane.[71] Subsequent studies with Gram-negative bacteria place the peptidoglycan near to the center of the periplasm in *C. crescentus*,[389] a *Prochlorococcus* cyanobacterium,[390] and the *Treponema pallidum* spirochete.[391] An exception may occur where peptidoglycan-protein complexation is critical to strengthening membrane-penetrating structures. For example, the tight binding by the stator proteins of its flagellum of *T. pallidum* to the peptidoglycan occurs adjacent to the inner leaflet of the outer membrane.[79] The outer membrane of the Gram-negative bacterium shares with the peptidoglycan the load-bearing task of the cell envelope.[392] The peptidoglycan that is found at the midsection of the *E. coli* bacterium has a uniform thickness of 6.4 nm, consistent with values from atomic force microscopy (AFM).[393] The thickness of the midsection peptidoglycan of the related Gram-negative bacterium *P. aeruginosa* is less (2.4 nm). In contrast, the thickness of the *E. coli* and *P. aeruginosa* cell envelopes are comparable as assessed by their outer membranes (approximately 7 nm), cytoplasmic membrane (approximately 6 nm), and periplasmic space (approximately 22 nm).[71] Cryoelectron tomography microscopy[72,77] of *E. coli* suggests similar dimensions for the cell envelope at the midsection and at the pole.[394,395] Although the exterior surface of Gram-negative bacteria is lipopolysaccharide, AFM detects the perturbation of their membrane surface, in the form of blebbing, as a result of β-lactam inactivation of peptidoglycan synthesis.[396] Cryo-EM also visualized the capsular layer of mycobacteria.[397]

In contrast to the mycobacterium and the Gram-negative bacterium, the exterior of the Gram-positive bacterium is composed of peptidoglycan, teichoic acids, and capsular saccharides.[164] AFM is useful for the imaging of this exterior.[398] The peptidoglycan of *B. subtilis* appears with AFM imaging as long (>1,000 NAG-NAM units) glycan strands, suggested to coincide with 50 nm-wide coiled "cables" running parallel to the short axis of the cell.[399] Subsequent AFM study of *B. subtilis* also found peptidoglycan cables and further demonstrated that the cables underwent remodeling during growth.[400] This observation is consistent with the observations that the peptidoglycan composition at the *B. subtilis* cell-division site is structurally distinctive (pentapeptide-rich),[401] and that the septal peptidoglycan of *S. pneumoniae* is extensively remodeled as its synthesis is completed.[402,403] Complementary observations with the cell shape dynamics of *S. aureus*,[404] and the D-amino acid remodeling of *S. aureus* during its infection of *Caenorhabditis elegans*, further emphasize the dynamic character of the bacterial peptidoglycan.[405] Swarming in two Gram-negative bacteria (*Proteus mirabilis* and *Vibrio parahaemolyticus*) coincides with reduced cell stiffness. This reduction in turn coincides with peptidoglycan that has shorter glycan strands (from approximately [NAG-NAM]$_{30}$ to [NAG-NAM]$_{20}$ and is thinner (from 1.5 nm to 1.0 nm), and with increased sensitivity of the bacteria to β-lactams.[406]

AFM study of *S. aureus* as a Gram-positive cocci is hampered by the orthogonal ribs of thickened septal peptidoglycan that mark past division planes.[407,408] AFM analysis of the wild-type ovococcoid bacterium *Lactococcus lactis* showed a featureless surface as a result of a uniform coat of exopolysaccharides.[78] Upon genetic deletion of wall-polysaccharide biosynthesis, the peptidoglycan was imaged as 25 nm-wide concentric bands perpendicular to the long axis of the cell. A detailed AFM analysis of polymeric Gram-negative peptidoglycan, in the form of isolated sacculus, confirms a mesh-like character comprising long (200 nm), circumferentially ordered glycan cables. This character is disrupted by inhibitors of MreB polymerization.[409] Exceptionally detailed deep-etch electron microscopy of *B. subtilis* shows distinction in appearance between the sidewall and septal surface. The former surface shows 13 nm circumferential peptidoglycan filaments that transition abruptly to the concentric 11 nm filaments of the septum. Visualization of the effect of lysozyme and β-lactams supports the structural assignment of these filaments as peptidoglycan.[410]

Solid-state NMR spectra of the sacculus support a defined three-dimensional structure for the peptidoglycan.[411,412] Nonetheless, the molecular-level structure (as a function of cell growth, shape, and subcellular location; and in response to stressors) is not known.[413] An important constraint is the total NAG-NAM mass in the bacterium. Wientjes *et al.* measured [^{3}H]-labeled *meso*-diaminopimelic acid (*m*-Dap) incorporation into the *E. coli* cell wall.[414] As *m*-Dap acid is exclusive to the peptidoglycan, the number of *m*-Dap molecules equals the number of NAG-NAM disaccharide segments. The value of 3.5×10^6 *m*-Dap molecules gives a mass for the *E. coli* sacculus of 2.3×10^9 Da. Using a value for the surface area of the *E. coli* bacterium of 10×10^6 nm^2, a thin (mostly single-layered with some triple-layered) peptidoglycan was calculated.[415] The alternative "scaffold model" (wherein the glycan strands are orthogonal to the membrane surface)[416] is less satisfactory.[133] Extensive NMR studies using isotopic labeling of the peptidoglycan of intact *S. aureus* implicate tightly packed glycan strands having both near-parallel and near-perpendicular stems as a repeating structural motif.[417,418] Structural models for the peptidoglycan polymer are discussed elsewhere.[101,419,420]

Cell wall-associated proteins have domains tasked with recognition of specific peptidoglycan sub-structure.

Amidase-dependent removal of the stem peptide at the septal peptidoglycan recruits to this "denuded" peptidoglycan the small number of bacterial proteins (in *E. coli*: FtsN, DamX, DedD, RlpA) that possess a sporulation-related repeat (SPOR domain).[287,421,422] SPOR-domain proteins are found in both Gram-negative and Gram-positive bacteria. Additional notable peptidoglycan-binding domains found in proteins are the SH3 anchor domains proteins of *S. aureus* that recognize the pentaglycine bridge motif of its peptidoglycan[423,424] and the ubiquitous LysM domain that recognizes the [GlcNAc–X–GlcNAc] substructure of chitin and the peptidoglycan. While there is an understanding of the substructure bound by these domains, the place and circumstance where this recognition occurs remains largely unknown.

The structure of the glycan strand of the peptidoglycan in solution is also important. This structure is encountered by the enzymes of peptidoglycan biosynthesis and the recognition proteins of innate immunity. In solution the solitary glycan strand shows a tertiary structure that is a right-handed helix.[125] This identical conformation is seen in the complex between a synthetic peptidoglycan strand and a human peptidoglycan-recognition protein[426] and numerous other peptidoglycan-binding proteins.

The Enzymes of Cell-Wall Biosynthesis as Targets for Antibiotics

The cell wall is the target of numerous antibiotics.[62–64] Exhaustive efforts to discover antibiotics defined the Golden Age of antibacterial discovery in the middle decades of the 20th century.[427–429] Many of the best antibiotics then discovered are still used in the clinic. Many of these structures target the enzymes of peptidoglycan biosynthesis.[430] While the discussion in this chapter has emphasized the β-lactam inhibitors of the PBP catalysts of peptidoglycan transpeptidation, other notable peptidoglycan-targeting antibiotics include fosfomycin, moenomycin, the glycopeptides (exemplified by vancomycin), and structures that inhibit peptidoglycan biosynthesis by complexing Lipid II. Current perspectives for each of these antibiotics are given.

Fosfomycin and moenomycin have (like the β-lactams) specific molecular targets. Fosfomycin is an oral, nontoxic, broad-spectrum, and bactericidal inhibitor of MurA, the first committed enzyme of peptidoglycan synthesis.[431–435] Although fosfomycin resistance develops easily *in vitro*, the *in vivo* development of resistance is much less frequent. As a consequence of this peculiar characteristic, fosfomycin continues in use worldwide (although less so in the United States) for the treatment of bacterial urinary tract infection. The identification of antibiotics that are synergistic when combined with fosfomycin has attracted considerable interest.[436–438] Moenomycin is an inhibitor of the glycan-synthesizing enzymes, including the high-mass PBPs, of peptidoglycan biosynthesis.[439] The gene cluster for its biosynthesis is known[440] and a presumed biosynthetic intermediate has antibiotic activity.[441] The structure of moenomycin bound to the transglycosylase domain of PBPs is known.[298,368,442,443] Moenomycin has validated assays to screen for new structures with a similar biological activity.[444–446] Given this understanding, and recognizing that initial studies assessing resistance mutation in *S. aureus* gave a low viability phenotype,[447] moenomycin presents opportunity for structure-based optimization.[448,449]

The glycopeptides and the Lipid II-binding antibiotics do not easily focus to a specific target. Vancomycin is the preeminent glycopeptide.[450,451] Its primary molecular mechanism as an antibiotic—based on its self-resistance mechanism and its Gram-positive activity—is steric interference with peptidoglycan biosynthesis as a result of its ability to non-covalently complex the -D-Ala-D-Ala terminus of the peptidoglycan stem.[452–454] Although vancomycin-resistant enterococci did not emerge as a clinically pervasive superbug,[455] the continuing erosion of the antibiotic efficacy of vancomycin[456,457] has stimulated continuing study of its mechanism, and continued structure-activity study. There is now evidence that the ability of vancomycin to block the incorporation of D-Ala into the teichoic acids of the cell wall is an important component of its mechansim.[458] A derivative of a glycopeptide structurally related to vancomycin, oritavancin, has the same ability.[459] The realization that the structure-activity landscape of the glycopeptides was underdeveloped has led to a resurgence of interest in this antibiotic class,[460–464] including the possibility of extending glycopeptide activity to Gram-negative bacteria.[465,466] Indeed, it is possible that notwithstanding the closeness of their structures, both teicoplanin[467–469] and oritavancin[470,471] each may have unique structure-activity character. Bacitracin is a mixture of structurally modified nonribosomal peptides produced by various *Bacillus* species, and used extensively as a topical antibiotic and as prophylaxis and therapy in food animals. The core structure of bacitracin is a cyclic dodecapeptide that contains both L- and D-amino acids. Bacitracin is active against many Gram-positive bacteria, but is inactive against MRSA. The complex of bacitracin with a divalent metal (most commonly zinc) tightly binds to the undecaprenol diphosphate lipid of the cytoplasmic membrane.[472,473] This complexation prevents the conversion of the undecaprenol diphosphate to undecaprenol phosphate, a transformation that is critical to undecaprenol recycling. Undecaprenol diphosphate is the lipid carrier of the structural bacterial glycosides. Its efficient recycling is critical to peptidoglycan biosynthesis (and in Gram-negative bacteria, also to lipopolysaccharide biosynthesis).[385,474] The determination of the crystal structure of a bacitracin-undecaprenol diphosphate mimic[475] opens the structure of this antibiotic to redesign so as to ameliorate its poor oral availability and toxicity. Resistance to bacitracin is seen in different bacterial species. It is caused either by ABC transporter (BcrAB) efflux pump or by a phosphatase that dephosphorylates the pyrophosphate to phosphate, reducing the concentration of the target molecule.[476–478] In this era of difficulty with respect to the discovery of new antibiotics, a resource is the natural products produced by the 99% of bacterial species that cannot be cultivated in laboratories.[479] The use of an "iChip" led to the isolation of *Eleftheria terrae*, a previously uncultivated Gram-negative bacterium. This bacterium produced the new antibiotic, teixobactin.[480] Teixobactin is a depsipeptide composed of 7 L-amino acids and 4 D-amino acids. Its unusual L-*allo*-enduracididine amino acid is part of the tetrapeptide lactone formed by an ester bond between D-Thr-8 and L-Ile-11. The lactone, but not the L-*allo*-enduracididine, is essential to its mechanism.[481,482] Teixobactin binds to the pyrophosphate moiety of Lipid II and Lipid III, precursors of peptidoglycan and teichoic acid biosynthesis, respectively, causing cell-wall damage and autolysin delocalization.[480,483,484] Teixobactin is active against Gram-positive bacteria (MRSA, VISA, *C. difficile*) and

mycobacteria. Its lack of activity against Gram-negative bacteria is due to its exclusion from access to its lipid targets in the periplasm, by the outer membrane. Indeed, teixobactin is active against an outer membrane-deficient *E. coli* strain, and when an outer membrane-disruptive antibiotic (such as colistin) is used in combination.[480,481] Lipid II is a new focus for antibiotic discovery and design.[206,485–487]

An emerging theme in antibiotic discovery is the use of genomics and proteomics for structure and for target identification.[488] For example, antibiotics that target the assembly of the cytoskeleton of *S. aureus*,[489–491] or interfere with teichoic-acid assembly,[492–494] are synergistic with β-lactam antibiotics. An inhibitor of the MurG enzyme of Lipid II biosynthesis is synergistic with the β-lactam antibiotics against *S. aureus* as a consequence of failed localization of PBP2 to the division septum when MurG is inhibited.[495] The possible necessity of combination antibiotic chemotherapy (as exemplified by the pairing of β-lactams with β-lactamases) is at the forefront of antibiotic research.[496–499]

Peptidoglycan Remodeling and Recycling

The processes of cell growth, division, and (in some bacteria) sporulation require not just the synthesis of new peptidoglycan, but the concurrent remodeling of existing cell wall. The presence of peptidoglycan-cleaving amidases affixed to the exterior of the Gram-positive cell wall is widely understood to reflect the necessity of enabling the outer diameter peptidoglycan to relax and expand as it is driven outward by the new peptidoglycan assembled at the membrane.[500] As the peptidoglycan of the Gram-negative is much thinner, the need for such a process is lower (or even, unnecessary). Accordingly, the discovery that *E. coli* remodels approximately 50% of its peptidoglycan per generation was a surprise.[501,502] The prescient suggestion that new peptidoglycan growth required excision of old peptidoglycan, to allow old peptidoglycan to act as a template for incorporation of new peptidoglycan, has the same credibility today decades after the suggestion was made.[9] How this task might be done at the molecular level is still unknown. Murein remodeling requires specific enzymes, called autolysins. This group of enzymes includes the nonhydrolytic lytic transglycosylases, and the hydrolytic DD-endopeptidase, carboxypeptidase, *N*-acetylmuramidase (lysozyme), acetylglucosaminidase, and *N*-acetylmuramyl-L-alanine amidase enzymes.[503,504] The estimate of 18 murein hydrolases in *E. coli* has expanded as a result of genetic and biochemical study,[505] as exemplified by the identification in *E. coli* of three redundantly essential hydrolases that cleave D-Ala-*m*-A$_2$pm crosslinks as an obligatory event in peptidoglycan growth[506] and the discovery of a quality control periplasmic enzyme (having both uncertain substrate and mechanism) that ensures peptidoglycan having a L-Ala as the position-1 amino acid of the peptidoglycan stem, and not either glycine or L-Ser.[507] Peptidoglycan with position-1 glycine or L-Ser is defective. Hyperstructures such as the flagellum and the secretion systems have dedicated peptidoglycan autolysins. The type VI secretion system (T6SS)[508] of *E. coli* requires adaptation of its peptidoglycan by the MltE lytic transglycosylase,[273] while the T6SS of *A. baumannii* requires adaptation of its peptidoglycan by the TagX L,D-endopeptidase. The redundancies in function among the breadth of enzymes has made the identification of their mechanistic role(s) exceedingly difficult. It is entirely possible that coordination of both the MltE and TagX enzymes is required. Extensive peptidoglycan remodeling occurs during spore formation, and subsequent spore resuscitation (germination) by the Gram-positive bacteria of the *Bacillales* and *Clostridiales* orders.[49,50,509,510] A structural sub-class of the cephalosporins, the cephamycins, inhibit specifically the sporulation PBPs of *Clostridioides difficile*. The most potent of the cephamycins, cefotetan, may represent chemotherapy for the prevention of recurrent *C. difficile* intestinal colonization.[511] Muropeptides stimulate growth resumption from stationary phase *E. coli*.[512]

Significant progress has been made, however, toward answers with respect to the biochemical pathways for peptidoglycan recycling. Under laboratory growth conditions peptidoglycan recycling by *E. coli* is not essential. Nonetheless, the efficiency with which *E. coli* accomplishes this recycling—90% of the released peptidoglycan is recovered and reused—can only be interpreted in terms of a competitive advantage to being able to do so.[513] Recycling in Gram-positive bacteria was once thought as insignificant (on the basis that peptidoglycan segments accumulated in the medium), but is now recognized to exist especially under conditions of limited nutrients, such as is necessary for stationary-phase survival.[514,515] Unique pathways are encountered for peptidoglycan recycling when different bacteria are examined.[516] This diversity may reflect an additional mechanism for competitive advantage. The direct relationship between peptidoglycan recycling and the expression of the AmpC β-lactamase by pathogenic Gram-negative *Enterobacteriales*, as a resistance response to the penicillins and cephalosporins, has generated intense interest. Figure 18.4 includes the pathways for peptidoglycan synthesis and peptidoglycan recycling partitioned among the cytoplasm, the cytoplasmic membrane, and the periplasm. At its center is the AmpR transcription factor that controls expression of the *ampC* gene. In the absence of a β-lactam antibiotic an "unperturbed" pool of recycled muropeptides enters the cytoplasm by transport through the PMF-dependent AmpG permease. In the absence of a β-lactam threat, the muropeptide structures (in partnership with key intermediates in peptidoglycan biosynthesis) that are bound to AmpR enforce repression of the synthesis of the AmpC β-lactamase. In the presence of β-lactams, the composition of the muropeptide pool is perturbed. The resulting different muropeptides displace the muropeptides bound to AmpR, leading to a conformational change in the AmpR-DNA complex that is now permissive for AmpC synthesis.[517–519] Understanding the molecular basis for the perturbation is central to preserving the clinical value of the cephalosporin β-lactams that are AmpC substrates.[520–523] The best-studied bacterium is *P. aeruginosa*.[104,524] Indeed, the potential therapeutic benefit of altering the peptidoglycan recycling so as to minimize the muropeptide perturbation (and prevent AmpC synthesis) is validated in a murine model of *P. aeruginosa* infection.[525,526] The relationship among muropeptide recycling, potential targets within the recycling pathway, and the breadth of AmpR regulation (beyond AmpC) are reviewed.[527–529] Although key aspects of the *P. aeruginosa* pathway hold for all AmpC-expressing *Enterobacteriaceae*, variation in this pathway is evident when these bacteria are compared.[289,290] In *Pseudomonas* the central enzyme defining the perturbation of the pool of

recycling muropeptides is the low-molecular-mass PBP4, acting as both a DD-carboxypeptidase with respect to the stem peptide and DD-endopeptidase with respect to the crosslinked stem of the peptidoglycan.[530–533] When this PBP4 is inactivated by a β-lactam, its catalytic activity is lost and the resulting replete stem structures, as distinct from the truncated stem structures, lead to derepression of AmpC synthesis.

Sensing and Preserving Peptidoglycan Integrity

Bacteria sense compromise in the structural integrity of their envelope. Not surprisingly, our understanding of how this sensing is done, and how it is responded to, derives largely from exposure of bacteria to cell envelope-targeting antibiotics. A notable example is the sophisticated monitoring of peptidoglycan recycling as a means to regulate production of the AmpC β-lactamase enzyme as a critical resistance mechanism in the *Enterobacteriaceae*, as just discussed. A second notable example is the adaptation of the self-resistance mechanism that enables vancomycin biosynthesis[534,535] by the *Enterococci* as a vancomycin resistance mechansim. Noncovalent engagement by vancomycin of the acyl-D-Ala-D-Ala stem terminus of the peptidoglycan by hydrogen bond formation is central to its mechanism. By alteration of the peptidoglycan biosynthetic pathway to use a acyl-D-Ala-D-Lac stem terminus, an energetically important hydrogen bond to the vancomycin interaction is lost, and with this loss its antibiotic activity. Although the use of this resistance mechanism by the *Enterococci* has not as led to pervasive infection caused by vancomycin-resistant *Enterococci* (VRE), persistent VRE infection remains clinically important.[536,537] Here it may be noted that intestinal commensals secrete a nisin-type, peptidoglycan-targeting lantibiotic in order to suppress VRE growth.[538,539] This observation emphasizes the broader perspective that the antibiotics should be understood to function as a means for ecological competitiveness within ceaseless internecine bacterial competition.[540,541]

The operation of these networks used by bacteria to detect and respond to cell-wall stress is complex. The SOS response is found in varying capacities among the bacteria (it is considered to be a robust response in *E. coli*, *P. aeruginosa*, and *B. subtilis*; and as less robust in *S. aureus*). As the SOS response is primarily a response to DNA damage, the ability of β-lactams to induce the SOS response is perhaps surprising. In *S. aureus*, β-lactams induce the LexA/RecA proteins, trigger prophage induction, and effect high-frequency horizontal transfer of pathogenicity islands.[542,543] Direct cell-wall damage in *S. aureus* is monitored by the VraSR two-component system, resulting in VraS auto-phosphorylation with subsequent VraR phosphorylation and dimerization.[544,545] The molecular identity stimulus effecting the autophosphorylation is not known. The effect of VraR dimerization is increased expression levels of genes relating to peptidoglycan and teichoic acid synthesis. In the mycobacteria and the Gram-positive Firmicutes and Actinobacteria, PASTA (for "penicillin-binding protein and serine-threonine kinase associated" domain)-containing proteins act as sensors. The PASTA domain interacts directly with the peptidoglycan,[546] and the PASTA domain-containing eukaryotic-type kinases interact directly with key PBPs and with complementary two-component systems.[547,548] While the importance of PASTA kinase activity to the Gram-positive bacteria with these activities is proven,[549] here also the molecular mechanisms of activation are not well understood. PASTA kinases are a particular focus as control points of the mycobacteria. Peptidoglycan fragments (muropeptides) resuscitate mycobacteria from host-evading dormancy[550,551] and resuscitate nonculturable mycobacteria.[552,553] In the example of the *M. tuberculosis* PknB kinase, a comprehensive evaluation of muropeptide structure implicated the co-presence of glycan strand length, a tetrapeptide stem, and a 1,6-anhydroMurNAc terminus as optimal for binding affinity.[554] The compromise of the peptidoglycan structure of one bacterium, as a result of the deliberate action of another, is exemplified by the synergy of the effectors injected into bacteria by the type VI secretion system of the Gram-negative bacterium *P. aeruginosa*.[555–557] One of the six effectors (Tse3) is a peptidoglycan-degrading enzyme that synergizes with loss of the proton motive force effected by the Tse4 effector.[558]

As bacterial regulation of their peptidoglycan is sophisticated, bacterial co-culture is a potentially valuable mechanism for unlocking cryptic antibiotic biosynthetic pathways. Notwithstanding the challenge of translating this conception to experiment, this approach led to the discovery of teixobactin, a Lipid II-binding antibiotic,[559] and darobactin, an inhibitor of the essential BamA chaperone of outer membrane protein insertion.[560] While the likelihood that these new antibiotics will have the practicality, efficacy, and safety required for clinical use is uncertain, there is no doubt that this approach defines an important path for new antibiotic discovery.[561–564]

Innate Immunity Detection of the Peptidoglycan

Muropeptide recognition is an essential component of the innate-immune system.[565,566] Deliberate muropeptide release to subvert the innate-immune system is a virulence mechanism of certain bacterial pathogens (notably the Gram-negative *Neisseria*).[567,568] The molecular basis for peptidoglycan recognition by the innate immunity-recognition proteins NOD1 (classic ligand is γ-D-glutamyl-*meso*-diaminopimelic acid, derived from degradation of the peptide stem of the peptidoglycan) and NOD2 (classic ligand is *N*-acetylmuramyl-L-Ala-D-isoglutamine, known as MDP, derived from the glycan strand of the peptidoglycan)[569,570] is summarized in a recent review.[571] The development of the chemistry for the synthesis of peptidoglycan structure has allowed evaluation of the structural basis for the long-recognized immunostimulatory capability of MDP.[572–575] The relevance of NOD2 to host defense and inflammation, and in the etiology of Crohn's disease (and other diseases), is reviewed.[576,577] The relationship of peptidoglycan structure to the immune response continues to be explored broadly. Five recent examples give a sense of the breadth of this exploration. Peptidoglycan photoaffinity reporters differentiate between the active and inactive states of NOD1 and NOD2 in the immune response pathway.[574] As noted previously, *B. burgdorferi* peptidoglycan is a persistent antigen in patients with Lyme disease.[174] Antibody neutralization of circulating peptidoglycan suppresses development of autoimmune arthritis and encephalomyelitis in murine pharmacological models.[578] The structurally different peptidoglycan of

stationary-phase *S. aureus* elicits a less robust immune response than that of exponential growth *S. aureus*.[579] The immune-stimulating property of the peptidoglycan is not limited to bacterial infection. A conjugate of MDP with docetaxel, a taxol derivative, is more active than docetaxel alone in the growth suppression of implanted tumors in mice.[580]

A notable recent development with respect to peptidoglycan and innate immunity is recognition of the ability of (many, if not all) cell-wall containing Gram-positive and Gram-negative bacteria to transform, in the presence of cell-wall-targeting antibiotics and other stressors, into a cell-wall-deficient "L-form" state.[581,582] In the L-form bacteria dispense with their peptidoglycan cell wall and use a membrane-only cell envelope. This change gives survival advantages. Without a cell wall the cell-wall-targeting antibiotic is impotent. Without a cell wall the L-form bacterium evades innate immunity. Thirdly, the metabolic alteration (from respiration to glycolysis) that accompanies the L-form transition is less stressful to the bacterium.[583] A closer relationship is observed between antibiotic lethality and the bacterial metabolic state, as opposed to the bacterial growth rate.[584] Compelling data are suggestive of the infection circumstances where L-form conversion confers a survival advantage.[585] These observations complement studies that correlate the suppression of peptidoglycan synthesis by cell-wall acting antibiotics[586] as an important tolerance resistance mechanism.[587–589]

Conclusion

The peptidoglycan is an essential structural polymer component of virtually all bacterial cell envelopes. It is not a static structure. Rather, it exists in perpetual states of remodeling, recycling, and regeneration with each process connected intimately to the bacterial cell cycle. The enzyme catalysts involved in this dynamic are the target of medical-critical antibiotics. Research vistas for the peptidoglycan include its pivotal role in the cohesion of the bacterial cell envelope; as a locus for signal transduction within the bacterium and by the host during bacterial infection; and as a frontier for new antibiotic discovery and the subversion of bacterial resistance.

REFERENCES

1. Pohlschroder, M., F. Pfeiffer, S. Schulze, and M. F. A. Halim. 2018. Archaeal cell surface biogenesis. *FEMS Microbiol. Rev.* 42:694–717. doi:10.1093/femsre/fuy027.

2. Gambelli, L., B. H. Meyer, M. McLaren, K. Sanders, T. E. F. Quax, V. A. M. Gold, S. V. Albers, and B. Daum. 2019. Architecture and modular assembly of *Sulfolobus* S-layers revealed by electron cryotomography. *Proc. Natl. Acad. Sci. USA.* 116:25278–25286. doi:10.1073/pnas.1911262116.

3. Klingl, A., C. Pickl, and J. Flechsler. 2019. Archaeal cell walls. *Subcell. Biochem.* 92:471–493. doi:10.1007/978-3-030-18768-2_14.

4. Silhavy, T. J., D. Kahne, and S. Walker. 2010. The bacterial cell envelope. *Cold Spring Harb. Perspect. Biol.* 2:a000414. doi:10.1101/cshperspect.a000414.

5. Dufresne, K., and C. Paradis-Bleau. 2015. Biology and assembly of the bacterial envelope. *Adv. Exp. Med. Biol.* 883:41–76. doi:10.1007/978-3-319-23603-2_3.

6. Kleanthous, C., and J. P. Armitage. 2015. The bacterial cell envelope. *Philos. Trans. R. Soc. Lond. B Biol. Sci.* 370:20150019. doi:10.1098/rstb.2015.0019.

7. Auer, G. K., and D. B. Weibel. 2017. Bacterial cell mechanics. *Biochemistry.* 56:3710–3724. doi:10.1021/acs.biochem.7b00346.

8. Sutcliffe, I. C. 2010. A phylum level perspective on bacterial cell envelope architecture. *Trends Microbiol.* 18:464–470. doi:10.1016/j.tim.2010.06.005.

9. Höltje, J. V. 1998. Growth of the stress-bearing and shape-maintaining murein sacculus of *Escherichia coli*. *Microbiol. Mol. Biol. Rev.* 62:181–203.

10. Typas, A., M. Banzhaf, C. A. Gross, and W. Vollmer. 2012. From the regulation of peptidoglycan synthesis to bacterial growth and morphology. *Nat. Rev. Microbiol.* 10:123–136. doi:10.1038/nrmicro2677.

11. Errington, J. 2013. L-form bacteria, cell walls and the origins of life. *Open Biol.* 3:120143. doi:10.1098/rsob.120143.

12. Braun, V. 2015. Bacterial cell wall research in Tübingen: a brief historical account. *Int. J. Med. Microbiol.* 305:178–182. doi:10.1016/j.ijmm.2014.12.013.

13. Zhao, H., V. Patel, J. D. Helmann, and T. Dörr. 2017. Don't let sleeping dogmas lie: new views of peptidoglycan synthesis and its regulation. *Mol. Microbiol.* 106:847–860. doi:10.1111/mmi.13853.

14. Pazos, M., and K. Peters. 2019. Peptidoglycan. *Subcell. Biochem.* 92:127–168. doi:10.1007/978-3-030-18768-2_5.

15. Brown, S., J. P. Santa Maria Jr., and S. Walker. 2013. Wall teichoic acids of Gram-positive bacteria. *Annu. Rev. Microbiol.* 67:313–336. doi:10.1146/annurev-micro-092412-155620.

16. Percy, M. G., and A. Gründling. 2014. Lipoteichoic acid synthesis and function in Gram-positive bacteria. *Annu. Rev. Microbiol.* 68:81–100. doi:10.1146/annurev-micro-091213-112949.

17. Gale, R. T., and E. D. Brown. 2015. New chemical tools to probe cell wall biosynthesis in bacteria. *Curr. Opin. Microbiol.* 27:69–77. doi:10.1016/j.mib.2015.07.013.

18. Keinhörster, D., S. E. George, C. Weidenmaier, and C. Wolz. 2019. Function and regulation of *Staphylococcus aureus* wall teichoic acids and capsular polysaccharides. *Int. J. Med. Microbiol.* 309:151333. doi:10.1016/j.ijmm.2019.151333.

19. Rajagopal, M., and S. Walker. 2017. Envelope structures of Gram-positive bacteria. *Curr. Top. Microbiol. Immunol.* 404:1–44. doi:10.1007/82_2015_5021.

20. Konovalova, A., D. E. Kahne, and T. J. Silhavy. 2017. Outer membrane biogenesis. *Annu. Rev. Microbiol.* 71:539–556. doi:10.1146/annurev-micro-090816-093754.

21. von Kügelgen, A., H. Tang, G. G. Hardy, D. Kureisaite-Ciziene, Y. V. Brun, P. J. Stansfeld, C. V. Robinson, and T. A. M. Bharat. 2020. *In situ* structure of an intact lipopolysaccharide-bound bacterial surface layer. *Cell.* 180:348–358. doi:10.1016/j.cell.2019.12.006.

22. Geisinger, E., W. Huo, J. Hernandez-Bird, and R. R. Isberg. 2019. *Acinetobacter baumannii*: envelope determinants that control drug resistance, virulence, and surface variability. *Annu. Rev. Microbiol.* 73:481–506. doi:10.1146/annurev-micro-020518-115714.

23. Grzegorzewicz, A. E., C. de Sousa-d'Auria, M. R. McNeil, E. Huc-Claustre, V. Jones, C. Petit, S. K. Angala, J. Zemanová, Q. Wang, J. M. Belardinelli, Q. Gao, K. Ishizaki, K. Mikušová, P. J. Brennan, D. R. Ronning, M. Chami, C. Houssin, and M. Jackson. 2016. Assembling of the *Mycobacterium tuberculosis* cell wall core. *J. Biol. Chem.* 291:18867–18879. doi:10.1074/jbc.M116.739227.

24. Puffal, J., A. García-Heredia, K. C. Rahlwes, M. S. Siegrist, and Y. S. Morita. 2018. Spatial control of cell envelope biosynthesis in mycobacteria. *Pathog. Dis.* 76:fty027. doi:10.1093/femspd/fty027.

25. Squeglia, F., A. Ruggiero, and R. Berisio. 2018. Chemistry of peptidoglycan in *Mycobacterium tuberculosis* life cycle: an off-the-wall balance of synthesis and degradation. *Chem.– Eur. J.* 24:2533–2546. doi:10.1002/chem.201702973.

26. Vincent, A. T., S. Nyongesa, I. Morneau, M. B. Reed, E. I. Tocheva, and F. J. Veyrier. 2018. The mycobacterial cell envelope: a relict from the past or the result of recent evolution? *Front. Microbiol.* 9:2341. doi:10.3389/fmicb.2018.02341.

27. Daffé, M., and H. Marrakchi. 2019. Unraveling the structure of the mycobacterial envelope. *Microbiol. Spectr.* 7. doi:10.1128/microbiolspec.GPP3-0027-2018.

28. Dulberger, C. L., E. J. Rubin, and C. C. Boutte. 2020. The mycobacterial cell envelope–a moving target. *Nat. Rev. Microbiol.* 18:47–59. doi:10.1038/s41579-019-0273-7.

29. Visweswaran, G. R., B. W. Dijkstra, and J. Kok. 2011. Murein and pseudomurein cell wall binding domains of bacteria and archaea—a comparative view. *Appl. Microbiol. Biotechnol.* 92:921–928. doi:10.1007/s00253-011-3637-0.

30. Waites, K. B., L. Xiao, Y. Liu, M. F. Balish, and T. P. Atkinson. 2017. *Mycoplasma pneumoniae* from the respiratory tract and beyond. *Clin. Microbiol. Rev.* 30:747–809. doi:10.1128/CMR.00114-16.

31. Liechti, G., E. Kuru, M. Packiam, Y. P. Hsu, S. Tekkam, E. Hall, J. T. Rittichier, M. VanNieuwenhze, Y. V. Brun, and A. T. Maurelli. 2016. Pathogenic chlamydia lack a classical sacculus but synthesize a narrow, mid-cell peptidoglycan ring, regulated by MreB, for cell division. *PLOS Pathog.* 12:e1005590. doi:10.1371/journal.ppat.1005590.

32. Otten, C., M. Brilli, W. Vollmer, P. H. Viollier, and J. Salje. 2017. Peptidoglycan in obligate intracellular bacteria. *Mol. Microbiol.* 107:142–163. doi:10.1111/mmi.13880.

33. Klöckner, A., H. Bühl, P. Viollier, and B. Henrichfreise. 2018. Deconstructing the chlamydial cell wall. *Curr. Top. Microbiol. Immunol.* 412:1–33. doi:10.1007/82_2016_34.

34. Bublitz, D. C., G. L. Chadwick, J. S. Magyar, K. M. Sandoz, D. M. Brooks, S. Mesnage, M. S. Ladinsky, A. I. Garber, P. J. Bjorkman, V. J. Orphan, and J. P. McCutcheon. 2019. Peptidoglycan production by an insect-bacterial mosaic. *Cell.* 179:703–712. doi:10.1016/j.cell.2019.08.054.

35. Radkov, A. D., and S. Chou. 2019. An affair to remember: how an endosymbiont partners with its host to build a cell envelope. *Cell.* 179:584–586. doi:10.1016/j.cell.2019.09.024.

36. Jacquier, N., P. H. Viollier, and G. Greub. 2015. The role of peptidoglycan in chlamydial cell division: towards resolving the chlamydial anomaly. *FEMS Microbiol. Rev.* 39:262–275. doi:10.1093/femsre/fuv001.

37. Pilhofer, M., K. Aistleitner, J. Biboy, J. Gray, E. Kuru, E. Hall, Y. V. Brun, M. S. Vannieuwenhze, W. Vollmer, M. Horn, and G. J. Jensen. 2013. Discovery of chlamydial peptidoglycan reveals bacteria with murein sacculi but without FtsZ. *Nat. Commun.* 4:2856. doi:10.1038/ncomms3856.

38. Packiam, M., B. Weinrick, W. R. Jacobs, and A. T. Maurelli. 2015. Structural characterization of muropeptides from *Chlamydia trachomatis* peptidoglycan by mass spectrometry resolves "chlamydial anomaly." *Proc. Natl. Acad. Sci. USA.* 112:11660–11665. doi:10.1073/pnas.1514026112.

39. Cabeen, M. T., and C. Jacobs-Wagner. 2005. Bacterial cell shape. *Nat. Rev. Microbiol.* 3:601–610.

40. Young, K. D. 2007. Bacterial morphology: why have different shapes? *Curr. Opin. Microbiol.* 10:596–600. doi:10.1016/j. mib.2007.09.009.

41. Wang, S., and J. W. Shaevitz. 2013. The mechanics of shape in prokaryotes. *Front. Biosci.* 5:564–574.

42. Peters, K., S. Kannan, V. A. Rao, J. Biboy, D. Vollmer, S. W. Erickson, R. J. Lewis, K. D. Young, and W. Vollmer. 2016. The redundancy of peptidoglycan carboxypeptidases ensures robust cell shape maintenance in *Escherichia coli. mBio.* 7:e00819–16. doi:10.1128/mBio.00819-16.

43. van Teeseling, M. C. F., M. A. de Pedro, and F. Cava. 2017. Determinants of bacterial morphology: from fundamentals to possibilities for antimicrobial targeting. *Front. Microbiol.* 8:1264. doi:10.3389/fmicb.2017.01264.

44. Taylor, J. A., S. R. Sichel, and N. R. Salama. 2019. Bent bacteria: a comparison of cell shape mechanisms in Proteobacteria. *Annu. Rev. Microbiol.* 73:457–480. doi:10.1146/annurev-micro-020518-115919.

45. Cava, F., and M. A. de Pedro. 2014. Peptidoglycan plasticity in bacteria: emerging variability of the murein sacculus and their associated biological functions. *Curr. Opin. Microbiol.* 18:46–53. doi:10.1016/j.mib.2014.01.004.

46. Bonnet, J., C. Durmort, M. Jacq, I. Mortier-Barrière, N. Campo, M. S. VanNieuwenhze, Y. V. Brun, C. Arthaud, B. Gallet, C. Moriscot, C. Morlot, T. Vernet, and A. M. Di Guilmi. 2017. Peptidoglycan *O*-acetylation is functionally related to cell wall biosynthesis and cell division in *Streptococcus pneumoniae. Mol. Microbiol.* 106:832–846. doi:10.1111/mmi.13849.

47. Sychantha, D., A. S. Brott, C. S. Jones, and A. J. Clarke. 2018. Mechanistic pathways for peptidoglycan *O*-acetylation and de-*O*-acetylation. *Front. Microbiol.* 9:2332. doi:10.3389/fmicb.2018.02332.

48. Brott, A. S., and A. J. Clarke. 2019. Peptidoglycan *O*-acetylation as a vrulence factor: its effect on lysozyme in the innate immune system. *Antibiotics.* 8:94. doi:10.3390/antibiotics8030094.

49. Moir, A., and G. Cooper. 2015. Spore germination. *Microbiol. Spectr.* 3:TBS–0014. doi:10.1128/microbiolspec. TBS-0014-2012.

50. Popham, D. L., and C. B. Bernhards. 2015. Spore peptidoglycan. *Microbiol. Spectr.* 3:TBS–0005. doi:10.1128/microbiolspec. TBS-0005-2012.

51. Coullon, H., A. Rifflet, R. Wheeler, C. Janoir, I. G. Boneca, and T. Candela. 2018. *N*-Deacetylases required for muramic-δ-lactam production are involved in *Clostridium difficile* sporulation, germination and heat resistance. *J. Biol. Chem.* 293:18040–18054. doi:10.1074/jbc.RA118.004273.

52. Sidarta, M., D. Li, L. Hederstedt, and E. Bukowska-Faniband. 2018. Forespore targeting of SpoVD in *Bacillus subtilis* is mediated by the N-terminal part of the protein. *J. Bacteriol.* 200:e00163–18. doi:10.1128/JB.00163-18.

53. Hsu, Y. -P., X. Meng, and M. S. VanNieuwenhze. 2016. Methods for visualization of peptidoglycan biosynthesis. *Meth. Microbiol.* 43:3–48. doi:10.1016/bs.mim.2016.10.004.

54. Hurtley, S. M. 2017. Coordinating cell wall synthesis and cell division. *Science.* 355:706. doi:10.1126/science.355.6326.706-c.

55. Osella, M., S. J. Tans, and M. C. Lagomarsino. 2017. Step by step, cell by cell: quantification of the bacterial cell cycle. *Trends Microbiol.* 25:250–256. doi:10.1016/j.tim.2016.12.005.

56. Willis, L., and K. C. Huang. 2017. Sizing up the bacterial cell cycle. *Nat. Rev. Microbiol.* 15:606–620. doi:10.1038/nrmicro.2017.79.

57. Dewachter, L., N. Verstraeten, M. Fauvart, and J. Michiels. 2018. An integrative view of cell cycle control in *Escherichia coli. FEMS Microbiol. Rev.* 42:116–136. doi:10.1093/femsre/fuy005.

58. Surovtsev, I. V., and C. Jacobs-Wagner. 2018. Subcellular organization: a critical feature of bacterial cell replication. *Cell.* 172:1271–1293. doi:10.1016/j.cell.2018.01.014.

59. Carballido-López, R. 2019. The bacterial cell cycle, chromosome inheritance and cell growth. *Nat. Microbiol.* 4:1246–1248. doi:10.1038/s41564-019-0528-0.

60. Reyes-Lamothe, R., and D. J. Sherratt. 2019. The bacterial cell cycle, chromosome inheritance and cell growth. *Nat. Rev. Microbiol.* 17:467–478. doi:10.1038/s41579-019-0212-7.

61. Booth, S., and R. J. Lewis. 2019. Structural basis for the coordination of cell division with the synthesis of the bacterial cell envelope. *Protein Sci.* 28:2042–2054. doi:10.1002/pro.3722.

62. Bugg, T. D. H., D. Braddick, C. G. Dowson, and D. I. Roper. 2011. Bacterial cell wall assembly: still an attractive antibacterial target. *Trends Biotechnol.* 29:167–173. doi:10.1016/j. tibtech.2010.12.006.

63. Silver, L. L. 2013. Viable screening targets related to the bacterial cell wall. *Ann. N. Y. Acad. Sci.* 1277:29–53. doi:10.1111/nyas.12006.

64. Nikolaidis, I., S. Favini-Stabile, and A. Dessen. 2014. Resistance to antibiotics targeted to the bacterial cell wall. *Protein Sci.* 23:243–259. doi:10.1002/pro.2414.

65. Walsh, C. T., and T. Wencewicz. 2016. *Antibiotics: Challenges, Mechanisms, Opportunities.* ASM Press, Washington D.C., pp. 477.

66. Wencewicz, T. A. 2016. New antibiotics from nature's chemical inventory. *Bioorg. Med. Chem.* 24:6227–6252. doi:10.1016/j.bmc.2016.09.014.

67. Beveridge, T. J. 2001. Use of the Gram stain in microbiology. *Biotech. Histochem.* 76:111–118.

68. Moyes, R. B., J. Reynolds, and D. P. Breakwell. 2009. Differential staining of bacteria: Gram stain. *Curr. Protoc. Microbiol.* 15:A.3C.1–A.3C.8. doi:10.1002/9780471729259.mca03cs15.

69. O'Toole, G. A. 2016. Classic spotlight: how the Gram stain works. *J. Bacteriol.* 198:3128. doi:10.1128/JB.00726-16.

70. Beveridge, T. J., and J. A. Davies. 1983. Cellular responses of *Bacillus subtilis* and *Escherichia coli* to the Gram stain. *J. Bacteriol.* 156:846–858.

71. Matias, V. R. F., A. Al-Amoudi, J. Dubochet, and T. J. Beveridge. 2003. Cryo-transmission electron microscopy of frozen-hydrated sections of *Escherichia coli* and *Pseudomonas aeruginosa. J. Bacteriol.* 185:6112–6118.

72. Jensen, G. J., and A. Briegel. 2007. How electron cryotomography is opening a new window onto prokaryotic ultrastructure. *Curr. Opin. Struct. Biol.* 17:260–267. doi:10.1016/j.sbi.2007.03.002.

73. Möll, A., S. Schlimpert, A. Briegel, G. J. Jensen, and M. Thanbichler. 2010. DipM, a new factor required for peptidoglycan remodelling during cell division in *Caulobacter crescentus. Mol. Microbiol.* 77:90–107. doi:10.1111/j.1365-2958.2010.07224.x.

74. Kishimoto-Okada, A., S. Murakami, Y. Ito, N. Horii, H. Furukawa, J. Takagi, and K. Iwasaki. 2010. Comparison of the envelope architecture of *E. coli* using two methods: CEMOVIS and cryo-electron tomography. *J. Electron. Microsc.* 59:419–426. doi:10.1093/jmicro/dfq056.

75. Matias, V. R. F., and T. J. Beveridge. 2005. Cryo-electron microscopy reveals native polymeric cell wall structure in *Bacillus subtilis* 168 and the existence of a periplasmic space. *Mol. Microbiol.* 56:240–251. doi:10.1111/j.1365-2958.2005.04535.x.

76. Matias, V. R. F., and T. J. Beveridge. 2007. Cryo-electron microscopy of cell division in *Staphylococcus aureus* reveals a mid-zone between nascent cross walls. *Mol. Microbiol.* 64:195–206. doi:10.1111/j.1365-2958.2007.05634.x.

77. Milne, J. L. S., and S. Subramaniam. 2009. Cryo-electron tomography of bacteria: progress, challenges and future prospects. *Nat. Rev. Microbiol.* 7:666–675. doi:10.1038/nrmicro2183.

78. Andre, G., S. Kulakauskas, M. -P. Chapot-Chartier, B. Navet, M. Deghorain, E. Bernard, P. Hols, and Y. F. Dufrêne. 2010. Imaging the nanoscale organization of peptidoglycan in living *Lactococcus lactis* cells. *Nat. Commun.* 1:1027. doi:10.1038/ncomms1027.

79. Liu, J., J. K. Howell, S. D. Bradley, Y. Zheng, Z. H. Zhou, and S. J. Norris. 2010. Cellular architecture of *Treponema pallidum*: novel flagellum, periplasmic cone, and cell envelope as revealed by cryo electron tomography. *J. Mol. Biol.* 403:546–561. doi:10.1016/j.jmb.2010.09.020.

80. Engholm, D. H., M. Kilian, D. S. Goodsell, E. S. Andersen, and R. S. Kjærgaard. 2017. A visual review of the human pathogen *Streptococcus pneumoniae*. *FEMS Microbiol. Rev.* 41:854–879. doi:10.1093/femsre/fux037.

81. Wilhelm, M. J., J. B. Sheffield, M. Sharifian Gh., Y. Wu, C. Spahr, G. Gonella, B. Xu, and H. -L. Dai. 2015. Gram's stain does not cross the bacterial cytoplasmic membrane. *ACS Chem. Biol.* 10:1711–1717.

82. Fuglsang-Damgaard, D., C. H. Nielsen, E. Mandrup, and K. Fuursted. 2011. The use of Gram stain and matrix-assisted laser desorption ionization time-of-flight mass spectrometry on positive blood culture: synergy between new and old technology. *Apmis.* 119:681–688. doi:10.1111/j.1600-0463.2011.02756.x.

83. Budin, G., H. J. Chung, H. Lee, and R. Weissleder. 2012. A magnetic Gram stain for bacterial detection. *Angew. Chem. Int. Ed.* 51:7752–7755. doi:10.1002/anie.201202982.

84. Váradi, L., M. Wang, R. R. Mamidi, J. L. Luo, J. D. Perry, D. E. Hibbs, and P. W. Groundwater. 2018. A latent green fluorescent styrylcoumarin probe for the selective growth and detection of Gram negative bacteria. *Bioorg. Med. Chem.* 26:4745–4750. doi:10.1016/j.bmc.2018.08.015.

85. Kwon, H. Y., X. Liu, E. G. Choi, J. Y. Lee, S. Y. Choi, J. Y. Kim, L. Wang, S. J. Park, B. Kim, Y. A. Lee, J. J. Kim, N. Y. Kang, and Y. T. Chang. 2019. Development of a universal fluorescent probe for Gram-positive bacteria. *Angew Chem Int Ed.* 58:8426–8431. doi:10.1002/anie.201902537.

86. Radkov, A. D., Y. P. Hsu, G. Booher, and M. S. VanNieuwenhze. 2018. Imaging bacterial cell wall biosynthesis. *Annu. Rev. Biochem.* 87:991–1014. doi:10.1146/annurev-biochem-062917-012921.

87. Hu, F., G. Qi, K. Kenry, D. Mao, S. Zhou, M. Wu, W. Wu, and B. Liu. 2020. Visualization and in-situ ablation of intracellular bacterial pathogen through metabolic labeling. *Angew. Chem. Int. Ed.* 59:9228-9292. doi:10.1002/anie.201910187.

88. Rood, I. G. H., and Q. Li. 2017. Review: molecular detection of extended spectrum-β-lactamase- and carbapenemase-producing Enterobacteriaceae in a clinical setting. *Diagn. Microbiol. Infect. Dis.* 89:245–250. doi:10.1016/j.diagmicrobio.2017.07.013.

89. Chan, H. L., L. Lyu, J. Aw, W. Zhang, J. Li, H. H. Yang, H. Hayashi, S. Chiba, and B. Xing. 2018. Unique fluorescent imaging probe for bacterial surface localization and resistant enzyme imaging. *ACS Chem. Biol.* 13:1890–1896. doi:10.1021/acschembio.8b00172.

90. deBoer, T., N. Murthy, N. Tarlton, R. Yamaji, S. Adams-Sapper, T. Wu, S. Maity, G. Vegesna, C. Sadlowski, P. DePaola, and L. W. Riley. 2018. An enzyme-mediated amplification strategy enables detection of β-lactamase activity directly in unprocessed clinical samples for phenotypic detection of β-lactam resistance. *ChemBioChem.* 19:2173–2177. doi:10.1002/cbic.201800443.

91. Neumann, K. D., J. E. Villanueva-Meyer, C. A. Mutch, R. R. Flavell, J. E. Blecha, T. Kwak, R. Sriram, H. F. VanBrocklin, O. S. Rosenberg, M. A. Ohliger, and D. M. Wilson. 2017. Imaging active infection in vivo using D-amino acid derived PET radiotracers. *Sci. Rep.* 7:7903. doi:10.1038/s41598-017-08415-x.

92. Stewart, M. N., M. F. L. Parker, S. Jivan, J. M. Luu, T. L. Huynh, B. Schulte, Y. Seo, J. E. Blecha, J. E. Villanueva-Meyer, R. R. Flavell, H. F. VanBrocklin, M. A. Ohliger, O. Rosenberg, and D. M. Wilson. 2020. High enantiomeric excess in-loop synthesis of D-[methyl-11C]-methionine for use as a diagnostic PET radiotracer in bacterial infection. *ACS Infect. Dis.* 6:43–49. doi:10.1021/acsinfecdis.9b00196.

93. Deibert, J., D. Kühner, M. Stahl, E. Koeksoy, and U. Bertsche. 2016. The peptidoglycan pattern of *Staphylococcus carnosus* TM300–detailed analysis and variations due to genetic and metabolic influences. *Antibiotics.* 5:33. doi:10.3390/antibiotics5040033.

94. Turner, R. D., W. Vollmer, and S. J. Foster. 2014. Different walls for rods and balls: the diversity of peptidoglycan. *Mol. Microbiol.* 91:862–874. doi:10.1111/mmi.12513.

95. Steenbakkers, P. J. M., W. J. Geerts, N. A. Ayman-Oz, and J. T. Keltjens. 2006. Identification of pseudomurein cell wall binding domains. *Mol. Microbiol.* 62:1618–1630. doi:10.1111/j.1365-2958.2006.05483.x.

96. Desmarais, S. M., M. A. De Pedro, F. Cava, and K. C. Huang. 2013. Peptidoglycan at its peaks: how chromatographic analyses can reveal bacterial cell wall structure and assembly. *Mol. Microbiol.* 89:1–13. doi:10.1111/mmi.12266.

97. Kraft, A. R., M. F. Templin, and J. V. Holtje. 1998. Membrane-bound lytic endotransglycosylase in *Escherichia coli*. *J. Bacteriol.* 180:3441–3447.

98. Byun, B., K. V. Mahasenan, D. A. Dik, D. R. Marous, E. Speri, M. Kumarasiri, J. F. Fisher, J. A. Hermoso, and S. Mobashery. 2018. Mechanism of the *Escherichia coli* MltE lytic transglycosylase, the cell-wall-penetrating enzyme for type VI secretion system assembly. *Sci. Rep.* 8:4110. doi:10.1038/s41598-018-22527-y.

99. Yunck, R., H. Cho, and T. G. Bernhardt. 2016. Identification of MltG as a potential terminase for peptidoglycan polymerization in bacteria. *Mol. Microbiol.* 99:700–718. doi:10.1111/mmi.13258.

100. Dik, D. A., D. R. Marous, J. F. Fisher, and S. Mobashery. 2017. Lytic transglycosylases: concinnity in concision of the bacterial cell wall. *Crit. Rev. Biochem. Mol. Biol.* 52:503–542. doi:10.1080/10409238.2017.1337705.

101. Vollmer, W., and S. J. Seligman. 2010. Architecture of peptidoglycan: more data and more models. *Trends Microbiol.* 18:59–66. doi:10.1016/j.tim.2009.12.004.

102. Desmarais, S. M., C. Tropini, A. Miguel, F. Cava, R. D. Monds, M. A. de Pedro, and K. C. Huang. 2015. High-throughput, highly sensitive analyses of bacterial morphogenesis using ultra performance liquid chromatography. *J. Biol. Chem.* 290:31090–31100. doi:10.1074/jbc.M115.661660.

103. Yulo, P. R. J., and H. L. Hendrickson. 2019. The evolution of spherical cell shape; progress and perspective. *Biochem. Soc. Trans.* 47:1621–1634. doi:10.1042/BST20180634.

104. Torrens, G., M. Pérez-Gallego, B. Moya, M. Munar-Bestard, L. Zamorano, G. Cabot, J. Blázquez, J. A. Ayala, A. Oliver, and C. Juan. 2017. Targeting the permeability barrier and peptidoglycan recycling pathways to disarm *Pseudomonas aeruginosa* against the innate immune system. *PLOS One.*12:e0181932. doi:10.1371/journal.pone.0181932.

105. Boneca, I. G., Z. -H. Huang, D. A. Gage, and A. Tomasz. 2000. Characterization of *Staphylococcus aureus* cell wall glycan strands, evidence for a new β-N-acetylglucosaminidase activity. *J. Biol. Chem.* 275:9910–9918.

106. Lee, M., D. Hesek, M. Suvorov, W. Lee, S. Vakulenko, and S. Mobashery. 2003. A mechanism-based inhibitor targeting the DD-transpeptidase activity of bacterial penicillin-binding proteins. *J. Am. Chem. Soc.* 125:16322–16326. doi:10.1021/ja038445l.

107. Kocaoglu, O., H. -C. T. Tsui, M. E. Winkler, and E. E. Carlson. 2015. Profiling of β-lactam selectivity for penicillin-binding proteins in *Streptococcus pneumoniae* D39. *Antimicrob. Agents Chemother.* 59:3548–3555. doi:10.1128/AAC.05142-14.

108. Kocaoglu, O., and E. E. Carlson. 2015. Profiling of β-lactam selectivity for penicillin-binding proteins in *Escherichia coli* strain DC2. *Antimicrob. Agents Chemother.* 59:2785–2790. doi:10.1128/AAC.04552-14.

109. Sutaria, D. S., B. Moya, K. B. Green, T. H. Kim, X. Tao, Y. Jiao, A. Louie, G. L. Drusano, and J. B. Bulitta. 2018. First penicillin-binding protein occupancy patterns of β-lactams and β-lactamase inhibitors in *Klebsiella pneumoniae. Antimicrob. Agents Chemother.* 62:e00282–18. doi:10.1128/AAC.00282-18.

110. Iannazzo, L., D. Sokora, S. Triboulet, M. Fonvielle, F. Compain, V. Dubée, J. -L. Mainardi, J. -E. Hugonnet, E. Braud, M. Arthur, and M. Etheve-Quelquejeu. 2016. Routes of synthesis of carbapenems for optimizing both the inactivation of L,D-transpeptidase LdtMt1 of *Mycobacterium tuberculosis* and the stability towards hydrolysis by β-lactamase BlaC. *J. Med. Chem.* 59:3427–3438. doi:10.1021/acs.jmedchem.6b00096.

111. Edoo, Z., M. Arthur, and J. E. Hugonnet. 2017. Reversible inactivation of a peptidoglycan transpeptidase by a β-lactam antibiotic mediated by β-lactam-ring recyclization in the enzyme active site. *Sci. Rep.* 7:9136. doi:10.1038/s41598-017-09341-8.

112. Bhattacharjee, N., S. Triboulet, V. Dubée, M. Fonvielle, Z. Edoo, J. -E. Hugonnet, M. Ethève-Quelquejeu, J. -P. Simorre, M. J. Field, M. Arthur, and C. M. Bougault. 2019. Negative impact of carbapenem methylation on the reactivity of β-lactams for cysteine acylation as revealed by quantum calculations and kinetic analyses. *Antimicrob. Agents Chemother.* 63:e02039–18. doi:10.1128/AAC.02039-18.

113. Triboulet, S., Z. Edoo, F. Compain, C. Ourghanlian, A. Dupuis, V. Dubée, L. Sütterlin, H. Atze, M. Etheve-Quelquejeu, J. E. Hugonnet, and M. Arthur. 2019. Tryptophan fluorescence quenching in β-lactam-interacting proteins is modulated by the structure of intermediates and final products of the acylation reaction. *ACS Infect. Dis.* 5:1169–1176. doi:10.1021/acsinfecdis.9b00023.

114. Lohans, C. T., H. T. H. Chan, T. R. Malla, K. Kumar, J. J. A. G. Kamps, D. J. B. McArdle, E. van Groesen, M. de Munnik, C. L. Tooke, J. Spencer, R. S. Paton, J. Brem, and C. Schofield. 2019. Non-hydrolytic β-lactam antibiotic fragmentation by L,D-transpeptidases and serine β-lactamase cysteine variants. *Angew. Chem. Int. Ed.* 58:1990–1994. doi:10.1002/anie.201809424.

115. Zhao, F., Y. J. Hou, Y. Zhang, D. C. Wang, and D. F. Li. 2019. The 1β-methyl group confers a lower affinity of l,d-transpeptidase Ldt$_{Mt2}$ for ertapenem than for imipenem. *Biochem. Biophys. Res. Commun.* 510:254–260. doi:10.1016/j.bbrc.2019.01.082.

116. Lavollay, M., M. Arthur, M. Fourgeaud, L. Dubost, A. Marie, N. Veziris, D. Blanot, L. Gutmann, and J. L. Mainardi. 2008. The peptidoglycan of stationary-phase *Mycobacterium tuberculosis* predominantly contains cross-links generated by L,D-transpeptidation. *J. Bacteriol.* 190:4360–4366. doi:10.1128/JB.00239-08.

117. Ngadjeua, F., E. Braud, S. Saidjalolov, L. Iannazzo, D. Schnappinger, S. Ehrt, J. E. Hugonnet, D. Mengin-Lecreulx, D. P. Patin, M. Ethève-Quelquejeu, M. Fonvielle, and M. Arthur. 2018. Critical impact of peptidoglycan precursor amidation on the activity of L,D-transpeptidases from *Enterococcus faecium* and *Mycobacterium tuberculosis. Chem.–Eur. J.* 24:5743–5747. doi:10.1002/chem.201706082.

118. Moré, N., A. M. Martorana, J. Biboy, C. Otten, M. Winkle, C. K. G. Serrano, A. Montón Silva, L. Atkinson, H. Yau, E. Breukink, T. den Blaauwen, W. Vollmer, and A. Polissi. 2019. Peptidoglycan remodeling enables *Escherichia coli* to survive severe outer membrane assembly defect. *mBio* 10:e02729–18. doi:10.1128/mBio.02729-18.

119. Hugonnet, J. E., D. Mengin-Lecreulx, A. Monton, T. den Blaauwen, E. Carbonnelle, C. Veckerlé, Y. V. Brun, M. van Nieuwenhze, C. Bouchier, K. Tu, L. B. Rice, and M. Arthur. 2016. Factors essential for L,D-transpeptidase-mediated peptidoglycan cross-linking and β-lactam resistance in *Escherichia coli. eLife.* 5:e19469. doi:10.7554/eLife.19469.

120. Caveney, N. A., G. Caballero, H. Voedts, A. Niciforovic, L. J. Worrall, M. Vuckovic, M. Fonvielle, J. -E. Hugonnet, M. Arthur, and N. C. J. Strynadka. 2019. Structural insight into YcbB-mediated β-lactam resistance in *Escherichia coli. Nat. Commun.* 10:1849. doi:10.1038/s41467-019-09507-0.

121. Kim, H. S., H. N. Im, D. R. An, J. Y. Yoon, J. Y. Jang, S. Mobashery, D. Hesek, M. Lee, J. Yoo, M. Cui, S. Choi, C. Kim, N. K. Lee, S. J. Kim, J. Y. Kim, G. Bang, B. W. Han, B. I. Lee, H. J. Yoon, and S. W. Suh. 2015. The cell shape-determining Csd6 protein from *Helicobacter pylori* constitutes a new family of L,D-carboxypeptidase. *J. Biol. Chem.* 290:25103–25117. doi:10.1074/jbc.M115.658781.

122. Yang, D. C., K. M. Blair, J. A. Taylor, T. W. Petersen, T. Sessler, C. M. Tull, C. Leverich, A. L. Collar, T. J. Wyckoff, J. Biboy, W. Vollmer, and N. R. Salama. 2019. A genome-wide *Helicobacter pylori* morphology screen uncovers a membrane spanning helical cell shape complex. *J. Bacteriol.* 201:e00724–18. doi:10.1128/JB.00724-18.

123. Schleifer, K. H., and O. Kandler. 1972. Peptidoglycan types of bacterial cell walls and their taxonomic implications. *Bacteriol. Rev.* 36:407–477.

124. Vollmer, W., D. Blanot, and M. A. de Pedro. 2008. Peptidoglycan structure and architecture. *FEMS Microbiol. Rev.* 32:149–167. doi:10.1111/j.1574-6976.2007.00094.x.

125. Heo, J., M. Hamada, H. Cho, H. Y. Weon, J. S. Kim, S. B. Hong, S. J. Kim, and S. W. Kwon. 2019. *Weissella cryptocerci* sp. nov., isolated from gut of the insect *Cryptocercus kyebangensis*. *Int. J. Syst. Evol. Microbiol.* 69:2801–2806. doi:10.1099/ijsem.0.003564.

126. Vollmer, W. 2008. Structural variation in the glycan strands of bacterial peptidoglycan. *FEMS Microbiol. Rev.* 32:287–306. doi:10.1111/j.1574-6976.2007.00088.x.

127. Schneider, T., M. M. Senn, B. Berger-Bachi, A. Tossi, H. G. Sahl, and I. Wiedemann. 2004. In vitro assembly of a complete, pentaglycine interpeptide bridge containing cell wall precursor (lipid II-Gly5) of *Staphylococcus aureus*. *Mol. Microbiol.* 53:675–685. doi:10.1111/j.1365-2958.2004.04149.x.

128. Ulrich, E. C., and W. A. van der Donk. 2016. Cameo appearances of aminoacyl-tRNA in natural product biosynthesis. *Curr. Opin. Chem. Biol.* 35:29–36. doi:10.1016/j.cbpa.2016.08.018.

129. Fuda, C., M. Suvorov, S. B. Vakulenko, and S. Mobashery. 2004. The basis for resistance to beta-lactam antibiotics by penicillin-binding protein 2a of methicillin-resistant *Staphylococcus aureus*. *J. Biol. Chem.* 279:40802–40806. doi:10.1074/jbc.M403589200.

130. Peacock, S. J., and G. K. Paterson. 2015. Mechanisms of methicillin resistance in *Staphylococcus aureus*. *Annu. Rev. Biochem.* 84:577–601. doi:10.1146/annurev-biochem-060614-034516.

131. Kim, C. K., C. Milheiriço, H. de Lencastre, and A. Tomasz. 2017. Antibiotic resistance as a stress response: recovery of high-level oxacillin resistance in methicillin-resistant *Staphylococcus aureus* "auxiliary" (*fem*) mutants by induction of the stringent stress response. *Antimicrob. Agents Chemother.* 61:e00313–17. doi:10.1128/AAC.00313-17.

132. Monteiro, J. M., G. Covas, D. Rausch, S. R. Filipe, T. Schneider, H. G. Sahl, and M. G. Pinho. 2019. The pentaglycine bridges of *Staphylococcus aureus* peptidoglycan are essential for cell integrity. *Sci. Rep.* 9:5010. doi:10.1038/s41598-019-41461-1.

133. Vollmer, W., and J. V. Holtje. 2004. The architecture of the murein (peptidoglycan) in Gram-negative bacteria: vertical scaffold or horizontal layer(s)? *J. Bacteriol.* 186:5978–5987. doi:10.1128/JB.186.18.5978–5987.2004.

134. Bradshaw, W. J., A. H. Davies, C. J. Chambers, A. K. Roberts, C. C. Shone, and K. R. Acharya. 2015. Molecular features of the sortase enzyme family. *FEBS J.* 282:2097–2114. doi:10.1111/febs.13288.

135. Siegel, S. D., M. E. Reardon, and H. Ton-That. 2017. Anchoring of LPXTG-like proteins to the Gram-positive cell wall envelope. *Curr. Top. Microbiol. Immunol.* 404:159–175. doi:10.1007/82_2016_8.

136. Malik, A., and S. B. Kim. 2019. A comprehensive in silico analysis of sortase superfamily. *J. Microbiol.* 57:431–443. doi:10.1007/s12275-019-8545-5.

137. Schneewind, O., and D. Missiakas. 2019. Sortases, surface proteins, and their roles in *Staphylococcus aureus* disease and vaccine development. *Microbiol. Spectr.* 7:PSIB–0004. doi:10.1128/microbiolspec.PSIB-0004-2018.

138. Schneewind, O., and D. M. Missiakas. 2019. Staphylococcal protein secretion and envelope assembly. *Microbiol. Spectr.* 7:10.1128/microbiolspec.GPP3-0070-2019. doi:10.1128/microbiolspec.GPP3-0070-2019.

139. Foster, T. J. 2019. Surface proteins of *Staphylococcus aureus*. *Microbiol. Spectr.* 7. doi:10.1128/microbiolspec.GPP3-0046-2018.

140. Pasquina, L. W., J. P. Santa Maria, and S. Walker. 2013. Teichoic acid biosynthesis as an antibiotic target. *Curr. Opin. Microbiol.* 16:531–537. doi:10.1016/j.mib.2013.06.014.

141. Siegel, S. D., J. Liu, and H. Ton-That. 2016. Biogenesis of the Gram-positive bacterial cell envelope. *Curr. Opin. Microbiol.* 34:31–37. doi:10.1016/j.mib.2016.07.015.

142. Pansegrau, W., and F. Bagnoli. 2017. Pilus assembly in Gram-positive bacteria. *Curr. Top. Microbiol. Immunol.* 404: 203–233. doi:10.1007/82_2015_5016.

143. Khare, B., and S. V. L. Narayana. 2017. Pilus biogenesis of Gram-positive bacteria: roles of sortases and implications for assembly. *Protein Sci.* 26:1458–1473. doi:10.1002/pro.3191.

144. Chang, C., B. R. Amer, J. Osipiuk, S. A. McConnell, I. H. Huang, V. Hsieh, J. Fu, H. H. Nguyen, J. Muroski, E. Flores, R. R. Ogorzalek Loo, J. A. Loo, J. A. Putkey, A. Joachimiak, A. Das, R. T. Clubb, and H. Ton-That. 2018. In vitro reconstitution of sortase-catalyzed pilus polymerization reveals structural elements involved in pilin cross-linking. *Proc. Natl. Acad. Sci. USA.* 115:E5477–E5486. doi:10.1073/pnas.1800954115.

145. Chang, C., C. Wu, J. Osipiuk, S. D. Siegel, S. Zhu, X. Liu, A. Joachimiak, R. T. Clubb, A. Das, and H. Ton-That. 2019. Cell-to-cell interaction requires optimal positioning of a pilus tip adhesin modulated by Gram-positive transpeptidase enzymes. *Proc. Natl. Acad. Sci. USA.* 116:18041–18049. doi:10.1073/pnas.1907733116.

146. Lukaszczyk, M., B. Pradhan, and H. Remaut. 2019. The biosynthesis and structures of bacterial pili. *Subcell. Biochem.* 92:369–413. doi:10.1007/978-3-030-18768-2_12.

147. Rebollo, I. R., S. McCallin, D. Bertoldo, J. M. Entenza, P. Moreillon, and C. Heinis. 2016. Development of potent and selective *S. aureus* sortase A inhibitors based on peptide macrocycles. *ACS Med. Chem. Lett.* 7:606–611. doi:10.1021/acsmedchemlett.6b00045.

148. He, W., Y. Zhang, J. Bao, X. Deng, J. Batara, S. Casey, Q. Guo, F. Jiang, and L. Fu. 2017. Synthesis, biological evaluation and molecular docking analysis of 2-phenyl-benzofuran-3-carboxamide derivatives as potential inhibitors of *Staphylococcus aureus* sortase A. *Bioorg. Med. Chem.* 25:1341–1351. doi:10.1016/j.bmc.2016.12.030.

149. Wang, G., X. Wang, L. Sun, Y. Gao, X. Niu, and H. Wang. 2018. Novel inhibitor discovery of *Staphylococcus aureus* sortase B and the mechanism confirmation via molecular modeling. *Molecules.* 23:977. doi:10.3390/molecules23040977.

150. Selvaraj, C., R. B. Priya, and S. K. Singh. 2018. Exploring the biology and structural architecture of sortase role on biofilm formation in Gram-positive pathogens. *Curr. Top. Med. Chem.* 18:2462–2480. doi:10.2174/1568026619666181130133916.

151. Wehrli, P. M., I. Uzelac, T. Olsson, T. Jacso, D. Tietze, and J. Gottfries. 2019. Discovery and development of substituted thiadiazoles as inhibitors of *Staphylococcus aureus* sortase A. *Bioorg. Med. Chem.* 27:115043. doi:10.1016/j.bmc.2019.115043.

152. Guttilla, I. K., A. H. Gaspar, A. Swierczynski, A. Swaminathan, P. Dwivedi, A. Das, and H. Ton-That. 2009. Acyl enzyme intermediates in sortase-catalyzed pilus morphogenesis in Gram-positive bacteria. *J. Bacteriol.* 191:5603–5612. doi:10.1128/JB.00627-09.

153. Hendrickx, A. P. A., R. J. L. Willems, M. J. M. Bonten, and W. van Schaik. 2009. LPxTG surface proteins of enterococci. *Trends Microbiol.* 17:423–430. doi:10.1016/j.tim.2009.06.004.

154. Manzano, C., T. Izoré, V. Job, A. M. Di Guilmi, and A. Dessen. 2009. Sortase activity is controlled by a flexible lid in the pilus biogenesis mechanism of Gram-positive pathogens. *Biochemistry*. 48:10549–10557. doi:10.1021/bi901261y.

155. Biswas, T., V. S. Pawale, D. Choudhury, and R. P. Roy. 2014. Sorting of LPXTG peptides by archetypal sortase A: role of invariant substrate residues in modulating the enzyme dynamics and conformational signature of a productive substrate. *Biochemistry*. 53:2515–2524. doi:10.1021/bi4016023.

156. Kühner, D., M. Stahl, D. D. Demircioglu, and U. Bertsche. 2014. From cells to muropeptide structures in 24 h: peptidoglycan mapping by UPLC-MS. *Sci. Rep.* 4:7494. doi:10.1038/srep07494.

157. Espaillat, A., O. Forsmo, K. El Biari, R. Björk, B. Lemaitre, J. Trygg, F. J. Cañada, M. A. de Pedro, and F. Cava. 2016. Chemometric analysis of bacterial peptidoglycan reveals atypical modifications that empower the cell wall against predatory enzymes and fly innate immunity. *J. Am. Chem. Soc.* 138:9193–9204. doi:10.1021/jacs.6b04430.

158. Alvarez, L., S. B. Hernandez, M. A. de Pedro, and F. Cava. 2016. Ultra-sensitive, high-resolution liquid chromatography methods for the high-throughput quantitative analysis of bacterial cell wall chemistry and structure. *Methods Mol. Biol.* 1440:11–27. doi:10.1007/978-1-4939-3676-2_2.

159. Bern, M., R. Beniston, and S. Mesnage. 2017. Towards an automated analysis of bacterial peptidoglycan structure. *Anal. Bioanal. Chem.* 409:551–560. doi:10.1007/s00216-016-9857-5.

160. Schaub, R. E., and J. P. Dillard. 2017. Digestion of peptidoglycan and analysis of soluble fragments. *Bio. Protoc.* 7:e2438. doi:10.21769/BioProtoc.2438.

161. Boulanger, M., C. Delvaux, L. Quinton, B. Joris, E. Pauw, and J. Far. 2019. *Bacillus licheniformis* peptidoglycan characterization by CZE-MS: assessment with the benchmark RP-HPLC-MS method. *Electrophoresis*. 40:2672–2682. doi:10.1002/elps.201900147.

162. Kumar, K., A. Espaillat, and F. Cava. 2017. PG-Metrics: a chemometric-based approach for classifying bacterial peptidoglycan data sets and uncovering their subjacent chemical variability. *PLOS One*. 12:e0186197. doi:10.1371/journal.pone.0186197.

163. Kumar, K., and F. Cava. 2018. Principal coordinate analysis assisted chromatographic analysis of bacterial cell wall collection: a robust classification approach. *Anal. Biochem.* 550:8–14. doi:10.1016/j.ab.2018.04.008.

164. Vollmer, W., O. Massidda, and A. Tomasz. 2019. The cell wall of *Streptococcus pneumoniae*. *Microbiol. Spectr.* 7:GPP3–0018. doi:10.1128/microbiolspec.GPP3-0018-2018.

165. Porfírio, S., R. W. Carlson, and P. Azadi. 2019. Elucidating peptidoglycan structure: an analytical toolset. *Trends Microbiol.* 27:607–622. doi:10.1016/j.tim.2019.01.009.

166. Chauhan, D., P. A. Srivastava, B. Ritzl, R. M. Yennamalli, F. Cava, and R. Priyadarshini. 2019. Amino acid-dependent alterations in cell wall and cell morphology of *Deinococcus indicus* DR1. *Front. Microbiol.* 10:1449. doi:10.3389/fmicb.2019.01449.

167. Bibek, G. C., G. S. Sahukhal, and M. O. Elasri. 2019. Role of the *msa*ABCR operon in cell wall biosynthesis, autolysis, integrity, and antibiotic resistance in *Staphylococcus aureus*. *Antimicrob. Agents Chemother.* 63:e00680–19. doi:10.1128/AAC.00680-19.

168. van der Aart, L. T., G. K. Spijksma, A. Harms, W. Vollmer, T. Hankemeier, and G. P. van Wezel. 2018. High-resolution analysis of the peptidoglycan composition in *Streptomyces coelicolor*. *J. Bacteriol.* 200:e00290–18. doi:10.1128/JB.00290-18.

169. Kim, B., Y. C. Wang, C. W. Hespen, J. Espinosa, J. Salje, K. J. Rangan, D. A. Oren, J. Y. Kang, V. A. Pedicord, and H. C. Hang. 2019. *Enterococcus faecium* secreted antigen a generates muropeptides to enhance host immunity and limit bacterial pathogenesis. *eLife*. 8:e45343. doi:10.7554/eLife.45343.

170. Schaub, R. E., and J. P. Dillard. 2019. The pathogenic *Neisseria* use a streamlined set of peptidoglycan degradation proteins for peptidoglycan remodeling, recycling, and toxic fragment release. *Front. Microbiol.* 9:73. doi:10.3389/fmicb.2019.00073.

171. Frirdich, E., J. Biboy, M. Pryjma, J. Lee, S. Huynh, C. T. Parker, S. E. Girardin, W. Vollmer, and E. C. Gaynor. 2019. The *Campylobacter jejuni* helical to coccoid transition involves changes to peptidoglycan and the ability to elicit an immune response. *Mol. Microbiol.* 112:280–301. doi:10.1111/mmi.14269.

172. Stankeviciute, G., A. V. Miguel, A. Radkov, S. Chou, K. C. Huang, and E. A. Klein. 2019. Differential modes of crosslinking establish spatially distinct regions of peptidoglycan in *Caulobacter crescentus*. *Mol. Microbiol.* 111:995–1008. doi:10.1111/mmi.14199.

173. Weber, B. S., S. W. Hennon, M. S. Wright, N. E. Scott, V. de Berardinis, L. J. Foster, J. A. Ayala, M. D. Adams, and M. F. Feldman. 2016. Genetic dissection of the type VI secretion system in *Acinetobacter* and identification of a novel peptidoglycan hydrolase, TagX, required for its biogenesis. *mBio*. 7:e01253–16. doi:10.1128/mBio.01253-16.

174. Jutras, B. L., R. B. Lochhead, Z. A. Kloos, J. Biboy, K. Strle, C. J. Booth, S. K. Govers, J. Gray, P. Schumann, W. Vollmer, L. K. Bockenstedt, A. C. Steere, and C. Jacobs-Wagner. 2019. *Borrelia burgdorferi* peptidoglycan is a persistent antigen in patients with lyme arthritis. *Proc. Natl. Acad. Sci. USA.* 116:13498–13507. doi:10.1073/pnas.1904170116.

175. Lee, M., D. Hesek, D. A. Dik, J. Fishovitz, E. Lastochkin, B. Boggess, J. F. Fisher, and S. Mobashery. 2017. From genome to proteome to eucidation of reactions for all eleven-known lytic transglycosylases from *Pseudomonas aeruginosa*. *Angew. Chem. Int. Ed.* 56:2735–2739. doi:10.1002/anie.201611279.

176. Ealand, C. S., E. E. Machowski, and B. D. Kana. 2018. β-Lactam resistance: the role of low molecular weight penicillin binding proteins, β-lactamases and LD-transpeptidases in bacteria associated with respiratory tract infections. *IUBMB Life*. 70:855–868. doi:10.1002/iub.1761.

177. Pidgeon, S. E., A. J. Apostolos, J. M. Nelson, M. Shaku, B. Rimal, M. N. Islam, D. C. Crick, S. J. Kim, M. S. Pavelka, B. D. Kana, and M. M. Pires. 2019. L,D-transpeptidase specific probe reveals spatial activity of peptidoglycan crosslinking. *ACS Chem. Biol.* 14:2185–2196. doi:10.1021/acschembio.9b00427.

178. Torrens, G., M. Escobar-Salom, E. Pol-Pol, C. Camps-Munar, G. Cabot, C. López-Causapé, E. Rojo-Molinero, A. Oliver, and C. Juan. 2019. Comparative analysis of peptidoglycans from *Pseudomonas aeruginosa* isolates recovered from chronic and acute infections. *Front. Microbiol.* 10:1868. doi:10.3389/fmicb.2019.01868.

179. Anderson, E. M., D. Sychantha, D. Brewer, A. J. Clarke, J. Geddes-McAlister, and C. M. Khursigara. 2020. Peptidoglycomics reveals compositional changes in peptidoglycan between biofilm- and planktonic-derived *Pseudomonas aeruginosa*. *J. Biol. Chem.* 295:504–516. doi:10.1074/jbc. RA119.010505.

180. de Pedro, M. A., J. C. Quintela, J. V. Holtje, and H. Schwarz. 1997. Murein segregation in *Escherichia coli*. *J. Bacteriol.* 179:2823–2834.

181. Cava, F., E. Kuru, Y. V. Brun, and M. A. de Pedro. 2013. Modes of cell wall growth differentiation in rod-shaped bacteria. *Curr. Opin. Microbiol.* 16:731–737. doi:10.1016/j.mib.2013.09.004.

182. Takacs, C. N., J. Hocking, M. T. Cabeen, N. K. Bui, S. Poggio, W. Vollmer, and C. Jacobs-Wagner. 2013. Growth medium-dependent glycine incorporation into the peptidoglycan of *Caulobacter crescentus*. *PLOS One*. 8:e57579. doi:10.1371/journal.pone.0057579.

183. Lam, H., D. C. Oh, F. Cava, C. N. Takacs, J. Clardy, M. A. de Pedro, and M. K. Waldor. 2009. D-Amino acids govern stationary phase cell wall remodeling in bacteria. *Science*. 325:1552–1555. doi:10.1126/science.1178123.

184. Cava, F., M. A. de Pedro, H. Lam, B. M. Davis, and M. K. Waldor. 2011. Distinct pathways for modification of the bacterial cell wall by non-canonical D-amino acids. *EMBO J.* 30:3442–3453. doi:10.1038/emboj.2011.246.

185. Cava, F., H. Lam, M. A. de Pedro, and M. K. Waldor. 2011. Emerging knowledge of regulatory roles of D-amino acids in bacteria. *Cell. Mol. Life Sci.* 68:817–831. doi:10.1007/s00018-010-0571-8.

186. Silva, A. M., C. Otten, J. Biboy, E. Breukink, M. VanNieuwenhze, W. Vollmer, and T. den Blaauwen. 2018. The fluorescent D-amino acid NADA as a tool to study the conditional activity of transpeptidases in *Escherichia coli*. *Front. Microbiol.* 9:2101. doi:10.3389/fmicb.2018.02101.

187. Hsu, Y. P., G. Booher, A. Egan, W. Vollmer, and M. S. VanNieuwenhze. 2019. D-Amino acid derivatives as in situ probes for visualizing bacterial peptidoglycan biosynthesis. *Acc. Chem. Res.* 52:2713–2722. doi:10.1021/acs.accounts.9b00311.

188. Botella, H., and J. Vaubourgeix. 2019. Building walls: work that never ends. *Trends Microbiol.* 27:4–7. doi:10.1016/j.tim.2018.11.006.

189. Taguchi, A., D. Kahne, and S. Walker. 2019. Chemical tools to characterize peptidoglycan synthases. *Curr. Opin. Chem. Biol.* 53:44–50. doi:10.1016/j.cbpa.2019.07.009.

190. Kuru, E., A. Radkov, X. Meng, A. Egan, L. Alvarez, A. Dowson, G. Booher, E. Breukink, D. I. Roper, F. Cava, W. Vollmer, Y. Brun, and M. S. VanNieuwenhze. 2019. Mechanisms of incorporation for D-amino acid probes that target peptidoglycan biosynthesis. *ACS Chem. Biol.* 14:2745–2756. doi:10.1021/acschembio.9b00664.

191. Zhang, Z. J., Y. C. Wang, X. Yang, and H. C. Hang. 2020. Chemical reporters for exploring microbiology and microbiota mechanisms. *ChemBioChem*. 21:19–32. doi:10.1002/cbic.201900535.

192. Kuru, E., S. Tekkam, E. Hall, Y. V. Brun, and M. S. VanNieuwenhze. 2015. Synthesis of fluorescent D-amino acids and their use for probing peptidoglycan synthesis and bacterial growth in situ. *Nat. Protoc.* 10:33–52. doi:10.1038/nprot.2014.197.

193. Hsu, Y. P., E. Hall, G. Booher, B. Murphy, A. D. Radkov, J. Yablonowski, C. Mulcahey, L. Alvarez, F. Cava, Y. V. Brun, E. Kuru, and M. S. VanNieuwenhze. 2019. Fluorogenic D-amino acids enable real-time monitoring of peptidoglycan biosynthesis and high-throughput transpeptidation assays. *Nat. Chem.* 11:335–341. doi:10.1038/s41557-019-0217-x.

194. DeMeester, K. E., H. Liang, J. Zhou, K. A. Wodzanowski, B. L. Prather, C. C. Santiago, and C. L. Grimes. 2019. Metabolic incorporation of N-acetyl muramic acid probes into bacterial peptidoglycan. *Curr. Protoc. Chem. Biol.* 11:e74. doi:10.1002/cpch.74.

195. Bisson-Filho, A. W., Y. P. Hsu, G. R. Squyres, E. Kuru, F. Wu, C. Jukes, Y. Sun, C. Dekker, S. Holden, M. S. VanNieuwenhze, Y. V. Brun, and E. C. Garner. 2017. Treadmilling by FtsZ filaments drives peptidoglycan synthesis and bacterial cell division. *Science*. 355:739–743. doi:10.1126/science.aak9973.

196. Yang, X., Z. Lyu, A. Miguel, R. McQuillen, K. C. Huang, and J. Xiao. 2017. GTPase activity-coupled treadmilling of the bacterial tubulin FtsZ organizes septal cell wall synthesis. *Science*. 355:744–747. doi:10.1126/science.aak9995.

197. Gautam, S., T. Kim, T. Shoda, S. Sen, D. Deep, R. Luthra, M. T. Ferreira, M. G. Pinho, and D. A. Spiegel. 2015. An activity-based probe for sudying crosslinking in live bacteria. *Angew. Chem. Int. Ed.* 54:10492–10496. doi:10.1002/anie.201503869.

198. Gautam, S., T. Kim, and D. A. Spiegel. 2015. Chemical probes reveal an extraseptal mode of cross-linking in *Staphylococcus aureus*. *J. Am. Chem. Soc.* 137:7441–7447.

199. Reichmann, N. T., A. C. Tavares, B. M. Saraiva, A. Jousselin, P. Reed, A. R. Pereira, J. M. Monteiro, R. G. Sobral, M. S. VanNieuwenhze, F. Fernandes, and M. G. Pinho. 2019. SEDS-bPBP pairs direct lateral and septal peptidoglycan synthesis in *Staphylococcus aureus*. *Nat. Microbiol.* 4:1368–1377. doi:10.1038/s41564-019-0437-2.

200. Botella, H., G. Yang, O. Ouerfelli, S. Ehrt, C. F. Nathan, and J. Vaubourgeix. 2017. Distinct spatiotemporal dynamics of peptidoglycan synthesis between *Mycobacterium smegmatis* and *Mycobacterium tuberculosis*. *mBio*. 8:e01183–17. doi:10.1128/mBio.01183-17.

201. Maitra, A., T. Munshi, J. Healy, L. T. Martin, W. Vollmer, N. H. Keep, and S. Bhakta. 2019. Cell wall peptidoglycan in *Mycobacterium tuberculosis*: an Achilles' heel for the TB-causing pathogen. *FEMS Microbiol. Rev.* 43:548–575. doi:10.1093/femsre/fuz016.

202. Bonnet, J., Y. S. Wong, T. Vernet, A. M. Di Guilmi, A. Zapun, and C. Durmort. 2018. One-pot two-step metabolic labeling of teichoic acids and direct labeling of peptidoglycan reveals the tight coordination of both polymers insertion in pneumococcus cell wall. *ACS Chem. Biol.* 13:2010–2016. doi:10.1021/acschembio.8b00559.

203. Nonejuie, P., M. Burkart, K. Pogliano, and J. Pogliano. 2013. Bacterial cytological profiling rapidly identifies the cellular pathways targeted by antibacterial molecules. *Proc. Natl. Acad. Sci. USA*. 110:16169–16174. doi:10.1073/pnas.1311066110.

204. Nonejuie, P., R. M. Trial, G. L. Newton, A. Lamsa, V. Ranmali Perera, J. Aguilar, W. T. Liu, P. C. Dorrestein, J. Pogliano, and K. Pogliano. 2016. Application of bacterial cytological profiling to crude natural product extracts reveals the antibacterial arsenal of *Bacillus subtilis*. *J. Antibiot.* 69:353–361. doi:10.1038/ja.2015.116.

205. Htoo, H. H., L. Brumage, V. Chaikeeratisak, H. Tsunemoto, J. Sugie, C. Tribuddharat, J. Pogliano, and P. Nonejuie. 2019. Bacterial cytological profiling as a tool to study mechanisms of action of antibiotics that are active against *Acinetobacter baumannii*. *Antimicrob. Agents Chemother.* 63:e02310–18. doi:10.1128/AAC.02310-18.

206. Tan, S., K. C. Ludwig, A. Müller, T. Schneider, and J. R. Nodwell. 2019. The lasso peptide siamycin-I targets lipid II at the Gram-positive cell surface. *ACS Chem. Biol.* 14:966–974. doi:10.1021/acschembio.9b00157.

207. Ojkic, N., J. López-Garrido, K. Pogliano, and R. G. Endres. 2016. Cell wall remodeling drives engulfment during *Bacillus subtilis* sporulation. *eLife.* 5:e18657. doi:10.7554/eLife.18657.

208. Vollmer, B., N. Steblau, N. Ladwig, C. Mayer, B. Macek, L. Mitousis, S. Sigle, A. Walter, W. Wohlleben, and G. Muth. 2019. Role of the *Streptomyces* spore wall synthesizing complex SSSC in differentiation of *Streptomyces coelicolor* A3(2). *Int. J. Med. Microbiol.* 309:151327. doi:10.1016/j.ijmm.2019.07.001.

209. Ono, D., T. Yamaguchi, M. Hamada, S. Sonoda, A. Sato, K. Aoki, C. Kajiwara, S. Kimura, M. Fujisaki, H. Tojo, M. Sasaki, H. Murakami, K. Kato, Y. Ishii, and K. Tateda. 2019. Analysis of synergy between beta-lactams and anti-methicillin-resistant *Staphylococcus aureus* agents from the standpoint of strain characteristics and binding action. *J. Infect. Chemother.* 25:273–280. doi:10.1016/j.jiac.2018.12.007.

210. Tiyanont, K., T. Doan, M. B. Lazarus, X. Fang, D. Z. Rudner, and S. Walker. 2006. Imaging peptidoglycan biosynthesis in *Bacillus subtilis* with fluorescent antibiotics. *Proc. Natl. Acad. Sci. USA* 103:11033–11038.

211. van Oosten, M., T. Schäfer, J. A. C. Gazendam, K. Ohlsen, E. Tsompanidou, M. C. de Goffau, H. J. M. Harmsen, L. M. A. Crane, E. Lim, K. P. Francis, L. Cheung, M. Olive, V. Ntziachristos, J. M. van Dijl, and G. M. van Dam. 2013. Real-time in vivo imaging of invasive- and biomaterial-associated bacterial infections using fluorescently labelled vancomycin. *Nat. Commun.* 4:2584. doi:10.1038/ncomms3584.

212. van Oosten, M., M. Hahn, L. M. A. Crane, R. G. Pleijhuis, K. P. Francis, J. M. van Dijl, and G. M. van Dam. 2015. Targeted imaging of bacterial infections: advances, hurdles and hopes. *FEMS Microbiol. Rev.* 39:892–916. doi:10.1093/femsre/fuv029.

213. Bernstein, H. D. 2011. The double life of a bacterial lipoprotein. *Mol. Microbiol.* 79:1128–1131. doi:10.1111/j.1365-2958.2011.07538.x.

214. Cowles, C. E., Y. Li, M. F. Semmelhack, I. M. Cristea, and T. J. Silhavy. 2011. The free and bound forms of Lpp occupy distinct subcellular locations in *Escherichia coli*. *Mol. Microbiol.* 79:1168–1181. doi:10.1111/j.1365-2958.2011.07539.x.

215. Asmar, A. T., and J. F. Collet. 2018. Lpp, the Braun lipoprotein, turns 50 - Major achievements and remaining issues. *FEMS Microbiol. Lett.* 365:fny199. doi:10.1093/femsle/fny199.

216. Braun, V. 2018. The outer membrane took center stage. *Annu. Rev. Microbiol.* 72:1–24. doi:10.1146/annurev-micro-090817-062156.

217. Shu, W., J. Liu, H. Ji, and M. Lu. 2000. Core structure of the outer membrane lipoprotein from *Escherichia coli* at 1.9 Å resolution. *J. Mol. Biol.* 299:1101–1112.

218. Liu, J., W. Cao, and M. Lu. 2002. Core side-chain packing and backbone conformation in Lpp-56 coiled-coil mutants. *J. Mol. Biol.* 318:877–888.

219. Magnet, S., S. Bellais, L. Dubost, M. Fourgeaud, J. L. Mainardi, S. Petit-Frère, A. Marie, D. Mengin-Lecreulx, M. Arthur, and L. Gutmann. 2007. Identification of the L,D-transpeptidases responsible for attachment of the Braun lipoprotein to *Escherichia coli* peptidoglycan. *J. Bacteriol.* 189:3927–3931. doi:10.1128/JB.00084-07.

220. Ni, Y., J. Reye, and R. R. Chen. 2007. *Lpp* deletion as a permeabilization method. *Biotechnol. Bioeng.* 97:1347–1356. doi:10.1002/bit.21375.

221. Sanders, A. N., and M. S. Pavelka. 2013. Phenotypic analysis of *Escherichia coli* mutants lacking L,D-transpeptidases. *Microbiology.* 159:1842–1852. doi:10.1099/mic.0.069211-0.

222. Asmar, A. T., J. L. Ferreira, E. J. Cohen, S. H. Cho, M. Beeby, K. T. Hughes, and J. F. Collet. 2017. Communication across the bacterial cell envelope depends on the size of the periplasm. *PLoS Biol.* 15:e2004303. doi:10.1371/journal.pbio.2004303.

223. Cohen, E. J., J. L. Ferreira, M. S. Ladinsky, M. Beeby, and K. T. Hughes. 2017. Nanoscale-length control of the flagellar driveshaft requires hitting the tethered outer membrane. *Science.* 356:197–200. doi:10.1126/science.aam6512.

224. Chang, T. W., Y. M. Lin, C. F. Wang, and Y. D. Liao. 2012. Outer membrane lipoprotein Lpp is Gram-negative bacterial cell surface receptor for cationic antimicrobial peptides. *J. Biol. Chem.* 287:418–428. doi:10.1074/jbc.M111.290361.

225. Ebbensgaard, A., H. Mordhorst, F. M. Aarestrup, and E. B. Hansen. 2018. The role of outer membrane proteins and lipopolysaccharides for the sensitivity of *Escherichia coli* to antimicrobial peptides. *Front. Microbiol.* 9:2153. doi:10.3389/fmicb.2018.02153.

226. Reusch, R. N. 2012. Insights into the structure and assembly of *Escherichia coli* outer membrane protein A. *FEBS J.* 279:894–909. doi:10.1111/j.1742-4658.2012.08484.x.

227. Tan, K., B. L. Deatherage Kaiser, R. Wu, M. Cuff, Y. Fan, L. Bigelow, R. P. Jedrzejczak, J. N. Adkins, J. R. Cort, G. Babnigg, and A. Joachimiak. 2017. Insights into PG-binding, conformational change, and dimerization of the OmpA C-terminal domains from *Salmonella enterica* serovar typhimurium and *Borrelia burgdorferi*. *Protein Sci.* 26:1738–1748. doi:10.1002/pro.3209.

228. Mushtaq, A. U., J. S. Park, S. H. Bae, H. Y. Kim, K. J. Yeo, E. Hwang, K. Y. Lee, J. G. Jee, H. K. Cheong, and Y. H. Jeon. 2017. Ligand-mediated folding of the OmpA periplasmic domain from *Acinetobacter baumannii*. *Biophys. J.* 112:2089–2098. doi:10.1016/j.bpj.2017.04.015.

229. Park, J. S., W. C. Lee, K. J. Yeo, K. S. Ryu, M. Kumarasiri, D. Hesek, M. Lee, S. Mobashery, J. H. Song, S. I. Kim, J. C. Lee, C. Cheong, Y. H. Jeon, and H. Y. Kim. 2012. Mechanism of anchoring of OmpA protein to the cell wall peptidoglycan of the Gram-negative bacterial outer membrane. *FASEB J.* 26:219–228. doi:10.1096/fj.11-188425.

230. Samsudin, F., M. L. Ortiz-Suarez, T. J. Piggot, P. J. Bond, and S. Khalid. 2016. OmpA: a flexible clamp for bacterial cell wall attachment. *Structure.* 24:2227–2235. doi:10.1016/j.str.2016.10.009.

231. Wojdyla, J. A., E. Cutts, R. Kaminska, G. Papadakos, J. T. Hopper, P. J. Stansfeld, D. Staunton, C. V. Robinson, and C. Kleanthous. 2015. Structure and function of the *Escherichia coli* Tol-Pal stator protein TolR. *J. Biol. Chem.* 290:26675–26687. doi:10.1074/jbc.M115.671586.

232. Boags, A. T., F. Samsudin, and S. Khalid. 2019. Binding from both sides: TolR and full-length OmpA bind and maintain the local structure of the *E.coli* cell wall. *Structure.* 27:713–724. doi:10.1016/j.str.2019.01.001.

233. Samsudin, F., A. Boags, T. J. Piggot, and S. Khalid. 2017. Braun's lipoprotein facilitates OmpA interaction with the *Escherichia coli* cell wall. *Biophys. J.* 113:1496–1504. doi:10.1016/j.bpj.2017.08.011.

234. Khalid, S., T. J. Piggot, and F. Samsudin. 2019. Atomistic and coarse grain simulations of the cell envelope of Gram-negative bacteria: what have we learned? *Acc. Chem. Res.* 52:180–188. doi:10.1021/acs.accounts.8b00377.

235. Skerniskyte, J., E. Karazijaitė, J. Deschamps, R. Krasauskas, R. Briandet, and E. Sužiedėlienė. 2019. The mutation of conservative Asp268 residue in the peptidoglycan-associated domain of the OmpA protein affects multiple *Acinetobacter baumannii* virulence characteristics. *Molecules.* 24:1972. doi:10.3390/molecules24101972.

236. Verhoeven, G. S., M. Dogterom, and T. den Blaauwen. 2013. Absence of long-range diffusion of OmpA in *E. coli* is not caused by its peptidoglycan binding domain. *BMC Microbiol.* 13:66. doi:10.1186/1471-2180-13-66.

237. Bosshart, P. D., I. Iordanov, C. Garzon-Coral, P. Demange, A. Engel, A. Milon, and D. J. Müller. 2012. The transmembrane protein *Kp*OmpA anchoring the outer membrane of *Klebsiella pneumoniae* unfolds and refolds in response to tensile load. *Structure.* 20:121–127. doi:10.1016/j.str.2011.11.002.

238. Godlewska, R., K. Wisniewska, Z. Pietras, and E. K. Jagusztyn-Krynicka. 2009. Peptidoglycan-associated lipoprotein (Pal) of Gram-negative bacteria: function, structure, role in pathogenesis and potential application in immunoprophylaxis. *FEMS Microbiol. Lett.* 298:1–11. doi:10.1111/j.1574-6968.2009.01659.x.

239. Egan, A. J., and W. Vollmer. 2013. The physiology of bacterial cell division. *Ann. N. Y. Acad. Sci.* 1277:8–28. doi:10.1111/j.1749-6632.2012.06818.x.

240. Zhang, X. Y., E. L. Goemaere, R. Thome, M. Gavioli, E. Cascales, and R. Lloubes. 2009. Mapping the interactions between *Escherichia coli* tol subunits: rotation of the TolR transmembrane helix. *J. Biol. Chem.* 284:4275–4282. doi:10.1074/jbc.M805257200.

241. Clavel, T., P. Germon, A. Vianney, R. Portalier, and J. C. Lazzaroni. 1998. TolB protein of *Escherichia coli* K-12 interacts with the outer membrane peptidoglycan-associated proteins Pal, Lpp and OmpA. *Mol. Microbiol.* 29:359–367.

242. Cascales, E., A. Bernadac, M. Gavioli, J. C. Lazzaroni, and R. Lloubes. 2002. Pal lipoprotein of *Escherichia coli* plays a major role in outer membrane integrity. *J. Bacteriol.* 184:754–759.

243. Gerding, M. A., Y. Ogata, N. D. Pecora, H. Niki, and P. A. de Boer. 2007. The trans-envelope Tol-Pal complex is part of the cell division machinery and required for proper outer-membrane invagination during cell constriction in *E. coli. Mol. Microbiol.* 63:1008–1025. doi:10.1111/j.1365-2958.2006.05571.x.

244. Goemaere, E. L., A. Devert, R. Lloubes, and E. Cascales. 2007. Movements of the TolR C-terminal domain depend on TolQR ionizable key residues and regulate activity of the Tol complex. *J. Biol. Chem.* 282:17749–17757. doi:10.1074/jbc.M701002200.

245. Yeh, Y. C., L. R. Comolli, K. H. Downing, L. Shapiro, and H. H. McAdams. 2010. The *Caulobacter* Tol-Pal complex is essential for outer membrane integrity and the positioning of a polar localization factor. *J. Bacteriol.* 192:4847–4858. doi:10.1128/JB.00607-10.

246. Petiti, M., B. Serrano, L. Faure, R. Lloubes, T. Mignot, and D. Duche. 2019. Tol energy-driven localization of *Pal* and anchoring to the peptidoglycan promote outer membrane constriction. *J. Mol. Biol.* 431:3275–3288. doi:10.1016/j.jmb.2019.05.039.

247. Pommier, S., M. Gavioli, E. Cascales, and R. Lloubes. 2005. Tol-dependent macromolecule import through the *Escherichia coli* cell envelope requires the presence of an exposed TolA binding motif. *J. Bacteriol.* 187:7526–7534. doi:10.1128/JB.187.21.7526-7534.2005.

248. Anwari, K., S. Poggio, A. Perry, X. Gatsos, S. H. Ramarathinam, N. A. Williamson, N. Noinaj, S. Buchanan, K. Gabriel, A. W. Purcell, C. Jacobs-Wagner, and T. Lithgow. 2010. A modular BAM complex in the outer membrane of the α-proteobacterium *Caulobacter crescentus*. *PLOS One.* 5:e8619. doi:10.1371/journal.pone.0008619.

249. Saaki, T. N. V., H. Strahl, and L. W. Hamoen. 2018. Membrane curvature and the Tol-Pal complex determine polar localization of the chemoreceptor Tar in *E. coli. J. Bacteriol.* 200:e00658–17. doi:10.1128/JB.00658-17.

250. Kowata, H., S. Tochigi, T. Kusano, and S. Kojima. 2016. Quantitative measurement of the outer membrane permeability in *Escherichia coli* lpp and *Tol-Pal* mutants defines the significance of Tol-Pal function for maintaining drug resistance. *J. Antibiot.* 69:863–870. doi:10.1038/ja.2016.50.

251. Bonsor, D. A., O. Hecht, M. Vankemmelbeke, A. Sharma, A. M. Krachler, N. G. Housden, K. J. Lilly, R. James, G. R. Moore, and C. Kleanthous. 2009. Allosteric β-propeller signalling in TolB and its manipulation by translocating colicins. *EMBO J.* 28:2846–2857. doi:10.1038/emboj.2009.224.

252. Li, C., Y. Zhang, M. Vankemmelbeke, O. Hecht, F. S. Aleanizy, C. Macdonald, G. R. Moore, R. James, and C. N. Penfold. 2012. Structural evidence that colicin A protein binds to a novel binding site of TolA protein in *Escherichia coli* periplasm. *J. Biol. Chem.* 287:19048–19057. doi:10.1074/jbc.M112.342246.

253. Rassam, P., K. R. Long, R. Kaminska, D. J. Williams, G. Papadakos, C. G. Baumann, and C. Kleanthous. 2018. Intermembrane crosstalk drives inner-membrane protein organization in *Escherichia coli. Nat. Commun.* 9:1082. doi:10.1038/s41467-018-03521-4.

254. Hirakawa, H., K. Suzue, K. Kurabayashi, and H. Tomita. 2019. The Tol-Pal system of uropathogenic *Escherichia coli* is responsible for optimal internalization into and aggregation within bladder epithelial cells, colonization of the urinary tract of mice, and bacterial motility. *Front. Microbiol.* 10:1827. doi:10.3389/fmicb.2019.01827.

255. Parsons, L. M., F. Lin, and J. Orban. 2006. Peptidoglycan recognition by Pal, an outer membrane lipoprotein. *Biochemistry.* 45:2122–2128. doi:10.1021/bi052227i.

256. Scheurwater, E. M., and L. L. Burrows. 2011. Maintaining network security: how macromolecular structures cross the peptidoglycan layer. *FEMS Microbiol. Lett.* 318:1–9. doi:10.1111/j.1574-6968.2011.02228.x.

257. Hosking, E. R., C. Vogt, E. P. Bakker, and M. D. Manson. 2006. The *Escherichia coli* MotAB proton channel unplugged. *J. Mol. Biol.* 364:921–937. doi:10.1016/j.jmb.2006.09.035.

258. Kojima, S., K. Imada, M. Sakuma, Y. Sudo, C. Kojima, T. Minamino, M. Homma, and K. Namba. 2009. Stator assembly and activation mechanism of the flagellar motor by the periplasmic region of MotB. *Mol. Microbiol.* 73:710–718. doi:10.1111/j.1365-2958.2009.06802.x.

259. Minamino, T., and K. Imada. 2015. The bacterial flagellar motor and its structural diversity. *Trends Microbiol.* 23:267–274. doi:http://dx.doi.org/10.1016/j.tim.2014.12.011.

260. Kojima, S., A. Shinohara, H. Terashima, T. Yakushi, M. Sakuma, M. Homma, K. Namba, and K. Imada. 2008. Insights into the stator assembly of the *Vibrio* flagellar motor from the crystal structure of MotY. *Proc. Natl. Acad. Sci. USA.* 105:7696–7701. doi:10.1073/pnas.0800308105.

261. Roujeinikova, A. 2008. Crystal structure of the cell wall anchor domain of MotB, a stator component of the bacterial flagellar motor: implications for peptidoglycan recognition. *Proc. Natl. Acad. Sci. USA.* 105:10348–10353. doi:10.1073/pnas.0803039105.

262. Reboul, C. F., D. A. Andrews, M. F. Nahar, A. M. Buckle, and A. Roujeinikova. 2011. Crystallographic and molecular dynamics analysis of loop motions unmasking the peptidoglycan-binding site in stator protein MotB of flagellar motor. *PLOS One.* 6:e18981. doi:10.1371/journal.pone.0018981.

263. Yonekura, K., S. Maki-Yonekura, and M. Homma. 2011. Structure of the flagellar motor protein complex PomAB: implications for the torque-generating conformation. *J. Bacteriol.* 193:3863–3870. doi:10.1128/JB.05021-11.

264. Andrews, D. A., Y. E. Nesmelov, M. C. Wilce, and A. Roujeinikova. 2017. Structural analysis of variant of *Helicobacter pylori* MotB in its activated form, engineered as chimera of MotB and leucine zipper. *Sci. Rep.* 7:13435. doi:10.1038/s41598-017-13421-0.

265. Liew, C. W., R. M. Hynson, L. A. Ganuelas, N. Shah-Mohammadi, A. P. Duff, S. Kojima, M. Homma, and L. K. Lee. 2018. Solution structure analysis of the periplasmic region of bacterial flagellar motor stators by small angle X-ray scattering. *Biochem. Biophys. Res. Commun.* 495:1614–1619. doi:10.1016/j.bbrc.2017.11.194.

266. Kojima, S., M. Takao, G. Almira, I. Kawahara, M. Sakuma, M. Homma, C. Kojima, and K. Imada. 2018. The helix rearrangement in the periplasmic domain of the flagellar stator B subunit activates peptidoglycan binding and ion influx. *Structure.* 26:590–598. doi:10.1016/j.str.2018.02.016.

267. Ishida, T., R. Ito, J. Clark, N. J. Matzke, Y. Sowa, and M. A. Baker. 2019. Sodium-powered stators of the bacterial flagellar motor can generate torque in the presence of phenamil with mutations near the peptidoglycan-binding region. *Mol. Microbiol.* 111:1689–1699. doi:10.1111/mmi.14246.

268. Hizukuri, Y., J. F. Morton, T. Yakushi, S. Kojima, and M. Homma. 2009. The peptidoglycan-binding (PGB) domain of the *Escherichia coli* pal protein can also function as the PGB domain in *E. coli* flagellar motor protein MotB. *J. Biochem.* 146:219–229. doi:10.1093/jb/mvp061.

269. Chou, S., N. K. Bui, A. B. Russell, K. W. Lexa, T. E. Gardiner, M. LeRoux, W. Vollmer, and J. D. Mougous. 2012. Structure of a peptidoglycan amidase effector targeted to Gram-negative bacteria by the type VI secretion system. *Cell Rep.* 1:656–664. doi:10.1016/j.celrep.2012.05.016.

270. Yang, L. C., Y. L. Gan, L. Y. Yang, B. L. Jiang, and J. L. Tang. 2017. Peptidoglycan hydrolysis mediated by the amidase AmiC and its LytM activator NlpD are critical for cell separation and virulence in the phytopathogen *Xanthomonas campestris. Mol. Plant Pathol.* 19:1705–1718. doi:10.1111/mpp.12653.

271. Bobrovskyy, M., S. E. Willing, O. Schneewind, and D. Missiakas. 2018. EssH peptidoglycan hydrolase enables *Staphylococcus aureus* type VII secretion across the bacterial cell wall envelope. *J. Bacteriol.* 200:e00268–18. doi:10.1128/JB.00268-18.

272. Burkinshaw, B. J., W. Deng, E. Lameignère, G. A. Wasney, H. Zhu, L. J. Worrall, B. B. Finlay, and N. C. Strynadka. 2015. Structural analysis of a specialized tpe III secretion system peptidoglycan-cleaving enzyme. *J. Biol. Chem.* 290:10406–10417. doi:10.1074/jbc.M115.639013.

273. Santin, Y. G., and E. Cascales. 2017. Domestication of a housekeeping transglycosylase for assembly of a type VI secretion system. *EMBO Rep.* 18:138–149. doi:10.15252/embr.201643206.

274. Wang, L., L. Y. Yang, Y. L. Gan, F. Yang, X. L. Liang, W. L. Li, and B. L. Jiang. 2019. Two lytic transglycosylases of *Xanthomonas campestris* pv. campestris associated with cell separation and type III secretion system respectively. *FEMS Microbiol. Lett.* 366:fnz073. doi:10.1093/femsle/fnz073.

275. Bateman, A., and M. Bycroft. 2000. The structure of a LysM domain from *E. coli* membrane-bound lytic murein transglycosylase D (MltD). *J. Mol. Biol.* 299:1113–1119.

276. Buist, G., A. Steen, J. Kok, and O. P. Kuipers. 2008. LysM, a widely distributed protein motif for binding to (peptido) glycans. *Mol. Microbiol.* 68:838–847. doi:10.1111/j.1365-2958.2008.06211.x.

277. Mesnage, S., M. Dellarole, N. J. Baxter, J. B. Rouget, J. D. Dimitrov, N. Wang, Y. Fujimoto, A. M. Hounslow, S. Lacroix-Desmazes, K. Fukase, S. J. Foster, and M. P. Williamson. 2014. Molecular basis for bacterial peptidoglycan recognition by LysM domains. *Nat. Commun.* 5:4269. doi:10.1038/ncomms5269.

278. Nakimbugwe, D., B. Masschalck, D. Deckers, L. Callewaert, A. Aertsen, and C. W. Michiels. 2006. Cell wall substrate specificity of six different lysozymes and lysozyme inhibitory activity of bacterial extracts. *FEMS Microbiol. Lett.* 259:41–46.

279. Callewaert, L., J. M. Van Herreweghe, L. Vanderkelen, S. Leysen, A. Voet, and C. W. Michiels. 2012. Guards of the great wall: bacterial lysozyme inhibitors. *Trends Microbiol.* 20:501–510. doi:10.1016/j.tim.2012.06.005.

280. Sharma, P., D. Dube, M. Sinha, S. Yadav, P. Kaur, S. Sharma, and T. P. Singh. 2013. Structural insights into the dual strategy of recognition by peptidoglycan recognition protein, PGRP-S: structure of the ternary complex of PGRP-S with lipopolysaccharide and stearic acid. *PLoS One.* 8:e53756. doi:10.1371/journal.pone.0053756.

281. Royet, J., and R. Dziarski. 2007. Peptidoglycan recognition proteins: pleiotropic sensors and effectors of antimicrobial defences. *Nat. Rev. Microbiol.* 5:264–277. doi:10.1038/nrmicro1620.

282. Dziarski, R., and D. Gupta. 2018. A balancing act: PGRPs preserve and protect. *Cell Host Microbe.* 23:149–151. doi:10.1016/j.chom.2018.01.010.

283. Wang, Q., M. Ren, X. Liu, H. Xia, and K. Chen. 2018. Peptidoglycan recognition proteins in insect immunity. *Mol. Immunol.* 106:69–76. doi:10.1016/j.molimm.2018.12.021.

284. Charroux, B., F. Capo, C. L. Kurz, S. Peslier, D. Chaduli, A. Viallat-Lieutaud, and J. Royet. 2018. Cytosolic and secreted peptidoglycan-degrading enzymes in *Drosophila* respectively control local and systemic immune responses to microbiota. *Cell Host Microbe.* 23:215–228. doi:10.1016/j.chom.2017.12.007.

285. van Heijenoort, J. 2007. Lipid intermediates in the biosynthesis of bacterial peptidoglycan. *Microbiol. Mol. Biol. Rev.* 71:620–635. doi:10.1128/MMBR.00016-07.

286. Woldemeskel, S. A., and E. D. Goley. 2017. Shapeshifting to survive: shape determination and regulation in *Caulobacter crescentus*. *Trends Microbiol.* 25:673–687. doi:10.1016/j.tim.2017.03.006.

287. Yahashiri, A., M. A. Jorgenson, and D. S. Weiss. 2017. The SPOR domain: a widely-conserved peptidoglycan binding domain that targets proteins to the site of cell division. *J. Bacteriol.* 199:e00118–17. doi:10.1128/JB.00118-17.

288. Caveney, N. A., F. K. K. Li, and N. C. Strynadka. 2018. Enzyme structures of the bacterial peptidoglycan and wall teichoic acid biogenesis pathways. *Curr. Opin. Struct. Biol.* 53:45–58. doi:10.1016/j.sbi.2018.05.002.

289. Borisova, M., J. Gisin, and C. Mayer. 2017. The *N*-acetylmuramic acid 6-phosphate phosphatase MupP completes the *Pseudomonas* peptidoglycan recycling pathway leading to intrinsic fosfomycin resistance. *mBio.* 8:e00092–17. doi:10.1128/mBio.00092-17.

290. Fumeaux, C., and T. G. Bernhardt. 2017. Identification of MupP as a new peptidoglycan recycling factor and antibiotic resistance determinant in *Pseudomonas aeruginosa*. *mBio.* 8:e00102–17. doi:10.1128/mBio.00102-17.

291. Smith, C. A. 2006. Structure, function and dynamics in the *mur* family of bacterial cell wall ligases. *J. Mol. Biol.* 362:640–655.

292. Kouidmi, I., R. C. Levesque, and C. Paradis-Bleau. 2014. The biology of Mur ligases as an antibacterial target. *Mol. Microbiol.* 94:242–253. doi:10.1111/mmi.12758.

293. Jukic, M., S. Gobec, and M. Sova. 2019. Reaching toward underexplored targets in antibacterial drug design. *Drug Dev. Res.* 80:6–10. doi:10.1002/ddr.21465.

294. Hartley, M. D., and B. Imperiali. 2012. At the membrane frontier: a prospectus on the remarkable evolutionary conservation of polyprenols and polyprenyl-phosphates. *Arch. Biochem. Biophys.* 517:83–97. doi:10.1016/j.abb.2011.10.018.

295. Liu, Y., and E. Breukink. 2016. The membrane steps of bacterial cell wall synthesis as antibiotic targets. *Antibiotics.* 5:28. doi:10.3390/antibiotics5030028.

296. Bouhss, A., A. E. Trunkfield, T. D. H. Bugg, and D. Mengin-Lecreulx. 2008. The biosynthesis of peptidoglycan lipid-linked intermediates. *FEMS Microbiol. Rev.* 32:208–233. doi:10.1111/j.1574-6976.2007.00089.x.

297. Koppermann, S., and C. Ducho. 2016. Natural products at work: structural insights into inhibition of the bacterial membrane protein MraY. *Angew. Chem. Int. Ed.* 55:11722–11724. doi:10.1002/anie.201606396.

298. Huang, C. Y., H. W. Shih, L. Y. Lin, Y. W. Tien, T. J. Cheng, W. C. Cheng, C. H. Wong, and C. Ma. 2012. Crystal structure of *Staphylococcus aureus* transglycosylase in complex with a lipid II analog and elucidation of peptidoglycan synthesis mechanism. *Proc. Natl. Acad. Sci. USA.* 109:6496–6501. doi:10.1073/pnas.1203900109.

299. Scheffers, D. J., and M. B. Tol. 2015. Lipid II: just another brick in the wall. *PLOS Pathog.* 11:e1005213. doi:10.1371/journal.ppat.1005213.

300. Laddomada, F., M. M. Miyachiro, M. Jessop, D. Patin, V. Job, D. Mengin-Lecreulx, A. Le Roy, C. Ebel, C. Breyton, I. Gutsche, and A. Dessen. 2019. The MurG glycosyltransferase provides an oligomeric scaffold for the cytoplasmic steps of peptidoglycan biosynthesis in the human pathogen *Bordetella pertussis*. *Sci. Rep.* 9:4656. doi:10.1038/s41598-019-40966-z.

301. Miyachiro, M. M., D. Granato, D. M. Trindade, C. Ebel, A. F. Paes Leme, and A. Dessen. 2019. Complex formation between Mur enzymes from *Streptococcus pneumoniae*. *Biochemistry.* 58:3314–3324. doi:10.1021/acs.biochem.9b00277.

302. Rubino, F. A., S. Kumar, N. Ruiz, S. Walker, and D. Kahne. 2018. Membrane potential is required for MurJ function. *J. Am. Chem. Soc.* 140:4481–4484. doi:10.1021/jacs.8b00942.

303. Sauvage, E., F. Kerff, M. Terrak, J. A. Ayala, and P. Charlier. 2008. The penicillin-binding proteins: structure and role in peptidoglycan biosynthesis. *FEMS Microbiol. Rev.* 32:234–258. doi:10.1111/j.1574-6976.2008.00105.x.

304. Pazos, M., K. Peters, and W. Vollmer. 2017. Robust peptidoglycan growth by dynamic and variable multi-protein complexes. *Curr. Opin. Microbiol.* 36:55–61. doi:10.1016/j.mib.2017.01.006.

305. Banzhaf, M., B. van den Berg van Saparoea, M. Terrak, C. Fraipont, A. Egan, J. Philippe, A. Zapun, E. Breukink, M. Nguyen-Disteche, T. den Blaauwen, and W. Vollmer. 2012. Cooperativity of peptidoglycan synthases active in bacterial cell elongation. *Mol. Microbiol.* 85:179–194. doi:10.1111/j.1365-2958.2012.08103.x.

306. Potluri, L., A. Karczmarek, J. Verheul, A. Piette, J. M. Wilkin, N. Werth, M. Banzhaf, W. Vollmer, K. D. Young, M. Nguyen-Disteche, and T. den Blaauwen. 2010. Septal and lateral wall localization of PBP5, the major D,D-carboxypeptidase of *Escherichia coli*, requires substrate recognition and membrane attachment. *Mol. Microbiol.* 77:300–323. doi:10.1111/j.1365-2958.2010.07205.x.

307. Potluri, L. P., M. A. de Pedro, and K. D. Young. 2012. *Escherichia coli* low-molecular-weight penicillin-binding proteins help orient septal FtsZ, and their absence leads to asymmetric cell division and branching. *Mol. Microbiol.* 84:203–224. doi:10.1111/j.1365-2958.2012.08023.x.

308. Wells, V. L., and W. Margolin. 2012. A new slant to the Z ring and bacterial cell branch formation. *Mol. Microbiol.* 84:199–202. doi:10.1111/j.1365-2958.2012.08029.x.

309. Sarkar, S. K., M. Dutta, C. Chowdhury, A. Kumar, and A. S. Ghosh. 2011. PBP5, PBP6 and DacD play different roles in intrinsic β-lactam resistance of *Escherichia coli*. *Microbiology.* 157:2702–2709. doi:10.1099/mic.0.046227-0.

310. Smith, J. D., M. Kumarasiri, W. Zhang, D. Hesek, M. Lee, M. Toth, S. Vakulenko, J. F. Fisher, S. Mobashery, and Y. Chen. 2013. Structural analysis of the role of *Pseudomonas aeruginosa* penicillin-binding protein 5 in β-lactam resistance. *Antimicrob. Agents Chemother.* 57:3137–3146. doi:10.1128/AAC.00505-13.

311. Dutta, M., D. Kar, A. Bansal, S. Chakraborty, and A. S. Ghosh. 2015. A single amino acid substitution in the Ω-like loop of *E. coli* PBP5 disrupts its ability to maintain cell shape and intrinsic β-lactam resistance. *Microbiology.* 161:895–902. doi:10.1099/mic.0.000052.

312. Schoenemann, K. M., and W. Margolin. 2017. Bacterial division: FtsZ treadmills to build a beautiful wall. *Curr. Biol.* 27:R301–R303. doi:10.1016/j.cub.2017.03.019.

313. Meeske, A. J., E. P. Riley, W. P. Robins, T. Uehara, J. J. Mekalanos, D. Kahne, S. Walker, A. C. Kruse, T. G. Bernhardt, and D. Z. Rudner. 2016. SEDS proteins are a widespread family of bacterial cell wall polymerases. *Nature.* 537:634–638. doi:10.1038/nature19331.

314. Leclercq, S., A. Derouaux, S. Olatunji, C. Fraipont, A. J. Egan, W. Vollmer, E. Breukink, and M. Terrak. 2017. Interplay between penicillin-binding proteins and SEDS proteins promotes bacterial cell wall synthesis. *Sci. Rep.* 7:43306. doi:10.1038/srep43306.

315. Ruiz, N. 2008. Bioinformatics identification of MurJ (MviN) as the peptidoglycan lipid II flippase in *Escherichia coli*. *Proc. Natl. Acad. Sci. USA.* 105:15553–15557. doi:10.1073/pnas.0808352105.

316. Sham, L. T., E. K. Butler, M. D. Lebar, D. Kahne, T. G. Bernhardt, and N. Ruiz. 2014. Bacterial cell wall. MurJ is the flippase of lipid-linked precursors for peptidoglycan biogenesis. *Science.* 345:220–222. doi:10.1126/science.1254522.

317. Young, K. D. 2014. Microbiology. A flipping cell wall ferry. *Science.* 345:139–140. doi:10.1126/science.1256585.

318. Kuk, A. C. Y., E. H. Mashalidis, and S. Y. Lee. 2017. Crystal structure of the MOP flippase MurJ in an inward-facing conformation. *Nat. Struct. Mol. Biol.* 24:171–176. doi:10.1038/nsmb.3346.

319. Zheng, S., L. T. Sham, F. A. Rubino, K. P. Brock, W. P. Robins, J. J. Mekalanos, D. S. Marks, T. G. Bernhardt, and A. C. Kruse. 2018. Structure and mutagenic analysis of the lipid II flippase MurJ from *Escherichia coli*. *Proc. Natl. Acad. Sci. USA.* 115:6709–6714. doi:10.1073/pnas.1802192115.

320. Kumar, S., F. A. Rubino, A. G. Mendoza, and N. Ruiz. 2019. The bacterial lipid II flippase MurJ functions by an alternating-access mechanism. *J. Biol. Chem.* 294:981–990. doi:10.1074/jbc.RA118.006099.

321. Kuk, A. C. Y., A. Hao, Z. Guan, and S. Y. Lee. 2019. Visualizing conformation transitions of the lipid II flippase MurJ. *Nat. Commun.* 10:1736. doi:10.1038/s41467-019-09658-0.

322. Wagstaff, J., and J. Löwe. 2018. Prokaryotic cytoskeletons: protein filaments organizing small cells. *Nature Rev. Microbiol.* 16:187.

323. Errington, J. 2015. Bacterial morphogenesis and the enigmatic MreB helix. *Nat. Rev. Microbiol.* 13:241–248. doi:10.1038/nrmicro3398.

324. de Boer, P. A. J. 2016. Classic spotlight: staying in shape and discovery of the *mrdAB* and *mreBCD* operons. *J. Bacteriol.* 198:1479. doi:10.1128/JB.00180-16.

325. Bratton, B. P., J. W. Shaevitz, Z. Gitai, and R. M. Morgenstein. 2018. MreB polymers and curvature localization are enhanced by RodZ and predict *E. coli*'s cylindrical uniformity. *Nat. Commun.* 9:2797. doi:10.1038/s41467-018-05186-5.

326. Hussain, S., C. N. Wivagg, P. Szwedziak, F. Wong, K. Schaefer, T. Izoré, L. D. Renner, M. J. Holmes, Y. Sun, A. W. Bisson-Filho, S. Walker, A. Amir, J. Löwe, and E. C. Garner. 2018. MreB filaments align along greatest principal membrane curvature to orient cell wall synthesis. *eLife.* 7:e32471. doi:10.7554/eLife.32471.

327. Wong, F., E. C. Garner, and A. Amir. 2019. Mechanics and dynamics of translocating MreB filaments on curved membranes. *eLife.* 8:e40472. doi:10.7554/eLife.40472.

328. Kawazura, T., K. Matsumoto, K. Kojima, F. Kato, T. Kanai, H. Niki, and D. Shiomi. 2017. Exclusion of assembled MreB by anionic phospholipids at cell poles confers cell polarity for bidirectional growth. *Mol. Microbiol.* 104:472–486. doi:10.1111/mmi.13639.

329. Shiomi, D. 2017. Polar localization of MreB actin is inhibited by anionic phospholipids in the rod-shaped bacterium *Escherichia coli*. *Curr. Gen.* 63:845–848. doi:10.1007/s00294-017-0696-5.

330. Sjodt, M., K. Brock, G. Dobihal, P. D. A. Rohs, A. G. Green, T. A. Hopf, A. J. Meeske, V. Srisuknimit, D. Kahne, S. Walker, D. S. Marks, T. G. Bernhardt, D. Z. Rudner, and A. C. Kruse. 2018. Structure of the peptidoglycan polymerase RodA resolved by evolutionary coupling analysis. *Nature.* 556: 118–121. doi:10.1038/nature25985.

331. Contreras-Martel, C., A. Martins, C. Ecobichon, D. M. Trindade, P. J. Matteï, S. Hicham, P. Hardouin, M. E. Ghachi, I. G. Boneca, and A. Dessen. 2017. Molecular architecture of the PBP2-MreC core bacterial cell wall synthesis complex. *Nat. Commun.* 8:776. doi:10.1038/s41467-017-00783-2.

332. Rohs, P. D. A., J. Buss, S. I. Sim, G. R. Squyres, V. Srisuknimit, M. Smith, H. Cho, M. Sjodt, A. C. Kruse, E. C. Garner, S. Walker, D. E. Kahne, and T. G. Bernhardt. 2018. A central role for PBP2 in the activation of peptidoglycan polymerization by the bacterial cell elongation machinery. *PLoS Genet.* 14:e1007726. doi:10.1371/journal.pgen.1007726.

333. White, C. L., A. Kitich, and J. W. Gober. 2010. Positioning cell wall synthetic complexes by the bacterial morphogenetic proteins MreB and MreD. *Mol. Microbiol.* 76:616–633. doi:10.1111/j.1365-2958.2010.07108.x.

334. Welsh, M. A., K. Schaefer, A. Taguchi, D. Kahne, and S. Walker. 2019. Direction of chain growth and substrate preferences of shape, elongation, division, and sporulation-family peptidoglycan glycosyltransferases. *J. Am. Chem. Soc.* 141:12994–12997. doi:10.1021/jacs.9b06358.

335. Garner, E. C., R. Bernard, W. Wang, X. Zhuang, D. Z. Rudner, and T. Mitchison. 2011. Coupled, circumferential motions of the cell wall synthesis machinery and MreB filaments in *B. subtilis*. *Science.* 333:222–225. doi:10.1126/science.1203285.

336. Domínguez-Escobar, J., A. Chastanet, A. H. Crevenna, V. Fromion, R. Wedlich-Söldner, and R. Carballido-López. 2011. Processive movement of MreB-associated cell wall biosynthetic complexes in bacteria. *Science.* 333:225–228. doi:10.1126/science.1203466.

337. Colavin, A., H. Shi, and K. C. Huang. 2018. RodZ modulates geometric localization of the bacterial actin MreB to regulate cell shape. *Nat. Commun.* 9:1280. doi:10.1038/s41467-018-03633-x.

338. Egan, A. J. F., and W. Vollmer. 2015. The stoichiometric divisome: a hypothesis. *Front. Microbiol.* 6:455.

339. den Blaauwen, T., L. W. Hamoen, and P. A. Levin. 2017. The divisome at 25: the road ahead. *Curr. Opin. Microbiol.* 36: 85–94. doi:10.1016/j.mib.2017.01.007.

340. Emami, K., A. Guyet, Y. Kawai, J. Devi, L. J. Wu, N. Allenby, R. A. Daniel, and J. Errington. 2017. RodA as the missing glycosyltransferase in *Bacillus subtilis* and antibiotic discovery for the peptidoglycan polymerase pathway. *Nat. Microbiol.* 2:16253. doi:10.1038/nmicrobiol.2016.253.

341. Rismondo, J., S. Halbedel, and A. Gründling. 2019. Cell shape and antibiotic resistance are maintained by the activity of multiple FtsW and RodA enzymes in *Listeria monocytogenes*. *mBio.* 10:e01448–19. doi:10.1128/mBio.01448-19.

342. Gamba, P., J. W. Veening, N. J. Saunders, L. W. Hamoen, and R. A. Daniel. 2009. Two-step assembly dynamics of the *Bacillus subtilis* divisome. *J. Bacteriol.* 191:4186–4194. doi:10.1128/JB.01758-08.

343. Lewis, R. J. 2017. The GpsB files: the truth is out there. *Mol. Microbiol.* 103:913–918. doi:10.1111/mmi.13612.

344. Eswara, P. J., R. S. Brzozowski, M. G. Viola, G. Graham, C. Spanoudis, C. Trebino, J. Jha, J. I. Aubee, K. M. Thompson, J. L. Camberg, and K. S. Ramamurthi. 2018. An essential *Staphylococcus aureus* cell division protein directly regulates FtsZ dynamics. *eLife.* 7:e38856. doi:10.7554/eLife.38856.

345. Cleverley, R. M., Z. J. Rutter, J. Rismondo, F. Corona, H. T. Tsui, F. A. Alatawi, R. A. Daniel, S. Halbedel, O. Massidda, M. E. Winkler, and R. J. Lewis. 2019. The cell cycle regulator GpsB functions as cytosolic adaptor for multiple cell wall enzymes. *Nat. Commun.* 10:261. doi:10.1038/s41467-018-08056-2.

346. Halbedel, S., and R. J. Lewis. 2019. Structural basis for interaction of DivIVA/GpsB proteins with their ligands. *Mol. Microbiol.* 111:1404–1415. doi:10.1111/mmi.14244.

347. Hammond, L. R., M. L. White, and P. J. Eswara. 2019. ¡vIVA la DivIVA. *J. Bacteriol.* 201:e00245–19. doi:10.1128/JB.00245-19.

348. Szwedziak, P., Q. Wang, T. A. M. Bharat, M. Tsim, and J. Löwe. 2014. Architecture of the ring formed by the tubulin homologue FtsZ in bacterial cell division. *eLife.* 3:e04601. doi:10.7554/eLife.04601.

349. Taguchi, A., M. A. Welsh, L. S. Marmont, W. Lee, M. Sjodt, A. C. Kruse, D. Kahne, T. G. Bernhardt, and S. Walker. 2019. FtsW is a peptidoglycan polymerase that is functional only in complex with its cognate penicillin-binding protein. *Nat. Microbiol.* 4:587–594. doi:10.1038/s41564-018-0345-x.

350. Söderström, B., K. Mirzadeh, S. Toddo, G. von Heijne, U. Skoglund, and D. O. Daley. 2016. Coordinated disassembly of the divisome complex in *Escherichia coli.* *Mol. Microbiol.* 101:425–438. doi:10.1111/mmi.13400.

351. Söderström, B., H. Chan, and D. O. Daley. 2019. Super-resolution images of peptidoglycan remodelling enzymes at the division site of *Escherichia coli.* *Curr. Genet.* 65:99–101. doi:10.1007/s00294-018-0869-x.

352. Söderström, B., K. Skoog, H. Blom, D. S. Weiss, G. von Heijne, and D. O. Daley. 2014. Disassembly of the divisome in *Escherichia coli*: evidence that FtsZ dissociates before compartmentalization. *Mol. Microbiol.* 92:1–9. doi:10.1111/mmi.12534.

353. Gray, A. N., A. J. F. Egan, I. L. van't Veer, J. Verheul, A. Colavin, A. Koumoutsi, J. Biboy, A. F. M. Altelaar, M. J. Damen, K. C. Huang, J. P. Simorre, E. Breukink, T. den Blaauwen, A. Typas, C. A. Gross, and W. Vollmer. 2015. Coordination of peptidoglycan synthesis and outer membrane constriction during *Escherichia coli* cell division. *eLife.* 4:e07118. doi:10.7554/eLife.07118.

354. Lee, T. K., K. Meng, H. Shi, and K. C. Huang. 2016. Single-molecule imaging reveals modulation of cell wall synthesis dynamics in live bacterial cells. *Nat. Commun.* 7:13170. doi:10.1038/ncomms13170.

355. Cesar, S., and K. C. Huang. 2017. Thinking big: the tunability of bacterial cell size. *FEMS Microbiol. Rev.* 41:672–678. doi:10.1093/femsre/fux026.

356. van der Ploeg, R., J. Verheul, N. O. E. Vischer, S. Alexeeva, E. Hoogendoorn, M. Postma, M. Banzhaf, W. Vollmer, and T. den Blaauwen. 2013. Colocalization and interaction between elongasome and divisome during a preparative cell division phase in *Escherichia coli.* *Mol. Microbiol.* 87:1074–1087. doi:10.1111/mmi.12150.

357. Yoshii, Y., H. Niki, and D. Shiomi. 2019. Division-site localization of RodZ is required for efficient Z ring formation in *Escherichia coli.* *Mol. Microbiol.* 111:1229–1244. doi:10.1111/mmi.14217.

358. Potluri, L. P., S. Kannan, and K. D. Young. 2012. ZipA is required for FtsZ-dependent preseptal peptidoglycan synthesis prior to invagination during cell division. *J. Bacteriol.* 194:5334–5342. doi:10.1128/JB.00859-12.

359. Pazos, M., K. Peters, M. Casanova, P. Palacios, M. VanNieuwenhze, E. Breukink, M. Vicente, and W. Vollmer. 2018. Z-ring membrane anchors associate with cell wall synthases to initiate bacterial cell division. *Nat. Commun.* 9:5090. doi:10.1038/s41467-018-07559-2.

360. Chung, H. S., Z. Yao, N. W. Goehring, R. Kishony, J. Beckwith, and D. Kahne. 2009. Rapid β-lactam-induced lysis requires successful assembly of the cell division machinery. *Proc. Natl. Acad. Sci. USA.* 106:21872–21877. doi:10.1073/pnas.0911674106.

361. Wong, F., and A. Amir. 2019. Mechanics and dynamics of bacterial cell lysis. *Biophys. J.* 116:2378–2389. doi:10.1016/j.bpj.2019.04.040.

362. Templin, M. F., D. H. Edwards, and J. V. Holtje. 1992. A murein hydrolase is the specific target of bulgecin in *Escherichia coli.* *J. Biol. Chem.* 267:20039–20043

363. Dik, D. A., C. S. Madukoma, S. Tomoshige, C. Kim, E. Lastochkin, W. C. Boggess, J. F. Fisher, J. D. Shrout, and S. Mobashery. 2019. Slt, MltD and MltG of *Pseudomonas aeruginosa* as targets of bulgecin A in potentiation of β-lactam antibiotics. *ACS Chem. Biol.* 14:296–303. doi:10.1021/acschembio.8b01025.

364. Cho, H., T. Uehara, and T. G. Bernhardt. 2014. β-Lactam antibiotics induce a lethal malfunctioning of the bacterial cell wall synthesis machinery. *Cell.* 159:1300–1311. doi:10.1016/j.cell.2014.11.017.

365. Jensen, C., K. T. Bæk, C. Gallay, I. Thalsø-Madsen, L. Xu, A. Jousselin, F. Ruiz Torrubia, W. Paulander, A. R. Pereira, J. W. Veening, M. G. Pinho, and D. Frees. 2019. The ClpX chaperone controls autolytic splitting of *Staphylococcus aureus* daughter cells, but is bypassed by β-lactam antibiotics or inhibitors of WTA biosynthesis. *PLOS Pathog.* 15:e1008044. doi:10.1371/journal.ppat.1008044.

366. Flores-Kim, J., G. S. Dobihal, A. Fenton, D. Z. Runder, and T. G. Bernhardt. 2019. A switch in surface polymer biogenesis triggers growth-phase-dependent and antibiotic-induced bacteriolysis. *eLife.* 8:e44912. doi:10.7554/eLife.44912.

367. Terrak, M., T. K. Ghosh, J. van Heijenoort, J. Van Beeumen, M. Lampilas, J. Aszodi, J. A. Ayala, J. M. Ghuysen, and M. Nguyen-Disteche. 1999. The catalytic, glycosyl transferase and acyl transferase modules of the cell wall peptidoglycan-polymerizing penicillin-binding protein 1b of *Escherichia coli.* *Mol. Microbiol.* 34:350–364.

368. King, D. T., G. A. Wasney, M. Nosella, A. Fong, and N. C. J. Strynadka. 2017. Structural insights into inhibition of *Escherichia coli* penicillin-binding protein 1B. *J. Biol. Chem.* 292:979–993. doi:10.1074/jbc.M116.718403.

369. Zijderveld, C. A., M. E. Aarsman, and N. Nanninga. 1995. Differences between inner membrane and peptidoglycan-associated PBP1B dimers of *Escherichia coli.* *J. Bacteriol.* 177:1860–1863.

370. Bertsche, U., E. Breukink, T. Kast, and W. Vollmer. 2005. In vitro murein peptidoglycan synthesis by dimers of the bifunctional transglycosylase-transpeptidase PBP1B from *Escherichia coli. J. Biol. Chem.* 280:38096–38101.

371. Boes, A., S. Olatunji, E. Breukink, and M. Terrak. 2019. Regulation of the peptidoglycan polymerase activity of PBP1b by antagonist actions of the core divisome proteins FtsBLQ and FtsN. *mBio.* 10:e01912–18. doi:10.1128/mBio.01912-18.

372. den Blaauwen, T., and J. Luirink. 2019. Checks and balances in bacterial cell division. *mBio.* 10:e00149–19. doi:10.1128/mBio.00149-19.

373. Fraipont, C., S. Alexeeva, B. Wolf, R. van der Ploeg, M. Schloesser, T. den Blaauwen, and M. Nguyen-Disteche. 2011. The integral membrane FtsW protein and peptidoglycan synthase PBP3 form a subcomplex in *Escherichia coli. Microbiology.* 157:251–259. doi:10.1099/mic.0.040071-0.

374. Vollmer, W., M. von Rechenberg, and J. V. Holtje. 1999. Demonstration of molecular interactions between the murein polymerase PBP1B, the lytic transglycosylase MltA, and the scaffolding protein MipA of *Escherichia coli. J. Biol. Chem.* 274:6726–6734.

375. von Rechenberg, M., A. Ursinus, and J. V. Holtje. 1996. Affinity chromatography as a means to study multienzyme complexes involved in murein synthesis. *Microb. Drug Resist.* 2:155–157.

376. Paradis-Bleau, C., M. Markovski, T. Uehara, T. J. Lupoli, S. Walker, D. E. Kahne, and T. G. Bernhardt. 2010. Lipoprotein cofactors located in the outer membrane activate bacterial cell wall polymerases. *Cell.* 143:1110–1120. doi:10.1016/j.cell.2010.11.037.

377. Typas, A., M. Banzhaf, B. van den Berg van Saparoea, J. Verheul, J. Biboy, R. J. Nichols, M. Zietek, K. Beilharz, K. Kannenberg, M. von Rechenberg, E. Breukink, T. den Blaauwen, C. A. Gross, and W. Vollmer. 2010. Regulation of peptidoglycan synthesis by outer-membrane proteins. *Cell.* 143:1097–1109. doi:10.1016/j.cell.2010.11.038.

378. Young, K. D. 2010. New ways to make old walls: bacterial surprises. *Cell.* 143:1042–1044. doi:10.1016/j.cell.2010.12.011.

379. Egan, A. J. F., R. Maya-Martinez, I. Ayala, C. M. Bougault, M. Banzhaf, E. Breukink, W. Vollmer, and J. P. Simorre. 2018. Induced conformational changes activate the peptidoglycan synthase PBP1B. *Mol. Microbiol.* 110:335–356. doi:10.1111/mmi.14082.

380. Thulin, E., and D. I. Andersson. 2019. Upregulation of PBP1B and LpoB in *cysB* mutants confers mecillinam resistance in *Escherichia coli. Antimicrob. Agents Chemother.* 63:e00612–19. doi:10.1128/AAC.00612-19.

381. Ranjit, D. K., and K. D. Young. 2013. The Rcs stress response and accessory envelope proteins are required for de novo generation of cell shape in *Escherichia coli. J. Bacteriol.* 195:2452–2462. doi:10.1128/JB.00160-13.

382. Weiss, D. S. 2013. *Escherichia coli* shapeshifters. *J. Bacteriol.* 195:2449–2451. doi:10.1128/JB.00306-13.

383. Ranjit, D. K., M. A. Jorgenson, and K. D. Young. 2017. PBP1B glycosyltransferase and transpeptidase activities play different essential roles during the de novo regeneration of rod-shaped morphology in *Escherichia coli. J. Bacteriol.* 199:e00612–16. doi:10.1128/JB.00612-16.

384. Meisel, U., J. V. Holtje, and W. Vollmer. 2003. Overproduction of inactive variants of the murein synthase PBP1B causes lysis in *Escherichia coli. J. Bacteriol.* 185:5342–5348.

385. Jorgenson, M. A., W. J. MacCain, B. M. Meberg, S. Kannan, J. C. Bryant, and K. D. Young. 2019. Simultaneously inhibiting undecaprenyl phosphate production and peptidoglycan synthases promotes rapid lysis in *Escherichia coli. Mol. Microbiol.* 112:233–248. doi:10.1111/mmi.14265.

386. Sarkar, S. K., M. Dutta, A. Kumar, D. Mallik, and A. S. Ghosh. 2012. Sub-Inhibitory cefsulodin sensitization of *E. coli* to β-lactams is mediated by PBP1b inhibition. *PLOS One.* 7:e48598. doi:10.1371/journal.pone.0048598.

387. Kumar, A., S. K. Sarkar, D. Ghosh, and A. S. Ghosh. 2012. Deletion of penicillin-binding protein 1b impairs biofilm formation and motility in *Escherichia coli. Res. Microbiol.* 163:254–257. doi:10.1016/j.resmic.2012.01.006.

388. Pepper, E. D., M. J. Farrell, and S. E. Finkel. 2006. Role of penicillin-binding protein 1b in competitive stationary-phase survival of *Escherichia coli. FEMS Microbiol. Lett.* 263:61–67.

389. Cabeen, M. T., M. A. Murolo, A. Briegel, N. K. Bui, W. Vollmer, N. Ausmees, G. J. Jensen, and C. Jacobs-Wagner. 2010. Mutations in the lipopolysaccharide biosynthesis pathway interfere with crescentin-mediated cell curvature in *Caulobacter crescentus. J. Bacteriol.* 192:3368–3378. doi:10.1128/JB.01371-09.

390. Ting, C. S., C. Hsieh, S. Sundararaman, C. Mannella, and M. Marko. 2007. Cryo-electron tomography reveals the comparative three-dimensional architecture of *Prochlorococcus*, a globally important marine cyanobacterium. *J. Bacteriol.* 189:4485–4493.

391. Izard, J., C. Renken, C. E. Hsieh, D. C. Desrosiers, S. Dunham-Ems, C. La Vake, L. L. Gebhardt, R. J. Limberger, D. L. Cox, M. Marko, and J. D. Radolf. 2009. Cryo-electron tomography elucidates the molecular architecture of *Treponema pallidum*, the syphilis spirochete. *J. Bacteriol.* 191:7566–7580. doi:10.1128/JB.01031-09.

392. Rojas, E. R., G. Billings, P. D. Odermatt, G. K. Auer, L. Zhu, A. Miguel, F. Chang, D. B. Weibel, J. A. Theriot, and K. C. Huang. 2018. The outer membrane is an essential load-bearing element in Gram-negative bacteria. *Nature.* 559:617–621. doi:10.1038/s41586-018-0344-3.

393. Yao, X., M. Jericho, D. Pink, and T. Beveridge. 1999. Thickness and elasticity of Gram-negative murein sacculi measured by atomic force microscopy. *J. Bacteriol.* 181:6865–6875.

394. Marko, M., C. Hsieh, R. Schalek, J. Frank, and C. Mannella. 2007. Focused-ion-beam thinning of frozen-hydrated biological specimens for cryo-electron microscopy. *Nat. Meth.* 4:215–217.

395. Zhang, P., C. M. Khursigara, L. M. Hartnell, and S. Subramaniam. 2007. Direct visualization of *Escherichia coli* chemotaxis receptor arrays using cryo-electron microscopy. *Proc. Natl. Acad. Sci. USA.* 104:3777–3781.

396. Ierardi, V., P. Domenichini, S. Reali, G. M. Chiappara, G. Devoto, and U. Valbusa. 2017. *Klebsiella pneumoniae* antibiotic resistance identified by atomic force microscopy. *J. Biosci.* 42:623–636.

397. Sani, M., E. N. Houben, J. Geurtsen, J. Pierson, K. de Punder, M. van Zon, B. Wever, S. R. Piersma, C. R. Jimenez, M. Daffe, B. J. Appelmelk, W. Bitter, N. van der Wel, and P. J. Peters. 2010. Direct visualization by cryo-EM of the mycobacterial capsular layer: a labile structure containing ESX-1-secreted proteins. *PLoS Pathog.* 6:e1000794. doi:10.1371/journal.ppat.1000794.

398. Dufrêne, Y. F. 2014. Atomic force microscopy in microbiology: new structural and functional insights into the microbial cell surface. *mBio.* 5:e01363–14. doi:10.1128/mBio.01363-14.

399. Hayhurst, E. J., L. Kailas, J. K. Hobbs, and S. J. Foster. 2008. Cell wall peptidoglycan architecture in *Bacillus subtilis. Proc. Natl. Acad. Sci. USA.* 105:14603–14608. doi:10.1073/pnas.0804138105.

400. Li, K., X. X. Yuan, H. M. Sun, L. S. Zhao, R. Tang, Z. H. Chen, Q. L. Qin, X. L. Chen, Y. Z. Zhang, and H. N. Su. 2018. Atomic force microscopy of side wall and septa peptidoglycan fom *Bacillus subtilis* reveals an architectural remodeling during growth. *Front. Microbiol.* 9:620. doi:10.3389/fmicb.2018.00620.

401. Angeles, D. M., Y. Liu, A. M. Hartman, M. Borisova, A. de Sousa Borges, N. de Kok, K. Beilharz, J.-W. Veening, C. Mayer, A. K. H. Hirsch, and D. J. Scheffers. 2017. Pentapeptide-rich peptidoglycan at the *Bacillus subtilis* cell-division site. *Mol. Microbiol.* 104:319–333. doi:10.1111/mmi.13629.

402. Tsui, H.-C. T., M. J. Boersma, S. A. Vella, O. Kocaoglu, E. Kuru, J. K. Peceny, E. E. Carlson, M. S. VanNieuwenhze, Y. V. Brun, S. L. Shaw, and M. E. Winkler. 2014. Pbp2x localizes separately from Pbp2b and other peptidoglycan synthesis proteins during later stages of cell division of *Streptococcus pneumoniae* D39. *Mol. Microbiol.* 94:21–40. doi:10.1111/mmi.12745.

403. Jacq, M., C. Arthaud, S. Manuse, C. Mercy, L. Bellard, K. Peters, B. Gallet, J. Galindo, T. Doan, W. Vollmer, Y. V. Brun, M. S. VanNieuwenhze, A. M. Di Guilmi, T. Vernet, C. Grangeasse, and C. Morlot. 2018. The cell wall hydrolase Pmp23 is important for assembly and stability of the division ring in *Streptococcus pneumoniae. Sci. Rep.* 8:7591. doi:10.1038/s41598-018-25882-y.

404. Monteiro, J. M., P. B. Fernandes, F. Vaz, A. R. Pereira, A. C. Tavares, M. T. Ferreira, P. M. Pereira, H. Veiga, E. Kuru, M. S. VanNieuwenhze, Y. V. Brun, S. R. Filipe, and M. G. Pinho. 2015. Cell shape dynamics during the staphylococcal cell cycle. *Nat. Commun.* 6:8055. doi:10.1038/ncomms9055.

405. Pidgeon, S. E., and M. M. Pires. 2017. Cell wall remodeling of *Staphylococcus aureus* in live *Caenorhabditis elegans. Bioconjug. Chem.* 28:2310–2315. doi:10.1021/acs.bioconjchem.7b00363.

406. Auer, G. K., P. M. Oliver, M. Rajendram, T. Y. Lin, Q. Yao, G. J. Jensen, and D. B. Weibel. 2019. Bacterial swarming reduces *Proteus mirabilis* and *Vibrio parahaemolyticus* cell stiffness and increases β-lactam susceptibility. *mBio.* 10:e00210–19. doi:10.1128/mBio.00210-19.

407. Touhami, A., M. H. Jericho, and T. J. Beveridge. 2004. Atomic force microscopy of cell growth and division in *Staphylococcus aureus. J. Bacteriol.* 186:3286–3295.

408. Turner, R. D., A. F. Hurd, A. Cadby, J. K. Hobbs, and S. J. Foster. 2013. Cell wall elongation mode in Gram-negative bacteria is determined by peptidoglycan architecture. *Nat. Commun.* 4:1496. doi:10.1038/ncomms2503.

409. Turner, R. D., S. Mesnage, J. K. Hobbs, and S. J. Foster. 2018. Molecular imaging of glycan chains couples cell-wall polysaccharide architecture to bacterial cell morphology. *Nat. Commun.* 9:1263. doi:10.1038/s41467-018-03551-y.

410. Tulum, I., Y. O. Tahara, and M. Miyata. 2019. Peptidoglycan layer and disruption processes in *Bacillus subtilis* cells visualized using quick-freeze, deep-etch electron microscopy. *Microscopy.* 68:441–449. doi:10.1093/jmicro/dfz033.

411. Romaniuk, J. A. H., and L. Cegelski. 2018. Peptidoglycan and teichoic acid levels and alterations in *S. aureus* by cell-wall and whole-cell NMR. *Biochemistry.* 57:3966–3975. doi:10.1021/acs.biochem.8b00495.

412. Bougault, C., I. Ayala, W. Vollmer, J. P. Simorre, and P. Schanda. 2019. Studying intact bacterial peptidoglycan by proton-detected NMR spectroscopy at 100 kHz MAS frequency. *J. Struct. Biol.* 206:66–72. doi:10.1016/j.jsb.2018.07.009.

413. Young, K. D. 2006. Too many strictures on structure. *Trends Microbiol.* 14:155–156. doi:10.1016/j.tim.2006.02.004.

414. Wientjes, F. B., C. L. Woldringh, and N. Nanninga. 1991. Amount of peptidoglycan in cell walls of Gram-negative bacteria. *J. Bacteriol.* 173:7684–7691.

415. Labischinski, H., E. W. Goodell, A. Goodell, and M. L. Hochberg. 1991. Direct proof of a "more-than-single-layered" peptidoglycan architecture of *Escherichia coli* W7: a neutron small-angle scattering study. *J. Bacteriol.* 173:751–756.

416. Dmitriev, B., F. Toukach, and S. Ehlers. 2005. Towards a comprehensive view of the bacterial cell wall. *Trends Microbiol.* 13:569–574.

417. Kim, S. J., M. Singh, M. Preobrazhenskaya, and J. Schaefer. 2013. *Staphylococcus aureus* peptidoglycan stem packing by rotational-echo double resonance NMR spectroscopy. *Biochemistry.* 52:3651–3659. doi:10.1021/bi4005039.

418. Kim, S. J., M. Singh, S. Sharif, and J. Schaefer. 2014. Cross-link formation and peptidoglycan lattice assembly in the FemA mutant of *Staphylococcus aureus. Biochemistry.* 53:1420–1427. doi:10.1021/bi4016742.

419. Kim, S. J., J. Chang, and M. Singh. 2015. Peptidoglycan architecture of Gram-positive bacteria by solid-state NMR. *Biochim. Biophys. Acta.* 1848:350–362. doi:10.1016/j.bbamem.2014.05.031.

420. Yang, H., M. Singh, S. J. Kim, and J. Schaefer. 2017. Characterization of the tertiary structure of the peptidoglycan of *Enterococcus faecalis. Biochim. Biophys. Acta.* 1859:2171–2180. doi:10.1016/j.bbamem.2017.08.003.

421. Yahashiri, A., M. A. Jorgenson, and D. S. Weiss. 2015. Bacterial SPOR domains are recruited to septal peptidoglycan by binding to glycan strands that lack stem peptides. *Proc. Natl. Acad. Sci. USA.* 112:11347–11352. doi:10.1073/pnas.1508536112.

422. Alcorlo, M., D. A. Dik, S. De Benedetti, K. V. Mahasenan, M. Lee, T. Domínguez-Gil, D. Hesek, E. Lastochkin, D. López, B. Boggess, S. Mobashery, and J. A. Hermoso. 2019. Structural basis of denuded glycan recognition by SPOR domains in bacterial cell division. *Nat. Commun.* 10:5567. doi:10.1038/s41467-019-13354-4.

423. Mitkowski, P., E. Jagielska, E. Nowak, J. M. Bujnicki, F. Stefaniak, D. Niedziałek, M. Bochtler, and I. Sabała. 2019. Structural bases of peptidoglycan recognition by lysostaphin SH3b domain. *Sci. Rep.* 9:5965. doi:10.1038/s41598-019-42435-z.

424. Gonzalez-Delgado, L. S., H. Walters-Morgan, B. Salamaga, A. J. Robertson, A. M. Hounslow, E. Jagielska, I. Sabała, M. P. Williamson, A. L. Lovering, and S. Mesnage. 2020. Two-site recognition of *Staphylococcus aureus* peptidoglycan by lysostaphin SH3b. *Nat. Chem. Biol.* 16:24–30. doi:10.1038/s41589-019-0393-4.

425. Meroueh, S. O., K. Z. Bencze, D. Hesek, M. Lee, J. F. Fisher, T. L. Stemmler, and S. Mobashery. 2006. Three-dimensional structure of the bacterial cell wall peptidoglycan. *Proc. Natl. Acad. Sci. USA.* 103:4404–4409.

426. Cho, S., Q. Wang, C. P. Swaminathan, D. Hesek, M. Lee, G. J. Boons, S. Mobashery, and R. A. Mariuzza. 2007. Structural insights into the bactericidal mechanism of human peptidoglycan recognition proteins. *Proc. Natl. Acad. Sci. USA.* 104:8761–8766. doi:10.1073pnas.0701453104.

427. Nicolaou, K. C., and S. Rigol. 2018. A brief history of antibiotics and select advances in their synthesis. *J. Antibiot.* 71: 153–184. doi:10.1038/ja.2017.62.

428. Durand, G. A., D. Raoult, and G. Dubourg. 2019. Antibiotic discovery: history, methods and perspectives. *Int. J. Antimicrob. Agents.* 53:371–382. doi:10.1016/j.ijantimicag.2018.11.010.

429. Hutchings, M. I., and A. W. Truman. 2019. Antibiotics: past, present and future. *Curr. Opin. Microbiol.* 51:72–80. doi:10.1016/j.mib.2019.10.008.

430. Sugimoto, A., A. Maeda, K. Itto, and H. Arimoto. 2017. Deciphering the mode of action of cell wall-inhibiting antibiotics using metabolic labeling of growing peptidoglycan in *Streptococcus pyogenes. Sci. Rep.* 7:1129. doi:10.1038/s41598-017-01267-5.

431. Falagas, M. E., E. K. Vouloumanou, G. Samonis, and K. Z. Vardakas. 2016. Fosfomycin. *Clin. Microbiol. Rev.* 29: 321–347. doi:10.1128/CMR.00068-15.

432. Dijkmans, A. C., N. V. O. Zacarías, J. Burggraaf, J. W. Mouton, E. B. Wilms, C. van Nieuwkoop, D. J. Touw, J. Stevens, and I. M. C. Kamerling. 2017. Fosfomycin: pharmacological, clinical and future perspectives. *Antibiotics.* 6:24. doi:10.3390/antibiotics6040024.

433. Silver, L. L. 2017. Fosfomycin: mechanism and resistance. *Cold Spring Harb. Perspect. Med.* 7:a025262. doi:10.1101/cshperspect.a025262.

434. Falagas, M. E., F. Athanasaki, G. L. Voulgaris, N. A. Triarides, and K. Z. Vardakas. 2019. Resistance to fosfomycin: mechanisms, frequency and clinical consequences. *Int. J. Antimicrob. Agents.* 53:22–28. doi:10.1016/j.ijantimicag.2018.09.013.

435. Aghamali, M., M. Sedighi, A. Z. Bialvaei, N. Mohammadzadeh, S. Abbasian, Z. Ghafouri, and E. Kouhsari. 2019. Fosfomycin: mechanisms and the increasing prevalence of resistance. *J. Med. Microbiol.* 68:11–15. doi:10.1099/jmm.0.000874.

436. Gil-Marqués, M. L., P. Moreno-Martínez, C. Costas, J. Pachón, J. Blázquez, and M. J. McConnell. 2018. Peptidoglycan recycling contributes to intrinsic resistance to fosfomycin in *Acinetobacter baumannii. J. Antimicrob. Chemother.* 73:2960–2968. doi:10.1093/jac/dky289.

437. Sherry, N., and B. Howden. 2018. Emerging Gram-negative resistance to last-line antimicrobial agents fosfomycin, colistin and ceftazidime-avibactam—epidemiology, laboratory detection and treatment implications. *Expert Rev. Anti-Infect. Ther.* 16:289–306. doi:10.1080/14787210.2018.1453807.

438. Flamm, R. K., P. R. Rhomberg, J. M. Lindley, K. Sweeney, E. J. Ellis-Grosse, and D. Shortridge. 2019. Evaluation of the bactericidal activity of fosfomycin in combination with selected antimicrobial comparison agents tested against Gram-negative bacterial strains by using time-kill curves. *Antimicrob. Agents Chemother.* 63:e02549–18. doi:10.1128/AAC.02549-18.

439. Graves-Woodward, K., and R. F. Pratt. 1999. Interactions of soluble penicillin-binding protein 2a of methicillin-resistant *Staphylococcus aureus* with moenomycin. *Biochemistry.* 38:10533–10542. doi:10.1021/bi982309p.

440. Ostash, B., A. Saghatelian, and S. Walker. 2007. A streamlined metabolic pathway for the biosynthesis of moenomycin A. *Chem. Biol.* 14:257–267.

441. Koyama, N., Y. Tokura, Y. Takahashi, and H. Tomoda. 2013. Discovery of nosokophic acid, a predicted intermediate of moenomycins, from nosokomycin-producing *Streptomyces* sp. K04-0144. *Bioorg. Med. Chem. Lett.* 23:860–863. doi:10.1016/j.bmcl.2012.11.044.

442. Lovering, A. L., L. H. de Castro, D. Lim, and N. C. Strynadka. 2007. Structural insight into the transglycosylation step of bacterial cell-wall biosynthesis. *Science.* 315:1402–1405. doi:10.1126/science.1136611.

443. Punekar, A. S., F. Samsudin, A. J. Lloyd, C. G. Dowson, D. J. Scott, S. Khalid, and D. I. Roper. 2018. The role of the jaw subdomain of peptidoglycan glycosyltransferases for lipid II polymerization. *Cell Surf.* 2:54–66. doi:10.1016/j.tcsw.2018.06.002.

444. Cheng, T. J., M. T. Sung, H. Y. Liao, Y. F. Chang, C. W. Chen, C. Y. Huang, L. Y. Chou, Y. D. Wu, Y. H. Chen, Y. S. Cheng, C. H. Wong, C. Ma, and W. C. Cheng. 2008. Domain requirement of moenomycin binding to bifunctional transglycosylases and development of high-throughput discovery of antibiotics. *Proc. Natl. Acad. Sci. USA.* 105:431–436. doi:10.1073/pnas.0710868105.

445. Huang, S. -H., W. -S. Wu, L. -Y. Huang, W. -F. Huang, W. -C. Fu, P. -T. Chen, J. -M. Fang, W. -C. Cheng, T. -J. R. Cheng, and C. -H. Wong. 2013. New continuous fluorometric assay for bacterial transglycosylase using Forster resonance energy transfer. *J. Am. Chem. Soc.* 135:17078–17089. doi:10.1021/ja407985m.

446. Wu, W. -S., W. -C. Cheng, T. -J. R. Cheng, and C. -H. Wong. 2018. Affinity-based screen for inhibitors of bacterial transglycosylase. *J. Am. Chem. Soc.* 140:2752–2755. doi:10.1021/jacs.7b13205.

447. Rebets, Y., T. Lupoli, Y. Qiao, K. Schirner, R. Villet, D. Hooper, D. Kahne, and S. Walker. 2014. Moenomycin resistance mutations in *Staphylococcus aureus* reduce peptidoglycan chain length and cause aberrant cell division. *ACS Chem. Biol.* 9:459–467. doi:10.1021/cb4006744.

448. Yu, J. Y., H. J. Cheng, H. R. Wu, W. S. Wu, J. W. Lu, T. J. Cheng, Y. T. Wu, and J. M. Fang. 2018. Structure-based design of bacterial transglycosylase inhibitors incorporating biphenyl, amine linker and 2-alkoxy-3-phosphorylpropanoate moieties. *Eur. J. Med. Chem.* 150:729–741. doi:10.1016/j.ejmech.2018.03.034.

449. Zhang, L., T. P. Ko, S. R. Malwal, W. Liu, S. Zhou, X. Yu, E. Oldfield, R. T. Guo, and C. C. Chen. 2019. Complex structures of MoeN5 with substrate analogues suggest sequential catalytic mechanism. *Biochem. Biophys. Res. Commun.* 511: 800–805. doi:10.1016/j.bbrc.2019.02.131.

450. Henson, K. E., M. T. Levine, E. A. Wong, and D. P. Levine. 2015. Glycopeptide antibiotics: evolving resistance, pharmacology and adverse event profile. *Expert Rev. Anti-Infect. Ther.* 13:1265–1278. doi:10.1586/14787210.2015.1068118.

451. McGuinness, W. A., N. Malachowa, and F. R. DeLeo. 2017. Vancomycin resistance in *Staphylococcus aureus. Yale J. Biol. Med.* 90:269–281.

452. Meziane-Cherif, D., P. J. Stogios, E. Evdokimova, A. Savchenko, and P. Courvalin. 2014. Structural basis for the evolution of vancomycin resistance D,D-peptidases. *Proc. Natl. Acad. Sci. USA.* 111:5872–5877. doi:10.1073/pnas.1402259111.

453. Binda, E., P. Cappelletti, F. Marinelli, and G. L. Marcone. 2018. Specificity of induction of glycopeptide antibiotic resistance in the producing *Actinomycetes*. *Antibiotics*. 7:36. doi:10.3390/antibiotics7020036.

454. Boger, D. L. 2017. The difference a single atom can make: synthesis and design at the chemistry-biology interface. *J. Org. Chem.* 82:11961–11980. doi:10.1021/acs.joc.7b02088.

455. Livermore, D. M. 2018. The 2018 garrod lecture: preparing for the black swans of resistance. *J. Antimicrob. Chemother.* 73:2907–2915. doi:10.1093/jac/dky265.

456. Randolph, A. G., R. Xu, T. Novak, M. M. Newhams, J. Bubeck Wardenburg, S. L. Weiss, R. C. Sanders, N. J. Thomas, M. W. Hall, K. M. Tarquinio, N. Cvijanovich, R. G. Gedeit, E. J. Truemper, B. Markovitz, M. E. Hartman, K. G. Ackerman, J. S. Giuliano, S. L. Shein, and K. L. Moffitt. 2019. Vancomycin monotherapy may be insufficient to treat methicillin-resistant *Staphylococcus aureus* coinfection in children with influenza-related critical illness. *Clin. Infect. Dis.* 68:365–372. doi:10.1093/cid/ciy495.

457. Thomsen, I. P. 2019. The concern for vancomycin failure in the treatment of pediatric *Staphylococcus aureus* disease. *Clin. Infect. Dis.* 68:373–374. doi:10.1093/cid/ciy497.

458. Singh, M., J. Chang, L. Coffman, and S. J. Kim. 2017. Hidden mode of action of glycopeptide antibiotics: inhibition of wall teichoic acid biosynthesis. *J. Phys. Chem. B.* 121:3925–3932. doi:10.1021/acs.jpcb.7b00324.

459. Chang, J., L. Coffman, and S. J. Kim. 2017. Inhibition of D-Ala incorporation into wall teichoic acid in *Staphylococcus aureus* by desleucyl-oritavancin. *Chem. Commun.* 53:5649–5652. doi:10.1039/c7cc02635h.

460. Sarkar, P., V. Yarlagadda, C. Ghosh, and J. Haldar. 2017. A review on cell wall synthesis inhibitors with an emphasis on glycopeptide antibiotics. *Med. Chem. Commun.* 8:516–533. doi:10.1039/C6MD00585C.

461. Blaskovich, M. A. T., K. A. Hansford, M. S. Butler, Z. Jia, A. E. Mark, and M. A. Cooper. 2018. Developments in glycopeptide antibiotics. *ACS Infect. Dis.* 4:715–735. doi:10.1021/acsinfecdis.7b00258.

462. Wu, Z. -C., N. A. Isley, and D. L. Boger. 2018. *N*-Terminus alkylation of vancomycin: ligand binding affinity, antimicrobial activity, and site-specific nature of quaternary trimethylammonium salt modification. *ACS Infect. Dis.* 4:1468–1474. doi:10.1021/acsinfecdis.8b00152.

463. Guan, D., F. Chen, L. Xiong, F. Tang, Faridoon, Y. Qiu, N. Zhang, L. Gong, J. Li, L. Lan, and W. Huang. 2018. Extra sugar on vancomycin: new analogues for combating multidrug-resistant *Staphylococcus aureus* and vancomycin-resistant enterococci. *J. Med. Chem.* 61:286–304. doi:10.1021/acs.jmedchem.7b01345.

464. Dhanda, G., P. Sarkar, S. Samaddar, and J. Haldar. 2019. Battle against vancomycin-resistant bacteria: recent developments in chemical strategies. *J. Med. Chem.* 62:3184–3205. doi:10.1021/acs.jmedchem.8b01093.

465. Yarlagadda, V., G. B. Manjunath, P. Sarkar, P. Akkapeddi, K. Paramanandham, B. R. Shome, R. Ravikumar, and J. Haldar. 2016. Glycopeptide antibiotic to overcome the intrinsic resistance of Gram-negative bacteria. *ACS Infect. Dis.* 2:132–139. doi:10.1021/acsinfecdis.5b00114.

466. Yarlagadda, V., P. Sarkar, S. Samaddar, G. B. Manjunath, S. D. Mitra, K. Paramanandham, B. R. Shome, and J. Haldar. 2018. Vancomycin analogue restores meropenem activity against NDM-1 Gram-negative pathogens. *ACS Infect. Dis.* 4:1093–1101. doi:10.1021/acsinfecdis.8b00011.

467. Szucs, Z., I. Bereczki, M. Csávás, E. Rőth, A. Borbás, G. Batta, E. Ostorházi, R. Szatmári, and P. Herczegh. 2017. Lipophilic teicoplanin pseudoaglycon derivatives are active against vancomycin- and teicoplanin-resistant enterococci. *J. Antibiot.* 70:664–670. doi:10.1038/ja.2017.2.

468. Yim, G., W. Wang, A. C. Pawlowski, and G. D. Wright. 2018. Trichlorination of a teicoplanin-type glycopeptide antibiotic by the halogenase StaI evades resistance. *Antimicrob. Agents Chemother.* 62:e01540–18. doi:10.1128/AAC.01540-18.

469. Huang, C. M., S. Y. Lyu, K. H. Lin, C. L. Chen, M. H. Chen, H. W. Shih, N. S. Hsu, I. W. Lo, Y. L. Wang, Y. S. Li, C. J. Wu, and T. L. Li. 2019. Teicoplanin reprogrammed with the *N*-acyl-glucosamine pharmacophore at the penultimate residue of aglycone acquires broad-spectrum antimicrobial activities effectively killing Gram-positive and -negative pathogens. *ACS Infect. Dis.* 5:430–442. doi:10.1021/acsinfecdis.8b00317.

470. Kaasch, A. J., and H. Seifert. 2016. Oritavancin: a long-acting antibacterial lipoglycopeptide. *Future Microbiol.* 11:843–855. doi:10.2217/fmb-2016-0003.

471. Bowden, S., C. Joseph, S. Tang, J. Cannon, E. Francis, M. Zhou, J. R. Baker, and S. K. Choi. 2018. Oritavancin retains a high affinity for a vancomycin-resistant cell-wall precursor via its bivalent motifs of interaction. *Biochemistry.* 57:2723–2732. doi:10.1021/acs.biochem.8b00187.

472. Kingston, A. W., H. Zhao, G. M. Cook, and J. D. Helmann. 2014. Accumulation of heptaprenyl diphosphate sensitizes *Bacillus subtilis* to bacitracin: implications for the mechanism of resistance mediated by the BceAB transporter. *Mol. Microbiol.* 93:37–49. doi:10.1111/mmi.12637.

473. Radeck, J., N. Lautenschläger, and T. Mascher. 2017. The essential UPP phosphatase pair BcrC and UppP connects cell wall homeostasis during growth and sporulation with cell envelope stress response in *Bacillus subtilis*. *Front Microbiol.* 8:2403. doi:10.3389/fmicb.2017.02403.

474. Zhao, J., J. An, D. Hwang, Q. Wu, S. Wang, R. A. Gillespie, E. G. Yang, Z. Guan, P. Zhou, and H. S. Chung. 2019. The lipid A 1-Phosphatase, LpxE, functionally connects multiple layers of bacterial envelope biogenesis. *mBio.* 10:e00886–19. doi:10.1128/mBio.00886-19.

475. Economou, N. J., S. Cocklin, and P. J. Loll. 2013. High-resolution crystal structure reveals molecular details of target recognition by bacitracin. *Proc. Natl. Acad. Sci. USA.* 110:14207–14212. doi:10.1073/pnas.1308268110.

476. Manson, J. M., S. Keis, J. M. Smith, and G. M. Cook. 2004. Acquired bacitracin resistance in *Enterococcus faecalis* is mediated by an ABC transporter and a novel regulatory protein, BcrR. *Antimicrob. Agents Chemother.* 48:3743–3748. doi:10.1128/AAC.48.10.3743-3748.2004.

477. Shaaly, A., F. Kalamorz, S. Gebhard, and G. M. Cook. 2013. Undecaprenyl pyrophosphate phosphatase confers low-level resistance to bacitracin in *Enterococcus faecalis*. *J. Antimicrob. Chemother.* 68:1583–1593. doi:10.1093/jac/dkt048.

478. Han, X., X. D. Du, L. Southey, D. M. Bulach, T. Seemann, X. X. Yan, T. L. Bannam, and J. I. Rood. 2015. Functional analysis of a bacitracin resistance determinant located on ICECp1, a novel Tn916-like element from a conjugative plasmid in *Clostridium perfringens. Antimicrob. Agents Chemother.* 59:6855–6865. doi:10.1128/AAC.01643-15.

479. Piddock, L. J. V. 2015. Teixobactin, the first of a new class of antibiotics discovered by iChip technology. *J. Antimicrob. Chemother.* 70:2679–2680. doi:10.1093/jac/dkv175.

480. Ling, L. L., T. Schneider, A. J. Peoples, A. L. Spoering, I. Engels, B. P. Conlon, A. Mueller, T. F. Schaberle, D. E. Hughes, S. Epstein, M. Jones, L. Lazarides, V. A. Steadman, D. R. Cohen, C. R. Felix, K. A. Fetterman, W. P. Millett, A. G. Nitti, A. M. Zullo, C. Chen, and K. Lewis. 2015. A new antibiotic kills pathogens without detectable resistance. *Nature.* 517:455–459. doi:10.1038/nature14098.

481. Ng, V., S. A. Kuehne, and W. C. Chan. 2018. Rational design and synthesis of modified teixobactin analogues: in vitro antibacterial activity against *Staphylococcus aureus, Propionibacterium acnes* and *Pseudomonas aeruginosa. Chem. Eur. J.* 24:9136–9147. doi:10.1002/chem.201801423.

482. Parmar, A., R. Lakshminarayanan, A. Iyer, V. Mayandi, E. T. Leng Goh, D. G. Lloyd, M. L. S. Chalasani, N. K. Verma, S. H. Prior, R. W. Beuerman, A. Madder, E. J. Taylor, and I. Singh. 2018. Design and syntheses of highly potent teixobactin analogues against *Staphylococcus aureus*, methicillin-resistant *Staphylococcus aureus* (MRSA), and vancomycin-resistant enterococci (VRE) in vitro and in vivo. *J. Med. Chem.* 61:2009–2017. doi:10.1021/acs.jmedchem.7b01634.

483. Homma, T., A. Nuxoll, A. B. Gandt, P. Ebner, I. Engels, T. Schneider, F. Götz, K. Lewis, and B. P. Conlon. 2016. Dual targeting of cell wall precursors by teixobactin leads to cell lysis. *Antimicrob. Agents Chemother.* 60:6510–6517. doi:10.1128/AAC.01050-16.

484. Wen, P. C., J. M. Vanegas, S. B. Rempe, and E. Tajkhorshid. 2018. Probing key elements of teixobactin-lipid II interactions in membranes. *Chem. Sci.* 9:6997–7008. doi:10.1039/c8sc02616e.

485. Ng, V., and W. C. Chan. 2016. New found hope for antibiotic discovery: lipid II inhibitors. *Chem. Eur. J.* 22:12606–12616. doi:10.1002/chem.201601315.

486. Grein, F., T. Schneider, and H. G. Sahl. 2019. Docking on lipid II — a widespread mechanism for potent bactericidal activities of antibiotic peptides. *J. Mol. Biol.* 431:3520–3530. doi:10.1016/j.jmb.2019.05.014.

487. Medeiros-Silva, J., S. Jekhmane, E. Breukink, and M. Weingarth. 2019. Towards the native binding modes of lipid II targeting antibiotics. *ChemBioChem.* 20:1731–1738. doi:10.1002/cbic.201800796.

488. Sutterlin, H. A., J. C. Malinverni, S. H. Lee, C. J. Balibar, and T. Roemer. 2018. Antibacterial new target discovery: sentinel examples, strategies, and surveying success. *Top. Med. Chem.* 25:1–30. doi:10.1007/7355_2016_31.

489. Tan, C. M., A. G. Therien, J. Lu, S. H. Lee, A. Caron, C. J. Gill, C. Lebeau-Jacob, L. Benton-Perdomo, J. M. Monteiro, P. M. Pereira, N. L. Elsen, J. Wu, K. Deschamps, M. Petcu, S. Wong, E. Daigneault, S. Kramer, L. Liang, E. Maxwell, D. Claveau, J. Vaillancourt, K. Skorey, J. Tam, H. Wang, T. C. Meredith, S. Sillaots, L. Wang-Jarantow, Y. Ramtohul, E. Langlois, F. Landry, J. C. Reid, G. Parthasarathy, S. Sharma, A. Baryshnikova, K. J. Lumb, M. G. Pinho, S. M. Soisson, and T. Roemer. 2012. Restoring methicillin-resistant *Staphylococcus aureus* susceptibility to β-lactam antibiotics. *Sci. Transl. Med.* 4:126ra35. doi:10.1126/scitranslmed.3003592.

490. Buss, J. A., V. Baidin, M. A. Welsh, J. Flores-Kim, H. Cho, B. M. Wood, T. Uehara, S. Walker, D. Kahne, and T. G. Bernhardt. 2019. A pathway-directed screen for inhibitors of the bacterial cell elongation machinery. *Antimicrob. Agents Chemother.* 63:e01530–18. doi:10.1128/AAC.01530-18.

491. Lui, H. K., W. Gao, K. C. Cheung, W. B. Jin, N. Sun, J. W. Y. Kan, I. L. K. Wong, J. Chiou, D. Lin, E. W. C. Chan, Y. C. Leung, T. H. Chan, S. Chen, K. F. Chan, and K. Y. Wong. 2018. Boosting the efficacy of anti-MRSA β-lactam antibiotics via an easily accessible, non-cytotoxic and orally bioavailable FtsZ inhibitor. *Eur. J. Med. Chem.* 163:95–115. doi:10.1016/j.ejmech.2018.11.052.

492. Lee, S. H., H. Wang, M. Labroli, S. Koseoglu, P. Zuck, T. Mayhood, C. Gill, P. Mann, X. Sher, S. Ha, S. W. Yang, M. Mandal, C. Yang, L. Liang, Z. Tan, P. Tawa, Y. Hou, R. Kuvelkar, K. DeVito, X. Wen, J. Xiao, C. Batchlett, C. J. Balibar, J. Liu, J. Xiao, N. Murgolo, C. G. Garlisi, P. R. Sheth, A. Flattery, J. Su, C. Tan, and T. Roemer. 2016. TarO-specific inhibitors of wall teichoic acid biosynthesis restore β-lactam efficacy against methicillin-resistant staphylococci. *Sci. Transl. Med.* 8:329ra32. doi:10.1126/scitranslmed.aad7364.

493. Foster, T. J. 2019. Can β-lactam antibiotics be resurrected to combat MRSA? *Trends Microbiol.* 27:26–38. doi:10.1016/j.tim.2018.06.005.

494. Coupri, D., A. Budin-Verneuil, A. Hartke, A. Benachour, L. Léger, T. Lequeux, E. Pfund, and N. Verneuil. 2019. Genetic and pharmacological inactivation of D-alanylation of teichoic acids sensitizes pathogenic enterococci to β-lactams. *J. Antimicrob. Chemother.* 74:3162–3169. doi:10.1093/jac/dkz322.

495. Mann, P. A., A. Muller, L. Xiao, P. M. Pereira, C. Yang, S. H. Lee, H. Wang, J. Trzeciak, J. Schneeweis, M. M. Dos Santos, N. Murgolo, X. She, C. Gill, C. J. Balibar, M. Labroli, J. Su, A. Flattery, B. Sherborne, R. Maier, C. M. Tan, T. Black, K. Onder, S. Kargman, F. J. Monsma, M. G. Pinho, T. Schneider, and T. Roemer. 2013. Murgocil is a highly bioactive staphylococcal-specific inhibitor of the peptidoglycan glycosyltransferase enzyme MurG. *ACS Chem. Biol.* 8:2442–2451. doi:10.1021/cb400487f.

496. Melander, R. J., and C. Melander. 2018. Antibiotic adjuvants. *Top. Med. Chem.* 25:89–118. doi:10.1007/7355_2017_10.

497. Tyers, M., and G. D. Wright. 2019. Drug combinations: a strategy to extend the life of antibiotics in the 21st century. *Nat. Rev. Microbiol.* 17:141–155. doi:10.1038/s41579-018-0141-x.

498. Douafer, H., V. Andrieu, O. Phanstiel, and J. M. Brunel. 2019. Antibiotic adjuvants: make antibiotics great again! *J. Med. Chem.* 62:8665–8880. doi:10.1021/acs.jmedchem.8b01781.

499. Mulani, M. S., E. E. Kamble, S. N. Kumkar, M. S. Tawre, and K. R. Pardesi. 2019. Emerging strategies to combat ESKAPE pathogens in the era of antimicrobial resistance: a review. *Front. Microbiol.* 10:539. doi:10.3389/fmicb.2019.00539.

500. Delauné, A., O. Poupel, A. Mallet, Y. M. Coic, T. Msadek, and S. Dubrac. 2011. Peptidoglycan crosslinking relaxation plays an important role in *Staphylococcus aureus* WalKR-dependent cell viability. *PLoS One.* 6:e17054. doi:10.1371/journal.pone.0017054.

501. Goodell, E. W. 1985. Recycling of murein by *Escherichia coli. J. Bacteriol.* 163:305–310.

502. Park, J. T., and T. Uehara. 2008. How bacteria consume their own exoskeletons (turnover and recycling of cell wall peptidoglycan). *Microbiol. Mol. Biol. Rev.* 72:211–227. doi:10.1128/MMBR.00027-07.

503. Vollmer, W. 2012. Bacterial growth does require peptidoglycan hydrolases. *Mol. Microbiol.* 86:1031–1035. doi:10.1111/mmi.12059.

504. Vermassen, A., S. Leroy, R. Talon, C. Provot, M. Popowska, and M. Desvaux. 2019. Cell wall hydrolases in bacteria: insight on the Diversity of cell wall amidases, glycosidases and peptidases toward peptidoglycan. *Front. Microbiol.* 10:331. doi:10.3389/fmicb.2019.00331.

505. Weiss, D. S. 2004. Bacterial cell division and the septal ring. *Mol. Microbiol.* 54:588–597. doi:10.1111/j.1365-2958.2004.04283.x.

506. Singh, S. K., L. SaiSree, R. N. Amrutha, and M. Reddy. 2012. Three redundant murein endopeptidases catalyze an essential cleavage step in peptidoglycan synthesis of *Escherichia coli* K12. *Mol. Microbiol.* 86:1036–1051. doi:10.1111/mmi.12058.

507. Parveen, S., and M. Reddy. 2017. Identification of YfiH (PgeF) as a factor contributing to the maintenance of bacterial peptidoglycan composition. *Mol. Microbiol.* 105:705–720. doi:10.1111/mmi.13730.

508. Cherrak, Y., N. Flaugnatti, E. Durand, L. Journet, and E. Cascales. 2019. Structure and activity of the type VI secretion system. *Microbiol. Spectr.* 7. doi:10.1128/microbiolspec.PSIB-0031-2019.

509. Setlow, P., S. Wang, and Y. Q. Li. 2017. Germination of spores of the orders *Bacillales* and *Clostridiales*. *Annu. Rev. Microbiol.* 71:459–477. doi:10.1146/annurev-micro-090816-093558.

510. Dembek, M., A. Kelly, A. Barwinska-Sendra, E. Tarrant, W. A. Stanley, D. Vollmer, J. Biboy, J. Gray, W. Vollmer, and P. S. Salgado. 2018. Peptidoglycan degradation machinery in *Clostridium difficile* forespore engulfment. *Mol. Microbiol.* 110:390–410. doi:10.1111/mmi.14091.

511. Srikhanta, Y. N., M. L. Hutton, M. M. Awad, N. Drinkwater, J. Singleton, S. L. Day, B. A. Cunningham, S. McGowan, and D. Lyras. 2019. Cephamycins inhibit pathogen sporulation and effectively treat recurrent *Clostridioides difficile* infection. *Nat. Microbiol.* 4:2237–2245. doi:10.1038/s41564-019-0519-1.

512. Joers, A., K. Vind, S. B. Hernández, R. Maruste, M. Pereira, A. Brauer, M. Remm, F. Cava, and T. Tenson. 2019. Muropeptides stimulate growth resumption from stationary phase in *Escherichia coli*. *Sci. Rep.* 9:18043. doi:10.1038/s41598-019-54646-5.

513. Uehara, T., K. Suefuji, N. Valbuena, B. Meehan, M. Donegan, and J. T. Park. 2005. Recycling of the anhydro-N-acetylmuramic acid derived from cell wall murein involves a two-step conversion to N-acetylglucosamine-phosphate. *J. Bacteriol.* 187:3643–3649.

514. Reith, J., and C. Mayer. 2011. Peptidoglycan turnover and recycling in Gram-positive bacteria. *Appl. Microbiol. Biotechnol.* 92:1–11. doi:10.1007/s00253-011-3486-x.

515. Borisova, M., R. Gaupp, A. Duckworth, A. Schneider, D. Dalügge, M. Mühleck, D. Deubel, S. Unsleber, W. Yu, G. Muth, M. Bischoff, F. Götz, and C. Mayer. 2016. Peptidoglycan recycling in Gram-positive bacteria is crucial for survival in stationary phase. *mBio.* 7:e00923-16. doi:10.1128/mBio.00923-16.

516. Mayer, C., R. M. Kluj, M. Mühleck, A. Walter, S. Unsleber, I. Hottmann, and M. Borisova. 2019. Bacteria's different ways to recycle their own cell wall. *Int. J. Med. Microbiol.* 309:151326. doi:10.1016/j.ijmm.2019.06.006.

517. Lindquist, S., F. Lindberg, and S. Normark. 1989. Binding of the *Citrobacter freundii* AmpR regulator to a single DNA site provides both autoregulation and activation of the inducible *ampC* β-lactamase gene. *J. Bacteriol.* 171:3746–3753.

518. Balcewich, M. D., T. M. Reeve, E. A. Orlikow, L. J. Donald, D. J. Vocadlo, and B. L. Mark. 2010. Crystal structure of the AmpR effector binding domain provides insight into the molecular regulation of inducible *ampC* β-lactamase. *J. Mol. Biol.* 400:998–1010. doi:10.1016/j.jmb.2010.05.040.

519. Dik, D. A., T. Domínguez-Gil, M. Lee, D. Hesek, B. Byun, J. Fishovitz, B. Boggess, L. M. Hellman, J. F. Fisher, J. A. Hermoso, and S. Mobashery. 2017. Muropeptide binding and the X-ray structure of the effector domain of the transcriptional regulator AmpR of *Pseudomonas aeruginosa*. *J. Am. Chem. Soc.* 137:1448–1451. doi:10.1021/jacs.6b12819.

520. Rodríguez-Baño, J., B. Gutiérrez-Gutiérrez, I. Machuca, and A. Pascual. 2018. Treatment of infections caused by extended-spectrum-β-lactamase-, AmpC-, and carbapenemase-producing *Enterobacteriaceae*. *Clin. Microbiol. Rev.* 31:e00079-17. doi:10.1128/CMR.00079-17.

521. Gutierrez-Gutiérrez, B., and J. Rodríguez-Baño. 2019. Current options for the treatment of infections due to extended-spectrum β-lactamase-producing Enterobacteriaceae in different groups of patients. *Clin. Microbiol. Infect.* 25:932–942. doi:10.1016/j.cmi.2019.03.030.

522. Kohlmann, R., T. Bähr, and S. G. Gatermann. 2019. Effect of *ampC* derepression on cefepime MIC in *Enterobacterales* with chromosomally encoded inducible AmpC β-lactamase. *Clin. Microbiol. Infect.* 25:1158. doi:10.1016/j.cmi.2019.05.007.

523. Tamma, P. D., Y. Doi, R. A. Bonomo, J. K. Johnson, and P. J. Simner. 2019. A primer on AmpC β-lactamases: necessary knowledge for an increasingly multidrug-resistant world. *Clin. Infect. Dis.* 69:1446–1455. doi:10.1093/cid/ciz173.

524. Pérez-Gallego, M., G. Torrens, J. Castillo-Vera, B. Moya, L. Zamorano, G. Cabot, K. Hultenby, S. Albertí, P. Mellroth, B. Henriques-Normark, S. Normark, A. Oliver, and C. Juan. 2016. Impact of AmpC derepression on fitness and virulence: the mechanism or the pathway? *mBio.* 7:e01783-16. doi:10.1128/mBio.01783-16.

525. Torrens, G., I. Sánchez-Diener, E. Jordana-Lluch, I. M. Barceló, L. Zamorano, C. Juan, and A. Oliver. 2019. In vivo validation of peptidoglycan recycling as target to disable AmpC-mediated resistance and reduce virulence enhancing the cell-wall-targeting immunity. *J. Infect. Dis.* 220:1729–1737. doi:10.1093/infdis/jiz377.

526. Mayer, C. 2019. Peptidoglycan recycling, a promising target for antibiotic adjuvants in antipseudomonal therapy. *J. Infect. Dis.* 220:1713–1714. doi:10.1093/infdis/jiz378.

527. Dhar, S., H. Kumari, D. Balasubramanian, and K. Mathee. 2018. Cell-wall recycling and synthesis in *Escherichia coli* and *Pseudomonas aeruginosa* - their role in the development of resistance. *J. Med. Microbiol.* 67:1–21. doi:10.1099/jmm.0.000636.

528. Dik, D. A., J. F. Fisher, and S. Mobashery. 2018. Cell-wall recycling of the Gram-negative bacteria and the nexus to antibiotic resistance. *Chem. Rev.* 118:5952–5984. doi:10.1021/acs.chemrev.8b00277.

529. Horcajada, J. P., M. Montero, A. Oliver, L. Sorlí, S. Luque, S. Gómez-Zorrilla, N. Benito, and S. Grau. 2019. Epidemiology and treatment of multidrug-resistant and extensively drug-resistant *Pseudomonas aeruginosa* infections. *Clin. Microbiol. Rev.* 32:e00031-19. doi:10.1128/CMR.00031-19.

530. Moyá, B., A. Beceiro, G. Cabot, C. Juan, L. Zamorano, S. Alberti, and A. Oliver. 2012. Pan-β-lactam resistance development in *Pseudomonas aeruginosa* clinical strains: molecular mechanisms, PBP profiles, and binding affinities. *Antimicrob. Agents Chemother.* 56:4771–4778. doi:10.1128/AAC.00680-12.

531. Cavallari, J. F., R. P. Lamers, E. M. Scheurwater, A. L. Matos, and L. L. Burrows. 2013. Changes to Its peptidoglycan-remodeling enzyme repertoire modulate β-lactam resistance in *Pseudomonas aeruginosa. Antimicrob. Agents Chemother.* 57:3078–3084. doi:10.1128/AAC.00268-13.

532. Lee, M., D. Hesek, B. Blázquez, E. Lastochkin, B. Boggess, J. F. Fisher, and S. Mobashery. 2015. Catalytic spectrum of the PBP4 of *Pseudomonas aeruginosa*, a nexus for the induction of β-lactam antibiotic resistance. *J. Am. Chem. Soc.* 137: 190–200. doi:10.1021/ja5111706.

533. Ropy, A., G. Cabot, I. Sánchez-Diener, C. Aguilera, B. Moya, J. A. Ayala, and A. Oliver. 2015. Role of *Pseudomonas aeruginosa* low-molecular-mass penicillin-binding proteins in AmpC expression, β-lactam resistance, and peptidoglycan structure. *Antimicrob. Agents Chemother.* 59:3925–3934. doi:10.1128/AAC.05150-14.

534. Schaberle, T. F., W. Vollmer, H. J. Frasch, S. Huttel, A. Kulik, M. Rottgen, A. K. von Thaler, W. Wohlleben, and E. Stegmann. 2011. Self-resistance and cell wall composition in the glycopeptide producer *Amycolatopsis balhimycina. Antimicrob. Agents Chemother.* 55:4283–4289. doi:10.1128/AAC.01372-10.

535. Hugonnet, J. -E., N. Haddache, C. Veckerlé, L. Dubost, A. Marie, N. Shikura, J. -L. Mainardi, L. B. Rice, and M. Arthur. 2014. Peptidoglycan cross-linking in glycopeptide-resistant *Actinomycetales. Antimicrob. Agents Chemother.* 58: 1749–1756. doi:10.1128/AAC.02329-13.

536. García-Solache, M., and L. B. Rice. 2019. The Enterococcus: a model of adaptability to its environment. *Clin. Microbiol. Rev.* 32:e00058–18. doi:10.1128/CMR.00058-18.

537. Rello, J., V. K. Eshwara, L. Lagunes, J. Alves, R. G. Wunderink, A. Conway-Morris, J. N. Rojas, E. Alp, and Z. Zhang. 2019. A global priority list of the TOp TEn resistant Microorganisms (TOTEM) study at intensive care: a prioritization exercise based on multi-criteria decision analysis. *Eur. J. Clin. Microbiol. Infect. Dis.* 38:319–323. doi:10.1007/s10096-018-3428-y.

538. Banla, L. I., N. H. Salzman, and C. J. Kristich. 2018. Colonization of the mammalian intestinal tract by enterococci. *Curr. Opin Microbiol.* 47:26–31. doi:10.1016/j.mib.2018.10.005.

539. Kim, S. G., S. Becattini, T. U. Moody, P. V. Shliaha, E. R. Littmann, R. Seok, M. Gjonbalaj, V. Eaton, E. Fontana, L. Amoretti, R. Wright, S. Caballero, Z. X. Wang, H. J. Jung, S. M. Morjaria, I. M. Leiner, W. Qin, R. J. J. F. Ramos, J. R. Cross, S. Narushima, K. Honda, J. U. Peled, R. C. Hendrickson, Y. Taur, M. R. M. van den Brink, and E. G. Pamer. 2019. Microbiota-derived lantibiotic restores resistance against vancomycin-resistant *Enterococcus. Nature.* 572:665–669. doi:10.1038/s41586-019-1501-z.

540. Strachan, C. R., and J. Davies. 2017. The whys and wherefores of antibiotic resistance. *Cold Spring Harb. Perspect. Med.* 7:a025171. doi:10.1101/cshperspect.a025171.

541. Granato, E. T., T. A. Meiller-Legrand, and K. R. Foster. 2019. The evolution and ecology of bacterial warfare. *Curr. Biol.* 29:R521–R537. doi:10.1016/j.cub.2019.04.024.

542. Cuirolo, A., K. Plata, and A. E. Rosato. 2009. Development of homogeneous expression of resistance in methicillin-resistant *Staphylococcus aureus* clinical strains is functionally associated with a β-lactam-mediated SOS response. *J. Antimicrob. Chemother.* 64:37–45. doi:10.1093/jac/dkp164.

543. Plata, K. B., S. Riosa, C. R. Singh, R. R. Rosato, and A. E. Rosato. 2013. Targeting of PBP1 by β-lactams determines *recA*/SOS response activation in heterogeneous MRSA clinical strains. *PLOS One.* 8:e61083. doi:10.1371/journal.pone.0061083.

544. Lee, H., S. Boyle-Vavra, J. Ren, J. A. Jarusiewicz, L. K. Sharma, D. T. Hoagland, S. Yin, T. Zhu, K. E. Hevener, I. Ojeda, R. E. Lee, R. S. Daum, and M. E. Johnson. 2019. Identification of small molecules exhibiting oxacillin synergy through a novel assay for inhibition of *vraTSR* expression in methicillin-resistant *Staphylococcus aureus. Antimicrob. Agents Chemother.* 63:e02593–18. doi:10.1128/AAC.02593-18.

545. Tajbakhsh, G., and D. Golemi-Kotra. 2019. The dimerization interface in VraR is essential for induction of the cell wall stress response in *Staphylococcus aureus*: a potential druggable target. *BMC Microbiol.* 19:153. doi:10.1186/s12866-019-1529-0.

546. Pensinger, D. A., A. J. Schaenzer, and J. -D. Sauer. 2018. Do shoot the messenger: PASTA kinases as virulence determinants and antibiotic targets. *Trends Microbiol.* 26:56–69. doi:10.1016/j.tim.2017.06.010.

547. Kellogg, S. L., and C. J. Kristich. 2018. Convergence of PASTA kinase and two-component signaling in response to cell wall stress in *Enterococcus faecalis. J. Bacteriol.* 200:e00086–18. doi:10.1128/JB.00086-18.

548. Labbe, B. D., C. L. Hall, S. L. Kellogg, Y. Chen, O. Koehn, A. M. Pickrum, S. P. Mirza, and C. J. Kristich. 2019. Reciprocal regulation of PASTA kinase signaling by differential modification. *J. Bacteriol.* 201:e00016–19. doi:10.1128/JB.00016-19.

549. Pensinger, D. A., K. M. Boldon, G. Y. Chen, W. J. Vincent, K. Sherman, M. Xiong, A. J. Schaenzer, E. R. Forster, J. Coers, R. Striker, and J. D. Sauer. 2016. The *Listeria monocytogenes* PASTA kinase PrkA and its substrate YvcK are required for cell wall homeostasis, metabolism, and virulence. *PLoS Pathog.* 12:e1006001. doi:10.1371/journal.ppat.1006001.

550. Kana, B. D., and V. Mizrahi. 2010. Resuscitation-promoting factors as lytic enzymes for bacterial growth and signaling. *FEMS Immunol. Med. Microbiol.* 58:39–50. doi:10.1111/j.1574-695X.2009.00606.x.

551. Veatch, A. V., and D. Kaushal. 2018. Opening Pandora's box: mechanisms of *Mycobacterium tuberculosis* resuscitation. *Trends Microbiol.* 26:145–157. doi:10.1016/j.tim.2017.08.001.

552. Nikitushkin, V. D., G. R. Demina, M. O. Shleeva, S. V. Guryanova, A. Ruggiero, R. Berisio, and A. S. Kaprelyants. 2015. A product of RpfB and RipA joint enzymatic action promotes the resuscitation of dormant mycobacteria. *FEBS J.* 282:2500–2511. doi:10.1111/febs.13292.

553. Rosser, A., C. Stover, M. Pareek, and G. V. Mukamolova. 2017. Resuscitation-promoting factors are important determinants of the pathophysiology in *Mycobacterium tuberculosis* infection. *Crit. Rev. Microbiol.* 43:621–630. doi:10.1080/1040841X.2017.1283485.

554. Wang, Q., R. Marchetti, S. Prisic, K. Ishii, Y. Arai, I. Ohta, S. Inuki, S. Uchiyama, A. Silipo, A. Molinaro, R. N. Husson, K. Fukase, and Y. Fujimoto. 2017. A comprehensive study of the interaction between peptidoglycan fragments and the extracellular domain of *Mycobacterium tuberculosis* Ser/Thr kinase PknB. *ChemBioChem*. 18:2094–2098. doi:10.1002/cbic.201700385.

555. Allsopp, L. P., T. E. Wood, S. A. Howard, F. Maggiorelli, L. M. Nolan, S. Wettstadt, and A. Filloux. 2017. RsmA and AmrZ orchestrate the assembly of all three type VI secretion systems in *Pseudomonas aeruginosa*. *Proc. Natl. Acad. Sci. USA*. 114:7707–7712. doi:10.1073/pnas.1700286114.

556. Bleves, S., and B. Berni. 2018. United we stand and divided we fall. *Nat. Microbiol*. 3:394–395. doi:10.1038/s41564-018-0130-x.

557. LaCourse, K. D., S. B. Peterson, H. D. Kulasekara, M. C. Radey, J. Kim, and J. D. Mougous. 2018. Conditional toxicity and synergy drive diversity among antibacterial effectors. *Nat. Microbiol*. 3:440–446. doi:10.1038/s41564-018-0113-y.

558. Lu, D., G. Shang, H. Zhang, Q. Yu, X. Cong, J. Yuan, F. He, C. Zhu, Y. Zhao, K. Yin, Y. Chen, J. Hu, X. Zhang, Z. Yuan, S. Xu, W. Hu, H. Cang, and L. Gu. 2014. Structural insights into the T6SS effector protein Tse3 and the Tse3-Tsi3 complex from *Pseudomonas aeruginosa* reveal a calcium-dependent membrane-binding mechanism. *Mol. Microbiol*. 92:1092–1112. doi:10.1111/mmi.12616.

559. Lewis, K. 2017. New approaches to antimicrobial discovery. *Biochem. Pharmacol*. 134:87–98. doi:https://doi.org/10.1016/j.bcp.2016.11.002.

560. Imai, Y., K. J. Meyer, A. Iinishi, Q. Favre-Godal, R. Green, S. Manuse, M. Caboni, M. Mori, S. Niles, M. Ghiglieri, C. Honrao, X. Ma, J. Guo, A. Makriyannis, L. Linares-Otoya, N. Böhringer, Z. G. Wuisan, H. Kaur, R. Wu, A. Mateus, A. Typas, M. M. Savitski, J. L. Espinoza, A. O'Rourke, K. E. Nelson, S. Hiller, N. Noinaj, T. F. Schäberle, A. D'Onofrio, and K. Lewis. 2019. A new antibiotic selectively kills Gram-negative pathogens. *Nature*. 576:459–464. doi:10.1038/s41586-019-1791-1.

561. Baltz, R. H. 2017. Molecular beacons to identify gifted microbes for genome mining. *J. Antibiot*. 70:639–646. doi:10.1038/ja.2017.1.

562. Netzker, T., M. Flak, M. K. Krespach, M. C. Stroe, J. Weber, V. Schroeckh, and A. A. Brakhage. 2018. Microbial interactions trigger the production of antibiotics. *Curr. Opin. Microbiol*. 45:117–123. doi:10.1016/j.mib.2018.04.002.

563. Zhang, C., and P. D. Straight. 2019. Antibiotic discovery through microbial interactions. *Curr. Opin. Microbiol*. 51:64–71. doi:10.1016/j.mib.2019.06.006.

564. Niu, G., and W. Li. 2019. Next-generation drug discovery to combat antimicrobial resistance. *Trends Biochem. Sci*. 44:961–972. doi:10.1016/j.tibs.2019.05.005.

565. Boudreau, M. A., J. F. Fisher, and S. Mobashery. 2012. Messenger functions of the bacterial cell wall-derived muropeptides. *Biochemistry*. 51:2974–2990. doi:10.1021/bi300174x.

566. Irazoki, O., S. B. Hernandez, and F. Cava. 2019. Peptidoglycan muropeptides: release, perception, and functions as signaling molecules. *Front. Microbiol*. 10:500. doi:10.3389/fmicb.2019.00500.

567. Chan, J. M., and J. P. Dillard. 2017. Attention seeker: production, modification, and release of inflammatory peptidoglycan fragments in *Neisseria* species. *J. Bacteriol*. 199:e00354–17. doi:10.1128/JB.00354-17.

568. Medina, K. M. P., and J. P. Dillard. 2018. Antibiotic targets in gonococcal cell wall metabolism. *Antibiotics*. 7:64. doi:10.3390/antibiotics7030064.

569. Chen, K.-T., D.-Y. Huang, C.-H. Chiu, W.-W. Lin, P.-H. Liang, and W.-C. Cheng. 2015. Synthesis of diverse *N*-substituted muramyl dipeptide derivatives and their use in a study of human NOD2 stimulation activity. *Chem. Eur. J*. 21:11984–11988.

570. Lauro, M. L., E. A. D'Ambrosio, B. J. Bahnson, and C. L. Grimes. 2017. Molecular recognition of muramyl dipeptide occurs in the leucine-rich repeat domain of Nod2. *ACS Infect. Dis*. 3:264–270. doi:10.1021/acsinfecdis.6b00154.

571. D'Ambrosio, E. A., W. R. Drake, S. Mashayekh, O. I. Ukaegbu, A. R. Brown, and C. L. Grimes. 2019. Modulation of the NOD-like receptors NOD1 and NOD2: a chemist's perspective. *Bioorg. Med. Chem. Lett*. 29:1153–1161. doi:10.1016/j.bmcl.2019.03.010.

572. Behr, M. A., and M. Divangahi. 2015. Freund's adjuvant, NOD2 and mycobacteria. *Curr. Opin. Microbiol*. 23:126–132. doi:10.1016/j.mib.2014.11.015.

573. Gobec, M., T. Tomašič, A. Štimac, R. Frkanec, J. Trontelj, M. Anderluh, I. Mlinaric-Rascan, and Ž. Jakopin. 2018. Discovery of nM desmuramylpeptide agonists of the innate immune receptor nucleotide-binding oligomerization domain-containing protein 2 (NOD2) possessing immunostimulatory properties. *J. Med. Chem*. 61:2707–2724. doi:10.1021/acs.jmedchem.7b01052.

574. Wang, Y. C., N. P. Westcott, M. E. Griffin, and H. C. Hang. 2019. Peptidoglycan metabolite photoaffinity reporters reveal direct binding to intracellular pattern recognition receptors and Arf GTPases. *ACS Chem. Biol*. 14:405–414. doi:10.1021/acschembio.8b01038.

575. Lazor, K. M., J. Zhou, K. E. DeMeester, E. A. D'Ambrosio, and C. L. Grimes. 2019. Synthesis and application of methyl *N,O*-hydroxylamine muramyl deptides. *ChemBioChem*. 20:1369–1375. doi:10.1002/cbic.201800731.

576. Al Nabhani, Z., G. Dietrich, J. P. Hugot, and F. Barreau. 2017. Nod2: the intestinal gate keeper. *PLOS Pathog*. 13:e1006177. doi:10.1371/journal.ppat.1006177.

577. Mukherjee, T., E. S. Hovingh, E. G. Foerster, M. Abdel-Nour, D. J. Philpott, and S. E. Girardin. 2019. NOD1 and NOD2 in inflammation, immunity and disease. *Arch. Biochem. Biophys*. 670:69–81. doi:10.1016/j.abb.2018.12.022.

578. Huang, Z., J. Wang, X. Xu, H. Wang, Y. Qiao, W. C. Chu, S. Xu, L. Chai, F. Cottier, N. Pavelka, M. Oosting, L. A. B. Joosten, M. Netea, C. Y. L. Ng, K. P. Leong, P. Kundu, K. P. Lam, S. Pettersson, and Y. Wang. 2019. Antibody neutralization of microbiota-derived circulating peptidoglycan dampens inflammation and ameliorates autoimmunity. *Nat. Microbiol*. 4:766–773. doi:10.1038/s41564-019-0381-1.

579. Balraadjsing, P. P., L. D. Lund, Y. Souwer, S. A. J. Zaat, H. Frøkiær, and E. C. de Jong. 2020. The nature of antibacterial adaptive immune responses against *Staphylococcus aureus* is dependent on the growth phase and extracellular peptidoglycan. *Infect. Immun*. 88:e00733–19. doi:10.1128/IAI.00733-19.

580. Wen, X., P. Zheng, Y. Ma, Y. Ou, W. Huang, S. Li, S. Liu, X. Zhang, Z. Wang, Q. Zhang, W. Cheng, R. Lin, H. Li, Y. Cai, C. Hu, N. Wu, L. Wan, T. Pan, J. Rao, X. Bei, W. Wu, J. Jin, J. Yan, and G. Liu. 2018. Salutaxel, a conjugate of docetaxel and a muramyl dipeptide (MDP) analogue, acts as multifunctional prodrug that inhibits tumor growth and metastasis. *J. Med. Chem.* 61:1519–1540. doi:10.1021/acs.jmedchem.7b01407.

581. Cambré, A., M. Zimmerman, U. Sauer, B. Vivijs, W. Cenens, C. W. Michiels, A. Aertsen, M. J. Loessner, J. -P. Noben, J. A. Ayala, R. Lavigne, and Y. Briers. 2015. Metabolite profiling and peptidoglycan analysis of transient cell wall-deficient bacteria in a new *Escherichia coli* model system. *Environ. Microbiol.* 17:1586–1599. doi:10.1111/1462-2920.12594.

582. Claessen, D., and J. Errington. 2019. Cell wall deficiency as a coping strategy for stress. *Trends Microbiol.* 27:1025–1033. doi:10.1016/j.tim.2019.07.008.

583. Kawai, Y., R. Mercier, K. Mickiewicz, A. Serafini, L. P. Sório de Carvalho, and J. Errington. 2019. Crucial role for central carbon metabolism in the bacterial L-form switch and killing by β-lactam antibiotics. *Nat. Microbiol.* 4:1716–1726. doi:10.1038/s41564-019-0497-3.

584. Lopatkin, A. J., J. M. Stokes, E. J. Zheng, J. H. Yang, M. K. Takahashi, L. You, and J. J. Collins. 2019. Bacterial metabolic state more accurately predicts antibiotic lethality than growth rate. *Nat. Microbiol.* 4:2109–2117. doi:10.1038/s41564-019-0536-0.

585. Mickiewicz, K. M., Y. Kawai, L. Drage, M. C. Gomes, F. Davison, R. Pickard, J. Hall, S. Mostowy, P. D. Aldridge, and J. Errington. 2019. Possible role of L-form switching in recurrent urinary tract infection. *Nat. Commun.* 10:4379. doi:10.1038/s41467-019-12359-3.

586. Cross, T., B. Ransegnola, J. H. Shin, A. Weaver, K. Fauntleroy, M. S. VanNieuwenhze, L. F. Westblade, and T. Dörr. 2019. Spheroplast-mediated carbapenem tolerance in Gram-negative pathogens. *Antimicrob. Agents Chemother.* 63:e00756–19. doi:10.1128/AAC.00756-19.

587. Lewis, K., and Y. Shan. 2017. Why tolerance invites resistance. *Science.* 355:796. doi:10.1126/science.aam7926.

588. Meylan, S., I. W. Andrews, and J. J. Collins. 2018. Targeting antibiotic tolerance, pathogen by pathogen. *Cell.* 172:1228–1238. doi:10.1016/j.cell.2018.01.037.

589. Windels, E. M., J. E. Michiels, B. Van den Bergh, M. Fauvart, and J. Michiels. 2019. Antibiotics: combatting tolerance to stop resistance. *mBio.* 10:e02095–19. doi:10.1128/mBio.02095-19.

19

The Human Microbiome in Health and Disease

Sandra B. Andersen, Menghan Liu, and Martin J. Blaser

CONTENTS

Introduction

Microbes, used collectively to describe bacteria, fungi, archaea, other unicellular organisms, and viruses, are everywhere. We are increasingly aware of their critical importance to human health: it is estimated that in the human body bacteria alone are as numerous as human cells (Sender et al. 2016), and in terms of gene content and function we are vastly outnumbered (Qin et al. 2010). Microbes contribute to the breakdown of food into accessible nutrients (Hooper et al. 2002; Smith et al. 2013), outcompete invading pathogens in the gut, on the skin and in the urogenital tract (Smits et al. 2016; Grice & Segre 2011; Hickey et al. 2012; Roubaud-Baudron et al. 2019), and affect our mood and behavior through metabolite signaling in the gut (Johnson & Foster 2018). Studying the human microbiome is not only of basic biological interest, but particularly important as we begin to identify associations between changes in composition and most metabolic and autoimmune disorders, including diabetes, obesity, asthma and allergies (Blaser & Falkow 2009). Correlative studies show that the human gut microbiome diversity is lower in populations in developed countries compared to those of hunter-gatherers and inhabitants of rural communities, with the loss suggested to be caused by consumption of less complex and more processed food, increased hygiene, and use of antibiotics (Clemente et al. 2015; Schnorr et al. 2014; Yatsunenko et al. 2012). Cause, effect, and mechanisms are studied *in vivo* in mouse models, and *in vitro* in cell cultures such as organoids, complemented by an increasing number of large-scale observational studies in human patients and the first randomized trials, as discussed later.

Advances in sequencing technologies have revolutionized our study, and understanding, of microbial communities. Circumventing culture-based approaches enables identification and quantitation of microbes that we are as of yet not able to grow in culture. The analysis is split between the lab bench, with, e.g., extraction of DNA, PCR amplification, and high-throughput sequencing, and the subsequent computational analysis by classifying sequences and comparing distributions between samples. Here, we aim to provide an introduction to this rapidly developing field of research, including core concepts, considerations, and pitfalls in the methodology and interpretation of data related to human health.

DOI: 10.1201/9781003099277-21

Methods in Microbiome Sampling and Analysis

Sample Collection and DNA Extraction

For collection of samples, timing and location should be considered in the study design to address biological variation. Microbiome composition differs vastly between body sites (Bouslimani et al. 2015), even between different section of the gut and fecal samples (Durbán et al. 2011; Gu et al. 2013), and vary with the time of day and season (Davenport et al. 2014; Zarrinpar et al. 2014). Technical variation may be introduced by sample storage conditions (Choo et al. 2015; Blekhman et al. 2016), and the complexity of the sample needs to be taken into account when extracting DNA; different kits and disruption methods favor different microbes so care should be taken in this step to minimize bias (Yuan et al. 2012). A step in which the sample is treated by bead-beating can facilitate both disruption of, e.g., tissues to dislodge microbes, and the release of DNA from cells. Specific filtration and extraction techniques can be employed to target DNA viruses (Minot et al. 2013; Kumar et al. 2017).

During extraction, and in the subsequent steps, it is critical to take precautions to avoid contamination and include negative controls, particularly when analyzing samples with a low abundance of microbes (Eisenhofer et al. 2019). Equipment and reagents contain traces of microbial DNA, colloquially referred to as the "kitome" (Salter et al. 2014). When including negative controls, the sequences retrieved can be subtracted from those obtained from the samples, as background noise (Segal et al. 2013; Goffau et al. 2019). An additional concern for samples with low levels of microbial DNA is whether it was present simply as that, i.e. a trace of dead cells, or from living organisms. As such, studies proclaiming characterization of microbiomes from organs usually believed to not contain living microbes, e.g., the placenta or brain, are increasingly met with skepticism (Eisenhofer et al. 2019).

Sequencing Platforms and Analysis Approaches

The development of high throughput sequencing technology is fast-paced, with competition on read length, error rate and type, run time, and cost. For a comparison of past and current technologies see Kchouk et al. (2017). The best choice will be determined by the research aims and the price available to the scientist. The short-read sequencing-by-synthesis approach of Illumina, with a low error rate, is used extensively for microbiome analysis. Single molecule, real-time (SMRT) sequencing by SMRT Pacific Biosciences (PacBio) and Oxford Nanopore Technologies (ONT) produce very long reads with a higher error rate. For whole genome assembly, technologies can be used in complement by using long reads to build scaffolds that are proof-read by short reads with high accuracy (Koren et al. 2012). Another approach, from 10X Genomics, is to partition long fragments in droplets for short read barcoded sequencing, where the barcodes are used to guide the assembly of contigs of the short reads (Bishara et al. 2018).

Estimating Diversity Using a Marker Gene

When targeting a specific genomic region for analysis, the DNA is amplified by PCR following extraction. Samples can be multiplexed in the sequencing step by adding short oligonucleotide barcodes to the primers, unique for each sample (Hamady et al. 2008). That allows for pooling of PCR products from multiple samples for sequencing, and separating the reads contributed from each sample in the analysis. Increasing the number of samples comes at the cost of the sequencing depth for each individual sample. PCR amplification intrinsically introduces bias, as some sequence variants are amplified at a higher rate than others. Such bias can be estimated, and computationally corrected for, by sequencing PCR products amplified at different cycle numbers (Silverman et al. 2019). Overall, one approach is to keep cycle number as low as possible. Another way to estimate technical bias is to include a mock community in analyses, with a known species composition, which also allows for comparisons between runs (Schloss et al. 2011).

Studies on the human microbiome usually focus on bacteria, which are considered the most important in terms of both biomass and function. The diversity of a bacterial community can be profiled by sequencing marker genes such as that for 16S rRNA. Ribosomal genes were early recognized as being ideal for this purpose, as they are sufficiently similar between species to amplify and align, yet enough different to confidently distinguish between organisms at least at the genus level and commonly considered to the species level. The 16S rRNA gene is on average 1525 bp long, and comprises variable and conserved regions. Primers targeting conserved regions ensure broad specificity, for amplification of hypervariable regions to estimate diversity. The majority of studies focus on a subset of the gene, such as the V3-V4 region, due to length limitations in read length in the sequencing step (Kumar et al. 2011, Yang et al. 2016, Rintala et al. 2017). Using the whole ribosomal RNA operon gives better taxonomic resolution to the strain level (de Oliveira Martins et al. 2020). High throughput sequencing of the full length 16S gene is possible with SMRT sequencing by constructing circular sequences; these are read multiple times and using the consensus output sequence compensates for the higher error rate of this technology (Callahan et al. 2019). Differences in copy-number of the 16S rRNA gene between species, and allele variation within genomes (Klappenbach et al. 2001), can skew the analyses of community composition. Alternatives to 16S are being explored, such as the single-copy highly conserved gene *rpoB*, which shows promise in terms of specificity and sensitivity in detecting diversity (Ogier et al. 2019). Eukaryotes can be targeted with 18S sequencing and the fungi in the mycobiome may be specifically targeted by sequencing of the Internal Transcribed Spacer (ITS) region (Nash et al. 2017). No general target region is available to analyze the virome, given the high diversity and frequently small genome size of viruses, so identification of DNA viruses requires a metagenomic approach.

Identifying Organisms in Data from Targeted Amplicon Sequencing

Analysis of the reads from high throughput sequencing aims to correctly identify unique biological variants, while limiting the noise introduced by sequencing errors. For 16S sequencing the clustering of sequences based on 97% similarity as operational taxonomic units (OTUs) has been the most widely used approach, using a database of reference sequences to group reads, or with *de novo* identification of OTUs. The most popular pipelines for this and subsequent steps are QIIME (Caporaso et al. 2010; Bolyen et al. 2019) and mothur (Westcott & Schloss, 2015), which are being continually improved. Here, the likelihood of missing potential strain level diversity is weighed against the error rate of the input. With the advent of more accurate sequencing, the use of Amplicon Sequence Variants (ASVs) may be replacing OTUs. Computational approaches such as DADA2 for identifying ASVs have the benefit of achieving a finer resolution of diversity, to single-nucleotide differences, and facilitate comparison across studies as they are not dependent on the other sequences in the dataset as is analysis of OTUs (Callahan et al. 2016, 2017). Following clustering of sequences, they are classified to the highest level possible, by comparison to a database (Balvocuite & Hudson 2017). This approach potentially permits resolution to the bacterial strain level.

Comparing Relative Abundances of Organisms

Amplicon-based diversity studies estimate the relative abundance of different organisms in a sample and should ideally be combined with an estimation of the absolute prevalence. Using only compositional data limits the choice of statistical tools and biases the impression of the similarity between samples: we would expect that microbiomes with the same diversity but a tenfold difference in actual numbers impact the host in different ways (Figure 19.1A; Morton et al. 2019). The actual abundance of microorganisms in a sample can be estimated by quantitative PCR targeting a region of a conserved indicator gene, e.g., the 16S rRNA gene (Nadkarni et al. 2002). A caveat, just as for 16S amplicon sequencing, is that different organisms have different copy-number of the 16S rRNA gene in their genome. An alternative approach is to stain bacteria with, e.g., a fluorescent nucleic dye or a probe with a broad specificity, and count the bacteria by flow cytometry (Props et al. 2017; Vandeputte et al. 2017). If an estimation of the absolute microbial prevalence is not feasible, Morton et al. (2019) suggest a computational approach to identify direction of community changes based on the differential ranking of microbes across samples.

Predicting Interactions from Relative Abundance Data

Comparing relative abundance data from multiple samples is used to predict interactions based on co-occurrence. The underlying hypothesis is that if a specific pattern in distribution is repeatedly observed, e.g., a positive or negative correlation between a consortium of species, this may be caused by interactions between those organisms (Faust & Raes 2012). Network analysis can identify pairwise and more complex correlations between organisms. Caution is advised when interpreting such results. Biologically, a positive correlation could arise if two organisms respond to availability of a specific food source, and network analysis may wrongly label such a competitive or neutral association as beneficial. Likewise, a strong negative correlation between the prevalence of organisms could represent the ability to use different resources, or be entirely spurious, rather than direct competition (Faust & Raes 2012). Computationally, there are limitations to comparing compositional data across samples. The measured abundances of organisms are inherently interdependent (Figure 19.1B), and this dependence is substantially more impactful in less-diverse communities, where, e.g., the importance of rare organisms, and their absence in some samples, may be inflated (Kurtz et al. 2015). This can be addressed in the choice of analytical approach, such as where the SPIEC-EASI method uses center-log transformation and assumes a sparse ecological network to decrease compositional bias (Kurtz et al. 2015).

Beyond 16S—Describing Lower Level Variation and Microbiome Function

While species composition and indices of diversity based on 16S rRNA gene sequencing are used to describe and compare communities, the overarching aim is to understand how composition reflects function. The PICRUSt software exploits the correlation between phylogeny and functional profile to extrapolate from 16S diversity to metabolic capabilities (Langille et al. 2013). This method maps 16S sequences to their closest relatives that have full genomes available and predicts whole genome level function based on that of related organisms. Even when 16S sequencing allows for identification of the correct species, an

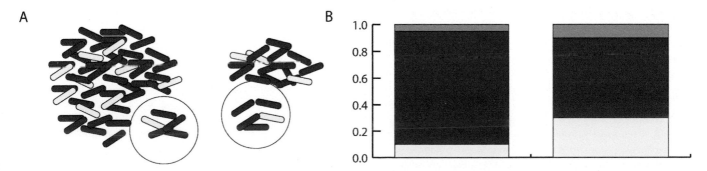

FIGURE 19.1 Challenges of using compositional data. A: An unbiased analysis of the diversity of the two microbiomes will show that the yellow bacteria comprise 20% in both, suggesting that the constituent populations are highly similar. However, quantifying the biomass will reveal a difference in the absolute number of bacteria that potentially can affect the host. B: The relative abundance of a species in a sample is dependent on the prevalence of other species as the sum is always 1. For two identical samples, in which extraction of DNA from the blue bacteria was more efficient in the first sample compared to the second, different measures of diversity are observed despite the proportion of DNA from the yellow bacteria being exactly the same in both samples.

obvious limitation of this approach is the inability to detect variation below the phylotype or species level, and the accuracy of microbiome function reconstruction is greatly restricted by the incompleteness of the database of genomes used. Metabolic profiles and production of exoproducts can vary extensively between strains and evolve over time within a microbiome (Zhang & Zhao 2016). Mutations such as single nucleotide polymorphisms (SNPs), and inserts and deletions can cause loss of function, while new traits can be acquired by recent or past horizontal transfer of genomic DNA or plasmids. Thus, functional analysis based on 16S is only a very crude approximation, but provides a well-established and cost-effective first-view into the biochemical attributes of a sample.

Metagenomics and metatranscriptomics with sequencing of DNA or cDNA associated with all organisms in a community, without prior PCR enrichment of a specific genomic region, provide more direct approaches to estimate function but are also both more expensive and computationally demanding. Depending on the study objective, sequences can be grouped by organisms or function. The reference database-based methods compare sequences to a library, and when a threshold is met, they are assigned to a corresponding organism or function (Franzosa et al. 2018; Segata et al. 2012; Wood & Salzberg, 2014). Here, the incompleteness of current libraries is an even greater challenge than for 16S sequencing. Alternatively, full genome *de novo* assembly can be considered, where sequences are stitched together based on overlapping patterns. It is challenging to assemble full genomes from a mixed community, but sequences can be "binned" based on patterns in oligonucleotide use, similarity to related genomes, and read prevalence. This approach is based on the assumption that different regions of a genome will have similar sequencing depth, so that contigs of sequences originating from the same organism are likely to have similar prevalence (Albertsen et al. 2013; Venter et al. 2004). The long reads of SMRT sequencing are particularly useful for such assembly, e.g., in resolving placement of repetitive elements, and this approach also identifies methylation patterns, which can be used to group sequences as they may be strain-specific (Zhu et al. 2018; Beaulaurier et al. 2019). The identification of functional pathways from genomic data can be done with or without full genome assembly and reveals the metabolic potential of a microbiome. Combining metagenomics and transcriptomics (the analysis of which genes are expressed by sequencing cDNA made from RNA) allows for inference of which members of the microbiome contribute to specific pathways and their relative importance (Liu et al. 2020).

Data Sharing

With the publication of research on sequencing data it should be, and is for most journals, a requirement to upload raw data to appropriate repositories. Care should be taken to include metadata for reproducibility, and so that data can be used for other studies (Navas-Molina et al. 2017). In addition to raw data, publication of the code used for analyses, including choice of parameters and tables with, e.g., OTUs or ASVs, will facilitate cross-study comparisons of the biological findings and the methodology used for sequencing and analysis. Sequencing data publication, however, also raises new questions about protection of privacy, as the microbiome is sufficiently unique to be used to identify individuals, and guidelines will surely need to be continuously developed (Franzosa et al. 2015; Wagner et al. 2016).

Interpreting Results

What Is a Healthy Microbiome?

A goal of human microbiome analyses is the identification of organisms that are associated with, and potentially contributing to, a specific host state, such as disease, geographic origin, diet, and medical treatment. Massive efforts such as the European Metagenomics of the Human Intestinal Tract project (MetaHIT) and the NIH Human Microbiome Project (HMP) have made great strides in developing computational methods and characterizing the microbiomes at different body sites (Qin et al. 2010; Peterson et al. 2009), and a wealth of metadata facilitates the finding of correlations between organisms and health factors. A major result is that there is no obvious composition of a core healthy microbiome, as taxonomic composition varies greatly with, e.g., diet and geographical location. However, as different species can overlap in their metabolic range, substantial variation may mask functional similarities (Bäckhed et al. 2012). Although such studies have shifted the focus from attempts to identify "good" and "bad" bacteria, there continue to be substantial efforts in this direction. In contrast, most research points to the importance of stability in composition and overall diversity, and that the loss of microbial diversity frequently is associated with poor outcomes.

The term dysbiosis has been used extensively to describe an unhealthy and unbalanced microbiome. However, the lack of a clear classification of what constitutes a healthy microbiome obviously poses a challenge to define what comprises a deviation from this, in particular if no "healthy" samples of a subject are available for comparison (Hooks & O'Malley 2017). Olesen & Alm (2016) argue that the way dysbiosis is used, without distinguishing whether an altered microbiome is the cause or effect of disease (or its treatment), is holding the field back from identifying the areas in which microbiome research can improve health. We do not need gut microbiome analyses to tell us that obese and healthy controls differ, so to be useful we would have to gain information on why they differ and how we can change it (Olesen & Alm 2016).

The Ecology and Evolution of Host-Microbiome Interactions

From a discovery-based research commencement, with the screening of many individuals, the microbiome field is entering a hypothesis-driven phase (Tripathi et al. 2018). Central to address questions such as why diversity is important and how it can be maintained or reconstituted, is the implementation of ecological and evolutionary theory. Ecological theory and experimental studies have dealt with how the diversity of an ecosystem affects function, resistance to change in the face of disturbance (Figure 19.2A), and resilience, as the likelihood to return to the starting point following perturbation (Figure 19.2B; Oliver et al 2015). Studies of aquatic systems have shown that the colonization success of a species is affected by arrival time through

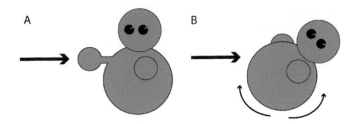

FIGURE 19.2 Characteristics of a stable ecosystem. Stable ecosystems can be characterized as both resistant and resilient in the face of a challenge. A: A resistant system will not be perturbed by a disturbance. An example is the resistance of a diverse gut microbiome to *Clostridium difficile* infection in which invaders are outcompeted. B: A resilient ecosystem bounces back to its initial state, like a roly-poly toy, following a disturbance. An example is the composition of the fecal microbiome of healthy volunteers treated with antibiotics for a short duration, in which diversity is largely restored after some time.

priority effects (Alford & Wilbur 1985; Shulman et al. 1983). The same dynamics matter for the human microbiome, where a perturbation can come in the form of dietary change or antibiotic treatment, and diversity affects the likelihood of pathogen invasion (Lozupone et al. 2012).

Evolutionary theory helps us consider the potential for host microbe-interaction (Foster et al. 2017). As we consider how an organism affects our health, it is easy to expect that beneficial traits were selected through co-evolution. The timescale that matters to a microbe with a short multiplication time, however, is if it can outcompete its neighbor, not whether it benefits the host. Can the host control microbiome composition, and if so, how? The most frequent interactions between microbes are expected to be competitive and weak, and those types of interactions, perhaps counter-intuitively, result in a more stable community (Coyte et al. 2015). The host may select for organisms with beneficial properties by providing resources that select for specific properties (Schluter & Foster 2012). In addition, the host may recognize groups of bacteria based on surface markers and can potentially remove or enhance them selectively (Foster et al. 2017).

With regard to the question of whether humans have co-evolved with our microbes, the term phylosymbiosis is used to describe congruency between host and symbiont phylogenies (Moran & Sloan 2015). There are obvious examples of this in obligate, vertically inherited, symbionts co-speciating with their host over time, such as many insects and their symbionts that provide essential nutrients (Douglas 2011). Yet such strict conditions do not have to be fulfilled to create patterns of matching phylogenies, which could arise also when related species acquire related microbes from shared or similar environmental niches. Groussin et al. 2017 teased apart the role of these mechanisms, and the effect of diet, in shaping mammalian microbiomes. A range of phylogenetically older microorganisms were positively correlated with diet, while showing a weaker association with host phylogeny. In contrast, other "younger" bacteria showed strong signals of host specificity and co-speciation.

Overinterpreting Results?

With a substantial decrease in sequencing prices, and ease of analyses facilitated by well-supported open-source bioinformatics tools, the number of microbiome studies has naturally increased. Careful studies have greatly increased our knowledge of the human microbiome, in addition to those of other organisms, and various environmental sites. With increased funding to the field (Proctor 2019), also comes the risk of overinterpreting results ("hype"). This may be in the form of equating correlation with causation, and explicitly translating patterns observed in mouse experiments to humans. Reduction of scientific findings to click-bait headlines by the press further exacerbates this issue. Critique of a study claiming to show a causative effect of microbiome composition on the development of autism (Sharon et al. 2019), highlight the challenges in this field of small sample sizes, translating mouse experiments to human conditions, selective collection of samples and analyses, and conflicting financial interests (Lowe 2019). As the field matures, such issues must be addressed.

How Close Are We to Direct Consumer Benefits?

A veritable industry has blossomed around direct-to-consumer microbiome sequencing. Several companies offer analyses of fecal samples, collected by the customer at home, for identification of "good and bad" microbes, and prediction of the optimal customized diet despite limited evidence to support their claims. The most promising results for designing dietary intervention come from using microbiome data as a component of a wider profiling of metabolic markers (Zeevi et al. 2015). In addition, there is a large consumer market for pre- and probiotics. The assumption has been that while perhaps not beneficial, probiotics would at least not be harmful. However, this may have been naive. Studies show that probiotics taken following a course of antibiotics can negatively affect the return to a pretreatment community (Suez et al. 2018), and probiotics given to ICU patients may lead to an increased rate of developing sepsis (Yelin et al. 2019). The problem is worsened by the multiplicity of "probiotic" products and the lack of transparency about actual compositions, or even their viability.

Clinical Applications of Microbiome Manipulation

Alleviating Malnourishment

Targeting gut microbiome function is showing potential in addressing the long-term consequences of early childhood malnutrition, where development is impaired despite subsequent increased caloric intake. In the first years of life, a healthy human gut microbiome changes in composition and shows an increase in complexity as it matures (Yatsunenko et al. 2012; Bokulich et al. 2016; Raman et al. 2019). Malnourished children were found to have immature microbiomes compared to healthy age-matched controls, possibly missing specific species (Subramanian et al. 2014; Blanton et al. 2016; Raman et al. 2019). Longitudinal sampling of fecal microbiomes and blood metabolites from malnourished children treated with a standard diet identified markers of incomplete recovery (Gehrig et al. 2019). Experiments with mice and piglets "humanized" by fecal transplants identified supplementary diet ingredients that benefitted microbiome maturation and host metabolite profile. The positive results were supported by a smaller-scale clinical study of malnourished Bangladeshi children (Gehrig et al. 2019), and shows that a supplementary diet targeting microbiome function, and not solely host nutrition, may be a constructive approach.

Outcompeting Pathogens by Fecal Microbiota Transplantation

Fecal microbiota transplantation (FMT) to treat recurrent infection with the bacterium *Clostridium difficile* represents a major success story in microbiome manipulation with clinical consequences (Smits et al. 2016; van Nood et al. 2013). *C. difficile* is a spore-forming, toxin-producing opportunistic pathogen that can establish infection in immunocompromised patients with low gut microbiome diversity. This increasingly common nosocomial infection causes severe, and potentially fatal, diarrhea and colitis, and unsuccessful treatment with antibiotics further diminishes competition in the gut and exacerbates problems with resistance (Leffler & Lamont, 2015). As a last resort, experimental treatment with a transplant of a healthy gut microbiome has proven highly effective (van Nood et al. 2013). Oral or rectal introduction of fecal material from a healthy donor contributes to establish a diverse microbiome capable of outcompeting *C. difficile*. It is still to be shown whether specific species, or the overall diversity, is most important to achieve this result, and even bacteria-free materials show beneficial effects, suggesting a role of metabolites (Ott et al. 2017). FMT has proven much less successful in the treatment of other conditions such as inflammatory bowel disease (Colman & Rubin 2014), and fatal outcomes following transplant of antibiotic resistant strains from a healthy donor highlights the importance of pretransplant screening, in particular as FMT recipients usually are immunosuppressed (DeFilipp et al. 2019). Development of a culture-based inoculum with a controlled composition may increase safety (Blaser 2019), and a clinical trial of autologous FMT, using material from the recipient acquired prior to antibiotic treatment shows promise for immunocompromised cancer patients (Taur et al. 2018).

Avoiding Side Effects of Drug Metabolism

Gut microbiome profiling can contribute to the development of personalized medicine. The effect of most drugs depends on how they are metabolized and the microbiome, with a wider array of biochemical pathways available than human cells, can play a significant role here. Large-scale screening of drug-microbe interactions revealed a broad range of drug modification by common gut microbes (Zimmermann et al. 2019a), and many widely used drugs have off-target effects on particular members of the gut microbiome (Maier et al. 2018). A mouse study aimed to disentangle the role of the host and microbes by comparing germ-free individuals with introduced microbiomes differing only in the ability of specific strains to metabolize two drugs (Zimmermann et al. 2019b). By modelling measured serum and blood metabolite concentrations, the host and microbial contributions to drug pharmacokinetics was accurately predicted.

For the drug irinotecan used against colorectal cancer, members of the gut microbiome can cause serious adverse reactions to treatment, by reactivating excreted metabolites (Wallace et al. 2010). Screening for strains with this enzymatic repertoire can identify patients that are better served with a different treatment (Guthrie et al. 2017).

Improving Response to Cancer Immunotherapy

Microbiome composition also influences the efficacy of cancer immunotherapy (CI). One of the most advanced CI strategies is to use immune checkpoint inhibitors such as anti-PD-1 blockade, which unleash the immune system to target cancer cells directly. However, fewer than 20% of CI patients show partial or complete response to treatment clinically (Carretero-González et al. 2018; Haslam & Prasad 2019). Recent studies demonstrated the microbiome may play a role in modulating CI efficacy and response. The baseline gut microbiome of CI responders is substantially different from CI nonresponders; and transferring of fecal samples from those patients to germ free mice recapitulated the phenotypes, suggesting the microbiome is sufficient to drive these different response (Gopalakrishnan et al. 2018; Matson et al. 2018; Routy et al. 2018). A consortium of 11 bacteria isolated from the human gut microbiota was shown to increase the population of CD8 T cells, which are crucial effector cells in CI, and increase the efficacy of immune checkpoint inhibitors (Tanoue et al. 2019). These studies suggest the potential for using microbiome screening to identify patients that will respond well to a specific treatment, or where manipulation of the gut microbiome prior to treatment can facilitate a better outcome, if the results can be recapitulated in clinical trials.

Bacteriophage Therapy to Eliminate Antibiotic Resistant Infections

While understudied, the human virome appears to be dynamic and dominated by viruses that target bacteria (bacteriophages) in a healthy host, which likely affect bacterial community composition (Minot et al. 2013). A study of IBD patients showed that both the bacterial and viral portions of the microbiomes differed from that of healthy controls (Clooney et al. 2019). Problems with antibiotic resistance has led to renewed interest in treatment with bacteriophages, which has been used particularly in parts of Eastern Europe and Russia (Chanishvili 2012; Chapter 54 in this volume). A high degree of host specialization of bacteriophages imply that they can be used for targeted treatment of specific pathogens, with limited collateral damage to the host microbiome, and may represent a last resort when antibiotic resistance rules out this first line of defense. A cystic fibrosis patient with an antibiotic resistant *Mycobacterium abscessus* infection was recently successfully treated with an engineered cocktail of bacteriophages (Dedrick et al. 2019), and a mouse study showed improvement of liver disease by targeting *Enterococcus faecalis* with bacteriophages (Duan et al. 2019). More research is required to inform us on the role of bacteriophages in healthy microbiomes, and how they can be used therapeutically.

Looking to the Future

As the above examples demonstrate, it is an exciting time to be working on the human microbiome. Technological advances expand the possibilities for studying host-microbe interactions on an ever-finer scale, and facilitate ways to experimentally test the selective pressures that have shaped them. Longitudinal sampling of individual patients or across a population can provide retrospective or near real-time tracking of natural community dynamics, or responses to treatment, and randomized controlled trials can test the promising findings of mouse studies in humans. With input from eco-evolutionary approaches, we should see the continued development of new strategies for ways to manipulate (to "harness") the microbiome for health benefits.

REFERENCES

Albertsen, M., Hugenholtz, P., Skarshewski, A., Nielsen, K. L., Tyson, G. W., & Nielsen, P. H. (2013). Genome sequences of rare, uncultured bacteria obtained by differential coverage binning of multiple metagenomes. *Nature Biotechnology*, 31(6), 533–538.

Alford, R. A., & Wilbur, H. M. (1985). Priority effects in experimental pond communities: Competition between Bufo and Rana. *Ecology*, 66(4), 1097–1105.

Bäckhed, F., Fraser, C. M., Ringel, Y., Sanders, M. E., Sartor, R. B., Sherman, P. M., … Finlay, B. B. (2012). Defining a healthy human gut microbiome: Current concepts, future directions, and clinical applications. *Cell Host & Microbe*, 12(5), 611–622.

Balvočiūtė, M., & Huson, D. H. (2017). SILVA, RDP, Greengenes, NCBI and OTT—How do these taxonomies compare? *BMC Genomics*, 18(S2), 114.

Beaulaurier, J., Schadt, E. E., & Fang, G. (2019). Deciphering bacterial epigenomes using modern sequencing technologies. *Nature Reviews Genetics*, 20(3), 157–172.

Bishara, A., Moss, E. L., Kolmogorov, M., Parada, A. E., Weng, Z., Sidow, A., … Bhatt, A. S. (2018). High-quality genome sequences of uncultured microbes by assembly of read clouds. *Nature Biotechnology*, 36(11), 1067–1075.

Blanton, L. V., Charbonneau, M. R., Salih, T., Barratt, M. J., Venkatesh, S., Ilkaveya, O., … Gordon, J. I. (2016). Gut bacteria that prevent growth impairments transmitted by microbiota from malnourished children. *Science*, 351(6275).

Blaser, M. J. (2019). Fecal microbiota transplantation for dysbiosis—Predictable risks. *New England Journal of Medicine*, 381(21), 2064–2066.

Blaser, M. J., & Falkow, S. (2009). What are the consequences of the disappearing human microbiota? *Nature Reviews. Microbiology*, 7(12), 887–894.

Blekhman, R., Tang, K., Archie, E. A., Barreiro, L. B., Johnson, Z. P., Wilson, M. E., … Tung, J. (2016). Common methods for fecal sample storage in field studies yield consistent signatures of individual identity in microbiome sequencing data. *Scientific Reports*, 6(1), 31519.

Bokulich, N. A., Chung, J., Battaglia, T., Henderson, N., Jay, M., Li, H., … Blaser, M. J. (2016). Antibiotics, birth mode, and diet shape microbiome maturation during early life. *Science Translational Medicine*, 8(343), 343ra82.

Bolyen, E., Rideout, J. R., Dillon, M. R., Bokulich, N. A., Abnet, C. C., Al-Ghalith, G. A., … Caporaso, J. G. (2019). Reproducible, interactive, scalable and extensible microbiome data science using QIIME 2. *Nature Biotechnology*, 37(8), 852–857.

Bouslimani, A., Porto, C., Rath, C. M., Wang, M., Guo, Y., Gonzalez, A., … Dorrestein P. C. (2015). Molecular cartography of the human skin surface in 3D. *Proceedings of the National Academy of Sciences*, 112(17), E2120–9.

Callahan, B. J., McMurdie, P. J., & Holmes, S. P. (2017). Exact sequence variants should replace operational taxonomic units in marker-gene data analysis. *The ISME Journal*, 11(12), 2639–2643.

Callahan, B. J., McMurdie, P. J., Rosen, M. J., Han, A. W., Johnson, A. J. A., & Holmes, S. P. (2016). DADA2: High-resolution sample inference from Illumina amplicon data. *Nature Methods*, 13(7), 581–583.

Callahan, B. J., Wong, J., Heiner, C., Oh, S., Theriot, C. M., Gulati, A. S., … Dougherty, M. K. (2019). High-throughput amplicon sequencing of the full-length 16S rRNA gene with single-nucleotide resolution. *Nucleic Acids Research*, 47(18), e103–e103.

Caporaso, J. G., Kuczynski, J., Stombaugh, J., Bittinger, K., Bushman, F. D., Costello, E. K., … Knight, R. (2010). QIIME allows analysis of high-throughput community sequencing data. *Nature Methods*, 7(5), 335–336.

Carretero-González, A., Lora, D., Ghanem, I., Zugazagoitia, J., Castellano, D., Sepúlveda, J. M., … de Velasco, G. (2018). Analysis of response rate with ANTI PD1/PD-L1 monoclonal antibodies in advanced solid tumors: A meta-analysis of randomized clinical trials. *Oncotarget*, 9(9), 8706–8715.

Chanishvili, N. (2012). Phage Therapy—History from Twort and d'Herelle through Soviet experience to current approaches. *Advances in Virus Research*, 83, 3–40.

Choo, J. M., Leong, L. E., & Rogers, G. B. (2015). Sample storage conditions significantly influence faecal microbiome profiles. *Scientific Reports*, 5(1), 16350. https://doi.org/10.1038/srep16350

Clemente, J. C., Pehrsson, E. C., Blaser, M. J., Sandhu, K., Gao, Z., Wang, B., … Dominguez-Bello, M. G. (2015). The microbiome of uncontacted Amerindians. *Science Advances*, 1(3), e1500183–e1500183.

Clooney, A. G., Sutton, T. D. S., Shkoporov, A. N., Holohan, R. K., Daly, K. M., O'Regan, O., … Hill, C. (2019). Whole-virome analysis sheds light on viral dark matter in inflammatory bowel disease. *Cell Host & Microbe*.

Colman, R. J., & Rubin, D. T. (2014). Fecal microbiota transplantation as therapy for inflammatory bowel disease: A systematic review and meta-analysis. *Journal of Crohn's and Colitis*, 8(12), 1569–1581.

Coyte, K. Z., Schluter, J., & Foster, K. R. (2015). The ecology of the microbiome: Networks, competition, and stability. *Science*, 350(6261), 663–666.

Davenport, E. R., Mizrahi-Man, O., Michelini, K., Barreiro, L. B., Ober, C., & Gilad, Y. (2014). Seasonal variation in human gut microbiome composition. *PLoS ONE*, 9(3), e90731.

de Oliveira Martins, L., Page, A. J., Mather, A. E., & Charles, I. G. (2020). Taxonomic resolution of the ribosomal RNA operon in bacteria: Implications for its use with long-read sequencing. *NAR Genomics and Bioinformatics*, 2(1).

Dedrick, R. M., Guerrero-Bustamante, C. A., Garlena, R. A., Russell, D. A., Ford, K., Harris, K., ... Spencer, H. (2019). Engineered bacteriophages for treatment of a patient with a disseminated drug-resistant *Mycobacterium abscessus*. *Nature Medicine*, *25*(5), 730–733.

DeFilipp, Z., Bloom, P. P., Torres Soto, M., Mansour, M. K., Sater, M. R. A., Huntley, M. H., ... Hohmann, E. L. (2019). Drug-resistant *E. coli* bacteremia transmitted by fecal microbiota transplant. *New England Journal of Medicine*, *381*(21):2043–2050.

Douglas, A. E. (2011). Lessons from studying insect symbioses. *Cell Host & Microbe*, *10*(4), 359–367.

Duan, Y., Llorente, C., Lang, S., Brandl, K., Chu, H., Jiang, L., ... Schnabl, B. (2019). Bacteriophage targeting of gut bacterium attenuates alcoholic liver disease. *Nature*, *575*(7783):505–511.

Durbán, A., Abellán, J. J., Jiménez-Hernández, N., Ponce, M., Ponce, J., Sala, T., ... Moya, A. (2011). Assessing gut microbial diversity from feces and rectal mucosa. *Microbial Ecology*, *61*(1), 123–133.

Eisenhofer, R., Minich, J. J., Marotz, C., Cooper, A., Knight, R., & Weyrich, L. S. (2019). Contamination in low microbial biomass microbiome studies: Issues and recommendations. *Trends in Microbiology*, *27*(2), 105–117.

Faust, K., & Raes, J. (2012). Microbial interactions: From networks to models. *Nature Reviews Microbiology*, *10*(8), 538–550.

Fischbach, M. A., & Segre, J. A. (2016). Signaling in host-associated microbial communities. *Cell*, *164*(6), 1288–1300.

Foster, K. R., Schluter, J., Coyte, K. Z., & Rakoff-Nahoum, S. (2017). The evolution of the host microbiome as an ecosystem on a leash. *Nature*, *548*(7665), 43–51.

Franzosa, E. A., Huang, K., Meadow, J. F., Gevers, D., Lemon, K. P., Bohannan, B. J. M., & Huttenhower, C. (2015). Identifying personal microbiomes using metagenomic codes. *Proceedings of the National Academy of Sciences*, *112*(22), E2930–E2938.

Franzosa, E. A., McIver, L. J., Rahnavard, G., Thompson, L. R., Schirmer, M., Weingart, G., ... Huttenhower, C. (2018). Species-level functional profiling of metagenomes and metatranscriptomes. *Nat Methods*, *15*(11), 962–8.

Gehrig, J. L., Venkatesh, S., Chang, H.-W., Hibberd, M. C., Kung, V. L., Cheng, J., ... Gordon, J. I. (2019). Effects of microbiota-directed foods in gnotobiotic animals and undernourished children. *Science (New York, N.Y.)*, *365*(6449).

Goffau, M. C. de, Lager, S., Sovio, U., Gaccioli, F., Cook, E., Peacock, S. J., ... Smith, G. C. S. (2019). Human placenta has no microbiome but can contain potential pathogens. *Nature 2019*, 1.

Gopalakrishnan, V., Spencer, C. N., Nezi, L., Reuben, A., Andrews, M. C., Karpinets, T. V., ... Wargo, J. A. (2018). Gut microbiome modulates response to anti-PD-1 immunotherapy in melanoma patients. *Science (New York, N.Y.)*, *359*(6371), 97–103.

Grice, E. A., & Segre, J. A. (2011). The skin microbiome. *Nature Reviews Microbiology*, *9*(4), 244–253.

Groussin, M., Mazel, F., Sanders, J. G., Smillie, C. S., Lavergne, S., Thuiller, W., & Alm, E. J. (2017). Unraveling the processes shaping mammalian gut microbiomes over evolutionary time. *Nature Communications*, *8*:14319.

Gu, S., Chen, D., Zhang, J.-N., Lv, X., Wang, K., Duan, L.-P., ... Wu, X.-L. (2013). Bacterial community mapping of the mouse gastrointestinal tract. *PLoS ONE*, *8*(10), e74957.

Guthrie, L., Gupta, S., Daily, J., & Kelly, L. (2017). Human microbiome signatures of differential colorectal cancer drug metabolism. *NPJ Biofilms and Microbiomes*, *3*(1), 27.

Hamady, M., Walker, J. J., Harris, J. K., Gold, N. J., & Knight, R. (2008). Error-correcting barcoded primers for pyrosequencing hundreds of samples in multiplex. *Nature Methods*, *5*(3), 235–237.

Haslam, A., & Prasad, V. (2019). Estimation of the percentage of US patients with cancer who are eligible for and respond to checkpoint inhibitor immunotherapy drugs. *JAMA Network Open*, *2*(5), e192535.

Hickey, R. J., Zhou, X., Pierson, J. D., Ravel, J., & Forney, L. J. (2012). Understanding vaginal microbiome complexity from an ecological perspective. *Translational Research*, *160*(4), 267–282.

Hooks, K. B., & O'Malley, M. A. (2017). Dysbiosis and its discontents. *MBio*, *8*(5):e01492–17.

Hooper, L. V., Midtvedt, T., & Gordon, J. I. (2002). How host-microbial interactions shape the nutrient environment of the mammalian intestine. *Annual Review of Nutrition*, *22*(1), 283–307.

Johnson, K. V.-A., & Foster, K. R. (2018). Why does the microbiome affect behaviour? *Nature Reviews Microbiology*, *16*(10), 647–655.

Kchouk, M., Gibrat, J.-F., & Elloumi, M. (2017). Generations of sequencing technologies: from first to next generation. *Biology and Medicine (Aligarh)*, *9*(3):100395.

Klappenbach, J. A., Saxman, P. R., Cole, J. R., & Schmidt, T. M. (2001). rrndb: the ribosomal RNA operon copy number database. *Nucleic Acids Research*, *29*(1), 181–184.

Koren, S., Schatz, M. C., Walenz, B. P., Martin, J., Howard, J. T., Ganapathy, G., ... Phillippy, A. M. (2012). Hybrid error correction and de novo assembly of single-molecule sequencing reads. *Nature Biotechnology*, *30*(7), 693–700.

Kumar, A., Murthy, S., & Kapoor, A. (2017). Evolution of selective-sequencing approaches for virus discovery and virome analysis. *Virus Research*, *239*, 172–179.

Kumar, P. S., Brooker, M. R., Dowd, S. E., & Camerlengo, T. (2011). Target region selection is a critical determinant of community fingerprints generated by 16S Pyrosequencing. *PLoS ONE*, *6*(6), e20956.

Kurtz, Z. D., Müller, C. L., Miraldi, E. R., Littman, D. R., Blaser, M. J., & Bonneau, R. A. (2015). Sparse and compositionally robust inference of microbial ecological networks. *PLOS Computational Biology*, *11*(5), e1004226.

Langille, M. G. I., Zaneveld, J., Caporaso, J. G., McDonald, D., Knights, D., Reyes, J. A., ... Huttenhower, C. (2013). Predictive functional profiling of microbial communities using 16S rRNA marker gene sequences. *Nature Biotechnology*, *31*(9), 814–821.

Leffler, D. A., & Lamont, J. T. (2015). *Clostridium difficile* infection. *New England Journal of Medicine*, *372*(16), 1539–1548.

Liu, M., Devlin, J. C., Hu, J., Volkova, A., Battaglia, T. W., Byrd, A., ... Nazzal, L. Microbial contributions to oxalate metabolism in health and disease. *medRxiv*. 2020. doi:10.1101/2020.01.27.20018770

Lowe, D. (2019). Autism mouse models for the microbiome? In the pipeline. Retrieved December 6, 2019, Available from https://blogs.sciencemag.org/pipeline/archives/2019/05/31/autism-mouse-models-for-the-microbiome

Lozupone, C. A., Stombaugh, J. I., Gordon, J. I., Jansson, J. K., & Knight, R. (2012). Diversity, stability and resilience of the human gut microbiota. *Nature*, *489*(7415), 220–230.

Maier, L., Pruteanu, M., Kuhn, M., Zeller, G., Telzerow, A., Anderson, E. E., ... Typas, A. (2018). Extensive impact of non-antibiotic drugs on human gut bacteria. *Nature*, *555*(7698), 623–628.

Matson, V., Fessler, J., Bao, R., Chongsuwat, T., Zha, Y., Alegre, M.-L., … Gajewski, T. F. (2018). The commensal microbiome is associated with anti–PD-1 efficacy in metastatic melanoma patients. *Science, 359*(6371), 104–108.

McLaren, M. R., Willis, A. D., & Callahan, B. J. (2019). Consistent and correctable bias in metagenomic sequencing experiments. *ELife, 8.*

Minot, S., Bryson, A., Chehoud, C., Wu, G. D., Lewis, J. D., & Bushman, F. D. (2013). Rapid evolution of the human gut virome. *Proceedings of the National Academy of Sciences of the United States of America, 110*(30), 12450–12455.

Moran, N. A., & Sloan, D. B. (2015). The hologenome concept: Helpful or hollow? *PLOS Biology, 13*(12), e1002311.

Morton, J. T., Marotz, C., Washburne, A., Silverman, J., Zaramela, L. S., Edlund, A., … Knight, R. (2019). Establishing microbial composition measurement standards with reference frames. *Nature Communications, 10*(1), 2719.

Nadkarni, M. A., Martin, F. E., Jacques, N. A., & Hunter, N. (2002). Determination of bacterial load by real-time PCR using a broad-range (universal) probe and primers set. *Microbiology, 148*(1), 257–266.

Nash, A. K., Auchtung, T. A., Wong, M. C., Smith, D. P., Gesell, J. R., Ross, M. C., … Petrosino, J. F. (2017). The gut mycobiome of the human microbiome project healthy cohort. *Microbiome, 5*(1), 153.

Navas-Molina, J. A., Hyde, E. R., Sanders, J. G., & Knight, R. (2017). The microbiome and big data. *Current Opinion in Systems Biology, 4*, 92–96.

Ogier, J.-C., Pagès, S., Galan, M., Barret, M., & Gaudriault, S. (2019). rpoB, a promising marker for analyzing the diversity of bacterial communities by amplicon sequencing. *BMC Microbiology, 19*(1), 171.

Olesen, S. W., & Alm, E. J. (2016). Dysbiosis is not an answer. *Nature Microbiology, 1*(12), 16228.

Oliver, T. H., Heard, M. S., Isaac, N. J. B., Roy, D. B., Procter, D., Eigenbrod, F., … Bullock, J. M. (2015). Biodiversity and resilience of ecosystem functions. *Trends in Ecology & Evolution, 30*(11), 673–684.

Ott, S. J., Waetzig, G. H., Rehman, A., Moltzau-Anderson, J., Bharti, R., Grasis, J. A., … Schreiber, S. (2017). Efficacy of sterile fecal filtrate transfer for treating patients with *Clostridium difficile* infection. *Gastroenterology, 152*(4), 799–811.

Peterson, J., Garges, S., Giovanni, M., McInnes, P., Wang, L., Schloss, J. A., … Guyer, M. (2009). The NIH human microbiome project. *Genome Research, 19*(12):2317–2323.

Proctor, L. (2019). Priorities for the next 10 years of human microbiome research. *Nature, 569*(7758), 623–625.

Props, R., Kerckhof, F.-M., Rubbens, P., De Vrieze, J., Hernandez Sanabria, E., Waegeman, W., … Boon, N. (2017). Absolute quantification of microbial taxon abundances. *The ISME Journal, 11*(2), 584–587.

Qin, J., Li, R., Raes, J., Arumugam, M., Burgdorf, K. S., Manichanh, C., … Zoetendal, E. (2010). A human gut microbial gene catalogue established by metagenomic sequencing. *Nature, 464*(7285):59–65.

Raman, A. S., Gehrig, J. L., Venkatesh, S., Chang, H.-W., Hibberd, M. C., Subramanian, S., … Gordon, J. I. (2019). A sparse covarying unit that describes healthy and impaired human gut microbiota development. *Science (New York, N.Y.), 365*(6449).

Rintala, A., Pietilä, S., Munukka, E., Eerola, E., Pursiheimo, J.-P., Laiho, A., … Huovinen, P. (2017). Gut microbiota analysis results are highly dependent on the 16S rRNA gene target region, whereas the impact of DNA extraction is minor. *Journal of Biomolecular Techniques : JBT, 28*(1), 19–30.

Robertson, R. C., Manges, A. R., Finlay, B. B., & Prendergast, A. J. (2019). The human microbiome and child growth – First 1000 days and beyond. *Trends in Microbiology, 27*(2), 131–147.

Roubaud-Baudron, C., Ruiz, V. E., Swan, A. M., Vallance, B. A., Ozkul, C., Pei, Z., …Blaser, M. J. (2019). Long-term effects of early-life antibiotic exposure on resistance to subsequent bacterial infection. *MBio, 10*(6).

Routy, B., Le Chatelier, E., Derosa, L., Duong, C. P. M., Alou, M. T., Daillère, R., …Zitvogel, L. (2018). Gut microbiome influences efficacy of PD-1-based immunotherapy against epithelial tumors. *Science (New York, N.Y.), 359*(6371), 91–97.

Salter, S. J., Cox, M. J., Turek, E. M., Calus, S. T., Cookson, W. O., Moffatt, M. F., …Walker, A. W. (2014). Reagent and laboratory contamination can critically impact sequence-based microbiome analyses. *BMC Biology, 12*(1), 87.

Schloss, P. D., Gevers, D., & Westcott, S. L. (2011). Reducing the effects of PCR amplification and sequencing artifacts on 16S rRNA-based studies. *PLoS ONE, 6*(12), e27310.

Schluter, J., & Foster, K. R. (2012). The evolution of mutualism in gut microbiota via host epithelial selection. *PLoS Biology, 10*(11), e1001424.

Schnorr, S. L., Candela, M., Rampelli, S., Centanni, M., Consolandi, C., Basaglia, G., … Crittenden, A. N. (2014). Gut microbiome of the Hadza hunter-gatherers. *Nature Communications, 5*(1), 3654.

Segal, L. N., Alekseyenko, A. V, Clemente, J. C., Kulkarni, R., Wu, B., Chen, H., … Weiden, M. D. (2013). Enrichment of lung microbiome with supraglottic taxa is associated with increased pulmonary inflammation. *Microbiome, 1*(1), 19.

Segata, N., Waldron, L., Ballarini, A., Narasimhan, V., Jousson, O., & Huttenhower, C. (2012). Metagenomic microbial community profiling using unique clade-specific marker genes. *Nat Methods, 9*(8), 811–4.

Sender, R., Fuchs, S., & Milo, R. (2016). Are we really vastly outnumbered? Revisiting the ratio of bacterial to host cells in humans. *Cell, 164*(3), 337–340.

Sharon, G., Cruz, N. J., Kang, D.-W., Gandal, M. J., Wang, B., Kim, Y.-M., …Mazmanian, S. K. (2019). Human gut microbiota from autism spectrum disorder promote behavioral symptoms in mice. *Cell, 177*(6), 1600–1618.e17.

Shulman, M. J., Ogden, J. C., Ebersole, J. P., McFarland, W. N., Miller, S. L., & Wolf, N. G. (1983). Priority effects in the recruitment of juvenile coral reef fishes. *Ecology, 64*(6), 1508–1513.

Silverman, J. D., Bloom, R. J., Jiang, S., Durand, H. K., Mukherjee, S., & David, L. A. (2019) Measuring and mitigating PCR bias in microbiome data. *bioRxiv.* https://doi.org/10.1101/604025

Smith, M. I., Yatsunenko, T., Manary, M. J., Trehan, I., Mkakosya, R., Cheng, J., …Gordon, J. I. (2013). Gut microbiomes of Malawian twin pairs discordant for kwashiorkor. *Science, 339*(6119), 548–554.

Smits, W. K., Lyras, D., Lacy, D. B., Wilcox, M. H., & Kuijper, E. J. (2016). *Clostridium difficile* infection. *Nature Reviews Disease Primers, 2*, 16020.

Subramanian, S., Huq, S., Yatsunenko, T., Haque, R., Mahfuz, M., Alam, M. A., ...Gordon, J. I. (2014). Persistent gut microbiota immaturity in malnourished Bangladeshi children. *Nature, 510*(7505), 417–421.

Suez, J., Zmora, N., Zilberman-Schapira, G., Mor, U., Dori-Bachash, M., Bashiardes, S., ...Elinav, E. (2018). Post-antibiotic gut mucosal microbiome reconstitution is impaired by probiotics and improved by autologous FMT. *Cell, 174*(6), 1406–1423.e16.

Tanoue, T., Morita, S., Plichta, D. R., Skelly, A. N., Suda, W., Sugiura, Y., ...Honda, K. (2019). A defined commensal consortium elicits CD8 T cells and anti-cancer immunity. *Nature, 565*(7741), 600–605.

Taur, Y., Coyte, K., Schluter, J., Robilotti, E., Figueroa, C., Gjonbalaj, M., ...Xavier, J. B. (2018). Reconstitution of the gut microbiota of antibiotic-treated patients by autologous fecal microbiota transplant. *Science Translational Medicine, 10*(460).

Tripathi, A., Marotz, C., Gonzalez, A., Vázquez-Baeza, Y., Song, S. J., Bouslimani, A., ...Knight, R. (2018). Are microbiome studies ready for hypothesis-driven research? *Current Opinion in Microbiology, 44*, 61–69.

van Nood, E., Vrieze, A., Nieuwdorp, M., Fuentes, S., Zoetendal, E. G., de Vos, W. M., ...Keller, J. J. (2013). Duodenal infusion of donor feces for recurrent *Clostridium difficile. New England Journal of Medicine, 368*(5), 407–415.

Vandeputte, D., Kathagen, G., D'hoe, K., Vieira-Silva, S., Valles-Colomer, M., Sabino, J., ... Raes, J. (2017). Quantitative microbiome profiling links gut community variation to microbial load. *Nature, 551*(7681), 507–511.

Venter, J. C., Remington, K., Heidelberg, J. F., Halpern, A. L., Rusch, D., Eisen, J. A., ... Smith, H. O. (2004). Environmental genome shotgun sequencing of the Sargasso Sea. *Science (New York, N.Y.), 304*(5667), 66–74.

Wagner, J., Paulson, J. N., Wang, X., Bhattacharjee, B., & Corrada Bravo, H. (2016). Privacy-preserving microbiome analysis using secure computation. *Bioinformatics, 32*(12), 1873–1879.

Wallace, B. D., Wang, H., Lane, K. T., Scott, J. E., Orans, J., Koo, J. S., ... Redinbo, M. R. (2010). Alleviating cancer drug toxicity by inhibiting a bacterial enzyme. *Science (New York, N.Y.), 330*(6005), 831–835.

Westcott, S. L., & Schloss, P. D. (2015). De novo clustering methods outperform reference-based methods for assigning 16S rRNA gene sequences to operational taxonomic units. *PeerJ, 3*, e1487.

Wood, D. E., & Salzberg, S. L. (2014). Kraken: Ultrafast metagenomic sequence classification using exact alignments. *Genome Biology, 15*(3), R46.

Yang, B., Wang, Y., & Qian, P.-Y. (2016). Sensitivity and correlation of hypervariable regions in 16S rRNA genes in phylogenetic analysis. *BMC Bioinformatics, 17*(1), 135.

Yatsunenko, T., Rey, F. E., Manary, M. J., Trehan, I., Dominguez-Bello, M. G., Contreras, M., & Gordon, J. I. (2012). Human gut microbiome viewed across age and geography. *Nature, 486*(7402), 222–227.

Yelin, I., Flett, K. B., Merakou, C., Mehrotra, P., Stam, J., Snesrud, E., & Priebe, G. P. (2019). Genomic and epidemiological evidence of bacterial transmission from probiotic capsule to blood in ICU patients. *Nature Medicine, 25*(11), 1728–1732.

Yuan, S., Cohen, D. B., Ravel, J., Abdo, Z., & Forney, L. J. (2012). Evaluation of methods for the extraction and purification of DNA from the human microbiome. *PLoS ONE, 7*(3), e33865.

Zarrinpar, A., Chaix, A., Yooseph, S., & Panda, S. (2014). Diet and feeding pattern affect the diurnal dynamics of the gut microbiome. *Cell Metabolism, 20*(6), 1006–1017.

Zeevi, D., Korem, T., Zmora, N., Israeli, D., Rothschild, D., Weinberger, A., & Segal, E. (2015). Personalized nutrition by prediction of glycemic responses. *Cell, 163*(5), 1079–1094.

Zhang, C., & Zhao, L. (2016). Strain-level dissection of the contribution of the gut microbiome to human metabolic disease. *Genome Medicine, 8*(1), 41.

Zhu, S., Beaulaurier, J., Deikus, G., Wu, T. P., Strahl, M., Hao, Z., & Fang, G. (2018). Mapping and characterizing N6-methyladenine in eukaryotic genomes using single-molecule real-time sequencing. *Genome Research, 28*(7), 1067–1078.

Zimmermann, M., Zimmermann-Kogadeeva, M., Wegmann, R., & Goodman, A. L. (2019a). Mapping human microbiome drug metabolism by gut bacteria and their genes. *Nature, 570*(7762), 462–467.

Zimmermann, M., Zimmermann-Kogadeeva, M., Wegmann, R., & Goodman, A. L. (2019b). Separating host and microbiome contributions to drug pharmacokinetics and toxicity. *Science, 363*(6427), eaat9931.

20

The Phylum Actinobacteria

Alan C. Ward, Nagamani Bora, Jenileima Devi, Alexander Escasinas, and Nicholas Allenby

CONTENTS

Introduction

The *Actinobacteria* are a major phylum in the bacterial domain. Many large-scale tools for visualizing taxonomic and phylogenetic relationships are emerging (de Vienne, 2016; Jovanovic and Mikheyev, 2019; Zhang et al. 2012) with a brilliant interactive visualization at http://lifemap-ncbi.univ-lyon1.fr/ (accessed February 2019) showing the relationship of this major clade to the Terrabacteria group, Bacteria, and the rest of the tree of life.

Their origin seems significantly more recent than the divergence of the Bacteria from Archaea and Eukaryota and has been linked to the colonization of the land (Battistuzzi et al., 2004).

The phylum encompasses a number of significant genera and species, including major human pathogens such as: *Mycobacterium tuberculosis* (Chai et al., 2018) and other members of the Mtb group; *Mycobacterium leprae* (Walker and Lockwood, 2006); and a number of other opportunistic pathogens, such as *M. avium, M. intracellulare, M. scrofulaceum, M. kansasii, M. xenopi, M. malmoense, M. fortuitum, M. marinum,*

DOI: 10.1201/9781003099277-22

M. ulcerans, and *M. abscessus* (Cowman et al., 2016). The genus *Corynebacterium* (Oliveira et al., 2017) includes human and animal pathogens (Bernard, 2012; Soares et al., 2013) and is a workhorse of industrial biotechnological production of biochemicals (Becker et al., 2018). The members of the genus *Bifidobacterium* are a major component of the gut microflora of adults and infants and crucial in immune development and immunomodulatory activity in a healthy gut (Ruiz et al., 2017).

Many actinobacteria are environmental including soil, rhizosphere, and endophytic organisms for example the genus *Frankia* responsible for symbiotic nitrogen fixation in nonleguminous plants (Nguyen et al., 2019; Ventura et al., 2007). Other actinobacteria species are cultured as endophytes from plants, especially *Streptomyces* but also *Microbispora*, *Micromonospora*, *Nocardioides*, *Nocardia*, and *Streptosporangium* are common (Dinesh et al., 2017); however, their role, for example, as symbiotic or commensal, is often less clear than in the case of *Frankia*. In the case of endophytic *Micromonospora* in nitrogen-fixing plants, plants with rhizobial nodules infected by *Micromonospora* do seem to show improved growth (Trujillo et al., 2015), though the basis for that improvement is unclear. In the soil, they play a role in degradation of biomass and in recycling, including recalcitrant molecules and pollutants (Alvarez et al., 2017). In activated sludge, as well as contributing to biodegradation (e.g., playing a positive role in enhanced biological phosphorus removal), filamentous Actinobacteria play a role in normal floc formation, as well as being major players in bulking and foaming, both significant problems in waste water treatment (Seviour et al., 2008).

Marine actinobacteria, once seen as terrestrial wash-in, are a rich source of novel diversity (Ward and Bora, 2006) and have come to prominence in search and discovery for bioactive metabolites, for example, *Salinispora* (Jensen et al., 2015). Both marine and terrestrial actinobacteria have been and continue to be the major source of bioactive natural products (Newman and Cragg, 2016), as have actinomycetes from diverse ecological niches (Bull et al., 2000), such as endophytes (Newman and Cragg, 2020).

The phylum *Actinobacteria* (Goodfellow, 2012) currently (List of Prokaryotic names with Standing in Nomenclature [LPSN], https://www.bacterio.net/ accessed February 2020) contains six classes with some deep-rooted clades (Table 20.1a). The major clades of diversity map roughly to the 31 orders (Table 20.1a and 20.1b). Understanding the diversity of organisms, their identity in the membership of functional biomes, and the ability to communicate that information is essential for our ability to interact with, manage, and exploit our relationships with the microbial world. That role should be played by systematics, the discovery, description, classification, and naming of organisms, enabling their identification. That reference classification consists of type

TABLE 20.1a

Phylum *Actinobacteria*—Deep-Rooted Orders

Class	Order	Families	Genera	Species	Reference Genera
Acidimicrobiia					
	Acidimicrobiales		8	11	
		Acidimicrobiaceae			
		Iamiaceae			
		Ilumatobacteraceae			*Illuminatobacter*
Coriobacteriia					
	Coriobacteriales		10	173	
		Atopobiaceae			*Atopobium*
		Coriobacteriaceae			*Collinsella*
	Eggerthellales		12	25	
		Eggerthellaceae			*Slackia*
Nitriliruptoria					
	Egibacterales		1	1	
		Egibacteraceae			
	Egicoccales		1	1	
		Egicoccaceae			
	Euzebyales		1	2	
		Euzebyaceae			
	Nitriliruptorales		1	1	
		Nitriliruptorales			
Rubrobacteria					
	Gaiellales		1	1	
		Gaiellaceae			
	Rubrobacterales		1	9	
		Rubrobacteraceae			*Rubrobacter*
Thermoleophilia					
	Solirubrobacterales		5	16	
		Baekduiaceae			
		Conexibacteraceae			
		Parviterribacteraceae			
		Patulibacteraceae			
		Solirubrobacteraceae			
	Thermoleophilales		1	2	
		Thermoleophilaceae			

TABLE 20.1b

Phylum and Class *Actinobacteria*

Class	Order	Families	Genera	Species	Reference Genera
Actinobacteria					
	Acidothermales		1	1	
		Acidothermaceae			*Acidithermus*
	Actinomycetales		20	87	
		Actinomycetaceae			*Actinomyces*
	Actinopolysporales		4	16	
		Actinopolysporaceae			*Actinopolyspora*
	Bifidobacteriales		10	89	
		Bifidobacteriaceae			*Bifidobacterium*
	Catenulisporales		3	10	
		Actinospicaceae			
		Catenulisporaceae			*Catenulispora*
	Corynebacteriales				
		Corynebacteriaceae	4	122	*Corynebacterium*
		Dietziaceae	1	11	*Detzia*
		Gordoniaceae	3	52	*Gordonia*
		Lawsonellaceae	1	1	*Lawsonella*
		Mycobacteriaceae	3	195	*Mycobacterium*
		Nocardiaceae	9	169	*Nocardia/Rhodococcus*
		Segniliparaceae	1	2	*Segniliparus*
		Tsukamurellaceae	1	13	*Tsukamurella*
	Cryptosporangiales		1	7	
		Cryptosporangiaceae			*Cryptosporangium*
	Frankiales		2	13	
		Frankiaceae			*Frankia*
		Motilibacteraceae			
	Geodermatophilales		6	45	
		Antricoccaceae			
		Geodermatophilaceae			*Geodermatophilus*
	Glycomycetales		6	32	
		Glycomycetaceae			*Glycomyces*
	Jiangellales		3	14	
		Jiangellaceae			*Jiangella*
	Kineosporiales		3	14	
		Kineosporiaceae			*Kineococcus*
	Micrococcales		147	1044	
		Beutenbergiaceae			
		Bogoriellaceae			*Georgenia*
		Brevibacteriaceae			*Brevibacterium*
		Cellulomonadaceae			*Cellulomonas*
		Demequinaceae			*Lysinimicrobium*
		Dermabacteraceae			*Brachybacterium*
		Dermacoccaceae			*Dermacoccus*
		Dermatophilaceae			*Dermatophilus*
		Intrasporangiaceae			*Intrasporangium*
		Jonesiaceae			
		Kytococcaceae			*Kytococcus*
		Microbacteriaceae			*Agromyces/Leucobacter*
		Micrococcaceae			*Arthrobacter*
		Ornithinimicrobiaceae			*Ornithininmicrobium*
		Promicromonosporaceae			*Isoptericola*
		Rarobacteraceae			*Rarobacter*
		Ruaniaceae			
		Tropherymataceae			*Tropheryma*
		Yaniellaceae			
	Micromonosporales		29	236	
		Micromonosporaceae			*Actinoplanes/Dactylosporangium*
					Micromonospora
	Nakamurellales		4	12	
		Nakamurellaceae			*Nakamurella*
	Propionibacteriales		33	256	
		Actinopolymorphaceae			*Actinopolyspora*
		Kribbellaceae			*Kribbella*
		Nocardioidaceae			*Nocardioides*
		Propionibacteriaceae			*Propionibacterium/Tessaracoccus*

(Continued)

TABLE 20.1b *(Continued)*

Phylum and Class *Actinobacteria*

Class	Order	Families	Genera	Species	Reference Genera
	Pseudonocardiales		35	339	
		Pseudonocardiaceae			*Amycolatopsis/Lentzea*
					Pseudonocardia/Sacharopolyspora
	Sporichthyales		3	4	
		Sporichthyaceae			
	Streptomycetales		6	709	
		Streptomycetaceae			*Streptomyces*
	Streptosporangiales		34	311	
		Nocardiopsaceae			*Nocardiopsis*
		Streptosporangiaceae			*Nonomuraea/Streptosporangium*
		Thermomonosporaceae			*Actinomadura*

species organized into a hierarchical set of nested groups—from strains, to species, genera, families, orders, classes, phyla, and domains.

Isolation and description of organisms now recognized as *Actinobacteria*, such as *Mycobacterium tuberculosis* (Koch, 1882; Taylor et al., 2003), *Streptomyces albus* (Rossi-Doria, 1891; Pridham and Lyons, 1961), and *Bifidobacterium bifidum* (Lee and O'sullivan, 2010; Tissier, 1899), have a long history. Descriptions and classifications were based on morphology, staining, growth, and physiology. Later, chemotaxonomic methods were added to the arsenal of techniques, as described in the first edition of the *CRC Practical Handbook of Microbiology* (Lechevalier and Lechevalier, 1989), just as molecular systematics—especially 16S gene sequencing—changed the field. Prior to 16S, taxonomy was explicitly an attempt to classify organisms into groups by similarity—not phylogeny—though evolutionary speculation was common (Stanier and van Niel, 1962). Subsequently, 16S revolutionized the field and 40 years of research have resulted in, probably, ~80,000–100,000 actinobacterial 16S sequences longer than 1,000 bp (the lower figure after removal of duplicate sequences) published in public databases. The integration of molecular systematics, chemotaxonomy, and phenotypic data in a polyphasic taxonomy did result in big changes in prokaryotic taxonomy (Zhulin, 2016). The chapter on actinobacteria in the second and third editions of the aforementioned handbook (Ward and Bora, 2008, 2015) was based upon 16S phylogeny. It has not taken 40 years for the next revolution, next generation, and third-generation sequencing (van Dijk et al., 2018); a revolution in biology has unfolded at tremendous speed, and the impact on the taxonomy and phylogeny of prokaryotes is illustrated by the contentious proposal (Bisgaard et al., 2019) to allow genome sequence data as type material for species descriptions (Whitman, 2015), as well as more than 20,000 actinobacterial genomes (though more than 6,700 genomes are for *Mycobacterium tuberculosis* strains) in the Genbank Genome database (https://www.ncbi.nlm.nih.gov/genome/—accessed February 2020). It is clear that genome phylogeny is the way forward, not only because of the sheer volume of phylogenetically relevant data, for inferring both phylogeny and taxonomic relationships, but also its integration with the further functional analysis for which phylogenetic analysis provides a framework. The Ad Hoc Committee on Reconciliation of Approaches to Bacterial Systematics (Wayne et al., 1987) agreed the complete DNA sequence should be the reference standard to determine phylogeny and that phylogeny should determine taxonomy.

Genomics

The advent of Sanger sequencing (Sanger et al., 1977) revolutionized biology over 25 years, evolving into a fast, accurate, low-cost, and high-throughput technology, in particular providing the backbone for the current prokaryotic and actinobacterial taxonomy from 16S sequencing. It culminated in the dramatic sequencing of the human genome (Lander et al., 2001; Venter et al., 2001)—taking 13 years and costing of the order of $3 billion. Sequencing a human genome drove the development of next-generation sequencing (Schloss, 2008)—454 (Margulies et al., 2005), Ion Torrent (Merriman and Rothberg, 2012), and Solexa (Davies, 2010). Massively parallel, high-throughput, accurate, short-read sequencing increased speed and minimized cost (van Dijk et al., 2014), but massive coverage cannot overcome the fundamental limitations of short reads (Salzberg and Yorke, 2005). Many of the actinobacterial genomes in the databases are highly fragmented with a few to thousands of contigs.

Third-generation, single-molecule, real-time, long-read sequencing followed rapidly (van Dijk et al., 2018) with PacBio (Eid et al., 2009) and Nanopore (Jain et al., 2015) sequencing. Long-read sequencing overcomes many of the limitations of short-read sequencing, but is significantly error prone (85–90% accuracy), though accuracy can be measured in different ways.

PacBio errors seem to be random and probably correctible by coverage (Chaisson et al., 2015), but computationally intensive. However, circular consensus sequencing (CCS), in which multiple rounds of sequencing around a single circularized fragment, removes the errors arising from mismapping of individual long reads. CCS initially came at the price of shorter long reads (<2 Kb), but recent developments (Wenger et al., 2019) promise long (average 13.5 Kb) accurate (96–99.9% depending how you measure) at the cost of sample preparation. The aim is a single sequencing technology to give comprehensive high-accuracy sequencing for human genomic DNA (with the complexity of large plant genomes an additional challenge). Although some actinobacterial genomes are relatively large (e.g., 10–14 Mb), they are simple compared to these eukaryotic genomes but nevertheless benefit from the sequencing technology developments.

Nanopore sequencing errors may contain a nonrandom element, though mostly correctible with coverage (Jain et al., 2018). However, to match PacBio CCS required polishing with short-read sequence data (Illumina Solexa). On-going software and hardware developments, such as neural network based Medaka (https://nanoporetech.github.io/medaka/) continue to improve accuracy.

Currently, combined long-read, PacBio sequencing with short-read Solexa data can be seen as state of the art (Gomez-Escribano et al., 2015). Initially, the PacBio RS II was a large floor-standing machine (though current Sequel II are still >300 kg) with complex sample preparation and data collection suited to generation of sequence data through a sequence provider, similar to Solexa data from MiSeq and HiSeq machines.

The nanopore Minion Mk1B sequencer is 90 g (3.3 cm × 10.5 cm × 2.3 cm) and much more directed to personal use.

Actinobacterial Genomics

There are 22,661 actinobacterial genomes in the Genbank Genome browser (17,893 in the validly described taxa listed in Table 20.1; see https://www.ncbi.nlm.nih.gov/genome/browse#!/prokaryotes/Actinobacteria—accessed February 2020). A majority of higher taxonomic levels are single species in a genus, which is the only member of a family and order (Figure 20.1). In Table 20.1, genera that constitute useful representatives of the major clades are highlighted. Genome sequencing is driven by research in the whole of biology but provides the data for comparing strains, organizing relationships and estimating phylogeny across diversity. There are about 4,479 actinobacterial species listed in validly described genera in the LPSN (Parte, 2018) (accessed February 2020), but the distribution of genomes across that reference set of type strains is not even (e.g., 6,710 genomes for *Mycobacterium tuberculosis*) and about 10% (1,860) are whole genomes (complete or single scaffold). The incentive for a whole genome phylogeny is the limited resolution of the 16S gene at the species level, as a single gene potentially influenced by lateral gene transfer, multiple copy number, and variable rates of evolution. Examples include strains with identical 16S sequences but significantly different genomic similarity (Stach et al., 2003); a study of 287 *Streptomyces* strains (Chevrette et al., 2019) with 100% 16S identity showed a genomic average nucleotide identity (ANI) as low as 84.4%. A limited number of examples are known in which multiple different 16S sequences occur in the same strain (Antony-Babu et al., 2017; Conville and Witebsky, 2007). Chevrette et al. (2019) show incongruence when 16S identity is plotted against genomic ANI, but 16S identity is not the optimum metric to determine phylogenetic relationships. Many studies show a good, if not exact, phylogenetic congruity between 16S phylogenetic trees and genome phylogenies (Hu et al., 2018; Jade et al., 2016; Ward and Allenby, 2018). Genomes are complex and determining metrics that best represent similarity between genomes is more of a challenge than for single-gene phylogenies. The first option is the ability to perform more single-gene phylogenies, for example, 23S rRNA. There are significant differences between multiple single-gene phylogenies so improvement on the single-gene 16S phylogeny depends upon generating a consensus from multiple single-gene phylogenies. However, being able to perform a multilocus sequence typing (MLST) analysis, but without the problems of finding conserved primer sites for polymerase chain reaction (PCR) amplification (Larsen et al., 2012; Miro et al., 2020), by alignment of concatenated gene sequences from whole genomes, is a better option. Classical MLST is linked to clinical identification and epidemiology of strains within a single pathogenic species, but has been extended to taxonomy (Rong and Huang, 2010; https://pubmlst.org/databases/). The equivalent phylogenomic procedure requires whole genome sequence data and has the aim of comparing organisms across the tree of life. It depends on the identification of single-copy number genes conserved across all the taxa to be analyzed, and the alignment, concatenation, and building of a tree, for example, with phylosift (Darling et al., 2014), though the 40 conserved genes for *Bacteria* are dominated by ribosomal proteins. These methods only use a fraction of the information encoded in the whole genome and exclude the pan-genome. Alignment of whole genomes for

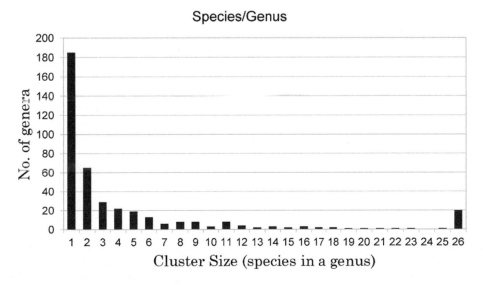

FIGURE 20.1 Numbers of species in genera.

comparative genomics is routine using software, including, for example, ProgressiveMauve (Darling et al., 2010), but multiple genome alignment at the scales needed for the current numbers of whole genomes is still challenging (Armstrong et al., 2019; Dewey, 2019) and an ongoing area of research, Dewey (2019) cites 40 whole genome alignment methods. New, large-scale phylogenomic trees use increasing numbers of marker genes—there are 381 marker genes in the phylogenomic tree of bacteria and archaea (Zhu et al., 2019)—requiring extensive analysis and massive computational resources. Alignment-free methods, such as CVTree3 (Zuo and Hao, 2015), use the whole genome and count K-peptides (peptide k-mers) in protein coding regions to measure genome similarity. Gene content phylogenies also use the whole genome, but need identification of homologous genes (Gupta and Sharma, 2015; Snel et al., 1999) as do gene order phylogenies (Zhou et al., 2017). DNA:DNA similarity was introduced in the 1970s and has been the fundamental basis for species identification in the era of 16S-based bacterial taxonomy (Wayne et al., 1987). The calculation of the ANI from genome sequences (Goris et al., 2007; Meier-Kolthoff et al., 2013) correlates well with DNA:DNA similarity. For large-scale phylogeny, the high-throughput version of the Genome BLAST Distance algorithm (Meier-Kolthoff et al., 2014) can be used to calculate genomic similarities.

Phylogenomic Studies in *Actinobacteria*

Studies of taxa in the *Actinobacteria* using phylogenomic methods provide models of the strategies and software to analyze genomic data, frequently illustrating the integration of both the phylogenomic analysis and addressing broader biological questions.

Acidimicrobiia

Phylogenomic studies have been done for *Acidimicrobiia* (Hu et al., 2018), a deep-rooted class with few cultivated members that are mostly isolated from extreme acidic environments, mine waste, and geothermal sites and able to oxidize ferrous ion (Jones and Johnson, 2015). However, they are abundant in marine and freshwater metagenomic studies and other uncultivatable members in the *Acidimicrobiia* are not acidophiles, for example *Candidatus Microthrix parvicella* from activated sludge and a related clade in the deep chlorophyll maximum (DCM) of the Sargasso Sea (Treusch et al., 2009). *Acidimicrobiia* were a major clade from the deep sea along the Southwest Indian Ridge (Chen et al., 2016), and four nearly complete genomes were assembled from samples from the Mediterranean DCM (Mizuno et al., 2015), including one encoding a novel actinorhodopsin (acidirhodopsin). The "acIV lineage" one of four major lineages detected in freshwater (Warnecke et al., 2004) is related to *Acidimicrobiales*.

Coriobacteriia

Coriobacterium glomerans was isolated from the gut of the red soldier bug (Haas and König, 1988) but is prevalent in the human distal gut. They metabolize drugs and bile acids, biotransform plant polyphenols to phytoestrogens, and have been associated with bacteremia, type 2 diabetes, cardiovascular disease, and rheumatoid arthritis. Bisanz et al. (2018) sequenced and annotated strains from

14 genera and performed genome-wide alignments with public sequences previously analyzed (Gupta et al., 2013).

Nitriliruptoria

The five members of the four genera in this class all have genomes, and although they are halophiles or halotolerant they have been isolated from other distinct ecological sites. Chen et al. (2020) have performed comparative genomics.

Thermoleophilia

Hu et al. (2019) performed a phylogenomic analysis of twelve genomes, six sequenced species, and six reconstructed genomes from metagenomic data. The genome from the one species in the order *Gaiellales*, *Gaiella occulta* F2-233, places it between *Thermoleophilia* and *Rubrobacteria* as indicated by 16S analysis. Members of the genus *Thermoleophilium*, the only genus in the order *Thermoleophilales*, are heat and oil loving, grow on n-alkanes, and are found largely in hot springs, while *Solirubrobacterales* contains species found in soil and grows on more diverse substrates, for example, part of the paddy soil microbiome (Li et al., 2018).

Bifidobacteriaceae

Members of the genus *Bifidobacterium* are facultatively anaerobic chemoorganotrophs found in the gut of humans, animals, and bees with diverse carbohydrate utilization abilities (Milani et al., 2016). They are linked to probiotics and development of the infant gut (Turroni et al., 2014. The other genera are more often present in the oral cavity and associated with organisms responsible for dental caries. Phylogenomic analysis was performed by Zhang et al. (2016) of the members of the genera *Bifidobacterium* and *Gardnerella*. Lugli et al. (2017) sequenced the members of the additional genera in *Bifidobacteriaceae* and identified acquisition of glycosyl hydrolase-encoding genes to enhance the utilization of carbohydrates, with the relative number of acquisitions linked to the diet of the host. *Gardnerella* was confirmed as more closely related to bifidobacteria rather than the other genera.

Corynebacteriales

Rhodococcus

The genus *Rhodococcus* contains *Rhodococcus equi*, an opportunistic pathogen for humans and animals, specifically foals; *R. fascians*, a pathogen forming leafy galls in dicotyledonous plants; and *R. erythropolis* used for industrial bioconversions. The genetic heterogeneity of the genus *Rhodococcus* is clear from 16S studies (Gürtler et al., 2004) and the analysis of some genomes (Creason et al., 2014). A more comprehensive phylogenomic study of more than 100 strains reveals the diverse groups, which they conclude are as similar to other *corynebacteria* as to one another (Sangal et al., 2016).

Rhodococcus equi

The phylogenomic analysis of 29 *R. equi* strains (Anastasi et al., 2016) confirmed that evolution of *R. equi* is by gene gain/loss in a clonal clade, while the lack of a phosphoenolpyruvate:carbohydrate transport system is shared only with the closely related *Rhodococcus defluvii*. The genomic relationships indicated that *R. equi* occupies a position within the *Rhodococcus* clade.

Tsukamurella

In describing *Tsukamurella hongkongensis*, Teng et al. (2015) observed high similarity in 16S across all described *Tsukamurella* species and anomalous similarity between pairs of validly described species (*Tsukamurella pulmonis/Tsukamurella spongiae*, *Tsukamurella tyrosinosolvens/Tsukamurella carboxydivorans*, and *Tsukamurella pseudospumae/Tsukamurella sunchonensis*). Resequencing of 16S revealed that these pairs had identical 16S sequences. Teng et al. (2016) performed a full phylogenomic and phenotypic analysis to reclassify these pairs of strains as the same species.

Geodermatophilales

Geodermatophilaceae

The family *Geodermatophilaceae* is distant from other actinobacterial clades and found in metagenomic studies of arid and hyperarid habitats. Initially problematic to study because of how difficult they are to cultivate, there are now many more isolates. They are clustered together by 16S because they are so different from their nearest neighbors. The phylogenomic analysis by Montero-Calasanz et al. (2017) clarifies the taxonomy and proposes a new genus *Klenkia*.

Micromonosporales

Micromonospora

Micromonospora have been isolated from soil, freshwater, and marine habitats, but also from insects, sponges, and as endophytes from plants. They are prolific producers of bioactive secondary metabolites, including antibiotics, antitumor, and biocontrol agents. A genome-based classification has been performed (Carro et al., 2018).

Salinispora

The three species of *Salinispora* are a well-studied clade of marine producers of bioactive metabolites. The three species are not definable as separate species by 16S, but are well-established species. They provide one of the best models for the integration of ecology, speciation, and the ecological role and search and discovery of bioactive metabolites (Patin et al., 2016). Millán-Aguiñaga et al. (2017) analyzed the genomes of 119 *Salinispora* genomes from diverse marine sediments and identified three major clades corresponding to the three identified species. They identified the core/pan-genomes and detected genes with evidence of recombination. Phylogenetic analysis was performed based on all core genes, core genes without evidence of recombination, and core genes with evidence of recombination. Phylogenies based on concatenation and coalescence of these individual gene trees all gave congruent phylogenies. The threshold to define species is not inherent in these tree distances.

An ANI analysis concluded that the data may encompass ten clusters that exhibit ANI similarity to less than the 95% species threshold. The paper provides a model for a wide-ranging theoretical and practical phylogenetic and systematic analysis of members of an actinobacterial clade.

Pseudonocardiales

Pseudonocardia

Pseudonocardia are famously associated with the leaf-cutter ant *Escovopsis-Pseudonocardia* assembly (Mueller et al., 2010). Caldera et al. (2019) analyzed the genomes of 29 ant-associated *Pseudonocardia* as well as their biosynthetic gene clusters (BGC) and identified significant variation of BGCs against a background of low diversity in the core genomes. They found elevated *Escovopsis* inhibition by *Pseudonocardia* from sites in the Panama Canal Zone compared to Costa Rica. That inhibition was locally adapted to the strains present at one site, but not the other two sites, in Panama. However, this pattern of inhibition was not correlated with the presence/absence of single BGCs. They assess these results against the geographic mosaic theory of coevolution.

Amycolatopsis

The genus *Amycolatopsis* is well-known for the prolific production of bioactive metabolites (Chen et al., 2016) including vancomycin and rifamycin. Adamek et al. (2018) sequenced two *Amycolatopsis* genomes and downloaded 41 others to analyze, generating a phylogenetic tree based upon an MLST analysis of gene sequences from the genomes, and mapped the BGCs identified by antiSMASH (Blin et al., 2019) onto the phylogenomic clades.

Streptomycetales

Streptomyces

The genus *Streptomyces* is the largest bacterial genus, the result of its predominance as a prolific producer of bioactive metabolites. The genus is extremely diverse and found across almost all ecological sites, the result perhaps of aerobic growth and profuse spore formation. *Streptomyces albidoflavus* is perhaps the most ubiquitous, with many independent genomes (30) defining a single species, and is found in diverse habitats (from air and soil to ants and marine cone snails) with a largely common set of BGCs (Ward and Allenby, 2018), though perhaps differentially expressed.

The phylogeny of the genus has been studied with 16S for all valid species (Labeda et al., 2012) and by MLST (Labeda et al., 2017). Whole genome phylogenies have been published (Alam et al., 2010; Chaves et al., 2018) but only including three and twenty-six genomes, respectively, at the time of writing there were 244 complete whole genomes and 1,891 genomes in total for the genus in Genbank.

Streptosporangiales

Nocardiopsis

Nocardiopsis has *N. dassonvillei* as the type strain of the genus. Both *N. dassonvillei* and *N. synnemataformans* are opportunistic pathogens and may cause mycetoma and abscesses. Otherwise, members of the genus are common in soil, though also found in marine and hyper-saline environments associated with sponges, coral, sea anemone, mollusks, pufferfish, earthworms, and mosquitos, and produce diverse bioactive metabolites. Li et al. (2013) analyzed 17 genomes and defined the core and pan-genomes.

Actinobacteria

Several studies have used whole genomes to look at the whole phylum. The first application of the analysis of whole genomes to the whole phylum (Sen et al., 2014) used 100 completely sequenced genomes representing 35 families and 17 orders. They evaluated eight different hypotheses for their phylogeny, including one based on a concatenation of 54 conserved proteins. They made proposals regarding a number of taxa, but recognized that formal proposals required more whole genomes. *Frankiales* and *Micrococcales* were recognized as Orders needing revision.

Salam et al. (2019) created a full 16S phylogenetic tree and resolved inconsistencies using gene sequences from whole genome data. A full phylogenomic analysis (Nouioui et al., 2018) was based upon 1,142 downloaded genomes and analyzed with the high-throughput version of the Genome BLAST Distance Phylogeny (Meier-Kolthoff et al., 2014) and presented as a multi-page phylogenetic tree and detailed taxonomic scrutiny, most of which is incorporated in the LPSN (Parte, 2018). Gupta (2019) provides a commentary on the analysis by Nouioui et al. (2018)..

Hands-on Whole Genome Sequencing and Analysis

The increasing number of publicly available actinobacterial whole genomes means that the sequencing of a lab strain under study is potentially leveraged by related strains already sequenced. If the project is based around a single strain, it probably makes sense to sequence with sequence providers for PacBio and Illumina MiSeq or HiSeq (Gomez-Escribano et al., 2015). Depending on the study objectives, the sequence data will still need extensive analysis and will involve significant bioinformatic analysis. Submitting the genome to the databases means that data can potentially be utilized to leverage other studies.

However, if the study objectives involve sequencing the genomes of multiple strains, the advances in sequencing make "in house" sequencing more accessible than ever. The portable nanopore sequencing device produced by Oxford Nanopore (https://nanoporetech.com/) is the most personal of personal sequencers producing significant data, perhaps 20 Gb sequenced bases per flow cell, as long reads (the record long read is about 2 Mb; hundreds of Kb are easily achieved and typically at least >7 Kb), though with relatively high error. Illumina has now produced the iSEQ (https://emea.illumina.com/systems/sequencing-platforms/iseq.html) as an inexpensive system with simple sample preparation producing 1.2 Gb or more of accurate short-read data. Alternatively, the error correcting short-read data can be provided by a low cost, relatively low coverage short read run from a service provider. Improved flow cells and base calling software continue to improve long read only assemblies.

The wet lab side is very accessible though DNA quality and preparation is much more important for long-read sequencing and beneficial for short-read sequencing too. The massive amounts of data generated are challenging. Barcoded samples can be run on both nanopore flow cells 12- and 24-barcode kits and the iSEQ 12-barcode kit, making sequencing costs per genome accessible, though not guaranteeing enough data for a complete genome for every sample in one run.

A $1,000 starter kit comes with the Oxford Nanopore MinION sequencer, two flow cells, and a library preparation kit. You will need a high-end computer with at least a core i7 processor, 16 Gb RAM, a 1 TB solid state drive, and USB 3.0 high-speed data connection to run the software to collect the data from a MinION run over 1 days.

DNA Preparation for Long-Read Sequencing

DNA preparation will depend on the organisms and sample, but long-read sequencing needs high molecular weight, high quality DNA that is completely free of contaminants, such as protein, polysaccharides, RNA, salt, phenol, chloroform, and SDS; it is essential to optimize this and minimize actions that shear the DNA. Growth of actinobacteria in an optimal medium supplemented with sucrose and glycine may aid lysis, though the concentrations may need to be optimized to prevent inhibition and premature lysis. Extraction of high-molecular-weight DNA using the salting out method (Hopwood et al., 2000) followed by a magnetic bead purification (Agencourt AMPure XP beads), or using the Qiagen Puregene Yeast/Bacteria kit or the MagAttract HMW DNA Kit has worked in our hands. Classic phenol:chloroform methods can work well, but it is essential that all traces of chloroform and phenol are removed. Size selection using magnetic beads can be used to deplete low-molecular-weight fragments, though short fragments can be utilized for error correction. Detailed protocols are on the Oxford Nanopore website (accessible after purchase of a starter kit). The same precautions apply for both nanopore sequencing and DNA preparation for PacBio sequencing—detailed guidelines are freely available online (https://www.pacb.com/wp-content/uploads/Technical-Note-Preparing-DNA-for-PacBio-HiFi-Sequencing-Extraction-and-Quality-Control.pdf—accessed September 2020). Although 260/280 nm and 260/230 nm values of 1.8–2.0 and 2.0–2.2 should be achieved, and can be determined by Nanodrop spectrophotometry, Nanodrop is not suitable for quantification (it typically overestimates DNA, perhaps twofold) and DNA quantification is essential and should be done fluorimetrically (Qubit or PicoGreen). Molecular weight should be checked by gel and TapeStation—running a gel is potentially diagnostic for other problems, such as the presence of RNA, phenol, or sheared DNA. Pipetting of high MW DNA needs wide-bore tips and avoidance of vortexing. More DNA is better, though low-input protocols are available (0.1–1 μg) HMW DNA is needed as input to the library preparation protocols.

Library Preparation

Single-molecule sequencing, PacBio (Wenger et al., 2019), and nanopore 2D sequencing revealed the extent of DNA damage, nicks, and abasic sites occurring in DNA preparations so that end-repair before attaching sequencing adapters and barcodes, followed by FFPE repair (e.g., NEB FFPE Repair Kit #M6630) improve DNA quality for long reads. Adding these steps, followed by DNA cleanup by magnetic beads, loses DNA at each step. The transposasome rapid sequencing kit is a fast, one-step library preparation method in which a transposasome adds the barcode and sequencing adapter in a brief 1-minute reaction. The ligation adapter kit adds adapter and barcode by ligation and is a multistep protocol (nominally 2–3 hours, but we allow longer). However, the ligation protocol tends to generate more data and longer average read length. Flow cells, kits, protocols, and software are all subject to updating.

Sequencing Run

Connect the sequencer to a USB port, insert the flow cell, and start the software as directed in the protocols. Next, run the QC check for the flow cell. Make up the priming buffer and transfer the priming buffer into the flow cell, then add the library preparation. In the protocol cited later, the priming buffer and library were pipetted directly into the flow cell port but later, flow cells use spot-on loading in which samples are added by dropping the sample onto the port where it is drawn in by capillary action, preventing bubbles entering the sequencing chamber. The sequencing can be followed on the software as single molecules attach to pores, transport through driven by the motor protein, and change the ion current flowing through the pore as the DNA bases translocate. The ion current traces (squiggles) are written to HDF5 files (https://www.hdfgroup.org/solutions/hdf5/) in nanopore fast5 format. They can be visualized with BulkVis (https://github.com/LooseLab/bulkvis).

A good overview of the steps, now slightly out of date, can be found in the protocol for ultra-long reads (Gong et al., 2019).

Basecalling

The next step, converting the raw signal data into FastQ base sequences, is the most demanding single step. The latest generation of basecallers are installed locally and run artificial neural networks. The Oxford Nanopore-supplied program, Guppy, comes in two formats: GuppyCPU and GuppyGPU. The neural network basecalling is computationally intensive and the Nvidia GPU-enabled Guppy requires a high-end GPU with a compute value >6.1—as in a high-end gaming computer. Basecalling the output from a single run can take several days, basecalling with the CPU version of Guppy is significantly slower.

Debarcoding

The Guppy software can debarcode the FastQ data relatively quickly and save the reads for each barcoded dataset into separate folders. Each folder then contains the reads as FastQ files, corresponding to the genome for the sample with that barcode.

Assembly

Assembling long reads with high errors is a different computational problem from assembling accurate short reads—different assemblers and assembly pipelines are needed (Wick and Holt, 2019). It make sense to assemble error-prone long reads with more than one assembler, but an initial assembly with Unicycler (Wick et al., 2017) means the same assembler can be used to assemble long read only, short read only, and do a hybrid assembly with both long and short reads. Then, other assemblers like Canu (Koren et al., 2017) and Flye (Kolmogorov et al., 2019) can be used to identify potential assembly mismatches.

Annotation

The NCBI's Prokaryote Genome Annotation Pipeline (PGAP —Haft et al., 2018) has been released as a standalone version (https://github.com/ncbi/pgap/wiki) with an ignore-all-errors parameter that will annotate draft assemblies. In particular, it will identify, count, and mark frameshifts and unexpected stop codons, which are hallmarks of the indel errors present in long-read assemblies

Conclusion

Whole genome sequencing has revolutionized biology and will undoubtedly change our understanding of the taxonomy and phylogeny of major clades, such as the *Actinobacteria*. But more particularly, it will integrate with broader research objectives, such as understanding evolution and identifying how pathogens function and can be targeted. The application of phylogenomics to the systematics of the actinobacteria is still in its infancy, but the research cited here not only provides more information about the relationships within the *Actinobacteria* studied, but also provides models for data analysis of both the organisms and how they function. These examples are important because we are in an era of big data, and massive data analysis with a plethora of software solutions.

The hands-on approach to genome sequencing identifies some major hurdles to the democratization of genome sequencing. Addressing these hurdles in the age of big data is going to be key for more than just genome sequencing or phylogenomics.

REFERENCES

Adamek M, Alanjary M, Sales-Ortells H, Goodfellow M, Bull AT, Winkler A, Wibberg D, Kalinowski J, Ziemert N (2018) Comparative genomics reveals phylogenetic distribution patterns of secondary metabolites in *Amycolatopsis* species. *BMC Genom.* 19, 426. doi: 10.1186/s12864-018-4809-4

Alam MT, Merlo ME, Takano E, Breitling R (2010) Genome-based phylogenetic analysis of streptomyces and its relatives. *Mol Phylogenet Evol.* 54(3), 763–772. doi: 10.1016/j.ympev.2009.11.019

Alvarcz A, Saez JM, Davila Costa JS, Colin VL, Fuentes MS, Cuozzo SA, Benimeli CS, Polti MA, Amoroso MJ (2017) Actinobacteria: current research and perspectives for bioremediation of pesticides and heavy metals. *Chemosphere.* 166, 41–62. doi: 10.1016/j.chemosphere.2016.09.070

Anastasi E, MacArthur I, Scortti M, Alvarez S, Giguère S, Vázquez-Boland JA (2016) Pangenome and phylogenomic analysis of the pathogenic actinobacterium *Rhodococcus equi*. *Gen Biol Evol.* 8(10), 3140–3148. doi: 10.1093/gbe/evw222

Antony-Babu S, Stien D, Eparvier V et al. (2017) Multiple *Streptomyces* species with distinct secondary metabolomes have identical 16S rRNA gene sequences. *Sci Rep.* 7, 11089. doi: 10.1038/s41598-017-11363-1

Armstrong J, Fiddes IT, Diekhans M, Paten B (2019) Whole-Genome alignment and comparative annotation. *Annu Rev Anim Biosci.* 7, 41–64. doi: 10.1146/annurev-animal-020518-115005

Battistuzzi FU, Feijao A, Hedges SB (2004) A genomic timescale of prokaryote evolution: insights into the origin of methanogenesis, phototrophy, and the colonization of land. *BMC Evol Biol.* 4, 44. doi: 10.1186/1471-2148-4-44

Becker J, Rohles CM, Wittmann C (2018) Metabolically engineered corynebacterium glutamicum for bio-based production of chemicals, fuels, materials, and healthcare products. *Metab Eng.* 50, 122–141. doi: 10.1016/j.ymben.2018.07.008

Bernard K (2012) The genus corynebacterium and other medically relevant coryneform-like bacteria. *J Clin Microbiol.* 50(10), 3152–3158.

Bisanz JE, Soto-Perez P, Lam KN, Bess EN, Haiser HJ, Allen-Vercoe E, Mekdal VM, Balskus EP, Turbaugh PJ (2018) Illuminating the microbiome's dark matter: a functional genomic toolkit for the study of human gut actinobacteria. *bioRXriv.* doi: 10.1101/304840

Bisgaard M, Christensen H, Clermont D, Dijkshoorn L, Janda JM, Moore ERB, Nemec A, Nørskov-Lauritsen N, Overmann J, Reubsaet FAG (2019) The use of genomic DNA sequences as typematerial for valid publication of bacterial species names will have severe implications for clinical microbiology and related discipline. *Diagn Microbiol Infect Dis.* 95, 102–103. doi: 10.1016/j.diagmicrobio.2019.03.007.

Blin K, Shaw S, Steinke K, Villebro R, Ziemert N, Lee SY, Medema MH, Weber T (2019) antiSMASH 5.0: updates to the secondary metabolite genome mining pipeline. *Nucleic Acids Res.* 47(W1), W81–W87. doi: 10.1093/nar/gkz310.

Bull AT, Ward AC, Goodfellow M (2000) Search and discovery strategies for biotechnology: the paradigm shift. *Microbiol Mol Bio Rev.* 64(3), 573–606. doi: 10.1128/MMBR.64.3.573-606.2000.

Caldera EJ, Chevrette MG, McDonald BR, Currie CR (2019) Local adaptation of bacterial symbionts within a geographic mosaic of antibiotic coevolution. *Appl Environ Microbiol.* 85(24), e01580–19. doi: 10.1128/AEM.01580-19.

Carro L, Nouioui I, Sangal V, Meier-Kolthoff JP, Trujillo ME, Montero-Calasanz M del C, Sahin N, Smith DL, Kim KE, Peluso P, Deshpande S, Woyke T, Shapiro N, Kyrpides NC, Klenk H-P, Göker M, Goodfellow M (2018) Genome-based classification of micromonosporae with a focus on their biotechnological and ecological potential. *Sci Rep.* 8, 525. doi: 10.1038/s41598-017-17392-0

Chai Q, Zhang Y, Liu CH (2018) *Mycobacterium tuberculosis*: an adaptable pathogen associated with multiple human diseases. *Front Cell Infect Microbiol.* 8, 158. doi: 10.3389/fcimb.2018.00158

Chaisson MJ, Huddleston J, Dennis MY, Sudmant PH, Malig M, Hormozdiari F, Antonacci F, Surti U, Sandstrom R, Boitano M, Landolin JM, Stamatoyannopoulos JA, Hunkapiller MW, Korlach J, Eichler EE2 (2015) Resolving the complexity of the human genome using single-molecule sequencing. *Nature.* 517, 608–611. doi: 10.1038/nature13907.

Chaves JV, Cinthya Ojeda CPO, da Silva IR, Procopio RE de L (2018) Identification and phylogeny of *Streptomyces* based on gene sequences. *Res J Microbiol.* 13, 13–20. doi: 10.3923/jm.2018.13.20.

Chen D, Tian Y, Jiao J et al. (2020) Comparative genomics analysis of *Nitriliruptoria* reveals the genomic differences and salt adaptation strategies. *Extremophiles.* 24, 249–264. doi: 10.1007/s00792-019-01150-3.

Chen P, Zhang L, Guo X, Dai X, Liu L, Xi L, et al. (2016) Diversity, biogeography, and biodegradation potential of actinobacteria in the deep-sea sediments along the southwest Indian ridge. *Front Microbiol.* 7, 1340. doi: 10.3389/fmicb.2016.01340.

Chen S, Wu Q, Shen Q, Wang H (2016) Progress in understanding the genetic information and biosynthetic pathways behind *Amycolatopsis* antibiotics, with implications for the continued discovery of novel drugs. *Chembiochem.* 17(2), 119–28. doi: 10.1002/cbic.201500542.

Chevrette MG, Carlos-Shanley C, Louie KB, Bowen BP, Northen TR, Currie CR (2019) Taxonomic and metabolic incongruence in the ancient genus *Streptomyces.Front Microbiol.* 10, 2170 doi: 10.3389/fmicb.2019.02170.

Conville PS, Witebsky FG (2007) Analysis of multiple differing copies of the 16S rRNA gene in five clinical isolates and three type strains of nocardia species and implications for species assignment. *J Clin Microbiol.* 45, 1146–1151.

Cowman S, Wilson R, Michael R. Loebinger MR (2016) Opportunistic mycobacterial diseases. *Infection.* 44(6), P390–392. doi: 10.1016/j.mpmed.2016.03.010.

Creason AL, Davis EW 2nd, Putnam ML, Vandeputte OM, Chang JH (2014) Use of whole genome sequences to develop a molecular phylogenetic framework for *Rhodococcus fascians* and the *Rhodococcus* genus. *Front Plant Sci.* 5, 406. doi: 10.3389/fpls.2014.00406.

Darling AE, Jospin G, Lowe E, Matsen FA IV, Bik HM, Eisen JA (2014) PhyloSift: phylogenetic analysis of genomes and metagenomes. *PeerJ.* 2, e243. doi: 10.7717/peerj.243.

Darling AE, Mau B, Perna NT (2010) ProgressiveMauve: multiple genome alignment with gene gain, loss and rearrangement. *PLOS ONE.* 5(6), e11147. doi: 10.1371/journal.pone.0011147.

Davies K (2010) 13 years ago, a beer summit in an English pub led to the birth of Solexa—and for now at least—the world's most popular second-generation sequencing technology. BioIT World. September 28, 2010. Available at www.bio-itworld.com/2010/issues/sept-oct/solexa.html.

de Vienne DM (2016) Lifemap: exploring the entire tree of life. *PLOS biology.* 14(12), e2001624. doi: 10.1371/journal.pbio.2001624.

Dewey CN (2019) Whole-Genome Alignment. In: Anisimova M. (eds) *Evolutionary Genomics. Methods in Molecular Biology*, vol 1910. Humana, New York, NY.

Dinesh R, Srinivasan V, Sheeja TE, Anandaraj M Srambikkal H (2017) Endophytic actinobacteria: diversity, secondary metabolism and mechanisms to unsilence biosynthetic gene clusters. *Crit Rev Microbiol.* 43(5), 546–566. doi: 10.1080/1040841X.2016.1270895.

Eid J, Fehr A, Gray J, Luong K, Lyle J, Otto G, Peluso P, Rank D, Baybayan P, Bettman B, Bibillo A, Bjornson K, Chaudhuri B, Christians F, Cicero R, Clark S, Dalal R, deWinter A, Dixon J, Foquet M, Gaertner A, Hardenbol P, Heiner C, Hester K, Holden D, Kearns G, Kong X, Kuse R, Lacroix Y, Lin S, Lundquist P, Ma C, Marks P, Maxham M, Murphy D, Park I, Pham T,

Phillips M, Roy J, Sebra R, Shen G, Sorenson J, Tomaney A, Travers K, Trulson M, Vieceli J, Wegener J, Wu D, Yang A, Zaccarin D, Zhao P, Zhong F, Korlach J, Turner S (2009) Real-time DNA sequencing from single polymerase molecules. *Science*. 323, 133–138. doi: 10.1126/science.1162986.

Gomez-Escribano JP, Castro JF, Razmilic V, Chandra G, Andrews B, Asenjo JA, Bibb MJ (2015) The *Streptomyces leeuwenhoekii* genome: de novo sequencing and assembly in single contigs of the chromosome, circular plasmid pSLE1 and linear plasmid pSLE2. *BMC Genom*. 16, 485. doi: 10.1186/s12864-015-1652-8.

Gong L, Wong C-H, Idol J, Ngan CY, Wei C-L (2019) Ultra-long read sequencing for whole genomic DNA analysis. *J Vis Exp*. 145, e58954. doi:10.3791/58954.

Goodfellow M (2012) Phylum XXVI. *Actinobacteria* phyl. nov. In: Goodfellow M., Kämpfer P., Busse, H.-J., Trujillo M.E., Suzuki K., Ludwig W., Whitman W.B. (eds) *Bergey's Manual of Systematic Bacteriology*, 2nd ed., vol. 5, p. 33–34. Springer, New York.

Goris J, Konstantinidis KT, Klappenbach JA, Coenye T, Vandamme P, Tiedje JM (2007) DNA–DNA hybridization values and their relationship to whole-genome sequence similarities. *Int J Syst Evol Microbiol*. 57(1), 81–91.

Gupta A, Sharma VK (2015) Using the taxon-specific genes for the taxonomic classification of bacterial genomes. *BMC Genom*. 16, 396. doi: 10.1186/s12864-015-1542-0.

Gupta RS (2019) Genome-Based taxonomic classification of the phylum *Actinobacteria*. *Front. Microbiol*. doi: 10.3389/fmicb.2019.00206.

Gupta RS, Chen WJ, Adeolu M, Chai Y (2013) Molecular signatures for the class *Coriobacteriia* and its different clades; proposal for division of the class *Coriobacteriia* into the emended order *Coriobacteriales*, containing the emended family *Coriobacteriaceae* and *Atopobiaceae* fam. nov., and *Eggerthellales* ord. nov., containing the family *Eggerthellaceae* fam. nov. *Int J Syst Evol Microbiol*. 63(9), 3379–3397. doi: 10.1099/ijs.0.048371-0.

Gürtler V, Mayall BC, Seviour R (2004) Can whole genome analysis refine the taxonomy of the genus *Rhodococcus*? *FEMS Microbiol Rev*. 28(3), 377–403. doi: 10.1016/j.femsre.2004.01.001.

Haas F, König H (1988) *Coriobacterium glomerans* gen. nov., sp. nov. from the intestinal tract of the red soldier bug. *Int J Syst Evol Microbiol*. 38(4), 382–384. doi: 10.1099/00207713-38-4-382.

Haft DH, DiCuccio M, Badretdin A, Brover V, Chetvernin V, O'Neill K, Li W, Chitsaz F, Derbyshire MK, Gonzales NR, Gwadz M, Lu F, Marchler GH, Song JS, Thanki N, Yamashita RA, Zheng C, Thibaud-Nissen F, Geer LY, Marchler-Bauer A, Pruitt KD (2018) RefSeq: an update on prokaryotic genome annotation and curation. *Nucleic Acids Res*. 46(D1), D851–D860. doi: 10.1093/nar/gkx1068.

Hopwood DA, Kieser T, Bibb M, Buttner M, Chater K (2000) *Practical Streptomycete Genetics*. John Innes Foundation, Norwich. ISBN: 9780708406236.

Hu D, Cha G, Gao B (2018) A phylogenomic and molecular markers based analysis of the class *Acidimicrobiia*. *Front Microbiol*. 9, 987. doi: 10.3389/fmicb.2018.00987.

Hu D, Zang Y, Mao Y, Gao B (2019) Identification of molecular markers that are specific to the class *Thermoleophilia*. *Front Microbiol*. 10, 1185. doi: 10.3389/fmicb.2019.01185.

Jain M, Fiddes IT, Miga KH, Olsen HE, Paten B, Akeson M (2015) Improved data analysis for the MinION nanopore sequencer. *Nat Methods*. 12, 351–356. doi: 10.1038/nmeth.3290.

Jain M, Koren S, Migal KH, Quick J, et al. (2018) Nanopore sequencing and assembly of a human genome with ultra-long reads. *Nat. Biotechnol*. 36, 338–345. doi: 10.1038/nbt.4060.

Jensen PR, Moore BS, Fenical W (2015) The marine actinomycete genus *Salinispora*: a model organism for secondary metabolite discovery. *Nat Prod Rep*. 32(5), 738–751. doi: 10.1039/c4np00167b.

Jones RM and Johnson DB (2015) *Acidithrix ferrooxidans* gen. nov., sp. nov.; a filamentous and obligately heterotrophic, acidophilic member of the *actinobacteria* that catalyzes dissimilatory oxido-reduction of iron. *Res Microbiol*. 166, 111–120. doi: 10.1016/j.resmic.2015.01.003.

Jovanovic N, Mikheyev AS (2019) Interactive web-based visualization and sharing of phylogenetic trees using phylogeny.IO. *Nucleic Acids Res*. 47(W1), W266–W269. doi:10.1093/nar/gkz356.

Koch R (1882) Die aetiologie der tuberculose. *Berl Klin Wochenschr*. 19, 221–230.

Kolmogorov M, Yuan J, Lin Y, et al. (2019) Assembly of long, error-prone reads using repeat graphs. *Nat Biotechnol*. 37, 540–546. doi: 10.1038/s41587-019-0072-8.

Koren S, Walenz BP, Berlin K, et al. (2017) Canu: scalable and accurate long-read assembly via adaptive *k*-mer weighting and repeat separation. *Genome Res*. 27(5), 722–736. doi: 10.1101/gr.215087.116.

Labeda DP, Dunlap CA, Rong X, Huang Y, Doroghazi JR, Ju KS, Metcalf WW (2017) Phylogenetic relationships in the family *Streptomycetaceae* using multi-locus sequence analysis. *Antonie Van Leeuwenhoek*. 110(4), 563–583. doi: 10.1007/s10482-016-0824-0.

Labeda DP, Goodfellow M, Brown R, Ward AC, Lanoot B, Vanncanneyt M, Swings J, Kim SB, Liu Z, Chun J, Tamura T, Oguchi A, Kikuchi T, Kikuchi H, Nishii T, Tsuji K, Yamaguchi Y, Tase A, Takahashi M, Sakane T, Suzuki KI, Hatano K (2012) Phylogenetic study of the species within the family *Streptomycetaceae*. *Antonie Van Leeuwenhoek*. 101(1), 73–104. doi: 10.1007/s10482-011-9656-0.

Lander ES, Linton LM, Birren B, Nusbaum C, Zody MC, Baldwin J et al. (2001) Initial sequencing and analysis of the human genome. *Nature*. 409, 860–921. doi: 10.1038/35057062.

Larsen MV, Consentino S, Rasmussen S, Friis C, Hasman H, Lykke Marvig R, et al. (2012) Multilocus sequence typing of total-genome-sequenced bacteria. *J Clin Microbiol*. 50, 1355–1361. doi: 10.1128/JCM.06094-11.

Lechevalier HA and Lechevalier MP (1989) *CRC Practical Handbook of Microbiology 1st Edition* In: O'Leary, WM (ed). CRC Press, Boca Raton, FL.

Lee J-H and O'sullivan DJ (2010) Genomic insights into bifidobacteria. *Microbiol Mol Biol Rev*. 74(3), 378–416. doi: 10.1128/MMBR.00004-10.

Li HW, Zhi XY, Yao JC, Zhou Y, Tang SK, Klenk HP, Zhao J, Li WJ (2013) Comparative genomic analysis of the genus *Nocardiopsis* provides new insights into its genetic mechanisms of environmental adaptability. *PLOS One*. 8(4), e61528. doi: 10.1371/journal.pone.0061528.

Li HY, Wang H, Wang HT, Xin PY, Xu XH, Ma Y, et al. (2018) The chemodiversity of paddy soil dissolved organic matter correlates with microbial community at continental scales. *Microbiome*. 6, 187. doi: 10.1186/s40168-018-0561-x.

Lugli GA, Milani C, Turroni F, Duranti S, Mancabelli L, Mangifesta M, Ferrario C, Modesto M, Mattarelli P, Jiří K, van Sinderen D, Ventura M (2017) Comparative genomic and phylogenomic analyses of the *Bifidobacteriaceae* family. *BMC Genom.* 18, 568. doi: 10.1186/s12864-017-3955-4.

Margulies M, Egholm M, Altman WE, Attiya S, Bader JS, Bemben LA, Berka J, et al. (2005) Genome sequencing in microfabricated high-density picolitre reactors. *Nature.* 437(7057), 376–380. doi: 10.1038/nature03959.

Meier-Kolthoff JP, Auch AF, Klenk H, et al. (2013) Genome sequence-based species delimitation with confidence intervals and improved distance functions. *BMC Bioinformatics.* 14, 60. doi: 10.1186/1471-2105-14-60.

Meier-Kolthoff, JP, Auch, AF., Klenk, H-P, Göker, M (2014) Highly parallelized inference of large genome-based phylogenies. *Concurr Comput Pract Exp.* 26, 1715–1729. doi: 10.1002/cpe.3112.

Merriman B, Rothberg JM (2012) Progress in ion torrent semiconductor chip based sequencing *Electrophoresis.* 33, 3397–3417. doi: 10.1002/elps.201200424.

Milani C, Turroni F, Duranti S, Lugli GA, Mancabelli L, Ferrario C, van Sinderen D, Ventura M (2016) Genomics of the genus *Bifidobacterium* reveals species-specific adaptation to the glycan-rich gut environment. *Appl Environ Microbiol.* 82(4), 980–991. doi: 10.1128/AEM.03500-15.

Millán-Aguiñaga N, Chavarria KL, Ugalde JA, Letzel A-C, Rouse GW, Jensen PR (2017) Phylogenomic insight into *Salinispora* (bacteria, actinobacteria) species designations. *Sci Rep.* 7, 3564. doi: 10.1038/s41598-017-02845-3.

Miro E, Rossen JWA, Chlebowicz MA, Harmsen D, Brisse S, Passet V, Navarro F, Friedrich AW, García-Cobos S (2020) Core/whole genome multilocus sequence typing and core genome SNP-based typing of OXA-48-producing *Klebsiella pneumoniae* clinical isolates from Spain. *Front Microbiol.* 10, 2961. doi: 10.3389/fmicb.2019.02961.

Mizuno CM, Rodriguez-Valera F, Ghai R (2015) Genomes of planktonic *Acidimicrobiales*: widening horizons for marine actinobacteria by metagenomics. *mBio.* 6, e2083–14. doi: 10.1128/mBio.02083-14.

Montero-Calasanz M del C, Meier-Kolthoff J-P, Zhang D-F, Yaramis A, Rohde M, Woyke T, Kyrpides NC, Schumann P, Li W-J, Göker M (2017) Genome-scale data call for a taxonomic rearrangement of *Geodermatophilaceae*. *Front Microbiol.* 8, 2501. doi: 10.3389/fmicb.2017.02501.

Mueller UG, Ishak H, Lee JC, Sen R, Gutell RR (2010) Placement of attine ant-associated *Pseudonocardia* in a global *Pseudonocardia* phylogeny (*Pseudonocardiaceae*, *Actinomycetales*): a test of two symbiont-association models. *Antonie van Leeuwenhoek.* 98(2), 195–212. doi: 10.1007/s10482-010-9427-3.

Newman DJ, Cragg GM (2016) Natural products as sources of new drugs from 1981 to 2014. *J Nat Prod.* 79(3), 629–661. doi: 10.1021/acs.jnatprod.5b01055.

Newman DJ, Cragg GM (2020) Plant endophytes and epiphytes: burgeoning sources of known and "unknown" cytotoxic and antibiotic agents? *Planta Med.* doi: 10.1055/a-1095-1111.

Nguyen TV, Wibberg D, Vigil-Stenman T, Berckx F, Battenberg K, Demchenko KN, Blom J, Fernandez MP, Yamanaka T, Berry AM, Kalinowski J, Brachmann A, Pawlowski K (2019) Frankia-enriched metagenomes from the earliest diverging symbiotic Frankia cluster: they come in teams. *Genome Biol Evol.* 11(8), 2273–2291. doi: 10.1093/gbe/evz153.

Norris PR (2012) Class II. *Acidimicrobiia* class. nov. In: Goodfellow M, Kämpfer P, H.-J. B, Trujillo ME, K.-I. S, Ludwig W, Whitman WB (eds), *Bergey's Manual of Systematic Bacteriology*, 2nd ed., vol. 5, part B, p. 1968. Springer, New York.

Nouioui I, Carrol L, García-López M, Meier-Kolthoff JP, Woyke T, Kyrpides NC, Pukall R, Klenk H-P, Goodfellow M, Göker M (2018) Genome-based taxonomic classification of the phylum *Actinobacteria*. *Front. Microbiol.* 9, 2007. doi: 10.3389/fmicb.2018.02007.

Oliveira A, Oliveira LC, Aburjaile F, Benevides L, Tiwari S, Jamal SB, Silva A, Figueiredo H, Ghosh P, Portela RW, De Carvalho Azevedo VA, Wattam AR (2017) Insight of genus *Corynebacterium*: ascertaining the role of pathogenic and non-pathogenic species. *Front Microbiol.* 8, 1937. doi: 10.3389/fmicb.2017.01937.

Parte AC (2018) LPSN — List of prokaryotic names with standing in nomenclature (bacterio.net), 20 years on. *Int J Syst Evol Microbiol.* 68, 1825–1829. doi: 10.1099/ijsem.0.002786.

Patin NV, Duncan KR, Dorrestein PC, Jensen PR (2016) Competitive strategies differentiate closely related species of marine actinobacteria. *ISME J.* 10(2), 478–490. doi: 10.1038/ismej.2015.128.

Pridham TG, Lyons AJ (1961) *Streptomyces albus* (Rossi-Doria) Waksman et Henrici: taxonomic study of strains labeled *Streptomyces albus*. *J. Bacteriol.* 81, 431–441.

Rong X, Huang Y (2010) Taxonomic evaluation of the *Streptomyces griseus* clade using multilocus sequence analysis and DNA-DNA hybridization, with proposal to combine 29 species and three subspecies as 11 genomic species. *Int J Syst Evol Microbiol.* 60, 696–703.

Rossl-Doria T (1891) Su di alcune specie di "Streptotrix" trovate nell'aria studiate in rapport a quelle giá note especialmente all' actinomyces. *Ann.igiene.* 1, 399–439.

Ruiz L, Delgado S, Ruas-Madiedo P, Sánchez B, Margolles A (2017) Bifidobacteria and their molecular communication with the immune system. *Front Microbiol.* 8, 2345. doi: 10.3389/fmicb.2017.02345.

Salam N, Jiao J, Zhang X, Li W (2019) Update in the classification of higher ranks in the phylum actinobacteria. *Int J Syst Evol Microbiol.* 70(2), 1331–1355. doi: 10.1099/ijsem.0.003920.

Salzberg SL, Yorke JA (2005) Beware of mis-assembled genomes. *Bioinformatics.* 21, 4320–4321. doi: 10.1093/bioinformatics/bti769.

Sangal V, Goodfellow M, Jones AL, Schwalbe EC, Blom J, Hoskisson PA, Sutcliffe, IC (2016) Next-generation systematics: an innovative approach to resolve the structure of complex prokaryotic taxa. *Sci Rep.* 6, 38392. doi: 10.1038/srep38392.

Sanger F, Nicklen S, Coulson AR (1977) DNA sequencing with chain-terminating inhibitors. *Proc Natl Acad Sci U S A.* 74, 5463–5467, doi: 10.1073/pnas.74.12.5463.

Schloss JA (2008) How to get genomes at one ten-thousandth the cost. *Nat. Biotechnol.* 26, 1113–1115. 10.1038/nbt1008-1113.

Sen A, Daubin V, Abrouk D, Gifford I, Berry AM, Normand P (2014) Phylogeny of the class *Actinobacteria* revisited in the light of complete genomes. The orders 'Frankiales' and *Micrococcales* should be split into coherent entities: proposal of *Frankiales* ord. nov., *Geodermatophilales* ord. nov., *Acidothermales* ord. nov. and *Nakamurellales* ord. nov. *Int J Syst Evol Microbiol.* 64(11), 3821–3832. doi: 10.1099/ijs.0.063966-0.

Seviour RJ, Kragelund C, Kong Y, Eales K, Nielsen JL, Nielsen PH (2008) Ecophysiology of the *Actinobacteria* in activated sludge systems. *Antonie Van Leeuwenhoek.* 94(1), 21–33. doi: 10.1007/s10482-008-9226-2.

Soares SC, Silva A, Trost E, Blom J, Ramos R, Carneiro A, Ali A, Santos AR, Pinto AC, Diniz C, Barbosa EG, Dorella FA, Aburjaile F, Rocha FS, Nascimento KK, Guimarães LC, Almeida S, Hassan SS, Bakhtiar SM, Pereira UP, Abreu VA, Schneider MP, Miyoshi A, Tauch A, Azevedo V (2013) The pan-genome of the animal pathogen *Corynebacterium pseudotuberculosis* reveals differences in genome plasticity between the biovar ovis and equi strains. *PLOS One.* 8(1), e53818. doi: 10.1371/journal.pone.0053818.

Stach JEM, Maldonado LA, Masson DG, Ward AC, Goodfellow M, Bull AT (2003) Statistical approaches for estimating actinobacterial diversity in marine sediments. *Appl Environ Microbiol.* 69(10), 6189–6200. doi: 10.1128/AEM.69.10.6189-6200.2003.

Stanier RY, van Niel CB (1962) The concept of a bacterium. *Arch Mikrobiol.* 42, 17–35. doi: 10.1007/BF00425185.

Taylor GM, Stewart GR, Cooke M, Chaplin S, Ladva S, Kirkup J, Palmer S, Young DB (2003) Koch's Bacillus – a look at the first isolate of *Mycobacterium tuberculosis* from a modern perspective. *Microbiology.* 149(11), 3213–3220. doi: 10.1099/mic.0.26654-0.

Teng JLL, Tang Y, Huang Y, Guo F-B, Wei W, Chen JHK, Wong SSY, Lau SKP, Woo PCY (2016) Phylogenomic analyses and reclassification of species within the genus *Tsukamurella*: insights to species definition in the post-genomic era. *Front Microbiol.* 7, 1137. doi: 10.3389/fmicb.2016.01137.

Teng JLL, Tang Y, Huang Y, Guo F-B, Wei W, Chen JHK, Wong SSY, Lau SKP, Woo PCY (2015) *Tsukamurella hongkongensis* sp. nov. and *Tsukamurella sinensis* sp. nov. isolated from patients with keratitis, catheter-related bacteraemia and conjunctivitis in Hong Kong. *Int J Syst Bacteriol.* 66, 391–397. 10.1099/ijsem.0.000733.

Tissier H (1899) Le bacterium coli et la reaction chromophile d'escherich. *Crit Rev Soc Biol.* 51, 943–945.

Tiwari K, Gupta RK. (2012) Rare actinomycetes: a potential storehouse for novel antibiotics. *Crit Rev Biotechnol.* 32, 108–132. doi: 10.3109/07388551.2011.562482.

Treusch AH, Vergin KL, Finlay LA, Donatz MG, Burton RM, Carlson CA, et al. (2009) Seasonality and vertical structure of microbial communities in an ocean gyre. *ISME J.* 3, 1148–1163. doi: 10.1038/ismej.2009.60

Trujillo ME, Riesco R, Benito P, Carro L (2015) Endophytic actinobacteria and the interaction of *Micromonospora* and nitrogen fixing plants. *Front Microbiol.* 6, 1341. doi: 10.3389/fmicb.2015.01341.

Turroni F, Ventura M, Butto LF, Duranti S, O'Toole PW, Motherway MO, van Sinderen D. (2014) Molecular dialogue between the human gut microbiota and the host: a *Lactobacillus* and *Bifidobacterium* perspective. *Cell Mol Life Sci.* 71, 183–203. doi: 10.1007/s00018-013-1318-0.

van Dijk EL, Auger H, Jaszczyszyn Y, Thermes C (2014) Ten years of next-generation sequencing technology. *Trends Genet.* 30, 418–426. 10.1016/j.tig.2014.07.001.

van Dijk EL, Jaszczyszyn Y, Naquin D, Thermes C (2018) The third revolution in sequencing technology. *Trends Genet.* 34(9), 666–681. doi: 10.1016/j.tig.2018.05.008.

Venter JC, Adams MD, Myers EW, Li PW, Mural RJ, Sutton GG et al. (2001) The sequence of the human genome. *Science.* 291, 1304–1351. doi: 10.1126/science.1058040.

Ventura M, Canchaya C, Tauch A, Chandra G, Fitzgerald GF, Chater KF, van Sinderen D (2007) Genomics of *Actinobacteria*: tracing the evolutionary history of an ancient phylum. *Microbiol Mol Biol R.* 71, 495–548. Doi: 10.1128/MMBR.00005-07.

Walker SL, Lockwood DN (2006) The clinical and immunological features of leprosy. *Br Med Bull.* 77–78, 103–121. doi:10.1093/bmb/ldl010.

Ward AC, Allenby NEE (2018) Genome mining for the search and discovery of bioactive compounds: the *Streptomyces* paradigm, *FEMS Microbiology Lett.* 365(24). doi: 10.1093/femsle/fny240.

Ward AC, Bora N (2006) Diversity and biogeography of marine actinobacteria. *Curr Opin Microbiol.* 9(3), 279–286. doi: 10.1016/j.mib.2006.04.004.

Ward AC, Bora N (2008) The Actinobacteria. In: Goldman E, Green LH (eds), *Practical Handbook of Microbiology*, 2nd ed., Chapter 27, pp. 375–444, CRC Press. doi: 10.1201/9781420009330

Ward AC, Bora N (2015) The Actinobacteria. In: Goldman E, Green LH (eds), *Practical Handbook of Microbiology*, 3rd ed., Chapter 30, pp. 505–547, CRC Press. doi: 10.1201/b17871

Warnecke F, Amann R, Pernthaler J (2004) Actinobacterial 16S rRNA genes from freshwater habitats cluster in four distinct lineages. *Environ. Microbiol.* 6, 242–253. doi: 10.1111/j.1462-2920.2004.00561.x.

Wayne LG, Brenner DJ, Colwell RR, Grimont PAD, Kandler O, Krichevsky MI, Moore LH, Moore WE, Murray RGE, Stackebrandt E, Starr MP, Truper HG (1987) Report of the ad hoc committee on reconciliation of approaches to bacterial systematics. *Int J Syst Bacteriol.* 37(4), 463–454. doi: 10.1099/00207713-37-4-463.

Wenger AM, Peluso P, Rowell WJ et al. (2019) Accurate circular consensus long-read sequencing improves variant detection and assembly of a human genome. *Nat Biotechnol.* 37, 1155–1162. doi: 10.1038/s41587-019-0217-9.

Whitman WB (2015) Genome sequences as the type material for taxonomic descriptions of prokaryotes. *Syst Appl Microbiol.* 38, 217–222. doi: 10.1016/j.syapm.2015.02.003.

Wick RR, Holt KE. (2019) Benchmarking of long-read assemblers for prokaryote whole genome sequencing. *F1000Res.* 8, 2138. doi: 10.12688/f1000research.21782.1

Wick RR, Judd LM, Gorrie CL, Holt KE (2017) Unicycler: resolving bacterial genome assemblies from short and long sequencing reads. *PLoS Comput Biol.* 13(6), e1005595. doi: 10.1371/journal.pcbi.1005595.

Zhang G, Gao B, Adeolu M, Khadka B, Gupta RS (2016) Phylogenomic analyses and comparative studies on genomes of the Bifidobacteriales: identification of molecular signatures specific for the order Bifidobacteriales and its different subclades. *Front Microbiol.* 7, 978. doi: 10.3389/fmicb.2016.00978.

Zhang H, Gao S, Lercher MJ, Hu S, Chen WH (2012) EvolView, an online tool for visualizing, annotating and managing phylogenetic trees. *Nucleic Acids Res.* (Web Server issue), W569–W572. doi: 10.1093/nar/gks576.

Zhou L, Lin Y, Feng B et al. (2017) Phylogeny analysis from gene-order data with massive duplications. *BMC Genom.* 18, 760. doi: 10.1186/s12864-017-4129-0.

Zhu, Q, Mai, U, Pfeiffer, W et al. (2019) Phylogenomics of 10,575 genomes reveals evolutionary proximity between domains bacteria and archaea. *Nat Commun.* 10, 5477. doi: 10.1038/s41467-019-13443-4.

Zhulin IB (2016) Classic spotlight: 16S rRNA redefines microbiology. *J Bact.* 198(20), 2764–2765. doi: 10.1128/JB.00616-16.

Zuo G, Hao B (2015) CVTree3 web server for whole-genome-based and alignment-free prokaryotic phylogeny and taxonomy. *Genom Proteom Bioinf.* 13, 321–331. doi: 10.1016/j.gpb.2015.08.004.

21

Archaea

Nina Dombrowski, Tara Mahendrarajah, Sarah T. Gross, Laura Eme, and Anja Spang

CONTENTS

Introduction

Just about half a century ago, all prokaryotes, i.e., cells without nucleus, were classified within one kingdom: *Monera*. However, in the late 1970s, scientists were starting to recognize that this classification system, based predominantly on morphological and metabolic traits, underestimated the vast diversity of prokaryotic life. Around the same time, the pioneering work of Carl Woese and George Fox led to the discovery that prokaryotes were, in fact, composed of two fundamentally different domains of life—the *Bacteria* and the *Archaea* (originally referred to as "Eubacteria" and "Archaebacteria," respectively) [1]. Woese and coworkers used the RNA components of the ribosome to reconstruct the first phylogenetic tree of life based on molecular data [2], which divided cellular organisms into three separate domains of life

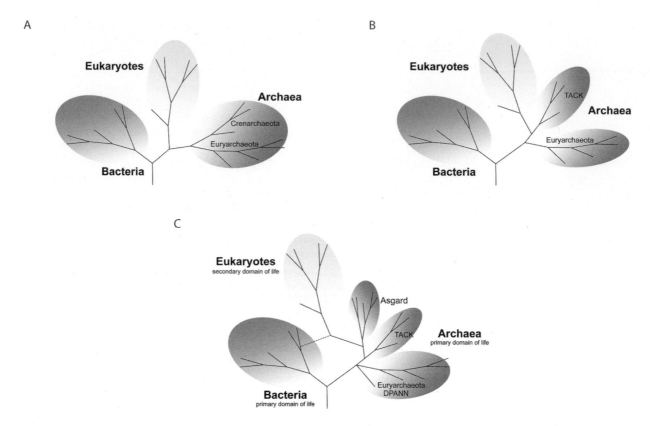

FIGURE 21.1 Schematic depictions of the relationship of *Archaea* with *Bacteria* and eukaryotes in the tree of life. A: Upon the discovery of *Archaea* as a separate domain, the tree of life was divided into three major domains. B: However, phylogenetic analyses of core informational proteins suggested later that eukaryotes may have evolved from within the *Archaea*, challenging the three-domain topology. C: Recent research, among others enabled by the discovery of the Asgard archaea, has shed further support on the branching of eukaryotes from within the *Archaea* (in terms of universal marker proteins). In turn, it has been suggested that the tree of life has two primary domains of life—the *Archaea* and *Bacteria*—and one secondary domain of life, which evolved from the former (see text for more details).

(Figure 21.1A)—the *Bacteria*, *Archaea*, and *Eukarya*, the latter of which comprised all organisms with a true nucleus [2]. At that time, it was suggested that *Archaea*, in spite of their superficial similarity to *Bacteria*, may be more closely related to eukaryotes than *Bacteria*. In fact, they seemed to harbor simplified versions of eukaryotic informational processing machineries (replication, transcription, translation, and cell division), in addition to unique characteristics such as ether-bound isoprenoids rather than ester-bound fatty acid-based lipids (Table 21.1). Subsequent research on *Archaea*, accompanied by extensive methodological developments in environmental microbiology, sequencing technologies, physiology, cell biology, and phylogenetics, has further changed our view on the diversity of life, the tree topology, as well as the ecological and evolutionary importance of *Archaea*. In particular, the use of cultivation-independent techniques, such as metagenomics and single-cell genomics, which allow us to obtain genomes of uncultivated organisms directly from environmental samples [3, 4], have been a key element leading to our changed perception of archaeal diversity and distribution. While *Archaea* have originally been viewed as comprising predominantly "extremophilic" organisms inhabiting environments with high temperature, salinity, and high or low pH, they are now known to be ubiquitous in all environments on Earth, including marine waters and freshwater lakes, sediments, soils (including plant roots), aquifers, and the human microbiome to name a few [5–7]. With their widespread ecological distribution and important

metabolic capabilities, *Archaea* are recognized as key players in a wide variety of biogeochemical processes, including the sulfur, nitrogen, and carbon cycles [8]. For instance, *Archaea* include the only known organisms able to conserve energy through the anaerobic production or consumption of methane in processes referred to as methanogenesis and anaerobic methane oxidation, respectively. Since methane is an extremely potent greenhouse gas, with a global-warming potential about 25 times greater than carbon dioxide, these *Archaea* have an essential role in the global carbon budget and consequently climate change [9]. Finally, the study of archaeal phylogenetic diversity and evolution has fundamentally changed our understanding of the eukaryotic cell (see below) [10].

Archaea and the Tree of Life

Since the discovery of the *Archaea* as a separate domain of life (Figure 21.1A), their relationship to *Bacteria* and eukaryotes has been a matter of debate and is regarded to be of fundamental importance for our understanding of the origin of eukaryotes. Eukaryotic cells are highly compartmentalized and it has long been recognized that eukaryotic compartments, such as mitochondria (the site of ATP generation via oxidative phosphorylation) and chloroplasts (the organelles in which photosynthesis occurs in plants), evolved as a result of endosymbiosis, i.e., mitochondria and chloroplasts seem to be derived from Alphaproteobacteria

TABLE 21.1

Comparison of Selected Characteristics of the Major Domains of Life

Characteristic	Bacteria	Archaea	Eukarya
Membrane-enclosed nucleus	No	No	Yes
Chromosomal structure	Circular	Circular	Linear
Peptidoglycan in cell wall	Yes	No	No
Membrane lipids	Ester-linked	Ether-linked	Ester-linked
Glycerol	Glycerol-3-phosphate	Glycerol-1-phosphate	Glycerol-3-phosphate
Ribosomes (mass)	70S	70S	80S
Initiator tRNA	formylmethionine	methionine	methionine
Introns	No	No	Yes
Operons	Yes	Yes	No
RNA polymerase	One (4 subunits)	One (8–12 subunits)	Three (12–14 subunits)
Transcription factors required	No	Yes	Yes
TATA box in promoter	No	Yes	Yes

and Cyanobacteria, respectively (e.g., reviewed in [11]). In contrast, the nature of the host cell taking up the progenitors of these compartments was unknown until recently; while some hypotheses suggested that this cell was a proto-eukaryote that already resembled extant eukaryotic cells, others point out that the host was an archaeon or even bacterium [11–13]. For a long time, the prevailing view was that *Archaea* and eukaryotes represent two independent sister lineages in the tree of life [2, 14], and it was unclear how the shared ancestor of *Archaea* and eukaryotes looked like. However, while certain phylogenetic analyses have supported this model, others have suggested alternative scenarios, in which eukaryotes evolved from within the *Archaea* ([15] and reference therein).

The use of cultivation-independent genomic approaches combined with improved phylogenetic methods and more realistic evolutionary models have recently led new insights into the evolutionary history of the Archaea, their placement in the tree of life and eukaryogenesis. In particular, these analyses have provided increasing support for eukaryotes branching from within the *Archaea* [10, 15] instead of as a sister lineage as originally assumed (Figure 21.1 B–C). Though eukaryotes initially appeared to branch close to the TACK superphylum (discussed later in this chapter) [16] (Figure 21.1B), it was challenging to pinpoint a specific archaeal lineage as being more closely related to eukaryotes than the others. The position of eukaryotes among *Archaea* became clearer with the recent discovery the Asgard archaea—a novel archaeal superphylum [17, 18] (discussed later in this chapter). Phylogenomic analyses revealed that Asgard archaea form a sister group of eukaryotes (Figure 21.1C) and harbor an extended set of proteins that were previously assumed to be specific to eukaryotes [17, 18] (discussed later in this chapter). Together, these findings indicate that eukaryotes may have evolved from a symbiosis between an archaeal host cell

and a bacterial endosymbiont, and also provide greater evidence in support of a two-domain tree of life [19–22], with *Archaea* and *Bacteria* representing two primary domains and eukaryotes being a secondary domain [15, 23]. Although the exact placement of eukaryotes with respect to the different members of the Asgard archaea remains to be elucidated, continued exploration of Asgard archaeal diversity will allow to further refine the position of Archaea and eukaryotes in the tree of life.

Archaeal Cell Biology and Eukaryotic Signature Proteins (ESPs)

In agreement with their close relationship to eukaryotes, *Archaea* encode informational processing machineries that closely resemble those of eukaryotic representatives. Although *Archaea* harbor a single circular chromosome like *Bacteria*, their replication machinery includes various components homologous (i.e., shared by common ancestry) to those of eukaryotes, while most functionally equivalent complexes in *Bacteria* are unrelated [24, 25]. For instance, *Archaea* and eukaryotes share homologous subunits comprising the origin of replication complex (ORC), a replicative helicase unit referred to as the CMG (Cdc45, MCM, GINS) complex, and the active replisome, which includes a two-subunit primase, a DNA polymerase sliding clamp and clamp loader, and DNA polymerases [24]. Yet, some *Archaea* also encode components that are absent from both eukaryotes and *Bacteria* and others that are shared with *Bacteria*. For example the two-subunit DNA polymerase D [24, 26] is unique to *Archaea* while the NAD+-dependent DNA ligase, the DNA gyrase, and the DNA primase DnaG are homologous to bacterial enzymes [25]. In many cases, archaeal complexes seem to represent a simplified version of their counterparts in eukaryotes [25], the latter of which often encode additional paralogous enzymes (i.e., those that evolved by gene duplication), whose evolution involved subfunctionalization [24]. For instance, while *Archaea* collectively encode three families of polymerase B, eukaryotes harbor the four polymerase B family enzymes referred to as Pol alpha, beta, gamma and delta [24, 26]. Notably, all of these eukaryotic enzymes seem to have evolved from two distinct archaeal polymerase B family homologs [18, 26]. Another interesting example represents the nucleosome: *Archaea* harbor histone-like proteins, which form a homodimeric histone complex in part homologous to the heterodimeric nucleosome of eukaryotes [27, 28].

Archaeal transcription also shares several features in common with eukaryotes. While many archaeal genomes encode gene clusters reminiscent of bacterial operons, the archaeal transcription machinery represents a simplified version of their eukaryotic counterparts [29]. For instance, the archaeal DNA-dependent RNA polymerase (RNAP) consists of 12-13 subunits, which are homologous to the subunits of the three eukaryotic RNA polymerases (RNAP I-III) [29]. In contrast, RNAP of Bacteria consists of only five subunits, two of which are distantly related to archaeal RNAP subunits 1 and 2 (i.e., RpoA and RpoB). Transcription initiation, which is based on the same molecular mechanisms across the domains, also involves homologous transcription factors in *Archaea* and eukaryotes [29].

Similarities in the translational machinery between *Archaea* and eukaryotes are also evident. Archaeal ribosomes are of comparable size to bacterial ribosomes (70S), but share various ribosomal subunits uniquely with eukaryotes [30]. Additionally, translation in *Archaea* is initiated by an initiator tRNA carrying methionine and several translation initiation factors, as is seen in eukaryotic organisms but contrasts with the use of formyl-methionine by bacteria. Further, a 22nd amino acid, pyrrolysine, has been identified uniquely in certain members of the *Archaea*, in particular methanogens [31].

Notably, *Archaea* not only share homologous replication, transcription, and translation machineries with eukaryotes, but have also been found to encode various so-called eukaryotic signature proteins (ESPs) [32], i.e., proteins that are generally absent from bacterial genomes while being central to the integrity and functioning of eukaryotic cells. These proteins include, for instance, components of the eukaryotic cytoskeleton (such as actin and tubulins), cell division and vesicle trafficking machineries, endosomal sorting complexes required for transport (ESCRT), as well as the proteasome and ubiquitin system [10].

In particular, members of the TACK archaea (discussed later in this chapter) including among others the *Cren-*, *Aig-* and *Thaumarchaeota* have early on been found to encode certain ESPs that were absent from *Euryarchaeota* [15, 16, 33–35]. For instance, while *Euryarchaeota* use FtsZ as major cell division protein, many *Cren-* and *Thaumarchaeota* harbor a cell division system (also referred to as cdvABC system) that includes homologs of eukaryotic ESCRT-III and an ATPase related to vacuolar protein sorting-associated protein 4 (Vps4) [36–39]. Furthermore, archaeal actin homologs referred to as crenactin, which are distantly related to eukaryotic actins, have been discovered in *Thermoproteales*, as well as in *Korarchaeota* [40]. Yutin and coworkers identified distant homologs of eukaryotic tubulins—the artubulins—in the genomes of two species of *Thaumarchaeota*, "*Candidatus Nitrosoarchaeum limnia*" and "*Candidatus Nitrosoarchaeum koreensis*" [41], and an analysis of the "*Candidatus Caldiarchaeaum subterraneum*" composite genome revealed the presence of a presumably fully functional ubiquitin-like protein modifier system [42].

The discovery of the Asgard archaea [17, 18], the currently most closely related archaeal sister lineage of eukaryotes, has recently revealed a variety of additional ESPs in *Archaea*. For instance, Asgard archaea not only encode additional homologs of eukaryotic informational processing machineries but also harbor simplified versions of the eukaryotic oligosaccharyl-transferase-complex and ubiquitin modifier system. Furthermore, they encode an extended set of small GTPases [17, 18], which are key regulators in eukaryotic cells with a central role in vesicle trafficking machineries [43]. Additional central components homologous to eukaryotic vesicle transport and tethering were identified in the genomes of the *Thorarchaeota* [18]. Further, Asgard archaea harbor protein domains homologous to the key domains of the three major eukaryotic ESCRT machinery complexes (ESCRT I-III) and a diversity of cytoskeleton-related proteins that are much more similar to their eukaryotic counterparts than those previously identified in *Archaea*. These include the lokiactins found across the Asgard representatives, as well as *bona fide* tubulins in *Odinarchaeota* [10, 17, 18]. Notably, Asgard archaea also encode actin-regulating proteins, such as the profilins [18], which were recently shown to be functionally equivalent to those of eukaryotes [44].

Altogether, archaeal information processing machineries as well as an extended set of ESPs in members of the TACK and in particular the Asgard archaea, further testify to the archaeal origin of the eukaryotic cell. Importantly, the study of these complexes in *Archaea* can help to provide a better understanding of eukaryotic cell biology and provide insight into the relative timing of the evolution of cellular complexity.

Archaeal Cell Membranes and Cells Walls

The composition of archaeal cell membranes differs fundamentally from those of *Bacteria* and eukaryotes [45]. For instance, the glycerol used to make archaeal phospholipids is a stereoisomer of the glycerol used to build bacterial and eukaryotic membranes, i.e., while *Archaea* use glycerol-1-phosphate, eukaryotes and bacteria have glycerol-3-phosphate. Furthermore, *Archaea* harbor isoprenoid side chains instead of the fatty acid side chains found in Bacteria and eukaryotes. These isoprenoids are bound to the glycerol backbone by ether linkages contrasting with the ester linkages formed between the bacterial and eukaryotic glycerol and fatty acid moieties. Archaeal isoprenoid side chains in the two monolayers of the lipid bilayer can be linked, thereby giving rise to transmembrane phospholipids. The isoprenes can also form five-carbon ring structures, which may function in the stabilization of the membranes of archaeal species that live in high temperature environments. More than 100 different ether-type polar lipids, such as phospholipids and glycolipids, have been identified in *Archaea* [46].

Different archaeal representatives differ with regard to their cell walls. In contrast to *Bacteria*, *Archaea* lack peptidoglycan and are thus naturally resistant to antibiotics that impair the synthesis of peptidoglycan, such as penicillins. Some species of methanogenic *Archaea* contain cell walls of pseudopeptidoglycan (pseudomurein) that superficially resemble bacterial peptidoglycan but contain different components (e.g., N-acetyltalosaminuronic acid instead of and N-acetylmuramic acid) and have β-1,3 instead of β-1,4-glycosidic bonds. Yet, most archaeal species lack pseudomurein and instead harbor cell walls made of proteins, glycoproteins, or polysaccharides [47]. For instance, a common cell wall structure found in *Archaea* is composed of a paracrystalline surface layer, termed S-layer, consisting of protein or glycoprotein moieties arranged in hexagonal patterns. Finally, some *Archaea*, such as certain members of the order *Thermoplasmatales*, lack cell walls altogether.

Taxonomic Diversity of *Archaea*

The *Archaea* were originally divided into two major phyla, termed *Crenarchaeota* and *Euryarchaeota* [2]. However, recent advances in culture-independent, high-throughput sequencing techniques have uncovered a large diversity of novel archaeal lineages, most of which remain uncultivated [5]. Many of these newly discovered archaeal lineages are only distantly related to established lineages within the *Cren-* and *Euryarchaeota*, which has led to the proposal of many additional archaeal phyla and superphyla during the past years [7]. Figure 21.2 summarizes the

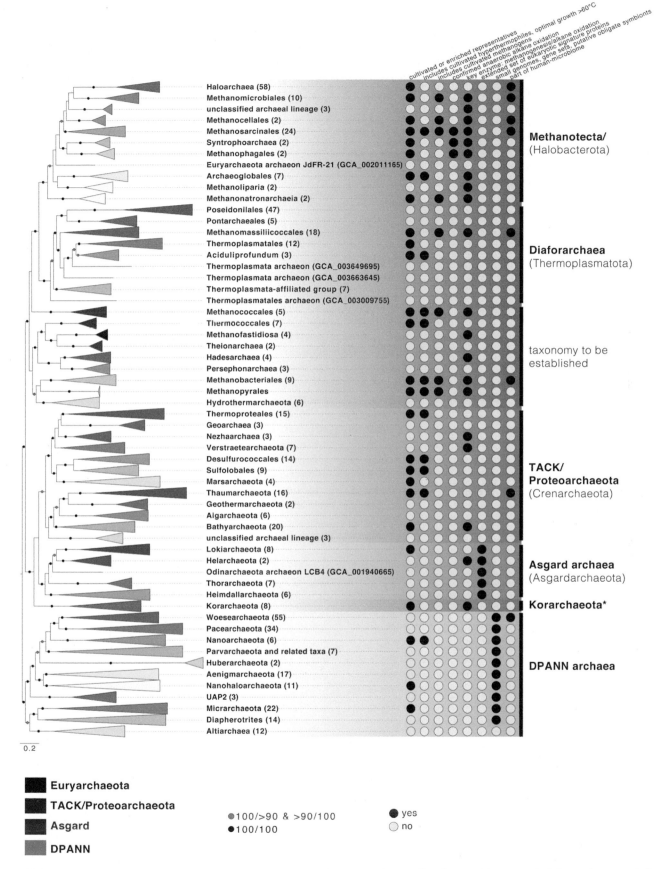

FIGURE 21.2 Depiction of the phylogenetic diversity of the *Archaea* and the presence/absence of key features. The unrooted phylogenetic tree was inferred with maximum likelihood using the IQ-tree software (with the C20+LG+F+R mixture model) and was based on an alignment of 11399 positions from a representative set of 569 archaeal taxa. Highly supported clusters (assessed by ultrafast bootstraps [220] and SH-like approximate-likelihood ratio test [221]) are indicated with gray or black dots based on their branch support values (see figure inlet). Taxa were predominantly named according to Adam et al. [7] (bold face), but alternative names suggested by GTDB (https://gtdb.ecogenomic.org/) are indicated in parenthesis when available. Numbers in brackets indicate the number of genomes/taxa per cluster. The presence and absence of certain features are shown with black and gray circles for each major taxonomic lineage (see figure inlet). Please note that the last column only reports those archaeal taxa that have been confirmed to be part of the human archaeome.

current understanding of the archaeal phylogeny, including established and proposed phyla, classes, and orders, as well as their general physiological grouping and certain features discussed below. However, please note that there is currently no consensus on how to best classify archaeal lineages. Therefore, a widely accepted taxonomy of the *Archaea* remains to be established [5]. In particular, there are currently two main classification schemes used: the classification suggested by Adam and coworkers that is implemented in NCBI [7] and the system introduced by the developers of the Genome Taxonomy Database (GTDB) (https://gtdb.ecogenomic.org/). The latter of these was suggested to provide a standardized and rank-normalized genome-based classification system, which was recently used to revise the bacterial taxonomy [48].

Euryarchaeota

The *Euryarchaeota* (Figure 21.2) comprise various cultivated and well-characterized archaeal species including the globally important methanogens (i.e., methane producers) as well as anaerobic methane-oxidizing Euryarchaeota (ANME) [49, 50]. Methanogens and ANME play a key role in the carbon cycle by anaerobically producing or consuming the potent climate gas, methane [8, 9, 50–52]. However, research during the past years has shown that the *Euryarchaeota* are a phylogenetically and physiologically much more diverse radiation than originally thought [5, 7]. Indeed, it remains to be elucidated whether *Euryarchaeota* comprise a monophyletic group or phylogenetically distinct divisions, some of which may be more closely related to the TACK and Asgard archaea [7, 53, 54]. In the following, we provide an overview of the major lineages comprising canonical and recently discovered lineages affiliating with the *Euryarchaeota*.

Methanotecta

The *Methanotecta* (Figure 21.2), a recently proposed superclass [7], comprise the so-called class II methanogens (*Methanosarcinales, Methanomicrobiales, Methanocellales*), several phylogenetically distinct ANME archaeal lineages, the *Haloarchaeota, Archaeoglobales*, as well as the more recently described archaeal orders referred to as *Methanonatronarchaeia, Syntrophoarchaeales, Methanoliparales*, and *Methanophagales*. We present major features of these different lineages below.

Methanomicrobiales

The order *Methanomicrobiales* comprises several families, such as the *Methanocalculaceae, Methanoregulaceae, Methanospirillaceae, Methanomicrobiaceae*, and *Methanocorpusculaceae* (e.g., reviewed in [55]), and can be found in a variety of anoxic habitats, including wetlands, soil, oceans and freshwater, landfills, rice paddies, as well as associated with animals [50]. Members of the *Methanomicrobiales* have diverse cell shapes, ranging from rods to cocci to plates, including motile and nonmotile species, and grow between 0°C and 60°C [55]. Cells are often surrounded by glycoprotein-containing S-layers. Many *Methanomicrobiales* use hydrogen and carbon dioxide to form methane and all species are obligate anaerobes. They can use formate and alcohol but not acetate and methylated C1-compounds as substrates for methanogenesis, distinguishing them from the *Methanosarcinales* [9, 55].

Methanosarcinales

The *Methanosarcinales* are closely related to the *Methanomicrobiales* and include families such as the *Methanosarcinaceae, Methanotrichaceae* (formerly *Methanosaetaceae*), and *Methermicoccaceae* (Table 21.1), as well as the *Methanoperedenaceae* (ANME-2d). While this order comprises diverse methanogenic organisms, it also includes representatives of the anaerobic methane-oxidizing *Euryarchaeota* ANME-2 and -3 lineages [50, 52, 56]. Similar to the *Methanomicrobiales*, representatives of the *Methanosarcinales* are found in a range of anoxic habitats [50]. Yet, in contrast to other methanogenic orders, the *Methanosarcinales* are known for their much wider substrate range for methanogenesis, i.e., members of this group not only use hydrogen and formate as substrates but also a variety of methylated compounds and acetate [8, 9]. Considering that methanogenesis based on acetate may contribute up to two-thirds of methane released to the atmosphere, members of this group have important roles in the global carbon cycle [8, 51]. Representatives of the ANME-2 and -3 lineages use the reverse methanogenesis pathway to anaerobically oxidize methane [52]. While some ANME-2 members can grow independently using nitrate, nitrite, or Fe(III) as electron acceptors [57–59], other ANME-2 grow in syntrophic consortia with bacterial partners (especially sulfate reducers) that serve as external electron sinks [52, 60]. Members of these groups are particularly abundant in the sulfate-methane transition zone in marine sediments and play an important role in the global carbon cycle by reoxidizing a large fraction of the methane produced in marine sediments before it can enter the atmosphere [52, 61].

Methanophagales (ANME-1)

The *Methanophagales* comprise another lineage of anaerobic methane-oxidizing archaea, also known as the ANME-1 lineage. While originally thought to affiliate with the *Methanosarcinales*, they were recently shown to represent a sister lineage of the *Syntrophoarchaeales* [7] (Figure 21.2 and later in this chapter). Similar to the ANME-2 and -3 lineages that belong to the *Methanosarcinales*, members of this group occur in diverse marine, terrestrial, and freshwater environments [62], are particularly abundant in the sulfur-methane transition zone [61], and use the reverse methanogenesis pathway for the anaerobic oxidation of methane (AOM) [63]. While ANME-1 has not been cultivated thus far, various lines of research have suggested that members are able to oxidize methane in syntrophy with sulfate-reducing bacteria (SRB) through direct electron transfer [52, 60, 64].

Methanocellales

Methanocellales represents a more recently described order of hydrogenotrophic methanogens that were originally referred to as Rice Cluster I (RC-I) [65] due to their initial discovery in rice paddy fields, where they are important producers of methane [66]. The first representative of this order, *Methanocella paludicola*, was isolated from an anaerobic, propionate-containing enrichment culture [65] and represents a nonmotile anaerobe with rod-shaped cells thriving at temperatures between 25°C and 40°C [65]. While the isolate performs methanogenesis using hydrogen, carbon dioxide, and formate, it uses acetate as a carbon source. Hydrogen is provided by its syntrophic partner, the bacterium *Syntrophobacter fumaroxidans* [67]. Similar metabolic features were found in other representatives of this order, including *M. arvoryzae* [68] and *M. conradii* [69].

Syntrophoarchaeales

Syntrophoarchaeales (sometimes assigned to the *Methanosarcinales*; Table 21.1) represent a recently discovered group of anaerobic, alkane-oxidizing archaea usually found in hydrocarbon-rich sediments [70, 71]. For example, the first two representatives of this lineage, *Syntrophoarchaeum butanivorans* and *Syntrophoarchaeum caldarius*, were originally isolated from hydrothermal- and hydrocarbon-rich marine sediments of the Guaymas Basin [71, 72]. Notably, *Syntrophoarchaeales* grow by the anaerobic oxidation of butane as well as propane, which are thought to be metabolized using the reverse methanogenesis pathway also operating in ANME archaea [73]. In particular, they encode subunits homologous to the Methyl-Coenzyme M Reductase (MCR) complex, which represents the key enzyme of methanogens catalyzing the demethylation of CH_3-S-CoM to methane [51]. In *Syntrophoarchaeales*, MCR is thought to be used in reverse and to mediate the first step of the breakdown of short-chain alkanes eventually yielding carbon dioxide as an end product [71]. As indicated by the names of members of this group, studied representatives grow syntrophically with the sulfate-reducing bacterium *Candidatus* Desulfofervidus auxilii.

Archaeoglobales

The *Archaeoglobales* comprises species belonging to the genera *Archaeoglobus*, *Ferroglobus*, and *Geoglobus* [74]. The *Archaeoglobus* sp. is believed to be predominantly composed of strictly anaerobic and hyperthermophilic members, growing optimally at 80°C and neutral pH. The best studied representatives are autotrophs and/or organotrophs and can reduce sulfate or sulfite during respiration [75]. Species of *Ferroglobus* grow by oxidation of $Fe(II)S^{2-}$ and H_2 [75], whereas *Geoglobus* grows anaerobically in the presence of acetate and ferric iron [74]. Recently, genomes of so far uncultivated members of the *Archaeoglobi* were reconstructed from environmental samples and shown to encode MCR-like protein complexes similar to those of methanogens and ANME archaea [76, 77]. Based on genomic inferences, it was suggested that the respective organisms may be able to grow by the oxidation of methane or alternative short-chain alkanes.

Methanoliparales

Methanoliparales is an uncultivated lineage within the *Methanotecta* that phylogenetically places between *Archaeoglobales* and a cluster comprising *Syntrophoarchaeales* and *Methanophagales*. *Methanoliparales* were first discovered in two metagenomes from a petroleum-enrichment culture and an oil seep and are represented by two metagenome-assembled genomes: *Candidatus Methanoliparum thermophilum* NM1a and *Candidatus Methanolliviera hydrocarbonicum* NM1b [78]. Genomic analyses suggest that *Methanolipirales* are methanogens that encode the Wood-Ljungdahl carbon fixation pathway and are capable of beta-oxidation. Interestingly, both genomes code for two distinct MCR complexes, which may be involved in methanogenesis and the oxidation of alkanes, respectively.

Haloarchaeota

Halobacteria, herein referred to as *Haloarchaea*, are a diverse group of *Archaea*, most of which are adapted to high salinity. Salt requirements of these species range from 1.5 to 5.2 M NaCl, although most strains grow best between 3.5 and 4.5 M NaCl, at or near the saturation point of salt (36% w/v salts). In order to maintain osmolarity of their cells in high-salt environments, haloarchaeal members accumulate up to 5 M intracellular levels of KCl to counterbalance high extracellular salt concentration. As a result, the entire intracellular machinery, including enzymes and structural proteins, must be adapted to high salt levels. The proteins of all haloarchaeal species have a very low isoelectric point and the genomes contain high GC contents that are well above 60% [79].

Some species of *Haloarchaea* are motile by means of tufts of flagella, although many species are nonmotile [75]. *Haloarchaea* comprise various aerobic or facultative anaerobes and show diverse morphologies and shapes, including rods, cocci, and a multitude of pleomorphic forms [75, 80]. The lack of turgor pressure within haloarchaeal cells enables the cells to tolerate the formation of corners, and as such, some species are even triangular or square-shaped [75, 80]. Cell envelopes of coccoid *Haloarchaea* are stable in the absence of salt, while, noncoccoid species maintain their integrity only in the presence of high concentrations of NaCl or KCl [75]. Non-coccoid species have a proteinaceous cell envelope with glycoprotein subunits forming a hexagonal pattern [75]. Species of *Haloarchaea* are abundant in salt lakes, inland seas, and evaporating ponds of seawater, such as the Dead Sea and solar salterns. *Haloarchaea* often tint the water column and sediments in bright colors due to the presence of retinal-based pigments. Some of these pigments are capable of the light-mediated translocation of ions across cell membranes. The best known halobacterial pigment is bacteriorhodopsin, which is an outwardly directed proton pump. Bacteriorhodopsin is involved in energy conservation and is the only nonchlorophyll-mediated light energy transducing system known to date [79]. Other retinal-based pigments found in *Haloarchaea* include halorhodopsin, which is an inward chloride pump involved in osmotic homeostasis, as well as sensory rhodopsin I and II (SRI and SRII, respectively). SRI and SRII can mediate positive and/or negative phototaxis [79].

Methanonatronarchaoeta

Another lineage of halophilic archaea are the *Methanonatronarchaeota*, which were first recovered from hypersaline anoxic lake sediments [81] and are currently represented by isolates from two distinct subgroups: the soda lake isolate *Methanonatronarchaeum thermophilum* AMET and the salt lake isolate *Candidatus Methanohalarchaeum thermophilum*

HMET [82]. *Cand. M. thermophilum* has motile, coccoid cells that are around 0.4 μM in diameter and are surrounded by an S-layer. These anaerobic organisms tolerate a range of pH (between 6.5 and 8 [HMET] and 9.5 and 9.8 [AMET]) and grow optimally at a temperature of 50°C and salt concentrations of 4 M by accumulating high concentrations of potassium inside their cells for osmotic balance ("salt-in strategy") [81, 82]. *Cand. M. thermophilum* is a heterotrophic methanogen that grows with C1-methylated compounds as electron acceptors, such as methanol or trimethylamine, and formate or hydrogen as electron donors [81]. The 16S rRNA gene analyses indicate that *Methanonatronarchaeota* are the first cultured representatives of the SA1 group, which is commonly found in hypersaline environments [81, 83]. Yet, the exact placement of *Methanonatronarchaeota* in the archaeal tree of life is still debated. While initial phylogenetic analyses placed this lineage sister to *Haloarchaea* [81], recent analyses have suggested that the *Methanonatronarchaeota* form an early diverging lineage of the *Methanotecta* [84].

Diaforarchaea

The *Diaforarchaea* comprise a recently suggested superclass [7] that includes the *Thermoplasmata* and related lineages, such as the diverse and abundant Marine Group II and III archaea [85, 86], now also known as the *Poseidoniales* and *Pontarchaeales*, respectively, as well as a recently discovered new order of methanogens, the *Methanomassiliicoccales* [87].

Thermoplasmatales

The *Thermoplasmatales* comprise the genera *Acidiplasma*, *Thermoplasma*, *Picrophilus*, *Cuniculiplasma*, and *Ferroplasma*. *Cunicuplasma*, *Thermoplasma*, and *Ferroplasma* are the only cultivated archaeal representatives that lack cell walls [75, 88]. Species of *Thermoplasma* are facultative anaerobes and obligate heterotrophs, using elemental sulfur for respiration. Most members of this group are thermoacidophiles and grow optimally at 60°C and pH 2 [75]. For instance, representatives may be found in self-heating coal refuse piles and in acidic solfatara fields [75].

Members of the *Picrophilus* are the most acidophilic organisms known so far [89]. They form irregular cocci that are 1–1.5 μm in diameter and contain S-layer cell walls [75]. *Picrophilus* are thermophilic and hyperacidophilic and grow at temperatures between 47°C and 60°C and pH ranges of 0–3.5 [75]. Their ability to grow at pH values near 0 and at high temperatures has shifted the physicochemical boundaries at which life was considered to exist.

In contrast to other members of the *Thermoplasmatales*, *Ferroplasma* are not thermophilic and can grow autotrophically using ferrous iron as energy and inorganic carbon as a carbon source. Representatives can be found in a variety of acidic environments with stable chemical conditions, such as ore deposits, mines, and acid mine drainage systems (natural or man-made), as well as in areas with geothermal activity [90, 91]. Representatives of this family are cell wall-lacking extreme and obligate acidophiles that are able to grow at pH values around 0. Together with members of *Picrophilum*, they comprise a group of the most extreme acidophilic organisms known, members of which tolerate high concentrations of iron, copper, zinc, and other metals [91].

Aciduliprofundales

Aciduliprofundales, formerly named the "deep-sea hydrothermal vent euryarchaeota 2" (DHVE2) lineage, is currently represented by the cultivated *Aciduliprofundum boonei* [92, 93]. As the original name suggests, *Aciduliprofundales* are predominantly found across hydrothermal vents, where they can represent up to 15% of the archaeal community [92–94]. *A. boonei* is an anaerobic heterotroph that ferments peptides and is able to reduce elemental sulfur or ferric iron at a pH between 3.3 and 5.8 (optimum pH 4.6) and an optimal growth temperature of 70°C [92]. This organism is motile with a single flagellum and has pleomorphic cells of about 0.6–1 μM in diameter that are surrounded by a single S-layer.

Methanomassiliicoccales

The order *Methanomassiliicoccales* represents the first lineage of the *Thermoplasmata* known to comprise methanogenic members [87], several of which have been isolated, such as *Methanomassiliicoccus luminyensis* [95, 96], *Candidatus Methanomethylophilus alvus* [97], and *Candidatus Methanoplasma termitum* [98]. *Methanomassiliicoccales* are widely distributed in wetlands and sediments as well as the gastrointestinal tracts of animals including those of humans and cows [87, 99, 100]. Members of this group comprise H_2-dependent methylotrophic methanogens, which are able to use methylated amines [100] including mono-, di-, and trimethylamines for methanogenesis. Considering that the latter compounds have been implicated in human disease, gut-associated members of the *Methanomassiliicoccales* may play an important role in human health [100].

Poseidoniales

The *Poseidoniales* [101], formerly Marine Group II (MG II), lack any cultured representatives and are mainly known from 16S rRNA gene diversity assays and genomic analyses. *Poseidoniales* are often found in the photic zone of marine waters and can present up to 15% of archaeal cells in the Atlantic ocean [102–104]. They are further divided into *Candidatus Poseidonaceae* (MGIIa) and *Candidatus Thalassarchaeacea* (MGIIb), whose abundances seasonally fluctuate, i.e., members of MGIIa and MGIIb are more abundant in the summer and winter, respectively [105]. Members of this group comprise aerobic heterotrophs with the potential to utilize a range of substrates such as proteins, peptides, amino acids, fatty acids, carbohydrates, xenobiotics, and agar [101, 106–110]. In addition, some representatives of the class *Ca. Poseidoniia*, found in the photic zone, encode proteorhodopsin indicative of a photoheterotrophic lifestyle [101, 107, 110].

Pontarchaeales

The order *Pontarchaeales*, or Marine Group III, are often found in the deep ocean, while being less abundant in the photic zone [102, 111]. Based on genomic data, it was inferred that deep-sea *Pontarchaeales* likely represent motile heterotrophs that might degrade proteins, carbohydrates, and lipids [112]. In contrast, surface dwelling members of the *Pontarchaeales* seem to encode photolyase and rhodopsin genes and in turn may be photoheterotrophs [111]. Notably, both the *Pontarchaeales* and the *Poseidoniales* lack the key archaeal lipid biosynthesis gene encoding glycerol-1 phosphate dehydrogenase, such that it is currently unclear whether members of these orders encode canonical archaeal lipids [45]. In particular, the presence of genes for glycerol-3 phosphate dehydrogenase, which is essential in the synthesis of bacterial lipids, has led to the suggestion that these *Archaea* may have mixed membranes [45, 101].

Other Euryarchaeota

The following section provides an overview of additional lineages affiliating with the *Euryarchaeota*, including methanogenic lineages that have been extensively studied in the past. However, some analyses indicate that at least some of these orders may be more closely related to the TACK and Asgard archaea [7, 53, 54].

Methanococcales

As the name implies, the *Methanococcales* include representatives with coccoid shapes and proteinous cell walls [75]. All members of this lineage are thought to be strict anaerobes that obtain energy by the reduction of CO_2 to methane [9] and comprise mesophilic (e.g., *Methanococcus*) to thermophilic (e.g., *Methanothermococcus*) to hyperthermophilic (e.g., *Methanocaldococcus*) taxa [75].

Thermococcales

Members of the *Thermococcales* represent anaerobic heterotrophs that utilize a wide range of organic compounds, including amino acids, a variety of sugars, and organic acids such as pyruvate. When available, they can use elemental sulfur as the terminal electron acceptor. Extensive research has been carried out on the metabolism of cultivated representatives and led to the discovery of unique enzymes and pathways [113]. Certain members of the *Thermococcales* represent important model organisms. For example, the hyperthermophilic *Pyrococcus furiosus*, which grows anaerobically at temperatures near 100°C using carbohydrates and peptides as carbon and energy sources [75], has been extensively used to study thermostable enzymes and adaptations to high-temperature environments [114].

Methanobacteriales

The *Methanobacteriales* comprise another lineage of methanogenic archaea that reduce CO_2 or methyl compounds with H_2, formate, or secondary alcohols as electron donors. They include rod-shaped, lancet-shaped, or coccoid members, which contain cell walls made of pseudopeptidoglycan. *Methanobacteriales* are widely distributed in nature and are found in anaerobic habitats such as aquatic sediments, soil, anaerobic sewage digesters, and the gastrointestinal tracts of animals to name a few [50, 75]

Methanopyrales

The *Methanopyrales* consists of a single genus, *Methanopyrus*, comprising rod-shaped members with cell walls made of pseudopeptidoglycan [75]. Known *Methanopyrus* are hyperthermophilic, and grow between 84°C and 110°C, with optimal growth at 98°C. Similar to other methanogenic lineages, members of this group have a chemolitoautotrophic lifestyle converting CO_2 and H_2 to methane [9, 75]. While it has proven difficult to resolve the exact phylogenetic placement of the *Methanopyrales* relative to other archaea, it has recently been suggested that this lineage forms a monophyletic clade together with the *Methanobacteriales* and the *Methanococcales* referred to as *Methanomada* [7]. However, it

remains to be determined whether these so-called group 1 methanogens [9] are indeed closely related phylogenetically (Figure 21.2).

Methanofastidiosales

Methanofastidiosales represent a recently discovered and thus far uncultivated archaeal lineage (also known as WSA2 or Arc I), whose members are present in diverse environments including sediments, groundwater, and bioreactors [115–117]. Metagenomic approaches have enabled the reconstruction of genomes of representatives of the *Methanofastidiosales* from wastewater-treatment bioreactors [117]. While members of this group encode key genes for methanogenesis, they lack genes related to carbon-fixation pathways and were suggested to solely use methylated thiols as substrates for methanogenesis [117].

Theionarchaeota

Theionarchaea (formerly Z7ME43) represents another clade of uncultivated archaea, which forms a sister lineage of the *Hadesarchaea* (see next) and was originally discovered in water-filled limestone sinkholes in northeastern Mexico [118]. This clade is currently represented by two genomes that were recovered from the White Oak River Estuary in North Carolina [119]. Genomic analyses indicated that *Theionarchaea* might conserve energy by peptide fermentation.

Hadesarchaea

The *Hadesarchaea*, which were originally referred to as the South-African Gold Mine Miscellaneous Euryarchaeal Group (SAGMEG), are distributed in a variety of anoxic environments, including the terrestrial subsurface as well as marine sediments, which cover a wide span of temperatures [120–123]. The first genomes of members of this clade were reconstructed from the water column of the White Oak River estuary [123] as well as Yellowstone National Park (YNP) hot spring sediments and indicated the capability of anaerobic CO oxidation potentially coupled to nitrite or H_2O reduction [123]. Notably, another genome of a member of the *Hadesarchaea* was recently obtained from a hot spring metagenome and shown to encode a *mcr*-like operon. Based on phylogenetic analyses of MCR subunits as well as genomic analyses, it was suggested that these *Hadesarchaea* may represent alkane-oxidizing archaea similar to members of the *Syntrophoarchaeales* [124] and perhaps some representatives of the *Bathyarchaeota* [50].

Persephonarchaea

The Mediterranean Sea Brine Lakes 1 (MSBL1) clade, now referred to as the *Persephonarchaea* [7], is another lineage of uncultivated archaea that is closely related to the *Hadesarchaea*. The *Persephonarchaea* are commonly found in marine hypersaline environments [125, 126] and comprise potential anaerobic mixotrophs that may conserve energy through sugar fermentation but may also be able to fix inorganic carbon [127]. Genomic inferences suggest that MSBL1 archaea synthesize trehalose as putative osmolyte to encounter the high salt conditions in their environment [127].

Hydrothermarchaeota

The *Hydrothermarchaeota* [7], also known as the Marine Benthic Group-E (MBG-E), were originally discovered in marine deep-sea sediments [128] and represent an uncultivated archaeal lineage widely distributed in deep subseafloor environments. Genomes from members of this group have been reconstructed from metagenomes of the Juan de Fuca Ridge flank, Guaymas Basin hydrothermal sediments, and the Mid-Atlantic Ridge of the South Atlantic Ocean [129–131]. Genomic analyses have indicated that *Hydrothermarchaea* are metabolically versatile [131] and include putative anaerobic chemolithoautotrophs that use carbon monoxide and/or hydrogen as electron donors as well as a variety of electron acceptors including nitrate and sulfate [132, 133].

The TACK Superphylum

The TACK superphylum was originally introduced to describe the *Crenarchaeota* and the related phyla referred to as the *Thaumarchaeota, Aigarchaeota*, and *Korarchaeota* [16]. During the past years, many additional lineages affiliating with the TACK archaea have been discovered through metagenomics and single cell genomics approaches and the TACK lineage has therefore been suggested to be referred to as the *Proteoarchaeota* [134]. However, a consensus has yet to be reached regarding both the naming as well as the validity of using a superphylum as a taxonomic level. In the following sections, we introduce canonical and recently discovered clades belonging to the TACK archaea.

Crenarchaeota

The *Crenarchaeota* includes a diversity of (hyper-) thermophilic archaeal species, many of which have been discovered through cultivation-based approaches before the onset of the genomics era in microbiology and now represent important model organisms. This taxon is composed of a single class, the *Thermoprotei*, which is subdivided into three to five subclades, the *Thermoproteales, Sulfolobales, Desulfurcoccales* as well as the *Fervidicoccales* and *Acidilobales*. However, the latter two may in fact belong to the *Desulfurococcales* (Figure 21.2). Cultured crenarchaeal species are morphologically diverse, and include rods, cocci, filamentous, and disk-shaped cells. Almost all cultured species are obligate (hyper-) thermophiles, with optimal growth temperatures ranging from 70°C to 113°C and many members are also acidophiles and capable of metabolizing sulfur. Representatives of the *Crenarchaeota* thrive in environments such as hot solfataras, volcanic areas, as well as hydrothermal vents at the bottom of the ocean. A variety of metabolic capabilities have been described in the different members of the *Crenarchaeota*. For instance, some *Thermoproteales* are chemolithoautotrophs, using carbon dioxide as a carbon source and conserving energy by the conversion of hydrogen and elemental sulfur to hydrogen sulfide. Others respire various organic substrates using oxygen, sulfur, nitrate, or nitrite as electron acceptors [75]. Many members of the *Desulfurococcales* are strict anaerobes and neutrophiles to weak acidophiles, growing optimally at pH 5.5–7.5 [135]. Representatives of the *Sulfolobales* are acidophilic hyperthermophiles, which can grow lithoautotrophically by oxidizing sulfur or chemoheterotrophically on simple reduced carbon compounds using sulfur derivatives as electron acceptors. Notably, the *Crenarchaeota* include several members that have been shown to be hosts of the small-celled *Nanoarchaeota* [136–140] (see later in this chapter). In particular, the biocoenosis between *Ignicoccus hospitalis*, a member of the *Desulfurococcales*, and its

nanoarchaeal ectosymbiont, *Nanoarchaeum equitans*, has been extensively studied and provides important insights into archaeal cell biology and cell-cell communication [141]. For instance, investigation of *I. hospitalis* has revealed remarkable cellular features including the presence of two outer membranes surrounding a large periplasmic space as well as an endomembrane system reminiscent of eukaryotic cells [142].

Thaumarchaeota

Environmental 16S rRNA-based surveys in the early 1990s have led to the discovery of uncultivated archaeal lineages distantly related to the *Crenarchaeota* in moderate marine and terrestrial ecosystems. The subsequent cultivation of the first representatives of these so-called mesophilic *Crenarchaeota* (also MG1) from marine and terrestrial environments [143] and the study of the first genomes of members of this group [144, 145], revealed that they form a separate phylum within the *Archaea* referred to as the *Thaumarchaeota* that distantly affiliates with the *Crenarchaeota*. Most cultivated *Thaumarchaeota* are chemolithoautotrophic ammonia-oxidizing archaea (AOA), which play an important role in the nitrogen and carbon cycles in both aquatic and terrestrial environments [146]. However, the reconstruction of genomes of deep-branching *Thaumarchaeota* has recently led to the suggestion that not all members of this group are AOA but instead represent chemoorganotrophs that may reduce oxygen, nitrate, or sulfur [147]. This notion was recently confirmed with the isolation of the thermoacidophilic, sulfur- and iron-reducing organoheterotrophic *Conexivisphaera calidus*, a potentially early diverging member of the *Thaumarchaeota* [148].

Aigarchaeota

The *Aigarchaeota* represent another proposed candidate phylum that comprises species of the Hot Water Crenarchaeotic Group I (HWCGI), members of which have not been cultivated so far. Genomic analyses of the first representatives of this group have suggested that the *Aigarchaeota* comprise both facultative and obligate anaerobes, which may respire a variety of organic substrates and perhaps also hydrogen and carbon monoxide using oxygen or oxidized sulfur or nitrogen compounds as electron acceptors [42, 149–152]. Furthermore, several representatives seem to have the ability to fix inorganic carbon. *Aigarchaeota* seem to predominantly inhabit thermally heated terrestrial and marine ecosystems, including hot springs, subsurface aquifers, and mine fracture waters [150, 152].

Korarchaeota

The *Korarchaeota* comprises a group of uncultivated *Archaea* that had already been discovered in the late 1990s in terrestrial and marine thermal environments [153]. The first member of this clade, referred to as "*Ca. Korarchaeum cryptofilum*," was shown to comprise ultra-thin, needle-shaped cells measuring up to 100 μm in length. Genomic analyses indicated that this organism represents a peptide fermenter with a unique set of informational processing genes, which early on indicated that it comprises the first member of a distinct archaeal phylum [154]. Recently, genomes of additional members of the *Korarchaeota* have been recovered from deep-sea hydrothermal vent sediments [130] and hot spring environments [18, 124, 155] providing novel insights into the metabolic features of this clade. Notably, genomic analyses revealed that certain members of the *Korarchaeota* harbor the key genes for methanogenesis, [155] which may for instance enable methanogenesis from methanol and hydrogen or the coupling of the anaerobic oxidation of methane with sulfite reduction [155].

Bathyarchaeota

Bathyarchaeota were originally discovered through 16S rRNA gene surveys in hot springs [153] and were referred to as Miscellaneous Crenarchaeota Group (MCG) [156] due to their distant affiliation with cultivated *Crenarchaeota*. This extremely diverse phylum is now subdivided into at least 25 subgroups, which are defined at family and order level [157]. Notably, members of this putative phylum-level lineage can be found in a diversity of anoxic marine, terrestrial, and hydrothermal environments including marine sediments and often represent the most abundant archaeal community members [157–159]. Based on genomic analyses, it is inferred that many *Bathyarchaeota* are heterotrophs with a wide substrate range including acetate, proteins, and aromatic compounds such as lignin [157, 160]. However, the *Bathyarchaeota* also includes putative acetogenic species [161] as well as organisms with *mcr* genes [162], which are closely related to those of *Syntrophoarchaea* [71]. In turn, it has been suggested that some members of the *Bathyarchaeota* may be able to mediate the anaerobic oxidation of short-chain alkanes [50].

Geoarchaeota

Geoarchaeota, also Novel Archaeal Group 1 (NAG1), are often found in hypoxic to oxic, hot, acidic, iron-rich springs [163–165] and represent a lineage of thus far uncultivated archaea which seem to be closely related to or part of the Crenarchaeota [164, 166]. Based on genomic inferences, it has been suggested that the *Geoarchaeota* are likely motile and might conserve energy through the oxidation of carbon monoxide, peptides, and/or carbohydrates using oxygen as a terminal electron acceptor [149, 164].

Verstraetearchaeota

Verstraetearchaeota were originally discovered in deep South-African Gold mine microbial communities through 16S rRNA gene surveys and were referred to as Terrestrial Miscellaneous Crenarchaeota Group (TMCG) [120]. Members of this group seem to be widely distributed and are also found in hydrocarbon-rich environments, sediments, soil, and wetlands [167]. First insights into the metabolic features of members of this group were derived from genomes assembled from anoxic digesters, named *Methanomethylicus sp.* and *Methanosuratus sp.* [167]. Subsequently, additional representatives were discovered and referred to as *Methanohydrogenales* and *Methanomediales* [168]. Notably, the *Verstraetearchaeota* comprise members with *mcr*-gene operons most similar to those found in methanogenic *Euryarchaeota*. In turn, based on genomic inferences, it was suggested that the *Verstraetearchaeota* likely include anaerobic methylotrophic as well as hydrogenotrophic methanogens [167–169].

Nezhaarchaeota

Nezhaarchaeota are a recent addition to the TACK superphylum represented solely by uncultivated members, whose genomes were assembled from hot spring metagenomes and hydrothermal sediments [77]. Notably, the *Nezhaarchaeota* encode a MCR protein cluster and are potential hydrogenotrophic methanogens [77].

Marsarchaeota

The *Marsarchaeota*, or "Novel Archaeal Group 2" (NAG2), are typically found in geothermal, iron oxide-rich mats [163]. The first genomes of members of this lineage were recently recovered from thermal (50–80°C) and acidic (pH 2.5–2.5) microbial mats from Yellowstone National Park [170] and led to the suggestions that the *Marsarchaeota* are aerobic chemoorganotrophs that degrade lipids, peptides, and carbohydrates and may be able to reduce ferric oxide.

Geothermarchaeota

The *Geothermarchaeota* represents one of the most recent additions to the *Archaea* and is thus far only represented by uncultivated members, whose genomes have been reconstructed from metagenomes from the Juan de Fuca Ridge subseafloor [129] and hydrothermal vent sediments in the Guaymas Basin [130]. Little is yet known about the lifestyle and ecological roles of *Geothermarchaeota*, and in-depth genomic analyses will be necessary to infer their metabolic potential.

The Asgard Superphylum

The Asgard superphylum is a recently described archaeal radiation, which comprises several different archaeal clades of high taxonomic rank (likely phylum-level) [17, 18]. Notably, phylogenetic and comparative genomic analyses have indicated that this archaeal clade includes the closest archaeal sister lineage of eukaryotes (discussed previously in this chapter). Members of this superphylum have originally been discovered in sediments all around the world, in which they can comprise a significant fraction of the microbial diversity. In the following, we briefly introduce the major metabolic features of the currently known members of the Asgard archaea, i.e., the *Loki-, Thor-, Odin-, Hel-,* and *Heimdallarchaeota*.

Lokiarchaeota

The *Lokiarchaeota* represents an archaeal lineage originally referred to as the Deep Sea Archaeal Group (DSAG) or Marine Benthic Group B (MBGB) archaea, which are abundant in diverse marine sediments [94, 171]. For example, the *Lokiarchaeota* comprise up to 10% of the microbial community in cold sediments near Loki's Castle hydrothermal vent field from which the first metagenomes were obtained [17, 18]. Members of the *Lokiarchaeota* might be autotrophs using the Wood-Ljungdahl pathway for carbon fixation [172]. However, genomic analyses also revealed the potential for the use of a variety of organic carbon substrates, suggesting that representatives of the *Lokiarchaeota* may predominantly rely on fermentative growth [20]. In fact, the successful cultivation of the first *Lokiarchaeote, Candidatus*

Prometheoarchaeum syntrophicum, revealed that this organism ferments amino acids enabled through a syntrophic interaction with hydrogen- or formate-consuming partner organisms [22].

Thorarchaeota

The *Thorarchaeota* share many metabolic features with the *Lokiarchaeota* [20, 173, 174]. Currently known representatives harbor a variety of genes likely encoding proteins involved in the usage of organic substrates. Furthermore, they encode the Wood-Ljungdahl pathway, which could be used for carbon fixation or serve as an electron sink during growth on organics. In contrast to currently known *Lokiarchaeota*, members of this group also harbor a putative NADH dehydrogenase that may enable respiratory growth in addition to fermentation [20]. Based on current environmental survey data, the *Thorarchaeota* seem less abundant than the *Lokiarchaeota* but occur in a wide variety of anoxic environments [18].

Heimdallarchaeota

Thus far known representatives of the *Heimdallarchaeota* are metabolically diverse and differ from other Asgard lineages [20]. While genomic analyses indicate that they are able to utilize a large variety of organic substrates similar to other members of the Asgards, they do not seem to be fermentative organisms and current members lack the Wood-Ljungdahl pathway. Instead, they encode a membrane-bound electron chain, which allows growth using oxygen and nitrite as electron acceptors [20, 21]. *Heimdallarchaeota* are currently thought to comprise the archaeal lineage most closely related to the archaeal ancestor of eukaryotes. However, though found in a variety of environmental samples including anoxic sediments and oxygenated waters, they are generally less abundant than the *Loki-* and *Thorarchaeota*.

Odinarchaeota

Odinarchaeota are currently represented by a single genome, which was obtained from a hot spring metagenome [18]. Similar to other members of the Asgard superphylum, *Odinarchaeum* encodes the ability to use organic compounds as growth substrates [20]. Yet, it lacks the key enzyme of the Wood-Ljungdahl pathway and instead encodes membrane-bound hydrogenases, which suggests that the thermophilic *Odinarchaeum* may conserve energy through fermentation of organic substrates to hydrogen, acetate, and carbon dioxide. Members of the *Odinarchaeota* are thought to predominantly inhabit thermal environments such as hot spring sediments and hydrothermal vents [18].

Helarchaeota

The *Helarchaeota* represent the most recently discovered clade within the Asgard archaea [175]. While they harbor similar gene sets as the *Loki-* and *Thorarchaeota*, currently known representatives of this lineage also contain *mcr*-gene clusters. Phylogenetic analyses of the encoded proteins revealed their close relationship with proteins of *Syntrophoarchaea* opening the possibility that certain members of the Asgard archaea have the potential to anaerobically oxidize short-chain alkanes, perhaps in syntrophy with microbial partners [175]. However, the environmental distribution of the *Helarchaeota* and the functional importance of this potential alkane metabolism in Asgard archaea remain to be determined.

The DPANN Superphylum

The DPANN superphylum is the fourth major radiation in the *Archaea*, besides the *Euryarchaeota*, TACK, and Asgard archaea [149, 176, 177]. Currently, this radiation is assumed to comprise a large diversity of distinct archaeal clades most of which seem to predominantly include members with extremely small genomes and cell sizes that are thought to depend on partner organisms for growth and survival [177]. While first defined in reference to the *Diapherotrites, Parvarchaeota, Aenigmarchaeota, Nanoarchaeota,* and *Nanohaloarchaeota* (DPANN) [149], additional lineages such as the *Micrarchaeota*, *Pacearchaeota*, *Woesearchaeota*, and *Huberarchaeota* are now also considered members of this group [177, 178]. Furthermore, the *Altiarchaeota* [179], representatives of which do not have reduced genomes, are sometimes considered to belong to the DPANN [177, 180]. However, the boundaries between certain clades of DPANN and other archaea (in particular the *Euryarchaeota*) are not well defined and it remains to be established which lineages indeed belong to a monophyletic (i.e., sharing a common ancestor) DPANN clade [180].

Nanoarchaeota

The first representative lineage of the DPANN archaea was already discovered in 2002, i.e., long before the DPANN radiation was known. In particular, Huber and coworkers discovered a small-celled organism in cultures of the crenarchaeaum, *I. hospitalis,* which they referred to as *N. equitans* [136]. Initially, it was suggested that this organism is the first representative of a novel phylum called *Nanoarchaeota* or may represent a highly derived member of the *Euryarchaeota* [136, 181]. However, upon the genomics-based discovery of additional archaeal lineages represented by organisms with small genomes, which affiliated with *Nanoarchaeota*, it was proposed that the *Nanoarchaeota* belong to the DPANN radiation [149].

Notably, the nanosized hyperthermophilic *N. equitans* is obligately host-dependent and grows as an ectoparasite on the surface of *I. hospitalis* [182, 183]. It lacks genes for many major metabolic pathways and in turn depends on its host for the acquisition of diverse metabolites likely including lipids, amino acids, and ATP. In line with this, the genome of *N. equitans* represents one of the smallest known genomes of any extracellular organism (480 kb) [184]. However, compared to the genomes of many bacterial endosymbionts, the genome of *N. equitans* does not show evidence of pseudogenes and contains a full complement of tightly packed genes encoding informational processing machineries [184]. Similar trends have recently been seen in other representatives of the DPANN radiation [177]. Members of the *Nanoarchaeota* have been found in a variety of thermal environments ranging from hydrothermal vents to hot springs and are now assumed to infect a variety of different crenarchaeal hosts. For instance, additional *Nanoarchaeota*, such as *Candidatus Nanobsidianus stetteri, Candidatus Nanopusillus acidilobi,* and *Candidatus Nanoclepta minutus* have recently been successfully co-cultivated with their crenarchaeal hosts referred to as

Acidolobus sp., Acidilobus sp. 7A, and *Zestosphaera tikiterensis*, respectively [138, 185, 186].

Overview of Other Putative DPANN Clades

Most of the other DPANN clades are represented by genomes reconstructed through metagenomics or single-cell genomics approaches. However, recent cultivation efforts have led to the enrichment of the first co-culture of a member of the *Nanohaloarchaeota* with its haloarchaeal host, i.e., *Ca. Nanohaloarchaeum antarcticus* and *Halorubrum lacusprofundi* [187], and of members of the *Micraarchaeota* members with their putative archaeal partners belonging to the *Thermoplasmatales* [188, 189]. Even though *Ca. N. antarcticus* has a larger genome and metabolic gene repertoire than *N. equitans*, it seems to obligately depend on its host for growth and survival [187].

Additional insights into the diversity and metabolic potential of members of the various DPANN clades predominantly derive from genomic inferences [6, 177, 180]. For instance, the *Woese-* and *Pacearchaeota* seemingly represent extremely diverse DPANN lineages whose members differ in the extent of genome reduction and metabolic capabilities. However, all representatives lack major and essential metabolic pathways indicating obligate host dependency. Few representatives of the DPANN, such as members of the *Diapherotrites* [190], may be able to conserve energy through fermentation. But the lack of some biosynthetic pathways indicates that they still depend on the external acquisition of certain amino acids, vitamins and/or cofactors [177, 180].

While much has to be learned about the DPANN archaea, the discovery of this large diversity of putative archaeal symbionts and the occurrence of certain representatives in almost all environments on Earth indicates that the future investigation of this radiation will be crucial for our understanding of both the evolution and ecology of *Archaea* and their impact on global nutrient cycles.

Altiarchaeota and its Symbiont—A Member of the Huberarchaeota

The *Altiarchaeota* represent a lineage variably affiliating either with the DPANN archaea or *Euryarchaeota* [6, 135, 177, 179, 180, 191] in phylogenetic analyses depending on the type of analysis (e.g., with regard to model choice) and data used. *Altiarchaeota* (formerly also referred to as SM1 *Euryarchaeota*) were originally discovered in a cold (~10°C), sulfurous Moor in Germany [135] but can also be found in sulfidic springs [192, 193], marine sediments, hot springs, and aquifers [191]. Notably, some members of the *Altiarchaeota* are found in microbial consortia that display a unique morphology described as a "string-of-pearls," which is several millimeters in length and consists of tiny white pearls (0.5–3 mm diameter) connected by thin threads [135]. The outer part of the pearl is composed of bacteria, such as the *Gammaproteobacterium Thiotrix uunzi* [194] or the *Epsilonproteobacterium* IMB1 [195], while the inside is dominated by *Altiarchaeota* [135]. The large size of the consortium allows for the effective enrichment of *Altiarchaeota* on polyethylene nets that can consist of ~98% of archaeal cells and ~2% bacteria [196]. Other representatives of the *Altiarchaeota* occur in almost single-member biofilms (~5% bacteria, ~95% *Altiarchaeota*) in sulfidic springs [192, 193].

Notably, *Altiarchaeota* have not only been found in symbiosis with bacteria but represent the hosts of members of the *Huberarchaeota*, a recent addition to the DPANN superphylum [178, 197]. Similar to other DPANN archaea, known members of the *Huberarchaeota* have reduced genomes and lack proteins related to energy metabolism, regeneration of redox equivalents and nucleotide biosynthesis indicating that they depend on a variety of compounds from their hosts.

The first insights into the metabolism of the Altiarchaeota came from the metagenome-assembled genome (MAG) of *Candidatus* Altiarchaeum hamiconexum, which was reconstructed from a cold, sulfidic spring in Germany [179]. Genomic analyses suggested this representative is an anaerobic autotroph, potentially capable of growth on carbon dioxide and possibly acetate, formate, and carbon monoxide [179]. *Ca. A.* hamiconexum is a biofilm-forming, nonmotile organism with coccoid cells (0.4–0.7 μM in diameter) and a double membrane [179]. Cells can be surrounded by up to 100 hair-like proteinaceous appendages of 2–3 μM length, so-called hami, which mediate adhesion to various surfaces [198]. However, representatives of the *Altiarchaeota* from sediments lack genes encoding proteins involved in hami formation and show specific adaptations to their respective environments [191].

Archaea as Part of the Human Microbiome

For a long time, it was thought that *Archaea* played minor roles in the microbiomes of humans and other mammals and true archaeal pathogens remain to be discovered. The first archaeon associated with humans was described in 1982, the methanogenic *Methanobrevibacter smithii*, which was isolated from human feces [199] suggesting that the methane exhaled by a certain proportion of humans may be produced by methanogens residing in the gastrointestinal tract [200]. Since then, several archaeal species have been identified to be associated with the intestinal, oral, gut, nasal, vaginal, lung, and skin microbiota of both humans and other animals [201–203]. However, their roles in human health and disease remain poorly understood [201–205]. In the following, we summarize current knowledge regarding the diversity and function of human-associated archaea.

Oral Archaeome

Methanogenic archaea are part of the oral archaeome with *Methanobrevibacter oralis* being the most frequently detected species [205, 206]. Notably, *M. oralis* seems to be correlated with periodontitis severity, supporting a potential pathogenic role of methanogenic archaea [206–208]. While no direct experimental evidence has demonstrated the virulence pattern of *M. oralis* and other oral archaeal species, the unique metabolism of methanogenic archaea provides insight into possible drivers of oral disease. For instance, methanogens in periodontal pockets may serve as an H_2 sink, which would favor the proliferation of syntrophic pathogenic microbes [206–209]. Recent investigation into microbial communities in the oral cavity has shown significant positive correlations between the abundance of methanogens with that of *Prevotella intermedia*, a known bacterial pathogen involved in periodontal infections such as periodontitis, gingivitis, and necrotizing ulcerative gingivitis [208].

The relationship between these two groups in periodontal pockets is still unknown, but indirect and direct associations between the methanogens and the local environment may be driving the proliferation of *P. intermedia* through a series of possible syntrophic interactions [208]. A current key research interest is to further determine the immediate role of archaea in the pathogenesis of periodontal infections [206, 210]

Gut Archaeome

To date, three species of methanogenic archaea have been cultivated and isolated from gut-derived samples, i.e., from human stool: *M. smithii*, *Methanosphaera stadtmanae*, and *Methanomassiliicoccus luminyensis* [95, 199, 211]. With the help of molecular tools, two candidate-species, *Candidatus Methanomassiliicoccus intestinalis* and *Candidatus Methanomethylophilus alvus*, in addition to several unknown members of the orders *Methanosarcinales*, *Methanobacteriales*, *Methanococcales*, *Methanomicrobiales*, and *Methanopyrales*, have been shown to inhabit the human gastrointestinal tract [202]. Further, the presence of methanogens in biopsy samples suggests that they may be associated with the mucosal lining in addition to their presumed presence in the lumen [202]. *M. smithii* is the major archaeal component in the human gut, while *M. stadtmanae* and *M. luminyensis* are less frequently detected species [201, 202] and appear to play an important role as H_2-consumers in the complex microbial ecosystem of the gut [201, 205, 209]. Fermentative microorganisms produce short-chain fatty acids and H_2, the former being consumed by the host and the latter being scavenged and consumed by the archaea. This removal of H_2 from the system by methanogens makes the fermentative processes kinetically more favorable and continuously drives this cyclical syntrophy [201, 202, 205]. Furthermore, there is evidence that methanogens may be involved in inflammatory bowel disease, irritable bowel syndrome, colorectal cancer, diverticulosis, and obesity [201, 205]. However, it is unclear whether methanogens directly or indirectly contribute to the development of gastrointestinal disorders and without doubt, more research is needed to unravel the role of archaea in intestinal disorders [204, 212]. For instance, it has also been suggested that some human-associated archaea may be mutualistic, providing health benefits and influencing host metabolism [202, 213].

Not all gut-associated archaea are methanogens however [202]. For instance, a halophilic archaeon belonging to the halobacteria, *Haloferax massiliensis*, was recently isolated from a human stool sample, reigniting the debate over whether halophiles may colonize the gut [214, 215]. Other studies have revealed a diversity of halobacteria-related sequences in fecal samples collected from healthy Korean people, with analyzed sequences representing *Halorubrum alimentarium* and *Halorubrum koreense*. Interestingly, both *H. alimentarium* and *H. koreense* had previously been isolated from salt-fermented sea foods suggesting native cuisine and eating habits may contribute to the propensity of these organisms in the gut environment [201, 216].

Global Human Archaeome

Technological advancements in high-throughput sequencing have further improved insights into the human microbiome and revealed unexpected diversity of representatives from archaeal phyla that had not previously been detected in human habitats, including members of the DPANN archaea. In particular, members of the *Woesearchaeota* appear to be present in the human lung, and while it is speculated that they may exhibit parasitic/symbiotic lifestyles, their environmental role remains unclear [202]. Analytical exploration into the distribution of archaeal signatures in the human body has revealed site-specific patterns that shape a preliminary biogeographical view of the human archaeome: (1) *Thaumarchaeota* on the skin, (2) methanogenic *Euryarchaeota* in the gut, (3) mixed skin-gut nasal archaeal communities, and (4) *Woesearchaeota* inhabiting the lungs [202].

While *M. smithii*, *M. oralis*, *M. stadtmanae*, *M. luminyensis*, and *H. massiliensis* are the only archaeal species that have been successfully isolated and cultivated from human habitats, efforts are underway to culture more archaeal species associated with humans in order to better understand their roles as potential pathogens or commensal members with potentially positive physiological impacts. For instance, a major step toward a better understanding of the function and dynamics of human-associated archaea may be gained through the Human Microbiome Project [209, 217].

Concurrent with efforts to culture archaeal species infecting humans and elucidate their potential roles in human pathogenesis, there are several initiatives aiming to identify antimicrobial agents that are effective against *Archaea*. Research shows that archaea are resistant to antibiotics used to treat bacterial infections, in part due to morphological and physiological features that impede the action of many bacterial-targeting antimicrobial agents. *In vivo* and *in vitro* experiments have indicated susceptibility of several archaeal groups to certain protein synthesis inhibitors, including fusidic acid and imidazole derivatives [218]. Nonetheless, antibiotic-resistant archaea may become indirectly susceptible to antimicrobial treatments when relying on chemically susceptible bacterial partners within their complex communities. To date, there are a limited number of antimicrobials that target archaea directly. Greater exploration into archaea as causative agents of human disease would also require further investigation into antiarchaeal compounds and treatments [210, 218].

Summary

Thought to be of limited ecological relevance originally, *Archaea* are now known to inhabit a wide range of ecosystems and to play a key role in major biogeochemical cycles [8]. Furthermore, *Archaea* have proven to be of fundamental importance for our understanding of the evolution of complex eukaryotic cells [10] and have emerged as important model systems. Notably, representatives of the *Archaea* are now known to form a stable part of the human microbiome and may even be involved in disease. Unique metabolic and cellular features of *Archaea* are being utilized in a variety of biotechnological applications as well as the development of novel adjuvants in the use of vaccines utilizing the unique membrane lipids of archaeal membranes [219]. Considering that a large fraction of *Archaea* of high-taxonomic rank likely still awaits discovery [5], the coming years will certainly witness further insights into the role of *Archaea* in ecological food webs, the evolution of life and human biology.

Funding

This work was supported by a grant of the Swedish Research Council (VR starting grant 2016-03559 to Anja Spang), the NWO-I foundation of the Netherlands Organisation for Scientific Research (WISE fellowship to Anja Spang). Furthermore, Laura Eme is currently supported by funding from the European Research Council (ERC) under the European Union's Horizon 2020 research and innovation program (grant agreement No 803151).

REFERENCES

1. Woese, C.R. and G.E. Fox, Phylogenetic structure of the prokaryotic domain: the primary kingdoms. *Proceedings of the National Academy of Sciences of the United States of America*, 1977. **74**(11): 5088–5090.

2. Woese, C.R., O. Kandler, and M.L. Wheelis, Towards a natural system of organisms: proposal for the domains archaea, bacteria, and eucarya. *Proceedings of the National Academy of Sciences of the United States of America*, 1990. **87**(12): 4576–4579.

3. Sangwan, N., F. Xia, and J.A. Gilbert, Recovering complete and draft population genomes from metagenome datasets. *Microbiome*, 2016. **4**: 8–8.

4. Woyke, T., D.F.R. Doud, and F. Schulz, The trajectory of microbial single-cell sequencing. *Nature Methods*, 2017. **14**(11): 1045–1054.

5. Spang, A., E.F. Caceres, and T.J.G. Ettema, Genomic exploration of the diversity, ecology, and evolution of the archaeal domain of life. *Science*, 2017. **357**(6351).

6. Castelle, C.J. and J.F. Banfield, Major new microbial groups expand diversity and alter our understanding of the tree of life. *Cell*, 2018. **172**(6): 1181–1197.

7. Adam, P.S., et al., The growing tree of archaea: new perspectives on their diversity, evolution and ecology. *ISME Journal*, 2017. **11**(11): 2407–2425.

8. Offre, P., A. Spang, and C. Schleper, Archaea in biogeochemical cycles. *Annual Review of Microbiology*, 2013.

9. Thauer, R.K., et al., Methanogenic archaea: ecologically relevant differences in energy conservation. *Nature Reviews Microbiology*, 2008. **6**(8): 579–591.

10. Eme, L., et al., Archaea and the origin of eukaryotes. *Nature Reviews Microbiology*, 2018. **16**(2): 120–120.

11. López García, and D. Moreira, Open questions on the origin of eukaryotes. *Trends in Ecology & Evolution*, 2015. **30**(11): 697–708.

12. Martin, W.F., S. Garg, and V. Zimorski, Endosymbiotic theories for eukaryote origin. *Philosophical Transactions of the Royal Society of London B: Biological Sciences*, 2015. **370**(1678): 20140330–20140330.

13. Guy, L., J.H. Saw, and T.J. Ettema, The archaeal legacy of eukaryotes: a phylogenomic perspective. *Cold Spring Harbor Perspectives in Biology*, 2014. **6**(10): a016022.

14. Iwabe, N., et al., Evolutionary relationship of Archaebacteria, eubacteria, and eukaryotes inferred from phylogenetic trees of duplicated genes. *Proceedings of the National Academy of Sciences of the United States of America*, 1989. **86**(23): 9355–9359.

15. Williams, T.A., et al., An archaeal origin of eukaryotes supports only two primary domains of life. *Nature*, 2013. **504**(7479): 231–236.

16. Guy, L. and T.J.G. Ettema, The archaeal 'TACK' superphylum and the origin of eukaryotes. *Trends in Microbiology*, 2011. **19**(12): 580–587.

17. Spang, A., et al., Complex archaea that bridge the gap between prokaryotes and eukaryotes. *Nature*, 2015. **521**(7551): 173–179.

18. Zaremba-Niedzwiedzka, K., et al., Asgard archaea illuminate the origin of eukaryotic cellular complexity. *Nature*, 2017. **541**(7637): 353–358.

19. Sousa, F.L., et al., Lokiarchaeon is hydrogen dependent. *Nature Microbiology*, 2016. **1**(5).

20. Spang, A., et al., Proposal of the reverse flow model for the origin of the eukaryotic cell based on comparative analyses of Asgard archaeal metabolism. *Nature Microbiology*, 2019.

21. Bulzu, P.A., et al., Casting light on Asgardarchaeota metabolism in a sunlit microoxic niche. *Nature Microbiology*, 2019. **4**(7): 1129–1137.

22. Imachi, H., et al., Isolation of an archaeon at the prokaryote-eukaryote interface. *Nature*, 2020.

23. Williams, T.A., et al., Phylogenomics provides robust support for a two-domains tree of life. *Nature Ecology and Evolution*, 2020. **4**(1): 138–147.

24. Makarova, K.S. and E.V. Koonin, Archaeology of eukaryotic DNA replication. *Cold Spring Harbor Perspectives in Biology*, 2013. **5**(11): a012963–a012963.

25. Raymann, K., et al., Global phylogenomic analysis disentangles the complex evolutionary history of DNA replication in archaea. *Genome Biology and Evolution*, 2014. **6**(1): 192–212.

26. Makarova, K.S., M. Krupovic, and E.V. Koonin, Evolution of replicative DNA polymerases in archaea and their contributions to the eukaryotic replication machinery. *Frontiers in Microbiology*, 2014. **5**: 354–354.

27. Mattiroli, F., et al., Structure of histone-based chromatin in archaea. *Science*, 2017. **357**(6351): 609–612.

28. Bhattacharyya, S., F. Mattiroli, and K. Luger, Archaeal DNA on the histone merry-go-round. *FEBS Journal*, 2018. **285**(17): 3168–3174.

29. Werner, F. and D. Grohmann, Evolution of multisubunit RNA polymerases in the three domains of life. *Nature Reviews Microbiology*, 2011. **9**(2): 85–98.

30. Yutin, N., et al., Phylogenomics of prokaryotic ribosomal proteins. *Genome Biology*, 2011. **12**(1): 1–1.

31. Krzycki, J.A., The direct genetic encoding of pyrrolysine. *Current Opinion in Microbiology*, 2005. **8**(6): 706–712.

32. Hartman, H. and A. Fedorov, The origin of the eukaryotic cell: a genomic investigation. *Proceedings of the National Academy of Sciences of the United States of America*, 2002. **99**(3): 1420–1425.

33. Brochier-Armanet, C., S. Gribaldo, and P. Forterre, A DNA topoisomerase IB in Thaumarchaeota testifies for the presence of this enzyme in the last common ancestor of Archaea and Eucarya. *Biology Direct*, 2008. **3**.

34. Koonin, E.V. and N. Yutin, The dispersed archaeal eukaryome and the complex archaeal ancestor of eukaryotes. *Cold Spring Harbor Perspectives in Biology*, 2014. **6**(4): a016188–a016188.

35. Saw, J.H., et al., Exploring microbial dark matter to resolve the deep archaeal ancestry of eukaryotes. *Philosophical Transactions of the Royal Society of London B: Biological Sciences*, 2015. **370**(1678): 20140328–20140328.

36. Samson, R.Y., et al., A role for the ESCRT system in cell division in archaea. *Science (New York, N.Y.)*, 2008. **322**(5908): 1710–1713.

37. Lindås, A.-C., et al., A unique cell division machinery in the archaea. *Proceedings of the National Academy of Sciences of the United States of America*, 2008. **105**(48): 18942–18946.

38. Ettema, T.J.G. and R. Bernander, Cell division and the ESCRT complex. *Communicative & Integrative Biology*, 2009. **2**(2): 86–88.

39. Pelve, E.A., et al., Cdv-based cell division and cell cycle organization in the thaumarchaeon Nitrosopumilus maritimus. *Molecular Microbiology*, 2011. **82**(3): 555–566.

40. Ettema, T.J., A.C. Lindas, and R. Bernander, An actin-based cytoskeleton in archaea. *Molecular Microbiology*, 2011. **80**(4): 1052–1061.

41. Yutin, N. and E.V. Koonin, Archaeal origin of tubulin. *Biology Direct*, 2012. **7**(1).

42. Nunoura, T., et al., Insights into the evolution of archaea and eukaryotic protein modifier systems revealed by the genome of a novel archaeal group. *Nucleic Acids Research*, 2011. **39**(8): 3204–3223.

43. Klinger, C.M., et al., Tracing the archaeal origins of eukaryotic membrane-trafficking system building blocks. *Molecular Biology and Evolution*, 2016. **33**(6): 1528–1541.

44. Akıl, C. and R.C. Robinson, Genomes of Asgard archaea encode profilins that regulate actin. *Nature*, 2018: 1.

45. Villanueva, L., S. Schouten, and J.S.S. Damsté, Phylogenomic analysis of lipid biosynthetic genes of Archaea shed light on the 'lipid divide.' *Environmental Microbiology*, 2017.

46. Koga, Y. and H. Morii, Recent advances in structural research on ether lipids from archaea including comparative and physiological aspects. *Bioscience, Biotechnology and Biochemistry*, 2005. **69**(11): 2019–2034.

47. Klingl, A., S-layer and cytoplasmic membrane - exceptions from the typical archaeal cell wall with a focus on double membranes. *Frontiers in Microbiology*, 2014. **5**: 624–624.

48. Parks, D.H., et al., A standardized bacterial taxonomy based on genome phylogeny substantially revises the tree of life. *Nature Biotechnology*, 2018.

49. Borrel, G., P.S. Adam, and S. Gribaldo, Methanogenesis and the Wood-Ljungdahl pathway: an ancient, versatile, and fragile association. *Genome Biology and Evolution*, 2016. **8**(6): 1706–1711.

50. Evans, P.N., et al., An evolving view of methane metabolism in the archaea. *Nature Reviews, Microbiology*, 2019. **17**(4): 219–232.

51. Ferry, J.G., How to make a living by exhaling methane. *Annual Review of Microbiology*, 2010. **64**: 453–473.

52. Knittel, K. and A. Boetius, Anaerobic oxidation of methane: progress with an unknown process. *Annual Review of Microbiology*, 2009. **63**(1): 311–334.

53. Raymann, K., C. Brochier-Armanet, and S. Gribaldo, The two-domain tree of life is linked to a new root for the Archaea. *Proceedings of the National Academy of Sciences of the United States of America*, 2015. **112**(21): 6670–6675.

54. Williams, T.A., et al., Integrative modeling of gene and genome evolution roots the archaeal tree of life. *Proc Natl Acad Sci U S A*, 2017. **114**(23): E4602–E4611.

55. Garcia, J.-L., B. Ollivier, and W.B. Whitman, The Order Methanomicrobiales, in *The Prokaryotes*, M. Dworkin, et al., Editors. 2006, Springer New York: New York, NY. 208–230.

56. Orsi, W.D., et al., Climate oscillations reflected within the microbiome of Arabian Sea sediments. *Scientific Reports*, 2017. **7**(1): 6040.

57. Haroon, M.F., et al., Anaerobic oxidation of methane coupled to nitrate reduction in a novel archaeal lineage. *Nature*, 2013. **500**(7464): 567–570.

58. Ettwig, K.F., et al., Nitrite-driven anaerobic methane oxidation by oxygenic bacteria. *Nature*, 2010. **464**(7288): 543–548.

59. Ettwig, K.F., et al., Archaea catalyze iron-dependent anaerobic oxidation of methane. *Proceedings of the National Academy of Sciences of the United States of America*, 2016. **113**(45): 12792–12796.

60. McGlynn, S.E., et al., Single cell activity reveals direct electron transfer in methanotrophic consortia. *Nature*, 2015. **526**(7574): 531–535.

61. Orsi, W.D., Ecology and evolution of seafloor and subseafloor microbial communities. *Nature Reviews Microbiology*, 2018: 1.

62. Timmers, P.H.A., et al., Reverse methanogenesis and respiration in methanotrophic archaea. *Archaea*, 2017.

63. Knittel, K. and A. Boetius, Anaerobic oxidation of methane: progress with an unknown process. *Annual Review of Microbiology*, 2009. **63**: 311–334.

64. Wegener, G., et al., Intercellular wiring enables electron transfer between methanotrophic archaea and bacteria. *Nature*, 2015. **526**(7574): 587–590.

65. Sakai, S., et al., Methanocella paludicola gen. nov., sp. nov., a methane-producing archaeon, the first isolate of the lineage 'Rice Cluster I', and proposal of the new archaeal order methanocellales ord. nov. *International Journal of Systematic and Evolutionary Microbiology*, 2008. **58**(4): 929–936.

66. Lu, Y. and R. Conrad, In Situ stable isotope probing of methanogenic archaea in the rice rhizosphere. *Science*, 2005. **309**(5737): 1088–1090.

67. Sakai, S., et al., Isolation of key methanogens for global methane emission from rice paddy fields: a novel isolate affiliated with the clone cluster rice cluster I. *Applied and Environmental Microbiology*, 2007. **73**(13): 4326–4331.

68. Sakai, S., et al., Methanocella arvoryzae sp. nov., a hydrogenotrophic methanogen isolated from rice field soil. *International Journal of Systematic and Evolutionary Microbiology*, 2010. **60**(12): 2918–2923.

69. Lü, Z. and Y. Lu, Methanocella conradii sp. nov., a thermophilic, obligate hydrogenotrophic methanogen, isolated from Chinese rice field soil. *PLOS ONE*, 2012. **7**(4).

70. Orcutt, B.N., et al., Impact of natural oil and higher hydrocarbons on microbial diversity, distribution, and activity in Gulf of Mexico cold-seep sediments. *Deep Sea Research Part II: Topical Studies in Oceanography*, 2010. **57**(21): 2008–2021.

71. Laso-Pérez, R., et al., Thermophilic archaea activate butane via alkyl-coenzyme M formation. *Nature*, 2016. **539**(7629): 396–401.

72. Laso-Perez, R., et al., Establishing anaerobic hydrocarbon-degrading enrichment cultures of microorganisms under strictly anoxic conditions. *Nature Protocols*, 2018. **13**(6): 1310–1330.

73. Laso-Pérez, R., et al., Anaerobic degradation of non-methane alkanes by "Candidatus Methanoliparia" in hydrocarbon seeps of the Gulf of Mexico. *mBio*, 2019. **10**(4).

74. Fedosov, D.V., et al., Investigation of the catabolism of acetate and peptides in the new anaerobic thermophilic bacterium Caldithrix abyssi. *Microbiology*, 2006. **75**(2): 119–124.

75. Boone, D.R., Castenholz, R. W., Garrity, G. M., ed., *Bergey's Manual of Systematic Bacteriology*. 2nd ed., vol. **1**. 2001, Springer-Verlag: New York. 721.

76. Boyd, J.A., et al., Divergent methyl-coenzyme M reductase genes in a deep-subseafloor Archaeoglobi. *The ISME Journal*, 2019: 1.

77. Wang, Y., et al., Diverse anaerobic methane- and multi-carbon alkane-metabolizing archaea coexist and show activity in Guaymas basin hydrothermal sediment. *Environmental Microbiology*, 2019. **0**(ja).

78. Borrel, G., et al., Wide diversity of methane and short-chain alkane metabolisms in uncultured archaea. *Nature Microbiology*, 2019. **4**(4): 603–613.

79. Soppa, J., From replication to cultivation: hot news from haloarchaea. *Current Opinion in Microbiology*, 2005. **8**(6): 737–44.

80. Walsby, A.E., Archaea with square cells. *Trends in Microbiology*, 2005. **13**(5): 193–195.

81. Sorokin, D.Y., et al., Discovery of extremely halophilic, methyl-reducing Euryarchaea provides insights into the evolutionary origin of methanogenesis. *Nature Microbiology*, 2017(in press).

82. Sorokin, D.Y., et al., Methanonatronarchaeum thermophilum gen. nov., sp. nov. and 'Candidatus Methanohalarchaeum thermophilum', extremely halo(natrono)philic methyl-reducing methanogens from hypersaline lakes comprising a new Euryarchaeal class Methanonatronarchaeia classis nov. *International Journal of Systematic and Evolutionary Microbiology*, 2018. **68**(7): 2199–2208.

83. Eder, W., et al., Prokaryotic phylogenetic diversity and corresponding geochemical data of the brine-seawater interface of the Shaban Deep, Red Sea. *Environmental Microbiology*, 2002. **4**(11): 758–763.

84. Aouad, M., et al., Evolutionary placement of methanonatronarchaeia. *Nature Microbiology*, 2019. **4**(4): 558.

85. Fuhrman, J.A., McCallum, K. and Davis, A.A., Novel major archaebacterial group from marine plankton. *Nature*, **356**(6365): 148–149.

86. DeLong, E.F., Archaea in coastal marine environments. *Proceedings of the National Academy of Sciences of the United States of America*, 1992. **89**(12): 5685–5689.

87. Borrel, G., et al., Comparative genomics highlights the unique biology of methanomassiliicoccales, a thermoplasmatales-related seventh order of methanogenic archaea that encodes pyrrolysine. *BMC Genomics*, 2014. **15**: 679–679.

88. Golyshina, O.V., et al., Biology of archaea from a novel family cuniculiplasmataceae (thermoplasmata) ubiquitous in hyperacidic environments. *Scientific Reports*, 2016. **6**: 39034–39034.

89. Schleper, C., et al., Picrophilus gen. nov., fam. nov.: a novel aerobic, heterotrophic, thermoacidophilic genus and family comprising archaea capable of growth around pH 0. *Journal of Bacteriology*, 1995. **177**(24): 7050–7059.

90. Edwards, R., D.P. Dixon, and V. Walbot, Plant glutathione S-transferases: enzymes with multiple functions in sickness and in health. *Trends in Plant Science*, 2000. **5**(5): 193–198.

91. Golyshina, O.V., Environmental, biogeographic, and biochemical patterns of archaea of the family ferroplasmaceae. *Applied and Environmental Microbiology*, 2011. **77**(15): 5071–5078.

92. Reysenbach, A.-L., et al., A ubiquitous thermoacidophilic archaeon from deep-sea hydrothermal vents. *Nature*, 2006. **442**(7101): 444–447.

93. Flores, G.E., et al., Distribution, abundance, and diversity patterns of the thermoacidophilic "Deep-Sea Hydrothermal Vent Euryarchaeota 2." *Frontiers in Microbiology*, 2012. **3**.

94. Takai, K. and K. Horikoshi, Genetic diversity of archaea in deep-sea hydrothermal vent environments. *Genetics*, 1999. **152**(4): 1285–1297.

95. Dridi, B., et al., Methanomassiliicoccus luminyensis gen. nov., sp. nov., a methanogenic archaeon isolated from human faeces. *International Journal of Systematic and Evolutionary Microbiology*, 2012. **62**(Pt 8): 1902–1907.

96. Kröninger, L., J. Gottschling, and U. Deppenmeier, Growth characteristics of methanomassiliicoccus luminyensis and expression of methyltransferase encoding genes. *Archaea*, 2017.

97. Borrel, G., et al., Genome sequence of "Candidatus Methanomethylophilus alvus" Mx1201, a methanogenic archaeon from the human gut belonging to a seventh order of methanogens. *Journal of Bacteriology*, 2012. **194**(24): 6944–6945.

98. Lang, K., et al., New mode of energy metabolism in the seventh order of methanogens as revealed by comparative genome analysis of "Candidatus Methanoplasma termitum". *Applied and Environmental Microbiology*, 2015. **81**(4): 1338–1352.

99. Poulsen, M., et al., Methylotrophic methanogenic Thermoplasmata implicated in reduced methane emissions from bovine rumen. *Nature Communications*, 2013. **4**.

100. Borrel, G., et al., Genome sequence of "Candidatus Methanomassiliicoccus intestinalis" issoire-Mx1, a third thermoplasmatales-related methanogenic archaeon from human feces. *Genome Announcements*, 2013. **1**(4).

101. Rinke, C., et al., A phylogenomic and ecological analysis of the globally abundant Marine Group II archaea (Ca. Poseidoniales ord. nov.). *ISME Journal*, 2019. **13**(3): 663–675.

102. Fuhrman, J. and A. Davis, Widespread archaea and novel bacteria from the deep sea as shown by 16S rRNA gene sequences. *Marine Ecology Progress Series*, 1997. **150**: 275–285.

103. Massana, R., et al., Vertical distribution and phylogenetic characterization of marine planktonic archaea in the Santa Barbara Channel. *Applied and Environmental Microbiology*, 1997. **63**(1): 50–56.

104. Teira, E., et al., Combining catalyzed reporter deposition-fluorescence in situ hybridization and Microautoradiography to detect substrate utilization by bacteria and archaea in the deep ocean. *Applied and Environmental Microbiology*, 2004. **70**(7): 4411–4414.

105. Orellana, L.H., et al., Niche differentiation among annually recurrent coastal marine group II euryarchaeota. *ISME Journal*, 2019.

106. Frigaard, N.-U., et al., Proteorhodopsin lateral gene transfer between marine planktonic bacteria and archaea. *Nature*, 2006. **439**(7078): 847–850.

107. Iverson, V., et al., Untangling genomes from metagenomes: revealing an uncultured class of marine euryarchaeota. *Science*, 2012. **335**(6068): 587–590.

108. Deschamps, P., et al., Pangenome evidence for extensive interdomain horizontal transfer affecting lineage core and shell genes in uncultured planktonic Thaumarchaeota and Euryarchaeota. *Genome Biology and Evolution*, 2014. **6**(7): 1549–1563.

109. Li, M., et al., Genomic and transcriptomic evidence for scavenging of diverse organic compounds by widespread deep-sea archaea. *Nature Communications*, 2015. **6**: 8933–8933.

110. Tully, B.J., Metabolic diversity within the globally abundant marine group II Euryarchaea offers insight into ecological patterns. *Nature Communications*, 2019. **10**(1): 271.

111. Haro-Moreno, J.M., et al., New insights into marine group III Euryarchaeota, from dark to light. *ISME Journal*, 2017. **11**(5): 1102–1117.

112. Li, M., et al., Genomic and transcriptomic evidence for scavenging of diverse organic compounds by widespread deep-sea archaea. *Nature Communications*, 2015. **6**: 8933.

113. Yokooji, Y., et al., Genetic examination of initial amino acid oxidation and glutamate catabolism in the hyperthermophilic archaeon Thermococcus kodakarensis. *Journal of Bacteriology*, 2013. **195**(9): 1940–8.

114. Poole, F.L., 2nd, et al., Defining genes in the genome of the hyperthermophilic archaeon pyrococcus furiosus: implications for all microbial genomes. *Journal of Bacteriology*, 2005. **187**(21): 7325–7332.

115. Chouari, R., et al., Novel predominant archaeal and bacterial groups revealed by molecular analysis of an anaerobic sludge digester. *Environmental Microbiology*, 2005. **7**(8): 1104–1115.

116. Dojka, M.A., et al., Microbial diversity in a hydrocarbon- and chlorinated-solvent-contaminated aquifer undergoing intrinsic bioremediation. *Applied Environmental Microbiology*, 1998. **64**(10): 3869–3877.

117. Nobu, M.K., et al., Chasing the elusive Euryarchaeota class WSA2: genomes reveal a uniquely fastidious methyl-reducing methanogen. *The ISME Journal*, 2016. **10**(10): 2478–2487.

118. Sahl, J.W., et al., A comparative molecular analysis of water-filled limestone sinkholes in north-eastern Mexico. *Environmental Microbiology*, 2011. **13**(1): 226–240.

119. Lazar, C.S., et al., Genomic reconstruction of multiple lineages of uncultured benthic archaea suggests distinct biogeochemical roles and ecological niches. *The ISME Journal*, 2017.

120. Takai, K., et al., Archaeal diversity in waters from Deep South African Gold Mines. *Applied and Environmental Microbiology*, 2001. **67**(12): 5750–5760.

121. Parkes, R.J., et al., Deep sub-seafloor prokaryotes stimulated at interfaces over geological time. *Nature*, 2005. **436**(7049): 390–394.

122. Biddle, J.F., et al., Heterotrophic archaea dominate sedimentary subsurface ecosystems off Peru. *Proceedings of the National Academy of Sciences of the United States of America*, 2006. **103**(10): 3846–3851.

123. Baker, B.J., et al., Genomic inference of the metabolism of cosmopolitan subsurface archaea, Hadesarchaea. *Nature Microbiology*, 2016: 16002.

124. Hua, Z.-S., et al., Insights into the ecological roles and evolution of methyl-coenzyme M reductase-containing hot spring Archaea. *Nature Communications*, 2019. **10**.

125. Wielen, P.W.J.J.v.d., et al., The enigma of prokaryotic life in deep hypersaline anoxic basins. *Science*, 2005. **307**(5706): 121–123.

126. Guan, Y., et al., Diversity of methanogens and sulfate-reducing bacteria in the interfaces of five deep-sea anoxic brines of the Red Sea. *Research in Microbiology*, 2015. **166**(9): 688–699.

127. Mwirichia, R., et al., Metabolic traits of an uncultured archaeal lineage -MSBL1- from brine pools of the Red Sea. *Scientific Reports*, 2016. **6**.

128. Vetriani, C., et al., Population structure and phylogenetic characterization of marine benthic archaea in deep-sea sediments. *Applied and Environmental Microbiology*, 1999. **65**(10): 4375–4384.

129. Jungbluth, S.P., J.P. Amend, and M.S. Rappé, Metagenome sequencing and 98 microbial genomes from Juan de Fuca Ridge flank subsurface fluids. *Scientific Data*, 2017. **4**: 170037.

130. Dombrowski, N., A.P. Teske, and B.J. Baker, Expansive microbial metabolic versatility and biodiversity in dynamic Guaymas basin hydrothermal sediments. *Nature Communications*, 2018. **9**(1): 4999.

131. Zhou, Z., et al., Genome- and community-level interaction insights into carbon utilization and element cycling functions of hydrothermarchaeota in hydrothermal sediment. *mSystems*, 2020. **5**(1): e00795–19.

132. Carr, S.A., et al., Carboxydotrophy potential of uncultivated hydrothermarchaeota from the subseafloor crustal biosphere. *ISME Journal*, 2019: 1.

133. Kato, S., et al., Metabolic potential of as-yet-uncultured archaeal lineages of Candidatus hydrothermarchaeota thriving in deep-sea metal sulfide deposits. *Microbes and Environments*, 2019. **34**(3): 293.

134. Petitjean, C., et al., Rooting the domain archaea by phylogenomic analysis supports the foundation of the new kingdom Proteoarchaeota. *Genome biology and evolution*, 2015. **7**(1): 191–204.

135. Rudolph, C., G. Wanner, and R. Huber, Natural communities of novel archaea and bacteria growing in cold sulfurous springs with a string-of-pearls-like morphology. *Applied Environmental Microbiology*, 2001. **67**(5): 2336–2344.

136. Huber, H., et al., A new phylum of Archaea represented by a nanosized hyperthermophilic symbiont. *Nature*, 2002. **417**(6884): 63–67.

137. Podar, M., et al., Insights into archaeal evolution and symbiosis from the genomes of a nanoarchaeon and its inferred crenarchaeal host from Obsidian Pool, Yellowstone National Park. *Biology Direct*, 2013. **8**(1).

138. Wurch, L., et al., Genomics-informed isolation and characterization of a symbiotic nanoarchaeota system from a terrestrial geothermal environment. *Nature Communications*, 2016. **7**: 12115.

139. Jarett, J.K., et al., Single-cell genomics of co-sorted Nanoarchaeota suggests novel putative host associations and diversification of proteins involved in symbiosis. *Microbiome*, 2018. **6**(1): 161.

140. John, E.S., et al., Deep-sea hydrothermal vent metagenome-assembled genomes provide insight into the phylum nanoarchaeota. *Environmental Microbiology Reports*, 2019. **11**(2): 262–270.

141. Huber, H., et al., The unusual cell biology of the hyperthermophilic Crenarchaeon Ignicoccus hospitalis. *Antonie Van Leeuwenhoek*, 2012. **102**(2): 203–219.

142. Heimerl, T., et al., A complex endomembrane system in the archaeon Ignicoccus hospitalis tapped by Nanoarchaeum equitans. *Frontiers in Microbiology*, 2017. **8**: 1072–1072.

143. Könneke, M., et al., Isolation of an autotrophic ammonia-oxidizing marine archaeon. *Nature*, 2005. **437**(7058): 543–546.

144. Brochier-Armanet, C., et al., Mesophilic Crenarchaeota: proposal for a third archaeal phylum, the Thaumarchaeota. *Nature Reviews. Microbiology*, 2008. **6**(3): 245–252.

145. Spang, A., et al., Distinct gene set in two different lineages of ammonia-oxidizing archaea supports the phylum Thaumarchaeota. *Trends in Microbiology*, 2010. **18**(8): 331–340.

146. Stahl, D.A. and J.R. de la Torre, Physiology and diversity of ammonia-oxidizing archaea. *Annual Review of Microbiology*, 2012. **66**(1): 83–101.

147. Beam, J.P., et al., Niche specialization of novel Thaumarchaeota to oxic and hypoxic acidic geothermal springs of Yellowstone National Park. *ISME Journal*, 2014. **8**(4): 938–951.

148. Kato, S., et al., Isolation and characterization of a thermophilic sulfur- and iron-reducing thaumarchaeote from a terrestrial acidic hot spring. *ISME Journal*, 2019. **13**(10): 2465–2474.

149. Rinke, C., et al., Insights into the phylogeny and coding potential of microbial dark matter. *Nature*, 2013. **499**(7459): 431–437.

150. Hedlund, B.P., et al., Uncultivated thermophiles: current status and spotlight on 'Aigarchaeota'. *Current Opinion in Microbiology*, 2015. **25**: 136–145.

151. Beam, J.P., et al., Ecophysiology of an uncultivated lineage of Aigarchaeota from an oxic, hot spring filamentous 'streamer' community. *ISME Journal*, 2016. **10**(1): 210–224.

152. Hua, Z.-S., et al., Genomic inference of the metabolism and evolution of the archaeal phylum Aigarchaeota. *Nature Communications*, 2018. **9**(1): 2832.

153. Barns, S.M., et al., Perspectives on archaeal diversity, thermophily and monophyly from environmental rRNA sequences. *Proceedings of the National Academy of Sciences of the United States of America*, 1996. **93**(17): 9188–9193.

154. Elkins, J.G., et al., A Korarchaeal genome reveals insights into the evolution of the archaea. *Proceedings of the National Academy of Sciences*, 2008. **105**(23): 8102–8107.

155. McKay, L.J., et al., Co-occurring genomic capacity for anaerobic methane and dissimilatory sulfur metabolisms discovered in the Korarchaeota. *Nature Microbiology*, 2019. **4**(4): 614–622.

156. Inagaki, F., et al., Microbial communities associated with geological horizons in coastal subseafloor sediments from the sea of Okhotsk. *Applied Environmental Microbiology*, 2003. **69**(12): 7224–7235.

157. Zhou, Z., et al., Bathyarchaeota: globally distributed metabolic generalists in anoxic environments. *FEMS Microbiology Reviews*, 2018. **42**(5): 639–655.

158. Teske, A.P., Microbial communities of deep marine subsurface sediments: molecular and cultivation surveys. *Geomicrobiology Journal*, 2006. **23**(6): 357–368.

159. Kubo, K., et al., Archaea of the miscellaneous Crenarchaeotal group are abundant, diverse and widespread in marine sediments. *ISME Journal*, 2012. **6**(10): 1949–1965.

160. Yu, T., et al., Growth of sedimentary Bathyarchaeota on lignin as an energy source. *Proceedings of the National Academy of Sciences*, 2018: 201718854.

161. He, Y., et al., Genomic and enzymatic evidence for acetogenesis among multiple lineages of the archaeal phylum Bathyarchaeota widespread in marine sediments. *Nature Microbiology*, 2016. **1**(6): 16035–16035.

162. Evans, P.N., et al., Methane metabolism in the archaeal phylum Bathyarchaeota revealed by genome-centric metagenomics. *Science*, 2015. **350**(6259): 434–438.

163. Kozubal, M.A., et al., Microbial iron cycling in acidic geothermal springs of Yellow Stone National Park: integrating molecular surveys, geochemical processes, and isolation of novel fe-active microorganisms. *Frontiers in Microbiology*, 2012. **3**.

164. Kozubal, M.A., et al., Geoarchaeota: a new candidate phylum in the Archaea from high-temperature acidic iron mats in Yellowstone National Park. *ISME Journal*, 2013. **7**(3): 622–634.

165. Beam, J.P., et al., Assembly and succession of iron oxide microbial mat communities in acidic geothermal springs. *Frontiers in Microbiology*, 2016. **7**.

166. Guy, L., et al., 'Geoarchaeote NAG1' is a deeply rooting lineage of the archaeal order Thermoproteales rather than a new phylum. *ISME Journal*, 2014. **8**(7): 1353–1357.

167. Vanwonterghem, I., et al., Methylotrophic methanogenesis discovered in the archaeal phylum verstraetearchaeota. *Nature Microbiology*, 2016. **1**: 16170–16170.

168. Berghuis, B.A., et al., Hydrogenotrophic methanogenesis in archaeal phylum verstraetearchaeota reveals the shared ancestry of all methanogens. *Proceedings of the National Academy of Sciences of the United States of America*, 2019. **116**(11): 5037–5044.

169. Kadnikov, V.V., et al., Genome of a member of the candidate archaeal phylum verstraetearchaeota from a subsurface thermal aquifer revealed pathways of methyl-reducing methanogenesis and fermentative metabolism. *Microbiology*, 2019. **88**(3): 316–323.

170. Jay, Z.J., et al., Marsarchaeota are an aerobic archaeal lineage abundant in geothermal iron oxide microbial mats. *Nature Microbiology*, 2018. **3**(6): 732–740.

171. Jorgensen, S.L., et al., Correlating microbial community profiles with geochemical data in highly stratified sediments from the Arctic Mid-Ocean Ridge. *Proceedings of the National Academy of Sciences of the United States of America*, 2012. **109**(42): E2846–E2855.

172. Wilke, A., et al., A RESTful API for accessing microbial community data for MG-RAST. *PLOS Computational Biology*, 2015. **11**(1): e1004008.

173. Seitz, K.W., et al., Genomic reconstruction of a novel, deeply branched sediment archaeal phylum with pathways for acetogenesis and sulfur reduction. *ISME Journal*, 2016.

174. Manoharan, L., et al., Metagenomes from coastal marine sediments give insights into the ecological role and cellular features of loki- and thorarchaeota. *mBio*, 2019. **10**(5): e02039–19.

175. Seitz, K.W., et al., Asgard archaea capable of anaerobic hydrocarbon cycling. *Nature Communications*, 2019. **10**(1): 1822.

176. Castelle, C.J., et al., Genomic expansion of domain archaea highlights roles for organisms from new phyla in anaerobic carbon cycling. *Current Biology*, 2015. **25**(6): 690–701.

177. Castelle, C.J., et al., Biosynthetic capacity, metabolic variety and unusual biology in the CPR and DPANN radiations. *Nature Reviews. Microbiology*, 2018. **16**(10): 629–645.

178. Probst, A.J., et al., Differential depth distribution of microbial function and putative symbionts through sediment-hosted aquifers in the deep terrestrial subsurface. *Nature Microbiology*, 2018. **3**(3): 328–336.

179. Probst, A.J., et al., Biology of a widespread uncultivated archaeon that contributes to carbon fixation in the subsurface. *Nature Communications*, 2014. **5**: 5497.

180. Dombrowski, N., et al., Genomic diversity, lifestyles and evolutionary origins of DPANN archaea. *FEMS Microbiology Letters*, 2019. **366**(2).

181. Brochier, C., et al., Nanoarchaea: representatives of a novel archaeal phylum or a fast-evolving euryarchaeal lineage related to thermococcales? *Genome Biology*, 2005. **6**(5).

182. Jahn, U., et al., Composition of the lipids of nanoarchaeum equitans and their origin from its host Ignicoccus sp. strain KIN4/I. *Archives of Microbiology*, 2004. **182**(5): 404–413.

183. Jahn, U., et al., Nanoarchaeum equitans and Ignicoccus hospitalis: new insights into a unique, intimate association of two archaea. *Journal of Bacteriology*, 2008. **190**(5): 1743–1750.

184. Waters, E., et al., The genome of Nanoarchaeum equitans: insights into early archaeal evolution and derived parasitism. *Proceedings of the National Academy of Sciences of the United States of America*, 2003. **100**(22): 12984–12988.

185. Munson-McGee, J.H., et al., Nanoarchaeota, their sulfolobales host, and nanoarchaeota virus distribution across Yellowstone National Park hot springs. *Applied and Environmental Microbiology*, 2015. **81**(22): 7860–7868.

186. St John, E., et al., A new symbiotic nanoarchaeote (candidatus nanoclepta minutus) and its host (zestosphaera tikiterensis gen. nov., sp. nov.) from a New Zealand hot spring. *Systematic and Applied Microbiology*, 2019. **42**(1): 94–106.

187. Hamm, J.N., et al., Unexpected host dependency of antarctic nanohaloarchaeota. *Proceedings of the National Academy of Sciences of the United States of America*, 2019. **116**(29): 14661–14670.

188. Krause, S., et al., Characterisation of a stable laboratory coculture of acidophilic nanoorganisms. *Scientific Reports*, 2017. **7**(1): 3289–3289.

189. Golyshina, O.V., et al., 'ARMAN' archaea depend on association with euryarchaeal host in culture and in situ. *Nature Communications*, 2017. **8**(1): 60.

190. Youssef, N.H., et al., Insights into the metabolism, lifestyle and putative evolutionary history of the novel archaeal phylum 'Diapherotrites.' *ISME Journal*, 2015. **9**(2): 447–460.

191. Bird, J.T., et al., Culture independent genomic comparisons reveal environmental adaptations for Altiarchaeales. *Frontiers in Microbiology*, 2016. **7**: 1221–1221.

192. Henneberger, R., et al., New insights into the lifestyle of the cold-loving SM1 euryarchaeon: natural growth as a monospecies biofilm in the subsurface. *Applied and Environmental Microbiology*, 2006. **72**(1): 192–199.

193. Probst, A.J., et al., Tackling the minority: sulfate-reducing bacteria in an archaea-dominated subsurface biofilm. *ISME Journal*, 2013. **7**(3): 635–651.

194. Moissl, C., C. Rudolph, and R. Huber, Natural communities of novel archaea and bacteria with a string-of-pearls-like morphology: molecular analysis of the bacterial partners. *Applied and Environmental Microbiology*, 2002. **68**(2): 933–937.

195. Rudolph, C., et al., Ecology and microbial structures of archaeal/bacterial strings-of-pearls communities and archaeal relatives thriving in cold sulfidic springs. *FEMS Microbiology Ecology*, 2004. **50**(1): 1–11.

196. Moissl, C., et al., In situ growth of the novel SM1 euryarchaeon from a string-of-pearls-like microbial community in its cold biotope, its physical separation and insights into its structure and physiology. *Archives of Microbiology*, 2003. **180**(3): 211–217.

197. Schwank, K., et al., An archaeal symbiont-host association from the deep terrestrial subsurface. *ISME Journal*, 2019. **13**(8): 2135–2139.

198. Moissl, C., et al., The unique structure of archaeal 'hami,' highly complex cell appendages with nano-grappling hooks. *Molecular Microbiology*, 2005. **56**(2): 361–370.

199. Miller, T.L., et al., Isolation of methanobrevibacter smithii from human feces. *Applied and Environmental Microbiology*, 1982. **43**(1): 227–232.

200. Levitt, M.D., et al., Stability of human methanogenic flora over 35 years and a review of insights obtained from breath methane measurements. *Clinical Gastroenterology and Hepatology*, 2006. **4**(2): 123–129.

201. Gaci, N., et al., Archaea and the human gut: new beginning of an old story. *World Journal of Gastroenterology*, 2014. **20**(43): 16062–16078.

202. Koskinen, K., et al., First insights into the diverse human archaeome: specific detection of archaea in the gastrointestinal tract, lung, and nose and on skin. *MBio*, 2017. **8**(6).

203. Pausan, M.R., et al., Exploring the Archaeome: Detection of Archaeal Signatures in the Human Body. *Frontiers in Microbiology*, 2019. **10**: 2796.

204. Conway de Macario, E. and A.J. Macario, Methanogenic archaea in health and disease: a novel paradigm of microbial pathogenesis. *International Journal of Medical Microbiology*, 2009. **299**(2): 99–108.

205. Sereme, Y., et al., Methanogenic archaea: emerging partners in the field of allergic diseases. *Clinical Review in Allergy and Immunology*, 2019. **57**(3): 456–466.

206. Nguyen-Hieu, T., et al., Methanogenic archaea in subgingival sites: a review. *APMIS*, 2013. **121**(6): 467–477.

207. Pérez-Chaparro, P.J., et al., Newly identified pathogens associated with periodontitis: a systematic review. *Journal of Dental Research*, 2014. **93**(9): 846–58.

208. Horz, H.P., et al., Relationship between methanogenic archaea and subgingival microbial complexes in human periodontitis. *Anaerobe*, 2015. **35**(Pt A): 10–12.

209. Horz, H.P. and G. Conrads, Methanogenic Archaea and oral infections-ways to unravel the black box. *Journal of Oral Microbiology*, 2011. 3.

210. Ramiro, F.S., et al., Effects of different periodontal treatments in changing the prevalence and levels of archaea present in the subgingival biofilm of subjects with periodontitis: A secondary analysis from a randomized controlled clinical trial. *International Journal of Dental Hygiene*, 2018. **16**(4): 569–575.

211. Miller, T.L. and M.J. Wolin, Methanosphaera stadtmaniae gen. nov., sp. nov.: a species that forms methane by reducing methanol with hydrogen. *Archives of Microbiology*, 1985. **141**(2): 116–122.

212. Horz, H.P. and G. Conrads, The discussion goes on: what is the role of Euryarchaeota in humans? *Archaea*, 2010. **2010**: 967271.

213. Jarrell, K.F., et al., Major players on the microbial stage: why archaea are important. *Microbiology*, 2011. **157**(Pt 4): 919–936.

214. Khelaifia, S. and D. Raoult, Haloferax massiliensis sp. nov., the first human-associated halophilic archaea. *New Microbes and New Infections*, 2016.

215. Khelaifia, S., et al., Genome sequence and description of haloferax massiliense sp. nov., a new halophilic archaeon isolated from the human gut. *Extremophiles*, 2018.

216. Nam, Y.D., et al., Bacterial, archaeal, and eukaryal diversity in the intestines of Korean people. *Journal of Microbiology*, 2008. **46**(5): 491–501.

217. Turnbaugh, P.J., et al., The human microbiome project: exploring the microbial part of ourselves in a changing world. *Nature*, 2007. **449**(7164): 804–810.

218. Khelaifia, S. and M. Drancourt, Susceptibility of archaea to antimicrobial agents: Applications to clinical microbiology. *Clinical Microbiology and Infection*, 2012. **18**(9): 841–848.

219. Jarrell, K.F., et al., Major players on the microbial stage: why archaea are important. *Microbiology*, 2011. **157**(4): 919–936.

220. Minh, B.Q., M.A. Nguyen, and A. von Haeseler, Ultrafast approximation for phylogenetic bootstrap. *Molecular Biology and Evolution*, 2013. **30**(5): 1188–1195.

221. Guindon, S., et al., New algorithms and methods to estimate maximum-likelihood phylogenies: assessing the performance of PhyML 3.0. *Systematic Biology*, 2010. **59**(3): 307–321.

22

The Genus Bacillus

Daniel R. Zeigler and John B. Perkins

CONTENTS

Introduction

The genus *Bacillus* comprises an expansive and diverse set of bacteria distinguished by a common feature: the ability to make dormant endospores aerobically when challenged with unfavorable growth conditions. A photograph of sporulating cells of *B. subtilis* is shown in Figure 22.1A. In 1989, Ruth Gordon authored a chapter for the *Practical Handbook of Microbiology* (CRC Press) that reviewed the systematics of the genus *Bacillus* (Gordon 1989). The various species of *Bacillus* were divided into three groups based on the morphologies of the sporangium (swollen or nonswollen) and the mature spore (spherical, cylindrical, or ellipsoidal) along with physiological and biochemical test results (such as catalase production and starch hydrolysis) and growth requirements (such as salinity, temperature, and pH). The species within these groups were highly heterogeneous. Since that time, however, there have been several major reclassifications of species within *Bacillus*, a reassessment necessitated by the introduction of more sophisticated analytical methods that rely on comparison at the genome level and by the constant influx of novel species isolated from unusual environments. *Bacillus* now represents just a small part of a larger taxonomical consortium of endospore-producing bacteria referred to as *Bacillus sensu lato*. Consequently, it is no longer possible in a review such as this one to describe each *Bacillus* species. On the contrary, the goal of this work is to provide the reader with an over-arching summary of *Bacillus sensu lato*, describing in broad brushstrokes the common characteristics of its current members. However, the review will also attempt to focus on those key species, especially *B. subtilis*, which play an increasingly important role in industry, medicine, and basic science. This two-tiered approach should provide the investigator with a greater appreciation of a fascinating and useful group of bacteria.

The Systematics of *Bacillus sensu lato*

Bacillus Systematics: The Classical Paradigm

The bacterial genus *Bacillus* has a long and rich history in the annals of microbiology. During the 1870s, Ferdinand Cohn, working at the University of Breslau, isolated a small, motile, aerobic bacterium from boiled hay infusions. He named it *Bacillus subtilis*, meaning "thin rod." Cohn provided the first accurate description of the *Bacillus* life cycle, demonstrating the formation, heat resistance, germination, and outgrowth of endospores (Cohn 1872). When Robert Koch established that a phenotypically similar organism, *B. anthracis*, was the causative agent of anthrax (Koch 1876), the scientific importance of this novel group of organisms became obvious. For the next 50 years, many new endospore-forming bacteria were identified. Due to limitations in the available tools, however, classification of these bacterial species was often a confusing and disorderly process.

The formidable challenge of bringing order to *Bacillus* taxonomy was met with admirable dedication by Nathan R. Smith,

DOI: 10.1201/9781003099277-24

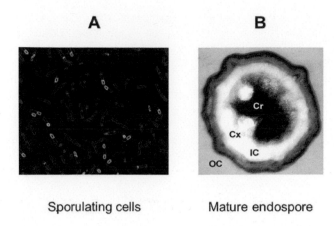

Sporulating cells   Mature endospore

FIGURE 22.1 Formation of endospores in representative *Bacillus* strains. (A) Fluorescence micrograph of *Bacillus subtilis* sporangia. The engulfment membrane (green) is visualized with the fluorescent protein fused to a sporulation protein that surrounds the growing spore (forespore). The cytoplasmic membrane is stained with the vital stain FM 4-64. (B) Cross-section showing the structure of a mature endospore of a wild-type derivative of *B. subtilis* 168. The endospore is approximately 1.2 μm in diameter and is composed of a thick spore coat that lies just underneath a small exosporium (not labeled). The spore coat contains up to 70 proteins, which are cross-linked to form two layers, a thick outer coat (OC) and a less dense layered inner coat (IC). The number and thickness of layers vary from spore to spore (Driks 2004). The spore coat encases the spore cortex (Cx) that consists of a large peptidoglycan layer with dipicolinic acid (DPA); this structure is thought to keep the interior of the endospore dehydrated, protecting the nucleoid from heat and radiation. The interior of the endospore, called the core (Cr), contains the ribosomes and the spore nucleoid. The highly condensed nucleoid is wrapped with small, acid-soluble spore proteins (SASPs) into a strongly protective structure not observed in growing cells. ([A] Micrograph taken by E. Angert and K. Price, Harvard University. Reprinted with permission of Prof. Richard Losick, Harvard University, and the American Society for Microbiology; [B] electron micrograph reproduced from Silvaggi et al., 2004 with permission from the American Society for Microbiology.)

Francis E. Clark, and Ruth E. Gordon in the 1930s and 1940s. They adopted a clear working definition of the genus as comprising "rod-shaped bacteria capable of aerobically forming refractile endospores that are more resistant than vegetative cells to heat, drying, and other destructive agencies" (Gordon, Haynes et al. 1973). Smith assembled a collection of 1,134 mesophilic *Bacillus* strains bearing 158 different species names. With his colleagues, he began to gather a meticulous set of morphological and physiological data on each strain. After analyzing the data—with a deliberate bias towards "lumping" rather than "splitting"—they assigned each isolate to one of 19 rigorously defined species (Smith, Gordon et al. 1946). Soon afterwards, Gordon and Smith repeated the process with a collection of 206 thermophilic *Bacillus* isolates, assigning them to only two additional species (Gordon and Smith 1949).

In the now classic 1973 monograph *The Genus Bacillus* (Gordon, Haynes et al. 1973), Gordon and her colleagues provided precise, detailed descriptions for 18 *Bacillus* species, together with a battery of 33 morphological and physiological tests and an identification key that could unambiguously assign most known isolates to one of those species. *Bacillus* taxonomy had grown into a logical, orderly, and lucid science. With this landmark publication, what one could call the classical period of *Bacillus* systematics, which largely succeeded in bringing order out of chaos, was drawing to a close.

Bacillus Systematics: The Molecular Paradigm

In the 1970s and early 1980s, Woese and his colleagues ushered in a dynamic new era of systematics, which led to the reordering of all known organisms into three domains of life, two of them prokaryotic (Fox, Pechman et al. 1977, Woese and Fox 1977). The core technology that underpinned this revolution was sequencing of the 16S ribosomal RNA. When the new tools of molecular taxonomy were applied to the genus *Bacillus*, the resulting data challenged the existing paradigm that the genus comprised a small number of rigorously defined species identifiable by certain key phenotypic characteristics. Ash utilized this new technology to scrutinize most of the known species of *Bacillus* and concluded that they fell into five distinct 16S sequence similarity groups, which they suggested could logically correspond to five novel genera (Ash, Farrow et al. 1991). *B. subtilis*, *B. licheniformis*, *B. amyloliquefaciens*, *B. megaterium*, *B. cereus*, *B. anthracis*, *Bacillus thuringiensis*, and many of the other best known and earliest studied species were placed in Group 1. This group can be termed *Bacillus sensu stricto*, since members of this group are considered to have the classic characteristics known for aerobic Gram-positive endospore-forming bacilli. It is important to note that additional species that were phenotypically similar to members of these five groups, yet were nonetheless sequence outliers, were thought to "represent the nuclei of other hitherto unrecognized genera."

An Overview of *Bacillus sensu lato*

This prediction was soon realized by a new generation of taxonomists who began mining the rich biodiversity of the earth's ecosystem in search for novel microorganisms. It is no exaggeration to say that virtually everywhere they have looked, microbiologists have uncovered Gram-positive endospore-formers closely related to the classical genus *Bacillus*. Such novel species have been recovered from ocean sediments thousands of meters below sea level (Rüger, Fritze et al. 2000, Lu, Nogi et al. 2001, Bae, Lee et al. 2005, Xu, Yu et al. 2019) and from stratospheric air samplers tens of kilometers above it (Shivaji, Chaturvedi et al. 2006). Other *Bacillus* species have been isolated from acidic geothermal pools and peat bogs (Simbahan, Drijber et al. 2004, Albert, Archambault et al. 2005) and from highly alkaline groundwater (Tiago, Chung et al. 2004). Several *Bacillus* species have been found in hypersaline terminal lakes, some contaminated with heavy metals (Switzer Blum, Burns Bindi et al. 1998, Arahal, Márquez et al. 1999, Vargas, Delgado et al. 2005, Lim, Jeon et al. 2006). Others have been discovered in human-created niches both ancient and modern, from Mexican shaft tombs and deteriorating Roman wall paintings (Heyrman, Balcaen et al. 2003, Gatson, Benz et al. 2006) to ultra-clean rooms in spacecraft assembly facilities (Venkateswaran, Kempf et al. 2003, La Duc, Satomi et al. 2004, Satomi, La Duc et al. 2006, Seuylemezian, Ott et al. 2019). With increasing frequency, novel *Bacillus* species have being isolated from the human gut (Tidjani Alou, Rathored et al. 2015, Tidjiani Alou, Rathored et al. 2015, Senghor, Seck et al. 2017). Plants continue to be a rich source of novel *Bacillus* species, some endophytic and others rhizosphere-associated (Reva, Smirnov et al. 2002, Olivera, Siñeriz et al. 2005). A wide diversity of *Bacillus* and related bacteria have even been detected by molecular methods in the subglacial Antarctic lake, Lake Vostok, buried under 3.7 km of ice (Shtarkman, Kocer et al. 2013).

As a result of these efforts, the systematics of the old genus *Bacillus* has been radically transformed. With the exponential influx of valid described endospore-forming species since the 1990s, the five *Bacillus* groups developed by Ash (Ash, Farrow et al. 1991) and later refined by Priest (1993) have not only been assigned novel genus names but have been subdivided into many additional genera, while the outliers from the original similarity groups have been assigned to novel genera as well. At the time of writing, there have been valid descriptions published for at least 1,070 *Bacillus*-like species comprising 102 genera (see the "List of Prokaryotic Names with Standing in Nomenclature," http://bacterio.net). This enlarged category of endospore formers is termed *Bacillus sensu lato* ("in the broad sense"), with the many of the original *Bacillus* species still defined as *Bacillus sensu stricto* ("in the narrow sense"). Moreover, during the decade from since 2004, novel species of *Bacillus sensu lato* have been described at an average rate of almost one per week. This trend shows no signs of abating. The interested reader is directed to several authoritative books and reviews that describe these new genera and their genotypic and phenotypic basis (Goto, Omura et al. 2000, Logan and Turnbull 2003, Fritze 2004, Logan and Rodriguez-Diaz 2006, Maughan and Van der Auwera 2011).

The oldest isolate—and therefore the nomenclatural type species for the genus *Bacillus*—is *B. subtilis* (Figure 22.1A). Descriptive studies have characterized *B. subtilis* as a mesophile, though a hardy one, growing at temperatures as low as 5–20°C and as high as 45–55°C and in salinities as great as 7% NaCl (w/v). Cells are moderate sized rods, about 0.7–0.8 µM by 2.0–3.0 µM, typically with many long peritrichous flagella and strong motility. Endospores are ellipsoidal and located near the end of the mother cell. The mature endospore, shown in Figure 22.1B, is released upon mother cell lysis. *B. subtilis* is one of the most thoroughly studied of all bacteria, both as an industrial powerhouse (see section entitled Industrial Applications of *Bacillus* later in this chapter) and a model system for cell development (see section entitled Cell Physiology and Development later in this chapter). A cluster of closely related species, labeled in Figure 22.2, is often referred to as the "*Bacillus subtilis* group." One of these species,

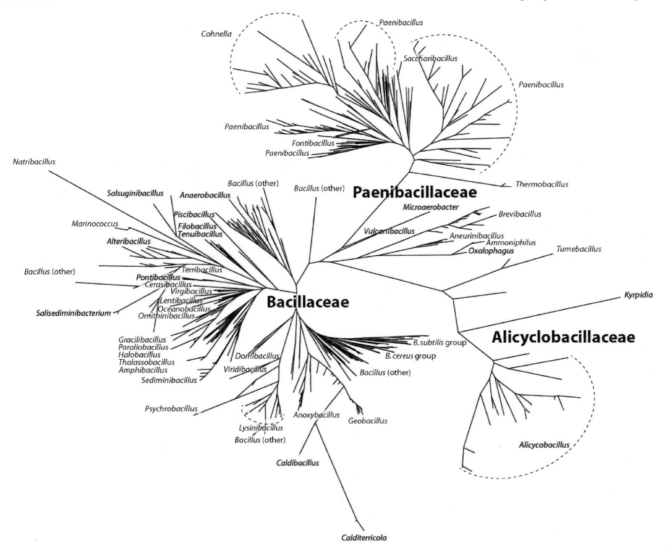

FIGURE 22.2 Phylogeny of *Bacillus sensu lato* from 16S rRNA gene sequences. GenBank DNA sequences for species type strains were aligned with ClustalW (Higgins, Thompson et al. 1994). An unrooted phylogenetic tree was constructed from the ClustalW distance matrix with the PHYLIP Neighbour application (Felsenstein 1989) and visualized with PhyloDraw. Dashed lines group sequences that have been assigned to the same taxon. A few species mentioned frequently in the text are indicated. The "*B. subtilis* group" consists of 14 species, including *B. licheniformis*, that are closely related to *B. subtilis*. The "*B. cereus* group" consists of six species, including *B. anthracis* and *B. thuringiensis*, which are closely related to *B. cereus*.

B. licheniformis, is also an important industrial microorganism. The species is named for its lichen-like appearance when grown on solid media, where it typically forms rough, dry, irregular colonies that adhere stubbornly to the agar surface (Gibson 1944).

Another very commonly studied cluster of species is known as the "*Bacillus cereus* group" (Figure 22.2). Descriptive studies of these species show that their cells are somewhat larger than those of *B. subtilis*, with widths of 1.0–1.2 µM and lengths of 3.0–5.0 µM. Most members of the group form colonies with a distinctive "ice crystal" surface texture, although strains of *B. mycoides* usually form complex rhizoid colonies. The *B. cereus* group of species is perhaps best known for its pathogenic members. *B. thuringiensis* forms large parasporal crystals composed of protoxins targeting specific invertebrate organisms, notably insect larvae (discussed later in this chapter). *B. anthracis* and certain strains of *B. cereus* are causative agents for human diseases (discussed later in this chapter). Some strains from the *B. cereus* group are known to produce spores with caps, spikes, filaments, and other bizarre appendages (Rampersad, Khan et al. 2003) (see Figure 22.3A and B). Unusual spore morphologies and complex developmental structures can also be found in isolates throughout the *Bacillus sensu lato* (Ajithkumar, Ajithkumar et al. 2001) (see Figure 22.3C).

Although these commonly studied species are illustrative of *Bacillus*, they are by no means fully representative of its diversity. It is now realized that the defining characteristic of the classical genus *Bacillus*—endospore formation—is an ancient trait that probably appeared a single time near the base of the bacterial phylum Firmicutes (Galperin, Mekhedov et al. 2012, Abecasis, Serrano et al. 2013, Traag, Pugliese et al. 2013). As a result, this characteristic circumscribes an extremely diverse set of bacteria that have adapted to life under a broad array of environmental conditions. Currently, these species have been grouped into three taxonomic families (*Bacillaceae, Paenibacillaceae, and Alicyclobacillaceae*), joining seven other non-*Bacillus* families in the order *Bacillales*. Many of these novel genera of *Bacillus sensu lato* contain organisms adapted to extreme conditions, bacteria largely unknown to the classical taxonomists of the 1930s and 1940s. Thermophiles, psychrophiles, acidophiles, alkaliphiles, and halophiles, as well as isolates that are neither rod-shaped nor spore-forming, all have a place within these newly described genera.

If we repeat the analysis of Ash, Farrow et al. (1991) and construct a phylogram from the 16S rRNA gene sequences of type strains within *Bacillus sensu lato*, some patterns begin to emerge (Figure 22.2). First, there is a reassuring—though by no means totally consistent—congruity between certain broad phenotypic traits and position on the phylogenetic tree. For example, two thermophilic genera, the aerobic *Geobacillus* and the anaerobic *Anoxybacillus*, cluster together on the phylogram. Several moderately alkalophilic and halophilic genera also tend to cluster together in another branch of the tree. There is some degree of convergence between classical phenotypic approaches and modern genotypic approaches to bacterial systematics after all. Secondly, species classified as *Bacillus sensu stricto* are nevertheless scattered in distantly related clusters all over the tree. Despite being subdivided multiple times, the genus is still phylogenetically incoherent. A similar situation exists for the genus *Paenibacillus*. It has recently been proposed that the genus *Bacillus* should be reserved only for those species clustering either with the *B. subtilis* or *B. cereus* groups in molecular sequence phylogenies, and that all other species should eventually be transferred to other genera (Bhandari, Ahmod et al. 2013). Very recently, six such genera have been proposed, based on comparative genomic approaches (Patel and Gupta 2019). There remains much work to be done, then, in reorganizing the taxonomy of these bacteria, even without taking into consideration the novel species that continue to be described. Careful systematic studies of *Bacillus* are still absolutely necessary if their taxonomy is to retain any significance and usefulness. A newly constituted Subcommittee on the Taxonomy of Order Bacillales under the oversight of the International Committee on Systematics of Prokaryotes has been charged with pursuing these goals (http://www.the-icsp.org/).

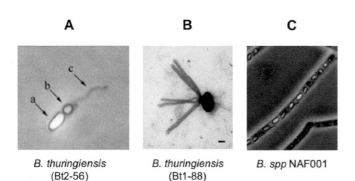

A	**B**	**C**
B. thuringiensis (Bt2-56)	*B. thuringiensis* (Bt1-88)	*B. spp* NAF001

FIGURE 22.3 Diversity of *Bacillus* endospores. (A) Phase contrast micrograph of a *B. thuringiensis* Bt2-56 cell, which produces a complex endospore assembly consisting of the spore, a spherical parasporal body, and a long filament, which upon germination remains attached to the cell wall of the outgrowing bacterium. This filament structure is unique and is not observed with other commonly known *B. thuringiensis* strains. (B) Electron micrograph of *B. thuringiensis* Bt1-88 spores with multiple filaments. This species is related to *B. thuringiensis* Bt2-56, but produces multiple filaments as opposed to singlets. Bar represents 500 nm. (C) Phase contrast micrograph of *Bacillus* spp. NAF001 filaments with endospores. Cells are also able to produce spore-like resting cells (SLRCs), which are heat resistant and can bud to form short filaments. ([A] The photograph is reprinted from Rampersad, Khan et al. 2003 with permission from Springer-Verlag New York Inc.; [B] the micrograph is reproduced with permission from Dra. Luz Irene Rojas Avelizapa, Departamento de Microbiología, Escuela National de Ciencias Biológicas, México; [C] the photograph is reprinted from Ajithkumar, Ajithkumar et al. (2001), with permission from the Society for General Microbiology.)

Characterization and Classification of Novel *Bacillus* Isolates

In today's era, what should a researcher do to characterize a *Bacillus* isolate adequately and to perhaps assign it to a novel species or even genus? In recent years, it has become increasingly practical simply to obtain a whole genome sequence (WGS) using one of the next generation (NextGen) sequencing technologies, a development that is rapidly changing nearly every aspect of microbiology (Loman, Constantinidou et al. 2012). Indeed, the existence of a publicly available genome sequence is a requirement for description of a novel species in most microbiology journals, and common standards have been widely adopted (Chun, Oren et al. 2018). Although further improvements in high throughput sequencing may render most other biochemical and molecular taxonomic methods obsolete in the near future, at this writing, some conventional methods can still provide an adequate level of characterization more rapidly and less expensively than genome sequencing. For example, a clinical microbiologist may need to determine quickly whether a certain *B. cereus* isolate from a patient with emetic food poisoning is identical with a previous isolate or represents a novel pathogenic strain. In that case—or in any number of applications where closely related members of the same species must be compared—a molecular fingerprinting technique using nucleic acid, peptide, or fatty acid profiles could still be an efficient approach. Several examples of these methods, together with references that illustrate their practical application to *Bacillus sensu lato*, are given in Table 22.1. In asking epidemiological questions that go deeper than mere strain

TABLE 22.1

Common Methods for Identifying and Characterizing Novel *Bacillus* Isolates

Method, with Selected Examples	Useful Range[a]	Application to *Bacillus Sensu Lato*
Physiological and morphological tests		
Microscopy and 26-test battery	species	Gordon, Haynes et al. 1973
Microscopy and 26-test battery	species	Priest and Alexander 1988
Microscopy and 20-test battery	species	Reva, Sorokulova et al. 2001
Carbohydrate utilization (commercial kit)	species	Logan and Berkeley 1984
DNA fingerprinting		
Restriction Fragment Length Polymorphism (RFLP)		
rRNA operons	subspecies-strain	Joung and Cote 2001; Coelho, von der Weid et al. 2003; Yi, Wu et al. 2012
amplified *secY* genes	genus-species	Palmisano, Nakamura et al. 2001
amplified *gyrB* genes	genus-species	Manzano, Cocolin et al. 2003
Pulsed Field Gel Electrophoresis (PFGE)	subspecies-strain	Gaviria Rivera and Priest 2003a
Conventional DNA sequencing		
Single locus sequence analysis		
16S rRNA	domain-genus	Ash, Farrow et al. 1991; Bavykin, Lysov et al. 2004; Zeigler 2005
23S rRNA	domain-genus	Bavykin, Lysov et al. 2004
groEL	species-subspecies	Chang, Shangkuan et al. 2003
gyrB	species-subspecies	Goto, Mochida et al. 2003; Bavykin, Lysov et al. 2004
recN	species-subspecies	Zeigler 2005
rpoB	species-subspecies	Palmisano, Nakamura et al. 2001
hag	subspecies	Xu and Cote 2006
panC	species-subspecies	Guinebretiere, Thompson et al. 2008
Multilocus Sequence Typing (MLST)		
adk, ccpA, ftsA, glpT, pyrE, recF, sucC	genus-subspecies	Helgason, Tourasse et al. 2004
glpF, gmk, ilvD, pta, pur, pycA, tpi	genus-subspecies	Didelot, Barker et al. 2009; Blackburn, Martin et al. 2013
rpoB, gyrB, pycA, mdh, mbl, mutS, plcR	genus-subspecies	Ko, Kim et al. 2004
rpoB, purH, gyrA, groEL, polC, 16S rRNA	genus-subspecies	Kubo, Rooney et al. 2011
adk, ccpA, pycA, glyA, glcK, glpF	genus-subspecies	Ge, Hu et al. 2011
Repetitive DNA PCR fingerprinting	subspecies-strain	Satomi, La Duc et al. 2006
Random Amplified Polymorphic DNA (RAPD)	subspecies-strain	Gaviria Rivera and Priest 2003b
Oligonucleotide microarray fingerprinting	subspecies-strain	Chandler, Alferov et al. 2006
Multilocus variable number tandem repeat (MLVA)	subspecies-strain	Durmaz, Doganay et al. 2012; Dhakal, Chauhan et al. 2013
Conserved signature indels in protein sequences	genus-species	Bhandari, Ahmod et al. 2013; Patel and Gupta 2019
Core genome sequence comparison		
Core genome protein sequence comparison	genus-subspecies	Patel and Gupta 2019
Core genome DNA sequence comparison	genus-subspecies	Bazinet 2017, Fan, Blom et al. 2017; Kim, Koh et al. 2017
Other molecular methods		
Fatty Acid profiling	species-subspecies	Kämpfer 1994; Schraft, Steele et al. 1996; Palmisano, Nakamura et al. 2001
MALDI-TOF Mass Spectroscopy	species-subspecies	Dickinson, La Duc et al. 2004; Farfour, Leto et al. 2012; Fernández-No, Böhme et al. 2013
Multilocus Enzyme Electrophoresis (MLEE)	subspecies-strain	Zahner, Momen et al. 1989; Zahner, Rabinovitch et al. 1994; Helgason, Økstad et al. 2000; Coelho, von der Weid et al. 2003

[a] Approximate range of taxonomic levels over which the method has maximum utility for both distinguishing and grouping related bacterial isolates.

identification, however, WGS analysis can provide characterization in unprecedented detail. A powerful example can be found in the forensic investigation into the "Amerithrax" incident in 2001, when comparative genomics combined with conventional analysis methodologies to reveal strain-specific differences in *B. anthracis* isolates (Rasko, Worsham et al. 2011).

When the aim is to address basic questions in environmental, evolutionary, or systematic microbiology, a broader analysis is generally required. Often, the very first information that a researcher will acquire is the 16S rRNA gene sequence of the isolate. While those data may be sufficient to place an isolate in a given genus—or perhaps to demonstrate that it is rather unlike any known genus—they may not be adequate for a reliable assignment at the species level or below. The 16S sequence is a powerful tool for tracing deeper phylogenies, but may show too little variability to differentiate among closely related species (Zeigler 2005). In fact, there are known examples where two *Bacillus* isolates may be clearly distinguishable by whole genome comparison techniques yet share essentially identical 16S rRNA genes (Fox, Wisotzkey et al. 1992). For this reason, a 16S sequence alone is not considered sufficient grounds to propose that a bacterial isolate belongs to a novel species. Unless a WGS is obtained, a 16S sequence should be supplemented with experiments to estimate whole genome relatedness between related organisms (Stackebrandt and Goebel 1994). Highly conserved protein-encoding genes may offer greater resolution when comparing bacteria at the species or subspecies level (Zeigler 2003). Comparing the concatenated set of gene sequences, a strategy known as multilocus sequence typing, has been a powerful tool for characterizing members of the *B. cereus* group, for example (Didelot, Barker et al. 2009). DNA-DNA cross-hybridization, together with %G + C genome content comparison, was once considered the "gold standard" in defining new species (Wayne, Brenner et al. 1987, Stackebrandt, Frederiksen et al. 2002), although a variety of other methods have also been useful (Gürtler and Mayall 2001). With the advent of low-cost DNA sequencing, however, WGS analysis is emerging as the new standard (Loman, Constantinidou et al. 2012). In any event, the ideal approach for systematics includes a polyphasic analysis, in which data gathered from complementary techniques are compared in order to draw a consensus conclusion (Brenner, Staley et al. 2001). Physiological and morphological phenotypic testing, together with 16S rRNA gene sequence comparison and comparative genomics using tools such as the Genome-to-Genome Distance Calculator (Meier-Kolthoff, Auch et al. 2013) and Average Nucleotide Identity (Yoon, Ha et al. 2017), provide a solid foundation for bacterial taxonomy (Ludwig and Klenk 2001, Maughan and Van der Auwera 2011, Chun, Oren et al. 2018). Table 22.1 also lists a variety of methods that have found application in *Bacillus* systematics.

Although these methods have increased our understanding of aerobic endospore former in *Bacillus sensu lato*, still only a handful of these species have been intensely studied for academic and commercial use. Nevertheless, as described in the following section, this small group of bacteria has led the way in various fields of science and industry, which has made a profound impact on our society.

Industrial Applications of *Bacillus*

Developing *Bacillus* Strains for Industry

Bacillus species have played a crucial role in establishing a wide range of sustainable industrial fermentation processes, which in some cases (e.g., riboflavin production) have supplanted older, less efficient chemical processes. Commercial products produced by *Bacillus* sp. range from specialty chemicals, surfactants, and antibiotics to vitamins and food enzymes (Hohmann and Stahmann 2010, Eggersdorfer, Laudert et al. 2012, Hohmann, van Dijl et al. 2017, Singh, Patil et al. 2019). Table 22.2 lists the key industrial fermentation processes that use *Bacillus* species.

In large part, many of these processes were established or improved using a combination of classical genetic methods and recombinant DNA technology and more recently using synthetic biology. Implementation of these latter techniques has been facilitated by the excellent natural competence and DNA recombination mechanisms exhibited by many of these species (Dubnau 1991, Hamoen, Venema et al. 2003). The process-development toolbox for *Bacillus* includes numerous plasmid cloning vectors (Errington 1988, Janniere, Bruand et al. 1990, Bruckner 1992, Kobayashi and Kawamura 1992, Vary 1992, Wang 1992, Choi, Wang et al. 2002), transposon mutagenesis systems (Youngman, Perkins et al. 1983, Perkins and Youngman 1986, Steinmetz and Richter 1994, Le Breton, Mohapatra et al. 2006), conjugative mobile elements (Lee, Thomas et al. 2012), antibiotic resistance markers to select for transformed cells (Horinouchi and Weisblum 1982, Perkins and Youngman 1983, Itaya, Kondo et al. 1989, Itaya, Yamaguchi et al. 1990, Dale, Langen et al. 1995, Guérout-Fleury, Shazand et al. 1995), constitutive and regulated promoters (Lee and Pero 1981, Udaka and Yamagata 1993–1994, Goldstein and Doi 1995, Nguyen, Nguyen et al. 2005, Terpe 2006), and single and multicopy gene expression cassettes, including those with promoter and signal sequence combinations to secrete high valued proteins and enzymes (see below). Introduction of genetic material into the cell can occur by natural competence, protoplast fusion, and electroporation transformation methods and in some instances by conjugation (Jensen, Andrup et al. 1996, Aquino de Muro and Priest 2000, Poluektova, Fedorina et al. 2004). Although most of these tools were first developed in *B. subtilis*, many have been adapted for use in related *Bacillus* species, facilitating the rapid development of new industrial processes. Moreover, the utility of these tools have been amplified by the availability of the complete genome sequences of *B. subtilis* and related species and the subsequent application of powerful omics technologies. The importance of genomics is addressed later in this review. So in many ways the development and ease of use of modern recombinant genetic tools in *Bacillus* rivals that of the other well-known production hosts, such as *E. coli* and *S. cerevisiae*. Extensive description of the key genetic modification and recombinant cloning techniques used for *Bacillus* has been well reviewed (Harwood and Cutting 1990, Meima, van Dijl et al. 2004, Perkins, Wyss et al. 2009, Dong and Zhang 2014, Gu, Xu et al. 2018, Liu, Yang et al. 2019).

TABLE 22.2

Industrial Fermentation Products Produced by *Bacillus* Species

Product	*Bacillus* Species	Industrial Application
Fine Chemicals		
Riboflavin	*B. subtilis*	Food and Pharma
Purine Nucleosides	*B. subtilis*	Food
Poly-γ-glutamic acid	*B. subtilis*	Food, Animal, and Pharma
D-ribose	*B. subtilis, B. pumilus*	Food, Animal, Cosmetics, and Pharma
Thaumatin	*B. subtilis*	Food and Pharma
Streptavidin	*B. subtilis*	Microarrays
2-actyl-1-pyrroline	*B. cereus*	Food
L-malate	*B. subtilis*	Food and Pharma
Biofuels and Bio-Based Chemicals		
Isobutanol	*B. subtilis*	Biofuel/Chemical
2-methyl-1-propanol	*B. subtilis*	Biofuel/Chemical
BDO	*B. subtilis*	Polymers
Acetoin	*B. subtilis*	Food and Beverage
H_2	*B. cereus, B. coagulans, B. licheniformis, B. megaterium, B. pumilus, B. sphaericus, B. subtilis, B. thuringiensis*	Biofuel
Polyhydroxybutyrate	*B. megaterium, B. cereus, B. licheniformis, B. mycoides*	Polymers
Surfactant	*B. subtilis*	Personal care, Detergents, Lubricants
Enzymes		
α-Amylase	*B. licheniformis, B. amyloliquefaciens, B. circulans, B. subtilis, G. stearothermophilus*[a]	Food, Paper, Starch, Textile, and Brewing
β-Amylase	*P. polymyxa*[b], *B. cereus, B. megaterium*	Textile
Alkaline phosphatase	*B. licheniformis*	Detergent
Cyclodextran Glucanotransferase	*P. macerans*[c], *B. megaterium, Bacillus* sp.	
β-Galactosidase	*G. stearothermophilus*[a], *B. circulans*	Food
β-Glucanase	*B. subtilis, B. circulans, B. amyloliquefaciens*	Beverage
β-Glucosidase	*Bacillus* sp.	
Glucose isomerase	*B. coagulans, B. acidipullulyticus, B. deramificans*	Starch
Glucosyl transferase	*B. megaterium*	
Glutaminase	*B. subtilis*	
Galactomannase	*B. subtilis*	
β-Lactamase	*B. licheniformis*	
Lipases	*Bacillus* sp.	Detergent
Neutral (metallo-) rotease	*B. lentus, P. polymyxa, B. subtilis, B. thermoproteolyticus, B. amyloliquefaciens*	
Penicillin acylase	*Bacillus* sp.	
Pullulanase	*Bacillus* sp., *B. acidopullulans*	Starch
Alkaline (serine-) protease	*B. amyloliquefaciens, B. amylosaccharicus, B. licheniformis, B. subtilis*	Detergent
Urease	*Bacillus* sp.	
Xylanases	*Bacillus* sp.	
Lignocellulolytic enzymes	*B. subtilis*	Food and biofuel
L-Asparaginase	*B. subtilis*	Food
Fibrinolytic enzymes	*Bacillus* sp.	Food and Pharma
Fermentation	*B. natto*	Food
Insect Pathogen	*B. thuringiensis, B. sphaericus, P. popilliae*[d] *P. lentimorbus*[e]	Agriculture
Plant Probiotic (Endophyte)	*B. subtilis, B. amyloliquefaciens, B. licheniformis, B. pasteurii, B. cereus, B. pumilus, B. mycoides,* and *B. sphaericus*	Agriculture

(Continued)

TABLE 22.2   *(Continued)*

Industrial Fermentation Products Produced by *Bacillus* Species

Product	*Bacillus* Species	Industrial Application
Animal Probiotic	*B. licheniformis*, *B. subtilis*, *B. cereus* var *toyoi*, *B. clausii*[f], *B. cereus*[f]	Animal feed
Aqua Probiotic	*B. subtilis*, *B. megaterium*, *B. licheniformis*, *P. polymyxa*[e], *Bacillus* sp.	Aquaculture feed
MEOR	*B. subtilis*	Enhanced oil recovery

Note: This table presents a condensed description of industrial products produced by *Bacillus* species. This table is by no means comprehensive since new products are continually being introduced into the market.

Sources: The data are compiled from previously published lists (enzymes from Meima et al. 2004 with permission from Horizon Scientific Press/Horizon Bioscience, UK; fine chemicals, insect pathogens, and fermentation from Schallmey et al. 2004, with permission from NRC Research Press, Canada and Gu et al. 2018; probiotics from Hong et al. 2005 with permission from Blackwell Publishing, UK) (Meima, van Dijl et al. 2004, Schallmey, Singh et al. 2004, Hong et al. 2005), and individual publications: β-galactosidase from (Kyoji 2011); L-malate from (Mu and Wen 2013); cellulosic enzymes from (Maki, Leung et al. 2009, Anderson, Robson et al. 2011, You, Zhang et al. 2012); proteases from (Contesini, de Melo et al. 2018); fibrinolytic enzymes from (Yogesh and Halami 2017); biofuels and chemicals from (Kumar, Patel et al. 2013 [isobutanol and H2]; Biswas, Yamaoka et al. 2012 [BDO]; Blombach and Eikmanns 2011[isobutanol]; Wang, Fu et al. 2012 and Zhang, Zhang et al. 2013 [acetoin]; and Li, Jia et al. 2012 [isobutanol]); plant endophytes from (Mongkolthanaruk 2012); and MEOR from (Gudiña, Pereira et al. 2013).

[a] Formerly *B. stearothermophilus*; [b]formerly *B. polymyxa*; [c]formerly *B. macerans*; [d]formerly *B. popilliae*; [e]formerly *B. lentimorbus*; [f]not approved for use in the EU.

Enzymes

By far the largest product group produced by *Bacillus* species is bulk and specialty enzymes, making up about 50% of the total market size. Applications include household detergents, starch hydrolysis, textile, baking, and beverages. Many of these enzymes have unique properties, such as tolerance to high temperatures (thermostability) or alkaline conditions, broad pH ranges, or inhibitory levels of by-products which make them the enzymes of choice in many industrial processes and household products (Meima, van Dijl et al. 2004). Moreover, with the availability of classical and recombinant methods, protein-engineered variants with novel or improved catalytic properties have been developed to expand their use into new commercial applications (Schmidt-Dannert, Pleiss et al. 1988, Takagi 1993, Nielsen and Borchert 2000, Olempska-Beer, Merker et al. 2006). Due to their propensity to secrete proteins into the extracellular medium, *Bacillus* species are also good hosts for the production of heterologous proteins; this product class will be addressed in later sections. Numerous plasmid-borne or chromosomal integration expression cassettes have been reported that combine transcriptional and translation controlling signals with secretion signal sequences that allow transport of the candidate proteins into the culture medium for subsequent isolation (Song, Nikoloff et al. 2015, Ozturk, Ergun et al. 2017, Cui, Han et al. 2018, Cai, Rao et al. 2019). In addition, a variety of protease-deficient mutants have been constructed to prevent degradation of the heterologous protein; strains with up to eight major and minor protease-encode genes have been constructed (Pero and Sloma 1993, Wu, Yeung et al. 2002).

More recently, the display of proteins on the surface of cells (rather than secretion and release from the cell) has shown great potential for use in the medical and agricultural fields. Such protein displays have been demonstrated for *B. subtilis* and *B. thuringiensis* using a variety of anchoring protein motifs found on the vegetative cell wall, such as *B. subtilis* LytE, (Chen, Wu et al. 2008). Anchoring is also possible using proteins involved in the process of sporulation, e.g., SpoIIIJ (Ding, Guan et al. 2019), or even on the spore surface itself via the *B. subtilis* spore coat proteins CotB (Isticato, Cangiano et al. 2001, Duc le, Hong et al. 2007, Hinc, Isticato et al. 2010), CotC (Ricca and Cutting 2003, Mauriello, Duc le et al. 2004, Hinc, Isticato et al. 2010), CotE (Hwang, Pan et al. 2013), CotG (Kim, Lee et al. 2005, Hinc, Isticato et al. 2010, Hwang, Pan et al. 2013, Liu, Liu et al. 2019), and CotA (Park, Kim et al. 2019), and on the *B. subtilis* inner-coat enzyme oxalate decarboxylase, OxdD (Potot, Serra et al. 2010). A full up-to-date accounting on spore-specific anchors and their displayed proteins can be found in research by Guoyan, Yingfang et al. (2019). In *B. thuringiensis*, proteins can be anchored to both surfaces via the spore cortex-lytic enzyme SceA (Shao, Ni et al. 2012). However, the success rate of displaying a functional enzyme is mixed (Potot, Serra et al. 2010). Future work is aimed to improve the technique to make success more predictable (Nguyen and Schumann 2014, Huang, Gosschalk et al. 2018).

Vitamins and Fine Chemicals

Another commercial area in which several *Bacillus* species play an important role as a fermentation host is the production of vitamins and fine chemicals. Commercialized processes include those for purine nucleotides, riboflavin, and poly-γ-glutamic acid (Ashiuchi, Shimanouchi et al. 2006), surfactants, D-ribose, and an array of minor products such as thaumatin, a low-calorie sugar substitute, polyhydroxybutyrate, streptavidin, and 2-acetyl-1-pyrroline. All but PHB (from *B. cereus*) utilize *B. subtilis* as the production host. *B. subtilis*-based processes for other vitamins, including biotin, pantothenate, thiamin and folic acid, and fine chemicals have as been reported, but none have been commercialized (Perkins, Wyss et al. 2009, Gu, Xu et al. 2018).

Products produced by these *Bacillus* species have a long history of safe use and as such have gained "generally regarded as safe" (GRAS) status in the United States (Silano and Silano 2008). GRAS status is based either on a recognized history of safe use in foods prior to 1958 or on scientific procedures similar to those required for a food additive regulation. An important criterion in determining GRAS status is the lack of pathogenicity of the production microorganism. In this regard, these *Bacillus* species are found not to be harmful to humans or animals.

A status similar to GRAS, called qualified presumption of safety (QPS), has been implemented by EU regulatory authorities. As of March 2019, the following *Bacillus* species have been granted QPS status: *Bacillus amyloliquefaciens, Bacillus atrophaeus, Bacillus clausii, Bacillus coagulans, Bacillus flexus, Bacillus fusiformis, Bacillus lentus, Bacillus licheniformis, Bacillus megaterium, Bacillus mojavensis, Bacillus pumilus, Bacillus smithii, Bacillus subtilis, Bacillus vallismortis,* and *Geobacillus stearothermophilus* (Koutsoumanis, Allende et al. 2019).

Another regulatory issue that affects the use of *Bacillus* in industrial processes that utilize GMO-derived strains is the physical containment of the engineered bacteria (EU Council Directive 90/219/EEC of April 23, 1990 on the contained use of genetically modified microorganisms). Since spores can persist in the environment for years due their unique resistance to heat, chemicals, and UV radiation, commercial strains are required to contain one or more mutations that arrest this terminal developmental stage. Consequently, most commercialized production strains contain a mutation that prevents the formation of a mature, resistant spore.

Biofuels and Bio-Based Chemicals

Replacing oil-based production of fuels and chemicals with sustainable production systems that use renewable (cellulosic) feedstocks is one of the major trends of the new millennium. In particular, many microbial metabolic pathways are currently being exploited to generate energy and produced valuable chemicals. The genus *Bacillus* plays a key role in this commercial endeavor due its wide use as an industrial fermentation host and the ease of metabolic engineering of its biosynthetic pathways as mentioned earlier. Examples include the production of biofuels, e.g., isobutanol and H_2 (Blombach and Eikmanns 2011, Li, Jia et al. 2012, Kumar, Patel et al. 2013), as well as the production of bio-based chemicals, e.g., BDO and PHA, that can be used in the formation of new polymer materials (Biswas, Yamaoka et al. 2012, Wang, Fu et al. 2012, Kumar, Patel et al. 2013, Zhang, Zhang et al. 2013, Rebecchi, Pinelli et al. 2018). Thermophilic members of *Bacillus sensu lato*, namely *Geobacillus* species, already serve as a production platform in the emerging second-generation biofuels industry (Taylor, Eley et al. 2009). In addition, more novel applications of *Bacillus*, such as in the production of biocement, biobricks, and other bio-based building materials, have been developed (Raut, Sarode et al. 2014, Wong 2015, Pungrasmi, Intarasoontron et al. 2019, Royne, Phua et al. 2019). The future goal will be to convert these lab-scale results to viable commercial processes.

A primary obstacle to fermentative production of bio-based fuels and chemicals is the high cost of degrading the cellulosic biomass into fermentable sugars, which is caused by the high recalcitrance of the plant biomass. However, production costs could be dramatically lowered if the biofuel or chemical production host could also synthesize cellulosic enzymes that solubilize the plant cell wall into fermentable sugars. This approach, referred to as consolidated bioprocessing (CBP), has been investigated in a number of naturally occurring cellulolytic microorganisms, such as *Clostridium thermocellum* (Maki, Leung et al. 2009, Singh, Mathur et al. 2018, Tian, Conway et al. 2019). However, since *C. thermocellum* is not easily engineered for biofuel production, an alternative approach has been to engineer microbes for both recombinant-product-forming and cellulose-utilizing abilities.

Bacillus and its relatives are another group of bacteria that has been successfully altered for CBP. Cellulolytic strains of *B. subtilis* have been engineered to display cellulase enzymes (e.g., *C. thermocellum* Cel8A endoglucanase) covalently anchored to the peptidoglycan cell wall (Anderson, Robson et al. 2011). A trifunctional cellulosome has also been displayed on the surface of *B. subtilis* using a cell-wall-binding module of LytE (You, Zhang et al. 2012). Cellulolytic activity can also occur if the minicellulosome is freely secreted into the medium and not attached to the cell surface. Recently, *Geobacillus* has been demonstrated to perform CBP for the production of ethanol on wheat straw (Bashir, Sheng et al. 2019). Future studies are aimed at generating better CBP strains by varying the cellulosic enzyme combination and improving cellulose hydrolysis by increasing cellulase synthesis and specific activity.

Conversely, *Bacillus* isolates (individually or in consortia) also play a role in the petroleum industry, as they are used in tertiary recovery of oil from mature wells (Malik and Ahmed 2012, Gudiña, Pereira et al. 2013). Microbial enhanced oil recovery (MEOR) is used to recover additional oil from older wells after primary and secondary methods have been used. Due to its high viscosity, up to two-thirds or the petroleum in a reservoir can be refractile to recovery by conventional methods. *Bacillus* species appear to be well suited for MEOR since they produce biosurfactants and oil-degrading enzymes that reduce viscosity, promoting oil flow and eventual recovery and thus extending the life of the existing oil reservoir. In recent years, sophisticated screens based on cyclic lipopeptide fingerprinting have been used to isolate new *Bacillus* isolates with improved oil-recovery properties (Farias, Hissa et al. 2018). In addition, application of classical mutagenesis has been used to obtain mutants with improved biosurfactant production (Bouassida, Ghazala et al. 2018).

Agriculture

Bacillus is an important commercial product in the agricultural sector. *B. thuringiensis, Lysinibacillus sphaericus, Paenibacillus popilliae,* and *P. lentimorbus* have long been recognized as natural biopesticides effective against a wide range of insect pests (Schnepf, Crickmore et al. 1998, Schallmey, Singh et al. 2004, el-Bendary 2006). By far, the most widely used is *B. thuringiensis,* which accumulates delta-endotoxin pesticides, or Cry proteins, in large, highly visible crystals during the process of sporulation. The species is equipped with a potent arsenal of highly specific pesticides that vary widely in target specificity from strain to strain. Cry proteins are classified according to their amino acid sequence, but the classifications frequently correspond to the insect order that the toxin is effective against. For example, Cry1 proteins of *B. thuringiensis* serovar. *kurstaki* are toxic mainly to *Lepidoptera* species, Cry3 proteins of *B. thuringiensis* biovar. *tenebrionis* are toxic to certain Coleoptera, and Cry4 and Cry10 proteins of *B. thuringiensis* serovar. *israelensis* are highly toxic against mosquitoes and black flies (Schnepf et al. 1998). As of 2019, there were 26 *B. thuringiensis*-based active ingredients and eight labeled products registered by the United States Environmental Protection Agency (EPA) (https://www.epa.gov/ingredients-used-pesticide-products/biopesticide-active-ingredients; https://iaspub.epa.gov/apex/pesticides/f?p=PPLS:1). Interestingly, an almost equal number of pesticide ingredients (18) and labeled products (nine) from other *Bacillus* species (e.g., *B. subtilis*, *B. firmus*, *B. amyloliquefaciens*, *B. mycoides*, *B. pumilus*, *Paenibacillus popilliae*, *B. cereus*, and *Lysinibacillus sphaericus*) have been registered by the EPA.

Moreover, with the advent of recombinant genetic techniques, the toxicity of these insecticides has been improved and the killing spectrum widened (Sanchis, Agaisse et al. 1996, Ghribj, Zouari et al. 2004, Liu and Dean 2006). The genes that encode these highly diverse toxins are now expressed directly into a variety of agricultural plants, such as corn, tobacco, and soybean, which further extends the usefulness of these *Bacillus* species in agriculture (Mehlo, Gahakwa et al. 2005, Christou, Capell et al. 2006). Between 1995 and 2018, 27 Bt *cry* genes have been approved for production in plants (i.e., plant incorporated protection, PIP): 12 for corn, 9 for cotton, 5 for soybean and 1 for potato (https://www.epa.gov/ingredients-used-pesticide-products/current-and-previously-registered-section-3-plant-incorporated).

The use of *Bacillus* for plant growth promotion (PGP) is an area with great potential for agriculture (Radhakrishnan, Hashem et al. 2017). Many species of *Bacillus* are found to be associated with plants as rhizosphere bacteria or endophytes (Mongkolthanaruk 2012). These species include *B. subtilis*, *B. amyloliquefaciens*, *B. licheniformis*, *B. pasteurii*, *B. cereus*, *B. pumilus*, *B. mycoides*, and *L. sphaericus*, *B. insolitus*, *B. velezensis*, and *B. megaterium*, and *B. mojavensis*. Application of *B. subtilis* on plant seeds has been shown to greatly enhance the survival and robustness of the seedlings (Cavaglieri, Orlando et al. 2005, Ugoji, Laing et al. 2005). *Bacillus sensu lato* species, such as *Paenibacillus polymyxa*, *B. cereus*, and *B. subtilis*, are also used to facilitate the establishment and durability of sport turf grass. The mechanisms by which *Bacillus* promotes plant growth are still being investigated, but include the synthesis of plant growth hormones and antimicrobial agents (Mongkolthanaruk 2012) and the production of enzymes, such as phytase (Tye, Siu et al. 2002), that increase nutrient availability (Gange and Hagley 2004). Interaction of beneficial bacteria may also induce plant stress genes, protecting the host from both pathogenic attach and abiotic stress, such as drought (Timmusk, Grantcharova et al. 2005). Finally, the ability of many *Bacillus* PGP strains to form biofilms on plant surfaces suggests that niche exclusion may be a key mechanism for controlling plant pathogens (Haggag and Timmusk 2008, Timmusk, van West et al. 2009, Beauregard, Chai et al. 2013, Chen, Yan et al. 2013).

Medical Applications of *Bacillus*

The importance of *Bacillus* in the field of medicine was established in the late 1800s by the identification of *B. anthracis* by Pasteur and Koch as the causative agent of anthrax. However, the genus as a whole is considered relatively benign in terms of human and animal disease when compared to other Gram-positive genera, like *Staphylococcus* and *Streptococcus*. *B. cereus* and the highly related species *B. weihenstephanensis* can be responsible for gastrointestinal illnesses caused by contaminated food or dairy products. Pathogenesis requires a common set of enterotoxin genes (Phelps and McKillip 2002). On rare occasions *B. cereus* can be an opportunistic human pathogen in immunocompromised patients (Kotiranta, Lounatmaa et al.

2000, Schoeni and Wong 2005). The general public's awareness of *Bacillus* has been heightened by the use of *B. anthracis* in acts and threats of bioterrorism (Schneider, Parish et al. 2004).

Despite this heightened awareness, what is not nearly so well-known are the numerous beneficial medical-related products produce by this genus (Table 22.3). By far, the most predominant product class is antibiotics. Most of these are peptide antibiotics that are synthesized either by nonribosomal enzymatic processes or by ribosomal synthesis of linear peptides that then undergo post-translational modification and proteolytic processing (Stein 2005, Ortiz and Sansinenea 2019). Another important product class is the synthesis of human proteins and vaccines, processes which take advantage of *Bacillus*' natural ability to produce and secrete enzymes as previously described for industrial and food enzymes. Almost all of these processes have been developed using *B. subtilis*, *B. megaterium*, and *Brevibacillus brevis* (Udaka and Yamagata 1993–1994, Vary 1994, Meima, van Dijl et al. 2004, Westers, Westers et al. 2004, Ferreira, Ferreira et al. 2005, Terpe 2006, Karauzum, Updegrove et al. 2018). The advantage of *B. brevis* is that extracellular protease production is low and heterologous protein expression is high. Although the number of implemented processes is not as extensive as for other better known protein production host such as *E. coli* or yeast, there are nevertheless several successful examples, such as the use of *B. megaterium* to produce HIV proteins for commercial diagnostic tests (Vary 1994). More recently, *B. subtilis* has been developed as a needle-free delivery vehicle of immunogens for intranasal, sublingual, and oral immunization by utilizing recombinant forms that display the antigen of the surface of the spore (Amuguni and Tzipori 2012, McKenney, Driks et al. 2013).

As mentioned in the preceding section, products produced by *B. subtilis* and related species have GRAS status. Moreover, the microorganisms themselves can also be of medical use as a probiotic. Two major human applications are as prophylactics and as health food supplements. Key species found in these preparations are *B. subtilis*, *B. licheniformis*, *B. coagulans*, *B. pumilus*, *B. clausii*, nonpathogenic *B. cereus* strains and other spore formers from *Bacillus sensu lato*. In Europe (excluding the UK), *Bacillus* probiotics are used prophylactically against gastrointestinal ailments. In southeast Asia, they are used as an adjunct to antibiotic treatment (Sanders, Morelli et al. 2003, Hong, Duc et al. 2005, Cutting 2011). Use in Western societies, especially in the United States, have increased over the last several years, perhaps due to use in developing countries as discussed above and with the realization that strains closely related to *B. subtilis* 168 are already consumed in *nattō*, a traditional Japanese fermented soy bean delicacy (Ueda 1989, Hosoi and Kiuchi 2004, Elshaghabee, Rokana et al. 2017, Jezewska-Frackowiak, Seroczynska et al. 2018).

B. cereus, *B. subtilis*, and *B. licheniformis* have also been shown to be a very potent animal probiotics, especially in the pig and poultry sectors and in aquaculture (Alexopoulos, Georgoulakis et al. 2004, Barbosa, Serra et al. 2004, Cartman and La Ragione 2004, Farzanfar 2006, Tam, Uyen et al. 2006, Mingmongkolchai and Panbangred 2018, Kuebutornye, Abarike et al. 2019). The effectiveness of these *Bacillus* species is thought to lie in their ability to be retained in the gut of the animal by undergoing repeated rounds of vegetative growth, sporulation, and germination, during which time the production of

TABLE 22.3

Medical and Human Health Products Produced by *Bacillus* Species

Product	*Bacillus* Species
Antibiotics	
Bacitracin	*B. licheniformis*
Gramicidin	*Brevibacillus brevis*
Tyrocidines	*B. brevis*
Subtilin	*B. subtilis*
Polymyxins	*B. polymyxa*
Edeines	*B. brevis*
Butirosins	*B. circulans*
Diagnostic test	
AIDS coat proteins	*B. megaterium*
Antigens for vaccines	
PA (*B. anthracis*)	*B. subtilis*
Pneumolysin (*Streptococcus pneumonia*)	*B. subtilis*
P1 (*Neisseria meningitides*)	*B. subtilis*
OmpP2 (*Haemophilus influenzae*)	*B. subtilis*
PT subunits (*Bordetella pertussis*)	*B. subtilis*
Omps and Hsp60 (*Chlamydia pneumonis*)	*B. subtilis*
Tetanus toxin fragment C (TTFC),	*B. subtilis*
Adjuvant	*B. subtilis*
Human proteins	
Growth hormone	*B. subtilis*
Interferon	*B. subtilis*
Proinsulin	*B. subtilis*
Tissue plasminogen activator	*B. subtilis*
scFv	*B. subtilis*
Insulin-like growth factor 1 (IGF-1)	*B. subtilis*
Human Probiotic	*B. subtilis, B. licheniformis, B. cereus, B. pumilus, B. polyfermenticus SCD, B. clausii, B. laterosporus, B. coagulans, B. polymyxa, B. bifidum*
Bioactive fine chemicals	
Hyaluronic acid	*B. subtilis*
Glucosamine	*B. subtilis*
Amorphadiene	*B. subtilis*
scyllo-inositol	*B. subtilis*
Heparosan	*B. subtilis*

Note: This table presents a condensed description of products produced by *Bacillus* species. This table is by no means comprehensive since new products are continually introduced into the market.

Sources: The data are compiled from previously published lists: antigens for vaccines from (Ferreira, Ferreira et al. 2005) with permission from Anais da Academia Brasileira de Ciências, Brazil and (Amuguni and Tzipori 2012); antibiotic and human proteins from (Schallmey, Singh et al. 2004), with permission from, NRC Research Press, Canada; probiotics from (Hong, Duc et al. 2005) with permission from Blackwell Publishing, UK (Schallmey, Singh et al. 2004, Ferreira, Ferreira et al. 2005, Hong, Duc et al. 2005), and individual publications (hyaluronic acid from (Jia, Zhu et al. 2013), glucosamine from (Liu, Liu et al. 2013), amorphadiene from (Zhou, Zou et al. 2013), vaccine adjuvant and delivery (Amuguni and Tzipori 2012), *scyllo*-Inositol and Heparosan from (Gu, Xu et al. 2018), and IGF-1 from (Wang, Yu et al. 2017).

bacteriocins, antibiotics, or other antimicrobial agents effectively prevents the establishment of pathogenic bacteria in the gut (Cutting 2011). Moreover, use of probiotics in combination with its complementary prebiotic (i.e., synbiotic) can further enhance the health benefit (Hamasalim 2016).

B. subtilis is also used to identify and validate novel antibacterial targets. *In silico* genome analysis and functional testing have been used to identify potential targets for new antimicrobial agents. Possible targets include proteases, biosynthetic enzymes for amino acids and vitamins (Gerdes, Scholle et al. 2002, Kobayashi, Ehrlich et al. 2003, Perkins, Goese et al. 2004, Yocum and Patterson 2005), and regulatory elements such as riboswitches (Blount and Breaker 2007). Due to its well-developed tools for genetics and recombinant engineering, *B. subtilis* has also been used as a screening host to identify new antimicrobial agents (Bunyapaiboonsri, Ramstroem et al. 2003, Lacriola, Falk et al. 2013).

Cell Physiology and Development

By virtue of its ability to make endospores through a primitive yet intricate developmental process, the genus *Bacillus* has been intensively studied in the academic arena. *B. subtilis* has received most of this attention (and research funding) due its nonpathogenicity and its user-friendly genetic system. Practically all *B. subtilis* strains used in academic research are derived from strain 168, a tryptophan-requiring derivative of wild-type *B. subtilis* Marburg isolated after x-ray mutagenesis (Burkholder and Giles 1947). It was with this strain that DNA transformation was first demonstrated for this species (Spizizen 1958). *B. subtilis* 168 has become a paradigm for Gram-positive molecular and developmental biology, second only to *E. coli* as a bacterial model organism.

The genetic analysis of sporulation mutations in *B. subtilis* 168 (Hoch 1971) was the first step in the ongoing endeavor to identify the proteins required for sporulation, to elucidate how these proteins function together to create a spore, and to understand how the cell regulates their expression in response to the environmental conditions that trigger the process. Figure 22.4 shows the well-defined developmental stages that lead from a vegetative cell to a mature spore. At least 239 genes are required for completing the intricate developmental program that converts the growing cell into two asymmetric compartments, into which are assembled the macromolecular structures that constitute the spore (Piggot and Losick 2001, Higgins and Dworkin 2012, Watabe 2013). To help explain the choreography of this process, concepts such as sigma-factor cascades (Losick and Pero 1981, Stragier and Losick 1996), multicomponent phosphorelays (Burbulys, Trach et al. 1991, Hoch 1993), inter-compartmental cross-talk (Bassler and Losick 2006), and cell cannibalism and fratricide (Gonzalez-Pastor, Hobbs et al. 2003, Ellermeier, Hobbs et al. 2006) have been introduced. More importantly these concepts can be applied to understand other more complex developmental processes, such as fruiting body formation in *Myxococcus* and swarmer-stalked cell differentiation in *Caulobacter* (Shapiro and Losick 2000, Ryan and Shapiro 2003).

Since sporulation is an energy-intensive process and requires a variety of molecular building blocks, substantial work has been

Key landmark
morphological stage

Appearance of
sigma factors

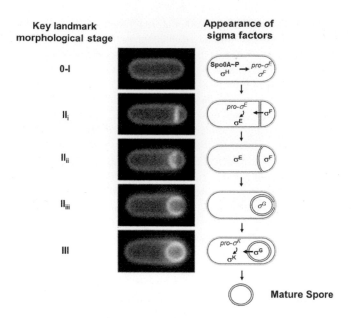

Mature Spore

FIGURE 22.4 Key landmark morphological stages in sporulation. Sporulating cells of *B. subtilis* stained with the membrane dye TMA-DPH is depicted in the middle panel. The sequence from top to bottom shows a cell at the predivisional stage (0-I) progressing though stages of polar division (II) to engulfment (III); stages (IV–VI) from spore maturation to final release of the mature spore from the mother cell are not shown. The right panel shows a schematic representation of the landmark morphological events and cascade-like appearance of sporulation-specific sigma factors. Activate and inactivate factors are highlighted in bold and italics, respectively. Spo0A~P and sigma H (σ^H) are first activated in the predivisional cell (stage 0) by external environmental signals (Piggot and Losick 2001). During this stage, terminal DNA replication of the genome takes place. Spo0A~P and σ^H direct the transcription of early sporulation genes, includes those involved in formation of the asymmetric septum and the synthesis of inactive sigma F (σ^F) and pro-σ^E (stage I). Formation of the sporulation septum triggers the activation of σ^F (stage II). This regulatory factor directs the transcription of forespore-specific genes, include those involved in eliciting an interdepartmental signal, which emanates from the forespore to the mother cell and activate sigma E (σ^E) through processing of inactive proprotein precursor (pro-σ^E). Translocation of the genome into the preforespore compartment occurs during this time. Inactive sigma G (σ^G), whose transcription is initiated by σ^F, is then synthesized during engulfment (stages II$_{i-iii}$), or shortly thereafter, but requires engulfment completion (stage III) for activation. Again, interdepartmental signal transduction, controlled by σ^G, is required to activate sigma K (σ^K) via processing of the inactive proprotein form (pro-σ^K). Final stages involve maturation of the spore cortex, including depositing of the spore-coat proteins, and eventual release of the mature spore from the mother cell. (Illustration reprinted from *Developmental Cell*, vol. 1, Rudner, D. Z., and Losick, R., "Morphological coupling in development: lessons from Prokaryotes," pp. 733–742. Copyright 2001, with permission from Elsevier.)

undertaken to elucidate central carbon metabolism, the pentose phosphate pathway, nitrogen and phosphate assimilation, and the biosynthesis of amino acids, vitamins, nucleotides, and macromolecules. Special emphasis has been placed on understanding carbon, nitrogen, and phosphate metabolism, since a limitation in any of these compounds induces sporulation (Hulett 1996, Fisher 1999, Sonenshein 2001).

Importantly, sporulation represents just one of several possible outcomes of late stationary growth phase. Considerable attention has been given to other phenomena that occur during this period

such as antibiotic production and genetic competence (Szurmant and Ordal 2004). Moreover, current thinking now places sporulation and these other cellular behaviors as part of an overarching developmental scheme that encompasses the bacterial population as a whole. This concept is referred to as bimodality (also referred to as bistability), a phenomenon in which a population of genetically identical cells can reversibly switch into different physiological states or cell types, such as one type that is geared to sporulate and a second that is not. The outcome of bimodality is to maximize the population's freedom of action to confront and overcome present and future adverse environmental conditions (Dubnau and Losick 2006), which has been referred to as bet-hedging. Within a stationary phase culture (or colony), cell types include individual motile and nonmotile cells, long-chained sessile cells, K-cells, and sporulating cells. The nonmotile cells secrete massive quantities of antibiotics and degradative enzymes (e.g., proteases, nucleases, and amylases), whereas some of the sessile cells release an extracellular matrix that is used to enable cells to adhere to a surface or to each other to form a biofilm (discussed next). These two cell types provide additional nutrients and protection to the cell population as it proceeds into late stationary growth phase. These cells can also induce cellular death and lysis and make the ultimate sacrifice to provide nutrients to the remainder of the cell population. The K-cells are arrested in cell growth and septum formation, but are active in DNA uptake and chromosomal integration, allowing these cells to potentially acquiring new functions to survive an increasingly nutrient-deficient environment. However, if nutrients are completely depleted, K-cells can transition into sporulation (Zeigler and Nicholson 2017).

Differentiation into multiple cell types is the first step to develop multicellular communities such as biofilms and fruiting bodies. It was originally thought that *B. subtilis* lacked this ability. However, studies of natural isolates (e.g., NCBI 3610) clearly show that this species can form such architecturally complex structures, in which sporulation, competence, and swimming are tightly interwoven (Branda, Gonzalez-Pastor et al. 2001, Branda, Gonzalez-Pastor et al. 2004, Shank and Kolter 2011, Vlamakis, Chai et al. 2013). Moreover, comparison of these natural isolates to domesticated counterparts, which have lost multicellularity, has led to the identification of master regulatory genes that control these processes (Kearns, Chu et al. 2005). Other controlling elements include extracellular peptides ("pheromones") that maintain cell-to-cell communication via quorum sensing and facilitate recruitment of cells for specific developmental roles. So far, peptide pheromones have been identified for sporulation, competence, and cell cannibalism/fratricide (Gonzalez-Pastor, Hobbs et al. 2003, Cormella and Grossman 2005).

To regulate the expression of genes necessary to carry out these developmental processes, several regulatory mechanisms are used in *B. subtilis*. One stratagem involves the deployment of alternate sigma subunits to temporally and spatially control the expression of specific groups of genes by altering the specificity of the RNA polymerase to recognize particular sets of promoters. *B. subtilis* contains 17 such sigma factors (Goldstein and Doi 1995, Haldenwang 1995, Missiakas and Raina 1998, Helman and Moran 2001, Helmann 2002). In most cases, the genes encoding these factors are themselves under control of a sigma subunit that is expressed at an earlier time in cell growth or in a preceding step in a developmental process. During sporulation, five sigma factors are expressed in a cascade-like manner to link the expression of specific gene sets to the landmark stages (see Figure 22.4). As an added layer of regulation, the activity of the sigma factors themselves are modulated through protein processing or by inhibition via antisigma factors (Hughes and Mathee 1998).

The two-component regulatory mechanism found in *E. coli* and other bacteria also play a major regulatory role in *B. subtilis* cell development, as described above, by permitting it to detect environmental signals and then to convert these stimuli into an actionable response at the cellular level. *B. subtilis* has 34 identified two-component systems that control key developmental activities, including competence, sporulation, protein secretion, and cell growth. Combination of several two-component systems result in formation of a gene regulatory network (GRN), which has the ability to dramatically shift gene expression profiles within a cell resulting in different cell type states. How is this switch-like behavior achieved? For the three-gene regulatory network, *sinI, sinR,* and *slrR,* that controls alternation between single motile cells and long connected chains of sessile cells, double feed-back loops are used (Piggot 2010). Microfluidic technology has been used to dissect the stochastic decision between formation of these two cell forms, indicating that that the motile state is "memory-less," exhibiting no autonomous control of the time spent in this state, whereas sessile cell chaining is strictly timed. (Norman, Lord et al. 2013). For the KinA-Spo0F-Spo0B-Spo0A phosphorelay, which is used to activate the onset of sporulation, a multi-layered feedback regulatory network is used, involving not only these four phospho-activated regulators, but an array of other kinases, phosphatases, and other stationary phase regulators (KinB, KinC, RapA, AbrB, etc.). Recent microfluidic studies of individual cells have shown that the process to enter sporulation is heterogeneous, stochastic, and memory-less. Moreover, the build-up of Spo0A~P, the ultimate regulatory protein that initiates sporulation, does not occur in a progressively higher oscillating manner, as initially thought (Narula, Fujita et al. 2016), but is activated abruptly within a single cell cycle, just before asymmetric division. It is postulated that the stochastic behavior of Spo0A activation is due to "noise" generated by the intrinsic structure of the phosphorelay itself (Russell, Cabeen et al. 2017). For biofilm formation, which is also controlled by Spo0A, intermediate levels of Spo0A~P result in biofilm matrix gene expression whereas higher levels induce sporulation genes. Although Spo0A~P does not accumulate in an oscillating manner, as mentioned above, biofilm formation does (Martinez-Corral, Liu et al. 2018). This is likely due to nitrogen (limitation) codependency in which cells in the biofilm periphery is driven by glutamate consumption and cells in the interior by ammonium production (Liu, Prindle et al. 2015).

A third regulatory mechanism used in *B. subtilis* is called the riboswitch. This mechanism involves the direct binding of metabolites to nascent mRNA molecules to regulate their expression. Depending on the secondary structure of the 5' leader, binding leads to termination of the nascent RNA transcript (or in Gram-negative bacteria the prevention of translation by blocking access of the ribosome binding site). In the case of regulation of the riboflavin biosynthetic genes, the riboswitch repressor has been identified as FMN. Riboswitch elements have also been detected to regulated gene expression involved in SAM, lysine, thiamine, purine, pyrimidine, and other metabolic pathways (Mandal and Breaker

2004, Vitreschak, Rodionov et al. 2004, Sudarsan, Hammond et al. 2006). Other riboswitches involving uncharged tRNA molecules have also been identified for amino acid metabolism genes (Henkin 1994, Henkin and Yanofsky 2002). Interestingly, it has been proposed that this mechanism is a vestige of an ancient RNA world, when living organisms used RNA as their main heredity material and source of enzymes (ribozymes).

A final regulatory mechanism has recently been described, by which *B. subtilis* can respond to physiological cues to regulate gene expression and hence development. This mechanism relies not on protein or RNA sensors, but on codon frequencies. High levels of serine in the growth medium suppress biofilm formation. In mRNA molecules, serine can be encoded either by any of six codons. Interestingly, the frequency of TCN serine codons is overrepresented in the *sinR* gene, which encodes the major repressor of biofilm formation genes, relative to the rest of the *B. subtilis* genome. Under serine starvation conditions, the rate of *sinR* transcription is therefore significantly reduced; this reduction is alleviated if the TCN codons are mutated to AGC or AGU, which also code for serine. Apparently the TCN codons act as a serine-sensing mechanism, bringing information on the nutritional state of the cell to bear on the decision to form biofilms during stationary growth phase (Subramaniam, Deloughery et al. 2013).

All the metabolic and developmental processes described so far occur during aerobic growth. Although originally thought to be a strict aerobe, *B. subtilis* is also capable of either anaerobic growth using nitrate or fumarate as the final electron accepter, or fermentative growth from glucose and pyruvate resulting in various end products such as ethanol, acetate, lactate, acetoin, and 2,3-butanediol (Nakano, Dailly et al. 1997, Nakano and Zuber 1998, Hartig and Jahn 2012). Other species able to grow or at least respire under anaerobic condition include *B. licheniformis* (Bulthuis, Rommens et al. 1991, Clements, Miller et al. 2002), *B. megaterium* (Freedman, Peet et al. 2018), *B. cereus,* and *B. mojavensis.* Extensive studies of *B. subtilis* have identified and characterized the two component regulatory mechanism (*resDE*) controlling the expression of a set of genes that permit growth under anaerobic conditions (Nakano, Yang et al. 1995, Nakano, Zuber et al. 1996, Sun, Sharkova et al. 1996, Hartig and Jahn 2012). Along with Fnr (the key regulator of genes involved with nitrate respiration), reduced oxygen tension is directly sensed by the membrane-bound ResDE. The ResDE response, in turn, is modified by nitric oxide and the NsrR repressor. Recently the genome-wide analysis of ResD, NsrR, and Fur regulation in *B. subtilis* has been elucidated during anaerobic fermentative growth, showing a complex arrangement of overlapping regulatory sites control genes within this regulon (Chumsakul, Anantsri et al. 2017). This regulatory mechanism is also conserved among *Bacillus* species that are facultative anaerobes. It will be interesting to see how this anaerobic behavior fits with the extensive repertoire of aerobic cell development displayed by *B. subtilis* in the overall ecology of this bacterium.

Future Technology and Applications

The authors have been involved in preparation of this chapter two times; first, 10 years ago for the second edition of *Practical Handbook of Microbiology,* and the update 4 years later for the third edition. Both times, we speculated on what the future held for the genus *Bacillus* in terms of the development of new technologies to further understand the genetics, physiology, and developmental biology of these species, and the application of this knowledge for new industrial and medical applications. We saw genomics and other "-omics" technologies as having a vital impact on *Bacillus* basic and applied research. This impact still applies today. We first noted that the determination of the *B. subtilis* genome sequence in 1997 (Kunst, Ogasawara et al. 1997) quickly led to the establishment of gene knock-out libraries (Schumann, Ehrlich et al. 2001, Kobayashi, Ehrlich et al. 2003), whole genome microarrays (Caldwell, Sapolsky et al. 2001, Lee, Zhang et al. 2001, Jürgen, Tobisch et al. 2005), regulatory element databases (Makita, Nakao et al. 2004), proteomic libraries (Bernhardt, Buttner et al. 1999, Hecker and Volker 2004), global mRNA half-life determinations (Hambraeus, von Wachenfeldt et al. 2003), minimal genome determinations (Westers, Dorenbos et al. 2003), and extensive metabolic flux determination (Fischer and Sauer 2005, Sauer and Elkmanns 2005). These approaches provided invaluable insights into *B. subtilis* as a developmental system and as an industrially important platform strain. In 2013, the *B. subtilis* genome sequence was updated to include toxin/antitoxin and small RNA genes, and a more complete metabolic network has been described (Belda, Sekowska et al. 2013). In addition, a more comprehensive inventory of proteins encoded by the *B. subtilis* genome was developed. More recent, the genome sequences of lab strains of *Bacillus subtilis* subsp. *subtilis* 168 and subspecies (e.g., *Bacillus subtilis* subsp. *spizizenii* strain W23; *Bacillus subtilis* subsp. *subtilis* KT3135, *Bacillus subtilis* "natto"), have been determined and used to elucidate the history of specific genetic lesions and ultimately the origin of 168 and W23 legacy strains (Zeigler, Pragai et al. 2008, Zeigler 2011, Kamada, Hase et al. 2014, Smith, Goldberg et al. 2014, Ahn, Jun et al. 2018). Finally, the introduction of microfluidics has taken place to analyze individual cells (Cabeen and Losick 2018), since it has been known for some time that average data obtained from microbial population-based studies may not be representative of the behavior, state, or phenotype of single cells. The ability to study single cells will make possible a more nuanced understanding of microbial development and physiology.

During this period, further improvements in the existing "-omics" platforms have also kept pace, such as metabolic flux (i.e., fluxome) and mathematical modeling (Dettwiler, Heinzle et al. 1993, Lee, Goel et al. 1997, Dasika, Gupta et al. 2004, Sauer 2006), identification of new global regulatory elements, such as sRNA (Silvaggi, Perkins et al. 2005, Silvaggi, Perkins et al. 2006), cataloging of global protein-protein interactions (Noirot and Noirot-Gros 2004), and establishment of metabolite compound libraries (i.e., metabolomics; Soga, Ohashi et al. 2003, Koek, Muilwijk et al. 2006). More recently, -omics platforms have expanded to include a two genome-scale deletion library (Koo, Kritikos et al. 2017), and attempts to combine -omics technology (i.e., mixomics) in single experiments to answer specific metabolic engineering questions such as riboflavin production (Hu, Lei et al. 2017).

Moreover, these data have been consolidated and organized into increasingly user-friendly "wiki"-style databases, facilitating the access and analysis of information on the genes, proteins, and regulatory elements of *B. subtilis*. Examples include *Subti*Wiki 1.0 and 2.0 (Michna, Zhu et al. 2016), *Subti*Pathway (Lammers, Florez et al. 2010), BacillOndex (Misirli, Wipat et al. 2013), and SporeWeb (Eijlander, de Jong et al. 2014).

More importantly, the establishment of these databases can be seen as part of a larger movement to dramatically improve the genetic toolbox of *B. subtilis* by the development of synthetic biology methodologies. This shift from traditional genetic engineering to synthetic biology-based technologies allows for faster, larger scale genetic changes. Over the last 10 years, development of such synthetic biology tools and methods have exploded by the development of newer gene expression systems based on the BioBrick concept, more efficient scar-less transformation systems and directed evolution platforms, counter-selection selection markers, marker-free knock-out and mutation delivery methods, large DNA cloning methods, cell-free and improved heterologous protein secretion, and genome size reduction to optimize *B. subtilis* as a production host (Itaya and Kaneko 2010, Toymentseva, Schrecke et al. 2012, Liu, Liu et al. 2013, Radeck, Kraft et al. 2013, Shi, Wang et al. 2013, Dong and Zhang 2014, Jeong, Park et al. 2015, Ogawa, Iwata et al. 2015, Tsuge, Sato et al. 2015, Ozturk, Ergun et al. 2017, Popp, Dotzler et al. 2017, Itaya, Sato et al. 2018, Liu, Zhang et al. 2019, Liu, Liu et al. 2019). More recently, the introduction of CRISPR-based tools has allowed faster and more precise alteration of the *B. subtilis* chromosome. These alterations include single base pair changes, insertions and deletions, including large DNA segments (So, Park et al. 2017). CRISPR/Cas systems that have been demonstrated to function in *Bacillus subtilis* are those from *Streptococcus pneumoniae* (Jiang, Bikard et al. 2013, Westbrook, Moo-Young et al. 2016, Burby and Simmons 2017a, Burby and Simmons 2017b) and *Streptococcus pyogenes* (Altenbuchner 2016, Zhang, Duan et al. 2016). Moreover, CRISPR/CasA gene editing has also been demonstrated in non-*Bacillus* model strains or for undomesticated strains with natural transformability: *B. licheniformis* and natural *Bacillus* spp. isolates found in the plant rhizosphere (Li, Cai et al. 2018, Yi, Li et al. 2018).

These new techniques have already borne fruit. For example, *B. subtilis* strains have been modified to contain the glyoxylate shunt, a pathway that allows for use of overflow metabolites as carbon source (Kabisch, Pratzka et al. 2013). These techniques have likewise facilitated the introduction of heterologous biosynthetic pathways for the production of commercially important compounds, such as amorphadiene (Zhou, Zou et al. 2013) and L-malate (Mu and Wen 2013). Finally, the first steps in construct a true chassis for SynBio engineering have taken place, such as reducing the genome to 2.68 Mb with 1,605 gene deletions, reducing carbon catabolite repression to simultaneously utilize of multiple carbon source from hydrolyzed biomass, engineering of the membrane to improve excretion of molecular compounds, and protecting the chassis against bacteriophage destruction (Tanaka, Henry et al. 2013, Ofir, Melamed et al. 2018, Liu, Liu et al. 2019).

Another trend we previously noted was the dramatically lowering cost of genomic (re)sequencing, allowing for the rapid determination of genetic differences between related species or industrial strain lineages (Albert, Dailidiene et al. 2005, Bennett, Barnes et al. 2005, Margulies, Egholm et al. 2005). These technologies include 454 pyrosequencing (Gharizadeh, Akhras et al. 2006), Illumina/Solexa, (Bentley, Balasubramanian et al. 2008)

SOLiD (Schuster 2008), and single molecule real time (SMRT) sequencing (Ardui, Ameur et al. 2018). Application of these sequencing technologies to improved strains recovered from classical mutagenesis programs can rapidly identify those critical mutations that result in higher metabolic function or enzyme overproduction. Moreover, this technology is being applied to rapidly sequence natural isolates of *B. subtilis* as a way to better understand genomic diversity (Earl, Losick et al. 2007, Zeigler, Pragai et al. 2008). Based on 2017 data, the number of sequenced *Bacillus* species and isolates has reached more than 13,000, an order of magnitude increase since our last review (https://gold.jgi-psf.org/; Reddy, Thomas et al. 2015, Mukherjee, Stamatis et al. 2017).

Of course, this wealth of genome sequence data opens the door to bring powerful -omics technologies to bear on the physiology and molecular biology of these species, perhaps uncovering novel developmental behaviors or industrial applications along the way. But in a broader sense, such studies also have a significant impact on how we perceive evolutionary changes at the genomic level. For example, based the high percentage of conserved genes among *Bacillus* species, one can consider genomes from highly related species to be actually composed of two components: a "core genome" shared by all members of the group, and a much larger "dispensable genome" consisting of partially shared or strain-specific genes. The core and dispensable genomes together comprise the group's "pangenome" (Tettelin, Masignani et al. 2005). An important implication of this concept is that bacteria are not completely clonal; that is, they do not inherit their genetic material (and therefore phenotypic traits) solely from a parental line, with a gradual accumulation of random mutations fixed by chance or natural selection. Instead, much of their genetic content is acquired through horizontal gene transfer (HGT) from nonrelated bacterial sources. One reasonable estimate, based on variations in G+C content in the DNA sequence, is that 552 of the roughly 4,112 protein-encoding genes in the *B. subtilis* chromosome entered the genome via HGT from non-*Bacillus* sources (Garcia-Vallve, Guzman et al. 2003). *B. cereus*, *B. thuringiensis*, and *B. anthracis* are known to share an extensive core genome while differing primarily in plasmid-borne pathogenicity genes (Rasko, Altherr et al. 2005). As the cost of DNA sequencing decreases through the introduction of new technologies (e.g., single-strand sequencing (Eid, Fehr et al. 2009), nanoball sequencing (Porreca 2010), and nanopore sequencing (Stoddart, Heron et al. 2009), genomes of a vast number of novel endospore-forming isolates will be determined, which future studies in *Bacillus* systematics will have to grapple with. Such approaches have already borne fruit. Using high-throughput sequencing technology, Kim et al. (2017) have identified core and strain-specific genes in 238 strains of five different *Bacillus* species from various food microbiomes; without resorting to culture-based whole genome sequencing, they were also able to reconstruct an unexpectedly large portion of the core genes, 98.22% of core genes in *B. amyloliquefaciens* and 97.77% of *B. subtilis*. Similar studies have already identified the number of harmful *Bacillus* species from food commodities procured from small and large retail outlets in a U.S. metropolitan area (Higgins, Pal et al. 2018), raw and pasteurized milk (Porcellato, Aspholm et al. 2019), and even in contaminated illegal drugs (Kalinowski, Ahrens et al. 2018). Alternative, we see similar

approaches being used to identify novel beneficial *Bacillus* probiotic strains from various microbiome sources (i.e., gastrointestinal tracts (GITs) and feces) of chickens, pigs, ruminants, and aquatic animals (Mingmongkolchai and Panbangred 2018).

As we have asked in the past, what next is in store for *Bacillus* in terms of new technologies and applications? We see a continuing trend of coupling synthetic biology and sequencing technology with miniaturization, allowing single cells to be studied on a molecular level. As mentioned above, single-cell analysis has been used to dissect the expression patterns of key regulatory genes during sporulation (e.g., Spo0A) and biofilm development, gaining a fuller understanding of these complex developmental processes. Microfluidics has also been used to decipher other *Bacillus* regulation circuits, such as germination (Zabrocka, Langer et al. 2015), the stress response controlled by sigma B (Cabeen, Russell et al. 2017), and more recently the *Bacillus subtilis* cell cycle (Lee, Wu et al. 2019). In the continuing years, we see this technology as the gold standard in studies of *Bacillus* developmental biology. Together with deep DNA sequencing, we see microfluidics and cell sorting also being applied to more commercial applications of *Bacillus*; to allow the isolation of commercially-important strains that would otherwise be hidden within a larger population or at least minimize the emergence of nonproducers (Rugbjerg and Sommer 2019). For example, Walser et al. were able to develop a high-throughput colony screening method using nanoliter-reactors. By use of alginate beads, this technique allowed single cells in 35 nl reaction compartments to be assessed for biological process properties (Walser, Pellaux et al. 2009). Use of cell sorters to isolate individual cells with favorable properties could potentially establish strains producing important metabolites. This "molecular" assessment can also be visualized by live single-cell imaging techniques allowing for the study of an entire lineage or history starting from a single common ancestor (de Jong, Beilharz et al. 2011). Overlaying this technology with mixomics (e.g., integrating ^{13}C metabolic flux analysis, metabolomics, and transcriptomics as described by Hu, Lei et al. 2017) should result in a powerful technology platform to improve protein and small molecule production in *Bacillus* species. In summary, further refinement of these technologies should foster new and exciting insights into the biology of *Bacillus* so that both our understanding of this intriguing endospore-forming taxon and our ability to harness its biotechnological potential will continue to grow.

Acknowledgements

We thank Richard Losick for critical reading of the manuscript.

REFERENCES

Abecasis, A. B., M. Serrano, R. Alves, L. Quintais, J. B. Pereira-Leal and A. O. Henriques (2013). A genomic signature and the identification of new sporulation genes. *J Bacteriol* **195**(9): 2101–2115.

Ahmed, I., A. Yokota, A. Yamazoe and T. Fujiwara (2007). Proposal of *Lysinibacillus boronitolerans* gen. nov. sp. nov., and transfer of *Bacillus fusiformis* to *Lysinibacillus fusiformis* comb. nov. and *Bacillus sphaericus* to *Lysinibacillus sphaericus* comb. nov. *Int J Syst Evol Microbiol* **57**(Pt 5): 1117–1125.

Ahn, S., S. Jun, H. J. Ro, J. H. Kim and S. Kim (2018). Complete genome of *Bacillus subtilis* subsp. *subtilis* KCTC 3135(T) and variation in cell wall genes of *B. subtilis* strains. *J Microbiol Biotechnol* **28**(10): 1760–1768.

Ajithkumar, V. P., B. Ajithkumar, K. Mori, K. Takamizawa, R. Iriye and S. Tabata (2001). A novel filamentous *Bacillus* sp., strain NAF001, forming endospores and budding cells. *Microbiology* **147**: 1415–1423.

Albert, R. A., J. Archambault, M. Lempa, B. Hurst, C. Richardson, S. Gruenloh, M. Duran, H. L. Worliczek, B. E. Huber, R. Rossello-Mora, P. Schumann and H. J. Busse (2007). Proposal of *Viridibacillus* gen. nov. and reclassification of *Bacillus arvi*, *Bacillus arenosi* and *Bacillus neidei* as *Viridibacillus arvi* gen. nov., comb. nov., *Viridibacillus arenosi* comb. nov. and *Viridibacillus neidei* comb. nov. *Int J Syst Evol Microbiol* **57**(Pt 12): 2729–2737.

Albert, R. A., J. Archambault, R. Rosselló-Mora, B. J. Tindall and M. Matheny (2005). *Bacillus acidicola* sp. nov., a novel mesophilic, acidophilic species isolated from acidic *Sphagnum* peat bogs in Wisconsin. *Int J Syst Evol Microbiol* **55**: 2125–2130.

Albert, T. J., D. Dailidiene, G. Dailide, J. E. Norton, A. Kalia, T. A. Richmond, M. Molla, J. Singh, R. D. Green and D. E. Berg (2005). Whole genome scanning for point mutations by hybridization-based "comparative genome resequencing": change associated with high-level metronidazole resistance in *Helicobacter pylori*. *Nat Meth* **2**: 951–953.

Albuquerque, L., I. Tiago, F. A. Rainey, M. Taborda, M. F. Nobre, A. Verissimo and M. S. da Costa (2007). *Salirhabdus euzebyi* gen. nov., sp. nov., a Gram-positive, halotolerant bacterium isolated from a sea salt evaporation pond. *Int J Syst Evol Microbiol* **57**(Pt 7): 1566–1571.

Alexopoulos, C., I. E. Georgoulakis, A. Tzivara, S. K. Kritas, A. Siochu and S. C. Kyriakis (2004). Field evaluation of the efficacy of a probiotic containing *Bacillus licheniformis* and *Bacillus subtilis* spores, on the health status and performance of sows and their litter. *J Anim Physiol Anim Nutr* **88**: 381–392.

Altenbuchner, J. (2016). Editing of the *Bacillus subtilis* genome by the CRISPR-Cas9 system. *Appl Environ Microbiol* **82**(17): 5421–5427.

Amoozegar, M. A., M. Bagheri, M. Didari, S. A. Shahzedeh Fazeli, P. Schumann, C. Sanchez-Porro and A. Ventosa (2013). *Saliterribacillus persicus* gen. nov., sp. nov., a moderately halophilic bacterium isolated from a hypersaline lake. *Int J Syst Evol Microbiol* **63**(Pt 1): 345–351.

Amuguni, H. and S. Tzipori (2012). *Bacillus subtilis*: a temperature resistant and needle free delivery system of immunogens. *Hum Vaccin Immunother* **8**(7): 979–986.

An, S. Y., M. Asahara, K. Goto, H. Kasai and A. Yokota (2007). *Terribacillus saccharophilus* gen. nov., sp. nov. and *Terribacillus halophilus* sp. nov., spore-forming bacteria isolated from field soil in Japan. *Int J Syst Evol Microbiol* **57**: 51–55.

Anderson, T. D., S. A. Robson, X. W. Jiang, G. R. Malmirchegini, H. P. Fierobe, B. A. Lazazzera and R. T. Clubb (2011). Assembly of minicellulosomes on the surface of *Bacillus subtilis*. *Appl Environ Microbiol* **77**(14): 4849–4858.

Aquino de Muro, M. and F. G. Priest (2000). Construction of chromosomal integrants of *Bacillus sphaericus* 2362 by conjugation with *Escherichia coli*. *Res Microbiol* **151**: 547–555.

Arahal, D. R., M. C. Márquez, B. E. Volcani, K. H. Schleifer and A. Ventosa (1999). *Bacillus marismortui* sp. nov., a new moderately halophilic species from the Dead Sea. *Int J Syst Evol Microbiol* **49**: 521–530.

Ardui, S., A. Ameur, J. R. Vermeesch and M. S. Hestand (2018). Single molecule real-time (SMRT) sequencing comes of age: applications and utilities for medical diagnostics. *Nucleic Acids Res* **46**(5): 2159–2168.

Ash, C., J. A. E. Farrow, S. Wallbanks and M. D. Collins (1991). Phylogenetic heterogeneity of the genus *Bacillus* revealed by comparative analysis of small-subunit-ribosomal RNA sequences. *Lett Appl Microbiol* **13**: 202–206.

Ashiuchi, M., K. Shimanouchi, T. Horiuchi, T. Kamei and H. Misono (2006). Genetically engineered poly-gamma-glutamate producer from *Bacillus subtilis* ISW1214. *Biosci Biotechnol Biochem* **70**: 1794–1797.

Bae, S. S., J. H. Lee and S. J. Kim (2005). *Bacillus alveayuensis* sp. nov., a thermophilic bacterium isolated from deep-sea sediments of the Ayu Trough. *Int J Syst Evol Microbiol* **55**: 1211–1215.

Barbosa, T. M., R. Serra and A. O. Henriques (2004). Gut sporeformers. *Bacterial spore formers: probiotics and emerging application*. E. Ricca, A. O. Henriques and S. M. Cutting. Norfolk, UK, Horizon Bioscience: 183–191.

Bashir, Z., L. Sheng, A. Anil, A. Lali, N. P. Minton and Y. Zhang (2019). Engineering *Geobacillus thermoglucosidasius* for direct utilisation of holocellulose from wheat straw. *Biotechnol Biofuels* **12**: 199.

Bassler, B. A. and R. Losick (2006). Bacterially speaking. *Cell* **125**: 237–246.

Bavykin, S. G., Y. P. Lysov, V. Zakhariev, J. J. Kelly, J. Jackman, D. A. Stahl and A. Cherni (2004). Use of 16S rRNA, 23S rRNA, and *gyrB* gene sequence analysis to determine phylogenetic relationships of *Bacillus cereus* group microorganisms. *J Clin Microbiol* **42**: 3711–3730.

Bazinet, A. L. (2017). Pan-genome and phylogeny of *Bacillus cereus* sensu lato. *BMC Evol Biol* **17**(1): 176.

Beauregard, P. B., Y. Chai, H. Vlamakis, R. Losick and R. Kolter (2013). *Bacillus subtilis* biofilm induction by plant polysaccharides. *Proc Natl Acad Sci U S A* **110**(17): E1621–1630.

Becher, D., K. Buttner, M. Moche, B. Hessling and M. Hecker (2011). From the genome sequence to the protein inventory of *Bacillus subtilis*. *Proteomics* **11**(15): 2971–2980.

Belda, E., A. Sekowska, F. Le Fevre, A. Morgat, D. Mornico, C. Ouzounis, D. Vallenet, C. Medigue and A. Danchin (2013). An updated metabolic view of the *Bacillus subtilis* 168 genome. *Microbiology* **159**(Pt 4): 757–770.

Bennett, S. T., C. Barnes, A. Cox, L. Davies and C. Brown (2005). Toward the $1,000 human genome. *Pharmacogenomics* **6**: 373–382.

Bentley, D. R., S. Balasubramanian, H. P. Swerdlow, G. P. Smith, J. Milton, C. G. Brown, K. P. Hall, D. J. Evers, C. L. Barnes, H. R. Bignell, J. M. Boutell, J. Bryant, et al. (2008). Accurate whole human genome sequencing using reversible terminator chemistry. *Nature* **456**(7218): 53–59.

Bernhardt, J., K. Buttner, C. Scharf and M. Hecker (1999). Dual channel imaging of two-dimensional electropherograms in *Bacillus subtilis*. *Electrophoresis* **20**: 2225–2240.

Bhandari, V., N. Z. Ahmod, H. N. Shah and R. S. Gupta (2013). Molecular signatures for *Bacillus* species: demarcation of the *Bacillus subtilis* and *Bacillus cereus* clades in molecular terms and proposal to limit the placement of new species into the genus *Bacillus*. *Int J Syst Evol Microbiol* **63**(Pt 7): 2712–2726.

Biswas, R., M. Yamaoka, H. Nakayama, T. Kondo, K. Yoshida, V. S. Bisaria and A. Kondo (2012). Enhanced production of 2,3-butanediol by engineered *Bacillus subtilis*. *Appl Microbiol Biotechnol* **94**(3): 651–658.

Blackburn, M. B., P. A. Martin, D. Kuhar, R. R. Farrar, Jr. and D. E. Gundersen-Rindal (2013). Phylogenetic distribution of phenotypic traits in *Bacillus thuringiensis* determined by multilocus sequence analysis. *PLOS One* **8**(6): e66061.

Blombach, B. and B. J. Eikmanns (2011). Current knowledge on iso-butanol production with *Escherichia coli*, *Bacillus subtilis* and *Corynebacterium glutamicum*. *Bioeng Bugs* **2**(6): 346–350.

Blount, K. F. and R. R. Breaker (2007). Riboswitches as antibacterial drug targets. *Nature Biotechnol* **24**: 1558–1564.

Bouassida, M., I. Ghazala, S. Ellouze-Chaabouni and D. Ghribi (2018). Improved biosurfactant production by *Bacillus subtilis* SPB1 mutant obtained by random mutagenesis and its application in enhanced oil recovery in a sand system. *J Microbiol Biotechnol* **28**(1): 95–104.

Branda, S. S., J. E. Gonzalez-Pastor, S. Ben-Yehuda, R. Losick and R. Kolter (2001). Fruiting body formation by *Bacillus subtilis*. *Proc Natl Acad Sci USA* **98**: 11621–11626.

Branda, S. S., J. E. Gonzalez-Pastor, E. Dervyn, S. D. Ehrlich, R. Losick and R. Kolter (2004). Genes involved in formation of structured multicellular communities by *Bacillus subtilis*. *J Bacteriol* **186**: 3970–3979.

Brenner, D. J., J. T. Staley and N. R. Krieg (2001). Classification of procaryotic organisms and the concept of bacterial speciation. *Bergey's manual of systematic bacteriology*, 2nd ed. G. M. Garrity. New York, NY, Springer: 27–31.

Bruckner, R. (1992). A series of shuttle vectors for *Bacillus subtilis* and *Escherichia coli*. *Gene* **122**: 187–192.

Bulthuis, B. A., C. Rommens, G. M. Koningstein, A. H. Stouthamer and H. W. van Verseveld (1991). Formation of fermentation products and extracellular protease during anaerobic growth of *Bacillus licheniformis* in chemostat and batch-culture. *Antonie Van Leeuwenhoek* **62**: 355–371.

Bunyapaiboonsri, T., H. Ramstroem, O. Ramstroem, J. Haiech and J.-M. Lehn (2003). Generation of bis-cationic heterocyclic inhibitors of *Bacillus subtilis* HPr kinase/phosphatase from a ditopic dynamic combinatorial library. *J Med Chem* **46**: 5803–5811.

Burbulys, D., K. A. Trach and J. A. Hoch (1991). Initiation of sporulation in *B. subtilis* is controlled by a multicomponent phosphorelay. *Cell* **64**: 545–552.

Burby, P. E. and L. A. Simmons (2017a). CRISPR/Cas9 editing of the *Bacillus subtilis* genome. *Bio Protoc* **7**(8).

Burby, P. E. and L. A. Simmons (2017b). MutS2 promotes homologous recombination in *Bacillus subtilis*. *J Bacteriol* **199**(2).

Burkholder, P. R. and N. H. Giles (1947). Induced biochemical mutations in *Bacillus subtilis*. *Am J Bot* **34**: 345–348.

Cabeen, M. T. and R. Losick (2018). Single-cell microfluidic analysis of *Bacillus subtilis*. *J Vis Exp* (131):56901.

Cabeen, M. T., J. R. Russell, J. Paulsson and R. Losick (2017). Use of a microfluidic platform to uncover basic features of energy and environmental stress responses in individual cells of *Bacillus subtilis*. *PLOS Genet* **13**(7): e1006901.

Cai, D., Y. Rao, Y. Zhan, Q. Wang and S. Chen (2019). Engineering *Bacillus* for efficient production of heterologous protein: current progress, challenge and prospect. *J Appl Microbiol* **126**(6): 1632–1642.

Caldwell, R., R. Sapolsky, W. Weyler, R. R. Maile, S. C. Causey and E. Ferrari (2001). Correlation between *Bacillus subtilis* ScoC phenotype and gene expression determined using microarrays for transcriptome analysis. *J Bacteriol* **183**: 7329–7340.

Carrasco, I. J., M. C. Marquez, Y. Xue, Y. Ma, D. A. Cowan, B. E. Jones, W. D. Grant and A. Ventosa (2007). *Salsuginibacillus kocurii* gen. nov., sp. nov., a moderately halophilic bacterium from soda-lake sediment. *Int J Syst Evol Microbiol* **57**(Pt 10): 2381–2386.

Carrasco, I. J., M. C. Marquez, Y. Xue, Y. Ma, D. A. Cowan, B. E. Jones, W. D. Grant and A. Ventosa (2008). *Sediminibacillus halophilus* gen. nov., sp. nov., a moderately halophilic, Gram-positive bacterium from a hypersaline lake. *Int J Syst Evol Microbiol* **58**(Pt 8): 1961–1967.

Cartman, S. T. and R. M. La Ragione (2004). Spore probiotics as animal feed supplements. *Bacterial spore formers: probiotics and emerging application*. E. Ricca, H. O. Henriques and S. M. Cutting. Norfolk, UK, Horizon Bioscience: 155–161.

Cavaglieri, L., J. Orlando, M. I. Rodríguez, S. Chulze and M. Etcheverry (2005). Biocontrol of *Bacillus subtilis* against *Fusarium verticillioides in vitro* and at the maize root level. *Res Microbiol* **156**: 748–754.

Chandler, D. P., O. Alferov, B. Chernov, D. S. Daly, J. Golova, A. Perov, M. Protic, R. Robison, M. Schipma, A. White and A. Willse (2006). Diagnostic oligonucleotide microarray fingerprinting of *Bacillus* isolates. *J Clin Microbiol* **44**: 244–250.

Chang, Y.-H., Y.-H. Shangkuan, H.-C. Lin and H.-W. Liu (2003). PCR assay of the *groEL* gene for detection and differentiation of *Bacillus cereus* group cells. *Appl Environ Microbiol* **69**: 4502–4510.

Chen, C. L., S. C. Wu, W. M. Tjia, C. L. Wang, M. J. Lohka and S. L. Wong (2008). Development of a LytE-based high-density surface display system in *Bacillus subtilis*. *Microb Biotechnol* **1**(2): 177–190.

Chen, Y., F. Yan, Y. Chai, H. Liu, R. Kolter, R. Losick and J. H. Guo (2013). Biocontrol of tomato wilt disease by *Bacillus subtilis* isolates from natural environments depends on conserved genes mediating biofilm formation. *Environ Microbiol* **15**(3): 848–864.

Choi, Y.-J., T.-T. Wang and B. H. Lee (2002). Positive selection vectors. *Crit. Rev. Biotechnol* **22**: 225–244.

Christou, P., T. Capell, A. Kohli, J. A. Gatehouse and A. M. Gatehouse (2006). Recent developments and future prospects in insect pest control in transgenic crops. *Trends Plant Sci* **11**: 302–308.

Chumsakul, O., D. P. Anantsri, T. Quirke, T. Oshima, K. Nakamura, S. Ishikawa and M. M. Nakano (2017). Genome-wide analysis of ResD, NsrR, and Fur binding in *Bacillus subtilis* during anaerobic fermentative growth by in vivo footprinting. *J Bacteriol* **199**(13).

Chun, J., A. Oren, A. Ventosa, H. Christensen, D. R. Arahal, M. S. da Costa, A. P. Rooney, H. Yi, X. W. Xu, S. De Meyer and M. E. Trujillo (2018). Proposed minimal standards for the use of genome data for the taxonomy of prokaryotes. *Int J Syst Evol Microbiol* **68**(1): 461–466.

Clements, L. D., B. S. Miller and U. N. Streips (2002). Comparative growth analysis of the facultative anaerobes *Bacillus subtilis*, *Bacillus licheniformis*, and *Escherichia coli*. *Syst Appl Microbiol* **25**: 284–286.

Coelho, M. R. R., I. von der Weid, V. Zahner and L. Seldin (2003). Characterization of nitrogen-fixing *Paenibacillus* species by polymerase chain reaction-restriction fragment length polymorphism analysis of part of genes encoding 16S rRNA and 23S rRNA and by multilocus enzyme electrophoresis. *FEMS Microbiol Letts* **222**: 243–250.

Cohn, F. (1872). Untersuchen über Bacterien. *Beitr z Biol Der Pflanz* **1**(Heft 2): 127–224.

Collins, M. D., P. A. Lawson, A. Willems, J. J. Cordoba, J. Fernandez-Garayzabal, P. Garcia, J. Cai, H. Hippe and J. A. E. Farrow (1994). The phylogeny of the genus *Clostridium*: proposal of five new genera and eleven new species combinations. *Int J Syst Bacteriol* **44**: 812–826.

Collins, M. D., B. M. Lund, J. A. E. Farrow and K. H. Schleifer (1983). Chemotaxonomic study of an alkalophilic bacterium, *Exiguobacterium aurantiacum* gen. nov., sp. nov. *J Gen Microbiol* **129**: 91–92.

Contesini, F. J., R. R. de Melo, and H. H. Sato (2018). An overview of *Bacillus* proteases: from production to application. *Critical Rev Biotechnol* **38**(3): 321–334.

Coorevits, A., A. E. Dinsdale, G. Halket, L. Lebbe, P. De Vos, A. Van Landschoot and N. A. Logan (2012). Taxonomic revision of the genus *Geobacillus*: emendation of *Geobacillus*, *G. stearothermophilus*, *G. jurassicus*, *G. toebii*, *G. thermodenitrificans* and *G. thermoglucosidans* (nom. corrig., formerly '*thermoglucosidasius*'); transfer of *Bacillus thermantarcticus* to the genus as *G. thermantarcticus* comb. nov.; proposal of *Caldibacillus debilis* gen. nov., comb. nov.; transfer of *G. tepidamans* to *Anoxybacillus* as *A. tepidamans* comb. nov.; and proposal of *Anoxybacillus caldiproteolyticus* sp. nov. *Int J Syst Evol Microbiol* **62**(Pt 7): 1470–1485.

Cormella, N. and A. D. Grossman (2005). Conservation of genes and processes controlled by the quorum response in bacteria: characterization of genes controlled by the quorum-sensing transcription factor ComA in *Bacillus subtilis*. *Mol Microbiol* **57**: 1159–1174.

Cui, W., L. Han, F. Suo, Z. Liu, L. Zhou and Z. Zhou (2018). Exploitation of *Bacillus subtilis* as a robust workhorse for production of heterologous proteins and beyond. *World J Microbiol Biotechnol* **34**(10): 145.

Cutting, S. M. (2011). *Bacillus* probiotics. *Food Microbiol* **28**(2): 214–220.

Dale, G. E., H. Langen, M. G. Page, R. L. Then and D. Stuber (1995). Cloning and characterization of a novel, plasmid-encoded trimethoprim- resistant dihydrofolate reductase from *Staphylococcus haemolyticus* MUR313. *Antimicrob Agents Chemother* **39**: 1920–1924.

Dasika, M. S., A. Gupta and C. D. Maranas (2004). A mixed integer linear programming (MLP) MILP framework for inferring time delay in gene regulatory networks. *Pac Symp Biocomput*, pp. 474–485.

de Jong, I. G., K. Beilharz, O. P. Kuipers and J. W. Veening (2011). Live cell imaging of *Bacillus subtilis* and *Streptococcus pneumoniae* using automated time-lapse microscopy. *J Vis Exp* (53).

Dettwiler, B. I., E. Heinzle and J. E. Prenosil (1993). A simulation model for the continuous production of acetoin and butanediol using *Bacillus subtilis* with integrated pervaporation separation. *Biotechnol Bioeng* **41**: 791–800.

Dhakal, R., K. Chauhan, R. B. Seale, H. C. Deeth, C. J. Pillidge, I. B. Powell, H. Craven and M. S. Turner (2013). Genotyping of dairy *Bacillus licheniformis* isolates by high resolution melt analysis of multiple variable number tandem repeat loci. *Food Microbiol* **34**(2): 344–351.

Dickinson, D. N., M. T. La Duc, M. Satomi, J. D. Winefordner, D. H. Powell and K. Venkateswaran (2004). MALDI-TOFMS compared with other polyphasic taxonomy approaches for the identification and classification of *Bacillus pumilus* spores. *J Microbiol Meth* **58**: 1–12.

Didari, M., M. A. Amoozegar, M. Bagheri, P. Schumann, C. Sproer, C. Sanchez-Porro and A. Ventosa (2012). *Alteribacillus bidgolensis* gen. nov., sp. nov., a moderately halophilic bacterium from a hypersaline lake, and reclassification of *Bacillus persepolensis* as *Alteribacillus persepolensis* comb. nov. *Int J Syst Evol Microbiol* **62**(Pt 11): 2691–2697.

Didelot, X., M. Barker, D. Falush and F. G. Priest (2009). Evolution of pathogenicity in the *Bacillus cereus* group. *Syst Appl Microbiol* **32**(2): 81–90.

Ding, Z., F. Guan, X. Yu, Q. Li, Q. Wang, J. Tian and N. Wu (2019). Identification of the anchoring protein SpoIIIJ for construction of the microbial cell surface display system in *Bacillus* spp. *Int J Biol Macromol* **133**: 614–623.

Dong, H. and D. Zhang (2014). Current development in genetic engineering strategies of *Bacillus* species. *Microb Cell Fact* **13**: 63.

Driks, A. (2004). The *Bacillus* spore coat. *Phytopathology* **94**: 1249–1251.

Dubnau, D. (1991). Genetic competence in *Bacillus subtilis*. *Microbiol Rev* **55**: 395–424.

Dubnau, D. and R. Losick (2006). Bistability in bacteria. *Mol Microbiol* **61**: 564–572.

Duc le, H., H. A. Hong, H. S. Atkins, H. C. Flick-Smith, Z. Durrani, S. Rijpkema, R. W. Titball and S. M. Cutting (2007). Immunization against anthrax using *Bacillus subtilis* spores expressing the anthrax protective antigen. *Vaccine* **25**(2): 346–355.

Durmaz, R., M. Doganay, M. Sahin, D. Percin, M. K. Karahocagil, U. Kayabas, B. Otlu, A. Karagoz, F. Buyuk, O. Celebi, Z. Ozturk and M. Ertek (2012). Molecular epidemiology of the *Bacillus anthracis* isolates collected throughout Turkey from 1983 to 2011. *Eur J Clin Microbiol Infect Dis* **31**(10): 2783–2790.

Earl, A. M., R. Losick and R. Kolter (2007). *Bacillus subtilis* genome diversity. *J Bacteriol* **189**: 1163–1170.

Echigo, A., T. Fukushima, T. Mizuki, M. Kamekura and R. Usami (2007). *Halalkalibacillus halophilus* gen. nov., sp. nov., a novel moderately halophilic and alkaliphilic bacterium isolated from a non-saline soil sample in Japan. *Int J Syst Evol Microbiol* **57**(Pt 5): 1081–1085.

Echigo, A., H. Minegishi, Y. Shimane, M. Kamekura and R. Usami (2012). *Natribacillus halophilus* gen. nov., sp. nov., a moderately halophilic and alkalitolerant bacterium isolated from soil. *Int J Syst Evol Microbiol* **62**(Pt 2): 289–294.

Eggersdorfer, M., D. Laudert, U. Letinois, T. McClymont, J. Medlock, T. Netscher and W. Bonrath (2012). One hundred years of vitamins-a success story of the natural sciences. *Angew Chem Int Ed Engl* **51**(52): 12960–12990.

Eid, J., A. Fehr, J. Gray, K. Luong, J. Lyle, G. Otto, P. Peluso, D. Rank, P. Baybayan, B. Bettman, A. Bibillo, K. Bjornson, B. Chaudhuri, F. Christians, R. Cicero, S. Clark, et al (2009). Real-time DNA sequencing from single polymerase molecules. *Science* **323**(5910): 133–138.

Eijlander, R. T., A. de Jong, A. O. Krawczyk, S. Holsappel and O. P. Kuipers (2014). SporeWeb: an interactive journey through the complete sporulation cycle of *Bacillus subtilis. Nucleic Acids Res* **42**(Database issue): D685–691.

el-Bendary, M. A. (2006). *Bacillus thuringiensis* and *Bacillus sphaericus* biopesticides production. *J Basic Microbiol* **46**: 158–170.

Ellermeier, C. D., E. C. Hobbs, J. E. Gonzalez-Pastor and R. Losick (2006). A three-protein signaling pathway governing immunity to a bacterial cannibalism toxin. *Cell* **124**: 549–559.

Elshaghabee, F. M. F., N. Rokana, R. D. Gulhane, C. Sharma and H. Panwar (2017). *Bacillus* as potential probiotics: Status, concerns, and future perspectives. *Front Microbiol* **8**: 1490.

Errington, J. (1988). Generalized cloning vectors for *Bacillus subtilis. Biotechnology* **10**: 345–362.

Euzéby, J. P. (1997). List of bacterial names with standing in nomenclature: a folder available on the Internet. *Int J Syst Bacteriol* **47**: 590–592.

Fan, B., J. Blom, H. P. Klenk and R. Borriss (2017). *Bacillus amyloliquefaciens*, *Bacillus velezensis*, and *Bacillus siamensis* form an "operational Group *B. amyloliquefaciens*" within the *B. subtilis* species complex. *Front Microbiol* **8**: 22.

Farfour, E., J. Leto, M. Barritault, C. Barberis, J. Meyer, B. Dauphin, A. S. Le Guern, A. Lefleche, E. Badell, N. Guiso, A. Leclercq, A. Le Monnier, M. Lecuit, V. Rodriguez-Nava, E. Bergeron, J. Raymond, S. Vimont, E. Bille, E. Carbonnelle, H. Guet-Revillet, H. Lecuyer, J. L. Beretti, C. Vay, P. Berche, A. Ferroni, X. Nassif and O. Join-Lambert (2012). Evaluation of the Andromas matrix-assisted laser desorption ionization-time of flight mass spectrometry system for identification of aerobically growing Gram-positive bacilli. *J Clin Microbiol* **50**(8): 2702–2707.

Farias, B. C. S., D. C. Hissa, C. T. M. do Nascimento, S. A. Oliveira, D. Zampieri, M. N. Eberlin, D. L. S. Migueleti, L. F. Martins, M. P. Sousa, D. N. Moyses and V. M. M. Melo (2018). Cyclic lipopeptide signature as fingerprinting for the screening of halotolerant *Bacillus* strains towards microbial enhanced oil recovery. *Appl Microbiol Biotechnol* **102**(3): 1179–1190.

Farzanfar, A. (2006). The use of probiotic in shrimp aquaculture. *FEMS Immunol Med Microbiol* **48**: 149–158.

Felsenstein, J. (1989). PHYLIP—Phylogeny inference package (version 3.2). *Cladistics* **5**: 164–166.

Fernández-No, I. C., K. Böhme, M. Díaz-Bao, A. Cepeda, J. Barros-Velázquez and P. Calo-Mata (2013). Characterisation and profiling of *Bacillus subtilis*, *Bacillus cereus* and *Bacillus licheniformis* by MALDI-TOF mass fingerprinting. *Food Microbiol* **33**(2): 235–242.

Ferreira, L. C. S., R. C. C. Ferreira and W. Schumann (2005). *Bacillus subtilis* as a tool for vaccine development: from antigen factories to delivery vectors. *An Acad Bras Ciênc* **77**: 113–124.

Fischer, E. and U. Sauer (2005). Large-scale in vivo flux analysis shows rigidity and suboptimal performance of *Bacillus subtilis* metabolism. *Nat Genet* **37**: 636–640.

Fisher, S. H. (1999). Regulation of nitrogen metabolism in *Bacillus subtilis*: vive la difference. *Mol Microbiol* **32**: 223–232.

Fortina, M. G., R. Pukall, P. Schumann, D. Mora, C. Parini, P. L. Manachini and E. Stackebrandt (2001). *Ureibacillus* gen. nov., a new genus to accommodate *Bacillus thermosphaericus* (Andersson *et al.* 1995), emendation of *Ureibacillus thermosphaericus* and description of *Ureibacillus terrenus* sp. nov. *Int J Syst Evol Microbiol* **51**: 447–455.

Fox, G. E., K. R. Pechman and C. R. Woese (1977). Comparative cataloging of 16S ribosomal ribonucleic acid: molecular approach to procaryotic systematics. *Int J System Bacteriol* **27**: 44–57.

Fox, G. E., J. D. Wisotzkey and P. Gurtshuk Jr. (1992). How close is close: 16S rRNA sequence identity may not be sufficient to guarantee species identity. *Int J Syst Bacteriol* **41**: 166–170.

Freedman, A. J. E., K. C. Peet, J. T. Boock, K. Penn, K. L. J. Prather and J. R. Thompson (2018). Isolation, development, and genomic analysis of *Bacillus megaterium* SR7 for growth and metabolite production under supercritical carbon dioxide. *Front Microbiol* **9**: 2152. https://pubmed.ncbi.nlm.nih.gov/30319556/.

Fritze, D. (2004). Taxonomy and systematics of the aerobic endospore forming bacteria: *Bacillus* and related genera. *Bacterial spore formers: Probiotics and emerging application.* E. Ricca, A. O. Henriques and S. M. Cutting. Norfolk, UK, Horizon Bioscience: 17–34.

Galperin, M. Y., S. L. Mekhedov, P. Puigbo, S. Smirnov, Y. I. Wolf and D. J. Rigden (2012). Genomic determinants of sporulation in Bacilli and Clostridia: towards the minimal set of sporulation-specific genes. *Environ Microbiol* **14**(11): 2870–2890.

Gange, A. C. and K. J. Hagley (2004). The potential for use of *Bacillus* spp. in sports turf management. *Bacterial spore formers: Probiotics and emerging application* E. Ricca, A. O. Henriques and S. M. Cutting. Norfolk, UK, Horizon Bioscience: 163–170.

Garcia-Vallve, S., E. Guzman, M. A. Montero and A. Romeu (2003). Horizontal gene transfer database (HGT-DB): a database of putative horizontally transferred genes in prokaryotic complete genomes. *Nucleic Acids Res* **31**(1):187–189. https://www.ncbi.nlm.nih.gov/pmc/articles/PMC165451/.

García, M. T., V. Gallego, A. Ventosa and E. Mellado (2005). *Thalassobacillus devorans* gen. nov., sp. nov., a moderately halophilic, phenol-degrading, Gram-positive bacterium. *Int J Syst Evol Microbiol* **55**: 1789–1795.

Gatson, J. W., B. F. Benz, C. Chandrasekaran, M. Satomi, K. Venkateswaran and M. E. Hart (2006). *Bacillus tequilensis* sp. nov., isolated from a 2000-year-old Mexican shaft-tomb, is closely related to *Bacillus subtilis. Int J Syst Evol Microbiol* **56**: 1475–1484.

Gaviria Rivera, A. M. and F. G. Priest (2003a). Pulsed field gel electrophoresis of chromosomal DNA reveals a clonal population structure to *Bacillus thuringiensis* that relates in general to crystal protein gene content. *FEMS Microbiol Lett* **223**: 61–66.

Gaviria Rivera, A. M. and F. G. Priest (2003b). Molecular Typing of *Bacillus thuringiensis* Serovars by RAPD-PCR. *Appl Environ Microbiol* **26**: 254–261.

Ge, Y., X. Hu, D. Zheng, Y. Wu and Z. Yuan (2011). Allelic diversity and population structure of *Bacillus sphaericus* as revealed by multilocus sequence typing. *Appl Environ Microbiol* **77**(15): 5553–5556.

Gerdes, S. Y., M. D. Scholle, M. D'Souza, A. Bernal, M. V. Baev, M. Farrell, O. V. Kurnasov, M. D. Daugherty, F. Mseeh, B. M. Polanuyer, J. W. Campbell, S. Anantha, K. Y. Shatalin, S. A. K. Chowdhury, M. Y. Fonstein and A. L. Osterman (2002). From genetic footprinting to antimicrobial drug targets: examples in cofactor biosynthetic pathways. *J Bacteriol* **184**: 4555–4572.

Gharizadeh, B., M. Akhras, N. Nourizad, M. Ghaderi, K. Yasuda, P. Nyren and N. Pourmand (2006). Methodological improvements of pyrosequencing technology. *J Biotechnol* **124**(3): 504–511.

Ghribj, D., N. Zouari and S. Jaoua (2004). Improvement of bioinsecticides production through mutagenesis of *Bacillus thuringiensis* by UV and nitrous acid affecting metabolic pathways and/or delta-endotoxin synthesis. *J Appl Microbiol* **97**: 338–346.

Gibson, T. (1944). A study of *Bacillus subtilis* and related organisms. *J Dairy Res* **13**: 248–260.

Goldstein, M. A. and R. H. Doi (1995). Prokaryotic promoters in biotechnology. *Biotechnol Annu Rev* **1**: 105–128.

Golovacheva, R. S. and G. I. Karavaiko (1978). *Sulfobacillus*, a new genus of thermophilic sporeforming bacteria. *Mikrobiologiya* **47**: 815–822.

Gonzalez-Pastor, J. E., E. C. Hobbs and R. Losick (2003). Cannibalism by sporulating bacteria. *Science* **301**: 510–513.

Gordon, R. E. (1989). The genus *Bacillus*. *Practical Handbook of Microbiology*. W. M. O'Leary. Orlando, CRC Press: 109–126.

Gordon, R. E., W. C. Haynes and C. H.-N. Pang (1973). *The Genus Bacillus*. Washington, DC, Agricultural Research Service, United States Department of Agriculture.

Gordon, R. E. and N. R. Smith (1949). Aerobic sporeforming bacteria capable of growth at high temperatures. *J Bacteriol* **58**: 327–341.

Gößner, A. S., R. Devereux, N. Ohnemüller, G. Acker, E. Stackebrandt and H. L. Drake (1999). *Thermicanus aegyptius* gen. nov., sp. nov., isolated from oxic soil, a fermentative microaerophile that grows commensally with the thermophilic acetogen *Moorella thermoacetica*. *Appl Environ Microbiol* **65**: 5124–5133.

Goto, K., K. Mochida, M. Asahara, M. Suzuki, H. Kasai and A. Yokota (2003). *Alicyclobacillus pomorum* sp. nov., a novel thermo-acidophilic, endospore-forming bacterium that does not possess omega-alicyclic fatty acids, and emended description of the genus *Alicyclobacillus*. *Int J Syst Evol Microbiol* **53**: 1537–1544.

Goto, K., T. Omura, Y. Hara and Y. Sadaie (2000). Application of the partial 16S rDNA sequence as a index for rapid identification of species in the genus *Bacillus*. *J Gen Appl Microbiol* **46**: 1–8.

Gu, Y., X. Xu, Y. Wu, T. Niu, Y. Liu, J. Li, G. Du and L. Liu (2018). Advances and prospects of *Bacillus subtilis* cellular factories: From rational design to industrial applications. *Metab Eng* **50**: 109–121.

Gudiña, E. J., J. F. Pereira, R. Costa, J. A. Coutinho, J. A. Teixeira and L. R. Rodrigues (2013). Biosurfactant-producing and oil-degrading *Bacillus subtilis* strains enhance oil recovery in laboratory sand-pack columns. *J Hazard Mater* **261**: 106–113.

Guérout-Fleury, A. M., K. Shazand, N. Frandsen and P. Stragier (1995). Antibiotic-resistance cassettes for *Bacillus subtilis*. *Gene* **167**: 335–336.

Guinebretiere, M. H., F. L. Thompson, A. Sorokin, P. Normand, P. Dawyndt, M. Ehling-Schulz, B. Svensson, V. Sanchis, C. Nguyen-The, M. Heyndrickx and P. De Vos (2008). Ecological diversification in the *Bacillus cereus* Group. *Environ Microbiol* **10**(4): 851–865.

Guoyan, Z., A. Yingfeng, H. Zabed, G. Qi, M. Yang, Y. Jiao, W. Li, S. Wenjing, Q. Xianghui (2019). *Bacillus subtilis* spore surface display technology: A review of its development and applications. *J Microbiol Biotechnol* **29**(2): 179–190.

Gürtler, V. and B. C. Mayall (2001). Genomic approaches to typing, taxonomy and evolution of bacterial isolates. *Int J Syst Evol Microbiol* **51**: 3–16.

Haggag, W. M. and S. Timmusk (2008). Colonization of peanut roots by biofilm-forming *Paenibacillus polymyxa* initiates biocontrol against crown rot disease. *J Appl Microbiol* **104**(4): 961–969.

Haldenwang, W. G. (1995). The sigma factors of *Bacillus subtilis*. *Microbiol Rev* **59**: 1–30.

Hamasalim, H. J. (2016). Synbiotic as feed additives relating to animal health and performance. *Adv Microbiology* **6**(04): 288.

Hambraeus, G., C. von Wachenfeldt and L. Hederstedt (2003). Genome-wide survey of mRNA half-lives in *Bacillus subtilis* identifies extremely stable mRNAs. *Mol. Genet Genomics* **269**: 706–714.

Hamoen, L. W., G. Venema and O. P. Kuipers (2003). Controlling competence in *Bacillus subtilis*: shared use of regulators. *Microbiology* **149**: 9–17.

Hartig, E. and D. Jahn (2012). Regulation of the anaerobic metabolism in *Bacillus subtilis*. *Adv Microb Physiol* **61**: 195–216.

Harwood, C. and S. M. Cutting, Eds. (1990). *Molecular biological methods for Bacillus*. Chichester, United Kingdom, John Wiley & Sons, Ltd.

Hatayama, K., H. Shoun, Y. Ueda and A. Nakamura (2006). *Tuberibacillus calidus* gen. nov., sp. nov., isolated from a compost pile and reclassification of *Bacillus naganoensis* Tomimura *et al.* 1990 as *Pullulanibacillus naganoensis* gen. nov., comb. nov. and *Bacillus laevolacticus* Andersch *et al.* 1994 as *Sporolactobacillus laevolacticus* comb. nov. *Int J Syst Evol Microbiol* **56**: 2545–2551.

Hecker, M. and U. Volker (2004). Towards a comprehensive understanding of *Bacillus subtilis* cell physiology by physiological proteomics. *Proteomics* **4**: 3727–3750.

Helgason, E., O. A. Økstad, D. A. Caugant, H. A. Johansen, D. A. Caugant, H. A. Johansen, A. Fouet, M. Mock, I. Hegna and A.-B. Kolstø (2000). *Bacillus anthracis*, *Bacillus cereus*, and *Bacillus thuringiensis*—one species on the basis of genetic evidence. *Appl Environ Microbiol* **66**: 2627–2630.

Helgason, E., N. Tourasse, R. Meisal, D. Caugant and A.-B. Kolstø (2004). Multilocus sequence typing scheme for bacteria of the *Bacillus cereus* group. *Appl Environ Microbiol* **70**: 191–201.

Helman, J. D. and C. P. Moran (2001). RNA polymerase and sigma factors. *Bacillus subtilis and its closest relatives: From genes to cells*. A. L. Sonenshein, J. A. Hoch and R. Losick. Washington, D.C., ASM Press: 289–312.

Helmann, J. D. (2002). The extracytoplasmic function (ECF) sigma factors. *Adv Microb Physio* **46**: 47–110.

Henkin, T. M. (1994). tRNA-directed transcription antitermination. *Mol Microbiol* **131**: 381–387.

Henkin, T. M. and C. Yanofsky (2002). Regulation by transcription attenuation in bacteria: how RNA provides instructions for transcription termination/antitermination decisions. *Bioessays* **24**: 700–707.

Heyndrickx, M., L. Lebbe, K. Kersters, P. De Vos, G. Forsyth and N. A. Logan (1998). *Virgibacillus*: a new genus to accommodate *Bacillus pantothenticus* (Proom and Knight 1950). Emended description of *Virgibacillus pantothenticus*. *Int J Syst Bacteriol* **48**: 99–106.

Heyrman, J., A. Balcaen, M. Rodriguez-Diaz, N. A. Logan, J. Swings and P. De Vos (2003). *Bacillus decolorationis* sp. nov., isolated from biodeteriorated parts of the mural paintings at the Servilia tomb, Roman necropolis of Carmona, Spain. *Int J Syst Evol Microbiol* **53**: 459–463.

Higgins, D. and J. Dworkin (2012). Recent progress in *Bacillus subtilis* sporulation. *FEMS Microbiol Rev* **36**(1): 131–148.

Higgins, D., C. Pal, I. M. Sulaiman, C. Jia, T. Zerwekh, S. E. Dowd and P. Banerjee. (2018). Application of high-throughput pyrosequencing in the analysis of microbiota of food commodities procured from small and large retail outlets in a U.S. metropolitan area - a pilot study. *Food Res Int* **105**: 29–40.

Higgins, D., J. Thompson, T. Gibson, J. D. Thompson, D. G. Higgins and T. J. Gibson (1994). CLUSTAL W: improving the sensitivity of progressive multiple sequence alignment through sequence weighting, position-specific gap penalties and weight matrix choice. *Nucl Acids Res* **22**: 4673–4680.

Hinc, K., R. Isticato, M. Dembek, J. Karczewska, A. Iwanicki, G. Peszynska-Sularz, M. De Felice, M. Obuchowski and E. Ricca (2010). Expression and display of UreA of *Helicobacter acinonychis* on the surface of *Bacillus subtilis* spores. *Microb Cell Fact* **9**: 2.

Hoch, J. A. (1971). Genetic analysis of pleiotropic negative sporulation mutants in *Bacillus subtilis*. *J Bacteriol* **105**: 896–901.

Hoch, J. A. (1993). Regulation of the phosphorelay and the initiation of sporulation in *Bacillus subtilis* " *Annu Rev Microbiol* **47**: 441–465.

Hohmann, H.-P. and K.-P. Stahmann (2010). Biotechnology of Riboflavin Production. *Comprehensive Natural Products II: Chemistry and Biology*. L. N. Mander and H. W. Liu. Oxford, Elsiever. **7**: 115–139.

Hohmann, H. P., J. M. van Dijl, L. Krishnappa and Z. Prágai (2017). Host organisms: *Bacillus subtilis*. *Industrial Biotechnology: Microorganisms*. C. Wittmann and J. C. Liao. Weinheim, Germany, Wiley-VCH Verlag GmbH & Co. KGaA. **1**: 221–297.

Hong, H. A., L. H. Duc and S. M. Cutting (2005). The use of bacterial spore formers as probiotics. *FEMS Microbiol Rev.* **29**: 813–835.

Horinouchi, S. and B. Weisblum (1982). Nucleotide sequence and functional map of pC194, a plasmid that specifies inducible chloramphenicol resistance. *J Bacteriol* **150**: 815–825.

Hosoi, T. and K. Kiuchi (2004). Production and probiotic effects of natto. *Bacterial Spore Formers: Probiotics and Emerging Applications*. E. Ricca, A. O. Henriques and S. M. Cutting. Norfolk, UK, Horizon Bioscience: 143–154.

Hu, J., P. Lei, A. Mohsin, X. Liu, M. Huang, L. Li, J. Hu, H. Hang, Y. Zhuang and M. Guo (2017). Mixomics analysis of *Bacillus subtilis*: effect of oxygen availability on riboflavin production. *Microb Cell Fact* **16**(1): 150.

Huang, G. L., J. E. Gosschalk, Y. S. Kim, R. R. Ogorzalek Loo and R. T. Clubb (2018). Stabilizing displayed proteins on vegetative *Bacillus subtilis* cells. *Appl Microbiol Biotechnol* **102**(15): 6547–6565.

Hughes, K. T. and K. Mathee (1998). The anti-sigma factors. *Annu Rev Microbiol* **52**: 231–286.

Hulett, F. M. (1996). The signal-transduction network for Pho regulation in *Bacillus subtilis*. *Mol Microbiol* **19**: 933–939.

Hwang, B. Y., J. G. Pan, B. G. Kim and J. H. Kim (2013). Functional display of active tetrameric beta-galactosidase using *Bacillus subtilis* spore display system. *J Nanosci Nanotechnol* **13**(3): 2313–2319.

Ishikawa, M., S. Ishizaki, Y. Yamamoto and K. Yamasato (2002). *Paraliobacillus ryukyuensis* gen. nov., sp. nov., a new Gram-positive, slightly halophilic, extremely halotolerant, facultative anaerobe isolated from a decomposing marine alga. *J Gen Appl Microbiol* **48**: 269–279.

Ishikawa, M., K. Nakajima, Y. Itamiya, S. Furukawa, Y. Yamamoto and K. Yamasato (2006). *Halolactibacillus halophilus* gen. nov., sp. nov. and *Halolactibacillus miurensis* sp. nov., halophilic and alkaliphilic marine lactic acid bacteria constituting a phylogenetic lineage in *Bacillus* rRNA group 1. *Int J Syst Evol Microbiol* **55**: 2427–2439.

Isticato, R., G. Cangiano, H. T. Tran, A. Ciabattini, D. Medaglini, M. R. Oggioni, M. De Felice, G. Pozzi and E. Ricca (2001). Surface display of recombinant proteins on *Bacillus subtilis* spores. *J Bacteriol* **183**(21): 6294–6301.

Itaya, M. and S. Kaneko (2010). Integration of stable extracellular DNA released from *Escherichia coli* into the *Bacillus subtilis* genome vector by culture mix method. *Nucleic Acids Res* **38**(8): 2551–2557.

Itaya, M., K. Kondo and T. Tanaka (1989). A neomycin resistance gene cassette selectable in a single copy state in the *Bacillus subtilis* chromosome. *Nucl Acids Res* **17**: 4410.

Itaya, M., M. Sato, M. Hasegawa, N. Kono, M. Tomita and S. Kaneko (2018). Far rapid synthesis of giant DNA in the *Bacillus subtilis* genome by a conjugation transfer system. *Sci Rep* **8**(1): 8792.

Itaya, M., I. Yamaguchi, K. Kobayashi, T. Endo and T. Tanaka (1990). The Blasticidin S resistance gene (*bsr*) selectable in a single copy state in the *Bacillus subtilis* chromosome. *J Biochem* **107**: 799–801.

Janniere, L., C. Bruand and S. D. Ehrlich (1990). Structurally stable *Bacillus subtilis* cloning vectors. *Gene* **87**: 53–61.

Jensen, G. B., L. Andrup, A. Wilcks, L. Smidt and O. M. Poulsen (1996). The aggregation-mediated conjugation system of *Bacillus thuringiensis* subsp. *israelensis*: host range and kinetics of transfer. *Curr Microbiol* **33**: 228–236.

Jeon, C. O., J. M. Lim, J. M. Lee, L. H. Xu, C. L. Jiang and C. J. Kim (2005). Reclassification of *Bacillus haloalkaliphilus* Fritze 1996 as *Alkalibacillus haloalkaliphilus* gen. nov., comb. nov. and the description of *Alkalibacillus salilacus* sp. nov., a novel halophilic bacterium isolated from a salt lake in China. *Int J Syst Evol Microbiol* **55**: 1891–1896.

Jeong, D. E., S. H. Park, J. G. Pan, E. J. Kim and S. K. Choi (2015). Genome engineering using a synthetic gene circuit in *Bacillus subtilis*. *Nucleic Acids Res* **43**(6): e42.

Jezewska-Frackowiak, J., K. Seroczynska, J. Banaszczyk, G. Jedrzejczak, A. Zylicz-Stachula and P. M. Skowron (2018). The promises and risks of probiotic *Bacillus* species. *Acta Biochim Pol* **65**(4): 509–519.

Jia, Y., J. Zhu, X. Chen, D. Tang, D. Su, W. Yao and X. Gao (2013). Metabolic engineering of *Bacillus subtilis* for the efficient biosynthesis of uniform hyaluronic acid with controlled molecular weights. *Bioresour Technol* **132**: 427–431.

Jiang, F., S. J. Cao, Z. H. Li, H. Fan, H. F. Li, W. J. Liu and H. L. Yuan (2012). *Salisediminibacterium halotolerans* gen. nov., sp. nov., a halophilic bacterium from soda lake sediment. *Int J Syst Evol Microbiol* **62**(Pt 9): 2127–2132.

Jiang, W., D. Bikard, D. Cox, F. Zhang and L. A. Marraffini (2013). RNA-guided editing of bacterial genomes using CRISPR-Cas systems. *Nat Biotechnol* **31**(3): 233–239.

Joung, K. B. and J. C. Cote (2001). A phylogenetic analysis of *Bacillus thuringiensis* serovars by RFLP-based ribotyping. *J Appl Microbiol* **91**: 279–289.

Jürgen, B., S. Tobisch, M. Wumpelmann, D. Gordes, A. Koch, K. Thurow, D. Albrecht, M. Hecker and T. Schweder (2005). Global expression profiling of *Bacillus subtilis* cells during industrial-close fed-batch fermentations with different nitrogen sources. *Biotechnol Bioeng* **92**: 277–298.

Kabisch, J., I. Pratzka, H. Meyer, D. Albrecht, M. Lalk, A. Ehrenreich and T. Schweder (2013). Metabolic engineering of *Bacillus subtilis* for growth on overflow metabolites. *Microb Cell Fact* **12**: 72.

Kalinowski, J., B. Ahrens, A. Al-Dilaimi, A. Winkler, D. Wibberg, U. Schleenbecker, C. Rückert and G. Grass (2018). Isolation and whole genome analysis of endospore-forming bacteria from heroin. *Forensic Sci Int Genet* **32**: 1–6.

Kamada, M., S. Hase, K. Sato, A. Toyoda, A. Fujiyama and Y. Sakakibara (2014). Whole genome complete resequencing of *Bacillus subtilis* natto by combining long reads with high-quality short reads. *PLOS One* **9**(10): e109999.

Kämpfer, P. (1994). Limits and possibilities of total fatty acid analysis for classification and identification of *Bacillus* species. *Syst Appl Microbiol* **17**: 86–98.

Kämpfer, P., S. P. Glaeser and H. J. Busse (2013). Transfer of *Bacillus schlegelii* to a novel genus and proposal of *Hydrogenibacillus schlegelii* gen. nov., comb. nov. *Int J Syst Evol Microbiol* **63**(Pt 5): 1723–1727.

Kämpfer, P., R. Rosselló-Mora, E. Falsen, H. J. Busse and B. J. Tindall (2006). *Cohnella thermotolerans* gen. nov., sp. nov., and classification of 'Paenibacillus hongkongensis' as *Cohnella hongkongensis* sp. nov. *Int J Syst Evol Microbiol* **56**: 781–786.

Karauzum, H., T. B. Updegrove, M. Kong, I. L. Wu, S. K. Datta and K. S. Ramamurthi (2018). Vaccine display on artificial bacterial spores enhances protective efficacy against *Staphylococcus aureus* infection. *FEMS Microbiol Lett* **365**(18).

Kearns, D. B., F. Chu, S. S. Branda, R. Kolter and R. Losick (2005). A master regulator for biofilm formation by *Bacillus subtilis*. *Mol Microbiol* **55**: 739–749.

Khelifi, N., E. Ben Romdhane, A. Hedi, A. Postec, M. L. Fardeau, M. Hamdi, J. L. Tholozan, B. Ollivier and A. Hirschler-Rea (2010). Characterization of *Microaerobacter geothermalis* gen. nov., sp. nov., a novel microaerophilic, nitrate- and nitrite-reducing thermophilic bacterium isolated from a terrestrial hot spring in Tunisia. *Extremophiles* **14**(3): 297–304.

Kim, J. H., C. S. Lee and B. G. Kim (2005). Spore-displayed streptavidin: a live diagnostic tool in biotechnology. *Biochem Biophys Res Commun* **331**(1): 210–214.

Kim, Y., I. Koh, M. Young Lim, W. H. Chung and M. Rho (2017). Pan-genome analysis of *Bacillus* for microbiome profiling. *Sci Rep* **7**(1): 10984.

Klenk, H. P., A. Lapidus, O. Chertkov, A. Copeland, T. G. Del Rio, M. Nolan, S. Lucas, F. Chen, H. Tice, J. F. Cheng, C. Han, D. Bruce, L. Goodwin, S. Pitluck, A. Pati, et al. (2011). Complete genome sequence of the thermophilic, hydrogen-oxidizing *Bacillus tusciae* type strain (T2) and reclassification in the new genus, *Kyrpidia* gen. nov. as *Kyrpidia tusciae* comb. nov. and emendation of the family Alicyclobacillaceae da Costa and Rainey,2010. *Stand Genomic Sci* **5**(1): 121–134.

Ko, K. S., J.-W. Kim, J.-M. Kim, W. Kim, S. Chung, I. J. Kim and Y.-H. Kook (2004). Population structure of the *Bacillus cereus* group as determined by sequence analysis of six housekeeping genes and the *plcR* gene. *Infect Immun* **72**: 5253–5261.

Kobayashi, K., S. D. Ehrlich, A. Albertini, G. Amati, K. K. Andersen, M. Arnaud, K. Asai, S. Ashikaga, S. Aymerich, P. Bessieres, F. Boland, S. C. Brignell, S. Bron, K. Bunai, et al. (2003). Essential *Bacillus subtilis* genes. *Proc Natl Acad Sci USA* **100**: 4678–4683.

Kobayashi, Y. and F. Kawamura (1992). Molecular cloning. *Biotechnology* **22**: 123–141.

Koch, R. (1876). Die Ätiologie der Milbrandkrankheit, begrüdet die Entwicklunsgesicht des Bacillus Anthracis. *Beitrag Biol Pflanzer* **2**: 277–310.

Koek, M. M., B. Muilwijk, M. J. van der Werf and T. Hankemeier (2006). Microbial metabolomics with gas chromatography/mass spectrometry. *Anal Chem* **78**: 1272–1281.

Koo, B. M., G. Kritikos, J. D. Farelli, H. Todor, K. Tong, H. Kimsey, I. Wapinski, M. Galardini, A. Cabal, J. M. Peters, A. B. Hachmann, D. Z. Rudner, K. N. Allen, A. Typas and C. A. Gross (2017). Construction and analysis of two genome-scale deletion libraries for *Bacillus subtilis*. *Cell Syst* **4**(3): 291–305.e7.

Kotiranta, A., K. Lounatmaa and M. Haapasalo (2000). Epidemiology and pathogenesis of *Bacillus cereus* infections. *Microbes Infect* **2**: 189–198.

Koutsoumanis, K., A. Allende, A. Álvarez-Ordóñez, D. Bolton, S. Bover-Cid, M. Chemaly, R. Davies, F. Hilbert and R. Lindqvist (2019). Update of the list of QPS-recommended biological agents intentionally added to food or feed as notified to EFSA 9: suitability of taxonomic units notified to EFSA until September 2018. *EFSA Journal* **17**(1): e05555.

Krishnamurthi, S., A. Ruckmani, R. Pukall and T. Chakrabarti (2010). *Psychrobacillus* gen. nov. and proposal for reclassification of *Bacillus insolitus* Larkin & Stokes, 1967, *B. psychrotolerans* Abd-El Rahman et al., 2002 and *B. psychrodurans* Abd-El Rahman et al., 2002 as *Psychrobacillus insolitus* comb. nov., *Psychrobacillus psychrotolerans* comb. nov. and *Psychrobacillus psychrodurans* comb. nov. *Syst Appl Microbiol* **33**(7): 367–373.

Kubo, Y., A. P. Rooney, Y. Tsukakoshi, R. Nakagawa, H. Hasegawa and K. Kimura (2011). Phylogenetic analysis of *Bacillus subtilis* strains applicable to natto (fermented soybean) production. *Appl Environ Microbiol* **77**(18): 6463–6469.

Kuebutornye, F. K. A., E. D. Abarike and Y. Lu (2019). A review on the application of *Bacillus* as probiotics in aquaculture. *Fish Shellfish Immunol* **87**: 820–828.

Kumar, P., S. K. Patel, J. K. Lee and V. C. Kalia (2013). Extending the limits of *Bacillus* for novel biotechnological applications. *Biotechnol Adv* **31**(8): 1543–1561.

Kunst, F., N. Ogasawara, I. Moszer, A. Albertini, G. Alloni, V. Azevedo, M. Bertero, P. Bessieres, A. Bolotin, S. Borchert, R. Borriss, L. Boursier, A. Brans, M. Braun, et al. (1997). The complete genome sequence of the gram-positive bacterium *Bacillus subtilis*. *Nature* **390**: 249–256.

Kyoji, G. (2011). The development and application of "Biolacta" and "Lactoles" that are β-galactosidase preparations derived from *Bacillus circulans*. *Miruku Saiensu* **60**: 105–109.

L'Haridon, S., M. L. Miroshnichenko, N. A. Kostrikina, B. J. Tindall, S. Spring, P. Schumann, E. Stackebrandt, E. A. Bonch-Osmolovskaya and C. Jeanthon (2006). *Vulcanibacillus modesticaldus* gen. nov., sp. nov., a strictly anaerobic, nitrate-reducing bacterium from deep-sea hydrothermal vents. *Int J Syst Evol Microbiol* **56**: 1047–1053.

La Duc, M. T., M. Satomi and K. Venkateswaran (2004). *Bacillus odysseyi* sp. nov., a round-spore-forming bacillus isolated from the Mars Odyssey spacecraft. *Int J Syst Evol Microbiol* **54**: 195–201.

Lacriola, C. J., S. P. Falk and B. Weisblum (2013). Screen for agents that induce autolysis in *Bacillus subtilis*. *Antimicrob Agents Chemother* **57**(1): 229–234.

Lammers, C. R., L. A. Florez, A. G. Schmeisky, S. F. Roppel, U. Mader, L. Hamoen and J. Stulke (2010). Connecting parts with processes: SubtiWiki and SubtiPathways integrate gene and pathway annotation for *Bacillus subtilis*. *Microbiology* **156**(Pt 3): 849–859.

Le Breton, Y., N. P. Mohapatra and W. G. Haldenwang (2006). In vivo random mutagenesis of *Bacillus subtilis* by use of TnYLB-1, a mariner-based transposon. *Appl Envir Microbiol* **72**: 327–333.

Lee, C. A., J. Thomas and A. D. Grossman (2012). The *Bacillus subtilis* conjugative transposon ICEBs1 mobilizes plasmids lacking dedicated mobilization functions. *J Bacteriol* **194**(12): 3165–3172.

Lee, G. and J. Pero (1981). Conserved nucleotide sequences in temporally controlled bacteriophage promoters. *J Mol Biol* **152**: 247–265.

Lee, J., A. Goel, M. M. Ataai and M. M. Domach (1997). Supply-side analysis of growth of *Bacillus subtilis* on glucose-citrate medium: feasible network alternatives and yield optimality. *Appl Environ Microbiol* **63**: 710–718.

Lee, J. M., S. Zhang, S. Saha, S. Santa Anna, C. Jiang, J. Perkins (2001). RNA expression analysis using an antisense *Bacillus subtilis* genome array. *J Bacteriol* **183**: 7371–7380.

Lee, S., L. J. Wu and J. Errington (2019). Microfluidic time-lapse analysis and reevaluation of the *Bacillus subtilis* cell cycle. *Microbiologyopen*: e876.

Li, K., D. Cai, Z. Wang, Z. He and S. Chen (2018). Development of an efficient genome editing tool in *Bacillus licheniformis* using CRISPR-Cas9 nickase. *Appl Environ Microbiol* **84**(6).

Li, S., X. Jia and J. Wen (2012). Improved 2-methyl-1-propanol production in an engineered *Bacillus subtilis* by constructing inducible pathways. *Biotechnol Lett* **34**(12): 2253–2258.

Lim, J. M., C. O. Jeon, S. M. Lee, J. C. Lee, L. H. Xu, C. L. Jiang and C. J. Kim (2006). *Bacillus salarius* sp. nov., a halophilic, spore-forming bacterium isolated from a salt lake in China. *Int J Syst Evol Microbiol* **56**: 373–377.

Lim, J. M., C. O. Jeon, S. M. Song and C. J. Kim (2005). *Pontibacillus chungwhensis* gen. nov., sp. nov., a moderately halophilic Gram-positive bacterium from a solar saltern in Korea. *Int J Syst Evol Microbiol* **55**: 165–170.

Liu, H., S. Yang, X. Wang and T. Wang (2019). Production of trehalose with trehalose synthase expressed and displayed on the surface of *Bacillus subtilis* spores. *Microb Cell Fact* **18**(1): 100.

Liu, J., A. Prindle, J. Humphries, M. Gabalda-Sagarra, M. Asally, D. Y. Lee, S. Ly, J. Garcia-Ojalvo and G. M. Suel (2015). Metabolic co-dependence gives rise to collective oscillations within biofilms. *Nature* **523**(7562): 550–554.

Liu, L., Y. Liu, H. D. Shin, R. R. Chen, N. S. Wang, J. Li, G. Du and J. Chen (2013). Developing *Bacillus* spp. as a cell factory for production of microbial enzymes and industrially important biochemicals in the context of systems and synthetic biology. *Appl Microbiol Biotechnol* **97**(14): 6113–6127.

Liu, W.-Q., L. Zhang, M. Chen and J. Li (2019). Cell-free protein synthesis: Recent advances in bacterial extract sources and expanded applications. *Biochem Eng J* **141**: 182–189.

Liu, X. S. and D. H. Dean (2006). Redesigning *Bacillus thuringiensis* Cry1Aa toxin into a mosquito toxin. *Protein Eng Des Sel* **19**: 107–111.

Liu, Y., L. Liu, J. Li, G. Du and J. Chen (2019). Synthetic biology toolbox and chassis development in *Bacillus subtilis*. *Trends Biotechnol* **37**(5): 548–562.

Logan, N. A. and R. C. Berkeley (1984). Identification of *Bacillus* strains using the API system. *J Gen Microbiol* **130**: 1871–1882.

Logan, N. A. and M. Rodriguez-Diaz (2006). *Bacillus, Alicyclobacillus* and *Paenibacillus. Principles and practice of clinical bacteriology*, 2nd ed. S. H. Gillespie and P. M. Hawkey. Chichester, UK, John Wiley: 139–158.

Logan, N. A. and P. C. B. Turnbull (2003). *Bacillus* and related genera. *Manual of clinical microbiology*, 8th ed. P. R. Murray, E. J. Baron, J. H. Jorgensen, M. A. Pfaller and R. H. Yolken. Washington, DC, American Society for Microbiology: 445–460.

Loman, N. J., C. Constantinidou, J. Z. Chan, M. Halachev, M. Sergeant, C. W. Penn, E. R. Robinson and M. J. Pallen (2012). High-throughput bacterial genome sequencing: an embarrassment of choice, a world of opportunity. *Nat Rev Microbiol* **10**(9): 599–606.

Losick, R. and J. Pero (1981). Cascades of sigma factors. *Cell* **25**: 582–584.

Lu, J., Y. Nogi and H. Takami (2001). *Oceanobacillus iheyensis* gen. nov., sp. nov., a deep-sea extremely halotolerant and alkaliphilic species isolated from a depth of 1050 m on the Iheya Ridge. *FEMS Microbiol Lett* **205**: 291–297.

Ludwig, W. and H.-P. Klenk (2001). Overview: a phylogenetic backbone and taxonomic framework for prokaryotic systematics. *Bergey's manual of systematic bacteriology*, 2nd ed. G. M. Garrity. New York, N.Y., Springer: 49–65.

Maki, M., K. T. Leung and W. Qin (2009). The prospects of cellulase-producing bacteria for the bioconversion of lignocellulosic biomass. *Int J Biol Sci* **5**(5): 500–516.

Makita, Y., M. Nakao, N. Ogasawara and K. Nakai (2004). DBTBS: database of transcriptional regulation in *Bacillus subtilis* and its contribution to comparative genomics. *Nucl Acids Res* **32**: D75–77.

Malik, Z. A. and S. Ahmed (2012). Degrading of petroleum hydrocarbons by oil field isolated bacterial consortia. *African J Biotechnol* **11**: 650–658.

Mandal, M. and R. R. Breaker (2004). Gene regulation by riboswitches. *Nat Rev Mol Cell Biol* **5**: 451–463.

Manzano, M., L. Cocolin, C. Cantoni and G. Comi (2003). *Bacillus cereus, Bacillus thuringiensis* and *Bacillus mycoides* differentiation using a PCR-RE technique. *Int J Food Microbiol* **81**: 249–254.

Margulies, M., M. Egholm, W. E. Altman, S. Attiya, J. S. Bader, L. A. Bemben, J. Berka, M. S. Braverman, Y. J. Chen, Z. Chen, S. B. Dewell, L. Du, J. M. Fierro, et al. (2005). Genome sequencing in microfabricated high-density picolitre reactors. *Nature* **437**: 376–380.

Marquez, M. C., I. J. Carrasco, Y. Xue, Y. Ma, D. A. Cowan, B. E. Jones, W. D. Grant and A. Ventosa (2008). *Aquisalibacillus elongatus* gen. nov., sp. nov., a moderately halophilic bacterium of the family Bacillaceae isolated from a saline lake. *Int J Syst Evol Microbiol* **58**(Pt 8): 1922–1926.

Martinez-Corral, R., J. Liu, G. M. Suel and J. Garcia-Ojalvo (2018). Bistable emergence of oscillations in growing *Bacillus subtilis* biofilms. *Proc Natl Acad Sci U S A* **115**(36): E8333–e8340.

Maughan, H. and G. Van der Auwera (2011). Bacillus taxonomy in the genomic era finds phenotypes to be essential though often misleading. *Infect Genet Evol* **11**(5): 789–797.

Mauriello, E. M., H. Duc le, R. Isticato, G. Cangiano, H. A. Hong, M. De Felice, E. Ricca and S. M. Cutting (2004). Display of heterologous antigens on the *Bacillus subtilis* spore coat using CotC as a fusion partner. *Vaccine* **22**(9–10): 1177–1187.

Mayr, R., H. J. Busse, H. L. Worliczek, M. Ehling-Schulz and S. Scherer (2006). *Ornithinibacillus* gen. nov., with the species *Ornithinibacillus bavariensis* sp. nov. and *Ornithinibacillus californiensis* sp. nov. *Int J Syst Evol Microbiol* **56**: 1383–1389.

McKenney, P. T., A. Driks and P. Eichenberger (2013). The *Bacillus subtilis* endospore: assembly and functions of the multilayered coat. *Nat Rev Microbiol* **11**(1): 33–44.

Mehlo, L., D. Gahakwa, P. T. Nghia, N. T. Loc, T. Capell, J. A. Gatehouse, A. M. Gatehouse and P. Christou (2005). An alternative strategy for sustainable pest resistance in genetically enhanced crops. *Proc Natl Acad Sci USA* **102**: 7812–7816.

Meier-Kolthoff, J. P., A. F. Auch, H. P. Klenk and M. Goker (2013). Genome sequence-based species delimitation with confidence intervals and improved distance functions. *BMC Bioinformatics* **14**: 60.

Meima, R. B., J. M. van Dijl, S. Holsappel and S. Bron (2004). Expression Systems in *Bacillus. Protein expression technologies: Current status and future trends.* F. Baneyx. Wymondham, UK, Horizon Bioscience: 199–252.

Michna, R. H., B. Zhu, U. Mader and J. Stulke (2016). SubtiWiki 2.0–an integrated database for the model organism *Bacillus subtilis. Nucleic Acids Res* **44**(D1): D654–662.

Miñana-Galbis, D., D. L. Pinzón, J. G. Lorén, A. Manresa and R. M. Oliart-Ros (2010). Reclassification of *Geobacillus pallidus* (Scholz et al. 1988) Banat et al. 2004 as *Aeribacillus pallidus* gen. nov., comb. nov. *Int J Syst Evol Microbiol* **60**(Pt 7): 1600–1604.

Mingmongkolchai, S. and W. Panbangred (2018). *Bacillus* probiotics: an alternative to antibiotics for livestock production. *J Appl Microbiol* **124**(6): 1334–1346.

Misirli, G., A. Wipat, J. Mullen, K. James, M. Pocock, W. Smith, N. Allenby and J. S. Hallinan (2013). BacillOndex: an integrated data resource for systems and synthetic biology. *J Integr Bioinform* **10**(2): 224.

Missiakas, D. and S. Raina (1998). The extracytoplasmic function sigma factors: role and regulation. *Mol Microbiol* **28**: 1059–1066.

Mongkolthanaruk, W. (2012). Classification of *Bacillus* beneficial substances related to plants, humans and animals. *J Microbiol Biotechnol* **22**(12): 1597–1604.

Moriya, T., T. Hikota, I. Yumoto, T. Ito, Y. Terui, A. Yamagishi and T. Oshima (2011). *Calditerricola satsumensis* gen. nov., sp. nov. and *Calditerricola yamamurae* sp. nov., extreme thermophiles isolated from a high-temperature compost. *Int J Syst Evol Microbiol* **61**(Pt 3): 631–636.

Mu, L. and J. Wen (2013). Engineered *Bacillus subtilis* 168 produces L-malate by heterologous biosynthesis pathway construction and lactate dehydrogenase deletion. *World J Microbiol Biotechnol* **29**(1): 33–41.

Mukherjee, S., D. Stamatis, J. Bertsch, G. Ovchinnikova, O. Verezemska, M. Isbandi, A. D. Thomas, R. Ali, K. Sharma, N. C. Kyrpides and T. B. Reddy (2017). Genomes OnLine Database (GOLD) v.6: data updates and feature enhancements. *Nucleic Acids Res* **45**(D1): D446–d456.

Nakamura, K., S. Haruta, S. Ueno, M. Ishii, A. Yokota and Y. Igarashi (2004). *Cerasibacillus quisquiliarum* gen. nov., sp. nov., isolated from a semi-continuous decomposing system of kitchen refuse. *Int J Syst Evol Microbiol* **54**: 1063–1069.

Nakano, M. M., Y. P. Dailly, P. Zuber and D. P. Clark (1997). Characterization of anaerobic fermentative growth of *Bacillus subtilis*: identification of fermentation end products and genes required for growth. *J Bacteriol* **179**: 6749–6755.

Nakano, M. M., F. Yang, P. Hardin and P. Zuber (1995). Nitrogen regulation of *nasA* and the *nasB* operon, which encode genes required for nitrate assimilation in *Bacillus subtilis. J Bacteriol* **177**: 573–579.

Nakano, M. M. and P. Zuber (1998). Anaerobic growth of a "strict aerobe." *Annu Rev Microbiol* **52**: 165–190.

Nakano, M. M., P. Zuber, P. Glaser, A. Danchin and F. M. Hulett (1996). Two-component regulatory proteins ResD-ResE are required for transcriptional activation of fnr upon oxygen limitation in *Bacillus subtilis. J Bacteriol* **178**: 3796–3802.

Narula, J., M. Fujita and O. A. Igoshin (2016). Functional requirements of cellular differentiation: lessons from *Bacillus subtilis. Curr Opin Microbiol* **34**: 38–46.

Nazina, T. N., T. P. Tourova, A. B. Poltaraus, E. V. Novikova, A. A. Grigoryan, A. E. Ivanova, A. M. Lysenko, V. V. Petrunyaka, G. A. Osipov, S. S. Belyaev and M. V. Ivanov (2001). Taxonomic study of aerobic thermophilic bacilli: descriptions of *Geobacillus subterraneus* gen. nov., sp. nov. and *Geobacillus uzenensis* sp. nov. from petroleum reservoirs and transfer of *Bacillus stearothermophilus, Bacillus thermo-catenulatus, Bacillus thermoleovorans, Bacillus kaustophilus, Bacillus thermoglucosidasius* and *Bacillus thermodenitrificans* to *Geobacillus* as the new combinations G. *stearothermophilus*, G. *thermocatenulatus*, G. *thermoleovorans*, G. *kaustophilus*, G. *thermoglucosidasius* and G. *thermodenitrificans. Int J System Evol Microbiol* **51**: 433–446.

Nguyen, H. D., Q. A. Nguyen, R. C. Ferreira, L. C. S. Ferreira, L. T. Tran and W. Schumann (2005). Construction of plasmid-based expression vectors for *Bacillus subtilis* exhibiting full structural stability. *Plasmid* **54**: 241–248.

Nguyen, Q. A. and W. Schumann (2014). Use of IPTG-inducible promoters for anchoring recombinant proteins on the *Bacillus subtilis* spore surface. *Protein Expr Purif* **95**: 67–76.

Nielsen, J. E. and T. V. Borchert (2000). Protein engineering of bacterial alpha-amylases. *Biochim Biophys Acta* **1543**: 253–274.

Niimura, Y., E. Koh, F. Yanagida, K.-I. Suzuki, K. Komagata and M. Kozaki (1990). *Amphibacillus xylanus* gen. nov., sp. nov., a facultatively anaerobic sporeforming xylan-digesting bacterium which lacks cytochrome, quinone, and catalase. *Int J Syst Bacteriol* **40**: 297–301.

Noirot, P. and M. F. Noirot-Gros (2004). Protein interaction networks in bacteria. *Curr Opin Microbiol* **7**: 505–512.

Norman, T. M., N. D. Lord, J. Paulsson and R. Losick (2013). Memory and modularity in cell-fate decision making. *Nature* **503**(7477): 481–486.

Nunes, I., I. Tiago, A. L. Pires, M. S. Da Costa and A. Veríssimo (2005). *Paucisalibacillus globulus* gen. nov., sp. nov., a Gram-positive bacterium isolated from potting soil. *Int J Syst Evol Microbiol* **56**: 1841–1845.

Nystrand, R. (1984). *Saccharococcus thermophilus* gen. nov., sp. nov., isolated from beet sugar extraction. *Syst Appl Microbiol* **5**: 204–219.

Ofir, G., S. Melamed, H. Sberro, Z. Mukamel, S. Silverman, G. Yaakov, S. Doron and R. Sorek (2018). DISARM is a widespread bacterial defence system with broad anti-phage activities. *Nat Microbiol* **3**(1): 90–98.

Ogawa, T., T. Iwata, S. Kaneko, M. Itaya and J. Hirota (2015). An inducible *recA* expression *Bacillus subtilis* genome vector for stable manipulation of large DNA fragments. *BMC Genomics* **16**: 209.

Olempska-Beer, Z. S., R. I. Merker, M. D. Ditto and M. J. DiNovi (2006). Food-processing enzymes from recombinant microorganisms — a review. *Regul Toxicol Pharmacol* **45**: 144–158.

Olivera, N., F. Siñeriz and J. D. Breccia (2005). *Bacillus patagoniensis* sp. nov., a novel alkalitolerant bacterium from the rhizosphere of *Atriplex lampa* in Patagonia, Argentina. *Int J Syst Evol Microbiol* **55**: 443–447.

Ortiz, A. and E. Sansinenea (2019). Chemical compounds produced by *Bacillus* sp. factories and their role in nature. *Mini Rev Med Chem* **19**(5): 373–380.

Ozturk, S., B. G. Ergun and P. Calik (2017). Double promoter expression systems for recombinant protein production by industrial microorganisms. *Appl Microbiol Biotechnol* **101**(20): 7459–7475.

Palmisano, M. M., L. K. Nakamura, K. E. Duncan, C. A. Istock and F. M. Cohan (2001). *Bacillus sonorensis* sp. nov., a close relative of *Bacillus licheniformis*, isolated from soil in the Sonoran Desert, Arizona. *Int J Syst Evol Microbiol* **51**: 1671–1679.

Park, J. H., W. Kim, Y. S. Lee and J. H. Kim (2019). Decolorization of acid green 25 by surface display of CotA laccase on *Bacillus subtilis* spores. *J Microbiol Biotechnol* **29**(9): 1383–1390.

Patel, S. and R. S. Gupta (2019). A phylogenomic and comparative genomic framework for resolving the polyphyly of the genus *Bacillus*: Proposal for six new genera of *Bacillus* species, *Peribacillus* gen. nov., *Cytobacillus* gen. nov., *Mesobacillus* gen. nov., *Neobacillus* gen. nov., *Metabacillus* gen. nov. and *Alkalihalobacillus* gen. nov. *Int J Syst Evol Microbiol* **70**(1):406–438. 10.1099/ijsem.1090.003775.

Perkins, J. B., M. Goese and G. Schyns (2004). Thiamin production by fermentation. European Patent EP1651780.

Perkins, J. B., M. Wyss, H.-P. Hohmann and U. Sauer (2009). Metabolic engineering in Bacillus subtilis. *The Metabolic Pathway Engineering Handbook*. C. D. Smolke. Boca Raton, CRC Press: 23-21–23-36.

Perkins, J. B. and P. J. Youngman (1983). *Streptococcus* plasmid pAMα1 is a composite of two separable replicons, one of which is closely related to *Bacillus* plasmid pBC16. *J Bacteriol* **155**: 607–615.

Perkins, J. B. and P. J. Youngman (1986). Construction and properties of Tn917–lac, a transposon derivative that mediates transcriptional gene fusions in *Bacillus subtilis*. *Proc Natl Acad Sci USA* **83**: 140–144.

Pero, J. and A. Sloma (1993). Proteases. *Bacillus subtilis and other gram-positive bacteria: Biochemistry, physiology, and molecular genetics*. A. L. Sonenshein, J. A. Hoch and R. Losick. Washington, DC, American Society for Microbiology: 939–952.

Phelps, R. J. and J. L. McKillip (2002). Enterotoxin production in natural isolates of Bacillaceae outside the *Bacillus cereus* group. *Appl Environ Microbiol* **68**: 3147–3151.

Piggot, P. (2010). Epigenetic switching: bacteria hedge bets about staying or moving. *Curr Biol* **20**(11): R480–482.

Piggot, P. and R. Losick (2001). Sporulation genes and intercompartmental regulation. *Bacillus subtilis and its closest relatives: From genes to cells*. A. L. Sonenshein, J. A. Hoch and R. Losick. Washington, DC, ASM Press: 483–517.

Pikuta, E., A. Lysenko, N. Chuvilskaya, U. Mendrock, H. Hippe, N. Suzina, D. Nikitin, G. Osipov and K. Laurinavichius (2000). *Anoxybacillus pushchinensis* gen. nov., sp. nov., a novel anaerobic, alkaliphilic, moderately thermophilic bacterium from manure, and description of *Anoxybacillus flavithermus* comb. nov. *Int J Syst Evol Microbiol* **50**: 2109–2117.

Poluektova, E. U., E. A. Fedorina, O. V. Lotareva and A. A. Prozorov (2004). Plasmid transfer in bacilli by a self-transmissible plasmid p19 from a *Bacillus subtilis* soil strain. *Plasmid* **52**: 212–217.

Popp, P. F., M. Dotzler, J. Radeck, J. Bartels and T. Mascher (2017). The *Bacillus* BioBrick Box 2.0: expanding the genetic toolbox for the standardized work with *Bacillus subtilis*. *Sci Rep* **7**(1): 15058.

Porcellato, D., M. Aspholm, S. B. Skeie, and H. Mellegård H. (2019). Application of a novel amplicon-based sequencing approach reveals the diversity of the *Bacillus* cereus group in stored raw and pasteurized milk. *Food Microbiol* **81**:32–39.

Porreca, G. J. (2010). Genome sequencing on nanoballs. *Nat Biotechnol* **28**(1): 43–44.

Potot, S., C. R. Serra, A. O. Henriques and G. Schyns (2010). Display of recombinant proteins on *Bacillus subtilis* spores, using a coat-associated enzyme as the carrier. *Appl Environ Microbiol* **76**(17): 5926–5933.

Priest, F. G. (1993). Systematics and ecology of *Bacillus*. *Bacillus subtilis and other gram-positive bacteria: Biochemistry, physiology, and molecular genetics*. A. L. Sonenshein, J. A. Hoch and R. Losick. Washington, DC, American Society for Microbiology: 1–16.

Priest, F. G. and B. Alexander (1988). A frequency matrix for probabilistic identification of some Bacilli. *J Gen Microbiol* **134**: 3001–3018.

Pungrasmi, W., J. Intarasoontron, P. Jongvivatsakul and S. Likitlersuang (2019). Evaluation of microencapsulation techniques for MICP bacterial spores applied in self-healing concrete. *Sci Rep* **9**(1): 12484.

Radeck, J., K. Kraft, J. Bartels, T. Cikovic, F. Durr, J. Emenegger, S. Kelterborn, C. Sauer, G. Fritz, S. Gebhard and T. Mascher (2013). The Bacillus BioBrick Box: generation and evaluation of essential genetic building blocks for standardized work with *Bacillus subtilis*. *J Biol Eng* **7**(1): 29.

Radhakrishnan, R., A. Hashem and E. F. Abd Allah (2017). *Bacillus*: A biological tool for crop improvement through bio-molecular changes in adverse environments. *Front Physiol* **8**: 667.

Rampersad, J., A. Khan and D. Ammons (2003). A *Bacillus thuringiensis* isolate possessing a spore-associated filament. *Curr Microbiol* **47**: 355–357.

Rasko, D. A., M. R. Altherr, C. S. Han and J. Ravel (2005). Genomic of the *Bacillus cereus* group of organisms. *FEMS Microbiol Rev.* **29**: 303–329.

Rasko, D. A., P. L. Worsham, T. G. Abshire, S. T. Stanley, J. D. Bannan, M. R. Wilson, R. J. Langham, R. S. Decker, L. Jiang, T. D. Read, A. M. Phillippy, S. L. Salzberg, et al. (2011). *Bacillus anthracis* comparative genome analysis in support of the Amerithrax investigation. *Proc Natl Acad Sci U S A* **108**(12): 5027–5032.

Raut, S. H., D. D. Sarode and S. S. Lele (2014). Biocalcification using *B. pasteurii* for strengthening brick masonry civil engineering structures. *World J Microbiol Biotechnol* **30**(1): 191–200.

Rebecchi, S., D. Pinelli, G. Zanaroli, F. Fava and D. Frascari (2018). Effect of oxygen mass transfer rate on the production of 2,3-butanediol from glucose and agro-industrial byproducts by *Bacillus licheniformis* ATCC9789. *Biotechnol Biofuels* **11**: 145.

Reddy, T. B., A. D. Thomas, D. Stamatis, J. Bertsch, M. Isbandi, J. Jansson, J. Mallajosyula, I. Pagani, E. A. Lobos and N. C. Kyrpides (2015). The Genomes OnLine Database (GOLD) v.5: a metadata management system based on a four level (meta) genome project classification. *Nucleic Acids Res* **43**(Database issue): D1099–1106.

Ren, P. G. and P. J. Zhou (2005). *Salinibacillus aidingensis* gen. nov., sp. nov. and *Salinibacillus kushneri* sp. nov., moderately halophilic bacteria isolated from a neutral saline lake in Xin-Jiang, China. *Int J Syst Evol Microbiol* **55**: 949–953.

Ren, P. G. and P. J. Zhou (2005). *Tenuibacillus multivorans* gen. nov., sp. nov., a moderately halophilic bacterium isolated from saline soil in Xin-Jiang, China. *Int J Syst Evol Microbiol* **55**: 95–99.

Reva, O. N., V. V. Smirnov, B. Pettersson and F. G. Priest (2002). *Bacillus endophyticus* sp. nov., isolated from the inner tissues of cotton plants, *Gossypium* sp. *Int J Syst Evol Microbiol* **52**.

Reva, O. N., I. B. Sorokulova and V. V. Smirnov (2001). Simplified technique for identification of the aerobic spore-forming bacteria by phenotype. *Int J Syst Evol Microbiol* **51**: 1361–1371.

Ricca, E. and S. M. Cutting (2003). Emerging applications of bacterial spores in nanobiotechnology. *J Nanobiotechnology* **1**(1): 6.

Rivas, R., P. Garcia-Fraile, J. L. Zurdo-Pineiro, P. F. Mateos, E. Martinez-Molina, E. J. Bedmar, J. Sanchez-Raya and E. Velazquez (2008). *Saccharibacillus sacchari* gen. nov., sp. nov., isolated from sugar cane. *Int J Syst Evol Microbiol* **58**(Pt 8): 1850–1854.

Royne, A., Y. J. Phua, S. Balzer Le, I. G. Eikjeland, K. D. Josefsen, S. Markussen, A. Myhr, H. Throne-Holst, P. Sikorski and A. Wentzel (2019). Towards a low CO_2 emission building material employing bacterial metabolism (1/2): The bacterial system and prototype production. *PLOS One* **14**(4): e0212990.

Rugbjerg, P. and M. O. A. Sommer (2019). Overcoming genetic heterogeneity in industrial fermentations. *Nat Biotechnol* **37**(8): 869–876.

Rüger, H. J., D. Fritze and C. Spröer (2000). New psychrophilic and psychrotolerant *Bacillus marinus* strains from tropical and polar deep-sea sediments and emended description of the species. *Int J Syst Evol Microbiol* **50**: 1305–1313.

Russell, J. R., M. T. Cabeen, P. A. Wiggins, J. Paulsson and R. Losick (2017). Noise in a phosphorelay drives stochastic entry into sporulation in *Bacillus subtilis*. *Embo j* **36**(19): 2856–2869.

Ryan, K. R. and L. Shapiro (2003). Temperal an spatial regulation in prokaryotic cell cycle progression and development. *Annu Rev Biochem* **72**: 367–394.

Saha, P., S. Krishnamurthi, A. Bhattacharya, R. Sharma and T. Chakrabarti (2010). *Fontibacillus aquaticus* gen. nov., sp. nov., isolated from a warm spring. *Int J Syst Evol Microbiol* **60**(Pt 2): 422–428.

Sanchis, V., H. Agaisse, J. Chaufaux and D. Lereclus (1996). Construction of new insecticidal *Bacillus thuringiensis* recombinant strains by using the sporulation non-dependent expression system of *cryIIIA* and a site specific recombination vector. *J Biotechnol* **48**: 81–96.

Sanders, M. E., L. Morelli and T. A. Tompkins (2003). Sporeformers as human probiotics: *Bacillus*, *Sporolactobacillus*, and *Brevibacillus*. *CRFSFS* **2**: 101–110.

Satomi, M., M. T. La Duc and K. Venkateswaran (2006). *Bacillus safensis* sp. nov., isolated from spacecraft and assembly-facility surfaces. *Int J Syst Evol Microbiol* **56**: 1735–1740.

Sauer, U. (2006). Metabolic networks in motion: 13 C-based flux analysis. *Mol Syst Biol* **2**: 1–10.

Sauer, U. and B. J. Elkmanns (2005). The PEP-pyruvate-oxaloacetate node as the switch point for carbon flux distribution in bacteria. *FEMS Microbiol Rev.* **29**: 765–794.

Schallmey, M., A. Singh and O. P. Ward (2004). Developments in the use of *Bacillus* species for industrial production. *Can J Microbiol* **50**: 1–17.

Schlesner, H., P. A. Lawson, M. D. Collins, N. Weiss, U. Wehmeyer, H. Völker and M. Thomm (2001). *Filobacillus milensis* gen. nov., sp. nov., a new halophilic spore-forming bacterium with Orn-D-Glu-type peptidoglycan. *Int J Syst Evol Microbiol* (51): 425–431.

Schmidt-Dannert, C., J. Pleiss and R. D. Schmid (1988). A toolbox of recombinant lipases for industrial applications. *Ann N Y Acad Sci* **864**: 14–22.

Schneider, K. R., M. E. Parish, R. M. Goodrich and T. Cookingham (2004). *Preventing foodborne illness: Bacillus cereus and Bacillus anthracis*. University of Florida, publication FSHN04-05. https://edis.ifas.ufl.edu/pdffiles/FS/FS10300.pdf.

Schnepf, E., N. Crickmore, J. Van Rie, D. Lereclus, J. Baum, J. Feitelson, D. R. Zeigler and D. H. Dean (1998). *Bacillus thuringiensis* and its pesticidal crystal proteins. *Microbiol Mol Biol Rev* **62**: 775–806.

Schoeni, J. L. and A. Wong (2005). *Bacillus cereus* food poisoning and its toxins. *J Food Prot* **68**: 636–648.

Schraft, H., M. Steele, B. McNab, J. Odumeru and M. W. Griffiths (1996). Epidemiological typing of *Bacillus* spp. isolated from food. *Appl Environ Microbiol* **62**: 4229–4232.

Schumann, W., S. D. Ehrlich and N. Ogasawara, Eds. (2001). *Functional analysis of bacterial genes: A practical manual*. West Sussex, UK, John Wiley & Sons Ltd.

Schuster, S. C. (2008). Next-generation sequencing transforms today's biology. *Nat Methods* **5**(1): 16–18.

Seiler, H., M. Wenning and S. Scherer (2013). *Domibacillus robiginosus* gen. nov., sp. nov., isolated from a pharmaceutical clean room. *Int J Syst Evol Microbiol* **63**(Pt 6): 2054–2061.

Senghor, B., E. H. Seck, S. Khelaifia, H. Bassene, C. Sokhna, P. E. Fournier, D. Raoult and J. C. Lagier (2017). Description of 'Bacillus dakarensis' sp. nov., 'Bacillus sinesaloumensis' sp. nov., 'Gracilibacillus timonensis' sp. nov., 'Halobacillus massiliensis' sp. nov., 'Lentibacillus massiliensis' sp. nov., 'Oceanobacillus senegalensis' sp. nov., 'Oceanobacillus *timonensis'* sp. nov., *'Virgibacillus dakarensis'* sp. nov. and *'Virgibacillus marseillensis'* sp. nov., nine halophilic new species isolated from human stool. *New Microbes New Infect* **17**: 45–51.

Seuylemezian, A., L. Ott, S. Wolf, J. Fragante, O. Yip, R. Pukall, P. Schumann and P. Vaishampayan (2019). *Bacillus glennii* sp. nov. and *Bacillus saganii* sp. nov., isolated from the vehicle assembly building at Kennedy Space Center where the Viking spacecraft were assembled. *Int J Syst Evol Microbiol* **70**(1): 71–76.

Shank, E. A. and R. Kolter (2011). Extracellular signaling and multicellularity in *Bacillus subtilis*. *Curr Opin Microbiol* **14**(6): 741–747.

Shao, X., H. Ni, T. Lu, M. Jiang, H. Li, X. Huang and L. Li (2012). An improved system for the surface immobilisation of proteins on *Bacillus thuringiensis* vegetative cells and spores through a new spore cortex-lytic enzyme anchor. *N Biotechnol* **29**(3): 302–310.

Shapiro, L. and R. Losick (2000). Dynamic spatial regulation in the bacterial cell. *Cell* **100**. 89–98.

Sheu, S. Y., A. B. Arun, S. R. Jiang, C. C. Young and W. M. Chen (2011). *Allobacillus halotolerans* gen. nov., sp. nov. isolated from shrimp paste. *Int J Syst Evol Microbiol* 61(Pt 5): 1023–1027.

Shi, T., G. Wang, Z. Wang, J. Fu, T. Chen and X. Zhao (2013). Establishment of a markerless mutation delivery system in *Bacillus subtilis* stimulated by a double-strand break in the chromosome. *PLOS One* **8**(11): e81370.

Shida, O., H. Takagi, K. Kadowaki and K. Komagata (1996). Proposal for two new genera, *Brevibacillus* gen. nov. and *Aneurinibacillus* gen. nov. *Int J Syst Bacteriol* **46**: 939–946.

Shida, O., H. Takagi, K. Kadowaki, L. K. Nakamura and K. Komagata (1997). Transfer of *Bacillus alginolyticus, Bacillus chondroitinus, Bacillus curdlanolyticus, Bacillus glucanolyticus, Bacillus kobensis,* and *Bacillus thiaminolyticus* to the Genus *Paenibacillus* and emended description of the genus *Paenibacillus*. *Int J Syst Bacteriol* **47**: 289–298.

Shivaji, S., P. Chaturvedi, K. Suresh, G. S. N. Reddy, C. B. S. Dutt, M. Wainwright, J. V. Narlikar and P. M. Bhargava (2006). *Bacillus aerius* sp. nov., *Bacillus aerophilus* sp. nov., *Bacillus stratosphericus* sp. nov. and *Bacillus altitudinis* sp. nov., isolated from cryogenic tubes used for collecting air samples from high altitudes. *Int J Syst Evol Microbiol* **56**: 1465–1473.

Shtarkman, Y. M., Z. A. Kocer, R. Edgar, R. S. Veerapaneni, T. D'Elia, P. F. Morris and S. O. Rogers (2013). Subglacial Lake Vostok (Antarctica) accretion ice contains a diverse set of sequences from aquatic, marine and sediment-inhabiting bacteria and eukarya. *PLOS One* **8**(7): e67221.

Silano, M. and V. Silano (2008). The fifth anniversary of the European Food Safety Authority (EFSA): Mission, organization, functioning and main results. *Fitoterapia* **79**(3): 149–160.

Silvaggi, J. M., J. B. Perkins and R. Losick (2005). Small untranslated RNA antitoxin in *Bacillus subtilis*. *J Bacteriol* **187**: 6641–6650.

Silvaggi, J. M., J. B. Perkins and R. Losick (2006). Genes for small, noncoding RNAs under sporulation control in *Bacillus subtilis*. *J Bacteriol* **188**: 532–541.

Silvaggi, J. M., D. L. Popham, A. Driks, P. Eichenberger and R. Losick (2004). Unmasking novel sporulation genes in *Bacillus subtilis*. *J Bacteriol* **186**(23): 8089–8095.

Simbahan, J., R. Drijber and P. Blum (2004). *Alicyclobacillus vulcanalis* sp. nov., a thermophilic, acidophilic bacterium isolated from Coso Hot Springs, California, USA. *Int J Syst Evol Microbiol* **54**: 1703–1707.

Singh, N., A. S. Mathur, R. P. Gupta, C. J. Barrow, D. Tuli and M. Puri (2018). Enhanced cellulosic ethanol production via consolidated bioprocessing by *Clostridium thermocellum* ATCC 31924. *Bioresour Technol* **250**: 860–867.

Singh, P., Y. Patil and V. Rale (2019). Biosurfactant production: emerging trends and promising strategies. *J Appl Microbiol* **126**(1): 2–13.

Skerman, V. B. D., V. McGowan and P. H. A. Sneath (1980). Approved lists of bacterial names. *Int J Syst Bacteriol* **30**: 225–420.

Smith, J. L., J. M. Goldberg and A. D. Grossman (2014). Complete genome sequences of *Bacillus subtilis* subsp. *subtilis* laboratory strains JH642 (AG174) and AG1839. *Genome Announc* **2**(4).

Smith, N. R., R. E. Gordon and F. E. Clark (1946). *Aerobic mesophilic sporeforming bacteria*. Washington, DC, United States Department of Agriculture.

So, Y., S. Y. Park, E. H. Park, S. H. Park, E. J. Kim, J. G. Pan and S. K. Choi (2017). A highly efficient CRISPR-Cas9-mediated large genomic deletion in *Bacillus subtilis*. *Front Microbiol* **8**: 1167.

Soga, T., Y. Ohashi, Y. Ueno, H. Naraoka, M. Tomita and T. Nishioka (2003). Quantitative metabolome analysis using capillary electrophoresis mass spectrometry. *J Proteome Res* **2**: 488–494.

Sonenshein, A. L. (2001). The Kreb citric acid cycle. *Bacillus subtilis and its closest relatives: From genes to cells*. A. L. Sonenshein, J. A. Hoch and R. Losick. Washington, DC, ASM Press: 151–162.

Song, Y., J. M. Nikoloff and D. Zhang (2015). Improving protein production on the level of regulation of both expression and secretion pathways in *Bacillus subtilis*. *J Microbiol Biotechnol* **25**(7): 963–977.

Sorokin, I. D., E. V. Zadorina, I. K. Kravchenko, E. S. Boulygina, T. P. Tourova and D. Y. Sorokin (2008). *Natronobacillus azotifigens* gen. nov., sp. nov., an anaerobic diazotrophic haloalkaliphile from soda-rich habitats. *Extremophiles* **12**(6): 819–827.

Spizizen, J. (1958). Transformation of biochemically deficient strains of *Bacillus subtilis* by deoxyribonucleate. *Proc Natl Acad Sci USA* **44**: 1072–1078.

Spring, S., W. Ludwig, M. C. Marquez, A. Ventosa and K. H. Schleifer (1996). *Halobacillus* gen. nov., with descriptions of *Halobacillus litoralis* sp. nov. and *Halobacillus trueperi* sp. nov., and transfer of *Sporosarcina halophila* to *Halobacillus halophilus* comb. nov. *Int J Syst Bacteriol* **46**: 492–496.

Stackebrandt, E., W. Frederiksen, G. M. Garrity, P. A. D. Grimont, P. Kampfer, M. C. J. Maiden, X. Nesme, R. Rossello-Mora, J. Swings, H. G. Truper, L. Vauterin, A. C. Ward and W. B. Whitman (2002). Report of the ad hoc committee for the re-evaluation of the species definition in bacteriology. *Int J Syst Evol Microbiol* **52**: 1043–1047.

Stackebrandt, E. and B. M. Goebel (1994). Taxonomic Note: a place for DNA-DNA reassociation and 16S rRNA sequence analysis in the present species definition in bacteriology. *Int J Syst Bacteriol* **44**: 846–849.

Stein, T. (2005). *Bacillus subtilis* antibiotics: structures, syntheses and specific functions. *Mol Microbiol* **56**: 845–857.

Steinmetz, M. and R. Richter (1994). Easy cloning of mini-Tn10 insertions from *Bacillus subtilis* chromosome. *J Bacteriol* **176**: 1761–1763.

Steven, B., M. Q. Chen, C. W. Greer, L. G. Whyte and T. D. Niederberger (2008). *Tumebacillus permanentifrigoris* gen. nov., sp. nov., an aerobic, spore-forming bacterium isolated from Canadian high Arctic permafrost. *Int J Syst Evol Microbiol* **58**(Pt 6): 1497–1501.

Stoddart, D., A. J. Heron, E. Mikhailova, G. Maglia and H. Bayley (2009). Single-nucleotide discrimination in immobilized DNA oligonucleotides with a biological nanopore. *Proc Natl Acad Sci U S A* **106**(19): 7702–7707.

Stragier, P. and R. Losick (1996). Molecular genetics of sporulation in *Bacillus subtilis*. *Annu Rev Genet* **30**: 297–341.

Subramaniam, A. R., A. Deloughery, N. Bradshaw, Y. Chen, E. O'Shea, R. Losick and Y. Chai (2013). A serine sensor for multicellularity in a bacterium. *Elife* **2**(0): e01501.

Sudarsan, N., M. C. Hammond, K. F. Block, R. Welz, J. E. Barrick, A. Roth and R. Breaker (2006). Tandem riboswitch architectures exhibit complex gene control functions. *Science* **314**: 300–304.

Sun, G., E. Sharkova, R. Chesnut, S. Birkey, M. F. Duggan, A. Sorokin, P. Pujic, S. D. Ehrlich and F. M. Hulett (1996). Regulators of aerobic and anaerobic respiration in *Bacillus subtilis*. *J Bacteriol* **178**: 1374–1385.

Switzer Blum, J., A. Burns Bindi, J. Buzzelli, J. F. Stolz and R. S. Oremland (1998). *Bacillus arsenicoselenatis*, sp. nov., and *Bacillus selenitireducens*, sp. nov. : two haloalkaliphiles from Mono Lake, California that respire oxyanions of selenium and arsenic. *Arch Microbiol* **171**: 19–30.

Szurmant, H. and G. W. Ordal (2004). Diversity in chemotaxis mechanisms among the bacteria and archaea. *Microbiol Mol Rev* **68**: 301–319.

Takagi, H. (1993). Protein engineering on subtilisin. *Int J Biochem* **25**: 307–312.

Tam, N. K. M., N. Q. Uyen, H. A. Hong, L. H. Duc, T. T. Hoa, C. R. Serra, A. O. Henriques and S. M. Cutting (2006). The intestinal life cycle of *Bacillus subtilis* and close relatives. *J Bacteriol* **188**: 2692–2700.

Tanaka, K., C. S. Henry, J. F. Zinner, E. Jolivet, M. P. Cohoon, F. Xia, V. Bidnenko, S. D. Ehrlich, R. L. Stevens and P. Noirot (2013). Building the repertoire of dispensable chromosome regions in *Bacillus subtilis* entails major refinement of cognate large-scale metabolic model. *Nucleic Acids Res* **41**(1): 687–699.

Tanasupawat, S., S. Namwong, T. Kudo and T. Itoh (2007). *Piscibacillus salipiscarius* gen. nov., sp. nov., a moderately halophilic bacterium from fermented fish (pla-ra) in Thailand. *Int J Syst Evol Microbiol* **57**(Pt 7): 1413–1417.

Taylor, M. P., K. L. Eley, S. Martin, M. I. Tuffin, S. G. Burton and D. A. Cowan (2009). Thermophilic ethanologenesis: future prospects for second-generation bioethanol production. *Trends Biotechnol* **27**(7): 398–405.

Terpe, K. (2006). Overview of bacterial expression systems for heterologous protein production: from molecular and biochemical fundamentals to commercial systems. *Appl Microbiol Biotechnol* **72**: 211–222.

Tettelin, H., V. Masignani, M. J. Cieslewicz, C. Donati, D. Medini, N. L. Ward, S. V. Angiuoli, J. Crabtree, A. L. Jones, A. S. Durkin, R. T. Deboy, T. M. Davidsen, et al. (2005). Genome analysis of multiple pathogenic isolates of *Streptococcus agalactiae*: implications for the microbial "pan-genome." *Proc Natl Acad Sci U S A*: 13950–13955.

Tiago, I., A. P. Chung and A. Veríssimo (2004). Bacterial diversity in a nonsaline alkaline environment: heterotrophic aerobic populations. *Appl Environ Microbiol* **70**: 7378–7387.

Tian, L., P. M. Conway, N. D. Cervenka, J. Cui, M. Maloney, D. G. Olson and L. R. Lynd (2019). Metabolic engineering of *Clostridium thermocellum* for n-butanol production from cellulose. *Biotechnol Biofuels* **12**: 186.

Tidjani Alou, M., J. Rathored, S. I. Traore, S. Khelaifia, C. Michelle, S. Brah, B. A. Diallo, D. Raoult and J. C. Lagier (2015). *Bacillus niameyensis* sp. nov., a new bacterial species isolated from human gut. *New Microbes New Infect* **8**: 61–69.

Tidjiani Alou, M., J. Rathored, S. Khelaifia, C. Michelle, S. Brah, B. A. Diallo, D. Raoult and J. C. Lagier (2015). *Bacillus rubiinfantis* sp. nov. strain mt2(T), a new bacterial species isolated from human gut. *New Microbes New Infect* **8**: 51–60.

Timmusk, S., N. Grantcharova and E. G. Wagner (2005). *Paenibacillus polymyxa* invades plant roots and forms biofilms. *Appl Environ Microbiol* **71**(11): 7292–7300.

Timmusk, S., P. van West, N. A. Gow and R. P. Huffstutler (2009). *Paenibacillus polymyxa* antagonizes oomycete plant pathogens *Phytophthora palmivora* and *Pythium aphanidermatum*. *J Appl Microbiol* **106**(5): 1473–1481.

Touzel, J. P., M. O'Donohue, P. Debeire, E. Samain and C. Breton (2000). *Thermobacillus xylanilyticus* gen. nov., sp. nov., a new aerobic thermophilic xylan-degrading bacterium isolated from farm soil. *Int J Syst Evol Microbiol* **50**: 315–320.

Toymentseva, A. A., K. Schrecke, M. R. Sharipova and T. Mascher (2012). The LIKE system, a novel protein expression toolbox for *Bacillus subtilis* based on the *liaI* promoter. *Microb Cell Fact* **11**: 143.

Traag, B. A., A. Pugliese, J. A. Eisen and R. Losick (2013). Gene conservation among endospore-forming bacteria reveals additional sporulation genes in *Bacillus subtilis*. *J Bacteriol* **195**(2): 253–260.

Tsuge, K., Y. Sato, Y. Kobayashi, M. Gondo, M. Hasebe, T. Togashi, M. Tomita and M. Itaya (2015). Method of preparing an equimolar DNA mixture for one-step DNA assembly of over 50 fragments. *Sci Rep* **5**: 10655.

Tye, A., F. Siu, T. Leung and B. Lim (2002). Molecular cloning and the biochemical characterization of two novel phytases from *B. subtilis* 168 and *B. licheniformis*." *Appl Microbiol Biotechnol* **59**: 190–197.

Udaka, S. and H. Yamagata (1993–1994). Protein secretion in *Bacillus brevis*. *Antonie van Leeuwenhoek* **64**: 137–143.

Ueda, S. (1989). Industrial applications of *B. subtilis*: utilization of soybean as natto, a traditional Japanese food. *Bacillus subtilis: Molecular biology and industrial applications*. B. Maruo and H. Yoshikawa. Amsterdam, Elsevier Science B.V.: 143–161.

Ugoji, E. O., M. D. Laing and C. H. Hunter (2005). Colonization of *Bacillus* spp. on seeds and in the plant rhizoplane. *J Environ Biol* **26**: 459–466.

van Dijl, J. M. and M. Hecker (2013). *Bacillus subtilis*: from soil bacterium to super-secreting cell factory. *Microb Cell Fact* **12**: 3.

Vargas, V. A., O. D. Delgado, R. Hatti-Kaul and B. Mattiasson (2005). *Bacillus bogoriensis* sp. nov., a novel alkaliphilic, halotolerant bacterium isolated from a Kenyan soda lake. *Int J Syst Evol Microbiol* **55**: 899–902.

Vary, P. (1992). Development of genetic engineering in *Bacillus megaterium*. *Biotechnology* **22**: 251–310.

Vary, P. (1994). Prime time for *Bacillus megaterium*. *Microbiology* **140**: 1001–1013.

Venkateswaran, K., M. Kempf, F. Chen, M. Satomi, W. Nicholson and R. Kern (2003). *Bacillus nealsonii* sp. nov., isolated from a spacecraft-assembly facility, whose spores are γ-radiation resistant. *Int J Syst Evol Microbiol* **53**: 165–172.

Vitreschak, A. G., D. A. Rodionov, A. A. Mironov and M. S. Gelfand (2004). Riboswitches: the oldest mechanism for the regulation of gene expression. *Trends Genet* **20**: 44–50.

Vlamakis, H., Y. Chai, P. Beauregard, R. Losick and R. Kolter (2013). Sticking together: building a biofilm the *Bacillus subtilis* way. *Nat Rev Microbiol* **11**(3): 157–168.

Wainø, M., B. J. Tindall, P. Schumann and K. Ingvorsen (1999). *Gracilibacillus* gen. nov., with description of *Gracilibacillus halotolerans* gen. nov., sp. nov.; transfer of *Bacillus dipsosauri* to *Gracilibacillus dipsosauri* comb. nov., and *Bacillus salexigens* to the genus *Salibacillus* gen. nov., as *Salibacillus salexigens* comb. nov. *Int J Syst Evol Microbiol* **49**: 821–831.

Walser, M., R. Pellaux, A. Meyer, M. Bechtold, H. Vanderschuren, R. Reinhardt, J. Magyar, S. Panke and M. Held (2009). Novel method for high-throughput colony PCR screening in nanoliter-reactors. *Nucleic Acids Res* **37**(8): e57.

Wang, L. F. (1992). Useful *Bacillus* strains and plasmids. *Biotechnology* **22**: 339–347.

Wang, J., H. Yu, S. Tian, H. Yang, J. Wang and W. Zhu (2017). Recombinant expression insulin-like growth factor 1 in *Bacillus subtilis* using a low-cost heat-purification technology. *Biochem* **63**:49-54.

Wang, M., J. Fu, X. Zhang and T. Chen (2012). Metabolic engineering of *Bacillus subtilis* for enhanced production of acetoin. *Biotechnol Lett* **34**(10): 1877–1885.

Wang, X., Y. Xue and Y. Ma (2011). *Streptohalobacillus salinus* gen. nov., sp. nov., a moderately halophilic, Gram-positive, facultative anaerobe isolated from subsurface saline soil. *Int J Syst Evol Microbiol* **61**(Pt 5): 1127–1132.

Watabe, K. (2013). [Overview of study on *Bacillus subtilis* spores]. *Yakugaku Zasshi* **133**(7): 783–797.

Wayne, L. G., D. J. Brenner, R. R. Colwell, P. A. D. Grimont, O. Kandler, M. I. Krichevsky, L. H. Moore, W. E. C. Moore, R. G. E. Murray, E. Stackebrandt, M. P. Starr and H. G. Truper. (1987). Report of the Ad-Hoc-Committee on Reconciliation of Approaches to Bacterial Systematics. *Int J Syst Bacteriol* **37**: 463–464.

Westbrook, A. W., M. Moo-Young and C. P. Chou (2016). Development of a CRISPR-Cas9 tool kit for comprehensive engineering of *Bacillus subtilis*. *Appl Environ Microbiol* **82**(16): 4876–4895.

Westers, H., R. Dorenbos, J. M. van Dijl, J. Kabel, T. Flanagan., K. M. Devine, F. Jude, S. J. Séror, A. C. Beekman, E. Darmon, C. Schevins, A. de Jong, et al. (2003). Genome engineering reveals large dispensable regions in *Bacillus subtilis*. *Mol Biol Evol* **20**: 2076–2090.

Westers, L., W. Westers and W. J. Quax (2004). *Bacillus subtilis* as a cell factory for pharmaceutical proteins: a biotechnological approach to optimize the host organism. *Biochim Biophys Acta* **1694**: 299–310.

Wisotzkey, J. D., P. Jurtshuk, Jr., G. E. Fox, G. Deinhard and K. Poralla (1992). Comparative sequences analyses on the 16S rRNA (rDNA) of *Bacillus acidocaldarius*, *Bacillus acidoterrestris*, and *Bacillus cycloheptanicus* and proposal for creation of a new genus, *Alicyclobacillus* gen. nov. *Int J Syst Bacteriol* **42**: 263–269.

Woese, C. R. and G. E. Fox (1977). Phylogenetic structure of the prokaryotic domain: the primary kingdoms. *Proc Natl Acad Sci USA* **74**: 5088–5090.

Wong, L. S. (2015). Microbial cementation of ureolytic bacteria from the genus *Bacillus*: a review of the bacterial application on cement-based materials for cleaner production. *J Cleaner Product* **93**: 5–17.

Wu, S. C., J. C. Yeung, Y. Duan, R. Ye, S. J. Szarka, H. R. Habibi and S. L. Wong (2002). Functional production and characterization of a fibrin-specific single-chain antibody fragment from *Bacillus subtilis*: effects of molecular chaperones and a wall-bound protease on antibody fragment production. *Appl Environ Microbiol* **68**: 3261–3269.

Xu, D. and J. Cote (2006). Sequence diversity of the *Bacillus thuringiensis* and *B. cereus* sensu lato flagellin (H antigen) protein: comparison with H serotype diversity. *Appl Environ Microbiol* **72**: 4653–4662.

Xu, X., L. Yu, G. Xu, Q. Wang, S. Wei and X. Tang (2019). *Bacillus yapensis* sp. nov., a novel piezotolerant bacterium isolated from deep-sea sediment of the Yap Trench, Pacific Ocean. *Antonie Van Leeuwenhoek* **113**(3):389–396.

Xue, Y., X. Zhang, C. Zhou, Y. Zhao, D. A. Cowan, S. Heaphy, W. D. Grant, B. E. Jones, A. Ventosa and Y. Ma (2006). *Caldalkalibacillus thermarum* gen. nov., sp. nov., a novel alkalithermophilic bacterium from a hot spring in China. *Int J Syst Evol Microbiol* **56**: 1217–1221.

Yi, J., H. Y. Wu, J. Wu, C. Y. Deng, R. Zheng and Z. Chao (2012). Molecular phylogenetic diversity of *Bacillus* community and its temporal-spatial distribution during the swine manure of composting. *Appl Microbiol Biotechnol* **93**(1): 411–421.

Yi, Y., Z. Li, C. Song and O. P. Kuipers (2018). Exploring plant-microbe interactions of the rhizobacteria *Bacillus subtilis* and *Bacillus mycoides* by use of the CRISPR-Cas9 system. *Environ Microbiol* **20**(12): 4245–4260.

Yocum, R. R. and T. A. Patterson (2005). Microorganisms and assays for the identification of antibiotics. U.S. Patent Application US2005158842.

Yogesh, D. and P. M. Halami (2017). Fibrinolytic enzymes of *Bacillus* spp.: an overview. *Int Food Res J* **24**(1): 35–47.

Yoon, J. H., K. H. Kang and Y. H. Park (2002). *Lentibacillus salicampi* gen. nov., sp. nov., a moderately halophilic bacterium isolated from a salt field in Korea. *Int J Syst Evol Microbiol* **52**: 2043–2048.

Yoon, J. H., S. J. Kang and T. K. Oh (2007). Reclassification of *Marinococcus albus* Hao et al. 1985 as *Salimicrobium album* gen. nov., comb. nov. and *Bacillus halophilus* Ventosa et al. 1990 as *Salimicrobium halophilum* comb. nov., and description of *Salimicrobium luteum* sp. nov. *Int J Syst Evol Microbiol* **57**(Pt 10): 2406–2411.

Yoon, J. H., N. Weiss, K. C. Lee, I. S. Lee, K. H. Kang and Y. H. Park (2001). *Jeotgalibacillus alimentarius* gen. nov., sp. nov., a novel bacterium isolated from jeotgal with L-lysine in the cell wall, and reclassification of *Bacillus marinus* Rüger 1983 as *Marinibacillus marinus* gen. nov., comb. nov. *Int J Syst Evol Microbiol* **51**: 2087–2093.

Yoon, S. H., S. M. Ha, J. Lim, S. Kwon and J. Chun (2017). A large-scale evaluation of algorithms to calculate average nucleotide identity. *Antonie Van Leeuwenhoek* **110**(10): 1281–1286.

You, C., X. Z. Zhang, N. Sathitsuksanoh, L. R. Lynd and Y. H. Zhang (2012). Enhanced microbial utilization of recalcitrant cellulose by an ex vivo cellulosome-microbe complex. *Appl Environ Microbiol* **78**(5): 1437–1444.

Youngman, P. J., J. B. Perkins and R. Losick (1983). Genetic transposition and insertional mutagenesis in *Bacillus subtilis* with *Streptococcus faecalis* transposon Tn917. *Proc Natl Acad Sci USA* **80**: 2305–2309.

Zabrocka, L., K. Langer, A. Michalski, J. Kocik and J. J. Langer (2015). A microfluidic device for real-time monitoring of *Bacillus subtilis* bacterial spores during germination based on non-specific physicochemical interactions on the nanoscale level. *Lab Chip* **15**(1): 274–282.

Zahner, V., H. Momen, C. A. Salles and L. Rabinovitch (1989). A comparative study of enzyme variation in *Bacillus cereus* and *Bacillus thuringiensis*. *J Appl Bacteriol* **67**: 275–282.

Zahner, V., L. Rabinovitch, C. F. Cavados and H. Momen (1994). Multilocus enzyme electrophoresis on agarose gel as an aid to the identification of entomopathogenic *Bacillus sphaericus* strains. *J Appl Bacteriol* **76**: 327–335.

Zaitsev, G. M., I. V. Tsitko, F. A. Rainey, Y. A. Trotsenko, J. S. Uotila, E. Stackebrandt and M. S. Salkinoja-Salonen (1998). New aerobic ammonium-dependent obligately oxalotrophic bacteria: description of *Ammoniphilus oxalaticus* gen. nov., sp. nov. and *Ammoniphilus oxalivorans* gen. nov., sp. nov. *Int J Syst Bacteriol* 48(Pt 1): 151–163.

Zavarzina, D. G., T. P. Tourova, T. V. Kolganova, E. S. Boulygina and T. N. Zhilina (2009). Description of *Anaerobacillus alkalilacustre* gen. nov., sp. nov.—Strictly anaerobic diazotrophic bacillus isolated from soda lake and transfer of *Bacillus arseniciselenatis*, *Bacillus macyae*, and *Bacillus alkalidiazotrophicus* to *Anaerobacillus* as the new combinations *A. arseniciselenatis* comb. nov., *A. macyae* comb. nov., and *A. alkalidiazotrophicus* comb. nov. *Microbiology* **78**(6): 723–731.

Zeigler, D. R. (2003). Gene sequences useful for predicting relatedness of whole genomes in bacteria. *Int J Syst Evol Microbiol* **53**: 1893–1900.

Zeigler, D. R. (2005). Application of a *recN* sequence similarity analysis to the identification of species within the bacterial genus *Geobacillus*. *Int J Syst Evol Microbiol* **55**: 1171–1179.

Zeigler, D. R. (2011). The genome sequence of *Bacillus subtilis* subsp. *spizizenii* W23: insights into speciation within the *B. subtilis* complex and into the history of *B. subtilis* genetics. *Microbiology* **157**: 2033–2041.

Zeigler, D. R. and W. L. Nicholson (2017). Experimental evolution of *Bacillus subtilis*. *Environ Microbiol* **19**(9): 3415–3422.

Zeigler, D. R., Z. Pragai, S. Rodriguez, B. Chevreux, A. Muffler, T. Albert, R. Bai, M. Wyss and J. B. Perkins (2008). The origins of 168, W23, and other *Bacillus subtilis* legacy strains. *J Bacteriol* 190: 6983–6995.

Zhang, K., X. Duan and J. Wu (2016). Multigene disruption in undomesticated *Bacillus subtilis* ATCC 6051a using the CRISPR/Cas9 system. *Sci Rep* 6: 27943.

Zhang, X., R. Zhang, T. Bao, T. Yang, M. Xu, H. Li, Z. Xu and Z. Rao (2013). Moderate expression of the transcriptional regulator ALsR enhances acetoin production by *Bacillus subtilis*. *J Ind Microbiol Biotechnol* **40**(9): 1067–1076.

Zhou, K., R. Zou, C. Zhang, G. Stephanopoulos and H. P. Too (2013). Optimization of amorphadiene synthesis in *Bacillus subtilis* via transcriptional, translational, and media modulation. *Biotechnol Bioeng* **110**(9): 2556–2561.

23

The Genus Bordetella*

Rita Austin and Tonya Shearin-Patterson

CONTENTS

Introduction and History

The members of the genus *Bordetella* are small non-spore-forming Gram-negative aerobic coccobacilli. There are now 16 named species. Among them are the more notable and familiar mammalian pathogens *Bordetella pertussis* and *B. parapertussis*$_{hu}$, which cause whooping cough in humans; *B. parapertussis*$_{ov}$, which causes a respiratory disease in sheep; and *B. bronchiseptica*, which causes respiratory diseases in numerous mammals.[1–7]

B. pertussis was the first *Bordetella* identified in the early 1900s by Jules Bordet at the Pasteur Institute.[8] It was originally named *Haemophilus pertussis*, based on some phenotypic similarities to members of the *Haemophilus* genus, but subsequent biochemical and genetic studies showed little similarity with *Haemophilus*. In 1952, the distinct genus *Bordetella* was named in honor of Bordet's foundational work.[9,10]

B. holmesii, B. trematum, and B. hinzii have also been associated with human infection. Although *B. hinzii* is more commonly isolated from the respiratory tracts of birds, all three species have caused bloodborne infections in immunocompromised individuals.[11–13] *B. holmesii* is also responsible for causing respiratory disease in humans,[14–19] whereas *B. trematum* more typically causes wound infections.[20,21] Recent additions to the genus *Bordetella*

include three organisms recovered from the respiratory tract of cystic fibrosis patients: *B. bronchialis, B. flabilis, and B. sputigena*.[22] *B. pseudohinzii* has been isolated from the respiratory system of mice.[23] *B. muralis, B. tumbae, and B. tumulicola* join *B. petrii* as organisms recovered in environmental samples.[24–27] *Bordetella avium*, more distantly related to the *B. bronchiseptica* cluster, is found in the respiratory tract of birds[28] and is a respiratory pathogen of poultry.[29] An overview of characteristics of the *Bordetella* can be found in Table 23.1.

Phylogenetic Relationships and Physiology

Genetic data suggest that four pathogenic species: *B. pertussis, B. parapertussis*$_{hu}$, *B. parapertussis*$_{ov}$, and *B. bronchiseptica* are related enough to actually be subspecies of a single ancestor of *B. bronchiseptica* and are sometimes referred to as the *B. bronchiseptica* cluster[2,3,30–32] An interesting aspect to note is that these organisms are capable of presenting markedly different antigenic profiles, depending on how they are grown. This is particularly important to be aware of if expression of desired proteins, such as those associated with virulence, is being sought for isolation or characterization.[1,33–40]

* We thank T. H. Stenson & M. S. Peppler for their contribution to this chapter in the second edition.

DOI: 10.1201/9781003099277-25

TABLE 23.1

The *Bordetella*[126,172,190,191,192,251-257]

Bordetella species	*pertussis*	*parapertussis*$_{hu/ov}$[a]	*bronchiseptica*	*avium*	*trematum*	*holmseii*	*hinzii*	*petrii*	*ansorpii*	*bronchialis*	*sputigena*	*flabilis*	*pseudohinzii*
Host Range	Human	Human/Sheep	Dog, Pig, Rodents Horse	Fowl	Humans	Humans	Birds, Humans	Environment, Humans	Human	Humans	Humans	Humans	Mouse
Disease or Specimen Source	Whooping Cough (typical)	Whooping Cough (atypical)	Kennel Cough, Atrophic Rhinitis, Snuffles	RTI	Wound	RTI (Whooping cough-like) Bacteremia	RTI in Birds, Bacteremia in Humans	upper RTI, Wound	Epidermal Cyst, Wound	RT of CF patients	RT of CF patient	RT of CF patient	NA
Visible colonies on agar (37°C)	3–5 days	2–3 days	1–2 days	24 hours	16–24 hours	3–5 days	2 days	2 days (30°C)	1–2 days	3 days (28°C)	3 days (28°C)	3 days (28°C)	48 hrs
Agar growth medium	BGA R-L	BGA R-L	BGA R-L SBA MAC	BHI SBA MAC	SBA	BGA SBA	TSA	BGA MAC SBA	SBA MAC	TSA SBA	TSA SBA	TSA SBA	BGA
Broth growth medium	SSB	SSB	SSB	BHI broth	BHI broth	BHI	BHI	LB broth	NA	MH	MH	MH	MH
Type strain	Tohama 1	12822	RB50	197N	LMG 13506	ATCC 51541	LMG 13501T	SE-111RT	SMC-8986T	LMG 28640T	LMG 28641T	LMG 28642T	8-296-03T
Genome size (basepairs)	4,086,186	4,773,551	5,338,400	3,732,255	4,485,537	3,699,674	4,885,897	5,287,950	6,210,412	5,878,756	ND	ND	4,490,371
Motility	–	–	+ (Bvg–)	+	+	–	+	–	+	+	+	+	+
LPS	No O side chain	O side chain	O side chain[b]	O side chain	O side chain	No O side chain	O side chain	No O side chain	Possible O side chain	ND	ND	ND	ND
Pertussis toxin	+ (Bvg+)	Not expressed	Delayed	Not encoded	Not encoded	Not encoded	Not encoded	Not encoded	Not encoded	ND	ND	ND	ND
bvg operon	*bvgASR*	*bvgASR*	*bvgASR*	*bvgAS*	*bvgAS''*	*bvgAS*	*bvgA'*	*bvgAS*	Neg	ND	ND	ND	ND
Targets for Molecular analysis (approximate copy number/genome)[126]	*IS481*(200) *IS1001*(200) *IS1002*(10) *IS1002*(<5) *ptxP*(1)	*IS1001*(20) *IS1002*(<5)	*IS481*(<5)[c] *IS1001*(<5)[d]	ND	ND	*IS481*(10) h*IS1001* *recA*(1)	*opmA*	ND	ND	ND	ND	ND	ND

a   *Bordetella parapertussis*$_{hu}$ and *B. parapertussis*$_{ov}$ are listed in one column. However, it should be remembered that they have distinct host specificities and that there is substantial data that result in them being classified as separate subspecies.

b   mostly identical to *B. parapertussis*$_{hu}$

c   for 1% of Isolates

d   for 50% of isolates

Abbreviations:: BGA = Bordet-Gengou agar; BHI = brain heart infusion; CF = cystic fibrosis; LB = Luria-Bertani broth; MAC = MacConkey agar; MH = Mueller Hinton; ND = not determined; R-L = Regan-Lowe agar; RT = respiratory tract; RTI = respiratory tract infection; SBA = sheep's blood agar; SSB = Stainer-Scholte broth; TSA = trypticase soy agar.

The regulation of virulence-associated antigens in the *Bordetella bronchiseptica* cluster is controlled by the *bvg* regulon.[1,33,41,42] "Bvg" is short for *Bordetella vir* (virulence) genes.[1,43] The Bvg regulon encodes three key gene products (BvgA, BvgS, and BvgR). BvgS and BvgA make up a classic two-component regulatory system involving a sensor and an activator, respectively. However, the BvgS sensor has more complexity than most common two-component sensors, including another phosphorylation site in what is known as the (cytoplasmic) linker domain.[1,38,39,44] The linker region and associated phosphorylation site add complexity to the signaling system that allows it to act more as a "rheostat switch" rather than an "on-off" signaler.[45–47] Evidence indicates that this allows the signaling system to control a more highly defined and complex phenotypic control than some other two-component global gene regulation systems. BvgS is a cytoplasmic membrane protein that senses environmental signals such as temperature, certain anions such as sulfate, and special molecules such as nicotinic acid.[44] At body temperatures and at *in vivo* conditions, BvgS is phosphorylated through a cascade relay system and transfers the phosphate group to the cytoplasmic protein BvgA. Phosphorylated BvgA can then bind to gene sequences that allow for transcription ("activation") of the virulence genes downstream from the promoter sites where BvgA binds.[48] The shorthand for genes that are transcribed this way is "*vags*," which stands for *vir*-activated genes. (Strictly speaking, these should be called "*bvg*"-activated genes" but the old name for the bvg locus was "*vir*" and the term *vags* was coined at that time).[48] Organisms expressing *vags* are said to be in the Bvg+ state. This is the fully virulent state of the organism as it is typically isolated from infected individuals.

One of the genes activated by BvgA is called BvgR, which is a repressor of a second family of genes called "*vrgs*" or *vir*-repressed genes.[40,49,50] When *vags* are highly expressed, *vrg* expression is repressed, and vice versa.[51] There are, however, a group of genes that can be expressed when environmental conditions result in an intermediate of BvgS and BvgA phosphorylation, and these are known as *bvg*-intermediate genes or "*vigs*."[52–56] The amount of phosphorylated BvgA has been shown to directly affect the positive or negative regulation of these *vigs*, independent of the BvgR repressor protein.[53,54,57] In sum, depending on the environment in which the organisms find themselves, certain gene products can be expressed or not, and this regulation is referred to as phenotypic modulation (or sometimes antigenic modulation).[34,35,46,54,58]

Environmental conditions are one way the organisms can change their phenotypes. Another mechanism is through mutation. In particular, a string of cytosines in the *bvgS* gene of *B. pertussis* is vulnerable to modification, causing *bvgS* to go out of frame and to not be expressed.[59] This results in a phenotype in which only *vrgs* are expressed and is therefore phenotypically Bvg–. These mutations are relatively rare (e.g., 1 in 10^6 colonies per subculture), depending on the strain, and are reversible.[38,59] The process has been called phase variation and importantly, in the pre-antibiotic era has been observed during the course of disease in *B. pertussis* human disease isolates.[60] When growing *Bordetella* organisms, one needs to keep in mind the potential for such alteration of phenotype, both spontaneous phase variation and the influence of media composition on antigenic modulation. Furthermore, despite relatively limited genetic variability, *B. pertussis* is recognized as having the ability to alter its genomic structure through accumulation of various mutations, including point mutations, gene deletions, and chromosomal inversions that are facilitated by insertion elements in its genome.[61–63] A gene-array study demonstrated alteration of gene expression in loci after as few as 12 culture passages.[64] When cultivating *B. pertussis*, it is of great importance to make a reserve stock culture of isolated strains and take care not to over-passage.

The remainder of this chapter will highlight the most important features of each species, with emphasis on *B. pertussis*.

Bordetella pertussis

Background

As discussed previously, *B. pertussis* was the first member of the *Bordetella* genus to be isolated and characterized. This priority reflects the negative impact the organism has had on human health as the agent of whooping cough in children. According to the Centers for Disease Control (CDC), prior to routine immunization in the United States in the 1940s, nearly 200,000 cases were recorded annually with approximately 9,000 fatalities.[65] Most affected were children under the age of 5 years, although whooping cough can occur at any age.[1,66,67] The incidence of pertussis in the United States for 2018 was reported to be 15,609 with five deaths.[68] This is substantially lower than the 2012 statistics mentioned in the previous edition, which was 48,277 with 20 deaths.[69]

Looking at the impact of pertussis disease globally is more challenging. Estimates of pertussis disease reported by the World Health Organization (WHO) in 1999 were 30.6 million cases and 390,000 deaths from pertussis in children younger than 5 years. In 2008, approximately 16 million cases of worldwide pertussis were estimated with 95% occurring in developing countries, resulting in approximately 195,000 child fatalities.[70] A more recent modeling study for 2014 estimated 24.1 million pertussis cases and 160,700 deaths in children younger than 5 years.[71]

Disease

B. pertussis is most often recognized for its role as the causative agent of whooping cough and the disease will be described in more detail in this section. As molecular detection of the organism has become more readily available, its possible implication in other respiratory disease such as community acquired pneumonia and asthma is being proposed and investigated.[72,73]

Whooping cough, also known as "pertussis," begins on average 7–10 days[74] after organisms are transferred from an infected individual to a susceptible person usually by aerosols generated through coughing or by direct contact with contaminated unwashed hands. *B. pertussis* was thought not to have any known environmental or animal reservoir nor was it traditionally thought to have a classical carriage state in people where it might be harbored in between outbreaks.[1,7,75–77] A recent systematic review of the literature, however, found that multiple studies have identified the presence of laboratory confirmed *B. pertussis* in asymptomatic household contacts and several studies provided evidence of asymptomatic transmission.[77]

Classic whooping cough, occurring primarily in unvaccinated children, will usually present with symptoms in a well-known pattern.[75] The first symptoms of classic pertussis disease will usually appear within 5–10 days of exposure and are much like those of a common cold, including runny nose, sneezing, and in some, a slight fever. This is known as the catarrhal stage. The second or paroxysmal stage, which begins during the second week of illness is named for the most dramatic and hallmark symptoms of the disease, the bouts of severe coughing fits, or paroxysms that occur. In a typical case, the patient coughs repeatedly and forcefully, multiple times in rapid succession. In a concentrated attempt to draw in air, the glottis is often narrowed, and the inspiration produces a whoop or crowing sound that gives the disease its name. Coughing is commonly so forceful that vomiting occurs. Episodes of paroxysmal coughing increase during this stage and tend to be most common at night. The lack of oxygen caused by the cough sequence is one of the main dangers of the disease and can lead to brain damage or even death. Coughing can also result in broken blood vessels in the conjunctivae, hernia, and prolapsed rectum. As the individual's immune defenses gain the upper hand in 1–4 weeks, the severity and frequency of the paroxysmal coughs subside slowly, taking up to 6 months to completely disappear. Even then, subsequent respiratory infections or irritations can trigger a cough reminiscent of the paroxysmal stage. The recovery period is called the convalescent stage; and if the individual has not experienced the more severe consequences of the paroxysmal stage, there usually are no other permanent side effects. Individuals who have recovered from pertussis can get the disease again.[1,78]

Infants demonstrate the highest mortality, presenting often only with initial symptoms of apnea and cyanosis.[70] Seizures in this age group are not uncommon. Infant fatalities are often a result of severe pulmonary hypertension leading to cardiac arrest.[1] Adults and adolescents can have a range of symptoms from persistent cough to those consistent with classic pertussis described above.[70] Infection in these age groups can occur in those that have had either a prior pertussis infection or the full series of immunization as immune response does not appear to offer long lasting protection.

Treatment

Susceptibility testing of *B. pertussis* is not routinely performed. Susceptibility to a wide range of antibiotics with the exception of the oral cephalosporins have been shown *in vitro*.[74] Antibiotic treatment for *B. pertussis* as recommended by the CDC (http://www.cdc.gov/pertussis/clinical/treatment.html) include the macrolides with trimethoprim sulfamethoxazole as an alternative for patients over 1-year old. Rare erythromycin resistance has been demonstrated and shown to be a result of a mutation in the macrolide binding site of tSe 23s rRNA.[79]

Treatment should be administered prior to the onset of the paroxysmal stage to have the benefit of a shorter symptomatic period and clearing of organism from the upper respiratory tract. This leads to a shorter period of possible transmission to other susceptible individuals lasting only 5 days versus 3 weeks for an untreated patient.[80] If treatment is not administered until after the paroxysmal symptoms start, clinical benefit is questionable. This is thought to be consistent with the pathogenesis of disease that is mediated through toxin production. Once the protein toxins are bound and internalized into their target human cells, the damage is done. Killing the infecting bacteria will prevent any further toxin production, but has no effect on the toxins already doing their job.[1,81,82]

Vaccination

The first vaccine for pertussis was implemented in 1943 in Canada and in 1944 in the United States. Most commonly, this consisted of formalin-treated, whole killed bacteria. The whole-cell pertussis vaccine is administered with diphtheria and tetanus toxoids in a single intramuscular injection and is known as DPT or DTP. A form of this vaccine is still used in developing countries and typically follows a three- to five-dose regimen starting at 2 or 3 months of age, with boosters at 2-month intervals for the first three doses, and subsequent boosters at different intervals depending on the jurisdiction administering the vaccine.[83]

In developing countries as the number of whooping cough cases declined, in large part due to vaccination, public concern was voiced over the side effects of the whole-cell vaccine. Concern peaked in the late 1970s and early 1980s when purported vaccine-associated encephalopathy and deaths occurred. Several countries suspended use of the vaccine. While no scientific study had ever established a clear cause-and-effect relationship between vaccination and brain damage or death, the scientific community responded by pursuing systematic studies of the molecular basis of disease and of protective components in the vaccine.[84–86] In 1981, Japan produced the first "acellular" vaccines, which were made from components released into broth during growth.[87] The Japanese vaccines were used with good effect in Japan; but in North America and Europe, further purification and quantification of individual antigens followed by extensive clinical trials, resulted in the first acellular vaccines approved for use in the United States in the late 1990s.[88] Because of the increased costs associated with production of the acellular vaccines, many countries continue to vaccinate with the whole-cell pertussis vacinne.[89] There are currently several combination vaccines offering an acellular pertussis component along with diphtheria and tetanus protections in use in the United States.

DTaP is recommended to be administered at 2, 4, 6, and 15–18 months with a fifth dose between 4 and 6 years of age. Additional recommendations include administering a dose of Tdap to adolescents and adults to help reduce the number of infections in these age groups. To improve protection of the most vulnerable in the population, the newborn, the CDC recommendations include vaccinating pregnant women between weeks 27 and 36 of each pregnancy as that should allow maternal antibodies to protect the infants until they begin to receive their own immunizations.[90] Research supports this with evidence of improved infant protection and by demonstrating newborns have statistically significant higher antibody titers against pertussis when mothers were vaccinated.[91–95] Despite this benefit, there is also some evidence that high maternal antibody titers could actually interfere with development of an infant's own immunity.[96,97] Additional protective strategy for newborns involves cocooning—urging all family members and potential close contacts of the infant to be up-to-date on their pertussis immunization.[98]

In the years that followed the administration of the acellular vaccines, there had been increased incidents of pertussis reported by the CDC with numbers reaching close to 50,000 cases in the United States in 2012.[99] Outbreaks had been documented in many states in the United States at that time, calling into question the effectiveness of the attenuated vaccines in preventing pertussis illness and brought attention to the apparent waning immunity.[98,100,101] In 2014, the pertussis vaccination was reportedly at 86% globally, but concern over increasing pertussis numbers existed.[102] This prompted the WHO to conduct a review of data from 19 countries. They concluded a global resurgence of the illness had not occurred, rather the dramatic increases in pertussis illness were limited to five countries: the United States, United Kingdom, Australia, Chile, and Portugal. The WHO further stated that several factors were contributing to the increase of pertussis in these countries, such as improvements in awareness, surveillance, and diagnosis in addition to the naturally occurring cycles of increased pertussis that happen every 2–5 years.[103] While these factors do, in fact, play an important role in the recent pertussis situation, it has been proposed that the real key to understanding the resurgence of pertussis disease lies in understanding the differences in immune response between the vaccine types, in particular the inability of the attenuated vaccine to provoke a mucosal immune response and prevent colonization with the organism, which appears to happen with the whole-cell vaccine or natural infection.[104–105]

To combat pertussis, work continues to develop and test new formulations of vaccine in hopes of achieving a more lasting protection. Researchers are investigating the use of live, attenuated intranasal or oral vaccines as an alternative to those currently in use.[106–112]

Clinical and Laboratory Diagnosis

Diagnosis of whooping cough can be accomplished by defined clinical criteria but is clearly strengthened by laboratory confirmation. Differing clinical definitions of whooping cough have resulted in different interpretations of vaccine efficacy and therefore need to be clearly stated and applied for meaningful results.[87] The ability to choose the most advantageous method of laboratory diagnosis for pertussis is greatly influenced by variability of several factors including: disease stage, vaccination and patient age. A recommendation for navigating these complexities is available in schematic form in the literature.[113]

Isolation and Growth

Isolation of the organism has been an important element in understanding infectious pertussis disease and is vital for future study. The "gold standard" for diagnosis of whooping cough is still the traditional isolation of the causative agent by growth on appropriate culture media. Isolation of living bacteria also provides a resource from which epidemiological studies of changes in genetic and antigenic structure can be monitored over time. Cultivation of living organisms from active cases continues to present a challenge. Culture is most successful within the first 2 weeks of the onset of cough.[114] Traditionally, collection of a sample involved using the cough plate technique: a Bordet-Gengou (BG) agar plate held in front of the patient during a coughing episode, to recover organism.[115]

Samples are typically taken by one of two means: a nasopharyngeal (NP) swab or an NP aspirate. *B. pertussis* can be found in ciliated epithelial cells which explains why the NP and not the throat, are acceptable for culture. The NP swab consists of a long (14 cm) flexible wire with a small swab. Dacron, calcium alginate, or nylon are acceptable materials for the swab, which is designed to be inserted through the nose to reach the back of the nasopharynx where the organisms reside. To increase the success of isolation, separate samples are often taken from each side of the nose. These are then either directly plated onto appropriate agar medium such as Regan-Lowe (RL), containing 40 g mL^{-1} cephalexin to suppress the growth of other intranasal bacteria, or are placed in a suitable transport medium (i.e., Stainer-Scholte broth (SSB) or RL transport medium) for delivery to a laboratory for further subculture. Likewise, the thin canula attached to a syringe containing a buffered wash solution is used to collect an NP aspirate from the same area.[116]

In successful culture attempts, white colonies about 1 mm in diameter appear within 3–7 days (may be up to 12 days) of incubation in a humidified environment at 36°C.[67] Recommended CO$_2$ level is less than 4%. Automated bacterial identification systems are not reliable for identification of *B. pertussis*. Biochemicals such as catalase, oxidase, urease, citrate, and nitrate were traditionally performed. *B. pertussis* is typically described as being nonmotile, but recent observations of motile strains from both laboratory-derived and clinical sources have been described.[117] More modern methods of identification are recommended over traditional biochemicals and include PCR and MALDI-TOF.[74]

Once viable organisms are recovered, they can be either lyophilized or frozen at −70°C for long-term storage. A Dacron swab is used to remove the pure growth from an entire 9-cm plate, and suspended in a suitable cryoprotective medium, such as 20% (w/v) skim milk.[118] For research purposes, growth of organisms from stock on BG medium containing 15% defibrinated sheep blood is recommended.[119] This is so the phenotype of the *Bordetella* organisms can be observed and sub-cultured faithfully. For example, wild-type (Bvg+) clinical isolates of *B. pertussis* grow as small (1–2 mm diameter) domed colonies with notable hemolysis on BG agar. Bvg– phase variants, in contrast, form larger (3–4 mm) flat, nonhemolytic colonies see Figure 23.1.[119] BG agar does not have the same shelf life as R-L agar but the charcoal base of R-L agar prevents the observation of hemolysis. For growth in broth, the preferred medium is SSB or modified SSB, which is aerated through shaking or by having a large surface-to-volume ratio for static liquid culture.[119,120] SSB is a synthetic medium developed for the commercial production of pertussis vaccine and has a good

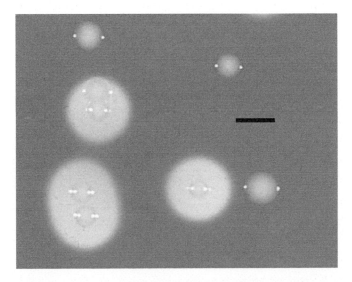

FIGURE 23.1 Colonies of *B. pertussis* phase variants on Bordet–Gengou agar (BGA). Bvg+ (domed, hemolytic) and Bvg– (flat, non-hemolytic). Bar = 3 mm.

track record. A drawback is that it must be made *de novo* in the laboratory because there is no commercial supplier. Other broth media have been used but the defined composition of SSB is a distinct advantage when it comes to reproducibility.

Non-culture Identification of Bordetella pertussis

Polymerase Chain Reaction (PCR).   One of the great advances in the diagnosis of whooping cough has been the introduction of PCR and its evolution from earlier conventional PCR to real-time PCR (RT-PCR), to its incorporation into multiplex respiratory panels. Many clinical labs in the United States have adopted these modern methodologies over traditional culture, and it has had a real impact on the number of reported cases.[121] PCR is an acceptable test method during the first 4 weeks of cough.[114] The technique is highly specific and is sensitive enough to detect very low numbers of organism. Unfortunately, the presence of nonviable DNA in a sample is also detectable, so care is needed to prevent contaminating DNA from giving a false positive and causing a pseudooutbreak.[122] Healthcare providers are cautioned to be aware of the ability of pertussis DNA, found in some vaccines, to contaminate patient samples. Best practices mandate that to prevent such false positives from occurring, specimen collection for pertussis testing should not take place in the same area as pertussis vaccination.[123] While both NP aspirate and NP swab are suitable for PCR analysis, an NP aspirate is preferred because the swab is easier to contaminate.[74,123,124] Due to the numerous commercially available or in-house developed PCR tests for pertussis testing, clinicians should consult the protocol issued by the testing laboratory for specifics about specimen collection and transport.

Several different targets are utilized in PCR testing of the *Bordetella*. Several insertion sequences (IS) were identified and used in early PCR testing but cross-reactivity had led to some diagnostic problems. IS*481* present in *B. pertussis*, has cross reactivity with both *B. holmesii* and less often *B. bronchiseptica*. IS*1002* is present in both *B. pertussis* and *B. parapertussis* and IS*1001* is found with *B. parapertussis* and *B. bronchiseptica*.[75,125]

One assay has been developed that targets all three in an effort to better identify the species of *Bordetella* present.[126] The pertussis toxin promotor (*ptxP*) gene has also been used to enhance detection of *B. pertussis*.[127–130] The IS*1001*-like (*IS1001Bhii*) target can be used to enhance discrimination of *B. holmesii*.[131] An additional target for *B. holmesii* is *recA*, an identified housekeeping gene.[132] A PCR test has been developed to detect Erythromycin resistant *B. pertussis* directly in nasal-pharyngeal swabs.[133]

Advances in technology have resulted in the automation of PCR assays that can be completed in a matter of hours. Such technology exists for assays designed strictly to identify *Bordetella* or for those multiplex assays that test for a host of additional potential viral and bacterial respiratory pathogens.[134] The latter are commercially available and serve as useful tools for clinical laboratories that lack a fully functional molecular laboratory.[113] The former can be commercially available or may be developed in house by a particular laboratory. Validation of all laboratory developed assays should be performed. This will assure quality. and guidelines for performing validations can be found in the literature.[113]

It is important to be aware of the limitations and variability of PCR result reporting from the many assays available for use in laboratory testing. PCR with a single-copy target may result in a false negative result due to gene loss as evidenced by the fluidity of the *Bordetella* genomes.[113] Diagnostic algorithms to aid in interpretation of molecular analysis for pertussis disease have been proposed and discussed in the literature.[135] It is predicted that the inclusion of *B. pertussis* in routine multiplex PCR used to diagnose respiratory disease may further implicate the organism in additional respiratory disease such as community acquired pneumonia.[72]

Fluorescence Microscopy.   Fluorescently labeled monoclonal antibody is commercially available to detect and distinguish between *B. pertussis* and *B. parapertussis*. The specificity is adequate, but it is no longer recommended by the CDC for clinical case definitions.[136,137] In settings where PCR equipment is not available, the monoclonal direct fluorescent antibody can be an adjunct to culture.[74]

Serological Diagnostics.   Problems with diagnosis of pertussis through serology exist and include: cross-reactivity, vaccine interference, and previous *Bordetella* infections. Despite these, it remains an important method of diagnosis while patients are in the latter stages of pertussis disease. The CDC recommends serology as a diagnostic tool to be used beginning 2 weeks after cough begins through week 12.

A number of studies have demonstrated the feasibility of using isolated antigens from *B. pertussis* as the foundation of antibody-based diagnosis by enzyme-linked immunosorbent assays (ELISAs).[113] Issues for standardization surround the number of serum samples and the number of antigens needed for reliability of the results.[138,139] Serologic tests are often used to retrospectively diagnose pertussis illness in adults, adolescents, and in older children.[75] This method should not be used, however, within a year of vaccination with acellular pertussis vaccine.[74] The CDC classifies a clinical case with only serological evidence of pertussis as a probable case.[137]

Epidemiology

Pertussis has changed from a disease of principally nonimmunized infants, aged newborn to 6 months, to a disease that includes a large proportion of cases in adolescents and adults.[140] Increased recognition and diagnosis of pertussis in older children, adolescents, and adults has been a result of several factors including: the clarification of case definitions, increased emphasis on reporting, and the availability of new molecular diagnostic technologies.[140]

CDC data on pertussis in the United States dates back to the 1920s. Passive reporting was done between 1922 and 1949 and cases were in the hundreds of thousands annually. After the introduction of the whole-cell vaccine in the 1940s, the number of pertussis cases trended downward to less than 10,000 by the mid-1960s. They remained about this level until the change to the attenuated pertussis vaccine was implemented. Since the change, the number of pertussis cases has gradually risen to greater than 15,000 annually with several years spiking much higher. The latest data shows a downward trend has once again been achieved in 2018.[68]

The early use of PCR for diagnosing pertussis was not without controversy as it contributed to misidentification of other members of the *Bordetella*, including *B. parapertussis* and *B. holmesii*, as *B. pertussis*. This is believed to be responsible for causing inaccuracies in reports of pertussis numbers resulting in false elevation. Another potential source of false positive diagnosis of pertussis has been shown to be the contamination of PCR specimens with DNA from pertussis vaccines. Vaccine DNA remains on environmental surfaces or the hands of workers in clinics where vaccines are administered and can contaminate specimens and lead to false reporting.[122]

It is also believed that the increasing numbers of unvaccinated children is a contributing factor to pertussis like illness in the United States.[142] Waning immunity helps to create vulnerable populations.[143,144]

In addition to the theories that have been proposed as contributing factors to the recent increase of pertussis infection seen in the United States between 2012 and 2014, emergence of vaccine resistant strains should be mentioned. Data suggests that evolutionary changes have led to the emergence of and persistence of clinical pertussis strains that differ from the vaccine strain. This difference appears to have emerged as a persisting evolutionary change in gene order.[145] It is important that the emergence and persistence of vaccine resistant strains should continue to be investigated.[146] Specifics of some of these changes are highlighted in the next few sections.

Pertactin Negative Pertussis

As the investigation into the biological cause of increasing numbers of pertussis disease began, documented clinical strains showing antigenic differences from vaccine components strains were found.[147,148] Pertactin is a major component in acellular vaccines and reports of pertactin negative strains of pertussis have been demonstrated in France,[149] Finland,[150] Japan,[151] Australia,[152] and the United States[153] prompting the attention of the CDC. The resulting CDC investigation of the genomic diversity of clinical isolates recovered in the United States between 2000 and 2013 demonstrated that pertactin-deficient strains increased markedly in number from 2010 to 2013.[154,155] Reports from Australian outbreaks that occurred between 2013 to 2017 indicate that pertactin-negative strains continued to persist.[152] Contrastingly, analysis of pertussis in Argentina, where vaccination with wP remains, showed low prevalence of pertactin-negative strains.[156] Monitoring of pertactin-negative strains is important for understanding changing pertussis disease and should continue.

Pertussis Toxin Production

The emergence of pertussis strains that produce higher levels of pertussis toxin associated with the *ptxP3* allele were first identified in the 1980s but continue to persist throughout the world.[148] Nearly 96% of isolates analyzed from a recent U.S. outbreak were *ptxP3* strains and produced higher levels of pertussis toxin.[155] Australia also saw an increase of *ptxP3* in its most recently described outbreak.[152] One European study looked at isolates from nine European countries between 2012 and 2015 and nearly 96% of the strains tested were found to carry the *ptxP3* allele.[157]

Research Methods

The characterization of different isolates of *B. pertussis* is an important adjunct in monitoring changing trends in pertussis disease.[1,81,158] An elaborate scheme for characterizing *B. pertussis* isolates by agglutination with adsorbed polyclonal antisera was developed by Eldering in the 1960s.[159,160] As antigens associated with virulence were characterized genetically and biochemically, assays had been developed to replace the traditional serological method with molecular techniques. These new methods provide much more sensitive and sophisticated tools for observing subtle changes among isolates.[161–163] Molecular analysis continues to be advanced and is applied by researchers to understand global pertussis patterns.[164,165]

Genomic Polymorphisms

Pulsed-field gel electrophoresis (PFGE) has provided a reliable means by which to fingerprint *B. pertussis* isolates.[166–170] This is a particularly useful tool for confirming the identity of known research strains, akin to karyotyping of tissue culture cells. For epidemiology of disease isolates, PFGE provides an entry point for comparisons. Other genomic techniques have been proposed for strain monitoring, but none has been uniformly adopted worldwide.[170,171]

Allelic Polymorphisms

Changes in gene sequences have been used to determine the genetic relatedness of individual strains of *B. pertussis* and for tracking changes in specific genes. Of special interest are genes associated with virulence that would amount to antigenic drift over time in disease isolates. Changes based on PCR and sequencing or single-strand polymorphisms have demonstrated insertions and deletions in pertussis toxin, pertactin, and fimbria genes that support a shift to new allelic types over time.[170,171,173–175]

Pathogenesis and Virulence

Overview

Because *B. pertussis* is only found in humans, the use of an animal model is necessary to study the definitive stepwise process of infection and disease. A commonly used animal model is the mouse aerosol challenge model. While mice provide insights into the disease process, it is important to note that the aerosol challenge causes a pneumonia and mice do not experience paroxysmal coughing.[176–178] The following scenario is pieced together from various animal sources and from human clinical data, but there are gaps that remain.

Once *B. pertussis* is inoculated into the human nose by aerosol or by hands contaminated with infected secretions, the organism colonizes the nasopharynx using adhesins, most likely fimbriae or filamentous hemagglutinin (FHA).[1,179,180] Organisms also colonize the bronchi where they adhere to the cilia of the respiratory epithelium by means of FHA, and likely other adhesins, and begin to grow. As part of their growth, several toxins are secreted that cause the cilia to stop beating and also affect the respiratory epithelium to cause an accumulation of mucus and trigger the paroxysmal cough.[1,74,181]

Organisms appear to be sequestered in the respiratory tree and are not found in the bloodstream, even in immunocompromised patients. Thus, *B. pertussis* does its damage in the extracellular compartment where in nonlethal cases it is eventually cleared by the host immune response and the respiratory epithelium returns to normal.[1,70] Post mortem examination of infant respiratory tissue samples showed histopathology characterized with necrotizing bronchiolitis, bronchopneumonia, pulmonary hemorrhage, pulmonary edema, and angiolymphatic leukocytosis.[182] Severe leukocytosis has been examined as an outcome predictor of severe infant disease.[183]

An impressive body of work has been devoted to the study of the molecular basis of pathogenesis in whooping cough. The goal of much of this work was to establish which proteins should be included in acellular vaccines to give the most complete immunity to *B. pertussis*. Not all the molecules known to contribute to virulence are found in an acellular vaccine, nor are all acellular vaccines composed of the same number of proteins. The following list will therefore be grouped based on whether the molecules are found in the acellular vaccine or not, and on the basis of how commonly they are found in the vaccine.

Virulence Factors of Bordetella pertussis

Virulence Factors Included in Acellular Vaccines.

1. **Pertussis toxin (Ptx):** The one protein found in all acellular vaccines is pertussis toxin. It has been shown to be necessary to produce disease in most animal models and has been attributed to many biological activities. This is due to its ability to enter cells and modify the inhibitory GTP binding protein of the mammalian cell adenylate cyclase system by covalently adding ADP-ribose derived from intracellular nicotinamide adenine dinucleotide (NAD). The consequence to the target cell is deregulation of normal cellular processes and, depending on the cell affected, can result in altered function that ultimately assists *B. pertussis* in its ability to infect in the lung. The crystal structure of Ptx has been determined; it is a unique A-B type toxin with an A (enzymatically active) subunit called S1 and a heteropentameric B (binding) subunit composed of one S2, one S3, two S4, and one S5 subunit.[164] Despite the clear association of Ptx with virulence, there is no definitive correlation between serum anti-Ptx levels and protection in humans. Ptx, along with cholera toxin, has also been widely used to study the cellular physiology of the 16 mammalian adenylate cyclase systems and other regulatory systems for which GTP-binding proteins play a key role in cellular homeostasis.[1,74,180,181] Worldwide emergence and persistence of various strains of *B. pertussis* that are found to contain the ptxP3 allele, which is associated with increased toxin production has been documented.[152,155] It is still unclear what exact role this emergence might play in recent increases in infection.[185]

2. **Filamentous hemagglutinin (FHA):** FHA is a unique adhesin with a rod-like structure and an affinity for sulfated glycolipids and cholesterol. It is presumed that antibody to FHA provides protection from colonization with virulent *B. pertussis*.[1,179,180]

3. **Pertactin (Prn):** Pertactin is an outer-membrane protein that has adhesive characteristics.[1,179,186,187] Its importance as a vaccine component was strongly supported by results of the early Japanese acellular vaccines that showed the presence of Prn in the most protective vaccines.[187] However the emergence of pertactin-negative strains along with pertactin polymorphism is a cause for concern and its impact on acellular vaccine protectiveness is unclear.[151] Research indicates that the majority of strains isolated in the United States during recent outbreaks were in fact, Prn negative.[188,189]

4. **Fimbriae (Fim):** Two major antigenic types of fimbriae recognized in *B. pertussis* are Fim2 and Fim3. Each is thought to play a role in the attachment of *B. pertussis* to respiratory epithelium. Subunits of each type have also been identified.[190,191] Epidemiological evidence supported the importance of having both Fim types in whole cell vaccines, and this observation was applied to the production of some of the licensed acellular vaccines.[1,180,187,192]

Virulence Factors Not Included in Acellular Vaccines.

1. **Adenylate cyclase-hemolysin or adenylate cyclase toxin (CyaA):** Early work by Alison Weiss showed that *cya* mutants were as avirulent in mice as mutants in pertussis toxin,[193,194] but the cloning and characterization of this complex molecule lagged behind that of the other virulence factors. It was never included in any of the acellular vaccines that are currently licensed. CyaA is responsible for the hemolysis produced by Bvg+ organisms on BGA and is also one of the most actively catalytic adenylate cyclases known in the bacterial world.[1,180,195,196] For this reason, it has been successfully used as a reporter molecule for bacterial two-hybrid systems.[197] In addition, the adenylate cyclase portion of this bifunctional protein is known to impede phagocytosis of *B. pertussis* by human neutrophils through its enzymatic activity on these host cells.[198–201]

2. **BrkA:** This protein is another *vag* product and gets its name from its ability to confer to *Bordetella* resistance to killing by serum.[1,202,203] The protein is a member of the autotransporter family of proteins and can mediate its own assembly into the outer membrane of *B. pertussis*.[1,204–206]

3. **Lipopolysaccharide (LPS):** The lipopolysaccharide of *B. pertussis* has the same endotoxic characteristics as other typical gram-negative bacteria but interestingly does not produce O side chains. On SDS-PAGE and oxidative silver staining, the great majority of clinical isolates produce two distinct bands.[37] The composition of these two species of short LPS is known, as are all of the genes responsible for its production.[207–209] Due to its endotoxicity, LPS was targeted as one of the components that should be absent from any acellular pertussis vaccine.[87]

4. **Tracheal cytotoxin (TCT):** TCT is a breakdown product of peptidoglycan synthesis in most *Bordetellae*, regardless of phase or phenotypic state. It has a profound effect on causing cilia to stop beating and is thought to be important in the pathogenesis of *Bordetella* disease. However, it is poorly immunogenic and has not been included in any acellular vaccine.[1,180,192,210]

5. **Dermonecrotic toxin (DNT):** DNT, a product of a *vag*, is another true toxin that is sensitive to denaturation by heat and is poorly immunogenic. It causes cell death in low concentrations and is thought to contribute to pertussis pathogenesis in humans although its precise role is still unknown.[1,180,192,211]

Bordetella parapertussis

B. parapertussis exists in two distinct lineages. One, designated *B. parapertussis*$_{hu}$, is a pathogen of humans and causes a milder form of whooping cough. The other lineage is designated *B. parapertussis*$_{ov}$ and is principally a respiratory pathogen of sheep, although human infections have been reported.[30] Each lineage is impressively homogeneous genetically with minor differences among strains, especially *B. parapertussis*$_{hu}$.[3]

One of the most intriguing aspects of *B. parapertussis*$_{hu}$ is its ability to cause a disease very similar to whooping cough without producing pertussis toxin. Whether or not *B. parapertussis*hu has caused true outbreaks of whooping cough on its own in the past remains open for debate.[212,213]

An interesting difference worth noting between *B. parapertussis*hu and *B. pertussis* includes the presence of a complete O antigen on *B. parapertussis*hu. It has been demonstrated that the O antigen plays a role in the inhibition of binding of *B. pertussis*-induced antibodies allowing the survival of *B. parapertussis* in immune hosts.[213]

*B. parapertussis*hu grows more rapidly than *B. pertussis* and is able to grow on blood-free media (such as nutrient agar), where it typically produces a pronounced browning of the medium due to the expression of an enzyme that cleaves tyrosine.[9] Colonies on RL appear larger and are not as shiny as *B. pertussis*. One routine biochemical difference between *B. pertussis* and *B. parapertussis* is the oxidase reaction, with *B. parapertussis* testing negative. Like *B. pertussis*, attempts at identification on commercial gram-negative identification systems are not recommended, while MALDI-TOF and molecular methods are. Molecular testing indicates the presence of *B. parapertussis* when positive results are reported for assays that target both IS1001 and IS1002.[125]

An important question that has been raised in the past is whether *B. parapertussis* might have had a resurgence with the advent of acellular pertussis vaccines. There is some evidence that whole-cell pertussis vaccines provided some cross-protection to *B. parapertussis*hu.[214] In contrast, the composition of even the five-component vaccines would not necessarily have enough similarity to antigens in *B. parapertussis*hu to have the same beneficial effect. Animal models show that there appears to be an enhancement of *B. parapertussis* infection following vaccination with aP vaccine allowing the development of more favorable hosts.[215] Reports on this question have come to conflicting conclusions.[216,217] Recent work, in the mouse model, with a new protein candidate for inclusion in pertussis vaccine, AfuA, indicated immune protection against both *B. pertussis* and *B. parapertussis* was achieved.[218]

Not nearly as much research has been focused on understanding the mechanisms of *B. parapertussis* pathogenesis. However, in the rat model of whooping cough, *B. pertussis* was able to cause rats to cough and *B. parapertussis*hu was not.[177] Infections with *B. parapertussis* outside of the respiratory tract are rare but have been reported.[217]

Bordetella bronchiseptica

B. bronchiseptica possesses genetic diversity. This makes some intuitive sense when one appreciates it has the second largest genome (only *B. ansorpii* is larger) and a broad host range. It has even been shown to survive in lake water.[219] *B. bronchiseptica* produce flagella as the product of a *vrg*. Its growth rate rivals that of *Escherichia coli* and it has full O side chains that are nearly identical to those of *B. parapertussis*hu.[207]

The regulation of virulence in *B. bronchiseptica* has been studied in detail; and because the Bvg locus is interchangeable with that of *B. pertussis*, *B. bronchiseptica* has been used as a model for virulence regulation in the *Bordetellae*. Rodents are part of the natural host range of *B. bronchiseptica*, so infection models are numerous and relevant. Its faster growth rate makes it an attractive organism to study. *B. bronchiseptica* has been shown to invade eukaryotic cells and has even been isolated from human blood.[220–222]

B. bronchiseptica is found in the respiratory tract of domestic mammals including dogs, cats, guinea pigs, and rabbits. It is the cause of kennel cough or tracheobronchitis in dogs.[223] Transmission occurs by close contact with animals or exposure of immunocompromised hosts to infectious materials.[224] The species can also be found as a commensal microbe of the upper respiratory tract of humans.

B. bronchiseptica is motile via peritrichous flagella and can be grown on routine culture media, such as 5% sheep blood agar (SBA) as small, convex colonies and on MacConkey as non-lactose fermenting rods[174] (Table 23.1) It has also been recovered from thioglycolate broth when incubated in 5% CO_2.[224] Identification of *B. bronchiseptica* was traditionally dependent on staining characteristics and biochemical reactions. The use of routine biochemical identification systems in clinical laboratories appears better than with *B. pertussis* and *B. parapertussis*. Supplemental testing through MALDI-TOF or molecular methods is recommended.

For all its capabilities, it is somewhat surprising that *B. bronchiseptica* does not cause more than rare zoonotic disease in humans, although these are becoming more common with the increased number of immunocompromised individuals.[225] *B. bronchiseptica*, along with *B. holmesii* and *B. hinzii*, can cause pertussis-like symptoms in immunocompromised hosts and those with underlying conditions, including HIV and cystic fibrosis.[74] Vaccines are available for veterinary use only and mainly for dogs and swine.

Other *Bordetella*

The other named *Bordetella* species will be discussed in this section. Most are easily recovered, within 1–3 days, on routine media such as sheep blood agar or trypticase soy agar. Most are considered strict aerobic organisms unless otherwise mentioned. These less common *Bordetella* species may become more commonplace in the future as more thorough identification methods are put into practice and increasing awareness of their potential role in illness is realized.

Bordetella avium

Of the classically studied *Bordetellae*, *B. avium* is genetically the least related to *B. pertussis* and yet it shares similar pathogenic mechanisms and some of the same virulence factors.[226,227] As a pathogen of poultry and other domestic birds, *B. avium* has an affinity for young turkeys and is therefore of economic consequence to the poultry industry.[29,74] *B. avium* most resembles *B. bronchiseptica*. Like *B. bronchiseptica*, *B. avium* produces flagella and can be recovered from sheep blood agar and MacConkey under aerobic conditions (see Table 23.1). Sequence data shows many differences in surface structures that likely contribute to the unique host range for *B. avium*.[226]

Both live attenuated and killed bacteria (bacterin) vaccines have been tested, but vaccination of turkeys is not generally practiced and, instead, good hygiene and biosafety practices are usually employed.[29] Fluorescent antibody and PCR tests are available for distinguishing isolates from *B. bronchiseptica* and *B. hinzii*.[29]

B. avium has demonstrated similar virulence factors to *B. pertussis* however, it is mainly a pathogen of poultry and other domestic birds.[74,227] This isolate has rarely caused human infections but is emerging as a potential pathogen of the respiratory tract.[228]

Bordetella holmesii

B. holmesii is an interesting member of the genus. In part, this is due to its similarity to *B. pertussis* in terms of its ability to cause human respiratory infection. *B. holmesii* causes a pertussis-like illness and should be differentiated from *B. pertussis* and *B. parapertussis*.[229,230] Misidentification occurred in the past through problems with the specificity of primers for RT-PCR contributing.[75,229] In the midst of pertussis outbreaks between 2010 and 2012 in both Japan and France, patients with pertussis-like illness were given an initial diagnosis of pertussis, when in fact *B. holmesii* was later found to be the causative agent.[231,232] In recent years, many other countries have reported the same after retrospective testing.

Because *B. holmesii* can be recovered from nasopharyngeal specimens and shares the insertion sequence IS481, testing for additional targets is important to avoid misdiagnosis of pertussis. Newer molecular assays have been developed with this in mind.[131,132]

Another case describes how bronchitis, presumptively thought to be caused by *Mycoplasma pneumonia*, was determined to be caused by *B. holmesii* through nucleic acid testing, reemphasizing the need for accurate diagnosis of this organism.[233]

In addition to the nasopharynx, *B. holmesii* has also been recovered in cases of bacteremia, endocarditis, pericarditis, and meningitis, highlighting its ability to cause invasive infections in the immunocompromised and otherwise healthy people.[234,235] It is possible for those with compromising disease or underlying conditions to be at greater risk of invasive disease if exposed. There is evidence of a prevalence of *B. holmesii* bacteremia among asplenic patients during increased pertussis cases in some regions of the United States.[236] Studies conducted to investigate possible development of immunity toward *B. holmesii* after a case of whooping cough, or vaccination with either wP or aP vaccine, determined protection against *B. holmesii* was inadequate.[237] *B. holmesii*, although causing similar respiratory disease, demonstrates a lack of the virulence factors of *B. pertussis* that causes more invasive disease contributing to this lack of protective immunity.

Susceptibility testing interpretations for the *Bordetella* species are not standardized and if necessary, an MIC method should be performed.[68]

B. hinzii, B. trematum, B. petrii, and B. ansorpii

The pathogenic potential of *B. hinzii* for humans has been more recently appreciated and its virulence properties continue to be assessed.[238] It remains an important concern in the avian population, especially in turkeys, where it is important to differentiate from *B. avium*. A specific PCR has been developed for *B. hinzii* identification, which targets the *ompA* gene as a result.[239] *B. trematum* has been isolated from a variety of human sources including an ear culture and various wound sources.[20,21] It has also been linked to chronic leg ulcers.[240] *B. trematum* has also been shown to progress from a soft-tissue infection into sepsis.[241] *B. petrii*, is a free-living environmental microorganism that grows under aerobic as well as anaerobic conditions.[26] It has been recovered from a case of chronic suppurative mastoiditis, and from the

respiratory tract.[24,25,242] *B. ansorpii* is predominantly an environmental organism, but possesses some of the virulence factors of other *Bordetella*, namely filamentous hemagglutinin and BvgAS virulence regulator. It lacks, however, the toxins known to this genus.[243] *B. ansorpii* has been isolated from an epidermal cyst and blood in addition to its environmental source.[74,244,245]

Newest Members of the *Bordetella*

B. bronchialis, B. sputigena, B. flabilis, B. pseudohinzii, B. muralis, B. tumbae and B. tumulicola

Of the recently named *Bordetella* species, *B. bronchialis*, *B. sputigena*, and *B. flabilis* were recovered from respiratory samples of cystic fibrosis patients and *B. pseudohinzii* was recovered from a cystic fibrosis mouse model.[246–248]

B. muralis, *B. tumbae*, and *B. tumulicola* have been found in environmental samples, such as plaster and murals, that date to more than 1,000 years old.[249] These species were not recovered from the stone that encased the art, however. It is proposed that these species may have some biodegradable capabilities as a result, but this is still being studied. Additional unnamed species have also been recovered from soil, rocks found in caves, plants, and a wide variety of bodies of water.[248] Some characteristics of the newer named species associated with human and animals are included in Table 23.1.

REFERENCES

1. Mattoo, S. and Cherry, J.D., Molecular pathogenesis, epidemiology, and clinical manifestations of respiratory infections due to *Bordetella pertussis* and other Bordetella subspecies. *Clin. Microbiol. Rev.*, 18(2), 326–382, 2005.
2. van der Zee, A., Mooi, F., Van Embden, J., and Musser, J., Molecular evolution and host adaptation of Bordetella spp.: phylogenetic analysis using multilocus enzyme 500 electrophoresis and typing with three insertion sequences. *J. Bacteriol.*, 179(21), 6609–6617, 1997.
3. Brinig, M.M., Register, K.B., Ackermann, M.R., and Relman, D.A., Genomic features of *Bordetella parapertussis* clades with distinct host species specificity. *Genome Biol.*, 7(9), R81, 2006.
4. Goodnow, R.A., Biology of Bordetella bronchiseptica. *Microbiol. Rev.*, 44(4), 722–738, 1980.
5. Heininger, U., Cotter, P.A., Fescemyer, H.W., Martinez de Tejada, G., Yuk, M.H., Miller, J.F., and Harvill, E.T., Comparative phenotypic analysis of the Bordetella parapertussis isolate chosen for genomic sequencing. *Infect. Immun.*, 70(7), 3777–3784, 2002.
6. Harvill, E.T., Cotter, P.A., and Miller, J.F., Pregenomic comparative analysis between Bordetella bronchiseptica RB50 and Bordetella pertussis Tohama I in murine models of respiratory tract infection. *Infect. Immun.*, 67(11), 6109–6118, 1999.
7. Diavatopoulos, D.A., Cummings, C.A., Schouls, L.M., Brinig, M.M., Relman, D.A., and Mooi, F.R., Bordetella pertussis, the causative agent of whooping cough, evolved from a distinct, human-associated lineage of B. bronchiseptica. *PLOS Pathog.*,1(4), e45, 2005.

8. Bordet, J. and Gengou, U., Le microbe de la coqueluche. *Ann. Inst. Pasteur.*, 20, 731–741, 1906.

9. Pittman, M., Bordetella, in *Bergey's Manual of Systemic Bacteriology*, 1st ed., Krieg, N.R. ed. Williams & Wilkins, Baltimore, MD, 338–393, 1984.

10. Moreno-López, M., El genero Bordetella. *Microbiologia Española.*, 5, 177–181, 1952.

11. Gadea, I., Cuenca-Estrella, M., Benito, N., Blanco, A., Fernandez-Guerrero, M.L., Valero-Guillen, P.L., and Soriano, F., Bordetella hinzii, a "new" opportunistic pathogen to think about. *J. Infect.*, 40(3), 298–299, 2000.

12. Vandamme, P., Hommez, J., Vancanneyt, M., Monsieurs, M., Hoste, B., Cookson, B., Wirsing von Konig, C.H., Kersters, K., and Blackall, P.J., Bordetella hinzii sp. nov., isolated from poultry and humans. *Int. J. Syst. Bacteriol.*, 45(1), 37–45, 1995.

13. Cookson, B.T., Vandamme, P., Carlson, L.C., Larson, A.M., Sheffield, J.V., Kersters, K., and Spach, D.H., Bacteremia caused by a novel Bordetella species, "B. hinzii". *J. Clin. Microbiol.*, 32(10), 2569–2571, 1994.

14. Shepard, C.W., Daneshvar, M.I., Kaiser, R.M., Ashford, D.A., Lonsway, D., Patel, J.B., Morey, R.E., Jordan, J.G., Weyant, R.S., and Fischer, M., Bordetella holmesii bacteremia: a newly recognized clinical entity among asplenic patients. *Clin. Infect. Dis.*, 38(6), 799–804, 2004.

15. Yih, W.K., Silva, E.A., Ida, J., Harrington, N., Lett, S.M., and George, H., Bordetella holmesii-like organisms isolated from Massachusetts patients with pertussis-like symptoms. *Emerg. Infect. Dis.*, 5(3), 441–443, 1999.

16. Tang, Y.W., Hopkins, M.K., Kolbert, C.P., Hartley, P.A., Severance, P.J., and Persing, D.H., Bordetella holmesii-like organisms associated with septicemia, endocarditis, and respiratory failure. *Clin. Infect. Dis.*, 26(2), 389–392, 1998.

17. Morris, J.T. and Myers, M., Bacteremia due to Bordetella holmesii. *Clin. Infect. Dis.*, (4), 912–913, 1998.

18. Weyant, R.S., Hollis, D.G., Weaver, R.E., Amin, M.F., Steigerwalt, A.G., O'Connor, S.P., Whitney, A.M., Daneshvar, M.I., Moss, C.W., and Brenner, D.J., Bordetella holmesii sp. nov., a new gram-negative species associated with septicemia.*J. Clin. Microbiol.*, 33(1), 1–7, 1995.

19. Lindquist, S.W., Weber, D.J., Mangum, M.E., Hollis, D G., and Jordan, J., Bordetella holmesii sepsis in an asplenic adolescent. *Pediatr. Infect. Dis. J.*, 14(9), 813–815, 1995.

20. Daxboeck, F., Goerzer, E., Apfalter, P., Nehr, M., and Krause, R., Isolation of Bordetella trematum from a diabetic leg ulcer. *Diabet Med.*, 21(11), 1247–1248, 2004.

21. Vandamme, P., Heyndrickx, M., Vancanneyt, M., Hoste, B., De Vos, P., Falsen, E., Kersters, K., and Hinz, K.H., Bordetella trematum sp. nov., isolated from wounds and ear infections in humans, and reassessment of Alcaligenes denitrificans Ruger and Tan 1983. *Int. J. Syst. Bacteriol.*, 46(4), 849–858, 1996.

22. Vandamme, P.A., Peeters, C., Cnockaert, M., Inganas, E., Falsen, E., Moore, E., Nunes, O., Manaia, C., Spilkers, T., and LiPuma, J., Bordetella bronchialis sp. nov., Bordetella flabilis sp. nov. and Bordetella sputigena sp. nov., isolated from human respiratory specimens, and reclassification of Achromobacter sediminum Zhang et al. 2014 as Verticia sediminum gen. nov., comb. nov. *Int. J. Syst. Evol. Microbiol.*, 65(10), 3674–3682, 2015.

23. Ivanov, Y.V., Linz, B., Register, K.B., Newman, J.D., Taylor, D.L., Boschert, K.R., LeGuyon, S., Wilson, E.F., Brinkac, L.M., Sanka, R., Greco, S.C., Klender, P.M., Losada, L., and Harvil,

E. Identification and taxonomic characterization of Bordetella pseudohinzii sp. nov. isolated from laboratory-raised mice. *Int. J. Syst. Evol. Microbiol.*, 66(12), 5452–5459, 2016.

24. Stark, D., Riley, L.A., Harkness, J., and Marriott, D., Bordetella petrii from a clinical sample in Australia: isolation and molecular identification. *J. Med. Microbiol.*, 56(Pt. 3), 435–437, 2007.

25. Fry, N.K., Duncan, J., Malnick, H., Warner, M., Smith, A.J., Jackson, M.S., and Ayoub, A., Bordetella petrii clinical isolate. *Emerg. Infect. Dis.*, 11(7), 1131–1133, 2005.

26. von Wintzingerode, F., Schattke, A., Siddiqui, R.A., Rosick, U., Gobel, U.B., and Gross, R., Bordetella petrii sp. nov., isolated from an anaerobic bioreactor, and emended description of the genus Bordetella. *Int. J. Syst. Evol. Microbiol.*, 51(Pt. 4), 1257–1265, 2001.

27. Nozomi Tazato, N., Handa, Y., Nishijima, M., Kigawa, R., Sano, C. and Sugiyama, J., Novel environmental species isolated from the plaster wall surface of mural paintings in the Takamatsuzuka Tumulus: Bordetella muralis sp. nov., Bordetella tumulicola sp. nov. and Bordetella tumbae sp. nov., *Int. J. Syst. Evol. Microbiol.*, 65(12), 4830–4838, 2015.

28. Whelen, C., Bordetella, in *Textbook of Diagnostic Microbiology*, 4th ed, Mahon, C., Lehman, D., Manuselis, G. eds. W. B. Saunders, Maryland Heights, MO, 423–426, 2011.

29. Jackwood, M.A., Bordetellosis, in *The Merck Veterinary Manual*, 9th ed., in Kahn, C.M. ed. Merck & Co., Inc., Whitehouse Station, NJ, 2006.

30. Gerlach, G., von Wintzingerode, F., Middendorf, B., and Gross, R., Evolutionary trends in the genus Bordetella. *Microbes Infect.*, 3(1), 61–72, 2001.

31. Aricò, B., Gross, R., Smida, J., and Rappuoli, R., Evolutionary relationships in the genus Bordetella. *Mol. Microbiol.*, 1(3), 301–308, 1987.

32. Parkhill, J., Sebaihia, M., Preston, A., Murphy, L.D., Thomson, N., Harris, D.E., Holden, M.T., Churcher, C.M., Bentley, S.D., Mungall, K.L., Cerdeno-Tarraga, A.M., Temple, L., et al., Comparative analysis of the genome sequences of Bordetella pertussis, Bordetella parapertussis and Bordetella bronchiseptica.*Nat. Genet.*, 35(1), 32–40, 2003.

33. Stibitz, S. and Miller, J.F., Coordinate regulation of virulence in Bordetella pertussis mediated by the vir (bvg) locus, in *Molecular Genetics of Bacterial Pathogenesis.*, 1st ed., ASM Press, Washington, DC, 407–422, 1994.

34. McPheat, W.L., Wardlaw, A.C., and Novotny, P., Modulation of Bordetella pertussis by nicotinic acid. *Infect. Immun.*, 41(2), 516–522, 1983.

35. Lacey, B.W., Antigenic modulation of Bordetella pertussis. *J. Hyg.*, 58, 57–93, 1960.

36. Peppler, M.S. and Schrumpf, M.E., Phenotypic variation and modulation in Bordetella bronchiseptica. *Infect. Immun.*, 44(3), 681–687, 1984.

37. Peppler, M.S., Two physically and serologically distinct lipopolysaccharide profiles in strains of Bordetella pertussis and their phenotype variants. *Infect. Immun.*, 43(1), 224–232, 1984.

38. Peppler, M.S., Isolation and characterization of isogenic pairs of domed hemolytic and flat nonhemolytic colony types of Bordetella pertussis. *Infect. Immun.*, 35(3), 840–851, 1982.

39. Stenson, T.H. and Peppler, M.S., Identification of two bvg-repressed surface proteins of Bordetella pertussis. *Infect. Immun.*, 63(10), 3780–3789, 1995.

40. Beattie, D.T., Knapp, S., and Mekalanos, J.J., Evidence that modulation requires sequences downstream of the promoters of two vir-repressed genes of Bordetella pertussis. *J. Bacteriol.*, 172(12), 6997–7004, 1990.

41. Uhl, M.A. and Miller, J.F., Bordetella pertussis BvgAS virulence control system, in *Two-Component Signal Transduction*, 1st ed., Hoch, J.A. and Silhavy, T.J. eds. ASM Press, Washington, DC, 1333–349, 1995.

42. Cotter, P.A. and DiRita, V.J., Bacterial virulence gene regulation: an evolutionary perspective, *Annu. Rev. Microbiol.*, 54, 519–565, 2000.

43. Aricò, B., Miller, J.F., Roy, C., Stibitz, S., Monack, D., Falkow, S., Gross, R., and Rappuoli, R., Sequences required for expression of Bordetella pertussis virulence factors share homology with prokaryotic signal transduction proteins. *Proc. Natl. Acad. Sci. U.S.A.*, 86(17), 6671–6675, 1989.

44. Uhl, M.A. and Miller, J.F., Autophosphorylation and phosphotransfer in the Bordetella pertussis BvgAS signal transduction cascade. *Proc. Natl. Acad. Sci. U.S.A.*, 91(3), 1163–1167, 1994.

45. Williams, C.L. and Cotter, P.A., Autoregulation is essential for precise temporal and steady-state regulation by the Bordetella BvgAS phosphorelay. *J. Bacteriol.*, 189, 1974–1982, 2007.

46. Veal-Carr, W.L. and Stibitz, S., Demonstration of differential virulence gene promoter activation in vivo in Bordetella pertussis using RIVET. *Mol. Microbiol.*, 55(3), 788–798, 2005.

47. Cotter, P.A. and Jones, A.M., Phosphorelay control of virulence gene expression in Bordetella. *Trends Microbiol.*, 11(8), 367–373, 2003.

48. Roy, C.R., Miller, J.F., and Falkow, S., The bvgA gene of Bordetella pertussis encodes a transcriptional activator required for coordinate regulation of several virulence genes. *J. Bacteriol.*, 171(11), 6338–6344, 1989.

49. Merkel, T.J., Barros, C., and Stibitz, S., Characterization of the bvgR locus of Bordetella pertussis. *J. Bacteriol.*, 180(7), 1682–1690, 1998.

50. Merkel, T.J. and Stibitz, S., Identification of a locus required for the regulation of bvg-repressed genes in Bordetella pertussis. *J. Bacteriol.*, 177(10), 2727–2736, 1995.

51. Knapp, S. and Mekalanos, J.J., Two trans-acting regulatory genes (*vir* and *mod*) control antigenic modulation in *Bordetella pertussis, J. Bacteriol.*, 170(11), 5059–5066, 1988.

52. Stockbauer, K.E., Fuchslocher, B., Miller, J.F., and Cotter, P.A., Identification and characterization of BipA, a Bordetella Bvg-intermediate phase protein. *Mol. Microbiol.*, 39(1), 65–78, 2001.

53. Williams, C.L., Boucher, P.E., Stibitz, S., and Cotter, P.A., BvgA functions as both an activator and a repressor to control Bvg phase expression of bipA in Bordetella pertussis. *Mol. Microbiol.*, 56(1), 175–188, 2005.

54. Vergara-Irigaray, N., Chavarri-Martinez, A., Rodriguez-Cuesta, J., Miller, J.F., Cotter, P.A., and Martinez de Tejada, G., Evaluation of the role of the Bvg intermediate phase in Bordetella pertussis during experimental respiratory infection. *Infect. Immun.*, 73(2), 748–760, 2005.

55. Cotter, P.A. and Miller, J.F., A mutation in the Bordetella bronchiseptica bvgS gene results in reduced virulence and increased resistance to starvation, and identifies a new class of Bvg-regulated antigens. *Mol. Microbiol.*, 24(4), 671–685, 1997.

56. Cummings, C.A., Bootsma, H.J., Relman, D.A., and Miller, J.F., Species- and strain-specific control of a complex, flexible regulon by Bordetella BvgAS. *J. Bacteriol.*, 188(5), 1775–1785, 2006.

57. Mishra, M. and Deora, R., Mode of action of the Bordetella BvgA protein: transcriptional activation and repression of the Bordetella bronchiseptica bipA promoter. *J. Bacteriol.*, 187(18), 6290–6299, 2005.

58. Melton, A.R. and Weiss, A.A., Characterization of environmental regulators of Bordetella pertussis. *Infect. Immun.*, 61(3), 807–815, 1993.

59. Stibitz, S., Aaronson, W., Monack, D., and Falkow, S., Phase variation in Bordetella pertussis by frameshift mutation in a gene for a novel two-component system. *Nature.*, 338(6212), 266–269, 1989.

60. Kasuga, T., Nakase, Y., Ukishima, K., and Takatsu, K., Studies on *Haemophilus pertussis*. V. Relation between the phase of bacilli and the progress of the whooping-cough. *Kitasato Arch. Exp. Med.*, 27(3), 57–62, 1954.

61. Stibitz, S. and Yang, M.S., Genomic plasticity in natural populations of Bordetella pertussis. *J. Bacteriol.*, 181(17), 5512–5515, 1999.

62. Caro, V., Hot, D., Guigon, G., Hubans, C., Arrive, M., Soubigou, G., Renauld-Mongenie, G., Antoine, R., Locht, C., Lemoine, Y., and Guiso, N., Temporal analysis of French Bordetella pertussis isolates by comparative whole-genome hybridization. *Microbes. Infect.*, 8(8), 2228–2235, 2006.

63. Cummings, C.A., Brinig, M.M., Lepp, P.W., van de Pas, S., and Relman, D.A., Bordetella species are distinguished by patterns of substantial gene loss and host adaptation. *J. Bacteriol.*, 186(5), 1484–1492, 2004.

64. Brinig, M.M., Cummings, C.A., Sanden, G.N., Stefanelli, P., Lawrence, A., and Relman, D.A., Significant gene order and expression differences in Bordetella pertussis despite limited gene content variation. *J. Bacteriol.*, 188(7), 2375–2382, 2006.

65. Centers for Disease Control. Pertussis: (Whooping Cough). http://www.cdc.gov/pertussis/about/faqs.html. Accessed in December 2013.

66. Hewlett, E.L., Bordetella species, in *Principles and Practice of Infectious Diseases*, 5th ed., Mandell, G.L., Bennett, J.E., and Dolin, R., eds. Churchill Livingstone, Philadelphia, PA, 2414–2422, 2000.

67. Olson, L.C., Pertussis. *Medicine*, 54(6), 427–468, 1975.

68. Centers for Disease Control, 2018. Final pertussis surveillance report. https://www.cdc.gov/pertussis/downloads/pertuss-surv-report-2018-508.pdf. Accessed on February 16, 2020.

69. Center for Disease Control. 2012. Final pertussis surveillance report. 62(33). http://www.cdc.gov/pertussis/downloads/pertussis-surveillance-report.pdf. Accessed on August 23, 2013.

70. World health organization. Pertussis vaccines: WHO position paper. *Weekly Epidemiological Record*. 40, 385–400, October 2010.

71. Yeung, K.H.T., Duclos, P., Nelson, E.A.S., and Hutubessy, R.C.W. An update of the global burden of pertussis in children younger than 5 years: a modelling study. *Lancet Infect. Dis.*, 17(9), 974–980.

72. López Luis, B.A., De Lourdes Guerrero Almeida, M., and Ruiz-Palacios, G.M. A place for Bordetella pertussis in PCR-based diagnosis of community-acquired pneumonia. *Infect. Dis.*, 2018.

73. Rubin, K., and Glazer, S. The pertussis hypothesis: Bordetella pertussis colonization in the pathogenesis of Alzheimer's disease. *Immunobiology*, 222(2), 228–240, 2017.

74. Heinz-Wirsing von Konig, C., Riffelmann, M., and Coenye, T. Bordetella and related genera, in *Manual of Clinical Microbiology*, 10th ed., Vol. 1,Jorgenson, J. ed. ASM Press,Washington, DC, 739–750, 2011.

75. Loeffelholz, M. Towards improved accuracy of Bordetella pertussis nucleic acid amplification tests. *J. Clin. Microbiol.*, 50(7), 2186–2190, 2012.

76. Wendelboe, A., Njamkepo, E., Bourillon, A., Floret, D., Gaudelus, D., Gerber, M., Grimprel, E. Greenberg, D., Halperin, S., Liese, J., Mun˜oz-Rivas, F., Teyssou, R., Guiso, N., and Van Rie, A. Transmission of Bordetella pertussis to young infants. *Pediatr. Infect. Dis. J.*, 26, 293–299, 2007.

77. Craig, R., Kunkel, E., Crowcroft, N.S., Fitzpatrick, M.C., de Melker, H., Althouse, B.M., Merkel, T. Scarpino, S.V., Koelle, K., Freidman, L., Arnold, C., and Bolotin, S., Asymptomatic infection and transmission of pertussis in households: a systematic review. *Clin. Infect. Dis.*, 70(1), 152–161, 2020.

78. Linnemann Jr., C.C., Host-parasite interactions in pertussis, in *International Symposium on Pertussis*, National Institutes of Health, Bethesda, MD, 1979.

79. Bartkus JM, Juni BA, Ehresmann K, Miller CA, Sanden GN, Cassiday PK, Saubolle M, Lee B, Long J, Harrison AR Jr, Besser JM., Identification of a mutation associated with erythromycin resistance in Bordetella pertussis: implications for surveillance of antimicrobial resistance. *J. Clin. Microbiol.*, 41(3), 1167–1172, 2003.

80. Center for Disease Control. Pertussis: (Whooping Cough) Treatment. http://www.cdc.gov/pertussis/clinical/treatment.html. Accessed January 2020.

81. Raguckas, S.E., VandenBussche, H.L., Jacobs, C., and Klepser, M.E., Pertussis resurgence: diagnosis, treatment, prevention, and beyond. *Pharmacotherapy*, 27(1), 41–52, 2007.

82. Carbonetti, N.H., Immunomodulation in the pathogenesis of Bordetella pertussis infection and disease. *Curr. Opin. Pharmacol.*, 7(3), 272–278, 2007.

83. World Health Organization. Pertussis vaccines. WHO position paper. *Weekly Epidemiolical Record*, 80(4), 31–39, 2005.

84. Edwards, K.M., Acellular pertussis vaccines—a solution to the pertussis problem? *J. Infect. Dis.*, 168(1), 15–20, 1993.

85. Decker, M.D. and Edwards, K.M., Acellular pertussis vaccines. *Pediatr. Clin. North. Am.*, 47(2), 309–335, 2000.

86. Pichichero, M.E., Edwards, K.M., Anderson, E.L., Rennels, M.B., Englund, J.A., Yerg, D.E., Blackwelder, W.C., Jansen, D.L., and Meade, B.D., Safety and immunogenicity of six acellular pertussis vaccines and one whole-cell pertussis vaccine given as a fifth dose in four- to six-year-old children. *Pediatrics*, 105(1), e11, 2000.

87. Edwards, K.M. and Decker, M.D., Pertussis Vaccine, in *Vaccines*, 4th ed., Plotkin, S.A. and Orenstein, W.A., eds. Elsevier, Philadelphia, PA, 471–528, 2004.

88. Cherry, J.D., The 112-year odyssey of pertussis and pertussis vaccines—mistakes made and implications for the future. *J. Ped. Infec. Dis. Soc.*, 8(4), 334–341, 2019.

89. World Health Organization. Pertussis. http://www.who.int/biologicals/vaccines/pertussis/en/ May 2015.

90. Centers for Disease Control. Pertussis Vaccination. https://www.cdc.gov/pertussis/vaccines.html. Accessed January 2020.

91. Gall, S.A., Myers, J., and Pichichero, M., Maternal immunization with tetanus-diphtheria-pertussis vaccine: effect on maternal and neonatal serum antibody levels. *Am. J. Obstet. Gynecol.*, 204(4), 334.e1–334.e5, 2011.

92. Gayatri, A., Andrews, N., Campbell, H., Riberio, S., Kara, E., Donegan, K., Fry, N.K., Miller, E., and Ramsay, M. Effectiveness of maternal pertussis vaccination in England: an observational study. *Lancet.*, 384(9953), 25–31, 2014.

93. Fernandes, E.G., Sato, A.P.S., Vaz-de-Lima, L.R.A., Rodrigues, M., Leite, D., deBrito, C.A., Luno, E.J.A., Carvalhanas, T.R.M.P., Ramos, M.L.B.N., Sato, H.K., and de Casttilho, E.A., The effectiveness of maternal pertussis vaccination in protecting newborn infants in Brazil: a case-control study. *Vaccine*, 37(36), 5481–5484, 2019.

94. Baxter, R., Bartlett, J., Fireman, B., Lewis, E., and Klein, N., Effectiveness of vaccination during pregnancy to prevent infant pertussis. *Pediatrics*, 139(5), 2017.

95. Skoff, T.H., Blain, A.E., Watt, J., Scherzinger, K., McMahon, M., Zansky, S.M., Kudish, K., Cieslak, P.R., Lewis, M., Shang, N., and Martin, S.W., Impact of the US maternal tetanus, diphtheria, and acellular pertussis vaccination program on preventing pertussis in infants <2 months of age: a case-control evaluation. *Clin. Infect. Dis.*, 65(12), 1977–1983, 2017.

96. Siegrist, C.A., Mechanisms by which maternal antibodies influence infant vaccine responses: review of hypotheses and definition of main determinants. *Vaccine.*, 21(24), 3406–3412, 2003.

97. Argondizo-Correia, C., Rodrigues, A.K.S., and DeBrito, C.A. Carolina., Neonatal immunity to Bordetella pertussis infection and current prevention strategies. *J. Immunol. Res.*, 2019, doi: 10.1155/2019/7134168.

98. Clark, T., Messonnier, N., and Hadler, S., Pertussis control: time for something new? *Trends Microbiol.*, 20(5), 211–213, 2012.

99. Centers for Disease Control. Pertussis Cases by Year. http://www.cdc.gov/pertussis/surv-reporting/cases-by-year.html. Accessed in August 2013.

100. Centers for Disease Control. Outbreaks of Pertussis associated with hospitals—Kentucky, Pennsylvania and Oregon, 2003. *MMWR*, 54(03), 67–71, 2005.

101. Poland, G., Editorial, Pertussis outbreaks and pertussis vaccines: new insights, new concerns, new recommendations? *Vaccine*, 30, 6957–6959, 2012.

102. World Health Organization. Pertussis vaccines: WHO position paper. *Weekly Epidemiological Record*, 35, 433–460, 2015.

103. WHO SAGE Pertussis Working Group. *Background Paper*. SAGE, April 2014, https://www.who.int/immunization/sage/meetings/2014/april/1_Pertussis_background_FINAL4_web.pdf)

104. Esposito, S., Stefanelli, P., Fry, N.K., Fedele, G., He, Q., Paterson, P., Tan, T., Knuf, M., Rodrigo, C., Weil-Olivier, C., Flanagan, K.L., Hung, I., Lutsar, I., Edwards, K., O'Ryan, M., and Principi, N., Pertussis prevention: reasons for resurgence, and differences in the current acellular pertussis vaccines. *Front. Immunol.*, 10, 1344, 2019.

105. Rotem, L. and Gill, C.J., The pertussis resurgence: putting together the pieces of the puzzle. *Trop. Dis. Travel Med. Vaccines*, 2, 26, 2016.

106. Lee, S.F., Halperin, S.A., Wang, H., and MacArthur, A., Oral colonization and immune responses to Streptococcus gordonii expressing a pertussis toxin S1 fragment in mice. *FEMS Microbiol. Lett.*, 208(2), 175–178, 2002.

107. Feunou, P. Kammoun, H., Debrie, A., Mielcarek, N., and Locht, C., Long term immunity against pertussis induced by a single nasal administration of live attenuated B. pertussis BPZE1. *Vaccine*, 28, 7047–7053, 2010.

108. Skerry, C. and Mahon, B., A live, attenuated Bordetella pertussis vaccine provides Long term protection against virulent challenge in a murine model. *Clin. Vac.cine Immunol.*, 18(2), 187–193, 2010.

109. Mielcarek, N., Debrie, A.S., Raze, D., Bertout, J., Rouanet, C., Younes, A.B., Creusy, C., Engle, J., Goldman, W.E., and Locht, C., A live attenuated B. pertussis as a single-dose nasal vaccine against whooping cough. *PLOS Pathog.*, 2(7), e65, 2006.

110. Debrie, A.S., Coutte, L., Raze, D., Mooi, F., Alexander, F., Gorringe, A., Mielcarek, N., and Locht, C., Construction and evaluation of Bordetella pertussis live attenuated vaccine strain BPZE1 producing Fim3. *Vaccine*, 36(11), 1345–1352, 2018.

111. Locht, C., Live pertussis vaccines: will they protect against carriage and spread of pertussis? *Clin. Microbiol. Infect.*, 22(Suppl 5):S96–S102, 2016.

112. Najminejad, H., Kalantar, S.M., Mokarram, A.R., Dabaghian, M., Abdollahpour-Alitappeh, M., Ebrahimi, S.M., Tebianian, M. Ramandi, M.F., and Sheikhha, M.H., Bordetella pertussis antigens encapsulated into N-trimethyl chitosan nanoparticulate systems as a novel intranasal pertussis vaccine, artificial cells. *Artif. Cells Nanomed. Biotechnol.*, 47(1), 2605–2611, 2019.

113. van der Zee, A., Schellekens, J.F.P., and Mooi, F.R. Laboratory diagnosis of pertussis. *Clin. Micro. Rev.*, 8(4), 1005–1026, 2015.

114. Centers for Disease Control. Pertussis (Whooping Cough). Diagnosis confirmation. https://www.cdc.gov/pertussis/clinical/diagnostic-testing/diagnosis-confirmation.html. Accessed January 2020.

115. Cherry, J.D. Historical perspective on pertussis and use of vaccines to prevent it: 100 years of pertussis (the cough of 100 days). *Microbe.*, 2(3), 139–144, 2007.

116. Centers for Disease Control. Pertussis (Whooping Cough). Specimen collection. https://www.cdc.gov/pertussis/clinical/diagnostic-testing/specimen-collection.html. Accessed January 2020.

117. Hoffman, C.L., Gonyar, L.A., Zacca, F., Sisti, F., Fernandez, J., Wong, T., Damron, F.H., and Hewlett, E., Bordetella pertussis can be motile and express flagellum-like structures. *Mol. Biol. Physiol..*, 2019, e00787–19.

118. Gherna, R.L., Preservation, in *Manual of Methods for General Bacteriology*, 1st ed., Gerhardt, P., Murray, R.G.E., Costilow, R.N., Nester, E.W., Wood, W.A., Krieg, N.R., and Phillips, G.B., eds. American Society for Microbiology, Washington, DC, 208–216, 1981.

119. Stenson, T. and Peppler, M., Bordetella, in *Practical Handbook of Microbiology*, 2nd ed., Goldman, E. and Green, H., eds. CRC Press. Boca Raton, FL, 2009.

120. Stainer, D.W. and Scholte, M.J., A simple chemically defined medium for the production of phase I Bordetella pertussis, *J. Gen. Microbiol.*, 63(2), 211–220, 1970.

121. Centers for Disease Control, *Pertussis. Incidence by year-United States, 1979–2004*, United States Public Health Service, Washington, DC, 2004.

122. Mandal S, Tatti KM, Woods-Stout D, Cassiday PK, Faulkner AE, Griffith MM, Jackson ML, Pawloski LC, Wagner B, Barnes M, Cohn AC, Gershman KA, Messonnier NE, Clark TA, Tondella ML, Martin SW., Pertussis pseudo-outbreak linked to specimens contaminated by Bordetella pertussis DNA from clinic surfaces. *Pediatrics*, 129(2), 424–30, 2012.

123. Centers for Disease Control. Pertussis (Whooping Cough). Diagnosis PCR best practices. https://www.cdc.gov/pertussis/clinical/diagnostic-testing/diagnosis-pcr-bestpractices.html. Accessed January 2020.

124. Halperin, S., Kasina, A., and Swift, M., Prolonged survival of Bordetella pertussis in a simple buffer after nasopharyngeal secretion aspiration. *Can. J. Microbiol.*, 38(11), 1210–1213, 1992.

125. Kilgore, P.E. and Coenye, T., Bordetella and related genera,in *Manual of Clinical Microbiology*, 12th ed., Vol. 1, Carrol, K.C., Pfaller, M.A., Landry, M.L., McAdam, A.J., Patel, R., Richter, S.S., and Warnock, D.W., eds. ASM Press, Washington, DC, 858–870, 2017.

126. Roorda, L., Buitenwerf, J., Ossewaarde, J.M., and van der Zee, A., A real-time PCR assay with improved specificity for detection and discrimination of all clinically relevant Bordetella species by the presence and distribution of three insertion sequence elements. *BMC Res. Notes.*, 4, 11. 2011.

127. Afonina, I., Metcalf, M., Mills, A., and Mahoney, W., Evaluation of 5′-MGB hybridization probes for detection of Bordetella pertussis and Bordetella parapertussis using different real-time PCR instruments. *Diagn. Microbiol. Infect. Dis.*, 60, 429–432, 2008.

128. He, Q., Schmidt Schläpfer, G., Just, M., Matter, H.C., Nikkari, S., Viljanen, M.K., and Mertsola, J., Impact of polymerase chain reaction on clinical pertussis: Finnish and Swiss experiences. *J. Infect. Dis.*, 174, 1288–1295, 1996.

129. Reizenstein, E., Johansson, B., Mardin, L., Abens, J., Möllby, R., and Hallander, H.O. Diagnostic evaluation of polymerase chain reaction discriminative for Bordetella pertussis, B. parapertussis, and B. bronchiseptica. *Diagn. Microbiol. Infect. Dis.*, 17, 185–191, 1993.

130. Schmidt-Schläpfer, G., Liese, J.G., Porter, F., Stojanov, S., Just, M., and Belohradsky, B.H. Polymerase chain reaction (PCR) compared with conventional identification in culture for detection of Bordetella pertussis in 7153 children. *Clin. Microbiol. Infect.*, 3, 462–467, 1997.

131. Tatti, K., Sparks, K., Boney, K., and Tondella, M., Novel multitarget real-time PCR assay for rapid detection of Bordetella species in clinical specimens. *J. Clin. Microbiol.*, 49(12), 4059–4066, 2011.

132. Vielemeyer, O., Crouch, J.Y., Edberg, S.C., and Howe, J.G., Identification of Bordetella pertussis in a Critically Ill human immunodeficiency virus-infected patient by direct genotypical analysis of gram-stained material and discrimination from B. Holmesii by using a unique RecA gene restriction enzyme site. *J. Clin. Microbiol.*, 42(2), 847–849, 2004.

133. Wang, Z., Han, R., Liu, Y., Ma, C., Li, H., and Yan, Y. Direct detection of erythromycin-resistant Bordetella pertussis in clinical Specimens by PCR. *J. Clin. Microbiol.*, 53(11), 2015.

134. Relich, R.F., Leber, A., Young, S., Schutzbank, T., Dunn, R., Farhang, J., and Uphoff, U., Multicenter clinical evaluation of the automated Aries Bordetella assay. *J. Clin. Microbiol.*, 57(2):e01471–18, 2019.

135. Valero-Rello, A., Henares, D., Acosta, L., Jane, M., Godoy, P., and Munoz-Almagro, C. Validation and implementation of a diagnostic algorithm for DNA detection of Bordetella

pertussis, B. parapertussis, and B. holmesii in a pediatric referral hospital in Barcelona. *Spain. J. Clin. Microbiol.*, 57(1), 2019.

136. She, R., Billetdeaux, I., Phansalkar, A., and Petti, C., Limited applicability of direct fluorescent-antibody testing for Bordetella sp. and Legionella sp. specimens for the clinical microbiology laboratory. *J. Clin. Microbiol.*, 45(7): 2212–2214, 2007.

137. Faulkner, A., Skoff, T., Martin, S., Cassiday, P., Tondella, M., Liang, J., Ejigiri, O., *Pertussis in Vaccine Preventable Disease Manuel*, 5th ed. CDC, Atlanta, GA, 2011.

138. Hallander, H.O., Microbiological and serological diagnosis of pertussis. *Clin. Infect. Dis.*, 28(Suppl. 2), S99–S106, 1999.

139. Watanabe, M., Connelly, B., and Weiss, A.A., Characterization of serological responses to pertussis. *Clin. Vaccine Immunol.*, 13(3), 341–348, 2006.

140. Ridda, I., Yin, J., King, C., MacIntyre, C., and McIntyre, P., The importance of pertussis in older adults: A growing case for reviewing vaccination strategy in the elderly. *Vaccine*, 30, 6745–6752, 2012.

141. Weigand, M.R., Williams, M.M., Peng, Y., Kania, D., Pawloski, L.C., and Tondella, M.L., Genomic survey of Bordetella pertussis diversity, United State, 2000-2013. *Emerg. Infect. Dis.*, 25(4), 2019.

142. Glanz, J., McClure, D., Magid, D., Daley, M., France, E., Salamon, D., and Hambridge, S., Parental refusal of pertussis vaccination is associated with an increased risk of pertussis infection in children. *Pediatrics*, 123(6), 1446–1451, 2009.

143. Klein, N., Bartlett, J., Rowhani-Rahbar, A., Fireman, B., and Baxter, R., Waning protection after fifth dose of acellular pertussis vaccine in children. *N. Engl. J. Med.*, 367, 1012–1019, 2012.

144. Sheridan, S., Ware, R., Grimwood, K., and Lambert, S., Number and order of whole cell pertussis vaccines in infancy and disease protection. *JAMA*, 308(5), 454–456, 2012.

145. Weigand, M.R., Peng, Y., Loparev, V., Batra, D., Bowden, K.E., Burroughs, M., Cassiday, P.K., Davis, J.K., Johnson, T., Juiebg, P., Knipe, K. Mathis, M., Pruitt, A.M., Rowe, L., Sheth, M., Tondella, M.L., and Williams, M.M., The history of Bordetella pertussis genome evolution includes structural rearrangement. *J. Bact.*, 199(8), 2017.

146. Bart, M.J., van der Heide, H., Zeddeman, A., Heuvelman, K., van Gent, M., and Mooj, F., Complete genome sequences of 11 Bordetella pertussis strains representing the pandemic PtxP3 lineage. *Genome Announc..*, 3(6), 2015.

147. Rohani, P and Drake, J., The decline and resurgence of pertussis in the US. *Epidemics*, 3, 183–188, 2011.

148. Mooi, F., VanLoo, I., and King, A., Adaptation of *Bordetella pertussis* to vaccination: a cause for reemergence? *Emerg. Infect. Dis..*, 7(3), 526–528, 2011.

149. Hegerle, N., Paris, A-S., Brun, D., et al. Evolution of French Bordetella pertussis and Bordetella parapertussis isolates: increase of Bordetellae not expressing pertactin. *Clin. Microbiol. Infect.*, 18, E40–E46, 2012.

150. Barkoff, A-M., Mertsola, J., Guillot, S., Guiso, N., Berbers, G., and He, Q., Appearance of Bordetella pertussis strains not expressing the vaccine antigen pertactin in Finland. *Clin. Vaccine Immunol.*, (19), 1703–1704, 2012.

151. Otsuka, N., Han, H.J., Toyoizumi-Ajisaka, H., Nakamura, Y., Arakawa, Y., Shibayama, K., and Kamach., K., Prevalence and genetic characterization of pertactin-deficient Bordetella pertussis in Japan. *PLOS One*, 7(2), e31985, 2012.

152. Xu, Z., Octavia, S., Luu, L., et al. Pertactin-negative and filamentous hemagglutinin-negative Bordetella pertussis, Australia, 2013–2017. *Emerg. Infect. Dis.*, 25(6), 1196–1199,2019.

153. Pawloski, L., Queenan, A., Cassiday, P., Lynch, A., Harrison, M., Shang, W., Williams, M., Bowden, K., Burgos Rivera, B., Qin, X., Messonnier, N., and Ton-della, M., Prevalence and molecular characterization of pertactin-deficient Bordetella pertussis in the United States. *Clin. Vaccine. Immunol.*, 21(2), 119–125, 2014.

154. Center for Disease Control, Pertussis: (Whooping Cough). Pertactin negative pertussis strains, 2013, http://www.cdc.gov/pertussis/pertactin-neg-strain.html. (accessed January 30, 2014.

155. Weigand, M.R., Williams, M.M., Peng, Y., et al. Genomic survey of Bordetella pertussis diversity, United States, 2000–2013. *Emerg. Infect. Dis.*, 25(4), 780–783, 2019.

156. Carriquiriborde, F., Regidor, V., Aispuro, P.M., et al. Rare detection of Bordetella pertussis pertactin-deficient strains in Argentina. *Emerg. Infect. Dis.*, 25(11), 2048–2054, 2019.

157. Barkoff, A.M., Mertsola, J., Pierard, D., Dalby, T., Hoegh, S.V., Guillot, S., Stefanelli, P., van Gent, M., Berbers, G., Vestrheim, D.F., Greve-Isdahl, M., Wehlin, L., Ljungman, M., Fry, N.K., Markay, K., Auranen, K., and He, Q. Surveillance of circulating Bordetella pertussis strains in Europe during 1998 to 2015. *J. Clin. Microbiol.*, 56(5):e01998–17, 2018.

158. Halperin, S.A., The control of pertussis—2007 and beyond. *N. Engl. J. Med.*, 356(2), 110–113, 2007.

159. Eldering, G., Holwerda, J., Davis, A., and Baker, J., Bordetella pertussis serotypes in the United States. *Appl. Microbiol.*, 18(4), 618–621, 1969.

160. Robinson, A., Ashworth, L.A. and Irons, L.I., Serotyping *Bordetella pertussis* strains, *Vaccine*, 7(6), 491–494, 1989.

161. de Melker, H.E., Schellekens, J.F., Neppelenbroek, S.E., Mooi, F.R., Rumke, H.C., and Conyn-van Spaendonck, M.A., Reemergence of pertussis in the highly vaccinated population of the Netherlands: observations on surveillance data. *Emerg Infect Dis.*, 6(4), 348–357, 2000.

162. Van Loo, I.H. and Mooi, F.R., Changes in the Dutch Bordetella pertussis population in the first 20 years after the introduction of whole-cell vaccines. *Microbiology*, 148(Pt. 7), 2011–2018, 2002.

163. Zeddeman, A., Witteveen, S., Bart, M.J., van Gent, M., van der Heide, H.G.J., Heuvelman, K.J., Schouls, L.M., and Mooi, F.R., Studying Bordetella pertussis populations by use of SNPeX, a simple high-throughput single nucleotide polymorphism typing method. *J. Clin. Microbiol.*, 53, 838–846, 2014.

164. Stein, P.E., Boodhoo, A., Armstrong, G.D., Cockle, S.A., Klein, M.H., and Read, R.J., The crystal structure of pertussis toxin. *Structure*, 2(1), 45–57, 1994.

165. Marchand-Austin, A., Tsang, R.S.W., Guthrie, J.L., Ma, J.H., Lim, G.H., Crowcroft, N.S., Deeks, S.L., Farrell, D.J., and Jamieson, F.B., Short-read whole-genome sequencing for laboratory-based surveillance of Bordetella pertussis. *J Clin Microbiol.* 55(5), 1446–1453, 2017.

166. Bisgard, K.M., Christie, C.D., Reising, S.F., Sanden, G.N., Cassiday, P.K., Gomersall, C., Wattigney, W.A., Roberts, N.E., and Strebel, P.M., Molecular epidemiology of Bordetella pertussis by pulsed-field gel electrophoresis profile: Cincinnati, 1989–1996. *J. Infect. Dis.*, 183(9), 1360–1367, 2001.

167. Cassiday, P., Sanden, G., Heuvelman, K., Mooi, F., Bisgard, K.M., and Popovic, T., Polymorphism in Bordetella pertussis pertactin and pertussis toxin virulence factors in the United States, 1935–1999. *J. Infect. Dis.*, 182(5), 1402–1408, 2000.

168. Hardwick, T.H., Cassiday, P., Weyant, R.S., Bisgard, K.M., and Sanden, G.N., Changes in predominance and diversity of genomic subtypes of Bordetella pertussis isolated in the United States, 1935 to 1999. *Emerg. Infect. Dis.*, 8(1), 44–49, 2002.

169. Hardwick, T.H., Plikaytis, B., Cassiday, P.K., Cage, G., Peppler, M.S., Shea, D., Boxrud, D., and Sanden, G.N., Reproducibility of Bordetella pertussis genomic DNA fragments generated by XbaI restriction and resolved by pulsed-field gel electrophoresis. *J. Clin. Microbiol.*, 40(3), 811–816, 2002.

170. Advani, A., Hallander, H.O., Dalby, T., Krogfelt, K.A., Guiso, N., Njamkepo, E., von König, C.H.W. et al., Pulsed-field gel electrophoresis analysis of Bordetella pertussis isolates circulating in Europe from 1998 to 2009. *J. Clin. Microbiol.*, 51(2), 422–428, 2013.

171. Schouls, L.M., van der Heide, H.G., Vauterin, L., Vauterin, P., and Mooi, F.R., Multiple-locus variable-number tandem repeat analysis of Dutch Bordetella pertussis strains reveals rapid genetic changes with clonal expansion during the late 1990s. *J. Bacteriol.*, 186(16), 5496–5505, 2004.

172. Forsyth, K.D., Wirsing von Konig, C.H., Tan, T., Caro, J., and Plotkin, S., Prevention of pertussis: recommendations derived from the second Global Pertussis Initiative roundtable meeting. *Vaccine*, 25(14), 2634–2642, 2007.

173. Lin, Y.C., Yao, S.M., Yan, J.J., Chen, Y.Y., Hsiao, M.J., Chou, C.Y., Su, H.P., Wu, H.S., and Li, S.Y., Molecular epidemiology of Bordetella pertussis in Taiwan, 1993–2004: suggests one possible explanation for the outbreak of pertussis in 1997. *Microbes Infect.*, 8(8), 2082–2087, 2006.

174. Tsang, R.S., Lau, A.K., Sill, M.L., Halperin, S.A., Van Caeseele, P., Jamieson, F., and Martin, I.E., Polymorphisms of the fimbria fim3 gene of Bordetella pertussis strains isolated in Canada. *J. Clin. Microbiol.*, 42(11), 5364–5367, 2004.

175. Tsang, R.S., Sill, M.L., Martin, I.E., and Jamieson, F., Genetic and antigenic analysis of Bordetella pertussis isolates recovered from clinical cases in Ontario, Canada, before and after the introduction of the acellular pertussis vaccine. *Can. J. Microbiol.*, 51(10), 887–892, 2005.

176. Halperin, S.A., Heifetz, S.A., and Kasina, A., Experimental respiratory infection with Bordetella pertussis in mice: comparison of two methods. *Clin. Invest. Med.*, 11(4), 297–303, 1988.

177. Hall, E., Parton, R., and Wardlaw, A.C., Time-course of infection and responses in a coughing rat model of pertussis, *J. Med. Microbiol.*, 48(1), 95–98, 1999.

178. Woods, D.E., Franklin, R., Cryz, S.J., Jr., Ganss, M., Peppler, M., and Ewanowich, C., Development of a rat model for respiratory infection with Bordetella pertussis. *Infect. Immun.*, 57(4), 1018–1024, 1989.

179. Locht, C., Bertin, P., Menozzi, F.D., and Renauld, G., The filamentous haemagglutinin, a multifaceted adhesion produced by virulent Bordetella spp. *Mol. Microbiol.*, 9(4), 653–660, 1993.

180. Mooi, F.R., Virulence factors of Bordetella pertussis. *Antonie Van Leeuwenhoek*, 54(5), 465–474, 1988.

181. Flak, T.A. and Goldman, W.E., Autotoxicity of nitric oxide in airway disease. *Am. J. Respir. Crit. Care Med.*, 154(4 Pt. 2), S202–S206, 1996.

182. Paddock, C.D., Sanden, G.N., Cherry, J.D., Gal, A.A., Langston, C., Tatti, K.M., Wu, K.H., Goldsmith, C.S., Greer, P.W., Montague, J.L., Eliason, M.T., Holman, R.C., Guarner, J., Shieh, W.J., and Zaki, S.R., Pathology and pathogenesis of fatal Bordetella pertussis infection in infants. *CID*, 47, 328–338, 2008.

183. Pierce, I., Klein, N., and Peters, M. Is leukocytosis a predictor of mortality in severe pertussis infection? *In. Care Med.*, 26(10), 1512–4, 2000.

184. Burnette, W.N., Is leukocytosis a predictor of mortality in severe pertussis infection? *Behring. Inst. Mitt.*, 98, 434–441, 1997.

185. Mooi, F, van Loo, I., van Gent, M., He, Q., Heuvelman, C., Bart, M., Diavatopoulos, D., Teunis, P., Nagelkerke, N., and Mertsola, J. Bordeltella pertussis strains with increased toxin production associated with pertussis resurgence. *Emerg. Infect. Dis.*, (5) 1206–1213, 2009.

186. Brennan, M.J. and Shahin, R.D., Pertussis antigens that abrogate bacterial adherence and elicit immunity. *Am. J. Respir. Crit. Care Med.*, 154(4 Pt. 2), S145–S149, 1996.

187. Poolman, J.T. and Hallander, H.O., Acellular pertussis vaccines and the role of pertactin and fimbriae. *Expert. Rev. Vaccines*, 6(1), 47–56, 2007.

188. Martin, S.W., Pawloski, L., Williams, M., Weening, K., DeBolt, C., Qin, X., Reynolds, L., Kenyon, C., Giambrone, G., Kudish, K., Miller, L., Selvage, D., Lee, A., Skoff, T.H., Kamiya, H., Cassidy, P.K., Tondella, M.L., and Clark, T.A., Pertactin-negative Bordetella pertussis strains: evidence for a possible selective advantage. *Clin. Infect. Dis.*, 60(2), 223–227, 2015.

189. Williams, M.M., Sen, K., Weigand, M.R., Skoff, T.H., Cunningham, V.A., Halse, T.A., Tondella, M.L., and CDC Pertussis Working Group. Bordetella pertussis strain lacking pertactin and pertussis toxin. *Emerg. Infect. Dis.*, 2(2), 2016.

190. Packard, E.R., Parton, R., Coote, J.G., and Fry, N.K., Sequence variation and conservation in virulence-related genes of Bordetella pertussis isolates from the UK. *J Med Microbiol.*, 53, 355–365, 2004.

191. Tsang, R.S., Lau, A.K., Sill, M.L., Halperin, S.A., Van Caeseele, P., Jamieson, F., and Martin, I.E., Polymorphisms of the fimbria fim3 gene of Bordetella pertussis strains isolated in Canada. *J Clin Microbiol.*, 42, 5364–5367, 2004.

192. Locht, C., Antoine, R., and Jacob-Dubuisson, F., Bordetella pertussis, molecular pathogenesis under multiple aspects. *Curr. Opin. Microbiol.*, 4(1), 82–89, 2001.

193. Weiss, A., Hewlett, E., Myers, G., and Falkow, S., Tn5-induced mutations affecting virulence factors of Bordetella pertussis. *Infect. Immun.*, 42(1), 33–41, 1983.

194. Weiss, A.A. and Hewlett, E.L., Virulence factors of Bordetella pertussis. *Annu. Rev. Microbiol.*, 40, 661–686, 1986.

195. Ladant, D. and Ullmann, A., Bordatella pertussis adenylate cyclase: a toxin with multiple talents. *Trends Microbiol.*, 7(4), 172–176, 1999.

196. Ahuja, N., Kumar, P., and Bhatnagar, R., The adenylate cyclase toxins. *Crit. Rev. Microbiol.*, 30(3).187–196, 2004.

197. Dautin, N., Karimova, G., and Ladant, D., Bordetella pertussis adenylate cyclase toxin: a versatile screening tool. *Toxicon*, 40(10), 1383–1387, 2002.

198. Mobberley-Schuman, P.S. and Weiss, A.A., Influence of CR3 (CD11b/CD18) expression on phagocytosis of Bordetella pertussis by human neutrophils. *Infect. Immun.*, 73(11), 7317–7323, 2005.

199. Mobberley-Schuman, P., Connelly, B., and Weiss, A., Phagocytosis of Bordetella pertussis incubated with convalescent serum. *J. Infect. Dis.*, 187(10), 1646–1653, 2003.

200. Weingart, C., Mobberley-Schuman, P., Hewlett, E., Gray, M., and Weiss, A., Neutralizing antibodies to adenylate cyclase toxin promote phagocytosis of Bordetella pertussis by human neutrophils. *Infect Immun.*, 68(12), 7152–7155, 2000.

201. Weingart, C. and Weiss, A., Bordetella pertussis virulence factors affect phagocytosis by human neutrophils. *Infect. Immun.*, 68(3), 1735–1739, 2000.

202. Barnes, M. and Weiss, A., BrkA protein of Bordetella pertussis inhibits the classical pathway of complement after C1 deposition. *Infect. Immun.*, 69(5), 3067–3072, 2001.

203. Fernandez, R. and Weiss, A., Cloning and sequencing of a Bordetella pertussis serum resistance locus. *Infect. Immun.*, 62(11), 4727–4738, 1994.

204. Oliver, D., Huang, G., Nodel, E., Pleasance, S., and Fernandez, R., A conserved region within the Bordetella pertussis autotransporter BrkA is necessary for folding of its passenger domain. *Mol. Microbiol.*, 47(5), 1367–1383, 2003.

205. Oliver, D., Huang, G., and Fernandez, R., Identification of secretion determinants of the Bordetella pertussis BrkA autotransporter. *J. Bacteriol.*, 185(2), 489–495, 2003.

206. Shannon, J. and Fernandez, R., The C-terminal domain of the Bordetella pertussis autotransporter BrkA forms a pore in lipid bilayer membranes. *J. Bacteriol.*, 181(18), 5838–5842, 1999.

207. Caroff, M., Aussel, L., Zarrouk, H., Martin, A., Richards, J.C., Therisod, H., Perry, M.B., and Karibian, D., Structural variability and originality of the Bordetella endotoxins. *J. Endotoxin. Res.*, 7(1), p. 63–68, 2001.

208. Caroff, M., Brisson, J., Martin, A., and Karibian, D., Structure of the Bordetella pertussis 1414 endotoxin. *FEBS Lett.*, 477(1–2), 8–14, 2000.

209. Preston, A., Thomas, R., and Maskell, D., Mutational analysis of the Bordetella pertussis wlb LPS biosynthesis locus. *Microb. Pathog.*, 33(3), 91–95, 2002.

210. Luker, K.E., Tyler, A.N., Marshall, G.R., and Goldman, W.E., Tracheal cytotoxin structural requirements for respiratory epithelial damage in pertussis. *Mol. Microbiol.*, 16(4) 733–743, 1995.

211. Matsuzawa, T., Fukui, A., Kashimoto, T., Nagao, K., Oka, K., Miyake, M., and Horiguchi, Y., Bordetella dermonecrotic toxin undergoes proteolytic processing to be translocated from a dynamin-related endosome into the cytoplasm in an acidification-independent manner. *J. Biol. Chem.*, 279(4), 2866–2872, 2004.

212. Cimolai, N. and Trombley, C., Molecular diagnostics confirm the paucity of parapertussis activity. *Eur J Pediatr.*, 160(8), 518, 2001.

213. Wolfe, D.N., Goebel, E.M., Bjornstad, O.N., Restif, O., and Harvill, E.T., The O antigen enables Bordetella parapertussis to avoid Bordetella pertussis-induced immunity. *Infect. Immun.*, 75(10), 4972–4979, 2007.

214. Stehr, K., Cherry, J., Heininger, U., Schmitt-Grohe, S., Uberall, M., Laussucq, S., Eckhardt, T., Meyer, M., Engelhardt, R., and Christenson, P., A comparative efficacy trial in Germany in infants who received either the Lederle/Takeda acellular pertussis component DTP (DTaP) vaccine, the Lederle whole-cell component DTP vaccine, or DT vaccine. *Pediatrics*, 101(1 Pt. 1), 1–11, 1998.

215. Long, G. Karanikas, A., Harvell, E., Read, A., and Peter, J. Hudson acellular pertussis vaccination facilitates Bordetella parapertussis infection in a rodent model of bordetellosis. *Proc. Biol. Sci.*, 277(1690), 2017–2025. 2010.

216. Liko, J., Robinson, S.G., and Cieslak, P.R., Do pertussis vaccines protect against Bordetella parapertussis? *Clin. Infect. Dis.*, 64(12), 1795–1797, 2017.

217. Toubiana, J., Azarnoush, S., Bouchez, V., Landier, A., Guillot, S., Matczak, S., Bonacorsi, S., and Brisse, S. Bordetella parapertussis bacteremia: clinical expression and bacterial genomics. *Open Forum Infect. Dis.*, 6(4), April 2019.

218. Hayes, A., Oviedo, J.M., Valdez, H., Laborde, J.M., Maschi, F., Ayala, M., Shah, R., Fernandez, L., and Rodriguez, M.E., A recombinant iron transport protein from Bordetella pertussis confers protection against Bordetella parapertussis. *Microbiol. Immunol.*, 61(10), 407–415, 2017.

219. Porter, J.F. and Wardlaw, A C., Long-term survival of Bordetella bronchiseptica in lakewater and in buffered saline without added nutrients, *FEMS Microbiol. Lett.*, 110(1), 33–36, 1993.

220. Dworkin, M.S., Sullivan, P.S., Buskin, S.E., Harrington, R.D., Olliffe, J., MacArthur, R.D., and Lopez, C.E., Bordetella bronchiseptica infection in human immunodeficiency virus-infected patients. *Clin. Infect. Dis.*, 28(5), 1095–1099, 1999.

221. Woolfrey, B.F. and Moody, J.A., Human infections associated with Bordetella bronchiseptica. *Clin Microbiol Rev.*, 4(3), 243–255, 1991.

222. Gueirard, P., Le Blay, K., Le Coustumier, A., Chaby, R., and Guiso, N., Variation in Bordetella bronchiseptica lipopolysaccharide during human infection. *FEMS Microbiol. Lett.*, 162(2), 331–337, 1998.

223. Winn Jr., W., Allen, S., Janda, W. Koneman, E., Procop, G., Schreckenberger, P., and Woods, G., Miscellaneous fastidious gram-negative bacilli, in *Koneman's Color Atlas and Textbook of Diagnostic Microbiology*, 6th ed. Lippincott Williams & Wilkins, Baltimore, MD, 510–522, 2006.

224. Forbes, B., Sahm, D., and Weissfeld, A.S., eds., Bordetella pertussis and parapertussis, in *Bailey and Scott's Diagnostic Microbiology*, 12th ed., Mosby, St. Louis, MO, 435–439, 2007.

225. Wernli, D. Emonet, S., Schrenzel, J., and Harbarth, S., Additional evaluation of eight cases of confirmed Bordetella bronchiseptica infection and colonization over a 15-year period. *Clin. Microbiol. Infect.*, 17, 201–203, 2011.

226. Sebaihia, M., Preston, A., Maskell, D.J., Kuzmiak, H., Connell, T.D., King, N.D., Orndorff, P.E., Miyamoto, D.M., Thomson, N.R., Harris, D., Goble, A., Lord, A., Murphy, L., et al., Comparison of the genome sequence of the poultry pathogen Bordetella avium with those of B. bronchiseptica, B. pertussis, and B. parapertussis reveals extensive diversity in surface structures associated with host interaction. *J. Bacteriol.*, 188(16), 6002–6015, 2006.

227. Spears, P.A., Temple, L.M., Miyamoto, D.M., Maskell, D.J., and Orndoroff, P.E., Unexpected similarities between Bordetella avium and other pathogenic Bordetellae. *Infect. Immunit.*, 71(5), 2591–2597, 2003.

228. Harrington, A.T., Castellanos, J.A., Ziedalski, T.M., Clarridge, J.E. III, and Cookson, B.T., Isolation of Bordetella avium and novel Bordetella strain from patients with respiratory disease. *Emerg. Infect. Dis.*, 15(1), 72–74, 2009.

229. Guthrie, J., Robertson, A., Tang, P., Jamieson, F., and Drews, J., Novel duplex real-time PCR assay detects Bordetella holmesii in specimens from patients with pertussis-like symptoms in Ontario.*Canada J. Clin. Microbiol.*, 48(4), 1435–1437, 2010.

230. Antila, M., He, Q., de Jong, C., Aarts, I., Verbakel, H., Bruisten, S., Keller, S., Haanpera, M., Makinen, J., Eerola, E., Viljanen, M.K., Mertsola, J., and van der Zee, A., Bordetella holmesii DNA is not detected in nasopharyngeal swabs from Finnish and Dutch patients with suspected pertussis. *J. Med. Microbiol.*, 55(Pt 8), 1043–1051, 2006.

231. Njamkepo, E., Bonacorsi, S., Debruyne, M., Gibaud, S., Guillot, S., and Guiso, N., Significant finding of B. holmesii DNA in nasopharyngeal samples from French patients with suspected pertussis. *J. Clin. Microbiol.*, 49(12), 4347–4348, 2011.

232. Kamiya, H., Otsuka, N., Ando, Y., Odaira, F., Yoshino, S., Kawano, K., Takahashi, H., Nishida, T., Hidaka, Y., Toyoizumi-Ajisaka, H., Shibayama, K., Kamachi, K., Sunagawa, T., Taniguchi, K., and Okabe, N., Transmission of Bordetella holmesii during pertussis outbreak, Japan. *Emerg. Infect. Dis.*,18(7), 1166–1169, 2012.

233. Katsukawa, C., Kushibiki, C., Nishito, A., Nishida, R., Kuwabara, N., Kawahara, R., Otsuka, N., Miyaji, Y., Toyoizumi-Ajisaka, T., and Kamachi, K., Bronchitis caused by Bordetella holmesii in a child with asthma misdiagnosed as mycoplasmal infection. *J. Infect. Chemother.*, 19(3), 534–357, 2013.

234. Tatti, K.M., Loparev, V.N., Ranganathan Ganakammal, S., Changayil, S., Frace, M., Weil, M.R., Sammons, S., MacCannell, D., Mayer, L.W., and Tondella, M.L., Draft genome sequences of Bordetella holmesii strains from blood (F627) and nasopharynx (H558). *Genome Announc.*, 1(2), e00056–13, 2013.

235. Pittet, L.F., Emonet, S., Schrenzel, J., Siegrist, C., and Posfay-Barbe, K. Bordetella holmesii: an underrecognized Bordetella species. *Lancet*, 14(6), 510–519, 2014.

236. Tartof, S., Gounder, P., Weiss, D., Lee, L., Cassiday, P., Clark, T., and Briere, E., Bordetella holmesii bacteremia cases in the United States, April 2010-January 2011. *CID*, 58, e39–e43, 2014.

237. Zhang, X., Weyrich, L.S., Lavine, J.S., Karanikas, A.T., and Harvill, E.T., Lack of cross-protection against Bordetella holmesii after pertussis vaccination. *Emerg. Infect. Dis.*, 18(11), 1771–1779, 2012.

238. Kattar, M.M., Chávez, J.F., Limaye, A.P., Rassoulian-Barrett, S.L., Yarfitz, S.L., Carlson, L.C., Houze, Y., Swanzy, S., Wood, B.L., and Cookson, B.T. Application of 16S rRNA gene sequencing to identify Bordetella hinzii as the causative agent of fatal septicemia. *J. Clin. Microbiol.*, 38(2), 789–794, 2000.

239. Register, K.B., Development of a PCR for identification of Bordetella hinzii. *Avian. Dis.*, 57(2), 307–310, 2013.

240. Almagro-Molto, M., Eder, W., and Schubert, S., Bordetella trematum in chronic ulcers: report on two cases and review of the literature. *Infection.*, 43(4), 489–494, 2015.

241. Majewski, L.L., Nogi, M., Bankowski, M.J., and Chung, H.H., Bordetella trematum sepsis with shock in a diabetic patient with rapidly developing soft tissue infection. *Diagn. Microbiol. Infect. Dis.*, 86(1), 112–114, 2016.

242. Le Coustumier, A., Njamkepo, E., Cattoir, V., Guillot, S., and Guiso, N. Bordetella petrii infection with long-lasting persistence in human. *Emerg. Infect. Dis.*, 17(4), 612–618, 2011.

243. Gross, R., Guzman, C.A., Sebaihia, M., dos Santos, V.A., Pieper, D.H., Koebnik, R., et al. The missing link: Bordetella petrii is endowed with both the metabolic versatility of environmental bacteria and virulence traits of pathogenic Bordetellae. *BMC Genom.*, 9, 449, 2008.

244. Ko, K.S., Peck, K.R., Oh, W.S., Lee, N.Y., Lee, J.H., and Song, J.H., New species of Bordetella, Bordetella ansorpii sp. nov., isolated from the purulent exudate of an epidermal cyst. *Clin. Microbiol*, 43(5), 2516–2519, 2005.

245. Fry, N., Duncan, J., Malnick, H., and Cockcroft, P., The first UK isolate of 'Bordetella ansorpii' from an immunocompromised patient. *J. Med. Microbiol.*, 56, 993–995, 2007.

246. Spilker, T., Darrah, R., and LiPuma, J.J., Complete genome sequences of Bordetella flabilis, Bordetella bronchialis, and "Bordetella pseudohinzii". *Genome Announc.*, 4(5), e01132–16, 2016.

247. Vandamme, P.A., Peeters, C., Cnockaert, M., Inganäs, E., Falsen, E., Moore, E.R., Nunes, O.C., Manaia, C.M., Spilker, T., and LiPuma, J.J., Bordetella bronchialis sp. nov., Bordetella flabilis sp. nov. and Bordetella sputigena sp. nov., isolated from human respiratory specimens, and reclassification of Achromobacter sediminum Zhang et al. 2014 as Verticia sediminum gen. nov., comb. nov. *Int J Syst Evol Microbiol.*, 65(10), 3674–3682, 2015.

248. Ivanov, Yury V. et al. Identification and taxonomic characterization of Bordetella pseudohinzii sp. nov. isolated from laboratory-raised mice. *Int. J. Syst. Evol. Microbiol.*, 66(12), 5452–5459, 2016.

249. Hamidou Soumana, I., Linz, B., and Harvill, E.T., Environmental origin of the genus Bordetella. *Front Microbiol.*, 8(28), 2017.

250. Vinogradov, E. and Caroff, M., Structure of the Bordetella trematum LPS O-chain subunit. *FEBS Lett.*, 579(1), 18–24, 2005.

251. Vinogradov, E., The structure of the core-O-chain linkage region of the lipopolysaccharide from Bordetella hinzii. *Carbohydr. Res.*, 342(3–4), 638–642, 2007.

252. Vinogradov, E., The structure of the carbohydrate backbone of the lipopolysaccharides from Bordetella hinzii and Bordetella bronchiseptica. *Eur. J. Biochem.*, 267(14), 4577–4582, 2000.

253. Gerlach, G., Janzen, S., Beier, D., and Gross, R., Functional characterization of the BvgAS two-component system of Bordetella holmesii. *Microbiology*, 150(11), 3715–3729, 2004.

254. Linz, B., Ivanov, Y.V., Preston, A., Brinkac, L., Parkhill, J., Kim, M., Harris, S.R., Goodfield, L.L., Fry, N.K., Gorringe, A.R., Nicholson, T.L., Register, K.B., Losada, L., and Harvill, E.T., Acquisition and loss of virulence-associated factors during genome evolution and speciation in three clades of Bordetella species. *BMC Genom.*, 17(1), 767, 2016.

255. LiPuma, J.J. and Spilker, T., Bordetella bronchialis strain AU3182, complete genome, 2016. (unpublished) Retrieved from https://www.ncbi.nlm.nih.gov/nuccore/CP016170.

256. Spilker, T., Darrah, R., and LiPuma, J.J., Bordetella pseudohinzii strain HI4681 chromosome, complete genome, 2016. (unpublished). Retrieved from https://www.ncbi.nlm.nih.gov/nuccore/CP016440.

24

The Genus Campylobacter

Collette Fitzgerald, Janet Pruckler, Maria Karlsson, and Patrick Kwan[*]
Updated 2021: Janet Pruckler, Lavin Joseph, Hayat Caidi, Mark Laughlin, Rachael D. Aubert

CONTENTS

Taxonomic History of Campylobacters

Spiral campylobacter-like organisms were first observed microscopically from the stool of children by Escherich in 1886. Between 1909 and 1944, there were a growing number of reports of similar "vibrio-like" organisms isolated from bovine and ovine sources, but they were not isolated from humans until 1938, in association with a milk-borne outbreak of gastroenteritis where blood cultures were positive for organisms resembling "Vibrio jejuni."[1] The microaerobic vibrios were assigned to the new genus Campylobacter in 1963,[2] and included just two

species: Campylobacter fetus and Campylobacter bubulus (now Campylobacter sputorum). Campylobacters were first successfully isolated from stool in the late 1960s using a filtration technique.[1] Later, the development of selective media brought the routine isolation of Campylobacter into the clinical microbiology setting and Campylobacter spp. rapidly became recognized as a common cause of bacterial gastroenteritis. The taxonomic structure of Campylobacter has changed substantially since its inception in 1963, first with a comprehensive study of the taxonomy of the genus in 1973.[3] By the late 1980s, there was a rapid increase in classification at the species level, and 14 species had been described (see review by Penner).[4]

---

[*] Disclaimer: Use of trade names is for identification only and does not imply endorsement by the Public Health Service or by the U.S. Department of Health and Human Services.

DOI: 10.1201/9781003099277-26

Advances in DNA technology, particularly 16S rRNA sequence analysis and DNA-DNA hybridization,[5] clarified the systematics of campylobacters and resulted in the extensive restructuring of the genus and assignment of campylobacters to the phylogenetic group rRNA superfamily VI.[6] The proposal that the genera *Campylobacter* and *Arcobacter* should be included in a separate family, the *Campylobacteraceae*, was also made in 1991 by Vandamme and Levy.[7] A detailed review of the taxonomy of *Campylobacteraceae* was published in 2008.[8] Thirty-one *Campylobacter* species are currently described within the family *Campylobacteraceae*. (Table 24.1). The disease spectrum of this genus ranges from well-established for *C. jejuni*, to emerging for

TABLE 24.1

Currently Described *Campylobacter* Species

Campylobacter Species	Known Sources	Human Disease Associated
C. jejuni subsp. *jejuni*	Poultry, cattle, sheep, wild birds, pigs	Gastroenteritis, meningitis, septicemia, Guillain-Barré syndrome
C. jejuni subsp. *doylei*	Humans	Gastroenteritis, septicemia
C. coli	Pigs, poultry, sheep / Wild birds, cattle	Gastroenteritis / Septicemia, meningitis
C. lari subsp. *lari*	Wild birds, poultry, dogs / Cats	Gastroenteritis / Septicemia
C. lari subsp. *concheus*	Shellfish	Gastroenteritis
C. fetus subsp. *fetus*	Cattle, sheep, reptiles	Gastroenteritis, septicemia
C. fetus subsp. *testudinum*	Humans, reptiles	Septicemia, chronic kidney disease
C. fetus subsp. *venerealis*	Cattle, sheep	Septicemia
C. upsaliensis	Dogs, cats	Gastroenteritis / Septicemia
C. helveticus	Cats, dogs	Gastroenteritis
C. insulaenigrae	Marine mammals	Gastroenteritis
C. peloridis	Shellfish	Gastroenteritis
C. hyointestinalis subsp. *hyointestinalis*	Pigs, cattle	Gastroenteritis
C. hyointestinalis subsp. *lawsonii*	Pigs	None at present
C. lanienae	Cattle, pigs	Gastroenteritis
C. sputorum bv. *sputorum*	Cattle, pigs	Abscesses, gastroenteritis
C. sputorum bv. *faecalis*	Sheep, bulls	None at present
C. sputorum bv. *paraureolyticus*	Cattle	Gastroenteritis
C. concisus	Humans, domestic pets	Gastroenteritis, periodontal disease, abscesses
C. curvus	Humans	Periodontal disease, gastroenteritis
C. rectus	Humans	Periodontal disease, abscesses
C. showae	Humans	Periodontal disease, abscesses
C. ureolyticus	Humans	Gastroenteritis, septicemia, soft tissue abscesses
C. gracilis	Humans	Periodontal disease, abscesses
C. hominis	Humans	None at present
C. mucosalis	Pigs	None at present
C. avium	Poultry	None at present
C. canadensis	Whooping cranes	None at present
C. cuniculorum	Rabbits	None at present
C. subantarcticus	Gray-headed albatrosses, black-browed albatrosses, gentoo penguins	None at present
C. volucris	Black-headed gulls	None at present
C. blaseri	Common seals	None at present
C. corcagiensis	Lion-tailed macaques	None at present
C. geochelonis	Hermann's tortoise	None at present
C. hepaticus	Chickens	None at present
C. iguaniorum	Reptiles	None at present
C. ornithocola	Wild birds	None at present
C. pinnipediorum subsp. *pinnipediorum*	Common seals, gray seals	None at present
C. pinnipediorum subsp. *caledonicus*	Common seals, gray seals	None at present

species such as *C. fetus* ssp. *fetus*, to speculative for species such as *C. concisus*. Since the last edition of this handbook, seven new *Campylobacter* species and one new subspecies have been described including *C. blaseri* from seals,[9] *C. corcagiensis* from lion-tailed macaques,[10] *C. geochelonis* from tortoises,[11] *C. hepaticus* from chickens,[12] *C. iguaniorum* from reptiles,[13] *C. ornithocola* from wild birds,[14] *C. pinnipediorum* from pinniped seals,[15] and *C. fetus* subsp. *testudinum* from humans and reptiles.[16]

The family *Campylobacteraceae* also contains 25 *Arcobacter* species and seven *Sulfurospirillum* species. Species belonging to the genus *Sulfurospirillum* have no known pathogenicity for humans or animals, and are considered environmental organisms.[17]

The most commonly recognized *Campylobacter* species, *C. jejuni*, is the leading cause of *Campylobacter* enteritis cases in humans; a number of other *Campylobacter* species, including *C. coli*, *C. upsaliensis*, *C. lari*, and *C. fetus*, have also been linked with enteritis (see Table 24.1). The true proportion of illness attributable to each of these species is not clear because campylobacters are difficult to differentiate, and many clinical laboratories do not identify *Campylobacter* isolates to the species level. *C. jejuni* is divided into two subspecies: *C. jejuni* ssp. *jejuni* (referred to as *C. jejuni* and the main focus of the chapter) and *C. jejuni* ssp. *doylei* (seldom found as the cause of human disease). *C. jejuni*, *C. coli*, *C. lari*, *C. upsaliensis*, and *C. helveticus* form a genetically close group as shown by 16S rRNA analysis[18] and are known as the thermotolerant or thermophilic campylobacters because they all grow optimally at 42°C.

Another group that is closely related genetically is the hydrogen-requiring species *Campylobacter sputorum*, *C. concisus*, *C. curvus*, *C. showae*, *C. rectus*, *C. gracilis*, and *C. hominis*. Most of these species have been associated with periodontal disease, although *C. hominis* has been isolated only from healthy human feces[19] and *C. sputorum* is also found in the reproductive and gastrointestinal tracts of various production animals. The taxonomic history of *C. sputorum* is complex and was revised in 1998 to include three biovars[20]: bv. *sputorum*, bv. *faecalis*, and bv. *paraureolyticus* based on the ability of a given strain to produce catalase or urease. Strains previously described as *C. bubulus* were reclassified as bv. *sputorum* as the tests that had been used to differentiate these two taxa were not reproducible.[21]

Campylobacter mucosalis and *C. sputorum* are closely related based on DNA-DNA hybridization studies, and they share a common source (porcine).[18] Because *C. mucosalis* strains require a hydrogen-enriched microaerobic atmosphere in order to grow, they will not be detected using culture methods used in most clinical laboratories, so their clinical significance is unknown. There are three *C. fetus* subspecies; *C. fetus* subsp. *fetus* is associated with both ovine/bovine reproductive disorders as well as human infections, *C. fetus* subsp. *testudinum* is associated with infections in both humans and reptiles[16] and *C. fetus* subsp. *venerealis* causes bovine-associated abortion and infertility.[3,22] This species is most similar phenotypically and genetically to the two *C. hyointestinalis* subspecies: *C. hyointestinalis* ssp. *hyointestinalis* and *C. hyointestinalis* ssp. *lawsonii*. Extensive 16S rRNA diversity has been reported for *C. hyointestinalis* strains.[23] *C. gracilis* was originally described as a *Bacteroides* species

associated with periodontal disease but formally assigned to the genus *Campylobacter* in 1995.[24] *C. gracilis* is often not recognized as being a *Campylobacter* because of its unusual morphology and metabolic characteristics; *C. gracilis* has an anaerobic growth requirement, a straight rod-shaped morphology, and is oxidase-negative.

General Characteristics

Campylobacter species are Gram-negative, nonspore forming rods. *Campylobacter* cells are typically curved, S-shaped, or spiral rods that are 0.2–0.9 μm wide and 0.5–5 μm long. Some species, such as *C. hominis* and *C. gracilis*, form straight rods. Organisms are usually motile by means of a single, polar, unsheathed flagellum at one or both ends. Cells of some species are nonmotile (*C. gracilis*) or have multiple flagella (*C. showae*). Most species require a microaerobic atmosphere for optimal growth; however, some species grow aerobically or anaerobically. An atmosphere containing increased hydrogen appears to be a growth requirement for other species such as *C. sputorum*, *C. concisus*, *C. mucosalis*, *C. curvus*, *C. showae*, *C. rectus*, *C. gracilis*, and *C. hominis*.[25]

Epidemiology

Campylobacters are a leading cause of bacterial enteritis worldwide and cause approximately 1.3 million illnesses annually in the United States.[26] Campylobacters can be transmitted directly from animal to person, through ingestion of fecally contaminated water, food, or by direct contact with animal feces or contaminated environmental surfaces. *Campylobacter* species are primarily zoonotic, with a variety of animals implicated as reservoirs of infection, including a diverse range of domestic and wild animals and birds.[1,27] In addition, water and the environment play a significant but poorly understood role in the epidemiology of campylobacteriosis.[28]

In the United States, requirements for reporting incidence of culture-confirmed infections vary by state. The Foodborne Diseases Active Surveillance network, FoodNet (Accessed August 5, 2019) provides uniform reporting from 10 sentinel sites, giving an accurate incidence of diagnosed infections. The reported incidence of *Campylobacter* infections in the United States in 2017 was 19.2 per 100,000 population, which represents a 10% increase from the 2014–2016 baseline.[29]

Most *Campylobacter* infections are acquired domestically, but *Campylobacter* is also a major cause of traveler's diarrhea.[30] While sporadic infections with *Campylobacter* are extremely common, *Campylobacter* outbreaks, defined as two or more persons epidemiologically linked by time and exposure, are relatively rare compared with other enteric pathogens. Outbreaks in the United States have been linked to a variety of foods, including poultry and other meats, and produce as well as contact with dogs obtained through the commercial dog industry. However, the most common single food implicated in *Campylobacter* outbreaks is unpasteurized milk.[31] Outbreaks have also been caused by contamination of untreated or poorly disinfected water supplies, and by drinking water from unprotected sources such as lakes or streams.[32,33] In contrast to other enteric pathogens, person-to-person transmission is unusual for outbreaks of campylobacteriosis despite the low infectious dose of *C. jejuni*. Handling or consumption of contaminated poultry meat is a major risk factor for human infection with *C. jejuni* and is thought to be responsible for making an important contribution to the number of sporadic *Campylobacter* infections seen annually.[34,35] While poultry is a principal source of campylobacteriosis, the contribution of other reservoirs to the overall burden of sporadic human disease has not been established.

In a typical case of diagnosed *Campylobacter* infection, illness lasts 5–7 days[36]; the infection resolves without antimicrobial treatment in the great majority of cases. Symptoms of infection typically begin 2–5 days after ingesting *C. jejuni* bacteria and include diarrhea with cramping abdominal pain, often with fever and some nausea. Extraintestinal *Campylobacter* infections are rare for people in good health, but are more likely to occur if the person is immunocompromised, elderly, or pregnant. Other complications of *Campylobacter* illness include bacteremia, Guillain-Barré syndrome (GBS), reactive arthritis, and irritable bowel syndrome.[37]

Isolation and Identification of *Campylobacter* spp.

Diagnosis of *Campylobacter* infection is usually made by isolating the organism from stool, blood, or other clinical specimens. Fastidious growth requirements can make isolation of *Campylobacter* from clinical specimens difficult. However, once isolated, presumptive identification of the genus *Campylobacter* is easily made based on morphology and darting motility. Further differentiation to species level is difficult because of the complex and rapidly evolving taxonomy and biochemical inertness of these bacteria. These problems have resulted in a proliferation of phenotypic and genotypic methods for identifying members of this group.

Direct Examination of Specimens

Microscopy

Presumptive diagnosis of a *Campylobacter* infection can be made directly by microscopic detection of the organism in a fresh stool specimen from patients with acute enteritis. Dark-field microscopy reveals the characteristic morphology and darting motility of campylobacters. Gram-stain analysis of stool specimens looking for organisms with typical *Campylobacter* morphology is also a highly sensitive and specific test that can be performed for rapid preliminary diagnosis of *Campylobacter* infection[38] and is currently underutilized.

Culture Independent Diagnostic Tests (CIDTs)

Culture independent methods for the direct detection of *Campylobacter* in stool specimens allow the detection of this fastidious organism without the specialized media and equipment that is needed for *Campylobacter* culture. Both immunologic and nucleic acid amplification tests (NAATs) are now commercially available. CIDTs are rapid, and provide same day

results, however, they do not yield an isolate for epidemiologic studies or antibiotic susceptibility testing.

Antigen Detection

Currently there are four immunoassays cleared by the U.S. Food and Drug Administration (FDA) for use in the United States that allow for the direct detection of Campylobacter antigen in stool specimens (ProSpecT™ Campylobacter Microplate Assay, Remel, Lenexa, KS; Premier™ CAMPY, Meridian Bioscience, Cincinnati, OH; ImmunoCard STAT! CAMPY®, Meridian Bioscience, Cincinnati, OH; and *Campylobacter* Quik Chek™, TechLab, Blacksburg, VA). These immunoassays detect both *C. jejuni* and *C. coli* in stool specimens, but cannot differentiate between them; the ProSpecT, Premier CAMPY and ImmunoCard STAT! tests have also been reported to cross react with *C. upsaliensis;* to date, a review or comparison of the performance of the *Campylobacter* Quik Chek has not been reported.[39,40] These methods have provided an attractive alternative to culture-based methods allowing rapid detection of *Campylobacter* within 2 hours.[41] However, inconsistent conclusions as to their sensitivity, specificity and false positivity rates have been reported[42,43] and cost is in excess of culture.

Nucleic Acid Detection

Many NAAT assays have been described to directly detect *Campylobacter* in stool specimens.[44] Currently there are six NAAT assays that detect *Campylobacter* and other gastrointestinal pathogens, FDA cleared for use in the United States. All of these tests detect a variety of gastrointestinal pathogens, including *Campylobacter*. The Prodesse Progastro SSCS (*Salmonella, Shigella, Campylobacter* and STEC) assay and BD MAX Enteric Bacterial Panel can detect the DNA from *C. jejuni* and *C. coli* but not differentiate between the species. The xTAG® GPP (gastrointestinal pathogen panel) and Verigene Enteric Pathogens assay can detect *C. jejuni, C. coli,* and *C. lari,* also without differentiating between the species. The BioFire Film Array and the Savyon GIP (gastrointestinal panel) both detects DNA from *C. jejuni, C. coli,* and *C. upsaliensis* but, again, does not differentiate between species. Advantages of using a nucleic acid detection approach include same-day detection and identification of *Campylobacter* to the species level and identification of the less-common *Campylobacter* species that are often missed by conventional culture.[45] However, this approach is more expensive than culture and does not provide an isolate for further characterization.

Transport Media

For the clinical microbiology laboratory whose main goal is to detect *Campylobacter* in the stool of patients with diarrhea, fresh stool (less than 2 hours old) is the ideal specimen. If a delay is anticipated, the specimen should be put in transport media such as modified Cary-Blair and kept at 4°C, not frozen. Transport media should also be used for rectal swab specimens. *Campylobacter* can be isolated from routine blood culture systems or wound specimens, so blood cultures should be done if the clinical features of the patient suggest campylobacteriosis.

Isolation from Human Specimens

Selective Media for Isolation

Diagnosis of *Campylobacter* gastroenteritis is commonly based on direct plating of stool onto selective media. No single medium will isolate all *Campylobacter* species. *C. jejuni* and *C. coli* can be isolated from stool samples by plating the stool on selective media and incubation in a microaerobic atmosphere at 42°C for 72 hours. A variety of media with different formulations, including both blood-based and charcoal-based selective media, are currently used for the isolation of these two species, most of which are commercially available. The history of the development of selective media for isolation of campylobacters was reviewed by Corry et al.[46] All the selective media contain a basal medium, either blood or other agents such as charcoal, to quench oxygen toxicity and a cocktail of antimicrobial agents to inhibit the normal flora associated with fecal specimens.

The use of antimicrobial-containing selective media and 42°C as the incubation temperature can inhibit the growth of the more fastidious emerging *Campylobacter* spp.[47] For example, the antimicrobial agent cephalothin that is present in some selective media formulations such as Campy-blood agar plates (BAPs) is inhibitory to *C. fetus* ssp. *fetus*. *C. jejuni* ssp. *doylei*, *C. upsaliensis*, and *C. concisus*. Therefore, these species generally do not grow on cephalothin-containing media. For primary isolation of *Campylobacter* from fecal samples, the use of cefoperazone-containing media instead of cephalothin-containing formulations is recommended.[25]

Enrichment Culture

A number of enrichment broths have been formulated to enhance the recovery of *Campylobacter* from stool samples.[46] Inclusion of an enrichment step may be beneficial in instances where low numbers of organisms are expected due to delayed transport to the laboratory, or after the acute stage of disease when the concentration of organisms may be low, such as in the investigation of GBS following acute *Campylobacter* infection.[25] Enrichment culture also is used to enhance the culture sensitivity of potentially environmentally stressed organisms. After selective enrichment, suspect *Campylobacter* colonies are subcultured onto *Campylobacter*-selective media such as mCCDA (modified charcoal cefoperazone deoxycholate agar).

Filtration Method for Isolation

A passive filtration technique that uses a nonselective medium developed by Steele and McDermott[48] is an effective approach to isolate campylobacters, particularly *Campylobacter* species that are susceptible to the antibiotics in selective agar isolation procedures.[49,50] The method is based on the principle that campylobacters can pass through membrane filters (0.45–0.65 µm) with relative ease (because of their small size and motility), while other stool flora are retained during the short processing time. Complete details of the procedure can be found in the *Manual of Clinical Microbiology*.[51] Incubation of nonselective media at 37°C under microaerobic conditions, preferably with an atmosphere containing increased hydrogen (for the hydrogen-requiring species) will allow isolation of the non-*C. jejuni*/*C. coli* species.

When this method is used, a variety of other *Campylobacter* taxa can be identified in diarrheic stool, although their pathogenic potential remains to be clarified.[52] Stool samples containing ~10^5 colony forming units (CFU) mL^{-1} *Campylobacter* will be detected with this method, and thus the filtration method is not as sensitive as primary culture with selective media.

Incubation Variables

Atmosphere

Most *Campylobacter* species require a microaerobic atmosphere containing approximately 5% O_2, 10% CO_2, and 85% N_2 for optimal recovery. There are a number of ways to generate microaerobic conditions; several manufacturers produce commercial microaerobic gas packs that are suitable for routine use in conjunction with an anaerobic jar, the evacuation-replacement method in conjunction with the approximate gas mixture, a tri-gas incubator, or the Anoxomat system (www.anoxomat.com. Accessed August 5, 2019) can also be used for routine cultures.[25] Some species of *Campylobacter*, such as *C. sputorum*, *C. concisus*, *C. mucosalis*, *C. curvus*, *C. rectus*, and *C. hyointestinalis* require an increased hydrogen concentration for isolation and growth. Typically, these species will not be recovered using commercial gas packs because the amount of hydrogen generated in properly used, commercial gas packs is less than 2%. A gas mixture containing at least 6% H_2 is sufficient for isolating hydrogen-requiring species.

Temperature and Length of Incubation

Species within the genus *Campylobacter* have different optimal temperatures for growth. Therefore, the choice of incubation temperature for routine stool cultures is critical in determining the spectrum of species that will be recovered.[25] It is common practice for most laboratories to use 42°C as the primary incubation temperature for *Campylobacter*, as this temperature allows growth of *C. jejuni* and *C. coli* on selective media. The nonthermophilic species will not grow at 42°C. In contrast, most *Campylobacter* species grow well at 37°C. Many of the selective media containing blood, such as Skirrow medium, were devised for use at 42°C and have poor selective properties at 37°C, whereas mCCDA and Karmali show good selective properties at 37°C.[46] Growth of *C. jejuni* and *C. coli* on selective media is typically visible within 24–48 hours. However, several studies report increased isolation rates when the incubation time is extended to 72 hours. Plates should be incubated a minimum of 72 hours before reporting as negative for *Campylobacter*.

Isolation of Campylobacters from Non-human Sources

Protocols for the isolation of *Campylobacter* species from food and water are available in the FDA's *Bacteriological Analytical Manual*[51] (https://www.fda.gov/food/laboratory-methods-food/bacteriological-analytical-manual-bam. Accessed August 5, 2019) and have recently been updated.[53] An international standard procedure for the detection of thermotolerant *Campylobacter* in food and animal foodstuff was first made by the ISO (International Organization for Standardization) in 1995.[54]

This method was updated in 2006 (ISO10272-1:2006),[55] in 2009[55], and more recently in 2017.[56] The recent validations of current methodologies to isolate *Campylobacter* spp. from foods were published.[57] Standard methods for the isolation of campylobacters from live animals are available from the World Organization of Animal Health (OIE) (www.oie.int. Accessed August 5, 2019).

Isolation of campylobacters from water can be difficult because campylobacters can be present in low numbers and may be stressed as a consequence of low temperature, osmotic stress, nutrient depletion, and/or competition from other organisms. For this reason, recovery of *Campylobacter* from water samples from contaminated surface and ground water sources implicated in *Campylobacter* outbreaks worldwide has not always been successful using conventional culture techniques.[58,59] ISO 17995 describes a standardized procedure for the isolation of *Campylobacter* from water.[60]

Identification

Identification of *Campylobacter* to the species level can be difficult because campylobacters are biochemically inert; few phenotypic tests are available to identify them to the species level. A presumptive identification of *Campylobacter* spp. can be made if oxidase-positive colonies exhibiting a characteristic Gram-stain appearance (curved Gram-negative rods) grow on selective media incubated at 42°C under microaerobic conditions. All *Campylobacter* species except *C. gracilis* are typically oxidase-positive. Campylobacters neither oxidize nor ferment glucose. Isolates that are hippurate-positive can be further identified as *C. jejuni* because it is the only *Campylobacter* sp. isolated from humans that produces the enzyme hippuricase. However, isolates of *C. jejuni* that are hippuricase-negative have been reported.[61] Hippuricase-negative *C. jejuni* strains cannot be differentiated from *C. coli* by phenotypic testing (see Figures 24.1A and B). Detection of species-specific sequences via PCR can be helpful in these instances to differentiate between these two closely related species.[62,63] The number of available PCR assays for species-specific identification of *C. jejuni* and *C. coli* continues to grow. Published evaluations of some of these assays highlight the importance of validating their sensitivity and specificity[62,63] and the continued need to revalidate protocols and certify their

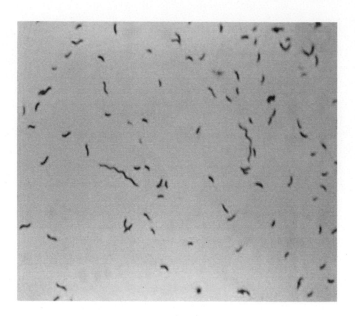

FIGURE 24.1B Gram stain image showing spiral morphology of *Campylobacter fetus* subsp.

accuracy as new species are described.[64] Figure 24.2 shows a flowchart for identification of *C. jejuni* and *C. coli*. Additional tests are needed to determine the species of hippurate-negative, indoxyl acetate-negative isolates.[25]

Nalidixic acid and cephalothin susceptibility testing have been used in species identification in the past.[65] A presumptive *Campylobacter* strain that was hippurate-positive, sensitive to nalidixic acid, and resistant to cephalothin was identified as a *C. jejuni*. However, fluoroquinolone resistant and cross-resistant to nalidixic acid *Campylobacter* species have become increasingly common, with rates reported to be as high as 80%.[66] Therefore, antimicrobial susceptibility tests can no longer be relied upon for the phenotypic identification of *Campylobacter* isolates.

The use of matrix-assisted laser desorption ionization time-of-flight (MALDI-TOF) mass spectrometry has more recently been shown to be an accurate, sensitive and rapid method for identification of *Campylobacter* spp.[67] and is becoming increasingly used in the routine clinical microbiology setting.[68]

Whole genome sequence (WGS) is an emerging technology that shows great potential for the identification of bacterial pathogens, including *Campylobacter,* in the clinical setting. WGS is already widely used in public health for subtyping and surveillance and by academia for research; however, there are challenges to the full implementation of WGS in the clinical setting.[69,70]

Antimicrobial Susceptibility Testing

The unique growth requirements of *Campylobacter* species present challenges for antimicrobial susceptibility determination. The Clinical and Laboratory Standards Institute (CLSI, formerly the National Committee for Clinical Laboratory Standards, NCCLS) and the European Committee for Antimicrobial Susceptibility Testing (EUCAST) are two widely used resources for standardized susceptibility testing methods,[71,72] and both committees provide specific guidance for *Campylobacter* testing. Before 2005, the CLSI described *Campylobacter* agar

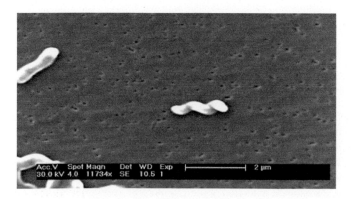

FIGURE 24.1A Electron micrograph of *Campylobacter jejuni*. (Courtesy of Public Health Image Library.)

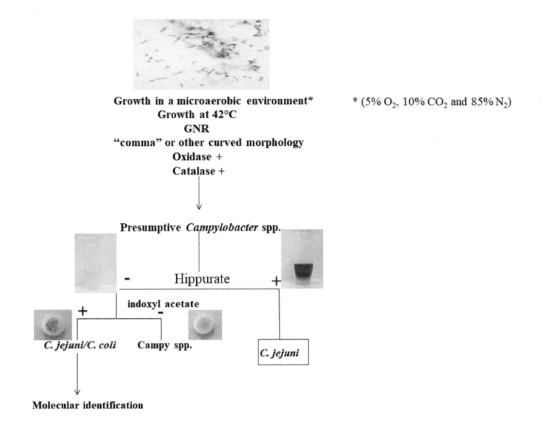

FIGURE 24.2 Identification of hippurate negative.

dilution testing as a standardized method for in *vitro* antimicrobial susceptibility testing of *Campylobacter* spp.[73] Given that agar dilution is cumbersome and can be costly to perform, broth microdilution and diffusion methods have also been used.[74–76] In 2004, broth microdilution minimum inhibitory concentration (MIC) was developed with quality control (QC) ranges for five antimicrobials: ciprofloxacin, doxycycline, gentamycin, erythromycin, and meropenem using *C. jejuni* ATCC 33560.[77] Methods for performing broth microdilution were also provided along with methodology for a variation of the standard disk diffusion method, allowing *Campylobacter* isolates to be screened for erythromycin and ciprofloxacin resistance. This approach uses growth up to the edge of a 6-mm disk (i.e., no zone of inhibition) as an indicator of acquired resistance (CLSI, M45AE). The CLSI agar dilution and disk diffusion methods require media supplemented with 5% sheep blood, whereas the CLSI broth microdilution tests require supplementation with 2.5–5% lysed horse blood. The tests can either be incubated at 36°C for 48 hours or at 42°C for 24 hours under microaerobic conditions. In the United States, if a laboratory chooses to apply EUCAST breakpoints, it is imperative that the FDA-cleared disk used contains the disk content assigned by EUCAST, and vice versa. Similarly, the disk quality control ranges are based on the disk content, and laboratories must ensure they are using the correct quality control range for the disk content employed. In 2017, CLSI and EUCAST formed a joint disk-diffusion working group endeavored to harmonize disk contents for new antimicrobial agents going forward.

Recently, EUCAST developed a standardized disk diffusion method for *Campylobacter* and used it to establish epidemiological cut off values (ECOFFs) for *C. jejuni* and *C. coli* with the same erythromycin and ciprofloxacin disk contents as those used by CLSI, as well as tetracycline (30 μg disk).[78] Compared to the CLSI method, EUCAST uses a different test medium (Mueller-Hinton agar with 5% mechanically defibrinated horse blood and 20 mg/L β-NAD). Based on the proposed ECOFFs, clinical breakpoints for disk diffusion were proposed beginning on January 1, 2013.[79]

Along with the disk diffusion guidelines, EUCAST also published ECOFFs and clinical breakpoints for MIC methods (http://www.eucast.org/mic_distributions_and_ecoffs/). Effective January 1, 2019, an updated correlation between MIC and zone distribution for ciprofloxacin, erythromycin, and tetracycline are available (http://www.eucast.org/ast_of_bacteria/calibration_and_validation/). As test medium for broth microdilution EUCAST recommends cation-adjusted Mueller-Hinton broth with 5% lysed horse blood and 20 mg/L β-NAD. The recommended incubation conditions for both disk diffusion and broth microdilution are 41 ± 1°C for 24 hours in a microaerobic environment.

Antimicrobial Resistance among Campylobacter

Antibiotic resistance in *Campylobacter* spp is an increasing concern, as it is in many other microorganisms. *Campylobacter* has evolved various mechanisms of resistance. These mechanisms include (i) restricting the access of antibiotics to their targets, which evolves reducing membrane permeability and increasing extrusion of antibiotics by efflux pumps; (ii) modification or protection of antibiotic targets; and (iii) modification or inactivation of antibiotics.[80] These emerging mechanisms threaten the usefulness of clinically important antibiotics used for treating human campylobacteriosis. As a foodborne pathogen, *Campylobacter* is constantly exposed to antimicrobials used for food production and has shown an amazing ability to adapt to antibiotic selection pressure.[81] The over-use of antibiotics in the human population and in animal husbandry has led to an increase in *Campylobacter* antibiotic-resistant infections, particularly with fluoroquinolones. Of greatest clinical concern would be the emergence of widespread macrolide resistance, since this class is the current treatment of choice of choice for campylobacteriosis. Also, the contribution of efflux pumps to antibiotic resistance warrants further studies, since this mechanism acts synergistically with other mechanisms of antibiotic resistance to confer high-level resistance in many instances.[81–83]

Antimicrobial resistance among *Campylobacter* varies by species, perhaps because *Campylobacter* species can be associated with different animal reservoirs. Resistance rates can vary between *C. jejuni* and *C. coli* because *C. jejuni* is more common in poultry and cattle while *C. coli* predominates in pigs[84] and antimicrobial use patterns vary by food animal species. The differences between *C. jejuni* and *C. coli* resistance are most notable in the case of the macrolides.

Macrolide Resistance

The increasing incidence of fluoroquinolone (FQ) resistance in *Campylobacter* has rendered macrolides such as erythromycin and azithromycin the drugs of choice for human campylobacteriosis.[85] Macrolide resistance among *Campylobacter jejuni* and *C. coli* was reviewed by Gibreel and Taylor.[86] This antimicrobial class includes erythromycin, azithromycin, and tylosin (a veterinary pharmaceutical), and resistance is attributed to chromosomal point mutations in the 23S rDNA that result in an altered ribosome binding site. These alterations most likely affect binding of the macrolide to the ribosome and are nontransmissible. The contribution of the CmeABC efflux systems to macrolide resistance in *Campylobacter* has been described in several recent studies.[87,88] The multidrug efflux pump CmeABC functions synergistically with 23S rRNA mutations to effect high-level macrolide resistance.[87–91] Almost all studies report higher macrolide resistance among *C. coli* than among *C. jejuni*. Because *C. coli* are associated with a swine reservoir, this raises questions about macrolide-selective pressure in those production settings.[66,92] The use of macrolides for growth promotion in animals was banned by the European Union in 1999 but they may still be used therapeutically.[93] In the United States, macrolide resistance was observed in 1.7% of *C. jejuni* and 2.7% of *C. coli* in 2011.[94]

Fluoroquinolone (FQ) Resistance

The fluoroquinolones such ciprofloxacin, enrofloxacin, and levofloxacin, are one of the drugs of choice to treat campylobacteriosis in humans as well as other bacterial diseases in both animals and humans.[81,95] In *Campylobacter*, the main resistance mechanism to FQ is mediated par points of mutations in the quinolone resistance determining region (QRDR) of *gyrA*.[96,97] One unique feature of FQ resistance in *Campylobacter* is that acquisition of high-level resistance does not stepwise accumulation of point of mutations in *gyrA*. Instead, a single mutation in *gyrA* can lead to clinically relevant levels of resistance in FQ antimicrobials.[98,99]

In addition to mutations in *gyrA*, the functional multidrug efflux pump, CmeABC, is also required for fluoroquinolone resistance in *Campylobacter*.[99,100] It is widely accepted that poultry is a common source of *C. jejuni* infections in humans, and fluoroquinolone-resistant *C. jejuni* has been isolated from retail poultry products.[101,102] Effective September 2005, enrofloxacin was withdrawn from use in U.S. poultry farms.[103]. However, it is not yet clear if the elevated rates of FQ resistance will decline after FQ restriction. Studies in several countries have shown that FQ resistance persists in poultry populations after the withdrawal of FQ, suggesting that there is a little/no fitness cost to FQ resistance in *Campylobacter*.[104,105]

Tetracycline Resistance

To date, resistance to tetracyclines in *Campylobacter* is conferred by a soluble protein factor called *tet*(O) and efflux pumps (CmeABC and *cme*G).[106] A recent study indicated that several critical residues located in the three loops of *tet*(O) disrupt the binding of tetracycline to the ribosome complex.[107] The *tet*(O) gene, which is widely present in *C. jejuni* and *C. coli*, can be located either in the chromosomal DNA or on a plasmid (e.g., pTet and pCC31).[108] Although high-level resistance to tetracyclines can be mediated by *tet*(O) alone, CmeABC and *cme*G multidrug efflux pumps contribute to both intrinsic and acquired resistance to tetracycline in *Campylobacter*, and function synergistically with *tet*(O) to confer high-level resistance to tetracycline.[109]

Aminoglycoside Resistance

Aminoglycosides are bactericidal antibiotics that bind to ribosomes and inhibit protein synthesis.[110] Aminoglycoside resistance was first detected in *C. coli* and was mediated by a 3'-aminoglycoside phosphotransferase (encoded by aphA-3) that has been previously described as conferring kanamycin resistance in *Streptococcus* and *Staphylococcus*.[83] Other *Campylobacter* strains harbor mosaic plasmids that contain various aminoglycoside resistance and insertion or transposon sequences from Gram-negative and Gram-positive bacteria, along with *tet(O)* genes[111–113] Other genes which confer kanamycin resistance include *aphA-1* and aphA-7, which were detected on plasmids in *C. jejuni*. Unlike *aphA-3* and *aphA-1*, which are thought have been horizontally acquired, *aph-7* has a similar G-C content to *C. jejuni* chromosomal DNA, suggesting it is intrinsic in *Campylobacter*.[112,113]

The gene conferring resistance to gentamycin in *Campylobacter* was initially annotated as *aac(6')-le/aph(2")-la* (also *aacA/aphD*), but it was found to encode the phosphotransferase only and thus was renamed *aph(2")-lf_*.[114] Very recently, several variants of 2"-phosphotransferase accounting for gentamycin resistance were identified in *Campylobacter* in the United States.[114,115] The contribution of efflux to aminoglycoside resistance in *Campylobacter* is not completely clear but it is likely to be less important than the plasmid-borne drug-modifying enzymes.[83]

Antimicrobial Resistance in Other Campylobacter Species

There is very little information available on the epidemiology of antimicrobial resistance among *Campylobacter* species other than *C. jejuni* and *C. coli*. Thwaites and Frost[116] reported susceptibility results for 25 isolates of *C. lari*. Resistance was observed for ampicillin (36%), kanamycin (60%), and tetracycline (12%); none were erythromycin-resistant. All the isolates were resistant to nalidixic acid, due to the long-recognized intrinsic resistance. They were also resistant to ciprofloxacin. In a study of 104 human isolates of *C. fetus* ssp. *fetus*, nonsusceptibility (intermediate or resistant) was found for cefotaxime (13%), erythromycin (71%), ciprofloxacin (5%), and tetracycline (34%).[117] In a 2006 study of 994 *C. fetus* isolated from cattle feces in Alberta, Canada, 39% were resistant to doxycycline and tetracycline, and 97% were resistant to nalidixic acid.[118] However, ciprofloxacin resistance was not found among the nalidixic acid-resistant *C. fetus* isolates, a phenotype typical for *C. fetus*. Ciprofloxacin resistance has been seen in *C. fetus* due to a *gyrA* mutation that has also been found in a ciprofloxacin-resistant *C. coli* strain.[119] Vandenberg et al.[50] found 100% nalidixic acid resistance among the eleven *C. fetus* strains isolated from humans in Belgium, but no resistance to the other five drugs tested, including ciprofloxacin. Of the 85 *C. upsaliensis* included in this study, 12% were nalidixic acid-resistant, 6% were ciprofloxacin-resistant, and 13% were erythromycin-resistant. Of 20 *C. concisus* isolates tested, 80% were nalidixic acid-resistant but only one isolate was ciprofloxacin-resistant.[81–83]

Epidemiologic Subtyping of *C. jejuni*

The ability to discriminate or subtype campylobacters below the level of species has been successfully applied to aid the epidemiological investigation of outbreaks of campylobacteriosis.[120] Subtyping provides information to recognize outbreaks of infection, match cases with potential vehicles of infection, and discriminate these from unrelated strains. In addition, these methods are important to identify potential reservoirs of strains that cause disease in humans, identify routes of transmission, and improve our understanding of *Campylobacter* epidemiology.[121]

A number of criteria are used to evaluate subtyping methods, including typability, reproducibility, discriminatory power, ease of use, ease of interpretation, and cost. The subtyping methods available to subtype *C. jejuni* vary considerably using these criteria, depending on the method used. The subtyping method of choice is ultimately determined by consideration of the basic microbiology of the organism in question, the nature of the microbiological question being asked, and, importantly, the ability of the typing method to detect significant epidemiological differences. Traditionally, phenotypic subtyping methods have been widely used to differentiate *C. jejuni* strains. There are two generally accepted, well-evaluated serotyping schemes that were developed in the 1980s for epidemiological characterization of *Campylobacter* isolates; the Penner scheme is based on heat-stable (HS) antigens using a passive hemagglutination technique,[122] and the Lior scheme is based on heat-labile (HL) antigens and a bacterial agglutination method.[123] The application of these techniques to *C. jejuni* isolates from human, farm animal, and environmental specimens has proven useful in outbreak investigations and for identification of potential reservoirs for human infection. However, the maintenance of a large panel of antisera that is not commercially available has prevented its

routine application for strain characterization and not all isolates are typable.[124]

The limitations associated with phenotypic subtyping methods and the rapid growth of molecular biology techniques led to the development of a wide range of molecular subtyping methods. A review of various molecular subtyping methods, including pulsed field gel electrophoresis (PFGE), and seven-gene multilocus sequence typing (MLST), used for characterization of *Campylobacter* isolates is described in Refs. [125, 126]. These methods demonstrated large genetic diversity among human *C. jejuni* isolates, indicative of the sporadic nature of human campylobacteriosis and the variety of different sources for *Campylobacter* infection. The implementation of standard PFGE protocols and data sharing through the PulseNet network provided additional ability to compare and exchange data and determine if the same *C. jejuni* strains are circulating in different places.[127–129]

Advances in DNA sequencing technology have provided a means to investigate genetic diversity at the nucleotide level. DNA sequence analysis is accurate, highly reproducible, does not rely on the interpretation of gel patterns and allows the inference of genetic relatedness between closely related strains. MLST characterizes bacterial isolates by indexing the variation present in short (400–550 nucleotide base pairs) gene fragments from seven housekeeping genes. MLST is not more discriminatory than PFGE, but also enables investigation of the population structure of the organism. In addition, MLST data generated is electronically portable and amenable to storage in web-accessible databases.[130] MLST schemes have been developed and databases are available for several *Campylobacter* species, including *C. jejuni,* at the pubMLST website (http://pubmlst.org/campylobacter). MLST studies have established the population structure for *C. jejuni*; strains have been confirmed to be genetically diverse, exhibit a high degree of plasticity, and have a weakly clonal population structure.[131] Subsequent studies have also begun to demonstrate associations of *C. jejuni* genotypes with particular animal hosts,[132–135] which greatly advanced the understanding of the molecular epidemiology of *C. jejuni*.

MLST data have been successfully applied in microbial subtype-based source attribution studies to understand the relative contribution of different sources to the burden of *C. jejuni* infections. These studies provided a mounting body of evidence that a large proportion of sporadic human infections originate from poultry sources, which stimulated and informed the implementation of intervention strategies that have led to sustained decline in the number of *Campylobacter* infections in New Zealand.[136,137]

The advent of next generation sequencing technology, accompanied by the rapidly expansion of sequencing platforms, has significantly increased the cost-effectiveness and throughput for WGS.[138] The sequencing of the first complete genome of a *C. jejuni* strain (NCTC11168) in 2000[139] and the subsequent sequencing of many more *Campylobacter* genomes since, has resulted in a new range of approaches to understand the genetic nature of pathogenicity, host association, environmental adaptation and evolution.[140] The vastly increased availability of full genome sequence data has paved the way for the emerging discipline of genomic epidemiology. The utility of WGS data to provide routine surveillance of *Campylobacter* infections has been shown in a real-time clinical setting to provide improved resolution among very closely related isolates and allowed the detection of single-strain clusters.[141,142] To facilitate this, web-accessible analysis tools and algorithms have been developed to compare various subsets of genetic loci, including MLST, core genome MLST (cgMLST), and whole genome MLST (wgMLST), without the need for a reference genome in a computationally nonintensive fashion.[143] Within PulseNet laboratories, foodborne outbreak clusters are detected using cgMLST. If further discrimination of isolates is needed, wgMLST or high-quality single-nucleotide polymorphism (hqSNP) analysis can be performed using the assembled sequences.[144–146]

The application of these powerful new molecular tools for comparative genomic studies of *C. jejuni* will facilitate the identification of phenotypic and molecular markers that may contribute to virulence, different disease outcomes, and host and ecological niche specificities. This, in turn, will greatly enhance source attribution studies to determine the fraction of *Campylobacter* infections attributable to different food sources, which will aid greatly the design of targeted control strategies to eliminate *C. jejuni* from the food chain so that food safety and public health risks are reduced.

REFERENCES

1. Butzler JP. 2004. Campylobacter, from obscurity to celebrity. *Clin Microbiol Infect.* 10:868–876.
2. Sebald M, Veron M. 1963. Teneur en bases de l'ADN et classification des vibrions. *Annales de l'Institut Pasteur.* 105:897–910.
3. Veron M, Chatelain R. 1973. Taxonomic study of the genus campylobacter Sebald and Véron and designation of the neotype strain for the type species, *Campylobacter fetus* (Smith and Taylor) Sebald and Veron). *Int J Syst Bacteriol.* 23:122–134.
4. Penner JL. 1988. The genus campylobacter: a decade of progress. *Clin Microbiol Rev.* 1:157–172.
5. Vandamme P, Pot B, Gillis M, de Vos P, Kersters K, Swings J. 1996. Polyphasic taxonomy, a consensus approach to bacterial systematics. *Microbiol Rev.* 60:407–438.
6. Vandamme P, Falsen E, Rossau R, Hoste B, Segers P, Tytgat R, De Ley J. 1991. Revision of campylobacter, helicobacter, and Wolinella taxonomy: emendation of generic descriptions and proposal of Arcobacter gen. nov. *Int J Syst Bacteriol.* 41:88–103.
7. Vandamme P, de Ley J. 1991. Proposal for a new family *Campylobacteraceae. Int J Syst Bacteriol.* 41:451–455.
8. Debruyne L, Gevers D, Vandamme P. 2008. Taxonomy of the family *Campylobacteraceae*, pp. 3–25, *in* Nachamkin I, C.M. Szymanski and M. J. Blaser (eds.), *Campylobacter*. ASM Press, Washington, DC.
9. Gilbert MJ, Zomer AL, Timmerman AJ, Spaninks MP, Rubio-Garcia A, Rossen JW, Duim B, Wagenaar JA. 2018. Campylobacter blaseri sp. nov., isolated from common seals (Phoca vitulina). *Int J Syst Evol Microbiol.* 68:1787–1794.
10. Koziel M, O'Doherty P, Vandamme P, Corcoran GD, Sleator RD, Lucey B. 2014. Campylobacter corcagiensis sp. nov., isolated from faeces of captive lion-tailed macaques (Macaca silenus). *Int J Syst Evol Microbiol.* 64:2878–2883.
11. Piccirillo A, Niero G, Calleros L, Perez R, Naya H, Iraola G. 2016. Campylobacter geochelonis sp. nov. isolated from the western Hermann's tortoise (testudo hermanni hermanni). *Int J Syst Evol Microbiol.* 66:3468–3476.
12. Van TTH, Elshagmani E, Gor MC, Scott PC, Moore RJ. 2016. Campylobacter hepaticus sp. nov., isolated from chickens with spotty liver disease. *Int J Syst Evol Microbiol.* 66:4518–4524.
13. Gilbert MJ, Kik M, Miller WG, Duim B, Wagenaar JA. 2015. Campylobacter iguaniorum sp. nov., isolated from reptiles. *Int J Syst Evol Microbiol.* 65:975–982.
14. Caceres A, Munoz I, Iraola G, Diaz-Viraque F, Collado L. 2017. Campylobacter ornithocola sp. nov., a novel member of the campylobacter lari group isolated from wild bird faecal samples. *Int J Syst Evol Microbiol.* 67:1643–1649.
15. Gilbert MJ, Miller WG, Leger JS, Chapman MH, Timmerman AJ, Duim B, Foster G, Wagenaar JA. 2017. Campylobacter pinnipediorum sp. nov., isolated from pinnipeds, comprising *Campylobacter pinnipediorum* subsp. pinnipediorum subsp. nov. and *Campylobacter pinnipediorum* subsp. caledonicus subsp. nov. *Int J Syst Evol Microbiol.* 67:1961–1968.
16. Fitzgerald C, Tu ZC, Patrick M, Stiles T, Lawson AJ, Santovenia M, Gilbert MJ, van Bergen M, Joyce K, Pruckler J, Stroika S, Duim B, Miller WG, Loparev V, Sinnige JC, Fields PI, Tauxe RV, Blaser MJ, Wagenaar JA.2014. *Campylobacter fetus* subsp. testudinum subsp. nov., isolated from humans and reptiles. *Int J Syst Evol Microbiol.* 64:2944–2948.
17. Schumacher W, Kroneck PMH, Pfennig N. 1992. Comparative systematic study on Spirillum 5175, *Campylobacter* and *Wolinella* species. Description of spirillum 5175 as Sulfurospirillum deleyanum gen. nov., sp. nov. *Arch Microbiol.* 158:287–293.
18. On SL. 2001. Taxonomy of campylobacter, Arcobacter, helicobacter and related bacteria: current status, future prospects and immediate concerns. *Symp Ser Soc Appl Microbiol.* 1S–15S.
19. Lawson AJ, On SL, Logan JM, Stanley J. 2001. Campylobacter hominis sp. nov., from the human gastrointestinal tract. *Int J Syst Evol Microbiol.* 51:651–660.
20. On SL, Atabay HI, Corry JE, Harrington CS, Vandamme P. 1998. Emended description of *Campylobacter sputorum* and revision of its infrasubspecific (biovar) divisions, including *C. sputorum* biovar paraureolyticus, a urease-producing variant from cattle and humans. *Int J Syst Bacteriol.* 48(Pt 1):195–206.
21. On SL. 2005. Taxonomy, phylogeny and methods for the identification of campylobacter species, pp. 14–42, *in Campylobacter Molecular and Cellular Microbiology.* Horizion Bioscience, Wymondham, UK.
22. Wagenaar JA, van Bergen MA, Blaser MJ, Tauxe RV, Newell DG, van Putten JP. 2014. *Campylobacter fetus* infections in humans: exposure and disease. *Clin Infect Dis.* doi:10.1093/cid/ciu085.
23. Harrington CS, On SL. 1999. Extensive 16S rRNA gene sequence diversity in campylobacter hyointestinalis strains: taxonomic and applied implications. *Int J Syst Bacteriol.* 49(Pt 3):1171–1175.
24. Vandamme P, Daneshvar MI, Dewhirst FE, Paster BJ, Kersters K, Goossens H, Moss CW. 1995. Chemotaxonomic analyses of bacteroides gracilis and bacteroides ureolyticus and reclassification of *B. gracilis* as campylobacter gracilis comb. nov. *Int J Syst Bacteriol.* 45:145–152.
25. Fitzgerald C, Nachamkin, I. 2011. Campylobacter and Arcobacter, pp. 885–899, *in Manual of Clinical Microbiology,* 10th ed. ASM Press, Washington, DC.

26. Scallan E, Hoekstra RM, Angulo FJ, Tauxe RV, Widdowson MA, Roy SL, Jones JL, Griffin PM. 2011. Foodborne illness acquired in the United States–major pathogens. *Emerg Infect Dis.* 17:7–15.

27. Altekruse SF, Stern NJ, Fields PI, Swerdlow DL. 1999. *Campylobacter jejuni*–an emerging foodborne pathogen. *Emerg Infect Dis.* 5:28–35.

28. Pitkanen T. 2013. Review of campylobacter spp. in drinking and environmental waters. *J Microbiol Methods.* 95:39–47.

29. Tack DM, Marder EP, Griffin PM, Cieslak PR, Dunn J, Hurd S, Scallan E, Lathrop S, Muse A, Ryan P, Smith K, Tobin-D'Angelo M, Vugia DJ, Holt KG, Wolpert BJ, Tauxe R, Geissler AL. 2019. Preliminary incidence and trends of infections with pathogens transmitted commonly through food - foodborne diseases active surveillance network, 10 U.S. Sites, 2015-2018. *MMWR Morb Mortal Wkly Rep.* 68:369–373.

30. Kendall ME, Crim S, Fullerton K, Han PV, Cronquist AB, Shiferaw B, Ingram LA, Rounds J, Mintz ED, Mahon BE. 2012. Travel-associated enteric infections diagnosed after return to the United States, foodborne diseases active surveillance network (FoodNet), 2004-2009. *Clin Infect Dis.*54(Suppl 5):S480–S487.

31. Taylor EV, Herman KM, Ailes EC, Fitzgerald C, Yoder JS, Mahon BE, Tauxe RV. 2013. Common source outbreaks of campylobacter infection in the USA, 1997-2008. *Epidemiol Infect.* 141:987–996.

32. O'Reilly CE, Bowen AB, Perez NE, Sarisky JP, Shepherd CA, Miller MD, Hubbard BC, Herring M, Buchanan SD, Fitzgerald CC, Hill V, Arrowood MJ, Xiao LX, Hoekstra RM, Mintz ED, Lynch MF. 2007. A waterborne outbreak of gastroenteritis with multiple etiologies among resort island visitors and residents: Ohio, 2004. *Clin Infect Dis.* 44:506–512.

33. Sacks JJ, Lieb S, Baldy LM, Berta S, Patton CM, White MC, Bigler WJ, Witte JJ. 1986. Epidemic campylobacteriosis associated with a community water supply. *Am J Public Health.* 76:424–428.

34. Friedman CR, Hoekstra RM, Samuel M, Marcus R, Bender J, Shiferaw B, Reddy S, Ahuja SD, Helfrick DL, Hardnett F, Carter M, Anderson B, Tauxe RV. 2004. Risk factors for sporadic *Campylobacter* infection in the United States: a case-control study in FoodNet sites. *Clin Infect Dis.* 38(Suppl 3):S285–S296.

35. Samuel MC, Vugia DJ, Shallow S, Marcus R, Segler S, McGivern T, Kassenborg H, Reilly K, Kennedy M, Angulo F, Tauxe RV. 2004. Epidemiology of sporadic campylobacter infection in the United States and declining trend in incidence, FoodNet 1996-1999. *Clin Infect Dis.* 38(Suppl 3).S165–S174.

36. Skirrow MaMB. 2000. Clinical aspects of *Campylobacter* infection. *In* Nachamkin MBaI (ed.), *Campylobacter.* American Society for Microbiology, Washington, DC.

37. Keithlin J, Sargeant J, Thomas MK, Fazil A. 2014. Systematic review and meta-analysis of the proportion of Campylobacter cases that develop chronic sequelae. *BMC Public Health.* 14:1203.

38. Wang H, Murdoch DR. 2004. Detection of campylobacter species in faecal samples by direct gram stain microscopy. *Pathology.* 36:343–344.

39. Couturier BA, Couturier MR, Kalp KJ, Fisher MA. 2013. Detection of non-jejuni and -coli campylobacter species from stool specimens with an immunochromatographic antigen detection assay. *J Clin Microbiol.* 51:1935–1937.

40. Hindiyeh M, Jense S, Hohmann S, Benett H, Edwards C, Aldeen W, Croft A, Daly J, Mottice S, Carroll KC. 2000. Rapid detection of *Campylobacter jejuni* in stool specimens by an enzyme immunoassay and surveillance for campylobacter upsaliensis in the greater Salt Lake City area. *J Clin Microbiol.* 38:3076–3079.

41. Granato PA, Chen L, Holiday I, Rawling RA, Novak-Weekley SM, Quinlan T, Musser KA. 2010. Comparison of premier CAMPY enzyme immunoassay (EIA), ProSpecT Campylobacter EIA, and ImmunoCard STAT! CAMPY tests with culture for laboratory diagnosis of campylobacter enteric infections. *J Clin Microbiol.* 48:4022–4027.

42. Bessede E, Delcamp A, Sifre E, Buissonniere A, Megraud F. 2011. New methods for detection of campylobacters in stool samples in comparison to culture. *J Clin Microbiol.* 49:941–944.

43. Myers AL, Jackson MA, Selvarangan R. 2011. False-positive results of campylobacter rapid antigen testing. *Pediatr Infect Dis J.* 30:542.

44. de Boer RF, Ott A, Guren P, van Zanten E, van Belkum A, Kooistra-Smid AM. 2013. Detection of campylobacter species and Arcobacter butzleri in stool samples by use of real-time multiplex PCR. *J Clin Microbiol.* 51:253–259.

45. Kulkarni SP, Lever S, Logan JM, Lawson AJ, Stanley J, Shafi MS. 2002. Detection of campylobacter species: a comparison of culture and polymerase chain reaction based methods. *J Clin Pathol.* 55:749–753.

46. Corry JE, Post DE, Colin P, Laisney MJ. 1995. Culture media for the isolation of campylobacters. *Int J Food Microbiol.* 26:43–76.

47. Man SM. 2011. The clinical importance of emerging campylobacter species. *Nat Rev Gastroenterol Hepatol.* 8:669–685.

48. Steele TW, McDermott SN. 1984. The use of membrane filters applied directly to the surface of agar plates for the isolation of *Campylobacter jejuni* from feces. *Pathology.* 16:263–265.

49. Lastovica AJ, Skirrow, MB. 2000. Clinical significance of campylobacter and related species other than *Campylobacter jejuni* and *C. coli*, pp. 89–120, *in Campylobacter*, 2nd ed. ASM Press, Washington, DC.

50. Vandenberg O, Houf K, Douat N, Vlaes L, Retore P, Butzler JP, Dediste A. 2006. Antimicrobial susceptibility of clinical isolates of non-jejuni/coli campylobacters and Arcobacters from Belgium. *J Antimicrob Chemother.* 57:908–913.

51. Nachamkin I. 2019. Campylobacter and Arcobacter, pp. 1028–1043, *in* Carroll KC, Phaller MA, Landry ML, McAdam AJ, Patel R, Richter SS, Warnock DW (eds.), *Manual of Clinical Microbiology*, 12th ed. ASM Press, Washington, DC.

52. Engberg J, On SL, Harrington CS, Gerner-Smidt P. 2000. Prevalence of campylobacter, Arcobacter, helicobacter, and Sutterella spp. in human fecal samples as estimated by a reevaluation of isolation methods for campylobacters. [see comments]. *J Clin Microbiol.* 38:286–291.

53. Gharst G, Bark DH, Newkirk R, Guillen L, Wang Q, Abeyta C, Jr. 2013. Evaluation and single-laboratory verification of a proposed modification to the U.S. Food and Drug Administration method for detection and identification of *Campylobacter jejuni* or campylobacter coli from raw silo milk. *J AOAC Int.* 96:1336–1342.

54. ANON. 1995. Microbiology of food and animal feeding stuffs—Horizontal method for detection of thermotolerant Campylobacter. International Organization for Standardization, Geneva.

55. ANON. 2006. Microbiology of food and animal feeding stuffs—Horizontal method for detection of Campylobacter spp. International Organization for Standardization, Geneva.

56. ANON. 2009. Microbiology of food and animal feeding stuffs—Horizontal method for detection of Campylobacter spp. International Organization for Standardization, Geneva.

57. Biesta-Peters EG, Jongenburger I, de Boer E, Jacobs-Reitsma WF. 2019. Validation by interlaboratory trials of EN ISO 10272—microbiology of the food chain—horizontal method for detection and enumeration of campylobacter spp. Part 1: detection method. *Int J Food Microbiol.* 288:39–46.

58. Bopp DJ, Sauders BD, Waring AL, Ackelsberg J, Dumas N, Braun-Howland E, Dziewulski D, Wallace BJ, Kelly M, Halse T, Musser KA, Smith PF, Morse DL, Limberger RJ. 2003. Detection, isolation, and molecular subtyping of *Escherichia coli* O157:H7 and *Campylobacter jejuni* associated with a large waterborne outbreak. *J Clin Microbiol.* 41:174–180.

59. Clark CG, Price L, Ahmed R, Woodward DL, Melito PL, Rodgers FG, Jamieson F, Ciebin B, Li A, Ellis A. 2003. Characterization of waterborne outbreak-associated *Campylobacter jejuni*, Walkerton, Ontario. *Emerg Infect Dis.* 9:1232–1241.

60. ANON. 2005. Water Quality—Detection and enumeration of thermotolerant Campylobacter species. International Organization for Standardization, Geneva.

61. Rautelin H, Jusufovic J, Hanninen ML. 1999. Identification of hippurate-negative thermophilic campylobacters. *Diagn Microbiol Infect Dis.* 35:9–12.

62. Burnett TA, Michael AH, Kuhnert P, Djordjevic SP. 2002. Speciating *Campylobacter jejuni* and campylobacter coli isolates from poultry and humans using six PCR-based assays. *FEMS Microbiol Lett.* 216:201–209.

63. On SL, Jordan PJ. 2003. Evaluation of 11 PCR assays for species-level identification of *Campylobacter jejuni* and campylobacter coli. *J Clin Microbiol.* 41:330–336.

64. On SL. 2013. Isolation, identification and subtyping of campylobacter: where to from here? *J Microbiol Methods.* 95:3–7.

65. Barrett TJ, Patton CM, Morris G K. 1988. Differentiation of campylobacter species using phenotypic characterization. *Lab Med.* 19:96–102.

66. Engberg J, Aarestrup FM, Taylor DE, Gerner-Smidt P, Nachamkin I. 2001. Quinolone and macrolide resistance in *Campylobacter jejuni* and C. coli: resistance mechanisms and trends in human isolates. *Emerg Infect Dis.* 7:24–34.

67. Mandrell RE, Harden LA, Bates A, Miller WG, Haddon WF, Fagerquist CK. 2005. Speciation of *Campylobacter coli, C. jejuni, C. helveticus, C. lari, C. sputorum,* and *C. upsaliensis* by matrix-assisted laser desorption ionization-time of flight mass spectrometry. *Appl Environ Microbiol.* 71:6292–6307.

68. Hsieh YH, Wang YF, Moura H, Miranda N, Simpson S, Gowrishankar R, Barr J, Kerdahi K, Sulaiman IM. 2018. Application of MALDI-TOF MS systems in the rapid identification of campylobacter spp. of public health importance. *J AOAC Int.* 101:761–768.

69. Mitchell SL, Simner PJ. 2019. Next-generation sequencing in clinical microbiology: are we there yet? *Clin Lab Med.* 39:405–418.

70. Tagini F, Greub G. 2017. Bacterial genome sequencing in clinical microbiology: a pathogen-oriented review. *Eur J Clin Microbiol Infect Dis.* 36:2007–2020.

71. ANON. 2000. EUCAST Definitive Document E.DEF 3.1, June 2000: determination of minimum inhibitory concentrations (mics) of antibacterial agents by agar dilution. *Clin Microbiol Infect.* 6:509–515.

72. NCCLS. 2003. Methods for Dilution Antimicrobial Susceptibility Tests for Bacteria that Grow Aerobically; Approved Standard-Sixth Edition. NCCLS, Wayne, Pennsylvania.

73. NCCLS. 2004. Performance Standards for Antimicrobial Susceptibility Testing; Fourteenth Informational Supplement. NCCLS, Wayne, PA.

74. Centers for Disease Control and Prevention. 2006. NARMS Human Isolates Final Report, 2003. *on* Division of Bacterial and Mycotic Diseases. http://www.cdc.gov/narms/reports/index.html. Accessed Nov. 11, 2006.

75. Danish Integrated Antimicrobial Resistance Monitoring and Research Programme. 2004. DANMAP 2003—Use of antimicrobial agents and occurrence of antimicrobial resistance in bacteria from food animals, foods and humans in Denmark. Søborg, Denmark.

76. Tenover FC, Baker CN, Fennell CL, Ryan CA. 1992. Antimicrobial resistance in campylobacter species, pp. 66–73, *in* Nachamkin I, Blaser MJ, Tompkins LS (eds.), *Campylobacter jejuni Current Status and Future Trends.* American Society of Microbiology, Washington, DC.

77. McDermott PF, Bodeis-Jones SM, Fritsche TR, Jones RN, Walker RD. 2005. Broth microdilution susceptibility testing of *Campylobacter jejuni* and the determination of quality control ranges for fourteen antimicrobial agents. *J Clin Microbiol.* 43:6136–6138.

78. (EUCAST) ECoASt. Clinical breakpoints, epidemiological cut-off (ECOFF) values and EUCAST disk diffusion methodology for *Campylobacter jejuni* and Campylobacter coli. Available at: http://www.eucast.org/fileadmin/src/media/PDFs/EUCAST_files/Consultation/Campylobacter_wide_consultation_August_2012.pdf

79. (EUCAST). ECoASt. Breakpoint tables for interpretation of MICs and zone diameters, version 4.0. Available at: http://www.eucast.org/fileadmin/src/media/PDFs/EUCAST_files/Breakpoint_tables/Breakpoint_table_v_4.0.pdf.

80. Efimochkina NR, Stetsenko VV, Bykova IV, Markova YM, Polyanina AS, Aleshkina AI, Sheveleva SA. 2018. Studying the phenotypic and genotypic expression of antibiotic resistance in *Campylobacter jejuni* under stress conditions. *Bull Exp Biol Med.* 164:466–472.

81. Shen Z, Wang Y, Zhang Q, Shen J. 2017. Antimicrobial resistance in campylobacter spp. *Microbiol Spectr.* 6.

82. Spizek, J. 2018. Fight against antimicrobial resistance. *Epidemiol Mikrobiol Imunol.* 67:74–80.

83. Iovine, NM. 2013. Resistance mechanisms in *Campylobacter jejuni*. *Virulence.* 4:230–240.

84. Aarestrup FM, Nielsen EM, Madsen M, Engberg J. 1997. Antimicrobial susceptibility patterns of thermophilic campylobacter spp. from humans, pigs, cattle, and broilers in Denmark. *Antimicrob Agents Chemother.* 41:2244–2250.

85. Bolinger H, Kathariou, S. 2017. The current state of macrolide resistance in campylobacter spp.: trends and impacts of resistance mechanisms. *Appl Environ Microbiol.* 83:e00416–e00417.

86. Gibreel A, Taylor DE. 2006. Macrolide resistance in *Campylobacter jejuni* and campylobacter coli. *J Antimicrob Chemother.* 58:243–255.

87. Liu D, Liu W, Lv Z, Xia J, Li X, Hao Y, Zhou Y, Yao H, Liu Z, Wang Y, Shen J, Ke Y, Shen Z. 2019. Emerging erm(B)-mediated macrolide resistance associated with novel multidrug resistance genomic islands in campylobacter. *Antimicrob Agents Chemother.* 63(7): e00153–19.

88. Pérez-Boto D, Paloma A, García-Peña FJ, Abad JC, Echeita MA, Amblar M. 2015. Isolation of a point mutation associated with altered expression of the CmeABC efflux pump in a multidrug-resistant *Campylobacter jejuni* population of poultry origin. *J Glob Antimicrob Resist.* 3:115–122.

89. Cagliero C, Mouline C, Payot S, Cloeckaert A. 2005. Involvement of the CmeABC efflux pump in the macrolide resistance of Campylobacter coli. *J Antimicrob Chemother.* 56:948–950.

90. Gibreel A, Wetsch NM, Taylor DE. 2007. Contribution of the CmeABC efflux pump to macrolide and tetracycline resistance in *Campylobacter jejuni. Antimicrob Agents Chemother.* 51:3212–3216.

91. Lin J, Cagliero C, Guo B, Barton YW, Maurel MC, Payot S, Zhang Q. 2005. Bile salts modulate expression of the CmeABC multidrug efflux pump in *Campylobacter jejuni. J Bacteriol.* 187:7417–7424.

92. Aarestrup FM. 2000. Occurrence, selection and spread of resistance to antimicrobial agents used for growth promotion for food animals in Denmark. *APMIS Suppl.* 101:1–48.

93. Council of the European Union. 1998. Council Regulation (EC) No 2821/98 of 17 December 1998 amending, as regards withdrawal of the authorisation of certain antibiotics, Directive 70/524/EED concerning additives in feedstuffs. *Official Journal of the European Communities.* L351:4–8.

94. Centers for Disease Control and Prevention. 2013. National Antimicrobial Resistance Monitoring System for Enteric Bacteria (NARMS): Human Isolates Final Report, 2011. http://www.cdc.gov/narms/pdf/2011-annual-report-narms-508c.pdf.

95. Hooper DC. 1998. Clinical applications of quinolones. *Biochim Biophys Acta.* 1400:45–61.

96. Luangtongkum T, Jeon B, Han J, Plummer P, Logue CM, Zhang Q. 2009. Antibiotic resistance in Campylobacter: emergence, transmission and persistence. *Future Microbiol.* 4:189–200.

97. Payot S, Bolla J-M, Corcoran D, Fanning S, Mégraud F, Zhang Q. 2006. Mechanisms of fluoroquinolone and macrolide resistance in campylobacter spp. *Microbes Infect.* 8:1967–1971.

98. Ge B, McDermott PF, White DG, Meng J. 2005. Role of efflux pumps and topoisomerase mutations in fluoroquinolone resistance in *Campylobacter jejuni* and campylobacter coli. *Antimicrob Agents Chemother.* 49:3347–3354.

99. Luo N, Sahin O, Lin J, Michel LO, Zhang Q. 2003. In vivo selection of campylobacter isolates with high levels of fluoroquinolone resistance associated with gyrA mutations and the function of the CmeABC efflux pump. *Antimicrob Agents Chemother.* 47:390–394.

100. Shen Z, Wang Y, Zhang Q, Shen J. 2018. Antimicrobial resistance in *Campylobacter* spp. *Microbiol Spectr.* 6.

101. Smith KE, Besser JM, Hedberg CW, Leano FT, Bender JB, Wicklund JH, Johnson BP, Moore KA, Osterholm MT. 1999. Quinolone-resistant *Campylobacter jejuni* infections in Minnesota, 1992-1998. Investigation Team. *N Engl J Med.* 340:1525–1532.

102. Ge B, White DG, McDermott PF, Girard W, Zhao S, Hubert S, Meng J. 2003. Antimicrobial-resistant *Campylobacter* species from retail raw meats. *Appl Environ Microbiol.* 69:3005–3007.

103. Nelson JM, Chiller TM, Powers JH, Angulo FJ. 2007. Fluoroquinolone-resistant campylobacter species and the withdrawal of fluoroquinolones from use in poultry: a public health success story. *Clin Infect Dis.* 44:977–980.

104. Luo N, Pereira S, Sahin O, Lin J, Huang S, Michel L, Zhang Q. 2005. Enhanced in vivo fitness of fluoroquinolone-resistant *Campylobacter jejuni* in the absence of antibiotic selection pressure. *Proc Natl Acad Sci U S A.* 102:541–546.

105. Zhang Q, Lin J, Pereira S. 2003. Fluoroquinolone-resistant campylobacter in animal reservoirs: dynamics of development, resistance mechanisms and ecological fitness. *Anim Health Res Rev.* 4:63–71.

106. Connell SR, Trieber CA, Dinos GP, Einfeldt E, Taylor DE, Nierhaus KH. 2003. Mechanism of Tet(O)-mediated tetracycline resistance. *EMBO J.* 22:945–953.

107. Li W, Atkinson GC, Thakor NS, Allas U, Lu CC, Chan KY, Tenson T, Schulten K, Wilson KS, Hauryliuk V, Frank J. 2013. Mechanism of tetracycline resistance by ribosomal protection protein Tet(O). *Nat Commun.* 4.

108. Batchelor RA, Pearson BM, Friis LM, Guerry P, Wells JM. 2004. Nucleotide sequences and comparison of two large conjugative plasmids from different campylobacter species. *Microbiology.* 150:3507–3517.

109. Mu Y, Shen Z, Jeon B, Dai L, Zhang Q. 2013. Synergistic effects of anti-CmeA and anti-CmeB peptide nucleic acids on sensitizing *Campylobacter jejuni* to antibiotics. *Antimicrob Agents Chemother.* 57:4575–4577.

110. Hormeño L, Ugarte-Ruiz M, Palomo G, Borge C4, Florez-Cuadrado D, Vadillo S, Píriz S, Domínguez L, Campos MJ, Quesada A. 2018. Ant(6)-I genes encoding aminoglycoside o-nucleotidyltransferases are widely spread among streptomycin resistant strains of *Campylobacter jejuni* and campylobacter coli. *Front Microbiol.* 2515:eCollection 2018.

111. Ansary A, Radu S. 1992. Conjugal transfer of antibiotic resistances and plasmids from *Campylobacter jejuni* clinical isolates. *FEMS Microbiol Lett.* 70:125–128.

112. Gibreel A, Sköld O, Taylor DE. 2004. Characterization of plasmid-mediated aphA-3 kanamycin resistance in *Campylobacter jejuni. Microb Drug Resist.* 10:98–105.

113. Gibreel A, Tracz DM, Nonaka L, Ngo TM, Connell SR, Taylor DE. 2002. Incidence of antibiotic resistance in *Campylobacter jejuni* isolated in Alberta, Canada, from 1999 to 2002, with special reference to tet(O)-mediated tetracycline resistance. *Antimicrob Agents Chemother.* 48:3442–3450.

114. Yao H, Liu D, Wang Y, Zhang Q, Shen Z. 2017. High prevalence and predominance of the aph(2″)-if gene conferring aminoglycoside resistance in campylobacter. *Antimicrob Agents Chemother.* 61:e00112– e00117.

115. Chen Y, Mukherjee S, Hoffmann M, Kotewicz ML, Young S, Abbott J, Luo Y, Davidson MK, Allard M, McDermott P, Zhao S. 2013. Whole-genome sequencing of gentamicin-resistant campylobacter coli isolated from U.S. retail meats reveals novel plasmid-mediated aminoglycoside resistance genes. *Antimicrob Agents Chemother.* 57:5398–5405.

116. Thwaites RT, Frost JA. 1999. Drug resistance in *Campylobacter jejuni*, C coli, and C lari isolated from humans in north west England and Wales, 1997. *J Clin Pathol.* 52:812–814.

117. Tremblay C, Gaudreau C, Lorange M. 2003. Epidemiology and antimicrobial susceptibilities of 111 *Campylobacter fetus* subsp. fetus strains isolated in Quebec, Canada, from 1983 to 2000. *J Clin Microbiol*. 41:463–466.

118. Inglis GD, Morck DW, McAllister TA, Entz T, Olson ME, Yanke LJ, Read RR. 2006. Temporal prevalence of antimicrobial resistance in Campylobacter spp. from beef cattle in Alberta feedlots. *Appl Environ Microbiol*. 72:4088–4095.

119. Engberg J, Keelan M, Gerner-Smidt P, Taylor DE. 2006. Antimicrobial resistance in *Campylobacter*, pp. 269–292, *in* Aarestrup FM (ed.), *Antimicrobial Resistance in Bacteria of Animal Origin*. ASM Press, Washington, DC.

120. Fitzgerald C, Helsel LO, Nicholson MA, Olsen SJ, Swerdlow DL, Flahart R, Sexton J, Fields PI. 2001. Evaluation of methods for subtyping *Campylobacter jejuni* during an outbreak involving a food handler. *J Clin Microbiol*. 39:2386–2390.

121. Fitzgerald C, Swaminathan, B, and Sails, AFitzgerald, C, Swaminathan, B, Sails, A. 2003. Detecting pathogens in food, pp. 271–293, *in Genetic Techniques: Molecular Subtyping Methods*. CRC Press, Boca Raton, FL.

122. Penner JL, Hennessy JN. 1980. Passive hemagglutination technique for serotyping *Campylobacter fetus* subsp. jejuni on the basis of soluble heat-stable antigens. *J Clin Microbiol*. 12:732–737.

123. Lior H, Woodward DL, Edgar JA, Laroche LJ, Gill P. 1982. Serotyping of *Campylobacter jejuni* by slide agglutination based on heat-labile antigenic factors. *J Clin Microbiol*. 15:761–768.

124. Fitzgerald C, Sails A, Fields PI. 2005. *Campylobacter jejuni* strain variation, pp. 59–77, *in Campylobacter Molecular and Cellular Biology*. Horizion Biosciences, Norfolk NR.

125. Wassenaar TM, Newell DG. 2000. Genotyping of campylobacter spp. *App Environ Microbiol*. 66:1–9.

126. Taboada EN, Clark CG, Sproston EL, Carrillo CD. 2013. Current methods for molecular typing of campylobacter species. *J Microbiol Methods*. 95:24–31.

127. Gardner TJ, Fitzgerald C, Xavier C, Klein R, Pruckler J, Stroika S, McLaughlin JB. 2011. Outbreak of campylobacteriosis associated with consumption of raw peas. *Clin Infect Dis*. 53:26–32.

128. Gerner-Smidt P, Hise K, Kincaid J, Hunter S, Rolando S, Hyytia-Trees E, Ribot EM, Swaminathan B. 2006. PulseNet USA: a five-year update. *Foodborne Pathog Dis*. 3:9–19.

129. Ribot EM, Fitzgerald C, Kubota K, Swaminathan B, Barrett TJ. 2001. Rapid pulsed-field gel electrophoresis protocol for subtyping of *Campylobacter jejuni*. *J Clin Microbiol*. 39:1889–1894.

130. Jolley KA, Chan MS, Maiden MC. 2004. mlstdbNet—distributed multi-locus sequence typing (MLST) databases. *BMC Bioinformatics*. 5:86.

131. Dingle KE, Colles FM, Wareing DR, Ure R, Fox AJ, Bolton FE, Bootsma HJ, Willems RJ, Urwin R, Maiden MC. 2001. Multilocus sequence typing system for *Campylobacter jejuni*. *J Clin Microbiol*. 39:14–23.

132. Colles FM, Jones K, Harding RM, Maiden MC. 2003. Genetic diversity of *Campylobacter jejuni* isolates from farm animals and the farm environment. *Appl Environ Microbiol*. 69:7409–7413.

133. Dingle KE, Colles FM, Ure R, Wagenaar JA, Duim B, Bolton FJ, Fox AJ, Wareing DR, Maiden MC. 2002. Molecular characterization of *Campylobacter jejuni* clones: a basis for epidemiologic investigation. *Emerg Infect Dis*. 8:949–955.

134. French N, Barrigas M, Brown P, Ribiero P, Williams N, Leatherbarrow H, Birtles R, Bolton E, Fearnhead P, Fox A. 2005. Spatial epidemiology and natural population structure of *Campylobacter jejuni* colonizing a farmland ecosystem. *Environ Microbiol*. 7:1116–1126.

135. Kwan PS, Birtles A, Bolton FJ, French NP, Robinson SE, Newbold LS, Upton M, Fox AJ. 2008. Longitudinal study of the molecular epidemiology of *Campylobacter jejuni* in cattle on dairy farms. *Appl Environ Microbiol*. 74:3626–3633.

136. Mullner P, Jones G, Noble A, Spencer SE, Hathaway S, French NP. 2009. Source attribution of food-borne zoonoses in New Zealand: a modified Hald model. *Risk Anal*. 29:970–984.

137. Sears A, Baker MG, Wilson N, Marshall J, Muellner P, Campbell DM, Lake RJ, French NP. 2011. Marked campylobacteriosis decline after interventions aimed at poultry, New Zealand. *Emerg Infect Dis*. 17:1007–1015.

138. Sboner A, Mu XJ, Greenbaum D, Auerbach RK, Gerstein MB. 2011. The real cost of sequencing: higher than you think! *Genome Biol*. 12:125.

139. Parkhill J, Wren BW, Mungall K, Ketley JM, Churcher C, Basham D, Chillingworth T, Davies RM, Feltwell T, Holroyd S, Jagels K, Karlyshev AV, Moule S, Pallen MJ, Penn CW, Quail MA, Rajandream MA, Rutherford KM, van Vliet AH, Whitehead S, Barrell BG. 2000. The genome sequence of the food-borne pathogen *Campylobacter jejuni* reveals hypervariable sequences. *Nature*. 403:665–668.

140. Miller WG, Parker CT. 2011. Campylobacter and Arcobacter, pp. 49–65, *in* Fratamico P, Liu Y, Kathariou S (eds.), *Genomes of Foodborne and Waterborne Pathogens*. ASM Press, Washington, DC.

141. Cody AJ, McCarthy ND, Jansen van Rensburg M, Isinkaye T, Bentley SD, Parkhill J, Dingle KE, Bowler IC, Jolley KA, Maiden MC. 2013. Real-time genomic epidemiological evaluation of human Campylobacter isolates by use of whole-genome multilocus sequence typing. *J Clin Microbiol*. 51:2526–2534.

142. Oakeson KF, Wagner JM, Rohrwasser A, Atkinson-Dunn R. 2018. Whole-genome sequencing and bioinformatic analysis of isolates from foodborne illness outbreaks of *Campylobacter jejuni* and salmonella enterica. *J Clin Microbiol*. 56.

143. Jolley KA, Bliss CM, Bennett JS, Bratcher HB, Brehony C, Colles FM, Wimalarathna H, Harrison OB, Sheppard SK, Cody AJ, Maiden MC. 2012. Ribosomal multilocus sequence typing: universal characterization of bacteria from domain to strain. *Microbiology*. 158:1005–1015.

144. Tolar B, Joseph LA, Schroeder MN, Stroika S, Ribot EM, Hise KB, Gerner-Smidt P. 2019. An Overview of PulseNct USA databases. *Foodborne Pathog Dis*. 16:457–462.

145. Gerner-Smidt P, Besser J, Concepcion-Acevedo J, Folster JP, Huffman J, Joseph LA, Kucerova Z, Nichols MC, Schwensohn CA, Tolar B. 2019. Whole genome sequencing: bridging one-health surveillance of foodborne diseases. *Front Public Health*. 7:172.

146. Ribot EM, Freeman M, Hise KB, Gerner-Smidt P. 2019. PulseNet: entering the age of next-generation sequencing. *Foodborne Pathog Dis*. 16:451–456.

25

Chlamydiae

Lourdes G. Bahamonde

CONTENTS

Introduction

The first CRC series chapter on Chlamydia was written by Leslie A. Page and published in 1989. Since then, significant advances in Chlamydiae diagnostics, genomics, pharmacotherapeutics, classification of new species, and bioinformatics have come to light. This chapter provides a review of our current understanding of Chlamydiia, while addressing how its classification has evolved to date.

With the advent of electron microscopy in the 1960s (Nunes and Gomes 2014), Chlamydiae were classified as bacteria, essentially because they possess DNA, RNA, ribosomes, and have a cell wall similar to that of Gram-negative bacteria (Nunes and Gomes 2014; Moulder 1966; Nunes et al. 2010). In 1966, Dr. Page first published evidence for the Chlamydiales order, the *Chlamydiaceae* family, and the *Chlamydia* genus (Page 1966). The classification system described at the time was based on a *Chlamydia*-specific developmental cycle that, ultimately,

DOI: 10.1201/9781003099277-27

separated the *Chlamydiae* from the *Rickettsiae* species (Ward 1983)—both obligate intracellular parasites. The identification of *C. trachomatis* and *C. psittaci* species were subsequently possible after the American Society of Microbiology adopted stricter numerical—plus and minus—taxonomic criteria for bacterial classification (Everett and Andersen 2001; Ward 1983). These numerical criteria were organized into matrices and only included phenotypic characteristics of bacteriological "morphology, biochemistry, culture composition, physiology, nutritional requirements, antigenic composition, and phage sensitivity" (Ward 1983). However, relying on phenotypic characteristics and numerical matrices is considered problematic because divergent organisms sharing the same environment typically have similar phenotypes, without evidence of evolutionary relatedness. Thus, phenotypic criteria may not necessarily imply relatedness, rather evolutionary convergence.

Chlamydiae's unique intracellular, biphasic developmental cycle has led to their historic taxonomic misclassification as protozoa or viruses (Nunes and Gomes 2014). The misnomer "*Chlamydia*," from the Greek word "Chlamys/Khlamus" to mean mantle or cloak, stemmed from a misconception dating back to 1907, when Halberstaedter and von Prowazek noted what they thought were bacteria with intracytoplasmic inclusions that appeared "draped/cloaked" around infected cell nuclei from humans with active trachoma (Longbottom and Coulter 2003; Budai 2007). Intracytoplasmic bodies or vacuoles were also noted in active human cases of urethritis, cervicitis, pneumonia (Nunes and Gomes 2014), conjunctivitis scrapings, as well as inoculated orangutans from trachoma cases (Longbottom and Coulter 2003). Yet, despite the high phenotypic homology, tissue tropism and host preference noted in the *Chlamydia* genus (Sachse et al. 2015), low homology by DNA-DNA hybridization was described across species (Nunes and Gomes 2014; Budai 2007; Sachse et al. 2015). In 1999, less than 70% hybridization homology was a common cutoff for the establishment of new species (Everett, Bush, and Andersen 1999), though homology by DNA-DNA hybridization within the phylum Chlamydiae, was 95% and was suggestive of a common origin (Nunes and Gomes 2014; Budai 2007).

In 1999, a controversial proposal was made to divide the *Chlamydiaceae* family into two genera and nine species (Clarke 2011; Everett, Bush, and Andersen 1999). In order to definitively establish the number and type of *Chlamydiae* species, researchers at the United States Department of Agriculture (USDA) National Animal Disease Center set about developing a DNA-based classification system based on 16S and 23S ribosomal RNA genes and the major outer membrane protein (MOMP)/ompA gene (Everett, Bush, and Andersen 1999). Based on gene analysis, the order Chlamydiales was redefined—the original family of *Chlamydiaceae* was analyzed by gene sequencing, which led to the establishment of four families, instead of one. The species that retained a >90% 16S rRNA sequence identity remained within the family *Chlamydiaceae*; whereas species with 80–90% sequence identity were separated into one of three new families—*Parachlamydiaceae*, *Simkaniaceae*, and *Waddliaceae* (Everett, Bush, and Andersen 1999).

Within this newly defined *Chlamydiaceae* family, the genus *Chlamydia* was expanded to include not just *C. trachomatis*, but also *C. suis*, and *C. muridarum* (Everett, Bush, and Andersen

1999). The species *C. psittaci* and its associated species, *C. pecorum* and *C. pneumoniae*, were moved to a new genus *Chlamydophila* in 1999. Three new species, *Chlamydophila felis*, *Chlamydophila caviae*, and *Chlamydophila abortus*, were added to *Chlamydophila* (Everett, Bush, and Andersen 1999). As of 2015, in light of newly identified species and more complete genomic sequence data, the taxonomy of *Chlamydia* species has undergone a second significant revision. The class of Chlamydiia was split into two orders: the retained, original order of Chlamydiales and a new order, Parachlamydiales (Gupta et al. 2015). The original four families classified in the original Chlamydiales order were subsequently further divided, while five additional *Candidatus* families were created (Gupta et al. 2015). Four of the five new candidate families were included under the Parachlamydiales order and the new candidate family, *Candidatus* Clavichlamydiaceae was retained within the original Chlamydiales order. The novel order of Parachlamydiales retained the *Parachlamydiaecae*, *Simkaniaceae*, and *Waddliaceae* families, with the added novel *Candidatus:* Criblamydiceae, Parailichlamydiceae, Piscichlamydiceae, and Rhabdochlamydiceae families (Gupta et al. 2015).

To date, the current Chlamydiia class taxonomy has been further diversified from one order with four families, to two orders (Chlamydiales and Parachlamydiales), with nine families—however, approval of the *Candidatus* families by the Subcommittee on Taxonomy of Chlamydiae of the International Committee on Systemics of Prokaryotes (ICSP) remains to be formalized as of 2019 (Borel and Greub 2019).

Significant controversy has existed within the *Chlamydiaceae* family as a result of inconsistencies in sequence-identity cutoff criteria in taxonomic classification. The 16S rRNA gene sequence identity threshold for the two genera (*Chlamydia* and *Chlamydophila*) proposed for *Chlamydiaceae* in 1999 was 95% (Everett, Bush, and Andersen 1999), however in 2008, more standardized and consistent evolutionary lineage criteria were recommended for boundary thresholds in prokaryotes (Sachse et al. 2015). The All-Species Living Tree Project, a collaborative of microbial taxonomists, created more consistent criteria for evolutionary lineage relationships (Yarza et al. 2008). To meet a standardized boundary threshold for a prokaryotic genus, a 94.5% sequence identity of 16S rRNA gene cut-off was instead recommended (Sachse et al. 2015). Newer analyses of 16S rRNA sequences generally showed higher sequence identities than the thresholds of 94.5–95%, which at the time, suggested that *Chlamydia* and *Chlamydophila* genera in separate phylogenetic branches were actually more similar to each other than were members of different genera in other bacterial groups (Stephens et al. 2009). Comparative pairwise protein sequencing has provided confirmation for emending two genres to a single genus (Stephens et al. 2009). In an analysis of 97 bacterial and archaeal genera, it was determined that a 50% threshold for the percentage of conserved proteins (POCP) was consistent with current levels of genus classification (Sachse et al. 2015), where *Chlamydia trachomatis* and *Chlamydophila psittaci* generated a POCP value of 82% (Sachse et al. 2015; Sachse and Laroucau 2015). In this review, we will be reverting to a single genus for all species.

Finally, a number of new species have been identified bringing the number of species from 9 to 13, and three additional *Candidatus* species (Cheong et al. 2019). Originally, the species

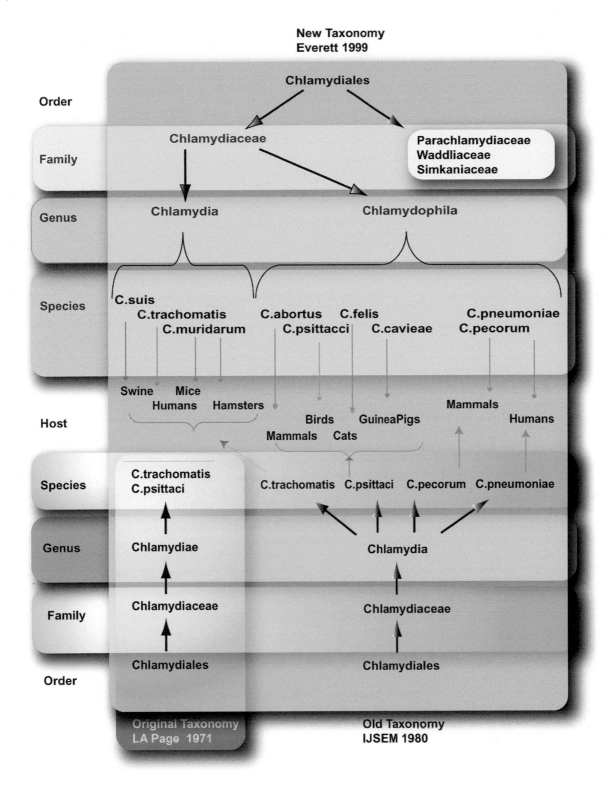

FIGURE 25.1A   Historical taxonomy overview.

C. psittaci was thought to be the only avian species. However, two new avian species were subsequently identified in pigeons and poultry—*Chlamydia avium* and *Chlamydia gallinacea*—which were *Chlamydiaceae*-positive species, but *Chlamydia-psittaci*-negative avian species (Cheong et al. 2019; Sachse et al. 2014). The species *Candidatus* Chlamydia ibidis has so far only been identified in feral populations of sacred ibis, wading birds

(Sachse et al. 2015). The rest of the new species, *Chlamydia serpentis*, *Chlamydia poikilothermis*, as well as the candidate species *Candidatus* Chlamydia corallus and *Candidatus* Chlamydia sanzinia, have all been found in captive snakes (Cheong et al. 2019; Sachse et al. 2014) (see Figures 25.1A and B).

As phylogenetically isolated, intracellular microbes, Chlamydiales are believed to have evolved differently than other

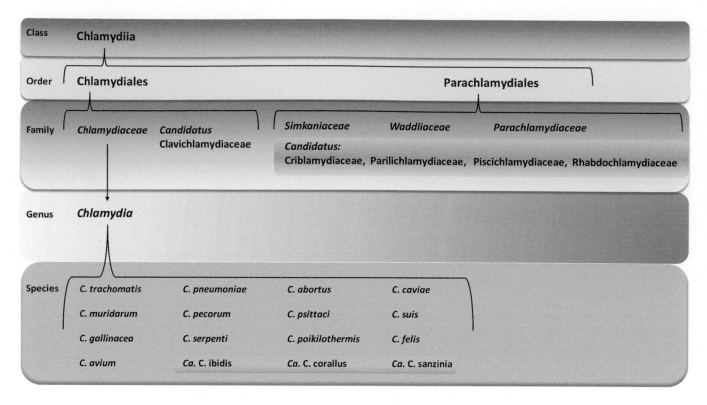

FIGURE 25.1B   Current taxonomic lineage.

well-characterized facultative and obligate intracellular eubacteria (Nunes et al. 2010). *Chlamydiae* species differ in host range, tissue tropism and pathogenicity (Nunes et al. 2010). Their interaction with multiple hosts can be elucidated from primate DNA analyses and from the isolation of plant-like genes; both suggest a divergence from a common ancestor approximately 700 million years ago (Clarke 2011) and an ancestral evolutionary relationship with cyanobacteria (Nunes et al. 2010). Understanding the mechanisms of genomic variation will enable us to identify potential genetic causes of pathogenicity (Nunes et al. 2010). Moreover, the suggestion that the species' pathogenicity could be partially mediated by an enhanced activation of the hosts' innate immune response could point to unique mechanisms underlying persistent infection (Rusconi and Greub 2011), and potentially lead to the development of innovative prophylactic and therapeutic treatments (Nunes et al. 2010).

Life Cycle

Chlamydiae are obligate intracellular bacteria that exist in two morphologically distinct forms and are characterized by a biphasic life cycle under cell culture conditions (Vandahl, Birkelund, and Christiansen 2004; Hogan et al. 2004). The bacteria alternate between a small infectious (300–400 nm) extracellular form—the elementary body (EB)—and a larger noninfectious (800–1000 nm) intracellular replicating form—the reticulate body (RB) (Murray 1998; Vandahl, Birkelund, and Christiansen 2004). Resistant to environmental factors, EB are spore-like and infectious, albeit unable to replicate outside of a host. EB are made up of DNA wrapped around histone-like proteins and

enveloped by an outer membrane of cross-linked proteins with disulfide bonds between cysteine residues for osmotic stability (Vandahl, Birkelund, and Christiansen 2004).

The EB life cycle begins when they attach to and stimulate phagocytosis by host cells (Figure 25.2). Within the phagosome, named an inclusion, the EB differentiate into metabolically active RB that later undergo multiple cycles of binary fission leading to their secondary differentiation back to an EB form (Hogan et al. 2004; Abdelrahman and Belland 2005). Lysis of the host cell releases a new generation of infectious EB that propagates

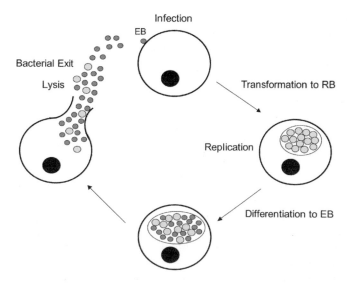

FIGURE 25.2   Chlamydiae life cycle.

among neighboring cells (Abdelrahman and Belland 2005). This life cycle often lasts 2–3 days, with some *Chlamydia*-like strains taking up to 14 days. The conditions under which this propagation occurs can lead to aberrant RB and can determine whether the host will continue to express infective EB forms with each bacterial cell division (Abdelrahman and Belland 2005). Propagation is linked to host immune responses *in vivo*, therefore pro-inflammatory chemokines, cytokines, nutrient availability and antibiotics are among some of the environmental factors that can disrupt the EB life cycle (Stephens 2003).

Characterization of *in vitro* persistent phase of infection and *in vivo* models suggest that Chlamydiae can persist in an altered form (Hogan et al. 2004). This form is now known as an aberrant body (AB) (Bayramova, Jacquier, and Greub 2018). These are persistent, viable nondividing, noninfectious forms. They are characterized as enlarged RBs due to continuous DNA replication without division, and their formation seems to be in response to stress with multiple stimuli (Bayramova, Jacquier, and Greub 2018). Antibiotics such as penicillin, clavulanic acid, and phosphomycin; depletion of nutrients such as iron, glucose, or amino acids; treatment with interferon; and coinfection with herpes, can all induce AB formation (Bayramova, Jacquier, and Greub 2018). These changes are reversible and continuation to the RB stage can occur upon removal of stressors, however, recurrent chlamydial infection may result from aberrant body formation (Hogan et al. 2004; Stephens 2003).

Identification

Phenotypic Markers

The genus *Chlamydia* is marked by the presence of intracellular glycogen, unfortunately those formally of the genus *Chlamydophila* are marked by neither the presence nor the absence of glycogen since some species show transient levels of glycogen during their life cycle. Originally, chlamydial species were distinguished by resistance to sulfadiazine, but this turned out to be highly variable.

Microimmunofluorescence, based on the immunoreactivity of the chlamydial major outer membrane protein (MOMP) and dominant antigen, is reported to be the first widely used methodology to discern differences in the molecular epidemiology between strains (Nunes et al. 2010). Initially, it employed polyclonal antibodies, later replaced by more specific and sensitive monoclonal antibodies (Nunes et al. 2010).

The *Chlamydiaceae* family is marked by detection with monoclonal antibodies to the group-specific lipopolysaccharide epitope: αKdo-(2→8)-αKdo-(2→4)-αKdo (originally: the genus-specific epitope) (Everett 1999). *Chlamydophilae* were originally differentiated from *Chlamydiae* by antibodies that recognize epitopes (NPTI, TLNPTI, LNPTIA, LNPTI) in the variable region - of the *Chlamydiae* MOMP, but not *Chlamydophilae*. Because these mAbs are routinely used to serotype human *C. trachomatis* isolates, cross-reaction may lead to errors in species identification (Everett 2000). It is worth noting that the classification of antibiotic resistance among species prior to 1999 is unreliable, since a DNA-typing, taxonomic classification system of species had not yet been adopted—e.g., data on antibiotic sensitivity and resistance for *C. trachomatis* prior to 1999 may have inadvertently included *C. trachomatis* and *C. suis* strains.

DNA Markers

Since phenotypic markers cannot reliably differentiate between Chlamydial species, DNA typing and sequencing is the best manner of identifying them. This makes sense since the various species were established via DNA sequence analysis in the first place. Real-time polymerase chain reaction (PCR) primers are available to amplify ompA (that encodes MOMP), 16S rRNA, and 23S rRNA genes (Nunes et al.; Everett 2000). The ompA primers are specific for all *Chlamydiaceae*, while the 16S and 23S primers are specific for all Chlamydiales (Everett 2000). Multilocus sequence typing using 400–500 base pair fragments of seven housekeeping genes (enoA, fumC, gatA, gidA, hemN, hlfX, oppa) have identified population genetic structure and diversity within *C. trachomatis* (Pannekoek et al. 2008) and *C. psittaci* (Pannekoek et al. 2010).

DNA Microarray technology is increasingly being deployed as a diagnostic technique capable of overcoming the limitations of PCR by increasing the number of targeted genomic loci used to identify an organism. This allows for rapid discrimination between species, strains, genotypes, serotypes, and resistance types. Validated protocols for 20 *Chlamydia* species are now available (Sachse and Ruettger 2015). Genomic sequence data is available for at least 20 *Chlamydiae* species within the phylum. These genomes are useful in developing phylogenetic trees and for identifying sequences valuable for typing of Chlamydiae during infection (Gupta et al. 2015). Conserved signature insertions/deletions (CSIs) which are insertions or deletions within protein sequences that are unique to a related group of organisms and conserved signature proteins (CSPs), which are proteins specific to certain lineages and found only in related groups (Gupta et al. 2015).

Thirty two CSIs have been identified that are specific to the phylum Chlamydiae with 17 CSIs being specific to the family *Chlamydiaceae* with three others being specific to the grouping of families *Candidatus* Criblamydiaceae, *Parachlamydiaceae*, *Simkaniaceae*, and *Waddliaceae*. In regard to the CSPs, 98 are specific to *Chlamydiaceae* and 15 are specific to the grouping of families *Candidatus* Criblamydiaceae, *Parachlamydiaceae*, *Simkaniaceae*, and *Waddliaceae*. Based on this finding, along with additional 16s rRNA gene sequence data, it was proposed that the class Chlamydiia be split into two orders. Chlamydiales would contain the family *Chlamydiaceae* and the *Candidatus* family Clavichlamydiacae. A novel order of Parachlamydiales was established to contain the families *Parachlamydiaceae*, *Simkaniaceae*, and *Waddliaceae* as well as the *Candidatus* families of Criblamydiaceae, Parilichlamydiaceae, Piscichlamydiaceae, and Rhabdochlamydiaceae (Gupta et al. 2015).

Pathogenicity of Chlamydiae

Chlamydiae were believed to exclusively cause infection and disease in humans and other animals (Everett 2000); however, as new and old species are genetically identified and classified, the epidemiology of Chlamydiae in a vast number of microbial, eukaryote hosts have come to light. There is also evidence of

Chlamydia-like strains found in soil, as well as *Chlamydia*-infected amoeba in marine water and anoxic, deep-marine sediment, without evidence of a host (Dharamshi et al. 2020). These recent findings may be of importance in understanding the comparative evolution and plausible tropism between *Chlamydia* species and specific hosts; as well as the potential mechanisms that may lead to their transmissibility to humans and other animals, some with greater pathogenicity than others (Fritsche et al. 2000; Everett 2000).

Many *Chlamydia* species carry a 7.5-kb plasmid containing reading frames not considered to be critical for its growth *in vitro;* however, it is demonstrated to play a role in species pathogenicity in certain species; e.g., *C. muridarum* (Zhong 2017). *Chlamydia* plasmids generally described encode eight open reading frames conventionally called plasmid glycoproteins 1–8 or pGP1–8, which are thought to promote species adaptation to various animal tissues and enable our further understanding of their pathogenicity (Zhong 2017).

Although host multitropism of some pathogens has rendered their manipulation possible in animal models, the human-restricted tropism of numerous viruses, bacteria, fungi, and parasites has seriously hampered our understanding of these pathogens. Uncovering the genetic basis underlying the narrow tropism of such pathogens is critical for understanding their mechanisms of infection and pathogenesis. Moreover, such genetic dissection is essential for the generation of permissive animal models that can serve as critical tools for the development of therapeutics or vaccines against challenging human pathogens. In this review, we describe different experimental approaches utilized to uncover the genetic foundation regulating pathogen host tropism as well as their relevance for studying the tropism of several important human pathogens (Douam et al. 2015).

Chlamydiales

The order Chlamydiales contains the *Chlamydiaceae* (Everett et al. 1999) and the *Candidatus* Clavichlamydiaceae families. Based on phylogenomic analyses of the phylum Chlamydiae by Gupta et al. (2015), members of this order can be distinguished from the Parachlamydiales by their branching in phylogenetic trees and distinguished from all other *Chlamydiae* species by Clade 1-specific CSIs; and typically showing >89% 16S rRNA gene sequence identity with each other (Gupta et al. 2015).

Chlamydiaceae

Chlamydiaceae, the original family described in previous editions of this book chapter, was formerly divided into two genera: *Chlamydia* and *Chlamydophila*. These are currently emended to one genus, *Chlamydia*, and now includes 13 phylogenctically classified species and three additional candidate species (Stephens et al. 2009). All are Gram-negative and express the family-specific lipopolysaccharide epitope αKdo-(2→8)-αKdo-(2→4)-αKdo. The extracellular osmotic stability of the *Chlamydiaceae* EBs is maintained by a complex of disulfide-cross-linked envelope proteins that include the 40 kDa major outer membrane protein, a hydrophilic cysteine-rich 60 kDa protein, and a low-molecular-mass cysteine-rich lipoprotein (Ward

1983). Phenotypic variability, host specificity and other molecular factors involved in chlamydial pathogenicity are thought to be the result of cumulative single nucleotide polymorphisms (SNPs) and small indels, as the species coevolves with their hosts (Nunes et al. 2010).

Chlamydia trachomatis

C. trachomatis comprises three geno-groups (formerly known as biovars) that are transmitted by sexual or other contact (Pannekoek et al. 2008; Stephens et al. 1998; Carlson et al. 2005). Their complete gene sequences have been identified (A/HAR-13 and D/UW-3/CX) (Stephens et al. 1998; Carlson et al. 2005) and these exhibit a genomic conservation of >98% (Nunes et al. 2010).

The *C. trachomatis* genome, under the influence of selective environmental factors and coexisting flora, gives the species an advantage in specific ecological niches (Nunes and Gomes 2014). *C. trachomatis* has only been isolated from humans. More recently however, Chlamydia-like microorganisms were detected in the gut microbiota of chimpanzees (Clarke 2011; Ochman et al. 2010). Whole genome sequence dissimilarities and phenotypic consequences of each colonizing species are believed to be under the direct influence of their host's innate immune response—their pattern-recognition receptors, signal molecules such as cytokines, and reactive oxygen species (Nunes and Gomes 2014; Rusconi and Greub 2011; Seth-Smith et al. 2013). Understanding the mechanisms underlying the coevolution of Chlamydiae and each host's innate immune response will enable the development of innovative treatment strategies, including effective vaccines (Rusconi and Greub 2011).

C. trachomatis is of global importance as the most common bacterial sexually transmitted infection and the leading cause of infectious blindness worldwide, according to the World Health Organization (Prevalence and Incidence of Sexually Transmitted Infections 2011; Rusconi and Greub 2011; Seth-Smith et al. 2013). *C. trachomatis* requires tissue culture for *in vitro* growth, a technically challenging and time-consuming technique (Seth-Smith et al. 2013).

Immunoreactivity of antibodies against epitopes on the *C. trachomatis* MOMP has enabled the identification of 19 distinct serovars, divided into three geno-groups—B class (serovars B, D, Da, E, L1, L2, L2a); C class (A, C, H, I, Ia, J, Ja, K, L3); and Intermediate class (F, G, Ga) (Everett, Bush, and Andersen 1999; Gomes et al. 2006; Pannekoek et al. 2008). Specifically, A, B, Ba, and C are implicated in chronic eye infections (trachoma) and serovars D, Da, E, F, G, Ga, H, I, Ia, J, and K in urogenital infections (Gomes et al. 2006). The lymphogranuloma venereum (LGV) serovars L1, L2, L2a, and L3 can cause more invasive urogenital disease, including painful buboes and proctatitis (Gomes et al. 2006; Manavi 2006). *C. trachomatis* has also been implicated in some forms of arthritis, neonatal conjunctivitis, and pneumonia (Everett, Bush, and Andersen 1999).

Although clinicians consider Chlamydiae culture methods to be the gold standard for identification of chlamydial infection, it has low specificity and it is laboratory expertise-dependent (Manavi 2006). Serotyping using direct fluorescent staining with monoclonal antibodies (DFA) is highly specific, but time and labor intensive (Manavi 2006). *C. trachomatis* routinely cross-reacts with *C. suis* epitopes with DFA serotyping, leading to errors in species identification (Everett 2000). Conversely, PCR has a sensitivity of 90% and specificity of 99% (Manavi 2006). Treatment of choice against *C. trachomatis* includes macrolides (azithromycin, erythromycin), tetracyclines, quinolones (ciprofloxacin, ofloxacin), and penicillins (amoxicillin) (Manavi 2006). Most *C. trachomatis* strains have extrachromosomal plasmids and, in general, therapeutic failure is uncommon (Nunes and Gomes 2014).

Chlamydia muridarum

Isolated from mice and hamsters, the two strains of *C. muridarum*—MoPn (Nigg) and SFPD (Weiss) had originally belonged to the *C. trachomatis* mouse pneumonitis biovar, due to their high relatedness (Cheong et al. 2019; Nunes and Gomes 2014; Everett, Bush, and Andersen 1999). SFPD is not known to cause disease, while MoPn may produce pneumonia in mice or cause an asymptomatic infection. The MoPn genome has been sequenced and the isolate used contained an extrachromosomal plasmid (Read et al. 2000). MoPn is sensitive to sulfadiazine.

Chlamydia suis

C. suis strains were originally referred to as *C. trachomatis* due to DNA sequence homology in the MOMP gene (*ompA*) (Everett 2000). *C. suis* had only been isolated from swine, where they caused conjunctivitis, enteritis, pneumonia, and asymptomatic infections (Everett 2000). *C. suis* DNA has also been more recently reported in conjunctival swabs of Nepalese trachoma patients and Belgian slaughterhouse workers by PCR and microarray, suggesting potential zoonotic transmission from pigs to humans (Borel, Polkinghorne, and Pospischil 2018). In the wild, *Chlamydiae* are commonly found in the intestine of pigs, in mostly endemic subclinical cases (Borel, Polkinghorne, and Pospischil 2018). However, *C. suis* might be opportunistic in nature; with variations in ocular, respiratory, and intestinal virulence influenced by the age, host immune status and/or the presence of a *C. suis* strain extrachromosomal plasmid (Borel, Polkinghorne, and Pospischil 2018).

Some strains of *C. suis* are resistant to sulfadiazine and/or tetracycline (Everett 2000); however, *C. suis* is the only chlamydial species known to have naturally acquired genes encoding for tetracycline resistance (Borel, Polkinghorne, and Pospischil 2018). Fears of horizontal transfer of the tetracycline resistant Tet(C) gene into human chlamydial pathogens is a growing fear (Bommana and Polkinghorne 2019).

Chlamydia abortus

C. abortus, formerly *Chlamydophila abortus*, was originally classified under *C. psittaci*, but is in fact its own species (Everett, Bush, and Andersen 1999). Associated primarily with cases of abortion and poor prognosis in neonate animals, *C. abortus* infections affect the placenta and has been reported to be endemic in ruminants—rabbits, guinea pigs, as well as pigs and horses (Borel, Polkinghorne, and Pospischil 2018). After aborting, most mammals have been reported to acquire immunity and rebreed successfully (Vretou et al. 2007). The potential to cause serious economic impacts to livestock have made research into *C. abortus* a priority (Bommana and Polkinghorne 2019).

In humans, upper respiratory infection due to *C. abortus* in a laboratory worker responsible for experimental, intranasal infections of sheep has also been reported as probable causative organism of human atypical pneumonia (Borel, Polkinghorne, and Pospischil 2018), which should compel public health officials to inform laboratory personnel of the potential for zoonotic chronic pulmonary infection by *C. abortus* via aerosol inhalation (Borel, Polkinghorne, and Pospischil 2018).

Documented cases of zoonotic abortion in women due to *C. abortus* are suspected to have resulted from close contact with sheep infected tissue (Everett 2000; Cheong et al. 2019). The complete genome has been sequenced; however, no strain of *C. abortus* has been identified to contain an extrachromosomal plasmid (Thomson et al. 2005). Chlamydiosis is a zoonosis that places pregnant women at a particularly high risk of abortion (Borel, Polkinghorne, and Pospischil 2018), with complicated cases of *C. abortus* septicemia also reported, carrying a fetal mortality of 94% and maternal mortality of 6.3% (Walder et al. 2005). Early treatment with erythromycin should be considered in suspected zoonotic exposure to *C. abortus*—especially since fetal abortions may be preventable with early administration of macrolides or tetracyclines (Walder et al. 2005).

Chlamydia caviae

C. caviae species, was taxonomically separated from *C. psittaci* in 1999 (Borel, Polkinghorne, and Pospischil 2018). *C. caviae*, formerly *Chlamydophila caviae*, has been exclusively isolated from conjunctivae of infected guinea pigs (Everett 2000), although it has also been identified in cats, dogs, rabbits, and horses as primary hosts with conjunctivitis and urogenital tract infections (Cheong et al. 2019). *C. caviae* can be transmitted sexually in humans, with clinical manifestations resembling active *C. trachomatis* disease (Cheong et al. 2019).

Guinea pig inclusion conjunctivitis has been reported in cytological smears stained with Giemsa, where intracytoplasmic inclusion bodies were evaluated from swabbed conjunctival epithelial cells confirmed to be *C. caviae* by PCR (Borel, Polkinghorne, and Pospischil 2018). The genome of *C. caviae* has been sequenced and it contains an extrachromosomal plasmid (Read et al. 2003) carrying virulence-associated chromosomal genes (Nunes and Gomes 2014).

Chlamydia felis

Formerly *Chlamydophila felis* it is endemic among domestic cats and dogs, its primary hosts (Cheong et al. 2019). *C. felis* primarily causes conjunctivitis, rhinitis, chronic salpingitis, and respiratory pathology in its primary host. Zoonotic infection of humans has also been reported, as a possible cause of conjunctivitis (Cheong et al. 2019). Strains differ in their pathogenesis, though the causes of these pathogenic differences are not clearly understood (Everett 2000). The genome of strain Fe/C-56, as well as that of other strains, have been sequenced (Azuma et al. 2006) and some have been identified to have an extrachromosomal plasmid, which may suggest evolutionary tropism (Douam et al. 2015; Zhong 2017).

Seroprevalence of *C. felis* amongst domestic cats is relatively high with rates among strays being >10% and among house cats at >3% in some countries. While cats are the primary host, dogs are now seen to be a significant reservoir of the bacterium, thus increasing the likelihood of zoonotic infection (Cheong et al. 2019). Chlamydial infection from a domestic cat with signs of conjunctivitis and rhinitis to a human-owner who developed acute follicular conjunctivitis was first demonstrated by Schachter et al. in 1969 (Paolillo et al. 2019). While direct evidence linking *C. felis* to human disease is ambiguous, a seroepidemiological study in Japan found a prevalence of 1.7% in the general population and 8.8% amongst small animal veterinarians (Cheong et al. 2019).

Zoonotic chlamydial infection is rare and it may be underestimated secondary to faulty isolation and detection practices (Di Francesco et al. 2004). Seroepidemiological data on *C. felis* reported sera tested for anti-*C. felis* antibodies by microimmunofluorescence test (MIF), using purified elementary bodies and fluorescein-conjugated goat anti-cat immunoglobulin G sera (Euroclone) (Di Francesco et al. 2004). However, sera positive for *C. felis* was also positive for *C. pneumoniae* and *C. psittaci*, demonstrating significant cross-reactivity between the species (Di Francesco et al. 2004). Given the higher seroprevalence of *C. pneumoniae* among humans, cross-reactivity is expected (Di Francesco et al. 2004), thus molecular classification techniques should be used in order to improve accuracy of evidence for zoonotic infection of *C. felis* (Longbottom and Livingstone 2006). Recently, *Parachlamydiaceae* have been identified in cats with neutrophilic and eosinophilic conjunctivitis (Sykes 2005) and the treatment of choice was found to be doxycycline, with azithromycin affording less efficacy (Sykes 2005).

Chlamydia pecorum

C. pecorum, formerly *Chlamydophila pecorum*, has been isolated only from mammals and is serologically and pathogenically diverse (Everett 2000). *C. pecorum*, was initially named *C. psittaci*, is reported to be more virulent in the koala than *C. pneumoniae* (Borel, Polkinghorne, and Pospischil 2018). Many animals, such as sheep, goats, pigs, and cattle, as well as wild ruminants, wild boars, and marsupials, such as koalas, can develop active clinical diseases (Borel, Polkinghorne, and Pospischil 2018). These include conjunctivitis, encephalomyelitis, enteritis, pneumonia, polyarthritis, infertility, and abortion, all of which have contributed to substantial population declines (Everett 2000).

C. pecorum infections in koalas have been associated with keratoconjunctivitis and aspermia with *C. pecorum* and *C. pneumoniae* (Borel, Polkinghorne, and Pospischil 2018). *C. pecorum* has been reported to have a prevalence of up to 100% in eastern states of Australia (Borel, Polkinghorne, and Pospischil 2018). Chlamydial inclusions were demonstrated in testis, Sertoli cells by IHC and TEM; and chlamydial DNA was also detected by PCR (Borel, Polkinghorne, and Pospischil 2018).

Chlamydia pneumoniae

Formerly *Chlamydophila pneumoniae* was previously synonymous with TWAR (a human biovar). *C. pneumoniae* primary hosts have included humans and a wide range of nonhuman animals, including mammals and reptiles (Cheong et al. 2019); however, zoonotic transmission has been linked to the discovery of nonhuman genotypes of *C. pneumoniae* in humans (Cheong et al. 2019). Pathology has been described in its primary host— respiratory disease not only in humans, but also in koalas, bandicoots, and horses, as well as a wide range of reptiles, such as snakes, iguanas, chameleons, frogs, turtles (Everett, Bush, and Andersen 1999; Cheong et al. 2019). Animal isolates have been shown to contain an extrachromosomal plasmid similar to that found in *C. trachomatis* and *C. muridarum*, but absent from human isolates (Everett, Bush, and Andersen 1999). Complete gene sequence of four strains, AR39, CWL029, J138, and TW183 have been cited in the literature (Read et al. 2000; Kalman et al. 1999; Shirai et al. 2000). The potential economic impact of *C. pneumoniae* transmission and the implication of pathogenic-association of disease onset and progression in asthma, primary biliary cirrhosis, atherosclerosis, and lung cancer have made research into *C. pneumoniae* a priority (Cheong et al. 2019).

The human *C. pneumoniae* strain TWAR is primarily a pathogen of the respiratory tract capable of causing acute or chronic bronchitis and pneumonia (Everett, Bush, and Andersen 1999). *C. pneumoniae* has been identified in asymptomatic and coinfected individuals (Cevenini, Donati, and Sambri 2002). It has been implicated in obstructive pulmonary disease, cystic fibrosis, Alzheimer's disease, asthma, erythema nodosum, reactive airway disease, Reiter's syndrome, and sarcoidosis (Everett, Bush, and Andersen 1999). There is evidence that *C. pneumoniae* may be involved in atherosclerosis and the acceleration of atherosclerotic-lesion development in humans and mice (de Kruif et al. 2005). Although *C. pneumoniae* primarily infects humans, mice have been found to be susceptible via intranasal inoculations, making them a useful animal model for the study of pneumonitis (Yang, Kuo, and Grayston 1993). The atherogenic properties of *C. pneumoniae* were not influenced by antibiotic therapy, unless the antibiotics were administered in the early, acute state of infection (de Kruif et al. 2005). Immunization against *C. pneumoniae* is the focus of current research (de Kruif et al. 2005; Longbottom and Livingstone 2006). At current, the most reliable assay for the detection of *C. pneumoniae* is real-time PCR; however, no well-standardized assay has been approved by the U.S. Food and Drug Administration (Kumar and Hammerschlag 2007).

Chlamydia psittaci

The eight formerly *Chlamydophila psittaci* serovars (A-F, M56, WC), and the three more recently described genotypes, have been identified in a wide range of bird groups, as its primary hosts (Heddema et al. 2006). *C. psittaci* strains (A-F) have been described as transmissible to birds, but have also been described as transmissible to mammals (WC and M56), as well as humans through possible inhalation (Cheong et al. 2019; Everett 2000) of feather dust or aerosolized fecal material (Vogler et al. 2019). Infection is often systemic, with or without apparent symptoms, including fever, headache, myalgia, and/or intermittent shedding (Cheong et al. 2019). *C. pneumoniae* has been reported to be passed through eggs and to chicks via regurgitation causing chronic Chlamydiosis (Cheong et al. 2019). Stress often triggers onset of severe symptoms, resulting in rapid deterioration and death. Serovar A is found among psittacine birds. Serovar B however, is described as endemic to pigeons; Serovar C among ducks and geese; Serovar D among turkeys; Serovar E among pigeons and other birds; and Serovar F, among parakeets and turkeys (Cheong et al. 2019). Genotyping of the *C. psittaci* ompA gene has revealed additional genotypes, including E/B, I, and J; with several strains containing an extrachromosomal plasmid (Everett 2000).

Infection by *C. psittaci* causes psittacosis (Heddema et al. 2006). In 1895, the term psittacosis derived from the Greek for parrot was first used to describe this disease, however a Gram-negative bacterium from parrots dying of psittacosis (*Bacillus psittacosis*) was instead identified belonging to the genus, *Salmonella* (Borel, Polkinghorne, and Pospischil 2018). Inconsistencies reported in classification of bacteriological isolates prompted the suspicion of a viral etiology of psittacosis, though in the late 1970s, *C. trachomatis* and *C. psittaci* were distinguished as separate species on the basis of sulfonamide susceptibility and glycogen accumulation by iodine staining (Borel, Polkinghorne, and Pospischil 2018).

C. psittaci has been particularly prevalent among bird slaughterhouse workers and owners of pet birds—with samples mainly of serovar A in psittacine birds and patients with psittacosis (Heddema et al. 2006; Longbottom et al. 2001; Everett, Bush, and Andersen 1999). Zoonotic infection is described as occurring horizontally by diseased or subclinically infected birds through contaminated aerosols and droppings (Chahota et al. 2006). Avian *C. psittaci* infections are often systemic, with bacteria infecting blood monocytes/macrophages leading to clinically unapparent manifestations of disease; although case reports of mucosal epithelial cells in the gastrointestinal tract, and other organs, leading to myocarditis, endocarditis, hepatitis, encephalitis, and meningitis have also been identified (Knittler and Sachse 2015; Longbottom and Livingstone 2006). Due to its airborne transmission, *C. psittaci* is classified as a Category B Crucial Biological agent as it could easily be misused as a biological warfare agent (Cheong et al. 2019). Currently, no vaccines have been developed against *C. psittaci* (Longbottom and Livingstone 2006). Innate immunity plays a key role in primary recognition of chlamydial infections; with adaptive immune response by lymphocytes, dendritic cells, antigen-presenting cells, enhanced major histocompatibility complex expression, and autophagic degradation aiding the elimination of the offending bacteria (Knittler and Sachse 2015).

Human cases of psittacosis are effectively treated using orally administered doxycycline and tetracycline hydrochloride for a period of 10–14 days in patients for whom tetracycline is contra-indicated (Bommana and Polkinghorne 2019). Availability of an effective treatment is reported as a main contributor for the general decline in psittacosis cases worldwide, particularly those with fatal outcome, in the past decades (Bommana and Polkinghorne 2019). Use of quinolones to treat chlamydia infection in humans has resulted in reports of treatment failure (Bommana and Polkinghorne 2019).

Treatment options for *C. psittaci* include bacteriostatic antibiotics effective against other *Chlamydiaceae*. In veterinary medicine, *C. psittaci* infections are treated with doxycycline or other tetracyclines and the fluoroquinolone, enrofloxacin, administered orally (feed/drinking water) or parenterally (intramuscular or subcutaneous routes) (Bommana and Polkinghorne

2019). Unfortunately, the widespread use of tetracycline in feed and/drinking water and long periods of treatment (21–25 days) in the poultry/bird industry is implicated in chronically subtherapeutic drug plasma levels that promote the emergence of drug-resistant *C. psittaci* strains (Bommana and Polkinghorne 2019). Despite available and effective antimicrobial treatments, vaccination remains at the forefront of best approaches to reduce chlamydial infections and prevalence in the population (Knittler and Sachse 2015). Antichlamydial T-cell vaccines induce cellular immune responses, which activate antigen-presenting cells effect long-lived T-cell memory and offer protective immunity against Chlamydiae (Knittler and Sachse 2015).

Chlamydia avium

C. avium, *Chlamydia gallinaceae*, and *Candidatus* Chlamydia ibidis are newly described avian species (Cheong et al. 2019). To date, *C. avium* has been identified in pigeons and psittacine birds, whereas *C. gallinaceae* is endemic in domestic poultry, such as chickens, ducks, turkeys, and guinea fowls (Sachse and Laroucau 2015; Sachse et al. 2014). Factors affecting infection rates of wild/captive and farmed fowl include the number of animals per farm, density of animals per stable, expanded feed, good on-farm hygiene management with hygienic storage of bedding material, and health monitoring of the animals (Borel, Polkinghorne, and Pospischil 2018). *C. avium* and *C. gallinacea* may be able to cause respiratory disease in psittacine birds and pigeons; neither species is known to be pathogenic in humans, but it has been suggested that *C. gallinacea* may be implicated in several cases of atypical pneumoniae in slaughterhouse workers (Cheong et al. 2019).

A complete genome of a *C. avium* strain from a feral pigeon in Italy was recently published (Floriano et al. 2020), describing significant variation in alignment and length of plasmid sequences that could shed light on the possible evolution of *C. avium* and differences in virulence of avian species (Floriano et al. 2020). To date, the only known hosts reported for *C. avium* are European feral pigeons and captive psittacine birds worldwide (Cheong et al. 2019), pointing to the potential *C. avium* spread in the wild by captive parakeet pets and other exotic pet birds. In light of more common reintroduction programs of captive birds into the wild worldwide, increased human contact (more commonly by people who feed birds, an evolving threat to sanitation and human health) warrants further investigation (Cheong et al. 2019).

Chlamydia gallinacea

The species had been previously described as "atypical chicken chlamydia," later named *Chlamydia gallinacea;* with reports of suspected non-*C. psittaci* pneumonia in poultry slaughterhouse workers. It was first identified in cloacal swabs and feces from clinically healthy chickens and turkeys from various European countries, and is not considered endemic in chickens (Vogler et al. 2019). Reports of bird-to-human transmission could not be demonstrated in recent studies by farmers caring for *C. gallinacea*-positive chicken flocks (Vogler et al. 2019). A review of zoonotic Chlamydiae described experimentally infected chickens that remained without overt clinical disease, except their showing signs of significant loss of body weight, with an isolated type strains of *C. gallinacea* containing a plasmid, suspected to be responsible for its virulence (Borel, Polkinghorne, and Pospischil 2018). Phylogenetic plasmids as well as polymorphic *ompA* sequences, have been typed in *C. gallinacea* genetic variant isolates from diverse geographic locations, with DNA detected in healthy dairy and beef cattle specimens from China, where both *C. gallinacea* and *C. psittaci* were detected in whole blood, milk, feces and vaginal swabs (Borel, Polkinghorne, and Pospischil 2018).

Chlamydia serpentis

The four newly identified species, *C. serpentis*, *C. poikilothermis*, *Ca.* C. corallus, and *Ca.* C. sanzinia, were all identified in captive snakes and little is understood of their pathogenicity or host range (Cheong et al. 2019). *C. serpentis* strains have also been reported in asymptomatic corn snakes, green bush vipers and other captive snakes, whose pathogenicity was believed to have been influenced and exacerbated by stress from their capture, transportation, high-density farming, or their hibernation state (Staub et al. 2018).

Comparative phylogenetic analyses of *Candidatus* C. sanzinia reported *C. pneumoniae* and *C. pecorum* forming a clade distinct from former "*Chlamydophila*" species (Bommana and Polkinghorne 2019). Culture-independent, full sequencing of genomic DNA obtained from the choana of a captive, healthy Madagascar tree boa found the novel *Candidatus* Chlamydia sanzinia, and similarly found from the choana of a captive Amazon basin emerald tree boa (Bommana and Polkinghorne 2019). Lack of suited *in vitro* culture systems remain a challenge to further the characterization of novel chlamydial strains (Staub et al. 2018). Molecular and phylogenetic characterization of free-living and captive turtle (semi-aquatic animal), as well as tortoise species (land animal) could help confirm these animals as *C. poikilothermis* host reservoirs (Staub et al. 2018). Implications of the increased occurrence of *Chlamydiaceae* in poikilothermic animals often found in captivity and recreational areas, such as zoos, parks, and lakes, as well as the zoonotic implications of reptiles should be further investigated (Staub et al. 2018).

Recent studies describing antibiotic sensitivity to tetracycline and moxifloxacin and phenotypic resistance to azithromycin in the newly described chlamydial species infecting snakes, *C. serpentis* and *C. poikilothermis* (Staub et al. 2018), demonstrate the potential for considerable variability in the antibiotic resistance profile of bacteria in the genus *Chlamydia*. As new chlamydial species continue to emerge in animals, the demonstration or prediction of antibiotic resistance to inform clinical treatment of these infections will become increasingly important (Bommana and Polkinghorne 2019).

Chlamydia poikilothermis

C. poikilothermis are reported in animals whose internal temperature can vary considerably; samples isolated from choanal and cloacal swabs from a captive corn snake were reported within the snake family *Colubridae*, although it is believed to also be present in other families of captive and free-ranging snakes (Staub et al. 2018). *C. poikilothermis* is reported susceptible to

tetracycline and moxifloxacin, but with intermediate resistance potential to azithromycin (Staub et al. 2018).

Candidatus Chlamydia ibidis

C. avium, *C. gallinacea*, and *Ca.* C. ibidis are three newly described avian chlamydial species; and like others of the family *Chlamydiaceae*, *Ca.* C. ibidis can be grown in cell culture and in embryonated chicken eggs (Vorimore et al. 2013). Morphology of this candidate species, as revealed by electron microscopy, demonstrated features typical of *Chlamydia* genus (Vorimore et al. 2013). *Ca.* C. ibidis has been described in cloacal swabs of healthy, feral African sacred ibis birds described as living along the west coast of France (Borel, Polkinghorne, and Pospischil 2018; Cheong et al. 2019), although its pathogenicity to avian species or other animals, is currently unknown (Borel, Polkinghorne, and Pospischil 2018). *Ca.* C. ibidis has also been described in the previously thought extinct crested ibis. In contrast to the African sacred ibis, the crested ibis is described as an endangered bird living in Sado Island, Japan after being introduced from China; however a few crested ibis were rediscovered in China in 1981 (Li et al. 2020) and harbored *C. psittaci* and *Ca.* C. ibidis (Li et al. 2020), suggestive of plausible evolution of *Chlamydiaceae* as potential threat to the endangered bird, which could warrant future protection practices (Li et al. 2020).

The genetic diversity in *Ca.* C. ibidis is suspected to have resulted from diverse host species not yet identified, as well as the geographic origin of the organisms (Li et al. 2020). To date, *Ca.* C. ibidis pathogenicity has not been confirmed in either feral African sacred ibises or crested ibises; for this reason, prevalence, transmission and pathogenicity of this *Candidatus* species have not been concluded (Li et al. 2020).

Candidatus Chlamydia sanzinia

Novel *Ca.* Chlamydia sanzinia was identified from culture-independent, full sequencing of genomic DNA obtained from the choana of a captive, healthy Madagascar tree boa in Switzerland (Bommana and Polkinghorne 2019).

Candidatus Chlamydia corallus

Novel *Ca.* Chlamydia corallus was identified from culture-independent, full sequencing of genomic DNA obtained from the choana of a captive, healthy Amazon basin emerald tree boa (Bommana and Polkinghorne 2019).

Chlamydiales-like Family Candidate

Candidatus Clavichlamydiaceae

The taxonomic family-level lineage *Candidatus* Clavichlamydiaceae was determined by phylogenetic relationships and genetic distance to be the closest relative to the original *Chlamydiaceae* (Greub 2013). This newest candidate family was originally reported as *Clavochlamydia* and was then renamed *Clavichlamydia* (Greub 2013). Chlamydiosis due to *Candidatus* Clavichlamydiaceae has been implicated in the bacterial pathology of marine and fresh water fish skin and gills (Borel, Polkinghorne, and Pospischil 2018).

Parachlamydiales

The Parachlamydiales order contains the *Parachlamydiaceae* family (Everett 1999), the *Simkaniaceae*, *Waddliaceae*, and four *Candidatus* families: Criblamydiaceae, Parilichlamydiaceae, Piscichlamydiaceae and Rhabdochlamydiaceae (Gupta et al. 2015). Based on phylogenomic analyses of the phylum Chlamydiae by Gupta et. al. (2015), Parachlamydiales can be distinguished from Chlamydiales by their branching in phylogenetic trees and three CSI variants in DNA topoisomerase I, TC_0635 and Phenylalanyl-tRNA synthetase beta subunit; they typically show >84% 16S rRNA gene sequence identity with each other (Gupta et al. 2015).

By immunohistochemistry using monoclonal antibody targeting of chlamydial lipopolysaccharides within the *Chlamydiaceae* family, antibodies directed against the group-specific LPS epitope αKdo-(2→8)-αKdo-(2→4)-αKdo, do not recognize members of the more recently classified *Parachlamydiaceae*, *Simkaniaceae*, or *Waddiaceae* families (Everett, Bush, and Andersen 1999; Corsaro and Greub 2006).

Parachlamydiaceae

Originally described as *Candidatus* Parachlamydia acanthamoebae, *Parachlamydiaceae* have been declared a new family (Everett, Bush, and Andersen 1999). Isolated from amoebae, the *Parachlamydiaceae* can also infect *Dictyostelium* and can be grown in cultures of African Green Monkey Kidney (Vero) cells (Everett, Bush, and Andersen 1999). Two genera have been isolated, *Parachlamydia* and *Neochlamydia* with other isolates being considered (Everett 2000). Among case reports, at least one of these isolates has been found to increase the cytotoxicity of the amoeba they infect and was isolated from humans during an outbreak of humidifier fever in Vermont, USA ("Hall's coccus") (Everett, Bush, and Andersen 1999). Other members of this family have been detected in cats, Australian marsupials, reptiles, and fishes (Corsaro and Greub 2006).

Simikaniaceae

The *Simikaniaceae* family originally contained only one strain: *Simikania negevensis*. First identified as a bacterial contaminant in cell cultures, the strain is associated with pneumonia in humans. The strain lacks any extrachromosomal plasmids and has a longer than normal developmental cycle in Vero cells (2 weeks versus the normal 2–3 days) (Everett 2000). Newly discovered members of this family, *Fritschea bemisiae* and *Fritschea eriococci*, have been found to infect invertebrates (Corsaro and Greub 2006).

Waddliaceae

The defining strain of this family, *Waddlia chondrophila* was identified in 1986 as an agent of bovine abortion. Serological studies have shown that titers of anti-*Waddlia* antibodies are statistically associated with cows that have aborted (Everett 2000). A second family member, *Waddlia malaysiensis* was identified in urine samples from Malaysian fruit bats (Corsaro and Greub 2006).

Parachlamydiales-like Family Candidates

Candidatus Parilichlamydiaceae, Rhabdochlamydiaceae, Criblamydiaceae and Piscichlamydiaceae

Currently, of these four candidate families, *Candidatus* Piscichlamydiaceae has been implicated in skin and gill pathology of diverse marine and fresh water fish (Borel, Polkinghorne, and Pospischil 2018). The reported pathology is described as local to multifocal epithelial hyperplasia with an interlamellar filling that can lead to fusion of the lamellae and cellular hypertrophy, with cyst-like inclusions within the affected gill epithelial cells (Borel, Polkinghorne, and Pospischil 2018; Stride et al. 2013). Histopathologically, these epithelial cysts have been reported to have intracytoplasmic inclusions that are sometimes basophilic and are referred to as epitheliocystis (Borel, Polkinghorne, and Pospischil 2018). Epitheliocystis in fish was first described in carp in 1920 and named in the 1960s in reference to gill cysts described in bluegill (Pawlikowska-Warych and Deptula 2016). Other histopathologic analyses of gill samples taken from striped trumpeter in Tasmania, Australia have identified epitheliocystis-like basophilic inclusions from distinct species of Chlamydiales as well as Parachlamydiales in yellowtail kingfish (Stride et al. 2013).

Highly diverse Chlamydiae have been described in >20 fish species (Borel, Polkinghorne, and Pospischil 2018); however, further phylogenetic differentiation will need to be supported by PCR and *in situ* hybridization (Borel, Polkinghorne, and Pospischil 2018). Common lesions observed in the form of epitheliocystis from infection of many of these *Candidatus* families, have been described in farmed fish, including barramundi, white sturgeon, silver perch, Atlantic salmon, red seabream, carp, and yellowtail amberjack (Pawlikowska-Warych and Deptula 2016). Common clinical signs that have been implicated include fish growth inhibition, spread to other tissues of fish, respiratory disorders and death of fish with high mortality rates (even up to 100%) (Pawlikowska-Warych and Deptula 2016).

Experimentally, epithelial cells in fish are also sensitive to infection and an immune response. (Pawlikowska-Warych and Deptula 2016). A common characteristic observed among Chlamydiales is their reported immune response by macrophage infiltrations, as well as their fastidious *in vitro* growth (Pawlikowska-Warych and Deptula 2016).

Conclusion

Due to being obligate pathogens coupled with their inability to be genetically manipulated like other bacteria, Chlamydiae are notoriously difficult to study (Subtil and Dautry-Varsat 2004). Yet the Chlamydiales has undergone tremendous reorganization in the last few years due to the application of whole genome sequencing and DNA-sequence comparison. These techniques continue to allow for greater understanding of the Chlamydia life cycle and its pathogenicity. The most recent taxonomy, as outlined here, is the subject of debate among Chlamydialogists, but is in line with a great number of taxonomic studies (Olsen, Woese, and Overbeek 1994; Everett and Andersen 2001). Continued genomic and proteomic studies will expand our understanding of the Chlamydiales order (Vandahl, Birkelund, and Christiansen 2004), leading to better treatment options that could potentially benefit both human and animal health.

Acknowledgments

Much appreciation to my husband, whose brilliance, endless support, and clear direction motivates me to challenge myself in ways unimaginable, and to our son, who is evidence that dreams do come true. We are grateful to our coeditor, Lorrence Green, for his saintly patience, virtuous guidance, and mentorship over the years.

REFERENCES

Abdelrahman, Y. M., and R. J. Belland. 2005. The Chlamydial developmental cycle. *FEMS Microbiol Rev* 29(5):949–59.

Azuma, Y., H. Hirakawa, A. Yamashita, et al. 2006. Genome sequence of the cat pathogen, chlamydophila felis. *DNA Res* 13(1):15–23.

Bayramova, F., N. Jacquier, and G. Greub. 2018. Insight in the biology of Chlamydia-related bacteria. *Microbes Infect* 20(7-8):432–40.

Bommana, S., and A. Polkinghorne. 2019. Mini review: antimicrobial control of Chlamydial infections in animals: current practices and issues. *Front Microbiol* 10:113.

Borel, N., and G. Greub. 2019. International committee on systematics of prokaryotes (ICSP) subcommittee on the taxonomy of Chlamydiae. Minutes of the closed meeting, 5 July 2018, Woudschoten, Zeist, The Netherlands. *Int J Syst Evol Microbiol* 69(8):2606–2608.

Borel, N., A. Polkinghorne, and A. Pospischil. 2018. A review on Chlamydial diseases in animals: Still a challenge for pathologists? *Vet Pathol* 55(3):374–90.

Budai, I. 2007. Chlamydia trachomatis: milestones in clinical and microbiological diagnostics in the last hundred years: a review. *Acta Microbiol Immunol Hung* 54(1):5–22.

Carlson, J. H., S. F. Porcella, G. McClarty, and H. D. Caldwell. 2005. Comparative genomic analysis of Chlamydia trachomatis oculotropic and genitotropic strains. *Infect Immun* 73(10):6407–18.

Cevenini, R., M. Donati, and V. Sambri. 2002. Chlamydia trachomatis — the agent. *Best Pract Res Clin Obstet Gynaecol* 16(6):761–73.

Chahota, R., H. Ogawa, Y. Mitsuhashi, K. Ohya, T. Yamaguchi, and H. Fukushi. 2006. Genetic diversity and epizootiology of Chlamydophila psittaci prevalent among the captive and feral avian species based on VD2 region of ompA gene. *Microbiol Immunol* 50(9):663–78.

Cheong, H. C., C. Y. Q. Lee, Y. Y. Cheok, G. M. Y. Tan, C. Y. Looi, and W. F. Wong. 2019. Diseases in primary hosts and zoonosis. *Microorganisms* 7(5).

Clarke, I. N. 2011. Evolution of Chlamydia trachomatis. *Ann N Y Acad Sci* 1230:E11–8.

Corsaro, D., and G. Greub. 2006. Pathogenic potential of novel Chlamydiae and diagnostic approaches to infections due to these obligate intracellular bacteria. *Clin Microbiol Rev* 19(2):283–97.

de Kruif, M. D., E. C. van Gorp, T. T. Keller, J. M. Ossewaarde, and H. T. Cate. 2005. Chlamydia pneumoniae infections in mouse models: relevance for atherosclerosis research. *Cardiovasc Res* 65(2):317–27.

Dharamshi, J. E., D. Tamarit, L. Eme, et al. 2020. Marine sediments illuminate Chlamydiae diversity and evolution. *Curr Biol* 30(6):1032–1048.e7.

Di Francesco, A., M. Donati, G. Battelli, R. Cevenini, and R. Baldelli. 2004. Seroepidemiological survey for Chlamydophila felis among household and feral cats in northern Italy. *Vet Rec* 155(13):399–400.

Douam, F., J. M. Gaska, B. Y. Winer, Q. Ding, M. von Schaewen, and A. Ploss. 2015. Genetic dissection of the host tropism of human-tropic pathogens. *Annu Rev Genet* 49:21–45.

Everett, K. D. 2000. Chlamydia and Chlamydiales: more than meets the eye. *Vet Microbiol* 75(2):109–26.

Everett, K. D., R. M. Bush, and A. A. Andersen. 1999. Emended description of the order Chlamydiales, proposal of Parachlamydiaceae fam. nov. and Simkaniaceae fam. nov., each containing one monotypic genus, revised taxonomy of the family Chlamydiaceae, including a new genus and five new species, and standards for the identification of organisms. *Int J Syst Bacteriol* 49(Pt 2):415–40.

Everett, K. D. E., and A. A. Andersen. 2001. Radical changes to chlamydial taxonomy are not necessary just yet – reply (Letter). *International Journal of Systematic and Evolutionary Microbiology* 51(1):251–3.

Floriano, A. M., S. Rigamonti, F. Comandatore, et al. 2020. Complete genome sequence of Chlamydia avium PV 4360/2, isolated from a feral pigeon in Italy. *Microbiol Resour Announc* 9(16).

Fritsche, T. R., M. Horn, M. Wagner, R. P. Herwig, K. H. Schleifer, and R. K. Gautom. 2000. Phylogenetic diversity among geographically dispersed Chlamydiales endosymbionts recovered from clinical and environmental isolates of Acanthamoeba spp. *Appl Environ Microbiol* 66(6):2613–9.

Gomes, J. P., A. Nunes, W. J. Bruno, M. J. Borrego, C. Florindo, and D. Dean. 2006. Polymorphisms in the nine polymorphic membrane proteins of Chlamydia trachomatis across all serovars: evidence for serovar Da recombination and correlation with tissue tropism. *J Bacteriol* 188(1):275–86.

Greub, Gilbert. 2013. International committee on systematics of prokaryotes subcommittee on the taxonomy of Chlamydiae. *International Journal of Systematic and Evolutionary Microbiology* 63(5):1934–5.

Gupta, R. S., S. Naushad, C. Chokshi, E. Griffiths, and M. Adeolu. 2015. A phylogenomic and molecular markers based analysis of the phylum Chlamydiae: proposal to divide the class Chlamydiia into two orders, Chlamydiales and Parachlamydiales ord. nov., and emended description of the class Chlamydiia. *Antonie Van Leeuwenhoek* 108(3):765–81.

Heddema, E. R., E. J. van Hannen, B. Duim, C. M. Vandenbroucke-Grauls, and Y. Pannekoek. 2006. Genotyping of Chlamydophila psittaci in human samples. *Emerg Infect Dis* 12(12):1989–90.

Hogan, R. J., S. A. Mathews, S. Mukhopadhyay, J. T. Summersgill, and P. Timms. 2004. Chlamydial persistence: beyond the biphasic paradigm. *Infect Immun* 72(4):1843–55.

Kalman, S., W. Mitchell, and R. Marathe, et al. 1999. Comparative genomes of Chlamydia pneumoniae and C. trachomatis. *Nat Genet* 21(4):385–9.

Knittler, M. R., and K. Sachse. 2015. Chlamydia psittaci: update on an underestimated zoonotic agent. *Pathog Dis* 73(1):1–15.

Kumar, S., and M. R. Hammerschlag. 2007. Acute respiratory infection due to Chlamydia pneumoniae: current status of diagnostic methods. *Clin Infect Dis* 44(4):568–76.

Li, Z., P. Liu, J. Hou, et al. 2020. Detection of Chlamydia psittaci and Chlamydia ibidis in the endangered crested ibis (Nipponia nippon). *Epidemiol Infect* 148:e1.

Longbottom, D., and L. J. Coulter. 2003. Animal chlamydioses and zoonotic implications. *J Comp Pathol* 128(4):217–44.

Longbottom, D., and M. Livingstone. 2006. Vaccination against chlamydial infections of man and animals. *Vet J* 171(2):263–75.

Longbottom, D., E. Psarrou, M. Livingstone, and E. Vretou. 2001. Diagnosis of ovine enzootic abortion using an indirect ELISA (rOMP91B iELISA) based on a recombinant protein fragment of the polymorphic outer membrane protein POMP91B of Chlamydophila abortus. *FEMS Microbiol Lett* 195(2):157–61.

Manavi, K. 2006. A review on infection with Chlamydia trachomatis. *Best Pract Res Clin Obstet Gynaecol* 20(6):941–51.

Moulder, J. W. 1966. The relation of the psittacosis group (Chlamydiae) to bacteria and viruses. *Annu Rev Microbiol* 20:107–30.

Murray, P.R. 1998. *Medical Microbiology*. 3rd ed. Mosby: St. Louis, MO.

Nunes, A., and J. P. Gomes. 2014. Evolution, phylogeny, and molecular epidemiology of Chlamydia. *Infect Genet Evol* 23:49–64.

Nunes, A., P. J. Nogueira, M. J. Borrego, and J. P. Gomes. 2010. Adaptive evolution of the Chlamydia trachomatis dominant antigen reveals distinct evolutionary scenarios for B- and T-cell epitopes: worldwide survey. *PLOS One* 5(10).

Ochman, H., M. Worobey, C. H. Kuo, et al. 2010. Evolutionary relationships of wild hominids recapitulated by gut microbial communities. *PLoS Biol* 8(11):e1000546.

Olsen, G. J., C. R. Woese, and R. Overbeek. 1994. The winds of (evolutionary) change: breathing new life into microbiology. *J Bacteriol* 176(1):1–6.

Page, L. A. 1966. Revision of the family Chlamydiaceae Rake (Rickettsiales): unification of the psittacosis-lymphogranuloma venereum-trachoma group of organisms in the genus Chlamydia Jones, Rake and Stearns, 19451. *Int J Syst Evol Microbiol* 16(2):223–52.

Pannekoek, Y., V. Dickx, D. S. Beeckman, et al. 2010. Multi locus sequence typing of Chlamydia reveals an association between Chlamydia psittaci genotypes and host species. *PLOS One* 5(12):e14179.

Pannckoek, Y., G. Morelli, B. Kusecek, et al. 2008. Multi locus sequence typing of Chlamydiales: clonal groupings within the obligate intracellular bacteria Chlamydia trachomatis. *BMC Microbiol* 8:42.

Paolillo, S., F. Veglia, E. Salvioni, et al. 2019. Heart failure prognosis over time: how the prognostic role of oxygen consumption and ventilatory efficiency during exercise has changed in the last 20 years. *Eur J Heart Fail* 21(2):208–17.

Pawlikowska-Warych, M., and W. Deptula. 2016. Characteristics of chlamydia-like organisms pathogenic to fish. *J Appl Genet* 57(1):135–41.

World Health Organization. 2011. Prevalence and incidence of selected sexually transmitted infections, *Chlamydia Trachomatis, Neisseria Gonorrhoeae*, syphilis, and *Trichomonas vaginalis*: methods and results used by WHO to generate 2005 estimates, Geneva.

Read, T. D., R. C. Brunham, C. Shen, et al. 2000. Genome sequences of Chlamydia trachomatis MoPn and Chlamydia pneumoniae AR39. *Nucleic Acids Res* 28(6):1397–406.

Read, T. D., G. S. Myers, R. C. Brunham, et al. 2003. Genome sequence of Chlamydophila caviae (Chlamydia psittaci GPIC): examining the role of niche-specific genes in the evolution of the Chlamydiaceae. *Nucleic Acids Res* 31(8):2134–47.

Rusconi, B., and G. Greub. 2011. Chlamydiales and the innate immune response: friend or foe? *FEMS Immunol Med Microbiol* 61(3):231–44.

Sachse, K., P. M. Bavoil, B. Kaltenboeck, et al. 2015. Emendation of the family Chlamydiaceae: proposal of a single genus, Chlamydia, to include all currently recognized species. *Syst Appl Microbiol* 38(2):99–103.

Sachse, K., and K. Laroucau. 2015. Two more species of Chlamydia-does it make a difference? *Pathog Dis* 73(1):1–3.

Sachse, K., K. Laroucau, K. Riege, et al. 2014. Evidence for the existence of two new members of the family Chlamydiaceae and proposal of Chlamydia avium sp. nov. and Chlamydia gallinacea sp. nov. *Syst Appl Microbiol* 37(2):79–88.

Sachse, K., and A. Ruettger. 2015. Rapid microarray-based genotyping of Chlamydia spp. strains from clinical tissue samples. *Methods Mol Biol* 1247:391–400.

Seth-Smith, H. M., S. R. Harris, R. J. Skilton, et al. 2013. Whole-genome sequences of Chlamydia trachomatis directly from clinical samples without culture. *Genome Res* 23(5):855–66.

Shirai, M., H. Hirakawa, M. Kimoto, et al. 2000. Comparison of whole genome sequences of Chlamydia pneumoniae J138 from Japan and CWL029 from USA. *Nucleic Acids Res* 28(12):2311–4.

Staub, E., H. Marti, R. Biondi, et al. 2018. Novel Chlamydia species isolated from snakes are temperature-sensitive and exhibit decreased susceptibility to azithromycin. *Sci Rep* 8(1):5660.

Stephens, R. S. 2003. The cellular paradigm of chlamydial pathogenesis. *Trends Microbiol* 11(1):44–51.

Stephens, R. S., S. Kalman, C. Lammel, et al. 1998. Genome sequence of an obligate intracellular pathogen of humans: Chlamydia trachomatis. *Science* 282(5389):754–9.

Stephens, R. S., G. Myers, M. Eppinger, and P. M. Bavoil. 2009. Divergence without difference: phylogenetics and taxonomy of Chlamydia resolved. *FEMS Immunol Med Microbiol* 55(2):115–9.

Stride, M. C., A. Polkinghorne, T. L. Miller, J. M. Groff, S. E. Lapatra, and B. F. Nowak. 2013. Molecular characterization of "Candidatus Parilichlamydia carangidicola," a novel Chlamydia-like epitheliocystis agent in yellowtail kingfish, Seriola lalandi (Valenciennes), and the proposal of a new family, "Candidatus Parilichlamydiaceae" fam. nov. (order Chlamydiales). *Appl Environ Microbiol* 79(5):1590–7.

Subtil, A., and A. Dautry-Varsat. 2004. Chlamydia: five years A.G. (after genome). *Curr Opin Microbiol* 7(1):85–92.

Sykes, J. E. 2005. Feline chlamydiosis. *Clin Tech Small Anim Pract* 20(2):129–34.

Thomson, N. R., C. Yeats, K. Bell, et al. 2005. The Chlamydophila abortus genome sequence reveals an array of variable proteins that contribute to interspecies variation. *Genome Res* 15(5):629–40.

Vandahl, B. B., S. Birkelund, and G. Christiansen. 2004. Genome and proteome analysis of Chlamydia. *Proteomics* 4(10):2831–42.

Vogler, B. R., M. Trinkler, H. Marti, et al. 2019. Survey on Chlamydiaceae in cloacal swabs from Swiss turkeys demonstrates absence of Chlamydia psittaci and low occurrence of Chlamydia *gallinacean*. *PLOS One* 14(12):e0226091.

Vorimore, F., R. C. Hsia, H. Huot-Creasy, et al. 2013. Isolation of a new Chlamydia species from the feral sacred Ibis (Threskiornis aethiopicus): Chlamydia ibidis. *PLOS One* 8(9):e74823.

Vretou, E., F. Radouani, E. Psarrou, I. Kritikos, E. Xylouri, and O. Mangana. 2007. Evaluation of two commercial assays for the detection of Chlamydophila abortus antibodies. *Vet Microbiol* 123(1–3):153–61.

Walder, G., H. Hotzel, C. Brezinka, et al. 2005. An unusual cause of sepsis during pregnancy: recognizing infection with chlamydophila abortus. *Obstet Gynecol* 106(5 Pt 2):1215–7.

Ward, M. E. 1983. Chlamydial classification, development and structure. *Br Med Bull* 39(2):109–15.

Yang, Z. P., C. C. Kuo, and J. T. Grayston. 1993. A mouse model of Chlamydia pneumoniae strain TWAR pneumonitis. *Infect Immun* 61 (5):2037–40.

Yarza, P., M. Richter, J. Peplies, et al. 2008. The all-species living tree project: a 16S rRNA-based phylogenetic tree of all sequenced type strains. *Syst Appl Microbiol* 31(4):241–50.

Zhong, G. 2017. Chlamydial plasmid-dependent pathogenicity. *Trends Microbiol* 25(2):141–52.

26

The Genus *Clostridium*

Peter Dürre

CONTENTS

Introduction

Clostridium is one of the largest bacterial genera, including more than 160 validly described species. Among these are several with enormous biotechnological potential (e.g., for production of biofuels, bulk chemicals, and important enzymes as well as for usage in cancer treatment) and also a few well-known pathogens. However, some of their toxins proved to be valuable in medical and cosmetic applications. Clostridia, thus, belong to the avant-garde of industrially useful microbes. Members of this genus stain, in general, Gram-positive, are more or less strictly anaerobic bacteria, employ an impressive number of varying fermentation routes, and are able to degrade numerous natural and artificial substances. Due to required precautions for excluding oxygen during handling, clostridia were, for a long time, virtually inaccessible at the genetic level. This situation has completely changed. Gene cloning, DNA transfer, gene expression modulation, and gene knock-out and knock-in systems have been successfully established. Thus, the road is paved for further elucidation and commercial exploitation of the enormous metabolic potential of the clostridia.

In the past three decades, six books have been published, completely devoted to clostridia (Minton and Clarke, 1989; Woods, 1993; Rood et al., 1997; Bahl and Dürre, 2001; Durre, 2005, Dürre, 2014). Due to the limited space of this chapter, the interested reader is referred to these references for additional and more detailed information.

Historical Background

Hippocrates (460–377 B.C.), a physician living on the Greek island of Kos, was probably the first who reported clostridia-effected diseases. His documentation describes gas gangrene (Sussman, 1958), caused by *C. histolyticum*, and lockjaw (also called opisthotonus or tetanus) (Kiple, 1993), caused by *C. tetani*. Until the Middle Ages, indigo dyeing in Europe was based on the woad plant. It contains two indigo precursors, isatan B (indoxyl-5-ketogluconate) and indican (indoxyl-β-D-glucoside), which are converted in a two-step process to indigo. Scientific research in the 1990s revealed that the responsible bacteria are *Enterobacter agglomerans* for the initial aerobic process and *C. isatidis* for the successive fermentation (Ewerdwalbesloh and Meyer, 1995; Padden et al., 1998; Padden et al., 1999). Similarly, it is now clear that flax and hemp retting, in use for thousands of years for clothing fabrication, are essentially dependent on clostridia (Bahl and Dürre, 1993; Tamburini et al., 2003). Microbiological investigations on these bacteria only started in 1861, when the famous French microbiologist Louis Pasteur discovered that microbes exist that can grow without oxygen (Pasteur, 1861). This represented a revolutionary finding at that time. It is usually concluded that the term "*Clostridium*" stems from Prazmowski (1880). However, before him, Trécul had already used this expression (compare Dürre, 2001). It is based on a Greek root, later became Latinized, and means "small spindle." Originally, it was used exclusively to describe the bacterial morphology, not any metabolic properties (Dürre, 2001). The first pure culture obtained was *C. butyricum* (Prazmowski, 1880), which now represents the type species of the genus *Clostridium*. Without a doubt, one of the world's most famous bacteria is *C. acetobutylicum*. It had both, enormous biotechnological as well as political impact. The acetone-butanol fermentation ranks second in size to ethanol production and was thus one of the largest bioprocesses ever performed. Until about 1950, two-thirds of the world's butanol production was obtained by fermentation, and the solvent was an important bulk chemical for industry (Jones and Woods, 1986; Dürre and Bahl, 1996). Originally, however, acetone was the desired product. Weizmann isolated the organism during World War I in the United Kingdom (Weizmann, 1915), when this chemical was in demand for production of ammunition. Despite his important contribution, he refused any honors, but made clear that he was in favor of a Jewish homeland in Palestine. There is no doubt that the Balfour Declaration of 1917 on this very subject was affected by Weizmann's achievements, who later became the first president of the State of Israel.

Metabolism

Clostridia in the public opinion are associated with production of a bad smell. In most cases, this is caused by butyric acid, one of the major fermentation products. Butyrate is produced by saccharolytic as well as proteolytic pathways (Table 26.1). For the former, the uptake of glucose (and other sugars) is performed by a phosphoenolpyruvate:phosphotransferase system. Glycolytic reactions then lead to pyruvate, which is split into acetyl-CoA, CO_2, and reduced ferredoxin by pyruvate:ferredoxin oxidoreductase. Ferredoxin is a small iron-sulfur protein with a very negative redox potential (~ –400 mV). This is of considerable advantage for the anaerobic clostridia, as they can "blow off" reducing equivalents as hydrogen (employing an additional hydrogenase) and do not need to sacrifice carbon compounds as acceptor molecules. Acetyl-CoA is the starting point for production of acetate (via phosphotransacetylase and acetate kinase, thereby also yielding ATP), ethanol (via acetaldehyde and alcohol dehydrogenases, thereby consuming reducing equivalents), and C_4 compounds such as butyrate and butanol. The latter step is accomplished by fusion of two acetyl-CoA to acetoacetyl-CoA and free coenzyme A. Acetoacetyl-CoA is reduced to

butyryl-CoA, which is transformed into butyrate or butanol by reactions analogous to those of the C_2 compounds. The ATP-yielding reaction might either be catalyzed by a butyrate kinase or by transfer of the CoA moiety to acetate and then conversion of acetyl-CoA via acetyl phosphate to acetate by acetate kinase. For caproate formation in case of *C. kluyveri*, a third molecule of acetyl-CoA is fused to butyryl-CoA. Except for the substrate combination ethanol/acetate, this organism can also grow on ethanol/succinate, thereby forming acetate and butyrate (in this case, directly from a C_4 compound). Acetone is made from acetoacetate after transfer of the CoA moiety to either acetate or butyrate, and isopropanol (e.g., by *C. beijerinckii*) by further reduction of acetone.

A landmark discovery in energy metabolism of anaerobes was the flavin-based and ferredoxin-dependent electron bifurcation that can be coupled to proton or Na^+ gradient generation and thus to ATP formation (Herrmann et al., 2008; Buckel and Thauer, 2018). The first such reaction found in clostridia was the reduction of crotonyl-CoA to butyryl-CoA in *C. kluyveri*, catalyzed by the butyryl-CoA dehydrogenase/EtfBC complex (Li et al., 2008). Etf stands for electron-transferring flavoprotein. This NADH-coupled reduction is highly exergonic, so that one electron of NADH is transferred to the more positive acceptor butyryl-CoA dehydrogenase and the second electron to the more negative acceptor ferredoxin. This bifurcation mechanism is repeated with a second NADH molecule. The total reaction sequence thus is:

$$\text{crotonyl-CoA} + 2 \text{ NADH} + \text{oxidized ferredoxin (Fd}_{ox})$$
$$\rightarrow \text{butyryl-CoA} + 2 \text{ NAD}^+ + \text{reduced ferredoxin (Fd}^{2-})$$

TABLE 26.1

Clostridial Fermentations

Fermentation Type	Typical Fermentation Balance	Example for Respective Species
Saccharolytic		
butyrate	glucose → 0.8 butyrate + 0.4 acetate + 2 CO_2 + 2.1 H_2	*C. butyricum*
butanol	glucose → 0.6 butanol + 0.2 acetone + 2.2 CO_2 + 1.4 H_2 + 0.04 butyrate + 0.14 acetate + 0.07 ethanol	*C. acetobutylicum*
homoacetate	fructose → 3 acetate	*C. aceticum*
Acidotrophic		
propionate	lactate → 0.66 propionate + 0.33 acetate + 0.33 CO_2	*C. propionicum*
propionate, Na^+-dependent	succinate → propionate + CO_2	probably *C. mayombei*
Alcoholotrophic	2 ethanol + acetate → butyrate + 0.33 caproate + 0.66 H_2	*C. kluyveri*
Autotrophic	2 CO_2 + 4 H_2 → acetate	*C. aceticum*
	10 CO → acetate + ethanol + 6 CO_2	*C. ljungdahlii*
	4 CO + 6 H_2 → acetate + ethanol	*C. ljungdahlii*
Proteolytic		
pairs of amino acids		
Stickland reaction	alanine + 2 glycine → 3 acetate + CO_2 + 3 NH_4^+	*C. sporogenes*
single amino acids		
2-, 3-elimination	threonine → 0.66 propionate + 0.33 butyrate + 0.66 CO_2 + NH_4^+	*C. propionicum*
B_{12}-dependent	glutamate → acetate + 0.5 butyrate + CO_2 + NH_4^+	*C. tetanomorphum*
SAM-dependent	lysine → butyrate + acetate + 2 NH_4^+	*C. subterminale*
Se-dependent	glycine → 0.75 acetate + 0.5 CO_2 + NH_4^+	*C. purinilyticum*
Heteroaromaticotrophic		
purine	adenine → acetate + formate + 2 CO_2 + 5 NH_4^+	*C. purinilyticum*
pyrimidine	orotic acid → aspartate + CO_2 + NH_4^+	*C. oroticum*

Note: H_2O is not indicated in fermentation reactions.

Abbreviation: SAM = S-adenosylmethionine.

TABLE 26.2

Flavin-Based and Ferredoxin (Flavodoxin)-Dependent Electron Bifurcation Enzyme Complexes

Enzyme Complex	Reaction Catalyzed	Organism
Butyryl-CoA dehydrogenase/electron-transfer flavoproteins (Bcd/Etf)	crotonyl-CoA + 2 NADH + Fd_{ox} → butyryl-CoA + 2 NAD^+ + Fd^{2-}	*C. kluyveri*
NAD^+-specific [FeFe]-hydrogenase (HydABC)	3 H^+ + NADH + Fd^{2-} ⇌ 2 H_2 + NAD^+ + Fd_{ox}	*Thermotoga maritima* *Acetobacterium woodii* *Moorella thermoacetica*
Ferredoxin:$NADP^+$ reductase (NfnAB)	2 $NADP^+$ + NADH + Fd^{2-} + H^+ ⇌ 2 NADPH + NAD^+ + Fd_{ox}	*C. kluyveri* *M. thermoacetica*
Heterodisulfide reductase (MvhADG-HdrABC)	CoM-S-S-CoB + 2 H_2 + Fd_{ox} → CoM-SH + CoB-SH + Fd^{2-} + 2 H^+	methanogenic Archaea
Caffeyl-CoA reductase (CarCDE)	caffeyl-CoA + 2 NADH + Fd_{ox} → dihydrocaffeyl-CoA + 2 NAD^+ + 2 H^+ + Fd^{2-}	*A. woodii*
Formate dehydrogenase (FdhF2-HylABC)	2 formate + NAD^+ + Fd_{ox} ⇌ 2 CO_2 + NADH + H^+ + Fd^{2-}	*C. acidurici*
$NADP^+$-specific [FeFe]-hydrogenase (HytA-E)	$NADP^+$ + Fd_{ox} + 2 H_2 ⇌ NADPH + Fd^{2-} + 3 H^+	*C. autoethanogenum*
Methylene-tetrahydrofolate reductase (MetF-HdrABCMvhD)	2 NADH + methylene-THF + Fd_{ox}(?) ⇌ 2 NAD^+ + methyl-THF + Fd^{2-} (?)	*M. thermoacetica*
Lactate dehydrogenase (LctBCD)	lactate + 2 NAD^+ + Fd^{2-} ⇌ pyruvate + 2 NADH + Fd_{ox}	*A. woodii*
F_{420}-specific heterodisulfide reductase (HdrA2B2C2)	CoM-S-S-CoB + 2 $F_{420}H_2$ + Fd_{ox} → CoM-SH + CoB-SH + Fd^{2-} + 2 F_{420} + 2 H^+	*Methanosarcina acetivorans*
NAD^+-specific ubiquinol reductase (FixABCX)	2 NADH + Q + Fld_{sq} → 2 NAD^+ + QH_2 + Fld_{hq}^{2-}	*Azotobacter vinelandii*

Abbreviations: CoM-SH, coenzyme M = 2-mercaptoethanesulfonate; CoB-SH, coenzyme B = 7-mercaptoheptanoylthreoninephosphate; F_{420}, factor 420 = deazaflavin adenine dinucleotide; Q = quinone; Fld_{sq} = flavodoxin semiquinone; Fld_{hq} = flavodoxin hydroquinone

Source: Data are taken from Buckel and Thauer (2018).

Reduced ferredoxin can be reoxidized at the membrane-bound Rnf complex (Rnf stems from *Rhodobacter* nitrogen fixation, in which organism this protein complex was discovered first), with concomitant NAD^+ reduction to NADH. This exergonic step is used by the Rnf complex to pump ions across the cytoplasmic membrane. This way, either a H^+ or a Na^+ gradient is formed, which can then be used for ATP formation by respective ATPases. So far, 11 flavin-based electron-bifurcating enzyme complexes have been discovered (Table 26.2), of which four are present in clostridia (Li et al., 2008). However, this might only be the tip of the iceberg, as several other reactions might be candidates for such a mechanism.

Proteolytic production of butyrate follows either the pathways described above starting with acetyl-CoA, or it is made directly from the C_4 compound threonine by a 2-,3-elimination reaction of the amino group (Table 26.1) (Buckel, 1990). For utilization of amino acids, fermentation in parallel oxidative and reductive branches is very common. Examples are the so-called Stickland reaction, in which pairs of amino acids are degraded, or the selenium-dependent glycine fermentation, in which one glycine is completely oxidized to CO_2 and the reducing equivalents are used to reduce three glycines to acetate via the selenoenzyme glycine reductase and producing ATP from acetyl-phosphate via acetate kinase (Dürre and Andreesen, 1982). In many cases, amino acids are transformed into the respective 2-oxo compounds, removing the amino group by 2-,3-elimination as already mentioned, oxidation, or transamination (Buckel, 1990). The 2-oxo acid can then be converted into a saturated fatty acid by a number of reduction, CoA activation, and dehydration steps. Other pathways employ B_{12}-dependent C-C or S-adenosylmethionine-dependent C-N rearrangements.

Another important route in some saccharolytic clostridia is homoacetate fermentation. By reutilization of the CO_2 formed in pyruvate degradation, three acetates are formed per hexose. Species such as *C. aceticum* are also able to grow autotrophically by converting CO_2 and H_2 into acetate. This metabolic route is called Wood-Ljungdahl pathway and depicted in Figure 26.1. It also allows the utilization of carbon monoxide as the sole carbon and energy source, e.g., by *C. ljungdahlii*. Energy conservation during autotrophic growth occurs by different mechanisms. The one molecule of ATP required for formyl-THF formation is compensated for by the ATP formed in the acetate kinase reaction. *C. ljungdahlii* builds up a proton gradient from reduced ferredoxin by action of the Rnf complex, with concomitant utilization of the produced NADH in formation of reduced THF-derivatives and formation of ethanol from acetyl-CoA (Köpke et al., 2010; Tremblay et al., 2012). A variation of this route is found in another anaerobe, *Acetobacterium woodii*, which contains a Na^+-pumping Rnf complex (Biegel and Müller, 2010). The situation is clearly different in acetogens such as *Moorella thermoacetica* (formerly *Clostridium thermoaceticum*) and *Thermoanaerobacter kivui* that employ an energy-conserving hydrogenase (Ech) for proton gradient formation (Schuchmann and Müller, 2014). An additional proton gradient might be generated in cytochrome- and quinone-harboring acetogens by a yet unknown electron transport chain. A possible model has been suggested by Das and Ljungdahl (2003).

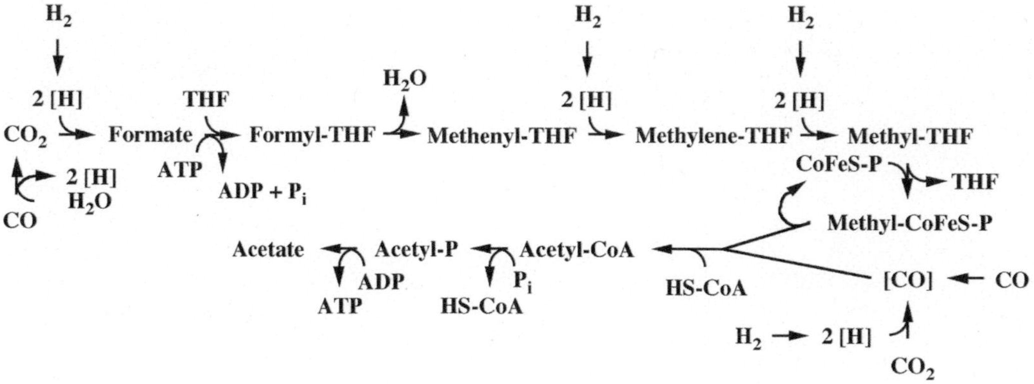

FIGURE 26.1 Wood-Ljungdahl pathway employed by clostridial acetogens. (*Abbreviations:* CoFeS-P, corrinoid iron-sulfur protein; Pi, inorganic phosphate; THF, tetrahydrofolate.)

A variety of organic acids, alcohols, polymers, aromatics, and halogenated substances can also be used as substrates by clostridia. For a detailed description, see Bahl and Dürre (1993). Several typical examples are listed in Table 26.1. Also, some clostridia such as *C. pasteurianum* are able to fix molecular nitrogen, thereby producing NH_4^+ and H_2.

Lactate is metabolized by several clostridia. *C. propionicum* converts this organic acid to propionate and acetate by branched pathways. 2 Lactate are activated with coenzyme A and dehydrated to yield acrylyl-CoA, which then is reduced to propionyl-CoA. Transfer of the CoA moiety to lactate leads to propionate formation. The reducing equivalents stem from oxidation of a third lactate to acetate and CO_2. The acetate kinase reaction of this branch generates the only ATP of this fermentation. It is thus clearly less efficient than the methylmalonyl-CoA pathway employed by *Propionibacterium*. Succinate decarboylation is another possibility to produce propionate from an organic acid. This way is employed by *Propionigenium modestum* and based on a membrane-located decarboxylation of methylmalonyl-CoA, which generates a Na^+-gradient. This gradient can be directly used for ATP generation by a Na^+-dependent F_1F_0-ATPase (as in *P. modestum*) or indirectly after conversion into a H^+-gradient by action of a Na^+/H^+ antiporter. The other decarboxylation product propionyl-CoA is used to activate succinate, and succinyl-CoA is converted into methylmalonyl-CoA. Thus, only three enzymes are required for this pathway. Within the clostridia, probably *C. mayombei* degrades succinate by Na^+-dependent decarboxylation (Bahl and Dürre, 1993). A dangerous organism for the cheese industry is *C. tyrobutyricum*, which forms butyrate from lactate. The so-called late blowing ruins structure and taste of contaminated cheese.

Heteroaromates stemming from the degradation of nucleic acids can also be used for growth by clostridia. *C. acidurici*, *C. cylindrosporum*, and *C. purinilyticum* are almost completely specialized on fermenting purines (*C. purinilyticum* grows also well on glycine). The pyrimidine moiety is split first, degraded to CO_2 and NH_4^+, and then the imidazole moiety is hydrolytically cleaved, yielding finally acetate (via glycine reductase and acetate kinase), formate, CO_2, and NH_4^+ (Dürre and Andreesen, 1983). Pyrimidine fermentation is accomplished by *C. glycolicum* and *C. oroticum* and is based on similar reactions (Vogels and van der Drift, 1976).

Clostridia belong to those bacteria, which can form endospores. These are the most resistant biological survival forms known and protect the genome against environmental danger such as desiccation, heat, radiation, and hazardous compounds. Retrieval of *C. aceticum*, a species considered to be lost from culture collections during World War II, provided evidence for successful germination even after four decades (Braun et al., 1981). Clostridia thus belong to the few prokaryotes able to perform cell differentiation. The sporulation process resembles very much that of *Bacillus* (Dürre and Hollergschwandner 2004), which is well understood. Initiation is achieved via a number of phosphate transfers from sensor kinases to the response regulator Spo0A, the so-called phosphorelay (Hoch 1993). Such a chain of transfer components is obviously not present in clostridia, as shown by the numerous genome sequencing projects, but the master regulator Spo0A is conserved and fulfills the same functions. In cooperation with the sporulation-specific sigma factor σ^H, Spo0A~P induces transcription of further forespore- and mother cell-specific sporulation sigma factors. In both compartments, pairs of these proteins become active successively, to guarantee a sequential number of steps, required for mature spore synthesis. In the forespore, σ^F and σ^G are responsible for this task and are regulated via anti- and anti-anti-sigma factors. In the mother cell, their companions are σ^E and σ^K, which are proteolytically activated. While most endospore-forming bacteria produce just one such compartment per cell, there are a few notable exceptions. *Anaerobacter polyendosporus* and *Metabacterium polyspora*, phylogenetically close relatives of *Clostridium*, produce up to five endospores in a single mother cell (Angert et al., 1996; Siunov et al., 1999). An evolutionary link to production of already living offspring might be represented by *Epulopiscium fishelsoni*. This is not only one of world's largest bacteria, surpassing many eukaryotic microorganisms in size, but also belongs phylogenetically to the clostridia and gives birth to several internal daughter cells (Angert et al., 1993).

An important task in food preservation is to prevent contamination with spores of pathogenic clostridia. For this purpose, preservatives can be added such as nitrite, which converts important redox-active iron-sulfur clusters (e.g., in ferredoxin) into inactive iron-nitric oxide complexes (Reddy et al., 1983; Carpenter et al., 1987), benzoic acid, sodium chloride, and sorbate. Acidification, desiccation, heating, ionizing radiation, and refrigeration are other methods to achieve extended shelf life (Lund and Peck, 2000). Hygienic conditions during preparation are of utmost importance, as failure in this respect might even affect vacuum-packed materials (production of gas and/or sulfides). In the future, high pressure treatment might become an important additional means to prevent clostridial spoilage of foods.

Phylogeny and Taxonomy

Phylogenetically, the clostridia belong to the low G + C Gram-positive phylum. This classification is based on 16 S rRNA gene sequence analysis. It reflects ordering of the species according to common ancestry, disregarding physiological and morphological features. Relatedness is usually displayed as a dendrogram. Historically, however, members of the genus *Clostridium* had been defined as performing an anaerobic life style, having a Gram-positive-type cell wall, producing endospores, and being unable of dissimilatory sulfate reduction. Thus, the genus represents a problem to taxonomists, and many bacteria, originally described as a *Clostridium*, have been reclassified. Taxonomy follows the scheme: order *Firmicutes*, class *Clostridia*, order *Clostridiales*, family *Clostridiaceae*, genus *Clostridium*. Compilations of validly described clostridia (currently more than 160 species, making *Clostridium* one of the largest bacterial genera) can be found in the internet (http://www.bacterio.net/clostridium.html, LPSN: List of Prokaryotic Names with Standing in Nomenclature—Genus Clostridium; http://www.taxonomicoutline.org/content/7/7/, Taxonomic Outline of the Bacteria and Archaea, Release 2007.07, Part 7—The Bacteria: Phylum Firmicutes, Class "Clostridia," by G. M. Garrity, T. G. Lilburn, J. R. Cole, S. H. Harrison, J. Euzéby, B. J. Tindall). There, and also in catalogs of culture collections such as ATCC (American Type Culture Collection) and DSMZ (Deutsche Sammlung von Mikroorganismen und Zellkulturen), are numerous indications of new genera with which former clostridia are now affiliated. These genera are *Caloramator, Dendrosporobacter, Eubacterium, Filifactor, Fusobacterium, Moorella, Oxalophagus, Oxobacter, Paenibacillus, Thermoanaerobacter, Thermoanaerobacterium, Sedimentibacter, Sporohalobacter,* and *Syntrophospora* (Stackebrandt, 2004; Wiegel et al., 2006). A further complication is that not all taxonomically validly described *Clostridium* species form phylogenetically coherent clusters. 16 S rDNA analyses of the *Bacillus/Clostridium* subphylum of the Gram-positives allowed identification of 20 clusters, designated I through XX. Clostridia are found in Clusters I (with a number of subdivisions and regarded as genus *Clostridium* sensu stricto [Stackebrandt and Hippe, 2001; Wiegel et al. 2006]), III, IV, XIa, XIb, XII, XIVa, XIVb, XVI, XVIII, and XIX. Cluster I contains the type species of the genus, *C. butyricum*. Different classifications of single species can be found (e.g., *C. limosum* in Cluster II; Stackebrandt and Hippe, 2001 or Cluster I; Wiegel et al., 2006). A compilation of genomes classified as *Clostridium* and *Clostrioides* revealed taxonomic inconsistencies even in Cluster I, arguably the authentic *Clostridium* genus (Cruz-Morales et al., 2019). Thus, the phylum *Firmicutes* will probably still remain the "greatest challenge for taxonomists" (Stackebrandt, 2004) for quite some time.

Pathogenicity

Despite the bad public reputation of clostridia caused by well-known species such as *C. botulinum* and *C. tetani*, only few members of this genus (less than 10%) form dangerous toxins. Some species contain more than one (outstanding is *C. perfringens* with 14 toxins), altogether 58 clostridial toxins have been identified thus far (Popoff and Stiles 2005). This number is slightly less than a fifth of all bacterial toxins. The clostridial toxins are proteins and can be classified as either pore-forming, zinc-dependent metalloproteases, glycosyltransferases, ADP-ribosyltransferases, or phospholipases (Table 26.3).

One of the most intensively studied representatives is the *C. perfringens* enterotoxin (CPE) (McClane, 2005). It is a common cause of foodborne disease. However, it should be kept in mind that the vast majority of strains of this species do not carry a *cpe* gene. While *in vitro* studies revealed that CPE can form pores in artificial lipid membranes, the *in vivo* effect relies on recruiting a number of host proteins to trigger massive changes in membrane permeability. Members of the claudin family are most probably the primary target and serve as receptors of CPE. With the help of additional membrane proteins, first a small complex of approximately 90 kDa and then large complexes of approximately 135, 155, and 200 kDa are formed. Interaction of a large complex with tight junction proteins leads to massive damage of the latter. CPE is one of the clostridial toxins, whose synthesis is tightly linked to sporulation. The protein is not secreted, but only set free upon lysis of the mother cell, together with the mature endospore.

The prototype of thiol-activated pore-forming toxins is perfringolysin O (Popoff and Stiles, 2005), which shows highest activity in the presence of reducing compounds. Elucidation of the structure revealed an elongated, rod-shaped molecule. Upon binding to its membrane target cholesterol, oligomerization of 40–50 molecules results in large pore formation. *C. septicum* α-toxin is secreted as a nontoxic precursor and becomes active by proteolysis. *C. perfringens* β-toxin does not destroy erythrocyte membranes, but rather forms cation-specific membrane pores. It is a major cause of necrotic enteritis in domesticated livestock and humans.

Most public attention achieves the clostridial neurotoxins (Johnson, 2005). Botulinum toxin is the most poisonous substance of biological origin known to date (for human beings a concentration of approximately 0.1 ng kg^{-1} has been calculated to be lethal) (Arnon, 1997). There are three forms of botulism: (1) infant (probably the most common with up to 100 cases per year in the United States, honey feeding before approximately 6 months of age is a risk factor); (2) foodborne (classical, ingestion of toxin); and (3) wound (very rare, bacterial spores infect a tissue with damaged blood and, thus, oxygen supply, germinate, and subsequently produce toxin). Botulinum toxin (BoNT) is on the list of bioterrorist agents, because of its extreme toxicity. In 1995, the Japanese Aum Shinrikyo sect wanted to use this compound to attack passengers on the Tokyo subway. Due to problems with aerosolization, they ultimately decided to use sarin gas. Also, following the first Gulf War in 1990 and during United Nations weapons inspections, Iraq conceded the production of botulinum toxin. About half of the total amount of 19,000 L had already been used for preparation of SCUD missile warheads and

TABLE 26.3

Clostridial Toxins

Class of Toxins	Type	Typical Representative	Pathogenic Species
Pore-forming	Enterotoxin	CPE	*C. perfringens*
	Thiol-activated	Perfringolysin O (Θ-toxin)	*C. perfringens*
	Aerolysin-like	α-toxin	*C. septicum*
	Hemolysin-like	β-toxin	*C. perfringens*
Zinc-dependent metalloprotease	Neurotoxin	Botulinum toxin	*C. botulinum A-F*
			C. baratii
			C. argentinense
		Tetanus toxin	*C. tetani*
	Collagenase	ColG, ColH	*C. histolyticum*
		ColA	*C. perfringens*
Glycosyltransferase	UDP-glucose-dependent (large clostridial toxins)	ToxA, ToxB	*C. difficile*
		Lethal toxin (LT)	*C. sordellii*
		Hemorrhagic toxin (HT)	*C. sordellii*
	UDP-N-acetyl-glucosamine-dependent	α-toxin	*C. novyi*
ADP-ribosyltransferase	Actin ADP-ribosylating (Binary toxins)	C2 toxin	*C. botulinum C, D*
		Iota toxin	*C. perfringens*
		Spiroforme toxin	*C. spiroforme*
		CDT	*C. difficile*
	Rho ADP-ribosylating	C3	*C. botulinum C, D*
		C3	*C. limosum*
Phospholipase	Phosphatidylcholine-, sphingomyelin-degrading	α-toxin	*C. perfringens*
		PLC	*C. absonum*
		PLC	*C. bifermentans*
		γ-toxin	*C. novyi*
		PLC	*C. haemolyticum*
	Additionally phosphatidylserine-degrading	α-toxin	*C. perfringens*
		γ-toxin	*C. novyi*
	Additionally phosphatidylethanol-amine-, phosphatidylinositol-degrading	γ-toxin	*C. novyi*

artillery shells (Arnon et al., 2001). However, even this extremely dangerous compound is now extensively used in medical therapy and cosmetic applications. Blepharospasm (unvoluntary eye lid contractions), cervical dystonia (involuntary contraction of neck and shoulder muscles), and strabismus (crossed eyes) are uncontrolled muscle spasms that can be treated by injecting sublethal doses of toxin. Economically even more important are cosmetic applications. Commercial preparations of botulinum toxin ("Botox", "Dysport") are used in treatment of face ageing (relaxation of muscle contractions leading to frown lines and wrinkles) as well as hyperhidrosis and hypersalivation (Keller and Vann, 2004). The injected toxin (in very diluted concentration) is active for only a few months, then injection must be repeated. Seven serotypes of botulinum toxin are known. The proteolytic group I of *C. botulinum* produces serotypes A, B, and F (whose genes are all located on the chromosome); the nonproteolytic group II produces B, E, and F (also from chromosomally located genes); and group III produces C and D (encoded by bacteriophages). Group IV is actually *C. argentinense* and produces serotype G (from a plasmid-located gene). In addition, strains of *C. barati* can produce F and strains of *C. butyricum* can produce E (both from chromosomally located genes). The molecular action starts by lysis of bacterial cells, release of the progenitor toxin,

and cleavage of the large precursor protein into a heavy and a light chain (approximately 100 and 50 kDa, respectively). The light chain carries the metalloprotease domain. Linkage of the chains is achieved by a disulfide bond. The heavy chain binds then to receptors at the nerve terminal and the complex of heavy and light chain is internalized as an endosomal vesicle. Acidification of the endosome leads to a conformational change of the heavy chain and subsequent translocation of the light chain into the cytosol. There, several SNARE (soluble N-ethylmaleimide sensitive factor attachment protein receptor) protein family members are proteolytically inactivated. Substrate specificity depends on the serotype of botulinum toxin. Serotypes A, C, and E cleave SNAP-25; C cleaves syntaxin; and B, D, F, and G cleave VAMP (vesicle-associated membrane protein). Due to the large number of serotypes, vaccination is a problem. Only a pentavalent toxoid (A-E) is available. Current treatment is injection of equine antiserum for absorbing circulating toxin. Medical countermeasure for infant botulism is an antitoxin-based therapy with a novel immune globulin.

With tetanus toxin, the situation is different. The genome of the causative agent, *C. tetani*, has been sequenced (discussed later in this chapter), and the gene encoding the toxin (*tetX*) is located on the 74-kb plasmid pE88. Thus, only one serotype is known and vaccination is the preferred medical countermeasure. It is performed with a toxoid originating from formaldehyde-inactivated 150-kDa toxin. Tetanus toxin (TeTx) is less toxic than BoNT. The human lethal dose is estimated at about 1 ng kg^{-1}. The disease is a spastic paralysis and almost exclusively caused by infections of deep wounds. Thereby, blood vessels are disrupted and the tissue becomes hypoxic, allowing the endospores to germinate and the arising vegetative cells to produce toxin. The mode of action is similar to that of BoNT, except that TeTx does not act on the peripheral, but rather on the central nervous system. TeTx also is cleaved into a light and a heavy chain, which are then linked by a disulfide bond. The complex binds to the neuromuscular junction of motorneurons. After internalization, it is transported to the spinal cord and migrates into inhibitory interneuron terminals. There, the catalytic activity of the light chain blocks release of inhibitory neurotransmitters by cleaving VAMP (also designated synaptobrevin).

Collagenases also belong to the class of zinc-dependent proteases. The causative agent of gas gangrene and myonecrosis, *C. histolyticum*, produces six such enzymes, which are now used in medical therapies (see following paragraph). They have molecular masses of 68–125 kDa, are also called α-clostripain, and specifically cleave collagen and gelatin at the glycine residue of a PXGP motif. *C. perfringens* also produces a collagenase (ColA), which resembles ColG of *C. histolyticum*.

The paradigm of a glycosyltransferase producer is *C. difficile*, a major contributor to healthcare-associated illnesses. Due to their molecular mass of up to 308 kDa, these proteins are also designated as large clostridial cytotoxins (Barth and Aktories, 2005). *C. difficile* produces ToxA and ToxB, which use UDP-glucose as a cofactor and glycosylate G-proteins such as Rho, Rac, Cdc42, Rap, and Ral. Similar enzymes are synthesized by *C. novyi* (α-toxin uses UDP-N-acetyl-glucosamine as a cofactor) and *C. sordellii* (hemorrhagic and lethal toxins, the latter ones act additionally on Ras, but not on Rho). The toxins are bound by a receptor of the host cell and taken up by endocytosis. After processing (the location of this process still awaits elucidation), acidification of the endosome leads to secretion of the enzyme domain into the cytosol of the host cell, where it modifies its target proteins. Glycosylated G-proteins become inactive, and signal chains important for proliferation, cell differentiation, organization of actin cytoskeleton, enzyme activation, and transcription are blocked. *C. difficile* is remarkable in another aspect: bacterial pathogens are generally considered to grow only under heterotrophic conditions. However, *C. difficile* has been shown to represent the first pathogen able of autotrophy, possibly adding to the persistence of this bacterium in the infected human intestinal tract (Köpke et al., 2013).

ADP-ribosyltransferases target either actin or Rho (Barth and Aktories, 2005). The best studied example of the first group is the C2 toxin of *C. botulinum*. It comprises two proteins, and the group is therefore referred to as binary toxins. In addition to type C and D *C. botulinum*, such proteins are produced by *C. difficile* (CDT), *C. perfringens* (iota toxin), and *C. spiroforme* (spiroforme toxin). In order to become cytotoxic, the binding component of C2 (C2II) is cleaved at its N terminus (removal of an app. 20-kDa peptide) to create C2IIa. This component oligomerizes into heptamers, binds to a receptor, and attaches to the host cell membrane. The other component, C2I, docks to this complex, which is then translocated into the host cell by endocytosis. Acidification of the endosome leads to a conformational change of C2IIa, its insertion into the membrane, and finally the release of C2I into the cytosol of the host cell, where it ADP-ribosylates G-actin monomers and thus prevents their polymerization.

The paradigm of the second group is C3 toxin of *C. botulinum*. It lacks a cell binding and transport domain. It ADP-ribosylates Rho GTPases, thereby inactivating them. A similar toxin is found in *C. limosum*. Since free toxin is poorly taken up by eukaryotic cells, invasion of pathogen into the host and subsequent release of toxin is a prerequisite for cytotoxic action.

Phospholipases produced by clostridia all belong to class C (Titball and Tweten, 2005). Such enzymes have been detected in *C. absonum*, *C. barati*, *C. bifermentans*, *C. haemolyticum*, *C. novyi*, *C. perfringens*, and *C. sordellii*. As yet to be characterized, all cleave phosphatidylcholine and sphingomyelin. Diseases caused by these toxins are gas gangrene and necrotic enteritis. The latter occurs in farmed fowl and increased significantly during the past decade.

Biotechnology

An outstanding historical application of biotechnology employing clostridia was the acetone-butanol fermentation by *C. acetobutylicum* (discussed earlier in this chapter). Plant sizes were enormous, the largest facilities in the United States were located in Terre Haute, Indiana (52 fermenters) and Peoria, Illinois (96 fermenters with a capacity of 189,250-liters each) (Gabriel, 1928). South Africa operated a plant until 1982 in Germiston, consisting of 12 production fermenters with a working volume of 90,000-liters each (Jones, 2001). Facilities were also located in Australia, Brazil, Canada, China, Egypt, India, Japan, the former Soviet Union, and the United Kingdom (Jones and Woods, 1986). Production in China was only stopped in 2004 (Chiao and Sun, 2007), but soon after restarted and later stopped again. In addition to acetone and 1-butanol, 2-propanol (or isopropanol) is another solvent produced by clostridia, namely *C. aurantibutyricum*, *C. beijerinckii*, *C. puniceum*, and *C. roseum* (Dürre and Bahl, 1996; Poehlein et al., 2017). Due to increasing crude oil prices, the fermentative production of butanol is again able to compete economically with the petrochemical process. Consequently, BP and DuPont formed a joint venture, Butamax Advanced Biofuels, aiming at producing biobutanol. A number of other companies are also actively pursuing the biotechnological production, especially focusing on substrates that do not compete for human nutrition (e. g., lignocellulosic hydrolysates) (Schiel-Bengelsdorf et al., 2013). In addition to its being an important bulk chemical for industrial purposes, butanol can also serve as a biofuel (Dürre, 2005; 2007b). Mixtures with gasoline are superior to those with ethanol, as butanol has a lower vapor pressure and a higher energy content, can be blended at higher concentrations than ethanol, and does not require specific adjustments of vehicle and engine technologies.

The clostridial metabolic pathways leading to solvent production are tightly coupled to regulation of endospore formation (Paredes et al., 2005, Dürre, 2005). This makes sense, as formation of acids during exponential growth leads to acidification of the surrounding environment. Anaerobic bacteria are unable to keep their internal pH constant, so it will drop in parallel with the external one, but being more alkaline by about 1 pH unit (Dürre et al., 1988). Reaching an external pH of about 4, undissociated acid from the outside will diffuse into the cytoplasm and will dissociate there. As a consequence, the proton gradient over the membrane will collapse, and the cell will die. Converting acids into neutral products will avoid this lethal effect, and thus respective species have an ecological advantage. However, since butanol in higher concentrations is also toxic, sporulation must be initiated at the same time as solvent formation to guarantee long-time survival. Overall, the cells thus manage to stay metabolically active for a longer period and to postpone entering a dormant stage.

Enzymes required for solventogenesis are coenzyme A transferase, butyraldehyde and butanol dehydrogenases for butanol formation, as well as coenzyme A transferase and acetoacetate decarboxylase for acetone production (an additional alcohol dehydrogenase allows 2-propanol synthesis from acetone in *C. beijerinckii*). In *C. acetobutylicum*, four dehydrogenases are known. Butanol dehydrogenases A and B are encoded by chromosomal genes, forming a monocistronic operon each (Walter et al., 1992). BdhA seems to function as an electron sink, removing excess reducing equivalents by alcohol formation, while BdhB plays a major role in massive butanol production (Sauer and Dürre, 1995). Initiation of solventogenesis is mediated by AdhE, a bifunctional butyraldehyde/butanol dehydrogenase, whose gene is located on a large plasmid (pSOL1, 192 kbp) and forms a common operon (*sol*) with two genes encoding the subunits of coenzyme A transferase. Directly downstream, but with convergent direction of transcription, lies a monocistronic operon encoding the gene for acetoacetate decarboxylase (*adc*). This arrangement is unique, as in other solventogenic species such as *C. beijerinckii*, *C. saccharobutylicum*, and *C. saccharoperbutylacetonicum* the *sol* operon consists of genes encoding butyraldehyde dehydrogenase, coenzyme A transferase, and acetoacetate decarboxylase (Dürre, 2005; Schiel-Bengelsdorf et al., 2013; Poehlein et al., 2017). Another *adhE* gene (*adhE2*) is also located on plasmid pSOL1. The signal(s) responsible for inducing solventogenesis (and also sporulation) have not been unequivocally identified, although pH, concentration of undissociated acids, metabolic intermediates, cofactors, and salts, as well as DNA supercoiling all play a role (Dürre, 1998; Girbal and Soucaille, 1998). The master regulator of sporulation, Spo0A, which in contrast to *Bacillus* species is directly phosphorylated and not by a phosphorelay (Steiner et al., 2011), also acts as a transcription factor during onset of solventogenesis. In addition to other regulatory proteins, RNA processing, a small noncoding regulatory RNA, and protein modification represent further levels of regulation (Dürre, 2005; 2011). The use of DNA microarrays allowed a new approach for studying this complex regulatory network (Tummala et al., 2005; Grimmler et al., 2011) and, consequently, construction of production strains with superior properties (Dürre, 2005; Tomas et al., 2005).

The emerging field of systems biology offers another possibility of improving performance of the acetone-butanol fermentation. Modelling of the metabolism of *C. acetobutylicum* can be based on enzymatic, transcriptome, proteome, and metabolome data or combinations thereof. Characterization of the quantitative dynamics of metabolic processes and regulatory networks within a cell will allow predictions of behavior under altered

conditions and thus improvement of productivity of butanol. Several such models have meanwhile been reported (Shinto et al., 2007; Senger and Papoutsakis, 2008a; 2008b; Jabbari et al., 2013; Millat et al., 2013a), but none does yet fully incorporate all important parameters such as pH-dependency of the shift and the whole repertoire of controlling mechanisms. An interesting aspect is the influence of population dynamics, which might result in a switch of an acidogenic to an solventogenic subpopulation (Millat et al., 2013b).

As already mentioned, industrial fermentation processes leading to large-value and low-price bulk chemicals are hampered, when using substrates that compete with human nutrition (e.g., sugar-or starch-containing materials). This raised the so-called food-versus-fuel debate. One way of avoiding this problem is the use of lignocellulosic hydrolysates. However, such hydrolysates require collection of biomass and costly pretreatments and may thus not always be applicable. Therefore, a very promising approach is the use of autotrophic clostridia that capture greenhouse gases such as CO or $CO_2 + H_2$ and use them as sole carbon and energy source, thus combining use of an abundant and cheap substrate with climate protection. Synthesis gas (Syngas) is a mixture of mostly CO and H_2 and often a waste product from the chemical industry or steel mills. It can also be easily generated from biomass. Syngas is already a major feedstock in the chemical industry, thus, its handling represents no problem. *C. autoethanogenum* or *C. ljungdahlii* form acetate and/or ethanol from syngas. Gas fermentation to ethanol has passed the pilot and demonstration scale (Köpke et al., 2011a) and the Chicago (U.S.)-based company LanzaTech has built a commercial size production plant (46,000 t/a) in China, which started operation in 2018. Instead of hydrogen, reducing power could also be provided by electricity. When electrons stem from photovoltaics, this could be even considered a new form of photosynthesis (Nevin et al., 2010). In addition to acetate and ethanol, 2,3-butanediol is also formed naturally by these acetogens (Köpke et al., 2011b). Recombinant strain construction allowed to expand the product repertoire, with *C. ljungdahlii* becoming a butanol producer (Köpke et al., 2010), *C. aceticum* an acetone producer (Schiel-Bengelsdorf and Dürre, 2012), and *C. autoethanogenum* and others an acetone and/or isopropanol producer (Chen et al., 2012).

Because some clostridia can degrade halogenated and hazardous compounds partially or completely, they may also be used in bioremediation processes. Examples are lindane (hexachlorocyclohexane, used as an insecticide), which is completely dechlorinated by *C. butyricum*; PCBs (polychlorinated biphenyls, used in insulating fluids), which is metabolized by *C. hydroxybenzoicum*, and TNT (trinitrotoluene, used as an explosive), which is reduced to triaminotoluene (Bahl and Dürre, 1993, Ahmad et al., 2005).

A medical application of a clostridial enzyme is wound debridement (Brett, 2005). Chronic, nonhealing wounds require removal of necrotic tissue, which is anchored by collagen to the wound surface, to achieve successful curing. Such treatment might become necessary in case of severe burns or dermal ulcer. One of the agents used for this goal is collagenase from *C. histolyticum*, of which six different such proteins are known. They cleave interstitial collagens into small peptides, mostly tripeptides. Collagen accounts for approximately 75% of the skin tissue (dry weight), and the enzymes act at physiological pH and temperature. Commercial preparations of collagenase (e.g., "Santyl," "Iruxol," or "Novuxol") are therefore particularly useful in the removal of detritus and contribute toward the formation of granulation tissue and subsequent epithelization. Collagen in healthy tissue or in newly formed granulation tissue is not attacked.

Oncolysis caused by clostridia can be used for treatment of solid cancers (Minton, 2003). It has been successfully tested at the laboratory stage with animals and is about to enter clinical trials (Brown and Liu, 2004). As clostridia perform an anaerobic lifestyle, their spores will only germinate in an anaerobic or at least hypoxic environment. This is a situation normally not occurring in mammals. Consequently, injected clostridial spores are cleared from blood within a few days (Brown and Liu, 2004). However, the environment of solid tumors in a mammal is hypoxic, thus allowing spores reaching such an area to germinate (only in this place, which is a very specific targeting) and to produce vegetative cells, which are able to multiply in that region. The bacteria proliferate at the expense of only necrotic tissue, as healthy, oxygenated tissue does not allow colonization. Recombinant spores will give rise to recombinant clostridia, which are able to produce specific proteins at the tumor. Examples are cytosine deaminase, carboxypeptidase G2, and nitroreductase that are all able to convert a harmless prodrug, injected into and delivered by the bloodstream, into an effective anticancer agent, which is generated only at the tumor and causes oncolysis (Minton, 2003). Another possibility is the transformation and expression of genes, which encode cytotoxic proteins. After secretion, the gene products will then be able to exert antitumor activity. Currently investigated are tumor necrosis factor α (TNFα) (van Mellaert et al., 2005) and, especially for treatment of pancreatic cancer, *C. perfringens* enterotoxin (CPE) (Dürre, 2007a). The direct use of *C. novyi*-NT (nontoxic) spores in combination with radiation therapy or as bacteriolytic therapy is another promising approach (Bettegowda et al., 2003; Maletzki et al., 2010).

Handling and Molecular Biology Techniques

Due to the large number of species in the genus *Clostridium*, almost the entire range of microbial metabolic properties is represented (Bahl and Dürre, 1993, Dürre, 2007a). Clostridia are found within acidophiles, neutrophiles, and alkaliphiles (thus covering the pH range from 4 to 10.5) as well as in psychrophiles, mesophiles, and thermophiles (thus covering most of the temperature range). Although in general being obligate anaerobes, some species are able to detoxify oxygen (to varying extent) by employing superoxide reductase, peroxidases, and/or superoxide dismutase (Dürre, 2001). In addition to using specific substrates already mentioned, clostridia can be grown in complex media commercially available (e.g., CMC, chopped meat-carbohydrate; PYG, peptone-yeast extract-glucose). Detailed media compositions are provided by culture collections distributing the species and in species-specific original publications.

The technique originally invented by Robert Hungate allows easy handling of clostridia by using anaerobic and prereduced media with help of titanium(III) citrate, a mixture of sodium sulfide and cysteine-HCl, or sodium thioglycolate (Breznak and Costilow, 1994). Media are boiled to remove oxygen and cooled under a constant flow of nitrogen. After addition of the reducing agents and a redox indicator (such as resorufin) for visual detection of potential oxygen contamination, the liquid is poured into tubes or bottles, and then a butyl rubber stopper is secured in place. After sterilization, inoculation or sampling can easily be performed with sterile syringes and needles. Solid media can be poured and streaked in an anaerobic chamber.

Many clostridial genomes have meanwhile been completely sequenced, many more are in various stages of completion. Mining of this information is in progress, one of the first general conclusions that can be drawn is that initiation of endospore formation is not triggered by a phosphorelay as in bacilli (Dürre and Hollergschwandner, 2004). Plasmids are quite common in clostridia, with about a third of the investigated species containing extrachromosomal elements (Lee et al., 1987). A number of shuttle vectors have been constructed, in general making use of erythromycin and chloramphenicol resistance genes (Tummala et al., 2001; Davis et al., 2005; Heap et al., 2009). However, in practice, these two antibiotics can be inactivated by clostridia so that clarithromycin (a pH-stable derivative of erythromycin) and thiamphenicol (a chloramphenicol derivative without the reducible nitro group) are employed. Transformation of free DNA into clostridia can be achieved by electroporation, and conjugation (direct cell contact with a donor cell) is also possible (Davis et al., 2005). Shelter against host restriction endonucleases can be artificially provided by a *Bacillus* phage methylase expressed in the intermediate host *E. coli* (Mermelstein and Papoutsakis, 1993).

Random mutagenesis has been tried with *C. acetobutylicum* and *C. tetani* using conjugative transposons from streptococci such as Tn*916*, Tn*925*, and Tn*1545* (for a review, see Dürre, 1993). Several mutants have been obtained, but because these elements prefer AT-rich regions found especially in promoters, structural gene inactivations are not frequent. However, based on the *mariner* transposon, an effective system for *in vivo* random mutagenesis of *C. difficile* has been developed (Cartman and Minton, 2010), which also might be transferable to other clostridia. Natural transposons are also found in clostridia. A conjugative element of 20.7 kb (Tn*5397*) is present in *C. difficile*, integrative mobilizable elements of 6.3 kb (Tn*4451* and Tn*4453*) have been detected in *C. perfringens* and *C. difficile*, a transmobilizable element of 9.6 kb (Tn*5398*) was found in *C. difficile*, and a compound transposon of 6.3 kb (Tn*5565*) has been described for *C. perfringens* (Lyras and Rood, 2005). Chemical mutagenesis by nitrite is not possible, as this compound destroys essential iron-sulfur centers, as mentioned before. However, a targeted gene knock out system with a high success rate has been developed, which is based on the use of a mobile group II intron from *Lactococcus lactis* ("ClosTron," Heap et al., 2010; Kuehne and Minton, 2012). The system can also be used for cargo DNA delivery (Kuehne et al., 2011). Integration of DNA into clostridial chromosomes could also be achieved without a counter-selection marker (Heap et al., 2012). Modulation of gene expression is possible by means of antisense RNA constructs (Tummala et al., 2005). In *C. saccharobutylicum*, a naturally expressed asRNA participates in regulation of nitrogen assimilation. The 43-bp transcript is complementary to a region at the start of the *glnA* gene (encoding glutamine synthetase) and thus inhibits translation during nitrogen rich conditions (Fierro-Monti et al., 1992). The toolbox for genetic manipulation of clostridia also contains several well-suited reporter gene systems. Successful use of *catP* (encoding chloramphenicol acetyltransferase), *eglA* (encoding β-1,4-endoglucanase), *gusA* (encoding β-glucuronidase), *lacZ* (encoding β-galactosidase), and *luxAB* or *lucB* (encoding luciferase) has been reported (Tummala et al., 2001; Feustel et al., 2004). The latter one is especially well suited, as no interfering metabolic equivalent is known from clostridia. In summary, all technologies required for genetic analysis and manipulation of members of this large bacterial genus are at hand.

Acknowledgements

Work in my laboratory was supported by grants from the BMBF GenoMikPlus program (Competence Network Göttingen), the transnational BMBF SysMO2 program (project COSMIC2), the ERA-Net IB3 project REACTIF, and the BMBF project Gasfermentation.

REFERENCES

Ahmad, F., J. B. Hughes, and G. N. Bennett. 2005. Biodegradation of hazardous materials by clostridia. In *Handbook on Clostridia*, ed. P. Dürre, 831–854, Boca Raton: CRC Press, Taylor & Francis Group.

Angert, E. R., K. D. Clements, and N. R. Pace. 1993. The largest bacterium. *Nature* 362:239–241.

Angert, E. R., A. E. Brooks, and N. R. Pace. 1996. Phylogenetic analysis of *Metabacterium polyspora*: clues to the evolutionary origin of daughter cell production in *Epulopiscium* species, the largest bacteria. *Journal of Bacteriology* 178:1451–1456.

Arnon, S. S. 1997. Human tetanus and human botulism. In *The Clostridia: Molecular Biology and Pathogenesis*, ed. J. I. Rood, B. A. McClane, J. G. Songer and R. W. Titball, 95–115, San Diego: Academic Press.

Arnon, S. S., R. Schechter, T. V. Inglesby, et al. 2001. Botulinum toxin as a biological weapon: medical and public health management. *The Journal of the American Medical Association* 285:1059–1070.

Bahl, H., and P. Dürre. 1993. Clostridia. In *Biotechnology*, 2nd edition, vol. 1, ed. H. Sahm, 285–323, Weinheim: VCH Verlagsgesellschaft mbH.

Bahl, H., and P. Dürre. 2001. Clostridia. *Biotechnology and Medical Applications*. Weinheim: Wiley-VCH.

Barth, H., and K. Aktories. 2005. Clostridial cytotoxins. In *Handbook on Clostridia*, ed. P. Dürre, 407–449, Boca Raton: CRC Press, Taylor & Francis Group.

Bettegowda, C., L. H. Dang, R. Abrams, D. L. Huso, L. Dillehay, I. Cheong, N. Agrawal, S. Borzillary, J. M. McCaffery, L. Watson, K.-S. Lin, F. Bunz, K. Baidoo, M. G. Pomper, K. W. Kinzler, B. Vogelstein, and S. Zhou. 2003. Overcoming the hypoxic barrier to radiation therapy with anaerobic bacteria. *Proceedings of the National Academy of Sciences United States of America* 100:15083–15088.

Biegel, E., and V. Müller. 2010. Bacterial Na$^+$-translocating ferredoxin:NAD$^+$ oxidoreductase. *Proceedings of the National Academy of Sciences United States of America* 107:18138–18142.

Braun, M., F. Mayer, and G. Gottschalk. 1981. *Clostridium aceticum* (Wieringa), a microorganism producing acetic acid from molecular hydrogen and carbon dioxide. *Archives of Microbiology* 128:288–293.

Brett, D. W. 2005. Clostridial collagenase in wound repair. In *Handbook on Clostridia*, ed. P. Dürre, 855–876, Boca Raton: CRC Press, Taylor & Francis Group.

Breznak, J. A., and R. N. Costilow. 1994. Physicochemical factors in growth. In *Methods for General and Molecular Bacteriology*, ed. P. Gerhardt, R. G. E. Murray, W. A. Wood and N. R. Krieg, 137–154, Washington, DC: American Society for Microbiology.

Brown, J. M., and S.-C. Liu. 2004. Use of anaerobic bacteria for cancer therapy. In *Strict and Facultative Anaerobes. Medical and Environmental Aspects*, ed. M. M. Nakano and P. Zuber, 211–219, Wymondham: Horizon Bioscience.

Buckel, W. 1990. Amino acid fermentations: coenzyme B$_{12}$-dependent and –independent pathways. In *The Molecular Basis of Bacterial Metabolism*, ed. G. Hauska and R. Thauer, 21–30, Heidelberg: Springer-Verlag.

Buckel, W., and R. K. Thauer. 2018. Flavin-based electron bifurcation, ferredoxin, flavodoxin, and anaerobic respiration with protons (Ech) or NAD$^+$ (Rnf) as electron acceptors: a historical review. *Frontiers in Microbiology* 9:401.

Carpenter, C. E., D. S. A. Reddy, and D. P. Cornforth. 1987. Inactivation of clostridial ferredoxin and pyruvate-ferredoxin oxidoreductase by sodium nitrite. *Applied and Environmental Microbiology* 53:549–552.

Cartman, S. T., and N. P. Minton. 2010. A *mariner*-based transposon system for *in vivo* random mutagenesis of *Clostridium difficile*. *Applied and Environmental Microbiology* 76:1103–1109.

Chen, W. Y., M. Koepke, F. Liew, and S. D. Simpson. 2012. Recombinant microorganisms and uses therefor. *Patent* WO 2012115527 A2.

Chiao, J.-S., and Z.-H. Sun. 2007. History of the acetone-butanol-ethanol fermentation industry in China: development of continuous production technology. *Journal of Molecular Microbiology and Biotechnology* 13:12–14.

Cruz-Morales, P., C. A. Orellana, G. Moutafis, G. Moonen, G. Rincon, L. K. Nielsen, and E. Marcellin. 2019. Revisiting the evolution and taxonomy of Clostridia, a phylogenomic update. *Genome Biology and Evolution* 11:2035–2044.

Das, A., and L. G. Ljungdahl. 2003. Electron-transport system in acetogens. In *Biochemistry and Physiology of Anaerobic Bacteria*, eds. L. G. Ljungdahl, M. Adams, L. Barton, J. Ferry and M. Johnson, 191–204, New York: Springer.

Davis, I. J., G. Carter, M. Young, and N. P. Minton. 2005. Gene cloning in clostridia. In *Handbook on Clostridia*, ed. P. Dürre, 37–52, Boca Raton: CRC Press, Taylor & Francis Group.

Dürre, P., and J. R. Andreesen. 1982. Selenium-dependent growth and glycine fermentation by *Clostridium purinolyticum*. *Journal of General Microbiology* 128:1457–1466.

Dürre, P., and J. R. Andreesen. 1983. Purine and glycine metabolism by purinolytic clostridia. *Journal of Bacteriology* 154:192–199.

Dürre, P., H. Bahl, and G. Gottschalk. 1988. Membrane processes and product formation in anaerobes. In *Handbook on Anaerobic Fermentations*, ed. L. E. Erickson and D. Y.-C. Fung. 187–206, New York: Marcel Dekker Inc.

Dürre, P. 1993. Transposons in clostridia. In *The Clostridia and Biotechnology*, ed. D. R. Woods, 227–246, Stoneham: Butterworth-Heinemann.

Dürre, P., and H. Bahl. 1996. Microbial production of acetone/butanol/isopropanol. In *Biotechnology*, 2nd ed, vol. 6, ed. M. Roehr, 229–268, Weinheim: VCH Verlagsgesellschaft mbH.

Dürre, P. 1998. New insights and novel developments in clostridial acetone/butanol/isopropanol fermentation. *Applied Microbiology and Biotechnology* 49:639–648.

Dürre, P. 2001. From Pandora's box to cornucopia: clostridia-a historical perspective. In *Clostridia. Biotechnology and Medical Applications*, ed. H. Bahl and P. Dürre, 1–17, Weinheim: Wiley-VCH.

Dürre, P., and C. Hollergschwandner. 2004. Initiation of endospore formation in *Clostridium acetobutylicum*. *Anaerobe* 10:69–74.

Dürre, P. 2005. *Handbook on Clostridia*. Boca Raton: CRC Press, Taylor & Francis Group.

Dürre, P. 2005. Formation of solvents in clostridia. In *Handbook on Clostridia*, ed. P. Dürre, 671–693, Boca Raton: CRC Press, Taylor & Francis Group.

Dürre, P. 2007a. Clostridia. *Encyclopedia of Life Sciences*, http://www.els.net/

Dürre, P. 2007b. Biobutanol: an attractive biofuel. *Biotechnological Journal* 2:1525–1534.

Dürre, P. 2011. Fermentative production of butanol — the academic perspective. *Current Opinion in Biotechnology* 22:331–336.

Dürre, P. 2014. *Systems Biology of Clostridium*. London: Imperial College Press.

Ewerdwalbesloh, I., and O. Meyer. 1995. Bacteriology and enzymology of the Woad fermentation. *Proceedings of the 2nd International Symposium on Woad, Indigo and other Natural Dyes: Past, Present and Future, Toulouse.*

Feustel, L., S. Nakotte, and P. Dürre. 2004. Characterization and development of two reporter gene systems for *Clostridium acetobutylicum*. *Applied and Environmental Microbiology* 70:798–803.

Fierro-Monti, I. P., S. J. Reid, and D. R. Woods. 1992. Differential expression of a *Clostridium acetobutylicum* antisense RNA: implications for regulation of glutamine synthetase. *Journal of Bacteriology* 174:7642–7647.

Gabriel, C. L. 1928. Butanol fermentation process. *Industrial and Engineering Chemistry* 28:1063–1067.

Girbal, L., and P. Soucaille. 1998. Regulation of solvent production in *Clostridium acetobutylicum*. *Trends in Biotechnology* 16:11–16.

Grimmler, C., H. Janssen, D. Krauße, R.-J. Fischer, H. Bahl, P. Dürre, W. Liebl, and A. Ehrenreich. 2011. Genome-wide gene expression analysis of the switch between acidogenesis and solventogenesis in continuous cultures of *Clostridium acetobutylicum*. *Journal of Molecular Microbiology and Biotechnology* 20:1–15.

Heap, J. T., O. J. Pennington, S. T. Cartman, and N. P. Minton. 2009. A modular system for *Clostridium* shuttle plasmids. *Journal of Microbiological Methods* 78:79–85.

Heap, J. T., S. A. Kuehne, M. Ehsaan, S. T. Cartman, C. M. Cooksley, J. C. Scott, and N. P. Minton. 2010. The ClosTron: mutagenesis in *Clostridium* refined and streamlined. *Journal of Microbiological Methods* 80:49–55.

Heap, J. T., M. Ehsaan, C. M. Cooksley, Y. K. Ng, S. T. Cartman, K. Winzer, and N. P. Minton. 2010. Integration of DNA into bacterial chromosomes from plasmids without a counter-selection marker. *Nucleic Acids Research* 40:e59.

Herrmann, G., E. Jayamani, G. Mai, and W. Buckel. 2008. Energy conservation via electron-transferring flavoprotein in anaerobic bacteria. *Journal of Bacteriology* 190:784–791.

Hoch, J. A. 1993. Regulation of the onset of the stationary phase and sporulation in *Bacillus subtilis*. *Advances in Microbial Physiology* 35:111–133.

Jabbari, S., E. Steiner, J. T. Heap, K. Winzer, N. P. Minton, and J. R. King. 2013. The putative influence of the *agr* operon upon survival mechanisms used by *Clostridium acetobutylicum*. *Mathematical Biosciences* 243:223–239.

Johnson, E. A. 2005. Clostridial neurotoxins. In *Handbook on Clostridia*, ed. P. Dürre, 491–525, Boca Raton: CRC Press, Taylor & Francis Group.

Jones, D. T., and D. R. Woods. 1986. Acetone-butanol fermentation revisited. *Microbiological Reviews* 50:484–524.

Jones, D. T. 2001. Applied acetone-butanol fermentation. In *Clostridia. Biotechnology and Medical Applications*, ed. H. Bahl and P. Dürre, 125–168, Weinheim: Wiley-VCH.

Keller, J. E., and W. F. Vann. 2004. Botulinum toxin. *Encyclopedia of Life Sciences*, http://www.els.net/.

Kiple, K. F. 1993. *The Cambridge World History of Human Disease*. Cambridge: Cambridge University Press.

Köpke, M., C. Held, S. Hujer, H. Liesegang, A. Wiezer, A. Wollherr, A. Ehrenreich, W. Liebl, G. Gottschalk, and P. Dürre. 2010. *Clostridium ljungdahlii* represents a microbial production platform based on syngas. *Proceedings of the National Academy of Sciences United States of America* 107:13087–13092.

Köpke, M., C. Mihalcea, J. C. Bromley, and S. D. Simpson. 2011a. Fermentative production of ethanol from carbon monoxide. *Current Opinion in Biotechnology* 22:320–325.

Köpke, M., C. Mihalcea, F. Liew, J. H. Tizard, M. S. Ali, J. J. Conolly, B. Al-Sinawi, and S. D. Simpson. 2011b. 2,3-Butanediol production by acetogenic bacteria, an alternative route to chemical synthesis, using industrial waste gas. *Applied and Environmental Microbiology* 77:5467–5475.

Köpke, M., M. Straub, and P. Dürre. 2013. *Clostridium difficile* is an autotrophic bacterial pathogen. *PLOS ONE* 8:e62157.

Kuehne, S. A., J. T. Heap, C. M. Cooksley, S. T. Cartman, and N. P. Minton. 2011. ClosTron-mediated engineering of *Clostridium*. *Methods in Molecular Biology* 765:389–407.

Kuehne, S. A., and N. P. Minton. 2012. ClosTron-mediated engineering of *Clostridium*. *Bioengineered* 3:247–254.

Lee, C.-K., P. Dürre, H. Hippe, and G. Gottschalk. 1987. Screening for plasmids in the genus *Clostridium*. *Archives of Microbiology* 148:107–114.

Li, F., J. Hinderberger, H. Seedorf, J. Zhang, W. Buckel, and R. K. Thauer. 2008. Coupled ferredoxin and crotonyl coenzyme A (CoA) reduction with NADH catalyzed by the butyryl-CoA dehydrogenase/Etf complex from *Clostridium kluyveri*. *Journal of Bacteriology* 190:843–850.

Lund, B. M., and M. W. Peck. 2000. *Clostridium botulinum*. In *The Microbiological Safety and Quality of Foods*, ed. B. M. Lund, A. C. Baird-Parker, and G. W. Gould, 1057–1109, Gaithersburg: Aspen Publ. Inc.

Lyras, D., and J. I. Rood. 2005. Transposable genetic elements of clostridia. In *Handbook on Clostridia*, ed. P. Dürre, 631–643, Boca Raton: CRC Press, Taylor & Francis Group.

Maletzki, C., M. Gock, U. Klier, E. Klar, and M. Linnebacher. 2010. Bacteriolytic therapy of experimental pancreatic carcinoma. *World Journal of Gastroenterology* 16:3546–3552.

McClane, B. A. 2005. Clostridial enterotoxins. In *Handbook on Clostridia*, ed. P. Dürre, 385–406, Boca Raton: CRC Press, Taylor & Francis Group.

Mermelstein, L. D., and E. T. Papoutsakis. 1993. *In vivo* methylation in *Escherichia coli* by the *Bacillus subtilis* phage ϕ3T I methyltransferase to protect plasmids from restriction upon transformation of *Clostridium acetobutylicum* ATCC 824. *Applied and Environmental Microbiology* 59:1077–1081.

Millat, T., H. Janssen, H. Bahl, R.-J. Fischer, and O. Wolkenhauer. 2013a. Integrative modelling of pH-dependent enzyme activity and transcriptomic regulation of the acetone-butanol-ethanol fermentation of *Clostridium acetobutylicum* in continuous culture. *Microbial Biotechnology* 6:526–539.

Millat, T., H. Janssen, G. T. Thorn, J. R. King, H. Bahl, R.-J. Fischer, and O. Wolkenhauer. 2013b. A shift in the dominant phenotype governs the pH-induced metabolic switch of *Clostridium acetobutylicum* in phosphate-limited continuous cultures. *Applied Microbiology Biotechnology* 97:6451–6466.

Minton, N. P., and D. J. Clarke. 1989. *Clostridia*. New York: Plenum Press.

Minton, N. P. 2003. Clostridia in cancer therapy. *Nature Reviews in Microbiology* 1:237–242.

Nevin, K. P., T. L. Woodard, A. E. Franks, Z. M. Summers, and D. R. Lovley. 2010. Microbial electrosynthesis: feeding microbes electricity to convert carbon dioxide and water to multicarbon extracellular organic compounds. *mBio* 1(2):e00103–10.

Padden, A. N., V. M. Dillon, P. John, J. Edmonds, M. D. Collins, and N. Alvarez. 1998. *Clostridium* used in medieval dyeing. *Nature* 396:225.

Padden, A. N., V. M. Dillon, J. Edmonds, M. D. Collins, N. Alvarez, and P. John. 1999. An indigo-reducing moderate thermophile from a woad vat, *Clostridium isatidis* sp. nov. *International Journal of Systematic Bacteriology* 49:1025–1031.

Paredes, C. J., K. V. Alsaker, and E. T. Papoutsakis. 2005. A comparative genomic view of clostridial sporulation and physiology. *Nature Reviews in Microbiology* 3:969–978.

Pasteur, L. 1861. Animacules infusoires vivant sans gaz oxygène libre et déterminant des fermentations. *Comptes Rendus Hebdomadaires des Séances de l'Académie des Sciences* 52:344–347.

Poehlein, A., J. D. Montoya Solano, S. K. Flitsch, P. Krabben, K. Winzer, S. R. Reid, D. T. Jones, E. Green, N. P. Minton, R. Daniel, and P. Dürre. 2017. Microbial solvent formation revisited by comparative genome analysis. *Biotechnology for Biofuels* 10:58.

Popoff, M. R., and B. G. Stiles. 2005. Clostridial toxins vs. other bacterial toxins. In *Handbook on Clostridia*, ed. P. Dürre, 323–383, Boca Raton: CRC Press, Taylor & Francis Group.

Prazmowski, A. 1880. Untersuchungen über die Entwickelungsgeschichte und Fermentwirkung einiger Bacterien-Arten. Ph. D. thesis, University of Leipzig.

Reddy, D., L. R. Lancaster, Jr., and D. P. Cornforth. 1983. Nitrite inhibition of *Clostridium botulinum*: electron spin resonance detection of iron-nitric oxide complexes. *Science* 221:769–770.

Rood, J. I., B. A. McClane, J. G. Songer, and R. W. Titball. 1997. *The Clostridia: Molecular Biology and Pathogenesis*. San Diego: Academic Press.

Sauer, U., and P. Dürre. 1995. Differential induction of genes related to solvent formation during the shift from acidogenesis to solventogenesis in continuous culture of *Clostridium acetobutylicum*. *FEMS Microbiology Letters* 125:115–120.

Schiel-Bengelsdorf, B., and P. Dürre. 2012. Pathway engineering and synthetic biology using acetogens. *Federation of European Biochemical Societies Letters* 586:2191–2198.

Schiel-Bengelsdorf, B., J. Montoya, S. Linder, and P. Dürre. 2013. Butanol fermentation. *Environmental Technology* 34(13–14):1691–710.

Schuchmann, K., and V. Müller. 2014. Autotrophy at the thermodynamic limit of life: a model for energy conservation in acetogenic bacteria. *Nature Reviews Microbiology* 12:809–821.

Senger, R. S., and E. T. Papoutsakis. 2008a. Genome-scale model for *Clostridium acetobutylicum*: Part I. Metabolic network resolution and analysis. *Biotechnology and Bioengineering* 101:1036–1052.

Senger, R. S., and E. T. Papoutsakis. 2008b. Genome-scale model for *Clostridium acetobutylicum*: Part II. Development of specific proton flux states and numerically determined sub-systems. *Biotechnology and Bioengineering* 101:1053–1071.

Shinto, H., Y. Tashiro, M. Yamashita, G. Kobayashi, T. Sekiguchi, T. Hanai, Y. Kuriya, M. Okamoto, and K. Sonomoto. 2007. Kinetic modeling and sensitivity analysis of acetone-butanol-ethanol production. *Journal of Biotechnology* 131:45–56.

Siunov, A. V., D. V. Nikitin, N. E. Suzina, V. V. Dimitriev, N. P. Kuzmin, and V. I. Duda. 1999. Phylogenetic status of *Anaerobacter polyendosporus*, an anaerobic, polysporic bacterium. *International Journal of Systematic Bacteriology* 49:1119–1124.

Stackebrandt, E., and H. Hippe. 2001. Taxonomy and systematics. In *Clostridia. Biotechnology and Medical Applications*, ed. H. Bahl and P. Dürre, 19–48, Weinheim: Wiley-VCH.

Stackebrandt, E. 2004. The phylogeny and classification of anaerobic bacteria. In *Strict and Facultative Anaerobes. Medical and Environmental Aspects*, ed. M. M. Nakano and P. Zuber, 1–25, Wymondham: Horizon Bioscience.

Steiner, E., A. E. Dago, D. I. Young, J. T. Heap, N. P. Minton, J. A. Hoch, and M. Young. 2011. Multiple orphan histidine kinases interact directly with Spo0A to control the initiation of endospore formation in *Clostridium acetobutylicum*. *Molecular Microbiology* 80:641–654.

Sussman, M. 1958. A description of *Clostridium histolyticum* gasgangrene in The Epidemics of Hippocrates. *Medical History* 2:226.

Tamburini, E., A. G. Leon, B. Perito, and G. Mastromei. 2003. Characterization of bacterial pectinolytic strains involved in the water retting process. *Environmental Microbiology* 5:730–736.

Titball, R. W., and R. K. Tweten. 2005. Membrane active toxins. In *Handbook on Clostridia*, ed. P. Dürre, 451–489, Boca Raton: CRC Press, Taylor & Francis Group.

Tomas, C. A., S. B. Tummala, and E. T. Papoutsakis. 2005. Metabolic engineering of solventogenic clostridia. In *Handbook on Clostridia*, ed. P. Dürre, 813–830, Boca Raton: CRC Press, Taylor & Francis Group.

Tremblay, P. L., T. Zhang, S. A. Dar, C. Leang, and D. R. Lovley. 2012. The Rnf complex of *Clostridium ljungdahlii* is a proton-translocating ferredoxin:NAD$^+$ oxidoreductase essential for autotrophic growth. *mBio* 4:e00406–12.

Tummala, S. B., C. Tomas, L. M. Harris et al. 2001. Genetic tools for solventogenic clostridia. In *Clostridia. Biotechnology and Medical Applications*, ed. H. Bahl and P. Dürre, 105–123, Weinheim: Wiley-VCH.

Tummala, S. B., C. A. Tomas, and E. T. Papoutsakis. 2005. Gene analysis of clostridia. In *Handbook on Clostridia*, ed. P. Dürre, 53–70, Boca Raton: CRC Press, Taylor & Francis Group.

van Mellaert, L., J. Theys, O. Pennington et al. 2005. Clostridia as production systems for prokaryotic and eukaryotic proteins of therapeutic value in tumor treatment. In *Handbook on Clostridia*, ed. P. Dürre, 877–893, Boca Raton: CRC Press, Taylor & Francis Group.

Vogels, G. D., and C. van der Drift. 1976. Degradation of purines and pyrimidines by microorganisms. *Bacteriological Reviews* 40:403–468.

Walter, K. A., G. N. Bennett, and E. T. Papoutsakis. 1992. Molecular characterization of two *Clostridium acetobutylicum* ATCC 824 butanol dehydrogenase isozyme genes. *Journal of Bacteriology* 174:7149–7158.

Weizmann, C. 1915. Improvements in the bacterial fermentation of carbohydrates and in bacterial cultures for the same. *British Patent* 4945.

Wiegel, J., R. Tanner, and F. A. Rainey. 2006. An introduction to the family Clostridiaceae. In *The Prokaryotes-A Handbook on the Biology of Bacteria*, vol. 4, ed. M. Dworkin, S. Falkow, E. Rosenberg, K.-H. Schleifer, E. Stackebrandt, 654–678, New York: Springer Science+Business Media, LLC.

Woods, D. R. 1993. *The Clostridia and Biotechnology*. Stoneham: Butterworth-Heinemann.

27

The Genus Corynebacterium

Lothar Eggeling and Michael Bott

CONTENTS

General Features

Corynebacterium was originally defined in 1896 by Lehmann and Neumann to accommodate nonmotile parasitic and pathogenic bacteria including diphtheroid bacilli [1]. "Diphther" is the Greek word for membrane and describes the fact that these latter bacteria typically can be isolated from a false membrane in the pharynx developed due to the necrotic action of the toxin made by *Corynebacterium diphtheriae*. Before the establishment of modern systematics the taxon *Corynebacterium* accommodated a number of heterogeneous bacteria. However, based on the extended use of chemotaxonomic markers, mainly in the last third of the last century, like cell wall chemistry and lipid composition, the taxon *Corynebacterium* was brought into a sharper focus and it was also recognized that it is a member of the so called CMN-group of bacteria, including besides *Corynebacterium* also bacteria like *Mycobacterium* and *Nocardia*. Based on 16S rRNA/rDNA sequence patterns and in particular whole genome sequences, it is now clear that *Corynebacterium* forms a robust and well-defined monophyletic group, which together with *Gordonia, Mycobacterium, Nocardia, Rhodococcus, Tsukamurella*, and *Dietzia* forms the order *Corynebacteriales* belonging to the large clade of *Actinobacteria* [2, 3]. Recent reviews and monographs covering several aspects of pathogenic and nonpathogenic *Corynebacterium* species are available [4–8].

The current number of validated *Corynebacterium* species in 2019 exceeds 132 [9]. A phylogenetic tree using a maximum likelihood analysis, based on approx. 2,500 sequences and evaluated among others by maximum parsimony and distance procedures is given in Figure 27.1. In addition, two type strains for the closely related species *Dietzia* and *Tsukamurella* are included as a reference. It can be seen, for instance, that the amino acid producing *Corynebacterium* species *C. glutamicum, C. efficiens*, and *C. callunae* cluster closely together.

Corynebacterium species can be isolated from a number of sources. Nonpathogenic species are found in a broad variety of habitats such as dairy products, animal fodder, rotting plant material, and also in soil as is the case with the biotechnologically important *C. glutamicum*. Pathogenic species are isolated from human or animal specimens, where they occur either as true pathogens or as cutaneous or mucocutaneous contaminants. Corynebacteria are part of the regular microbiome of the skin, as is the case with *C. xerosis*, which however in immunocompromised hosts may cause infections, such as endocarditis, pneumonia, empyema, wound infection, and peritonitis. Although there is no selective medium or enrichment procedure known specifically suited for *Corynebacterium* species, a number of rapid identification and differentiation methods are commercially available that are based on physiological traits. These include the API CORYNE (bioMerieux), API ZYM (bioMerieux), BIOLOG (Biolog), MINITEK (Becton Dickinson), and RapID CB Plus (Remel) identification systems. They are possibly biased toward the identification of clinically relevant species.

For the in-depth identification of *Corynebacterium* species and the study of their phylogenetic relationship, DNA and 16S rRNA analysis is used. Besides total DNA characterization by either thermal denaturation, or hybridization, most informative and convenient is 16S rDNA sequence analysis. For this purpose, a large fragment of the 16S rRNA gene is amplified via PCR, using the universal primers pA 5′-AGAGTTTGATCCTGGCTCAG (positions 8 to 27 of the *Escherichia coli* sequence) and pH* 5′-AAGGAGGTGATCCAGCCGCA (positions 1,541 to 1,522). These bind near the 5' and 3' ends of the rRNA gene and the PCR product is subsequently sequenced and analyzed by comparison with known 16S rDNA sequences [10], as has been done for the *Corynebacterium* species compiled in Figure 27.1.

Corynebacterium cells appear rod-shaped with a somewhat irregular ("coryneform") morphology. Two cells may be arranged in V-formation as a consequence of cell division ("snapping mode"). Also packages of several cells in parallel arrangement ("pallisades") occur. Cells are nonmotile, and mycelia, capsules, and spores are absent. Some species show piliation, like *C. diphtheriae*. Corynebacteria are catalase-positive,

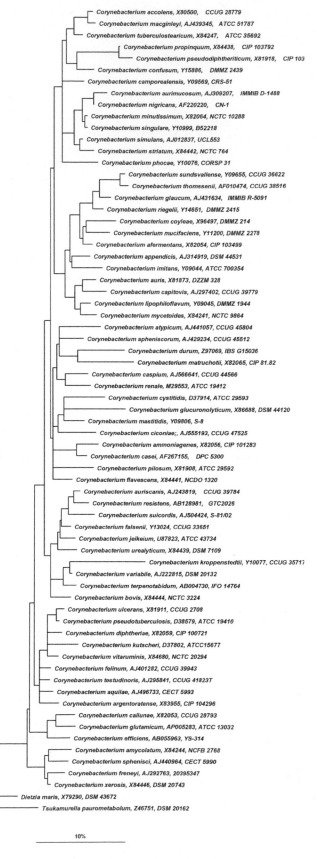

10%

FIGURE 27.1 Clustering of *Corynebacterium* species. The scale bar indicates 10% estimated sequence divergence. (Figure kindly provided by Dr. W. Ludwig, Technical University Munich.)

aerobic or facultatively anaerobic bacteria. They grow in the range of 15–45°C.

Cell Wall

Interestingly, in some properties, the architecture and function of the cell wall of the Gram-positive *Corynebacterium* and related genera like *Mycobacterium* resembles those of the Gram-negative cell envelope, although the molecular details are strikingly different. *Corynebacterium* possesses an outer lipid bilayer membrane [11]. However, this is made up of α-branched, β-hydroxylated fatty acids (mycolic acids, see below) instead of linear fatty acids [12]. In addition *Corynebacterium* possesses unusual sugar polymers like arabinogalactan (AG), which contains D-arabinose, which rarely occurs in nature as the "D" enantiomer [13]. The major macromolecular cell wall components are AG, peptidoglycan, lipomannan, and lipoarabinomannan. Whereas the peptidoglycan is similar to that found in other bacteria, like *E. coli* for instance [14], the sugar polymers and liposugar Alderw polymers deserve closer attention since they are common to *Corynebacterium* and other members of the order *Corynebacteriales* within the *Actinobacteria*, giving them their unique position.

The sugar polymers are of relevance since they might come in direct contact with the environment, respectively, the host, and therefore decide on immunological properties and persistence of the bacterium [15]. Much progress has been made in recent years on the enzymology and structure of these polymers. AG is covalently linked via its galactan part to peptidoglycan via a unique α-L-rhamnopyranose–(1→3)-α-D-GlcNAc-(1→P) linker unit. To the arabinose part of AG mycolic acids are bound. Thus AG is part of a large mycolyl-AG-peptidoglycan (mAGP) complex. It can be considered as the major scaffold within the cell wall, whose absence or disorder causes serve growth defects [16–18], and existing antibiotics, like ethambutol, have AG as target as might have future antibiotics hoped for. For isolation of the mAGP complex cells are suspended in phosphate-buffered saline containing 2% Triton X-100 (pH 7.2), disrupted by sonication and centrifuged. The pelleted material is then extracted with 2% SDS in phosphate-buffered saline at 95°C to remove associated proteins, successively washed with water, 80% (v/v) acetone in water, and acetone, and finally lyophilized to yield a highly purified cell wall preparation [19].

Access to the glycosyl composition of AG is by the further analysis of the isolated cell wall preparation. This is hydrolyzed with 2 M trifluoroacetic acid (TFA) and reduced with $NaBH_4$ to result in alditols, which are per-*O*-acetylated and examined by gas chromatography [20–22]. An example for such a cell wall analysis for *C. glutamicum* is shown in Figure 27.2, left panel. A recent comparison of *C. glutamicum*, *C. diphtheriae*, *C. xerosis*, and *C. amycolatum* by such methodology confirmed that all species have as expected primarily arabinose (Ara) and galactose (Gal) in their arabinogalactan. In addition, there are small but significant amounts of glucose linked to AG of *C. xerosis* and *C. amycolatum*, whereas AG of *C. diphtheriae* contains additionally mannose [28] and AG of *C. glutamicum* rhamnose [23].

Besides the total composition of AG, further information concerning the structure of this sugar polymer comes from a linkage

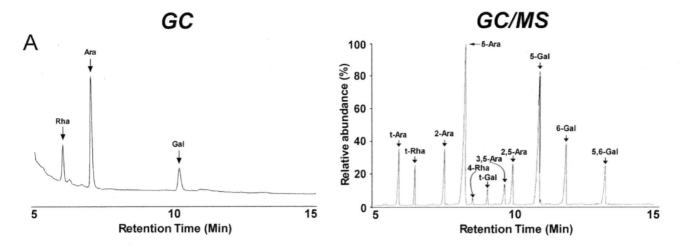

FIGURE 27.2 On the left is shown a GC analysis of arabinogalactan of *C. glutamicum* showing the basic building blocks arabinose, galactose, and rhamnose of this heterosugarpolymer, whereas on the right, the individually linked sugars are shown as analyzed by GC/MS.

analysis of the sugars. This is done by per-*O*-methylation of cell wall preparations using dimethyl sulfinyl carbanion [19, 22, 24], followed by hydrolysis using 2 M TFA and reduction with NaBH₄. Subsequently, the free hydroxyl groups of per-*O*-methylated sugar derivatives are per-*O*-acetylated and examined by gas chromatography/mass spectrometry. Such an analysis for *C. glutamicum* is shown in Figure 27.2, right panel. It reveals that 5-Gal*f*, 6-Gal*f* and 5,6-Gal*f* residues are present. They form a linear chain of alternating β(1→6) linked β(1→5) Gal*f* residues [25, 26]. To this backbone, arabinose chains are attached at the 8th, 10th, and 12th β(1→6) Gal*f* residue [24]. Thus the latter Gal*f* residue with arabinose attached has three linkages, and results in the 5,6-Gal*f* residue (for convention the 1 position is not given). The arabinan domain is made up of α(1→5), α(1→3), and β(1→2) arabinofuranosyl linkages to form a branched polymer of approximately 25 residues in size [19].

To the *t*-Ara*f* and 2-Ara*f* residues mycolic acids are bound representing the unique lipophilic components of the cell envelopes of *Corynebacterium*. In *Corynebacterium*, the mycolic acids are of rather simple structure and consist of two condensed linear fatty acid chains, whereas in *Mycobacterium*, the basic structure is retained, but one of the two fatty acids is much more elaborated. Interestingly, there are a few atypical *Corynebacterium* species not possessing mycolic acids. These are *C. amycolatum*, *C. kroppenstedtii*, and *C. atypicum*. The mycolic acids in *Corynebacterium* species are collectively called corynomycolic acids and consist of up to 38 carbon atoms. The most prominent mycolic acid in *C. glutamicum* has the formula 34:1 followed by 32:0 [21, 27], the latter apparently the result of a condensation of two palmitic acid molecules, and the former of one palmitic plus one oleic acid molecule. The mycolic acids formed may depend on the culture conditions. Thus addition of oleic acid to *C. glutamicum* results in a dominant content of more than 90% of the mycolic acid of the type 36:2.

In addition to the mycolic acids bound to arabinogalactan mycolic acids exist bound to trehalose to form trehalose mono- and dimycolates, respectively. These latter mycolic acid conjugates are solvent extractable [12]. Together with the mycolic acids bound to AG, they form the outer lipid bilayer of *Corynebacterium*, which is clearly visible by electronmicroscopic techniques in addition to the cytoplasmic membrane [11, 28]. The existence of an outer lipid layer is entirely consistent with the presence of a range of pore-forming proteins, porins, within the cell walls of these Gram-positive bacteria [29].

The condensation of two linear fatty acids to the mycolic acids requires the activity of a specific enzyme complex as evident from a genetic analysis of *C. glutamicum*. Whereas the carboxylase α-subunit D1 is a component of the ubiquitous acetyl-CoA carboxylase required for linear fatty acid synthesis, the paralogous carboxylase α-subunits D2 and D3 present in all *Corynebacterium* species are essential for the mycolic acid-forming enzyme complex [21, 30, 31]. The acetyl-CoA carboxylase consists of the carboxylase α-subunit D1, the biotin-carrying β-subunit and an additional ε-subunit. It carboxylates acetyl-CoA in *Corynebacterium* to provide malonyl CoA necessary for linear fatty acid synthesis, as is similar the case in Gram-negative bacteria and other organisms. However, the carboxylase involved in mycolic acid synthesis is apparently a specific evolutionary incidence, involving gene duplications, to derive an enzyme complex structurally and mechanistically related to the acetyl-CoA carboxylase. The acyl carboxylase consists of the carboxylase α-subunits D2 and D3 forming a complex together with the same β- and ε-subunit as required for acetyl-CoA carboxylase complex formation. Although the details are not yet fully understood, it is clear that one linear fatty acid is carboxylated forming an intermediate during condensation of the two linear fatty acid chains by the single polyketide synthase present in *Corynebacterium* [21, 31, 32]. Due to the importance of pathogenic *Corynebacterium* species like *C. diphtheriae* or *C. jeikeium* and the related *Mycobacterium* species, which also possess mycolic acids, considerable efforts are currently made to understand the biochemical details of mycolic acid synthesis. Structural information on one of the carboxylase α-subunits involved is available [32, 33]. In this respect, the ease of handling *C. glutamicum* and its rather small genome structure just containing the genes required for core reactions of cell wall synthesis, and not overloaded with paralogous genes as is the case in *Mycobacterium*, offers a great advantage to understand the basic principles of mycolic acid synthesis.

Genomes

At present, approximately 60 genome sequences of different *Corynebacterium* species are deposited. Many of them are at the level of contigs or scaffolds. The majority of the sequences are derived from pathogenic species. Of the facultative intracellular pathogen *C. pseudotuberculosis,* 15 isolates from different hosts and countries were sequenced and their genomes compared [34, 35]. Analysis of pathogenic islands enabled differentiation of *C. pseudotuberculosis* biovar *ovis* from *C. pseudotuberculosis* biovar *equi*. In the biovar *ovis* strains, a high similarity of the pilus genes is present, whereas these genes show great variability in the biovar *equi* strains. This may be responsible for the ability of the biovar *ovis* strains to spread throughout host tissues and penetrate cells to live intracellularly, in contrast to biovar *equi* strains, which rarely attack visceral organs. Similarly, the genomes of 13 different *C. diphtheriae* isolates were compared to identify a core genome consisting of 1,632 conserved genes. With each newly sequenced strain, an average 65 unique new genes are found. From genome comparisons, it appears that variation in the genome is due to horizontal gene transfer between species and that this is a strategy of *C. diphtheriae* to establish differences in host-pathogen interactions [36].

Nonpathogenic *Corynebacterium* species from which genome sequences are available include, among others, that of *C. glutamicum*. Indeed, the *C. glutamicum* sequence was the first *Corynebacterium* sequence published due to the importance of *C. glutamicum* in industry. Mutants of this bacterium have been used for amino acid production since more than 50 years. Currently, L-glutamate and L-lysine are produced annually in amounts exceeding 3.5 million and 2.5 million tons, respectively. This is done in fermenters of about 500 m³ in size with mutants of *C. glutamicum* [4, 37]. It is, therefore, not surprising that even two sequences of the type strain *C. glutamicum* ATCC 13032 were released at almost the same time, reflecting the commercial interest in this organism [38, 39]. They are not perfectly identical, but differ roughly by 27 kb, which is mainly due to additional copies of insertion elements and an additional prophage region present in the larger genome. The prophage regions can be deleted leading to a 6% reduced genome, which

does not influence growth [40]. There is a publically available genome sequence of an L-lysine-producing *C. glutamicum* mutant, which was isolated from the type strain *C. glutamicum* ATCC13032 by a new selection procedure [41]. This mutant is characterized by 268 single nucleotide polymorphisms with the mutation *murE*-G81E identified as a causative mutation leading to increased L-lysine synthesis [42]. Other nonpathogenic *Corynebacterium* species for which whole genome sequences are available include *C. ammoniagenes, C. flavescens,* and *C. variabile;* all are involved in smear-ripening of cheeses where they contribute to the specific flavor and textural properties of the cheese [43]. The genome sequence is also known for *C. halotolerans* [44], which tolerates salt concentrations up to 25%, a trait that could be of industrial interest. The GC content of 68.4% is exceptionally high for a *Corynebacterium* species. Also *C. efficiens* has a high GC content (63.4%), which is attributed to its high growth temperature of up to 45°C [45]. This trait is of also interest for the large-scale production of amino acids, since cooling costs in fermentations are reduced [4].

Not unexpectedly, the overall genome sequence organization of the *Corynebacterium* species shows that genes are largely syntenic. The comparison of *C. glutamicum* versus *C. diphtheriae* is shown in Figure 27.3, left panel. It reveals one insertion in *C. glutamicum* compared to *C. diphtheriae*, in the range from 1,775,332 to 1,973,456 bp. Nothing in that ~200 kb range of the *C. glutamicum* genome is found in *C. diphtheriae*. In contrast, the comparison of *C. diphtheriae* versus *C. jeikeium* or *C. glutamicum* versus *C. jeikeium* (Figure 27.3, middle and right panels) reveals one major large inversion in *C. jeikeium*, which is interestingly symmetric about the origin of replication creating an "X" pattern (or the beginning of one). Such features of genome structures are known for moderately related bacteria and were also shown to occur in *E. coli* and *Salmonella typhimurium* under laboratory conditions [38]. In *C. jeikeium*, there are also some smaller breakpoints of synteny present, indicating that this species is the most distant of the three analyzed. *C. jeikeium* is characterized by a restricted carbohydrate utilization pattern as compared to the other *Corynebacterium* species, and it lacks phosphoenolpyruvate:carbohydrate phosphotransferase systems (PTSs), which might be one reason for the smaller genome and might also reflect an adaptation of

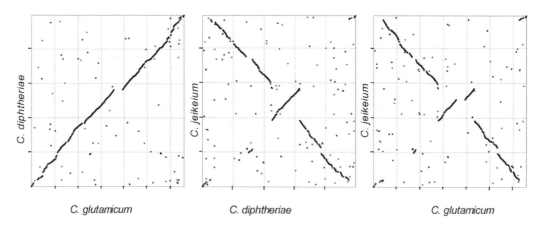

FIGURE 27.3 Synteny between the genomes of *C. glutamicum* ATCC 13032, *C. jeikeium* K411, and *C. diphtheriae* NCTC 13129. The diagram shows scatterplots of the conserved protein sequences between each pair of species, forming syntenic regions between the corynebacterial genomes. The x and y axes are scaled in 0.5 Mbp. (The alignments were kindly provided by S. Salzberg, Bioinformatics and Computational Biology, University of Maryland.)

C. jeikeium to the availability of nutrients in the predominantly colonized areas of the human skin.

Transcriptional Regulators

There are structural and functional types of transcription factors (TRs). Most of them present in *Corynebacterium* consist of one protein, as is also the case in other bacteria [46]. These one-component systems directly bind to DNA, typically in the promoter region, to induce or repress downstream genes. In contrast, two-component systems consist of one protein that is usually membrane-bound with kinase activity and a second protein that functions in response to phosphorylation by the kinase as transcriptional regulator [47]. This particular setting enables transmission of extracytoplasmic signals to the genome. A third type of TRs is the sigma factors, which are bound to the core RNA polymerase to initiate transcription of genes. The transcriptional regulatory repertoire is similar between the closely related nonpathogenic *Corynebacterium* species, but differs remarkable when comparing the nonpathogenic species with their pathogenic relatives [43]. This is related with the complexity and size of the genomes, a trend that has been observed in many other microbial families and can be interpreted with the necessity to integrate the additional genetic information into the regulatory network of the cell [48].

Genome comparisons enabled to define the core set of TRs in *Corynebacterium* [49, 50]. It includes 22 DNA-binding TRs, two two-component systems, and six sigma factors (Table 27.1). The high degree of conservation indicates that the respective TRs fulfill important functions in *Corynebacterium*. Among the core TRs is DtxR, which was first identified in *Corynebacterium diphtheria* as a repressor of the diphtheria toxin gene *tox,* located on the corynebacteriophage ß [51]. It is now clear that DtxR has a much broader function [52, 53], as revealed by DNA microarray analysis and proteome analysis with *C. glutamicum* [54] and other species [19, 55]. It is involved in iron homeostasis and represses in *C. glutamicum* more than 50 genes under iron excess, including those of three other transcriptional regulators, RipA, HrrA, and the ArsR-type regulator GlyR. The AraC-type regulator RipA was shown to repress a set of prominent iron proteins under iron limitation, e.g. aconitase, succinate dehydrogenase, nitrate reductase, isopropylmalate dehydratase, and others [56]. The anticipated function of RipA is therefore to reduce the cell's iron demand under iron-limiting conditions. HrrA mediates heme-dependent gene regulation when hemin is used as an iron source [57]. DtxR is not only a repressor, but presumably also an activator of about a dozen genes that encode the iron storage proteins ferritin and Dps and proteins required for the assembly of iron-sulfur clusters in proteins. Another TR belonging to the core set of regulators is ClgR (Table 27.1), which is involved in proteolysis and DNA repair [58]. The TRs RamA and RamB of *C. glutamicum* are involved in carbon flux control [59–62], with RamA belonging to the core set of regulators. Among the two-component systems, MtrAB is retained in all *Corynebacterium* species studied so far (Table 27.1). It is involved in osmoregulation and cell wall homeostasis [63]. Also, PhoSR and CgtSR4 are highly conserved two-component systems, with the former involved in phosphate homeostasis [64] and the latter with unknown function, but probably essential for growth [47].

In *C. glutamicum,* a minimum of 159 TRs can be regarded as the regulatory repertoire required to coordinate the expression of about 3,000 predicted protein-coding genes under varying environmental conditions [49, 50, 65]. These are organized in a hierarchical and modular transcriptional regulatory network of global, master, or local regulators [66]. Most prominent is GlxR as global regulator, which is a cAMP-sensing TR. In *C. glutamicum,* it represses or activates 25 regulators and thereby controls either directly or indirectly by hierarchical cross-regulations the expression of a large number of genes. Almost 200 genes are directly regulated by GlxR. It can be regarded as a major regulatory hub to adapt *C. glutamicum* to fluctuations in the environment influencing the growth conditions and in particular to

TABLE 27.1

General Genome Features of Selected *Corynebacterium* Species

	***Corynebacterium* Species**				
Feature	***glutamicum*** **ATCC 13032**	***efficiens*** **YS-314**	***diphtheriae*** **NCTC13129**	***jeikeium*** **K411**	***pseudotuberculosis*** **PAT10**
Reason of interest	Industrial amino acid producer	Potential amino acid producer	Human pathogen, causative agent of diphtheria	Multiresistant nosocomial pathogen	Animal pathogen, causative agent of caseous lymphadenitis
Genome size (bp)	3,309,401	3,147,090	2,488,635	2,462,499	2,335,323
Number of coding regions	3,073	3,064	2,388	2,181	2,200
Number of protein coding regions	2,993	2,994	2,272	2,120	2,079
G+C content	53.8	62.9	53.5	61.4	52.2
Plasmids	0	2			0
Refseq	NC_003450.3	GCA_000011305.1	GCF_000195815.1	GCA_000006605.1	NC_017305.1
Reference	Ikeda and Nakagawa [39]; Kalinowski [38]	Nishio et al. [45]	Cerdeño-Tárraga, et al. [67]	Tauch et al. [68]	Cerdeira et al. [93]

TABLE 27.2

Core Set of Transcriptional Regulators in *Corynebacterium*

C. glutamicum CDS	*C. efficiens* CDS	Gene	Regulator Family	Regulatory Role[a]	Regulated Targets/physiological Function
cg0337	*ce0283*	*whiB4, whcA*	WhiB	R	Oxidative stress response genes
cg0350	*ce0287*	*glxR*	CRP	D	cAMP-sensing global regulator
cg0850	*ce0758*	*whiB2, whcD*	WhiB	enhances TR WhiA	Cell division and fatty acid synthesis genes
cg0878	*ce0783*	*whcE*	WhiB	A	Thioredoxin genes
cg1486	*ce1426*	*ltbR*	IclR	R	Leucine and tryptophan synthesis genes
cg1585	*ce1531*	*argR*	ArgR	R	Arginine and glutamine synthesis genes
cg1631	*ce1574*	*ftsR*	MerR	D	Cell division genes including *ftsZ*
cg1633	*ce1576*	*merR2*	MerR	-	Targets and function unknown
cg1738	*ce1663*	*acnR*	TetR	R	Aconitase gene
cg1765	*ce1687*	*sufR*	ArsR	R	Iron-sulfur cluster biogenesis genes
cg2109	*ce1817*	*oxyR*	LysR	R	Oxidative stress response genes
cg2103	*ce1812*	*dtxR*	DtxR	D	Iron metabolism genes
cg2112	*ce1820*	*nrdR*	NrdR	R	Ribonucleotide reductase genes
cg2114	*ce1823*	*lexA*	LexA	R	SOS and DNA repair genes
cg2152	*ce1885*	*clgR*	HtH_3	A	Proteolysis and DNA repair genes
cg2502	*ce2180*	*zur*	FUR	R	Zinc metabolism genes
cg2516	*ce2190*	*hrcA*	HrcA	R	Heat-shock response genes
cg2831	*ce2445*	*ramA*	LuxR	D	Carbon metabolism genes
cg2910	*ce2511*	*ipsA*	LacI	R	Inositol-derived lipid synthesis genes
cg3097	*ce2626*	*hspR*	MeR	R	Heat-shock response genes
cg3253	*ce2788*	*mcbR*	TetR	R	Sulfur metabolism genes
cg3315	*ce2826*	*uspR*	MarR	R	Universal stress response gene cg3316
cg0862	*ce0796*	*mtrA*	Response[b]	D	Cell envelope and osmoprotection genes
cg2888	*ce2494*	*phoR*	Response[b]	A	Phosphate metabolism genes
cg0309	*ce0223*	*sigC*	Sigma factor	n/a	Respiratory chain genes
cg0876	*ce0782*	*sigH*	Sigma factor	n/a	Heat and oxidative stress response genes
cg1271	*ce1177*	*sigE*	Sigma factor	n/a	Cell surface stress response genes
cg2092	*ce1804*	*sigA*	Sigma factor	n/a	Primary (housekeeping) sigma factor
cg2102	*ce1811*	*sigB*	Sigma factor	n/a	Nonessential primary-like sigma factor
cg3420	*ce2932*	*sigM*	Sigma factor	n/a	Disulfide stress-related genes

[a] A = activator; D = dual regulator; R = repressor; n/a = not applicable.
[b] Response, response regulator of a two-component signal transduction system.
Source: Ref. [50] and recent literature.

the availability of alternative carbon and energy sources. Since GlxR belongs to the core set of regulators (Table 27.2), it might exert a similar global function in all corynebacteria. The TRs SigH, SigB, DtxR, LexA, PhoR, and McbR operate at a lower hierarchical level (Table 27.2). They control, at least in *C. glutamicum*, the expression of a number of functionally related genes, including TRs. In contrast to the extended regulons of these global TRs, local TRs act on a single or a few transcription units and tend to be encoded next to the regulated gene(s). A typical example is the first TR identified in *C. glutamicum*, which is LysG, controlling the transcription of the adjacent *lysE* gene encoding the basic amino acid exporter LysE, which has made possible the industrial use of *C. glutamicum* for L-lysine production [4, 69, 70].

Posttranscriptional Regulation

The genomes of *C. glutamicum*, *C. efficiens*, *C. diphtheriae*, and *C. jeikeium* contain four or five (*C. jeikeium*) genes for serine/threonine protein kinases. Those common to all are designated PknA, PknB, PknL, and PknG, the one specific for *C. jeikeium* has been named PknJ (*jk1732*). In addition, a gene for a phosphoserine/phosphothreonine-specific phosphoprotein phosphatase (*ppp*) is present in these four species. The genes for PknA, PknB, and Ppp are part of a gene cluster that is highly conserved in the order *Corynebacteriales* and implicated in cell division. Whereas PknA, PknB, PknL, and PknJ are predicted to contain a single transmembrane helix, and thus are membrane-integral proteins, PknG appears to be a cytosolic protein. Recent proteome studies with *C. glutamicum* led to the identification of the first target of PknG, a 15-kDa protein with a forkhead-associated domain that was designated OdhI [71]. In its unphosphorylated state, OdhI binds to the E1 subunit OdhA of the 2-oxoglutarate dehydrogenase complex and inhibits the activity of this enzyme complex [72, 73], which catalyzes the oxidative decarboxylation of 2-oxoglutarate to succinyl-CoA in the citric acid cycle. Phosphorylation of OdhI at threonine residue 14 by PknG relieves the inhibitory function of OdhI, and thus, allows for a high activity of 2-oxoglutarate dehydrogenase. This happens for example when glutamine is used as carbon source, which is catabolized via glutamate and 2-oxoglutarate. However, there might be many other stimuli that affect the phosphorylation status of OdhI. Most interestingly, not only PknG, but also one or several of the other serine/threonine protein kinases can phosphorylate OdhI, as even in a *pknG* mutant part of OdhI is present in a phosphorylated state [74]. Thus, OdhI might be able to integrate diverse stimuli to optimally control the flux at the 2-oxoglutarate node of metabolism. This novel control mechanism for 2-oxoglutarate dehydrogenase via PknG and OdhI, probably operating in many corynebacteria and mycobacteria, is of prime importance for glutamate production [75, 76], as a reduced 2-oxoglutarate dehydrogenase activity was shown to be the key factor in the metabolic network leading to glutamate. A *C. glutamicum* mutant lacking *odhI* was severely inhibited (90–100%) in its ability to secrete glutamate under different conditions. On the other hand, a mutant lacking *pknG* showed significantly improved glutamate production under certain conditions [75]. It is now clear that the ability of *C. glutamicum* to excrete L-glutamate under specific conditions, which led to its isolation, is dependent on the mechanosensitive channel MscCG [77]. Regulation of the activity of MscCG is a topic of current studies.

Plasmids and Transposon Use

Based on the strong demand to improve the amino acid production properties of *C. glutamicum,* a broad set of vectors has been developed. They include shuttle vectors, shuttle expression vectors, vectors for integration into the chromosome, promoter- and terminator-probe vectors, vectors for site specific integration, and vectors for self-cloning. They are compiled in the *Handbook of Corynebacterium glutamicum,* which is also a comprehensive source for additional information on *Corynebacterium* [4]. The plasmids are mostly based on the two different replicons pBL1 and pCG1 and their close relatives occurring in *Corynebacterium,* each of these plasmids replicating by a rolling circle mechanism. For more details on the plasmids of amino acid-producing coryneform bacteria, excellent reviews are available [76–80]. It has already been demonstrated that derivatives of pBL1 and pCG1 also replicate in other *Corynebacterium* species, including pathogenic ones, thus presuming that they might be useful for molecular work in probably all *Corynebacterium* strains. There are also a few examples on vectors originally developed for use in other species than *C. glutamicum* and shown to replicate in *C. glutamicum* (Table 27.3). To these vectors belongs pNG2, the archetype of vectors like

TABLE 27.3

Vectors of Different Origin Shown to Replicate in an Other *Corynebacterium* Species

Original Isolation	Plasmid Name	Plasmid Family	Demonstrated Replication	Reference
C. glutamicum	pBL1	pBL1	*C. callunae*	[81]
			C. diphtheriae	[82]
C. glutamicum	pCG1	pCG1	*C. ammoniagenes*	[83]
			C. callunae	[83]
			C. diphtheriae	[82]
			C. pilosum	[83]
C. glutamicum	pGA1	pCG1	*C. callunae*	[84]
			C. diphtheriae	[82]
Brevibacterium stationis	pBY503	pCG1	*C. glutamicum*	[85]
C. callunae	pCC1	pBL1	*C. glutamicum*	[86]
C. diphtheriae	pNG2	pCG1	*C. ulcerans*	[87]
			C. glutamicum	[87]
C. renale	pCR1		*C. glutamicum*	[88]
			C. lilium	[88]
			C. acetoacidophilum	[88]

pCG1, and originally isolated from *C. diphtheriae.* The smallest plasmid isolated to date from *Corynebacterium* is the cryptic plasmid pCR1 isolated from *C. renale,* which has a size of 1,488 bp and was shown to replicate also in *C. glutamicum, C. lilium,* and *C. acetoacidophilum* [88]. These data corroborate the belief that the molecular tools developed for *C. glutamicum* are of broader use.

It is important to realize that the developed vectors might be present in different copy numbers in the host. This is of particular relevance in case they encode proteins that, due to structure or localization, are deleterious to the cells, as can be the case for membrane proteins or for deregulated biosynthesis enzymes. For representative examples, their relative copy numbers can be compared in Figure 27.4. The lowest copy numbers in *C. glutamicum* have pWK0 [89] and pBHK18 [90], both using the replicon of pNG2 [87], and pKW0 using the replicon of pGA1 [84]. Based on homoserine dehydrogenase and threonine dehydratase genes cloned in pWK0 and enzyme activity determinations, pWK0 is present in three to four copies in *C. glutamicum* [89]. Plasmid pEKEx2 might be of intermediate copy number, whereas the highest copy numbers are present for pJC1 and pECT18mob2. An assessment of the band intensities from the gels and transformations with aliquots of plasmids isolated from *C. glutamicum* suggests that the copy numbers of pJC1 and pECT18mob2 are at least ten-fold increased relative to pKW0. As can be seen from Figure 27.4, the copy numbers in *E. coli* are very much different. As a control, pUC18 was isolated in an identical manner as the other vectors from *E. coli* and shown on the far right. Plasmid pKW0 is also present in *E. coli* at a low copy number, but since replication of this vector relies in *E. coli* on *repA,* which is under control of the temperature sensitive repressor cI[857] [91], a shift to higher cultivation temperatures allows "run-away replication" and, therefore, easy preparation of sufficient vector for further use in recombinant work.

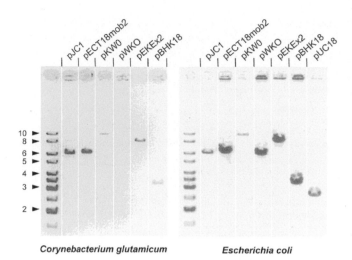

FIGURE 27.4 Comparison of relative copy numbers. Plasmids were isolated from recombinant strains of 100 mL LB cultures and isolated via the QIAfilter Plasmid Midi Kit according to the protocol of the supplier (QIAGEN, Hilden, Germany). 2.5 µl plasmid preparation digested with EcoRI was applied in each lane for plasmids isolated from *C. glutamicum,* and 0.5 µl plasmid preparation digested with EcoRI was applied for plasmids isolated from *E. coli.* With plasmid pWKO isolated from *C. glutamicum,* no band is visible in the reproduction, but is on the original gel.

For transformation of *Corynebacterium,* various methods have been developed and used over time. The most efficient protocol is also characterized by its simplicity. It uses brain-heart infusion medium containing 91 g sorbitol/L (BHIS) for cultivation of cells up to a low optical density at 578 nm (OD_{578}) of 1.75 [82]. Cells are carefully washed and transformed via electroporation. Applying the same protocol to *C. diphtheriae,* a very high transformation efficiency of up to 5.6×10^5 colony forming units per µg DNA is obtained when DNA directly prepared from *E. coli* is used. Such high transformation efficiency is particular of relevance when vectors are integrated into the chromosome of the recipient, since additional host-specific recombination processes reduce the chance of plasmid, once it is present in the cytosol, to be integrated in the chromosome. Also for *C. pseudotuberculosis,* similar transformation protocols have been described [92–94].

For a genome-scale analysis, application of transposon mutagenesis is a valuable tool. It is established and used in a number of applications for *C. glutamicum* with Tn*5531* [95–97]. This transposon is composed of *aph3* conferring kanamycin resistance flanked by two IS*1207* sequences encoding the transposase. Since Tn*5531* relies on the IS*1207* sequences, only strains without IS*1207* present in the chromosome are appropriate hosts to generate mutants. Whereas the type strain *C. glutamicum* ATCC 13032 contains seven copies of highly similar sequences related to IS*1207,* hosts like *C. glutamicum* ATCC 14752, *C. efficiens* YS-314, and *Corynebacterium* sp. 2262 are appropriate. Another transposon operating in *C. glutamicum* is based on IS*6100,* which is present on plasmid pTET3 isolated from strain *C. glutamicum* LP-6. The transposon is useful to isolate mutants from the type strain ATCC 13032 [98]. Also, other transposons were isolated and shown to operate in *C. glutamicum* [99, 100]. There is little information on the application of these transposon delivery vectors in other *Corynebacterium*

species. However, a transposon derivative based on Tn*4001* and originally developed for *Streptococcus pyogenes* was profitably used in *C. pseudotuberculosis* to identify genes encoding extracytosolic genes, like fimbriae or transporter subunits [101]. A number of techniques for genome-scale deletions and analysis were developed for *C. glutamicum* R, which are compiled in a recent review [102].

Whereas the *in vivo* systems described depend on nonreplicative delivery vectors, a recent *in vitro* system has been developed [103], which might represent an attractive alternative in particular due to its ease of handling, its general applicability, and its commercial availability (Epicentre Biotechnologies, Madison, WI, USA). It involves the *in vitro* formation of released Tn5 transposition complexes, called transposomes, followed by introduction of the complexes into the target cell of choice by electroporation. This simple technology has been used for construction and characterization of transposon insertion mutations in *C. diphtheriae* and the isolation of mutations that affect expression of the diphtheria toxin repressor (DtxR) [104]. It has similarly been used in *C. diphtheriae* to isolate a gene involved in adherence to epithelial cells [105], and to identify iron-regulated promoters [106]. The transposome technology was also successfully applied to *C. matruchotii,* leading to the identification of an integral membrane transporter probably linked to corynomycolic acid synthesis [107], and leading to mutants that synthesize unnatural hybrid fatty acids that functionally replace corynomycolic acid. Application of the transposome system to *C. glutamicum* R enabled to construct a library of 2,300 different mutants covering 75% of the genes in that strain [108]. The transposome-mediated insertions in selected *Corynebacterium* species, and their established function in a wide range of organisms, including *Acinetobacter, Mycobacterium, Proteus,* and even *Saccharomyces,* and *Trypanosoma,* reveals that this molecular device will be instrumental for the analysis of other *Corynebacterium* species, too.

A recent tool introduced for *Corynebacteriales* like *C. glutamicum* or *Mycobacterium smegmatis* uses the CRISPR/Cas system [109, 110]. Simplicity and programmability of this system has enabled its straightforward application to specifically target any genomic location. In *C. glutamicum* the cleavage activity of the nuclease Cas12a was directed by its crRNA to a sensitive region in the γ-glutamyl kinase, which together with single-stranded DNA (ssDNA) recombineering, enabled to relieve allosteric feedback inhibition of the γ-glutamyl kinase by L-proline, thus converting the wild type of *C. glutamicum* to an L-proline producer [109]. Similarly, the system was used to perform codon saturation mutagenesis at critical positions in the mechanosensitive channel MscCG to identify new gain-of-function mutations leading to enhanced L-glutamate export of *C. glutamicum* [111].

Conclusion

Nonpathogenic *Corynebacterium* species find many applications, like amino acid production with *C. glutamicum, C. efficiens,* and *C. callunae* [4, 50], or nucleotide production with *C. ammoniagenes* [112]. This strong industrial interest explains the interest of companies and associated research groups to understand *Corynebacterium* at the physiological, molecular, and genomic level. A recent vision is to use the ability of *C. glutamicum* to make, besides the "bulk" products already mentioned, additional metabolites that are cell inherent or whose synthesis capacity is artificially expanded, summarized under the term "white biotechnology" [113, 114]. Fortunately, the fast development of standard techniques, such as genome sequencing and RNA-Seq, has also generated the basic blueprint information for a number of pathogenic *Corynebacterium* species. The fact that *Corynebacterium* species share basic molecular features helps to deepen the knowledge on pathogenic *Corynebacterium* species, with *C. diphtheriae* probably representing the most important species with respect to its epidemiological properties [115, 116]. However, *C. xerosis*, *C. urealyticum*, *C. ulcerans*, *C. pseudotuberculosis*, *C. striatum*, and further species pose severe problems as nosocomial pathogens, for instance, where a molecular understanding on physiology and antibiotic resistance mechanisms might assist to have control on these bacteria.

REFERENCES

1. Lehmann, K.B. and Neumann, R.O. (1896). *Atlas und Grundriss der Bakteriologie und Lehrbuch der speziellen bakteriologischen Diagnostik*, J.F. Lehmann, München, Germany.
2. Stackebrandt, E., Rainey, F.A. and Rainey, N.L. (1997). Proposal for a new hierarchic classification system, *Actinobacteria* classis nov. *Int J Syst Bacteriol* 47, 479–491.
3. Gao, B.L. and Gupta, R.S. (2012). Phylogenetic framework and molecular signatures for the main clades of the phylum *Actinobacteria*. *Microbiol Mol Bio Rev* 76, 66–112.
4. Eggeling, L. and Bott, M. (2005). *Handbook of Corynebacterium glutamicum*, CRC Press, Boca Raton, FL.
5. Vázquez-Boland, J.A. and Meijer, W.G. (2019). The pathogenic actinobacterium *Rhodococcus equi*: what's in a name? *Mol Microbiol* 112, 1–15.
6. Bernard, K. (2012). The genus *Corynebacterium* and other medically relevant coryneform-like bacteria. *J Clin Microbiol* 50, 3152–3158.
7. Jankute, M., Cox, J.A., Harrison, J. and Besra, G.S. (2015). Assembly of the mycobacterial cell wall. *Annu Rev Microbiol* 69, 405–423.
8. Hacker, E., Antunes, C.A., Mattos-Guarald, A.L., Burkovski, A. and Tauch, A. (2016). *Corynebacterium ulcerans*, an emerging human pathogen. *Future Microbiol* 11, 1191–1208.
9. List of procaryotic names (http://www.bacterio.net/).
10. Pascual, C., Lawson, P.A., Farrow, J.A., Gimenez, M.N. and Collins, M.D. (1995). Phylogenetic analysis of the genus *Corynebacterium* based on 16S rRNA gene sequences. *Int J Syst Bacteriol* 45, 724–728.
11. Hoffmann, C., Leis, A., Niederweis, M., Plitzko, J.M. and Engelhardt, H. (2008). Disclosure of the mycobacterial outer membrane: cryo-electron tomography and vitreous sections reveal the lipid bilayer structure. *Proc Natl Acad Sci U S A* 105, 3963–3967.
12. Bansal-Mutalik, R. and Nikaido, H. (2011). Quantitative lipid composition of cell envelopes of *Corynebacterium glutamicum* elucidated through reverse micelle extraction. *Proc Natl Acad Sci U S A* 108, 15360–15365.
13. Mishra, A.K., Driessen, N.N., Appelmelk, B.J. and Besra, G.S. (2011). Lipoarabinomannan and related glycoconjugates: structure, biogenesis and role in *Mycobacterium tuberculosis* physiology and host-pathogen interaction. *FEMS Microbiol Rev* 35, 1126–1157.
14. Janczura, E., Leyh-Bouille, M., Cocito, C. and Ghuysen, J.M. (1981). Primary structure of the wall peptidoglycan of leprosy-derived corynebacteria. *J Bacteriol* 145, 775–759.
15. Abrahams, K.A. and Besra, G.S. (2018). Mycobacterial cell wall biosynthesis: a multifaceted antibiotic target. *Parasitology* 145, 116–133.
16. Birch, H.L. et al. (2008). Biosynthesis of mycobacterial arabinogalactan: identification of a novel $\alpha(1 \rightarrow 3)$ arabinofuranosyltransferase. *Mol Microbiol* 69, 1191–1206.
17. Mishra, A.K. (2012). Differential arabinan capping of lipoarabinomannan modulates innate immune responses and impacts T helper cell differentiation. *J Biol Chem* 287, 44173–44183.
18. Jankute, M. et al. (2018). The singular *Corynebacterium glutamicum* Emb arabinofuranosyltransferase polymerises the $\alpha(1 \rightarrow 5)$ arabinan backbone in the early stages of cell wall arabinan biosynthesis. *Cell Surf* 2, 38–53.
19. Besra, G.S., Khoo, K.H., Mcneil, M.R., Dell, A., Morris, H.R. and Brennan, P.J. (1995). A new interpretation of the structure of the mycolyl arabinogalactan complex of mycobacterium-tuberculosis as revealed through characterization of oligoglycosylalditol fragments by fast-atom-bombardment mass-spectrometry and H-1 nuclear-magnetic-resonance spectroscopy. *Biochemistry* 34, 4257–4266.
20. Alderwick, L.J. (2005). Deletion of Cg-*emb* in *Corynebacterianeae* leads to a novel truncated cell wall arabinogalactan, whereas inactivation of Cg-*ubiA* results in an arabinan-deficient mutant with a cell wall galactan core. *J Biol Chem* 280, 32362–32371.
21. Portevin, D., Sousa-D'Auria, C.C., Houssin, C., Grimaldi, C., Chami, M., Daffe, M. and Guilhot, C. (2004). A polyketide synthase catalyzes the last condensation step of mycolic acid biosynthesis in mycobacteria and related organisms. *Proc Natl Acad Sci U S A* 101, 314–319.
22. Daffe, M., Brennan, P.J. and McNeil, M. (1990). Predominant structural features of the cell wall arabinogalactan of *Mycobacterium tuberculosis* as revealed through characterization of oligoglycosyl alditol fragments by gas chromatography/mass spectrometry and by 1H and 13C NMR analyses. *J Biol Chem* 265, 6734–6743.
23. Birch, H.L. et al. (2009). Identification of a terminal rhamnopyranosyltransferase (RptA) involved in *Corynebacterium glutamicum* cell wall biosynthesis. *J Bacteriol* 191, 4879–4887.
24. Alderwick, L.J., Seidel, M., Sahm, H., Besra, G.S. and Eggeling, L. (2006). Identification of a novel arabinofuranosyltransferase (AftA) involved in cell wall Arabinan biosynthesis in *Mycobacterium tuberculosis*. *J Biol Chem* 281, 15653–15661.
25. Kremer, L. (2001). Galactan biosynthesis in *Mycobacterium tuberculosis* - identification of a bifunctional UDP-galactofuranosyltransferase. *J Biol Chem* 276, 26430–26440.
26. Mikusova, K., Yagi, T., Stern, R., McNeil, M.R., Besra, G.S., Crick, D.C. and Brennan, P.J. (2000). Biosynthesis of the galactan component of the mycobacterial cell wall. *J Biol Chem* 275, 33890–33897.

27. Radmacher, E., Alderwick, L.J., Besra, G.S., Brown, A.K., Gibson, K.J.C., Sahm, H. and Eggeling, L. (2005). Two functional FAS-I type fatty acid synthases in *Corynebacterium glutamicum*. *Microbiology* 151, 2421–2427.

28. Puech, V. (2001). Structure of the cell envelope of corynebacteria: importance of the non-covalently bound lipids in the formation of the cell wall permeability barrier and fracture plane. *Microbiology* 147, 1365–1382.

29. Abdali, N., et al. (2018). Identification and characterization of smallest pore-forming protein in the cell wall of pathogenic *Corynebacterium urealyticum* DSM 7109. *BMC Biochem* 19, 3–21.

30. Gande, R. et al. (2004). Acyl-CoA carboxylases (accD2 and accD3), together with a unique polyketide synthase (Cg-pks), are key to mycolic acid biosynthesis in *Corynebacterianeae* such as *Corynebacterium glutamicum* and *Mycobacterium tuberculosis*. *J Biol Chem* 279, 44847–44857.

31. Gande, R., Dover, L.G., Krumbach, K., Besra, G.S., Sahm, H., Oikawa, T. and Eggeling, L. (2007). The two carboxylases of *Corynebacterium glutamicum* essential for fatty acid and mycolic acid synthesis. *J Bacteriol* 189, 5257–5264.

32. Gago, G., Diacovich, L., Arabolaza, A., Tsai, S.C. and Gramajo, H. (2011). Fatty acid biosynthesis in actinomycetes. *FEMS Microbiol Rev* 35, 475–497.

33. Pawelczyk, J., Viljoen, A., Kremers, L. and Dziadek, J. (2017). The influence of AccD5 on AccD6 carboxyltransferase essentiality in pathogenic and non-pathogenic mycobacterium. *Sci Rep.* 7, 42692.

34. Soares, S.C. (2013). The pan-genome of the animal pathogen *Corynebacterium pseudotuberculosis* reveals differences in genome plasticity between the biovar *ovis* and *equi* strains. *PLOS One* 8, e53818.

35. Silva, A. (2012). Complete genome sequence of *Corynebacterium pseudotuberculosis* Cp31, isolated from an Egyptian buffalo. *J Bacteriol* 194, 6663–4.

36. Trost, E. (2012). Pangenomic study of *Corynebacterium diphtheriae* that provides insights into the genomic diversity of pathogenic isolates from cases of classical diphtheria, endocarditis, and pneumonia. *J Bacteriol* 194, 3199–31215.

37. Eggeling, L. and Bott, M. (2015). A giant market and a powerful metabolism: L-lysine provided by *Corynebacterium glutamicum*. *Appl Microbiol Biotechnol* 99, 3387–3389.

38. Kalinowski, J. (2003). The complete *Corynebacterium glutamicum* ATCC 13032 genome sequence and its impact on the production of L-aspartate-derived amino acids and vitamins. *J Biotechnol* 104, 5–25.

39. Ikeda, M. and Nakagawa, S. (2003). The *Corynebacterium glutamicum* genome: features and impacts on biotechnological processes. *Appl Microbiol Biotechnol* 62, 99–109.

40. Baumgart, M., Unthan, S., Rückert, C., Sivalingam, J., Grünberger, A., Kalinowski, J., Bott, M., Noack, S. and Frunzke, J. (2013). Construction of a prophage-free variant of *Corynebacterium glutamicum* ATCC 13032 for use as a platform strain for basic research and industrial biotechnology. *Appl Environ Microbiol* 79, 6006–6015.

41. Binder, S., Schendzielorz, G., Stabler, N., Krumbach, K., Hoffmann, K., Bott, M. and Eggeling, L. (2012). A high-throughput approach to identify genomic variants of bacterial metabolite producers at the single-cell level. *Genome Biol* 13, R40.

42. Binder, S., Siedler, S., Marienhagen, J., Bott, M. and Eggeling, L. (2013). Recombineering in *Corynebacterium glutamicum* combined with optical nanosensors: a general strategy for fast producer strain generation. *Nucleic Acids Res* 41, 6360–6369.

43. Schröder, J., Maus, I., Trost, E. and Tauch, A. (2011). Complete genome sequence of *Corynebacterium variabile* DSM 44702 isolated from the surface of smear-ripened cheeses and insights into cheese ripening and flavor generation. *BMC Genomics* 12, 545.

44. Rückert, C., Albersmeier, A., Al-Dilaimi, A., Niehaus, K., Szczepanowski, R. and Kalinowski, J. (2012). Genome sequence of the halotolerant bacterium *Corynebacterium halotolerans* type strain YIM 70093(T) (= DSM 44683(T)). *Stand Genomic Sci* 7, 284–293.

45. Nishio, Y., Nakamura, Y., Kawarabayasi, Y., Usuda, Y., Kimura, E., Sugimoto, S., Matsui, K., Yamagishi, A., Kikuchi, H., Ikeo, K., Gojobori, T. (2003). Comparative complete genome sequence analysis of the amino acid replacements responsible for the thermostability of *Corynebacterium efficiens*. *Genome Res* 13, 1572–1579.

46. Ulrich, L.E., Koonin, E.V. and Zhulin, I.B. (2005). One-component systems dominate signal transduction in prokaryotes. *Trends Microbiol* 13, 52–56.

47. Bott, M. and Brocker, M. (2012). Two-component signal transduction in *Corynebacterium glutamicum* and other corynebacteria: on the way towards stimuli and targets. *Appl Microbiol Biotechnol* 94, 1131–1150.

48. Rodionov, D.A. (2007). Comparative genomic reconstruction of transcriptional regulatory networks in bacteria. *Chem Rev* 107, 3467–3697.

49. Brinkrolf, K., Brune, I. and Tauch, A. (2007). The transcriptional regulatory network of the amino acid producer *Corynebacterium glutamicum*. *J Biotechnol* 129, 191–211.

50. Schröder, J. and Tauch, A. (2013). The transcriptional regulatory network of *Corynebacterium glutamicum*. In *Corynebacterium glutamicum: biology and biotechnology*, Yukawa, H. and Inui, M. eds., pp. 239–261, Springer, Heidelberg, New York.

51. Boyd, J., Oza, M.N. and Murphy, J.R. (1990). Molecular cloning and DNA sequence analysis of a diphtheria tox iron-dependent regulatory element (*dtxR*) from *Corynebacterium diphtheriae*. *Proc Natl Acad Sci U S A* 87, 5968–5972.

52. Brune, I., Werner, H., Huser, A.T., Kalinowski, J., Puhler, A. and Tauch, A. (2006). The DtxR protein acting as dual transcriptional regulator directs a global regulatory network involved in iron metabolism of *Corynebacterium glutamicum*. *BMC Genomics* 7, 21.

53. Wennerhold, J. and Bott, M. (2006). The DtxR regulon of *Corynebacterium glutamicum*. *J Bacteriol* 188, 2907–2918.

54. Küberl, A., Mengus-Kaya, A., Polen, T. and Bott, M. (2020). The iron deficiency response of Corynebacterium glutamicum and a link to thiamine biosynthesis. *Appl Environm Microbiol* 86: e00065–20.

55. Wittchen, M., Busche, T., Gaspar, A.H., Lee, J.H., Ton-That, H., Kalinowski, J. and Tauch, A. (2018). Transcriptome sequencing of the human pathogen Corynebacterium diphtheriae NCTC 13129 provides detailed insights into its

transcriptional landscape and into DtxR-mediated transcriptional regulation. *BMC Genomics* 19, 82.

56. Wennerhold, J., Krug, A., and Bott, M. (2005). The AraC-type regulator RipA represses aconitase and other iron proteins from Corynebacterium under iron limitation and is itself repressed by DtxR. *J Biol Chem* 280, 40500–40508.

57. Frunzke, J., Gätgens, C., Brocker, M. and Bott, M. (2011). Control of heme homeostasis in *Corynebacterium glutamicum* by the two-component system HrrSA. *J Bacteriol* 193, 1212–1221.

58. Engels, S., Schweitzer, J.E., Ludwig, C., Bott, M. and Schaffer, S. (2004). *clpC* and *clpP1P2* gene expression in *Corynebacterium glutamicum* is controlled by a regulatory network involving the transcriptional regulators ClgR and HspR as well as the ECF sigma factor sigma(H). *Mol Microbiol* 52, 285–302.

59. Shah, A., Blombach, B., Gauttam, R. and Eikmanns, B.J. (2018). The RamA regulon: complex regulatory interactions in relation to central metabolism in *Corynebacterium glutamicum*. *Appl Microbiol Biotechnol* 102, 5901–5910.

60. van Ooyen, J., Noack, S., Bott, M., Reth, A. and Eggeling, L. (2012). Improved L-lysine production with *Corynebacterium glutamicum* and systemic insight into citrate synthase flux and activity. *Biotech Bioeng* 109, 2070–2081.

61. Bussmann, M., Emer, D., Hasenbein, S., Degraf, S., Eikmanns, B.J. and Bott, M. (2009). Transcriptional control of the succinate dehydrogenase operon *sdhCAB* of *Corynebacterium glutamicum* by the cAMP-dependent regulator GlxR and the LuxR-type regulator RamA. *J Biotech* 143, 173–182.

62. Bott, M. (2007). Offering surprises: TCA cycle regulation in *Corynebacterium glutamicum*. *Trends Microbiol* 15, 417–425.

63. Möker, N., Brocker, M., Schaffer, S., Krämer, R., Morbach, S. and Bott, M. (2004). Deletion of the genes encoding the MtrA-MtrB two-component system of *Corynebacterium glutamicum* has a strong influence on cell morphology, antibiotics susceptibility and expression of genes involved in osmoprotection. *Mol Microbiol* 54, 420–438.

64. Kocan, M., Schaffer, S., Ishige, T., Sorger-Herrmann, U., Wendisch, V.F., and Bott, M. (2006). Two-component systems of Corynebacterium glutamicum: Deletion analysis and involvement of the PhoS-PhoR system in the phosphate starvation response. *J Bacteriol* 188, 724–732.

65. Toyoda, K. and Inui, M. (2016). Regulons of global transcription factors in *Corynebacterium glutamicum*. *Appl Microbiol Biotechnol* 100, 45–60.

66. Schröder, J. and Tauch, A. (2010). Transcriptional regulation of gene expression in *Corynebacterium glutamicum*: the role of global, master and local regulators in the modular and hierarchical gene regulatory network. *FEMS Microbiol Rev* 34, 685–737.

67. Cerdeno-Tarraga, A.M., Efstratiou, A., Dover, L.G., Holden, M.T.G., Pallen, M., Bentley, S.D., et al. (2003). The complete genome sequence and analysis of *Corynebacterium diphtheriae* NCTC13129. *Nucleic Acids Res.* 31, 6516–6523.

68. auch, A., Kaiser, O., Hain, T., Goesmann, A., Weisshaar, B., Albersmeier, A., et al. (2005). Complete genome sequence and analysis of the multiresistant nosocomial pathogen *Corynebacterium jeikeium* K411, a lipid-requiring bacterium of the human skin flora. *J Bacteriol* 187, 4671–4682.

69. Vrljic, M., Sahm, H. and Eggeling, L. (1996). A new type of transporter with a new type of cellular function: L-lysine export from *Corynebacterium glutamicum*. *Mol Microbiol* 22, 815–826.

70. Bellmann, A., Vrljic, M., Pátek, M., Sahm, H., Krämer, R. and Eggeling, L. (2001). Expression control and specificity of the basic amino acid exporter LysE of *Corynebacterium glutamicum*. *Microbiology* 147, 1765–74.

71. Niebisch, A., Kabus, A., Schultz, C., Weil, B. and Bott, M. (2006). Corynebacterial protein kinase G controls 2-oxoglutarate dehydrogenase activity via the phosphorylation status of the OdhI protein. *J Biol Chem* 281, 12300–12307.

72. Krawczyk, S., Raasch, K., Schultz, C., Hoffelder, M., Eggeling, L. and Bott, M. (2010). The FHA domain of OdhI interacts with the carboxyterminal 2-oxoglutarate dehydrogenase domain of OdhA in *Corynebacterium glutamicum*. *FEBS Lett* 584, 1463–1468.

73. Hoffelder, M., Raasch, K., van Ooyen, J. and Eggeling, L. (2010). The E2 domain of OdhA of *Corynebacterium glutamicum* has succinyltransferase activity dependent on lipoyl residues of the acetyltransferase AceF. *J Bacteriol* 192, 5203–5211.

74. Schultz, C., Niebisch, A., Schwaiger, A., Viets, U., Metzger, S., Bramkamp, M. and Bott, M. (2009). Genetic and biochemical analysis of the serine/threonine protein kinases PknA, PknB, PknG and PknL of *Corynebacterium glutamicum*: evidence for non-essentiality and for phosphorylation of OdhI and FtsZ by multiple kinases. *Mol Microbiol* 74, 724–741.

75. Schultz, C., Niebisch, A., Gebel, L., and Bott, M. (2007). Glutamate production by Corynebacterium glutamicum: dependence on the oxoglutarate dehydrogenase inhibitor protein OdhI and protein kinase PknG. *Appl Microbiol Biotechnol* 76, 691–700.

76. Komine-Abe, A., Nagano-Shoji, M., Kubo, S., Kawasaki, H., Yoshida, M., Nishiyama, M., and Kosono, S. (2017). Effect of lysine succinylation on the regulation of 2-oxoglutarate dehydrogenase inhibitor, OdhI, involved in glutamate production in *Corynebacterium glutamicum*. *Biosci Biotechnol Biochem* 81, 2130–2138.

77. Nakayama, Y., Hashimoto, K.I., Kawasaki, H. and Martinac, B. (2019). "Force-From-Lipids" mechanosensation in *Corynebacterium glutamicum*. *Biophys Rev* 11, 327–333.

78. Tauch, A., Puhler, A., Kalinowski, J. and Thierbach, G. (2003). Plasmids in *Corynebacterium glutamicum* and their molecular classification by comparative genomics. *J Biotechnol* 104, 27–40.

79. Nesvera, J. and Patek, M. (2011). Tools for genetic manipulations in *Corynebacterium glutamicum* and their applications. *Appl Microbiol Biotechnol* 90, 1641–1654.

80. Eggeling, L. and Reyes, O. (2005). Experiments. In *Handbook of Corynebacterium glutamicum*, Eggeling, L. and Bott, M., eds., CRC Press, Boca Raton, FL.

81. Sandoval, H., del Real, G., Mateos, L.M., Aguilar, A. and Martín, J.F. (1985). Screening of plasmids in non-pathogenic corynebacteria. *FEMS Microbiol Lett* 27, 93–98.

82. Tauch, A., Kirchner, O., Löffler, B., Gotker, S., Pühler, A. and Kalinowski, J. (2002). Efficient electro transformation of *Corynebacterium diphtheriae* with a mini-replicon derived from the *Corynebacterium glutamicum* plasmid pGA1. *Curr Microbiol* 45, 362–367.

83. Schäfer, A., Kalinowski, J., Simon, R., Seep-Feldhaus, A.H. and Pühler, A. (1990). High-frequency conjugal plasmid transfer from gram-negative *Escherichia coli* to various gram-positive coryneform bacteria. *J Bacteriol* 172, 1663–1666.

84. Sonnen, H., Thierbach, G., Kautz, S., Kalinowski, J., Schneider, J., Puhler, A. and Kutzner, H.J. (1991). Characterization of pGA1, a new plasmid from *Corynebacterium glutamicum* LP-6. *Gene* 107, 69–74.

85. Satoh, Y., Hatakeyama, K., Kohama, K., Kobayashi, M., Kurusu, Y. and Yukawa, H. (1990). Electrotransformation of intact cells of *Brevibacterium flavum* MJ-233. *J Ind Microbiol* 5, 159–165.

86. Sandoval, H., Aguilar, A., Paniagua, C. and Martín, J.F. (1984). Isolation and physical characterization of plasmid pCC1 from Corynebacterium callunae and construction of hybrid derivatives. *Appl Microbiol Biotechnol*. 19, 409–413.

87. Serwold-Davis, T.M., Groman, N. and Rabin, M. (1987). Transformation of *Corynebacterium diphtheriae*, *Corynebacterium ulcerans*, *Corynebacterium glutamicum*, and *Escherichia coli* with the *C. diphtheriae* plasmid pNG2. *Proc Natl Acad Sci U S A* 84, 4964–4968.

88. Srivastava, P., Nath, N. and Deb, J.K. (2006). Characterization of broad host range cryptic plasmid pCR1 from *Corynebacterium renale*. *Plasmid* 56, 24–34.

89. Reinscheid, D.J., Kronemeyer, W., Eggeling, L., Eikmanns, B.J. and Sahm, H. (1994). Stable expression of *hom-1-thrB* in *Corynebacterium glutamicum* and its effect on the carbon flux to threonine and related amino acids. *Appl Environ Microbiol* 60, 126–132.

90. Kirchner, O. and Tauch, A. (2003). Tools for genetic engineering in the amino acid-producing bacterium *Corynebacterium glutamicum*. *J Biotechnol* 104, 287–299.

91. Larsen, J.E., Gerdes, K., Light, J. and Molin, S. (1984). Low-copy-number plasmid-cloning vectors amplifiable by derepression of an inserted foreign promoter. *Gene* 28, 45–54.

92. Songer, J.G., Hilwig, R.W., Leeming, M.N., Iandolo, J.J. and Libby, S.J. (1991). Transformation of *Corynebacterium pseudotuberculosis* by electroporation. *Am J Vet Res* 52, 1258–1261.

93. Cerdeira, L.T. et al. (2011). Whole-genome sequence of *Corynebacterium pseudotuberculosis* PAT10 strain isolated from sheep in Patagonia, Argentina. *J Bacteriol* 193, 6420–6421.

94. Dorella, F.A., Estevam, E.M., Cardoso, P.G., Savassi, B.M., Oliveira, S.C., Azevedo, V. and Miyoshi, A. (2006). An improved protocol for electrotransformation of *Corynebacterium pseudotuberculosis*. *Vet Microbiol* 114, 298–303.

95. Ankri, S., Serebrijski, I., Reyes, O. and Leblon, G. (1996). Mutations in the *Corynebacterium glutamicum* proline biosynthetic pathway: a natural bypass of the *proA* step. *J Bacteriol* 178, 4412–4419.

96. Bonamy, C., Labarre, J., Cazaubon, L., Jacob, C., Le Bohec, F., Reyes, O. and Leblon, G. (2003). The mobile element IS*1207* of *Brevibacterium lactofermentum* ATCC 21086: isolation and use in the construction of Tn*5531*, a versatile transposon for insertional mutagenesis of *Corynebacterium glutamicum*. *J Biotechnol* 104, 301–309.

97. Simic, P., Sahm, H. and Eggeling, L. (2001). L-threonine export: use of peptides to identify a new translocator from *Corynebacterium glutamicum*. *J Bacteriol* 183, 5317–5324.

98. Mormann, S., Lömker, A., Rückert, C., Gaigalat, L., Tauch, A., Pühler, A. and Kalinowski, J. (2006). Random mutagenesis in *Corynebacterium glutamicum* ATCC 13032 using an IS*6100*-based transposon vector identified the last unknown gene in the histidine biosynthesis pathway. *BMC Genomics* 7, 205.

99. Inui, M., Tsuge, Y., Suzuki, N., Vertes, A.A. and Yukawa, H. (2005). Isolation and characterization of a native composite transposon, Tn*14751*, carrying 17.4 kilobases of *Corynebacterium glutamicum* chromosomal DNA. *Appl Environ Microbiol* 71, 407–416.

100. Tsuge, Y., Ninomiya, K., Suzuki, N., Inui, M. and Yukawa, H. (2005). A new insertion sequence, IS*14999*, from *Corynebacterium glutamicum*. *Microbiology* 151, 501–508.

101. Dorella, F.A., Estevam, E.M., Pacheco, L.G., Guimaraes, C.T., Lana, U.G., Gomes, E.A., Barsante, M.M., Oliveira, S.C., Meyer, R., Miyoshi, A. and Azevedo, V. (2006). In vivo insertional mutagenesis in Corynebacterium pseudotuberculosis: an efficient means to identify DNA sequences encoding exported proteins. *Appl Environ Microbiol*. 72, 7368–7372.

102. Suzuki, N. and Inui, M. (2013). Genome engineering of *Corynebacterium glutamicum*. In *Corynebacterium glutamicum: biology and biotechnology*, Yukawa H., Inui M., (eds.), vol. 23, pp. 89–105, Springer-Verlag, Berlin, Heidelberg.

103. Goryshin, I.Y., Jendrisak, J., Hoffman, L.M., Meis, R. and Reznikoff, W.S. (2000). Insertional transposon mutagenesis by electroporation of released Tn*5* transposition complexes. *Nat Biotechnol* 18, 97–100.

104. Oram, D.M., Avdalovic, A. and Holmes, R.K. (2002). Construction and characterization of transposon insertion mutations in *Corynebacterium diphtheriae* that affect expression of the diphtheria toxin repressor (DtxR). *J Bacteriol* 184, 5723–5732.

105. Kolodkina, V., Denisevich, T. and Titov, L. (2011). Identification of *Corynebacterium diphtheriae* gene involved in adherence to epithelial cells. *Infect Genet Evol* 11, 518–521.

106. Spinler, J.K., Zajdowicz, S.L.W., Haller, J.C., Oram, D.M., Gill, R.E. and Holmes, R.K. (2009). Development and use of a selectable, broad-host-range reporter transposon for identifying environmentally regulated promoters in bacteria. *FEMS Microbiol Lett* 291, 143–150.

107. Wang, C., Hayes, B., Vestling, M.M. and Takayama, K. (2006). Transposome mutagenesis of an integral membrane transporter in *Corynebacterium matruchotii*. *Biochem Biophys Res Commun* 340, 953–960.

108. Suzuki, N., Okai, N., Nonaka, H., Tsuge, Y., Inui, M. and Yukawa, H. (2006). High-throughput transposon mutagenesis of *Corynebacterium glutamicum* and construction of a single-gene disruptant mutant library. *Appl Environ Microbiol* 72, 3750–3755.

109. Jiang, Y., et al. (1917). CRISPR-Cpf1 assisted genome editing of *Corynebacterium glutamicum*. *Nat Commun* 8, 15179.

110. Yan, M.Y., Yan, H.Q., Ren, G.X., Zhao, J.P., Guo, X.P. and Sun. Y.C. (2017). CRISPR-Cas12a-assisted recombineering in bacteria. *Appl Environ Microbiol* 83, e00947–17.

111. Krumbach K. Sonntag, C.K., Eggeling, L., and Mariehagen, J. (2019). CRISPR-Cas12a mediated genome editing to introduce amino acid substitutions into the mechanosensitive channel MscCG of *Corynebacterium glutamicum. ACS Synth Biol* 8, 2726–2734.

112. Abbouni, B., Elhariry, H.M. and Auling, G. (2004). Overproduction of NAD$^+$ and 5'-inosine monophosphate in the presence of 10 μM Mn^{2+} by a mutant of *Corynebacterium ammoniagenes* with thermosensitive nucleotide reduction (*nrd*^{ts}) after temperature shift. *Arch Microbiol* 182, 119–125.

113. Becker, J., Rohles, C.M. and Wittmann, C. (2018). Metabolically engineered *Corynebacterium glutamicum* for bio-based production of chemicals, fuels, materials, and healthcare products. *Metab Eng* 50, 122–141.

114. Yukawa, H. and Inui, M. (2013). *Corynebacterium glutamicum: biology and biotechnology.* Springer, Heidelberg, New York.

115. Holmes, R.K. (2000). Biology and molecular epidemiology of diphtheria toxin and the tox gene. *J Infect Dis.* 181 (Suppl 1), S156–167.

116. Truelove, S.A., Keegan, L.T., Moss, W.J., Chaisson, L.H., Macher, E., Azman, A.S. and Lessler, J. (2020) Clinical and epidemiological aspects of diphtheria: a systematic review and pooled analysis. *Clin Infect Dis* 71, 89–97.

28

The Family Enterobacteriaceae

J. Michael Janda and Denise L. Lopez

CONTENTS

Introduction

There is no single family in the δ class of the phylum Proteobacteria that has had a greater impact on medicine and public health, molecular genetics and phylogeny, pathogenesis, gene structure, regulation, and function, or microbial ecology than the Enterobacteriaceae. The family currently consists of more than 65 validated genera including over 350 species (http://www.bacterio.net). About 25% of current genera consist of a single species (e.g., Plesiomonas) while other genera contain more than 20 species (e.g., Erwinia, Xenorhabdus). The metamorphosis of the family over the past few decades and the logarithmic explosion in the number of recognized taxa is due to pioneering studies conducted by the Centers for Disease Control and Prevention and the Pasteur Institute (French: Institut Pasteur) and the upsurge in phylogenetic investigations employing molecular chronometers such as digital DNA-DNA hybridization, 16S rDNA and rpoB gene sequencing, matrix-assisted laser desorption/ionization time of flight mass spectrometry (MALDI-TOF), amino acid sequence identity, multilocus sequence analysis (MLST), average nucleotide identity, and genome-to-genome distance calculator.[1] These investigations have led to dramatic classification changes within some important members of the family with some genera, such as Enterobacter, experiencing marked attrition in the number of valid species within the taxon. The latest proposed additions to the family Enterobacteriaceae include the genera Metakosakonia, Pluralibacter,[2] and Mixta.[3]

In contrast to the common misconception that enterobacteria are solely inhabitants of the gastrointestinal tract of vertebrates, the family Enterobacteriaceae is widely dispersed in nature and many species exist in free-living states in diverse niches in the biosphere. These ecosystems can be broadly broken down into four major groups, namely strains (1) principally associated in commensal or saprophytic states in the alimentary tract of humans and other vertebrates or at extraintestinal sites; (2) intimately associated as epi- or endophytic symbionts with plants or plant-associated diseases; (3) chiefly found in water, soil, invertebrate species, and industrial processes; and (4) obligate endosymbionts or commensals of insects.[4] Some species have developed such stable symbiotic relationships with their hosts that no genomic rearrangements or gene acquisitions have occurred over the past 50 million years.[4]

Due to the "molecular taxonomy" revolution, the classic biochemical features that previously defined membership in this family are no longer used routinely, but instead rely almost exclusively on genetic relatedness (Table 28.1). Biochemical traits are still valid for the identification of most common genera and species in clinical laboratory settings but rare species or new genera and species most recently proposed for inclusion in the family defy traditional phenotypic identification. For bacteria that cannot be grown on routine media or that lack the required gene arrays to be cultured in vitro, inclusion in the family Enterobacteriaceae is based on a genetic rather than phenetic definition. Accurate biochemical identification of less commonly encountered groups of bacteria in this family requires reliance on a large battery of phenotypic tests, many of which are not readily available on commercial panels. Current clinical laboratory alternatives include 16S rDNA gene sequencing and MALDI-TOF although even proprietary databases associated with these commercial products may not achieve the discriminatory level needed without in-house matrix supplementation. Whole genome sequencing (WGS) may be relevant in particular settings. Although traditionally most members of this family have been susceptible to a wide variety of antimicrobial agents, including extended-spectrum cephalosporins, the exceptionally high usage of these drugs in hospital settings has led to the development of resistance to many β-lactam antibiotics via extended-spectrum β-lactamases (ESBLs) and carbapenemases.[5] Of particular concern is the recent emergence of carbapenemases

DOI: 10.1201/9781003099277-30

TABLE 28.1

Defining Phenotypic Traits of the Family *Enterobacteriaceae*

Character	Marker for Family	Notable Exceptions
Metabolism		
Facultatively anaerobic rods	+	*Alterococcus agarolyticus*
Morphology		
Gram reaction	–	
Structure		
Possession of ECA	+	*Dickeya chrysanthemi*
Spore formation	–	*Serratia marcescens (rare)*
Biochemical		
Oxidase	–	*Plesiomonas*
Catalase	+	*Shigella dysenteriae 1, Xenorhabdus*
Nitrate reductase	+	*Erwinia, Lonsdalea, Yersinia (some)*
D-glucose	+	
D-xylose	+	*Cedecea (some), Cosenzea, Edwardsiella, Lonsdalea, Morganella*

Note: ECA, enterobacterial common antigen.

and colistin resistance in *Klebsiella pneumoniae* (KPC, NDM-1, MRC), *Enterobacter cloacae* (KPC, VIM, OXA-48), and *Escherichia coli* (KPC, NDM-1, OXA-48, MRC), which are a major concern in healthcare settings.[5] Carbapenems-resistant *Enterobacteriaceae* are commonly found in wildlife, pets, and livestock presenting a significant public health risk to the general population.[6]

Many species in the family *Enterobacteriaceae* serve as prototypes or models for studying global processes thought to play important roles in microbial communities, cell-to-cell communication, and gene expression with regard to virulence. *E. coli* has long been an established vector system to clone and express genes and their gene products. Other high-profile processes that enterobacteria are thought to be intimately involved in include quorum sensing (*E. coli*, *Salmonella*) and biofilm formation (*E. coli*, *Klebsiella*, *Salmonella*).[7]

Klebsiella and *Enterobacter*

Klebsiella

Members of the genus *Klebsiella* are widely distributed in nature and are ubiquitous in forest environments, soil, vegetation, and water.[4] They can exist in a free-living state in the environment for prolonged periods of time. Klebsiellae are also found as normal commensals in the gastrointestinal tract of many vertebrates and mammals, including birds, reptiles, and even insects. As animal pathogens they cause diseases such as mastitis and septicemia in bovine and porcine species respectively.

It is presently difficult attempting to define biochemical characteristics for inclusion in the genus *Klebsiella* because the genus continues to undergo major taxonomic revisions. Most traditional klebsiellae are nonmotile, Voges-Proskauer (VP) positive (acetylmethylcarbinol production), and indole-negative; however,

there are exceptions to each of these traits. Other common traits that are associated with many *Klebsiella* species include urea hydrolysis and fermentation of *m*-inositol, a carbohydrate-like compound that is not commonly utilized by many other enterobacteria. With the revised taxonomy within the genus, many isolates are difficult or impossible to identify to species without the use of PCR, MALDI-TOF, or specific gene sequencing.[8] Many klebsiellae isolates, particularly those of *K. pneumoniae*, can produce copious capsular material (K antigen) that can often be observed on primary plating of isolates recovered from clinical specimens.

With current revisions the genus *Klebsiella* is composed of 13 species. Several former species have been transferred to the genus, *Raoultella*,[9] while others relegated to subspecies status within *K. pneumoniae* (ssp. *ozaenae* and ssp. *rhinoscleromatis*). The species names *Enterobacter aerogenes* and *Klebsiella mobilis* are homotypic synonyms and share the same type strain ATCC 13048[T]. The type strain is clearly a klebsiellae. Based upon rules of nomenclature the proposal has been made to use the name *Klebsiella aerogenes* to resolve this taxonomic quagmire.[10]

Klebsiella, and in particular *K. pneumoniae* and *Klebsiella oxytoca*, are extremely important human pathogens in the healthcare setting. These species are often responsible for devastating healthcare-associated illnesses such as septicemia and endocarditis, pneumonia, infections of the face (orbital cellulitis and endophthalmitis), bones (osteomyelitis), and urinary tract.[4] Of particular concern with regard to the treatment of such infections is the rapid emergence of ESBLs in these species and the difficulty in eradicating such systemic infections with the currently available arsenal of drugs.[5,11] Furthermore, hypervirulent *K. pneumoniae* infections, originally reported from Asia, are now being seen globally. These strains can cause pyogenic liver and lung abscesses with possible secondary complications such as endophthalmitis and meningitis. Infections are primarily caused by *K. pneumoniae* strains carrying the K1 or K2 capsular antigens plus specific other virulence determinants.[12,13] Persons with underlying conditions such as diabetes, or who are of Asian ancestry, appear to be especially prone to developing such infections. Hypervirulent *K. pneumoniae* are also being reported as significant causes of monomicrobial necrotizing fasciitis[14] and a call for consensus definition and international collaboration has been made.[15] Outbreaks associated with high neonatal mortality rates have also been recently linked to hypervirulent *K. variicola*.[16] Another evolving human infection associated with klebsiellae concerns *Klebsiella oxytoca* antibiotic-associated hemorrhagic colitis.[17] This hospital-acquired infection is most often observed in patients receiving penicillin or nonsteroidal anti-inflammatory drugs and is uniquely associated with certain cytotoxin-producing *Klebsiella oxytoca* isolates.[17,18]

Enterobacter

Most of the statements previously made regarding the genus *Klebsiella* equally apply to the genus *Enterobacter*. This includes their ubiquitous environmental distribution (trees, plants, crops, soil, water, foods), an incredibly complicated taxonomy, a lack of definitive phenotypic traits to characterize the genus, enhanced drug resistance mediated via ESBLs, and their overall importance as healthcare-associated pathogens.[4,19,20]

This genus has numerous complex taxonomic issues. From a peak of more than 30 validated species within this genus, less than 20 are now considered true members (http://www.bacterio.net/enterobacter.html). Former members of this genus have been transferred to newly created genera including *Lelliottia*, *Kosakonia*, and *Pluralibacter*, while two prominent species, *Enterobacter sakazakii* and *Enterobacter agglomerans*, have been transferred to the genera *Cronobacter* and *Pantoea*, respectively. *E. cloacae*, the preeminent pathogen of this genus, as currently defined, is genetically heterogeneous, being composed of from 5 to 12 distinct genomic groups within the complex that cannot be unambiguously separated from one another on a biochemical basis.[4,21,22] The second most common species in the genus, *E. aerogenes*, belongs in the genus *Klebsiella* (as discussed previously). Most members of the genus are VP and ornithine decarboxylase-positive, utilize citrate, and ferment L-arabinose, L-rhamnose, raffinose, and D-xylose; exceptions to each of these characteristics exist.[4,21] Some species such as *Enterobacter sakazakii* (now *Cronobacter sakazakii*) often produce yellow-pigmented colonies upon subculture.

E. cloacae causes many serious, life-threatening medical complications associated with hospitalization, among which are pediatric and adult bacteremia, central nervous system (CNS) disease and meningitis, and respiratory and urinary tract infections.[4,19,20] A less common but equally devastating pathogen is *Cronobacter* (*Enterobacter*) *sakazakii*, which causes serious invasive disease (meningitis, brain abscess) in infants and is linked to contaminated powered infant formula.[22,23] *E. cloacae* and *Enterobacter asburiae* produce saxitoxins in the rumina of cattle associated with bovine paraplegic syndrome and can cause coliform mastitis in dairy cows.[4]

Proteus, Morganella, **and** Providencia

In the 1980s, these three genera were grouped together in the tribe Proteeae, based on similar phenotypic and morphologic features. In many ways these associations still remain valid. Phylogenetically, these genera appear at the periphery of the *Enterobacteriaceae* based on 16S rDNA gene sequencing, and they are only 20% or less related to core members of this family (such as *E. coli*, *Salmonella*, *Enterobacter*, and *Citrobacter*).

A number of biochemical features are almost exclusively associated with these three genera in the family *Enterobacteriaceae* and include production of phenylalanine deaminase, elaboration of a reddish-brown pigment on media containing DL-tryptophan, and the ability to degrade L-tryosine crystals via a tyrosine phenol-lyase. The most remarkable distinguishing feature within this tribe is the ability of many *Proteus* strains belonging to the species *Proteus mirabilis* and *Proteus vulgaris* to form concentric rings of outgrowth on solid media such as MacConkey agar.[4,24] This swarming phenomenon is due to conversion of swimmer cells in broth to swarmer cells in agar and is associated with cellular elongation and increased flagellin synthesis. Other features that help separate these three genera from one another include formation of H_2S on Triple Sugar Iron (TSI) agar slants and degradation of gelatin by *Proteus* and the fermentation of mannose by both *Morganella* and *Providencia*.

The singular disease principally associated with *Proteus*, *Morganella*, and *Providencia* is urinary tract infections (UTIs), and the preeminent pathogen of this group is *P. mirabilis* (up to 5% of all UTIs).[4,24,25] *P. mirabilis* typically produces uncomplicated UTIs in healthy women and less frequently in young children. Less often, *P. mirabilis* can cause complicated UTIs in persons with structural, physiologic, or neurologic disorders of the urinary tract necessitating catheterization.[24] The virulence factors thought to be operative in *P. mirabilis* UTIs have been extensively studied and include MR/P fimbriae, a urease, cytotoxins, an IgA degrading protease, an hemolysin, and flagella.[24] Proteeae are also common causes of bacteremia[26] and a variety of other less frequently encountered clinical conditions including CNS disease, wound infections, and rheumatoid arthritis.[4] *Providencia alcalifaciens* and related species are associated with travelers' diarrhea and foodborne outbreaks of gastroenteritis in Japan and Kenya.[27,28] Recent studies suggest that *Proteus* spp. may play a role in various gastrointestinal disturbances including gastroenteritis and Crohn's disease.[29]

Members of the Proteeae are often isolated from homeothermic and poikilothermic species, including dogs, cows, birds, snakes, and fish.[4] Some studies have found clonal relatedness between strains recovered from humans and their companion animals.[30] *Morganella morganii*, a copious histamine producer, is associated with the spoilage of fish and the clinical condition known as scombroid poisoning.[31] A number of infections in animals have been attributed to these groups, including equine abortions, hoof canker, stomatitis, neonatal diarrhea in calves, and a large variety of additional systemic infections.

Salmonella

The genus *Salmonella* has two species, *S. bongori* (previously subspecies V) and *S. enterica*, with the latter species subdivided into six subspecies (I, II, IIIA, IIIB, IV, and VI).[32] All of the commonly encountered serotypes belong in subspecies I of *S. enterica* (subsp. *enterica*) while organisms previously known as *Arizona* fall within subspecies IIIA and IIIB. *S. bongori* and serotypes from all six *S. enterica* subspecies have been isolated from humans although *S. bongori* and *S. enterica* subspecies VI isolates are rare.[33] *Salmonella* is part of the intestinal microbiome of many reptiles, which serve as the most common source for human infections caused by subspecies II, IIIA, IIIB, and IV.[34,35] As the practice of maintaining reptiles as pets has increased, so has the risk of *Salmonella* transmission to humans and other animals.[35] Several serotypes are host-adapted (but can cause disease in other hosts, e.g., Dublin in cattle) or host-restricted (Gallinarum in poultry, Typhi in humans) but most serotypes cause disease in both warm- and cold-blooded animals. The gastrointestinal tract is the natural habitat for salmonellae but unlike *E. coli* it can survive for weeks to years in water and soil. Any food contaminated by fecal matter, including fresh fruits, vegetables, and shellfish, may serve as a source of salmonella infection; however, meats, raw milk, and eggs are the most frequently involved causes.[4] Poultry farms can have high prevalence of low-virulence *Salmonella* serotypes, which has important public health implications because horizontal gene transmission has been documented in this genus, as has antibiotic resistance.[35] Infections may be acquired by direct transmission from both domesticated and wild animals and birds, although infrequently.

Nontyphoidal salmonellae (NTS) are a major cause of diarrheal disease in both industrialized and developing countries. In the United States in 2015, the annual incidence rate was estimated to be 15.7 cases/100,000 population.[36] *Salmonella* Enteritidis can be found in the egg yolks from seemingly healthy chickens. This serotype traveled, with the transport of infected chickens, down the east coast in the late 1980s and is now the most common serotype of *Salmonella* associated with gastroenteritis in the United States.[37] In 2015, it made up 19.2% of serotypes isolated, followed by Newport (11.4%), and Typhimurium (10.4%). The emergence of multidrug-resistant strains have raised the importance of more precise source tracking methods, using molecular tools such as WGS.[38] Multidrug-resistant strains related to *Salmonella* Enteritidis and another subspecies *enterica* serotype, *S.* Typhimurium, have been associated with outbreaks worldwide. Drug-resistance is a major public health concern and *Salmonella* have demonstrated resistance to multiple classes of antimicrobials, including penicillin, sulfonamides, tetracyclines, fluoroquinolones, aminoglycosides, and cephalosporins.[35] For typhoid fever (caused by *Salmonella enterica* serotype Typhi) there are more than 20 million new cases per year.[39] Highest incidence rates occur in sub-Saharan Africa (>700 cases/100,000 population), and, of particular concern, multidrug-resistant (MDR) strains are on the rise. There have even been cases of extensively drug-resistant (XDR) strains, which demonstrate resistance to cephalosporins and can result in treatment failure.[40] The next highest incidence rates occur in other areas of Asia, Africa, Latin America, the Caribbean, and Oceania (exclusive of Australia and New Zealand) with rates of 15 to >400.[41] The typhoid vaccine is an important public health control measure in areas with high incidence rates and antimicrobial resistance. In industrialized countries (Europe, North America, etc.), the rate is less than <1 per 100,000 population.[33] Mortality rates are difficult to determine for typhoid fever. Current estimates suggest a 1% overall case fatality rate.[4,41] Gastrointestinal illness, the most common infection caused by NTS, is characterized by abdominal pain and watery diarrhea (mild to cholera-like) with mucus and, in some cases, blood (occult or visible). Patients with NTS may develop a temporary carrier state usually lasting 1–2 months (occasionally up to a year with infants). Permanent carriers are rare (1%) with NTS infection; however, 2–5% of typhoid patients will become chronic carriers, shedding high levels of *S.* Typhi while outwardly asymptomatic; 25% of typhoid carriers have no history of disease.[4] NTS cause a variety of extraintestinal illnesses, the most common and/or severe of which include sepsis, UTI, meningitis, and osteomyelitis. Enteric fever is caused by serotypes Typhi and Paratyphi A and C, all of which are host-adapted to humans, and tartrate-negative strains of serotype Paratyphi B. Untreated typhoid fever has three stages and persists approximately 3–4 weeks.[42] During the first week, symptoms are consistent with gastroenteritis, and stool cultures will be transiently positive while blood is usually negative. The second stage includes bacteremia and other extraintestinal symptoms, and cultures from blood, stool, bone marrow, and rose spots (if present) are generally positive. If complications arise, they usually occur in the third week; otherwise, the patient begins convalescence;

cultures are not usually positive at this time, with the possible exception of stool.

Both NTS and serotypes causing enteric fever are easily isolated from stool in the acute phase of disease on traditional enteric isolation media such as MacConkey (MAC), xylose-lysine-desoxycholate (XLD), Hektoen (HE), and/or Salmonella-Shigella (SS) agars. Other media specific for salmonellae are also available (bismuth sulfite, brilliant green) as are a variety of chromogenic agars. To detect carriers, preincubation in enrichment broth is recommended. Selenite F with cystine and Hajna's Gram-negative broth are generally used; tetrathionate does not recover serotype Typhi and should not be used for human specimens. Most standard blood culture methods are satisfactory for recovery of both NTS and agents of typhoid fever; however, for the latter group, use of 10% Oxgall in distilled water in lieu of tryptic soy broth and using a 1:10 dilution of blood (5 mL) to medium (25 mL), or culture of bone marrow, will increase the chance of recovery of these agents.[43,44] Serological assays are not useful for diagnosis. However, PCR is sensitive and specific and the faster turnaround time may be a useful complement to culture. All commercial systems approved for clinical diagnostic use in the United States identify *Salmonella* satisfactorily for phenotypically typical strains;[4] however, they fail to identify lactose-, sucrose-, or indol-positive strains of salmonellae. Strains of serotype Typhi are anaerogenic, produce little or no H_2S, and give negative reactions for citrate and ornithine. All subspecies of NTS, including serotype Typhi, are easily identified with a minimal number of conventional media, including lactose- and sucrose-positive strains that are not correctly identified using commercial systems. Most subgroup I strains of NTS are indol- and *o*-nitrophenyl-β-D-galactopyranoside (ONPG)-negative and H_2S-, LDC-, and citrate-positive and ferment dulcitol, a combination distinctive for salmonellae. Subgroup III strains are ONPG- and malonate-positive and dulcitol-negative. DNA- and protein-based methods of identification are increasingly common in clinical laboratories, in particular WGS and MALDI-TOF MS.[45]

For epidemiological purposes, salmonellae have classically been characterized to serotype using the Kauffmann-White schema.[46] Cell surface (somatic) antigens place the organism within a group while the type is determined by a particular combination of flagellar antigens. There are more than 2,000 serotypes of salmonellae distributed in more than 40 groups, although approximately 95% of salmonellae isolated from humans belong to groups A, B, C_1, C_2, D, and E. Typhi strains have a Vi (virulence) capsule that masks the somatic antigen; full identification of these strains requires biochemical and serological testing, including testing for Vi antigen. There are commercial molecular serotyping kits available as a high-throughput alternative to traditional serotyping.[47] Because of their frequent involvement in outbreaks, salmonellae are often subjected to additional genetic characterization. Whereas pulsed-field gel electrophoresis used to be the method of choice, WGS is now preferred, offering additional data on antimicrobial resistance genes and other virulence factors.

As with most enteric pathogens, salmonellae virulence is multifactorial, and both chromosomal and plasmid determinants are involved. Twelve pathogenicity islands (PIs) have been identified in salmonellae, two of which (*Salmonella* PI-1 and PI-2) contain type III secretion systems that are involved in invasion and intracellular multiplication.[48] A number of other chromosomal genes are involved in adhesion and colonization of the bacterium to the intestine and resistance to gastric acidity. Determinants on virulence plasmids mediate colonization, resistance to complement-mediated lysis, and enhance the ability of strains to multiply in extraintestinal tissue. Biofilm formation, which is enhanced in *Salmonella* using a quorum sensing mechanism, is an important virulence factor. Biofilms allow *Salmonella*, subspecies *enterica* in particular, to persist in the environment and throughout the food processing chain.

Shigella and Escherichia Coli

The genera *Shigella* and *Escherichia* genetically reside in a single genus by DNA hybridization, although the genus name, *Shigella*, is presently retained because of the immense medical importance of shigellae in comparison to the commensal nature of most *E. coli*.[49] Pathogenic *E. coli* can be subdivided into two major groups, intestinal pathogenic *E. coli* (IPEC) and extraintestinal pathogenic *E. coli* or ExPEC (see Table 28.2 for groupings and acronyms).[50] ExPEC strains primarily affect humans and the poultry industry (APEC). Hybrids of both groups containing both IPEC and ExPEC virulence genes exist and appear to be increasing.[50] *E. coli* can be serotyped, and specific types are associated with each of the above groups, the most notable of which is *E. coli* O157:H7 with the STEC group. Shigellae are broken down into four groups based on their cell surface (somatic) antigens and are designated as A (*Shigella dysenteriae*), B (*Shigella flexneri*), C (*Shigella boydii*), and D (*Shigella sonnei*).[51] Three groups (A, B, and C) are divided into types and *S. flexneri* types are further subdivided into subtypes. Currently subgroup A (*S. dysenteriae*) contains 17 serotypes, subgroup B (*S. flexneri*) 14 serotypes, and subgroup C (*S. boydii*) 19 serotypes.[51] *S. boydii* 13 has been reclassified as *Escherichia albertii*.

E. coli makes their way into the food chain via fecal contamination of animal carcasses during slaughter, with subsequent improper handling (storage and/or cooking) of the product or by fecal adulteration of agricultural crops in the field or packing sheds. A study of retail foods at 10 commercial markets in the Minneapolis-St. Paul area sampled between 2001 and 2003 found 24% of the 1,648 food items contaminated with *E. coli*.[52] Of these food items, poultry (92%) and meats (69%) predominated with only 9% of produce and other nonmeat items culture-positive.[52] Shigellae are host-adapted to the human intestine and, outside of humans, have only rarely been found to infect dogs or primates; isolations of shigellae from food and water only occur following contamination with human feces.[4]

E. coli can cause a variety of extraintestinal infections, and has been isolated from virtually every organ and anatomical site. In 2000, in the United States alone, the estimate of mortality from *E. coli* bloodstream infections was 40,000 deaths, with $1.1 to $2.8 billion in healthcare costs.[53] Similar values for *E. coli* UTIs included 6–8 million cases of uncomplicated cystitis, 250,000 cases of pyelonephritis, and a quarter- to half-million cases of catheter-associated UTI.[53] The World Health Organization and CDC estimate that there are between 80 and 165 million cases of shigella per year on a worldwide basis with >90% occurring in developing countries.[51] In comparison, the CDC estimates that approximately 500,000 cases of shigellosis with 38 deaths occur in the United States yearly.[51,54,55] The mortality rate in developing countries exceeds 1 million, and 61% of deaths occur in children under 5 years of age. In industrialized nations, the estimate of infections is 1.5 million per year with a 0.2% mortality rate. The predominant two species causing shigellosis are *S. flexneri* and *Shigella sonnei*. *Shigella sonnei* appears to be replacing *S. flexneri* as the predominant species in developing countries and in regions undergoing transitional environments.[55]

Cases of *E. coli* sepsis usually arise following UTI and occur more frequently in women than in men. Approximately 85–95% of patients who develop *E. coli* sepsis have underlying conditions with liver disease, hematologic malignancies, and solid tumors or other cancers being the most common.[4] *E. coli* meningitis can occur in any age group; neonates primarily acquire infection by vertical transmission from colonized mothers during or immediately following delivery and less frequently via the nosocomial route.[56] Approximately 90% of these strains carry a K1 capsule (which is identical to the type B polysaccharide of *Neisseria meningitidis*) and an increasing number of strains carry the CTX-M type or TEM-type extended spectrum β-lactamase.[4,56]

TABLE 28.2

Pathogenic Groups of *Escherichia*

	Group Designation	Disease, Site of Infection	Pathogenic Mechanism
Diarrheagenic	Enteropathogenic (EPEC)	Watery diarrhea, small intestine	Adherence
	Enteroinvasive (EIEC)	Watery diarrhea followed by bloody diarrhea, colon	Invasion
	Enterotoxigenic (ETEC)	Watery diarrhea, small intestine	Heat-stable and/or heat-labile toxin
	Shiga-toxin producing (STEC) Also known as: Enterohemorrhagic (EHEC) Verotoxigenic (VTEC)	Watery diarrhea, sometimes bloody, especially for O157:H7, colon	Shiga toxins, type 1 and/or type 2
	Enteroadherent (EAEC)	Watery diarrhea, small intestine and colon	Adherence
	Diffuse adhering (DAEC)	Watery diarrhea, small intestine	Adherence
Extraintestinal (ExPEC)	Uropathogenic *E. coli* (UPEC)	Urinary tract, bloodstream, meninges	Adherence, toxins, biofilm formation, siderophore
	Neonatal Meningitis *E. coli* (NMEC)	Meninges, cerebrospinal fluid, blood	Cell invasion, complement inhibition, meningitis-associated fimbriae

E. coli cystitis is characterized by acute dysuria and lower back pain, while pyelonephritis involves the upper urinary tract (kidneys and/or pelvis) and may be of a mild nature or cause vomiting and nausea with concomitant sepsis.[57] Gastrointestinal symptoms for the diarrheagenic groups of *E. coli* are given in Table 28.2. *Shigella* infections are typically characterized by generalized symptoms of fever, fatigue, anorexia, and malaise at onset, followed by watery diarrhea and abdominal pain.[58] Blood and mucus are present in approximately 40% and 50% of stools, respectively, with bloody stools occurring more commonly in children than in adults.[4] Although such isolations are infrequent, shigella can be isolated from sources other than stool, including urine, blood, wounds, sputum, gallbladder, and ears. *Shigella* bacteremia is rare, occurring primarily in developing countries in malnourished children with decreased serum bactericidal activity.[4] Both STEC and shigellae, especially *S. dysenteriae* type 1, can cause hemolytic uremic syndrome (HUS), a disease characterized by microangiopathic hemolytic anemia, thrombocytopenia, and acute renal failure.[59] *Shigella*- and *E. coli*-associated HUS is linked to the production of Shiga-toxin (Stx) by these organisms. In the United States, the most common serotype associated with HUS is O157:H7; however, other serotypes of STEC and shigellae are capable of causing the disease. Recently two clusters of *Shigella sonnei* producing Stx1 have caused more than 50 illnesses in California.[60] Although no cases of HUS were detected, 71% of the case-patients presented with bloody diarrhea.

E. coli can be easily isolated from blood by routine blood culture methods, but if a single bottle is used, aerobic incubation recovers more isolates than anaerobic (nonvented) incubation.[4] Chromogenic agars are equal or superior to routine media (blood or MacConkey agars) used for UTI isolation in terms of numbers recovered, and have the added advantage that *E. coli* can be readily distinguished by their colony appearance and easily confirmed by a spot indole test.[4] Isolation of *E. coli* strains involved in diarrheal disease is a much more difficult task because most are biochemically indistinguishable from commensal strains.[61] *E. coli* O157:H7 strains, which are delayed D-sorbitol fermenters, are the exception to this rule, and this trait is exploited in media used for their isolation. Sorbitol MacConkey (SMAC) agar contains D-sorbitol as a substitute for lactose and is the primary isolation medium for O157 strains that present as colorless colonies.[47] Other formulations exist, such as SMAC and CT-SMAC with cefixime and tellurite incorporated to increase selectivity. A variety of chromogenic agars have been developed that detect both O157:H7 STEC as well as non-O157 STEC strains.[62,63] Immunomagnetic separation is a sensitive isolation technique for O157 and for a limited number of other serotypes of STEC, in which somatic antibody is coated on magnetic beads. Enzyme immunoassays (EIAs), which are commercially available and FDA approved, can be used to detect the somatic O157 antigen or Shiga toxin (Meridian Bioscience, Cincinnati, OH; Remel, Lenexa, KS). Stx can be detected directly from stool as well as from an isolate of STEC, but there is a 3-log difference between EIA Stx detection from stool and the more sensitive tissue culture technique using Vero cells. An Oxoid latex agglutination assay for Stx (E. coli Latex Test) is also commercially available (ThermoFisher Scientific, Waltham MA). Both the latex assay and the EIA detect the presence of Stx but unlike the

Vero cell assay with neutralizing antibody to type 1 and 2, neither can detect which toxin type(s), 1 or 2, are present. Multiplex PCR assays are now widely available and FDA-approved, which detect a variety of syndromic diseases including gastroenteritis. Almost all such platforms include both STEC and *Shigella* targets on them such as Luminex GPP panel (Austin TX), Verigene EP (Luminex Corporation), and BioFire GIP (Salt Lake City, UT).[64] Panels detect from 9 to 22 bacterial, viral, and parasitic targets simultaneously.[64] Isolates of *E. coli* from other types of diarrheal diseases not included on syndromic panels or with hybrid characteristics are detected by determining the presence of the virulence factor(s) causing the infection. As with most specialized assays, these tests are available in a limited number of laboratories. Direct plating of stool is the method of choice for the isolation of shigellae for public health reasons; if culture is not performed within 2–4 hours of taking the stool, a transport medium must be used because shigellae are extremely sensitive to metabolic end products of other bacteria, phages, etc. present in stool. MAC, XLD, and HE agars work well for isolation of shigella, as does SS agar, but the latter may be too inhibitory for some strains; colonies of shigella will be colorless on these media (colonies on HE may appear green because of the color of the medium). *Shigella sonnei* is easily recognizable on plating media as it throws off large, rough colonies as well as smaller convex, smooth colonies; laboratories should select the smooth colonies for further testing.

Biochemically, typical strains of *E. coli* present no challenge to commercial identification systems.[4] However, *E. coli* are phenotypically diverse organisms and may be biochemically inactive (indol-, lysine- or ornithine decarboxylase-, and ONPG-negative) or conversely can give positive reactions for tests (H_2S, urea) that are usually negative; such strains are not accurately identified by commercial kits.[4] Shigellae do not fare as well as *E. coli* in commercial identification systems. In data derived from various studies, of 198 strains tested collectively, 14 different systems correctly identified only 162 *Shigella*.[4] There is considerable biochemical variation between species and types within species; distinctive traits include lack of mannitol fermentation for *S. dysenteriae* strains, ODC- and ONPG-positive reactions for *Shigella sonnei*, and gas-production from D-glucose for some strains of *S. flexneri* type 6 and *S. boydii* type 14.

Virulence characteristics are specific for each of the diarrheagenic *E. coli* groups and for strains of ExPEC. Numerous genes, located on both the chromosome and on plasmids, are involved for each group but the primary virulence mechanism is given in Table 28.2. Shigellae are invasive pathogens; a 220-kb plasmid encodes the genes involved in invasion but genes that regulate plasmid-borne virulence factors are located on the chromosome along with other virulence genes.

Yersinia

The genus *Yersinia* presently consists of 18 validly published species, although the legitimacy of *Yersinia wautersii* ("Korean strains") has been questioned. The disease spectrum and pathogenicity of taxa within the genus ranges from humans (*Y. pestis*) to fish (*Yersinia ruckeri*) to insects (*Yersinia entomophaga*). Many of the more recently described environmental species have only been rarely isolated, and then from samples such as soil, vegetation, and water.

Fourteen of the 18 *Yersinia* species have been recovered from clinical samples, which includes the recent isolation of *Y. entomophaga* from the urine of an 85-year-old man.[65] The preeminent human pathogens of the genus are *Yersinia pestis* (bubonic plague), *Yersinia enterocolitica* (enterocolitis, mesenteric lymphadenitis), and *Yersinia pseudotuberculosis* (mesenteric lymphadenitis, septicemia). Environmentally, *Y. ruckeri* is an increasingly important pathogen of fish, causing enteric redmouth disease in salmonids.[66]

While most strains of *Y. pestis* and *Y. pseudotuberculosis* are virulent, only certain *Y. enterocolitica* strains with specific serotype/biotype combinations are thought to cause disease. Currently there are two subspecies of *Y. enterocolitica*: subspecies *enterocolitica* and *paleartica*.[67] In general, bioserotypes of subspecies *enterocolitica* are considered to be more virulent with the exception of subspecies *paleartica* serotype O:3 biotype 4, which is the dominant bioserotype worldwide.[68] Since these subspecies can only be identified by molecular techniques such as 16S rDNA gene sequencing or MLST, the subspecies designations aren't currently used.[69] *Y. pestis* is a zoonotic organism causing epidemic plague in a number of rodents as well as humans.[70] Rats (*Rattus rattus* and *R. norvegicus*) play a major role in the spread of plague but are not significant in maintaining *Y. pestis* in the wild; enzootic rodents highly resistant to the organism (carriers) are believed to be one reservoir as well as soil.[70] Unlike domesticated dogs, ferrets, coyotes, skunks, and raccoons that are highly resistant to infection, cats can develop clinical plague.[71] Lagomorphs, birds, boars, and pigs, and rodents and aquatic environments are major reservoirs for *Y. pseudotuberculosis* and *Y. enterocolitica*, respectively, although both can be isolated from a wide variety of sources.[4] In rare cases, inpatients with underlying disorders may present with infections caused by *Yersinia frederiksensii, Yersinia kristensenii, Yersinia intermedia, Yersinia bercovieri, Yersinia mollaretii,* or *Yersinia rohdei*.

Plague (*Y. pestis*) is a vector-borne (flea) zoonotic infection that can occur in one of three forms (bubonic, septicemic, pneumonic).[72] The organism migrates via the bloodstream to the lymphatic system, causing inflammation of the lymph nodes (buboes) in the groin, axilla, or neck[71] with an incubation period of 2–8 days following the fleabite. Plague is characterized by fever, chills, headache, and weakness, and, if present, gastrointestinal symptoms of abdominal pain, nausea, vomiting, and diarrhea. Secondary pneumonic plaque follows spread from the bloodstream to the respiratory tract and is distinguished by bronchopneumonia, cavitation, or consolidation with bloody or purulent sputum.[71] Patients with secondary pneumonic plague can infect close contacts with primary pneumonic plague in 1–3 days; these cases will be fatal if not treated within 1 day. In primary septicemic plague, which occurs in 25% of cases, symptoms resemble those of any Gram-negative infection (there are no buboes) but the mortality rate is 30–50%. Plague now primarily occurs as sporadic cases associated with rodent contact[70]; however, outbreaks do occur, most notably in Madagascar in 2017 where over 2,000 suspected cases of pneumonic plague were reported.[72] In endemic areas, plague may occur as a mild or unapparent disease, detected primarily in seroprevalence studies.[71] In the United States, squirrels, rabbits, and prairie dogs are the most common vectors involved in plague cases but sources cannot be identified for a third of all infections. Infected cats have transmitted plague to humans by scratches (bubonic) and aerosolized droplets (pneumonic).

Y. pseudotuberculosis is a relatively uncommon zoonotic pathogen and infections are more often seen in Europe where a variety of animals including voles, hares, fowl, and pigs harbor these microbes as pathogens or commensals. While *Y. pseudotuberculosis* is most often acquired by consumption of contaminated foods, the chief clinical manifestations are often not enteric (diarrhea) per se but rather as mesenteric lymphadenitis (mimicking appendicitis), granulomatous disease, or upon dissemination as septicemia. As with *Y. enterocolitica, Y. pseudotuberculosis* occurs in patients with underlying medical complications such as iron overload conditions.[4] Most infections occur in older children and young adults, especially in males.

Enteritis is the most frequently encountered gastrointestinal illness associated with *Y. enterocolitica* but it can also cause mesenteric lymphadenitis and terminal ileitis, conditions that mimic appendicitis. *Y. enterocolitica* is also a major cause of transfusion-related sepsis.[73] *Y. enterocolitica* serogroups of biotype 1A are regarded as nonpathogenic because they lack the virulence characteristics of pathogenic strains. However recent studies suggest some or many biotype 1A strains may cause disease.[74] Patients who carry the HLA-B27 allele are at increased risk of developing reactive arthritis (RA) following gastrointestinal illness; approximately 10% of patients with yersinia-associated RA develop chronic arthritis.[4]

Y. pestis present in blood, bubo aspirates, sputum, or cerebrospinal fluid exhibit a characteristic bipolar appearance when stained with Wayson's or Giemsa stain. Colonies on blood agar are opaque with a fried egg appearance appearing within 48 hours; however, plates should be held 7 days. *Y. pseudotuberculosis* is most often isolated from blood, and routine blood culture techniques are satisfactory. Like, *Y. enterocolitica, Y. pseudotuberculosis* can be isolated from MAC or SS agars although recovery on these media is variable for both organisms and colonies are often small (<1 mm) or pinpoint. On Cefsulodin-Irgasan-Novobiocin (CIN), a selective media designed specifically for isolation of *Y. enterocolitica*, colonies have a bull's-eye appearance with a red center and colorless apron. Both *Aeromonas* and *Plesiomonas shigelloides* also grow on CIN, appearing as bull's-eyes and colorless colonies, respectively, which allow this medium to be used to isolate pathogens other than yersinia. CIN, as prepared for *Y. enterocolitica*, could be inhibitory for strains of *Y. pestis*.

With respect to the identification of yersinia, the performance of commercial identification systems, overall, is mediocre, with *Y. pseudotuberculosis* faring the worst (correctly identified only 83% of the time).[4] Only three systems (API 20E, MicroScan Walk/Away, and VITEK) licensed for clinical use in the United States include *Y. pestis* in their database (Tier 1 Select Agent). All isolates of this organism should be sent to an appropriate public health reference laboratory immediately if the isolate is suspicious for plague. Most kits identify *Y. enterocolitica* as "*Y. enterocolitica* group," a designation that includes *Yersinia frederiksenii, Y. intermedia,* and *Y. kristensenii*, species generally considered nonpathogenic; systems that identify *Y. enterocolitica* as a species (API 20E and RapID) do not distinguish between virulent and nonvirulent biotypes. Using conventional media, *Y. pestis* and *Y. pseudotuberculosis* are nonmotile at 35°C, do not ferment sucrose, and are LDC-, ODC-, and ADH-negative; unlike *Y. pestis*, *Y. pseudotuberculosis* is urea-positive and L-rhamnose-negative. *Y. enterocolitica* hydrolyzes urea, ferments sucrose but not L-rhamnose, and is ODC-positive.

All three pathogenic species have a type III secretion system, also called the Yop virulon, located on a 70-kb plasmid. The Yop virulon is composed of a Ysc secretion apparatus (protein pump), Yop effector (causes pore formation in the host cell), and Yop translocator (blocks the host cell's ability to respond to infection) proteins.[72,75] Furthermore, a chromosomal high-pathogenicity island, containing genes encoding for a siderophore with a high affinity for iron, occurs in *Y. enterocolitica* pathogenic serogroup IB, *Y. pseudotuberculosis* serogroups I and III, and *Y. pestis*.[76]

Additional Members of the Family *Enterobacteriaceae*

There are many other genera and species in the family *Enterobacteriaceae* that cause serious diseases or illnesses in humans, animals, plants, and other living creatures. Many of these diseases are not specific to a single entity; that is, many different species can cause bloodborne disease in humans to varying degrees. However, a number of syndromes are either species-specific or are intimately linked to the genus and/or species, and some of these are presented in Table 28.3.

The genus *Citrobacter* consists of a large number of species (*n* = 15), most of which are lysine decarboxylase-negative (LDC-negative) and citrate-positive. Members of the *Citrobacter freundii* complex are also H$_2$S-positive on TSI. The most common

species involved in human disease belong to the *Citrobacter freundii* complex, where they primarily cause bacteremia and UTIs.[4] A less frequently isolated indole-positive species, *Citrobacter koseri* causes a variety of CNS diseases (ventriculitis, meningitis, brain abscesses) in neonates less than 2 months of age. Risk factors for developing CNS disease caused by *Citrobacter koseri* include a gestation age of less than 37 weeks and low birth weight.[77] *Citrobacter rodentium*, a citrobacter species that does not cause disease in humans, can cause massive outbreaks of morbidity and mortality in mouse and gerbil colonies.[78]

Several other enterobacteria typically cause diarrhea as their most common clinical presentation. *Edwardsiella tarda*, an H$_2$S-, indole-, and LDC-positive species, causes watery to bloody diarrhea, particularly in subtropical regions of the world.[79] Infection is often associated with close contact with reptiles. *P. shigelloides*, the only oxidase-positive member of the family, causes enteritis and is associated with consumption of seafood or shellfish or contaminated water.[80]

At least six genera are considered phytopathogens. Not all isolates of these individual species cause disease under all circumstances, and many exist as commensals or part of the microbial rhizosphere. Many of these strains exhibit varying degrees of pigmentation (e.g., yellow) upon subculture in the laboratory. Specific tests have been designed to measure pathogenicity, such as the onion maceration test.

TABLE 28.3

Selected Examples of Major Diseases Associated with Other Members of the Family *Enterobacteriaceae*

Category	Disease	Agent(s)
Humans	Gastroenteritis	*Edwardsiella tarda, Plesiomonas shigelloides*
	Neonatal meningitis, brain abscess	*Citrobacter koseri Cronobacter sakazakii*
Animals	Murine colonic hyperplasia	*Citrobacter rodentium*
	Amber disease (grub)	*Serratia entomophila*
	Insect sepsis	*Photorhabdus*
	Sepsis in tilapia, ducks, and egrets	*Edwardsiella* spp.
	Sepsis in penguins	*Plesiomonas shigelloides*
Plants	Bark cankers	*Brenneria species*
	Stem canker, fire blight	*Erwinia species*
	Black spot necrosis, pink disease, seed and boil rot, onion rot, brown apical necrosis of walnut, leaf blight and vascular wilt of maize	*Pantoea species*
	Soft rot, black leg disease	*Dickeya species Pectobacterium carotovorum*
	Bark necrosis	*Samsonia*

REFERENCES

1. Mahato, NK, et al. Microbial taxonomy in the era of OMICS: application of DNA sequences, computational tools and techniques. *Antonie van Leeuwenhoek* 110, 1357, 2017.
2. Alnajar, S. and Gupta, R.S. Phylogenomics and comparative genomic studies delineate six main clades within the family *Enterobacteriaceae* and support the reclassification of several polyphyletic members of the family. *Infect. Genet. Evol.* 54, 108, 2017.
3. Palmer, M., et al. *Mixta* gen. nov., a new genus in the *Erwiniaceae. Int. J. Syst. Evol. Microbiol.* 68, 1396, 2018.
4. Janda, J.M., and Abbott, S.L. *The Enterobacteria*, 2nd ed., ASM Press, Washington, DC, 2006.
5. Rodriguez-Baño, J., et al. Treatment of infections caused by extended-spectrum-beta-lactamase-, AmpC-, and carbapenemase-producing *Enterobacteriaceae. Clin. Microbiol. Rev.* 31, e00079-17, 2018.
6. Köck, R, et al. Carbapenem-resistant Enterobacteriaceae in wildlife, food-producing, and companion animals: a systematic review. *Clin. Microbiol. Infect.* 24, 1241, 2018.
7. Wolska, K.I., et al. Genetic control of bacterial biofilms. *J. Appl. Genetics* 57, 225, 2016.
8. Rodrigues, C., et al. Identification of *Klebsiella pneumoniae, Klebsiella quasipneumoniae, Klebsiella variicola* and related phylogroups by MALDI-TOF mass spectrometry. *Front. Microbiol.* 9, 3000, 2018.
9. Drancourt, M., et al. Phylogenetic analyses of *Klebsiella* species delineate *Klebsiella* and *Raoultella* gen. nov., with description of *Raoultella ornithinolytica* comb. nov., *Raoultella terrigena* comb. nov. and *Raoultella planticola* comb. nov.. *Int. J. Syst. Evol. Microbiol.* 51, 925, 2001.

10. Tindall, B.J., Sutton, G. and Garrity, G. *Enterobacter aerogenes* Hormaeche and Edwards 1960 (Approved Lists 1980) and *Klebsiella mobilis* Bascomb et al. 1971 (Approved Lists 1980) share the same nomenclatural type (ATCC 13048) on the Approved Lists and are homotypic synonyms, with consequences for the name *Klebsiella mobilis* Bascomb et al. 1971 (Approved Lists 1980). *Int. J. Syst. Evol. Microbiol.* 67, 502, 2017.

11. Chong, Y., Shimoda, S. and Shimono, N. Current epidemiology, genetic evolution and clinical impact of extended-spectrum β-lactamase-producing *Escherichia coli* and *Klebsiella pneumoniae*. *Infect. Genet. Evol.* 61, 185, 2018.

12. Shon, A.S., Bajwa, R.P.S. and Russo, T.A. Hypervirulent (hypermucoviscous) *Klebsiella pneumoniae*: a new and dangerous breed. *Virulence* 4, 107, 2013.

13. Siu, L.K., et al. *Klebsiella pneumoniae* liver abscess: a new Invasive syndrome. *Lancet Infect. Dis.* 12, 881, 2012.

14. Rahim, G.R., et al. Monomicrobial *Klebsiella pneumoniae* necrotizing fasciitis: an emerging life-threatening entity. *Clin. Microbiol. Infect.* 25, 316, 2019.

15. Harada, S., and Doi, Y. Hypervirulent *Klebsiella pneumoniae*: a call for consensus definition and international collaboration. *J. Clin. Microbiol.* 56, e00959–18, 2018.

16. Farzana, R., et al. Outbreak of hypervirulent multi-drug resistant *Klebsiella variicola* causing high mortality in neonates in Bangladesh. *Clin. Infect. Dis.* 68, 1225, 2019.

17. Högenauer, C., et al. *Klebsiella oxytoca* as a causative organism of antibiotic-associated hemorrhagic colitis. *N. Engl. J. Med.* 355, 2418, 2006.

18. Fisher, A. and Halalau, A. A case report and literature review of *Clostridium difficile* negative antibiotic associated hemorrhagic colitis caused by *Klebsiella oxytoca*. *Case Rep. Gastrointestinal. Med.* 2018, 7264613, 2018.

19. Messatesta, M.L., Gona, F. and Stefani, S. *Enterobacter cloacae* complex: clinical impact and emerging antibiotic resistance. *Future Microbiol.* 7, 887, 2012.

20. Davin-Regli, A. and Pagès, J.-M. *Enterobacter aerogenes* and *Enterobacter cloacae*; versatile bacterial pathogens confronting antibiotic treatment. *Front. Microbiol.* 6, 392, 2015.

21. Kremer, A. and Hoffmann, H. Prevalences of the *Enterobacter cloacae* complex and its phylogenetic derivatives in the nosocomial environment. *Eur. J. Clin. Microbiol. Infect. Dis.* 31, 2951, 2012.

22. Farmer III, J.J. My 40-year history with *Cronobacter/Enterobacter sakazakii*—lessons learned, myths debunked, and recommendations. *Front Pediatr.* 3, 84, 2015.

23. Healy, B., et al. *Cronobacter* (*Enterobacter sakazakii*): an opportunistic foodborne pathogen. *Foodborne Path. Dis.* 7, 339, 2010.

24. Armbruster, C.E., Mobley, H.L.T., and Pearson M.M. Pathogenesis of *Proteus mirabilis* infection. *EcoSal Plus.* 8, 10.1128/ecosalplus.ESP-0009-2017, 2018.

25. Liu, H. et al. *Morganella morganii*, a non-negligent opportunistic pathogen. *Int, J. Infect. Dis.* 50, 10, 2016.

26. Kim, B.-N. et al. Bacteraemia due to tribe Proteeae: a review of 132 cases during a decade (1991–2000), *Scand. J. Infect. Dis.* 35, 98, 2003.

27. Yoh, M. et al. Importance of *Providencia* species as a major cause of travellers' diarrhea. *J. Med. Microbiol.* 54, 1077, 2005.

28. Shah, M., et al. First report of a foodborne *Providencia alcalifaciens* outbreak in Kenya. *Am. J. Trop. Med. Hyg.* 93, 497, 2015.

29. Hamilton, A.L., et al. *Proteus* spp. as putative gastrointestinal pathogens. *Clin. Microbiol. Rev.* 31, e00085-17, 2018.

30. Mrques, C., et al. Clonal relatedness of *Proteus mirabilis* strains causing urinary tract infections in companion animals and humans. *Vet. Microbiol.* 228, 77, 2019.

31. Lorca, T.A. et al. Growth and histamine formation of *Morganella morganii* in determining the safety and quality of inoculated and uninoculated bluefish (*Pomatomus saltatrix*). *J. Food Prot.* 64, 2015, 2001.

32. Desai, P.T. et al. Evolutionary genomics of *Salmonella enterica* subspecies. *mBio* 4, e00579-12, 2013.

33. Aleksic, S., Heinzerling, F. and Bockemuhl, J. Human infection caused by salmonellae of subspecies II to VI in Germany, 1977–1992. *Zentbl. Bakteriol.* 283, 391, 1996.

34. Mermin, J., et al. Reptiles, amphibians and human *Salmonella* infection: a population-based, case-control study. *Clin. Infect. Dis.* 38 (Suppl. 3), S253, 2004.

35. Lamas, A., et al. A comprehensive review of non-*enterica* subspecies of *Salmonella enterica*. *Microbiologic Res.* 206, 60, 2018.

36. Centers for Disease Control. Foodborne Diseases Active Surveillance Network (FoodNet): FoodNet 2015 Surveillance Report (Final Data). Atlanta, GA. U.S. Dept Health and Human Services, 2017.

37. Centers for Disease Control and Prevention. An atlas of *Salmonella* in the United States, 1968-2011. Laboratory-based Enteric Disease Surveillance. Atlanta, GA. U.S. Dept of Health and Human Services, 2013.

38. Ferrari, R., Panzenhagen, P.H.N. and Conte-Junior, C.A. Phenotypes and genotypic eligible methods for *Salmonella* Typhimurium source tracking. *Front. Microbiol.* 8, 2587, 2017.

39. Dougan, G. and Baker, S. *Salmonella enterica* serovar Typhi and the pathogenesis of typhoid fever. *Annu. Rev. Microbiol.* 68, 317, 2014.

40. Britto, C.D., et al. A systematic review of antimicrobial resistance in *Salmonella enterica* serovar Typhi, the etiological agent of typhoid. *PLOS Negl. Trop. Dis.* 12(10), e0006779, 2018.

41. Buckle, G.C., Fischer Walker, C.L. and Black, R.E. Typhoid fever and paratyphoid fever: systematic review to estimate global morbidity and mortality for 2010. *J. Glob. Health* 2, 1, 2012.

42. Goldberg, M.B. and Rubin, R.H. The spectrum of Salmonella infection. In *Infectious Diarrhea*. R.C. Moellering and Gorbach, S.L., eds. The W.B. Saunders Co., Philadelphia, PA, 1988.

43. Wain, J., et al. Quantitation of bacteria in bone marrow from patients with typhoid fever; relationships between counts and clinical features. *J. Clin. Microbiol.* 39, 1571, 1998.

44. Escamilla, J., Florez-Ugarte, H. and Kilpatrick, M.E. Evaluation of blood clot cultures for isolation of *Salmonella typhi*, *Salmonella paratyphi-A*, and *Brucella melitensis*. *J. Clin. Microbiol.* 24, 388, 1986.

45. Nørskov-Lauritsen, N. and Ridderberg, W. Approaches to the identification of aerobic gram-negative bacteria. *Manual of Clinical Microbiology*, 12th ed., ASM Press, Washington, DC, 2019.

46. Popoff, M.Y. Antigenic Formulas of the Salmonella Serovars. 8th ed. *WHO Collaborating Centre for Reference and Research on Salmonella*, Institut Pasteur, Paris, France, 2001.

47. Yoshida, C., et al. Evaluation of molecular methods for the identification of *Salmonella* serovars. *J. Clin. Microbiol.* 54, 1992, 2016.

48. Tomljenovic-Berube, A.M., et al. Mapping and regulation of genes within *Salmonella* pathogenicity island 12 that contribute to in vitro fitness of *Salmonella enterica* serovar Typhimurium. *Infect. Immun.* 81, 2394, 2013.

49. Chattaway, M.A., et al. Identification of *Escherichia coli* and *Shigella* species from whole-genome sequences. *J. Clin. Microbiol.* 55, 616, 2017.

50. Lindstedt, B.-A., et al. High frequency of hybrid *Escherichia coli* strains with combined intestinal pathogenic *Escherichia coli* (IPEC) and extraintestinal pathogenic *Escherichia coli* (ExPEC) virulence factors isolated from human faecal samples. *BMC Infect. Dis.* 18, doi.org/10.1186/s12879-018-3449-2, 2018.

51. Dekker, J.P. and Frank, K.M. *Salmonella, Shigella*, and *Yersinia*. In *Diagnostic Testing for Enteric Pathogens, An Issue of Clinics in laboratory Medicine*, Vol 35-2. A.J. McAdam, ed. Elsevier, Philadelphia, USA, 2015.

52. Johnson, J.R., et al. Antimicrobial-resistant and extraintestinal pathogenic *Escherichia coli* in retail foods. *J. Infect. Dis.* 191, 1040, 2005.

53. Russo, T.A. and Johnson, J.R. Medical and economic impact of extraintestinal infections due to *Escherichia coli*: focus on an increasingly important endemic problem. *Microbe Infection* 5, 449, 2003.

54. Kotloff, K.L. et al. Global burden of *Shigella* infections: implications for vaccine development and implementation of control strategies. *Bull. WHO* 77, 651, 1999.

55. Baker, S. and The, H.C. Recent insights into *Shigella*: a major contributor to the global diarrhoeal disease burden. *Curr. Opin. Infect. Dis.* 31, 449, 2018.

56. Kim, K.S. Human meningitis-associated *Escherichia coli*. *EcoSal Plus* 7, 10.1128/ecosalplus.ESP-0015-2015, 2016.

57. Sarowska, J., et al. Virulence factors, prevalence and potential transmission of extraintestinal pathogenic *Escherichia coli* isolated from different sources: recent reports. *Gut Pathog.* 11, doi.org/10.1186/s13099-019-290-0, 2019.

58. Kotloff, K.L., et al. Shigellosis. *Lancet* 391, 801, 2019.

59. Cody, E.M. and Dixon, B.P. Hemolytic uremic syndrome. *Pediatr, Clin. N. Am.* 66, 235, 2019.

60. Lamba, K., et al. Shiga toxin 1-producing *Shigella sonnei* infections, California, United States, 2014-2015. *Emerg. Infect. Dis.* 22, 679, 2016.

61. Bryan, A., Yioungster, I. and McAdam, A.J. Shiga toxin producing Escherichia coli. In *Diagnostic Testing for Enteric Pathogens, An Issue of Clinics in laboratory medicine*, Vol 35-2. A.J. McAdam, ed. Elsevier, Philadelphia, USA, 2015.

62. Gouali, M., et al. Evaluation of CHROMagar STEC and STEC 104 chromogenic agar media for detection of Shiga toxin-producing *Escherichia coli. J. Clin. Microbiol.* 51, 894, 2013.

63. Zelyas, N., et al. Assessment of commercial chromogenic solid media for the detection of non-O157 Shiga toxin-producing *Escherichia coli. Diagn. Microbiol, Infect. Dis.* 85, 302, 2016.

64. Ramanan, P., et al. Syndromic panel-based testing in clinical microbiology. *Clin. Microbiol. Rev.* 31, e00024-17,2018.

65. Le Guern, A.-S., et al. First isolation of *Yersinia entomophaga* in human urine. *New Microbe New Infect.* 26, 3, 2018.

66. Kumar, G., et al. *Yersinia ruckeri*, the causative agent of enteric redmouth disease in fish. *Vet. Res.* 46, 103, 2015.

67. Neubauer, H., et al. *Yersinia enterocolitica* 16S rRNA gene types belong to the same genospecies but form three homology groups. *Int. J. Med. Microbiol.* 290, 61, 2000.

68. Batzilla, J., et al. *Yersinia enterocolitica palearctica* serobiotype O:3/4 – a successful group of emerging zoonotic pathogens. *BioMed. Genomics* 12, 348, 2011.

69. Hall, M., et al. Use of whole-genome sequence data to develop a multilocus sequence typing tool that accurately identifies *Yersinia* Isolates to the species and subspecies levels. *J. Clin. Microbiol.* 53, 35, 2015.

70. Raoult, D., et al. Plague: history and contemporary analysis. *J. Infect.* 66, 18, 2013.

71. Smego, R.A., Frean, J. and Koornhof, H.J. Yersiniois I: microbiological and clinicoepidemiological aspects of plague and non-plague *Yersinia* infection. *Eur. J. Clin. Microbiol. Infect. Dis.* 18, 1, 1991.

72. Demeure, C.E., et al. *Yersinia pestis* and plague: an updated view on evolution, virulence determinants, immune subversion, vaccination, and diagnostics. *Gene Immun.* 20, 357, 2019.

73. Guinet, F., Carniel, E. and Leclercq, A. Transfusion-transmitted *Yersinia enterocolitica* sepsis. *Clin. Infect. Dis.* 53, 583, 2011.

74. Stephan, R., et al. Characteristics of *Yersinia enterocolitica* biotype 1A strains isolated from patients and asymptomatic carriers. *Eur. J. Clin. Microbiol. Infect. Dis.* 32, 869, 2013.

75. Ruckdeschel, K. *Yersinia* species disrupt immune responses to subdue the host. *ASM News* 66, 470, 2000.

76. Bancerz-Kisiel, A., et al. The most important virulence markers of *Yersinia enterocolitica* and their role during infection. *Genes* 9, 235, 2018.

77. Doran, T.I. The role of *Citrobacter* in clinical disease of children: review. *Clin. Infect. Dis.* 28, 384, 1999.

78. Collins, J.V.V., et al. *Citrobacter rodentium*: infection, inflammation and the microbiota. *Nat. Rev. Microbiol.* 12, 612, 2014.

79. Janda, J.M. and Abbott, S.L. Infections associated with the genus *Edwardsiella*: the role of *Edwardsiella tarda* in human disease. *Clin. Infect. Dis.* 17, 742, 1993.

80. Janda, J.M., Abbott, S.L. and McIver, C.J. *Plesiomonas shigelloides* revisited. *Clin. Microbiol. Rev.* 29, 349, 2016.

29

Haemophilus *Species*

Elisabeth Adderson

CONTENTS

Organization and Characteristics

The genus *Haemophilus* is a member of the family *Pasteurellaceae*. The taxonomy of the genus has evolved rapidly with the availability of sensitive genotyping and protein analysis (Kuhnert and Christensen 2019). Of the current 13 species, nine with host specificity for humans are divided into three phenotypic groups: the *H. influenzae* group (*H. influenzae*, *H. aegyptius*, and *H. haemolyticus*), the *H. parainfluenzae* group (*H. parainfluenzae*, *H. parahaemolyticus*, *H. paraphrohaemolyticus*, *H. pittnamiae*, and *H. sputorum*), and a group comprising the single species *H. ducreyi* (Kuhnert and Christensen 2019; Norskov-Lauritsen 2014). Significant phenotypic and genotypic differences suggest that, while it currently remains a member of the genus, *H. ducreyi* would be more properly recognized as a distinct member of Pasteurella (Sturm 1981; Lagergard et al. 2011). Four species, *H. felis*, *H. haemoglobulinophilus*, *H. paracuniculus*, and *H. parasuis*, colonize and may be responsible for disease in animals. *H. parasuis*, however, will undergo reclassification and validation of as a new genus, *Glaesserella parasuis*, in the near future (Christensen and Bisgaard 2018; Dickerman et al. 2019).

Haemophilus species constitute part of the normal flora of the upper respiratory track of mammalian hosts. All are facultatively anaerobic, nonspore forming, pleomorphic, coccobacilli that are fastidious, requiring rich media supplemented with V factor (nicotinamide adenine dinucleotide [NAD], nicotinamide adenine dinucleotide phosphate [NADP], or nicatinamine mononucleotide) and/or X factor (protoporphyrin IX). *Haemophilus*, from the Greek *blood-loving*, reflects that these growth factors are most commonly provided by erythrocytes–X factor by hemin or hemoglobin and Y factor by NAD. Most *Haemophilus* species grow best when incubated at 35–37°C in a moist environment supplemented with 5–10% CO_2 and have an optimal pH of 7.6 (Ledeboer and Doern 2015). *H. ducreyi* grows better at 33°C and a pH of 6.5–7.0, and *H. felis*, *H. paracuniculus*, and *H. paraphrohaemolyticus* have a strict requirement for CO_2 enrichment. All species except *H. ducreyi* ferment carbohydrates, producing acid and, in some cases, gas; these characteristics are useful in species identification. Some *H. influenzae* elaborate a polysaccharide capsule.

General Methods for the Identification and Differentiation of *Haemophilus* Species

Collection and Processing

Clinical specimens should be processed promptly since most species are highly susceptible to drying and other environmental changes. The highest recovery rate is obtained if samples

DOI: 10.1201/9781003099277-31

are inoculated directly onto appropriate media for isolation. Automated blood culture systems readily isolate *Haemophilus* species from blood specimens. These should be inoculated into an appropriate blood culture bottle and transported promptly to the laboratory. Cerebrospinal fluid (CSF) specimens should be processed within an hour after collection or, if this is not feasible, inoculated into Trans-Isolate medium and transported at 20–25°C (Castillo et al. 2011). For *H. ducreyi*, swabs from the undermined edge of genital and cutaneous ulcers are the preferred specimen. If immediate inoculation onto culture media is not possible, the use of thioglycollate/hemin-based transport media containing selenium dioxide, L-glutamine and albumin held at 4°C is recommended (Dangor et al. 1993; Lewis and Ison 2006).

Growth of *Haemophilus* requires a source of X and V factors. X factor is immediately available to bacteria from erythrocytes, but V factor must be released from cells by heating (*chocolating*) or peptic digestion (Fildes enrichment). The most common medium used is chocolate agar prepared with 5% horse or sheep blood. Selective media, such as chocolate agar with vancomycin, with or without bacitracin and clindamycin, or cefsulodin is useful in optimizing yields from respiratory specimens. For optimal detection of *H. ducreyi*, the use of more than one culture medium containing vancomycin as a selective agent is recommended, although sensitivity remains low (30–80%) even with these techniques (Lewis and Ison 2006).

Identification

Most *Haemophilus* species grow to 1–2 mm diameter colonies after 24–48 hours of incubation. *H. influenzae* form grayish, semiopaque, smooth, convex colonies. Encapsulated strains may become confluent and demonstrate iridescence, particularly on clear agar. Colonies of *H. parainfluenzae* influenza may be rough and wrinkled. Those of *H. aegypticus* rarely grow larger than 1 mm in diameter and typically require more prolonged incubation than less fastidious species. Colonies of *H. ducreyi* are smooth, cohesive, and grow to a diameter of 0.1–0.5 mm after incubation for three to four days. Indole-producing *Haemophilus* strains have a characteristic pungent smell, like that of *Escherichia coli*. The odor of other strains is described as resembling that of a mouse nest. *Haemophilus* isolates can be stored at ambient temperature after lyophilization in skim milk and by freezing broth cultures or bacterial suspensions in 10% glycerol at –70°C or heavily inoculated cotton swabs at –135°C.

In Gram-stained smears, *Haemophilus* species are small (0.2–0.3 µm × 0.5–1.0 µm) pleomorphic, gram-negative rods. *H. ducreyi* often forms characteristic parallel chains described as *railroad tracks*, *school of fish*, or *fingerprints*, particularly in direct smears of clinical specimens and in broth culture. Identification of *Haemophilus* species is confirmed by the demonstration of X and V factor dependence. X-factor independent species will grow on sheep blood agar, whereas those species requiring both V and X factors cannot. ß-Hemolytic *S. aureus* both secrete NAD and release hemin from erythrocytes, permitting the growth of satellite colonies of *Haemophilus* strains requiring these factors for growth. Some laboratories, therefore, determine X and Y factor dependency by observing satellite colonies of the unknown bacteria around a streak of *S. aureus* (*Staph streak*) on blood agar. The satellite phenomena, however,

is not exclusive to *Haemophilus*, and it is sometimes difficult to detect small satellite colonies. The need for X and V factors can also be demonstrated by growth around factor-impregnated filter paper disks or strips. *Quad* and *tri* plates, in which unsupplemented blood agar and media incorporating one or more factors are separated into individual compartments, are also commercially available. Care must be taken to avoid carryover of X factor from the primary isolation media.

The porphyrin test is the most reliable method to detect X-factor dependence and has the additional advantage of being independent of bacterial growth. δ-Aminolevulinic acid hydrochloride is mixed with bacteria from an agar plate culture. X factor-independent strains are able to convert this substrate to porphyrin and porphobilinogen, which are detected after 4–24 hours of incubation by the demonstration of red fluorescence in ultraviolet light or a red color change when the suspension is mixed with Kovac's reagent.

Speciation

Carbohydrate fermentation and the expression of catalase, urease and ß-lactamase are key reactions for further differentiation of *Haemophilus* species (Table 29.1). The observation of hemolysis on blood agar is chiefly useful to distinguish pathogenic *H. influenzae* from the commensal species *H. haemolyticus*, as classical biochemical identification fails to distinguish between these bacteria. Semiautomated and automated biochemical differentiation systems accurately identify clinical strains of *H. influenzae* and *H. parainfluenzae*, but identification failures and misidentifications of other *Haemophilus* species are not uncommon (Rennie et al. 2008; Valenza et al. 2007).

The accuracy of identification of *Haemophilus* species by matrix-assisted laser desorption/ionization time of flight mass spectrometry (MALDI-TOF MS) is comparable to that of phenotypic methods for some species, but that of less commonly encountered species remains challenging (Powell et al. 2013). While increasingly utilized by clinical laboratories, the utility of this method is currently limited by small numbers of reference spectra for some species and the inability to distinguish between some closely related species (Norskov-Lauritsen 2014; Powell et al. 2013).

Nucleic acid amplification tests (NAATs) have been developed for the identification and speciation of cultured bacteria and to directly detect bacteria in blood, other normally sterile body fluids, and respiratory secretions. Methods employing 16S rRNA, housekeeping, or species-specific gene sequencing have been developed. Currently, however, clinical experience with NAATs for *Haemophilus* species other than *H. influenzae* and *H. ducreyi* is limited, and these tests are not generally available.

Genomes and Genetic Diversity

The genetic diversity and population structure of *Haemophilus* have been studied by a variety of techniques, including multilocus enzyme electrophoresis, ribotyping, pulsed-field gel electrophoresis and whole genome, 16S rRNA, and multilocus sequence typing. The first bacterial genome to be completely sequenced was that of *H. influenzae* Rd KW20 (Rd), a nonencapsulated variant of a serotype d strain (Fleischmann et al. 1995). Many complete and draft genomes of encapsulated *H. influenzae*,

TABLE 29.1

Major Differential Characteristics of *Haemophilus* Species

Species	Hemolysis	Factor Requirement		Catalase	ß-galactosidase	Urease	CO₂ᵃ	Fermentation of		
		X	V					Sucrose	Lactose	Mannose
H. influenzae	+	+	+	+	–	V^b	–	+	–	–
H. aegypticus	+	+	+	+	–	+	–	+	–	–
H. hemolyticus	+	+	+	+	–	+	–	+	–	–
H. parainfluenzae	–	–	+	V	+	V	–	V	–	+
H. parahaemolyticus	–	–	+	V	+	+	–	+	–	–
H. paraphrohaemolyticus	–	–	+	V	V	+	V	+	–	–
H. pittmaniae	+	–	+	V	+	–	–	+	–	+
H. sputorum	+	–	+	V	+	+	–	+	–	–
H. felis	+	+	V	–	+	–	–	+	+	+
H. haemoglobinophilus	+	–	+	–	+	–	+	+	–	+
H. paracuniculus	+	+	+	–	+	–	–	+	+	NK^c
H. ducreyi	+	+	–	–	–	–	–	–	–	–

ᵃ Growth enhanced by CO_2.
ᵇ V, variable.
ᶜ NK, not known.

NTHi, and other *Haemophilus* species have since been reported and have contributed to the understanding of the population structure and pathogenicity. The size of these genomes varies between 1.7 and 2.1 Mb.

Haemophilus influenzae

H. influenzae are divided into encapsulated and nonencapsulated (nontypeable, NTHi) strains. Encapsulated strains express one of six antigenically distinct polysaccharide capsules (types a–f). The capsulation (*cap*) locus consists of two highly conserved regions common to all capsular types and a region containing type-specific genes responsible for capsular biosynthesis. Serotyping of *H. influenzae* is most commonly performed by slide agglutination, counterimmunoelectrophoresis, or enzyme-linked immunoassay using antisera specific for individual capsular polysaccharides. MALDI-TOF MS may also have utility in capsular serotyping of *H. influenzae* (Takeuchi et al. 2018; Mansson et al. 2018). Typing systems based on polymerase chain reaction (PCR) amplification of serotype-specific segments of the capsulation locus and serotype-specific gene probes are more sensitive and specific than serological assays and are available through reference laboratories.

H. influenzae can also be subdivided into eight biotypes on the basis of indole production and the expression of urease and ornithine decarboxylase (Table 29.2) (Kilian 1976). Each of these activities is readily determined by simple colorimetric assets. Biotype I strains include encapsulated serotype a, b and f strains, while serotype a and c strains are frequent in biotype II and serotype d and e strains in biotype IV. Several commercial systems for identification and biotyping are available. These, however, may require additional tests to speciate isolates other than *H. influenzae* and *H. parainfluenzae*.

The population structure of encapsulated serotypes of *H. influenzae* is relatively clonal, whereas NTHi isolates are genetically distinct from encapsulated isolates and more diverse (Meats et al. 2003). It is hypothesized that this genetic heterogeneity results from the acquisition of different members of a complete complement of *contingency genes* available from the population supergenome (Shen et al. 2005). Such genetic exchange is facilitated by the ability of *H. influenzae*, which is naturally transformable, to take up DNA from highly related organisms by recognition of uptake signal sequences that are widely distributed throughout the genome (Danner et al. 1980; Smith et al. 1995). Examples of horizontal gene transfer between NTHi strains and between *H. influenzae* and other respiratory pathogens have been reported (Hiltke et al. 2003; Kroll et al. 1998). The rate of recombination in *H. influenzae* is not known, but it appears to be higher in NTHi than among encapsulated isolates. The simultaneous presence of multiple distinct strains of NTHi in the respiratory tract and the ability to undergo continuous reassortment may be a strategy that allows these bacteria to adapt to changing environmental factors.

H. influenzae has few two-component and global regulatory systems. An alternative strategy for gene regulation is provided by simple contingency loci, which contain di- or tetranucleotide tandem sequence repeats within or 5' to coding regions (Bayliss et al. 2001). These repeats undergo rapid, reversible, recombinase-independent mutation, resulting in alterations in promoter activity or shifts in translational reading frames (*slipped-strand mismatch repair*). *H. influenzae* has multiple such loci, including genes involved in LPS biosynthesis, adhesion, Fe acquisition, and restriction-modification enzyme systems (Bayliss et al. 2001). This phase variation increases fitness by permitting bacteria to very rapidly alter their immunogenic surface structures and nutritional requirements to survive in different ecological niches (Poole et al. 2013).

Infections Caused by *H. influenzae*

Robert Pfeiffer first described the types of species of *H. influenza* during the 1889–1892 influenza pandemic, and until 1918,

TABLE 29.2

Biochemical Classification of *H. influenzae* and *H. parainfluenzae*

Species	Biotype	Indole	Urease	Ornithine Decarboxylase	Associated Capsular Type	Clinical Association
H. influenzae	I	+	+	+	a, b, f	Respiratory and invasive infection
	II	+	+	–	a, c	Respiratory
	III	–	+	–	a	Respiratory
	IV	–	+	+	d, e	Genitourinary, obstetrical, neonatal infection
	V	+	–	+		Invasive
	VI	–	–	+		
	VII	+	–	–		
	VIII	–	–	–		
H. influenzae Biogroup aegypticus		–	+	–		Brazilian purpuric fever
H. parainfluenzae	I	–	–	–		
	II	–	+	+		
	III	–	+	-		
	IV	+	+	+		
	V	–	–	–		
	VI	+	–	–		
	VII	+	+	–		
	VIII	+	–	–		

it was assumed to be the etiologic agent of these infections. It is now recognized as an important cause of respiratory, invasive, and, less commonly, genitourinary and obstetrical infections. *H. influenzae* colonizes the nasopharyngeal tract and, occasionally, the female genital tract. NTHi may be found in the respiratory tract of up to 80% of adults and children, whereas less than 5% of healthy persons are colonized by encapsulated strains, particularly since the introduction of *H. influenzae* serotype b (Hib) vaccines in the 1980s. Bacteria are transmitted by inhalation of airborne droplets or by direct contact with infected respiratory tract secretions. Detailed epidemiologic studies have revealed that *H. influenzae* colonization is a dynamic process. Individual strains are carried for variable periods of time before loss or replacement by new strains, and several distinct strains may simultaneously cocolonize individuals (Murphy et al. 1999; Smith-Vaughan et al. 1996). A single clonal group may predominate during infection, however. The vast majority of colonized persons are asymptomatic.

Infectious Caused by Encapsulated *H. influenzae*

Serotype b *H. influenzae* (Hib) is a cause of invasive disease, including bacteremia, meningitis, epiglottitis, periorbital cellulitis, and soft-tissue and skeletal infections (Wilfert 1990). Serum antibody concentrations of antitype b capsular antibody correlate inversely with the risk of infection. Children between 3 months and 4 years of age are most susceptible to infection because of loss of maternally-acquired immunoglobulin, coupled with an age-dependent impairment in the acquisition of specific antibody against Hib and other T-independent bacterial polysaccharides (Anderson et al. 1977). Hib infection is also more frequent

in certain ethnic populations, including Australian and some North American indigenous peoples. Highly immunogenic polysaccharide-protein conjugate vaccines introduced in the 1980s have dramatically reduced the incidence of Hib infection among vaccinated children (MacNeil et al. 2011).

Nonserotype b encapsulated *H. influenzae*, particularly serotypes a and f, are occasional causes of bacteremia and other invasive disease (Soeters et al. 2018). Molecular typing has demonstrated that these invasive strains are clonally related and possess virulence factors that are also found in serotype b strains (Ogilvie et al. 2001; Rodriguez et al. 2003; Ulanova and Tsang 2014; Omikunle et al. 2002; Potts et al. 2019; Singh et al. 2014).

Pathogenesis

The pathogenesis of Hib infections can be divided into four stages: colonization or respiratory epithelium, invasion of epithelial and endothelial cells, persistence in the bloodstream, and central nervous system invasion. Bacterial factors contributing to pharyngeal colonization include hemagglutinating pili, which adhere to mucin, extracellular matrix proteins, and epithelial cells, the major nonpilus adhesin, Hsf, and phosphocholine (Farley et al. 1990; Geme and Cutter 1995; Weiser et al. 1998). The persistence of bacteria at mucosal surfaces is promoted by bacterial products that interfere with the host's physical and immune defenses and damage and impair ciliary function, including protein D and Hsf surface fibril, which are unique to encapsulate strains, and peptidoglycan (Janson et al. 1999; Male 1979; Wilson and Cole 1988; Singh et al. 2014). Adherence of *H. influenzae* to epithelial cells results in the loss of integrity of tight junctions and sloughing of cells. This epithelial damage, which is attributable in part to the bacterial lipooligosaccharide (LOS), exposes nonluminal epithelial cells and basement membrane, to which bacteria adhere in large numbers. The genome of most *H. influenzae* strains contains *igaA*, a gene that encodes a type I IgA$_1$ protease (Kilian et al. 1979). This enzyme cleaves the hinge region of both serum and secreted IgA1, eliminating the agglutination activity of free and antigen-bound immunoglobulin.

Endothelial invasion is an active process in which bacteria are taken up within vacuoles and translocated to the basal cell surface and bloodstream. The type b capsule plays a critical role in the persistence of bacteria in the bloodstream (Moxon and Kroll 1988). The hydrophilic capsule may provide a physical barrier to phagocytosis, preventing clearance of Hib from the bloodstream by complement-mediated phagocytosis. Supporting this hypothesis is the observation that the opsonophagocytic killing of Hib is inversely proportional to the amount of capsule expressed by the bacteria. ∝-D-Galactose[1,4]-ß-D-galactose and sialic acid components of lipopolysaccharide (LPS), the major component of the Gram-negative outer membrane, also contribute to resistance to bacterial clearance (Hood et al. 1999; Weiser et al. 1998). After gaining access to the bloodstream, Hib adhere to and damage the blood-brain barrier, allowing bacteria to translocate across intercellular tight junctions to the CSF. LPS increases blood-brain barrier permeability and also contributes to the systemic symptoms of Hib infection.

Diagnosis of Hib Infections

Several growth-independent tools have been developed for the diagnosis of Hib infection. While not intended to replace traditional culture, these are useful in situations where the yield of conventional culture may be low, such as patients who have received antibiotics before diagnostic tests are obtained, and in settings, such as resource-limited countries, where conventional microbiology has limitations. Encapsulated *Haemophilus* species shed capsular polysaccharide during growth, and rapid detection of this antigen by latex agglutination or enzyme-linked immunoassay is possible in up to 90% of cases of Hib meningitis. Antigen may be detectable in CSF for more than a week after bacteria cease to be cultivatable. Antigen detection can be performed on other normally sterile body fluids and concentrated urine; however, the sensitivity and specificity of the assay in these settings is not established. Bacterial nucleic acids also may persistent at sites of infection or in clinical specimens longer than viable bacteria. Single and multiplex NAAT have been developed for the direct detection of Hib and other common meningeal pathogens in cerebrospinal fluid specimens. While results show good agreement with conventional testing and, relative to culture, they are rapid and sensitive, most reported studies have included only small numbers of cases caused by *H. influenzae* (Liesman et al. 2018).

Disease Caused by NTHi

NTHi are common causes of upper respiratory tract infections and conjunctivitis, especially among children, and important causes of lower respiratory tract infections, particularly in young children, adults with chronic obstructive pulmonary disease (COPD) and patients with cystic fibrosis (Van Eldere et al. 2014). Infections result from direct spread from colonized sites in the upper respiratory tract or by contact with infected respiratory secretions. Risk factors for infection include factors that impair normal mucociliary function or result in inflammation, including antecedent viral infection, smoking, alterations in mucous (such as those seen in patients with cystic fibrosis), and anatomic abnormalities. Rare cases of bacteremia and meningitis are also reported, primarily in neonates, the elderly, and immunocompromised persons (Soeters et al. 2018).

The ability of NTHi to adhere to respiratory epithelium and to resist clearance by host immune defenses is critical to their survival. NTHi adhere to and invade bronchial epithelial cells by interactions between phosphocholine and platelet-activating factor receptor (Swords et al. 2000). Multiple and redundant pilus and nonpilus adhesins have evolved to facilitate NTHi persistence in the upper respiratory tract. The P2 and P5 porins and pili mediate binding to mucin (Gilsdorf et al. 1996; Kubiet et al. 2000; Reddy et al. 1996). P5 also binds intracellular adhesion molecule 1 (ICAM1) and carcinoembryonic antigen-related adhesion molecule 1 (Avadhanula et al. 2006; Bookwalter et al. 2008). Other adhesins include P4 (vitronectin, fibronectin), the HMW1 and HMW2 high-molecular-weight surface proteins (extracellular matrix proteoglycans), Hia (a homolog of the Hsf of encapsulated strains that is generally mutually exclusive with HMW1/HMW2), and opacity-associated protein A (Barenkamp and St Geme 1996; Jurcisek et al. 2007; St Geme et al. 1993; Weiser et al. 1995; Su et al. 2016). *Haemophilus* adhesion and penetration protein (Hap) binds fibronectin, laminin, and collagen IV, mediating attachment, bacterial aggregation, and invasion of respiratory epithelial cells (Fink et al. 2002). A second ubiquitous multifunctional adhesin, protein E, binds both vitronectin and, with Hap, laminin (Hallstrom et al. 2011). The ABC transporter protein F binds laminin and human pulmonary epithelial cells (Jalalvand et al. 2013). While considered a nonmotile organism, some *H. influenzae* strains express a homolog of type IV pili (Tfp) and exhibit twitching motility (Bakaletz et al. 2005). Tfp promotes bacterial adherence to respiratory epithelial cells by binding ICAM1 and mucin and plays an important role in the development of biofilm (Jurcisek et al. 2007). Epithelial adherence and other virulence factors of NTHi may be regulated by the DNA methylase gene *mod10A*, with loss of expression caused by polymerase slippage at a long tetrameric repeat tract leading to the upregulation of adhesins (VanWagoner et al. 2016).

Although considered an extracellular pathogen, NTHi, like Hib, may invade and multiply in epithelial cells. Inside cells, these bacterial may be relatively protected from killing by many host immune defenses and antimicrobial agents (Forsgren et al. 1994). Invasion of respiratory epithelial cells by NTHi is mediated by Hap, protein E, HMW1/HMW2, the ABC transporter sensitivity to antimicrobial peptides A (SapA), protein D, IgA1 proteases, and LOS (Kenjale et al. 2009; Ikeda et al. 2015; Mell et al. 2016; Raffel et al. 2013; Ruan et al. 1990; Swords et al. 2001; Clementi et al. 2014). NTHi may also, independent of phagocytosis, be internalized by macrophages.

Lacking a polysaccharide capsule to block host immune factors, NTHi have evolved diverse mechanisms to persist in the respiratory tract. Protein D is a highly conserved surface lipoprotein that both impairs ciliary function and, like LOS, induces morphological damage to cilia (Forsgren et al. 2008; Janson et al. 1999). While most NTHi express type 1 IgA$_1$ protease, a clonally related group of strains associated with COPD also possesses a second gene, *igaB*, which encodes a type 2 IgA$_1$ protease (Murphy et al. 2011). Multiple factors, including phase variation in LOS and the acquisition of human complement inhibitors, such as plasminogen, C4B-binding protein, factor H (P5), factor H-like protein and vitronectin (P4, proteins E and F), contribute to NTHi serum resistance (Barthel et al. 2012; Hallstrom et al. 2006; Hallstrom et al. 2007; Hallstrom et al. 2008; Su et al. 2013). The Sap ABC transporter permits NTHi to evade the bacterial effects of host-derived antimicrobial peptides such as defensins and cathelicidins by transporting these to the bacterial cytoplasm, where they are degraded by peptidase activity (Shelton et al. 2011). Certain modifications to LOS structure may decrease complement deposition and activation and inflammatory responses (Clark et al. 2012). Outer membrane vesicles, blebs separating from the bacterial outer membrane, are able to exploit host immune responses to promote colonization and invasion (Kaparakis-Liaskos and Ferrero 2015).

The formation of biofilm, comprised chiefly of LOS and bacterial and eukaryotic DNA, in a process that requires Hfp and may involve P2, P5, and P6, permits NTHi to persist at mucosal sites (Jurcisek and Bakaletz 2007; Langereis and Hermans 2013; Murphy and Kirkham 2002). In an animal model of middle ear infection, antibody against the DNA-binding protein DNABII, which plays an important role in the structural stability of NTHi biofilm, reduced biofilm mass and resulted in more rapid clearance of bacteria (Goodman et al. 2011).

Antigenic diversity is another strategy by which NTHi avoids immune recognition by the host. Genomic alteration critical virulence factors may occur through slipped-strand misrepair during colonization (Pettigrew et al. 2018). Structural diversity may also arise through point mutations in surface-exposed structures or genetic exchange between NTHi and closely related bacteria (Duim et al. 1994; Shen et al. 2005). Some NTHi strains with defects in components of the methyl-directed mismatch DNA repair system have exceptionally high rates of mutation; this may provide these bacteria with a survival advantage (Watson et al. 2004). The structure of LOS may be modified independently of phase variation in response to environmental cues and growth conditions (Oerlemans et al. 2019). Finally, the structure of some forms of LOS is shared by host glycolipids and glycosphingolipids, and these may fail to be recognized as foreign by the host (Harvey et al. 2001; Ng et al. 2019).

Other NTHi Infections

A recently reported cryptic biotype IV genospecies, *Haemophilus quentini*, is associated with urogenital infections, serious infections of parturient women and neonates (Kus et al. 2019). A distinct NTHi, biogroup *aegyptius*, causes seasonal epidemics of conjunctivitis and a clonal strain of this biogroup caused outbreaks of Brazilian purpuric fever (BPF), a syndrome characterized by purulent conjunctivitis, shock, and hemorrhagic skin lesions, in young children in the 1980s and 1990s (Harrison et al. 2008). BPF-associated strains possessed genes that encode unique adhesins and fimbrial operons that may be responsible for their unusual pathogenicity (Strouts et al. 2012).

Treatment of *H. influenzae* Infections

Serious invasive *H. influenzae* infections are typically treated with parenteral third generation cephalosporins, carbapenems, or ampicillin. Less critical infections may be treated with oral ampicillin or cephalosporins. A quarter to a half of isolates are ampicillin resistant. While beta-lactamase elaboration, predominantly plasmid-encoded class A serine enzymes, is the most prevalent resistance mechanism in NTHi, strains exhibiting beta-lactamase negative ampicillin resistance (BLNAR) and, in some cases, cephalosporin and carbapenem resistance, have emerged rapidly in some countries (Van Eldere et al. 2014). Reduced susceptibility of BLNAR isolates is a consequence of one or more alterations in *ftsI*, the gene encoding penicillin binding protein 3 (Ubukata et al. 2001). Most of these strains remain susceptible to quinolone and carbapenem antimicrobials.

Haemophilus parainfluenzae

H. parainfluenzae are the most prevalent *Haemophilus* species in normal flora of the human oropharynx and are a common commensal of the genitourinary tract (Kuklinska and Kilian 1984). Like *H. influenzae*, *H. parainfluenzae* may be subdivided into 8 biotypes (Table 29.2). Interestingly, while *H. parainfluenzae* has a lipid A and inner core LPS very similar to that of *H. influenzae*, it has reduced outer core LPS repertoire and, consistent with the observation that the genome does not contain lengthy tetranucleotide repeat tracts, there is no evidence for phase variation of cell surface components (Young and Hood 2013).

These bacteria occasionally cause respiratory and invasive diseases, most notably endocarditis and biliary tract infections (Darras-Joly et al. 1997). They are frequently isolated from sputum of adults with COPD, but their contribution to exacerbations of pulmonary disease remains controversial (Sethi and Murphy 2001). Antigen and antibody-directed against *H. parainfluenzae* outer membrane antigen preparations have also been detected in the glomeruli of patients with IgA nephropathy, and administration of outer membrane antigens induces glomerular deposition of IgA and mesangial proliferation in an experimental animal model, suggesting *H. parainfluenzae* colonization or infection may trigger immune-complex mediated renal disease (Yamamoto et al. 2002).

Haemophilus ducreyi

First described by Auguste Ducrey in 1889, *H. ducreyi* is the causative agent of the sexually-transmitted genital ulcer disease chancroid (soft chancre) and, more recently, has been recognized as a cause of chronic cutaneous ulcers in children in low-income countries (Lewis and Mitja 2016). Whole genome and multilocus sequencing suggest *H. ducreyi* is composed of two clonal populations that diverged almost 2 million years ago (White et al. 2005; Ricotta et al. 2011; Marks et al. 2018). Strains causing cutaneous ulceration are genetically related to class 1 strains causing genital ulcers and may have evolved through a more recent divergence from these bacteria (Gangaiah et al. 2015; Gaston et al. 2015). The *H. ducreyi* chromosome has no lengthy regions of synteny with that of Rd or the NTHi 86-028NP (Challacombe et al. 2007; Harrison et al. 2005). Observed differences in genes involved in LOS synthesis, carbohydrate and amino acid utilization, nucleotide biosynthesis, iron acquisition and oxidative stress may account for the phenotypic differences between these *Haemophilus* strains and their adaptation to particular environments.

The pathogenesis of *H. ducreyi* infection has been studied using human and animal models of cutaneous inoculation. The ability of bacteria to cause disease is dependent on the ability to form aggregates, adhere to keratinocytes, extracellular matrix and serum proteins, to damage host cells, and to resist killing by human antimicrobial peptides and phagocytosis by polymorphonuclear leukocytes and macrophages (Alfa and DeGagne 1997; Bauer et al. 2001). The expression of high-molecular-weight LOS may be important for adherence to mucosal surfaces. Three fimbria-like proteins, Flp1, Flp2, and Flp3, contribute to the ability of bacteria to form microcolonies and adhere to fibroblasts—the expression of these adhesins is required for virulence in humans (Nika et al. 2002; Spinola et al. 2003; Janowicz et al. 2011). *H. ducreyi* serum resistance protein A (DrsA), a trimeric autotransporter adhesion involved in serum resistance, is the major fibronectin-binding determinant of the organism and also mediates binding to vitronectin and fibrinogen (Cole et al. 2002; Fusco et al. 2013; Leduc et al. 2009). The *H. ducreyi* homolog of the GroEL heath shock protein binds glycosphingolipids exposed on the surface of epithelial cells, NcaA is a type I collagen-binding outer membrane protein, and fibrinogen binder A (FgbA) contributes to the virulence of *H. ducreyi* in experimental human infection (Alfa and DeGagne 1997; Fulcher et al. 2006; Pantzar et al. 2006).

H. ducreyi expresses two extracellular toxins. Hemolysin (HhdA) is cytotoxic to keratinocytes, fibroblasts, macrophages, and lymphocytes. The cytolethal distending toxin (CTD) causes cell cycle arrest and apoptosis in fibroblasts and lymphocytes; the formation of ulcers is likely attributable to these virulence factors (Cope et al. 1997).

The sensitive to antimicrobial peptides (Sap) influx transporter, proton motive force multiple transferable resistance transporter, and modifications of lipopolysaccharide or LOS mediate resistance to human antimicrobial peptides (Mount et al. 2010; Rinker et al. 2011; Trombley et al. 2015). *H. ducreyi* is highly resistant to complement-mediated killing by human serum. At least four outer membrane proteins contribute to serum resistance, DsrA, DltA (*H. ducreyi* lectin A), LspA1, and LspA2 (Leduc et al. 2004; Mock et al. 2005). LspA1 and LspA2 (large supranuclear proteins A1 and A2) are highly homologous secreted proteins that interfere with protein tyrosine kinase signaling in the early phases of Fc receptor-mediated phagocytosis (Janowicz et al. 2004). Fibrinogen binder A (FgbA) may contribute to virulence by promoting the deposition of fibrin on the bacterial cell surface (Bauer et al. 2009). Lysis of immune cells by HdtA and CTD may also contribute to the ability of *H. ducreyi* to evade host immune responses.

Infectious Caused by *H. ducreyi*

Historically, chancroid has been primarily a disease of resource-limited countries in Africa, Asia, Latin America, and the Caribbean. It is more commonly seen in men, particularly uncircumcised men, and female sex workers, and is a major cofactor for the transmission of HIV (Greenblatt et al. 1988). The global epidemiology of chancroid is difficult to estimate because sensitive and specific laboratory tests are often not available in regions with the highest endemicity. While the precise incidence is unclear, reductions in the proportion of GUD attributable to *H. ducreyi* in many countries have been reported (Gonzalez-Beiras et al. 2016). This has been attributed to a 1991 World Health Organization recommendation for syndromic management of genital ulcer disease (i.e. without microbiological confirmation of etiology), improvements in health care delivery, and with behavioral changes, such as condom use, intended primarily to reduce the risk of HIV prevention.

The inoculum required to cause chancroid is low, 1–100 organisms (Al-Tawfiq and Spinola 2002). In males, cutaneous or mucosal inoculation through minor trauma results in formation of a painful papule following an incubation period of 3–10 days. This becomes pustular and then ulcerates, forming a superficial ulcer with a friable base and ragged undermined edge, characteristically covered by a gray or yellow necrotic exudate. Autoinoculation may cause "kissing ulcers" on opposing skin. Painful unilateral inguinal lymph nodes (buboes) occur in 10–70% of patients and may suppurate and rupture. Most infected women are asymptomatic. Bacteria remain localized to the superficial tissues; invasive infection has not been described.

Recent studies have identified *H. ducreyi* as a previously unrecognized cause of chronic skin ulcers in Africa and the Pacific islands, predominantly in children and sometimes in association with coinfection by *Treponema pallidum* subsp. *pertenue*, the etiologic agent of the cutaneous ulcer disease, yaws (Mitja et al. 2014; Marks et al. 2014). The precise pathogenesis is unclear, but *H. ducreyi* DNA has been detected on the skin of well children, flies, and fomites such as bed sheets, suggesting an environmental reservoir for the bacteria exists (Houinei et al. 2017). No clinical features appear to reliably distinguish ulcerations caused by *H. ducreyi* from yaws.

Diagnosis and Treatment of *H. ducreyi* Infections

Laboratory confirmation of *H. ducreyi* is challenging because of its fastidious growth conditions. Although not routinely available in some countries, NAATs are preferred for the diagnosis of chancroid. Multiplex PCR assays for the simultaneous detection of *H. ducreyi* and other etiologic agents of genital ulcer disease have a sensitivity and specificity of over 95%. While the definitive diagnosis of chancroid requires isolation from genital lesions, the sensitivity of culture, even under ideal conditions, is estimated to be less than 75% when NAATs are considered to be the gold standard (Lewis 2003). Positive serology may persist for months after infection and is useful only for epidemiological investigations.

H. ducreyi is generally susceptible to fluoroquinolones, erythromycin, azithromycin, and ceftriaxone. Ulcers typically improve within a week of initiation of therapy. Treatment failures are more common in persons with HIV infection and may reflect coinfection with other pathogens, particularly herpes simplex virus. Fluctuant buboes may be treated by aspiration or surgical excision.

Other *Haemophilus* Species

H. aegyptius (Koch-Weeks bacillus) is a frequent cause of acute bacterial conjunctivitis in humans and a rare agent of human and veterinary infections (Pittman and Davis 1950). *H. haemolyticus* and *H. paraphrohaemolyticus* are human respiratory commensals and anecdotal causes of invasive infection (Murphy et al. 2007). *H. pittmaniae* and *H. sputorum* colonize the oropharynx of humans and are also rare pathogens (Norskov-Lauritsen et al. 2005; Norskov-Lauritsen et al. 2012).

REFERENCES

Al-Tawfiq, J. A., and S. M. Spinola. 2002. *Haemophilus ducreyi*: clinical disease and pathogenesis. *Curr Opin Infect Dis* 15:43–47.

Alfa, M. J., and P. DeGagne. 1997. Attachment of *Haemophilus ducreyi* to human foreskin fibroblasts involves LOS and fibronectin. *Microb Pathog* 22:39–46.

Anderson, P., D. H. Smith, D. L. Ingram, J. Wilkins, P. F. Wehrle, and V. M. Howie. 1977. Antibody of polyribophate of *Haemophilus influenzae* type b in infants and children: effect of immunization with polyribophosphate. *J Infect Dis* 136 (Suppl):S57–62.

Avadhanula, V., C. A. Rodriguez, G. C. Ulett, L. O. Bakaletz, and E. E. Adderson. 2006. Nontypeable *Haemophilus influenzae* adheres to intercellular adhesion molecule 1 (ICAM-1) on respiratory epithelial cells and upregulates ICAM-1 expression. *Infect Immun* 74:830–838.

Bakaletz, L. O., B. D. Baker, J. A. Jurcisek, A. Harrison, L. A. Novotny, J. E. Bookwalter, R. Mungur, and Jr., R. S. Munson. 2005. Demonstration of type IV pilus expression and a twitching phenotype by *Haemophilus influenzae*. *Infect Immun* 73:1635–1643.

Barenkamp, S. J., and J. W. St Geme, 3rd. 1996. Identification of a second family of high-molecular-weight adhesion proteins expressed by non-typable *Haemophilus influenzae*. *Mol Microbiol* 19:1215–1223.

Barthel, D., B. Singh, K. Riesbeck, and P. F. Zipfel. 2012. *Haemophilus influenzae* uses the surface protein E to acquire human plasminogen and to evade innate immunity. *J Immunol* 188:379–385.

Bauer, M. E., M. P. Goheen, C. A. Townsend, and S. M. Spinola. 2001. *Haemophilus ducreyi* associates with phagocytes, collagen, and fibrin and remains extracellular throughout infection of human volunteers. *Infect Immun* 69:2549–2557.

Bauer, M. E., C. A. Townsend, R. S. Doster, K. R. Fortney, B. W. Zwickl, B. P. Katz, S. M. Spinola, and D. M. Janowicz. 2009. A fibrinogen-binding lipoprotein contributes to the virulence of *Haemophilus ducreyi* in humans. *J Infect Dis* 199:684–692.

Bayliss, C. D., D. Field, and E. R. Moxon. 2001. The simple sequence contingency loci of *Haemophilus influenzae* and *Neisseria meningitidis*. *J Clin Invest* 107:657–662.

Bookwalter, J. E., J. A. Jurcisek, S. D. Gray-Owen, S. Fernandez, G. McGillivary, and L. O. Bakaletz. 2008. A carcinoembryonic antigen-related cell adhesion molecule 1 homologue plays a pivotal role in nontypeable *Haemophilus influenzae* colonization of the chinchilla nasopharynx via the outer membrane protein P5-homologous adhesin. *Infect Immun* 76:48–55.

Castillo, D., B. Harcourt, C. Hatcher, M. Jackson, L. Katz, R. Mair, L. Mayer, et al. 2011. Laboratory methods for the diagnosis of meningitis. Chapter 5: Collection and transport of clinical specimens, Centers for Disease Control and Prevention. Available at: https://www.cdc.gov/meningitis/lab-manual/chpt05-collect-transport-specimens.html.

Challacombe, J. F., A. J. Duncan, T. S. Brettin, D. Bruce, O. Chertkov, J. C. Detter, C. S. Han, M. Misra, P. Richardson, R. Tapia, N. Thayer, G. Xie, and T. J. Inzana. 2007. Complete genome sequence of *Haemophilus somnus* (*Histophilus somni*) strain 129Pt and comparison to *Haemophilus ducreyi* 35000HP and *Haemophilus influenzae* Rd. *J Bacteriol* 189:1890–1898.

Christensen, H., and M. Bisgaard. 2018. Classification of genera of Pasteurellaceae using conserved predicted protein sequences. *Int J Syst Evol Microbiol* 68:2692–2696.

Clark, S. E., J. Snow, J. Li, T. A. Zola, and J. N. Weiser. 2012. Phosphorylcholine allows for evasion of bactericidal antibody by *Haemophilus influenzae*. *PLOS Pathog* 8:e1002521.

Clementi, C. F., A. P. Hakansson, and T. F. Murphy. 2014. Internalization and trafficking of nontypeable *Haemophilus influenzae* in human respiratory epithelial cells and roles of IgA1 proteases for optimal invasion and persistence. *Infect Immun* 82:433–444.

Cole, L. E., T. H. Kawula, K. L. Toffer, and C. Elkins. 2002. The *Haemophilus ducreyi* serum resistance antigen DsrA confers attachment to human keratinocytes. *Infect Immun* 70:6158–6165.

Cope, L. D., S. Lumbley, J. L. Latimer, J. Klesney-Tait, M. K. Stevens, L. S. Johnson, M. Purven, Jr., R. S. Munson, T. Lagergard, J. D. Radolf, and E. J. Hansen. 1997. A diffusible cytotoxin of *Haemophilus ducreyi*. *Proc Natl Acad Sci U S A* 94:4056–4061.

Dangor, Y., F. Radebe, and R. C. Ballard. 1993. Transport media for *Haemophilus ducreyi*. *Sex Transm Dis* 20:5–9.

Danner, D. B., R. A. Deich, K. L. Sisco, and H. O. Smith. 1980. An eleven-base-pair sequence determines the specificity of DNA uptake in *Haemophilus* transformation. *Gene* 11:311–318.

Darras-Joly, C., O. Lortholary, J. L. Mainardi, J. Etienne, L. Guillevin, and J. Acar. 1997. *Haemophilus* endocarditis: report of 42 cases in adults and review. *Haemophilus* endocarditis study group. *Clin Infect Dis* 24:1087–1094.

Dickerman, A., A. B. Bandara, and T. J. Inzana. 2019. Phylogenomic analysis of *Haemophilus parasuis* and proposed reclassification to *Glaesserella parasuis*, gen. nov., comb. nov. *Int J Syst Evol Microbiol*.

Duim, B., L. van Alphen, P. Eijk, H. M. Jansen, and J. Dankert. 1994. Antigenic drift of non-encapsulated *Haemophilus influenzae* major outer membrane protein P2 in patients with chronic bronchitis is caused by point mutations. *Mol Microbiol* 11:1181–1189.

Farley, M. M., D. S. Stephens, S. L. Kaplan, and Jr., E. O. Mason. 1990. Pilus- and non-pilus-mediated interactions of *Haemophilus influenzae* type b with human erythrocytes and human nasopharyngeal mucosa. *J Infect Dis* 161:274–280.

Fink, D. L., B. A. Green, and J. W. St Geme, 3rd. 2002. The *Haemophilus influenzae* hap autotransporter binds to fibronectin, laminin, and collagen IV. *Infect Immun* 70:4902–4907.

Fleischmann, R. D., M. D. Adams, O. White, R. A. Clayton, E. F. Kirkness, A. R. Kerlavage, C. J. Bult, J. F. Tomb, B. A. Dougherty, J. M. Merrick, et al. 1995. Whole-genome random sequencing and assembly of *Haemophilus influenzae* Rd. *Science* 269:496–512.

Forsgren, A., K. Riesbeck, and H. Janson. 2008. Protein D of *Haemophilus influenzae*: a protective nontypeable *H. influenzae* antigen and a carrier for pneumococcal conjugate vaccines. *Clin Infect Dis* 46:726–731.

Forsgren, J., A. Samuelson, A. Ahlin, J. Jonasson, B. Rynnel-Dagoo, and A. Lindberg. 1994. *Haemophilus influenzae* resides and multiplies intracellularly in human adenoid tissue as demonstrated by in situ hybridization and bacterial viability assay. *Infect Immun* 62:673–679.

Fulcher, R. A., L. E. Cole, D. M. Janowicz, K. L. Toffer, K. R. Fortney, B. P. Katz, P. E. Orndorff, S. M. Spinola, and T. H. Kawula. 2006. Expression of *Haemophilus ducreyi* collagen binding outer membrane protein NcaA is required for virulence in swine and human challenge models of chancroid. *Infect Immun* 74:2651–2658.

Fusco, W. G., C. Elkins, and I. Leduc. 2013. Trimeric autotransporter DsrA is a major mediator of fibrinogen binding in *Haemophilus ducreyi*. *Infect Immun* 81:4443–4452.

Gangaiah, D., K. M. Webb, T. L. Humphreys, K. R. Fortney, E. Toh, A. Tai, S. S. Katz, A. Pillay, C. Y. Chen, S. A. Roberts, Jr., R. S. Munson, and S. M. Spinola. 2015. *Haemophilus ducreyi* cutaneous ulcer strains are nearly identical to class I genital ulcer strains. *PLOS Negl Trop Dis* 9:e0003918.

Gaston, J. R., S. A. Roberts, and T. L. Humphreys. 2015. Molecular phylogenetic analysis of non-sexually transmitted strains of *Haemophilus ducreyi*. *PLOS One* 10:e0118613.

Geme, J. W., 3rd, and D. Cutter. 1995. Evidence that surface fibrils expressed by *Haemophilus influenzae* type b promote attachment to human epithelial cells. *Mol Microbiol* 15:77–85.

Gilsdorf, J. R., M. Tucci, and C. F. Marrs. 1996. Role of pili in *Haemophilus influenzae* adherence to, and internalization by, respiratory cells. *Pediatr Res* 39:343–348.

Gonzalez-Beiras, C., M. Marks, C. Y. Chen, S. Roberts, and O. Mitja. 2016. Epidemiology of *Haemophilus ducreyi* Infections. *Emerg Infect Dis* 22:1–8.

Goodman, S. D., K. P. Obergfell, J. A. Jurcisek, L. A. Novotny, J. S. Downey, E. A. Ayala, N. Tjokro, B. Li, S. S. Justice, and L. O. Bakaletz. 2011. Biofilms can be dispersed by focusing the immune system on a common family of bacterial nucleoid-associated proteins. *Mucosal Immunol* 4:625–637.

Greenblatt, R. M., S. A. Lukehart, F. A. Plummer, T. C. Quinn, C. W. Critchlow, R. L. Ashley, L. J. D'Costa, J. O. Ndinya-Achola, L. Corey, A. R. Ronald, et al. 1988. Genital ulceration as a risk factor for human immunodeficiency virus infection. *AIDS* 2:47–50.

Hallstrom, T., H. Jarva, K. Riesbeck, and A. M. Blom. 2007. Interaction with C4b-binding protein contributes to nontypeable *Haemophilus influenzae* serum resistance. *J Immunol* 178:6359–6366.

Hallstrom, T., B. Singh, F. Resman, A. M. Blom, M. Morgelin, and K. Riesbeck. 2011. *Haemophilus influenzae* protein E binds to the extracellular matrix by concurrently interacting with laminin and vitronectin. *J Infect Dis* 204:1065–1074.

Hallstrom, T., E. Trajkovska, A. Forsgren, and K. Riesbeck. 2006. *Haemophilus influenzae* surface fibrils contribute to serum resistance by interacting with vitronectin. *J Immunol* 177:430–436.

Hallstrom, T., P. F. Zipfel, A. M. Blom, N. Lauer, A. Forsgren, and K. Riesbeck. 2008. *Haemophilus influenzae* interacts with the human complement inhibitor factor H. *J Immunol* 181:537–545.

Harrison, A., D. W. Dyer, A. Gillaspy, W. C. Ray, R. Mungur, M. B. Carson, H. Zhong, J. Gipson, M. Gipson, L. S. Johnson, L. Lewis, L. O. Bakaletz, and Jr., R. S. Munson. 2005. Genomic sequence of an otitis media isolate of nontypeable *Haemophilus influenzae*: comparative study with *H. influenzae* serotype d, strain KW20. *J Bacteriol* 187:4627–4636.

Harrison, L. H., V. Simonsen, and E. A. Waldman. 2008. Emergence and disappearance of a virulent clone of *Haemophilus influenzae* biogroup *aegyptius*, cause of Brazilian purpuric fever. *Clin Microbiol Rev* 21:594–605.

Harvey, H. A., W. E. Swords, and M. A. Apicella. 2001. The mimicry of human glycolipids and glycosphingolipids by the lipooligosaccharides of pathogenic *Neisseria* and *Haemophilus*. *J Autoimmun* 16:257–262.

Hiltke, T. J., A. T. Schiffmacher, A. J. Dagonese, S. Sethi, and T. F. Murphy. 2003. Horizontal transfer of the gene encoding outer membrane protein P2 of nontypeable *Haemophilus influenzae*, in a patient with chronic obstructive pulmonary disease. *J Infect Dis* 188:114–117.

Hood, D. W., K. Makepeace, M. E. Deadman, R. F. Rest, P. Thibault, A. Martin, J. C. Richards, and E. R. Moxon. 1999. Sialic acid in the lipopolysaccharide of *Haemophilus influenzae*: strain distribution, influence on serum resistance and structural characterization. *Mol Microbiol* 33:679–692.

Houinei, W., C. Godornes, A. Kapa, S. Knauf, E. Q. Mooring, C. Gonzalez-Beiras, R. Watup, R. Paru, P. Advent, S. Bieb, S. Sanz, Q. Bassat, S. M. Spinola, S. A. Lukehart, and O. Mitja. 2017. *Haemophilus ducreyi* DNA is detectable on the skin of asymptomatic children, flies and fomites in villages of Papua New Guinea. *PLOS Negl Trop Dis* 11:e0004958.

Ikeda, M., N. Enomoto, D. Hashimoto, T. Fujisawa, N. Inui, Y. Nakamura, T. Suda, and T. Nagata. 2015. Nontypeable *Haemophilus influenzae* exploits the interaction between protein-E and vitronectin for the adherence and invasion to bronchial epithelial cells. *BMC Microbiol* 15:263.

Jalalvand, F., Y. C. Su, M. Morgelin, M. Brant, O. Hallgren, G. Westergren-Thorsson, B. Singh, and K. Riesbeck. 2013. *Haemophilus influenzae* protein F mediates binding to laminin and human pulmonary epithelial cells. *J Infect Dis* 207:803–813.

Janowicz, D. M., S. A. Cooney, J. Walsh, B. Baker, B. P. Katz, K. R. Fortney, B. W. Zwickl, S. Ellinger, and Jr., R. S. Munson. 2011. Expression of the Flp proteins by *Haemophilus ducreyi* is necessary for virulence in human volunteers. *BMC Microbiol* 11:208.

Janowicz, D. M., K. R. Fortney, B. P. Katz, J. L. Latimer, K. Deng, E. J. Hansen, and S. M. Spinola. 2004. Expression of the LspA1 and LspA2 proteins by *Haemophilus ducreyi* is required for virulence in human volunteers. *Infect Immun* 72:4528–4533.

Janson, H., B. Carlen, A. Cervin, A. Forsgren, A. B. Magnusdottir, S. Lindberg, and T. Runer. 1999. Effects on the ciliated epithelium of protein D-producing and -nonproducing nontypeable *Haemophilus influenzae* in nasopharyngeal tissue cultures. *J Infect Dis* 180:737–746.

Jurcisek, J. A., and L. O. Bakaletz. 2007. Biofilms formed by nontypeable *Haemophilus influenzae* in vivo contain both double-stranded DNA and type IV pilin protein. *J Bacteriol* 189:3868–3875.

Jurcisek, J. A., J. E. Bookwalter, B. D. Baker, S. Fernandez, L. A. Novotny, Jr., R. S. Munson, and L. O. Bakaletz. 2007. The PilA protein of non-typeable *Haemophilus influenzae* plays a role in biofilm formation, adherence to epithelial cells and colonization of the mammalian upper respiratory tract. *Mol Microbiol* 65:1288–1299.

Kaparakis-Liaskos, M., and R. L. Ferrero. 2015. Immune modulation by bacterial outer membrane vesicles. *Nat Rev Immunol* 15:375–387.

Kenjale, R., G. Meng, D. L. Fink, T. Juehne, T. Ohashi, H. P. Erickson, G. Waksman, and J. W. St Geme, 3rd. 2009. Structural determinants of autoproteolysis of the *Haemophilus influenzae* hap autotransporter. *Infect Immun* 77:4704–4713.

Kilian, M. 1976. A taxonomic study of the genus *Haemophilus*, with the proposal of a new species. *J Gen Microbiol* 93:9–62.

Kilian, M., J. Mestecky, and R. E. Schrohenloher. 1979. Pathogenic species of the genus *Haemophilus* and *Streptococcus pneumoniae* produce immunoglobulin A1 protease. *Infect Immun* 26:143–149.

Kroll, J. S., K. E. Wilks, J. L. Farrant, and P. R. Langford. 1998. Natural genetic exchange between *Haemophilus* and *Neisseria*: intergeneric transfer of chromosomal genes between major human pathogens. *Proc Natl Acad Sci U S A* 95:12381–12385.

Kubiet, M., R. Ramphal, A. Weber, and A. Smith. 2000. Pilus-mediated adherence of *Haemophilus influenzae* to human respiratory mucins. *Infect Immun* 68:3362–3367.

Kuhnert, P., and H. Christensen. 2019. International committee on systematics of prokaryotes subcommittee on the taxonomy of Pasteurellaceae minutes of the meetings, 9 October 2018, Prato, Italy. *Int J Syst Evol Microbiol* 69:871–872.

Kuklinska, D., and M. Kilian. 1984. Relative proportions of *Haemophilus* species in the throat of healthy children and adults. *Eur J Clin Microbiol* 3:249–252.

Kus, J. V., M. Shuel, D. Soares, W. Hoang, D. Law, and R. S. W. Tsang. 2019. Identification and characterization of "*Haemophilus quentini*" strains causing invasive disease in Ontario, Canada (2016-2018). *J Clin Microbiol* 57:e01254–19.

Lagergard, T., I. Bolin, and L. Lindholm. 2011. On the evolution of the sexually transmitted bacteria *Haemophilus ducreyi* and *Klebsiella granulomatis*. *Ann N Y Acad Sci* 1230:E1–E10.

Langereis, J. D., and P. W. Hermans. 2013. Novel concepts in nontypeable *Haemophilus influenzae* biofilm formation. *FEMS Microbiol Lett* 346:81–89.

Ledeboer, N. A., and G. V. Doern. 2015. Haemophilus. In: J. H. Jorgenson, M. A. Pfaller, K. C. Carroll, G. Funke, M. L. Landry, S. S. Richter and D. W. Warnock (eds.), *Manual of Clinical Microbiology* (ASM Press: Sterling, VA).

Leduc, I., B. Olsen, and C. Elkins. 2009. Localization of the domains of the *Haemophilus ducreyi* trimeric autotransporter DsrA involved in serum resistance and binding to the extracellular matrix proteins fibronectin and vitronectin. *Infect Immun* 77:657–666.

Leduc, I., P. Richards, C. Davis, B. Schilling, and C. Elkins. 2004. A novel lectin, DltA, is required for expression of a full serum resistance phenotype in *Haemophilus ducreyi*. *Infect Immun* 72:3418–3428.

Lewis, D. A. 2003. Chancroid: clinical manifestations, diagnosis, and management. *Sex Transm Infect* 79:68–71.

Lewis, D. A., and C. A. Ison. 2006. Chancroid. *Sex Transm Infect* 82 (Suppl 4):iv19–20.

Lewis, D. A., and O. Mitja. 2016. *Haemophilus ducreyi*: from sexually transmitted infection to skin ulcer pathogen. *Curr Opin Infect Dis* 29:52–57.

Liesman, R. M., A. P. Strasburg, A. K. Heitman, E. S. Theel, R. Patel, and M. J. Binnicker. 2018. Evaluation of a commercial multiplex molecular panel for diagnosis of infectious meningitis and encephalitis. *J Clin Microbiol* 56:e01927–17.

MacNeil, J. R., A. C. Cohn, M. Farley, R. Mair, J. Baumbach, N. Bennett, K. Gershman, L. H. Harrison, R. Lynfield, S. Petit, A. Reingold, W. Schaffner, A. Thomas, F. Coronado, E. R. Zell, L. W. Mayer, T. A. Clark, and N. E. Messonnier. 2011. Current epidemiology and trends in invasive *Haemophilus influenzae* disease–United States, 1989-2008. *Clin Infect Dis* 53:1230–1236.

Male, C. J. 1979. Immunoglobulin A1 protease production by *Haemophilus influenzae* and *Streptococcus pneumoniae*. *Infect Immun* 26:254–261.

Mansson, V., J. R. Gilsdorf, G. Kahlmeter, M. Kilian, J. S. Kroll, K. Riesbeck, and F. Resman. 2018. Capsule typing of *Haemophilus influenzae* by matrix-assisted laser desorption/ionization time-of-flight mass spectrometry. *Emerg Infect Dis* 24:443–452.

Marks, M., K. H. Chi, V. Vahi, A. Pillay, O. Sokana, A. Pavluck, D. C. Mabey, C. Y. Chen, and A. W. Solomon. 2014. *Haemophilus ducreyi* associated with skin ulcers among children, Solomon Islands. *Emerg Infect Dis* 20:1705–1707.

Marks, M., M. Fookes, J. Wagner, R. Ghinai, O. Sokana, Y. A. Sarkodie, A. W. Solomon, D. C. W. Mabey, and N. R. Thomson. 2018. Direct whole-genome sequencing of cutaneous strains of *Haemophilus ducreyi*. *Emerg Infect Dis* 24:786–789.

Meats, E., E. J. Feil, S. Stringer, A. J. Cody, R. Goldstein, J. S. Kroll, T. Popovic, and B. G. Spratt. 2003. Characterization of encapsulated and noncapsulated *Haemophilus influenzae* and determination of phylogenetic relationships by multilocus sequence typing. *J Clin Microbiol* 41:1623–1636.

Mell, J. C., C. Viadas, J. Moleres, S. Sinha, A. Fernandez-Calvet, E. A. Porsch, J. W. St Geme, 3rd, C. Nislow, R. J. Redfield, and J. Garmendia. 2016. Transformed recombinant enrichment profiling rapidly identifies HMW1 as an intracellular invasion locus in *Haemophilus influenzae*. *PLOS Pathog* 12:e1005576.

Mitja, O., S. A. Lukehart, G. Pokowas, P. Moses, A. Kapa, C. Godornes, J. Robson, S. Cherian, W. Houinei, W. Kazadi, P. Siba, E. de Lazzari, and Q. Bassat. 2014. *Haemophilus ducreyi* as a cause of skin ulcers in children from a yaws-endemic area of Papua New Guinea: a prospective cohort study. *Lancet Glob Health* 2:e235–241.

Mock, J. R., M. Vakevainen, K. Deng, J. L. Latimer, J. A. Young, N. S. van Oers, S. Greenberg, and E. J. Hansen. 2005. *Haemophilus ducreyi* targets src family protein tyrosine kinases to inhibit phagocytic signaling. *Infect Immun* 73:7808–7816.

Mount, K. L., C. A. Townsend, S. D. Rinker, X. Gu, K. R. Fortney, B. W. Zwickl, D. M. Janowicz, S. M. Spinola, B. P. Katz, and M. E. Bauer. 2010. *Haemophilus ducreyi* SapA contributes to cathelicidin resistance and virulence in humans. *Infect Immun* 78:1176–1184.

Moxon, E. R., and J. S. Kroll. 1988. Type b capsular polysaccharide as a virulence factor of *Haemophilus influenzae*. *Vaccine* 6:113–115.

Murphy, T. F., A. L. Brauer, S. Sethi, M. Kilian, X. Cai, and A. J. Lesse. 2007. *Haemophilus haemolyticus*: a human respiratory tract commensal to be distinguished from *Haemophilus influenzae*. *J Infect Dis* 195:81–89.

Murphy, T. F., and C. Kirkham. 2002. Biofilm formation by nontypeable *Haemophilus influenzae*: strain variability, outer membrane antigen expression and role of pili. *BMC Microbiol* 2:7.

Murphy, T. F., A. J. Lesse, C. Kirkham, H. Zhong, S. Sethi, and Jr., R. S. Munson. 2011. A clonal group of nontypeable *Haemophilus influenzae* with two IgA proteases is adapted to infection in chronic obstructive pulmonary disease. *PLOS One* 6:e25923.

Murphy, T. F., S. Sethi, K. L. Klingman, A. B. Brueggemann, and G. V. Doern. 1999. Simultaneous respiratory tract colonization by multiple strains of nontypeable *Haemophilus influenzae* in chronic obstructive pulmonary disease: implications for antibiotic therapy. *J Infect Dis* 180:404–409.

Ng, P. S. K., C. J. Day, J. M. Atack, L. E. Hartley-Tassell, L. E. Winter, T. Marshanski, V. Padler-Karavani, A. Varki, S. J. Barenkamp, M. A. Apicella, and M. P. Jennings. 2019. Nontypeable *Haemophilus influenzae* has evolved preferential use of N-acetylneuraminic acid as a host adaptation. *MBio* 10.

Nika, J. R., J. L. Latimer, C. K. Ward, R. J. Blick, N. J. Wagner, L. D. Cope, G. G. Mahairas, Jr., R. S. Munson, and E. J. Hansen. 2002. *Haemophilus ducreyi* requires the flp gene cluster for microcolony formation in vitro. *Infect Immun* 70:2965–2975.

Norskov-Lauritsen, N. 2014. Classification, identification, and clinical significance of *Haemophilus* and *aggregatibacter* species with host specificity for humans. *Clin Microbiol Rev* 27:214–240.

Norskov-Lauritsen, N., B. Bruun, C. Andersen, and M. Kilian. 2012. Identification of haemolytic *Haemophilus* species isolated from human clinical specimens and description of *Haemophilus sputorum* sp. nov. *Int J Med Microbiol* 302:78–83.

Norskov-Lauritsen, N., B. Bruun, and M. Kilian. 2005. Multilocus sequence phylogenetic study of the genus *Haemophilus* with description of *Haemophilus pittmaniae* sp. nov. *Int J Syst Evol Microbiol* 55:449–456.

Oerlemans, M. M. P., S. J. Moons, J. J. A. Heming, T. J. Boltje, M. I. de Jonge, and J. D. Langereis. 2019. Uptake of sialic acid by nontypeable *Haemophilus influenzae* increases complement resistance through decreasing IgM-dependent complement activation. *Infect Immun* 87:e00077-19.

Ogilvie, C., A. Omikunle, Y. Wang, 3rd, J.W St Geme, C. A. Rodriguez, and E. E. Adderson. 2001. Capsulation loci of non-serotype b encapsulated *Haemophilus influenzae*. *J Infect Dis* 184:144–149.

Omikunle, A., S. Takahashi, C. L. Ogilvie, Y. Wang, C. A. Rodriguez, 3rd, J. W. St Geme, and E. E. Adderson. 2002. Limited genetic diversity of recent invasive isolates of non-serotype b encapsulated *Haemophilus influenzae*. *J Clin Microbiol* 40:1264–1270.

Pantzar, M., S. Teneberg, and T. Lagergard. 2006. Binding of *Haemophilus ducreyi* to carbohydrate receptors is mediated by the 58.5-kDa GroEL heat shock protein. *Microbes Infect* 8:2452–2458.

Pettigrew, M. M., C. P. Ahearn, J. F. Gent, Y. Kong, M. C. Gallo, J. B. Munro, A. D'Mello, S. Sethi, H. Tettelin, and T. F. Murphy. 2018. *Haemophilus influenzae* genome evolution during persistence in the human airways in chronic obstructive pulmonary disease. *Proc Natl Acad Sci U S A* 115:E3256–E3265.

Pittman, M., and D. J. Davis. 1950. Identification of the Koch-Weeks bacillus (*Hemophilus aegyptius*). *J Bacteriol* 59:413–426.

Poole, J., E. Foster, K. Chaloner, J. Hunt, M. P. Jennings, T. Bair, K. Knudtson, E. Christensen, Jr., R. S. Munson, P. L. Winokur, and M. A. Apicella. 2013. Analysis of nontypeable *Haemophilus influenzae* phase-variable genes during experimental human nasopharyngeal colonization. *J Infect Dis* 208:720–727.

Potts, C. C., N. Topaz, L. D. Rodriguez-Rivera, F. Hu, H. Y. Chang, M. J. Whaley, S. Schmink, A. C. Retchless, A. Chen, E. Ramos, G. H. Doho, and X. Wang. 2019. Genomic characterization of *Haemophilus influenzae*: a focus on the capsule locus. *BMC Genomics* 20:733.

Powell, E. A., D. Blecker-Shelly, S. Montgomery, and J. E. Mortensen. 2013. Application of matrix-assisted laser desorption ionization-time of flight mass spectrometry for identification of the fastidious pediatric pathogens *Aggregatibacter*, *Eikenella*, *Haemophilus*, and *Kingella*. *J Clin Microbiol* 51:3862–3864.

Raffel, F. K., B. R. Szelestey, W. L. Beatty, and K. M. Mason. 2013. The *Haemophilus influenzae* sap transporter mediates bacterium-epithelial cell homeostasis. *Infect Immun* 81:43–54.

Reddy, M. S., J. M. Bernstein, T. F. Murphy, and H. S. Faden. 1996. Binding between outer membrane proteins of nontypeable *Haemophilus influenzae* and human nasopharyngeal mucin. *Infect Immun* 64:1477–1479.

Rennie, R. P., C. Brosnikoff, S. Shokoples, L. B. Reller, S. Mirrett, W. Janda, K. Ristow, and A. Krilcich. 2008. Multicenter evaluation of the new vitek 2 neisseria-haemophilus identification card. *J Clin Microbiol* 46:2681–2685.

Ricotta, E. E., N. Wang, R. Cutler, J. G. Lawrence, and T. L. Humphreys. 2011. Rapid divergence of two classes of *Haemophilus ducreyi*. *J Bacteriol* 193:2941–2947.

Rinker, S. D., M. P. Trombley, X. Gu, K. R. Fortney, and M. E. Bauer. 2011. Deletion of mtrC in *Haemophilus ducreyi* increases sensitivity to human antimicrobial peptides and activates the CpxRA regulon. *Infect Immun* 79:2324–2334.

Rodriguez, C. A., V. Avadhanula, A. Buscher, A. L. Smith, 3rd, J. W. St Geme, and E. E. Adderson. 2003. Prevalence and distribution of adhesins in invasive non-type b encapsulated *Haemophilus influenzae*. *Infect Immun* 71:1635–1642.

Ruan, M. R., M. Akkoyunlu, A. Grubb, and A. Forsgren. 1990. Protein D of *Haemophilus influenzae*. A novel bacterial surface protein with affinity for human IgD. *J Immunol* 145:3379–3384.

Sethi, S., and T. F. Murphy. 2001. Bacterial infection in chronic obstructive pulmonary disease in 2000: a state-of-the-art review. *Clin Microbiol Rev* 14:336–363.

Shelton, C. L., F. K. Raffel, W. L. Beatty, S. M. Johnson, and K. M. Mason. 2011. Sap transporter mediated import and subsequent degradation of antimicrobial peptides in *Haemophilus*. *PLOS Pathog* 7:e1002360.

Shen, K., P. Antalis, J. Gladitz, S. Sayeed, A. Ahmed, S. Yu, J. Hayes, S. Johnson, B. Dice, R. Dopico, R. Keefe, B. Janto, W. Chong, J. Goodwin, R. M. Wadowsky, G. Erdos, J. C. Post, G. D. Ehrlich, and F. Z. Hu. 2005. Identification, distribution, and expression of novel genes in 10 clinical isolates of nontypeable *Haemophilus influenzae*. *Infect Immun* 73:3479–3491.

Singh, B., Y. C. Su, T. Al-Jubair, O. Mukherjee, T. Hallstrom, M. Morgelin, A. M. Blom, and K. Riesbeck. 2014. A fine-tuned interaction between trimeric autotransporter *Haemophilus* surface fibrils and vitronectin leads to serum resistance and adherence to respiratory epithelial cells. *Infect Immun* 82:2378–2389.

Smith, H. O., J. F. Tomb, B. A. Dougherty, R. D. Fleischmann, and J. C. Venter. 1995. Frequency and distribution of DNA uptake signal sequences in the *Haemophilus influenzae* Rd genome. *Science* 269:538–540.

Smith-Vaughan, H. C., A. J. Leach, T. M. Shelby-James, K. Kemp, D. J. Kemp, and J. D. Mathews. 1996. Carriage of multiple ribotypes of non-encapsulated *Haemophilus influenzae* in aboriginal infants with otitis media. *Epidemiol Infect* 116:177–183.

Soeters, H. M., A. Blain, T. Pondo, B. Doman, M. M. Farley, L. H. Harrison, R. Lynfield, L. Miller, S. Petit, A. Reingold, W. Schaffner, A. Thomas, S. M. Zansky, X. Wang, and E. C. Briere. 2018. Current epidemiology and trends in invasive *Haemophilus influenzae* disease-United States, 2009-2015. *Clin Infect Dis* 67:881–889.

Spinola, S. M., K. R. Fortney, B. P. Katz, J. L. Latimer, J. R. Mock, M. Vakevainen, and E. J. Hansen. 2003. *Haemophilus ducreyi* requires an intact flp gene cluster for virulence in humans. *Infect Immun* 71:7178–7182.

St Geme, J. W., 3rd, S. Falkow, and S. J. Barenkamp. 1993. High-molecular-weight proteins of nontypable *Haemophilus influenzae* mediate attachment to human epithelial cells. *Proc Natl Acad Sci U S A* 90:2875–2879.

Strouts, F. R., P. Power, N. J. Croucher, N. Corton, A. van Tonder, M. A. Quail, P. R. Langford, M. J. Hudson, J. Parkhill, J. S. Kroll, and S. D. Bentley. 2012. Lineage-specific virulence determinants of *Haemophilus influenzae* biogroup *aegyptius*. *Emerg Infect Dis* 18:449–457.

Sturm, A. W. 1981. Identification of *Haemophilus ducreyi*. *Antonie Van Leeuwenhoek* 47:89–90.

Su, Y. C., F. Jalalvand, M. Morgelin, A. M. Blom, B. Singh, and K. Riesbeck. 2013. *Haemophilus influenzae* acquires vitronectin via the ubiquitous protein F to subvert host innate immunity. *Mol Microbiol* 87:1245–1266.

Su, Y. C., O. Mukherjee, B. Singh, O. Hallgren, G. Westergren-Thorsson, D. Hood, and K. Riesbeck. 2016. *Haemophilus influenzae* P4 interacts with extracellular matrix proteins promoting adhesion and serum resistance. *J Infect Dis* 213:314–323.

Swords, W. E., B. A. Buscher, K. Ver Steeg Ii, A. Preston, W. A. Nichols, J. N. Weiser, B. W. Gibson, and M. A. Apicella. 2000. Non-typeable *Haemophilus influenzae* adhere to and invade human bronchial epithelial cells via an interaction of lipooligosaccharide with the PAF receptor. *Mol Microbiol* 37:13–27.

Swords, W. E., M. R. Ketterer, J. Shao, C. A. Campbell, J. N. Weiser, and M. A. Apicella. 2001. Binding of the non-typeable *Haemophilus influenzae* lipooligosaccharide to the PAF receptor initiates host cell signalling. *Cell Microbiol* 3:525–536.

Takeuchi, N., S. Segawa, N. Ishiwada, M. Ohkusu, S. Tsuchida, M. Satoh, K. Matsushita, and F. Nomura. 2018. Capsular serotyping of *Haemophilus influenzae* by using matrix-associated laser desorption ionization-time of flight mass spectrometry. *J Infect Chemother* 24:510–514.

Trombley, M. P., D. M. Post, S. D. Rinker, L. M. Reinders, K. R. Fortney, B. W. Zwickl, D. M. Janowicz, F. M. Baye, B. P. Katz, S. M. Spinola, and M. E. Bauer. 2015. Phosphoethanolamine transferase LptA in *Haemophilus ducreyi* modifies lipid A and contributes to human defensin resistance in vitro. *PLOS One* 10:e0124373.

Ubukata, K., Y. Shibasaki, K. Yamamoto, N. Chiba, K. Hasegawa, Y. Takeuchi, K. Sunakawa, M. Inoue, and M. Konno. 2001. Association of amino acid substitutions in penicillin-binding protein 3 with beta-lactam resistance in beta-lactamase-negative ampicillin-resistant *Haemophilus influenzae*. *Antimicrob Agents Chemother* 45:1693–1699.

Ulanova, M., and R. S. W. Tsang. 2014. *Haemophilus influenzae* serotype a as a cause of serious invasive infections. *Lancet Infect Dis* 14:70–82.

Valenza, G., C. Ruoff, U. Vogel, M. Frosch, and M. Abele-Horn. 2007. Microbiological evaluation of the new VITEK 2 Neisseria-Haemophilus identification card. *J Clin Microbiol* 45:3493–3497.

Van Eldere, J., M. P. Slack, S. Ladhani, and A. W. Cripps. 2014. Non-typeable *Haemophilus influenzae*, an under-recognised pathogen. *Lancet Infect Dis* 14:1281–1292.

VanWagoner, T. M., J. M. Atack, K. L. Nelson, H. K. Smith, K. L. Fox, M. P. Jennings, T. L. Stull, and A. L. Smith. 2016. The modA10 phasevarion of nontypeable *Haemophilus influenzae* R2866 regulates multiple virulence-associated traits. *Microb Pathog* 92:60–67.

Watson, M. E., Jr., J. L. Burns, and A. L. Smith. 2004. Hypermutable *Haemophilus influenzae* with mutations in mutS are found in cystic fibrosis sputum. *Microbiology* 150:2947–2958.

Weiser, J. N., S. T. Chong, D. Greenberg, and W. Fong. 1995. Identification and characterization of a cell envelope protein of *Haemophilus influenzae* contributing to phase variation in colony opacity and nasopharyngeal colonization. *Mol Microbiol* 17:555–564.

Weiser, J. N., N. Pan, K. L. McGowan, D. Musher, A. Martin, and J. Richards. 1998. Phosphorylcholine on the lipopolysaccharide of *Haemophilus influenzae* contributes to persistence in the respiratory tract and sensitivity to serum killing mediated by C-reactive protein. *J Exp Med* 187:631–640.

White, C. D., I. Leduc, B. Olsen, C. Jeter, C. Harris, and C. Elkins. 2005. *Haemophilus ducreyi* outer membrane determinants, including DsrA, define two clonal populations. *Infect Immun* 73:2387–2399.

Wilfert, C. M. 1990. Epidemiology of *Haemophilus influenzae* type b infections. *Pediatrics* 85:631–635.

Wilson, R., and P. J. Cole. 1988. The effect of bacterial products on ciliary function. *Am Rev Respir Dis* 138:S49–53.

Yamamoto, C., S. Suzuki, H. Kimura, H. Yoshida, and F. Gejyo. 2002. Experimental nephropathy induced by *Haemophilus parainfluenzae* antigens. *Nephron* 90:320–327.

Young, R. E., and D. W. Hood. 2013. *Haemophilus parainfluenzae* has a limited core lipopolysaccharide repertoire with no phase variation. *Glycoconj J* 30:561–576.

30

The Genus Helicobacter

Ernestine M. Vellozzi and Edmund R. Giugliano

CONTENTS

Introduction

Many observations of spiral bacteria colonizing the gastric mucosa of animals and humans have been noted throughout the 20th century, some of which suggested their role in gastric pathology.[1–3] Despite the presence of these microorganisms, the stomach was still considered a sterile environment. It was the discovery of *Helicobacter pylori* that forever changed our perception of the stomach, and made us view this organ as a habitat for specialized bacteria. Recent evidence supports the theory that humans and *H. pylori* have coevolved for millennia. Genetic studies indicate that humans have been colonized with the organism for at least 58,000 years and that *H. pylori* most likely post-dated the evolution of humans.[4–7] Strain types that prevail within certain regions of the world have also been found to correlate with human migration patterns.[5,8]

Isolated from the human stomach in 1982, this organism was first categorized as a spiral-shaped bacterium resembling *Campylobacter*.[9] The organism bore many similarities to the campylobacters, including morphology, growth under microaerophilic conditions, G + C content, among others. As a result, the organism first referred to as a *Campylobacter*-like organism was later named *Campylobacter pylordis* in 1985.[10] The species name was subsequently revised to "*pylori*" in 1987.[11] Electron microscopy and fatty acid analysis of *C. pylori* soon demonstrated that it showed marked differences from the other *Campylobacter* species.[12] *C. pylori* was later reclassified into the new genus *Helicobacter*. This new classification was supported by sequencing studies of 16S rRNA.[13] Early studies described this bacterium as a probable cause of gastritis and peptic ulcer disease, and this etiology would later be validated.[14] Now, there is much evidence to support the role of *H. pylori* in various clinical disorders of the upper gastrointestinal tract, such as chronic gastritis, peptic ulcer disease, mucosa-associated lymphoid tissue (MALT) lymphoma, and gastric adenocarcinoma.[15–17] The clinical significance of *H. pylori* has been emphasized by the World Health Organization, via the classification of this organism as a Class I carcinogen of gastric cancer.[18] This bacterium is currently considered the most common etiologic agent of infection-related cancers and it is believed that up to 5.5% of the global cancer burden is represented by this organism.[19,20] In more recent years, *H. pylori* infection has been implicated in some extradigestive diseases, such as iron-deficiency anemia, idiopathic thrombocytopenic purpura (ITP),

DOI: 10.1201/9781003099277-32

cardiovascular diseases, and diabetes mellitus, among others.[21,22] The isolation, characterization, and recognition of *H. pylori* have generated a renewed interest in the possible association of other spiral mucus-associated bacteria with disease.

Many animals are now known to be colonized by a wide range of highly adapted bacteria that belong to the genus *Helicobacter*.[23] The genus has grown from the initial named two species in 1989 (*H. pylori* and *H. mustelae*) to approximately 46 species.[24,25] The *Helicobacter* species as a group can be broadly categorized according to their site of colonization in the host. Distinct groups of species within the genus have been identified using 16S rRNA gene sequencing. These groups are broadly categorized as gastric (stomach), enterohepatic (intestine, liver or biliary tract) and a subgroup of enterohepatic species which lack a characteristic sheathed flagellum (Table 30.1). Gastric *Helicobacter* species are widely distributed in mammalian hosts and are nearly universally present in the gastric mucosa. This group of organisms can cause an inflammatory response resembling that seen with *H. pylori* in humans.[28] The gastric helicobacters often serve as good models of human disease. Enterohepatic *Helicobacter* species are a diverse group of organisms that have been identified in the intestinal and hepatobiliary tracts of various mammal and bird hosts, with several species associated with human infection (Table 30.2). These organisms have been linked with inflammation in immunocompromised humans where more severe clinical disease is observed.[29] Among the unsheathed subgroup of enterohepatic helicobacters are species associated with gastroenteritis, and possible liver disease in humans.[23,30–32]

This chapter describes the genus *Helicobacter* and characterizes the role of these organisms in the pathogenesis of gastric and enterohepatic disease. Epidemiology, transmission, pathogenesis, virulence factors, along with methods in culture, identification, detection, and treatment, are reviewed. Subsequent discussion of animal models in this chapter illustrates how these models have helped investigators gather pertinent information to further our understanding of this fascinating group of organisms comprising the genus *Helicobacter*.

Genus Description

The genus *Helicobacter* comprises more than 40 species, with additional provisionally named species waiting to be validated. Others have been isolated but await classification. New and more sophisticated techniques, especially in the field of molecular biology, have helped in the further delineation of the genus.[33–36] These Gram-negative, nonspore-forming rods typically have spiral, curved, or fusiform morphology ranging from 0.3 μm to 1.0 μm in width to 1.5–10.0 μm in length. The morphology of the spiral-shaped body varies greatly among the species. For example, *H. acinonychis* is an "S"-shaped organism similar to *H. pylori*, while *H. felis* and *H. heilmannii* have tightly spiraled bodies that resemble a corkscrew. Motility in the genus *Helicobacter* is due to the presence of flagella, which give them a characteristic rapid corkscrew-like or slower wave-like motion[33] Most species have bundles of multiple sheathed flagella, having either a polar or bipolar distribution, while others have a single polar or bipolar

flagellum. *H. pylori* isolates, in contrast to other *Helicobacter* species, have multiple monopolar sheathed flagella (Table 30.1). Flagellar distribution can also be peritrichous (*H. mustelae*) or nonsheathed (*H. pullorum*, *H. rodentium*, and *H. mesocricetorum*).[33,34] These basic morphology and motility characteristics are thought to play an important role in the ability of the organism to maneuver successfully in the mucous layer of their host gastrointestinal tracts.

The host range of colonization varies among the gastric *Helicobacter* species, with some being more restrictive than others. For example, *H. pylori* is associated with humans, *H. felis* with cats and dogs, *H. mustelae* with ferrets and minks, *H. salomonis* with dogs, and *H. acinonychis*, which specifically colonizes the cheetah. Other gastric species, for example, *H. bizzozeronii*, *H. heilmannii*, and related bacteria, exhibit a much broader host specificity, naturally colonizing, cats, dogs, and primates, including humans.[37,38] The presence of gastric *Helicobacter* species is associated with inflammation and/or ulceration, and gastric mucosa-associated lymphoid tissue lymphoma in humans, like *H. heilmannii*, which is known to colonize the gastric mucosa of animals. There is also some evidence to suggest that *H. heilmannii* infection may be an example of a zoonosis.[38–41]The gastric helicobacters are recognized as important pathogens in gastroduodenal disease. In contrast, the enterohepatic helicobacters inhabit the intestinal and hepatobiliary tracts of mammals and birds. In humans, enterohepatic helicobacters, such as *H. canadensis*, *H. canis*, *H. cinaedi*, *H. fennelliae*, *H. pullorum*, *H. winghamensis*, and *Helicobacter* sp. *strain flexispira taxon 8*, have been isolated from rectal swabs and feces. Enterohepatic helicobacters have been linked to severe inflammatory lesions in the lower bowel, gall bladders, and livers of infected mammalian hosts including humans.[38–46] Furthermore, *H. mustelae*, *H. hepaticus*, and *H. bilis*, which are known to have numerous virulence factors, have been found to display carcinogenic potential in animals and have been implicated in causing disease in humans as well.[41]

In the laboratory, most species of *Helicobacter* will grow at 37°C under microaerophilic conditions in an atmosphere of reduced oxygen concentration (5–10% O_2), and increased humidity and carbon dioxide levels (5–12% CO_2). The role of atmospheric hydrogen in the microaerophilic incubation of *Helicobacter* is not fully understood. The requirement appears to be strain dependent, being required by some strains and having a growth stimulating effect in others.[33,34] Routine aerobic atmospheres do not support the growth of most *Helicobacter* species. Some species grow poorly, at best, in these conditions whereas several *Helicobacter* species, such as *H. rodentium*, can grow both in microaerobic and anaerobic conditions.[47] Another rodent enteric species, *H. ganmani* grows best in an anaerobic cabinet, although strict anaerobic conditions can be lethal to the organism.[48] When helicobacters are cultured for a prolonged period of time, or if the growth conditions are suboptimal, cell morphology may become spheroid or coccoid in nature. These unusual forms are not successfully subcultured.[24,49] As shown in Table 30.1, there exists some variation in the key biochemical traits within the genus *Helicobacter* which can therefore aid in species identification. Biochemically, helicobacters display a respiratory type of metabolism. Neither oxidation nor

TABLE 30.1

Key Characteristics for Identification of *Helicobacter* Species[a]

Helicobacter Species	Catalase	Urease	Nitrate	Indoxyl Acetate Hydrolysis	Alkaline Phosphatase	ɣ-Glutamyl Transpeptidase	Growth: At 42°C	with 1% glycine	Resistance to: Nal	Ceph	Mol% G+C	Flagellum:[b] type (no.)	sheath
Gastric													
H. acinonychis	+	+	−	−	+	+	−	−	R	S	30	B (2-5)	+
H. baculiformis	+	+	+	−	+	+	−	−	I	R	ND	B (11)	+
H. bizzozeronii	+	+	+	+	+	+	+	−	R	S	ND	B (10-20)	+
"Candidatus H. bovis"	ND	+	ND	ND	ND	ND	ND	ND	ND	ND	ND	M/B (1-4)	?
H. cetorum	+	+	−	−	−	+	+	−	I	S	ND	B (2)	+
H. cynogastricus	+	+	+	−	+	+	−	−	ND	ND	ND	B (6-12)	+
H. felis	+	+	+	−	+	+	+	−	R	S	42	B (14-20)	+
"Candidatus H. heilmannii"	ND	+	ND	ND	ND	ND	ND	ND	ND	ND	ND	B (10-20)	+
H pylori	+	+	−	−	+	+	−	−	R	S	35-37	B (4-8)	+
H. salomonis	+	+	+	+	+	+	−	−	R	S	ND	B (10-23)	+
H. suis	+	+	−	−	+	+	−	−	ND	ND	ND	B (4-10)	+
Enterohepatic													
H. anseris	−	+	−	+	−	−	+	W	S	R	ND	B (2)	+
H. aurati	+	+	−	+	+	+	+	−	S	R	ND	B (7-10)	+
H. bilis	+	+	+	−	−	+	−	+	R	R	ND	B (3-14)	+
H. brantae	−	−	−	+	−	−	+	W	S	R	ND	B (2)	+
H. canis	−	−	−	+	+	+	+	−	S	I	48	B (2)	+
H. cholecystus	+	−	+	−	+	−	+	+	I	R	ND	M (1)	+
H. cinaedi	+	−	+	−	−	−	−	+	S	I	37-38	B (1-2)	+
H. equorum	+	−	+	−	+	−	−	−	R	R	ND	M (1)	+
H. fennelliae	+	−	−	+	+	−	−	+	S	S	35	M/B (1-2)	+
H. hepaticus	+	+	+	+	−	−	−	+	R	R	35	B (2)	+
H. marmotae	+	+	−	−	+	−	+	+	R	R	ND	B (2)	+
H. mastomyrinus	+	+	−	−	−	−	+	+	R	R	ND	B (2)	+
H. muridarum	+	+	−	−	+	+	−	−	R	R	34	B (10-14)	+
H. mustelae	+	+	+	+	+	+	+	−	S	R	36	P (4-8)	+
H. pametensis	+	−	+	−	+	−	+	+	S	S	38	B (2)	+
H. trogontum	+	+	+	−	−	+	+	ND	R	R	ND	B (4-7)	+
H. typhlonius	+	−	+	−	−	−	+	+	S	R	ND	B (2)	+
Unsheathed													
H. canadensis	+	−	W	+	−	−	+	+	R	R	33	B (2)	−
H. ganmani	−	−	+	−	−	ND	+	−	R	S	ND	B (2)	−
H. mesocricetorum	+	−	+	ND	+	−	+	−	S	R	ND	B (2)	−
H. pullorum	+	−	+	−	−	ND	+	−	R	S	34-35	M (1)	−
H. rodentium	+	−	+	−	−	−	+	+	R	R	ND	B (2)	−
"H. winghamensis"	−	−	−	+	−	ND	−	+	R	R	ND	B (2)	−

[a] Test results from Baele et al.[26] and Fox and Megraud[27] and formal species descriptions. +, positive; −, negative; W, weakly positive; ND, not determined; Nal, nalidixic acid; Ceph, cephalexin; R, resistant; S, susceptible.

[b] B, bipolar; M, monopolar, P, peritrichous.

Source: See Refs. [24], [26], and [27]. (©2011. Used with permission of the American Society for Microbiology. No further reproduction or distribution is permitted without their prior written permission.)

TABLE 30.2

Helicobacter Species, Hosts, and Disease Spectrum

Helicobacter Species	Primary Hosts	Primary Site(s)	Diseases in Humans
Gastric			
H. acinonychis	Large felines (cheetahs)	Stomach	Not reported
H. baculiformis	Cats	Stomach	Not reported
H. bizzozeronii	Dogs, cats	Stomach	Gastritis, ulcer
"*Candidatus* H. bovis"	Cattle	Stomach	Gastritis
H. cetorum	Dolphins, whales	Stomach	Not reported
H. cynogastricus	Dogs	Stomach	Not reported
H. felis	Cats, dogs	Stomach	Gastritis, ulcer
"*Candidatus* H. heilmannii"	Humans	Stomach	Gastritis, ulcer
H. pylori	Humans	Stomach	Gastritis, ulcer, MALT, lymphoma, gastric cancer
H. salomonis	Dogs	Stomach	Gastritis, ulcer
H. suis	Pigs	Stomach	Gastritis, ulcer
Enterohepatic			
H. anseris	Geese	Intestine	Not reported
H. aurati	Rodents (hamsters)	Intestine	Not reported
H. bilis	Rodents, dogs	Intestine, liver	Sepsis
H. brantae	Geese	Intestine	Not reported
H. canis	Dogs	Intestine	Not reported
H. cholecystus	Rodents (hamsters)	Liver	Not reported
H. cinaedi	Rodents, dogs, primates	Intestine	Colitis, sepsis, cellulitis
H. equorum	Horses	Intestine	Not reported
H. fennelliae	Dogs	Intestine	Colitis, sepsis
H. hepaticus	Rodents	Intestine, liver	Not reported
H. marmotae	Rodents (woodchucks), cats	Intestine, liver	Not reported
H. mastomyrinus	Rodents (*Mastomys*)	Intestine, liver	Not reported
H. muridarum	Rodents	Intestine	Not reported
H. mustelae	Ferrets, minks	Stomach	Not reported
H. pametensis	Birds (terns), pigs	Intestine	Not reported
H. trogontum	Rodents	Intestine	Not reported
H. typhlonius	Rodents	Intestine	Not reported
Unsheathed			
H. canadensis	Chickens, geese	Intestine	Gastroenteritis
H. ganmani	Rodents (mice)	Intestine	Liver disease?
H. mesocricetorum	Rodents (hamsters)	Intestine	Not reported
H. pullorum	Chickens	Intestine	Gastroenteritis
H. rodentium	Rodents	Intestine	Not reported
"*H. winghamensis*"	Rodents	Intestine	Gastroenteritis

Source: ©2011. Used with permission of the American Society for Microbiology. No further reproduction or distribution is permitted without their prior written permission.

fermentation is observed when examining standard methods of carbohydrate utilization. However, glucose oxidation appears to occur in *H. pylori,* and oxidase activity is found in all species.[50] Nearly all *H. pylori* strains require arginine, histidine, leucine, methionine, phenylalanine, and valine for growth.[51,52] These amino acid requirement are similar to those required by humans.[53] The principal means of energy production and biosynthesis is believed to occur via glycolysis and gluconeogenesis, respectively. Other catabolic pathways, such as the glyoxylate shunt, is absent in these organisms.[50,54]

Epidemiology and Transmission of *Helicobacter pylori* and Other Helicobacters

Preventing the spread of infectious pathogenic agents has historically relied on public health measures that were based on sound epidemiological data. In the case of *Helicobacter*, and in particular *H. pylori*, the epidemiology and especially the transmission mechanisms are incompletely understood. A thorough understanding of these processes must be first elucidated prior to the possible implementation of control measures for this group of organisms. *H. pylori* in particular, has emerged as one of the most common chronic bacterial infections in the world, affecting half the global population.[55] Although the human stomach is the natural niche for *H. pylori*, there have been reports implicating other sources for the bacterium.[56] Zoonotic transmission routes have been proposed, such as transmission by houseflies, and some domestic animals such as dogs, cats, and sheep, as well as iatrogenic sources.[41,57–59] However, evidence suggests direct person-to-person transmission of the organism, especially from mothers to children and among siblings, is the most probable source. There also appears to be a greater chance of acquiring infection in the early years of life.[6,24,56] This mode of transmission is supported by the clustering of cases within families, and the similarity of genotypes found among isolates of related persons.[60–63] Direct person-to-person transmission is also supported by the fact that *H. pylori* can be found in regurgitated gastric juice, which would allow the organism to temporarily colonize the oral cavity. *H. pylori* has also been isolated from dental plaque, saliva, and vomitus where the organism can remain viable for hours.[56]

Although, the mechanism by which *H. pylori* is transmitted from one host to another remains speculative, the gastro-oral, oral-oral, and fecal-oral routes are the most likely means of transmission.[56,64,65] In developing countries, fecal/oral transmission of *H. pylori* appears to be a likely means of spread of *Helicobacter* where lack of sanitation may pose a greater risk.[64,65] Water has been proposed as a source of *Helicobacter* infection, since water biofilms have been suggested to provide the bacteria with a protective habitat enabling them to endure the water handling process.[66,67] This route of transmission has been supported by both epidemiologic studies which have shown *H. pylori* prevalence to be associated with water related sources, as well as with more rapid acquisition rate, and those that have detected or isolated *H. pylori* from water sources by using autoradiographic, molecular, culture and other methods.[68–79] Water may serve as a secondary transmission source in the fecal-oral transmission of *H. pylori*, acting as a reservoir for the organism, as well as a source of infection.[79]

Human infections with *H. heilmannii* and related species of gastric origin are generally considered uncommon, with prevalence rates ranging from less than 0.3–6%.[41,80] There have been reports of human infection with one or more species of zoonotic origin, notably, *H. salomonis*, *H. felis*, *H. suis*, and "*Candidatus H. bovis.*" This indicates that cats, dogs and swine may be possible sources of infection, although the mode of transmission is not yet known.[80–83]

Among the enterohepatic helicobacters, species such as *H. bilis*, *H. canadensis*, *H. canis*, *H. cinaedi*, *H. fennelliae*, *H. pullorum*, and *H. winghamensis*, are known to produce clinical symptoms in humans. See Table 30.2 The enterohepatic *H. mustelae*, which colonizes ferrets and minks, has a similar relationship of age-associated prevalence that mimics that of *H. pylori* in humans. Studies have shown that *H. mustelae* infection ranges from 0% in ferrets less than 1 month of age to 100% in adult animals over 1 year of age.[84,85] Fecal-oral spread of *H. mustelae* in the ferret population has likewise been proposed, as well as the possible zoonotic transmission of these gastric helicobacters to humans.[86] Naturally infected cats and sheep have been found to have *H. pylori* in their secretions and feces, also raising the possibility of transmission to humans.[87,88] Likewise, the zoonotic food-borne transmission of enterohepatic helicobacters to humans, such as *H. pullorum* and *H. canadensis,* has been reported.[45,89] Enterohepatic *H. pullorum*, is considered an emergent human pathogen and has been implicated in gastroenteritis, inflammatory bowel disease, and chronic liver diseases.[41] A high prevalence of *H. pullorum* has been found in chicken flocks, with strains of the bacterium isolated from chicken meat, which suggests this organism may be a foodborne pathogen.[42,90,91] The significance of enterohepatic helicobacters in transmission of disease and the true prevalence of these organisms in human and animal populations are yet to be determined. What has become apparent over the years is that *Helicobacter* species are commonly present as part of the gastric, enteric, and hepatobiliary microbiota in humans and other animals.[24,92] See Table 30.2.

Pathogenesis

The gastric helicobacters have evolved and acquired a number of traits that enable them to survive in a harsh environment, and the complex mechanisms used by these organisms to establish infection and maintain colonization is just beginning to be understood. Knowledge of the prevalence and natural history of *Helicobacter* infections as they occur in nature in nonhumans is likewise limited.[93] Gastric *Helicobacter* species exhibit different degrees of host specificity; hence, manifestations of disease are host specific in most cases. High levels of host specificity in gastric helicobacters are most evident in those species that adhere tightly to host tissues and are able to agglutinate red blood cells.[94,95] The wide range of *Helicobacter* infections seems to indicate that these bacteria may not always have a true pathogenic relationship with their host.[96] Although some *Helicobacter* strains clearly cause more damage to cells than others, much of the pathogenesis seen in *Helicobacter* infections such as *H. pylori* may be due to the host response to the bacterium, rather than any significant toxicity mediated by the bacterium itself. Some studies suggest that colonization of the stomach by this organism serves as a protective factor against gastroesophageal reflux disease and adenocarcinoma of the esophagus.[96] These types of pathology have increased at an alarming rate over the years and coincide with the dramatic decrease in gastric cancer and increased eradication of *H. pylori*.[97] This view of the role of *Helicobacter* is one of a commensal displaying a symbiotic host-parasite relationship.[6,98] Most of the pathology seen with gastric helicobacters such as *H. acinonychis*, and *H. felis* occurs when the organism is placed in an unnatural host or when the infection becomes chronic over time. Of course there are exceptions; for example, *H. cinaedi* and *H. fennelliae* clearly produce disease in the primate model and have also been found to be associated with human disease.[41,99] Unlike *H. pylori*, *H. cinaedi* has been cultured from human blood and joint fluid, as has *H. fennelliae* from human blood.[100] It is believed that helicobacters that invade the bloodstream do so via colonization of the lower gastrointestinal tract.[101] *H. cinaedi* infection has been linked to possible animal human transmission, having been isolated for the first time in an adult who developed bacterial meningitis, after prolonged contact with her cat.[102] *H. heilmannii* and related gastric species have been associated with mild to moderate gastritis, peptic ulcer disease and gastric mucosa-associated lymphoid tissue (MALT) lymphomas in human adults.[41,103,104–106] Among the enterohepatic species of *Helicobacter*, *H. hepaticus*, *H. mustelae*, and *H. bilis*, exhibit carcinogenic potential in animals, and have been associated with numerous virulence genes, which might result in human disease as well.[39–41] For example, in the naturally infected ferret, infection with *H. mustelae* can also lead to precancerous lesions and has also been linked to gastric adenocarcinoma and MALT lymphoma in these animals.[107] Other enterohepatic species, such as *H. hepaticus*, *H. bilis*, and *H. ganmani*, have been detected by PCR in specimens from patients with hepatobiliary diseases.[30,31,108] However, such organisms have not yet been isolated from these patients. In addition, these organisms may be associated with Crohn's disease, inflammatory bowel disease and ulcerative colitis.[109,110] Similar helicobacters have also been found to colonize and cause inflammation in other mammalian species. This includes, for example, *H. cinaedi* in the rhesus monkey and *H. hepaticus* in mice.[111–113] *H. rodentium*, *H. typhlonius*, *H. bilis*, and *H. hepaticus* were recently found to be the most frequently detected enterohepatic species in wild house mice, suggesting that these animals serve as a potential reservoir for infection of both humans and other vertebrates.[114] *H. cinaedi*, *H. bilis*, and *H. canis* can cause severe infections, mostly in immunocompromised patients, such as bacteremia, cellulitis, cutaneous diseases, and fever of unknown origin.[115] Additional species of enterohepatic helicobacters have been implicated as causes of human gastroenteritis, in compromised patients such as *H. canadensis*, *H. pullorum*, and *H. winghamensis*.[44,45,62,89,116–119] *Helicobacter* species, such as *H. bilis*, *H. pullorum*, and *H. hepaticus* have also been detected in the human hepatobiliary tract.[109,120]

The role of enterohepatic helicobacters in veterinary and human medicine is increasingly recognized. However, more studies are still required to prove this association.[115]

In humans, natural infection with *H. pylori* is likely acquired quite differently from animal infection models, which require large and often repeated inocula.[121,122] It is widely believed that in the absence of treatment, *H. pylori* infection, once established, persists for life. In the elderly, however, it is likely that infection can disappear. This is most likely due to the stomach mucosa becoming increasingly atrophic with age, rendering it inhospitable to colonization.[123] The proportion of acute infection that persists is not known. However, in studies that have followed the natural history of disease progression in populations, spontaneous elimination has been reported.[123,124]

Helicobacter can cause direct damage to epithelial cells in the gastric mucosa as well as induce an inflammatory response in the host. Both host factors and organism factors determine the phenotypic expression of the infection over time. Essentially, all persons colonized with *H. pylori* develop gastric inflammation,

known as chronic superficial gastritis. The condition can develop into chronic inflammation of the gastric mucosa if the bacteria are not eradicated. However, this chronic inflammation is clinically silent in most infected persons.[38] This type of gastritis usually predisposes the host to hyperacidity and duodenal ulcer disease. In cases where the gastritis results from long-standing infection, it becomes more generalized and affects the corpus, leading to glandular atrophy and intestinal metaplasia. This is a recognized precursor state for gastric ulcer disease, gastric carcinoma, and MALT lymphoma.[125–128]

The role of *H. pylori* is less clear in chronic inflammation in coronary disease, inflammatory bowel disease, liver and biliary tract diseases and pancreatic cancer.[21,129–131] However other helicobacter species, such as *H. hepaticus*, *H. bilis*, and *H. cinaedi* are known to infect the intestines and biliary tract in humans and *H. bilis* has been found to be highly prevalent in patients with pancreaticobiliary maljunction.[21,132] Hence, additional studies are needed to determine the prevalence and capacity of non-*pylori* helicobacters to cause disease. Other medical conditions in which there is a suggested role for *H. pylori* include nonulcer dyspepsia, nonsteroidal anti-inflammatory drug-induced (NSAID) ulcer, gastroesophageal reflux disease (GERD), iron-deficiency anemia, and idiopathic thrombocytopenic purpura. There has also been a growing interest in the association of *H. pylori* with cardiovascular, neurologic, hematologic, dermatologic, head and neck, urogynecology diseases, diabetes mellitus and metabolic syndrome.[132–138]

Virulence Factors

Many virulence factors have been described in *Helicobacter* over the years, as summarized in Table 30.3. These factors enable the organism to survive the extreme acidic environment of the gastric tract, to reach the more neutral environment of the mucus layer, to colonize the gastric epithelium, and to evade the human immune response. This progression of events ultimately results in persistence of the organism, leading to possible infection and subsequent disease. Flagella, urease, and adhesins are essential colonizing factors. Among the other unique adaptations of *H. pylori* that enable the organism to survive in the hostile environment of the stomach are the presence of vacuolating cytotoxin (VacA) and oncogenic cytotoxin associated antigen A (cagA) protein. These virulence factors are discussed in greater detail in this section.[97,139–207,210]

Flagella and Motility

Movement is an important function of the gastric colonizers. First, it prevents the organisms from being washed out of the stomach and, second, allows the organisms to find suitable nutrients, while evading the local hostile environment. A fully functioning flagellum structure is essential for optimal survival. Studies involving the genes responsible for the flagella filament structure reveal that successful colonization and organism survival became compromised when these genes were disrupted.[210] Most of the gastric colonizers possess a spiral-shaped body with one exception, *H. mustelae*, which is a short rod. The morphology of the bacterium and the type of flagella dictate how the organism moves

TABLE 30.3

Virulence Factors of *Helicobacter*

Virulence Factor	Type	Proposed Function	Ref.
Flagella	Organelle	Promotes colonization; enhances bacterial penetration into the mucus layer and enables movement away from acidic environment.	[214–216]
Urease	Enzyme, immunogen	Colonization, neutralization, tissue damage (tight junctions). Causes gastric cancer progression	[143, 148–151]
Lewis Blood Group Antigen Adhesins: BabA (Binds to B and H-1 antigens) SabA (Binds with sialyl-Lex antigen)	ABO antigen-binding adhesins	Mediate bacterial attachment and colonization of gastric epithelium; Molecular mimicry and autoimmunity	[151, 210 220–223, 237–238]
Adhesins: OipA HopQ	Outer Membrane Protein (OMP)	Bacterial adherence to gastric epithelium; OipA also promotes a proinflammatory environment and damages mucosal layer. HopQ inhibits immune cell activities and facilitates CagA translocation.	[151, 210, 220, 221, 224–229]
Superoxide dismutase	Enzyme	Prevents phagocytosis and bacterial cell death	[151]
Catalase	Enzyme	Prevents phagocytosis and bacterial cell death	[151, 210]
Glycosulfatase	Proteolytic enzyme	Mucin degradation	[151, 210]
Phospholipase A	Enzyme	Tissue damage (cell membrane disruption)	[151]
Lipopolysaccharide (LPS)	Endotoxin	Affects macrophage-lymphocyte interactions; May contribute to *H. pylori* permanence; Epithelial cell apoptosis; molecular mimicry	[151, 153–155, 159, 164, 234–239]
Vacuolating cytotoxin A (VacA)	Cytotoxin	Causes cell vacuolization, cell necrosis and apoptosis.	[151, 166–168, 207–208, 210]
Cytolethal distending toxin (CDT)	Cytotoxin	Chronic inflammation and tissue damage (found in enterohepatic species of *Helicobacter*)	[151, 171–179]
Cytotoxin associated antigen (CagA)	Immunodominant cell surface antigen	Oncoprotein, which alters cell-signaling pathways; Induces chromosomal Instability, cell proliferation, and apoptosis; Causes IL-8 release and proto-oncogene activation.	[151, 188–194]
Coccoid forms	Morphological adaptation	Possible bacterial cell survival; May be associated with transmission and virulence.	[195–198, 201, 202]

and ultimately where in the gastric mucosa the organism will be located. For example, the movement of *H. felis* and *H. heilmannii* differs completely from that of *H. mustelae*, and, therefore, these bacteria are found in different regions of the stomach. These spiral organisms do not adhere to the cell surface, but instead are found free, tracking backward and forward along the mucus strands with their corkscrew motion.[209] The curved morphology of *Helicobacter* and the four to eight polar-sheathed flagella found in species such as *H. pylori,* create a screw-like motion, which may enable the organism to penetrate the mucin layer. The organism moves through the gastric mucosa epithelium layer to the basal layer where the pH value is close to 7.0.[210] Motility in helicobacters has been found to be pH dependent and becomes impaired at a pH below 4.0.[139] Mutations in any of the genes associated with the motility and the chemotaxis systems of *H. pylori* have been found to eliminate the bacterium's ability to infect the stomach and establish colonization.[210,211] Patients infected with *H. pylori* demonstrating increased motility were shown to also have increased bacterial density and a more intense inflammatory response in the upper stomach, and therefore potentially manifesting more severe pathological outcomes.[212] Therefore, flagella may be considered an early stage colonization virulence factor.[210] In addition to motility, flagella have been shown to play a role in bacterial adherence to mammalian host cells in a number of hosts.[213,214] However, a study by Clyne et al. indicated that flagella did not play a direct role in adhesion of both *H. pylori* and *H. mustelae* to gastric epithelial cells.[215] However, the genes involved in flagellar biosynthesis in *H. pylori* may also regulate

the expression of adhesins, which are also important virulence factors associated with the bacterium.[215] Although many components of *H. pylori* flagella have been characterized and data regarding flagellar function and regulation are rapidly increasing, further studies are required to confirm the association between flagella an pathogenicity in *H. pylori*.[216]

Urease

Urease is a major virulence factor of gastric *Helicobacter* species. Among the proposed functions for the presence of urease in *Helicobacter* is the ability of the enzyme to protect the bacterium from the hostile acid environment of the stomach. Urease is also strongly immunogenic and chemotactic for phagocytes.[140] The enzyme catalyzes the breakdown of urea to alkaline ammonia and bicarbonate, which is further broken down to carbon dioxide and water (H_2O) molecules. These by-products provide the necessary environment for *H. pylori*'s gastric survival, and continuous urease expression has been found to be necessary for colonization even after the organism's successful adaptation in the stomach.[142] Support for this theory has been demonstrated in studies with urease-negative mutants, which were shown to lack the ability to colonize the gastric mucosa at normal physiological pH.[141,142,217] *H. pylori* produces approximately 10% of its protein as urease enzyme.[143] The urease produced by *H. pylori* is of special interest as it is 10–100 times more active than other bacterial ureases and is located in the membrane, rather than the cytoplasm, of the organism. In addition, the Michaelis-Menten constant (K_m)

of the urease is much lower than that of other bacterial ureases, a finding consistent with the very low urea levels in the stomach.[144] Several *in vitro* studies have also linked urease to the associated pathology resulting from *H. pylori* infection. Ammonia production has been found to disrupt the tight cell junctions, breach the cellular integrity, and ultimately cause damage to the gastric epithelium.[145,146] Carbon dioxide protects *H. pylori* from the bactericidal activity of metabolic products like nitric oxide as well as intracellular killing of phagocytic cells.[147] Urease has also been reported to possibly contribute to tumor growth and metastatic dissemination by inducing new blood vessel formation from pre-existing vasculature, or the process of angiogenesis. This process could play a role in gastric cancer progression.[148] Induction of hypoxia-inducible factor (HIF) has been associated with chronic *H. pylori* infection, and this transcription factor contributes to the development and progression of certain cancers.[149] *H. pylori* urease has also been shown to bind to major histocompatibility complex class II molecules and induce cell apoptosis.[150] The ureases of *H. pylori*, *H. felis*, and *H. heilmannii* are more closely related to one another than they are to the urease of *H. mustelae*.[218] Studies have shown that the enzyme of *H. heilmannii* might be different from those of other gastric *Helicobacter* species. However, there is also evidence to suggest that the *H. heilmannii* enzyme may be more similar to the urease of *H. felis*.[219] Unlike helicobacters that colonize the stomach, expression of urease in enterohepatic *Helicobacter* species is variable. See Table 30.1.

Adhesins

Adhesins are bacterial proteins, glycoconjugates, or lipids that are involved in the initial stages of colonization by mediating the interaction between the bacterium and the host cell surface.[151] Host cell receptors are composed of lipids, proteins, glycolipids, or glycoproteins. Adherence of bacteria to host cell receptors triggers cellular changes that include signal transduction cascades, which can lead to infiltration of inflammatory cells and to the local persistence of the organism. When the mucosal layer lining the gastric epithelium is colonized by *H. pylori*, the interaction between bacterial adhesins with cellular receptors ultimately protects the organism from displacement from the stomach due to peristalsis, gastric emptying, and mucus shedding. The bacterium also benefits by having access to nutrients for improved growth and enables bacterial toxin release and other effector molecules to gain entrance to host cells.[151,152] *H. pylori* adhesins are considered important virulence factors involved in many processes that occur both during the early and chronic phase of infection.[152] These adhesins belong to the outer membrane protein (OMP) family of *H. pylori*. The most well-known adhesins in this group include the blood group antigen-binding adhesin (BabA) and sialic acid-binding adhesin (SabA), the outer inflammatory protein (OipA) and *H. pylori* outer membrane protein (HopQ).[151] The OMP adhesins not only provide the bacteria attachment to host cells, but also augments the pathogenicity of *H. pylori*. BabA plays an important role in bacterial attachment and is associated with cellular pathogenicity. Two closely related paralogs, BabB and BabC have also been identified but their functions have not yet been elucidated.[151,220,221] BabA adhesin interacts with the difucosylated ABO/Lewis b (LeB) antigen present on red blood cells and gastrointestinal mucosa epithelial cells. Infection with BabA producing strains of *H. pylori* have been associated with an increased risk for peptic ulcer disease.[221] Also because of the preference of BabA *H. pylori* strains to bind to ABO blood group antigens, and especially to blood group O, infected individuals with this blood type are at higher risk for developing duodenal ulcer disease.[221] The SabA adhesin binds to Lewis X (sLex) and Lewis a (sLea) antigens, which are commonly found in the infected and inflamed gastric mucosa.[222] SabA expression has been found to respond to changing conditions in different regions of the stomach. This gives *H. pylori* the ability to quickly adapt to varying microenvironments or host immune responses, and results in long-term colonization and infection.[223] SabA production has also been reported to be associated with severe intestinal metaplasia, gastric atrophy, and development of gastric cancer.[222] Outer inflammatory protein (OipA) is another virulence protein comprising the *H. pylori* OMP family and is encoded by the *oipA* gene.[151] When the gene is expressed, OipA enables *H. pylori* to adhere to gastric epithelial cells and is associated with mucosal damage, host cell apoptosis, Interleukin (IL)-8 induction, duodenal ulcers, and development of gastric cancer.[151,224] *H. pylori* HopQ protein also contributes to bacterial adherence and interacts with the family of the carcinoembryonic antigen-related cell adhesion molecules (CEACAM), which are expressed on the gastric epithelium during gastritis and gastric cancer development.[151,224] In addition to the role played by HopQ in gastric colonization, recent studies have also shown its association with gastric ulcers, gastric cancer development, inhibition of immune cell activity, and the development of B-cell non-Hodgkin lymphoma.[151,225,226]

The attachment mechanisms of *H. pylori* to gastric epithelial cells that are associated with cytoskeletal rearrangements, and modification of host proteins, are also known to occur in enteropathogenic *Escherichia coli* adherence processes.[227] Similar structures have also been observed in ultrastructural studies of *H. mustelae* in its adherence to the ferret gastric mucosa.[228] The outer membrane proteins required for *in vitro* attachment of *H. pylori* to gastric epithelium have not been demonstrated in other gastric helicobacters such as *H. mustelae* and *H. felis*.[229] However, *H. mustelae* and *H. pylori* appear to bind common lipid receptors such as phosphatidylethanolamine (PE). The adhesion to intact eukaryotic cells has been found to correlate with the amount of PE present in the membrane.[230,231] Other cell surface properties, such as hydrophobicity, may also contribute to the gastric epithelial binding of helicobacters, as in the case of *H. mustelae*.[232] *H. heilmannii*, which is typically found in the mucus layer above surface epithelial cells, does not show the closely associated adherence structures seen in *H. pylori* adherence.[233]

Other Enzymes

Additional enzymes have been identified in the genus *Helicobacter* that enhance survival in the host environment. The action of superoxide dismutase, for example, breaks down the superoxide produced in polymorphonuclear leukocytes and macrophages, preventing their destruction of organisms. This enzyme has been isolated from *H. pylori*, for example, and is believed to play a role in its survival.[148] In a similar fashion, catalase is believed to protect *H. pylori* and other helicobacters from the damaging effects of hydrogen peroxide released from phagocytes.[149] Also, both urease and catalase, when excreted into the surrounding environment, are believed to have a somewhat protective role against the humoral immune response.[148] Proteolytic enzymes such as glycosulfatase may cause degradation of gastric mucin, and the production of phospholipase A may contribute to the degradation of cell membranes by helicobacters.[150,151]

Lipopolysaccharide

The cell wall of most Gram-negative bacteria contains lipopolysaccharide (LPS) and which is typically composed of three layers: a hydrophobic lipid A (or endotoxin) that anchors the molecule to the outer membrane; a variable O-antigen extending from the cell to the external environment; and a core-oligosaccharide that links the O-antigen to the lipid A.[153] Each component of the LPS layer has its own unique function contributing to the bacteria's survival in the host. The LPS is known to play an essential role in inducing host activation of cytokine release and increased inflammatory response, resulting in a strong host immune response. The O-antigen of *H. pylori* LPS has been reported to contribute to the organism's virulence. *H. pylori* can imitate carbohydrate structures present within human blood cells, secretions and particularly epithelial cells by incorporating Lewis antigens into their O-chains.[154,159] Lewis antigens have been found to be expressed in over 80% of *H. pylori* strains tested.[237] To evade host immune surveillance, *H. pylori* constitutively modifies its Lipid A, allowing the organism to persistently colonize its host.[156] The unique lipid A structure of *H. pylori* enables the organism to resist cationic antimicrobial peptides (CAMPs), and to evade Toll-like receptor 4 (TLR-4) recognition.[154,156] Previous reports have shown that the LPS of *H. pylori* has a lower immunological activity when compared to other helicobacters and to other Gram-negative bacteria.[198] For example, *H. felis*-associated LPS appears to exert an inflammatory effect on host cells, whereas *H. pylori* endotoxin is a poor inducer of cytokine production by inflammatory cells *in vitro*.[235] This observation is explained in part by the unusual structure of lipid A in *H. pylori*.[153,154,156] The lower biological activity of the LPS, combined with host antigens on its surface, may be a mechanism for *H. pylori* to evade the host inflammatory response, and may explain why *H. pylori* infection is chronic and can last for decades.[236] Autoantibodies targeted against the bacterial Lewis antigens may also play a role in the pathogenesis of gastric disease.[237] This role is supported by studies that show that *Helicobacter* strains isolated from patients with ulcer disease express an increased number of Lewis antigens as compared to isolates cultured from patients with dyspepsia.[157,158] In contrast, the LPS of the gastric helicobacter *H. mustelae* does not express Lewis antigens, nor are the antigens expressed on ferret gastric epithelial cells. However, both ferret gastric epithelium and *H. mustelae* LPS express blood group antigen A. This antigenic expression may be considered a mechanism of molecular mimicry similar to the expression of Lewis antigens by *H. pylori*.[238,239]

Vacuolating Cytotoxin A

The toxins of *Helicobacter* species have been implicated in lesion formation associated with gastric and enterohepatic infections. The *H. pylori* VacA protein was the first of these toxins to be identified and characterized. This protein was initially described as inducing a vacuolating cytopathic effect in HeLa cell tissue cultures and is now known to play a crucial role in the pathogenicity of *H. pylori* due to its interaction with gastric epithelial cells.[151,161,162] VacA is an 88 kDa pore-forming toxin whose action on host cells is characterized by the formation of large vacuoles, which give the toxin its name.[151] The vacuoles formed are also believed to create an intracellular space for the survival of the bacterium.[205] The cellular morphologic changes induced by the protein as well as the resulting gastric epithelial cell necrosis, appear to create a nutrient-rich environment that optimizes organism survival.[163–165] The gene encoding VacA is present in all *H. pylori* strains and displays allelic diversity, which explains the variation in cytotoxic activity between strains.[163,164] The *H. pylori* strains most frequently associated with duodenal ulcer disease were found to produce toxigenic forms of VacA.[151] Both purified VacA protein and sonicated cell extracts of VacA-producing *H. pylori* were found to induce ulcer-like lesions when administered to mice, which supports the possible role of VacA in mediation of ulcer formation in the host.[240] Peptic ulcers have also been observed in gnotobiotic piglets and Mongolian gerbils that were colonized with *H. pylori*. Interestingly, these effects were also observed in mice, ferrets, and swine that were infected with non-*H. pylori* species.[94,241]

VacA has been found to promote the formation of acidic vacuoles in the cytoplasm of gastric epithelial cells.[163] This is believed to be caused by cellular collapse, resulting from the integrity of mitochondria, cytoplasmic membrane, and endomembranous structures, becoming destabilized.[163,165] In addition, the VacA protein is believed to be involved in the activation and suppression of the immune response through its effects on T-cells and antigen-presenting cells, which can result in immune tolerance and persistence of *H. pylori* infection.[163,166] The effects of this virulence factor have also been found to enhance *H. pylori* gastritis as well as ulcer and cancer development. This suggests that VacA could possibly be used as a biomarker for the prediction of disease development.[163–167,203,204] This is supported by studies that have shown that antibody produced against VacA is associated with increased risk for gastric cancer development and extra-gastric diseases, such as colorectal cancer, especially in African Americans.[203–208] Immunization of mice with VacA has been shown to prevent infection when mice were subsequently challenged with *H. pylori*.[208]

The *vacA* gene appears to be absent from other gastric helicobacters such as *H. felis*, and therefore, is not directly associated with damage to the gastric epithelium. Rather, the host T cell response, initiated by *Helicobacter* infection, is believed to be responsible for the resulting gastric pathology in *H. felis*.[242] Similarly, *H. mustelae* does not appear to produce vacuolating cytotoxin activity *in vitro*.[243] This observation suggests that there may be other factors involved in ulcer disease production among the other *Helicobacter* species.

Cytolethal Distending Toxin

The cytolethal distending toxin (CDT) is a bacterial virulence factor produced by several Gram-negative pathogenic bacteria, including some of the enterohepatic *Helicobacter* species implicated in several chronic infections.[171] The CDTs associated with these many organisms have been found to differ in their sequence and structure, and are associated with the organism's pathogenicity, primarily by promoting persistent infection in their hosts. At the host level, CDTs cause cell distension, cell cycle block, and DNA damage, eventually leading to cell death.[171] In addition to these known *in vitro* effects of CDTs, there is also evidence that this toxin is involved in chronic inflammatory processes.[172,173] Among the enterohepatic helicobacters, CDT was first demonstrated in *H. hepaticus,* which produced a toxic effect in a murine liver cell line resulting in a granular appearance of the affected cells.[174] This latter CDT activity of *H. hepaticus* was found to closely resemble the CDT of *Campylobacter* species.[174,175] CDT-producing *H. hepaticus* closely resembles the human pathogen *H. pylori*, and is considered the prototype of enterohepatic helicobacters.[175] *H. hepaticus* infection induces chronic active hepatitis, hepatocellular tumors, typhlocolitis, and lower bowel cancer in susceptible mice.[175,178,179] This finding implicates CDT as having carcinogenic potential *in vivo*, although further studies using *in vivo* animal models are needed. The expression of CDT activity has also been observed in other enterohepatic *Helicobacter* species, such as *H. canis, H. bilis, H. cinaedi,* and in *H. pullorum*, which is considered an emerging human pathogen and implicated in several digestive pathologies.[41,42,177–179]

Cytotoxin-Associated Gene Antigen

One of the most intensely studied proteins of *H. pylori* is Cytotoxin-associated gene antigen (CagA). The *cagA* gene that encodes CagA is present within the *cag* pathogenicity island (*cag* PAI). The *cag* PAI is a 40 kb DNA segment that encodes for approximately 32 genes including components of the type IV secretion system.[164,180] The type IV secretion system (T4SS) is one of several types of secretion systems, which microorganisms use for the transport of macromolecules such as proteins and DNA across the cell envelope. Some T4SSs are used by pathogenic Gram-negative bacteria to translocate a wide variety of virulence factors into the host cell.[181] Upon infection, *H. pylori* utilizes this latter system to translocate CagA into host epithelial cells, which then localizes in the host cell plasma membrane. Once incorporated, a series of events occur, which starts with the kinase-mediated tyrosine phosphorylation of CagA.[164] In addition to CagA, *H. pylori* peptidoglycan is delivered into host cells, by the T4SS, which ultimately leads to the stimulation of pro-inflammatory cytokines such as interleukin-8.[151,182] Ultimately, the interaction between *cag* PAI positive strains and host cells results in the alteration of cell signaling pathways that regulate the cellular oncogenic response, induction of morphologic changes, chromosomal instability, cell proliferation, and apoptosis. This increases the risk for transformation and ultimately progression to gastric cancer.[151,183–187]

Both *cagA*-positive and *cagA*-negative *H. pylori* strains exist in nature. All *H. pylori* strains induce gastritis, however, those strains that are *cag* PAI positive, increase the risk for severe gastritis, atrophy, dysplasia, and gastric adenocarcinoma when compared to negative strains. The *cag* PAI is present in approximately 60% – 70% of Western *H. pylori* strains and close to 100% of East Asian *H. pylori* strains, which helps to explain the high gastric cancer incidence in that region.[187–193]

Negative *cag* PAI strains are located predominantly in the mucus gel layer of the stomach whereas positive strains are found adjacent or adherent to gastric epithelial cells, indicating the *cag* PAI genotype influences the colonization and distribution of the organism in the stomach.[194]

In animal hosts infected with non-*H. pylori* gastric helicobacters, such as *H. felis*, development of severe, chronic gastritis is known to occur. However, the presence of a *cagA* equivalent in these bacteria has not yet been definitively demonstrated. The induction of chronic gastritis in *Helicobacter*-infected animal hosts may therefore be independent of this cytotoxin.[242]

Coccoid Forms

H. pylori is typically recognized as a seagull-shaped or curved Gram-negative rod. However, when growing in either *in vitro* conditions or in suboptimal environments, the organism is capable of morphological transformation into an intermediate or transitional U-shaped and C-shaped cell (resting or dormant state), and a spherical or coccoid form (viable but unculturable form).[195] In addition, filamentous (elongated) forms of *H. pylori* have also been described.[195] This pleomorphism is not unique to *H. pylori* as this ability has been observed in many other Gram-negative rods. The coccoid form of *H. pylori* has been speculated to represent a dormant or stressed form of the organism, a resulting sign of cell adaptation, or a process of aging.[196] Since observing them in *H. pylori*, the coccoid forms and their viability, have been and continue to be a topic of debate. However, when the first observation of *H. pylori* coccoid forms were made in biopsy specimens, more emphasis and interest was generated in knowing their clinical importance and activity.[197] Coccoid forms are thought to play a potential role in relapsed or recurrent gastric disease and may even represent a form of antibiotic resistance.[195] Currently, it is believed that eradication therapies administered to patients, should aim at eliminating both bacilli and coccoid forms of *H. pylori*.[195,196] Other studies have indicated that adaptational changes in cell morphology are related to the virulence profile of *H. pylori*. Less virulent *H. pylori* strains have been found to have a lower potential for coccoid formation, than highly pathogenic strains, whereas strains of lower pathogenicity were more apt to form filamentous or elongated forms of the bacterium.[198–200] Reduction in cell size among *H. pylori* strains exposed to suboptimal conditions has been also found to be positively correlated with the presence of certain virulence genes, such as *vacA, cagA*, and *babA*.[198] However, controversy still exists as to the infectivity of coccoid forms and their role in transmission, colonization, infection, and pathogenesis, and further studies are needed to better understand the importance of this pathogen morphotype.[201–202]

Culture of *Helicobacter*

The culture and isolation of *Helicobacter* in the laboratory is technically demanding and requires proper specimen transport, specific media, and growth conditions.[33,244] Given the effectiveness of currently available *Helicobacter* detection methods, it could be argued that culture is not an essential routine procedure in the diagnosis of *H. pylori* infections. However, there are many instances in which culture remains a valuable tool. For example, culture is indicated when evaluating antimicrobial resistance to prevent antibiotic treatment failures. There is also an increasing need to fingerprint isolates of *H. pylori* and other *Helicobacter* species for epidemiological purposes. Many of these methods of molecular typing are relatively simple, well validated, and readily available, requiring pure cultures of newly isolated organisms.[60,61] Culture is also a basic research tool in the study of virulence factors such as urease and cytotoxin, among others. It is important to note that continuous passage on artificial media may cause isolates to lose properties essential for growth in the host and induction of disease. For this reason, those undertaking research with laboratory strains of *Helicobacter* isolates should repeat their studies using newly isolated organisms.[247]

Specimens for Culture of *Helicobacter*

Blood, feces, and gastric and tissue biopsies are sources for *Helicobacter* isolation. Although, *H. pylori* and other gastric helicobacters are rarely isolated from human blood, the enterohepatic helicobacters can often cause invasive infections. If bacteremia is suspected, peripheral venous blood should be collected as recommended by the manufacturer, in commercially available aerobic and anaerobic blood culture bottles.[247] The gastric helicobacters *H. pylori* and *H. heilmannii* are rarely isolated from human fecal specimens. However, enterohepatic helicobacters, like campylobacters, can be cultured routinely by modified stool culture, preferably within 4 hours of collection.[33] Glycerol-containing media can serve as both transport and storage media for biopsy specimens. Brucella broth with 20% glycerol is recommended as it is readily available commercially. Biopsy material should be minced or homogenized in 0.9% saline or 20% glucose prior to plating.[247]

Culture Techniques

In the laboratory, *Helicobacter* strains will typically grow under microaerophilic conditions (4% O_2, 5% CO_2, 5% H_2, and 86% N_2) at 37°C with increased humidity.[33,247] Microaerophilic environments can be created using a variable-atmosphere incubator, partially evacuated anaerobic jars with defined gas mixtures, or commercial gas-generating packets.[26] Growth does not occur in aerobic conditions, with the exception of species like the enterohepatic *H. rodentium*, which grows both in microaerophilic and anaerobic conditions. Although some species may take up to 10 days to grow, many others produce colonies in 3–4 days of incubation. *Helicobacter* can be grown on a variety of rich agar bases supplemented with 5–10% whole blood or serum.[33] These media include brain heart infusion agar, brucella agar, Wilkins-Chalgren agar, Trypticase soy agar, chocolate agar, Skirrow *Campylobacter* medium and Mueller Hinton agar.[33] Vancomycin, nalidixic acid, and amphotericin are common additives when a selective medium is required. When isolating *H. pylori*, the antibiotic cephalothin should be omitted, due to its susceptibility to this agent.[247,248] There are no recommended culture methods in the routine clinical laboratory for the isolation of *H. heilmannii* and other non-*H. pylori* gastric helicobacters, however, most media that will support the growth of *H. pylori* can be used to grow the gastric helicobacter *H. felis*.

Recently, a novel isolation method using high acidity and modified gaseous conditions has been developed for the isolation of *H. suis* from pig gastric tissue, but this medium's ability to isolate helicobacter from human gastric biopsy specimens has not been evaluated.[26] Fastidious species such as *H. bizzozeronii* and *H. salomonis* prefer very moist environments, and a thin broth layer can be poured on the top of the agar surface to stimulate growth.[249] Successful isolation of most enterohepatic helicobacters from feces requires a defined microaerophilic atmosphere with some strains requiring 5%–10% H_2 for optimum growth. Commercially available gas-generating packets may be inadequate in supplying the increased atmospheric hydrogen required by this group of helicobacters.[33] Certain species such as *H. bilis* and *H. hepaticus* have other specific growth requirements.[244]

Identification

Helicobacters produce varying colonial phenotypes on blood agar. *H. pylori* and other gastric helicobacters tend to produce entire gray and translucent colonies, whereas the intestinal helicobacters, such as *H. cinaedi*, *H. fennelliae*, *H. pullorum*, and *H. canadensis*, produce various swarming phenotypes as most isolates are motile. *H. pylori* cells have a curved or helical morphology, which is often lost on subculture, where the organism may subsequently appear as curved, "U" shaped or even as straight rods. Other gastric species are usually larger in size with a helical morphology that may be more pronounced.[33,245] The enterohepatic species morphologically resemble other Gram-negative spiral or curved bacteria.

Helicobacters cultured from blood may require special staining techniques to visualize the organisms. Such is the case when culturing *H. cinaedi*, which is generally not seen in a Gram stain preparation, but instead requires dark-field microscopy, Giemsa, or acridine orange staining for the best visualization of the bacterial cells.[100] When microscopically identifying helicobacters from biopsy material, Gram stains of histologic sections or smears of gastric mucus often can reveal the presence of the organisms. A modified Gram stain using carbol-fuchsin (0.5% wt/vol) counterstain has been found to augment detection of most helicobacters.[33] The Warthin-Starry and modified Giemsa stains can be used to identify the presence of the bacterial cells in tissue (Figure 30.1). Other non-*H. pylori* gastric helicobacters can be visualized by Giemsa stain and can be distinguished from *H. pylori* by their distinct, tightly spiral morphology.[245] Triple stains, which combine hematoxylin and eosin, Alcian blue, and a third stain such as the Genta or the El-Zimaity triple stain, have been found to be superior for identification of *H. pylori* and allow for easier delincation of gastric morophology.[250,251]

Helicobacters are routinely tested for cytochrome oxidase, catalase, and urease activity. All are oxidase-positive and most species, including *H. pylori*, are catalase-positive and urease-positive.[248] Susceptibility to nalidixic acid and cephalothin are also used as phenotypic markers of identification (Table 30.1). The enterohepatic species have several unique biochemical characteristics (Table 30.1). Biochemical and other testing should be carried out using recommended procedures.[246] Species such as *H. candidiasis*, *H. canis*, *H. fennelliae*, and *H. pullorum* that lack urease activity, resemble enteric campylobacters and therefore, definitive identification cannot be based on phenotype alone. Growth at 42°C and indoxyl acetate hydrolysis are examples of the useful tests that can be used to distinguish these species

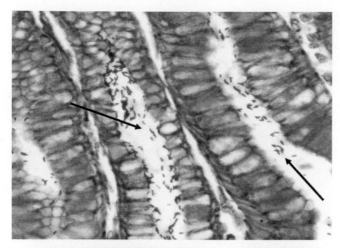

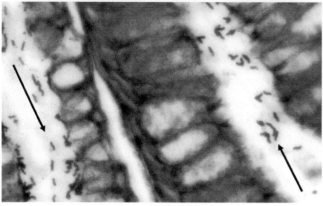

FIGURE 30.1 Panel A shows a Warthin Starry stain of *Helicobacter pylori* in the gastric mucosa demonstrating typical curved cell morphology. Panel B shows an enlarged portion of photomicrograph from Panel A. (Photomicrograph courtesy of Long Island Jewish Medical Center, Northwell Health System, Department of Pathology, New Hyde Park, New York.)

(Table 30.1). When culturing fecal specimens, it is important to note that samples may be co-colonized with multiple *Helicobacter* and *Campylobacter* species making organism identification challenging.[33]

There are currently several methods that can be used in the definitive identification of helicobacters, which include: nucleic acid amplification tests (NAATs) and matrix-assisted laser desorption/ionization time of flight (MALDI-TOF) mass spectrometry. NAATs are usually considered to be a good identification method due to their great sensitivity and specificity. PCR has been used for the rapid identification of *Helicobacter* species in clinical specimens, and has been used to identify *H. cinaedi* infections, as well as and for screening of carriers.[252–254] The advantage of MALDI-TOF when compared to other identification methods is the availability of results within a few hours.[255] This technique has also been used for the identification of *Helicobacter* species (*H. pullorum and H. pametensis*) and has helped to distinguish them from phenotypically similar *Campylobacter* species in clinical specimens.[256] However, this method has not been as successful for the identification of *H. pylori* because of its high intraspecies variability.[257] A variety of rapid typing methods have also been developed for clinical isolates, such as 16S rRNA

gene sequencing, polymerase chain reaction-restriction fragment length polymorphism analysis (PCR-RFLP), and pulse field gel electrophoresis (PFGE). All of these methods enable subtyping of unrelated strains, but have differing accuracy, discriminatory power, and reproducibility.[258,259]

Antibiotic Susceptibility and Eradication Therapy

Various phenotypic and genotypic susceptibility testing methods are used to assess antimicrobial resistance in *Helicobacter*. Broth microdilution, disk diffusion, the Epsilometer test (E-test®), and agar dilution are among the phenotypic methods available used specifically for the susceptibility testing of *H. pylori*.[260–262] Several methods involving gene amplification, rapid sequencing by pyrosequencing, or fluorescent *in situ* hybridization, real-time PCR, PCR-restriction fragment length polymorphism among others have been developed for the rapid detection of mutations associated with antibiotic resistance in *H. pylori*.[263,264] The Clinical and Laboratory Standards Institute (CLSI) recommendations for antimicrobial susceptibility testing are currently available for *H. pylori* only. The CLSI recommends agar dilution susceptibility testing of clarithromycin when testing *H. pylori*. The treatment of *H. pylori* continues to be a much studied and debated topic. Increasing antibiotic resistance trends have become a stumbling block to current therapies and makes universal consensus on best treatment options challenging.[265] All patients who test positive for active infection should be treated, and the treatment that is chosen should provide an eradication rate of at least 90%.[266] Current *H. pylori* treatment regimens include: triple therapy, sequential therapy (where a patient is given one treatment followed by another), quadruple therapy, and levofloxacin-based triple therapy.[266,267] When choosing an appropriate treatment, other factors must be taken into consideration, which include: previous antibiotic exposure, regional antibiotic resistance patterns, and as noted above, eradication rates. It is believed that all of these factors may influence treatment outcome.[266,268] There are two suggested first-line triple therapy protocols, and they consist of clarithromycin, a proton pump inhibitor (PPI), and either amoxicillin or metronidazole.[266,267,269] Antacid medications such as proton pump inhibitors or H_2 antagonists are usually added to the regimen to reduce acid output and to encourage ulcer healing. Proton pump inhibitors suppress acid production by halting the mechanism that pumps acid into the stomach. H_2 blockers work by blocking histamine, which stimulates acid secretion. Historically, clarithromycin-based therapies were considered to be the most efficacious, best tolerated and safest to administer.[266] However, clarithromycin use has been threatened by increasing rates of antibiotic resistance, which have caused the eradication rate to fall below 80%.[263,264,266,270,271] Therefore it is recommended that this therapy choice be used in geographic areas where clarithromycin resistance rates are less than 15%, and for patients who have not had previous macrolide exposure.[263,264,266,267,269] Bismuth quadruple therapy has been found to result in eradication rates that are similar to clarithromycin-based regimens.[262,264,266,269] Bismuth-containing compounds, such as bismuth subsalicylate, can be added to the regimen to protect the stomach lining from acid damage. The compound also has some inhibitory effect on the organism. The first line bismuth quadruple therapy is recommended in areas of high antibiotic resistance, in cases of prior macrolide exposure, and in patients where initial clarithromycin-based therapy has failed.[262,264,269] In addition, concomitant nonbismuth quadruple therapy administered for a recommended 14 days, is another first-line treatment choice for patients.[262,264,269] This regimen can also be used as salvage therapy in patients with persistent *H. pylori* infection, and where primary therapy has failed.[264,266] The College of American Gastroenterology (ACG) has recommended this latter approach based on a meta-analysis of 19 trials (>2,000 patients), which showed a mean eradication rate of 88%.[264,270] This eradication rate was also achieved in areas having high clarithromycin and metronidazole resistance. In the United States, levels of metronidazole resistance have been documented at 20–40% with similar levels seen in Europe.[264,268,270]

In cases of treatment failure using standard therapies, two other classes of antibiotics are recommended as a third choice for organism eradication. Levofloxacin and the antituberculosis agent, rifabutin, may be combined with a proton pump inhibitor and amoxicillin. This has been recognized as a promising regimen by the ACG based on results of international studies.[264,269] However, although levofloxacin-based therapy has been found to be comparable to clarithromycin triple therapy, *H. pylori* resistance rates to levofloxacin in the United States was found to be greater than 30% when isolates were tested from 2009 through 2011.[264] The use of rifabutin either in a triple regimen with a PPI and amoxicillin, or high dose dual therapy with a PPI and amoxicillin, are two other reported treatment options. However, there is insufficient efficacy data currently available for the rifabutin treatment options.[264,266,269] The limited antibiotic regimens used to treat *H. pylori* combined with the growing resistance of *H. pylori* to these antibiotics, is a cause for concern. Improved local surveillance networks are needed to better select appropriate eradication regimens for each geographic region, in addition to long-term solutions for preventing infection with this pathogen.[263,264,266]

Optimal treatment regimens have not yet been established for other gastric helicobacters. However, there is evidence that treatment regimens used for *H. pylori* can resolve gastritis and peptic ulcer disease caused by these organisms as well as *H. heilmannii*-associated low-grade MALT lymphoma. There is no information regarding frequency of resistance to clarithromycin and other antibiotics and no PCR assays have yet been developed to determine antibiotic resistance in this group of organisms.[24,103,245] Although, *H. heilmannii* appears to be susceptible to triple therapy in humans, long-term eradication from animal hosts such as cats, dogs, and primates has not been as effective.[265,272] However, triple therapy has been shown to significantly reduce the burden of gastric helicobacters in experimentally infected mouse stomachs.[273] *H. felis*, which has been isolated from cats, dogs, and occasionally, the human stomach, is sensitive to a variety of antimicrobial agents. These compounds include metronidazole, cephalothin, erythromycin, ampicillin, and the antimicrobial agents used in triple therapy. Clearance of *H. felis* from mice using the triple therapy combination has led to the development of an animal model that can be used for the study of new antimicrobial regimens against *H. pylori*.[274] The antimicrobial therapy that is required to clear *H. mustelae* infection from ferrets is similar to that used for *H. pylori* in humans.[275] *H. acinonychis* is

resistant to nalidixic acid and susceptible to ampicillin, penicillin, erythromycin, gentamicin, and chloramphenicol. *In vivo* triple antibiotic therapy has been found to be unsuccessful in totally clearing this organism from the cheetah host. This observation is also true for *H. salomonis* and *H. bizzozeronii,* which can be suppressed but not be totally eradicated from the stomach of dogs.[276]

There are currently no guidelines for the antimicrobial susceptibility testing of and recommended interpretive criteria for the enterohepatic helicobacters.[24,25] These organisms, which can be causative agents of bacteremia, often require combination antibiotic therapies that include aminoglycosides, such as amikacin or gentamicin. However, prolonged therapy is often necessary to eradicate infection. Treatment for *H. cinaedi*, for example, can include antibiotics such as erythromycin, ciprofloxacin, levofloxacin, gentamicin, tetracycline, and beta-lactams administered for at least a 2- to 3-week period.[80,100,276,277] Susceptibility testing of this latter species is probably warranted, as *in vitro* resistance has been correlated with treatment failures.[24,100] Although data are limited, gentamicin and ampicillin-sulbactam have been found to successfully treat bacteremia caused by *H. fennelliae*.[102,278] Likewise, limited information is available for treatment regimens for gastroenteritis caused by *H. canis, H. bilis,* and *H. pullorum*. However, bacteremia caused by *H. bilis* and *H. canis* were recently shown to be successfully treated with ertapenem, azithromycin, levofloxacin, doxycycline, and ceftriaxone.[279,280]

In addition to antibiotics, there is now evidence indicating that some natural compounds may have an inhibitory effect against *Helicobacter*. In particular, studies have shown that regular consumption of broccoli sprouts may eradicate *H. pylori*. Similarly, polyphenolics found in green and other teas have been shown to suppress *H. pylori* growth *in vitro* and in animal models.[281,282] Green tea extract has also been found to inhibit *H. pylori* adhesion to human epithelial cells *in vitro*.[283] Other nonantibiotic agents that have been reported to have anti-*H. pylori* activity include quince and cranberry juice, black caraway, garlic, and turmeric.[284,285] Recent interest has also been generated in the use of probiotic therapy, used as an adjunct to *H. pylori* eradication regimens. Reports and current meta-analyses have suggested that probiotic adjunct therapy resulted in an increased cure rate of approximately 10% – 4%.[265] Further studies are needed, however, to determine if these and other agents could potentially be used for prophylaxis or adjunct therapy for *Helicobacter* infection, and possibly help control the rising bacterial resistance of these organisms to antibiotics.[286]

Detection Methods

Historically, *H. pylori* testing was indicated for patients having clinical manifestations of infection, a history of peptic ulcer disease, or MALT lymphoma.[262,264,287,289] Recently, use of diagnostic testing has been expanded beyond gastrointestinal manifestations of *H. pylori* infection. Evidence now supports testing patients with nongastric diseases, such as unexplained iron-deficiency anemia or idiopathic thrombocytopenic purpura.[262,264,266,289]

There are several invasive and noninvasive approaches for the detection of *Helicobacter,* including *H. pylori*. Invasive methods require endoscopic biopsies and include rapid urease testing, histological examination, culture, and polymerase chain reaction.

TABLE 30.4

Methods for Detection of *Helicobacter*

Test Method	Principle	Comments	References
Histological examination	Direct observation of organism in tissue	Requires invasive procedures, such as endoscopy. Sensitivity is dependent on bacterial load present.	[250–251, 287, 289, 293]
Culture	Isolation of *Helicobacter* from specimens	100% specificity. Requires invasive procedures; technically demanding; Allows antibiotic susceptibility testing of isolates.	[244, 247–248, 257–258, 293]
Rapid urease	Indirect detection of *Helicobacter* by demonstration of urea breakdown products	Rapid results and easy to perform, False-positive results possible; endoscopy required.	[245–247, 293]
Radiolabeled C13 or C14 urea breath tests (UBT)	Measurement of labeled CO_2 in the breath; derived from urease hydrolysis of ingested C-labeled urea	Considered most reliable noninvasive test. Well tolerated; high sensitivity and specificity. Used in diagnosis of *H. pylori* infection and monitoring post treatment eradication.	[293, 300]
Serology	Measurement of *Helicobacter*-specific immunoglobulins IgG, IgA, and IgM by ELISA and latex agglutination	Noninvasive; good epidemiologic study tool; may not differentiate active vs. past infection. Widely available for *H. pylori* antibodies; not commercially available for *H. cinaedi* or other enterohepatic helicobacter; Method of choice for initial *H. pylori* detection due to cost efficiency and ease of performance.	[293, 297–298]
Fecal Antigen Test	Direct detection of *Helicobacter* antigen in feces using monoclonal antibodies by ELISA or immunochromatography assay (ICA) and antibodies.	Noninvasive; rapid assays available; high sensitivity and specificity; used in diagnosis and as an alternative method to UBT for eradication confirmation of *H. pylori* infection.	[292, 293]
DNA amplification	PCR assays are available for speciation, determination of resistance genes, 16S rRNA relatedness and for direct detection of *H. pylori* in stool and other specimens.	PCR is not available for all species; no suitable PCR assays available for fecal detection of enterohepatic *Helicobacter* species of clinical relevance.	[252–254, 259, 293, 295–296]

Noninvasive testing methods include serologic testing, the urea breath test, stool antigen test, and stool-based PCR assays.[288,289] Although the details of these test methods are beyond the scope of this chapter, additional information can be found in Table 30.4.[33,247–251,288–301]

Animal Models for the Study of *Helicobacter*

Much of the current understanding of *Helicobacter* pathogenesis, and especially that of *H. pylori*, has derived from *in vitro* studies. However, these methods often fail to demonstrate the true complexities of the host-pathogen relationship. The availability of small animal models has augmented and improved our understanding of the *in vivo* mechanisms of host adaptation by these organisms. Many different animal models, such as mice, ferrets, dogs, and cats, have been used over the years for studying *Helicobacter* organisms.[23] These animal models have generated important information that has also been physiologically relevant to human disease. Mice, rats, and rabbits were some of the first animal models used for the establishment of *H. pylori* infection, but they did not prove successful.[302] Animal models were then sought based on their natural infectivity with closely related *Helicobacter* species, such as the ferret or nonhuman primates, or by their ability to be experimentally infected with *H. pylori*, such as gnotobiotic piglets and dogs.[303–306] Experimental *H. pylori* infections have also been developed in Mongolian gerbils, guinea pigs, cats, and macaque monkeys.[307–310] Likewise, several animal models have been developed to study other gastric helicobacters, as well as the enterohepatic helicobacters such as *H. hepaticus*, *H. cinaedi*, *H. bilis*, and *H. fennelliae*.[242,311–318]

The ferret, having a stomach with anatomic and physiological similarities to that of humans, was also found to develop naturally occurring gastritis and gastric ulcer disease caused by *Helicobacter mustelae*.[303] As a result, the *H. mustelae* ferret model became one of the best-studied animal models of gastric *Helicobacter* infection in the natural host.[319,320] This model was instrumental in solidifying the role of *H. pylori* pathogenicity in humans. Although the ferret model is an excellent research tool, it is not always readily accessible.[320] Other animal models, such as the gnotobiotic piglets and primates, are likewise expensive and impractical for widespread use although they have been successfully utilized.[197,217,321]

Although there is no one model that is perfect for all aspects of research applications, the mouse model, based on cost and practicality, is the animal model of choice for *Helicobacter* studies. The use of mice models has been able to demonstrate the immune system modulation effects caused by gastric and enterohepatic *Helicobacter* species.[322–325] These models have also helped elucidate the importance of both intestinal and extraintestinal coinfection with regard to patient outcomes especially in terms of the degree of inflammation, and tumor development.[323,326,327] *H. felis*, which normally colonizes the gastric mucosa of cats and dogs, was found to readily colonize mice, particularly the gastric antrum.[328] This was an important finding because *H. pylori* could not be made to colonize early mouse models.[302] The first demonstration of the potential use of the *H. felis* mouse model was in studies with germ-free mice, where the bacterium was found to induce an active chronic gastritis and inflammatory effects similar to those induced by *H. pylori* in humans.[329] Subsequently, the *H. felis* mouse model provided an opportunity to study and understand an infectious process that resembled gastric pathological changes in human

H. pylori infection.[330,331] One limitation of the *H. felis* model is that it lacks the cag PAI found in *H. pylori*.[332] In spite of this, the ability of *H. felis* to induce severe gastritis and carcinoma in certain mice strains, makes this a valuable model to study inflammation-associated gastric carcinogenesis. With the availability of the *H. pylori* genome and recent advances in genetic engineering, *in vivo* imaging, and system-wide genomics and proteomics, the *H. felis* mouse model is being replaced with newer genetically engineered mouse study models.[54,160] The first published report of induced gastric carcinoma by *H. pylori* was reported using transgenic insulin-gastrin (INS-GAS) mice.[333,334] These mice overexpress pancreatic gastrin making them useful in the study of gastric cancer. This mouse model was also recently used to demonstrate the link between exposure to Swedish smokeless tobacco snuff and gastric cancer.[335] INS-GAS mice have a unique and important feature, i.e. male-predominant cancer expression, which mimics that seen in humans. Studies using these animals have shown that estrogen has a protective effect against cancer in both males and females infected with *H. pylori*.[334,336]

Other mouse models are being used to study the role of coinfection with other bacteria, viruses, and eukaryotic parasites and the modulation of *Helicobacter* disease. Such infections may augment or ameliorate gastric disease depending on location and the type of immune response elicited.[323,327] These findings help explain the high degree of regional differences seen in *H. pylori* disease outcomes among human populations, and which appear to depend on the incidence of endemic coinfections.[337] For example, in mice, helminth and protozoal parasites have been shown to impact the progression of gastritis and preneoplasia induced by *H. felis* and enterohepatic species of *Helicobacter* in the lower bowel. Such coinfections can impact the phenotypic disease induced by *H. pylori*.[322–324] In humans, infection with *Schistosoma mansoni* has been proposed to be one factor in the protection of individuals against concomitant *Helicobacter* gastritis in low-lying tropical regions.[338] Generally speaking, infectious agents that provoke a Th1 type immune response, regardless of tissue site, have been shown to augment disease induced by gastric helicobacters, whereas a Th2 type immune response may ameliorate disease.[337,339–341] This supports other animal studies that have shown that proinflammatory gut microbes such as *H. hepaticus* can promote extraintestinal cancers including mammary and liver carcinomas.[339]

Animal models and the use of sophisticated molecular tools will continue to yield new and pertinent information in our understanding of *Helicobacter* disease. In particular, new insights will continue to be unraveled regarding the complex host-pathogen mechanisms that influence clinical disease and its possible progression to cancer. This information will more importantly shed light on how these mechanisms might be therapeutically modulated to improve patient outcomes.

REFERENCES

1. Blaser, M.J. and Atherton, J.C., *Helicobacter pylori* persistence: biology and disease, *J. Clin. Invest.*, 113, 321, 2004.
2. Doenges, J.L., Spirochetes in the gastric glands of *Macacus rhesus* and of man without related diseases, *Arch. Pathol.*, 27, 469, 1939.
3. Freedberg, A.S. and Baron, L.E., The presence of spirochetes in human gastric mucosa, *Am. J. Dig. Dis.*, 7, 443, 1940.
4. Covacci, A. and Rappuoli, R., *Helicobacter pylori*: molecular evolution of a bacterial quasi-species, *Curr. Opin. Microbiol.*, 1, 96, 1998.
5. Linz, B. et al., An African origin for the intimate associate between humans and *Helicobacter pylori*, *Nature*, 445, 915, 2007.
6. Li, J. and Perez-Perez, G.I., *Helicobacter pylori* the latent human pathogen or an ancestral commensal organism, *Front. Microbiol.*, 9, 609, 2018.
7. Moodley, Y., et al., Age of the association between *Helicobacter pylori* and man, *PLOS Path.*, 8, 2012.
8. Suzuki, R.et al., Molecular epidemiology, population genetics, and pathogenic role of *Helicobacter pylori*, *Infect. Genet. Evolu.*, 12, 203, 2012.
9. Warren, J.R. and Marshall, B., Unidentified curved bacilli on gastric epithelium in active chronic gastritis, *Lancet*, 321, 1273, 1983.
10. Anonymous, Validation of the publication of new names and new combinations previously effectively published outside the IJSB: list no. 17, *Int. J. Syst. Bacteriol.*, 35, 223, 1985.
11. Marshall, B.J. and C.S. Goodwin, Revised nomenclature of *C. pyloridis*, *Int. J. Syst. Bacteriol.*, 37, 68, 1987.
12. Goodwin, C.S. et al., Unusual cellular fatty acids and distinctive ultrastructure in a new spiral bacterium (*Campylobacter pyloridis*) from the human gastric mucosa, *J. Med. Microbiol.*, 19, 257, 1985.
13. Romaniuk, P.J. et al., *Campylobacter pylori*, the spiral bacterium associated with human gastritis is not a true *Campylobacter* sp., *J. Bacteriol.*, 169, 2137, 1987.
14. Marshall, B.J. and Warren, J.R., Unidentified curved bacilli in the stomach of patients with gastritis and peptic ulceration, *Lancet*, 323, 1311, 1984.
15. Kusters, J.G. et al., Pathogenesis of *Helicobacter pylori* infection, *Clin. Microbiol. Rev.*, 19, 449, 2006.
16. Diaconu, S. et al., *Helicobacter pylori* infection: old and new, *J. of Med. and Life.*, 10(2), 112, 2017.
17. Chey, W.D., et al., ACG clinical guideline: treatment of *Helicobacter pylori* infection, *Am. J. Gastroenterol.*, 112, 212, 2017.
18. Anonymous, Schistosomes, liver flukes and *Helicobacter pylori*. IARC Working Group on the Evaluation of Carcinogenic Risks to Humans, *IARC Monogr. Eval. Carcinog. Risks Hum.*, 61(5), 1, 1994.
19. Mbulaiteye, S.M. et al., *Helicobacter pylori* associated global gastric cancer burden, *Front. Biosci. (Landmark Ed).*, 14, 1490, 2009.
20. Malfertheiner, P., et al., *Helicobacter pylori* infection: new facts in clinical management, *Curr. Treat. Options Gastroenterol*, 16, 605, 2018.
21. Testerman, T.L. and Morris, J., Beyond the stomach: an updated view of *Helicobacter pylori* pathogenesis, diagnosis, and treatment, *World J. Gastroenterol.*, 20(36), 12781, 2014.
22. Zhou, X., et al., Association between *Helicobacter pylori* infection and diabetes mellitus: a meta-analysis of observational studies, *Diabetes Res. Clin. Pract.*, 99, 200, 2013.
23. Menard, A. and Smet, A., Review: Other *Helicobacter* species, *Helicobacter.*, 24, (Suppl. 1): e126451, 2019.

24. Lawson, A.J., *Helicobacter*, in *Manual of Clinical Microbiology*, 10th ed., Vol. I, Versoalovic, J. Ed., ASM Press, Washington, DC, 2011, chap 54.

25. Smet, A, et al., Macroevolution of gastric *Helicobacter* species unveils interspecies admixture and time of divergence, *The ISME J.*, 12, 2518, 2018.

26. Baele, M. et al., Isolation and characterization of *Helicobacter suis* sp. nov from pig stomachs, *Int. J. Syst. Evol. Microbiol.*, 58, 1350, 2008.

27. Fox, J.G. and Megraud, F., *Helicobacter*, in *Manual of Clinical Microbiology*, 9th ed., Vol. I, Murray, P. R. Ed., ASM Press, Washington, DC, 2007, Chap. 60.

28. Fox, J.G. and Lee, A., The role of *Helicobacter* species in newly recognized gastrointestinal diseases of animals, *Lab Anim. Sci.*, 47, 222, 1997.

29. Solnick, J.V. and Schauer, D.B., Emergency of the *Helicobacter* genus in the pathogenesis of gastrointestinal disease, *Clin. Microbiol. Rev.*, 14, 59, 2001.

30. Avenaud, P. et al., Detection of *Helicobacter* species in the liver of patients with and without primary liver carcinoma, *Cancer*, 89, 1431, 2000.

31. Cindoruk, M. et al., Identification of *Helicobacter* species by 16S rDNA PCR and sequence analysis in human liver samples from patients with various etiologies of benign liver diseases, *Eur. J. Gastroenterol.*, 20, 33, 2008.

32. Mateos-Munoz, B. et al., Enterohepatic *Helicobacter* other than *Helicobacter pylori*, *Revista Espanola De Enfermedades Digestivas*, 105, 8, 477, 2013.

33. Versalovic, J. and Fox, J.G., Helicobacter, in *Manual of Clinical Microbiology*, 8th ed., Vol. I, Murray, P.R., Ed., ASM Press, Washington, DC, 2011, chap. 58.

34. Lawson, A.J. Helicobacter, in *Manual of Clinical Microbiology*, 11th ed., Vol. II, Carroll, C. and Funke, G., Eds., ASM Press, Washington, DC, 2015, chap. 57.

35. Dewhirst, F.E. et al., '*Flexispira rappini*' strains represent at least 10 *Helicobacter* taxa, *Int. J. Syst. Evol. Microbiol.*, 50, 1781, 2000.

36. Franco-Duarte, R. et al., Advances in chemical and biological methods to identify microorganisms-from past to present, *Microorganisms*, 7, 130, 2019.

37. Happonen, I. et al., Occurrence and topological mapping of gastric *Helicobacter*-like organisms and their association with histological changes in apparently healthy dogs and cats, *J. Vet. Med.*, 43, 305, 1996.

38. Flahou, B. et al., Non-*Helicobacter pylori Helicobacter* infections in humans and animals, *Helicobacter pylori Res.*, 233, 2016.

39. Haesebrouck, F. et al., Gastric helicobacters in domestic animals and nonhuman primates and their significance for human health, *Clin. Microbiol. Rev.*, 22, 202, 2009.

40. De Groote, D. et al., Detection of non-pylori *Helicobacter* species in "*Helicobacter heilmannii*" infected humans, *Helicobacter*, 10, 398, 2005.

41. Mladenova-Hristova, I. et al., Zoonotic potential of *Helicobacter* species, *J. Microbiol. Immunol. Infect.*, 50, 265, 2017.

42. Borges, V. et al., *Hellocbacter pullorum* isolated from fresh chicken meat: antibiotic resistance and genomic traits of an emerging foodborne pathogen, *Appl. Environ. Microbiol.*, 81, 8155, 2015.

43. Burnens, A P. et al., Novel *Campylobacter*-like organism resembling *Helicobacter fennelliae* isolated from a boy with gastroenteritis and from dogs, *J. Clin. Microbiol.*, 31, 1916, 1993.

44. Stanley, J. et al., *Helicobacter canis* sp. nov., a new species from dogs: an integrated study of phenotype and genotype, *J. Gen. Microbiol.*, 139, 2495, 1993.

45. Stanley, J. et al., *Helicobacter pullorum* sp. nov. genotype and phenotype of a new species isolated from poultry and from human patients with gastroenteritis, *Microbiology*, 140, 3441, 1994.

46. Melito, P.L. et al., *Helicobacter winghamensis* sp. nov., a novel *Helicobacter* sp. isolated from patients with gastroenteritis, *J. Clin. Microbiol.*, 39, 2412, 2001.

47. Shen, Z. et al., *Helicobacter rodentium* sp. nov., a urease negative *Helicobacter* species isolated from laboratory mice. *Int. J. Syst. Bacteriol.*, 47, 627, 1997.

48. Robertson, B.R. et al., *Helicobacter ganmani* sp. nov., a urease negative anaerobe isolated from the intestines of laboratory mice, *Int. J. Syst. Evol. Microbiol.*, 51, 1881, 2001.

49. Garrity, G.M. et al., Helicobacteriaceae, in *Bergey's Manual of Systematic Bacteriology, Volume Two: The Proteobacteria (Part C)*, 2nd ed., Garrity, G.M. et al., Ed., Springer-Verlag, New York, 2005, p.1169.

50. Chalk, P.A., Roberts, A.D., and Blows, W.M., Metabolism of pyruvate and glucose by intact cells of *Helicobacter pylori* studied by 13C NMR spectroscopy, *Microbiology*, 140, 2085, 1994.

51. Reynolds, D.J. and Penn, C.W., Characteristics of *Helicobacter pylori* growth in a defined medium and determination of its amino acid requirements, *Microbiology*, 140, 2649, 1994.

52. Testerman, T.L. et al., *Helicobacter pylori* growth in urease detection in a chemically defined medium Ham's F-12 nutrient mixture, *J. Clin. Microbiol.*, 39, 3842, 2001.

53. Schilling, C.H. et al., Genome-scale metabolic model of *Helicobacter pylori* 26695, *J. Bacteriol.*,184, 4582, 2002.

54. Tomb, J.F. et al., The complete genome sequence of the gastric pathogen *Helicobacter pylori*, *Nature*, 388, 539, 1997.

55. Percival, S.I., et al., Biofilms and *Helicobacter pylori*: dissemination and persistence within the environment and host, *World J. Gastroenterol. Pathophysiol*, 5, 122, 2014.

56. Payão, S.L.M. and Rasmussen, L.T., *Helicobacter pylori* and its reservoirs: a correlation with the gastric infection, *World J. Gastrointest. Pharmacol. Ther.* 7(1), 126, 2016.

57. Junqueira, A.C.M., et al., The microbes of bowflies and houseflies as bacterial transmission reservoirs, *Sci. Rep.*, 7, 16324, 2017.

58. Momtaz, H., et al., Study of *Helicobacter pylori* genotype status in cows sheep, goats and human beings, *BMC Gastroenterol.*, 14, 61, 2014.

59. Peters, C., et al., The occupational risk of *Helicobacter pylori* infection among gastroenterologists and their assistants, *BMC Infect. Dis.*, 11, 154, 2011.

60. Khalifa, M.M., et al., *Helicobacter pylori* a poor man's gut pathogen? *Gut Pathogens*, 2, 2, 2010.

61. Mamishi, S., et al., Intrafamilial transmission of *Helicobacter pylori*: genotyping of faecal samples, *Br. J. Biomed. Sci.*, 73, 38, 2016.

62. Weyermann, M., et al., Acquisition of *Helicobacter pylori* infection in early childhood: independent contributions of infected mother, father and siblings, *Am J. Gastroenterol.*, 104, 182, 2009.

63. Goh, K.L. et al., Epidemiology of *Helicobacter pylori* infection and public health implications, *Helicobacter*, 16, 1, 2011.

64. Thomas, J E. et al., Isolation of *Helicobacter pylori* from human feces, *Lancet*, 341, 380, 1993.

65. Bruce, M.G and Maaroos, H.I., Epidemiology of *Helicobacter pylori* infections, *Helicobacter*, 13 (Suppl. 1), 1, 2008.

66. Bellack, N.R., et al., A conceptual model of water's role as a reservoir in Helicobacter pylori transmission: a review of the evidence, *Epidemiol. Infect.*, 134,439, 2006.

67. Gião, M.S., et al., Persistence of *Helicobacter pylori* in heterotrophic drinking water biofilms, *Appl. Environ. Microbiol.*, 74, 5898, 2008.

68. Moreno, Y. et al., Survival and viability of *Helicobacter pylori* after inoculation into chlorinated drinking water, *Water Res.*, 41, 3490, 2007.

69. Hulten, K. et al., *Helicobacter pylori* in the drinking water in Peru, *Gastroenterology*, 110, 1031, 1996.

70. Graham, D.Y., et al., Seroepidemiology of *Helicobacter pylori* infection in India. Comparison of developing and developed countries, *Dig. Dis. Sci.*, 36, 1084, 1991.

71. Lee, Y.Y., et al., Sociocultural and dietary practices among Malay subjects in the north-eastern region of Peninsular Malaysia: a region of low prevalence of *Helicobacter pylori* infection, *Helicobacter*, 17, 54, 2012.

72. Porras, C., et al., Epidemiology of Helicobacter pylori infection in six Latin American countries (SWOG Trial S0701), *Cancer Causes Control*, 24, 209, 2013.

73. Degnan, A.J., et al., Development of a plating medium for selection of *Helicobacter pylori* from water samples, *Appl. Environ. Microbiol.*, 69, 29, 2003.

74. Azevedo, N.F., et al., Nutrient shock and incubation atmosphere influence recovery of culturable *Helicobacter pylori* from water, *Appl. Environ. Microbiol.*, 70, 490, 2004.

75. Moreno, Y., and Ferrus, M.A., Specific detection of cultivable *Helicobacter pylori* cells from wastewater treatment plants, *Helicobacter* 17, 327, 2012.

76. Al-Sulami, A.A., et al., Isolation and identification of Helicobacter pylori from drinking water in Basra governorate, Iraq, *East Mediterr. Health J.*, 16, 920, 2011.

77. Bahrami, A.R., et al., Detection of *Helicobacter pylori* in city water, dental units' water, and bottled mineral water in Isfahan, Iran, *Scientific World J.*, 2013, 280510, 2013.

78. Shahamat, M., et al., Use of autoradiography to assess viability of *Helicobacter pylori* in water, *Appl. Environ. Microbiol.*, 59, 1231, 1993.

79. Aziz, R.K., et al., Contaminated water as a source of *Helicobacter pylori* infection: a review, *J. Advanced Res.*, 6, 539, 2015.

80. O'Rourke, J.L., Grehan, M. and Lee, A., Non-pylori *Helicobacter* species in humans, *Gut*, 49, 601, 2001.

81. De Groote, D. et al., Detection of non-pylori *Helicobacter* species in "*Helicobacter heilmannii*"-infected humans, *Helicobacter*, 10(5), 398, 2005.

82. Haesebrouck, F. et al., Gastric helicobacters in domestic animals and non-human primates and their significance in human health, *Clin. Microbiol. Rev.*, 22, 202, 2009.

83. Van den Bulck, K. et al., Identification of non-*Helicobacter pylori* spiral organisms in gastric samples from humans, dogs and cats, *J. Clin. Microbiol.*, 43, 2256, 2005.

84. Fox, J.G. et al., Gastric colonization *by Campylobacter pylori* subsp. *mustelae* in ferrets, *Infect. Immun.*, 56, 2994, 1988.

85. Fox, J.G. et al., Gastric colonization of the ferret with *Helicobacter* species: natural and experimental infections, *Rev. Infect. Dis.*, 13(Suppl. 8), S671, 1991.

86. Fox, J.G. et al., *Helicobacter mustelae* isolation from feces of ferrets: evidence to support fecal-oral transmission of gastric *Helicobacter*, *Infect. Immun.*, 60, 606, 1992.

87. Dore, M.P. et al., Isolation of *Helicobacter pylori* from sheep—implications for transmission to humans, *Am. J. Gastroenterol.*, 96, 1396, 2001.

88. Fox, J.G. et al., Local immune response in *Helicobacter pylori* infected cats and identification of *H. pylori* in saliva, gastric fluid and feces, *Immunology*, 88, 400, 1996.

89. Fox, J.G. et al., *Helicobacter canadensis* sp. nov. isolated from humans with diarrhea as an example of an emerging pathogen, *J. Clin. Microbiol.*, 38, 2546, 2000.

90. Zanoni, R.G. et al., Occurrence and antibiotic susceptibility of *H. pullorum* from broiler chickens and commercially laying hens in Italy, *Int. J. Food Microbiol.*, 116, 168, 2007.

91. Gonzalez, A. et al., A novel real time PCR assay for the detection of *Helicobacter pullorum*-like organisms in chicken products, *Int. Microbiol.*, 11, 203, 2008.

92. Blaser, M.J. In a world of black and white, *Helicobacter pylori* is gray, *Ann. Intern. Med.*, 130, 695, 1999.

93. Go, M.F., Natural history and epidemiology of *Helicobacter pylori* infection, *Aliment. Pharacol. Ther.*, 16 (Suppl. 1), 3, 2002.

94. O'Rourke, J.L., Lee, A., and Fox, J.G., *Helicobacter* infection in animals: a clue to the role of adhesion in the pathogenesis of gastroduodenal disease, *Eur. J. Gastroenterol. Hepatol.*, 4, (Suppl. 1), S31, 1992.

95. Taylor, N.S. et al., Haemagglutination profiles of *Helicobacter* species that cause gastritis in man and animals, *J. Med. Microbiol.*, 37, 299, 1992.

96. Malnick, S.D.H., et al., *Helicobacter pylori*: friend or foe? *World J. Gastroenterol.*, 20, 8979, 2014.

97. Uemura, N. et al., *Helicobacter pylori* infection and the development of gastric cancer, *N. Engl. J. Med.*, 345, 784, 2001.

98. Blaser, M.J., An endangered species in the stomach, *Sci. Am.*, 292(2), 38, 2005.

99. Vandamme, P. et al., Identification of *Campylobacter cinaedi* isolated from blood and feces from children and adult females, *J. Clin. Microbiol.*, 28, 1016, 1990.

100. Kiehlbauch, J A. et al., *Helicobacter cinaedi*-associated bacteremia and cellulitis in immunocompromised patients, *Ann. Intern. Med.*, 121, 90, 1994.

101. Hsueh, P R. et al., Septic shock due to *Helicobacter fennelliae* in a non-human immunodeficiency virus-infected heterosexual patient, *J. Clin. Microbiol.*, 37, 2084, 1999.

102. Sugiyama, A., et al., First adult case of *Helicobacter cinaedi* meningitis, *J. Neurol. Sci.*, 16, 2011.

103. Morgner, A. et al., *Helicobacter heilmannii*-associated primary gastric low-grade MALT lymphoma: complete remission after curing the infection, *Gastroenterology*, 118, 821, 2000.

104. Ali, B., et al., Presence of gastric *Helicobacter* species in children suffering from gastric disorders in southern Turkey, *Helicobacter*, 23, 2018.

105. Schott, T. et al., Comparative genomics of Helicobacter pylori and the human-derived *Helicobacter bizzozeronii* CIII-1 strain reveal the molecular basis of the zoonotic nature of non-pylori gastric *Helicobacter* infections in humans, *BMC Genomics*, 12, 534, 2011.

106. Bento-Miranda, M. and Figueiredo, C., *Helicobacter heilmannii* senu lato: an overview of the infections in humans, *World J. Gastroenterol.*, 20, 17779, 2014.

107. Erdman, S E. et al., *Helicobacter mustelae*-associated gastric MALT lymphoma in ferrets, *Am. J. Pathol.*, 151, 273, 1997.

108. Rocha, M. et al., Association of *Helicobacter* species with hepatitis C cirrhosis with or without hepatocellular carcinoma, *Gut*, 54, 396, 2005.

109. Fox, J.G. et al., Hepatic *Helicobacter* species identified in bile and gall bladder tissue from Chileans with chronic cholecystitis, *Gastroenterology*, 114, 755, 1998.

110. Laharie, D. et al., Association between enterohepatic *Helicobacter* species and Crohn's disease: a prospective cross-sectional study, *Aliment. Pharmacol. Ther.*, 30, 283, 2009.

111. Fox, J.G. et al., *Helicobacter hepaticus* sp. nov., a microaerophilic bacterium isolated from livers and intestinal mucosal scrapings from mice, *J. Clin. Microbiol.*, 32, 1238, 1994.

112. Fox, J.G. et al., Isolation of *Helicobacter cinaedi* from the colon, liver, and mesenteric lymph node of a rhesus monkey with chronic colitis and hepatitis, *J. Clin. Microbiol.*, 39, 1580, 2001.

113. Fox, J.G. et al., Persistent hepatitis and enterocolitis in germfree mice infected with *Helicobacter hepaticus*, *Infect. Immun.*, 64, 3673, 1996.

114. Wasimuddin, W. et al., High prevalence and species diversity of *Helicobacter spp.* detected in wild house mice, *Appl. Environ. Microbiol.*, 78, 8158, 2012.

115. Flahou, B. et al., Gastric and enterohepatic non-*Helicobacter pylori* helicobacters, *Helicobacter*, 18(Suppl. 1), 66, 2013.

116. Ménard, A. and Smet, A., Review: other *Helicobacter* species, *Heliocobacter*, 24 (Suppl. 1), 214, 2019.

117. Ménard, A, et al., Gastric and enterohepatic helicobacters other than *Helicobacter pylori*, *Helicobacter*, 19(Suppl. 1), 2014.

118. Ceelen, L. et al., Prevalence of *Helicobacter pullorum* among patients with gastric disease and clinically healthy persons, *J. Clin. Microbiol.*, 43(6), 2984, 2005.

119. Solnick, J V. et al., Extragastric manifestations of *Helicobacter pylori* infection—other *Helicobacter* species, *Helicobacter*, 11(Suppl. 1), 46, 2006.

120. Nilsson, H O. et al., Identification of *Helicobacter pylori* and other *Helicobacter* species by PCR, hybridization, and partial DNA sequencing in human liver samples from patients with primary sclerosing cholangitis or primary biliary cirrhosis, *J. Clin. Microbiol.*, 38, 1072, 2000.

121. Whary, M.T. and Fox, J.G., Detection, eradication, and research implications of *Helicobacter* infection in laboratory rodents, *Lab Animal*, 35(7), 25, 2006.

122. Lee, A. et al., A standardized mouse model of *Helicobacter pylori* infection. Introducing the Sydney strain, *Gastroenterology*, 112, 1386, 1997.

123. Goodman, K. and Cockburn, M., The role of epidemiology in understanding the health effect of *Helicobacter pylori*, *Epidemiology*, 12(2), 266, 2001.

124. Goodman, K. et al., Dynamics of *Helicobacter pylori* infection in a US-Mexico cohort during the first two years of life, *Int. J. Epidemiol.*, 34(6), 1348, 2005.

125. McCarthy, C. et al., Long-term prospective study of *Helicobacter pylori* in nonulcer dyspepsia, *Dig. Dis. Sci.*, 40, 114, 1995.

126. Hansson, L.E. et al., The risk of stomach cancer in patients with gastric or duodenal ulcer disease, *N. Engl. J. Med.*, 335, 242, 1996.

127. Ahmad, A., Govil, Y., and Frank, B.B., Gastric mucosa-associated lymphoid tissue lymphoma, *Am. J. Gastroenterol.*, 98, 975, 2003.

128. Mager, D.L., Bacteria and cancer: cause, coincidence or cure? A review, *J. Transl. Med.*, 4, 14, 2006.

129. Manolakis, A., et al., A review of the postulated mechanisms concerning the association of *Helicobacter pylori* with ischemic heart disease, *Helicobacter*, 12, 287, 2007.

130. Stolzenberg-Solomon, R Z. et al., *Helicobacter pylori* seropositivity as a risk factor for pancreatic cancer, *J. Natl. Cancer Inst.*, 93, 937, 2001.

131. de Martel, C. et al., *Helicobacter* species in cancers of the gallbladder and extrahepatic billiary tract, *Br. J. Cancer*, 100, 194, 2009.

132. Kosaka, T., et al., *Helicobacter bilis* colonization of the biliary system in patients with pancreaticobiliary maljunction, *Br. J. Surg.*, 97, 544, 2010.

133. Goni, E., and Franceschi, F., *Helicobacter pylori* and extragastric diseases, *Helicobacter*, 21 (Suppl 1), 45, 2016.

134. Diaconu, S., et al., *Helicobacter pylori* infection: old and new, *J. of Med. and Life*, 10(2), 112, 2017.

135. Bazmamoun, H., et al., *Helicobacter* infection in children with type 1 diabetes mellitus: a case-control study, *J. Res. Health Sci.*, 16(2), 68, 2016.

136. Di Simone, N., et al., *Helicobacter pylori* infection contributes to placental impairment in preeclampsia: basic and clinical evidences, *Helicobacter.*, 92, 1153, 2016.

137. Hudak, L., et al., An updated systemic review and meta-analysis on the association between *Helicobacter pylori* infection and iron deficiency anemia, *Helicobacter*, 22(1), 39, 2017.

138. Upsala, S., et al., Association between *Helicobacter pylori* infection and metabolic syndrome: a systemic review and meta-analysis, *J. Digest. Dis.*, 17(7), 433, 2016.

139. Hazell, S L. et al., *Campylobacter pyloridis* and gastritis: association with intercellular spaces and adaptation to an environment of mucus as important factors in colonization of the gastric epithelium, *J. Infect. Dis.*, 153, 658, 1986.

140. Hawtin, P.R., Stacey, A.R., and Newell, D.G., Investigation of the structure and location of the urease of *Helicobacter pylori* using monoclonal antibodies, *J. Gen. Microbiol.*, 136, 1995, 1990.

141. Marcus, E.A., et al., The periplasmic alpha-carbonic anhydrase activity of *Helicobacter pylori* is essential for acid acclimation, *J. Bacteriol.*, 187, 729, 2005.

142. Debowski, A.W., et al., *Helicobacter pylori* gene silencing in vivo demonstrates urease is essential for chronic infection, *PLOS Pathog.*, 13, e10066464, 2017.

143. Schoep, T.D., et al., Surface properties of *Helicobacter pylori* urease complex are essential for persistence, *PLOS ONE*, 5, e15042, 2010.

144. Mobley, H.L. et al., *Helicobacter pylori* urease: properties and role in pathogenesis, *Scand. J. Gastroenterol. Suppl.*, 187, 39, 1991.

145. Lytton, S.D., et al., Production of ammonium by Helicobacter pylori mediates occluding processing and disruption of tight junctions in Caco-2 cells, *Microbiology*, 151, 3267, 2005.

146. Wroblewski, L.E., et al., *Helicobacter pylori* dysregulation of gastric epithelial tight junctions by urease-mediated myosin II activation, *Gastroenterology*, 136, 236, 2009.

147. Kuwahara, H., et al., *Helicobacter pylori* urease suppresses bactericidal activity of peroxynitrite via carbon dioxide production, *Infect. Immun.*, 68, 4378, 2000.

148. Olivera-Severo, D., et al., A new role for *Helicobacter pylori* urease: contributions to angiogenesis, *Front. Microbiol.*, 8, 1883, 2017.

149. Valenzuela-Valderrama, M., et al., The *Helicobacter pylori* urease virulence factor is required for induction of hypoxia-induced Factor-1 α in gastric cells, *Cancers*, 11, 799, 2019.

150. Fan, X., et al., *Helicobacter pylori* urease binds to class II MHC on gastric epithelial cells and induces their apoptosis, *J. Immunol.*, 165, 1918, 2000.

151. Ansari, S. and Yamaoka, Y., Helicobacter pylori virulence factors exploiting gastric colonization and its pathogenicity, *Toxins*, 11, 677, 2019.

152. Aspholm, M., et al., *Helicobacter pylori* adhesion to carbohydrates, *Methods Enzymol.*, 417, 293, 2006.

153. Leker, K., et al., Comparison of lipopolysaccharide composition of two different strains of *Helicobacter pylori*, *BMC Microbiology*, 17, 226, 2017.

154. Li, H., et al., The redefinition of *Helicobacter pylori* lipopolysaccharide O-antigen and core-oligosaccharide domains, *PLOS Pathogens*, 13(3), e1006280, 2017.

155. Heneghan, M.A., McCarthy, C.F., and Moran, A.P., Relationship of blood group determinants on *Helicobacter pylori* lipopolysaccharide with host Lewis phenotype and inflammatory response, *Infect. Immun.*, 68, 937, 2000.

156. Cullen, T.W., et al., *Helicobacter pylori* versus the host: remodeling of the bacterial outer membrane is required for survival in the gastric mucosa, *PLOS Pathogen*, 7, e1002454, 2011.

157. Wirth, H P. et al., Expression of Lewis X and Y blood group antigens by *Helicobacter pylori* strains is related to cagA status, *Gastroenterology*, 110, A296, 1996.

158. Zheng, P Y. et al., Association of peptic ulcer with increased expression of Lewis antigens but not cagA, iceA and vacA in *Helicobacter pylori* isolates in an Asian population, *Gut*, 47, 18, 2000.

159. Moran, A P. et al., Molecular mimicry of host structures by lipopolysaccharides of *Campylobacter* and *Helicobacter* spp.: implications in pathogenesis, *J. Endotox. Res.*, 3, 521, 1996.

160. Marchetti, M. et al., Development of a mouse model of *Helicobacter pylori* infection that mimics human disease, *Science*, 267, 1655, 1995.

161. Cover, T.L. and Blaser, M.J., Purification and characterization of the vacuolating toxin from *Helicobacter pylori*, *J. Biol. Chem.*, 267, 10570, 1992.

162. Phadais, S H. et al., Pathological significance and molecular characterization of the vacuolating toxin gene of *Helicobacter pylori*, *Infect. Immun.*, 62, 1557, 1994.

163. Bittencourt de Brito, B. et al., Pathogenesis and clinical management of *Helicobacter pylori* gastric infection, *World J. Gastroenterol.*, 25(37), 5578, 2019.

164. Roesler, B M. et al., Virulence factors of *Helicobacter pylori*: A review, *Clinical Medicine Insights: Gastroenterol.*, 7, 9, 2014.

165. Atherton, J C. et al., Clinical and pathological importance of heterogeneity in vacA, the vacuolating cytotoxin gene of *Helicobacter pylori*, *Gastroenterol.*, 112, 97, 1997.

166. Djekic, A. and Möller, A, The immunomodulator VacA promotes immune tolerance and persistent *Helicobacter pylori* infection through its activities on T-cells and antigen-presenting cells, *Toxins.*, 8, 187, 2016.

167. Boquet, P. and Ricci, V, Intoxication strategy of helicobacter pylori VacA toxin, *Trends Microbiol.*, 20, 165, 2012.

168. Shirasaka, D. et al., Analysis of *Helicobacter pylori* VacA gene and serum antibodies to VacA in Japan, *Dig. Dis. Sci.*, 45, 789, 2000.

169. Jang, T.J. and Kim, J.R., Proliferation and apoptosis in gastric antral epithelial cells in patients infected with *Helicobacter pylori*, *J. Gastroenterol.*, 35, 265, 2000.

170. Houghton, J. et al., Tumor necrosis factor alpha and interleukin 1 up-regulate gastric mucosal fas antigen expression in *Helicobacter pylori* infection, *Infect. Immun.*, 68, 1189, 2000.

171. Pons, B. J. et al., Cytolethal distending toxin subunit B: a review of structure-function relationship, *Toxins*, 11, 595, 2019.

172. Guerra, L.et al., Do bacterial genotoxins contribute to chronic inflammation, genomic instability and tumor progression? *FEBS J.*, 278(23), 4577, 2011.

173. Belibasakis, G.N. and Bostanci, N., Inflammatory and bone remodeling responses to the cytolethal distending toxins, *Cells*, 3, 236, 2014.

174. Taylor, N.S., Fox, J.G. and Yan, L., *In vitro* hepatotoxic factor in *Helicobacter hepaticus*, *H. pylori* and other *Helicobacter* species, *J. Med. Microbiol.*, 42, 48, 1995.

175. Ge, Z. et al., *Helicobacter hepaticus* cytolethal distending toxin promotes intestinal carcinogenesis in 129Rag2-deficient mice, *Cell Microbiol*, 19, 1, 2017.

176. Zhongming, G., Shauer, D.B. and Fox, J.G., *In-vivo* virulence properties of bacterial cytolethal distending toxin, *Cell. Microbiol.*, 10, 1599, 2008.

177. Chien, C.C. et al., Identification of cdtB homologues and cytolethal distending toxin activity in enterohepatic *Helicobacter* spp., *J. Med. Microbiol.*, 49, 525, 2000.

178. Péré-Védrenne, C. et al., The cytolethal distending toxin subunit CdtB of *Helicobacter* induces a Th17-related and antimicrobial signature in intestinal and hepatic cells *in vitro*, *J. Infect. Dis.*, 213, 1979, 2016.

179. Shen, Z. et al., Cytolethal distending toxin promotes *Helicobacter cinaedi*-associated typhlocolitis in interleukin-10 deficient mice, *Infect. Immun.*, 77, 2508, 2009.

180. Covacci, A. and Rappuoli, R., Tyrosine-phosphorylated bacterial proteins: Trojan horses for the host cell, *J. Exp. Med.*, 191, 587, 2000.

181. Rego, A.T., Chandra, V. and Waksman, G., Two-step and one-step secretion mechanisms in Gram-negative bacteria: contrasting the type IV secretion system and the chaperone-usher pathway of pilus biogenesis, *Biochem. J.*, 425, 475, 2010.

182. Kaparakis, M. et al., Bacterial membrane vesicles deliver peptidoglycan to NOD1 in epithelial cells, *Cell. Microbiol*, 12, 372, 2010.

183. Higashi, H. et al., Biological activity of the *Helicobacter pylori* virulence factor CagA is determined by variation in tyrosine phosphorylation sites, *Proc. Natl. Acad. Sci. USA*, 99, 14428, 2002.

184. Higashi, H. et al., *Helicobacter pylori* CagA induces Ras-independent morphogenetic response through SHP-2 recruitment and activation, *J. Biol. Chem.*, 279, 17205, 2004.

185. Amieva, M R. et al., Disruption of the epithelial apical-junctional complex by *Helicobacter pylori* CagA, *Science*, 300, 1430, 2003.

186. Brandt, S. et al., NF-kappaB activation and potentiation of proinflammatory responses by the *Helicobacter pylori* CagA protein, *Proc. Natl. Acad. Sci. USA*, 102, 9300, 2005.

187. Butt, J. et al., Serologic response to *Helicobacter pylori* proteins associated with risk of colorectal cancer among diverse populations in the United States, *Gastroenterology*, 156, 175, 2019.

188. Noto, J.M. and Peek Jr., M., The *Helicobacter pylori* cag pathogenicity island, in *Helicobacter Species: Methods and Protocols, Methods in Molecular Biology*, Houghton, J. Ed., Humana Press, Springer, New York, 2012, chap 7.

189. Wroblewski, L.E. and Peek, R. M., Jr., *Helicobacter pylori*: Pathogenic enablers-toxic relationships in the stomach, *Nat. Rev. Gastroenterol. Hepatol.*, 13, 317, 2016.

190. Ahmadzadeh, A, et al., Association of CagPAI integrity with severeness of *Helicobacter pylori* infection in patients with gastritis, *Pathol. Biol.*, 63, 252, 2015.

191. Markovska, R. et al., Status of *Helicobacter pylori* cag pathogenicity island (cagPAI) integrity and significance of its individual genes, *Infect. Genet. Evol.*, 59, 167, 2018.

192. Yadegar, A. et al., Genetic diversity and amino acid sequence polymorphism in *Helicobacter pylori* CagL hypervariable motif and its association with virulence markers and gastroduodenal diseases, *Cancer Med.*, 8, 1619, 2019.

193. Choi, S.I. et al., CDX1 expression induced by CagA-expressing *Helicobacter pylori* promotes gastric tumorigenesis, *Mol. Cancer Res.*, 17, 2169, 2019.

194. Camorlinga-Ponce, M. et al., Topographical localisation of *cag*A postive and *cag*A negative *Helicobacter pylori* strains in the gastric mucosa; an in situ hybridisation study, *J. Clin. Pathol.*, 57, 822, 2004.

195. Reshetnyak, V.I. and Reshnetnyak, T.M., Significance of dormant forms of *Helicobacter pylori* in ulcerogenesis, *World J. Gastroenterol.*, 23, 4867, 2017.

196. Reinhard, H. et al., Coccoid *Helicobacter pylori*: an uncommon form of a common pathogen, *Ibnosina J. Med. Biomed. Sci.*, 11, 120, 2019.

197. Balakrishna, J.P, and Filatov, A., Coccoid forms of *Helicobacter pylori* causing active gastritis, *Am. J. Clin. Pathol.*, 140(Suppl 1), A101, 2013.

198. Krzyzek, P., et al., Intensive formation of coccoid forms as a feature strongly associated with highly pathogenic *Helicobacter pylori* strains, *Folia Microbiologica*, 64, 273, 2019.

199. Vitoriano, I. et al., Ulcerogenic *Helicobacter pylori* strains isolated from children: a contribution to get insight into the virulence of the bacteria, *PLOS ONE.*, 6(10), e26265, 2011.

200. Loke, M.F. et al., Understanding the dimorphic lifestyles of human gastric pathogen *Helicobacter pylori* using the SWATH-based proteomics approach, *Sci. Rep.*, 6, 26784, 2016.

201. Sarem, M. and Corti, R., Role of *Helicobacter pylori* coccoid forms in infection and recrudescence, *Gastroenterol. Hepatol.*, 39, 28, 2015.

202. Assadi, M.M. et al., Methods for detecting the environmental coccoid form of *Helicobacter pylori*, *Front. Public Health.*, 3, 147, 2015.

203. Li, Q., et al., Serum VacA antibody is associated with risks of peptic ulcer and gastric cancer: A meta-analysis, *Microb. Pathog.*, 99, 220, 2016.

204. Butt, J., et al., Serologic response to *Helicobacter pylori* proteins associated with risk of colorectal cancer among diverse populations in the United States, *Gastroenterology*, 156, 175, 2019.

205. Cover, T L. et al., Induction of gastric epithelial cell apoptosis by *Helicobacter pylori* vacuolating cytotoxin, *Cancer Res.*, 63, 951, 2003.

206. Terebiznik, M R. et al., Effect of *Helicobacter pylori's* vacuolating cytotoxin on the autophagy pathway in gastric epithelial cells, *Autophagy*, 5, 370, 2009.

207. Liu, X. et al., A systemic review on the association between the *Helicobacter pylori vac*A I genotype and gastric disease, *FEBS Open Bio.*, 6, 409, 2016.

208. Sezikli, M. et al., Frequencies of serum antibodies to *Helicobacter pylori* CagA and VacA in a Turkish population with various gastroduodenal diseases, *Int. J. Clin. Prac.*, 60, 1239, 2006.

209. Lee, A. et al., Isolation of a spiral-shaped bacterium from a cat stomach, *Infect. Immun.*, 56, 2843, 1988.

210. Cheng-Yen, K., et al., *Helicobacter pylori* infection: an overview of bacterial virulence factors and pathogenesis, *Biomed. J.*, 39, 14, 2016.

211. Howitt, M.R., et al., ChePep controls *Helicobacter pylori* infection of the gastric glands and chemotaxis in the *Epsilonproteobacteria*, *M. Bio.*, 2, 5, 2011.

212. Kao, CY., et al., Higher motility enhances bacterial density and inflammatory response in dyspeptic patients infected with *Helicobacter pylori*, *Helicobacter*, 17, 411, 2012.

213. Merino, S., et al., Role of *Aeromonas hydrophila* flagella glycosylation in adhesion to Hep-2 cells, biofilm formation and immune stimulation, *Int. J. Mol. Sci.*, 15, 21935, 2014.

214. Bucior, I., et al., *Pseudomonas aeruginosa* pili and flagella mediate distinct binding and signaling events at the apical and basolateral surface of airway epithelium, *PLOS. Pathol.*, 8, 2012.

215. Clyne, M., et al., Adherence of isogenic flagellum-negative mutants of *Helicobacter pylori* and *Helicobacter mustelae* to human and ferret gastric epithelial cells, *Infect. Immun.*, 68, 4335, 2000.

216. Gu, H., Role of flagella in the pathogenesis of *Helicobacter pylori*, *Curr. Microbiol.*, 74, 863, 2017.

217. Eaton, K A. et al., Essential role of urease in pathogenesis in gastritis induced by *Helicobacter pylori* in gnotobiotic piglets, *Infect. Immun.*, 59, 2470, 1991.

218. Solnick, J V. et al., Construction and characterization of an isogenic urease-negative mutant of *H. mustelae*, *Infect. Immun.*, 63, 3718, 1995.

219. Solnick, J V. et al., Molecular analysis of urease genes from a newly identified uncultured species of *Helicobacter*, *Infect. Immun.*, 62, 1631, 1994.

220. Ansari, S., and Yamaoka, Y., *Helicobacter pylori* BabA in adaptation for gastric colonization, *World J. Gastroenterol.*, 23, 4158, 2017.

221. Aspholm-Hurtig, M., et al., Functional adaptation of BabA, the *H. pylori* ABO blood group antigen binding adhesin, *Science*, 305, 519, 2004.

222. Huang, Y., et al., Adhesin and invasion of gastric mucosa epithelial cells by *Helicobacter pylori, Front. Cell. Infect. Microbiol.*, 6, 159, 2016.

223. Goodwin, A.C., et al., Expression of the *Helicobacter pylori* adhesin SabA is controlled via phase variation and the ArsRS signal transduction system, *Microbiology*, 154, 2231, 2008.

224. Sallas, M.L., et al., Status (on/off) of *oip*A gene: their associations with gastritis and gastric cancer and geographic origins, *Arch. Microbiol.*, 201, 93, 2019.

225. Gur, C., et al., The *Helicobacter pylori* HopQ outer membrane protein inhibits immune cell activities, *OncoImmunology*, 8, e1553487, 2019.

226. Yakoob, J., et al., Gastric lymphoma: association with *Helicobacter pylori* outer membrane protein Q (HopQ) and cytotoxic-pathogenicity activity island (CPAI) genes, *Epidemiol. Infect.*, 145, 3468, 2017.

227. Segal, E.D., Falkow, S., and Thompkins, L.S., *Helicobacter pylori* attachment to gastric cells induces cytoskeletal rearrangements and tyrosine phosphorylation of host proteins, *Proc. Natl. Acad. Sci. USA.*, 93, 1259, 1996.

228. O'Rourke, J., Lee, A., and Fox, J.G., An ultrastructural study of *Helicobacter mustelae* and evidence of a specific association with gastric mucosa, *J. Med. Microbiol.*, 36, 420, 1992.

229. Odenbreit, S. et al., Genetic and functional characterization of the alpAB gene locus for the adhesion of *Helicobacter pylori* to human gastric tissue, *Mol. Microbiol.*, 31, 1537, 1999.

230. Gold, B.D. et al., Comparison of *Helicobacter mustelae* and *Helicobacter pylori* adhesion to eukaryotic cells *in vitro*, *Gastroenterology*, 109, 692, 1995.

231. Gold, B.D., Sherman, P.M., and Lingwood, C.A., *Helicobacter mustelae* and *Helicobacter pylori* bind to common lipid receptors *in vitro*, *Infect. Immun.*, 61, 2632, 1993.

232. Gold, B.D. et al., Surface properties of *Helicobacter mustelae* and ferret gastrointestinal mucosa, *Clin. Investig. Med.*, 19, 92, 1996.

233. Stolte, M. et al., A comparison of *Helicobacter pylori* and *H. heilmannii* gastritis. A matched control study involving 404 patients, *Scand. J. Gastroenterol.*, 32, 28, 1997.

234. Muotiala, A. et al., Low biological activity of *Helicobacter pylori* lipopolysaccharide, *Infect. Immun.*, 60, 1714, 1992.

235. Bliss, C.M. Jr, et al., *Helicobacter pylori* lipopolysaccharide binds to CD14 and stimulates release of interleukin-8, epithelial neutrophil-activating peptide 78, and monocyte chemotactic protein 1 by human monocytes, *Infect. Immun.*, 66, 5357, 1998.

236. Blaser, M.J. and Parsonnet, J., Parasitism by the "slow" bacterium *Helicobacter pylori* leads to altered homeostasis and neoplasia, *J. Clin. Investi.*, 94, 4, 1994.

237. Appelmelk, B J. et al., Potential role of molecular mimicry between *Helicobacter pylori* lipopolysaccharide and host Lewis blood group antigens in autoimmunity, *Infect. Immun.*, 64, 2031, 1996.

238. O. Cróinin, T. et al., Molecular mimicry of ferret gastric epithelial blood group antigen A by *Helicobacter mustelae*, *Gastroenterology*, 11, 690, 1998.

239. Monteiro, M.A. et al., The lipopolysaccharide of *Helicobacter mustelae* type strain ATCC 43772 expresses the monofucosyl A type 1 histo-blood group epitope, *FEMS Microbiol. Lett.*, 154, 103, 1997.

240. Telford, J L. et al., Gene structure of the *Helicobacter pylori* cytotoxin and evidence of its key role in gastric disease, *J. Exp. Med.*, 179, 1653, 1994.

241. Ikeno, T. et al., *Helicobacter pylori* induced chronic active gastritis, intestinal metaplasia and gastric ulcer in Mongolian gerbils, *Am. J. Pathol.*, 154, 951, 1999.

242. Roth, K.A. et al., Cellular Immune response are essential for the development of *Helicobacter felis*-associated gastric pathology, *J. Immunol.*, 163, 1490, 1999.

243. Morgan, D. R., Fox, J.G., and Leunk, R.D., Comparison of isolates of *Helicobacter pylori* and *Helicobacter mustelae*, *J. Clin. Microbiol.*, 29, 395, 1991.

244. Whitmire, J.M. and Merrell, D.S., Successful culture techniques for Helicobacter Species: General culture techniques for *Helicobacter pylori*, Methods and Protocols, in *Methods in Molecular Biology*, Houghton, J. Ed., Humana Press, Springer, New York, 2012, chap 4.

245. Jothimani, D K. et al., A rare case of gastric erosions, *Gut*, 58, 1669, 2009.

246. Dewhirst, F.E., Fox J.G. and On, S.L., Recommended minimal standards for describing new species of the genus *Helicobacter*, *Int. J. Syst. Evol. Microbiol.*, 50, 2231, 2000.

247. Han, S.W. et al., Transport and storage of *Helicobacter pylori* from gastric mucosal biopsies and clinical isolates, *Eur. J. Clin. Microbiol. Infect. Dis.*, 14, 349, 1995.

248. Schrader, J A. et al., A role for culture and diagnosis of *Helicobacter pylori*–related gastric disease, *Am. J. Gastroenterol.*, 88, 1729, 1993.

249. Vakil, N. et al., The cost-effectiveness of diagnostic testing strategies for *Helicobacter pylori*, *Am. J. Gastroenterol.*, 95, 1691, 2000.

250. El-Zimaity, H.M., et al., Histologic assessment of *Helicobacter pylori* status after therapy: comparison of Giemsa, Diff-Quik, and Genta stains, *Mod. Pathol.*, 11, 288, 1998.

251. Genta, R.M. and Graham, D.Y., Comparison of biopsy sites for the histopathologic diagnosis of *Helicobacter pylori*: a topographic study of *H. pylori* density and distribution, *Gastrointest. Endosc.*, 40, 342, 1994.

252. Gonzalez, A. et al., A novel real-time PCR assay for the detection of *Helicobacter pullorum*-like organisms in chicken products, *Int. Microbiol.*, 11, 203, 2008.

253. Moyaert, H. et al., Evaluation of 16S rRNA gene-based PCT assays for genus-level identification of *Helicobacter* species, *J. Clin. Microbiol.*, 46, 1867, 2008.

254. Oyama, K. et al., Identification odf and screening for *Helicobacter cinaedi* infections and carriers via nested PCR, *J. Clin. Microbiol.*, 50, 3893, 2012.

255. Alispahic, M. et al., Species-specific identification and differentiation of *Arcobacter, Helicobacter* and *Campylobacter* by full-spectral matrix-associated laser desorption/ionization time of flight mass spectrometry analysis, *J. Med. Microbiol.*, 59, 295, 2010.

256. Melito, P L. et al., Differentiation of clinical Helicobacter pullorum isolates from related *Helicobacter* and *Campylobacter* species, *Helicobacter*, 5, 142, 2000.

257. Ilina, E.N., Direct matrix-assisted laser desorption-ionization(MALDI) mass-spectrometry bacteria profiling for identifying and characterizing pathogens, *Acta Naturae*, 1, 115, 2009.

258. Welker, M. and Moore, E. R., Applications of whole-cell matrix-assisted laser-desorption/ionization time-of-flight mass spectrometry in systemic microbiology, *Syst. Appl. Microbiol.*, 34, 2, 2011.

259. Owen, R.J. et al., Heterogeneity and subtyping, in *Helicobacter Pylori: Physiology and Genetics*, Mobley, H.L.T., Mendez. G.L. and Hazell. S.L., Eds., ASM Press, Washington, DC, 2001.

260. Hachem, C.Y. et al., Antimicrobial susceptibility testing of *Helicobacter pylori*. Comparison of E-test, broth microdilution, and disk diffusion for ampicillin, clarithromycin, and metronidazole, *Diagn. Microbiol. Infect. Dis.*, 24, 37, 1996.

261. Clinical and Laboratory Standards Institute (CLSI), *Performance Standard for Antimicrobial Susceptibility Testing, M100*, 29th ed., CLSI, Wayne, PA, 2019.

262. Malfertheiner, P. et al., Management of *Helicobacter pylori* infection-the Maastricht V/Florence Consensus Report, *Gut.*, 66(1), 6, 2017.

263. Abadi, A.T.B., Resistance to clarithromycin and gastroenterologist's persistence roles in nomination for *Helicobacter pylori* as high priority pathogen by World Health Organization, *World J. Gastroenterol.*, 23, 6379, 2017.

264. Chey, W.D. et al., ACG clinical guideline: treatment of *Helicobacter pylori* infection, *Am. J. Gastroenterol.*, 112, 212, 2017.

265. O' Morain, N.R. et al., Treatment of *Helicobacter pylori* infection in 2018, *Helicobacter*, 23, e12519, 2019.

266. Myran, L. and Zarbock, S.D., Management of *Helicobacter pylori* infection, *US Pharm.*, 43(4), 27, 2018.

267. De Francesco, V. et al., First-line therapies for *Helicobacter pylori* eradication: a critical reappraisal of updated guidelines, *Ann. Gastroenterol.*, 30, 373, 2017.

268. Savoldi, A. et al., Prevalence of antibiotic resistance in *Helicobacter pylori*: a systematic review and meta-analysis in world health organization regions, *Gastroenterology*, 155, 1372, 2018.

269. Fallone, C A. et al., The Toronto Consensus for the treatment of *Helicobacter pylori* infection in adults, *Gastroenterology.*, 151, 51, 2016.

270. Georgopoulos, S D. et al., Randomized clinical trial comparing ten day concomitant and sequential therapies for *Helicobacter pylori* eradication in a high clarithromycin resistance area, *Eur. J. Intern. Med.*, 32, 84, 2016.

271. Goddard, A F. et al., Healing of duodenal ulcer after eradication of *H. heilmannii*, *Lancet*, 349, 1815, 1997.

272. Neiger, R., Seiler, G., and, Schmassmann, A., Use of a urea breath test to evaluate short term treatments for cats naturally infected with *Helicobacter heilmannii*, *Am. J. Vet. Res.*, 60, 880, 1999.

273. Matsui, H. et al., Evaluation of antibiotic therapy for the eradication of "*Candidatus* Helicobacter heilmannii," *Antimicrob. Agents Chemother.*, 52, 2988, 2008.

274. Dick-Hegedus, E. and Lee, A., Use of a mouse model to examine anti-*Helicobacter pylori* agents, *Scand. J. Gastroenterol.*, 26, 909, 1991.

275. Marini, R P. et al., Ranitidine bismuth citrate and clarithromycin, alone or in combination, for eradication of *Helicobacter mustelae* infection in ferrets, *Am. J. Vet. Res.*, 60, 1280, 1999.

276. Jalava, K. et al., *Helicobacter salomonis* sp. nov., A canine gastric *Helicobacter* sp. related to *Helicobacter felis* and *Helicobacter bizzozeronii*, *Int. J. Syst. Bacteriol.*, 47, 975, 1997.

277. Matsumoto, T. et al., Phylogeny of a novel "*Helicobacter heilmannii*" organism from a Japanese patient with chronic gastritis based on DNA sequence analysis or 16S rRNA and urease genes, *J. Microbiol.*, 47, 201, 2009.

278. Ng, V.L. et al., Successive bacteremias with "*Campylobacter cinaedi*" and "*Campylobacter fennelliae*" in a bisexual male, *J. Clin. Microbiol.*, 25, 2008, 1987.

279. Turvey, S E. et al., Successful approach to treatment of *Helicobacter bilis* infection in X-linked agammaglobulinemia, *J. Clin. Immunol.*, 32, 1404, 2012.

280. Abidi, M Z. et al., *Helicobacter canis* bacteremia in a patient with fever of unknown origin, *J. Clin. Microbiol.*, 51, 1046, 2012.

281. Galan, M.V., Kishan, A.A., and Silverman, A.L., Oral broccoli sprouts for the treatment of *Helicobacter pylori* infection: a preliminary report, *Dig. Dis. Sci.*, 49, 1088, 2004.

282. Ankolekar, C. et al., Inhibitory potential of tea polyphenolics and influence of extraction time against *Helicobacter pylori* and lack of inhibition of beneficial lactic acid bacteria, *J. Medicinal Food*, 14(11), 1321, 2011.

283. Lee, J. et al., Inhibition of pathogenic bacterial adhesion by acidic polysaccharide from green tea (*Cameilia sinensis*), *J. Agric. Food Chem.*, 54(23), 871, 2006.

284. Babarikina, A., Nikolajeva, V. and Babarykind, D., Anti-*Helicobacter* activity of certain food plant extract juices and their composition *in vitro*, *Food Nutrition Sci.*, 2(8), 868, 2011.

285. Sadeghian, S. et al., Bacteriostatic effect of dill, fennel, caraway and cinnamon extracts against *Helicobacter pylori*, *J. Nutrition. Environ. Med.*, 15, 47, 2005.

286. Keenan, J I. et al., Individual and combined effects of food on *Helicobacter pylori* growth, *Phytother. Res.*, 24(8), 1229, 2010.

287. Benoit, A. et al., Diagnosis of Helicobacter pylori in infection on gastric biopsies: standard stain, special stain or immunochemistry? *Ann. Pathol.*, 38(6), 363, 2018.

288. Chey, W.D. and Wong, B.C.Y., American college of gastroenterology guideline on the management of *Helicobacter pylori* infection, *Am. J. Gastroenterol.*, 102, 1808, 2007.

289. Chen, F. et al., Hit or a miss: concordance between histopathologic-endoscopic findings in gastric mucosal biopsies, *Ann. Diagn. Pathol.*, 38, 106, 2019.

290. El-Serag. H.B. et al., Houston consensus conference on testing for *Helicobacter pylori* infection in the United States, *Clin. Gastroenterol. Hepatol.*, 16, 992, 2018.

291. Moayyedi, P M. et al., ACG and CAG clinical guideline: management of dyspepsia, *Am. J. Gastroenterol.*, 112, 988, 2017.

292. Mounsey, A. and Leonard, E.A., Noninvasive diagnostic tests for *Helicobacter pylori* infection, *Am. Fam. Physician*, 100(1), 16, 2019.

293. Makristathis, A. et al., Review: diagnosis of *Helicobacter pylori* infection, *Helicobacter.*, 24 (Suppl. 1), e12641, 2019.

294. Dechant, F X. et al., Accuracy of different rapid urease tests in comparison with histopathology in patients with endoscopic signs of gastritis, *Digestion*, 28, 1, 2019.

295. Bénéjat, L., Real-time PCR for *Helicobacter* diagnosis, the best tools available, *Helicobacter*, 23, e12512, 2018.

296. Shafaie, E. et al., Multiplex serology of *Helicobacter pylori* antigens in detection of current infection and atrophic gastritis- a simple and cost-efficient method, *Microb. Pathog.*, 119, 137, 2018.

297. Wilcox, M.H. et al., Accuracy of serology for the diagnosis of *Helicobacter pylori* infection — a comparison of eight kits, *J. Clin. Pathol.*, 49, 373, 2018.

298. Syrjänen, K. et al., GastroPanel® biomarker assay: the most comprehensive test for *Helicobacter pylori* infection and its clinical sequelae, a critical review, *Anticancer Res.*, 39, 1091, 2019.

299. Graham, D.Y. and Miftahussurur, M., *Helicobacter* urease for diagnosis of *Helicobacter* infection: a minireview, *J Adv. Res.*, 13, 51, 2018.

300. Perets, T.T. et al., Optimization of ^{13}C-urea breath test threshold levels for the detection of *Helicobacter pylori* in a national reference laboratory, *J. Clin. Lab. Anal.*, 33, e22674, 2019.

301. Sevin, E. et al., Codetection of *Helicobacter pylori* and of its resistance to clarithromycin by PCR, *FEMS Microbiol. Lett.*, 165, 369, 1998.

302. Cantorna, M.T. and Balishe, E., Inability of human clinical strains of *Helicobacter pylori* to colonize the alimentary tract of germ free rodents, *Can. J. Microbiol.*, 36, 237, 1990.

303. Fox, J.G. et al., *Helicobacter mustelae* associated gastritis in ferrets: an animal model of *Helicobacter pylori* gastritis in humans, *Gastroenterology*, 99, 352, 1990.

304. Dubois, A. et al., Natural gastric infection with *Helicobacter pylori* in monkeys: a model for spiral bacteria infection in humans, *Gastroenterology*, 106, 1405, 1994.

305. Krakowka, S. et al., Establishment of gastric *Campylobacter pylori* infection in the neonatal gnotobiotic piglet, *Infect. Immun.*, 55, 2789, 1987.

306. Radin, J.M. et. al., *Helicobacter pylori* infection in gnotobiotic beagle dogs, *Infect. Immun.*, 58, 2606, 1990.

307. Matsumoto, S. et al., Induction of ulceration and severe gastritis in Mongolian gerbils by *Helicobacter pylori* infection, *J. Med. Microbiol.*, 46, 391, 1997.

308. Shomer, N H. et al., Experimental *Helicobacter pylori* infection induces antral gastritis and gastric mucosa-associated lymphoid tissue in guinea pigs, *Infect. Immun.*, 65, 4858, 1998.

309. Fox, J.G. et al., *Helicobacter pylori*–induced gastritis in the domestic cat, *Infect. Immun.*, 63, 2674, 1995.

310. Dubois, A. et al., Host specificity of *Helicobacter pylori* strains and host responses in experimentally challenged non-human primates, *Gastroenterology*, 116, 90, 1999.

311. Czinn, S.J., Cai, A., and Nedrud, J.G., Protection of germ-free mice from infection by *Helicobacter felis* after active oral or passive IgA immunization, *Vaccine*, 11, 637, 1993.

312. Foltz, C J. et al., Spontaneous inflammatory bowel disease in multiple mutant mouse lines: association with colonization by *Helicobacter hepaticus*, *Helicobacter*, 3, 69, 1998.

313. Wang, T. et al., Synergistic interaction between hypergastrinemia and *Helicobacter* infection in a mouse model of gastric carcinoma, *Gastroenterology*, 118, 36, 2000.

314. Chin, E.Y. et al., *Helicobacter hepaticus* infection triggers inflammatory bowel disease in T cell receptor αβ-deficient mice, *Comp. Med.*, 50, 586, 2000.

315. Flores, B M. et al., Experimental infection of pig-tailed macaques (*Macaca nemestrina*) with *Campylobacter cinaedi* and *Campylobacter fennelliae*, *Infect. Immun.*, 58, 3947, 1990.

316. Fox, J.G. et al., Chronic proliferative hepatitis in A/JCr mice associated with persistent *Helicobacter hepaticus* infection: a model of *Helicobacter*-induced carcinogenesis, *Infect. Immun.*, 64, 1548, 1996.

317. Haines, D C. et al., Inflammatory large bowel disease in immunodeficient rats naturally and experimentally infected with *Helicobacter bilis*, *Vet. Pathol.*, 35, 202, 1998.

318. Franklin, C L. et al., Enterohepatic lesions in SCID mice infected with *Helicobacter bilis*, *Lab. Anim. Sci.*, 48, 334, 1998.

319. Fox, J.G., Anatomy of the ferret, in *Biology and Diseases of the Ferret*, Fox, J.G., Ed., The Williams and Wilkins Company, Baltimore, Maryland, 1998, p. 19.

320. Fox, J.G. and Lee, A., Gastric *Helicobacter* infection in animals: natural and experimental infections, in *Helicobacter pylori: Biology and Clinical Practice*, Goodwin, C.S. and Worsley, B.W., Eds., CRC Press, Boca Raton, FL, 1993, p. 407.

321. Euler, A.R. et al., Evaluation of two monkey species (*Macaca mulatta* and *Mucaca fascicularis*) as possible models for human *Helicobacter pylori* disease, *J. Clin. Microbiol.*, 28, 2285, 1990.

322. Fox, J.G. et al., Concurrent enteric helminth infection modulates inflammation and gastric immune responses and reduces *Helicobacter*-induced gastric atrophy, *Nat. Med.*, 6, 536, 2000.

323. Stoicov, C., et al., Coinfection modulates inflammatory responses and clinical outcome of *Helicobacter felis*, and *Toxoplasma gondii* infections, *J. Immunol.*, 173, 3329, 2004.

324. Lemke, L B. et al., Concurrent *Helicobacter bilis* infection in C57BL/6 mice attenuates proinflammatory *H. pylori*-induced gastric pathology, *Infect. Immun.*, 77, 2147, 2009.

325. Wrobleuski, L. E., Peek, R.M. Jr. and Wilson, K.T., *Helicobacter pylori* and gastric cancer: factors that modulate disease risk, *Clin. Microbiol. Rev.*, 23, 713, 2010.

326. De Sablet, T. et al., Phylogeographic origin of *Helicobacter pylori* is a determinant of gastric cancer risk, *Gut*, 60(9), 1189, 2010.

327. Ghoshal, U.C., Chaturvedi, R. and Correa, P., The enigma of *Helicobacter pylori* infection and gastric cancer, *Indian J. Gastroenterol.*, 29, 95, 2010.

328. Dick, E. et al., Use of the mouse for the isolation and investigation of stomach-associated, spiral-helical shaped bacteria from man and other animals, *J. Med. Microbiol.*, 29, 55, 1989.

329. Lee, A. et al., A small animal model of human *Helicobacter pylori* active chronic gastritis, *Gastroenterology*, 99, 1315, 1990.

330. Taylor, N.S. and Fox J.G., Animal models of *Helicobacter*-induced diseases: methods to successfully infect the mouse, *Methods Mol. Biol.*, 921, 131, 2012.

331. Roth, K.A. et al., Cellular immune responses are essential for the development of *Helicobacter felis*-associated gastric pathology, *J. Immunol.*, 163, 1490, 1999.

332. Gasbarrini, A. et al., *Helicobacter pylori* and extragastric diseases-other helicobacters, *Helicobacter* 8 (Suppl. 1), 68, 2003.

333. Fox, J.G. et al., Host and microbial constituents influence *Helicobacter pylori*-induced cancer in a murine model of hypergastrinemia, *Gastroenterol*, 124, 1879, 2003.

334. Fox, J.G. et al., *Helicobacter pylori*-associated gastric cancer in INS-GAS mice is gender specific, *Cancer Res.*, 63, 942, 2003.

335. Stenstrom, B. et al., Swedish moist snuff accelerates gastric cancer development in *Helicobacter pylori*-infected wild-type and gastrin transgenic mice, *Carcinogenesis*, 28(9), 2041, 2007.

336. Ohtani, M. et al., Protective role of 17{beta}-estradiol against the development of *Helicobacter pylori* - induced gastric cancer in INS-GAS mice, *Carcinogenesis*, 28, 2597, 2007.

337. Whary, M T. et al., Intestinal helminthiasis in Columbian children promotes a Th2 response to *Helicobacter pylori*: possible implications for gastric carcinogenesis, *Cancer Epidemiol. Biomarkers Prev.*, 14, 1464, 2005.

338. Abou Holw, S.A. et al., Impact of coinfection with *Schistosoma mansoni* on *Helicobacter pylori* induced disease, *J. Egypt. Soc. Parasitol.*, 38, 73, 2008.

339. Rao, V.P. et al., Proinflammatory CD4+CD45RBhi lymphocytes promote mammary and intestinal carcinogenesis in ApcMin/+ mice, *Cancer Res.*, 66, 57, 2006.

340. Fox, J.G. et al., Gut microbes define liver cancer risk in mice exposed to chemical and viral transgenic hepatocarcinogens, *Gut*, 59, 88, 2010.

341. Brenner, H., Rothenbacher, D. and Arndt, V., Epidemiology of stomach cancer, *Methods Mol. Biol.*, 472, 467, 2009.

31

The Genus Legionella

Ashley M. Joseph and Stephanie R. Shames

CONTENTS

The Genus *Legionella*

Members of the genus *Legionella* are Gram-negative, aerobic bacilli that belong to the gamma-subgroup of proteobacteria. *Legionella* are facultative, intracellular bacteria that are ubiquitously found in freshwater ponds and streams where they colonize phagocytic protozoa. The relevance of these organisms to human health was first recognized following the eponymous 1976 outbreak at a meeting of the American Legion in Philadelphia. During the weeks following the convention, 221 attendees fell ill with unexplained, pneumonia-like symptoms; 34 of these patients later died (Garrett 1994). Legionnaires' disease was ultimately linked to infection by a bacterium that was named *Legionella pneumophila*. An examination of the air conditioning system at the convention site revealed high levels of *L. pneumophila*-laden protozoa, indicating that exposure to these bacteria via inhalation of contaminated aerosols could result in severe respiratory disease.

Subsequent outbreaks of *Legionella* infection have been traced to several types of man-made water reservoirs, including cooling towers, industrial air conditioners, hot water systems and fountains. If improperly maintained, biofilms may develop on the accumulated sediment in these structures, creating an ecosystem that supports both protozoa and *Legionella* growth. Members of the genus *Legionella* are found in numerous natural ecosystems proximal to human populations, but their ability to colonize human hosts depends upon inhalation of aerosolized water-borne bacteria. There has been one recorded case of *Legionella* human-to-human transmission stemming from the 2014 Vila Franca de Xira, Portugal outbreak that resulted in 400 cases and 14 deaths, one of the largest out breaks to date (Borges *et al.* 2016).

The diseases resulting from exposure to *Legionella*, collectively termed legionellosis, include Legionnaires' disease and Pontiac fever. Legionnaires' disease is a severe infection of the respiratory tract that is fatal in approximately 10% of cases (CDC 2018). Pontiac fever is a milder illness, with flu-like symptoms that may resolve within a few hours to a few days of onset. Whereas Legionnaires' disease is thought to result from an infection where bacteria are actively replicating in the lung, Pontiac fever may result from an acute exposure to high levels of bacteria without bacterial multiplication. All forms of legionellosis can be successfully treated with a range of antibiotics, including macrolides (erythromycin, azithromycin, clarithromycin), tetracycline (doxycycline), or broad-spectrum fluoroquinolones (Amsden 2005).

Legionella infection is one of the most common causes of community-acquired pneumonia (CAP) and is second only to bacteremic pneumococcal pneumonia in causing severe CAP that requires treatment in intensive care. *Legionella* has also emerged as a common nosocomial acquired pneumonia, a phenomenon of particular concern since mortality resulting from legionellosis is highest in elderly and immunocompromised patients (Amsden 2005). In the United States between 8,000 and 18,000 people are diagnosed with legionellosis each year; however, this number is estimated to be higher as many cases go undiagnosed (Amsden 2005).

Thus far, 65 *Legionella* species, consisting of more than 80 strains, have been identified (Gomez-Valero *et al.* 2019).

DOI: 10.1201/9781003099277-33

While many of these species can occasionally cause disease in humans, analysis of culture-confirmed cases of legionellosis indicates that *L. pneumophila* is responsible for greater than 90% of disease, with the exception of Australia and New Zealand, where 30–55% of cases are caused by *L. longbeachae* (Mercante and Winchell 2015). Other strains periodically isolated from immunocompromised patients include *L. bozemanii*, *L. micdadei*, *L. feeleii*, *L. dumoffii*, *L. wadsworthii*, and *L. anisa* (Yu *et al.* 2002). Of the 15 serogroups identified for *L. pneumophila*, the vast majority of disease results from infection by serogroup 1, with serogroups 4 and 6 periodically found in nosocomial outbreaks (Amsden 2005). Due to the marked clinical prevalence of *L. pneumophila* serogroup 1, the majority of scientific research has focused on these, commonly Philadelphia 1 and Paris strains.

Despite this historical focus on serogroup 1, the emergence of high-throughput sequencing technologies promises to shed new light on *Legionella* diversity and the evolution of disease. Early microarray-based comparisons between different genetic backgrounds of *L. pneumophila* have suggested that serogroup 1 clinical prevalence might derive from a unique lipopolysaccharide (Cazalet *et al.* 2008). Whole genome comparisons between a serogroup 6 strain *L. pneumophila* str. Thunder Bay and *L. pneumophila* serogroup 1 identified genetic differences within the serogroup 6 O-antigen biosynthetic cluster (Khan *et al.* 2013). Compared to serogroup 1 strains, Thunder Bay displayed reduced O-antigen decoration, increased susceptibility to complement-mediated killing, and reduced dissemination in a mouse model of disease. Whole genome sequencing revealed that the strain responsible for the first human-to-human transmission, *L. pneumophila* subspecies Fraseri, contained a rare 65 kb pathogenicity island carrying enzymes that mediate oxidative stress, an efflux pump-like homolog, as well as the unique lipopolysaccharides present in clinical strains of serogroup 1 (Borges *et al.* 2016). Additionally, some strains have developed the ability to infect human hosts through the acquirement of a diverse effector protein repertoire that will be discussed later.

L. pneumophila Growth in Nature and in the Laboratory

In freshwater ecosystems, *L. pneumophila* colonizes biofilms where they are phagocytized and replicate within free-living protozoa including *Acanthamoeba*, *Hartmonella*, and the model amoeba *Dictyostelium* (Moffat and Tompkins 1992; Fields *et al.* 1993; Solomon *et al.* 2000). To study the basis of human infection in the laboratory, *L. pneumophila* can be cultured within a range of mammalian cell lines. Primary macrophages derived from the bone marrow of laboratory mice, such as A/J and C57BL/6, have been used in numerous studies (Yamamoto, Klein, and Friedman 1991; Saito *et al.* 2001), in addition to macrophage-like cell lines such as THP1, U937, RAW, and J774 (King *et al.* 1991; Husmann and Johnson 1992; Cirillo, Falkow, and Tompkins 1994; Diez *et al.* 2000). *L. pneumophila* can also replicate within cells that are not inherently phagocytic, provided they have a means of internalization. For example, *L. pneumophila* that have been opsonized with a specific antibody can be internalized by Chinese hamster ovary (CHO) cells expressing the FcγRII receptor. Cell biological assays indicate that the intracellular life cycle

events that unfold in this cell line parallel those seen in protozoan and phagocyte hosts (Kagan *et al.* 2004).

The nutrients required for *L. pneumophila* to establish and maintain its intracellular niche are yet to be fully described. Microarray analysis of differential gene expression in *L. pneumophila* growing extracellularly in broth culture or intracellularly in *Acanthamoeba castellanii* indicate that intracellular growth involves scavenging for amino acids by upregulation of factors governing amino acid uptake and catabolism. In addition, *L. pneumophila* may also acquire carbohydrate from its host, as intracellular growth also leads to an upregulation of the gene cluster encoding the Entner-Douderoff pathway, a putative glucokinase and a sugar transporter (Brüggemann, Cazalet, and Buchrieser 2006).

The growth medium commonly used for extracellular culture of *Legionella* spp. consists of yeast extract and ACES buffer, with potassium hydroxide added to adjust the pH to 6.9 (Feeley *et al.* 1979). This medium is then supplemented with cysteine and iron, and for solid medium, activated charcoal is added to chelate an unknown factor in agar that is toxic to *Legionella*. Bacterial growth is optimal at 37°C, but lower temperatures (20–25°C) are often required for studies involving protozoan host species. Streptomycin, chloramphenicol, gentamicin, and kanamycin are frequently used for antibiotic selection; however, *L. pneumophila* is resistant to even high doses of ampicillin.

Biphasic Life Cycle of *L. pneumophila*

Studies of *L. pneumophila* grown in broth or in protozoa have shown that this bacterium has (at least) two different stages in its life cycle. While growing exponentially in broth culture, and within the first several hours following host cell entry, *L. pneumophila* adopts a nonmotile, nonflagellated state, which is competent to replicate to high levels, but shows a low level of infectivity of new host cells. When cultured in broth to high density in post-exponential phase, *L. pneumophila* become motile due to the production of flagella and these bacteria are highly infectious. This phase roughly correlates with late time points of *in vivo* infection of host cells when *L. pneumophila* have completed several rounds of replication and exist within densely populated replicative compartments. With nutrient concentrations and host cell viability on the decline, these virulent, motile bacteria are primed to carry out new rounds of infection in neighboring host cells. In addition, the transition of *L. pneumophila* from its replicative form to infectious form is associated with increased resistance to osmotic and heat stress, sensitivity to sodium chloride, increased pigmentation and a reduction in bacterial cell size (Byrne and Swanson 1998).

These broadly defined stages of *L. pneumophila*'s life cycle have been termed the replicative and transmissive phases, and the regulatory mechanisms underlying this phase switch are beginning to be described. A screen for mutants that expressed low levels of flagellin during the during post-exponential growth conditions revealed that a two-component regulatory system, LetA/LetS, the sigma factor FliA, and the stationary phase sigma factor RpoS function in concert to regulate *L. pneumophila*'s transition from its replicative to transmissive form (Hammer, Tateda, and Swanson 2002; Lynch *et al.* 2003) . The LetA/LetS two-component system is thought to respond to ppGpp, an alarmone produced as part of a stringent response under conditions of nutrient starvation such as low amino acid levels (Hammer and Swanson 1999; Jain, Kumar, and Chatterji 2006). In turn, LetA/LetS induces transition to the transmissive phase by relieving repression by CsrA (Molofsky and Swanson 2003). Csr regulatory systems are found in numerous bacterial species and are responsible for post-transcriptional regulation of gene expression on a global level (Romeo 1998; Majdalani, Vanderpool, and Gottesman 2008). However, the requirement for LetA/LetS signaling in *L. pneumophila* virulence appears to be host cell type-dependent. *LetA* mutants that showed a replication defect in the protozoan host *Acanthamoeba castellanii* were able to grow efficiently in A/J mouse bone marrow-derived macrophages (Hammer, Tateda, and Swanson 2002; Lynch *et al.* 2003).

The role of the sigma factor RpoS in phase switching is no yet fully understood. Due to the requirement for this transcription factor for intracellular survival and replication, RpoS function has been placed amidst a cascade of replicative phase regulatory events (McNealy *et al.* 2005; Abu-Zant *et al.* 2006). However, seemingly contradictory data reveal an upregulation of RpoS expression during the stationary phase as *L. pneumophila* prepares to transform into the transmissive phase (Hales and Shuman 1999; Bachman and Swanson 2001; Gal-Mor and Segal 2003; Molofsky and Swanson 2003).

Finally, the successful transformation of replicating *L. pneumophila* to virulent, motile, bacteria requires the signaling cascade of the flagella regulon. Upon entering the transmissive phase, the alternative sigma factor RpoN and the transcriptional activator FleQ positively regulate expression of *fleN* (FleN may be an antiactivator of FleQ), *fliM*, *fleSR*, and numerous other genes whose products are required for flagellum biosynthesis (Dasgupta and Ramphal 2001; Jacobi, Schade, and Heuner 2004; Brüggemann *et al.* 2006). RpoN and the FleS/FleR two-component system act in concert to trigger expression of the alternative sigma factor FliA, which controls the last stages of flagellum biosynthesis, including production of the FlaA flagellin subunit (Heuner and Steinert 2003; Brüggemann *et al.* 2006).

Intracellular Life Cycle of *L. pneumophila*

Bacterial Entry

Since *L. pneumophila* was identified as the causative agent of the fatal pneumonia outbreak in 1976, the intracellular life cycle events of this bacterium have been extensively dissected. Early work by Marcus Horwitz and colleagues provided an essential descriptive model of *L. pneumophila*'s intracellular trafficking within both protozoan and macrophage hosts (Horwitz 1983a; 1983b). Subsequent studies have revealed that *L. pneumophila* is able to evade the degradative endocytic compartments of the host cell (Berger, Merriam and Isberg 1994) and establish a replicative niche by remodeling its phagosome into a ribosome-studded vacuole that resembles the host cell endoplasmic reticulum (ER) (Tilney *et al.* 2001; Kagan and Roy 2002; Robinson and Roy 2006) (Figure 31.1).

L. pneumophila enter host cells by phagocytosis, engaging host cell surface receptors that appear to vary with host cell type. In mammalian monocytes, bacterial entry can be partially inhibited

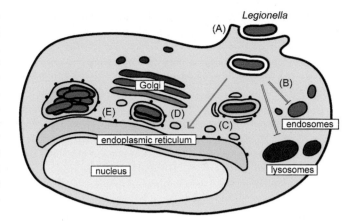

FIGURE 31.1 Intracellular life cycle of *Legionella pneumophila*. (A) *L. pneumophila* engage protozoan or mammalian host cell surface receptors and are internalized by phagocytosis. (B) The bacterium inhibits transport of its phagosome down the endocytic pathway to degradative lysosomes and (C) associates with membrane vesicles and mitochondria within minutes of internalization. (D) *L. pneumophila* completes remodeling of its vacuole and begins to replicate in a compartment that is morphologically indistinguishable from the ER. (E) Bacterial replication proceeds for several hours until the LCV is lysed and *L. pneumophila* exit host cells to initiate new rounds of infection. (Figure adapted from Roy and Tilney [2002].)

by the presence of antibodies specific for the C1 and C3 complement receptors on the host cell surface (Payne and Horwitz 1987). It has also been shown that opsonization of *L. pneumophila* with a bacteria-specific antibody allows for Fc-receptor mediated phagocytosis (Horwitz and Silverstein 1981) though the absence of *L. pneumophila*-specific IgG in the lungs of naïve human hosts renders this pathway of uncertain clinical relevance. In addition, levels of complement in the lung are normally low (Reynolds and Newball 1974), suggesting that *L. pneumophila* may also engage as yet unidentified surface receptors to induce uptake into mammalian host cells.

L. pneumophila's natural hosts—freshwater protozoa—lack both Fc and complement receptors. The bacteria infiltrate various protozoan host through predatory grazing of biofilms that *Legionella* has colonized (Boamah *et al.* 2017). The fate of the bacteria once it enters the cell is determined by the host infected. Members of the Cercozoa and Amoebozoa phylums such as *Cashia limocoides*, are efficient at digesting *Legionella* pathogens, while many protozoa such as the aforementioned *Acanthamoeba*, provide useful replication reservoirs (Boamah *et al.* 2017). Following adherence to host cells, internalization of *L. pneumophila* proceeds via conventional zippering phagocytosis that can be potently blocked by depolymerization of the actin cytoskeleton with Cytochalasin D (Elliott and Winn 1986). There is also evidence that *L. pneumophila* can induce a novel form of entry termed "coiling phagocytosis," in which the bacterium is enveloped in multiple layers of host cell membrane during internalization (Horwitz 1984). This phenomenon has been observed when professional phagocytes are infected with relatively high doses of live or heat-killed *L. pneumophila* and can be disrupted if the bacteria are pretreated with anti-*L. pneumophila* antibody. This suggests that high concentrations of a stable bacterial surface component may be required to perturb phagocyte membrane and induce coiling, a model supported by the observation that only certain *L. pneumophila* serotypes induce coiling phagocytosis (Rechnitzer and Blom 1989).

The role of coiling phagocytosis during *L. pneumophila*'s life cycle is unclear. Bacterial viability has been shown to be required for all *L. pneumophila* virulence factors identified to date, including Type IV secretion (discussed below) and changes in morphology, motility, and gene expression associated with *L. pneumophila*'s transition from a replicating, noninfectious state to a stationary and transmissive one. Since heat-killed bacteria can induce coiling phagocytosis but are quickly delivered to lysosomes and degraded (Horwitz 1984), it is unlikely that this novel mode of entry plays a significant role in the later stages of *L. pneumophila*'s unique pattern of intracellular trafficking.

Intracellular Trafficking and Replicative Vacuole Formation

Once inside its host cell, *L. pneumophila* must quickly execute two essential tasks to successfully establish its intracellular niche. The bacterium must avoid the fate of most cargo internalized by phagocytic cells by preventing its rapid delivery to lysosomes, the degradative terminal compartment of the endocytic pathway. Concurrently, *L. pneumophila* must establish a subcellular niche that supports bacterial replication.

Early experiments by Horwitz and colleagues showed that phagosomes containing live *L. pneumophila* acquire lysosomal markers at greatly reduced levels compared to phagosomes containing fixed *L. pneumophila* or *E. coli* and this separation from the endocytic pathway is maintained for several hours following infection (Horwitz 1983a; 1983b). Subsequently, various lines of experimentation have generated conflicting data regarding late-stage fusion events of *L. pneumophila*'s compartment with acidic lysosomes and the importance of acidic pH for efficient bacterial growth.

Evidence supporting a need for acidification of the *L. pneumophila*-containing vacuole (LCV) is provided by the reduced numbers of bacteria recovered from host cells treated with the vacuolar ATPase inhibitor bafilomycin-A (BFA), which prevents acidification of endocytic compartments (Sturgill-Koszycki and Swanson 2000). However, BFA treatment has been shown to dysregulate a number of cellular pathways (Clague *et al.* 1994; Reaves and Banting 1994; van, Geuze and Stoorvogel 1997; Yamamoto *et al.* 1998), calling into question the ability to directly interpret these data as proof of *L. pneumophila*'s requirement for an acidified compartment. In addition, bacterial growth assays conducted in monocyte cell lines with differing capacities to acidify their endocytic compartments yielded indistinguishable levels of bacterial growth (Wieland, Goetz, and Neumeister 2004). Therefore, it is likely that the only prudent action that *L. pneumophila* may take relative to the endocytic pathway is to avoid it.

With the degradative effects of the lysosome held at bay, *L. pneumophila* conducts an extensive remodeling of its plasma membrane-derived phagosome. Within 15 minutes of internalization, the LCV is surrounded by numerous smooth vesicles and mitochondria (Horwitz 1983a). As the infection proceeds, these vesicles flatten and fuse with the limiting membrane of the LCV, which also becomes studded with ribosomes (Horwitz 1983a; Tilney *et al.* 2001; Robinson and Roy 2006). Localization of various ER proteins to this compartment

shortly after infection, including calnexin, BiP, and KDEL epitope-containing proteins, indicate that these vesicles are derived from the ER (Swanson and Isberg 1995; Kagan and Roy 2002; Derre and Isberg 2004). Conversely, LCVs do not accumulate the Golgi protein Giantin (Kagan and Roy 2002), suggesting that *L. pneumophila* gains access to secretory membrane traffic directly, rather than circuitously through the Golgi as is seen with the trafficking of the Cholera and Shiga bacterial toxins (Sandvig and van Deurs 2002).

Disruption of early secretory traffic using chemical inhibitors has allowed for a further narrowing of the window in which *L. pneumophila* intercepts the membrane traffic used to remodel its phagosome. Brefeldin A inhibits ADP ribosylation factor 1 (ARF1)-dependent vesicle budding from the ER. The microtubule-based transport of these vesicles from the ER through the intermediate compartment (ERGIC) to the Golgi can be blocked with the microtubule disrupting agent nocodazole. The LCV is able to recruit the ERGIC protein p58 in the presence of nocodazole but not when cells are treated with Brefeldin A (Kagan and Roy 2002). Thus, following internalization, *L. pneumophila* remodels its phagosome into a ribosome-studded, ER-like vacuole by intercepting and fusing with ER vesicles before they are transported through the ERGIC to the Golgi.

The previously proposed model to describe the generation of *L. pneumophila*'s unique vacuole invoked the involvement of host autophagic machinery to convert the LCV into an autophagosomal compartment. Common features shared by autophagosomes and LCVs include a double membrane and the presence of ER proteins such as calnexin and BiP (Horwitz 1983a; Swanson and Isberg 1995; Tilney *et al.* 2001). The induction of autophagy in macrophages by either amino acid starvation or rapamycin treatment results in both the recruitment of autophagy proteins Atg7 and Atg8 to LCVs and an increase in colony-forming units recovered from autophagy-induced cells compared to untreated conditions (Amer and Swanson 2005). However, the discovery of the *Legionella* translocated protein, RavZ has complicated this model. RavZ directly inactivates Atg8 proteins of autophagosomal membranes through hydrolyzation of C-terminal amide bonds, preventing conjugation to Atg7 and Atg3, and similarly restricting the recruitment of ubiquitin, thus inhibiting autophagy (Choy *et al.* 2012; Kubori *et al.* 2017).

It has also been shown that *L. pneumophila* grown in wild type and autophagy-deficient *Dictyostelium discoideum* were similarly able to form normal vacuoles and replicate to comparable levels (Otto *et al.* 2004). Additionally, the *L. pneumophila* effector *Lp*SPL targets the host cell's sphingolipid metabolism through lyase activity. Sphingolipids are eukaryotic bioactive molecules that are central regulators of several intracellular processes, including autophagy and cell death. By preventing the accumulation of the sphingolipids, *Lp*SPL plays a role in *L. pneumophila* evasion of host detection (Rolando *et al.* 2016). A third effector protein, LegA9, has been shown to promote clearance of *L. pneumophila* infections in BMDMs and wild type mice. LCVs of mutants lacking LegA9 have decreased ubiquitin labeling, promoting the evasion of autophagy, although the exact function of LegA9 is currently unknown (Khweek *et al.* 2013). These findings taken together contradict the notion of autophagy's role in LCV formation and suggest that evasion of host defenses plays a larger part in *Legionella* intracellular replication.

Egress

Little is known about the final phase of *L. pneumophila*'s intracellular existence. At 12 to 24 hours after infection, vacuoles containing multiplying *L. pneumophila* have grown substantially, housing dozens of bacteria. Bacterial cells now in the transmissive phase of the biphasic life cycle egress through the formation of pores in the LCV membrane. The *icmT* gene appears to play a role in pore-mediated egress, as mutants defective in host cell release all feature identical point mutations on the 3'- end following a stretch of poly(T). Electron microscopy studies have revealed what appear to be cytosolic *L. pneumophila* at late stages of infection indicating that there may be a delay between rupture of the LCV and egress of bacteria from the host cell to go on to further rounds of infection (Molmeret *et al.* 2004).

Identification of the Dot/Icm Secretion System as a Key Component of *L. pneumophila*'s Intracellular Life Cycle

To determine the mechanisms underlying *L. pneumophila*'s unique intracellular life cycle, numerous screens have been conducted to identify the bacterial genes required for intracellular growth. Various mutagenesis strategies were used to generate *L. pneumophila* strains unable to replicate within and lyse host cells, to prevent fusion of bacteria-containing phagosomes with lysosomes, or to recruit host cell vesicles and organelles.

One such set of mutants was generated by repeated passage of *L. pneumophila* on suboptimal laboratory media to allow for the loss of genes that are not essential for extracellular replication (Marra *et al.* 1992). One mutant strain isolated by this method, named 25D, was unable to grow inside a macrophage-like cell line, to avoid delivery to lysosomes, to recruit host vesicle traffic, and to kill host cells—a read-out for host cell lysis. By transforming this mutant strain with a cosmid library of *L. pneumophila* genomic DNA, a locus of genes capable of complimenting all of these defects were defined and subsequently named the *intracellular multiplication* (*icm*) locus.

Another strategy seeking a similar category of intracellular growth-deficient mutants was devised using *L. pneumophila*

thymine auxotrophs that are only viable if they are not replicating when cultured in the absence of thymine (Berger and Isberg 1993). Mutagenesis of this *thy* strain followed by several rounds of infection and bacterial recovery produced a pool of mutants enriched for strains that were deficient when tested in intracellular growth and phagosome trafficking assays. Complementation by transformation of these mutants with a *L. pneumophila* genomic library also identified a locus of relevant genes, which was subsequently called the *defect in organelle trafficking* (*dot*) locus.

The *icm* and *dot* loci defined by these screening techniques encompass similar though not identical sets of *L. pneumophila* genes (Figure 31.2). Sequencing of nearby open reading frames and further analysis of salt-sensitive mutants produced a list of 27 *dot/icm* genes that are essential for *L. pneumophila*'s intracellular growth and trafficking (Andrews, Vogel and Isberg 1998; Segal and Suman 1998). Homology found within *dotB*, *dotG*, *dotL*, and *dotM* to known components of *Agrobacterium tumefaciens* conjugation machinery led to the discovery that the *dot/icm* genes collectively encode the components of a type IVb secretion system (Vogel *et al.* 1998). Members of this family of specialized secretion systems are ancestrally related to bacterial conjugation systems and are found across a wide range of pathogenic bacterial species (Christie 2001). Indeed, the *L. pneumophila dot/icm* system can support conjugative plasmid DNA transfer (Vogel *et al.* 1998), though it remains unclear whether this activity serves a present role in the life-cycle of *L. pneumophila* or is merely a vestige of the Dot/Icm apparatus' evolutionary past. *Helicobacter pylori*, the causative agent of stomach ulcers, uses a type IV secretion system to deliver the effector protein CagA into mammalian cells (Odenbreit *et al.* 2000) via its type IV secretion system, the plant pathogen *Agrobacterium tumefaciens* injects both protein effectors as well as oncogenic T-DNA to produce the tumorous growth characteristic of infected plant tissues (Christie 2001).

Further studies have found that in addition to the Dot/Icm apparatus, *L. pneumophila* encodes a veritable laundry list of other secretion systems, including a second type IV system called *lvh*, a type II system called *lsp*, a type I system called *lss*, and numerous secretion systems with homology to the *tra* conjugation machinery. Analysis of the Paris strain

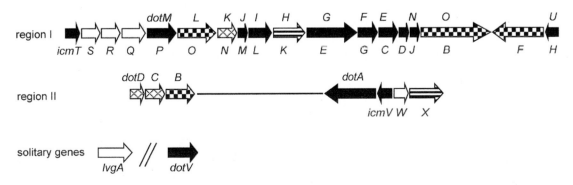

FIGURE 31.2 The *Legionella pneumophila dot/icm* locus encodes a type IVb secretion system. The 27 *dot/icm* genes are located in two regions of the *L. pneumophila* chromosome. Subcellular localization predictions for each gene product are indicated as follows: inner membrane (black), outer membrane (hatched), periplasmic (striped) or soluble (white). Proteins with predicted ATP/GTP binding sites are checkered. *Dot/icm* genes bearing homology to components of the *Agrobacterium tumefaciens* Ti plasmid Tra/Trb conjugation machinery include: *dotB* (*trbB*), *dotG* (*trbI*), *dotL* (*trbC*), and *dotM* (*trbA*). (Figure adapted from Segal and Suman (1998) and Nagai and Kubori [2011].)

of *L. pneumophila* has also revealed a type V secretion that serves as an autotransporter (Cazalet *et al.* 2004; Brüggemann *et al.* 2006). The *lsp* type II system has been shown to be important for efficient growth of *L. pneumophila* in protozoan and mammalian cells and is thought to control secretion of factors with protease, RNaseA and lipase activities, among others (Rossier, Starkenburg, and Cianciotto 2003). The role of the other secretion systems in *L. pneumophila*'s intracellular growth has yet to be studied in depth.

Identification of Dot/Icm Effector Substrates

Identification of a Large Arsenal

Proper functioning of the Dot/Icm secretion apparatus has been shown to be absolutely required for *L. pneumophila*'s intracellular replication. Therefore, it follows that the identification of the substrates of this transporter is essential for the understanding of *L. pneumophila* pathogenesis. However, the screens that successfully identified the Dot/Icm transporter as the predominant tool required for *L. pneumophila*'s intracellular growth failed to isolate mutants lacking individual Dot/Icm substrates. As an answer to this challenge, numerous novel strategies to hunt for Dot/Icm secreted proteins have been developed.

As described above, these original mutagenesis screens employed both gross intracellular growth and phagosome mistrafficking as phenotypic read-outs. Though the mutants isolated in this manner only contained lesions within individual *dot/icm* genes, it is probable that the absence of any one component of the apparatus results in a Dot/Icm complex that is misassembled, improperly functioning, or both. Therefore, the strong intracellular growth and trafficking defects displayed by these mutants most likely reflect their inability to secrete most if not all of *L. pneumophila*'s effector proteins into the host cell cytosol. Current research utilizes mutants deficient in the *dotA* component of the secretion system as a control as theses mutants lack the ability to secrete substrates into the host cell. Although the specific function of the DotA protein remains elusive, studies have found that the membrane protein is itself translocated by the Dot/Icm system (Nagai and Roy 2001).

Given the varied and complex aspects of its intracellular life cycle, it is probable that *L. pneumophila* requires the coordinate function of numerous effector proteins to establish its replicative vacuole. Within this repertoire, there may be proteins of redundant or overlapping function and proteins that mimic host cell activities. A mutation in any one of these genes may not cripple the bacterium sufficiently for it to be selected using such severe defects as overall intracellular growth or trafficking as a marker. Therefore, novel strategies to identify Dot/Icm substrates have allowed for the isolation of *L. pneumophila* genes whose products are responsible for more subtle phenotypes such as acquisition of a single cellular marker or minor perturbations of host cell pathways. In addition, strategies that permit the analysis of candidate genes in isolation have provided a means by which to sidestep the complications posed by functionally redundant effector proteins.

This more targeted approach was used to identify the *L. pneumophila* protein RalF as a secreted substrate of the Dot/Icm secretion apparatus (Nagai *et al.* 2002). This protein was identified using an *in silico* screen of the *L. pneumophila* genome based on the presence of a Sec7 domain, a catalytic motif conserved throughout the family of guanine nucleotide exchange factors for eukaryotic ARF proteins. Dot/Icm-mediated translocation of RalF into host cells was found to be required for the recruitment of detectable levels of host ARF1 to the *L. pneumophila* phagosome. However, when tested for growth in multiple host cell types, *L. pneumophila* lacking RalF showed no defect in their ability to grow compared to wild type bacteria, supporting the hypothesis that single effector mutants are unlikely to elicit gross defects in such assays.

Since the discovery of RalF, several additional strategies have been used to identify a growing list of Dot/Icm substrates. Bacterial two-hybrid screening using IcmG/DotF (predicted to be an inner membrane protein of the Dot/Icm complex that interacts with the carboxy terminus of RalF) was conducted and candidates were tested for translocation using the Cre/loxP system. These experiments yielded a large number of substrate of Dot/Icm transporter (Sid) proteins (Luo and Isberg 2004). *Saccharomyces cerevisiae* has been used in two different studies to identify *L. pneumophila* effectors based on their presumed ability to affect (and interfere with) host membrane transport when present in high levels. The VPS inhibitor proteins (Vip) were isolated using a well-established assay that measures perturbation of membrane transport via missorting of vacuolar cargo to the cell surface (Shohdy *et al.* 2005). Yeast lethal factors (Ylf) A and B are two Dot/Icm substrates that were identified based upon their ability to inhibit yeast growth when conditionally overexpressed (Campodonico, Chesnel, and Roy 2005). Yeast two-hybrid analysis using IcmW (a soluble protein predicted to mediate association of translocated effectors with the Dot/Icm apparatus) as bait isolated the IcmW interacting proteins (Wip) that were subsequently shown to be translocated into host cells in a Dot/Icm-dependent manner (Ninio *et al.* 2005).

In recent years, direct screens for translocation efficiency have been performed on large libraries of *L. pneumophila* proteins fused to reporter sequences, dramatically increasing the number of known Dot/Icm effector substrates. In one instance, *Legionella*-specific or eukaryotic-like genes were fused to a translocation-deficient N-terminal fragment of the substrate, sidC. A library of *L. pneumophila* strains expressing these fusions was then used to infect host cells and localization of each fusion protein to the *Legionella*-containing vacuole was measured by anti-sidC immunofluorescence (Huang *et al.* 2011). In another screen, a beta-lactamase reporter system was used to systematically measure the translocation of approximately 800 open reading frames during infection (Zhu *et al.* 2011). To date, more than 300 Dot/Icm translocated substrates have been identified in *L. pneumophila* str. Philadelphia-1, presenting researchers with the challenge of assigning function to each protein or family of proteins in the context of *L. pneumophila*'s intracellular life cycle.

The identification of such a large arsenal of Dot/Icm substrates in *L. pneumophila* str. Philadelphia-1 has also led to a greater understanding of the translocation signal required by the Dot/Icm system. Translocation of Dot/Icm substrates has been shown to be facilitated by glutamate-rich amino acid sequence located near the C-terminus of substrate proteins (Burstein *et al.* 2009;

Huang *et al.* 2011). This signal, together with other common features of translocated substrates, has been used to predict translocated substrates *in silico* in other species of *Legionella* and the closely related species, *Coxiella burnetii* (Lifshitz *et al.* 2013). Given that most large-scale screens for translocation efficiency to-date have been performed in only one strain of *L. pneumophila* (Philadelphia-1), the ability to predict effector proteins from primary sequence information will likely prove critical as access to affordable sequencing technologies continues to increase the number of genomes available for analysis.

Collectively Essential, Individually Dispensable

The extensive collection of Dot/Icm substrates in *L. pneumophila* and the essentiality of the translocation apparatus itself for intracellular replication speak to the importance of this arsenal to virulence. Nevertheless, the vast majority of translocated substrates are themselves are dispensable for intracellular replication in the standard host cells (mammalian and amoebal) used in the lab to model virulence. Current models suggest that the dispensability of individual translocated substrates derives from functional redundancy between these genes. The most striking example of substrate redundancy comes from a study in which the *L. pneumophila* genome was minimized through directed deletion of clustered nonessential genes (O'Connor *et al.* 2011). In this study, the simultaneous deletion of 31% of the known Dot/Icm substrates in one strain had little impact on its ability to replicate within primary mouse macrophages. Other multilocus deletions, spanning as few as 4% of the known Dot/Icm substrates, did display defects in one or more amoebal hosts. Some researchers have utilized transposon insertion sequencing (InSeq) to generate mutant libraries allowing for the screening of loss-of-function mutations that lead to enhanced or detrimental to *Legionella* intracellular growth in various hosts (Shames *et al.* 2017). The effector protein, LegC4, is one example of an effector protein that can promote replication in one host, while being detrimental in another. In the presence of a host immune system, LegC4 hinders intracellular replication by increasing cytokine-mediated host defenses (Ngwaga *et al.* 2019); however, LegC4 is required for wild-type replication of *L. pneumophila* within the natural host *Acanthamoeba castellanii* (Shames *et al.* 2017). This suggests that substrate function may be less redundant in some environmental hosts, which provides one possible explanation for the evolutionary maintenance of redundancy.

Evolution of *L. pneumophila* as an Intracellular Pathogen

Despite the threat the *L. pneumophila* poses to human health, only one human-to-human transmission of disease has ever been observed, and occurred between subjects with close interaction (Borges *et al.* 2016). In contrast, *L. pneumophila* persistence in nature depends upon its capacity to avoid predation by and replicate inside a diverse set of protozoan hosts (Fields 1996). Taken together, these observations suggest a critical role for bacteria-protist interactions in the evolution of *L. pneumophila* into an intracellular pathogen (Molmeret *et al.* 2005). Extended passage of *L. pneumophila* in a single host cell type has been shown to

lead to specialization, suggesting that cycling through a diverse set of environmental hosts may play a critical role in maintaining the pathogen's broad host range (Ensminger *et al.* 2012). Given the importance of protozoa to the evolutionary trajectory of *L. pneumophila* and other intracellular pathogens, relatively little is understood about how the intracellular niche within individual hosts can vary; for instance, with respect to nutrient availability or vacuolar integrity.

A recent study uncovered over 18,000 effector proteins from 58 *Legionella* species that suggested horizontal gene transfer (HGT) and multiple eukaryotic crossover events have provided the bacterium with its impressive repertoire of Dot/Icm substrates (Gomez-Valero *et al.* 2019). Analysis of the sequenced genomes of three serogroup 1 species, Philadelphia-1, Paris and Lens, revealed that *L. pneumophila* displays a high degree of genome plasticity between species (Cazalet *et al.* 2004; de Felipe et al. 2005). Paris and Lens alone differ by three plasmids and approximately 13% genomic sequence (Cazalet *et al.* 2004). G-C content calculations for all three species indicate that *L. pneumophila* has likely acquired a large number of its predicted effector proteins from other organisms (Cazalet *et al.* 2004; de Felipe et al. 2005). The identification of a large group of *L. pneumophila* G-C biased open reading frames that encode proteins containing eukaryotic-like motifs strongly suggests that HGT from eukaryotic host cells themselves have supplied *L. pneumophila* with a number of these open reading frames.

De Felipe and colleagues have described a group of 46 genes termed *L. pneumophila* eukaryotic-like genes (*leg*) by screening the Philadelphia-1 genome for open reading frames baring eukaryotic sequences or motifs with a low prevalence in other prokaryotes. This strategy yielded a diverse collection of proteins containing common protein-protein interaction domains, including Ankyrin repeats, coiled coils, and Leucine-rich repeats. Several other genes encode proteins with motif or active site homology suggesting they may serve as serine-threonine kinases, lipid signaling factors, GDP-GTP exchange proteins and factors involved in ubiquitination. A number of these proteins are translocated into host cells in a Dot/Icm-dependent manner, providing evidence that *L. pneumophila*'s ability to manipulate its host cell is partially dependent upon effector proteins built using components of the host cell itself (de Felipe *et al.* 2005). Since this initial discovery, comparative genomics has been used to identify Dot/Icm substrates that display orthology to one or more amoebal proteins, providing further evidence that interdomain horizontal gene transfer has played an important role in expanding the arsenal of Dot/Icm substrates (Urbanus *et al.* 2016). The full potential of these approaches, however, remains largely unrealized given the limited number of sequenced protozoan genomes.

Affecting the Effectors

Somewhat exclusively to *Legionella* is the presence of meta-effectors. This class of effector proteins regulates the function of another effector or group of effectors. Many bacterial genuses produce protein antagonists that regulate the function of a proceeding protein in a toxin-antitoxin manner (Melderen and Bast 2009). *Legionella* meta-effectors are unique in they directly regulate their target effector through inactivation and/or modification (Urbanus *et al.* 2016). The first observed case of this was the discovery of LubX and SidH in *L. pneumophila*. LubX is an E3 Ubiquitin Ligase that is secreted into the host cytosol in the late stages of infection and marks SidH for degradation (Quaile *et al.* 2015). The meta-effector, LupA, acts as a deubiquitinase to inactivate its cognate effector LegC3 (Urbanus *et al.* 2016). Interestingly, some meta-effectors have been found to be individually important for *Legionella* virulence (Liu and Luo 2007; Shames *et al.* 2017). *L. pneumophila* cells lacking the meta-effectors MesI or SidJ undergo attenuated growth due to unsuppressed toxicity of their effector targets. SidJ is a glutamylase that modulates the SidE family of ubiquitin ligases (Bhogaraju *et al.* 2019) while MesI specifically regulates the translation inhibitor SidI through unknown mechanisms (Shames *et al.* 2017; Joseph *et al.* 2020). In all, 143 effector pairs have been identified in *Legionella*, 19 of which have been found to have coevolved (Burstein *et al.* 2016). However, these analyses focused on pairs found in two or more species, and individual species of *Legionella* may harbor more effector pairs then currently recognized.

Conclusions

L. pneumophila is one of several facultative intracellular bacterial species that survive by subverting host defenses while engaging and co-opting other host pathways to create a replicative niche. With few exceptions, studies of *L. pneumophila* in both protozoan and mammalian host-cell systems have revealed remarkable parallels in the intracellular life cycle, indicating that the host proteins and pathways targeted by *L. pneumophila* to accomplish this feat are well conserved across Eukarya. In recent years, several groups have generated a long list of proteins that can be translocated by the Dot/Icm system into host cells. Characterization of these proteins will provide the mechanistic details underlying the unique phenomena observed during *L. pneumophila* infection, trafficking, replication, and egress. Additionally, the sheer number and diversity of *Legionella* effector proteins and their ability to infect and colonize a variety of eukaryotic hosts make the genus a valuable tool in the study of host-pathogen interactions.

Acknowledgments

This article is based on Chapter 35, *The Genus Legionella*, by Alexander W. Ensminger, Eva M. Campodonico and Craig R. Roy, in the third edition of this series. The previous contribution of these authors is very much appreciated and acknowledged.

REFERENCES

Abu-Zant A, Asare R, Graham J *et al.* Role for RpoS but Not RelA of *Legionella pneumophila* in modulation of phagosome biogenesis and adaptation to the phagosomal microenvironment. *Infect Immun* 2006;**74**:3021–6.

Amer AO, Swanson MS. Autophagy is an immediate macrophage response to *Legionella pneumophila*. *Cell Microbiol* 2005;**7**:765–78.

Amsden GW. Treatment of Legionnaires' disease. *Drugs* 2005;**65**: 605–14.

Andrews H, Vogel J, Isberg R. Identification of linked *Legionella pneumophila* genes essential for intracellular growth and evasion of the endocytic pathway. *Infect Immun* 1998;**66**: 950–8.

Bachman MA, Swanson MS. RpoS co-operates with other factors to induce Legionella pneumophila virulence in the stationary phase. *Mol Microbiol* 2001;**40**:1201–14.

Berger KH, Isberg RR. Two distinct defects in intracellular growth complemented by a single genetic locus in legionella pneumophila. *Mol Microbiol* 1993;**7**:7–19.

Berger KH, Merriam JJ, Isberg RR. Altered intracellular targeting properties associated with mutations in the legionella pneumophila dotA gene. *Mol Microbiol* 1994;**14**:809–22.

Bhogaraju S, Bonn F, Mukherjee R *et al.* Inhibition of bacterial ubiquitin ligases by SidJ–calmodulin catalysed glutamylation. *Nature* 2019;**572**:382–6.

Boamah D, Zhou G, Ensminger A, O'Connor T. From many hosts, one accidental pathogen: the diverse protozoan hosts of Legionella. *Front Cell Infect Microbiol.* 2017;**30**:477.

Borges V, Nunes A, Sampaio DA *et al.* Legionella pneumophila strain associated with the first evidence of person-to-person transmission of Legionnaires' disease: a unique mosaic genetic backbone. *Sci Rep-uk* 2016;**6**:26261.

Brüggemann H, Cazalet C, Buchrieser C. Adaptation of Legionella pneumophila to the host environment: role of protein secretion, effectors and eukaryotic-like proteins. *Curr Opin Microbiol* 2006;**9**:86–94.

Brüggemann H, Hagman A, Jules M *et al.* Virulence strategies for infecting phagocytes deduced from the in vivo transcriptional program of Legionella pneumophila. *Cell Microbiol* 2006;**8**:1228–40.

Burstein D, Amaro F, Zusman T *et al.* Genomic analysis of 38 *Legionella* species identifies large and diverse effector repertoires. *Nat Genet* 2016;**48**:167–75.

Burstein D, Zusman T, Degtyar E *et al.* Genome-scale identification of *Legionella pneumophila* effectors using a machine learning approach. *Plos Pathog* 2009;**5**:e1000508.

Byrne B, Swanson, MS. Expression of *Legionella pneumophila* virulence traits in response to growth conditions. *Infect Immun* 1998;**66**:3029–34.

Campodonico EM, Chesnel L, Roy CR. A yeast genetic system for the identification and characterization of substrate proteins transferred into host cells by the *Legionella pneumophila* Dot/Icm system. *Mol Microbiol* 2005;**56**:918–33.

Cazalet C, Jarraud S, Ghavi-Helm Y *et al.* Multigenome analysis identifies a worldwide distributed epidemic *Legionella pneumophila* clone that emerged within a highly diverse species. *Genome Res* 2008;**18**:431–41.

Cazalet C, Rusniok C, Brüggemann H *et al.* Evidence in the *Legionella pneumophila* genome for exploitation of host cell functions and high genome plasticity. *Nat Genet* 2004;**36**:1165–73.

CDC. Legionella (Legionnaires' Disease and Pontiac Fever). 2018.

Choy A, Dancourt J, Mugo B *et al.* The *Legionella* effector RavZ inhibits host autophagy through irreversible atg8 deconjugation. *Science* 2012;**338**:1072–6.

Christie PJ. Type IV secretion: intercellular transfer of macromolecules by systems ancestrally related to conjugation machines. *Mol Microbiol* 2001;**40**:294–305.

Cirillo J, Falkow S, Tompkins L. Growth of *Legionella pneumophila* in acanthamoeba castellanii enhances invasion. *Infect Immun* 1994;**62**:3254–61.

Clague M, Urbé S, Anieto F *et al.* Vacuolar ATPase activity is required for endosomal carrier vesicle formation. *J Biological Chem* 1994;**269**:21–4.

Dasgupta N, Ramphal R. Interaction of the antiactivator FleN with the transcriptional activator FleQ regulates flagellar number in pseudomonas aeruginosa. *J Bacteriol* 2001;**183**:6636–44.

Derre I, Isberg R. *Legionella pneumophila* replication vacuole formation involves rapid recruitment of proteins of the early secretory system. *Infect Immun* 2004;**72**:3048–53.

Diez E, Yaraghi Z, MacKenzie A *et al.* The neuronal apoptosis inhibitory protein (Naip) is expressed in macrophages and is modulated after phagocytosis and during intracellular infection with *Legionella pneumophila*. *J Immunol* 2000;**164**: 1470–7.

Elliott J, Winn W. Treatment of alveolar macrophages with cytochalasin D inhibits uptake and subsequent growth of *Legionella pneumophila*. *Infect Immun* 1986;**51**:31–6.

Ensminger A, Yassin Y, Miron A, Isberg R. Experimental evolution of *Legionella pneumophila* in mouse macrophages leads to strains with altered determinants of environmental survival. *PLoS Pathog.* 2012;**8**:e1002731.

Feeley J, Gibson R, Gorman G *et al.* Charcoal-yeast extract agar: primary isolation medium for *Legionella pneumophila*. *J Clin Microbiol* 1979;**10**:437–41.

de Felipe K, Pampou S, Jovanovic O *et al.* Evidence for acquisition of Legionella type IV secretion substrates via interdomain horizontal gene transfer. *J Bacteriol* 2005;**187**:7716–26.

Fields BS. The molecular ecology of legionellae. *Trends in Microbiology* 1996;**4**:286–90.

Fields B, Fields S, Loy J *et al.* Attachment and entry of *Legionella pneumophila* in *Hartmannella vermiformis*. *J Infect Dis* 1993;**167**:1146–50.

Gal-Mor O, Segal G. The *Legionella pneumophila* GacA homolog (LetA) is involved in the regulation of icm virulence genes and is required for intracellular multiplication in *Acanthamoeba castellanii*. *Microb Pathogenesis* 2003;**34**:187–94.

Garrett L. *The Coming Plague: Newly Emerging Diseases in a World Out Of Balance*. Farrar Straus & Giroux, New York, NY, 1994.

Gomez-Valero L, Rusniok C, Carson D *et al.* More than 18,000 effectors in the legionella genus genome provide multiple, independent combinations for replication in human cells. *Proc National Acad Sci* 2019;**116**:201808016.

Hales L, Shuman H. The *Legionella pneumophila* rpoS gene is required for growth within acanthamoeba castellanii. *J Bacteriol* 1999;**181**:4879–89.

Hammer BK, Swanson MS. Co-ordination of *Legionella pneumophila* virulence with entry into stationary phase by ppGpp. *Mol Microbiol* 1999;**33**:721–31.

Hammer BK, Tateda ES, Swanson MS. A two-component regulator induces the transmission phenotype of stationary-phase *Legionella pneumophila*. *Mol Microbiol* 2002;**44**:107–18.

Heuner K, Steinert M. The flagellum of *Legionella pneumophila* and its link to the expression of the virulent phenotype. *Int J Med Microbiol* 2003;**293**:133–43.

Horwitz M. Formation of a novel phagosome by the Legionnaires' disease bacterium (*Legionella pneumophila*) in human monocytes. *J Exp Medicine* 1983a;**158**:1319–31.

Horwitz M. The Legionnaires' disease bacterium (*Legionella pneumophila*) inhibits phagosome-lysosome fusion in human monocytes. *J Exp Medicine* 1983b;**158**:2108–26.

Horwitz M. Phagocytosis of the Legionnaires' disease bacterium (*Legionella pneumophila*) occurs by a novel mechanism: engulfment within a pseudopod coil. *Cell* 1984;**36**:27–33.

Horwitz M, Silverstein S. Interaction of the Legionnaires' disease bacterium (*Legionella pneumophila*) with human phagocytes. I. L. pneumophila resists killing by polymorphonuclearleukocytes, antibody, and complement. *J Exp Medicine* 1981;**153**:386–97.

Huang L, Boyd D, Amyot WM *et al.* The E block motif is associated with *Legionella pneumophila* translocated substrates. *Cell Microbiol* 2011;**13**:227–45.

Husmann L, Johnson W. Adherence of *Legionella pneumophila* to guinea pig peritoneal macrophages, J774 mouse macrophages, and undifferentiated U937 human monocytes: role of Fc and complement receptors. *Infect Immun* 1992;**60**: 5212–8.

Jacobi S, Schade R, Heuner K. Characterization of the alternative sigma factor 54 and the transcriptional regulator FleQ of *Legionella pneumophila*, which are both involved in the regulation cascade of flagellar gene expression. *J Bacteriol* 2004;**186**:2540–7.

Jain V, Kumar M, Chatterji D. ppGpp: stringent response and survival. *J Microbiol Seoul Korea* 2006;**44**:1–10.

Joseph A, Pohl A, Ball, T.J. *et al.* The Legionella *pneumophila* metaeffector Lpg2505 (MesI) regulates SidI-mediated translation inhibition and novel glycosyl hydrolase activity. *Infect Immun.* 2020;**88**.

Kagan JC, Roy CR. Legionella phagosomes intercept vesicular traffic from endoplasmic reticulum exit sites. *Nat Cell Biol* 2002;**4**:945–54.

Kagan JC, Stein M-P, Pypaert M *et al.* Legionella subvert the functions of Rab1 and Sec22b to create a replicative organelle. *J Exp Medicine* 2004;**199**:1201–11.

Khan M, Knox N, Prashar A *et al.* Comparative genomics reveal that host-innate immune responses influence the clinical prevalence of *Legionella pneumophila* Serogroups. *Plos One* 2013;**8**:e67298.

Khweek AA, Caution K, Akhter A *et al.* A bacterial protein promotes the recognition of the *legionella pneumophila* vacuole by autophagy. *Eur J Immunol* 2013;**43**:1333–44.

King C, Fields B, Shotts E *et al.* Effects of cytochalasin D and methylamine on intracellular growth of *Legionella pneumophila* in amoebae and human monocyte-like cells. *Infect Immun* 1991;**59**:758–63.

Kubori T, Bui XT, Hubber A *et al.* Legionella RavZ plays a role in preventing ubiquitin recruitment to bacteria-containing vacuoles. *Front Cell Infect Microbiol*, 2017;**7**:384.

Lifshitz Z, Burstein D, Peeri M *et al.* Computational modeling and experimental validation of the Legionella and Coxiella virulence-related type-IVB secretion signal. *Proc National Acad Sci* 2013;**110**:E707–15.

Liu Y, Luo Z-Q. The *Legionella pneumophila* effector SidJ Is required for efficient recruitment of endoplasmic reticulum proteins to the bacterial phagosome. *Infect Immun* 2007;**75**:592–603.

Luo Z-Q, Isberg R. Multiple substrates of the *Legionella pneumophila* Dot/Icm system identified by interbacterial protein transfer. *Proc National Acad Sci* 2004;**101**:841–6.

Lynch D, Fieser N, Glöggler K *et al*. The response regulator LetA regulates the stationary-phase stress response in *Legionella pneumophila* and is required for efficient infection of Acanthamoeba castellanii. *FEMS Microbiol Lett* 2003;**219**:241–8.

Majdalani N, Vanderpool CK, Gottesman S. Bacterial small RNA regulators. *Crit Rev Biochem Mol* 2008;**40**:93–113.

Marra A, Blander S, Horwitz M *et al*. Identification of a *legionella pneumophila* locus required for intracellular multiplication in human macrophages. *Proc National Acad Sci* 1992;**89**:9607–11.

McNealy T, Forsbach-Birk V, Shi C *et al*. The Hfq homolog in *legionella pneumophila* demonstrates regulation by LetA and RpoS and interacts with the global regulator CsrA. *J Bacteriol* 2005;**187**:1527–32.

Melderen L, Bast M. Bacterial toxin–antitoxin systems: more than selfish entities? *PLOS Genet* 2009;**5**:e1000437.

Mercante J, Winchell J. Current and emerging Legionella diagnostics for laboratory and outbreak investigations. *Clin Microbiol Rev*. 2015;**28**:95–133.

Moffat J, Tompkins L. A quantitative model of intracellular growth of *Legionella pneumophila* in Acanthamoeba castellanii. *Infect Immun* 1992;**60**:296–301.

Molmeret M, Bitar D, Han L *et al*. Disruption of the phagosomal membrane and egress of *Legionella pneumophila* into the cytoplasm during the last stages of intracellular infection of macrophages and Acanthamoeba polyphaga. *Infect Immun* 2004;**72**:4040–51.

Molmeret M, Horn M, Wagner M *et al*. Amoebae as training grounds for intracellular bacterial pathogens. *Appl Environ Microb* 2005;**71**:20–8.

Molofsky AB, Swanson MS. *Legionella pneumophila* CsrA is a pivotal repressor of transmission traits and activator of replication. *Mol Microbiol* 2003;**50**:445–61.

Nagai H, Kagan JC, Zhu X *et al*. A bacterial guanine nucleotide exchange factor activates ARF on Legionella phagosomes. *Science* 2002;**295**:679–82.

Nagai H, Kubori T. Type IVB secretion systems of Legionella and other gram-negative bacteria. *Front Microbiol* 2011;**2**:136.

Nagai H, Roy CR. The DotA protein from *Legionella pneumophila* is secreted by a novel process that requires the Dot/Icm transporter. *Embo J* 2001;**20**:5962–70.

Ngwaga T, Hydock AJ, Ganesan S *et al*. Potentiation of cytokine-mediated restriction of Legionella intracellular replication by a Dot/icm-translocated effector. *J Bacteriol* 2019;**201**, DOI: 10.1128/jb.00755-18

Ninio S, Zuckman-Cholon DM, Cambronne ED *et al*. The Legionella IcmS–IcmW protein complex is important for Dot/Icm-mediated protein translocation. *Mol Microbiol* 2005;**55**:912–26.

O'Connor TJ, Adepoju Y, Boyd D *et al*. Minimization of the *Legionella pneumophila* genome reveals chromosomal regions involved in host range expansion. *Proc National Acad Sci* 2011;**108**:14733–40.

Odenbreit S, Püls J, Sedlmaier B *et al*. Translocation of helicobacter pylori CagA into Gastric epithelial cells by type IV Secretion. *Science* 2000;**287**:1497–500.

Otto GP, Wu MY, Clarke M *et al*. Macroautophagy is dispensable for intracellular replication of *Legionella pneumophila* in Dictyostelium discoideum. *Mol Microbiol* 2004;**51**: 63–72.

Payne N, Horwitz M. Phagocytosis of *Legionella pneumophila* is mediated by human monocyte complement receptors. *J Exp Medicine* 1987;**166**:1377–89.

Quaile AT, Urbanus ML, Stogios PJ *et al*. Molecular characterization of LubX: functional divergence of the U-Box fold by *Legionella pneumophila*. *Struct Lond Engl 1993* 2015;**23**:1459–69.

Reaves B, Banting G. Vacuolar ATPase inactivation blocks recycling to the trans-golgi network from the plasma membrane. *Febs Lett* 1994;**345**:61–6.

Rechnitzer C, Blom J. Engulfment of the Philadelphia strain of *Legionella pneumophila* within pseudopod coils in human phagocytes. Comparison with other Legionella strains and species. *APMIS* 1989;**97**:105–14.

Reynolds H, Newball H. Analysis of proteins and respiratory cells obtained from human lungs by bronchial lavage. *J Lab Clin Med* 1974;**84**:559–73.

Robinson CG, Roy CR. Attachment and fusion of endoplasmic reticulum with vacuoles containing *Legionella pneumophila*. *Cell Microbiol* 2006;**8**:793–805.

Rolando M, Escoll P, Nora T *et al*. *Legionella pneumophila* S1P-lyase targets host sphingolipid metabolism and restrains autophagy. *Proc National Acad Sci* 2016;**113**:1901–6.

Romeo T. Global regulation by the small RNA-binding protein CsrA and the non-coding RNA molecule CsrB. *Mol Microbiol* 1998;**29**:1321–30.

Rossier O, Starkenburg S, Cianciotto N. *Legionella pneumophila* type II protein secretion promotes virulence in the A/J mouse model of Legionnaires' disease pneumonia. *Infect Immun* 2003;**72**:310–21.

Roy CR, Tilney LG. The road less traveled. *J Cell Biology* 2002;**158**:415–9.

Saito M, Kajiwara H, Miyamoto H *et al*. Fate of *Legionella pneumophila* in macrophages of C57BL/6 chronic granulomatous disease mice. *Microbiol Immunol* 2001;**45**:539–41.

Sandvig K, van Deurs B. Membrane traffic exploited by protein toxins. *Annu Rev Cell Dev Bi* 2002;**18**:1–24.

Segal G, Suman HA. How is the intracellular fate of the *Legionella pneumophila* phagosome determined? *Trends Microbiol* 1998;**6**:253–5.

Shames SR, Liu L, Havey JC *et al*. Multiple *Legionella pneumophila* effector virulence phenotypes revealed through high-throughput analysis of targeted mutant libraries. *Proc Nat Acad Sci USA* 2017;**114**:E10446–54.

Shohdy N, Efe J, Emr S *et al*. Pathogen effector protein screening in yeast identifies Legionella factors that interfere with membrane trafficking. *Proc Nat Acad Sci USA* 2005;**102**:4866–71.

Solomon JM, Rupper A, Cardelli JA *et al*. Intracellular growth of *Legionella pneumophila* in Dictyostelium discoideum, a system for genetic analysis of host-pathogen interactions. *Infect Immun* 2000;**68**:2939–47.

Sturgill-Koszycki S, Swanson MS. *Legionella pneumophila* replication vacuoles mature into acidic, endocytic organelles. *J Exp Medicine* 2000;**192**:1261–72.

Swanson, MS, Isberg RR. Association of *Legionella pneumophila* with the macrophage endoplasmic reticulum. *Infect Immun* 1995;**63**:3609–20.

Tilney L, Harb O, Connelly P *et al*. How the parasitic bacterium *Legionella pneumophila* modifies its phagosome and transforms it into rough ER: implications for conversion of plasma membrane to the ER membrane. *J Cell Sci* 2001;**114**:4637–50.

Urbanus ML, Quaile AT, Stogios PJ *et al.* Diverse mechanisms of metaeffector activity in an intracellular bacterial pathogen, *Legionella pneumophila. Mol Syst Biol* 2016;**12**:893.

van WA, Geuze H, Stoorvogel W. Heterogeneous behavior of cells with respect to induction of retrograde transport from the trans-Golgi network to the Golgi upon inhibition of the vacuolar proton pump. *Eur J Cell Biol* 1997;**74**:417–23.

Vogel JP, Andrews HL, Wong S *et al.* Conjugative transfer by the virulence system of *Legionella pneumophila. Science* 1998;**279**:873–6.

Wieland H, Goetz F, Neumeister B. Phagosomal acidification is not a prerequisite for intracellular multiplication of *Legionella pneumophila* in human monocytes. *J Infect Dis* 2004;**189**:1610–4.

Yamamoto A, Tagawa Y, Yoshimori T *et al.* Bafilomycin A1 prevents maturation of autophagic vacuoles by inhibiting fusion between autophagosomes and lysosomes in rat hepatoma cell line, H-4-II-E Cells. *Cell Struct Funct* 1998;**23**:33–42.

Yamamoto Y, Klein T, Friedman H. *Legionella pneumophila* growth in macrophages from susceptible mice is genetically controlled. *Proc Soc Exp Biol Med* 1991;**196**:405–9.

Yu VL, Plouffe JF, Pastoris M *et al.* Distribution of Legionella species and serogroups isolated by culture in patients with sporadic community-acquired legionellosis: an international collaborative survey. *J Infect Dis* 2002;**186**:127–8.

Zhu W, Banga S, Tan Y *et al.* Comprehensive identification of protein substrates of the Dot/Icm type IV transporter of *Legionella pneumophila. PLOS One* 2011;**6**:e17638.

32

The Genus Listeria

Sukhadeo Barbuddhe, Torsten Hain, Swapnil P. Doijad, and Trinad Chakraborty

CONTENTS

DOI: 10.1201/9781003099277-34

Introduction

The genus *Listeria* belongs to the phylum Firmicute, the order Bacillales, the class Bacilli, and the family *Listeriaceae* together with the genus *Brochothrix*. The *Listeria* is Gram-positive bacteria with low G+C content closely related to *Bacillus*, *Clostridium*, *Enterococcus*, *Streptococcus*, and *Staphylococcus*. The natural habitat of these bacteria is thought to be decomposing plant matter, in which they live as saprophytes. The potential primary sources for the introduction of *L. monocytogenes* pathogenic to humans into the food chain and food-processing plants include livestock and produce farms. Isolation of *L. monocytogenes* has been reported from cattle, silage, animal feeds, manure, and growing grass, among others [1, 2] as well as from soil, water, and vegetation of natural and urban environments [3, 4].

Historically, Hülpers isolated bacteria from a liver necrosis in a rabbit in 1911 that were pathogenic for mice and called it *Bacillus hepatica* according to the isolation site [5]. In 1926, a bacterium was isolated from dead laboratory rabbits and guinea pigs exhibiting monocytosis by Murray, Webb and Swann who called it *Bacterium monocytogenes* [6]. Later, a similar bacterium was isolated from wild gerbils with "Tiger River Disease" in South Africa and named as *Listerella hepatolytica* to honor Lord Joseph Lister [7]. Nyfeldt reported the first confirmed isolation from humans in Denmark [8]. These two similar species were named *Listerella hepatolyticus* and finally renamed as *Listeria* in 1940 for taxonomic reasons [9].

Listeria monocytogenes contamination is one of the leading microbiological causes of food recalls, mainly of meat, poultry, seafood, and dairy products. More recently contamination of fresh fruit and produce has been associated with outbreaks of listeriosis. Prevention and control measures are based on hazard analysis and critical control point programs throughout the food industry, and on specific recommendations for high-risk groups.

Trends of listeriosis incidence differ around the world. This is related to dietary preferences and food safety policies [10–12]. High hospitalization and mortality rates worldwide, especially in high-risk groups such as the elderly, the newborns, and immuno-compromised individuals as well as pregnant women have been associated with *L. monocytogenes* foodborne outbreaks. There were more foodborne outbreaks associated with *L. monocytogenes* between 2005 and 2015 in the EU (83) than in the United States (47), resulting in 757 and 491 cases, respectively [13, 14]. A new and significant source of infection is related to the use of antibiologicals to treat immune-related disease such as rheumatoid arthritis [15].

L. monocytogenes is morphologically indistinguishable from other *Listeria* species, and often causes nonspecific clinical symptoms; therefore, laboratory testing is required to differentiate *L. monocytogenes* from other *Listeria* species, and to diagnose listeriosis. Following recent advances in molecular genetic techniques, methods targeting unique genes in *Listeria* have been designed for the specific differentiation of *L. monocytogenes* from other *Listeria* species. These methods are intrinsically more precise and less affected by natural variation than the phenotypic methods. A great opportunity to use these tools toward the development of molecular subtyping techniques has been provided by the recent developments in the field of genomics to demonstrate occurrence and pinpoint circulation of strains. Whole genome sequence (WGS)-based typing provides higher discriminatory power, better data accuracy, and reproducibility thus allowing higher laboratory through-put at lower cost and has become indispensable to support a successful outbreak investigation by appropriate public health analysts and epidemiologists [16]. Classical epidemiological approach coupled with molecular subtyping and WGS techniques were used to identify and confirm a *Listeria* gastroenteritis outbreak associated with consumption of sliced cold beef ham in Italy [17].

Extensive research in recent decades has led to significant insights regarding *Listeria* species and listeriosis [18]. The establishment of animal models and *in vitro* cell culture systems for listeriosis has helped the delineation of key stages in *L. monocytogenes* infection and pathogenesis [19].

In-depth reviews on the taxonomy, isolation, identification, epidemiology, pathogenesis, virulence determinants and genomics of *L. monocytogenes* are available [20–26].

The Genus *Listeria*

With the description of several novel species, the genus *Listeria* includes 20 species now, *L. monocytogenes, L. seeligeri, L. ivanovii, L. welshimeri, L. marthii, L. innocua, L. grayi, L. fleischmannii, L. floridensis, L. aquatica, L. newyorkensis, L. cornellensis, L. rocourtiae, L. weihenstephanensis, L. grandensis, L. riparia, L. booriae* [27], *L. costaricensis* [28], *L. goaensis* [29], and *L. thailandensis* [30]. In 2012, *L. weihenstephanensis* was isolated from the water plant *Lemna trisulca* sampled from a fresh water pond [31] and *L. fleischmannii* isolated from cheese from a ripening cellar [32]. Sampling of seafood and dairy processing facilities in the north-eastern United States revealed two novel species, *L. booriae* and *L. newyorkensis* [33], while *L. floridensis, L. aquatica, L. cornellensis, L. riparia,* and *L. grandensis* were isolated from agricultural and natural environments [34]. *L. costaricensis* was isolated from a food processing drainage system in Costa Rica [28]. *L. goaensis* [29] was isolated from mangrove swamps in Goa, India, while *L. thailandensis* from fried chicken [30]. These novel species represent an important branching point in *Listeria* phylogeny. They also present a diagnostic challenge for clinical, food and public health microbiologists seeking to identify *L. monocytogenes*. *L. monocytogenes* is an opportunistic pathogen in human beings and various animal species, whereas *L. ivanovii* mainly affects ruminants, causing abortion, and only occasionally occurring in man. Phylogenetic analyses, based on the *16S* and *23S rRNA* genes and the *iap, prs, vclA, vclB,* and *ldh* genes, indicate that *L. innocua* is highly related to *L. monocytogenes*, while *L. welshimeri* is more distant, exhibiting the deepest branching of this group. The second group has *L. ivanovii*, together with *L. seeligeri, L. grayi* seems to be very distant from these two groups [35]. The 16S rRNA gene sequence analysis confirmed close phylogenetic relatedness of *L. marthii* to *L. monocytogenes* and *L. innocua* and more distant relatedness to *L. welshimeri, L. seeligeri, L. ivanovii* and *L. grayi* [36]. Phylogenetic analysis of partial sequences of *L. marthii* for *sig*B, *gap,* and *prs* showed formation of a well-supported sistergroup to *L. monocytogenes* [36]. The 16S rRNA gene sequence analysis revealed a close phylogenetic similarity of *L. weihenstephanensis* to *L. rocourtiae* and a more distant relationship to other *Listeria* species [31] (Figure 32.1).

Characteristics

Listeria species are a small, Gram-positive, nonsporulating, facultatively anaerobic rod that measures 1–2 × 0.5 microns and shows characteristic tumbling motility at or around 25°C. *Listeria* is able to multiply at high salt concentrations (10% NaCl) and at a broad range of pH (pH 4.5 to 9) and temperature (0–45°C, optimum 30–37°C)[37]. The morphology of *Listeria* as seen by Gram staining and electron microscopy is depicted in Figures 32.2A and 32.2B, respectively.

Listeria species are closely related bacteria that share many morphological and biochemical characteristics. *Listeria* species are catalase and Voges-Proskauer reaction positive, and indole and oxidase negative. *Listeria* species can hydrolyze aesculin, but not urea or reduce nitrates. A rapid test strip for identification has been developed [38]. *Listeria* species show variations in their ability to hemolyze horse or sheep red blood cells, and in their ability to produce acid from L-rhamnose, D-xylose and a-methyl-D-mannoside [39].

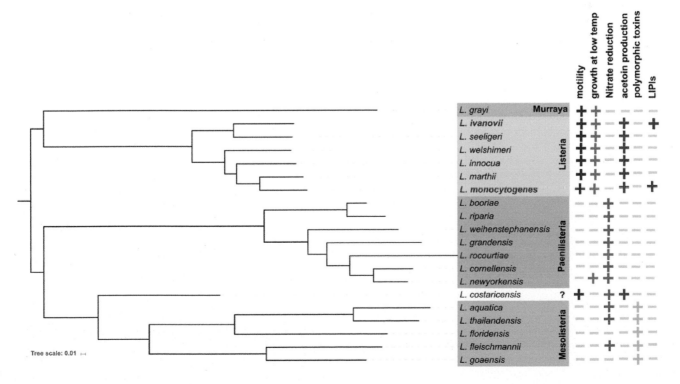

FIGURE 32.1 Phylogenetic tree of currently identified species of genus *Listeria*.

FIGURE 32.2A   Gram-stained *Listeria monocytogenes* (magnification 100×).

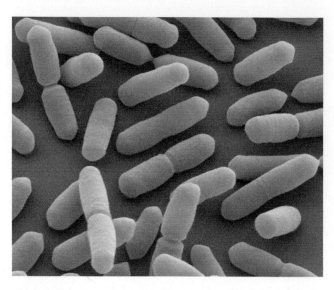

FIGURE 32.2B   Scanning electron microscopy of *Listeria monocytogenes* (magnification 16,000×).

Methods for Confirmation/Differentiation of *Listeria*

As *L. monocytogenes* is morphologically indistinguishable from other *Listeria* species additional, laboratory testing is required to differentiate *L. monocytogenes* from other *Listeria* species. Table 32.1 shows a scheme for identification of *Listeria* spp. *L. ivanovii* is differentiated biochemically from *L. monocytogenes* and other *Listeria* species by its production of a wide, clear or double zone of hemolysis on sheep or horse blood agar, strong lecithinase reaction with or without charcoal in the medium, a positive Christie-Atkins-Munch-Petersen (CAMP) reaction with *Rhodococcus equi* but not with hemolytic *Staphylococcus aureus*, and fermentation of D-xylose but not L-rhamnose [40]. *L. monocytogenes* requires charcoal for its lecithinase reaction [41]. *L. innocua* is distinguished from *L. monocytogenes* on the basis of its negative CAMP reaction and its failure to cause α-hemolysis or to show PI-PLC activity on chromogenic media [39].

An assay based on phosphatidylinositol-specific phospholipase C activity (PI-PLC) has been described for the discrimination of pathogenic and nonpathogenic *Listeria* species based on which hemolytic but nonpathogenic species, i.e., *L. seeligeri*, can be separated from the hemolytic and pathogenic species, i.e., *L. monocytogenes* and *L. ivanovii* [42]. The application of a multiplex polymerase chain reaction (PCR) assay that selectively amplifies a commonly shared region of the *iap* gene that facilitates the differentiation of all six *Listeria* species in a single test has been developed [43].

Biology of *Listeria*

L. monocytogenes is a remarkable bacterium that has acquired a diverse collection of virulence factors, each with unique properties and functions. Its life cycle reflects its remarkable adaptation to intracellular survival and multiplication in professional phagocytic and nonphagocytic cells of vertebrates and invertebrates.

Pathogenic *Listeria* species are able to breach endothelial and epithelial barriers of the infected hosts including the intestinal, blood-brain and fetoplacental barrier, are able to invade and replicate in phagocytic and nonphagocytic cells [18]. Entry of the pathogen is mediated by internalins A and B, which are expressed on the surface of the bacterium. Two virulence associated molecules are responsible for lysis of the primary single-membraned vacuoles and subsequent escape by *L. monocytogenes*: listeriolysin O (LLO) and phosphatidylinositol-phospholipase C (PI-PLC). After lysis of the primary single-membraned vacuoles, *L. monocytogenes* is released to the cytosol, where it undergoes intracellular growth and multiplication. The bacterial surface protein ActA leads to the intracellular mobility and cell-to-cell spread, which is cotranscribed with PC-PLC and mediates the formation of polarized actin tails that propel the bacteria toward the cytoplasmic membrane, which enables the bacterium to infect the second cell [18]. Immunofluorescence staining of both bacteria and the host cell actin in infected cells reveals the unique ability of these bacteria to move intracellularly (Figure 32.2C).

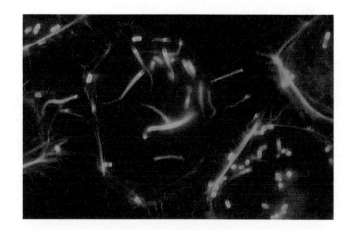

FIGURE 32.2C   Immunofluorescence staining of intracellular motile *Listeria* (bacteria are stained red, while the actin tail of *Listeria monocytogenes* is stained green with FITC-labeled phalloidin).

TABLE 32.1

A Scheme for the Phenotypic Identification of *Listeria*

	L. monocytogenes	L. innocua	L. seeligeri	L. ivanovii	L. welshimeri	L. marthii	L. grayi	L. rocourtiae	L. weihenstephanensis	L. cornellensis	L. riparia	L. grandensis	L. fleischmannii	L. aquatica	L. floridensis	L. newyorkensis	L. booriae	L. costaricensis	L. goaensis	L. thailandensis
Gram characters	+	+	+	+	+	+	+	+	+	+	+	+	+	+	+	+	+	+	+	+
Catalase	+	+	+	+	+	+	+	+	+	+	+	+	+	+	+	+	+	−	+	+
Oxidase	−	−	−	−	−	−	−	−	−	−	−	−	−	−	−	−	−	−	−	−
D-Mannitol	−	−	−	−	−	−	+	+	+	−	v	−	v	−	−	+	+	−	−	−
L-Rhamnose	+	v	−	−	v	−	−	+	+	−	+	−	+	+	+	v	+	+	+	+
D-Xylose	−	−	+	+	+	−	+	+	+	+	+	+	+	+	+	+	+	+	+	+
α-Methyl D-mannosidase	+	+	−	−	+	+	v	+	−	−	+	−	−	+	−	−	+	+	−	−
Voges-Proskauer	+	+	+	+	+	+	+	−	−	−	−	−	−	v	−	−	−	−	+	+
Nitrate Reduction	−	−	−	−	−	−	v	+	+	+	+	+	+	+	−	+	+	+	−	+
Motility	+	+	+	+	+	+	+	−	−	−	−	−	−	−	−	−	−	+	−	−
Methyl Red	+	+	+	+	+	+	+	v	v	v	+	v	+	+	+	+	+	NA	+	NA
Hemolysis	+	−	+	+	−	−	−	−	−	−	−	−	−	−	−	−	−	−	+(*)	−
D-Arylamidase	−	+	+	v	v	−	+	−	−	−	−	−	−	−	−	−	−	−	−	−
PI-PLC	+	−	−	+	−	−	−	−	−	−	−	−	−	−	−	−	−	−	−	−
D-Tagatose	−	−	−	−	+	−	−	−	−	−	−	−	+	−	−	−	−	+	−	+
D-Ribose	−	−	−	+	−	−	+	+	−	+	V	+	+	+	−	+	V	+	−	+
Sucrose	+	+	+	+	+	−	−	−	−	−	−	v	−	−	−	−	−	+	−	−
D-Turanose	−	v	−	−	−	+	+	−	−	−	−	v	−	−	−	−	−	−	−	−
Glycerol	v	+	+	+	+	−	v	+	+	v	v	−	+	v	−	+	+	+	(+)	(+)
D-Galactose	v	−	−	v	−	−	+	+	−	−	+	−	−	−	+	+	+	+	−	−
L-arabinose	−	−	−	−	−	−	−	−	−	v	+	−	−	+	+	+	+	−	−	−
Inositol	−	−	−	−	−	−	−	−	−	−	v	−	v	v	−	−	−	−	−	+
Methyl α-D-mannose	−	−	NA	−	NA	−	+	−	−	−	−	−	v	−	−	−	−	+	(+)	−
D-Maltose	+	+	+	+	+	+	+	+	+	+	+	+	+	−	+	+	+	+	+	−
D-Lactose	+	+	+	+	+	+	+	+	v!	(+)	+	−	+	−	+	+	+	+	+	−
D-Melibiose	v!	v	−	−	−	v	−	+	−	v	v	v	−	−	−	−	+	−	−	−
Inulin	v!	v!	−	−	−	−	−	−	−	−	−	−	−	−	−	−	−	−	−	−
D-Melezitose	v	v	v	v	v	−	−	−	−	−	−	−	v	−	−	−	−	−	−	−
D-Lyxose	v	v	−	−	v	−	v	−	−	−	−	−	−	−	v	+	−	−	−	−
D-glucose	v!	v!	+	v!	+	v!	+	+	+	+	+	+	+	+	+	+	+	+	+	+
D-Arabitol	+	+	+	+	+	+	+	−	+	−	−	v	+	−	−	−	+	+	+	+
L-Fucose	−	−	−	−	−	−	−	−	−	−	−	−	−	−	−	−	−	+	−	−
potassium 5-ketogluconate	−	−	−	−	−	−	−	−	−	−	−	−	−	−	−	−	−	+	−	−

Notes: +, positive reaction; (+), weak positive; −, negative reaction; v, variable between isolates or replicates; v!, variable between the studies (possible due to difference in incubation time and temperature); NA, not available; (*) α-hemolysis.

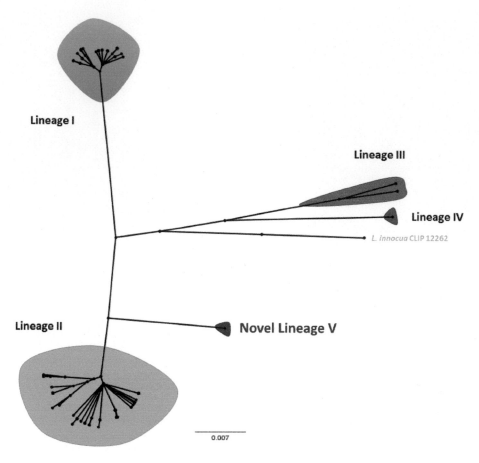

FIGURE 32.3   Phylogenomic lineages of *L. monocytogenes*.

The genes encoding the virulence-associated proteins PIPLC, LLO, Mpl, ActA and PC-PLC are located in a 9.6-kb virulence gene cluster [44], which is principally regulated by the pleiotropic virulence regulator, PrfA (a 27-kDa protein encoded by *prf*A). The various pathogenicity islands in *Listeria* species are shown in Figure 32.3. In addition to these virulence-associated genes and proteins, several other genes, such as *iap*, *bsh*, *vip*, *inlJ*, *auto*, *ami*, and *bilA*, are also involved in *L. monocytogenes* virulence and pathogenicity. The resistance of *L. monocytogenes* to acidic conditions and to bile salts makes it particularly adept at infecting the gastrointestinal tract [45].

Intracellular gene expression profiling of *L. monocytogenes* indicated that many of the genes are upregulated in the cytosol and allowed the identification of activation of several *L. monocytogenes* genes, including the *plc*A and *prf*A after infection of the host cells. Indeed, *L. monocytogenes* modulates the expression of 500 genes for its survival in this cellular compartment [46, 47].

A hybrid sublineage of the major lineage II (HSL-II) and serotype 4h of *L. monocytogenes* were discovered while characterizing the isolates from severe ovine listeriosis outbreaks from China [48]. HSL-II isolates were highly virulent and exhibited higher organ colonization capacities than well-characterized hypervirulent strains of *L. monocytogenes* in an orogastric mouse infection model. HSL-II isolates are the only known *L. monocytogenes* to carry virulence factors from a related species in the animal pathogen *L. ivanovii*. These isolates harbor a unique wall teichoic acid (WTA) structure

essential for resistance to antimicrobial peptides and responsible for bacterial invasion and overall virulence phenotype [48] (Figure 32.4).

General Ecology of *Listeria*

Some *Listeria* strains have the ability to survive for long times under adverse environmental conditions and to persist in niches in food processing equipment, and associated drains, walls, and ceilings. Various compounds that are used to inhibit growth of the pathogen include organic acids, fatty acids, antioxidants, sodium nitrite, smoke, spices, and herbs.

Due to its environmental persistence *L. monocytogenes* is recognized as a problem for the food industry, because of its ability to form biofilms. Biofilms are microbial communities adhered to biotic or abiotic surfaces including polystyrene, polypropylene, glass, stainless steel, quartz, marble, and granite [49] coated by self-produced extracellular polymers, thereby conferring protection to bacterial cells from a variety of environmental factors, such as ultraviolet rays, toxic metals, acids, desiccation, salinity, and antimicrobials [50, 51] and decreasing the efficiency of cleaning and disinfection procedures [52].

Listeria spp. showed similar prevalence in samples from natural (23.4%) and urban (22.3%) environments. *L. seeligeri* and *L. welshimeri* were significantly associated with natural environments, while *L. innocua* and *L. monocytogenes* were significantly associated with urban environments [53]. Members of the genus *Listeria* were common in urban and natural environments

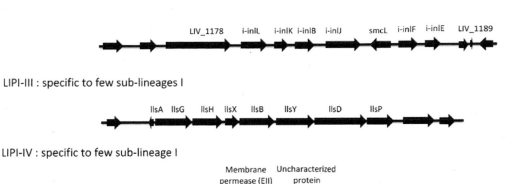

FIGURE 32.4 *Listeria* pathogenicity islands.

and showed species- and subtype-specific associations with different environments and areas (Figure 32.5). The study also indicated that *Listeria* species and subtypes within these species may show distinct ecological preferences suggesting development of molecular approaches for source tracking [53]. Sources of transmission of Listeria species between the natural environment, food industry, and humans are shown in Figure 32.6.

A study of the caged laying hens in France has shown that 29.5% are contaminated by *L. monocytogenes* [54]. Another study in five egg-breaking plants in Western France revealed the presence of *Listeria* spp. and *L. monocytogenes* in the environment of egg-breaking plants with 65.1% and 8.0% of contaminated samples, respectively. The typing of 253 isolates of *L. monocytogenes* by PFGE analysis using *Apa*I and *Asc*I enzymes identified a single dominant *L. monocytogenes* cluster that was better

adapted to egg-breaking plants [55]. Analysis of strains revealed a high diversity with 46 different pulsotypes, and the raw material was implicated as a source of contamination of egg-breaking plants. One was dominant in the five egg-breaking plants during the four seasons studied.

Rodents have been reported to be capable of carrying pathogenic bacteria in their intestines, such as *Listeria*, and disseminating those pathogens into the natural environment and possibly transmit the pathogen between wild animals and humans. In a study, intestinal fecal samples of rodents were found to be positive for *L. monocytogenes* (3.22%) and for *L. ivanovii* (2.05%). Other species identified included *L. innocua*, *L. fleischmannii*, and *L. floridensis* [56]. In another study, *L. ivanovii* (3.7%) was isolated from wild rodents and the genetic diversity was relatively low [57]. Isolation of *L. monocytogenes* was reported

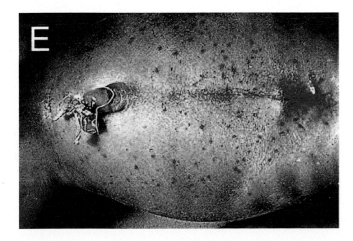

FIGURE 32.5A Stillborn with granulomatosis infantiseptica characterized by the presence of miliary-disseminated pyogranulomatous lesions on body surface.

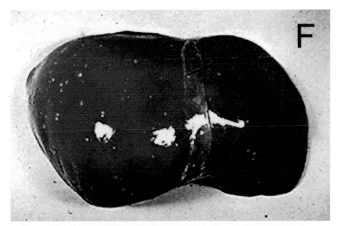

FIGURE 32.5B Hematoxylin-eosin-stained section from the liver showing two subcapsular pyogranulomes.

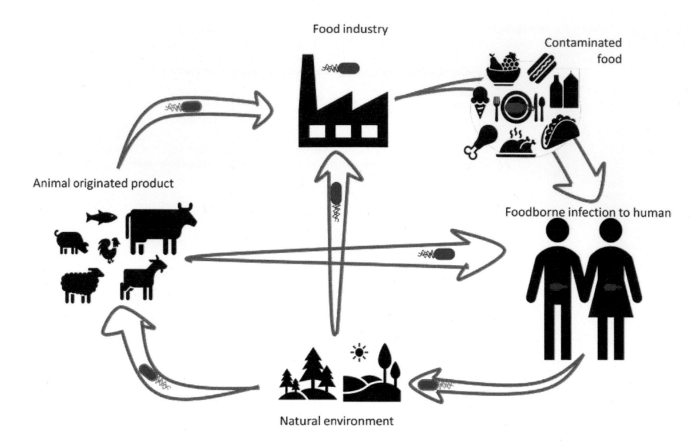

FIGURE 32.6   Sources of transmission of *Listeria* species between the natural environment, food industry, and humans.

from farm environment (soil) and invertebrates (fly and tick) [58]. Characterization of *L. monocytogenes* isolated from black-headed gulls in Dianchi Lake and other regions revealed potential public health risks in regions where the migratory birds passage and reside [59]. The occurrence of *L. monocytogenes* in black-headed gull feces in 2016, 2017, and 2018 was found to be 1.0%, 1.0%, and 0.6%, respectively.

Listeria and Biofilms

Biofilm formation by foodborne pathogens constitutes a grave concern in food industry. *L. monocytogenes* has the ability to form biofilms thereby persisting for a long-time favoring cross-contamination of foods. A number of studies have been conducted to better characterize the listerial biofilm extracellular matrix that represents the major component of a biofilm.

The biofilms represent structured communities of bacterial cells embedded in a self-produced matrix of extracellular polymeric substances (EPSs) leading to persistence of the pathogen in food processing environment. Exopolysaccharides, proteins, and eDNA are the main molecules composing the biofilm matrix of several bacteria [60]. Listerial biofilms were found to be abundant in proteins [61].

The formation of biofilms alters the phenotype and gene expression. The biofilm formation occurs on several surfaces employed in the food industry including stainless steel, polystyrene, and glass [62], signifying a serious concern for food safety. Upon maturity, microbial cells detach from the biofilm and scatter into the environment (in their planktonic form), thereby

acting as a potential source of contamination [63]. Various areas in processing facilities, such as floors, waste water pipes, bends in pipes, conveyor belts, rubber seals, and stainless steel surfaces, are some of the most common sources involved in biofilm accumulation. As evidenced by electron microscopic studies, strong biofilm-forming isolates could synthesize a biofilm within 24 hours on surfaces important in food industries, such as stainless steel, ceramic tiles, high-density polyethylene plastics, polyvinyl chloride pipes, and glass [64]. Abundance of certain fatty acids such as iso-C15:0, anteiso-C15:0, and anteiso-C17:0 correlated with the biofilm-forming capability of *L. monocytogenes*. [64]. Cells within biofilms are reported to be more resistant to biocides and stress conditions [65, 66].

Temperature, the nature of adhesion surface, and its hydrophobicity play an important role in biofilm formation [62, 67]. Complex organization of *L. monocytogenes* biofilms was observed at both 22°C and 37°C in terms of cell number and EPSs produced [62]. *L. monocytogenes* form biofilms at 4°C and 12°C with higher levels on glass compared to stainless steel and polystyrene. Biofilm formation at low temperatures used during food processing and storage increases the likelihood of cross-contamination. Reports on a correlation between lineages and biofilm forming ability (with lineage II strains presenting higher levels of biofilm production) [68] have been reported.

In natural environments, such as in food industry, biofilms may be composed of multiple bacterial species, which are more resistant to disinfectants and sanitizers. An increase of number of listerial cells within biofilms was reported when cocultivated with *Flavobacterium* spp. [69]. Despite the great advances made over the last few decades in the field of food safety, several outbreaks of foodborne listeriosis with high rates of morbidity and mortality have been reported [70, 71]. *L. monocytogenes-Escherichia coli* biofilms were effectively removed using DNase I followed by pronase and cellulose [72]. The antibiofilm strategies are focused on use of the enzymes [73], bacteriophages [74], or combined strategies [75]. Bacteriocinogenic lactic acid bacteria (LAB), their bacteriocins, and bioengineered bacteriocin derivatives have been proposed for controlling *L. monocytogenes* biofilms on different surfaces through inhibition, competition, exclusion, and displacement [76].

Disease in Humans

Listeriosis is considered an emerging health problem, especially among the elderly, including those with no underlying medical conditions [12]. In recent years, the trends of listeriosis incidence have been found to differ around the world, probably related to dietary preferences and food safety policies [11, 77–79]. Hence both an increase in incidence [80, 81] and a reduction in the number of cases of listeriosis have been reported [82, 83]. A recent report from the Centers for Disease Control and Prevention (CDC) estimates 1,600 cases and 255 deaths per year in the United States [84]. *L. monocytogenes* is the foodborne pathogen responsible for the majority of deaths in the United States after nontyphoid *Salmonella* spp. In Europe, statistically significant increasing trends in listeriosis notification rates from 2005 to 2009 were noted in Austria, Denmark, Hungary, Italy, Spain, and Sweden [85].

L. monocytogenes has great public health significance, and it is currently considered as one of the leading emerging pathogens associated with foodborne diseases [86].

The majority of cases in adults and juveniles occur among the immunosuppressed, i.e., patients receiving steroid or cytotoxic therapy or with malignant neoplasms. Other "at risk" groups include patients with AIDS, diabetics, elderly people, kidney dialysis patients, individuals with prosthetic heart valves, or replacement joints, and individuals with alcoholism or alcoholic liver disease. Approximately one-third of patients with listerial meningitis and around 10% with primary bacteremia are immuno-competent [87]. Rarer presentations in this patient group include meningoencephalitis and encephalitis, together with infections with identifiable foci, i.e., endocarditis, pneumonia, peritonitis, and deep-seated abscess formation. During the early stages of infection, human listeriosis often displays non-specific flu-like symptoms (e.g., chills, fatigue, headache, and muscular and joint pain) and gastroenteritis. However, without appropriate antibiotic treatment, it can develop into septicemia, meningitis, encephalitis, abortion and, in some cases, death [18]. Minor skin infections, particularly in farmers or veterinarians after handling bovine calves or abortions, have been recorded. The symptoms such as vomiting and conjunctivitis have also been seen. A case of *L. monocytogenes* septic arthritis of the right hip in a 66-year-old male treated with mycophenolate mofetil for polyarteritis nodosa was reported [88]. *L. monocytogenes* was reported as a cause of fatal spontaneous bacterial peritonitis in a 51-year-old man with alcoholic cirrhosis with jaundice, ascites, and lower limbs edema [89]. Anti-TNF-α drugs blocking

the host's response against various microorganisms, particularly intracellular agents like *L. monocytogenes*, increase the risk of disease. Two cases of *L. monocytogenes* meningitis in ulcerative colitis patients under infliximab plus steroids were reported [90].

Cases of *L. monocytogenes* endophthalmitis in human adults have also been reported [91]. A review of 38 consecutive patients with pleural infection, pneumonia, or both revealed *L. monocytogenes* as a rare but severe cause of pneumonia and pleural infection in older immunocompromised patients [92]. Endovascular listeriosis, manifested as vascular infections and endocarditis, mostly in older patients with vascular or cardiac valve prosthetic devices and co-morbidities, is a rare but severe infection. In addition, 71 cases with vascular aneurysms/prosthetic infections, endocarditis, or both were reviewed [93].

In South Africa, the world's largest listeriosis outbreak attributed to ready-to-eat processed meat products with a total of 1,060 confirmed cases was reported during 2017–2018 [94]. The clinical isolates were largely (91%) of sequence type 6 (ST6). Higher mortality was recorded in neonatal cohort probably due to more severe associated with ST6 and the predominance of early-onset disease [95]. An outbreak with *L. monocytogenes* ST2 clone lacking chitinase ChiB causing 163 cases of noninvasive listeriosis was reported in Germany in 2015 [96].

Interestingly in Europe, the incidence of listeriosis appears to have increased since 2000 [97]; in particular, the incidence among those >65 years old appears to have increased over the last years. Because of its high fatality rate, listeriosis ranks among the most frequent causes of death due to foodborne illness. *L. monocytogenes* infections are responsible for the highest hospitalization rates (91%) among known foodborne pathogens and have been linked to sporadic episodes and large outbreaks of human illness worldwide. Figures 32.5A and 32.5B show a stillborn with granulomatosis infantiseptica characterized by the presence of military-disseminated pyogranulomatous lesion on body surface and cut from the liver showing two subcapsular pyogranulomes, respectively.

Disease in Animals

The spectrum of the disease in animals is broad ranging from asymptomatic infection and carriage to uncommon cutaneous lesions or various focal infections such as conjunctivitis, urethritis, endocarditis, and severe disturbance of the gait, followed by death. Farm animals can be asymptomatic or cause encephalitis, septicemia, and abortions and may be a source of *L. monocytogenes* in the farm environment. *L. monocytogenes* causes invasive and often fatal disease including CNS infection in numerous animal species, including farm ruminants, horses, dogs, pigs, deer, South American camelids, and cats [98].

Abortions and perinatal deaths are common in cattle and sheep. *L. monocytogenes* is a well-recognized cause of mastitis, abortion, repeat breeding, infertility, encephalitis, and septicemia in cattle [99–100]. *L. ivanovii* has also been implicated as a cause of abortion in cattle and sheep but occurs less frequently than *L. monocytogenes*. Listeric infections and abortions usually develop in the late winter or early spring. Abortions are most commonly recognized in the last trimester of pregnancy, frequently in the absence of other clinical signs.

Immunity to *Listeria*

Listeria infection in mice has been exploited extensively as a model to study molecular mechanisms of early innate as well as adaptive immunity. Restriction of parenchymal and systemic spreading of *L. monocytogenes* is achieved by containment in granulomatous structures or abscesses containing cellular and acellular components. The major role in this process has been attributed to an intricate cooperation between macrophages, neutrophils, and T cells. During systemic *L. monocytogenes* infection in mice macrophages, neutrophils, natural killer (NK) cells, and dendritic cells are central to the early innate immune response, acting as both the host for and forming a major line of defense against this pathogen [101].

The development of a protective immune response to *L. monocytogenes* has been ascribed to the involvement of CD4+ and CD8+ T-cell populations. In this regard, class I MHC and class II MHC-restricted epitopes of LLO for CD8+ and CD4+ T-cells, respectively, have been identified [102]. CD8+ cells have been shown to lyse macrophages infected with *L. monocytogenes* or pulsed with *L. monocytogenes* peptides, latter requiring additional *in vitro* stimulation for expressing significant cytolytic activity [103]. It is also now known that CD4+ T-cell help is required during priming of CD8+ T-cells to generate a life-long protective memory response to *L. monocytogenes* infection [104]. The increased levels of non-MHC-restricted cytotoxic activity, mediated by NK cells, have also been observed following *in vivo* challenge with *L. monocytogenes* in a mouse model [105].

Live, replicating bacteria that grow in the host cell cytosol induce acquired cell-mediated immunity to *L. monocytogenes*, whereas killed bacteria, or those trapped in a phagosome, fail to induce protective immunity [106]. It has been shown that three innate immune pathways are induced by infection with *L. monocytogenes*: expression of inflammatory cytokines through an MyD88-dependent pathway emanating from a phagosome; a STING/IRF3-dependent pathway; and low levels of a Caspase-1-dependent, AIM2-dependent inflammasome pathway, which results in proteolytic activation and secretion of IL-1β and IL-18 and pyroptotic cell death [106].

Activation of TH₁ cells, subset of CD4+ T helper lymphocytes by various antigens, has been found to produce gamma-interferon [IFN γ] that stimulates macrophages [102]. Tumor necrosis factor and interferon gamma are essential for *in vivo* immune defense against primary infection with *L. monocytogenes* [107]. Full induction of protective immunity requires infection with live, cytosol-invasive *L. monocytogenes*. *L. monocytogenes* possesses many attributes that would make it a good antigen delivery system for vaccine strategies aimed at the induction of CD8+ T-cell responses [104].

Surveillance of Listeriosis

Listeria spp., including *L. monocytogenes*, is ubiquitous. The presence of this organism has been reported in the environment including soil, water, sewage, vegetation (e.g., grass, meadows, forests, silage), wild animal feces, as well as on the farm and in food processing facilities [108]. Contamination of food along the food chain has always been shown as the cause of listeriosis. Outbreaks from *L. monocytogenes* are rare; however, they receive considerable attention due to high case fatality rate and seriously affected cases. The outbreaks are often the result of errors made by workers in manufacturing plants leading to major economic consequences, especially if the products affect international trade. In the EU in 2007, the notification rate was the highest in the elderly (53.1% of cases,1.0 case per population of 100,000) followed by children under the age of 5 (0.51 cases per population of 100,000) [109]. Surveillance programs are largely lacking in developing countries. There, it is assumed that listeriosis cases rarely are reported and traditional food storage and preparation situations preclude opportunities for growth of the pathogen in RTE foods. Cold foods are very popular in the Middle East with a tendency to keep the food refrigerated for a longer time. In, India and Africa, there is little refrigeration in homes. Therefore, a comprehensive study for contamination of RTE foods by *L. monocytogenes* in developing countries should be considered by WHO and FAO. The lessons learned from the study and earlier outbreak investigations can then be shared with food control agencies in developing countries. Surveillance of listeriosis and monitoring of *L. monocytogenes* over many decades in developed countries has demonstrated that the pathogen has become a major contributor to human foodborne disease because of its high hospitalization and death rates.

The global estimates of *L. monocytogenes* prevalence have been reported to be at 2.9% in deli meat, at 2.4% in soft cheese, and in packaged salad at 2.0% [110]. However, global summary prevalence estimates for risk assessments are not advisable owing to the high heterogeneity between studies.

Identification of the global trends in *L. monocytogenes* epidemiology using ProMED reports from 1996 to2018 revealed 123 *Listeria* events [111]. A total of 65% events were associated with two or more human cases, 11% with only one human case, and 24% were precautionary food product recalls due to the presence of bacterial contamination without associated human cases. The food vehicle implicated was identified in 85% outbreak events and in 77% sporadic case events. Educating high-risk individual, such as pregnant women and immunocompromised individuals, about safe food-handling practices is required [111].

The economic cost estimates associated with foodborne disease are important to inform public health decision-making. Costs associated with the cases and those incurred by the implicated plant and federal agencies responding to the outbreak were estimated to be nearly $242 million Canadian dollars [112], demonstrating the considerable economic burden at both the individual and population levels associated with foodborne disease and foodborne outbreaks in particular.

Antigens of *Listeria*

Conventional Antigens

Somatic (O) and flagellar (H) antigens of *L. monocytogenes* have been described and used for serological groupings. The conventional antigens employed in various serological assays include heat-killed suspensions to detect antibodies to somatic antigens and formalin-treated organisms for flagellar antigens [113–114].

These antigens exhibit cross-reactions with *Staphylococcus aureus*, *Corynebacterium pyogenes*, *Escherichia coli*, *Streptococcus faecalis*, and *Bacillus* species [115], when employed in serological tests like agglutination, complement fixation, and immunoprecipitation. Trypsinization of heat-killed antigen of *L. monocytogenes* has been reported to increase its specificity and sensitivity besides avoiding cross-reactions [113].

Listeriolysin O (LLO)

Of the various factors incriminated for the virulence of *L. monocytogenes*, listeriolysin O (LLO) is highly immunogenic and can serve as an important target for innate and adaptive immune responses in different animal models and humans [116].

LLO is a secreted 58-kDa protein belonging to the group of cholesterol-dependent cytolysins and is antigenically related to streptolysin-O [SLO], pneumolysin, and perfringolysin [117]. It contained 529 amino acids and exhibits strong regional homologies to SLO and pneumolysin [118]. The purification of LLO by column chromatography with cross-linked polydextrans and carboxymethyl cellulose has been described [119]. When expressed in *L. innocua*, expression of LLO had been found to increase 500-fold and purification of LLO by ion-exchange chromatography yielded a homogeneous protein free of p60 [120].

The demonstration that LLO could be detected in serum samples after absorption with SLO [121] has been suggestive of specific antigenic epitopes in LLO; therefore, an immunoassay based upon purified LLO would be a reliable indicator of listeric infection [122–123]. LLO has been shown to be a reliable indicator for serodiagnosis of listeriosis in sheep by enzyme-linked immunosorbent assay (ELISA) [40] and in human beings by dot-blot assay, especially when bacteria could not be isolated [121, 124].

Protein p60 (cwhA)

Antibodies directed against the major secreted protein of *L. monocytogenes*, termed p60, have been found to be more frequently than antilisteriolysin antibodies in sera of listeriosis patients [125], which were also found to recognize the native and denatured secreted p60 protein of all serotypes of this species [126]. Antibodies raised against synthetic peptides derived from p60 could be useful for the development of an immunological assay specific for the foodborne pathogen *L. monocytogenes* [126].

Surface Proteins/Outer Membrane Proteins

An approach in immunological detection of pathogenic *L. monocytogenes* has been the use of antibodies directed against specific cell surface proteins/outer membrane proteins of the microorganisms. Certain bacterial outer membrane proteins/antigens have been reported to be specific [127]. Characteristic banding patterns of the cell surface proteins of various *Listeria* spp. by SDS-PAGE and immunoblotting have been reported [128]. The cell surface proteins are specific for species and serovars with molecular weights of 64 and 68 kDa. Monoclonal antibodies associated with 66 kDa cell surface antigen specific for *L. monocytogenes* have been produced [129], which could be useful for detecting the organism in clinical samples and foods.

Phosphatidylinositol-specific phospholipase C (PI-PLC) is an essential determinant of *L. monocytogenes* pathogenesis and secreted by only pathogenic species of *Listeria* [42]. It has been purified to homogeneity [130]. It is a 33-kDa protein with isoelectric point of 9.4. Its secretion can be enhanced by the addition of divalent cations to the culture medium. Expression of PI-PLC in the host tissues may be studied by detection of specific antibodies against it, which in turn may be useful in the diagnosis of *L. monocytogenes* infection. The serodiagnostic efficacy of two recombinant antigens namely, rLLO and rPI-PLC, revealed rLLO to be a better antigen than rPI-PLC in ELISA for the serodiagnosis of listeriosis in animals [131]

A ≈77-kDa antigen that is associated with the cell surface of live *L. monocytogenes* serotype 4b cells was identified by mass spectrometry and N-terminal sequencing to be IspC [132]. The presence of IspC was shown to be highly conserved within *L. monocytogenes* serotype 4b. The monoclonal antibodies developed had high fidelity and affinity for the IspC protein and serotype 4b isolates making it a promising candidate for use in the development of a specific *L. monocytogenes* serotype 4b diagnostic test.

Diagnosis of Listeriosis

Listeriosis can be tentatively diagnosed on the basis of clinical symptoms and demonstration of the organisms in smear by Gram staining, or by immunofluorescence. The organism can be isolated from clinical specimens, such as blood, cerebrospinal fluid (CSF), and meconium of newborns or fetus in abortion cases, and feces, vomitus, food stuffs/animal feed, and vaginal secretions of infected individuals or animals.

Detection of soluble antigen in CSF especially in meningitis cases of humans may be useful, but it is not reliable. Serodiagnostic methods, such as serum agglutination, complement fixation test, hemagglutination test, hemagglutination inhibition test, antibody precipitation test, growth inhibition test, and ELISA exist, but crude antigens employed in the tests show cross-reactivity. Listeriosis should be differentiated from influenza, tuberculous meningitis (in humans), and rabies, brucellosis, pasteurellosis, toxoplasmosis (in animals), especially in abortion cases. In view of the high cost, more time and skills required by isolation procedures along with their impracticability for screening large numbers of samples, the serological, novel surface antigens, and pathogenicity marker-based methods play an important role in its rapid and reliable diagnosis.

Isolation of *Listeria*

Attempts to isolate and identify *L. monocytogenes* and to limit its proliferation in foods have been the focus of a significant international effort [133]. Historically, it has been challenging to isolate *Listeria* from food or other samples and this explains why it remained unnoticed as a major food pathogen until recently. The earliest method available was the cold enrichment technique [134], which remained the only available method for many years. This required inoculation of the sample into a nutrient broth lacking selective agents followed by incubation at 4°C for long periods primarily to isolate the pathogen from infected animal or human tissue.

The key issues, enrichment/isolation time and the recovery of stressed *Listeria* cells, were addressed when further methods of enrichment and isolation were developed. Owing to many foodborne outbreaks of listeriosis, a zero tolerance level (absence in 25 g of food) has been implemented. Therefore, the tests must be able to detect one *Listeria* organism per 25g of food if to be approved by the regulatory agencies. *Listeria* cells are fastidious and can be rapidly outgrown by competitors. The recovery of *L. monocytogenes* from food, animals/human beings, and environmental samples requires the use of enrichment cultures followed by selective plating and where injured organisms are likely to be present, a pre-enrichment step is required [135]. These methods are sensitive but often time consuming, laborious, and may take 5–6 days before the result is available. Two of the most widely-used culture reference methods for detection of *Listeria* in all foods are the FDA bacteriological and analytical method (BAM) and the International Organization of Standards (ISO) 11290 method.

In the FDA BAM method, a sample (25 g) is enriched for 48 hours at 30°C in *Listeria* enrichment broth (LEB)[136] containing the selective agents acriflavine, nalidixic acid, and the antifungal agent cycloheximide, followed by plating onto selective agar (Oxford, PALCAM, MOX, or LPM) [10]. The ISO 11290 method employs a two-stage enrichment process: the first enrichment in half Fraser broth [137] for 24 hours, then an aliquot is transferred to full-strength Fraser broth for further enrichment followed by plating on Oxford and PALCAM agars. Fraser broth also contains the selective agents, acriflavine and nalidixic acid, as well as esculin, which allow detection of β-D-glucosidase activity by *Listeria*, observed as blackening of the growth medium [138]. Primary or pre-enrichment broths usually contain lower amounts of the selective agents to aid the resuscitation of possibly injured cells. Both primary and secondary broths contain a phosphate-buffering system.

The USDA and Association of Analytical Chemists (AOAC/IDF) methods use a modification of University of Vermont Medium (UVM) [139] containing acriflavine and nalidixic acid for primary enrichment, followed by secondary enrichment in Fraser broth and plating onto Modified Oxford (MOX) agar containing the selective agents, moxalactam and colistin sulfate [138].

The FDA method was designed for processing dairy products, whereas the USDA method was designed and has been officially recommended primarily for meat and poultry products [140], the latter being slightly superior for detection of *L. monocytogenes* in foods and environmental samples. The detection rate of *L. monocytogenes* in milk using the FDA method by changing the final Oxford and PALCAM plating agar to *L. monocytogenes* Blood Agar (LMBA) has been improved [141].

Selective or Differential Plating Media

Use of potassium tellurite to inhibit Gram-negative bacteria was the first significant step in producing a *Listeria* selective agar (LSA)[142]. Later on another LSA referred to as McBride *Listeria* agar (MLA) was formulated in 1960 by substituting phenyl ethanol agar containing lithium chloride, glycine, and blood [143].

Later, modified McBride *Listeria* agar (MMLA) [144] and LiCl-phenylethanol-moxalactam agar (LPMA) [135] were developed. The FDA method employed MMLA, while the USDA method used LiCl-phenylethanol-moxalactam agar (LPMA) as isolation agars [140].

A range of media has been developed, which include Oxford agar [145] LiCl-ceftazidime agar, modified (LCAM) [146], polymyxin-acriflavine-lithium chloride-ceftazidime-aesculin-mannitol (PALCAM), polymyxin-acriflavine-lithium chloride-ceftazidime-aesculin-mannitol egg yolk [147], Dominguez-Rodriguez isolation agar [148], Dominguez-Rodriguez *Listeria* selective agar medium, modified [149]; enhanced hemolysis agar (EHA) [145], modified Vogel Johnson agar (MVJ), [150],and MVJ modified further [151].

Chromogenic Media

Phosphatidylinositol-specific phospholipase C is an enzyme produced only by *L. monocytogenes* and *L. ivanovii*, which hydrolyses a specific substrate added to the medium, producing an opaque halo around the colonies. Substrates for the detection of β-glucosidase, common to all *Listeria* spp., may be used as elective feature for the detection of all *Listeria* colonies. Lithium chloride, nalidixic acid, and/or cyclohexamide are added to the media to obtain sufficient selectivity [152]. The chromogenic media that are commercially available include "Agar *Listeria* according to Ottaviani and Agosti," BCM *L. monocytogenes* detection system, CHROM agar, and Rapid'L.mono. Chromogenic media are simple, cost-effective, and easy to interpret [138]. Chromogenic media have been found to be a good supplementation to PALCAM agar [153].

Virulence Determination

Many *L. monocytogenes* strains are highly pathogenic, others are relatively avirulent and cause little harm in the host. A variety of methods have been developed to gauge the virulence of *L. monocytogenes* strains. Besides, hemolysin, mouse pathogenicity, tissue culture systems, and the detection of virulence-associated proteins and genes have been used to identify virulent *Listeria*.

Animal Inoculation

The mouse virulence assay has been frequently used for assessing *L. monocytogenes* virulence. This assay is done by inoculating groups of mice with varying doses of *L. monocytogenes* bacteria via the oral, nasal, intraperitoneal, intravenous, or subcutaneous routes [154]. Tests using oral inoculation in mice must be interpreted with caution since the main bacterial invasion protein internalin A does not bind to the mouse E-cadherin gene, resulting in inefficient translocation of these bacteria from the gastrointestinal tract to deeper tissues and organs [155]. The virulence of *L. monocytogenes* for man has been correlated with pathogenicity in mice [156], particularly in those made immunocompromised by treatment with carragecnan [157]. However, a large difference in the 50% of lethal doses (LD_{50}) between virulent and less virulent *L. monocytogenes* strains have been observed in immunocompromised model unlike the normal mice, which allows a clear and rapid means of distinguishing between the strains with a single dose inoculum and with the sole criterion being death of mice [158]. In some instances, the relatively high numbers of animals required have rendered LD_{50} determination impracticable; therefore, it has been emphasized that criterion for assessing virulence of *Listeria* strains on the basis of death is wrong and should be replaced more appropriately with persistence of microorganisms in liver or spleen following inoculation [159]. Recently, relative virulence (%) has been described as an alternative to LD50 measurement of the mouse virulence assay for *L. monocytogenes* [160]. The relative virulence (%) is obtained by dividing the number of dead mice by the total number of mice tested for a particular strain, using a known virulent strain (e.g., *L. monocytogenes* EGD) as a reference. All the nonhemolytic species of *Listeria* (*L. innocua*, *L. welshimeri*, and *L. grayi*) and the weakly hemolytic *L. seeligeri* are avirulent in mouse pathogenicity tests [156].

Besides mice, Sprague Dawley rats pre-treated with cimetidine have also been developed as an experimental model for gastric intubation [161]. The chick embryo test has been reported to agree with the mouse bioassay for assessment of the pathogenicity of *Listeria* species [162] as embryos inoculated with pathogenic strains through chorioallantoic membrane (CAM) route died within 72 hours, whereas those inoculated with nonpathogenic strains survived [163]. The LD$_{50}$ has been reported to be less than 6×10^2 cells for virulent strains [164]. Yolk sac challenge of 7-day-old chick embryos has been found to be less suitable than the CAM challenge for assessing virulence because of nonspecific deaths encountered in the former route [162].

Esophageal inoculation of juvenile rats with 10^6 cfu *L. monocytogenes* showed about a 50% infection rate in the liver or spleen [161]. An experimental keratoconjunctivitis test (Anton's eye test) can be performed in either guinea pigs or rabbits by inoculating a live bacterial suspension onto the eye [165].

Approximation of the infectious dose of *L. monocytogenes* in foods using cynomolgus monkey as a nonhuman primate model indicated that animals receiving 10^9 cells of *L. monocytogenes* became noticeably ill with symptoms of septicemia, irritability, loss of appetite, and occasional diarrhea [166].

The possibility of addressing aspects of mammalian innate immunity in invertebrates has opened a new arena for developing invertebrate models to study human infections [167]. The greater wax moth, *Galleria mellonella*, has been studied as a reliable model to investigate the *Listeria* pathogenesis [167, 168] and as a source of peptides exhibiting anti-*Listeria*-activity [167]. The *Galleria* model has been used to investigate the differences between infections caused by strains with different virulence potentials in the mouse infection model and revealed a strong correlation with virulence previously determined by the mouse model [167]. Using *Galleria* as a model, bacterial infection can be studied at 37°C as most of the human pathogens induce their virulence factors at this temperature. It has been shown that *G. mellonella* can be used to study brain infection and its impact on larval development as well as the activation of stress responses and neuronal repair mechanisms [169].

In vitro *Cell Assays*

L. monocytogenes is able to infect and grow intracellularly in a range of mammalian cell types growing *in vitro*, including enterocytes, macrophages, hepatocytes, neuronal cells, and fibroblasts [18, 170] as well as invertebrate, *Drosophila*

melanogaster-derived cells [171]. These methods measure the ability of *monocyto genes* to cause cytopathogenic effects in the enterocyte-like cell line Caco-2 [172] to form plaques in the human adenocarcinoma cell line HT-29 [173] and L929 mouse fibroblast [174] and also help to study the heterogeneity of virulence factors. *L. monocytogenes*, *L. ivanovii*, and *L. seeligeri* show properties of invasion and spreading, but other *Listeria* species do not [172, 175].

The main advantages of *in vitro* cell assays include their relatively low cost and ease of use. However, these tests are time-consuming and occasionally variable, which has prevented them from being adopted in clinical laboratories for determining *L. monocytogenes* virulence and pathogenic potential [19]

Expression of Virulence-Associated Factors

In vitro demonstration of LLO, PC-PLC, and PI-PLC activities often provides general guidance on the pathogenic potential of *L. monocytogenes* strains. However, its reliability as a virulence indicator is not satisfactory. Detection of *L. monocytogenes* virulence-associated genes by PCR and genetic lineage analysis has not resulted in a clear correlation between these genes and the underlying virulence of *L. monocytogenes* [99, 176, 177]. Also, mutations in virulence-associated genes, resulting in the expression of truncated or nonfunctional proteins [178–179], can occur. Targeting these gene mutations as a means of differentiating *L. monocytogenes* virulence often experience difficulties. An optimal strategy for *L. monocytogenes* virulence testing should be the detection of virulence-specific gene[s] that are present only in virulent strains, but absent in avirulent strains [19].

Immunological Tests

Conventional Tests

Serological methods have been reported to be nonspecific because of antigenic cross-reactivity between *L. monocytogenes* and other Gram-positive bacteria, such as *Staphylococci*, *Enterococci*, and *Bacillus* spp. [115]. Such methods also lack sensitivity, and, therefore, cannot be used for reliable diagnosis of listeriosis [121].

Enzyme-Linked Immunosorbent Assay (ELISA)

ELISAs for *L. monocytogenes* detection are either based on polyclonal antibody [180] or on monoclonal antibodies (mAbs) [181]. There are a number of ELISA formats including direct ELISAs, sandwich ELISAs, and competitive ELISAs. MAbs based-ELISA and dot-blot assays have been reported to identify *Listeria* spp. in food [182–183] and clinical samples [184]. However, some of these mAbs have been shown to react with all species of *Listeria* [182–183].

The *Listeria*-Tek assay, an ELISA-based method, has been reported to be a rapid and simple procedure for determination of *Listeria* spp. in foods [185]. The ELISA and the USDA procedure have been proved to be equally sensitive for processing the food samples having *L. monocytogenes* count >3 cfu/g and had detection limits of approximately 10^6 cfu/ml and 10^4 cfu/mL in pure cultures, respectively [186]. The ELISA for detection of

L. monocytogenes in dairy, seafood, and meat products has been adopted first action by AOAC International [187].

The most commonly used immunoassays for the detection of the pathogen are based on the use of whole cells. Often for detection of *L. monocytogenes* from foods, the sample is enriched, heat-killed [188–189], or formalin-fixed [190]. However, many of the cell-surface antigens are genus-specific rather than *L. monocytogenes*-specific [191–192]. Detection of the flagella of the bacterium has also been attempted [193–194].

ELISAs using the O and H antigens [195], as well as whole-cell protein extracts [196], have also been used to detect the pathogen. Monoclonal antibodies are used in the Vitek Immunodiagnostic Assay System (VIDAS)-LMO (bioMerieux Vitek, Hazeltown, MO) and Lister test (Vicam, Watertown, MA) [188, 197]. For field application, a point-of-care-testing (POCT) version of ELISA was employed using the concept of cross-flow chromatography [198] and applied for the detection of *L. monocytogenes* [199].

Lateral Flow Assay

Recently, a lateral-flow enzyme immunochromatography coupled with an immunomagnetic step was developed for rapid detection of *L. monocytogenes* in food matrices [200] and was demonstrated to be highly sensitive, specific, and rapid compared with spectroscopic methods [201, 202]. The assay yielded a limit of detection of 95 and 97 ± 19.5 cfu/mL in buffer solution and 2% milk sample, respectively [200] with the separation and detection time within 2 hours. The system did not show cross-reactivity with other bacteria (e.g., *Escherichia coli* O157:H7, *Salmonella typhimurium*, and *Salmonella enteritidis*). The analytical procedure could be useful in the field for rapid screening for pathogenic and as a preventive measure to contain disease outbreak.

Immunomagnetic Separation (IMS)

Paramagnetic polystyrene beads possessing pathogen-specific antibodies covalently linked to the bead surface are used, to which the pathogen antigens binds in the presence of a magnetic field [203]. Immunomagnetic nanoparticles coated with rabbit anti-*Listeria monocytogenes* antibodies were used for the rapid detection of *L. monocytogenes* from milk samples in combination with real-time PCR with the limit of detection greater than 10^2 cfu/0.5 mL [204]. In another study, paramagnetic beads coated with recombinant *Listeria* phage endolysin-derived cell-wall-binding domain proteins specific for *Listeria* spp. were used to detect bacteria in artificially contaminated raw milk with a sensitivity of 10^2 to 10^3 cfu/mL [205].

Immunofluorescence Test

The fluorescent antibody technique for the detection of *L. monocytogenes* in smear—impression smears from dead tissues, meat, and milk—was described by Khan *et al.* [206]. However, they reported cross-reaction with micrococci and streptococci. Later, immunofluorescence tests based on monoclonal antibodies for detection of *L. monocytogenes* from the clinical specimens, such as necropsy tissue [184, 207] and specific-protein p60, have been developed [208].

Cellular Assays

Cell-mediated immune (CMI) effector functions have been shown to play important role in the host defense against different facultative intracellular bacterial pathogens through cytokine production and perhaps by recognition and lysis of macrophages and other cells infected with these pathogens including *L. monocytogenes* [209]. The measurement of CMI can be accomplished by several methods such as delayed type hypersensitivity, lymphocyte proliferation assay, cytotoxicity assay, and cytokine assay, which are either generalized or specialized in nature [210].

Delayed-type hypersensitivity (DTH) is characterized by a localized swelling occurring 48–72 hours after intradermal injection with the microbial antigen under test. DTH carried out with soluble crude antigen fractions of nonpathogenic *L. innocua* (serotype 6a) or *L. monocytogenes* (serotype 4 b) has been shown to be useful in detecting listeriosis at an early stage [211] using experimental mouse model.

Cytotoxicity assays are based on the measurement of the ability of lymphocytes to kill target cells. Cytotoxic T-lymphocytes have been found to play an important role in the immunity of hosts to facultative intracellular bacteria [212].

Cytokine released by activated lymphocytes have been assayed to serve as a measure of CMI. Cytokine assays might be in the form of bioassays or a direct concentration measurement as in ELISA, radio-immunoassays, or precipitation assays or detection of specific mRNA in lymphocytes [210]. IFNγ in the blood streams and spleens of mice has been detected from Day 1 to Day 4 postinfection with *L. monocytogenes* by double sandwich ELISA as well as immuno-histochemical techniques [213]. IFNγ appeared as early as Day 7 of an oral infection in bovines [214].

Molecular Methods

DNA Probes

The presence of a target sequence is detected using a single-stranded nucleic acid that is enzyme- or radiolabeled. Datta *et al.* [215] reported the first DNA probe wherein a *Hind*III-*Hind*II fragment of a presumptive hemolysin gene was used in a trial for specific detection of *L. monocytogenes*. Nonisotopic colorimetric detection has been applied in a rapid nucleic acid dipstick hybridization assay for detection of *Listeria* sp. in food and environmental samples [216]. Different DNA probes such as digoxigenin-labelled synthetic oligonucleotide probe [217] on a *16S rRNA* sequence [218] have been developed. DNA probes targeting the inlA and *plc*A [219–220], and the *prf*A [221] genes have also been developed. As this procedure exploits differences among *Listeria* species at the genetic level, it is more specific than biochemical and serological methods that are phenotype based.

Fluorescence in situ Hybridization

Fluorescence *in situ* hybridization (FISH) is used to study the presence and distribution of specific strains in microbial communities. FISH can be used in phylogenetic studies, and in assessing the spatial distribution of target microbes in communities such as biofilms [222]. FISH using oligonucleotide probes specific for the virulence gene transcript, *iap*-mRNA has been carried out which allowed to analyze virulence gene expression of *L. monocytogenes* within a mixed microbial community [222].

Nucleic Acid Amplification

In the first reported PCR for identification of *L. monocytogenes*, the *hly* sequence published by Mengaud *et al.*[223] was used. This PCR was used to detect *L. monocytogenes* in water, whole milk, and human cerebrospinal fluid. Standard PCR followed by dot-blot hybridization has come out as an encouraging approach for the diagnosis of *Listeria* meningitis [224].

PCRs have been developed for the identification of *L. monocytogenes* in food samples [225–226] and in vegetables [227]. The minimum detection limits of *L. monocytogenes* in a 25-g sample after 48 hours incubation have been reported to be 10 cells [228] and 4–10 cells after overnight incubation [229]. Using a two-step PCR with nested primers, one colony-forming unit (CFU) *L. monocytogenes* could be directly detected in 25 mL of raw milk [230]. However, false-negative results have also been reported with PCR while analyzing foods containing high populations of *L. monocytogenes* [229].

As PCR has the ability to selectively amplify specific targets present in low concentrations, it offers exquisite specificity, unsurpassed sensitivity, rapid turnover, and ease of automation for laboratory detection of *L. monocytogenes* from clinical specimens [19]. The molecular differences within *16S* and *23S rRNA* genes, intergenic spacer regions, *hly*, *inlA*, *inlB*, *iap*, and other genes can be used to differentiate *L. monocytogenes* from other *Listeria* species and common bacteria [231]. Table 32.2 lists the genes employed for PCR-based identification of *Listeria*. The nested PCR protocol to detect the *hly*A

TABLE 32.2

Identification of *Listeria* Species by PCR-Based Procedures

Target Gene	References
16S rRNA gene	[344–345]
23S rRNA gene	[291,346]
16S/23S rRNA intergenic regions	[347–348]
hly	[99,349–353]
plcA, plcB	[42,99,177,219]
actA	[99,177,354]
prfA	[42,355–356]
inlA, inlB, inl AB	[177,220,350,357–358]
iap	[344,359–360]
lma/dth18	[361–363]
fbp	[364]
flaA	[365]
pepC	[366]
clpE	[219]
sigB	[367–368]
Vip	[369]
lse24-315	[370]
lin2483	[371]
liv22-228	[372]

gene has been used as a screening test to detect *L. monocytogenes* in Minas Frescal cheese, allowing earlier detection of the pathogen that can later be confirmed by the ISO standardized reference method [232].

Reverse Transcription (RT)-PCR

Testing of food or environmental samples for pathogenic *Listeria* should only target living organisms since only live *Listeria* cells can cause disease. The choice of RNA or mRNA as a target for food pathogen testing has gathered increasing favor since the presence of mRNA is an indication of the living state of the cell [233–234].

L. monocytogenes has been detected using RT-PCR in artificially inoculated meat samples [235] by targeting mRNA transcripts of the *hly*, *prfA*, and *iap* genes, in waste samples [236] by targeting the transcripts for *rRNA* genes, and also for the detection of heat-injured *monocytogenes* by targeting the *hly* transcript [237].

Real-Time PCR

Real-time PCR has been used to identify and quantify *L. monocytogenes* in foods and clinical samples in several studies [238–241]. Real-time PCR is quantitative, which is a significant advantage over other molecular methods, and so this technology is extremely attractive for food testing and epidemiological investigations. A combination of four qualitative SYBR®Green qPCR screening assays targeting two levels of discrimination, *Listeria* genus (except *Listeria grayi*) and *Listeria monocytogenes*, has been described [242] with the limit of detection of the SYBR®Green qPCR assays determined between two and five copies of target genes per qPCR reaction.

Other Techniques for Detection of *Listeria*

Biosensor-Based Techniques

The combined use of micro- and nanofabrication techniques in the area of biosensors holds a great promise in the area of detection of foodborne pathogens [243]. Aptamer-A8, specific for internalin A, was used in the fiberoptic sensor together with antibody in a sandwich format for detection of *L. monocytogenes* from food [244]. A mammalian cell-based biosensor, wherein the B-lymphocyte Ped-2E9 cell line embedded in a collagen matrix served as a sensing element, was developed and tested for detection of *L. monocytogenes* and its toxin, listeriolysin O, from artificially inoculated food samples [245]. It has a detection level of 10^2 to 10^4 cfu/g in artificially inoculated food and beverages. An internalin A probe-based genosensor that discriminates whole DNA samples of *L. monocytogenes* strains from other nonpathogenic *Listeria* species DNA has been developed [246].

Gold nanoparticles/horseradish peroxidase-encapsulated polyelectrolyte nanocapsule was developed for signal amplification in *L. monocytogenes* detection and demonstrated that the bioconjugate nanocapsules showed 30 times greater sensitivity and a shorter assay time (5 minutes) when compared to conventional ELISA using an HRP-labeled antibody, for a given quantity of antibody [247].

Loop-Mediated Isothermal Amplification (LAMP)

Loop-mediated isothermal amplification (LAMP) allows a rapid amplification of nucleic acids under isothermal conditions. LAMP can be combined with a rhodamine-based dual chemosensor for efficient, field-friendly detection of *L. monocytogenes*. The LAMP-chemosensor method was used to detect low levels of *L. monocytogenes* DNA. LAMP has the advantages of better sensitivity and speed and less dependence on equipment than the standard PCR and this can be useful in the field as a routine diagnostic tool [248].

LAMP assay targeting the *hly*A gene was developed and the amplification products were visualized by calcein and manganous ion and agarose gel electrophoresis. It has been opined that the LAMP assay can facilitate the surveillance for contamination of *L. monocytogenes* in food [249].

Spectroscopy-Based Techniques

Various spectrophotometric methods like Fourier transform infrared (FT-IR) and matrix-assisted laser desorption/ionization time-of-flight mass spectroscopy (MALDI-TOF MS) have been developed to detect foodborne pathogens [250]. A rapid method involving MALDI-TOF MS showed promise for identification of *Listeria* species and typing and even allowed for differentiation at the level of clonal lineages among pathogenic strains of *L. monocytogenes* [251]. The method examines the chemistry of major proteins, yielding profile spectra consisting of a series of peaks, a characteristic "fingerprint" mainly derived from ribosomal proteins. Mass spectra derived from *Listeria* isolates showed characteristic peaks, conserved at both the species and lineage levels.

The availability of reference databases such as MALDI Biotyper (Bruker Corporation) and SARAMIS (bioMérieux), which include reference spectra for numerous microorganisms, has made detection of microbes possible in a time-saving manner compared to the currently available screening and confirmation techniques for *Listeria* [252].

Other Methods

Other rapid methods for detecting *Listeria* spp. include the use of flow cytometry [95], nucleic acid sequence-based amplification [253], and electrical impedance [254].

Immunomagnetic separation used primarily to isolate strains of *L. monocytogenes* from pure cultures as well as from heterogenous suspensions has been viewed as a new approach for extraction and isolation of pathogenic bacteria directly from foods [255].

A fluorescent *in situ* hybridization (FISH] method in conjunction with fluorescein-labeled peptide nucleic acid (PNA) probes (PNA-FISH) for detection of *Listeria* species was developed [256]. Three PNA probes Lis-16S-1, Lm-16S-2, and Liv-16S-5 were suitable for specific identification of *Listeria* genus, *L. monocytogenes*, and *L. ivanovii*, respectively.

The Roka *Listeria* Detection Assay, in conjunction with a new media formulation (R2 Medium), allowed for the early detection of *Listeria* in ice cream and may be applied in other food matrixes and environmental samples [257].

Methods for Subtyping of *Listeria*

Subtyping plays an important role in epidemiology during outbreaks of *L. monocytogenes* and trace back investigations and also for understanding the pathophysiology of the organisms [258–259] and source attribution [260].

It is also useful to examine the epidemiology and population genetics of *L. monocytogenes* and is integral to control and prevention programs aimed at limiting listeriosis. A variety of conventional, phenotypic, and DNA-based subtyping methods have been described for differentiation of *L. monocytogenes* beyond the species and subspecies levels [261]. While phenotype-based methods have been used for many years to subtype *L. monocytogenes* and other foodborne pathogens, DNA-based subtyping methods are generally more discriminatory and amenable to inter laboratory standardization and are thus increasingly replacing phenotype-based subtyping methods [262]. Commonly used phenotype-based subtyping methods for *L. monocytogenes* and other foodborne pathogens include serotyping, phage typing, and multilocus enzyme electrophoresis (MLEE)[261, 263]. The genetic subtyping approach encompasses PCR-based approaches (e.g., random amplified polymorphic DNA and amplified fragment length polymorphism), PCR-restriction fragment length polymorphism (PCR-RFLP), ribotyping, pulsed-field gel electrophoresis, and DNA sequencing-based subtyping techniques, e.g., multilocus sequence typing (MLST) [264–267]. The phenotypic subtyping approach occasionally suffers from low discrimination and reproducibility, the genetic subtyping approach is highly sensitive, discriminatory, and reproducible. For improved subtyping discrimination, a combination of two or more subtyping techniques, be they gene or phenotype based, is often used in practice for epidemiologic investigation of *L. monocytogenes* outbreaks. The methods for subtyping of *Listeria* are listed in Table 32.3.

Phenotyping Typing Methods

Serotyping

Serotyping is a universally accepted subtyping method for *L. monocytogenes*. Identification of the strain serotype permits differentiation between important foodborne strains. *Listeria* species possess group-specific surface proteins, such as somatic (O) and flagellar (H) antigens that are useful targets for serological detection with corresponding monoclonal and polyclonal antibodies. While there are 15 *Listeria* somatic (O) antigen subtypes (I–XV), flagellar (H) antigens comprise four subtypes (A–D) [268–269] with the serotypes of individual *Listeria* strains being determined by their unique combinations of O and H antigens. An ELISA has recently been developed to improve efficiency of serotyping [195].

Serotyping may potentially be useful for tracking *L. monocytogenes* strains involved in disease outbreaks. Indeed, it has been observed that *L. monocytogenes* serotypes 1/2a, 1/2b, and 4b are responsible for 98% of documented human listeriosis cases, whereas serotypes 4a and 4c are rarely associated with outbreaks of the disease [270–271]. However, due mainly to the high cost of acquiring subtype-specific antisera, serotyping methods are

TABLE 32.3

Methods for Subtyping of *Listeria* Species

Methods	References
Phenotypic Methods	
Serotyping	[195,265]
Phage typing	[275,373–374]
MLEE	[275]
Esterase typing	[375]
Genotypic Methods	
PFGE	[266,287,376]
Ribotyping	[282–283,377]
RFLP	[282]
RAPD	[285–286]
AFLP	[288,290,378]
PCR-RFLP	[291]
REP-PCR	[292]
DNA microarray	[301–303]
DNA sequencing based	[265,282,379]
MLST	[378,380]
MVLA	[296,381]
Multilocus TR sequence analysis (MLTSA)	[382]
Whole genome sequencing	[278, 383, 384]

not routinely performed in clinical laboratories. Serotyping does not correlate with the species distinctions. Serotyping methods have largely been superseded by molecular procedures that are intrinsically more specific and sensitive for the identification and differentiation of *Listeria* species. The development of PCR-based serotyping procedures has provided additional tools for the identification and grouping of *L. monocytogenes* [272–273]. The multiplex PCR method, in combination with a simple slide agglutination assay, can be very useful in quickly identifying serotypes of major *L. monocytogenes* serotypes [274].

Phage Typing

Phage typing of *L. monocytogenes* has been used as a discriminatory typing system. Bacteriophages have the capacity to lyse closely related *Listeria* because of their host specificity independently of the species and serovar identities. *Listeria* strains can be separated into distinct phage groups and phagovars, which are useful for tracking the origin and course of listeriosis outbreaks [275]. With close to 10% of *Listeria* strains being untypable (especially serovar 3 and *L. grayi* strains), the usefulness of phage typing as an independent tool for epidemiological investigations is severely constrained.

Multilocus Enzyme Electrophoresis (MLEE)

MLEE is a protein-based, isoenzyme typing method that correlates specific protein-band patterns with genotypes. MLEE differentiates bacterial strains by detecting variations in the patterns of the electrophoretic mobilities of various constitutive enzymes. Based on the similar electrophoretic types detected in

MLEE, *L. monocytogenes* serovars 1/2b, 3b, and 4b are classified into one distinct division, and serovars 1/2a, 1/2c, and 3a in another division [276–277]. The detection of a large number of electrophoretic types in *L. monocytogenes* strains by MLEE necessitates careful optimization and standardization of the test procedure so that run-to-run variations are minimized.

Genetic Typing Methods

DNA-based typing of *L. monocytogenes* can be divided into two broad categories. In the first approach, different fragments of genome generated either by cleaving with restriction enzymes are compared or amplified segments of DNA are visualized after gel electrophoresis separation [278]. The techniques include ribotyping (RT), pulsed field gel electrophoresis (PFGE), rep-PCR, and random amplification of polymorphic DNA. The second approach involves the sequence-based subtyping, which is faster and cheaper. The techniques include whole genomic sequencing (WGS), next-generation sequencing (NGS), multilocus variable tandem repeats analysis (MLVA), and DNA-based microarrays.

Pulse Field Gel Electrophoresis (PFGE)

PFGE is a highly reproducible, discriminatory and effective molecular typing method that is based on restriction fragment length polymorphisms (RFLPs) of bacterial DNA [12] and provides a "gold standard" for comparing isolates analyzed in different labs with different techniques. PFGE uses selected restriction enzymes to yield fewer and larger fragments of DNA, resulting in a higher level of fragment resolution. PFGE provides sensitive subtype discrimination and is often considered the standard subtyping method for *L. monocytogenes* [279]. However, this method is not automated and is labor intensive.

PulseNet, a national network of public health and food regulatory laboratories, is established in the United States to detect clusters of foodborne disease and respond quickly to foodborne outbreak investigations [265]. The laboratories use a highly standardized 1-day PFGE to subtype the bacteria and exchange normalized DNA fingerprint patterns via the internet. PFGE has become the mainstay in molecular subtyping activity. Using standard protocols, the CDC has initiated PulseNet USA to store, analyze, and share digital PFGE images on a real-time basis. This assists in outbreak and trace back investigations of outbreaks caused by foodborne bacterial pathogens [280–281].

Ribotyping

Ribotyping is similar to RFLP in that it uses restriction endonuclease digestion of DNA to create a pattern that can be analyzed. It involves the restriction enzyme digestion of chromosomal DNA followed by DNA hybridization using an *rRNA* gene probe. Weidmann et al. [282] used ribotyping on different isolates of *L. monocytogenes* to group different lineages and relate the pathogenicity of the organism. *Eco*RI ribotyping was used to demonstrate the level of genetic diversity among *L. monocytogenes* contamination of Gorgonzola cheeses [283]. Although several attempts have been made to use ribotyping as a standard tool for strain characterization, its usefulness was found to be limited because of its low discriminatory power [284].

RAPD and Arbitrarily Primed PCR (AP-PCR)

Randomly amplified polymorphic DNA (RAPD) and AP-PCR analysis make use of a short arbitrary primer (usually 10 bases long for AP-PCR, and 10–15 bases long for RAPD) that anneals randomly along genomic DNA to amplify a number of fragments within the genome at a relatively low temperature (around 36°C). RAPD was used to follow the incidence and typing of *L. monocytogenes* in both poultry [285] and vegetable [286] processing plants. RAPD is more economical and faster than other typing methods and is particularly suitable for testing fewer than 50 strains.

Amplified Fragment Length Polymorphism (AFLP)

AFLP is a modification of RFLP through the addition of adaptors to restriction enzyme-digested DNA, followed by PCR amplification and electrophoretic separation of PCR products [19]. AFLP can be used to differentiate strains of *L. monocytogenes* on a more discriminating basis than serotyping [242–243, 287–288]. AFLP is sensitive and reproducible, thus representing a valuable tool in the characterization of *L. monocytogenes* strains, and also in the identification of *Listeria* species [289–290].

PCR-RFLP

PCR-RFLP involves the PCR amplification of one or more *L. monocytogenes* housekeeping or virulence-associated genes (e.g., *hly*, *act*A, and *inl*A), digestion with selected restriction enzymes and separation by agarose gel electrophoresis. Subsequent examination of the distinct band patterns permits differentiation of *L. monocytogenes* subtypes [164]. Paillard et al. [291] used PCR–RFLP on the 23S rRNA gene to determine the species of *Listeria* in sludge samples. PCR-RFLP provides a sensitive, discriminatory, and reproducible method for tracking and epidemiological investigation of *L. monocytogenes* bacteria, if used in combination with other subtyping procedures.

REP-PCR

Repetitive extragenic palindromic (REP) and enterobacterial repetitive intergenic consensus (ERIC) sequences represent useful primer-binding sites for PCR amplification of the *L. monocytogenes* genome to achieve species and strain discrimination and divide *L. monocytogenes* strains into four clusters that match the origin of isolation, each consisting of multiple subtypes [292]. REP-PCR showed a higher discriminative potential than ERIC-PCR and a comparable discriminative potential as RAPD combining three to four primers [292].

MLST

DNA sequencing-based methods are being developed and increasingly used for subtyping and characterizing bacterial isolates. In these methods, complete or partial nucleotide sequences are determined for one or more bacterial genes or chromosomal regions, thus providing unambiguous and discrete data. The advantages of sequencing methods over DNA fragment size-based typing methods include their ability to generate unambiguous data that are portable through web-based databases and that can be used for phylogenetic analyses [293–294]. While a variety of DNA sequence-based subtyping strategies targeting virulence genes, housekeeping genes, or other chromosomal genes and regions are feasible, MLST, which is an extension of MLEE, represents a widely used strategy [295].

Multilocus sequence typing (MLST) is a reference method for global epidemiology and population biology of bacteria. The application of MLST to *L. monocytogenes* effectively allows isolate comparisons across laboratories (www.pasteur.fr/mlst). The existence of few prevalent and globally distributed clones has been shown following, genotyping of 300 isolates from the five continents and diverse sources, some of which include previously described epidemic clones [296].

The implementation of DNA sequence-based subtyping approaches for routine characterization of human, animal, and food *L. monocytogenes* isolates will not only allow for sensitive and standardized subtyping for outbreak detection, but it will also provide an opportunity for using subtyping data to probe the evolution of this foodborne pathogen and to track the spread of clonal groups [262]. With the cost of DNA sequencing decreasing rapidly, MLST is poised to play a more important role in *L. monocytogenes* subtyping and phylogenetic studies.

MLVA

Multilocus variable tandem repeats analysis (MLVA), based upon identifying a number of small repeats (5–10bp) at different genetic loci, is a sequence-based method. Each locus is PCR amplified by specific primers resulting in different fragment lengths and translated into to the number of repeats. These repeats are arranged in a string to form a pattern representing a strain. MLVA could be automated for high throughput and high reproducibility has been found to be useful in discriminating strains from a single outbreak from the unrelated strains. MLVA could be a very useful tool for discriminating pure cultures as well as food enrichment broths directly, thereby reducing the time for detection and subtyping [297]. A modified protocol using five loci has been evaluated and found to be slightly more discriminatory than single-enzyme PFGE [298]. A website, MLVA-NET (http://www.pasteur.fr/recherche/genopole/PF8/mlva/) proposing the five-loci method as the standard method for MLVA analysis of *L. monocytogenes*, is maintained by the Institute Pasteur [278].

Using various subtyping procedures, *L. monocytogenes* strains have been grouped into three genetic lineages (or divisions), with lineage I consisting of serotypes 1/2b, 3b, 4b, 4d, and 4e; lineage II of serotypes 1/2a, 1/2c, 3a, and 3c; and lineage III of serotypes 4a and 4c [299]. A recent classification describes four lineages of *L. monocytogenes* with coincident niches: lineage I encompasses serotypes 1/2b, 3b, 4b, and 3c; lineage II includes serotypes 1/2a, 1/2c, and 3a; lineage III comprises serotypes 4a, 4b, and 4c; and lineage IV comprises 4a, 4b, and 4c [300]. Most isolates belong to lineages I and II, comprising the serotypes most commonly associated with human clinical cases. Lineage II strains are commonly isolated from food and animals with listeriosis, whereas outbreaks of human listeriosis are associated with lineage I. Lineages III and IV strains are rare and mostly isolated from animal sources [300].

DNA Microarrays

Microarrays are composed of a number of discreetly located DNA probes corresponding to an oligonucleotide specific to a target DNA sequence fixed on a solid substrate, such as glass. DNA microarrays are used to investigate microbial evolution and epidemiology and can serve as a diagnostic tool for clinical, environmental, or food testing. Two approaches are used, one is based on sequence-specific oligonucleotides and the other employs specific PCR products.

Several microarray-based strategies were developed to differentiate between the six listerial species [219] and for discrimination among *L. monocytogenes* serovars [301–302] and phylogenetic lineages [303–304]. Microarray-based assays were also applied to investigate the genome evolution within the genus *Listeria* [299] and for identification of natural atypical *L. innocua* strains harboring genes of the LIPI-1 [305].

Oligonucleotide microarrays based on the *iap* and *hly* genes have been used simultaneously to detect and discriminate between *Listeria* species [219]. Borucki et al. [301] developed a mixed genome microarray to identify gene sequences that differentiate different serotypes of *L. monocytogenes*. The most attractive feature of microarrays is the capability of simultaneous identification and typing of several *Listeria* strains in one test.

Genome Sequencing

Since the 1990s, the genome sequencing methods have been developed for strain identification, strain characterization for epidemiological use and for establishing phylogenetic relationships and genetic evolution of organisms [278]. New generation sequencing techniques that are relatively cheap and fast have revolutionized the sequence-based subtyping and our understanding of the genome structure. More than 50,000 whole genome sequences are available as of the year 2020 in public databases representing several outbreak and nonoutbreak-associated strains [214].

The availability of genome sequences of *L. monocytogenes* serovars, *L. innocua*, *L. welshimeri*, and *L. seeligeri* [306–309], has provided insight into the molecular basis of the pathogenesis determinants of *Listeria* species. Further, sequencing of the strains from other species of this genus will provide a rich resource for understanding the sources of variation and evolutionary history. Comparative genomics could reveal genetic loci that confer specific pathogenic traits to epidemic strains. The information provides access to niche-specific candidate genes and proteins that will need to be evaluated by traditional physiological, biochemical, and genetic approaches, in turn enabling to develop novel ways to interrupt transmission and prevent foodborne transmission.

Prevention and Control

Listeriosis can be prevented in humans by taking care during handling of the abortion cases in both humans and animals, avoiding consumption of contaminated foodstuffs and cross-infections, especially among infants in hospitals. Culling infected animals should be advocated as they secrete the organisms in secretions and excretions, especially in the cases of mastitis. Care in the use and preparation of silage is important as the pathogen grows luxuriantly at a pH greater than 5, particularly when fermentation is ineffective and molds grow. So far, no vaccine is available against listeriosis.

The government, industry, and consumer are the most important parties in ensuring food safety. The government should establish standards and develop codes of practice and ensure that these will be followed. Food producers and food preparers should produce safe food, achieved by adequate hygiene standards, good manufacturing practices (GMP), and implementation of hazard analysis and critical control points (HACCPs). Consumers, in particular those who buy food and prepare meals, should have a basic knowledge of safe food preparation.

Treatment

The organism is, so far, usually sensitive to a wide range of antibiotics [310]. Ampicillin, amoxicillin, tetracyclines, chloramphenicol, β-lactam antibiotics, together with an aminoglycoside, trimethoprim, and sulfamethoxazole are recommended. The use of ampicillin plus gentamicin is recommended to treat CNS infections, endocarditis, and infections in immunocompromised patients [311–312]. As per the IDSA guidelines of *Listeria* encephalitis, the use of ampicillin plus gentamicin or SXT (evidence A-III) is recommended [313]. Based on the delayed *in vitro* bactericidal activity of ampicillin, the synergistic effect of both agents *in vitro*, and on results of animal models [314], an aminoglycoside was added. However, it has been opined that the potential benefit attained with the addition of an aminoglycoside may decrease due to the risk of nephrotoxicity and the possible teratogenic effect in pregnant women [315]. Use of combination of cotrimoxazole has been suggested as it is effective *in vitro* [316] and bactericidal, and when combined with ampicillin than the combination of ampicillin plus gentamicin (93.3% vs. 43%) [317].

Recently, it has been found that fosfomycin is extremely effective in controlling listerial growth in a mouse model of infection. The basis of this exquisite sensitivity has been elucidated by the finding that following entry of the bacteria to the cytoplasm, intracellularly growing *Listeria* express a permease for the uptake of hexose phosphates (UhpT), and that fosfomycin is preferentially translocated into growing bacteria via this transporter [318].

L. monocytogenes infections are usually treated with a single antimicrobial agent and drugs are only combined for treatment of immunocompromised patients [319]. Recently, reports have described clinical strains resistant to chloramphenicol, erythromycin, streptomycin, tetracycline, vancomycin, and trimethoprim [320–322]. Resistant *L. monocytogenes* strains have also been found in food samples [323]. Use of the disinfectants, Virkon®S 2.0%, H_2O_2 5.0%, and TH^{4+} 1.0%, loaded on silver nanoparticles composite had the strong bactericidal effect against *L. monocytogenes* [324]. Currently, for treatment of *L. monocytogenes* endocarditis penicillin G or ampicillin (AMP) monotherapy, or AMP + gentamicin combination therapy is often favored. In an investigation AMP + ceftriaxone (CRO) and AMP + daptomycin (DAP) were each found to have synergistic activity against a *L. monocytogenes* endocarditis isolate [325].

Biotechnological Applications

L. monocytogenes has become an important paradigm for immunological investigation and also an important model system for analysis of the molecular mechanisms of intracellular parasitism and pathophysiology [45]. For many years *L. monocytogenes* has been used as a model of host-disease immunology,

As a facultative intracellular bacterium, *L. monocytogenes* survive within cells after phagocytosis, it is an ideal vector for the delivery of antigens to be processed and presented through both the class I and II antigen-processing pathways. It is possible that the virulence factors released in cytosol can enhance the immunogenicity of tumor-associated antigens, which are poorly immunogenic [326]. In several mouse models, *Listeria*-based vaccines have been demonstrated to be an effective method of influencing tumor growth and eliciting potent antitumor immune responses [327]. Active immunotherapy targeting dendritic cells (DCs) have shown great promise in preclinical models and in human clinical trials for the treatment of malignant disease [328]. Live-attenuated *L. monocytogenes* found to target naturally the DCs *in vivo* and stimulate both innate and adaptive cellular immunity. *L. monocytogenes*-based vaccines engineered to express cancer antigens have demonstrated striking efficacy in several animal models and have resulted in encouraging anecdotal survival benefit in early human clinical trials [328]. Many synchronous and disparate action mechanisms stimulating innate and cell-mediated adaptive immunity and reducing immunosuppressive influences in the tumor microenvironment are utilized by *L. monocytogenes* immunotherapy [329].

LLO incorporated into liposomes can be used as an efficient vaccine delivery system carrying a viral antigenic protein to generate protective antiviral immunity [330].

Therapeutic vaccines of cancer including live vaccination strategies, having the capacity of breaking the immune tolerance and invoking long-term immune response targeting cancer cells without autoimmunity, have been emerged as an attractive approach [331]. Several bacterial strains including *L. monocytogenes* effectively colonize solid tumors and act as antitumor therapeutics. *Listeria* has been engineered to express a number of HIV/SIV antigens and has been tried as a live bacterial vaccine vector for the delivery of HIV antigens [332].

Listeria monocytogenes as a Vector for Cancer Immunotherapy

L. monocytogenes has been found to selectively infect antigen-presenting cells. It activates tumor-targeting CTL-mediated immunity through efficient delivery of tumor antigens to both the MHC Class I and II pathways [333–334]. *L. monocytogenes* is a versatile bacterial vector and has ability to induce therapeutic immunity against a wide-array of tumor-associated antigens (TAAs). Therefore, *L. monocytogenes*-based immunotherapies have delivered impressive therapeutic efficacy in preclinical models of cancer for two decades and are now showing promise clinically. The history leading up to the development of current Lm-based immunotherapies is extensively reviewed [333].

Incorporation of multiple epitopes from several antigens into a polyvalent vaccine has been suggested as an effective strategy going forward for *L. monocytogenes*-based immunotherapies to mitigate immune escape [335]. It has been suggested that in addition to the development of polyvalent *L. monocytogenes*-based vaccines, the new version could explore tumor tropism of their attenuated *L. monocytogenes* vaccines [336]. Treatment of stubborn malignancies such as pancreatic ductal carcinoma could be harnessed through the primary and metastatic tumor tropism and killing by *Listeria* [336]. Owing to this tumor tropism, a possible synergistic strategy of *L. monocytogenes*-based vaccines may go forward to deliver therapeutic proteins or expression plasmids to the tumor microenvironment [337]. This synergism of *L. monocytogenes*-based immunotherapies in combination with radiation, adjuvants, and therapeutic antibodies suggests versatile use [338–339].

L. monocytogenes infection has been known to induce robust $CD8^+$ T-cell responses having a critical role in resolving primary infection and provide protective immunity to reinfections [340]. The listeriolysin O peptide 91-99 ($GNP-LLO_{91-99}$) having antimetastatic properties, coupled with gold glyconanoparticles, are vaccine delivery platforms that facilitate immune-cell targeting and increase antigen loading [341] and have been proposed to function as immune stimulators and immune effectors and as safe cancer therapies, alone or in combination with other immunotherapies. *L. monocytogenes* immunotherapy with a broad effect on the immune system, axalimogene filolisbac (ADXS11-001), had shown promising results in Phase 2 studies for treatment of human papillomavirus-associated cancers, including cervical cancer [342].

Immunotherapy with $CD8^+$ T-cells activated by *Listeria* in young and old mice had shown an antitumor effect at young but not at old age; however, the metastases at old age were shown to be eliminated through different mechanism(s) [343].

Outlook and Perspective

Listeria remains among the deadliest known foodborne pathogen worldwide. Great progress has been made in eliciting the mechanisms leading to pathogenesis of this bacterium and a large battery of tests are now available to detect and diagnostically classify *Listeria* both at the species level and even permit discrimination between strains associated with human and animal infections or even sporadic from epidemic strains. The advent of functional listerial genomics is providing new possibilities to comprehensively catalogue genes and their products involved in the transition from life in the environment to life within infected cells. The availability of genome sequences will provide us with an opportunity to define the evolutionary paths taken to pathogenesis and help us understand the emergence of *Listeria* spp. as major foodborne pathogens. Recombinant *Listeria* is reported to be effective delivery vectors for the development of cancer vaccines, which needs to be explored further.

The discovery of isolates harboring pan-species virulence genes of the genus *Listeria* and novel species has opened new areas of research in genus *Listeria*.

We now have a good understanding of those listerial factors that predispose to biofilm formation and persistence. Thus, ability to determine hypervirulence properties of emerging *L. monocytogenes* variants has increased our understanding of virulence potential that can now be translated to probiotic health and food safety measures. The increase of incidence causing both sporadic and epidemic disease worldwide suggests that increased vigilance must now be implemented, particularly in food-processing facilities.

Acknowledgements

We thank M. Rohde (Hemholtz Institute for Infectious Disease Research, Braunschweig) for electron microscopy. The work reported in this manuscript was made possible by grants from the Bundesministerium fuer Bildung und Forschung (BMBF) through the German Centre for Infection and the Deutsche Forschungs Gemeinschaft (DFG) to T.H. and T.C., and from the Department of Biotechnology, Government of India to S.B. [BT/01/CEIB/11/VI/13].

REFERENCES

1. Nightingale, K.K. et al. Ecology and transmission of *Listeria monocytogenes* infecting ruminants and in the farm environment. *Appl. Environ. Microbiol.*, 70, 4458, 2004.
2. Castro, H.K. and Lindström, M. Ecology and surveillance of *Listeria monocytogenes* on dairy cattle farms. *Int. J. Infect. Dis.*, 53, 68, 2016.
3. Sauders, B.D. et al. Diversity of *Listeria* species in urban and natural environments. *Appl. Environ. Microbiol.*, 78, 4420, 2012.
4. Linke, K. et al. Reservoirs of *Listeria* species in three environmental ecosystems. *Appl. Environ. Microbiol.*, 80, 5583, 2014.
5. Hülpers, G. Lefvernekros hos kanin orsakad af en ej förut beskrifven bakterie. (Liver necrosis in rabbit caused by an hitherto unknown bacterium). *Svensk. Vet. Tidskrift.*, 2, 265, 1911.
6. Murray, E.G.D., Webb, R.A. and Swann, M.B.R. A disease of rabbits characterised by a large mononuclear leukocytosis, caused by a hitherto undescribed bacillus *Bacterium monocytogenes* (n. sp.). *J. Pathol. Bacteriol.*, 29, 407, 1926.
7. Pirie, J.H.H. A new disease of veld rodents. 'Tiger river disease'. *Publ. S. Afr. Inst. Med. Res.*, 3, 163, 1927.
8. Nyfelt, A. Etiologie de la mononucle´ose infectieuse. *C.R. Soc. Biol.*, 101, 590, 1929.
9. Pirie, J.H. The genus *Listerella* pirie. *Science*, 91, 383, 1940.
10. Koch, J. and Stark, K. Significant increase of listeriosis in Germany: epidemiological patterns 2001-2005. *Euro. Surveill.*, 11, 85, 2006.
11. Kasper, S. et al. Epidemiology of listeriosis in Austria. *Wien. Klin. Wochenschr.*, 121, 113, 2009.
12. Muñoz, P. et al. Listeriosis: an emerging public health problem especially among the elderly. *J. Infect.*, 64, 19, 2012.
13. Rodríguez-López, P. et al. Quantifying the combined effects of pronase and benzalkonium chloride in removing late-stage *Listeria monocytogenes–Escherichia coli* dual-species biofilms. *Biofouling.*, 33, 690, 2018.
14. EFSA. The European Union summary report on trends and sources of zoonoses, zoonotic agents and food-borne outbreaks in 2014. *EFSA J.*, 13, 4329, 2015.
15. Leclercq, A., Charlier, C. and Lecuit, M. Global burden of listeriosis: the tip of the iceberg. *Lancet Infect Dis.*, 14, 1027, 2014.
16. Pietzka, A. et al. Whole genome sequencing based surveillance of *L. monocytogenes* for early detection and investigations of listeriosis outbreaks. *Front. Public Health.*, 7, 139. 2019.
17. Maurella, C. et al. Outbreak of febrile gastroenteritis caused by *Listeria monocytogenes* 1/2a in sliced cold beef ham, Italy, May 2016. *Euro Surveill.*, 23, 2018.
18. Vazquez-Boland, J.A. et al. *Listeria* pathogenesis and molecular virulence determinants. *Clin. Microbiol. Rev.*, 14, 584, 2001.
19. Liu, D. Identification, subtyping and virulence determination of *Listeria monocytogenes*, an important foodborne pathogen. *J. Med. Microbiol.*, 55, 645, 2006.
20. Pizarro-Cerdá, J. and Cossart, P. *Listeria monocytogenes*: cell biology of invasion and intracellular growth. *Microbiol Spectr.*, 6, 2018.
21. Lomonaco, S., Nucera, D. and Filipello, V. The evolution and epidemiology of *Listeria monocytogenes* in Europe and the United States. *Infect. Genet. Evol.*, 35, 172, 2015.
22. de Noordhout, C.M. et al. The global burden of listeriosis: a systematic review and meta-analysis. *Lancet Infect. Dis.*, 14, 1073, 2014.
23. Churchill, R.L., Lee, H. and Hall, J.C. Detection of *Listeria monocytogenes* and the toxin listeriolysin O in food. *J. Microbiol. Methods.*, 64, 141, 2006.
24. Hain, T., Steinweg, C. and Chakraborty, T. Comparative and functional genomics of *Listeria* spp. *J. Biotechnol.*, 126, 37, 2006.
25. Lamont, R.F. et al. Listeriosis in human pregnancy: a systematic review. *J. Perinat. Med.*, 39, 227, 2011.
26. Milillo, S.R. et al. A review of the ecology, genomics and stress response of *Listeria innocua* and *Listeria monocytogenes*. *Crit. Rev. Food Sci. Nutr.*, 52, 712, 2012
27. Orsi, R.H. and Wiedmann, M. Characteristics and distribution of *Listeria* spp., including *Listeria* species newly described since 2009. *Appl. Microbiol. Biotechnol.*, 100, 5273, 2016.
28. Núñez-Montero, K. et al. *Listeria costaricensis* sp. nov. *Int. J. Syst. Evol. Microbiol.*, 68, 844, 2018.
29. Doijad, S.P.et al. *Listeria goaensis* sp. nov. *Int. J. Syst. Evol. Microbiol.*, 68, 3285, 2018.
30. Leclercq, A. et al. *Listeria thailandensis* sp. nov. *Int. J. Syst. Evol. Microbiol.*, 69, 74, 2019.
31. Lang Halter, E., Neuhaus, K. and Scherer, S. *Listeria weihenstephanensis* sp.nov., isolated from the water plant Lemna trisulca of a German fresh water pond. *Int. J. Syst. Evol. Microbiol.*, 63, 641, 2013.
32. Bertsch, D. et al. *Listeria fleischmannii* sp. nov., isolated from cheese. *Int. J. Syst. Evol. Microbiol.*, 63, 526, 2013.
33. Weller, D. et al. *Listeria booriae* sp. nov. and *Listeria newyorkensis* sp. nov., from food processing environments in the USA. *Int. J. Syst. Evol. Microbiol.*, 65, 286, 2015.
34. den Bakker, H.C. et al. *Listeria floridensis* sp. nov., *Listeria aquatica* sp. nov., *Listeria cornellensis* sp. nov., *Listeria riparia* sp. nov. and *Listeria grandensis* sp. nov., from agricultural and natural environments. *Int. J. Syst. Evol. Microbiol.*, 64, 1882, 2014.
35. Schmid, M.W. et al. Evolutionary history of the genus *Listeria* and its virulence genes *Syst. Appl. Microbiol.*, 28, 1, 2005.
36. Graves, L.M. et al. *Listeria marthii* sp. nov., isolated from the natural environment, Finger Lakes National Forest. *Int. J. Syst. Evol. Microbiol.*, 60, 1280, 2010.

37. Grau, F.H. and Vanderlinde, P.B. Growth of *Listeria monocyto-genes* on vacuum packaged beef. *J. Food Prot.*, 53, 739, 1990.

38. Bille, J. et al. API *Listeria*, a new and promising one-day system to identify *Listeria* isolates. *Appl. Environ. Microbiol.*, 58, 1857, 1992.

39. Robinson, R.K., Batt, C.A. and Patel, P.D., eds. *Encyclopedia of Food Microbiology*, Academic Press, San Diego, CA, 2000.

40. Rocourt, J. and Catimel, B. Biochemical characterization of species in the genus *Listeria*. *Zentralbl. Bakteriol. Mikrobiol. Hyg. [A]*, 260, 221, 1985.

41. Ermolaeva, S. et al. A simple method for the differentiation of *Listeria monocytogenes* based on induction of lecithinase activity by charcoal. *Int. J. Food Microbiol.*, 82, 87, 2003.

42. Notermans, S.H. et al. Phosphatidylinositol-specific phospholipase C activity as a marker to distinguish between pathogenic and nonpathogenic *Listeria* species. *Appl. Environ. Microbiol.*, 57, 2666, 1991.

43. Bubert, A. et al. Differential expression of *Listeria monocytogenes* virulence genes in mammalian host cells. *Mol. Gen. Genet.*, 261, 323, 1999.

44. Gouin, E., Mengaud, J. and Cossart, P. The virulence gene cluster of *Listeria monocytogenes* is also present in *Listeria ivanovii*, an animal pathogen, and *Listeria seeligeri*, a non-pathogenic species. *Infect. Immun.*, 62, 3550, 1994.

45. Hamon, M., Bierne, H. and Cossart, P. *Listeria monocytogenes*: a multifaceted model *Nat. Rev. Microbiol.*, 4, 423, 2006.

46. Chatterjee, S.S. et al. Intracellular gene expression profile of *Listeria monocytogenesInfect. Immun.*, 74, 1323, 2006.

47. Joseph, B. et al. Identification of *Listeria monocytogenes* genes contributing to intracellular replication by expression profiling and mutant screening. *J. Bacteriol.*, 188, 556, 2006.

48. Yin, Y. et al. A hybrid sub-lineage of *Listeria monocytogenes* comprising hypervirulent isolates. *Nat Commun.*, 10, 4283, 2019.

49. Silva, S. et al. Adhesion to and viability of *Listeria monocytogenes* on food contact surfaces. *J. Food Prot.*, 71, 1379, 2008.

50. Hall-Stoodley, L., Costerton, J.W. and Stoodley, P. Bacterial biofilms: from the natural environment to infectious diseases. *Nature Rev. Microbiol.*, 2, 95, 2004.

51. Carpentier, B. and Cerf, P. Persistence of Listeria monocytogenes in food industry equipment and premises. *Int. J. Food Microbiol.*, 145, 1, 2011.

52. da Silva, E.P. and De Martinis, E.C. Current knowledge and perspectives on biofilm formation: the case of *Listeria monocytogenes*. *Appl. Microbiol. Biotechnol.*, 97, 957, 2013.

53. Sauders, B.D. et al. Diversity of *Listeria* species in urban and natural environments. *Appl. Environ. Microbiol.*, 78, 4420, 2012.

54. Chemaly, M. et al. Prevalence of *Listeria monocytogenes* in poultry production in France. *J. Food Prot.*, 71, 1996, 2008.

55. Rivoal, K. et al. Detection of *Listeria* spp. in liquid egg products and in the egg breaking plants environment and tracking of *Listeria monocytogenes* by PFGE. *Int. J. Food Microbiol.*, 166, 109, 2013.

56. Wang, Y. et al. Isolation and characterization of *Listeria* species from rodents in natural environments in China. *Emerg. Microbes. Infect.*, 6, e44, 2017.

57. Cao, X. et al. Prevalence and characteristics of *Listeria ivanovii* strains in wild rodents in China. *Vector Borne Zoonotic Dis.*, 19, 8, 2019.

58. Kulesh, R. et al. The occurrence of *Listeria monocytogenes* in goats, farm environment and invertebrates, *Biolog. Rhythm Res.*, 2019, doi: 10.1080/09291016.2019.1660836,

59. Gan, L. et al. Carriage and potential long distance transmission of *Listeria monocytogenes* by migratory black-headed gulls in Dianchi Lake, Kunming. *Emerg. Microbes Infect.*, 8, 1195, 2019.

60. Flemming, H. and Wingender, J. The biofilm matrix. *Nat. Rev. Microbiol.*, 8, 623, 2010

61. Combrouse, T. et al. Quantification of the extracellular matrix of the *Listeria monocytogenes* biofilms of different phylogenic lineages with optimization of culture conditions. *J. Appl. Microbiol.*, 114, 1120, 2013.

62. Di Bonaventura, G. et al. Influence of temperature on biofilm formation by *Listeria monocytogenes* on various food-contact surfaces: relationship with motility and cell surface hydrophobicity. *J. Appl. Microbiol.*, 104, 1552, 2008.

63. Reis-Teixeira, F.B.D., Alves, V.F. and de Martinis, E.C.P. Growth, viability and architecture of biofilms of *Listeria monocytogenes* formed on abiotic surfaces. *Brazilian J. Microbiol.*, 48, 587, 2017.

64. Doijad, S.P. et al. Biofilm-forming abilities of *Listeria monocytogenes* Serotypes isolated from different sources. *PLOS One*, 10, e0137046, 2015.

65. Wang, J. Persistent and transient *Listeria monocytogenes* strains from retail deli environments vary in their ability to adhere and form biofilms and rarely have *inl*A premature stop codons. *Foodborne Pathog. Dis.*, 12, 1, 2015.

66. Nowak, J. Persistent *Listeria monocytogenes* strains isolated from mussel production facilities form more biofilm but are not linked to specific genetic markers. *Int. J. Food Microbiol.*, 256, 45, 2017.

67. Midelet, G., Kobilinsky, A. and Carpentier, B. Construction and analysis of fractional multifactorial designs to study attachment strength and transfer of Listeria monocytogenes from pure or mixed biofilms after contact with a solid model food. *Appl. Environ. Microbiol.*, 72, 2313, 2006.

68. Borucki, M.K. et al. Variation in biofilm formation among strains of *Listeria monocytogenes*. *Appl. Environ. Microbiol.*, 69, 7336, 2003.

69. Bremer, P.J., Monk, I. and Osborne, C.M. Survival of *Listeria monocytogenes* attached to stainless steel surfaces in the presence or absence of *Flavobacterium* spp. *J. Food Prot.*, 64, 1369, 2001.

70. EFSA and ECDC. The European Union summary report on trends and sources of zoonoses, zoonotic agents and foodborne outbreaks in 2016. *EFSA J.*, 15, 5077, 2017.

71. Marder, E.P. et al. Preliminary incidence and trends of infection with pathogens transmitted commonly through food—foodborne diseases active surveillance network, 10 U.S. Sites, 2006–2017. *Morb. Mortal. Wkly. Rep.*, 64, 324, 2018.

72. Rodríguez-López, P. et al. Removal of *Listeria monocytogenes* dual-species biofilms using combined enzyme-benzalkonium chloride treatments. *Biofouling.*, 33, 45, 2017.

73. Nguyen, U.T. and Burrows, L.L. DNase I and proteinase K impair *Listeria monocytogenes* biofilm formation and induce dispersal of pre-existing biofilms. *Int. J. Food Microbiol.*, 187, 26, 2014.

74. Gray, J.A.et al. Novel biocontrol methods for *Listeria monocytogenes* biofilms in food production facilities. *Front. Microbiol.*, 9, 605, 2018.

75. Rodríguez-López, P. et al. Quantifying the combined effects of pronase and benzalkonium chloride in removing late-stage *Listeria monocytogenes–Escherichia coli* dual-species biofilms. *Biofouling.*, 33, 690, 2018.

76. Camargo, A.C. et al. Lactic acid bacteria (LAB) and their bacteriocins as alternative biotechnological tools to control *Listeria monocytogenes* biofilms in food processing facilities. *Mol. Biotechnol.*, 60, 712, 2018.

77. Mylonakis, E., Hohmann, E.L. and Calderwood, S.B. Central nervous system infection with *Listeria monocytogenes*. 33 years' experience at a general hospital and review of 776 episodes from the literature. *Medicine (Baltimore)*, 77, 313, 1998.

78. Doorduyn, Y. et al. Invasive *Listeria monocytogenes* infections in the Netherlands, 1995-2003. *Eur. J. Clin. Microbiol. Infect. Dis.*, 25, 433, 2006.

79. Cartwright, E.J. et al. Listeriosis outbreaks and associated food vehicle, United States, 1998-2008. *Emerg. Infect. Dis.*, 19, 1, 2013.

80. Gillespie, I.A. et al. Changing pattern of human listeriosis, England and Wales, 2001-2004. *Emerg. Infect Dis.*, 12, 1361, 2006.

81. Goulet, V. et al. Increasing incidence of listeriosis in France and other European countries. *Emerg. Infect. Dis.*, 14, 734, 2008.

82. Gerner-Smidt, P. et al. Invasive listeriosis in Denmark 1994-2003: A review of 299 cases with special emphasis on risk factors for mortality. *Clin. Microbiol. Infect.*, 11, 618, 2005.

83. Voetsch, A.C. et al. Reduction in the incidence of invasive listeriosis in foodborne diseases active surveillance network sites, 1996-2003. *Clin. Infect. Dis.*, 44, 513, 2007.

84. Scallan, E. et al. Foodborne illness acquired in the United States—major pathogens. *Emerg. Infect. Dis.*, 17, 7, 2011.

85. Pontello, M. et al. *Listeria monocytogenes* serotypes in human infections (Italy, 2000-2010). *Ann. Ist. Super Sanita.*, 48, 146, 2012.

86. World Health Organization [WHO] Risk assessment of *Listeria monocytogenes* in ready-to-eat foods: interpretative summary. Available at https://www.who.int/publications/i/item/risk-assessment-of-listeria-monocytogenes-in-ready-to-eat-foods-interpretive-summary, 2004. Assessed on 24 September, 2020.

87. Smerdon, W.J. et al. Surveillance of listeriosis in England and Wales, 1995-1999. *Commun. Dis. Public Health*, 4, 188, 2001.

88. Del Pozo, J.L. et al., *Listeria monocytogenes* septic arthritis in a patient treated with mycophenolate mofetil for polyarteritis nodosa: a case report and review of the literature. *Int. J. Infect. Dis.*, 17, 132, 2013.

89. Cardoso, C.et al. Spontaneous bacterial peritonitis caused by *Listeria monocytogenes*: a case report and literature review. *Anal. Hepatol.*, 11, 955, 2012.

90. Abreu, C. et al. *Listeria* infection in patients on anti-TNF treatment: report of two cases and review of the literature. *J. Crohns Colitis.*, 7, 175, 2013.

91. Chersich, M.F. et al. Diagnosis and treatment of *Listeria monocytogenes* endophthalmitis: a systematic review. *Ocul. Immunol. Inflamm.*, 26, 508, 2018.

92. Morgand, M. et al. *Listeria monocytogenes*-associated respiratory infections: a study of 38 consecutive cases. *Clin. Microbiol. Infect.*, 24, 1339.e1–1339.e5, 2018.

93. Shoai-Tehrani, M. et al. *Listeria* endovascular infections study group. 2019. *Listeria monocytogenes*-associated endovascular infections: a study of 71 consecutive cases. *J. Infect.*, 79, 322, 2019.

94. Smith, A.M. et al. Outbreak of *Listeria monocytogenes* in South Africa, 2017-2018: Laboratory activities and experiences associated with whole-genome sequencing analysis of isolates. *Foodborne Pathog. Dis.*, 16, 524, 2019.

95. Dramowski, A. et al. Neonatal listeriosis during a countrywide epidemic in South Africa: a tertiary hospital's experience. *S. Afr. Med. J.*, 108, 818, 2018.

96. Halbedel, S.et al. A *Listeria monocytogenes* ST2 clone lacking chitinase ChiB from an outbreak of non-invasive gastroenteritis. *Emerg. Microbes Infect.*, 8, 17, 2019.

97. Allerberger, F. and Wagner, M. Listeriosis: a resurgent foodborne infection. *Clin. Microbiol. Infect.*, 16, 16, 2010.

98. Oevermann, A. et al. Neuropathogenesis of naturally occurring encephalitis caused by *Listeria monocytogenes* in ruminants. *Brain Pathol.*, 20, 378, 2010.

99. Rawool, D.B. et al. Detection of multiple virulence-associated genes in *Listeria monocytogenes* isolated from bovine mastitis cases. *Int. J. Food Microbiol.*, 113, 201, 2007.

100. Shakuntala, I. et al. Isolation of *Listeria monocytogenes* from buffaloes with reproductive disorders and its confirmation by polymerase chain reaction. *Vet. Microbiol.*, 117, 229, 2006.

101. Popov, A. et al. Indoleamine 2,3-dioxygenase-expressing dendritic cells form suppurative granulomas following Listeria monocytogenes infection. *J. Clin. Invest.*116(12): 3160–3170.

102. Safley, S.A. et al. Role of listeriolysin-O [LLO] in the T lymphocyte response to infection with *Listeria monocytogenes*. Identification of T cell epitopes of LLO. *J. Immunol.*, 146, 3604, 1991.

103. Kaufmann, S.H. Immunity to intracellular bacteria. *Annu. Rev. Immunol.*, 11, 129, 1993.

104. Lara-Tejero, M. and Pamer, E.G. T cell responses to *Listeria monocytogenes*. *Curr Opin Microbiol.*, 7, 45, 2004.

105. Kearns, R.J. and Leu, R.W. Modulation of natural killer activity in mice following infection with *Listeria monocytogenes*. *Cell Immunol.*, 84, 361, 1984.

106. Witte, C.E. et al. Innate immune pathways triggered by *Listeria monocytogenes* and their role in the induction of cell mediated immunity. *Adv. Immunol.*, 113, 135, 2012.

107. Pamer, E.G. Immune responses to *Listeria monocytogenes*. *Nat. Rev. Immunol.*, 4, 812, 2004.

108. Gianfranceschi, M. et al. Incidence of *Listeria monocytogenes* in food and environmental samples in Italy between 1990 and 1999: Serotype distribution in food, environmental and clinical samples. *Eur. J. Epidemiol.*, 18, 1001, 2003.

109. European Food Safety Authority. The community summary report on trends and sources of zoonoses and zoonotic agents in the European Union in 2007. *EFSA J.*, 223, 18, 2009.

110. Churchill, K.J. et al. Prevalence of *Listeria monocytogenes* in select ready-to-eat foods-deli meat, soft cheese, and packaged salad: a systematic review and meta-analysis. *J. Food Prot.*, 82, 344, 2019.

111. Desai, A.N. et al. Changing epidemiology of *Listeria monocytogenes* outbreaks, sporadic cases, and recalls globally: a review of ProMED reports from 1996 to 2018. *Int J Infect Dis.*, 84, 48, 2019.

112. Thomas, M.K. et al. Economic cost of a *Listeria monocytogenes* outbreak in Canada, 2008. *Foodborne Pathog. Dis.*, 12, 966, 2015.

113. Osebold, J.W., Aalund, O. and Crisp, C.E. Chemical and immunological composition of surface structures of *Listeria monocytogenes*. *J. Bacteriol.*, 89, 84, 1965.

114. Larsen, S.A., Wiggins, G.A. and Albritton, W.L. *Immune response to Listeria*. In *Manual of Clinical Immunology*, Rose, N.R. and Friedman, H. eds., 2nd ed., American Society for Microbiology, Washington, DC, 1980, 506.

115. Gray, M.L. and Killinger, A.H. *Listeria monocytogenes* and listeric infections *Bacteriol. Rev.*, 30, 309, 1966.

116. Hernández-Flores, K.G. and Vivanco-Cid, H. Biological effects of listeriolysin O: implications for vaccination. *Biomed Res. Int.*, 2015: 360741, 2015, doi: 10.1155/2015/360741.

117. Geoffroy, C. et al. Purification, characterization, and toxicity of the sulfhydryl-activated hemolysin listeriolysin O from *Listeria monocytogenes. Infect. Immun.*, 55, 1641, 1987.

118. Leimeister-Wachter, M. and Chakraborty, T. Detection of listeriolysin, the thiol-dependent hemolysin in *Listeria monocytogenes, Listeria ivanovii*, and *Listeria seeligeri. Infect. Immun.*, 57, 2350, 1989.

119. Mengaud, J. et al. A genetic approach to demonstrate the role of listeriolysin O in the virulence of *Listeria monocytogenes. Acta Microbiol. Hung.*, 36, 177, 1989.

120. Darji, A. et al. Hyperexpression of listeriolysin in the non-pathogenic species *Listeria innocua* and high yield purification. *J Biotechnol.*, 43, 205, 1995.

121. Berche, P. et al. Detection of anti-listeriolysin O for serodiagnosis of human listeriosis. *Lancet*, 335, 624, 1990.

122. Low, J.C., Davies, R.C. and Donachie, W. Purification of listeriolysin O and development of an immunoassay for diagnosis of listeric infections in sheep. *J. Clin. Microbiol.*, 30, 2705, 1992.

123. Barbuddhe, S.B., Malik, S.V. and Gupta, L.K. Kinetics of antibody production and clinical profiles of calves experimentally infected with *Listeria monocytogenes. J. Vet. Med. B.*, 47, 497, 2000.

124. Barbuddhe, S.B., Malik, S.V. and Kumar, P. High seropositivity against listeriolysin O in humans. *Ann. Trop. Med. Parasitol.*, 93, 537, 1999.

125. Gentschev, I. et al. Identification of p60 antibodies in human sera and presentation of this listerial antigen on the surface of attenuated salmonellae by the HlyB-HlyD secretion system. *Infect. Immun.*, 60, 5091, 1992.

126. Bubert, A. et al. Synthetic peptides derived from the *Listeria monocytogenes* p60 protein as antigens for the generation of polyclonal antibodies specific for secreted cell-free *L. monocytogenes* p60 proteins. *Appl. Environ. Microbiol.*, 60, 3120, 1994.

127. Chang, T.C., Chen, C.H. and Chen, H.C. Development of a latex agglutination test for the rapid identification of *Vibrio parahaemolyticus. J Food Prot.*, 57, 31, 1994.

128. Tabouret, M., de Rycke, J. and Dubray, G. Analysis of surface proteins of *Listeria* in relation to species, serovar and pathogenicity. *J. Gen. Microbiol.*, 138, 743, 1992.

129. Bhunia, A.K. and Johnson, M.G. Monoclonal antibody specific for Listeria monocytogenes associated with a 66-kilodalton cell surface antigen. *Appl. Environ. Microbiol.*, 58, 1924, 1992.

130. Goldfine, H. and Knob, C. Purification and characterization of *Listeria monocytogenes* phosphatidylinositol-specific phospholipase C. *Infect. Immun.*, 60, 4059, 1992.

131. Suryawanshi, R.D. et al. Comparative diagnostic efficacy of recombinant LLO and PI-PLC-based ELISAs for detection of listeriosis in animals. *J. Microbiol. Methods.*, 137, 40, 2017.

132. Ronholm, J. et al. Monoclonal antibodies recognizing the surface autolysin IspC of *Listeria monocytogenes* serotype 4b: epitope localization, kinetic characterization, and cross-reaction studies. *PLOS One*, 8, e55098, 2013.

133. Farber, J.M. and Peterkin, P.I. *Listeria monocytogenes*, a food-borne pathogen. *Microbiol. Rev.*, 55, 476, 1991.

134. Gray, M.L. et al. A new technique for isolating listerelle from the bovine brain. *J Bacteriol.*, 55, 471, 1948.

135. Curtis, G.D. and Lee, W.H. Culture media and methods for the isolation of *Listeria monocytogenes. Int. J. Food Microbiol.*, 26, 1, 1995.

136. Lovett, J., Francis, D.W. and Hunt, J.M. *Listeria monocytogenes* in raw milk: detection, incidence and pathogenicity. *J Food Prot.*, 50, 185, 1987.

137. Fraser, J.A. and Sperber, W.H. Rapid detection of *Listeria* spp. in food and environmental samples by esculin hydrolysis. *J. Food Prot.*, 51, 762, 1988.

138. Gasanov, U., Hughes, D. and Hansbro, P.M. Methods for the isolation and identification of *Listeria* spp. and *Listeria monocytogenes:* a review. *FEMS Microbiol. Rev.*, 29, 851, 2005.

139. Donnelly, C.W. and Baigent, G.J. Method for flow cytometric detection of *Listeria monocytogenes* in milk. *Appl. Environ. Microbiol.*, 52, 689, 1986.

140. Brackett, R.E. and Beuchat, L.R. Methods and media for the isolation and cultivation of *Listeria monocytogenes* from various foods. *Int. J. Food Microbiol.*, 8, 219, 1989.

141. Kells, J. and Gilmour, A. Incidence of *Listeria monocytogenes* in two milk processing environments, and assessment of *Listeria monocytogenes* blood agar for isolation. *Int. J. Food Microbiol.*, 91, 167, 2004.

142. Gray, M.L., Stafseth, H.J. and Thorp, F. Jr. The use of potassium tellurite, sodium azide and acetic acid in a selective medium for the isolation of *Listeria monocytogenes. J Bacteriol.*, 59, 443, 1950.

143. McBride, M.E. and Girard. K.F. A selective method for the isolation of *Listeria monocytogenes* from mixed bacterial populations. *J Lab Clin Med.*, 55, 153, 1960.

144. Lee, W.H. and McClain, D. Improved *Listeria monocytogenes* selective agar. *Appl. Environ. Microbiol.*, 52, 1215, 1986.

145. Curtis, G.D.W., Nicholas, W.W. and Falla, T.J. Selective agents for *Listeria* can inhibit their growth. *Lett Appl Microbiol.*, 8, 169, 1989.

146. Lachica, R.V. Selective plating medium for quantitative recovery of food-borne *Listeria monocytogenes. Appl. Environ. Microbiol.*, 56, 167, 1990.

147. van Netten, P. et al. Liquid and solid selective differential media for the detection and enumeration of *L. monocytogenes* and other *Listeria* spp. *Int. J. Food Microbiol.*, 8, 299, 1989.

148. Dominguez-Rodriguez, L. et al. New methodology for the isolation of *Listeria* microorganisms from heavily contaminated environments. *Appl. Environ. Microbiol.*, 47, 1188, 1984.

149. Blanco, M, et al. A technique for the direct identification of haemolytic pathogenic *Listeria* on selective plating media. *Lett. Appl. Microbiol.*, 9, 125, 1989.

150. Buchanan, R.L. et al. Comparison of lithium chloride-phenylethanol-moxalactam and modified Vogel Johnson agars for detection of *Listeria* spp. in retail-level meats, poultry, and seafood. *Appl. Environ. Microbiol.*, 55, 599, 1989.

151. Smith, J.L. and Buchanan, R.L. Identification of supplements that enhance the recovery of *Listeria monocytogenes* on modified Vogel Johnson agar. *J Food Safety*, 10, 155, 2007.

152. Vlaemynck, G., Lafarge, V. and Scotter, S. Improvement of the detection of *Listeria monocytogenes* by the application of ALOA, a diagnostic, chromogenic isolation medium. *J. Appl. Microbiol.*, 88, 430, 2000.

153. Stessl, B. et al. Performance testing of six chromogenic ALOA-type media for the detection of *Listeria monocytogenes*. *J. Appl. Microbiol.* 106, 651, 2009.

154. Menudier, A., Bosiraud, C. and Nicolas, J.A. Virulence of *Listeria monocytogenes* serovars and *Listeria* spp. in experimental infection in mice. *J. Food Prot.*, 54, 917, 1991.

155. Lecuit, M. et al. A single amino acid in E-cadherin responsible for host specificity towards the human pathogen *Listeria monocytogenes*. *EMBO J.*, 18, 3956, 1999.

156. Mainou-Fowler, T., MacGowan, A.P. and Postlethwaite, R. Virulence of *Listeria* spp.: course of infection in resistant and susceptible mice. *J. Med. Microbiol.*, 27, 131, 1988.

157. Stelma, G.N., Jr. et al. Pathogenicity test for *Listeria monocytogenes* using immunocompromised mice. *J. Clin. Microbiol.*, 25, 2085, 1987.

158. Tabouret, M. et al. Pathogenicity of *Listeria monocytogenes* isolates in immunocompromised mice in relation to listeriolysin production. *J. Med. Microbiol.*, 34, 13, 1991.

159. Bracegirdle, P. et al. A comparison of aerosol and intragastric routes of infection with *Listeria* spp *Epidemiol. Infect.*, 112, 69, 1994.

160. Liu, D. *Listeria monocytogenes*: comparative interpretation of mouse virulence assay. *FEMS Microbiol. Lett.*, 233, 159, 2004.

161. Schlech, W.F., III, Chase, D.P. and Badley, A. A model of food-borne *Listeria monocytogenes* infection in the Sprague-Dawley rat using gastric inoculation: development and effect of gastric acidity on infective dose. *Int. J. Food Microbiol.*, 18, 15, 1993.

162. Notermans, S. et al. The chick embryo test agrees with the mouse bioassay for assessment of the pathogenicity of *Listeria* species. *Lett. Appl. Microbiol.*, 13, 161, 1991.

163. Terplan, G. and Steinmeyer, S. Investigations on the pathogenicity of *Listeria* spp. by experimental infection of the chick embryo. *Int. J. Food Microbiol.*, 8, 277, 1989.

164. Schonberg, A. Method to determine virulence of *Listeria* strains. *Int. J. Food Microbiol.*, 8, 281, 1989.

165. Anton, W. Kritisch-experimenteller beitrag zur biologie des *Bacterium monocytogenes*. *Zentralbl. Bakteriol. Parasitenkd. Infektionskr.*, 131, 89, 1934.

166. Farber, J.M. et al. Feeding trials of *Listeria monocytogenes* with a nonhuman primate model. *J. Clin. Microbiol.*, 29, 2606, 1991.

167. Mukherjee, K. et al. *Galleria mellonella* as a model system for studying *Listeria pathogenesis*. *Appl. Environ. Microbiol.*, 76, 310, 2010.

168. Joyce, S.A. and Gahan, C.G. Molecular pathogenesis of *Listeria monocytogenes* in the alternative model host *Galleria mellonella*. *Microbiology.*, 156, 3456, 2010.

169. Mukherjee, K. et al. Brain infection and activation of neuronal repair mechanisms by the human pathogen *Listeria monocytogenes* in the lepidopteran model host *Galleria mellonella*. *Virulence*, 4, 324, 2013.

170. Dons, L. et al. Rat dorsal root ganglia neurons as a model for *Listeria monocytogenes* infections in culture. *Med Microbiol Immunol [Berl].*, 188, 15, 1999.

171. Mansfield, B.E. et al. Exploration of host-pathogen interactions using *Listeria monocytogenes* and *Drosophila melanogaster*. *Cell Microbiol.*, 5, 901, 2003.

172. Pine, L. et al. Cytopathogenic effects in enterocyte like Caco-2 cells differentiate virulent from avirulent *Listeria* strains. *J. Clin. Microbiol.*, 29, 990, 1991.

173. Roche, S.M. et al. Assessment of the virulence of *Listeria monocytogenes*: agreement between a plaque-forming assay with HT-29 cells and infection of immunocompetent mice. *Int. J. Food Microbiol.*, 68, 33, 2001.

174. Chatterjee, S.S. et al. Invasiveness is a variable and heterogeneous phenotype in *Listeria monocytogenes* serotypestrains. *Int. J. Med. Microbiol.*, 296, 277, 2006.

175. Van Langendonck, N. et al. Tissue culture assays using Caco-2 cell line differentiate virulent from non-virulent *Listeria monocytogenes* strains. *J. Appl. Microbiol.*, 85, 337, 1998.

176. Nishibori, T. et al. Correlation between the presence of virulence-associated genes as determined by PCR and actual virulence to mice in various strains of *Listeria* spp *Microbiol. Immunol.*, 39, 343, 1995.

177. Jaradat, Z.W., Schutze, G.E. and Bhunia, A.K. Genetic homogeneity among *Listeria monocytogenes* strains from infected patients and meat products from two geographic locations determined by phenotyping, ribotyping and PCR analysis of virulence genes. *Int. J. Food Microbiol.*, 76, 1, 2002.

178. Roberts, A., Chan, Y. and Wiedmann, M. Definition of genetically distinct attenuation mechanisms in naturally virulence-attenuated *Listeria monocytogenes* by comparative cell culture and molecular characterization. *Appl. Environ. Microbiol.*, 71, 3900, 2005.

179. Roche, S.M. et al. Investigation of specific substitutions in virulence genes characterizing phenotypic groups of low-virulence field strains of *Listeria monocytogenes*. *Appl. Environ. Microbiol.*, 71, 6039, 2005.

180. Olapedo, D.K. Detection of *Listeria monocytogenes* using polyclonal antibody. *Lett Appl Microbiol.*, 14, 26, 1992.

181. Beumer, R.R. Detection of *Listeria* spp. with a monoclonal antibody based ELISA. *Food Microbiology*, 6, 171, 1989.

182. Farber, J.M. and Speirs, J.I. Monoclonal antibodies directed against the flagellar antigens of *Listeria* spp. and their potential in EIA based methods. *J Food Prot.*, 50, 479, 1987.

183. Mattingly, J.A. et al. Rapid monoclonal antibody-based enzyme-linked immunosorbent assay for detection of *Listeria* in food products. *J. Assoc. Off. Anal. Chem.*, 71, 679, 1988.

184. McLauchlin, J. et al. Monoclonal antibodies show *Listeria monocytogenes* in necropsy tissue samples. *J. Clin. Pathol.*, 41, 983, 1988.

185. Walker, S.J., Archer, P. and Aleyard, J. Comparison of the Listeria Tek ELISA kit with cultural procedures for the detection of *Listeria* species in foods. *Food Microbiol.*, 7, 335, 1990.

186. Norrung, B., Solve, M., Ovesen, M., and Skovgaard, N. Evaluation of an ELISA test for the detection of *Listeria* spp. *J Food Prot.*, 54, 752, 1991.

187. Curiale, M.S., Lepper, W. and Robison, B. Enzyme-linked immunoassay for detection of *Listeria monocytogenes* in dairy products, sea foods, and meats: collaborative study. *J. AOAC Int.*, 77, 1472, 1994.

188. Sewell, A.M. et al. The development of an efficient and rapid enzyme linked fluorescent assay method for the detection of *Listeria* spp. from foods. *Int. J. Food Microbiol.*, 81, 123, 2003.

189. Silbernagel, K.M. et al. Evaluation of the VIDAS *Listeria* [LIS] immunoassay for the detection of *Listeria* in foods using demi-Fraser and Fraser enrichment broths, as modification of AOAC Official Method 999.06 [AOAC Official Method 2004.06]. *J. AOAC Int.*, 88, 750, 2005.

190. Solve, M., Boel, J. and Norrung, B. Evaluation of a monoclonal antibody able to detect live *Listeria monocytogenes* and *Listeria innocua*. *Int. J. Food Microbiol.*, 57, 219, 2000.

191. Durham, R.J. et al. A monoclonal antibody enzyme immunoassay [ELISA] for the detection of *Listeria* in foods and environmental samples. In *Foodborne Listeriosis*, Miller, A.J., Smith, J.L., and Somkuti, G.A., eds., Elsevier Science Publishers, New York, 1990, 105.

192. Knight, M.T. et al. TECRA *Listeria* Visual Immunoassay [TLVIA] for detection of *Listeria* in foods: collaborative study. *J. AOAC Int.*, 79, 1083, 1996.

193. Kim, S.H. et al. Development of a sandwich ELISA for the detection of *Listeria* spp. using specific flagella antibodies. *J. Vet. Sci.*, 6, 41, 2005.

194. Skjerve, E., Bos, W. and van der, G.B. Evaluation of monoclonal antibodies to *Listeria monocytogenes* flagella by checkerboard ELISA and cluster analysis. *J. Immunol. Methods*, 144, 11, 1991.

195. Palumbo, J. D. et al. Serotyping of *Listeria monocytogenes* by enzyme-linked immunosorbent assay and identification of mixed-serotype cultures by colony immunoblotting. *J. Clin. Microbiol.*, 41, 564, 2003.

196. Bourry, A., Cochard, T. and Poutrel, B. Serological diagnosis of bovine, caprine, and ovine mastitis caused by *Listeria monocytogenes* by using an enzyme-linked immunosorbent assay. *J. Clin. Microbiol.*, 35, 1606, 1997.

197. Allerberger, F. *Listeria*: growth, phenotypic differentiation and molecular microbiology. *FEMS Immunol. Med. Microbiol.*, 35, 183, 2003.

198. Cho, J.H. et al. An enzyme immunoanalytical system based on sequential cross-flow chromatography. *Anal. Chem.*, 77, 4091, 2005.

199. Seo, S.M. et al. An ELISA-on-a Chip biosensor system for early screening of *Listeria monocytogenes* in contaminated food products. *Bull. Korean Chem. Soc.*, 30, 2993, 2009.

200. Cho, I.H. and Irudayaraj, J. Lateral-flow enzyme immunoconcentration for rapid detection of *Listeria monocytogenes*. *Anal. Bioanal. Chem.*, 405, 3313, 2013.

201. Gupta, M. et al. Quantification of foodborne pathogens in different food matrices using FTIR and artificial neural networks. *Trans. ASABE*, 49, 1249, 2006.

202. Ravindranath, S., Wang, Y. and Irudayaraj, J. SERS driven Cross-platform based multiplex pathogen detection. *Sens. Actuat. B. Chem.*, 152, 183, 2011.

203. Wadud, S. et al. Evaluation of immunomagnetic separation in combination with ALOA Listeria chromogenic agar for the isolation and identification of *Listeria monocytogenes* in ready-to-eat foods. *J. Microbiol. Methods*, 81, 153, 2010.

204. Yang, H. et al. Rapid detection of Listeria monocytogenes by nanoparticle-based immunomagnetic separation and real-time PCR. *Int. J. Food Microbiol.*, 118, 132, 2007.

205. Walcher, G. et al. Evaluation of paramagnetic beads coated with recombinant *Listeria* phage endolysin-derived cell-wall-binding domain proteins for separation of *Listeria monocytogenes* from raw milk in combination with culture-based and real-time polymerase chain reaction-based quantification. *Foodborne Pathog. Dis.*, 7, 1019, 2010.

206. Khan, M.A., Seaman, A. and Woodbine, M. Immunofluorescent identification of *Listeria monocytogenes*. *Zentralbl. Bakteriol. [Orig.A]*, 239, 62, 1977.

207. McLauchlin, J., Ridley, A.M. and Taylor, A.G. The use of monoclonal antibodies against *Listeria monocytogenes* in a direct immunofluorescence technique for the rapid presumptive identification and direct demonstration of *Listeria* in food. *Acta Microbiol. Hung.*, 36, 467, 1989.

208. Ruhland, G.J. et al. Cell-surface location of *Listeria*-specific protein p60–detection of *Listeria* cells by indirect immunofluorescence. *J. Gen. Microbiol.*, 139, 609, 1993.

209. Czuprynski, C.J. Host defence against *Listeria monocytogenes*: implications for food safety. *Food Microbiol.*, 11, 131, 1994.

210. Clough, N.E. and Roth, J.A. Methods for assessing cell-mediated immunity in infectious disease resistance and in the development of vaccines. *J. Am. Vet. Med. Assoc.*, 206, 1208, 1995.

211. Klink, M. et al. Specific cellular and humoral reactions as markers of *Listeria monocytogenes* infections. *Acta Microbiol. Pol.*, 43, 335, 1994.

212. Kaufmann, S.H.E. CD8+ lymphocytes in intracellular microbial infections. *Immunol Today*, 9, 168, 1988.

213. Nakane, A. et al. Evidence that endogenous gamma interferon is produced early in *Listeria monocytogenes* infection *Infect. Immun.*, 58, 2386, 1990.

214. Barbuddhe, S.B. et al. Kinetics of interferon-gamma production and its comparison with anti-listeriolysin O detection in experimental bovine listeriosis. *Vet. Res. Commun.*, 22, 505, 1998.

215. Datta, A.R., Wentz, B.A. and Hill, W.E. Identification and enumeration of beta-hemolytic *Listeria monocytogenes* in naturally contaminated dairy products. *J. Assoc. Off Anal. Chem.*, 71, 673, 1988.

216. King, W. et al. A new colorimetric nucleic acid hybridization assay for *Listeria* in foods. *Int. J. Food Microbiol.*, 8, 225, 1989.

217. Kim, C. et al. Rapid confirmation of *Listeria monocytogenes* isolated from foods by a colony blot assay using a digoxigenin-labeled synthetic oligonucleotide probe. *Appl. Environ. Microbiol.*, 57, 1609, 1991.

218. Bobbit, J.A. and Betts, R.P. Confirmation of *Listeria monocytogenes* using a commercially available nucleic acid probe. *Food Microbiol.*, 9, 311, 1992.

219. Volokhov, D. et al. Identification of *Listeria* species by microarray-based assay. *J. Clin. Microbiol.*, 40, 4720, 2002.

220. Ingianni, A. et al. Rapid detection of *Listeria monocytogenes* in foods, by a combination of PCR and DNA probe. *Mol. Cell Probes.*, 15, 275, 2001.

221. Wernars, K. et al. Suitability of the *prfA* gene, which encodes a regulator of virulence genes in *Listeria monocytogenes*, in the identification of pathogenic *Listeria* spp. *Appl. Environ. Microbiol.*, 58, 765, 1992.

222. Wagner, M. et al. *In situ* detection of a virulence factor mRNA and *16S rRNA* in *Listeria monocytogenes*. *FEMS Microbiol. Lett.*, 160, 159, 1998.

223. Mengaud, J. et al. Expression in *Escherichia coli* and sequence analysis of the listeriolysin O determinant of *Listeria monocytogenesInfect. Immun.*, 56, 766, 1988.

224. Jaton, K., Sahli, R. and Bille, J. Development of polymerase chain reaction assays for detection of *Listeria monocytogenes* in clinical cerebrospinal fluid samples. *J. Clin. Microbiol.*, 30, 1931, 1992.

225. Rossen, L.et al. A rapid polymerase chain reaction [PCR]-based assay for the identification of *Listeria monocytogenes* in food samples. *Int. J. Food Microbiol.*, 14, 145, 1991.

226. Niederhauser, C. et al. Use of polymerase chain reaction for detection of *Listeria monocytogenes* in food. *Appl. Environ. Microbiol.*, 58, 1564, 1992.

227. Torriani, S. and Pallotta, M.L. Use of polymerase chain reaction to detect *Listeria monocytogenes in silages. Biotechnol Tech*, 8, 157, 1994.

228. Bohnert, M. et al. Use of specific oligonucleotides for direct enumeration of *Listeria monocytogenes* in food samples by colony hybridization and rapid detection by PCR. *Res. Microbiol.*, 143, 271, 1992.

229. Wang, R.F., Cao, W.W. and Johnson, M.G. *16S rRNA*-based probes and polymerase chain reaction method to detect *Listeria monocytogenes* cells added to foods. *Appl. Environ. Microbiol.*, 58, 2827, 1992.

230. Herman, L.M., De Block, J.H. and Moermans, R.J. Direct detection of *Listeria monocytogenes* in 25 milliliters of raw milk by a two-step PCR with nested primers. *Appl. Environ. Microbiol.*, 61, 817, 1995.

231. Aznar, R. and Alarcon, B. On the specificity of PCR detection of *Listeria monocytogenes* in food: a comparison of published primers. *Syst. Appl. Microbiol.*, 25, 109, 2002.

232. Souza, S.M. et al. Using nested PCR to detect the hlyA gene of *Listeria monocytogenes* in Minas Frescal cow's milk cheese. *J. Food Prot.*, 75, 1324, 2012.

233. Novak, J.S. and Juneja, V.K. Detection of heat injury in *Listeria monocytogenes* Scott A. *J. Food Prot.*, 64, 1739, 2001.

234. Keer, J.T. and Birch, L. Molecular methods for the assessment of bacterial viability. *J. Microbiol. Methods*, 53, 175, 2003.

235. Klein, P.G. and Juneja, V.K. Sensitive detection of viable *Listeria monocytogenes* by reverse transcription-PCR. *Appl. Environ. Microbiol.*, 63, 4441, 1997.

236. Burtscher, C. and Wuertz, S. Evaluation of the use of PCR and reverse transcriptase PCR for detection of pathogenic bacteria in biosolids from anaerobic digestors and aerobic composters. *Appl. Environ. Microbiol.*, 69, 4618, 2003.

237. Koo, K. and Jaykus, L.A. Selective amplification of bacterial RNA: use of a DNA primer containing mismatched bases near its 30 terminus to reduce false-positive signals. *Lett. Appl. Microbiol.*, 31, 187, 2000.

238. Hein, I. et al. Detection and quantification of the *iap* gene of *Listeria monocytogenes* and *Listeria innocua* by a new real-time quantitative PCR assay. *Res. Microbiol.*, 152, 37, 2001.

239. Hough, A.J. et al. Rapid enumeration of *Listeria monocytogenes* in artificially contaminated cabbage using real-time polymerase chain reaction. *J. Food Prot.*, 65, 1329, 2002.

240. Bhagwat, A.A. Simultaneous detection of *Escherichia coli O157:H7*, *Listeria monocytogenes* and *Salmonella* strains by real-time PCR. *Int. J. Food Microbiol.*, 84, 217, 2003.

241. Guilbaud, M. et al. Quantitative detection of *Listeria monocytogenes* in biofilms by real-time PCR. *Appl. Environ. Microbiol.*, 71, 2190, 2005.

242. Barbau-Piednoir, E. et al. Development and validation of qualitative SYBR®Green real-time PCR for detection and discrimination of *Listeria* spp. and *Listeria monocytogenes*. *Appl. Microbiol. Biotechnol.*, 97, 4021, 2013.

243. Lazcka, O., Campo, F.J. and Munoz, F.X. Pathogen detection: a perspective of traditional methods and biosensors. *Biosens. Bioelectron*, 22, 1205, 2007.

244. Ohk, S.H. et al. Antibody-aptamer functionalized fibre-optic biosensor for specific detection of *Listeria monocytogenes* from food. *J. Appl. Microbiol.*, 109, 808, 2010.

245. Banerjee, P. and Bhunia, A.K. Cell-based biosensor for rapid screening of pathogens and toxins. *Biosens. Bioelectron.*, 26, 99, 2010.

246. Bifulco, L., Ingianni, A. and Pompei, R. An internalin a probe-based genosensor for *Listeria monocytogenes* detection and differentiation. *Biomed. Res. Int.*, 2013, 6401, 2013.

247. Oaew, S. et al. Gold nanoparticles/horseradish peroxidase encapsulated polyelectrolyte nanocapsule for signal amplification in *Listeria monocytogenes* detection. *Biosens. Bioelectron.*, 34, 238, 2012.

248. Wang, D. et al. Rapid detection of *Listeria monocytogenes* in raw milk with loop-mediated isothermal amplification and chemosensor. *J. Food Sci.*, 76, M611, 2011.

249. Tang, M. et al. Rapid and sensitive detection of *Listeria monocytogenes* by loop-mediated- isothermal amplification. *Curr. Microbiol.*, 63, 511, 2011.

250. Jadhav, S. et al. Methods used for the detection and subtyping of *Listeria monocytogenes*. *J. Microbiol. Methods*, 88, 327, 2012.

251. Barbuddhe, S.B.et al. Rapid identification and typing of *Listeria* species using matrix assisted laser desorption ionization-time of flight mass spectrometry. *Appl.Environ. Microbiol.*,74, 5402, 2008.

252. Welker, M. and Moore, E.R.B. Applications of whole-cell matrix-assisted laser desorption/ionization time-of-flight mass spectrometry in systematic microbiology. *Syst. Appl. Microbiol.*, 34, 2, 2011.

253. Uyttendaele, M. et al. Development of NASBA, a nucleic acid amplification system, for identification of *Listeria monocytogenes* and comparison to ELISA and a modified FDA method. *Int. J. Food Microbiol.*, 27, 77, 1995.

254. Hancock, I., Bointon, B. M. and McAthey, P. Rapid detection of *Listeria* spp. by selective impedimetric assay. *Lett. Appl. Microbiol.*, 16, 311, 1993.

255. Skjerve, E., Rorvik, L.M. and Olsvik, O. Detection of *Listeria monocytogenes* in foods by immunomagnetic separation. *Appl. Environ. Microbiol.*, 56, 3478, 1990.

256. Zhang, X. et al. Peptide nucleic acid fluorescence in situ hybridization for identification of *Listeria* genus, *Listeria monocytogenes* and *Listeria ivanovii*. *Int. J. Food Microbiol.*, 157, 309, 2012.

257. Arias-Rios, E.V. et al. Rapid detection of *Listeria* in ice cream in 13 hours using the Roka *Listeria* detection assay. *J. AOAC Int.*, 101, 1806, 2018.

258. Roberts, A.J. et al. Some *Listeria monocytogenes* outbreak strains demonstrate significantly reduced invasion, inlA transcript levels, and swarming motility in vitro. *Appl. Environ. Microbiol.*, 75, 5647, 2009.

259. Laksanalamai, P. et al. High density microarray analysis reveals new insights into genetic footprints of *Listeria monocytogenes* strains involved in listeriosis outbreaks. *PLOS One*, 7, e32896, 2012.

260. Ferreira, V. et al. Diverse geno- and phenotypes of persistent *Listeria monocytogenes* isolates from fermented meat sausage production facilities in Portugal. *Appl. Environ. Microbiol.*, 77, 2701, 2011.

261. Graves, L.M., Swaminathan, B. and Hunter, S.B. Subtyping *Listeria monocytogenes*. In *Listeria, Listeriosis, and Food Safety*, Ryser, E. and Marth, E., eds., Marcel Dekker, New York, 1999, 279.

262. Wiedmann, M. Molecular subtyping methods for *Listeria monocytogenes*. *J. AOAC Int.*, 85, 524, 2002.

263. Schonberg, A. et al. Serotyping of 80 strains from the WHO multicentre international typing study of *Listeria monocytogenes*. *Int. J. Food Microbiol.*, 32, 279, 1996.

264. Bruce, J.L. et al. Sets of *EcoRI* fragments containing ribosomal RNA sequences are conserved among different strains of *Listeria monocytogenes*. *Proc. Natl. Acad. Sci. U.S.A.*, 92, 5229, 1995.

265. Graves, L.M. and Swaminathan, B. PulseNet standardized protocol for subtyping *Listeria monocytogenes* by macrorestriction and pulsed-field gel electrophoresis. *Int. J. Food Microbiol.*, 65, 55, 2001.

266. Jeffers, G.T.et al. Comparative genetic characterization of *Listeria monocytogenes* isolates from human and animal listeriosis cases. *Microbiology*, 147, 1095, 2001.

267. Katzav, M. et al. Pulsed-field gel electrophoresis typing of *Listeria monocytogenes* isolated in two Finnish fish farms. *J. Food Prot.*, 69, 1443, 2006.

268. Seeliger, H.P.R. and Hohne, K. Serotyping of *Listeria monocytogenes* and related species. *Methods Microbiol.*, 13, 31, 1979.

269. Seeliger, H.P.R. and Jones, D. *Listeria*. In *Bergey's Manual of Systematic Bacteriology*, vol. 2, Sneath, P. H.A. et al., eds. Baltimore, MD: Williams and Wilkins, 1986, 1235.

270. Jacquet, C.et al. Expression of *ActA*, *Ami*, *InlB*, and listeriolysin O in *Listeria monocytogenes* of human and food origin. *Appl. Environ. Microbiol.*, 68, 616, 2002.

271. Wiedmann, M, et al. Free in PMC ribotype diversity of *Listeria monocytogenes* strains associated with outbreaks of listeriosis in ruminants. *J Clin Microbiol.*, 34, 1086, 1996.

272. Borucki, M.K. and Call, D.R. *Listeria monocytogenes* serotype identification by PCR *J. Clin. Microbiol.*, 41, 5537, 2003.

273. Doumith, M. et al. Differentiation of the major *Listeria monocytogenes* serovars by multiplex PCR. *J. Clin. Microbiol.*, 42, 3819, 2004.

274. Burall, L.S., Simpson, A.C. and Datta, A.R. Evaluation of a serotyping scheme using a combination of an antibody based serogrouping method and a multiplex PCR assay for identifying the major serotypes of *Listeria monocytogenes*. *J. Food Prot.*, 74, 403, 2011.

275. Audurier, A. et al. A phage typing system for *Listeria monocytogenes* and its use in epidemiological studies. *Clin. Invest Med.*, 7, 229, 1984.

276. Bibb, W.F. et al. Analysis of clinical and food-borne isolates of *Listeria monocytogenes* in the United States by multilocus enzyme electrophoresis and application of the method to epidemiologic investigations. *Appl. Environ. Microbiol.*, 56, 2133, 1990.

277. Piffaretti, J.C. et al. Genetic characterization of clones of the bacterium *Listeria monocytogenes* causing epidemic disease. *Proc. Natl. Acad. Sci. U.S.A*, 86, 3818, 1989.

278. Datta, A.R., Laksanalamai, P. and Solomotis, M. Recent developments in molecular sub-typing of *Listeria monocytogenes*. *Food Addit. Contam. Part A Chem. Anal. Control Expo. Risk Assess*, 30, 1437, 2013.

279. Graves, L.M. et al. Comparison of ribotyping and multilocus enzyme electrophoresis for subtyping of *Listeria monocytogenes* isolates. *J. Clin. Microbiol.*, 32, 2936, 1994.

280. Gerner-Smidt, P. et al. PulseNet USA: a five-year update. *Foodborne Pathog. Dis.*, 3, 9, 2006.

281. Halpin, J.L. et al. Re-evaluation, optimization, and multi-laboratory validation of the PulseNet- standardized pulsed-field gel electrophoresis protocol for *Listeria monocytogenes*. *Foodborne Pathog. Dis.*, 7, 293, 2010.

282. Wiedmann, M. et al. Ribotypes and virulence gene polymorphisms suggest three distinct *Listeria monocytogenes* lineages with differences in pathogenic potential. *Infect. Immun.*, 65, 2707, 1997.

283. Manfreda, G. et al. Occurrence and ribotypes of *Listeria monocytogenes* in Gorgonzola cheeses. *Int. J. Food Microbiol.*, 102, 287, 2005.

284. Chen, Y., Zhang, W. and Knabel, S.J. Multi-virulence-locus sequence typing identifies single nucleotide polymorphisms which differentiate epidemic clones and outbreak strains of *Listeria monocytogenes*. *J. Clin. Microbiol.*, 45, 835, 2007.

285. Lawrence, L.M. and Gilmour, A. Characterization of *Listeria monocytogenes* isolated from poultry products and from the poultry-processing environment by random amplification of polymorphic DNA and multilocus enzyme electrophoresis. *Appl. Environ. Microbiol.*, 61, 2139, 1995.

286. Aguado, V., Vitas, A.I. and Garcia-Jalon, I. Characterization of Listeria monocytogenes and *Listeria innocua* from a vegetable processing plant by RAPD and REA. *Int. J. Food Microbiol.*, 90, 341, 2004.

287. Aarts, H.J., Hakemulder, L.E. and Van Hoef, A.M. Genomic typing of *Listeria monocytogenes* strains by automated laser fluorescence analysis of amplified fragment length polymorphism fingerprint patterns. *Int. J. Food Microbiol.*, 49, 95, 1999.

288. Vogel, B.F. et al. High-resolution genotyping of *Listeria monocytogenes* by fluorescent amplified fragment length polymorphism analysis compared to pulsed-field gel electrophoresis, random amplified polymorphic DNA analysis, ribotyping, and PCR-restriction fragment length polymorphism analysis. *J. Food Prot.*, 67, 1656, 2004.

289. Guerra, M.M., Bernardo, F. and McLauchlin, J. Amplified fragment length polymorphism (AFLP) analysis of *Listeria monocytogenes*. *Syst. Appl. Microbiol.*, 25, 456, 2002.

290. Keto-Timonen, R.O., Autio, T.J. and Korkeala, H.J. An improved amplified fragment length polymorphism [AFLP] protocol for discrimination of *Listeria* isolates. *Syst. Appl. Microbiol.*, 26, 236, 2003.

291. Paillard, D. et al. Rapid identification of *Listeria* species by using restriction fragment length polymorphism of PCR-amplified *23S rRNA* gene fragments. *Appl. Environ. Microbiol.*, 69, 6386, 2003.

292. Jersek, B. et al. Typing of *Listeria monocytogenes* strains by repetitive element sequence-based PCR. *J. Clin. Microbiol.*, 37, 103, 1999.

293. Chan, M.S., Maiden, M.C. and Spratt, B.G. Database-driven multi locus sequence typing [MLST] of bacterial pathogens. *Bioinformatics.*, 17, 1077, 2001.

294. Enright, M.C. et al. Multilocus sequence typing of *Streptococcus pyogenes* and the relationships between emm type and clone. *Infect. Immun.*, 69, 2416, 2001.

295. Spratt, B.G. Multilocus sequence typing: molecular typing of bacterial pathogens in an era of rapid DNA sequencing and the internet. *Curr. Opin. Microbiol.*, 2, 312, 1999.

296. Chenal-Francisque, V. et al. Optimized Multilocus variable-number tandem-repeat analysis assay and its complementarity with pulsed-field gel electrophoresis and multilocus sequence typing for *Listeria monocytogenes* clone identification and surveillance. *J. Clin. Microbiol.*, 51, 1868, 2013.

297. Chen, S. et al. Multiple-locus variable number of tandem repeat analysis [MLVA] of *Listeria monocytogenes* directly in food samples. *Int. J. Food Microbiol.*, 148, 8, 2011.

298. Lindstedt, B.A. et al. Multiple-locus-variable-number tandem-repeat analysis of *Listeria monocytogenes* using multicolour capillary electrophoresis and comparison with pulsed field gel electrophoresis. *J. Microbiol. Methods*, 72, 141, 2008.

299. Doumith, M. et al. New aspects regarding evolution and virulence of *Listeria monocytogenes* revealed by comparative genomics and DNA arrays. *Infect. Immun.*, 72, 1072, 2004.

300. Orsi, R.H., Den Bakker, H.C. and Wiedmann, M. *Listeria monocytogenes* lineages: genomics, evolution, ecology, and phenotypic characteristics. *Int. J. Med. Microbiol.*, 301, 79, 2011.

301. Borucki, M.K. et al. Discrimination among *Listeria monocytogenes* isolates using a mixed genome DNA microarray. *Vet. Microbiol.*, 92, 351, 2003.

302. Rudi, K., Katla, T. and Naterstad, K. Multi locus fingerprinting of *Listeria monocytogenes* by sequence-specific labeling of DNA probes combined with array hybridization. *FEMS Microbiol. Lett.*, 220, 9, 2003.

303. Call, D.R., Borucki, M.K. and Besser, T.E. Mixed-genome microarrays reveal multiple serotype and lineage-specific differences among strains of *Listeria monocytogenes*. *J. Clin. Microbiol.*, 41, 632, 2003.

304. Zhang, C. et al. Genome diversification in phylogenetic lineages I and II of *Listeria monocytogenes*: identification of segments unique to lineage II populations. *J. Bacteriol.*, 185, 5573, 2003.

305. Johnson, J. et al. Natural atypical *Listeria innocua* strains with *Listeria monocytogenes* pathogenicity island 1 genes. *Appl. Environ. Microbiol.*, 70, 4256, 2004.

306. Glaser, P. et al. Comparative genomics of *Listeria* species. *Science*, 294, 849, 2001.

307. Nelson, K.E. et al. Whole genome comparisons of serotype 4b and 1/2a strains of the food-borne pathogen *Listeria monocytogenes* reveal new insights into the core genome components of this species. *Nucleic Acids Res.*, 32, 2386, 2004.

308. Hain, T. et al. Whole-genome sequence of *Listeria welshimeri* reveals common steps in genome reduction with *Listeria innocua* as compared to *Listeria monocytogenes*. *J. Bacteriol.*, 188, 7405, 2006.

309. Steinweg, C. et al. The complete genome sequence of *L. seeligeri*, a non-pathogenic member of the genus *Listeria*. *J. Bacteriol.*, 192, 1473, 2010.

310. Jones, E.M. and MacGowan, A.P. Antimicrobial chemotherapy of human infection due to *Listeria monocytogenes*. *Eur. J. Clin. Microbiol. Infect. Dis.*, 14, 165, 1995.

311. Janakiraman, V. Listeriosis in pregnancy: diagnosis, treatment, and prevention. *Rev. Obstet. Gynecol.*, 1, 179, 2008.

312. Lorber, B. ed. *Mandell, Douglas, and Bennett's Principles and Practice of Infectious Diseases*, 7th ed., Philadelphia, PA: Churchill Livingstone Elsevier, 2009.

313. Tunkel, A.R. et al. The management of encephalitis: clinical practice guidelines by the Infectious Diseases Society of America. *Clin. Infect. Dis.*, 47, 303, 2008.

314. Safdar, A. and Armstrong, D. Antimicrobial activities against 84 *Listeria monocytogenes* isolates from patients with systemic listeriosis at a comprehensive cancer center [1955-1997]. *J. Clin. Microbiol.*, 41, 483, 2003.

315. Hof, H. Listeriosis: therapeutic options. *FEMS Immunol. Med. Microbiol.*, 35, 203, 2003.

316. Blazquez, R. et al. Activity in vitro of 22 antimicrobial agents against clinical isolates of *Listeria monocytogenes*. *Clin. Microbiol. Infect.*, 2, 63, 1996.

317. Merle-Melet, M. et al. Is amoxicillin- cotrimoxazole the most appropriate antibiotic regimen for *Listeria* meningoencephalitis? Review of 22 cases and the literature. *J. Infect.*, 33, 79, 1996.

318. Scortti, M. et al. Coexpression of virulence and fosfomycin susceptibility in *Listeria*: molecular basis of an antimicrobial *in vitro-in vivo* paradox. *Nat. Med.*, 12, 515, 2006.

319. Moellering Jr., R.C., Wennersten, C. and Weinberg, A.N. Synergy of penicillin and gentamicin against *Enterococci*. *J. Infect. Dis.*, 124, S207, 1971.

320. Biavasco, F. et al. In vitro conjugative transfer of *VanA* vancomycin resistance between *Enterococci* and Listeriae of different species. *Eur. J. Clin. Microbiol. Infect. Dis.*, 15, 50, 1996.

321. Charpentier, E. and Courvalin, P. Emergence of the trimethoprim resistance gene *dfrD* in *Listeria monocytogenes* BM4293. *Antimicrob. Agents Chemother.*, 41, 1134, 1997.

322. Poyart-Salmeron, C. et al. Transferable plasmid-mediated antibiotic resistance in *Listeria monocytogenes*. *Lancet*, 335, 1422, 1990.

323. Roberts, M.C. et al. Transferable erythromycin resistance in *Listeria* spp. isolated from food. *Appl. Environ. Microbiol.*, 62, 269, 1996.

324. Mohammed, A.N. and Abdel Aziz, S.A.A. Novel approach for controlling resistant *Listeria monocytogenes* to antimicrobials using different disinfectants types loaded on silver nanoparticles (AgNPs). *Environ. Sci. Pollut. Res. Int.*, 26, 1954, 2019.

325. Kumaraswamy, M. et al. *Listeria monocytogenes* endocarditis: case report, review of the literature, and laboratory evaluation of potential novel antibiotic synergies. *Int. J. Antimicrob. Agents*, 51, 468, 2018.

326. Singh, R. and Paterson, Y. *Listeria monocytogenes* as a vector for tumor-associated antigens for cancer immunotherapy. *Expert. Rev. Vaccines*, 5, 541, 2006.

327. Singh, R. et al. Fusion to listeriolysin O and delivery by *Listeria monocytogenes* enhances the immunogenicity of HER-2/neu and reveals subdominant epitopes in the FVB/N mouse. *J. Immunol.*, 175, 3663, 2005.

328. Le, D.T., Dubenksy, T.W. and Brockstedt, D.G. Clinical development of *Listeria monocytogenes*-based immunotherapies. *Semin. Oncol.*, 39, 311, 2012.

329. Rothman, J. and Paterson, Y. Live attenuated *Listeria*-based immunotherapy. *Expert Rev. Vaccines.*, 12, 493, 2013.

330. Mandal, M. et al. Cytosolic delivery of viral nucleoprotein by listeriolysin O-liposome induces enhanced specific cytotoxic T lymphocyte response and protective immunity. *Mol. Pharm.*, 1, 2, 2004.

331. Bolhassani, A. and Zahedifard, F. Therapeutic live vaccines as a potential anticancer strategy. *Int. J. Cancer*, 131, 1733, 2012.

332. Paterson, Y. and Johnson, R.S. Progress towards the use of *Listeria monocytogenes* as a live bacterial vaccine vector for the delivery of HIV antigens. *Expert. Rev. Vaccines*, 3, S119, 2004.

333. Wood, L.M. and Paterson Y. Attenuated *Listeria monocytogenes*: a powerful and versatile vector for the future of tumor immunotherapy. *Front. Cell Infect. Microbiol.*, 4, 51, 2014.

334. Bolhassani, A., Naderi, N. and Soleymani, S. Prospects and progress of *Listeria*-based cancer vaccines. *Expert Opin. Biol. Ther.*, 17, 1389, 2017.

335. Chen, Y. et al. Development of a *Listeria monocytogenes*-based vaccine against hepatocellular carcinoma. *Oncogene*, 31, 2140, 2012.

336. Quispe-Tintaya, W. et al. Nontoxic radioactive *Listeria* (at) is a highly effective therapy against metastatic pancreatic cancer. *Proc. Natl. Acad. Sci. U.S.A.*, 110, 8668, 2013.

337. Stritzker, J. et al. Prodrugcon-verting enzyme gene delivery by *L. monocytogenes*. *BMC Cancer*, 8, 94, 2008.

338. Hannan, R. et al. Combined immunotherapy with *Listeria monocytogenes*-based PSA vaccine and radiation therapy leads to a therapeutic response in a murine model of prostate cancer. *Cancer Immunol. Immunother.*, 61, 2227, 2012.

339. Mkrtichyan, M. et al. Anti-PD-1 antibody significantly increases therapeutic efficacy of *Listeria monocytogenes* (Lm)-LLO immunotherapy. *J. Immunother. Cancer*, 1, 15, 2013.

340. Qiu, Z., Khairallah, C. and Sheridan, B.S. *Listeria monocytogenes*: a model pathogen continues to refine our knowledge of the CD8 T cell response. *Pathogens*, 7, 55, 2018.

341. Terán-Navarro, H. et al., Pre-clinical development of *Listeria*-based nanovaccines as immunotherapies for solid tumours: insights from melanoma. *Oncoimmunology*, 8, e1541534, 2018.

342. Basu, P. et al. A randomized phase 2 study of ADXS11-001 *Listeria monocytogenes*-Listeriolysin O immunotherapy with or without cisplatin in treatment of advanced cervical cancer. *Int. J. Gynecol. Cancer*, 28, 764, 2018.

343. Jahangir, A. et al. Immunotherapy with *Listeria* reduces metastatic breast cancer in young and old mice through different mechanisms. *Oncoimmunology*, 6, e1342025, 2017.

344. Wesley, I.V. et al. Application of a multiplex polymerase chain reaction assay for the simultaneous confirmation of *Listeria monocytogenes* and other *Listeria* species in turkey sample surveillance. *J. Food Prot.*, 65, 780, 2002.

345. Somer, L. and Kashi, Y. A PCR method based on 16S rRNA sequence for simultaneous detection of the genus *Listeria* and the species *Listeria monocytogenes* in food products. *J. Food Prot.*, 66, 1658, 2003.

346. Sallen, B. et al.Comparative analysis of *16S* and *23S rRNA* sequences of *Listeria* species. *Int. J. Syst. Bacteriol.*, 46, 669, 1996.

347. Graham, T.A. et al. Inter- and intraspecies comparison of the *16S-23S rRNA* operon intergenic spacer regions of six *Listeria* spp. *Int. J. Syst. Bacteriol.*, 47, 863, 1997.

348. O'Connor, L. et al. Rapid polymerase chain reaction/DNA probe membrane-based assay for the detection of *Listeria* and *Listeria monocytogenes* in food. *J. Food Prot.*, 63, 337, 2000.

349. Furrer, B. et al. Detection and identification of *Listeria monocytogenes* in cooked sausage products and in milk by *in vitro* amplification of haemolysin gene fragments. *J. Appl. Bacteriol.*, 70, 372, 1991.

350. Lunge, V.R. et al. Factors affecting the performance of 5' nuclease PCR assays for *Listeria monocytogenes* detection. *J. Microbiol. Methods*, 51, 361, 2002.

351. Thimothe, J. et al. Tracking of *Listeria monocytogenes* in smoked fish processing plants. *J. Food Prot.*, 67, 328, 2004.

352. Paziak-Domanska, B. et al. Evaluation of the API test, phosphatidylinositol-specific phospholipase C activity and PCR method in identification of *Listeria monocytogenes* in meat foods. *FEMS Microbiol. Lett.*, 171, 209, 1999.

353. Greer, G.G. Bacterial contamination of recirculating brine used in the commercial production of moisture-enhanced pork. *J Food Prot.*, 67, 185, 2004.

354. Longhi, C. et al. Detection of *Listeria monocytogenes* in Italian-style soft cheeses. *J. Appl. Microbiol.*, 94, 879, 2003.

355. D'Agostino, M. A validated PCR-based method to detect *Listeria monocytogenes* using raw milk as a food model–towards an international standard. *J. Food Prot.*, 67, 1646, 2004.

356. Liu, D. et al. Characterization of virulent and avirulent *Listeria monocytogenes* strains by PCR amplification of putative transcriptional regulator and internalin genes. *J. Med. Microbiol.*, 52, 1065, 2003.

357. Jung, Y.S. et al. Polymerase chain reaction detection of *Listeria monocytogenes* on frankfurters using oligonucleotide primers targeting the genes encoding internalin AB. *J. Food Prot.*, 66, 237, 2003.

358. Pangallo, D. et al. Detection of *Listeria monocytogenes* by polymerase chain reaction oriented to *inlB* gene. *New Microbiol.*, 24, 333, 2001.

359. Cocolin, L. et al. Direct identification in food samples of *Listeria* spp and Listeria monocytogenes by molecular methods. *Appl. Environ. Microbiol.*, 68, 6273, 2002.

360. Schmid, M. et al. Nucleic acid-based, cultivation-independent detection of *Listeria* spp and genotypes of *L monocytogenes*. *FEMS Immunol. Med. Microbiol.*, 35, 215, 2003.

361. Johnson, W.M. et al. Detection of genes coding for listeriolysin and *Listeria monocytogenes* antigen A [*ImaA*] in *Listeria* spp. by the polymerase chain reaction. *Microb. Pathog.*, 12, 79, 1992.

362. Wernars, K. Use of the polymerase chain reaction for direct detection of *Listeria monocytogenes* in soft cheese. *J. Appl. Bacteriol.*, 70, 121, 1991.

363. Fluit, A.C. et al. Detection of *Listeria monocytogenes* in cheese with the magnetic immuno-polymerase chain reaction assay. *Appl. Environ. Microbiol.*, 59, 1289, 1993.

364. Gilot, P. and Content, J. Specific identification of *Listeria welshimeri* and *Listeria monocytogenes* by PCR assays targeting a gene encoding a fibronectin-binding protein. *J. Clin. Microbiol.*, 40, 698, 2002.

365. Gray, D.I. and Kroll, R.G. Polymerase chain reaction amplification of the *flaA* gene for the rapid identification of *Listeria* spp. *Lett. Appl. Microbiol.*, 20, 65, 1995.

366. Winters, D.K., Maloney, T.P. and Johnson, M.G. Rapid detection of *Listeria monocytogenes* by a PCR assay specific for an aminopeptidase. *Mol. Cell Probes*, 13, 127, 1999.

367. Moorhead, S.M., Dykes, G.A. and Cursons, R.T. An SNP-based PCR assay to differentiate between *Listeria monocytogenes* lineages derived from phylogenetic analysis of the sigB gene. *J. Microbiol. Methods*, 55, 425, 2003.

368. Nightingale, K. et al. Combined *sigB* allelic typing and multiplex PCR provide improved discriminatory power and reliability for *Listeria monocytogenes* molecular serotyping. *J. Microbiol. Methods*, 68, 52, 2007.

369. Cabanes, D. et al. Gp96 is a receptor for a novel *Listeria monocytogenes* virulence factor, Vip, a surface protein. *EMBO J.*, 24, 2827, 2005.

370. Liu, D. et al. Species-specific PCR determination of *Listeria seeligeri*. *Res. Microbiol.*, 155, 741, 2004.

371. Rodriguez-Lazaro, D. et al. Quantitative detection of *Listeria monocytogenes* and *Listeria innocua* by real-time PCR: assessment of *hly*, *iap*, and *lin02483* targets and AmpliFluor technology. *Appl. Environ. Microbiol.*, 70, 1366, 2004.

372. Liu, D. et al. PCR detection of a putative N-acetylmuramidase gene from *Listeria ivanovii* facilitates its rapid identification. *Vet. Microbiol.*, 101, 83, 2004.

373. Loessner, M.J. and Busse, M. Bacteriophage typing of *Listeria* species *Appl. Environ. Microbiol.*, 56, 1912, 1990.

374. Capita, R. et al. Evaluation of the international phage typing set and some experimental phages for typing of *Listeria monocytogenes* from poultry in Spain. *J Appl Microbiol.*, 92, 90, 2002.

375. Harvey, J. and Gilmour, A. Characterization of *Listeria monocytogenes* isolates by esterase electrophoresis. *Appl. Environ. Microbiol.*, 62, 1461, 1996.

376. Aarnisalo, K. et al. Typing of *Listeria monocytogenes* isolates originating from the food processing industry with automated ribotyping and pulsed-field gel electrophoresis. *J. Food Protect.*, 66, 249, 2003.

377. Suihko, M.L. et al. Characterization of *Listeria monocytogenes* isolates from the meat, poultry and seafood industries by automated ribotyping. *Int J Food Microbiol.*, 30, 137, 2002.

378. Parisi, A. et al. Amplified fragment length polymorphism and multi-locus sequence typing for high-resolution genotyping of *Listeria monocytogenes* from foods and the environment. *Food Microbiol.*, 27, 101, 2010.

379. Nadon, C.A. et al. Correlations between molecular subtyping and serotyping of *Listeria monocytogenes*. *J. Clin. Microbiol.*, 39, 2704, 2001.

380. Cantinelli, T. et al. "Epidemic clones" of *Listeria monocytogenes* are widespread and ancient clonal groups. *J. Clin. Microbiol.*, 51, 3770, 2013.

381. Miya S. et al. Development of a multilocus variable-number of tandem repeat typing method for *Listeria monocytogenes* serotype 4b strains. *Int. J. Food Microbiol.*, 124, 239, 2008.

382. Miya, S. et al. Highly discriminatory typing method for *Listeria monocytogenes* using polymorphic tandem repeat regions. *J Microbiol. Methods.*, 90, 285, 2012.

383. den Bakker, H.C. et al. Genome sequencing identifies *Listeria fleischmannii* subsp. *coloradonensis* subsp. nov., isolated from a ranch. *Int. J. Syst. Evol. Microbiol.*, 63, 3257, 2013.

384. Gilmour, M.W. et al. High-throughput genome sequencing of two *Listeria monocytogenes* clinical isolates during a large foodborne outbreak. *BMC Genomics.*, 11, 120, 2010.

33

The Genus Mycobacterium

Leen Rigouts and Sari Cogneau

CONTENTS

Introduction

The genus *Mycobacterium* consists of more than 190 species and belongs to the family of *Mycobacteriaceae*, order Corynebacteriales, phylum/class Actinobacteria, and kingdom Bacteria (http://www.dsmz.de/bacterial-diversity/prokaryotic-nomenclature-up-to-date).[1] Mycobacteria are nonmotile, rod-shaped bacilli, characterized by an extremely lipid-rich cell wall comprising mycolic acids. These long-chain (C-60 to C-90) fatty acids make all mycobacterial species "acid-fast" as they resist acid-alcohol decolorization following staining with phenicated dyes like fuchsine, enabling their microscopic detection. Mycobacteria should not be confused with members of the closely related genera *Nocardia, Corynebacterium*, or *Rhodococcus*, which have shorter-chain (C-22 to C-64) mycolic acids and can be partially acid-fast.[1]

Based on phenotypic characters, the genus was proposed in 1896 by Lehmann and Neuman.[2] With the advent of DNA-based analysis, mycobacterial species were initially differentiated at <70% similarity by DNA-DNA hybridization,[3] and later by <98.8% 16SrRNA Sanger sequence similarity[4] with or without

DOI: 10.1201/9781003099277-35

inclusion of additional housekeeping genes like *hsp65* or *rpoB*.[5] These phenotypic and genotypic analyses are all in agreement with the initially described genus *Mycobacterium,* phylogenetically distinguishing five major groups.

Recently, the taxonomy of the genus *Mycobacterium* has been challenged based on a core protein analysis—identifying insertions/deletions of amino acids, or proteins exclusively found in evolutionarily-related groups of species—suggesting to split the genus in five monophyletic groups designated as the "Tuberculosis-Simiae," "Terrae," "Triviale," "Fortuitum-Vaccae," and "Abscessus-Chelonae" clades.[6] The five proposed genera exactly overlap the major clades of the classical taxonomy. Gupta et al. propose to keep the genus *"Mycobacterium"* for species belonging to the "Tuberculosis-Simiae" clade and to introduce four novel genera hosting the nontuberculous mycobacteria (NTM) of the above mentioned clades in *"Mycolicibacter," "Mycolicibacillus," "Mycolicibacterium,"* and *"Mycobacteroides,"* respectively. This reclassification has been debated and questioned for reasons of methodology and clinical relevance.[7,8] Tortoli and colleagues argue:

> NTM may be diverse in their genetics and biology, but they produce remarkably similar disease manifestations in distinct populations at risk. This uniformity and the lessons to be learned from it are important for both clinicians as well as the affected patients.

The newly proposed reallocation risks to cause confusion among clinicians and clinical microbiologists, while not being beneficial for the patient. Also, the novel genus *"Mycobacterium"* is said to include all of the major human pathogens[6] but does not comprise clinically important species like *M. abscessus*.[9] Regarding the methodological approach for this reclassification, while Tortoli and colleagues identified minor technical errors (mislabeling for some of the sequences used),[10] Meehan and colleagues questioned the correctness of the analysis in a more fundamental way.[8] Analyzing the percentage of conserved proteins (POCP) revealed that all 145 tested species in the original *Mycobacterium* genus fell within the 50% boundary of all others, while all 223 tested non-*Mycobacterium* species from the *Corynebacteriaceae* family had <50% POCP to all *Mycobacterium* species, strongly supporting retention of the classical genus *Mycobacterium*.[8] Given the equal taxonomic validity of both classification systems, the newly proposed genus and species names can be used as synonyms.[7] For sake of clarity in this handbook on clinical practice, we will stick to the classical genus *"Mycobacterium"* and related species names in this chapter.

With their generation time varying from 2 to 48 hours, mycobacterial species can be classified as slow growers—requiring more than 7 days to form visible colonies on solid medium—and rapid growers exhibiting visible growth in less than 7 days. This growth-rate-based separation is supported by phylogenetic trees constructed from 16S rRNA or whole genome sequences.[11] Combining growth rate with ability to produce pigmentation and clinical significance, Runyon classified members of the *Mycobacterium* genus into five groups,[12,13] a classification that is barely used nowadays.

According to their infectiousness, mycobacteria are classified in three categories: (1) strict pathogens, with members of the *M. tuberculosis* complex (MTBc) and *M. leprae* having the biggest public health impact; (2) opportunistic pathogens like *M. avium*, that are found in the environment and may cause disease in persons with predisposing conditions or compromised immunity; and (3) saprophytes like *M. gordonae*, that are ubiquitous and can be found in environmental sources, yet rarely or never cause infection and are considered overall nonpathogenic. Most (opportunistic) pathogenic mycobacteria are slow growers.

Clinical Importance of Mycobacteria

The clinical importance of mycobacteria cannot be overemphasized, especially for members of the MTBc, *M. leprae*, *M. ulcerans*, and some opportunistic NTM in specific vulnerable populations.

Mycobacterium tuberculosis Complex (MTBC)

Tuberculosis (TB) is caused by members of the *M. tuberculosis* complex (MTBC) comprising *M. tuberculosis sensu stricto*, *M. africanum*, *M. bovis*, *M. canettii*, *M. caprae*, *M. microti*, *M. mungi*, *M. pinnipedii*, *M. orygis*, and *M. surricatae*.[14] While *M. tuberculosis* and *M. africanum* are human-adapted species, the other species prefer (specific) animal hosts with possible zoonotic spill over to humans, mostly from *M. bovis* via cattle.[14] An attenuated strain of *M. bovis*, *M. bovis* BCG (Bacille Calmette-Guérin), is used as a vaccine in many high-burden countries to protect against acquiring TB, though its efficacy is questionable.[15]

The two human-adapted species comprise eight phylogenetic lineages (L1–L8),[16] with L1–L4 and L7–L8 being *M. tuberculosis sensu stricto*, and L5 and L6 *M. africanum*. As for their geographic repartition, L2 and L4 are the most widespread, while L1 and L3 are intermediately distributed, L5 and L6 are restricted to West Africa, and L7 is found only in Ethiopia.[17] *M. africanum* is responsible for up to 40% of TB in West Africa.[18]

Worldwide, TB is among the top 10 causes of death, and the first cause of death by a single infectious agent (above HIV/AIDS).[19] It was estimated that in 2018, worldwide, 10 million people developed TB, of whom 1.2 million died. Two-thirds of cases occurred in eight high-burden countries, with India (27%) and China (9%) ranking first. Only 6% of global cases were in the World Health Organization (WHO) European Region (3%) and the WHO Region of the Americas (3%). In terms of TB incidence, Africa ranked first with 231 per 100,000 population, which is almost twofold the worldwide incidence (132 per 100,000 population).[19] Likewise, the highest TB mortality rate was reported from Africa, where the TB epidemic is fueled by the HIV/AIDS epidemic. People living with HIV are 20–30 times more likely to develop active TB disease than people without HIV. Following huge TB control efforts, increased accessibility to antiretroviral treatment and availability of new anti-TB drugs, the worldwide TB incidence rate is falling at about 2% per year, with the fastest regional declines between 2013 and 2017 in the WHO European (on average, 5% per year) and African (3.8% per year) regions.[19] Nevertheless, TB continues to be a major public health problem in many countries.

Of added importance is the number of MTBC strains that are found resistant to one or more of the antitubercular agents.

TB disease is always treated with a combination therapy. The preferred regimen for treating adults with TB consists of an intensive phase of 2 months of isoniazid (INH), rifampin (RIF), pyrazinamide (PZA), and ethambutol (EMB) followed by a continuation phase of 4 months of INH and RIF. RIF, a strong bactericidal and sterilizing drug, is considered the core drug for standard TB treatment.[20] Patients infected with MTBC strains resistant to (at least) RIF or RIF and INH—defined as multidrug resistant (MDR)—should be treated with alternative drugs in so-called MDR-TB regimens. While these drugs are commonly known as "second-line" drugs, recent guidelines from the WHO classify these drugs in group A, B, and C.[21] Individualized MDR-TB regimens using three drugs from group A, complemented by one or two drugs from group B, and one or more from C, have been proposed for a duration of 18 months or more.[22] Alternatively, a standardized short MDR-TB regimen (9–12 months) has been shown to be noninferior compared to the long regimen,[22] and was successfully implemented under programmatic conditions in multiple African and Asian countries.[23,24] The short regimen includes a fourth-generation fluoroquinolone (gatifloxacin, moxifloxacin, or levofloxacin) as core drug.[20] Recently, the use of injectable aminoglycosides (amikacin or kanamycin) or polypeptides (capreomycin) has been discouraged, due to the irreversible ototoxicity.[21]

An MDR patient may further develop resistance to other drugs during the course of treatment. MDR combined with resistance to a fluoroquinolone and an injectable aminoglycoside or polypeptide is defined as extreme drug resistance (XDR). XDR-TB treatment regimens require further inclusion of less efficient, more toxic group C drugs[21] or new drugs like bedaquiline and delamanid. XDR-TB treatment is difficult and costly.

MDR- and XDR-TB are a major concern worldwide, with cases reported in all countries. According to the WHO, in 2018, globally 3.4% of new cases and 18% of previously treated TB cases had RIF-resistant/MDR disease, with three countries accounting for half of the world's cases of RIF-resistant/MDR-TB: India (27%), China (14%), and the Russian Federation (9%).[19] Among MDR cases 6.2% were estimated to be XDR, numbers that were slightly lower compared to 2018 data.[19]

Non-tuberculous Mycobacteria (NTM)

Mycobacterial species different from MTBC are grouped as NTM, also referred to as atypical or anonymous mycobacteria, or mycobacteria other than tuberculosis (MOTT).

Humans can acquire NTM infection from various sources in the environment such as natural and engineered water systems, mist, aerosols, dust, and soil,[25] or from contact with plants, birds, fish, and other animals.[26,27] In humans, NTM can be found as colonizers without clinical relevance—with chronic lung disease as predisposing factor—or they can be seen as mostly intracellular pathogens. Human-to-human transmission of NTM was not known to be a common route of transmission, till recent evidence documented human-to-human transmission in cystic fibrosis patients.[28–30]

While the frequency of clinical isolation of NTM is increasing in many countries with decreasing MTBC prevalence,[27,31–33] and even in countries endemic for TB,[34,35] their clinical relevance remains often unclear.

The diagnosis of NTM disease remains a clinical dilemma for clinicians as NTM exist in the environment, and mere isolation from an unsterile site could be a reflection of environmental contamination. Correctly identifying and determining the clinical relevance are however paramount, as the time-consuming and complicated treatment of NTM disease will vary depending on the species. NTM infections most commonly manifest as pulmonary disease (especially in adults), followed by skin and soft tissue infections, lymphadenitis (especially in children), or disseminated disease in severely compromised patients.[36]

General criteria for the diagnosis of pulmonary NTM disease as established by the American Thoracic Society (ATS) may guide on clinical decision-making.[37] One of the criteria states that two positive cultures should be obtained from subsequent sputum specimens OR a single positive culture from a bronchial aspirate. Nevertheless, single-sputum NTM isolation can provide evidence of true NTM-lung-disease, taking into account patient's characteristics like bronchiectasis, younger age, and more severe radiographic pulmonary lesions.[38]

It is to be noted that only around 20–25% of patients with pulmonary NTM isolates meet the ATS criteria,[32,35,39,40] which can increase up to 60% among patients at higher risk like cystic fibrosis patients.[41] Of note, a considerable proportion of patients with NTM pulmonary disease in Europe have no detected underlying lung disease or immunodeficiency.[42] In the United States, the annual rates of pulmonary NTM disease range from 1.4 to 13.9 per 100,000 inhabitants,[43] while in Europe rates seem to be lower ranging from 0.2 to 2.9 per 100,000 population.[42]

The probability of clinical relevance varies across species. NTM commonly found in patients with pulmonary symptoms are *M. avium*, *M. intracellulare*, *M. kansasii*, *M. xenopi*, *M. abscessus*, *M. fortuitum*, and *M. scrofulaceum*.[32,36,40,44] *M. abscessus* is increasingly seen as pulmonary pathogen.[39,40] While the less common *M. genavense* has a very high likelihood of being clinically relevant in pulmonary disease, *M. avium* and *M. kansasii* have a moderate odds, and *M. gordonae* is almost never involved in disease though frequently being isolated.[32,40] The frequency of isolated NTM species may vary across geographical regions and settings.[45] Also, the relative frequency of NTM species may change over time, as has been noticed in Europe with shifts of *M. kansasii* and *M. malmoense* frequency.[42]

Regarding extra-pulmonary NTM disease, the majority of lymphadenitis cases are seen in young children, and most often caused by *M. avium*, *M. scrofulaceum*, *M. simiae*, *M. haemophilum*, *M. malmoense*, or *M. marinum*.[32,46-48]

NTM species most commonly reported from skin and soft tissue infections are *M. fortuitum*, *M. abscessus*, *M. chelonae* among the rapidly growing mycobacteria, and *M. marinum*, *M. ulcerans*, *M. chimaera*, and *M. haemophilum* as slowly growing species.[27] Using inadequately disinfected medical equipment has been associated with outbreaks of NTM skin and soft tissue infections. Because of their lipophilic character, environmental mycobacteria are frequently found in biofilms, making them even more resistant to decontamination with standard antiseptics and biocides, including chlorhexidine and glutaraldehyde.[42,49] Recently, some compounds have been described to be active against NTM in biofilms.[50,51] Potable water systems have been proposed as an advantageous pre-infection niche for bacteria—including NTM-colonizing lungs of CF patients.[52] In a recent comparison of Norwegian chlorinated and non-chlorinated drinking water distribution systems, it was found that residual chloramine may increase mycobacterial biomass, while it may also decrease mycobacterial diversity.[53] As a result, mycobacterial species have been associated with both pseudo-outbreaks and real outbreaks of nosocomial infections following surgery or cosmetic procedures. In the United States, cosmetic surgery tourism constitutes an increased risk for NTM soft skin tissue disease.[54,55] Molecular typing of the isolated strains during an outbreak with *M. abscessus* after breast augmentation surgery in Venezuela undoubtedly identified tap water as the source of infection.[56] Also from nonhospital/cosmetic settings, outbreaks have been reported, such as *M. marinum* infections following fish handling[57] or use of premixed ink as the common source of infection with *M. chelonae*.[58]

The NTM commonly found as causative agent in patients with disseminated disease are *M. avium, M. intracellulare, M. kansasii, M. haemophilum,* and *M. chelonae*.[36,59]

M. gordonae and *M. engbaekii* are among the NTM most frequently found as contaminants with no or very rare clinical implication.

Trying to ascertain the prevalence of NTM disease may be complicated by several factors. As opposed to TB, NTM disease is not notifiable to public health authorities at a national or an international level, with a few exceptions such as Oregon State (the U.S.A). Therefore, underreporting might occur, which can prohibit collection of systematic data and potentially lead to an underestimation of NTM disease. Conversely, overdiagnosis may occur when patients are diagnosed with (and treated for) non-pathogenic mycobacteria, and in those cases in which the clinical relevance of the potential pathogenic mycobacterium isolated has been misjudged. Finally, care must be taken with regard to the trend analysis of certain species, as the degree of taxonomical diversification is much higher in contemporary studies using molecular identification techniques, than it is in older studies based on phenotypic characteristics.[42]

Clinically Significant Species of Slowly Growing NTM

The following are some of the most important slow-growing NTM species that have been associated with clinical disease. An overview of the clinical relevance for additional slowly growing NTM species can be consulted at https://bccm.belspo.be/about-us/bccm-itm.

Mycobacterium leprae

This mycobacterium is an obligatory intracellular pathogen and the causative agent of leprosy or Hansen's disease. The disease mainly affects the skin, the peripheral nerves, mucosa of the upper respiratory tract, and the eyes. It can range from mild and self-limiting (tuberculoid type) to severely debilitating (lepromatous type). Leprosy is also classified as paucibacillary (PB) or multibacillary (MB), based on the number of skin lesions and the presence of nerve involvement. It is an ancient disease with cases depicted around 600 BC. There has always been a large social stigma associated with leprosy, which resulted in victims being segregated to leper colonies.

In humans, *M. leprae* is a very slow-growing organism with a long incubation period, it can take from 6 months to 20 years for an infected individual to develop symptoms after exposure (WHO fact sheet). *M. leprae* does not grow on laboratory media, likely due to the loss of important genes, with a reduced genome size of 3.3 Mbp compared to the 4.1 Mbp of MTBC.[60]

In its efforts to eliminate leprosy, the WHO recently introduced the concept of "one report for leprosy" by launching an online data collection tool to facilitate the gathering of epidemiological and programmatic indicators from member states. In 2018, there were 208,619 new leprosy cases registered globally, according to official figures from 159 countries from the six WHO Regions (https://www.who.int/news-room/fact-sheets/detail/leprosy). The majority of these cases were found in nine countries in Asia, Africa, and Latin America. In the United States, leprosy is seen in about 200 new cases every year, with nine-banded armadillos being an important reservoir.[61]

Current diagnosis of leprosy relies often solely on clinical findings, without microbiological confirmation. Microscopy for detection of acid-fast bacilli can confirm the diagnosis of more advanced leprosy, but it is less effective in the early stages or in paucibacillary cases. Nucleic acid-based detection techniques for *M. leprae* are becoming a standard of care for supporting the diagnosis of leprosy, with the repetitive sequence RLEP as a promising target for qPCR, allowing the detection of approximately one *M. leprae* genome in a sample (https://internationaltextbookofleprosy.org/chapter/pathogen-detection). No commercial molecular tests for *M. leprae* detection are available.

Clinical diagnosis of leprosy is based on the presence of at least one of three cardinal signs: (1) definite loss of sensation in a pale (hypopigmented) or reddish skin patch; (2) thickened or enlarged peripheral nerve with loss of sensation and/or weakness of the muscles supplied by that nerve; or (3) presence of acid-fast bacilli.[62]

Leprosy is now totally treatable with a free-of-cost multidrug therapy including dapsone, RIF, and clofazimine.

Mycobacterium ulcerans

This species is a close relative to *M. marinum,* but differs in its ability to produce a mycolactone toxin. It is the causative agent of Buruli ulcer (BU), a disease which occurs in at least 33 countries with mostly tropical and subtropical or temperate climates. Most cases are reported in West and Central Africa, while Australia remains an important non-African endemic country. In 2018, 2,713 suspected BU cases were reported from 14 countries, confirming the recent increase after years of decline from 2010 (5,000 cases) to 2016 (1,961 cases). The reasons for the initial decline and recent increase remain unclear https://www.who.int/news-room/fact-sheets/detail/buruli-ulcer-(mycobacterium-ulcerans-infection).

The disease begins as a painless nodule, but if left untreated, can progress to a debilitating skin ulcer or osteomyelitis, and can lead to permanent disfigurement and long-term disability.[63] It is the mycolactone that causes subcutaneous tissue destruction and local immunosuppression, aggravating the symptoms. Most of the lesions appear on the limbs, especially on the lower extremities.[63] *M. ulcerans* is an environmental bacterium, and the exact means of transmission remain unknown, but most infections appear to be found adjacent to bodies of (stagnant) water. There is some indication that mosquitoes may be responsible for transmission of the disease in Australia,[64,65] although evidence of *M. ulcerans* positive mosquitoes remains limited.[66] In Africa, where BU is most prevalent, no environmental source has been identified so far, with one exception of a *M. ulcerans* isolate grown from a water strider in Benin.[67]

Although the disease can often be diagnosed on clinical presentation, this organism can grow on standard media suitable for mycobacterial growth, with a growth temperature optimum of 30°C. The most sensitive diagnosis for BU, however, is done by PCR targeting the insertion sequence IS*2404*.

Buruli ulcer is treatable with combination therapy using RIF and clarithromycin for 8 weeks. Surgical debridement is often required as well, especially for larger ulcers or advanced disease.[63]

Mycobacterium avium Complex (MAC)

Originally, *M. avium* and *M. intracellulare* were the only two species belonging to the *Mycobacterium avium* complex (MAC).[37] Sequencing studies of the 16S rRNA gene and 16S–23S internal transcribed spacer (ITS) confirmed the MAC as a diverse, yet relatively clearly defined and clustered group among the slowly growing NTM, comprising the species *M. avium*, *M. intracellulare** (including *M. chimaera**, *M. yongonense**, *M. paraintracellulare**), *M. colombiense*, *M. arosiense*, *M. vulneris*, *M. bouchedurhonense*, *M. timonense*, *M. marseillense*, and *M. lepraemurium*.[68–70]

M. intracellulare cannot be easily differentiated from *M. avium* through traditional physical and biochemical tests;[37] it requires genetic analysis.

Members of the MAC are ubiquitous in soil and many natural and man-made water systems.[25,71,72] Although usually not considered pathogenic in the general community, these organisms are the most commonly isolated causative agents of NTM-mediated lung disease with TB-like symptoms,[37,70] and they may cause lymphadenitis in children and/or other extrapulmonary and disseminated infections in immunocompromised patients.[37,68,73] Pulmonary disease attributed to the MAC is mainly caused by *M. avium* and *M. intracellulare*. Other MAC species have been increasingly reported as potential causative agents of NTM lung infections, but their full spectrum of diseases is not yet entirely known.[68] One member of the MAC, namely *M. lepraemurium*, is known as the etiologic agent of feline and murine leprosy.[74,75]

Infections are most probably acquired by inhalation or ingestion of the MAC species. Although frequently isolated from respiratory specimens, isolation of MAC does not necessarily imply clinical relevance in pulmonary disease.[37]

To date, there is still an ongoing debate on whether or not specific species should be assigned within the complex or if the MAC should be considered as a single entity. Moreover, the American Thoracic Society and Infectious Diseases Society of America (ATS/IDSA) stated that although the differentiation between *M. avium* and *M. intracellulare* may be important epidemiologically for research purposes and in the future for therapeutic use, the distinction is not yet clinically significant enough

to routinely identify MAC isolates to the species level.[37] Also, disease caused by all members of the MAC is treated in a similar manner using a regimen specific to the MAC.[68]

Treatment regimens for MAC infections include combination therapy, typically with a macrolide (clarithromycin or azithromycin) as core drug, with RIF (or another companion drug), EMB and occasionally amikacin in cases of disseminated disease when a more aggressive therapy is required.[37,76] ATS recommends clarithromycin susceptibility testing for new, previously untreated MAC isolates.[37]

Mycobacterium avium

M. avium was originally divided into three subspecies corresponding to pathogenicity and host-range characteristics: *M. avium avium*, *M. avium paratuberculosis*, and *M. avium silvaticum*.[77,78] To elucidate the epidemiology of *M. avium*-associated disease, Mijs et al. clustered the isolates originating from humans and pigs together as a suggested fourth subspecies called *M. avium hominissuis*,[79] the most abundant etiological agent of *M. avium*-related disease in humans.[80,81] Additionally, Mijs and colleagues stated that the designation *M. avium avium* should be reserved for bird-type strains because of its association with avian TB.[79] *M. avium paratuberculosis* on the other hand is the causative agent of Johne's disease resulting in chronic granulomatous enteritis of ruminant livestock and wildlife and may have a similar—yet not proven—role in Crohn's diseases in humans.[82-84] Finally, *M. avium silvaticum* is highly related to *M. avium avium* and has been isolated primarily from wood pigeons wherein it can cause TB-like lesions.[78]

Tortoli and colleagues suggest combining these *M. avium* subspecies within a single taxon, and to regard them as synonyms based on their very high average nucleotide identity (ANI) and genome-to-genome distances (GG) values well exceeding the subspecies thresholds.[85] In consideration of their adaptation to specific hosts, the previously named subspecies should be reclassified as variants.[85] Subspecies status of *M. avium paratuberculosis* is proposed to be maintained due to the significant differences in clinical presentation between this subspecies and other variants of the MAC.[77,85]

This would result in *M. avium variant avium*, *M. avium variant hominissuis*, *M. avium variant silvaticum*, and *M. avium paratuberculosis*

M. avium grows at 30–37°C on media suited for mycobacterial growth. Primary isolation of both *M. avium paratuberculosis* and *M. silvaticum* requires adding mycobactin to the culture medium.[86] Besides, *M. silvaticum* is unable to grow on egg media and needs the stimulation of growth by pyruvate and at pH 5.5, gradually losing this phenotype upon subculture.[78]

Mycobacterium intracellulare and Closely Related Species

This mycobacterium is mainly a respiratory pathogen in humans, responsible for most MAC lung diseases in the United States, while an association with disseminated disease is less common.[35,77]

M. chimaera became well known for its association with contaminated heater-cooler devices used in open heart surgery during a worldwide outbreak.[87,88]

M. yongonense and *M. chimaera* were recently reclassified as subspecies of *M. intracellulare*, resulting in the designation of three new taxa: *M. intracellulare intracellulare*, *M. intracellulare yongonense*,[89] and *M. intracellulare chimaera*.[69] In 2019, *M. intracellulare chimaera* has even been proposed as synonym of *M. intracellulare yongonense*.[85]

Likewise, *M. paraintracellulare*[90] was proposed to represent a fourth subspecies or even a synonym of *M. intracellulare* rather than a separate species based on its average nucleotide identity (ANI) value being well within intraspecies values.[68,69] Also, the need to distinguish *M. paraintracellulare* from *M. intracellulare* is questioned as there is no reason of clinical relevance to do so.[68]

This would result in a new taxonomy to recognize only a single species, *M. intracellulare*, with two subspecies: *M. intracellulare intracellulare* and *M. intracellulare chimaera*.[85]

Mycobacterium haemophilum

This organism was first isolated in 1978 from a patient with Hodgkin's disease, and it was shown to cause skin lesions and disseminated infections in patients with AIDS in the early- to mid-1990s. Indeed, *M. haemophilum* has been described in immunosuppressed patients, including those with lymphoma, those who undergo bone marrow or cardiac transplantation, and patients receiving steroid therapy.[91] It has also been described as a cause of lymphadenitis in immunocompetent children.[92]

The growth requirement of these organisms is rather unique among mycobacteria as they require hemin or ferric ammonium citrate for growth, at an optimal temperature of 30°C.[93] This probably accounts for the relative infrequency of its isolation, especially in laboratories solely relying on culture at 37°C. It is advisable that all patients presenting with a skin lesion have their specimen cultured onto medium containing a source of hemin with incubation at 30°C.

There is no standard therapy for infections with *M. haemophilum* but regimens including clarithromycin, doxycycline, ciprofloxacin, amikacin, INH, and RIF have led to resolution of the disease,[94] while surgical intervention may be favored in case of cervicofacial lymphadenitis.[95]

Mycobacterium kansasii

This mycobacterium causes a disease that resembles pulmonary TB. The patients present with cough, fever, and night sweats as is seen in TB.[96] Typically, the patient is an older male with a prior history of TB, chronic obstructive pulmonary disease or other chronic lung diseases. Cavitary disease may be seen on x-ray. Extra-pulmonary disease caused by *M. kansasii* has also been described occasionally.[59,97] The odds of clinical significance for a positive *M. kansasii* culture is relatively high.[32] In a Korean study, 52% of all patients with *M. kansasii* respiratory isolates exhibited clinically significant disease.[98]

In the past, seven subtypes of *M. kansasii* have been reported based on target sequences such as *hsp65*. Recent WGS analyses support a reclassification of this polyphyletic group in separate species rather than subtypes,[99,100] with clearer separation of their clinical relevance (Table 33.1).

M. kansasii has similar growth conditions as *M. tuberculosis*. Standard treatment involves the use of first-line anti-TB drugs

TABLE 33.1

Reclassification Species Within the *Mycobacterium kansasii* Group

(New) Species Name[99,100]	Former *M. kansasii* Subtype	Pathogenicity	Isolation from Clinical Specimens
M. kansasii	1	Most pathogenic	Frequent
M. persicum	2	Pathogenic	Limited data
M. pseudokansasii	3	Rarely pathogenic	Frequent
M. species[a]	4		
M. innocens	5	Rarely pathogenic	Limited data
M. attenuatum	6	Nonpathogenic	Very rarely
No suggestion[b]	7		

[a] To be defined as soon as a Type strain is available; [b] Lack of Type strain and full genome.

RIF, INH, and RMB, while RIF-resistant *M. kansasii* treatment can be based on drug-susceptibility testing data.[29]

Mycobacterium malmoense

This mycobacterium is rarely seen in the United States but is recognized as causing pulmonary disease in Northern Europe, with a frequency exceeding that of *M. kansasii*-isolation in some UK settings. In the United Kingdom, *M. malmoense* isolates had a moderate chance of being clinically significance,[40] while in the Netherlands *M. malmoense* was found to be clinically relevant at much higher (80% of cases) levels. These usually cause pulmonary disease with serious morbidity.[101] The typical patient is a male over his 50s with prior or underlying lung disease.[101,102] There have also been reports of—mostly pediatric—lymphadenitis caused by *M. malmoense*.[101,103]

M. malmoense has no specific growth requirements—even though growth on solid medium can remain very slight, with good growth at 37°C.[104]

Therapy includes the use of RIF and EMB, supplemented with a macrolide,[29] although some studies showed no additional benefit of adjunctive clarithromycin.[101,105]

Mycobacterium xenopi

Like most of the other slowly growing opportunistic pathogenic NTM, this species causes pulmonary disease, with various presentations (cavitary, nodular, or diffuse infiltrate).[106] *M. xenopi* disease has become prominent in Northern European and American countries with a low TB prevalence, yet is rarely or not seen in TB high prevalent countries (China,[107] Tanzania,[34] and Mozambique[35]). Nevertheless, *M. xenopi* is a fairly ubiquitous environmental bacterium found mostly in water, including the water supplies of hospitals. The typical patient is an older male with a prior history of TB or other lung disease. The patient often presents with cavitary lung disease. The probability of being clinically relevant was found to be moderate 40–60%.[32,39,40]

This organism is easily grown in culture but prefers an elevated growth temperature (42°C) and forms typical "bird's nest" colonies when grown on solid media.[108]

M. xenopi disease is hard to treat with around 30% success rate.[109] Variable success has been reported using a four-drug regimen comprising RIF, EMB, a macrolide (clarithromycin or azithromycin), and a fluoroquinolone (cipro- or moxifloxacin) or INH.[29,106] For severe disease an injectable aminoglycoside (amikacin or streptomycin) can be added.[29]

Mycobacterium celatum

This mycobacterium was first described in the early 1990s as a causative agent of disseminated disease in AIDS patients with a very low CD4 count.[110] Subsequently, *M. celatum* has been identified as a cause of pulmonary disease in immunocompetent individuals as well, with disease resembling that of MTBC and other NTM species.[111] *M. celatum* isolates can be easily misidentified as *M. xenopi*, sharing many biochemical/phenotypic characteristics including a typical fatty acid pattern.[112] *M. celatum* can be differentiated from *M. xenopi* by its poor growth at 45°C, and production of larger colonies on 7H10 agar.[112] Using the AccuProbe assay—which detects MTBC-specific nucleic acids—*M. celatum* isolates can be falsely positive.[113]

There is no standard therapy for infections with *M. celatum*. Isolates of *M. celatum* are generally resistant to RIF. Successful treatment has been achieved by combining clarithromycin, EMB, and ciprofloxacin.[114]

Mycobacterium marinum

Mycobacterium marinum was first observed in 1926 causing infections in saltwater fish in the Philadelphia aquarium.[115] The first reported cases in humans were from an epidemic in Sweden in 1954. These patients, who all swam in the same swimming pool, developed lesions, which began as papules at the site of infection and mostly became ulcerated and/or exudative. *M. marinum* disease, often referred to as "swimming pool granuloma," has also been associated with fish tank ownership.[116] The infection usually occurs on the upper extremities, including the hands, where the organism gains entrance through scratches and abrasions, and generally remains localized at the site of infection.

M. marinum can be cultured on standard media used for mycobacterial growth, but with incubation at 30°C, the optimum growth temperature for this species.

Therapy with regimens containing EMB, RIF, clarithromycin, amikacin, or doxycycline is usually effective. In some instances, surgical intervention is required.[117]

Clinically Significant Species of Rapidly Growing NTM

At present, rapidly growing mycobacteria comprise approximately 50% of all recognized and validated species belonging to the genus *Mycobacterium*.[118,119] Although ubiquitous in environmental reservoirs, some rapidly growing NTM are frequently isolated in healthcare facilities and have emerged as important human pathogens.[120] They are commonly found in abscesses formed after an unintentional inoculation or puncture wound, through contamination of surgical wounds or via traumatic injuries.[27] *M. abscessus* is increasingly seen as pulmonary pathogen.[39,40] *M. abscessus*, *M. chelonae*, and *M. fortuitum* together embody more than 80% of clinical isolates of rapidly growing NTM.[119,121]

First-line TB drugs are ineffective against NTM disease due to rapid growers.

An overview of the clinical relevance for additional rapidly growing NTM species can be consulted at https://bccm.belspo.be/about-us/bccm-itm.

Mycobacterium abscessus

This organism was first identified in 1952 by Moore and Frerichs, who isolated it from a patient's knee abscess.[122] Species name designation was based on the mycobacterium's ability to cause human skin and soft tissue infection, producing deep subcutaneous abscesses.[123]

Over time, *M. abscessus*' taxonomy, particularly the identification of species and subspecies, has faced many challenges resulting in a cumbersome cascade of nomenclature changes. *M. abscessus* and *M. chelonae* were originally considered to be the same species as they present almost identical biochemical features.[10,121,123] After first being designated subspecies status in 1972,[124] Kusunoki and Ezaki re-elevated *M. abscessus* to species status in 1992 based on genomic DNA hybridization studies.[125] As a result, *M. chelonae* subspecies *chelonae* was also reinstated again to its former name *M. chelonae*. After *M. abscessus* was recognized as a distinct species, the mycobacterium has been further categorized into three distinct subspecies: *M. abscessus abscessus*, *M. abscessus bolletii*, and *M. abscessus massiliense*.[126-128]

Precise identification to subspecies level of the disease-causing strain in *M. abscessus* pulmonary infections is essential for diagnosis and management as the subspecies differ in clinical disease presentation as well as in susceptibility to drugs and in optimal therapeutic regimens.[129] *M. abscessus abscessus* and *M. abscessus bolletii* have an *erm* (erythromycin ribosomal methylase) gene that confers inducible macrolide resistance, whereas most *M. abscessus massiliense* strains have a dysfunctional *erm* gene, resulting in macrolide susceptibility.[130] Consequently, better treatment outcomes have been reported in patients with *M. abscessus massiliense* pulmonary infection than those with *M. abscessus abscessus* disease.[131]

M. abscessus abscessus is the most common pathogen of the three *M. abscessus* subspecies,[132] and typically it is associated with a wide range of skin and soft tissue infections in human hosts,[37,123,133] while being increasingly the cause of pulmonary infections in patients with underlying structural lung disease such as cystic fibrosis and bronchiectasis.[134]

Skin and soft tissue infections often develop after the use of equipment or injected substances contaminated with these rapid growing mycobacteria during medical, cosmetic, and plastic surgery (such as liposuction,[135] mammaplasty,[56] and mesotherapy[136]), pedicures,[137] and after trauma.[138,139] Also pseudoinfections due to contaminated bronchoscopes and endoscopes have been ascribed to *M. abscessus*[140] and *M. abscessus bolletii*.[141]

M. abscessus is a nonchromogenic rapidly growing NTM characterized by its inability to grow at 42°C.

Current guidelines recommend an initial treatment for 2–4 months consisting of two parenteral intravenous agents (such as amikacin, cefoxitin, imipenem, and tigecycline) and a macrolide antibiotic if the species is macrolide susceptible.[37,133] If not, the macrolide should be replaced by another agent based on *in vitro* drug susceptibility testing data. Secondarily, an additional inhaled/oral antibiotic (fluoroquinolone, linezolid, or clofazimine) can be added during the continuation phase.[133]

Mycobacterium chelonae

As stated before, *M. chelonae* is closely related to *M. abscessus* but lacks the *erm* gene that confers macrolide resistance in *M. abscessus*.[142] *M. chelonae* is also highly resistant to cefoxitin, while *M. abscessus* is intermediately susceptible. Additionally, *M. chelonae* is only a rare cause of pulmonary infections and instead has a greater preference for skin and soft tissue infections.[27,143] Furthermore, *M. chelonae* has been associated with catheter-related, disseminated and bone infections, mainly in immunocompromised patients with risk factors such as organ transplantation, renal failure, surgery, long-term corticosteroid therapy, or chemotherapy.[144,145] Infections of *M. chelonae* can also be linked to ophthalmic surgery such as LASIK (laser in situ keratomileusis)[146] and cosmetic procedures (inoculations, plastic surgery, mesotherapy).[147]

This species of rapidly growing NTM is abundant in man-made environments, such as tap water, but has also been retrieved from natural environments, such as lakes, freshwater rivers, and seas.[148] Given its ubiquitous presence in the environment, exposure to *M. chelonae* is common and is often seen in clinical labs, contaminating solutions, medical equipment, and surgical wounds.

M. chelonae grows optimally at 30–32°C.

Treatment regimens of *M. chelonae* infections may include tobramycin, clarithromycin, linezolid, imipenem, or clofazimine. To avoid the development of macrolide resistance, clarithromycin will usually be combined with a second agent.[37]

Mycobacterium fortuitum Group

In 1938, da Costa Cruz (1938) proposed the name *M. fortuitum* for a strain isolated from a patient with a postinjection skin abscess.[121] More than 30 years later, Stanford and Gunthorpe showed that the isolate was identical to a formerly recognized species named *M. ranae*, isolated from a frog by Küster in 1905.[149,150] In 1972, the request of Runyon to maintain the species designation *M. fortuitum* was accepted, which to date still stands in nomenclature as such.[121]

The *M. fortuitum* group comprises the species *M. fortuitum, M. peregrinum, M. senegalense, M. alvei, M. houstonense, M. setense, M. neworleansense, M. boenickei, M. septicum, M. brisbanense,* and *M. porcinum,*[151,152] and its species are commonly found in drinking water, water distribution systems, and in a variety of soil worldwide.[148]

In a series of cases of extrapulmonary disease caused by the *M. fortuitum* group, Wallace et al.[153] reported that almost 80% of the infections were due to the single species *M. fortuitum.*

M. fortuitum has also been associated with a number of nosocomial pseudo-outbreaks,[154,155] and it has been linked to surgical site infections, skin lesions, postinjection abscesses, otitis media, and catheter-related infections.[156–160] Whirlpool pedicure footbaths in nail salons have also been identified as a potential source of *M. fortuitum*-associated furunculosis.[118,161]

M. fortuitum grows under the same conditions as MTBC.

Treatment of *M. fortuitum* infections consists of minimal 4 months of therapy with at least two agents with *in vitro* activity against the clinical isolate.[76] This regimen may include a combination of amikacin, fluoroquinolones, some tetracyclines, cefoxitin, imipenem, or sulfonamides.[153,162,163] In some cases where drug therapy is difficult, surgical intervention is required.[118]

Diagnosis

Clinical suspicion of mycobacterial disease (including x-ray or other imaging techniques) can be confirmed bacteriologically.

For decades, smear microscopy to detect acid-fast bacilli (AFB) and isolation of mycobacteria by culture were the most important diagnostic tools for mycobacterial disease, especially in low-resource settings where microscopy often was the only available and affordable method. Microscopy however lacks sensitivity (for pulmonary TB overall around 60% and only 22–43% in HIV-positive individuals; 10–60% for clinically significant NTM disease),[40] while culture is significantly more sensitive (80–95%) but can take weeks for *M. tuberculosis* complex and slowly growing NTM to become positive. With the advent of nucleic-acid amplification techniques in the 1990s, rapid, specific, and sensitive diagnosis became conceivable, but it was mostly restricted to well-equipped laboratories in high-income countries. Only with the introduction and roll out of the cartridge-based GeneXpert MTB/RIF assay (Cepheid, U.S.A), molecular diagnosis for TB—with simultaneous detection of rifampicin resistance—has become widely available and increasingly is the first test of choice in TB high prevalent settings.[164,165,166] TB-LAMP, another relatively easy isothermal DNA amplification assay to detect MTBC, has been endorsed in 2016,[167] but it has been rolled out in only a limited number of settings.[168,169]

The antigen-detection-based lateral flow-lipoarabinomannan (LF-LAM) assay has been recommended to assist in TB diagnosis in specific populations, namely adults, adolescents, and children living with HIV with signs and symptoms of TB and CD4 cell count under 200 in inpatient settings as well as in adults, adolescents, and children living with HIV with signs and symptoms of TB irrespective of the CD4 cell count in outpatient settings.[170,171] LAM, a lipopolysaccharide, is present in mycobacterial cell walls and is released from metabolically active or degenerating bacterial cells. An LF-LAM test is to be done manually on a urine sample.[170]

Regarding serodiagnostics, the WHO strongly recommended that commercially available assays should not be used for the diagnosis of pulmonary or extra-pulmonary TB.[172]

Despite progress in molecular diagnosis, culture remains important for subsequent phenotypic drug-susceptibility testing (especially for new drugs like bedaquiline), monitoring of patients during treatment, and as a gold standard test for most mycobacterial diseases. Once a culture becomes positive and is confirmed to yield AFB, the isolated bacteria can be identified by phenotypic or genotypic means prior to phenotypic drug-susceptibility testing.

Of note, two test systems are available for the identification of latent TB infection: the tuberculosis skin test (TST) and interferon-gamma assays (IGRAs). The TST consists of an intradermal injection of purified protein derivatives (PPD, tuberculin), with a read-out by qualified staff after 48–72 hours. This assay, albeit cheap and not requiring any laboratory facility, is not specific because of cross-reaction due to prior BCG vaccination or exposure to NTM.[173] Efforts are ongoing to validate new skin tests using more specific RD1-antigens, not shared by *M. bovis* BCG, showing similar sensitivity.[174,175] IGRAs are *in vitro* blood tests to measure cell-mediated immune response, and make use of the same RD1-antigens.[173] Both test systems have limited value to predict progress to active TB.

Specimen Selection

Respiratory specimens constitute the most common type of specimen submitted for the culture and isolation of mycobacteria, with only 10–20% of clinical specimens originating from nonpulmonary sites.[176] For respiratory infections, sputum is to be obtained by preference early in the morning. The patient should be instructed to rinse his/her mouth with bottle water (tap water is not recommended), and cough up expectorate into a sterile sputum collection device specifically designed for mycobacterial culture. Multiple early morning sputum samples improve the chances to grow mycobacteria, especially NTM. If the patient can't produce sputum, which is often the case for children and very sick persons, a sputum induction can be performed. In this instance, the patient is either placed inside a special chamber or a hood is placed over the patient's head, and nebulized saline is introduced, causing the patient to cough up sputum.

In children or in presumptive patients having difficulty expectorating, specimens can also be obtained by the use of a bronchoscope, yielding a bronchial wash, bronchial alveolar lavage, or biopsy.[177] Bronchoscopy can assist in specifically sampling the affected area of the lung identified by x-ray. Alternatively, gastric lavage is used in patients who are not able to expectorate sputum and, hence, swallow it. Gastric lavage specimens need to be processed within 4 hours after collection.[178] Gastric lavage had higher positivity rates compared to induced sputum for TB diagnosis in children and adults.[177,179] Data on the isolation of MTBC and especially NTM from gastric lavage are scarce.[179] Recently, examination of stool by GeneXpert MTB/RIF (Cepheid, U.S.A) yielded good results and was superior compared to gastric lavage in individuals unable to produce sputum.[180] This application has however not yet been recommended by WHO.

For extrapulmonary disease, biopsies, or body fluids can be used for diagnosis. Biopsy specimens can be obtained from any site suspected of being infected with mycobacteria. Biopsies of an organ can be sampled either by ultrasound guidance or through general surgery. Skin lesions can be sampled by punch biopsy, scraping the margin of the lesion, or swabbing the ulcerated region. Drying of swabs should be avoided. It is rarely productive to collect samples from the necrotic center of the lesions. Body fluids can be obtained through aseptic fine needle aspiration. Urine samples are rarely positive for growth of mycobacteria. If indicated, early morning midstream urine should be collected. For the LF-LAM assay, any urine sample may do. Blood is not commonly used for detection of mycobacteria except for disseminated disease, especially in AIDS patients with low CD4 counts. Blood can be inoculated directly into special blood culture bottles of some of the automated mycobacterial culture systems.

Specimen Storage and Transport

Although mycobacteria can be recovered several days after sample collection, non-delayed processing will increase chances of successful isolation. Long delays in processing the sample increase the risk of overgrowth by microorganisms from the common microflora or environmental contamination. This risk can be reduced by ensuring refrigerated storage and transport (2–8°C), or in case of expected long delays from remote places (>7 days) and/or in the absence of an ensured cold-chain, transport medium can be added. To this end, a commercial medium like OMNIgene.SPUTUM (DNA Genotek, Canada) or the noncommercial cetylpyridinium chloride (CPC) can be added, allowing storage and transport at ambient temperature for up to 8 days or 28 days, respectively, prior to culture. Both reagents are also suited to conduct subsequent molecular analysis.[181] If only molecular analysis are to be done, addition of ethanol (final concentration ≥50%) allows for long-term storage (years) and shipping at ambient temperature.[181]

Isolation of Mycobacteria from Clinical Specimens (Culture)

Specimen Processing

Sputa and other specimens collected from the respiratory tract are "contaminated" with normal microflora. In addition, sputum is a viscous nonhomogenous specimen. Hence the need to process these specimens with a mucolytic agent and a decontaminating agent prior to inoculation on culture medium. If not, the normal flora will overgrow the more slowly growing species of mycobacteria, making their isolation from clinical samples impossible.

Several mucolytic and bactericidal/fungicidal reagents are used for this purpose.[178] The most commonly used are 4% NaOH or equal quantities of 4% NaOH solution and sodium citrate plus N-acetyl L-cysteine (NALC) solution. The 4% NaOH helps in liquefying the specimen as well as killing the normal flora of bacteria/fungi. NALC at concentrations of 0.5–2.0%, when combined with NaOH, facilitates decontamination by further digesting mucopurulent specimens, which allows the NaOH to penetrate; sodium citrate aids in the liquification by binding heavy metals, thus stabilizing NALC and allowing it to work properly. NALC can be used with reduced concentrations of NaOH (final concentration of 1% in sputum). For isolation of mycobacteria from specimens that are likely contaminated with Gram-negative organisms such as *Pseudomonas aeruginosa*, as observed in patients of cystic fibrosis and bronchiectasis, a combination of Nalc-NaOH processing with 5% oxalic acid treatment may be more successful.[178]

The optimal volume of sputum to be processed is 5–10 mL. An equal volume of the digesting/decontaminating solution is added to the specimen, vortexed, and allowed to stand for 15–20 minutes at room temperature. Then, the alkali are either neutralized by adding an equal volume of phosphate buffered saline (PBS; pH 6.8, 0.067 M) or HCl buffer (1N) till neutral pH as indicated by the phenol red indicator, or "washed" by adding an excess (usually till 45 mL total volume) of PBS buffer.[178] PBS lowers the specific gravity of the specimen and gently neutralizes the specimen. The amount of NaOH used, the time allowed for decontamination, and the addition of neutralizing or washing buffer to the digested/decontaminated specimen are critical because both the common microflora and mycobacteria are affected by the exposure to the high pH of sodium hydroxide.

The specimen is then centrifuged at 3,000 × g for 15–20 minutes to concentrate the bacilli at the bottom.[182] The obtained sediment can be used to inoculate appropriate culture media, and—in case not yet done directly from the nonprocessed specimen—to prepare a smear for microscopic examination, or to perform a DNA extraction and subsequent molecular analysis.

Urine and other large volume samples should be concentrated (15 minutes at 3,000 × g), and nonsterile biopsies minced (sterile mortar and pestle) prior to processing. Specimens aseptically collected from normally sterile body sites do not need to be processed before inoculation into the appropriate media.

Smear Preparation

Smears can be prepared directly from clinical specimens (direct) or from the sediment of processed specimens in case culture is done as well (indirect).

Staining and Reading of Smears

Since the detection of TB bacilli by Robert Koch, the most commonly used dye for staining mycobacteria is basic fuchsine. This dye can stain all microorganisms, but only few will retain the dye in the cell wall after discoloration with an acid-alcohol solution. The composition of the mycobacterial cell wall (mycolic acids) of these organisms renders them resistant to discoloration of the primary dye, hence the term "acid-fastness."

The classic method, the Ziehl-Neelsen stain, uses carbolfuchsin (0.3%) as the primary dye. In this method, the stain is heated until steam arises (three times at intervals of 3–5 minutes) to allow optimal penetration of the cell wall. The modified or cold Kinyoun stain also uses carbolfuchsin as the primary dye but without heating and a contact time of 20–25 minutes. In both methods, acid-fast organisms will appear red against the background stain, which is usually methylene blue. Reading of the slides requires objective 100× and the use immersion oil.

Alternatively, fluorescent stains are used that consist of the fluorochrome stain auramine, alone or in combination with another fluorochrome, rhodamine. These stained smears must be read using a fluorescent microscope, where mycobacteria will fluoresce a bright yellow to orange. Fluorescence microscopy allows reading at a lower magnification (objective 25× or 40×), speeding up analysis. However, care must be taken to distinguish fluorescing artifacts from true organisms. The introduction of LED-based microscopes has enabled the roll out of fluorescence microscopy, even to remote places. In 2011, the WHO recommended that conventional fluorescence microscopy be replaced by LED microscopy, and that LED microscopy be phased in as an alternative for conventional Ziehl-Neelsen light microscopy.[183]

A minimum of 5,000–10,000 bacilli per mL specimen is needed to be observable in microscopy.

Isolation of Mycobacteria by Culture

For an optimal diagnosis of mycobacterial infection, their isolation from clinical specimens is considered as a gold standard. Digested, decontaminated, and concentrated specimen is inoculated on culture medium and incubated at 30°C and/or 37°C for 6–8 weeks to allow growth of mycobacteria. As a rule, media inoculated with specimens obtained from skin lesions should always be incubated at the regular 37 ± 1°C as well as at the lower 30 ± 1°C temperature. Once growth of AFB shows up and is confirmed, it is reported as positive. Traditionally one or more pieces of a solid medium or one solid and one liquid medium are inoculated.

Solid media used for mycobacterial culture can be divided into two main categories: egg-based and agar-based media. The egg-based media consist of (part of the) eggs, salts, potato flour, amino acids, glycerol, and malachite green as an inhibitor of bacterial growth from the common microflora persisting prior specimen processing. The media are solidified by heating the mixture to a temperature of 85°C for approximately 45 minutes in a process called inspissation. This process also sterilizes the medium. The most common form of egg-based medium used in the United States and many other countries is Löwenstein-Jensen (LJ) medium. Modified egg-based media may be more suited for the isolation of specific species. For example, *M. bovis* (BCG) grows better on medium without glycerol and supplemented with pyruvate (Coletsos or Stonebrink medium), while *M. haemophilum* needs addition of ferric-ammonium citrate for its growth. Agar-based media commonly employed are Middlebrook 7H10 and 7H11, both containing defined salt concentrations, growth factors such as oleic acid, albumin, catalase, glycerol, and a low concentration of malachite green. Medium 7H11 differs from 7H10 by the addition of pancreatic digest of casein to facilitate the growth of fastidious MTBC cultures.[184] Recently a new selective agar-based medium has been developed ("rapid growing mycobacteria" RGM), which enabled sensitive (98%) detection of NTM from CF patient samples.[185] Prolonged incubation up to 26 days further increased growth of MAC, with higher recovery compared to the classical medium for isolation of mycobacteria.[186,187]

Solid media can be prepared in house or purchased as ready-to-use from commercial companies. They may be supplemented with antibiotic mixtures to further reduce culture contamination rates. As, the optimal growth temperature for most NTM ranges between 28°C and 37°C, it is important to incubate at least one tube at 37°C and one at 30°C, especially in case of biopsies or pus from skin lesions.

In resource-poor settings, specimens can be cultured using only LJ medium, but for optimal isolation of mycobacteria from clinical specimens, both liquid and solid media should be employed to decrease the time to positivity and to increase the probability of a positive result (>10%).[188] The number of positive cultures for liquid is significantly higher than for any solid medium. Several automated systems using liquid medium for isolation of mycobacteria have been developed and are now used routinely in many laboratories, with the BACTEC 460 (BD Diagnostic Systems, U.S.) being the first introduced in the 1980s. This semiautomated system utilized a modified Middlebrook

7H9 broth containing a radiolabeled substrate that would be hydrolyzed by microorganisms growing in the medium, releasing radiolabeled carbon dioxide that was measured as an indicator of growth. The BACTEC 460 system has been discontinued due to concerns of radioactivity and its disposal and has been replaced by the nonradiometric BACTEC MGIT 960 system (BD Diagnostic Systems, Sparks, MD). This system builds on the principle of oxygen consumption by growing organisms. Oxygen present in the medium quenches the fluorescent indicator embedded in silicone at the bottom of the tube. When the oxygen is consumed by growing bacilli, the tubes will begin to fluoresce in the presence of UV light. Automated reading each hour, will flag these tubes as positive. The instrument provides the time to positivity and a growth-unit value (GU). Prior to specimen inoculation, a growth supplement including an antibiotic mixture (PANTA) must be added. MGIT 960 can be used for all types of specimens except for blood. Blood must be inoculated into a different type of bottle, the MYCO/F Lytic F, and incubated in a different instrument, the BACTEC 9240.

Bacterial contamination, which survives decontamination and grows in the presence of added antibiotics, can lead to a false-positive result. Unlike solid medium, liquid medium does not allow for the observation of colony characteristics, colony counts or mixed cultures. Any positive culture must therefore be confirmed by acid-fast staining, and, if possible, by purity check on a blood agar.

Scores of publications have verified MGIT 960 superior performance over solid media.

A couple of other commercial automated liquid systems have been introduced, such as VersaTREK Myco System (Trek Diagnostic Systems, U.S.A) and BacT/Alert (Biomérieux, France), but these systems are less commonly used.

Identification of Mycobacteria

It is imperative that MTBC can be immediately distinguished from NTM.

Upon isolation, initial mycobacterial species identification may rely on the observed growth rate (rapid versus slow), pigmentation, and colony appearance, the latter mainly observable from solid medium. They are described as either buff (white- to cream-colored colony) or a chromogen (yellow to orange pigmentation). The chromogens are classified as scoto- or photochromogenic, with scotochromogens being pigmented regardless of growing in the light or dark, whereas photochromogens will only develop pigment if exposed to light.[183] Most mycobacteria can be speciated by their biochemical profile,[183] which can be very tedious, requiring up to several weeks, and may not always allow for reliable (sub) species differentiation. Susceptibility to paranitrobenzoic acid (PNB; 500 µg/mL) has for long been a marker to differentiate MTBC from PNB-resistant NTM.[189] In most laboratories, however, molecular assays have displaced biochemical tests.

To rapidly differentiate MTBC from NTM growth, a lateral flow immunochromatographic assay detecting the presence of the MTBc-associated MPT 64 antigen has been developed. Three commercial tests are available: Capilia TB NEO (Tauns, Japan), BD MGIT™ TBc ID test (Becton Dickinson, U.S.), and SD Bioline TB Ag MPT64 Rapid (Standard Diagnostics, South Korea).

In 15 minutes, the tests differentiate MTBC species from NTM in a positive culture. In a multicenter study, all three tests showed 100% sensitivity for detection of MTB from MGIT on the day of positivity. While the BD assay is not indicated for use with solid media cultures, the sensitivity of Capilia and SD Bioline assays was also 100% when applied on isolates grown on LJ cultures.[190] The SD Bioline assay may yield a weak false-positive result with *M. gastri*.[190] In a comparative study in Uganda, the SD Bioline assay performed equally well as the PNB assay.[191] Of note, not all MTBC lineages may be detected with a similar sensitivity, as evidenced by false-negative reactions for *M. africanum* in Benin and the Gambia.[192,193]

In case no MTBC is detected by an MPT64-based assay or an MTBC-specific molecular assay, most laboratories rely now on molecular testing for further NTM speciation. In house PCR- and sequencing-based identification of conserved genes is the reference method for the identification of mycobacteria.[5] This technology commonly targets a number of housekeeping genes, such as the ones encoding the 16S rRNA (*rrs*), the 65-kDa heat shock protein (*hsp65*), the RNA polymerase β-subunit (*rpoB*), bus is mostly restricted to reference laboratories. The vast majority of clinical laboratories will make use of commercially available assays.[5] While these can reliably differentiate MTBC from other mycobacterial species with very high sensitivity and specificity, they can identify only a limited number of NTM species. A comprehensive overview of various (molecular) assays for NTM species identification, discussing advantages, and shortcomings is provided by Misch and colleagues.[27] Some examples are discussed next.

The FDA-approved AccuProbe test (Gen-Probe, U.S.A) has single-stranded DNA probes that hybridize with ribosomal RNA from several mycobacterial species. It has probes specific for MTBC, *M. avium* complex, *M. kansasii*, and *M. gordonae*. The test can be used directly from positive cultures, within clinically relevant time frames (24–48 hours), with a sensitivity and specificity comparable to the lateral flow assay, yet requiring more complex manipulations.

Numerous assays based on the reversed hybridization line probe technology have been developed for the identification of mycobacteria: the InnoLiPA Mycobacteria (Innogenetics, Belgium) identifying 13 NTM species, and various GenoType assays from Hain Lifescience (Germany). The GenoType *Mycobacterium* CM and GenoType Cm*direct* both differentiate >20 common NTM species and MTBc, with the former requiring a positive culture, while the latter can also be used directly on decontaminated sputum. The GenoType NTM-DR allows for the same species identification as "CM," but in addition detects resistance mutations associated with clarithromycin and amikacin. For further identification of 19 less common NTM species, the GenoType *Mycobacterium* AS can be used on positive cultures.

In addition, molecular tests used for direct detection of MTBC in clinical specimens can also be used for identification of MTBC in grown cultures.

The cumbersome analysis of mycolic acid profiles using high performance (or pressure) liquid chromatography (HPLC), is nowadays rarely used in routine practice. The commercialized Sherlock Midi system (Midi Labs, Newark, DE), contains a database to which the profile of the unknown organism can be matched to speciate the isolate.

In clinical practice, matrix-assisted laser desorption/ionization time-of-flight (MALDI-TOF) mass spectrometry has shown value in replacing the more costly and time-consuming 16S rRNA gene PCR and sequencing-based approach as the primary method for identification of microorganisms.[194] In a recent multicenter study, two commercial databases (Bruker Biotyper with Mycobacterial Library v5.0.0 and bioMérieux VITEK MS with v3.0 database) showed comparable performance, with 92% and 95% of NTM strains identified at least to the complex/group level, and 62% and 57% to the highest taxonomic level.[195] Differentiation between members of *M. abscessus, M. fortuitum, M. mucogenicum, M. avium,* and *M. terrae* complexes/groups was problematic for both systems, as was identification of *M. chelonae* for the Bruker system.[195] Similarly, a seeded-culture study that incorporated clinically prevalent respiratory microbiota, achieved 99% correct identification of mycobacteria, indicating that residual patient microbiota in a positive liquid culture was reduced by the Vitek MS.[196] In another study, the MAC subtyper module using the MALDI Biotyper algorithm, allowed for correct identification of 100% of the *M. intracellulare* and 82% of the *M. chimaera,* two closely related species.[197] Forero-Morales and colleagues have described a novel mycobacterial inactivation and protein extraction protocol (MIPE), which provides reliable MALDI-TOF results from solid and liquid media, while ensuring laboratory safety.[198] Finally, MALDI-TOF MS is intended for the identification of pure cultures; mixed *Mycobacterium* cultures present a challenge, as both species may not be distinguishable, especially in case of closely related taxa.[199]

Finally, the next-generation sequencing based Deeplex Myc-TB assay (Genoscreen, Lille, France), can detect 156 mycobacterial species either directly from smear-positive clinical specimens or from positive cultures, using nucleotide identity at the *hsp65* gene.[200]

Direct Testing of Specimens for *Mycobacterium Tuberculosis* Using Molecular Methods

Molecular diagnostics can be used to test directly on clinical specimens. Several commercial assays for detection of MTBC are being used since the 1990s (Table 33.2), such as the isothermal MTD amplification test by Gen-Probe (San Diego, U.S.A), the ProbeTec ET *Mycobacterium tuberculosis* complex Direct Test (DTB) (Becton-Dickinson), or the COBAS Taq-Man MTB Test (Roche Molecular Systems, Indianapolis, U.S.A). These assays were approved for AFB smear-positive respiratory specimens, yet require large instruments, are expensive or can be cumbersome.

Among the most common newer commercial molecular tests are the GenoType MTBC (Hain Lifesciences, GmbH, Nehren, Germany) and the less commonly used INNO-LiPA Mycobacteria (LiPA Innogenetics, Ghent, Belgium). These tests detect MTBc in a clinical specimen. The sensitivity and specificity are good in smear-positive sputum specimens but low in smear-negative specimens.

The most recently introduced is the Gene Xpert MTB/RIF (Cepheid, California, U.S.A) automated test. It is the most simple and rapid test with results available within 2-hours' time. Specimen is mixed with a reagent in a measured quantity and

TABLE 33.2

Overview of FDA-Approved Nucleic-Acid Amplification Techniques (NAAT) for the Detection and Identification of Mycobacteria

Target	Assay	Company
Mycobacterium tuberculosis	Xpert MTB/RIF Assay	Cepheid
	BDProbetec ET Mycobacterium tuberculosis complex culture identification kit	Becton, Dickinson, & Co.
	Amplified Mycobacterium tuberculosis Direct Test	Gen-Probe, Inc.
	Amplicor Mycobacterium tuberculosis test	Roche Molecular Systems, Inc.
	SNAP M. tuberculosis complex	Syngene, Inc.
	Accuprobe Mycobacterium tuberculosis complex Test	Gen-Probe, Inc.
	Rapid Diagnostic System for Mycobacterium tuberculosis	Gen-Probe, Inc.
	Rapid Identification Test for Mycobacterium tuberculosis complex	Gen-Probe, Inc.
Mycobacterium species	Accuprobe *Mycobacterium avium* complex culture	Gen-Probe, Inc.
	Accuprobe Mycobacterium kansasii Identification Test	Gen-Probe, Inc.
	SNAP *Mycobacterium avium* complex	Syngene, Inc.
	Accuprobe Mycobacterium intracellulare Culture Identification Test	Gen-Probe, Inc.
	Accuprobe Mycobacterium gordonae culture identification Test	Gen-Probe, Inc.
	Rapid Diagnostic System for Mycobacterium gordonae	Gen-Probe, Inc.
	Rapid Diagnostic System for Mycobacteria	Gen-Probe, Inc.
	Rapid Identification Test for *Mycobacterium avium*	Gen-Probe, Inc.
	Gen-Probe Mycobacterium Rapid Confirmation System	Gen-Probe, Inc.

Source: https://www.fda.gov/medical-devices/vitro-diagnostics/nucleic-acid-based-tests (update September 8, 2020).

is placed in the instrument. All the necessary molecular procedures are automatically carried out inside the instrument, and results for the presence or absence of the MTBc and presence of rifampicin resistance is displayed within two hours. Sensitivity and specificity for smear-positive is very high, but for smear-negative specimens is rather low. This system has been tried in the peripheral sites with encouraging results.[40] Detection from smear-negative and culture-positive specimens needs improvement while detection of rifampicin resistance at a low level may be a concern.[41]

Other less commonly used commercial CE-labeled real-time PCR systems are available:

• The artus *Mycobacterium tuberculosis* RG PCR (artus MTB) kit (QIAGEN, Germany) has also been approved by the FDA and detects MTBc and/or MAC.

- The Geno-Sen's MTC/MOTT Real-Time PCR (Corbett Research, Australia) detects the genus *Mycobacterium* and MTBC.
- The Abbott RealTime MTB (Abbott Diagnostics, U.S.A) detects MTBC, while the variant Abbott RealTime MTB RIF/INH allows for simultaneous detection of resistance to rifampicin and/or isoniazid.
- The Speed-oligo Direct *Mycobacterium tuberculosis* assay (Vircell, Spain) includes an automated or manual post-PCR strip hybridization (dipstick format).
- The Truenat MTB Plus (MolBio, India) allows for detection of MTBC, while the Truenat MTB-RIF Dx subsequently detects rifampicin resistance in *M. tuberculosis* in Truenat MTB/MTB Plus positive specimens.

More sophisticated micro-array or lab-on chip-based CE-approved systems comprise:

- The CapitalBio Mycobacteria Real-Time PCR Detection Kit (CapitalBio Corporation, China) detects and differentiates MTBC from NTM, while the CapitalBio M. tuberculosis Drug Resistance Array Kit in addition detects 14 of the most frequent mutations associated with resistance to isoniazid and rifampicin.
- The VereMTB™ Detection Kit (Veredus Laboratories, Singapore) simultaneously detects MTBC, resistance to rifampicin and isoniazid, as well as 14 NTM species.

A non-CE-IVD approved test includes the REBA Myco ID (YD Diagnostics, South Korea).

Drug-susceptibility Testing of Mycobacteria Species

Details of drugs effective against clinically important mycobacteria and procedures for drug-susceptibility testing (DST) can be found in the Clinical Laboratory Standard Institutes (CLSI) guidelines.[201]

Susceptibility testing of MTBc isolates is advisable even in newly diagnosed infections, as an individual may have been infected with a resistant strain (primary resistance),[202] and drug resistance rates vary widely across global regions and settings. The WHO therefore recommends "universal DST," i.e., the search and identification of RR-TB among new TB patients, wherever this is

logistically feasible.[203] Universal DST in all presumptive TB cases in the GeneXpert-based algorithm in South Africa resulted in a higher overall proportion of MDR-TB cases being diagnosed.[164]

Regarding susceptibility testing of NTM, the U.S. CLSI provides clear guidance on when and how to test and how to interpret minimal inhibitory concentrations (MICs),[201] while the British Thoracic Society recommends to follow CLSI, admitting that evidence is poor and DST should only be carried out on isolates where there is clinical suspicion of disease.[29] In the absence of clear *in vitro/in vivo* associations and clear-cut test concentrations for most NTM and drugs, the British guidelines distilled some evidence statements that could guide patient management (Table 33.3).

Overall, there are two major approaches for testing drug resistance, using culture-based or phenotypic and molecular-based or genotypic methods, with NTM-DST being mostly limited to phenotypic DST, with preference for MIC determination and reporting.

Phenotypic Susceptibility Testing

Culture-based or phenotypic DST can be performed on solid or in liquid medium, the latter having the preference for MTBC in terms of speed, and for NTM in terms of reliability. TB-DST should always be carried out in a confined laboratory following at least the safety precautions recommended by WHO.[204]

Classically, TB drug susceptibility testing is performed on the same solid media as used for primary isolation, such as LJ or Middlebrook 7H10 and 7H11 agar medium. The proportion method is the most applied, establishing the proportion of bacilli growing on culture-containing medium as compared to drug-free slants, considering an isolate as clinically resistant if >1% of the bacterial population resists the drug. To this end, an appropriate bacterial suspension (e.g., 10^{-3} dilution MacFarland #1) of the isolated culture is inoculated on the drug-containing slants, while a 1:100 dilution of the same suspension is inoculated on ad drug-free control. Solid media for DST can be obtained commercially or prepared in house. The drugs are added to the medium prior to coagulation. With exception of pyrazinamide (PZA) all TB drugs can be tested on solid medium, with clearly defined critical concentrations, specific for the drug and medium used.[205] Of note, LJ medium should not be used for bedaquiline susceptibility testing, because of its high protein-binding capacity. Solid medium DST, while being reliable, requires at least 3–4 weeks of incubation, with preferably a second reading at 6 weeks,[205] which is not in keeping with current U.S. Centers for Disease Control recommendations.[202]

TABLE 33.3

Evidence-Based Impact of Drug Resistance on NTM Treatment Outcome

Species	Disease Presentation	*In Vitro* Resistance[a] to	Impact	Evidence Level[b]
MAC	Not specified	Macrolides & amikacin	Treatment failure	2++
M. kansasii	Not specified	Rifampicin	Treatment failure	2++
M. abscessus	Pulmonary	Macrolides	Treatment failure	2++
M. abscessus	Extrapulmonary	Cefoxitin, amikacin and co-trimoxazole	Treatment failure	2++

Source: Adapted from Ref. [28].

[a] determined by broth dilution.

[b] 2++ = High-quality. Systematic reviews of case–control or cohort studies, or high quality case–control or cohort studies with a very low risk of confounding, bias or chance, and a high probability that the relationship is causal.

The use of liquid medium for detection and drug susceptibility testing has been recommended by the U.S. Centers for Disease Control and WHO,[202,204] MGIT 960 is the most widespread used liquid culture system for susceptibility testing of isolated MTBC against first- and second-line drugs, as well as new drugs like bedaquiline[208] and delamanid.[209,210] The system also relies on the proportion method principle, and results can be reported as early as 5–7 days after primary isolation. Details of the MGIT DST procedure are covered in the MGIT Manual published by FIND.[211] First-line and some of the second-line drugs are available in ready-to-use lyophilized form, while others need to be purchased and dissolved/diluted in house. Although recommended by WHO and CDC, MGIT960 testing risks to miss some cases of rifampicin resistance.[212,213]

PZA testing requires a low pH (pH 5.0) which can be achieved using adapted MGIT medium (BACTEC™ MGIT™ 960 PZA Kit), the only recommended phenotypic assay, despite previous association with a high rate of false-resistance results.[201,207] Several studies have shown good agreement of MGIT-PZA testing with *pncA* sequencing, if done meticulously; careful inoculum preparation is essential for correctly performing PZA testing reliably.[214]

Regarding NTM susceptibility testing, a variety of methods have been proposed, including broth dilution tests, the E-test, and proportional- and absolute-concentration methods.[42] E-tests suffer from high intra- and interlaboratory variability, and the proportional method on solid media or in MGIT has its breakpoints not yet defined for NTM. Hence the preference for the broth microdilution method.[201]

Molecular-Based Drug-susceptibility Testing

In the last decade, several genotypic DST procedures have been introduced, detecting mutations in one or more specific genes associated with resistance to (a) particular (class of) drugs. Genotypic testing is significantly faster than phenotypic testing. Among the first-line drugs, molecular testing overall has an excellent sensitivity for rifampicin and a good sensitivity for isoniazid, while both having excellent specificity. Sensitivity and specificity for ethambutol and PZA are debatable. Regarding second-line drugs, both for injectables and fluoroquinolones a relatively high sensitivity is observed. Details of the likelihood of association with drug resistance can be found on the Relational Sequencing TB Knowledgebase (ReSeqTB) platform (https://platform.reseqtb.org/).[215]

The FDA-approved GeneXpert MTB/RIF is the most widely used assay, with a pooled sensitivity and specificity of 96.8% and 98.4%, respectively, to correctly identify rifampicin-resistance in clinical specimens.[216] False-resistant results have been associated with a low bacillary load,[217,218,219] or the presence of synonymous mutations not associated with resistance.[220] Therefore, it is recommended to repeat testing in case of a low pretest probability for rifampicin resistance or a very low bacillary load. The new version Xpert Ultra may be more specific as it includes melting temperatures to identify the mutation type.[221] Further roll out of this version will provide data on its performance as an initial diagnostic test in field settings.

Results for the clinical evaluation of the chip-based real-time micro-PCR Truenat MTB-RIF Dx (MolBio, India) are expected by 2020.

While not being FDA approved, the WHO has endorsed the GenoType MTBDR*plus* (Hain Lifesciences, Germany) and the Genoscholar NTM-MDR TB (Nipro, Japan) line probe assays for detection of resistance to rifampicin and isoniazid in cultures and smear-positive samples.[222] Likewise, the GenoType MTBDR*sl* (Hain Lifesciences, Germany) can detect resistance to fluoroquinolones and second-line injectables.[223] The only assay for rapid detection of PZA resistance (Genoscholar PZA-TB II) has not been endorsed yet.[224]

Finally, next-generation sequencing-based Deeplex Myc-TB assay (Genoscreen, Lille, France) can predict resistance to 13 anti-TB drugs/drug classes, directly from smear-positive sputum or from culture isolates.[225] This assay has not yet been CE or FDA approved. Likewise, whole genome sequencing can predict TB drug resistance with a good sensitivity and specificity for most drugs,[226] albeit with yet unclear genotypic-phenotypic associations for new drugs like bedaquiline and delamanid.[227,228]

REFERENCES

1. Goodfellow, M. Corynebacteriales ord. nov. In: *Bergey's Manual of Systematics of Archaea and Bacteria*, Eds.: Whitman, W.B., Rainey, F., Kämpfer, P., Trujillo, M., Chun, J., DeVos, P., et al. (Wiley, Hoboken, NJ) (2012).
2. Lehmann, K.B. & Neumann, R. Anhang I. 2. Mycobacterium. In: *Lehmann's Medicinische Handatlanten. Band X. Atlas und Grundriss der Bakteriologie und Lehrbuch der Speciellen Baketeriologischen Diagnostik. Teil II: Text*, Vol. 2, Eds.: Lehmann, K.B. & Neumann, R. 363–375 (Verlag von J. F. Lehmann, München) (1896).
3. Wayne, L.G. & Kubica, G.P. Family mycobacteriaceae, genus *Mycobacterium*. In: *Bergey's Manual of Systematic Bacteriology*, Vol. 2, Eds.: Holt, J.G., Sneath, P.H.A., Mair, N.S. & Sharpe, M.E. 1436–1445 (Williams & Wilkins, Baltimore, London, Los Angeles, Sydney) (1986).
4. Stackebrandt, E., Rainey, F.A. & Ward-Rainey, N.L. Proposal for a new hierarchic classification system, actinobacteria classic nov. *Int J Syst Bacteriol* **47**(**2**), 479–491 (1997).
5. Tortoli, E. Microbiological features and clinical relevance of new species of the genus *Mycobacterium*. *Clin Microbiol Rev* **27**, 727–752 (2014).
6. Gupta, R.S., Lo, B. & Son, J. Phylogenomics and comparative genomic studies robustly support division of the genus *Mycobacterium* into an emended genus *Mycobacterium* and four novel genera. *Front Microbiol* **9**, 67 (2018).
7. Tortoli, E. *et al.* Same meat, different gravy: ignore the new names of mycobacteria. (Editorial). *Eur Respir J* **54**, 1900795 (2019).
8. Meehan, C.J., Cogneau, S., Avakimyan, A., Diels, M. & Rigouts, L. Reconstituting the genus *Mycobacterium*. *Abstract Book 40th Annual Congress of the European Society of Mycobacteriology* **OR17**, 21 (2019).
9. Lee, M.R. *et al. Mycobacterium abscessus* complex infections in humans. *Emerg Infect Dis* **21**, 1638–1646 (2015).
10. Tortoli, E. Phylogenomics and comparative genomic studies robustly support division of the genus mycobacterium into an emended genus Mycobacterium and four novel genera. *Front Microbiol* **9**, (2018).

11. Tortoli, E. Chapter 1 The Taxonomy of the Genus Mycobacterium. In: *Nontuberculous Mycobacteria (NTM): Microbiological, Clinical and Geographical Distribution*, Eds.: Velayati, A.A. & Farnia, P. (Academic Press, Cambridge, MA) (2019).

12. Runyon, E.H. Anonymous mycobacteria in pulmonary disease. *Med Clin North Am* **43**, 273–290 (1959).

13. Runyon, E.H. Pathogenic mycobacteria. *Adv Tuberc Res* **14**, 235–287 (1965).

14. Gagneux, S. Ecology and evolution of *Mycobacterium tuberculosis*. *Nature Rev Microbiol* **16**, 202–213 (2018).

15. Colditz, G.A. *et al.* Efficacy of BCG vaccine in the prevention of tuberculosis: meta-analysis of the published literature. *JAMA* **271**, 698–702 (1994).

16. Ngabonziza, J.C.S. *et al.* A sister lineage of the Mycobacterium tuberculosis complex discovered in the African Great Lakes region. *Nat Commun* **11**(1), 2917 (2020 Jun). doi: 10.1038/s41467-020-16626-6.

17. Coscolla, M. & Gagneux, S. Consequences of genomic diversity in *Mycobacterium tuberculosis*. *Sem Immunol* **26**, 431–444 (2014).

18. de Jong, B.C., Antonio, M. & Gagneux, S. *Mycobacterium africanum*—review of an important cause of human tuberculosis in West Africa. *PLOS Negl Trop Dis* **4**(9), e744, (2010).

19. World Health Organization. Global tuberculosis report 2019. *World Health Organization Document WHO/CDS/TB/2019*.15, 1–284 (2019).

20. Van Deun, A. *et al.* Principles for constructing a tuberculosis treatment regimen: the role and definition of core and companion drugs. (Perspective). *Int J Tuberc Lung Dis* **22**, 239–245 (2018).

21. World Health Organization. Rapid communication: key changes to treatment of multidrug- and rifampicin-resistant tuberculosis (MDR/RR-TB). *World Health Organization Document WHO/CDS/TB/2018.18*, 1–9 (2018).

22. Nunn, A.J. *et al.* A trial of a shorter regimen for rifampin-resistant tuberculosis. *N Engl J Med* **380**, 1201–1213 (2019).

23. Aung, K.J.M. *et al.* Successful '9-month Bangladesh regimen' for multidrug-resistant tuberculosis among over 500 consecutive patients. *Int J Tuberc Lung Dis* **18**, 1180–1187 (2014).

24. Trébucq, A. *et al.* Treatment outcome with a short multidrug-resistant tuberculosis regimen in nine African countries. *Int J Tuberc Lung Dis* **22**, 17–25 (2018).

25. Falkinham III, J.O. Surrounded by mycobacteria: nontuberculous mycobacteria in the human environment. *J. Appl. Microbiol* (2009).

26. Lake, M.A., Ambrose, L.R., Lipman, M.C.I. & Lowe, D.M. "Why me, why now?" Using clinical immunology and epidemiology to explain who gets nontuberculous mycobacterial infection. *BMC Medicine* **14**, 54 (2016).

27. Misch, E.A., Saddler, C. & Davis, J.M. Skin and soft tissue infections due to nontuberculous mycobacteria. *Curr Infect Dis Rep* **20**, (2018).

28. Aitken, M.L. *et al.* Respiratory outbreak of *Mycobacterium abscessus* subspecies *massiliense* in a lung transplant and cystic fibrosis center. (Correspondence). *Am J Respir Crit Care Med* **185**, 231–232 (2012).

29. Haworth, C.S. & Floto, R.A. Introducing the new BTS guideline: management of non-tuberculous mycobacterial pulmonary disease (NTM-PD). (Editorial). *Thorax* **72**, 969–970 (2017).

30. Bryant, J.M., Grogono, D.M., Parkhill, J. & Floto, R.A. Transmission of M abscessus in patients with cystic fibrosis reply. *Lancet* **382**, 504–504 (2013).

31. Martín-Casabona, N. *et al.* Non-tuberculous mycobacteria: patterns of isolation. A multi-country retrospective survey. *Int J Tuberc Lung Dis* **8**, 1186–1193 (2004).

32. van Ingen, J. *et al.* Clinical relevance of non-tuberculous mycobacteria isolated in the Nijmegen-Arnhem region, The Netherlands. *Thorax* **64**, 502–506 (2009).

33. Moore, J.E., Kruijshaar, M.E., Ormerod, L.P., Drobniewski, F. & Abubakar, I. Increasing reports of non-tuberculous mycobacteria in England, Wales and Northern Ireland, 1995–2006. *BMC Public Health*, **10**, 612, (2010).

34. Hoza, A.S., Lupindu, A.M., Mfinanga, S.G.M., Moser, I. & König, B. The role of nontuberculous mycobacteria in the diagnosis, management and quantifying risks of tuberculosis in Tanga, Tanzania. *Tanzania J Health Res* **18**(2) (2016).

35. López-Varela, E. *et al.* High rates of non-tuberculous mycobacteria isolation in Mozambican children with presumptive tuberculosis. *PLOS One* **12**, e0169757 (2017).

36. Blanc, P. *et al.* Nontuberculous mycobacterial infections in a French hospital: a 12-year retrospective study. *PLOS One* **11**, e0168290 (2016).

37. Griffith, D.E. *et al.* An official ATS/IDSA statement: diagnosis, treatment, and prevention of nontuberculous mycobacterial diseases. [Erratum appears in: *Am J Respir Crit Care Med* 2007; 175:744–5]. *Am J Respir Crit Care Med* **175**, 367–416 (2007).

38. Lee, M.R. *et al.* Factors associated with subsequent nontuberculous mycobacterial lung disease in patients with a single sputum isolate on initial examination. *Clin Microbiol Infect* **21**(2015).

39. Dakić, I. *et al.* Pulmonary isolation and clinical relevance of nontuberculous mycobacteria during nationwide survey in Serbia, 2010–2015. *PLOS One* **13**, e0207751 (2018).

40. Schiff, H.F. *et al.* Clinical relevance of non-tuberculous mycobacteria isolated from respiratory specimens: seven year experience in a UK hospital. *Sci Rep* **9**, (2019).

41. Perez, J.J.B. *et al.* Clinical significance of environmental mycobacteria isolated from respiratory specimens of patients with and without silicosis. *Arch Bronconeumol* **52**, 145–150 (2016).

42. Wassilew, N., Hoffmann, H., Andrejak, C. & Lange, C. Pulmonary disease caused by Non-Tuberculous Mycobacteria. *Respiration* **91**, 386–402 (2016).

43. Adjemian, J. *et al.* Epidemiology of nontuberculous mycobacterial lung disease and tuberculosis, Hawaii, USA. *Emerg Infect Dis* **23**, 439–447 (2017).

44. Garima, K. *et al.* Are we overlooking infections owing to nontuberculous mycobacteria during routine conventional laboratory investigations? *Int J Mycobacteriol* **1**, 207–211 (2012).

45. Varghese, B. *et al.* The first Saudi Arabian national inventory study revealed the upcoming challenges of highly diverse nontuberculous mycobacterial diseases. *PLOS Negl Trop Dis* **12**, (2018).

46. Tortoli, E. Epidemiology of cervico-facial pediatric lymphadenitis as a result of nontuberculous mycobacteria. *Int J Mycobacteriol* **1**, 165–169 (2012).

47. Heraud, D., Carr, R.D., McKee, J. & Dehority, W. Nontuberculous mycobacterial adenitis outside of the head and neck region in children: a case report and systematic review of the literature. *Int J Mycobacteriol* **5**, 351–353 (2016).

48. Shih, D.C. *et al.* Extrapulmonary nontuberculous mycobacterial disease surveillance–Oregon, 2014-2016. *Morb Mortal Wkly Rep* **67**, 854–857 (2018).

49. Kaestli, M. *et al.* Opportunistic pathogens and large microbial diversity detected in source-to-distribution drinking water of three remote communities in Northern Australia. *PLOS Negl Trop Dis* **13**, e0007672 (2019).

50. Marini, E. *et al.* Efficacy of carvacrol against resistant rapidly growing mycobacteria in the planktonic and biofilm growth mode. *PLOS One* **14**, (2019).

51. Cox, K.E. & Melander, C. Anti-biofilm activity of quinazoline derivatives against Mycobacterium smegmatis. *MedChemComm* **10**, 1177–1179 (2019).

52. Wargo, M.J. Is the potable water system an advantageous preinfection niche for bacteria colonizing the cystic fibrosis lung? *mBio* **10**, (2019).

53. Waak, M.B., LaPara, T.M., Halle, C. & Hozalski, R.M. Nontuberculous Mycobacteria in two drinking water distribution systems and the role of residual disinfection. *Environ Sci Tech* **53**, 8563–8573 (2019).

54. Cusumano, L.R. *et al.* Rapidly growing Mycobacterium infections after cosmetic surgery in medical tourists: the Bronx experience and a review of the literature. *Int J Infect Dis* **63**, 1–6 (2017).

55. Avanzi, A., Bierbauer, K., Vales-Kennedy, G. & Covino, J. Nontuberculous mycobacteria infection risk in medical tourism. *JAAPA* **31**, 45–47 (2018).

56. Torres-Coy, J.A., Rodriguez-Castillo, B.A., Perez-Alfonzo, R. & de Waard, J.H. Source investigation of two outbreaks of skin and soft tissue infection by Mycobacterium abscessus subsp abscessus in Venezuela. *Epidemiol Infect* **144**, 1117–1120 (2016).

57. Yacisin, K. *et al.* Outbreak of non-tuberculous mycobacteria skin or soft tissue infections associated with handling fish - New York City, 2013-2014. *Epidemiol Infect* **145**, 2269–2279 (2017).

58. Kennedy, B.S. *et al.* Outbreak of *Mycobacterium chelonae* infection associated with Tattoo Ink. *N Eng J Med* **367**, 1020–1024 (2012).

59. Henkle, E., Hedberg, K., Schafer, S.D. & Winthrop, K.L. Surveillance of extrapulmonary nontuberculous mycobacteria infections, Oregon, USA, 2007–2012. *Emerg Infect Dis* **23**, 1627–1630 (2017).

60. Chavarro-Portillo, B., Soto, C.Y. & Guerrero, M.I. Mycobacterium leprae's evolution and environmental adaptation. *Acta Trop* **197**, 105041 (2019).

61. Logas, C.M. & Holloway, K.B. Cutaneous leprosy in Central Florida man with significant armadillo exposure. *BMJ Case Rep* **12**, (2019), doi: 10.1136/bcr-2019-229287.

62. World Health Organization. WHO Guidelines for the diagnosis, treatment and prevention of leprosy. World Health Organization Document, ISBN: 978 92 9022 638 3 (2018).

63. World Health Organization. Treatment of *Mycobacterium ulcerans* disease (Buruli ulcer). World Health Organization Document, WHO/HTM/NTD/IDM/2012.1, 1–66 (2012).

64. Johnson, P.D.R. *et al.* *Mycobacterium ulcerans* in mosquitoes captured during outbreak of Buruli ulcer, southeastern Australia. *Emerg Infect Dis* **13**, 1653–1660 (2007).

65. Lavender, C.J. *et al.* Risk of Buruli ulcer and detection of *Mycobacterium ulcerans* in mosquitoes in southeastern Australia. *PLOS Negl Trop Dis* **5**, e1305 (2011).

66. Singh, A., McBride, J.H.W., Govan, B. & Pearson, M. Survey of local fauna from endemic areas of northern Queensland, Australia for the presence of *Mycobacterium ulcerans*. *Int J Mycobacteriol* **8**, 48–52 (2019).

67. Portaels, F. *et al.* First cultivation and characterization of *Mycobacterium ulcerans* from the Environment. *PLOS Neg Trop Dis* **2**, (2008).

68. van Ingen, J., Turenne, C.Y., Tortoli, E., Wallace, R.J., Jr. & Brown-Elliott, B.A. A definition of the *Mycobacterium avium* complex for taxonomical and clinical purposes, a review. *Int J Syst Evol Microbiol* **68**, 3666–3677 (2018).

69. Tortoli, E. *et al.* The new phylogeny of the genus Mycobacterium: The old and the news. *Infect Genet Evol* **56**, 19–25 (2017).

70. Kim, S.Y. *et al.* Distribution and clinical significance of *Mycobacterium avium* complex species isolated from respiratory specimens. *Diagn Microbiol Infect Dis* **88**, 125–137 (2017).

71. Honda, J.R., Virdi, R. & Chan, E.D. Global environmental nontuberculous mycobacteria and their contemporaneous man-made and natural niches. *Front Microbiol* **9**, (2018).

72. Hamilton, K.A., Ahmed, W., Toze, S. & Haas, C.N. Human health risks for *Legionella* and *Mycobacterium avium* complex (MAC) from potable and non-potable uses of roof-harvested rainwater. *Water Res* **119**, 288–303 (2017).

73. Horsburgh, C.R. *et al.* Geographic and seasonal variation in *Mycobacterium avium* bacteremia among North American patients with AIDS. *A J Med Sci* **313**, 341–345 (1997).

74. Benjak, A. *et al.* Insights from the genome sequence of Mycobacterium lepraemurium: massive gene decay and reductive evolution. *mBio* **8**, (2017).

75. O'Brien, C.R. *et al.* Feline leprosy due to Mycobacterium lepraemurium: Further clinical and molecular characterisation of 23 previously reported cases and an additional 42 cases. *J Feline Med Surgery* **19**, 737–746 (2017).

76. Brown-Elliott, B.A., Nash, K.A. & Wallace, R.J., Jr. Antimicrobial susceptibility testing, drug resistance mechanisms, and therapy of infections with nontuberculous mycobacteria. *Clin Microbiol Rev* **25**, 545–82 (2012).

77. Turenne, C.Y., Wallace, R., Jr. & Behr, M.A. *Mycobacterium avium* in the postgenomic era. *Clin Microbiol Rev* **20**, 205–29 (2007).

78. Thorel, M.F., Krichevsky, M. & Levy-Frebault, V.V. Numerical taxonomy of mycobactin-dependent mycobacteria, emended description of *Mycobacterium avium*, and description of *Mycobacterium avium* subsp. avium subsp. nov., *Mycobacterium avium* subsp. paratuberculosis subsp. nov., and *Mycobacterium avium* subsp. silvaticum subsp. nov. *Int J Syst Bacteriol* **40**, 254–60 (1990).

79. Mijs, W. *et al.* Molecular evidence to support a proposal to reserve the designation *Mycobacterium avium* subsp. avium for bird-type isolates and 'M. avium subsp. hominissuis' for the human/porcine type of M. avium. *Int J Syst Evol Microbiol* **52**, 1505–18 (2002).

80. Shin, S.J. *et al.* Efficient differentiation of *Mycobacterium avium* complex species and subspecies by use of five-target multiplex PCR. *J Clin Microbiol* **48**, 4057–62 (2010).

81. Alvarez, J. *et al.* Genetic diversity of *Mycobacterium avium* isolates recovered from clinical samples and from the environment: molecular characterization for diagnostic purposes. *J Clin Microbiol* **46**, 1246–51 (2008).

82. Feller, M. *et al. Mycobacterium avium* subspecies *paratuberculosis* and Crohn's disease: a systematic review and meta-analysis. *Lancet Infect Dis* **7**, 607–613 (2007).

83. McNees, A.L., Markesich, D., Zayyani, N.R. & Graham, D.Y. Mycobacterium paratuberculosis as a cause of Crohn's disease. *Expert Rev Gastroenterol Hepatol* **9**, 1523–34 (2015).

84. Harris, J.E. & Lammerding, A.M. Crohn's disease and *Mycobacterium avium* subsp. paratuberculosis: current issues. *J Food Prot* **64**, 2103–10 (2001).

85. Tortoli, E. *et al.* Genome-based taxonomic revision detects a number of synonymous taxa in the genus Mycobacterium. *Infect Genet Evol* **75**, 103983 (2019).

86. Wells, S.J. *et al.* Evaluation of a rapid fecal PCR test for detection of *Mycobacterium avium* subsp. paratuberculosis in dairy cattle. *Clin Vaccine Immunol* **13**, 1125–30 (2006).

87. van Ingen, J. *et al.* Global outbreak of severe *Mycobacterium chimaera* disease after cardiac surgery: a molecular epidemiological study. *Lancet Infect Dis* **17**, 1033–1041 (2017).

88. Sax, H. *et al.* Prolonged outbreak of *Mycobacterium chimaera* infection after open-chest heart surgery. *Clin Infect Dis* **61**, 67–75 (2015).

89. Castejon, M., Menendez, M.C., Comas, I., Vicente, A. & Garcia, M.J. Whole-genome sequence analysis of the *Mycobacterium avium* complex and proposal of the transfer of *Mycobacterium yongonense* to *Mycobacterium intracellulare* subsp. *yongonense* subsp. nov. *Int J Syst Evol Microbiol* **68**, 1998–2005 (2018).

90. Lee, S.Y. *et al.* Mycobacterium paraintracellulare sp. nov., for the genotype INT-1 of Mycobacterium intracellulare. *Int J Syst Evol Microbiol* **66**, 3132–41 (2016).

91. LaBombardi, V.J. & Nord, J.A. Clinical and laboratory aspects of *Mycobacterium haemophilum* infections. *Rev Med Microbiol* **9**, 49–54 (1998).

92. Armstrong, K.L., James, R.W., Dawson, D.J., Francis, P.W. & Masters, B. *Mycobacterium-haemophilum* causing perihilar or cervical lymphadenitis in healthy-children. *J Pediatr* **121**, 202–205 (1992).

93. Sompolinsky, D., Lagziel, A., Naveh, D. & Yankilevitz, T. *Mycobacterium haemophilum* sp-nov, a new pathogen of humans. *Int J System Bacteriol* **28**, 67–75 (1978).

94. Nookeu, P., Angkasekwinai, N., Foongladda, S. & Phoompoung, P. Clinical characteristics and treatment outcomes for patients infected with *Mycobacterium haemophilum*. *Emerg Infect Dis* **25**, 1648–1652 (2019).

95. Lindeboom, J.A., Bruijnesteijn van Coppenraet, L.E.S., van Soolingen, D., Prins, J.M. & Kuijper, E.J. Clinical manifestations, diagnosis, and treatment of *Mycobacterium haemophilum* infections. *Clin Microbiol Rev* **24**, 701–717 (2011).

96. Maliwan, N. & Zvetina, J.R. Clinical features and follow up of 302 patients with *Mycobacterium kansasii* pulmonary infection: a 50 year experience. *Postgrad Med J* **81**, 530–533 (2005).

97. Nakamura, T. *et al. Mycobacterium kansasii* arthritis of the foot in a patient with systemic lupus erythematosus. *Intern Med* **40**, 1045–1049 (2001).

98. Moon, S.M. *et al.* Clinical significance of *Mycobacterium kansasii* isolates from respiratory specimens. *PLOS One* **10**, e0139621 (2015).

99. Shahraki, A.H. *et al. Mycobacterium persicum* sp. nov., a novel species closely related to *Mycobacterium kansasii* and *Mycobacterium gastri*. *Int J Syst Evol Microbiol* **67**, 1766–1770 (2017).

100. Tagini, F. *et al.* Phylogenomics reveal that *Mycobacterium kansasii* subtypes are species-level lineages. Description of *Mycobacterium pseudokansasii* sp. nov., *Mycobacterium innocens* sp. nov. and *Mycobacterium attenuatum* sp. nov. *Int J Syst Evol Microbiol* **69**, 1696–1704 (2019).

101. Hoefsloot, W. *et al.* Clinical relevance of *Mycobacterium malmoense* isolation in the Netherlands. *Eur Respir J* **34**, 926–931 (2009).

102. Petrie, G. *et al.* Pulmonary disease caused by M-malmoense in HIV negative patients: 5-yr follow-up of patients receiving standardised treatment. *Euro Respir Jl* **21**, 478–482 (2003).

103. Lopez-Calleja, A.I., Lezcano, M.A., Samper, S., de Juan, F. & Revillo, M.J. Mycobacterium malmoense lymphadenitis in Spain: first two cases in immunocompetent patients. *Eur J Clin Microbiol Infect Dis* **23**, 567–9 (2004).

104. Schroder, K.H. & Juhlin, I. Mycobacterium-malmoense sp-nov. *Int J System Bacteriol* **27**, 241–246 (1977).

105. Jenkins, P.A. *et al.* Clarithromycin vs ciprofloxacin as adjuncts to rifampicin and ethambutol in treating opportunist mycobacterial lung diseases and an assessment of *Mycobacterium vaccae* immunotherapy. *Thorax* **63**, 627–634 (2008).

106. Andréjak, C. *et al. Mycobacterium xenopi* pulmonary infections: a multicentric retrospective study of 136 cases in northeast France. *Thorax* **64**, 291–296 (2009).

107. Duan, H. *et al.* Clinical significance of nontuberculous mycobacteria isolated from respiratory specimens in a Chinese Tuberculosis Tertiary Care Center. *Sci Rep* **6**, 36299 (2016).

108. Schwabacher, H. A strain of mycobacterium isolated from skin lesions of a cold-blooded animal, Xenopus laevis, and its relation to atypical acid-fast bacilli occurring in man. *J Hyg (Lond)* **57**, 57–67 (1959).

109. Diel, R. *et al.* Microbiological and clinical outcomes of treating non-*Mycobacterium avium* complex nontuberculous mycobacterial pulmonary disease. A systematic review and meta-analysis. *Chest* **152**, 120–142 (2017).

110. Tortoli, E. *et al.* Isolation of the newly described species Mycobacterium celatum from AIDS patients. *J Clin Microbiol* **33**, 137–40 (1995).

111. Piersimoni, C., Zitti, P.G., Nista, D. & Bornigia, S. *Mycobacterium celatum* pulmonary infection in the immunocompetent: case report and review. *Emerg Infect Dis* **9**, 399–402 (2003).

112. Butler, W.R. *et al.* Mycobacterium celatum sp. nov. *Int J Syst Bacteriol* **43**, 539–48 (1993).

113. Butler, W.R., O'Connor, S.P., Yakrus, M.A. & Gross, W.M. Cross-reactivity of genetic probe for detection of Mycobacterium tuberculosis with newly described species Mycobacterium celatum. *J Clin Microbiol* **32**, 536–8 (1994).

114. Jun, H.J., Lee, N.Y., Kim, J. & Koh, W.J. Successful treatment of Mycobacterium celatum pulmonary disease in an immunocompetent patient using antimicobacterial chemotherapy and combined pulmonary resection. *Yonsei Med J* **51**, 980–983 (2010).

115. Aronson, J.D. Spontaneous tuberculosis in salt water fish. *J Infect Dis* **39**, 315–320 (1922).

116. Zeligman, I. Mycobacterium marinum granuloma. A disease acquired in the tributaries of Chesapeake Bay. *Arch Dermatol* **106**, 26–31 (1972).

117. Edelstein, H. *Mycobacterium marinum* skin infections. Report of 31 cases and review of the literature. *Arch Intern Med* **154**, 1359–1364 (1994).

118. Forbes, B.A. *et al.* Practice guidelines for clinical microbiology laboratories: mycobacteria. *Clin Microbiol Rev* **31**, e00038–17 (2018).

119. Brown-Elliott, B.A. & Philley, J.V. Rapidly growing mycobacteria. *Microbiol Spectr* **5**, (2017).

120. Griffith, D.E. & Aksamit, T.R. Nontuberculous mycobacterial disease therapy. Take it to the limit one more time. (Editorial). *Chest* **150**, 1177–1178 (2016).

121. Brown-Elliott, B.A. & Wallace, R.J., Jr. Clinical and taxonomic status of pathogenic nonpigmented or late-pigmenting rapidly growing mycobacteria. *Clin Microbiol Rev* **15**, 716–46 (2002).

122. Moore, M. & Frerichs, J.B. An unusual acid-fast infection of the knee with subcutaneous, abscess-like lesions of the gluteal region; report of a case with a study of the organism, Mycobacterium abscessus, n. sp. *J Invest Dermatol* **20**, 133–69 (1953).

123. Lopeman, R.C., Harrison, J., Desai, M. & Cox, J.A.G. Mycobacterium abscessus: environmental bacterium turned clinical nightmare. *Microorganisms* **7**, (2019).

124. Kubica, G.P. *et al.* Cooperative numerical-analysis of rapidly growing mycobacteria. *J Gen Microbiol* **73**, 55-+ (1972).

125. Levyfrebault, X., Grimont, F., Grimont, P.A.D. & David, H.L. Deoxyribonucleic-acid relatedness study of the mycobacterium-fortuitum-mycobacterium-chelonae complex. *Int J System Bacteriol* **36**, 458–460 (1986).

126. Cho, Y.J. *et al.* The genome sequence of 'Mycobacterium massiliense' strain CIP 108297 suggests the independent taxonomic status of the *Mycobacterium abscessus* complex at the subspecies level. *PLOS One* **8(11)**, e81560, (2013).

127. Adekambi, T., Sassi, M., van Ingen, J. & Drancourt, M. Reinstating Mycobacterium massiliense and Mycobacterium bolletii as species of the Mycobacterium abscessus complex. *Int J System Evol Microbiol* **67**, 2726–2730 (2017).

128. Tortoli, E. *et al.* Emended description of *Mycobacterium abscessus*, *Mycobacterium abscessus* subsp abscessus and *Mycobacterium abscessus* subsp bolletii and designation of *Mycobacterium abscessus* subsp massiliense comb. nov. *Int J System Evol Microbiol* **66**, 4471–4479 (2016).

129. Koh, W.J. *et al.* Clinical significance of differentiation of *Mycobacterium massiliense* from *Mycobacterium abscessus*. *Am J Respir Crit Care Med* **183**, 405–410 (2011).

130. Ahmed, I., Hasan, R. & Shakoor, S. Chapter 3 Susceptibility Testing of Nontuberculous Mycobacteria. In: *Nontuberculous Mycobacteria (NTM): Microbiological, Clinical and Geographical Distribution*, Eds: Velayati, A.A. & Farnia, P. (Academic Press, Cambridge, MA) (2019).

131. Lyu, J. *et al.* A shorter treatment duration may be sufficient for patients with Mycobacterium massiliense lung disease than with Mycobacterium abscessus lung disease. *Respir Med* **108**, 1706–1712 (2014).

132. Koh, W.J., Stout, J.E. & Yew, W.W. Advances in the management of pulmonary disease due to *Mycobacterium abscessus* complex. *Int J Tuberc Lung Dis* **18**, 1141–1148 (2014).

133. Kasperbauer, S.H. & De Groote, M.A. The treatment of rapidly growing mycobacterial infections. *Clin Chest Med* **36**, 67–78 (2015).

134. Sanguinetti, M. *et al.* Fatal pulmonary infection due to multidrug-resistant *Mycobacterium abscessus* in a patient with cystic fibrosis. *J Clin Microbiol* **39**, 816–819 (2001).

135. Meyers, H. *et al.* An outbreak of *Mycobacterium chelonae* following liposuction. *Clin Infect Dis* **34**, 1500–1507 (2002).

136. Galmes-Truyols, A. *et al.* An outbreak of cutaneous infection due to *Mycobacterium abscessus* associated to mesotherapy. *Enferm Infecc Microbiol Clin* **29**, 510–514 (2011).

137. Stout, J.E. *et al.* Pedicure-associated rapidly growing mycobacterial infection: an endemic disease. *Clin Infect Dis* **53**, 787–792 (2011).

138. Schnabel, D. *et al.* Multistate US outbreak of rapidly growing mycobacterial infections associated with medical tourism to the Dominican Republic, 2013-2014. *Emerg Infect Dis* **22**, 1340–1347 (2016).

139. Sfeir, M. *et al.* Mycobacterium abscessus complex infections: a retrospective cohort study. *Open Forum Infect Dis* **5**, ofy022 (2018).

140. Weber, D.J. & Rutala, W.A. Lessons from outbreaks associated with bronchoscopy. *Infect Control Hosp Epidemiol* **22**, 403–8 (2001).

141. Guimaraes, T. *et al.* Pseudooutbreak of rapidly growing mycobacteria due to *Mycobacterium abscessus* subsp bolletii in a digestive and respiratory endoscopy unit caused by the same clone as that of a countrywide outbreak. *Am J Infect Control* **44**, E221–E226 (2016).

142. Nash, K.A., Brown-Elliott, B.A. & Wallace, R.J. A novel gene, erm(41), confers inducible macrolide resistance to clinical isolates of *Mycobacterium abscessus* but is absent from *Mycobacterium chelonae*. *Antimicrob Agents Chemother* **53**, 1367–1376 (2009).

143. Griffith, D.E., Girard, W.M. & Wallace, R.J. Clinical features of pulmonary disease caused by rapidly growing mycobacteria. An analysis of 154 patients. *Am Rev Respir Dis* **147**, 1271–1278 (1993).

144. Wallace, R.J., Brown, B.A. & Onyi, G.O. Skin, soft-tissue, and bone-infections due to mycobacterium-chelonae-chelonae: importance of prior corticosteroid-therapy, frequency of disseminated infections, and resistance to oral antimicrobials other than clarithromycin. *J Infect Dis* **166**, 405–412 (1992).

145. Merlin, T.L. & Tzamaloukas, A.H. Mycobacterium chelonae peritonitis associated with continuous ambulatory peritoneal dialysis. *Am J Clin Pathol* **91**, 717–20 (1989).

146. Kheir, W.J., Sheheitli, H., Fattah, M.A. & Hamam, R.N. Nontuberculous Mycobacterial ocular infections: a systematic review of the literature. *Biomed Res Int* (2015).

147. Meyers, H. *et al.* An outbreak of Mycobacterium chelonae infection following liposuction. *Clin Infect Dis* **34**, 1500–1507 (2002).

148. Leao, S.C. *et al. Practical Handbook for the Phenotypic and Genotypic Identification of Mycobacteria* (Vanden Broele, Brugge, Belgium) (2004).

149. Stanford, J.L. & Gunthorpe, W.J. Serological and bacteriological investigation of Mycobacterium ranae (fortuitum). *J Bacteriol* **98**, 375–83 (1969).

150. Kuster, E. Ueber kaltblutertuberkulose. *Muenchener Medizinische Wochenschrift* **57**, 3 (1905).

151. Adekambi, T. & Drancourt, M. Dissection of phylogenetic relationships among 19 rapidly growing *Mycobacterium* species by 16S rRNA, hsp65, sodA, recA and rpoB gene sequencing. *Int J Syst Evol Microbiol* **54**, 2095–105 (2004).

152. Schinsky, M.F. *et al.* Taxonomic variation in the *Mycobacterium fortuitum* third biovariant complex: description of *Mycobacterium boenickei* sp. nov., *Mycobacterium*

houstonense sp. nov., *Mycobacterium neworleansense* sp. nov. and *Mycobacterium brisbanense* sp. nov. and recognition of *Mycobacterium porcinum* from human clinical isolates. *Int J Syst Evol Microbiol* **54**, 1653–67 (2004).

153. Wallace, R.J., Jr. *et al.* Clinical disease, drug susceptibility, and biochemical patterns of the unnamed third biovariant complex of *Mycobacterium fortuitum*. *J Infect Dis* **163**, 598–603 (1991).

154. Wallace, R.J., Jr., Brown, B.A. & Griffith, D.E. Nosocomial outbreaks/pseudo-outbreaks caused by nontuberculous mycobacteria. *Annu Rev Microbiol* **52**, 453–490 (1998).

155. Labombardi, V.J., O'Brien A, M. & Kislak, J.W. Pseudo-outbreak of *Mycobacterium fortuitum* due to contaminated ice machines. *Am J Infect Control* **30**, 184–6 (2002).

156. Nolan, C.M., Hashisaki, P.A. & Dundas, D.F. An outbreak of soft-tissue infections due to *Mycobacterium fortuitum* associated with electromyography. *J Infect Dis* **163**, 1150–1153 (1991).

157. Hoy, J.F., Rolston, K.V.I., Hopfer, R.L. & Bodey, G.P. *Mycobacterium fortuitum* bacteremia in patients with cancer and long-term venous catheters. *Am J Med* **83**, 213–217 (1987).

158. Plemmons, R.M., McAllister, C.K., Liening, D.A. & Garces, M.C. Otitis media and mastoiditis due to *Mycobacterium fortuitum*: case report, review of four cases, and a cautionary note. *Clin Infect Dis* **22**, 1105–6 (1996).

159. Hector, J.S. *et al.* Large restriction fragment patterns of genomic *Mycobacterium fortuitum* DNA as strain-specific markers and their use in epidemiologic investigation of four nosocomial outbreaks. *J Clin Microbiol* **30**, 1250–5 (1992).

160. Hoffman, P.C., Fraser, D.W., Robicsek, F., O'Bar, P.R. & Mauney, C.U. Two outbreaks of sternal wound infection due to organisms of the *Mycobacterium fortuitum* complex. *J Infect Dis* **143**, 533–42 (1981).

161. Winthrop, K.L. *et al.* The clinical management and outcome of nail salon-acquired Mycobacterium fortuitum skin infection. *Clin Infect Dis* **38**, 38–44 (2004).

162. Wallace, R.J., Swenson, J.M., Silcox, V.A. & Bullen, M.G. Treatment of nonpulmonary infections due to *Mycobacterium fortuitum* and *Mycobacterium chelonei* on the basis of in vitro susceptibilities. *J Infect Dis* **152**, 500–514 (1985).

163. Swenson, J.M., Wallace, R.J., Silcox, V.A. & Thornsberry, C. Antimicrobial susceptibility of five subgroups of *Mycobacterium fortuitum* and *Mycobacterium chelonae*. *Antimicrob Agents Chemother* **28**, 807–811 (1985).

164. Naidoo, P., Dunbar, R., Caldwell, J., Lombard, C. & Beyers, N. Has universal screening with Xpert® MTB/RIF increased the proportion of multidrug-resistant tuberculosis cases diagnosed in a routine operational setting? *PLOS One* **12**, e0172143 (2017).

165. Churchyard, G.J. *et al.* Xpert MTB/RIF versus sputum microscopy as the initial diagnostic test for tuberculosis: a cluster-randomised trial embedded in South African rollout of Xpert MTB/RIF. *Lancet Glob Health* **3**, e450–e457 (2015).

166. Albert, H. *et al.* Development, roll-out and impact of Xpert MTB/RIF for tuberculosis: what lessons have we learnt and how can we do better? *Eur Respir J* **48**, 516–525 (2016).

167. World Health Organization. The use of loop-mediated isothermal amplification (TB-LAMP) for the diagnosis of pulmonary tuberculosis. Policy guidance. *World Health Organization Document WHO/HTM/TB/2016*.11, 1–39 (2016).

168. Shete, P.B., Farr, K., Strnad, L., Gray, C.M. & Cattamanchi, A. Diagnostic accuracy of TB-LAMP for pulmonary tuberculosis: a systematic review and meta-analysis. *BMC Infect Dis* **19**, 268 (2019).

169. Wang, G.R. *et al.* Xpert MTB/RIF Ultra improved the diagnosis of paucibacillary tuberculosis: a prospective cohort study. *J Infect* **78**, 311–316 (2019).

170. World Health Organization. The use of lateral flow urine lipoarabinomannan assay (LF-LAM) for the diagnosis and screening of active tuberculosis in people living with HIV. *World Health Organization Document WHO/HTM/TB/2015*.25, 1–62 (2015).

171. World Health Organization. Lateral flow urine lipoarabinomannan assay (LF-LAM) for the diagnosis of active tuberculosis in people living with HIV. Policy update. World Health Organization Document WHO/CDS/TB/2019.16 (2019).

172. World Health Organization. Commercial serodiagnostic tests for diagnosis of tuberculosis. Expert group meeting report 22 July 2010. *World Health Organization Document WHO/HTM/TB/2011*.14, 1–66 (2011).

173. Pai, M. & Behr, M. Latent *Mycobacterium tuberculosis* infection and interferon-gamma release assays. *Microbiol Spectrum* **4**, TBTB2-0023-2016 (2016).

174. Aggerbeck, H. *et al.* C-Tb skin test to diagnose *Mycobacterium tuberculosis* infection in children and HIV-infected adults: a phase 3 trial. *PLOS One* **13**, e0204554 (2018).

175. Slogotskaya, L., Bogorodskaya, E., Ivanova, D. & Sevostyanova, T. Comparative sensitivity of the test with tuberculosis recombinant allergen, containing ESAT6-CFP10 protein, and Mantoux test with 2 TU PPD-L in newly diagnosed tuberculosis children and adolescents in Moscow. *PLOS One* **13**, (2018).

176. Smith, G.S. *et al.* Epidemiology of nontuberculous mycobacteria isolations among central North Carolina residents, 2006–2010. *J Infect* **72**, 678–686 (2016).

177. Brown, M. *et al.* Prospective study of sputum induction, gastric washing, and bronchoalveolar lavage for the diagnosis of pulmonary tuberculosis in patients who are unable to expectorate. *Clin Infect Dis* **44**, 1415–1420 (2007).

178. GLI. Global Laboratory Initiative advancing TB diagnosis: GUIDE to TB Specimen Referral Systems and Integrated Networks. *Global Laboratory Initiative Document* (2018).

179. Kordy, F. *et al.* Utility of gastric aspirates for diagnosing tuberculosis in children in a low prevalence area predictors of positive cultures and significance of non-tuberculous Mycobacteria. *Pediatr Infect Dis J* **34**, 91–93 (2015).

180. Liu, R.M. *et al.* GeneXpert of stool versus gastric lavage fluid for the diagnosis of pulmonary tuberculosis in severely ill adults. *Infection* **47**, 611–616 (2019).

181. Sanoussi, C.N. *et al.* Storage of sputum in cetylpyridinium chloride, OMNIgene.SPUTUM, and ethanol is compatible with molecular tuberculosis diagnostic testing. *J Clin Microbiol* **57**, e00275–19 (2019).

182. Kent, P.T. & Kubica, G.P. Public health mycobacteriology. A guide for the level III laboratory. Ed: U.S. Department of Health and Human Services, 1–207 (Centers for Disease Control) (1985).

183. World Health Organization. Fluorescent light-emitting diode (LED) microscopy for diagnosis of tuberculosis. Policy statement. *World Health Organization Document WHO/HTM/TB/2011*.8, 1–12 (2011).

184. Cohn, M.L., Waggoner, R.F. & McClatchy, J.K. The 7H11 medium for the cultivation of mycobacteria. *Am Rev Respir Dis* **98**, 295–296 (1968).

185. Preece, C.L. *et al.* A novel culture medium for isolation of rapidly-growing mycobacteria from the sputum of patients with cystic fibrosis. *J Cyst Fibros.* **15**(2), 186–191 (2016 Mar). doi: 10.1016/j.jcf.2015.05.002.

186. Stephenson, D. *et al.* An evaluation of methods for the isolation of nontuberculous mycobacteria from patients with cystic fibrosis, bronchiectasis and patients assessed for lung transplantation. *BMC Pulm Med* **19**, 19 (2019).

187. Plongla, R., Preece, C.L., Perry, J.D. & Gilligan, P.H. Evaluation of RGM medium for isolation of nontuberculous mycobacteria from respiratory samples from patients with cystic fibrosis in the United States. *J Clin Microbiol* **55**, 1469–1477 (2017).

188. Sorlozano, A. *et al.* Comparative evaluation of three culture methods for the isolation of Mycobacteria from clinical samples. *J Microbiol Biotechnol* **19**, 1259–1264 (2009).

189. Rastogi, N., Goh, K.S. & David, H.L. Selective-inhibition of the mycobacterium-tuberculosis complex by P-nitro-alpha-acetylamino-beta-hydroxypropio phenone (Nap) and P-nitrobenzoic acid (Pnb) used in 7h11 agar medium. *Res Microbiol* **140**, 419–423 (1989).

190. Chikamatsu, K. *et al.* Comparative evaluation of three immunochromatographic identification tests for culture confirmation of *Mycobacterium tuberculosis* complex. *BMC Infect Dis* 14(1), 54 (2014).

191. Orikiriza, P. *et al.* Evaluation of the SD Bioline TB Ag MPT64 test for identification of Mycobacterium tuberculosis complex from liquid cultures in Southwestern Uganda. *Af J Lab Med* **6**, (2017).

192. Sanoussi, C.N. *et al.* Low sensitivity of the MPT64 identification test to detect lineage 5 of the *Mycobacterium tuberculosis* complex. *J Med Microbiol* **67**, 1718–1727 (2018).

193. Ofori-Anyinam, B. *et al.* Impact of the *Mycobacterium africanum* West Africa 2 Lineage on TB diagnostics in West Africa: decreased sensitivity of rapid identification tests in the Gambia. *PLOS Negl Trop Dis* **10**, e0004801 (2016).

194. Greco, V. *et al.* Applications of MALDI-TOF mass spectrometry in clinical proteomics. *Expert Rev Proteomics* **15**, 683–696 (2018).

195. Brown-Elliott, B.A. *et al.* Comparison of two commercial matrix-assisted laser desorption/ionization-time of flight mass spectrometry (MALDI-TOF MS) systems for identification of nontuberculous mycobacteria. *Am J Clin Pathol* **152**, 527–536 (2019).

196. Miller, E. *et al.* Performance of Vitek MS v3.0 for identification of *Mycobacterium* species from patient samples by use of automated liquid systems. *J Clin Microbiol* **56**, e00219–18 (2018).

197. Epperson, L.E. *et al.* Evaluation of a novel MALDI biotyper algorithm to distinguish *Mycobacterium intracellulare* from *Mycobacterium chimaera*. *Front Microbiol* **9**, 3140 (2018).

198. Forero Morales, M.P., Lim, C.K., Shephard, L. & Weldhagen, G.F. Mycobacterial inactivation protein extraction protocol for matrix-assisted laser desorption ionization time-of-flight characterization of clinical isolates. *Int J Mycobacteriol* **7**, 217–221 (2018).

199. Drancourt, M. Detection of microorganisms in blood specimens using matrix-assisted laser desorption ionization time-of-flight mass spectrometry: a review. *Clin Microbiol Infect* **16**, 1620–1625 (2010).

200. Jouet A. *et al.* Deep amplicon sequencing for culture-free prediction of susceptibility or resistance to 13 anti-tuberculous drugs. *Eur Respir J*, 2002338 (2020). doi: 10.1183/13993003.02338-2020.

201. Woods, G.L. *et al.* Susceptibility testing of mycobacteria, Nocardia spp and other aerobic Actinomycetes. *Clinical and Laboratory Standard Institute document* M24–A3 (2018).

202. Lewinsohn, D.M. *et al.* Official American Thoracic Society/Infectious Diseases Society of America/Centers for Disease Control and Prevention clinical practice guidelines: diagnosis of tuberculosis in adults and children. *Clin Infect Dis* **64**, 111–115 (2017).

203. Gilpin, C., Korobitsyn, A. & Weyer, K. Current tools available for the diagnosis of drug-resistant tuberculosis. *Ther Adv Infect Dis* **3**, 145–151 (2016).

204. World Health Organization. Tuberculosis laboratory biosafety manual. *World Health Organization Document WHO/HTM/TB/2012.*11, 1–50 (2012).

205. World Health Organization. Technical report on critical concentrations for drug susceptibility testing of medicines used in the treatment of drug-resistant tuberculosis. *World Health Organization Document WHO/CDS/TB/2018.*5, 1–106 (2018).

206. Van Deun, A. *et al.* Rifampin drug resistance tests for tuberculosis: challenging the gold standard. *J Clin Microbiol* **51**, 2633–2640 (2013).

207. World Health Organization. Technical manual for drug susceptibility testing of medicines used in the treatment of tuberculosis. *World Health Organization Document* WHO/CDS/TB/2018.24, 1–39 (2018).

208. Torrea, G. *et al.* Bedaquiline susceptibility testing of Mycobacterium tuberculosis in an automated liquid culture system. *J Antimicrob Chemother* **70**, 2300–5 (2015).

209. Keller, P.M. *et al.* Determination of MIC distribution and epidemiological cutoff values for bedaquiline and delamanid in *Mycobacterium tuberculosis* using MGIT 960 system equipped with TB eXiST. *Antimicrob Agents Chemother* **59**, 4352–4355 (2015).

210. Schena, E. *et al.* Delamanid susceptibility testing of Mycobacterium tuberculosis using the resazurin microtitre assay and the BACTEC MGIT 960 system. *J Antimicrob Chemother* **71**, 1532–9 (2016).

211. Siddiqi, S. & Ruesch Gerdes, S. MGIT Procedure Manual—Mycobacteria Growth Indicator Tube (MGIT) Culture and Drug Susceptibility Demonstration Projects. *Foundation for Innovative New Diagnostics (FIND) document*(2006).

212. Rigouts, L. *et al.* Rifampin resistance missed in automated liquid culture systems for *Mycobacterium tuberculosis* isolates with specific *rpoB* mutations. *J Clin Microbiol* **51**, 2641–2645 (2013).

213. Miotto, P., Cabibbe, A.M., Borroni, E., Degano, M. & Cirillo, D.M. Role of disputed mutations in the *rpoB* gene in interpretation of automated liquid MGIT culture results for rifampin susceptibility tesing of *Mycobacterium tuberculosis*. *J Clin Microbiol* **56**, e01599–17 (2018).

214. Hoffner, S. *et al.* Proficiency of drug susceptibility testing of *Mycobacterium tuberculosis* against pyrazinamide: the Swedish experience. *Int J Tuberc Lung Dis* **17**, 1486–1490 (2013).

215. Ezewudo, M. *et al.* Integrating standardized whole genome sequence analysis with a global *Mycobacterium tuberculosis* antibiotic resistance knowledgebase. *Sci Rep* **8**, 15382 (2018).

216. Drobniewski, F. *et al.* Systematic review, meta-analysis and economic modelling of molecular diagnostic tests for anti-biotic resistance in tuberculosis. *Health Technol Assess* **19**, 1–188, vii–viii (2015).

217. Sahrin, M. *et al.* Discordance in Xpert® MTB/RIF assay results among low bacterial load specimens in Bangladesh. *Int J Tuberc Lung Dis* **22**, 1056–1062 (2018).

218. Ocheretina, O. *et al.* False-positive rifampin resistant results with Xpert MTB/RIF version 4 assay in clinical samples with a low bacterial load. *Diagn Microbiol Infect Dis* **85**, 53–55 (2016).

219. Semuto Ngabonziza, J.C. *et al.* Prevalence and drivers of false-positive rifampicin-resistant Xpert MTB/RIF results: a prospective observational study in Rwanda. *Lancet Microbe*, 1(2), e74–e83 (2020), https://doi.org/10.1016/S2666-5247(20)30007-0

220. Mathys, V., van de Vyvere, M., de Droogh, E., Soetaert, K. & Groenen, G. False-positive rifampicin resistance on Xpert® MTB/RIF caused by a silent mutation in the *rpoB* gene. *Int J Tuberc Lung Dis* **18**, 1255–1257 (2014).

221. Chakravorty, S. *et al.* The new Xpert MTB/RIF Ultra: improving detection of *Mycobacterium tuberculosis* and resistance to rifampin in an assay suitable for point-of care testing. *mBio* **8**, e00812–17 (2017).

222. World Health Organization. The use of molecular line probe assays for the detection of resistance to isoniazid and rifampicin.World Health Organization Document WHO/HTM/TB/2016.12 (2016).

223. World Health Organization. The use of molecular line probe assays for the detection of resistance to second-line anti-tuberculosis drugs.World Health Organization Document WHO/HTM/TB/2016.07 (2016).

224. Driesen, M. *et al.* Evaluation of a novel line probe assay to detect resistance to pyrazinamide, a key drug used for tuberculosis treatment. *Clin Microbiol Infect* **24**, 60–64 (2018).

225. Jouet, A., *et al.* Deep amplicon sequencing for culture-free prediction of susceptibility or resistance to 13 anti-tuberculous drugs. *Eur Respir J*, 2002338 (2020 Sep). doi: 10.1183/13993003.02338-2020.

226. World Health Organization. The use of next-generation sequencing technologies for the detection of mutations associated with drug resistance in Mycobacterium tuberculosis complex: technical guide. *World Health Organization Document WHO/CDS/TB/2018*.19 (2018).

227. Villellas, C. *et al.* Unexpected high prevalence of resistance-associated *Rv0678* variants in MDR-TB patients without documented prior use of clofazimine or bedaquiline. *J Antimicrob Chemother* **72**, 684–690 (2017).

228. Yang, J.S., Kim, K.J., Choi, H. & Lee, S.H. Delamanid, Bedaquiline, and Linezolid Minimum Inhibitory concentration distributions and resistance-related gene mutations in multidrug-resistant and extensively drug-resistant Tuberculosis in Korea. *Ann Lab Med* **38**, 563–568 (2018).

34

Mycoplasma and Related Organisms

Bahman Rostama and Meghan A. May

CONTENTS

Introduction

Mycoplasma spp. are members of the class *Mollicutes*, have the smallest cells among free-living eubacteria, and have genomes presumed to approach the minimal essential information for independent cellular life (1). Mycoplasmas and the other mollicutes evolved from Gram-positive ancestors (2) by reductive processes that resulted in obligate association with eukaryotic host cells. The distinguishing characteristics of mollicutes include absence of a cell wall, small cell size (200–500 nm), small genome size (580–2200 kbp), low G + C content (typically in the range of 23–34 mol%, but 40 mol% in *Mycoplasma pneumoniae*), 16S rDNA sequences clearly affiliated with the class (3), unique codon usage (e.g., UGA as a tryptophan codon) in some lineages, and minimal metabolic capabilities. Cells are bound only by a single membrane, resulting in a general cellular pleomorphism; however, cytoskeletal elements confer helicity or polarity in a small number of species. Some species exhibit rotatory, flexional, or gliding motility. The best-studied mollicutes are significant pathogens of humans, domesticated animals, or plants. Additionally, a small number of species are common contaminants of *in vitro* eukaryotic cell cultures (4). Scientific organizations devoted to the study of these unique organisms include The International Organization for Mycoplasmology (IOM-Online.org) and its International Research Programme on Comparative Mycoplasmology, The United States Organization for Mycoplasmology (USOMycoplasmology.org), The Asian Organization for Mycoplasmology (square.umin. ac.jp/aom/), European Society for Clinical Microbiology and Infectious Disease Study Group for Mycoplasma Infections, the International Committee for the Systematics of Prokaryotes subcommittee for the Taxonomy of *Mollicutes*, Australian Society for Microbiology (Division of Medical and Veterinary Microbiology, Special Interest Group for *Mycoplasmatales*), and Division G of the American Society for Microbiology (ASM.org). The purpose of this chapter is to introduce scientists who have been trained in other disciplines to Mycoplasma and related organisms, and to provide an entry to the literature of practical mycoplasmology with emphasis on vaccinology and methods of genetic manipulation for microbiologists who specialize in other species.

Species Concept for Mycoplasma and Related Organisms

The species concept for Mycoplasma and related organisms is similar to the general species concept for other bacteria (5–7). As with other bacteria, in Mycoplasma, the correlation between 16S rDNA

DOI: 10.1201/9781003099277-36

sequence similarity and gold standard DNA-DNA hybridization (DDH) values is practically useful although imprecise (7–9) for a few notable species. For example, *Mycoplasma gallisepticum* and *Mycoplasma imitans* have >99% 16S rDNA similarity, yet show only 40% DDH (10). Within-species variability among *Mycoplasma hominis* strains can generate as little as 50% intraspecific DDH, while maintaining >99% 16S rDNA similarity (11). Stackebrandt and Goebel (12) proposed an upper limit of 97% 16S rDNA similarity as a threshold, which if exceeded, would indicate a need for additional tests to determine whether strains should be regarded as a single or separate species. A contemporary interpretation of this guideline indicates that complete genome sequences should be derived to resolve ambiguity between closely related taxa. For Mycoplasma and related organisms, experience suggests that a frontier of about 94% 16S rDNA sequence similarity indicates either a spectrum of related species from which an isolate would have to be distinguished by additional means. Current mycoplasmology employing at minimum a combination of 16S rDNA sequence analyses and supplementary phenotypic data to identify species is consistent with standards established by the microbial systematics community for a species concept (13). However, novel species descriptions proactively avoid ambiguous taxonomic assignment and are greatly strengthened by the generation of a complete genome sequence and analyses of the average nucleotide identity (ANI) between pairs of genome sequences of closely related species (14). The accepted boundary between species of 70% DDH empirically equates to about 95–96% ANI also for species circumscription of Mycoplasma and related organisms (7, 15), but like DDH, the ANI is not suitable for the determination of supraspecific relationships among mollicutes (16).

Taxonomic Ambiguities

The order Mycoplasmatales presently contains the single family *Mycoplasmataceae*. About 130 species have been named. Its polyphyletic genus *Mycoplasma* is divided into two groups based on ribosomal RNA sequence, named for their most prominent human pathogens: hominis and pneumoniae (17). These groups are further subdivided into phylogenetic clusters, and this strategy was extended to include the anomalous mycoides cluster. This group consists of a few species, regrettably including the genus *Mycoplasma* type species (*Mycoplasma mycoides* subsp. *mycoides*), whose 16S rDNA sequences are affiliated unambiguously with those of the family *Entomoplasmataceae* rather than with the *Mycoplasmataceae*. The nomenclature of those species is a matter of continuing controversy. Renaming of this small number of species is not possible, as *M. mycoides* has taxonomic precedent over later-named species. Furthermore, multiple members of the mycoides cluster cause World Organization for Animal Health (OIE)-listed diseases, and their cultivation, transport, and reporting are tightly controlled. Alterations in nomenclature would introduce potentially catastrophic confusion.

Despite these practical challenges, several recent revisions were proposed to resolve the polyphyly within the class. Affected species include numerous organisms of clinical and/or regulatory significance. The proposed revisions are based on combinations of characteristic sequence polymorphisms and indels, which in some species' cases are based on as few as a single strain genome. For these logistic and technological reasons, the International Committee for the Systematics of Prokaryotes Subcommittee for the Taxonomy of *Mollicutes* has published a formal request for opinion on the preservation of the original names (18). In so doing, the committee has soundly rejected the inclusion of the proposed order name *Mycoplasmoidales*, the proposed family names *Metamycoplasmataceae* and *Mycoplasmoidaceae*, the proposed genus names *Malacoplasma*, *Mesomycoplasma*, *Metamycoplasma*, *Mycoplasmoides*, and *Mycoplasmopsis*, and all species names included therein (19, 20). It is therefore strongly recommended that the revised names not be adopted at this time.

Uncultivated *Mollicutes*

The International Code of Nomenclature of Prokaryotes (ICNP, also referred to as "the Bacteriological Code") states that valid publication of a novel taxon requires isolation in axenic culture. The indefinite rank *Candidatus* is used for those organisms that clearly represent extant taxa, yet are prohibited from fulfilling the standards for valid publication for technological reasons (20, 21). Major plant pathogens in the class *Mollicutes* have been described and most fall under the heading of "*Candidatus* Phytoplasma" spp. The phytoplasmas have not been readily cultivated *in vitro*, in part because among other biochemical lesions, they lack the capacity for ATP synthesis (22). A small number of reports of axenic phytoplasma cultivation using undefined medium are promising, but remain to be independently verified (23). At this time, these taxa retain the rank *Candidatus*, and infected plants are surveilled and detected using molecular techniques (24).

Hemotrophic mycoplasmas (colloquially referred to as "hemoplasmas") such as "*Candidatus Mycoplasma haemofelis*," which were historically affiliated to the rickettsial genera *Haemobartonella* or *Eperythrozoon* based on phenotypic traits, also remain uncultivated in artificial media. About a dozen different hemoplasmas are distinguishable by host origin and 16S rDNA sequences. A small number of hemoplasmas whose detection and description predate the introduction of the "*Candidatus*," have full genus-species name combinations despite being uncultivatable. These include *Mycoplasma wenyonii*, *Mycoplasma coccoides*, *Mycoplasma ovis*, and *Mycoplasma suis* (1).

The rapid expansion of metagenomics datasets has allowed for the detection of numerous uncultivated *Mollicutes* through global 16S rDNA amplicon library sequencing. For example, Eckburg et al. (25) determined the 16S rRNA gene sequences of at least 10 novel phylotypes among the human intestinal microbiota that cluster distinctly enough to suggest the existence of a previously undiscovered order within the class *Mollicutes*. Similarly, the isolation or detection of 16S ribotypes in invertebrates (insects, shellfish) that appear to affiliate with the genus *Mycoplasma* question the previous genus description featuring vertebrate host habitats (26–29). As metagenomics continues to grow as a field, the number of discreet but uncultivated taxa belonging to the *Mollicutes* will undoubtedly expand.

Biology and Biochemistry of the Class *Mollicutes*

Mollicutes have been observed in close association with plants, invertebrates, and all classes of vertebrate hosts except amphibians. A flask-shaped cellular morphology, terminal adhesion-related structures, and a complex cytoskeleton are common in the lineages of the *Mycoplasma pneumoniae* and *Mycoplasma sualvi* clusters, and helical morphology is distinct to members of the genus *Spiroplasma*. The general morphological and metabolic simplicity of mycoplasmas and related organisms is consistent with their limited genome size, and leads to few unique biochemical traits that can be exploited for rapid differentiation between species (30). The lack of anabolic pathways in mycoplasmas is compensated for by an overrepresentation of transport-associated proteins in their proteomes. Pathways for energy generation are also minimal, as mycoplasmas lack the tricarboxylic acid cycle and other mediators of oxidative phosphorylation (31). Sugar fermenting mycoplasmas generate much of their energy from glycolysis and the pyruvate dehydrogenase pathway. The acidic byproducts of these pathways cause a characteristic downward pH shift of the culture medium following growth of those species. Some mycoplasmas hydrolyze arginine as a means of energy generation through the relatively simple arginine dihydrolase pathway, which requires only three enzymes. Hydrolysis of arginine results in the accumulation of ammonia, resulting in the characteristic upward pH shift of the culture medium seen following growth. A small number of species (e.g., *Mycoplasma bovis*) do not cause any shift in pH of the culture medium following growth. Some mycoplasmas catabolize organic acids and alcohols, such as lactic acid, pyruvic acid, oxobutyric acid, ethanol, glycerol, or isopropanol (31–33). Catabolism of glycerol leads to the excretion of hydrogen peroxide, which is a critical mediator of pathogenicity for some species (34–36). The hydrolysis of urea for ATP synthesis is the defining trait of the genus *Ureaplasma*; however, it cannot be used to differentiate between species within the genus.

Growth Characteristics

Mollicutes spp. are aerobic or facultatively anaerobic (with the exception of *Anaeroplasma* spp.), and have been isolated from vertebrate, invertebrate, or plant hosts. Excluding the genera *Acholeplasma* and *Asteroleplasma*, they require cholesterol and/or other sterols for growth, a need that is usually met by adding serum to the culture medium. A variety of culture media have been described (37). Nonmotile species or species with only gliding motility, tend to form umbonate ("fried egg") colonies on solid media. Satellite colonies are often observed for motile species on agar. Most *Mycoplasma* spp. grow best at temperatures reflecting their usual habitat. Optimum growth at 37°C is a common feature of species isolated from homeothermic vertebrates, and the lower part of the permissive temperature range even of *Mycoplasma* spp. isolated from poikilothermic fish and reptiles is only 20–25°C. The differential utilization of glucose and arginine and the ability to hydrolyze urea (for *Ureaplasma* spp.) and esculin (for *Acholeplasma* spp.) are important diagnostic features of *Mollicutes* spp.

The nutritional requirements of mollicutes were reviewed extensively by Miles (38). Tully (37) described in detail the most commonly used complex media formulations for cultivation of *Mycoplasma* spp. and related organisms. American Type Culture Collection medium 988 (SP-4) is a rich undefined medium that has proven to be superior for meeting their comparatively fastidious nutritional requirements during primary isolation, and for the maintenance of many different mollicutes. It is prepared

as an aqueous base, cooled to 56°C after autoclaving, to which is then added aseptically a mixture of serum, yeast extract, and other supplements that have been sterilized by sequential passage through 0.80 μm, 0.45 μm, and 0.22 μm filters. Penicillin G or other cell wall-targeting antibiotics are included to discourage the growth of other bacteria. The final pH of the complete medium, 7.6–7.8, facilitates phenol red-based detection of a pH shift caused by growth of mollicutes that excrete acidic endproducts of fermentation, although not all species do so. Isolation and growth of arginine-hydrolyzing species can be enhanced by including 5 mL/L of a 42% w/v arginine solution in the supplements, in which case the final pH of the complete medium should be 7.0–7.2 to facilitate detection of an alkaline shift. Evaluation of clinical specimens wherein a novel *Mollicutes* species is suspected can be cultivated in SP-4 medium supplemented with both glucose and arginine. Some species, such as *Mycoplasma fermentans*, are capable of utilizing both sugars and arginine as carbon sources. An initial decrease in pH may reverse later during their course of growth in a complex medium (31). Quality control tests for the serum and yeast extract components and for complete SP-4 media are particularly important because serum quality is susceptible to variation and yeast extract contains labile components (37). Commonly used alternatives to SP-4, such as Frey's, Hayflick, and Friis media, differ mainly in the proportions of inorganic salts, amino acids, serum sources, and types of antibiotic supplements included. Urea must be supplied for primary isolation and cultivation of ureaplasmas, and they require a medium with pH 5.5–6.5. A variety of liquid and solid media formulations for *Ureaplasma* spp. culture from human and animal sources were described in detail by Shepard (39). Phenol red and bromothymol blue indicator broths have been developed to detect urease activity. Incorporation of calcium chloride in the differential agar medium A8 distinguishes *Ureaplasma* spp. colonies, which appear deep brown or brown-black in color (as a result of ammonia production through urease activity) from *Mycoplasma* spp., which remain colorless on A8 agar. Liquid and solid culture media for Spiroplasma spp. were described in detail by Whitcomb (40) and Hackett and Whitcomb (41). SP-4 or M1D media, which substitutes *Drosophila* tissue culture medium for CMRL 1066 and also includes fructose, sucrose, and sorbitol, are suggested for routine culture of a wide array of spiroplasmas from various serogroups. Sorbitol is used to adjust the osmolality to approximately 500 mOsm. Spiroplasmas have also been cultured in defined media or cocultured with mammalian or insect cells (42). Few *Mollicutes* species grow reliably well by 24 hours of incubation, especially at primary isolation from an infected host. The majority require multiple days before positive indicators of growth are observed, and clinical specimens should be left for up to one month without growth before being categorized as negative for growth.

Major Clinical Presentations

General Features of Mycoplasmosis and Phytoplasmosis

Mycoplasma and *Ureaplasma* species most commonly infect epithelial surfaces, and a small number of species can spread between distal tissues. Mycoplasmosis is a highly inflammatory condition, and lesions in naturally and experimentally infected animals exhibit hallmarks of chronic inflammatory responses. Most manifestations cause significant morbidity and low mortality, with the notable exceptions of *M. mycoides* subsp. *mycoides*, *Mycoplasma capricolum*

subsp. *capripneumoniae*, and *Mycoplasma alligatoris* (28). Phytoplasmosis similarly causes a general failure to thrive along with numerous species-specific clinical presentations. Disease states associated with specific *Mollicutes* species are detailed in Tables 34.1–34.3.

TABLE 34.1

Mollicutes Diseases of Animals[a]

Species	Host	Disease	Clinical Signs/Notes
Livestock Pathogens			
M. alkalescens	Cattle	URTD	Nasal discharge, lameness, decreased milk production
M. arginini	Cattle	RTI	Infertility
M. bovigenitalium	Cattle	LRTD, URTD, RTI	Nasal discharge, coughing, infertility
M. leachii	Cattle	Mastitis, arthritis	Lameness, decreased milk production
M. bovis	Cattle	URTD, LRTD	Coughing, head tilt, purulent discharge, lameness, decreased milk production
M. bovoculi	Cattle	Conjunctivitis	Ocular discharge
M. californicum	Cattle	Mastitis	Decreased milk production
M. canadense	Cattle	Mastitis	Decreased milk production
M. dispar	Cattle	LRTD	Coughing
M. leachii	Cattle	RTI	APO
M. mycoides mycoides	Cattle	CBPP	Coughing, fever, torticollis
M. verecundum	Cattle	Conjunctivitis	Ocular discharge
M. wenyonii	Cattle	Hemolytic anemia	Malaise, jaundice
M. gallisepticum	Galliform	CRD, IS, neurological disease, arthritis	Rales, coughing, fever, sinus swelling, ataxia, ocular discharge, lameness
M. synoviae	Galliform	Arthritis, LRTD, RTI	Rales, coughing, sinus swelling, PEQ, reduced hatchability, lameness
M. anatis	Duck	URTD	Nasal discharge
M. adleri	Goat	Arthritis	Lameness
M. agalactiae	Goat	LRTD, RTI, contagious agalactia	Coughing, ocular discharge, decreased milk production, lameness, APO
M. capricolum	Goat	Contagious agalactia, LRTD, encephalitis	Lameness, decreased milk production, coughing, sudden death
M. capricolum capripneumoniae	Goat	CCPP	Coughing, fever, torticollis
M. mycoides capri	Cattle	Contagious agalactia, LRTD	Lameness, decreased milk production, ocular discharge, coughing
M. putrefaciens	Goat	Septicemia, contagious agalactia	Fever, malaise, cataracts, blindness, lameness, decreased milk production
M. subdolum	Horse	RTI	APO, infertility
M. haemolamae	Llama	Infectious anemia	Malaise
M. haemosuis	Pig	Hemolytic anemia	Malaise, jaundice
M. hyopneumoniae	Pig	PEP	Coughing, malaise, arrhythmia
M. hyorhinis	Pig	URTD, LRTD	Nasal discharge, coughing, arrhythmia, lameness
M. hyosynoviae	Pig	Arthritis	Lameness
M. arginini	Sheep	LRTD, URTD	Nasal discharge, coughing, ocular discharge
M. conjunctivae	Sheep, chamois, ibex	Conjunctivitis	Cataracts
M. ovipneumoniae	Sheep	URTD, LRTD	Nasal discharge, coughing
M. ovis	Sheep	Hemolytic anemia	Malaise, jaundice
M. meleagridis	Turkey	LRTD, RTI	PEQ, reduced hatchability, Rickets, skeletal deformity
M. iowae	Turkey	Arthritis, osteitis, RTI	PEQ, reduced hatchability, ocular discharge, lameness, skeletal deformity
M. pullorum	Turkey	RTI	Reduced hatchability
S. eriocheiris	Chinese mitten crab	Tremor disease	Tremor, death
S. apis	Honeybees	May disease	Tremor, paralysis
S. melliferum	Honeybees	Colony loss	Sudden death of bees

TABLE 34.1 *(Continued)*

Mollicutes Diseases of Animals

Species	Host	Disease	Clinical Signs/Notes
Wildlife Pathogens			
M. alligatoris	Alligator	LRTD, nephritis, ME, fulminant inflammatory disease	Rapid death, lethargy, rapid death, lameness
M. felifaucium	Cheetah, puma	Gastroenteritis	Emesis, wasting
M. agassizii	Desert tortoise	URTD	Nasal exudates
M. buteonis	Falcon	LRTD, unclear	Uncoordinated movement, lameness, skeletal deformity
M. gallsepticum	Passerines (primarily finches)	Conjunctivitis	Ocular discharge
M. pneumoniae	Hominids (chimpanzee, rhesus monkeys)	PAP, URTD	Coughing, malaise
M. iguanae	Iguana	Osteitis	Lameness
M. sphenisci	Jackass penguin	URTD	Halitosis, choanal discharge
M. crocodyli	Nile crocodile	LRTD	Lameness
M. haemodidelphidis	Opossum	Infectious anemia	Malaise
M. orale	Orangutan	URTD	Rales, nasal discharge
M. columborale	Pigeon	LRTD	Rales, coughing, lethargy
M. phocicerebrale	Seal, Human	Eye infection, Seal finger (hu)	Cataracts, blindness
M. phocirhinis	Seal	URTD	Rhinitis
M. zalophi	Sea lion	Necrotizing LRTD	Lameness, coughing
M. conjunctivae	Sheep, chamois, ibex	Conjunctivitis	Ocular discharge, cataracts
M. kahanei	Squirrel monkey	Hemolytic anemia	Malaise, jaundice, arrythmia
M. sturni	Starling	Conjunctivitis	Ocular discharge
M. mobile	Tench	Red gill disease	Ulceration, labored breathing
M. testudineum	Tortoise	URTD	Nasal discharge
M. corogypsi	Vulture	Abscess	Chronic exudate
M. vulturii	Vulture	LRTD	Rales, coughing
Companion and Laboratory Animal Pathogens			
M. oxoniensis	Chinese hamster	Conjunctivitis	Ocular discharge
M. cricetuli	Chinese hamster	Conjunctivitis	Ocular discharge
M. canis	Dog	RTI, UTI, URTD	Frequency, coughing
M. cynos	Dog	pneumonia	Coughing, malaise
M. edwardii	Dog	Septicemia	Fever, lameness
M. haemocanis	Dog	Infectious anemia	Malaise
M. maculosum	Dog	LRTD, UTI	Coughing, frequency
M. spumans	Dog	arthritis	Lameness
M. haemominutum	Domestic cat	Infectious anemia	Malaise
M. felis	Cat, Horse	URTD (Fe), LRTD (h)	Nasal discharge, labored breathing, ocular discharge
M. gateae	Cat	URTD	Anorexia, lethargy, ocular discharge, lameness
M. haemofelis	Cat	Infectious anemia	Malaise
M. equigenitalium	Horse	RTI	APO, infertility
M. equirhinis	Horse	Inflammatory airway disease	Nasal discharge, coughing, rales
M. coccoides	Mouse	Infectious anemia	Malaise
M. haemomuris	Mouse	Infectious anemia	Malaise
M. neurolyticum	Mouse	Rolling disease	Uncoordinated movement
M. pulmonis	Mouse, rat	Murine respiratory mycoplasmosis	Nasal discharge, coughing, anorexia, lameness
M. ravipulmonis	Mouse	Grey lung disease	Coughing
M. arthriditis	Rat	Septicemia, arthritis	Lameness
S. poulsonii	Drosophila spp.	Sex ratio abnormalities	Overabundance of males due to lethality in females

Abbreviations: URTD = upper respiratory tract disease; LRTD = lower respiratory tract disease; RTI = reproductive tract infection; CBPP = contagious bovine pleuropneumonia; CCPP = contagious caprine pleuropneumonia; CRD = chronic respiratory disease; IS = infectious sinusitis; PEP = porcine enzootic pneumonia; APO = adverse pregnancy outcomes; ME = meningoencephalitis; RTI = reproductive tract infection; UTI = urinary tract infection.

TABLE 34.2

Mollicutes Diseases of Humans

Species	Host Tissue	Disease	Clinical Signs/Notes
M. pneumoniae	Respiratory tract	PAP, encephalitis, meningitis, conjunctivitis	Postinfections inflammatory complications can occur; additional body sites are rarely implicated
M. genitalium	Urogenital tract	NGU, PID, cervicitis, orchitis	Infertility, preterm labor, APOs
M. hominis	Urogenital tract	BV, cerebral abscess, NGU	Infertility, APOs
M. amphoriforme	Respiratory tract	LRTI	Diagnostic differentiation from *M. pneumoniae* historically challenging
M. phocicerebrale	Skin	"seal finger" (ulcerative keratitis)	Occurs secondary to seal/marine mammal contact
Ureaplasma spp. (adult)	Urogenital tract	NGU, chorioamnionitis, endometritis, hyperammonemia	Strain-dependent pathogenicity, APOs, hyperammonemia occurs in immunocompromised patients only
Ureaplasma spp. (neonate)	Respiratory tract, CNS	BPD, encephalitis, meningitis	Premature infants born to infected mothers

Abbreviations: APO = adverse pregnancy outcome; BPD = bronchopulmonary dysplasia; BV = bacterial vaginosis; CNS = central nervous system; LRTI = lower respiratory tract infection; NGU = nongonococcal urethritis; PAP = primary atypical pneumonia; PID = pelvic inflammatory disease.

TABLE 34.3

Mollicutes Diseases of Plants

Species	Plant Host	Disease
S. citri	Citrus (orange; grapefruit)	Citrus stubborn
S. kunkelii	Corn/maize	Corn stunt
Candidatus P. asteris	Aster family plants	Aster yellows disease, corn stunt, cabbage proliferation disease, apple sessile leaf, strawberry green petal, blueberry stunt, tomato big bud, potato purple top, primrose virescence, hydrangea phyllody
Candidatus P. americanum	Potato	Potato purple top
Candidatus P. autralasia	Papaya	Papaya yellow crinkle disease
Candidatus P. allocasuarinae	Slaty sheoak	Sheoak (allocasuarina) yellows
Candidatus P. australiense	Grapevine; papaya; strawberry, peanut, apple, alfalfa, sweet potato	Australian grapevine yellows, papaya die-back, strawberry lethal yellows, peanut witches' broom, cocky apple witches' broom, sweet potato little leaf, alfalfa witches' broom
Candidatus P. mali	Apple	Apple proliferation disease
Candidatus P. vitis	Grapevine	Grapevine flavescence doree
Candidatus P. prunorum	Stone-containing fruits	European stone fruit yellows
Candidatus P. fraxini	Ash, lilac, alfalfa	Ash yellows, lilac witches' broom, alfalfa witches' broom
Candidatus P. lycopersici	Tomato	Hoja de perejil
Candidatus P. trifolii	Potato	Colombian potato purple top, potato witches' broom
Candidatus P. ziziphi	Cherry, nectarine	Nectarine yellows, cheery lethal yellows
Candidatus P. aurantifolia	Lime	Witches' broom disease of lime
Candidatus P. brasiliense	Alfalfa, hibiscus	Alfalfa witches' broom, hibiscus witches' broom
Candidatus P. fragariae	Strawberry	Strawberry lethal yellows
Candidatus P. japonicum	Coconut, sugarcane, strawberry	Coconut lethal yellows, sugarcane yellows, strawberry green petal
Candidatus P. pyri	Pear, peach	Peach yellow leafroll, pear decline
Candidatus P. phoenicum	Almond	Almond witches' broom
Candidatus P. oryzae	Rice, sugarcane	Ricc yellow dwarf; sugarcane white leaf
Candidatus P. rhamni	Buckthorn	Buckthorn witches' broom
Candidatus P. spartii	Sweet potato	Sweet potato little leaf
Candidatus P. solani	Grapevine	Grapevine yellows
Candidatus P. omaniense	Blueberry, peach, sugarcane, grapevine, walnut, cherry, tea	Blueberry proliferation; Canadian peach X; little peach; sugarcane yellows; Virginia grapevine yellows; walnut witches' broom; cherry buckskin disease ("Western X"); weeping tea disease
Candidatus P. ulmi	Elm, Indian hemp	Elm yellows; hemp dogbane yellows

Transmission between hosts occurs primarily by direct contact (includes nose-to-nose contact, sexual contact, or milk feeding) (mycoplasmas, ureaplasmas, acholeplasmas, anaeroplasmas), fomite transmission (veterinary mycoplasmas, spiroplasmas [via plant surfaces]), or by insect vector (phytoplasmas, spiroplasmas, entomoplasmas, hemotropic mycoplasmas) (1, 28). Aerosol transmission of respiratory mycoplasmosis between humans or animals housed in high density (e.g., dormitory settings, military barracks, or livestock and poultry rearing) has also been observed (43–45). Vertical and perinatal transmission of mycoplasmas, ureaplasmas, and spiroplasmas has been reported (28). The clinical signs of vertically transmitted mycoplasmosis are distinct from those displayed by the infected mother, as congenital infection with *M. hominis* and *Ureaplasma* species can lead to pneumonia, bronchopulmonary dysplasia, meningitis, or encephalitis in (frequently premature) newborns born to mothers with either clinical or subclinical urogenital tract infections (46–48).

Impacts and Outcomes of Mycoplasmosis

The chronic nature of veterinary mycoplasmoses makes them diseases of concern for livestock, who often exhibit decreased production. Coinfections of mycoplasmas with additional pathogens notoriously exacerbate of disease over and above those of either pathogen alone (28). Mycoplasmosis of livestock species leads to billions of dollars of economic loss each year. Several mycoplasmal diseases are listed by OIE and the Food and Agriculture Organization of the United Nations (FAO) because of the potentially devastating impact on local and national economies and food security. Morbidity associated with mycoplasmosis results in decreased production for poultry, swine, cattle, and small ruminants in the forms of downgraded carcasses, lowered feed conversion, loss of egg or milk production, eggs or milk not suitable for sale, and infertility. The most common mycoplasmal livestock diseases worldwide include porcine enzootic pneumonia due to *Mycoplasma hyopneumoniae*, chronic respiratory disease/avian mycoplasmosis due to *M. gallisepticum* and/or *Mycoplasma synoviae*, bovine respiratory disease/mastitis/arthritis due to *M. bovis*, and bovine mastitis/contagious agalactia due to *Mycoplasma agalactiae*, *Mycoplasma mycoides* subsp. *capri*, *Mycoplasma capricolum* subsp. *capricolum*, and *Mycoplasma putrefaciens*. Over and above these classic manifestations of mycoplasmosis, contagious bovine pleuropneumonia (CBPP) and contagious caprine pleuropneumonia (CCPP) caused by *M. mycoides* subsp. *mycoides* and *M. capricolum* subsp. *capripneumoniae*, respectively, feature both high mortality and high morbidity in survivors. Outbreaks can thus cause catastrophic economic losses. Both diseases have been eradicated from the Americas, Australia, and Western Europe, and their import and cultivation are tightly regulated and highly restricted. The impact of CBPP and CCPP on the livelihood of sub-Saharan Africans, in particular, is widespread and multifaceted. Cattle with CBPP frequently die of disease or are culled to prevent the spread of infection. Survivors provide less meat, 90% less milk, and are far less able to perform production work (49–51). Quarantine is often implemented during CBPP and/or CCPP outbreaks, which severely restricts movement of infected herds and potentially prohibits animals from accessing grazing grounds or water sources.

Both the health of uninfected animals in the herd and the ability of farmers to access the value of their cattle or goats (52) are thus negatively affected by quarantine. All mycoplasmal diseases of animals are described in Table 34.1.

Urogenital mycoplasmosis of humans has been linked to pelvic inflammatory disease, infertility, spontaneous abortion, and preterm labor. *M. genitalium*, a causative agent of nongonococcal urethritis, in particular is strongly associated with pelvic inflammatory disease, spontaneous abortion, and infertility (28, 46). Active infection is required to induce spontaneous abortion and preterm labor, but pelvic inflammatory disease and infertility persist even if infection is resolved. Preterm labor and/or premature rupture of membranes during ureaplasmosis of pregnant women is a significant risk factor neonatal respiratory tract or central nervous system infections, and chronic lung conditions can persist in infants following resolution of infection (28, 47). Respiratory mycoplasmosis caused by *M. pneumoniae* has been associated with asthma exacerbation, postinfectious reactive airway disease, and adult-onset asthma, and these conditions persist after successful elimination of the bacteria (53, 54). Impaired fertility, chronic pain, reactive airway disease, and lifelong complications from premature delivery can have significant impacts on economic livelihood and quality of life. All mycoplasmal and ureaplasmal diseases of animals are described in Table 34.1.

Several grain crops, fruit trees/bushes, nut trees, vegetables, and ornamental plants are affected by phytoplasmosis. General disease features of phytoplasmosis are variable and species-specific, but can include phyllody, greening (virescence), witches' broom formation, stunting, floral sterility, yellowing, leaf curling, purpling, cupping, vein clearing, and failure to thrive (28). Numerous diseases of important crop species are caused by phytoplasmas (Table 34.3). As with veterinary mycoplasmosis, numerous phytoplasmal diseases are listed by the International Plant Protection Convention (IPPC), and the import and reporting of infected plants is tightly controlled.

Detection and Surveillance

Growth in culture is often considered the gold standard for positive detection of an organism for most bacterial infections. In addition to detection by culture, nucleic acid amplification tests (NAATs) and metagenomics techniques have been used widely to detect mollicutes in clinical material, cell cultures, and environmental samples (55). Molecular techniques avoid the limitations of culturing, which include sometimes very slow growth, false-negative results due to the complex nutritional requirements of many species, or uninformative results due to overgrowth with fungal or other bacterial contaminants during primary isolation. Commercial NAATs have been developed for the human pathogens *M. pneumoniae* and *M. genitalium*, with the latter having the additional benefit of simultaneously detecting single nucleotide polymorphisms (SNPs) conferring macrolide resistance (Table 34.4). Serological methods of diagnosis were once preferred; however, NAATs have rapidly become the diagnostic method of choice for the major human mycoplasmoses (i.e., primary atypical pneumonia and nongonococcal urethritis). A qualitative method of detection in cell culture systems, based on mycoplasma-specific ATP synthesis activity present in tissue culture supernatant medium,

TABLE 34.4

Commercial Diagnostic NAATs Detecting Mycoplasmosis

Disease[a]	Host	Organism	Test	Manufacturer
WP/CAP	Human	*M. pneumoniae*	Alethia™	Meridian Biosciences
WP/CAP	Human	*M. pneumoniae*	BioFire® FilmArray® (RP)[c]	bioMérieux
NGU	Human	*M. genitalium*	Aptima® MG	Hologic
NGU	Human	*M. genitalium*	*ResistancePlus®* MG[d]	SpeeDX
BoM	Cattle	*M. bovis*	*VetMax™ MastiType bovis*[c]	Thermo Scientific™
BoM	Cattle	Multiple[b]	*VetMax™ MastiType Series*[c]	Thermo Scientific™
BoM	Cattle	Multiple[b]	*Mastit4 4B, 4BD*[c]	*DNA-Diagnostic*
BoM	Cattle	*M. bovis*	*Bactotype® Mastitis HP3*[c]	QIAGEN
BoM	Cattle	*M. bovis*	*Bovicheck® Mbovis*	BioVet®
BRD	Cattle	*M. bovis*	Pneumo4 PN4B, 4BC, 4BVC55	*DNA-Diagnostic*
PEP	Swine	*M. hyopneumoniae*	*Swinecheck® Mhyopneumoniae*	BioVet®
PEP	Swine	*M. hyopneumoniae*	*VetMax™ Mhyopneumoniae*	Thermo Scientific™
PEP	Swine	*M. hyopneumoniae*	*RealPCR Mhyo*	IDEXX
RTI	Swine	*M. hyosynoviae*	*Swinecheck® Mhyosynoviae*	BioVet®
Arthritis	Swine	*M. hyorhinis*	*Swinecheck® Mhyorhinis*	BioVet®
AM/CRD	Poultry	*M. gallisepticum*	*VetMax™ Mg and Ms*[b]	Thermo Scientific™
AM	Poultry	*M. synoviae*	*VetMax™ Mg and Ms*[b]	Thermo Scientific™
AM/CRD	Poultry	*M. gallisepticum*	*RealPCR MG/MS*[b]	IDEXX
AM	Poultry	*M. synoviae*	*RealPCR MG/MS*[b]	IDEXX
Phytoplasmosis	Plants	Multiple[e]	*Phytoplasma Universal Detection*	Nippon Gene

[a] *Abbreviations:* WP/CAP = "walking pneumonia"/community acquired pneumonia; NGU = nongonococcal urethritis; BoM = bovine mastitis; BRD = bovine respiratory disease; PEP = porcine enzootic pneumonia; RTI = respiratory tract infection; AM = avian mycoplasmosis; CRD = chronic respiratory disease.

[b] Detects Mycoplasma alkalescens, Mycoplasma bovis, Mycoplasma bovigenitalium, Mycoplasma canadense, Mycoplasma californicum and/or contains genus-wide primers.

[c] Part of a multiplex panel.

[d] Detects M. genitalium presence and macrolide susceptibility simultaneously.

[e] Detects all members of the genus "Candidatus Phytoplasma."

is also available commercially with a standardized protocol for research purposes (MycoAlert®, Lonza Group Ltd.).

Because of the technical challenges to culturing certain species and the severely restricted usefulness of conventional metabolic diagnostic tests of isolates, diagnosis of veterinary mycoplasmosis often relies on enzyme-linked immunosorbent assays (ELISA) or NAATs. ELISA techniques have been described in detail for most major veterinary pathogens, and many are commercially available. Commercial NAATs are also available for most major agricultural mycoplasmas (Table 34.4), with the notable exceptions of *M. mycoides mycoides* and *M. capricolum capripneumoniae*. Additionally, many veterinary diagnostic laboratories have developed and validated in-house ELISA and/or NAATs for agricultural and companion animals available on a fee-for-service basis.

Treatment and Control

Antimicrobial Treatment

Mycoplasmas are intrinsically resistant to many antibiotics. Lacking a cell wall, mycoplasmas are not affected by β-lactams, vancomycin, or fosfomycin. In addition, mycoplasmas as a whole

are resistant to sulfonamides and trimethoprim because they do not synthesize their own nucleotides; resistant to rifampicin because they express a form of RNA polymerase unrecognized by the antibiotic; and resistant to polymyxins (56). Individual species can exhibit even a broader spectrum of antibiotic resistance. Examples include the resistance to erythromycin and azithromycin exhibited by *M. hominis*, *M. fermentans*, *Mycoplasma hyopneumoniae*, *Mycoplasma flocculare*, *Mycoplasma pulmonis*, *Mycoplasma hyorhinis*, *Mycoplasma hyosynoviae*, *Mycoplasma meleagridis*, and *M. bovis*, which is apparently mediated by a change in sequence of the 23S RNA (57). Antibiotic treatment options for mycoplasmosis are therefore limited due to their intrinsic resistance and the toxicity of some of the antibiotics that are effective. Treatment of mycoplasmosis often involves the use of antibiotics that inhibit DNA replication or protein synthesis, such as tetracyclines, fluoroquinolones, aminoglycosides, certain macrolides or ketolides, and under rare circumstances chloramphenicol. Mycoplasmal infections require long-term antimicrobial therapy, as short-term treatment often results in recurrence of the infection. This is speculated to be due to sequestration of a small number of bacteria in privileged sites, potentially including inside host cells (28).

Macrolides are the treatment of choice for human mycoplasmosis based on their effectiveness and their minimal side effect profile relative to tetracyclines. Macrolides less readily penetrate the blood-brain barrier, however, making them suboptimal treatments for disseminated mycoplasmosis. Fluoroquinolones exhibit a bactericidal effect on Mycoplasma spp. (12, 58, 59). Fluoroquinolones were historically less widely used to treat human mycoplasmosis, but their use is becoming more common due to increasing levels of macrolide resistance in both *M. pneumoniae* and *M. genitalium*. Factors leading to macrolide resistance differ for the two organisms. *M. pneumoniae* resistance is being driven largely by over-the-counter availability in many part of the world, and the emergence of resistant strains tends to occur in these areas (60, 61). In contrast, the single-dose azithromycin treatment given at the point-of-care (POC) recommended by the United States Centers for Disease Control and Prevention as of 2010 to combat nongonococcal urethritis is effective against *C. trachomatis*, but is insufficient to resolve *M. genitalium* infection. This treatment thus contributes to antibiotic resistance by *M. genitalium* in patients who are presumptively treated for chlamydia (62). The newly available near-POC diagnostic tests for *M. genitalium* should be an effective counter-measure against this source of developed macrolide resistance. Human mycoplasmosis is also responsive to tetracyclines, but the use of doxycycline is now preferred only in treatment-refractory cases due to its side effect profile. Chloramphenicol is used rarely, and is only recommended for central nervous system infections in neonates that are unresponsive to other antibiotics (37).

Tetracyclines and macrolides are used commonly to treat animal mycoplasmoses, based on their low cost and the fact that resistance is not intrinsic (56). Preference for tylosin and o-tetracycline is standard due to their limited use in human disease. The use of fluoroquinolones in veterinary medicine is growing, although the threat of generating resistant, nonmycoplasma zoonotic agents may dictate against their use (63). Aminoglycosides, pleuromutilins, lincosamides, and phenicols are also occasionally employed in veterinary medicine. As with human medicine, increasing reports of antimicrobial resistance among livestock pathogens emphasize the need to employ nonchemotherapeutic disease control approaches to veterinary mycoplasmosis (discussed later) (56, 64–66).

Antibiotic use to control phytoplasmosis has been carefully examined and found to be largely ineffective. Treatment with tetracyclines, phenicols, and thiazolides fails to transiently or permanently resolve signs of infection (67, 68). Application of potentially antimicrobial plant oils such as rose oil, clove oil, and eucalyptus oil were also ineffective (68). However, some efficacy toward alleviation of disease has been reported following treatment with organic acids such as salicylic acid and gibberellic acid (69, 70).

Infection Control Measures

Human mycoplasmosis is typically spread via droplet aerosol or by sexual contact. Accordingly, measures to control transmission are consistent with other respiratory or sexually transmitted pathogens. Patients exhibiting signs of community-acquired pneumonia associated with *M. pneumoniae* or *Mycoplasma amphoriforme* in an outpatient setting should be masked upon entry to a clinical setting to prevent spread to other patients in common areas. Prevention of *M. genitalium* transmission can be achieved

through barrier protection such as condoms (71). There are no vaccines available to protect against human mycoplasmosis.

Control of phytoplasmosis can be an extraordinary challenge, as treatment is largely ineffective and plants cannot be vaccinated. Because phytoplasmas are vector-borne pathogens, the strongest preventative measures involve pest control, with a particular focus on eliminating leafhoppers and psyllids. Addition disease prevention can be achieved through management practices such as surveillance, import control, and roughing of infected plants (69).

Control of veterinary mycoplasmas is approached by a combination of vaccination and ancillary animal husbandry measures. Universal hygiene measures that prevent the spread of infection (e.g., disinfection of ear tagging equipment, milking equipment, and shearing tools; restriction of immediate movement between isolated groups of animals by farm personnel; and pest control) can reduce the incidence of mycoplasmosis. Additional management practices such as all in/all out production, use of artificial insemination over external breeders, minimizing herd/flock sizes, earlier weaning, periodic or continuous antibiotic use, and maintaining appropriate air quality within production facilities have all been shown to have additional benefits toward the prevention of mycoplasmosis (65, 66, 72, 73).

Vaccination is the most promising approach to the control of mycoplasma infections in animals, and several experimental vaccines have been described (reviewed in reference [74]). A number of vaccines are commercially available for the protection of livestock and poultry against infection by Mycoplasma spp. (Table 34.5). No subunit or nucleic acid-based vaccines against mycoplasmosis have been approved for use as of this writing, and only a single recombinant vaccine against *M. gallisepticum* employing a fowlpox vector has been approved (Vectormune FP MG, Ceva). Bacterins are the simplest vaccines to produce, and although many Mycoplasma spp. are capable of antigenic variation, potentially lowering the efficacy of a bacterin, many are commercially available. All other licensed vaccines are live, attenuated strains, despite their potential for residual virulence or the threat of reversion to pathogenic capabilities.

Several vaccines have been described for mycoplasmoses of small and large ruminants. Sonicated *Mycoplasma capricolum* subsp. *capripneumoniae* adjuvanted with saponin dramatically reduced endemic CCPP in goat herds (75), and a similar bacterin composed of lyophilized *M. capricolum* subsp. *capripneumoniae* adjuvanted with saponin was both completely protective and stable at ambient temperature, an important factor in the design of vaccines for diseases prevalent in developing countries (76). Commercial bacterins are also available against *M. bovis* and *M. agalactiae*, though their success is somewhat equivocal. Live, attenuated vaccines are available for CBPP and contagious agalactia due to *M. agalactiae*. The vaccine strain T1/44 of *M. mycoides* subsp. *mycoides* was generated by serial passage *in ovo*. The attenuating mutations of this strain are not clear, despite its recent genome sequence (77). T1/44 appears moderately efficacious and is commonly used in *M. mycoides* subsp. *mycoides* (SC)-endemic regions, despite the common occurrence of injection-site reactions ("Willem's reaction" and/or subdermal abscess) (51, 54, 78). Outbreaks of CBPP occurring in vaccinated herds, as well as T1/44 challenge studies, suggest that residual pathogenicity exists (79, 80). The vaccine strain of

TABLE 34.5

Commercially Available Vaccines Against Mycoplasmosis[a]

Trade Name	Agent	Strain	Vaccine Type	Manufacturer
Pulmo-Guard™ MpB	*M. bovis*	ND	Bacterin	Boehringer Ingelheim
Mycomune®	*M. bovis*	ND	Bacterin	AgriLabs
Myco-Bac® B	*M. bovis*	ND	Bacterin	Texas Vet Lab, Inc.
Contravax	*M. mycoides mycoides*	$T_1/44$	Live	KeVeVaPI
PERIBOV	*M. mycoides mycoides*	$T_1/44$	Live	BVI
Caprivax	*M. mycoides capri*	F38	Bacterin	KeVeVaPI
Capripen	*M. mycoides capri*	F38	Bacterin	PVCI
AGALAXIPRA®	*M. agalactiae*	ND	Bacterin	HIPRA
Agalaksipen	*M. agalactiae*	AIK	Live	PVCI
Agalaksipen-Inactive	*M. agalactiae*	AIK	Bacterin	PVCI
RespiSure1	*M. hyopneumoniae*	P-5722	Bacterin	Zoetis
Ingelvac® MycoFLEX™	*M. hyopneumoniae*	J	Bacterin	Boehringer Ingelheim
MycoSilencer ONCE	*M. hyopneumoniae*	ND	Bacterin	Merck
M. hyopneumoniae Live Vaccine	*M. hyopneumoniae*	168	Live	Tianbang Bio-Industry Co
M+Pac®	*M. hyopneumoniae*	J	Bacterin	Intervet
Mypravac® Suis	*M. hyopneumoniae*	J	Bacterin	Hipra
Resprotek™ One Shot	*M. hyopneumoniae*	ND	Bacterin	Bayer
Serkel Pneumo	*M. hyopneumoniae*	ND	Bacterin	Vencofarma
MG-Bac	*M. gallisepticum*	ND	Bacterin	Zoetis
Gallivac MG TS11	*M. gallisepticum*	ts-11	Live	Sanofi Russia
MycoVax® TS-11	*M. gallisepticum*	ts-11	Live	Boehringer Ingelheim (Merial)
Vaxsafe® TS/11	*M. gallisepticum*	ts-11	Live	Bioproperties
PoulVac® Myco F	*M. gallisepticum*	F	Live	Zoetis
CEVAC® MGF	*M. gallisepticum*	F	Live	Ceva
Nobilis MG	*M. gallisepticum*	6/85	Live	MSD Animal Health
Vectormune® FP MG	*M. gallisepticum*	N/A	Vectored	Ceva
Myco-Galli MG-70	*M. gallisepticum*	ND	Live	BioVet Vaxxinova
Vaxsafe® MS-H	*M. synoviae*	MS-H	Live	Bioproperties
MycoVax® MS-H	*M. synoviae*	MS-H	Live	Boehringer Ingelheim (Merial)
M. synoviae MS	*M. synoviae*	MS-H	Live	PharmaSure LTD

[a] *Abbreviations:* ND = not described; N/A = not applicable; KeVeVaPI = Kenya Veterinary Vaccine Production Institute; BVI = Botswana Vaccine Institute; PVCI = Pendik Veterinary Control Institute.

M. agalactiae, AIK40, was generated by serial passage *in vitro*. Though not widely used, AIK40 appears to protect vaccinated sheep and goats against arthritis and keratoconjunctivitis (74). It was not completely protective against agalactia, however, as mild mastitis was observed in some challenged dams (81).

Many inactivated vaccines are available for the swine pathogen *M. hyopneumoniae*. Despite this, porcine enzootic pneumonia continues to be problematic in many countries, calling the efficacy of these preparations into question. Multiple studies have demonstrated their poor immunogenicity, and their inability to prevent colonization with *M. hyopneumoniae* (66, 82). Given the difficulty of eradicating *M. hyopneumoniae* from production facilities once introduced, there is understandable hesitation to hastily adopt live, attenuated vaccines. One such vaccine, Mycoplasma Hyopneumoniae Live Vaccine (strain 168, Tianbang Bio-Industry Co) has been described and is being used with apparent success in China (83, 84).

Many vaccines are commercially available targeting avian mycoplasmosis due to *M. gallisepticum*. A single bacterin (MG-Bac, Zoetis) is available, and the lone vectored vaccine against a member of the genus (Vectormune FP MG, Ceva) also targets this organism. However, both MG-Bac and Vectormune FP MG fail to fully prevent respiratory lesions and impacts on egg production, respectively (85, 86). Numerous live, attenuated vaccines are also available against *M. gallipseticum*, and are almost exclusively composed of either strains ts-11, 6/85, or F. Though these vaccines are largely successful, balancing the stronger safety versus poorer efficacy of 6/85, as opposed to the high efficacy but higher potential for residual pathology of F, can be challenging. A single live, attenuated vaccine strain (MS-H) has been described as protective against avian mycoplasmosis due to *M. synoviae*. Vaccination with MS-H does not prevent colonization with *M. synoviae*, but it does protect against clinical signs and minimize horizontal transmission between infected animals (44, 87, 88).

Molecular Biology and Genetic Manipulation

Mollicute Genome Characteristics

With their small genome size and diverse range of specialized hosts, mollicutes constitute a unique model organism for studying comparative genomics, and as economically important human, animal, and plant pathogens, understanding their molecular mechanisms is important for controlling their virulence (14, 89). Mollicutes have undergone reductive evolution, shedding nonessential genes, leading to small, high-density genomes with a core pool of genes and few extrachromosomal elements. The use of operon systems has also helped reduce the number of gene regulatory regions, but despite the shortage of space many duplicated and repeating sequences, and various nonchromosomal elements can also be found in their genomes. The 0.5–2 Mbp circular dsDNA chromosomes have low GC content (35% average, the lowest at 25%), and low number of rRNAs relative to other bacteria (89, 90).

While mollicute evolution has been influenced by gene loss, there is also increasing evidence for episodes of expansion via horizontal gene transfer. Indeed, comparative genomic methods show large chromosomal exchanges, high genome plasticity, and mosaics of mobile genetic elements (MGEs). The noncore accessory genomes of mollicutes are mostly contained in insertion sequence (IS) elements, integrative conjugative elements (ICEs), and a few phages and some plasmids (mostly in the spiroplasmas, phytoplasmas, and ruminant pathogens) (91). Phytoplasmas also contain sequence-variable mosaics (SVMs, evolutionary remnants of phages and other MGEs), and potential mobile units (PMUs), putative conjugative transposons that contain genes for recombination, replication, transposases, and effector virulence proteins that increase in expression in insect vectors (92, 93). Some mollicutes also have CRISPR elements in their genomes, as well as CRISPR-like structures that are smaller and less conserved (89). Reductive evolution and gene decay results in formation of pseudogenes (genes with coding sequence loss or change >20%), but these are found more often in the mobilomes than in the core genomes which contain the essential genes (89).

Genome Plasticity

Mollicute genomes are dynamic and their plasticity is evident in the variety of genetic elements they contain. Horizontal gene transfers (HGT) between related species and strains create the mosaicism and heterogeneity that is important for evolving with host defenses, avoiding genome stagnation, and adapting to new hosts, niches, and environments. While relative to other bacteria mollicutes have fewer MGEs, knowing the impacts of the mobilome on physiology, adaptation, evolution, and virulence are of critical importance, especially among pathogens of interest (28). MGEs, including ICEs, IS, phages, plasmids, integrative mobilizable elements (IMEs), PMUs, and Tra islands are highly heterogeneous and often restricted to specific species (94).

Mollicute ICEs (described in greater detail below) are important elements for HGT, in that they provide the conjugation machinery necessary for genetic transfer between cells. IMEs, also known as Genomic Islands, encode excision and integration machinery and depend on ICEs for conjugative transfer

(89, 91, 95). A few mycoplasma species have Tra islands (one to two copies per genome), containing coding sequences for proteins with transmembrane domains, as well as phage- and plasmid-related gene homologs, such as helicases, methylases, machinery for replication-initiation and plasmid partition, and ssDNA binding proteins (ssb). Tra islands also contain a copy of *IS3*, and some ICE genes such as *traE*. Tra islands represent integrative elements, and although they are distinct from ICE, they do contribute to genome mosaicism by differential integration and genome rearrangement (89, 96, 97).

Compared to other bacteria, plasmids are rare in mollicutes, and mostly found in ruminant- and plant-pathogens, such as spiroplasmas, phytoplasmas, and mycoplasmas of the *mycoides* clade (98–112). However, the plasmids in these genera are different, despite their shared niche in plants. Mollicute plasmids are smaller than those of other bacteria, with 1–25 coding sequences within the 1 kbp–35 kbp plasmids (113). Although many of the plasmids are cryptic, some affect the host phenotype by expression of adhesins for transmission, conjugation machinery similar to ICE, and ssb for stabilizing the plasmids during rolling circle replication (99, 100, 105, 109, 114, 115). Plasmids can also be dynamic and undergo rearrangements and recombinations, but they retain their sequences and architecture distinct from mollicute chromosomes (103, 115, 116). As in comparative genomics, plasmids from different species can be compared for overall sequence, gene orders, and coding sequence similarities for evidence of shared lineage and HGT.

As with plasmids, phages are mostly found in spiroplasmas and phytoplasmas, and can be both ssDNA or dsDNA, (+) or (−) stranded, linear or circular free-phage-plasmids, and chromosome-integrated with truncated coding sequences (89, 117). Some phages resemble IS sequences in that they can encode excisionases and integrases, disrupting genes and reorganizing genomes, potentially altering the mycoplasma's virulence, antigen expression, lipoprotein expression, or encode their own novel surface proteins (117–122). Phages, combined with other mobile elements, can drive the formation of phytoplasma SVMs, increasing the differences between related strains (94).

Integrative Conjugative Elements (ICEs)

Relative to other bacteria, there is a lower frequency of conjugal DNA transfers in mollicutes (90). However, those species capable of HGT have greater phenotypic variation, with increased options for potential genes that may increase fitness (and gene decay for those that do not) (91, 113). HGT is important for evolution, adaptation, and acquiring new traits. The highest levels of HGT occur in ruminant mollicutes, with comparative genomic evidence of ~18% HGT between *M. agalactiae* and the *mycoides* cluster (28). Evidence of HGT is often observed in phenotypic changes, such as resistance to antimicrobial agents or degradative enzymes (91). HGT and mosaicism in the pathogens make diagnostics and identification of serovars difficult and may influence gain or loss of virulence (123–126). There is little evidence of HGT between mollicutes and nonmollicutes. Some sequence similarities can be found in *M. mobile*, *M. arthriditis*, and *M. hominis* with Gram(+) and proteobacteria, but these are likely due to shared ancestry. In ureaplasmas and mycoplasmas, there is evidence of evolutionary acquisition of the SOS system's *ruvB* gene from Gram(-) bacteria in the same operon as *ruvA* that was inherited from a Gram(+) ancestor (127). There is evidence of HGT in Spiroplasmas, especially *S. mirum* and *S. citri*, but their mechanisms are still not yet elucidated (91, 128, 129). Spiroplasmas do have homologs of *Tra* genes for conjugation, and viral components for transduction (109, 130–132). The viral sequences found in Spiroplasmas can also cause genome rearrangement and facilitate incorporation of novel fragments of DNA (92).

In mycoplasmas, ICEs (also referred to as MICE or conjugative transposons) are large (20–30 kbp), modular, self-transmissible families of genetic elements that contain the genes for excision of the ICE out of the chromosome, the proteins for formation of the conjugative channel, proteins for replication of the complementary strand in the donor and the recipient, and for integration into the chromosomes at random (89–91). This conjugative HGT requires one of the two cells to contain at least one ICE, and the process is cell-cell contact dependent. ICEs are mainly found in the hominis and spiroplasma groups, and ICE homologs are found in the pneumoniae and ureaplasma groups. Some mycoplasmas species do not have ICEs, such as *M. genitalium* and *M. pneumoniae* in the pneumoniae clade, and *M. synoviae* and *M. pulmonis* in the hominis clade. These species have relatively stable genomes, both over time and over geographic distribution (91, 133). ICE in the hominis group have lower GC% than the rest of their chromosomes, while ICE in the mycoides group have the same GC% as their chromosomes (89). The same ICEs can be found in distant mycoplasma species, also making them useful tools for comparative genomics (134, 135). Vestigial ICE that have been disrupted by other elements (e.g., IS or transposons) also exist in mollicute genomes, and are often degraded over time, with smaller footprints and decayed pseudogenes (91, 97, 135).

The trigger for activating chromosomal ICE excision and gene expression is yet unknown, but ICE are often silent and expressed at low frequency in most mycoplasmas (91, 136). Upon excision, the ICE becomes a circular ssDNA, and the copy that is transferred is then replicated in the recipient cell, as is the ssDNA copy that remained in the donor (89, 96, 123, 137, 138). Large chromosomal transfers often occur during ICE conjugation, and interestingly, as the ICE transfers from ICE(+) to ICE(−), the chromosome fragment transfers from the ICE(−) to the ICE(+), and they have equal frequencies of transfer (139). The DDE recombinases encoded by ICE integrate the element randomly into the genome, but the chromosome fragments transferred to the ICE-donor integrate via homologous recombination, displacing that region from the host's chromosome (96, 140).

The types of DDE recombinase and inverted repeats that help the excised ICE circularize are unique to the type (family) of ICE. Multiple ICE can be found in the same organism, with either multiple copies of the same ICE or two difference ICEs (91). Separate but distinct ICE types exist, corresponding to the type of mycoplasma, often even if they share the same niche in the same host (96, 97, 123, 136, 138, 141–145). It is difficult to experiment with ICE for HGT because in the lab ICE are activated less frequently and may require specific stressors or host factors that are not yet known. Mycoplasma chromosomes can have their own non-ICE associated conjugative elements, but all observed conjugative transfers have required either the donor or recipient to have at least one ICE (91).

Insertion sequences (IS) are small and simple mobile genetic elements within mollicute genomes, which can be transposed within a genome, replicated via vertical transfer or potentially transferred between two cells during HGT (146–148). IS elements encode transposases (sometimes multiple), and often contain pseudogenes that have undergone gene decay (89, 97, 149). IS element genetic structures are used to classify them into families, and their copy number patterns and locations in the genome are used to identify mycoplasma strains (97, 123–125, 150–152). IS elements can disrupt genes and cause gene rearrangements, carry out tandem duplications or inversions, and carry restriction-modification systems (126, 145, 150, 153–156). While some mycoplasma species can contain few or no IS elements, the IS elements in *M. mycoides* subs. *mycoides* strain PG1 occupy 13% of the genome (89, 97, 149, 157). Some IS elements can also carry with them virulence genes, but their own distribution, copy number, and integration sites can directly contribute to the mollicute genome plasticity and dynamics (126).

Antigenic Variation

Mollicutes rely on structures on their surface to mediate targeted or nonspecific binding and motility on environmental surfaces, especially molecules on host cells required for successful infection and survival. These adhesion structures can thus provide ideal targets for immune recognition and response. Antigenic variations of these surface molecules by mollicutes can expand the repertoire of host targets to which they can bind, and to help evade the host's immune response (158). Although immune evasion is critical for survival, antigenic variation is not the function of the surface molecules, and the consequence of the variation (the cell's phenotypic change) is what is observed when the protein is absent or not in its primary form (159). Continued stress from the host can make the change in the new form of the antigen permanent, creating resistance and a new instance of the chronic infection (160). Sometimes, the protein that undergoes variability is not the target antigen, but the one performing the role of masking (161). Successful antigenic variations can result in worsened susceptibility to the immune system for an individual cell, but greater fitness for survival in a different way, such as the ability to bind new substrates for escape, or evasion via biofilm formation (162–165). Survival of the antigenic switch usually means either the variable protein or antigen is temporarily dispensable, or the new alternative form is still functional enough (166). Some antigenic variations that results from genetic switches provide protection due to more system-wide effects, such as genome-wide modifiers, or global changes in translation as ribosomes are altered (167). One antigen variation target can, itself, have two or more genetic switches associated with it, and one genetic switch can have effects on two different antigenic variation targets, attesting to the complexity of strategies available to organisms with limited coding capacities (158, 168–170).

Antigenic variations are stochastic events, even in a clonal population, and as they occur randomly, the individuals with the most optimal variation survive and thrive (171). The frequency of the antigenic variation likely corresponds to the frequency of changes in the host environment and the urgency of the mollicute to adapt. Due to selection pressures, the antigenic variations rely on the genetic switches to be fairly high-frequency to provide the right combination in time for evasion. As such, the genomic elements that provide switching must be less rigid, such as areas of rapid point mutations, slipped-strand mispairings, and DNA rearrangements. In addition to genetic switches, epigenetic and postgenetic (post-transcriptional or post-translational) variation strategies can also occur. The resulting variations can be phase variations, where the genes are turned on or off, domain variations, where the gene is still expressed but in a new form, and epitope masking, where the susceptible area of the protein is hidden from threat (172).

Genetic switches can involve the coding sequence of a gene, the promoter, or both. Some switches can be simple, such as point mutations, additions, or deletions, resulting from template or synthesis strand slippage, usually at hot spots where there are numerous short nucleotide repeats. Switches can be inversion recombinations, caused by site-specific recombinations that require small, conserved inverted repeats around a target, a recombinase enzyme, and the proximity of the recombinase gene to its target can affect the frequency of the switch. Switches can be homologous recombination events that require larger areas of sequence homology, and the resulting crossover exchanges can happen across greater distances in the genome. Switches can also affect larger swaths of the genome caused by excisions or duplication events. Switches that result in epitope masking occur in the gene of a partner protein of the actual target antigen (172).

Specific examples of these antigenic variation strategies undertaken by different mollicute species have been reported. Genes with point mutations or strand slippage resulting from nucleotide repeats are common. In the *vlp* gene of *M. hyorhinis*, *vmc* genes of *M. mycoides* subs. *mycoides* SC and *M. capricolum* subs. *capricolum*, *vlhA* gene of *M. gallisepticum*, and *maa2* gene of *M. arthritidis*, the promoters all contain different variations of polynucleotide repeats, where if a mutation changes the number of repeats, the promoter stops functioning and the genes are turned off (173–176). These simple mutations can be seen in the coding sequences as well, for example in the *vaa* gene of *M. hominis*, and the *gapA* gene of *M. gallisepticum*, wherein short repeat sequences cause frameshift or nonsense mutation that form truncated transcripts (177, 178). The *gapA* gene of *M. gallisepticum* is also part of an operon system, and the mutation also inhibits the expression of the downstream gene (*pvpA*) (169).

In the *vpma* cluster of *M. agalactiae*, six downstream genes in a locus share one promoter, and an area of the 5' region of each gene is subject to inversion by the xerI recombinase, determining which of those genes is acted upon by the promoter and turned on at any one time (179, 180). In the *vsa* cluster of *M. pulmonis*, only one gene out of many in the cluster are transcriptionally active, determined by repetitive domains in each copy, that are inverted by HvsR recombinase and made available to a 5' signal region and the promoter (181, 182). Inverted repeat recombinations also affect the *mpl* cluster of *M. penetrans* and the *vlhA* cluster of *M. synoviae*, where recombinase-mediated inversions in the promoter or coding sequence, respectively, cause phase variation by turning the genes off (120, 183, 184). The *mba* locus of *U. parvum* undergoes inversions that switch genes adjacent to a promoter to turn them on and off, and in the *mpl* clusters of *M. penetrans*, each gene's own promoter can invert to turn the gene on or off (120, 185).

Homologous recombinations occur across greater distances in the genome and require larger homologous sequences for crossing over events. The *mgp* genes of *M. genitalium*, containing three copies of a three-gene operon, with two of them inactivated by truncation, can undergo homologous recombination to turn on different gene clusters by filling in the gaps of the truncation (186–188). The *vsp* loci of *M. bovis* undergo intrachromosomal homologous recombination between different copies to create new chimeric genes (158). *MgPa/MgPar* regions of *M. genitalium* and *M. pneumoniae* also undergo homologous recombination to create new antigen domains (186–188). Some of the masking events that have been reported include the P56 protein of *M. hominis* being masked by the phase variable P120 protein, and the maa1 protein of *M. arthritidis* being masked by a phase- and size-variable maa2 protein (161, 174).

There are also examples of multiswitching strategies. In the *mba* locus of *U. parvum*, a two-switch control combines strand-slippage in a poly-nucleotide repeat in the promoter for phase variation, and regional inversion events that change the number of open reading frames of the coding sequence downstream, leading to size variation (185, 189). A four-switch strategy exists in the *vsp* cluster of *M. bovis*, including: an inversion in a 5' UTR cassette that changes ribosomal binding; an inversion in a second 5' UTR region that prohibits the promoter's activity; tandem repeats in the coding sequence that create size variants, and homologous recombinations between the different loci, creating new chimeric proteins (158, 168, 170, 190). Finally, an epigenetic strategy was reported in the *hsd* locus of *M. pulmonis*, where inversions in the coding sequence of a methyltransferase enzyme in one of the restriction-modification system changes the DNA sequence affinity of the enzyme's binding site, potentially changing the global methylation landscape (167).

The functions of most of the antigens may be adhesion, but the larger consequence of their variation can be much more dynamic, such as growth alterations, quorum sensing, biofilm formation, or heme adsorption. It is also important to remember that the experimental laboratory environment has its own environmental pressures on mollicutes. Within the same population, there can be differences in metabolic conditions for growth, fitness advantages, regional differences in toxicity, population pressures, or nutrition pressures, all of which can influence antigenic variation (172).

Molecular Toolbox for Mollicutes

It is challenging to study mollicutes because many are difficult to cultivate, and sequence assemblies are complicated by the genomes' repeats and duplications. Endogenous restriction-modification systems often obstruct plasmids, phages, and other gene transfers, complicating the genetic manipulation of mycoplasmas. There are also a limited number of naturally occurring plasmids in mollicutes, complicating efforts to design vectors that can successfully and stably transform multiple species. Analysis of mollicute genes as clones in other model organisms (e.g., *E. coli*) is also hampered by the altered usage of the UGA codon for encoding tryptophan (all mollicutes except phytoplasmas and acholeplasmas), whereas otherwise UGA would be interpreted as a universal stop codon (89, 102, 135). Mollicutes have also lost their mismatch repair mechanisms, further complicating attempts at molecular manipulation (90, 191). The use of antibiotic selection markers in mollicutes is also difficult due to the lack of a cell wall, limiting the options of both antibiotics and resistance marker genes. Nevertheless, a variety of genetic manipulation tools and methods have been developed to study mollicutes. For example, the UGA codon usage problem can be resolved by genetic construct synthesis or site-directed mutagenesis of the cloned gene, converting each instance of UGA to UGG (192, 193). The restriction-modification roadblock, too, can potentially be resolved, either by inactivation of the existing enzymes (if there are only a few), or by *in vitro* methylation of the introduced DNA to prevent restriction enzyme degradation (194). New plasmids and transposons are also being made for better targeting and replication of various mollicute species. Next-generation sequencing methods have also improved the understanding of mollicute genomes. This includes the potential for discovering novel extrachromosomal elements, better understanding of evolutionary and phenotypic traits, designing novel genetic tools corresponding to the right parts for the right species, and uncovering the pathogenic mechanisms for better molecular targeting.

An emerging useful tool in genetic manipulation of mollicutes has been yeast (194, 195). Many studies have used yeast (usually *Saccharomyces cerevisiae*) to replicate mollicute genomes, and this mechanism was indispensable for the generation of the multiple iterations of the synthetic mycoplasmas developed at the J. Craig Venter Institute (discussed in the next section). Expansion of synthetic and naturally occurring mycoplasma genomes has also facilitated the application of other molecular tools, such as cloning in *Cre/LoxP* cassettes, and manipulations of genes using the CRISPR-Cas system (196, 197).

The use of transposons to directly manipulate mollicutes has seen great improvement and yielded great results. The Tn916 conjugative transposon is a large transposon with site-specific conjugative properties, capable of being used as a delivery vector. Tn916 can be delivered using electroporation, polyethylene glycol, or transformation with the pAM120 plasmid (198–200). Tn916 also naturally encodes a *tetM* gene for tetracycline resistance. Another widely used transposon is Tn4001, often used to randomly transpose into mycoplasmas, creating mutant libraries for phenotypic analysis of interrupted genes. Modified versions of Tn4001 have improved upon its transformation efficiencies, and engineering in restriction sites to expand its function for gene delivery (201). Minitransposons have also been used as delivery vehicles, with the caveat that their transposase genes are outside of the transposable element, and excision from the primary site occurs only once (202–204).

Despite the lack of a cell wall, there are antibiotics to which mollicutes are sensitive, and for which resistance and selection options exist. As mentioned before, mollicutes are sensitive to tetracycline, and the Tn916 *tetM* gene confers resistance, with the added benefit that, as mollicutes are often intracellular microbes, tetracycline can penetrate eukaryotic membranes for *in vivo* experimentation (196, 205). Erythromycin resistance can be conferred via the *pAMb1* gene, chloramphenicol resistance via the *cat* gene (chloramphenicol acyltransferase from *E-coli* Tn9), puromycin resistance via *pac* (puromycin n-acetyl transferase), and gentamycin/tobramycin/kanamycin resistance via *aacA-aphD* (part of Tn4001). Resistance to fluoroquinolones

can be generated by targeted mutations of the endogenous gyrase and topoisomerase genes, on the sites bound by the antibiotic (56).

Albeit transformation of mollicutes with plasmids has not been very stable or successful, improvements have been made in plasmids to increase efficiency. As new, naturally occurring plasmids are discovered, some have been found to be capable of acting as expression vectors in other species, such as pMyBK1, derived from *M. yeatsii*, capable of expressing in mycoplasmas of the *mycoides* cluster (206). Similarly, pKMK1 obtained from *M. mycoides* has successfully transformed *M. mycoides* subs. *capri*, and *M. capricolum* (207, 208). The efficiency of suicide plasmids in allele exchange via double crossover events have significantly improved by inclusion of the *recA* (recombinase) gene in the plasmid (209). Some of the best improvements in plasmid use in mollicutes have been seen in oriC plasmids. Extrachromosomal replication and stability of oriC plasmids had been a problem, and genomic integration of oriC improved its stability and expression. The oriC plasmid has been further improved by incorporation of DNA replication boxes from different species of mycoplasmas around the actual oriC site in the plasmid, making it extrachromosomally stable and replication competent when transformed into the mycoplasma with the homologous DNA box. OriC plasmids had also previously been used to transform spiroplasmas (with very few crossover events), but it is not known if the stability and efficiency of oriC can be improved by the use of cloning in natural spiroplasma plasmid DNA boxes into the oriC site, as was seen with mycoplasmas (209). However, some spiroplasmas naturally have multiple small extrachromosomal plasmids, each with a different replication region, and naturally occurring spiroplasma plasmids have been successfully used to transform other spiroplasma species that are resistant to artificial plasmids and oriC (109, 113).

Reporter transposons and plasmids have also been developed to determine the functions and activity of regulatory and promoter regions of mollicute genes, often using the *lacZ* gene as the reporter (211–213). While mutagenesis and gene knockout experiments can identify essential genes and phenotypes, the inability to positively regulate gene expression in mollicutes has been a roadblock. However, new plasmids and combination regulatory strategies have helped push these boundaries. One strategy combined the expression of the Cre enzyme, under the control of the tetracycline repressor promoter (*tetR*), regulating Cre expression (and targeting of desired *LoxP* targets in *M. genitalium*) via application of tetracycline (197). A newer inducible mechanism has utilized a CRISPR-Cas-mediated interference strategy for repression of endogenous gene expression in *M. pneumoniae* and *M. mycoides*. Once again, under the control of a tetracycline (*tetON*) promoter, a nuclease-defective Cas9 protein is guided to the gene of interest, tuneably suppressing its transcription (196).

Synthetic Biology

Mollicutes, as the smallest free-living organisms, are inherently good subjects for genome science. In 1995, owing to its size, the genome of *M. genitalium* was fully sequenced by J. Craig Venter's genome research group, uncovering 470 genes in the 580 kbp genome. In pursuit of identifying the minimal cell and the minimal essential genome, targeted deletions of the

470 genes helped identify ~100 genes that could be inactivated while maintaining life, leaving 350–400 genes (in this model) as the minimal essential genome.

To manufacture their own *M. genitalium* minimal essential genome, Venter's group designed a cassette-based artificial chromosome and took a strategy of iterative deletions of the nonessential genes. In 2008, his group chemically synthesized, ligated, and assembled a 583 kbp artificial version of *M. genitalium* G37 in yeast, with the addition of incorporated genetic barcodes for validation. This organism was dubbed JCVI 1.0 (214). Having proven the concept, and due to the slow rate of growth of *M. genitalium*, Venter's group switched to using the faster-growing *M. mycoides* subsp. *capri*, and in 2010, they chemically synthesized its 1.08 Mbp genome starting only with a digital sequence, assembling it as before in yeast, with a genetic barcode, protected it from restriction enzymes by methylation, and transplanted the *M. mycoides* subsp. *capri* genome into *M. capricolum* subsp. *capricolum* cells to prove the ability of this synthetic genome to maintain life (215). This organism was dubbed JCVI Syn1.0. Using transposon-mutagenesis methods, Venter's group subsequently identified essential gene pairs and essential genome regulatory regions, and further minimized the essential genome from JCVI Syn1.0 down to 531 kbp and 473 genes, capable of maintaining free-living and reproductive life as long as all of its nutrients were provided. This organism was dubbed JCVI Syn3.0 (216).

Although it is much easier to edit the genomes of existing organisms rather than synthesizing them *de-novo*, going forward, synthetic biology can serve many purposes, including genome recording and preservation, testing novel synthetic amino acids, creation of synthetic cells for industrial products and services, biodefense, production of medicines, electronics, biofuels, biomaterials, as well as creation of organisms for bioremediation, and biosensing and reporting (217, 218). As synthetic biology advances and new tools are developed, the possibilities are only limited by the imagination.

Acknowledgments

The authors wish to acknowledge salary support from the National Science Foundation.

REFERENCES

1. Brown DR, May M, Bradbury JaM, Johansson K-E. *Mollicutes.* In: Whitman WB, editor. *Bergey's Manual of Systematic of Archaea and Bacteria.* 4. 3rd ed. Hoboken, NJ: John Wiley & Sons, Inc.; 2018.
2. Maniloff Jack. Phylogeny and evolution. In: Razin S, Herrmann R, editors. *Molecular Biology and Pathogenicity of Mycoplasmas.* New York, NY: Kluwer Academic/Plenum Publishers; 2002. pp. 31–43.
3. Weisburg WG, Tully JG, Rose DL, Petzel JP, Oyaizu H, Yang D, et al. A phylogenetic analysis of the mycoplasmas: Basis for their classification. *J Bacteriol.* 1989;171(12):6455–67.
4. Baseman JB, Tully JG. Mycoplasmas: Sophisticated, reemerging, and burdened by their notoriety. *Emerg Infect Dis.* 1997;3(1):21–32.
5. Rosselló-Mora R. Opinion: The species problem, can we achieve a universal concept? *Syst Appl Microbiol.* 2003;26(3):323–6.
6. Rosselló-Mora R. Updating prokaryotic taxonomy. *J Bacteriol.* 2005;187(18):6255–7.
7. Rosselló-Mora R, Amann R. The species concept for prokaryotes. *FEMS Microbiol Rev.* 2001;25(1):39–67.
8. Keswani J, Whitman WB. Relationship of 16S rRNA sequence similarity to DNA hybridization in prokaryotes. *Int J Syst Evol Microbiol.* 2001;51(Pt 2):667–78.
9. Stackebrandt E, Goebel B. International journal for the systematics of bacteriology taxonomic Note: A place for DNA-DNA reassociation and 16S rRNA sequence analysis in the present species definition in bacteriology. *Int J Syst Bacteriol.* 1994(44):846–9.
10. Bradbury JM, Abdul-Wahab OM, Yavari CA, Dupiellet JP, Bové JM. Mycoplasma imitans sp. nov. is related to Mycoplasma gallisepticum and found in birds. *Int J Syst Bacteriol.* 1993;43(4):721–8.
11. Blanchard A, Yáñez A, Dybvig K, Watson HL, Griffiths G, Cassell GH. Evaluation of intraspecies genetic variation within the 16S rRNA gene of Mycoplasma hominis and detection by polymerase chain reaction. *J Clin Microbiol.* 1993;31(5):1358–61.
12. Bebear CM, Bové JM, Bebear C, Renaudin J. Characterization of Mycoplasma hominis mutations involved in resistance to fluoroquinolones. *Antimicrob Agents Chemother.* 1997;41(2):269–73.
13. Stackebrandt E, Frederiksen W, Garrity GM, Grimont PA, Kämpfer P, Maiden MC, et al. Report of the ad hoc committee for the re-evaluation of the species definition in bacteriology. *Int J Syst Evol Microbiol.* 2002;52(Pt 3):1043–7.
14. Brown DR, Bradbury JM. The contentious taxonomy of *Mollicutes.* In: Browning GF, Citti C, editors. *Mollicutes: Molecular Biology and Pathogenesis.* Norfolk, UK: Horizon Bioscience; 2014.
15. Firrao G, Martini M, Ermacora P, Loi N, Torelli E, Foissac X, et al. Genome wide sequence analysis grants unbiased definition of species boundaries in "Candidatus Phytoplasma." *Syst Appl Microbiol.* 2013;36(8):539–48.
16. Firrao G, Brown DR. International Committee on Systematics of Prokaryotes. Subcommittee on the taxonomy of Mollicutes: Minutes of the meetings, July 15th and 19th 2012, Toulouse, France. *Int J Syst Evol Microbiol.* 2013;63(Pt 6):2361–4.
17. Johansson KE, Pettersson B. Taxonomy of *Mollicutes.* In: Razin S, Herrmann R, editors. *Molecular Biology and Pathogenicity of Mycoplasmas.* New York, NY: Kluwer Academic/Plenum Publishers; 2002. pp. 1–29.
18. Balish M, Bertaccini A, Blanchard A, Brown D, Browning G, Chalker V, et al. Recommended rejection of the names Malacoplasma gen. nov., Mesomycoplasma gen. nov., Metamycoplasma gen. nov., Metamycoplasmataceae fam. nov., Mycoplasmoidaceae fam. nov., Mycoplasmoidales ord. nov., Mycoplasmoides gen. nov., Mycoplasmopsis gen. nov. [Gupta, Sawnani, Adeolu, Alnajar and Oren 2018] and all proposed species comb. nov. placed therein. *Int J Syst Evol Microbiol.* 2019; 69(11):3650–53.
19. Gupta RS, Sawnani S, Adeolu M, Alnajar S, Oren A. Correction to: Phylogenetic framework for the phylum Tenericutes based on genome sequence data: Proposal for the creation of a new order Mycoplasmoidales ord. nov.,

containing two new families Mycoplasmoidaceae fam. nov. and Metamycoplasmataceae fam. nov. harbouring Eperythrozoon, Ureaplasma and five novel genera. *Antonie Van Leeuwenhoek*. 2018;111(12):2485–6.

20. Gupta RS, Sawnani S, Adeolu M, Alnajar S, Oren A. Phylogenetic framework for the phylum Tenericutes based on genome sequence data: Proposal for the creation of a new order Mycoplasmoidales ord. nov., containing two new families Mycoplasmoidaceae fam. nov. and Metamycoplasmataceae fam. nov. harbouring Eperythrozoon, Ureaplasma and five novel genera. *Antonie Van Leeuwenhoek*. 2018;111(9):1583–630.

21. International code of nomenclature of prokaryotes. *Int J Syst Evol Microbiol*. 2019;69(1A):S1–S111.

22. Oshima K, Kakizawa S, Nishigawa H, Jung HY, Wei W, Suzuki S, et al. Reductive evolution suggested from the complete genome sequence of a plant-pathogenic phytoplasma. *Nat Genet*. 2004;36(1):27–9.

23. Contaldo N, Satta E, Zambon Y, Paltrinieri S, Bertaccini A. Development and evaluation of different complex media for phytoplasma isolation and growth. *J Microbiol Methods*. 2016;127:105–10.

24. Namba S. Molecular and biological properties of phytoplasmas. *Proc Jpn Acad Ser B Phys Biol Sci*. 2019;95(7):401–18.

25. Eckburg PB, Bik EM, Bernstein CN, Purdom E, Dethlefsen L, Sargent M, et al. Diversity of the human intestinal microbial flora. *Science*. 2005;308(5728):1635–8.

26. Hulcr J, Rountree NR, Diamond SE, Stelinski LL, Fierer N, Dunn RR. Mycangia of ambrosia beetles host communities of bacteria. *Microb Ecol*. 2012;64(3):784–93.

27. King GM, Judd C, Kuske CR, Smith C. Analysis of stomach and gut microbiomes of the eastern oyster (Crassostrea virginica) from coastal Louisiana, USA. *PLOS One*. 2012;7(12):e51475.

28. May M, Balish MF, Blanchard A. The Order *Mycoplasmatales*. In: Rosenberg E, DeLong EF, Lory S, Stackebrandt E, Thompson F, editors. *The Prokaryotes*. 4th ed., Berlin/Heidelberg: Springer; 2014, pp. 515–550.

29. Ramírez AS, Vega-Orellana OM, Viver T, Poveda JB, Rosales RS, Poveda CG, et al. First description of two moderately halophilic and psychrotolerant Mycoplasma species isolated from cephalopods and proposal of Mycoplasma marinum sp. nov. and Mycoplasma todarodis sp. nov. *Syst Appl Microbiol*. 2019;42(4):457–67.

30. Poveda JB. Biochemical characteristics in mycoplasma identification. *Methods Mol Biol*. 1998;104:69–78.

31. Razin S, Yogev D, Naot Y. Molecular biology and pathogenicity of Mycoplasmas. *Microbiol Mol Biol Rev*. 1998;62(4):1094–156.

32. Khan LA, Miles RJ, Nicholas RA. Hydrogen peroxide production by Mycoplasma bovis and Mycoplasma agalactiae and effect of in vitro passage on a Mycoplasma bovis strain producing high levels of H2O2. *Vet Res Commun*. 2005;29(3):181–8.

33. Taylor RR, Varsani H, Miles RJ. Alternatives to arginine as energy sources for the non-fermentative Mycoplasma gallinarum. *FEMS Microbiol Lett*. 1994;115(2–3):163–7.

34. Bischof DF, Janis C, Vilei EM, Bertoni G, Frey J. Cytotoxicity of Mycoplasma mycoides subsp. mycoides small colony type to bovine epithelial cells. *Infect Immun*. 2008;76(1):263–9.

35. Pilo P, Frey J, Vilei EM. Molecular mechanisms of pathogenicity of Mycoplasma mycoides subsp. mycoides SC. *Vet J*. 2007;174(3):513–21.

36. Pilo P, Vilei EM, Peterhans E, Bonvin-Klotz L, Stoffel MH, Dobbelaere D, et al. A metabolic enzyme as a primary virulence factor of Mycoplasma mycoides subsp. mycoides small colony. *J Bacteriol*. 2005;187(19):6824–31.

37. Tully JG. Culture medium formulation for primary isolation and maintenance of *Mollicutes*. In: Razin S, Tully JR, editors. *Molecular and 50 agnostic Procedures in Mycoplasmology*. Vol. 1. San Diego, CA: Academic Press; 1995, pp. 33–9.

38. Miles RJ. Catabolism in mollicutes. *J Gen Microbiol*. 1992;138(9):1773–83.

39. Shepard MC. Culture media for ureaplasmas. In: Razin S, Tully JG, editors. *Methods in Mycoplasmology*. Vol. 1. New York, NY: Academic Press; 1983, pp. 137–46.

40. Whitcomb RF. Culture Media for Spiroplasmas. In: Razin S, Tully JG, editors. *Methods in Mycoplasmology*. Vol. 1. New York, NY: Academic Press; 1983, pp. 147–58.

41. Hackett KJ, Whitcomb RF. Cultivation of spiroplasmas in undefined and defined media. In: Razin S, Tully JG, editors. *Molecular and Diagnostic Procedures in Mycoplasmology*. Vol. 1. San Diego, CA: Academic Press; 1995, pp. 41–54.

42. Hackett KJ, Lynn DE. Cell-assisted growth of a fastidious spiroplasma. *Science*. 1985;230(4727):825–7.

43. Atkinson TP, Balish MF, Waites KB. Epidemiology, clinical manifestations, pathogenesis and laboratory detection of Mycoplasma pneumoniae infections. *FEMS Microbiol Rev*. 2008;32(6):956–73.

44. Feberwee A, Dijkman R, Klinkenberg D, Landman WJM. Quantification of the horizontal transmission of Mycoplasma synoviae in non-vaccinated and MS-H-vaccinated layers. *Avian Pathol*. 2017;46(4):346–58.

45. Tanskanen R. Transmission of Mycoplasma dispar among a succession of newborn calves on a dairy farm. *Acta Vet Scand*. 1987;28(3-4):349–60.

46. Viscardi RM. Ureaplasma species: Role in diseases of prematurity. *Clin Perinatol*. 2010;37(2):393–409.

47. Waites KB, Katz B, Schelonka RL. Mycoplasmas and ureaplasmas as neonatal pathogens. *Clin Microbiol Rev*. 2005;18(4):757–89.

48. Waites KB, Crouse DT, Cassell GH. Therapeutic considerations for Ureaplasma urealyticum infections in neonates. *Clin Infect Dis*. 1993;17 Suppl 1:S208–14.

49. Mariner JC, Catley A. Towards sustainable CBPP control programmes for Africa. *The dynamics of CBPP endemism and development of effective control strategies*. Rome: Food and Agriculture Organization of the United Nations Corporate Document Repository; 2003.

50. Tambi NE, Maina WO, Ndi C. An estimation of the economic impact of contagious bovine pleuropneumonia in Africa. *Rev Sci Tech*. 2006;25(3):999–1011.

51. Thiaucourt F. 2.4.9 Contagious bovine pleuropneumonia. *World Organisation for Animal Health (OIE) Terrestrial Manual: World Organisation for Animal Health (OIE)*; 2008. pp. 713–24.

52. Barrett C, Osterloh S, Little PD, McPeak JG. Constraints limiting marketed off-take rates among pastoralists. Research Brief 04-06-PARIMA. Davis, CA: GL-CRSP (Global Livestock Collaborative Research Support Program), University of California-Davis; 2004.

53. Darveaux JI, Lemanske RF. Infection-related asthma. *J Allergy Clin Immunol Pract.* 2014;2(6):658–63.

54. Watanabe H, Uruma T, Nakamura H, Aoshiba K. The role of Mycoplasma pneumoniae infection in the initial onset and exacerbations of asthma. *Allergy Asthma Proc.* 2014;35(3):204–10.

55. Razin S. Diagnosis of Mycoplasmal infections. In: Razin S, Herrmann R, editors. *Molecular Biology and Pathogenicity of Mycoplasmas.* London, UK: Kluwer; 2002, pp. 531–44.

56. Bebear C, Kempf I. Antimicrobial therapy and antimicrobial resistance. Blanchard A, Browning G, editors. *Norfolk*, NR United Kingdom: Horizon Biosciences; 2005.

57. Pereyre S, Gonzalez P, De Barbeyrac B, Darnige A, Renaudin H, Charron A, et al. Mutations in 23S rRNA account for intrinsic resistance to macrolides in Mycoplasma hominis and Mycoplasma fermentans and for acquired resistance to macrolides in M. hominis. *Antimicrob Agents Chemother.* 2002;46(10):3142–50.

58. Bebear CM, Renaudin H, Bryskier A, Bebear C. Comparative activities of telithromycin (HMR 3647), levofloxacin, and other antimicrobial agents against human mycoplasmas. *Antimicrob Agents Chemother.* 2000;44(7):1980–2.

59. Duffy LB, Crabb DM, Bing X, Waites KB. Bactericidal activity of levofloxacin against Mycoplasma pneumoniae. *J Antimicrob Chemother.* 2003;52(3):527–8.

60. Ho PL, Law PY, Chan BW, Wong CW, To KK, Chiu SS, et al. Emergence of macrolide-resistant Mycoplasma pneumoniae in Hong Kong is linked to increasing macrolide resistance in multilocus variable-number tandem-repeat analysis type 4-5-7-2. *J Clin Microbiol.* 2015;53(11):3560–4.

61. Zhao S, Musa SS, Qin J, He D. Phase-shifting of the transmissibility of macrolide-sensitive and resistant Mycoplasma pneumoniae epidemics in Hong Kong, from 2015 to 2018. *Int J Infect Dis.* 2019;81:251–3.

62. Manhart LE. Mycoplasma genitalium: An emergent sexually transmitted disease? *Infect Dis Clin North Am.* 2013;27(4):779–92.

63. Teuber M. Spread of antibiotic resistance with food-borne pathogens. *Cell Mol Life Sci.* 1999;56(9-10):755–63.

64. Calcutt MJ, Lysnyansky I, Sachse K, Fox LK, Nicholas RAJ, Ayling RD. Gap analysis of Mycoplasma bovis disease, diagnosis and control: An aid to identify future development requirements. *Transbound Emerg Dis.* 2018;65 (Suppl 1):91–109.

65. Kleven SH. Control of avian mycoplasma infections in commercial poultry. *Avian Dis.* 2008;52(3):367–74.

66. Maes D, Segales J, Meyns T, Sibila M, Pieters M, Haesebrouck F. Control of Mycoplasma hyopneumoniae infections in pigs. *Vet Microbiol.* 2008;126(4):297–309.

67. Aldaghi M, Massart S, Druart P, Bertaccini A, Jijakli MII, Lepoivre P. Preliminary evaluation of antimicrobial activity of some chemicals on in vitro apple shoots infected by 'Candidatus Phytoplasma mali'. *Commun Agric Appl Biol Sci.* 2008;73(2):335–41.

68. Upadhyay R. Varietal Susceptibility and Effect of Antibiotics on Little Leaf Phytoplasma of Brinjal (Solanummelongena L). *Int J Emerg Trends Sci Technol.* 2016;3:3911–4.

69. Kumari S, Nagendran K, Rai AB, Singh B, Rao GP, Bertaccini A. Global Status of Phytoplasma Diseases in Vegetable Crops. *Front Microbiol.* 2019;10:1349.

70. Sánchez-Rojo S, López-Delgado HA, Mora-Herrera ME, Almeyda-Leon HI, Zavaleta-Mancera HA, Espinosa-Victoria D. Salicylic acid protects potato plants-from Phytoplasma-associated stress and improves tuber photosynthate assimilation. *American Journal of Potato Research.* 2011;88(2):175–83.

71. Couldwell DL, Jalocon D, Power M, Jeoffreys NJ, Chen SC, Lewis DA. High prevalence of resistance to macrolides and frequent anorectal infection in men who have sex with men in western Sydney. *Sex Transm Infect.* 2018;94(6):406–10.

72. Gille L, Callens J, Supré K, Boyen F, Haesebrouck F, Van Driessche L, et al. Use of a breeding bull and absence of a calving pen as risk factors for the presence of Mycoplasma bovis in dairy herds. *J Dairy Sci.* 2018;101(9): 8284–90.

73. Maunsell FP, Woolums AR, Francoz D, Rosenbusch RF, Step DL, Wilson DJ, et al. Mycoplasma bovis infections in cattle. *J Vet Intern Med.* 2011;25(4):772–83.

74. Browning GF, Whithear KG, Geary SJ. Vaccines to control Mycoplasmosis. In: Blanchard A, Browning GF, editors. *Mycoplasmas: Molecular Biology, Pathogenicity, and Strategies for Control.* Norfolk, NR, United Kingdom: Horizon Bioscience; 2005, pp. 569–99.

75. King GJ, Kagumba M, Kariuki DP. Trial of the efficacy and immunological response to an inactivated mycoplasma F38 vaccine. *Vet Rec.* 1992;131(20):461–4.

76. Rurangirwa FR, McGuire TC, Kibor A, Chema S. An inactivated vaccine for contagious caprine pleuropneumonia. *Vet Rec.* 1987;121(17):397–400.

77. Gourgues G, Barré A, Beaudoing E, Weber J, Magdelenat G, Barbe V, et al. Complete Genome Sequence of Mycoplasma mycoides subsp. mycoides T1/44, a Vaccine Strain against Contagious Bovine Pleuropneumonia. *Genome Announc.* 2016;4(2).

78. Nkando I, Ndinda J, Kuria J, Naessens J, Mbithi F, Schnier C, et al. Efficacy of two vaccine formulations against contagious bovine pleuropneumonia (CBPP) in Kenyan indigenous cattle. *Res Vet Sci.* 2012;93(2):568–73.

79. Hubschle O, Lelli R, Frey J, Nicholas R. Contagious bovine pleuropneumonia and vaccine strain T1/44. *Vet Rec.* 2002;150(19):615.

80. Mbulu RS, Tjipura-Zaire G, Lelli R, Frey J, Pilo P, Vilei EM, et al. Contagious bovine pleuropneumonia (CBPP) caused by vaccine strain T1/44 of Mycoplasma mycoides subsp. mycoides SC. *Vet Microbiol.* 2004;98(3–4):229–34.

81. Foggie A, Etheridge JR, Erdag O, Arisoy F. Contagious agalactia of sheep and goats studies on live and dead vaccines in lactating sheep. *J Comp Pathol.* 1971;81(1):165–72.

82. Fisch A, Beutinger Marchioro S, Klazer Gomes C, Galli V, Rodrigues de Oliveira N, Simionatto S, et al. Commercial Bacterins did not Induce Detectable Levels of Antibodies in Mice against Mycoplasma hyopneumoniae Antigens Strongly Recognized by Swine Immune System. *Trials in Vaccinology.* 2016;5(32–37).

83. Martelli P, Saleri R, Cavalli V, De Angelis E, Ferrari L, Benetti M, et al. Systemic and local immune response in pigs intradermally and intramuscularly injected with inactivated Mycoplasma hyopneumoniae vaccines. *Vet Microbiol.* 2014;168(2–4):357–64.

84. Xiong Q, Wei Y, Xie H, Feng Z, Gan Y, Wang C, et al. Effect of different adjuvant formulations on the immunogenicity and protective effect of a live Mycoplasma hyopneumoniae vaccine after intramuscular inoculation. *Vaccine.* 2014;32(27):3445–51.

85. Ferguson-Noel N, Cookson K, Laibinis VA, Kleven SH. The efficacy of three commercial Mycoplasma gallisepticum vaccines in laying hens. *Avian Dis.* 2012;56(2):272–5.

86. Karaca K, Lam KM. Efficacy of commercial Mycoplasma gallisepticum bacterin (MG-Bac) in preventing air-sac lesions in chickens. *Avian Dis.* 1987;31(1):202–3.

87. Jones JF, Whithear KG, Scott PC, Noormohammadi AH. Determination of the effective dose of the live Mycoplasma synoviae vaccine, Vaxsafe MS (strain MS-H) by protection against experimental challenge. *Avian Dis.* 2006;50(1):88–91.

88. Feberwee A, Landman WJ, von Banniseht-Wysmuller T, Klinkenberg D, Vernooij JC, Gielkens AL, et al. The effect of a live vaccine on the horizontal transmission of Mycoplasma gallisepticum. *Avian Pathol.* 2006;35(5):359–66.

89. Marenda M. Genome Mosaics. In: Browning GF, Citti C, editors. *Mollicutes: Molecular Biology and Pathogenesis.* Norfolk, UK: Caistor Academic Press; 2014.

90. Cordova CM, Hoeltgebaum DL, Machado LD, Santos LD. Molecular biology of mycoplasmas: From the minimum cell concept to the artificial cell. *An Acad Bras Cienc.* 2016;88 Suppl 1:599–607.

91. Citti C, Dordet-Frisoni E, Nouvel LX, Kuo CH, Baranowski E. Horizontal gene transfers in Mycoplasmas (Mollicutes). *Curr Issues Mol Biol.* 2018;29:3–22.

92. Ku C, Lo WS, Chen LL, Kuo CH. Complete genomes of two dipteran-associated spiroplasmas provided insights into the origin, dynamics, and impacts of viral invasion in spiroplasma. *Genome Biol Evol.* 2013;5(6):1151–64.

93. Sugio A, Hogenhout SA. The genome biology of phytoplasma: Modulators of plants and insects. *Curr Opin Microbiol.* 2012;15(3):247–54.

94. Wei W, Davis RE, Jomantiene R, Zhao Y. Ancient, recurrent phage attacks and recombination shaped dynamic sequence-variable mosaics at the root of phytoplasma genome evolution. *Proc Natl Acad Sci U S A.* 2008;105(33):11827–32.

95. Daccord A, Ceccarelli D, Burrus V. Integrating conjugative elements of the SXT/R391 family trigger the excision and drive the mobilization of a new class of Vibrio genomic islands. *Mol Microbiol.* 2010;78(3):576–88.

96. Calcutt MJ, Lewis MS, Wise KS. Molecular genetic analysis of ICEF, an integrative conjugal element that is present as a repetitive sequence in the chromosome of Mycoplasma fermentans PG18. *J Bacteriol.* 2002;184(24):6929–41.

97. Nouvel LX, Sirand-Pugnet P, Marenda MS, Sagné E, Barbe V, Mangenot S, et al. Comparative genomic and proteomic analyses of two Mycoplasma agalactiae strains: Clues to the macro- and micro-events that are shaping mycoplasma diversity. *BMC Genomics.* 2010;11:86.

98. Bergemann AD, Finch LR. Isolation and restriction endonuclease analysis of a mycoplasma plasmid. *Plasmid.* 1988;19(1):68–70.

99. Berho N, Duret S, Danet JL, Renaudin J. Plasmid pSci6 from Spiroplasma citri GII-3 confers insect transmissibility to the non-transmissible strain S. citri 44. *Microbiology.* 2006; 152(Pt 9):2703–16.

100. Davis RE, Dally EL, Jomantiene R, Zhao Y, Roe B, Lin S, et al. Cryptic plasmid pSKU146 from the wall-less plant pathogen Spiroplasma kunkelii encodes an adhesin and components of a type IV translocation-related conjugation system. *Plasmid.* 2005;53(2):179–90.

101. Djordjevic SR, Forbes WA, Forbes-Faulkner J, Kuhnert P, Hum S, Hornitzky MA, et al. Genetic diversity among Mycoplasma species bovine group 7: Clonal isolates from an outbreak of polyarthritis, mastitis, and abortion in dairy cattle. *Electrophoresis.* 2001;22(16):3551–61.

102. Dybvig K, Voelker LL. Molecular biology of mycoplasmas. *Annu Rev Microbiol.* 1996;50:25–57.

103. Joshi BD, Berg M, Rogers J, Fletcher J, Melcher U. Sequence comparisons of plasmids pBJS-O of Spiroplasma citri and pSKU146 of S. kunkelii: Implications for plasmid evolution. *BMC Genomics.* 2005;6:175.

104. Liefting LW, Andersen MT, Lough TJ, Beever RE. Comparative analysis of the plasmids from two isolates of "Candidatus Phytoplasma australiense." *Plasmid.* 2006;56(2):138–44.

105. Nishigawa H, Miyata S, Oshima K, Sawayanagi T, Komoto A, Kuboyama T, et al. In planta expression of a protein encoded by the extrachromosomal DNA of a phytoplasma and related to geminivirus replication proteins. *Microbiology.* 2001; 147(Pt 2):507–13.

106. Nishigawa H, Oshima K, Kakizawa S, Jung HY, Kuboyama T, Miyata S, et al. A plasmid from a non-insect-transmissible line of a phytoplasma lacks two open reading frames that exist in the plasmid from the wild-type line. *Gene.* 2002;298(2):195–201.

107. Oshima K, Kakizawa S, Nishigawa H, Kuboyama T, Miyata S, Ugaki M, et al. A plasmid of phytoplasma encodes a unique replication protein having both plasmid- and virus-like domains: Clue to viral ancestry or result of virus/plasmid recombination? *Virology.* 2001;285(2):270–7.

108. Petrzik K, Krawczyk K, Zwolinska A. Two high-copy plasmids found in plants associated with strains of "Candidatus Phytoplasma asteris." *Plasmid.* 2011;66(2):122–7.

109. Saillard C, Carle P, Duret-Nurbel S, Henri R, Killiny N, Carrère S, et al. The abundant DNA content of the Spiroplasma citri GII3-3X genome. *BMC Genomics.* 2008;9:195.

110. Thiaucourt F, Manso-Silvan L, Salah W, Barbe V, Vacherie B, Jacob D, et al. Mycoplasma mycoides, from "mycoides Small Colony" to "capri." A microevolutionary perspective. *BMC Genomics.* 2011;12:114.

111. Tran-Nguyen LT, Gibb KS. Extrachromosomal DNA isolated from tomato big bud and Candidatus Phytoplasma australiense phytoplasma strains. *Plasmid.* 2006;56(3):153–66.

112. Berho N, Duret S, Renaudin J. Absence of plasmids encoding adhesion-related proteins in non-insect-transmissible strains of Spiroplasma citri. *Microbiology.* 2006;152(Pt 3):873–86.

113. Marenda M, Barbe V, Gourgues G, Mangenot S, Sagne E, Citti C. A new integrative conjugative element occurs in Mycoplasma agalactiae as chromosomal and free circular forms. *J Bacteriol.* 2006;188(11):4137–41.

114. Bai X, Fazzolari T, Hogenhout SA. Identification and characterization of traE genes of Spiroplasma kunkelii. *Gene.* 2004;336(1):81–91.

115. Breton M, Duret S, Arricau-Bouvery N, Béven L, Renaudin J. Characterizing the replication and stability regions of Spiroplasma citri plasmids identifies a novel replication protein and expands the genetic toolbox for plant-pathogenic spiroplasmas. *Microbiology.* 2008;154(Pt 10):3232–44.

116. Nishigawa H, Oshima K, Kakizawa S, Jung HY, Kuboyama T, Miyata S, et al. Evidence of intermolecular recombination between extrachromosomal DNAs in phytoplasma: A trigger

for the biological diversity of phytoplasma? *Microbiology.* 2002;148(Pt 5):1389–96.

117. Maniloff J, Kampo GJ, Dascher CC. Sequence analysis of a unique temperature phage: Mycoplasma virus L2. *Gene.* 1994;141(1):1–8.

118. Clapper B, Tu AH, Elgavish A, Dybvig K. The vir gene of bacteriophage MAV1 confers resistance to phage infection on Mycoplasma arthritidis. *J Bacteriol.* 2004;186(17):5715–20.

119. Jomantiene R, Zhao Y, Davis RE. Sequence-variable mosaics: Composites of recurrent transposition characterizing the genomes of phylogenetically diverse phytoplasmas. *DNA Cell Biol.* 2007;26(8):557–64.

120. Röske K, Blanchard A, Chambaud I, Citti C, Helbig JH, Prevost MC, et al. Phase variation among major surface antigens of Mycoplasma penetrans. *Infect Immun.* 2001;69(12):7642–51.

121. Röske K, Calcutt MJ, Wise KS. The Mycoplasma fermentans prophage phiMFV1: Genome organization, mobility and variable expression of an encoded surface protein. *Mol Microbiol.* 2004;52(6):1703–20.

122. Melcher U, Sha Y, Ye F, Fletcher J. Mechanisms of spiroplasma genome variation associated with SpV1-like viral DNA inferred from sequence comparisons. *Microb Comp Genomics.* 1999;4(1):29–46.

123. Hu WS, Hayes MM, Wang RY, Shih JW, Lo SC. High-frequency DNA rearrangements in the chromosomes of clinically isolated Mycoplasma fermentans. *Curr Microbiol.* 1998;37(1):1–5.

124. March JB, Clark J, Brodlie M. Characterization of strains of Mycoplasma mycoides subsp. mycoides small colony type isolated from recent outbreaks of contagious bovine pleuropneumonia in Botswana and Tanzania: Evidence for a new biotype. *J Clin Microbiol.* 2000;38(4):1419–25.

125. Pitcher D, Hilbocus J. Variability in the distribution and composition of insertion sequence-like elements in strains of Mycoplasma fermentans. *FEMS Microbiol Lett.* 1998;160(1):101–9.

126. Szczepanek SM, Tulman ER, Gorton TS, Liao X, Lu Z, Zinski J, et al. Comparative genomic analyses of attenuated strains of Mycoplasma gallisepticum. *Infect Immun.* 2010;78(4):1760–71.

127. Omelchenko MV, Makarova KS, Wolf YI, Rogozin IB, Koonin EV. Evolution of mosaic operons by horizontal gene transfer and gene displacement in situ. *Genome Biol.* 2003;4(9):R55.

128. Lo WS, Gasparich GE, Kuo CH. Found and Lost: The fates of horizontally acquired genes in arthropod-symbiotic spiroplasma. *Genome Biol Evol.* 2015;7(9):2458–72.

129. Lo WS, Huang YY, Kuo CH. Winding paths to simplicity: Genome evolution in facultative insect symbionts. *FEMS Microbiol Rev.* 2016;40(6):855–74.

130. Ku C, Lo WS, Kuo CH. Horizontal transfer of potential mobile units in phytoplasmas. *Mob Genet Elements.* 2013;3(5):e26145.

131. Lo WS, Ku C, Chen LL, Chang TH, Kuo CH. Comparison of metabolic capacities and inference of gene content evolution in mosquito-associated Spiroplasma diminutum and S. taiwanense. *Genome Biol Evol.* 2013;5(8):1512–23.

132. Paredes JC, Herren JK, Schüpfer F, Marin R, Claverol S, Kuo CH, et al. Genome sequence of the Drosophila melanogaster male-killing Spiroplasma strain MSRO endosymbiont. *MBio.* 2015;6(2).

133. Xiao L, Ptacek T, Osborne JD, Crabb DM, Simmons WL, Lefkowitz EJ, et al. Comparative genome analysis of Mycoplasma pneumoniae. *BMC Genomics.* 2015;16:610.

134. Pereyre S, Sirand-Pugnet P, Beven L, Charron A, Renaudin H, Barré A, et al. Life on arginine for Mycoplasma hominis: Clues from its minimal genome and comparison with other human urogenital mycoplasmas. *PLOS Genet.* 2009;5(10):e1000677.

135. Sirand-Pugnet P, Lartigue C, Marenda M, Jacob D, Barré A, Barbe V, et al. Being pathogenic, plastic, and sexual while living with a nearly minimal bacterial genome. *PLOS Genet.* 2007;3(5):e75.

136. Dordet Frisoni E, Marenda MS, Sagné E, Nouvel LX, Guérillot R, Glaser P, et al. ICEA of Mycoplasma agalactiae: A new family of self-transmissible integrative elements that confers conjugative properties to the recipient strain. *Mol Microbiol.* 2013;89(6):1226–39.

137. Pinto PM, Oliveira de Carvalho M, Alves-Junior L, Brocchi M, Schrank IS. Molecular Analysis of an integrative conjugative element, ICEH, present in the chromosome of different strains of *Mycoplasma hyopneumoniae. Genet Mol Biol.* 2007;30(1).

138. Vasconcelos AT, Ferreira HB, Bizarro CV, Bonatto SL, Carvalho MO, Pinto PM, et al. Swine and poultry pathogens: the complete genome sequences of two strains of Mycoplasma hyopneumoniae and a strain of Mycoplasma synoviae. *J Bacteriol.* 2005;187(16):5568–77.

139. Dordet-Frisoni E, Sagné E, Baranowski E, Breton M, Nouvel LX, Blanchard A, et al. Chromosomal transfers in mycoplasmas: When minimal genomes go mobile. *mBio.* 2014;5(6):e01958.

140. Brochet M, Da Cunha V, Couvé E, Rusniok C, Trieu-Cuot P, Glaser P. Atypical association of DDE transposition with conjugation specifies a new family of mobile elements. *Mol Microbiol.* 2009;71(4):948–59.

141. Calderon-Copete SP, Wigger G, Wunderlin C, Schmidheini T, Frey J, Quail MA, et al. The Mycoplasma conjunctivae genome sequencing, annotation and analysis. *BMC Bioinformatics.* 2009;10 Suppl 6:S7.

142. Liu W, Feng Z, Fang L, Zhou Z, Li Q, Li S, et al. Complete genome sequence of Mycoplasma hyopneumoniae strain 168. *J Bacteriol.* 2011;193(4):1016–7.

143. Manso-Silván L, Tardy F, Baranowski E, Barré A, Blanchard A, Breton M, et al. Draft Genome Sequences of Mycoplasma alkalescens, Mycoplasma arginini, and Mycoplasma bovigenitalium, Three species with equivocal pathogenic status for cattle. *Genome Announc.* 2013;1(3):e00348–13.

144. Minion FC, Lefkowitz EJ, Madsen ML, Cleary BJ, Swartzell SM, Mahairas GG. The genome sequence of Mycoplasma hyopneumoniae strain 232, the agent of swine mycoplasmosis. *J Bacteriol.* 2004;186(21):7123–33.

145. Wise KS, Calcutt MJ, Foecking MF, Röske K, Madupu R, Methé BA. Complete genome sequence of Mycoplasma bovis type strain PG45 (ATCC 25523). *Infect Immun.* 2011;79(2):982–3.

146. Lee IM, Zhao Y, Bottner KD. Novel insertion sequence-like elements in phytoplasma strains of the aster yellows group are putative new members of the IS3 family. *FEMS Microbiol Lett.* 2005;242(2):353–60.

147. Mahillon J, Chandler M. Insertion sequences. *Microbiol Mol Biol Rev.* 1998;62(3):725–74.

148. Thomas A, Linden A, Mainil J, Bischof DF, Frey J, Vilei EM. Mycoplasma bovis shares insertion sequences with Mycoplasma agalactiae and Mycoplasma mycoides subsp. mycoides SC: Evolutionary and developmental aspects. *FEMS Microbiol Lett.* 2005;245(2):249–55.

149. Pilo P, Fleury B, Marenda M, Frey J, Vilei EM. Prevalence and distribution of the insertion element ISMag1 in Mycoplasma agalactiae. *Vet Microbiol.* 2003;92(1-2):37–48.

150. Bischof DF, Vilei EM, Frey J. Genomic differences between type strain PG1 and field strains of Mycoplasma mycoides subsp. mycoides small-colony type. *Genomics.* 2006;88(5):633–41.

151. Miles K, Churchward CP, McAuliffe L, Ayling RD, Nicholas RA. Identification and differentiation of European and African/Australian strains of Mycoplasma mycoides subspecies mycoides small-colony type using polymerase chain reaction analysis. *J Vet Diagn Invest.* 2006;18(2):168–71.

152. Miles K, McAuliffe L, Persson A, Ayling RD, Nicholas RA. Insertion sequence profiling of UK Mycoplasma bovis field isolates. *Vet Microbiol.* 2005;107(3-4):301–6.

153. Dybvig K, Cao Z, French CT, Yu H. Evidence for type III restriction and modification systems in Mycoplasma pulmonis. *J Bacteriol.* 2007;189(6):2197–202.

154. Li Y, Zheng H, Liu Y, Jiang Y, Xin J, Chen W, et al. The complete genome sequence of Mycoplasma bovis strain Hubei-1. *PLoS One.* 2011;6(6):e20999.

155. Papazisi L, Gorton TS, Kutish G, Markham PF, Browning GF, Nguyen DK, et al. The complete genome sequence of the avian pathogen Mycoplasma gallisepticum strain R(low). *Microbiology.* 2003;149(Pt 9):2307–16.

156. Sasaki Y, Ishikawa J, Yamashita A, Oshima K, Kenri T, Furuya K, et al. The complete genomic sequence of Mycoplasma penetrans, an intracellular bacterial pathogen in humans. *Nucleic Acids Res.* 2002;30(23):5293–300.

157. Westberg J, Persson A, Holmberg A, Goesmann A, Lundeberg J, Johansson KE, et al. The genome sequence of Mycoplasma mycoides subsp. mycoides SC type strain PG1T, the causative agent of contagious bovine pleuropneumonia (CBPP). *Genome Res.* 2004;14(2):221–7.

158. Lysnyansky I, Ron Y, Sachse K, Yogev D. Intrachromosomal recombination within the vsp locus of Mycoplasma bovis generates a chimeric variable surface lipoprotein antigen. *Infect Immun.* 2001;69(6):3703–12.

159. Neyrolles O, Chambaud I, Ferris S, Prevost MC, Sasaki T, Montagnier L, et al. Phase variations of the Mycoplasma penetrans main surface lipoprotein increase antigenic diversity. *Infect Immun.* 1999;67(4):1569–78.

160. Citti C, Kim MF, Wise KS. Elongated versions of Vlp surface lipoproteins protect Mycoplasma hyorhinis escape variants from growth-inhibiting host antibodies. *Infect Immun.* 1997;65(5):1773–85.

161. Zhang Q, Wise KS. Coupled phase-variable expression and epitope masking of selective surface lipoproteins increase surface phenotypic diversity in Mycoplasma hominis. *Infect Immun.* 2001;69(8):5177–81.

162. Simmons WL, Bolland JR, Daubenspeck JM, Dybvig K. A stochastic mechanism for biofilm formation by Mycoplasma pulmonis. *J Bacteriol.* 2007;189(5):1905–13.

163. Simmons WL, Denison AM, Dybvig K. Resistance of Mycoplasma pulmonis to complement lysis is dependent on the number of Vsa tandem repeats: Shield hypothesis. *Infect Immun.* 2004;72(12):6846–51.

164. Simmons WL, Dybvig K. The Vsa proteins modulate susceptibility of Mycoplasma pulmonis to complement killing, hemadsorption, and adherence to polystyrene. *Infect Immun.* 2003;71(10):5733–8.

165. Simmons WL, Dybvig K. Biofilms protect Mycoplasma pulmonis cells from lytic effects of complement and gramicidin. *Infect Immun.* 2007;75(8):3696–9.

166. Khiari AB, Guériri I, Mohammed RB, Mardassi BB. Characterization of a variant vlhA gene of Mycoplasma synoviae, strain WVU 1853, with a highly divergent haemagglutinin region. *BMC Microbiol.* 2010;10:6.

167. Dybvig K, Sitaraman R, French CT. A family of phase-variable restriction enzymes with differing specificities generated by high-frequency gene rearrangements. *Proc Natl Acad Sci U S A.* 1998;95(23):13923–8.

168. Behrens A, Heller M, Kirchhoff H, Yogev D, Rosengarten R. A family of phase- and size-variant membrane surface lipoprotein antigens (Vsps) of Mycoplasma bovis. *Infect Immun.* 1994;62(11):5075–84.

169. Boguslavsky S, Menaker D, Lysnyansky I, Liu T, Levisohn S, Rosengarten R, et al. Molecular characterization of the Mycoplasma gallisepticum pvpA gene which encodes a putative variable cytadhesin protein. *Infect Immun.* 2000;68(7):3956–64.

170. Lysnyansky I, Ron Y, Yogev D. Juxtaposition of an active promoter to vsp genes via site-specific DNA inversions generates antigenic variation in Mycoplasma bovis. *J Bacteriol.* 2001;183(19):5698–708.

171. Glew MD, Browning GF, Markham PF, Walker ID. pMGA phenotypic variation in Mycoplasma gallisepticum occurs in vivo and is mediated by trinucleotide repeat length variation. *Infect Immun.* 2000;68(10):6027–33.

172. Zimmerman CU. Current insights into phase and antigenic variation in Mycoplasmas. In: Browning GF, Citti C, editors. *Mollicutes: Molecular Biology and Pathogenesis.* Norfolk, UK: Caister Academic Press; 2014.

173. Markham PF, Glew MD, Sykes JE, Bowden TR, Pollocks TD, Browning GF, et al. The organisation of the multigene family which encodes the major cell surface protein, pMGA, of Mycoplasma gallisepticum. *FEBS Lett.* 1994;352(3):347–52.

174. Washburn LR, Weaver KE, Weaver EJ, Donelan W, Al-Sheboul S. Molecular characterization of Mycoplasma arthritidis variable surface protein MAA2. *Infect Immun.* 1998;66(6):2576–86.

175. Wise KS, Foecking MF, Röske K, Lee YJ, Lee YM, Madan A, et al. Distinctive repertoire of contingency genes conferring mutation-based phase variation and combinatorial expression of surface lipoproteins in Mycoplasma capricolum subsp. capricolum of the Mycoplasma mycoides phylogenetic cluster. *J Bacteriol.* 2006;188(13):4926–41.

176. Yogev D, Rosengarten R, Watson-McKown R, Wise KS. Molecular basis of Mycoplasma surface antigenic variation: A novel set of divergent genes undergo spontaneous mutation of periodic coding regions and 5' regulatory sequences. *EMBO J.* 1991;10(13):4069–79.

177. Winner F, Markovà I, Much P, Lugmair A, Siebert-Gulle K, Vogl G, et al. Phenotypic switching in Mycoplasma gallisepticum hemadsorption is governed by a high-frequency, reversible point mutation. *Infect Immun.* 2003;71(3):1265–73.

178. Zhang Q, Wise KS. Localized reversible frameshift mutation in an adhesin gene confers a phase-variable adherence phenotype in mycoplasma. *Mol Microbiol.* 1997;25(5):859–69.

179. Chopra-Dewasthaly R, Citti C, Glew MD, Zimmermann M, Rosengarten R, Jechlinger W. Phase-locked mutants of Mycoplasma agalactiae: Defining the molecular switch of

high-frequency Vpma antigenic variation. *Mol Microbiol.* 2008;67(6):1196–210.

180. Glew MD, Marenda M, Rosengarten R, Citti C. Surface diversity in Mycoplasma agalactiae is driven by site-specific DNA inversions within the vpma multigene locus. *J Bacteriol.* 2002;184(21):5987–98.

181. Bhugra B, Voelker LL, Zou N, Yu H, Dybvig K. Mechanism of antigenic variation in Mycoplasma pulmonis: Interwoven, site-specific DNA inversions. *Mol Microbiol.* 1995;18(4):703–14.

182. Shen X, Gumulak J, Yu H, French CT, Zou N, Dybvig K. Gene rearrangements in the vsa locus of Mycoplasma pulmonis. *J Bacteriol.* 2000;182(10):2900–8.

183. Horino A, Sasaki Y, Sasaki T, Kenri T. Multiple promoter inversions generate surface antigenic variation in Mycoplasma penetrans. *J Bacteriol.* 2003;185(1):231–42.

184. Noormohammadi AH, Markham PF, Kanci A, Whithear KG, Browning GF. A novel mechanism for control of antigenic variation in the haemagglutinin gene family of mycoplasma synoviae. *Mol Microbiol.* 2000;35(4):911–23.

185. Zimmerman CU, Stiedl T, Rosengarten R, Spergser J. Alternate phase variation in expression of two major surface membrane proteins (MBA and UU376) of Ureaplasma parvum serovar 3. *FEMS Microbiol Lett.* 2009;292(2):187–93.

186. Iverson-Cabral SL, Astete SG, Cohen CR, Rocha EP, Totten PA. Intrastrain heterogeneity of the mgpB gene in Mycoplasma genitalium is extensive in vitro and in vivo and suggests that variation is generated via recombination with repetitive chromosomal sequences. *Infect Immun.* 2006;74(7): 3715–26.

187. Iverson-Cabral SL, Astete SG, Cohen CR, Totten PA. mgpB and mgpC sequence diversity in Mycoplasma genitalium is generated by segmental reciprocal recombination with repetitive chromosomal sequences. *Mol Microbiol.* 2007;66(1):55–73.

188. Ma L, Jensen JS, Myers L, Burnett J, Welch M, Jia Q, et al. Mycoplasma genitalium: An efficient strategy to generate genetic variation from a minimal genome. *Mol Microbiol.* 2007;66(1):220–36.

189. Zimmerman CU, Rosengarten R, Spergser J. Ureaplasma antigenic variation beyond MBA phase variation: DNA inversions generating chimeric structures and switching in expression of the MBA N-terminal paralogue UU172. *Mol Microbiol.* 2011;79(3):663–76.

190. Lysnyansky I, Sachse K, Rosenbusch R, Levisohn S, Yogev D. The vsp locus of Mycoplasma bovis: Gene organization and structural features. *J Bacteriol.* 1999;181(18):5734–41.

191. Kamashev D, Oberto J, Serebryakova M, Gorbachev A, Zhukova Y, Levitskii S, et al. Mycoplasma gallisepticum produces a histone-like protein that recognizes base mismatches in DNA. *Biochemistry.* 2011;50(40):8692–702.

192. Knudtson KL, Manohar M, Joyner DE, Ahmed EA, Cole BC. Expression of the superantigen Mycoplasma arthritidis mitogen in Escherichia coli and characterization of the recombinant protein. *Infect Immun.* 1997;65(12):4965–71.

193. Simionatto S, Marchioro SB, Galli V, Luerce TD, Hartwig DD, Moreira AN, et al. Efficient site-directed mutagenesis using an overlap extension-PCR method for expressing Mycoplasma hyopneumoniae genes in Escherichia coli. *J Microbiol Methods.* 2009;79(1):101–5.

194. Lartigue C, Vashee S, Algire MA, Chuang RY, Benders GA, Ma L, et al. Creating bacterial strains from genomes that have been cloned and engineered in yeast. *Science.* 2009;325(5948):1693–6.

195. Benders GA, Noskov VN, Denisova EA, Lartigue C, Gibson DG, Assad-Garcia N, et al. Cloning whole bacterial genomes in yeast. *Nucleic Acids Res.* 2010;38(8):2558–69.

196. Mariscal AM, Kakizawa S, Hsu JY, Tanaka K, González-González L, Broto A, et al. Tuning gene activity by inducible and targeted regulation of gene expression in minimal bacterial cells. *ACS Synth Biol.* 2018;7(6):1538–52.

197. Noskov VN, Ma L, Chen S, Chuang RY. Recombinase-mediated cassette exchange (RMCE) system for functional genomics studies in Mycoplasma mycoides. *Biol Proced Online.* 2015;17:6.

198. Halbedel S, Stülke J. Tools for the genetic analysis of Mycoplasma. *Int J Med Microbiol.* 2007;297(1):37–44.

199. Roberts AP, Hennequin C, Elmore M, Collignon A, Karjalainen T, Minton N, et al. Development of an integrative vector for the expression of antisense RNA in Clostridium difficile. *J Microbiol Methods.* 2003;55(3):617–24.

200. Roberts AP, Mullany P. A modular master on the move: The Tn916 family of mobile genetic elements. *Trends Microbiol.* 2009;17(6):251–8.

201. Montero-Blay A, Miravet-Verde S, Lluch-Senar M, Piñero-Lambea C, Serrano L. SynMyco transposon: Engineering transposon vectors for efficient transformation of minimal genomes. *DNA Res.* 2019;26(4):327–39.

202. Pich OQ, Burgos R, Planell R, Querol E, Piñol J. Comparative analysis of antibiotic resistance gene markers in Mycoplasma genitalium: Application to studies of the minimal gene complement. *Microbiology.* 2006;152(Pt 2):519–27.

203. Pour-El I, Adams C, Minion FC. Construction of mini-Tn4001tet and its use in Mycoplasma gallisepticum. *Plasmid.* 2002;47(2):129–37.

204. Zimmerman CU, Herrmann R. Synthesis of a small, cysteine-rich, 29 amino acids long peptide in Mycoplasma pneumoniae. *FEMS Microbiol Lett.* 2005;253(2):315–21.

205. Yao F, Svensjö T, Winkler T, Lu M, Eriksson C, Eriksson E. Tetracycline repressor, tetR, rather than the tetR-mammalian cell transcription factor fusion derivatives, regulates inducible gene expression in mammalian cells. *Hum Gene Ther.* 1998;9(13):1939–50.

206. Breton M, Tardy F, Dordet-Frisoni E, Sagne E, Mick V, Renaudin J, et al. Distribution and diversity of mycoplasma plasmids: Lessons from cryptic genetic elements. *BMC Microbiol.* 2012;12:257.

207. King KW, Dybvig K. Mycoplasmal cloning vectors derived from plasmid pKMK1. *Plasmid.* 1994;31(1):49–59.

208. King KW, Dybvig K. Transformation of Mycoplasma capricolum and examination of DNA restriction modification in M. capricolum and Mycoplasma mycoides subsp. mycoides. *Plasmid.* 1994;31(3):308–11.

209. Allam AB, Reyes L, Assad-Garcia N, Glass JI, Brown MB. Enhancement of targeted homologous recombination in Mycoplasma mycoides subsp. capri by inclusion of heterologous recA. *Appl Environ Microbiol.* 2010;76(20): 6951–4.

210. Duret S, Berho N, Danet JL, Garnier M, Renaudin J. Spiralin is not essential for helicity, motility, or pathogenicity but is required for efficient transmission of Spiroplasma citri by its leafhopper vector Circulifer haematoceps. *Appl Environ Microbiol.* 2003;69(10):6225–34.

211. Knudtson KL, Minion FC. Construction of Tn4001lac derivatives to be used as promoter probe vectors in mycoplasmas. *Gene.* 1993;137(2):217–22.

212. Liu L, Dybvig K, Panangala VS, van Santen VL, French CT. GAA trinucleotide repeat region regulates M9/pMGA gene expression in Mycoplasma gallisepticum. *Infect Immun.* 2000;68(2):871–6.

213. Liu L, Panangala VS, Dybvig K. Trinucleotide GAA repeats dictate pMGA gene expression in Mycoplasma gallisepticum by affecting spacing between flanking regions. *J Bacteriol.* 2002;184(5):1335–9.

214. Gibson DG, Benders GA, Andrews-Pfannkoch C, Denisova EA, Baden-Tillson H, Zaveri J, et al. Complete chemical synthesis, assembly, and cloning of a Mycoplasma genitalium genome. *Science.* 2008;319(5867):1215–20.

215. Gibson DG, Glass JI, Lartigue C, Noskov VN, Chuang RY, Algire MA, et al. Creation of a bacterial cell controlled by a chemically synthesized genome. *Science.* 2010;329(5987):52–6.

216. Hutchison CA, Chuang RY, Noskov VN, Assad-Garcia N, Deerinck TJ, Ellisman MH, et al. Design and synthesis of a minimal bacterial genome. *Science.* 2016;351(6280):aad6253.

217. Callaway E. 'Minimal' cell raises stakes in race to harness synthetic life. *Nature.* 2016;531(7596):557–8.

218. Rothschild LJ. Synthetic biology meets bioprinting: Enabling technologies for humans on Mars (and Earth). *Biochem Soc Trans.* 2016;44(4):1158–64.

35

The Family Neisseriaceae

Yvonne A. Lue

CONTENTS

Taxonomy

The family of *Neisseriaceae* as defined by *Bergey's Manual of Systematic Bacteriology* consists of two separate rRNA super families that are not related. The true *Neisseriaceae* and two *Kingella* ssp. belong to the β-subclass of Proteobacteria, while false *Neisseria*, *Moraxella*, *Branhamella*, and the *Acinetobacter* genera belong to the γ-subclass of the Proteobacteria, which have been removed from the family of *Neisseriaceae*. In addition, genetic studies have shown that *Eikenella*, *Simonsiella*, *Alyseilla* EF-4a, EF-4b, M-5, and M-6 (CDC groups) are related to the true Neisseriaceae and two *Kinegella* ssp. The present classification of the family of Neisseriaceae contains *Neisseria*, *Kingella*, *Eikenella*, *Simonsiella*, *Alyseilla* EF-4a, EF-4b, M-5, and M-6 (1, 2). This chapter will only discuss *Neisseria gonorrhoeae* and *Neisseria meningitidis* because of their pathogenicity in humans. Species of the genus *Neisseria* found in humans and animals inhabiting mucous membranes are listed in Table 35.1 (3).

The *Neisseria* are Gram-negative diplococci, with adjacent sides that are flattened, giving the organisms a "coffee-bean" appearance. These organisms are aerobic, nonmotile, grow best in an atmosphere of 5–10% CO_2, produce cytochrome oxidase, and do not form spores. Acid from carbohydrates, nitrate reduction, polysaccharide from sucrose, presence of deoxyribonuclease (DNase), superoxol (30% hydrogen peroxide), pigment production, and the resistance to Colistin (10 µg) are used to differentiate the species (Table 35.2). Nongrowth-dependent rapid carbohydrate tests are available for the detection of acid production from carbohydrates for the speciation of *Neisseria* (3, 4).

Neisseria gonorrhoeae

N. gonorrhoeae (the gonococcus) is the causative agent of gonorrhea, a sexually transmitted infection, and for several other conditions that are sequelae to the infection. The organism infects only humans, and transmission is almost always by person-to-person contact. It is the second most commonly reported bacteria causing an infectious disease.

Microscopically, the organism is indistinguishable from other *Neisseria,* but, *N. gonorrhoeae* is the most fastidious in its growth requirements. The gonococcus requires the amino acid cysteine and a usable energy source (i.e., glucose, pyruvate, or lactose), and does not grow on blood agar. It will grow on supplemented chocolate agar. *N. gonorrhoeae* ferments glucose but not maltose, fructose, or sucrose.

Laboratory Diagnosis

Diagnosis of gonococcal infection can be made at two levels, presumptive and confirmed. The isolation of the organism and the diagnosis of the infection must be reported to the local public health authorities. There are serious medicolegal consequences for misdiagnosing gonorrhea or misidentifying *N. gonorrhoeae*.

Gram stain showing Gram-negative intracellular microorganisms is presumptive for the diagnosis of gonorrhea. This test has a sensitivity of 50–70% and generally used for symptomatic males. For symptomatic females, an endocervical smear has 40–60% sensitivity and its use discouraged. A definitive diagnosis for females must be made by the recovery of *N. gonorrhoeae* in culture from an endocervical specimen or by detection using a nucleic acid amplification procedure.

TABLE 35.1

Host Range of *Neisseria* Species in Humans and Animals

Humans	Species
Urogenital tract	N. gonorrhoeae, N. meningitidis
Oropharynx	N. meningitidis, N. lactamica, N. cinerea, N. flavescens, N. subflava, N. sicca, N. mucosa, N. elongate, N. polysaccharea
Animals	
Oropharynx	
Cat	N. canis
Dog	N. weaveri
Guinea pig	N. denitrificans and N. caviae
Iguana	N. iguanae
Rabbit	N. cuniculi
Monkey	N. macacae
Sheep and cattle	N. ovis

Source: Modified from *Identification of N. gonorrhoeae and Related Species*, Centers for Disease Control and Prevention, Sexually Transmitted Disease, Atlanta, GA, 2004.

The Gram stain has low predictive value and is not useful for the asymptomatic female (1, 2, 5).

The site of specimen collection depends on the gender and the patient's sexual practices. Sites for obtaining specimens include urethra, endocervix, rectum, oropharynx, joint fluid, blood, skin lesions, and conjunctiva, depending on the clinical impression. The gonococcus is the most sensitive of the *Neisseria* species to adverse conditions, such as drying and extremes in temperature (4).

The correct transportation of the specimen is very important in maintaining the viability of the gonococcus for definitive identification. The specimen must be collected with a Dacron or rayon swab that must be transported in Stuart's or Amies buffered semisolid transport media at room temperature. The specimen can also be inoculated directly onto culture medium immediately after collection. Culture media, such as Martin-Lewis, Thayer-Martin, and New York City medium in JEMBEC plates, can be used to transport the specimens (2). The medium, at room temperature, is inoculated and the CO_2-generating tablet is placed in the area provided. The inoculated agar is placed in a zip lock plastic bag and sent to the laboratory at room temperature. Upon receipt, the inoculated plate is placed in the incubator at 35–37°C with 3–5% CO_2.

Specimens submitted on swabs must be inoculated on an appropriate medium within 24 hours of collection. Specimens collected from sites with normal bacteria flora must be inoculated on a selection medium such as Thayer-Martin, Martin-Lewis, or New York City agar. Chocolate agar medium is used for specimens from normally sterile body sites. Inoculated media must be incubated in the atmosphere or 3–5% CO_2 at 35–37°C and examined every 24 hours for 72 hours. A higher concentration of CO_2 can inhibit the growth of some strains (2).

The colonies of *N. gonorrhoeae* are about 0.5 mm in diameter, glistening, and raised. A pinkish brown pigment is visible when sweeping the colonies with a cotton swab. For definitive identification, the organism must be Gram-negative diplococci, oxidase-positive, and ferment glucose, but not maltose, lactose, sucrose, or fructose. The test systems RapID™ NH System (Remel), API NH (bioMérieux) and Vitek 2 (bioMérieux) NHI

TABLE 35.2

Differential Characteristics of Human *Neisseria* Species (3)

Species	*Acid from					Nitrate Reduction	Polysaccharide from Sucrose	DNase	Superoxol	Pigment	Colistin Resistance 10 μg
	G	M	S	F	L						
N. gonorrhoeae	+	–	–	–	–	–	–	–	Strong (4+) positive (explosive)	–	R
N. gonorrhoeae ssp. kochii	+	–	–	–	–	–	–	–	Strong (4+) positive (explosive	–	R
N. meningitidis	+	+	–	–	–	–	–	–	Weak (1+) to strong (4+) positive	–	R
N. lactamica	+	+	–	–	+	–	–	–	Weak (1+) to strong (3+) positive	+	R
N. polysaccharea	+	+	–	–	–	–	+	–	Weak (1+) to strong (3+) positive	–	(R)
N. cinerea	–	–	–	–	–	–	–	–	Weak (2+) positive	–	(R)
N. flavescens	–	–	–	–	–	+	–	–	Weak (2+) positive	+	S
N. mucosa	+	+	+	+	–	+	+	–	Weak (2+) positive	d	S
N. subflava biovar subflava	+	+	–	–	–	–	–	–	Weak (2+) positive	+	S
N. subflava biovar flava	+	+	–	+	–	–	–	–	Weak (2+) positive	+	S
N. subflava biovar perflava	+	+	+	+	–	–	+	–	Weak (2+) positive	+	(R)
N. sicca	+	+	+	+	–	–	+	–	Weak (2+) positive	–	S
N. elongata	–	–	–	–	–	–	–	+	–	–	S

have replaced the slower carbohydrate tests. Definitive identification can also be made with direct fluorescent monoclonal antibodies, coagulation (Phadebact Monoclonal GC) and colorimetric test (GonoGen II).

Matrix-assisted laser desorption/ionization time of flight mass spectrometry (MALDI-TOF MS) is being used by numerous laboratories because it is fast and gives accurate identification for clinically significant organisms. There is limited data on the accuracy of MALDI-TOF MS for the identification of *Neisseria* species (6).

Specimens from patients suspected of sexual abuse must be collected from the site of abuse for the recovery and definitive identification of *N. gonorrhoeae*. The identification of *N. gonorrhoeae* recovered from specimens in sexual abuse cases should be determined by at least two methods. The antimicrobial sensitivity pattern of the isolates from the alleged abuser and the patient are valuable in linking the isolate recovered from the victim and the alleged offender.

Nucleic acid amplification tests (NAAT) have a greater sensitivity than culture and almost the same specificity for the detection of *N. gonorrhoeae*. The specimens for the NAAT include urine (preferable for males), endocervical, urethral, self-collected vaginal, rectal, and pharyngeal specimens. Vaginal swabs have a greater sensitivity than endocervical swabs that are more sensitive than urine specimens. All FDA-approved NAAT for *N. gonorrhoeae* also include the target for *Chlamydia trachomatis* since infections with each of these organisms have similar clinical presentations. The advantages of the NAAT are longer specimen stability after collection, decreased turnaround time for results, and greater sensitivity than culture. Each of the NAAT has a unique system for the specimen transport and, except for the liquid Pap medium, they are not interchangeable. See Table 35.3 for a list of current FDA-approved NAAT with the specimen types (7).

There have been reports of cross-reactivity with genital saprophytic *Neisseria* species resulting in false NAAT-positive results. The Centers for Disease Control and Prevention (CDC) issued guidelines for NAAT when the positive predictive value is <90%. Repeat testing for positives specimens should be performed using a test that does not detect commensal *Neisseria* species. The limitations of NAAT include the high cost and the inability for antibiotic testing to monitor for resistance (7).

Because NAATs are significantly more sensitive with specificity almost equal to culture, they are recommended for rape and sexual abuse cases. Positive results should be confirmed by another method (7).

Gonococcal Disease

In the United States, gonorrhea is the second most commonly reported notifiable infection. There are 820,000 new gonococcal infections occurring each year based on the CDC estimates. The rate of reported gonorrhea cases increased 18.6% during 2016–2017, a 75.2% increased since the historic low in 2009 (3).

Acute anterior urethritis is the most common manifestation of gonorrhea. This usually appears in 90% of males 2–8 days after exposure and is manifested by a purulent discharge. Acute epididymitis is the most common local complication of the infection. In females, the endocervix is the primary site of infection. Urethral infection and infection of the Bartholin's glands are seen in women who have gonococcal cervicitis. Ascending infection occurs in 10–20% of females with a gonococcal infection. Asymptomatic infection can lead to pelvic inflammatory disease (PID), which can result in obstruction of the fallopian tubes and resulting sterility. Gonorrhea in women can facilitate the transmission of human immunodeficiency virus (HIV) (1, 8).

Manifestation of disseminated gonococcal disease in males and females includes dermatitis, arthritis-tenosynovitis syndrome, monoarticular septic arthritis, perihepatitis, endocarditis, and meningitis. Ophthalmia neonatorum, a disease of newborns, is prevented by prophylactic instilling in the eyes of the infant a solution of penicillin or silver nitrate immediately after birth.

Sulfonamides were used successful in the treatment of gonorrhea. During World War II, strains of *N. gonorrhoeae* resistant to sulfonamides appeared. Later, penicillin became the drug of choice; but with the acquisition of plasmid-mediated resistance, this treatment is now ineffective (9).

The CDC issued guidelines for treatment in 2015. For patients with uncomplicated genital, rectal, or pharyngeal gonorrhea, the CDC now recommends combination therapy with ceftriaxone, 250 mg as a single intramuscular dose, plus either azithromycin, 1 gram orally in a single dose, or doxycycline, 100 mg orally twice daily for 7 days. If ceftriaxone is not available, CDC recommends cefixime, 400 mg orally, plus either azithromycin, 1 gram orally, or doxycycline, 100 mg orally twice daily for 7 days. For patients with a severe allergy to cephalosporins, the recommendation is a single 2 g dose of azithromycin orally. In both circumstances, when ceftriaxone is not used, CDC recommends a test of cure for these patients one week after treatment. This is an important change in the treatment guidelines. Clinicians are encouraged to be vigilant for cephalosporin resistance (10). If a patient fails treatment with CDC recommended therapy, the clinician should collect a specimen for culture and sensitivity, re-treat the patient,

TABLE 35.3

Nucleic Acid Amplification Tests

Manufacturer		Instrumentation	Specimen
Roche Cobas® CT/NG	Roche Molecular Systems	Cobas 6800/8800	Vaginal (clinician or self-collected), urine in males only
Abbott Real-time CT/NG	Abbott Molecular	m2000sp and m2000rt	Endocervix, vaginal (clinician or self-collected), urethra, and urine
Becton Dickinson	BD MAX™ CT/GC/TV	BD MAX	Endocervix, vaginal (clinician or self-collected), urethra, and urine
Hologic	Aptima Combo 2 CT/GC	Tigris and Panther	Endocervix, vaginal (clinician or self-collected), urethra, urine, rectal, and pharynx
Cepheid	Xpert CT/NG	GeneXpert®	Endocervix, vaginal (clinician or self-collected), urethra, urine, rectal, and pharynx

ensure the patient's recent partners are treated, and notify CDC through the local STD program within 24 hours (8).

To date, there are no documented treatment failures in the United States. However, Morbidity and Mortality Weekly Reports (MMWR), show that there are a number of international reports of cefixime treatment failures, and the bacteria's history of becoming resistant to antibiotics used for treatment, points to the increasing likelihood that gonococcal cephalosporin resistance and treatment failures are on the horizon in the United States. Refer to the current edition of Clinical Laboratory Standards Institute (CLSI) M100 Performance Standards for Antimicrobial Susceptibility for the recommended antimicrobials to be tested by the Bauer-Kirby method.

Test of cure is not recommended for individuals who receive the recommended therapeutic regimen. Culture for the recovery of *N. gonorrhoeae* should be performed if symptoms persist after therapy with an established regimen and antimicrobial susceptibility performed for the recovered N. gonorrhoeae isolate.

Neisseria meningitidis

Colonies of this species grow to be 1 mm in diameter after 18–24 hours of incubation and are usually gray and mucoid on primary isolation media. The organism grows very well on sheep blood or chocolate agar at 35–37°C and is enhanced by incubation in an atmosphere of 3–5% CO_2. Identification can be made by observing acid production from glucose and maltose, but not from sucrose, fructose, or lactose. The RapID NH System (Remel) and API NH (bioMérieux) have replaced the individual carbohydrate substrates (5, 3).

N. meningitidis is separated into 13 serologic groups (A, B, C, D, H, I, K, L, X, Y, Z, W135, and 29E) based on capsular polysaccharide. Encapsulated cells of freshly isolated organisms demonstrate the quellung reaction when mixed with homologous antiserum. Serogroups A, B, and C are usually isolated from epidemic and endemic cases (1, 11).

The pathogenicity of meningococcus is linked to a potent endotoxin, which comprises approximately 50% of the outer membrane of the organism. The endotoxin has been characterized as a lipooligosaccharide because it lacks the multiple repeating sugars of a lipopolysaccharide. The endotoxin has been implicated as the cause of septic shock seen in patients with meningococcemia (12).

Meningococcal Disease

Humans are the only natural host for *N. meningitidis*, and the disease is maintained and spread by carriers harboring the organism in their nasopharynx and pharynx. Infection commonly results from asymptomatic nasopharyngeal carriage. There is a 10–20% carriage rate in the population at any given time. The organism evades the immune system and enters the bloodstream to cause disease. Diseases caused by *N. meningitidis* include meningoencephalitis, meningitis with or without meningococcemia, meningococcemia without meningitis, and bacteremia without complications. Hematogenous spread of the organism can lead to endocarditis, pericarditis, arthritis, endophthalmitis,

osteomyelitis, and peritonitis. Risk factors of meningococcal disease include individuals with complement or properdin component deficiencies, hepatic failure, systemic lupus erythematous, multiple myeloma, and asplenia. Meningococci have been implicated as the etiological agent of approximately 2–4% of community-acquired pneumonia. Pharyngitis is associated with recent contact with carriers of *N. meningitidis* and can be a prior symptom and sign of serious disease (1).

Laboratory Diagnosis

For the diagnosis of meningococcal disease, the organism should be isolated from cerebrospinal fluid (CSF), blood, or aspirates of petechiae. Synovial fluid, sputum, conjunctival and nasopharyngeal swabs should be obtained when clinically indicated. These specimens must be inoculated onto blood and chocolate media as quickly as possible. Gram-stained smears showing Gram-negative diplococci of CSF, aspirates, and other fluids can give the presumptive diagnosis. Latex agglutination and coagglutination tests are direct antigen tests to detect meningococcal capsular polysaccharide in CSF, urine, and serum, but have fallen into disuse (1, 2).

The significance of the recovery of *N. meningitidis* from rectal and genital specimens is unknown. The recovery of this organism from pharyngeal specimen identifies the individual as a carrier. Prophylaxis is considered during outbreaks of meningococcal disease.

Chemotherapy and Prophylaxis

With the introduction of sulfonamides, chemotherapy was the main method of treating meningococcal disease. The appearance of resistance to this agent in 1963, however, has made this treatment ineffective. Benzyl penicillin is the preferred agent for treatment of meningococcal meningitis as long as the strain does not produce β-lactamase with cefotaxime and ceftriaxone also being effective alternate agents.

Vaccines have been successful in reducing the incidence of disease in the military, containing epidemics and in individuals with complement disease. This protection is not long-lasting. Vaccines against *N. meningitidis* were available in the United States since 1981. Meningococcal polysaccharide vaccine (MPSV4 or Menomune®) and meningococcal conjugate vaccine (MCV4 or MenactraT) prevent four types of meningococcal disease. These vaccines prevent disease caused by two of the three most common serotypes in the United States (serogroups C, Y, and W-135) and serotype A that causes epidemics in Africa. Vaccines against serotype B were introduced in 2018. Meningococcal vaccines cannot prevent all types of the disease, but they can protect many people who might become sick if they do not receive the vaccine (12).

Meningitis due to *N. meningitidis* must be reported to state and/or local health departments to assure follow-up of close contacts and recognize outbreaks.

Laboratory Safety

N. meningitidis is now classified as a biosafety level 2 organism, which requires that all manipulations of specimens suspected of containing this organism, and meningococcal culture, be conducted in a biological safety cabinet. It is recommended that laboratory personnel be offered the meningococcal vaccine. The vaccine will not decrease the risk of infection but can reduce the potential risk of laboratory-acquired disease (2).

Other *Neisseria* Species

The other *Neisseria* species listed in Table 35.1 are members of the microbiota of the pharyngeal and female urogenital tracts and are rarely associated with disease. *N. cinerea, N. elongate, N. flavescens, N. mucosa, N. sicca,* and *N. subflava* have been reported as the causative agents of endocarditis. *Neisseria lactamica, N. polysaccharea,* and *N. subflava* have been reported to be the etiological agent for septicemia and meningitidis.

REFERENCES

1. Armstrong, D. et al. *Infectious Diseases*. Mosby, London, UK, 1999.
2. Murray, P.R. et al. *Manual of Clinical Microbiology*, 8th ed. American Society for Microbiology, Washington, DC, 2003.
3. Identification of N. Gonorrhoeae and Related Species. Centers for Disease Control and Prevention, Sexually Transmitted Disease, Atlanta, GA, 2004.
4. Goldman, E. and Green, L. H. *Practical Handbook of Microbiology*, 2nd ed. CRC Press, Boca Raton, FL, 2009.
5. Versalovic, J. et al. *Manual of Clinical Microbiology*, 10th ed. American Society for Microbiology, Washington, DC, 2011.
6. Morel, F. et al. Use of Andromas and Bruker MALDI-TOF MS in the identification of Neisseria. *Eur. J. Clin. Microbial. Infect. Dis.* 2018, 37:2273–2277.
7. Screening tests to detect Chlamydia trachomatis and Neisseria gonorrhoeae infections. Morbidity and Mortality Weekly Report, Recommendations and Reports 51, 15, 2002.
8. Centers for Disease Control (CDC) Recommendations and Reports/Vol. 64/No. 3 June 5, 2015 Sexually Transmitted Diseases Treatment Guidelines, 2015
9. Centers for Disease Control (CDC). Sexually transmitted diseases treatment guidelines 2006. *Morbidity and Mortality Weekly Report, Recommendation and Reports.* 2006, 55(RR11):1–94.
10. Centers for Disease Control (CDC). Oral cephalosporins no longer a recommended treatment for gonococcal infections. *Morbidity and Mortality Weekly Report, Recommendation and Reports.* 2012, 61(31): 590–594.
11. Harrison, O. B. et al. Description and nomenclature of Neisseria meningitidis capsule locus. *Emerg. Infect. Dis.* 2013, 19:566–573.
12. Centers for Disease Control (CDC)/Division of Bacterial and Mycotic Diseases. Meningococcal Disease, 2005.

36

The Genus Pseudomonas

Layla Ramos-Hegazy, Shubham Chakravarty, and Gregory G. Anderson

CONTENTS

Introduction

Members of the Gram-negative bacterial genus *Pseudomonas* are ubiquitous in soil and water ecosystems [1]. Their genetic flexibility, versatility in optimizing the usage of a vast array of organic and inorganic compounds, and innate capability to survive under diverse, often trying, environmental conditions are the key factors behind their successful inhabitation of most environmental niches found on our planet [2]. From the point of view of humankind, this genus is intriguing indeed. On the one hand, the ability of *Pseudomonas* to produce an arsenal of degradative factors makes it one of the most feared pathogens of plants and animals [3, 4]. On the other hand, this same feature, combined with biochemical proficiency for metabolizing toxic hydrocarbons and other compounds, have rendered it one of the most prized organisms for bioremediation of recalcitrant wastes [5–7]. This chapter briefly describes the current taxonomical position and general features of *Pseudomonas*. This discussion is followed by a section describing some of the prominent members of this genus, and their respective roles in pathogenesis, bioremediation, and plant symbiosis.

Taxonomy

Pseudomonas is part of the bacterial family *Pseudomonadaceae*, which is also composed of *Azotobacter*, *Azomonas*, *Cellvibrio*, and *Azorhizophilus*, among others [2]. *Pseudomonadaceae*

DOI: 10.1201/9781003099277-38

are commonly characterized as chemoorganotrophic, aerobic, and incapable of photosynthesis. There are also able to survive under a myriad of nutritional environments [8]. Classification and taxonomy of this genus has undergone numerous transitions since its inclusion in the first edition of *Bergey's Manual of Determinative Bacteriology* in 1923 [9]. Specifically, numerous species that were once classified in *Pseudomonas* are now recognized as distinct genera, through advances in molecular techniques. While the common methods of taxonomical organization of the *Pseudomonas* genus include cell morphology, pigment types, nutritional behaviors, and genetic organization, the widespread phenotypic characteristics mostly studied are carbon utilization patterns, antibiotic resistance, and antibiotic and amino acid synthesis [10].

There are currently more than 100 species in the genus *Pseudomonas* [11], and 13 species (and numerous strains thereof) are sequenced and annotated. Table 36.1 describes the various approaches used in the current era for *Pseudomonas* taxonomic studies. While the benefits of chemotaxonomic studies have been widely published [12, 13], gene sequencing and nucleic acid fingerprinting hold the greatest promise for solving *Pseudomonas* taxonomic issues [11]. In the past decade, clinical isolate identification has primarily been achieved based on protein analysis using matrix-assisted laser desorption/ionization time of flight mass spectrometry (MALDI-TOF MS) [11]. Inclusion of new species in the genus *Pseudomonas* requires 16S rRNA sequencing, DNA-DNA hybridization, phenotypic analysis, fatty acid characterization, and multilocus sequencing analysis [11].

The Life of *Pseudomonas*

Cellular Characteristics

Pseudomonas species are Gram-negative, catalase positive, and either oxidase positive or negative, depending on the species. They are slightly curved or linear rod-shaped cells, having a maximum length of approximately 4 µm. *Pseudomonas* species sometimes exist as single cells and occasionally in pairs, or even short chains [2]. They are often motile via polar flagella, although lateral flagella are occasionally found, usually associated with swarming [2, 14, 15]. The flagellar number is decided by FleN, a putative ATP-GTP binding protein [2, 16].

Pili are also commonly reported and are particularly studied for their role in pathogenesis of clinically relevant species such as *P. aeruginosa* [17]; nevertheless, other typical nonpathogenic *Pseudomonas* species like *P. fluorescens* also possess pili [18]. Intriguingly, low G + C content of pilin genes, in relation to the higher average chromosomal G + C, indicates horizontal acquisition by *Pseudomonas* species [19].

There are several colored pigments produced by *Pseudomonas* species. For instance, under limiting iron conditions, some *Pseudomonas* species produce pigments that fluoresce under UV light, and these pigments facilitate bacterial iron uptake [20]. Others are nonfluorescent and participate in other processes. Common pigments characteristic of *Pseudomonas* are listed in Table 36.2.

Intracellularly, *Pseudomonas* species form inclusion bodies of various substances. For instance, under nitrogen deficient

TABLE 36.1

Current Methods Used in *Pseudomonas* Classification

Approach	Illustrative Citation(s)	Comments/Description
Chemotaxonomic		
Whole fatty acid analysis	[256]	Has been used to examine the taxonomic position of certain *P. marginalis* strains included under the *P. fluorescens* group
SDS-PAGE profiling	[257]	Fingerprinting *Pseudomonas* strains (isolates from disparate phylogenetic groups) at the species level
Siderotyping and pyoverdine analysis	[20]	Involves isoelectrophoretic characterization of siderophores and often species-specific pyoverdines; mass spectrometry has greatly enhanced this approach
Use of stable low–molecular-weight RNA	[258, 259]	5S rRNA and tRNA are studied; staircase electrophoresis [260] has improved this technique
Fluorescent spectroscopy fingerprinting	[261]	Emission spectral analysis of intrinsic fluorophores like NADH, tryptophan
Matrix-assisted laser desorption/ionization time of flight mass spectrometry (MALDI-TOF MS)	[262]	A type of bacterial fingerprinting that uses mass spectrometry to analyze protein composition.
Gene Sequencing Studies		
16S rRNA sequencing	[10, 263]	16S rRNA is widely used to determine phylogenetic relationships between organisms
Housekeeping genes as phylogenetic markers	[264]	Phylogenetic resolution of this method (using *rpoB* for instance) is often much more than can be obtained using 16S rRNA [265]
16S-23S rRNA intergenic spacer (ITS) marker	[266]	ITS has high sequence and size variability, and has phylogenetic resolution high enough to differentiate bacteria even at the strain level; the productivity of this method has been reported for *Pseudomonas* classification [267]

TABLE 36.2

Common Pigments in *Pseudomonas*

Pigment	Bacterial Strain	Color
Pyoverdine	*Pseudomonas sp.*	Yellow-green, fluorescent
Pyocyanin	*P. aeruginosa*	Blue
Pyorubin	*P. aeruginosa*	Red
Oxychlororaphin	*P. aureofaciens*	Orange
Chlororaphin	*P. chlororaphis*	Green
Oxychlororaphin	*P. chlororaphis*	Orange

Source: Data compiled from refs. [2, 268, 269].

conditions, *Pseudomonas* accumulates poly-β-hydroxybutyrate (PHB) granules. When growing on gluconate or alkanes, *Pseudomonas* assembles polyhydroxyalkanoates.

Physiology and Metabolism

As mentioned, *Pseudomonas* exhibits great metabolic plasticity [2, 21], and extensive analysis has been performed to elucidate *Pseudomonas* metabolism [2, 22, 23]. The TCA cycle exists in all species of *Pseudomonas* and it is important for metabolism. The Entner-Doudoroff glycolytic pathway often prevails over the Embden-Meyerhof pathway due to the lack of 6-phosphofructokinase in many species. The pentose phosphate pathway, the glyoxylate shunt, and other important metabolic pathways, are also widespread in this genus.

Catabolite repression is also reported in *Pseudomonas* [24, 25], which allows *Pseudomonas* to utilize a medley of available carbon sources in a preferential and orderly manner [26, 27]. Organic and amino acids are preferred to glucose [28], followed by mannitol and histidine. Crc and other signal transduction proteins facilitate catabolite repression [29–32]. Intriguingly, in some cases, like during repression of phenol metabolism in *P. putida* [33], cells can mediate catabolite repression by discerning the redox state of respiratory chains.

Pseudomonas has a propensity to use amino acids as carbon and nitrogen sources. Specific membrane permeases translocate amino acids available in the environment into the cell cytoplasm [34]. The ability to use amino acids as nutrients poses a great advantage to *Pseudomonas*, because it can readily assimilate amino acids into the cell biomass with minimal processing. Alternatively, additional catabolic activity can funnel amino acids into central metabolism. Amongst amino acids, arginine metabolism, especially, is key in *Pseudomonas* biochemistry, with several catabolic pathways identified for breaking down this substrate [35, 36]. In *P. aeruginosa*, for instance, numerous arginine-responsive genes are controlled by the arginine regulatory protein ArgR [37].

Analysis of amino acid metabolism in *Pseudomonas* is important not only for biochemical elucidation, but also for taxonomical classification of the genus. For example, analysis of L-phenylalanine and L-tyrosine pathways has been used to classify members of *Pseudomonas* [38]. Furthermore, the multibranched pathways of aromatic amino acid synthesis have elucidated phylogenetic relationships in *Pseudomonas* [39–41].

Though typically considered a strict aerobe, particular *Pseudomonas* species perform fermentations and/or anaerobic respiration. Specifically, arginine [42, 43] and pyruvate [43] fermentation pathways have been reported. Additionally, *Pseudomonas* utilizes nitrogenous substances, like nitrate, as terminal electron acceptors for anaerobic respiration [44]. In addition to denitrification activities of some species, nitrogen fixing species like *P. stutzeri* have also been reported [45]. A few *Pseudomonas* species oxidize NO to nitrate [2], though generally considered as a detoxification or co-oxidation process rather than an energy yielding one. Another notable exception to the aerobic *Pseudomonas* species is *P. chloritidismutans*, which employs chlorate as an electron acceptor in conjunction with oxygen [46].

Habitat

Pseudomonas is ubiquitous in the environment, which is attributed to its metabolic versatility [47]. It can be found in soil and water environments, as well as plant and animal tissues and many other niches. *Pseudomonas* species have even been found in hospital saline solutions [48], Antarctic cyanobacterial mats [49], and water from plumbing fixtures [50]. *Pseudomonas* species grow over a temperature range of 4–42°C, and pH 4–8 is considered favorable. As mentioned, *Pseudomonas* species are generally strict aerobes, although *P. aeruginosa* and other denitrifying species exhibit fermentation and anaerobic respiration [51].

Though culture-dependent studies indicate *Pseudomonas* as an important soil microorganism [52], culture-independent methods, such as metagenomics, suggest a relative scarcity of *Pseudomonas* in the soil environment [53, 54]. Nevertheless, when growing in soil, *Pseudomonas* does so in conjunction with other bacteria, like *Streptomyces*, that supply monomeric carbon compounds [2].

Pseudomonas is considered extremely suited to thriving in the rhizosphere [55, 56], the territory impacted by vegetation [57]. *Pseudomonas* possesses numerous attributes necessary for optimizing its survival in this niche, like chemotaxis, adherence factors, and mechanism for evasion of plant defenses [2]. Importantly, *Pseudomonas* outcompetes other microorganisms to enhance its hold on the rhizosphere [58]. Different *Pseudomonas* species exert different effects on plants. Some species, like *P. syringae*, are epiphytic in nature and effectively colonize plant leaves and other plant surfaces to cause disease [59], while others benefit the plant by producing phytohormones [60]. However, many species neither harm nor hurt plants [2]. Generally, though, *Pseudomonas* species live a saprophytic or pathogenic lifestyle, scavenging organic matter [61, 62].

Laboratory Culture of *Pseudomonas*

Under laboratory conditions, *Pseudomonas* is generally grown in media containing organic compounds, at neutral pH and mesophilic temperatures (~30°C) [63–65]. These conditions yield a high growth rate. Media replete in peptides, like tryptic soy broth/agar, is preferred, especially by the saprophytic species of the genus; the peptide concentration of the media may be changed to optimize the amount and variety of species maintained in culture [66]. *Pseudomonas* generally requires media containing 0.1–1% (w/v) organic matter to maintain good growth. No separate mineral, vitamin, or growth factor is required in the media. Denitrifying *Pseudomonas* bacteria need nitrate, anaerobic conditions, and temperatures between 30 and 40°C for their enrichment [67].

As mentioned above, many *Pseudomonas* species produce fluorescent molecules. Differentiation of these species on laboratory media requires iron depletion to promote secretion of the fluorescent siderophores [2]. King's A and B media are commonly used for fluorescent *Pseudomonas* enrichment [68]. The antibiotic trimethoprim is often used to discern and isolate fluorescent species from nonfluorescent ones [69].

Although a few selective/differential media exist for *Pseudomonas* isolation (e.g., Pseudosel agar medium), highly selective media generally prohibit *Pseudomonas* growth [2, 70], and it is often suggested not to use such media when isolating *Pseudomonas* from niches already exerting strong selection [2, 3]. Cetrimide is recommended [2] to inhibit accompanying microorganisms, enhance pigment production, and enrich for *P. aeruginosa*. A few unconventional methods of *Pseudomonas* isolation utilize toxic resin acids [71] or unorthodox terminal electron donors and acceptors [46].

Control of Gene Expression

Pseudomonas species have some of the largest genomes among sequenced bacteria, often containing more than 6 Mbp, and more than 5,000 genes. A complex network of transcriptional and post-transcriptional regulatory mechanisms controls expression of these numerous genes [72, 73]. Several of these regulatory mechanisms are discussed next.

Quorum Sensing (QS)

Pseudomonas species produce many different QS molecules to regulate gene expression. QS facilitates coordinated gene expression in response to environmental signals and bacterial population density. Thus, pseudomonads use QS to coordinate population-wide gene expression for pathogenesis, metabolism, and many other processes [74].

Two-Component Systems (TCSs)

All *Pseudomonas* species produce TCSs. These regulatory systems mediate a variety of functions for the microbes (Table 36.3).

Sigma Factors

Many sigma factors have been reported in *Pseudomonas* species, and these sigma factors are critical for optimal RNA polymerase-mediated gene transcription activity. Some important ones are described next.

σ^{70}

This sigma factor controls housekeeping gene transcription [75]. The consensus sequence found in σ^{70}-dependent promoters in *Pseudomonas* has similarity to that found in *Escherichia coli* [2].

Extracytoplasmic Function (ECF) Sigma Factors

This subgroup of σ^{70} family factors is important for control of extracytoplasmic stress [76]. An important ECF sigma factor is AlgU, which regulates biosynthesis of the secreted polysaccharide alginate [77]. It also influences expression of heat shock sigma factor σ^{32}, and hence serves as a global gene regulator [78]. PvdS, another prominent ECF sigma factor, regulates pyoverdine siderophore biosynthesis. A sequence at the −35 position, known as the "IS box," is critical for PvdS-dependent promoter functioning [79]. Orthologues of the *P. aeruginosa* PvdS exist in *P. putida* (strain KT2440) and *P. fluorescens* [80].

σ^{54}

This sigma factor is encoded by *rpoN* and is distinct from the σ^{70} family [81]. The number and function of σ^{54}-dependent genes varies greatly amongst various *Pseudomonas* species [82–84]. Some of the major functions of RpoN include regulation of glutamine synthetase and urease in *P. aeruginosa* and *P. putida*, as well as some virulence factors in *P. aeruginosa*, and regulation of the phytotoxin coronatine in *P. syringae*. In *P. aeruginosa*, RpoN regulates over 700 genes, and blocking this sigma factor can increase antibiotic sensitivity [85].

Other Transcriptional Regulators

Pseudomonas contains quite a few transcriptional regulator, including members of the AsnC, GntR, LacI, LuxR, and MarR families [2, 86]. For catabolism of aromatic hydrocarbons, various species use σ^{54}-dependent regulator families such as XylR, TouR, and DmpR [81, 87–90]. Table 36.4 summarizes some of the most common transcriptional regulators.

Post-transcriptional Regulation

Post-transcription regulation has also been reported in *Pseudomonas*. For example, in *P. aeruginosa*, the TCS GacAS and the response regulator RetS regulate transcription of two small noncoding regulatory RNAs: RsmY and RsmZ [91]. These RNAs control the intracellular concentration of free RsmA and RsmF, prominent members of the CsrA family of RNA-binding proteins [92]. Rsm regulons control almost a tenth of *P. aeruginosa* transcriptome and are important for activating the phenotypes

---

TABLE 36.3

Common Two-Component Systems in *Pseudomonas* and their Functions

HK[a]	RR[b]	Function	References
PilS	PilR	Regulate pilin gene (*pilA*) expression	[270]
NtrB	NtrC	Nitrogen metabolism	[2]
PhoR	PhoB	Phosphate assimilation	[2]
FleS	FleR	Regulate motility and adhesion properties	[271]
GacS	GacA	Pathogenesis	[110]
PfeS	PfeR	Iron uptake	[272]

[a] Sensor histidine kinase (HK)
[b] Response regulator (RR)

TABLE 36.4

Families of Transcriptional Regulators that are Highly Represented in *Pseudomonas*

Regulator Family	No. of Members	Function	References
LysR	>100	Expression and activation of factors involved in metabolism of aromatic compounds (PcaQ and CatR); diverse functions especially in soil and plant-related environments	[2]
AraC	>30	Pathogenesis, stress response regulation, carbon metabolism	[2]
NtrC	~25	Phenylalanine metabolism (PhhR), lipase synthesis	[273, 274]

associated with acute infections [92]. RsmA and RsmF control translation of certain transcripts by binding to conserved 5'-ANGGAN-3' motifs present within stem-loop secondary structures of target mRNAs, such as those for type three secretion system (T3SS) and type six secretion system (T6SS) [92, 93]. Such binding sites often overlap or are proximal to the Shine-Dalgarno sequence of these target mRNAs [94–98], and thus RsmA and RsmF binding blocks ribosome recognition and translation initiation [92]. RsmA can also act as a post-transcriptional stimulator of gene expression by affecting mRNA stability and secondary structure [97, 99]. RsmY and RsmZ each have 4-6 RsmA binding sites. Thus, presence of these regulatory RNAs controls the free concentration of RsmA and its ability to bind target mRNAs [91]. Recent studies uncovered two additional small noncoding regulatory RNAs, RsmV, and RsmW, that work in a GacAS-independent manner to sequester RsmA [100, 101]. Similarly, in *P. fluorescens*, the *hcnABC* operon (involved in expression of hydrogen cyanide synthase expression) is under control of the small noncoding RNAs RsmA and RsmE [102–105].

Examples of Prominent *Pseudomonas* Species

This section highlights four major *Pseudomonas* species: *P. aeruginosa*, *P. syringae*, *P. putida*, and *P. fluorescens*. These species are examples of *Pseudomonas* pathogenesis, bioremediation, or plant growth promotion.

Pathogenesis of *P. aeruginosa*

P. aeruginosa causes infection in a wide range of different host organisms, including humans. The importance of *P. aeruginosa* as an opportunistic pathogen stems from three major concerns: high incidence of infection, high morbidity and mortality, and recalcitrance to immune or antibiotic clearance [106]. Moreover, this pathogen can cause both acute infections and chronic infections (Figure 36.1). For instance, *P. aeruginosa* is the leading

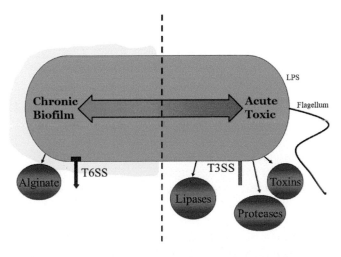

FIGURE 36.1 *P. aeruginosa* exhibits an acute infection lifestyle and a chronic infection lifestyle by alternate regulation of different sets of virulence factors. The dashed line separates these two lifestyles in this drawing, but, in reality, a continuum of phenotypes exists between these two extremes.

pathogen causing acute respiratory infections in mechanically ventilated and immunocompromised individuals and chronic infections in the airways of individuals with cystic fibrosis (CF) [107–109]. *P. aeruginosa* has also been implicated in other opportunistic diseases such as those of the urinary tract, eye, skin, and burn wounds [110].

General Pathogenicity Attributes

During infection, *P. aeruginosa* is thought to be acquired from the environment, including, importantly, the hospital. Bacterial cell surface virulence factors, such as flagella, pili, and lipopolysaccharide (LPS), aid in initial adhesion to the host epithelial layer [106]. Attachment is followed by deployment of the T3SS by *P. aeruginosa* to inject effector molecules (ExoU, ExoS, ExoT, ExoY) into the host epithelium. These effectors facilitate host cell cytoskeleton rearrangement, escape from host cell phagocytic engulfment, and cytotoxicity leading to tissue necrosis [110]. Other secreted virulence factors involved in host cell cytotoxicity and tissue destruction include elastase, phospholipase C, and pigments such as pyoverdine and pyocyanin [111–119]. Production of this arsenal of virulence factors is tightly regulated by an intricate network of QS, TCSs, stress response systems, and other factors [110, 120–123]. This regulatory network controls the virulence phenotype of *P. aeruginosa*, producing an acute infection characterized by toxin production and other classical pathogenic activities, a chronic infection characterized by biofilm formation and exopolysaccharide secretion, or a continuum or phenotypes in between (depending upon the type and level of inputs the regulators receive) (Figure 36.1) [124].

Cell Surface Virulence Factors

As the name suggests, these factors are localized on the cell surface. They mediate bacterial motility and the early stages of infection, such as initial adhesion to the host cell surface.

Flagella: In addition to providing motility for *P. aeruginosa*, flagella bind to asialoGM1, a normal constituent of the host epithelial cell membrane [125]. They are also known to prompt an NFκB-mediated inflammatory reaction, such as production of interleukin (IL)-8, by interacting with TLR2 and TLR5 [126, 127]. Because flagella are immunogenic, they often are down-regulated during infection, especially in chronic infections like those in the CF airways (discussed later in this chapter). Immunogenicity of flagellar proteins has been exploited by researchers to develop immunological products targeted at them [128–130].

Pili: Pili are also critical for *P. aeruginosa* virulence [131–133], particularly for adherence to the host cell surface by binding to asialoGM1 [134, 135]. They are also involved in twitching motility, which is dependent upon retractile movement of the pili. This activity leads to "spreading," rather than "swimming," of the bacterial population on the host tissue surface. This kind of locomotion is especially prevalent in respiratory tract infections [136]. Because of the importance of pili for virulence, immunological targeting of pili is a major area of research [137–140].

LPS: LPS is crucial in *P. aeruginosa* pathogenesis [141]. LPS assists *P. aeruginosa* in binding to asialoGM1 [135] and it elicits pro-inflammatory cascades through interaction of lipid A with TLR4/CD14 [142, 143].

Type III Secretion System (T3SS)

T3SS is a complex molecular syringe that *P. aeruginosa* uses to inject four effector toxins (ExoU, ExoS, ExoT, and ExoY) into the host cell cytosol. Different strains contain different combinations of these effectors. Intriguingly, mutants that lack effectors, but still express the T3SS structure are still virulent [144], indicating that the T3SS needle complex by itself might be important in *P. aeruginosa* pathogenesis. PcrV is an important structural protein making up the tip of the T3SS "needle" [145], and immuno-therapeutics targeted at PcrV have achieved some success [146–150].

ExoU: ExoU is the most potent of the *P. aeruginosa* T3SS effectors [144, 151] and it is the predominant cytotoxin injected by the T3SS [151–153]. It destroys host cell membranes through its phospholipase/lysophospholipase activities [154–156]. Anti-ExoU immunotherapy has been generally unsatisfactory [106], but some success has been achieved with phospholipase A2 inhibitors in vitro [157]. As the mechanism of ExoU activity becomes clearer, newer therapeutic strategies will emerge.

ExoS: The ExoS cytotoxin [144, 158] disrupts the host cell cytoskeleton through two different activities [106]. The C-terminus, which requires a 14-3-3 cofactor protein, contains ADP-ribosyltransferase activity [151, 159, 160] and the N-terminal domain acts as a Rho GTPase-activating protein (GAP). ExoS provokes inflammatory responses through TLR2 and TLR4 [161].

ExoT: ExoT is regarded a minor effector [151]. This toxin contains domains similar to ExoS, although the ADP-ribosyltransferase domain in ExoT affects a more restricted subset of host cellular proteins, including Crk [162]. ExoT induces cytoskeletal rearrangements, leading to inhibition of *P. aeruginosa* internalization and stunted wound healing [163, 164]. Some literature sources have reported that ExoT production decreases ExoU-mediated cell cytotoxicity [144].

ExoY: ExoY functions as an adenylate cyclase [110], and upon being injected into the host cell cytoplasm, increases intracellular cAMP levels [165]. Increased host intracellular cAMP leads to increased tissue porosity, especially in lung infections [166].

Type VI Secretion System (T6SS)

T6SS acts as a contractile molecular syringe, consisting of a sheath and a needle-like structure anchored to the inner and outer membrane via a baseplate complex. *P. aeruginosa* contains three different T6SS loci named H1-, H2-, and H3-T6SS [167]. The main proteins that make up this structure including the needle, baseplate, and spike all show significant homology to the T4 phage tail structure [168]. T6SS primarily acts in bacterial defense but can also mediate intracellular host invasion [167]. T6SS enhances *P. aeruginosa* ecological fitness by reducing colonization of other bacteria present in the same ecological niche [168]. Once *P. aeruginosa* comes in contact with a neighboring bacterial cell, the tube inside the sheath will extend, penetrate through the prey's membrane, and inject effectors into the cell [168].

H1-T6SS: H1-T6SS primarily serves as an antibacterial weapon to outcompete other bacteria that may colonize the same ecological niche as *P. aeruginosa* [167]. Once in contact with other bacteria, it will inject the effectors Tse1-7 that cause a wide range of damage including dormancy and degradation of peptidoglycan and dinucleotides [167]. To reduce damage against

TABLE 36.5

Secreted Virulence Factors of *P. aeruginosa*

Factor	Description/Function
Pyocyanin	Blue pigment; induces IL-8 expression and apoptosis of neutrophils; causes oxidative damage in host cells
Pyoverdine	Siderophore for iron chelation and uptake; virulence regulator controlling secretion of Exotoxin A and itself
Alkaline protease	Lyses fibrin in host tissue; secreted by type I secretion system; prevalent in corneal infections and in lung disease
Protease IV	Important in keratitis caused by *P. aeruginosa*; also implicated in airway infections due to destruction of lung surfactant proteins A, B, and D
Elastase	Metalloproteinase secreted by the type II secretion system (T2SS) into the extracellular space; destroys tight-junctions leading to porosity of airway epithelial tissue; induces inflammation by recruiting neutrophils and eliciting IL-8 response; also reported to disrupt lung surfactant proteins A and D
Phospholipase C	Secreted into the extracellular space by the T2SS; disrupts host cell membrane phospholipids, especially in acute lung disease; participates in surfactant destruction, inflammation induction, and inhibition of neutrophil oxidative burst
Exotoxin A (ExoA)	Toxin secreted by the T2SS; functions as an ADP-ribosyltransferase that blocks elongation factor 2, leading to impaired protein synthesis and host cell death; also diminishes host response to infection

Source: Data obtained from ref. [106].

other *P. aeruginosa* cells, immunity proteins Tsi1-6 directly interact with the effectors to render them inactive [167].

H2- and H3-T6SS: H2- and H3-T6SS mediate invasion of *P. aeruginosa* into pulmonary epithelial cells, likely via microtubules and actin [167]. Effector toxins released by H2- and H3-T6SS include phospholipases PldA and PldB that degrade the host cell membrane [167, 169].

Other Secreted Virulence Elements

P. aeruginosa secretes several other factors that contribute to disease progression. These factors, such as secreted pigments, enzymes, and proteases, destroy host tissues, increase tissue porosity, and induce host inflammatory reaction, among other functions (Table 36.5) [74, 106].

Quorum Sensing (QS)

In *P. aeruginosa*, there are three overlapping QS systems. Two of these systems, Las and Rhl, produce acyl homoserine lactone (AHL) autoinducers, while the third involves the quinolone molecule PQS [74]. QS controls expression of Type II Secretion (T2SS), T6SS, elastase, pyocyanin, and pyoverdine in *P. aeruginosa*, among other factors [74, 167].

P. aeruginosa Pathogenesis in the CF Lung Environment

CF is a congenital genetic disease, marked by improper chloride secretion across cell membranes [170]. Particularly in the airway epithelium, this defect leads to unusually high amounts of mucus building up [171], which provides an excellent niche for microbes to proliferate. Though there are multiple pathogens that colonize the CF lung, *P. aeruginosa* predominates from adolescence though adulthood [171].

P. aeruginosa causes a persistent, lifelong infection in the airways of individuals with CF [63, 77, 109]. These chronic infections initiate as *P. aeruginosa* adjusts to the CF airway environment and transitions to a biofilm lifestyle [63, 74, 172, 173]. This biofilm infection in CF lungs is strikingly dissimilar to acute *P. aeruginosa* infections. Selective pressure from the host's immune system and complex genetic networks activate the switch from an initial acute (planktonic) infection to a chronic (biofilm) infection lifestyle (Figure 36.1) [124]. T3SS and expression of toxins like elastase are diminished. Likewise, virulence features that trigger host immunity, like flagella and pili, are down-regulated once the pathogen has established adherent contact with the host respiratory epithelium. On the other hand, biofilm bacteria produce vast amounts of the exopolysaccharide alginate, leading to a highly mucoid phenotype and upregulation of T6SS activity [124, 174].

P. aeruginosa biofilms in the context of CF airways hold serious clinical implications [77]. Firstly, CF biofilms are nearly impossible to eradicate due to elevated antibiotic resistance [172]. Secondly, despite intense host immune infiltration, the biofilm structure protects resident bacteria from destruction. Moreover, inflammation leads to damage of surrounding host tissues [175]. Thus, *P. aeruginosa* persists for the life of the individual, and it is recognized as the leading cause of morbidity and mortality in CF patients [171, 176]. Understanding the molecular cascades that enable *P. aeruginosa* to form biofilms has been the focus of much research [110].

Infrequent Pseudomonas *Species as Human Pathogens*

P. aeruginosa has been implicated as the foremost pathogenic strain for human infections. Nevertheless, other *Pseudomonas* species have been reported in human infection, including

P. putida, *P. fluorescens*, *P. pseudoalcaligenes*, and *P. stutzeri* [177]. While very few infections involving *P. putida* bacteremia and *P. stutzeri* endocarditis have been reported, many of them involved immunocompromised patients and/or patients who had recently received invasive surgery prior to infection [178, 179].

The Phytopathogen *P. syringae*

P. syringae is the predominant plant pathogen in the *Pseudomonas* genus [180]. There are more than 60 defined pathovars of this bacterium [180, 181], and they infect a variety of plants, including lilac, maple, saucer magnolia, cherry, plum, apple, basswood, dogwood, and forsythia [182, 183]. Diseases caused by this pathogen range from bacterial specks on tomatoes, halo blight on beans, and bleeding cankers on horse-chestnuts [184]. *P. syringae* has been isolated from *Primula officinalis*, *P. grandiflora*, *P. elatior*, *Dodecatheon pulchellum*, *Silene acaulis*, *Hutchinsia alpina*, *Geranium sylvaticum*, and *Onobrychis montana*. Bacterial populations are minimal on emerging and developing plant tissues, but under favorable conditions, large CFUs have been reported [185–187]. In perennial plants, *P. syringae* manifestations are higher in spring compared to summer [186, 188–192].

Mechanism of Spread

P. syringae can exist in clouds and can be spread through snow and rain (Figure 36.2) [193, 194]. Water-borne *P. syringae* can then reach wild and cultivated plants, and eventually return to clouds as aerosols. At each step of this ecological cycle, strains diversify as selective pressures select for genetic modifications.

Pathogenic Properties and Genes

P. syringae is associated with ice nucleation [180, 195], which can lead to frost injury and destruction of plant tissues [182, 183]. *P. syringae* also regulates activities to enhance its survival on

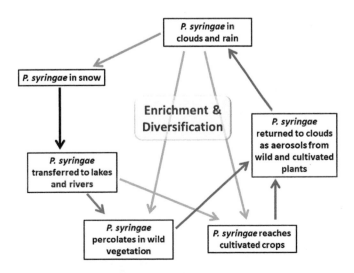

FIGURE 36.2 The transmission cycle of phytopathogenic *P. syringae*. See reference [195] for more information about this subject.

plants, including UV tolerance and production of extracellular polysaccharide to cope with dryness [180]. Alginate has specifically been reported as a virulence factor for *P. syringae* [196]. In some *P. syringae* pathovars (e.g., tomato), pili have been described as important virulence factors [197–202]. Additionally, genes of the *hrp* (hypersensitivity reaction (HR) and pathogenicity) regulon encode a T3SS involved in rapid crumpling and plant death [203].

Another important gene important for *P. syringae* pathogenesis is *gacS*, encoding a sensor kinase [180]. The utility and significance of *gacS* in *P. syringae* seems to be pathovar specific [204]. For example, it is associated with production of syringopeptin toxin in pathovar (pv.) *syringae*, while in pv. *coronafaciens*, it leads to expression of tabtoxin [205]. In addition to toxin regulation, *gacS* also affects production of proteases, QS molecules, alginate, and swarming motility [206–210]; multiple genes under the *gacS* regulon have been reported [207].

Bioremediation by *P. putida*

P. putida has been quite useful for environmental biotechnological applications. *P. putida* strain KT2440 [211, 212] is particularly utilized in environmental remediation applications. Additionally, it populates the rhizosphere of cultivated plants, and thus enhances plant growth [213].

Metabolic Potential for Bioremediation

P. putida can catabolize a wide array of compounds [213]. For this activity, it possesses a vast repertoire of degradative enzymes (mostly oxygenases and oxidoreductases) [214] to break down environmental pollutants (Table 36.6). Genetic modification has created *P. putida* strains capable of degrading many environmental pollutants such as hydrocarbons [215], naphthalene [216], and 1,2,3-trichloropropane (TCP) [217]. Some of its catabolic substrates are lignin derivatives produced during plant disintegration [218]. *P. putida* breaks these compounds down to precursors that are then fed into central metabolism [219, 220]. Importantly, this species contains excellent chemosensory mechanisms [213].

TABLE 36.6

Common Catabolic Enzymes Chromosomally Encoded in *P. putida*

Enzyme Type	Putative Number	References
Dioxygenase	18	[275, 276]
Monooxygenase	15	[213]
Oxidoreductase	80	[213]
XenB (PP0920: metabolizes trinitrotoluene (TNT))	1	[277, 278]
Hydrolase	51	[213]
Transferase	62	[213]
Dehydrogenase	40	[213]
Enzymes encoded by the *ssuFBCDAE* operon	6	[213]
Glutathione-S-transferase	12	[279]
Chlorohydrolase	3	[213]

In addition to its metabolic flexibility, *P. putida* has extensive transporter systems to facilitate transport of substrates and end products into and out of the cell [213], such as PcaK [221] for 4-hydroxybenzoate transport and members of the BenF/PhaK/OprD porin family [222, 223]. The presence of 11 LysE family amino acid efflux transporter proteins [213] helps alleviate deleterious levels of amino acids or their analogues.

P. putida *in the Rhizosphere*

ABC transporters facilitate *P. putida* survival in the rhizosphere [213]. As in other root associated microbes, *P. putida* KT2440 has a putative ABC family opine transporter [224] and other enzymes for metabolizing opine, which is a plant produced molecule. Additionally, *P. putida* can protect its host by expressing a type VI secretion system that injects toxins into a wide range of plant pathogens, thereby inhibiting their growth [225]. Additionally, it has an efflux mechanism for removal of fusaric acid, a toxin produced by plant pathogens like *Fusarium oxysporum* [226], suggesting that *P. putida* could be used as an effective biopesticide [213].

P. fluorescens, the Phyto-Benefactor

P. fluorescens is widely accepted as a promoter of plant health and productivity [227]. This microbe is metabolically versatile and adaptive [228], and it aids in crop growth and expansion [229, 230] by producing metabolites like phytohormones [231–233]. In some cases, it has been reported to produce proteins that limit frost injury in plants [234, 235], while in other instances, this species boosts plants' nutrition [236] by increasing the availability of key nutrients [237, 238]. Additionally, rhizosphere-dwelling *P. fluorescens* produce secondary metabolites like antibiotics [227] that limit the growth of deleterious populations of pathogenic organisms around the plant. Several phenazine compounds produced by *P. fluorescens* act as antimicrobial compounds against fungal, bacterial, and oomycete plant pathogens [239]. Altogether, the extraordinary ability of *P. fluorescens* to thrive in the rhizosphere yields great benefits to agriculture [227]. *P. fluorescens* strain Pf-5 has been well characterized for its role in aiding plant growth [227].

Fitness to Exist in the Rhizosphere

Pf-5 expresses a variety of factors that promotes its growth in the rhizosphere. For instance, it can metabolize a wide range of compounds commonly found in this environment, many of which are organic molecules exuded by roots or other plant tissues [240, 241]. Additionally, Pf-5 can outcompete other microbes in the environment, and this feature has rendered it a promising biocontrol agent [227]. Pf-5 exhibits resistance to a range of antibiotics [227] and toxins produced by plant pathogens [242], and it has a large repertoire of efflux pumps for removal of toxic compounds [243]. Furthermore, it expresses ten peroxidases, six catalases, and several superoxide dismutases to cope with oxidative stress [244].

Regulatory Mechanisms

P. fluorescens contains a multitude of regulatory genes in its large genome (7.06 Mb) to adjust to various environmental niches [227].

Among these are 68 putative sensor histidine kinases and 113 response regulators; also it has the highest count of ECF sigma factors (27) amongst all *Proteobacteria* [227]. Several antisigma factors [245] have also been reported in this species.

Role of Pf-5 in Biocontrol

P. fluorescens strain Pf-5 promotes plant growth because of its propensity to produce numerous antimicrobials [246] and other secondary metabolites [247]. Almost 6% of the Pf-5 genome encodes secondary metabolites [227]. Antibiotics produced by Pf-5 include pyoluteorin [248–250], 2,4-diacetylphloroglucinol (DAPG), and pyrrolnitrin [251]. HCN is also produced by Pf-5 [252]. The surfactant lipopeptide is effective against the phytopathogen *Phytophthora ramorum* [253]. Pf-5 also produces chitinase [254] and can thus serve to degrade chitin in fungal cell walls. Other prominent antimicrobials include bacteriocins [247] and exoenzymes [227]. To further strengthen its role as a biocontrol agent, Pf-5 has a homolog of the *mcf* (*makes* caterpillars *f*loppy) insect toxin gene also found in the bacterium *Photorhabdus luminescens* [255]. Moreover, Pf-5 is devoid of numerous virulence elements found in other *Pseudomonas* species, such as T3SS, toxins, and exoenzymes [227]. This lack of plant destructive factors enhances the agricultural suitability of *P. fluorescens*.

Conclusion

Pseudomonas is a genus that continues to fascinate microbiologists with its phenotypic and genotypic diversity. In some cases, as with the human and plant pathogens *P. aeruginosa* and *P. syringae*, respectively, this diversity and is highly detrimental, while in others, like *P. putida* and *P. fluorescens*, metabolic versatility can be exploited to our advantage. As additional genomic and transcriptomic data is revealed, we will likely discover novel properties of this extraordinary genus.

Acknowledgments

G. Anderson was supported by RSFG from IUPUI.

REFERENCES

1. Ashish, J., G.N. Warghane, B.B. Wagh, S.P. Nag, M.L. Jisnani, R. Thaware, and H. S. Kitey, *Isolation and characterization of Pseudomonas species from Godavari river sample. Asiatic J Biotechnol Res*, 2011. **2**(7): p. 862–866.
2. Moore, E.B., et al., *Nonmedical: Pseudomonas*, in *The Prokaryotes*, M. Dworkin, et al., Editors. 2006, Springer: New York. p. 646–703.
3. Schroth, M., D. C. Hildebrand, and N. Panopoulos, *Phytopathogenic pseudomonads and plant-associated pseudomonads*, in *The Prokaryotes*, 2nd ed., H.G.T. A. Balows, M. Dworkin, W. Harder, and K. H. Schleifer, Editor. 1992, Springer-Verlag: New York, NY, p. 3104–3131.
4. Palleroni, N.J., *Present situation of the taxonomy of aerobic pseudomonads.*, in *Pseudomonas: Molecular Biology and Biotechnology*, S.S. E. Galli and B. Witholt, Editors. 1992, ASM Press: Washington, DC, p. 105–115.
5. Dorn, E., et al., *Isolation and characterization of a 3-chlorobenzoate degrading pseudomonad. Arch Microbiol*, 1974. **99**(1): p. 61–70.
6. Zeyer, J., P.R. Lehrbach, and K.N. Timmis, *Use of cloned genes of Pseudomonas TOL plasmid to effect biotransformation of benzoates to cis-dihydrodiols and catechols by Escherichia coli cells. Appl Environ Microbiol*, 1985. **50**(6): p. 1409–13.
7. Olivera, E.R., et al., *Genetically engineered Pseudomonas: a factory of new bioplastics with broad applications. Environ Microbiol*, 2001. **3**(10): p. 612–8.
8. Palleroni, N.J., D.H. Pieper, and E.R.B. Moore, *Microbiology of Hydrocarbon-Degrading Pseudomonas*, in *Handbook of Hydrocarbon and Lipid Microbioloyg*, K. Timmis, Editor. 2010, Springer: Berlin Heidelberg, p. 1787–1798.
9. Bergey, D.H., F.C. Harrison, R.S. Breed, B.W. Hammer, and F.M. Huntoon, *Bergey's Manual of Determinative Bacteriology*, 1st ed. 1923, Williams and Wilkins: Baltimore, MD.
10. Palleroni, N.J., *Pseudomonas*, in Bergey's Manual of Systematic Bacteriology. *Part B: The Proteobacteria*, D.J. Brenner, Krieg, N.R., and Staley, J.T., Editors. 2005, Springer: New York, NY, p. 323–379.
11. Peix, A., M.H. Ramirez-Bahena, and E. Velazquez, *The current status on the taxonomy of Pseudomonas revisited: An update. Infect Genet Evol*, 2018. **57**: p. 106–116.
12. Denner, E.B., et al., *Reclassification of Pseudomonas echinoides Heumann 1962, 343AL, in the genus Sphingomonas as Sphingomonas echinoides comb. nov. Int J Syst Bacteriol*, 1999. **49**(Pt 3): p. 1103–9.
13. Kampfer, P., E. Falsen, and H.J. Busse, *Reclassification of Pseudomonas mephitica Claydon and Hammer 1939 as a later heterotypic synonym of Janthinobacterium lividum (Eisenberg 1891) De Ley et al. 1978. Int J Syst Evol Microbiol*, 2008. **58**(Pt 1): p. 136–8.
14. Palleroni, N.J., et al., *Taxonomy of the aerobic pseudomonads: the properties of the Pseudomonas stutzeri group. J Gen Microbiol*, 1970. **60**(2): p. 215–31.
15. Shinoda, S. and K. Okamoto, *Formation and function of Vibrio parahaemolyticus lateral flagella. J Bacteriol*, 1977. **129**(3): p. 1266–71.
16. Dasgupta, N., S.K. Arora, and R. Ramphal, *fleN, a gene that regulates flagellar number in Pseudomonas aeruginosa. J Bacteriol*, 2000. **182**(2): p. 357–64.
17. Doig, P., et al., *Role of pili in adhesion of Pseudomonas aeruginosa to human respiratory epithelial cells. Infect Immun*, 1988. **56**(6): p. 1641–6.
18. Vesper, S.J., *Production of Pili (Fimbriae) by Pseudomonas fluorescens and correlation with attachment to corn roots. Appl Environ Microbiol*, 1987. **53**(7): p. 1397–405.
19. West, S.E. and B.H. Iglewski, *Codon usage in Pseudomonas aeruginosa. Nucleic Acids Res*, 1988. **16**(19): p. 9323–35.
20. Meyer, J.M., et al., *Siderophore typing, a powerful tool for the identification of fluorescent and nonfluorescent pseudomonads. Appl Environ Microbiol*, 2002. **68**(6): p. 2745–53.
21. Den Dooren de Jong, L.E., *Bijdrage tot de kennis vanhet mineralisatieproces: Thesis*. 1926, Technische Hogeschool, Delft. Nijgh & Van Ditmar.: Rotterdam, the Netherlands. p. 1–199.
22. Stanier, R.Y., N.J. Palleroni, and M. Doudoroff, *The aerobic pseudomonads: a taxonomic study. J Gen Microbiol*, 1966. **43**(2): p. 159–271.

23. Palleroni, N.J. and M. Doudoroff, *Some properties and subdivisions of the genus Pseudomonas. Ann Rev Phytopathol,* 1972. **10**: p. 73–100.

24. Sze, C.C., T. Moore, and V. Shingler, *Growth phase-dependent transcription of the sigma(54)-dependent Po promoter controlling the Pseudomonas-derived (methyl)phenol dmp operon of pVI150. J Bacteriol,* 1996. **178**(13): p. 3727–35.

25. Sze, C.C., L.M. Bernardo, and V. Shingler, *Integration of global regulation of two aromatic-responsive sigma(54)-dependent systems: a common phenotype by different mechanisms. J Bacteriol,* 2002. **184**(3): p. 760–70.

26. Cases, I., V. de Lorenzo, and J. Perez-Martin, *Involvement of sigma 54 in exponential silencing of the Pseudomonas putida TOL plasmid Pu promoter. Mol Microbiol,* 1996. **19**(1): p. 7–17.

27. Cases, I., F. Velazquez, and V. de Lorenzo, *Role of ptsO in carbon-mediated inhibition of the Pu promoter belonging to the pWW0 Pseudomonas putida plasmid. J Bacteriol,* 2001. **183**(17): p. 5128–33.

28. Collier, D.N., P.W. Hager, and P.V. Phibbs, Jr., *Catabolite repression control in the Pseudomonads. Res Microbiol,* 1996. **147**(6-7): p. 551–61.

29. Rojo, F., and A. Dinamarca, *Catabolite Repression and Physiological Control,* in *Pseudomonas,* J.L. Ramos, Editor. 2004, Kluwer Academic/Plenum Publishers: New York, NY, p. 365–387.

30. MacGregor, C.H., et al., *The nucleotide sequence of the Pseudomonas aeruginosa pyrE-crc-rph region and the purification of the crc gene product. J Bacteriol,* 1996. **178**(19): p. 5627–35.

31. Canosa, I., et al., *A positive feedback mechanism controls expression of AlkS, the transcriptional regulator of the Pseudomonas oleovorans alkane degradation pathway. Mol Microbiol,* 2000. **35**(4): p. 791–9.

32. Yuste, L. and F. Rojo, *Role of the crc gene in catabolic repression of the Pseudomonas putida GPol alkane degradation pathway. J Bacteriol,* 2001. **183**(21): p. 6197–206.

33. Petruschka, L., et al., *The cyo operon of Pseudomonas putida is involved in carbon catabolite repression of phenol degradation. Mol Genet Genomics,* 2001. **266**(2): p. 199–206.

34. Hechtman, P. and C.R. Scriver, *Neutral amino acid transport in Pseudomonas fluorescens. J Bacteriol,* 1970. **104**(2): p. 857–63.

35. Jann, A., H. Matsumoto, and D. Haas, *The fourth arginine catabolic pathway of Pseudomonas aeruginosa. J Gen Microbiol,* 1988. **134**(4): p. 1043–53.

36. Stalon, V., et al., *Catabolism of arginine, citrulline and ornithine by Pseudomonas and related bacteria. J Gen Microbiol,* 1987. **133**(9): p. 2487–95.

37. Lu, C.D., Z. Yang, and W. Li, *Transcriptome analysis of the ArgR regulon in Pseudomonas aeruginosa. J Bacteriol,* 2004. **186**(12): p. 3855–61.

38. Byng, G.S., et al., *The evolutionary pattern of aromatic amino acid biosynthesis and the emerging phylogeny of pseudomonad bacteria. J Mol Evol,* 1983. **19**(3-4): p. 272–82.

39. Byng, G.S., et al., *Variable enzymological patterning in tyrosine biosynthesis as a means of determining natural relatedness among the Pseudomonadaceae. J Bacteriol,* 1980. **144**(1): p. 247–57.

40. Byng, G.S., J.F. Kane, and R.A. Jensen, *Diversity in the routing and regulation of complex biochemical pathways as indicators of microbial relatedness. Crit Rev Microbiol,* 1982. **9**(4): p. 227–52.

41. Whitaker, R.J., et al., *Diverse enzymological patterns of phenylalanine biosynthesis in pseudomonads are conserved in parallel with deoxyribonucleic acid homology groupings. J Bacteriol,* 1981. **147**(2): p. 526–34.

42. Miller, D.L. and V.W. Rodwell, *Metabolism of basic amino acids in Pseudomonas putida. Intermediates in L-arginine catabolism. J Biol Chem,* 1971. **246**(16): p. 5053–8.

43. Eschbach, M., et al., *Long-term anaerobic survival of the opportunistic pathogen Pseudomonas aeruginosa via pyruvate fermentation. J Bacteriol,* 2004. **186**(14): p. 4596–604.

44. van Hartingsveldt, J. and A.H. Stouthamer, *Mapping and characterization of mutants of Pseudomonas aeruginosa affected in nitrate respiration in aerobic or anaerobic growth. J Gen Microbiol,* 1973. **74**(1): p. 97–106.

45. Yan, Y., et al., *Nitrogen fixation island and rhizosphere competence traits in the genome of root-associated Pseudomonas stutzeri A1501. Proc Natl Acad Sci USA,* 2008. **105**(21): p. 7564–9.

46. Wolterink, A.F., et al., *Pseudomonas chloritidismutans sp. nov., a non-denitrifying, chlorate-reducing bacterium. Int J Syst Evol Microbiol,* 2002. **52**(Pt 6): p. 2183–90.

47. Rojo, F., *Carbon catabolite repression in Pseudomonas: optimizing metabolic versatility and interactions with the environment. FEMS Microbiol Rev,* 2010. **34**(5): p. 658–84.

48. van der Kooij, D., *The occurrence of Pseudomonas spp. in surface water and in tap water as determined on citrate media. Antonie Van Leeuwenhoek,* 1977. **43**(2): p. 187–97.

49. Reddy, G.S., et al., *Psychrophilic pseudomonads from Antarctica: Pseudomonas antarctica sp. nov., Pseudomonas meridiana sp. nov. and Pseudomonas proteolytica sp. nov. Int J Syst Evol Microbiol,* 2004. **54**(Pt 3): p. 713–9.

50. Mena, K.D. and C.P. Gerba, *Risk assessment of Pseudomonas aeruginosa in water. Rev Environ Contam Toxicol,* 2009. **201**: p. 71–115.

51. Schreiber, K., et al., *Anaerobic survival of Pseudomonas aeruginosa by pyruvate fermentation requires an Usp-type stress protein. J Bacteriol,* 2006. **188**(2): p. 659–68.

52. Green, S.K., et al., *Agricultural plants and soil as a reservoir for Pseudomonas aeruginosa. Appl Microbiol,* 1974. **28**(6): p. 987–91.

53. Nogales, B., et al., *Identification of the metabolically active members of a bacterial community in a polychlorinated biphenyl-polluted moorland soil. Environ Microbiol,* 1999. **1**(3): p. 199–212.

54. Nogales, B., et al., *Combined use of 16S ribosomal DNA and 16S rRNA to study the bacterial community of polychlorinated biphenyl-polluted soil. Appl Environ Microbiol,* 2001. **67**(4): p. 1874–84.

55. Rainey, P.B., *Adaptation of Pseudomonas fluorescens to the plant rhizosphere. Environ Microbiol,* 1999. **1**(3): p. 243–57.

56. Lugtenberg, B.J., L. Dekkers, and G.V. Bloemberg, *Molecular determinants of rhizosphere colonization by Pseudomonas. Annu Rev Phytopathol,* 2001. **39**: p. 461–90.

57. Hiltner, L., *Über neuer Erfahrungen und Probleme auf dem Gebiete der Bodenbakteriologie unter bessonderer Berüksichtigung der Gründung und Brache. Arb Dtsch Landwirtsch Ges Berl,* 1904. **98**: p. 59–78.

58. Faraji, M., M. Ahmadzadeh, K. Behboudi, S.M. Okhovvat, M. Ruocco, M. Lorito, M. Rezaei-Tavirani, and H. Zali, *Study of proteome pattern of Pseudomonas fluorescens strain UTPF68 in Interaction with Trichoderma atroviride strain P1 and tomato. J Paramed Sci,* 2013. **4**(1).

59. Xin, X.F. and S.Y. He, *Pseudomonas syringae pv. tomato DC3000: a model pathogen for probing disease susceptibility and hormone signaling in plants. Annu Rev Phytopathol*, 2013. **51**: p. 473–98.

60. Brandl, M.T., B. Quinones, and S.E. Lindow, *Heterogeneous transcription of an indoleacetic acid biosynthetic gene in Erwinia herbicola on plant surfaces. Proc Natl Acad Sci USA*, 2001. **98**(6): p. 3454–9.

61. Adetuyi, F.C., et al., *Saprophytic Pseudomonas syringae strain M1 of wheat produces cyclic lipodepsipeptides. FEMS Microbiol Lett*, 1995. **131**(1): p. 63–7.

62. Abboud, M.M., et al., *Copper uptake by Pseudomonas aeruginosa isolated from infected burn patients. Curr Microbiol*, 2009. **59**(3): p. 282–7.

63. Anderson, G.G., et al., *The Pseudomonas aeruginosa magnesium transporter MgtE inhibits transcription of the type III secretion system. Infect Immun*, 2010. **78**(3): p. 1239–49.

64. Bao, Z., et al., *Genomic plasticity enables phenotypic variation of Pseudomonas syringae pv. tomato DC3000. PLOS One*, 2014. **9**(2): p. e86628.

65. Decoin, V., et al., *A type VI secretion system is involved in Pseudomonas fluorescens bacterial competition. PLOS One*, 2014. **9**(2): p. e89411.

66. Aagot, N., et al., *An altered Pseudomonas diversity is recovered from soil by using nutrient-poor Pseudomonas-selective soil extract media. Appl Environ Microbiol*, 2001. **67**(11): p. 5233–9.

67. Pichinoty, F., et al., *[Study of 14 denitrifying soil bacteria of the "pseudomonas stutzeri" group isolated by enrichment culture in the presence of nitrous oxide (author's transl)]. Ann Microbiol (Paris)*, 1977. **128a**(1): p. 75–87.

68. King, E.O., M.K. Ward, and D.E. Raney, *Two simple media for the demonstration of pyocyanin and fluorescin. J Lab Clin Med*, 1954. **44**(2): p. 301–7.

69. Gould, W.D., et al., *New selective media for enumeration and recovery of fluorescent pseudomonads from various habitats. Appl Environ Microbiol*, 1985. **49**(1): p. 28–32.

70. Gilardi, G.L., Pseudomonas, in *Manual of Clinical Microbiology*, 4th ed., E.H. Lennette, A. Balows, W.J. Hausler Jr., and H.J. Shadomy, Editors. 1985, ASM Press: Washington, DC, p. 350–372.

71. Mohn, W.W., et al., *Physiological and phylogenetic diversity of bacteria growing on resin acids. Syst Appl Microbiol*, 1999. **22**(1): p. 68–78.

72. Williams, P. and M. Camara, *Quorum sensing and environmental adaptation in Pseudomonas aeruginosa: a tale of regulatory networks and multifunctional signal molecules. Curr Opin Microbiol*, 2009. **12**(2): p. 182–91.

73. Moreno, R., et al., *The Pseudomonas putida Crc global regulator is an RNA binding protein that inhibits translation of the AlkS transcriptional regulator. Mol Microbiol*, 2007. **64**(3): p. 665–75.

74. Balasubramanian, D., et al., *A dynamic and intricate regulatory network determines Pseudomonas aeruginosa virulence. Nucleic Acids Res*, 2013. **41**(1): p. 1–20.

75. Marques, S., et al., *Transcriptional induction kinetics from the promoters of the catabolic pathways of TOL plasmid pWW0 of Pseudomonas putida for metabolism of aromatics. J Bacteriol*, 1994. **176**(9): p. 2517–24.

76. Martinez-Bueno, M.A., et al., *Detection of multiple extra-cytoplasmic function (ECF) sigma factors in the genome of Pseudomonas putida KT2440 and their counterparts in Pseudomonas aeruginosa PA01. Environ Microbiol*, 2002. **4**(12): p. 842–55.

77. Govan, J.R. and V. Deretic, *Microbial pathogenesis in cystic fibrosis: mucoid Pseudomonas aeruginosa and Burkholderia cepacia. Microbiol Rev*, 1996. **60**(3): p. 539–74.

78. Schurr, M.J. and V. Deretic, *Microbial pathogenesis in cystic fibrosis: co-ordinate regulation of heat-shock response and conversion to mucoidy in Pseudomonas aeruginosa. Mol Microbiol*, 1997. **24**(2): p. 411–20.

79. Wilson, M.J., B.J. McMorran, and I.L. Lamont, *Analysis of promoters recognized by PvdS, an extracytoplasmic-function sigma factor protein from Pseudomonas aeruginosa. J Bacteriol*, 2001. **183**(6): p. 2151–5.

80. Sexton, R., et al., *Iron-responsive gene expression in Pseudomonas fluorescens M114: cloning and characterization of a transcription-activating factor, PbrA. Mol Microbiol*, 1995. **15**(2): p. 297–306.

81. Valls, M., I. Cases, and V. de Lorenzo., Transcription Mediated by rpoN-dependent Promoters, in *Pseudomonas*, J.L. Ramos, Editor. 2004, Kluwer Academic/Plenum Publishers: New York, NY, p. 289–318.

82. Kohler, T., et al., *Involvement of Pseudomonas putida RpoN sigma factor in regulation of various metabolic functions. J Bacteriol*, 1989. **171**(8): p. 4326–33.

83. Totten, P.A., J.C. Lara, and S. Lory, *The rpoN gene product of Pseudomonas aeruginosa is required for expression of diverse genes, including the flagellin gene. J Bacteriol*, 1990. **172**(1): p. 389–96.

84. Hendrickson, E.L., P. Guevera, and F.M. Ausubel, *The alternative sigma factor RpoN is required for hrp activity in Pseudomonas syringae pv. maculicola and acts at the level of hrpL transcription. J Bacteriol*, 2000. **182**(12): p. 3508–16.

85. Lloyd, M.G., et al., *Blocking RpoN reduces virulence of Pseudomonas aeruginosa isolated from cystic fibrosis patients and increases antibiotic sensitivity in a laboratory strain. Sci Rep*, 2019. **9**(1): p. 6677–6677.

86. Knoten, C.A., et al., *KynR, a Lrp/AsnC-type transcriptional regulator, directly controls the kynurenine pathway in Pseudomonas aeruginosa. J Bacteriol*, 2011. **193**(23): p. 6567–75.

87. Ramos, J.L., S. Marques, and K.N. Timmis, *Transcriptional control of the Pseudomonas TOL plasmid catabolic operons is achieved through an interplay of host factors and plasmid-encoded regulators. Annu Rev Microbiol*, 1997. **51**: p. 341–73.

88. Arenghi, F.L., et al., *New insights into the activation of o-xylene biodegradation in Pseudomonas stutzeri OX1 by pathway substrates. EMBO Rep*, 2001. **2**(5): p. 409–14.

89. Jaspers, M.C., et al., *Transcriptional organization and dynamic expression of the hbpCAD genes, which encode the first three enzymes for 2-hydroxybiphenyl degradation in Pseudomonas azelaica HBP1. J Bacteriol*, 2001. **183**(1): p. 270–9.

90. Shingler, V., Transcriptional Regulation and Catabolic Strategies of Phenol Degradative Pathways, in *Pseudomonas*, J.L. Ramos, Editor. 2004, Kluwer Academic/Plenum Publishers: New York, NY, p. 451–477.

91. Intile, P.J., et al., *The AlgZR two-component system recalibrates the RsmAYZ posttranscriptional regulatory system to inhibit expression of the Pseudomonas aeruginosa type III secretion system. J Bacteriol*, 2014. **196**(2): p. 357–66.

92. Chakravarty, S. and E. Masse, *RNA-dependent regulation of virulence in pathogenic bacteria. Front Cell Infect Microbiol*, 2019. **9**: p. 337.

93. Allsopp, L.P., et al., *RsmA and AmrZ orchestrate the assembly of all three type VI secretion systems in Pseudomonas aeruginosa. Proc Natl Acad Sci USA*, 2017. **114**(29): p. 7707–7712.

94. Baker, C.S., et al., *CsrA inhibits translation initiation of Escherichia coli hfq by binding to a single site overlapping the Shine-Dalgarno sequence. J Bacteriol*, 2007. **189**(15): p. 5472–81.

95. Irie, Y., et al., *Pseudomonas aeruginosa biofilm matrix polysaccharide Psl is regulated transcriptionally by RpoS and post-transcriptionally by RsmA. Mol Microbiol*, 2010. **78**(1): p. 158–72.

96. Dubey, A.K., et al., *RNA sequence and secondary structure participate in high-affinity CsrA-RNA interaction. RNA*, 2005. **11**(10): p. 1579–87.

97. Wei, B.L., et al., *Positive regulation of motility and flhDC expression by the RNA-binding protein CsrA of Escherichia coli. Mol Microbiol*, 2001. **40**(1): p. 245–56.

98. Wang, X., et al., *CsrA post-transcriptionally represses pgaABCD, responsible for synthesis of a biofilm polysaccharide adhesin of Escherichia coli. Mol Microbiol*, 2005. **56**(6): p. 1648–63.

99. Patterson-Fortin, L.M., et al., *Dual posttranscriptional regulation via a cofactor-responsive mRNA leader. J Mol Biol*, 2013. **425**(19): p. 3662–77.

100. Janssen, K.H., et al., *RsmV, a small noncoding regulatory RNA in Pseudomonas aeruginosa that sequesters RsmA and RsmF from target mRNAs. J Bacteriol*, 2018. **200**(16).

101. Miller, C.L., et al., *RsmW, Pseudomonas aeruginosa small non-coding RsmA-binding RNA upregulated in biofilm versus planktonic growth conditions. BMC Microbiol*, 2016. **16**(1): p. 155.

102. Blumer, C., et al., *Global GacA-steered control of cyanide and exoprotease production in Pseudomonas fluorescens involves specific ribosome binding sites. Proc Natl Acad Sci USA*, 1999. **96**(24): p. 14073–8.

103. Haas, D., C. Keel, and C. Reimmann, *Signal transduction in plant-beneficial rhizobacteria with biocontrol properties. Antonie Van Leeuwenhoek*, 2002. **81**(1-4): p. 385–95.

104. Heeb, S., C. Blumer, and D. Haas, *Regulatory RNA as mediator in GacA/RsmA-dependent global control of exoproduct formation in Pseudomonas fluorescens CHA0. J Bacteriol*, 2002. **184**(4): p. 1046–56.

105. Pessi, G., and D. Haas, Cyanogenesis, in *Pseudomonas*, J.L. Ramos, Editor. 2004, Kluwer Academic/Plenum Publishers: New York, NY, p. 671–687.

106. Kipnis, E., T. Sawa, and J. Wiener-Kronish, *Targeting mechanisms of Pseudomonas aeruginosa pathogenesis. Med Mal Infect*, 2006. **36**(2): p. 78–91.

107. Chastre, J. and J.Y. Fagon, *Ventilator-associated pneumonia. Am J Respir Crit Care Med*, 2002. **165**(7): p. 867–903.

108. Chastre, J. and J.L. Trouillet, *Problem pathogens (Pseudomonas aeruginosa and Acinetobacter). Semin Respir Infect*, 2000. **15**(4): p. 287–98.

109. Gomez, M.I. and A. Prince, *Opportunistic infections in lung disease: Pseudomonas infections in cystic fibrosis. Curr Opin Pharmacol*, 2007. **7**(3): p. 244–51.

110. Diaz, M.R., J.M. King, and T.L. Yahr, *Intrinsic and extrinsic regulation of type III secretion gene expression in Pseudomonas aeruginosa. Front Microbiol*, 2011. **2**: p. 89.

111. Leidal, K.G., K.L. Munson, and G.M. Denning, *Small molecular weight secretory factors from Pseudomonas aeruginosa have opposite effects on IL-8 and RANTES expression by human airway epithelial cells. Am J Respir Cell Mol Biol*, 2001. **25**(2): p. 186–95.

112. Denning, G.M., et al., *Pseudomonas pyocyanin increases interleukin-8 expression by human airway epithelial cells. Infect Immun*, 1998. **66**(12): p. 5777–84.

113. Meyer, J.M., et al., *Pyoverdin is essential for virulence of Pseudomonas aeruginosa. Infect Immun*, 1996. **64**(2): p. 518–23.

114. Takase, H., et al., *Impact of siderophore production on Pseudomonas aeruginosa infections in immunosuppressed mice. Infect Immun*, 2000. **68**(4): p. 1834–9.

115. Matsumoto, K., *Role of bacterial proteases in pseudomonal and serratial keratitis. Biol Chem*, 2004. **385**(11): p. 1007–16.

116. Azghani, A.O., *Pseudomonas aeruginosa and epithelial permeability: role of virulence factors elastase and exotoxin A. Am J Respir Cell Mol Biol*, 1996. **15**(1): p. 132–40.

117. Azghani, A.O., L.D. Gray, and A.R. Johnson, *A bacterial protease perturbs the paracellular barrier function of transporting epithelial monolayers in culture. Infect Immun*, 1993. **61**(6): p. 2681–6.

118. Azghani, A.O., E.J. Miller, and B.T. Peterson, *Virulence factors from Pseudomonas aeruginosa increase lung epithelial permeability. Lung*, 2000. **178**(5): p. 261–9.

119. Wiener-Kronish, J.P., et al., *Alveolar epithelial injury and pleural empyema in acute P. aeruginosa pneumonia in anesthetized rabbits. J Appl Physiol (1985)*, 1993. **75**(4): p. 1661–9.

120. Telford, G., et al., *The Pseudomonas aeruginosa quorum-sensing signal molecule N-(3-oxododecanoyl)-L-homoserine lactone has immunomodulatory activity. Infect Immun*, 1998. **66**(1): p. 36–42.

121. Smith, R.S., et al., *The Pseudomonas aeruginosa quorum-sensing molecule N-(3-oxododecanoyl)homoserine lactone contributes to virulence and induces inflammation in vivo. J Bacteriol*, 2002. **184**(4): p. 1132–9.

122. Smith, R.S., et al., *IL-8 production in human lung fibroblasts and epithelial cells activated by the Pseudomonas autoinducer N-3-oxododecanoyl homoserine lactone is transcriptionally regulated by NF-kappa B and activator protein-2. J Immunol*, 2001. **167**(1): p. 366–74.

123. Smith, R.S., et al., *The Pseudomonas autoinducer N-(3-oxododecanoyl) homoserine lactone induces cyclooxygenase-2 and prostaglandin E2 production in human lung fibroblasts: implications for inflammation. J Immunol*, 2002. **169**(5): p. 2636–42.

124. Valentini, M., et al., *Lifestyle transitions and adaptive pathogenesis of Pseudomonas aeruginosa. Curr Opin Microbiol*, 2018. **41**: p. 15–20.

125. Feldman, M., et al., *Role of flagella in pathogenesis of Pseudomonas aeruginosa pulmonary infection. Infect Immun*, 1998. **66**(1): p. 43–51.

126. Adamo, R., et al., *Pseudomonas aeruginosa flagella activate airway epithelial cells through asialoGM1 and toll-like receptor 2 as well as toll-like receptor 5. Am J Respir Cell Mol Biol*, 2004. **30**(5): p. 627–34.

127. DiMango, E., et al., *Diverse Pseudomonas aeruginosa gene products stimulate respiratory epithelial cells to produce interleukin-8. J Clin Invest*, 1995. **96**(5): p. 2204–10.

128. Landsperger, W.J., et al., *Inhibition of bacterial motility with human antiflagellar monoclonal antibodies attenuates Pseudomonas aeruginosa-induced pneumonia in the immunocompetent rat. Infect Immun*, 1994. **62**(11): p. 4825–30.

129. Doring, G., et al., *Parenteral application of a Pseudomonas aeruginosa flagella vaccine elicits specific anti-flagella antibodies in the airways of healthy individuals. Am J Respir Crit Care Med*, 1995. **151**(4): p. 983–5.

130. Doring, G. and F. Dorner, *A multicenter vaccine trial using the Pseudomonas aeruginosa flagella vaccine IMMUNO in patients with cystic fibrosis. Behring Inst Mitt*, 1997(98): p. 338–44.

131. Farinha, M.A., et al., *Alteration of the pilin adhesin of Pseudomonas aeruginosa PAO results in normal pilus biogenesis but a loss of adherence to human pneumocyte cells and decreased virulence in mice. Infect Immun*, 1994. **62**(10): p. 4118–23.

132. Comolli, J.C., et al., *Pseudomonas aeruginosa gene products PilT and PilU are required for cytotoxicity in vitro and virulence in a mouse model of acute pneumonia. Infect Immun*, 1999. **67**(7): p. 3625–30.

133. Tang, H., M. Kays, and A. Prince, *Role of Pseudomonas aeruginosa pili in acute pulmonary infection. Infect Immun*, 1995. **63**(4): p. 1278–85.

134. Hahn, H.P., *The type-4 pilus is the major virulence-associated adhesin of Pseudomonas aeruginosa–a review. Gene*, 1997. **192**(1): p. 99–108.

135. Gupta, S.K., et al., *Pili and lipopolysaccharide of Pseudomonas aeruginosa bind to the glycolipid asialo GM1. Infect Immun*, 1994. **62**(10): p. 4572–9.

136. Mattick, J.S., *Type IV pili and twitching motility. Annu Rev Microbiol*, 2002. **56**: p. 289–314.

137. Hahn, H., et al., *Pilin-based anti-Pseudomonas vaccines: latest developments and perspectives. Behring Inst Mitt*, 1997(98): p. 315–25.

138. Sheth, H.B., et al., *Development of an anti-adhesive vaccine for Pseudomonas aeruginosa targeting the C-terminal region of the pilin structural protein. Biomed Pept Proteins Nucleic Acids*, 1995. **1**(3): p. 141–8.

139. Cachia, P.J., D.J. Kao, and R.S. Hodges, *Synthetic peptide vaccine development: measurement of polyclonal antibody affinity and cross-reactivity using a new peptide capture and release system for surface plasmon resonance spectroscopy. J Mol Recognit*, 2004. **17**(6): p. 540–57.

140. Hertle, R., R. Mrsny, and D.J. Fitzgerald, *Dual-function vaccine for Pseudomonas aeruginosa: characterization of chimeric exotoxin A-pilin protein. Infect Immun*, 2001. **69**(11): p. 6962–9.

141. Tang, H.B., et al., *Contribution of specific Pseudomonas aeruginosa virulence factors to pathogenesis of pneumonia in a neonatal mouse model of infection. Infect Immun*, 1996. **64**(1): p. 37–43.

142. Backhed, F., et al., *Structural requirements for TLR4-mediated LPS signalling: a biological role for LPS modifications. Microbes Infect*, 2003. **5**(12): p. 1057–63.

143. Hajjar, A.M., et al., *Human Toll-like receptor 4 recognizes host-specific LPS modifications. Nat Immunol*, 2002. **3**(4): p. 354–9.

144. Lee, V.T., et al., *Activities of Pseudomonas aeruginosa effectors secreted by the Type III secretion system in vitro and during infection. Infect Immun*, 2005. **73**(3): p. 1695–705.

145. Goure, J., et al., *The V antigen of Pseudomonas aeruginosa is required for assembly of the functional PopB/PopD translocation pore in host cell membranes. Infect Immun*, 2004. **72**(8): p. 4741–50.

146. Sawa, T., et al., *Active and passive immunization with the Pseudomonas V antigen protects against type III intoxication and lung injury. Nat Med*, 1999. **5**(4): p. 392–8.

147. Neely, A.N., et al., *Passive anti-PcrV treatment protects burned mice against Pseudomonas aeruginosa challenge. Burns*, 2005. **31**(2): p. 153–8.

148. Faure, K., et al., *Effects of monoclonal anti-PcrV antibody on Pseudomonas aeruginosa-induced acute lung injury in a rat model. J Immune Based Ther Vaccines*, 2003. **1**(1): p. 2.

149. Frank, D.W., et al., *Generation and characterization of a protective monoclonal antibody to Pseudomonas aeruginosa PcrV. J Infect Dis*, 2002. **186**(1): p. 64–73.

150. Shime, N., et al., *Therapeutic administration of anti-PcrV F(ab')(2) in sepsis associated with Pseudomonas aeruginosa. J Immunol*, 2001. **167**(10): p. 5880–6.

151. Shaver, C.M. and A.R. Hauser, *Relative contributions of Pseudomonas aeruginosa ExoU, ExoS, and ExoT to virulence in the lung. Infect Immun*, 2004. **72**(12): p. 6969–77.

152. Finck-Barbancon, V., et al., *ExoU expression by Pseudomonas aeruginosa correlates with acute cytotoxicity and epithelial injury. Mol Microbiol*, 1997. **25**(3): p. 547–57.

153. Allewelt, M., et al., *Acquisition of expression of the Pseudomonas aeruginosa ExoU cytotoxin leads to increased bacterial virulence in a murine model of acute pneumonia and systemic spread. Infect Immun*, 2000. **68**(7): p. 3998–4004.

154. Sato, H., et al., *The mechanism of action of the Pseudomonas aeruginosa-encoded type III cytotoxin, ExoU. Embo J*, 2003. **22**(12): p. 2959–69.

155. Tamura, M., et al., *Lysophospholipase A activity of Pseudomonas aeruginosa type III secretory toxin ExoU. Biochem Biophys Res Commun*, 2004. **316**(2): p. 323–31.

156. Pankhaniya, R.R., et al., *Pseudomonas aeruginosa causes acute lung injury via the catalytic activity of the patatin-like phospholipase domain of ExoU. Crit Care Med*, 2004. **32**(11): p. 2293–9.

157. Phillips, R.M., et al., *In vivo phospholipase activity of the Pseudomonas aeruginosa cytotoxin ExoU and protection of mammalian cells with phospholipase A2 inhibitors. J Biol Chem*, 2003. **278**(42): p. 41326–32.

158. Goehring, U.M., et al., *The N-terminal domain of Pseudomonas aeruginosa exoenzyme S is a GTPase-activating protein for Rho GTPases. J Biol Chem*, 1999. **274**(51): p. 36369–72.

159. Fu, H., J. Coburn, and R.J. Collier, *The eukaryotic host factor that activates exoenzyme S of Pseudomonas aeruginosa is a member of the 14-3-3 protein family. Proc Natl Acad Sci USA*, 1993. **90**(6): p. 2320–4.

160. Maresso, A.W., M.R. Baldwin, and J.T. Barbieri, *Ezrin/ radixin/moesin proteins are high affinity targets for ADP-ribosylation by Pseudomonas aeruginosa ExoS. J Biol Chem*, 2004. **279**(37): p. 38402–8.

161. Epelman, S., et al., *Different domains of Pseudomonas aeruginosa exoenzyme S activate distinct TLRs. J Immunol*, 2004. **173**(3): p. 2031–40.

162. J. T. Barbieri, J. Sun, *Pseudomonas aeruginosa ExoS and ExoT. Rev Physiol Biochem Pharmacol*, 2004. **152**: p. 79–92.

163. Garrity-Ryan, L., et al., *The arginine finger domain of ExoT contributes to actin cytoskeleton disruption and inhibition of internalization of Pseudomonas aeruginosa by epithelial cells and macrophages. Infect Immun*, 2000. **68**(12): p. 7100–13.

164. Geiser, T.K., et al., *Pseudomonas aeruginosa ExoT inhibits in vitro lung epithelial wound repair. Cell Microbiol*, 2001. **3**(4): p. 223–36.

165. Yahr, T.L., et al., *ExoY, an adenylate cyclase secreted by the Pseudomonas aeruginosa type III system. Proc Natl Acad Sci USA*, 1998. **95**(23): p. 13899–904.

166. Sayner, S.L., et al., *Paradoxical cAMP-induced lung endothelial hyperpermeability revealed by Pseudomonas aeruginosa ExoY. Circ Res*, 2004. **95**(2): p. 196–203.

167. Sana, T.G., B. Berni, and S. Bleves, *The T6SSs of Pseudomonas aeruginosa strain PAO1 and their effectors: beyond bacterial-cell targeting. Front Cell Infect Microbiol*, 2016. **6**: p. 61.

168. Ho, Brian T., Tao G. Dong, and John J. Mekalanos, *A view to a kill: the bacterial type VI secretion system. Cell Host Microbe*, 2014. **15**(1): p. 9–21.

169. Russell, A.B., et al., *Diverse type VI secretion phospholipases are functionally plastic antibacterial effectors. Nature*, 2013. **496**(7446): p. 508–12.

170. Ko, Y.H. and P.L. Pedersen, *Cystic fibrosis: a brief look at some highlights of a decade of research focused on elucidating and correcting the molecular basis of the disease. J Bioenerg Biomembr*, 2001. **33**(6): p. 513–21.

171. Gibson, R.L., J.L. Burns, and B.W. Ramsey, *Pathophysiology and management of pulmonary infections in cystic fibrosis. Am J Respir Crit Care Med*, 2003. **168**(8): p. 918–51.

172. Redelman, C.V., S. Chakravarty, and G.G. Anderson, *Antibiotic treatment of Pseudomonas aeruginosa biofilms stimulates expression of the magnesium transporter gene mgtE. Microbiology*, 2014. **160**(Pt 1): p. 165–78.

173. Dasgupta, N., et al., *Transcriptional induction of the Pseudomonas aeruginosa type III secretion system by low Ca2+ and host cell contact proceeds through two distinct signaling pathways. Infect Immun*, 2006. **74**(6): p. 3334–41.

174. Jones, A.K., et al., *Activation of the Pseudomonas aeruginosa AlgU regulon through mucA mutation inhibits cyclic AMP/ Vfr signaling. J Bacteriol*, 2010. **192**(21): p. 5709–17.

175. Bjarnsholt, T., et al., *Pseudomonas aeruginosa biofilms in the respiratory tract of cystic fibrosis patients. Pediatr Pulmonol*, 2009. **44**(6): p. 547–58.

176. Lyczak, J.B., C.L. Cannon, and G.B. Pier, *Establishment of Pseudomonas aeruginosa infection: lessons from a versatile opportunist. Microbes Infect*, 2000. **2**(9): p. 1051–60.

177. Gilardi, G.L., *Infrequently encountered Pseudomonas species causing infection in humans. Ann Intern Med*, 1972. **77**(2): p. 211–5.

178. Yoshino, Y., et al., *Pseudomonas putida bacteremia in adult patients: five case reports and a review of the literature. J Infect Chemother*, 2011. **17**(2): p. 278–82.

179. Halabi, Z., et al., *Pseudomonas stutzeri prosthetic valve endocarditis: A case report and review of the literature. J Infect Public Health*, 2019. **12**(3): p. 434–437.

180. Hirano, S.S. and C.D. Upper, *Bacteria in the leaf ecosystem with emphasis on Pseudomonas syringae-a pathogen, ice nucleus, and epiphyte. Microbiol Mol Biol Rev*, 2000. **64**(3): p. 624–53.

181. Baltrus, D.A., H.C. McCann, and D.S. Guttman, *Evolution, genomics and epidemiology of Pseudomonas syringae: challenges in bacterial molecular plant pathology. Mol Plant Pathol*, 2017. **18**(1): p. 152–168.

182. Jones, R.K. and D.M. Benson. *Diseases of Woody Ornamentals and Trees in Nurseries*. 2001, St. Paul, MN: APS Press.

183. Sinclair, W.A., H. H. Lyon, and W. T. Johnson, *Diseases of trees and shrubs*. 1987, Ithaca, NY: Cornell University Press.

184. Preston, G.M., *Profiling the extended phenotype of plant pathogens: challenges in bacterial molecular plant pathology. Mol Plant Pathol*, 2017. **18**(3): p. 443–456.

185. Hirano, S.S., M. K. Clayton, and C. D. Upper, *Estimation of and temporal changes in means and variances of populations of Pseudomonas syringae on snap bean leaflets. Phytopathology*, 1994. **84**: p. 934–940.

186. Lindow, S.E., Population dynamics of epiphytic ice nucleation active bacteria on frost sensitive plants and frost control by means of antagonistic bacteria, in *Plant Cold Hardiness and Freezing Stress—Mechanisms and Crop Implications*, P.H. Li and A. Sakai, Editors. 1982, Academic Press: New York, NY, p. 395–416.

187. Smitley, D.R., and S. M. McCarter, *Spread of Pseudomonas syringae pv. tomato and role of epiphytic populations and environmental conditions in disease development. Plant Dis*, 1982. **66**: p. 713–717.

188. Gross, D.C., et al., *Distribution, population dynamics, and characteristics of ice nucleation-active bacteria in deciduous fruit tree orchards. Appl Environ Microbiol*, 1983. **46**(6): p. 1370–9.

189. Ercolani, G.L., *Occurrence of Pseudomonas savastanoi (E. F. Smith) Stevens as an epiphyte of olive trees, in Apulia. Phytopathol Mediterr*, 1971. **10**: p. 130–132.

190. Ercolani, G.L., *Pseudomonas savastanoi and other bacteria colonizing the surface of olive leaves in the field. J. Gen. Microbiol.*, 1978. **109**: p. 245–257.

191. Martin, J.M.S., *Characteristics and population densities of fluorescent pseudomonads from cherry and apricot leaf surfaces in Portugal. Garcia De Orta Ser. Estud. Agron*, 1982. **9**: p. 249–254.

192. Roos, I.M.M., and M. J. Hattingh, *Resident populations of Pseudomonas syringae on stone fruit tree leaves in South Africa. Phytophylactica*, 1986. **18**: p. 55–58.

193. Sands, DC, V.E. Langhans, A.L. Scharen, and G. de Smet, *The association between bacteria and rain and possible resultant meteorological implications. J Hungarian Meteorol Serv*, 1982. **86**: p. 148–152.

194. Amato, P., et al., *Microorganisms isolated from the water phase of tropospheric clouds at the Puy de Dome: major groups and growth abilities at low temperatures. FEMS Microbiol Ecol*, 2007. **59**(2): p. 242–54.

195. Morris, C.E., D.C. Sands, B.A. Vinatzer, C. Glaux, C. Guilbaud, A. Buffière, S. Yan, H. Dominguez, and B.M. Thompson, *The life history of the plant pathogen Pseudomonas syringae is linked to the water cycle. ISME J*, 2008. **2**: p. 321–334.

196. Yu, J., et al., *Involvement of the exopolysaccharide alginate in the virulence and epiphytic fitness of Pseudomonas syringae pv. syringae. Mol Microbiol*, 1999. **33**(4): p. 712–20.

197. Romantschuk, M., *Attachment of plant pathogenic bacteria to plant surfaces. Annu Rev Phytopathol*, 1992. **30**: p. 225–43.

198. Romantschuk, M. and D.H. Bamford, *The causal agent of halo blight in bean, Pseudomonas syringae pv. phaseolicola, attaches to stomata via its pili. Microb Pathog*, 1986. **1**(2): p. 139–48.

199. Bjorklof, K., et al., *High frequency of conjugation versus plasmid segregation of RP1 in epiphytic Pseudomonas syringae populations. Microbiology*, 1995. **141**(Pt 10): p. 2719–27.

200. Roine, E., et al., *Characterization of type IV pilus genes in Pseudomonas syringae pv. tomato DC3000. Mol Plant Microbe Interact*, 1998. **11**(11): p. 1048–56.

201. Romantschuk, M., E.-L. Nurmiaho-Lassila, E. Roine, and A. Suoniemi., *Pilus-mediated adsorption of Pseudomonas syringae to the surface of host and non-host plant leaves. J. Gen. Microbiol.*, 1993. **139**: p. 2251–2260.

202. Romantschuk, M., E. Roine, K. Björklöf, T. Ojanen, E.-L. Nurmiaho-Lassila, and K. Haahtela, *Microbial attachment to plant aerial surfaces, in Aerial plant surface microbiology*, P.C.N. C. E. Morris, and C. Nguyen-The, Editor. 1996, Plenum Press: New York, NY, p. 43–57.

203. Goodman, R.N., and A. J. Novacky., *The hypersensitive reaction in plants to pathogens—a resistance phenomenon.* 1994, American Phytopathological Society: St. Paul, MN.

204. Rich, J.J., S.S. Hirano, and D.K. Willis, *Pathovar-specific requirement for the Pseudomonas syringae lemA gene in disease lesion formation. Appl Environ Microbiol*, 1992. **58**(5): p. 1440–6.

205. Barta, T.M., T.G. Kinscherf, and D.K. Willis, *Regulation of tabtoxin production by the lemA gene in Pseudomonas syringae. J Bacteriol*, 1992. **174**(9): p. 3021–9.

206. Kinscherf, T.G. and D.K. Willis, *Swarming by Pseudomonas syringae B728a requires gacS (lemA) and gacA but not the acyl-homoserine lactone biosynthetic gene ahlI. J Bacteriol*, 1999. **181**(13): p. 4133–6.

207. Kitten, T., et al., *A newly identified regulator is required for virulence and toxin production in Pseudomonas syringae. Mol Microbiol*, 1998. **28**(5): p. 917–29.

208. Klement, Z., *Rapid detection of the pathogenicity of phytopathogenic Pseudomonads. Nature*, 1963. **199**: p. 299–300.

209. Dumenyo, C.K., A. Mukherjee, W. Chun, and A. K. Chatterjee, *Genetic and physiological evidence for the production of N-acyl homoserine lactones by Pseudomonas syringae pv. syringae and other fluorescent plant pathogenic Pseudomonas species. Eur. J. Plant Pathol*, 1998. **104**: p. 569–582.

210. Hrabak, E.M., and D. K. Willis, *Involvement of the lemA gene in production of syringomycin and protease by Pseudomonas syringae pv. syringae. Mol Plant-Microbe Interact*, 1993. **6**: p. 368–375.

211. Bagdasarian, M., et al., *Specific-purpose plasmid cloning vectors. II. Broad host range, high copy number, RSF1010-derived vectors, and a host-vector system for gene cloning in Pseudomonas. Gene*, 1981. **16**(1-3): p. 237–47.

212. Regenhardt, D., et al., *Pedigree and taxonomic credentials of Pseudomonas putida strain KT2440. Environ Microbiol*, 2002. **4**(12): p. 912–5.

213. Nelson, K.E., et al., *Complete genome sequence and comparative analysis of the metabolically versatile Pseudomonas putida KT2440. Environ Microbiol*, 2002. **4**(12): p. 799–808.

214. Resnick, S.M., and D.T. Gibson, *Diverse reactions catalyzed by naphthalene dioxygenase from Pseudomonas sp. Strain. J Ind Microbiol*, 1996. **17**: p. 438–457.

215. Tahseen, R., et al., *Enhanced degradation of hydrocarbons by gamma ray induced mutant strain of Pseudomonas putida. Biotechnol Lett*, 2019. **41**(3): p. 391–399.

216. Kim, J. and W. Park, *Genome analysis of naphthalene-degrading Pseudomonas sp. AS1 harboring the megaplasmid pAS1. J Microbiol Biotechnol*, 2018. **28**(2): p. 330–337.

217. Gong, T., et al., *Combinatorial metabolic engineering of Pseudomonas putida KT2440 for efficient mineralization of 1, 2, 3-trichloropropane. Sci Rep*, 2017. **7**(1): p. 7064.

218. Dagley, S., *Catabolism of aromatic compounds by microorganisms. Adv Microb Physiol*, 1971. **6**(0): p. 1–46.

219. Luengo, J.M., J.L. Garcia, and E.R. Olivera, *The phenylacetyl-CoA catabolon: a complex catabolic unit with broad biotechnological applications. Mol Microbiol*, 2001. **39**(6): p. 1434–42.

220. Harwood, C.S. and R.E. Parales, *The beta-ketoadipate pathway and the biology of self-identity. Annu Rev Microbiol*, 1996. **50**: p. 553–90.

221. Nichols, N.N. and C.S. Harwood, *PcaK, a high-affinity permease for the aromatic compounds 4-hydroxybenzoate and protocatechuate from Pseudomonas putida. J Bacteriol*, 1997. **179**(16): p. 5056–61.

222. Olivera, E.R., et al., *Molecular characterization of the phenylacetic acid catabolic pathway in Pseudomonas putida U: the phenylacetyl-CoA catabolon. Proc Natl Acad Sci USA*, 1998. **95**(11): p. 6419–24.

223. Cowles, C.E., N.N. Nichols, and C.S. Harwood, *BenR, a XylS homologue, regulates three different pathways of aromatic acid degradation in Pseudomonas putida. J Bacteriol*, 2000. **182**(22): p. 6339–46.

224. Lyi, S.M., S. Jafri, and S.C. Winans, *Mannopinic acid and agropinic acid catabolism region of the octopine-type Ti plasmid pTi15955. Mol Microbiol*, 1999. **31**(1): p. 339–47.

225. Bernal, P., et al., *The Pseudomonas putida T6SS is a plant warden against phytopathogens. ISME J*, 2017. **11**(4): p. 972–987.

226. Schnider-Keel, U., et al., *Autoinduction of 2,4-diacetylphloroglucinol biosynthesis in the biocontrol agent Pseudomonas fluorescens CHA0 and repression by the bacterial metabolites salicylate and pyoluteorin. J Bacteriol*, 2000. **182**(5): p. 1215–25.

227. Mavrodi, Dmitri V., I.T Paulsen, Q. Ren, and J.E. Loper, Genomics of Pseudomonas Fluorescens PF-5, in *Pseudomonas*, J-L Ramos, A. Filloux, Editors. 2007, Springer: New York, NY, p. 3–30.

228. Leisinger, T. and R. Margraff, *Secondary metabolites of the fluorescent pseudomonads. Microbiol Rev*, 1979. **43**(3): p. 422–42.

229. Arshad, M. and W.T. Frankenberger, *Microbial biosynthesis of ethylene and its influence on plant growth. Adv Microb Ecol*, 1992. **12**: p. 69–111.

230. Arshad, M. and W.T. Frankenberger, *Plant growth-regulating substances in the rhizosphere: microbial production and functions. Adv Agron*, 1998. **62**: p. 45–151.

231. Patten, C.L. and B.R. Glick, *Bacterial biosynthesis of indole-3-acetic acid. Can J Microbiol*, 1996. **42**(3): p. 207–20.

232. Dubeikovsky, A.N., E.A. Mordukhova, V.V. Kochetkov, F.Y. Polikarpova, and A.M. Boronin, *Growth promotion of black currant softwood cuttings by recombinant strain Pseudomonas fluorescens Bsp53a synthesizing an increased amount of indole-3-acetic acid. Soil Biol Biochem*, 1993. **25**: p. 1277–1281.

233. Loper, J.E. and M.N. Schroth, *Influence of bacterial sources of indole-3-acetic acid on root elongation of sugar beet. Phytopathology*, 1986. **76**: p. 386–389.

234. Lee, R.E., G.J. Warren, and L.V. Gusta, *Biological Ice Nucleation and its Applications*. 1995: APS Press: St. Paul, MN.

235. Lindow, S.E., *Control of epiphytic ice nucleation-active bacteria for management of plant frost injury*, in *Biological Ice Nucleation and its Applications*, R.E. Lee, G.J. Warren, and L.V. Gusta, Editor. 1995, APS Press: St. Paul, MN, p. 239–256.

236. Lifshitz, R., et al., *Nitrogen-fixing pseudomonads isolated from roots of plants grown in the canadian high arctic. Appl Environ Microbiol*, 1986. **51**(2): p. 251–5.

237. Goldstein, A.H., *Recent progress in understanding the molecular genetics and biochemistry of calcium phosphate solubilization by Gram negative bacteria. Biol Agr Hort*, 1995. **12**: p. 185–193.

238. Torriani-Gorini, A., E, Yagil, and S. Silver, *Phosphate in Microorganisms: Cellular and Molecular Biology*. 1994, Washington, D.C.: ASM Press.

239. Biessy, A. and M. Filion, *Phenazines in plant-beneficial Pseudomonas spp.: biosynthesis, regulation, function and genomics. Environ Microbiol*, 2018. **20**(11): p. 3905–3917.

240. Paulsen, I.T., et al., *Complete genome sequence of the plant commensal Pseudomonas fluorescens Pf-5. Nat Biontechnol*, 2005. **23**(7): p. 873–8.

241. Loper, J.E., D.Y. Kobayashi, and I.T. Paulsen, *The genomic sequence of Pseudomonas fluorescens Pf-5: insights into biological control. Phytopathology*, 2007. **97**(2): p. 233–8.

242. Notz, R., et al., *Fusaric acid-producing strains of Fusarium oxysporum alter 2,4-diacetylphloroglucinol biosynthetic gene expression in Pseudomonas fluorescens CHA0 in vitro and in the rhizosphere of wheat. Appl Environ Microbiol*, 2002. **68**(5): p. 2229–35.

243. Paulsen, I.T., *Multidrug efflux pumps and resistance: regulation and evolution. Curr Opin Microbiol*, 2003. **6**(5): p. 446–51.

244. Kim, Y.C., C.D. Miller, and A.J. Anderson, *Superoxide dismutase activity in Pseudomonas putida affects utilization of sugars and growth on root surfaces. Appl Environ Microbiol*, 2000. **66**(4): p. 1460–7.

245. Schnider-Keel, U., et al., *The sigma factor AlgU (AlgT) controls exopolysaccharide production and tolerance towards desiccation and osmotic stress in the biocontrol agent Pseudomonas fluorescens CHA0. Appl Environ Microbiol*, 2001. **67**(12): p. 5683–93.

246. Kraus, J. and J.E. Loper, *Characterization of a genomic region required for production of the antibiotic pyoluteorin by the biological control agent Pseudomonas fluorescens Pf-5. Appl Environ Microbiol*, 1995. **61**(3): p. 849–54.

247. Parret, A.H., K. Temmerman, and R. De Mot, *Novel lectin-like bacteriocins of biocontrol strain Pseudomonas fluorescens Pf-5. Appl Environ Microbiol*, 2005. **71**(9): p. 5197–207.

248. Nowak-Thompson, B., S.J. Gould, and J.E. Loper, *Identification and sequence analysis of the genes encoding a polyketide synthase required for pyoluteorin biosynthesis in Pseudomonas fluorescens Pf-5. Gene*, 1997. **204**(1-2): p. 17–24.

249. Nowak-Thompson, B., et al., *Characterization of the pyoluteorin biosynthetic gene cluster of Pseudomonas fluorescens Pf-5. J Bacteriol*, 1999. **181**(7): p. 2166–74.

250. Howell, C.R., and R.D. Stipanovic, *Suppression of Pythium ultimum induced damping-off of cotton seedlings by Pseudomonas fluorescens and its antibiotic pyoluteorin. Phytopathology*, 1980. **70**: p. 712–715.

251. Howell, C.R., and R.D. Stipanovic, *Control of Rhizoctonia solani in cotton seedlings with Pseudomonas fluorescens and with an antibiotic produced by the bacterium. Phytopathology*, 1979. **69**: p. 480–482.

252. Kraus, J., and J.E. Loper, *Lack of evidence for a role of antifungal metabolite production by Pseudomonas fluorescens Pf-5 in biological control of Pythium damping off of cucumber. Phytopathology*, 1992. **82**: p. 264–271.

253. Gross, H., et al., *The genomisotopic approach: a systematic method to isolate products of orphan biosynthetic gene clusters. Chem Biol*, 2007. **14**(1): p. 53–63.

254. Folders, J., et al., *Characterization of Pseudomonas aeruginosa chitinase, a gradually secreted protein. J Bacteriol*, 2001. **183**(24): p. 7044–52.

255. Daborn, P.J., et al., *A single Photorhabdus gene, makes caterpillars floppy (mcf), allows Escherichia coli to persist within and kill insects. Proc Natl Acad Sci USA*, 2002. **99**(16): p. 10742–7.

256. Janse, J.D., J.H.J. Derks, B.E. Spit and W.R. Van der Tuin, *Classification of fluorescent soft rot Pseudomonas bacteria, including P. marginalis strains, using whole cell fatty acid analysis. Syst Appl Microbiol*, 1992. **15**: p. 538–553.

257. Vancanneyt, M., U. Torck, D. Dewettinck, M. Vaerewijck and K. Kersters, *Grouping of pseudomonads by SDS-PAGE of whole-cell proteins. Syst Appl Microbiol*, 1996. **19**: p. 556–568.

258. Höfle, M.G., *Identification of bacteria by low molecular weight RNA profiles: a new chemotaxonomic approach. J Microbiol Methods*, 1988. **8**: p. 235–248.

259. Höfle, M.G., *Transfer RNAs as genotypic fingerprints of eubacteria. Arch Microbiol*, 1990. **153**: p. 299–304.

260. Cruz-Sanchez, J.M., et al., *Enhancement of resolution of low molecular weight RNA profiles by staircase electrophoresis. Electrophoresis*, 1997. **18**(11): p. 1909–11.

261. Tourkya, B., et al., *Fluorescence spectroscopy as a promising tool for a polyphasic approach to pseudomonad taxonomy. Curr Microbiol*, 2009. **58**(1): p. 39–46.

262. Angeletti, S., *Matrix assisted laser desorption time of flight mass spectrometry (MALDI-TOF MS) in clinical microbiology. J Microbiol Methods*, 2017. **138**: p. 20–29.

263. Anzai, Y., et al., *Phylogenetic affiliation of the pseudomonads based on 16S rRNA sequence. Int J Syst Evol Microbiol*, 2000. **50**(Pt 4): p. 1563–89.

264. Hilario, E., T.R. Buckley, and J.M. Young, *Improved resolution on the phylogenetic relationships among Pseudomonas by the combined analysis of atp D, car A, rec A and 16S rDNA. Antonie Van Leeuwenhoek*, 2004. **86**(1): p. 51–64.

265. Ait Tayeb, L., et al., *Molecular phylogeny of the genus Pseudomonas based on rpoB sequences and application for the identification of isolates. Res Microbiol*, 2005. **156**(5-6): p. 763–73.

266. Gurtler, V. and V.A. Stanisich, *New approaches to typing and identification of bacteria using the 16S-23S rDNA spacer region. Microbiology*, 1996. **142**(Pt 1): p. 3–16.

267. Guasp, C., et al., *Utility of internally transcribed 16S-23S rDNA spacer regions for the definition of Pseudomonas stutzeri genomovars and other Pseudomonas species. Int J Syst Evol Microbiol*, 2000. **50**(Pt 4): p. 1629–39.

268. Hugh, R., and G. L. Gilardi., *Pseudomonas in Manual of Clinical Microbiology*, 3rd ed. E. H. Lennette, A. Balows, W.J. Hausler Jr., and J.P. Truant, Editors. 1980, ASM Press: Washington, DC, p. 288–317.

269. Meyer, J.M., and J-M. Hornsperger., *Iron metabolism and siderophores in Pseudomonas and related species.*, in *Biotechnology Handbooks, Vol. 10: Pseudomonas*, T.C. Monte, Editor. 1998, Plenum Publishing: New York, NY, p. 201–243.

270. Hobbs, M., et al., *PilS and PilR, a two-component transcriptional regulatory system controlling expression of type 4 fimbriae in Pseudomonas aeruginosa. Mol Microbiol*, 1993. **7**(5): p. 669–82.

271. Ritchings, B.W., et al., *Cloning and phenotypic characterization of fleS and fleR, new response regulators of Pseudomonas aeruginosa which regulate motility and adhesion to mucin. Infect Immun*, 1995. **63**(12): p. 4868–76.

272. Dean, C.R. and K. Poole, *Expression of the ferric enterobactin receptor (PfeA) of Pseudomonas aeruginosa: involvement of a two-component regulatory system. Mol Microbiol*, 1993. **8**(6): p. 1095–103.

273. Song, J. and R.A. Jensen, *PhhR, a divergently transcribed activator of the phenylalanine hydroxylase gene cluster of Pseudomonas aeruginosa. Mol Microbiol*, 1996. **22**(3): p. 497–507.

274. Krzeslak, J., et al., *Lipase expression in Pseudomonas alcaligenes is under the control of a two-component regulatory system. Appl Environ Microbiol*, 2008. **74**(5): p. 1402–1411.

275. Wolfe, M.D., et al., *Benzoate 1,2-dioxygenase from Pseudomonas putida: single turnover kinetics and regulation of a two-component Rieske dioxygenase. Biochemistry*, 2002. **41**(30): p. 9611–26.

276. Kita, A., et al., *An archetypical extradiol-cleaving catecholic dioxygenase: the crystal structure of catechol 2,3-dioxygenase (metapyrocatechase) from Pseudomonas putida mt-2. Structure*, 1999. **7**(1): p. 25–34.

277. Blehert, D.S., B.G. Fox, and G.H. Chambliss, *Cloning and sequence analysis of two Pseudomonas flavoprotein xenobiotic reductases. J Bacteriol*, 1999. **181**(20): p. 6254–63.

278. Pak, J.W., et al., *Transformation of 2, 4, 6-trinitrotoluene by purified xenobiotic reductase B from Pseudomonas fluorescens I-C. Appl Environ Microbiol*, 2000. **66**(11): p. 4742–50.

279. Vuilleumier, S. and M. Pagni, *The elusive roles of bacterial glutathione S-transferases: new lessons from genomes. Appl Microbiol Biotechnol*, 2002. **58**(2): p. 138–46.

37

The Family Rickettsiaceae

Timothy P. Driscoll, Victoria I. Verhoeve, Magda Beier-Sexton, Abdu F. Azad, and Joseph J. Gillespie

CONTENTS

Introduction

Bacteria of the order Rickettsiales (*Alphaproteobacteria*) are Gram-negative, small, rod-shaped, and coccoid, with all described species existing as obligate intracellular parasites of a wide range of eukaryotic organisms (Gillespie et al., 2012a). Before the "DNA revolution," bacteria were assigned to Rickettsiales based primarily on chemical composition and morphology. Intraordinal classification employed a taxonomic system (i.e., generic characteristics) based on five major biological properties: (1) human disease and geographic distribution, (2) natural vertebrate hosts and other animal vectors, (3) experimental infections and serology reactions and cross-reactions, (4) strain cultivation and stability, and (5) energy production and biosynthesis. This pioneering systematic work resulted in a Rickettsiales hierarchy of three families containing nine obligate and facultative intracellular pathogenic genera: (1) *Rickettsiaceae*: genera *Rickettsia, Coxiella, Rochalima,* and *Ehrlichia*; (2) *Bartonellaceae*: genera *Bartonella, Haemobartonella, Eperythrozoon,* and *Grahamella*; and (3) *Anaplasmataceae*: genus *Anaplasma*. However, rickettsial classification has been substantially revised since the turn of the millennium, owing to the technological advances in molecular sequence generation and the advent of several new fields of study, including molecular systematics, phylogenomics, and bioinformatics. Contemporary Rickettsiales taxonomy

is radically different from the traditional system, with such tremendous changes as the reassignment of the Q-fever agent (*Coxiella burnetii*) to *Gammaproteobacteria* (Williams et al., 2010) and the placement of the causative agents of several human diseases such as endocarditis, Trench fever, and Cat-scratch disease (*Bartonella* spp.) to the Order Rhizobiales (Williams et al., 2007).

This chapter primarily focuses on the family *Rickettsiaceae*, which is currently comprised well-studied species in the genera *Rickettsia* and *Orientia*, as well as candidate species in several provisional genera that are poorly understood (Castelli et al., 2019; Driscoll et al., 2013; Schrallhammer et al., 2013; Szokoli et al., 2016). *Rickettsia* and *Orientia* contain many known and potential pathogens of great concern as the causative agents of emerging and re-emerging human diseases (Table 37.1). However, this work also reviews other families that are either currently included in the order Rickettsiales (*Holosporaceae, Anaplasmataceae,* and *Midichloriaceae*) or were previously classified as Rickettsiales (e.g., *Bartonellaceae*). A summary of the current knowledge of rickettsial phenotypic and genotypic characteristics is presented, as well as a synopsis of the factors governing rickettsial virulence and pathogenicity. A perspective on the changing ecology of rickettsial pathogens is provided, illustrating the emerging and re-emerging properties of many pathogenic species. Finally, a discussion is given on the utilization of rickettsial pathogens as agents of bioterrorism.

DOI: 10.1201/9781003099277-39

TABLE 37.1

Some Epidemiological Features of Rickettsial Diseases

	Agent	Disease/dominant symptoms	Vector/Reservoir Host	Geographic Distribution
Typhus fevers	R. prowazekii	Epidemic typhus	Human body louse/Humans	Africa, Asia, America
		Sylvatic typhus	Flea and louse/Flying squirrels	USA (only)
	R. typhi	Murine (endemic) typhus	Rodent and cat fleas/Rats, mice, opossums (U.S.)	Worldwide
Transitional group	R. akari	Rickettsialpox	Mite/house mice	Worldwide
	R. felis	Cat flea rickettsiosis	Fleas/Domestic cats, opossums (USA)	Worldwide
	R. australis	Queensland tick typhus	Tick/Rodents	Australia, Tasmania
Tick-transmitted spotted fevers	R. rickettsia	Rocky Mountain spotted fever	Tick/Rodents, rabbits	North, Central, and South America
	R. parkeri	Mild spotted fever	Tick/Rodents?	USA, Brazil, Uruguay
	R. conorii	Mediterranean Spotted fever	Tick/Rodents, hedgehogs	Europe, Asia, Africa,
	R. sibirica	North Asian tick typhus, Siberian tick typhus	Tick/Rodents	Russia, China, Mongolia, Europe
	R. africae	African tick-bite fever	Tick/Rodents	Sub-Saharan Africa, Caribbean
	R. japonica	Oriental spotted fever	Tick/?	Japan
	R. slovaca	Necrosis, erythema	Tick/Rodents and lagomorphs	Europe
	R. helvetica	Aneruptive fever	Tick/Rodents	Old World
	R. honei	Finders Island spotted fever, Thai tick typhus	Ticks/?	Australia, Thailand
Scrub typhus	O. tsutsugamushi	Scrub typhus	Mites/Rodents	Indian subcontinent, Asia, Australia

Rickettsial Classification

Rickettsiaceae

Within *Rickettsiaceae*, only species of the genus *Rickettsia*, unique bacteria with highly reduced genomes (Andersson and Andersson, 1999; Andersson et al., 1998) and a close evolutionary relationship with the mitochondrial ancestor (Emelyanov, 2003), remain from the original four genera (Table 37.2). Of particular note, one former species of *Rickettsia*, the scrub typhus agent *R. tsutsugamushi*, is now placed in a novel genus, *Orientia* (Tamura et al., 1995). For the taxon *Coxiella*, a monotypic genus originally considered a sister lineage to *Rickettsia* species, reassignment to the *Gammaproteobacteria* was suggested based on phylogenetic analysis of 16S ribosomal DNA sequences (Weisburg et al., 1989). Robust phylogenomic analysis has confirmed the close relationship of *Coxiella* and *Legionella* species within the *Gammaproteobacteria* (Williams et al., 2010). Phylogenetic analysis of the genus *Rochalima* resulted in its unification with *Bartonella* species in the *Bartonellaceae*, with all species renamed as *Bartonella* spp. (Brenner et al., 1993). Finally, the genus *Ehrlichia* has been moved to the *Anaplasmataceae*, as its current described species share greater affinity with *Anaplasma* spp. than to species of *Rickettsia* (Dumler et al., 2001).

Species of the genus *Rickettsia* have traditionally been classified into either the Typhus Group (TG) or the Spotted Fever Group (SFG) rickettsiae. However, this classification is no longer valid for several reasons. First, many uncharacterized species of *Rickettsia*, particularly those with unknown pathogenicity and no known association with blood-feeding arthropods or vertebrates, are ancestral to the more commonly known pathogenic strains (Perlman et al., 2006). Second, several characterized species of *Rickettsia*, such as *R. bellii* (Stothard and Fuerst, 1995), *R. canadensis* (Gillespie et al., 2007) and *R. helvetica* (Driscoll et al., 2013), do not group within the TG or SFG based on robust phylogenetic analysis. Finally, a monophyletic clade that lies between TG and SFG rickettsiae, containing *R. felis*, *R. akari*, and *R. australis*, has been shown to share many traits similar to non-SFG rickettsiae (Gillespie et al., 2007, 2015a). This group was named Transitional Group (TRG) rickettsiae, with all other early diverging *Rickettsia* lineages assigned to provisional groups that need further classificatory attention (Murray et al., 2016). Recent advances in phylogenetic methods that accommodate base compositional biases in nucleotide and amino acid sequences, which are critical to employ for the highly AT-rich genomes of Rickettsiales, suggest that TRG and TG rickettsiae share a common origin relative to the other rickettsial lineages (Driscoll et al., 2013) (Figure 37.1).

TABLE 37.2

Pathogens That Are No Longer Associated with Rickettsiales and/or Rickettsiaceae

	Agent	Disease/dominant symptoms	Vector/Reservoir Host	Geographic Distribution
Coxiella	C. burnetii	Q-fever	Tick/Goats, sheep, cattle, domestic cats	Worldwide
Bartonella	B. henselae	Cat-scratch disease	Cat flea/Domestic cat	Worldwide
	B. quintana	Trench fever	Human body louse/ Humans	Worldwide
	B. bacilliformis	Oroya fever	Sand fly/?	Peru, Ecuador, Colombia
Ehrlichia	E. chaffeensis	Monocytic Ehrlichiosis	Tick/Mammals (deer, rodents)	Worldwide
	E. ewingii	?	Tick/Deer?	USA
Neorickettsia	N. sennetsu	Sennetsu fever	Snail/Fish	Japan, Malaysia
Anaplasma	A. phagocytophilum	Anaplasmosis	Tick/Rodents, other small mammals	USA, Europe, Asia, Africa

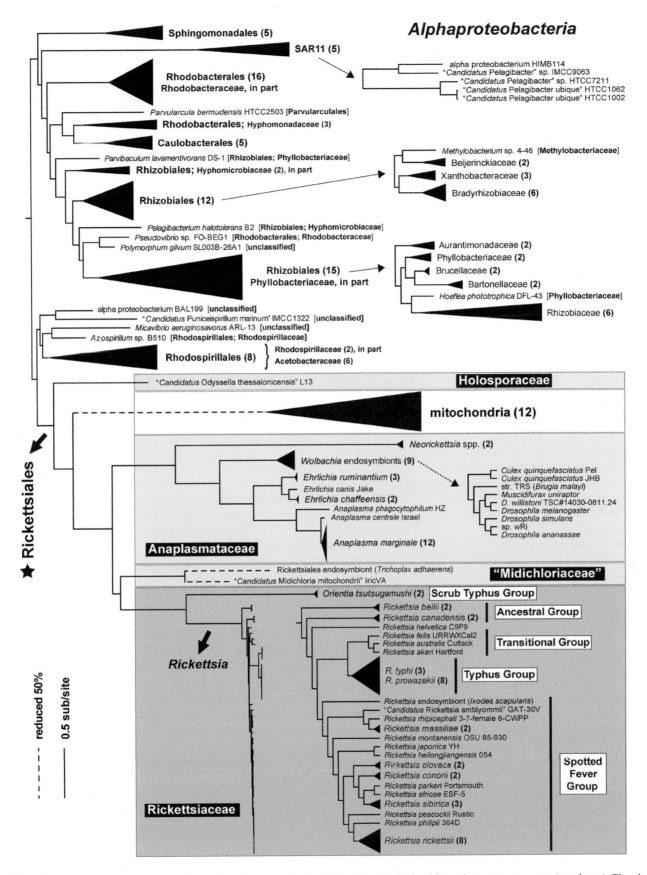

FIGURE 37.1 Genome-based phylogeny estimated for 164 alphaproteobacterial taxa, 12 mitochondria, and two outgroup taxa (not shown). The phylogenetic pipeline that entails orthologous protein group (OG) generation, OG alignment (and masking of less conserved positions), and concatenation of aligned OGs is described elsewhere (Driscoll et al., 2013). Tree was estimated using the CAT-GTR model of substitution as implemented in PhyloBayes v3.3 (Lartillot and Philippe, 2004, 2006). Tree is a consensus of 1,522 trees (post burn-in) pooled from two independent Markov chains run in parallel. Branch support was measured via posterior probabilities, which reflect frequencies of clades among the pooled trees (all branches were recovered at 100%). Rickettsiales is noted with a star. Mitochondria, boxed; rickettsial families *Holosporaceae*, *Anaplasmataceae* and *Midichloriaceae* boxed and shaded light gray; *Rickettsiaceae*, boxed and shaded dark gray. Classification scheme for *Rickettsia* spp. follows previous studies (Gillespie et al., 2007, 2008). More details such as taxon names, genome accession numbers, etc., are provided elsewhere (Driscoll et al., 2013).

An assemblage of 1) Scrub Typhus Group rickettsiae (STG), which currently comprises *Orientia tsutsugamushi* and *O. chuto* (Izzard et al., 2010), 2) *Occidentia massiliensis* (Mediannikov et al., 2014), 3) "*Candidatus* Megaira polyxenophila" (Lanzoni et al., 2019), and 4) many other uncharacterized environmental strains typically isolated from aquatic ciliates, hydra, etc. (Boscaro et al., 2019) form sister lineages to *Rickettsia* species or early branching lineages of *Rickettsiaceae*. The recently sequenced genome of "*Candidatus* Arcanobacter lacustris" indicates it is the most ancient lineage of *Rickettsiaceae*, as it contains chemotaxis genes and vertically inherited flagellar genes reflective of a once facultative obligate intracellular lifestyle (Martijn et al., 2015). Collectively, this reclassification of *Rickettsiaceae* has been continually supported in various robust phylogenomic analyses (Gillespie et al., 2008, 2009a, 2012a, 2015a; Kaur et al., 2012). The monophyly (common ancestry) of the reclassified *Rickettsiaceae* is indisputable, thus providing a final stability to the classification of this natural rickettsial group.

Rickettsiella

The genus *Rickettsiella*, originally described as a member of Rickettsiales (Weiss et al., 1984), comprises a group of intracellular pathogens of diverse arthropod species. Based on characteristics divergent from typical rickettsiae, such as unique cell morphology (Weiser and Zizka, 1968), mode of reproduction (Götz, 1971, 1972), and an unusual morphogenic cycle most similar to *Chlamydia* spp. (Federici, 1980), the placement of *Rickettsiella* within Rickettsiales was always considered tenuous. Phylogenetic analysis confirmed the erroneous taxonomic assignment of *Rickettsiella* (Roux et al., 1997), with an eventual reassignment to the *Coxiellaceae* (*Gammaproteobacteria*: Legionellales) (Garrity et al., 2005). Robust phylogenomic analysis has confirmed the close relationship of *Coxiella* spp. with *Rickettsiella* within the *Gammaproteobacteria* (Leclerque, 2008; Williams et al., 2010).

Anaplasmataceae

Molecular phylogenetic analysis revealed that the once monotypic *Anaplasmataceae* was comprised of four lineages containing five described genera: *Anaplasma*, *Cowdria*, *Ehrlichia*, *Neorickettsia*, and *Wolbachia*. In a landmark study, the majority of the species in these five genera were meticulously placed into the genera *Neorickettsia*, *Wolbachia*, *Anaplasma*, and *Ehrlichia*, with substantial revision of the latter two genera (Dumler et al., 2001). Robust phylogeny estimations have supported the monophyly of the contemporary *Anaplasmataceae*, and its sister relationship to the *Rickettsiaceae* within Rickettsiales (Driscoll et al., 2013, 2017; Gillespie et al., 2012a; Williams et al., 2007). A lineage containing *Neorickettsia* spp. and the recently discovered "*Candidatus* Xenolissoclinum pacificiensis," an endoparasite of the tunicate *Lissoclinum patella* (Kwan and Schmidt, 2013), is currently the most ancestral lineage within the *Anaplasmataceae* (Driscoll et al., 2017), followed by the *Wolbachia* endosymbionts and parasites, then the sister clade containing *Anaplasma* and *Ehrlichia* (Figure 37.1). The latter two genera are much less divergent from one another than any other generic comparisons

within the Rickettsiales, suggesting their relatively recent divergence from one another (Driscoll et al., 2013).

Midichloriaceae

A novel rickettsial family, *Midichloriaceae*, was recently proposed based on the discovery of a large clade of species that is distinct from the *Rickettsiaceae* and *Anaplasmataceae* (Driscoll et al., 2013; Gillespie et al., 2012a; Montagna et al., 2013). This family includes many species with diverse eukaryotic hosts, which are predominantly aquatic (i.e., corals, sponges, placozoans, Hydra spp., and protists). *Midichloriaceae* was named after the tick symbiont "*Candidatus* Midichloria mitochondrii," which is known to invade and devour mitochondria (Sassera et al., 2006). Unique genomic traits define "*Candidatus* M. mitochondrii," particularly genes encoding flagella and cbb(3) oxidase (Sassera et al., 2011). However, the phylogenetic position of the "Midichloriaceae" within the Rickettsiales is still tenuous. Genome-based phylogenies have supported the sister relationship of *Midichloriaceae* with either *Rickettsiaceae* (Driscoll et al., 2013; Sassera et al., 2011), as depicted in Figure 37.1, or *Anaplasmataceae* (Driscoll et al., 2017). The latter is probably more accurate, provided that 16S rDNA-based phylogenies including far more environmental isolates suggest a closer relationship of "Midichloriaceae" with *Anaplasmataceae* (Driscoll et al., 2013; Gillespie et al., 2012a). The analysis of 16S rDNA sequences illustrates the extraordinary diversity within the *Midichloriaceae* and paves the way for future genome sequencing of additional taxa to better understand the proximate position of this natural group within the Rickettsiales (Figure 37.2). The position of the recently described "*Candidatus* Deianiraea vastatrix," a presumed obligate exoparasite of ciliates, is currently tenuous but stands to inform on the origin of obligate intracellularity in Rickettsiales (Castelli et al., 2019).

Holosporaceae

Family *Holosporaceae* was first added to the Rickettsiales in 2001 (Boone et al., 2001). This group is composed primarily of endosymbionts of protists, particularly amoebae, and was shown to form a distinct lineage basal to the derived rickettsial lineages (Baker et al., 2003). Much of the knowledge pertaining to these organisms comes from the described species of *Holospora*, particularly the paramecia associated *H. obtusa* (Gromov and Ossipov, 1981). Robust phylogenomics analysis of the first sequenced genomes for *Holosporaceae*, *H. undulata* (Dohra et al., 2013) and "*Candidatus* Odyssella thessalonicensis" (Birtles et al., 2000), placed *Holosporaceae* as a basal lineage to the three other rickettsial families and mitochondria (Driscoll et al., 2013) (Figure 37.1). While broader analysis of 16S rDNA sequences confirmed this arrangement of rickettsial families (Driscoll et al., 2013) (Figure 37.2), more recent studies argue that *Holosporaceae* is not a rickettsial family (Hess et al., 2016; Muñoz-Gómez et al., 2019; Szokoli et al., 2016). Like *Midichloriaceae*, the diversity encompassed within the *Holosporaceae* is extraordinary, necessitating the collection of more genome sequences from other species to bolster the taxonomic position at the root of the rickettsial phylogeny. This information will prove invaluable for further stabilization of

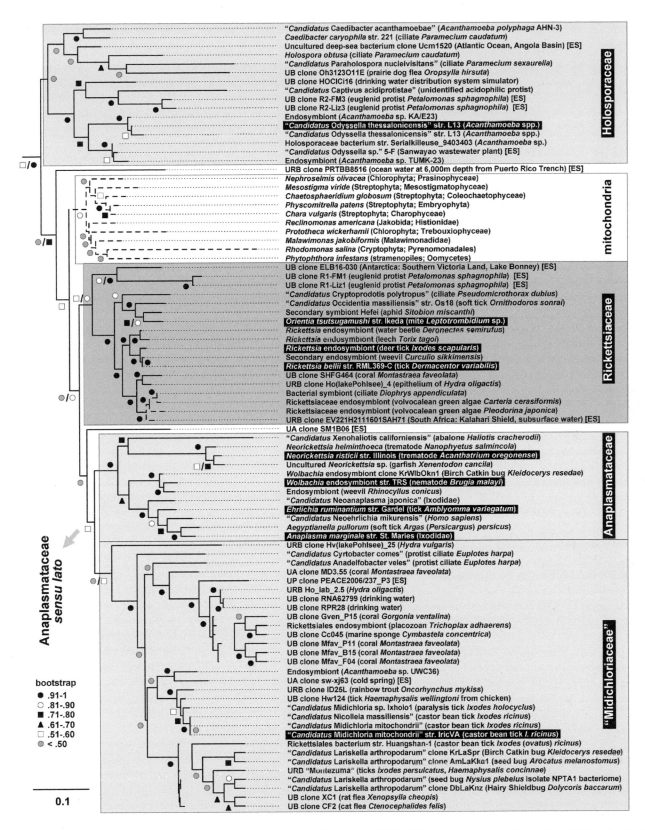

FIGURE 37.2 16S rDNA-based phylogeny estimation for Rickettsiales. Phylogeny of SSU rDNA sequences was estimated for 78 Rickettsiales taxa, 10 mitochondria, and five outgroup taxa (not shown). The phylogenetic pipeline that entails alignment and tree-building methods is described elsewhere (Driscoll et al., 2013). Tree is final optimization likelihood: (−22042.321923) using GTR substitution model with GAMMA and proportion of invariant sites estimated. Brach support is from 1,000 bootstrap pseudoreplications. For nodes represented by two bootstrap values, the left is from the analysis that included 10 mitochondrial sequences, with the right from the analysis without the mitochondrial sequences. All nodes with single bootstrap values had similar support in both analyses. Dashed branches are reduced 75% (mitochondria) or increased 50% (within *Rickettsiaceae*). Cladograms depict minimally resolved lineages within the "Midichloriaceae." For each taxon, associated hosts are within parentheses, with ES depicting an environmental sample.) Taxa within black boxes have available genome sequence data. More details such as taxon names, sequence accession numbers, etc., are provided elsewhere (Driscoll et al., 2013). (*Abbreviations:* UB, uncultured bacterium; UP, uncultured proteobacterium; UA, uncultured alphaproteobacterium; URB, uncultured Rickettsiales bacterium.

rickettsial classification and will also shed light on the evolution of diverse strategies of obligate intracellular lifestyle.

Rickettsial Genotypic and Phenotypic Characteristics

The first sequenced genome of a rickettsial species, *R. prowazekii*, revealed numerous pseudogenes and a highly reduced genome relative to facultative intracellular and free-living bacterial species (Andersson et al., 1998). Since then, dozens of additional genomes from species within all major genera have been sequenced, all being highly reductive yet encoding slightly different metabolic capacities (Driscoll et al., 2017; Gillespie et al., 2012a). Regarding Rickettsiaceae, by April 2014 three genome sequences of STG rickettsiae were available, coupled with 59 genome sequences from species of *Rickettsia* (Table 37.3). The genomes of two strains of *O. tsutsugamushi* possess a staggering proliferation of mobile genetic elements (MGEs), on par with that of the human genome (Cho et al., 2007; Nakayama et al., 2008). Still, these genomes contain several hundred core genes in common with *Rickettsia* genomes, though with slightly fewer encoding metabolic enzymes (e.g., an incomplete TCA cycle) (Driscoll et al., 2017; Nakayama et al., 2010). Analysis of the sequenced genomes of *O. chuto* (Izzard et al., 2010) and *Occidentia massiliensis* (Mediannikov et al., 2014) placed these species together with *O. tsutsugamushi* strains in a sister clade to all described *Rickettsia* species (Driscoll et al., 2017).

Importantly, the genome of *Occidentia massiliensis* does not encode the proliferated MGEs that characterize the *O. tsutsugamushi* genomes, suggesting the latter acquired these elements after diverging from the former (the *O. chuto* genome has not yet been analyzed for MGEs). It is tempting to speculate that these proliferated MGEs in *O. tsutsugamushi* genomes play a role in pathogenesis, possibly in providing the rickettsiae with a dynamic arsenal of surface antigens for avoidance of the host immune response (Cho et al., 2007; Gillespie et al., 2012b). Aside from several known host factors (i.e., fibronectin, α5β1 integrins and syndecan-4 receptors), three proteins have been shown to mediate attachment and entry of *O. tsutsugamushi* into host cells through a clathrin-dependent endosomal pathway (Ge and Rikihisa, 2011). Two of these proteins, TSA (47 kDa surface protein) and surface antigen ScaC, are unknown from *Occidentia massiliensis* (the *O. chuto* genome has not yet been scrutinized). The other protein, HtrA (47 kDa surface protein), is highly conserved across *Rickettsiaceae* genomes, with its function as a surface protein in *O. tsutsugamushi* possibly a moonlighting role that needs to be determined for other rickettsial species. Collectively, it is apparent that substantial genotypic differences underlay the STG rickettsiae genomes, illustrating the need to employ phylogenomics as a means to correlate pending characterized phenotypic traits with underlying genotype.

Substantial variation in size and gene number is seen across the 59 sequenced *Rickettsia* genomes (Table 37.3). The TG rickettsiae genomes are the smallest and encode the fewest number of genes, consequences of being the most degraded and rapidly evolving genomes of *Rickettsia* (Blanc et al., 2007). The TG rickettsiae genomes also contain a slightly higher AT-bias within their genomes. Outside of TG rickettsiae, such genome metrics do not consistently define other monophyletic groups. This is exemplified by the larger *Rickettsia* genomes, which are found in ancestral taxa (e.g., *R. bellii*), TRG rickettsiae (*R. felis*) and SFG rickettsiae (REIS, or now referred to as *R. buchneri* (Kurtti et al., 2015)). In fact, the larger *Rickettsia* genomes tend to have higher numbers of transposases and other MGEs relative to the smaller genomes, illustrating the role of lateral gene transfer in the evolution of *Rickettsia* species (Gillespie et al., 2012b; Hagen et al., 2018). The presence of plasmid(s) is also variable across species (and sometimes strains) (Gillespie et al., 2015a), yet it is likely that all species are able to harbor plasmids given that *R. prowazekii*, which does not encode plasmids, can stably maintain a transfected SFG-like plasmid (Wood et al., 2012). Notwithstanding these variable features across *Rickettsia* genomes, mapping of protein families over generated genome-based phylogenies strongly supports the three-group classification scheme (Gillespie et al., 2007), with each group having a distinct genetic profile that is predominantly of vertical descent (Gillespie et al., 2008, 2012a).

Phylogenomics has provided some clues for why certain species of *Rickettsia* do not fit well into the traditional TG and SFG rickettsiae. For instance, *R. canadensis* was previously associated with either TG or SFG rickettsiae until a robust phylogeny estimation suggested its closer affiliation with *R. bellii* (Gillespie et al., 2007). This result was consistent with only a few other reports (Stothard and Fuerst, 1995; Vitorino et al., 2007), as most other phylogenetic studies grouped *R. canadensis* erroneously based on the molecular marker being employed for phylogeny estimation. This instability of *R. canadensis* in analysis of different genes was clearly illustrated (Vitorino et al., 2007), and correlates with phenotypic traits that do not clearly delineate this species into TG or SFG rickettsiae. For example, like SFG rickettsiae, *R. canadensis* (1) infects ticks and is maintained transstadially and transovarially (Brinton and Burgdorfer, 1971; Burgdorfer, 1968); (2) grows within the nuclei of its host (Burgdorfer, 1968; Burgdorfer and Brinton, 1970); and (3) contains genes encoding both Sca0 and Sca5 surface antigens (Ching et al., 1990; Dasch and Bourgeois, 1981) (Figure 37.3). However, similar to TG rickettsiae, *R. canadensis* grows abundantly in yolk sac, lyses red blood cells, is susceptible to erythromycin, and forms smaller plaques relative to SFG rickettsiae (Myers and Wisseman, 1981). Genomic characteristics are just as ambiguous, as despite sharing roughly the same G + C% as TG rickettsiae (Eremeeva et al., 2005; Myers and Wisseman, 1981) and being more similar to TG rickettsiae in the percentage of coding sequence per genome and number of predicted genes (Table 37.3), *R. canadensis* shares more small repetitive elements with SFG rickettsiae than TG rickettsiae (Eremeeva et al., 2005). Collectively, these attributes for *R. canadensis* illustrate how a rigorous phylogenomic analysis is imperative for generating a robust systematic position for rickettsial species that defy older paradigms for classification into TG and SFG rickettsiae (Gillespie et al., 2007, 2009a).

Phylogenomics has also been employed to characterize the TRG rickettsiae (Gillespie et al., 2007, 2008, 2015a), as well as identify that *R. helvetica* does not currently fit within the four-group classification scheme (Driscoll et al., 2013, 2017). The emerging diversity of genus *Rickettsia* (Gillespie et al., 2012a; Murray et al., 2016; Perlman et al., 2006; Weinert et al.,

TABLE 37.3

Characteristics of Rickettsiaceae Genome Sequences[a]

Genome	Contig	%GC	Length	Genes	Plasmids
Scrub Typhus Group (STG)					
Rickettsiaceae bacterium str. Os18	301	31.25	1469247	1310	0
Orientia tsutsugamushi str. Ikeda	1	30.51	2008987	2197	0
Orientia tsutsugamushi str. Boryong	1	30.53	2127051	2364	0
Ancestral Lineages					
Rickettsia sp. MEAM1 (Bemisia tabaci)	607	32.40	1105415	1036	0
Rickettsia bellii str. RML369-C	1	31.65	1522076	1612	0
Rickettsia bellii str. OSU 85-389	1	31.63	1528980	1657	0
Rickettsia canadensis str. CA410	1	31.01	1150228	1130	0
Rickettsia canadensis str. McKiel	1	31.05	1159772	1230	0
Incertae sedis					
Rickettsia helvetica str. C9P9	1	32.20	1369827	1739	1
Transitional Group (TRG)					
Rickettsia felis str. LSU-Lb	35	32.44	1467654	1831	2
Rickettsia felis str. URRWXCal2	1	32.45	1485148	1810	2
Rickettsia felis str. LSU	21	32.02	1483097	2194	1
Rickettsia akari str. Hartford	1	32.33	1231060	1437	0
Rickettsia australis str. Cutlack	1	32.25	1296670	1565	1
Rickettsia australis str. Phillips	45	31.88	1274508	1525	0
Typhus Group (TG)					
Rickettsia typhi str. TH1527	1	28.92	1112372	875	0
Rickettsia typhi str. B9991CWPP	1	28.92	1112957	877	0
Rickettsia typhi str. Wilmington	1	28.92	1111496	892	0
Rickettsia prowazekii str. GvV257	1	28.99	1111969	902	0
Rickettsia prowazekii str. RpGvF24	1	28.99	1112101	897	0
Rickettsia prowazekii str. GvF12	10	27.94	1109257	966	0
Rickettsia prowazekii str. BuV67-CWPP	1	29.00	1111445	901	0
Rickettsia prowazekii str. Katsinyian	1	29.00	1111454	902	0
Rickettsia prowazekii str. NMRC Madrid E	1	29.00	1111520	926	0
Rickettsia prowazekii str. Madrid E	1	29.00	1111523	924	0
Rickettsia prowazekii str. Rp22	1	29.00	1111612	910	0
Rickettsia prowazekii str. Cairo 3	20	36.12	1113960	900	0
Rickettsia prowazekii str. Chernikova	1	29.01	1109804	892	0
Rickettsia prowazekii str. Dachau	1	29.01	1108946	907	0
Rickettsia prowazekii str. Breinl	1	29.01	1109301	914	0
Spotted Fever Group (SFG)					
Rickettsia monacensis str. IrR/Munich	88	32.09	1268385	1684	1
Rickettsia endosymbiont of *Ixodes scapularis*	16	38.05	1776032	2404	4
Rickettsia montanensis str. OSU 85-930	1	32.57	1279798	1513	0
"*Candidatus* Rickettsia amblyommii" str. GAT-30V	1	32.43	1407796	1846	3
Rickettsia rhipicephali str. 3-7-female6-CWPP	1	32.41	1290368	1621	1

(Continued)

TABLE 37.3 *(Continued)*

Characteristics of Rickettsiaceae Genome Sequences[a]

Genome	Contig	%GC	Length	Genes	Plasmids
Rickettsia massiliae str. AZT80	1	32.61	1263659	1601	1
Rickettsia massiliae str. MTU5	1	32.54	1360898	1721	1
"*Candidatus* Rickettsia gravesii" str. BWI-1	28	32.58	1327625	1687	0
Rickettsia japonica str. YH	1	32.35	1283087	1575	0
Rickettsia heilongjiangensis str. 054	1	32.32	1278468	1562	0
Rickettsia honei str. RB	11	32.54	1268758	1614	0
Rickettsia peacockii str. Rustic	1	32.56	1288492	1558	1
Rickettsia philipii str. 364D	1	32.47	1287740	1570	0
Rickettsia rickettsii str. Hlp#2	1	32.47	1270751	1574	0
Rickettsia rickettsii str. Colombia	1	32.46	1270083	1560	0
Rickettsia rickettsii str. Brazil	1	32.45	1255681	1547	0
Rickettsia rickettsii str. Sheila Smith	1	32.47	1257710	1577	0
Rickettsia rickettsii str. Arizona	1	32.44	1267197	1580	0
Rickettsia rickettsii str. Hauke	1	32.45	1269721	1570	0
Rickettsia rickettsii str. Iowa	1	32.45	1268175	1595	0
Rickettsia rickettsii str. Hino	1	32.45	1269837	1593	1
Rickettsia slovaca str. D-CWPP	1	32.50	1275720	1598	0
Rickettsia slovaca 13-B	1	32.50	1275089	1611	0
Rickettsia parkeri str. Portsmouth	1	32.43	1300386	1604	0
Rickettsia africae str. ESF-5	1	32.40	1278530	1545	1
Rickettsia sibirica str. 246	1	32.47	1250020	1554	0
Rickettsia sibirica subsp. *sibirica* BJ-90	8	32.51	1254734	1588	0
Rickettsia sibirica subsp. *mongolitimonae* HA-91	21	31.92	1252337	1616	0
Rickettsia conorii str. Malish 7	1	32.44	1268755	1578	0
Rickettsia conorii subsp. *indica* ITTR	10	32.49	1249482	1601	0
Rickettsia conorii subsp. *israelensis* ISTT CDC1	33	31.83	1252815	1640	0
Rickettsia conorii subsp. *caspia* A-167	25	32.38	1260331	1657	0

[a] All sequences were annotated with RAST (Aziz et al., 2008).

2009) suggests that additional groups of rickettsiae will likely be needed to further stabilize *Rickettsia* classification, though such endeavors must employ rigorous phylogenomic analysis. Additionally, phenotypic data is critical for understanding the underlining genetic attributes of *Rickettsia* genomes. Without such data, the results of phylogenomics cannot be effectively interpreted. For example, the analysis of 19 secreted proteins across 59 *Rickettsia* genomes shows some patterns that define the four groups of rickettsiae (Gillespie et al., 2015b) (Figure 37.3B). However, it is difficult to associate phenotypic traits with these genotypic maps, for several reasons. First, thorough and rigorously tested phenotypic traits are not available in the literature for all rickettsial species. Second, presence alone of a gene suspected to contribute to a phenotype is not enough; expression and other post-genomic data are necessary to determine if the gene is functional. Finally, and most concerning, is the poor quality of many genomes sequenced with short read sequencing technologies. It is apparent that most genomes sequenced since 2010 have not been manually

curated for confirmation of automated annotation, and many genome projects have not published an assembly. The WGS "raw" data may contain areas of low sequence coverage that, unfortunately, result in misassembly or, even worse, errors in gene prediction and annotation. This results in apparently truncated genes, split genes, or pseudogenes; however, the probable full-length of these genes is masked by poor quality of data and a lack of manual evaluation.

Collectively, a vast treasure trove of genomic data is amassing for species of *Rickettsiaceae*. Phylogenomics is providing an invaluable tool for improving rickettsial classification and identifying genotypic features that define specific lineages of rickettsial species. Undoubtedly, the challenges highlighted above will be met with the careful fusion of experimental data and genomic data, with the latter manually processed to provide an accurate genotype for all species. Thus, the framework for scaffolding pathogenicity factors over rickettsial lineages, correlating phenotype with genotype, is set and provides an invaluable resource for future research.

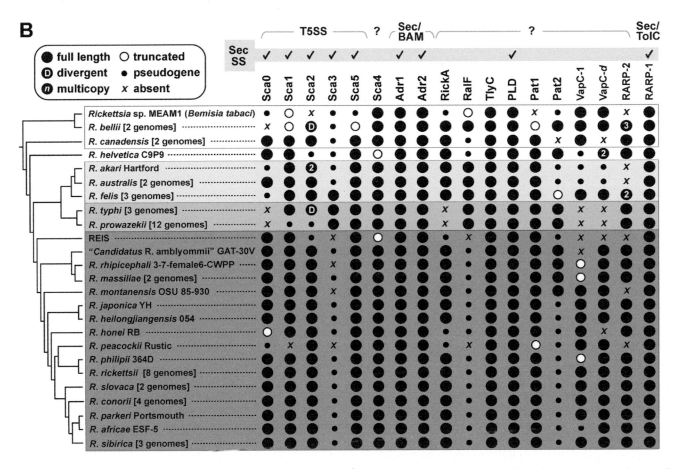

FIGURE 37.3 *Rickettsia* phenotypic and genotypic variability. (A) Variation in pathogenicity (decreasing left to right) and several phenotypic traits for nine well-studied *Rickettsia* species. For transmission: V, vertical (transovarial); H, horizontal (transstadial only). For *R. typhi*, "min" reflects the small actin tails responsible for erratic rickettsia movement (Heinzen et al., 1993; Van Kirk et al., 2000; Teysseire and Raoult, 1992). For *R. felis*, "C" denotes contradictory data. (B) Conservation and distribution across 59 *Rickettsia* genomes of genes encoding 18 secretory proteins. Phylogeny at left was estimated as previously described (Driscoll et al., 2013), with additional genomes annotated using RAST (Aziz et al., 2008). Classification scheme for *Rickettsia* spp. follows previous studies: white, ancestral group; light gray, transitional group; gray, typhus group; dark gray, spotted fever group (Gillespie et al., 2007, 2008). *Rickettsia helvetica* is unclassified (*incertae sedis*) following recent recommendations (Driscoll et al., 2013). Protein names are listed at top. Secretion pathway information is above protein names, with checkmarks depicting proteins with NT Sec secretion signals (Ammerman et al., 2008). Inset: large black circles, full-length proteins; medium-sized open circles, truncated proteins that may have additional fragments encoded by separate genes; small black circles, probable pseudogenes with one or more fragments that do not span the complete protein; Xs, absent genes with zero significant matches using BLASTP. For Sca2, D denotes divergent passenger domains fused to conserved Sca2 autotransporter domains. Numbers on other circles depict proteins encoded by multiple genes.

Rickettsial Virulence and Pathogenesis

The majority of extant, described species of *Rickettsia* and *Orientia* are associated with arthropods during some phase of their life cycle. Vector maintenance of rickettsiae involves either horizontal (arthropod-vertebrate cycle) or vertical (trans-ovarial) transmission, or both, as is the case for many of the well-studied *Rickettsia* pathogens (Figure 37.3A). Aside from the spread of *R. prowazekii* via lice-feeding among tightly associated communities, humans are considered a dead-end host for rickettsiae, the result of accidental infection (Sahni et al., 2013). The rickettsiae transmitted via blood-feeding arthropods are found within all four phylogenetic groups (STG, TRG, TG, SFG rickettsiae) as well as certain ancestral *Rickettsia* species (e.g., *R. bellii* and *R. canadensis*) and include pathogens vectored by various species of ticks, fleas, lice, and mites. These rickettsial species have varying degrees of pathogenicity for eukaryotic hosts, including mammals and arthropods. For example, *R. prowazekii* and *R. rickettsii* produce severe and often fatal disease in humans, and their maintenance in their vectors (human body lice and tick, respectively) is lethal to the arthropod host (Azad and Beard, 1998). However, infection with *R. typhi*, while symptomatic and even fatal in humans, has no effect on the vector fleas. While both *R. prowazekii* and *R. typhi* infection are lethal to human body lice, infection with two pathogenic SFG rickettsiae (*R. rickettsii* and *R. conorii*) had no effect on louse fitness. In contrast to all described species of TG and TRG rickettsiae, not all species of SFG rickettsiae cause disease in vertebrate hosts; for example, *R. montanensis*, *R. peacockii*, and *R. rhipicephali* are proposed to be nonpathogenic for mammals and lack any apparent adverse effects on the survival and fitness of their tick hosts (Azad and Beard, 1998). Furthermore, the pathogenicity of species of ancestral *Rickettsia* species remains unknown, as does the ability of other poorly described species not associated with blood-feeding arthropod vectors (Gillespie et al., 2012a).

Over the past 20 years, the application of molecular biology tools, especially recombinant DNA technology for rickettsial diagnosis, resulted in the discovery of several new species of pathogenic rickettsiae. Of paramount importance was the sequencing of over 60 genomes of the members of *Rickettsiaceae* (Table 37.3). While this information revealed some intriguing information (e.g., close phylogenetic relationship to mitochondria, extensive numbers of pseudogenes in varying states of degradation, genome reduction and loss of many genes needed for free living life style), it also confirmed many features of rickettsial pathogens that were already known (e.g., ATP/ADP transporter system, absence of rickettsial bacteriophages, lack of genes for glycolysis and the biosynthesis of most amino acids, vitamins and cofactors) (Gillespie et al., 2012a). Aside from evolutionary studies and comparative metabolic profiling (Driscoll et al., 2017), more recent studies utilizing genomic data tend to focus on specific genes that potentially play a direct role in host pathogenicity. To this end, studies focusing on genes that define particular rickettsial groups are invaluable for understanding lineage-specific pathogenicity factors (Ammerman et al., 2009; Lehman et al., 2018; Rahman et al., 2010, 2013; Rennoll-Bankert et al., 2015).

For *Rickettsia* spp., genomics has driven the characterization of genes encoding secreted proteins involved in (or predicted to be involved in) adhesion, phagosomal escape, host actin polymerization, toxin secretion, and hemolysis (Gillespie et al., 2015b) (Figure 37.3B). The secretion systems themselves (Gillespie et al., 2009b, 2010, 2015c, 2016; Kaur et al., 2012; Rahman et al., 2003, 2005, 2007; Sears et al., 2012) as well as the factors mediating protein secretion (Ammerman et al., 2009, 2008; Lehman et al., 2018; Rennoll-Bankert et al., 2015), are also becoming better understood. Collectively, these components of the rickettsial secretome are highly targeted by researchers interested in identifying the mechanisms of rickettsial pathogenicity. Although rickettsial invasion and destruction of eukaryotic host cells is considered the basis for rickettsial pathogenicity, the precise role of presumed "virulence factors" requires further elucidation. The question that remains to be answered is: what are the underlying mechanisms that make some strains of *Rickettsia* so virulent and others avirulent? While the relevance of these findings to virulence is still an open question, a blend of advances in genetic manipulation, post genomic approaches, and phylogenomics (including more sequenced genomes) will undoubtedly provide insights into the underlying molecular basis of rickettsial virulence. It is only a matter of time before the genetic basis of rickettsial pathogenicity is illuminated.

Changing Ecology of Rickettsial Pathogens

Throughout modern history, the morbidity and mortality associated with rickettsial infections has been underestimated, primarily due to misdiagnosis. This unfortunate truth remains at present, with the significance and impact of rickettsioses greatly underappreciated worldwide, including in the United States. Fatalities associated with rickettsial infections continue to decline worldwide due to the effectiveness of antibiotics and physician awareness and improved rapid diagnosis. However, the "flu-like" symptoms presented by patients during early infection often delay the administration of appropriate antibiotics. Furthermore, the long incubation time (often greater than one week) allows rickettsiae to grow to large numbers and disseminate throughout the vascular system, compromising vascular integrity and setting the stages for more serious complications, including noncardiogenic pulmonary edema, acute respiratory distress, compromised central nervous system (CNS), and failure of multiple organ systems (Sahni et al., 2013). However, perhaps most important to the underestimation of rickettsial diseases is the growing number of emerging pathogens that were previously considered nonpathogenic, such as *R. massiliae*, *R. felis*, *R. parkeri*, *R. aeschlimannii*, *R. slovaca*, and *R. helvetica*. This occurrence, in conjunction with reemerging pathogens in novel focal areas, which in some cases are vectored by previously unrecognized arthropod species (Dumler and Walker, 2005), present a dire need to better understand the ecological landscapes of rickettsial pathogens and their ability to adapt to changing environments and maintain their virulence.

In recent decades, the improvement of molecular diagnostics and the importance of selected rickettsial pathogens as biothreat agents have rekindled interest in rickettsioses. Recent "sero-surveys" have demonstrated a high prevalence of rickettsial diseases worldwide, particularly in warm and humid climates. Additionally, new molecular diagnostic technologies helped to narrow the gap of the continental divide in rickettsial distributions. For example, several rickettsial species that were known only in the United States are now reported in Central and South America. The substantial diversity of rickettsial pathogens throughout the world, coupled with variable clinical manifestations, provides a daunting task to review the ecology of all rickettsial diseases. Thus, further description below is limited to four important rickettsial pathogens: 1) the louse-borne epidemic typhus agent *R. prowazekii*, 2) the flea-borne endemic typhus agent *R. typhi*, 3) the tick-borne Rocky Mountain Spotted Fever agent *R. rickettsii*, and 4) the mite-borne scrub typhus agent, *O. tsutsugamushi*.

Classic Epidemic Typhus

Epidemic typhus is one of the most virulent diseases known to humanity. Symptoms appear about 10 days after an infected body louse has bitten a person and include a high fever of about 42°C, extreme pain in the muscle and joints, stiffness, and cerebral impairment. During the second week, the patient may become delirious with neurological symptoms and may experience stupor. Gangrene and necrosis may occur due to thrombosis of the small vessels in the extremities. Mortality rate in untreated patients is approximately 20%. In severe epidemics, the mortality rate is often as high as 40%. The human body louse, *Pediculus humanus corporis*, is the principal vector for *R. prowazekii*. Although the head louse, *P. h. capitis*, is capable of maintaining *R. prowazekii* experimentally, its role in the transmission of this rickettsiosis is not well established. Body lice feed only on humans, although the laboratory-adapted colony can be maintained on rabbit or artificially fed on defibrinated blood, and all three stages of the louse life cycle (i.e., eggs, nymphs, and adults) may occur on the same host. The lice prefer a lower temperature (20°C) and are normally found in the folds of clothing. The body louse will abandon a patient with a temperature greater than 40°C to seek another host. This attribute is a major factor in the transmission of typhus within the susceptible population. Humans serve as host to *R. prowazekii* and lice and are reservoirs of the rickettsiae. In addition, humans serve as a mobile component of the louse-borne typhus cycle, such that behavior influences the pattern of typhus transmission. The conditions that allow for the coexistence of body lice and a susceptible population could be the starting point for an epidemic of *R. prowazekii* to flare in refugee camps.

Louse-borne epidemic typhus is still endemic in highlands and cold areas of Africa, Asia, Central and South America, and in parts of Eastern Europe. Another face of typhus, hardly studied in depth, is recrudescent typhus (Brill-Zinsser disease), in which the symptoms are less pronounced, and the mortality rate is less than 1%. However, these patients could serve as a long-range source of *R. prowazekii*, permitting transmission of rickettsiae to occur months to years after the primary infection. In the United Sates, *R. prowazekii* is maintained in the sylvatic form involving the flying squirrel and its fleas and lice. Fortunately, immunity to typhus rickettsiae develops after recovery from infection. Furthermore, successful treatment with tetracycline and doxycycline is the approach of choice to reduce morbidity and eliminate mortality in susceptible populations. Although vaccination against *R. prowazekii* has been partially achieved with inoculation of inactivated rickettsiae or attenuated strains (e.g., str. Madrid E), these vaccine approaches have unfortunately been accompanied with undesirable toxic reactions and difficulties in standardization.

Endemic Typhus

In the United States, a major concern is the changing ecology of endemic typhus, hereafter referred to as murine typhus. For instance, in both south Texas and southern California, the classic cycle of *R. typhi*, which involves commensal rats and primarily the rat flea (*Xenopsylla cheopis*), has been replaced by the Virginia opossum (*Didelphis virginiana*)/cat flea (*Ctenocephalides felis*) cycle. Curiously, however, infected rats and their fleas are difficult to document within Texas and California's murine typhus foci (Azad and Beard, 1998; Azad et al., 1997). Similarly, the association of 33 cases of locally acquired murine typhus in Los Angeles County with seropositive domestic cats and opossums was also confirmed. Additionally, based on serological surveys, *R. typhi* infections also occur in inland cities in Oklahoma, Kansas, and Kentucky, where urban and rural-dwelling opossums thrive. Thus, the maintenance of *R. typhi* in the cat flea/opossum cycle is of potential public health importance and a major health risk considering the distribution of opossum, which spans the United States, Mexico, Central America, and Canada. A previous search for the rural reservoir of murine typhus also resulted in the discovery of *R. felis*, the second flea-borne associated rickettsial species in opossums (Azad and Beard, 1998; Azad et al., 1997). Urban rat/flea populations are still the main reservoir of *R. typhi* worldwide and particularly in many cities where urban settings provide a constellation of factors for the perpetuation of the *R. typhi* cycle, including declining infrastructures, increased immunocompromised populations, homelessness, and high population density of rats and fleas.

Rocky Mountain Spotted Fever (RMSF)

RMSF is one of the most virulent human infections in the United States. *R. rickettsii*, the causative agent of RMSF, is a true zoonotic bacterium that cycles between ixodid ticks and wildlife populations, not only in the United States, but also in Mexico and in Central and South America. RMSF, like all rickettsial infections, is classified as a zoonosis requiring a biological vector such as tick to be transmitted between animal hosts (and accidentally to humans). Human infection occurs via the bite of an infected tick. Initial signs and symptoms of the disease include fever, headache, and muscle pain. This is followed by rash and organ-specific symptoms such as nausea, vomiting, and abdominal pain. Delayed treatment results in hospitalization and sequelae, such as amputation, deafness, and permanent learning impairment. The disease can be difficult to diagnose in the early stages due to nonspecific presentations and, unfortunately, in the absence of prompt and appropriate treatment, it often kills the infected patient. Mortality of up to 75% had been reported before antibiotic discovery, and even today 5–10% of children and adults die from RMSF, with many more requiring intensive care. RMSF is a reportable disease in the United States, although the number of reported cases annually ranges between 250 and 1,200.

All SFG rickettsiae are transmitted by ixodid ticks (Azad and Beard, 1998; Parola et al., 2005). In addition to *R. rickettsii*, the etiologic agent of RMSF, several other tick-borne rickettsial species are also human pathogens (Table 37.1). In the United States, at least five other SFG rickettsiae, namely *R. amblyommii*, *R. montanensis*, *R. peacockii*, *R. buchneri*, and *R. rhipicephali*, are routinely isolated only from ticks yet cause limited or no known pathogenicity to humans or certain laboratory animals. *R. parkeri*, which was originally considered nonpathogenic, is now widely considered a human pathogen (Paddock et al., 2004); thus, the other five rickettsial species could also be etiologic agents of as-yet-undiscovered, less severe rickettsioses. However, the genome sequences of *R. peacockii* (Felsheim et al., 2009) and *R. buchneri* (Gillespie et al., 2012b) reveal the proliferation of mobile elements that have interrupted many genes, some of which are candidate virulence factors, suggesting that these species may be restricted to their arthropod vectors.

Distribution of SFG rickettsiae is limited to that of their tick vectors. In the United States, a high prevalence of SFG species in ticks cannot be explained without the extensive contributions of transovarial transmission. The transovarial and transstadial passage of SFG rickettsiae within tick vectors in nature ensures rickettsial survival without requiring the complexity inherent in an obligate multi-host reservoir system (Azad and Beard, 1998). Although many genera and species of ixodid ticks are naturally infected with rickettsiae, *Dermacentor andersoni* (Rocky Mountain wood tick) and *D. variabilis* (American dog tick or Wood tick) are the major vectors of *R. rickettsii* in the United States.

Scrub Typhus

Scrub typhus is an acute, febrile, infectious illness caused by *O. tsutsugamushi* (formerly *Rickettsia tsutsugamushi*, as previously discussed) (Salje, 2017). Long considered a monotypic genus, it is now apparent that substantial diversity exists within the sister clade to genus *Rickettsia* (Figure 37.2), with a recently described novel species of *Orientia*, *O. chuto* (Izzard et al., 2010). Additionally, novel strains of *O. tsutsugamushi* continue to be discovered in geographically isolated regions worldwide (Jiang et al., 2013; Kelly et al., 2009; Odorico et al., 1998; Yang et al., 2012). Like *Rickettsia* spp., *O. tsutsugamushi* is an obligate intracellular Gram-negative bacterium. However, as compared to *Rickettsia* spp., *O. tsutsugamushi* possesses a different cell wall structure, lacking canonical peptidoglycan and any component of lipopolysaccharide (Atwal et al., 2017; Driscoll et al., 2017), and also has a slightly different composition of metabolic genes (Driscoll et al., 2017; Gillespie et al., 2012a; Min et al., 2008). Humans acquire scrub typhus via the bite of infected larval stages of numerous species of trombiculid mites (*Leptotrombidium* spp.). Scrub typhus is endemic in regions of eastern Asia and the southwestern Pacific (Korea to Australia), and also from Japan to India and Pakistan. While no significant morbidity or mortality occurs in patients who receive appropriate treatment, pneumonia, myocarditis, disseminated intravascular coagulation, and death can occur in 0–30% of untreated patients or those infected with antibiotic resistant *O. tsutsugamushi* strains.

The contemporary urban and rural cycles of rickettsial pathogens present a potential for future outbreaks amid the drastic increase in housing construction, urban sprawl and related developmental expansion, all of which provide mammalian reservoirs and their blood-sucking ectoparasites ample harborage and proximity to human habitations. Considering the existing trends in global population expansion, rickettsial agents will continue to be introduced into human populations, with susceptible populations the primary targets for emerging and re-emerging pathogens. To this end, the advancements in classification and diagnostics brought about by the molecular technologies described above have a broader impact, reshaping the knowledge of the eco-dynamics and compositional patterns of rickettsial species worldwide.

Rickettsial Pathogens as Biothreat Agents

Throughout modern history, the epidemics of louse-borne typhus have been important in the molding of human destiny, being credited with causing more deaths than all the wars in history (Azad, 2007; Azad and Radulovic, 2003). For example, more than 30 million cases of louse-borne typhus occurred during and immediately after World War I, causing an estimated 3 million deaths. In the wake of war, famine, flood, and other disasters, the explosive spread of the brutal epidemic of louse-borne typhus within crowded human populations made a deep impression on the commanders of the Russian Red Army, and by 1928, *R. prowazekii* was transformed into a battlefield weapon. Many years later, the Japanese Army successfully tested "biobombs" containing *R. prowazekii*, causing outbreaks of typhus. Thus, the precedent exists for utilizing pathogenic rickettsiae in warfare, and recent increased risk of misuse of bacterial pathogens as weapons of terror is no longer fiction (Azad, 2007; Azad and Radulovic, 2003).

The rickettsial diseases vary from mild to very severe clinical presentations, with case fatality ranging from 2% to more than 30% (Table 37.4). The severity of rickettsioses has been associated with age, delayed diagnosis, hepatic and renal dysfunction, CNS abnormalities, and pulmonary compromise. Despite the variability in clinical presentations, pathogenic rickettsiae cause debilitating diseases and sometimes death, and several rickettsial pathogens could be used as a biological weapon. Realistically, only *R. prowazekii* and *R. rickettsii* pose serious problems, especially when some salient features of pathogenic rickettsial species are compared with those from several category A agents (Table 37.4). Despite the fact that effective chemotherapy is available and effective control measures are known, rickettsial diseases continue to be a problem in the United States and many other parts of the world.

Acknowledgments

The work presented in this chapter has been supported with funds from the National Institute of Health (NIH)/National Institute of Allergy and Infectious Diseases (NIAID) grants R01AI017828 and R01AI126853 to AFA, and R21AI26108 and R21AI146773 to JJG. TPD was supported by start-up funding by West Virginia

TABLE 37.4

Comparison of Selected Rickettsial Pathogens to Examples of Biological Warfare Bacteria

Bacteria	Disease/Incubation[a]	Natural Hosts	Treatment/ Vaccine	Rapid Diagnosis	Transmission[b]	Mortality Rate[c]
Bacillus anthracis	Anthrax/variable	+	+/+	+	Inh, Ing	5–80%
Clostridium botulinum	Botulism/12–36h	–	+/–	+	Ing	70%
Yersinia pestis	Plague/2–7d	+	+/±	+	V, Inh	30–90%
Burkholderia mallei	Glanders/3–6d	+	+/–	+	Inh, C	95%
Francisella tularensis	Tularemia/2–10d	+	+/–	+	C, V	5%
Coxiella burnetii	Q-fever/7–14d	+	+/–	+	Inh, V	2–3%
Rickettsia prowazekii	Epidemic Typhus/6–14d	+	+/–	+	V, inh	30%
Rickettsia rickettsii	RMSF/3–15d	+	+/–	+	V, inh	20–25%
Rickettsia typhi	Endemic Typhus/6–14d	+	+/–	+	V, Inh	4%

[a] Incubation period: dependent upon bacterial dose and mode of exposure.

[b] Mode of exposure: C, cutaneous; Inh, inhalation (aerosol); Ing, ingestion; V, vector-borne.

[c] Fatality rate (dependent upon the mode of exposure).

University. The content is solely the responsibility of the authors and does not necessarily represent the official views of the funding agencies. The funders had no role in study design, data collection and analysis, decision to publish, or preparation of the manuscript.

REFERENCES

Ammerman, N., Gillespie, J., Neuwald, A., Sobral, B., and Azad, A. (2009). A typhus group-specific protease defies reductive evolution in rickettsiae. *J. Bacteriol. 191*, 7609–7613.

Ammerman, N.C., Rahman, M.S., and Azad, A.F. (2008). Characterization of sec-translocon-dependent extracytoplasmic proteins of rickettsia typhi. *J. Bacteriol. 190*, 6234–6242.

Andersson, J.O., and Andersson, S.G. (1999). Genome degradation is an ongoing process in rickettsia. *Mol. Biol. Evol. 16*, 1178–1191.

Andersson, S.G., Zomorodipour, A., Andersson, J.O., Sicheritz-Pontén, T., Alsmark, U.C., Podowski, R.M., Näslund, A.K., Eriksson, A.S., Winkler, H.H., and Kurland, C.G. (1998). The genome sequence of rickettsia prowazekii and the origin of mitochondria. *Nature. 396*, 133–140.

Atwal, S., Giengkam, S., Chaemchuen, S., Dorling, J., Kosaisawe, N., Vannieuwenhze, M., Sampattavanich, S., Schumann, P., and Salje, J. (2017). Evidence for a peptidoglycan-like structure in *Orientia tsutsugamushi. Mol. Microbiol. 105*, 440–452.

Azad, A.F. (2007). Pathogenic rickettsiae as bioterrorism agents. *Clin. Infect. Dis. 45* Suppl 1, S52–5.

Azad, A.F., and Beard, C.B. (1998). Rickettsial pathogens and their arthropod vectors. *Emerg. Infect. Dis. 4*, 179–186.

Azad, A.F., and Radulovic, S. (2003). Pathogenic rickettsiae as bioterrorism agents. *Ann. N. Y. Acad. Sci. 990*, 734–738.

Azad, A.F., Radulovic, S., Higgins, J.A., Noden, B.H., and Troyer, J.M. (1997). Flea-borne rickettsioses: Ecologic considerations. *Emerg. Infect. Dis. 3*, 319–327.

Aziz, R.K., Bartels, D., Best, A.A., DeJongh, M., Disz, T., Edwards, R.A., Formsma, K., Gerdes, S., Glass, E.M., Kubal, M., et al. (2008). The RAST server: Rapid annotations using subsystems technology. *BMC Genomics. 9*, 75.

Baker, B.J., Hugenholtz, P., Dawson, S.C., and Banfield, J.F. (2003). Extremely acidophilic protists from acid mine drainage host Rickettsiales-lineage endosymbionts that have intervening sequences in their 16S rRNA genes. *Appl. Environ. Microbiol. 69*, 5512–5518.

Birtles, R.J., Rowbotham, T.J., Michel, R., Pitcher, D.G., Lascola, B., Alexiou-Daniel, S., and Raoult, D. (2000). "Candidatus Odyssella thessalonicensis" gen. nov., sp. nov., an obligate intracellular parasite of acanthamoeba species. *Int. J. Syst. Evol. Microbiol. 50 Pt. 1*, 63–72.

Blanc, G., Ogata, H., Robert, C., Audic, S., Suhre, K., Vestris, G., Claverie, J.-M., and Raoult, D. (2007). Reductive genome evolution from the mother of rickettsia. *PLOS Genet. 3*, e14.

Boone, D., Castenholz, R., and Garrity, G. (2001). *Bergey's Manual of Systematic Bacteriology*, New York, NY: Springer.

Boscaro, V., Husnik, F., Vannini, C., and Keeling, P.J. (2019). Symbionts of the ciliate Euplotes: Diversity, patterns and potential as models for bacteria-eukaryote endosymbioses. *Proc. Biol. Sci. 286*, 20190693.

Brenner, D.J., O'Connor, S.P., Winkler, H.H., and Steigerwalt, A.G. (1993). Proposals to unify the genera Bartonella and Rochalimaea, with descriptions of Bartonella quintana comb. nov., Bartonella vinsonii comb. nov., Bartonella henselae comb. nov., and Bartonella elizabethae comb. nov., and to remove the family Bartonellaceae from the order Rickettsiales. *Int. J. Syst. Bacteriol. 43*, 777–786.

Brinton, L.P., and Burgdorfer, W. (1971). Fine structure of rickettsia Canada in tissues of Dermacentor andersoni stiles. *J. Bacteriol. 105*, 1149–1159.

Burgdorfer, W. (1968). Observations on rickettsia Canada a recently described member of the typhus group rickettsiae. *J. Hyg. Epidemiol. Microbiol. Immunol. 12*, 26–31.

Burgdorfer, W., and Brinton, L.P. (1970). Intranuclear growth of rickettsia Canada, a member of the typhus group. *Infect. Immun. 2*, 112–114.

Castelli, M., Sabaneyeva, E., Lanzoni, O., Lebedeva, N., Floriano, A.M., Gaiarsa, S., Benken, K., Modeo, L., Bandi, C., Potekhin, A., et al. (2019). Deianiraea, an extracellular bacterium associated with the ciliate paramecium, suggests an alternative scenario for the evolution of Rickettsiales. *ISME J. 3*, 1–15.

Ching, W.M., Dasch, G.A., Carl, M., and Dobson, M.E. (1990). Structural analyses of the 120-kDa serotype protein antigens of typhus group rickettsiae. Comparison with other S-layer proteins. *Ann. N. Y. Acad. Sci. 590*, 334–351.

Cho, N.-H., Kim, H.-R., Lee, J.-H., Kim, S.-Y., Kim, J., Cha, S., Kim, S.-Y., Darby, A.C., Fuxelius, H.-H., Yin, J., et al. (2007). The Orientia tsutsugamushi genome reveals massive proliferation of conjugative type IV secretion system and host cell interaction genes. *Proc. Natl. Acad. Sci. 104*, 7981–7986.

Dasch, G., and Bourgeois, A. (1981). Antigens of the typhus group of rickettsiae: Importance of the species-specific surface protein antigens in eliciting immunity. In *Rickettsiae and Rickettsial Diseases*, W. Burgdorfer and R. Anacker, eds., New York, NY: Academic Press, pp. 61–70.

Dohra, H., Suzuki, H., Suzuki, T., Tanaka, K., and Fujishima, M. (2013). Draft genome sequence of *Holospora undulata* strain HU1, a micronucleus-specific symbiont of the ciliate *Paramecium caudatum*. *Genome Announc. 1*, e00664–13.

Driscoll, T.P., Gillespie, J.J., Nordberg, E.K., Azad, A.F., and Sobral, B.W. (2013). Bacterial DNA sifted from the *Trichoplax adhaerens* (Animalia: Placozoa) genome project reveals a putative rickettsial endosymbiont. *Genome Biol. Evol. 5*, 621–645.

Driscoll, T.P., Verhoeve, V.I., Guillotte, M.L., Lehman, S.S., Rennoll, S.A., Beier-Sexton, M., Rahman, M.S., Azad, A.F., and Gillespie, J.J. (2017). Wholly *rickettsia*! reconstructed metabolic profile of the quintessential bacterial parasite of eukaryotic cells. *mBio. 8*, e00859–17.

Dumler, J.S., and Walker, D.H. (2005). Rocky mountain spotted fever–changing ecology and persisting virulence. *N. Engl. J. Med. 353*, 551–553.

Dumler, J.S., Barbet, A.F., Bekker, C.P.J., Dasch, G.A., Palmer, G.H., Ray, S.C., Rikihisa, Y., and Rurangirwa, F.R. (2001). Reorganization of genera in the families Rickettsiaceae and Anaplasmataceae in the order Rickettsiales: Unification of some species of ehrlichia with Anaplasma, Cowdria with ehrlichia and ehrlichia with Neorickettsia, descriptions of six new species combinations and designation of ehrlichia equi and "HGE agent" as subjective synonyms of ehrlichia phagocytophila. *Int. J. Syst. Evol. Microbiol. 51*, 2145–2165.

Emelyanov, V. V (2003). Mitochondrial connection to the origin of the eukaryotic cell. *Eur. J. Biochem. 270*, 1599–1618.

Eremeeva, M.E., Madan, A., Shaw, C.D., Tang, K., and Dasch, G.A. (2005). New perspectives on rickettsial evolution from new genome sequences of rickettsia, particularly R. canadensis, and *Orientia tsutsugamushi*. *Ann. N. Y. Acad. Sci. 1063*, 47–63.

Federici, B.A. (1980). Reproduction and morphogenesis of Rickettsiella chironomi, an unusual intracellular procaryotic parasite of midge larvae. *J. Bacteriol. 143*, 995–1002.

Felsheim, R.F., Kurtti, T.J., and Munderloh, U.G. (2009). Genome sequence of the endosymbiont Rickettsia peacockii and comparison with virulent Rickettsia rickettsii: Identification of virulence factors. *PLOS One. 4*, e8361.

Garrity, G.M., Bell, J.A., and Lilburn, T. (2005). Family II. Coxiellaceae fam. nov. In *Bergey's Manual of Systematic Bacteriology*, G. Garrity, D. Brenner, N. Krieg, and J. Staley, eds., New York, NY: Springer, pp. 237–247.

Ge, Y., and Rikihisa, Y. (2011). Subversion of host cell signaling by Orientia *tsutsugamushi*. *Microbes Infect. 13*, 638–648.

Gillespie, J.J., Beier, M.S., Rahman, M.S., Ammerman, N.C., Shallom, J.M., Purkayastha, A., Sobral, B.S., and Azad, A.F. (2007). Plasmids and rickettsial evolution: Insight from *Rickettsia felis*. *PLOS One. 2*, e266.

Gillespie, J.J., Williams, K., Shukla, M., Snyder, E.E., Nordberg, E.K., Ceraul, S.M., Dharmanolía, C., Rainey, D., Soneja, J., Shallom, J.M., et al. (2008). Rickettsia phylogenomics: Unwinding the intricacies of obligate intracellular life. *PLOS One. 3*.

Gillespie, J.J., Ammerman, N.C., Beier-Sexton, M., Sobral, B.S., and Azad, A.F. (2009a). Louse- and flea-borne rickettsioses: Biological and genomic analyses. *Vet. Res. 40*, 12.

Gillespie, J.J., Ammerman, N.C., Dreher-Lesnick, S.M., Rahman, M.S., Worley, M.J., Setubal, J.C., Sobral, B.S., and Azad, A.F. (2009b). An anomalous type IV secretion system in Rickettsia is evolutionarily conserved. *PLOS One. 4*, e4833.

Gillespie, J.J., Brayton, K.A., Williams, K.P., Diaz, M.A., Brown, W.C., Azad, A.F., and Sobral, B.W. (2010). Phylogenomics reveals a diverse Rickettsiales type IV secretion system. *Infect Immun. 78*, 1809–1823.

Gillespie, J.J., Nordberg, E.K., Azad, A.A., and Sobral, B.W. (2012a). Phylogeny and comparative genomics: The shifting landscape in the genomics era. In *Intracellular Pathogens II: Rickettsiales*, A.F. Azad and G.H. Palmer, eds., Boston, MA: American Society of Microbiology, pp. 84–141.

Gillespie, J.J., Joardar, V., Williams, K.P., Driscoll, T.P., Hostetler, J.B., Nordberg, E., Shukla, M., Walenz, B., Hill, C.A., Nene, V.M., et al. (2012b). A rickettsia genome overrun by mobile genetic elements provides insight into the acquisition of genes characteristic of an obligate intracellular lifestyle. *J. Bacteriol. 194*, 376–394.

Gillespie, J.J., Driscoll, T.P., Verhoeve, V.I., Utsuki, T., Husseneder, C., Chouljenko, V.N., Azad, A.F., and Macaluso, K.R. (2015a). Genomic diversification in strains of rickettsia felis isolated from different arthropods. *Genome Biol. Evol. 7*, 35–56.

Gillespie, J.J., Kaur, S.J., Sayeedur Rahman, M., Rennoll-Bankert, K., Sears, K.T., Beier-Sexton, M., and Azad, A.F. (2015b). Secretome of obligate intracellular rickettsia. *FEMS Microbiol. Rev. 39*.

Gillespie, J.J., Phan, I.Q.H., Scheib, H., Subramanian, S., Edwards, T.E., Lehman, S.S., Piitulainen, H., Sayeedur Rahman, M., Rennoll-Bankert, K.E., Staker, B.L., et al. (2015c). Structural insight into how bacteria prevent interference between multiple divergent type iv secretion systems. *MBio. 6*, e01867–15.

Gillespie, J.J., Phan, I.Q.H., Driscoll, T.P., Guillotte, M.L., Lehman, S.S., Rennoll-Bankert, K.E., Subramanian, S., Beier-Sexton, M., Myler, P.J., Rahman, M.S., et al. (2016). The *rickettsia* type IV secretion system: Unrealized complexity mired by gene family expansion. *Pathog. Dis. 74*, ftw058.

Götz, P. (1971). Multiple cell division as a mode of reproduction of a cell-parasitic bacterium. *Naturwissenchaften. 58*, 569–570.

Götz, P. (1972). "Rickettsiella chironomi": An unusual bacterial pathogen which reproduces by multiple cell division. *J Invertebr Pathol. 20*, 22–30.

Gromov, B. V., and Ossipov, D. V. (1981). Holospora (ex Hafkine 1890) nom. rev., a genus of bacteria inhabiting the nuclei of paramecia. *Int. J. Syst. Bacteriol. 31*, 348–352.

Hagen, R., Verhoeve, V.I., Gillespie, J.J., and Driscoll, T.P. (2018). Conjugative transposons and their cargo genes vary across natural populations of Rickettsia buchneri infecting the tick Ixodes scapularis. *Genome Biol. Evol. 10*, 3218–3229.

Heinzen, R.A., Hayes, S.F., Peacock, M.G., and Hackstadt, T. (1993). Directional actin polymerization associated with spotted fever group rickettsia infection of Vero cells. *Infect. Immun. 61*, 1926–1935.

Hess, S., Suthaus, A., and Melkonian, M. (2016). "*Candidatus* Finniella" (*Rickettsiales, Alphaproteobacteria*), novel endosymbionts of Viridiraptorid amoeboflagellates (Cercozoa, Rhizaria). *Appl. Environ. Microbiol. 82*, 659–670.

Izzard, L., Fuller, A., Blacksell, S.D., Paris, D.H., Richards, A.L., Aukkanit, N., Nguyen, C., Jiang, J., Fenwick, S., Day, N.P.J., et al. (2010). Isolation of a novel *Orientia* species (*O. chuto* sp. nov.) from a patient infected in Dubai. *J. Clin. Microbiol. 48*, 4404–4409.

Jiang, J., Paris, D.H., Blacksell, S.D., Aukkanit, N., Newton, P.N., Phetsouvanh, R., Izzard, L., Stenos, J., Graves, S.R., Day, N.P.J., et al. (2013). Diversity of the 47-kD HtrA nucleic acid and translated amino acid sequences from 17 recent human isolates of Orientia. *Vector Borne Zoonotic Dis. 13*, 367–375.

Kaur, S.J., Sayeedur Rahman, M., Ammerman, N.C., Beier-Sexton, M., Ceraul, S.M., Gillespie, J.J., and Azada, A.F. (2012). TolC-dependent secretion of an ankyrin repeat-containing protein of *Rickettsia typhi. J. Bacteriol. 194*.

Kelly, D.J., Fuerst, P.A., Ching, W.-M., and Richards, A.L. (2009). Scrub typhus: The geographic distribution of phenotypic and genotypic variants of Orientia tsutsugamushi. *Clin. Infect. Dis. 48 Suppl 3*, S203–30.

Van Kirk, L.S., Hayes, S.F., and Heinzen, R.A. (2000). Ultrastructure of rickettsia rickettsii actin tails and localization of cytoskeletal proteins. *Infect. Immun. 68*, 4706–4713.

Kurtti, T.J., Felsheim, R.F., Burkhardt, N.Y., Oliver, J.D., Heu, C.C., and Munderloh, U.G. (2015). Rickettsia buchneri sp. nov., a rickettsial endosymbiont of the blacklegged tick Ixodes scapularis. *Int. J. Syst. Evol. Microbiol. 65*, 965–970.

Kwan, J.C., and Schmidt, E.W. (2013). Bacterial endosymbiosis in a chordate host: Long-term co-evolution and conservation of secondary metabolism. *PLOS One. 8*, e80822.

Lanzoni, O., Sabaneyeva, E., Modeo, L., Castelli, M., Lebedeva, N., Verni, F., Schrallhammer, M., Potekhin, A., and Petroni, G. (2019). Diversity and environmental distribution of the cosmopolitan endosymbiont "Candidatus megaira." *Sci. Rep. 9*.

Lartillot, N., and Philippe, H. (2004). A bayesian mixture model for across-site heterogeneities in the amino-acid replacement process. *Mol. Biol. Evol. 21*, 1095–1109.

Lartillot, N., and Philippe, H. (2006). Computing Bayes factors using thermodynamic integration. *Syst. Biol. 55*, 195–207.

Leclerque, A. (2008). Whole genome-based assessment of the taxonomic position of the arthropod pathogenic bacterium Rickettsiella grylli. *FEMS Microbiol. Lett. 283*, 117–127.

Lehman, S.S., Noriea, N.F., Aistleitner, K., Clark, T.R., Dooley, C.A., Nair, V., Kaur, S.J., Rahman, M.S., Gillespie, J.J., Azad, A.F., et al. (2018). The rickettsial ankyrin repeat protein 2 is a type IV secreted effector that associates with the endoplasmic reticulum. *mBio. 9*.

Martijn, J., Schulz, F., Zaremba-Niedzwiedzka, K., Viklund, J., Stepanauskas, R., Andersson, S.G.E., Horn, M., Guy, L., and Ettema, T.J.G. (2015). Single-cell genomics of a rare environmental Alphaproteobacterium provides unique insights into Rickettsiaceae evolution. *ISME J. 9*, 2373–2385.

Mediannikov, O., Nguyen, T.-T., Bell-Sakyi, L., Padmanabhan, R., Fournier, P.-E., and Raoult, D. (2014). High quality draft genome sequence and description of Occidentia massiliensis gen. nov., sp. nov., a new member of the family Rickettsiaceae. *Stand. Genomic Sci. 9*, 9.

Min, C.-K., Yang, J.-S., Kim, S., Choi, M.-S., Kim, I.-S., and Cho, N.-H. (2008). Genome-based construction of the metabolic pathways of *Orientia tsutsugamushi* and comparative analysis within the Rickettsiales order. *Comp. Funct. Genomics. 2008*, 1–14.

Montagna, M., Sassera, D., Epis, S., Bazzocchi, C., Vannini, C., Lo, N., Sacchi, L., Fukatsu, T., Petroni, G., and Bandi, C. (2013). "Candidatus Midichloriaceae" fam. Nov. (Rickettsiales), an ecologically: Widespread clade of intracellular Alphaproteobacteria. *Appl. Environ. Microbiol. 79*, 3241–3248.

Muñoz-Gómez, S.A., Hess, S., Burger, G., Franz Lang, B., Susko, E., Slamovits, C.H., and Roger, A.J. (2019). An updated phylogeny of the Alphaproteobacteria reveals that the parasitic Rickettsiales and Holosporales have independent origins. *Elife. 8*.

Murray, G.G.R., Weinert, L.A., Rhule, E.L., and Welch, J.J. (2016). The phylogeny of rickettsia using different evolutionary signatures: How tree-like is bacterial evolution? *Syst. Biol. 65*, 265–279.

Myers, W., and Wisseman, C. (1981). The taxonomic relationship of Rickettsia Canada to the typhus and spotted fever groups of the genus Rickettsia. In *Rickettsiae and Rickettsial Diseases*, W. Burgdorfer and R. Anacker, eds., New York, NY: Academic Press, pp. 313–325.

Nakayama, K., Yamashita, A., Kurokawa, K., Morimoto, T., Ogawa, M., Fukuhara, M., Urakami, H., Ohnishi, M., Uchiyama, I., Ogura, Y., et al. (2008). The Whole-genome sequencing of the obligate intracellular bacterium Orientia tsutsugamushi revealed massive gene amplification during reductive genome evolution. *DNA Res. 15*, 185–199.

Nakayama, K., Kurokawa, K., Fukuhara, M., Urakami, H., Yamamoto, S., Yamazaki, K., Ogura, Y., Ooka, T., and Hayashi, T. (2010). Genome comparison and phylogenetic analysis of Orientia tsutsugamushi strains. *DNA Res. 17*, 281–291.

Odorico, D.M., Graves, S.R., Currie, B., Catmull, J., Nack, Z., Ellis, S., Wang, L., and Miller, D.J. (1998). New Orientia tsutsugamushi strain from scrub typhus in Australia. *Emerg. Infect. Dis. 4*, 641–644.

Paddock, C.D., Sumner, J.W., Comer, J.A., Zaki, S.R., Goldsmith, C.S., Goddard, J., McLellan, S.L.F., Tamminga, C.L., and Ohl, C.A. (2004). Rickettsia parkeri: A newly recognized cause of spotted fever rickettsiosis in the United States. *Clin. Infect. Dis. 38*, 805–811.

Parola, P., Davoust, B., and Raoult, D. (2005). Tick- and flea-borne rickettsial emerging zoonoses. *Vet. Res. 36*, 469–492.

Perlman, S.J., Hunter, M.S., and Zchori-Fein, E. (2006). The emerging diversity of Rickettsia. *Proc. Biol. Sci. 273*, 2097–2106.

Rahman, M.S., Simser, J.A., Macaluso, K.R., and Azad, A.F. (2003). Molecular and functional analysis of the lepB gene, encoding a type I signal peptidase from rickettsia rickettsii and rickettsia typhi. *J. Bacteriol. 185*, 4578–4584.

Rahman, M.S., Simser, J.A., Macaluso, K.R., and Azad, A.F. (2005). Functional analysis of secA homologues from rickettsiae. *Microbiology. 151*, 589–596.

Rahman, M.S., Ceraul, S.M., Dreher-Lesnick, S.M., Beier, M.S., and Azad, A.F. (2007). The lspA gene, encoding the type II signal peptidase of rickettsia typhi: Transcriptional and functional analysis. *J. Bacteriol. 189*, 336–341.

Rahman, M.S., Ammerman, N.C., Sears, K.T., Ceraul, S.M., and Azad, A.F. (2010). Functional characterization of a phospholipase A2 homolog from rickettsia typhi. *J. Bacteriol. 192*, 3294–3303.

Rahman, M.S., Gillespie, J.J., Kaur, S.J., Sears, K.T., Ceraul, S.M., Beier-Sexton, M., and Azad, A.F. (2013). Rickettsia typhi possesses phospholipase A2 enzymes that are involved in infection of host cells. *PLOS Pathog. 9*.

Rennoll-Bankert, K.E., Rahman, M.S., Gillespie, J.J., Guillotte, M.L., Kaur, S.J., Lehman, S.S., Beier-Sexton, M., and Azad, A.F. (2015). Which way in? The RalF Arf-GEF orchestrates rickettsia host cell invasion. *PLOS Pathog. 11*, e1005115.

Roux, V., Bergoin, M., Lamaze, N., and Raoult, D. (1997). Reassessment of the taxonomic position of Rickettsiella grylli. *Int. J. Syst. Bacteriol. 47*, 1255–1257.

Sahni, S.K., Narra, H.P., Sahni, A., and Walker, D.H. (2013). Recent molecular insights into rickettsial pathogenesis and immunity. *Future Microbiol. 8*, 1265–1288.

Salje, J. (2017). Orientia tsutsugamushi: A neglected but fascinating obligate intracellular bacterial pathogen. *PLOS Pathog. 13*.

Sassera, D., Beninati, T., Bandi, C., Bouman, E.A.P., Sacchi, L., Fabbi, M., and Lo, N. (2006). "*Candidatus* Midichloria mitochondrii", an endosymbiont of the tick *Ixodes ricinus* with a unique intramitochondrial lifestyle. *Int. J. Syst. Evol. Microbiol. 56*, 2535–2540.

Sassera, D., Lo, N., Epis, S., D'Auria, G., Montagna, M., Comandatore, F., Horner, D., Peretó, J., Luciano, A.M., Franciosi, F., et al. (2011). Phylogenomic evidence for the presence of a flagellum and cbb(3) oxidase in the free-living mitochondrial ancestor. *Mol. Biol. Evol. 28*, 3285–3296.

Schrallhammer, M., Ferrantini, F., Vannini, C., Galati, S., Schweikert, M., Görtz, H.-D., Verni, F., and Petroni, G. (2013). "Candidatus megaira polyxenophila" gen. nov., sp. nov.: Considerations on evolutionary history, host range and shift of early divergent rickettsiae. *PLOS One. 8*, e72581.

Sears, K.T., Ceraul, S.M., Gillespie, J.J., Allen, E.D., Popov, V.L., Ammerman, N.C., Rahman, M.S., and Azad, A.F. (2012). Surface proteome analysis and characterization of surface cell antigen (sca) or autotransporter family of rickettsia typhi. *PLOS Pathog. 8*.

Stothard, D.R., and Fuerst, P.A. (1995). Evolutionary analysis of the spotted fever and thyphus groups of rickettsia using 16s rRNA gene sequences. *Syst. Appl. Microbiol. 18*, 52–61.

Szokoli, F., Castelli, M., Sabaneyeva, E., Schrallhammer, M., Krenek, S., Doak, T.G., Berendonk, T.U., and Petroni, G. (2016). Disentangling the taxonomy of Rickettsiales and description of two novel symbionts ("Candidatus bealeia paramacronuclearis" and "Candidatus fokinia cryptica") sharing the cytoplasm of the ciliate protist paramecium biaurelia. *Appl. Environ. Microbiol. 82*, 7236–7247.

Tamura, A., Ohashi, N., Urakami, H., and Miyamura, S. (1995). Classification of rickettsia tsutsugamushi in a new genus, Orientia gen. nov., as Orientia tsutsugamushi comb. nov. *Int. J. Syst. Bacteriol. 45*, 589–591.

Teysseire, N., and Raoult, D. (1992). Comparison of western immunoblotting and microimmunofluorescence for diagnosis of Mediterranean spotted fever. *J. Clin. Microbiol. 30*, 455–460.

Vitorino, L., Chelo, I.M., Bacellar, F., and Zé-Zé, L. (2007). Rickettsiae phylogeny: A multigenic approach. *Microbiology. 153*, 160–168.

Weinert, L.A., Werren, J.H., Aebi, A., Stone, G.N., and Jiggins, F.M. (2009). Evolution and diversity of rickettsia bacteria. *BMC Biol. 7*.

Weisburg, W.G., Dobson, M.E., Samuel, J.E., Dasch, G.A., Mallavia, L.P., Baca, O., Mandelco, L., Sechrest, J.E., Weiss, E., and Woese, C.R. (1989). Phylogenetic diversity of the rickettsiae. *J. Bacteriol. 171*, 4202–4206.

Weiser, J., and Zizka, Z. (1968). Electron microscope studies of Rickettsiella chironomi in the midge Camptochironomous tentans. *J. Invertebr. Pathol. 12*, 222–230.

Weiss, E., Dasch, G.A., and Chang, K.-P. (1984). Genus VIII. Rickettsiella Philip 1956. In *Bergey's Manual of Systematic Bacteriology.*, N. Krieg and J. Holt, eds., Baltimore, MD: Williams & Wilkins, pp. 713–717.

Williams, K.P., Sobral, B.W., and Dickerman, A.W. (2007). A robust species tree for the Alphaproteobacteria. *J. Bacteriol. 189*, 4578–4586.

Williams, K.P., Gillespie, J.J., Sobral, B.W.S., Nordberg, E.K., Snyder, E.E., Shallom, J.M., and Dickerman, A.W. (2010). Phylogeny of Gammaproteobacteria. *J. Bacteriol. 192*.

Wood, D.O., Hines, A., Tucker, A.M., Woodard, A., Driskell, L.O., Burkhardt, N.Y., Kurtti, T.J., Baldridge, G.D., and Munderloh, U.G. (2012). Establishment of a replicating plasmid in rickettsia prowazekii. *PLOS One. 7*, e34715.

Yang, H.-H., Huang, I.-T., Lin, C.-H., Chen, T.-Y., and Chen, L.-K. (2012). New genotypes of Orientia tsutsugamushi isolated from humans in eastern Taiwan. *PLOS One. 7*, e46997.

38

Microbiological and Clinical Aspects of the Pathogenic Spirochetes

Charles S. Pavia

CONTENTS

Introduction

This chapter focuses on the pathogenic spirochetes, highlighting their unique basic biological properties and the important epidemiologic, clinical, pathologic, and diagnostic entities and immune phenomena associated with the diseases caused by them. Major emphasis is placed on the causative agents of Lyme disease (*Borrelia burgdorferi* and related strains) and syphilis (*Treponema pallidum*), because these represent the most common spirochetal diseases worldwide, including North America, and have generated the most interest and discussions among clinicians, scientists, patients, and other members of the lay public. In addition, their social, behavioral, as well as biomedical implications have long been recognized and these have led to considerable scientific and political debate, and unfounded misconceptions.

Basic Biology of the Spirochetal Bacteria

Spirochetes are a highly specialized group of motile Gram-negative spiral-shaped bacteria (Table 38.1), usually having a slender and tightly helically coiled structure. They range from 7 μm to 50 μm in length and about 0.5 μm in width. One of the unique features of spirochetes is their motility by a rapidly drifting rotation, often associated with a flexing or undulating movement along the helical path. Such locomotion is due to the presence of axial fibrils, also known as flagella, that are wound around the main body (protoplasmic cylinder) and enclosed by the outer cell wall or sheath of these organisms. These bacteria belong to the order Spirochaetales, which includes two families: Spirochaetaceae and Leptospiraceae. Important members of these groups include the genera *Borrelia, Leptospira,* and *Treponema.*

The spirochetes generally comprise a fastidious group of bacteria; that is, they can be difficult to grow (and therefore to study) in the laboratory, often requiring highly specialized media and culture conditions (such as low oxygen tension) to optimize their replicating capabilities. Some, such as *Treponema pallidum,* can only be maintained consistently in a replicating state by *in vivo* passage in rabbits, although a recent breakthrough in culture techniques may change the need for this long-standing procedure. The spirochetes live primarily as extracellular pathogens, rarely (if ever) growing within a host cell. A few reports have claimed finding spirochetes inside host cells but there is no evidence indicating that these are live or replicating organisms. Unlike most bacteria, spirochetes do not stain well with aniline dyes such as those used in the Gram stain procedure. However, their cell walls do resemble,

DOI: 10.1201/9781003099277-40

TABLE 38.1

Unusual Features of the Pathogenic Spirochetes

- Large bacteria: up to 50 μ in length, but very thin in diameter compared with other bacteria (e.g., cocci and rods are 1–3 μ in length), red blood cell diameter is 6–8 μ; they also exhibit a unique spiral, helical shape (Figure 38.1).
- Most of them require special staining techniques and microscopy for visualization, such as silver stain, fluorescence, or dark-field microscopy.
- They are nutritionally fastidious and exhibit a slow rate of growth: estimated 24–33 hour division time *in vivo* and slightly less *in vitro*; compare with *Escherichia coli*: 20 minutes.
- They are extremely sensitive to elevated temperatures (≥38°C). Except for one recently reported successful attempt, pathogenic treponemes *cannot* be cultivated on artificial media; other spirochetes, such as the *Borrelia* and *Leptospira* can be grown with some difficulty or with special media.
- They cause chronic, stage-related, and sometimes extremely debilitating or crippling disease in the untreated host.
- They do *not* seem to produce toxins.
- The interplay or interrelationship between the invading spirochetes and the subsequent host response as factors in the disease process have yet to be clearly or fully defined.
- They have endoflagella intertwined between the cell wall and protoplasmic cylinder—also called axial fibrils. Most bacterial flagella are extracellular.
- Most pathogenic spirochetes (*Borrelia, Treponema*) are microaerophilic (once thought to be anaerobes).
- *Borrelia burgdorferi* and its related strains are perhaps the most unique of the spirochetes, by having linear plasmids that code for outer-surface proteins.

both structurally and biochemically, those of other Gram-negative bacteria and are thus classified within this very large group of bacteria. Other staining procedures, such as silver, Giemsa, or the Wright stain, may be useful. The best way to visualize spirochetes, especially when they are alive and motile, is through the use of dark-field or phase-contrast microscopy or after staining with a fluorochrome dye, such as acridine orange, and then viewing under a microscope equipped for fluorescence microscopy (Figure 38.1). When present, their appearance in tissue specimens can often be revealed best by the silver-staining technique.

The infections caused by the spirochetes are important public health problems throughout the world, leading to such diseases as Lyme and the relapsing fever borrelioses, syphilis and the other treponematoses, and leptospirosis (Table 38.2). A better understanding of the molecular biology, pathogenesis, and immunobiology of the disease-causing spirochetes has become crucial in efforts to try to develop effective vaccines because there has been no significant modification in excessive sexual activity or in consistently practicing "safe sex" (for syphilis), personal

hygiene practices (for some of the treponemal diseases), or vector or reservoir control (for Lyme disease and the other related borrelioses). Further knowledge of immune responses to spirochetes is essential for their eventual control by immunization, and studies of the host-spirochete relationship have led to important new insights related to the immunobiology of these pathogens. Serologic techniques have long been used as indispensable diagnostic tools for the detection of many of the spirochetal diseases, especially Lyme disease and syphilis. Unfortunately, as may occur in other infectious processes, the host response to the spirochetes, as part of the normal protective mechanisms, may paradoxically cause an immunologically induced disease in the affected individual, leading to the complications of carditis, arthritis and the neuropathies of Lyme disease, as well as aortitis, immune-complex glomerulonephritis, neurologic anomalies and the gummatous lesions of syphilis.

Lyme Disease

General Features

In the mid-1970s, a geographic clustering of an unusual rheumatoid arthritis-like condition, involving mostly children and young adults, occurred in northeastern Connecticut. This condition proved to be a newly discovered disease named Lyme disease, after the town of its origin.[1] The arthritis was characterized by intermittent attacks of asymmetric pain and swelling, primarily in the large joints (especially the knees) over a period of a few years. Epidemiological and clinical research showed that the onset of symptoms was preceded by an insect bite and unique skin rash probably identical to that of an illness following a tick bite, first described in Europe in the early 1900s.[2]

The beneficial effects of penicillin or tetracycline in early cases suggested a microbial origin (likely bacterial) for what was initially called Lyme arthritis. According to the U.S. Centers for Disease Control and Prevention (CDC), each year, approximately 25,000 to 30,000 cases of Lyme disease are reported to the CDC by state health departments and the District of Columbia (Figure 38.2). However, this number is believed to be a sizable

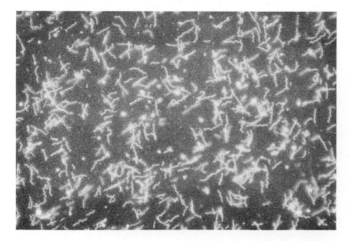

FIGURE 38.1 Photomicrograph of culture-grown *Borrelia burgdorferi* (sensu stricto) stained with the fluorochrome dye, acridine orange, and visualized through a microscope equipped for fluorescence microscopy; magnification 500×.

TABLE 38.2

Epidemiology of Spirochetal Infections

Pathogenic Spirochete	Human Disease	Vector or Source of Infection
Borrelia		
B. burgdorferi (sensu lato)[a]	Lyme disease or	Ixodid ticks
B. burgdorferi (sensu stricto)	Lyme borreliosis	*I. scapularis* or *pacificus*
B. afzelii		*I. ricinus*
B. garinii		
B. recurrentis	Epidemic relapsing fever	Body louse, *Pediculus humanus*
B. hermsii		
B. turicatae	Endemic relapsing fever	Ornithodoros ticks
B. parkeri		
B. miyamotoi		Various species of Ixodid ticks
Leptospira		
Leptospira interrogans	Leptospirosis (Weil's disease)	Exposure to contaminated animal urine
Treponema		
T. pallidum subspecies *pallidum*	Syphilis	Sexual contact, transplacental
T. pallidum ssp. *endemicum*	Bejel (endemic syphilis)	Direct contact with contaminated eating utensils
T. pallidum ssp. *pertenue*	Yaws	Direct contact with infected skin lesions
T. carateum	Pinta	

[a] Sensu stricto strains are found almost exclusively in North America, while *B. afzelii* and *B. garinii* are found only in Europe.

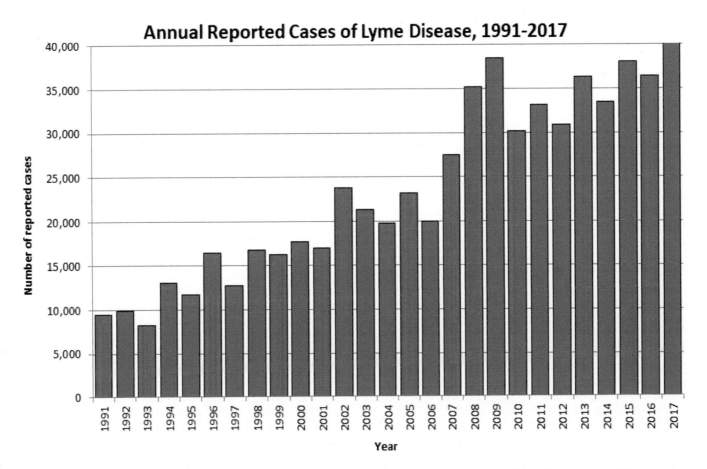

FIGURE 38.2 The graph displays the number of cases of Lyme disease reported to CDC which have increased steadily over the past 26 years. (From Centers for Disease Control and Prevention website: https://www.cdc.gov/lyme/why-is-cdc-concerned-about-lyme-disease.html.)

underestimate of the actual cases of Lyme disease that occur in the United States every year. It was announced in the summer of 2013, by the CDC, that the number of Americans diagnosed with Lyme disease each year is probably around 300,000. This preliminary estimate is based on studies conducted by the CDC and published a year later[3] that analyzed medical claim information, test results from clinical laboratories, and self-reported cases derived from a survey of the general public. The validity of such findings, however, still require scrutiny and further analysis, especially in terms of delineating laboratory-confirmed cases in relation to only probable cases having little or no confirmatory laboratory test results.

Lyme disease is now the most common tick-transmitted illness, and it has been reported in almost all the contiguous states of the United States and in numerous countries worldwide. However, it occurs primarily in three geographic regions in North America: (1) the coastal areas of the Northeast from Maine to Maryland, (2) the Midwest in Wisconsin and Minnesota, and (3) the far West in parts of California and Oregon. These geographic areas parallel the location of the primary tick vector of Lyme disease in the United States—*Ixodes scapularis* (formerly called *I. dammini*) in the east and Midwest and *I. pacificus* in the far west. Outside North America, Lyme disease is most prevalent in western Europe, especially in Austria, Germany, and the Scandinavian countries, corresponding to the distribution of *I. ricinus* ticks. The greatest concentration of cases is in the northeastern United States, particularly in Connecticut, Pennsylvania, and New York state, where the disease is endemic on eastern Long Island and just north of New York City in neighboring Westchester and Putnam counties.

In the early 1980s, spirochetal organisms were isolated and cultured from the midguts of *Ixodes* ticks taken from Shelter Island, NY (an endemic focus), and shortly thereafter they were cultured from the skin rash site, blood, and cerebrospinal fluid of patients with Lyme disease. This newly discovered spirochete called *B. burgdorferi* is microaerophilic, resembles other spirochetes morphologically, and is larger and not as tightly coiled than the treponemes. Recently, the Lyme disease spirochetes have been given updated taxonomic designations under the broad category of *B. burgdorferi* (sensu lato) (Table 38.2). Unlike most of the pathogenic treponemes, *B. burgdorferi* and related strains can be readily cultivated *in vitro* in a highly fortified growth media.[4]

Protection to *B. burgdorferi* may develop slowly, and it is unclear whether resistance to reinfection occurs. Experimental animal studies have shown that immune sera, that was derived from untreated patients having high titer antiborrelial antibodies, can transfer protection to normal recipients challenged with *B. burgdorferi*.[5] Monoclonal antibodies to borrelial outer surface proteins are also protective,[6] and these were the major target antigens for the first vaccine that became available for human use. It should be noted that, for a variety of reasons, the vaccine was withdrawn from the market by the manufacturer in 2002, after its initial approval in 1998.

Clinical Aspects

Lyme disease is an illness having protean manifestations with various signs and symptoms that include the following: (1) an erythematous-expanding red annular rash with central clearing; (2) fever, headache, stiff neck, nausea, and vomiting; (3) neurologic complications such as facial nerve (Bell's) palsy and meningitis; and (4) arthritis in about 50% of untreated patients.[7] These symptoms occur most frequently from May to November when ticks are active and numerous, and people are engaged in many outdoor activities. The most characteristic feature of early Lyme disease is a skin rash, often referred to as erythema migrans (EM), which appears shortly (3–32 days) after a bite from an infected tick. The lesion typically expands almost uniformly from the center of the bite and is usually flat or slightly indurated with central clearing and reddening at the periphery. It is noteworthy, however, that many Lyme disease victims either do not recall being bitten by a tick, or they do not notice a rash developing at an unrecognized tick bite, or possibly they do not go on to develop a classic EM at all. On the other hand, at various intervals after the initial rash, some patients develop similar but smaller multiple secondary annular skin lesions that last for several weeks to months. Due to the expanding nature of this rash and its originally described chronic form in European patients, it was initially referred to as erythema chronicum migrans. However, this term is rarely used now, with the term EM or multiple EM being the current descriptor. Biopsy of these skin lesions reveals a lymphocytic and plasmacytic infiltrate, and *Borrelia* can be cultured from them but not always. Various flu-like symptoms such as malaise, fever, headache, stiff neck, and arthralgias are often associated with EM. The serious extracutaneous manifestations of Lyme disease may include migratory and polyarticular arthritis, neurologic and cardiac involvement with cranial nerve palsies and radiculopathy, myocarditis, and arrhythmias. Lyme arthritis typically involves a knee or other large joint. It may enter a chronic phase, leading to destruction of bone and joints if left untreated. Interestingly, Lyme arthritis is less common in Europe than in the United States, but neurologic complications are more prevalent in Europe. Another geographic consideration involves the later-stage chronic skin condition known as acrodermatitis chronicum atrophicans, which can occur in Europe but has rarely been observed in North America. Unique strain variations expressing antigenic sub-types between European and North American isolates of *B. burgdorferi*[8] probably explain these dissimilarities, and this has led to additional species designations for other related Lyme disease-causing spirochetes, such as *B. afzelii* and *B. garinii,* which are not found in North America. Also, several other species within the senso-lato geno-complex, such as *B. lusitaniae* (found in Portugal) and *B. japonica* (found in certain parts of the Far East), have been identified but their pathogenic capabilities, as possible causes of Lyme disease, have yet to be well characterized.

In most cases, humoral and cell-mediated immune responses are activated during borrelial infection.[7,9,10] Antibody, mostly of the IgM class, can be detected shortly after the appearance of EM; thereafter, a gradual increase in overall titer and a switch to a predominant IgG antibody response occurs for the duration of an untreated infection. Most notably, very high levels of antibody have been found in serum and joint fluid taken from patients with moderate to severe arthritis, making this the most actively invasive and debilitating form of Lyme disease. Although the presence of such high antibody titers against *B. burgdorferi* may reduce the spirochete load somewhat, they appear not to ameliorate the disease process completely and, indeed, may actually contribute to some of the pathologic changes. These consistent serologic responses have led to the development of laboratory tests designed to aid in the diagnosis of Lyme borreliosis. For cellular immunity, lymphocyte transformation assays have shown that peripheral blood T cells from Lyme disease patients respond to borrelial antigens primarily after early infection and following successful treatment.[9,10] Also, addition of antigens to synovial cells *in vitro* from infected patients triggers the production of interleukin-1, which could account for many of the harmful inflammatory reactions associated with this disease.[11] Other *in vitro* studies[12,13] have shown that human mononuclear and polymorphonuclear phagocytes can both ingest and presumably destroy *Borrelia*. Thus, borrelial antigen-stimulated T cells or their products may activate macrophages, limiting dissemination and resulting in enhanced phagocytic activity and the eventual clearance of spirochetes from the primary lesion.

Diagnosis

Clinically, Lyme disease mimics other disorders, many of which are not infectious and therefore would ordinarily not be responsive to antibiotic therapy. Because *B. burgdorferi* is the causative agent, finding the organism, or any of its key detectable components, in suspected cases is the most definitive diagnosis. However, in the vast majority of cases, the Lyme spirochete cannot be consistently isolated or identified due to its transient presence in blood and other body fluids and various tissues, coupled with its highly fastidious growth requirements for in vitro culture. Accordingly, immune responses (antibody production) specific for *B. burgdorferi* must be used to confirm the diagnosis.[14] Unfortunately, the antibody response is often not detectable in the early treatable stage, and the clinical impression cannot always be confirmed.

Isolation of the spirochete unambiguously confirms the diagnosis of Lyme borreliosis. Recovery of *Borrelia burgdorferi* is possible but the frequency of isolation from the blood or other body fluids of acutely ill patients is very low.[15,16] Better success rates in isolating *Borrelia* have been achieved after culturing skin biopsy specimens of clear-cut EM rashes.[17] Despite such success, borrelial cultivation can usually only be done in a few laboratories or institutions because the medium is expensive and very complex, and cultures can take up to several weeks of incubation before spirochetes are detected,[15,16] but most are positive within 4 weeks.

Visualization of the spirochetes in tissue or body fluids also has been used to diagnose Lyme borreliosis.[18] In the early stage of the disease, when EM is present, the Warthin-Starry or modified Dieterle silver stain can identify spirochetes in half or more of skin biopsies obtained from the outer portion of the lesion.[19] However, few microorganisms are present and they can be confused with normal skin structures or tissue breakdown products by inexperienced laboratory personnel. Immunohistologic examination of tissue has rarely been successful in determining the presence of *Borrelia* and, in chronic Lyme disease, spirochetes are rarely detectable by any microscopic technique.[20]

Serologic tests are, for all practical purposes, the only detection systems routinely available for the confirmation of Lyme borreliosis. One of the standard serological tests, either an enzyme-linked immunosorbent assay (ELISA) or an indirect immunofluorescence assay (IFA),[14] is available in many public and private laboratories. Blood samples obtained within 3 weeks of the onset of EM are frequently serologically negative in both assays.[21] In the past, these assays were not standardized, with laboratories using different antigen preparations and *cutoff* values. Workers, using the same set of sera, have reported interlaboratory variation in results and interpretations.[22] There is also considerable variability in the serologic response pattern of patients with Lyme disease. Nonetheless, over the past several years there has been considerable progress in development of better antibody detection systems. Currently, there are several commercially available serologic test kits that claim to have improved sensitivity (e.g., the C6 ELISA) without sacrificing specificity. Finally, if antibiotics are administered during early illness, antibody production can be aborted or severely curtailed,[7,14] although this is not always the case, particularly when convalescent sera are analyzed.[23]

The existence of antigenically different strains of *B. burgdorferi* sensu lato throughout the world[8] may account for some of the variability in antibody response patterns. In addition, assays currently either use the whole spirochete or a crude bacterial sonicate as antigen, or a mixture of recombinant-derived proteins. While the latter results in greater specificity, they tend to be less sensitive. With these assays, cross-reactions have been observed with other spirochetes, in particular *T. pallidum* and the relapsing fever *Borrelia* species.[24] For all of these assay systems, false-positive reactions are relatively rare except possibly in hyperendemic areas, where considerably more testing (often unnecessary) would be expected because of heightened awareness of Lyme disease within the general community. The cause of these positives is not always clear but they can occur if a patient has certain other disorders, such as syphilis (active or inactive), infectious mononucleosis, lupus, or rheumatoid arthritis. In the absence of a clear-cut EM rash, laboratory testing plays a vital role in establishing or confirming the diagnosis.

Attempts to improve antibody detection have used Western (immunoblot) analysis for the detection of IgM and IgG antibodies and have used purified flagellin antigen or the VlsE (or C6) peptides in the ELISA. Immunoblots are more sensitive and more specific than ELISAs. Although not standardized, commercial immunoblot test kits are now being offered to further verify a routine ELISA-based serologic test result, especially in troublesome cases, in conjunction with the two-tiered testing algorithm recommended by the CDC,[25] as well as for further confirmation of a true serologic response to specific borrelial antigens and possible infection. Some studies have shown, however, that immunoblotting could not overcome the inability to detect antibody during the first 3 weeks of infection.[26] The performance of the ELISA can also be improved using purified flagellin protein as antigen.[27] Antibodies to the 41-kDA, flagellum-associated component peak at 6–8 weeks, and often appear very early (<4 weeks) during an initial infection. Unfortunately, epitopes on this antigen are shared by many other spirochetes, and neither IgM nor IgG antibodies to this antigen are specific for *B. burgdorferi*.[26] It is noteworthy, however, that the initial development of an ELISA using recombinant VlsE (or C6) peptides showed considerable improvement in both specificity and sensitivity over prior assay systems.[28]

The prevailing sentiment within (as well as outside) the Lyme disease research and diagnostic community is that serological verification of this disorder is still fraught with difficulties. False-negative results are likely to occur if serum is obtained within 4 weeks of initial infection, or if the patient has been treated with antibiotics soon after a tick bite or a recognized EM rash. False-positives occur if large numbers of patients with a low *a priori* probability of having Lyme borreliosis are examined. Interlaboratory agreement on what constitutes a positive result varies, in part because of the diversity of detection systems being used by diagnostic testing facilities, which contain different antigen preparations and cut-off values developed by various manufacturers for serodiagnostic purposes. Compounding this problem, diagnosis of initial infection can be clinically difficult because only 60–75% of patients with Lyme borreliosis present with or recall having developed an EM rash, or have a clear and consistent epidemiologic history.[7]

The initial attempts in developing nonserologic based assays that were meant to address these apparent shortcomings in serologic testing have led to the development of alternative or supplemental diagnostic procedures such as the polymerase chain reaction (PCR) and the lymphocyte proliferation assay. Using the PCR and selective probes, it is possible to detect a single organism in a serum or tissue sample, and such gene amplification procedures show great promise for the early detection of *B. burgdorferi*.[29] In this regard, using primers directed at the rRNA genes of *B. burgdorferi*, it is possible to find the Lyme disease spirochete directly from skin biopsy material[30] as well as from short-term cultures of tissue extracts.[31] From a realistic standpoint, however, and because it may be technically demanding and not cost-effective for most diagnostic labs handling just a few specimens, PCR may continue to be primarily a research tool rather than a routine diagnostic procedure. Other limitations include its relatively poor sensitivity beyond the infected skin site and its inability to distinguish live organisms from remnants of dying organisms that may be present in a patient sample.

Additionally, some attention had previously been given to a cell-mediated immune (CMI)-based assay system[9,10] designed to measure past or current exposure to *B. burgdorferi* by virtue of the patient's lymphocytes to respond *in vitro* to undergo DNA synthesis (aka lymphocyte transformation or proliferation) in the presence of specific borrelial antigens. This laboratory procedure is generally considered a good *in vitro* correlate of the classic delayed type hypersensitivity (DTH) reaction,[32] analogous to what is found when measuring *in vivo* tuberculin sensitivity following a Mantoux skin test using a purified protein derivative (PPD) of *Mycobacterium tuberculosis*. For purposes related to Lyme disease, it was found that these lymphocyte proliferation assays have been helpful in identifying patients with active disease in the absence of detectable antibody (serologic) responses. However, as was the case with the early development of serologic tests, this assay has yet to be standardized and there is concern over the evidence[9] for elevated responses occurring in some healthy controls, thereby limiting the usefulness of this technique. Another *in vitro* correlate of the DTH reaction, which is currently being developed and is an extension of the lymphocyte proliferation assay, is based on Quantiferon technology.[33] This assay system relies on the ability of a patient's T cells to be stimulated to produce interferon-gamma after *in vitro* culture

with peptide antigens of *B. burgdorferi*. So far there is only preliminary data showing that this assay system would be a significant alternative or supplement to serologic testing.

Finally, apart from PCR- and CMI-based techniques, another nonserologic detection system has used the newly developed metabolomics approach for the purpose of identifying unique biomarkers that appear in the blood of patients with early Lyme disease. Using this approach, it has been shown[34,35] that certain serum and urine small molecule metabolites can accurately distinguish patients with early Lyme disease from healthy controls and from patients with other diseases. The results of such analyses involved the use of liquid chromatography-mass spectrometry instrumentation which could be a serious limiting factor in the routine implementation of this diagnostic approach.

While PCR- and metabolomic-based procedures seem promising as exquisite and novel diagnostic tools for selective stages of Lyme disease, serological testing, for a variety of reasons (cost, standardization, need for sophisticated equipment and technical skill), will continue to be the mainstay for the laboratory detection of the vast majority of Lyme disease cases, as it currently is for syphilis (caused by the related spirochete, *T. pallidum*) and for certain other infectious and noninfectious disorders. Nonetheless, continued refinements along these lines will be needed and should be geared toward developing as economical a system as possible combined with one having optimal sensitivity and specificity.

Prophylactic Measures

Avoiding *Borrelia*-infected ticks or tick-infested areas will guarantee protection against Lyme infection. For those living in endemic areas, a few simple precautions will help minimize possible exposure. These include wearing clothing that fully protects the body and using repellents that contain DEET (diethyltoluamide). If a tick does attach to the skin, careful removal with tweezers shortly after it attaches, followed by application of alcohol or another suitable disinfectant, will make borrelial transmission unlikely.

In the early 1990s, considerable attention turned toward the development of a vaccine for Lyme disease. For a few years already, a canine vaccine consisting of whole inactivated organisms (Bacterin) was available, and still is, for veterinary purposes,[36] primarily for preventing *B. burgdorferi* infection in dogs. Those that had been developed for humans consisted of recombinant outer surface proteins (Osp) of *Borrelia burgdorferi*. In 1998, the FDA gave final approval to the first Lyme disease vaccine (called LYMErix®), which was shown to be safe and effective in extensively conducted clinical trials. It consisted of DNA-derived OspA of B31—a well-characterized tick isolate of *B. burgdorferi*—incorporated with an adjuvant.[37] However, in less than 4 years, the vaccine was withdrawn by the manufacturer, mostly due to concerns over its safety and unfavorable publicity resulting from claims of purported serious side effects in a select group of vaccines. Although studies continue with other vaccine candidates (especially in Europe), it is unclear when they may be available for widespread use.

Since the Lyme vaccine is no longer marketed and a number of substantive limitations exist to the use of oral doxycycline for prophylaxis shortly after a documented tick bite (cited below, under treatment) including the fact that the drug is contraindicated in young children (<8 years of age) and pregnant or lactating women, consideration has been given to using topical antibiotic preparations instead that might have antiborrelial activity. In this regard, a variety of topical antimicrobials are in wide general usage for treatment of common conditions, both infectious and noninfectious. These preparations are well accepted and safe and because *B. burgdorferi* remains at the site of deposition in the skin after an *I. scapularis* tick bite for ≥2 days, perhaps the timely application of topical formulations of active antibiotics at the tick bite site could eliminate the infection. In the first published study[38] to test this hypothesis, several topical antimicrobial preparations (tetracycline hydrochloride, doxycycline, penicillin G, amoxicillin, ceftriaxone, and erythromycin) were shown to be highly effective (nearly 100%) in preventing *B. burgdorferi* infection in a mouse model when these preparations were applied within 48 hours after the spirochete-infected ticks had detached and the drug was given in the appropriate concentration and for the appropriate duration of time. However, none of the antibiotic preparations studied by these investigators is commercially available for use topically in the United States. Also, these investigators dissolved all the antibiotics studied, except for erythromycin, in dimethyl sulfoxide for improved absorption, but this compound is unsatisfactory, based on the formulation used, for routine use in humans. Unfortunately, as a follow-up to these initial seemingly promising results, in carefully controlled studies[39] using the murine model for experimental Lyme disease, topical applications of 2% erythromycin and 3% tetracycline preparations, that are acceptable for use in humans, were found to be ineffective in eliminating *B. burgdorferi* from the tick bite site or in preventing dissemination to other tissues.

Treatment

In general, early Lyme disease is readily treatable with a 2–3 week course of antibiotics, such as amoxicillin and doxycycline.[40] Later complications, such as arthritis and the neuropathies, may require more intense and prolonged antibiotic therapy, including intravenous treatment, with ceftriaxone being the preferred parenteral drug. Based on an extensive and well controlled study, showing favorable outcomes in treating patients with a single dose of doxycycline soon after a documented tick bite,[41] consideration should be given to following this treatment strategy, especially in highly endemic areas and this practice has recently been endorsed by the CDC, after meeting certain criteria, for the prevention of Lyme disease but not for other tick-borne diseases.[42]

Relapsing Fever Borreliosis

Relapsing fever is an acute febrile disease of worldwide distribution and is caused by arthropod-borne spirochetes belonging to the genus *Borrelia*.[43,44] Two major forms of this illness are louse-borne relapsing fever (for which humans are the reservoir and the body louse *Pediculus humanus* is the vector) and tick-borne relapsing fever (TBRF, for which rodents and other small animals are the major reservoirs and ticks of the genus *Ornithodoros* are the vectors). *B. recurrentis* causes louse-borne relapsing fever and is transmitted from human to human following the ingestion of infected human blood by the louse and subsequent transmission of spirochetes onto the skin or mucous membranes of a new host when the body louse is crushed. The disease is endemic in parts of central and east Africa and South America. The causative organisms of tick-borne relapsing fever are numerous and include *B. hermsii*, *B. turicatae*, and *B. parkeri* in North America; *B. hispanica* in Spain; *B. duttonii* in eastern Africa; and *B. persica* in Asia. Ticks become infected by biting and sucking blood from a spirochetemic animal. The infection is transmitted to humans or animals when saliva is released by a feeding tick through bites or penetration of intact skin.

After an individual has been exposed to an infected louse or tick, *Borrelia* penetrate the skin and enter the bloodstream and lymphatic system. After a 1- to 3-week incubation period, spirochetes replicate in the blood and there is an acute onset of shaking chills, fever, headache, and fatigue. Concentrations of *Borrelia* can reach as high as 10^8 spirochetes per milliliter blood, and these are clearly visible after staining blood smears with Giemsa or Wright's stain. During the febrile disease, *Borrelia* are present in the patient's blood but disappear prior to afebrile episodes and subsequently return to the bloodstream during the next febrile period. Jaundice can develop in some severely ill patients as a result of intrahepatic obstruction of bile flow and hepatocellular inflammation; if left untreated, patients can die from damage to the liver, spleen, or brain. The majority of untreated patients, however, recover spontaneously. They produce borrelial antibodies that have agglutinating, complement-fixing, borreliacidal, and immobilizing capabilities, and that render patients' immune to reinfection with the same *Borrelia* serotype. Serologic tests designed to measure these antibodies are of limited diagnostic value because of antigenic variation among strains and the coexistence of mixed populations of *Borrelia* within a given host during the course of a single infection. Diagnosis in the majority of cases requires demonstration of spirochetemia in febrile patients.

A partial account of an interesting clinical case pertaining to relapsing fever borreliosis was published recently in the *New England Journal of Medicine*[45]. It discusses a 13-year-old girl who presented with recurrent fevers, with temperatures up to 104°F (40°C), over a 3-week period, with associated headache, chills, and malaise. She and her family had stayed in a cabin in the eastern Sierra Nevada Mountains 10 days before she first had a fever. The patient's brother had similar symptoms. The siblings were febrile for 5 days and then symptom-free for 7 days. During a relapse of the girl's fever, blood was drawn; on examination of a blood smear, a *B. hermsii* spirochete was detected (Figure 38.3; arrow). Because of their symptoms and recent travel to an area where tickborne relapsing fever is endemic, the siblings were treated with doxycycline for presumed *B. hermsii* infection. Tests for serum IgG and IgM antibodies to *B. hermsii* were subsequently positive. After the first dose of doxycycline, defervescence occurred in both patients and associated symptoms resolved.

Borrelia miyamotoi is a newly recognized, culturable agent of human disease, and may be prevalent in areas where Lyme disease is endemic in the United States.[46] Using specialized media, *B. miyamotoi* isolates from certain parts of Japan and North America have been cultured but the reported yields of cultures for a North American isolate were lower than that expected for *B. burgdorferi* and several other *Borrelia* species. Based on recent DNA sequencing studies, it has been established that *B. miyamotoi* is a relapsing-fever group species. It is a spiral-shaped

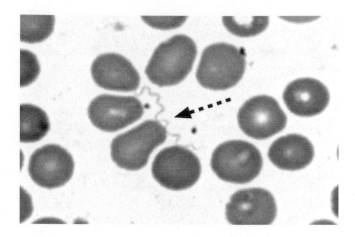

FIGURE 38.3 Giemsa-stained peripheral blood smear confirming a diagnosis of relapsing fever borreliosis.

bacterium that is only distantly related to the spirochete that causes Lyme disease. First identified in 1995 in ticks from Japan, it has since been detected in two species of North American ticks, the deer tick that transmits Lyme disease and the Western black-legged tick (*I. pacificus*), which is also known to transmit several diseases, including Lyme disease, anaplasmosis, and babesiosis. Human infections with *B. miyamotoi* were first described in 2011 in a report from Russia. Most of the patients had fever, headache, and muscle aches—the symptoms typical of TBRF. Symptoms similar to those of Lyme disease, such as the erythema migrans rash (bull's-eye rash), arthritis, or facial palsy, were uncommon.

Since the first three cases of human infection with *B. miyamotoi* were identified in the United States,[46] subsequent studies have identified several more cases. Examples of these patients include an elderly, immunocompromised woman with confusion and an unsteady gate. The bacteria were seen in samples of the patient's spinal fluid, and she recovered when treated with antibiotics. The two other patients had fever, chills, and muscle aches, similar to the symptoms of the patients in Russia.

Blood tests based on detection of antibodies appear to be useful indicators of active infection,[43] but still require further validation. Diagnosis currently relies on the use of tests to detect DNA of the organism in whole blood using the polymerase chain reaction or by serology. Serologic testing for *B. miyamotoi* is done using a recombinant glycerophosphodiester phosphodiesterase (rGlpQ) protein.[47] These tests are under continuous development and refinement, and only a few laboratories perform these tests. Culturing *B. miyamotoi* from whole blood of suspect patients is not routinely done due to the need for a highly specialized growth media, and the limited number of laboratories having the capability to isolate this difficult-to-culture spirochete on a routine basis. It should be noted that blood tests for Lyme disease are unlikely to be helpful in diagnosing *B. miyamotoi* infections. Physicians have successfully treated patients infected with *B. miyamotoi* with a 2-week course of doxycycline. The key signs and symptoms of this infection have not been clearly defined and some of them could easily be confused with other abnormal conditions. Most studies have focused on hospitalized patients and may have overestimated the severity of a typical infection. It is also unclear how common this infection is in the United States.

Leptospirosis (Weil's Disease)

Leptospiral infections are zoonoses widely distributed throughout the world. Human disease is somewhat more common in tropical than in temperate areas, and infection occurs more in the summer or autumn in temperate regions.[48] Most humans are infected by leptospires in water used for drinking, washing, bathing, or swimming primarily after contact with contaminated urine, from leptospiruric animals, that has entered the water supply. The most common sources of organisms are chronically infected rats and other rodents, cattle, horses, pigs, and dogs. Numerous wild animals and even reptiles and amphibians may be carriers. In North America, unvaccinated dogs are the major reservoir for exposure of humans to this disease. The routine vaccination of dogs against *Leptospira* is probably an important preventive measure. Leptospirosis is an occupational hazard for rice and sugar cane field workers, military personnel during field exercises, people who swim in contaminated ponds or streams, farmers, veterinarians, and sewer and abattoir workers. In tropical rainforests, infected rodents contaminate stream banks as well as the water. Rainstorms raise the level of the water, which washes leptospires into streams and makes swimming very hazardous. Related to this concern, large outbreaks of leptospirosis were reported among athletes participating in triathlon events held within the past 20 years in both the United States in 1998[49] and Malaysia in 2000.[50] More recently in the winter of 2018, a very small and brief outbreak occurred in the Bronx section of New York City (a large urban area) that involved three people who contracted leptospirosis. Although New York City typically sees only about one to three cases of leptospirosis per year, this is the first time that health officials identified a cluster of leptospirosis cases (meaning more than a single case occurring in the same place around the same time) in the city. All three people were hospitalized with kidney and liver failure, and one person died as a result of complications from their infection. It is believed that all three victims were exposed to infected rat urine. Such an outbreak points to the sometimes unpredictable epidemiologic nature of this disease.

Leptospirosis is an acute, febrile disease caused by various serotypes of *Leptospira* (there exist about 170 serovars in 20 serologic groups). Often referred to as Weil's disease, infection with *Leptospira interrogans* causes diseases that are extremely varied in their clinical presentations and that are also found in a variety of wild and domestic animals. After entering the body through the mucosal surface or breaks in the skin, leptospiral bacteria cause an acute illness characterized by fever, chills, myalgias, severe headaches, aseptic meningitis, rash, hemolytic anemia, uveitis, conjunctival suffuseness, and gastrointestinal problems. Most human infections are mild and anicteric, although in a small proportion of victims, severe icteric disease can occur and be fatal, primarily as a result of renal failure, myocarditis, and damage to small blood vessels. After infection of the kidneys, leptospiras are excreted in the urine. Liver dysfunction with hepatocellular damage and jaundice is common. Antibiotic treatment is curative if begun during early disease, but its value, thereafter, is questionable.

Diagnosis of leptospirosis depends on either seroconversion or the demonstration of spirochetes in clinical specimens. The macroscopic slide agglutination test, which uses formalized antigen, offers safe and rapid antibody screening. Measurement of antibody for a specific serotype, however, is performed with the very sensitive microscopic agglutination test involving live organisms. This method provides the most specific reaction with the highest titer and fewer cross-reactions. Agglutinating IgM-class-specific antibodies are produced during early infection and persist in high titers for many months. Protective and agglutinating antibodies often persist in sera of convalescent patients and may be associated with resistance to future infections.

Syphilis

General Features

The origins and history of syphilis are filled with many mysteries and hypotheses. Biblical references suggest its presence in early civilization. Other evidence points to its prevalence primarily after Columbus and his crew returned to Europe from the New World in 1493. From that point on, syphilis spread throughout Europe, affecting all levels of society including political and religious leaders. Indeed, the dementia and insanity associated with late-stage syphilis that may have occurred in certain afflicted rulers or monarchs probably changed the course of history during the 16th and 17th centuries. Before antibiotics, many toxic drugs were used for treatment; as early as 1905, a blood test was developed for the diagnosis of syphilis—the so-called Wasserman test—the prototype for the current nonspecific serologic tests designed to measure antibodies to cardiolipin.

Syphilis is still a significant worldwide problem, and after herpes, gonorrhea, and chlamydial infections, it is the fourth or fifth most commonly reported sexually transmitted disease in the United States. The most recent rise in both same-sex and heterosexual infection occurred in the mid-1980s to the early 1990s (Figure 38.4), which coincided and co-factored with the huge number of newly diagnosed HIV cases. At the same time, this also paralleled an alarming increase in congenital syphilis in many urban areas where drug abuse and the frequent exchange of sexual services for drugs were common practices among those using illicit drugs.

Treponema pallidum subspecies *pallidum* is the spirochetal bacterium that causes syphilis and was once referred to as "The Great Imitator" (or "Mimicker"), due primarily to the myriad clinical presentations often associated with this disease.[51,52] If left untreated, syphilis can have severe pathologic effects, leading to irreversible damage to the cardiovascular, central nervous, and musculoskeletal systems.[51,52] This treponeme's genome was sequenced in 1998, yet up until very recently,[53] it had remained nonculturable on artificial media for more than a century. Despite this extraordinary achievement, it remains to be seen if other laboratories are able to grow this pathogen on a consistent basis. Prior to the successful *in vitro* cultivation of *T. pallidum*,[53] it had to be passaged *in vivo* using rabbits to maintain sufficient numbers of live organisms needed for conducting meaningful laboratory studies. It is also highly motile, infectious, and replicates extracellularly and very slowly *in vivo*. Because of its uniqueness, it has taken many years of research to acquire our current understanding of the treponemal bacteria, such as *T. pallidum* and this has probably delayed efforts toward any possible vaccine development. *T. pallidum*, unlike some of the other spirochetes, is shorter and more tightly coiled than the *Borrelia* and *Leptospiras*.

With the institution of antibiotic therapy in the mid-1940s, the incidence of syphilis fell sharply from a high of 72 cases per 100,000 population in 1943 to about 4 per 100,000 in 1956. During the late 1970s and early 1980s, syphilis increased rapidly within the homosexual community. Such findings can be attributed, in part, to changing lifestyles, sexual practices, and other factors such as an unusually high prevalence and reduced efficacy of antibiotics in patients with acquired immunodeficiency syndrome (AIDS). Syphilis continues to rank annually in the top five of the most frequently reported communicable disease in the United States, although its overall prevalence pattern has remained almost unchanged for the past several years (Figure 38.4).

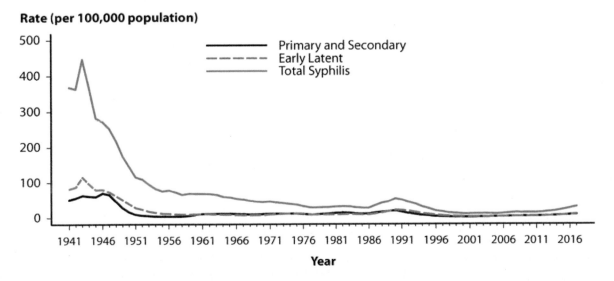

FIGURE 38.4 Rates of reported syphilis cases by stage of infection, United States, 1941–2017. (Based on data from Centers for Disease Control and Prevention. https://www.cdc.gov/std/stats17/syphilis.htm; accessed on July 8, 2019.)

The course of syphilis in humans is marked by several interesting phenomena. Without treatment, the disease will usually progress through several well-defined stages (somewhat resembling Lyme disease). This is unlike most other infectious diseases, which are ultimately eliminated by the host's immune system or, in severe cases, result in death. The relatively slow generation time of treponemes, which is estimated at 30–33 hours, contributes to this unique course. During the first two stages (primary and secondary syphilis), there is almost unimpeded rapid growth of *T. pallidum*, leading to an early infectious spirochetemic phase of the disease. The third stage (tertiary syphilis) occurs much later, following a prolonged latency period. Alterations in this stage are due primarily to tissue damaging immune responses elicited by small numbers of previously deposited or disseminated spirochetes.

Syphilis activates both humoral and cell-mediated immunity, but this protection is only partial. The relative importance of each type of immune response is not fully known. Protective immunity against re-exposure is incomplete, especially during early stages when it develops relatively slowly.

Clinical Aspects

The severe late manifestations or complications of syphilis occur in the blood vessels and perivascular areas. However, sexual contact is the common mode of transmission that follows inoculation of infectious treponemes onto the mucous membranes of genital organs or of the oropharynx.

The first clinically apparent manifestation of syphilis (primary syphilis) is an indurated, circumscribed, relatively avascular and painless ulcer (chancre) at the site of treponemal inoculation. Spirochetemia with secondary metastatic distribution of microorganisms occurs within a few days after onset of local infection but clinically apparent secondary lesions may not be observed for 2–4 weeks. The chancre lasts 10–14 days before healing spontaneously.

The presence of metastatic infection (secondary syphilis) is manifested by highly infectious mucocutaneous lesions of extraordinarily diverse description as well as headache, low-grade fever, diffuse lymphadenopathy, and a variety of more sporadic phenomena. The lesions of secondary syphilis ordinarily go on to apparent spontaneous resolution in the absence of treatment. However, until solid immunity develops (a matter of about 4 years), 25% of untreated syphilitic patients may be susceptible to repeated episodes of spirochetemia and metastatic infection.

Following the resolution of secondary syphilis, the disease enters a period of latency with only abnormal serologic tests to indicate the presence of infection. During this time, persistent or progressive focal infection is presumably taking place, but the precise site remains unknown in the absence of specific symptoms and signs. One site of potential latency, the central nervous system, can be evaluated by examining the cerebrospinal fluid in which pleocytosis, elevated protein levels, and a positive serologic test for syphilis are indicative of asymptomatic neurosyphilis.

Only about 15% of patients with untreated latent syphilis go on to develop symptomatic tertiary syphilis. Serious or fatal tertiary syphilis in adults is virtually limited to disease of the aorta (aortitis with aneurysm formation and secondary aortic valve

insufficiency), the central nervous system (tabes dorsalis, general paresis), the eyes (interstitial keratitis), or the ears (nerve deafness). Less frequently, the disease becomes apparent as localized single or multiple granulomas known as *gummas*. These lesions are typically found in the skin, bones, liver, testes, or larynx. The histopathologic features of the gumma resemble those of earlier syphilitic lesions, except that the vasculitis is associated with increased tissue necrosis and often frank caseation. With its myriad organ-system involvement and symptomatology, syphilis, it's not surprising to have long been called the *The Great Imitator*.

Congenital syphilis is the direct result of treponemes crossing the placenta and fetal membranes, especially during midpregnancy, leading to spirochetemia and widespread dissemination after entering the fetal circulation. Fetal death and abortion can occur. Surviving babies with the disease have prominent early symptoms of hepatosplenomegaly, multiple long bone involvement, mouth and facial anomalies (e.g., saddle nose), and skin lesions. Treponemal antibodies (especially IgM) found in the newborn's blood is highly diagnostic.

Diagnosis

In its primary and secondary stages, syphilis can be diagnosed by dark-field microscopic examination of material taken from suspected lesions. This rather delicate and cumbersome procedure is now rarely done in most clinical settings and confirmatory tests rely solely on serology. Diagnostic serologic changes do not begin to occur until 14–21 days following acquisition of infection. Serologic tests provide important confirmatory evidence for secondary syphilis but are the only means of diagnosing latent infection. Many forms of tertiary syphilis can be suspected on clinical grounds, but serologic tests are important in confirming the diagnosis. Spirochetes are notoriously difficult to demonstrate in the late stages of syphilis.

Two main categories of serologic tests for syphilis are available: (1) tests for nonspecific reaginic (or anticardiolipin) antibody and (2) tests for specific treponemal antibody. In the United States, testing for syphilis traditionally has consisted of initial screening with the relatively inexpensive nonspecific test, then retesting reactive specimens with the specific, and slightly more expensive, treponemal test (both types of tests are explained next). When both test results are reactive, they indicate present or past infection. However, for economic reasons and because of variable response patterns exhibited by some suspect patients (such as with false-positives associated with the nontreponemal tests), some high-volume clinical laboratories have begun using automated treponemal tests, which has led to reversing the previously established and more traditional testing sequence by first screening with a treponemal test and then retesting reactive results with a nontreponemal test. This approach has introduced complexities in test interpretation that did not exist with the traditional sequence. Specifically, screening with a treponemal test sometimes identifies persons who are reactive to the treponemal test but nonreactive to the nontreponemal test. No formal recommendations exist regarding how such results derived from this new testing sequence should be interpreted, or how patients with such results should be managed thereby making the attending physician to have to rely on using sound judgement and expertise in deciding how to deal with a suspect patient in the context of the clinical picture. If they have not been previously treated, patients with reactive results from treponemal tests and nonreactive results from nontreponemal tests should be treated for late latent syphilis.

Tests for Reaginic or Anticardiolipin Antibody

This is an unfortunate and confusing designation; there is no relationship or connection between *T. pallidum*-related reaginic antibody and IgE reaginic antibody (the antibody associated with immediate hypersensitivity reactions or anaphylaxis). Patients with syphilis develop an antibody response to a tissue-derived substance (from beef heart) that is thought to be a component of mitochondrial membranes and is called *cardiolipin*. Antibody to cardiolipin antigen is known as the Wassermann, or reaginic antibody. Numerous variations (and names) are associated with tests for this antigen. The simplest and most practical of these are the VDRL test (a test originally developed, many years ago, by the Venereal Disease Research Laboratory of the U.S. Public Health Service), which involves a slide microflocculation technique and can provide qualitative and quantitative data, and the rapid plasma reagin (RPR) circle card test (a simplified version of the VDRL test). Positive tests are considered to be diagnostic of syphilis when there is a high or increasing titer or when the medical history is compatible with primary or secondary syphilis. The tests may also be of prognostic aid in following response to therapy because the antibody titer will revert to negative within 1 year of treatment for seropositive primary syphilis or within 2 years of that for secondary syphilis. Because cardiolipin antigen is found in the mitochondrial membranes of many mammalian tissues as well as in diverse microorganisms, it is not surprising that antibody to this antigen should appear during other diseases, leading to a false-positive test result. A positive VDRL or RPR test may be encountered, for example, in patients with infectious mononucleosis, leprosy, hepatitis, or systemic lupus erythematosus, and various other conditions. Although the VDRL test lacks specificity for syphilis, its great sensitivity and simplicity, along with being cost-efficient, makes it extremely useful.

Tests for Treponemal Antibody

The first test used for detecting specific antitreponemal antibody was the *Treponema pallidum* immobilization (TPI) test. Although highly reliable, it proved too cumbersome for routine use. In its place and used initially as a major test, until recently, was the fluorescent *T. pallidum* antibody (FTA) test (Figure 38.5). If virulent *T. pallidum* grown from an infected rabbit testicle is placed and fixed on a slide and then overlaid with serum from a patient with antibody to treponemes, an antigen-antibody reaction will occur. The bound antibody can then be detected by means of a fluoresceinated antihuman immunoglobulin antibody. The specificity of the test for *T. pallidum* is enhanced by first absorbing the serum with nonpathogenic treponemal strains. This modification is referred to as the FTA-ABS test. (If specific fluoresceinated anti-IgM antibody to human gamma globulin is used, the acuteness of the infection or the occurrence of congenital syphilis can be assessed. However, this test may sometimes be falsely positive or negative in babies born of mothers with syphilis.)

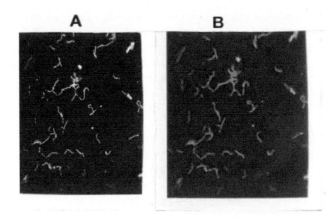

FIGURE 38.5 Indirect fluorescent antibody (IFA) method for detecting antibodies against *T. pallidum*. Image A shows *T. pallidum* organisms viewed under darkfield microscopy. Image B shows the same *T. pallidum* organisms reacting with serum from an immune patient, followed by treatment with fluorescein-tagged goat antihuman immunoglobulin, and then viewed using fluorescence microscopy.

The FTA-ABS test is reactive in approximately 80% of patients with primary syphilis (vs. 50% for the VDRL test). Both tests are positive in virtually 100% of patients with secondary syphilis. Whereas the VDRL test shows a tendency to decline in titer after successful treatment, the FTA-ABS test may remain positive for years. It is especially useful in confirming or ruling out a diagnosis of syphilis in patients with suspected biologic false-positive reactions to the VDRL test. However, even the FTA-ABS test may be susceptible to false-positive reactions, especially in the presence of lupus erythematosus (an autoimmune disorder of unknown etiology).

The microhemagglutination-*Treponema pallidum* (MHA-TP) test, a simple passive hemagglutination test, has been used as a satisfactory substitute for the FTA-ABS test. Treponemal antigens are coupled onto animal erythrocytes and, in the presence of specific antibody, agglutination or clumping occurs, indicative of a positive reaction. Its principal advantages are economy of technician time, low cost and it can be read visually without the need for any sophisticated instrumentation. Its results correlate closely with those of the FTA-ABS test, except during primary and early secondary syphilis when both the VDRL and FTA-ABS tests are more likely to show reactivity. An alternative version of the MHA-TP test, that is in current use, is the *Treponema pallidum* particle agglutination assay which is also an indirect agglutination assay used for detection and titration of specific antibodies directed against a collection of *T. pallidum* antigens. In the test, gelatin particles are sensitized with *T. pallidum* antigens. Patient serum is mixed with the reagent containing the sensitized gelatin particles. The particles aggregate to form clumps when the patient serum is positive for syphilis. A negative test shows no clumping of gelatin particles. More recently, commercially available ELISA detection systems for measuring specific treponemal antibodies have been developed that can also be used for aiding in the diagnosis of syphilis. In addition to serum, cerebrospinal fluid can also be analyzed for treponemal antibodies, and be used to confirm a case of neurosyphilis. It should be noted, however, that the VDRL test is the only one that can be used with reliability in the evaluation of cerebrospinal fluid.

As noted previously, the interpretation of serologic data from patients with syphilis may be extremely complex in some cases. For example, a prozone phenomenon may be encountered in secondary syphilis; serofastness may characterize late syphilis; and the VDRL test may be negative in up to a third of patients with late latent syphilis.

Prophylactic Measures

As is the case with other sexually transmitted infections, prevention of syphilis requires the practice of safe sex techniques such as the use of condoms. These, if used properly, can be an effective barrier against the sexual transmission of *T. pallidum*. Early treatment with antibiotics is the only known way to prevent the later ravages of syphilis. Experimental vaccines have proven impractical or fail to afford complete protection, based primarily on animal studies involving the use of the experimental rabbit model for syphilis.

Treatment

Penicillin is the drug of choice for syphilis in all its stages. For those allergic to penicillin, alternative antibiotics such as doxycycline and erythromycin (azithromycin) can be used. Because the lesions of tertiary syphilis may be irreversible, it is crucial to identify and treat the disease before tertiary lesions or damage to key organ sites begin to occur. AIDS patients with syphilis must be treated more intensively with penicillin.[54] This reinforces the notion that curing syphilis depends on interactions between an intact immune system and the treponemicidal effects of antibiotics. So far there is no evidence that drug-resistant strains exist, unlike what has been found many other nonspirochetal bacteria.

Nonvenereal Treponematoses

The causes of yaws (*Treponema pallidum* ssp. *pertenue*), pinta (*T. carateum*), and bejel (*T. pallidum* and *endemicum*) are human pathogens responsible for this group of contagious diseases, which are endemic among rural populations in tropical and subtropical countries.[55] Unlike syphilis, these diseases are not transmitted by sexual activity but arise when treponemes are transmitted primarily by direct contact, mostly among children living under poor sanitary conditions. These three treponemal species are morphologically and antigenically similar to *T. pallidum* but give rise to slightly different disease manifestations. Pinta causes skin lesions only, yaws causes skin and bone lesions, and bejel (so-called endemic syphilis) affects the mucous membranes, skin, and bones. They do resemble venereal syphilis by virtue of the self-limiting primary and secondary lesions, a latency period with clinically dormant disease, and late lesions that are frequently highly destructive. The serologic responses for all three diseases are indistinguishable from one another and from that of venereal syphilis, and there is the same degree of slow development of protective immunity associated with prolonged untreated infection.

The spirochetes that cause yaws and pinta are transmitted through broken skin after close contact with infected people.

The diseases are most prevalent in rural tropical areas where hygiene is poor and living conditions are crowded. Yaws occurs most often in 2- to 5-year-olds, and pinta occurs most often in older children and adolescents. During global campaigns in the 1950s and 1960s, both diseases were nearly eradicated, but their prevalence has increased in the absence of continuous control measures. Indeed there is now much concern over the most recent resurgence of yaws primarily in the Papua New Guinea/Melanesia areas.[56] Yaws could easily be included in that long list of neglected tropical diseases that are currently getting significant worldwide attention by the World Health Organization and a few nongovernmental organizations including certain philanthropic foundations for improved methods of control and access to treatment. However, except for the WHO and one pharmaceutical company that plans to donate massive quantities of azithromycin as part of an eradication effort, no other response against yaws has occurred.

With yaws, there is usually an incubation period of 2 weeks to 3 months before a papular-type lesion appears on the face, arms, and legs and persists for weeks or months. This papilloma expands slowly to form multiple papular or raspberry-like lesions that may ulcerate in the center. The primary papillomas heal spontaneously within months. Similar secondary lesions may erupt, heal, and spread in successive crops. Also, painful papillomas and hyperkeratosis may appear on the palms and soles, along with lymphadenopathy, and destructive osteitis and periosteitis of the long bones. A variety of skin lesions can appear in the late stage of the disease, with gummatous lesions at several sites. Yaws cripples and deforms but is rarely fatal.

With pinta, 1–8 weeks after infection, a scaling papule appears on the hands, legs, face, or dorsum of the feet. The papule may be accompanied by lymphadenitis. In 3–12 months, a maculopapular erythematous rash appears at the sites of the primary lesions. It may change color from blue to violet to brown and finally become depigmented. The rash and pigmented lesions can recur for years, and the depigmented lesions can cause considerable disfigurement.

Spirochetes can be found in early skin lesions in patients having either yaws or pinta. Serologic tests for syphilis become positive during the early stages of either disease. Small dose of an antibiotic such as penicillin cures the infection. Improved sanitation, including increased use of soap and water, will prevent transmission. Case monitoring and antibiotic treatment control these diseases.

Spirochetal Infections During Pregnancy and Transfusions

Except for syphilis, there is little evidence or few well-documented cases for other maternally acquired spirochetal infections that lead to congenital disease.[57] Nonetheless, syphilis was probably the first infectious disease where transplacental transmission was first recognized, and which could lead to serious outcomes to the mother or the developing fetus, involving miscarriages, stillbirths, or neonatal disease. Syphilis is now a relatively rare occurrence in the newborn, except in certain large urban settings such as New York City. *T. pallidum* bacteria are usually acquired transplacentally, although intrapartum acquisition is possible. A mother with untreated primary or secondary syphilis is unlikely to have normal children; about half will be premature or suffer perinatal death, and the remainder will have congenital disease. These proportions decrease to 10% perinatal death and 10% congenital syphilis when maternal infection is acquired during late gestation. About 50% of infected newborns are initially asymptomatic. Babies with congenital infection may have a vesicular or bullous rash that includes the palms and soles; they may have rhinitis (sniffles), a maculopapular rash, pneumonia, condylomata lata, hepatosplenomegaly, generalized lymphadenopathy, and/or abnormal CSF. These manifestations may not be apparent at birth but may develop during the first few weeks of life. Radiographs often reveal bony lesions at 1–3 months postnatally, and these are characteristic for symmetric metaphyseal involvement with elevation of the periosteum, and osteomyelitic lesions, most often involving the humerus and tibia. These bone changes occur in 90% of infants who manifest with congenital syphilis. The osteochondritis and periostitis may be painful and may be manifested by the pseudoparalysis of a limb due to pain. Diagnosis may be problematic and usually depends on serologic confirmation of clinically based suspicions.

Infected infants are often asymptomatic at birth; and if infection was late in the pregnancy, they may be seronegative. Either the VDRL/RPR tests or the specific treponemal tests may be used, but it should be noted that any maternal serologic reactivity, in the form of IgG antibodies, will be transmitted to the newborn. All these tests measure both IgG and IgM. However, for measuring IgM only (produced by the newborn), recently developed ELISAs and the FTA-IgM test can be used. This would be helpful in identifying an actual congenital infection because IgM is not transferred across the placenta. Problems exist, however, in the overall sensitivity and specificity of these newer assays. The infant's passively acquired VDRL reactivity gradually decreases by 3 months, and passive TPPA/ELISA reactivity should decrease significantly by 6 months.

All infants suspected of having congenital syphilis on the basis of maternal history or physical findings should be carefully examined, investigated by serology, and have a lumbar puncture to collect cerebral spinal fluid for routine analysis, a possible dark-field exam, and a VDRL/RPR test. The placenta should be examined for focal villitis and other possible abnormalities. If the serologically positive mother did not receive adequate therapy or received nonpenicillin therapy, or if adequate follow-up is not assured, the child should be treated with penicillin at birth. Infants with neurosyphilis should be seen at follow-up at 3 months, then at 6-month intervals for repeat serology, CSF exam, and clinical reevaluation, until it is clear that VDRL/RPR titers are falling.

Prior to there being sophisticated screening techniques, various pathogens were known to cause infections following a blood transfusion. These included the spirochetes, *T. pallidum* and *B. recurrentis*, the protozoans *Babesia* and *Plasmodia*, and several viruses. Currently, transfusion-associated infections with spirochetes are practically nonexistent, due to the use of up to date screening procedures, but this was not always the case, as was the situation, even fairly recently with other infectious diseases, such as hepatitis B and HIV infections. Historically, there had always been a concern that the spirochetemic phase of an infection with

T. pallidum, resulting from extracutaneous spread of the organism, could be a source of transfusion-associated syphilis, especially if a prospective donor was asymptomatic. Likewise, with Lyme disease, there is ample evidence, based primarily on blood culture studies,[16,17] that the spread of *B. burgdorferi* organisms from the EM rash to the various extracutaneous sites occurs mainly through the hematogenous route. Currently however, the CDC does not include *B. burgdorferi* as one of the pathogens to be tested for its possible presence in donated blood. Fortunately, it is highly unlikely that Lyme disease patients would want to donate blood during their symptomatic, and presumably spirochetemic phase, and thus it would be highly remote that blood containing *Borrelia* bacteria would be unwittingly transferred to a prospective recipient.[58] Still, it is important that blood collecting agencies nationwide and throughout the world, such as the American Red Cross, remain vigilant and continue to use the most sensitive test procedures in the screening for these diseases in potential blood donors.

Summary

The spirochetes are a unique group of bacteria that can be distinguished morphologically from most other bacteria based on their large size and helical or corkscrew-shaped appearance. They also possess flagella that are internal rather than extracellular, which is more characteristic of most other motile organisms. They include numerous species belonging to the *Borrelia*, *Leptospira*, and *Treponema* genera. Infection with most of the spirochetes occurs initially in the skin before they disseminate to extracutaneous sites. Spirochetes usually cause diseases that are nonacute upon initial exposure of a susceptible host to the infectious agent, but they can have devastating consequences on the human body later on, in the absence of curative antibiotic therapy, or when treatment is delayed.

The sexually transmitted treponemal disease, syphilis, has been with us for many centuries, perhaps as far back as Biblical times, whereas the tick-borne borrelial-caused illness, Lyme disease, was initially described as a serious clinical entity only about 40 years ago. Both of these pathogens can cause a variety of multisystem disorders that can be easily confused with other infectious or noninfectious illnesses. Early diagnosis of syphilis and Lyme disease is essential so that effective antibiotic treatment can be given leading to the prevention of chronic debilitating sequelae. Other spirochetal infections caused by various other *Borrelia*, *Leptospira*, and *Treponema* are uncommon in developed countries but still can cause significant morbidity in various other parts of the world, especially in poor, underdeveloped and tropical regions.

All spirochetal infections rely heavily on serologic techniques for verifying or establishing the diagnosis, although molecular techniques and other nonserologic laboratory tests can be helpful under certain circumstances. Nonmedical preventive measures include avoiding contact with (1) the appropriate insect vector, (2) an infected sex partner or body fluid, and (3) contaminated fomites or skin lesions. Although vaccines for human use are currently unavailable, there is hope that continuing research along these lines will lead to the development of safe and effective ones.

REFERENCES

1. Hinckley, A.F. et al. Lyme disease testing by large commercial laboratories in the United States. *Clin. Infect. Dis.* 59:676, 2014.
2. Steere, A.C. et al. Lyme arthritis: an epidemic of oligoarticular arthritis in children and adults in three connecticut communities. *Arthritis Rheumat.* 20:7, 1977.
3. Afzelius, A. Report to verhandlungen der dermatologischen gesellshaft zu stockholm. *Acta Derm. Venereol.* 2:120, 1921.
4. Barbour, A. Isolation and cultivation of Lyme disease spirochetes. *Yale J. Biol. Med.* 57:521, 1984.
5. Pavia, C.S. et al. Activity of sera from patients with Lyme disease against *Borrelia burgdorferi*. *Clin Infect. Dis.* 25(Suppl. 1):S25, 1997.
6. Simon, M.M. et al. Recombinant outer surface protein a from *Borrelia burgdorferi* induces antibodies protective against spirochetal infection in mice. *J. Infect. Dis.* 164:123, 1991.
7. Steere, A.C. Lyme borreliosis in 2005, 30 years after initial observations in Lyme, CT. *Wien. Klin. Wochenschr.* 118:625, 2006.
8. Barbour, A. et al. Heterogeneity of major proteins of Lyme disease Borreliae: a molecular analysis of American and European isolates. *J. Infect. Dis.* 152:478, 1985.
9. Horowitz, H.W. and Pavia, C.S. et al. Sustained cellular immune responses to *Borrelia burgdorferi*: lack of correlation with clinical presentation and serology. *Clin. Diagn. Lab. Immunol.* 1:373, 1994.
10. Dattwyler, R.J. et al. Seronegative Lyme disease: dissociation of specific T- and B-lymphocyte responses to *Borrelia burgdorferi*. *N. Engl J. Med.* 319:1441, 1988.
11. Habicht, G.S. et al. Lyme disease spirochetes induce human and mouse interleukin-I production. *J. Immunol.* 134:3147, 1985.
12. Benach, J.L. et al. Interaction of phagocytes with the Lyme disease spirochete; role of the Fc receptor. *J. Infect. Dis.* 150:497, 1984.
13. Georgilis, K. et al. Infectivity of *Borrelia burgdorferi* correlates with resistance to elimination by phagocytic cells. *J. Infect. Dis.* 163:150, 1991.
14. Aguero-Rosenfeld, M.E. et al. Diagnosis of Lyme borreliosis. *Clin. Microbiol. Rev.* 18:484, 2005.
15. Steere, A.C. et al. The spirochetal etiology of Lyme disease. *N. Engl. J. Med.* 308:733, 1983.
16. Wormser, G.P. et al. Improving the yield of blood cultures in early Lyme disease. *J. Clin. Microbiol.* 38:1648, 1998.
17. Wormser, G.P. et al. Use of a novel technique of cutaneous lavage for diagnosis of Lyme disease associated with *erythema migrans*. *JAMA.* 268:1311, 1992.
18. Park, H.J. et al. *Erythema chronicum migrans* of Lyme disease: diagnosis by monoclonal antibodies. *J. Am. Acad. Dermatol.* 15:111, 1986.
19. Duray, P.H. et al. Demonstration of the Lyme disease spirochetes by modified Dieterle stain method. *Lab. Med.* 16:685, 1985.
20. Duray, P.H. and Steere, A.C. Clinical pathology correlates of Lyme disease by stage. *Ann. N.Y. Acad. Sci.* 539:65, 1988.
21. Hedberg, C.M. et al. An interlaboratory study of antibody to *Borrelia burgdorferi*. *J. Infect. Dis.* 155:1325, 1987.
22. Bakken, L.L. et al. Performance of 45 laboratories participating in a proficiency testing program for Lyme disease serology. *JAMA.* 268:891, 1992.
23. Pavia, C.S. et al. An indirect hemagglutination antibody test to detect antibodies to *Borrelia burgdorferi* in patients with Lyme disease. *J. Microbiol. Methods.* 40:163, 2000.
24. Grodzicki, R.L. and Steere, A.C. Comparison of immunoblotting and indirect ELISA using different antigen preparations for diagnosing early Lyme disease. *J. Infect. Dis.* 157:790, 1988.
25. Centers for Disease Control and Prevention. Case definitions for infectious conditions under public health surveillance: Lyme disease. *MMWR.* 46:20, 1997.
26. Dattwyler, R.J. et al. Immunological aspects of Lyme borreliosis. *Rev. Infect. Dis.* 11(S6):1494, 1989.
27. Hanson, L. et al. Measurement of antibodies to the *Borrelia burgdorferi* flagellum improves serodiagnosis in Lyme disease. *J. Clin. Microbiol.* 26:338, 1988.
28. Bacon, R.M. et al. Serodiagnosis of Lyme disease by kinetic enzyme-linked immunosorbent assay using recombinant VlsE1 or peptide antigens of *Borrelia burgdorferi* compared with 2-tiered testing using whole-cell lysates. *J. Infect. Dis.* 187:1187, 2003.
29. Rosa, P.A. and Schwann, T.G. A specific and sensitive assay for the Lyme disease spirochete *B. burgdorferi* using the polymerase chain reaction. *J. Infect. Dis.* 160:1018, 1989.
30. Schwartz, I. et al. Diagnosis of early Lyme disease by polymerase chain reaction amplification or culture of skin biopsies from erythema migrans. *J. Clin. Microbiol.* 30:3082, 1992.
31. Schwartz, I. et al. Polymerase chain reaction amplification of culture supernatants for rapid detection of *Borrelia burgdorferi*. *Eur. J. Clin. Microbiol. Infect. Dis.* 12:879, 1993.
32. Oppenheim, J. Relationship of *in vitro* lymphocyte transformation to delayed hypersensitivity in guinea pigs and man. *Fed. Proc.* 27:21, 1968.
33. Callister, S.M. et al. detection of IFN-γ secretion by T cells collected before and after successful treatment of early Lyme disease. *Clin. Infect. Dis.* 62:1235, 2016.
34. Mollins, C.R. et al. Development of a metabolic biosignature for detection of early Lyme disease. *Clin. Infect. Dis.* 60(12):1767, 2015.
35. Pegalajar-Jurado, A. et al. Identification of urine metabolites as biomarkers of early Lyme disease. *Sci. Rep.* 8(1):12204, 2018.
36. Chu, H.J. et al. Immunogenicity and efficacy study of a commercial *B. burgdorferi* bacterin. *J. Am. Vet. Med. Assoc.* 201:403, 1992.
37. Steere, A.C. et al. Vaccination against Lyme disease with recombinant *Borrelia burgdorferi* outer surface lipoprotein A with adjuvant. *N. Engl. J. Med.* 339:209, 1998.
38. Shih, C-M. and Spielman, A. Topical prophylaxis for Lyme disease after tick bite in a rodent model. *J. Infect. Dis.* 168:1042, 1993.
39. Wormser, G.P., Daniels, T.J., Bittker, S., Cooper, D., Wang, G. and Pavia C.S. Failure of topical antibiotics to prevent disseminated *Borrelia burgdorferi* infection following a tick bite in C3H/HeJ mice. *J. Infect. Dis.* 205(6):991–4, 2012.
40. Wormser, G.P. et al. The clinical assessment, treatment, and prevention of Lyme disease, human granulocytic anaplasmosis and babesiosis: clinical practice guidelines by the infectious diseases society of America. *Clin Infect. Dis.* 43:1089, 2006.
41. Nadelman, R.B. et al. Prophylaxis with single-dose doxycycline for the prevention of Lyme disease after an *Ixodes scapularis* tick bite. *N. Engl. J. Med.* 345:79, 2001.
42. CDC. Tick-bite prophylaxis. https://www.cdc.gov/ticks/tickborne-diseases/tick-bite-prophylaxis.html. Accessed on July 16, 2019.
43. Roscoe, C. and Epperly, T. Tick-borne relapsing fever. *Am. J. Physician.* 72:2039, 2005.

44. Rebaudat, S. and Parola, P. Epidemiology of relapsing fever borreliosis in Europe. *FEMS Immunol. Med. Microbiol.* 48:11, 2006.

45. Gholkar, N. and Lehman D. Images in clinical medicine. *Borrelia hermsii* (relapsing fever). *N. Engl. J. Med.* 368(3):266. 2013.

46. Krause, P.J. et al. Human *Borrelia miyamotoi* infection in the United States. *N. Engl. J Med.* 368(3):291–3, 2013.

47. Koetsveld. J. et al. Serodiagnosis of *Borrelia miyamotoi* disease by measuring antibodies against GlpQ and variable major proteins. *Clin Microbiol Infect.* 24:1338.e1–1338.e7, 2018.

48. Lecour, H. et al. Human leptospirosis—a review of 50 cases. *Infection.* 17:8–12, 1989.

49. Morgan, J. et al. Outbreak of leptospirosis among triathlon participants and community residents in Springfield, Illinois, 1998. *Clin. Infect. Dis.* 34:1593, 2002.

50. Sejuar, J. et al. Leptospirosis in "Eco-Challenge" athletes, Malaysian Borneo, 2000. *Emerg. Infect. Dis.* 9:702, 2003.

51. Peeling, R.W. and Hook, E.W. The pathogenesis of syphilis: the great mimicker, revisited. *J. Pathol.* 208:224, 2006.

52. Lafond, R.E. and Lukehart, S.A. Biological basis for syphilis. *Clin. Microbiol. Rev.* 19:24, 2006.

53. Edmondson, D. et al. Long-term in vitro culture of the syphilis spirochete treponema pallidum subsp. pallidum. *mBio.* 9(3):e01153–18, 2018.

54. Zellan, R.E. and Augenbraun, M. Syphilis in the HIV-infected patient: an update on epidemiology, diagnosis and treatment. *Curr. HIV/AIDS Rep.* 1:142, 2004.

55. Antel, G.M. et al. The endemic treponematoses. *Microbes Infect.* 4:83, 2002.

56. Enserink, M. A second chance. On a remote Melanesian island, a Spanish doctor has revived the 60-year-old quest to eradicate a disfiguring disease. *Science.* 361:216, 2018.

57. Remington, J.S. and Klein, J.O. *Infectious Diseases of the Fetus and Newborn Infant*, 7th edition, Saunders/Elsevier, Amsterdam, The Netherlands, 2016.

58. Pavia, C.S. and Plummer M.M. Transfusion-associated Lyme disease—although unlikely, it is still a concern worth considering. *Front. Microbiol.* 9:2070, 2018.

39

Staphylococcus aureus *and Related Staphylococci*

Volker Winstel, Olaf Schneewind,† and Dominique Missiakas

CONTENTS

Introduction

Staphylococcus aureus was first characterized as a human isolate in the late 19th century. Sir Alexander Ogston in his lecture at the Ninth Surgical Congress in Berlin in 1880 reported the presence of "Micrococci" associated with pus in surgical wound infections. Later, Sir Ogston would use the word staphylococci to refer to these organisms (1). He used eggs to isolate pure cultures of staphylococci and showed that rabbits inoculated with these cultures developed abscesses thereby fulfilling Koch's postulates for the identification of the etiological agent of suppurative abscesses (1). *S. aureus* is a physiological commensalism of the human skin, nares, and mucosal surfaces, and bacteriological culture of the nose and skin of healthy humans invariably reveals staphylococci. In 1884, Rosenbach isolated two colony types of staphylococci found on humans and, based on their pigmentation, proposed the nomenclature *S. aureus* and *S. albus* for the

yellow and white isolates, respectively (2). The latter species is now named *S. epidermidis* and until the early 1970s, *S. aureus*, *S. epidermidis*, and *S. saprophyticus* were the only three species in the genus *Staphylococcus*. Genotypic properties and refined taxonomy have led to the distinction of over 40 species during the last four decades (3, 4). This review will describe *Staphylococcus* species briefly and provide a more detailed account on laboratory manipulation of *S. aureus* and *S. epidermidis*.

Classification, Natural Habitat, and Epidemiology

Classification and Natural Habitat

The genus *Staphylococcus* is in the bacterial family *Staphylococcaceae*, which includes the lesser known genera, *Gemella*, *Jeotgalicoccus*, *Macrococcus*, and *Salinicoccus* (5). Nearby phylogenetic relatives include members of the genus *Bacillus* and *Listeria* in the family *Bacillaceae* and *Listeriaceae*,

†Deceased May 26, 2019

DOI: 10.1201/9781003099277-41

respectively (5). More than 60 species and subspecies of the genus *Staphylococcus* are distinguished based on biochemical analysis, DNA-DNA hybridization studies and sequencing analyses of the conserved genes *gap*, 16S rRNA, *hsp60*, *rpoB*, *sodA*, and *tuf* (3–8). Various species are often distinguished as coagulase-positive and -negative, and novobiocin susceptible or resistant (4).

Members of the genus *Staphylococcus* inhabit the skin, skin glands, and mucous membranes of humans, animals, and birds. Staphylococci found on the skin are either resident or transient. Transient bacteria are encountered upon exposure with exogenous source and upon contact between different host species. Such transient organisms are mostly eliminated unless the normal defense barriers of the host are compromised. Yet, host switching represents an important mechanism in the evolution of *Staphylococcus* (*vide infra*) (7). *S. epidermidis* is the most prevalent and persistent *Staphylococcus* species on human skin. Although it is present all over the body surface, it prefers hydrated sites such as the anterior nares and apocrine glands. Similarly, *S. aureus*, *S. hominis*, and *S. haemolyticus* and to a lesser extent *S. warneri* share the same habitat although as a result of its adaptability *S. hominis* colonizes drier regions on the human skin. *S. aureus* is found in the nares and other anatomical locales of humans and can exist as a resident or transient member of the natural flora. *S. aureus* can also colonize domestic animals and birds. *S. capitis* colonizes the human scalp following puberty and large populations may be found in areas with numerous sebaceous glands. *S. auricularis* demonstrates a strong preference for the adult human external auditory meatus. Some subspecies of *S. auricularis* preferentially colonize the ears of nonhuman primates. *S. intermedius* and *S. felis* are major species of the domestic dog and cat, respectively. The host range of these staphylococci is however not restricted to the skin of dogs and cats and may encompass other animals or even humans. Other staphylococci including *S. xylosus*, *S. saprophyticus*, and *S. cohnii* are more prevalent on lower primates and mammals while *S. delphini*, *S. caprae*, and *S. lutrae* have been specifically isolated from dolphins and domestic goat with suppurative skin abscesses, and dead otters, respectively. Interestingly, pigmented *S. xylosus* strains are used as fermenting agents in production facilities, and the activity of nitrate reductase contributes to the development of the red color characteristic of dried sausages or the orange color that coats some cheeses. General taxonomic, environmental, and genetic information about staphylococci can be found at https://bacdive.dsmz.de/.

Epidemiology of Staphylococcus Infections

S. aureus and *S. epidermidis* are always found associated with humans and bear the greatest pathogenic potential. Likewise, they are the best and most studied species. Among the other human commensals, *S. haemolyticus*, *S. simulans*, *S. cohnii*, *S. capitis*, *S. warneri*, and *S. lugdunensis* are occasional causative agents of infections in humans and *S. saprophyticus* is often associated with urinary tract infection (reviewed in Refs. [9, 10]). Similarly, commensals of animals such as *S. intermedius*, *S. hyicus*, or *S. sciuri* display virulence traits in infected animals. Because they colonize the human skin continuously, staphylococcal strains are exposed to all antibiotic therapies

(11). *S. aureus* is a major cause of nosocomial and community-acquired infections. Whenever drug-resistant microbes emerge, these strains can spread by direct contact very rapidly among human populations, as exemplified by the threat of methicillin-resistant *S. aureus* (MRSA) worldwide and vancomycin-resistant *S. aureus* (VRSA) (12–14). New antibiotics (daptomycin, linezolid, mupirocin) have been developed with the concomitant emergence of resistance mechanisms with dire consequences for the clinical management of such infections (15). *S. aureus* isolates collected during outbreaks were traditionally characterized by a phage-typing method based on phenotypic markers in the envelope. This typing method requires maintenance of many phage stocks and is technically challenging. Various molecular typing methods are used for the epidemiological analysis of MRSA strains since such strains encode defined genetic markers. Molecular typing relying on plasmid analysis is often used but often lacks accuracy, as plasmids are easily lost, especially when strains are grown in laboratory media. Pulsed field gel electrophoresis, whereby genomic DNA is cut with a restriction enzyme that generates large fragments of 50–700 kbp, represents a reliable method for genetic characterization. Recently, whole-genome sequencing has been used to characterize MRSA outbreaks and to define transmission pathways (16–18). In these approaches, high-throughput sequencing technology is used to construct a phylogenetic tree by comparing single-nucleotide polymorphisms (SNPs) in the core genome to a reference genome (an epidemic MRSA clone versus a sequenced MRSA clone). It is expected that the time and cost associated with sequencing analyses will eventually be reduced to levels where the technology will represent the standard of clinical care (17, 19).

Staphylococci should always be considered potential pathogens as the difference between commensal and pathogen is tenuous, or at least difficult to define at the molecular or genetic level. The spectrum of diseases caused by staphylococci ranges from soft tissue infections, abscesses in organ tissues, osteomyelitis, endocarditis, toxic shock syndrome, to necrotizing pneumonia (20–22). The ability to replicate in different pathological-anatomical sites (23) is enabled by a long list of virulence factors, including cell-wall anchored proteins, secreted toxins such as hemolysins, leukocidins and enterotoxins, capsular- and exopolysaccharides, iron-transport systems, and modulators of host immune functions in addition to antibiotic-resistance genes (20, 24–26). Production of toxins is not limited to *S. aureus*. For example, some strains of the fermenting *S. xylosus* isolated from milk of cows with mastitis produce several enterotoxins (27). Many staphylococci encode all the attributes to form biofilms, attaching effectively to surfaces and developing into antibiotic recalcitrant community structures (28). Biofilm-based infections caused by *S. aureus* and coagulase-negative staphylococci represent a significant burden on the healthcare system, in particular infections of medical devices and surgical wounds (29). An appreciation of staphylococci-host interactions can be gained through the exploitation of animal models of infections (see reviews Refs. [30–35]). In these models, the ability to colonize, replicate, disseminate and either persist or precipitate lethal disease can be compared between wild-type bacteria and isogenic mutants or in the presence of candidate therapeutics. The mouse remains the preferred model for studying *S. aureus* (31). However, some *S. aureus* virulence factors have a restricted human tropism. Examples include staphylococcal superantigen-like (SSL) proteins, phage-encoded immune evasion molecules staphylokinase (SAK), chemotaxis inhibitory protein (CHIPS) and staphylococcal complement inhibitor (SCIN), as well as Panton-Valentine leukocidin (PVL) and γ-hemolysin CB (HlgCB) (36–38). Thus, the contribution of these factors to disease cannot be evaluated in animals. It is interesting, for example, that epidemiological studies attribute the predominance of the highly virulent community-associated MRSA strain USA300 to PVL (39, 40). PVL and HlgCB belong to the family of bicomponent pore-forming toxins that target and kill phagocytes. PVL is a prophage-encoded, pore-forming exotoxin composed of subunits LukS-PV and LukF-PV (41, 42). Each subunit binds a different receptor on human phagocytes; LukS-PV binds human complement C5a receptor 1 (C5aR1), while LukF-PV binds human CD45 independently of the S-component. HlgC, the S component of HlgCB, also binds hC5aR1 (43–45). The identification of these receptors explains the cellular tropism and human specificity of these two leukocidins; the development of a human C5aR1 knock-in mouse model substantiates the contribution of HlgCB during infection (45).

Biochemical Properties

General Properties: Appearance and Metabolic Properties

Members of the genus *Staphylococcus* are Gram-positive cocci with sizes ranging between 0.5 and 1.5 µm in diameter. Staphylococci divide in two perpendicular planes and new daughter cells may remain associated in irregular grape-like clusters, a feature that distinguishes them from the slightly oblong streptococci that divide in one plane and usually grow in chains. Cocci also occur singly, in pairs, tetrads and short chains. Staphylococci are nonmotile, nonspore-forming, and although genes encoding for capsule formation are found in the genus, usually staphylococci grow nonencapsulated. Most species are facultative anaerobes that grow by aerobic respiration or by fermentation yielding principally lactic acid. Staphylococci grow more rapidly under aerobic conditions with the exception of *S. saccharolyticus* and *S. aureus* subsp. *anaerobius*. This latter subspecies is the etiological agent of abscess disease in sheep, also known as Morel's disease, which was first described by Morel in 1911 (46, 47). Aerobic respiration is achieved by the presence of a- and b-type cytochromes and electron transfer in the membrane is mediated by menaquinones. Thus far, c-type cytochromes, where heme is covalently attached to cysteine residues, have only been identified in *S. lentus*, *S. sciuri*, and *S. vitulus*.

Carbohydrate uptake and metabolism have been best characterized for *S. aureus* and *S. xylosus* or may be deduced from genome analysis of *S. aureus* and *S. epidermidis* strains. Two types of carbohydrate transport systems have been investigated, the phosphoenolpyruvate (PEP):sugar phosphotransferase system (PTS) and the PTS-independent carbohydrate transport whereby sugars taken up by a dedicated permease are subsequently phosphorylated by an ATP-dependent kinase (instead of PEP-dependent phosphorylation). Both systems have evolved to take up glucose as demonstrated in *S. aureus* and *S. xylosus* (48).

Glucose oxidation is mediated by the fructose 1,6-bisphosphate and the oxidative pentose phosphate pathways. There is no evidence for the existence of the Entner-Doudoroff pathway and, in the absence of oxygen, staphylococci will behave like facultative anaerobes. Enzymes of the entire tricarboxylic acid (TCA) cycle are encoded in the genome of *S. aureus* and reduced cofactors generated by the TCA cycle are re-oxidized by an electron transport chain containing a- and b-type cytochromes. A typical F0F1-ATPase is encoded in the genome of staphylococci. In addition to glucose, *S. aureus* may use D-galactose in a pathway that converts it to D-galactose 6-P and next tagatose. Tagatose-6-P kinase, the key enzyme of the tagatose pathway is inducible by growth on galactose or lactose and is found in *S. aureus*, *S. epidermidis*, and *S. hominis*, but absent in *S. intermedius*, *S. saprophyticus*, and *S. xylosus*. In these organisms, galactose is oxidized via the Leloir pathway (49). Other sugars can be metabolized by staphylococci and a summary of these other metabolic pathways can be found in reference (1). Upon glucose depletion, *S. aureus* cells growing in aerobic conditions will oxidize acetate, succinate, and malate (50). Most strains examined thus far encode a lactate dehydrogenase that is activated by fructose-1,6-bisphosphate, itself upregulated during anaerobic conditions in the presence of glucose. Staphylococci may use nitrate as an electron acceptor in the absence of oxygen, a metabolic activity that has broad application in food technology, however little is known about the biochemistry and genetic requirement for nitrate reduction with the exception of *S. carnosus* (51).

Most *Staphylococcus* species are oxidase-negative (exceptions include *S. lentus*, *S. sciuri*, and *S. vitulus*) and catalase-positive (exceptions include the anaerobic *S. saccharolyticus* and *S. aureus* subsp. *anaerobius*). The catalase test helps distinguish the catalase-negative streptococci from staphylococci. The test is performed by adding 3% hydrogen peroxide to a colony on an agar plate or slant. Catalase-positive cultures produce O_2 and bubble. Most staphylococci are susceptible to lysostaphin (*vide infra*). Most staphylococci are also susceptible to furazolidone (100 µg on disk; see Ref. [52]), display low resistance to erythromycin (0.04 µg/mL) and produce acid from glycerol oxidation (53).

Major Differences Between S. aureus and Other Staphylococci

Strains of *S. aureus* are classified into sequence type (ST) following a multilocus sequence typing (MLST) scheme based on sequences of internal fragments of seven housekeeping genes (*arcC*, *aroE*, *glpF*, *gmk*, *pta*, *tpi*, and *yqiL*) (54). Strains that are only different at one or two of the seven MLST sites are grouped into a so-called clonal complex (19, 54). *S. aureus* forms fairly large, yellow colonies on rich medium due to the production of the orange carotenoid staphyloxanthin (55, 56). Production of this membrane-bound pigment helps scavenge reactive oxygen species and protects *S. aureus* from neutrophil killing (57). *S. aureus* clonal complex 75 (CC75) has been described as the distinct species *S. argenteus*; both species share the same virulence factors and antibiotic resistance genes but *S. argenteus* conspicuously lacks the genes for staphyloxanthin production (reviewed by Ref. [58]). Surprisingly, *S. argenteus* exhibits comparable morbidity and health care-associated infection rates as *S. aureus* (reviewed by Ref. [58]).

Hallmark of *S. aureus* is its ability to clot human or animal blood even in the presence of coagulation inhibitors (warfarin, heparin, hirudin, or calcium chelators) (59, 60). Coagulation is caused by two secreted proteins, Coa and vWbp, that bind to prothrombin to promote its non-proteolytic activation (reviewed by Ref. [61]). The coagulation test is used extensively as a diagnostic tool to differentiate *S. aureus* from other staphylococci (22). Prothrombin in complex with Coa or vWbp cleaves fibrinogen but not with other thrombin substrates (clotting factors FV, FVIII, FXI, protein C, antithrombin, and plasmin) (reviewed by Ref. [61]). Thus, by secreting Coa and vWbp, *S. aureus* usurps the zymogen of one clotting factor (FII) to modify the coagulation cascade of human or animal blood, causing exuberant, uncontrolled polymerization of fibrin without activating the other clotting and inflammatory factors. Another diagnostic tool, the slide agglutination test, monitors the agglutination of *S. aureus* immersed in calcium-chelated plasma (62). Recently, the biochemical attributes and physiological relevance of staphylococcal agglutination have been shown to require Coa and vWbp as well as clumping factor A (ClfA) (63). ClfA is a surface protein that promotes precipitation of staphylococci through association with soluble fibrinogen (clumping reaction) (64, 65) but can also bind to coagulases-polymerized fibrin cables, thereby tethering fibrin cables to the staphylococcal surface (63). Importantly, Coa, vWbp, and ClfA have been shown to be important virulence factors during the pathogenesis of *S. aureus* infections, enabling the formation of abscesses for staphylococcal replication and the depletion of clotting factors from blood (63, 66). Most other staphylococci and thus cannot manipulate blood hemostasis for virulence. Protein-ligand interaction studies have shown that surface proteins Fbl, SdrG, and SpsD of *S. lugdunensis*, *S. epidermidis*, and *S. pseudintermedius*, respectively, bind human fibrinogen *in vitro*. *S. pseudintermedius* also displays SpsL in its envelope (67). SpsL binds canine fibrinogen with high affinity and may promote bacterial aggregation, biofilm formation, and resistance to neutrophil phagocytosis, suggesting a role similar to *S. aureus* ClfA during pathogenesis (68).

S. aureus secretes α-hemolysin, which causes complete lysis of red blood cells (this lysis is referred to as β-hemolysis). Lack of hemolytic activity around colonies on blood agar is referred to as γ-hemolysis. *Staphylococcus* species are either β- or γ-hemolytic. Most *S. aureus* strains encode *hla*, the structural gene for α-hemolysin (α-toxin), on their chromosome (69, 70). Occasionally *S. aureus* appears nonhemolytic on blood agar. This has been attributed to mutations in the *agr* or *sae* loci (*vide infra*) or in other regulatory loci that control globally the expression of secreted proteins. The *hla* gene encodes a 293 amino acid long protein secreted as a water-soluble monomer (71), which represents the founding member of bacterial pore-forming β-barrel toxins (72). Hla engages surface receptors of sensitive host cells, thereby promoting its oligomerization into a heptameric prepore and insertion of a β-barrel with 2 nm pore diameter into the plasma membrane (73). Hla pores form in lymphocytes, macrophages, type I alveolar cells, pulmonary endothelium, and erythrocytes, whereas granulocytes and fibroblasts are resistant to lysis (71, 74). Instillation of purified Hla into rabbit or rat lung tissue triggers vascular leakage and pulmonary hypertension (74–76). ADAM10 serves as the receptor for α-toxin (77).

ADAM10 contains a zinc-dependent metalloprotease domain displayed on the surface of many host cells. This extracellular domain cleaves the ectodomain of several host proteins including members of the cadherin family (reviewed in Ref. [78]). E-cadherin of epithelial cells, such as pneumocytes and keratinocytes, is a principal substrate for ADAM10. Epithelial cells are also key targets of α-toxin (78). Cleavage of E-cadherin by ADAM10 is enhanced in the presence of α-toxin and results in a loss of the homotypic interaction of the cadherin molecules on adjacent cells at the adherens junction (79), thereby injuring the epithelial tissue barrier function and promoting *S. aureus* virulence (80, 81).

S. aureus is the only staphylococcal species that displays Staphylococcal protein A (SpA; Protein A) on its surface. Protein A has been exploited commercially for its ability to bind antibodies. Protein A binds the Fcγ region of immunoglobulins as well the V_H3 region of IgM molecules of B lymphocytes, i.e., B cell receptors (82–84). Binding to Fcγ results in the coating of bacteria with immunoglobulins, which blocks phagocytosis by immune cells. Protein A released by *S. aureus* binds to Fab and crosslinks V_H3 clan IgM promoting B cell superantigen activity and the futile production of V_H3 antibodies that cannot neutralize *S. aureus* (84, 85). Consequently, mutants that no longer produce protein A display reduced virulence in animals (86–88).

Cell Wall Envelope

The Gram stain broadly differentiates bacteria into Gram-positive and Gram-negative groups. Staphylococci have a thick (60–80 nm) continuous cell wall also called the murein sacculus. The cell wall is composed largely of peptidoglycan also known as muropeptide or murein. Peptidoglycan also functions as a scaffold for the immobilization wall teichoic acids (WTA) (89, 90), capsular polysaccharides (91, 92), and proteins (93) (*vide infra*). Eighty percent of all capsular types found in clinical isolates of *S. aureus* strains are serotypes 5 and 8 (92). Polysaccharide intercellular adhesin (PIA) may also be found in the envelope (94). The genes responsible for PIA synthesis were first revealed in *S. epidermidis* and designated *icaABC* (94). PIA is a polymer of *N*-acetylglucosamine, a surface carbohydrate involved in biofilm formation (94) that contributes to the pathogenesis of biomaterial-associated infections (95). The *ica* genes are also found in other staphylococci including *S. aureus* where it also contributes to formation of biofilms on the surfaces of biomedical implants (96). Detailed information on capsular polysaccharides and PIA can be found in the following reviews (4, 92, 97).

The cell wall presents a unique problem in cellular architecture, as new incoming units must be inserted into a peptidoglycan sheet that is held together by covalent bonds. Random openings in the wall would otherwise result in cell lysis. Further, the cell wall must grow into a specific shape. This process is achieved by a regulated pattern of enzymatic activities that controls opening and closure of covalent bonds in particular during cell division. Distinct morphological features of the cell wall have been revealed by electron microscopy of the newly emerging cross walls of dividing cells, revealing "murosomes," i.e., small puncture sites along the division plane of future daughter cells (98, 99).

Peptidoglycan

Composition

Most bacterial peptidoglycans, with the exception of archae-bacteria, have a very similar structure. The peptidoglycan consists of a backbone of glycan chains of the repeating disac-charides N-acetylglucosamine (NAG) and its lactic ether N-acetylmuramic acid (NAM). Each disaccharide carries a tetra-peptide substituent of alternating L and D amino acids. A peptide bridge links the terminal COOH of D-alanine of one tetrapeptide to an NH_2 group of a tetrapeptide on a neighboring glycan chain (Figure 39.1). In *S. aureus*, each interbridge peptide consists of a pentaglycine (Figure 39.1) and peptidoglycan crosslinking is extensive (100). In rapidly growing cells, more than 95% of the subunits are crosslinked and bear almost no free carboxyl groups, since the D-glutamic acid α-carboxyl group is amidated (101). *S. aureus* peptidoglycan is extensively modified on the

N-acetylmuramic acid residues with O-acetyl substituents such as the 4-*N*, 6-*O*-diacetyl derivative (102). This modification ren-ders the peptidoglycan resistant to egg-white lysozyme and other muramidases (103). In *S. aureus*, the enzyme O-acetyltransferase A is responsible for modification of peptidoglycan and resistance to lysozyme (104). During infection, phagocytosis and lysozyme-based cell-wall degradation of *S. aureus* favors inflammasome activation and IL-1β secretion (105). Thus, acetylation of pepti-doglycan endows pathogenic staphylococci with a specific sub-version mechanism against host defenses.

Biosynthesis

Peptidoglycan synthesis is initiated in the cytoplasm by modify-ing the uridine nucleotide of NAM (UDP-NAM) with the pen-tapeptide at its lactic carboxyl end. This product is also named UDP-NAM-pentapeptide or Park's nucleotide and accumulates in *S. aureus* inhibited by penicillin (106, 107). It is formed by

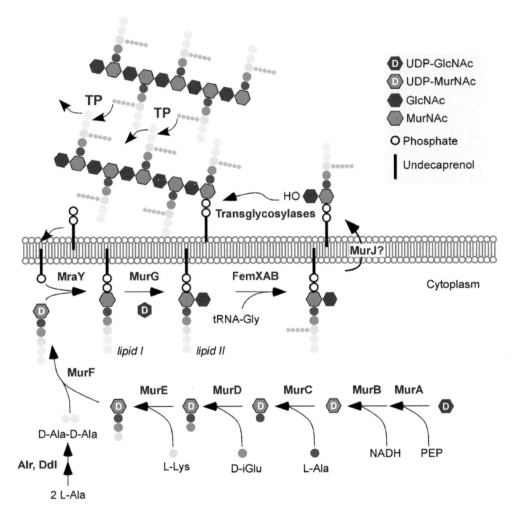

FIGURE 39.1 Model of peptidoglycan biogenesis. The generation of cell wall precursors begins in the cytoplasm, resulting in the synthesis of the soluble intermediate UDP-MurNAc-L-Ala-D-*i*Glu-L-Lys-D-Ala-D-Ala (Park's nucleotide). The peptidoglycan precursor subunit is transferred to the bactoprenol lipid carrier in the membrane to generate lipid I and further converted to the lipid II intermediate following addition of GlcNAc and the penta-Gly cross-bridge in *S. aureus*. Lipid II is translocated to the outer leaflet of the plasma membrane and transglycosylases link peptidoglycan precursors in a processive manner possibly onto bactoprenol. A murein hydrolase (SgtB) cleaves glycan strands to physiological sizes (not shown). Transglycosylation causes the concomitant displacement of the lipid carrier which must be dephosphorylates and retransported to the cis side of the membrane (not shown). The terminal D-alanine of the wall pentapeptide is subject to substitution by cross-linking to other wall subunits catalyzed by transpeptidases or can be removed by the action of carboxypeptidases (not shown). Enzymes known to be involved in distinct steps of this pathway are shown in blue. (*Abbreviations:* PEP, phospho-enolpyruvate; TP: transglycosylase.)

the sequential addition of L-alanine, D-glutamic acid, L-lysine and D-alanyl-D-alanine to the D-lactyl group of UDP-NAM and each step is catalyzed by a specific synthetase (106, 108) (see Ref. [109] for a review). Further, as for *Escherichia coli,* the enzyme MurI converts the L-Glu to D-Glu (110), while L-Ala is converted to D-Ala by the alanine-racemase Alr in *S. aureus* (111). A dedicated ligase generates a D-Alanyl-D-alanine dimer (108). The final soluble intermediate is UDP-NAM- L-Ala-D-*iso*-Glu-L-Lys-D-Ala-D-Ala. This intermediate is transferred onto the membrane carrier lipid, undecaprenol-P (also called bactoprenol-P or C55-P) by the conserved enzyme MraY to yield lipid I (112–115). Lipid I faces the cytoplasm and UDP-NAG is linked to the muramoyl moiety at the expense of UPD by the conserved enzyme MurG to generate the disaccharide lipid II precursor (Figure 39.1) (116, 117). Lipid II precursors are modified by sequential addition of glycine to the ε-amino group of lysine, using glycyl-tRNA as donor, in a ribosome independent fashion (118). *S. aureus* encodes three species of glycyl-tRNA for cell-wall biosynthesis and one for protein synthesis (119). The factors essential for methicillin (Fem) resistance (120) catalyze the incorporation of glycyl residues in the cross bridge. FemX links the first glycine residue to the ε-amino group of lysine, FemA incorporates glycyl residues 2 and 3 and FemB, glycyl residues 4 and 5 (100, 121, 122). The pentaglycine bridges is essential for cell integrity of *S. aureus* (123) and is also the target of anti-staphylococcal endopeptidases such as lysostaphin, secreted by *S. simulans* biovar *staphylolyticus* (124), and Ale-1, secreted by *S. capitis* EPK1 (125). To protect themselves from the lytic activity of lysostaphin and Ale-1, *S. simulans* and *S. capitis* encode the lysostaphin immunity factors (Lif) and endopeptidase resistance (Epr), respectively (126, 127). In *S. epidermidis,* four different pentapeptide cross-bridges have been isolated (128). Each contains glycine and L-serine residues in the ratio 3:2 in a characteristic sequence where glycine is always the initial substituent of the ε-amino group of L-lysine.

The completed lipid II-pentaglycine intermediate is flipped across the membrane by a mechanism that is not completely resolved, possibly involving MurJ or FtsW (129, 130) (Figure 39.1). MurJ is a member of the MOP (multidrug/oligosaccharidyl-lipid/polysaccharide) exporter superfamily and has been suggested to be the lipid II flippase in *E. coli* (131). In *S. aureus,* the essential gene SAV1754 (SACOL1804) has been proposed to serve as a functional MurJ ortholog (132) and microscopy studies suggest that MurJ is recruited to the septum by divisome proteins, driving peptidoglycan incorporation to midcell (133). FtsW is a member of the SEDS (sporulation, elongation, division, and synthesis) protein family, also suggested to be a lipid II flippase (134). More recently, SEDS proteins were shown to be peptidoglycan transglycosylases that may work together with a cognate penicillin-binding protein (PBP; *vide infra*) (135, 136). Following translocation of the lipid II-pentaglycine intermediate, glycan polymerization is carried by highly processive transglycosylases. The presence of uncrosslinked peptidoglycan in the cell wall suggests that transglycosylation reaction precedes transpeptidation reaction. The released carrier undecaprenol-PP then loses a P, thereby regenerating the form that can accept a precursor. Lastly, peptide cross-linking between adjacent glycan chains is catalyzed by transpeptidases that are sensitive to β-lactams and thus referred to as penicillin binding proteins

(PBP) (137). Most organisms encode multiple PBPs that are involved in peptidoglycan growth, cell shape and septum formation. Some PBPs encode D,D-carboxypeptidases that release the terminal D-Ala from a pentapeptide and thus generate a tetrapeptide with a free terminus (that cannot enter a cross-link). These enzymes also have endopeptidase activity on existing cross-links. They function to guide morphogenesis and can be eliminated without loss of viability. *S. aureus* contains only four PBPs, PBP1-4, and MRSA strains contain a fifth known as PBP2A. The low affinity of PBP2A for β-lactams makes it the determinant factor for resistance (138). PBP1 is an essential transpeptidase involved in cell division and separation (139, 140). PBP2 is the only bifunctional PBP with both transpeptidase and transglycosylase activities. This protein is essential in MSSA strains, but dispensable in MRSA strains owing to the presence of PBP2A, unless β-lactams are present (141). Interaction between the SEDS protein FtsW (transglycosylase) and PBP1 (transpeptidase) may be critical for initiating septum formation upon FtsZ assembly, while later processive peptidoglycan synthesis is carried out by PBP2 (136). PBP3 and PBP4 are nonessential proteins. The exact function of PBP3 is unknown but loss of PBP4 results in highly cross-linked peptidoglycan (142). *S. aureus* also encodes two proteins with monofunctional transglycosylase domains, and two transpeptidase-like proteins, but none are essential for growth (142).

The so-called autolysins play an important role during cell division when new cell wall synthesis must take place and be added to the preexisting wall (143). Autolysins catalyze amidase and glycosidase activities that split bonds between tetrapeptides and glycan chains, respectively. In *S. aureus,* and *S. epidermidis*, the major murein hydrolases involved in cell separation are named AtlA (144) and AtlE (145), respectively. Both proteins bear a conserved modular organization encompassing a signal peptide for secretion, followed by propeptide, amidase, three repeat (R1-R3) and glucosaminidase domains. Secreted autolysins are proteolytically cleaved after the propeptide and repeat R2, leading to the formation of two processed hydrolases that bind to the septum sites of dividing cells (146–148). The repeat sequences are responsible for the targeting of the enzymes to the septum of dividing cells (149, 150). Presumably, wall teichoic acid (WTA) present in the old cell wall but not yet accessible in the septum region prevents binding of repeat-containing peptides (150), whereas lipoteichoic acid (LTA) present at the septum favors binding of the hydrolases (151) thereby restricting the activity of these enzymes.

Glycan chain length of peptidoglycan is constant in staphylococci (152, 153). In *S. aureus,* glycan chain length is regulated by SagB, a membrane-associated N-acetylglucosaminidase, that cleaves polymerized glycan strands to their physiological length (154). Loss of *sagB* perturbs growth, cell morphology, and protein trafficking into and across the envelope (154). All these defects can be explained by the exaggerated length of peptidoglycan strands (154, 155).

Protoplasts

The cytoplasmic membrane is pressed against the wall by an internal osmotic pressure (turgor pressure) that exceeds that of the external environment. Digestion of the wall or inhibition of its synthesis in growing cells leads to cell lysis. Cell wall of Gram-positive bacteria can be destroyed by digestion of the peptidoglycan using lysozyme, a protein that breaks the 1,4-glycosidic bonds between N-acetylglucosamine and N-acetylmuramic acid. Because of extensive peptidoglycan modification, staphylococci are insensitive to lysozyme (102, 103) and the glycyl-glycine endopeptidase lysostaphin is used to digest the cell wall. This procedure is necessary for the extraction of cytoplasmic components such as DNA, RNA, and proteins. Cell lysis will not occur in the presence of lysostaphin when the medium is sufficiently hypertonic. Typically, a solute such as sucrose 0.5 M in the buffer solution, will balance the concentration of solutes inside the cell. This leads to the formation of an osmotically sensitive sphere called a protoplast. Such sucrose-stabilized bodies can grow briefly without cell division. However, if placed in hypotonic solution, lysis occurs immediately.

Peptidoglycan Biosynthesis Steps Inhibited by Antibiotics

Penicillin interferes with wall formation by inhibiting the transpeptidation reaction. Electron microscopy reveals that penicillin-treated cells accumulate amorphous material between the membrane and the wall (98). There is a close structural analogy between the D-Ala-D-Ala and the β-lactam ring of penicillin or other β-lactam antibiotics. This ring reacts with the transpeptidase to forms a stable covalent bond that inactivates the enzyme. Cycloserine, a structural analog of D-Ala is a competitive inhibitor of both racemization of L-Ala to D-Ala and conversion of the latter to the dipeptide D-Ala-D-Ala. Fosfomycin, an analog of P-enolpyruvate, irreversibly inhibits the conversion of P-enolpyruvate to the ether-linked lactyl group of UDP-NAM. Neomycin blocks transfer of muramyl peptide to the growing chain. Vancomycin blocks transpeptidation by binding to D-Ala-D-Ala in several intermediates and in the peptidoglycan form. Bacitracin also binds to a substrate instead of an enzyme. It binds undecaprenol-PP blocking its dephosphorylation and thereby preventing its recycling to undecaprenol-P. In response to vancomycin selective pressure, some staphylococcal strains acquired mutations that trigger altered cell wall envelope structures and low-level vancomycin resistance (14, 156). Other strains such as VRSA acquired the *vanA* resistance gene from enterococci (13, 157). VanA ligates β-hydroxy carboxylic acids, for example D-lactate, to D-Ala and the product, D-Ala-D-Lactate is then incorporated into wall peptides (158). The net result of staphylococcal *vanA* gene acquisition, vancomycin resistance, stems from the reduced affinity of the glycopeptide antibiotic to the depsipeptide bond between D-Ala-D-Lactate (159).

Resistance to methicillin by MRSA results from the acquisition of the oxacillin/methicillin-resistance gene *mecA* that encodes PBP2A. *mecA* was possibly acquired from *S. sciuri* by an unknown mechanism (160). *mecA* is located on a mobile genetic element called staphylococcal cassette chromosome (SCC) in *S. aureus* (21–67 kb fragment) (161). Because MRSA and VRSA strains have also acquired macrolide, tetracycline, aminoglycoside, chloramphenicol and fluoroquinolone resistance mechanisms (162), pharmaceutical, scientific, and medical communities have suddenly begun to realize that the therapeutic arsenals are close to depletion causing a threat for infectious disease therapy (163).

Teichoic Acids

Cell wall teichoic acid (WTA) and lipoteichoic acid (LTA) are secondary wall polymers that traverse the envelope of Gram-positive bacteria (164). WTA is tethered to peptidoglycan whereas LTA is retained in the outer leaflet of the plasma membrane through attachment to a glycolipid. *S. aureus* WTA is a polymer of 30 to 50 ribitol-phosphate (Rbo-P) subunits connected via 1,5-phosphodiester bonds (165). The repeating ribitol units may be further substituted with D-alanine and GlcNAc via α- or β-glycosidic linkages (166, 167). $Rbo-P_n$ is tethered to peptidoglycan via the murein linkage unit, $GlcNAc-ManNAc-(Gro-P)_{2-3}$ (168). Synthesis of the murein linkage unit is initiated by TagO (also referred to as TarO), which links UDP-GlcNAc onto undecaprenol-P (Figure 39.2) (169–171). Next, TagA adds ManNAc (168, 172), TagBDF adds Gro-P, and TarILJ adds Rbo-P (173–175). The product of this pathway, $C_{55}-PP-GlcNAc-ManNAc-(Gro-P)_{2-3}-(Rbo-P)_{30-50}$, is flipped across the plasma membrane by the TagGH transporter (176). Attachment of WTA polymers to the C6-hydroxyl of *N*-acetylmuramic acid within peptidoglycan requires genes encoding LytR-CpsA-Psr (LCP) proteins (177, 178). *S. aureus* encodes three LCP enzymes and a mutant lacking all three genes does not harbor WTA in the cell wall envelope (179) and instead releases these polymers in the medium (89).

The first two genes of the WTA pathway (*tagO tagA*) can be deleted without abolishing staphylococcal growth (172, 180, 181). In contrast, *tagBDFGHtarIJL* cannot be deleted unless staphylococci carry inactivating mutations in *tagA* or *tagO* (182). This synthetic viable phenotype is explained as the limited availability of bactoprenol and its undecaprenyl-phosphate derivatives ($C_{55}-PP$ and $C_{55}-P$) for peptidoglycan cell wall biosynthesis (182).

Unlike WTA, LTA is composed of 1,3-polyglycerolphosphate (poly[Gro-P]) linked to the C-6 of the non-reducing glycosyl within the glycolipid anchor ß-gentiobiosyldiacylglycerol (Glc[α1-2]Glc (α1-3)-diacylglycerol (Glc_2-DAG) (328). Assembly of LTA can be divided into three steps: (1) synthesis of the glycolipid anchor diglucosyl-diacylglycerol (Glc_2-DAG) by PgcA, GtaB, and YpfP; (2) polymerization of the poly(Gro-P) chain by LtaS; and (3) D-alanylation of poly(Gro-P) by Dlt enzymes (183–190). When Glc_2-DAG cannot be synthesized because of mutations in the *pgcA*, *gtaB*, or *ypfP* genes, the poly(Gro-P) is tethered to DAG, instead of Glc_2-DAG (187, 191). In staphylococci, *ypfP* is located in an operon with *ltaA*, a gene that encodes a polytopic membrane protein and has been proposed to translocate Glc_2-DAG (Figure 39.3) (191). Polymerization of the poly(Gro-P) chain, is catalyzed by LtaS, a polytopic membrane protein with a large C-terminal domain on the outer surface of the bacterial membrane (Figure 39.3). LtaS transfers *sn*-glycerol 1-phosphate from phosphatidylglycerol onto Glc_2-DAG with subsequent stepwise addition of Gro-P at the distal end of the polymerizing chain (183). *S. aureus ltaS* mutants display a severe cell division defects and grow poorly, suggesting that LTA synthesis is required for growth (192, 193). Incorporation of D-alanine esters confers positive charges onto otherwise negatively charged polymers (194) and requires genes of the *dltXABCD* operon (195). Two models have been proposed to account for the biochemical activity of DltABCD proteins (the *dltX* gene is not conserved in

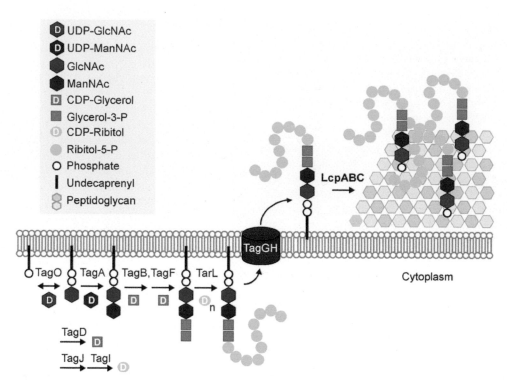

FIGURE 39.2 Model of wall teichoic acid biogenesis in *S. aureus*. The complete synthesis of WTA is catalyzed on bactoprenol P on the *cis* side of the plasma membrane. The assembled molecule is translocated by the membrane transporter (TagG, TagH), and ligated to peptidoglycan by Lcp enzymes. The WTA can be further glycosylated and modified with d-alanyl residues (not shown, see text for details).

bacteria and it has no assigned function so far). In the first model, DltB transfers D-alanine from DltC to undecaprenol-phosphate and flips the lipid-linked intermediate across the membrane, whereas DltD, acting on the *trans* side of the membrane, transfers D-alanine to LTA (196, 197). In the second model, DltD

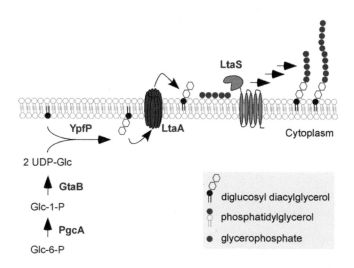

FIGURE 39.3 Model of lipoteichoic acid biogenesis in *S. aureus*. PgcA, GtaB, and YpfP synthesize the glycolipid anchor diglycosyl-diacylglycerol (Glc$_2$DAG). The 12-transmembrane-domain permease LtaA (brown) transports the glycolipid anchor to the *trans* side of the membrane. Next, LtaS (red) cleaves glycerophosphate (Gro-P) subunits from phosphatidyl glycerol (PG) and polymerizes Gro-P onto Glc$_2$DAG. Polymerization of Gro-P by LtaS occurs at the distal end of the extending polymer, up to 50 PG molecules may be added.

promotes transfer of D-alanine between DltA and DltC in the cytoplasm. Alanylated-DltC is translocated across the membrane by DltB and may then transfer D-alanine directly onto LTA (194). The Dlt enzymes transfer D-alanine onto LTA but not onto WTA. D-alanyl esters are transferred from LTA to WTA in a manner that may not involve any further catalysis (198, 199).

S. aureus mutants lacking WTA such as *tagO* mutants have been shown to be unable to colonize cotton rat nares (181); the exact function of LTA is unknown although the presence of LTA at the dividing septum may favor the recruitment of autolysins for cell separation (151). In *S. aureus,* the growth defect caused by loss of *ltaS* can be rescued by inactivation of *gdpP* (200). GdpP is a phosphodiesterase that hydrolyzes the signaling molecule cyclic-di-AMP (327). Increased cyclic-di-AMP in the *gdpP* mutant was shown to be associated with increased peptidoglycan crosslinking that could somehow compensate for the lack of LTA (200). Cyclic-di-AMP was also found to regulate the production of the potassium transporters Kdp and Ktr (201, 202). Neuhaus and Baddiley referred to LTA as a polyelectrolyte that forms a "continuum of anionic charges" (194). Thus, it is tempting to speculate that osmotic homeostasis in *S. aureus* caused by loss of LtaS may be compensated by the regulatory activity of cyclic-di-AMP (201, 202). The contributions of D-alanyl esterification and glycosylation of TAs are better appreciated. *Dlt* mutants of *S. aureus* are more susceptible to cationic antimicrobial host factors, antibiotics, bacteriocins and neutrophils (203, 204). In contrast, several glycosyltransferases modify WTA to generate anomeric heterogeneity that results in altered ability of WTA to elicit specific antibody responses, bind bacteriophages, and resist methicillin (205, 206).

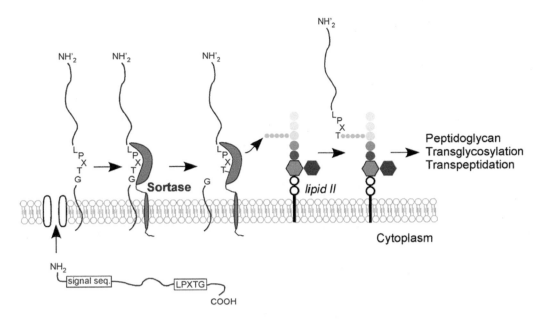

FIGURE 39.4 Sortase-mediated anchoring of proteins to the cell wall envelope of *S. aureus*. Precursor proteins bear an N-terminal signal sequence and C-terminal sorting signal with the conserved LPXTG motif. Precursor proteins are first secreted across the plasma membrane (Sec machinery) and the signal sequence is cleaved to yield a new N-terminus (NH'$_2$). The C-terminal is retained in the membrane. Sortase A cleaves the LPXTG motif to form an acyl-enzyme intermediate that is resolved by lipid II. The cell-wall anchored protein-linked to lipid II is then incorporated into the cell wall via the trans-peptidation and transglycosylation reactions as shown in Figure 39.1.

Cell-Wall Anchored Surface Proteins

Staphylococci display several surface proteins in the envelope, some of which are covalently attached to peptidoglycan following secretion through the canonical secretory pathway (207). The mechanism responsible for protein attachment (anchoring) was first solved for *S. aureus* protein A. Protein A can be released as a homogeneous molecule following lysostaphin treatment (208). However, mechanical treatment of cells releases Protein A with incremental mass increases that can be attributed to the amino sugars *N*-acetylglucosamine and *N*-acetylmuramic acid, suggesting that protein A is linked to peptidoglycan when immobilized in the envelope (208). Protein A contains two hydrophobic domains, one at the N-terminus and one at the C-terminus. The C-terminus carries the LPXTG sequence that is an almost invariant motif shared by most cell wall anchored proteins, where X is any amino acid, followed by a C-terminal hydrophobic domain and a tail of mostly positively charged residues (209). The LPXTG motif and C-terminal tail are necessary for cell wall anchoring (210). Following translocation, the LPXTG motif sorting signal is cleaved between the threonine (T) and the glycine (G) residues (211) by the enzyme sortase A (Figure 39.4) (93). Sortase A also catalyzes the transpeptidation reaction between the carboxyl of threonine and the free amino group of pentaglycine cross-bridges in the staphylococcal cell wall (212) by using lipid II peptidoglycan intermediates as a substrate (213). A second sortase, SrtB is responsible for the anchoring of Isd factors (iron-regulated surface determinant) of *S. aureus* but recognizes a slightly different sorting signal. Isd factors are responsible for hemoglobin binding and passage of heme-iron to the cytoplasm and act as an import apparatus that uses cell wall-anchored proteins to relay heme-iron across the bacterial envelope (214).

Sortase A-anchored surface proteins can be identified following a genome wide search for candidate polypeptides bearing a predicted signal sequence for secretion at the N-terminus and the LPXTG motif at the C-terminus. Between 18 and 22 genes encoding substrates of Sortase-A can be identified in the genomes of *S. aureus* (25). The function of cell wall anchored proteins is not exhaustively known, nonetheless many of them function as microbial surface components recognizing adhesive matrix molecules (MSCRAMMs) and thus represent bacterial elements of tissue adhesion and immune evasion (67, 86, 207, 215, 216).

Virulence Factors and Regulation

S. aureus is the leading cause of hospital acquired infections (22). The spectrum of human diseases caused by staphylococci ranges from soft tissue infections, deep-seated abscesses, osteomyelitis, endocarditis, toxic shock syndrome to necrotizing pneumonia all of which can be tested using small animal model of infection (20, 22, 23, 31). Research over the past several decades identified *S. aureus* cell wall anchored proteins, secreted toxins, capsular- and exo-polysaccharides, iron-transport systems, and modulators of host immune functions in addition to antibiotic-resistance genes as important virulence factors (20, 25, 26, 84). Such versatility is enabled by complex regulatory pathways that include sigma factors, two-component systems, metabolite-responsive regulatory proteins, RNA-binding proteins, and regulatory RNAs (217, 218).

Many staphylococci perform a bacterial census via the secretion of auto-inducing peptides known as AgrD. Auto-inducing peptides are processed and secreted via AgrB, and bind to a cognate histidine kinase (AgrC) at threshold concentration, thereby activating phospho-relay reactions and activation of the AgrA transcription factor (217, 219). The so-called accessory gene regulator (*agr*) is one of 16 two-component systems in *S. aureus*. Activation of AgrA leads to the production of RNAIII, the main intracellular effector of the quorum-sensing system (217, 219). RNAIII encodes the small δ-hemolysin in *S. aureus*, as well as antisense RNAs that regulate the translation or stability of mRNAs encoding transcriptional regulators, major virulence factors, and cell-wall metabolism enzymes (217). During infection, Agr-mediated bacterial census ensures massive secretion of virulence factors when staphylococcal counts are high, presumably increasing the likelihood of bacterial spread in infected tissues and/or systemic dissemination (217, 220, 221).

Virulence factor synthesis is also metabolically controlled. Three major metabolic regulators, CodY, CcpA, and Rex proteins, regulate dozens of metabolism and virulence genes (222). CodY activity is controlled by the branched-chain amino acids isoleucine, leucine, and valine; CcpA, a member of the LacI family, responds to the availability of preferred carbon sources; Rex is a global regulator of genes whose products interconvert NADH and NAD$^+$ (222). More directly, the tricarboxylic acid (TCA) exerts a metabolic regulation over the synthesis of capsular polysaccharide by providing phosphoenolpyruvate for gluconeogenesis (223). However, the level of activity of the TCA cycle may lead to transcriptional repression of the *icaADBC* operon encoding the enzymes of polysaccharide intercellular adhesin (PIA) (224). The interconnection between virulence and metabolism is particularly intriguing when staphylococci switch from planktonic to biofilm lifestyles and develop "persister" cells in a metabolically quiescent state (see Ref. [225] for a review on this subject).

Genome and Genomic Variability

Comparative and pan-genomic analyses of staphylococcal genomes revealed interesting genomic variabilities and strain-specific features linked to virulence, hospital adaptation, or commensalism (226–229). Most extensive studies have been performed for *S. aureus* and *S. epidermidis*. Their pan-genome is subclassified into three major sections: the core genome, the accessory genome, and the unique genome (227). While the unique genome encompasses genes that can be assigned to a single genome, the core genome comprises a shared set of genes found in all isolates (227). In *S. aureus*, the core genome (approximately 1,400 genes) is composed of essential genes involved in DNA and RNA biosynthesis, replication, and metabolism along with genes required for virulence, including those encoding for surface proteins, sortase, α-hemolysin, several exotoxins and the *agr* regulon. On the contrary, the accessory genome is highly diverse and contains an array of varying mobile genetic elements (MGE) such as pathogenicity islands, prophages, plasmids, transposons, integrative and conjugative elements, and the staphylococcal cassette chromosome (SCC) (230, 231). Most of these elements have been acquired by horizontal gene transfer events and typically encode for proteins affecting virulence, antibiotic resistance, host-adaptation, and specificity (230, 231). For example, all MRSA strains carry the *mecA*-encoding SCC*mec* element always inserts into *orfX* (231). However, not all SCC elements harbor *mecA*. Non-*mec* SCC elements contain fitness and resistance determinants such as resistance to fusidic acid (SCC476) or mercury chloride (SCC*mercury*) (232, 233). Strains with SCC*mec* elements such as the hypervirulent and pandemic isolate USA300 (ST8) or the emerging MRSA lineage ST239 carry the arginine catabolic

mobile element (ACME) gene cluster (234, 235). ACME affects skin colonization, survival in acidic environments, and evasion from host immune cell responses (236–238).

Earlier work identified bacteriophages as important contributors to pathogenesis and the evolution of staphylococcal genomes (231, 239, 240). For example, virtually all sequenced *S. aureus* genomes carry at least one and up to four prophages (241). Certain *S. aureus* bacteriophages encode the previously mentioned innate immune modulators chemotaxis inhibitory protein (*chp*), staphylokinase (*sak*), and staphylococcal complement inhibitor (*scn*) (239, 242). Other bacteriophages bear coding sequences for potent toxins including enterotoxins, exfoliative toxin, or PVL (231, 240), and are associated with severe soft tissue infections and fulminant necrotizing pneumonia (243–245). The φSPβ-like prophage found in widespread Asian MRSA isolates, harbors SasX, a sortase-anchored surface protein speculated to promote nasal colonization and immune evasion (246). In addition, clinical strains bear bacteriophage-related *S. aureus* pathogenicity island (SaPI) (231, 247). These elements are 14–27 kbp DNA in size and occupy constant positions in the chromosomes of toxigenic strains (231, 247). SaPIs encode for various enterotoxins, multidrug resistance transporters, toxic-shock-syndrome toxin-1, and an array of T-cell superantigens (247, 248). Secreted T-cell superantigens cause massive cytokine release that leads to fatal toxic shock (249). SaPIs rely on helper bacteriophages for mobilization as they lack genes encoding for structural phage elements (247, 250, 251). In this manner, SaPIs can be transferred across species boundaries thereby substantially contributing to bacterial evolution and the emergence of new epidemic clones (252–254). Although uncommon, host switching also contributes to the evolution of *Staphylococcus*. For example, human-to-poultry and bovine-to-human host switches have been observed for *S. aureus* (255, 256). Host switching may occur through single nucleotide change (*S. aureus* adaptation to rabbits) as well as horizontal gene transfer (257).

Practical Growth and Manipulations

Growth in Laboratory Media

Under laboratory conditions, most staphylococci can be grown in nonselective and rich media including, but not limited to, trypticase soy broth (TSB), brain heart infusion, or Luria Bertani broth. For optimal growth, a temperature of 37°C should be considered although most staphylococci replicate at temperatures ranging from 15°C to 45°C. In addition, *S. aureus* and related species can be cultivated in high-salt media (up to 20% NaCl) or synthetic media mimicking natural niches such as human nares (258–260). In routine processes, specific media are often used to select for *S. aureus*. For example, direct enumeration of *S. aureus* from human swabs can be made on Baird-Parker agar containing rabbit plasma-fibrinogen tellurite, lithium chloride, and egg yolk (261). Clear halos develop around gray-black colonies of *S. aureus*. Coagulase-negative staphylococci typically do not develop the characteristic halo zones. Because the presence or absence of coagulase is a distinctive feature of staphylococci, several coagulase tests have also been developed. Here, isolated colonies are grown in 0.2 mL of brain heart infusion for 18–24 hours at a temperature of 37°C. After incubation, 0.5 mL of reconstituted coagulase plasma with ethylenediaminetetraacetic acid (plasma is commercially available and reconstituted according to manufacturer's directions) is added to the culture, followed by another subsequent incubation step at 37°C. Samples are examined for clot formation in 6-hour intervals. Other media, often derived from Baird-Parker medium, are used for the nonselective isolation and enumeration of coagulase-positive staphylococci from food sources. Moreover, Schleifer-Krämer agar can be used for enrichment of staphylococci from foods. Methods and recipes for specific media can also be found in (259, 262).

Biosafety Measures

S. aureus is an opportunistic and adaptable pathogen with the ability to infect, invade, persist, and replicate in any human tissue including skin, bone, visceral organs, or vasculature. Manipulation with most *Staphylococcus* species should be performed following biosafety level 2 measures. General guidelines for BSL-2 practice can be obtained from the latest edition (5th edition) of Biosafety in Microbiological and Biomedical Laboratories available via the following web link: https://www.cdc.gov/labs/BMBL.html.

Molecular Biology and Genetic Manipulation

Staphylococcal Cloning Plasmids

Many staphylococcal plasmids have been developed for genetic manipulation and cloning purposes. These plasmids are grouped in families based on replication initiation mechanisms. Representative prototype vectors, which are copied via a rolling circle mechanism, are pC194, pT181, pSN2, and pE194 (263–265). Other vectors such as pSK1 or pI258 utilize a theta (θ)-type replication mechanism (263, 265). Derivatives of these vectors carrying mutations with conditional replication phenotypes have been isolated. For example, pTS1, a temperature-sensitive (*ts*) *E.coli/S. aureus* shuttle plasmid, carries a chloramphenicol resistance determinant for plasmid selection and the pE194*ts ori* (origin) for propagation in staphylococci (266). PTS1 is frequently used for allelic replacement (*vide infra*). A set of inducible expression systems, some of which are derived from pI258, have also been developed (265). Integrative vectors with an *E. coli* replicon for cloning, but unable to replicate in staphylococci, have been developed taking advantage of staphylococcal phages. Genes carried by these plasmids can only be expressed following site-specific recombination onto the bacterial chromosome. Of note, stable integration does not require selection and facilitates single copy gene-dosage analysis. Lee and colleagues developed an array of prototype single-copy integration vector systems, most of which are based on the site-specific recombination of bacteriophage L54a (267–269). In *S. aureus*, these vectors integrate into the *attB* attachment site located within the *geh* gene encoding for lipase (267–269). Vectors pLL29 and pLL39 contain a second attachment site derived from lysogenic bacteriophage φ11 that facilitates integration into an alternative genomic locus and can be used to preserve an intact lipase (270). In addition, numerous pCL83/84-based integration plasmids have been engineered (271–273).

Allelic Replacement

Allelic replacement is frequently used to generate chromosomal mutations (Figure 39.5). Generally, shuttle vector systems such

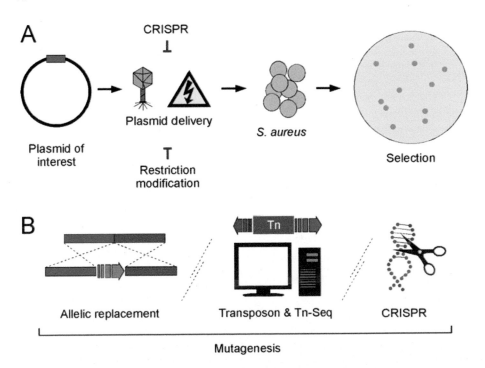

FIGURE 39.5 Genetic engineering of staphylococci. Panel A: A plasmid of interest carried by *S. aureus* RN4220 is typically transferred to clinical staphylococci via electroporation or phage-mediated transduction. Restriction modification systems or CRISPR/Cas9 loci may interfere at this stage. Panel B: Following transformation (or transduction), various technologies can be applied for subsequent mutagenesis.

as pTS1 carrying replication-defective conditional mutations are used. Under nonpermissive condition, i.e., temperature shift, plasmid integration into the bacterial chromosome occurs via homologous recombination (274). As a result of single cross-over recombination events, merodiploid cells bearing both wild-type and mutant alleles may develop and can be resolved under permissive conditions, in which recombination and vector excision are thought to be triggered by a rolling-circle mechanism. In case of double cross-over recombination, wild-type or mutant alleles reside in the genome while the reciprocal alleles replicate with the excised vector.

Cloning design for successful allelic replacement should include genomic sequences of 0.4–1.0 kbp flanking the gene of interest or target sequence to be deleted or modified. When desired, an antibiotic resistance marker (i.e., *ermC*) may be inserted between the flanking sequences to accelerate the identification of recombinants (Table 39.1). Selection takes place at the permissive temperature of 30°C on agar plates containing chloramphenicol. Further selection with chloramphenicol (and erythromycin) and a shift to a nonpermissive temperature, typically 43°C, ensures integration of the vector into the staphylococcal chromosome. Next, variants are grown under the permissive temperature (30°C) and plated on agar supplemented with erythromycin. Several colonies are patched onto non- and antibiotic-containing agar plates (chloramphenicol). Colonies that grow in the absence of chloramphenicol, but not in its presence, should be further analyzed for plasmid loss and the presence of the mutated allele (274).

Arnaud and coworkers developed pMAD for allelic replacement in Gram-positive bacteria (275). pMAD contains sequence fragments of pBR322 and the pE194*ts ori* along with *bgaB* derived from *Bacillus stearothermophilus* (275). *bgaB* encodes for a thermostable β-galactosidase which facilitates cleavage of the chromogenic substrate X-gal (5-bromo-4-chloro-3-indoyl-β-d-galactopyranoside) in *S. aureus*. Depending on the integration status of pMAD, *S. aureus* colonies may appear blue (episomal), light-blue (single cross over), or white (double cross over) thereby providing a rapid screening tool for excision and loss of the plasmid (275). In addition, Bae and Schneewind designed pKOR1 (another pTS1 derivative) taking advantage of antisense *secY* RNA expression for counter-selection during the allelic replacement procedure (276). In pKOR1, expression of the antisense *secY* RNA is controlled by $P_{xyl/tetO}$, an anhydrotetracycline (ATc)-inducible module. Since SecY is essential, expression of *secY* antisense RNA efficiently prevents colony formation on solid media (277). Following plasmid transfer to *S. aureus* via electroporation, replication of pKOR1 requires a permissive temperature of 30°C. Similar to pTS1, a thermal shift to 43°C (nonpermissive

TABLE 39.1

Preparation of Electrocompetent *S. aureus* and Their Transformation with DNA

1. Streak *S. aureus* strain to be transformed from a frozen stock on a tryptic soy agar (TSA) plate and incubate overnight at 37°C.

2. Pick an isolated colony with a sterile loop and inoculate 25 mL tryptic soy broth (TSB) in a 100-mL flask.

3. Incubate overnight at 37°C with shaking (160–200 rpm)

4. Dilute the overnight culture in 200 mL TSB to a final $OD_{600} = 0.1$; the use of a large 2-L flask is recommended.

5. Grow cells at 37°C to early-log phase ($OD_{600} = 0.5$) with vigorous shaking (approximate incubation time is 90–120 minutes).

6. Transfer the culture into sterile spin bottles and collect cells by centrifugation at 5,000 x g for 10 minutes and 4°C.

7. Discard the supernatant and suspend the cell pellet in 40 mL of ice-cold sterile 0.5 M sucrose in deionized water.

8. Transfer the suspension to a prechilled 50 mL sterile centrifuge tube; keep sample on ice.

9. Collect cells by centrifugation at 8,000 x g (10 minutes, 4°C).

10. Discard supernatant and suspend the cell pellet in 20 mL of the ice-cold 0.5 M sucrose solution as above.

11. Collect cells by centrifugation at 8,000 x g (10 minutes, 4°C).

12. Repeat steps 10 and 11 once more.

13. Resuspend the pellet in 2 mL ice-cold 0.5 M sucrose solution. Keep samples on ice.

14. Transfer 100 μl aliquots of the prepared electrocompetent cells into microcentrifuge tubes chilled on ice.

15. Freeze tubes by plunging them in a dry ice-ethanol bath; store cells at –80°C until use.

16. For electroporation with plasmid DNA, retrieve a tube of competent cells from the freezer and place tube on ice.

17. Add 500–1,000 ng of purified plasmid to thawed cells.

18. Incubate the DNA/cell mixture for 30 minutes at room temperature.

19. Transfer the DNA/cell mixture into a 0.1-cm prechilled electroporation cuvette

20. Electroporate the cells using the following parameters: voltage = 2.5 kV, resistance = 100 Ω, capacity = 25 μF.

21. Subsequently, add 1 mL TSB kept at room temperature and transfer the entire content to sterile 2 mL reactions tubes.

22. Incubate for 1 hour at 37°C (no shaking required). In case temperature-sensitive plasmids are transformed, incubate sample at 30°C accordingly.

23. Plate 50–200 μl aliquots on multiple TSA plates containing the appropriate antibiotic for plasmid selection.

24. Incubate plates at 37°C (or 30°C) for at least 16 hours (or until colonies are visible).

25. Extract plasmid DNA from potential *S. aureus* clones using lysostaphin digestion and a commercially available plasmid isolation kit

26. Validate plasmid via agarose gel electrophoresis.

temperature) and homologous recombination facilitate integration of the vector into the staphylococcal chromosome. Next, plasmid eviction is triggered by growing cells to permissive temperature (30°C) and subsequent plating of 10^4-fold dilutions (100 µl) onto plates containing 0, 1, or 2 µg/mL ATc. ATc-induced expression of the *secY* antisense RNA suppresses growth of cells containing pKOR1 thereby selecting against co-integrants that are not resolved (refer to Ref. [276] for a more detailed protocol). *SecY* antisense RNA has also been used for counterselection in other vector systems. The $P_{xyl/tetO}$-inducible *secY* antisense element of pKOR1 was transferred into pIMAY with the temperature-sensitive replicon of the lactococcal plasmid pWV01ts (278). Elements of pKOR1 and pBT2 have also been combined to generate pBASE6, another vector system that employs *secY* antisense RNA for counterselection (279). All plasmids, pKOR1, pIMAY and pBASE6 have been successfully used for allelic replacement procedures in *S. aureus* and coagulase negative staphylococci such as *S. epidermidis* and *S. lugdunensis*.

Transposon Mutagenesis and Tn-Seq

Transposons can be adapted for random genetic mutagenesis of bacteria including staphylococci (Figure 39.5). Most transposons are delivered using vectors with temperature sensitive replicons. In some instances, two plasmids are used to deliver the transposon and its cognate transposase, other technologies rely on single vector systems. For example, pTV1 harbors a transposon Tn917 (280) that has been used for insertional mutagenesis of various Gram-positive microbes including *S. aureus* (281–283). Another single vector system is pBTn, which takes advantage of a *himarI* transposase driven by a xylose-inducible promotor (284). The transposon system *bursa aurealis* developed by Bae *et al.* consists of two plasmids with temperature-sensitive replicons, pBursa (transposon) and pFA545 (transposase *himarI*) (281). *Bursa aurealis* was used to generate a large collection of *S. aureus* mutants, the ΦNΞ (Phoenix) library (284), as well as the publicly available Nebraska Tn Mutant Library, a transposon library generated in USA300-derived MRSA strain (285). Detailed instructions on the use of the *bursa aurealis* technology can be found in (281, 286).

Transposons insertional mutagenesis has also been combined with next-generation sequencing technologies in staphylococcal research (Figure 39.5). This experimental approach, known as Tn-Seq, relies on genome-wide transposon mutagenesis coupled to high-throughput sequencing to define the contribution of crucial genes to certain phenotypes. In general, *himarI mariner* based transposon insertion systems are used to generate a saturated transposon mutant pool. Following library generation, the mutant pool is subjected to a strong selection procedure designed with specific goals. Ultimately, PCR, massive sequencing, and bioinformatics are used to identify transposon insertion sites and mutational lesion frequencies. For example, such a pooled *S. aureus* transposon library was used to isolate genes for replication in whole human blood, aqueous fluid, and vitreous fluid to discover 426 genes presumably required for *S. aureus* survival and growth under infection-associated conditions (287). Several other studies employed similar approaches to identify staphylococcal genes contributing to antibiotic resistance (288), temperature stress (289), or virulence (290, 291).

Genome Editing Via CRISPR/Cas9 Mutagenesis

Clustered regularly interspaced short palindromic repeat (CRISPR)/Cas9-powered genome editing is a powerful genetic tool that facilitates rapid and efficient disruption of genomic loci. While initially designed to edit mammalian genomes, an array of vector systems has been developed for manipulation of bacterial and staphylococcal chromosomes (Figure 39.5). For example, Chen and colleagues designed pCasSA suitable for gene deletion, gene knock-down, insertion, and single-base substitution mutation in *S. aureus* (292). PCasSA harbors the *cas9* gene from *Streptococcus pyogenes* (SpCas9), which is driven by the *S. aureus rpsL* promoter (292). Expression of the single guide RNA (sgRNA; sequence targeted for mutagenesis) is controlled by the *capIA* promotor (292). Moreover, pCasSA contains a temperature-sensitive origin (*repF*), thus facilitating efficient curing of the vector after successful mutagenesis (292). Based on pCasSA, Gu *et al.* developed pnCasSA-BEC, an elegant base-editing system that can be used for gene inactivation or introduction of point mutations in *S. aureus* (293). The system relies on a *S. pyogenes*-derived Cas9 nickase fused to a cytidine deaminase (rat APOBEC1) thereby enabling highly efficient C to T conversion in staphylococcal genomes (293). The genome editing toolbox for staphylococci has further been expanded by Liu *et al.* who designed another single vector system (pLQ-$P_{xyl/tet}$-*cas9*-P_{spac}-sgRNA) (294). This plasmid carries the sgRNA under the control of the P_{spac} promotor while *cas9* is placed under the constitutive promotor $P_{xyl/tet}$ (294). pLQ-$P_{xyl/tet}$-*cas9*-P_{spac}-sgRNA also includes a temperature-sensitive origin and target gene-flanking elements required to repair Cas9-mediated double-strand breaks (294). Detailed protocols on the use of these tools can be found in references (292–294).

Bacteriophage-Mediated Transduction

Many naturally occurring *S. aureus* bacteriophages facilitate rapid exchange of mobile genetic elements between clonal complexes thereby representing excellent tools for genetic manipulations. All known *S. aureus* bacteriophages have double-stranded DNA genomes and belong to the order of *Caudovirales* (295, 240). According to their morphology, staphylococcal bacteriophages are further subdivided into three major families: *Siphoviridae* (long, noncontractile tails), *Myoviridae* (contractile tails), and *Podoviridae* (short, noncontractile tails) (241, 240). Lytic *Myoviridae* and *Podoviridae* cannot be used for genetic manipulations. However, *Siphoviridae*, with three main serological groups (A, B, F), are temperate phages that have the capacity to transduce genetic material between staphylococci (240, 296, 297). For example, φ11, a serogroup B transducing phage has a burst size of approximately 250 plaque-forming units and exhibits a low lysogenization frequency (1–10%) (298). Use of φ11 is ideal for transducing alleles linked or marked with a selectable trait between staphylococcal isolates. Here, phage lysates are generated by infecting a donor strain grown in TSB containing 5 mM $CaCl_2$ at a multiplicity of infection of 1. Addition of $CaCl_2$ favors phage adsorption. Lysis, which can be performed on both soft agar and liquid broth, is achieved after 16 hours of incubation at 37°C. To remove cellular debris and potential bacterial survivors, recovered phage lysates must be centrifuged and sterilized by filtration. For transduction experiments, a multiplicity of infection of 1 phage particle to 10 bacteria is recommended. Samples should be incubated at 37°C for 30 minutes, followed by a centrifugation step to separate bacteria and phage particles. Supernatants containing phages are discarded and bacterial pellets washed twice with TSB containing the calcium chelating agent, sodium citrate (40 mM). Selection of transductants is performed by plating bacteria are onto media with the appropriate antibiotic and sodium citrate (40 mM). Transductants typically arise within 48 hours of incubation at 37°C and should be streaked at least twice on sodium citrate-containing plates to eliminate residual lytic phage. Likewise, phages may also be used to transfer plasmid DNA among staphylococci (Figure 39.5) (299, 300). Not all strains are susceptible to φ11 or can be transduced by other serogroup B phages as the host range of staphylococcal phages largely depends on their cognate cell surface receptor: WTA (254, 301). Differences in phage susceptibility can also be accounted for by postadsorption events, such as genetic barriers that limit or prevent phage replication (*vide infra*).

Barriers to Genetic Manipulation in Staphylococci and Technical Solutions

Genetic barriers such as restriction modification systems (R-M systems) or CRISPR/Cas9 loci substantially contribute to bacterial and staphylococcal evolution (230). Such barriers also interfere with laboratory procedures employed for genetic manipulation of *S. aureus* and related staphylococci (Figure 39.5) (230). For example, R-M systems limit or completely prevent transfer of plasmid DNA to staphylococci from exogenous sources (302–304). Three major types of R-M systems (types I, II, and IV) have been identified in *S. aureus* (302–306), where type I and IV systems mainly impede plasmid transfer events (302–304). Thus, *E. coli*-derived plasmid DNA is often passaged in the *S. aureus* laboratory strain RN4220 (307). RN4220 accepts *E. coli*-propagated plasmid DNA due to nitrosoguanidine-induced mutation(s) in its R-M systems (308), recently mapped to the *hsdR* and *sauUSI* genes (302, 303). Plasmids extracted from strain RN4220 can be electroporated into most *S. aureus* isolates (Table 39.1). Alternatively, a set of specifically engineered *E. coli* strains lacking the *dcm* gene required for cytosine methylation of DNA may be used to overcome the type IV restriction barrier (278). Plasmid DNA derived from these *E. coli* strains can be electroporated directly into some clinical isolates without using strain RN4220 (278). To bypass both type I and IV R-M systems, Jones *et al.* generated *E. coli dcm* mutants with a plasmid artificial modification (PAM) system, which ensures protection of the vector from the restriction endonuclease in the target *S. aureus* strain (309). In this manner, clinical isolates belonging to clonal complexes 1, 5, and 8 can be transformed with plasmid DNA derived from *E. coli* (309).

Exogenous DNA is also rapidly recognized and destroyed by CRISPR/Cas9 systems (310). These loci facilitate adaptive immunity against bacteriophages and other foreign DNA elements such as plasmids, thus interfering with genetic engineering of bacteria (Figure 39.5) (310–312). However, CRISPR/Cas9 systems are typically not found in *S. aureus* genomes as only a few isolates carry CRISPR/Cas9 modules (313–315). These systems may play a more important role in coagulase-negative staphylococci and

hamper their genetic manipulation (316). Accordingly, genetic engineering of most coagulase-negative staphylococci continues to be laborious and complex (278, 317–319).

The Challenge in Staphylococcal Research

Invasive infections caused by Gram-positive pathogens such as Staphylococci, Streptococci, or Enterococci represent a serious public health threat (320–322). Specifically, the emergence of rapidly spreading hyper-virulent and multidrug-resistant clones is a daunting task for clinicians, pharmaceutical companies, and the research community (321–323). While warranted, the development of new antimicrobials to combat the emergence of drug-resistant strains is laborious, time- and cost-intensive, and such efforts are often deemed unprofitable. To circumvent this problem, many well-conceived approaches have been initiated to develop a vaccine, or alternative anti-infective therapeutics against *S. aureus* (324). A large body of studies has been conducted to understand the composition and biogenesis of staphylococcal cell wall envelope and to identify valuable targets for vaccines and therapeutics. However, attempts to design vaccines that target cell envelope structures such as capsular polysaccharide, or LTA, or the IsdB protein have failed in various clinical trials (325). Clearly, the wide array of virulence and immune evasion factors, the genetic variability of clinical *S. aureus* isolates have prevented linear progress toward vaccine development (84, 324–326). Yet, this genetic variability, versatility to adapt to new hosts and environment, and expandable immune subversion attributes make staphylococci an endless treasure-trove for the avid microbiologists.

REFERENCES

1. Ogston A. 1883. Micrococcus poisoning. *J Anat Physiol.* 17:24–58.
2. Rosenbach FJ. 1884. *Microorganismen Bei Den Wund-Infections-Krankheiten Des Menschen*, ed. JF Bergmann. Wiesbaden, Germany, pp. 1–122.
3. Ghebremedhin B, Layer F, Konig W, Konig B. 2008. Genetic classification and distinguishing of *Staphylococcus* species based on different partial gap, 16S rRNA, hsp60, rpoB, sodA, and tuf gene sequences. *J Clin Microbiol.* 46:1019–25.
4. Götz F, Bannerman T, Schleifer K-H. 2006. The Genera *Staphylococcus* and *Macrococcus*. In *The Prokaryotes*, eds. M Dworkin, S Falkow, E Rosenberg, K-H Schleifer, E Stackebrandt, 4th ed. Springer, New York, pp. 5–75.
5. Garrity GM, Holt JG. 2001. The Road Map to the Manual. In *Bergey's Manual of Systematic Bacteriology*, eds. RD Boone, RW Castenholz, GM Garrity. Springer-Verlag, New York, pp. 119–66.
6. Kloos W, Schleifer K-H, Götz F. 1992. The Genus *Staphylococcus*. In *The Prokaryotes*, eds. A Balows, HG Truper, M Dworkin, W Harder, K-H Schleifer. Springer-Verlag, New-York, NY, pp. 1369–420.
7. Lamers RP, Muthukrishnan G, Castoe TA, Tafur S, Cole AM, Parkinson CL. 2012. Phylogenetic relationships among staphylococcus species and refinement of cluster groups based on multilocus data. *BMC Evol Biol.* 12:171.
8. Mellmann A, Becker K, von Eiff C, Keckevoet U, Schumann P, Harmsen D. 2006. Sequencing and staphylococci identification. *Emerg Infect Dis.* 12:333–36.

9. Argemi X, Hansmann Y, Prola K, Prevost G. 2019. Coagulase-negative Staphylococci pathogenomics. *Int J Mol Sci.* 20.

10. Rosenstein R, Gotz F. 2013. What distinguishes highly pathogenic staphylococci from medium- and non-pathogenic? *Curr Top Microbiol Immunol.* 358:33–89.

11. Neu HC. 1992. The crisis in antibiotic resistance. *Science.* 257:1064–73.

12. Brumfitt W, Hamilton-Miller J. 1989. Methicillin-resistant *Staphylococcus aureus*. *N Engl J Med.* 320:1188–99.

13. Chang S, Sievert DM, Hageman JC, Boulton ML, Tenover FC, et al. 2003. Infection with vancomycin-resistant *Staphylococcus aureus* containing the *vanA* resistance gene. *N Engl J Med.* 348:1342–47.

14. Hiramatsu K, Hanaki H, Ino T, Yabuta K, Oguri T, Tenover FC. 1997. Methicillin-resistant *Staphylococcus aureus* clinical strain with reduced vancomycin susceptibility. *J Antimicrob Chemother.* 40:135–36.

15. Liu C, Bayer A, Cosgrove SE, Daum RS, Fridkin SK, et al. 2011. Clinical practice guidelines by the infectious diseases society of America for the treatment of methicillin-resistant *Staphylococcus aureus* infections in adults and children: executive summary. *Clin Infect Dis.* 52:285–92.

16. Harris SR, Feil EJ, Holden MT, Quail MA, Nickerson EK, et al. 2010. Evolution of MRSA during hospital transmission and intercontinental spread. *Science.* 327:469–74.

17. Harrison EM, Paterson GK, Holden MT, Larsen J, Stegger M, et al. 2013. Whole genome sequencing identifies zoonotic transmission of MRSA isolates with the novel mecA homologue mecC. *EMBO Mol Med.* 5:509–15.

18. Koser CU, Holden MT, Ellington MJ, Cartwright EJ, Brown NM, et al. 2012. Rapid whole-genome sequencing for investigation of a neonatal MRSA outbreak. *N Engl J Med.* 366:2267–75.

19. Planet PJ, Narechania A, Chen L, Mathema B, Boundy S, et al. 2017. Architecture of a species: phylogenomics of *Staphylococcus aureus*. *Trends Microbiol.* 25:153–66.

20. Archer GL. 1998. *Staphylococcus aureus*: a well-armed pathogen. *Clin Infect Dis.* 26:1179–81.

21. David MZ, Daum RS. 2017. Treatment of *Staphylococcus aureus* infections. *Curr Top Microbiol Immunol.* 409:325–83.

22. Lowy FD. 1998. *Staphylococcus aureus* infections. *New Engl J Med.* 339:520–32.

23. Kuehnert MJ, Kruszon-Moran D, Hill HA, McQuillan G, McAllister SK, et al. 2006. Prevalence of *Staphylococcus aureus* nasal colonization in the United States, 2001-2002. *J Infect Dis.* 193:172–79.

24. Dinges MM, Orwin PM, Schlievert PM. 2000. Exotoxins of *Staphylococcus aureus*. *Clin Microbiol Rev.* 13:16–34.

25. Marraffini LA, DeDent AC, Schneewind O. 2006. Sortases and the art of anchoring proteins to the envelopes of gram-positive bacteria. *Microbiol Mol Biol Rev.* 70:192–221.

26. Novick RP. 2003. Mobile genetic elements and bacterial toxinoses: the superantigen-encoding pathogenicity islands of *Staphylococcus aureus*. *Plasmid.* 49:93–105.

27. Fijalkowski K, Struk M, Karakulska J, Paszkowska A, Giedrys-Kalemba S, et al. 2014. Comparative analysis of superantigen genes in *Staphylococcus xylosus* and *Staphylococcus aureus* isolates collected from a single mammary quarter of cows with mastitis. *J Microbiol.* 52:366–72.

28. Paharik AE, Horswill AR. 2016. The staphylococcal biofilm: adhesins, regulation, and host response. *Microbiol Spect.* 4(2).

29. Sievert DM, Ricks P, Edwards JR, Schneider A, Patel J, et al. 2013. Antimicrobial-resistant pathogens associated with healthcare-associated infections: summary of data reported to the national healthcare safety network at the centers for disease control and prevention, 2009-2010. *Infect Control Hosp Epidemiol.* 34:1–14.

30. Cheng AG, DeDent AC, Schneewind O, Missiakas DM. 2011. A play in four acts: *Staphylococcus aureus* abscess formation. *Trends Microbiol.* 19:225–32.

31. Kim HK, Missiakas D, Schneewind O. 2014. Mouse models for infectious diseases caused by *Staphylococcus aureus*. *J Immunol Methods.* 410:88–99.

32. Malachowa N, Kobayashi SD, Braughton KR, DeLeo FR. 2013. Mouse model of *Staphylococcus aureus* skin infection. *Methods Mol Biol.* 1031:109–16.

33. Marra A. 2014. Animal models in drug development for MRSA. *Methods Mol Biol.* 1085:333–45.

34. Sun Y, Emolo C, Holtfreter S, Wiles S, Kreiswirth B, et al. 2018. Staphylococcal protein A contributes to persistent colonization of mice with *Staphylococcus aureus*. *J Bacteriol.* 200.

35. Tarkowski A, Collins LV, Gjertsson I, Hultgren OH, Jonsson IM, et al. 2001. Model systems: modeling human staphylococcal arthritis and sepsis in the mouse. *Trends Microbiol.* 9:321–26.

36. Langley R, Wines B, Willoughby N, Basu I, Proft T, Fraser JD. 2005. The staphylococcal superantigen-like protein 7 binds IgA and complement C5 and inhibits IgA-Fc alpha RI binding and serum killing of bacteria. *J Immunol.* 174:2926–33.

37. Loffler B, Hussain M, Grundmeier M, Bruck M, Holzinger D, et al. 2010. *Staphylococcus aureus* Panton-Valentine leukocidin is a very potent cytotoxic factor for human neutrophils. *PLOS Pathog.* 6:e1000715.

38. Rooijakkers SH, van Kessel KP, van Strijp JA. 2005. Staphylococcal innate immune evasion. *Trends Microbiol.* 13:596–601.

39. DeLeo FR, Otto M, Kreiswirth BN, Chambers HF. 2010. Community-associated meticillin-resistant *Staphylococcus aureus*. *Lancet.* 375:1557–68.

40. Vandenesch F, Naimi T, Enright MC, Lina G, Nimmo GR, et al. 2003. Community-acquired methicillin-resistant *Staphylococcus aureus* carrying Panton-Valentine leukocidin genes: worldwide emergence. *Emerg Infect Dis.* 9:978–84.

41. Diep BA, Gill SR, Chang RF, Phan TH, Chen JH, et al. 2006. Complete genome sequence of USA300, an epidemic clone of community-acquired meticillin-resistant *Staphylococcus aureus*. *Lancet.* 367:731–39.

42. Panton PN, Valentine FCO. 1932. Staphylococcal toxin. *Lancet.* 222:506–08.

43. Reyes Robles T, Alonzo F, 3rd, Kozhaya L, Lacy DB, Unutmaz D, Torres VJ. 2013. *Staphylococcus aureus* leukotoxin ED targets the chemokine receptors CXCR1 and CXCR2 to kill leukocytes and promote infection. *Cell Host Microbe.* 14:453–59.

44. Spaan AN, Henry T, van Rooijen WJ, Perret M, Badiou C, et al. 2013. The staphylococcal toxin Panton-Valentine leukocidin targets human C5a receptors. *Cell Host Microbe.* 13:584–94.

45. Tromp AT, Van Gent M, Abrial P, Martin A, Jansen JP, et al. 2018. Human CD45 is an F-component-specific receptor for the staphylococcal toxin Panton-Valentine leukocidin. *Nat Microbiol.* 3:708–17.

46. De La Fuente R, Suarez IG, Schleifer KH. 1985. *Staphylococcus aureus* subsp. *anaerobius* subsp. nov., the

causal agent of abscess disease of sheep. *Int J Syst Bacteriol.* 35:99–102.

47. Elbir H, Feil EJ, Drancourt M, Roux V, El Sanousi SM, et al. 2010. Ovine clone ST1464: a predominant genotype of *Staphylococcus aureus* subsp. anaerobius isolated from sheep in Sudan. *J Infect Dev Ctries.* 4:235–38.

48. Reizer J, Hoischen C, Titgemeyer F, Rivolta C, Rabus R, et al. 1998. A novel protein kinase that controls carbon catabolite repression in bacteria. *Mol Microbiol.* 27:1157–69.

49. Schleifer K-H, Hartinger A, Götz F. 1978. Occurrence of D-tagatose-6-phosphate pathway of D-galactose metabolism among *staphylococci*. *FEMS Microbiol Lett.* 3:9–11.

50. Strasters KC, Winkler KC. 1963. Carbohydrate metabolism of *Staphylococcus aureus*. *J Gen Microbiol.* 33:213–29.

51. Neubauer H, Gotz F. 1996. Physiology and interaction of nitrate and nitrite reduction in *Staphylococcus carnosus*. *J Bacteriol.* 178:2005–09.

52. Baker JS. 1984. Comparison of various methods for differentiation of staphylococci and micrococci. *J Clin Microbiol.* 19:875–79.

53. Schleifer KH, Kloos WE. 1975. A simple test system for the separation of staphylococci from micrococci. *J Clin Microbiol.* 1:337–38.

54. Enright MC, Day NP, Davies CE, Peacock SJ, Spratt BG. 2000. Multilocus sequence typing for characterization of methicillin-resistant and methicillin-susceptible clones of *Staphylococcus aureus*. *J Clin Microbiol.* 38:1008–15.

55. Marshall JH, Wilmoth GJ. 1981. Proposed pathway of triterpenoid carotenoid biosynthesis in *Staphylococcus aureus*: evidence from a study of mutants. *J Bacteriol.* 147:914–19.

56. Pelz A, Wieland KP, Putzbach K, Hentschel P, Albert K, Götz F. 2005. Structure and biosynthesis of staphyloxanthin from *Staphylococcus aureus*. *J Biol Chem.* 280:32493–98.

57. Clauditz A, Resch A, Wieland KP, Peschel A, Götz F. 2006. Staphyloxanthin plays a role in the fitness of *Staphylococcus aureus* and its ability to cope with oxidative stress. *Infect Immun.* 74:4950–53.

58. Kaden R, Engstrand L, Rautelin H, Johansson C. 2018. Which methods are appropriate for the detection of *Staphylococcus argenteus* and is it worthwhile to distinguish *S. argenteus* from *S. aureus*? *Infect Drug Resist.* 11:2335–44.

59. Loeb L. 1903. The influence of certain bacteria on the coagulation of blood. *J Med Res.* 10:407.

60. Much H. 1908. Über eine vorstufe des fibrinfermentes in kulturen von *staphylokokkus aureus*. *Biochem Z.* 14:143–55.

61. Thomer L, Schneewind O, Missiakas D. 2016. Pathogenesis of *Staphylococcus aureus* bloodstream infections. *Annu Rev Pathol.* 11:343–64.

62. Kolle W, Otto R. 1902. Die differenzierung der staphylokokken mittelst der agglutination. *Z Hygiene.* 41:369–79.

63. McAdow M, Kim HK, DeDenta AC, Hendrickx APA, Schneewind O, Missiakas DM. 2011. Preventing *Staphylococcus aureus* sepsis through the inhibition of its agglutination in blood. *PLOS Pathog.* 7:e1002307.

64. Hawiger J, Timmons S, Strong DD, Cottrell BA, Riley M, Doolittle RF. 1982. Identification of a region of human fibrinogen interacting with staphylococcal clumping factor. *Biochemistry.* 21:1407–13.

65. McDevitt D, Francois P, Vaudaux P, Foster TJ. 1995. Identification of the ligand-binding domain of the surface-located fibrinogen receptor (clumping factor) of *Staphylococcus aureus*. *Mol Microbiol.* 16:895–907.

66. Cheng AG, McAdow M, Kim HK, Bae T, Missiakas DM, Schneewind O. 2010. Contribution of coagulases towards *Staphylococcus aureus* disease and protective immunity. *PLOS Pathog.* 6:e1001036.

67. Foster TJ. 2019. The MSCRAMM Family of cell-wall-anchored surface proteins of gram-positive cocci. *Trends Microbiol.* 27:927–41.

68. Pickering AC, Vitry P, Prystopiuk V, Garcia B, Hook M, et al. 2019. Host-specialized fibrinogen-binding by a bacterial surface protein promotes biofilm formation and innate immune evasion. *PLOS Pathog.* 15:e1007816.

69. O'Reilly M, de Azavedo JC, Kennedy S, Foster TJ. 1986. Inactivation of the alpha-haemolysin gene of *Staphylococcus aureus* 8325-4 by site-directed mutagenesis and studies on the expression of its haemolysins. *Microb Pathog.* 1:125–38.

70. O'Reilly M, Kreiswirth BN, Foster TJ. 1990. Cryptic alpha-toxin gene in toxic shock syndrome and septicemia strains of *Staphylococcus aureus*. *Mol Microbiol.* 4:1947–55.

71. Bhakdi S, Tranum-Jensen J. 1991. Alpha-toxin of *Staphylococcus aureus*. *Microbiol Rev.* 55:733–51.

72. Song L, Hobaugh MR, Shustak C, Cheley S, Bayley H, Gouaux JE. 1996. Structure of staphylococcal alpha-hemolysin, a heptameric transmembrane pore. *Science.* 274:1859–66.

73. Gouaux E, Hobaugh M, Song L. 1997. Alpha-hemolysin, gamma-hemolysin, and leukocidin from *Staphylococcus aureus*: distant in sequence but similar in structure. *Protein Sci.* 6:2631–35.

74. McElroy MC, Harty HR, Hosford GE, Boylan GM, Pittet JF, Foster TJ. 1999. Alpha-toxin damages the air-blood barrier of the lung in a rat model of *Staphylococcus aureus*-induced pneumonia. *Infect Immun.* 67:5541–44.

75. Seeger W, Bauer M, Bhakdi S. 1984. Staphylococcal alpha-toxin elicits hypertension in isolated rabbit lungs. Evidence for thromboxane formation and the role of extracellular calcium. *J Clin Invest.* 74:849–58.

76. Seeger W, Birkemeyer RG, Ermert L, Suttorp N, Bhakdi S, Duncker HR. 1990. Staphylococcal alpha-toxin-induced vascular leakage in isolated perfused rabbit lungs. *Lab Invest.* 63:341–49.

77. Wilke GA, Bubeck Wardenburg J. 2010. Role of a disintegrin and metalloprotease 10 in *Staphylococcus aureus* alpha-hemolysin-mediated cellular injury. *Proc Natl Acad Sci USA.* 107:13473–78.

78. Berube BJ, Bubeck Wardenburg J. 2013. *Staphylococcus aureus* alpha-toxin: nearly a century of intrigue. *Toxins (Basel).* 5:1140–66.

79. Powers ME, Kim HK, Wang Y-T, Bubeck-Wardenburg J. 2012. ADAM10 mediates vascular injury induced by *Staphylococcus aureus* α-hemolysin. *J Infect Dis.* 206:352–356.

80. Inoshima I, Inoshima N, Wilke GA, Powers ME, Frank KM, et al. 2011. A *Staphylococcus aureus* pore-forming toxin subverts the activity of ADAM10 to cause lethal infection in mice. *Nat Med.* 17:1310–14.

81. Inoshima N, Wang Y, Bubeck Wardenburg J. 2012. Genetic requirement for ADAM10 in severe *Staphylococcus aureus* skin infection. *J Invest Dermatol.* 132:1513–16.

82. Goodyear CS, Silverman GJ. 2004. Staphylococcal toxin induced preferential and prolonged *in vivo* deletion of innate-like B lymphocytes. *Proc Natl Acad Sci USA.* 101:11392–27.

83. Sasso EH, Silverman GJ, Mannik M. 1989. Human IgM molecules that bind staphylococcal protein A contain VHIII H chains. *J Immun.* 142:2778–83.

84. Thammavongsa V, Kim HK, Missiakas D, Schneewind O. 2015. Staphylococcal manipulation of host immune responses. *Nat Rev Microbiol.* 13:529–43.

85. Kim HK, Cheng AG, Kim H-Y, Missiakas DM, Schneewind O. 2010. Non-toxigenic protein A vaccine for methicillin-resistant *Staphylococcus aureus* infections. *J Exp Med.* 207:1863–70.

86. Cheng AG, Kim HK, Burts ML, Krausz T, Schneewind O, Missiakas DM. 2009. Genetic requirements for *Staphylococcus aureus* abscess formation and persistence in host tissues. *FASEB J.* 23:3393–404.

87. Jonsson P, Lindberg M, Haraldsson I, Wadstrom T. 1985. Virulence of *Staphylococcus aureus* in a mouse mastitis model: studies of hemolysin, coagulase, and protein A as possible virulence determinants with protoplast fusion and gene cloning. *Infect Immun.* 49:765–69.

88. Patel AH, Nowlan P, Weavers ED, Foster T. 1987. Virulence of protein A-deficient and alpha-toxin-deficient mutants of *Staphylococcus aureus* isolated by allele replacement. *Infect Immun.* 55:3103–10.

89. Chan YG, Frankel MB, Dengler V, Schneewind O, Missiakas D. 2013. *Staphylococcus aureus* mutants lacking the LytR-CpsA-Psr family of enzymes release cell wall teichoic acids into the extracellular medium. *J Bacteriol.* 195:4650–59.

90. Coley J, Archibald AR, Baddiley J. 1976. A linkage unit joining peptidoglycan to teichoic acid in *Staphylococcus aureus* H. *FEBS Lett.* 61:240–42.

91. Chan YG, Kim HK, Schneewind O, Missiakas D. 2014. The capsular polysaccharide of *Staphylococcus aureus* is attached to peptidoglycan by the LytR-CpsA-Psr (LCP) family of enzymes. *J Biol Chem.* 289:15680–90.

92. O'Riordan K, Lee JC. 2004. *Staphylococcus aureus* capsular polysaccharides. *Clin Microbiol Rev.* 17:218–34.

93. Mazmanian SK, Liu G, Ton-That H, Schneewind O. 1999. *Staphylococcus aureus* sortase, an enzyme that anchors surface proteins to the cell wall. *Science.* 285:760–63.

94. Heilmann C, Schweitzer O, Gerke C, Vanittanakom N, Mack D, Gotz F. 1996. Molecular basis of intercellular adhesion in the biofilm-forming *Staphylococcus epidermidis*. *Mol Microbiol.* 20:1083–91.

95. Rupp ME, Fey PD, Heilmann C, Gotz F. 2001. Characterization of the importance of *Staphylococcus epidermidis* autolysin and polysaccharide intercellular adhesin in the pathogenesis of intravascular catheter-associated infection in a rat model. *J Infect Dis.* 183:1038–42.

96. Cramton SE, Gerke C, Schnell NF, Nichols WW, Götz F. 1999. The intercellular adhesion (*ica*) locus is present in *Staphylococcus aureus* and is required for biofilm formation. *Infect Immun* 67:5427–33.

97. Joo HS, Otto M. 2012. Molecular basis of in vivo biofilm formation by bacterial pathogens. *Chem Biol.* 19:1503–13.

98. Giesbrecht P, Kersten T, Maidhof H, Wecke J. 1998. Staphylococcal cell wall: morphogenesis and fatal variations in the presence of penicillin. *Microbiol Mol Biol Rev.* 62:1371–414.

99. Giesbrecht P, Kersten T, Wecke J. 1992. Fan-shaped ejections of regularly arranged murosomes involved in penicillin-induced death of staphylococci. *J Bacteriol.* 174:2241–52.

100. Henze U, Sidow T, Wecke J, Labischinski H, Berger-Bächi B. 1993. Influence of *femB* on methicillin resistance and peptidoglycan metabolism in *Staphylococcus aureus*. *J Bacteriol.* 175:1612–20.

101. Tipper DJ, Strominger JL, Ensign JC. 1967. Structure of the cell wall of *Staphylococcus aureus*, strain Copenhagen. VII. Mode of action of the bacteriolytic peptidase from myxobacter and the isolation of intact cell wall polysaccharides. *Biochemistry.* 6:906–20.

102. Tipper DJ, Tomoeda M, Strominger JL. 1971. Isolation and characterization of -1,4-N-acetylmuramyl-N-acetylglucosamine and its O-acetyl derivative. *Biochemistry.* 10:4683–90.

103. Warren GH, Gray J. 1965. Effect of sublethal concentrations of penicillins on the lysis of bacteria by lysozyme and trypsin. *Proc Soc Exp Biol Med.* 120:504–11.

104. Bera A, Herbert S, Jakob A, Vollmer W, Götz F. 2005. Why are pathogenic staphylococci so lysozyme resistant? The peptidoglycan *O*-acetyltransferase OatA is the major determinant for lysozyme resistance of *Staphylococcus aureus*. *Mol Microbiol.* 55:778–87.

105. Shimada T, Park BG, Wolf AJ, Brikos C, Goodridge HS, et al. 2010. *Staphylococcus aureus* evades lysozyme-based peptidoglycan digestion that links phagocytosis, inflammasome activation, and IL-1beta secretion. *Cell Host Microbe.* 7:38–49.

106. Ito E, Strominger JL. 1960. Enzymatic synthesis of the peptide in a uridine nucleotide from *Staphylococcus aureus*. *J Biol Chem.* 235:PC5–6.

107. Park JT. 1952. Uridine-5'-pyrophosphate derivatives. III. Amino acid-containing derivatives. *J Biol Chem.* 194:897–904.

108. Ito E, Strominger JL. 1973. Enzymatic synthesis of the peptide in bacterial uridine nucleotides. VII. Comparative biochemistry. *J Biol Chem.* 248:3131–36.

109. Barreteau H, Kovac A, Boniface A, Sova M, Gobec S, Blanot D. 2008. Cytoplasmic steps of peptidoglycan biosynthesis. *FEMS Microbiol Rev.* 32:168–207.

110. Doublet P, van Heijenoort J, Mengin-Lecreulx D. 1992. Identification of the *Escherichia coli* murI gene, which is required for the biosynthesis of D-glutamic acid, a specific component of bacterial peptidoglycan. *J Bacteriol.* 174:5772–79.

111. Kullik I, Giachino P, Fuchs T. 1998. Deletion of the alternative sigma factor sigmaB in *Staphylococcus aureus* reveals its function as a global regulator of virulence genes. *J Bacteriol.* 180:4814–20.

112. Chung BC, Zhao J, Gillespie RA, Kwon DY, Guan Z, et al. 2013. Crystal structure of MraY, an essential membrane enzyme for bacterial cell wall synthesis. *Science.* 341:1012–16.

113. Higashi Y, Strominger JL, Sweeley CC. 1967. Structure of a lipid intermediate in cell wall peptidoglycan synthesis: a derivative of C55 isoprenoid alcohol. *Proc Natl Acad Sci USA.* 57:1878–84.

114. Ikeda M, Wachi M, Jung HK, Ishino F, Matsuhashi M. 1991. The *Escherichia coli* mraY gene encoding UDP-N-acetylmuramoyl-pentapeptide: undecaprenyl-phosphate phospho-N-acetylmuramoyl-pentapeptide transferase. *J Bacteriol.* 173:1021–26.

115. Pless DD, Neuhaus FC. 1973. Initial membrane reaction in peptidoglycan synthesis. Lipid dependence of

phospho-n-acetylmuramyl-pentapeptide translocase (exchange reaction). *J Biol Chem.* 248:1568–76.

116. Bouhss A, Trunkfield AE, Bugg TD, Mengin-Lecreulx D. 2008. The biosynthesis of peptidoglycan lipid-linked intermediates. *FEMS Microbiol Rev.* 32:208–33.

117. Mengin-Lecreulx D, Texier L, Rousseau M, van Heijenoort J. 1991. The *murG* gene of *Escherichia coli* codes for the UDP-N-acetylglucosamine: N-acetylmuramyl-(pentapeptide) pyrophosphoryl-undecaprenol N-acetylglucosamine transferase involved in the membrane steps of peptidoglycan synthesis. *J Bacteriol.* 173:4625–36.

118. Kamiryo T, Matsuhashi M. 1972. The biosynthesis of the cross-linking peptides in the cell wall peptidoglycan of *Staphylococcus aureus. J Bacteriol.* 247:6306–11.

119. Green CI, Vold BS. 1993. *Staphylococcus aureus* has clustered tRNA genes. *J Bacteriol.* 175:5091–96.

120. Berger-Bächi B. 1994. Expression of resistance to methicillin. *Trends Microbiol.* 2:389–09.

121. Rohrer S, Berger-Bachi B. 2003. FemABX peptidyl transferases: a link between branched-chain cell wall peptide formation and beta-lactam resistance in gram-positive cocci. *Antimicrob Agents Chemother.* 47:837–46.

122. Rohrer S, Ehlert K, Tschierske M, Labischinski H, Berger-Bächi B. 1999. The essential *Staphylococcus aureus* gene *fmhB* is involved in the first step of peptidoglycan pentaglycine interpeptide formation. *Proc Natl Acad Sci USA.* 96:9351–56.

123. Monteiro JM, Covas G, Rausch D, Filipe SR, Schneider T, et al. 2019. The pentaglycine bridges of *Staphylococcus aureus* peptidoglycan are essential for cell integrity. *Sci Rep.* 9:5010.

124. Schindler CA, Schuhardt VT. 1964. Lysostaphin: a new bacteriolytic agent for the *Staphylococcus. Proc Natl Acad Sci USA.* 51:414–21.

125. Sugai M, Fujiwara T, Akiyama T, Ohara M, Komatsuzawa H, et al. 1997. Purification and molecular characterization of glycylglycine endopeptidase produced by *Staphylococcus capitis* EPK1. *J Bacteriol.* 179:1193–202.

126. Sugai M, Fujiwara T, Ohta K, Komatsuzawa H, Ohara M, Suginaka H. 1997. *epr*, which encodes glycylglycine endopeptidase resistance, is homologous to *femAB* and affects serine content of peptidoglycan cross bridges in *Staphylococcus capitis* and *Staphylococcus aureus. J Bacteriol.* 179:4311–18.

127. Thumm G, Gotz F. 1997. Studies on prolysostaphin processing and characterization of the lysostaphin immunity factor (Lif) of *Staphylococcus simulans* biovar staphylolyticus. *Mol Microbiol.* 23:1251–65.

128. Tipper DJ. 1969. Structures of the cell wall peptidoglycans of *Staphylococcus epidermidis* Texas 26 and *Staphylococcus aureus* Copenhagen. II. Structure of neutral and basic peptides from hydrolysis with the myxobacter al-1 peptidase. *Biochemistry.* 8:2192–202.

129. Fay A, Dworkin J. 2009. Bacillus subtilis homologs of MviN (MurJ), the putative *Escherichia coli* lipid II flippase, are not essential for growth. *J Bacteriol.* 191:6020–28.

130. Ruiz N. 2008. Bioinformatics identification of MurJ (MviN) as the peptidoglycan lipid II flippase in *Escherichia coli. Proc Natl Acad Sci USA.* 105:15553–57.

131. Sham LT, Butler EK, Lebar MD, Kahne D, Bernhardt TG, Ruiz N. 2014. Bacterial cell wall. MurJ is the flippase of lipid-linked precursors for peptidoglycan biogenesis. *Science.* 345:220–22.

132. Huber J, Donald RG, Lee SH, Jarantow LW, Salvatore MJ, et al. 2009. Chemical genetic identification of peptidoglycan inhibitors potentiating carbapenem activity against methicillin-resistant *Staphylococcus aureus. Chem Biol.* 16:837–48.

133. Monteiro JM, Pereira AR, Reichmann NT, Saraiva BM, Fernandes PB, et al. 2018. Peptidoglycan synthesis drives an FtsZ-treadmilling-independent step of cytokinesis. *Nature.* 554:528–32.

134. Mohammadi T, van Dam V, Sijbrandi R, Vernet T, Zapun A, et al. 2011. Identification of FtsW as a transporter of lipid-linked cell wall precursors across the membrane. *EMBO J.* 30:1425–32.

135. Meeske AJ, Riley EP, Robins WP, Uehara T, Mekalanos JJ, et al. 2016. SEDS proteins are a widespread family of bacterial cell wall polymerases. *Nature.* 537:634–38.

136. Taguchi A, Welsh MA, Marmont LS, Lee W, Sjodt M, et al. 2019. FtsW is a peptidoglycan polymerase that is functional only in complex with its cognate penicillin-binding protein. *Nat Microbiol.* 4:587–94.

137. Tipper DJ, Strominger JL. 1968. Biosynthesis of the peptidoglycan of bacterial cell walls. XII. Inhibition of cross-linking by penicillins and cephalosporins: studies in *Staphylococcus aureus in vivo. J Biol Chem.* 243:3169–79.

138. Hartman BJ, Tomasz A. 1984. Low affinity penicillin binding protein associated with b-lactam resistance in *Staphylococcus aureus. J Bacteriol.* 158:513–16.

139. Pereira SF, Henriques AO, Pinho MG, de Lencastre H, Tomasz A. 2007. Role of PBP1 in cell division of *Staphylococcus aureus. J Bacteriol.* 189:3525–31.

140. Pereira SF, Henriques AO, Pinho MG, de Lencastre H, Tomasz A. 2009. Evidence for a dual role of PBP1 in the cell division and cell separation of *Staphylococcus aureus. Mol Microbiol.* 72:895–904.

141. Pinho MG, Filipe SR, de Lencastre H, Tomasz A. 2001. Complementation of the essential peptidoglycan transpeptidase function of penicillin-binding protein 2 (PBP2) by the drug resistance protein PBP2A in *Staphylococcus aureus. J Bacteriol.* 183:6525–31.

142. Reed P, Atilano ML, Alves R, Hoiczyk E, Sher X, et al. 2015. *Staphylococcus aureus* survives with a minimal peptidoglycan synthesis machine but sacrifices virulence and antibiotic resistance. *PLOS Pathog.* 11:e1004891.

143. Sugai M, Yamada S, Nakashima S, Komatsuzawa H, Matsumoto A, et al. 1997. Localized perforation of the cell wall by a major autolysin: *atl* gene products and the onset of penicillin-induced lysis of *Staphylococcus aureus. J Bacteriol.* 179:2958–062.

144. Oshida T, Sugai M, Komatsuzawa H, Hong Y-M, Suginaka H, Tomasz A. 1995. A *Staphylococcus aureus* autolysin that has an N-acetylmuramoyl-L-alanine amidase domain and an endo-b-N-acetylglucosaminidase domain: cloning, sequence analysis, and characterization. *Proc Natl Acad Sci USA.* 92:285–89.

145. Heilmann C, Hussain M, Peters G, Gotz F. 1997. Evidence for autolysin-mediated primary attachment of *Staphylococcus epidermidis* to a polystyrene surface. *Mol Microbiol.* 24:1013–24.

146. Biswas R, Voggu L, Simon UK, Hentschel P, Thumm G, Gotz F. 2006. Activity of the major staphylococcal autolysin Atl. *FEMS Microbiol Lett.* 259:260–68.

147. Komatsuzawa H, Sugai M, Nakashima S, Yamada S, Matsumoto A, et al. 1997. Subcellular localization of the major autolysin, ATL and its processed proteins in *Staphylococcus aureus*. *Microbiol Immunol*. 41:469–79.

148. Yamada S, Sugai M, Komatsuzawa H, Nakashima S, Oshida T, et al. 1996. An autolysin ring associated with cell separation of *Staphylococcus aureus*. *J Bacteriol*. 178:1565–71.

149. Baba T, Schneewind O. 1998. Targeting of muralytic enzymes to the cell division site of gram-positive bacteria: repeat domains direct autolysin to the equatorial surface ring of *Staphylococcus aureus*. *EMBO J*. 17:4639–46.

150. Schlag M, Biswas R, Krismer B, Kohler T, Zoll S, et al. 2010. Role of staphylococcal cell wall teichoic acid in targeting the major autolysin Atl. *Mol Microbiol*. 75:864–73.

151. Zoll S, Schlag M, Shkumatov AV, Rautenberg M, Svergun DI, et al. 2012. Ligand-binding properties and conformational dynamics of autolysin repeat domains in staphylococcal cell wall recognition. *J Bacteriol*. 194:3789–802.

152. Ward JB. 1973. The chain length of the glycans in bacterial cell walls. *Biochem J*. 133:395–98.

153. Ward JB, Perkins HR. 1973. The direction of glycan synthesis in a bacterial peptidoglycan. *Biochem J*. 135:721–28.

154. Chan YG, Frankel MB, Missiakas D, Schneewind O. 2016. SagB glucosaminidase is a determinant of *Staphylococcus aureus* glycan chain length, antibiotic susceptibility, and protein secretion. *J Bacteriol*. 198:1123–36.

155. Wheeler R, Turner RD, Bailey RG, Salamaga B, Mesnage S, et al. 2015. Bacterial cell enlargement requires control of cell wall stiffness mediated by peptidoglycan hydrolases. *MBio*. 6:e00660.

156. Weigel LM, Clewell DB, Gill SR, Clark NC, McDougal LK, et al. 2003. Genetic analysis of a high-level vancomycin-resistant isolate of *Staphylococcus aureus*. *Science*. 302:1569–71.

157. Tenover FC, Biddle JW, Lancaster MV. 2001. Increasing resistance to vancomycin and other glycopeptides in *Staphylococcus aureus*. *Emerg Infect Dis*. 7:327–32.

158. Bugg TDH, Wright GD, Dutka-Malen S, Arthur M, Courvalin P, Walsh CT. 1991. Molecular basis for vancomycin resistance in *enterococcus faecium* BM4147: biosynthesis of a depsipeptide peptidoglycan precursor by vancomycin resistance proteins VanH and VanA. *Biochemistry*. 30:10408–15.

159. Walsh CT. 1993. Vancomycin resistance: decoding the molecular logic. *Science*. 261:308–09.

160. Wu H, Fives-Taylor PM. 2001. Molecular strategies for fimbrial expression and assembly. *Crit. Rev Oral Biol Med*. 12:101–15.

161. Katayama Y, Ito T, Hiramatsu K. 2000. A new class of genetic element, staphylococcus cassette chromosome mec, encodes methicillin resistance in *Staphylococcus aureus*. *Antimicrob Agents Chemother*. 44:1549–55.

162. Kuroda M, Ohta T, Uchiyama I, Baba T, Yuzawa H, et al. 2001. Whole genome sequencing of meticillin-resistant *Staphylococcus aureus*. *Lancet*. 357:1225–40.

163. Projan SJ. 2003. Why is big pharma getting out of antibacterial drug discovery? *Cur Opin Microbiol*. 6:427–30.

164. Archibald AR, Baddiley J, Heckels JE. 1973. Molecular arrangement of teichoic acid in the cell wall of *Staphylococcus aureus*. *Nature New Biol*. 241:29–31.

165. Armstrong JJ, Baddiley J, Buchanan JG, Davison AL, Kelemen MV, Neuhaus FC. 1959. Composition of teichoic acids from a number of bacterial cell walls. *Nature*. 184:247–49.

166. Sanderson AR, Strominger JL, Nathenson SG. 1962. Chemical structure of teichoic acid from *Staphylococcus aureus*, strain Copenhagen. *J Biol Chem*. 237:3603–13.

167. Vinogradov E, Sadovskaya I, Li J, Jabbouri S. 2006. Structural elucidation of the extracellular and cell-wall teichoic acids of *Staphylococcus aureus* MN8m, a biofilm forming strain. *Carbohydr Res*. 341:738–43.

168. Yokohama K, Miyashita T, Arakai Y, Ito E. 1986. Structure and functions of linkage unit intermediates in the biosynthesis of ribitol teichoic acids in *Staphylococcus aureus* H and Bacillus subtilis W23. *Eur J Biochem*. 161:479–89.

169. Soldo B, Lazarevic V, Karamata D. 2002. *tagO* is involved in the synthesis of all anionic cell-wall polymers in Bacillus subtilis 168. *Microbiology*. 148:2079–87.

170. Wyke AW, Ward JB. 1977. Biosynthesis of wall polymers in Bacillus subtilis. *J Bacteriol*. 130:1055–63.

171. Xia G, Maier L, Sanchez-Carballo P, Li M, Holst O, Peschel A. 2010. Glycosylation of wall teichoic acid in *Staphylococcus aureus* by TarM. *J Biol Chem*. 285.

172. D'Elia MA, Henderson JA, Beveridge TJ, Heinrichs DE, Brown ED. 2009. The *N*-acetylmannosamine transferase catalyzes the first committed step of teichoic acid assembly in Bacillus subtilis and *Staphylococcus aureus*. *J Bacteriol*. 191:4030–34.

173. Lazarevic V, Abellan FX, Möller SB, Karamata D, Mauël C. 2002. Comparison of ribitol and glycerol teichoic acid genes in Bacillus subtilis W23 and 168: identical function, similar divergent organization, but different regulation. *Microbiology*. 148:815–24.

174. Mauel C, Young M, Margot P, Karamata D. 1989. The essential nature of teichoic acids in Bacillus subtilis as revealed by insertional mutagenesis. *Mol Gen Genet*. 215:388–94.

175. Pooley HM, Abellan FX, Karamata D. 1991. A conditional-lethal mutant of Bacillus subtilis 168 with a thermosensitive glycerol-3-phosphate cytidylyltransferase, an enzyme specific for the synthesis of the major wall teichoic acid. *J Gen Microbiol*. 137:921–28.

176. Lazarevic V, Karamata D. 1995. The *tagGH* operon of Bacillus subtilis 168 encodes a two-component ABC transporter involved in the metabolism of two wall teichoic acids. *Mol Microbiol*. 16:345–55.

177. Kawai Y, Marles-Wright J, Cleverley RM, Emmins R, Ishikawa S, et al. 2011. A widespread family of bacterial cell wall assembly proteins. *EMBO J*, 30:4931–41.

178. Kojima N, Araki Y, Ito E. 1983. Structure of linkage region between ribitol teichoic acid and peptidoglycan in cell walls of *Staphylococcus aureus* H. *J Biol Chem*. 258:9043–45.

179. Dengler V, Meier PS, Heusser R, Kupferschmied P, Fazekas J, et al. 2012. Deletion of hypothetical wall teichoic acid ligases in *Staphylococcus aureus* activates the cell wall stress response. *FEMS Microbiol Lett*. 333:109–20.

180. D'Elia MA, Millar KE, Beveridge TJ, Brown ED. 2006. Wall teichoic acid polymers are dispensable for cell viability in Bacillus subtilis. *J Bacteriol*. 188:8313–16.

181. Weidenmaier C, Kokai-Kun JF, Kristian SA, Chanturiya T, Kalbacher H, et al. 2004. Role of teichoic acids in *Staphylococcus aureus* nasal colonization, a major risk factor in nosocomial infections. *Nat Med*. 10:243–45.

182. D'Elia MA, Pereira MP, Chung YS, Zhao W, Chau A, et al. 2006. Lesions in teichoic acid biosynthesis in *Staphylococcus*

aureus lead to a lethal gain of function in the otherwise dispensable pathway. *J Bacteriol.* 188:4183–89.

183. Fischer W. 1994. Lipoteichoic acid and lipids in the membrane of *Staphylococcus aureus. Med. Microbiol Immunol.* 183:61–76.

184. Forsberg CW, Wyrick PB, Ward JB, Rogers HJ. 1973. Effect of phosphate limitation on the morphology and wall composition of Bacillus licheniformis and its phosphoglucomutase-deficient mutants. *J Bacteriol.* 113:969–84.

185. Jorasch P, Warnecke DC, Lindner B, Zahringer U, Heinz E. 2000. Novel processive and nonprocessive glycosyltransferases from *Staphylococcus aureus* and *Arabidopsis thaliana* synthesize glycolipids, glycophospholipids, glycosphingolipids, and glycosylsterols. *Eur J Biochem.* 267:3770–83.

186. Jorasch P, Wolter FP, Zahringer U, Heinz E. 1998. A UDP glucosyltransferase from Bacillus subtilis successively transfers up to four glucose residues to 1,2-diacylglycerol: expression of *ypfP* in *Escherichia coli* and structural analysis of its reaction products. *Mol Microbiol.* 29:419–30.

187. Kiriukhin MY, Debabov DV, Shinabarger DL, Neuhaus FC. 2001. Biosynthesis of the glycolipid anchor in lipoteichoic acid of *Staphylococcus aureus* RN4220: role of YpfP, the diglucosyldiacylglycerol synthase. *J Bacteriol.* 183:3506–14.

188. Lazarevic V, Soldo B, Medico N, Pooley H, Bron S, Karamata D. 2005. Bacillus subtilis alpha-phosphoglucomutase is required for normal cell morphology and biofilm formation. *Appl Environ Microbiol.* 71:39–45.

189. Pooley HM, Paschoud D, Karamata D. 1987. The gtaB marker in Bacillus subtilis 168 is associated with a deficiency in UDPglucose pyrophosphorylase. *J Gen Microbiol.* 133:3481–93.

190. Soldo B, Lazarevic V, Margot P, Karamata D. 1993. Sequencing and analysis of the divergon comprising gtaB, the structural gene of UDP-glucose pyrophosphorylase of Bacillus subtilis 168. *J Gen Microbiol.* 139:3185–95.

191. Grundling A, Schneewind O. 2007. Genes required for glycolipid synthesis and lipoteichoic acid anchoring in *Staphylococcus aureus. J Bacteriol.* 189:2521–30.

192. Grundling A, Schneewind O. 2007. Synthesis of glycerolphosphate lipoteichoic acid in *Staphylococcus aureus. Pro Natl Acad Sci USA,* 104:8478–83.

193. Oku Y, Kurokawa K, Matsuo M, Yamada S, Lee B, Sekimizu K. 2009. Pleiotropic roles of polyglycerolphosphate synthase of lipoteichoic acid in growth of *Staphylococcus aureus* cells. *J Bacteriol.* 191:141–51.

194. Neuhaus FC, Baddiley J. 2003. A continuum of anionic charge: structures and functions of D-alanyl-teichoic acids in gram-positive bacteria. *Microbiol Mol Biol Rev.* 67:686–723.

195. Koprivnjak T, Mlakar V, Swanson L, Fournier B, Peschel A, Weiss JP. 2006. Cation-induced transcriptional regulation of the dlt operon of *Staphylococcus aureus. J Bacteriol.* 188:3622–30.

196. Perego M, Glaser P, Minutello A, Strauch MA, Leopold K, Fischer W. 1995. Incorporation of D-alanine into lipoteichoic acid and wall teichoic acid in Bacillus subtilis. Identification of genes and regulation. *J Biol Chem.* 270:15598–606.

197. Reichmann NT, Cassona CP, Grundling A. 2013. Revised mechanism of D-alanine incorporation into cell wall polymers in gram-positive bacteria. *Microbiology.* 159:1868–77.

198. Haas R, Koch HU, Fischer W. 1984. Alanyl turnover from lipoteichoic acid to teichoic acid in *Staphylococcus aureus. FEMS Microbiol Lett.* 21:27–31.

199. Koch HU, Doker R, Fischer W. 1985. Maintenance of D-alanine ester substitution of lipoteichoic acid by reesterification in *Staphylococcus aureus. J Bacteriol.* 164:1211–17.

200. Corrigan RM, Abbott JC, Burhenne H, Kaever V, Grundling A. 2011. c-di-AMP is a new second messenger in *Staphylococcus aureus* with a role in controlling cell size and envelope stress. *PLOS Pathog.* 7:e1002217.

201. Corrigan RM, Campeotto I, Jeganathan T, Roelofs KG, Lee VT, Grundling A. 2013. Systematic identification of conserved bacterial c-di-AMP receptor proteins. *Proc Natl Acad Sci USA.* 110:9084–89.

202. Price-Whelan A, Poon CK, Benson MA, Eidem TT, Roux CM, et al. 2013. Transcriptional profiling of *Staphylococcus aureus* during growth in 2 M NaCl leads to clarification of physiological roles for Kdp and Ktr K+ uptake systems. *mBio.* 4:e00407–13.

203. Peschel A, Otto M, Jack RW, Kalbacher H, Jung G, Götz F. 1999. Inactivation of the *dlt* operon in *Staphylococcus aureus* confers sensitivity to defensins, protegrins, and other antimicrobial peptides. *J Biol Chem.* 274:8405–10.

204. Peschel A, Vuong C, Otto M, Götz F. 2000. The D-alanine residues of *Staphylococcus aureus* teichoic acids alter the susceptibility to vancomycin and the activity of autolytic enzymes. *Antimicrob Agents Chemother.* 44:2845–47.

205. Brown S, Santa Maria JP, Jr., Walker S. 2013. Wall teichoic acids of gram-positive bacteria. *Annu Rev Microbiol.* 67:313–36.

206. Missiakas D. 2019. *Staphylococcus aureus* TarP: a brick in the wall or Rosetta Stone? *Cell Host Microbe.* 25:182–83.

207. Navarre WW, Schneewind O. 1999. Surface proteins of gram-positive bacteria and the mechanisms of their targeting to the cell wall envelope. *Microbiol Mol Biol Rev.* 63:174–229.

208. Sjöquist J, Meloun B, Hjelm H. 1972. Protein A isolated from *Staphylococcus aureus* after digestion with lysostaphin. *Eur J Biochem.* 29:572–78.

209. Fischetti VA, Pancholi V, Schneewind O. 1990. Conservation of a hexapeptide sequence in the anchor region of surface proteins from gram-positive cocci. *Mol Microbiol.* 4:1603–05.

210. Schneewind O, Model P, Fischetti VA. 1992. Sorting of protein A to the staphylococcal cell wall. *Cell.* 70:267–81.

211. Navarre WW, Schneewind O. 1994. Proteolytic cleavage and cell wall anchoring at the LPXTG motif of surface proteins in gram-positive bacteria. *Mol Microbiol.* 14:115–21.

212. Schneewind O, Fowler A, Faull KF. 1995. Structure of the cell wall anchor of surface proteins in *Staphylococcus aureus. Science.* 268:103–06.

213. Perry AM, Ton-That H, Mazmanian SK, Schneewind O. 2002. Anchoring of surface proteins to the cell wall of *Staphylococcus aureus.* III. Lipid II is an *in vivo* peptidoglycan substrate for sortase-catalyzed surface protein anchoring. *J Biol Chem.* 277:16241–48.

214. Mazmanian SK, Skaar EP, Gaspar AH, Humayun M, Gornicki P, et al. 2003. Passage of heme-iron across the envelope of *Staphylococcus aureus. Science.* 299:906–09.

215. DeDent A, Kim HK, Missiakas DM, Schneewind O. 2012. Exploring *Staphylococcus aureus* pathways to disease for vaccine development. *Semin Immunopathol.* 34:317–33.

216. Foster TJ, Höök M. 1998. Surface protein adhesins of *Staphylococcus aureus*. *Trends Microbiol*. 6:484–88.

217. Bronesky D, Wu Z, Marzi S, Walter P, Geissmann T, et al. 2016. *Staphylococcus aureus* RNAIII and its regulon link quorum sensing, stress responses, metabolic adaptation, and regulation of virulence gene expression. *Annu Rev Microbiol*. 70:299–316.

218. Ibarra JA, Perez-Rueda E, Carroll RK, Shaw LN. 2013. Global analysis of transcriptional regulators in *Staphylococcus aureus*. *BMC Genomics*. 14:126.

219. Novick RP. 2003. Autoinduction and signal transduction in the regulation of staphylococcal virulence. *Mol Microbiol*. 48:1429–49.

220. Cheung AL, Nishina KA, Trotonda MP, Tamber S. 2008. The SarA protein family of *Staphylococcus aureus*. *Int J Biochem Cell Biol*. 40:355–61.

221. Romilly C, Caldelari I, Parmentier D, Lioliou E, Romby P, Fechter P. 2012. Current knowledge on regulatory RNAs and their machineries in *Staphylococcus aureus*. *RNA Biol*. 9:402–13.

222. Richardson AR, Somerville GA, Sonenshein AL. 2015. Regulating the intersection of metabolism and pathogenesis in gram-positive bacteria. *Microbiol Spect*. 3, doi: 10.1128/microbiolspec.MBP-0004-2014.

223. Sadykov MR, Mattes TA, Luong TT, Zhu Y, Day SR, et al. 2010. Tricarboxylic acid cycle-dependent synthesis of *Staphylococcus aureus* type 5 and 8 capsular polysaccharides. *J Bacteriol*. 192:1459–62.

224. Vuong C, Kidder JB, Jacobson ER, Otto M, Proctor RA, Somerville GA. 2005. *Staphylococcus epidermidis* polysaccharide intercellular adhesin production significantly increases during tricarboxylic acid cycle stress. *J Bacteriol*. 187:2967–73.

225. Archer NK, Mazaitis MJ, Costerton JW, Leid JG, Powers ME, Shirtliff ME. 2011. *Staphylococcus aureus* biofilms: properties, regulation, and roles in human disease. *Virulence*. 2:445–59.

226. Argemi X, Matelska D, Ginalski K, Riegel P, Hansmann Y, et al. 2018. Comparative genomic analysis of *Staphylococcus lugdunensis* shows a closed pan-genome and multiple barriers to horizontal gene transfer. *BMC Genomics*. 19:621.

227. Bosi E, Monk JM, Aziz RK, Fondi M, Nizet V, Palsson BO. 2016. Comparative genome-scale modelling of *Staphylococcus aureus* strains identifies strain-specific metabolic capabilities linked to pathogenicity. *Proc Natl Acad Sci USA*. 113:E3801–09.

228. Conlan S, Mijares LA, Program NCS, Becker J, Blakesley RW, et al. 2012. *Staphylococcus epidermidis* pan-genome sequence analysis reveals diversity of skin commensal and hospital infection-associated isolates. *Genome Biol*. 13:R64.

229. Pain M, Hjerde E, Klingenberg C, Cavanagh JP. 2019. Comparative genomic analysis of staphylococcus haemolyticus reveals key to hospital adaptation and pathogenicity. *Front Microbiol*. 10:2096.

230. Lindsay JA. 2014. *Staphylococcus aureus* genomics and the impact of horizontal gene transfer. *Int J Med Microbiol*. 304:103–09.

231. Malachowa N, DeLeo FR. 2010. Mobile genetic elements of *Staphylococcus aureus*. *Cell Mol Life Sci: CMLS*. 67:3057–71.

232. Chongtrakool P, Ito T, Ma XX, Kondo Y, Trakulsomboon S, et al. 2006. Staphylococcal cassette chromosome mec (SCCmec) typing of methicillin-resistant *Staphylococcus aureus* strains isolated in 11 Asian countries: a proposal for a new nomenclature for SCCmec elements. *Antimicrob Agents Chemother*. 50:1001–12.

233. Holden MT, Feil EJ, Lindsay JA, Peacock SJ, Day NP, et al. 2004. Complete genomes of two clinical *Staphylococcus aureus* strains: evidence for the rapid evolution of virulence and drug resistance. *Proc Natl Acad Sci USA*. 101:9786–91.

234. Diep BA, Gill SR, Chang RF, Phan TH, Chen JH, et al. 2006. Complete genome sequence of USA300, an epidemic clone of community-acquired meticillin-resistant *Staphylococcus aureus*. *Lancet*. 367:731–39.

235. Espedido BA, Steen JA, Barbagiannakos T, Mercer J, Paterson DL, et al. 2012. Carriage of an ACME II variant may have contributed to methicillin-resistant *Staphylococcus aureus* sequence type 239-like strain replacement in Liverpool Hospital, Sydney, Australia. *Antimicrob Agents Chemother*. 56:3380–83.

236. Diep BA, Stone GG, Basuino L, Graber CJ, Miller A, et al. 2008. The arginine catabolic mobile element and staphylococcal chromosomal cassette mec linkage: convergence of virulence and resistance in the USA300 clone of methicillin-resistant *Staphylococcus aureus*. *J Infect Dis*. 197:1523–30.

237. Planet PJ, LaRussa SJ, Dana A, Smith H, Xu A, et al. 2013. Emergence of the epidemic methicillin-resistant *Staphylococcus aureus* strain USA300 coincides with horizontal transfer of the arginine catabolic mobile element and speG-mediated adaptations for survival on skin. *MBio*. 4:e00889–13.

238. Thurlow LR, Joshi GS, Clark JR, Spontak JS, Neely CJ, et al. 2013. Functional modularity of the arginine catabolic mobile element contributes to the success of USA300 methicillin-resistant *Staphylococcus aureus*. *Cell Host Microbe*. 13:100–07.

239. van Wamel WJ, Rooijakkers SH, Ruyken M, van Kessel KP, van Strijp JA. 2006. The innate immune modulators staphylococcal complement inhibitor and chemotaxis inhibitory protein of *Staphylococcus aureus* are located on beta-hemolysin-converting bacteriophages. *J Bacteriol*. 188:1310–15.

240. Xia G, Wolz C. 2014. Phages of *Staphylococcus aureus* and their impact on host evolution. *Infect, Genet Evol*. 21:593–601.

241. Moller AG, Lindsay JA, Read TD. 2019. Determinants of phage host range in staphylococcus species. *Appl Environ Microbiol*. 85.

242. Coleman DC, Sullivan DJ, Russell RJ, Arbuthnott JP, Carey BF, Pomeroy HM. 1989. *Staphylococcus aureus* bacteriophages mediating the simultaneous lysogenic conversion of beta-lysin, staphylokinase and enterotoxin A: molecular mechanism of triple conversion. *J Gen Microbiol*. 135:1679–97.

243. Diep BA, Sensabaugh GF, Somboonna N, Carleton HA, Perdreau-Remington F. 2004. Widespread skin and soft-tissue infections due to two methicillin-resistant *Staphylococcus aureus* strains harboring the genes for Panton-Valentine leucocidin. *J Clin Microbiol*. 42:2080–84.

244. Francis JS, Doherty MC, Lopatin U, Johnston CP, Sinha G, et al. 2005. Severe community-onset pneumonia in healthy adults caused by methicillin-resistant *Staphylococcus aureus* carrying the Panton-Valentine leukocidin genes. *Clin Infect Dis*. 40:100–07.

245. Gillet Y, Issartel B, Vanhems P, Fournet JC, Lina G, et al. 2002. Association between *Staphylococcus aureus* strains

carrying gene for Panton-Valentine leukocidin and highly lethal necrotising pneumonia in young immunocompetent patients. *Lancet.* 359:753–59.

246. Li M, Du X, Villaruz AE, Diep BA, Wang D, et al. 2012. MRSA epidemic linked to a quickly spreading colonization and virulence determinant. *Nat Med.* 18:816–19.

247. Novick RP, Christie GE, Penades JR. 2010. The phage-related chromosomal islands of gram-positive bacteria. *Nat Rev Microbiol.* 8:541–51.

248. Novick RP. 2019. Pathogenicity Islands and their role in staphylococcal biology. *Microbiol Spect.* 7(3), doi: 10.1128/microbiolspec.GPP3-0062-2019.

249. Tuffs SW, Haeryfar SMM, McCormick JK. 2018. Manipulation of innate and adaptive immunity by staphylococcal superantigens. *Pathogens.* 7:53.

250. Tormo MA, Ferrer MD, Maiques E, Ubeda C, Selva L, et al. 2008. *Staphylococcus aureus* pathogenicity island DNA is packaged in particles composed of phage proteins. *J Bacteriol.* 190:2434–40.

251. Ubeda C, Maiques E, Knecht E, Lasa I, Novick RP, Penades JR. 2005. Antibiotic-induced SOS response promotes horizontal dissemination of pathogenicity island-encoded virulence factors in staphylococci. *Mol Microbiol.* 56:836–44.

252. Chen J, Carpena N, Quiles-Puchalt N, Ram G, Novick RP, Penades JR. 2015. Intra- and inter-generic transfer of pathogenicity island-encoded virulence genes by cos phages. *ISME J.* 9:1260–63.

253. Chen J, Novick RP. 2009. Phage-mediated intergeneric transfer of toxin genes. *Science.* 323:139–41.

254. Winstel V, Liang C, Sanchez-Carballo P, Steglich M, Munar M, et al. 2013. Wall teichoic acid structure governs horizontal gene transfer between major bacterial pathogens. *Nature Communications.* 4:2345.

255. Lowder BV, Guinane CM, Ben Zakour NL, Weinert LA, Conway-Morris A, et al. 2009. Recent human-to-poultry host jump, adaptation, and pandemic spread of *Staphylococcus aureus. Proc Natl Acad Sci USA.* 106:19545–50.

256. Sakwinska O, Giddey M, Moreillon M, Morisset D, Waldvogel A, Moreillon P. 2011. *Staphylococcus aureus* host range and human-bovine host shift. *Appl Environ Microbiol.* 77:5908–15.

257. Viana D, Comos M, McAdam PR, Ward MJ, Selva L, et al. 2015. A single natural nucleotide mutation alters bacterial pathogen host tropism. *Nat Genet.* 47:361–66.

258. Bruins MJ, Juffer P, Wolfhagen MJ, Ruijs GJ. 2007. Salt tolerance of methicillin-resistant and methicillin-susceptible *Staphylococcus aureus. J Clin Microbiol.* 45:682–83.

259. Krismer B, Liebeke M, Janek D, Nega M, Rautenberg M, et al. 2014. Nutrient limitation governs *Staphylococcus aureus* metabolism and niche adaptation in the human nose. *PLOS Pathog.* 10:e1003862.

260. Parfentjev IA, Catelli AR. 1964. Tolerance of *Staphylococcus aureus* to sodium chloride. *J Bacteriol.* 88:1–3.

261. Boothby D, Daneo-Moore L, Shockman GD. 1971. A rapid, quantitative, and selective estimation of radioactively labeled peptidoglycan in gram-positive bacteria. *Anal. Biochem.* 44:645–53.

262. Gotz F, Bannerman T, Schleifer K-H. 2006. The Genera *Staphylococcus* and *Macrococcus.* In *The Prokaryotes*, eds. M Dworkin, S Falkow, E Rosenberg, K-H Schleifer, E Stackebrandt, 4. Springer, New York, pp. 5–75.

263. Firth N, Jensen SO, Kwong SM, Skurray RA, Ramsay JP. 2018. Staphylococcal plasmids, transposable and integrative elements. *Microbiol Spectr.* 6, doi: 10.1128/microbiolspec. GPP3-0030-2018.

264. Novick RP. 1989. Staphylococcal plasmids and their replication. *Annu Rev Microbiol.* 43:537–65.

265. Prax M, Lee CY, Bertram R. 2013. An update on the molecular genetics toolbox for staphylococci. *Microbiology.* 159:421–35.

266. O'Connell D, Pattee PA, Foster TJ. 1993. Sequence and mapping of the *aroA* gene of *Staphylococcus aureus* 8325-4. *J. Gen. Microbiol.* 139:1449–60.

267. Lee CY, Buranen SL, Ye ZH. 1991. Construction of single-copy integration vectors for *Staphylococcus aureus. Gene.* 103:101–05.

268. Lee CY, Iandolo JJ. 1986. Integration of staphylococcal phage L54a occurs by site-specific recombination: structural analysis of the attachment sites. *Proc Natl Acad Sci USA.* 83:5474–78.

269. Lee CY, Iandolo JJ. 1986. Lysogenic conversion of staphylococcal lipase is caused by insertion of the bacteriophage L54a genome into the lipase structural gene. *J Bacteriol.* 166:385–91.

270. Luong TT, Lee CY. 2007. Improved single-copy integration vectors for *Staphylococcus aureus. J Microbiol Methods.* 70:186–90.

271. Benton BM, Zhang JP, Bond S, Pope C, Christian T, et al. 2004. Large-scale identification of genes required for full virulence of *Staphylococcus aureus. J Bacteriol.* 186:8478–89.

272. Bottomley AL, Kabli AF, Hurd AF, Turner RD, Garcia-Lara J, Foster SJ. 2014. *Staphylococcus aureus* DivIB is a peptidoglycan-binding protein that is required for a morphological checkpoint in cell division. *Mol Microbiol.* 94(5), doi: 10.1111/mmi.12813

273. Mainiero M, Goerke C, Geiger T, Gonser C, Herbert S, Wolz C. 2010. Differential target gene activation by the *Staphylococcus aureus* two-component system saeRS. *J Bacteriol.* 192:613–23.

274. Foster TJ. 1998. Molecular genetic analysis of staphylococcal virulence. *Methods Microbiol.* 27:432–54.

275. Arnaud M, Chastanet A, Debarbouille M. 2004. New vector for efficient allelic replacement in naturally nontransformable, low-GC-content, gram-positive bacteria. *Appl Environ Microbiol.* 70:6887–91.

276. Bae T, Schneewind O. 2006. Allelic replacement in *Staphylococcus aureus* with inducible counter-selection. *Plasmid.* 55:58–63.

277. Ji Y, Zhang B, Van Horn SF, Warren P, et al. 2001. Identification of critical staphylococcal genes using conditional phenotypes generated by antisense RNA. *Science.* 293:2266–69.

278. Monk IR, Shah IM, Xu M, Tan MW, Foster TJ. 2012. Transforming the untransformable: application of direct transformation to manipulate genetically *Staphylococcus aureus* and *Staphylococcus epidermidis. mBio.* 3:e00277-11.

279. Geiger T, Francois P, Liebeke M, Fraunholz M, Goerke C, et al. 2012. The stringent response of *Staphylococcus aureus* and its impact on survival after phagocytosis through the induction of intracellular PSMs expression. *PLOS Pathog.* 8:e1003016.

280. Hartley RW, Paddon CJ. 1986. Use of plasmid pTV1 in transposon mutagenesis and gene cloning in *Bacillus amyloliquefaciens. Plasmid.* 16:45–51.

281. Bae T, Banger AK, Wallace A, Glass EM, Aslund F, et al. 2004. *Staphylococcus aureus* virulence genes identified by

bursa aurealis mutagenesis and nematode killing. *Proc Nat Acad Sci USA.* 101:12312–17.

282. Mei JM, Nourbakhsh F, Ford CW, Holden DW. 1997. Identification of *Staphylococcus aureus* virulence genes in a murine model of bacteraemia using signature-tagged mutagenesis. *Mol Microbiol.* 26:399–407.

283. Schwan WR, Coulter SN, Ng EY, Lnghorne MH, Ritchie HD, et al. 1998. Identification and characterization of the PutP proline permease that contributes to *in vivo* survival of *Staphylococcus aureus* in animal models. *Infect. Immun.* 66:567–72.

284. Li M, Rigby K, Lai Y, Nair V, Peschel A, et al. 2009. *Staphylococcus aureus* mutant screen reveals interaction of the human antimicrobial peptide dermcidin with membrane phospholipids. *Antimicrob Agents Chemother.* 53:4200–10.

285. Fey PD, Endres JL, Yajjala VK, Widhelm TJ, Boissy RJ, et al. 2013. A genetic resource for rapid and comprehensive phenotype screening of nonessential *Staphylococcus aureus* genes. *MBio.* 4:e00537–12.

286. Bae T, Glass EM, Schneewind O, Missiakas D. 2008. Generating a collection of insertion mutations in the *Staphylococcus aureus* genome using bursa aurealis. *Methods Mol Biol.* 416:103–16.

287. Valentino MD, Foulston L, Sadaka A, Kos VN, Villet RA, et al. 2014. Genes contributing to *Staphylococcus aureus* fitness in abscess- and infection-related ecologies. *mBio.* 5:e01729–14.

288. Rajagopal M, Martin MJ, Santiago M, Lee W, Kos VN, et al. 2016. Multidrug intrinsic resistance factors in *Staphylococcus aureus* identified by profiling fitness within high-diversity transposon libraries. *mBio.* 7, doi: 10.1128/mBio.00950-16.

289. Santiago M, Matano LM, Moussa SH, Gilmore MS, Walker S, Meredith TC. 2015. A new platform for ultra-high density *Staphylococcus aureus* transposon libraries. *BMC Genom.* 16:252.

290. Grosser MR, Paluscio E, Thurlow LR, Dillon MM, Cooper VS, et al. 2018. Genetic requirements for *Staphylococcus aureus* nitric oxide resistance and virulence. *PLoS Pathog.* 14:e1006907.

291. Wilde AD, Snyder DJ, Putnam NE, Valentino MD, Hammer ND, et al. 2015. Bacterial hypoxic responses revealed as critical determinants of the host-pathogen outcome by TnSeq analysis of *Staphylococcus aureus* invasive infection. *PLOS Pathog.* 11:e1005341.

292. Chen W, Zhang Y, Yeo WS, Bae T, Ji Q. 2017. Rapid and efficient genome editing in *Staphylococcus aureus* by using an engineered CRISPR/Cas9 system. *J Am Chem Soc.* 139:3790–95.

293. Gu T, Zhao S, Pi Y, Chen W, Chen C, et al. 2018. Highly efficient base editing in *Staphylococcus aureus* using an engineered CRISPR RNA-guided cytidine deaminase. *Chem Sci.* 9:3248–53.

294. Liu Q, Jiang Y, Shao L, Yang P, Sun B, et al. 2017. CRISPR/Cas9-based efficient genome editing in *Staphylococcus aureus*. *Acta Biochim Biophys Sin.* 49:764–70.

295. Deghorain M, Van Melderen L. 2012. The staphylococci phages family: an overview. *Viruses.* 4:3316–35.

296. Dowell CE, Rosenblum ED. 1962. Serology and transduction in staphylococcal phage. *J Bacteriol.* 84:1071–75.

297. Maslanova I, Doskar J, Varga M, Kuntova L, Muzik J, et al. 2013. Bacteriophages of *Staphylococcus aureus* efficiently package various bacterial genes and mobile genetic elements including SCCmec with different frequencies. *Environ Microbiol Rep.* 5:66–73.

298. Novick RP. 1967. Properties of a cryptic high-frequency transducing phage in *Staphylococcus aureus*. *Virology.* 33:155–66.

299. Krausz KL, Bose JL. 2016. Bacteriophage transduction in *Staphylococcus aureus*: broth-based method. *Methods Mol Biol.* 1373:63–68.

300. Maslanova I, Stribna S, Doskar J, Pantucek R. 2016. Efficient plasmid transduction to *Staphylococcus aureus* strains insensitive to the lytic action of transducing phage. *FEMS Microbiol Lett.* 363(19).

301. Xia G, Corrigan RM, Winstel V, Goerke C, Grundling A, Peschel A. 2011. Wall teichoic acid-dependent adsorption of staphylococcal siphovirus and myovirus. *J Bacteriol.* 193:4006–09.

302. Corvaglia AR, Francois P, Hernandez D, Perron K, Linder P, Schrenzel J. 2010. A type III-like restriction endonuclease functions as a major barrier to horizontal gene transfer in clinical *Staphylococcus aureus* strains. *Proc Natl Acad Sci USA.* 107:11954–58.

303. Waldron DE, Lindsay JA. 2006. Sau1: a novel lineage-specific type I restriction-modification system that blocks horizontal gene transfer into *Staphylococcus aureus* and between *S. aureus* isolates of different lineages. *J Bacteriol.* 188:5578–85.

304. Xu SY, Corvaglia AR, Chan SH, Zheng Y, Linder P. 2011. A type IV modification-dependent restriction enzyme SauUSI from *Staphylococcus aureus* subsp. aureus USA300. *Nucleic Acids Res.* 39:5597–610.

305. Stobberingh EE, Schiphof R, Sussenbach JS. 1977. Occurrence of a class II restriction endonuclease in *Staphylococcus aureus*. *J Bacteriol.* 131:645–49.

306. Szilak L, Venetianer P, Kiss A. 1990. Cloning and nucleotide sequence of the genes coding for the Sau96I restriction and modification enzymes. *Nucleic Acids Res.* 18:4659–64.

307. Kreiswirth BN, Lofdahl S, Betley MJ, O'Reilly M, Schlievert PM, et al. 1983. The toxic shock syndrome exotoxin structural gene is not detectably transmitted by a prophage. *Nature.* 305:709–12.

308. Peng HL, Novick RP, Kreiswirth B, Kornblum J, Schlievert P. 1988. Cloning, characterization, and sequencing of an accessory gene regulator (*agr*) in *Staphylococcus aureus*. *J Bacteriol.* 170:4365–72.

309. Jones MJ, Donegan NP, Mikheyeva IV, Cheung AL. 2015. Improving transformation of *Staphylococcus aureus* belonging to the CC1, CC5 and CC8 clonal complexes. *PLOS One.* 10:e0119487.

310. Barrangou R, Marraffini LA. 2014. CRISPR-Cas systems: prokaryotes upgrade to adaptive immunity. *Mol Cell.* 54:234–44.

311. Barrangou R, Fremaux C, Deveau H, Richards M, Boyaval P, et al. 2007. CRISPR provides acquired resistance against viruses in prokaryotes. *Science.* 315:1709–12.

312. Goldberg GW, Marraffini LA. 2015. Resistance and tolerance to foreign elements by prokaryotic immune systems—curating the genome. *Nat Rev Immunol.* 15:717–24.

313. Golding GR, Bryden L, Levett PN, McDonald RR, Wong A, et al. 2012. Whole-genome sequence of livestock-associated st398 methicillin-resistant *Staphylococcus aureus* isolated from humans in Canada. *J Bacteriol.* 194:6627–28.

314. Kinnevey PM, Shore AC, Brennan GI, Sullivan DJ, Ehricht R, et al. 2013. Emergence of sequence type 779

methicillin-resistant *Staphylococcus aureus* harboring a novel pseudo staphylococcal cassette chromosome mec (SCCmec)-SCC-SCCCRISPR composite element in Irish hospitals. *Antimicrob Agents Chemother.* 57:524–31.

315. Larsen J, Andersen PS, Winstel V, Peschel A. 2017. *Staphylococcus aureus* CC395 harbours a novel composite staphylococcal cassette chromosome mec element. *J Antimicrob Chemother.* 72:1002–05.

316. Marraffini LA, Sontheimer EJ. 2008. CRISPR interference limits horizontal gene transfer in staphylococci by targeting DNA. *Science.* 322:1843–45.

317. Costa SK, Donegan NP, Corvaglia AR, Francois P, Cheung AL. 2017. Bypassing the restriction system to improve transformation of *Staphylococcus epidermidis. J Bacteriol.* 199.

318. Winstel V, Kuhner P, Krismer B, Peschel A, Rohde H. 2015. Transfer of plasmid DNA to clinical coagulase-negative staphylococcal pathogens by using a unique bacteriophage. *Appl Environ Microbiol.* 81:2481–88.

319. Winstel V, Kuhner P, Rohde H, Peschel A. 2016. Genetic engineering of untransformable coagulase-negative staphylococcal pathogens. *Nature Protocols.* 11:949–59.

320. Menichetti F. 2005. Current and emerging serious gram-positive infections. *Clin Microbiol Infect.* 11 Suppl 3:22–28.

321. Munita JM, Bayer AS, Arias CA. 2015. Evolving resistance among gram-positive pathogens. *Clin Infect Dis.* 61 Suppl 2:S48–57.

322. Woodford N, Livermore DM. 2009. Infections caused by gram-positive bacteria: a review of the global challenge. *J Infect.* 59 Suppl 1:S4–16.

323. Chambers HF, Deleo FR. 2009. Waves of resistance: *Staphylococcus aureus* in the antibiotic era. *Nat Rev Microbiol.* 7:629–41.

324. Missiakas D, Schneewind O. 2016. *Staphylococcus aureus* vaccines: deviating from the carol. *J Exp Med.* 213:1645–53.

325. Proctor RA. 2012. Challenges for a universal *Staphylococcus aureus* vaccine. *Clin Infect Dis.* 54:1179–86.

326. McCarthy AJ, Lindsay JA. 2010. Genetic variation in *Staphylococcus aureus* surface and immune evasion genes is lineage associated: implications for vaccine design and host-pathogen interactions. *BMC Microbiol.* 10:173.

327. Rao F, See RY, Zhang D, Toh DC, Ji Q, Liang ZX. 2010. YybT is a signaling protein that contains a cyclic dinucleotide phosphodiesterase domain and a GGDEF domain with ATPase activity. *J Biol Chem.* 285:473–82.

328. Duckworth M, Archibald AR, Baddiley J. 1975. Lipoteichoic acid and lipoteichoic acid carrier in *Staphylococcus aureus* H. *FEBS Lett.* 53:176–79.

40

Streptococcus

Vincent A. Fischetti and Patricia Ryan

CONTENTS

Introduction

Streptococci will appear under the microscope as round bacteria arranged in pairs or in chains. By the Gram-staining technique, they will be Gram-positive. The chaining characteristic is best observed when organisms are grown in liquid media or isolated from infected body fluids such as blood. Streptococci are differentiated as alpha, beta, and gamma types based on their activity on the surface of blood agar, which could differ somewhat based on the species and age of the red blood cells. Alpha-hemolytic streptococcal colonies are surrounded by a narrow zone of hemolysis that shows green discoloration based on the hemolysin's action on the hemoglobin; beta-hemolytic streptococci show a well-defined clear zone of hemolysis around the colony; gamma-hemolytic streptococci have no effect on the red blood cells. *Streptococcus pyogenes* are nearly always beta-hemolytic, whereas closely related Groups B and C streptococci usually appear as beta-hemolytic colonies, but different strains can vary in their hemolytic activity. Nearly all strains of *Streptococcus pneumoniae* are alpha-hemolytic but have been shown to exhibit beta-hemolysis during anaerobic incubation.

Most oral streptococci and enterococci are nonhemolytic, and thus considered gamma types. The property of hemolysis is used in rapid screens for identification of *S. pyogenes* and *S. pneumoniae* but may be unreliable for general differentiation of other streptococci. In general, strains that produce a hemolysin are professional pathogens, responsible for direct human and animal diseases, whereas nonhemolytic streptococci are mostly opportunistic pathogens.

Hemolytic Streptococci

Hemolysins are considered virulence determinants, and as such, hemolytic streptococci are found to be human and animal pathogens. Most clinical isolates appear as beta-hemolytic streptococci when grown on blood agar. *S. pyogenes*, groups B and C streptococci and *Streptococcus pneumoniae* (described below in more detail), are the most prevalent streptococcal pathogens isolated from humans. The human oropharynx is the sole known natural reservoir for *S. pyogenes* in the environment. These organisms are maintained in the throats of humans (usually children) in a

DOI: 10.1201/9781003099277-42

"carrier state," with up to a 30% frequency [1]. Group B streptococci are predominantly found colonizing the human vagina, but it is also associated with animal infections. Although resistance to penicillin-based antibiotics has yet to emerge in *S. pyogenes*, infections caused by this organism continue to occur with occasional widespread outbreaks.

Classification

The classification of hemolytic streptococci was considerably simplified when Lancefield showed that surface antigens (carbohydrates) extracted with heat from the cell wall of streptococci could react with carbohydrate-specific antisera prepared in rabbits. Lancefield's group A, for example, are the *S. pyogenes* strains responsible for human diseases such as scarlet fever, streptococcal pharyngitis, erysipelas, puerperal sepsis, and wound infections. N-acetylglucosamine is the group A-specific carbohydrate, whereas N-acetylgalactosamine is the group C streptococcal determinant. Extended studies of the cell wall carbohydrates of streptococci from several sources have shown at least 13 different serologic groups encompassing human and animal pathogens and commensals.

The group A streptococci may be subdivided into more than 150 different types based on a highly variable region of a fibrillar surface protein, called M protein. While M protein was used as a serologic determinant for early classification of group A strains (M serotypes), recently, the sequence of the variable region of the M protein gene (*emm*) has been used instead (emm-type) with identical results. M and *emm* typing are generally used for epidemiologic purposes in studying the spread of streptococcal strains and the diseases they cause.

Capsular serotyping has been used to differentiate different strains of group B streptococci (*S. agalactiae*). To date, 10 capsular serotypes have been described (Ia, Ib, II-IX). Worldwide, serotype III strains are of particular importance, since they are responsible for the majority of infections, including neonatal meningitis.

S. pyogenes Extracellular Products

S. pyogenes produce a wide array of extracellular products, many of which are considered virulence factors (reviewed in refs. [2, 3]).

Streptococcal pyrogenic exotoxins (Spes) are produced by most *S. pyogenes* strains. The most potent of which, SpeA, is responsible for the rash in cases of scarlet fever and its gene is carried by a prophage [4]. Currently, up to 13 different Spes have been reported [5]. In all cases, Spes are superantigens, or mitogenic proteins with the capacity to crosslink the V_b domain of the T-cell receptor and the major histocompatibility complex of class II molecules on the surface of an antigen-presenting cell. This T-cell stimulation results in high systemic levels of proinflammatory cytokines and T-cell mediators, causing hypotension, fever, and shock. As a result, Spes have been suggested to be the prime mediators in streptococcal toxic shock syndrome. Interestingly, except for SpeB (a cysteine protease), Spes are not part of the bacterial genome but are carried on prophages widely found in streptococci.

Streptolysin O (SLO) is an oxygen-labile protein with the capacity to lyse red blood cells under anaerobic conditions. It is highly antigenic such that the antibodies produced during infection could be quantified in relation to their capacity to neutralize the hemolytic activity of SLO. It has been shown that the titer of antibodies to streptolysin O (ASO) is related to a recent *S. pyogenes* infection. Streptolysin S (SLS) on the other hand is stable in air and is the molecule responsible for the hemolytic zone around streptococcal colonies on blood plates. SLS is a nonantigenic 2.8 kD peptide that is tightly associated with the bacterial cell surface with potent toxic effects on experimental animals in vivo and on leukocytes in vitro [6].

Streptokinase is a secreted molecule with no intrinsic enzymatic activity, however when bound to plasminogen it initiates its conversion to plasmin, the active protease with fibrinolytic activity. Because of this activity, streptokinase has had some success in human medicine as an agent to lyse fibrin clots in coronary arterial thrombosis, acute pulmonary embolism, and deep venous thrombosis. However, because of its antigenicity, it has appreciated limited use.

Hyaluronidase is an enzyme that acts on hyaluronic acid, the ground substance in connective tissue. Because of this activity, hyaluronidase is considered the "spreading factor" for these streptococci. Surprisingly, the structurally identical hyaluronic acid is also the composition of the capsule found on many strains of *S. pyogenes*, a capsule that enables the organism to resist phagocytic attack by human leukocytes [3]. While the capsule is found on the organism during log-phase growth, it quickly falls off during stationary phase.

S. pyogenes also produce four enzymes that can cleave nucleic acids, specifically DNA. These DNases, or sometimes referred to as streptodornase A through D, all possess deoxyribonuclease activity while streptodornase B and D possess ribonuclease activity as well. It has been suggested that by digesting the DNA released from dead mammalian cells at an infected site, the enzyme reduces the local viscosity allowing the organism to spread more easily through the tissues and as such are considered to be virulence factors. As such, DNases have also been referred to as "spreading factors."

Surface Proteins on *S. pyogenes*

In addition to secreted molecules, *S. pyogenes* express a variety of molecules that are displayed on the cell surface, which are critical for colonization and infectivity (reviewed in refs. [2, 7, 8]). Among these are M protein, IgA binding protein, IgG binding protein, C5a peptidase, serum opacity factor, fibronectin binding protein, plasmin binding protein, and five glycolytic enzymes (glyceraldehyde-3-phosphate dehydrogenase, alpha-enolase, phosphoglycerate mutase, phosphoglycerate kinase, trios phosphate isomerase) to name a few. All these proteins have been implicated in some way with streptococcal pathogenesis; however, here we will discuss the major virulence determinant, the M protein, in detail.

M Protein

The streptococcal M protein is probably one of the best-defined molecules among the known bacterial virulence determinants. It was discovered more than 90 years ago by Rebecca Lancefield [9], and it is clear that protective immunity to

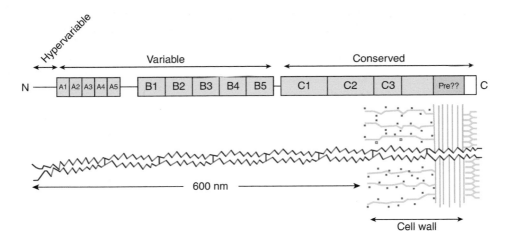

FIGURE 40.1 Sequence repeat and other regions of the M protein molecule (top) and the corresponding coiled-coil M protein situated in the bacterial cell wall (bottom).

group A streptococcal infection is achieved through antibodies directed to the M protein [10]. The M molecule is a fibrillar coiled-coil molecule composed of sequence repeat regions extending from the cell wall surface. The A- and B-repeats located within the N-terminal half of the molecule are antigenically variable among the >150 known streptococcal types with the N-terminal nonrepetitive region and A-repeats exhibiting hypervariability (Figure 40.1). The C-terminal C-repeats, the majority of which are still surface exposed, contain epitopes that are highly conserved among the identified M proteins [11]. Because of its antigenically variable N-terminal region, the M protein provides the basis for the Lancefield serological typing scheme for group A streptococci [10].

The M protein is considered the major virulence determinant because of its ability to prevent phagocytosis when present on the streptococcal surface and thus, by this definition all clinical isolates express M protein. This function may in part be attributed to the specific binding of complement factor H to both the conserved C-repeat domain [12] and the fibrinogen bound to the B-repeats [13], thus preventing the deposition of C3b on the streptococcal surface. It is proposed that when the streptococcus contacts serum, serum factor H binds to the M molecule and inhibits or reverses the formation of C3b, Bb complexes, helping to convert C3b to its inactive form (iC3b) on the bacterial surface, thus preventing C3b-dependent phagocytosis. This is supported by studies showing that antibodies directed to the B- and C-repeat regions of the M protein are unable to promote phagocytosis [14] while antibodies to the N-terminal type-specific region opsonize the organism. This may be the result of the ability of M-bound factor H to also control the binding of C3b to the Fc receptors on these antibodies, resulting in inefficient phagocytosis [15]. Antibodies directed to the hypervariable N-terminal region are opsonic, perhaps because they cannot be controlled by the factor H bound to the distantly located B- and C-repeat regions. Thus, it appears that the streptococcus has devised a method to protect its conserved region from being used against itself by binding factor H to regulate the potentially opsonic antibodies that bind to these regions. Both the N-terminal hypervariable region and the conserved region are targets for vaccine development (discussed later) [16].

Streptococcal Diseases

S. pyogenes are responsible for several diseases including, pharyngitis, otitis media, impetigo, meningitis, necrotizing fasciitis, scarlet fever, erysipelas, rheumatic fever, and acute glomerulonephritis, to name a few [2, 17]. The latter two, rheumatic fever, and acute glomerulonephritis, are sequelae of an *S. pyogenes* infection. Although a great deal of research has focused on these diseases, little is known regarding their etiology except that an *S. pyogenes* infection preceded the disease, suggesting that an immune component may be responsible for the observed symptoms.

The latent period for rheumatic fever is about three weeks after a streptococcal pharyngitis [17]. Curiously, none of the other *S. pyogenes* infections culminates in this disease, only acute pharyngitis ultimately results in damage to the mitral heart valve. It has been proven that aggressive treatment of the streptococcal pharyngitis with penicillin within 10 days of its onset will significantly reduce and even prevent cardiac damage. However, if untreated, recurrent rheumatic attacks following streptococcal pharyngitis may occur resulting in further cardiac damage ultimately necessitating valvular replacement surgery. It should be emphasized that *S. pyogenes* are never isolated from the damaged heart, strongly suggesting that based on the timing of the disease, an immunological component is responsible for the observed damage.

Acute glomerulonephritis may follow a *S. pyogenes* infection by about one week, resulting in malfunction of the kidney glomeruli. However, in this disease the patients usually undergo spontaneous recovery. Treatment is usually managing the reduced kidney function until the condition resolves itself. Whether this is a toxic or an immunological disease has not yet been resolved.

In recent years, streptococcal toxic shock syndrome (STSS) has been a concern worldwide. It is a severe illness associated with invasive or noninvasive *S. pyogenes* infection. The disease may result from an infection at any site but most often in association with infection of a cutaneous lesion. In some cases, a traumatic injury without any external breaks has been the focus, suggesting that organisms circulating in the blood of carriers could bind to the site of the trauma to initiate the localized infection. Toxicity and a rapid progression of the clinical course (24–48 hours) are characteristic, having a case-fatality rate of >50%. In most instances, patients present with shock-like symptoms and a necrotic lesion or painful abscess. However, in cases where no injury is obvious, these are more difficult to diagnose and rapidly become serious. Treatment is usually management of the shock, debridement of the lesion, and intravenous antibiotics. Fortunately, *S. pyogenes* are not resistant to penicillin-based antibiotics, making this the drug of choice. In more severe cases, amputations may be necessary due to the manifestations of septic shock.

More than 100 years ago, scarlet fever was considered a deadly disease, and a severe complication of a streptococcal infection, usually pharyngitis. Today, perhaps due to changes in the bacteria and treatment options, scarlet fever is not as serious, but still results in a body rash during a pharyngitis. Erysipelas (cellulitis with fever and toxicity) is also less common today again perhaps due to changes in the bacteria, the host, and the availability of antibiotics. Neonatal meningitis is a disease that has been increasing in the past few decades. This disease, caused by group B streptococci (GBS), results from the contamination of newborns of mothers vaginally colonized with GBS during natural childbirth. The organism and its pathogenic program are discussed later in this chapter.

Vaccine Development

To date, no vaccine is available to prevent infections by *S. pyogenes*, making pharyngitis one of the only childhood diseases for which no vaccine exists. This could be due to several factors. The first being that the M protein is the major virulence determinant for which type-specific antibodies were shown to be neutralizing. However, since there are >150 different serotypes identified, a polyvalent vaccine directed to the most prevalent types would need to be developed. This was accomplished after years of testing [18, 19]. In a second approach, it was discovered that antibodies directed to the conserved C-repeat region of the M protein while not opsonic, could protect mice from colonization by multiple *S. pyogenes* serotypes when delivered to the oral mucosa [20, 21]. Unfortunately, both apparently effective vaccine approaches have yet to be licensed for human testing and use. The major reason for this could be that the streptococcal antigen/s responsible for the observed heart-reactive antibodies seen in rheumatic fever patients has yet to be identified, so companies are reluctant to administer a streptococcal product as a vaccine for fear of inducing the disease it is designed to prevent.

Laboratory Diagnosis

The hemolytic reaction and colonial morphology exhibited on blood agar has been used as the standard to classify streptococci. *S. pyogenes* is associated with beta-hemolysis, or complete lysis of blood cells around the colony. *S. pyogenes* are nearly always beta-hemolytic as opposed to group *B streptococci*, which can exhibit alpha-, beta- or gamma-hemolysis. Serological

identification is the most accurate method of streptococcal identification; however, these reagents are for research purposes and are unavailable for routine use. Because *S. pyogenes* are sensitive to bacitracin, and most other beta-hemolytic streptococci are not, using a bacitracin disk on a blood agar plate can be used for identification.

Rapid Tests

There are several rapid test kits for *S. pyogenes* in use today that rely on the group-specific carbohydrate found in the cell wall of this organism [22]. These test kits, which are routinely employed in doctor's offices and hospitals, use acid to extract the carbohydrate determinant (N-acetylglucosamine) from the streptococci on a swab and antibodies to that released sugar are used for positive identification. Since *S. pyogenes* is the only streptococcus with this determinant, most of these rapid tests are >95% specific and about 90% sensitive for these bacteria.

Other Beta-Hemolytic Streptococci

Groups C and G Streptococci

According to recent taxonomic studies, large colony-forming groups C and G streptococci that infect humans are classified as *Streptococcus dysgalactiae* subsp. *equisimilis* [23], the others infect animals. These human and animal species have been isolated from severe systemic infections in humans that are usually associated with food born contamination (primarily from milk and cheese). More commonly however, groups C and G streptococci may be found colonizing the human nasopharynx and have been associated with outbreaks of acute pharyngitis. Because of the specificity of antigen-specific tests for the group A streptococci, group C and G streptococci colonizing and infecting the nasopharynx are usually missed. Recently, group G streptococci have been identified as an important opportunistic and nosocomial pathogen. In certain tropical climates (Caribbean and northern Australia), group G as well as group C streptococci have been associated with cases of rheumatic fever, a disease that for decades has only been associated with group A streptococcal pharyngitis [24]. Since both groups C and G streptococci have surface M protein, they can be subtyped based on the antigenic and sequence differences in this molecule.

Group B Streptococci (GBS)

GBS are Gram-positive encapsulated organisms found predominantly in pairs or short chains. They not only share the Lancefield group B polysaccharide antigen, but they are discriminated based on their type-specific capsular polysaccharides which separate them into ten antigenically unique types (Ia, Ib, II-IX) [25]. The capsule is a major virulence factor for GBS. It is an antiphagocytic molecule, which helps the organism evade phagocytic attack except when type specific antibodies are present [25]. "Nontypeable" strains have also been identified, which account for ~1% of isolates from neonatal infections. On blood agar, GBS appear beta-hemolytic with a narrow zone of hemolysis, however ~2% of isolates are nonhemolytic. All beta-hemolytic isolates produce a red-orange pigment, granadaene, when cultured under certain conditions [26]. An additional important surface protein is the immunogenic protein Sip, which is shared by all GBS isolates [16]. Pilus-like structures have also been described which are composed of three subunits and used by the organism as an adhesin [27].

As mentioned above, this human pathogen is responsible for neonatal meningitis and is a leading cause of neonatal sepsis, which presents within the first 24–48 hours after birth. The organism, which colonizes the genitourinary or gastrointestinal tracts of 10–30% of pregnant women, is acquired vertically after the amniotic sac ruptures, but can also invade through intact amniotic membranes. The bacterium is aspirated into the lungs of the neonate, resulting in sepsis, pneumonia, and, less commonly, meningitis. The resulting infection is a leading cause of early neonatal death [28].

It is essential, prior to giving birth, that maternal carriers be identified by culture-based screening and treated with intravenous antibiotic prophylaxis during labor and delivery [26]. Past studies have shown that intrapartum antibiotic prophylaxis is ~88% effective in preventing these neonatal infections [29, 30]. A more recent study [31] has shown that intrapartum treatment with β-lactam antibiotics (penicillin or ampicillin) for 4 hours or more before delivery was most effective at preventing neonatal infections. Although screening for potential carriers has become standard practice, for those mothers who have not been identified prior to presenting in labor, risk-based screening is essential, which evaluates a number of other factors that correlate with increased risk of neonatal infections including intrapartum temperatures >100.4°C, amniotic rupture >18 hours, and delivery <37 weeks gestation [32].

Identification

Vaginal and rectal cultures from pregnant women (35–37 weeks gestation) are now the standard practice for screening potential carriers of GBS. Specimens are collected and transported in commercially available media [33] and held at 4°C until processed. Within 24 hours of collection, specimens are inoculated into enrichment media, and this step, rather than direct plating, has been shown to decrease the incidence of false-negative results. Enrichment broths include TransVag broth (ToddHewitt supplemented with 8 μg/mL gentamicin and 15 μg/mL nalidixic acid) and Lim broth (essentially TransVag broth in which gentamicin is replaced by colistin [10 μg/mL]) [34].

Enrichment is followed by traditional subculturing on blood agar or chromogenic agar, followed by the CAMP test, hippurate hydrolysis assays, or latex agglutination with antisera against group B–specific carbohydrates. While DNA probes and PCR-based assays are commercially available, the necessity of performing enrichment culture to decrease false-negative results makes these assays less feasible in a hospital setting to screen intrapartum women who were not screened earlier in their pregnancy. In those instances, postpartum cultures or risk-based screening may prove more beneficial. While the CDC does not recommend any of the rapid nucleic acid-based techniques on nonenriched samples in the intrapartum setting, nucleic acid amplification techniques on enriched samples have been shown to be 92.5–100% sensitive [32].

Other Streptococci

Viridans Streptococci

The viridans streptococci are a heterogeneous group, which includes both alpha-hemolytic and nonhemolytic streptococci. Many species are components of the normal flora of humans and animals; however, the members of this group can cause human disease if given access to normally sterile sites in the body. These opportunistic pathogens are primarily associated with the formation of dental caries (*Streptococcus mutans*), infective endocarditis (*S. sanguis*) and bone, joint, and liver abscesses (*S. anginosus*) [35].

Viridans streptococci include approximately 30 species that belong to one of five general groups: Mutans, Salivarius, Sanguinis, Mitis, and Anginosus [36]. The members of the viridans group are historically considered difficult to differentiate on the species level, as they do not contain C-carbohydrate antigens in their cell walls, they cannot be serologically identified using the traditional methods described for the beta-hemolytic streptococci. It is generally accepted that the viridans group are first differentiated from other streptococci by exclusion: all members are bile insoluble, nonbeta-hemolytic, pyrrolidonyl arylamidase (PYR) negative, cannot grow in 6.5% NaCl, and with the exception of the Salivarius group, are bile-esculin negative.

The differentiation of the viridans streptococci into groups or species requires a combination of molecular and biochemical tools. Several species-specific biochemical characteristics may prove useful (https://www.cdc.gov/streplab/other-strep/general-methods-section2.html#table2). For example, the Voges-Proskauer reaction assists in the differentiation of the viridans streptococcal species and is a key reaction for the identification of the Anginosus group. Members of the *S. mutans* species/group produce acid in nearly all carbohydrate broths, and the strains belonging to the *S. salivarius* group are the only viridans to hydrolyze urea. The *S. mitis* group members are the only viridans that are negative for the following biochemical reactions: mannitol or sorbitol fermentation, arginine or urea hydrolysis, and Voges-Proskauer and bile esculin tests. The species within each of the major groupings can be further differentiated by additional biochemical characteristics [36, 37]

The clinical significance of viridans group members is generally species-dependent and as such, accurate species differentiation is the first step in properly diagnosing the etiologic agent of an infection to prescribe the proper treatment course. Commercially available kits for biochemical identification of streptococci are available, such as the API 20 Strep kit and Rapid ID 32 Strep kit (both from bioMérieux, Hazelwood, MO). These tests, however, are generally more reliable for the nonviridans streptococci, and do not always provide accurate results for the identification of individual viridans group species [38, 39]. In addition to biochemical analyses, several molecular assays have been developed for viridans identification, including PCR-based techniques, ribotyping, and sequence comparisons of various target genes including 16s rRNA [40], manganese-dependent superoxide dismutase, and *groESL* genes [41].

Various automated systems are now used to identify clinical isolates of the viridans group, including VITEK 2 (bioMérieux) and Phoenix systems (Becton Dickinson Diagnostic Systems,

Sparks, MD). The VITEK system, however, has recently been reported [40, 42] to often be unreliable for the viridans group (particularly *S. mitis* and *S. sanguinis*), with only about 40% of clinical isolates correctly identified at the species level. The performance of the Phoenix system for identification of *Streptococci* and *Enterococci* was recently evaluated for reproducibility and reliability in comparison to the API system [43]. Phoenix correctly identified 67% of clinical isolates of viridans streptococci at the species level, with most accurate results obtained for *S. anginosus* (100%) and *S. sanguinis* groups (75%), and the least correct results for the *S. mitis* group (53%). The discrepancies resulting from all automated systems, however, can generally be resolved with further testing using molecular techniques or other biochemical tests.

Streptococcus pneumoniae

S. pneumoniae are Gram-positive cocci often arranged in pairs (diplococci) or in short chains that commonly asymptomatically colonize the human nasopharynx. Like other streptococci, pneumococci are nonmotile, catalase-negative, facultative anaerobes that ferment glucose to lactic acid. *S. pneumoniae* produce a large polysaccharide capsule, which is considered to be one of the major virulence factors (reviewed in [44]), as it confers antiphagocytic properties on the organism, and was recently shown to limit mucus-mediated clearance of the organism from human mucosal surfaces [45]. The composition of the polysaccharide capsule forms the basis for pneumococcal serotyping by the Neufeld Quellung reaction, which involves agglutination of the organism with type-specific capsular antibodies [46]. To date, at least 90 serotypes have been identified [44].

Pneumococci (and other streptococci in the mitis group [47]) contain phosphorylcholine that is covalently linked to teichoic acid in the cell wall and to the lipoteichoic acid component of the cell membrane [48]; thus, choline is an absolute growth requirement of the organism. A group of surface-exposed choline-binding proteins (as many as 15 in some strains [49]) interact noncovalently with the choline residues in the cell wall, and exhibits a number of different functions essential to the overall biology of the organism, including host tissue adherence and autolysis sink [50].

Pneumococcal Diseases

S. pneumoniae cause a wide variety of illness including lobar pneumonia, meningitis, otitis media, bacteremia, and sinusitis. Of great importance to human health, it is worth noting that in 2016, *S. pneumoniae* was the leading cause of global mortality from pneumonia and was responsible for more deaths than all other causes combined [51]. Most deaths occurred in African and Asian countries, where vaccination programs (discussed later) are less robust or altogether absent.

The elderly and children under the age of 2 years carry the major disease burden and the World Health Organization estimates that nearly 500,000 children (most of whom are from underdeveloped nations) die each year of pneumococcal infections (https://www.cdc.gov/pneumococcal/global.html). Recently, the Centers for Disease Control (www.cdc.gov) estimated that *S. pneumoniae*

caused over 175,000 cases of pneumonia requiring hospitalization, 6–7 million cases of otitis media, 50,000 cases of bacteremia (13,000 of which are children) and 3,000–6,000 cases of bacterial meningitis (700 of whom were children) (http://www.cdc.gov/vaccines/pubs/pinkbook/pneumo.html).

Pneumococci produce several virulence determinants that contribute to the overall pathogenicity of the organism, and the antiphagocytic polysaccharide capsule has long been considered one of the most important. There are however, numerous cell-surface associated and secreted proteins involved in various disease processes including (but not limited to) hyaluronate lyase, pneumolysin, choline binding proteins, two neuraminidases, autolysin LytA, pneumococcal surface antigen PsaA, and pneumococcal surface proteins PspA and PspC (reviewed in [52, 53]). Also contributing to virulence is the ability of pneumococci to undergo spontaneous phase variation between an opaque and a transparent colony phenotype. Although the mechanisms of this process are still not fully understood, opaque variants have more capsular polysaccharide than the transparent phenotype and are more commonly associated with invasive infections. The transparent phenotype is generally better able to colonize the nasopharynx than the opaque variants [54, 55].

The pathogenic program of pneumococci is a complicated, multifaceted process that depends on the interaction between host and bacterium, as well as the genetic and physiological characteristics of both. Recent findings [56, 57] indicate that the capsular types, as well as the overall genetic background of the pneumococci located at specific infection sites, contribute to the degree to which isolates are virulent. Furthermore, a relationship has been shown to exist between capsular type and risk for, and outcome of, invasive pneumococcal disease.

Identification

S. pneumoniae are fragile organisms, lysing within 18–24 hours after the start of growth due to the production of autolysin, a cell-wall degrading lytic enzyme [58, 59]. Pneumococci are fastidious, growing best in 5% CO_2 on agar containing blood as a source of catalase. On blood agar, colonies are round and about 1 mm in diameter. Colonies initially appear raised, but over time, become flattened and shiny (or mucoid) in appearance, with depressed centers (resulting from the action of autolysin) [60]. In 5% CO_2, colonies are surrounded by large zones of alpha-hemolysis (resulting from the production of pneumolysin); a characteristic that helps to distinguish pneumococci from the beta-hemolytic group A streptococci.

Further tests are necessary to distinguish *S. pneumoniae* from the alpha-hemolytic viridans streptococci, and classically include differentiation based on colony morphology, bile (deoxycholate) solubility [61] and optochin (cuprein hydrochloride) sensitivity [62]. The latter two tests are still considered generally reliable for initial identification, with reported sensitivities of >98% and 90–100%, respectively [60]. The bile solubility test is based on the observation that broth cultures of *S. pneumoniae* lyse (solubilize) when treated with a 2% solution of the bile acid salt, sodium deoxycholate. This assay remains one of the most sensitive and specific tests for the identification of pneumococci [60]. Optochin (ethylhydrocupreine hydrochloride) sensitivity is determined using disks impregnated with the compound that are placed on a freshly streaked agar plate of bacteria. After incubation, a zone of inhibition (>14 mm) surrounding the disk indicates that the organism is a pneumococcus [36]. Growth of other alpha-hemolytic streptococci in the area surrounding the disk will not be inhibited. For cases in which discrepancies in either test occur, other identification methods should be included, as there have been reports of optochin-susceptible viridans group streptococci as well as bile-insoluble *S. pneumoniae* [63]. The traditional Quellung (capsular swelling) reaction for identification is cumbersome and labor-intensive and has been generally restricted to specialized laboratories.

Newer diagnostic tests, including identification and typing methods are now available for *S. pneumoniae* (reviewed in [52]) that facilitate rapid identification of pneumococci. An overview of several such tests is provided here.

Detection of the polysaccharide capsule by latex agglutination forms the basis of a number of the currently available serological tests, including Pneumoslide (Becton-Dickinson Diagnostic Systems, Sparks, MD), Directigen (Becton-Dickinson), and Pneumotest-Latex (Statens Serum Institut, Copenhagen, Denmark); however, many false positives involving alpha- and nonhemolytic streptococci have been reported with both Pneumoslide and Directigen [60]. In contrast, the Pneumotest-Latex method has recently been shown to detect 90 different pneumococcal serotypes with 95% accuracy [64]. Serological tests that detect capsular antigens by coagglutination techniques are also available, including Phadebact (Bactus AB, Huddinge, Sweden), which, in independent tests, correctly identified 98% of pneumococcal isolates [60]. Obviously, the major limiting factor for all these serological tests is that they can only be used on encapsulated isolates.

Several diagnostic methods have been developed that are based on specific virulence factors of *S. pneumoniae*. Pneumolysin-derived peptides can be detected immunologically (enzyme-linked immunosorbent assays [ELISA] and agglutination) and genetically (nested-PCR, real-time PCR and multilocus sequence typing [MLST]). PCR-based assays targeting autolysin are also available. These methods (reviewed in [53]) have facilitated the identification of the organism from cultured specimens and clinical samples such as urine, sputum, and pleural fluid. More recently, Pimenta et al. [65] have developed a real-time PCR-based assay for detection of 21 different capsular polysaccharides.

Importantly for diagnostics, the CDC has recently noted that the serotypes of *S. pneumoniae* involved in invasive diseases have changed greatly since the heptavalent vaccine was introduced in 2001 and provides up-to-date primer sequences for PCR assays and MLST analysis on their website. Recent studies [66] to evaluate such methods have shown that real-time PCR assays to detect the *lytA* gene in cerebrospinal fluid, together with Gram stains of the CSF, were sensitive and specific (>90% for both) in diagnosing pneumococcal meningitis.

For clinical cases in which it is difficult to obtain isolates of *S. pneumoniae* (low recovery after antibiotic treatment, negative subcultures, sampling procedures that are too invasive), an immunochromatographic membrane assay (Binax NOW, Binax, Portland, ME) is available that, with great rapidity, sensitivity, and specificity (80–95%), detects the presence of the C polysaccharide cell wall antigen (common to all serotypes) in urine [67, 68],

blood culture bottles [69], bronchoalveolar lavage fluid [70], and cerebrospinal fluid [71]. The main drawback of this methodology is reduced accuracy in children, as it also detects pneumococcal carriage [72].

Treatment and Vaccines

For most hospital and outpatient settings, penicillin, erythromycin, and tetracycline have been recommended for the treatment of pneumococcal infections; however, in 2017, the CDC estimated that nearly one-third of pneumococcal strains from pneumonia patients are resistant to common antibiotics [73]. Specifically, data from the Active Bacterial Core Surveillance project at the CDC (http://www.cdc.gov/abcs/reports-findings/survreports/spneu17.html) revealed that 2.4% of *S. pneumoniae* strains were resistant to penicillin (a decline from 2012 when approximately 6.8% of isolates were resistant and from 2005 when 34% were resistant) and 29% were resistant to erythromycin (no change from 2012 reports). The 2017 prevalence of resistance to tetracycline was reportedly 11.3% and trimethoprim-sulfamethoxazole resistance was 6.9% (a decrease from the 20% reported in 2012).

There are currently two pneumococcal vaccines, both based on capsular polysaccharides, which are licensed for use in the United States. The 13-valent pneumococcal conjugate vaccine (PCV13, Prevnar 13®) was introduced in the United States in 2010 for the prevention of systemic infections in children and replaced the original heptavalent pneumococcal conjugate vaccine (PCV7, Prevnar®) approved in 2000. The 13-valent vaccine is recommended for children aged 2–59 months and for children aged 60–71 months who are at increased risk for pneumococcal disease [74, 75]. In 2012, it was approved for use in adults aged 50 years and older [76].

The second vaccine, PPSV23, a 23-valent polysaccharide vaccine was initially recommended only for adults aged 65 or older; however, in 2010, the Advisory Committee on Immunization Practices (ACIP) at the CDC published new recommendations for the PPSV23 vaccine to include adult asthmatics, adult cigarette smokers, and adults with asplenia or other immunocompromising conditions [74]. Although this vaccine is not particularly immunogenic in children under 2 years of age, it is recommended for children over 2 who have underlying immunocompromising medical conditions. In 2014, the ACIP further recommended that PCV13 be given in series with PPSV23 for all adults over 65 years old [77].

It is of great significance that after the introduction of the heptavalent vaccine in 2001 the CDC's Bacterial Surveillance Program reported a significant decrease in invasive pneumococcal diseases (IPD) in the United States (https://www.cdc.gov/pneumococcal/surveillance.html). In 1998, there were an estimated 95 cases of IPD per 100,000 children less than 2 years old, while in 2008, the reported numbers of infected children dropped to 21 per 100,000 children. After the replacement of the heptavalent vaccine with PCV13, the reported cases of IPD have dropped even further to seven children per 100,000 in 2017 (https://www.cdc.gov/abcs/reports-findings/survreports/spneu17.html). For adults less than 65 years of age, the reported number of cases of IPD has been reduced by 50% (16 per 100,000 in 1998 to seven per 100,000 in 2016). Adults over 65 years of age demonstrated an even greater reduction in cases, falling from 61 cases to 26 cases per 100,000 (https://www.cdc.gov/abcs/reports-findings/survreports/spneu17.html). Furthermore, Tomczyk et al. recently reported two interesting findings after tracking the occurrences of antibiotic-resistance cases of IPD in the 3 years following the introduction of PCV13 [78]. First, they noted large decreases in the rates of antibiotic-nonsusceptible IPD caused by the serotypes included in PCV13 vaccine (but not in PCV7 vaccine). Second, they found a decrease in cases of multidrug-nonsusceptible IPD across all age groups, particularly among children younger than 5 years old and adults aged 65 years or older [78]. The role that conjugate vaccines have on decreasing antimicrobial resistance, while indirectly allowing resistant non-vaccine-serotype clones to continue to emerge and expand, was recently extensively reviewed [73].

S. pneumoniae is a naturally transformable organism, and as such, during a physiological state known as competence [79–81], the bacterium can readily acquire exogenous DNA, thereby increasing its genetic plasticity. This trait is one of the underlying factors that contribute to capsular switching, increased antibiotic resistance and virulence factor exchange [82]. Furthermore, pneumococci harbor bacteriophage, and recent reports indicate that at least 70%, and as many as 90%, of the pneumococcal genomes from clinical isolates contain prophages or phage remnants. Therefore, through both transformation and transduction, pneumococci can readily exchange and acquire genetic material, including novel capsular genes, antibiotic resistance markers and encoded virulence determinants: factors that may potentially compound our ability to treat and to prevent infections caused by this organism.

Enterococci

Members of the genus *Enterococcus,* once included in genus *Streptococcus* [83, 84], are catalase-negative Gram-positive cocci that appear in short chains, in pairs or as single cocci. Most strains react with Lancefield group D typing serum, grow in 6.5% NaCl and produce pyrrolidonyl arylamidase. Colonies on agar plates are generally larger than streptococci, growing to 2–3 mm in diameter within 2 days, and appear raised and gray/white in color.

Enterococci share a few characteristics with other Gram-positive cocci and as such, a battery of biochemical tests (reviewed in Murray [85]) can be performed to differentiate enterococcal isolates from less commonly encountered Gram-positive cocci. Examples of such tests include growth on bile-esculin media (most enterococci are positive), production of gas from glucose (most enterococci are negative), and the ability to grow at both 45°C and 10°C (enterococci generally grow at both temperatures).

Enterococci are components of the flora of the intestinal tract, oral cavity and vaginal canal of humans and animals, and have historically been considered commensals with low pathogenic potential. The past two decades, however, have seen an increase in enterococcal infections, including those of the urinary tract, burn wounds, surgical incisions, and heart valves. Enterococci colonize catheters and implanted medical devices, which often results in endocarditis and septicemia. At least 12 different species have been associated with various illness; however, two

species, *E. faecalis* and *E. faecium,* have emerged in recent decades as major causes of nosocomial infections (reviewed in Mundy [86], with *E. faecalis* responsible for the vast majority (80–90%) of the infections caused by this genus.

As most human infections are caused by either *E. faecalis* or *E. faecium,* it is often necessary to specifically differentiate these two species. Clinical laboratories use automated and rapid identification systems such as the VITEK II system (bioMérieux), BBL Crystal kits (Becton Dickinson, BD) and more recently, the Phoenix system (Becton Dickinson, BD). Given reported concerns about the reliability of such systems [83], molecular techniques and species-specific biochemical differences (e.g., using API®20 Strep tests, bioMérieux) can facilitate the identification process. In terms of simple biochemical differences, *E. faecalis,* but not *E. faecium,* will grow on medium containing 0.4% telluride, reduce tetrazolium to formazan and most isolates will produce acid from sorbitol, glycerol, and D-tagatose. On the other hand, most strains of *E. faecium,* but not *faecalis* will produce acid from L-arabinose and melibiose [85].

PCR-based techniques for the amplification and downstream sequencing of the 16S rRNA genes have been used for several years to distinguish enterococci from lactococci and to identify specific enterococcal species [84]. Similar techniques involving the sequencing of species-specific regions of other genes have been described (e.g., heat shock protein 60 and manganese-dependent superoxide dismutase, *sodA,* genes) [58]. Recently, a novel multiplex PCR technique, incorporating both 16S rRNA and *sodA* specific primers, has been developed that allows for the simultaneous identification of both genus and species from a variety of sample types (feces, retail foods, animal carcasses). Additionally, peptide nucleic acid fluorescent in situ hybridization (PNA FISH®) tools have been introduced to differentiate *E. faecalis* from other enterococcal species in blood cultures within 3 hours (AdvanDX, Inc.).

E. faecalis and *E. faecium* elaborate virulence factors that contribute to the infection process. Although both species have acquired antibiotic resistance determinants (discussed below), the best-studied virulence factors are predominantly associated only with *E. faecalis.* For example, this species elaborates a cytolytic toxin (hemolysin), gelatinase, aggregation substance, and an enterococcal surface protein, Esp; each of these enhance virulence, but they are not (or very rarely) produced by *E. faecium* [86]. The enterococcal cytolysin is a proinflammatory, acutely lethal, hemolytic toxin, which also functions as a bacteriocin against other Gram-positive bacteria. The aggregation substance is a surface protein that is multifunctional: it not only promotes aggregate formation during bacterial conjugation to facilitate efficient contact between donor and recipient strains for plasmid transfer, it is also associated with binding to, and subsequent intracellular survival within, human neutrophils. The role of this virulence factor as well as the numerous others elaborated by *E. faecalis* are reviewed extensively elsewhere [86].

The organism is spread through direct contact with stool, contaminated urine, and blood, and indirectly via healthcare workers and contaminated hospital surfaces. The clinical relevance of enterococci has increased due to the emergence of vancomycin resistant strains (VREs) and multidrug resistant isolates, and it is now the second most isolated bacteria in nosocomial infections (second to *Staphylococci*). Data from the CDC report entitled *Antibiotic resistance threats in the United States, 2019* (https://www.cdc.gov/drugresistance/pdf/threats-report/2019-ar-threats-report-508.pdf) indicate that there are an estimated 54,500 healthcare associated VRE infections in the United States. In the years prior to 1990, VRE accounted for less than 0.5% of hospital acquired infections (www.cdc.gov). In 2017, the WHO ranked VRE (more specifically, vancomycin-resistant *E. faecium*) as a "high priority" due to the dearth of antibiotics to treat such infections and the economic impact of health care costs associated with such infections (https://www.who.int/medicines/publications/global-priority-list-antibiotic-resistant-bacteria/en/).

The proportion of VRE infections differs by species, with over 77% attributed to *E. faecium* and 9% caused by *E. faecalis*. The numbers from the SENTRY antimicrobial surveillance program agree with those from the CDC, as SENTRY reports that 80.7% of all bloodstream VRE infections are caused by *E. faecium* (a marked increase from the 57% in 2000) and 5.3% are attributed to *E. faecalis*. SENTRY further reported that the vancomycin-resistant *E. faecalis* strains also showed resistance to amoxicillin (close to 95%), levofloxacin (94%), and erythromycin (94%). The CDC has determined that vancomycin-resistant *E. faecalis* strains are a *serious threat* and calls for prompt action (compliance to CDC guidelines for detection, prevention, tracking, and reporting) to ensure the problem does not escalate.

The explanation for the rapid emergence of VRE is markedly different in different parts of the world. In Europe, for example, the emergence of vancomycin resistance has been associated with avoparcin, a vancomycin-related antibiotic, used as a growth promoter in various agricultural processes and added to animal feed until it was banned in 1997 [84]. Colonization of humans with vancomycin resistant strains likely occurred after ingestion of meat products resulting from such practices. In the United States, where avoparcin was never approved for use in animal feed, the major reservoirs for VRE are hospital staff and patients [86]. A root cause of vancomycin resistance appears, at least in part, to be the improper use of antimicrobial agents, and VRE spread seems to be directly linked to clinical settings where the organism is most prevalent.

The recent emergence of multidrug resistance in enterococci presents an even bigger challenge to the health care profession. Multidrug-resistant isolates generally emerge from intrinsic baseline resistance to antibiotics or by acquiring resistance through mobile genetic element transfers (transposons and plasmids) [86, 87]. Studies have found a link between ampicillin and vancomycin resistance, particularly in *E. faecium*, and it has been suggested that β-lactam exposure is a predisposing factor for multidrug resistance. Never has the need to control the spread of VRE in nosocomial settings been more apparent than after a recent report detailing the horizontal transfer of one such vancomycin resistance gene (*vanA*) from *E. faecalis* to methicillin-resistant strains of *Staphylococcus aureus* (MRSA) [87]. Not only are the reports of multidrug-resistant VRE increasing, but the danger of such strains is no longer restricted to other *Enterococci*, as the passage of resistance markers to other genera is now a documented reality.

Lactococci

Lactococci are Gram-positive, catalase-negative cocci that grow in chains and produce lactic acid from the fermentation of lactose. The two major species, *L. lactis* subsp. *lactis* and *L. lactis* subsp. *cremoris* were previously categorized as lactic acid streptococci until they were moved to their own genus in 1985 [88]. They can be distinguished from other similar cocci by specific phenotypic characteristics: they do not produce gas from glucose fermentation, most strains can grow in 6.5% NaCl, grow poorly at 45°C (distinguishing them from enterococci) but well at 10°C, and exhibit positive reactions on bile-esculin media. Further differentiation from the enterococci relies on additional biochemical tests (reviewed by Facklam and Elliott [89]): lactococcal strains do not produce acid in arabinose broth (as do *E. faecalis*), but will produce acid in mannitol broth (unlike *E. durans* and *E. hirae*).

Lactococci are common environmental bacteria that are most notable for their use in the food industry for food preservation and flavor. The biochemical process that produces lactic acid during carbohydrate fermentation has several specific applications: the byproducts of the fermentative pathway can enhance food flavor; the decrease in pH during lactic acid production can precipitate proteins that help to change or improve food texture. The increased acidity (which can be as low as 4.0) inhibits the growth of many microorganisms, thereby increasing the shelf life of such fermented foods. *L. lactis* is used in the production of fermented milk products (such as buttermilk) and is the major starter bacterium in the cheese industry.

Beyond the food industry, *L. lactis* has an important place in the biotechnology industry due to the production of nisin, a 34 amino acid potent bacteriocin with broad-spectrum activity against several bacterial species (reviewed in [90, 91]). The lactococci are also involved in other applications, including the expression of antigens for the development of mucosal vaccines, gene expression systems (NICE), and the production of human proteins such as cytokines for in situ applications [92].

REFERENCES

1. Stromberg, A., A. Schwan and O. Cars, Throat carrier rates of beta-hemolytic streptococci among healthy adults and children. *Scand. J. Infect. Dis.*, 1988. **20**: p. 411–417.

2. Ferretti, J.J., D.L. Stevens and V.A. Fischetti (eds.), 2016. *Streptococcus pyogenes: Basic biology to clinical manifestations.* University of Oklahoma Health Sciences Center: Oklahoma City, OK. Available from: https://www.ncbi.nlm.nih.gov/books/NBK333424/

3. Bisno, A.L., M.O. Brito and C.M. Collins, Molecular basis of group A streptococcal virulence. *Lancet Infect. Dis.*, 2003. **3**: p. 191–200.

4. Zabriskie, J.B., The role of temperate bacteriophage in the production of erythrogenic toxin by group A streptococci. *J. Exp. Med.*, 1964. **119**: p. 761–779.

5. Spaulding, A.R., et al., Staphylococcal and streptococcal superantigen exotoxins. *Clin. Microbiol. Rev.*, 2013. **26**(3): p. 422–447.

6. Alouf, J.E., Streptococcal toxins (streptolysin O, streptolysin S, erythrogenic toxin). *Pharmacol. Ther.*, 1980. **11**: p. 617–717.

7. Fischetti V.A., 2016. M Protein and other surface proteins on streptococci. In: Ferretti J.J., Stevens D.L., Fischetti V.A. (eds.), *Streptococcus pyogenes: Basic Biology to Clinical Manifestations*. University of Oklahoma Health Sciences Center: Oklahoma City, OK. Available from: https://www.ncbi.nlm.nih.gov/books/NBK333431/

8. Ryan P.A. and B. Juncosa, 2016. Group A streptococcal adherence. In: Ferretti J.J., Stevens D.L., Fischetti V.A. (eds.), *Streptococcus pyogenes: Basic Biology to Clinical Manifestations*. University of Oklahoma Health Sciences Center: Oklahoma City (OK). Available from: https://www.ncbi.nlm.nih.gov/books/NBK333427/

9. Todd, E.W. and R.C. Lancefield, Variants of hemolytic streptococci; their relation to type-specific substance, virulence, and toxin. *J. Exp. Med.*, 1928. **48**: p. 751–767.

10. Fischetti, V.A., Streptococcal M protein: molecular design and biological behavior. *Clin. Microbiol. Rev.*, 1989. **2**: p. 285–314.

11. Jones, K.F., et al., *Location of variable and conserved epitopes among the multiple serotypes of streptococcal M protein. J. Exp. Med.*, 1985. **161**: p. 623–628.

12. Fischetti, V.A., R.D. Horstmann and V. Pancholi, Location of the complement factor H binding site on streptococcal M6 protein. *Infect. Immun.*, 1995. **63**(1): p. 149–153.

13. Horstmann, R.K., et al., Role of fibrinogen in complement inhibition by streptococcal M protein. *Infect. Immun.*, 1992. **60**: p. 5036–5041.

14. Fischetti, V.A., et al., Structure, function, and genetics of streptococcal M protein. *Rev. Infect. Dis.*, 1988. **10 Suppl 2**: p. S356–S359.

15. Ehlenberger, A.G. and V. Nussenzweig, Role of C3b and C3d receptors in phagocytosis. *J.Exp.Med.*, 1977. **145**: p. 357–371.

16. Brodeur, B.R., et al., Identification of group B streptococcal Sip protein, which elicits cross-protective immunity. *Infect. Immun.*, 2000. **68**(10): p. 5610–5618.

17. Carapetis, J.R., M. McDonald and N.J. Wilson, Acute rheumatic fever. *Lancet*, 2005. **366**: p. 155–168.

18. Steer, A.C., et al., *Status of research and development of vaccines for Streptococcus pyogenes. Vaccine*, 2016. **34**(26): p. 2953–2958.

19. Dale, J.B., et al., *Potential coverage of a multivalent M protein-based group A streptococcal vaccine. Vaccine*, 2013. **31**(12): p. 1576–1581.

20. Bessen, D. and V.A. Fischetti, Passive acquired mucosal immunity to group A streptococci by secretory immunoglobulin A. *J. Exp. Med.*, 1988. **167**: p. 1945–1950.

21. Bessen, D. and V.A. Fischetti, Synthetic peptide vaccine against mucosal colonization by group A streptococci. I. Protection against a heterologous M serotype with shared C repeat region epitopes. *J. Immunol.*, 1990. **145**: p. 1251–1256.

22. Leung, A.K., et al., Rapid antigen detection testing in diagnosing group A beta-hemolytic streptococcal pharyngitis. *Expert Rev. Mol. Diagn.*, 2006. **6**(5): p. 761–766.

23. Facklam, R., What happened to the streptococci: overview of taxonomic and nomenclature changes. *Clin. Microbiol.*, 2002. **15**: p. 613–630.

24. McDonald, M.I., et al., Low rates of streptococcal pharyngitis and high rates of pyoderma in Australian aboriginal communities where acute rheumatic fever is hyperendemic. *Clin. Infect. Dis.*, 2006. **43**(6): p. 683–689.

25. Madrid, L., et al., Infant Group B streptococcal disease incidence and serotypes worldwide: systematic review and meta-analyses. *Clin. Infect Dis.*, 2017. **65**(suppl_2): p. S160–S172.

26. Rosa-Fraile, M., et al., Granadaene: proposed structure of the group B streptococcus polyenic pigment. *Appl. Environ. Microbiol.*, 2006. **72**(9): p. 6367–6370.

27. Lauer, P., et al., Genome analysis reveals pili in Group B streptococcus. *Science*, 2005. **309**(5731): p. 105.

28. Franciosi R.A., J.D. Knostman and R.A. Zimmerman, Group B streptococcal neonatal and infant infections. *J. Pediatr.*, 1973. **82**: 707–718.

29. Schrag S.J., E.R. Zell, R. Lynfield, A. Roome, K.E. Arnold, A.S. Craig, L.H. Harrison, et al., A populationbased comparison of strategies to prevent early-onset group B streptococcal disease in neonates. *N. Engl. J. Med.*, 2002. **347**: 233–239.

30. Van Dyke M.K., C.R. Phares, R. Lynfield, A.R. Thomas, K.E. Arnold, A.S. Craig and J. Mohle-Boetani, et al., Evaluation of universal antenatal screening for group B streptococcus. *N. Engl. J. Med.*, 2009. **360**: 2626–2636.

31. Verani J.R., L. McGee, and S.J. Schrag, Division of Bacterial Diseases NCfI, Respiratory Diseases CfDC, Prevention, 2010. *Prevention of perinatal group B streptococcal disease—revised guidelines from CDC, 2010. MMWR*, **59**: 1–36 (Recommendations and Reports).

32. Teese N., D. Henessey, C. Pearce, N. Kelly and S. Garland, *Screening protocols for group B streptococcus: are transport media appropriate? Infect. Dis. Obstet. Gynecol.*, 2003. **11**: 199–202.

33. Fenton L.J. and M.H. Harper, *Evaluation of colistin and nalidixic acid in Todd-Hewitt broth for selective isolation of group B streptococci. J. Clin. Microbiol.*, 1979. **9**: 167–169.

34. Fairlie T., E.R. Zell and S. Schrag, Effectiveness of intrapartum antibiotic prophylaxis for prevention of early-onset group B streptococcal disease. *Obstet. Gynecol.*, 2013. **121**: 570–577.

35. Coykendall A.L., Classification and identification of the viridans. *Clin. Microbiol.*, 1989. **2**: 315–328.

36. Facklam R.R. and Washington JA, II. 1991. Streptococcus and related catalase-negative gram-positive cocci, p. 238–257. In Balows A., Hausler W.J., Jr., Hermann K.L., Isenberg H.D., Shadomy H.J. (eds.), *Manual of Clinical Microbiology*, 5th edn., vol. **5**. American Society for Microbiology, Washington, DC.

37. Facklam R., 2002. What happened to the streptococci: Overview of taxonomic and nomenclature changes. *Clin. Microbiol.*, **15**: 613–630.

38. Gorm Jensen T., H. Bossen Konradsen and B. Brunn, Evaluation of the rapid ID 32 strep system. *Clin. Microbiol. Infect.*, 1999. **5**: 417–423.

39. Bosshard P.P., S. Abels, M. Altwegg and E.C.Z.R. Bottger, Comparison of conventional and molecular methods for identification of aerobic catalase-negative gram-positive cocci in the clinical laboratory. *J. Clin. Microbiol.*, 2004. **42**: 2065–2073.

40. Haanpera M., J. Jalava, P.M.O. Huovinen and K. Rantakokko-Jalava, Identification of alpha-hemolytic streptococci by pyrosequencing the 16S rRNA gene and by use of VITEK 2. *J. Clin. Microbiol.*, 2007. **45**: 762–770.

41. Teng L.J., P.R. Hsueh, J.C. Tsai, P.W.H.J.C. Chen, H.C. Lai, C.N. Chun-Nan Lee and S.W. Ho, groESL sequence determination, phylogenetic analysis, and species differentiation for viridans group streptococci. *J. Clin. Microbiol.*, 2002. **40**: 3172–3178.

42. Gavin P.J., J.R. Warren, A.A. Obias, S.M. Collins and L.R. Peterson, Evaluation of the VITEK 2 for rapid identification of clinical isolates of gram-negative bacilli and members of the family Streptococcaceae. *Eur. J. Clin. Microbiol. Infect. Dis.*, 2002. **21**: 869–874.

43. Brigante G., F. Luzzaro, A. Bettacini, G. Lombardi, F. Meacci, B. Pini, S. Stefani and A. Toniolo, Use of the phoenix automated system for identification of Streptococcus and Enterococcus spp. *J. Clin. Microbiol.*, 2006. **44**: 3263–3267.

44. Lopez R., *Pneumococcus: The sugar coated-bacteria. Int. Microbiol.*, 2006. **9**: 176–190.

45. Nelson A.L., A.M. Roche, J.M. Gould, K. Chim and A.J.W.J.N. Ratner, Capsule enhances pneumococcal colonization by limiting mucus-mediated clearance. *Infect. Immun.*, 2007. **75**: 83–90.

46. Sorensen U.B., Typing of pneumococci by using 12 pooled antisera. *J. Clin. Microbiol.*, 1993. **8**: 2097–2100.

47. Kolberg J., E.A. Höiby and E. Jantzen, Detection of phosphorylcholine epitope in streptococci, Haemophilus and pathogenic Neisseriae by immunoblotting. *Microb. Pathog.*, 1997. **22**: 321–329.

48. Tomasz A., Choline in the cell wall of a bacterium: novel type of polymer-linked choline in pneumococcus. *Science*, 1967. **157**: 694–697.

49. Tettelin H., Complete genome sequence of a virulent isolate of Streptococcus pneumoniae. *Science*, 2001. **293**: 498–506.

50. Gosink K.K., F.R. Mann, C. Guglielmo, E.I. Tuomanen and H.R. Masure, Role of novel choline binding proteins in virulence of Streptococcus pneumoniae. *Infect. Immun.*, 2000. **68**: 5690–5695.

51. GBD 2016, Lower Respiratory Infections Collaborators. Estimates of the global, regional, and national morbidity, mortality, and aetiologies of lower respiratory infections in 195 countries, 1990-2016: A systematic analysis for the Global Burden of Disease Study 2016. *Lancet. Infect. Dis.* 2018. **18**: 1191–1210.

52. Jedrzejas M.J., *Pneumococcal virulence factors: structure and function. Microbiol. Mol. Biol. Rev.*, 2001. **65**: 187–207.

53. Garcia-Suarez M.M., F. Vaquez and F.J. Mendez, Streptococcus pneumoniae virulence factors and their clinical impact: An update. *Enferm. Infect. Microbiol. Clin.*, 2006. **24**: 512–517.

54. Weiser J.N., R. Austrian, P.K. Sreenivasan and H.R. Masure, Phase variation in pneumococcal opacity: relationship between colonial morphology and nasopharyngeal colonization. 1994. *Infect. Immun.*, **62**: 2582–2589.

55. Kim J.O., S. Romero-Steiner, U.B.S. Sorensen, J. Blom, M. Carvalho, S. Barnard, G. Carlone and J.N. Weiser, Relationship between cell surface carbohydrates and intrastrain variation on opsonophagocytosis of Streptococcus pneumoniae. *Infect. Immun.*, 1999. **67**: 2327–2333.

56. Crook D.W., Capsular type and the pneumococcal human host—Parasite relationship. *Clin. Infect. Dis.*, 2006. **42**: 460–462.

57. Sjostrom K., C. Spindler, A. Ortqvist, M. Kalin, A. Sandgren and B. Henriques-Normark. S. Kuhlmann-Berenzen, Clonal and capsular types decide whether pneumococci will act as a primary or opportunistic pathogen. *Clin. Infect. Dis.*, 2006. **42**: 451–456.

58. Mosser J.L. and A. Tomasz, Choline-containing teichoic acid as a structural component of pneumococcal cell wall and its role in sensitivity to lysis by an autolytic enzyme. *J. Biol. Chem.*, 1970. **245**: 287–298.

59. Garcia E., J.L. Garcia, C. Ronda, P. Garcia and R. Lopez, Cloning and expression of the pneumococcal autolysin gene in Escherichia coli. *Mol. Gen. Genet.*, 1985. **201**: 225–230.

60. Kellogg J.A., D.A. Bankert, C.J. Elder, J.L. Gibbs and M.C. Smith, Identification of Streptococcus pneumoniae revisited. *J. Clin. Microbiol.*, 2001. **39**: 3373–3375.

61. Howden R., A rapid bile solubility test for pneumococci. *J. Clin. Pathol.*, 1979. **12**: 1293–1294.

62. Wasilauskas B.L. and K.D. Hampton, An analysis of Streptococcus pneumoniae identification using biochemical and serological procedures. *Diagn. Microbiol. Infect. Dis.* 1984. **2**: 301–307.

63. Carvalho M.G.S., A.G. Steigerwalt, T. Thompson, D. Jackson and R.R. Facklam, Confirmation of nontypeable Streptococcus pneumoniae-like organisms isolated from outbreaks of epidemic conjunctivitis as Streptococcus pneumoniae. *J. Clin. Microbiol.*, 2003. **41**:4415–4417.

64. Slotved H.C., N. Kaltoft, I.C. Skovsted, M.B. Kerrn and F. Espersen, Simple, rapid latex agglutination test for serotyping of pneumococci (Pneumotest-Latex). *Clin. Microbiol.*, 2004. **42**: 2518–2522.

65. Pimenta F.C., A. Roundtree, A. Soysal, M. Bakir, M. du Plessis, N. Wolter, A. von Gottberg, L. McGee, G. Carvalho Mda and B. Beall, Sequential triplex real-time PCR assay for detecting 21 pneumococcal capsular serotypes that account for a high global disease burden. *J. Clin. Microbiol.*, 2013. **51**: 647–652.

66. Wu H.M., S.M. Cordeiro, B.H. Harcourt, M. Carvalho, J. Azevedo, T.Q. Oliveira and M.C. Leite, et al., Accuracy of real-time PCR, Gram stain and culture for Streptococcus pneumoniae, Neisseria meningitidis and Haemophilus influenzae meningitis diagnosis. *BMC Infect. Dis.*, 2013. **13**: 26.

67. Murdoch D.R., R.T.R. Laing, G.D. Mills, N.C. Karalus, G.I. Town, M.S. Reller and L.B. Reller, Evaluation of a rapid immunochromatogenic test for detection of Streptococcus pneumoniae antigen in urine samples from adults with community-acquired pneumonia. *J. Clin. Microbiol.*, 2001. **39**: 3495–3498.

68. Briones M.L., J. Blanquer, D. Ferrando, M.L. Blasco, C. Gimeno and J. Marin, Assessment of analysis of urinary pneumococcal antigen by immunochromatography for etiologic diagnosis of community acquired pneumonia in adults. *Clin. Vaccine Immunol.*, 2006. **13**: 1092–1097.

69. Petti C.A., C.W. Woods and L.B. Reller, Streptococcus pneumoniae antigen test using positive blood culture bottles as an alternative method to diagnose pneumococcal bacteremia. *J. Clin. Microbiol.*, 2005. **43**: 2510–2512.

70. Jacobs J.A., E.E. Stobberingh, E.I.M. Cornelissen and M. Drent, Detection of Streptococcus pneumoniae antigen in bronchoalveolar lavage fluid samples by a rapid immunochromatographic membrane assay. *J. Clin. Microbiol.*, 2005. **43**: 4037–4040.

71. Marcos M.A., E. Martinez, M. Almela, J. Mensa, and M.T. Jimenez de Anta, New rapid antigen test for diagnosis of Pneumococcal meningitis. *Lancet.*, 2001. **357**:1499–1500.

72. Murdoch D.R., K.L. O'Brien, A.J. Driscoll, R.A. Karron and N. Bhat, Pneumonia methods working G, team PC. 2012. Laboratory methods for determining pneumonia etiology in children. *Clin. Infect. Dis.*, **54** (Suppl. 2): S146–S152.

73. Kim L., L. McGee, S. Tomczyk and B. Beall, Biological and epidemiological features of antibiotic-resistance

Streptococcus pneumoniae in pre- and post-conjugate vaccine eras: A United States perspective. *Clin Microbiol Rev.*, 2016. **29**: 525–52

74. Centers for Disease Control, Prevention, Advisory Committee on Immunization P. Updated recommendations for prevention of invasive pneumococcal disease among adults using the 23-valent pneumococcal polysaccharide vaccine (PPSV23). *MMWR.*, 2010. **59**: 1102–1106.

75. Nuorti J.P. and C.G. Whitney, Centers for Disease Control, Prevention, Prevention of pneumococcal disease among infants and children—Use of 13-valent pneumococcal conjugate vaccine and 23-valent pneumococcal polysaccharide vaccine—Recommendations of the Advisory Committee on Immunization Practices (ACIP). *MMWR.*, 2010. **59**: 1–18

76. Centers for Disease Control, Prevention. Licensure of 13-valent pneumococcal conjugate vaccine for adults aged 50 years and older. *MMWR.*, 2012. **61**: 394–395.

77. Matanock A., G. Lee, R. Gierke, M. Kobayashi, A. Leidner and T. Pilishvili, Use of 13-Valent Pneumococcal Conjugate Vaccine and 23-Valent Pneumococcal Polysaccharide Vaccine Among Adults Aged ≥65 Years: Updated Recommendations of the Advisory Committee on Immunization Practices. *MMWR*, 2019. **68**: 1069–1075.

78. Tomczyk S., R. Lynfield, W. Schaffner, A. Reingold, L. Miller, S. Petit, C. Holtzman, S.M. Zansky, A. Thomas, J. Baumbach, L.H. Harrison, M.M. Farley, B. Beall, L. McGee, R. Gierke, T. Pondo and L. Kim, Prevention of antibiotic-nonsusceptible invasive pneumococcal disease with the 13-valent pneumococcal conjugate vaccine. *Clin. Infect. Dis.* 2016. **62**: 1119–1125.

79. Tomasz A., Control of the competent state in Pneumococcus by a hormone-like cell product: an example for a new type of regulatory mechanism in bacteria. *Nature*, 1965. **208**: 155–159.

80. Morrison D.A., Streptococcal competence for genetic transformation: regulation by peptide pheromones. *Microb. Drug Resist.*, 1997. **3**: 27–37.

81. Claverys J.P., M. Prudhomme and B. Martin, Induction of competence Regulons as a general response to stress in gram-positive bacteria. *Ann. Rev. Microbiol.*, 2006. **60**: 451–475.

82. Jeffries J.M., A. Smith, S.C. Clarke, C. Dowson and T.J. Mitchell, Genetic analysis of diverse disease-causing pneumococci indicates high levels of diversity within serotypes and capsule switching. *J. Clin. Microbiol.*, 2004. **42**: 5681–5688.

83. Jackson C.R., P.J. Fedorka-Cray and J.B. Barrett, Use of genus-and species-specific multiplex PCR for identification of Enterococci. *J. Clin. Microbiol.*, 2004. **42**: 3558–3565.

84. Deasy B.M., M.C. Rea, G.F. Fitzgerald, T.M. Cogan and T.P. Beresford, A rapid PCR based method to distinguish between Lactococcus and Enterococcus. *Syst. Appl. Microbiol.*, 2000. **23**: 510–522.

85. Murray B.E., The life and times of the Enterococcus. *Clin. Microbiol. Rev.*, 1990. **3**: 46–65.

86. Mundy L.M., D.F. Sahm and M. Gilmore, Relationships between enterococcal virulence and antimicrobial susceptibility. *Clin. Microbiol. Rev.*, 2006. **13**: 513–522.

87. Willems R.J.L., J. Top, M. van Santen, D.A. Robinson, T.M. Coque, F. Baquero, H. Grundmann and M.J.M. Bontem, Global spread of vancomycin-resistant Enterococcus faecium from distinct nosocomial genetic complex. *Emerg. Infect. Dis.*, 2005. **11**: 821–828.

88. Schleifer K.H., J. Kraus, C. Dvorak, R. Kilpper-Balz, M.D. Collins and W. Fischer, Transfer of Streptococcus lactis and related streptococci to the genus Lactococcus gen. nov. *Syst. Appl. Microbiol.*, 1985. **6**: 183–195.

89. Facklam R.and J.A. Elliot, Identification, classification, and clinical relevance of catalase negative, gram positive Cocci, excluding the Streptococci and Enterococci. *Clin. Microbiol. Rev.*, 1995. **8**: 479–495.

90. Klaenhammer T.R., Genetics of bacteriocins produced by lactic acid bacteria. *FEMS Microbiol. Rev.*, 1993. **12**: 39–85.

91. Jack R.W., J.R. Tagg, and B. Ray, Bacteriocins of gram-positive bacteria. *Microbiol. Rev.*, 1995. **59**: 171–200.

92. Mierau I. and M. Kleerebezem, 10 years of the nisin-controlled gene expression system (NICE) in Lactococcus lactis. *Appl. Microbiol. Biotechnol.*, 2005. **68**: 705–717.

41

The Genus Vibrio *and Related Genera*

Seon Young Choi, Anwar Huq, and Rita R. Colwell

CONTENTS

Introduction

Vibrios are short curved or straight cells, which are single or united into spirals. They grow well and rapidly on the surfaces of standard culture media and can be readily isolated from estuarine, marine, and fresh-water samples. These heterotrophic organisms vary in their nutritional requirements; some occur as parasites and pathogens for animals and for humans. The short curved, asporogenous, Gram-negative rods that are members of the genus *Vibrio* are most commonly encountered in the estuarine, marine, or fresh-water habitat. Distinguishing species of the genus *Vibrio* from other related genera can be presumptively accomplished by examining Gram stains of carefully prepared specimens, followed by electron microscopy to confirm morphology. *Vibrio* species are short rods with a curved axis, motile by means of a single polar flagellum. *Vibrio* species may be short, straight rods (1.5–3.0 μm × 0.5 μm), or they may be S-shaped or spiral-shaped when individual cells are joined. Possession of two or more flagella in a polar tuft has also been demonstrated in *Vibrio* species, as have lateral flagella in *Vibrio parahaemolyticus*. Thus, to identify and classify vibrios, physiological and biochemical taxonomic tests are done.[1] *Vibrio* species are facultatively anaerobic, with both a respiratory (oxygen-utilizing) and a fermentative metabolism. Related genera may be aerobic or microaerophilic, with a strictly respiratory metabolism (oxygen is the terminal electron acceptor).

Since the eighth edition of *Bergey's Manual of Determinative Bacteriology*,[1] the genus *Vibrio* was placed among the Gram-negative facultatively anaerobic rods in the family *Vibrionaceae*. The family *Vibrionaceae* comprises several genera including *Vibrio*, *Photobacterium*, and *Salinivibrio*. *Vibrio* species are fermentative, producing acid in carbohydrate-containing media without formation of gas, and possess a DNA with G + C in the range from 39 to 49 moles %. Vibrios require 1–3% NaCl for growth and the salt requirement is absolute. Vibrios, including *Vibrio cholerae*, are autochthonous to the aquatic environment. Molecular genetic and genomic sequence data have proved useful in establishing relationships among the genera *Plesiomonas*, *Aeromonas*, and *Vibrio*.[2] In the latest edition of *Bergey's Manual*, *Plesiomonas* and *Aeromonas* are classified in other families, although they are oxidase positive genera of fermentative bacteria as are *Vibrio* and *Photobacterium*.[3]

Genus *Vibrio* Pacini, 1854

The type species for the genus *Vibrio* is *Vibrio cholerae* Pacini, 1854. The genus *Vibrio* has been expanded to include many more species than the 40 species of *Vibrio* listed in the latest edition of *Bergey's Manual*.[3] The intense studies of the genus *Vibrio* performed over the years resulted in considerable changes. Many *Vibrio* species have been reclassified and transferred to other or new genera. Strictly anaerobic species have been removed from the genus and no cultures of obligately anaerobic *Vibrio* species are extant. Microaerophilic species, including *Vibrio fetus* and *Vibrio bubulus*, have been transferred to the genus *Campylobacter* and another microaerophilic species, *Vibrio succinogenes,* is now the type species of the genus *Wolinella*.[4,5] Also the pink-pigmented methanol-oxidizing bacterium, *Vibrio extorquens*, is reclassified in the genus *Methylobacterium*.[6]

The number of characterized and defined species resident in the genus *Vibrio* has been significantly increased, with many novel species having been described.

Description of the genus *Vibrio*, as amended by the Subcommittee on Taxonomy of Vibrios, International Committee on Nomenclature of Bacteria, is concise and remains a reasonably good working definition:

> Gram-negative, asporogenous rods which have a single, rigid curve or which are straight. Motile by means of a single, polar flagellum. Produce indophenol oxidase and catalase. Ferment glucose without gas production. Acidity is produced from glucose by the Embden-Meyerhof glycolytic pathway. The guanine plus cytosine in the DNA of *Vibrio* species is within the range of 40 to 50 moles per cent.

The genus *Vibrio* has been a subject of many polyphasic studies. The genus *Listonella,* proposed by MacDonell and Colwell,[7] is polyphyletic and belongs to the genus *Vibrio* on the basis of 16S rRNA phylogeny (Figure 41.1).[8] *Listonella anguillarum* and

DOI: 10.1201/9781003099277-43

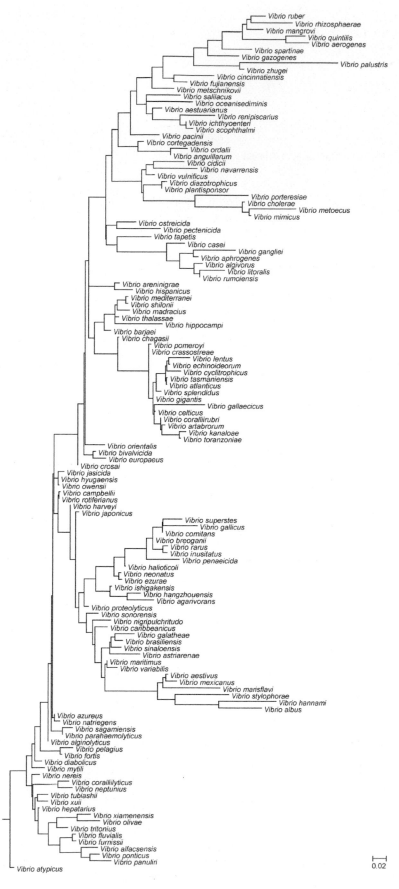

FIGURE 41.1 *Listonella anguillarum.*

Listonella pelagius are included in the genus *Vibrio* in the latest edition of the *Bergey's Manual*. Most of the present members of the genus *Vibrio* form a monophyletic group except for *Vibrio calviensis,* which should be transferred to the genus *Enterovibrio*.[9]

The type species of the genus, *Vibrio cholerae* Pacini 1854, can be succinctly described as follows: producing L-lysine and L-ornithine decarboxylases; not producing L-arginine dihydrolase and hydrogen sulfide (Kligler iron agar). The G + C content in the DNA of *Vibrio cholerae* is approximately 48 ± 1%. *Vibrio cholerae* includes strains that may or may not elicit the cholera-red (nitroso-indole) reaction, may or may not be hemolytic, may or may not be agglutinated by Gardner and Venkatraman O group I antiserum, and may or may not be lysed by Mukherjee *Vibrio cholerae* bacteriophages I, II, III, IV, and V.[10]

Vibrio cholerae strains possess a common H antigen and can be serologically grouped into many serotypes according to their O antigens. Strains agglutinated by Gardner and Venkatraman O group I antiserum are in serotype I and were the principal cause of cholera in man until the emergence of *Vibrio cholerae* serotype O139. An M antigen can obscure the agglutinability of mucoid strains of *Vibrio cholerae*.

In the past, many cholerae-like vibrios were given separate species status because they had been isolated from patients suffering diarrhea, not cholera, or from water and foods. Among these, *Vibrio albensis* can be considered to be a biotype of *Vibrio cholerae*. The so-called nonagglutinable (NAG) or noncholera (NCV) vibrios, including *Vibrio eltor*, were shown many years ago by Citarella and Colwell,[11] using DNA-DNA hybridization, to be *Vibrio cholerae* and this conclusion has been substantiated.

Since its initial isolation by Fujino and Fukumi[12] in 1953, *Vibrio parahaemolyticus* has received considerable attention and rightfully, as it is a major cause of seafood borne illness. *Vibrio parahaemolyticus* (syn. *Pasteurella parahaemolytica, Pseudomonas enteritis, Oceanomonas parahaemolytica*) is the causative agent of food poisoning arising from ingestion of contaminated seafood and is autochthonous to the marine and estuarine environment.

The species *Vibrio alginolyticus*, although previously considered by some investigators to be synonymous with *Vibrio parahaemolyticus*, is a separate species.

According to results of DNA-DNA hybridization and other techniques, *Vibrio mimicus* is highly related to *Vibrio cholerae*. With *Vibrio cholerae*, the species *Vibrio mimicus* will be the basis for the definition of the genus *Vibrio*.[3,13]

Other species of *Vibrio* that should be considered are listed in Figure 41.1 and Table 41.1. Vibrios are naturally occurring in the aquatic environment and are found in sea water and associated with human and marine animals.[14,15]

Distinctions can be made among the species of the genus *Vibrio* and useful differentiating characteristics are given in Table 41.1. Some species of *Vibrio* may demonstrate sheathed flagella—i.e., a flagellum with a central core and an outer sheath—and, under certain conditions of growth, peritrichous flagella.[16] "Round bodies" or spheroplasts are commonly present during various stages of growth,[17,18] and fimbriae (pili) have been observed in strains of *Vibrio cholerae*.[19]

Vibrios as Human Pathogens

Many species of the genus *Vibrio* are commensals or pathogenic for fish, shellfish, mollusks, Crustacea, and other aquatic animals. Among all the major bacterial foodborne pathogens (e.g., Salmonella, Listeria, Shiga toxin–producing *Escherichia coli* (STEC), and Campylobacter), vibrios are a distinct group that is increasing in incidence in the United States.[20–22] Many are associated with foodborne outbreaks and of significant importance are *Vibrio cholerae, Vibrio parahaemolyticus*, and *Vibrio vulnificus,* three well-characterized human pathogenic bacterial species.[23]

V. cholerae is responsible for seven historical cholera pandemics since 1817. Cholera outbreaks occur throughout the world, from Southeast Asia, Africa, the Middle East, to South and Central America. In the past decade, a cholera outbreak in Haiti occurred after the earthquake of 2010 and in October 2016, another outbreak was reported in Yemen after war. Also many countries in Sub-Saharan Africa suffer recurrent cholera outbreaks.[24–28] This bacterium can cause severe, dehydrating diarrhea in otherwise healthy people, resulting in approximately 100,000 deaths annually.[29] The major virulence factors of *V. cholerae* are cholera toxin (CT, *ctxAB*) and type IV pilus toxin coregulated pilus (TCP) encoded by a lysogenic bacteriophage (CTXφ), and the *tcpA-F* operon located in the Vibrio pathogenicity island 1 (VPI-1). In regard to antimicrobial resistance, ICE (integrative conjugative element) of the SXT/R391 family was originally reported in *V. cholerae* O139 strain MO10 and it encodes resistance to sulfamethoxazole (*sul2*), trimethoprim (*dfr*), chloramphenicol (*floR*), and streptomycin (*strA* and *strB*).[30] Since the report of O139 strain MO10, the number of reports of SXT-related ICEs is increasing in Asia and Africa. Also, various types with a difference in antibiotic-resistant genes have been reported and ICEs play a critical role in acquiring antimicrobial resistance genes by vibrios.[31–36]

Human-illness-associated strains, harboring CT, are generally associated with the *V. cholerae* O1 or O139 serogroups. *V. cholerae* O1 serogroup is further classified into two biotypes, classical and El Tor, based on a number of phenotypic characteristics, including Voges-Proskauer reaction and sensitivity to polymyxin B. Classical biotype strains were responsible for the first six cholera pandemics in the early 19th century, while El Tor biotype strains producing El Tor type CT are responsible for the current seventh cholera pandemic since 1961. Beginning in the early 1990s, new variants of the *V. cholerae* O1 serogroup, hybrid and altered strains have emerged and have replaced El Tor strains in Asian countries, but also have been reported in Africa.[37–39] Comparative genomics and genome sequencing analysis show VPI-1, VPI-2, Vibrio seventh pandemic island 1 (VSP-1) and VSP-2 to be present in the *V. cholerae* genome, findings reported after the complete genome sequence of *V. cholerae* O1 El Tor N16961 was published.[40,41] Serotype O139 first emerged in South India and Bangladesh during 1992 and can cause epidemics of cholera. Since first emergence in India, it has been reported in the Southeast Asia and China subsequently.[42–45]

In the 1950s, *V. parahaemolyticus* was isolated and described as an important seafood-borne bacterium and subsequently shown to be the primary cause of human acute gastroenteritis worldwide. *V. parahaemolyticus* also can cause wound

TABLE 41.1

Differential Characteristics among Important *Vibrio* Spp.

	V. alginolyticus	*V. carchariae*	*V. cholerae*	*V. cincinnatiensis*	*V. camsela*	*V. fluvialis*	*V. furnissii*	*V. hollisae*	*V. metschnikovii*	*V. minicus*	*V. parahaemolyticus*	*V. vulnificus*
Swarming (marine agar, 25°C)	+	+	−	+	−	−	−	−	−	−	+	−
Yellow pigment at 25°C	−	−	−	−	−	−	−	−	−	−	−	−
Indole production	v	+	+	−	−	v	v	+	v	+	+	+
Voges–Proskauer reaction	+	v	v	−	+	−	−	−	+	−	−	−
Arginine dehydrogenase	−	−	−	−	+	+	+	−	v	−	−	−
Lysine dehydrogenase	+	+	+	v	v	−	−	−	v	+	+	+
Ornithine dehydrogenase	v	−	+	−	−	−	−	−	−	+	+	v
D-Glucose, acid production	+	v	+	+	+	+	+	+	+	+	+	+
D-Glucose, gas production	−	−	−	−	−	−	+	−	−	−	−	−
Nitrate→nitrite	+	+	+	+	+	+	+	+	−	+	+	+
Oxidase	+	+	+	+	+	+	+	+	−	+	+	+
Lipase (corn oil)	v	−	+	v	−	+	v	−	+	v	+	+
ONPG test	−	−	+	v	−	v	v	−	v	+	−	v
Tyrosine clearing	v	−	v	−	−	v	v	−	−	v	v	v
Acid Production From												
L-Arabinose	−	−	−	+	−	+	+	+	−	−	v	
D-Arabitol	−	−	−	−	−	v	v	−	−	−	−	
Cellobiose	−	v	−	+	−	v	v	−	−	−	−	+
Lactose	−	−	−	−	−	−	−	v	v	v	−	v
Maltose	+	+	+	+	+	+	+	−	+	+	+	+
D-Mannitol	+	v	+	+	−	+	+	−	+	+	+	v
Salicin	−	−	−	+	−	−	−	−	−	−	−	+
Sucrose	+	v	+	+	−	+	+	−	+	−	−	v
Growth in Nutrient Broth												
0% NaCl	−	−	+	−	−	−	−	−	−	+	−	−
1% NaCl	+	+	+	+	+	+	+	+	+	+	+	+
6% NaCl	+	s	v	+	+	+	+	v	v	v	+	v
8% NaCl	+	−	−	v	−	v	v	−	v	−	v	−
10% NaCl	v	−	−	−	−	−	−	−	−	−	−	−
Sensitivity To:												
O/129	v	+	+	v	+	v	−	v	+	+	v	+
Polymyxin B	v	+	v	+	v	+	v	+	+	v	−	−

Notes: +, 90% to 00% positive; −, 0% to 10% positive; v, 11% to 89% positive. Data obtained from Farmer and Hickman-Brenner.[27]

infections and septicemia. With *V. vulnificus*, *V. parahaemolyticus* is the major cause of seafood-borne infections and mortality in the United States.[22,46,47] Until 1996, *V. parahaemolyticus* infections were associated with diverse serovars of the pathogen, but after 1996 the most common serotype has been *V. parahaemolyticus* O3:K6, associated with the pandemic clonal complex (CC) 3 and the multilocus sequence typing (MLST) sequence type (ST) 3, with specific genetic markers (*tdh* positive, *toxRS/*new, and/or *orf8* from the f237 filamentous phage, *trh* and urease negative).[48–50] An outbreak of *V. parahaemolyticus* O3:K6 that began in Kolkata, India in 1996, rapidly spread throughout Southeast Asia, the Atlantic and Gulf coasts of the United States, as well as in Europe, Africa, and Central and South America.[51–54] Subsequently, there have been reports of more than 20 serovariants of O3:K6, including O4:K68, O1:K25, O6:K18, O4:K12, and O1:KUT (untypable).[53–56] Potentially virulent strains of *V.*

parahaemolyticus are commonly distinguished from virulent strains by the presence of two hemolysins, namely thermostable direct hemolysin (*tdh*) and Tdh-related hemolysin (*trh*).[57–59] Pathogenic strains of *V. parahaemolyticus* can produce either TDH, TRH, or both. The finding of *tdh* gene was also reported in several other vibrios, including *V. hollisae, V. mimicus,* and *V. cholerae* non-O1, suggesting horizontal transfer of that gene to other species.[60] Genomic sequencing of *V. parahaemolyticus* RIMD 2210633 revealed two pathogenicity islands encoded T3SS1 and some cognate effectors and Vp-PAI, also referred to as VPaI-7, encoding T3SS2, its known effectors and the *tdh* genes.[61–64] Comparative genomics analysis revealed additional genomic islands.[65–67] The complete genome of *V. parahaemolyticus* RIMD 2210633 O3:K6 serotype strains and many other strains of *V. parahaemolyticus* have been sequenced, including *V. parahaemolyticus* AQ3810 serotype O3:K6, strain 10329 serotype O4:K12, and the serotype O4:K68 strains AN-5034, K5030, and Peru-466. The *V. parahaemolyticus* isolate PCV08-7 *tdh+/trh-* and prepandemic strain BB22OP have also been sequenced.[51,61,64,68,69]

V. vulnificus is a dangerous pathogen, with a case-fatality rate exceeding 50% and now recognized as the most virulent of the pathogenic vibrios in the United States. Indeed, this single bacterial species is responsible for 95% of all seafood-borne deaths and is capable of causing gastroenteritis, wound infections, and fatal septicemia in humans.[23,70,71] Cases where skin or wounds have been contaminated with *V. vulnificus* have been described in Belgium, Denmark, Germany, Spain, Sweden, and the Netherlands.[23,72,73] Especially, the infection of *V. vulnificus* shows an obvious gender disparity with males making up around 85% of reported infections.[74] Strains of *V. vulnificus* are classified into three biotypes, based on their biochemical and serological characteristics, as well as host range. Strains of biotype 1 are responsible for most human infections and biotype 2 is primarily pathogenic for eels (genus *Anguilla*), but may also act as an opportunistic human pathogen. *V. vulnificus* biotype 3 possesses biochemical properties of both biotypes 1 and 2 and recently emerged in Israel, associated with an outbreak of human wound infections.[75–78] Strains of *V. vulnificus* can be divided into two groups based on alleles of the virulence correlated gene *vcg* (environmental and clinical types *vcgE* and *vcgC*), the *pilF* gene that encodes a protein required for pilus type IV assembly, and by comparison of 16S rRNA sequences (A-type and B-type).[79–81] Following genome sequencing of two clinical strains of *V. vulnificus* C-type, and utilizing comparative genomics based on results of genome sequencing projects, many more strains have been described, including E-type strains, providing interesting genomic diversity.[82–87] *V. vulnificus* possesses many virulence determinants, one of which is a polysaccharide that confers resistance to phagocytosis and serum complement. An important characteristic of *V. vulnificus* is that it produces a variety of iron uptake systems, including siderophore (vulnibactin), and utilizes hemoglobin and hemin as iron sources via the heme receptor HupA. *V. vulnificus* possesses exoenzymes and exotoxins, including metalloprotease (VvpE), phospholipase A, cytolytic hemolysin (VvhA), and repeats-in-toxin (RtxA1). Flagella and pili have been described in *V. vulnificus* to be important virulence factors.[88–98]

These three *Vibrio* spp. are the dominant human pathogenic species of the genus Vibrio. However, it is also reported in the Cholera and Other Vibrio Illness Surveillance (COVIS) system of the Centers for Disease Control and Prevention that more than 20% of patients with diagnosed illness are concluded to have their disease condition caused by *V. alginolyticus, V. fluvialis, V. mimicus, V. hollisae* (formerly *Grimontia hollisae*) and related species, including *Photobacterium damselae* subsp. *damselae* (formerly *V. damsela*), *V. furnissii,* or *V. metschnikovii*.[74] In human clinical specimens worldwide, the cases of *Vibrio cincinnatiensis* and *Vibrio harveyi* (*V. carchariae*) are shown rarely as well.[3,99–101] The status of the latter as human enteric or wound pathogens is less clear, but nevertheless significant. Among noncholera vibrios, *V. alginolyticus,* formerly regarded as biotype 2 of *V. parahaemolyticus,* is gradually recognized as an emerging human pathogen responsible for more than 10% of patient cases. *V. alginolyticus* is reported the causative agent of wound and ear infections, intracranial infection, peritonitis, and osteomyelitis.[74,102] The incidence of *V. alginolyticus* infection significantly increases during summer months and most reports of *V. alginolyticus* wound infections result from exposure of cuts or abrasions to contaminated seawater.[22,103,104]

Including *V. cholerae, V. parahaemolyticus, V. vulnificus,* and *V. alginolyticus,* vibrios are ubiquitous bacteria in estuarine and marine environments and an important cause of diseases in human and aquatic animals. Their abundance in the natural environment mirrors the ambient environmental temperature and, with climate change, especially warming of the sea surface, which is strongly associated with spread of vibrio pathogens in the environment. In an aquatic environment, most notably infectious disease outbreaks will depend on climate change, including warm air with low river flows, heavy rainfall, and also extreme weather events.[47,105–111] Discovering the link between growth and spread of Vibrio spp. in the oceans and climate change should be widely investigated as a potential emerging issue.

REFERENCES

1. Buchanan, R.E., and Gibbons, N.E., *Bergey's Manual of Determinative Bacteriology*, 8th ed., Williams & Wilkins, Baltimore, MD, 1974.
2. Ruimy, R., Breittmayer, V., Elbaze, P., Lafay, B., Boussemart, O., Gauthier, M., and Christen, R., *Int J Syst Bacteriol*, 44, 416, 1994.
3. Bergey, D.H., Brenner, D.J, Krieg, N.R., and Staley, J.T., *Bergey's Manual of Systematic Bacteriology*, 2nd ed. Vol. 2 Proteobacteria Part B: The Gammaproteobacteria, Springer, New York, NY, 2005.
4. Veron, M., and Chatelain, R., *Int J Syst Evol Microbiol*, 23, 122, 1973.
5. Tanner, A.C.R., Badger, S., Lai, C.H., Listgarten, M.A., Visconti, R.A., and Socransky, S.S., *Int J Syst Evol Microbiol*, 31, 432, 1981.
6. Green, Peter N., and Ardley, Julie K., *Int J Syst Evol Microbiol*, 68, 2727, 2018.
7. MacDonell, M.T., and Colwell, R.R., *Syst Appl Microbiol*, 6, 171, 1985.
8. Yoon, S.H., Ha, S.M., Kwon, S., Lim, J., Kim, Y., Seo, H., and Chun, J., *Int J Syst Evol Microbiol*, 67, 1613, 2017.
9. Thompson, F.L., Thompson, C.C., Naser, S., Hoste, B., Vandemeulebroecke, K., Munn, C., Bourne, D., and Swings, J., *Int J Syst Evol Microbiol*, 55, 913, 2005.

10. Hugh, R., and Feeley, J.C., *Int J Syst Bacteriol*, 22, 123, 1972.

11. Citarella, R.V., and Colwell, R.R., *J Bacteriol*, 104, 434, 1970.

12. Fujino, T., and Fukumi, H.E., *Vibrio Parahaemolyticus*, 2nd ed. (in Japanese), Naya Shoten, Tokyo, 1967.

13. Davis, B.R., Fanning, G.R., Madden, J.M., Steigerwalt, A.G., Bradford, H.B., Smith, H.L., and Brenner, D.J., *J Clin Microbiol*, 14, 631, 1981.

14. Colwell, R.R., and Morita, R.Y., *J Bacteriol*, 88, 831, 1964.

15. Bianchi, Micheline A.G., *Archiv Mikrobiol*, 77, 127, 1971.

16. Baumann, P., Baumann, L., and Mandel, M., *J Bacteriol*, 107, 268, 1971.

17. Felter, R.A., Kennedy, S.F., Colwell, R.R., and Chapman, G.B., *J Bacteriol*, 102, 552, 1970.

18. Kennedy, S.F., Colwell, R.R., and Chapman, G.B., *Can J Microbiol*, 16, 1027, 1970.

19. Tweedy, J.M., Park, R.W., and Hodgkiss, W., *J Gen Microbiol*, 51, 235, 1968.

20. Tack, D.M., Marder, E.P., Griffin, P.M., Cieslak, P.R., Dunn, J., Hurd, S., Scallan, E., Lathrop, S., Muse, A., Ryan, P., Smith, K., Tobin-D'Angelo, M., Vugia, D.J., Holt, K.G., Wolpert, B.J., Tauxe, R., and Geissler, A.L., *Am J Transplant*, 19, 1859, 2019.

21. Crim, S.M., Iwamoto, M., Huang, J.Y., Griffin, P.M., Gilliss, D., Cronquist, A.B., Cartter, M., Tobin-D'Angelo, M., Blythe, D., Smith, K., Lathrop, S., Zansky, S., Cieslak, P.R., Dunn, J., Holt, K.G., Lance, S., Tauxe, R., and Henao, O.L., *MMWR Morb Mortal Wkly Rep*, 63, 328, 2014.

22. Newton, A., Kendall, M., Vugia, D.J., Henao, O.L., and Mahon, B.E., *Clin infect dis*, 54 *Suppl*, 5, S391, 2012.

23. Feldhusen, F., *Microbes Infect*, 2, 1651, 2000.

24. Centers for Disease Control and Prevention (CDC), *MMWR Morb Mortal Wkly Rep*, 59, 1411, 2010.

25. Ingelbeen, B., Hendrickx, D., Miwanda, B., van der Sande, M.A.B., Mossoko, M., Vochten, H., Riems, B., Nyakio, J.P., Vanlerberghe, V., Lunguya, O., Jacobs, J., Boelaert, M., Kebela, B. I., Bompangue, D., and Muyembe, J.J., *Emerg Infect Dis*, 25, 856, 2019.

26. Sema Baltazar, C., Langa, J.P., Dengo Baloi, L., Wood, R., Ouedraogo, I., Njanpop-Lafourcade, B.M., Inguane, D., Elias Chitio, J., Mhlanga, T., Gujral, L., Gessner B.D., Munier, A., and Mengel M.A., *PLOS Negl Trop Dis*, 11, e0005941, 2017.

27. Cartwright, E.J., Patel, M.K., Mbopi-Keou, F.X., Ayers, T., Haenke, B., Wagenaar, B.H., Mintz, E., and Quick, R., *Epidemiol Infect*, 141, 2083, 2013.

28. World Health Organization (WHO) Weekly Epidemiological Record, Cholera. Available at: https://www.who.int/wer/en/

29. Mutreja, A., Kim, D.W., Thomson, N.R., Connor, T.R., Lee, J.H., Kariuki, S., Croucher, N. J., Choi, S.Y., Harris, S.R., Lebens, M., Niyogi, S.K., Kim, E.J., Ramamurthy, T., Chun, J., Wood, J.L., Clemens, J.D., Czerkinsky, C., Nair, G.B., Holmgren, J., Parkhill, J., and Dougan, G., *Nature*, 477, 462, 2011.

30. Hochhut, B., and Waldor, M.K., *Mol Microbiol*, 32, 99, 1999.

31. Pande, K., Mendiratta, D.K., Vijayashri, D., Thamke, D.C., and Narang, P., *Indian J Med Res*, 135, 346, 2012.

32. Dalsgaard, A., Forslund, A., Sandvang, D., Arntzen, L., and Keddy, K., *J. Antimicrob Chemother*, 48, 827, 2001.

33. Ceccarelli, D., Spagnoletti, M., Bacciu, D., Danin-Poleg, Y., Mendiratta, D.K., Kashi, Y., Cappuccinelli, P., Burrus, V., and Colombo, M.M., *Int J Med Microbiol*, 301, 318, 2011.

34. Spagnoletti, M., Ceccarelli, D., Rieux, A., Fondi, M., Taviani, E., Fani, R., Colombo, M.M., Colwell, R.R., and Balloux, F., *mBio*, 5, 2014.

35. Wozniak, R.A., Fouts, D.E., Spagnoletti, M., Colombo, M.M., Ceccarelli, D., Garriss, G., Dery, C., Burrus, V., and Waldor, M.K., *PLOS Genet*, 5, e1000786, 2009.

36. Dalia, A.B., Seed, K.D., Calderwood, S.B., and Camilli, A., *Proc Natl Acad Sci U S A*, 112, 10485, 2015.

37. Nair, G.B., Faruque, S.M., Bhuiyan, N.A., Kamruzzaman, M., Siddique, A.K., and Sack, D.A., *J Clin Microbiol*, 40, 3296, 2002.

38. Nair, G.B., Qadri, F., Holmgren, J., Svennerholm, A.M., Safa, A., Bhuiyan, N.A., Ahmad, Q.S., Faruque, S.M., Faruque, A.S., Takeda, Y., and Sack, D.A., *J Clin Microbiol*, 44, 4211, 2006.

39. Lee, J.H., Han, K.H., Choi, S.Y., Lucas, M.E., Mondlane, C., Ansaruzzaman, M., Nair, G. B., Sack, D.A., von Seidlein, L., Clemens, J.D., Song, M., Chun, J., and Kim, D.W., *J Med Microbiol*, 55, 165, 2006.

40. Dziejman, M., Balon, E., Boyd, D., Fraser, C.M., Heidelberg, J.F., and Mekalanos, J.J., *Proc Natl Acad Sci U S A*, 99, 1556, 2002.

41. Heidelberg, J.F., Eisen, J.A., Nelson, W.C., Clayton, R.A., Gwinn, M.L., Dodson, R.J., Haft, D.H., Hickey, E.K., Peterson, J.D., Umayam, L., Gill, S.R., Nelson, K.E., Read, T.D., Tettelin, H., Richardson, D., Ermolaeva, M. D., Vamathevan, J., Bass, S., Qin, H., Dragoi, I., Sellers, P., McDonald, L., Utterback, T., Fleishmann, R.D., Nierman, W.C., White, O., Salzberg, S.L., Smith, H.O., Colwell, R.R., Mekalanos, J.J., Venter, J.C., and Fraser, C.M., *Nature*, 406, 477, 2000.

42. Ramamurthy, T., Garg, S., Sharma, R., Bhattacharya, S. K., Nair, G.B., Shimada, T., Takeda, T., Karasawa, T., Kurazano, H., Pal, A., and et al, *Lancet*, 341, 703, 1993.

43. Albert, M.J., Siddique, A.K., Islam, M.S., Faruque, A.S., Ansaruzzaman, M., Faruque, S. M., and Sack, R.B., *Lancet*, 341, 704, 1993.

44. Li, B.S., Xiao, Y., Wang, D.C., Tan, H.L., Ke, B.X., He, D.M., Ke, C.W., and Zhang, Y.H., *Epidemiol infect*, 144, 2679, 2016.

45. Bodhidatta, L., Echeverria, P., Hoge, C.W., Pitarangsi, C., Serichantalergs, O., Henprasert-Tae, N., Harikul, S., and Kitpoka, P., *Epidemiol Infect*, 114, 71, 1995.

46. Haendiges, J., Rock, M., Myers, R.A., Brown, E.W., Evans, P., and Gonzalez-Escalona, N., *Emerg Infect Dis*, 20, 718, 2014.

47. Martinez-Urtaza, J., Baker-Austin, C., Jones, J.L., Newton, A.E., Gonzalez-Aviles, G.D., and DePaola, A., *N Engl J Med*, 369, 1573, 2013.

48. Okuda, J., Ishibashi, M., Hayakawa, E., Nishino, T., Takeda, Y., Mukhopadhyay, A.K., Garg, S., Bhattacharya, S.K., Nair, G.B., and Nishibuchi, M., *J Clin Microbiol*, 35, 3150, 1997.

49. Okura, M., Osawa, R., Iguchi, A., Arakawa, E., Terajima, J., and Watanabe, H., *J Clin Microbiol*, 41, 4676, 2003.

50. Gonzalez-Escalona, N., Martinez-Urtaza, J., Romero, J., Espejo, R.T., Jaykus, L.A., and DePaola, A., *J Bacteriol*, 190, 2831, 2008.

51. Gonzalez-Escalona, N., Strain, E.A., De Jesus, A.J., Jones, J.L., and Depaola, A., *J Bacteriol*, 193, 3405, 2011.

52. Chowdhury, N.R., Stine, O.C., Morris, J.G., and Nair, G.B., *J Clin Microbiol*, 42, 1280, 2004.

53. Velazquez-Roman, J., Leon-Sicairos, N., de Jesus Hernandez-Diaz, L., and Canizalez-Roman, A., *Front Cell Infect Microbiol*, 3, 110, 2014.

54. Nair, G.B., Ramamurthy, T., Bhattacharya, S.K., Dutta, B., Takeda, Y., and Sack, D. A., *Clin Microbiol Rev*, 20, 39, 2007.

55. Morris, J.G., *Clin Infect Dis*, 37, 272, 2003.

56. Chowdhury, N.R., Chakraborty, S., Ramamurthy, T., Nishibuchi, M., Yamasaki, S., Takeda, Y., and Nair, G.B., *Emerg Infect Dis*, 6, 631, 2000.

57. Honda, T., Ni, Y.X., and Miwatani, T., *Infect Immun*, 56, 961, 1988.

58. Tsunasawa, S., Sugihara, A., Masaki, T., Sakiyama, F., Takeda, Y., Miwatani, T., and Narita, K., *J Biochem*, 101, 111, 1987.

59. Zhang, X.H., and Austin, B., *J Appl Microbiol*, 98, 1011, 2005.

60. Nishibuchi, M., Janda, J.M., and Ezaki, T., *Microbiol Immunol*, 40, 59, 1996.

61. Makino, K., Oshima, K., Kurokawa, K., Yokoyama, K., Uda, T., Tagomori, K., Iijima, Y., Najima, M., Nakano, M., Yamashita, A., Kubota, Y., Kimura, S., Yasunaga, T., Honda, T., Shinagawa, H., Hattori, M., and Iida, T., *Lancet*, 361, 743, 2003.

62. Panina, E.M., Mattoo, S., Griffith, N., Kozak, N.A., Yuk, M.H., and Miller, J.F., *Mol Microbiol*, 58, 267, 2005.

63. Ono, T., Park, K.S., Ueta, M., Iida, T., and Honda, T., *Infect Immun*, 74, 1032, 2006.

64. Boyd, E.F., Cohen, A.L., Naughton, L.M., Ussery, D.W., Binnewies, T.T., Stine, O.C., and Parent, M.A., *BMC Microbiol*, 8, 110, 2008.

65. Hurley, C.C., Quirke, A., Reen, F.J., and Boyd, E.F., *BMC Genomics*, 7, 104, 2006.

66. Wang, H.Z., Wong, M.M., O'Toole, D., Mak, M.M., Wu, R.S., and Kong, R.Y., *Appl Environ Microbiol*, 72, 4455, 2006.

67. Okura, M., Osawa, R., Arakawa, E., Terajima, J., and Watanabe, H., *J Clin Microbiol*, 43, 3533, 2005.

68. Tiruvayipati, S., Bhassu, S., Kumar, N., Baddam, R., Shaik, S., Gurindapalli, A.K., Thong, K.L., and Ahmed, N., *Gut Pathog*, 5, 37, 2013.

69. Jensen, R.V., Depasquale, S.M., Harbolick, E.A., Hong, T., Kernell, A.L., Kruchko, D.H., Modise, T., Smith, C.E., McCarter, L.L., and Stevens, A.M., *Genome Announce*, 1, 2013.

70. Hlady, W.G., and Klontz, K.C., *J Infect Dis*, 173, 1176, 1996.

71. Oliver, J. D., *Microb Ecol*, 65, 793, 2013.

72. Said, R., Volpin, G., Grimberg, B., Friedenstrom, S.R., Lefler, E., and Stahl, S., *J Hand Surg Br*, 23, 808, 1998.

73. Hoi, L., Larsen, J.L., Dalsgaard, I., and Dalsgaard, A., *Appl Environ Microbiol*, 64, 7, 1998.

74. Centers for Disease Control and Prevention (CDC), the Cholera and Other Vibrio Illness Surveillance (COVIS) system, Available at: https://www.cdc.gov/vibrio/surveillance.html

75. Bisharat, N., Agmon, V., Finkelstein, R., Raz, R., Ben-Dror, G., Lerner, L., Soboh, S., Colodner, R., Cameron, D.N., Wykstra, D.L., Swerdlow, D.L., and Farmer, J.J., *Lancet*, 354, 1421, 1999.

76. Tison, D.L., Nishibuchi, M., Greenwood, J.D., and Seidler, R.J., *Appl Environ Microbiol*, 44, 640, 1982.

77. Amaro, C., and Biosca, E.G., *Appl Environ Microbiol*, 62, 1454, 1996.

78. Zaidenstein, R., Sadik, C., Lerner, L., Valinsky, L., Kopelowitz, J., Yishai, R., Agmon, V., Parsons, M., Bopp, C., and Weinberger, M., *Emerg Infect Dis*, 14, 1875, 2008.

79. Rosche, T.M., Yano, Y., and Oliver, J.D., *Microbiol Immunol*, 49, 381, 2005.

80. Nilsson, W.B., Paranjype, R.N., DePaola, A., and Strom, M.S., *J Clin Microbiol*, 41, 442, 2003.

81. Roig, F.J., Sanjuan, E., Llorens, A., and Amaro, C., *Appl Environ Microbiol*, 76, 1328, 2010.

82. Froelich, B.A., Williams, T.C., Noble, R.T., and Oliver, J.D., *Appl Environ Microbiol*, 78, 3885, 2012.

83. Kim, Y.R., Lee, S.E., Kim, C.M., Kim, S.Y., Shin, E.K., Shin, D.H., Chung, S.S., Choy, H.E., Progulske-Fox, A., Hillman, J.D., Handfield, M., and Rhee, J.H., *Infect Immun*, 71, 5461, 2003.

84. Park, J.H., Cho, Y.J., Chun, J., Seok, Y.J., Lee, J.K., Kim, K.S., Lee, K.H., Park, S.J., and Choi, S.H., *J Bacteriol*, 193, 2062, 2001.

85. Morrison, S.S., Williams, T., Cain, A., Froelich, B., Taylor, C., Baker-Austin, C., Verner-Jeffreys, D., Hartnell, R., Oliver, J.D., and Gibas, C.J., *PLOS one*, 7, e37553, 2012.

86. Gulig, P.A., de Crecy-Lagard, V., Wright, A.C., Walts, B., Telonis-Scott, M., and McIntyre, L.M., *BMC genomics*, 11, 512, 2010.

87. Cohen, A.L., Oliver, J.D., DePaola, A., Feil, E.J., and Boyd, E.F., *Appl Environ Microbiol*, 73, 5553, 2007.

88. Ran Kim, Y., and Haeng Rhee, J., *Biochem Biophys Res Commun*, 304, 405, 2003.

89. Lee, J.H., Rho, J.B., Park, K.J., Kim, C.B., Han, Y.S., Choi, S.H., Lee, K.H., and Park, S.J., *Infect Immun*, 72, 4905, 2004.

90. Simpson, L.M., White, V.K., Zane, S.F., and Oliver, J.D., *Infect Immun*, 55, 269, 1987.

91. Kreger, A., and Lockwood, D., *Infect Immun*, 33, 583, 1981.

92. Lee, J.H., Kim, M.W., Kim, B.S., Kim, S.M., Lee, B.C., Kim, T.S., and Choi, S.H., *J Microbiol*, 45, 146, 2007.

93. Koo, B.S., Lee, J.H., Kim, S.C., Yoon, H.Y., Kim, K.A., Kwon, K.B., Kim, H.R., Park, J.W., and Park, B.H., *Int J Mol Med*, 20, 913, 2007.

94. Oliver, J.D., Wear, J.E., Thomas, M.B., Warner, M., and Linder, K., *Diagn Microbiol Infect Dis*, 5, 99, 1986.

95. Yoshida, S., Ogawa, M., and Mizuguchi, Y., *Infect Immun*, 47, 446, 1985.

96. Helms, S.D., Oliver, J.D., and Travis, J.C., *Infect Immun*, 45, 345, 1984.

97. Simpson, L.M., and Oliver, J.D., *Infect Immun*, 41, 644, 1983.

98. Wang, J., Sasaki, T., Maehara, Y., Nakao, H., Tsuchiya, T., and Miyoshi, S., *Microb Pathog*, 44, 494, 2008.

99. Brayton, P.R., Bode, R.B., Colwell, R.R., MacDonell, M.T., Hall, H.L., Grimes, D.J., West, P.A., and Bryant, T.N., *J Clin Microbiol*, 23, 104, 1986.

100. Pavia, A.T., Bryan, J.A., Maher, K.L., Hester, T.R., and Farmer, J.J., *Ann Intern Med*, 111, 85, 1989.

101. Hundenborn, J., Thurig, S., Kommerell, M., Haag, H., and Nolte, O., *Case Rep Med*, 2013, 610632, 2013.

102. Sakazaki, R., *Jpn J Med Sci Biol*, 21, 359, 1968.

103. Weis, K.E., Hammond, R.M., Hutchinson, R., and Blackmore, C.G., *Epidemiol Infect*, 139, 591, 2011.

104. Baker-Austin, C., Trinanes, J.A., Salmenlinna, S., Lofdahl, M., Siitonen, A., Taylor, N.G., and Martinez-Urtaza, J., *Emerg Infect Dis*, 22, 1216, 2016.

105. Baker-Austin, C., Trinanes, J., Gonzalez-Escalona, N., and Martinez-Urtaza, J., *Trends Microbiol*, 25, 76, 2017.

106. Vezzulli, L., Grande, C., Reid, P.C., Helaouet, P., Edwards, M., Hofle, M.G., Brettar, I., Colwell, R.R., and Pruzzo, C., *Proc Natl Acad Sci U S A*, 113, E5062, 2016.

107. McLaughlin, J.B., DePaola, A., Bopp, C.A., Martinek, K.A., Napolilli, N.P., Allison, C.G., Murray, S.L., Thompson, E.C., Bird, M.M., and Middaugh, J.P., *Engl J Med*, 353, 1463, 2005.

108. Centers for Disease Control and Prevention (CDC), *MMWR Morb Mortal Wkly Rep*, 54, 928, 2005.

109. Jutla, A., Whitcombe, E., Hasan, N., Haley, B., Akanda, A., Huq, A., Alam, M., Sack, R. B., and Colwell, R., *Am J Trop Med Hyg*, 89, 597, 2013.

110. Vezzulli, L., Colwell, R.R., and Pruzzo, C., *Microb Ecol*, 65, 817, 2013.

111. European Food Safety Authority (EFSA), Annual report of the emerging risks exchange networks 2015, 13, 1067E, 2016.

42

Yersinia

Ryan F. Relich and Meghan A. May

CONTENTS

Introduction

The genus *Yersinia* comprises Gram-negative rod-shaped bacteria that inhabit diverse ecosystems and includes several pathogens of humans and other animals. The most notorious species, *Yersinia pestis*, is the causative agent of plague, which is one of the oldest recorded infectious diseases. Plague has caused at least three major pandemics at approximately 600-year intervals starting in the 6th century AD. This pandemic began in Egypt and spread through the Middle East to Europe. The second pandemic occurred during the 14th century AD and was called The Black Death. It began in the area of the Black Sea, spread to Europe, and was responsible for the death of more than a quarter of Europe's population. The third plague pandemic, termed the Third Pandemic, began in 1855 in China and spread to India, Egypt, Portugal, Japan, Paraguay, Eastern Africa, Manila, Scotland, Australia, the United States, and several other countries. This infection is believed to have caused more than 150 epidemics of varying degrees until the 1950s.[1–3]

Because The Black Death was, according to historical accounts, characterized by swollen lesions (buboes) in the groin and elsewhere on the body, as was the Third Pandemic, scientists and historians have assumed that the earlier pandemic was an outbreak of the same disease. This led to the conclusion that The Black Death was caused by the bacterium *Y. pestis*, which is spread by fleas with the help of rodent reservoirs. However, because buboes are features of other diseases, too, this view has been questioned by some.

Taxonomy

The genus name *Yersinia* was coined by van Loghem in 1944 in honor of the French bacteriologist Alexandre Yersin who, in 1894, first isolated the causative agent of plague, "*Pasteurella pestis*."[4] This organism was subsequently renamed *Yersinia pestis*. In 2016, the genus *Yersinia*, along with the genera *Chania*, *Ewingella*, *Rhanella*, and *Serratia*, was assigned to the new family *Yersiniaceae*.[5] Later, the genera *Chimaeribacter*, *Nissabacter*, *Rouxiella*, *Samsonia*, and *Candidatus* Fukatsuia were also included in this family. Previously, these organisms were classified among the *Enterobacteriaceae*. This systematics change does not impact the binomial species names of any members of the genus *Yersinia*, and is not expected to affect diagnostic testing (e.g., interpretation of antimicrobial susceptibility test data) or regulations concerning disease reporting, biological containment, isolate transportation, biological safety, and biological security.

DOI: 10.1201/9781003099277-44

To date, there are 19 named *Yersinia* spp. and numerous as-of-yet unnamed isolates. Of the former, three species are important human pathogens: *Yersinia enterocolitica*, *Y. pestis*, and *Yersinia pseudotuberculosis*. The extremely high ribosomal RNA sequence identity and disparate clinical presentations (and consequent biocontainment controls) between *Y. pestis* and *Y. pseudotuberculosis* are a frustrating taxonomic paradox. The logistical challenges associated with altering the nomenclature of *Y. pestis* make resolving this instance of polyphyly impractical, as first measured and articulated by Bercovier *et al.*[6] *Y. pseudotuberculosis* has taxonomic precedent over *Y. pestis*, and therefore the informal designation "*Y. pseudotuberculosis* complex" was created to address this ambiguity. This group consists of *Y. pseudotuberculosis*, *Y. pestis*, *Yersinia similis*, and *Yersinia wautersii*.[7,8]

Other species in the genus include *Yersinia frederiksenii*, *Yersinia intermedia*, *Yersinia kristensenii*, *Yersinia aldovae*, *Yersinia bercovieri*, *Yersinia mollaretti*, *Yersinia aleksiciae*, *Yersinia hibernica*, *Yersinia entomophaga*, *Yersinia massiliensis*, *Yersinia nurmii*, *Yersinia pekkanenii*, *Yersinia ruckeri*, *Yersinia rohdei*, *Y. similis*, and *Y. wautersii*.[9,10] The status of these organisms as primary or opportunistic pathogens is equivocal, with the exception of *Y. ruckeri*, the unambiguous etiologic agent of enteric redmouth disease (ERM) of farmed fish.[11] Significant polyphyly still exists within *Y. enterocolitica* and *Y. kristensenii*, suggesting that further resolution of informal groups into extant novel species is likely.

Epidemiology

The natural reservoirs of *Y. pestis* are rodents (e.g., rats, mice, rabbits, prairie dogs, and ground squirrels). Organisms are usually spread from a reservoir to humans by fleas. Much less frequently, the organisms can be spread to humans through direct contact with infected animals. The Oriental rat flea, *Xenopsylla cheopis* (Figure 42.1), is considered the classic vector of plague; however, other flea species can be major vectors of endemic plague in the western parts of the United States. For example, ground squirrel fleas (*Oropsylla montana*) have been found to be vectors in California, and prairie dog and rock squirrel fleas (*Oropsylla hirsuta*, *O. montana*, and *Hoplopsyllus anomalus*) have been found to be vectors in New Mexico.[12,13]

Flea-mediated transmission of *Y. pestis* depends on masses of the organism blocking the flea's proventriculus foregut. This blockage causes the flea to be unable to feed, and the effect of which is the flea repeatedly biting a host in an attempt to obtain a blood meal, which optimizes transmission of *Y. pestis* from the flea to the host (e.g., a human). Should infection progress to the lungs (pneumonic plague; discussed later), aerosol-based transmission between humans is possible. Direct transmission from infected companion animals or wildlife to humans has also been reported.[14,15] The cat flea, *Ctenocephalides felis*, is a poor vector for plague, and cats usually become infected by eating an infected rodent rather than by a flea bite. Infected cats can transmit plague to humans through bites, scratches, and aerosols. Dogs rarely develop clinical signs during *Y. pestis* infection, but have played a role in transmission by bringing infected fleas into the proximity of their owners.[16] Because *Y. pestis* is often low on

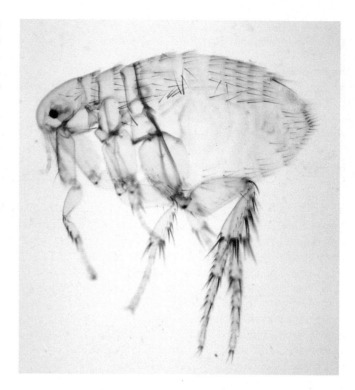

FIGURE 42.1 *Xenopsylla cheopis*, the Oriental rat flea, the primary vector of *Y. pestis*. Original magnification, 4x.

the diagnostic index of suspicion for dogs with severe pneumonia, untreated canine pneumonic plague has, in rare cases, lead to widespread human exposure.[17]

Prairie dogs have also emerged as important sources of human plague cases, both as wildlife reservoirs and as pets.[18,19] Plague is still endemic in many countries, most of which report at least some human cases each year. Madagascar has had notably high levels of seasonal plague each year since its introduction in 1990, with most cases occurring between September and April. The plague season of 2017–2018 was dramatically intense, with that year's suspected and confirmed case total exceeding 2,400.[20] This epidemic consisted largely of urban cases and a higher proportion of pneumonic plague, whereas most of the world's cases occur in rural settings and are bubonic plague. These differences likely led to the higher incidence and are evidence of the potential for higher-than-expected plague case-loads under certain circumstances.

Y. enterocolitica is also found in a wide variety of animals, including humans, but the primary reservoir for this species is swine.[3] In addition, as *Y. enterocolitica* can survive for a time in the environment, contaminated soil, water, and vegetation can be a part of the chain of transmission to animals and humans.[3] *Y. enterocolitica* is an infrequent cause of disease in the United States, but is common in northern Europe.[21,22]

Y. pseudotuberculosis is the most uncommon of these three human pathogens. It is found in a wide variety of domestic and wild animals, including fowl. The main reservoirs for this organism are rodents, rabbits, and wild birds. Most cases of disease associated with *Y. pseudotuberculosis* are reported in European children. Although *Y. enterocolitica* and *Y. pseudotuberculosis* are found worldwide, they are primarily seen in the warmer areas of Europe. They can also be found in Asia, South Africa, Australia, and the United States.[21,22]

Characteristics of Pathogenic *Yersinia* spp.

Cellular Morphology

Yersinia spp. are nonspore-forming Gram-negative bacilli measuring 0.5–0.8 μm wide by 1–3 μm long.[3] They appear in Gram-stained smears as small, individual cells when cultures on solid media are used to prepare smears (Figure 42.2). In smears from broth (e.g., blood culture broth), they can form short chains.[3] Staining with either Wright or Wayson stains can demonstrate bipolar staining, a phenomenon in which the cell poles stain, but the center of the cell does not. Note that bipolar staining is not a defining characteristic of *Yersinia* and it is not always obvious. *Y. pestis* is nonmotile, but all other species are motile by peritrichous flagella at 22–30°C; all species are nonmotile at 37°C.[3]

Culture

Yersinia spp. will grow on a variety of nonselective and selective media under aerobic conditions. Optimal growth temperatures range from 25–28°C, but many species can grow at temperatures as low as 0°C to as high as 45°C. In clinical microbiology laboratories, cultures for *Y. enterocolitica* are often incubated at room temperature to optimize recovery of this organism; however, growth will still occur at 35–37°C.[3] Unlike other closely related bacteria, *Yersinia* spp. grow slower, so colonies are often tiny (<1 mm) after 24 hours of incubation at their optimal growth temperature.

On sheep blood agar, colonies are nonhemolytic and tiny (<1 mm) after 24 hours of incubation. Following 48 hours of incubation, colonies are larger (~2 mm in size), have an irregular edge, and are often gray-white and opaque (Figure 42.3A). Older cultures (>48 hours) of *Y. pestis* exhibit a fried-egg morphology; the central mass of the colony is more opaque than the peripheral growth. On MacConkey agar, colonies are small and colorless, as lactose is not fermented. Cefsulodin-Irgasan®-novobiocin (CIN) agar is a medium developed specifically to selectively isolate *Yersinia* spp. from stool and other contaminated specimen types. The growth of most other Gram-negative bacilli is inhibited by CIN agar, but some, including

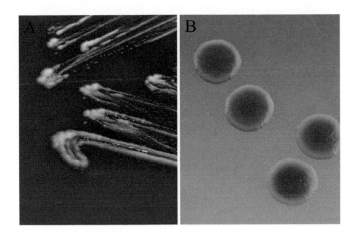

FIGURE 42.3 Colonial characteristics of *Y. pestis* grown on blood agar after 48 hours of incubation (A) and *Y. enterocolitica* growing on CIN agar (B). On CIN agar, isolates of *Y. enterocolitica* often display the characteristic "bull's eye" appearance in which the center of the colony is more red than its lighter periphery. (Image courtesy of Hardy Diagnostics, www.HardyDiagnostics.com.)

Morganella spp. and *Serratia* spp., can grow and appear similar to *Yersinia* colonies.[3] Although colonies of these organisms are often larger than those of *Yersinia*. On CIN agar, *Y. enterocolitica* often develops characteristic "bull's eye" colonies that have a dark red center surrounded by a lighter periphery (Figure 42.3B).

Identification of Isolates

Physiochemical identification of isolates requires the use of a variety of biochemical media or multisubstrate panels, such as those used with manual (e.g., analytical profile index [API]) and automated (e.g., VITEK 2, MicroScan WalkAway) phenotypic identification systems. In general, all *Yersinia* spp. ferment glucose, are cytochrome oxidase-negative, and reduce nitrate to nitrite. Table 42.1 lists distinguishing biochemical test results for human-pathogenic yersiniae.

Note that *Y. pestis* is often misidentified as *Y. pseudotuberculosis* by automated phenotypic identification systems, so all isolates identified as such should be tested further to rule out *Y. pestis* as the true identity. Sentinel-level laboratories (e.g., clinical microbiology laboratories) should perform a minimal amount of testing on suspected *Y. pestis* isolates to avoid excessive amplification and manipulation of cultures, which could inadvertently lead to laboratory-acquired infections. If laboratories cannot rule out *Y. pestis* as the identity of an unknown isolate, it should be referred to a Laboratory Response Network (LRN) Reference Laboratory (e.g., a state public health laboratory) for additional testing and then to a National Laboratory for further characterization. Guidelines for ruling out are available from the American Society for Microbiology's LRN Sentinel Level Clinical Laboratory Protocols website: https://www.asm.org/Articles/Policy/Laboratory-Response-Network-LRN-Sentinel-Level-C.

Newer identification technologies, including nucleic acid amplification tests that enable detection of *Yersinia* spp. DNA directly from clinical specimens (e.g., *Y. enterocolitica* DNA from stool specimens) and matrix-assisted laser desorption/ionization time-of-flight mass spectrometry that permits rapid

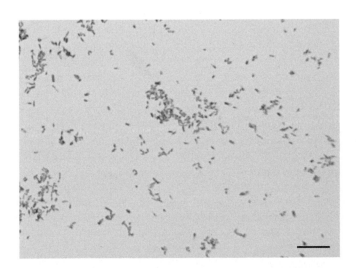

FIGURE 42.2 Gram stain of *Y. pestis* prepared from a blood agar culture that was incubated for 48 hours. Cells are small, plump, and stain Gram-negative. Scale bar, 10 μm.

TABLE 42.1

Biochemical Test Results Helpful for Identification of Human-Pathogenic *Yersinia* Species[a]

Biochemical Results[b]	*Yersinia enterocolitica*	*Yersinia pestis*	*Yersinia pseudotuberculosis*
Indole production	V	–	–
Voges-Proskauer	V	–	–
Motility	+	–	+
Ornithine decarboxylase	+	–	–
Urease	+	–	+
Esculin hydrolysis	–	+	+
Carbohydrate utilization			
Arabinose	+	+	–
Cellobiose	+	–	–
Inositol	+	–	–
Melibiose	–	–	+
Rhamnose	–	–	weak +
Sorbitol	+	–	–
Sorbose	+	–	–
Sucrose	+	–	–
Xylose	V	+	+

[a] Table adapted from reference [3].

[b] Reactions were determined by incubation of biochemical media at 25–28°C. Abbreviations are as follows: (+), positive reaction; (–), negative reaction; (V), variable reaction

identification of cultured isolates, have supplanted traditional biochemical testing methods in many diagnostic laboratories. In doing so, these technologies have increased the accuracy and precision of *Yersinia* spp. detection and identification and, in the majority of cases, have substantially decreased the turnaround time for disease diagnosis.[23] A number of publications describing the performance characteristics of these technologies applied to the identification of *Yersinia* spp. are available.[24–28]

Laboratory Safety

Extreme care should be used when handling cultures suspected of being, or confirmed to be, *Y. pestis*. Currently, *Y. pestis* is classified as a CDC/USDA Animal and Plant Health Inspection Service Tier 1 Select Agent that requires biosafety level (BSL)-3 containment, security, and work practices for handling, storage, and inactivation of specimens and cultures containing virulent *Y. pestis*. All manipulations of virulent *Y. pestis* strains, including, but not limited to, cultivation, storage, transfer, inactivation, animal inoculation, and disposal, requires documentation to ensure compliance with Select Agent rules, which are available on the Federal Select Agent Program website, https://www.selectagents.gov/index.html. *Y. pestis* strains containing deletions in the *pgm* locus, which are designated Pgm⁻, including strain EV and its various substrains, are excluded from Select Agent rules and regulations.[29] For these strains, standard BSL-2 containment, security, and work practices are appropriate for manipulation of cultures.

Other *Yersinia* spp., including *Y. enterocolitica* and *Y. pseudotuberculosis*, can be safely handled at BSL-2; however, a thorough laboratory risk assessment must be conducted prior to any diagnostic or experimental work to ensure that all appropriate procedures (e.g., spill cleanup and emergency response procedures), engineering controls (e.g., biological safety cabinet [BSC]), and personal protective equipment (PPE; e.g., gloves and laboratory coats) are in place prior to commencement of work.

Parcels containing cultures of virulent strains of *Y. pestis* must be shipped in accordance with the International Air Transport Association's (IATA) Dangerous Goods Regulations and, if cultures are shipped within the United States, the U.S. Department of Transportation's (U.S. DOT) Hazardous Materials Regulations.[30,31] According to both IATA and the U.S. DOT, *Y. pestis* cultures must be shipped according to Category A Infectious Substance Affecting Humans (UN 2814) packaging and shipping regulations.[30,31]

Diseases

Plague

Bubonic Plague

Following a bite from an infected flea, *Y. pestis* travels to the nearest lymph node where they are ingested by macrophages. Once inside, they rapidly multiply to high levels. The ensuing inflammatory reaction causes the formation of buboes, which are, although not specific to bubonic plague, a hallmark of this disease. Buboes are hot, red, and painful oval swellings of the skin and surrounding tissue that range in size from 1 cm to 10 cm in length.[32] Patients also usually present with fever.

Septicemic Plague

Y. pestis can spread to the bloodstream and cause septicemic plague. Bloodstream infections often result in the dissemination of *Y. pestis* to various organs and, as a result, patients can develop disseminated intravascular coagulation, bleeding, organ failure, and irreversible shock. Infected patients often develop necrotic lesions on the skin, which turn black, which is why this disease was named "The Black Death." Blockage of vessels in the fingers, toes, ears, and nose can lead to gangrene. In cases of primary septicemic plague, buboes may be absent, leading to clinical misdiagnosis.[3]

Pneumonic Plague

Primary pneumonic plague is seen when a person inhales *Y. pestis* and the organisms proliferate within the lungs. Persons with either the primary or secondary form of pneumonic plague are very contagious and can easily spread the infection from person to person via aerosols. Pneumonic plague is rapidly fatal, which makes rapid diagnosis critical.[3] These patients normally present with cough, hemoptysis, and may or may not have buboes.[3]

Miscellaneous Plague Manifestations

Plague meningitis is a rare complication of infection that occurs as acute early disease or as a result of inadequate treatment for bubonic plague. This syndrome is characterized by headache, fever, meningismus, and cerebrospinal fluid (CSF) pleocytosis. *Y. pestis* can also manifest as pharyngeal plague, which is characterized by a painful, inflamed pharynx and associated local lymphadenopathy, which may mimic acute tonsillitis. This form of disease can rarely follow either ingestion or inhalation of the bacteria. Plague can also rarely manifest as gastrointestinal symptoms, including nausea, vomiting, diarrhea, and abdominal pain, which may precede the formation of buboes.[3]

Yersinia enterocolitica Infections

Yersiniosis

Gastrointestinal infections caused by *Y. enterocolitica* are associated with consuming contaminated food or water. Yersiniosis, which is caused by *Y. enterocolitica*, is most often transmitted by consumption of raw or undercooked pork products (e.g., chitterlings) and unpasteurized milk, but other meats and beverages are also implicated in transmission.[3] Symptoms of yersiniosis in children include watery and occasionally bloody diarrhea, fever, and abdominal pain. Mesenteric lymphadenitis is a potential complication of yersiniosis, and its symptoms (fever and right-sided abdominal pain) mimic those of appendicitis.[3] These symptoms can last for 2–3 weeks and normally occur in children. A reactive polyarthritis may also develop in patients after 2 to 14 days of infection. This process involves two to four joints (knees, ankles, toes, fingers, wrists), and symptoms can persist for up to 4 months. Less commonly seen complications associated with *Y. enterocolitica* infections include ankylosing spondylitis, reactive arthritis, erythema nodosum, exudative pharyngitis, pneumonia, empyema, lung abscess, septicemia, endocarditis, and mycotic aneurysms.[3]

Yersinia pseudotuberculosis Infections

Far East Scarlet-Like Fever

Like *Y. enterocolitica*, *Y. pseudotuberculosis* causes self-limited mesenteric lymphadenitis; however, in Japan and Russia, *Y. tuberculosis* is associated with sporadic outbreaks of a disease characterized by severe systemic inflammation. This condition is called Far East scarlet-like fever. Signs and symptoms often include hyperemic skin covered in a punctate rash, submandibular lymphadenopathy, hyperemic tongue, pale nasolabial triangle, and tonsillitis, mimicking those of scarlet fever.[33]

Diagnosis, Treatment, and Prevention

For diagnosis of plague, aspirates of buboes and blood should be cultured.[3] Additional specimens can include sputum, bronchial washes, swabs from skin lesions or the pharyngeal mucosa, and CSF. Aspirates of buboes are obtained using sterile saline and a small needle and syringe. A small amount (up to 2 mL) of sterile saline is infused into the bubo and is then aspirated. Gram-stained or Wayson-stained direct smears of bubo aspirates often reveal an acute inflammatory infiltrate, characterized by the presence of abundant polymorphonuclear neutrophils, and small, plump Gram-negative bacilli. Wayson stain will demonstrate bipolar staining; however, as mentioned before, this attribute is not always apparent and is not specific to *Y. pestis* or other *Yersinia* spp. Refer to laboratory safety and federal Select Agent mandates for ruling out and reporting *Y. pestis*.

Serological diagnosis of plague is also possible, and in the United States, a number of commercial reference laboratories offer testing. The passive hemagglutination test using F1 antigens from the organism is performed on acute and/or convalescent serum samples. A four-fold rise in IgG titer from acute to convalescent samples, or a single titer of greater than 1:16, is presumptive evidence for disease, and a titer of 1:128 is diagnostic for infection.[3] Most patients will have antibodies by the second or third week after infection. Additional diagnostic techniques include polymerase chain reaction (PCR) for detecting the nucleic acids of the organisms, and enzyme immunoassay for the detection of antigens associated with the organisms.[4]

Diagnosis of *Y. enterocolitica* and *Y. pseudotuberculosis* infections normally depends on the isolation of the organisms from stool, mesenteric lymph nodes, peritoneal fluid, blood, or abscesses. Rapid, multiplex real-time PCR assays are now commercially available to clinical laboratories for detecting *Y. enterocolitica* in stool specimens. These syndromic panels permit the simultaneous detection of several pathogens that are frequently associated with specific disease syndromes, including enteritis.

If indicated, antimicrobial susceptibility testing (AST) of isolates can be performed by disk diffusion or methods that determine the minimal inhibitory concentration of a drug or drugs (e.g., microbroth dilution testing). AST of *Y. pestis* should not be attempted by routine clinical microbiology laboratories; instead, isolates should be referred to an LRN Reference Laboratory or National Laboratory for this testing. Frequently updated breakpoints for interpretation of AST results are available from the Clinical and Laboratory Standards Institute and other sources.

If left untreated, plague can result in a 50% mortality rate. Streptomycin is the drug of choice for treatment of all types of plague.[3] Alternatively, gentamicin or gentamicin in combination with doxycycline can be used.[3] Chloramphenicol can be used when high tissue concentrations are necessary, as with cases of meningitis. Resistant strains of *Y. pestis* have rarely been encountered, and treatment does not seem to select for a resistant population of organisms, as relapses do not occur.

Infections caused by *Y. enterocolitica* can be treated with aminoglycosides, chloramphenicol, tetracycline, trimethoprim-sulfamethoxazole, piperacillin, ciprofloxacin, and the third-generation cephalosporins.[3,22] *Y. pseudotuberculosis* is normally susceptible to the aminoglycosides, chloramphenicol, tetracycline, and cephalosporins; although, therapy is usually not warranted in patients with mesenteric adenitis.[3] Septicemia caused by either of these organisms is usually greater than 50% fatal despite appropriate therapy.

A formalin-killed plague vaccine exists, but is only available currently from a manufacturer in Australia. It is not protective against the pulmonary form of the disease, and immunity requires several doses. It is rarely used today except for personnel in research laboratories who work with *Y. pestis*. Research is currently underway to produce a vaccine that will be effective against the pulmonary form of the disease, as concerns mount over this organism potentially being used as a biological weapon.

Persons living in endemic areas should take precautions against becoming infected with the plague bacteria. Protective measures against exposure to rodents and their fleas are of utmost importance. Public health measures aimed at curbing the transmission of infections caused by *Y. enterocolitica* and *Y. pseudotuberculosis* include appropriate sanitation measures in food preparation and not eating undercooked pork.[3]

REFERENCES

1. Demeure C, Dussurget O, Fiol GM, Le Guern AS, Savin C, Pizarro-Cerdá J. 2019. *Yersinia pestis* and plague: An updated view on evolution, virulence determinants, immune subversion, vaccination and diagnostics. *Microbes Infect.* 21(5–6):202–212.
2. Bramanti B, Dean KR, WallØe L, Chr Stenseth N. 2019. The third plague pandemic in Europe. *Proc Biol Sci.* 286(1901):20182429.
3. Kingry LC, Tarr CL, Petersen JM. 2019. Yersinia, p. 751–764, in *Manual of Clinical Microbiology*, 12th ed. Carroll KC, Pfaller MA, Landry ML, McAdam AL, Patel R, Richter SS, Warnock DW, eds. ASM Press, Washington, D.C.
4. van Loghem JJ. 1944. The classification of the plague bacillus. *Antonie van Leeuwenhoek.* 10:15–16.
5. Adeolu M, Alnajar S, Naushad S, S Gupta R. 2016. Genome-based phylogeny and taxonomy of the 'Enterobacteriales': Proposal for *Enterobacterales* ord. nov. divided into the families *Enterobacteriaceae*, *Erwiniaceae* fam. nov., *Pectobacteriaceae* fam. nov., *Yersiniaceae* fam. nov., *Hafniaceae* fam. nov., *Morganellaceae* fam. nov., and *Budviciaceae* fam. nov. *Int J Syst Evol Microbiol.* 66(12):5575–5599.
6. Bercovier H, Mollaret HH, Alonso JM, Brault J, Fanning GR, Steigerwalt AG, Brenner DJ. 1980. Intra- and interspecies relatedness of *Yersinia pestis* by DNA hybridization and its relationship to *Yersinia pseudotuberculosis*. *Curr Microbiol.* 4:225–229.
7. Laukkanen-Ninios R, Didelot X, Jolley KA, Morelli G, Sangal V, Kristo P, Brehony C, Imori PF, Fukushima H, Siitonen A, Tseneva G, Voskressenskaya E, Falcao JP, Korkeala H, Maiden MC, Mazzoni C, Carniel E, Skurnik M, Achtman M. 2011. Population structure of the *Yersinia pseudotuberculosis* complex according to multilocus sequence typing. *Environ Microbiol.* 13(12):3114–3127.
8. Savin C, Martin L, Bouchier C, Filali S, Chenau J, Zhou Z, Becher F, Fukushima H, Thomson NR, Scholz HC, Carniel E. 2014. The *Yersinia pseudotuberculosis* complex: Characterization and delineation of a new species, *Yersinia wautersii*. *Int J Med Microbiol.* 30(3-4):452–463.
9. Murros-Kontiainen A, Johansson P, Niskanen T, Fredriksson-Ahomaa M, Korkeala H, Björkroth J. 2011. *Yersinia pekkanenii* sp. nov. *Int J Syst Evol Microbiol.* 61(Pt 10):2363–2367.
10. Hurst MR, Becher SA, Young SD, Nelson TL, Glare TR. 2011. *Yersinia entomophaga* sp. nov., isolated from the New Zealand grass grub *Costelytra zealandica*. *Int J Syst Evol Microbiol.* 61(Pt 4):844–849.
11. Kumar G, Menanteau-Ledouble S, Saleh M, El-Matbouli M. 2015. *Yersinia ruckeri*, the causative agent of enteric redmouth disease in fish. *Vet Res.* 46:103.
12. Maupin GO, Beard ML, Hinkson G, Barnes AM, Craven RB. 1991. Studies on the control of plague in the western United States: Laboratory trials of six insecticide dust formulations applied to soil for the control of the plague vector *Oropsylla montana* (Siphonaptera: Ceratophyllidae). *J Med Entomol.* 28(6):770–775.
13. Davis RM, Smith RT, Madon MB, Sitko-Cleugh E. 2002. Flea, rodent, and plague ecology at Chuchupate Campground, Ventura County, California. *J Vector Ecol.* 27(1):107–127.
14. Oyston PC, Williamson D. 2011. Plague: Infections of companion animals and opportunities for intervention. *Animals (Basel).* 1(2):242–255.
15. Bevins SN, Tracey JA, Franklin SP, Schmit VL, Macmillan ML, Gage KL, Schriefer ME, Logan KA, Sweanor LL, Alldredge MW, Krumm C, Boyce WM, Vickers W, Riley SP, Lyren LM, Boydston EE, Fisher RN, Roelke ME, Salman M, Crooks KR, Vandewoude S. 2009. Wild felids as hosts for human plague, Western United States. *Emerg Infect Dis.* 15(12):2021–2024.
16. Centers for Disease Control and Prevention (CDC). 2011. Notes from the field: Two cases of human plague—Oregon, 2010. *MMWR Morb Mortal Wkly Rep.* 60(7):214.
17. Schaffer PA, Brault SA, Hershkowitz C, Harris L, Dowers K, House J, Aboellail TA, Morley PS, Daniels JB. 2019. Pneumonic plague in a dog and widespread potential human exposure in a veterinary hospital, United States. *Emerg Infect Dis.* 25(4):800–803.
18. Roth JD. 2019. Sylvatic plague management and prairie dogs—A meta-analysis. *J Vector Ecol.* 44(1):1–10.
19. Chomel BB. 2015. Diseases transmitted by less common house pets. *Microbiol Spectr.* 3(6):1–23.
20. Randremanana R, Andrianaivoarimanana V, Nikolay B, Ramasindrazana B, Paireau J, Ten Bosch QA, Rakotondramanga JM, Rahajandraibe S, Rahelinirina S, Rakotomanana F, Rakotoarimanana FM, Randriamampionona LB, Razafimbia V, De Dieu Randria MJ, Raberahona M, Mikaty G, Le Guern AS, Rakotonjanabelo LA, Ndiaye CF, Rasolofo V, Bertherat E, Ratsitorahina M, Cauchemez S, Baril L, Spiegel A, Rajerison M. 2019. Epidemiological

characteristics of an urban plague epidemic in Madagascar, August-November, 2017: An outbreak report. *Lancet Infect Dis.* 19(5):537–545.

21. Chlebicz A, Śliżewska K. 2018. Campylobacteriosis, salmonellosis, yersiniosis, and listeriosis as zoonotic foodborne diseases: A review. *Int J Environ Res Public Health.* 15(5):1–25.

22. Drummond N, Murphy BP, Ringwood T, Prentice MB, Buckley JF, Fanning S. 2012. *Yersinia enterocolitica*: A brief review of the issues relating to the zoonotic pathogen, public health challenges, and the pork production chain. *Foodborne Pathog Dis.* 9(3):179–189.

23. Buchan BW, Ledeboer NA. 2014. Emerging technologies for the clinical microbiology laboratory. *Clin Microbiology Rev.* 27(4):783–822.

24. Ayyadurai S, Flaudrops C, Raoult D, Drancourt M. 2010. Rapid identification and typing of *Yersinia pestis* and other *Yersinia* species by matrix-assisted laser desorption/ionization time-of-flight (MALDI-TOF) mass spectrometry. *BMC Microbiol.* 10:285.

25. Lasch P, Wahab T, Weil S, Pályi B, Tomaso H, Zange S, Kiland Granerud B, Drevinek M, Kokotovic B, Wittwer M, Pflüger V, Di Caro A, Stämmler M, Grunow R, Jacob D. 2015. Identification of highly pathogenic microorganisms by matrix-assisted laser desorption ionization-time of flight mass spectrometry: Results of an interlaboratory ring trial. *J Clin Microbiol.* 53(8):2632–2640.

26. Morka K, Bystron J, Bania J, Korzeniowska-Kowal A, Korzekwa K, Guz-Regner K, Bugla-Ploskonska G. 2018. Identification of *Yersinia enterocolitica* isolates from humans, pigs and wild boars by MALDI-TOF MS. *BMC Microbiol.* 18(1):86.

27. Khare R, Espy MJ, Cebelinski E, Boxrud D, Sloan LM, Cunningham SA, Pritt BS, Patel R, Binnicker MJ. 2014. Comparative evaluation of two commercial multiplex panels for detection of gastrointestinal pathogens by use of clinical stool specimens. *J Clin Microbiol.* 52(10):3667–3673.

28. Tilmanne A, Martiny D, Quach C, Wautier M, Vandenberg O, Lepage P, Hallin M. 2019. Enteropathogens in paediatric gastroenteritis: Comparison of routine diagnostic and molecular methods. *Clin Microbiol Infect.* 25(12):1519–1524.

29. CDC Division of Select Agents and Toxins, APHIS Agriculture Select Agent Services. Select Agents and Toxins Exclusions. Federal Select Agent Program. https://www.selectagents.gov/SelectAgentsandToxinsExclusions.html. Last accessed, January 17, 2020.

30. CDC Division of Select Agents and Toxins, APHIS Agriculture Select Agent Services. Guidance on the transfer of Select Agents and Toxins. https://www.selectagents.gov/tgd-intro.html. Last accessed, January 17, 2020.

31. IATA. Dangerous Goods Regulations. https://www.iata.org/en/publications/dgr/. Last accessed, January 17, 2020.

32. Southwick, F. 2007. Bioterrorism, p. 357–358, in *Infectious Diseases, a Clinical Short Course,* 2nd ed. Southwick F, ed. McGraw Hill, New York, NY.

33. Amphlett A. 2015. Far East scarlet-like fever: A review of the epidemiology, symptomatology, and role of superantigenic toxin, *Yersinia pseudotuberculosis*-derived mitogen A. *Open Forum Infec Dis.* 3(1):1–7.

43

Other Anaerobic Bacteria: Bacteroides, Porphyromonas, Prevotella, Tannerella, Fusobacterium, and Gram-positive Anaerobic Cocci

Joseph J. Zambon and Violet I. Haraszthy

CONTENTS

Introduction

Anaerobic bacteria live in environments with reduced oxygen levels and include both commensals microorganisms and microbial pathogens. Anaerobic pathogens are frequently components of the resident (normal) microflora. They are opportunistic pathogens that can cause disease if, for example, they are traumatically inoculated into adjacent sites or by elimination of microbial competitors following antibiotic therapy. Resident anaerobic bacteria can also become pathogenic by gaining access to the bloodstream and being transported from their normal ecologic niche to distant sites in the body. Anaerobic bacteria that normally reside in the human oral cavity—members of the genera *Porphyromonas* and *Prevotella*— can be found in atherosclerotic plaques (Haraszthy et al., 2000; Kozarov et al., 2006; Zhong et al., 2008; Chhibber-Goel et al., 2016). A number of oral species have been associated with atherosclerotic plaques including *A. actinomycetemcomitans, C. pneumoniae, C. rectus, E. corrodens, E. hormaechei, F. necrophorum, F. nucleatum, H. pylori, M. pneumoniae, P. aeruginosa, P. endodontalis, P. gingivalis, P. intermedia, P. luteola, P. nigrescens, S. gordonii, S. mitis, S. mutans, S. oralis, S. sanguinis, T. denticola, T. forsythia*, and *Veillonella* (Chhibber-Goel et al., 2016). Many species, including *Porphyromonas gingivalis* and *Prevotella intermedia*, have been identified by polymerase chain reaction (PCR) amplification of bacterial 16S rDNA and species-specific DNA probes. Consequently, anaerobic species such as these that have their primary ecologic niche in dental plaque in the human oral cavity are thought to gain access to the blood stream and become part of atherosclerotic plaques, possibly playing a role in their formation.

As in the case of *P. gingivalis* and *P. intermedia* in atherosclerotic plaques, anaerobic infections are typically mixed infections comprising several species almost evenly divided between anaerobes and facultative species. The microbial ecology of these mixed anaerobic infections is mutually sustaining with facultative species reducing oxygen levels sufficient to permit the growth of the anaerobic species, while anaerobic species inhibit phagocytosis by host immune cells and provide metabolic end products that are necessary for the growth of the facultative species (Styrt and Gorbach, 1989). The importance of mixed infections in human disease is well known and was shown in studies by Smith in 1930. He found that individual bacterial species isolated from what at the time was known as Vincent's infection, and is now called necrotizing gingivitis, were individually unable to cause disease in animal models, but could cause disease when two or more species were used.

Environments with reduced oxygen tension favor anaerobes. Certain diseases or traumatic injuries can reduce oxygen levels in tissues favoring anaerobic infection. Certain disease processes or traumatic injuries can diminish blood flow to a tissue and consequently reduce tissue oxygen levels favoring anaerobic infection. For example, microvascular changes associated with diabetes mellitus can diminish blood flow, particularly to the extremities, and make them susceptible to anaerobic infections as frequently seen in diabetic foot ulcers. In a study of 900 clinical isolates

DOI: 10.1201/9781003099277-45

collected from patients with intra-abdominal and diabetic foot infections, 61% were anaerobes, of which 56% were *Bacteroides* (Edmiston et al., 2004). Similarly, a prominent feature of chronic periodontitis is the development of epithelial-lined soft tissue pockets around the teeth, sometimes approaching 1 cm in depth. In the depth of these pockets, oxygen is decreased and the proportion of anaerobes in the biofilm is increased (Mombelli et al., 1996; Tanaka et al., 1998). Certain anatomic sites have decreased vascularity with aging favoring the growth of anaerobic bacteria. This occurs in the dental pulp in teeth where blood flow is reduced over time as the pulp chamber narrows with the deposition of secondary dentin. Carious lesions in the dental pulp can become infected with strict anaerobes such as *Porphyromonas endodontalis*, leading to odontogenic abscess (Siqueira et al., 2001; Narayanan and Vaishnavi, 2010).

Culture and identification of anaerobic microorganisms is expensive and limited in clinical settings (Ortiz and Sande, 2000). It is often reserved for situations in which empirical therapy has failed to produce a satisfactory result. Clinical microbiological tests for anaerobes often have limited effect on therapy. Less than 1% of anaerobic blood cultures are associated with clinically significant anaerobic bacteremia. Although mortality for these infections is high, patients frequently receive presumptive treatment with appropriate antibiotics based not on the results of anaerobic culture, but on clinical presentation suggestive of an anaerobic infection (Byrd and Roy, 2003). The results of blood cultures of patients with community-acquired pneumonia rarely alter the empiric choice of antibiotics (Ramanujam and Rathlev, 2006).

DNA-based technologies have largely supplanted traditional methods for the identification of bacterial pathogens, including anaerobic bacteria. These methods depend on the determination and use of nucleic acid sequences specific for the target microorganism or microbial taxon. They generally require traditional microbial culture and sequencing of the microbial nucleic acids. That is, the microorganism must be cultivable in order to isolate and sequence the nucleic acids. The species or group-specific sequence can then be employed in a number of different formats such as real time PCR, cloned, sequenced and compared to a database of nucleic acid sequences such as GenBank of the National Center for Biotechnology Information (United States), European Molecular Biology Laboratory Nucleotide Sequence Database, or the DNA Databank of Japan. The direct amplification of microbial nucleic acids, sometimes referred to as broad-range PCR, can identify both cultivable and noncultivable bacteria (Relman, 2002; Johnson et al., 2019). This method is based on the highly conserved nature of 16S ribosomal RNA within species and genera. Analysis of the 16S rRNA sequence facilitates the classification of bacteria and defines phylogenetic relationships between species. For example, *Bacteroides splanchnicus*, which can be phenotypically distinguished from other members of the genera by its production of *n*-butyric acid, is distinct from the genus *Bacteroides* as shown by analysis of 16S rRNA (Olsen and Shah, 2003). Conversely, bacteria that are phenotypically indistinguishable from other species are identified as novel species based on 16S rRNA sequence. Thus, nucleic acid sequencing has greatly increased the number of bacterial species, including many previously never-cultured anaerobic bacteria among the *Bacteroides* and *Fusobacteria*. Consequently, while the identification and characterization of cultivable anaerobic bacteria,

including *Bacteroides* and *Fusobacterium*, have previously relied on phenotypic features, biochemical tests, % G + C, and DNA–DNA hybridization, the identification of both cultivable and noncultivable microorganisms is based on nucleic acid analysis (Clarridge, 2004).

Bacteroides, Porphyromonas, and Prevotella

Of the genera within the family *Bacteroidaceae* (class: Bacteroidetes; order: Bacteroidales), only a few genera have been associated with significant disease in humans. These include *Bacteroides, Porphyromonas, Prevotella*, and *Fusobacteria*, the latter genus to be considered separately.

Bacterial species within the genera *Bacteroides, Porphyromonas*, and *Prevotella* are Gram-negative, obligately anaerobic (sometimes strictly anaerobic, sometimes exhibiting various degrees of aerotolerance), nonmotile (some with twitching motility), and nonspore-forming. The bacterial cells generally range in size from coccobacillus to long rods. The species grow at 37°C, particularly on blood agar. Colonies are generally 1–3 mm in diameter, circular, with an entire edge, low convex, smooth, and shiny. Strains are generally not hemolytic, but some strains produce small amounts of hemolysin and some are beta hemolytic. Strains generally ferment carbohydrates although, as described later, sugar fermentation has been a major feature in distinguishing species and in the taxonomy of this group of microorganisms. They can produce gas and acid, including volatile fatty acids such as acetic and succinic acids. The genera vary in the guanine and cytosine contents of the DNA. For the genus *Bacteroides*, the G + C content is 40–48 mol%. For the *Porphyromonas*, the G + C content is 40–55 mol%; and for the *Prevotella*, the G + C content is 40–60 mol%.

Both *Bacteroides* and *Prevotella* have a cell-wall peptidoglycan based on mesodiaminopimelic acid, similar to other Gram-negative bacteria, and as distinguished from the *Porphyromonas*, which have a cell-wall peptidoglycan based on lysine (Shah et al., 1976).

In the human body, species in these three genera colonize a variety of mucosal surfaces—oral, pharyngeal, gastrointestinal, and genitourinary—where they exist as part of the normal (resident) microflora and where they can cause opportunistic infections. As resident species, they are important in stabilizing the microbial ecology on mucosal surfaces and in preventing colonization by exogenous pathogens, a process referred to as colonization resistance (van der Waaij et al., 1972). These species, particularly the *Bacteroides*, are major constituents of the microflora throughout the human gastrointestinal tract where they exist in close contact with the mucosa of the small and large intestines. They outnumber aerobic bacteria in the intestine by 1,000:1 and are associated with diseases of the colon. These genera, particularly the *Porphyromonas* and the *Prevotella*, rarely the *Bacteroides*, are also found in the upper gastrointestinal tract. Like *Bacteroides* in feces in the intestine, *Porphyromonas* and *Prevotella* are major components of supragingival and, especially, subgingival dental plaque in the human oral cavity. These genera are associated with oral infectious diseases such as chronic periodontitis, periodontal abscess, endodontic infections, periapical abscess, and sometimes life-threatening soft tissue infections of the head and neck region. In the oropharynx, *Bacteroides, Porphyromonas*, and *Prevotella* can be found in tonsillar crypts where they can cause peritonsillar abscess (Kuhn et al., 1995) and in the fissures sometimes found on the dorsal surface of the tongue (Faveri et al., 2006) where they have can cause oral halitosis. Infection with these bacteria in the oropharynx can also be associated with chronic (but not acute) sinusitis and chronic (but not acute) otitis media. These bacteria are present in the upper (but not lower) respiratory tract where they are associated with community-acquired aspiration pneumonia, necrotizing pneumonia, lung abscess, and empyema (Pryor et al., 2001; Johnson and Hirsch, 2003; Brook, 2004a). They are present in the urogenital tract, particularly the vagina, where they can cause bacterial vaginosis, a disease that significantly increases the risk for preterm, low-birth-weight babies. Bacterial vaginosis is characterized by an alteration in the microbial ecology with decreased numbers of lactobacilli and increased numbers of *Prevotella*. Specific bacteria implicated in bacterial vaginosis include *Bacteroides ureolyticus* (reclassified as *Campylobacter ureolyticus* [Vandamme et al., 2010]), and *Prevotella bivia* (Boggess et al., 2005). Finally, *Bacteroides*, *Porphyromonas*, and *Prevotella* are not usually found on the human skin except in the case of human and dog bite wounds (Talan et al., 1999).

Taxonomy of *Bacteroides*, *Porphyromonas*, and *Prevotella*

The taxonomy of the *Bacteroides*, *Porphyromonas*, and *Prevotella* has undergone numerous revisions reflective of changing methodology. Originally, all three genera were classified within a single genera—*Bacteroides*—based on phenotypic and biochemical tests. The bacterial cells were described as Gram-negative, nonspore-forming, anaerobic, often pleomorphic bacilli (usually with rounded or pointed ends, sometimes fusiform or filamentous). Growth is stimulated by bile, hemin, and vitamin K. Steroid hormones, such as estradiol and progesterone, can also stimulate growth. This is important in that people with elevated serum hormone levels, such as pregnant women or children during the time of puberty, can have oral infections with these species as a result of steroid hormone- stimulated growth (Kornman and Loesche, 1982; Rawlinson et al., 1998).

These bacteria are chemoorganotrophs, which ferment carbohydrates to produce mixed acid fermentation end products, including acetate, succinate, lactate, formate, or propionate. Distinct in this regard is *B. splanchnicus*, which produces butyric acid, including isobutyrate and isovalerate. Comparative nucleic acid sequence analysis has shown that *B. splanchnicus* is not a member of the genus (Olsen and Shah, 2003). The absence of butyric acid production differentiates *Bacteroides*, *Porphyromonas*, and *Prevotella* from *Fusobacterium*, which do produce butyric acid.

Prominent among the *Bacteroides* is the species *Bacteroides fragilis*, which at one time consisted of five subspecies: *distasonis*, *fragilis*, *ovatus*, *thetaiotaomicron*, and *vulgatus*. These were subsequently shown by DNA–DNA homology to be poorly related and were then defined as individual species:

B. distasonis, *B. fragilis*, *B. ovatus*, *B. thetaiotaomicron*, and *B. vulgatus* (Cato and Johnson, 1976). *Bacteroides merdae*, *B. splanchnicus*, and *Bacteroides stercoris* were identified as closely related species. Recently, 16S rRNA sequence analysis has shown that *B. distasonis*, *B. merdae*, and *Bacteroides goldsteinii* should not be classified within *Bacteroides*.

Bacteroides that produce darkly pigmented colonies constituted an easily distinguishable group that provided the impetus for further taxonomic revisions as well as a relatively easy method for assessing their importance in various diseases, such as human chronic periodontitis and, most recently, canine periodontitis (Hardham et al., 2005). Species within the original definition of *Bacteroides* included those within the human oral cavity that produce black or brown pigmented colonies when cultured on blood agar media (sometimes referred to as *black-pigmenting Bacteroides*). Colonies from some species exhibit brick-red fluorescence when exposed to UV light. Like the darkly pigmented colonies, fluorescence has been used as a way of differentiating these species.

In 1921, Oliver and Wherry isolated bacteria producing black or beige pigmented colonies and named them *Bacterium melaninogenicum* (*melanin producing*), mistakenly believing that the pigment in the colonies was melanin. Chemical analysis has since demonstrated that, rather than melanin, the pigments responsible for the dark coloration of the bacterial colonies are protohemin and protoporphyrin, the latter responsible for the brick-red fluorescence of colonies exposed to UV light (Shah et al., 1979; Slots and Reynolds, 1982).

The pigmented *Bacteroides* originally classified as *B. melaninogenicum* were subdivided by Moore and Holdeman into three subspecies based on sugar fermentation: *Bacteroides melaninogenicus* ssp. *melaninogenicus* for strong fermenters, *Bacteroides melaninogenicus* ssp. *intermedius* for bacteria with intermediate fermentative ability, and *Bacteroides melaninogenicus* ssp. *asaccharolyticus* for nonfermenting species. Subsequently, these subspecies were elevated to species level as *B. melaninogenicus*, *B. intermedius*, and *B. asaccharolyticus* (Moore and Holdeman, 1974). The sugar-fermenting species *B. melaninogenicus* was divided into *B. loescheii*, *B. denticola*, and *B. melaninogenicus* (Holdeman and Johnson, 1982). The intermediately sugar-fermenting species *B. intermedius* became two species: *B. intermedius* and *B. corporis* (Johnson and Holdeman, 1983). The sugar nonfermenting strains from the gastrointestinal tract could be distinguished from the sugar nonfermenting strains from the oral cavity both by DNA analysis and by phenotypic traits. Oral strains produced phenylacetic acid, while the gastrointestinal strains did not (Kaczmarek and Coykendall, 1980). The asaccharolytic species found in dental plaque around the gingiva were first defined as *Bacteroides gingivalis* (Coykendall et al., 1980) and then placed into a new genus *Porphyromonas* as *P. gingivalis* (Shah and Collins, 1988). Asaccharolytic species from the oral cavity associated with the root canals of teeth were likewise first defined as *Bacteroides endodontalis* (van Steenbergen et al., 1984) and then placed into the genus *Porphyromonas* as *P. endodontalis* (Shah and Collins, 1988). Two weakly fermentative species originally categorized within *B. intermedius* were later categorized in the genus *Porphyromonas* as *P. levii* (Shah et al., 1995) and *Porphyromonas macacae*.

Bacteroides

Bacteroides fragilis

B. fragilis is the type species and the most important species in the genus *Bacteroides* (Table 43.1) by virtue of its distribution

TABLE 43.1

Classification of *Bacteroidetes*

Phylum XIV Bacteroidetes

    Class I. Bacteroidia

        Order I. Bacteroidales

            Family I. Bacteroidaceae

                Genus I. Bacteroides
                Species

                    B. acidifaciens

                    B. caccae

                    B. cellulosolvens

                    B. capillosus

                    B. coprocola

                    B. coprosuis

                    B. distasonis

                    B. eggerthii

                    B. finegoldii

                    B. forsythus

                    B. fragilis

                    B. goldsteinii

                    B. heparinolytica

                    B. helcogenes

                    B. intestinalis

                    B. merdae

                    B. massiliensis

                    B. nordii

                    B. ovatus

                    B. prius

                    B. pyogenes

                    B. salyersiae

                    B. splanchnicus

                    B. stercoris

                    B. tectus

                    B. thetaiotaomicron

                    B. uniformis

                    B. vulgatus

                    B. xylanilyticus

                Genus II. Acetofilamentum

                Genus III. Acetomicrobium

                Genus IV. Acetothermus

                Genus V. Anaerorhabdus

            Family II. "Marinilabiaceae"

                Genus I. Marinilabilia

                Genus II. Alkaliflexus

                Genus III. Anaerophaga

            Family III. "Rikenellaceae"

                Genus I. Rikenella

                Genus II Alistipes

TABLE 43.1 *(Continued)*

Classification of *Bacteroidetes*

Family IV. "Porphyromonadaceae"	
	Genus I. Porphyromonas
	Genus II. Barnesiella
	Genus III. Dysgonomonas
	Genus IV. Paludibacter
	Genus V. Parabacteroides
	Genus VI. Petrimonas
	Genus VII. Proteiniphilum
	Genus VIII. Tannerella
Family V. "Prevotellaceae"	
	Genus I. Prevotella
	Genus II. Xylanibacter

Source: Adapted from *Bergey's Manual of Systematic Bacteriology*, Volume 4.

and its pathogenic potential. It is the leading human anaerobic pathogen. The cells are 0.5–0.8 μm in diameter, 1.5–4.5 μm long with rounded ends, and most strains are encapsulated. It is present as part of the normal gastrointestinal microflora (i.e., a commensal) in proportions of about 0.5–2% of the gastrointestinal bacteria, and it constitutes about 10% of all fecal *Bacteroides*. However, it causes clinical problems in a disproportionately high number of cases relative to its numbers, pointing to the presence of virulence factors in some strains. *B. fragilis* is the most frequently isolated anaerobe from clinical specimens of the gastrointestinal tract of both humans and animals. *B. fragilis* recently has been shown to evolve within individuals in response to factors, such as diet, enabling the bacterium to persist in the intestine (Zhao et al., 2019). It is isolated particularly from intra-abdominal abscesses and from cases of peritonitis usually following perforation of the intestines due to pathological processes, trauma, or abdominal surgery. As in other mixed anaerobic infections, it is often isolated together with other anaerobes and facultative species. In contrast to its presence mainly in the gastrointestinal tract, two other members of the genus—*Bacteroides bivius* and *Bacteroides disiens*—are found mainly in the female genital tract. *B. fragilis* shares the same cellular and colonial morphology as the *Porphyromonas* and *Prevotella*, but it can be distinguished in that its growth is stimulated by the presence of 20% bile.

B. fragilis is typically nonhemolytic, but 10 hemolysin genes have been described (Robertson et al., 2006). *B. fragilis* is unusual among *Bacteroides* in possessing a polysaccharide capsule, seen especially in fresh isolates, but which is lost on repeated subculture (Kasper, 1976). Like capsules in other bacterial species, the *B. fragilis* capsule can inhibit phagocytosis by host immune cells and can also play a role in abscess formation as demonstrated in animal models (Nakatani et al., 1996). Two types of capsular polysaccharides have been described (Tzianabos et al., 1992). *B. fragilis* also has a lipopolysaccharide notable for a

lipid A component chemically distinct from other species; the glucosamine lacks attached phosphate groups and the amino sugars have fewer attached fatty acids. The lipopolysaccharide also lacks beta-hydroxymyristic acid. This accounts for its much lower biological activity as compared to other lipopolysaccharide endotoxins (Weintraub et al., 1989; Hofstad, 1992).

Strains of *B. fragilis* can produce an enterotoxin (enterotoxigenic *B. fragilis* [ETBF]) causing watery diarrhea, especially in children between 1 and 5 years of age (Pantosti et al., 1997). While it is present in 6.5% of otherwise healthy people, ETBF accounts for up to 20% of cases of diarrhea. ETBF has also been associated with diarrheal disease in livestock, with inflammatory bowel disease, and with colorectal cancer, the latter thought to occur through modulation of mucosal immune responses and induction of changes in epithelial cells. The *B. fragilis* enterotoxin (*B. fragilis* toxin [BFT]) is a 20 kDa zinc metalloprotease. There are three subtypes encoded by the *B. fragilis* toxin (bft) gene (Kato et al., 2000). The enterotoxin binds to intestinal epithelial cells facilitated by pili present on the cell surface of some *B. fragilis* strains. The enterotoxin cleaves the protein E-cadherin, which is responsible for intercellular adhesion (Sears, 2001; Wu et al., 2007). ETBF-infected mice develop colitis and colonic tumors while nontoxin-producing *B. fragilis*-infected mice do not (Wu et al., 2007) and ETBF is found more frequently in colon-cancer patients (Toprak et al., 2006). The frequency of *B. fragilis* in patients with colorectal cancer is about twice that in control subjects (Haghi et al., 2019).

B. fragilis together with other species in the *B. fragilis* group—*B. distasonis*, *B. ovatus*, *B. thetaiotaomicron*, and *B. vulgatus*—frequently produce beta-lactamase and are resistant to penicillins, but they are generally susceptible to metronidazole, carbapenems, and beta-lactam antibiotics. There has been a marked increase in antibiotic resistance among *Bacteroides* as well as *Porphyromonas* and *Prevotella* species in recent years. Multidrug-resistant *B. fragilis* has been reported including an isolate resistant to resistance to penicillin, clindamycin, metronidazole, cefoxitin, meropenem, imipenem, piperacillin/tazobactam, and tigecycline (Sherwood et al., 2011).

Bacteroides thetaiotaomicron

B. thetaiotaomicron is another important member of the genus *Bacteroides*. It is a nonspore- forming, Gram-negative anaerobic bacillus, which may be motile or nonmotile. It is the dominant bacteria in the distal human gastrointestinal tract (Moore and Holdeman, 1974). Next to *B. fragilis*, it is the most important cause of subdiaphragmatic infections in humans including intra-abdominal sepsis as well as bacteremia. It also has an important symbiotic relationship with its human host. It stimulates both the development of the human gastrointestinal tract and host immune functions (Hooper et al., 2002). Studies of *B. thetaiotaomicron* have been key to understanding the role of the gut microflora in carbohydrate metabolism. *B. thetaiotaomicron* metabolizes the contents of the intestine—otherwise indigestible, mainly fibrous, dietary carbohydrates as well as carbohydrates produced by humans, such as mucins—releasing sugars for its own metabolism and for other gut microorganisms (Gilmore and Ferretti, 2003).

B. thetaiotaomicron is also the first bacterium in this group to be completely sequenced (Xu et al., 2003). The type strain of *B. thetaiotaomicron* (VPI-5482 = ATCC 29148) originally isolated from feces taken from a healthy adult has a 6.26 Mb genome. The genome encodes 4,776 proteins and has a G+C content of 42.8%. Consistent with its role in metabolizing carbohydrates in the gut, a large portion of the *B. thetaiotaomicron* genome encodes enzymes involved in polysaccharide uptake and degradation such as glycosyl hydrolases, carbohydrate-binding proteins, and glycosyltransferases involved in capsular polysaccharide formation.

Porphyromonas

Porphyromonas are part of the phylum Bacteroidetes (previously the Cytophaga-Flavobacteria- Bacteroides group [Boone and Castenholtz, 2001]), order Bacteroidales, family *Porphyromonadaceae*. *Porphyromonas* are named after the porphyrin pigment—protoheme—that results in the colonies becoming darkly pigmented. The *Porphyromonas* include Gram-negative nonspore-forming anaerobic nonmotile rods, which are sometimes short or coccobacillary rods that produce smooth, shiny, circular, convex colonies with an entire edge. Growth is optimum at 37°C, and most species, but not all, produce the protoheme pigment that causes the colonies on blood agar to darken from edge to center over a period of 3–6 days. They are asaccharolytic species. The G + C content of the DNA ranges from 40 mol% to 55 mol%. They are primarily found in the oral cavity of humans, nonhuman primates, and other animals. Species of *Porphyromonas* are shown in Table 43.2. *Porphyromonas asaccharolytica* is the type species of the genera and, like *Bacteroides*, is found in the normal human intestinal tract and the human vagina. It is found in mixed infections, including superficial abscesses and ulcers of the genitalia and perineum (Duerden, 1993).

Porphyromonas gingivalis

P. gingivalis (formerly *Bacteroides gingivalis*) is an anaerobic, nonfermenting, Gram-negative short rod frequently found in the human oral cavity where it is part of the normal microflora. *P. gingivalis* is nonclonal with variable pathogenicity between strains (Tribble et al., 2013). For example, *P. gingivalis* strains W50 and W83 induce alveolar bone loss in mice, while strain A7A1-28 does not (Marchesan et al., 2012). *P. gingivalis* virulence factors include capsule, gingipains (cell surface cysteine proteinases that regulate host inflammatory responses (Huq et al. (2013), hemagglutinins, hemolysin, iron uptake transporters, and toxic outer membrane blebs. The genome of *P. gingivalis* strain W83 has been sequenced (Nelson et al., 2003) and is closely related to *B. fragilis* and *B. thetaiotaomicron*. It contains six putative hemagglutinin-like genes, 36 previously unidentified peptidases, and genes to produce toxic metabolic end products that may contribute to periodontal disease.

P. gingivalis can be isolated from dental plaque, particularly subgingival dental plaque in adults. An anaerobe, *P. gingivalis*, is found in higher numbers in the depths of periodontal pockets where the oxygen tension is lower and, which, consequently, favors its growth. *P. gingivalis* is less frequently found in dental plaque in children, and it appears to become established in the human oral cavity following puberty. *P. gingivalis* can also be found, but in fewer numbers, in parts of the oral cavity contacting dental plaque, such as in saliva, the buccal mucosa, the lateral borders of the tongue, and the tonsils, particularly if the tonsils are cryptic.

P. gingivalis is one of the major etiological agents in the pathogenesis and progression of the inflammatory events of periodontal disease (Hajishengallis et al., 2014). As described next, there are several factors in its pathogenesis including the fact that it is found in high prevalence in dental plaque in humans with chronic periodontitis. That is, when *P. gingivalis* is present in the dental plaque, periodontal pockets will often become deeper and there will be greater loss of the connective tissue attachment of the gingiva to the teeth and loss of alveolar bone. *P. gingivalis* is considered one of a group of "red complex" bacteria, each of which is found in cases of progressing periodontitis (Socransky et al., 1998). Conversely, healthy gingiva is associated with the absence of *P. gingivalis*. Interestingly, bacteria similar to *P. gingivalis* that infect the oral cavities of other animals, such as monkey, dogs, and cats, are also associated with periodontitis in those animals. *P. gingivalis* produces a number of virulence factors including:

1. Cell surface fimbriae that facilitate bacterial adherence to tooth surfaces and to epithelial cells, such as those present in the gingival sulcus or periodontal pocket. *P. gingivalis* can agglutinate red blood cells, bind and invade gingival epithelial cells via integrin host cell receptors, and induce cytokine expression resulting in alveolar bone resorption. *P. gingivalis* produces two types of fimbriae, Mfa1 and FimA, the latter long fimbriae from the *fimA* gene. There are six genotypic variants based on nucleotide sequence differences in the *fimA* genes—types I–V and Ib. Type II fimA exhibit greater adhesive and invasive

TABLE 43.2

Species of *Porphyromonas*

Species	Subclusters	Features	G + C mol%	Growth Characteristics	Type Strain	References
Found mainly in humans						
P. asaccharolytica	1	Type species of the genus (1970) Intestinal flora Vaginal flora Diabetic ulcers and gangrene Genital ulcers	52–54	Darkly pigmented colonies Inhibited by bile and bile salts Fluorescence with UV light	ATCC 25260	Moore and Holdeman (1974); Shah and Collins (1988); Duerden (1993)
P. bennonis		Human wounds and abscesses in the buttocks and groin	58	Weakly pigmented	ATCC BAA-1629	Summanen et al. (2005)
P. catoniae	4	Gingiva	49–51	The only saccharolytic and nonpigmented species in the genus	ATCC 51270	Moore et al. (1994); Willems and Collins (1995b)
P. endodontalis	1	Infected dental root canals (1984) Periapical abscess	49–51	Darkly pigmented colonies	ATCC 35406	van Steenbergen et al. (1984); Shah and Collins (1988)
P. gingivalis	3	Dental plaque Human periodontal disease Feline periodontal disease		Fluorescence with UV light Produces collagenase Has trypsin-like activity	ATCC 33277	Coykendall et al. (1980); Shah and Collins (1988); Norris and Love (1999)
P. somerae	2	Human infections Formerly *P. levii-like organisms* (PLLOs)		Darkly pigmented colonies	ATCC BAA-1230	Summanen et al. (2005)
P. uenonis	1	Human infections, likely of intestinal origin	52.5	Darkly pigmented colonies Fluorescence with UV light	ATCC BAA-906	Finegold et al. (2004)
Found mainly in animals						
P. cangingivalis	2	Dog subgingival dental plaque Periodontal disease in dogs	49–51	Darkly pigmented colonies	NCTC 12856	Collins et al. (1994)
P. canoris	2	Gingiva in dogs with periodontal disease Dog bite wounds in humans;	49–51	Darkly pigmented colonies Fluorescence with UV light	NCTC 12835	Love et al. (1994) Citron et al. (1996)
P. cansulci	5	Dog subgingival dental plaque	49–51	Darkly pigmented colonies	NCTC 12858	Collins et al. (1994)
P. circumdentaria	1	Periodontal disease in dogs Cat gingiva	40–42	Fluorescence with UV light Darkly pigmented colonies	NCTC 12469	Love et al. (1992)
P. crevioricanis	3	Cat dental plaque Feline periodontal disease Dog gingiva	44–45	Fluorescence with UV light Darkly pigmented colonies	ATCC 55563	Norris and Love (1999); Hirasawa and Takada (1994)
P. denticanis *P. gingivicanis*	1	Periodontal disease in dogs Most closely related to *B. splanchnicus* by 16S rRNA Dog gingiva	41–42	Darkly pigmented colonies Darkly pigmented colonies	ATCC 55562	Hardham et al. (2005); Hirasawa and Takada (1994)
P. gulae	3	Periodontal disease in dogs Dental plaque in animals	51	Darkly pigmented colonies	ATCC 51700	Fournier et al. (2001)
P. levii	2	Bear, cat, coyote, dog, wolf, and monkey Canine periodontal disease Intestinal tract of cattle	45–48	Darkly pigmented colonies	ATCC 29147	Hardham et al. (2005); Johnson and Holdeman (1983)
		Cow rumen		Weakly fermentative		Shah et al. (1995)
P. macacae	4	Periodontitis in monkeys	43–44	Fluorescence with UV light Differentiable from *P. macacae* based on phenotypic differences	ATCC 33141	Slots and Genco (1980)
P. salivosa		Cat bite wounds in humans Feline periodontal disease	42–44	Darkly pigmented colonies Weakly fermentative Differentiable from *P. salivosa* based on phenotypic differences	NCTC 11362	Love (1995); Love et al. (1987, 1992); Love (1995)
		Canine periodontal disease				Citron et al. (1996); Norris and Love (1999)

capabilities (Nagano et al., 2013) and predominate in chronic periodontitis (Amano et al., 2004; Moon et al., 2012; Fabrizi et al., 2013). *P. gingivalis* fimbriae degrade integrin-related signaling molecules, thereby disrupting cellular proliferation, which enables it to persist in the gingival epithelium (Amano, 2007) where it becomes resistant to antibiotic treatment (Irshad et al., 2012).

2. A bacterial collagenase capable of degrading type I collagen and thought to be responsible for the connective tissue destruction seen in periodontitis (Houle et al., 2003). *P. gingivalis* collagenase is a therapeutic target for low-dose doxycycline therapy used to treat chronic periodontitis (Vernillo et al., 1994), and it is the basis for a diagnostic test for *P. gingivalis* and oral bacteria with similar trypsin-like activities able to hydrolyze *N*-benzoyl-dl- arginine-2-naphthyl-amide (Lee et al., 2006).

3. A lipopolysaccharide endotoxin associated with alveolar bone loss (Millar et al., 1986), but, which like the LPS from *B. fragilis*, has relatively weak biological activity based on the chemical structure of the lipid A moiety (Ogawa et al., 2007).

4. A capsular polysaccharide encoded by a glycosyltransferase (Davey and Duncan, 2006), which, like the capsule in *B. fragilis*, can inhibit phagocytosis by host immune cells and complement-mediated killing (Slaney et al., 2006) as well as a hemagglutinin, hemolysin, hyaluronidases, phospholipase, alkaline phosphatase, and acid phosphatase (Kantarci and Van Dyke, 2002).

5. Bacterial proteases including trypsin-, thiol-, and caseinolytic proteinases (Curtis et al., 2001). The *P. gingivalis* proteases are important factors in colonization of the periodontal pocket (Dubin et al., 2013). *P. gingivalis* proteases fall into two general classes; the cysteine proteinase (also known as "trypsin-like" enzymes) and the serine proteinase (Bostanci and Belibasakis, 2012). The cysteine proteinases are commonly knowns as gingipains and cleave polypeptides at the C-terminal after arginine-gingipain R, of which there are two types, RgpA and RgpB, or lysine-gingipain K, of which there is one type, Kgp. Gingipain R degrades integrin-fibronectin-binding cytokine, immunoglobulin, and complement factors.

The complete bacterial genome from *P. gingivalis* strain W83, originally isolated from a human oral infection by H. Werner (Loos et al., 1993), has been sequenced and found to exhibit similarities to other *Bacteroides* species, such as *B. fragilis* and *B. thetaiotaomicron*, to the Cytophaga- Flavobacteria- Bacteroides phylum, and to the green-sulfur bacteria (Nelson et al., 2003). It exhibits 2,343,476 base pairs and 2,053 genes, of which 96.83% are protein encoding. It also exhibits a large number, at least 96, mobile insertion sequences and genes similar to those found in other *Bacteroides*, including immunoreactive surface proteins and proteins for aerotolerance (Duncan, 2003). Whole-genome comparison between virulent *P. gingivalis* strain W83 and avirulent *P. gingivalis* strain ATCC 33277 shows 93% similarity, with 7% composed of variant or missing genes, including insertion sequences, the immunoreactive surface protein RagA, enzymes involved in polysaccharide capsule synthesis, and pathogenicity islands possibly acquired by lateral gene transfer (Chen et al., 2004).

P. gingivalis has been recently associated with extraoral diseases including rheumatoid arthritis, Alzheimer's disease, and pancreatic cancer. Patients with rheumatoid arthritis have a higher prevalence of severe periodontitis than patients without rheumatoid arthritis (Ogrendik, 2012; Totaro et al., 2013). *P. gingivalis* DNA, as well as that of *Prevotella intermedia*, *Prevotella melaninogenica*, and *Tannerella forsythia*, can be detected in the synovial fluid of arthritic joints and correlated with the number of missing teeth, which is a measure of the severity of periodontal disease (Reichert et al., 2013). Patients with both severe periodontitis and rheumatoid arthritis have higher antibody titers to *P. gingivalis* than patients with severe periodontitis who did not have rheumatoid arthritis (Smit et al., 2012). Autoimmunity to citrullinated proteins is a feature of rheumatoid arthritis, and anticitrullinated antibodies are found in gingival crevicular fluid from inflamed gingiva. *P. gingivalis* is unique among oral bacteria in having an enzyme, peptidyl-arginine deiminase, capable of citrullinating proteins (Janssen et al., 2013; Quirke et al., 2013). It can also be shown to cause rheumatoid arthritis in animals by other mechanisms including induction of NETosis, osteoclastogenesis, and Th17 proinflammatory response (Perricone et al., 2019)

Postmortem examination of the brains of patients with Alzheimer's disease demonstrates lipopolysaccharide from *P. gingivalis* (Poole et al., 2013), and Alzheimer's patients have elevated levels of serum antibodies to *P. gingivalis* compared to control subjects (Sparks Stein et al., 2012). Similarly, subjects exhibiting elevated levels of *P. gingivalis* antibodies were at twice the risk of pancreatic cancer compared to controls (Michaud et al., 2012).

Porphyromonas endodontalis

P. endodontalis is a bacterium that is closely related to *P. gingivalis* and causes dental root canal infections and endodontic abscesses together with other species (van Steenbergen et al., 1984). *P. endodontalis* infections are characterized by acute symptoms, such as pain, swelling, and purulence. As opposed to the aerotolerance exhibited by *P. gingivalis*, *P. endodontalis* is a strict anaerobe whose intolerance for atmospheric oxygen has made it difficult to detect on routine culture. For example, *P. endodontalis* can be detected more than twice as frequently in chronically infected dental root canals by the PCR than it can be detected by anaerobic culture (Tomazinho and Avila-Campos 2007). The lipopolysaccharide from *P. endodontalis* can stimulate cytokine production, and the bacterium can induce dental pulp to produce matrix metalloproteinases (Chang et al., 2002).

Prevotella

The genus *Prevotella* (Shah and Collins, 1990; family: *Prevotellaceae*), named after the French anaerobic microbiologist Prevot, includes species previously classified as *Bacteroides*.

As described earlier, these species are obligately anaerobic, nonspore-forming, nonmotile, Gram-negative rods. They produce gray, brown, or black colonies that are shiny, circular, and convex with an entire edge, although there are also nonpigmenting species within the genus. They *moderately* ferment sugars, require hemin and menadione for growth, and are bile sensitive. The cell walls contain mesodiaminopimelic acid. The G+C content ranges from 40 mol% to 60 mol%. There are 45 cultured species of which 38 have been isolated from humans (Alauzet et al., 2010, Table 43.3) and of which 31 have been isolated from the oral cavities of humans and other species. Only one species, *Prevotella paludivivens*, has been isolated from the environment of rice plants.

In humans, *Prevotella* are predominantly oral species (Table 43.3) that can be found in infections of the head and neck regions, including the oral cavity where they can cause periodontal diseases and oral halitosis (Krespi et al., 2006), sinus infections (Brook, 2005a), and infections of the tonsils and middle ear (Brook, 2005b). They can also be found in the gastrointestinal tract (Brook, 2004b) and the urogenital tract, particularly *P. bivia* and *Prevotella disiens*, which can cause gynecologic infections, including bacterial vaginosis—a risk factor for preterm low birth weight (Marrazzo, 2004). In animals, *Prevotella* comprise the most numerous Gram-negative species in the gastrointestinal tract (Peterka et al., 2003). *P. melaninogenica* has been associated with oral cancer as specific for tumorigenic tissues, and together with *Capnocytophaga gingivalis* and *Streptococcus mitis* as a salivary marker for the early detection of oral squamous cell carcinoma (Mager et al., 2005). Other species of *Prevotella* include *Prevotella heparinolytica* and *Prevotella zoogleoformans*, which are closely related to *Bacteroides* species and may represent a new genus within the family *Prevotellaceae*.

Prevotella intermedia

Prior to reclassification in the genus *Prevotella*, this species of intermediately fermentative *Bacteroides* was known as *B. intermedius*. It can be distinguished on primary culture plates from other darkly pigmented colonies by the brick-red fluorescence observed in the colonies exposed to UV light. *P. intermedia* can also be distinguished from other darkly pigmented colonies by its production of galactosidase. *P. intermedia* is associated with different forms of human periodontal disease, including moderate to severe gingivitis, chronic periodontitis, and necrotizing gingivitis.

Tannerella

Most oral species once categorized as *Bacteroides* have been placed into the genus *Porphyromonas* or *Prevotella* as described earlier. One notable exception is an organism originally isolated by Anne Tanner of the Forsyth Institute from dental plaques in adults with progressing advanced periodontitis (Tanner et al., 1986). It has subsequently been isolated from gingivitis, periodontitis, endodontic infections, and infections around dental implants. This microorganism was first described as *fusiform Bacteroides* based on its cellular appearance—tapered (fusiform) ends with central swellings. The species, originally named *Bacteroides forsythus*, was found to be distinct from *Bacteroides*—it is not resistant to bile—and from *Porphyromonas* based on 16S rRNA phylogenetic analysis. It was placed into a new genus as *Tannerella*, first as *Tannerella forsythus* (Sakamoto et al., 2002) and then as *T. forsythia* (Maiden et al., 2003) It is a slow-growing, strict anaerobe with unique growth requirements. The colonies are tiny and opaque, and appear on primary culture often as satellites of *Fusobacterium nucleatum* colonies, where they appear as speckled pale pink, circular, entire, convex, and sometimes with a depressed center. *T. forsythia* requires media containing *N*-acetylmuramic acid (Wyss, 1989), and it is one of a few oral species from among the up to 1,200 bacterial species that can inhabit the human oral cavity, such as *Aggregatibacter actinomycetemcomitans* and *P. gingivalis,* for which there is significant evidence for its role in the etiology of human periodontitis.

T. forsythia along with *P. gingivalis* and *T. denticola* are routinely isolated from subgingival plaques in patients with chronic periodontitis. Although *T. forsythia* is the only member of the genus cultured to date, other members of the genus have been detected by their 16S rRNA sequence. *Tannerella* sequences have been identified in the gut flora from insects, such as termites and scarab beetle (Tanner and Izard, 2006). The full sequence of the *T. forsythia* type strain ATCC 43037 has been determined. It contains 3,405,543 base pairs, 3,034 predicted open reading frames, 15 pathogenicity islands, and a G+C content of 46.8 mol% (Chen et al., 2005).

T. forsythia produces several virulence factors, including trypsin-like and PrtH proteases, SiaH and NanH sialidases, a hemagglutinin, alpha-d-glucosidase and *N*-acetyl-beta-glucosaminidase, epithelial cell-adhering and invasion-promoting S-layer, and an apoptosis-inducing activity (Sharma, 2010). It has demonstrated virulence in animal models causing skin abscesses and alveolar bone loss enhanced by coinfection with *F. nucleatum* or *P. gingivalis* (Kesavalu et al., 2007). Along with *P. gingivalis*, it has been associated with esophageal cancer (Malinowski et al., 2019). *Tannerella forsythia* can be detected in the synovial fluid of arthritic joints.

Fusobacterium

Fusobacterium (Table 43.4) is a genus of Gram-negative nonspore-forming anaerobic bacteria that has a distribution similar to *Bacteroides*. Similar to the species described earlier, *Fusobacterium* lives on mucous membranes in both humans and other animals. Like *Bacteroides* and related species, some species of *Fusobacterium* are pathogenic and are found in purulent infections and in gangrene. Fusobacterium cells are filamentous or spindle-shaped, varying in size and motility, and its major metabolic end product is butyric acid.

Fusobacterium has been implicated in the etiology of colorectal cancer as have some other microorganisms in the gastrointestinal tract. Studies in animal models have pointed to potential mechanisms. Fusobacterium could promote oncogene transcription and induce immune suppression. Fusobacterium is associated with epigenetic changes of malignant epithelial cells (Zhou et al., 2018).

TABLE 43.3

Prevotella Species

Species	Isolated From	G + C mol%	Growth Characteristics	Type Strain	References
P. albensis	Rumen. Previously *Prevotella ruminicola* subsp. *ruminicola*	39–43	Nonpigmented	NCTC 13060	Avgustin et al. (1997)
P. amnii	Human amniotic fluid		Nonpigmented	JCM 14753	Lawson et al. (2008)
P. aurantiaca	Human oral cavity	39–41	Pigmented	JCM 15754	Sakamoto et al. (2010)
P. baroniae	Endodontic and periodontal infections, dentoalveolar abscesses, and dental plaque	52	Nonpigmented	DSM 16972	Downes et al. (2005)
P. bergensis	Skin and soft tissues	48		DSM 17361	Downes et al. (2006)
P. bivia	Gynecologic and obstetrical infections		Nonpigmented	NCTC 11156	Shah and Collins (1990)
P. brevis	Rumen	45–51	Nonpigmented	NCTC 13061	Avgustin et al. (1997)
P. bryantii	Rumen	39–43	Abundant extracellular DNase Nonpigmented	NCTC 13062	Avgustin et al. (1997)
P. buccae	Human oral cavity		Nonpigmented	ATCC 33574	Shah and Collins (1990)
P. buccalis	Human oral cavity		Nonpigmented	ATCC 35310	Shah and Collins (1990)
P. copri	Human feces	45	Nonpigmented	DSM 18205	Hayashi et al. (2007)
P. corporis	Human oral cavity		Pigmented	ATCC 33547	Shah and Collins (1990)
P. dentalis	Human oral cavity. Formerly *Mitsuokella dentalis* and *Hallella seregens*	56–60	Nonpigmented	ATCC 49559	Willems and Collins (1995a)
P. dentasini	Donkey oral cavity	49–51	Pigmented	JCM 15908	Takada et al. (2010)
P. denticola	Human oral cavity	49–51	Pigmented	ATCC 35308	Shah and Collins (1990)
P. disiens	Human oral cavity		Nonpigmented	ATCC 29426	Shah and Collins (1990)
P. enoeca	Human oral cavity		Nonpigmented	ATCC 51261	Moore et al. (1994)
P. falsenii	Monkey oral cavity	44–48	Pigmented	JCM 15124	Sakamoto et al. (2009)
P. fusca	Human oral cavity	43	Pigmented	DSM 22504	Downes and Wade (2011)
P. histicola	Human oral cavity				Downes et al. (2008)
P. intermedia	Human oral cavity	40–42	Pigmented	ATCC 25611	Shah and Collins (1990)
P. loescheii	Human oral cavity	46–48	Pigmented	ATCC 15930	Holdeman and Johnson (1982)
P. maculosa	Human oral cavity	48	Nonpigmented	DSM 19339	Downes et al. (2007)
P. marshii	Human oral cavity	51	Nonpigmented	DSM 16973	Downes et al. (2005)
P. melaninogenica	Type species of the genus Human oral cavity and vagina	40–42	Pigmented	ATCC 25845	Downes et al. (1990)
P. micans	Human oral cavity	46	Pigmented	DSM 21469	Downes et al. (2009)
P. multiformis	Human oral cavity		Nonpigmented	JCM 12541	Sakamoto et al. (2005a)
P. multisaccharivorax	Human oral cavity		Nonpigmented	DSM 17128	Sakamoto et al. (2005b)
P. nanceiensis	Human blood cultures, lung abscess, and bronchoalveolar lavage fluid	39.4	Nonpigmented	CCUG 54409	Alauzet et al. (2007)
P. nigrescens	Human oral cavity	40–42	Pigmented, can be distinguished from *P. intermedia* only by oligonucleotide probes	ATCC 33563	Shah and Gharbia (1992) and Matto et al. (1997)
P. oralis	Human oral cavity		Nonpigmented	NCTC 11459	Shah and Collins (1990)
P. oris	Human oral cavity		Nonpigmented	ATCC 33573	Shah and Collins (1990)
P. oulorum	Human oral cavity	45	Nonpigmented	ATCC 43324	Shah and Collins (1990)
P. pallens	Human oral cavity		Pigmented	NCTC 13042	Kononen et al. (1998)
P. paludivivens	Plant residue and rice roots in irrigated rice-field soil	39.2	Nonpigmented	DSM 17968	Ueki et al. (2007)
P. pleuritidis	Human pleural fluid	45.4	Nonpigmented	JCM 14110	Sakamoto et al. (2007)
P. ruminicola	Rumen	45–52	Nonpigmented	ATCC 19189	Shah and Collins (1990)
P. saccharolytica	Human oral cavity	44	Nonpigmented	DSM 22473	Downes et al. (2010)
P. salivae	Human oral cavity	41.3	Nonpigmented	DSM 15606	Sakamoto et al. (2004)
P. scopos	Human oral cavity	41	Pigmented	DSM 22613	Downes and Wade (2011)
P. shahii	Human oral cavity	44.3	Pigmented	DSM 15611	Sakamoto et al. (2004)
P. stercorea	Human feces	48	Nonpigmented	DSM 18206	Hayashi et al. (2007)
P. tannerae	Human oral cavity		Pigmented	ATCC 51259	Moore et al. (1994)
P. timonensis	Human breast abscess		Nonpigmented	CCUG 50105	Glazunova et al. (2007)
P. veroralis	Human oral cavity	42	Nonpigmented	ATCC 33779	Wu et al. (1992)

TABLE 43.4

Taxonomy of *Fusobacteria*

Phylum XIX:
Fusobacteria

Class I: Fusobacteriia

Order I: Fusobacteriales

Family I: Fusobacteriaceae

Genus: Fusobacterium

Species:	Subspecies:
	F. nucleatum
	F. nucleatum subsp. nucleatum
	F. nucleatum subsp. animalis
	F. nucleatum subsp. fusiforme
	F. nucleatum subsp. polymorphum
	F. nucleatum subsp. vincentii
F. canifelinum	
F. equinum	
F. gonidiaformans	
F. mortiferum	
F. naviforme	
F. necrogenes	
F. necrophorum	
	F. necrophorum subsp. necrophorum
	F. necrophorum subsp. funduliforme
F. perfoetens	
F. periodonticum	
F. russii	
F. simiae	
F. ulcerans	
F. varium	

Source: Adapted from Bergey's Manual of Systematic Bacteriology, Volume 4.

F. nucleatum is the type species of the genus. It is indigenous to the human oral cavity (where it can cause periodontal disease) and can be found in dental plaque in association with other Gram-positive and Gram-negative species. The cells are spindle shaped, 5–10 μm long, and often paired end to end. *F. nucleatum* can be found in mixed infections from the head and neck, as well as infections in the chest, lung, liver, and abdomen. It is associated with colorectal cancer, pancreatic cancer, oral cancer, and premature and term stillbirths (Shang and Liu, 2018). The complete genome of *F. nucleatum* strain ATCC 25586 has been sequenced (Kapatral et al., 2002). It has a 27 mol% G+C content and 2.17 Mb encoding 2,067 open reading frames.

Fusobacterium necrophorum is also a component of the normal human oropharyngeal, gastrointestinal, and urogenital tract flora. The otogenic variant can cause abscesses, otitis, sinusitis, sinus thrombosis, mastoiditis, and meningitis (Creemers-Schild et al., 2014). There is some evidence that *F. necrophorum* disease is increasing (Brazier et al., 2002). The cells are filamentous,

curved, with spherical enlargements. *F. necrophorum* is responsible for about 10% of sore throats, second only to group A streptococci (Aliyu et al., 2004), and can cause a particularly serious disease in healthy young adults known as Lemierre's syndrome—a life-threatening infection that follows an initial sore throat. *F. necrophorum* can also cause meningitis, gastrointestinal and urogenital infections, and a number of clinical syndromes known as necrobacillosis (Hagelskjaer Kristensen and Prag, 2000). *F. necrophorum* causes disease through the production of several virulence factors, including a leukotoxin, proteolytic enzymes, lipopolysaccharide, and hemagglutinin.

Gram-positive Anaerobic Cocci

Previously classified as *Peptostreptococcus* (phylum: Firmicutes—Gram-positive bacteria; class: Clostridia; order: Clostridiales), the Gram-positive anaerobic cocci (GPAC) are slow-growing and nonspore-forming cocci that are 0.3–1.8 μm in diameter and usually arranged in chains, pairs, tetrads, or clumps. They were a poorly described group of bacteria previously referred to as *anaerobic cocci, anaerobic streptococci, anaerobic Gram-positive cocci,* and *Peptococcus* although they have been the subject of increasing study in recent years. The genera and species of the GPAC are listed in Table 43.5. In addition to *Peptostreptococcus* (species include *P. anaerobius, P. stomatis,* and *P. russellii*), the GPAC include the genera *Finegoldia (F. magna), Parvimonas* (species include *P. olsenii* and *P. gorbachii*), *Anaerococcus, Peptococcus, Peptoniphilus* (species include *P. ivorii, P. harei,* and *P. niger*), *Gallicola* (species include *G. barnesae*), *Murdochiella (M. asaccharolytica), Atopobium, Anaerosphaera (A. aminiphila), Coprococcus, Sarcina, Ruminococcus,* and *Blautia (B. product)* (Murphy and Frick, 2013). Studies of the *Peptostreptococcus* like the larger GPAC group have been hampered by taxonomic uncertainties, slow bacterial-cell growth, and oxygen sensitivity. They are part of the normal flora on mucosal surfaces of the oral cavity, gastrointestinal tract, and urogenital tract, and account for 25–30% of all anaerobic bacteria isolated from clinical specimens. They are the most frequently isolated group of anaerobic microorganisms isolated from clinical infections followed by *Prevotella* species and the *B. fragilis* group (Mikamo et al., 2011). They are commensals and opportunistic pathogens recovered from oral infections, deep organ abscesses, obstetric and gynecological sepsis, and chronic wounds, such as leg ulcers. They are components of mixed anaerobic infections but may also be found as the sole infecting microorganism. As more is known about the GPAC, their importance in certain infections such as bacteremias and infected joints has increased.

The most commonly isolated GPAC from infections are (using the revised taxonomy) *F. magna, M. asaccharolytica, Anaerococcus prevotii,* and *Parvimonas micra. F. magna* (formerly *Peptococcus magnus* and *Peptostreptococcus magnus*) is the most pathogenic of the GPAC and is isolated from skin lesions, soft tissue abscesses, joint infections, pleural empyema, vaginosis, and infective endocarditis. *F. magna* has been isolated from infected orthopedic joints (Söderquist et al., 2017). *F. magna* virulence factors include protein L, which binds to immunoglobulin light chains. Protein L is thought to enhance *F. magna*'s ability to colonize host tissues (Ricci et al., 2001), and it induces the synthesis and release of IL-4 and IL-13, making protein L a bacterial superantigen (Genovese et al., 2003). *F. magna* also produces a peptostreptococcal albumin-binding protein (de Château and Björck, 1994) that facilitates nutrient access, increases growth rate, and inactivates antibacterial peptides (Egesten et al., 2011). *Anaerococcus sp. A20* was isolated from the skin of the human axilla and is responsible for axillary odor as it releases 3-hydroxy-3-metyl-hexanoic acid, the main component of axillary odor (Fujii et al., 2014). *M. asaccharolytica* is the sole species in the genus *Murdochiella* and has been isolated from human wounds (Ulger-Toprak et al., 2010). *A. prevotii* is a component of the resident flora of the skin, oral cavity, and gut but is also isolated from vaginal discharges, ovarian, peritoneal, sacral, and lung abscesses (Labutti et al., 2009), diabetic foot ulcers, and pressure ulcers (Dowd et al., 2008). *P. micra* (formerly *Peptostreptococcus micros* and *Micromonas micros*) is part of the resident gastrointestinal flora including the oral cavity and is a pathogen in oral infections including chronic periodontitis and endodontic lesions as well as extraoral chronic wounds, burns, and surgical infections. *P. micra* virulence factors include proteases, gelatinases, collagenase, and hemolysins (Grenier and Bouclin, 2006; Ota-Tsuzuki and Alves Mayer, 2010).

TABLE 43.5

Gram-Positive Anaerobic Cocci

Anaerococcus	
	A. hydrogenalis
	A. lactolyticus
	A. murdochii
	A. octavius
	A. prevotii
	A. senegalensis
	A. tetradius
	A. vaginalis
Anaerosphaera	
	A. aminiphila
Blautia	
	B. product
Finegoldia	
	F. magna
Gallicola	
	G. barnesae
Murdochiella	
	M. asaccharolytica
Parvimonas	
	P. micra
Peptoanaerobacter	
	Peptoanaerobacter stomatis
Peptoniphilus	
	P. asaccharolyticus
	P. coxii
	P. duerdenii
	P. gorbachii
	P. grossensis
	P. harei
	P. indolicus
	P. ivorii
	P. koenoeneniae
	P. lacrimalis
	P. methioninivorax
	P. olsenii
	P. timonensis
	P. tyrrelliae
Peptococcus	
	P. niger
Peptostreptococcus	
	P. anaerobius
	P. russellii
	P. stomatis

REFERENCES

Alauzet, C., H. Marchandin, and A. Lozniewski. 2010. New insights into *Prevotella* diversity and medical microbiology. *Future Microbiology*, 5:1695–1718.

Alauzet, C., F. Mory, J. P. Carlier, H. Marchandin, E. Jumas-Bilak, and A. Lozniewski. 2007. *Prevotella nanceiensis* sp. nov., isolated from human clinical samples. *International Journal of Systematic Bacteriology*, 57:2216–2220.

Aliyu, S. H., R. K. Marriott, M. D. Curran, S. Parmar, N. Bentley, N. M. Brown, J. S. Brazier, and H. Ludlam. 2004. Real-time PCR investigation into the importance of *Fusobacterium necrophorum* as a cause of acute pharyngitis in general practice. *Journal of Medical Microbiology*, 53:1029–1035.

Amano, A. 2007. Disruption of epithelial barrier and impairment of cellular function by *Porphyromonas gingivalis. Frontiers in Bioscience*, 12:3965–3974.

Amano, A., I. Nakagawa, N. Okahashi, and N. Hamada. 2004. Variations of *Porphyromonas gingivalis* fimbriae in relation to microbial pathogenesis. *Journal of Periodontal Research*, 39:136–142.

Avgustin, G., R. J. Wallace, and H. J. Flint. 1997. Phenotypic diversity among ruminal isolates of *Prevotella ruminicola*: proposal of *Prevotella brevis* sp. nov., *Prevotella bryantii* sp. nov., and *Prevotella albensis* sp. nov. and redefinition of *Prevotella ruminicola. International Journal of Systematic Bacteriology*, 47:284–288.

Boggess, K. A., T. N. Trevett, P. N. Madianos, L. Rabe, S. L. Hillier, J. Beck, and S. Offenbacher. 2005. Use of DNA hybridization to detect vaginal pathogens associated with bacterial vaginosis among asymptom- atic pregnant women. *American Journal of Obstetrics and Gynecology*, 193:752–756.

Boone, D.R. and R. W. Castenholtz (Eds.). 2001. *Bergey's Manual of Systematic Bacteriology*, 2nd ed., Vol. 1. Springer-Verlag, New York.

Bostanci, N., and G. N. Belibasakis. 2012. Porphyromonas gingivalis: an invasive and evasive opportunistic oral pathogen. *FEMS Microbiology Letters*, 333(1): 1–9. doi: 10.1111/j.1574-6968.2012.02579.x.

Brazier, J. S., V. Hall, E. Yusuf, and B. I. Duerden. 2002. *Fusobacterium necrophorum* infections in England and Wales 1990–2000. *Journal of Medical Microbiology*, 51:269–272.

Brook, I. 2004a. Anaerobic pulmonary infections in children. *Pediatric Emergency Care*, 20:636–640.

Brook, I. 2004b. Urinary tract and genito-urinary suppurative infections due to anaerobic bacteria. *International Journal of Urology*, 11:133–141.

Brook, I. 2005a. Microbiology of intracranial abscesses and their associated sinusitis. *Archives of Otolaryngology - Head and Neck Surgery*, 131:1017–1019.

Brook, I. 2005b. The role of bacterial interference in otitis, sinusitis and tonsillitis. *Journal of Otolaryngology - Head and Neck Surgery*, 133:139–146.

Byrd, R. P. Jr. and T. M. Roy. 2003. Anaerobic blood cultures: useful in the ICU? *Chest*, 123:2158–2159.

Cato, E. P. and J. L. Johnson. 1976. Reinstatement of species rank for *Bacteroides fragilis, B. ovatus, B. distasonis, B. thetaiotaomicron* and *B. vulgatus*. Designation of neotype strains for *B. fragilis* (Veillon and Zuber) Castellani and Chalmers and *B. thetaiotaomicron* (Distaso) Castellani and Chalmers. *International Journal of Systematic Bacteriology*, 26:230–237.

Chang, Y. C., C. C. Lai, S. F. Yang, Y. Chan, and Y. S. Hsieh. 2002. Stimulation of matrix metalloproteinases by black-pigmented *Bacteroides* in human pulp and periodontal ligament cell cultures. *Journal of Endodontics*, 28:90–93.

Chen, T., K. Abbey, W. J. Deng, and M. C. Cheng. 2005. The bioinformatics resource for oral pathogens. *Nucleic Acids Research*, 33(Web Server issue):W734–W740.

Chen, T., Y. Hosogi, K. Nishikawa, K. Abbey, R. D. Fleischmann, J. Walling, and M. J. Duncan. 2004. Comparative whole-genome analysis of virulent and avirulent strains of *Porphyromonas gingivalis. Journal of Bacteriology*, 186:5473–5479.

Chhibber-Goel, J., V. Singhal, D. Bhowmik, et al. 2016. Linkages between oral commensal bacteria and atherosclerotic plaques in coronary artery disease patients. *NPJ Biofilms Microbiomes* 2, 7.

Citron, D. M., S. Hunt Gerardo, M. C. Claros, F. Abrahamian, D. Talan, and E. J. Goldstein. 1996. Frequency of isolation of *Porphyromonas* species from infected dog and cat bite wounds in humans and their characterization by biochemical tests and arbitrarily primed-polymerase chain reaction fingerprinting. *Clinical Infectious Disease*, 23(Suppl. 1):S78–S82.

Clarridge, J. E. 3rd. 2004. Impact of 16S rRNA gene sequence analysis for identification of bacteria on clinical microbiology and infectious diseases. *Clinical Microbiology Reviews*, 17:840–862.

Cole, J. R., B. Chai, R. J. Farris, Q. Wang, A. S. Kulam-Syed-Mohideen, D. M. McGarrell, A. M. Bandela, E. Cardenas, G. M. Garrity, and J. M. Tiedje. 2007. The ribosomal database project (RDP-II): introducing myRDP space and quality controlled public data. *Nucleic Acids Research*, 35:D169–D172.

Cardenas, E., G. M. Garrity, and J. M. Tiedje. 2007. The ribosomal database project (RDP-II): Introducing myRDP space and quality controlled public data. *Nucleic Acids Research*, 35:D169–D172.

Collins, M. D., D. N. Love, J. Karjalainen, A. Kanervo, B. Forsblom, A. Willems, S. Stubbs, E, Sarkiala, G. D. Bailey, D. I. Wigney, et al. 1994. Phylogenetic analysis of members of the genus *Porphyromonas* and description of *Porphyromonas cangingivalis* sp. nov. and *Porphyromonas cansulci* sp. nov. *International Journal of Systematic Bacteriology*, 44:674–679.

Coykendall A. L., F. S. Kaczmarek, and J. Slots. 1980. Genetic heterogeneity in *Bacteroides asaccharolyticus* (Holdeman and Moore, 1970) Finegold and Barnes 1977 (Approved List 1980) and proposal of *Bactroides gingivalis* sp. nov. and *Bacteroides macacae* (Slots and Genco) comb nov. *International Journal of Systematic Bacteriology*, 30:559–564.

Creemers-Schild, D., F., Gronthoud, L. Spanjaard, L. G. Visser, C. N. M. Brouwer, and E. J. Kuijper. 2014. *Fusobacterium necrophorum*, an emerging pathogen of otogenic and paranasal infections? *New Microbes New Infections*, 2: 52–57.

Curtis M. A., J. Aduse-Opoku, and M. Rangarajan. 2001. Cysteine proteases of *Porphyromonas gingivalis. Critical Reviews in Oral Biology and Medicine*, 12:192–216.

Davey M. E. and M. J. Duncan. 2006. Enhanced biofilm formation and loss of capsule synthesis: deletion of a putative glycosyltransferase in *Porphyromonas gingivalis. Journal of Bacteriology*, 188: 5510–5523.

de Château, M. and L. Björck. 1994. Protein PAB, a mosaic albumin-binding bacterial protein representing the first contemporary example of module shuffling. *Journal of Biological Chemistry*, 269:12147–12151.

Dowd, S. E., R. D. Wolcott, Y. Sun, T. McKeehan, E. Smith, and D. Rhoads. 2008. Polymicrobial nature of chronic diabetic foot ulcer biofilm infections determined using bacterial tag encoded FLX amplicon pyrosequencing (bTEFAP). *PLOS ONE*, 3:e3326.

Downes, J., F. E. Dewhirst, A. C. R. Tanner, and W. G. Wade. 2013. Description of *Alloprevotella rava* gen. nov., sp. nov., isolated from the human oral cavity, and reclassification of Prevotellatannerae Moore et al. 1994 as *Alloprevotella tannerae* gen. nov., comb. nov. *International Journal of Systematic and Evolutionary Bacteriology*, 63:1214–1218.

Downes, J., S. J. Hooper, M. J. Wilson, and W. G. Wade. 2008. *Prevotella histicola* sp. nov., isolated from the human oral cavity. *International Journal of Systematic and Evolutionary Bacteriology*, 58:1788–1791.

Downes, J., M. Liu, E. Kononen, and W. G. Wade. 2009. *Prevotella micans* sp. nov., isolated from the human oral cavity.

International Journal of Systematic and Evolutionary Bacteriology, 59:771–774.

Downes, J., I. Sutcliffe, A. C. R. Tanner, and W. G. Wade. 2005. *Prevotella marshii* sp. nov. and *Prevotella baroniae* sp. nov., isolated from the human oral cavity. *International Journal of Systematic and Evolutionary Bacteriology*, 55:1551–1555.

Downes, J., A. C. R. Tanner, F. E. Dewhirst, and W. G. Wade. 2010. *Prevotella saccharolytica* sp. nov., isolated from the human oral cavity *International Journal of Systematic Bacteriology*, 60:2458–2461.

Downes, J., I. C. Sutcliffe, V. Booth, and W. G. Wade. 2007. Prevotella maculosa sp. nov., isolated from the human oral cavity. *International Journal of Systematic and Evolutionary Microbiology*, 57(Pt 12):2936–2939.

Downes, J., and W. G. Wade. 2011. Prevotella fusca sp. nov. and Prevotella scopos sp. nov., isolated from the human oral cavity. *International Journal of Systematic and Evolutionary Microbiology*. 61(Pt 4):854–858.

Dubin, G., J. Koziel, K. Pyrc, B. Wladyka, and J. Potempa J. 2013. Bacterial proteases in disease – role in intracellular survival, evasion of coagulation/ fibrinolysis innate defenses, toxicoses and viral infections. *Current Pharmaceutical Design*, 19(6):1090–1113.

Duerden, B. I. 1993. Black-pigmented Gram-negative anaerobes in genitourinary tract and pelvic infections. *FEMS Immunology and Medical Microbiology*, 6:223–227.

Duncan, M. J. 2003. Genomics of oral bacteria. *Critical Reviews in Oral Biology and Medicine*, 14:175–187.

Edmiston, C. E., C. J. Krepel, G. R. Seabrook, L. R. Somberg, A. Nakeeb, R. A. Cambria, and J. B. Towne. 2004. *In vitro* activities of moxifloxacin against 900 aerobic and anaerobic surgical isolates from patients with intra-abdominal and diabetic foot infections. *Antimicrobial Agents and Chemotherapy*, 48:1012–1016.

Egesten, A., I. M. Frick, M. Morgelin, A. I. Olin, and L. Bjorck. 2011. Binding of albumin promotes bacterial survival at the epithelial surface. *Journal of Biological Chemistry*, 286:2469–2476.

Fabrizi, S., R. León, V. Blanc, D. Herrera, and M. Sanz. 2013. Variability of the *fimA* gene in *Porphyromonas gingivalis* isolated from periodontitis and non-periodontitis patients. *Medicina Oral Patologia Oral y Cirugia Bucal*, 18:e100–5.

Faveri, M., M. Feres, J. A. Shibli, R. F. Hayacibara, M. M. Hayacibara, and L. C. de Figueiredo. 2006. Microbiota of the dorsum of the tongue after plaque accumulation: an experimental study in humans. *Journal of Periodontology*, 77:1539–1546.

Finegold S. M., M. L. Vaisanen, M. Rautio, E. Eerola, P. Summanen, et al. 2004. *Porphyromonas uenonis* sp. nov., a pathogen for humans distinct from *P. asaccharolytica* and *P. endodontalis*. *J Clinical Microbiolology*, 42:5298–5301.

Fournier, D., C. Mouton, P. Lapierre, T. Kato, K. Okuda, and C. Menard. 2001. *Porphyromonas gulae* sp. nov., an anaerobic, Gram-negative coccobacillus from the gingival sulcus of various animal hosts. *International Journal of Systematic and Evolutionary Microbiology*, 51:1179–1189.

Fujii, J. S., K. Takayuki, I. Keiji, and F. Ryosuke. 2014. A newly discovered *Anaerococcus* strain responsible for axillary odor and a new axillary odor inhibitor, pentagalloyl glucose *FEMS Microbiology Ecology*, 89:198–207.

Genovese, A., G. Borgia, L. Bjorck, A. Petraroli, A. de Paulis, M. Piazza, and G. Marone. 2003. Immunoglobulin superantigen protein L induces IL-4 and IL-13 secretion from human Fc epsilon RI+ cells through interaction with the kappa light chains of IgE. *Journal of Immunology*, 170:1854–1861.

Gharbia, S. E., H. N. Shah, and K. Bernard. 2012. International Committee on Systematics of Prokaryotes Subcommittee on the taxonomy of Gram-negative anaerobic rods: Minutes of the open meeting, 1–2 February 2011, Health Protection Agency, Colindale, London, UK. *International Journal of Systematic and Evolutionary Bacteriology*, 62:467–471.

Gilmore, M. S. and J. J. Ferretti. 2003. Microbiology. The thin line between gut commensal and pathogen. *Science*, 299:1999–2002.

Glazunova, O. O., T. Launay, D. Raoult, and V. Roux. 2007. *Prevotella timonensis* sp. nov., isolated from a human breast abscess. *International Journal of Systematic and Evolutionary Microbiology*, 57:883–886.

Grenier, D. and R. Bouclin. 2006. Contribution of proteases and plasmin-acquired activity in migration of *Peptostreptococcus micros* through a reconstituted basement membrane. *Oral Microbiology and Immunology*, 21:319–325.

Hagelskjaer Kristensen, L. and J. Prag. 2000. Human necrobacillosis, with emphasis on Lemierre's syndrome. *Clinical Infectious Disease*, 31: 524–532.

Haghi, F., E. Goli, B. Mirzaei, and H. Zeighami. 2019. The association between fecal enterotoxigenic *B. fragilis* with colorectal cancer. *BMC Cancer*, 19:879.

Hajishengallis G. and R. J. Lamont. 2014. Breaking bad: manipulation of the host response by *Porphyromonas gingivalis*. *European Journal of Immunology*, 44:328–338.

Haraszthy, V. I., J. J. Zambon, M. Trevisan, M. Zeid, and R. J. Genco. 2000. Identification of periodontal pathogens in atheromatous plaques. *Journal of Periodontology*, 71:1554–1560.

Hardham, J., K. Dreier, J. Wong, C. Sfintescu, and R. T. Evans. 2005. Pigmented-anaerobic bacteria associated with canine periodontitis. *Veterinary Microbiology*, 106:119–128.

Hayashi, H., K. Shibata, M. Sakamoto, S. Tomita, and Y. Benno. 2007. *Prevotella copri* sp. nov. and *Prevotella stercorea* sp. nov., isolated from human faeces. *International Journal of Systematic and Evolutionary Microbiology*, 57:941–946.

Hirasawa, M. and K. Takada. 1994. *Porphyromonas gingivicanis* sp. nov. and *Porphyromonas crevioricanis* sp. nov., isolated from beagles. *International Journal of Systematic Bacteriology*, 44:637–640.

Hofstad, T. 1992. Virulence factors in anaerobic bacteria. *European Journal of Clinical Microbiology and Infectious Disease*, 11:1044–1048.

Holdeman, L. V. and J. L. Johnson. 1982. Description of *Bacteroides loeschii* sp. nov. and emendation of the descriptions of *Bacteroides melaninogenicus* (Oliver and Wherry) Roy and Kelly 1939 and *Bacteroides denticola* Shah and Collins 1981. *International Journal of Systematic Bacteriology*, 32:399.

Hooper, L.V., T. Midtvedt, and J. I. Gordon. 2002. How host-microbial interactions shape the nutrient environment of the mammalian intestine. *Annual Review of Nutrition*, 22:283–307.

Houle, M. A., D. Grenier, P. Plamondon, and K. Nakayama. 2003. The collagenase activity of *Porphyromonas gingivalis* is due to Arg-gingipain. *FEMS Microbiological Letters*, 221:181–185.

Huq, N. L., C. A. Seers, E. C. Toh, S. G. Dashper, N. Slakeski, L. Zhang, B. R. Ward, et al. 2013. Propeptide mediated inhibition of cognate gingipain proteinases. *PLOS ONE*, 10(8):e65447.

Irshad M., W. A. van der Reijden, W. Crielaard, and M. L. Laine. 2012. In vitro invasion and survival of *Porphyromonas gingivalis* in gingival fibroblasts, role of the capsule. *Archivum Immunologiae et Therapiae Experimentalis (Warsz)*, 60:469–476.

Janssen, K. M., A. Vissink, M. J. de Smit, J. Westra, and E. Brouwer. 2013. Lessons to be learned from periodontitis. *Current Opinion in Rheumatology*, 25(2):241–247.

Johnson, J. L. and C. S. Hirsch. 2003. Aspiration pneumonia. Recognizing and managing a potentially growing disorder. *Postgraduate Medicine*, 113:99–102.

Johnson J. L. and L. V. Holdeman. 1983. *Bacteroides intermedius* comb. nov., and descriptions of *B. corporis* sp. nov. and *Bacteroides levii* sp. nov. *International Journal of Systematic Bacteriology*, 33:15–25.

Johnson, J. S., D. J. Spakowicz, B. Hong, et al. 2019. Evaluation of 16S rRNA gene sequencing for species and strain-level microbiome analysis. *Nature Communications*, 10:5029.

Kaczmarek, F. S. and A. L. Coykendall. 1980. Production of phenylacetic acid by strains of *Bacteroides asaccharolyticus* and *Bacteroides gingivalis* (sp. nov.). *Journal of Clinical Microbiology*, 12:288–290.

Kantarci, A. and T. E. Van Dyke. 2002. Neutrophil-mediated host response to *Porphyromonas gingivalis*. *Journal of the International Academy of Periodontology*, 4:119–125.

Kapatral, V., I. Anderson, N. Ivanova, G. Reznik, T. Los, A. Lykidis, A. Bhattacharyya, et al. 2002. Genome sequence and analysis of the oral bacterium *Fusobacterium nucleatum* strain ATCC 25586. *Journal of Bacteriology*, 184:2005–2018.

Kasper, D. L. 1976. The polysaccharide capsule of *Bacteroides fragilis* subspecies *fragilis*: immunochemical and morphologic definition. *Journal of Infectious Disease*, 133:79–87.

Kato, N., C. X. Liu, H. Kato, K. Watanabe, Y. Tanaka, T. Yamamoto, K. Suzuki, and K. Ueno. 2000. A new subtype of the metalloprotease toxin gene and the incidence of the three bft subtypes among *Bacteroi- des fragilis* isolates in Japan. *FEMS Microbiology Letters*, 182:171–176.

Kesavalu, L., S. Sathishkumar, V. Bakthavatchalu, C. Matthews, D. Dawson, M. Steffen, and J. L. Ebersole. 2007. Rat model of polymicrobial infection, immunity, and alveolar bone resorption in periodontal disease. *Infection and Immunity*, 75:1704–1712.

Kononen, E., E. Eerola, E. V. Frandsen, J. Jalava, J. Matto, S. Salmenlinna, and H. Jousimies-Somer. 1998. Phylogenetic characterization and proposal of a new pigmented species to the genus Prevotella: *Prevotella pallens* sp. nov. *International Journal of Systematic Bacteriology*, 48:47–51.

Kornman, K. S. and W. J. Loesche. 1982. Effects of estradiol and progesterone on *Bacteroides melaninogenicus* and *Bacteroides gingivalis*. *Infection and Immunity*, 35:256–263.

Kozarov, E., D. Sweier, C. Shelburne, A. Progulske-Fox, and D. Lopatin. 2006. Detection of bacterial DNA in atheromatous plaques by quantitative PCR. *Microbes and Infection*, 8:687–689.

Krespi, Y. P., M. G. Shrime, and A. Kacker. 2006. The relationship between oral malodor and volatile sulfur compound-producing bacteria. *Otolaryngology — Head and Neck Surgery*, 135:671–676.

Kuhn, J. J., I. Brook, C. L. Waters, L. W. Church, D. A. Bianchi, and D. H. Thompson. 1995. Quantitative bacteriology of tonsils removed from children with tonsillitis hypertrophy and recurrent tonsillitis with and without hypertrophy. *Annals of Otology, Rhinology, and Laryngology*, 104:646–652.

Labutti, K., R. Pukall, K. Steenblock, T. Glavina Del Rio, H. Tice, A. Copeland, J.-F. Cheng et al. 2009. Complete genome sequence of *Anaerococcus prevotii* type strain (PC1). *Standards in Genomic Sciences*, 1:159–165.

Lee, Y., W. S. Tchaou, K. B. Welch, and W. J. Loesche. 2006. The transmission of BANA-positive periodontal bacterial species from caregivers to children. *Journal of the American Dental Association*, 137:1539–1546.

Loos, B. G., D. W. Dyer, T. S. Whittam, and R. K. Selander. 1993. Genetic structure of populations of *Porphyromonas gingivalis* associated with periodontitis and other oral infections. *Infection and Immunity*, 61:204–212.

Love, D. N., G. D. Bailey, S. Collings, and D. A. Briscoe. 1992. Description of *Porphyromonas circumdentaria* sp. nov. and reassignment of *Bacteroides salivosus* (Love, Johnson, Jones, and Calverley 1987) as *Porphyromonas* (Shah and Collins 1988) *salivosa* comb. nov. *International Journal of Systematic Bacteriology*, 42:434–438.

Love, D. N. 1995. *Porphyromonas macacae* comb. nov., a consequence of *Bacteroides macacae* being a senior synonym of *Porphyromonas salivosa*. *International Journal of Systematic Bacteriology*, 45:90–92.

Love, D. N., J. Karjalainen, A. Kanervo, B. Forsblom, E. Sarkiala, G. D. Bailey, D. I. Wigney, and H. Jousimies-Somer. 1994. *Porphyromonas canoris* sp. nov., an asaccharolytic, black-pigmented species from the gingival sulcus of dogs. *International Journal of Systematic Bacteriology*, 44:204–208.

Maiden, M. F. J., P. Cohee, and A. C. Tanner. 2003. Proposal to conserve the adjectival form of the specific epi- thet in the reclassification of *Bacteroides forsythus* Tanner et al. 1986 to the genus *Tannerella* Sakamoto et al. 2002 as *Tannerella forsythia* corrig., gen. nov., comb. nov. Request for an Opinion. *International Journal of Systematic and Evolutionary Microbiology*, 53:2111–2112.

Mager, D. L., A. D. Haffajee, P. M. Devlin, C. M. Norris, M. R. Posner, and J. M. Goodson. 2005. The salivary microbiota as a diagnostic indicator of oral cancer: A descriptive, non-randomized study of cancer-free and oral squamous cell carcinoma subjects. *Journal of Translational Medicine*, 3:27.

Malinowski, B., A. Węsierska, K. Zalewska, M. Maya, M. Sokołowska, W. Bursiewicz, S. Maciej, M. Ozorowski, K, Pawlak-Osińska, and M, Wiciński. 2019. The role of *Tannerella forsythia* and *Porphyromonas gingivalis* in pathogenesis of esophageal cancer. *Infectious Agents Cancer*, 14:3.

Marchesan, J. T., T. Morelli, S. K. Lundy, Y. Jiao, S. Lim, N. Inohara, G. Nunez, D. A. Fox, and W. V. Giannobile. 2012. Divergence of the systemic immune response following oral infection with distinct strains of *Porphyromonas gingivalis*. *Molecular Oral Microbiology*, 27:483–495.

Marrazzo, J. M. 2004. Evolving issues in understanding and treating bacterial vaginosis. *Expert Review of Anti-Infective Therapy*, 2:913–922.

Matto, J, S. Asikainen, M. L. Vaisanen, M. Rautio, M. Doarela, M. Summanen, S. Finegold, and H. Jousimies-Somer. 1997. Role of *Porphyromonas gingivalis*, *Prevotella intermedia*, and *Prevotella nigrescens* in extraoral and some odontogenic infections. *Clinical Infectious Disease*, 25(Suppl. 2):S194–S198.

Michaud, D. S., J. Izard, C. S. Wilhelm-Benartzi, D. H. You, V. A. Grote, A. Tjønneland, C. C. Dahm, et al. 2012. Plasma antibodies to oral bacteria and risk of pancreatic cancer in a large European prospective cohort study. *Gut*, doi: 10.1136/gutjnl-2012-303006.

Millar, S. J., E. G. Goldstein, M. J. Levine, and E. Hausmann. 1986. Modulation of bone metabolism by two chemically distinct lipopolysaccharide fractions from *Bacteroides gingivalis*. *Infection and Immunity*, 51:302–306.

Mikamo, H., S. Arakawa, M. Fujiwara, H. Funada, T. Inamatsu, S. Iwata, A. Kaneko, et al. 2011. Chapter 1-1. Anaerobic infections (General): Epidemiology of anaerobic infections. *Journal of Infection and Chemotherapy*, 17(Suppl. 1):4–12.

Mombelli. A., M. Tonetti, B. Lehmann, and N. P. Lang. 1996. Topographic distribution of black-pigmenting anaerobes before and after periodontal treatment by local delivery of tetracycline. *Journal of Clinical Periodontology*, 23:906–913.

Moon, J. H., Y. Herr, H. W. Lee, S. I. Shin, C. Kim, A. Amano, J. Y. Lee. 2012. Genotype analysis of *Porphyromonas gingivalis* fimA in Korean adults using new primers. *Journal of Medical Microbiology*, Dec 21.

Moore, L. V., J. L. Johnson, and W. E. Moore. 1994. Descriptions of *Prevotella tannerae* sp. nov. and *Prevotella enoeca* sp. nov. from the human gingival crevice and emendation of the description of *Prevotella zoogleoformans*. *International Journal of Systematic Bacteriology*, 44:599–602.

Moore, W. E. C. and L. V. Holdeman. 1974. Human fecal flora: The normal flora of 20 Japanese-Hawaiians. *Applied Microbiology*, 27:961–979.

Murphy, E. C. and I. M. Frick. 2013. Gram-positive anaerobic cocci—Commensals and opportunistic pathogens. *FEMS Microbiological Reviews*, 37:520–553.

Nagano, K., Y. Abiko, Y. Yoshida, and F. Yoshimura. 2013. Genetic and antigenic analyses of *Porphyromonas gingivalis* FimA fimbriae. *Molecular Oral Microbiology*, 28:392–403.

Nakatani, T., T. Sato, B. F. Trump, J. H. Siegel, and K. Kobayashi. 1996. Manipulation of the size and clone of an intra-abdominal abscess in rats. *Research in Experimental Medicine (Berlin)*, 196:117–126.

Narayanan, L. L., and C. Vaishnavi. 2010. Endodontic microbiology. *Journal of Conservative Dentistry*, 13(4):233–239. doi:10.4103/0972-0707.73386. PMID: 21217951; PMCID: PMC3010028.

Nelson, K. E., R. D. Fleischmann, R. T. DeBoy, I. T. Paulsen, D. E. Fouts, J. A. Eisen, S. C. Daugherty, R. J. Dodson, A. S. Durkin, M. Gwinn, D. H. Haft, J. F. Kolonay, W. C. Nelson, T. Mason, L. Tallon, J. Gray, D. Granger, H. Tettelin, H. Dong, J. L. Galvin, M. J. Duncan, F. E. Dewhirst, and C. M. Fraser. 2003. Complete genome sequence of the oral pathogenic bacterium *Porphyromonas gingivalis* strain W83. *Journal of Bacteriology*, 185:5591–5601.

Norris, J. M. and D. N. Love. 1999. Associations amongst three feline *Porphyromonas* species from the gingi- val margin of cats during periodontal health and disease. *Veterinary Microbiology*, 65:195–207.

Ogrendik, M. 2012. Does periodontopathic bacterial infection contribute to the etiopathogenesis of the autoimmune disease rheumatoid arthritis? *Discovery Medicine*, 13:349–355.

Ogawa, T., Y. Asai, Y. Makimura, and R. Tamai. 2007. Chemical structure and immunobiological activity of *Porphyromonas gingivalis* lipid A. *Frontiers in Bioscience*, 12:3795–3812.

Oliver, W. W. and W. B. Wherry. 1921. Notes on some bacteria parasites of the human mucous membranes. *Journal of Infectious Disease*, 28:341–344.

Olsen, I. and H. N. Shah. 2001. International committee on systematics of prokaryotes subcommittee on the taxonomy of gram-negative anaerobic rods. *International Journal of Systematic and Evolutionary Microbiology*, 51:1943–1944.

Olsen, I. and H. N. Shah. 2003. International committee on systematics of prokaryotes subcommittee on the taxonomy of gram-negative anaerobic rods. *International Journal of Systematic and Evolutionary Microbiology*, 53:923–924.

Ortiz, E. and M. A. Sande. 2000. Routine use of anaerobic blood cultures: are they still indicated? *American Journal of Medicine*, 108:445–447.

Ota-Tsuzuki, C., and M. P. Alves Mayer. 2010. Collagenase production and hemolytic activity related to 16S rRNA variability among Parvimonas micra oral isolates. *Anaerobe*, 16(1):38–42. doi: 10.1016/j.anaerobe.2009.03.008.

Pantosti, A., M. G. Menozzi, A. Frate, L. Sanfilippo, F. D'Ambrosio, and M. Malpeli. 1997. Detection of enterotoxigenic *Bacteroides fragilis* and its toxin in stool samples from adults and children in Italy. *Clinical Infectious Diseases*, 24:12–16.

Paul, A., P. A. Lawson, E. Moore, and E. Falsen. 2008. *Prevotella amnii* sp. nov., isolated from human amniotic fluid. *International Journal of Systematic Bacteriology*, 58:89–92.

Perricone C., F. Ceccarelli, M. Saccucci, G. Di Carlo, D. P. Bogdanos et al. 2019. *Porphyromonas gingivalis* and rheumatoid arthritis. *Current Opinion in Rheumatology*, 31:517–524.

Peterka, M., K. Tepsic, T. Accetto, A. Kostanjsek, A. Ramsak, L. Lipoglavsek, and G. Avgustin. 2003. Molecular microbiology of gut bacteria: genetic diversity and community structure analysis. *Acta Microbiologica et Immunologica Hungarica*, 50:395–406.

Poole, S., S. K. Singhrao, L. Kesavalu, M. A. Curtis, and S. J. Crean. 2013. Determining the presence of periodontopathic virulence factors in short-term postmortem Alzheimer's disease brain tissue. *Journal of Alzheimers Disease*, 36:665–677.

Pryor, J. P., E. Piotrowski, C. W. Seltzer, and V. H. Gracias. 2001. Early diagnosis of retroperitoneal necrotizing fasciitis. *Critical Care Medicine*, 29:1071–1073.

Quirke, A. M., E. B. Lugli, N. Wegner, B. C. Hamilton, P. Charles, M. Chowdhury, A. J. Ytterberg, et al. 2013. Heightened immune response to autocitrullinated *Porphyromonas gingivalis* peptidylarginine deiminase: A potential mechanism for breaching immunologic tolerance in rheumatoid arthritis. *Annals of Rheumatic Diseases*.doi: 10.1136/annrheumdis-2012-202726

Ramanujam, P. and N. K. Rathlev. 2006. Blood cultures do not change management in hospitalized patients with community-acquired pneumonia. *Academic Emergency Medicine*, 13:740–745.

Rawlinson, A., T. F. Walsh, A. Lee, and S. J. Hodges. 1998. Phylloquinone in gingival crevicular fluid in adult periodontitis. *Journal of Clinical Periodontology*, 25:662–665.

Reichert, S., M. Haffner, G. Keyßer, C. Schäfer, J. M. Stein, H. G. Schaller, A. Wienke, H. Strauss, S. Heide, S. Schulz. 2013. Detection of oral bacterial DNA in synovial fluid. *Journal of Clinical Periodontology*, 40:591–598.

Relman, D. A. 2002. New technologies, human-microbe interactions, and the search for previously unrecognized pathogens. *Journal of Infectious Disease*, 186(Suppl. 2):S254–S258.

Ricci, S., D. Medaglini, H. Marcotte, A. Ols.n, G. Pozzi, and L. Bj. rck. 2001. Immunoglobulin-binding domains of peptostreptococcal protein L enhance vaginal colonization of mice by *Streptococcus gordonii*. *Microbial Pathogenesis*, 30:229–235.

Robertson, K. P., C. J. Smith, A. M. Gough, and E. R. Rocha. 2006. Characterization of *Bacteroides fragilis* hemolysins and regulation and synergistic interactions of HlyA and HlyB. *Infection and Immunity*, 74:2304–2316.

Sakamoto M., Y. Huang, M. Umeda, I. Ishikawa, and Y. Benno. 2005. *Prevotella multiformis* sp. nov., isolated from human subgingival plaque. *International Journal of Systematic and Evolutionary Microbiology*, 55:815–819.

Sakamoto M., H. Kumada, N. Hamada, Y. Takahashi, M. Okamoto, M. A. Bakir, and Y. Benno. 2009. *Prevotella falsenii* sp. nov., a *Prevotella intermedia*-like organism isolated from monkey dental plaque. *International Journal of Systematic and Evolutionary Microbiology*, 59:319–322.

Sakamoto M. and Ohkuma, M. 2012. Reclassification of *Xylanibacter oryzae* Ueki et al. 2006 as *Prevotella oryzae* comb. nov., with an emended description of the genus *Prevotella. International Journal of Systematic and Evolutionary Microbiology*, 62:2637–2642.

Sakamoto, M. and M. Ohkuma. 2013. Porphyromonas crevioricanis is an earlier heterotypic synonym of *Porphyromonas cansulci* and has priority. *International Journal of Systematic and Evolutionary Microbiology*, 63:454–457.

Sakamoto, M., K. Ohkusu, T. Masaki, H. Kako, T. Ezaki, and Y. Benno. 2007. *Prevotella pleuritidis* sp. nov., isolated from pleural fluid. *International Journal of Systematic and Evolutionary Microbiology*, Aug;57(Pt 8):1725–1728.

Sakamoto, M., M. Suzuki, Y. Huang, M. Umeda, I. Ishikawa, and Y. Benno. 2004. *Prevotella shahii* sp. nov. and *Prevotella salivae* sp. nov., isolated from the human oral cavity. *International Journal of Systematic and Evolutionary Microbiology*, 54:877–883.

Sakamoto, M., N. Suzuki, and M. Okamoto. 2010. *Prevotella aurantiaca* sp. nov., isolated from the human oral cavity. *International Journal of Systematic and Evolutionary Microbiology*, 60:500–503.

Sakamoto, M., M. Suzuki, M. Umeda, L. Ishikawa, and Y. Benno. 2002. Reclassification of *Bacteroides forsythus* (Tanner et al. 1986) as *Tannerella forsythensis* corrig., gen. nov., comb. nov. *International Journal of Systematic and Evolutionary Microbiology*, 52:841–849.

Sakamoto, M., M. Umeda, I. Ishikawa, and Y. Benno. 2005. *Prevotella multisaccharivorax* sp. nov., isolated from human subgingival plaque. *International Journal of Systematic and Evolutionary Microbiology*, 55:1839–1843.

Sears, C. L. 2001. The toxins of *Bacteroides fragilis. Toxicon*, 39:1737–1746.

Shah, H. N., R. Bonnett, B. Mateen, and R. A. Williams. 1979. The porphyrin pigmentation of subspecies of *Bacteroides melaninogenicus. Biochemical Journal*, 180:45–50.

Shah, H. N. and M. D. Collins. 1988. Proposal for reclassification of *Bacteroides asaccharolyticus, Bacteroides gingivalis* and *Bacteroides endodontalis* in a new genus, *Porphyromonas. International Journal of Systematic Bacteriology*, 38:128–131.

Shah, H. N. and D. M. Collins. 1990. *Prevotella*, a new genus to include *Bacteroides melaninogenicus* and related species formerly classified in the genus *Bacteroides. International Journal of Systematic Bacteriology*, 40:205–208.

Shah, H. N., M. D. Collins, I. Olsen, B. J. Paster, and F. E. Dewhirst. 1995. Reclassification of *Bacteroides levii* (Holdeman, Cato, and Moore) in the genus *Porphyromonas* as *Porphyromonas levii* comb. nov. *International Journal of Systematic Bacteriology*, 45:586–588.

Shah, H. N. and S. E. Gharbia. 1992. Biochemical and chemical studies on strains designated *Prevotella inter- media* and proposal of a new pigmented species, *Prevotella nigrescens* sp. nov. *International Journal of Systematic Bacteriology*, 42:542–546.

Shah, H. N., R. A. Williams, G. H. Bowden, and J. M. Hardie. 1976. Comparison of the biochemical properties of *Bacteroides melaninogenicus* from human dental plaque and other sites. *Journal of Applied Bacteriology*, 41:473–495.

Shang, F. M., and H. L. Liu. 2018. Fusobacterium nucleatum and colorectal cancer: A review. *World Journal of Gastrointestinal Oncology*, 10(3):71–81.

Sharma, A. 2010. Virulence factors of *Tannerella forsythia. Periodontology 2000*, 54:106–116.

Sherwood, J. E., S. Fraser, D. M. Citron, H. Wexler, G. Blakely, K. Jobling, and S. Patrick. 2011. Multi-drug resistant *Bacteroides fragilis* recovered from blood and severe leg wounds caused by an improvised explosive device (IED) in Afghanistan. *Anaerobe*, 17:152–155.

Siqueira, J. F. Jr, I. N. Rocas, J. C. Oliveira, and K. R. Santos. 2001. Molecular detection of black-pigmented bacteria in infections of endodontic origin. *Journal of Endodontology*, 27:563–566.

Slaney, J. M., A. Gallagher, J. Aduse-Opoku, K. Pell, and M. A. Curtis. 2006. Mechanisms of resistance of *Porphyromonas gingivalis* to killing by serum complement. *Infection and Immunity*, 74:5352–5361.

Slots, J. and R. J. Genco. 1980. *Bacteroides melaninogenicus* subsp. *macacae*, a new subspecies from mon- key periodontopathogenic indigenous microflora. *International Journal of Systematic Bacteriology*, 30:82–85.

Slots J. and H. S. Reynolds. 1982. Long-wave UV light fluorescence for identification of black-pigmented *Bac- teroides* spp. *Journal of Clinical Microbiology*, 16:1148–1151.

Smit, M. D., J. Westra, A. Vissink, B. Doornbos-van der Meer, E. Brouwer, and A. J. van Winkelhoff. 2012. Periodontitis in established rheumatoid arthritis patients: a cross-sectional clinical, microbiological and serological study. *Arthritis Research and Therapeutics*, 14:R222.

Smith, D. T. 1930. Fusospirochetal disease of the lungs produced with cultures from Vincent's angina. *Journal of Infectious Disease*, 46:303–310.

Socransky, S. S., A. D. Haffajee, M. A. Cugini, C. Smith, and R. L. Kent, Jr. 1998. Microbial complexes in subgingival plaque. *Journal of Clinical Periodontology*, 25:134–144.

Söderquist, B., S. Björklund, B. Hellmark, A. Jensen, and H. Brüggemann. 2017. *Finegoldia magna* isolated from orthopedic joint implant-associated infection. *Journal of Clinical Microbiology*, 55:3283–3291.

Sparks Stein, P., M. J. Steffen, C. Smith, G. Jicha, J. L. Ebersole, E. Abner, and D. Dawson 3rd. 2012. Serum antibodies to periodontal pathogens are a risk factor for Alzheimer's disease. *Alzheimers Dementia*, 8:196–203.

Styrt, B. and S. L. Gorbach. 1989. Recent developments in the understanding of the pathogenesis and treatment of anaerobic infections (2). *New England Journal of Medicine*, 321:240–246.

Summanen, P. H., B. Durmaz, M. L. Väisänen, C. Liu, D. Molitoris, E. Eerola, I. M. Helander, and S. M. Finegold. 2005 *Porphyromonas somerae* sp. nov., a pathogen isolated from humans and distinct from *Porphyromonas levii. Journal of Clinical Microbiology*, 43:4455–4459

Takada, K., K. Hayashi, Y. Sato, and M. Hirasawa. 2010. *Prevotella dentasini* sp. nov., a black-pigmented species isolated from the oral cavity of donkeys. *International Journal of Systematic and Evolutionary Microbiology*, 60:1637–1639.

Talan, D. A., D. M. Citron, F. M. Abrahamian, G. J. Moran, and E. J. Goldstein. 1999. Bacteriologic analysis of infected dog and cat bites. Emergency Medicine Animal Bite Infection Study Group. *New England Journal of Medicine*, 340:85–92.

Tanaka, M., T. Hanioka, K. Takaya, and S. Shizukuishi. 1998. Association of oxygen tension in human periodontal pockets with gingival inflammation. *Journal of Periodontology*, 69:1127–1130.

Tanner, A. C., and J. M. Goodson. 1986. Sampling of microorganisms associated with periodontal disease. *Oral Microbiology and Immunology.* 1(1):15–22. doi: 10.1111/j.1399-302x.1986. tb00310.x.

Tanner, A. C. and J. Izard. 2006. *Tannerella forsythia*, a periodontal pathogen entering the genomic era. *Periodontology 2000*, 42:88–113.

Tomazinho, L. F. and M. J. Avila-Campos. 2007. Detection of *Porphyromonas gingivalis*, *Porphyromonas endodontalis*, *Prevotella intermedia*, and *Prevotella nigrescens* in chronic endodontic infection. *Oral Surgery, Oral Medicine, Oral Pathology, Oral Radiology and Endodontology*, 103:285–288.

Toprak, N. U., A. Yagci, B. M. Gulluoglu, M. L. Akin, P. Demirkalem, T. Celenk, and G. Soyletir. 2006. A possible role of *Bacteroides fragilis* enterotoxin in the aetiology of colorectal cancer. *Clinical Microbiology and Infection*, 12:782–786.

Totaro, M. C., P. Cattani, F. Ria, B. Tolusso, E. Gremese, A. L. Fedele, S. D'Onghia, et al. 2013. *Porphyromonas gingivalis* and the pathogenesis of rheumatoid arthritis: Analysis of various compartments including the synovial tissue. *Arthritis Research and Therapy*, 18(15):R66.

Tribble, G. D., J. E. Kerr, and B. Y. Wang. 2013. Genetic diversity in the oral pathogen Porphyromonas gingivalis: Molecular mechanisms and biological consequences. *Future Microbiology*, 8:607–620.

Tzianabos, A. O., A. Pantosti, H. Baumann, J. R. Brisson, H. J. Jennings, and D. L. Kasper. 1992. The capsular polysaccharide of *Bacteroides fragilis* comprises two ionically linked polysaccharides. *Journal of Biological Chemistry*, 267:18230–18235.

Ueki, A., H. Akasaka, A. Satoh, D. Suzuki, and K Ueki. 2007. Prevotella paludivens sp. nov., a novel strictly anaerobic, Gram-negative, hemicellulosedecomposing bacterium isolated from plant residue and rice roots in irrigated rice-field soil. *International Journal of Systematic and Evolutionary Microbiology*, 57(Pt 8):1803–1809. doi: 10.1099/ijs.0.64914-0.

Ulger-Toprak, N., C. Liu, P. H. Summanen, and S. M. Finegold. 2010. *Murdochiella asaccharolytica* gen. nov., sp. nov., a Gram-stain-positive, anaerobic coccus isolated from human wound specimens. *International Journal of Systematic and Evolutionary Microbiology*, 60:1013–1016.

van der Waaij, D., J. M. Berghuis-de Vries, and J. E. C. Lekkerkerk-van der Wees. 1972. Colonization resistance of the digestive tract and the spread of bacteria to the lymphatic organs in mice. *Journal of Hygiene (London)*, 70:335–342.

van Steenbergen, T. J. M., A. J. van Winkelhoff, D. Mayrand, D. Grenier, and J. de Graaff. 1984. *Bacteroides endodontalis* sp. nov., an asaccharolytic black pigmented *Bacteroides* species from infected dental root canals. *International Journal of Systematic Bacteriology*, 34:118–120.

Vandamme, P., L. Debruyne, E. De Brandt, and E. Falsen. 2010. Reclassification of *Bacteroides ureolyticus* as *Campylobacter ureolyticus* comb. nov., and emended description of the genus Campylobacter. *International Journal of Systematic and Evolutionary Microbiology*, 60:2016–2022.

Vernillo, A. T., N. S. Ramamurthy, L. M. Golub, and B. R. Rifkin. 1994. The nonantimicrobial properties of tetracycline for the treatment of periodontal disease. *Current Opinion in Periodontology*, 2:111–118.

Vlek, A. L. M., M. J. M. Bonten, and C. H. E. Boel. 2012. Direct matrix-assisted laser desorption ionization time-of-flight mass spectrometry improves appropriateness of antibiotic treatment of bacteremia. *PLOS ONE*, 7:e32589.

Watabe, J., Y. Benno, and T. Mitsuoka. 1983. Taxonomic study of *Bacteroides oralis* and related organisms and proposal of *Bacteroides veroralis* sp. nov. *International Journal of Systematic Bacteriology*, 33:57–64.

Weintraub, A., U. Zähringer, H. W. Wollenweber, U. Seydel, and E. T. Rietschel. 1989. Structural characteriza- tion of the lipid A component of *Bacteroides fragilis* strain NCTC 9343 lipopolysaccharide. *European Journal of Biochemistry*, 183:425–431.

Willems, A. and M. D. Collins. 1995a. 16S rRNA gene similarities indicate that *Hallella seregens* (Moore and Moore) and *Mitsuokella dentalis* (Haapasalo et al.) are genealogically highly related and are members of the genus *Prevotella*: emended description of the genus *Prevotella* (Shah and Collins) and description of *Prevotella dentalis* comb. Nov. *International Journal of Systematic Bacteriology*, 45: 832–836.

Willems, A. and M. D. Collins. 1995b. Reclassification of *Oribaculum catoniae* (Moore and Moore 1994) as *Porphyromonas catoniae* comb. nov., and emendation of the genus *Porphyromonas*. *International Journal of Systematic Bacteriology*, 45:578–581.

Wu, C. C., J. L. Johnson, W. E. C. Moore, and L. V. H. Moore. 1992. Emended descriptions of *Prevotella denticola*, *Prevotella loescheii*, *Prevotella veroralis*, and *Prevotella melaninogenica*. *International Journal of Systematic Bacteriology*, 42:536–541.

Wu, S., K. J. Rhee, M. Zhang, A. Franco, and C. L. Sears. 2007. *Bacteroides fragilis* toxin stimulates intestinal epithelial cell shedding and {gamma}-secretase-dependent E-cadherin cleavage. *Journal of Cell Science*, 120:1944–1952.

Wyss, C. 1989. Dependence of proliferation of *Bacteroides forsythus* on exogenous N-acetylmuramic acid. *Infection and Immunity*, 57:1757–1759.

Xu, J., M. K. Bjursell, J. Himrod, S. Deng, L. K. Carmichael, H. C. Chiang, L. V. Hooper, and J. I. Gordon. 2003. A genomic view of the human-*Bacteroides thetaiotaomicron* symbiosis. *Science*, 299:2074–2076.

Zhao, S., T. D. Lieberman, P. Mathilde, K. F. Kauffman, S. M. Gibbons, M. Groussin, J. X. Ramnik, and E. J. Alm. 2019. Adaptive evolution within gut Microbiomes of healthy people. *Cell Host and Microbe*; 25:656–667.

Zhong, L. I., Y. M. Zhang, L. H. Liu, P. Liang, A. R. Murat, and S. Askar. 2008. Detection of periodontal pathogens in coronary atherosclerotic plaques. *Zhonghua Kou Qiang Yi Xue Za Zhi*, 43:4–7.

Zhou, Z., J. Chen, H. Yao, and H. Hu. 2018. Fusobacterium and colorectal cancer. *Frontiers of Oncology*, 8:371.

44

Other Gram-Negative Bacteria: Acinetobacter, Burkholderia, *and* Moraxella

Rebecca E. Colman and Jason W. Sahl

CONTENTS

Introduction

Gram-negative bacteria do not retain the crystal violet stain during classical Gram staining methods to differentiate between bacterial pathogens. Gram-negative bacteria do not follow a monophyletic, phylogenetic pattern, and thus, the designation is primarily of importance in the determination of which antimicrobial to use when treating bacterial infections. For example, the Gram-negative outer cell membrane makes them resistant to several classes of antibiotics. This chapter briefly discusses species of the following Gram-negative genera in the phylum Proteobacteria: *Acinetobacter*, *Burkholderia*, and *Moraxella*.

Genus *Acinetobacter*

The genus *Acinetobacter* is a heterogeneous group of organisms, including strains associated with both the natural environment and human infection. *Acinetobacter* is in the class Gammaproteobacteria and the family *Moraxellaceae*, where there are 63 named species and several unnamed species (http://www.bacterio.net/acinetobacter.html) (1).

While organisms in *Acinetobacter* generally inhabit soil and water environments, some are nosocomial pathogens and others are opportunistic human pathogens. In the clinical setting, *Acinetobacter* infections are classified as either nosocomial or community acquired, although the majority of cases are community acquired (2). *Acinetobacter baumannii* is the primary organism identified in nosocomial infections, particularly in intensive care units, where it has been associated with 20% of all infections (3). However, other species have also been associated with nosocomial infections, including genomic species 13 (*Acinetobacter nosocomialis* [4]) and 3 (*Acinetobacter pittii* [4]). These three species are closely related according to DNA-DNA hybridization studies and are difficult to distinguish phenotypically and thus are often grouped together into the *Acinetobacter*

calcoaceticus-baumannii (Acb) complex. The complication with this complex designation is that a fourth species, *A. calcoaceticus*, is included in the complex, but is a soil bacterium not associated with human disease (1). Also included in the Acb complex is *Acinetobacter oleivorans*, a species not associated with pathogenesis that can grow solely on diesel fuel (5). A recent clinical study in the United States demonstrated that most *Acinetobacter* bloodstream infections were caused by *A. baumannii* (63% of cases), *A. nosocomialis* (21% of cases), and *A. pittii* (8% of cases) (6). Species outside of the Acb complex, including *Acinetobacter ursingii*, have also been associated with nosocomial infection (7).

Acinetobacter spp. are difficult to differentiate experimentally, even with commercial typing systems (8, 9). A range of methods have been investigated to classify and discriminate between *Acinetobacter* species. One study assessed a variety of methods and found pulsed-field gel electrophoresis (PFGE) to be the most discriminating method in the differentiation of *Acinetobacter* spp (10). However, the PFGE protocol must be strictly standardized (11), and even then, the results are not always reproducible (12). Additional methods, including optimal growth temperature, cannot accurately distinguish between species within the Acb complex (13). Additional analyses applied to the identification of *Acinetobacter* spp. include amplified 16S rDNA restriction analysis (14), amplified fragment length polymorphism (AFLP) (15), and *rpoB* gene sequence analysis (16). However, many of these methods can either not be compared between studies or cannot differentiate between related species in the Acb complex. Core genome phylogenetics has recently been used to differentiate between *Acinetobacter* spp. and will likely be the most accurate method moving forward to differentiate between species (17, 18).

A. baumannii is the most clinically relevant species of *Acinetobacter* (9) and causes a range of negative interactions with the human host, from asymptomatic colonization (19) to septicemia and death (20–22). One of the primary human health concerns of *A. baumannii* infection is the organism's potential resistance to all known antibiotics (9, 21). Multidrug-resistant (resistant to at least three classes of drugs specifically

DOI: 10.1201/9781003099277-46

cephalosporin, fluoroquinolones, and aminoglycosides) (23) and extensively-drug-resistant (MDR plus resistant to carbapenems) (24) isolates of *A. baumannii* have been identified in hospitals around the world (25). The combination of desiccation resistance (26), antibiotic resistance, and attachment to abiotic surfaces (27), including surfaces used in joint replacements (e.g., titanium), has made *A. baumannii* a pathogen of great concern in the hospital environment.

Genus *Burkholderia*

Originally, a member of the genus *Pseudomonas*, the genus *Burkholderia* has species that occupy a wide range of ecological niches, including soil, water, the rhizosphere, animals, and humans. *Burkholderia* is a member of the class Betaproteobacteria, family *Burkholderiaceae*, and consists of 122 named species (http://www.bacterio.net/burkholderia.html).

The *Burkholderia cepacia* complex (Bcc) consisting of at least 17 species, is a group of organisms consisting of extraordinary metabolic diversity that enables the inhabitants to colonize a wide range of environments (28). Some members of this complex have been leveraged for bioremediation applications, because of their ability to degrade pollutants in water and soils (29). There are also several strains in the Bcc that produce antifungal compounds and fix atmospheric nitrogen (28, 30). While the Bcc was originally identified as a plant pathogen (31), species in this complex generally have ecologically beneficial interactions with plants (32). Because of the ability of several species within the Bcc to colonize the rhizosphere on economically important crops, there has been interest in using Bcc isolates for plant-growth promotion (30). Issues with their use have arisen due to the fact that many species in *Burkholderia* have emerged as important human pathogens. These environmental organisms in the Bcc have been implicated as opportunistic pathogens for cystic fibrosis (CF) and immunocompromised patients.

Bcc species can cause severe respiratory infections in CF patients (33) and currently there is no distinction between environmental and clinical isolates, implying that the natural environment is the source of infections (34). Identified in 1984 as an increasing problem among CF patients and noting multiresistance to antibiotics, Bcc species were concluded to cause the *cepacia syndrome* (35). Although this syndrome is not common, it presents a high fatality rate and affects the ability of the patient to qualify for a lung transplant (33). Several strains within the Bcc are shown to be highly transmissible among patients through social contact (36, 37). Because these are ecologically diverse species, they can often survive in a variety of environments that can put human health in danger. Contaminated pharmaceuticals, cosmetics, disinfectants, and preservative products are a major source of Bcc bacteria. Furthermore, in the clinical setting, bacteria from this complex have been isolated from tap water, dialysis machines, nebulizers, catheters, blood gas analyzers, thermometers, ventilators, temperature sensors, solutions, and intravenous fluids (38, 39). One issue associated with the Bcc complex is their intrinsic resistance to most of the clinically available antimicrobials including aminoglycosides, quinolones, polymyxins, and β-lactams (40).

Another complex within *Burkholderia* is the *Burkholderia pseudomallei* complex (Bpc), which contains *B. singularis*, *B. oklahomensis*, *B. humptydooensis*, two putative novel species (41), *B. thailandensis*, and the Select Agents: *B. pseudomallei* and *B. mallei*, which can cause serious disease in humans and animals (42). *B. pseudomallei*, an organism found naturally in soil and water, causes the disease melioidosis in Southeast Asia, northern Australia, the Middle East, China, India, and South America (43, 44). The highest number of human cases is in Thailand, where it accounts for up to 20% of community-acquired septicemias. Melioidosis mostly affects adults, often with underlying predisposing conditions that result in an impaired immune response (42). There are a variety of clinical manifestations, both acute and chronic, although asymptomatic colonization has also been observed (45). *B. pseudomallei* is also recognized as a major veterinary pathogen, causing disease in many animal species, and is especially a concern for zoos (42). The other Select Agent, *B. mallei* causes the disease glanders (42), mainly a disease of equines, but can be transmitted to humans. The disease can be either acute or chronic, and the organism cannot persist outside its host. *Burkholderia mallei* is a clone of *B. pseudomallei*, which has gone through significant genomic reduction (46, 47). Both *B. mallei* and *B. pseudomallei* have been identified as pathogens of interest for many years, and knowledge on the mechanisms underlying their virulence has recently advanced (48).

Genus *Moraxella*

The genus *Moraxella* contains species that are coccoid or rod-shaped, but are all genetically related (49–51). *Moraxella* is in the class Gammaproteobacteria and the family *Moraxellaceae*, where there are 22 named species (http://www.bacterio.net/moraxella.html).

Several species within *Moraxella* cause clinical disease, while others are found as part of the natural human flora. *Moraxella lacunata* (52), *M. nonliquefaciens*, and *bovis* (53) have been associated with eye infections. Additionally, *M. nonliquefaciens* can cause septicemia in leukemic patients (54), but may also be found as part of normal airway flora (50, 55). The most clinically important species of this genus is *Moraxella catarrhalis*, which originally was thought of as a commensal of the upper respiratory tract but, as of late, has been identified to cause disease of the eye, ear, and sinus in children (56).

Only recently has there been clinical interest in *M. catarrhalis*, and as such, not much is known about the pathogenesis of the species. Phenotyping strategies have been developed for epidemiological typing, although none have been accepted internationally. Electrophoretic profiling of outer membrane proteins (57), typing of lipopolysaccharide (58), and isoelectric focusing of β-lactamase proteins (59) have all been proposed for typing this species. DNA techniques have also been developed, but there are no universally adopted techniques. Restriction endonuclease analysis (60, 61), macrorestriction enzyme (62), PFGE (63), strain-specific DNA probes (64), AFLP, and ribotyping (65, 66) have all been typing procedures discussed in the literature. As genetic studies are being conducted, the possibility of frequent horizontal gene transfer has become evident (67) allowing for the sharing of important genetic material potentially linked to antibiotic resistance. There are greater than 100 whole-genome sequences for *M. catarrhalis* (68) and additional sequencing may lead to the identification of an appropriate and comprehensive typing system.

Historically, there is an inverse relationship between patient age and colonization of *M. catarrhalis* that is still present today (69, 70). While carriage rates are high in children, *M. catarrhalis* is also an opportunistic pathogen causing sinusitis, otitis media, tracheitis, bronchitis, and pneumonia, and can cause systemic disease (e.g., meningitis and sepsis) (51). *M. catarrhalis* is one of the top three pathogens in acute otitis media, which is a major cause of morbidity in early childhood (51). Infections in adults include laryngitis, bronchitis, and pneumonia, and a rare but serious infection is endocarditis. *Moraxella catarrhalis* can also manifest as a nosocomial pathogen, but without a reliable typing system, the conformation of the spread of the organism is difficult (51, 71). This species produces β-lactamase and thus has almost universal β-lactamase-mediated resistance to penicillins. *Moraxella catarrhalis* is also inherently resistant to trimethoprim; however, it is universally sensitive to most antibiotics used in respiratory infection treatments (72–74).

Conclusions

The *Acinetobacter, Burkholderia*, and *Moraxella* genera contain a mix of important human pathogens and environmental bacteria. The study of pathogenesis in these organisms is important not only to control the spread of infection but also in understanding the genetic effects that allow soil bacteria to transition to the colonization of the human host. A comprehensive study of environmental and clinical isolates will allow researchers to better understand the evolution of these clinically important genera. These studies will help guide more accurate diagnostics for human pathogens and will also help to understand the acquisition of antimicrobial resistance mechanisms, which will focus targeted treatment therapeutics.

REFERENCES

1. Visca P, Seifert H, Towner KJ. 2011. Acinetobacter infection–an emerging threat to human health. *IUBMB Life.* 63:1048–54.
2. Ong CW, Lye DC, Khoo KL, Chua GS, Yeoh SF, Leo YS, Tambyah PA, Chua AC. 2009. Severe community-acquired Acinetobacter baumannii pneumonia: an emerging highly lethal infectious disease in the Asia-Pacific. *Respirology.* 14:1200–5.
3. Hood MI, Mortensen BL, Moore JL, Zhang Y, Kehl-Fie TE, Sugitani N, Chazin WJ, Caprioli RM, Skaar EP. 2012. Identification of an Acinetobacter baumannii zinc acquisition system that facilitates resistance to calprotectin-mediated zinc sequestration. *PLOS Pathog.* 8:e1003068.
4. Nemec A, Krizova L, Maixnerova M, van der Reijden TJ, Deschaght P, Passet V, Vaneechoutte M, Brisse S, Dijkshoorn L. 2011. Genotypic and phenotypic characterization of the Acinetobacter calcoaceticus-Acinetobacter baumannii complex with the proposal of Acinetobacter pittii sp. nov. (formerly Acinetobacter genomic species 3) and Acinetobacter nosocomialis sp. nov. (formerly Acinetobacter genomic species 13TU). *Res Microbiol.* 162:393–404.
5. Kang YS, Jung J, Jeon CO, Park W. 2011. Acinetobacter oleivorans sp. nov. is capable of adhering to and growing on diesel-oil. *J Microbiol.* 49:29–34.
6. Wisplinghoff H, Paulus T, Lugenheim M, Stefanik D, Higgins PG, Edmond MB, Wenzel RP, Seifert H. 2012. Nosocomial bloodstream infections due to Acinetobacter baumannii, Acinetobacter pittii and Acinetobacter nosocomialis in the United States. *J Infect.* 64:282–90.
7. Horii T, Tamai K, Mitsui M, Notake S, Yanagisawa H. 2011. Blood stream infections caused by Acinetobacter ursingii in an obstetrics ward. *Infect Genet Evol.* 11:52–6.
8. Bernards AT, van der Toorn J, van Boven CP, Dijkshoorn L. 1996. Evaluation of the ability of a commercial system to identify Acinetobacter genomic species. *Eur J Clin Microbiol Infect Dis.* 15:303–8.
9. Peleg AY, Seifert H, Paterson DL. 2008. Acinetobacter baumannii: emergence of a successful pathogen. *Clin Microbiol Rev.* 21:538–82.
10. Marcos MA, Jimenez de Anta MT, Vila J. 1995. Correlation of six methods for typing nosocomial isolates of Acinetobacter baumannii. *J Med Microbiol.* 42:328–35.
11. Seifert H, Dolzani L, Bressan R, van der Reijden T, van Strijen B, Stefanik D, Heersma H, Dijkshoorn L. 2005. Standardization and interlaboratory reproducibility assessment of pulsed-field gel electrophoresis-generated fingerprints of Acinetobacter baumannii. *J Clin Microbiol.* 43:4328–35.
12. Dijrshoorn L, Nemec A. 2008. The diversity of the genus Acinetobacter, in Gerischer U (ed), *Acinetobacter Molecular Biology.* Caister Academic Press, University of Ulm, Germany.
13. Gerner-Smidt P, Tjernberg I, Ursing J. 1991. Reliability of phenotypic tests for identification of Acinetobacter species. *J Clin Microbiol.* 29:277–82.
14. Vaneechoutte M, Dijkshoorn L, Tjernberg I, Elaichouni A, de Vos P, Claeys G, Verschraegen G. 1995. Identification of Acinetobacter genomic species by amplified ribosomal DNA restriction analysis. *J Clin Microbiol.* 33:11–5.
15. Janssen P, Maquelin K, Coopman R, Tjernberg I, Bouvet P, Kersters K, Dijkshoorn L. 1997. Discrimination of Acinetobacter genomic species by AFLP fingerprinting. *Int J Syst Bacteriol.* 47:1179–87.
16. La Scola B, Gundi VA, Khamis A, Raoult D. 2006. Sequencing of the rpoB gene and flanking spacers for molecular identification of Acinetobacter species. *J Clin Microbiol.* 44:827–32.
17. Chan JZ, Halachev MR, Loman NJ, Constantinidou C, Pallen MJ. 2012. Defining bacterial species in the genomic era: insights from the genus Acinetobacter. *BMC Microbiol.* 12:302.
18. Sahl JW, Gillece JD, Schupp JM, Waddell VG, Driebe EM, Engelthaler DM, Keim P. 2013. Evolution of a pathogen: a comparative genomics analysis identifies a genetic pathway to pathogenesis in Acinetobacter. *PLOS One.* 8:e54287.
19. Marchaim D, Navon-Venezia S, Schwartz D, Tarabeia J, Fefer I, Schwaber MJ, Carmeli Y. 2007. Surveillance cultures and duration of carriage of multidrug-resistant Acinetobacter baumannii. *J Clin Microbiol.* 45:1551–5.
20. Cisneros JM, Reyes MJ, Pachon J, Becerril B, Caballero FJ, Garcia-Garmendia JL, Ortiz C, Cobacho AR. 1996. Bacteremia due to Acinetobacter baumannii: epidemiology, clinical findings, and prognostic features. *Clin Infect Dis.* 22:1026–32.

21. Van Looveren M, Goossens H, Group AS. 2004. Antimicrobial resistance of Acinetobacter spp. in Europe. *Clin Microbiol Infect.* 10:684–704.

22. Zurawski DV, Thompson MG, McQueary CN, Matalka MN, Sahl JW, Craft DW, Rasko DA. 2012. Genome sequences of four divergent multidrug-resistant Acinetobacter baumannii strains isolated from patients with sepsis or osteomyelitis. *J Bacteriol.* 194:1619–20.

23. Manchanda V, Sanchaita S, Singh N. 2010. Multidrug resistant acinetobacter. *J Glob Infect Dis.* 2:291–304.

24. Park YK, Peck KR, Cheong HS, Chung DR, Song JH, Ko KS. 2009. Extreme drug resistance in Acinetobacter baumannii infections in intensive care units, South Korea. *Emerg Infect Dis.* 15:1325–7.

25. Perez F, Hujer AM, Hujer KM, Decker BK, Rather PN, Bonomo RA. 2007. Global challenge of multidrug-resistant Acinetobacter baumannii. *Antimicrob Agents Chemother.* 51:3471–84.

26. Jawad A, Heritage J, Snelling AM, Gascoyne-Binzi DM, Hawkey PM. 1996. Influence of relative humidity and suspending menstrua on survival of Acinetobacter spp. on dry surfaces. *J Clin Microbiol.* 34:2881–7.

27. Brossard KA, Campagnari AA. 2012. The Acinetobacter baumannii biofilm-associated protein plays a role in adherence to human epithelial cells. *Infect Immun.* 80:228–33.

28. Sousa SA, Ramos CG, Leitao JH. 2011. Burkholderia cepacia complex: emerging multihost pathogens equipped with a wide range of virulence factors and determinants. *Int J Microbiol.* 2011.

29. Mars AE, Houwing J, Dolfing J, Janssen DB. 1996. Degradation of toluene and trichloroethylene by Burkholderia cepacia G4 in growth-limited fed-batch culture. *Appl Environ Microbiol.* 62:886–91.

30. Chiarini L, Bevivino A, Dalmastri C, Tabacchioni S, Visca P. 2006. Burkholderia cepacia complex species: health hazards and biotechnological potential. *Trends Microbiol.* 14:277–86.

31. Burkloder WH. 1950. Sour skin, a bacterial rot of Onion bulbs. *Phytopathology.* 40:115–117.

32. Parke JL, Gurian-Sherman D. 2001. Diversity of the Burkholderia cepacia complex and implications for risk assessment of biological control strains. *Annu Rev Phytopathol.* 39:225–58.

33. Vial L, Chapalain A, Groleau MC, Deziel E. 2011. The various lifestyles of the Burkholderia cepacia complex species: a tribute to adaptation. *Environ Microbiol.* 13:1–12.

34. Mahenthiralingam E, Baldwin A, Dowson CG. 2008. Burkholderia cepacia complex bacteria: opportunistic pathogens with important natural biology. *J Appl Microbiol.* 104:1539–51.

35. Isles A, Maclusky I, Corey M, Gold R, Prober C, Fleming P, Levison H. 1984. Pseudomonas cepacia infection in cystic fibrosis: an emerging problem. *J Pediatr.* 104:206–10.

36. Biddick R, Spilker T, Martin A, LiPuma JJ. 2003. Evidence of transmission of Burkholderia cepacia, Burkholderia multivorans and Burkholderia dolosa among persons with cystic fibrosis. *FEMS Microbiol Lett.* 228:57–62.

37. LiPuma JJ, Dasen SE, Nielson DW, Stern RC, Stull TL. 1990. Person-to-person transmission of Pseudomonas cepacia between patients with cystic fibrosis. *Lancet.* 336:1094–6.

38. Geftic SG, Heymann H, Adair FW. 1979. Fourteen-year survival of Pseudomonas cepacia in a salts solution preserved with benzalkonium chloride. *Appl Environ Microbiol.* 37:505–10.

39. Marioni G, Rinaldi R, Ottaviano G, Marchese-Ragona R, Savastano M, Staffieri A. 2006. Cervical necrotizing fasciitis: a novel clinical presentation of Burkholderia cepacia infection. *J Infect.* 53:e219–22.

40. Leitao JH, Sousa SA, Cunha MV, Salgado MJ, Melo-Cristino J, Barreto MC, Sa-Correia I. 2008. Variation of the antimicrobial susceptibility profiles of Burkholderia cepacia complex clonal isolates obtained from chronically infected cystic fibrosis patients: a five-year survey in the major Portuguese treatment center. *Eur J Clin Microbiol Infect Dis.* 27:1101–11.

41. Sahl JW, Vazquez AJ, Hall CM, Busch JD, Tuanyok A, Mayo M, Schupp JM, Lummis M, Pearson T, Shippy K, Colman RE, Allender CJ, Theobald V, Sarovich DS, Price EP, Hutcheson A, Korlach J, LiPuma JJ, Ladner J, Lovett S, Koroleva G, Palacios G, Limmathurotsakul D, Wuthiekanun V, Wongsuwan G, Currie BJ, Keim P, Wagner DM. 2016. The effects of signal erosion and core genome reduction on the identification of diagnostic markers. *mBio.* 7.

42. Galyov EE, Brett PJ, DeShazer D. 2010. Molecular insights into *Burkholderia pseudomallei* and *Burkholderia mallei* pathogenesis. *Annu Rev Microbiol.* 64:495–517.

43. Cheng AC, Currie BJ. 2005. Melioidosis: epidemiology, pathophysiology, and management. *Clin Microbiol Rev.* 18:383–416.

44. White NJ. 2003. Melioidosis. *Lancet.* 361:1715–22.

45. Pearson T, Sahl JW, Hepp CM, Handady K, Hornstra H, Vazquez AJ, Settles E, Mayo M, Kaestli M, Williamson CHD, Price EP, Sarovich DS, Cook JM, Wolken SR, Bowen RA, Tuanyok A, Foster JT, Drees KP, Kidd TJ, Bell SC, Currie BJ, Keim P. 2019. Pathogen to commensal: longitudinal within-host population dynamics, evolution, and adaptation during a chronic >16-year Burkholderia pseudomallei infection. *bioRxiv*, doi: https://doi.org/10.1101/552109.

46. Godoy D, Randle G, Simpson AJ, Aanensen DM, Pitt TL, Kinoshita R, Spratt BG. 2003. Multilocus sequence typing and evolutionary relationships among the causative agents of melioidosis and glanders, Burkholderia pseudomallei and Burkholderia mallei. *J Clin Microbiol.* 41:2068–79.

47. Nierman WC, DeShazer D, Kim HS, Tettelin H, Nelson KE, Feldblyum T, Ulrich RL, Ronning CM, Brinkac LM, Daugherty SC, Davidsen TD, Deboy RT, Dimitrov G, Dodson RJ, Durkin AS, Gwinn ML, Haft DH, Khouri H, Kolonay JF, Madupu R, Mohammoud Y, Nelson WC, Radune D, Romero CM, Sarria S, Selengut J, Shamblin C, Sullivan SA, White O, Yu Y, Zafar N, Zhou L, Fraser CM. 2004. Structural flexibility in the Burkholderia mallei genome. *Proc Natl Acad Sci U S A.* 101:14246–51.

48. Stone JK, DeShazer D, Brett PJ, Burtnick MN. 2014. Melioidosis: molecular aspects of pathogenesis. *Expert Rev Anti Infect Ther.* 12:1487–99.

49. Enright MC, Carter PE, MacLean IA, McKenzie H. 1994. Phylogenetic relationships between some members of the genera NEISSERIA, Acinetobacter, Moraxella, and Kingella based on partial 16S ribosomal DNA sequence analysis. *Int J Syst Bacteriol.* 44:387–91.

50. Pettersson B, Kodjo A, Ronaghi M, Uhlen M, Tonjum T. 1998. Phylogeny of the family Moraxellaceae by 16S rDNA sequence analysis, with special emphasis on differentiation of Moraxella species. *Int J Syst Bacteriol.* 48 Pt 1:75–89.

51. Verduin CM, Hol C, Fleer A, van Dijk H, van Belkum A. 2002. Moraxella catarrhalis: from emerging to established pathogen. *Clin Microbiol Rev.* 15:125–44.

52. Ringvold A, Vik E, Bevanger LS. 1985. Moraxella lacunata isolated from epidemic conjunctivitis among teen-aged females. *Acta Ophthalmol (Copenh).* 63:427–31.

53. Hughes DE, Pugh GW, Jr. 1970. A five-year study of infectious bovine keratoconjunctivitis in a beef herd. *J Am Vet Med Assoc.* 157:443–51.

54. Brorson JE, Falsen E, Nilsson-Ehle H, Rodjer S, Westin J. 1983. Septicemia due to Moraxella nonliquefaciens in a patient with multiple myeloma. *Scand J Infect Dis.* 15:221–3.

55. Tonjum T, Caugant DA, Bovre K. 1992. Differentiation of Moraxella nonliquefaciens, M. lacunata, and M. bovis by using multilocus enzyme electrophoresis and hybridization with pilin-specific DNA probes. *J Clin Microbiol.* 30:3099–107.

56. Catlin BW. 1990. Branhamella catarrhalis: an organism gaining respect as a pathogen. *Clin Microbiol Rev.* 3:293–320.

57. Bartos LC, Murphy TF. 1988. Comparison of the outer membrane proteins of 50 strains of Branhamella catarrhalis. *J Infect Dis.* 158:761–5.

58. Van Hare GF, Shurin PA, Marchant CD, Cartelli NA, Johnson CE, Fulton D, Carlin S, Kim CH. 1987. Acute otitis media caused by Branhamella catarrhalis: biology and therapy. *Rev Infect Dis.* 9:16–27.

59. Nash DR, Wallace RJ, Jr., Steingrube VA, Shurin PA. 1986. Isoelectric focusing of beta-lactamases from sputum and middle ear isolates of Branhamella catarrhalis recovered in the United States. *Drugs.* 31 Suppl 3:48–54.

60. Kawakami Y, Ueno I, Katsuyama T, Furihata K, Matsumoto H. 1994. Restriction fragment length polymorphism (RFLP) of genomic DNA of Moraxella (Branhamella) catarrhalis isolates in a hospital. *Microbiol Immunol.* 38:891–5.

61. Patterson TF, Patterson JE, Masecar BL, Barden GE, Hierholzer WJ, Jr., Zervos MJ. 1988. A nosocomial outbreak of Branhamella catarrhalis confirmed by restriction endonuclease analysis. *J Infect Dis.* 157:996–1001.

62. Vu-Thien H, Dulot C, Moissenet D, Fauroux B, Garbarg-Chenon A. 1999. Comparison of randomly amplified polymorphic DNA analysis and pulsed-field gel electrophoresis for typing of Moraxella catarrhalis strains. *J Clin Microbiol.* 37:450–2.

63. Yano H, Suetake M, Kuga A, Irinoda K, Okamoto R, Kobayashi T, Inoue M. 2000. Pulsed-field gel electrophoresis analysis of nasopharyngeal flora in children attending a day care center. *J Clin Microbiol.* 38:625–9.

64. Beaulieu D, Scriver S, Bergeron MG, Low DE, Parr TR, Jr., Patterson JE, Matlow A, Roy PH. 1993. Epidemiological typing of Moraxella catarrhalis by using DNA probes. *J Clin Microbiol.* 31:736–9.

65. Bootsma HJ, van der Heide HG, van de Pas S, Schouls LM, Mooi FR. 2000. Analysis of Moraxella catarrhalis by DNA typing: evidence for a distinct subpopulation associated with virulence traits. *J Infect Dis.* 181:1376–87.

66. Verduin CM, Kools-Sijmons M, van der Plas J, Vlooswijk J, Tromp M, van Dijk H, Banks J, Verbrugh H, van Belkum A. 2000. Complement-resistant Moraxella catarrhalis forms a genetically distinct lineage within the species. *FEMS Microbiol Lett.* 184:1–8.

67. Bootsma HJ, Aerts PC, Posthuma G, Harmsen T, Verhoef J, van Dijk H, Mooi FR. 1999. Moraxella (Branhamella) catarrhalis BRO beta-lactamase: a lipoprotein of gram-positive origin? *J Bacteriol.* 181:5090–3.

68. Zomer A, de Vries SP, Riesbeck K, Meinke AL, Hermans PW, Bootsma HJ. 2012. Genome sequence of Moraxella catarrhalis RH4, an isolate of seroresistant lineage. *J Bacteriol.* 194:6969.

69. Arkwright JA. 1907. On the Occurrence of the Micrococcus catarrhalis in normal and catarrhal noses and its differentiation from other gram-negative Cocci. *J Hyg (Lond).* 7:145–54.

70. Ejlertsen T, Thisted E, Ebbesen F, Olesen B, Renneberg J. 1994. Branhamella catarrhalis in children and adults. A study of prevalence, time of colonisation, and association with upper and lower respiratory tract infections. *J Infect.* 29:23–31.

71. Ikram RB, Nixon M, Aitken J, Wells E. 1993. A prospective study of isolation of Moraxella catarrhalis in a hospital during the winter months. *J Hosp Infect.* 25:7–14.

72. Berk SL, Kalbfleisch JH. 1996. Antibiotic susceptibility patterns of community-acquired respiratory isolates of Moraxella catarrhalis in Western Europe and in the USA. The Alexander project collaborative group. *J Antimicrob Chemother.* 38 Suppl A:85–96.

73. Hoogkamp-Korstanje JA, Dirks-Go SI, Kabel P, Manson WL, Stobberingh EE, Vreede RW, Davies BI. 1997. Multicentre in-vitro evaluation of the susceptibility of Streptococcus *pneumoniae,* Haemophilus influenzae and Moraxella catarrhalis to ciprofloxacin, clarithromycin, co-amoxiclav and sparfloxacin. *J Antimicrob Chemother.* 39:411–4.

74. McGregor K, Chang BJ, Mee BJ, Riley TV. 1998. Moraxella catarrhalis: clinical significance, antimicrobial susceptibility and BRO beta-lactamases. *Eur J Clin Microbiol Infect Dis.* 17:219–34.

45

Selected Zoonotic Pathogens

Sanjay K. Shukla and Steven L. Foley

CONTENTS

Introduction

Zoonosis is described as an infectious disease that is transmitted primarily from animals to humans. More than 950 of some 1,500 known pathogens have been categorized as zoonotic[1], and, therefore, present a significant public health risk. Consequently, the World Health Organization is engaged to address the diseases caused by existing and emerging zoonoses. In addition, One Health Initiative has been launched to emphasize that both human and animal health are inextricably linked.[2] Zoonotic bacteria have assumed additional importance since more than two-thirds of all emerging pathogens during the last three decades are zoonotic in origin. These pathogens are represented by diverse taxa and are not restricted to any particular class or group of bacteria. Data from the European Union member states have shown zoonotic bacteria with high levels of antimicrobial resistance in humans, animals, and food, which is a concern for European food safety authorities.[3] Climate change may also affect the emergence of new zoonotic diseases particularly caused by pathogens whose life cycle exists partly outside of their human hosts.[4] Furthermore, effects of climate and climate changes, particularly on the vector-borne diseases, including the ones caused by the zoonotic bacteria, is a concern since rise in temperatures, precipitation, and climate-change driven changes in the animal hosts could help establish and spread zoonotic pathogens to newer geographical areas.[5] Zoonotic pathogens maintain either an ongoing reservoir life cycle in animals or arthropod vectors without maintaining a permanent life cycle in humans; in some cases the pathogens can jump the species barrier and maintain a permanent life cycle in humans, possibly not needing the animal reservoir. Figure 45.1 shows the potential for emerging new zoonotic pathogens as a consequence of increased interactions between humans, vectors, and companion and wild animals.

Climate change, including higher precipitation and global warning, may also contribute to the emergence and/or enhanced virulence of zoonotic pathogens. These interactions allow species to jump hosts, get adapted and become opportunistic pathogens. The parasitic vectors and wild animals further aid in the maintenance and spread of potential zoonotic microbial agents. Zoonotic pathogens' ability to colonize human organs and their disease-causing capability varies considerably. Because these pathogens have evolved to live in multiple hosts including both vertebrates and invertebrates to maintain their enzootic lifestyle, their genomes have evolved to help them sustain life in different host environments. Some of them have a diminished ability to synthesize essential proteins because they can be acquired from the host. However, many of them retain their ability to make specific cofactors and vitamins to sustain efficient growth. These pathogens probably exploit receptors that are common in multiple hosts to colonize and/or gain entry into human hosts. Several groups of pathogens with representatives that can cause zoonotic disease are described in their own chapters in this book. In this chapter, we briefly discuss the pathogens of the following zoonotic genera: *Anaplasma*, *Bartonella*, *Borrelia*, *Brucella*, *Coxiella*, *Francisella*, and *Pasteurella*.

Anaplasma

The family *Anaplasmataceae* includes several tick-borne zoonotic pathogens. The genus *Anaplasma* is one of four genera placed in the newly organized family *Anaplasmataceae*. The other three genera in this family are *Ehrlichia*, *Neorickettsia*, and *Wolbachia*. However, this section will primarily focus on *Anaplasma phagocytophilum*. In the newly proposed taxonomic reclassification of the order *Anaplasmatales* by Dumler et al. in 2001, the three previously described species, *Ehrlichia equi*,

DOI: 10.1201/9781003099277-47

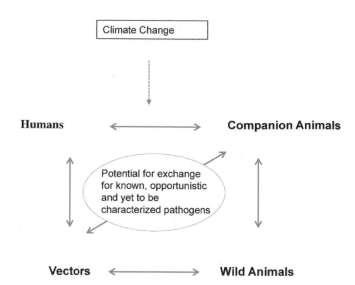

FIGURE 45.1 A schematic diagram of potential for emerging new zoonotic pathogens as a consequence of increased interaction among humans, companion animals, disease vectors, and wild animals. The relative likelihood of transmission and bacterial species jump in these hosts may vary.

Ehrlichia phagocytophila, and human granulocytic agent were combined and named as *A. phagocytophilum*.[6] This regrouping was justified by Dumler and colleagues following a molecular taxonomic approach that showed at least 99.1% nucleotide similarity in their 16S rRNA genes and identical GroESL amino acid sequences.[6]

A. phagocytophilum are pleomorphic, Gram-negative, small, obligate intracellular, zoonotic bacterium that are 0.5–1.5 μm in size. They reside and replicate in membrane-bound vacuoles of neutrophils of eukaryotic cells but can also infect bone marrow progenitors and endothelial cells.[7,8] The other rickettsia that infects granulocytes is *Ehrlichia ewingii*, but it is primarily confined to dogs even though it has been reported to cause infection in humans as well.[9] Some of the common natural hosts for *A. phagocytophilum* are humans, deer, horses, sheep, cattle, bison, and wild rodents. This pathogen cannot be transmitted transovarially in the invertebrate hosts.[10] *A. phagocytophilum* exhibit leukocyte-specific tropisms and infect granulocytes; hence the disease it causes was named granulocytic anaplasmosis. A wide range of vertebrate animals including horses, sheep, and humans can develop granulocytic anaplasmosis. It is a relatively uncommon zoonotic disease in humans and was first recognized in the upper midwestern United States in 1994.[11,12] However, equine granulocytic anaplasmosis has been described as early as 1968 in the western United States and Canada[13] and in dogs in 1982.[14] *A. phagocytophilum* infections have been described in California, the upper midwestern and northeastern regions of the United States, British Columbia, Great Britain, and Europe.

This pathogen multiplies in granulocytes and forms membrane-bound, intracytoplasmic colonies known as morulae,[15] which could be seen in the peripheral smear of the blood upon staining with Giemsa stain. The symptoms of granulocytic anaplasmosis are not remarkable and overlap with other tick-borne diseases. Symptoms include undifferentiated febrile illness, rigors, headache, myalgia, and malaise often accompanied by leucopenia and thrombocytopenia. Elevation of liver enzymes such

as alanine aminotransferase and aspartate aminotransferase has also been reported but the presence of skin rash is not common.

Since granulocytic anaplasmosis is a tick-borne disease, it is seasonal and often coincides with tick and human outdoor activities (hiking, camping, etc.) in tick-endemic wooded areas. Due to the seasonality of the disease, its transmission and dissemination is intertwined with the 2-year life cycle of the tick vectors, specifically their activities in spring and summer. The disease could be acquired when an *A. phagocytophilum* infected tick of the *Ixodes ricinus* complex transmits the pathogen to the host during a blood meal. Tick vectors differ in different geographic areas. It is transmitted by *Ixodes pacificus* in the western United States, *Ixodes scapularis* in the midwestern and eastern United States,[16,17] *I. ricinus* in Europe, and *Ixodes persulcatus* in Asia. The tick life cycle could be considered starting with a newly hatched tick larvae parasitizing on white-footed mice in spring. The larvae feed on the mice over the summer and then molt to the nymphal stage. At the nymphal stage, ticks typically acquire the pathogen from the blood meal of their hosts, which could be rodents, deer, or humans. Over winter, nymphs engage in another blood meal before falling off from the host and molting into an adult stage. Adults, both males and females, then search for another host, such as white-tailed deer, and mate there. Subsequently, females drop off and lay eggs. The eggs hatch and repeat the life cycle next spring. The risk of human infestation is higher when the nymphal tick is searching for a new host, although larvae and adults are capable of feeding on humans as well. Humans can get infected via a blood meal by an infected tick. Studies on mice suggested that infected nymphs may transmit the pathogen within 24 hours of attachment to the host.[18] The main mammalian reservoir for *A. phagocytophilum* is the white-footed mouse (*Peromyscus leucopus*), although white tailed deer (*Odocoileus virginianus*) also play a role in the maintenance of the pathogen. Other reported vertebrate hosts included humans, cats, sheep, cattle, horses, llamas, bison, etc., whereas ticks are the invertebrate hosts. Granulocytic anaplasmosis is rarely a fatal disease, although it could be fatal in immunocompromised individuals.

A probable diagnosis for granulocytic anaplasmosis includes a febrile illness with a history of tick exposure/bite, the presence of intracytoplasmic morulae in a blood smear (Figure 45.2), more than four times *A. phagocytophilum* antibodies than the cutoff, or a positive polymerase chain reaction (PCR) result for the pathogen using *Anaplasma*-specific primers. Since *A. phagocytophilum* is an obligate intracellular pathogen, it is quite difficult to culture this pathogen in the laboratory routinely. However, it could be cultured in HeLa cells[19] and has been successfully

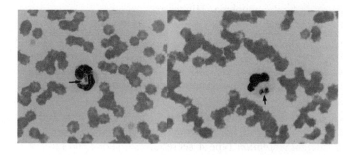

FIGURE 45.2 *Anaplasma phagocytophilum* morulae within the cytoplasm of a cat granulocyte.

cultured in a tick cell line as well.[20] Availability of the culture grown *Anaplasma* cells have advanced our knowledge of the pathogen's physiology and genomics. The genome size of *A. phagocytophilum* strains has been found to be remarkably consistent in size, conserved in genomic content, and ranges between 1.47 Mb and 1.51 Mb.[10] The genome, despite numerous repeats, shows flexibility and harbor genes that allow them to have a dual existence (as invertebrate symbionts and human/animal pathogens).[6] The genome contains 1,369 open reading frames including genes for major vitamins, co-factors and nucleotides, but with a limited ability to make amino acids.[10] Since this pathogen lacks the genes for biosynthesis of lipopolysaccharide and most of the genes of peptidoglycan biosynthesis,[21] it does not trigger a robust immune response.[22] A significant percent of its genome is dedicated to nucleotide, co-factor and vitamin biosynthesis, and protein synthesis, but less for the regulatory and metabolic functions.[10] Limited genotypic studies on the pathogen suggests that not all *A. phagocytophilum* strains are equal. Disparities exist in disease incidence in different endemic areas, clinical severity, seroprevalences and disease manifestation in humans, horses, cattle, ruminants, etc. from both North America and Europe, suggesting there may be yet-to-be-understood genotypic differences that exist among strains of this species.[16,23,24] Several 16S rRNA gene-based variants of *A. phagocytophilum* have been described from animal isolates.[16,25–27] Based on the genetic variants found in Rhode Island and Connecticut, Massung et al. have hypothesized that some variants are incapable of infecting humans. Further, in areas where these variants are predominant in ticks, a lower incidence of human infection causing strains would be expected in ticks in these areas and consequently a lower number of human cases.[16,25–27] Although 16S rRNA gene of the human isolates are found to be identical, genetic divergence in *A. phagocytophilum* has been noted when comparing their ankyrin gene, *ankA* or membrane surface protein 4 gene, *msp4*.[28,29] Cows with granulocytic anaplasmosis could be infected with multiple variants of *A. phagocytophilum* at the same time.[30] High levels of IFN-γ, IL-10, IL-12, and serum ferritin have been observed in patients with *A. phagocytophilum* infection. However, the severity of infection could be correlated with the level of IL-12 and ferritin and the ratio of IL-10 to IFN-γ in serum.[31,32] A pathophysiological role for proinflammatory response in human anaplasmosis have been proposed recently. In a case control study that measured select cytokines in sera from patients with human anaplasmosis (HA) and healthy control subjects, it was observed that proinflammatory cytokines, particularly, IFN-γ, on average were 7.8 times higher in HA patients. In addition, IL-1β, IL-8, IL-6, TNF-α, and IL-10 were significantly associated with severity of thrombocytopenia. Higher concentrations of IL-13 and IL-5 were associated with diarrhea and vomiting symptoms.[33] In laboratory study, sarolaner (Simparica™) prevented the transmission of *A. phagocytophilum* in dogs after being challenged with infected ticks.[34]

Additional immune-response studies in humans are needed to assess their complete biological significance, specifically with respect to in the pathogen's virulence potential and differences in host susceptibility. A conserved membrane protein, VirB10 of *A. phagocytophilum* type 4 secretion has shown promise as a vaccine candidate in mouse model. This vaccine showed partial protection against *A. phagocytophilum* challenge through the production of IFN-γ CD $^+$ T cells.[35] Doxycycline and rifampin are the most effective drugs based on the susceptibility testing[28]; most people do recover if a timely therapy is instituted. Since this pathogen is present in blood, there is a potential risk for anaplasmosis if a recipient received *A. phagocytophilum*-infected blood from a donor. Interestingly, mechanisms used by the *A. phagocytophilum* to infect and multiply in ticks, and in vertebrate host cells, appear to be conserved across all tick species and vertebrate hosts.[36]

Bartonella

Members of the genus *Bartonella* are small, Gram-negative, aerobic α-proteobacteria, which also includes genera such as *Agrobacterium*, *Brucella*, *Rhizobium*, and *Rickettsia*.[37] Many *Bartonella* species are potential zoonotic pathogens including *B. bacilliformis*, *B. henselae*, *B. quintana*, *B. elizabethae*, *B. grahamii*, *B. koehlerae*, *B. vinsonii*, and *B. washoensis*.[38,39] The first species characterized was *B. bacilliformis* that was associated with infections in the Peruvian Andes.[40] *B. quintana* was isolated in 1960 and identified as the causative agent of trench fever, a disease that infected over a million soldiers during World War I.[41] The more extensive recognition of *Bartonella* species causing a wider range of infections occurred following the beginning of the AIDS epidemic, where immunocompromised individuals developed various diseases manifestations that were tied to *Bartonella* infections.[42] One of the challenges to earlier recognition of *Bartonella* serving as a source of human infections was the challenge in culturing the fastidious pathogens that often required weeks to grow for primary culture. With the advent of improved molecular detection (PCR, DNA sequencing, etc.) and serological methods there was a wider understanding of the diseases associated with the genus.[43] In tissues, the organisms are visible with Warthin-Starry silver impregnation stain as tightly compacted clumps. In red blood cells, the organisms can be identified by May-Grünwald-Giemsa staining.[44]

The current, most common species associated with human infections is *B. henselae*, the causative agent of cat-scratch disease (CSD).[45] Based on insurance claim data, there are approximately 12,500 CSD infections diagnosed in the United States each year, with the majority of the cases being relatively mild, producing a lesion at the bite/scratch site and regional lymphadenopathy; however, ~500 cases required hospitalization.[46] Zoonotic transmission of *B. henselae* to humans generally occurs through a cat scratch, cat flea bite, or possibly less likely through a tick bite.[39] The organisms are present in flea feces and are transmitted to humans following disruption of the epithelial barrier during scratching or biting.

The prevalence of *B. henselae* in cats is generally higher in areas of warmer and more humid climates and lower in colder climates; for example, in the Philippines the prevalence was 68% and 0% in Norway.[44] Likewise in the United States, the highest incidence of infections is in the southern states and lower in the northern and western mountain states. This distribution is likely due to the numbers of cat fleas present throughout the year in the different locations. Cats were generally found to be asymptomatic carriers of the organism in experimentally infected conditions, although the bacterium was able to persist for up to

32 weeks.[47] The organism can persist in the feline bloodstream within erythrocytes and potentially invade blood vessel endothelial cells. Endothelial invasion has been noted *in vitro*, but not yet identified *in vivo*.[48] The vast majority of cases of bartonellosis in immunocompetent individuals present as typical CSD with limited regional lymphadenopathy and red-to-brown papules at the site of infection. However, 10–20% of patients may have multiple lymph nodes affected during the disease course. In addition to the skin lesions and lymphadenopathy, many patients report low-grade fever, malaise, headache, or sore throat. Atypical CSD occurs in 11–12% of patients and involves additional disease manifestations including granulomatous conjunctivitis, hepatitis and splenitis, osteitis, pneumonitis, and neurological pathologies.[37] In addition, patients infected with *B. henselae* or *B. quintana* may develop endocarditis or bacillary angiomatosis (BA), a proliferative disease of the vascular tissue that can lead to papular lesions of the skin or impact a number of different internal organs. BA is most commonly seen in immunocompromised patients, such as those with HIV infections or who have had a transplant.[48] *B. quintana*, which is carried by fleas and lice, is the causative agent of trench fever that impacted several troops during World War I. People who develop trench fever often have high fevers, severe headache, especially behind the eyes, and pain in the legs and back.[49]

Typical CSD infections are primarily diagnosed clinically based on the history of a cat bite or scratch and the detection of enlarged lymph nodes and the granulomatous lesion at the infection site. The detection of atypical *B. henselae* infections relies on diagnostic testing including serology, PCR-based detection, or direct culture of the pathogen, which is often problematic due to the fastidious nature of the organism and the long incubation period for growth.[44] Direct culture is important however since some patients with BA do not generate a detectable antibody response. Antimicrobial therapy for *B. henselae* infections generally includes macrolides such as erythromycin, azithromycin or clarithromycin, doxycycline, and rifampin.[48] A meta-analysis of the treatment for CSD, found that antimicrobial therapy did not reduce the cure rate or duration of infection in most patients.[50] Prevention of infection appears to be the most effective way to limit the impact of *B. henselae* on human health. For example, immunocompromised individuals that choose to have pets should adopt those that are serologically negative for *Bartonella* exposure and work to limit flea infestation on the animals to break the chain of transmission of the pathogen.[44]

Borrelia

Although our understanding of the genetics and pathophysiology of *Borrelia* has been significantly better now due to tools available in genomics, proteomics, and bioinformatics, an effective vaccine for humans remains elusive. The genus *Borrelia*, which presently consists of 11 genospecies, belongs to the family *Spirochaetaceae*. However, not all species are present in all endemic locations and of the 11 genospecies only *Borrelia andersonii*, *Borrelia bissettii*, and *Borrelia burgdorferi* sensu stricto have been identified in North America. Seven species, including *Borrelia garinii*, *Borrelia afzelii*, *Borrelia valaisiana*, *Borrelia japonica*, *Borrelia tanukii*, *Borrelia turdi*, and *Borrelia*

sinica have been identified from Asia. The five European species are *B. burgdorferi* sensu stricto, *B. garinii*, *B. afzelii*, *B. valaisiana*, and *Borrelia lusitaniae*.[51] This pathogen is transmitted by vectors of the *I. persulcatus* complex (also called the *I. ricinus* complex). *I. scapularis* is the principal vector for this pathogen in the northeastern and north-central United States, whereas *I. pacificus* is the primary vector in the western United States. *I. ricinus* and *I. persulcatus* are the principal vectors for Europe and Asia, respectively. Interestingly, not all *Borrelia* species are known to cause infections. Based on several genetic and immunologic considerations, all *Borrelia* species could be placed into three phylogenetic distinct genomic clusters: *B. burgdorferi* sensu stricto, *B. garinii*, and *B. afzelii*. In the United States, *B. burgdorferi* sensu stricto used to be the only species that caused infection in humans but in 2014 another species, named as *B. mayonii* has been reported to cause Lyme disease and it can be detected by a PCR test from blood.[52-54]

The full syndrome of Lyme disease was initially recognized as juvenile rheumatoid arthritis in Connecticut, United States in 1975. However, this spirochetal bacterium was discovered by Willy Burgdorferi and colleagues in 1981 from the nymph of the tick, *I. scapularis*[55] that was subsequently linked to Lyme disease as recognized by Steere et al. and others.[56,57] *Borrelia* are highly motile, corkscrew shaped, endoflagellated bacterium with cells 10–30 μm long and 0.2–0.5 μm wide. It has a relatively small genome of ~1.5 Mb, which consists of a linear chromosome (~0.91 Mb) and 21 plasmids (nine circular and 12 linear) with a G + C content of 28.6%. The genome contains ~1,701 predicted genes. However, a whole genome sequence comparison of 13 *B. burgdorferi* isolates suggested the range of genome size between 1.35 Mb and 1.52 Mb and the number of plasmids in strains can vary between 13 and 21.[58] Some plasmids are essential and are considered as mini chromosomes. Unlike other prokaryotic genomes, the *B. burgdorferi* genome is unique because of its linear chromosome and multiple linear plasmids. A 16S rRNA gene-based PCR assay based on the two highly conserved segments of the gene has been developed to detect *Borrelia*.[59] Its genomic novelty is also due to the presence of over 150 lipoprotein-encoding genes, including a surface exposed lipoprotein called VlsE that undergoes extensive antigenic variation.[60] Analysis of the plasmid sequence suggests evidence of recent genomic rearrangement.[53] *B. burgdorferi* genome encodes for a limited number of biosynthetic proteins and therefore has to depend on its host for many of its nutritional needs. Since this pathogen makes no known toxins, it causes infection by attaching to the host cells and migrating through the tissues.

B. burgdorferi sensu lato maintains a complicated enzootic life cycle in nature, which involves one of ticks of the *I. persulcatus* complex and a range of animal hosts. Both the transmission vectors and the animal hosts vary depending upon geographic locations. Its incidence in 2012 in the United States was 7.0 cases per 100,000 persons. However, the incidence is higher in the eastern coastal states, i.e., New Hampshire: 75.9, Maine: 66.6, Vermont: 61.7 per 100,000 persons.[61] Lyme disease can be acquired after being bitten by a *B. burgdorferi*-infected tick that engages in a blood meal from the host. Small mammals serve as the reservoir for this pathogen for the vector's larva and nymphs, whereas the white-tailed deer is the reservoir for the adults. Once infected with the pathogen, larvae can maintain the agent during molting to nymph and adult stages. Lyme disease has multiple phases and typically starts with a spreading skin lesion (*erythema migrans*) at the tick bite site (phase 1). This typically happens in spring and/or summer months when tick activity is usually high. In phase 2, which could occur within days to weeks of phase 1, the pathogen may cause additional skin lesions or infect other organs such as the heart, joints or even the nervous system. In phase 3, which may begin within months to years after the initial bite, is a stage where manifestation of the latent infections takes place. Recently, cases of cardiac bradydysrhythmia have been reported in cats from Lyme borreliosis[62] and this disease was consistent with what has been reported in humans and dogs.

As expected for vector-borne pathogens, *B. burgdorferi* sensu lato genome expresses different genes in response to different microenvironments as part of its multiple-host associated life cycle. Not all genes that are expressed in mammalian hosts are expressed in tick vectors or vice versa.[63] Outer surface proteins (Osp), which are lipoproteins, are among the major antigens of this pathogen. OspA and OspB are expressed in the midgut of the tick vector, whereas OspC is expressed in the salivary gland of the tick once the pathogen migrates there.

The presence of *B. burgdorferi* in a clinical sample could be determined in a laboratory by several approaches, including the direct observation of the pathogen under a microscope. Biopsy material from skin lesions or cerebrospinal fluid or blood samples may be used to isolate the agent in the original Kelly medium[64] or its modified versions such as Barbour-Stoenner-Kelly II medium (BSK-II), BSK-H, Kelly–Pettenkofer (MKP) medium.[53] *B. burgdorferi* sensu lato has a long generation time of 7–20 hours, which necessitates a long incubation period, often up to 12 weeks to detect a positive culture. However, daily microscopic examination of the culture supernatant could facilitate detection within a week in many cases.[65] The time to infer a positive culture could be further reduced by testing the culture supernatant by PCR even before it is suspected to be positive by microscopic examination.[66] In general, recovery of the *B. burgdorferi* was best achieved from skin lesion biopsy followed by plasma, serum and whole blood, respectively. The latter three are poor clinical samples for a Lyme PCR assay probably because of the low number of the organisms present or the presence of PCR inhibitors. A number of qualitative and quantitative PCR-based diagnostic assays have been developed and these assays could target either chromosomally encoded genes (such as the 16S and 23S rRNA genes, 5S-23S rRNA intergenic region, *recA*, *flab*, *p66*, etc.) or the plasmid encoded gene *ospA*, and not each target gene has the same level of sensitivity.[67] To confirm Lyme arthritis, PCR-based assays rather than the culture method are recommended from the synovial fluids of the affected joints.[51]

Immunologically, Lyme borreliosis is frequently diagnosed by detecting antibodies in the host's serum against a panel of *B. burgdorferi* antigens that are available as part of the Lyme borreliosis detection kits. Both IgG and IgM antibodies could be detected against *B. burgdorferi* FlaB (41 kDa) and FlaA (37 kDa) antigens within days of exposure to the Lyme agent.[68] Plasmid encoded OspC, which is heterogeneous in nature (21–25 kDa), is another immunodominant antigen during the early stage of the disease.[39] Some of the other commercially available Lyme antibody detection kits use indirect immunofluorescent-antibody assay, enzyme-linked immunosorbent assay, or Western immunoblots or their variations. For the immunofluorescent-antibody assay, diluted serum titers of 128 or 256 that react with fixed Lyme antigen on the glass slide are considered a positive reaction.[69]

Western immunoblots are useful in determining the different stages of Lyme borreliosis as antibody response to different antigens varies. This assay is somewhat variable due to the different strain types and different geographic sources of strains used in the preparation of antigens. Newer enzyme immunoassays, which use recombinant antigens of OspC, vlsE, FlaA, P66, etc., have also been developed but their relative sensitivity and reactivity to corresponding IgM and IgG vary.[53] Borreliacidal antibody assays have also been developed to detect Lyme borreliosis by exploiting the killing properties of OspC in combination with a complex of complements that act upon the pathogen's outer surface proteins. The borreliacidal assay makes use of flow cytometry and/or staining dyes to distinguish the dead bacterial cells from the viable ones.[70] Overall, all diagnostic tests for Lyme borreliosis should be interpreted in the context of the endemicity of the pathogen in that geographic area, relevant travel history, clinical samples tested, duration of the infection following tick-bite and the type of assay used. *Borrelia*-induced monocyte and T-cell derived cytokines profile showed that they were influenced by both genetic and nongenetic factors.[71] Notably, age strongly impaired the IL-22 responses; activation of HIF-1 alpha was observed along with the increase in glycolysis derived lactate in Lyme disease patients.[71]

Lyme disease can be effectively treated with appropriate antibiotics during the early stage of infection whereas Lyme arthritis and neuroborreliosis can be effectively treated with third generation cephalosporins. However, prevention of exposure and early treatment should be the strategy for an effective management.[72] Vaccine development for Lyme disease has been fraught with frustration and technical challenges. Vaccines made from outer surface proteins, OspA, OspC, and OspA and OspC in combination has met with limited success. The recombinant OspA vaccine LYMErix had promise, but was withdrawn due to poor sales, need for a multiple booster shots, and the potential for an autoimmune response against human lymphocyte function associated antigen -1.[73] Using the combination of omics and bioinformatics approaches, an interactome of *Borrelia* proteins suggested two proteins, ErpY and ErpX, as potential targets for vaccine development.[74]

Brucella

Brucella species are important zoonotic pathogens that cause infections that manifest in different disease syndromes.[75] Several different species of livestock and other animals are normal hosts for members of the genus *Brucella*. Members of the genus have a high degree of genetic similarity, but are currently separated into multiple species, based largely on the animal hosts where they normally reside.[76] *Brucella abortus* is predominantly associated with cattle, *B. canis* with dogs, *B. ceti* and *B. pinnipedialis* with marine mammals, *B. melitensis* with goats, *B. neotomae* with rats, *B. ovis* with sheep, and *B. suis* with swine.[77,78] The most commonly associated species with zoonotic disease are *B. melitensis*, *B. abortus*, and *B. suis*.[79] The number of cases of brucellosis is hard to determine, but it is likely that there are nearly 500,000 new cases each year worldwide.[80] Part of the challenge in determining the infection is that the symptoms of the acute phase of disease are the fairly nondescript "flu-like symptoms" that are associated with multiple infections.[78] The chronic phase of infection is more severe, in that after a 2- to 4-week latency period postinfection, the organism can be disseminated throughout the body where they infect tissues leading to endocarditis, meningitis, spondylitis, arthritis, or potential reproductive-system problems, such as inflammation of the epididymis and testes or spontaneous abortion if the disease occurs during pregnancy.[81] In more developed countries, the number of cases is relatively small, e.g., in the United States, the average number of infections reported annually is ~100.[81]

Brucella are Gram-negative, facultative, intracellular bacteria that are members of the α-proteobacteria along with organisms such as *Agrobacterium, Anaplasma, Bartonella, Rhizobium*, and *Rickettsia*. A distinctive feature of some *Brucella* species is the presence of two chromosomes in their genome. Species with two chromosomes include *B. abortus, B. melitensis*, and some biotypes of *B. suis* (others contain a single larger chromosome)[62] and there appears to be some duplication of genes between each chromosome. *Brucella* also lack several of the virulence factors that are characteristic of other Gram-negative pathogens, including many toxins and type 3 secretion systems. Due to this genetic reductionism in prototypical virulence factors, *Brucella* have evolved other characteristics that facilitate intracellular pathogenesis. The bacteria have acquired genes associated with acid tolerance that allow for transversion of the upper gastrointestinal and within phagocytic vesicles and they have modified lipopolysaccharide (LPS) antigens that make *Brucella* less immunogenic than that of most Gram-negative organisms.[79,82] The modified LPS layer limits the activation of the alternative complement pathway and production of certain cytokines, as well as decreasing the susceptibility to bactericidal substances produced by the host. The mild immune response and resistance to toxic substance likely allows for the persistence of the organism in a host.[77] Additionally, *Brucella* express a VirB type 4 secretion system (T4SS) that allows for the transfer of effector molecules to the host cells allowing for protection during the intracellular phase of infection.[78,83]

The zoonotic transmission of *Brucella* occurs primarily through direct contact with or consumption of contaminated animal products or inhalation of aerosolized particles containing infectious bacteria. The most common cause of infection in the United States is associated with the consumption unpasteurized dairy products originating from *Brucella* endemic areas.[81] However, infections associated with occupational exposure of animal caretakers, veterinarians, and laboratory personnel have also been reported. A recent systemic review of the literature has shown that there have been isolated reports of human-to-human transmission of *Brucella*, most commonly through breast feeding, as well as through sexual contact, aerosol exposure, and via blood transfusion or bone marrow transplantation.[84] During the infectious process, the organisms penetrate the nasal, oral, or pharyngeal mucosa and enter phagocytes that are transferred to regional lymph nodes.[77] The exact mechanism by which *Brucella* enter the cells is not fully understood. However, the organisms that enter phagocytes are able to survive in a *Brucella*-containing

vacuole (BCV) and avoid fusion with the lysosome[79] and a key contributor to the ability to survive and persist in the BCV is the expression of the VirB T4SS factors that impact intracellular trafficking and persistence.[79,83] After the postinfection latency period, the organism may be released from the phagocytes and widely disseminated throughout the body leading to debilitating disease.[80]

The most effective way to limit brucellosis is to prevent human infections by preventing infection in livestock and through the pasteurization of dairy products.[85] Currently there are effective veterinary vaccines against *B. melitensis* and *B. abortus*, which have helped to limit the number of human infections in the United States. Consequently, many of the human brucellosis cases are likely associated with the consumption of unpasteurized dairy products illegally brought in from countries where brucellosis in animals remains a problem. The recommended treatment of brucellosis is a multidrug approach that includes doxycycline, streptomycin, and rifampin. Other agents effective against *Brucella* include gentamicin, trimethoprim/sulfamethoxazole, and the fluoroquinolones.[81,86]

Coxiella

Coxiella burnetii is the causative agent of Q fever, a disease found throughout the world. The microorganisms are Gram-negative coccobacilli that are obligate intracellular pathogens. Genetically, the organisms are closely related to *Legionella pneumophila* and are members of the γ-proteobacteria.[87] *C. burnetii* is a clonal pathogen and its genetic variation is achieved through single nucleotide polymorphisms, and many of them are nonsynonymous.[88] Some *C. burnetii* cells exist as relatively dormant small colony variants. These organisms tend to be more resistant to chemical agents and heating. Small colony variants are more resistant to the environmental stresses of the external environments and are the primary form of inhaled microorganisms.[89] Other cells are transformed into large cell variants that are able to multiply within macrophages and monocytes. As *C. burnetii* are taken up by the phagocytes, they enter the phagosome. As the small colony variants enter the phagosome, they appear to delay lysosomal fusion and are transformed into the large cell variant state prior to the formation of the phagolysosome. The low pH environment of the phagolysosome serves as the site of cellular reproduction.[90] *C. burnetii* also can undergo antigenic phase variation following isolation. Organisms recently isolated from humans or other animals display a phase I type antigenic profile consisting of a thick lipopolysaccharide layer capable of blocking antibodies from binding to the surface of the organism. The enhanced lipopolysaccharide layer is one of the major virulence factors employed by *C. burnetii*. Upon repeated passage of the microorganisms in cell culture or embryonated eggs, *C. burnetii* can shift to displaying the phase II antigenic structure, which is characterized by having an altered lipopolysaccharide structure and reduced infectivity.[91]

Several different animal species may be infected with *C. burnetii*, including many mammals, birds, fish, and arthropods such as ticks. Among mammals, cattle, sheep, goats, cats, and dogs are potential sources of human infections. Spread to humans is generally through direct contact with infected animals, their byproducts, or through the inhalation of microorganisms that are aerosolized or associated with dust. In infected animals, such as sheep, the placenta may have a very high level of *C. burnetii* that can be aerosolized following birth and serve as a source of infection. Additionally, the hides of infected animals are also a potential source of infection, especially for abattoir workers.[87] Multiple genotypic tools including comparative genome hybridization have identified eight genogroups (genogroup I–VIII),[92] but not all genomic groups of *C. burnetii* are equally virulent.[93]

C. burnetii infections can present as either an acute illness or as a chronic disease. Many patients with an acute form of Q fever present with a fever that may be associated with pneumonia and hepatitis. However, in some cases, fever may not be present, while in others, it can last a week or more. In one preantibiotic era study, the length of fever ranged from 5 to 57 days with a median of 10 days.[87,94] There is a high level of variation in patient presentation, which can confound the diagnosis. The mortality rate of acute infections is around 1–2% and is often associated with more severe manifestations of Q fever such as myocarditis. In addition to fever, many patients with acute Q fever also have severe headaches, night sweats, muscle and joint pain, and anorexia, leading to rapid weight loss. The early treatment of patients with an antibiotic, such as doxycycline, appears to reduce the length of fever and decreases the recovery time in patients with pneumonia. Patients who develop pneumonia are characterized by having fever, a nonproductive cough that often expels blood and a severe headache. Radiographic findings are often nonspecific and of limited value for diagnosing Q fever.[95]

In patients who develop chronic disease, endocarditis is the major manifestation, accounting for 60–70% of the cases and the detection of endocarditis is often delayed up to a year or more from onset to symptoms, which is associated with significant morbidity. Patients who do not receive antimicrobial therapy for the endocarditis have a high level of mortality. Primary risk factors for patients to develop endocarditis include having an underlying condition, such as a heart valve abnormality, or being immunocompromised.[87] In addition to endocarditis, other presentations can occur, including bone and joint infections, infections of the kidney, liver, or lungs, or a recrudescence of disease during pregnancy. Pregnancy related infections have been associated with premature birth, miscarriage, and neonatal mortality.[94]

The clinical diagnosis of Q fever can be difficult due to the variable nature of disease presentation. Due to the zoonotic nature, a history of animal exposure to animals known to be associated with Q fever may be helpful. Laboratory testing for *C. burnetii* antibodies is the current best diagnostic tool and the primary detection method is through indirect fluorescence detection.[91] Other serological tests based on complement fixation and enzyme immunoassays are available, but appear to lack the desired specificity and sensitivity for detection. Interestingly, the antibodies against the phase II antigens are generally higher in acute infections and phase I antigens are elevated in chronic infections. Other detection methods are available that utilize PCR detection of the organisms or rely on histological changes in infected tissue. Presently, neither PCR, nor histological testing appears to be as valuable as serology for the detection of Q fever.[87]

The treatment of Q fever is dependent on the type of infection. For acute infections, doxycycline treatment for 2 weeks is valuable in reducing the length of fever and/or pneumonia.[91] In children who are infected, treatment with trimethoprim/sulfamethoxazole is preferred. Treatment for patients with chronic endocarditis is controversial with some advocating lifelong antibiotic treatment while others recommending shorter courses of 1.5–2 years with combination therapy.[87] Combination of doxycycline with hydroxychloroquine appears to be effective in the treatment of chronic endocarditis. Hydroxychloroquine raises the pH of the phagolysosome, allowing the doxycycline to function with greater efficiency in clearing the infection. The use of trimethoprim/sulfamethoxazole to treat *C. burnetii* infections during pregnancy limits the risk for spontaneous abortion and neonatal infections and may prevent recrudescence.[87,96] In some countries, there are effective vaccines available to prevent Q fever. However, in the United States, there is only an investigational vaccine available for laboratory workers who handle live microorganisms. When available, people who are at greatest risk for exposure (abattoir workers, laboratory personnel, animal handlers) should receive the vaccine to limit infection.[94,96]

Francisella

The genus *Francisella* is part of the γ-proteobacteria along with genera such as *Escherichia*, *Pseudomonas*, *Salmonella*, *Yersinia*, and *Vibrio*. *Francisella tularensis* is a small Gram-negative coccobacillus-shaped organism that is the causative agent of the disease tularemia.[97] Multiple animal species are known to host *F. tularensis*, including rabbits, ground squirrels, muskrats and other rodents.[98] In addition, the bacterium can be carried by many arthropod vectors, including several species of ticks, mosquitoes, and biting flies.[99] *F. tularensis* is divided into three subspecies, *tularensis*, *holarctica*, and *mediasiatica*, each having varying degrees of virulence in rabbits and humans, with *tularensis* and *holarctica* being most virulent to humans.[100,101] A second species, *F. novicida* is a the cause of a limited number of human infections, predominantly in immunocompromised patients.[101]

While the incidence of tularemia in the United States has been declining since the 1930s, there are concerns about the potential use of *F. tularensis* as a bioterrorism agent.[100,102] Cases of tularemia are likely under diagnosed, since many strains lack the significant virulence factors that are associated with severe disease. In highly virulent strains, an infectious dose of less than 10 colony-forming units can cause significant illness with mortality rates of up to 5–10%, without appropriate antimicrobial therapy.[103] The spectrum of disease associated with *F. tularensis* infections is quite varied depending on the route of infection. The most common manifestation is ulceroglandular tularemia, which is characterized initially by chills, fever, headache, and body pain that develop 3–6 days postinfection. Transmission of the pathogen to humans can occur through handling infected wild game, being bitten by an insect vector that recently fed on an infected mammal, or through contaminated water.[99] With ulceroglandular infections, an ulcer often develops at the site of infection that can persist for months. The bacteria can also enter the lymphatic system and migrate to the regional lymph nodes, causing swelling of the nodes. Glandular infections can progress without the development of the ulcerative skin lesion. In some cases, the bacteria can infect other tissues of the body leading to increased morbidity.

Other routes of infection include the eye, pharynx, gastrointestinal (GI) tract, and lungs leading to different disease syndromes and clinical outcomes. Oculoglandular infections, or Parinaud's syndrome, originate in the conjunctiva of the eye and progress with ulceration of the cornea and regional lymph node involvement.[99] Oropharyngeal and GI tularemia typically occurs after the consumption of contaminated food or water and may result in a painful infection of the throat, with tonsillitis and enlargement of the cervical and mesenteric lymph nodes that can drain into the GI tract.[100,103] Patients may also develop bowel ulceration and accompanied by GI bleeding, diarrhea, and vomiting that can lead to shock. Consequently, the case fatality rate may be as high as 60% with GI involvement.[99] Patients can also develop typhoidal tularemia, which can be problematic to diagnose due to the lack of distinct signs and/or apparent routes of infection common to other types of tularemia infections. Most patients have a high-grade fever, but other signs may be more variable.[97,100] Pneumonic infections can also develop through hematogenous spread during the progression of ulceroglandular tularemia[97] and the fatality rate of pneumonic infections is nearly 40%.[99] Pneumonic tularemia is typically the most acutely severe form of tularemia occurring following inhalation of the bacteria. Those with the greatest risk of pneumonic tularemia include farmers, landscapers, and laboratory workers who come in contact with aerosols or dust containing *F. tularensis*.[99]

Tularemia diagnosis is often dependent on the signs and symptoms associated with the different types of disease manifestation. For example, ulceroglandular tularemia is characterized by the ulcerative lesion that develop at the site of infection and the corresponding enlargement and tenderness of the adjacent lymph nodes to the site of infection.[97,100] Culturing of the infecting *F. tularensis* is the most definitive method of confirming an infection, however there are challenges as the culture media needs enrichment with cysteine to facilitate growth, however blood culture is often negative for growth of the organisms.[99] Serological testing has also been shown to have diagnostic utility for the detection of tularemia, however there is the potential for cross reactivity with other bacterial pathogens that can confound the interpretation of results.[99] Another drawback of serological testing is the delay of 2–4 weeks postinfection required to develop a high enough antibody titer to be detected efficiently.[98,100] PCR has also been shown to have utility in the detection of *F. tularensis* in certain samples, for example PCR has shown to be effective in the detection of *F. tularensis* from ulceroglandular primary lesions, blood, and respiratory specimens.[100,104]

Early and appropriate antimicrobial therapy is important for the treatment of severe cases of tularemia to avoid potential septic shock and respiratory distress. The primary antibiotic recommended includes the aminoglycosides, including streptomycin (first line), gentamicin, or kanamycin for 7–14 days. Other agents with efficacy include the tetracyclines and the fluoroquinolones. Relapse of infection is possible in some cases, especially if the antimicrobial prescribed is bacteriostatic, thus a longer course of therapy may be required.[97,100] Work continues on the development of an effective vaccine against *F. tularensis*. Such a vaccine would be most beneficial for people living in highly endemic areas involved in activities that put them at elevated risk for exposure, including laboratory personnel, hunters, outdoors people, and people working with potentially contaminated soils, such as landscapers and farmers.[97,105]

Pasteurella

Members of the genus *Pasteurella* are Gram-negative facultative anaerobes that are part of the γ-proteobacteria that are carried in the oral and gastrointestinal tract of many animal species.[106] The most common species associated with zoonotic disease is *P. multocida*.[107] *P. multocida* was first identified by Louis Pasteur as the causative agent of the poultry disease, fowl cholera, and has subsequently found to be a potential pathogen to a variety of animal species.[108] Other *Pasteurella* species associated with human disease include *P. canis*, *P. stomatis*, and *P. dagmatis*.[109] As many as 90% of cats and 66% of dogs carry *Pasteurella* asymptomatically and most of the human cases of pasteurellosis are due to animal bites or scratches. However, transmission can occur following contact with saliva from infected animals or inhalation of the agent.[110,111] It is estimated that between 20% and 50% of cat and dog bites become infected by *Pasteurella*.[111] *Pasteurella* can also be isolated from the upper respiratory tract of many livestock species as well and serve as a source of zoonotic transmission.[109,112] Many of these nonbite infections can lead to severe manifestations, including infections of the bone and joints, skin, soft tissue, meningitis, peritonitis, pneumonia, endocarditis, and septicemia. The initial exposure is through the oral/respiratory route leading to an initial colonization of the upper respiratory tract, followed by hematogenous spread to the affected organ system.[107,109,110,113,114] Between 16% and 31% of *Pasteurella* infections occur without apparent direct animal contact, which may be indicative of human-to-human transmission of the pathogens.[115]

The initial symptoms of skin and soft tissue infections, such as from animal bites, involves swelling and tenderness at the site of infection, followed by regional lymphadenopathy, wound discharge, and fever, which occur in approximately 20–40% of cases.[109] More severe manifestations may include septic arthritis and/or osteomyelitis, especially when the bite wound involves the periosteum where there is spread from soft tissue adjacent to the bone.[116] Additionally, patients receiving corticosteroids due to underlying arthritis are at increased risk of developing septic arthritis.[109] In rare cases, patients can develop *P. multocida*-associated septicemia that can lead to development of endocarditis and meningitis.[114,116,117] Most patients that develop septicemia often have pre-existing health conditions that predisposes them to the invasive disease.[113]

In the majority of non soft-tissue infections, the initial *Pasteurella* colonization often occurs through fimbriae-associated adherence in the upper respiratory tract to the mucosal epithelial cells. After skin and soft tissue, the respiratory tract is the most common site of infection, with the tonsils likely the primary site of colonization.[109] *P. multocida* cause upper respiratory disease such as sinusitis and bronchitis and lower respiratory tract infections, including empyema and pneumonia. Patients that develop lower respiratory illnesses typically have underlying pathologies.[116] There have also been reports of *Pasteurella*-associated urinary tract infections, endocarditis, and peritonitis.

Urinary tract infections are likely due to direct infection of urinary tract colonization rather than hematogenous spread.[110] A number of virulence factors have been identified in *Pasteurella* strains including those that encode the production of fimbriae, toxins, polysaccharide capsules, lipopolysaccharide, and iron acquisition proteins.[108] Toxins detected within virulent strains likely aid in the invasion of different host tissues,[109] while the capsule aids in resistance to complement and phagocyte-mediated killing, adherence and resistance to drying.[108]

The typical diagnosis of *Pasteurella* infections involves the clinical presentation and culture of the organisms from the site of infection or from the patient's blood.[112] Additionally, diagnostic technologies, such as the matrix-assisted laser desorption/ionization time of flight (MALDI-TOF) can be used to identify *P. multocida*.[117] Penicillin or amoxicillin-clavulanic acid are the primary recommended therapies for most forms of *Pasteurella* infections.[109,118] If penicillins are contraindicated, macrolides, such as clindamycin, are recommended for use.[119] The treatment regimen is dependent on the severity of disease, with mild soft tissue infections treated with antibiotics on an outpatient basis, whereas more severe infections with deeper tissue and/or bone and joint involvement may require hospitalization and more aggressive antimicrobial and surgical interventional therapy.[118] The use of antimicrobial prophylaxis following animal bites remains somewhat controversial, however many clinicians recommend a short course of antimicrobials to reduce the risk of infection of the wound.[109]

Conclusions

The taxonomic diversity of zoonotic pathogens is quite substantial and ranges from mycobacteria, to Gram-positive to Gram-negative organisms. Some are obligate intracellular bacteria while others are not. Many zoonotic pathogens are spread by direct contact with another vertebrate host through contact or bite wounds, while others are transferred through an arthropod vector, such as a tick or mosquito. The number of potential zoonotic pathogens is also quite substantial but our understanding of many of them is quite limited. In this chapter, we focused on seven genera of zoonotic pathogens that were not covered in other chapters in this book. Some zoonotic pathogens, such as *Bacillus anthracis*, *F. tularensis*, and *Brucella abortus* have garnered renewed interests because of their potential role as bioterror agents. The list of new zoonotic pathogens is likely to grow as integration of MALDI-TOF mass spectrometry for bacterial identification in clinical microbiology laboratories has the potential to identify new and clinically relevant zoonotic pathogens from animal sources. Additionally, the "One Health" laboratory practice where clinical microbiologists overseeing both human and animal microbiological specimens may play a greater role in making zoonotic connections.

While a number of zoonotic pathogens are likely to increase, their control and prevention will be complicated due to high degree of diversity among these pathogens and many modes of transmission. Steady surveillance of infectious diseases in wild and domesticated animals, animal reservoirs, known vectors, and monitoring human exposure will not only enhance our understanding of the ecology of many of the zoonotic pathogens, but also zoonotic pathogen prevalence and their transmission dynamics. Knowledge of the ecological factors that promote the emergence of zoonotic pathogens will help us create ways to reduce or break the transmission cycles. Zoonotic pathogens and diseases remain a significant cause of morbidity and mortality throughout the world and an increased understanding of these pathogens and their ecology is warranted.

Acknowledgment

Authors would like to acknowledge Krishna Ganta for proofreading the manuscript. SKS would like to acknowledge the support from Marshfield Clinic Research Institute. The material described in this manuscript is not a formal dissemination of information by FDA and does not represent agency position or policy.

REFERENCES

1. Taylor LH, Latham SM, Woolhouse ME. Risk factors for human disease emergence. *Philos Trans R Soc Lond B Biol Sci.* 2001 Jul 29;356(1411):983–9.
2. http://www.onehealthinitiative.com/mission-statement
3. Friedrich M. Antimicrobial resistance on the rise in zoonotic bacteria in Europe. *JAMA.* 2019;321(15):1448.
4. Baylis M. Potential impact of climate change on emerging vector-borne and other infections in the UK. *Environ Health.* 2017;16(Suppl 1):112.
5. Ogden NH, Lindsay LR. Effects of climate and climate change on vectors and vector-borne diseases: ticks are different. *Trends Parasitol.* 2016 Aug;32(8):646–656.
6. Dumler JS, Barbet AF, Bekker CP, Dasch GA, Palmer GH, Ray SC, Rikihisa Y, Rurangirwa FR. Reorganization of genera in the families *Rickettsiaceae* and *Anaplasmataceae* in the order *Rickettsiales*: unification of some species of *Ehrlichia* with *Anaplasma*, *Cowdria* with *Ehrlichia* and *Ehrlichia* with *Neorickettsia*, descriptions of six new species combinations and designation of *Ehrlichia equi* and "HGE agent" as subjective synonyms of *Ehrlichia phagocytophila*. *Int J Syst Evol Microbiol.* 2001 Nov;51(Pt 6):2145–65.
7. Herron MJ, Ericson ME, Kurtti TJ, Munderloh UG. The interactions of *Anaplasma phagocytophilum*, endothelial cells, and human neutrophils. *Ann N Y Acad Sci.* 2005 Dec;1063:374–82.
8. Klein MB, Miller JS, Nelson CM, Goodman JL. Primary bone marrow progenitors of both granulocytic and monocytic lineages are susceptible to infection with the agent of human granulocytic ehrlichiosis. *J Infect Dis.* 1997 Nov;176(5):1405–9.
9. Buller RS, Arens M, Hmiel SP, Paddock CD, Sumner JW, Rikihisa Y, Unver A, Gaudreault-Keener M, Manian FA, Liddell AM, Schmulewitz N, Storch GA. *Ehrlichia ewingii*, a newly recognized agent of human ehrlichiosis. *N Engl J Med.* 1999 Jul 15;341(3):148–55.
10. Dunning Hotopp JC, Lin M, Madupu R, Crabtree J, Angiuoli SV, Eisen JA, Seshadri R, Ren Q, Wu M, Utterback TR, Smith S, Lewis M, Khouri H, Zhang C, Niu H, Lin Q, Ohashi N, Zhi N, Nelson W, Brinkac LM, Dodson RJ, Rosovitz MJ, Sundaram J, Daugherty SC, Davidsen T, Durkin AS, Gwinn M, Haft DH, Selengut JD, Sullivan SA, Zafar N, Zhou L, Benahmed F, Forberger H, Halpin R, Mulligan S, Robinson J, White O, Rikihisa Y, Tettelin H. Comparative genomics of emerging human ehrlichiosis agents. *PLOS Genet.* 2006 Feb;2(2):e21. Epub 2006 Feb 17.

11. Bakken JS, Dumpler JS, Chen SM, Eckman MR, Van Etta LL, Walker DH. Human granulocytic ehrlichiosis in upper Midwest United States. A new species emerging? *JAMA*. 1994 Jul 20;272(3):212–8.

12. Chen SM, Dumler JS, Bakken JS, Walker DH. Identification of a granulocytotropic *Ehrlichia* species as the etiologic agent of human disease. *J Clin Microbiol*. 1994 Mar;32(3):589–95.

13. Madigan JE, Gribble D. Equine ehrlichiosis in northern California: 49 cases (1968-1981). *J Am Vet Med Assoc*. 1987 Feb 15;190(4):445–8.

14. Madewell BR, Gribble DH. Infection in two dogs with an agent resembling *Ehrlichia equi*. *J Am Vet Med Assoc*. 1982 Mar 1;180(5):512–4.

15. Rikihisa Y. The tribe Ehrlichieae and ehrlichial diseases. *Clin Microbiol Rev*. 1991 Jul;4(3):286–308.

16. Massung RF, Mauel MJ, Owens JH, Allan N, Courtney JW, Stafford KC 3rd, Mather TN. Genetic variants of *ehrlichia phagocytophila*, Rhode Island and Connecticut. *Emerg Infect Dis*. 2002 May;8(5):467–72.

17. Richter PJ, Kimsey RB, Madigan JE, Barlough JE, Dumler JS, Brooks DL. *Ixodes pacificus* (acari: ixodidae) as a vector of *ehrlichia equi* (*Rickettsiales: Ehrlichieae*). *J Med Entomol*. 1996 Jan;33(1):1–5.

18. des Vignes F, Piesman J, Heffernan R, Schulze TL, Stafford KC 3rd, Fish D. Effect of tick removal on transmission of *Borrelia burgdorferi* and *ehrlichia phagocytophila* by *Ixodes scapularis* nymphs. *J Infect Dis*. 2001 Mar 1;183(5):773–8. Epub 2001 Feb 1.

19. Goodman JL, Nelson C, Vitale B, Madigan JE, Dumler JS, Kurtti TJ, Munderloh UG. Direct cultivation of the causative agent of human granulocytic ehrlichiosis. *N Engl J Med*. 1996 Jan 25;334(4):209–15.

20. Munderloh UG, Jauron SD, Fingerle V, Leitritz L, Hayes SF, Hautman JM, Nelson CM, Huberty BW, Kurtti TJ, Ahlstrand GG, Greig B, Mellencamp MA, Goodman JL. Invasion and intracellular development of the human granulocytic ehrlichiosis agent in tick cell culture. *J Clin Microbiol*. 1999 Aug;37(8):2518–24.

21. Lin M, Rikihisa Y. *Ehrlichia chaffeensis* and *Anaplasma phagocytophilum* lack genes for lipid A biosynthesis and incorporate cholesterol for their survival. *Infect Immun*. 2003 Sep;71(9):5324–31.

22. Rikihisa Y. Molecular events involved in cellular invasion by *Ehrlichia chaffeensis* and *Anaplasma phagocytophilum*. *Vet Parasitol*. 2010 Feb 10;167(2-4):155–66.

23. Asanovich KM, Bakken JS, Madigan JE, Aguero-Rosenfeld M, Wormser GP, Dumler JS. Antigenic diversity of granulocytic *Ehrlichia* isolates from humans in Wisconsin and New York and a horse in California. *J Infect Dis*. 1997 Oct;176(4):1029–34.

24. Dumler JS, Asanovich KM, Bakken JS. Analysis of genetic identity of North American *Anaplasma phagocytophilum* strains by pulsed-field gel electrophoresis. *J Clin Microbiol*. 2003 Jul;41(7):3392–4.

25. Belongia EA, Reed KD, Mitchell PD, Kolbert CP, Persing DH, Gill JS, Kazmierczack JJ. Prevalence of granulocytic *Ehrlichia* infection among white-tailed deer in Wisconsin. *J Clin Microbiol*. 1997 Jun;35(6):1465–8.

26. Foley JE, Crawford-Miksza L, Dumler JS, Glaser C, Chae JS, Yeh E, Schurr D, Hood R, Hunter W, Madigan JE. Human granulocytic ehrlichiosis in Northern California: two case descriptions with genetic analysis of the *Ehrlichiae*. *Clin Infect Dis*. 1999 Aug;29(2):388–92.

27. Poitout FM, Shinozaki JK, Stockwell PJ, Holland CJ, Shukla SK. Genetic variants of *Anaplasma phagocytophilum* infecting dogs in western Washington State. *J Clin Microbiol*. 2005 Feb;43(2):796–801.

28. Massung RF, Owens JH, Ross D, Reed KD, Petrovec M, Bjoersdorff A, Coughlin RT, Beltz GA, Murphy CI. Sequence analysis of the ank gene of granulocytic *Ehrlichiae*. *J Clin Microbiol*. 2000 Aug;38(8):2917–22.

29. de la Fuente J, Massung RF, Wong SJ, Chu FK, Lutz H, Meli M, von Loewenich FD, Grzeszczuk A, Torina A, Caracappa S, Mangold AJ, Naranjo V, Stuen S, Kocan KM. Sequence analysis of the msp4 gene of *Anaplasma phagocytophilum* strains. *J Clin Microbiol*. 2005 Mar;43(3):1309–17.

30. Tegtmeyer P, Ganter M, von Loewenich FD. Simultaneous infection of cattle with different *Anaplasma phagocytophilum* variants. *Ticks Tick Borne Dis*. 2019 Aug;10(5):1051–1056.

31. Dumler JS, Barat NC, Barat CE, Bakken JS. Human granulocytic anaplasmosis and macrophage activation. *Clin Infect Dis*. 2007 Jul 15;45(2):199–204. Epub 2007 Jun 5.

32. Dumler JS. The biological basis of severe outcomes in Anaplasma phagocytophilum infection. *FEMS Immunol Med Microbiol*. 2012 Feb;64(1):13–20.

33. Schotthoefer AM, Schrodi SJ, Meece JK, Fritsche TR, Shukla SK. Pro-inflammatory immune responses are associated with clinical signs and symptoms of human anaplasmosis. *PLOS One*. 2017 Jun 19;12(6):e0179655.

34. Honsberger NA, Six RH, Heinz TJ, Weber A, Mahabir SP, Berg TC. Efficacy of sarolaner in the prevention of *Borrelia burgdorferi* and *Anaplasma phagocytophilum* transmission from infected *Ixodes scapularis* to dogs. *Vet Parasitol*. 2016 May 30;222:67–72.

35. Crosby FL, Lundgren AM, Hoffman C, Pascual DW, Barbet AF. VirB10 vaccination for protection against Anaplasma phagocytophilum. *BMC Microbiol*. 2018 Dec 18;18(1):217.

36. de la Fuente J, Estrada-Peña A, Cabezas-Cruz A, Kocan KM. Anaplasma phagocytophilum uses common strategies for infection of ticks and vertebrate hosts. *Trends Microbiol*. 2016 Mar;24(3):173–180.

37. Slater LN, Welch DF. *Bartonella*, including cat-scratch disease. In: Mandell GL, Bennett JE, Dolin R, editors. *Principles and Practice of Infectious Diseases*. 6th ed. Philadelphia, PA: Elsevier Churchill Livingstone; 2005, pp. 2733–2748.

38. Pulliainen AT, Dehio C. Persistence of *Bartonella* spp. stealth pathogens: from subclinical infections to vasoproliferative tumor formation. *FEMS Microbiol Rev*. 2012 May;36(3):563–99.

39. Chomel BB, Kasten RW. Bartonellosis, an increasingly recognized zoonosis. *J Appl Microbiol*. 2010 Sep;109(3):743–50.

40. Sanchez Clemente N, Ugarte-Gil CA, Solorzano N, Maguina C, Pachas P, Blazes D, Bailey R, Mabey D, Moore D. Bartonella bacilliformis: a systematic review of the literature to guide the research agenda for elimination. *PLOS Negl Trop Dis*. 2012;6(10):e1819.

41. Brouqui P, Raoult D. Arthropod-borne diseases in homeless. *Ann N Y Acad Sci*. 2006;1078:223–235.

42. Breitschwerdt EB. Bartonellosis: one health perspectives for an emerging infectious disease. *ILAR J*. 2014;55(1):46–58.

43. Guptill L. Bartonellosis. *Vet Microbiol*. 2010;140(3-4):347–359.

44. Boulouis HJ, Chang CC, Henn JB, Kasten RW, Chomel BB. Factors associated with the rapid emergence of zoonotic *Bartonella* infections. *Vet Res.* 2005 May-Jun;36(3):383–410.

45. Cheslock MA, Embers ME. Human bartonellosis: an under-appreciated public health problem? *Trop Med Infect Dis.* 2019;4(2):69.

46. Nelson CA, Saha S, Mead PS. Cat-Scratch disease in the United States, 2005-2013. *Emerging Infectious Diseases.* 2016;22(10):1741–1746.

47. Massei F, Gori L, Macchia P, Maggiore G. The expanded spectrum of bartonellosis in children. *Infect Dis Clin North Am.* 2005 Sep;19(3):691–711.

48. Rolain JM, Brouqui P, Koehler JE, Maguina C, Dolan MJ, Raoult D. Recommendations for treatment of human infections caused by *Bartonella* species. *Antimicrob Agents Chemother.* 2004 Jun;48(6):1921–33.

49. Maurin M, Raoult D. *Bartonella (Rochalimaea) quintana* infections. *Clin Microbiol Rev.* 1996 Jul;9(3):273–92.

50. Prutsky G, Domecq JP, Mori L, Bebko S, Matzumura M, Sabouni A, Shahrour A, Erwin PJ, Boyce TG, Montori VM, Malaga G, Murad MH. Treatment outcomes of human bartonellosis: a systematic review and meta-analysis. *Int J Infect Dis.* Oct;17(10):e811–9. Epub 2013 Apr 18.

51. Aguero-Rosenfield ME, Wang G, Schwartz I, Wormser GP. Diagnosis of Lyme borreliosis, *Clin Microbiol Rev.* 2005 Jul;18(3):484–509.

52. Pritt BS, Mead PS, Johnson DKH, Neitzel DF, Respicio-Kingry LB, Davis JP, Schiffman E, Sloan LM, Schriefer ME, Replogle AJ, Paskewitz SM, Ray JA, Bjork J, Steward CR, Deedon A, Lee X, Kingry LC, Miller TK, Feist MA, Theel ES, Patel R, Irish CL, Petersen JM. Identification of a novel pathogenic Borrelia species causing Lyme borreliosis with unusually high spirochaetaemia: a descriptive study. *Lancet Infect Dis.* 2016 May;16(5):556–564.

53. Aguero-Rosenfield ME, Nowakowlski J, McKenna DF, Carbonaro CA, Worsmer GP. Serodiagnosis of early Lyme diseases, *J Clin Microbiol.* 1993 Dec;31(12):3090–5.

54. Mathiesen DA, Oliver JH Jr, Kolbert CP, Tullson ED, Johnson BJ, Campbell GL, Mitchell PD, Reed KD, Telford SR 3rd, Anderson JF, Lane RS, Persing DH. Genetic heterogeneity of *Borrelia burgdorferi* in the United States, *J Infect Dis.* 1997 Jan;175(1):98–107.

55. Burgdorfer W, Barbour AG, Hayes SF, Benach JL, Grunwaldt E, Davis JP. Lyme disease — a tick-borne spirochetosis? *Science.* 1982 Jun 18;216(4552):1317–9.

56. Benach JL, Bosler EM, Hanrahan JP, Coleman JL, Habicht GS, Bast TF, Cameron DJ, Ziegler JL, Barbour AG, Burgdorfer W, Edelman R, Kaslow RA. Spirochetes isolated from two patients with Lyme diseases. *N Engl J Med.* 1983 Mar 31;308(13):740–2.

57. Steere AC, Grodzicki RL, Kornblatt AN, Craft JE, Barbour AG, Burgdorfer W, Schmid GP, Johnson E, Malawista SE. The spirochetal etiology of Lyme disease. *N Engl J Med.* 1983 Mar 31;308(13):733–40.

58. Schutzer SE, Fraser-Liggett CM, Casjens SR, Qiu WG, Dunn JJ, Mongodin EF, Luft BJ. Whole-genome sequences of thirteen isolates of *Borrelia burgdorferi. J Bacteriol.* 2011 Feb;193(4):1018–20.

59. Lee SH, Healy JE, Lambert JS. Single core genome sequencing for detection of both *Borrelia burgdorferi* sensu lato and relapsing fever Borrelia species. *Int J Environ Res Public Health.* 2019 May 20;16(10): pii: E1779.

60. Zhang JR, Norris SJ. Genetic variation of the *Borrelia burgdorferi* gene vlsE involves cassette-specific, segmental gene conversion. *Infect Immun.* 1998 Aug;66(8):3698–704.

61. http://www.cdc.gov/lyme/

62. Tørnqvist-Johnsen C, Dickson SA, Rolph K, Palermo V, Hodgkiss-Geere H, Gilmore P, Gunn-Moore DA. First report of Lyme borreliosis leading to cardiac bradydysrhythmia in two cats. *JFMS Open Rep.* 2020 Jan 2;6(1):2055116919898292.

63. Gilmore RD Jr, Mbow ML, Stevenson B. Analysis of *Borrelia burgdorferi* gene expression during life cycle phases of the tick vector, *Ixodes scapularis. Microbes Infect.* 2001 Aug;3(10):799–808.

64. Kelly R. Cultivation of *Borrelia hermsi. Science.* 1971 Jul 30;173(3995):443–4.

65. Reed KD. Laboratory testing for Lyme disease: possibilities and practicalities. *J Clin Microbiol.* 2002 Feb;40(2):319–24.

66. Schwartz I, Bittker S, Bowen SL, Cooper D, Pavia C, Wormser GP. Polymerase chain reaction amplification of culture supernatants for rapid detection of *Borrelia burgdorferi. Eur J Clin Microbiol Infect.* 1993 Nov;12(11):879–82.

67. Schmidt BL. PCR in laboratory diagnosis of human *Borrelia burgdorferi* infections. *Clin Microbiol Rev.* 1997 Jan;10(1):185–201.

68. Grodzicki RL, Steere AL. Comparison of immunoblotting and indirect enzyme-linked immunosorbent assay using different antigen preparations for diagnosing early Lyme disease. *J Infect Dis.* 1988 Apr;157(4):790–7.

69. Magnarelli LA, Meegan JM, Anderson JF, Chappell WA. Comparison of an indirect fluorescent-antibody test with an enzyme-linked immunosorbent assay for serological studies of Lyme disease. *J Clin Microbiol.* 1984 Aug;20(2):181–4.

70. Callister SM, Jobe DA, Agger WA, Schell RF, Kowalski TJ, Lovrich SD, Marks JA. Ability of the borreliacidal antibody test to confirm Lyme disease in clinical practice. *Clin Diagn Lab Immunol.* 2002 Jul;9(4):908–12.

71. Oosting M, Kerstholt M, Ter Horst R, Li Y, Deelen P, Smeekens S, Jaeger M, Lachmandas E, Vrijmoeth H, Lupse M, Flonta M, Cramer RA, Kullberg BJ, Kumar V, Xavier R, Wijmenga C, Netea MG, Joosten LA. Functional and genomic architecture of *Borrelia burgdorferi*-induced cytokine responses in humans. *Cell Host Microbe.* 2016 Dec 14;20(6):822–833.

72. Nathwani D, Hamlet N, Walker E. Lyme disease: a review. *Br J Gen Pract.* 1990 Feb;40(331):72–4.

73. Embers ME, Narasimhan S. Vaccination against Lyme disease: past, present, and future. *Front Cell Infect Microbiol.* 2013 Feb 12;3:6. eCollection 2013.

74. Bencurova E, Gupta SK, Oskoueian E, Bhide M, Dandekar T. Omics and bioinformatics applied to vaccine development against Borrelia. *Mol Omics.* 2018 Oct 8;14(5):330–340.

75. Ustun I, Ozcakar L, Arda N, Duranay M, Bayrak E, Duman K, Atabay M, Cakal BE, Altundag K, Guler S. *Brucella* glomerulonephritis: case report and review of the literature. *South Med J.* 2005 Dec;98(12):1216–7.

76. Cutler SJ, Whatmore AM, Commander NJ. Brucellosis–new aspects of an old disease. *J Appl Microbiol.* 2005;98(6):1270–81.

77. Ko J, Splitter GA. Molecular host-pathogen interaction in brucellosis: current understanding and future approaches to vaccine development for mice and humans. *Clin Microbiol Rev.* 2003 Jan;16(1):65–78.

78. Celli, J. The intracellular life cycle of *Brucella* spp. *Microbiol Spectr.* 2019;7(2): 10.1128/microbiolspec.BAI-0006-2019.

79. Martirosyan A, Moreno E, Gorvel JP. An evolutionary strategy for a stealthy intracellular *Brucella* pathogen. *Immunol Rev.* 2011 Mar;240(1):211–34.

80. Pappas G, Papadimitriou P, Akritidis N, Christou L, Tsianos EV. The new global map of human brucellosis. *Lancet Infect Dis.* 2006 Feb;6(2):91–9.

81. Pappas G, Akritidis N, Bosilkovski M, Tsianos E. Brucellosis. *N Engl J Med.* 2005 Jun 2;352(22):2325–36.

82. El-Sayed A, Awad W. 2018. Brucellosis: Evolution and expected comeback. *Int J Vet Sci Med.* 6(Suppl):S31–S35.

83. de Jong MF, Rolan HG, Tsolis RM. Innate immune encounters of the (Type) 4th kind: *brucella*. *Cell Microbiol.* 2010 Sep 1;12(9):1195–202. Epub 2010 Jul 28.

84. Tuon FF, Gondolfo RB, Cerchiari N. Human-to-human transmission of Brucella - a systematic review. *Trop Med Int Health.* 2017;22(5):539–546.

85. Franc KA, Krecek RC, Häsler BN, Arenas-Gamboa AM. Brucellosis remains a neglected disease in the developing world: a call for interdisciplinary action. *BMC Public Health.* 2018 Jan 11;18(1):125.

86. Young EJ. *Brucella* species. In: Mandell GL, Bennett JE, Dolin R, editors. *Principles and Practice of Infectious Diseases.* 6th ed. Philadelphia, PA: Elsevier Churchill Livingstone; 2005, pp. 2669–2674.

87. Parker NR, Barralet JH, Bell AM. Q fever. *Lancet.* 2006;367:679–688.

88. Hemsley CM, O'Neill PA, Essex-Lopresti A, Norville IH, Atkins TP, Titball RW. Extensive genome analysis of *Coxiella burnetii* reveals limited evolution within genomic groups. *BMC Genomics.* 2019 Jun 5;20(1):441.

89. Arricau-Bouvery N, Rodolakis A. Is Q fever an emerging or re-emerging zoonosis? *Vet Res.* 2005 May-Jun;36(3):327–49.

90. Howe D, Mallavia LP. *Coxiella burnetii* exhibits morphological change and delays phagolysosomal fusion after internalization by J774A.1 cells. *Infect Immun.* 2000 Jul;68(7):3815–21.

91. Fournier PE, Marrie TJ, Raoult D. Diagnosis of Q fever. *J Clin Microbiol.* 1998 Jul;36(7):1823–34.

92. Beare PA[1], Samuel JE, Howe D, Virtaneva K, Porcella SF, Heinzen RA. Genetic diversity of the Q fever agent, *Coxiella burnetii*, assessed by microarray-based whole-genome comparisons. *J Bacteriol.* 2006 Apr;188(7):2309–24.

93. Long CM, Beare PA, Cockrell DC, Larson CL, Heinzen RA. Comparative virulence of diverse *Coxiella burnetii* strains. *Virulence.* 2019 Dec;10(1):133–150.

94. Marrie TJ, Raoult D. Coxiella burnetii (Q Fever). In: Mandell, GL, Bennett JE, and Dolin R, editors. *Principles and Practice of Infectious Diseases*, 6th ed., vol. 2. Philadelphia, PA: Elsevier Churchill Livingstone; 2005, pp. 2296–2303.

95. Marrie TJ. *Coxiella burnetii* pneumonia. *Eur Respir J.* 2003 Apr;21(4):713–9.

96. Karakousis PC, Trucksis M, Dumler JS. Chronic Q fever in the United States. *J Clin Microbiol.* 2006 Jun;44(6):2283–7.

97. Penn RL. *Francisella tularensis* (Tularemia). In: Mandell GL, Bennett JE, Dolin R, editors. *Principles and Practice of Infectious Diseases.* 6th ed. Philadelphia, PA: Elsevier Churchill Livingstone; 2005, pp. 2674–2685.

98. Petersen JM, Schriefer ME. Tularemia: emergence/re-emergence. *Vet Res.* 2005 May-Jun;36(3):455–67.

99. Foley JE, Nieto NC. Tularemia. *Vet Microbiol.* 2010 Jan 27;140(3-4):332–8. Epub 2009 Aug 8.

100. Eliasson H, Broman T, Forsman M, Back E. Tularemia: current epidemiology and disease management. *Infect Dis Clin North Am.* 2006 Jun;20(2):289–311, ix.

101. Zellner B, Huntley JF. Ticks and tularemia: do we know what we don't know? *Front Cell Infect Microbiol.* 2019;9:146.

102. Farlow J, Wagner DM, Dukerich M, Stanley M, Chu M, Kubota K, Petersen J, Keim P. *Francisella tularensis* in the United States. *Emerg Infect Dis.* 2005 Dec;11(12):1835–41.

103. Tarnvik A, Berglund L. Tularaemia. *Eur Respir J.* 2003 Feb;21(2):361–73.

104. Keim P, Johansson A, Wagner DM. Molecular epidemiology, evolution, and ecology of *Francisella*. *Ann N Y Acad Sci.* 2007 Jun;1105:30–66. Epub 2007 Apr 13.

105. Roberts LM, Powell DA, Frelinger JA. Adaptive immunity to *Francisella tularensis* and considerations for vaccine development. *Front Cell Infect Microbiol.* 2018;8:115.

106. Prescott LM, Harley JP, Klein DA. *Microbiology.* New York, NY: McGraw-Hill; 2005.

107. Holst E, Rollof J, Larsson L, Nielsen JP. Characterization and distribution of *Pasteurella* species recovered from infected humans. *J Clin Microbiol.* 1992 Nov;30(11):2984–7.

108. Harper M, Boyce JD, Adler B. *Pasteurella multocida* pathogenesis: 125 years after Pasteur. *FEMS Microbiol Lett.* 2006 Dec;265(1):1–10.

109. Zurlo JJ. *Pasteurella* Species. In: Mandell GL, Bennett JE, Dolin R, editors. *Principles and Practice of Infectious Diseases.* 6th ed. Philadelphia, PA: Elsevier Churchill Livingstone; 2005, pp. 2687–2690.

110. Liu W, Chemaly RF, Tuohy MJ, LaSalvia MM, Procop GW. *Pasteurella multocida* urinary tract infection with molecular evidence of zoonotic transmission. *Clin Infect Dis.* 2003 Feb 15;36(4):E58–60. Epub 2003 Jan 31.

111. Iaria C, Cascio A. Please, do not forget *Pasteurella multocida*. *Clin Infect Dis.* 2007 Oct 1;45(7):940.

112. Tan JS. Human zoonotic infections transmitted by dogs and cats. *Arch Intern Med.* 1997 Sep 22;157(17):1933–43.

113. Ashley BD, Noone M, Dwarakanath AD, Malnick H. Fatal *Pasteurella dagmatis* peritonitis and septicaemia in a patient with cirrhosis: a case report and review of the literature. *J Clin Pathol.* 2004 Feb;57(2):210–2.

114. Boerlin P, Siegrist HH, Burnens AP, Kuhnert P, Mendez P, Prétat G, Lienhard R, Nicolet J. Molecular identification and epidemiological tracing of *Pasteurella multocida* meningitis in a baby. *J Clin Microbiol.* 2000 Mar;38(3):1235–7.

115. Hubbert WT, Rosen MN. Pasteurella multocida infections. II. Pasteurella multocida infection in man unrelated to animal bite. *Am J Public Health Nations Health.* 1970 Jun;60(6):1109–17.

116. Weber DJ, Wolfson JS, Swartz MN, Hooper DC. *Pasteurella multocida* infections. Report of 34 cases and review of the literature. *Medicine (Baltimore).* 1984 May;63(3):133–54.

117. Martin TCS, Abdelmalek J, Yee B, Lavergne S, Ritter M. *Pasteurella multocida* line infection: a case report and review of literature. *BMC Infect Dis.* 2018;18(1):420.

118. Mogilner L, Katz C. Pasteurella multocida. *Pediatr Rev.* 2019;40(2):90–92.

119. Gilbert DN, Moellering RC, Sande MA. *The Sanford Guide to Antimicrobial Therapy.* Hyde Park, VT: Antimicrobial Therapy, Inc.; 2003.

46

Fungi

Charles Adair

CONTENTS

DOI: 10.1201/9781003099277-48

The Fungus Kingdom Mycota

The Mycota is one of the taxonomic kingdoms of life, and all members of the kingdom share a set of characteristics that separates them from the other kingdoms. Genomic evidence shows that this kingdom has a closer evolutionary relationship to the animal kingdom than to any other eukaryotic kingdom, which is a factor with clinical implications that will be addressed later in this chapter [1–4].

Based on genomic and ultrastructural evidence, various Mycologists have subdivided the Mycota into at least a dozen phyla, and this chapter describes the 11 that appear to be the most valid (some very briefly) [2–5]. All 11 are treated as separate phyla for consistency, even though there is uncertainty about the status of some of them. The names of these phyla refer to some unique feature of the phylum or to a representative member of the phylum, and (with one exception) these names end with the suffix -mycota.

The Mycota include four phyla with swimming zoospores, a primitive feature that reflects the aquatic ancestry of this kingdom: the Chytridiomycota, Blastocladiomycota, Neocallimastigomycota, and Olpidiomycota [2, 3, 5, 6]. These four phyla are given the informal name "Lower Fungi with Zoospores" in this chapter, and some of the more notable activities of organisms in this group are described within the section with that name. There are another five phyla given the informal designation "Lower Fungi without Zoospores," since they appear to be relatively primitive as well, but they do not produce motile zoospores: the Glomeromycota, the Mucoromycota, the Zoopagomycota, the Entomophthoromycota, and the Microsporidia [2, 3, 5, 7]. Several of these phyla had originally been classified within a phylum known as the Zygomycota because of similarities in their reproductive cycles that included the formation of a zygospore, but more recently they have been recognized as different phyla based on genomic data, making that earlier phylum name invalid [7, 8]. Several species within these phyla are of clinical importance, while others are significant for economic or ecological reasons. Finally, the Ascomycota and Basidiomycota are the most highly evolved and specialized phyla in many respects, identified collectively by some authors as the Dikarya (to be explained later) but more commonly referred to as the Higher Fungi, and these are covered in detail in their own section of this chapter with that title [9, 10].

Each of these phyla is subdivided into fungal classes, orders, families, genera, and species following the standard taxonomic system of classification. (Note that some Mycological texts refer to each subdivision of the fungus Kingdom as a Division, but an increasing number of publications utilize the term Phylum, so that term is employed in this chapter for consistency.)

Within this standard taxonomic system the species name is a Latin binomial (capitalized genus and lower-case specific epithet, written as a Latin name), with priority given to the first valid binomial that was assigned to any given organism. However, there are certain challenges regarding the naming of fungi that have complicated the definition of the first valid binomial, as explained in more detail below.

Biology of Fungi

The Mycota consists of heterotrophic, eukaryotic organisms that are functionally single-celled, absorbing nutrients directly into their protoplasm through a chitinous cell wall and a cell membrane, that have a relatively simple growth form known as a thallus (never forming true tissues or organs), and that typically reproduce by means of spores [2, 4, 11, 12].

Fungal Cell Structure

The cells of all fungi are enclosed in a cell wall that is permeable to gases and dissolved solutes. This cell wall consists of molecules of chitin, chitosan, glucans (glucose polymers), glycoproteins, and mannans and/or galactomannans. Each of these macromolecules is extruded into the cell wall space by plasma membrane-associated synthases. These macromolecules are cross-linked by extracellular enzymes to create the wall matrix that provides a protective layer against osmotic and other stresses, as shown in Figure 46.1 [13, 14].

Additional components are deposited into the wall structure that creates an amorphous layer that covers and encloses the cross-linked matrix, and this layer is quite variable depending on the adaptations of the fungus. Two examples of functions of this layer are to provide adhesion to a substrate or to mask antigenic wall components in the case of species adapted to infect humans and other animals [13].

The protoplasm of fungi contains nuclei and other eukaryotic organelles, but plastids (the category of organelles that include chloroplasts) are absent from the protoplasm of the members of this kingdom. The nuclei in the growing and metabolizing cells of many of the fungi are haploid, meaning that the only time when a diploid nucleus occurs in such fungi is at one specific point in its life cycle, after which that diploid nucleus divides by meiosis. However, diploid nuclei are common in other fungi (especially the yeasts), and polyploid and aneuploid chromosome numbers can occur as well. The haploid nuclei are approximately 2 microns in diameter, which are relatively small among eukaryotic organisms (and difficult to study microscopically), and the number of chromosomes in those that have been studied ranges from 3 to 15 [11, 12, 15].

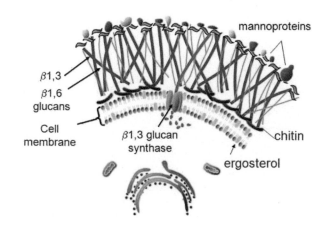

FIGURE 46.1   Ultrastructure of fungal cell walls.

The selectively permeable fungal plasma membrane has the typical phospholipid bilayer with membrane-bound proteins that is characteristic of all eukaryotic organisms. However, the sterol component of the membrane is ergosterol, contrasting with the cholesterol in animal plasma membranes and phytosterols in plant plasma membranes. This difference in membrane sterol composition has been targeted in the development of some antifungal therapeutic agents. One of these is amphotericin B, which binds with ergosterol and creates a polar pore in the plasma membrane that has lethal effects on the fungus, but this agent unfortunately causes significant side effects in treated patients. The azoles, another type of antifungal therapeutic agent, inhibit the synthesis of ergosterol, and include fluconazole and miconazole. Ergosterol is also found in the plasma membranes of certain Protista, such as *Trypanosoma* (the cause of African sleeping sickness), so the same therapeutic agents may be utilized to treat infections by these organisms [13, 16].

Fungal Physiology and Nutrition

Most fungi are capable of aerobic respiration at atmospheric concentrations of oxygen, but many are facultatively anaerobic, utilizing various fermentative pathways at very low oxygen levels or in the complete absence of oxygen; very few are obligate anaerobes. There are a few species of fungi that are thermophilic (capable of growth at temperatures of 45–60°C), many that are psychrotrophic (capable of growth at 0°C but with optimum growth above 20°C), and very few that are psychrophilic (capable of growth at 0°C but unable to grow above 20°C) [17, 18]. Many fungi are osmophilic, and some of these are associated with spoilage of sugary food products, while a small number are halophilic, capable of growth in hypersaline environments. Other extremophiles among fungi include acid-tolerant species, many of which can grow in environments with a pH of 3–4, while a small number have been detected at a pH below 3.0 [17–21].

The nutrients for the heterotrophic nutrition of fungi must be absorbed in solution through the permeable cell wall, so fungi are dependent on a damp or moist environment that provides such solutions. Fungi typically release enzymes through the cell wall into their surroundings that digest complex organic compounds into soluble molecules, and they also release waste products including various metabolites through their cell walls into their surroundings [22].

Sources of Fungal Nutrition

The nutritional requirements of fungi are met through several different kinds of interactions with their surroundings and other organisms. Many fungi are saprotrophs, absorbing their nutrients from nonliving organic matter, thus functioning ecologically as decomposers, and their impact on human affairs can be beneficial or harmful. Fungi of medical and agricultural importance are often necrotrophic, invading a living host with the release of toxins and enzymes that kill the host tissues, releasing soluble nutrients that are absorbed by the fungus, and typically creating a disease condition in the host. A few fungi are biotrophic (often referred to as obligate parasites), absorbing nutrients while maintaining an intimate relationship with host tissues that does not cause tissue death immediately, or at all, and the effect of this interaction on the host can be harmful or beneficial. Those necrotrophic and biotrophic fungi that cause disease in a host organism are identified as pathogens infecting a host to create that diseased condition. The disease cycle leading to such an infection is closely related to the life cycle and means of dispersal of the pathogenic fungus [22, 23].

Symbiotic Associations of Fungi

A beneficial biotrophic relationship with plants (known as a symbiosis) has evolved in several different phyla of fungi in the form of mycorrhiza (literally "fungus-root"), which occurs as endomycorrhizae or ectomycorrhizae. Endomycorrhizae consist of an invasion of the outer cortex layer of a plant root by the fungus, leading to penetration of individual plant cell walls by the hyphae without killing the host protoplast, since the hyphae neither penetrate the host plasma membrane nor release any products that injure the protoplast. The most commonly occurring endomycorrhiza is the arbuscular form involving the Glomeromycota, which is described later in the section about that phylum, but there are also other types of endomycorrhiza associated with the Ericaceae (the heath family of plants) and with the Orchidaceae. In an ectomycorrhiza, the fungal hyphae do not penetrate the host plant cell walls, but instead grow in the intercellular spaces between the cortex root cells. In all types of mycorrhizae, the hyphae of the fungus extend outside the root into the surrounding soil or other substrate, and in the case of ectomycorrhizae the hyphae create a mantle of mycelium around the root known as a "hartig net." The diagram in Figure 46.2 illustrates the relationship between fungus and root in each of these kinds of mycorrhizae. In most cases, the symbiotic relationship involves enhanced absorption of inorganic nutrients (and in some cases water) by the fungus, which are delivered to the plant, and transfer of carbohydrate food from the plant to the fungus. However, in the case of orchid endomycorrhizae, the fungus typically transfers carbohydrates to the plant [24–28].

Lichens consist of a biotrophic relationship between a fungus (almost always a member of the Ascomycota, although a very small number of Basidiomycota are also involved) and an "alga" that is either one of the Cyanobacteria or one of the Chlorophyta. Although this relationship is traditionally included among examples of symbiosis, it is actually closer to a case of cultivation of an autotroph by a heterotroph, with the "alga" performing photosynthesis (and nitrogen fixation in some Cyanobacteria) and thus providing food to the fungus, while it gets mineral nutrients, exposure to light, and a chance to grow in unusual locations in return. All the algae in this relationship are capable of growing independently in aqueous environments, whereas the fungal component is highly dependent on its autotrophic partner. The fungal component makes up 90–95% of the mass of a lichen, determines is shape, and reproduces with its typical sexual or asexual spore stages. The algal component is in the upper layers of the structure, with its cells essentially caged in by fungal hyphae and supplying either glucose or sugar alcohols to the fungus. A type of vegetative propagation occurs in some lichens by the release of powdery soredia (clusters of a few algal cells surrounded by fungal hyphae) or other comparable structures [29, 30].

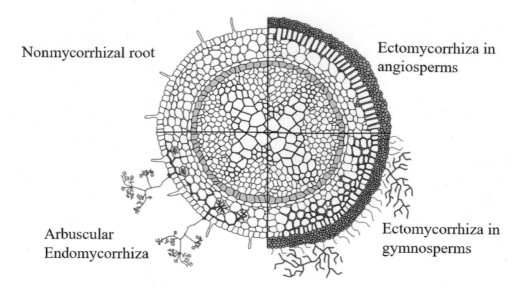

FIGURE 46.2   Comparisons of mycorrhizal types.

Fungal Growth Forms

The two most common growth forms of fungi are branching hyphae and single-celled yeasts. Branching hyphae occur in nearly all the major fungal phyla, while the yeast form of growth occurs in several of them, especially among the Ascomycota and to a lesser extent in the Basidiomycota.

The Branching Hyphae Growth Form

This growth form is created as a result of the tubular extension and branching of the growing hypha as the turgor pressure within the protoplasm pushes against the cell wall, stretching those regions of the cell wall at the tips where there is the least cross-linking [14]. The branching hyphae grow continuously and function in nutrient absorption, while a fungus is metabolically active, except when they function in the formation of sexual or asexual spores, and for convenience this growing and feeding stage is described as "vegetative." Mitosis of the hyphal nuclei occurs simultaneously with hyphal growth so that nuclei are distributed throughout the growing protoplasm. The diameters of most fungal hyphae range from 2 to 5 µm, although in some of the Mucoromycota they can be as much as 15 µm in diameter, as shown in Figure 46.3 [12].

In several of the fungal phyla—primarily the Ascomycota and Basidiomycota—there is a development of septa (internal cross-walls) at regular intervals within the hypha, and the presence or absence of septa in various groups of fungi can be readily observed by light microscopy, as shown in Figure 46.4. There are usually either one or two nuclei within each segment of the hypha created by septa, depending on various factors described in a later section of this chapter, and a function of these septa in some fungi is to prevent nuclear migration within the hypha that would alter this number. A complete septum without a pore is produced in all fungal phyla in the process of forming reproductive structures. The septa in actively functioning hyphae are perforated by various types of pores that allow the passage of ribosomes and other smaller cytoplasmic components, and the plasma membrane of adjacent cells is continuous through the margins of the pore. The septa of most Ascomycota have a round pore in the center and associated with each septum is a type of cytoplasmic microbody known as a Woronin body on either side of the pore. The Woronin body quickly plugs the pore if the hypha is damaged, preventing leakage of protoplasm from adjacent sections of the hypha. The septal pores of Basidiomycota are more complex, with either a dolipore septum (consisting of a barrel-shaped structure in the center of the pore with a portion of the endoplasmic reticulum described as a "parenthesome" on either side of it), or a "pulley wheel occlusion" suspended in the middle of the pore, nearly blocking it. The ultrastructure of hyphal pores and associated structures is illustrated in Figure 46.5 [12, 31, 32].

Hyphae can extend for considerable lengths and develop into macroscopic arrays that are easily visible without a microscope,

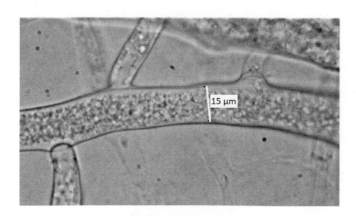

FIGURE 46.3   Nonseptate hypha of *Mucor plumbeus*.

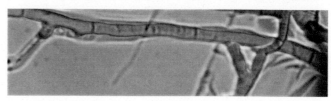

FIGURE 46.4   Septate hypha of *Helminthosporium solani*.

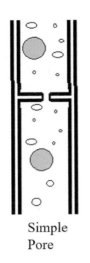
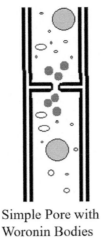

| Simple
Pore | Simple Pore with
Woronin Bodies | Dolipore Septum with
Continuous Parenthesomes |

FIGURE 46.5 Simplified diagrams of septal pore types occurring in fungal hyphae.

in which case the growth is described as "mycelium." Such growth is seen in the familiar forms known as molds and in the interwoven structure of such fruiting bodies as mushrooms. Because of the growth of hyphae into the substrate from which the fungus obtains its nutrients, a fungal colony is strongly adherent to that surface. When grown in culture on solid medium, fungal colonies display colors and textures that are characteristic of the species involved [12].

The Yeast Growth Form

The yeasts are an alternate growth form that occurs in some fungi that are essentially metabolically active spores (compared to the temporarily dormant state that is typical of most fungal spores). Many of the familiar yeasts are fairly small (4–6 μm), have round to elliptical shapes and have relatively thin cell walls. Most yeasts are dependent on outside forces for their dispersal beyond the site of a local population, and enter a metabolically inactive state until they are exposed to a wet or moist environment containing dissolved nutrients to support their metabolism [12, 33].

Population increase in most yeasts occurs by the formation of a blastospore, commonly described as "budding," which is essentially a smaller sister cell created when one of the nuclei formed by mitosis migrates into a ballooning portion of the cell wall until a separate cell is created, isolated from the original cell by a septum, as illustrated in Figure 46.6. However, there are other modes of reproduction in certain yeasts (such as the "medial fission" of fission yeasts described later in this chapter) [33, 34].

Rapid population growth of yeast cells in a favorable environment creates the characteristic macroscopic appearance of a yeast culture, which can resemble a bacterial colony on solid media. There are some fungi (described as dimorphic) that can transition between hyphal and yeast forms depending on their environmental conditions, such as *Histoplasma capsulatum* (causal agent of the human infection known as histoplasmosis), which is filamentous when growing saprotrophically at room temperature, but has a yeast growth form when growing necrotrophically inside the human body at 37°C [35].

Blastospores can also be formed directly on hyphae of certain fungi, which can create the macroscopic appearance of a yeast even though the organism is filamentous, and there are some yeasts that take on a hypha-like shape known as pseudohyphae (sometimes described as "pseudomycelium") when individual cells become unusually elongated or cling together in a linear arrangement, shown in Figure 46.7 [36].

Sexual and Asexual Reproductive Stages of Fungi

The traditional name for the sexual spore stage in the field of Mycology had originally been the "perfect stage," but this has been replaced by the term <u>teleomorph</u> ("end form"). Parallel with this terminology, the traditional name for the asexual spore stage had been the "imperfect stage," and this has been replaced by the term <u>anamorph,</u> which describes all structures (including hyphae) that are not involved in the sexual stage of the organism. Finally, the term <u>holomorph</u> describes the complete picture of

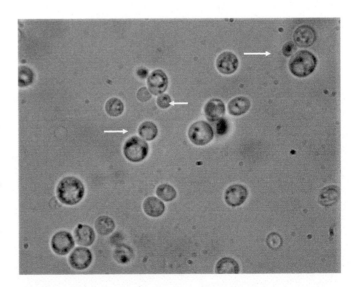

FIGURE 46.6 Blastospore formation ("budding") in *Saccharomyces cerevisiae*.

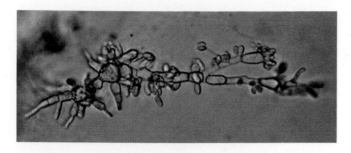

FIGURE 46.7 Pseudohyphae of *Aureobasidium pullulans.*

both the sexual and asexual stages of a fungus, so it is a combination of the teleomorph and the anamorph [9, 10].

Sexual Spore Stage

The sexual stage of a fungus life cycle consists of the events leading up to the fusion of two compatible haploid nuclei to form a diploid zygote and its subsequent division by meiosis to produce haploid nuclei typically packaged within spores. The fusion of the compatible haploid nuclei is more accurately known as karyogamy ("nuclear marriage") than "fertilization," since there are no equivalents to male or female gametes among fungi [37–39]. Some fungi employ elaborate methods for bringing together the two compatible haploid nuclei in this process, often having mating types that prevent karyogamy of nuclei within the same colony or between closely related strains. A fusion of fungal cells is therefore necessary to bring these two compatible haploid nuclei together, whether these are two hyphae, a hypha and a specialized spore, two spores, or other specialized structures, and this cell fusion is known as plasmogamy ("cell marriage"). Following meiosis of the diploid zygote (or comparable cell), the resulting haploid nuclei, or their mitotic products, are immediately or eventually packaged into characteristic spores that are associated with the sexual stage of the fungus. The events of plasmogamy, karyogamy and meiosis are so unique to each phylum of Mycota, and to each of the various taxonomic subgroups within each of those phyla, that they are a dependable basis on which to identify and classify these organisms [37–39].

Asexual Spore Stage

Those spores that are produced by mitosis from various hyphal structures are known as zoospores, sporangiospores, or conidia depending on their properties and the structure in which or on which they are formed. There is a tremendous variety of asexual spore morphology, and also in the modes of asexual spore formation and dispersal, all of which are utilized in recognizing fungal species. However, the asexual spore stage is much less reliable than the sexual spore stage for establishing the identity and taxonomic placement of an unknown fungus. Since their mitotic origin results in cloning, the reproductive function of zoospores, sporangiospores and conidia is asexual [40–42].

Practical Challenges Caused by the Teleomorph/ Anamorph Duality of Many Fungi

A recurring theme in the field of Mycology has been the difficulty in making the connection between the teleomorph and

the anamorph of a large number of fungi in several of the phyla of the Mycota, especially those in the Ascomycota. One of the reasons for this is that the sexual and asexual spore stages are typically quite dissimilar, and often produced at different times during the development of a fungus in its natural habitat. Also, the successful initiation and outcome of the sexual spore stage in many fungi requires the interaction of two compatible mating types which may not always be present in the same location at the same time, so the teleomorph may occur rarely or never. The mitotically produced spores associated with the anamorph of most fungi function effectively as a means of asexual reproduction, and they can also function as an effective means of initiating host infections by necrotrophic and biotrophic fungi of clinical or agricultural importance. Because of this characteristic of many fungi that have an impact on human affairs, these fungi have been historically recognized by their anamorph stage, have been given Latin binomials based on that stage, and are described in the scientific literature by those binomials [5, 9].

At the same time, the teleomorphs of many of these same fungi have been identified as unique species and given Latin binomials, but the connection with the corresponding anamorph may or may not have been recognized at the time they were first described. As a result, there are many fungi that have two Latin binomials, one for the anamorph and one for the teleomorph. Advances in mycological techniques, including both the availability of genomic data and methods of ultrastructural analysis, have resulted in the publication of increasing numbers of teleomorph binomials that are linked to their corresponding anamorph binomial [9].

While the publication of two different binomials for the same fungal species would appear to be cause for confusion, most scientists investigating these organisms in the context of agricultural, clinical, and industrial applications have recognized the basis for these binomials and have cited them appropriately in the scientific literature. However, there is a debate arising from the principle of "one fungus one name" that is consistent with the current interpretation of the International Code of Botanical Nomenclature, the recognized authority on scientific naming of fungi. While the use of both the teleomorph binomial and anamorph binomial was originally considered to be acceptable, the current standard being advocated is to utilize only the teleomorph binomial, even in cases when the anamorph binomial was published first. In practical terms, much of the scientific literature regarding clinical, agricultural, and industrial investigations of fungi has identified these organisms by the anamorph binomial, since that is the stage that is relevant to these fields. The exclusion of the anamorph binomial from future publications in these fields would appear to be a more significant source of confusion than following the earlier convention of identifying the organism with both binomials when appropriate [43–45].

Spore Structure and Germination

Whether produced sexually or asexually, each fungal spore is basically a package (or set of packages) of protoplasm containing at least one nucleus, and is typically capable of initiating the growth of the organism in a location or at a time that is separate from its point of origin.

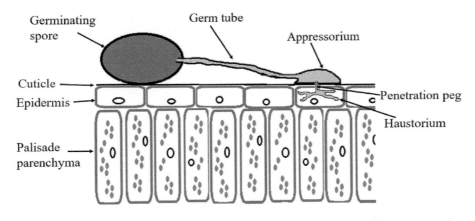

FIGURE 46.8   Germ tube with appressorium penetrating the host leaf epidermis.

Spore Germination

In most cases, the contents of the spore have a low level of metabolic activity until they are stimulated to germinate by conditions that are favorable for the growth of the organism that they can potentially develop into. Most fungal spores produce at least one germ tube that emerges through the spore wall and that essentially functions as the first hypha of the organism. The fungus at the time of spore germination is highly vulnerable to conditions that are unfavorable for its continued growth and development, and in the case of plant pathogenic fungi this is the stage that is most susceptible to disease control measures such as fungicide application. For a germ tube of a plant pathogenic fungus to be successful in initiating an infection, it must penetrate the surface of the plant, which in a few groups is accomplished by growing into a stoma (gas exchange pore on the surface of leaves and young stems). However, in most cases the penetration involves the expansion of the hyphal tip into an appressorium that adheres to the plant cuticle, while a thin infection peg pushes through the cell wall with a combination of mechanical pressure and enzymatic action, as shown in Figure 46.8. Once inside the plant cell or tissue, the hypha expands to its typical diameter and begins the internal invasion of the plant by whatever mechanism is characteristic for that kind of fungus [46–49].

There are some exceptions to the above description of spore production and function, including the blastospores ("buds") produced by yeasts as described above that are metabolically active at the time of formation, the behavior of flagellated zoospores produced by several small phyla of Lower Fungi with Zoospores described below, and certain specialized spores that only function in bringing compatible mating types together, as in the case of "spermatia" produced at one point in the life cycle of the Pucciniales (the rust fungi), also described in more detail within the Higher Fungi later in this chapter [34, 50].

Lower Fungi with Zoospores

The production of zoospores by the four phyla in this category reflects the aquatic evolutionary origins of the fungus kingdom. Although they have relatively less impact on human affairs than the fungi in the next sections of this chapter, the activities of some members of the zoospore-producing fungi are associated with economic losses as well as ecological damage [51].

Phylum Chytridiomycota

The fungi in this phylum are primarily aquatic, occurring in fresh water, salt water, and moist soil. Many of the species are saprotrophic, functioning as decomposers in the ecosystem. However, there are several biotrophic species (obligate parasites) that can cause economic loss, and there is also a significant necrotrophic pathogen of amphibians in this phylum [51].

Characteristics of the Phylum

Hyphae in the Chytridiomycota (known as "chytrids") are non-septate when present, but some members of the phylum absorb nutrients directly into a thallus consisting only of a developing zoosporangium, and some absorb nutrients with rhizoids (thin, branching extensions that do not contain nuclei). A typical chytrid thallus with rhizoids is shown in Figure 46.9. Cleavage of multinucleate protoplasm inside a zoosporangium results in the formation of zoospores, and these zoospores may escape through an opened operculum (lid) or through pores or slits in the zoosporangium. Zoospores display chemotactic behavior that favors their migration toward a suitable substrate, and once contact is made, they form a single-celled cyst (with or without retraction of the flagellum), which develops a chitinous cell wall and then germinates with a germ tube that penetrates the substrate [6, 51].

In those Chytridiomycota in which the sexual stage has been observed, there are several methods by which compatible haploid nuclei are brought together, including fusion of two zoospores,

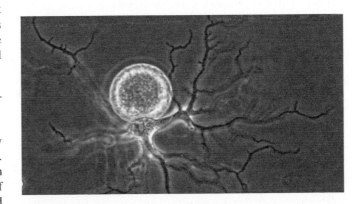

FIGURE 46.9   Chytrid thallus with rhizoids.

fusion of a zoospore with a nonmotile cell, and fusion of rhizoids of compatible strains. The zygote often functions as a survival structure during periods when the nutrient source is unavailable or conditions are unfavorable for growth. Meiosis of the zygote is typically followed by the development of haploid zoospores within a zoosporangium [51].

Class Chytridiomycetes

This class within the phylum Chytridiomycota is notable because of agricultural losses caused by some of its species and because one genus in this class has had a highly negative ecological impact in recent years.

One of the genera within this class is *Synchytrium*, a biotrophic parasite of flowering plants, ferns, mosses, and algae, and the species *Synchytrium endobioticum* is a well-known plant pathogen in this genus. This species infects epidermal cells of developing potato tubers, stimulating them to overgrow, resulting in the symptoms of potato scab or potato wart. When the fungus produces its zygotes toward the end of the growing season, additional disruptive host cell divisions occur, leading to decay of the potato tuber [52].

Also included in the class Chytridiomycetes is the genus *Batrachochytrium* causing an infection known as chytridiomycosis in amphibians that has come to be recognized as a significant cause of population decline of many species in this class of animals. This fungus is a necrotrophic pathogen, capable of releasing enzymes that digest the epidermis of various frogs and salamanders, leading to the death of the infected animal. The zoosporangia of the fungus develop in the invaded epidermis, leading to release of additional generations of zoospores that initiate infections of increasing numbers of amphibians. This chytridiomycosis of amphibians has resulted in the extinction of several species, although some species are naturally resistant (including bullfrogs, that appear to be help disperse the zoospores of the parasite through their activities) [6, 53, 54].

Phylum Blastocladiomycota

An economically important species within this phylum is the plant pathogen *Physoderma maydis*, which causes brown spot and stalk rot of maize; other species in this genus cause similar diseases of other crop plants. Plant disease caused by these pathogens is more severe during periods of high rainfall during the growing season, since free water favors the dispersal of their zoospores to above-ground portions of the plant [6, 55].

Phylum Neocallimastigomycota

The organisms within this phylum are unusual among fungi in that they are anaerobic, lacking mitochondria. These organisms are part of the microbiome of the digestive tracts of herbivorous animals that consume a high-fiber diet. The polysaccharide-degrading enzymes released by these fungi function in hydrolyzing resistant plant polymers, making the resulting carbohydrates available for the nutrition of their host animal [6, 56, 57].

Phylum Olpidiomycota

An economically important member of this phylum is the genus *Olpidium*, which infects a number of plants, animals, protists, and other fungi. Plant pathogens in this genus initiate an infection by means of zoospores that are released from zoosporangia on roots of infected plants and that migrate to roots of neighboring plants. Although the infection symptoms caused by *Olpidium* are relatively minor, it is the migration of its zoospores from one plant to another that can cause economic loss in crop plants. If a plant is suffering a virus infection in addition to an *Olpidium* infection, the migration of *Olpidium* zoospores from the double-infected plant can serve as a virus vector, introducing the virus into a neighboring healthy plant as the zoospores encyst, germinate, and penetrate the roots of that plant with their germ tubes [6, 58, 59].

Lower Fungi Without Zoospores

This informal collection of phyla includes fungi that may be entirely microscopic, or if they have visible mycelium it is never organized into complex structures like mushrooms. The hyphae are nonseptate when present, and the structures associated with their sexual stages (if known) are microscopic [7].

Phylum Glomeromycota

The Glomeromycota are a unique phylum of fungi with a relatively small number of species, but this phylum is vast in terms of its distribution and presence among nearly all plant populations of the world. These are the arbuscular mycorrhizal fungi that have a mutualistic biotrophic relationship with most photosynthetic organisms growing on land—primarily members of the Angiosperms (flowering plants), but also with the Bryophyta (mosses and their relatives), and even some Cyanobacteria, in which case the Cyanobacteria are inside the fungus. The symbiotic relationship with land plants began at least 400 million years ago, as observed in fossil evidence, and appears to have been a major factor that contributed to the survival and evolution of plants on land into the highly successful kingdom that now supports much of the animal life on the planet [7, 24, 26].

The Glomeromycota reproduce asexually with multinucleate spores that can be as large as 0.5 mm in diameter. The asexual spores in this phylum may be formed individually by expansion of the hyphal tip followed by cutting off of the spore or, in some cases, by cleavage of multinucleate protoplasm to form several spores within a type of sporangium. No sexual stage has been identified in this phylum [7, 25].

Spore germination in this phylum may occur when conditions are favorable, but is often stimulated by root exudates of compatible plant species. A germ tube develops that penetrates through the root epidermis into the root cortex (the outer layer of tissue that surrounds the central vascular cylinder). As the hyphae grow into this tissue, they penetrate the walls of living host cells and grow and branch inside each cell space but outside the host cell plasma membrane, developing into an arbuscule (named for the treelike appearance of their branching arrangement), as shown in Figure 46.10 [7, 26].

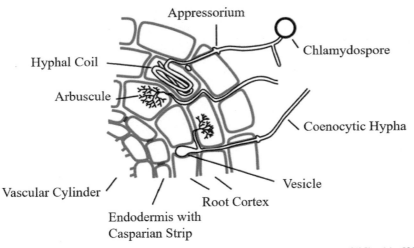

FIGURE 46.10   Glomeromycota infection on a host root with development of arbuscules.

At the same time, the hyphae of this fungus grow into the soil surrounding the root, creating a large absorbing surface area that functions in taking up various inorganic ions, including phosphate and nitrate ions, and transporting these into the root-invading hyphae that connect with the arbuscules. This intimate contact between fungus plasma membrane and host plasma membrane facilitates the transfer of carbohydrate from the plant to the fungus and the transfer of essential nutrients (inorganic ions) from the fungus to the plant [7, 27, 28].

Phylum Mucoromycota

This is one of three phyla of fungi that had previously been included in a larger phylum known as the Zygomycota. That identity is now considered invalid because it had included three similar-looking groups of fungi that are now recognized as a three separate phyla based on genomic and ultrastructural data [7, 8].

Characteristics of the Phylum

Many of the members of the Mucoromycota develop into macroscopically visible colonies and conspicuous spore-producing structures, so these are more familiar than the preceding phyla of fungi described in this chapter.

In some but not all members of the phylum, colony expansion may be augmented by the formation of long horizontal aerial hyphae known as "stolons" that arch over the surface until they make contact with the substrate. In those species producing such stolons, the hypha is anchored to the substrate at the point of contact by sprouting rhizoids (thin, branching extensions that do not contain nuclei) that penetrate the surface and function in the saprotrophic nutrition of the fungus. The macroscopically visible growth of mycelium in this phylum has a woolly appearance because of the profuse development of aerial hyphae [7].

Sexual Spore Stage

The characteristic structure associated with the sexual stage of fungi in this phylum is the zygospore. The development of the zygospore begins with the fusion of compatible haploid nuclei that are brought together by modified hyphal branches, known as zygophores, that arise from hyphae of compatible mating types of the organism, illustrated in Figure 46.11. The zygophores in most of the genera in this phylum are aerial hyphae that

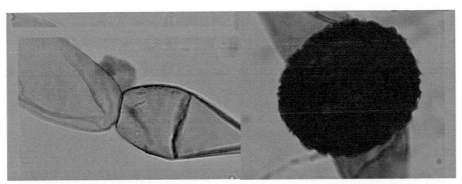

Zygophores making contact and   Mature zygospore
gametangium development

FIGURE 46.11   *Rhizopus oryzae*: Representative images of the sexual stage.

release pheromones into the air that attract the growth of the corresponding structure of the opposite mating type. Once the two zygophores make contact, they flatten and adhere in the region of contact, then each one becomes enlarged by expansion of its surrounding cell wall. A septum cuts off the terminal portion of each zygophore, resulting in the creation of two adhering cells that contain one or more of the haploid nuclei of that mating type, known as gametangia (literally "gamete vessels"). The cell wall substance in the region of contact between these two gametangia becomes hydrolyzed, resulting in the plasmogamy of the two gametangia, representing the next stage of zygospore development. That portion of the zygophore below each gametangium is known as a suspensor, and each one continues to maintain a functional connection to the gametangia and to the developing zygospore, enabling the transfer of metabolites to these structures from the supporting hyphae of the "parent" organisms [7, 39].

A notable feature of zygospore development in this phylum is the formation of a thick wall with conspicuous warty projections surrounding the nucleus or nuclei that are contained inside, and this protective wall has been described as a zygosporangium with a zygospore inside. There are dense deposits of melanin in this thick zygosporangium wall, making the contents invisible to both light microscopy and electron microscopy, but cryofracturing of the zygosporangium wall has allowed the visualization of its contents in some studies [7, 39].

The zygospore is the site of karyogamy and meiosis, leading to the formation of haploid nuclei. There is some variation among different members of this phylum in terms of the timing of these events and whether all potential gamete nuclei participate in the process, or if some remain haploid (therefore remaining part of the population of haploid nuclei prior to germination of the zygosporangium). As would be expected for this type of structure, the zygospore functions as a survival structure, remaining dormant until it is stimulated to germinate by conditions favorable for the growth of the fungus. Zygosporangium germination typically results in the development of a sporangium containing sporangiospores, although, in some cases, the germ tube from the zygosporangium just continues growth as a new hypha. The sporangium that develops following the germination of the zygosporangium is indistinguishable from a sporangium of the type that the same organism produces in the process of asexual reproduction, so the characteristics of these sporangia are described in the following section dealing with asexual reproduction in this phylum [7, 39].

Asexual Spore Stage

As in the case of sporangia in other phyla described previously, sporangiospores in the Mucoromycota are typically created by cleavage of multinucleate protoplasm within a sporangium. Most sporangia range in size from 40 μm to 350 μm in diameter, and the sporangiospores can be from 4 μm to 11 μm in diameter with globose or angular shapes. The wall of the sporangium becomes fragile and easily disrupted at maturity in most species, allowing the sporangiospores to be released, as shown in Figure 46.12. Depending on the species, the sporangiospores may be dry and released into the surrounding air, or they may be slimy or become suspended in a spore droplet. Following the disintegration of a sporangium wall, the basal portion may persist in the form of a "collarette." Some sporangia with air-borne spores are produced on relatively tall sporangiophores (as much as several centimeters in some species) that raise the released spores above the level of surface laminar air flow and into a more turbulent zone, increasing the probability of wide dispersal of the sporangiospores [7, 8].

The sporangiophore terminates in a dome-shaped or globose columella that extends into the sporangium and persists after the sporangium wall disintegrates. In some species, the sporangiophore is expanded in the region below the columella, a feature known as an apophysis. The presence and appearance of the columella, apophysis, and the collarette described previously can be useful identifying features of some species [7, 8].

The sporangiophores in some members of the phylum are branched, producing a sporangium on each branch as in *Mucor plumbeus* shown in Figure 46.13, while others are unbranched. There is also wide variation in the arrangement and number of sporangiospores within each sporangium, including some (known as sporangioles) that contain only one sporangiospore or a few sporangiospores, and some (known as merosporangia) that contain 5–10 merosporangiospores in a row [7, 8].

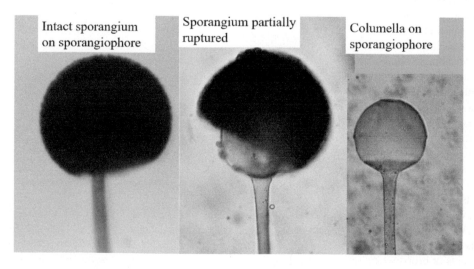

Intact sporangium on sporangiophore | Sporangium partially ruptured | Columella on sporangiophore

FIGURE 46.12 *Rhizopus oryzae*: Representative images of the asexual stage.

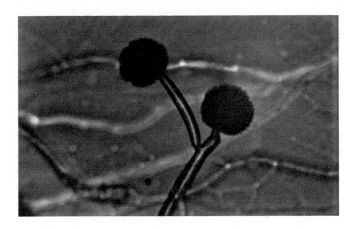

FIGURE 46.13 *Mucor plumbeus*: Sporangia borne on branching sporangiophores.

Impact of Mucoromycota on Human Affairs

The members of this phylum are all saprotrophic, but several also function as necrotrophs when susceptible host tissue is available.

Mucormycosis in Humans

Several species of Mucoromycota are associated with mucormycosis of humans (also known in earlier literature as zygomycosis because of the previous use of "Zygomycota" as the name for this phylum). The genera most commonly associated with human infections are *Rhizopus* and *Mucor*, with *Rhizopus oryzae* and *Rhizopus microsporus* being the two most common causal agents, but there are species in several other genera that may also be involved [60].

The spores of these organisms are common in the environment, especially in soil, but most people are resistant to mucormycotic infections. These infections are mostly limited to individuals with uncontrolled diabetes or certain cancers, those who are taking immune-suppressing treatments following organ transplant, those with immune deficiency diseases, those taking broad-spectrum antibiotics or corticosteroids for an extended period of time, and in those suffering from severe burns and other injuries that become contaminated with the fungal spores [60, 61].

Types of mucormycosis that occur include rhinocerebral (moving from infected sinuses into the brain), pulmonary, gastrointestinal, cutaneous (developing after skin trauma in otherwise healthy individuals), and disseminated when spread through the bloodstream. Available antifungal treatments, such as amphotericin B and the azoles (which target the ergosterol component of the fungal plasma membrane as described previously) have very limited effectiveness, so the mortality rate from these infections is approximately 50% [61].

Mucormycosis in Domesticated Animals

Mucormycosis has also been reported in domesticated animals including cats, dogs, cows, and pigs, commonly causing pneumonia and enteritis. As in the case of humans, healthy animals are generally not susceptible, while those with extended treatment with corticosteroids or broad-spectrum antibiotics, or with certain virus diseases, are more susceptible [62].

Economic Loss and Plant Disease Caused by Mucoromycota

Familiar examples of economic loss caused by the activity of species in this phylum include "black bread mold" growth on baked goods (saprotrophic) and the damage to harvested fruits sometimes observed on overripe or bruised produce. Various species of both *Rhizopus* and *Mucor* have been reported as the causal agents of postharvest "soft rot" of a wide range of fruits and certain vegetables, such as sweet potatoes and white potatoes. "Head rot"—when flower heads are destroyed before they are ready for harvesting—is a significant loss in sunflower cultivation caused by several species of *Rhizopus* [63, 64].

Economically Beneficial Mucoromycota

There are beneficial uses of fungi in this phylum, including the involvement of *Rhizopus oligosporus* and *Rhizopus oryzae* in the aerobic fermentation of the Indonesian soy product tempeh. Besides creating a desirable texture and flavor in this product, the enzymatic action of the fungal growth hydrolyzes the oligosaccharides that are associated with undesirable side effects of eating soybeans, and also reduces the level of phytic acid that tends to interfere with absorption of phosphorus and inositol by the human digestive system [65–67].

Rhizopus oryzae is also widely utilized in the field of industrial microbiology as a source of enzymes, organic acids, and ethanol that are useful in the pharmaceutical and food processing industries and as a source of industrial raw material [68].

Phylum Zoopagomycota

This is the second of the phyla of fungi that had previously been considered a member of the now-invalid phylum Zygomycota and consists of microscopic fungi without any clinical or economic impact on human affairs. All of its members are parasitic, with most of them obtaining their nutrition biotrophically. Their hosts include soil nematodes, rotifers, protozoa (such as amoebae), and some Mucoromycota [69].

Phylum Entomophthoromycota

This is the third of the phyla of fungi that had previously been considered a member of the now-invalid phylum Zygomycota, and—as reflected in its name ("insect destroyer")—is a group known for parasitizing insects. This phylum also includes some fungi with clinical significance as parasites of humans and some domesticated animals [70].

Although it has nonseptate hyphae and zygospores that look similar to those of the Mucoromycota, its asexual spores are quite different, consisting of sporangia that are forcibly discharged from the sporangiophore on which they are formed. According to some reports, the mechanism involves the formation of a gas bubble at the point of contact of sporangium and sporangiophore, contained within the outer cell wall that surrounds both structures, leading to sufficient pressure to break the outer cell wall and to propel the sporangium several millimeters away from the sporangiophore. In most cases, the sticky sporangium surface favors contact with the exoskeleton of a susceptible host insect. Germination of a sporangium can lead to penetration of the insect exoskeleton and invasion of the insect hemocoel (inner fluid-filled body cavity). The resulting internal infection often includes production of additional sporangiophores growing out of the body of the insect while it is still alive, favoring the dissemination of sporangia to other potential hosts [70].

Most of the other species within this phylum are saprotrophic, utilizing dead plant material as a food source, or in some cases living in the intestinal tract of reptiles, amphibians, and some fish. However, *Basidiobolus ranarum* is a species that can parasitize human and other animal tissues as a necrotrophic pathogen, causing a granuloma [71]. Most human infections are associated with tropical climates, while reported animal hosts include horses, dogs, and sheep. The closely related *Conidiobolus coronatus* is associated with rhinofacial "zygomycosis" (another case of naming according to the original name of the phylum in which it had been classified). This infection begins in the nasal passages and leads to destructive subcutaneous involvement. As in the case of *Basidiobolus*, most human infections occur in the tropics, and this species causes similar infections in some domesticated animals [62].

Phylum Microsporidia

The atypical name of this fungal phylum reflects the historically uncertain classification of these organisms, which had been considered to be protozoa of one type or another prior to their recognition as fungi. These organisms are essentially single-celled intracellular biotrophs that lack mitochondria, having an anaerobic metabolism. They are capable of forming resistant spores that have a chitinous layer, a feature demonstrating that they are fungi. Parasitism of a host by these organisms is generally initiated by the germinating spore by means of a penetration apparatus that allows the organism to pass through the intestinal lining to initiate an infection, illustrated in Figure 46.14. Many of the species in this phylum cause a chronic infection of insects that disrupts the host physiology. This phylum is of particular clinical importance because some species can infect the intestinal lining of immunocompromised humans, causing a chronic wasting condition [72, 73].

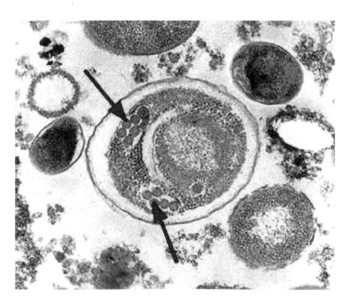

FIGURE 46.14 A mature spore of a member of the Microsporidia. Arrows indicate the double rows of polar tubule coils in a cross-section, which characterize a mature *E. bieneusi* spore.

The Higher Fungi

Both phyla in this category, the Ascomycota and the Basidiomycota, have septate hyphae. Almost all holomorphs in these two phyla exist as a dikaryon for at least a portion of their life cycles, this being the state in which two compatible haploid nuclei coexist within the same cell (defined in this case as the segment of a hypha between two septa) without fusing to become a zygote. To create this dikaryotic hypha, there must be plasmogamy without karyogamy, resulting in two haploid nuclei sharing the same body of cytoplasm and potentially providing the same genetic function as a diploid nucleus. Because of this shared nuclear condition, which is not found in any other phyla of fungi, these two phyla are described by some authors as being within the subkingdom Dikarya [9, 10].

Phylum Ascomycota

This is an extremely large phylum of fungi that includes saprotrophic, necrotrophic, and biotrophic members, among which are a number of adaptations for these modes of nutrition. Most of the organisms in this phylum occur as septate hyphae containing one haploid nucleus per cell. There are also a number of yeasts in this phylum, and in some of these yeasts the nuclear condition in nonreproductive cells may be diploid. As described above, there is a small round pore in the middle of each septum of most filamentous Ascomycota and there is a Woronin body on either side of the pore that functions in plugging the pore if a hypha is damaged. Many species in this phylum produce macroscopic structures that house the phases of the sexual stage, some of which are comparable to mushrooms in their size, complexity, and potential edibility by humans. There is also a wide assortment of conidial shapes and means of production of the asexual spores of these fungi, with some species being known only by their anamorph because of the rarity or absence of the sexual stage [74].

Sexual Spore Stage

Although there are some variations among different classes within this phylum, the typical sexual spore stage in the Ascomycota results in the formation of eight ascospores inside an ascus (plural asci). The process begins with plasmogamy achieved by specialized hyphal structures known as gametangia, identified as such because they contain haploid nuclei of compatible mating types that could potentially function as gametes in karyogamy. Two kinds of gametangia are formed: the ascogonium, a receiver of gamete nuclei, and the antheridium, a donator of gamete nuclei. The process of plasmogamy begins in many Ascomycota by the expansion of the ascogonium cell wall on the surface facing the antheridium, creating an extension of this cell into a thread like (or hair like) tube known as the trichogyne. This structure grows toward and penetrates the antheridium, after which haploid nuclei migrate from the antheridium into the ascogonium through the trichogyne. In some Ascomycota, a conidium (asexually-produced spore) can take the place of an antheridium as the source of compatible haploid nuclei that migrate into the ascogonium [9, 37].

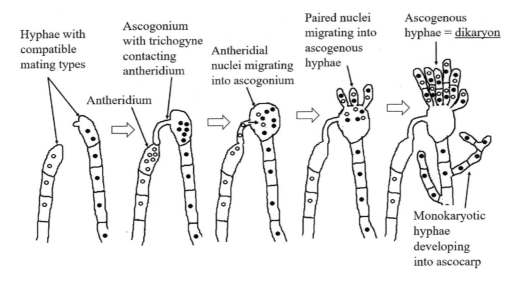

FIGURE 46.15 The Ascomycota sexual stage from ascogonium/antheridium to ascogenous hyphae.

Once the haploid nuclei from the antheridium are inside the ascogonium, the haploid nuclei of compatible mating types pair up, but do not fuse, thus, no zygote is formed yet in this process. These pairs of nuclei divide synchronously by mitosis, resulting in many pairs of compatible haploid nuclei within the ascogonium. Numerous narrow septate hyphae containing these nuclei then grow out of the ascogonium with a pair of compatible nuclei contained within each cell; these nuclei divide by mitosis as each new cell is formed, resulting in dikaryotic <u>ascogenous</u> hyphae. Figure 46.15 illustrates the stages that occur up to this point. The formation of these dikaryotic hyphae is the basis for the inclusion of the Ascomycota in the subkingdom Dikarya [9, 37].

The terminal portion of each of these narrow hyphae bends back on itself forming what is known as a crozier. The two nuclei located within this crozier divide by mitosis synchronously with parallel division figures, resulting in one set of daughter nuclei arriving at the bend of the crozier and the other set of daughter nuclei located away from the bend. Two septa typically form across the crozier, one cutting off the "apical cell" containing one haploid nucleus and the other cutting off the basal "stalk cell" containing the other haploid nucleus, leaving the pair of compatible haploid nuclei alone in the "penultimate cell," as shown in Figure 46.16 [9, 37].

It is at this point when this pair of haploid nuclei finally forms a diploid zygote by the process of karyogamy. Meiosis of this zygote occurs immediately, resulting in four haploid nuclei, and in nearly all Ascomycota each of these haploid nuclei divides by mitosis, resulting in the formation of eight nuclei inside of what is now the ascus. Cleavage of the cytoplasm around each nucleus creates eight ascospores inside the expanding ascus, a feature that occurs in the sexual stage of nearly every member of the Ascomycota (the only exceptions being some that form just four ascospores out of the products of meiosis, as in the case of the yeasts, or that have different structural arrangements leading to larger numbers of ascospores) [9, 37].

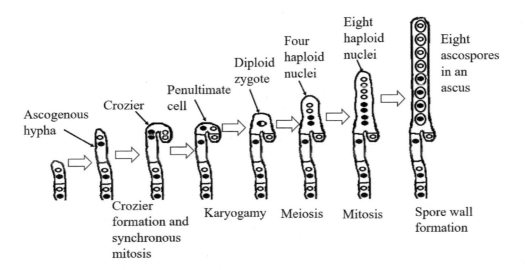

FIGURE 46.16 The Ascomycota sexual stage from crozier to ascus.

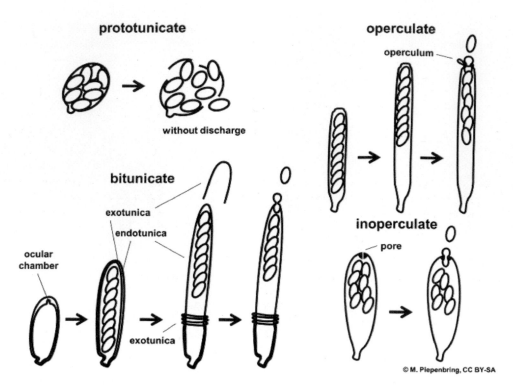

FIGURE 46.17 Types of ascus morphology.

Variation in Ascus Morphology

In many of the Ascomycota, the ascospores are ejected through the tip of the ascus by the sudden release of osmotic pressure that has built up because of a high concentration of glycerol and other solutes within the ascus, although there are some asci that have more passive ascospore release mechanisms. The process of ascus formation in various classes of Ascomycota results in one of four distinctly different ascus structures related to the mode of ascospore release. Terms used to describe these forms include "tunicate" referring to the nature of the wall of the ascus and "operculate" referring to the nature of the opening (if any) of the ascus. As indicated by the labels in Figure 46.17, these four types of ascus structure are as follows:

a. Unitunicate-operculate asci: Elongated asci having a single functional wall and a pre-existing operculum at the tip that opens at a sufficient pressure to allow the ascospores to be ejected.

b. Unitunicate-inoperculate asci: Elongated asci having a single functional wall but no pre-existing operculum, but instead having a ring of cell-wall material at the tip that is stretched open at a sufficient pressure to allow the ascospores to be ejected.

c. Bitunicate asci: Elongated asci with a thin inelastic outer wall and a thick elastic inner wall. The inner wall absorbs water by imbibition while the ascus takes up water by osmosis, resulting in expansion of the ascus leading to rupture of the thin outer wall, and continued lengthwise expansion until the ascus extends above its surrounding structures and finally bursts at the tip to allow the ascospores to be ejected.

d. Prototunicate asci: More or less spherical asci that do not eject their ascospores under pressure, with a wall that may dissolve at maturity to allow ascospores to ooze out, or that may depend on outside forces to rupture the wall and disperse the ascospores. This type of ascus has evidently evolved independently in several different classes of Ascomycota from one of the three ascus categories described above; therefore, it represents a reduction of a previously more complex structure typical for each group [9, 75].

Variation in Ascocarp Morphology

In most of the Ascomycota, there is extensive growth of the surrounding "parent" monokaryotic hyphae (having only one haploid nucleus per cell) that originally gave rise to the ascogonium and the resulting asci of one of the above categories. This hyphal growth produces the supporting and more or less enclosing ascocarp ("ascus fruit") of various forms, and of which there are also four commonly observed categories (correlated with one or more of the four categories of asci described previously). The four commonly observed ascocarp categories are:

a. Apothecium: An open fruiting body in the form of a cup- or bowl-shaped supporting structure that forms under a developing layer of upward-pointing asci that are packed together on the exposed surface as shown in Figure 46.18. This fertile layer of asci is known as a "hymenium," and the species that produce this type of ascocarp are commonly known as cup fungi. Most of the asci in this layer mature simultaneously and release all their ascospores at once. This is the only category of ascocarp formed

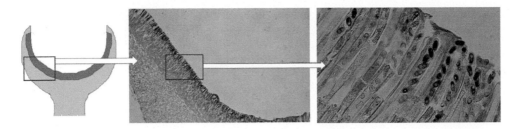

FIGURE 46.18 *Peziza cerea* apothecium diagram (left), magnified detail of hymenium (center) and asci with ascospores (right).

by species that have unitunicate-operculate asci, while some of the species that produce unitunicate-inoperculate asci also produce an apothecial ascocarp.

b. Perithecium: A flask-shaped fruiting body that develops <u>around</u> a developing group of asci, with a narrow aperture that allows only one or a few asci at a time to get through the opening to release their ascospores, shown in Figure 46.19. The only kind of Ascomycota that form this kind of ascocarp are those that produce unitunicate-inoperculate asci.

c. Pseudothecium: This as also a flask-shaped fruiting body that requires its asci to take their turns in releasing their ascospores. However, it differs from the true perithecium because its hyphal structure begins growing <u>before</u> the asci develop, so that the developing asci must find their way through pre-existing hyphae as they expand, rather than being surrounded as they develop (thus the reasoning for the "pseudo-" prefix for its name). Figure 46.19 also shows an example of this structure. Only those species that produce bitunicate asci form this type of ascocarp.

d. Cleistothecium: These fruiting bodies form <u>around</u> a developing group of asci, but they have no opening, as shown in Figure 46.19. In some groups, the asci inside are not adapted for ejecting their ascospores into the air and would meet the definition of prototunicate asci described previously; in others, the cleistothecium opens at a certain point, allowing the asci inside to eject their ascospores [9].

Asexual Spore Stage

The majority of species in the Ascomycota produce conidia as the asexual spore stage, and these are typically adapted for rapid and efficient dispersal of the organism to new substrates and sources of nutrients. After the conidia separate from the hypha of origin, they are capable of germinating with a germ tube in a different location or at a different time from their place and time of origin. In a few species the anamorphic stage may involve the formation of survival structures such as chlamydospores or sclerotia. Chlamydospores are a type of asexual spore with a greatly thickened wall, while sclerotia are dense masses of hyphae with an outer layer consisting of dead cells with thickened walls and an inner core of viable but dormant cells. Melanin is commonly deposited in the walls of both structures as an additional protective feature [76–78].

In several species of Ascomycota with highly successful asexual reproduction, the teleomorph is rarely encountered or is still unknown. In the case of these species, it is difficult or impossible to identify the corresponding holomorph, given the need for the teleomorph for a definitive answer. Many economically important fungi—both beneficial and harmful—are known only by their anamorph, therefor it has been necessary to construct an artificial classification system based largely on the mode of conidial formation or other anamorphic features. Because of the earlier tradition of referring to the asexual spore stage as the "imperfect stage," the fungi that are studied in this artificial classification system were known as the Fungi Imperfecti. A more authentic-sounding title for the same category of fungi is "Deuteromycetes," but this has no more validity than Fungi Imperfecti. [76, 78, 79].

Classification of Anamorphic Ascomycota Based on Organization of Conidiophores

An example of a classification system that is employed in the identification of anamorphic fungi that continues to be useful (even though it may not reflect the "true" holomorphic identity

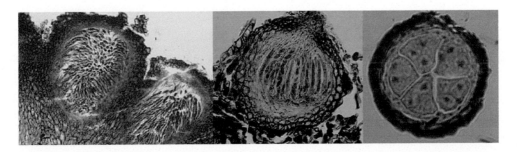

FIGURE 46.19 Comparisons of perithecium, pseudothecium, and cleistothecium structure. Left: *Nectria cinnabarina* perithecia showing ascospores in rows within asci. Middle: *Venturia inaequalis* pseudothecium showing rows of ascospores within asci. Right: *Erysiphe* cleistothecium containing several asci and ascospores in sectional view.

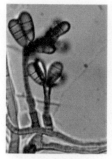

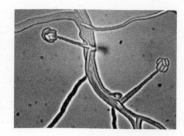

Helminthosporium solani: simple conidiophores

Fusarium roseum: developing sporodochium

Acremonium strictum: conidia in slime droplets on conidiophores prior to formation of synnemata

FIGURE 46.20 Representative modes of organization of conidiophores among Ascomycota hyphomycete anamorphs. Left: *Helminthosporium solani*, simple conidiophores. Middle: *Fusarium roseum*, developing sporodochium. Right: *Acremonium strictum*, conidia in slime droplets on conidiophores prior to formation of synnemata.

of the fungi that are organized within it) relates to the organization of the conidia-forming hyphae (conidiophores). This system is based on whether the fungus produces its conidia on hyphae that are uncovered and exposed to the air as they form conidia (in which case it is referred to as a hyphomycete) or if its conidia are produced in a complex more or less closed structure (referred to as a coelomycete) [9, 74, 76].

Those fungi commonly known as molds (in which individual exposed conidiophores are distributed throughout the growing colony) are familiar examples of hyphomycetes. However, this category also includes some fungi in which the conidiophores are clustered together into a tight bundle (a synnema) that produces a mass of conidia at the top (slimy and adapted for insect dispersal), and there are other fungi in which the conidiophores are packed together on top of a cushion of supporting hyphae (a sporodochium) adapted for rain-splashing of the conidia [9].

Coelomycetes include plant parasitic fungi that form closely packed conidiophores within host plant tissue (at the epidermal layer or deeper) in a disc-shaped structure known as an acervulus that has a superficial similarity to an apothecium; large numbers of conidia arising from these conidiophores create sufficient pressure to rupture the plant tissues, allowing the conidia to be dispersed (often by rain-splashing). There are other coelomycetes that form tightly-packed conidiophores within a flask-shaped

structure that has a superficial similarity to a perithecium but is known as a pycnidium; conidia arising from these conidiophores are extruded through the opening of the pycnidium in a slimy mass in wet weather, also adapted for rain-splashing of conidia. Representative hyphomycete organization of conidiophores is illustrated in Figure 46.20, and coelomycete organization of conidiophores is shown in Figure 46.21 [74, 78].

Classification of Anamorphic Ascomycota Based on Conidial Morphology and Details of Conidiogenesis

Most conidia are globose to elliptical in shape or, in some cases are short cylinders, and these are known as amerospores, while a two-celled conidium divided by one septum is a didymospore. A conidium consisting of several cells in a row separated by two or more transverse septa is a phragmospore, while a conidium in which there are both transverse and longitudinal septa creating multiple cells is a dictyospore. Variations in conidial shape include scolecospores, which are long and thin (with the length more than 15 times the width), staurospores that have radiating arms, and helicospores that are curved through more than a half-circle. Conidia are often hyaline, but some of the larger conidia with septations are dark because of melanin deposition in the wall. Some varieties of conidial morphology are shown in Figure 46.22 [9, 80].

In most of the Ascomycota with a hyphal growth form, conidia develop by one of two basic mechanisms known as thallic or

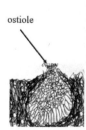
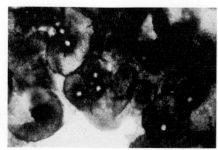

ostiole

FIGURE 46.21 Representative modes of organization of conidiophores among Ascomycota coelomycete anamorphs. Left: *Nectria cinnabarina*, acervulus erupting through three bark. Middle: A diagram of pycnidium indicating the ostiole. Right: *Phoma herbarum*, pycnidia whole mount from culture. Note the ostioles and conidial masses (orange in color).

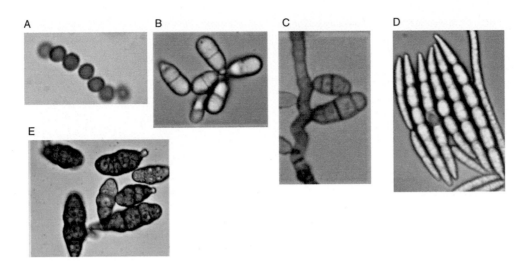

FIGURE 46.22 Types of conidial morphology. A: Amerospores of *Aspergillus niger.* B: Didymospores of *Trichothecium roseum.* C: Phragmospores of *Curvularia inaequalis.* D: Scolecospores of *Fusarium roseum.* E: Dictyospores of *Alternaria alternata.*

blastic. Thallic conidiogenesis occurs by the conversion of sections of hyphae that were located between adjacent septa (or at the end of a hypha) into the resulting conidia, with more or less modification of the shape of the original hyphal sections as they mature, and the resulting amerospores are likely to have a short cylindrical shape rather than an elliptical or globose shape. Blastic conidiogenesis occurs by budding of a conidium from its hypha of origin and is comparable to the blastospore formation described in the introductory section of this chapter in which the yeast growth form was explained. As the bud develops into the conidium, it takes on the shape that is characteristic for the conidia in that species as genetically determined regions of the cell wall balloon out [74, 76].

In most species of Ascomycota with blastic conidiogenesis, the hypha of origin is modified for conidial production and is recognized as a conidiophore. Both thallic conidia and blastic conidia are separated from the hypha of origin or from each other by the formation of either a double septum or by the formation of a separating cell between the conidium and its hypha of origin, followed by the breakdown of cell wall material to release the conidium. The separation of a conidium by wall breakdown at a double septum is known as schizolytic, while the release of a conidium by breakdown of a separating cell is known as rhexolytic [74, 78, 81, 82].

Compared with thallic conidiogenesis, there is much more variety in the modes of blastic conidiogenesis, depending on whether there is a chain of conidia or individual conidia, whether the newest conidia in a chain are at the base or at the tip of the chain, whether or not the hypha of origin is modified into a conidiophore, what shapes, septations, and colors occur in the conidia that are produced, in addition to the distinction between hyphomycetes and coelomycetes described previously [9, 74].

Several commonly observed modes of conidiogenesis in anamorphic Ascomycota are listed next and illustrated with examples in Figure 46.23.

a. Arthric: Thallic conidiogenesis in which the hyphal cells develop into conidia that separate schizolytically into a chain that appears jointed, thus the basis for the name.

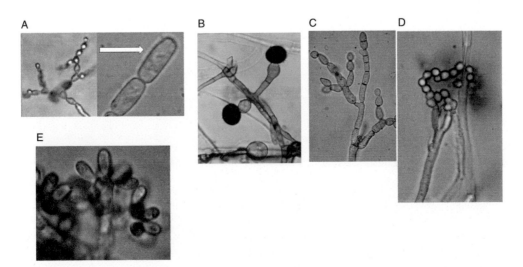

FIGURE 46.23 Modes of conidiogenesis among anamorphic Ascomycota. A: Arthric conidia of *Neurospora intermedia.* B: Solitary conidia of *Nigrospora sphaerica.* C: Acropetal conidia of *Cladosporium cladosporioides.* D: Phialidic conidia of *Penicillium notatum.* E: Synchronous conidia of *Botrytis cinerea.*

b. Solitary: Thallic conidiogenesis in which the terminal cell of the hypha becomes modified into a conidium, often becoming enlarged before it separates schizolytically.

c. Acropetal: Blastic conidiogenesis occurs at the tip of a hypha, and is followed by blastic conidiogenesis of an additional conidium arising from the first conidium, then from the second and third conidium, with the process repeating (with double conidia formed on some of the conidia, resulting in branching) until there is a branching chain of conidia, with the oldest at the base and the youngest at the tips of the branches.

d. Phialidic: Blastic conidiogenesis occurs from within an elongated flask-shaped cell at the tip of the conidiophore, with each new conidium pushing the previous one farther away from the phialide as it emerges. In some fungi, each conidium may cling to the previous conidium so that a chain develops with the oldest conidium at the tip and the youngest at the base, while in others, the conidia may accumulate around the phialide. Since new cell-wall material is laid down around each conidium as it is budded from the phialide, this may result in an internal thickening of the phialide until it no longer functions, at which time a new phialide develops next to the original phialide.

e. Synchronous: Blastic conidiogenesis occurs at several points over the surface of an enlarged terminal cell of the conidiophore, resulting in simultaneous formation of many conidia [9, 74, 78].

Classes and Representative Orders of Ascomycota of Importance in Human Affairs

The holomorphs of many of the Ascomycota have been identified and their relationships have been analyzed on the basis of genomic and ultrastructural evidence, leading to the recognition of three subphyla: the Taphrinomycotina, the Saccharomycotina, and the Pezizomycotina [3, 9, 74].

Taphrinomycotina

The members of this subphylum do not produce ascogenous hyphae or ascocarps and occur as a yeast form for part or all of their life cycle. This subphylum is distinct from the "true" yeast subphylum Saccharomycotina that includes the more familiar species utilized as leavening agents and brewing and is apparently descended from the ancestors of the entire phylum. The three classes of significance in this subphylum are the Schizosaccharomycetes, Taphrinomycetes, and Pneumocystidomycetes [9].

Schizosaccharomycetes

This is the class of "fission yeasts" in which the organism consists of rod-shaped cells 7–14 μm in length that elongate at the cell tips until they divide by mitosis at their medial region to produce two daughter cells of equal size (a process known as "medial fission," which is quite different from the budding process in other yeast-forming fungi). Typical cells of this group are shown in Figure 46.24. The economic significance of the genus *Schizosaccharomyces* is that of a fermenting organism utilized in the production of some alcoholic beverages, but the species *Schizosaccharomyces pombe* is also of importance as a model organism that has been the subject of extensive research into cell cycle regulation and genetics, and its genome was fully sequenced in 2002 [9, 83].

Taphrinomycetes

This class includes the one order Taphrinales and is unusual among Ascomycota in that the growing hyphae are diploid rather than haploid or dikaryotic. Although these organisms form ascogenous hyphae, the asci are produced at the surface of the tissues of a host organism without any supporting ascocarp, and the mature asci release their ascospores by splitting open at the tip rather than having an operculum or expanding ring of cell wall material, as illustrated in Figure 46.24. The ascospores bud to form yeast cells that form additional cells by budding, representing the yeast stage in this order. The best-known species in this class is *Taphrina deformans*, the causal organism of a leaf curl disorder in peaches and almonds. This disease can cause significant damage to the leaves of infected plants and the loss

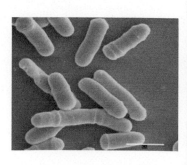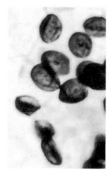

FIGURE 46.24 Three classes within the Taphrinomycotina. Left: Schizosaccharomycetes, *Schizosaccharomyces pombe*. Middle: Taphrinomycetes, *Taphrina deformans*. Right: Pneumocystidomycetes, *Pneumocystis jirovecii*.

of developing fruit. The various species in this class are fairly host-specific and cause abnormal developmental symptoms in the plants that are infected [9, 84].

Pneumocystidomycetes

This class includes the one order Pneumocystidales that contains the one family *Pneumocystidaceae* that contains the one genus *Pneumocystis* in which the species now known as *Pneumocystis jirovecii* is the causal organism of pneumocystis pneumonia. This organism was previously considered to be a protozoan with the name *Pneumocystis carinii*, but it was subsequently recognized as a member of the Ascomycota, and as such the species name was officially changed to *P. jirovecii*. However, it should be noted that the well-established name *P. carinii* has been retained in much of the clinical literature on this disease. This organism appears to be fairly ubiquitous and is readily cleared by a competent immune system if there is an exposure to it, but it has been recognized in recent years as a biotrophic parasite, causing disease in immunocompromised individuals. The organism occurs as an extracellular parasite in lung tissue, as shown in Figure 46.24, but it has not yet been grown in culture. An increase in the number of parasitic cells by mitosis during infection is comparable to the asexual reproduction of other Ascomycota, and the formation of globular cysts with a thick cell wall, followed by their rupture to release eight spores, is seen as evidence that this is a type of ascus in which karyogamy and meiosis take place. Until the development of therapies to control this infection, the disease caused by this organism contributed to the death of many individuals infected by the human immunodeficiency virus [85–87].

Saccharomycotina

These are the true yeasts that reproduce asexually by budding, and that produce neither ascogenous hyphae nor an ascocarp. Their growth forms include septate hyphae, pseudohyphae, and typically budding yeast cells. The hyphal septa in this subphylum differ from those of other Ascomycota in that they are perforated by many micropores, and there are no Woronin bodies associated with them. This subphylum includes the one class Saccharomycetes that contains the one order Saccharomycetales. Of the 13 families described in this order, the *Saccharomycetaceae* is most familiar in that it includes both the well-known "baker's/brewer's yeast" genus *Saccharomyces* and the diverse genus *Candida* that includes species of clinical importance.

The Genus Saccharomyces

There is considerable variation in the chromosome number in *Saccharomyces* populations, including the common occurrence of metabolically active diploid cells arising through the plasmogamy ("mating") of compatible haploid cells and of aneuploid and polyploid forms. These cells reproduce asexually by budding, but in *Saccharomyces cerevisiae* certain conditions may trigger the meiosis of the diploid nucleus of a cell, resulting in the formation of four ascospores in an ascus derived from the mother cell in which meiosis took place. Such ascospores typically have a thick resistant wall, and function in survival rather than in dispersal of the organism [15, 88, 89].

The commercial and biopharmaceutical utilization of *Saccharomyces* species is based on technology developed in the production of beer and wine. *Saccharomyces cerevisiae* is utilized in the production of ale, while *S. carlsbergensis* is involved in the production of lager. *S. cerevisiae* var. *ellipsoideus* is the most common wild yeast from grape skins that is associated with the production of grape wines. With the development of techniques for genetic recombination, *S. cerevisiae* has been successfully utilized in the production of enzymes, biomass, and a number of pharmaceutical products such as hepatitis B subunit vaccine, Leukine, platelet derived growth factor, and Gardasil [88, 90–93].

Although the teleomorph of *Candida albicans* is *Saccharomycopsis*, this species and other pathogenic *Candida* species are primarily known by their anamorph name and morphology in clinical practice. *Candida albicans* is normally a harmless part of the intestinal microbiome, but under certain conditions it can infect the mucous membranes (causing thrush or vaginal candidiasis) and skin (cutaneous candidiasis). In individuals who are immunocompromised or otherwise compromised by metabolic disorders or long-term utilization of catheters or intravenous delivery of nutrients, candidiasis can extend throughout the alimentary canal or become disseminated as a septicemia. *Candida auris* (named for its original isolation from the ear) has been recognized in recent years as the causal agent in increasing numbers of systemic infections, complicated by its resistance to most antifungal treatments [94–97].

Candida species display dimorphism by switching from a yeast form to a filamentous form (pseudohyphae) at 37°C, neutral pH, elevated CO_2 levels, and serum presence, then switching back to the yeast form at lower temperatures, acidic pH, and rich nutritional conditions. Both forms occur in culture as shown in Figure 46.25. This property evidently enhances the ability of pathogenic species to pass through tissue barriers until they reach circulatory pathways leading to systemic invasion [95, 97].

Pezizomycotina

This is the largest of the subphyla of Ascomycota, and the teleomorph (when present) forms its asci within some type of ascocarp. The holomorphic species are classified into six classes, 18 orders, and a great number of families and genera. In addition to the known holomorphs, there remain many anamorphs that

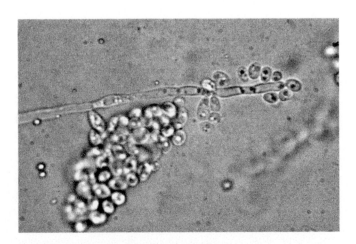

FIGURE 46.25 Candida albicans growth form.

can only be indirectly correlated with the holomorphic species. Representative examples of some of the holomorphs and anamorphs of importance in human affairs are described here.

Class Pezizomycetes

This class includes the one order Pezizales, the group in which the asci are unitunicate-operculate, and in most of the genera the asci appear on above-ground apothecia, as represented in Figure 46.18. Most of the members of this order are saprotrophs, although some are biotrophs that have an ectomycorrhizal relationship with tree roots. The apothecia of these fungi are often seen in wooded areas, including those of the genus *Morchella* in which a large edible ascocarp known as the "morel" is produced. The asci of the morel are on multiple apothecia that form in pits on the surface of a ridged conical head supported by a hollow stalk. A second type of edible ascocarp is produced by the genus *Tuber*, which has evolved into an entirely closed underground structure containing prototunicate asci surrounded by a dense mass of hyphae. There are various colors, flavors, and fragrances of these ascocarps in different species of *Tuber*, which are well known as the edible truffles. The fragrance and edibility of truffles function in attracting animals (primarily mammals, such as pigs) that disperse the ascospores during their feeding activity. *Tuber* is also notable because its mycelium grows in association with roots of trees with which they have an ectomycorrhizal relationship [98–101].

Class Dothideomycetes

All members of this class produce bitunicate asci, and the ascospores typically become septate internally to form either phragmospores or dictyospores. Although most of the fungi within this class are saprotrophic, many them are also necrotrophic plant pathogens. The order Dothidiales includes the family Venturiaceae, which includes the causal organism of apple scab disease, *Venturia inaequalis*. The major damage in this disease results from necrotic spots on the leaves and disfiguring scabs on the fruit. During the growing season, its anamorph *Spilocaea pomi* produces blastic conidia on exposed conidiophores (thus is a hyphomycete), and these conidia initiate new infections during the growing season. The teleomorph consists of pseudothecia that develop from mycelium in dead leaves on the ground in the spring, as shown in Figure 46.19, leading to the release of ascospores that begin the next disease cycle [9, 102].

The Capnodiales are another order in the Dothideomycetes of importance as plant pathogens, and they include the family Mycosphaerellaceae containing the genus *Pseudocercospora*, which is the anamorph of *Mycosphaerella*. The conidia are elongated phragmospores produced in an acervulus on short conidiophores. Species in this genus cause of variety leaf and fruit spots and blights of many tropical fruit crops, including sigatoka disease of bananas, something that represents a significant threat to banana production in many regions [103].

The Pleosporales are a third order of economic significance in the class Dothideomycetes and include the families Pleosporaceae and Didymellaceae that include teleomorphs of several plant pathogenic species. However, these plant pathogens are better known by the names assigned to their anamorphic forms (some of which have no known teleomorph), represented by the following examples.

a. *Alternaria* is a hyphomycete anamorph in the family Pleosporaceae, and it produces branching acropetal chains of conidia that become darkened dictyospores through internal septations and accumulation of melanin as they mature, as shown in Figure 46.26. The numerous species of this genus are typically necrotrophic, causing leaf spots as well as lesions on fruits and stems, and they attack a very wide host range of cultivated plants. The conidia in this genus are also important as an airborne allergen [9, 104].

b. *Stemphylium* is another hyphomycete anamorph in the family Pleosporaceae, associated with the holomorph *Pleospora*, but the sexual stage has not been observed in nature. Its conidia are produced blastically on short conidiophores and become darkened dictyospores through internal septations and accumulation of melanin as they mature, similar to those of *Alternaria*, except that its conidia are solitary, also shown in Figure 46.26. It is a necrotrophic plant pathogen often attacking the leaves of vegetable and field crops [105].

c. A third example of a destructive plant pathogen in the Pleosporaceae is the genus *Bipolaris*, a hyphomycete producing long elliptical (15–20 μm × 70–160 μm) phragmospores blastically on simple conidiophores that have a geniculate appearance, bending back and forth at the points where the conidia arise. There are several plant pathogenic species in this genus, many of them identified in the older literature as *Drechslera* or *Helminthosporium*. The species *Bipolaris maydis* (of which the teleomorph is *Cochliobolus heterostrophus*) is the cause of southern corn blight, a disease that caused major losses of the corn crop during 1969–1970 because most of the hybrid corn at that time shared a genotype that made it susceptible to this infection [106].

d. *Phoma* is a coelomycete anamorph in the family Didymellaceae, and the holomorph associated with most species in this genus is *Didymella*. The conidia of *Phoma* are produced on short phialides lining the interior of a pycnidium and are extruded in a slimy mass that is adapted for rain splashing of conidia, as illustrated in Figure 46.21. Species of *Phoma* are necrotrophic pathogens of numerous cultivated plants, causing leaf spots, blights, and stem cankers [48, 107, 108].

Class Eurotiomycetes

Most of the members of this class have prototunicate asci enclosed in a cleistothecial ascocarp. One of the orders in this class is the Onygenales that is of considerable clinical importance. One of the medically important families in this order is the Arthrodermataceae, known for the ability of many of its members to digest keratin. As with the Pleosporaceae described above, the fungi represented by teleomorphs in this family are better known by the names assigned to their anamorphic forms (some of which have no known teleomorph). Only nonliving tissues of the skin, hair, and nails are invaded by these dermatophytes, but the metabolites of the fungi create the familiar symptoms and host tissue responses, including shedding of thickened keratinized epithelium that tends to favor dispersal of

 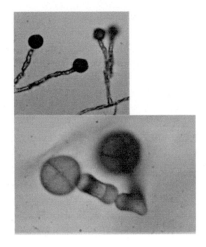

FIGURE 46.26 Plant pathogenic members of the Pleosporaceae known only by their anamorphs. Left: *Alternaria alternata*. Right: *Stemphylium sarcinaefome*.

conidia of these fungi to additional hosts. The anamorphic characteristics of three commonly encountered dermatophytes are as follows:

a. *Microsporum* is identified based on its large septate macroconidia with echinulate to verrucose (more or less spiny and roughened) cell walls, formed by thallic conidiogenesis. No microconidia are produced in this genus.

b. Several species of *Trichophyton* produce pyriform/globose microconidia by blastic conidiogenesis directly on growing hyphae, as shown in Figure 46.27, and some produce macroconidia that are similar to those produced by *Microsporum* except they are smooth-walled.

c. *Epidermophyton* produces both septate macroconidia and arthric conidia by thallic conidiogenesis, but it does not form blastic microconidia, in contrast with *Trichophyton* [94, 95, 109].

Two other families of clinical significance in the Onygenales are the Onygenaceae and the Ajellomycetaceae. The Onygenaceae includes the clinically important anamorphic genus *Coccidioides*, of which the species *Coccidioides immitis* is the cause of coccidiosis in humans. This organism is placed into this holomorphic family based on genomic and ultrastructural evidence, even though its teleomorph is unknown. It grows as a filamentous fungus in nitrogen-rich soil and produces small (3–5 μm) arthric conidia by thallic conidiogenesis as shown in Figure 46.27. If the conidia are inhaled by a human, they parasitize the lung tissue as a necrotrophic pathogen, switching to a yeast form under physiological conditions. During the infection, the fungal cells become enlarged into spherules in which large numbers of nuclei are formed by mitosis, after which cytoplasmic cleavage results in the formation of numerous endospores. The endospores can continue the lung infection, or if released into the environment can germinate to form the hyphae of the filamentous saprotrophic stage [94, 110].

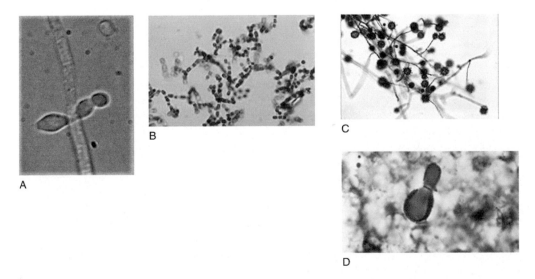

FIGURE 46.27 Human pathogens in three of the families in the order Onygenales. A: *Trichophyton mentagrophytes* in the Arthrodermataceae. B: *Coccidioides immitis* in the Onygenaceae. C: *Ajellomyces capsulatus* (*Histoplasma capsulatum*) in the Ajellomycetaceae. D: *Ajellomyces dermatitidis* (*Blastomyces dermatitidis*) in the Ajellomycetaceae.

The Ajellomycetaceae includes the clinically important genus *Ajellomyces* in which there are two species causing human infections, but much of the literature regarding these infections refers to the anamorph names, and the clinical terms for these infections are also based on the anamorph names. The anamorph of *Ajellomyces dermatitidis* is *Blastomyces dermatitidis*, which forms short conidiophores that produce small (2–10 μm) blastic conidia that are readily airborne and inhaled, shown in Figure 46.27. The infection is known as blastomycosis, and typically begins as granulomas in the lungs showing up as pneumonia, but can become disseminated to other organs, especially the skin, bones, joints, and the genitourinary tract. This disease is endemic to eastern North America, including Eastern Canada and the United States, and is apparently associated with high organic-matter soil with an acidic pH and high moisture levels created by nearby waterways [94, 111–113].

The anamorph of *Ajellomyces capsulatus* is *Histoplasma capsulatum,* which forms short conidiophores that produce two types of conidia blastically: ovoid microconidia (2–4 μm) and globose macroconidia (8–15 μm) with finger-like projections on the cell wall, also shown in Figure 46.27. This species occurs widely in nitrogen-rich soil and is common in central and eastern North America.

The microconidia are readily airborne and inhaled, but when phagocytized they remain active within the phagocytes, leading to a condition known as histoplasmosis. This organism is dimorphic, switching to a yeast form when growing necrotrophically inside the human body at 37°C [35].

The Eurotiomycetes are also of considerable importance because of the inclusion in this class of the order Eurotiales, in which the known teleomorphs produce a thick-walled cleistothecium surrounding several spherical prototunicate asci (shown in Figure 46.28) containing distinctive ascospores (many of them oblate with an equatorial ridge, or "pulley-wheel" shape). The family of most importance to humans within this order is the Trichocomaceae, given that it includes the teleomorphs of the highly significant anamorphic genera *Aspergillus* and *Penicillium* (although some anamorphic species in these genera have no known teleomorph). Some of the holomorphic genera that are associated with *Aspergillus* species are *Emericella,*

Eurotium, and *Neosartorya*; holomorphic genera associated with *Penicillium* species include *Carpentales, Eupenicillium,* and *Talaromyces* [9, 91, 114–116].

The Trichocomaceae also includes the holomorphic genus *Byssochlamys,* which is primarily known as a genus containing heat-resistant spores that germinate leading to growth within containers of canned fruits and bottled fruit juices resulting in their spoilage. A common anamorph of *Byssochlamys* is *Paecilomyces,* a very diverse genus found on a wide variety of substrates, and the necrotrophic infections of fruit crops that it causes may explain its presence in canned fruit products [117–120].

The asexual spores of *Aspergillus, Penicillium,* and *Paecilomyces* are formed blastically on phialides from which they are produced in basipetal succession, and they tend to cling together in chains until they are dislodged and become airborne. The conidiophores of *Aspergillus* species arise from a vesicle supported by an unbranched conidiophore. The phialides in *Penicillium* species develop in parallel bundles supported by hyphae with various levels of branching, creating the familiar "paint brush" form. Conidiophores of *Aspergillus* and *Penicillium* are shown in Figure 46.28. The phialides of *Paecilomyces* species may be solitary, in pairs, as verticils, or in penicillate heads [9].

Several *Aspergillus* species have a beneficial impact on human affairs, and examples include the use of *Aspergillus oryzae* in the production of soy sauce, miso, and saké, the use of *Aspergillus niger* in the production of citric acid, and the production by this and other *Aspergillus* species of enzymes and organic acids utilized in processing of food and animal feed [121, 122].

Growth of *Aspergillus* species can also have a negative impact, including the production of carcinogenic aflatoxins in stored grains and nuts by *Aspergillus flavus* and *Aspergillus parasiticus,* as well as spoilage of stored foods. *Aspergillus fumigatus* (with *Neosartorya* identified as its teleomorph) is a species of clinical importance because of its ability to colonize the pulmonary system as an opportunistic pathogen, which can include growth within a lung cavity created by a previous occurrence of Mycobacterial tuberculosis (known as an aspergilloma), leading to breakdown of the cavity wall [94, 123].

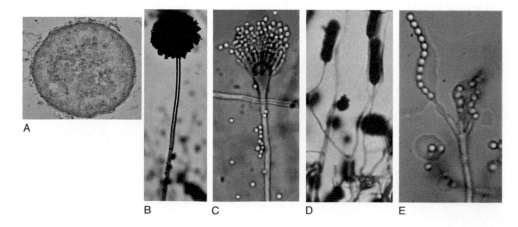

FIGURE 46.28 Teleomorph and anamorph stages of significant genera in the Trichocomaceae. A: Cleistothecium of teleomorph stage of *Aspergillus* sp. Note the ascospores within prototunicate asci. B: *Aspergillus niger* conidiophore. C: *Aspergillus oryzae* conidiophore. D: *Aspergillus fumigatus* conidiophores. E: *Penicillium notatum* conidiophore.

Penicillium is the genus well-known for the production of penicillin by some of its species, as discovered and published by Sir Alexander Fleming in 1929 (but not produced in large quantities until the 1940s). Fleming originally described the antibiotic from *Penicillium notatum*, but *Penicillium chrysogenum* became the major source of this antibiotic during the years in which it was first utilized for treating bacterial infections. Griseofulvin is an antifungal antibiotic produced by *Penicillium griseofulvum*. Various *Penicillium* species are also utilized for their production of enzymes and organic acids for industrial use, and the growth of species such as *Penicillium caseiolum*, *Penicillium camemberti*, and *Penicillium roqueforti* on certain cheeses creates desirable modifications of those foods. However, there are also species of *Penicillium* that produce mycotoxins in stored food products, including patulin formation in stored apples by *Penicillium expansum* [91, 121, 122].

Class Leotiomycetes

All members of this class have unitunicate-inoperculate asci formed within various types of ascocarps, including apothecia and cleistothecia. The holomorph is well known for most of the fungi in this class, although different names continue to be used for the holomorph and anamorph stages of most of these organisms since a number of plant diseases are associated with the anamorph stage [9].

The order Helotiales within this class include several necrotrophic plant pathogens that cause significant economic loss, most of them in the family Sclerotiniaceae, but there are destructive pathogens in other families as well. Three examples of plant pathogens in the Sclerotiniaceae are:

a. *Sclerotinia sclerotiorum*, which is known for producing sclerotia (but not conidia) and causing such diseases as stem rot, crown rot, and blossom blight of field crops. Apothecia develop from overwintering sclerotia at the beginning of the next growing season, releasing ascospores that are the inoculum for the next disease cycle [48].

b. *Monilinia fructicola*, which causes soft rot of peaches and other stone fruits. Apothecia develop from mummified fruit containing mycelium from the previous season, releasing ascospores at the time fruit trees are flowering to initiate new infections. Blastic acropetal conidia are produced on the infected plant tissues during the growing season as inoculum for additional infections, and this anamorphic stage has the name *Monilia fructicola* [48].

c. *Botryotinia cinerea*, which represents the teleomorph of the anamorph *Botrytis cinerea*; the anamorph name of this organism is most widely used, given that the blastic synchronous conidia are the only commonly observed stage of this fungus, and the primary overwintering structures are sclerotia that germinate to produce new mycelium. This organism is a necrotrophic pathogen with an extremely wide host range, commonly described as "gray mold." It frequently causes a soft rot of mature plant structures such as fruits, storage organs, and flowers, but it can also attack susceptible leaves and seedlings. Its conidiophores are branched with multiple denticles on a terminal cell (shown in Figure 46.23) on which many individual conidia are produced by synchronous blastic conidiogenesis [9, 48].

Another order in the Leotiomycetes that includes significant plant pathogens is the Erysiphales—the powdery mildews that attack a wide range of cultivated plants. The members of this order are biotrophic pathogens, dependent on a parasitic relationship with a host plant. The hyphae of a powdery mildew remain outside the host plant except for the penetration of the cell wall of individual epidermal cells by an appressorium and the development inside each cell of a fungal haustorium that typically branches within the host cell and absorb nutrients through the plasma membrane of the living host cell. The anamorph genus for most powdery mildews is *Oidium*, which develops unbranched conidiophores on the external hyphae on the surface of the host as shown in Figure 46.29. These conidiophores produce basauxic chains of conidia (with the youngest conidium at the base of the chain) that function as the inoculum during the growing season. The conidia have a powdery appearance on the surface of the plant (the basis for the common name), and are airborne to new host plants; they are unique in their specific requirement for a dry surface but high humidity for successful germination on the surface of the new host. Cleistothecia typically develop on the surface of the infected plant late in the season (as shown in Figure 46.29) and serve as overwintering structures. The cleistothecia often have distinctive appendages that may play a role in the dispersal of the cleistothecia themselves and serve as an identifying feature.

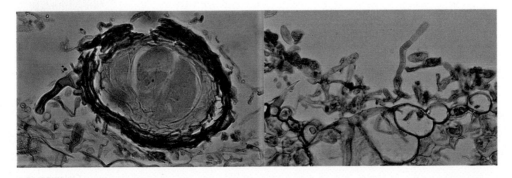

FIGURE 46.29 Teleomorph and anamorph of powdery mildew. Left: Cleistothecium; note the broken appendage. Right: Conidiophores with conidia nearby; note haustoria penetrating epidermal cells.

They open to allow the asci to eject ascospores that have the same germination requirements as the conidia [9, 124].

Class Sordariomycetes

Members of this class with a known teleomorph have unitunicate-inoperculate asci that develop inside perithecial ascocarps. Most of the seven orders in this class are saprotrophic, although they also include necrotrophic or biotrophic plant pathogens, some parasites on arthropod hosts, and one species that causes a necrotrophic infection of humans. Selected examples are described next:

a. Hypocreales: This order includes several holomorphic families of significance along with one related anamorphic genus for which the teleomorph has not yet been identified.

The Nectriaceae include the holomorphic genus *Nectria* that is known as a necrotrophic pathogen of tree species, causing cankers and dieback, as illustrated in Figures 46.19 and 46.21. The anamorph of *Nectria haematococca* is *Fusarium solani*, a significant necrotrophic pathogen of numerous commercially important plants. Fusarium species produce canoe-shaped phragmospores with a distinctive "foot cell" at one end (macroconidia) as shown in Figure 46.30, as well as single-celled microconidia on short slender phialides; chlamydospores often develop from one or more cells of a macroconidium. *Fusarium solani* and other species in this genus attack seedlings and roots, and also cause a wilt disease when they invade the vascular tissue of a plant host, releasing a toxin into the vascular stream that causes necrotic symptoms above the infection site. A mycoprotein food product known as "Quorn" has been developed in recent years through the culturing of *Fusarium venenatum* [125–127].

The Hypocreaceae include the holomorphic genus *Hypocrea* which is known as a saprotroph on wood, but species in its anamorphic genus *Trichoderma* (which forms conidia on phialides on branching hyphae as seen in Figure 46.30) are often mycoparasitic (attacking other fungi). This trait makes *Trichoderma* of interest as a candidate for biological control of plant pathogenic fungi, and the cellulases produced by these fungi are also of value in digesting cellulose [9, 128, 129].

The small anamorphic family Stachybotryaceae includes the saprotrophic species *Stachybotrys chartarum* (in addition to some plant pathogenic species). A corresponding teleomorph of *Stachybotrys* has been reported to be *Melanopsamma* in the holomorphic family *Niessliaceae* in the Hypocreales, but the anamorph binomial *Stachybotrys chartarum* is the name most associated with the fungus that produces trichothecenes that have been reported as contributing to sick building conditions. These mycotoxins can be released into the air from growth of colonies of *Stachybotrys chartarum* on interior surfaces of cellulose-containing building material (such as wallboard), which has become wet from flooding damage. The conidia of this species are dark, elliptical amerospores, produced singly on swollen phialides that develop in a cluster at the tip of the conidiophore, as shown in Figure 46.30. The conidia tend to collect into a slimy mass and would only be released into the air if the interior surfaces of the colonized building material were exposed through demolition activity, thus these conidia would not normally be found at high levels in indoor air [130–134].

The family *Clavicipitaceae* includes the holomorphic species *Claviceps purpurea* that is a biotrophic parasite of rye. The rye flowers are inoculated by ascospores released by this species in the spring, leading to fungal growth in place of the normal rye ovary tissue. The anamorphic stage known as *Sphacelia* forms conidia in phialides on a mycelial mat that produces a sugary nectar, and this nectar attracts insects that disseminate the conidia to new hosts. Later in the season the mycelial mats develop into an ergot, which is a type of sclerotium formed on the rye plant. The ergots typically fall to the ground as an overwintering structure, and in the spring each one sprouts several

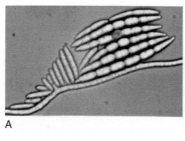
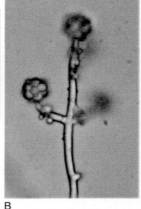

FIGURE 46.30 Anamorphic stages of members of three families in the Hypocreales. A: Nectriaceae: *Fusarium roseum* conidia from pure culture. B: Hypocreaceae: *Trichoderma viride*. C: Stachybotryaceae: *Stachybotrys chartarum*.

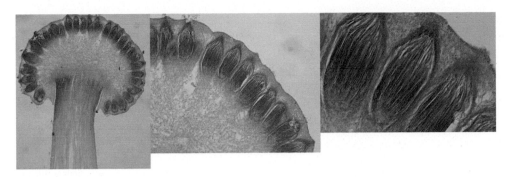

FIGURE 46.31 *Claviceps purpurea* stroma with multiple perithecia, magnified to show needle-shaped ascospores. Left: Terminal knob on stroma with imbedded perithecia. Middle: Detail of perithecia showing openings to the surface of the terminal knob. Right: Needle-shaped ascospores within asci contained within perithecia.

upright stalks made up of interwoven hyphae, each with a terminal knob. Perithecia develop on the surface these knobs, leading to the formation and release of the ascospores that begin the disease cycle again, as shown in Figure 46.31. If the developing ergots become harvested with rye grain instead of falling on the ground, the alkaloids in the ergots can produce the symptoms of ergotism if edible products made from the rye grain are consumed. Ingestion of ergot alkaloids leads to extreme smooth muscle contractions, which result in interrupted circulation and gangrene in affected tissues. This property of the alkaloids has been used pharmaceutically to induce uterine contractions and to control postpartum bleeding [9, 135].

Cordyceps is another genus in the family *Clavicipitaceae* that is of interest because of a highly beneficial pharmaceutical product derived from one of its anamorphs. While the holomorphic species *Cordyceps subsessilis* parasitizes beetle larvae, its anamorph *Tolypocladium inflatum* is the source of cyclosporine, the immunosuppressant that has become of great importance in minimizing organ transplant rejection as well as controlling some autoimmune disorders [91, 136].

Verticillium is apparently a member of the Hypocreales based on its characteristics, but its teleomorph is unknown at this time. The fungi in this genus invade plant hosts through natural wounds in the roots and enter the vascular tissue where they create a severe wilt disease because of xylem blockage. *Verticillium* produces blastic conidia on phialides that are arranged in whorls ("verticils") around an upright conidiophore, and the conidia tend to accumulate in a slimy cluster around each phialide in culture, but are carried by the xylem stream in a host plant to initiate infections at higher locations in the stem or trunk. Some *Verticillium* species produce microsclerotia as survival structures in the soil, some produce chlamydospores, and some survive in the soil as mycelium [137].

b. Sordariales. The species in this order are typically saprotrophic, which can lead to economic loss as in the case of the genus *Chaetomium* in the *Chaetomiaceae*, known for is cellulose-decomposing activity affecting canvas and other fabrics. The species *Neurospora crassa* in the *Sordariaceae* has been of particular importance in genetics research, and one of the factors that made it suitable for such research is its ability to grow on minimal media, making it possible to connect biochemical changes to genetic factors [138].

c. Ophiostomatales. Several genera and species in the family *Ophiostomataceae* are associated with beetles that tunnel through tree bark and function as vectors of fungi that invade the sapwood of trees, often causing a wilt disease or staining of the wood. The asci in this order break down to release their ascospores in a slimy mass that oozes out of the long tubular neck of the perithecium where they are picked up by tunneling beetles. *Ophiostoma ulmi* has caused the death of many American elms by the misnamed Dutch elm disease since its introduction to North America from Asia by way of the Netherlands. The conidiophores of its anamorph *Graphium* are bundled into a synnema that retains the conidia in a slimy droplet accessible to passing beetles. Another species in this genus that causes sapwood staining is *Ophiostoma stenoceras*, and its anamorph *Sporothrix schenckii* is associated with a subcutaneous infection in humans known as sporotrichosis when its conidia in soil or peat moss are introduced into a superficial wound. It responds to the higher temperature of a human host by transforming into a yeast that can become disseminated, leading to a significant medical condition in immunocompromised patients [94, 139, 140].

d. Glomerellales. Several species of necrotrophic plant pathogens attacking a wide range of hosts are within the holomorphic family *Glomerellaceae*, primarily in the holomorphic genus *Glomerella*, but the causal organisms are primarily known as the anamorphic genus *Colletotrichum*. The members of this coelomycete genus produce single-celled conidia (amerospores) within an acervulus (a disc-shaped mass of conidiophores originating under the surface of the host plant). When the cuticle or epidermis of the host ruptures from the pressure of the accumulating conidia, the resulting lesions are recognized as an anthracnose. [141].

Phylum Basidiomycota

Like most of the Ascomycota, the Basidiomycota typically have septate hyphae containing haploid nuclei, although there are a few members of the phylum that occur in a yeast form. Many Basidiomycota produce macroscopic basidiocarps that house the phases of the sexual stage, and many of them have anamorphs that produce conidia or sclerotia. A significant feature of the Basidiomycota is the common occurrence of dikaryotic vegetative hyphae, compared with the brief dikaryotic phase of most Ascomycota that is represented only by the ascogenous hyphae developing from an ascogonium. A hyphal structure associated with these dikaryotic hyphae in many of the Basidiomycota is the "clamp connection," which consists of the growth of a narrow tubular hypha extending from the new terminal cell to the adjacent cell concurrently with the simultaneous mitosis of the two nuclei in that cell. The clamp connection delivers one of the two resulting nuclei back to the adjacent cell so that both cells have one of each the two original nuclei, as shown in Figure 46.32. There are three subphyla in the Basidiomycota—Agaricomycotina, Pucciniomycotina, and Ustilaginomycotina—all of which produce basidiospores at the conclusion of their sexual stage, but the processes associated with this stage are unique to each of the three subphyla [10].

Agaricomycotina

Most of the members of this subphylum produce readily visible basidiocarps, comparable to the ascocarps described for the Ascomycota, and among these are the familiar mushrooms, bracket fungi, and puffballs along with the other less commonly seen earthstars, stinkhorns, and bird's nest fungi. The nutrition of most of the fungi in this subphylum is saprotrophic, with only a small number parasitic toward plants. Although the holomorph basidiocarp structure is most often recognized in this subphylum, conidia are produced by the anamorphs of several members of this group. The class Agaricomycetes includes most of the members of the Agaricomycotina subphylum, with a smaller number of species in two smaller classes known as the Dacrymycetes and Tremellomycetes [10].

The dikaryotic condition of fungi in this class originates through anastomosis of monokaryotic hyphae with compatible mating types and is maintained by the formation of clamp connections described earlier in this section. Dolipore septa (having a round pore with a barrel-shaped structure in the center, and bracketed by endoplasmic reticulum "parenthesomes" on either side, as shown in Figure 46.5) are also common in the Agaricomycotina [10, 142].

The two compatible nuclei of the dikaryotic hyphae typically form a diploid zygote by karyogamy within the club-shaped basidium (known as a holobasidium) that is characteristic of this subphylum. This is followed immediately in most members of this subphylum by meiosis, and the four resulting haploid nuclei are typically packaged into four basidiospores that form outside the basidium, created blastically on sterigmata, to which the basidiospores are attached asymmetrically. This sequence is illustrated in the series of diagrams in Figure 46.33. The basidiospores are propelled off the sterigmata by a mechanism that depends on the condensation of water vapor and osmosis of that water into a droplet containing mannitol and monosaccharides that is located at the base of each basidiospore. The basidiospores also accumulate a film of water by condensation on their hygroscopic surfaces, and at a certain point the basal droplets shift and merge with the water film on each spore, resulting in a force that separates the basidiospores from their sterigmata. This force is sufficient to carry the basidiospores far enough horizontally to reach a vertical space within the basidiocarp, allowing them to fall until they reach the turbulent air zone underneath and become dispersed by air currents. The basidia are typically packed together into a hymenium on the surface of the basidiocarp structure (commonly on gills or within pores), which are maintained in the necessary vertical orientation for basidiospores to escape from the structure [143].

There are exceptions to the typical dikaryotic state of the mycelium of the Agaricomycotina, such as the genus *Armillaria* in which karyogamy occurs immediately after plasmogamy, resulting in diploid vegetative hyphae. Basidiospores are produced as usual by meiosis of the diploid nuclei in the basidia that are formed on basidiocarps of such fungi [143].

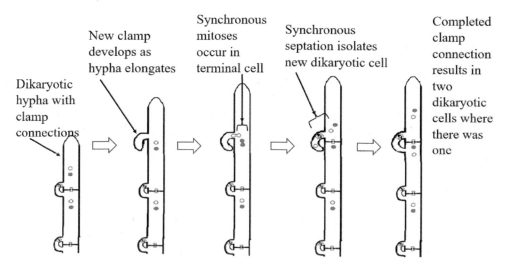

FIGURE 46.32 Clamp connection function in maintaining dikaryotic condition in Basidiomycota.

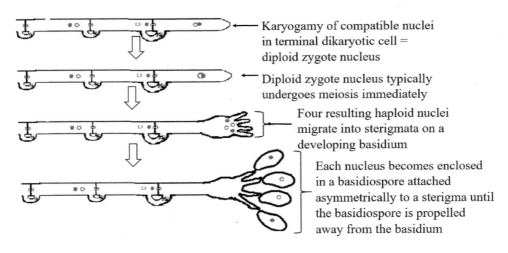

→ Karyogamy of compatible nuclei in terminal dikaryotic cell = diploid zygote nucleus

← Diploid zygote nucleus typically undergoes meiosis immediately

Four resulting haploid nuclei migrate into sterigmata on a developing basidium

Each nucleus becomes enclosed in a basidiospore attached asymmetrically to a sterigma until the basidiospore is propelled away from the basidium

FIGURE 46.33   The sexual stage associated with the Agaricomycotina subphylum of Basidiomycota.

Based on molecular data, many of the Agaricomycetes have been classified into one of two subclasses—the Agaricomycetidae and the Phallomycetidae—but there are a number of other members of this class that are given the Latin designation "*Incertae sedis*" (uncertain position). The Agaricomycetidae is the group in which the members produce the mushrooms, brackets, and puffballs as their basidiocarps, developing from mycelium that obtains nutrients from a substrate (or host) [10].

The morphology of most basidiocarps in this subclass is adapted for the type of basidiospore dispersal described above, with a central stipe (stalk) supporting the pileus (cap) that commonly has its hymenium on the surface of gills or a porous structure underneath. A protective membrane-like layer known as a veil sometimes encloses all are part of the basidiocarp during its development, and remnants of this veil may persist on the fully opened structure in the form of a volva (surrounding the base of the stipe), patches on the surface of the cap, an annulus ringing the stipe, or a cortina (curtain-like remnants) hanging down from the edge of the pileus [144].

The two major orders within this subclass are the Agaricales and the Boletales. Although the Agaricales include the more familiar mushrooms with their basidia on gills, this order also includes the puffball *Lycoperdon* in which basidiospores are not propelled off of sterigmata, but are retained within the basidiocarp until it ripens to release the basidiospores in the powdery

puff that occurs when raindrops (or other forces) strike the outer wall. There are a vast number of families, genera, and species within this subclass, and some representative examples are the edible *Agaricus* (the familiar grocery store mushroom) and *Lentinula* (the shiitake mushroom), the hallucinogenic *Psilocybe*, the tree pathogen *Armillaria* that attacks the roots and kills the tree, and numerous poisonous genera and species such as *Amanita*. Figure 46.34 shows cross sections of the pileus of a representative basidiocarp at different magnifications, revealing basidiospores arising from basidia lining the hymenium of the gills. Many members of this subclass are ectomycorrhizal on tree roots, primarily those of conifers. The Boletales produce fleshy basidiocarps similar to those of the Agaricales, but the hymenium is more often located within a porous structure than on gills, although gills also occur in some genera. Most of the Boletales are associated with ectomycorrhizae with tree roots [9, 28, 145–147].

Of the orders in the *Incertae sedis*, the Polyporales are significant as a large, economically destructive order since it includes numerous wood-rotting fungi. As the name indicates, the basidiocarps of the Polyporales typically have pores on the underside lined with basidia, as shown in the series of cross sections through the hymenium on the gills of a porous basidiocarp in Figure 46.35. Although some are fleshy and temporary like those of the Agaricales, many of them are corky or woody

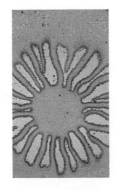
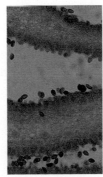
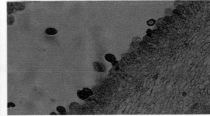

FIGURE 46.34   *Agaricus campestris* basidiospores on basidia in a hymenium on gills of basidiocarp. Left: Basidiocarp cross-section with gills radiating from the central stipe. Middle: Hymenium of basidia on the gill surface with basidiospores. Right: Magnified view of middle panel.

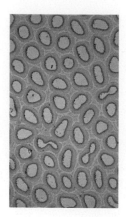
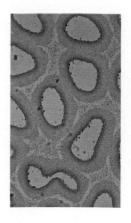

FIGURE 46.35 *Polyporus sp.* Basidiospores on basidia in a hymenium on surface of pores of basidiocarp. Left: Basidiocarp cross-section with pores on lower surface. Middle: Hymenium of basidia on pore surface with basidiospores. Right: Magnified view of middle panel.

perennial structures in which a new layer of pores is added to previous layers each year. Some members of this order (such as *Heterobasidion* and *Fomes*) are pathogenic, causing the weakening or death of host trees, while others such as *Polyporus* and *Ganoderma* are saprotrophic, growing on exposed heartwood of trees or on structural lumber. Fungi in this order are differentiated by the type of enzymatic digestion of wood they cause (brown rot fungi digest cellulose but not lignin, while white rot fungi digest lignin but not cellulose), the kinds of hyphae that make up the basidiocarp (whether they consist of only generative hyphae that create basidia or also contain thick-walled skeletal hyphae that create a hardened basidiocarp or thin-walled binding hyphae or both, along with other microscopic features). Anamorphs of these fungi may have thallic arthric conidia or blastic conidia produced on a vesicle [9, 48].

The subclass Phallomycetidae contains a relatively small number of orders, including the Phallales that rely mostly on flies for basidiospore dispersal. The members of this group release their basidiospores into a slimy mass with an odor of decaying animal flesh or feces, typically presented on upright basidiocarps that are often colorful and conspicuous [9, 80].

The class Dacrymycetes represents the jelly fungi that grow on decaying wood and that produce a gelatinous matrix through which their basidia elongate until they reach the surface where the basidiospores are propelled into the air.

The class Tremellomycetes also produces gelatinous basidiocarps, but differ from the Dacrymycetes in having vertically septate basidia with long sterigmata on which the basidiospores develop. Most of the anamorphs in this class are budding yeasts, and the basidiospores also germinate as budding yeast cells in most species. All teleomorphs in this class are mycoparasites, attacking both Ascomycota and Basidiomycota that occur on wood. The family Tremellaceae within the order Tremellales contains the genus *Filobasidiella neoformans* that is better known by its anamorph name *Cryptococcus neoformans*, which is of clinical importance as the cause of cryptococcosis. This species grows saprotrophically as a yeast on pigeon droppings and similar high-nitrogen substrates, and the cells are coated with a polysaccharide capsule that has a gelatinous consistency, as illustrated in Figure 46.36. This capsule functions as a virulence factor when the cells are inhaled and become established in a necrotrophic lung infection in individuals who are immunocompromised. The

Cryptococcus yeast cells can become disseminated and initiate infections in multiple organs, including a meningitis with a very high mortality rate [148].

Pucciniomycotina

This subphylum consists of several classes and orders of fungi that are related to each other according to genomic data, but there is just one order in just one class that is of most importance as a category of plant pathogens: the Pucciniales in the Pucciniomycetes that cause rust diseases of plants. These fungi are biotrophic pathogens, with an obligately parasitic relationship with their hosts, and many of the rust fungi alternate between two host plant species, in which case they are described as being heteroecious; those that only occur on one host species are autoecious. The entry of the parasite into the host plant is initiated by spores that germinate on the surface and penetrate the epidermis with an appressorium. This leads to mostly localized infections of the host by the growth of hyphae within the tissues but between the cells, except for haustoria that penetrate individual cells and absorb nutrients through the plasma membrane of the host cells that the haustoria are in contact with. An overwintering spore known as the teliospore takes the place of the basidiocarp in the

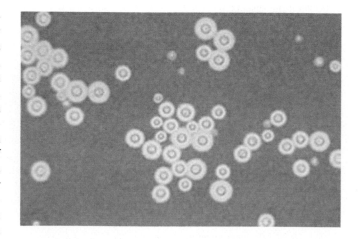

FIGURE 46.36 *Cryptococcus neoformans* suspended in India ink to reveal the capsule around each cell.

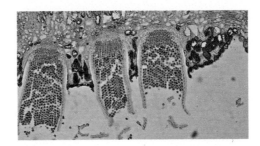

FIGURE 46.37 *Puccinia graminis* spore stages on *Berberis vulgaris*. Left: Spermatia with receptive hyphae extending above the surface of an infected leaf. Right: Aecia breaking through the lower epidermis and releasing aeciospores that function in inoculating wheat *Triticum aestivum*.

teleomorphic stage of the members of this order, and the anamorph of these fungi is represented by a series of conidial stages that are distinctive and characteristic of members of this order. If the teleomorphic teliospores are included, there can be as many as five spore stages in the life cycle of macrocyclic rust fungi, but microcyclic rust fungi have fewer than five spore stages. Both monokaryotic and dikaryotic phases occur in the life cycle, and the dikaryotic hyphae have a unique septal pore structure among Basidiomycota: a simple round pore in the center of which is suspended a "pulley wheel occlusion." The dikaryotic hyphae in this subphylum do not have the clamp connections that are common in the other two subphyla [10, 149].

The stem rust of wheat caused by *Puccinia graminis* subspecies *tritici* is widely cited as a representative macrocyclic heteroecious rust because of its economic impact as well as its well-defined spore stages, so it is described here for the same reasons. The annual disease cycle of this pathogen begins with airborne basidiospores that function as the inoculum that initiates the infection of the leaves of barberry (*Berberis*) by monokaryotic hyphae, leading to a fairly minor leaf spot disease. As the barberry leaf infection progresses, the fungus produces a specialized type of single-celled conidia known as spermatia, by phialidic conidiogenesis, within a droplet of nectar, along with several receptive hyphae, all of which are formed within a pycnidium-like structure known as a spermagonium, as shown in Figure 46.37. These conidia do not function as additional inoculum leading to additional barberry leaf infections, but instead are adapted for dispersal to other spermagonia by nectar-feeding insects where they function in the delivery of

nuclei to receptive hyphae if they are of the opposite mating type, resulting in the establishment of dikaryotic hyphae. On the underside of the same leaf, the resulting dikaryotic hyphae create an acervulus-like structure under the epidermis known as an aecium in which unbranched conidiophores produce dikaryotic conidia known as aeciospores by arthrosporic conidiogenesis, also shown in Figure 46.37. Once the host epidermis ruptures to expose the aeciospores, they become airborne as the inoculum that initiates an infection of wheat plants (*Triticum aestivum*). As in the barberry host, the wheat infection is biotrophic, with an intercellular invasion that involves penetration of host cells with haustoria. The wheat infection leads to the formation of acervulus-like structures on wheat stems known as uredinia, within which conidiophores produce dikaryotic urediniospores on unbranched conidiophore in "pustules" under the epidermis that have a rust-red color, shown in Figure 46.38. The color of these urediniospores is the basis for the name of the disease (and for the ancient Roman festival of Robigalia that was held to appease the Roman god Robigus to protect the wheat fields against this disease). Once the host plant epidermis ruptures to release the urediniospores, they become airborne to other wheat plants and function as the inoculum for new stem rust infections, which can rapidly become widespread in a wheat-growing region, leading to considerable economic loss. Both the physiological effects of the infection and the damage to the wheat stems by this spore production causes reduced yields, excessive water loss, disruption of water transport, and weakening of stem tissue leading to stem breakage known as lodging. Toward the end of the growing season, the same stem lesions mature into telia in which the

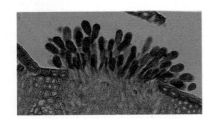
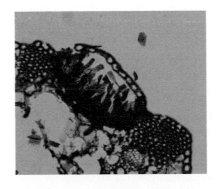

FIGURE 46.38 *Puccinia graminis* spore stages on *Triticum aestivum*. Left: Uredinia breaking through the epidermis of wheat and releasing urediniospores that function in inoculating wheat plants. Right: Telia developing on dead wheat stems and producing teliospores that overwinter until they germinate in the spring, leading to basidiospore production.

conidiophores produce terminal dikaryotic, pigmented, thick-walled, two-celled teliospores that remain attached to the stem tissue, shown in Figure 46.38. Karyogamy occurs within each cell of the teliospore while it functions as an overwintering structure very much like a sclerotium. At the beginning of the next growing season, at the time when barberry leaves have opened, each cell of the teliospore germinates with a short germ tube that develops into a septate four-celled basidium with a sterigma on each cell bearing a basidiospore. The basidiospores are propelled into the air by the same mechanism described above for other basidia, resulting in their being dispersed by wind, and those that land on barberry leaves may germinate and penetrate the leaf tissue to initiate a new infection [50, 150, 151].

Another macrocyclic heteroecious rust pathogen causing economic loss is *Cronartium ribicola*, causing blister rust of eastern white pine (*Pinus strobus*) and western white pine (*Pinus monticola*). The disease cycle begins with the infection of pine needles and twigs by basidiospores in the spring, followed by development on the twigs of spermagonia that form spermatia and receptive hyphae in a nectar droplet leading to dikaryotization as nectar-feeding insects serve as vectors for the spermatia, corresponding to the same events described above for stem rust of wheat. The aecia that subsequently develop on the bark of the pine twigs release aeciospores that function as the inoculum for the infection of wild currant (*Ribes*), but the persistence and expansion of these aecia causes permanent damage to the bark of these trees. When this damage extends to the main trunk of the tree, its bark becomes girdled (the term for the destruction of the bark around the entire circumference of the tree), which kills both the food-conducting phloem and the meristematic vascular cambium that functions in creating new layers of wood and bark, and results in the death of the tree. Once the wild currant plants are infected, uredinia are formed during the growing season that release urediniospores that can inoculate other wild currants, followed by the formation of telia that produce the overwintering teliospores in which karyogamy and meiosis lead to the release of basidiospores in the spring [152, 153].

Coffee rust is a microcyclic autoecious rust disease caused by *Hemileia vastatrix*, which produces only three spore stages in the disease cycle: Urediniospores that function as the inoculum that initiates infections of leaves of coffee trees, teliospores in which karyogamy and meiosis occur, and the basidiospores that form on septate basidia following meiosis in the teliospore. The urediniospores function in inoculating leaves of coffee trees and are produced by conidiophores that emerge through leaf stomata. The infection causes severe defoliation of the coffee trees, leading to the decline and death of the trees. Although teliospores and their resulting basidiospores are produced in the same lesions, there is no known alternate host that is inoculated by these basidiospores.

There are numerous rust diseases of economically important plants that display various disease cycles such as the ones described above. An example of another variation is represented by the rust diseases of members of the bean family (Fabaceae) caused by *Uromyces* species which are both macrocyclic and autoecious. All five spore stages are produced on the same host plant by fungi in this genus. The disease primarily causes a leaf spot that may cause defoliation but may also extend to a blight of young plants [154, 155].

Ustilaginomycotina

As in the case of the Pucciniomycotina, no basidiocarp is produced in this subphylum, and the teleomorph is represented by either a basidium developing directly from dikaryotic hyphae or by a basidium developing from a germinating teliospore. Nearly all of the members of this subphylum are plant parasites, primarily attacking flowering plants, and the commonly recognized diseases they cause are known as smuts (a reference to the "dirty" appearance of a black mass of teliospores) and bunts (a term reportedly arising from a contraction of the original "burnt ears" describing the affected grain). In most of the orders within this subphylum the organism is dimorphic, growing as saprotrophic monokaryotic budding yeasts and as biotrophic dikaryotic septate hyphae at different stages of the life cycle. Two classes have been identified within this subphylum, the Exobasidiomycetes and the Ustilaginomycetes [10, 156].

Exobasidiomycetes

Of the eight orders in this class, the Tilletiales (containing the one family *Tilletiaceae*) contains several destructive plant pathogens. It is the only order in this class that is not dimorphic, occurring entirely as dikaryotic hyphae, and it has dolipore septa similar to those of the Agaricomycotina. The teleomorph in the *Tilletiaceae* consists of overwintering dikaryotic teliospores in which karyogamy and meiosis leads to the formation of a type of holobasidium (nonseptate but elongated rather than club shaped) that forms 8–16 terminal basidiospores. These basidiospores immediately undergo dikaryotization through plasmogamy between compatible basidiospores by the connection of outgrowths from the spores themselves that results in an H-shaped arrangement of connected spores. The resulting dikaryotic hyphae may function in the inoculation of seedlings of the host plant (often members of the *Poaceae* (grass family), or they may produce ballistoconidia (propelled into the air by various mechanisms) that function as the inoculum. The *Tilletiaceae* grow intercellularly within the apical meristem of the plant until it produces an inflorescence, at which time the fungus typically replaces the normal ovary tissue of the flower with its own hyphae. These hyphae become converted into a mass of black or brown teliospores known as a sorus when the structure is mature. A representative pathogen in this family is *Tilletia caries*, causing common bunt of wheat, also known as stinking smut because of the release of trimethylamine with the odor of decayed fish from the sorus [156–158].

Ustilaginomycetes

Members of this class are dimorphic, with a budding yeast anamorph, while the teleomorph consists of an overwintering teliospore in which karyogamy and meiosis lead to the formation of a septate basidium, producing four basidiospores on sterigmata. The basidiospores give rise to the yeast anamorph that grows saprotrophically, and neither the basidiospores nor the yeast function as inoculum in the disease cycle. After dikaryotization occurs through anastomosis of yeast cells of compatible mating types, the resulting dikaryotic hyphae are capable of initiating an infection of a host plant (primarily members of the *Poaceae*) with an appressorium, leading to an intracellular invasion of the host tissue. This infection has a limited impact on the physiology of the host plant until an inflorescence develops, at which time the fungus typically occupies the normal ovary tissue of the flower with its own hyphae. When the structure is mature, a gelatinization of hyphal cell walls occurs, resulting in an arthritic formation of black teliospores when the structure is mature, thus following a developmental pattern similar to that of the *Tilletiaceae* described previously. An unusual feature of the Ustilaginales is the absence of septal pores in physiologically inactive hyphae. A representative pathogen in the *Ustilaginaceae* is *Ustilago maydis* causing corn smut, in which some of the kernels on a developing ear of corn become converted to smut sori that become much larger than normal corn kernels, eventually disintegrating to release the black smut teliospores. Prior to the maturation of a sorus into teliospores, the enlarged sori consisting of fungal hyphae growing within the original ovary tissue are utilized as an edible product known locally as "cuitlacoche" or "huitlacoche." While corn smut creates the enlargement of kernels described above, smut diseases of a wide range of hosts in the *Poaceae* are caused by various genera and species in the *Ustilaginaceae*, typically involving replacement of the reproductive structures of the infected plant by teliospores without such a conspicuous sign of the disease [10, 156, 159, 160].

REFERENCES

1. Simpson, A.G. and A.J. Roger, *The real 'kingdoms' of eukaryotes. Curr Biol*, 2004. **14**(17): p. R693–696.
2. Kendrick, B., *The fifth kingdom*, 4th ed. 2017, Hackett Publishing Company, Inc: Indianapolis, IN.
3. Money, N.P., *Fungal diversity*, in *The fungi*, 3rd ed., S.C. Watkinson, L. Boddy, and N.P. Money, Editors. 2016, Academic Press: Boston, MA. p. 1–36.
4. Kendrick, B., *Kingdoms, classification, nomenclature, and biodiversity*, in *The fifth kingdom*, 2017, Hackett Publishing Company: Indianapolis, IN. p. 1–15.
5. Hibbett, D.S., et al., *A higher-level phylogenetic classification of the fungi. Mycol Res*, 2007. **111**(Pt 5): p. 509–547.
6. Kendrick, B., *A mixed bag: protozoan 'Pseudofungi'; kingdom Eumycota phylum 1 Chytridiomycota, phylum 2 Blastocladiomycota, phylum 3 Neocallimastigomycota*, in *The Fifth Kingdom*, 2017, Hackett Publishing Company: Indianapolis, IN. p. 16–37.
7. Kendrick, B., *Eumycotan fungi; phylum 4 Zygomycota, phylum 5 Glomeromycota, phylum 6 Microsporidia*, in *The fifth kingdom*. 2017, Hackett Publishing Company: Indianapolis, IN. p. 38–54.
8. Spatafora, J.W., et al., *A phylum-level phylogenetic classification of Zygomycete fungi based on genome-scale data. Mycologia*, 2016. **108**(5): p. 1028–1046.
9. Kendrick, B., *Kingdom Eumycota; Phylum 7 Ascomycota*, in *The fifth kingdom*. 2017, Hackett Publishing Company: Indianapolis, IN. p. 55–104.
10. Kendrick, B., *Kingdom Eumycota; phylum 8 Basidiomycota*, in *The fifth kingdom*. 2017, Hackett Publishing Company: Indianapolis, IN. p. 105–138.
11. Watkinson, S., Boddy, L., Money, N., in *The fungi*, 3rd ed., S.C. Watkinson, L. Boddy, and N.P. Money, Editors. 2016, Academic Press: Boston, MA. p. 466.
12. Money, N.P., *Fungal cell biology and development*, in *The fungi*, 3rd ed., S.C. Watkinson, L. Boddy, and N.P. Money, Editors. 2016, Academic Press: Boston, MA. p. 37–66.
13. Gow, N.A.R., J.P. Latge and C.A. Munro, *The fungal cell wall: structure, biosynthesis, and function. Microbiol Spectr*, 2017. **5**(3).
14. Takeshita, N., *Coordinated process of polarized growth in filamentous fungi. Biosci Biotechnol Biochem*, 2016. **80**(9): p. 1693–1699.
15. Todd, R.T., A. Forche and A. Selmecki, *Ploidy variation in fungi: polyploidy, aneuploidy, and genome evolution. Microbiology Spectr*, 2017. **5**(4). 10.1128/microbiolspec. FUNK-0051-2016.
16. Boddy, L., *Interactions with humans and other animals*, in *The Fungi*, 3rd ed., S.C. Watkinson, L. Boddy, and N.P. Money, Editors. 2016, Academic Press: Boston, MA. p. 293–336.
17. de Oliveira, T.B., E. Gomes and A. Rodrigues, *Thermophilic fungi in the new age of fungal taxonomy. Extremophiles*, 2015. **19**(1): p. 31–37.
18. Gostincar, C., et al., *Extremotolerance in fungi: evolution on the edge. FEMS Microbiol Ecol*, 2010. **71**(1): p. 2–11.
19. Aguilera, A., *Eukaryotic organisms in extreme acidic environments, the Rio Tinto case. Life (Basel)*, 2013. **3**(3): p. 363–374.
20. Gunde-Cimerman, N., J. Ramos and A. Plemenitas, *Halotolerant and halophilic fungi. Mycol Res*, 2009. **113**(Pt 11): p. 1231–1241.
21. Gounot, A.M., *Psychrophilic and psychrotrophic microorganisms. Experientia*, 1986. **42**(11-12): p. 1192–1197.
22. Watkinson, S.C., *Physiology and Adaptation*, in *The fungi*, 3rd ed., S.C. Watkinson, L. Boddy, and N.P. Money, Editors. 2016, Academic Press: Boston, MA. p. 141–187.
23. Watkinson, S.C., *Mutualistic symbiosis between fungi and autotrophs*, in *The Fungi*, 3rd ed., S.C. Watkinson, L. Boddy, and N.P. Money, Editors. 2016, Academic Press: Boston, MA. p. 205–243.
24. Buscot, F., *Implication of evolution and diversity in arbuscular and ectomycorrhizal symbioses. J Plant Physiol*, 2015. **172**: p. 55–61.
25. Gutjahr, C. and M. Parniske, *Cell and developmental biology of arbuscular mycorrhiza symbiosis. Annu Rev Cell Dev Biol*, 2013. **29**: p. 593–617.
26. Lanfranco, L., P. Bonfante and A. Genre, *The mutualistic interaction between plants and arbuscular mycorrhizal fungi. Microbiol Spectr*, 2016. 4(6).
27. Roth, R. and U. Paszkowski, *Plant carbon nourishment of arbuscular mycorrhizal fungi. Curr Opin Plant Biol*, 2017. **39**: p. 50–56.
28. Kendrick, B., *Mycorrhizas*, in *The fifth kingdom*, 2017. Hackett Publishing Company: Indianapolis, IN. p. 323–344.
29. Kendrick, B., *Lichens–dual extremophile organisms*, in *The fifth kingdom*. 2017, Hackett Publishing Company: Indianapolis, IN. p. 145–154.
30. Spribille, T., *Relative symbiont input and the lichen symbiotic outcome. Curr Opin Plant Biol*, 2018. **44**: p. 57–63.
31. Markham, P. and A.J. Collinge, *Woronin bodies of filamentous fungi. FEMS Microbiol Letters*, 1987. **46**(1): p. 1–11.
32. Markham, P., *Occlusions of septal pores in filamentous fungi. Mycolog Res*, 1994. **98**(10): p. 1089–1106.
33. Kendrick, B., *Yeasts–compact polyphyletic extremophile fungi*, in *The fifth kingdom*. 2017, Hackett Publishing Company: Indianapolis, IN. p. 139–144.
34. Bhavsar-Jog, Y.P. and E. Bi, *Mechanics and regulation of cytokinesis in budding yeast. Semin Cell Dev Biol*, 2017. **66**: p. 107–118.
35. Boyce, K.J. and A. Andrianopoulos, *Fungal dimorphism: the switch from hyphae to yeast is a specialized morphogenetic*

adaptation allowing colonization of a host. FEMS Microbiol Rev, 2015. **39**(6): p. 797–811.

36. Sudbery, P., N. Gow and J. Berman, *The distinct morphogenic states of Candida albicans. Trends Microbiol*, 2004. **12**(7): p. 317–24.

37. Bennett, R.J. and B.G. Turgeon, *Fungal sex: the Ascomycota. Microbiol Spectr*, 2016. **4**(5).

38. Coelho, M.A., et al., *Fungal sex: the Basidiomycota. Microbiol Spectr*, 2017. **5**(3).

39. Lee, S.C. and A. Idnurm, *Fungal Sex: the Mucoromycota. Microbiol Spectr*, 2017. **5**(2).

40. Anderson, J.L., B.P.S. Nieuwenhuis, and H. Johannesson, *Asexual reproduction and growth rate: independent and plastic life history traits in Neurospora crassa. ISME J*, 2019. **13**(3): p. 780–788.

41. Ashu, E.E. and J. Xu, *The roles of sexual and asexual reproduction in the origin and dissemination of strains causing fungal infectious disease outbreaks. Infect Genet Evol*, 2015. **36**: p. 199–209.

42. Dyer, P.S. and U. Kuck, *Sex and the imperfect fungi. Microbiol Spectr*, 2017. **5**(3).

43. Tsang, C.-C., et al., *Taxonomy and evolution of Aspergillus, Penicillium and Talaromyces in the omics era — past, present and future. Comput Struct Biotechnol J*, 2018. **16**: p. 197–210.

44. Houbraken, J., R.P. de Vries and R.A. Samson, *Modern taxonomy of biotechnologically important aspergillus and Penicillium species*, in *Advances in Applied Microbiology*, S. Sariaslani and G.M. Gadd, Editors. 2014, Academic Press: San Diego, CA. p. 199–249.

45. Money, N.P., *Against the naming of fungi. Fungal Biol*, 2013. **117**(7): p. 463–465.

46. Gougouli, M. and K.P. Koutsoumanis, *Relation between germination and mycelium growth of individual fungal spores. Int J Food Microbiol*, 2013. **161**(3): p. 231–9.

47. Kendrick, B., *Fungicides*, in *The fifth kingdom*. 2017, Hackett Publishing Company: Indianapolis, IN. p. 265–275.

48. Kendrick, B., *Fungal diseases of crops and trees*, in *The fifth kingdom*. 2017, Hackett Publishing Company: Indianapolis, IN. p. 247–264.

49. Ryder, L.S. and N.J. Talbot, *Regulation of appressorium development in pathogenic fungi. Curr Opin Plant Biol*, 2015. **26**: p. 8–13.

50. Bruckart, W.L., 3rd, et al., *Life cycle of Puccinia acroptili on Rhaponticum (= Acroptilon) repens. Mycologia*, 2010. **102**(1): p. 62–8.

51. James, T.Y., et al., *A molecular phylogeny of the flagellated fungi (Chytridiomycota) and description of a new phylum (Blastocladiomycota). Mycologia*, 2006. **98**(6): p. 860–71.

52. Przetakiewicz, J., *First report of Synchytrium endobioticum (Potato Wart Disease) pathotype 18(T1) in Poland. Plant Dis*, 2014. **98**(5): p. 688.

53. Voyles, J., et al., *Pathogenesis of chytridiomycosis, a cause of catastrophic amphibian declines. Science*, 2009. **326**(5952): p. 582–5.

54. Voyles, J., E.B. Rosenblum and L. Berger, *Interactions between Batrachochytrium dendrobatidis and its amphibian hosts: a review of pathogenesis and immunity. Microbes Infect*, 2011. **13**(1): p. 25–32.

55. Wu, S.H., et al., *Yield loss model and yield loss mechanism of high-yielding summer maize infected by Physoderma maydis. Ying Yong Sheng Tai Xue Bao*, 2011. **22**(3): p. 720–6.

56. Griffith, G.W., et al., *Anaerobic fungi: Neocallimastigomycota. IMA Fungus*, 2010. **1**(2): p. 181–5.

57. Gruninger, R.J., et al., *Anaerobic fungi (phylum Neocallimastigomycota): advances in understanding their taxonomy, life cycle, ecology, role and biotechnological potential. FEMS Microbiol Ecol*, 2014. **90**(1): p. 1–17.

58. Lay, C.Y., C. Hamel and M. St-Arnaud, *Taxonomy and pathogenicity of Olpidium brassicae and its allied species. Fungal Biol*, 2018. **122**(9): p. 837–846.

59. Rochon, D., et al., *Molecular aspects of plant virus transmission by Olpidium and Plasmodiophorid vectors. Annu Rev Phytopathol*, 2004. **42**: p. 211–41.

60. Mendoza, L., et al., *Human fungal pathogens of Mucorales and Entomophthorales. Cold Spring Harb Perspect Med*, 2014. **5**(4).

61. Farmakiotis, D. and D.P. Kontoyiannis, *Mucormycoses. Infect Dis Clin North Am*, 2016. **30**(1): p. 143–63.

62. Seyedmousavi, S., et al., *Fungal infections in animals: a patchwork of different situations. Med Mycol*, 2018. **56**(suppl_1): p. 165–187.

63. Rico-Munoz, E., R.A. Samson and J. Houbraken, *Mould spoilage of foods and beverages: using the right methodology. Food Microbiol*, 2019. **81**: p. 51–62.

64. Shtienberg, D., *Rhizopus head rot of confectionery sunflower: effects on yield quantity and quality and implications for disease management. Phytopathology*, 1997. **87**(12): p. 1226–32.

65. Hachmeister, K.A. and D.Y. Fung, *Tempeh: a mold-modified indigenous fermented food made from soybeans and/or cereal grains. Crit Rev Microbiol*, 1993. **19**(3): p. 137–88.

66. Jelen, H., et al., *Determination of compounds responsible for tempeh aroma. Food Chem*, 2013. **141**(1): p. 459–65.

67. Samson, R.A., J.A. Van Kooij and E. De Boer, *Microbiological quality of commercial tempeh in the Netherlands. J Food Prot*, 1987. **50**(2): p. 92–94.

68. Karatay, S.E. and G. Donmez, *Evaluation of biotechnological potentials of some industrial fungi in economical lipid accumulation and biofuel production as a field of use. Prep Biochem Biotechnol*, 2014. **44**(4): p. 332–41.

69. Corsaro, D., et al., *New insights from molecular phylogenetics of Amoebophagous fungi (Zoopagomycota, Zoopagales). Parasitol Res*, 2018. **117**(1): p. 157–167.

70. Gryganskyi, A.P., et al., *Phylogenetic lineages in Entomophthoromycota. Persoonia*, 2013. **30**: p. 94–105.

71. Shaikh, N., et al., *Entomophthoramycosis: a neglected tropical mycosis. Clin Microbiol Infect*, 2016. **22**(8): p. 688–94.

72. Field, A.S. and D.A. Milner, Jr., *Intestinal microsporidiosis. Clin Lab Med*, 2015. **35**(2): p. 445–59.

73. Han, B. and L.M. Weiss, *Microsporidia: obligate intracellular pathogens within the fungal kingdom. Microbiol Spectr*, 2017. **5**(2):10.1128/microbiolspec.FUNK-0018-2016.

74. Moore, D., G.D. Robson and P.J. Trinci, *21st century guidebook to fungi*. 2011, Cambridge University Press: New York. p. 217–225.

75. Tedersoo, L., A.E. Arnold and K. Hansen, *Novel aspects in the life cycle and biotrophic interactions in Pezizomycetes (Ascomycota, Fungi). Mol Ecol*, 2013. **22**(6): p. 1488–93.

76. Dijksterhuis, J., *Fungal spores: highly variable and stress-resistant vehicles for distribution and spoilage. Food Microbiol*, 2019. **81**: p. 2–11.

77. Jung, B., S. Kim and J. Lee, *Microcyle conidiation in filamentous fungi. Mycobiology*, 2014. **42**(1): p. 1–5.

78. Campbell, C.K., *Conidiogenesis of fungi pathogenic for man.* Crit Rev Microbiol, 1986. **12**(4): p. 321–41.

79. Druzhinina, I. and C.P. Kubicek, *Species concepts and biodiversity in Trichoderma and Hypocrea: from aggregate species to species clusters?* J Zhejiang Univ Sci B, 2005. **6**(2): p. 100–12.

80. Kendrick, B., *Spore Dispersal in Fungi*, in *The fifth kingdom*. 2017, Hackett Publishing Company: Indianapolis, IN. p. 156–174.

81. Money, N.P., *Spore production, discharge, and dispersal*, in *The fungi*, 3rd ed., S.C. Watkinson, L. Boddy, and N.P. Money, Editors. 2016, Academic Press: Boston, MA. p. 67–97.

82. Wyatt, T.T., H.A.B. Wösten and J. Dijksterhuis, *Fungal spores for dispersal in space and time*, in *Advances in Applied Microbiology*, S. Sariaslani and G.M. Gadd, Editors. 2013, Academic Press, San Diego, CA. p. 43–91.

83. Hoffman, C.S., V. Wood and P.A. Fantes, *An ancient yeast for young geneticists: a primer on the Schizosaccharomyces pombe model system.* Genetics, 2015. **201**(2): p. 403.

84. Rossi, V. and L. Languasco, *Influence of environmental conditions on spore production and budding in Taphrina deformans, the causal agent of peach leaf curl.* Phytopathology, 2007. **97**(3): p. 359–65.

85. Gigliotti, F., A.H. Limper and T. Wright, *Pneumocystis.* Cold Spring Harb Perspect Med, 2014. **4**(12): p. a019828.

86. Morris, A. and K.A. Norris, *Colonization by Pneumocystis jirovecii and its role in disease.* Clin Microbiol Rev, 2012. **25**(2): p. 297–317.

87. White, P.L., M. Backx and R.A. Barnes, *Diagnosis and management of Pneumocystis jirovecii infection.* Expert Rev Anti Infect Ther, 2017. **15**(5): p. 435–447.

88. Neiman, A.M., *Ascospore formation in the yeast Saccharomyces cerevisiae.* Microbiol Mol Biol Rev: MMBR, 2005. **69**(4): p. 565–584.

89. Coluccio, A., R.K. Rodriguez, M.J. Kernan and A.M. Neiman, *The yeast spore wall enables spores to survive passage through the digestive tract of Drosophila.* PLOS One, 2008. 3.

90. Kurtzman, C.P. and J.W. Fell, *The yeasts, a taxonomic study*, 4th ed. 1998, Elsevier Science: Amsterdam.

91. Kendrick, B., *Commercial exploitation of fungal metabolites and mycelia*, in *The fifth kingdom*. 2017, Hackett Publishing Company: Indianapolis, IN. p. 420–428.

92. Stewart, G.G., Review: genetic manipulation of Saccharomyces sp. that produce ethanol, related metabolites/enzymes and biomass*, in *Reference Module in Food Science*, 2019, Elsevier.

93. Meehl, M.A. and T.A. Stadheim, *Biopharmaceutical discovery and production in yeast.* Curr Opin Biotechnol, 2014. **30**: p. 120–127.

94. Kendrick, B., *Medical Mycology*, in *The Fifth Kingdom*. 2017, Hackett Publishing Company: Indianapolis, IN. p. 409–419.

95. Weitzman, I. and R.C. Summerbell, *The dermatophytes.* Clin Microbiol Rev, 1995. **8**(2): p. 240–59.

96. Brandt, M.E. and S.R. Lockhart, *Recent taxonomic developments with Candida and other opportunistic yeasts.* Curr Fungal Infect Rep, 2012. **6**(3): p. 170–177.

97. Yue, H., et al., *Filamentation in Candida auris, an emerging fungal pathogen of humans: passage through the mammalian body induces a heritable phenotypic switch.* Emerg Microbes Infect, 2018. **7**(1): p. 188–188.

98. Pinto, S., et al., *Does soil fauna like truffles just as humans do? One-year study of biodiversity in natural Brules of Tuber Aestivum vittad.* Sci Total Environ, 2017. **584-585**: p. 1175–1184.

99. Alvarez-Lafuente, A., et al., *Multi-cropping edible truffles and sweet chestnuts: production of high-quality Castanea sativa seedlings inoculated with Tuber aestivum. Its ecotype T. uncinatum, T. brumale, and T. macrosporum.* Mycorrhiza, 2018. **28**(1): p. 29–38.

100. Buntgen, U., et al., *New insights into the complex relationship between weight and maturity of Burgundy Truffles (Tuber aestivum).* PLOS One, 2017. **12**(1): p. e0170375.

101. Tietel, Z. and S. Masaphy, *True morels (Morchella)-nutritional and phytochemical composition, health benefits and flavor: a review.* Crit Rev Food Sci Nutr, 2018. **58**(11): p. 1888–1901.

102. Bowen, J.K., et al., *Venturia inaequalis: the causal agent of apple scab.* Mol Plant Pathol, 2011. **12**(2): p. 105–22.

103. Crous, P.W., et al., *Phylogenetic lineages in Pseudocercospora.* Stud Mycol, 2013. **75**: p. 37–114.

104. Woudenberg, J.H.C., et al., *Alternaria section Alternaria: Species, Formae speciales or pathotypes?* Stud Mycol, 2015. **82**: p. 1–21.

105. Woudenberg, J.H.C., et al., *Stemphylium revisited.* Stud Mycol, 2017. **87**: p. 77–103.

106. Mubeen, S., et al., *Study of southern corn leaf blight (SCLB) on maize genotypes and its effect on yield.* J Saudi Soc Agricultur Sci, 2017. **16**(3): p. 210–217.

107. Chen, Q., et al., *Resolving the Phoma enigma.* Stud Mycol, 2015. **82**: p. 137–217.

108. Bennett, A., M.M. Ponder and J. Garcia-Diaz, *Phoma infections: classification, potential food sources, and its clinical impact.* Microorganisms, 2018. **6**(3): p. 58.

109. Ogawa, H., et al., *Dermatophytes and host defence in cutaneous mycoses.* Med Mycol, 1998. **36 Suppl 1**: p. 166–73.

110. Kirkland, T.N. and J. Fierer, *Coccidioides immitis and posadasii; a review of their biology, genomics, pathogenesis, and host immunity.* Virulence, 2018. **9**(1): p. 1426–1435.

111. Castillo, C.G., C.A. Kauffman and M.H. Miceli, *Blastomycosis.* Infect Dis Clin North Am, 2016. **30**(1): p. 247–64.

112. Kane, J., et al., *Blastomycosis: a new endemic focus in Canada.* Can Med Assoc J, 1983. **129**(7): p. 728–731.

113. Baumgardner, D.J., et al., *Epidemiology of blastomycosis in a region of high endemicity in north central Wisconsin.* Clin Infect Dis, 1992. **15**(4): p. 629–35.

114. Kendrick, B., *Food Spoilage by Fungi*, in *The fifth kingdom*, 2017, Hackett Publishing Company: Indianapolis, IN. p. 365–371.

115. Money, N.P., *Chapter 12—Fungi and Biotechnology*, in *The fungi*, 3rd ed., S.C. Watkinson, L. Boddy, and N.P. Money, Editors. 2016, Academic Press: Boston, MA. p. 401–424.

116. Samson, R.A. and J.I. Pitt, *Integration of Modern Taxonomic Methods for Penicillium and Aspergillus Classification*, 2000, Harwood Academic Publishers: Amsterdam.

117. Houbraken, J. and R.A. Samson, *Current taxonomy and identification of foodborne fungi.* Curr Opin Food Sci, 2017. **17**: p. 84–88.

118. King, A.D., H.D. Michener and K.A. Ito, *Control of Byssochlamys and related heat-resistant fungi in grape products.* App Microbiol, 1969. **18**(2): p. 166–173.

119. Samson, R.A., et al., *Polyphasic taxonomy of the heat resistant ascomycete genus Byssochlamys and its Paecilomyces anamorphs.* Persoonia, 2009. **22**: p. 14–27.

120. Biango-Daniels, M.N., et al., *Fruit infected with Paecilomyces niveus: a source of spoilage inoculum and patulin in apple juice concentrate? Food Control*, 2019. **97**: p. 81–86.

121. Max, B., et al., *Biotechnological production of citric acid. Brazilian Journal of Microbiology: [publication of the Brazilian Society for Microbiology]*, 2010. **41**(4): p. 862–875.

122. Gupta, V., editor, *New and future developments in microbial biotechnology and bioengineering: aspergillus system properties and applications*, 2016, Elsevier: Amsterdam.

123. Pitt, J.I. and R.A. Samson, *Nomenclatural considerations in naming species of Aspergillus and its teleomorphs. Stud Mycol*, 2007. **59**: p. 67–70.

124. Glawe, D.A., *The powdery mildews: a review of the world's most familiar (yet poorly known) plant pathogens. Annu Rev Phytopathol*, 2008. **46**: p. 27–51.

125. Hirooka, Y., A.Y. Rossman and P. Chaverri, *A morphological and phylogenetic revision of the Nectria cinnabarina species complex. Stud Mycol*, 2011. **68**: p. 35–56.

126. Coleman, J.J., *The Fusarium solani species complex: ubiquitous pathogens of agricultural importance. Mol Plant Pathol*, 2016. **17**(2): p. 146–158.

127. O'Donnell, K., E. Cigelnik, and H.H. Casper, *Molecular phylogenetic, morphological, and mycotoxin data support reidentification of the Quorn mycoprotein fungus as Fusarium venenatum. Fungal Genet Biol*, 1998. **23**(1): p. 57–67.

128. Schuster, A. and M. Schmoll, *Biology and biotechnology of trichoderma. App Microbiol Biotechnol*, 2010. **87**(3): p. 787–799.

129. Kendrick, B., *Fungi as agents of biological control*, in *The fifth kingdom*. 2017, Hackett Publishing Company: Indianapolis, IN. p. 276–294.

130. Foladi, S., et al., *Study on fungi in archives of offices, with a particular focus on Stachybotrys chartarum. J de Mycologie Médicale*, 2013. **23**(4): p. 242–246.

131. Hintikka, E.-L., *The role of stachybotrys in the phenomenon known as sick building syndrome*, in *Advances in applied microbiology*, 2004, Academic Press: San Diego, CA. p. 155–173.

132. Jarvis, B.B., *Stachybotrys chartarum: a fungus for our time. Phytochemistry*, 2003. **64**(1): p. 53–60.

133. Shariat, C. and H.R. Collard, *Acute lung injury after exposure to Stachybotrys chartarum. Respir Med Extra*, 2007. **3**(2): p. 74–75.

134. Castlebury, L.A., et al., *Multigene phylogeny reveals new lineage for Stachybotrys chartarum, the indoor air fungus. Mention of trade names or commercial products in this article is solely for the purpose of providing specific information and does not imply recommendation or endorsement by the USDA. Mycolog Res*, 2004, **108**(8): p. 864–872.

135. Kendrick, B., *Mycotoxins in food and feed*, in *The fifth kingdom*. 2017, Hackett Publishing Company: Indianapolis, IN. p. 372–391.

136. Jegorov, A., et al., *Cyclosporins from Tolypocladium terricola. Phytochemistry*, 1995. **38**(2): p. 403–407.

137. Inderbitzin, P., et al., *Phylogenetics and taxonomy of the fungal vascular wilt pathogen Verticillium, with the descriptions of five new species. PLOS One*, 2011. **6**(12): p. e28341–e28341.

138. Ebbole, D., *Neurospora: a new (?) model system for microbial genetics: Neurospora 2000, Asilomar, CA, USA, 9–12 March 2000. Trends in Genetics*, 2000. **16**(7): p. 291–292.

139. De Beer, Z.W., et al., *Phylogeny of the Ophiostoma stenoceras: Sporothrix schenckii complex. Mycologia*, 2003. **95**: p. 434–41.

140. Spatafora, J.W. and M. Blackwell, *The polyphyletic origins of Ophiostomatoid fungi. Mycol Res*, 1994. **98**(1): p. 1–9.

141. Weir, B.S., P.R. Johnston and U. Damm, *The Colletotrichum gloeosporioides species complex. Stud Mycol*, 2012. **73**: p. 115–180.

142. Walther, G., S. Garnica, and M. Weiß, *The systematic relevance of conidiogenesis modes in the gilled Agaricales1 1Dedicated to John Webster on the occasion of his 80th birthday. Mycol Res*, 2005. **109**(5): p. 525–544.

143. Banuett, F., *From dikaryon to diploid. Fungal Biol Rev*, 2015. **29**(3): p. 194–208.

144. Halbwachs, H., J. Simmel, and C. Bässler, *Tales and mysteries of fungal fruiting: how morphological and physiological traits affect a pileate lifestyle. Fungal Biol Rev*, 2016. **30**(2): p. 36–61.

145. Kendrick, B., *Fungi as food*, in *The fifth kingdom*. 2017, Hackett Publishing Company: Indianapolis, IN. p. 345–358.

146. Kendrick, B., *Poisonous and Hallucinogenic mushrooms*, in *The fifth kingdom*. 2017, Hackett Publishing Company: Indianapolis. p. 392–408.

147. Pegler, D.N. and T.W.K. Young, *A natural arrangement of the Boletales, with reference to spore morphology. Trans Brit Mycol Soc*, 1981. **76**(1): p. 103–146.

148. Maziarz, E.K. and J.R. Perfect, *Cryptococcosis. Infect Dis Clin North Am*, 2016. **30**(1): p. 179–206.

149. Bennell, A.P. and D.M. Henderson, *Urediniospore and teliospore development in Tranzschelia (Uredinales). Trans Brit Mycol Soc*, 1978. **71**(2): p. 271–278.

150. Abbasi, M., S.B. Goodwin and M. Scholler, *Taxonomy, phylogeny, and distribution of Puccinia graminis, the black stem rust: new insights based on rDNA sequence data. Mycoscience*, 2005. **46**(4): p. 241–247.

151. Tiburzy, R., E.M.F. Martins and H.J. Reisener, *Isolation of haustoria of Puccinia graminis f.sp.tritici from wheat leaves. Experiment Mycol*, 1992. **16**(4): p. 324–328.

152. Kaitera, J. and H. Nuorteva, *Inoculations of eight Pinus species with Cronartium and Peridermium stem rusts. Forest Ecol Manage*, 2008. **255**(3): p. 973–981.

153. Lu, P., et al., *Seedling survival of Pinus strobus and its interspecific hybrids after artificial inoculation of Cronartium ribicola. Forest Ecol Manage*, 2005. **214**(1): p. 344–357.

154. Lapwood, D.H., et al., *An effect of rust (Uromyces viciae-fabae) on the yield of spring-sown field beans (Vicia faba) in the UK. Crop Protect*, 1984. **3**(2): p. 193–198.

155. Groth, J.V., *Uromyces appendiculatus, rust of Phaseolus beans*, in *Advances in plant pathology*, G.S. Sidhu, Editor. 1988, Academic Press. p. 389–400.

156. Weiss M., R. Bauer, J.P. Sampaio, F. Oberwinkler. *Tremellomycetes and Related Groups*, in *Systematics and Evolution. The Mycota (A Comprehensive Treatise on Fungi as Experimental Systems for Basic and Applied Research)*, vol 7A, D. McLaughlin, J. Spatafora, Editors. 2014, Springer, Berlin, Heidelberg. p. 331–355.

157. Olson, Å. and J. Stenlid, *Pathogenic fungal species hybrids infecting plants. Microbes Infect*, 2002. **4**(13): p. 1353–1359.

158. Marshall, G.M., *The incidence of certain seed-borne diseases in commercial seed samples. Annal Appl Biol*, 1960. **48**(1): p. 34–38.

159. McKenzie, E.H.C. and K. Vánky, *Smut fungi of New Zealand: an introduction, and list of recorded species. New Zealand J Botany*, 2001. **39**(3): p. 501–515.

160. Piepenbring, M., M. Stoll and F. Oberwinkler, *The generic position of Ustilago maydis, Ustilago scitaminea*, and *Ustilago esculenta* (Ustilaginales). *Mycolog Prog*, 2002. **1**(1): p. 71–80.

47

Introduction to Parasites

Purnima Bhanot

CONTENTS

DOI: 10.1201/9781003099277-49

Introduction[*]

The focus of this chapter is on human parasites, particularly those that are major pathogens. The word "parasite" is derived from the Greek *para*, meaning eating and *sitos* meaning by the side of. Parasitism is a biological niche in which one organism derives benefit at the expense of another.

Parasites can be unicellular or multicellular organisms, ranging in size from protozoa of microscopic dimensions to tapeworms that can attain a length of 20–30 feet. They can be classified based on where they live: ectoparasitic if they live on the body, e.g., lice, or endoparasitic if they live inside the body.

This chapter describes a limited number of parasitic diseases as representative of major groups of parasitic organisms. The major groups of parasites included in the chapter are the protists (protozoan) and helminths. Insects (mosquitoes, flies, etc.) and acarids (ticks), while human parasites, are dealt with only in the context of their role as disease vectors.

Global Dissemination of Parasites

The numbers of people suffering and dying from parasitic diseases is staggering. It is estimated that annually, there are more than 200 million cases of malaria [1], almost 1 billion cases of soil-transmitted helminths [2], and 8 million infections of *Trypanosoma cruzi* [3]. These diseases have mortalities ranging from the hundreds of thousands to thousands annually, with children representing a large number of deaths. While the vast majority of the disease burden lies in developing countries, increased international travel and climate change mean that the developed world is not immune.

Adaptations for Parasitism as a Way of Life

Parasite Life Cycles: Definitions

The *definitive* host of a parasite is the one in which the parasite becomes sexually mature. Thus, for the malaria parasite, the mosquito is the definitive host, and the human is the intermediate host. The *intermediate* host (e.g., snails, fish) is one that harbors the developmental stages of the parasite and serves as a food source for the definitive host, thus completing the cycle

from human to human. Some parasites may have a primary, secondary, and even a tertiary intermediate host. The *vector* is the means by which a parasite is spread. In African sleeping sickness, the tsetse fly is the vector and, in its quest for blood, transmits the parasite from animal to animal, animal to human, or from human to human. Water may also serve as a vector, transmitting water-borne infections. A *mechanical* vector is one that can transmit a parasite, but in which no parasite development occurs. Flies and roaches as purveyors of filth may serve as *mechanical* vectors by transmitting disease organisms from fecal waste to food waiting to be consumed. A *reservoir* is an animal, other than human, that harbors the parasite and can serve as a source of human infections.

Mechanisms of Pathogenesis

A parasite inflicts damage upon the host (Table 47.1). Competition with the parasite for nutrients may lead to malnutrition, slowed development of the host, or vulnerability to other infections. Some parasites, such as *P. falciparum* and *T. brucei*, cause death if left untreated. In well-fed, otherwise healthy individuals, parasites may cause illness but only rarely death. A tapeworm may even have an advantageous effect on its host. It has been suggested that a benign tapeworm can, by its presence in the intestine, present a target for the immune system, making it less likely that lymphocytes (see later) would cause autoimmune diseases [4]. In a mixed malarial infection, the presence of *Plasmodium malariae* may moderate symptoms caused by *P. falciparum* [5]. Parasites living in the intestinal tract, such as tapeworms, compete with the host for nutrients passing through the gut. Physical damage can result from attachment by hooks to the intestinal wall (some tapeworms) or migration of helminth larvae through body tissues and organs. Some parasitic worms (e.g., *Ascaris*) can, by their numbers, physically obstruct the intestine. Destruction of cells or tissues can be a consequence of feeding by the parasite such as when *Entamoeba histolytica* attaches to the intestine and ingests epithelial cells. The malarial parasite, *Plasmodium* develops within the host's erythrocytes, causing them to lyse and leading to anemia. Toxins or waste products of the parasite can also cause damage to the host, as well as activating an intense immune response from the host that can have harmful consequences, akin to what occurs in toxic shock syndrome. The presence of a parasite or its secretions may cause an allergic response. In many helminth infections, a large part of the disease pathology is caused by the human immune response. Parasites produce a variety of enzymes, including proteases, peptidases, hydrolases, lipases, and phospholipases that can damage the integrity of host tissues and cell membranes, providing nutrients and shelter for the parasite and avenues for parasite expansion.

---

[*] This article is based on Chapter 45 in the second edition of this series, and Chapter 49 in the third edition, "Introduction to Parasites" by Fred Schuster (second edition), and Purnima Bhanot and Fred Shuster (third edition). The previous contributions of Dr. Shuster (deceased) are very much appreciated and acknowledged.

TABLE 47.1

Protistan Parasites of Humans

Parasite Genus	Species	Disease	Area Affected	Geographic Distribution	Transmission	Diagnosis
				Amoebozoa (Amoebae)		
Acanthamoeba	Several species: *A. castellanii*, *A. culbertsoni*, *A. healyi*, *A. polyphaga*, *A. hatchetti*	Granulomatous amoebic encephalitis	Cutaneous lesions, CNS	Cosmopolitan	Contact with soil, water, air-blown cysts, immunocompromised status	IFA, brain/skin biopsy, culture, PCR
		Amoebic keratitis	Cornea	Cosmopolitan	Corneal trauma, contact lens use	Corneal scraping, serology, culture, PCR
Balamuthia	*B. mandrillaris*	Granulomatous amoebic encephalitis	Cutaneous lesions, CNS	Cosmopolitan	Contact with soil, water, air-blown cysts	Brain/skin biopsy, culture, serology, PCR
Entamoeba	*E. histolytica*	Amoebiasis; amoebic dysentery	Colon, liver, brain	Cosmopolitan	Fecal-oral route, sewage-contaminated uncooked foods	Stool samples for trophic amoebae or cysts, immunoassay, serology for tissue stages
				Opisthokonta (Protists with fungal affinities/origins)		
Microsporidia	*Encephalitozoon cuniculi* *Enterocytozoon bieneusi*	Microsporidiosis	Intestinal tract, brain	Cosmopolitan	Fecal-oral transmission, immunocompromised status	Tissue biopsy, H&E, trichrome, and IFA staining, ELISA
Pneumocystis	*P. jirovecii*	*Pneumocystis (carinii) jirovecii* pneumonia (PCP)	Lungs	Cosmopolitan	Airborne transmission, immunocompromised status	Bronchoalveolar lavage
				Chromalveolata (Protists with apicoplasts; ciliates)		
Babesia	*B. divergens* *B. microti*	Babesiosis	Blood	Europe, United States, Africa, Asia	Zoonotic disease, insect transmission (tick), blood transfusions	Blood films, PCR, ELISA, animal inoculation
Balantidium	*B. coli*	Balantidiosis	Intestine; mostly in pigs	Tropical to subtropical	Fecal-oral route	Trophic ciliates or cysts in stool sample
Cryptosporidium	*C. hominis*	Cryptosporidiosis	Small intestine	Cosmopolitan	Fecal-oral route, water, contaminated foods, immunocompromised status	Stool, sputum, urine samples for oocysts, modified acid-fast (positive), IFA, PCR
Cyclospora	*C. cayetanensis*	Cyclosporiasis	Small intestine	Cosmopolitan	As for *Cryptosporidium*	As for *Cryptosporidium*
Plasmodium	*P. falciparum*	Malignant tertian malaria (48-hour periodicity)	Blood, brain	Africa, Asia, South America	Mosquitoes (*Anopheles* spp.)	Blood films, PCR, IFA, ELISA
	P. vivax	Benign tertian malaria	Blood	Asia, North Africa		Blood films, PCR, IFA
	P. malariae	Quartan malaria (72 periodicity)	Blood	SE Asia, Central America, Pacific area		Blood films, PCR, IFA
	P. ovale	Mild tertian malaria	Blood	West Africa, Pacific area, India		Blood films, PCR, IFA
Toxoplasma	*T. gondii*	Toxoplasmosis	Fetus	Cosmopolitan	Fecal-oral route, infected meat, cat feces, immunocompromised status	ELISA, PCR, IFA, culture, commercial kits available

(Continued)

TABLE 47.1 (Continued)

Protistan Parasites of Humans

Parasite Genus	Species	Disease	Area Affected	Geographic Distribution	Transmission	Diagnosis
			Excavata (Flagellate or amoeba-flagellate protists)			
Giardia	G. duodenalis	Giardiasis	Small intestine	Cosmopolitan	Fecal-oral route	Stool sample, ELISA, IFA, commercial kits available
Leishmania	L. tropica L. major	Old World leishmaniasis (Oriental sore)	Cutaneous ulcerated lesions	Africa, Middle East, Asia Minor, India	Sandfly (Phlebotomus et al.)	Presence of Leishman-Donovan bodies in aspirates (non-flagellate [amastigote] stages in tissue aspirates). ELISA, PCR, clinical ID
	L. donovani	Visceral leishmaniasis (Kala azar)	Enlarged liver and spleen cutaneous lesions	India, Africa, Mediterranean area		
	L. braziliensis	Mucocutaneous leishmaniasis (Espundia)	Cutaneous lesions	Latin America north to Mexico		
	L. mexicana	New World leishmaniasis (Chiclero ulcer)	Cutaneous lesions	South and Central Americas		
Naegleria	N. fowleri	Primary amoebic meningoen-cephalitis	CNS	Cosmopolitan	Swimming, water sports (fresh water)	Microscopic examination of CSF, culture, PCR
Trichomonas	T. vaginalis	Trichomoniasis	Urogenital tract of females and males	Cosmopolitan	Sexually transmitted disease	Vaginal discharge (wet-mount), commercial kits, PCR
Trypanosoma	T. brucei rhodesiense, T. brucei gambiense	East and West African sleeping sickness	CNS	Africa	Tsetse fly (Glossina spp.)	Blood films, CSF, culture, lymph node aspirate, ELISA, card agglutination test
	T. cruzi	American trypanosomiasis (Chagas' disease)	Cardiovascular, gastrointestinal systems, brain	Argentina north to Mexican border	Reduviid bug (Triatoma spp.), blood transfusions	Blood films, IFA, culture, clinical ID, PCR, xenodiagnosis (in triatomid bug)

Abbreviations: CSF = cerebrospinal fluid; ELISA = enzyme-linked immunosorbent assay; H&E = hematoxylin-eosin stain (tissue); IFA = immunofluorescent antibody staining; IIF = indirect immunofluorescent staining (tissue); PCR = polymerase chain reaction.

The course of parasitic infections can be influenced by factors such as the infective dose of the parasite, the virulence of the particular strain or species, the ability of the parasite to circumvent the host's immune system, and the immune status and/or health of the host. Immunosuppressed individuals, the elderly, and the young are more vulnerable to many diseases, parasitic or otherwise.

Host Factors

The Immune Response

The Humoral Pathway

There are two major immune responses of the host. One is the *humoral* pathway resulting in antibody production by B-lymphocytes specifically targeted at the invading parasite, also called adaptive or inducible immunity. Immunity can be conferred to a neonate by transplacental transfer of maternal antibodies and provide protection for several months after birth before it begins to fade and before the child's own immune system becomes functionally mature. This offers a window of opportunity for parasites to infect and cause disease or death. Falciparum malaria has its greatest impact on children between 1 and 5 years of age. Antibodies produced during parasitic disease remain for varying lengths of time even after the disease has been cleared, conferring weak or strong immune protection against reinfection. Immunoglobulin M (IgM) is associated with the initial stage of infection, and IgG with postacute to convalescent or later stages of infection, protection against recurrent infection, and placental transfer. IgA is a secretory antibody released from mucosal surfaces (intestinal, pulmonary and urogenital tracts, corneal surface), and has a role in antiparasite activity. IgE is linked to allergic responses and parasitic infections and causes degranulation of eosinophils, with release of hydrolytic enzymes, peroxidases, and a core cationic protein granule. Eosinophils can adhere via antibodies to the surface of intestinal helminths and release granule contents that aid in attacking the worm's surface, allowing other immune molecules to penetrate and destroy the worm.

Cell-Mediated Immunity

The second type is the *cell-mediated* immune response (innate or nonspecific immunity) leading to the activation of T-lymphocytes, activated macrophages, killer T-cells, and other cell types that are functionally equipped to destroy the parasite, particularly those parasites that are intracellular and sheltered from contact with antibodies. The cell-mediated response can activate the *complement system* (alternative and classical pathways) that may or may not damage and/or lyse the parasite. Cell-cell communication between immune response components is the function of interleukins, cytokines, and other molecules such as tumor necrosis factor and interferon. *Plasmodium* toxins stimulate T-cell proliferation, which leads to overproduction of interferon and tumor necrosis factor. An excess of these proinflammatory cytokines can be as damaging to the host as the disease itself, affecting erythrocyte formation and a possible cause of malaria-associated anemia [6]. Other parasites have evolved mechanisms

to neutralize these activators. Tapeworms of various species can inhibit production of interferon and interleukins, thus muting the immune response [4].

Parasite Evasion of the Immune Response

Parasites have intrinsic and extrinsic ways of avoiding the host's immune response. Chief among these is antigenic variation that is well developed in several protozoa. *T. brucei*, *P. falciparum* and *Giardia lamblia* are all capable of shedding their antigenic coat to which the host has developed antibody sensitivity, and replacing it with a new coat requiring a build-up of new antibody by the host. This delay enables the parasites to remain within the host and prolong periods of transmission.

Leishmania spp. and *Trypanosoma cruzi* develop safely within the host's macrophages, activist components of the antiparasite response. Other parasites (e.g., schistosome worms) cloak themselves with host molecules to escape detection as foreign antigen. Components of the complement pathway can be neutralized by some parasites by preventing complement activation and subsequent lysis. Still other parasites may shed complement activation components as they bind to the parasite surface, or produce substances that activate and deplete complement, but at a safe distance from the surface of the parasite.

Human Parasitic Diseases

Protists and helminths are two major groups of organisms that are parasites of humans (Tables 47.1 and 47.2). The insects and ticks include human parasites but they will be treated in this review with regard to their role as vectors of parasitic diseases (Table 47.3). These tables are not comprehensive and include the more familiar human parasites as well as those responsible for the more devastating diseases, causing both morbidity and mortality in large numbers of individuals. There are less than 400 species of protists and helminths that are parasitic in humans: approximately 300 species of worms and greater than 70 protist species [7].

Protists

Disease agents belonging to this group are responsible for some of the most widespread and fatal diseases of humans and other mammals. Malaria is the deadliest parasitic disease. The World Health Organization (WHO) estimates that there were about 220 million cases of malaria in 2010 causing almost 600,000 deaths [1]. Most of these deaths are of children in sub-Saharan Africa. African sleeping sickness or trypanosomiasis is another major killer disease, not only of humans but also cattle and wild game—which serve as reservoirs of disease—preventing the raising and grazing of cattle in vast areas of Africa.

Falciparum Malaria

An abundant literature on falciparum malaria exists and this capsule summary is insufficient to convey all that is known of the disease. *Plasmodium falciparum* is the most lethal of the human malarial parasites, and the only one to affect the CNS, causing

TABLE 47.2

Helminth Parasites of Humans

Parasite Genus	Disease	Body Area Affected	Distribution	Transmission
		Trematodes (Flukes)		
Clonorchis sinensis	Clonorchiasis	Liver, bile duct	Asia	Eating raw or undercooked fish
Schistosoma haematobium	Schistosomiasis (bilharziasis)	Bladder	Africa, Middle East	Aquatic larvae (cercariae) penetrate through skin
S. japonicum		Veins draining small intestine	Japan, China, SE Asia	
S. mansoni		Veins draining large intestine	Middle East, Africa	
		Cestodes (Tapeworms)		
Diphyllobothrium latum	Diphyllobothriasis (fish tapeworm)	Intestine	United States, Europe	Eating raw or undercooked fish
Echinococcus granulosus	Hydatid cyst (larval stage)	Brain	Cosmopolitan	Acquiring ova from infected dogs
saginatus	Taeniasis (beef tapeworm)	Intestine	Cosmopolitan	Eating raw or undercooked beef
Taenia solium	Taeniasis (pork tapeworm)	Intestine	Cosmopolitan	
	Cysticercosis (larval stage)			
		Brain		
		Nematodes (Round Worms)		
Ancylostoma duodenale	Hookworm	Small intestine	SE Asia, northern Africa, Europe	Terrestrial larvae penetrate through skin
Ascaris lumbricoides	Ascariasis	Intestine	Cosmopolitan	Contact with worm ova
Baylisascaris procyonis	Baylisascariasis	Migrating larvae	United States	Transmitted via raccoon feces (geophagia)
Dracunculus medinensis	Dracunculiasis (guinea worm)	Females release ova in subcutaneous sites (legs, feet)	Middle East	Intermediate host (crustacean) present in drinking water
Enterobius vermicularis	Enterobiasis	Colon, perianal area	Cosmopolitan	Contact with worm ova
Gnathostoma spinigerum	Gnathostomiasis	Migrating larvae	Asia, SW Pacific	Eating raw or undercooked fish
Necator americanus	Hookworm	Small intestine	SE Asia, United States, SW Pacific	Terrestrial larvae penetrate through skin
Onchocerca volvulus	Onchocerciasis (river blindness)	Eyes	Africa, Central and South Americas	Transmitted by black fly (*Simulium*)
Trichinella spiralis	Trichinellosis	Muscle tissues, migrating larvae	Cosmopolitan	Eating raw or undercooked pork
Wuchereria bancrofti	Filariasis (elephantiasis)	Lymphatic system	SE Asia, Pacific area, Central Africa	Transmitted by mosquitoes (Anopholes Culex, Aedes)

TABLE 47.3

Insects and Acarids Important as Vectors of Human Parasitic Disease

Insect	Genus	Disease	Transmission
Black fly	*Simulium*	Onchocerciasis	Injection from salivary glands
Cockroach	All species	None	Mechanical transmission
House fly	*Musca domestica*	None	Mechanical transmission
Mosquito	*Anopheles*	Malaria	Injection from salivary glands
Sandfly	*Phlebotomus*	Leishmaniasis	Injection
Tick	*Ixodes*	Babesiosis	Injection
Triatomid bug (kissing bug)	*Triatoma*	Chagas' disease	Defecation at site of insect bite
Tsetse fly	*Glossina*	Trypanosomiasis	Injection

cerebral malaria [8]. The disease is actually two infections: malaria as the blood disease and as a fulminant cerebral infection. Cerebral malaria is more likely to develop in naive children than in adults with some immunity to the disease. Following injection of the infective stage (sporozoite) from *Anopheles* spp., the organism invades hepatic cells where asexual reproduction occurs. Infected liver cells release merozoites, which penetrate erythrocytes, reproduce asexually, and cause their hemolysis at approximately 48-hour intervals. The cycle of infection and lysis is synchronous. At lysis, there is release of toxins and wastes from the erythrocytes that causes the chills and fever typifying the disease. Infected erythrocytes become sequestered in the microvasculature of the brain. They bind to the endothelial walls of the blood vessels by parasite-induced knobs on the erythrocyte surface, and form rosettes of erythrocytes facilitating vessel occlusion [8]. The consequences of this blockage are oxygen insufficiency, necrosis, edema, hemorrhage, and inflammation in the brain. Overproduction of proinflammatory cytokines stimulated by *Plasmodium* antigens may aggravate the condition [9, 10]. Prevention includes screens, netting, mosquito repellent, and prophylactic drugs if traveling in malaria-endemic areas.

Plasmodium vivax

P. vivax is most geographically wide-spread human malaria species [11]. It is most common in Asia and South and Central America. In these areas, *P. falciparum* and *P. vivax* malaria have roughly equal prevalence. Importantly, intrahepatic stages of *P. vivax* can form dormant stages known as *hypnozoites*. Hypnozoites can stay dormant for weeks up to a year before reactivating to cause the relapses characteristic of *P. vivax* infection. The primary receptor for *P vivax* on the erythrocyte surface is the Duffy antigen receptor for chemokines (DARC). West Africans carry the Duffy-negative phenotype [12] and are resistant to *P. vivax* malaria. *P vivax* preferentially invades immature red blood cells and parasitemias are usually lower than in *P. falciparum* infections [11].

Diagnosis and Assessment

The gold standard for malaria diagnosis is the microscopic examination of blood smears. However, simple and sensitive antibody tests that detect specific parasite proteins are now widely used. Some these tests can distinguish between different *Plasmodium* species.

African Sleeping Sickness

The subspecies *Trypanosoma brucei gambiense* is an etiologic agent of African trypanosomiasis, also known as Gambian sleeping sickness in West and Central Africa. The tsetse fly, *Glossina* spp., is the vector of the disease and, once infected, remains infected for its life. The fly acquires the parasite while obtaining blood from a human and passes it on to another host after the transmissive stage of the parasite enters the fly's salivary glands. In both the fly and the human, the parasite passes through several pleomorphic stages, ranging from a slender swimming form with an elongate undulating membrane and free flagellum (Figure 47.1) to a shorter stage lacking the free flagellum that is found in the salivary glands

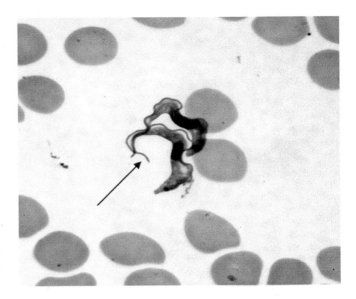

FIGURE 47.1 Trypanosomes seen in a stained blood film. The two organisms (trypomastigote stage) have a wavy outline due to undulating membrane that runs the entire length of the body. A free flagellum projects beyond the undulating membrane. Giemsa stain. (Courtesy of DPDx Parasite Image Library at the Centers for Disease Control and Prevention.)

of the tsetse fly. *T. b. gambiense* causes a chronic form of sleeping sickness affecting the central nervous system, while a related subspecies, *T. b. rhodesiense*, causes a similar but fulminant infection in humans. Of diagnostic importance is elevated protein (and IgM) and leukocyte levels in the central nervous system (CNS). *T. b. rhodesiense* occurs in eastern and Central Africa and is also found in wild game animals that serve as reservoirs for the parasite. *T. brucei brucei* infects cattle and wild animals (nagana), but not humans. Sleeping sickness is biphasic: an early blood phase and a later CNS phase, and different drugs are used for the different stages (Table 47.4).

American Trypanosomiasis

American trypanosomiasis is caused by *Trypanosoma cruzi*, and transmitted by a triatomid ("kissing") bug. Transmission from insect to human occurs when the bug defecates on the skin while feeding. The deposited fecal material contains the infective parasites and is usually rubbed into the break in the skin caused by the bite or into the eyes.

Babesiosis

This is a zoonotic blood infection transmitted by the bite of an infected tick, and similar to malaria in intraerythrocytic location and malaria-like symptoms (fever, chills, aches, and anemia). Two species are involved in human infections: *Babesia microti* in North America and *B. divergens* in Europe, with the latter species being the more virulent (5% vs. 42% mortality, respectively) [13, 14]. Coinfected ticks can spread both babesiosis and Lyme borreliosis disease (caused by the bacterium *Borrelia burgdorferi*) simultaneously. Splenectomized and immunosuppressed individuals are at high risk for developing fulminating disease if infected. Rodents, deer, elk, and cattle serve as reservoir hosts. Diagnosis can be difficult, and hamster inoculation (xenodiagnosis) is useful to boost parasitemia

TABLE 47.4

Drug Therapy for Protistan and Helminthic Infections

Disease	Recommended Drug Therapy
Protistan Diseases	
Acanthamoebiasis	Keratitis: Polyhexamethylene biguanide, chlorhexidine gluconate with or without Brolene® Amoebic encephalitis: Empirical therapy—pentamidine isethionate, sulfadiazine, flucytosine, fluconazole, or itraconazole
Balamuthiasis	Encephalitis: Empirical therapy — pentamidine isethionate, sulfadiazine, clarithromycin (or azithromycin), fluconazole, flucytosine
African sleeping sickness, west (*Trypanosoma b. gambiense*)	Human infection Early stage: Pentamidine isethionate or suramin Late stage (CNS involvement): Melarsoprol, eflornithine
African sleeping sickness, east (*Trypanosoma b. rhodesiense*)	Human infection Early stage: Suramin Late stage: melarsoprol
Entamoebiasis (amoebic dysentery)	Iodoquinol or metronidazole or tinidazole
Chagas' disease	Benznidazole or nifurtimox
Giardiasis	Metronidazole or tinidazole
Leishmaniasis	Cutaneous, mucosal, and visceral disease: Sodium stibogluconate
Microsporidiosis	Albendazole plus fumagillin
Naegleriasis (primary amoebic meningoen-cephalitis)	Amphotericin B plus miconazole
Plasmodium falciparum (malaria)	Malaria acquired in chloroquine-resistant area: Atovaquone-proguanil or quinine sulfate plus doxycycline or mefloquine or artesunate plus mefloquine
Plasmodium vivax (malaria)	Malaria acquired in chloroquine-resistant area: Quinine sulfate plus doxycycline or mefloquine
Plasmodium spp. (other human malarias)	Malaria acquired in chloroquine-sensitive area: Chloroquine phosphate or quinidine gluconate (oral), quinidine dihydrochloride (parenteral) or artemether For prevention of relapse: Primaquine phosphate General prevention: Chloroquine-sensitive areas—mefloquine, doxycycline, chloroquine-resistant areas—chloroquine phosphate or atovaquone proguanil
Pneumocystis pneumonia (PCP)	Trimethoprim-sulfamethoxazole
Toxoplasmosis	Pyrimethamine plus sulfadiazine
Trichomoniasis (*Trichomonas vaginalis*)	Metronidazole or tinidazole

to make a diagnosis. *In vitro* cultivation is possible but exacting [15]. Prevention includes wearing protective clothing (long sleeves, full-length pants) and applying tick repellent when working or recreating in areas where ticks are abundant (particularly in the Northeast, upper Midwest, West, and Northwest coast of the United States).

Cryptosporidiosis and Cyclosporiasis

Caused in humans by *Cryptosporidium hominis*, *C. parvum*, and *Cyclospora cayetanensis*; spread by fecal-oral contamination, usually of drinking water but also food (raw shellfish, uncooked vegetables, salad greens). Person-to-person spread can occur among children (nurseries and day care) and among the elderly in nursing facilities. Infective oocysts are shed in stools; they are visible with modified acid-fast staining. Oocysts of *Cryptosporidium* are smaller (~5 µm) than those of *Cyclospora* (~9 µm) [16]. A major outbreak of cryptosporidiosis occurred in Milwaukee, Wisconsin, in 1993 that resulted in more than 400,000 infections was caused by a contaminated water supply.

Cyclospora outbreaks have been traced to imported produce. In 2013, a cyclosporiasis outbreak sickened more than 600 people in 25 states. Some cases were traced to salad mixes and fresh cilantro. Symptoms are typical of gastrointestinal infection

and can be particularly serious in HIV/AIDS patients. Diarrhea is self-limiting (~2 weeks) in immunocompetent individuals. Antibiotics are not helpful in treatment except, perhaps, in HIV/AIDS patients. Prevention: avoid possibly contaminated drinking water (bottled water for HIV/AIDS patients), uncooked fruits and vegetables, and contact with farm animals.

Toxoplasmosis

Caused by *Toxoplasma gondii* and transmitted to humans in undercooked meat (horse meat, pork, lamb, beef) and by cat feces containing oocysts [17]. The domestic cat is the definitive host. In humans, the main pathway of infection is via meat, and the cat is unnecessary in transmission of *Toxoplasma* to humans. Symptoms are flu-like and pass quickly, but congenital infections (leading to hydrocephalus, microcephaly, chorioretinitis, cerebral calcification, and mental retardation in the neonate and juveniles) can develop in pregnant women who are not immune from earlier exposure; the risk of fetal infection increases with the gestation period. The organism is also an opportunistic pathogen of HIV/AIDS patients and other immunosuppressed individuals. Reactivation of latent infections can occur in individuals who secondarily become immunosuppressed. Toxoplasmosis is more common abroad (20–40% of adult Americans are seropositive for *Toxoplasma* antibodies vs. approximately 70% in France and about 70–90% in Latin America) and in immigrants to the United States [18]. Prevention includes avoiding raw or undercooked meats that may transmit parasite. Pregnant women in particular should avoid eating undercooked meats or contact with cat litter boxes.

Leishmaniasis

Leishmaniasis is caused by different species of the genus, *Leishmania*, each with a distinct geographic distribution. They are transmitted by certain species of the sandfly. It occurs primarily as either cutaneous or visceral leishmaniasis. Cutaneous leishmaniasis is caused mostly by *L. major*, *L. tropica*, and *L. Mexicana*. Visceral leishmaniasis is caused mostly by *L. donovani* and *L. chagasi*. The parasites infect macrophages and undergo asexual development within the cell. In cutaneous leishmaniasis, the parasites remain at the site of infection causing a skin lesion. The skin ulcer is often self-healing and leads to long-term immunity against cutaneous leishmaniasis. In visceral leishmaniasis, the parasites infect the reticuloendothelial system compromising the patient's immune system. Domestic dogs and rodents are the major reservoirs of human disease. Prevention relies on control of the sandfly.

Giardiasis and Amoebiasis

Giardiasis is caused by *Giardia lamblia*, an amitochondriate, flagellated protozoan. It is transmitted through water contaminated with environmentally resistant cyst forms. Beavers are the major reservoir. After encountering low pH in the stomach and pancreatic enzymes, the ingested cysts excyst to form two motile trophozoites in the small intestine. The trophozoite has two nuclei, eight flagella and an adhesive disc on its ventral surface. The adhesive disc allows the trophozoites to attach to epithelial cells of the intestinal villi. The trophozoite then divide through binary fission. Some trophozoites eventually encyst in the small intestine and pass out with feces.

Infection with *G. lamblia* is often symptomatic. Clinical disease consists of severe diarrhea. Flattening of the villi as a result of infection also leads to malabsorption. The parasite is able to evade the host's immune reaction through variation of its major surface protein, the cysteine-rich variant surface protein. Prevention of Giardiasis consists of proper disposal of fecal matter and filtering drinking water.

Amoebiasis is caused by the pseudopod forming protozoan, *Entamoeba histolytica*. Infection can lead to liver abscess and amebic colitis. *E. histolytica* life cycle consists of the infectious cyst form and the invasive trophozoite form. Cysts are generally ingested by eating fecally infected food or water. In the intestine, the cyst forms eight trophozoites. Trophozoites can either invade the intestinal mucosa forming flask-shaped ulcers or colonize the liver.

Helminths

The helminths are a group of parasitic organisms that includes flatworms and round worms. Flatworms have a dorsoventrally flattened body and include cestodes such as tapeworms and trematodes (also known as flukes) such schistosomes. Roundworms include soil-borne nematodes and filarial worms such as those that cause lymphatic filariasis.

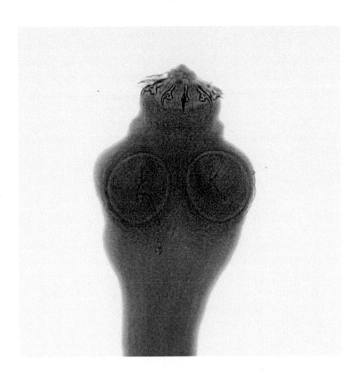

FIGURE 47.2 Scolex ("head") of an adult tapeworm, *Taenia solium*. The scolex has a ring of hooks (rostellum) and below that are four suckers (acetabula). The scolex tapers to form the neck, which is the generative region where proglottids are formed. (Courtesy of DPDx Parasite Image Library at the Centers for Disease Control and Prevention.)

Cestodes (Tapeworms)

Tapeworms are best known as intestinal parasites but, as larval stages, can penetrate the intestinal wall and migrate through tissues and organs. Damage by the migration of the larvae and the final site of the larvae can cause life-threatening conditions. In a "normal" infection of the human, the larval stage (in beef, pork, or fish) is digested free in the intestine, grows, and reaches sexual maturity (Figure 47.2).

Pork Tapeworm

The human, as the definitive host, contracts the disease by eating undercooked pork. The pig, as the intermediate host, becomes infected with ova when it has access to human fecal matter containing proglottids and/or eggs. This occurs in areas of the world where human waste is used as fertilizer or in rural areas where pigs forage on refuse. In the pig intestine, the eggs hatch and produce larvae that penetrate the intestinal wall and migrate to muscle tissue where they become encapsulated. Pork flesh containing these larvae (bladder worms), which can be seen as lustrous beads, is called measly pork and can be identified with the naked eye. In the human small intestine, the bladder worm is digested free from its capsule, attaches to the intestinal wall, and develops into the adult worm. The human can inadvertently become a dead-end intermediate host for the larva if the tapeworm egg itself is ingested. The freed larva (cysticercus) then migrates through host tissue, coming to rest in muscle or organs and causing a condition known as cysticercosis, or neurocysticercosis if the traveling larva ends its journey in the brain. The latter condition which is accompanied by seizures can be fatal. In the

United States, most cases of cysticercosis occur in foreign-born Hispanics (Mexico and Latin America), while it is relatively rare in native-born citizens [19]. An outbreak of cysticercosis among Orthodox Jews—proscribed from eating pork or pork products—was traced to Hispanic household employees. Prevention includes avoiding undercooked pork products, particularly in developing countries where swine have access to garbage and other waste materials. Emphasize importance of hand washing to food handlers.

Trematodes (Flukes)

The fluke body is unsegmented. The adult worm may have one or two suckers, one ventrally located and the second a sucker that surrounds the mouth leading to a sac-like gut but lacking an anal opening. Most flukes are monoecious (have male and female sex organs in the same individual) although *Schistosoma* spp. are dioecious (the sexes in separate individuals). Intestinal flukes can be identified by the shapes of their ova as seen in stool samples (Figure 47.3).

Schistosomiasis

Schistosomiasis (also called bilharzia or snail fever), caused by *Schistosoma* spp., is transmitted by water contaminated with the parasite's larval forms. Chronic schistosomiasis causes delays in physical growth and cognitive development in children. The schistosomes are found in the circulatory system of humans. Three of the medically important species recognized in this chapter are *S. haematobium;* which is found in veins associated with the urinary bladder and lower abdomen; *S. japonicum;*

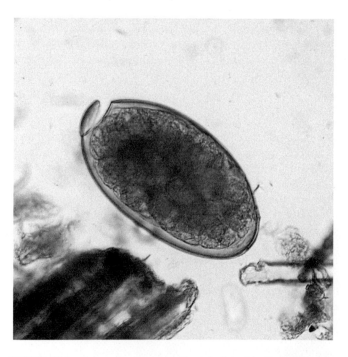

FIGURE 47.3 Ovum of a flatworm (*Fasciola hepatica,* liver fluke). The eggs are shed in the stool and identifying them is probably the chief way in which diagnosis is made. The egg has a distinctive operculum at one end through which the larva (miracidium) will emerge. (Courtesy of DPDx Parasite Image Library at the Centers for Disease Control and Prevention.)

which is found in veins about the small intestine; and *S. mansoni,* which is found in veins associated with the large intestine.

The female is harbored within a groove of the male worm, ensuring close contact and fertilization of ova. Depending on the site of the adults in the body, eggs can be found in the urine or in the stool. Ova that do not pass out of the body can penetrate the venule walls, causing granuloma formation. Once in water, the egg hatches to release a ciliated miracidium that infects a snail as secondary host. The snail host is specific for each of the three different species. After undergoing development in the snail, a cercarial stage is released which can infect humans by penetrating the skin. They then enter the circulatory system and are carried to the lungs and liver, and finally to their specific sites in the venous circulation. Prevention calls for avoidance of streams, lakes, and ponds in endemic areas for washing and swimming because of cercariae in the water.

Helminth Infections: Nematodes

Nematode worms are among the most numerous organisms in the world. The group includes many free-living forms as well as medically significant parasites.

Onchocerciasis

The disease, also called river blindness, is caused by the filarial worm *Onchocerca volvulus,* and is transmitted by the black fly (*Simulium* spp.). The fly becomes infected with larvae (microfilariae) when it feeds on the blood of an individual with onchocerciasis. The microfilariae migrate from the fly's stomach to its proboscis. Once inoculated into the human host, the male and female worms accumulate in cutaneous nodules. Following mating, the larvae appear in the circulatory system. The disease is found in Central and South America, Mexico, the Middle East, and Africa, with distribution to the Americas a consequence of the slave trade. Microfilariae can also enter the eye where damage is done in large part by the host's immune response to the larvae. Prevention includes water filtration in endemic areas, insecticide use to eliminate black flies.

Lymphatic Filariasis

The disease, also called elephantiasis, is caused by the filarial worms *Wuchereria bancrofti* and *Brugia malayi,* transmitted to humans via mosquito bite. Infected mosquitoes deposit worm larvae on the skin. The larvae enter the lymphatics and mature into adults. Adult living worms can suppress the host's immune response. Dead or dying worms can elicit an immunopathology that can lead to an occlusion of lymphatics resulting in lymph accumulation in tissues or lymphedema in the affected region of the body. Elephantiasis is a consequence of long-term infection. Prevention requires the use of netting and mosquito repellant.

Nematode Infections

Ascariasis

One of the most common nematode infections, it is caused by *Ascaris lumbricoides.* It is acquired by ingesting food contaminated with Ascaris eggs. The adult worm is found in the human

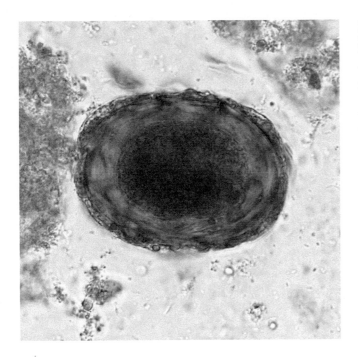

FIGURE 47.4 Ovum of the nematode worm *Ascaris lumbricoides*. The egg with an embryo inside has a thick, protective, and mammilate wall that can withstand long-term environmental exposure. (Courtesy of DPDx Parasite Image Library at the Centers for Disease Control and Prevention.)

intestine and eggs are shed into the feces (Figure 47.4). The thick-walled eggs are highly resistant to environmental extremes and can remain viable for months after being shed. Large numbers of worms can cause blockage of the intestinal lumen, requiring surgical removal. Ascariasis is more likely to occur in rural farming areas where there is close contact between humans and pigs, both of which share the worm infection (*A. lumbricoides* in humans and *A. suum* in swine). Prevention requires proper disposal of human waste and avoiding soil containing human or swine excrement, particularly for children who are more likely to transfer eggs on their fingers after playing in or eating soil.

Trichuriasis

Trichuriasis is caused by *Trichuris trichiura*, commonly known as the whipworm. It is transmitted to humans through ingestion of *T. trichiura* eggs. Ingested eggs hatch into larvae in the small intestines, where they develop into adults. Adults *T. trichiura* migrate into the colon and permanently attach to the mucosa by embedding their anterior portion in the tissue. Adult females release eggs that are passed out in feces. Heavy infections can lead to dysentery and colitis in children. Severe cases may lead to rectal prolapse. Prevention calls for proper disposal of human waste and avoiding soil containing human excrement.

Parasite Epidemiology

Epidemiology: Ways That Parasites Spread

Parasite epidemiology attempts to explain how diseases are spread from person to person. Epidemiology has moved well beyond its beginnings in hand washing and the fundamental principles of hygiene. Modern epidemiology focuses more on the parasite genome, enabling identification of parasite strains and the presence of genes that confer drug resistance. The tools of the trade include evolutionary analyses and construction of phylogenetic trees, mapping the genomic composition of parasite populations, and, not forgetting the vectors, the genomes of the insects that transmit disease [20]. Molecular studies of parasite distribution and spread are the counterpart of studies tracing outbreaks of bacterial infections through plasmids and antimicrobial resistance genes that have revolutionized prokaryote epidemiology. In this chapter, however, more attention is given to the classic approach to epidemiology. Categories recognized below are not mutually exclusive and there much is overlap between them.

Insect Transmission

Elimination of insects as vectors—a desirable goal but not always possible—can reduce incidence of disease. Controlling populations of *Plasmodium*-carrying mosquitoes using insecticides or protecting susceptible children in malarious areas from being bitten by use of insecticide-impregnated bed netting, can reduce deaths due to malaria. Mosquitoes, however, have developed resistance to some insecticides (e.g., DDT), reducing their efficacy in controlling insect populations. As noted earlier, insects such as roaches and flies can transmit protistan cysts and helminth ova on mouthparts, legs, and feet as well as within their digestive tract. Regurgitation of gut contents prior to feeding and fecal deposition are other ways in which infectious material can be transferred [21].

Fecal-Oral Transmission

Human or animal feces transmit many parasitic diseases by parasite ova and protistan cysts. Such transmission can occur where feces are used as fertilizer ("night soil") or contaminate water used for bathing, drinking or irrigation. This is particularly the case with cysts of *Entamoeba histolytica*, *Giardia duodenalis*, and *Balantidium coli* that can contaminate the water supply in countries lacking water treatment standards. Feces-contaminated vegetables, eaten without cooking, are another result of fecal-oral transmission. This has become increasingly common in recent years as such foods, sometimes from abroad but also from "safe" farms, come on the market.

Direct Blood Transmission

This route is perhaps a consequence of human ingenuity that allows for blood collection for transfusions. Although the blood supply is monitored for HIV and prion diseases (bovine spongiform encephalopathy), etiologic agents of parasitic diseases are usually not looked for. Malaria and babesiosis, both blood infections, have been transmitted by human blood transfusions, as has Chagas' disease and toxoplasmosis.

Zoonotic Transmission

A zoonotic disease is one that is spread from animal to human. Some of the zoonotic diseases include babesiosis (from rodents, cattle, deer, elk, and sheep) to humans via ticks; trypanosomiasis (from game animals and cattle via the tsetse fly); toxoplasmosis (from cats, pigs, lambs, and horses to humans), ascariasis, and echinococcosis. The transition from a mobile hunter-gatherer society to a settled life on farms, and ultimately communities, encouraged domestication of cattle, pigs, and other animals, offering opportunities for animal parasites to jump from one host to a new and virgin host. Nomadic people leave their wastes behind them but as communities increased in numbers, human and animal excreta would accumulate. Some was undoubtedly used as fertilizer as in areas of Asia today, but the cumulative build-up of waste would also contaminate water supplies, affording a means of disease transmission.

Sexual Transmission

Two protistan parasites can be transmitted via sexual contact. *Trichomonas vaginalis* can be spread as a sexually transmitted disease by intercourse, with transmission from male to female or female to male. *Entamoeba histolytica* has been spread by anal intercourse among male homosexuals.

Person-to-Person Transmission

Most parasitic diseases require an intermediate host before cycling to the human host. Pinworms, highly resilient intestinal parasites, can be spread directly from an infected child to other children and adults as a result of poor sanitary habits (careful hand washing). *Cryptosporidium* can be spread directly in high-risk groups such as children in daycare centers and institutionalized individuals.

Food and Food Handler Transmission

Food may serve as a vehicle for transmission of many parasites, including those found in feces. Food contaminated with human or animal waste, particularly if not cooked (e.g., salad greens, fruits), can transmit *Entamoeba histolytica*, *Cryptosporidium*, and *Cyclospora*. Food handlers indifferent to washing hands after toileting have also been implicated in transmission.

Undercooked pork, beef, or fish can carry larval stages of tapeworms or *Trichinella* (pork) (Figure 47.5). In Asia, intermediate hosts of some parasites (e.g., fish, crabs) may be eaten raw or lightly cooked and pass on larval stages to humans and other mammals as definitive hosts.

Water Transmission

Perhaps the simplest and most common vehicle for disease transmission is water, either directly or indirectly. Contaminated drinking water can transmit amoebic dysentery and giardiasis (discussed previously: fecal-oral transmission). Numerous efforts have been made to provide water to water-scarce areas of Africa for drinking, irrigation, and hydroelectric power through the construction of dams. The resultant lakes and irrigation canals, however, have also attracted snails, in particular those that are intermediate hosts for schistosome worms, resulting in a sharp increase in schistosomiasis [22]. The water also provides a breeding ground for mosquitoes that spread malaria.

FIGURE 47.5 Section through muscle tissue showing an encysted larva of *Trichinella spiralis*. If not killed by cooking, the larvae are digested free of their capsules and develop into sexually mature adults. Larvae produced in the lumen penetrate the intestinal wall to encyst in muscle tissues. (Courtesy of DPDx Parasite Image Library at the Centers for Disease Control and Prevention.)

Diagnostic Techniques

As with all infectious diseases, a critical factor in initiating appropriate therapy is early identification of the etiologic agent. After decades of testing, the protocols for detection of many parasitic disease agents are well established. Cultivation of protist parasites from a clinical sample is helpful in some diagnoses. Helminth parasites, however, are not cultured.

Clinical Diagnosis

Some parasitic infections can be recognized based on the patient's symptoms, recent travel history, occupation, contact with domestic and wild animals, unusual diets, immune status, insect bites, and place of birth (i.e., immigrant or native-born). Amoebic dysentery would be a logical diagnosis for a patient with diarrhea who had recently returned from a vacation in a developing country. Examination of a stool sample for amoebic cysts or trophozoites could verify the diagnosis and treatment could be started. Visceral leishmaniasis (kala-azar) would be a differential diagnosis for a child with an enlarged spleen and liver who emigrated from the Middle East (Table 47.1). Many other parasitic diseases have vague symptoms, are difficult to identify, and may require extensive testing. The clinician also has information from the standard panel of diagnostic tests, including the complete blood count (CBC); analysis of CSF (protein, white blood cells, and glucose in the CSF); and radiographic imaging (CT scans, MRIs). Neurocysticercosis, in which the larval stage of the pork tapeworm develops in the brain and eventually becomes encapsulated, will show up in an MRI. Other parasitic infections where calcification facilitates diagnosis are dracunculiasis, balamuthiasis, and baylisascariasis. Infection by

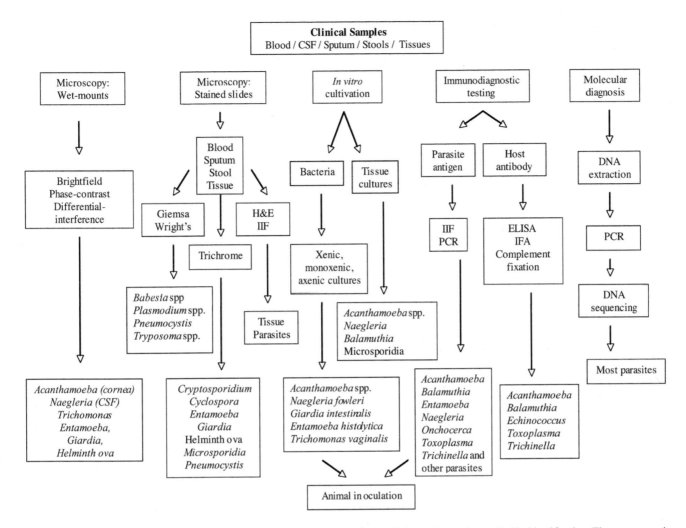

FIGURE 47.6 An algorithm based on the types of clinical samples and the various techniques that can be applied in identification. The same organisms can appear in several pathways, indicating that there is more than one way for identification of the parasite. It is assumed in this algorithm that most parasitic organisms can be identified by the PCR technique.

parasites can lead to elevated IgE levels and eosinophilia. If the blood sample is counted electronically instead of using a stained blood film, eosinophils may not be differentiated from other leukocytes and might be missed. In malaria infections, a drop in erythrocyte numbers and anemia are to be expected and, coupled with a stained blood film, can confirm a diagnosis.

Laboratory Diagnosis

Where testing is based on established criteria, such as cysts in a stool sample or *Plasmodium*-infected erythrocytes in a blood film, a rapid diagnosis is possible. Many parasitic diseases, however, are more difficult to identify. Symptoms may be vague or nonspecific and may mimic a number of other conditions.

Clinical Samples

The most common samples for testing include whole blood (malaria, trypanosomiasis), serum for serological testing (toxoplasmosis), stool samples for gastrointestinal parasites (worm ova), cysts of amoebae (dysentery), tissue samples for hematoxylin-eosin or other types of staining, sputum for lung parasites

(*Paragonimus, Pneumocystis*), vaginal smears (trichomoniasis), and sigmoidoscopic (pinworms), or lymphatic aspirates (trypanosomes). Based on the type of specimen, special handling may be required. Although serum samples for antibody testing can be frozen and stored for prolonged periods of time, whole blood for blood film staining must be processed immediately. With other types of specimens, the integrity of the parasite might be damaged should the sample be kept at or near 37°C for prolonged time periods or, at the other extreme, placed in a refrigerator or freezer. Delayed transit of a sample to the laboratory can also reduce the chance of detecting more fragile parasites. Figure 47.6 presents an algorithm for the testing of clinical samples.

Specific Techniques

Direct and Indirect Testing for Parasites

Testing for the presence of parasites covers a broad range of techniques. Direct testing aims to visualize the parasite with a microscope in a blood film, stool, urine, or bronchoalveolar sample or biopsied tissue. Even with these long-established techniques, however, the parasite can elude the technician. For many parasites, multiple samples must be taken over several days because of periodicity in the presence of parasite stages in the clinical sample. The names characterizing the different malarial diseases (Table 47.1) denote differences in the time intervals at which the acute phase of disease occurs (i.e., approximately 48 hours in tertiary malaria; approximately 72 hours in quartan malaria). Early erythrocytic ("ring") stages of *Plasmodium falciparum* are infrequently seen in circulating blood or blood films because these infected erythrocytes are likely to bind to the walls of blood vessels in the brain. The filarial worm *Wuchereria bancrofti* (cause of elephantiasis) appears in circulating blood at night, corresponding to the time that mosquito vectors would be most actively feeding.

In indirect testing, where testing for antibodies is involved, samples from the acute as well as the convalescent phases of the disease are needed to detect changes in antibody titer or in types of antibodies present (IgM vs. IgG). Antibodies in the patients' blood samples may cross-react with the "wrong" antigens, leading to misdiagnosis and unnecessary or inappropriate therapy. Some immunoserology tests may give problematic results (e.g., for neurocysticercosis) and may need corroboration by some other technique to get a definitive diagnosis.

Cultivation of Parasites

In vitro cultivation of parasites is useful in confirming identification, obtaining yields of parasites for molecular and biochemical studies, for assessing virulence, for electron microscopy, and for testing antimicrobial efficacy in treating infections. Many protist parasites can be cultured, although often requiring complex media, demanding growth conditions, and inconvenient transfer schedules to keep cultures from dying out. Some media are available commercially and the American Type Culture Collection (www.ATCC.org), in addition to having active cultures or frozen strains of parasitic protists available for purchase, can also provide media formulations and prepared media. Bacteria generally accompany protists established in culture from clinical samples, the number depending on the site in the body from which they were taken. CSF, for example, should be sterile and anything isolated (e.g., *Naegleria fowleri*) from it would be significant. Cultures with undefined bacteria are referred to as xenic or polyxenic cultures (from *xenos* [Greek], meaning stranger). The next step would be to establish the protist in monoxenic (single "stranger") culture with a defined bacterial species using antibiotics to eliminate unknown or unwanted microbes. The goal is eliminating all bacteria to obtain an axenic culture (without "strangers," or bacteria-free). Antibiotics frequently used for eliminating bacteria are penicillin-streptomycin, gentamycin, and amphotericin B for fungal contaminants.

Naegleria fowleri and *Acanthamoeba* can be cultured by transfer of clinical samples (*Naegleria*: brain tissue, CSF; *Acanthamoeba* spp.: brain or cutaneous tissues, corneal scrapings for amoebic keratitis) to nonnutrient agar plates with a film of bacteria as a food source (any of the following: *Escherichia coli, Klebsiella pneumoniae, Enterobacter* spp.) spread over the agar surface and incubated at 30–37°C. Amoebae will grow out within 24–72 hours [23]. *Balamuthia*, another free-living

amoeba causing cutaneous and CNS infections, can be cultured from brain tissue (often obtained at autopsy) on tissue culture cells. Monkey kidney cells are a good source for growth, but other tissue culture lines have been used [24]. *Entamoeba histolytica*, *Giardia duodenalis*, and *Trichomonas vaginalis* require complex media and are often accompanied by a mixed population of indigenous bacteria that can overgrow in the enriched medium [25]. An *Entamoeba histolytica* look-alike, *E. dispar*, may also be present in cultured stool [26], as well as several other commensal amoebae from the intestine (*Entamoeba coli*, *Endolimax nana*).

The hemoflagellates (trypanosomes, *Leishmania*) are medically important protists that can be isolated from blood and tissues samples, and needle aspirates of cutaneous lesions. The media used for these organisms are varied and complex [27]. *Plasmodium* and *Babesia*, both intraerythrocytic blood parasites, can be grown in culture but are especially challenging in terms of media and growth conditions, requiring skills that are usually restricted to research laboratories [28, 29]. Microsporidia, intracellular parasites that can be isolated from urine, corneal scrapings, and bronchoalveolar lavage, can be maintained in a variety of tissue culture lines [30]. *Toxoplasma* can be identified by inoculation of clinical samples (e.g., CSF) into tissue cultures. *Pneumocystis jirovecii*, although an important opportunistic parasite, has yet to be cultured *in vitro*, beyond mere maintenance.

Animal Inoculations (Xenodiagnosis)

Mice, rats, hamsters, or guinea pigs can be inoculated with infectious material from a clinical sample or with isolated etiologic agents themselves if in culture and observed for signs of disease. An advantage of this technique is that if parasitemia is low in the host and diagnosis is uncertain (as in *Babesia* infections), animal inoculation can allow for an increase under *in vivo* conditions and aid in diagnosis. The disadvantage with the technique is that it may take days or even weeks for results to develop. Some isolates such as amoebae (*Balamuthia mandrillaris*, *Acanthamoeba* spp.) from culture can be inoculated in mice via an intranasal or intracranial route and cause death of the animal within a week [31]. Insect vectors for parasitic diseases (*Anopheles* mosquitoes, triatomid bugs, etc.) can be inoculated by allowing them to feed on an infected human or animal. Some pathogenic strains lose virulence with prolonged *in vitro* culture and may require animal passage to maintain or up-regulate their virulence.

Immunofluorescent Testing for Antibodies (IFA)

Patient serum is prepared in a series of dilutions. Antibodies in the serum bind to the antigen (the parasite) that is fixed and dried on a slide. This, in turn, is treated with anti-human fluorochrome-conjugated antibody raised in goats or rabbits. Fluorescence resulting from binding of the fluorescent antibody to the human antiparasite antibody can be visualized using a fluorescence microscope. The antibody titer is that serum dilution (e.g., 1:64), beyond which fluorescence can no longer be detected. The technique is useful for determining antibody titers at different stages of disease.

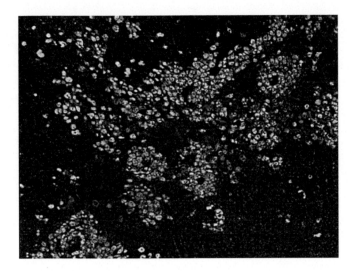

FIGURE 47.7 Section through brain stained with immunofluorescent antibody specific for *Balamuthia mandrillaris*. The amoebae fluoresce brightly against unstained parenchymal background. (Courtesy of Dr. G. S. Visvesvara.)

Immunofluorescent Staining for Parasites in Tissue (IIF)

IIF (indirect immunofluorescence) is a qualitative test for the presence of the parasite in tissue. Biopsy or necropsy tissue sections on slides are exposed to goat or rabbit antiparasite serum, followed by fluorochrome-conjugated antirabbit or antigoat serum. Parasites in tissue will fluoresce when the slide is examined with a fluorescence microscope, but not surrounding host tissue (Figure 47.7). The immunostaining is highly specific, making this a particularly reliable diagnostic test.

Enzyme-Linked Immunosorbent Assay (ELISA)

Enzyme immunoassay for antibody is widely used for diagnosis of many parasitic diseases. Commercial ELISA kits are available, precluding the do-it-yourself, time-consuming preparation of plates and assembling necessary reagents, thus simplifying the procedure. Because of considerations such as cost of the kits, the number of tests being performed in the laboratory, and the shelf life of the kit, it may not be feasible to perform ELISA testing, but rather to send samples to state or national reference laboratories.

Microscopic Diagnosis

For many diagnostic procedures, the microscope remains an important and affordable tool, especially in laboratories where more sophisticated testing and equipment are not available. A major use is in the examination of blood smears in conjunction with staining, using Giemsa or Wright's stains for blood smears. Stained blood films are useful in determining parasitemia, the percentage of infected erythrocytes. Changes in parasitemia can be used to check response to drug therapy or determine if a particular therapy is effective. Stained films are also effective in distinguishing between infection with *Plasmodium* or *Babesia* (pear-shaped in erythrocytes).

The microscope is used with a variety of clinical samples: Fixed and stained with hematoxylin-eosin (tissue sections), or iron hematoxylin or trichrome for stool samples. Lugol's iodine, used in moderation, can also be a useful stain, enhancing the appearance of ova in stools and the presence of flagella on protists, although more conventional staining should be used to corroborate observations. A compendium of stains, formulas for preparation, and directions for their use can be found in Garcia [16]. In some instances, unstained material can be examined directly (e.g., CSF, lymph node aspirates) as wet-mounts.

A phase-contrast or differential-inference microscope can enhance the appearance of unstained specimens in wet-mounts and facilitate parasite identification. For extended viewing of wet-mounts, a combination of petroleum jelly and paraffin can be used to seal the coverslip to the slide, thus slowing drying. Fresh, unfixed wet-mounts are particularly useful for observing protist motility. Examination of CSF wet-mounts for the presence of active, rapidly moving amoebae is essential in the diagnosis of amoebic meningoencephalitis caused by *Naegleria fowleri*. In old or refrigerated CSF samples, leukocytes may be mistaken for amoebae, causing misdiagnosis. Flagellated trichomonads from vaginal discharge in a wet-mount typically show a wobbly movement, similar to that of a spinning top that is about to topple.

Concentrating the clinical sample is often useful for increasing the chance of detecting parasites or parasitic stages that may be sparse in samples. Thick versus thin blood films can be helpful in detecting and identifying malaria parasites. A thick film offers a better chance of seeing parasites in the blood film, while a thin film is better able to identify the different species of *Plasmodium*. Helminth ova in stools may be obscured by fecal debris, and separating them using flotation (zinc floatation) or sedimentation (formal-ethyl acetate), the former being more reliable and easier to use than the latter, is frequently done [16].

Histopathology

Sectioned tissue from biopsies, whether brain, liver, or intestine, stained with hematoxylin-eosin stain can provide definitive evidence for the identification of parasites (Figure 47.8).

Molecular Diagnostic Techniques: PCR

Once a highly specialized technique used only in research and reference laboratories, the polymerase chain reaction (PCR) is now available to a wider audience of diagnosticians. Its basis is the use of primer sets containing base pairs unique for strands of parasite nuclear or mitochondrial DNA, giving an amplified product that can be visualized on a gel. The resulting bands identify the DNA as coming from a particular parasite whose DNA is present in host tissues or body fluids. Many, if not most, parasites can be identified by the PCR. In conventional PCR, the amplified DNA extract from a clinical sample is run on a gel and identified by the number of base pairs in the amplification product and by comparison with positive and negative controls. The identification can be further confirmed by DNA sequencing. Nested PCR is a modification of the conventional PCR technique in which sequential primer sets are used in producing an amplification product. Real-time PCR is an alternative technique that is more sensitive than conventional PCR and has additional advantages.

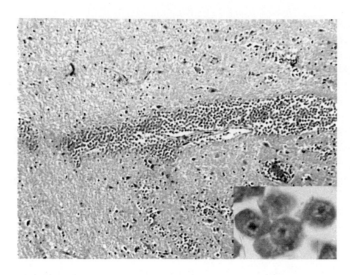

FIGURE 47.8 Section through brain with primary amoebic meningoencephalitis caused by *Naegleria fowleri*. A heavy concentration of amoeba is seen in the perivascular space. Hematoxylin-eosin stain. Inset: high-magnification image of *Naegleria* amoebae showing nucleus with large centrally located nucleolus in each amoeba. (Courtesy of Dr. G.S. Visvesvara.)

The amplification and detection (fluorescent probe) steps occur in a closed tube, reducing the chance of contamination and environmental spread of DNA [32]. Additionally, the products are analyzed in real-time using a thermal cycler. The technique is faster (hours vs. overnight) and requires no post-amplification handling. Real-time PCR can identify DNA from different organisms simultaneously as, for example, *Naegleria fowleri*, *Acanthamoeba*, and *Balamuthia mandrillaris*, free-living amoebae responsible for amoebic encephalitis [33]. It has also been used for differentiating between species of *Plasmodium* more effectively than stained blood films and has shown evidence of mixed human infections [34].

PCR cannot distinguish between DNA from living versus dead organisms, which can make a difference in deciding to treat or not treat a disease with drugs. PCR can be useful in retrospective studies of disease by analysis of archival slides and formalin-preserved tissues, but is less reliable in identifying DNA from parasite tissue samples because of fragmentation of the DNA caused by formalin fixation.

Drug Therapy

As with all infectious diseases, the ideal drug for use against an etiologic agent is one that will specifically target a particular function or metabolic pathway in the parasite that is either absent from the host's cells or less sensitive to the drug. Drug treatment is dynamic; new drugs come on the market, new variations of existing drugs become available, the U.S. Food and Drug Administration may approve drugs that were previously unavailable for use in the United States, or resistance to a particular drug therapy (e.g., use of chloroquine in areas where chloroquine resistance is common) may preclude its use in treatment and the cost of the drug, particularly when considering widespread use in endemic regions. Additionally, there are other considerations that will affect drug choice and dosage: dosage adjusted for age,

oral versus parenteral administration, stage of disease, drug use for prophylaxis, drug allergies or negative interactions with other medications, drug safety during pregnancy, disseminated or localized infections, and patient compliance with the recommended drug therapy.

A consequence of drug therapy is death of parasites. Although this is a desired effect, massive release of antigens from disintegrating parasites may present more of a problem to the immune system than the presence of the living organisms. Large-scale death of helminths in tissue can elicit an allergic reaction from the host caused by an immune response to the dying parasites. Antihistamines and corticosteroids can mute the drug-initiated immune reaction, but in some instances palliative measures may be the preferred course of action. Table 47.4 presents general information about drugs for protistan and helminthic diseases. The reader should seek medical advice regarding current guidelines [35, 36] for drug use against specific diseases.

Mechanisms of Action of Selected Drugs

Several drugs that are used against a number of different parasites are examined for their mechanisms of action. The general effects include interference with parasite metabolic pathways, inhibition of microtubule assembly preventing mitosis or disorganizing cell structure, alteration of permeability of the parasite tegument, and damage to parasite DNA.

Selected Drugs Effective in Treating Protistan Diseases

1. *Artemisinin and derivatives.* These antimalarial compounds are derived from the medicinal herb *Artemisia annua* that has been used for centuries in China as a treatment for malaria [37]. Derivatives of the compound include artemether and artesunate. The drugs themselves are effective for relatively short time periods and, to extend their duration of efficacy, they have been coupled with other drugs (e.g., artesunate-mefloquine).

2. *Chloroquine.* Chloroquine was used successfully for many years in the treatment of malaria until drug resistance began to show up in malarial parasites. Now its use is restricted to parts of the world where malarial strains remain sensitive to the drug. The drug is a weak base and is taken up into acidic food vacuoles of the intracellular parasite. Several mechanisms of action have been reported for the drug: prevention of detoxification of internalized heme (from host erythrocyte hemoglobin) accumulated by the malarial parasite, formation of drug-heme complexes, or interference with nucleic acid biosyntheses. Chloroquine-sensitive strains accumulate the drug, while chloroquine-resistant strains do not.

3. *Metronidazole (Flagyl®) and tinidazole (Tindamax®; Fasigyn®).* Metronidazole is a synthetic nitroimidazole antiprotistan drug. When taken up, it is reduced by the parasite's electron transport system, resulting in production of free radicals. A gradient is established across the parasite membrane between the reduced molecule

within the parasite and its normal oxidized molecular state outside the membrane, enhancing uptake of the drug. It has also been shown to damage the parasite's DNA and is a mutagenic agent (in rodents). It is particularly useful in treating infections caused by anaerobic or microaerophilic protists such as *Entamoeba histolytica, Giardia intestinalis,* and *Trichomonas vaginalis.* Tinidazole is similar to metronidazole in its action and the organisms effected, but is reportedly faster acting.

4. *Miltefosine (hexadecylphosphocholine).* Miltefosine is an alkylphosphocholine developed as an anticancer drug but has been shown effective in the treatment of leishmaniasis, including visceral [38], cutaneous, and mucocutaneous [39] forms. It is available as a pill for oral administration and has limited side effects (diarrhea, vomiting) in patients. The drug produces an apoptosis-like effect on the parasite, resulting in membrane breakdown and lysis.

Selected Drugs Effective in Treating Helminth Diseases

1. *Praziquantel (Biltricide®).* Praziquantel alters the intracellular calcium levels of sensitive flat worms [40]. Among its effects on helminths, at low concentrations it causes spastic paralysis leading to expulsion of the worm from the host. At high concentrations, it leads to damage to the worm's tegument, alters permeability, and causes its disintegration, opening the tegument to the host's immune system. Both these events are triggered by an increase in calcium ions. It has also been used in treating neurocysticercosis. Although praziquantel is used in treating many fluke and cestode diseases, nematodes do not respond to treatment with this drug.

2. *Albendazole (Albenza®) and Mebendazole (Vermox®).* Both drugs are benzimidazole broad-spectrum antihelminthics. Their effects include preventing glucose absorption across the worm tegument, interfering with mitochondrial activity, and inhibiting microtubule formation by binding to tubulin. A concentration differential exists between sensitivity of nematode tubulin and host tubulin to benzimidazoles, the former being more sensitive to lower drug concentrations. Their use is primarily against nematodes: pinworm, hookworm, and mixed infections. They are used in the treatment of inoperable neurocysticercosis in conjunction with corticosteroids, the latter to prevent an inflammatory response to the disintegrating cysticercoid larvae.

Conclusions and Perspectives

Parasites are well adapted to their ecologic niche, and in many diseases have achieved détente with their hosts. The parasites have had millennia to perfect their way of life; the host has decades to adjust. As the result of living in a disease-endemic region, the host may develop some immunity against the parasite (e.g., falciparum malaria in individuals older than 5 years), allowing survival.

Although major efforts are underway to eliminate specific parasitic diseases, problems remain:

1. Prohibitive cost of medication in developing countries.
2. Patient compliance in taking necessary antiparasite medications despite unpleasant side effects or prolonged use
3. Development of resistance to drugs of choice
4. Prevention of reinfection in endemic areas

Progress on vaccines against the most devastating of the parasitic diseases, malaria, has been slow and the successes once envisioned have yet to be realized, with antigenic variation as a major obstacle to overcome; any vaccine, to be successful, has to protect against all stages of the parasite life cycle. With continued sequencing studies of parasite DNA and identification of gene products, new drugs aimed at specific functions in parasites may prove more effective in combating diseases than those in the current pharmacopoeia.

Looking ahead to the generally accepted effects of global climate change, conditions may become more favorable for survival and dissemination of human parasites. Resource-starved developing nations will likely bear the burden of increased parasite loads. As a result of global warming, water may become scarcer in areas of the world, further concentrating peoples around water sources leading to their contamination with enhanced development of larval stages and vectors.

REFERENCES

1. World Malaria Report 2019, World Health Organization.
2. World Health Organization Fact Sheet 2019, Soil-transmitted helminth infections.
3. Pan American Health Organization Fact Sheet 2019, Chagas in the Americas.
4. McKay, D.M. and Webb, R.A., Cestode infection: immunological considerations from host and tapeworm perspectives, in *Parasitic Flatworms: Molecular Biology, Biochemistry, Immunology and Physiology*, Maule, A.G. and Marks, N.J., Eds., Cabi, Oxfordshire, UK, 2006, chap. 9.
5. Black, J., Hommel, M., Snounou, G., and Pinder, M., Mixed infections with *Plasmodium falciparum* and *P. malariae* and fever in malaria. *Lancet*, 343, 1095, 1994.
6. Miller, L.H., Good, M.F., and Milon, G., Malaria pathogenesis. *Science*, 264, 1878, 1994.
7. Cox, F.E.G., History of human parasitology. *Clin. Microbiol. Rev.*, 15, 595, 2002.
8. Chen, Q.Q., Schlichtherle, M., and Wahlgren, M., Molecular aspects of severe malaria. *Clin. Microbiol. Rev.*, 13, 439, 2000.
9. Birbeck, G.L., Cerebral malaria. *Curr. Treat. Options Neurol.*, 6, 125, 2004.
10. Clark, I.A., Alleva, L.M., Mills, A.C., and Cowden, W.B., Pathogenesis of malaria and clinically similar conditions. *Clin. Microbiol. Rev.*, 17, 509, 2004.
11. Gething, P.W. et al., A long neglected world malaria map: Plasmodium vivax endemicity in 2010. *PLOS Negl. Trop. Dis.*, 6, e1814, 2012.

12. Howes R.E., Patil A.P., Piel, F.B., Nyangiri, O.A., Kabaria, C.W., et al., The global distribution of the Duffy blood group. *Nat Commun*, 2, 266, 2011.

13. Homer, M.J., Aguilar-Delfin, I., Teleford, S.R. III, et al., Babesiosis. *Clin. Microbiol. Rev.*, 13, 451, 2000.

14. Zintl, A., Mulcahy, G., Skerrett, H.E., et al., *Babesia divergens*, a bovine blood parasite of veterinary and zoonotic importance. *Clin. Microbiol. Rev.*, 16, 622, 2003.

15. Schuster, F.L., Cultivation of *Babesia* and *Babesia*-like blood parasites: agents of an emerging zoonotic disease. *Clin. Microbiol. Rev.*, 15, 365, 2002.

16. Garcia, L.S., *Diagnostic Medical Parasitology*, 5th ed., ASM Press, Washington, DC, 2007.

17. Montoya, J.G. and Liesenfeld, O., Toxoplasmosis. *Lancet*, 363, 1965, 2004.

18. Jones, J.L., Kruszon-Moran, D., Wilson, M., et al., *Toxoplasma gondii* infection in the United States: seroprevalence and risk factors. *Am. J. Epidemiol.*, 154, 357, 2001.

19. Sorvillo, F.J., DiGiorgio, C., and Waterman, S.H., Deaths from cysticercosis, United States. *Emerg. Inf. Dis.*, 13, 230, 2007.

20. Conway, D.J., Molecular epidemiology of malaria. *Clin. Microbiol. Rev.*, 20, 188, 2007.

21. Graczyk, T.K., Knight, R., and Tamang, L., Mechanical transmission of human protozoan parasites by insects. *Clin. Microbiol. Rev.*, 18, 128, 2005.

22. Fenwick, A., Waterborne infectious diseases—could they be consigned to history? *Science*, 313, 1077, 2006.

23. Schuster, F.L., Cultivation of pathogenic and opportunistic free-living amebas. *Clin. Microbiol. Rev.*, 15, 342, 2002.

24. Visvesvara, G.S., Moura, H., and Schuster, F.L., Pathogenic and opportunistic free-living amoebae: *Acanthamoeba* spp., *Balamuthia mandrillaris*, *Naegleria fowleri*, and *Sappinia diploidea*. *FEMS Immunol Med. Microbiol.*, 50, 1, 2007.

25. Clark, C.G. and Diamond, L.S., Methods for cultivation of luminal protists of clinical importance. *Clin. Microbiol. Rev.*, 15, 329, 2002.

26. Diamond, L.S. and Clark, C.G., *Enatmaoeba histolytica*, Schaudinn 1903 (emended Walker, 1911) separating it from *Entamoeba dispar*, Brumpt 1925. *J. Eukaryot. Microbiol.*, 40, 340, 1993.

27. Schuster, F.L. and Sullivan, J.J., Cultivation of clinically significant hemoflagellates. *Clin. Microbiol. Rev.*, 15, 374, 2002.

28. Schuster, F.L., Cultivation of *Plasmodium* spp. *Clin. Microbiol. Rev.*, 15, 355, 2002.

29. Schuster, F.L., Cultivation of *Babesia* and *Babesia*-like blood parasites: agents of an emerging zoonotic disease. *Clin. Microbiol. Rev.*, 15, 365. 2002.

30. Visvesvara, G.S., *In vitro* cultivation of microsporidia of clinical importance. *Clin. Microbiol. Rev.*, 15, 401, 2002.

31. Visvesvara, G.S., Schuster, F.L., and Martinez, A.J., *Balamuthia mandrillaris*, N.G., N. Sp., agent of meningoencephalitis in humans and other animals. *J. Eukaryot. Microbiol.*, 40, 504, 1993.

32. Espy, M.J., Uhl, J.R., Sloan, M., et al., Real-time PCR in clinical microbiology: applications for routine laboratory testing. *Clin. Microbiol. Rev.*, 19, 165, 2006.

33. Qvarnstrom, Y., Visvesvara, G.S., Sriram, R., and da Silva, A.J., A multiplex real-time PCR assay for simultaneous detection of *Acanthamoeba* spp., *Balamuthia mandrillaris* and *Naegleria fowleri*. *J. Clin. Microbiol.*, 44, 3589, 2006.

34. Perandin, F., Manca, A., Calderaro, A., et al., Development of a real-time PCR assay for detection of *Plasmodium falciparum*, *Plasmodium vivax*, and *Plasmodium ovale* for routine clinical diagnosis. *J. Clin. Microbiol.*, 42, 1214, 2004.

35. Gilbert, D.N., Moellering, R.C., Elliopoulos, G.M., and Sande, M.A., Eds, *The Sanford Guide to Antimicrobial Therapy*, Antimicrobial Therapy, Inc., Sperryville, VA, 2006.

36. Abramowicz, M. and Zuccotti, G., Eds., *Medical Letter on Drugs and Therapeutics. Drugs for Parasitic Infections*, The Medical Letter, Inc., New Rochelle, NY, 2004.

37. Adjuik, M., Agnamey, P. Babiker, A., et al., The International Artemisinin Study Group. Artesunate combinations for treatment of malaria: meta-analysis. *Lancet*, 363, 9, 2004.

38. Sundar, S., Jha, T.K., Thakur, C.P., et al., Oral miltefosine for Indian visceral leishmaniasis. *N. Eng. J. Med.*, 347, 1739, 2002.

39. Soto, J. and Soto, P., Miltefosine: oral treatment of leishmaniasis. *Expert Rev. Anti. Infect. Ther.*, 4, 177, 2006.

40. Greenberg, R.M., Praziquantel: mechanism of action, in *Parasitic Flatworms: Molecular Biology, Biochemistry, Immunology and Physiology*, Maule, A.G. and Marks, N.J., Eds., Cabi, Oxfordshire, UK, 2006.

48

Introduction to Bacteriophages

Elizabeth Kutter and Emanuel Goldman

CONTENTS

Introduction

The Nature of Bacteriophages

Bacteriophages are viruses that only infect bacteria. They are like complex spaceships (Figure 48.1), each carrying its genome from one susceptible bacterial cell to another in which it can direct the production of more phage. Each phage particle (virion) contains its nucleic acid genome (DNA or RNA) enclosed in a protein or lipoprotein coat, or capsid; the combined nucleic acid and capsid form the nucleocapsid. The target host for each phage is a specific group of bacteria. This group is often some subset of one species, but several related species can sometimes be infected by the same phage.

Phages, like all viruses, are absolute parasites. Although they carry all the information to direct their own reproduction in an appropriate host, they have no machinery for generating energy and no ribosomes for making proteins. They are the most abundant living entities on earth, found in very large numbers wherever their hosts live—in sewage and feces, in the soil, in deep thermal vents, and in natural bodies of water. Their high level of specificity, long-term survivability, and ability to reproduce rapidly in appropriate hosts contribute to their maintenance of a dynamic balance among the wide variety of bacterial species in any natural ecosystem. When no appropriate hosts are present, many phages can maintain their ability to infect for decades, unless damaged by external agents.

Some phages have only a few thousand bases in their genome, whereas phage G, the largest sequenced to date, has 497,513 base pairs—as much as an average bacterium, though still lacking the genes for such essential bacterial machinery as ribosomes. Over 95% of the phages described in the literature to date belong to the Caudovirales (tailed phages). Their virions are approximately half double-stranded DNA and half protein by mass, with icosahedral heads assembled from many copies of a specific protein or two. The corners are generally made up of pentamers of a protein, and the rest of each side is made up of hexamers of the same or a similar

DOI: 10.1201/9781003099277-50

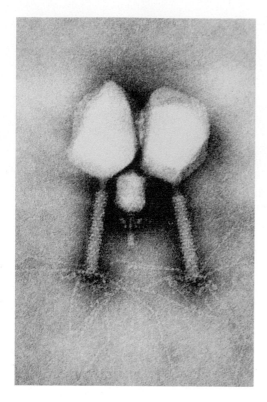

FIGURE 48.1 A "family portrait": Bacteriophages φ29 and T2. (Electron micrograph provided by Dwight Anderson.)

protein. The three families are defined by their very distinct tail morphologies: 60% of the characterized phages are *Siphoviridae*, with long, flexible tails; 25% are *Myoviridae*, with double-layered, contractile tails; and 15% are *Podoviridae*, with short, stubby, tails. The latter may have some key infection proteins enclosed inside the head that can form a sort of extensible tail upon contact with the host, as shown most clearly for coliphage T7.[3]

Bacteriophage T4, a myovirus, is the most extensively studied[2b] of the larger phages at the genetic, molecular, and functional levels (Figure 48.2). T4 and its close relative T2 played key roles in the development of molecular biology: The identification of DNA as the genetic material, the discovery of messenger RNA (mRNA), the triplet nature of the genetic code and the deciphering of that code. Their use of a noncanonical base, 5-hydroxymethylcytosine, rather than cytosine, in their DNA was crucial for establishing that viruses encoded enzymes, and for totally shutting off all host transcription, including that of ribosomal RNA (rRNA). Until then, the enormous amounts of rRNA had masked the existence of the labile messenger RNAs (mRNAs), which mediate the transfer of the information from the nucleotide sequence of each gene in DNA to the protein it encodes.

Archaea have their own set of infecting viruses, sometimes called *archaephages*. Many of these have unusual, often pleomorphic shapes, that are unique to the Archaea. However, many viruses identified to date for the Crenarchaeota kingdom of Archaea look like typical tailed bacteriophages.[4] See Chapter 21 in this book for more information on Archaea.

The 10 families of tailless phages described to date each have very few members. They are differentiated by shape (rods, spherical, lemon-shaped, or pleomorphic); by whether they are enveloped in a lipid coat; by having double- or single-stranded DNA or RNA genomes, segmented or not; and by whether they are released by lysis of their host cell or are continually extruded from the cell surface. Their general structures, sizes, nucleic acids, adsorption sites, and modes of release are described in detail in recent books,[1,5] as well as the basic infection processes of the *Inoviridae*, *Leviviridae*, *Microviridae*, and *Tectiviridae*. Relatively little is known about most of the others, which generally have been isolated under extremes of pH, temperature, or salinity, and have only been observed in Archaea.

Phages can also be divided into two classes based on lifestyle: obligatorily/professionally *virulent* and *temperate*. Professionally virulent phages can only multiply by means of a lytic cycle; the phage virion adsorbs to the surface of a host cell and injects its genome, which takes over much of host metabolism and sets up molecular machinery for making more phages. The host cell then lyses minutes or hours later, liberating many new phages. *Temperate phages*, in contrast, have a choice of reproductive modes when they infect a new host cell. Sometimes the infecting phage initiates a lytic cycle, resulting after a given short time in lysis of the cell and release of new phage, as just described. The infecting phage may alternatively initiate a *lysogenic* cycle; instead of replicating, the phage genome assumes a quiescent state called a prophage, generally integrated into the host genome but sometimes maintained as a plasmid. It remains in this condition indefinitely, being replicated as its host cell reproduces to make a clone of cells all containing *prophages*; these cells are said to be lysogenized or lysogenic (i.e., capable of producing lysis) because one of these prophages occasionally comes out of its quiescent condition and enters the lytic cycle. The factors affecting the choice to lysogenize or to reenter into a lytic cycle are described below. As discussed by Levin and Lenski,[6] the lysogenic state is highly evolved, requiring co-evolution of virus and host that presumably reflects various advantages to both. Temperate phages can help protect their hosts from infection by other similar phages and can lead to significant changes in the properties of their hosts, including restriction systems and resistance to antibiotics and other environmental insults. As discussed later in this chapter, they may even convert the host to a pathogenic phenotype, as in diphtheria or enterohemorrhagic *Escherichia coli* (EHEC) strains. Bacteriophages lambda (λ), P1, Mu, and various dairy phages are among the best-studied temperate phages. (Note that mutation of certain genes can create virulent derivatives of temperate phages; these are still considered members of their temperate phage families, especially since they can pick up various genes through recombination with their integrated relatives.)

The larger virulent phages generally encode many host-lethal proteins. Some of them disrupt host replication, transcription, or translation; they may also degrade the host genome, destroy, or redirect certain host enzymes, and/or alter the bacterial membrane. There have also been found many small host-lethal early phage genes that could be exploited to develop new antimicrobials or other modulators of bacterial physiology.[6b] The temperate phages, in contrast, generally do much less restructuring of the host, and they carry few if any host-lethal proteins that would need to be kept under tight control during long-term lysogeny. They always encode a *repressor* protein, which acts at a few *operator* sites to block transcription of other phage genes. This

GENOMIC MAP OF BACTERIOPHAGE T4

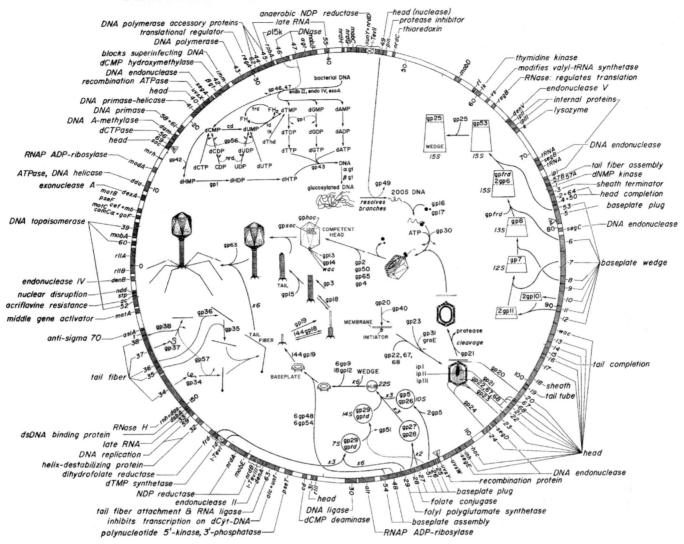

FIGURE 48.2 This genomic map of bacteriophage T4, incorporating detailed illustrations of the pathways of DNA biosynthesis and phage particle construction, has been updated multiple times by Burton Guttman and Elizabeth Kutter from the version originally drawn as the frontispiece for the 1983 book "Bacteriophage T4." That book, which grew out of the 1981 biennial Evergreen International Phage Meeting and involved 71 members of the T4 community, was the first monograph ever published by ASM press.

repressor may be the only phage-encoded protein produced during the lysogenic state, but often a few other genes that may be beneficial to host survival are also expressed from the prophages. The repressor also blocks lytic infection by other phages of the same immunity group—that is, other phages whose genes can be regulated by the same repressor. In this way, a temperate phage generally protects its host bacterium from infection by several kinds of phages.

Antigenic Properties of Phages

Phages, like other protein structures, can elicit an antibody response, and antiphage antibodies have been used in a variety of ways since early phage researchers first showed that rabbits injected with phage lysates produced phage-neutralizing antibody; in 1926, d'Herelle,[7] and in 1959 Adams,[8] reviewed early

studies in the field. For most phages, the neutralization follows first-order kinetics to an inactivation of about 99%, with the survivors then generally being more resistant and often having smaller plaque size. The activity of a given preparation at a standard temperature and salt concentration is denoted by a velocity constant, K, which is related to the log of the percent of phage still infectious after a given exposure time t. As emphasized by Adams, such factors as salt concentration are very important in titering antibody activity. Some phages require specific divalent cations but monovalent cations above 10^{-2} M and divalent cations above 10^{-3} M are generally inhibitory. He cites an example of anti-T4 serum with an inactivation constant (K value) of over 105 min^{-1} under optimal conditions but only 600 min^{-1} in broth. Also, somewhat surprisingly, different assay hosts may give different estimates of the fraction of nonneutralized phage. Adams suggested this may be due to minor differences in the receptors recognized on various hosts, leading to differences in bonding strength. We also know now that some phages can use totally different receptors on different hosts, as described for T4 later in this chapter. Once determined for a given set of conditions, the *K* value can be used to calculate the appropriate antibody dilution for any given experiment.

With the aid of noted Russian electron microscopist Tamara Tikhonenko,[9] such techniques as immunoelectron microscopy were used to carry out particularly detailed studies of the process of inactivation by antibodies.[10] Implications of phage antigenicity for human therapy are discussed in chapter 54.

Recent experiments have shown that phage can elicit antiviral immune responses in mammalian cells, which turn out to protect a bacterial pathogen, *Pseudomonas aeruginosa*, from being cleared from infected patients.[10b]

Susceptibility of Phages to Chemical and Physical Agents

Phages vary greatly in their sensitivity to various chemical and physical agents, in ways that are generally unpredictable and need to be determined experimentally in each case. For example, no one has yet explained the stability of phage T1 to drying, a stability that has led to much grief in the biotech industry as well as in phage labs from contamination by ubiquitous T1-like phages. There are, however, certain general principles. For example, almost all phages are very susceptible to UV light in the range of 260 nm as well as in the far UV; in addition to the general effects of sunlight, there are multiple reports of phage collections being lost to fluorescent-lighted refrigerators. Other factors potentially affecting phage infectivity include pH, ascorbic acid, urea, urethane, detergents, chelating agents, mustard gas, alcohols, and heat inactivation. The most extensive summary of the known effects of chemical and physical agents on phages is found in Adams[8]; there is also very good information in Ackermann and Dubow.[11]

Phages generally are stable at pH 5–8, and many are stable down to pH 3 or 4; each phage needs to be specifically characterized. Phage are often quite sensitive to protein-denaturing agents such as urea and urethane, but the level of inactivation depends on both concentration and temperature and differs for different phages. Not surprisingly, detergents generally have far less effect on phage than they do on bacteria, although the few phages that are enveloped in membranes are quite susceptible, while chelating agents have strong effects on some phages but not on others; this apparently depends mainly on ionic cofactor requirements for adsorption. Chloroform has little or no effect on nonenveloped phages; in fact, one needs to be very careful not to get phage contaminants into the chloroform being used to produce phage lysis. However, contaminants toxic to phage in high concentrations are found in some commercial chloroform. Mutagenic agents such as mustard gas, nitric oxide, and ultraviolet light inactivate phage and can induce the lytic cycle in many lysogens. Lytic phages can generally still infect cells recently inactivated with UV or other mutagens, because they do not require ongoing synthesis of host proteins after infection. In fact, UV inactivation of the host just prior to infection is sometimes used to virtually eliminate residual synthesis of host proteins for experiments using pulse labeling to examine patterns of phage protein synthesis.

Interaction of Phage with the Host Cell

Virtually every structure on the outer surface of a bacterial cell is likely to serve as a receptor for some phage. See Chapter 18 in this book.

The properties of the distinctive outer membrane of Gram-negative cells have been reviewed in detail by Nikaido[12,13] and Ghuysen[14] and are shown in Figure 48.3a. The inner face of the outer membrane has a phospholipid composition similar to that of the cytoplasmic membrane, to which is attached the peptidoglycan layer. In contrast, the lipid part of the outer face is mainly a unique substance called lipopolysaccharide (LPS), seen nowhere else in nature. LPS is composed of three parts: a hydrophobic lipid A membrane anchor, connected by a core polysaccharide, to a complex distal polysaccharide O-antigen. The O-antigens play important roles in bacterial interactions with mammalian hosts and in virulence; they are often indicated in strain names, such as *E. coli* O157. They can also serve as efficient phage receptors; obviously, such phages are likely to have very narrow host ranges. Many common lab strains, including *E. coli* K-12, lack any O-antigen; *E. coli* B, used extensively for phage work, even lacks the more distal part of the LPS core, but they are found in most clinical strains. The outer membrane also contains several families of general porins—proteins that form large β-barrel channels with charged central restrictions that support nonspecific rapid passage of small hydrophilic molecules but exclude large and lipophilic molecules, and are generally present in high quantity. It also contains various high-affinity receptor proteins that catalyze specific transport of solutes such as vitamin B12, catechols, fatty acids, and different iron derivatives. There are about 3 million molecules of LPS on the surface of each cell, plus 700,000 molecules of lipoproteins, and 200,000 molecules of the porins. The concentrations of the specific other outer membrane proteins used as receptors vary considerably, depending on environmental conditions and the general need for the particular compound each one transports; sex pili and other such structures are also used as sites of initial attachment by some phages, leading, for example, to phages that are male-specific. Some filamentous phages actually inject through the pili.

Appropriate positioning of the phage tail on the cell surface[14b] triggers irreversible events leading to DNA delivery into the host. There is some evidence supporting that phage T4 can actually "walk" along the surface of the bacterium to find its receptor (Liefting W, Dreesens L, Aubin-Tam ME, Beaumont H, personal communication). In Gram-negative cells, DNA delivery includes getting its DNA safely across the periplasmic space (periplasm), which appears to be a sort of viscous gel that contains many nucleases and proteases, in addition to transport across the outer and inner membranes. In some phages, it has been shown that lipopolysaccharide alone is capable of triggering DNA release from phage particles.[14c] There are few phages where much is known about the mechanism or energetics of DNA delivery, but it is clear that it can differ substantially from phage to phage.

Proteins TonB and TolA, which are anchored to the inner membrane but span the periplasmic space and are particularly important to various uptake systems, are also crucial to infection by such phages as T5. A detailed review by Letellier et al.[15] focuses primarily on three coliphages that have been best studied

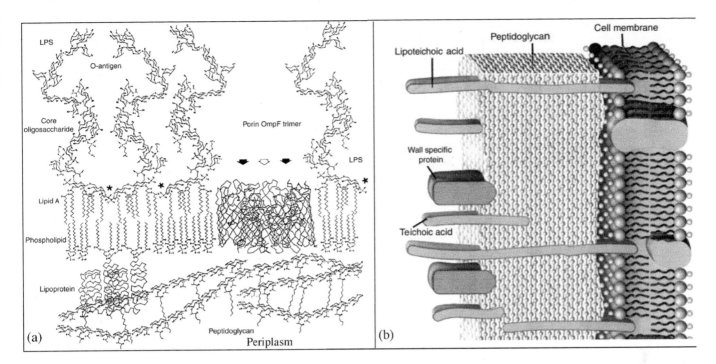

FIGURE 48.3 (a) Details of the outer membrane structures of Gram-negative bacteria. (With permission, from Hancock, R.E.W., Karunaratne, D.N., and Bernegger-Egli, C. Molecular Organization and Structural role of Outer Membrane Macromolecules, *The Bacterial Cell Wall*, J.M. Ghuysen, Hackenbeck, R. (Eds.). Elsevier, Amsterdam, 1994, 263–279.) (b) General structure of the cell wall of Gram-positive bacteria (from Kutter et al.[2]). Recent work indicates that there is actually an inner wall zone between the peptidoglycan and the cell membrane in *S. aureus*.[72]

regarding the DNA transfer process: myovirus T4, siphovirus T5, and podovirus T7. For each, a very long polyanionic genome—50 times the cell length for T4—must penetrate unscathed across two hydrophobic barriers, the outer and inner membranes, as well as the peptidoglycan sheath and the nuclease-containing periplasmic space. As they discuss, the rate of transfer can be as high as 3,000–4,000 base-pairs/second, in contrast to the 100 base-pairs/second seen for conjugation and natural transformation. Furthermore, the very high efficiency of infection for phages like T4 indicates that the process almost always occurs without significant damage to the infecting DNA molecule. The mechanisms are highly varied; often, ATP, a membrane potential, or enzyme action is involved, but some phages, such as T5, can enter cells in the absence of metabolic energy sources—it is here that TolA and TonB are involved for T5, but not used for T4 or T7.

The volume of the rigid phage capsids does not change as the DNA is injected into the cell, and the simple release of energy from metastable packing in the particle does not provide sufficient energy to transport a DNA molecule several micrometers long through the very narrow channel inside the tail tube. For *B. subtilis* phage SP82G, McAllister[16] showed that the rate of DNA transfer into the cell is constant over the whole genome and is highly temperature dependent, and the second-step transfer of the major part of coliphage T5 DNA proceeds normally even in experiments where the DNA has already been released from its capsid.[15] Though the myoviruses have contractile tails, it does not appear that their tail tubes actually pierce the inner membrane and a potential gradient is still required for DNA entry into the cell. All in all, as discussed by Molineux,[3] it is clear that the widely invoked "hypodermic syringe" injection metaphor is generally inaccurate and the mechanisms providing energy for the transfer differ from phage to phage. In the case of T7, for example, the energy for transfer is largely supplied by transcription of the incoming DNA, starting from a promoter near the unique beginning of the molecule and depending on the fact that all of the transcription is in the same direction. The relatively slow speed of this process allows time for T7-encoded mechanisms to block damage to the DNA by host nucleases.

The interactions of phage with Gram-positive hosts have been much less well studied. Here, the phage recognize parts of the peptidoglycan layer and/or other molecules embedded in it (Figure 48.3b); for example, particular teichoic acids are involved for myoviruses like *B. subtilis* phage SPO1, listeria phage A511, and *Staphylococcus aureus* phages ISP and Twort. Specific interactions with the cell membrane appear to also be involved in the final, irreversible step. Little is yet known about the specific adhesins of most phages infecting Gram-positive bacteria or about the various receptors to which they bind. Phage specificity may relate to the substantial variations in the amino acids used for the cross-linking peptides in the exposed thick peptidoglycan structure; there are also variations in the teichoic acids protruding from it. A variety of cell-wall-associated proteins are also bound to the cell either via a specific C-terminal anchor sequence or as N-terminal lipoproteins, and interact in various ways with the environment; staph protein A, which binds the constant region of IgG and has been implicated in pathogenesis, is the most famous of these.

Duplessis and Moineau[17] published the first identification and characterization of a phage gene involved in recognition of Gram-positive bacteria, for the sequenced phage DT1 and six related virulent phages with different host ranges on *Streptococcus thermophilus.* They confirmed that *orf18* encodes the adhesin, which has a conserved N-terminal domain of nearly 500 amino acids, a collagen-like sequence, and a largely conserved C-terminal domain with an internal 145 amino acid variable region (VR2). Using DT1 to infect a host in which MD4 genes *17–19* had been cloned, they generated five recombinant phages that now had the MD4 host range; all five contained the MD4 version of the VR2 region, but were otherwise largely like DT1.

Ravin et al.[18] showed that the specificity of adsorption of the isometric-headed phage LL-H and the prolate-headed JCL1032 to their host *Lactobacillus delbrueckii* involves a conserved C-terminal region in LL-H Gp71 and JCL 1032 ORF 474. Stuer-Lauridsen et al.[19] showed that four virulent prolate-headed phages (c2 species) of *Lactococcus lactis* with different host ranges first reversibly recognize a specific combination of carbohydrates in the outer cell wall. They then bind irreversibly with a host surface protein called PIP (phage infection protein), with no other known function, triggering DNA release from the capsid. Using the fully sequenced phages φc2 and φbIL67 and the sequenced least-conserved late regions of the other phages, the authors showed that the host-range-determining element is in the central 462 bp segment of a gene designated *115, 35,* and *2* in three different phages, with 65–71% pairwise similarity between them. While most genes in this region show no homology with unrelated phages, parts of this host-range-determining element resemble genes from phages of *S. thermophilus* and *L. lactis* that have also been implicated in host range determination processes. The product of a neighboring gene, *110/31/5,* with 83–95% similarity, had earlier been shown to bind to the tail spike and apparently to be involved in the infection process; they could not clone this second gene in their shuttle vector, perhaps because it is the one responsible for binding to PIP and helping transfer the DNA into the cell.

The phage infection process is very much dependent on the host metabolic machinery, so in most cases it is highly affected by what the host was experiencing shortly prior to infection, as well as by the energetic state and what nutrients and other conditions are present during the infection process itself.[20] Note that the phages that efficiently turn off host gene function, such as coliphages T4 and T5 and *B. subtilis* phage SPO1, generally render their hosts incapable of responding to substantial environmental changes after infection. For temperate phages, the energetic state of the cell also generally has significant effects on the probability of establishing lysogeny and, in some cases, on the reactivation of phages from lysogens.

Most bacteria used in phage studies use two basic kinds of metabolic pathways to get the energy they need: Fermentation and respiration. In both cases, electrons are withdrawn from an oxidizable molecule, freeing energy which is then stored as ATP, NADH, or similar coenzymes. These electrons, now at a lower potential, must, in turn, be removed by reducing some final acceptor. In fermentation, electrons removed from substrates such as glucose are transferred to breakdown products of the glucose to produce wastes such as ethanol or lactic acid, and only a small fraction of the energy in the glucose can be used. In respiration, the electrons are eventually passed through a membrane-bound electron transport system (ETS) and are carried away by reducing oxygen to water (in ordinary aerobic respiration), or, in anaerobic respiration, by reducing ferric to ferrous iron, sulfate

to sulfide, fumarate to succinate, or nitrate to nitrite or N2. The electron transport system is oriented in the cell membrane so as to transport protons, H+, to the outside of the cell membrane. This builds up an electrochemical potential, known as the proton motive force (PMF), across the cell membrane. This PMF is critical for infection by most phages. As discussed by Goldberg,[21] the PMF is separable into two parts: a pH gradient and a membrane potential that is due to the general separation of charges. For many (but not all) phages, DNA transfer into the cell requires the membrane potential.

The Infection Process

Overview

The general lytic infection process for tailed phages is presented here, with some examples from model phages. More detail on these subjects can be found in Abedon and Calendar's *The Bacteriophages,*[5] in Kutter and Sulakvelidze's *Bacteriophage: Biology and Applications,* or in Granoff and Webster's *Encyclopedia of Virology,*[22] as well as in books and articles on the individual phages.

Since the early studies of d'Herelle, the details of phage-host interactions have been studied by use of the single-step growth curve (Figure 48.4a), as systematized in 1939 by Ellis and Delbrück.[23] Phage are mixed with appropriate host bacteria at a low multiplicity of infection. After a few minutes for adsorption, the infected cells are diluted (to avoid attachment of released phage to uninfected cells or bacterial debris) and samples are plated at various times to determine infective centers. An infective center is either a single phage particle or an infected cell that bursts on the plate to produce a single plaque (see Chapter 7 for a discussion of bacteriophage plaque assays). The number of plaques generally remains constant at the number of infected cells for a characteristic time, called the *latent period,* and then rises sharply, leveling off at many times its initial value as each cell lyses and liberates the completed phage. The ratio between the numbers of plaques obtained before and after lysis is called the burst size. Both the burst size and the latent period are characteristic of each phage strain under particular conditions, but are affected by the host used, medium, and temperature. As first shown by Doermann,[24] if the infected cells are broken open at various times after infection, the phage seem to have all disappeared for a certain period. This eclipse period was a mystery until the nature of the phage particle and of the infection process was determined and it was realized that during this period only naked phage DNA is present in the cell. Using chloroform-induced lysis followed by plating, both the eclipse period and the subsequent rate of intracellular synthesis of viable phage particles are now routinely measured. Infecting at a high multiplicity (5–10 phage/cell) allows one to also measure the effectiveness of killing (i.e., the number of bacterial survivors) and the impact of the phage infection in terms of continued expansion of cell mass and eventual cell lysis, as reflected in the absorbance or optical density (Figure 48.4b).

The infection process involves several tightly programmed steps, as diagramed in Figure 48.5. The efficiency, timing, and other aspects of the process may be very much affected by the

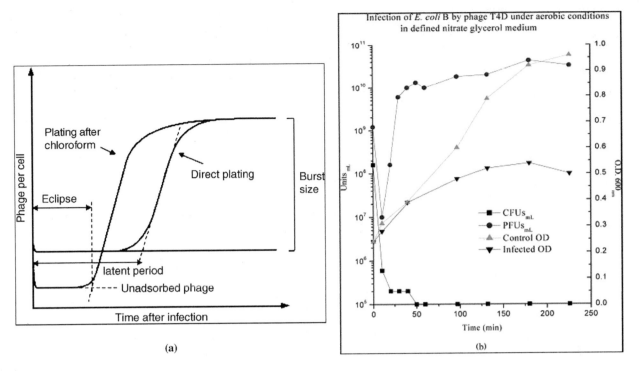

FIGURE 48.4 (a) Classical single-step growth analysis of phage infection. (b) Infection of *E. coli* B by T4D at an MOI of 9 in a minimal medium. Note the very rapid initial drop in both surviving bacteria and unadsorbed phage, the doubling of the cell mass even after infection, as measured by absorbance (OD600), and the fact that there is a long delay before lysis actually occurs under these conditions; this phenomenon of "lysis inhibition" is discussed in the text.

host metabolic state, and in many cases the host can no longer adapt to major metabolic changes once the phage has taken over the cell. Thus, for example, studies on anaerobic infection can only be carried out in host cells that were themselves grown anaerobically.

Most research on phage have utilized bacteria growing exponentially. However, in nature, phage may be more likely to encounter bacteria in the stationary phase. This condition has not been studied very much. Research has revealed that phage T4 can infect and kill stationary phase hosts but enter what has been termed a "hibernation" mode between middle and late gene expression. Phage morphogenesis is delayed until additional nutrients are added to the medium, after which infection resumes and goes to completion. Not all phage-infected stationary cells

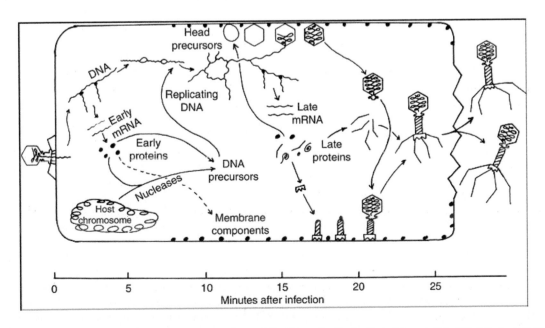

FIGURE 48.5 An overview of the T4 infection cycle. (From Mathews, C.K., Kutter, E., Mosig, G. and Berget, P.B. [Eds.] Bacteriophage T4. American Society for Microbiology, Washington, D.C., 1983, as reproduced in Kutter et al.[1])

follow this prototype—there is also an alternate "scavenger" response, where the phage uses whatever nutrients are available to generate a very low burst of progeny, only about one phage per infected cell.[24b]

Adsorption

Infection by tailed phages starts when specialized adsorption structures, such as fibers or spikes, bind to specific surface molecules or capsules on their target bacteria. In Gram-negative bacteria, virtually any of the proteins, oligosaccharides, and lipopolysaccharides can act as receptors for some phage. The more complex murein of Gram-positive bacteria offers a very different set of potential binding sites. Many phages require clusters of one specific kind of molecule that is present in high concentration to properly position the phage tail for surface penetration. However, coliphage N4 manages to use a receptor called NfrA, which is only present in a few copies per cell. The attachment of phages often involves two separate stages and two different receptors. T4-like phages, for instance, must bind at least three of their six long tail fibers to primary receptor molecules to trigger rearrangements of baseplate components, which then bind irreversibly to a second outer-membrane receptor that is involved in actual cell-surface penetration, eventually leading to the triggering of DNA release into the cell. Different members of the T4 family use very different primary receptors, but they seem to all use the heptose residue of the LPS inner core for the secondary receptor.

Adsorption velocity and efficiency are important parameters that may vary for a given phage-host system depending on external factors and host physiological state. For example, the lambda receptor is only expressed in the presence of the sugar maltose. Many phages require specific cofactors, such as Ca^{2+}, Mg^{2+}, or simply any divalent cation. T4B strains (but not T4D strains) require tryptophan for binding, but can bypass this requirement for a short "nascent" period when they are first released from the prior host and still attached to the inside of the membrane by their baseplates.[25-27] A nascent-phage period with broader adsorption properties was also observed for *Staphylococcus* phage in early work, and described for a *Streptococcus* phage by Evans,[28] but the mechanism(s) of nascent-phage interaction to give the extended host range is not known.

Bacteria commonly develop resistance to a particular phage through mutational loss or alteration of receptors used by that phage. However, losing some receptors offers no protection against the many other kinds of phage that use different cell-surface molecules as receptors. Furthermore, in most cases, the phage can acquire a compensating adaptation through appropriate host range mutations, which alter the tail fibers so they can recognize the altered cell-surface protein or bind to a different receptor. This is likely less efficient than bacteria acquiring resistance, since phage adaptation to new receptors requires establishment of a new functional interaction, while development of host resistance is a "negative" event, requiring loss of function.[29] Still, some such mutants seem to be present in every substantial phage population. Furthermore, some phages, such as P1 and Mu, encode multiple versions of the tail fibers. Others can recognize multiple receptors; for example, T4 tail fibers bind efficiently to an *E. coli* B-specific lipopolysaccharide, to the outer membrane

protein OmpC found on K strains, and to OmpF.[30] The adhesion regions of the tail fibers are highly variable between related phages, with high rates of recombination that facilitate the formation of new, chimeric adhesins. Not surprisingly, there is much interest in engineering new receptor-recognition elements into the tail fibers of well-characterized phages so they can infect taxonomically distant hosts.

The physiological state of a cell can substantially change the concentration of particular cell-surface molecules and thus the efficiency of infection by certain phages. Many of the surface molecules that particular phages use as receptors are crucial to the bacterial cell, at least under some environmental conditions, so resistance may lead to loss of important functions and reduction in competitiveness. Such phages are particularly useful for phage therapy applications.

Penetration

After irreversible attachment, the phage genome passes through the tail into the host cell. This is not actually an "injection" process, as has often been depicted, but involves mechanisms of DNA transfer specific for each phage. In general, the tail tip has an enzymatic mechanism for penetrating the peptidoglycan layer and then touching or penetrating the inner membrane to release the DNA directly into the cell; the binding of the tail also releases a mechanism that has been blocking exit of the DNA from the capsid until properly positioned on a potential host. The DNA is then drawn into the cell by processes that generally depend on cellular energetics but are poorly understood except for a few phages. For example, as noted earlier, in T7 the entry of the DNA is mediated by the process of its transcription, but that is an unusual case. No host mutants have yet been identified that specifically block phage DNA transport across the cell membrane for any phage. That process thus probably involves host components, which are indispensable to the viability of the host, at least for those phages like T4 where it is very rapid and generally complete.

Once inside the cell, the phage DNA is potentially susceptible to host exonucleases and restriction enzymes. Therefore, many phage circularize their DNA rapidly by means of sticky ends or terminal redundancies, or have the linear ends protected; T4 encodes a special protein, gp2, for this purpose. Many also have methods to inhibit host nucleases (T7, T4) or use an odd nucleotide in their DNA such as hydroxymethyldeoxycytidine (hmdC: T4-related coliphages) or hydroxy methyldeoxyuridine (hmdU: SPO1-related phages infecting Gram-positive bacteria and ViI-like phages infecting enteric bacteria). In other cases, their genomes have been selected over evolutionary time to eliminate sites that would be recognized by the restriction enzymes present in their common hosts (staphylococcal phage Sb-1, coliphage N4).

The Transition from Host to Phage-Directed Metabolism

The initial step generally involves recognition by the host RNA polymerase of very strong phage promoters, leading to the transcription of immediate early genes. The products of these genes may protect the phage genome and restructure the host appropriately for the needs of the phage; they may inactivate host proteases and block restriction enzymes, directly terminate various host macromolecular biosyntheses, and/or inactivate some host proteins. The large virulent phages such as T4, SPO1, and Sb-1 encode many proteins sometimes referred to as "monkey wrenches," which are lethal to the host even when cloned individually and which appear to participate in this process of host takeover. A set of middle genes is often then transcribed, producing products that synthesize the new phage DNA, followed by expression of a set of late genes that encode the components of the phage particle and factors needed for assembly. For some phages, these transitions involve the synthesis of new sigma factors or DNA-binding proteins to reprogram the host RNA polymerase; other phages encode their own RNA polymerase(s) for one or more of these stages. Degradation of host DNA, inhibition of host replication, and inhibition of transcription and translation of host mRNAs are other mechanisms that can contribute to reprogramming the cell for the synthesis of new phage. For lytic development of temperate phages and for some lytic phages, the programmed synthesis of phage proteins is superimposed on the host pattern. At the other extreme, 2-D gel analysis of proteins radioactively pulse-labeled after T4 infection shows that synthesis of host proteins is totally inhibited during the first couple of minutes after infection.[31]

New techniques, such as RNA-seq, are now being used to look more carefully at the details of host and phage gene expression after infection by a variety of phages, with very different results being seen for different phages.[32,33] Unexpectedly, RNA-seq technology has revealed that host transcription, and even phage mediated transcription of host genes, occurs during lytic infections with many *Pseudomonas aeruginosa* lytic phages.[33b,33c] This may function to allow the bacterium to respond to stresses during the infection.

Morphogenesis

The DNA is packaged into preassembled icosahedral protein shells called procapsids. In most phages, their assembly involves complex interactions between specific scaffolding proteins and the major head structural proteins, followed by proteolytic cleavage of both the scaffolding and the N-terminus of the main head proteins. Before or during packaging, the head expands and becomes more stable, with increased internal volume for the DNA. Located at one vertex of the head is a portal complex that serves as the starting point for head assembly, the docking site for the DNA packaging enzymes, a conduit for the passage of DNA, and, for myoviruses and siphoviruses, a binding site for the phage tail, which is assembled separately. The assembly of key model phages provides the best-understood examples in biology of morphogenesis at the molecular level.[33] It also has provided the models and components for several key developments in nanotechnology.

Cell Lysis

The final step—lysis of the host cell—is a precipitous event whose timing is tightly controlled.[34] If lysis happens too quickly, too few new phages will have been made to effectively carry on the cycle; if lysis is delayed too long, opportunities for infection and a new explosive cycle of reproduction will have been lost.[35,36]

As discussed in Chapter 18, the shape-determining rigid wall that protects virtually all bacteria from rupture in media of low osmolarity (and must be ruptured to release phage) is actually one enormous molecule, forming a peptidoglycan or murein sack that completely encloses the cell; the sack grows and divides without losing its structural integrity and is the site of attack of antibiotics like penicillin (see Chapter 51 in this book). The glycan warp of this sack consists of chains of alternating sugars—*N*-acetylglucosamine (NAG) and *N*-acetyl muramic acid (NAM)—wrapped around the axis of the cylinder for rod-shaped bacteria. These chains are cross-linked by short peptides to make a two-dimensional sheet. The tailed phages all use two components for lysis: a lysin (an enzyme capable of cleaving one of the key bonds in the peptidoglycan matrix) and a holing (a protein that assembles pores in the inner membrane at the appropriate time to allow the lysin to reach the peptidoglycan layer and precipitate lysis). The timing is affected by growth conditions and genetics; mutants with altered lysis times can be selected. The T4-like phages often have an additional mechanism for adapting to host availability. If other phages of the same species try to enter the cell more than a few minutes after T4 infection, the infected cell interprets that as a sign that there is more phage around than available hosts, and substantially delays lysis. This phenomenon, termed "lysis inhibition," may last many hours; the precise length is characteristic of the specific phage involved and the environmental conditions.[35] For T4 grown aerobically at 37°C, it typically lasts about 6 hours and can be instigated even when the superinfection by the secondary phages occurs just a minute or two before the normal time of lysis. The general mechanisms and wide variety of holins and lysins are discussed by Young and Wang[37] and by Kutter et al.[2].

Mechanisms of Host Survival of Phage Attack

Phage are the most abundant biological form on earth and have been estimated to outnumber their bacterial hosts by at least an order of magnitude. Nevertheless, bacterial species survive and even flourish despite being so heavily outnumbered by their predators. Obviously, it is essential for phage to not obliterate their bacterial hosts since they would then not have a vehicle for their own growth. The evolution of lysogeny and temperate phages (discussed later) are one major solution to this issue, along with immunity to superinfection conferred by many prophages (a result of several alternate mechanisms, including the lysogenic repressor protein repressing expression of the superinfecting phage genome). Those bacterial species capable of forming spores represent another solution to this issue, as spores are resistant to many environmental insults, including phage attack. In addition, bacteria have themselves evolved several mechanisms for surviving in an environment heavily populated with phage predators.[38] As mentioned above, one common mechanism for bacterial survival to phage attack is mutation of a receptor for phage adsorption, where the altered receptor won't support phage attachment and penetration. Given the large populations involved, it is virtually to be expected that in a susceptible bacterial population subjected to phage attack, there will reside one or more random variants to which the phage cannot attach. If the phage kills most of their wild-type brethren, these variants will now take over the population in what could be described as a textbook exercise in Darwinian evolution. Some bacteria also produce molecules that hinder access to, alter, or block phage receptors when challenged with a phage predator.

A potential variation on this theme is that bacteria can, in some small proportion of the population, enter a state of quiescence, where the cell is not growing but not dead. This is believed to be the mechanism by which some bacteria survive exposure to antibiotics.[39] It is also possible (though not yet demonstrated, to our knowledge) that if infecting phage require growing cells to support their own reproduction, some quiescent cells will survive the phage attack and later be able to regenerate the bacterial population when conditions are suitable for survival. In the case of T4, phage entering stationary-phase cells show the option of going into a so-called hibernation mode, capable of delaying completion of the infection process until needed nutrients become available, while others follow a scavenger pathway, using the limited nutrients available to make a few phage but unable then to respond to nutrient addition. The balance between the two pathways is affected by both host and phage genetics as well as environmental conditions.

Also mentioned above, many phage have evolved mechanisms to protect their DNA from bacterial host restriction enzymes that would otherwise destroy the invading DNA. Nevertheless, the existence of these restriction enzymes thus represents another mechanism by which bacteria are able to survive despite being so outnumbered by phage. Restriction enzymes are not without risk: they are capable of destroying host bacterial DNA as well as foreign (invading) DNA. The risk is tolerated because most restriction enzymes co-evolved with modification enzymes, generally referred to as "restriction/modification" systems. In these systems, modification enzymes chemically alter the host DNA (e.g., by methylation), and the altered host DNA is refractory to degradation by the co-evolved restriction enzyme.[40] Therefore, this restriction/modification pair confers an epigenetic alteration of the cellular DNA that allows the DNA to remain intact in the presence of the restriction enzyme, whereas any foreign DNA that is not decorated with the same chemical modification will be destroyed.

A bacterial population may also survive a phage attack if the specific cells infected by phage undergo an abortive infection; that is, the infected cells die before the phage have had the time to reproduce. This could be considered "altruistic" on the part of the cells that die, since they take the phage down with them before the phage have emerged to kill their bacterial brethren. While there are several known mechanisms that accomplish this (including drastic reduction of ATP levels, or cleavage of protein synthesis components such as EF-Tu or tRNA^{Lys}), one interesting basis for abortive infection has recently been analyzed to result from a toxin-antitoxin (TA) system. Originally identified as a mechanism for plasmid maintenance, and later shown to exist in bacterial genomes as well, TA systems comprise pairs of elements, a stable toxin that can kill the cell harboring it, and an unstable antitoxin that protects the cell from the lethality of the toxin. While these systems are tightly controlled, the underlying principle is that if the antitoxin, or the gene encoding the antitoxin, is lost, the toxin kills the cell because it is stable and the unstable antitoxin is not regenerated to protect the cell. In one reported case of cells using this mechanism to abort phage infection,[41] the antitoxin is an RNA, while the toxin is a protein.

The phytopathogen *Erwinia carotovora* subspecies *atroseptica* encodes the toxin, triggering abortive infections of many different phages. The phage infection is thought to prevent replenishment of the antitoxin, so the toxin then kills the cell before the phage have had the time to reproduce. The consequence of this bacterial apoptosis is the death of both the cells that were infected, and the phage that infected them, while the rest of the bacterial population survives.

With the advent of widespread DNA sequencing of many genomes, another mechanism of bacterial resistance to phage infection has been recognized. This is a kind of adaptive immunity, in which the bacteria acquire specific sequence information from their predators and incorporate this sequence information as spacers into a specialized genetic element named CRISPR (Clustered Regularly Interspaced Short Palindromic Repeats). CRISPR sequences are transcribed and processed by CRISPR-associated (Cas) systems, leading to production of short interfering RNAs (crRNAs) that then function in a manner similar to that of eukaryotic RNA interference (RNAi). The interfering RNAs target foreign nucleic acid, preventing expression and causing degradation of the foreign genetic material.[42] Presently, about half of eubacterial species, and 85% of archaebacterial species, have been identified as harboring CRISPR-Cas systems, though these numbers may increase in the future. CRISPR-Cas systems have been developed as general tools for editing genomes. See Chapter 16 in this volume for details.

Identified sequences in the systems studied to date stem from temperate phages (which are discussed in detail later) rather than from the obligatorily/professionally lytic phages, which is not surprising since bacteria infected by such lytic phages generally don't survive to pass on any such sequence information to their progeny. In the case of temperate phages, the CRISPR-Cas systems may help control both infection by new phages and the activation of prophages carrying the target sequences.

An interesting question about CRISPR systems is how do bacteria obtain the pieces of phage nucleic acid that are then incorporated into CRISPR elements? One possibility is that defective phage particles produced during a lytic infection contain portions of the phage genome that are insufficient to support productive infection in a recipient host. This would be similar to the principle of transduction, described later. The defective phage particle is then the vehicle for delivering phage genetic material that becomes incorporated into CRISPR elements. Another possibility is that acquisition of phage sequences for CRISPR spacers may be accomplished by some bacterial cells managing to survive a phage attack, even though the phage genetic material has entered the cell, that is, a kind of abortive infection that does not result in the death of the cell. Perhaps quiescent cells[39] (as mentioned previously) could provide a mechanism for this kind of abortive infection, allowing cell survival even in the presence of a phage genome. Upon leaving the quiescent state, portions of that phage genome could then be incorporated as spacers in CRISPR elements. This would probably not apply to the professionally lytic phages, whose genomes contain host-lethal genes that would kill the cell upon leaving the quiescent state.

At the end of the day, phage have to balance their need to reproduce with their need to allow the vehicles for their reproduction to remain available. A summary of the major mechanisms discussed here is in Chart 48.1. The balancing act between bacteria and their phage predators is a fascinating area of study that continues to reveal sometimes surprising new information.

CHART 48.1
Major Mechanisms of Bacterial Survival to Phage Attack

1. Blocking phage adsorption
   This can be accomplished by alteration of the phage's receptor, hindering phage access to the receptor, or by mutation of the receptor to a resistant phenotype.

2. *Restriction/modification systems*
   Host DNA is chemically modified to be resistant to a restriction enzyme, while foreign DNA not decorated with that modification will be destroyed by the restriction enzyme.

3. *Abortive infection*
   A kind of "altruistic" action on the part of some cells that die from the phage infection before the phage can reproduce. Mechanisms include membrane alteration, loss of ATP, degradation of components of the protein synthesis apparatus, and toxin-antitoxin systems.

4. *CRISPR adaptive immunity*
   Cells incorporate DNA sequences of the phages that attack them into CRISPR genetic elements and transcribe these sequences into interfering RNAs that block phage gene expression and lead to degradation of the invading phage genome.

Lysogeny

The concept of lysogeny has had a checkered history. Early phage investigators in the 1920s and 1930s claimed to find phages irregularly associated with their bacterial stocks and believed that bacteria were able to spontaneously generate phage, which were thus long considered by many to be some sort of "ferment" or enzyme rather than a living virus. When Max Delbrück and his companions began their work, they confined themselves to the classical set of coliphages designated T1–T7, none of which showed this property, and Delbrück attributed the earlier reports to methodological sloppiness. However, the phenomenon could no longer be denied after the careful work of Lwoff and Gutmann[43]; through microscopic observation of individual cells of *Bacillus megaterium* in microdrops, they demonstrated that cells could continue to divide in phage-free medium with no sign of phage production, but that an occasional cell would lyse spontaneously and liberate phage. Lwoff named the hypothetical intracellular state of the phage genome a prophage and showed later that by treating lysogenic cells with agents such as ultraviolet light, the prophage could be uniformly induced to come out of its quiescent state and initiate lytic growth. The prophage carried by a bacterial strain is given in parentheses; thus, K(P1) means bacterial strain K carrying prophage P1.

Esther Lederberg[44] showed that strains of *E. coli* K-12 carried such a phage, which she named lambda (λ). Meanwhile,

Jacob and Wollman[45] had been investigating the phenomenon of conjugation between donor (Hfr) and recipient (F⁻) strains of *E. coli*. They found that matings of F⁻ strains carrying λ and non-lysogenic Hfrs proceeded normally, but that reciprocal matings yielded no recombinants and, in fact, produced a burst of lambda phage. A mating of Hfr(λ) with nonlysogenic F⁻ would proceed normally if it were stopped before transfer of the *gal* (galactose metabolism) genes, but if conjugation proceeded long enough for the *gal* genes to enter the recipient cell, the prophage would be induced (zygotic induction). These experiments indicated that the λ prophage occupies a specific location near the *gal* genes; that a lysogenic cell maintains the prophage state by expression of one λ gene, encoding a specific repressor protein, which represses expression of all other λ genes; and that if the *gal* genes—and thus the λ prophage—are transferred into a nonlysogenic cell during mating, the prophage finds itself in a cytoplasm lacking repressor and therefore expresses its other genes and enters the lytic cycle.

Typical plaques made by phage λ are turbid, due to lysogenization of some bacteria within the plaque. Mutants of λ producing clear plaques are unable to lysogenize. Analysis of these mutants revealed three genes, designated cI, cII, and cIII, whose products are required for lysogeny. The cI gene encodes the repressor protein. Allen Campbell demonstrated that the sequence of genes in the prophage is a circular permutation of their sequence in the phage genome. He therefore postulated that the prophage is physically inserted into the host genome by circularization of the infecting genome followed by crossing-over between this genome and the (circular) bacterial genome. We now know that the genome in a λ virion has short, complementary single-stranded ends; as one step in lysogenization, it circularizes through internal binding of these ends and is then integrated (by means of a specific integrase) at a point between the *gal* and *bio* (biotin biosynthesis) genes.

The mechanism of the molecular decision between lytic and lysogenic growth has been worked out in most detail for phage λ,[46] but Dodd and Egan[47] found great similarities for the unrelated temperate phage 186. In each case, lysogeny is governed by a repressor protein, CI, which binds to a set of operators in the lysogenic state and represses the expression of all genes except its own (Figure 48.6). Both phages have critical promoter regions, one promoting lysogeny and the other lysis, which are closely associated, and lysogeny is promoted by binding of the CI protein to these sites in such a way as to inhibit lytic growth. DNA looping contributes to stability of the repression,[47b] similar to the role of DNA looping in repression of several catabolic operons. The CI proteins of both phages are strongly cooperative, forming tetramers or octamers. In the λ case, CI is involved in a molecular competition with another protein, Cro, which promotes the lytic cycle. The transition toward lysogeny in λ is also promoted by two other proteins, CII and CIII, which bind to critical promoters and stimulate transcription of the *cI* gene and others. The stability of CII is determined by factors that measure the cell's energy level. A cell with sufficient energy has little cyclic AMP (cAMP); when the cell is energy-starved (leading to low intracellular glucose), the cAMP concentration is high. A high level of cAMP promotes CII stabilization and thus lysogeny. It is clearly adaptive for a phage genome entering a new cell to sense whether there is sufficient energy to make a large burst of phage or whether the energy level is low, so its best strategy for survival is to go into a prophage state.

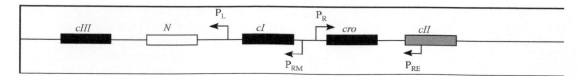

FIGURE 48.6   The regulatory region of the phage λ genome where the decision is made between lysis and lysogeny. The heart of the matter is competition between two proteins, CI and Cro. Initially, lysogeny is promoted by the combined action of the cII and cIII proteins, which act at P_{RE} (promoter right for establishment) and promote transcription of the *cI* gene; the cI protein then binds to various operator/promoter sites and represses them. In particular, cI prevents transcription of the *N* gene whose product acts at other sites in the genome to promote expression of genes required for lytic growth. Also, cI binds to P_{RM} (promoter right for maintenance) and promotes further transcription of its own gene. These lysogeny-directed processes are in competition with transcription from P_R of the *cro* gene and other genes to the right needed for lytic growth. Cro and cI compete for binding sites in the complex promoter/operator region that includes both P_{RM} and P_R where the lysis/lysogeny decision is ultimately made.

Notice that the critical event in establishing lysogeny is regulation of the phage genes. Integration of the phage genome into the host is secondary, and so it is understandable that other temperate phages, such as P1, can establish lysogeny with their prophages functioning as plasmids in the cytoplasm. Many interesting variations on lysogeny and temperate phages have been discovered.

Transduction

Generalized Transduction

As noted previously, bacteriophage package their genomes into phage capsids. Also noted above, the host's bacterial DNA is often subject to degradation during phage infection. The consequence of these two phenomena is that capsids of many phages occasionally package pieces of host DNA instead of phage DNA, making a transducing particle, that is, a vehicle that transmits a piece of DNA from one bacterium to another. This is because the phage particle containing the piece of bacterial DNA can attach to an uninfected bacterium and have its DNA contents taken up by the recipient cell. If the introduced donor DNA recombines with the chromosome of the recipient cell, then genes carried on the donor DNA will be incorporated into that recipient.

Originally discovered in 1952 by Zinder and Lederberg,[48] this process is called generalized transduction, because any portion of the genome can, in principle, be packaged and transferred to a recipient cell. In practice, some phages are much better than others in carrying out transduction, which is a consequence of the characteristics of the specific phage life cycle. Phage T4, for example, is not a very good transducing phage because it is too effective in degrading host DNA, whereas some integrity of host DNA is required for effective transduction, especially by a phage designed to package as long a DNA molecule as T4 does (about 180 kb). By contrast, phage P1 is a workhorse of transduction, very widely used by researchers, because it generates capsid-size chunks of host DNA and is capable of effectively packaging them in transducing particles. The frequency of P1 transduction of any given gene is generally taken to be about 10^{-6}. Multiplicity of infection (MOI) needs to be considered in transduction experiments, because too high an MOI will make it unlikely that a recipient cell will only receive a transducing particle and not be killed by a co-infecting virulent phage particle as well. An illustration of the principle of generalized transduction is shown in Figure 48.7.

Specialized Transduction

Temperate phages have the alternate lifestyle choice of lysogenizing the hosts they infect, integrating into the host chromosome or, like phage P1, forming a plasmid and keeping their genomes largely dormant. But when adverse conditions activate a lysogen, the phage enter a lytic cycle. Rarely, excision of temperate phage DNA from the chromosome is imprecise, resulting in an excised piece of DNA that contains most of the phage genome but also part of the host chromosome that was physically adjacent to the prophage DNA on the chromosome (Figure 48.8). When this excised piece of DNA is packaged in a phage capsid, it becomes a vehicle for transmitting these adjacent host gene(s) to a recipient cell. Because of the imprecise excision, the partial phage genome is usually defective, so the recipient cell will not be subject to a lytic cycle of the phage; instead, the defective phage DNA, carrying the adjacent host DNA, will recombine with the recipient cell's chromosome and thereby transfer genetic material from donor to recipient.

First characterized in 1956 by Morse et al.,[49] this process is called specialized transduction because it can transfer only genes that were physically adjacent to the prophage when it resided in the chromosome of the donor (prior to induction). The host genes transferred can be from either side of the prophage DNA in the lysogen and, depending on the extent of imprecise excision, can include one or a few genes.

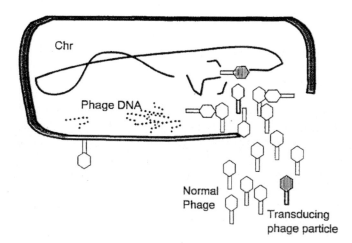

FIGURE 48.7   Generalized transduction. During a lytic phage infection, host chromosomal DNA is degraded. Occasionally, a fragment of the host chromosome is inadvertently packaged into a phage capsid. This "transducing phage particle" can transfer the packaged part of the donor chromosome to a recipient cell. (Illustration courtesy of M. Zafri Humayun.)

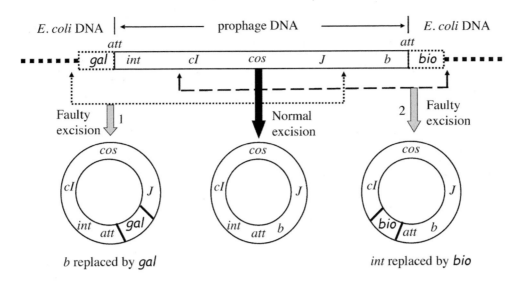

FIGURE 48.8   Specialized transduction results from faulty excision of the lambda prophage. The linear lambda phage DNA (not shown) has cohesive ends (*cos*) that lead to formation of circular lambda DNA. The attachment site (*att*) in the circular lambda DNA recombines with the *E. coli* chromosome between the *gal* and *bio* genes, integrating the lambda prophage DNA to make a lysogen. When the lysogen is induced to a lytic cycle, prophage DNA is excised, and the restored circular lambda DNA will act as a template for virus growth. As a rare event, excision of the prophage is faulty, generating circular lambda DNA containing either the adjacent *E. coli gal* gene (excision 1 in the figure) or the adjacent *E. coli bio* gene (excision 2), in place of lambda genes. Transfer of these *E. coli* genes to recipients in subsequent infections constitute examples of specialized transduction. Representative lambda genes: *int*, integrase; *cI*, repressor; *J*, tail fiber; *b*, region of lambda chromosome that can be deleted with phage retaining viability. (Based in part on ideas described by Campbell.[109])

Many temperate phages have their own specific sites of integration, so the genes they transfer are characteristic of the phage and are a consequence of the specific integration sites. However, some phages, like Mu (for "mutator"), integrate randomly into the genome and a piece longer than the phage genome is packaged directly from the host, so they always carry some neighboring host DNA with them during the packaging process when they are induced. Thus, they are able to affect a generalized transduction of short segments from throughout the genome, and the transducing phage is still fully capable of carrying out a normal infection cycle.

Transduction plays important roles in bacterial genetics, both in the lab and in the wild, and is one of the three major mechanisms for gene transfer between bacteria (the other two being conjugation and transformation). Briani et al.[50] have emphasized the importance of temperate phages as "replicons endowed with horizontal transfer capabilities," and they have probably been major factors in bacterial evolution by moving segments of genomes into new organisms. Broudy and Fischetti[51] have reported *in vivo* lysogenic conversion of Tox− *S. pyogenes* to Tox+ with lysogenic streptococci or free phage in a mammalian host. Thus, this process presumably goes on all the time in nature.

In a variation on transduction, some phages do not degrade host DNA, leaving intact plasmids capable of altering other bacteria by transformation. These phages have been termed "superspreaders",[51b] and they could be significant contributors to the spread of antibiotic resistance as well as other properties.

Role of Bacteriophage in Causing Pathogenesis

Although most phage genes are dormant in a lysogen, particularly genes responsible for lytic growth of the phage, there are a few genes that continue to be expressed from the integrated phage genome. These include, of course, genes like the lambda repressor, which prevents expression of the lytic cycle. However, other phage-specific genes may also be expressed that direct synthesis of one or more products to change the cell's phenotype; the new product(s) did not exist in the cell prior to lysogeny and are often beneficial to host survival in some environments, giving the lysogenized host a selective advantage. This phenomenon is known as phage conversion.

Bacterial toxins, such as diphtheria, cholera, and Strep A toxin, and new antigens, such as *Salmonella* O antigen, are medically important examples of phage conversion, which led a famous bacterial geneticist, William Hayes (1918–1994), to remark "We incriminate the bacteria for the sins of its viruses." Table 48.1 lists a few prominent examples of phage-encoded pathogenesis factors. As an introductory chapter to the field, we will only briefly describe three of the very many known instances of temperate phage-determined virulence. Extensive coverage of this topic may be found in recent publications.[52,53]

Diphtheria Toxin

For an excellent review on diphtheria toxin (DT), see Holmes.[54] When *Corynebacterium diphtheriae* are lysogenized by phage β, the prophage expresses the *tox* gene, leading to the production of a precursor form of DT, which is then processed to the mature toxin. DT inhibits eukaryotic protein synthesis by inactivating elongation factor EF-2. It accomplishes this by covalent attachment of an ADP-ribosyl group to a unique residue in EF-2, a modified form of histidine called diphthamide.

Expression of the *tox* gene is controlled by an iron-regulated repressor encoded by the *dtxR* gene on the *C. diphtheriae* chromosome. In the presence of iron, the DtxR protein is active as a repressor and prevents transcription of the *tox* gene. In the

TABLE 48.1

Examples of Bacteriophage-Encoded Pathogenesis Factors

Pathogenesis Factor	Bacteria	Phage
O antigen[72–76] (immune system evasion)	*Salmonella* species	ε15, ε34
SopE[77] (invasion, type III secretion)	*Salmonella typhimurium*	SopEφ
SodC[78] (superoxide dismutase)	*Salmonella typhimurium*	Gifsy-2
Cholera toxin[55]	*Vibrio cholerae*	CTXφ
Diphtheria toxin[79]	*Corynebacterium diphtheriae*	Converting β–phage
Shiga-like toxin–I and –II[80,81]	Enterotoxigenic *E. coli*	Stx converting phage
Serum resistance[82]	*Escherichia coli*	λ
Enterotoxin A and staphylokinase[83]	*Staphylococcus aureus*	φ13
Streptococcal pyrogenic exotoxins A&C[84]	Group A *Streptococci*	T12 and 3GL16

absence of iron, the repressor is inactive and there is no repression of *tox*, leading to production of DT (Figure 48.9). Since iron concentrations tend to be low in mammalian hosts, pathogenesis occurs when β-phage-carrying *C. diphtheriae* colonize the throat of an affected individual. See Chapter 27 of this volume for a full discussion of *Corynebacteria*.

Cholera Toxin

Recognition that cholera toxin (CT) is encoded by a lysogenic phage was relatively recent,[55] perhaps in part because the phage involved, CTXφ, is a filamentous phage that does not kill the host bacterial cell during productive growth. CT, comprised of one "A" subunit and five "B" subunits, inactivates a GTP-binding protein needed to control cellular cyclic-AMP synthesis, causing

overproduction of cAMP and massive fluid and electrolyte loss from intestinal cells.[56]

Vibrio cholerae ordinarily occupy a salt-water habitat, which is not the environment of a mammalian gut. Thus, one can argue that a bacterial sensor system of a low-salt environment activates synthesis of CT, and the profuse resultant diarrhea facilitates the release and spread of the bacteria from the affected individual. It is known that environmental conditions affect *V. cholerae* virulence.[57] Chapter 41 in this volume presents a full discussion of *Vibrio* species.

The sensor system for activation of CT synthesis is complex, with many cellular components[58] as well as a recently discovered lysogenic satellite phage, RS1.[59] The regulator ToxT activates transcription of the *ctxA* and *ctxB* toxin genes from a single promoter; there is much greater synthesis of the B subunit compared to the A subunit, which is probably controlled at the translational level. The *toxT* gene, as well as other virulence genes activated by ToxT (toxin coregulated pilus [tcp] and accessory colonization factors [acf]), are located on the Vibrio pathogenicity island (VPI) region of the *V. cholerae* chromosome. This region was also thought to represent a lysogenic phage genome, but evidence for this was not compelling and this view is generally no longer current.[60]

Synthesis of ToxT itself is activated by other *V. cholerae* genes, *toxR, toxS, tcpP,* and *tcpH.* The products of these genes are located in the cell membrane, where they may play a role in sensing a low salt level in the environment, activating ToxT synthesis, which in turn, activates CT production from the lysogen.

Recently, the satellite phage RS1 was shown to be another lysogen of *V. cholerae* and to play a role in CT production. RS1 itself is defective and only grows during growth of CTXφ. The *ctxAB* genes can also be transcribed in the lysogen from an upstream promoter for phage growth, which is repressed by the RstR repressor. Phage RS1 encodes an antirepressor, RstC, which inactivates repression by RstR, leading to increased synthesis of the CT genes. RstC itself is also repressed by the RstR repressor; therefore induction of the prophage by DNA damage (the "SOS"

Diphtheria Toxin Regulation

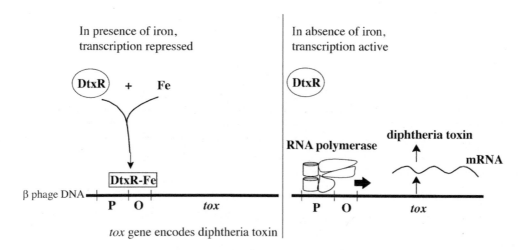

FIGURE 48.9 Negative control of diphtheria toxin by a repressor protein. The toxin is encoded by a phage gene in a lysogen. The DtxR repressor protein is encoded in the *C. diphtheria* genome. (*Abbreviations:* P, promoter; O, operator; Fe, iron.)

Regulation of Cholera Virulence

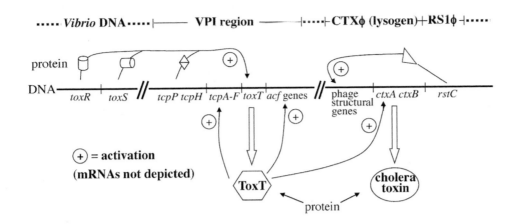

tcp=toxin coregulated pilus; acf=accessory colonization factor;
VPI=Vibrio pathogenicity island; φ = lysogenic phage genome.

FIGURE 48.10 Regulation of virulence gene expression in *Vibrio cholerae*. TcpP/tcpH and toxR/toxS are located in the cell membrane and cooperate in activating transcription of *toxT*. ToxT activates transcription of *tcpA-F*, *ctxAB*, and *acf* genes. Genes for toxR/toxS are encoded on the ancestral *V. cholerae* chromosome and regulate genes other than those on CTXφ and VPI (formerly thought to be a lysogen). Satellite phage RS1φ encodes the RstC antirepressor, which inactivates the RstR repressor (not shown) of CTXφ growth. Inactivating the RstR repressor also leads to transcription of the *ctxAB* genes. Synthesis of RstC follows an SOS response.

response), which inactivates RstR, is necessary to obtain production of RstC.[60] Thus, there is also a second mechanism available, mediated by satellite phage RS1, to stimulate synthesis of CT.

A schematic diagram of the regulatory circuits for production of CT is shown in Figure 48.10. For simplicity, the effect of RstC is shown only as activation of transcription upstream of the *ctxAB* genes (whereas the actual mechanism involves the intermediate step of inactivation of the RstR repressor).

Shiga-Like Toxins

Escherichia coli, ordinarily a commensal universally found in the gut, can become highly pathogenic following lysogeny with Stx-converting phages. Pathogenicity is generally associated with the serotype O157:H7, although there are some other serotypes as well, resulting in hemorrhagic colitis, hemolytic uremic syndrome, and kidney failure. See Chapter 28 of this volume for a discussion of Enterobacteriaceae, including *Escherichia coli*.

The Shiga-like toxins, so-named due to similarity to toxins from *Shigella dysenteriae*, are also comprised of one "A" subunit and five "B" subunits. They inhibit protein synthesis by attacking eukaryotic ribosomal RNA.

The phages responsible for this are of the lambda family, and carry genes, *stxA* and *stxB*, that encode the two toxin subunits. Of the two prominent toxin types, type 1 is regulated by iron, while type 2 is not. Type 2 is repressed in the lysogen, and expression follows induction of the lysogen to a lytic cycle.[61] Expression of Type 1 in the lysogen, under conditions of low iron concentration, is enhanced following induction of the lysogen to a lytic cycle.[62] The involvement of lambdoid phages in pathogenesis is why they are not good choices for phage therapy (see Chapter 54).

A simplified schematic diagram illustrating regulation of expression of Shiga toxin-type 2 is shown in Figure 48.11.

Practical Information about Selected Phages

Tailed Phages

Table 48.2 lists, in alphabetical order, selected tailed phages that have been extensively studied. Pathogenesis-causing phages previously listed in Table 48.1 are not included. All of these phage have linear double-stranded DNA genomes. Many exhibit terminal redundancy with (e.g., P1, P22, T4, ViI, SPO1, K,) or without (e.g., λ, P2, P4, T5, T7) circular permutation.[63] Some key characteristics of the phages are presented, and the "comments" column highlights especially interesting or unique features.

More in-depth general resources include Granoff and Webster's *Encyclopedia of Virology*,[22] both editions of *The Bacteriophages*,[64,65] Ackermann and DuBow[11] and Adams.[8] The complete genomic sequences of hundreds of phages have now been determined and are available in the bacteriophage section of the genome sites at the National Center for Biotechnology Information <http://www.ncbi.nlm.nih.gov>, as are the sequences of many of the prophages determined in the course of microbial genome projects.

Phages of *E. coli* and *B. subtilis* have generally been the best studied regarding genetics and the physiology of the infection process, taking advantage of and contributing to the broad knowledge about these two key model organisms. Coliphage T4 in particular has provided the data used to identify the probable functions of many apparently related genes in genetically unstudied phages (Figure 48.2).

Shiga toxin 2 expression in *E. coli* O157:H7

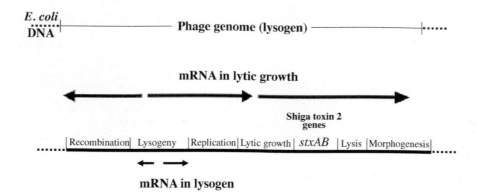

FIGURE 48.11   Shiga toxin 2 expressed following λ-family prophage induction in *E. coli* O157:H7. Putative genome organization (not to scale) and transcription of Stx2-encoding phage is based on known lambda-family phage. Shown are relevant functional genes clusters, as indicated. *stxAB* represents the genes encoding the pathogenic Shiga toxin 2. Below is shown the pattern of transcription initiating at the early promoters in the lysogen. Above is shown the patterns of transcription initiating at the early and late promoters during lytic growth.

TABLE 48.2

Survey of Selected DNA-Containing Tailed Phages (Caudovirales)*

Phage	Family[a]	Host(s)	Size[b]	Temperate[c]	Transduction	Comments
		Gram-Negative Bacteria				
Lambda (λ)	Sipho	*Escherichia coli*	48.5[85]	+	specialized	Genetic engineering vector
Mu	Myo	*E. coli, Salmonella,* others	36.7[86]	+	generalized	Random integration in host DNA
N4	Podo	*E. coli*	70.2[87]			RNA polymerase used in biotech
N15	Sipho	*E. coli*	46.3[88]	+		Prophage a linear plasmid
P1	Myo	broad host range	93.6[89]	+	generalized	Prophage a circular plasmid
P2	Myo	*E. coli*	33.6[90]	+		DNA packaged from monomeric circles
P4	Myo	*E. coli*	11.6[91]	+		Defective; satellite of phage P2
P22	Podo	*Salmonella typhimurium*	41.7[92]	+	generalized	Transduction via lytic cycle; modifies O antigen
phiKZ(φKZ)	Myo	*Pseudomonas*	280[93]		generalized	Transduction at very low frequency
T1	Sipho	*E. coli*	48.8[94]			Viable even after dried out
T4[d]	Myo	*E. coli*	169[95]			5-hmdC replaces C in DNA
T5	Sipho	*E. coli*	122[96]			Encodes tRNAs for all 20 amino acids
T7[e]	Podo	*E. coli* (F⁻ only)	39.9[97]			Many biotech applications
		Gram-Positive Bacteria				
C₁	Podo	group C *Streptococci*	16.7[98]			Potential medical use of "lysin"
G	Myo	*Bacillus megaterium*	497.5[99]			Largest characterized virus
L5	Sipho	*Mycobacteria*	52.3[100]	+	specialized	Useful for *Mycobacterial* genetics
MM1	Sipho	*Streptococcus pneumoniae*	40.2[101]	+		NOT coliphage MM1; many unknown genes
phi11(φ11)	Sipho	*Staphylococcus aureus*	43.6[102]	+	generalized	Closely related to pathogenic phage φ13
phi29(φ29)	Podo	*Bacillus subtilis*	19.3[103]			Biotech use of DNA polymerase
phi105(φ105)	Sipho	*B. subtilis*	39.3[104]	+	specialized	Similar to lambda
Sb-1	Myo	*S. aureus*	120[105]			Resistance rare; therapeutic use
SPβ	Sipho	*B. subtilis, B. globigii*	134[106]	+	specialized	Secretes betacin, kills nearby cells
SPO1	Myo	*B. subtilis*	140[107]			Transcriptional sigma cascade; hmdU replaces T

*   For more detailed discussion of most of the phages in this table (and related phages), see Chapter 3 in *Bacteriophages: Biology and Applications*[1]
[a]   Virus type[108]: Myoviridae, viruses with contractile tails; Siphoviridae, viruses with long noncontractile tails; Podoviridae, viruses with short noncontractile tails.
[b]   genome size in kbp.
[c]   + = known to be capable of lysogeny.
[d]   T2 and T6 are very closely related to T4.
[e]   T3 is very closely related to T7.

Note that, in general, the designation of phage names has been quite arbitrary, giving little or no information about phage properties or relationships.

Nontailed Phages

phiX174 (φX174): The prototype member of the *Microviridae*, with a 5.4-kb circular single-stranded DNA genome[66] and an icosahedral virion. The DNA is a positive (+) strand (i.e., sequence is in the same orientation as its mRNA) with overlapping genes. Infects *Enterobacteriaceae* such as *E. coli, Salmonella* and *Shigella,* and has been extensively used in studies of DNA replication.

M13: The prototype member of the *Inoviridae*, with a 6.4 kb circular positive single-stranded DNA genome,[67] and an amorphous, filamentous virion. Infects male *E. coli* by adsorbing to the tip of the pilus; nonlytic, but turbid plaques can be discerned due to slower growth of infected bacteria. Has been extensively used for cloning and DNA sequencing, and in "phage display" technology, described in detail in Chapter 9 in this book.

MS2: The prototype member of the *Leviviridae,* with a 3.6 kb linear single-stranded highly structured RNA genome[68] and an icosahedral virion. Infects male *E. coli* by adsorbing to the sides of the pilus; lytic virus but does not shut off host metabolism. Positive-strand RNA genome functions as an mRNA. High affinity of coat protein to specific sequences in RNA has been exploited in biotechnology.

phi6 (φ6): The prototype member of the *Cystoviridae,* with a genome of three double-stranded RNAs of 6.4, 4.1, and 2.9 kb (reviewed in Mindich[69]) and an enveloped virion. Infects the plant pathogen *Pseudomonas syringae* by initially adsorbing to the pilus, though other family members do not use the pilus for entry.

PRD1: The prototype member of the *Tectiviridae,* with a 14.9-kb linear double-stranded DNA genome,[70] and an icosahedral virion coating an internal lipid membrane. Infects Gram-negative bacteria including *Enterobacteriaceae (E. coli, Salmonella, etc.)* and *Pseudomonas.* Certain conjugative plasmids required for phage adsorption to the cell wall (not pili).[71] Has garnered attention as a model for human adenovirus.

REFERENCES

1. Guttman, B., Raya, R., and Kutter, E. Basic Phage Biology (chapter 3). *Bacteriophages: Biology and Applications,* Kutter, E., and Sulakvelidze A. (Eds.), CRC Press, Boca Raton, FL, 2005, 29–66.
2. Kutter, E., Raya, R., and Carlson, K. Molecular Mechanisms of Phage Infection (chapter 7). *Bacteriophages: Biology and Applications,* Kutter, E., and Sulakvelidze A. (Eds.), CRC Press, Boca Raton, FL, 2005, 165–222.
2b. Kutter, E., Bryan, D., Ray, G., Brewster, E., Blasdel, B., and Guttman, B. From host to phage metabolism: hot tales of phage t4's takeover of *E. coli. Viruses,* 10, 387, 2018.
3. Molineux, I.J. No syringes please, ejection of phage T7 DNA from the virion is enzyme driven. *Mol. Microbiol.,* 40, 1–8, 2001.
4. Prangishvili, D. Evolutionary insights from studies on viruses of hyperthermophilic archaea. *Res. Microbiol.,* 154, 289–294, 2003.
5. Abedon, S.T., and Calendar, R.L. (Eds.) *The Bacteriophages,* second edition, Oxford University Press, New York, NY, 2005, part III, 129–210.
6. Levin, B.R., and Lenski, R.E. Bacteria and phage: a model system for the study of the ecology and co-evolution of hosts and parasites. *Ecology and Genetics of Host-Parasite Interactions.* The Linnean Society of London, London, 1985, 227–241.
6b. De Smet, J., Zimmermann, M., Kogadeeva, M., Ceyssens, P.J., Vermaelen, W., Blasdel, B., Jang, H.B., Sauer, U., and Lavigne, R. High coverage metabolomics analysis reveals phage-specific alterations to *pseudomonas aeruginosa* physiology during infection. *ISME J.,* 10, 1823–1835, 2016.
7. d'Herelle, F. *The Bacteriophage and Its Behavior,* The Williams & Wilkins Company, Baltimore, MD, 1926.
8. Adams, M.H. *Bacteriophages,* Interscience Publishers, New York, NY, 1959.
9. Tikhonenko, A.S., Gachechiladze, K.K., Bespalova, I.A., Kretova, A.F., and Chanishvili, T.G. Electron-microscopic study of the serological affinity between the antigenic components of phages T4 and DDVI. *Mol. Biol. (Mosk),* 10, 667–673, 1976.
10. Gachechiladze, K. Antigenicity of Phages. *Bacteriophages: Biology and Applications,* Kutter, E., and Sulakvelidze A. (Eds.), CRC Press, Boca Raton, FL, 2005, 33–37.
10b. Sweere, J.M., Van Belleghem, J.D., Ishak, H., Bach, M.S., Popescu, M., Sunkari, V., Kaber, G., Manasherob, R., Suh, G.A., Cao, X., de Vries, C.R., Lam, D.N., Marshall, P.L., Birukova, M., Katznelson, E., Lazzareschi, D.V., Balaji, S., Keswani, S.G., Hawn, T.R., Secor, P.R., and Bollyky, P.L. Bacteriophage trigger antiviral immunity and prevent clearance of bacterial infection. *Science,* 363, eaat9691, 2019.
11. Ackermann, H.W., and Dubow, M. *Viruses of Prokaryotes,* CRC Press, Boca Raton, FL, 1987.
12. Nikaido, H. Outer Membrane. *Escherichia coli and Salmonella,* Neidhardt, F.C., et al. (Eds.), American Society for Microbiology, Washington, D.C., 1996, 29–47.
13. Nikaido, H. Molecular basis of bacterial outer membrane permeability revisited. *Microbiol. Mol. Biol. Rev.,* 67, 593–656, 2003.
14. Ghuysen, J.M., and Hackenbeck, R. *Bacterial Cell Wall,* Elsevier, Amsterdam & New York, 1994.
14b. Nobrega, F.L., Vlot, M., de Jonge, P.A., Dreesens, L.L., Beaumont, H.J.E., Lavigne, R., Dutilh, B.E., and Brouns, S.J.J. Targeting mechanisms of tailed bacteriophages. *Nat Rev Microbiol,* 16, 760–773, 2018.
14c. Andres, D., Roske, Y., Doering, C., Heinemann, U., Seckler, R., and Barbirz, S. Tail morphology controls DNA release in two *salmonella* phages with one lipopolysaccharide receptor recognition system. *Mol. Microbiol.,* 83, 1244–1253, 2012.
15. Letellier, L., Boulanger, P., Plancon, L., Jacquot, P., and Santamaria, M. Main features on tailed phage, host recognition and DNA uptake. *Front. Biosci.,* 9, 1228–1339, 2004.
16. McAllister, W.T. Bacteriophage infection: which end of the SP82G genome goes in first? *J. Virol.,* 5, 194–198, 1970.

17. Duplessis, M., and Moineau, S. Identification of a genetic determinant responsible for host specificity in Streptococcus thermophilus bacteriophages. *Mol. Microbiol.*, 41, 325–336, 2001.

18. Ravin, V., Raisanen, L., and Alatossava, T. A conserved C-terminal region in Gp71 of the small isometric-head phage LL-H and ORF474 of the prolate-head phage JCL1032 is implicated in specificity of adsorption of phage to its host, *lactobacillus delbrueckii. J. Bacteriol.*, 184, 2455–2459, 2002.

19. Stuer-Lauridsen, B., Janzen, T., Schnabl, J., and Johansen, E. Identification of the host determinant of two prolate-headed phages infecting *Lactococcus lactis. Virology*, 309, 10–17, 2003.

20. Kutter, E., Stidham, T., Guttman, B., Kutter, E., Batts, D., Peterson, S., Djavakhishvili, T., et al. Genomic Map of Bacteriophage T4. *Molecular Biology of Bacteriophage T4*, Karam, J., Drake, J.W., Kreuzer, K.N., Mosig, G., Hall, D.H., Eiserling, F.A., Black, L.W., et al. (Eds.), American Society for Microbiology, Washington, D.C., 1994, 491–519.

21. Goldberg, E., Grinius, L., and Letellier, L. Recognition, Attachment, and Injection. *Molecular Biology of Bacteriophage T4*, Karam, J.D., Drake, J.W., Kreuzer, K.N., Mosig, G., Hall, D.H., Eiserling, F.A., Black, L.W., et al. (Eds.), American Society for Microbiology, Washington, D.C., 1994, 347–356.

22. Granoff, A., and Webster, R.G. (Eds.) *Encyclopedia of Virology*, Academic Press, New York, NY, 1999.

23. Ellis, E.L., and Delbrück, M. The growth of bacteriophage. *J. Gen. Physiol.*, 22, 365, 1939.

24. Doermann, A.H., The vegetative state in the life cycle of bacteriophage: Evidence for its occurrence and its genetic characterization. *Cold Spring Harbor Symp. Quant. Biol.*, 18, 3–11, 1953.

24b. Bryan, D., El-Shibiny, A., Hobbs, Z., Porter, J., and Kutter, E.M. Bacteriophage T4 infection of stationary phase *E. coli*: life after log from a phage perspective. *Front Microbiol.*, 7, 1391, 2016.

25. Wollman, E.L., and Stent, G.S. Studies on activation of T4 bacteriophage by cofactor. IV. Nascent activity. *Biochem. Biophys. Acta*, 9, 538–550, 1952.

26. Brown, D.T., and Anderson, T.F., Effect of host cell wall material on the adsorbability of cofactor-requiring T4. *J. Virol.*, 4, 94–108, 1969.

27. Simon, L.D. The infection of *Escherichia coli* by T2 and T4 bacteriophages as seen in the electron microscope. III. Membrane-associated intracellular bacteriophages. *Virology*, 38, 285–296, 1969.

28. Evans, A.C. The potency of nascent *Streptococcus* bacteriophage B. *J Bacteriol*, 39, 597–604, 1940.

29. Lenski, R.E. Coevolution of bacteria and phage: are there endless cycles of bacterial defenses and phage counterdefenses? *J. Theor. Biol.*, 108, 319–325, 1984.

30. Tetart, F., Desplats, C., and Krisch, H.M. Genome plasticity in the distal tail fiber locus of the T-even bacteriophage: Recombination between conserved motifs swaps adhesin specificity. *J Mol. Biol.*, 282, 543–556, 1998.

31. Cowan, J., d'Acci, K., Guttman, B., and Kutter, E. Gel Analysis of T4 Prereplicative Proteins. *Molecular Biology of Bacteriophage T4*, Karam, J., et al. (Eds.), American Society for Microbiology, Washington, D.C., 1994, 520–527.

32. Létoffé, S., Audrain, B., Bernier, S.P., Delepierre, M., and Ghigo, J.-M. Aerial exposure to the bacterial volatile compound trimethylamine modifies antibiotic resistance of physically separated bacteria by raising culture medium pH. *mBio*, 5(1), e00944–13, 2014.

33. Leiman, P.G., Arisaka, F., van Raaij, M.J., Kostyuchenko, V.A., Aksyuk, A.A., Kanamaru, S., and Rossman, M.G. Morphogenesis of the T4 tail and tail fibers. *Virol J*, 7, 355–(28 pages), 2010.

33b. Blasdel, B.G., Chevallereau, A., Monot, M., Lavigne, R., and Debarbieux, L. Comparative transcriptomics analyses reveal the conservation of an ancestral infectious strategy in two bacteriophage genera. *ISME J.*, 11, 1988–1996, 2017.

33c. Chevallereau, A., Blasdel, B.G., De Smet, J., Monot, M., Zimmermann, M., Kogadeeva, M., and Lavigne, R. Next-Generation "-omics" approaches reveal a massive alteration of host RNA metabolism during bacteriophage infection of pseudomonas aeruginosa. *PLOS Genet.*, 12, e1006134, 2016.

34. Wang, I.N., Deaton, J., and Young, R. Sizing the holin lesion with an endolysin-beta-galactosidase fusion. *J. Bacteriol.*, 185, 779–787, 2003.

35. Abedon, S.T. Selection for lysis inhibition in bacteriophage. *J. Theor. Biol.*, 146, 501–511, 1990.

36. Abedon, S.T. Phage Ecology (chapter 5). *The Bacteriophages*, second edition, Abedon, S.T., and Calendar, R.L. (Eds.), Oxford University Press, New York, NY, 2005, 37–48.

37. Young, R., and Wang, I.N. Phage Lysis (chapter 10). *The Bacteriophages*, second edition, Abedon, S.T., and Calendar, R.L. (Eds.), Oxford University Press, New York, NY, 2005, 104–128.

38. Labrie, S.J., Samson, J.E., and Moineau, S. Bacteriophage resistance mechanisms. *Nat. Rev., Microbiol.*, 8, 317–327, 2010.

39. Rittershaus, E.S., Baek, S.H., and Sassetti, C.M. The normalcy of dormancy: common themes in microbial quiescence. *Cell Host Microbe.*, 13, 643–651, 2013.

40. Wilson, G.G., and Murray, N.E. Restriction and modification systems. *Annu. Rev. Genet.*, 25, 585–627, 1991.

41. Fineran, P.C., Blower, T.R., Foulds, I.J., Humphreys, D.P., Lilley, K.S., and Salmond, G.P. The phage abortive infection system, ToxIN, functions as a protein-RNA toxin-antitoxin pair. *Proc. Natl. Acad. Sci. U. S. A.*, 106, 894–899, 2009.

42. Richter, C., Chang, J.T., and Fineran, P.C. Function and regulation of clustered regularly interspaced short palindromic repeats (CRISPR)/CRISPR associated (Cas) systems. *Viruses*, 4, 2291–2311, 2012.

43. Lwoff, A., and Gutmann, A. Recherches sur un *bacillus megaterium* lysogene. *Ann Inst Pasteur*, 78, 711–739, 1950.

44. Lederberg, E.M., and Lederberg, J. Genetic studies of lysogenicity in *Escherichia coli. Genetics*, 38, 51–64, 1953.

45. Jacob, F., and Wollman, E.L. *Sexuality and the Genetics of Bacteria*, Academic Press, New York, NY, 1961.

46. Little, J.W. Gene Regulatory Circuitry of Phage λ (chapter 8). *The Bacteriophages*, Abedon, S.T, and Calendar, R.L. (Eds.), Oxford University Press, New York, NY, 2005, 74–82.

47. Dodd, I.B., and Egan, J.B. Action at a distance in CI repressor regulation of the bacteriophage 186 genetic switch. *Mol. Microbiol.*, 45, 697–710, 2002.

47b. Lewis, D.E.A., Gussin, G.N., and Adhya, S. New Insights into the phage genetic switch: Effects of bacteriophage lambda operator mutations on DNA looping and regulation of PR, PL, and PRM. *J Mol Biol.*, 428, 4438–4456, 2016.

48. Zinder, N.D., and Lederberg, J. Genetic exchange in *salmonella. J. Bacteriol.*, 64, 679–699, 1952.

49. Morse, M.L., Lederberg, E.M., and Lederberg, J. Transduction in *Escherichia coli* K-12. *Genetics*, 41, 142–156, 1956.

50. Briani, F., Deho, G., Forti, F., and Ghisotti, D. The plasmid status of satellite bacteriophage P4. *Plasmid*, 45, 1–17, 2001.

51. Broudy, T.B., and Fischctti, V.A. In vivo lysogenic conversion of Tox(-) *Streptococcus pyogenes* to Tox(+) with lysogenic streptococci or free phage. *Infect. Immun.*, 71, 3782–3786, 2003.

51b. Keen, E.C., Bliskovsky, V.V., Malagon, F., Baker, J.D., Prince, J.S., Klaus, J.S., and Adhya, S.L. Novel "Superspreader" bacteriophages promote horizontal gene transfer by transformation. *mBio.*, 8, e02115–e02116, 2017.

52. Waldor, M.K., Friedman, D.I., and Adhya, S.L. (Eds.) *Phages: Their Role in Bacterial Pathogenesis and Biotechnology*, ASM Press, Washington, D.C., 2005.

53. Boyd, E.F. Bacteriophage-encoded bacterial virulence factors and phage-pathogenicity island interactions. *Adv Virus Res*, 82, 91–118, 2012.

54. Holmes, R.K. Biology and molecular epidemiology of diphtheria toxin and the *tox* gene. *J. Infect. Dis.*, 181(Suppl 1), S156–167, 2000.

55. Waldor, M.K., and Mekalanos, J.J. Lysogenic conversion by a filamentous phage encoding cholera toxin. *Science*, 272, 1910–1914, 1996.

56. Spangler, B.D. Structure and function of cholera toxin and the related *Escherichia coli* heat-labile enterotoxin. *Microbiol. Rev.*, 56, 622–647, 1992.

57. Skorupski, K., and Taylor, R.K. Control of the ToxR virulence regulon in *vibrio cholerae* by environmental stimuli. *Mol. Microbiol.*, 25, 1003–1009, 1997.

58. Cotter, P.A., and DiRita, V.J. Bacterial virulence gene regulation: an evolutionary perspective. *Annu. Rev. Microbiol.*, 54, 519–565, 2000.

59. Davis, B.M., Kimsey, H.H., Kane, A.V., and Waldor, M.K. A satellite phage-encoded antirepressor induces repressor aggregation and cholera toxin gene transfer. *EMBO J.*, 21, 4240–4249, 2002.

60. Davis, B.M., and Waldor, M.K. Filamentous phages linked to virulence of *vibrio cholerae*. *Curr. Opin. Microbiol.*, 6, 35–42, 2003.

61. Wagner, P.L., Neely, M.N., Zhang, X., Acheson, D.W., Waldor, M.K., and Friedman, D.I. Role for a phage promoter in Shiga toxin 2 expression from a pathogenic *Escherichia coli* strain. *J. Bacteriol.*, 183, 2081–2085, 2001.

62. Wagner, P.L., Livny, J., Neely, M.N., Acheson, D.W., Friedman, D.I., and Waldor, M.K. Bacteriophage control of Shiga toxin 1 production and release by *Escherichia coli*. *Mol. Microbiol.*, 44, 957–970, 2002.

63. Fujisawa, H., and Morita, M. Phage DNA packaging. *Genes Cells*, 2, 537–545, 1997.

64. Calendar, R. (Ed.) *The Bacteriophages*, first edition, Plenum Press, New York, NY, 1988.

65. Abedon, S.T., and Calendar, R. L. (Eds.). *The Bacteriophages*, second edition, Oxford University Press, New York, NY, 2005.

66. Sanger, F., Coulson, A.R., Friedmann, C.T., Air, G.M., Barrell, B.G., Brown, N.L., Fiddes, J.C., Hutchison III, C.A., Slocombe, P.M., and Smith, M. The nucleotide sequence of bacteriophage {phi}X174. *J. Mol. Biol.*, 125, 225–246, 1978.

67. van Wezenbeek, P.M., Hulsebos, T.J., and Schoenmakers, J.G. Nucleotide sequence of the filamentous bacteriophage M13 DNA genome: comparison with phage fd. *Gene*, 11, 129–148, 1980.

68. Fiers, W., Contreras, R., Duerinck, F., Haegeman, G., Iserentant, D., Merregaert, J., Min Jou, W., Molemans, F., Raeymaekers, A., Van den Berghe, A., Volckaert, G., and Ysebaert, M. Complete nucleotide sequence of bacteriophage MS2 RNA: primary and secondary structure of the replicase gene. *Nature*, 260, 500–507, 1976.

69. Mindich, L. Precise packaging of the three genomic segments of the double-stranded-RNA bacteriophage phi6. *Microbiol Mol Biol Rev.*, 63, 149–160, 1999.

70. Bamford, J.K.H., Hänninen, A.-L., Pakula, T.M., Ojala, P.M., Kalkkinen, N., Frilander, M., and Bamford, D.H. Genome organization of membrane-containing bacteriophage PRD1. *Virology*, 183, 658–676, 1991.

71. Olsen, R.H., Siak, J.-S., and Gray, R.H. Characteristics of PRD1, a plasmid-dependent broad host range DNA bacteriophage. *J. Virol.*, 14, 689–699, 1974.

72. Matias, V.R., and Beveridge, T.J. Native cell wall organization shown by cryo-electron microscopy confirms the existence of a periplasmic space in *Staphylococcus aureus*. *J. Bacteriol.* 188, 1011–1021, 2006.

73. Robbins, P.W., and Uchida, T. Determinants of specificity in *Salmonella*: changes in antigenic structure mediated by bacteriophage. *Fed. Proc.*, 21, 702, 1962.

74. Losick, R. Isolation of a trypsin-sensitive inhibitor of O-antigen synthesis involved in lysogenic conversion by bacteriophage epsilon-15. *J. Mol. Biol.*, 42, 237, 1969.

75. Wright, A. Mechanism of conversion of the *Salmonella* O antigen by bacteriophage epsilon 34. *J. Bacteriol.*, 105, 927, 1971.

76. Smith, H.W., and Parsell, Z. The effect of virulence of converting the O antigen of *Salmonella cholerae-suis* from 627 to 617 by phage. *J. Gen. Microbiol.*, 81, 217, 1974.

77. Mirold, S., Rabsch, W., Rohde, M., Stender, S., Tschape, H., Russmann, H., Igwe, E., and Hardt, W.D. Isolation of a temperate bacteriophage encoding the type III effector protein SopE from an epidemic *Salmonella typhimurium* strain. *Proc. Natl. Acad. Sci. U. S. A.*, 96, 9845, 1999.

78. Figueroa-Bossi, N., and Bossi, L. Inducible prophages contribute to *salmonella* virulence in mice. *Mol. Microbiol.*, 33, 167, 1999.

79. Groman, N.B. Conversion by Corynephages and its role in the natural history of diphtheria. *J Hyg (Lond).*, 93, 405, 1984.

80. Muniesa, M., de Simon, M., Prats, G., Ferrer, D., Panella, H., and Jofre, J. Shiga toxin 2-converting bacteriophages associated with clonal variability in *Escherichia coli* O157:H7 strains of human origin isolated from a single outbreak. *Infect. Immun.*, 71, 4554, 2003.

81. Huang, A., Friesen, J., and Brunton, J.L. Characterization of a bacteriophage that carries the genes for production of Shiga-like toxin 1 in *Escherichia coli*. *J. Bacteriol.*, 169, 4308, 1987.

82. Barondess, J.J., and Beckwith, J. *Bor* gene of phage λ, involved in serum resistance, encodes a widely conserved outer membrane lipoprotein. *J. Bacteriol.*, 177, 1247, 1995.

83. Coleman, D.C., Sullivan, D.J., Russell, R.J., Arbuthnott, J.P., Carey, B.F., and Pomeroy, H.M. *Staphylococcus aureus* bacteriophages mediating the simultaneous lysogenic conversion of beta-lysin, staphylokinase and enterotoxin A: molecular mechanism of triple conversion. *J. Gen. Microbiol.*, 135, 1679, 1989.

84. Johnson, L.P., Schlievert, P.M., and Watson, D.W. Transfer of group A streptococcal pyrogenic exotoxin production to nontoxigenic strains of lysogenic conversion. *Infect. Immun.*, 28, 254, 1980.

85. Sanger, F., Coulson, A.R., Hong, G.F., Hill, D.F., and Petersen, G.B., Nucleotide sequence of bacteriophage lambda DNA. *J Mol Biol*, 162, 729–773, 1982.

86. Morgan, G.J., Hatfull, G.F., Casjens, S., and Hendrix R.W. Bacteriophage Mu genome sequence: analysis and comparison with Mu-like prophages in Haemophilus, Neisseria and Deinococcus. *J Mol Biol.*, 317, 337–359, 2002.

87. GenBank ACCESSION # NC_008720.

88. Ravin, V., Ravin, N., Casjens, S., Ford, M.E., Hatfull, G.F., and Hendrix, R.W. Genomic sequence and analysis of the atypical temperate bacteriophage N15. *J Mol Biol.*, 299, 53–73, 2000.

89. Lobocka, M.B., Rose, D.J., Plunkett, G. 3rd, Rusin, M., Samojedny, A., Lehnherr, H., Yarmolinsky, M.B., and Blattner, F.R. Genome of bacteriophage P1. *J Bacteriol.*, 186, 7032–7068, 2004.

90. GenBank ACCESSION # NC_001895.

91. Halling, C., Calendar, R., Christie, G.E., Dale, E.C., Dehò, G., Finkel, S., Flensburg, J., Ghisotti, D., Kahn, M.L., and Lane, K.B. DNA sequence of satellite bacteriophage P4. *Nucleic Acids Res.*, 18, 1649, 1990.

92. Pedulla, M.L., Ford, M.E., Karthikeyan, T., Houtz, J.M., Hendrix, R.W., Hatfull, G.F., Poteete, A.R., Gilcrease, E.B., Winn-Stapley, D.A., and Casjens, S.R. Corrected sequence of the bacteriophage p22 genome. *J Bacteriol.*, 185, 1475–1477, 2003.

93. Mesyanzhinov, V.V., Robben, J., Grymonprez, B., Kostyuchenko, V.A., Bourkaltseva, M.V., Sykilinda, N.N., Krylov, V.N., et al. The genome of bacteriophage fKZ of *Pseudomonas aeruginosa. J Mol Biol*, 317, 1B19, 2002.

94. Roberts, M.D., Martin, N.L., and Kropinski, A.M. The genome and proteome of coliphage T1. *Virology*, 318, 245–266, 2004.

95. Miller, E.S., Kutter, E., Mosig, G., Arisaka, F., Kunisawa, T., and Ruger, W. Bacteriophage T4 genome. *Microbiol. Mol. Biol. Rev.*, 67, 86–156, 2003.

96. GenBank ACCESSION # NC_005859.

97. Dunn, J.J., and Studier, F.W. Complete nucleotide sequence of bacteriophage T7 DNA and the locations of T7 genetic elements. *J Mol Biol.*, 166, 477–535, 1983.

98. Nelson, D., Schuch, R., Zhu, S., Tscherne, M., and Fischetti, V.A., Genomic sequence of C1, the streptococcal phage. *J Bacteriol*, 185, 3325–3332, 2003.

99. Dore, E., Frontali, C., and Grignoli, M. The molecular complexity of G DNA. *Virology*, 79, 442–445, 1977. Sequence in Genbank: http://www.ncbi.nlm.nih.gov/nuccore/593777468.

100. Hatfull, G.F., and Sarkis, G.J. DNA sequence, structure and gene expression of mycobacteriophage L5: a phage system for mycobacterial genetics. *Mol. Microbiol.*, 7, 395–405, 1993.

101. Obregon, V., Garcia, J.L., Garcia, E., Lopez, R., and Garcia, P., Genome organization and molecular analysis of the temperate bacteriophage MM1 of *streptococcus pneumoniae. J Bacteriol*, 185, 2362–2368, 2003.

102. Iandolo, J.J., Worrell, V., Groicher, K.H., Qian, Y., Tian, R., Kenton, S., Dorman, A., Ji, H., Lin, S., Loh, P., Qi, S., Zhu, H., and Roe, B.A. Comparative analysis of the genomes of the temperate bacteriophages phi 11, phi 12 and phi 13 of *Staphylococcus aureus* 8325. *Gene*, 289, 109–118, 2002.

103. Vlcek, C., and Paces, V. Nucleotide sequence of the late region of Bacillus phage phi 29 completes the 19,285-bp sequence of phi 29 genome. Comparison with the homologous sequence of phage PZA. *Gene*, 46, 215–25, 1986.

104. GenBank ACCESSION # AB016282.

105. GenBank ACCESSION # NC_023009

106. GenBank ACCESSION # AF020713.

107. Parker, M.L., Ralston, E.J., and Eiserling, F.A. Bacteriophage SPO1 structure and morphogenesis. II. Head structure and DNA size. *J. Virol.*, 46, 250–259, 1983.

108. Maniloff, J., and Ackermann, H.W. Taxonomy of bacterial viruses: establishment of tailed virus genera and the order caudovirales. *Arch Virol.*, 143, 2051–2063, 1998.

109. Campbell, A. Episomes. *Adv Genet*, 11, 101–145, 1962

49

Introduction to Virology

Ken S. Rosenthal

CONTENTS

DOI: 10.1201/9781003099277-51

Introduction

Viruses are simple infectious agents and yet, as they parasitize our bodies, they can cause devastating disease. Viral diseases range from the common cold and diarrhea to life-threatening encephalitis, hemorrhagic fever, and smallpox. Several different viruses may cause a single disease, as determined by the tissue or organ that is affected, or one virus may cause different diseases due to the nature of its interaction with the body. Ultimately, the nature of the disease caused by the virus is a function of the tissue that is infected, the nature of the virus, and the host immune response to the virus. Interestingly, each of these parameters is determined by the structure and mode of replication of the virus. This chapter introduces the basic viral structures and modes of replication and presents the consequences of these properties for the different virus families. *In many cases, knowing the structure of the virus allows prediction of many of the other properties of the virus.* This introduction also includes a discussion of general mechanisms of viral pathogenesis and disease production, epidemiology, antiviral drugs, vaccines, and viral detection before a synopsis of each of the virus families is presented. References to other basic reviews on virology by the author (1–4) are provided.

Definition of a Virus

Unlike bacteria, fungi, or parasites, a virus does not replicate by binary fission but rather by synthesis and assembly. Viruses neither have organelles nor can they make energy. Viruses are obligate intracellular parasites that depend on the biochemical machinery of the infected cell for replication. As obligate parasites, viruses must continue to find new host cells and individuals and be able to use the biochemical machinery of the cell to replicate or the virus will disappear. The virus uses different molecular tricks to manipulate the cell and its host to replicate and spread.

Structure

Although we may think of a virus as having characteristics of a living microbe, it is not alive. The virion particle is a very large biochemical complex consisting of a nucleic acid genome protected and carried within either a protein shell, termed a capsid, or within a membrane envelope. (Although prions were considered atypical viruses at one time, they are rogue, infectious proteins and not viruses). The basic virion particle consists of a genome of either RNA or DNA (but only one or the other) that is packaged for protection in a protein capsid or a membrane envelope. Some viruses may carry auxiliary RNA or proteins within the virion particle to facilitate replication in the host.

Human viruses range in size and complexity from the tiny Norwalk virus (22 nm), which causes outbreaks of diarrhea, to the poxviruses, which are almost visible under a light microscope (Tables 49.1 and 49.2). They may be as simple as the rabies virus, which has only five proteins, or as complex as the cytomegalovirus, which encodes approximately 110 proteins. The larger viruses have more space to hold a larger genome and encode more proteins. These "extra" proteins may facilitate the replication of the virus; allow replication in difficult cells, such as neurons; or facilitate escape from host protective innate and immune responses. Larger viruses also require larger aerosol droplets for transmission through the air.

The viral genome is the core of the virus and acts like a computer memory to store the genetic information for the virus. Like computer memory (e.g., a hard drive, CD-ROM, or USB memory device), the viral genome can take different forms. It can consist of DNA—either linear, circular, single or double stranded—or double- or single-stranded RNA of positive or negative sense. Most DNA genomes can establish themselves in the nucleus and utilize the nuclear machinery for mRNA synthesis and replication of the genome, whereas RNA genomes are more temporary. Most RNA viral genomes replicate in the cytoplasm (where RNA belongs) and must provide and encode the enzymes to make mRNA and replicate their genome. DNA and positive strand RNA (other than retroviruses) genomes are sufficient to infect cells when microinjected into the cells.

The outer structures of the virus are a package that protects the genome, delivers it from host to host, and provides the means for interacting with and entering the appropriate target cell (Figure 49.1). Capsids are protein shells usually in symmetrical icosahedral or icosadeltahedral shapes (soccer balls)

TABLE 49.1

DNA Viruses

Family	Genome	Virion	Virion Size (nm)	Example(s)
Parvoviridae	ss, linear	Capsid	18–26	B19, Boca
Papillomaviridae	ss, circular	Capsid	45–55	Human papilloma
Polyomaviridae	ss, circular	Capsid	45–55	JC, BK, Merkel
Hepadnaviridae	ds, circular	Enveloped	42	Hepatitis B
Adenoviridae	ds, linear	Capsid with fibers	70–90	Adenovirus
Herpesviridae	ds, linear	Enveloped	120–130	HSV, VZV, EBV, CMV, HHV6-8
Poxviridae	ds, linear	Enveloped, brick	$300 \times 200 \times 100$	Smallpox, Molluscum contagiosum

Abbreviations: ss = single stranded; ds = double stranded; HSV = herpes simplex virus; VZV = varicella zoster virus; EBV = Epstein Barr virus; CMV = cytomegalovirus; HHV = human herpes virus.

TABLE 49.2

RNA Viruses

Family	Genome	Virion	Virion Size (nm)	Example(s)
Caliciviridae	ss, positive	Capsid	35–40	Norovirus
Hepeviridae	ss, positive	Capsid	27-34	Hepatitis E virus
Picornaviridae	ss, positive	Capsid	25–30	Polio, echo, parecho, entero, HAV, rhino
Coronaviridae	ss, positive	Enveloped	80–130	Corona, SARS, MERS, SARS-CoV-2
Paramyxoviridae	ss, negative	Enveloped	150–300	Measles, mumps, parainfluenza, respiratory syncytial virus, metapneumovirus
Orthomyxoviridae	ss, negative	Enveloped	80–120	Influenza A, B, C
Rhabdoviridae	ss, negative	Enveloped	180 × 75 bullet	Rabies
Filoviridae	ss, negative	Enveloped	800 × 80 fiber	Ebola, Marburg
Reoviridae	Double stranded	Double capsid	60–80	Rotavirus, Colorado tick fever
Togaviridae	ss, positive	Enveloped	60–70	Rubella, EEV
Flaviviridae	ss, positive	Enveloped	40–50	West Nile encephalitis, Yellow fever, dengue, zika
Bunyaviridae	ss, negative	Enveloped	90–100	Ca. encephalitis, hantavirus
Arenaviridae	ss, ambisense	Enveloped	50–300	Lassa fever, LCMV
Retroviridae	ss, positive	Enveloped	80–110	HIV, HTLV

Abbreviations: ss = single stranded; ds = double stranded; SARS = severe acute respiratory syndrome; MERS = middle eastern respiratory syndrome; EEV = eastern equine encephalitis virus; LCMV = lymphocytic choriomeningitis virus; HIV = human immunodeficiency virus; HTLV = human T lymphotropic virus.

(Table 49.3). Like hardboiled eggs, the protein of the capsid forms a shell that is impervious to drying, acids, and detergents. In contrast, enveloped viruses are surrounded by a membrane consisting of phospholipids and proteins, which must stay wet and can be disrupted by detergents and harsh environments, including the acids and bile of the gut. Of course, there are exceptions to these generalizations and they will be identified throughout the text.

The nature of the viral disease, the means by which it spreads, the type of immune response necessary to control the viral infection, and the types of antiviral drugs that can be used to combat the infection are determined to a large extent by the viral structure and mechanism of replication (discussed later). The impervious shell of capsid viruses allows them to be transmitted by the fecal-oral route and resist drying and disinfection with detergents and

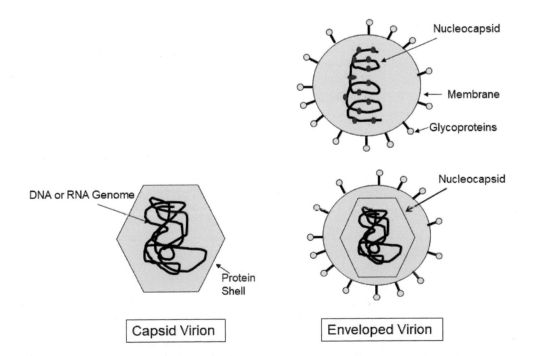

FIGURE 49.1 Virion structure. The virion particle consists of a DNA or RNA genome surrounded by either a protein shell (called a capsid) or a cell-like membrane (called an envelope). The genome may be associated with or enclosed by protein to form the nucleocapsid. The virion may also contain other structures and proteins that will facilitate the replication of the virus within the cell. The outside structures of the capsid or the glycoproteins of the envelope mediate the interaction of the virus with the cell and are the primary targets of protective neutralizing antibodies.

TABLE 49.3

Implications of Virus Structure

	Capsid	*Envelope*
Components	Protein shell	Phospholipids, lipids, glycoproteins, proteins
Implications	*Hard-boiled egg-like shell*	*Membrane structure*
	Resistant to acid, dryness detergents[a]	*Labile to acid, dryness, detergents[b]*
	Can be spread by fecal-oral and other means	*Must stay wet, cannot spread by fecal-oral route[b]*
		Envelope antigens incorporated into cell membranes
	Released by cell death and lysis[c]	*Released from cell without killing cell or by other means*
	Antibody often sufficient to prevent and resolve infection	*Cell-mediated immunity (CMI) necessary to eliminate virus producing cell*
		CMI induces more systemic response and inflammation

[a] Rhinovirus capsid is acid labile.
[b] Coronavirus glycoprotein forms an exoshell around its envelope to resist detergents, acid, and allow fecal-oral spread.
[c] Some capsid viruses (hepatitis A) can be released through exocytosis.

solvents. This is not so for enveloped viruses. Rhinovirus (common cold) is an exception and is susceptible to acid pH. This is the basis for incorporating citric acid into tissues to kill these common cold viruses. As discussed later, antibody is usually sufficient to control most naked capsid viruses. This is why inactivated virus vaccines, which generate primarily an antibody response, are usually effective for capsid viruses but may not be effective for enveloped viruses. In contrast, enveloped viruses cannot withstand the harsh conditions of the gastrointestinal (GI) tract; they are readily disinfected by soap or solvents and are more sensitive to heat and dryness. Enveloped viruses are transmitted in fluids (e.g., aerosols, saliva, blood, semen, and tissue transplants). The exceptions are coronaviruses and hepatitis B virus, which can withstand the GI tract. A combination of cell-mediated and antibody immune responses are important to control enveloped viruses.

Replication and Consequences

Every virus must replicate and produce more viral progeny or disappear. In many respects, a virus utilizes host cells as a factory that provides the raw materials, energy, and most of the synthetic machinery to produce the genome and structural components of new viruses. For every infecting virus, the cell can produce 100 to 100,000 new infectious virions and many times more noninfectious (defective) particles. Viruses are notorious for making mistakes, but when there are a million siblings, even 10% infectious virus is a lot. The virus can kill the cell quickly, slowly, or not at all. The outcomes depend on the type of virus and how it interacts with the cell.

All viruses follow a similar set of steps to generate new viruses (Figure 49.2). The basic steps for replicating the virus include:

1. Recognition and attachment
2. Entry into the cell
3. Uncoating
4. Synthesis of the macromolecular components, including the genome and the structural parts of the virion particle
5. Assembly of the new virion particles
6. Exit from the cell

Each of these steps is performed differently by different viruses, depending on the structure of the viral genome and whether the virion has a naked capsid or an envelope.

Recognition, Attachment, and Entry

Finding and binding to the appropriate cell is mediated by the surface structures of the naked capsid virus or by the glycoproteins that decorate the envelope. These viral attachment proteins (VAP) are the keys that interact with cell surface receptors and open the doors of entry into the cell. Some of these structures are like skeleton keys and bind to many different cell types, whereas others are very specific for certain cell types. A canyon on the surface of the rhino or polio virus (picornaviruses) acts as a keyhole that can be filled by specific receptors from the cell which are members of the immunoglobulin family of glycoproteins. The glycoproteins on the Epstein-Barr virus bind to the receptor for the C3d component of complement (CD21), identifying the B lymphocyte as the principal target for the virus. The HIV gp120 glycoprotein binds to the CD4 molecule to target subsets of T cells and myeloid cells (dendritic cells and macrophages), but can switch target cells by mutating and changing the portion of the gp120 that binds to a chemokine receptor present on either myeloid cells and activated T cells (CCR5) or one that is present on other T cells (CXCR4).

The mechanism of virus entry into the cell depends on the virion structure. Binding of the picornavirus receptor to the canyon on the virion surface opens the capsid structure and creates a hydrophobic portal for injection of the viral genome into the cytoplasm. Some enveloped viruses, including herpes, retro-, and paramyxoviruses will fuse their membrane envelope with the plasma membrane of the cell to deliver their genome into the cell. Many viruses bind to receptors on the cell surface that trigger the internalization of the virus by receptor-mediated endocytosis, the same pathway that cells use to take up transferrin or low-density lipoprotein. Although one would think that the normal acidification of the endosomal vesicle would be detrimental to a virus, they use this as a trigger to enter the cytoplasm. Acidification activates the fiber protein of the adenovirus capsid to lyse the vesicle and break into the cytoplasm. Acidification changes the conformation of the viral attachment glycoproteins of enveloped viruses, such as influenza or rabies virus, to expose regions that promote fusion of their membrane

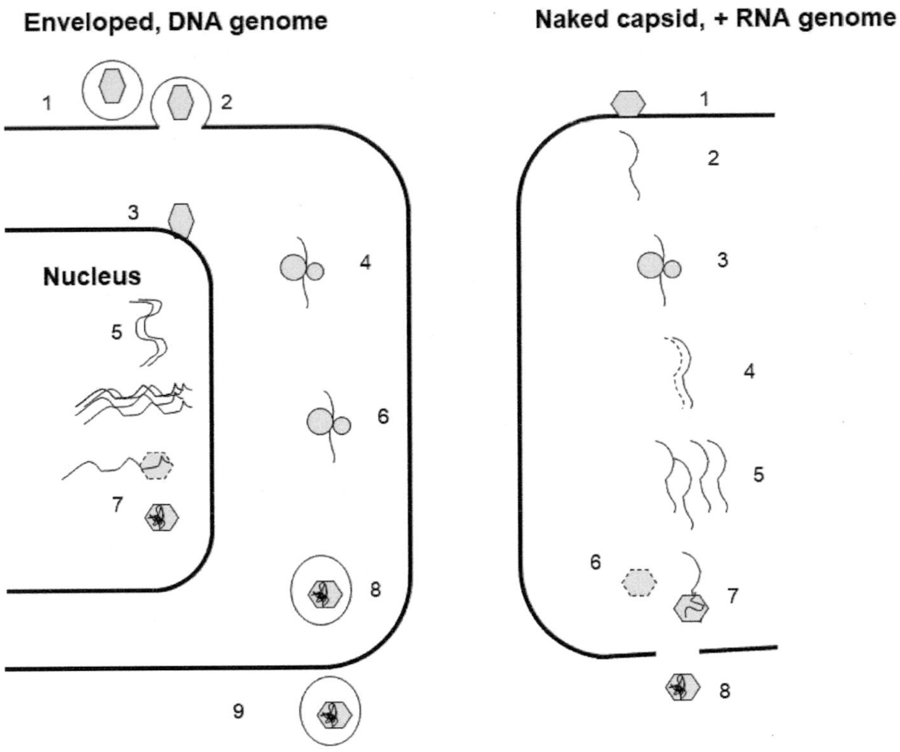

Enveloped, DNA genome

Naked capsid, + RNA genome

FIGURE 49.2 Virus replication. Virus replication proceeds in an ordered, sequential fashion. Left: The steps in virus replication for an enveloped virus with a DNA genome, like a herpes virus, are (1) binding, (2) fusion and entry, (3) delivery of genome to nucleus, (4) immediate early and early mRNA and protein synthesis, (5) replication of the genome, (6) late mRNA and protein synthesis, (7) capsid assembly and filling, (8) envelopment of genome-filled capsid, and (9) exit of cell by exocytosis or cell lysis. Right: The steps in virus replication for a capsid virus with a positive-sense RNA genome, like a picornavirus (polio or rhino), are (I) binding, (2) entry and uncoating, (3) protein synthesis, (4) production of an RNA template, (5) genome replication, (6) capsid assembly, (7) filling capsid with genome, and (8) cell lysis and release.

with the vesicular membrane of the cell to deliver their genome into the cytoplasm.

Viral mRNA and Protein Synthesis

Once inside the cell, the viral genome inserts into the cell's machinery to promote the synthesis of viral mRNA and protein (Table 49.4). DNA viruses, except poxviruses, deliver their genome to the nucleus where it will be mistaken for cellular DNA. The sequences of the viral genes must contain the same genetic passwords as cellular DNA so that they can use the cell's machinery for making and processing mRNA. RNA will be made by the host DNA-dependent RNA polymerase and modified to make the viral mRNAs look and act like cellular mRNAs. The RNA will be processed into mRNA by removal of introns (if present), become stabilized by the addition of a 3'-polyadenosine tail, and pick up the recognition structure for attachment to the ribosome and protein synthesis with the addition of a 5'-methylated (7-methylguanosine) cap. Poxviruses are the DNA virus exception because they replicate in the cytoplasm. The poxviruses pay the price for being more independent of the nucleus by having to provide their own molecular tools for mRNA synthesis.

Most DNA viruses divide protein synthesis into early and late phases. The proteins made during the early phase convert the cell into a more efficient factory for replicating virus. Once the cell has been subverted and new viral genomes are being produced, then the late genes are turned on and the cell churns out large amounts of viral structural proteins and glycoproteins to make the capsid or envelope proteins. Activation of the late genes and synthesis of the structural proteins of a DNA virus is usually a death sentence for the cell.

Most RNA viruses replicate in the cytoplasm, which provides easy access to the ribosome for protein synthesis. However, using RNA as the genetic information carrier is foreign to the cell because the cell has no polymerases to copy RNA from RNA, and cannot make mRNA or replicate the genome from an RNA template. Hence, most RNA viruses must encode and/or carry their own RNA-dependent RNA polymerases within the virion particle. In addition, these viruses must develop a way to acquire a 5' cap or an alternative way to bind to the ribosome for protein synthesis.

The viruses with a genome that looks like mRNA (positive-sense), except retroviruses, can immediately bind to ribosomes and initiate protein synthesis. The resulting protein, called a polyprotein, includes proteases to cleave itself or is cleaved by other proteases into the individual viral proteins. Viruses with a negative-sense RNA (like a photographic negative) cannot be deciphered by the cell's ribosome. The genome of these viruses is decorated by enzymes (RNA-dependent RNA polymerase), as part of the nucleocapsid, that will convert the genome into individual mRNAs for each of the viral proteins when released from the virion particle. The orthomyxoviruses (influenza) are exceptional because their proteins are encoded by individual segments of negative-stranded RNA, and these segments are converted

TABLE 49.4

Implications of Viral Genome

	Petite DNA Genome[a]	Large DNA Genome[b]	RNA Genome	Retrovirus
Genome structure	Small DNA	Large DNA	(+)RNA, (−)RNA, (+/−)RNA	(+)RNA
Viruses	Polyoma, Papilloma, Parvo	Adeno, Herpes	See Table 49.2	HIV, HTLV
Implications	*Must infect growing cells or stimulate growth in cells to replicate DNA*	*Encodes polymerase and manipulates cell growth to replicate DNA*	*MUST ENCODE RNA dep RNA POLYMERASE. (−) and (+/−)RNA viruses must carry polymerase in virion*	*Encodes and carries RNA dep DNA polymerase (reverse transcriptase)*
	DNA goes to nucleus and may integrate	*DNA goes to nucleus and remains in cell*	*RNA is in the cytoplasm and transient species[c]*	*cDNA of genome integrates into chromosome*
	Encodes proteins to direct viral gene transcription.	*Encodes proteins to direct viral gene transcription.*	*Viral polymerases are very error prone and cause mutations*	*Viral polymerases are very error prone and cause mutations*
	Uses cell enzymes in nucleus to generate mRNA	*Uses cell enzymes in nucleus to generate mRNA*	*Encodes enzymes to generate mRNA*	*Uses cell enzymes in nucleus to generate mRNA and genome*
	Stimulation of cell growth may lead to transformation	*Encodes proteins to manipulate host response*	*Encodes proteins to manipulate host response*	*Encodes proteins to manipulate host response*

[a] *Memory aid*: P for petite, parvo-, polyoma-, and papilloma viruses.

[b] Poxviruses are the exception to most rules since they replicate in the cytoplasm, hepadnaviruses replicate through an RNA intermediate and are not included in table.

[c] Influenza mRNA and replication occur in the nucleus.

into mRNA in the nucleus. Influenza steals the 5' cap from cellular mRNA, and the 3'-polyadenosine is attached before leaving the nucleus so that their viral mRNA will look like cellular messenger RNA. The reo- and rotaviruses have a segmented double-stranded RNA genome surrounded by components of the polymerase (an inner capsid) that will convert the negative strands of the genome into individual mRNAs.

Although the retroviruses, including HIV, have a positive sense RNA genome, they carry a reverse transcriptase enzyme that converts the RNA genome into DNA and then another viral protein promotes the integration of the viral DNA into the host chromosome. The integrated retroviral DNA becomes a very active gene within the host chromosome to produce viral mRNAs and new copies of the viral genome are transcribed as if it were a cellular gene.

Once the virus has manipulated the cell into making mRNA, the viral proteins are synthesized by the cell's ribosome and glycoproteins are processed by the same machinery as cellular glycoproteins. Different viruses have different ways to regulate when early and late proteins are made. The herpesviruses (DNA viruses) have three phases of protein synthesis, the immediate early genes are activated by viral and cellular proteins and the early and late genes are activated by viral proteins. The early proteins of the togaviruses and coronaviruses (positive RNA genome) are translated directly from the infecting genome while the mRNA for the late proteins is made from the negative RNA template that is used to replicate the genome. During the late phase of protein synthesis, the cellular machinery is likely to be devoted to production of viral proteins to the exclusion of cellular proteins.

Genome Replication

The new genome of DNA viruses is replicated like cellular DNA, in a semiconservative manner. The amounts of deoxyribonucleotides and DNA polymerase in a resting cell are limited and this

can prevent or retard the replication of a virus. The smaller DNA viruses get around this problem by replicating in growing cells (B19 parvovirus in erythrocyte precursors) or by removing or blocking the action of inherent cellular growth inhibitors (p53 or RB growth suppressor proteins). The larger viruses encode their own tools for viral replication, a viral DNA polymerase and even scavenging enzymes for deoxyribonucleotides, which provide more independence from the cell and speed up the replication of the genome.

Most RNA viruses replicate their genome using an RNA-dependent RNA polymerase that is carried into the cell or encoded by the virus and made soon after infection. This polymerase makes an RNA template from the genome and uses it to synthesize positive- or negative-sense RNA genomes. During some point during the production of an mRNA or a new genome, a double-stranded RNA structure will be generated between the template and the new copy. The double-stranded RNA replicative intermediate is the most powerful activator of Type 1 interferons and activates other innate responses through Toll-like (TLR) and other receptors.

The retroviruses and hepatitis B virus utilize both RNA and DNA in their replication. During the early steps in retrovirus infection, the incoming RNA genome is converted into DNA by the viral encoded and virion carried reverse transcriptase enzyme (RNA-dependent DNA polymerase) and another viral enzyme (integrase) facilitates the integration of the genome into the host chromosome. Viral mRNA and new genomes are produced by the cell's DNA-dependent RNA polymerase as if it were making mRNA. For hepatitis B virus, the cell makes different sizes of mRNA, including a longer than full genomic length RNA copy from the DNA genome. The RNA circle that results is transcribed into the circular hepatitis B virus DNA genome using the viral reverse transcriptase.

The viral polymerase is the major target for antiviral drugs. Nucleotide analogs inhibit virus replication or promote hypermutation.

Virus Assembly

Ultimately, all parts of the virus must be assembled into the virion particle so that it can move to new target cells and hosts. At first thought, the process seems analogous to assembling a jigsaw puzzle by shaking the parts in a box. However, the process is not as random as the shaken puzzle box. For some capsid viruses, individual proteins associate into larger and larger complexes that eventually become empty capsid shells. The large number of viral proteins within the cell promotes enough collisions between the proteins to drive the process to completion. Many of these structures are built on scaffolds made of cellular membranes or other viral proteins. Some proteins must be cleaved before they become fully functional and fit into the assembly.

To finish the process, the genome can be inserted into an assembled capsid shell, or the capsid proteins can assemble around or associate with the genome. For enveloped viruses, the protein-coated genomes (nucleocapsid) associate with proteins on the inside of the cellular membrane and the structure wraps itself in the membrane to bud from the cell. The viral encoded proteases of retroviruses finalize the assembly process after envelopment and are a target for antiviral drugs.

Virus Release

Most capsid viruses accumulate in the cytoplasm and must cause the cell to lyse in order to be released. Lysis of the cell may be a consequence of toxic viral proteins, the build-up of viral material, an inability to rebuild cellular structures, or induction of apoptosis or necrosis. Most enveloped viruses are more subtle and acquire their envelope from the plasma membrane as they push their way out of the cell without killing the cell. Herpesviruses are released by vesicular transport mechanisms, similar to those used to release proteins as well as by cell lysis.

Upon release, the viruses can spread to other cells and expand the infection. Some viruses (e.g., herpes simplex virus (HSV)) can move between cells (intercellular spaces) to infect neighboring cells without exposing themselves to antibody detection. Other viruses (HSV, varicella zoster, measles and other paramyxoviruses, and HIV) recruit large numbers of cells into their factory by fusing them together into multinucleated giant cells (syncytia). Syncytia formation is also a diagnostic characteristic of these viruses.

Host Protective Responses

Antiviral immune responses are a part of viral disease. They are essential for control of infection and yet they contribute to the pathology of the disease. Most viruses encode proteins and mechanisms to escape or shut off the host protections. Antiviral host protective responses are based on the following military-like rules:

1. Keep them out and they can cause no harm.
2. Prevent them from establishing a toehold.
3. If they acquire a toehold, then prevent them from spreading.
4. Eliminate the invaders and their factories at all cost.

The initial antiviral protection consists of barriers such as skin and mucus; unfavorable conditions such as the acid and detergent-like bile found in the gut; and antimicrobial peptides such as the defensins, which are found in many parts of the body and especially the mouth. Body temperature and fever are deterrents to many viruses; for example, the rhinoviruses are restricted to the upper respiratory tract because they cannot replicate at body temperature in the lower lung.

Innate responses react quickly in response to microbial components and prevent the establishment of a toehold. These responses include the type 1 interferons, natural killer (NK) cells, macrophages, and dendritic cells. Neutrophils and complement have less important protective roles in antiviral responses than antibacterial responses but contribute to pathogenesis.

The cell has sensors for cytoplasmic RNA, double stranded RNA and cytoplasmic DNA produced during viral replication that activate antiviral responses, including type 1 interferon production. Enzymes, such as APOBEC3 (apolipoprotein B mRNA editing enzyme), promote mutagenesis of single stranded DNA in the cytoplasm, as would be produced during retrovirus infection. HIV counteracts this enzyme with its vif protein.

Type 1 interferons (IFN) include more than 20 different IFN alphas, and one type of IFN beta. They received their name because they "interfere" with virus replication. The type 1 IFNs are produced in response to viral DNA, RNA, and especially the double-stranded RNA replicative intermediate generated during RNA virus replication. Plasmacytoid dendritic cells produce large amounts of type 1 IFNs in response to viruses in the blood. Type 1 IFNs are distinguished from interferon-γ (IFN-γ), the only type 2 IFN, which is produced by NK and T cells in response to activation of immune responses (discussed later).

IFN produced and released from a virus-infected cell acts as an early warning system to surrounding cells. The IFN binds to cell surface receptors and induces the production of antiviral proteins, which become activated upon virus infection of the cell. These antiviral proteins are part of the "antiviral state" and include a protein kinase (PKR), which when activated by double-stranded RNA will phosphorylate a key controlling protein in the ribosome assembly to prevent protein synthesis. Another protein indirectly activates an RNase in response to virus infection. Both of these interferon-induced and virus-activated responses work together to prevent protein synthesis and hence block virus replication. IFN also activates NK cells and other body defenses. These actions act locally to stop the progression of the infection but also provide an early warning system in case the virus does spread. Unfortunately, these protections cause the flu-like symptoms associated with many viral infections.

NK cells can recognize and kill viral-infected cells as well as produce IFN-γ. IFN-γ acts as a cytokine to activate macrophages and dendritic cells. Activated macrophages also have killer cell activity. Activated dendritic cells initiate the antigen-specific antiviral immune response.

The goal of the protective innate and immune responses is to completely eliminate the virus (resolve the infection) by eliminating infectious virus and virus-producing cells. Some virus infections can be controlled by antibody, whereas others require both antibody and cell-mediated responses. The rule is that if the virus kills the cellular factory as part of virus replication, then antibody is sufficient to resolve the infection. If the virus does not kill the infected cell, then cell-mediated immune responses must kill the cell to resolve the infection.

Antibody-mediated protection acts primarily to limit the spread of the virus. It is directed against the proteins and structures that bind to the cell surface receptor, the viral attachment proteins, or structures. The antibody coats the virus and prevents it from being able to bind and infect cells and promotes the uptake and destruction of the virus by macrophages (opsonization). Antibodies to other viral proteins are not protective but provide serological evidence of the infection.

T-cell responses help the antiviral response with cytokines and provide killer cells to destroy the viral infected cell factories. The killer CD8-T cell responses are directed against viral peptides that decorate the major histocompatibility antigens (MHC-I) on the surface of the infected cell. CD8-T cells and CD-4 T cells can also make IFN-γ to maintain the cytolytic response. Cell-mediated responses usually cause flu-like symptoms, inflammation, and tissue damage contributing to the disease presentation. More description of the antimicrobial immune response can be obtained in Ref. [1–4] and other sources.

Mobilization of the antigen-specific immune response takes longer to activate than the local militia of innate responses, especially if it is the first challenge with this strain of virus. Maturation of new antigen-specific immune responses takes at least 7–10 days, sufficient time for the virus to spread, reach the target tissue for the disease (e.g., for polio virus, to the central nervous system), and cause damage. As a result, there is more virus and infected cells and a much stronger immune response is necessary to resolve the infection of an immunonaive individual and this may result in greater immunopathology (discussed later). Prior immunization, either by natural infection or vaccination, shortens the time to approximately three days and the infection is usually stopped before signs of disease begin.

Pathogenesis and Disease

Most viral diseases are caused by a combination of viral pathogenesis and immunopathogenesis.

For some viruses, such as polio, the viral pathogenesis is the primary cause of disease as the virus kills the infected cell and targets motor neurons of the anterior horn and brainstem. For rhinoviruses (common cold) and the more serious influenza and herpes viral diseases, disease results from a combination of viral and immunopathogenesis whereas for measles, mumps, hepatitis viruses, and hantaviruses, immunopathogenesis is the primary cause of disease.

Viral Pathogenesis

Viral pathogenesis is caused by changes in the cell that are induced by viral replication and viral replication-induced cell death. The type and extent of viral pathogenesis is determined by the nature of the virus-cell interaction. Many viruses, especially the larger viruses, encode proteins that can manipulate the cell and the host to promote virus replication. Only half of the approximately 80 genes encoded by HSV are necessary to replicate the virus in tissue culture; the other genes support virus

replication in difficult cells, like neurons, and help the virus spread or escape the immune response.

Viral disease is primarily determined by the cell type and tissue (tropism) that are infected by the virus and the consequences of the virus-cell interaction (e.g., hepatitis, encephalitis). The expression of virus receptor molecules on the target cell is the initial and most important determinant of the nature of the viral disease.

Once the virus has picked, attached, and entered the target cell, there are several different possible outcomes: replication and viral-induced cell death, replication without cell death (chronic infection), latency, or transformation. Most naked capsid viruses are cytolytic because they must kill the infected cell to be released and spread to other cells, whereas many enveloped viruses leave the cell without killing the cell as they acquire their envelope from the cell membrane and bud away from the cell. For a DNA virus, the outcome can be determined by the ability of the cell to make the viral proteins. For example, HSV establishes a latent infection without apparent effect on neurons because only one RNA (latency-associated transcript (LAT) is transcribed in these cells, and this RNA does not encode a detectable protein. For simian virus 40 (SV40), the early gene, T antigen, stimulates cell growth, which facilitates replication of the DNA viral genome; but if the late genes for the capsid proteins cannot be transcribed, as in rodent cells, the stimulation of cell growth becomes uncontrolled, and can lead to oncogenic transformation. The T antigen protein prevents the function of the p53 and RB cellular growth suppressors and cell growth and mutation ensues. SV40 replicates and kills monkey cells, a situation that is inconsistent with oncogenic transformation. Similarly, the E6 and E7 proteins of the high risk for cancer human papilloma virus types, including Types 16 and 18, also prevent the function of the p53 and RB. When the HPV DNA genome becomes circularized or integrated into the host chromatin, a gene important for virus replication is inactivated and the cell grows out of control. HPV 16, 18 and several other types are associated with human cervical carcinomas and other cancers.

Although the virus may not kill the cell, it may cause characteristic changes to cellular structures within the cell. These changes may be seen as histological staining aberrations, including inclusion bodies, or syncytia formation (multinucleated giant cells due to cell-cell fusion). For example, herpes simplex virus causes syncytia formation, margination of the chromatin, and a Cowdry Type A intranuclear inclusion body. More examples are presented in Table 49.5.

TABLE 49.5

Visual Examples of Viral Pathogenesis

Example	Virus
Cytolysis	Most viruses, especially herpes, picorna, adeno, pox
Syncytia formation	HSV, VZV, paramyxoviruses, HIV
Inclusion bodies:	
Cowdry type A intranuclear	HSV
Owl's eye intranuclear	CMV
Intranuclear basophilic	Adenovirus
Intracytoplasmic acidophilic	Poxvirus
Negri bodies intracytoplasmic	Rabies
Perinuclear cytoplasmic acidophilic	Reoviruses

The incubation period for disease will be short if the primary site of infection is also the target tissue. The incubation period is longer if the virus must be amplified and then spread to the target tissue through the blood stream (viremia) or the disease is due to immunopathogenesis (e.g., hepatitis). Oftentimes, a prodrome of flu-like symptoms occurs while the virus spreads by viremia to the disease related target tissue.

Immunopathogenesis

Although the immune response is essential to control the virus infection, it is also a major contributor to the disease signs and symptoms. It takes energy to combat a viral infection, and many protective immune responses also cause cell damage. Interferon and cytokines not only mobilize the body's defenses against the virus, but also cause the flu-like symptoms of fever, malaise, headache, muscle ache, and lack of appetite. Inflammation, rashes, and tissue destruction may be a consequence of cell-mediated immune killing of infected cells to combat the virus. However, insufficient immune responses can result in chronic infections, as with hepatitis B virus infection of children ("no pain, no gain"). Immune complexes between viral antigen and antibody can lead to hypersensitivity reactions as with the rash following infection of children and the arthritis that accompanies adult infection by the B19 parvovirus. Similarly, the large amount of viral antigen particles (HBsAg) during hepatitis B infection can cause glomerulonephritis.

Laboratory Detection

This section is not a replacement for a chapter on the subject but provides a summary. An overview of laboratory diagnosis of microbial diseases, including viruses, was developed by the IDSA (5).

Choice of the appropriate sample for analysis is the first step in viral laboratory analysis. The patient's symptoms and the source and nature of the clinical sample are the first clues to the proper sample and to identification of the virus. Virus isolation was the gold standard for detection of viruses but this is not possible in all cases and is being replaced by genome amplification techniques, such as polymerase chain reaction (PCR) for DNA and reverse transcriptase PCR (RT-PCR) for RNA. Growth of the virus isolate in appropriate cell cultures and the cytopathological effects induced by virus replication (see Table 49.5) can provide important clues to the trained eye. Due to the expense, difficulty, and risk, most laboratories have minimized the use of virus isolation and are identifying viruses by detection of the viral genome or antigens using molecular biological or immunological techniques. Enzyme-linked immunosorbent assays (ELISA), immunofluorescence, and related tests are useful for detecting viral antigen in serum or other sample and in infected tissue culture cells. *In situ* hybridization, PCR, RT-PCR, and related approaches can detect and quantitate the viral genome. Newer gene amplification techniques and DNA sequence analysis have made quantitation of the virus load for HIV, Hepatitis C (HCV), and other viruses, the standard analytical tool to follow the patient's disease progress and antiviral drug compliance. Multiplex PCR techniques are available for different diseases to allow rapid screening for multiple agents. This approach can provide rapid distinction of a treatable bacterial from an untreatable viral cause of pneumonia as well as the type and strain of the virus (e.g., influenza type). New, cheaper, and more rapid DNA sequencing techniques allow rapid definitive identification of the agent and can distinguish the mutations that may be associated with antiviral drug resistance or immune escape mechanisms. (A full description of these techniques can be found in Chapters 10 and 11).

Serological analysis provides a history of a patient's infections. The presence of IgM (immunoglobulin M) to viral antigen indicates ongoing disease, and a fourfold increase in antibody titer between acute and convalescent serum indicates recent infection. Analysis of antibody responses to specific viral antigens can also be used to follow the course of some viral diseases, especially for Hepatitis B virus and Epstein-Barr virus. ELISA tests are important for screening the blood supply for HIV, HCV, HBV, and HTLV-1. In addition to the ELISA and immunofluorescence tests mentioned earlier, antibody neutralization of virus infection and inhibition of hemagglutination (for those viruses that bind to erythrocyte structures and agglutinate the cells), termed hemagglutination inhibition (HI), can be used to evaluate serum for antiviral activity. HI is used to validate the annual influenza vaccine activity.

Epidemiology

Similar to infectious disease, which studies the consequence of viral infection of the body and its control by the immune response, epidemiology studies the viral infection of the population and the control of its spread by immunization of the populace. Specifically, viral epidemiology studies the mechanisms and modes of transmission of the virus, including vectors; the risk factors; the at-risk groups; seasonal, geographical, and environmental considerations for infection; as well as modes of controlling virus spread. The personal history provides epidemiologic data and is a gold mine of information for making the diagnosis of a viral disease.

Transmission

The most common means of virus transmission are the respiratory and fecal-oral routes. Other transmission routes include contact with fomites carrying the virus (e.g., paper tissues), contaminated secretions (e.g., saliva, semen) or blood upon sex, at birth, by transfusion or organ transplant, and from insect or animal vectors.

Some viruses are zoonoses, capable of infecting insects or animals other than humans. A bird, rodent, or monkey may be a reservoir for maintaining the virus and an arthropod, such as a mosquito or tick, may be the vector for transmitting the virus. The arthropod would acquire the virus while taking a blood meal and transmit the virus to its next host. To be a good reservoir, the animal must maintain a sufficient viremia for the vector to acquire the virus. Humans are primarily at risk for infection when they encroach on the environment of the reservoir or vector.

The route of transmission is determined by several factors, but primarily what the virion structure can withstand and the tissue that replicates the virus. As indicated earlier, capsid viruses can be transmitted by most routes, but enveloped viruses must remain wet during transmission to be functional, and most enveloped viruses are destroyed in the GI tract. Viruses released in respiratory aerosols replicate in the lung. For fecal-oral transmission, the virus can replicate in mucosal epithelium or lymphoid tissue of the intestines or the pharynx; most of these viruses are capsid viruses. Transmission in blood, by a mosquito or in a transfusion, requires that the virus infect blood cells or the cells lining the blood vessels, establish a sufficient viremia and be in the bloodstream when the blood bank or the arthropod takes its blood.

Sexual transmission requires an additional criterion. Sexually transmitted viruses replicate on the genitalia or within secretory cells and virus transmission must occur prior to, or in the absence of symptoms. Aches, pains, lesions, or other symptoms are a deterrent to the transmission of STDs (sexually transmitted diseases).

Geography and Season

Many viral diseases have worldwide distribution. The geographic limitations to virus spread have dissolved due to global transportation. Passengers and freight can carry a virus or an infected mosquito to new geographic regions within a day. A perfect example of this was the outbreak of severe acute respiratory syndrome (SARS) in 2002. SARS spread from Vietnam to China to Hong Kong and then to Toronto, Canada. The spread of, and outbreak of measles in the United States is a more recent example. Vector-borne viral diseases are endemic to the habitat of the insect or animal vector. This is especially true for many of the mosquito-borne encephalitis and hemorrhagic fever viruses. Influenza spreads throughout the world in infected birds as well as infected people.

The reasons for the seasonal occurrence of viral infections are sometimes obvious—for example, outbreaks of arboviral (arthropod-borne virus) encephalitis occur during the summer when mosquitos are prevalent. The seasonal nature of influenza outbreaks may be due to the increased proximity of people during the fall and winter, or that the temperature and humidity of the air may stabilize or promote transmission of the virus-loaded aerosol.

Who is at Risk

There are many different risk factors in addition to geography that determine whether an individual is likely to come into contact with a virus and the severity of the resulting infection. These include the age, health, and immune status of the individual, as well as occupation, travel history, lifestyle, sexual activity, and interactions with children, especially daycare facilities.

Vaccines and Antiviral Drugs

Vaccines

Vaccination is probably the most beneficial treatment that a physician can provide because the immune response is the only therapy for most viral diseases (6, 7). Passive immunization can be used as a temporary therapy to prevent or ameliorate the course of viral disease. The use of human antibodies is preferable to avoid the possibility of developing serum sickness upon a second administration of animal antibody. For example, rabies immunoglobulin is injected around the site of an animal bite to neutralize any potential infecting virus. Varicella zoster immunoglobulin (VZIG) can be given to children with leukemia because they are at risk for serious outcomes of infection.

Active immunization with an attenuated (nondisease causing) virus, inactivated, or subunit vaccine is preferable to passive immunization because it initiates a longer-term immune response in the individual (Table 49.6). Live attenuated vaccines utilize virus variants that cause mild or no disease in humans. The vaccine viruses cannot grow efficiently at human body temperature, they have been adapted to animal or tissue culture cells rather than human cells, or they are a related animal virus that does not grow well in humans but shares antigens with the human virus (e.g., smallpox and rotavirus vaccines). These live vaccines elicit a natural, long-lived immunity with effective cellular and antibody responses. For example, the live attenuated vaccine for influenza is dripped into the nose, grows efficiently at 25°C but very poorly at 37°C and elicits IgG, secretory IgA, T cell responses, and longer immune memory. However, there is a minor risk of disease with some live vaccines in the immunocompromised individual or upon mutation of the vaccine virus (e.g., poliovirus). Inactivated vaccines consist of purified viral components or virion particles that are heat, radiation, or chemically treated to destroy the infectivity of the virus. The hepatitis B virus and the human papillomavirus vaccines are produced by genetic engineering and consist of virus-like particles (VLPs) containing one of the viral proteins. Inactivated vaccines produce a more short-lived and mainly antibody, not cell-mediated immunity. Adjuvants, larger amounts, and booster immunizations enhance the response to inactivated vaccines.

New approaches to vaccines being developed for dengue, HIV, HCV, and other very difficult viruses include the use of new adjuvants, immunization with hybrid viruses, DNA vaccines, and even self-replicating RNA vaccines, which resemble RNA viruses. Newer adjuvants enhance the potency and stimulate cell-mediated immunity to the vaccine proteins. Hybrid virus vaccines for HIV incorporate HIV genes into adenovirus, poxvirus, or other vectors. DNA vaccines incorporate genes for cell mediated and antibody responses into plasmids that can be utilized by human cells. Safe and more efficient methods of DNA and RNA vaccine delivery have been developed to make this approach viable.

Vaccine immunization of the community reduces the number of virus susceptible individuals and is termed "herd immunity." Routine immunization programs are available for infants and children and for individuals at risk of infection due to their vocation or travel into an endemic region. Immunization programs have led to the elimination of smallpox throughout the world; natural polio in most of the world; measles, mumps, and rubella except in the Third World; and reduction of rotavirus and influenza infections and disease. Complacency and misinformation have compromised the delivery of these programs in many parts of the developed world as evidenced by outbreaks of measles. Measles is so contagious that outbreaks will occur if less than 10% of the population is unimmunized, which includes

TABLE 49.6

Most Common Viral Vaccines

Vaccine	Type of Vaccine	Who Should Receive Vaccine[a]
Measles	Live, attenuated	1-year-olds, booster at 4–6 years
Mumps	Live, attenuated	1-year-olds, booster at 4–6 years
Rubella	Live, attenuated	1-year-olds, booster at 4–6 years
Varicella Zoster Virus *Varicella*	Live, attenuated	1-year-olds, booster at 4–6 years
Varicella Zoster Virus *Zoster*	• Stronger version of live attenuated[b] • Adjuvanted subunit	>60-year-olds
Polio	• Killed, intramuscular • Live, attenuated, oral[b]	2-, 4-, and 6-month-olds 2-month-olds
Influenza	• Inactivated • Live, attenuated, aerosol	Adults, annually Ages 2–49 year-olds
Hepatitis A	Inactivated	12-month-olds, Travelers and at-risk population
Hepatitis B	Virus-like particle	2-, 4-, and 8 months old and at-risk population
Rabies	Inactivated	Infected individuals, veterinarians, and others at risk
Rotavirus	• Live attenuated, oral • Live reassortant, oral	2-, 4-, 6 months old
Human papillomavirus	Virus-like particle	Teen and young adult women Teen boys

[a] As per "Recommended Immunization Schedules for Persons Aged 0 Through 18 Years."
[b] No longer recommended in the United States.
Source: https://www.cdc.gov/vaccines/schedules/downloads/child/0-18yrs-child-combined-schedule.pdf.

individuals who are less than 1 year of age, the minimal age for measles vaccination.

Antiviral Drugs

A summary of antiviral drugs will follow, but the subject is discussed in depth in Chapter 53.

The best antiviral drug targets are the DNA polymerase of herpes viruses, the reverse transcriptase polymerase of retroviruses and hepatitis B virus, and other polymerases. Other targets include the protease of HIV and HCV, the integrase, fusion and attachment steps of HIV, the neuraminidase of influenza A and B, and the M2 protein of influenza A. Inhibitors of viral polymerases include nucleotide/nucleoside analogs such as acyclovir, penciclovir, and azidothymidine (AZT). The anti-HSV and varicella zoster virus drugs (acyclovir and penciclovir) and their prodrugs (valacyclovir and famciclovir) are very safe drugs because they require activation by a viral enzyme called thymidine kinase. There are also nonnucleos(t)ide inhibitors that bind outside of the active site of the enzyme, such as for the HIV reverse transcriptase. The HIV and HCV proteases are also good targets for antiviral drugs because they are essential for the proper assembly of an infectious virion and activation of the viral proteins. Inhibitors of the influenza neuraminidase, such as zanamivir and oseltamivir, cause the virus to clump and stick to the cell surface and this prevents virus release. Blockage of the M2 protein channel of influenza A by amantadine and rimantadine prevent the release of the genome into the cytoplasm. Ribavirin is a guanosine analog that induces hypermutation and inactivation of viral genomes as well as inhibiting other important processes.

Description of Virus Families

DNA Viruses

Parvoviruses

The parvoviruses are the smallest DNA viruses and include the B19, bocavirus, and adeno-associated viruses (8–10). Parvoviruses have a single-stranded DNA genome contained in a naked capsid. They require growing cells to provide DNA nucleotides and polymerase for replication. B19 targets erythroid precursor cells. During replication, the single-stranded genome is converted into double-stranded DNA.

B19 causes erythema infectiosum (fifth disease—one of the five childhood exanthems) in children and polyarthritis without rash in adults (11). In children, a period of high fever is followed several days later by a maculopapular rash caused by immune complexes. B19's predilection for erythroid precursor cells can cause an aplastic crisis in a person with chronic hemolytic anemia (e.g., sickle cell anemia) and neonatal infection may be fatal due to hydrops fetalis. There are neither vaccines nor antiviral drugs. Bocavirus may cause acute respiratory disease (12). Adeno associated viruses do not cause disease but are good vectors for genetic engineering (13).

Papillomaviruses

There are at least 100 types of the wart-causing human papillomavirus (HPV) that infect cutaneous or mucosal tissues (14, 15). HPVs are naked capsid viruses with a double-stranded DNA genome. These viruses can stimulate the cell to grow but utilize cellular polymerases for replication of the genome. The virus is transmitted by contact. In addition to the benign common and other warts of the skin, other HPV types can cause growths on mucosal epithelial surfaces of the throat, conjunctiva, and anogenital regions. High risk HPV strains, including HPV 16, 18, 31, 33, 45, 52, and 58, inhibit cellular growth suppressor proteins p53 and RB105, integrate into the host chromosome, and can promote cell progression to cancer. HPV 16, 18, and some other types are common sexually transmitted diseases that can lead to dysplasia and cervical carcinoma. These high-risk HPV types are also associated with anal, penile, and oropharyngeal cancers. Genome detection methods such as PCR and *in situ* DNA analysis are used to detect and identify the virus. Surgical removal of the wart or growth is the most common therapy. A nucleotide analog, cidofovir, and an immunomodulator, imiquimod, are FDA-approved therapies. A virus-like particle vaccine formulation including HPV 6, 11, 16, 18, 31, 33, 45, 52, and 58 is recommended for young women and boys to prevent infection and cancer.

Polyomaviruses

The three human polyomaviruses, JC, BK, and Merkel Cell polyomavirus (MCV), are close cousins of the papillomaviruses. JC and BK are ubiquitous but only cause disease in immunocompromised individuals (16). The JC virus causes progressive multifocal leukoencephalopathy (PML), a disease of multiple neurologic symptoms; and the BK virus causes hemorrhagic cystitis. MCV is associated with Merkel cell carcinoma, a highly aggressive skin cancer. There are neither vaccines nor antiviral drugs.

Adenoviruses

The adenoviruses are larger double-stranded DNA viruses (17, 18). Their naked icosadeltahedral capsid has a fiber at each of the 12 vertices that serves as the viral attachment protein. These viruses encode a DNA polymerase and replication is divided into early and late phases. Respiratory diseases and pharyngoconjunctival fever (pink eye with a sore throat) are most common but adenovirus can cause other syndromes. Immunocompromised patients are at risk to serious adenovirus disease. The virus is spread in aerosols, in contaminated water (e.g., swimming pools), fecal contact, and by contact with fomites. There are live attenuated vaccines to adenovirus types 4 and 7 for military (but not civilian) use. Adenovirus infections may be treated with cidofovir. These and other adenovirus types are also used as gene delivery vehicles for gene replacement therapy and hybrid vaccines.

Herpesviruses

The herpesviruses are large viruses with a double-stranded DNA genome packaged in an icosadeltahedral capsid that is surrounded by an envelope. Cell-mediated immune responses are essential for control of the herpesviruses. Herpesvirus replication progresses through immediate early, early, and late phases of mRNA and protein expression but transition through these phases depends on the type and state of the infected cell. The herpesviruses establish latent infection within specific cell types

of the host, different for each of the viruses. All the herpesviruses can reactivate upon immunosuppression and HSV reactivates upon stress—emotional as well as physical.

There are eight different human herpesviruses, all but HHV8 are ubiquitous. They include herpes simplex virus (HSV) types 1 and 2 (19, 20), both of which cause oral and genital herpes, encephalitis, keratoconjunctivitis, and severe neonatal infections; varicella zoster virus (VZV) (21), which causes varicella (chicken pox) and upon recurrence, zoster (shingles); Epstein-Barr virus (EBV) (22, 23), which causes heterophile-positive infectious mononucleosis and is associated with Burkitt's, other lymphomas and nasopharyngeal carcinoma; cytomegalovirus (CMV) (24–26), which can cause a mononucleosis-like disease, causes several different opportunistic diseases in immunocompromised individuals and is the most common viral cause of congenital disease; HHV6 and 7 (27, 28), which cause exanthema subitum (roseola), a rash in children, and is integrated into the chromosome of 1% of the population with neurologic consequences upon recurrence; and HHV8, which causes Kaposi's sarcoma, multicentric Castleman disease, and primary effusion lymphoma (29, 30). Kaposi's sarcoma can affect normal individuals, but is usually an opportunistic disease associated with AIDS. Immune and inflammatory responses to silent recurrences of HSV, EBV, CMV, and HHV6 and 7 have been implicated to trigger different neurologic and inflammatory diseases.

HSV and CMV can be isolated in tissue culture but these methods are being replaced by immunological and genome detection methods that are used for the other herpesviruses. Serology is commonly used to detect and follow the progression of EBV-associated infectious mononucleosis. EBV infection of B cells induces production of heterophile antibodies, which distinguish the resulting mononucleosis from those caused by CMV, HIV, or toxoplasma. Detection of antibodies to the EBV MA (membrane antigen) and VCA (viral capsid antigen) are present during disease and antibody to EBNA (Epstein-Barr nuclear antigen) indicates convalescence from infectious mononucleosis or prior infection.

Acyclovir, valaciclovir, penciclovir, and famciclovir are available for treatment of HSV and VZV. Ganciclovir, valganciclovir, cidofovir, and foscarnet are available for CMV. Passive immune protection with VZIG and prophylactic protection with a live vaccine is available for varicella and an adjuvanted subunit vaccine is available for zoster.

Poxviruses

Poxviruses are the largest viruses that cause human disease. Poxviruses have a double-stranded DNA genome that is enclosed in an ovoid to brick-shaped enveloped virion with a complex structure. The poxviruses are the exception to most of the rules for virus replication because they are DNA viruses that replicate and assemble in the cytoplasm. As such, their genome encodes the polymerase and other enzymes necessary to transcribe mRNA as well as replicate its genome.

This virus family includes the deadly smallpox, animal pox viruses (31) that cause similar vesicular diseases, and molluscum contagiosum (32). Through a very effective vaccination program and careful epidemiology, the smallpox virus was eradicated from all parts of the Earth except for the biowarfare factories of the former Soviet Union and the deep freezers of the CDC in the United States. Molluscum contagiosum is a wart-like skin disease. The smallpox vaccine consists of a related virus, vaccinia, which shares immunogenicity with smallpox but usually causes benign disease in immunocompetent humans. The vaccine is only administered to military and at-risk personnel. Vaccinia is a good vector for genetic engineering and has been used to develop hybrid vaccines for HIV and other viruses.

RNA Viruses

Caliciviruses and Hepeviridae

The caliciviruses are small naked capsid viruses with a positive-sense RNA genome. The most well-known of this genus are the noroviruses (33–35). Most caliciviruses cause outbreaks of gastroenteritis. Norovirus is the most common cause of foodborne gastroenteritis, is very resistant to inactivation and infected individuals shed large amounts of virus in their stool and vomit even after recovery. Norovirus has a predilection to infect and cause more serious disease in individuals with cell surface carbohydrates of the type O (not A or B) blood group.

Hepatitis E virus (HEV) was classified as a calicivirus but now has its own family, *Hepeviridae* (36). HEV causes acute hepatitis similar to hepatitis A virus (HAV) but the mortality rate is 1–2% (10 times greater than for HAV) and up to 20% for pregnant women. HEV is detected by ELISA.

Picornaviruses

The picornaviruses (pico-small, RNA viruses) are the prototypical naked capsid viruses with a positive-sense RNA genome (37–39). This family includes the enteroviruses, which cause all kinds of different diseases except gastroenteritis, and the rhinoviruses, which cause the common cold. A canyon on the surface of these viruses binds to a receptor on the target cell and the RNA genome is injected into the cell. The positive-sense RNA genome binds to a ribosome and a polyprotein is made that becomes proteolyzed into individual proteins including an RNA-dependent RNA polymerase, capsid subunits, and other proteins. A negative-sense RNA template is made, from which more mRNA and new genomes are produced by the viral-encoded polymerase. An empty procapsid shell is filled with the RNA genome, and the virus kills and then leaves the cell.

The most famous member of the enteroviruses is poliovirus, infamous for causing paralytic disease. Fortunately, effective immunization programs with the attenuated-oral or the inactivated vaccines have led to the elimination of natural polio in most parts of the world. Other members of this family include the Coxsackie A viruses, which cause hand, foot and mouth disease, herpangina, common cold-like, polio-like, and other diseases; Coxsackie B virus (B is for body), which causes pleurodynia and myocardial infections; echoviruses; parechoviruses; and some of the higher-numbered enteroviruses, which normally are benign, but can cause viral meningitis in children younger than 1 year. Coxsackie B virus infection can lead to type 1 diabetes. Hepatitis A virus, which causes acute hepatitis, is also an enterovirus. The enteroviruses are transmitted by the fecal-oral route, but some are also transmitted in aerosols and on fomites.

The rhinoviruses cause the common cold and are distinguished from the enteroviruses because they are disrupted by acids, cannot be transmitted by the fecal-oral route, and cannot grow efficiently at temperatures above 34°C. These conditions restrict the virus to the upper respiratory tract and aerosol dispersion.

Live attenuated and killed polio vaccines are composed of two or three poliovirus types. Although the live vaccine is cheaper, easier to administer, and has other benefits, the killed vaccine is safer and currently preferred where risk of exposure is minimal. The hepatitis A virus vaccine used in most parts of the world is also inactivated, but a live vaccine is used in China. Pleconaril is an antiviral drug that binds to the cell receptor binding canyon on the virion surface, prevents uncoating of the virus but is no longer available.

Coronaviruses

Coronaviruses are enveloped viruses with a positive-sense RNA genome. The virus derives its name from the "corona" seen in electron micrographs around the virion created by the viral glycoproteins (40, 41). This structure also protects the virion from harsh conditions and allows this enveloped virus to be transmitted by the fecal-oral route in addition to the respiratory route. Like rhinoviruses, most of these viruses do not replicate well at 37°C, are restricted to the upper respiratory tract and cause the common cold. The severe acute respiratory syndrome coronavirus (SARS-CoV-1 and SARS-CoV-2) and Middle Eastern respiratory syndrome virus (MERS-CoV) can grow at 37°C and cause severe lung disease. SARS-CoV, MERS-CoV, and SARS-CoV-2 originated as animal (bats) viruses and have animal reservoirs. SARS-CoV-1 caused an outbreak of deadly severe acute respiratory syndrome, which started in Vietnam and China and spread to Toronto, Canada (42). An outbreak of MERS-CoV occurred in 2013 in Saudi Arabia. As this book goes to press, December 2020, the world is in the grips of a new coronavirus pandemic (COVID-19) caused by SARS-CoV-2. Based on the extensive tissue expression of the angiotensin converting enzyme 2 (ACE2) receptor for SARS-CoV-2 and inflammatory responses, disease can affect many organs and range from asymptomatic to lung and multisystem organ failure (42a). Careful quarantine limited the spread and disease caused by SARS-CoV-1. Vaccines and antiviral drugs for SARS-CoV-2 are available.

Paramyxoviruses

The paramyxoviruses include the well-known measles and mumps viruses, as well as very common parainfluenza, respiratory syncytial, and metapneumo respiratory disease viruses (43). The paramyxoviruses are large enveloped viruses with a negative-sense RNA genome. As negative-strand RNA viruses, they carry their polymerase in the virion wrapped around the genome as part of the nucleocapsid. Individual mRNAs and a genomic-length positive-sense RNA are transcribed from the genome and the latter form of RNA is used as a template to make new genomes. The paramyxoviruses express a fusion protein that allows them to enter the cell by fusion of the envelope with the plasma membrane, and the fusion protein also promotes cell-cell fusion into multinucleated giant cells (syncytia). These viruses exit the cell without killing the cell.

Paramyxoviruses are good inducers and responsive to type 1 interferons. Cell-mediated immunity is essential for controlling paramyxoviruses but the resulting "flu-like" symptoms and inflammation contribute to the production of disease.

As respiratory viruses, the paramyxoviruses are very contagious. Parainfluenza and respiratory syncytial viruses (RSVs) (44–46) are ubiquitous, causing anything from a serious cold to bronchiolitis or pneumonia. In young children, parainfluenza causes croup and RSV causes a serious pneumonia. The severity of RSV infections in newborns, especially those born prematurely, prompted the development of antiviral therapy with aerosolized ribavirin or treatment with anti-RSV immunoglobulin.

The classic symptoms of measles (47–50) are cough, coryza (runny nose), conjunctivitis, rash, and Koplik spots (small vesicular lesions in the mouth). The more serious outcomes include pneumonia, encephalitis, and post-infectious encephalitis. Most of the disease symptoms are due to immune pathogenesis. Measles is very contagious and remains a major killer especially for vitamin A deficient individuals. Sequelae of measles may include post measles encephalitis and subacute sclerosing panencephalitis (SSPE). Mumps infects glandular tissue and the immune response causes swelling that can result in parotitis, orchitis, and meningoencephalitis. Fortunately, there are very effective live, attenuated vaccines for protection against measles and mumps that are administered as part of the MMR (measles, mumps, rubella) or MMRV (with varicella) vaccines to children after 1 year of age and again at 4-6 years of age. Unfortunately, increasing numbers of individuals are opting out of vaccination programs and large outbreaks of measles and mumps have returned.

Orthomyxoviruses

The orthomyxoviruses include influenza A, B, and C (51–56). These are medium sized enveloped viruses that have a segmented, negative-strand RNA genome. Most of the RNA genome segments encode individual viral proteins. The virions are decorated with hemagglutinin (HA or H) and neuraminidase (NA or N) glycoproteins. The HA binds to a sialic acid cellular receptor, which is on epithelial cells and on erythrocytes. The NA cleaves sialic acid from virion glycoproteins and mucus to prevent the virus from binding to itself and facilitates exit from the infected cell and spread. Having a segmented genome allows the influenza to reassort their segments when a cell is infected with more than one strain of influenza to create hybrid viruses.

Influenza A is a zoonose, a virus that infects animals and humans. Reassortment of gene segments during a mixed animal-human influenza virus infection or mutation can create new human viruses that can spread worldwide. These greatly feared pandemics usually originate in the Far East, although the most recent "swine flu" originated in Mexico. Influenza viruses are named for their type, place of original isolation, date of original isolation, and type of hemagglutinin and neuraminidase (for example, A/Bangkok/1/79(H3N2) virus). Influenza B is strictly a human virus. Influenza B can also change, but primarily by mutation and less drastically than Influenza A. Influenza C causes mild disease. Influenza can be assayed by its ability to agglutinate erythrocytes (hemagglutination) but detection of antigen by ELISA tests or the genome by RT-PCR are preferred procedures. Anti-influenza antibody generated by the annual vaccines is validated by hemagglutination inhibition as a surrogate of protection.

Oseltamivir (Tamiflu), zanamivir (Relenza) and peramivir (Rapivab) inhibit the neuraminidase of influenza A and B. Amantadine and rimantadine are antiviral drugs that inhibit a channel formed by the M2 protein of influenza A (not B or C) to prevent the initial release of the genome from the virion particle within the cell to block replication, but these drugs are no longer recommended due to the prevalence of resistant strains. An annual vaccine program utilizes a mixture of three or four inactivated viruses or their HA and NA produced in fertilized eggs or tissue culture or the HA produced by genetic engineering comprising the influenza A and B types that are predicted to occur during that year. A live attenuated inhaled vaccine (LAIV) (FluMist) is also available for persons aged 2–49 years.

Rhabdoviruses, Filoviruses, and Bornaviruses

These viruses have similar structures and were once considered to be in the same viral family. These are enveloped viruses with single-stranded, negative-sense RNA genomes. The rhabdoviruses have a characteristic bullet shape, whereas the filo- and bornaviruses are filamentous. The most famous and only rhabdovirus to affect humans is rabies (57). Reservoirs of this virus are present in raccoons, foxes, skunks, other forest wildlife, bats, and even farm animals. Infected dogs are still a common vector in countries that lack pet immunization programs. Infection of humans is always fatal unless the individual is treated with anti-rabies immunoglobulin and immunized with an inactivated vaccine prepared in human diploid fibroblasts or chick embryo cells. Rabies takes a long time to progress from the site of the bite to the brain and does not cause signs of disease until it reaches the salivary glands and the brain. The characteristic fear of drinking water (hydrophobia) is due to the pain upon swallowing due to infection of the salivary glands. Infection of the highly innervated salivary glands and mental changes in animals (e.g., mad dog) provides the source of virus and the attitude (madness) for transmitting the virus. Analysis of brain tissue from an infected animal will show Negri inclusion bodies in the cytoplasm, but virus infection is usually detected by immunofluorescence or genome detection. As mentioned, there is an inactivated vaccine that is administered to people at risk for infection, such as veterinarians, and to people who have been bitten by an animal that is a possible carrier of rabies. A hybrid vaccine consisting of the rabies virus glycoprotein gene genetically inserted into the vaccinia virus is infused into blocks of food and airdropped into the forest to immunize wild animals.

The Marburg and Ebola filoviruses cause severe or fatal hemorrhagic fevers and are endemic in regions of Africa (58). These viruses are so virulent that they often burn out their susceptible population by killing their human cohort before they can spread to another community. Recent outbreaks of Ebola have occurred in the Democratic Republic of the Congo. These infections may be responsive to ribavirin or immunoglobulin from survivors. A hybrid virus vaccine has been developed for Ebola. Bornaviruses have been implicated to trigger psychiatric diseases.

Reoviruses

The reoviruses derive their name as the acronym for "respiratory enteric orphan" viruses because no human disease was associated with them until the rotaviruses were identified. The Colorado tick fever virus is also a reovirus. Reoviruses have a double capsid and a double-stranded segmented RNA genome. Partial protease digestion exposes the viral attachment proteins and allows the virion to disassemble when inside the target cells of the GI tract. Like the orthomyxoviruses, the individual negative-strand segments of the double stranded genome are transcribed into positive-sense mRNA. An early set of proteins are made that corral the positive-sense segments into an inner capsid, late proteins are made, the double-stranded RNA is recreated, the outer virion capsid acquired, and the virus is released.

Rotaviruses are the most common cause of infantile diarrhea and a major cause of death due to dehydration of infants in underdeveloped countries (59–61). Rotaviruses are usually detected in stool by ELISA. A live, attenuated human virus or a bovine reassorted vaccine is administered at 2, 4, and 6 months of age to prevent this disease.

Togaviruses

The togaviruses consist of the alphaviruses, which are zoonoses that are transmitted by mosquitoes and considered arboviruses (arthropod-borne viruses), and rubella virus, which is transmitted by aerosols. The togaviruses are small with an envelope that is tightly wrapped around a nucleocapsid containing a positive-strand RNA genome. After entering the cell by receptor-mediated endocytosis, the genome is released, binds to ribosomes, and a polymerase and other early proteins are cleaved from the initially translated polyprotein. After a negative-strand RNA template is made for new genomes, smaller mRNAs are also made that encode a polyprotein for the capsid and envelope glycoproteins.

The alphaviruses are zoonoses, which usually have a bird or rodent reservoir and a mosquito vector (62). These viruses, which include Western, Eastern, and Venezuelan equine encephalitis viruses, cause either systemic flu-like symptoms or encephalitis in humans. Chikungunya causes systemic, arthralgia and arthritic disease (63).

Rubella virus is not a zoonose, is transmitted in aerosols, only infects humans, has one serotype, and causes rubella (German measles), a childhood rash with possible arthritic symptoms due to immune pathogenesis. Rubella was a major cause of congenital defects (deafness, mental retardation, and cataracts) until vaccine programs limited the spread of and infection by this virus. The vaccine is live and attenuated and administered with the measles and mumps (MMR) and potentially VZV (MMRV) vaccines.

Flaviviruses

The flaviviruses are arboviruses, except for the hepatitis C virus (HCV). These viruses are small and enveloped with a positive-strand RNA genome that replicates like the picornaviruses, except with an envelope surrounds its nucleocapsid. The arbo-flaviviruses (64–70) cause mild systemic flu-like symptoms, systemic disease (zika), hemorrhagic disease (dengue, yellow fever), or encephalitis (St. Louis, West Nile, Japanese encephalitis viruses). Zika virus remains in the blood for long periods after infection and can be transmitted in semen and in utero to the fetus where it can cause congenital anomalies. St. Louis and West Nile encephalitis and dengue and yellow fever viruses establish urban cycles of human-mosquito-human infection based on the habitat of the mosquito vector. Vaccines are available for the yellow fever and Japanese encephalitis viruses.

Hepatitis C (HCV) virus is transmitted in blood, semen, and vaginal fluids (71–76). HCV usually causes chronic hepatitis (70%), which can progress to cirrhosis (20%), liver failure (6%), or hepatocellular carcinoma (4%). There are 6 different HCV types, type 1 being the most common. They differ in their sensitivity to antiviral drugs. ELISA is used for serological screening to protect the blood supply and RT-PCR is used to follow the progression of disease and treatment. HCV disease is treated with a combination of drugs that inhibit the polymerase, polymerase/virulence enhancer, or protease. Treatment has the potential to cure the HCV infection. Previous therapy utilized ribavirin and interferon alpha.

Bunyaviruses

The bunyaviruses are enveloped viruses with a negative-strand RNA genome. There are many different bunyaviruses with exotic names that are indicative of where in the world they were discovered. Most bunyaviruses are transmitted by arthropods, including mosquitoes, flies, or ticks, depending on the virus (77–79). The hantaviruses infect rodents and are spread in their urine and stool (80, 81). The California encephalitis viruses and hantaviruses occur in the United States. These viruses cause encephalitis and hemorrhagic disease, respectively. There is no treatment or vaccine for these viruses.

Arenaviruses

The arenaviruses (*arena* is Greek term for "sandy") get their name because they incorporate ribosomes into the virion particle and this gives them a sandy appearance in electron micrographs (82). These enveloped viruses contain two circular RNA genome segments; the larger segment is a negative-sense RNA while the other is ambisense. Early mRNA from the ambisense segment is copied from the entering genomic negative-sense RNA. Later, when positive RNA templates are made to replicate the genome, another mRNA is also made for the late structural proteins. The arenaviruses are transmitted in urine from mice and other rodents and Lassa fever is also transmitted in secretions and fluids from infected humans. Lymphocytic choriomeningitis virus is seen in rodent infested areas around the world (83). Lassa fever (84), Machupo, and Junin viruses, in Western Africa, Argentina and Bolivia, respectively, cause deadly hemorrhagic fever. Lassa fever has been treated with immunoglobulin and ribavirin.

Retroviruses

The retroviruses include two human disease viruses—human immunodeficiency virus (HIV) (85–95) and human T-leukemia virus (HTLV) (96)—as well as oncogenic animal viruses, some nonpathogenic viruses, and many endogenous viruses that are carried in our chromosomes. The retrovirus envelope encloses a nucleocapsid that contains two copies of the positive-sense RNA, two transfer RNAs, an RNA-dependent RNA polymerase (reverse transcriptase), and other enzymes. The basic retrovirus genome consists of three genes: The gag (capsid proteins), pol (polymerase, integrase and protease), and env (glycoprotein) genes. HTLV and HIV encode other genes to facilitate the replication and pathogenesis of the virus. HIV binds to CD4 molecules and a chemokine receptor as a co-receptor on T-cells, macrophages, dendritic and other cells. Initially, the virus uses the CCR5 chemokine receptor and then later in the disease, some of the virus mutates to utilize the CXCR4 receptor. Persons lacking the CCR5 protein are naturally resistant to HIV. The virus fuses its envelope with the cell membrane to enter the cell, releasing the genome and proteins into the cytoplasm. The positive-strand RNA is converted into a negative-strand cDNA and then a circular double-stranded DNA by the reverse transcriptase enzyme carried in the virion. The HIV polymerase is extremely error prone and readily generates new versions of the virus through mutation. An integrase enzyme, carried into the cell, facilitates the integration of the genome into the host chromosome of growing cells and several mRNAs and a genomic-length, positive-sense RNA are transcribed by cellular enzymes as if from very active genes. A polyprotein, consisting of the capsid and enzymatic proteins, associates with viral glycoprotein modified plasma membranes. A pair of RNA genomes and tRNAs then associates with the protein-modified membrane and the virion buds out of the cell. After leaving the cell, the virion protease cleaves the polypeptide into functional proteins. Eventually, the cell will die. HTLV-1 also binds to CD4 molecules on T cells, its cDNA gets integrated into the host chromosome, but the virus stimulates the growth of the cell. There is a long latent period prior to the onset of acute T cell lymphocytic leukemia in infected individuals.

Many articles and books have been written about HIV and its disease, acquired immunodeficiency syndrome (AIDS). In summary, HIV infects the controlling cells of the immune response, the CD4 expressing macrophage, dendritic cell, and CD4 T-cell, to eventually deplete the person's immune response and cause the individual to die of opportunistic infections. Soon after HIV infection, the activated CD4 T cells of the intestinal and other systems are eliminated. Initial symptoms of acute disease may resemble mononucleosis. After a variable period of subclinical immune dysfunction and a switch in viral specificity to the CXCR4 chemokine receptor, CD4 T cell numbers drop precipitously followed by a drop in CD8 T cell numbers, resulting in significant immunodeficiency. Individuals are plagued by opportunistic diseases caused by mycobacteria and other bacteria, fungi and parasites, and recurrences of herpesviruses.

Virus infection can be detected by genome analysis and serology and the blood supply is screened by ELISA. Western blot analysis was necessary for confirmation but has been replaced by newer ELISA assays that detect an early viral protein (p24) as well as anti-HIV antibodies. Virus load, the number of viral genomes in the blood, is quantitated by real-time reverse transcriptase PCR analysis or other genome detection methods. These tests are more sensitive than serology and allow early detection and quantification of the infection and the ability to follow the course of disease and compliance with antiviral therapy.

The high mutation rate of the virus promotes escape from antibody control and single antiviral drug treatment. A cocktail of antiviral drugs (highly active antiretroviral therapy [HAART]) consists of nucleotide and/or non-nucleotide analog inhibitors of the reverse transcriptase, inhibitors of the protease, integrase, binding to CCR5, and/or virus fusion. Prophylactic therapy (PrEP (pre-exposure prophylaxis)) or (PEP [post-exposure prophylaxis]) are suggested for individuals at risk for infection.

HTLV infection is usually asymptomatic but can progress to adult acute T-cell lymphocytic leukemia (ATLL) over the course of many years. HTLV can also cause tropical spastic paraparesis, a nononcogenic neurologic disease.

Up to 8% of the human chromosome is made up of retroviral sequences. Human endogenous retroviruses (HERV) are quiet but may influence body functions, immune and other responses. A retroviral protein, syncytin, is induced during pregnancy and promotes placental formation.

Hepadnaviruses and Hepatitis D Virus

Hepatitis B virus (HBV) is the only member of the hepadnaviruses that infects humans (97–100). The hepadnaviruses are enveloped DNA viruses but, like the retroviruses, encode a reverse transcriptase and replicate through an RNA intermediate. HBV replicates in hepatocytes without killing the cell. The DNA genome of the virus is transcribed in the nucleus into several mRNAs and one RNA that is larger than the genome is used as a template for making the genomic DNA. The negative DNA strand is synthesized from the template RNA by the viral encoded reverse transcriptase and as the positive DNA strand is being made, the RNA is degraded and the genome is enclosed in a capsid structure that causes synthesis to stop prior to completion. The virus acquires its envelope by budding from the plasma membrane without killing the cell. Large amounts of the glycoproteins of the virus are produced, and these form particles (HBsAg) that are present in the blood of infected individuals.

Hepatitis B, also known as serum hepatitis, has an insidious onset and usually will resolve, but it can also cause chronic disease in 10% of patients. Interestingly, the symptoms of disease are caused by the immune resolution of the infection and a mild case of HBV is likely to become chronic. This is especially true for infants and children. Chronic HBV can result in primary hepatocellular carcinoma (PHC) after 20 or more years. The course of the disease is followed by detection of viral proteins and serology. The presence of the HBsAg with another viral antigen, HBeAg, is indicative of active disease and virus production, while just HBsAg can indicate chronic disease. Antibodies to HBsAg (anti-HBsAg) are detectable only after HBsAg disappears from the blood and indicates resolution of infection or a vaccinated individual. IgM for the HBcAg (capsid antigen) indicates recent infection and may be present after HBsAg ceases to be produced but before anti-HBsAg can be detected ("the window"). HBV is transmitted in blood, semen, vaginal secretions, and saliva. Antiviral drugs target the viral reverse transcriptase polymerase and include lamivudine, entecavir, tenofovir, adefovir dipivoxil, and telbivudine. The HBV vaccine consists of virus-like particles that form from the HBsAg. Originally, the vaccine was purified from the blood of chronic carriers of the disease but is now prepared by genetic engineering. In addition to reducing the incidence of hepatitis, vaccination also reduces the risk for PHC.

Hepatitis D virus (HDV) is somewhat of a parasite of hepatitis B virus. HDV can only replicate in HBV-infected cells and the HDV RNA genome is packaged in HBV envelopes. HDV exacerbates HBV disease, promoting fulminant hepatitis and cirrhosis.

Prions

Prions, by definition, are infectious proteins and not really viruses. The prion protein, called PrP^{SC}, is a structural variant of a normal and natural cell surface protein (101). Binding of the prion protein to the natural analog changes its configuration to resemble the prion protein, adds it to the aggregate to extend the prion structure. Prions are very resistant to inactivation by detergents, disinfectants, heat, autoclaving, and irradiation. Prions cause spongiform encephalopathies, including Creutzfeldt-Jakob disease (CJD), Kuru, Gerstmann-Straussler-Scheinker syndrome, and fatal familial insomnia, which can be infectious or inheritable diseases. CJD has an incubation period that could last for decades prior to the onset of a progressive degenerative neurologic disease. A variant CJD (vCJD) has a more rapid onset. After a long incubation period, the spongiform encephalopathies are characterized by a loss of muscle control, shivering, myoclonic jerks, and tremors, rapidly progressive dementia, and death. Prions are very difficult to detect, but an assay has been developed (102). There is no vaccine or treatment other than supportive care for these diseases.

REFERENCES

1. Murray P., Rosenthal KS, and Pfaller M. 2021. *Medical Microbiology*, 9th ed., Elsevier-Saunders, Philadelphia, PA.
2. Rosenthal KS. 2005. Are microbial symptoms "self-inflicted"? The consequences of immunopathology. *Infect. Dis. Clin. Prac.* 13:306–310.
3. Rosenthal KS. 2006. Vaccines make good immune theater: immunization as described in a three act play. *Infect. Dis. Clin. Prac.* 14:35–45.
4. Rosenthal KS. 2006. Viruses: microbial spies and saboteurs. *Infect. Dis. Clin. Prac.* 14:97–106.
5. Miller JM, et al. 2018. A guide to utilization of the microbiology laboratory for diagnosis of infectious diseases: 2018 update by the Infectious Diseases Society of America and the American Society for Microbiology. *Clin Infect Dis.* 67:e1–e94.
6. Rosenthal KS, Sikon J, Kuntz, A. 2015. Why don't we have a vaccine against? Part 1. Viruses. *Infect. Dis. Clin. Pract.* 23:202–210.
7. Centers for Disease Control and Prevention, Hamborsky J, Kroger A, Wolfe S, eds. 2015. *Epidemiology and Prevention of Vaccine-Preventable Diseases.* 13th ed. Public Health Foundation, Washington, DC. Available at: https://www.cdc.gov/vaccines/pubs/pinkbook/index.html
8. Qiu J, Söderlund-Venermo M, Young NS. 2016. Human parvoviruses. *Clin Microbiol Rev.* 30 (1):43–113.
9. Young NS, Brown KE. 2004. Parvovirus B19. *N. Engl. J. Med.* 350:586–97.
10. Qiu J, Söderlund-Venermo M, Young NS. 2016. Human parvoviruses. *Clin. Micro. Rev.* 30:43–113.
11. Centers for Disease Control. Parvovirus B19 and Fifth Disease. Available at: https://www.cdc.gov/parvovirusb19/fifth-disease.html
12. Guido M, Tumolo MR., Verri T, Romano A, Serio F, De Giorgi M, De Donno A, Bagordo F, Zizza A. 2016. Human bocavirus: current knowledge and future challenges. *World J. Gastroenterol.* 22:8684–8697.
13. Naso MF, Tomkowicz B, Perry WL, Strohl WR. 2017. Adeno-associated virus (AAV) as a vector for gene therapy. *BioDrugs.* 31:317–334.
14. Moody CA, Laimins LA. 2010. Human papillomavirus oncoproteins: pathways to transformation. *Nat Rev Cancer.* 10:550–560.
15. Gearhart PA, et al. 2019. Human papillomavirus (HPV). Available at: http://emedicine.medscape.com/article/219110-overview.
16. Rinaldo CH, Hirsch HH. 2013. The human polyomaviruses: from orphans and mutants to patchwork family. *APMIS.* 121:681–684.
17. Tebruegge M, Curtis NJ. 2012. Adenovirus: an overview for pediatric infectious diseases specialists. *Ped Infect Dis J.* 31:626–627.
18. Ghebremedhin B. 2014. Human adenovirus: viral pathogen with increasing importance. *Eur J Microbiol Immunol.* 4(1):26–33.
19. Corey L, Wald A. 2009. Maternal and neonatal herpes simplex virus infections. *N. Engl. J. Med.* 361:1376–1385.
20. Kinchington PR, St Leger AJ, Guedon JG, Hendricks RL. 2012. Herpes simplex virus and varicella zoster virus, the house guests who never leave. *Herpesviridae.* 3:5. Available at: http://www.herpesviridae.org/content/3/1/5
21. Arvin AM, Moffat JF, Sommer M. et al. 2010. Varicella-zoster virus T cell tropism and the pathogenesis of skin infection. *Curr Top Microbiol Immunol.* 342:189–209.
22. Luzuriaga K., Sullivan JL. 2010. Infectious mononucleosis. *N. Engl. J. Med.* 362:1993–2000.
23. Odumade OA, Hogquist KA, Balfour HH Jr. 2011. Progress and problems in understanding and managing primary Epstein-Barr virus infections. *Clin. Microbiol. Rev.* 24:193–209.
24. Cannon MJ, Hyde TB, Schmid DS. 2011. Review of cytomegalovirus shedding in bodily fluids and relevance to congenital cytomegalovirus infection. *Rev Med Virol* 21:240–255.
25. Britt WJ. 2018. Maternal immunity and the natural history of congenital human cytomegalovirus infection. *Viruses.* 10(8):405.
26. Manicklal S, Emery VC, Lazzarotto T, Boppana SB, Gupta RK. 2013. The "silent" global burden of congenital cytomegalovirus. *Clin. Microbiol. Rev.* 26:86–102.
27. Blumberg BM. 2017. The secret lives of human herpes virus 6 (HHV-6). *J. Neurol. Disord.* 5: 352.
28. HHV-6 Foundation (website): http://hhv-6foundation.org/
29. Edelman DC. 2005. Human herpesvirus 8—a novel human pathogen, *Virol J.* 2:78–110.
30. Dittmer DP, Damania B. 2016. Kaposi sarcoma–associated herpesvirus: immunobiology, oncogenesis, and therapy. *J. Clin. Invest.* 126:3165–3175.
31. Frey SE, Belshe RB. 2004. Poxvirus zoonoses — putting pocks into context. *N. Engl. J. Med.* 350:324–327.
32. Chen X, Anstey AV, Bugert JJ. 2013. Molluscum contagiosum virus infection. *Lancet Infect Dis.* 13:877–888.
33. Hall AJ, Eisenbart VG, Etingüe A, Gould L, Lopman BA, Parashar, UD. 2012. Epidemiology of foodborne

norovirus outbreaks, united states, 2001–2008. *Emerg Infect Dis.* 18:1566–1573.

34. Patel MM, Hall AJ, Vinje J, et al. 2009. Noroviruses: a comprehensive review, *J. Clin. Virol.*44:1–8.

35. Bok K, Green KY. 2012. Norovirus gastroenteritis in immunocompromised patients. *N. Engl. J. Med.* 367:2126–2132.

36. Hoofnagle JH, Nelson KE, Purcell RH. 2012. Hepatitis E. *N. Engl. J. Med.*367:1237–1244.

37. Racaniello VR. 2006. One hundred years of poliovirus pathogenesis. *Virology.* 344: 9–16.

38. Muir P. 2017. Enteroviruses. *Medicine.* 45:12:794–797.

39. Shah S. 2018. Picornavirus infections. Available at: www.emedicine.com/med/topic1831.htm

40. Weiss SR, Leibowitz JL. 2011. Coronavirus pathogenesis. *Adv. Virus Res.* 81:85–164.

41. Weiss SR, Navas-Martin S. 2005. Coronavirus pathogenesis and the emerging pathogen severe acute respiratory syndrome coronavirus. *Microbiol. Mol. Biol. Rev.* 69:635–64.

42. Graham RL, Donaldson EF, Baric RS. 2013. A decade after SARS: strategies for controlling emerging coronaviruses. *Nat. Rev. Microbiol.* 11:836–848.

42 a. Petrosillo N. et al. 2020. COVID-19, SARS and MERS: are they closely related? *Clin Microbiol Infect.* 26(6):729–734.

43. Parks GD, Alexander-Miller MA. 2013. Paramyxovirus activation and inhibition of innate immune responses. *J Mol Biol.* 425(24):4872–92.

44. Griffiths C, Drews SJ, Marchanta DJ. 2017. Respiratory syncytial virus: infection, detection, and new options for prevention and treatment. *Clin. Micro. Rev.* 30:277–319.

45. Tregoning JS, Schwarze J. 2010. Respiratory viral infections in infants: causes, clinical symptoms, virology, and immunology. *Clin. Microbiol. Rev.* 23:74–98.

46. González PA, Bueno SM, Carreño LJ, Riedel CA, Kalergis AM. 2012. Respiratory syncytial virus infection and immunity. *Rev Med Virol.* 22A:230–244.

47. Griffin DE, Oldstone MM. 2009. Measles: pathogenesis and control. *Curr. Top Microbiol Immunol.* 330:1.

48. Perry, R.T., Halsey, N.A. 2004. The clinical significance of measles: a review. *J. Infect. Dis.* 189:S4–S16.

49. Ludlow M, McQuaid S, Milner D, de Swart RL, Duprex WP. 2015. Pathological consequences of systemic measles virus infection. *J. Pathol.* 235:253–265.

50. Strebel PM, Orenstein WA. 2019. Measles. *N. Engl. J. Med.* 381:349–357.

51. Stevens J, Blixt O, Paulson JC, Wilson IA. 2006. Glycan microarray technologies: tools to survey host specificity of influenza viruses. *Nat Rev Microbiol* 4(11): 857–864. (Issue devoted to Influenza Pandemics).

52. Medina RA, García-Sastre, A. 2011. Influenza a viruses: new research developments. *Nat Rev Microbiol.* 9:590–603.

53. Tosh PK, Jacobson RM, Poland GA. 2010. Influenza vaccines: from surveillance through production to protection. *Mayo. Clinic. Proceedings.* 85:257–273.

54. Oldstone MBA, Compans RW. 2014. *Influenza Pathogenesis and Control—Volume I and II. Current Topics in Microbiology and Immunology.* 385–386, Springer, New York, NY.

55. Centers for Disease Control and Prevention. Influenza (flu). Available at: www.cdc.gov/flu/

56. Derlet RW, Nguyen HH. Sandrock CE. 2018. Influenza. Available at: http://emedicine.medscape.com/article/219557-overview

57. Schnell MJ, McGettigan JP, Wirblich C, Papaneri A. 2009. The cell biology of rabies virus: using stealth to reach the brain. *Nat Rev Microbiol.* 8:51–61.

58. Sullivan N, Yang Z-H, Nabel GJ. 2003. Ebola virus pathogenesis: implications for vaccines and therapies. *J. Virol.* 77:9733–9737.

59. Hu L, Crawford SE, Hyser JM, Estes MK, Venkataram Prasad, BV. 2012. Rotavirus non-structural proteins: structure and function. *Curr Opin Virol.* 2:380–388.

60. Greenberg HB, Estes MK. 2009. Rotaviruses: from pathogenesis to vaccination, *Gastroenterology.* 136:1939–1951.

61. Nguyen DD, Henin SS, King BR. 2018. Rotavirus. Available at: http://emedicine.medscape.com/article/803885-overview

62. Weaver SC, Barrett AD. 2004. Transmission cycles, host range, evolution and emergence of arboviral disease. *Nat Rev Microbiol.* 2:789–801.

63. Centers for Disease Control and Prevention. Chikungunya Virus. Available at: https://www.cdc.gov/chikungunya/. Accessed September 3, 2018

64. Suthar MS, Diamond MS, Gale M Jr. 2013. West Nile virus infection and immunity. *Nat Rev Microbiol.* 11:115–128.

65. Guzman MG, Halstead SB, Artsob H, Buchy P, Farrar J, et al. 2010. Dengue: a continuing global threat. *Nat Rev Microbiol.* 8:S7–S16.

66. Centers for Disease Control and Prevention. Division of Vector Borne Diseases. Available at: www.cdc.gov/ncezid/dvbd/.

67. Centers for Disease Control and Prevention. Dengue. Available at: www.cdc.gov/dengue/.

68. Centers for Disease Control and Prevention. Dengue training modules. Available at: https://www.cdc.gov/dengue/training/cme/ccm/page32415.html

69. Centers for Disease Control and Prevention. West Nile virus. Available at: www.cdc.gov/ncidod/dvbid/westnile/index.htm.

70. Centers for Disease Control and Prevention. Zika Virus. Available at: https://www.cdc.gov/zika/index.html

71. Lindenbach BD, Rice CM. 2013. The ins and outs of hepatitis C virus entry and assembly. *Nat Rev Microbiol.* 11:688–700.

72. Dhawan VK. 2019. Hepatitis C. Available at: https://emedicine.medscape.com/article/177792-overview

72a. Hepatitis C Online. 2018. HCV Medications. Available at: https://www.hepatitisc.uw.edu/page/treatment/drugs. Accessed September 21, 2018.

73. Infectious Disease Society of America. HCV guidelines. Available at: https://www.hcvguidelines.org/

74. Mauss S, Berg T, Rockstroh J, et al. #11: The 2012 short guide to hepatitis C. www.flyingpublisher.com/0011.php.

75. Dhawan VK. Hepatitis C. Available at: http://emedicine.medscape.com/article/177792-overview.

76. Hepatitis C. *Nature.* 474(Supp. S1–S48): Available at: www.nature.com/nature/outlook/hepatitis-c/index.html.

77. Ayoade FO, Anderson WE, Soliman E, Perez N. 2016. California encephalitis. Available at: http://emedicine.medscape.com/article/234159-overview.

78. Centers for Disease Control and Prevention. Heartland virus disease. Available at: www.cdc.gov/ncezid/dvbd/heartland/index.html.

79. Centers for Disease Control and Prevention. La Crosse encephalitis. Available at: www.cdc.gov/lac/.

80. Vaheri A, Strandin T, Hepojoki J, Sironen T, Henttonen H, Mäkelä S, Mustonen J. 2013. Uncovering the mysteries of hantavirus infections. *Nat Rev Microbiol.* 11:539–551.

81. Avšič-Županc T, Saksida A, Korva M. 2016. Hantavirus infections. *Clin Microbiol Infect.* 21S:e6–e16.
82. Gompf SG, Smith KM, Choe, U. Arenaviruses. 2018. Available at: http://emedicine.medscape.com/article/212356-overview. September 4, 2018.
83. Centers for Disease Control and Prevention. Lymphocytic choriomeningitis (LCMV). Available at: www.cdc.gov/vhf/lcm/.
84. Centers for Disease Control and Prevention. Lassa fever. Available at: www.cdc.gov/vhf/lassa/.
85. Han Y, Wind-Rotolo M, Yang H-C, Siliciano JD, Siliciano RF. 2007. Experimental approaches to the study of HIV-1 latency. *Nat Rev Microbiol.* 5: 95–106.
86. AIDS/HIV. Available at: http://aids.about.com/.
87. AIDS Education Global Information System: Homepage. Available at: www.aegis.com.
88. Global information and education on HIV and AIDS. Available at: https://www.avert.org/public-hub
89. Gillroy SA, Faragon JJ. HIV infection and AIDS. Available at: http://emedicine.medscape.com/article/211316-overview.
90. HIV/AIDS statistics and surveillance. Available at: https://www.cdc.gov/hiv/library/reports/hiv-surveillance.html.
91. AIDS information. National Institutes of Health. Available at: http://aidsinfo.nih.gov/.
92. Updated U.S. Public Health Service guidelines for the management of exposures to HIV and recommendations for post-exposure prophylaxis. Available at: https://stacks.cdc.gov/view/cdc/20711
93. HIV AIDS Treatment. Federal Drug Administration. Available at: https://www.fda.gov/ForPatients/Illness/HIVAIDS/Treatment/ucm118915.htm
94. Cohen MS, Shaw GM, McMichael AJ, Haynes BF. 2011. Acute HIV-1 infection. *N Engl. J. Med.* 364:1943–1954
95. Piot P, Quinn TC. 2013. Response to the AIDS pandemic—a global health model *N. Engl. J. Med.* 368:2210–8.
96. Yabes JM. 2019. Human T-cell lymphotropic viruses. Available at: http://emedicine.medscape.com/article/219285-overview.
97. Ganem D., Prince AM. 2004. Hepatitis B virus infection—natural history and clinical consequences. *N. Engl. J. Med.* 350:1118–1129.
98. Centers for Disease Control and Prevention: Viral hepatitis. Available at: www.cdc.gov/hepatitis/.
99. Hepatitis B Foundation: Hepatitis B Statistics. www.hepb.org/hepb/statistics.htm.
100. National Institute of Allergy and Infectious Diseases. Hepatitis. Available at: https://www.niaid.nih.gov/diseases-conditions/hepatitis
101. Head MW, Ironside JW. 2012. The contribution of different prion protein types and host polymorphisms to clinicopathological variations in Creutzfeldt–Jakob disease. *Rev Med Virol.* 22: 214–229.
102. Orrú CD, Groveman BR, Hughson AG, et al. 2015 Jan-Feb. Rapid and sensitive RT-QuIC detection of human Creutzfeldt-Jakob disease using cerebrospinal fluid. *mBio.* 6(1):e02451–14.

50

Emerging Viruses

Meghan A. May and Ryan F. Relich

CONTENTS

Introduction

Emerging viruses are classified as those viruses that are new to human hosts, rising significantly in incidence, or have established endemicity in a new geographic location. In contrast, re-emerging viruses are those that were largely controlled and whose incidence is rising again. Emerging viruses that are of primary concern to human health, as of this writing, belong to a relatively small number of viral taxa, and almost exclusively feature RNA genomes. This chapter aims to introduce general principles that drive viral emergence, identify several prominent emerging viruses (Table 50.1) and note their clinical presentations (Table 50.2), recommended treatment and control (Table 50.3), diagnostic approaches and recommendations, competent transmission vectors (Table 50.4), synonyms and diagnostic codes (Table 50.5), and geographic distributions (Figures 50.1–50.13). This chapter will only briefly discuss re-emerging diseases (Table 50.6) and will defer discussions of novel pandemic influenza virus strains and human immunodeficiency virus (HIV) to more specialized resources due to the depth and broad scope of these topics. We have also declined to include additional viruses that are at times included in the category of emerging viruses due to either their current global ubiquity and disease burden following re-emergence (e.g., dengue virus) or the ambiguity surrounding their causal role in disease outbreaks (e.g., Bas-Congo virus, Bornavirus) [1, 2].

Principles and Drivers of Virus Emergence

New, emerging, and re-emerging viruses are a perpetual threat to human health. Since the dawn of recorded history, there have been records of sudden and unexpected plagues, some of which recede just as suddenly and some of which continue to circulate quietly, with sudden outbreaks recurring every so often [3, 4]. The overwhelming majority of emerging viruses are zoonotic, meaning they are existing viruses of animals that are introduced into the human population or develop the capacity to infect humans [5]. The circumstances of emergence for each virus are somewhat unique, but there are several features that are common to the phenomenon.

Disease ecology, a collective consideration of the intersections between host, pathogen, and environment, has shown to be a very useful framework around which to consider emerging pathogens. Emergence or re-emergence of viruses often feature changes to the environment of humans, viruses and their reservoirs and/or vectors, or some combination thereof [6]. Notable examples of such changes include the expansion of building and construction, shifting ecologies due to climate change, population density, wildlife and livestock densities, and sanitation. These factors often intersect and amplify each other. For example, rapid development of a country on the edges of historical climate zones would lead to extensive construction, which can create sources of standing water, lead to deforestation, or disrupt watersheds or habitats in ways that displace wildlife that reservoir and/or transmit human-pathogenic viruses. The subsequent increase in the human and livestock densities, perhaps in the absence of strict regulatory controls, can lead to increased travel and transit to other parts of the world, which, in turn, can promote the introduction of foreign arthropod vectors from neighboring climate zone. These actions can lead to increases in populations of arthropod vectors, allow the introduction of vectors previously not present to the area, bring humans into closer contact with animals, or a combination thereof. Surveillance for, and prediction of, disease emergence is best performed by integrating wildlife carriage, vector distribution, wildlife/livestock interactions, and human infection incidence.

Many viruses are biologically primed to evolve more rapidly than their hosts, and viruses that emerge in new hosts often undergo a period of rapid diversification as they adapt to their new habitat. Such viruses often encode their own polymerases with poor or absent proofreading capacity, resulting in high rates of nucleic acid substitutions. While these error-prone polymerases often result in replication inefficiencies because lethal mutations can occur, their overall contribution to fitness for many

DOI: 10.1201/9781003099277-52

TABLE 50.1

Notable Emerging Viruses

Virus	Abbrev.	Genus	Family	Reservoir Species[a]
Alkhurma (aka Alkhumra) virus	ALKV	*Flavivirus*	*Flaviviridae*	Ruminants
Andes virus	ANDV	*Orthohantavirus*	*Hantaviridae*	Rodents
Banna virus	BFV	*Coltivirus*	*Reoviridae*	Livestock
Barmah Forest virus	BAV	*Alphavirus*	*Togaviridae*	Marsupials
Bayou virus	BAYV	*Orthohantavirus*	*Hantaviridae*	Rodents
Black Creek Canal virus	BCCV	*Orthohantavirus*	*Hantaviridae*	Rodents
Bourbon virus	BRBV	*Thogotovirus*	*Orthomyxoviridae*	Cervids; procyons
Bundibugyo virus	BDBV	*Ebolavirus*	*Filoviridae*	Bats (?)
Chikungunya virus	CHIKV	*Alphavirus*	*Togaviridae*	NHPs; bats
Colorado tick fever virus	CTFV	*Coltivirus*	*Reoviridae*	Small mammals
Crimean-Congo hemorrhagic fever virus	CCHFV	*Orthonairovirus*	*Nairoviridae*	Livestock; artiodactyls
Dobrava virus[b]	DOBV	*Orthohantavirus*	*Hantaviridae*	Rodents
Eastern equine encephalitis virus	EEEV	*Alphavirus*	*Togaviridae*	Birds
Ebola virus	EBOV	*Filovirus*	*Filoviridae*	Bats (?)
Eyach virus	EYAV	*Coltivirus*	*Reoviridae*	Rodents
Hantaan virus	HTNV	*Orthohantavirus*	*Hantaviridae*	Rodents
Heartland virus	HRTV	*Phlebovirus*	*Phenuiviridae*	Cervids; procyons
Hendra virus	HEV	*Henipavirus*	*Paramyxoviridae*	Bats
Jamestown Canyon virus	JCV	*Orthobunyavirus*	*Peribunyaviridae*	Cervids
Japanese encephalitis virus	JEV	*Flavivirus*	*Flaviviridae*	Birds
Junin virus[b]	JUNV	*Mammarenavirus*	*Arenaviridae*	Rodents
Kyasanur Forest disease virus	KFDV	*Flavivirus*	*Flaviviridae*	NHPs
La Crosse encephalitis virus	LACV	*Orthobunyavirus*	*Peribunyaviridae*	Sciurids
Lassa virus	LASV	*Mammarenavirus*	*Arenaviridae*	Rodents
Lujo virus	LUJV	*Mammarenavirus*	*Arenaviridae*	Unconfirmed
Machupo virus	MACV	*Mammarenavirus*	*Arenaviridae*	Rodents
Marburg virus	MARV	*Marburgvirus*	*Filoviridae*	Bats
Mayaro virus	MAYV	*Alphavirus*	*Togaviridae*	NHPs
Middle East respiratory syndrome coronavirus	MERS-CoV	*Betacoronavirus*	*Coronaviridae*	Bats
Monkeypox virus	MPV	*Orthopoxvirus*	*Poxviridae*	Mammals
Monongahela virus	MGLV	*Orthohantavirus*	*Hantaviridae*	Rodents
Murray Valley encephalitis virus	MVEV	*Flavivirus*	*Flaviviridae*	Birds
Nipah virus	NIV	*Henipavirus*	*Paramyxoviridae*	Bats
O'nyong'nyong virus	ONNV	*Alphavirus*	*Togaviridae*	Unknown
Orf virus	ORFV	*Parapoxvirus*	*Poxviridae*	Ruminants
Oropouche virus	OROV	*Orthobunyavirus*	*Peribunyaviridae*	NHPs; birds; rodents
Powassan virus	POWV	*Flavivirus*	*Flaviviridae*	Birds
Puumala virus	PUUV	*Orthohantavirus*	*Hantaviridae*	Rodents
Rift Valley fever virus	RVFV	*Phlebovirus*	*Phenuiviridae*	Mammals
Ross River encephalitis virus	RRV	*Alphavirus*	*Togaviridae*	Mammals; birds
Saaremaa virus	SAAV	*Orthohantavirus*	*Hantaviridae*	Rodents
Saint Louis encephalitis virus	SLEV	*Flavivirus*	*Flaviviridae*	Birds
Seoul virus	SEOV	*Orthohantavirus*	*Hantaviridae*	Rodents
Severe acute respiratory syndrome coronavirus	SARS-CoV	*Betacoronavirus*	*Coronaviridae*	Bats
Sever acute respiratory syndrome coronavirus 2	SARS-CoV-2	Betacoronavirus	Coronaviridae	Bats
Severe fever with thrombocytopenia syndrome virus.[b]	SFTSV	*Banyangvirus*	*Phenuiviridae*	Livestock; rodents
Sin Nombre virus[b]	SNV	*Orthohantavirus*	*Hantaviridae*	Rodents

TABLE 50.1 *(Continued)*

Notable Emerging Viruses

Virus	Abbrev.	Genus	Family	Reservoir Species[a]
Sudan virus	SUDV	*Ebolavirus*	*Filoviridae*	Bats
Tai Forest virus	TAFV	*Ebolavirus*	*Filoviridae*	Bats
Tick-borne encephalitis virus	TBEV	*Flavivirus*	*Flaviviridae*	Rodents
Usutu virus	USUV	*Flavivirus*	*Flaviviridae*	Birds
Venezuelan equine encephalitis virus	VEEV	*Alphavirus*	*Togaviridae*	Rodents
West Nile virus	WNV	*Flavivirus*	*Flaviviridae*	Birds
Western equine encephalitis virus	WEEV	*Alphavirus*	*Togaviridae*	Birds
Yellow fever virus[c]	YFV	*Flavivirus*	*Flaviviridae*	NHPs
Zika virus	ZIKV	*Flavivirus*	*Flaviviridae*	NHPs

[a] Reservoir species listed are the primary source of infection; "NHP" = nonhuman primate.

[b] Taxonomic revisions are as follows: New York virus has been reclassified as a variant of Sin Nombre virus (SNV).

[c] YFV is endemic in several parts of Africa, but is considered an emerging/re-emerging disease in parts of the Western Hemisphere.

TABLE 50.2

Clinical Presentations Associated with Emerging Viruses[a]

Virus	Rash	Fever	Hemorrhage	Pain	Encephalitis	Resp. Syndrome	GI/Kidney	Mortality[b]
ALKV	MBFM	YES	TTP, VHF	------	YES	-------	H, HFRS	HIGH
ANDV	-------	YES	-------	------	-------	HPS; HCPS	-------	HIGH
BAV	-------	YES	-------	-------	YES	-------	N, V	HIGH
BAYV	-------	YES	-------	-------	-------	HPS	-------	HIGH
BCCV	-------	YES	-------	------	-------	HPS	-------	HIGH
BDBV	-------	YES	VHF	------	-------	-------	V,D	HIGH
BFV	P, VES	YES	------	YES	------	------	N	LOW
BRBV	-------	YES	TTP	------	-------	------	-------	HIGH
CHIKV	MP	YES	-------	YES	YES	-------	N	LOW
CCHFV	PET	YES	VHF	YES	-------	-------	V,J,D	HIGH
CTFV	-------	YES	-------	YES	-------	-------	V	LOW
DOBV	-------	YES	HFRS	------	-------	-------	HFRS	MODERATE
EBOV	-------	YES	VHF	------	-------	-------	V,D,J,E	HIGH
EEEV	-------	YES	-------	------	YES	-------	N,V,D	HIGH
EYAV	-------	YES	-------	------	YES	-------		???
HTNV	-------	YES	HFRS	------	-------	-------	HFRS	HIGH
HRTV	-------	YES	TTP	YES	-------	-------	E,D	HIGH
HEV	-------	YES	-------	------	YES	C	-------	HIGH
JCV	-------	YES	-------	------	YES	-------	-------	LOW
JEV	-------	YES	-------	------	YES	-------	V	HIGH[c]
JUNV	PET	YES	VHF	------	-------	-------	N,V	HIGH
KFDV	-------	YES	VHF	YES	-------	-------	V	MODERATE
LACV	-------	YES	-------	------	YES	-------	N,V	LOW
LASV	-------	YES	VHF	YES	-------	RDS	V	HIGH
LUJV	MBFM	YES	TTP	YES	-------	RDS	D,E	HIGH
MACV	-------	YES	VHF	------	YES	-------	V,D	HIGH
MARV	MP	YES	VHF	------	-------	-------	V,D,J,E	HIGH
MAYV	MP	YES	-------	YES	-------	-------	-------	LOW

(Continued)

TABLE 50.2 *(Continued)*

Clinical Presentations Associated with Emerging Viruses[a]

Virus	Rash	Fever	Hemorrhage	Pain	Encephalitis	Resp. Syndrome	GI/Kidney	Mortality[b]
MERS-CoV	-------	YES	-------	------	-------	C, RDS	-------	HIGH
MGLV	-------	YES	-------	-------	-------	HPS	-------	HIGH
MPV	VES	YES	------	------	------	------	------	MODERATE
MVEV	-------	YES	-------	------	YES	-------	N,V	HIGH[c]
NiV	-------	YES	-------	-------	YES	C, RDS	-------	HIGH
ORFV	VES	NO	-------	------	-------	------	------	LOW
ONNV	MP	YES	-------	YES; ITCH	-------	-------	-------	LOW
OROV	PET	YES	PH (rare)	YES	YES	-------	V	LOW
POWV	-------	YES	-------	------	YES	-------	V	MODERATE
PUUV	-------	YES	HFRS	------	-------	-------	HFRS	LOW
RVFV	-------	YES	VHF (rare)	------	YES	-------	J,V	HIGH[c]
RRV	MP	YES	-------	YES	YES (rare)	-------	-------	LOW
SAAV	-------	YES	HFRS	------	-------	-------	HFRS	LOW
SFTSV	-------	YES	TTP	------	-------	-------	V,D,E	HIGH
SLEV	-------	YES	-------	------	YES	-------	N	MODERATE
SEOV	-------	YES	HFRS	------	-------	-------	HFRS	LOW
SARS-CoV	-------	YES	-------	------	-------	C, RDS	D	MODERATE
SARS-CoV-2	-------	YES	-------	------	-------	C, RDS	D	MODERATE
SNV	-------	YES	-------	------	-------	HPS	-------	HIGH
SUDV	-------	YES	VHF	------	-------	-------	V,D	HIGH
TAFV	-------	YES	VHF	------	-------	-------	V,D,J,E	HIGH
TBEV	-------	YES	-------	------	YES	-------	N	HIGH[c]
USUV	MP	YES	-------	------	YES	-------	-------	LOW
VEEV	------	YES	-------	------	YES	-------	N,V	LOW HIGH[d]
WNV	MP	YES	-------	YES	YES	-------	N,V	HIGH[c]
WEEV	-------	YES	-------	------	YES	-------	N,V	HIGH[c]
YFV	-------	YES	VHF	------	YES	-------	V,J,E	HIGH
ZIKV	MP	YES	-------	ITCH	YES	-------	-------	LOW HIGH[d]

[a] *Abbreviations*: C – cough; D – diarrhea; E – elevated liver enzymes; H – hematuria; HCPS – Hantavirus cardiopulmonary syndrome; HFRS – hemorrhagic fever with renal syndrome; HPS – hantavirus pulmonary syndrome; J – jaundice; MBFM – morbilliform; MP – maculopapular; N – nausea; P – popular; PET – petechial; RDS – respiratory distress syndrome; TTP – thrombocytopenia; V – vomiting; VES – vesicular; VHF – viral hemorrhagic fever

[b] Mortality classifications: LOW = 0–1%; MODERATE = 2–10%; HIGH = >10%.

[c] Mortality is high in patients presenting with encephalitis, but low in patients presenting subclinically or with mild febrile illness.

[d] Mortality is low in pregnant women but high in developing fetuses, leading to stillbirth or spontaneous abortion.

TABLE 50.3

Emerging Viruses, Disease Names, and Clinical Interventions

Virus	Common Disease Name	Treatment	Vaccine Available?
ALKV	Alkhurma hemorrhagic fever	Supportive care	NO
ANDV	Hantavirus pulmonary syndrome	Supportive care	NO
BAV	Banna virus encephalitis	Supportive care	NO
BAYV	Hantavirus pulmonary syndrome	Supportive care	NO
BCCV	Hantavirus pulmonary syndrome	Supportive care	NO
BDBV	Ebola virus disease	Supportive care; mABs	NO[a]
BFD	Barmah Forest disease	Supportive care	NO
BRBV	Bourbon virus disease	Supportive care	NO
CHIKV	Chikungunya	Supportive care	NO
CCHFV	Crimean-Congo hemorrhagic fever	Supportive care, ribavirin?	NO

TABLE 50.3 *(Continued)*

Emerging Viruses, Disease Names, and Clinical Interventions

Virus	Common Disease Name	Treatment	Vaccine Available?
CTFV	Colorado tick fever	Supportive care	NO
DOBV	Hemorrhagic fever with renal syndrome	Supportive care, ribavirin?	NO
EBOV	Ebola virus disease	Supportive care; mABs	YES (VSV-EBOV)
EEEV	Eastern equine encephalomyelitis	Supportive care	NO
EYAC	Eyach virus encephalitis	Supportive care	NO
HTNV	Hemorrhagic fever with renal syndrome	Supportive care, ribavirin?	NO
HRTV	Heartland virus disease	Supportive care	NO
HEV	Hendra virus disease	Supportive care	NO
JCV	Jamestown Canyon virus disease	Supportive care	NO
JEV	Japanese encephalitis	Supportive care	YES (IXIARO)
JUNV	Argentinian hemorrhagic fever	Ribavirin	YES (Candid#1)
KFDV	Kyasanur Forest disease	Supportive care	YES ("KFD vaccine")
LACV	La Crosse encephalitis	Supportive care	NO
LASV	Lassa fever	Ribavirin	NO
LUJV	Lujo hemorrhagic fever	Supportive care, ribavirin?	NO
MACV	Bolivian hemorrhagic fever	Ribavirin	NO[b]
MARV	Marburg hemorrhagic fever	Supportive care	NO
MAYV	Mayaro virus disease	Supportive care	NO
MERScoV	Middle East respiratory syndrome	Supportive care	NO
MGLV	Hantavirus pulmonary syndrome	Supportive care	NO
MPV	Monkeypox	Supportive care	NO
MVEV	Murray Valley encephalitis	Supportive care	NO
NIV	Nipah virus disease	Supportive care	NO
ONNV	O'Nyong Nyong	Supportive care	NO
ORFV	Contagious pustular dermatitis	Supportive care	NO
OROV	Oropouche fever	Supportive care	NO
POWV	Powasson encephalitis	Supportive care	NO
PUUV	Hemorrhagic fever with renal syndrome	Supportive care, ribavirin?	NO
RVFV	Rift Valley fever	Supportive care	NO[c]
RRV	Ross River fever	Supportive care	NO
SAAV	Hemorrhagic fever with renal syndrome	Supportive care	NO
SFTSV	Severe fever with thrombocytopenia syndrome	Supportive care	NO
SLEV	Saint Louis encephalitis	Supportive care	NO
SEOV	Hemorrhagic fever with renal syndrome	Supportive care	NO
SARS-CoV	Severe acute respiratory syndrome	Supportive care	NO
SARS-CoV-2	coronavirus disease 2019 (COVID-19)	Supportive care; some antiviral drugs	YES
SNV	Hantavirus pulmonary syndrome	Supportive care	NO
SUDV	Ebola virus disease	Supportive care; mABs	NO[a]
TAFV	Ebola virus disease	Supportive care; mABs	NO[a]
TBEV	Tick-borne encephalitis	Supportive care	NO
USUV	Usutu fever	Supportive care	NO
VEEV	Venezuelan equine encephalomyelitis	Supportive care	NO
WNV	West Nile fever; West Nile encephalitis	Supportive care	NO
WEEV	Western equine encephalomyelitis	Supportive care	NO
YFV	Yellow fever	Supportive care	YES (17D; YF-VAX)
ZIKV	Zika virus disease; Zika congenital syndrome	Supportive care*	NO

[a] The VSV-EBOV vaccine may provide cross-protection.

[b] The Candid#1 JUNV vaccine may provide cross-protection.

[c] An efficacious veterinary vaccine is available against RVFV, which has not be utilized in humans.

TABLE 50.4

Major[a] Competent Vectors for Emerging Viruses

Arthropod Vector	Type[b]	Viruses Transmitted	Geographic Distribution
Aedes aegypti	M	CHIKV, ZIKV, MAYV, YFV	Tropics worldwide
Aedes albopictus	M	CHIKV, ZIKV, WNV, USUV, YFV	Rethink
Aedes africanus	M	CHIKV, YFV, WNV, RVFV	Africa
Aedes camptorhynchus	M	RRV, BFV	Australia
Aedes furcifer-taylori	M	CHIKV	Africa
Aedes luteocephalus	M	CHIKV, YFV	Africa
Aedes mcintoshi	M	RVFV	Africa
Aedes notoscriptus	M	RRV	Australia, USA (West Coast)
Aedes normanensis	M	BFV	Australia
Aedes ochraceus	M	RVFV	Africa
Aedes triseriatus	M	LACV	USA, Canada
Aedes vigilax	M	RRV, BFV	Australia
Amblyomma americanum	T	BRBV, HRTV	USA (East, Midwest)
Anopheles funestus	M	ONNV	Africa
Anopheles gambiae	M	ONNV	Africa
Anopheles punctipennis	M	JCV	North America
Coquillettidia perturbans	M	JCV, EEEV, WNV, BFV	Worldwide
Coquillettidia venezuelensis	M	OROV	South America, Central America
Culex annulirostris	M	RRV, MVEV	Australia
Culex annulus	M	BAV	Southeast Asia
Culex pipiens	M	WNV, LACV, RVFV, USUV	Worldwide
Culex quinquefasciatus	M	OROV, WNV, SLEV, WEEV	Tropics/subtropics worldwide
Culex tarsalis	M	WNV, WEEV,	North America
Culex tritaeniorhynchus	M	JEV, BAV	Eastern Hemisphere
Culicoides paraensis	Mi	OROV	Western Hemisphere
Culiseta melanura	M	EEEV, WEEV	North America
Dermacentor andersoni	T	CTFV	Western North America
Haemagogus janthinomys	M	MAYV, YFV	South America
Haemaphysalis spinigera	T	KFDV	India, Sri Lanka, Vietnam
Haemaphysalis longicornis	T	POWV	Central/East Asia, Australia
Hyalomma anatolicum	T	CCHFV	Central Asia, Southern Europe
Hyalomma asiaticum	T	CCHFV	Middle East
Hyalomma marginatum	T	CCHFV	Southern Europe, North Africa
Hyalomma rufipes	T	CCHFV, ALKV	Southern Europe, Africa
Ixodes cookie	T	POWV	USA, Eastern Canada
Ixodes ovatus	T	TBEV	Southeast Asia
Ixodes persulcatus	T	TBEV	Europe, Asia
Ixodes ricinus	T	TBEV, EYAV	Europe
Ixodes scapularis	T	POWV	USA (East), Canada (Southeast)
Ochlerotatus abserratus	M	JCV	USA
Ochlerotatus canadensis	M	JCV, EEEV	Eastern North America
Ochlerotatus cantator	M	JCV	Canada (East), USA (Northeast)
Ornithodoros savignyi	T	ALKV	Africa, Asia
Sabethes chloropterus	M	YFV, SLEV	South America, Central America

[a] Vectors showing transmission under experimental settings are not included.

[b] *Vector abbreviations:* M = mosquito; Mi = mite; T = tick.

TABLE 50.5

Diagnostic Codes and Synonyms

Virus	ICD9	ICD10	Synonyms
ALKV	063.8	A98.8	Alkhurma disease; Fakeeh; Kadam
ANDV	078.89	B33.4	Hantavirus respiratory distress syndrome; Hantavirus cardiopulmonary syndrome
BAYV	078.89	B33.4	Hantavirus respiratory distress syndrome
BAV	063.9	A83.8	Balaton; Kadipiro; Liao Ning
BCCV	078.89	B33.4	Hantavirus respiratory distress syndrome
BFV	066.8	A92.8	Gan Gan; Kokobera; Trubanaman
BRBV	078.89	A85.2	None
BDBV	065.8	A98.4	Ebola Bundibugyo; Ebola Uganda
CHIKV	066.2	A92.1	Buggy Creek; Getah; Kidenga pepo; Knuckle fever; Me Tri
CCHFV	065.0	A98.0	Congo Fever; Dugbe; Erve; Ganjam; Kemerovo; Orungo; Xinjiang hemorrhagic fever
DOBV	078.6	A98.5	Dobrava/Belgrade hemorrhagic fever; Bosnian hemorrhagic fever; Sochi virus
CTFV	066.1	A93.2	American mountain fever; Mountain fever; Mountain tick fever
EEEV	062.2	A83.2	Caaingua; Madariaga virus
EYAV	063.9	A83.8	Balaton; Kadipiro; Liao Ning
HTNV	078.6	A98.5	Korean hemorrhagic fever; Churilov disease
HRTV	078.89	A93.8	None
HEV	078.89	B33.8	Bat paramyxovirus; Equine morbillivirus; Menangle
JCV	062.5	A83.5	California encephalitis
JEV	062.0	A83.0	Alfuy; Nam Dinh; Russian Autumnal encephalitis; Summer encephalitis
JUNV	078.7	A96.0	Junin; Junin hemorrhagic fever
KFDV	065.2	A98.2	Monkey fever; Nanjianyin
LACV	062.5	A83.5	California encephalitis
LASV	078.89	A96.2	Lassa-Fieber; Luna virus
LUJV	078.89	A96.2	None
MACV	078.7	A96.1	Bolivian hemorrhagic fever; Chapare; Machupo hemorrhagic fever
MARV	078.89	A98.3	Durba syndrome; Green monkey disease
MAYV	066.3	A92.8	Una
MERScoV	079.82	U04.9	Novel CoV 2012; Human betacoronavirus 2c England-Qatar; HCoV-EMC
MGLV	078.89	B33.4	Hantavirus respiratory distress syndrome
MPV	057.8	B04	None
MVEV	062.4	A83.4	Australian encephalitis; Australian X disease; Kunjin
NIV	078.89	B33.8	Hendra-Like virus
ONNV	066.3	A92.8	Igbo Ora
ORFV	078.89	B08.0	Contagious ecthyma; Ecthyma contagiosum; Ovine pustular dermatitis; Scabby mouth
OROV	066.3	A93.0	Iquitos Virus
POWV	063.8	A84.8	Deer Tick Virus
PUUV	078.6	A98.5	Scandinavian hemorrhagic fever; Nephropathia Epidemica
RVFV	066.3	A92.4	Arumowot; Enzootic hepatitis, Ntepes; Rift Valley Koorts; Zinga
RRV	066.3	B33.1	Edge Hill disease; Stratford disease
SAAV	078.6	A98.5	Nephropathia Epidemica
SFTSV	065.9	A99	Guertu virus; Huaiyangshan virus
SLEV	062.3	A83.3	None
SEOV	078.6	A98.5	Korean hemorrhagic fever; Churilov disease
SARScoV	079.82	U04.9	SARG; SRA; SRAS
SNV	078.89	B33.4	Hantavirus respiratory distress syndrome; New York Orthohantavirus
SUDV	065.8	A98.4	Ebola Sudan

(Continued)

TABLE 50.5 *(Continued)*

Diagnostic Codes and Synonyms

Virus	ICD9	ICD10	Synonyms
TFAV	065.8	A98.4	Ebola Tai Forest
TBEV	063.2	A84.1	Diphasic meningoencephalitis; Diphasic milk fever; European tick-borne encephalitis
USUV	073.5	A83.8	SAAr 1776
VEEV	066.2	A92.2	Everglades Virus; Rio Negro Virus; Tonante virus
WNV	066.4	A92.3	Near Eastern equine encephalitis; Ntaya virus
WEEV	062.1	A83.1	None
YFV	060	A95	Bulan fever; Magdalena Fever; Pest of Havana; Stranger's fever
EBOV	065.8	A98.4	Ebola Zaire
ZIKV	078.89	A92.8	None

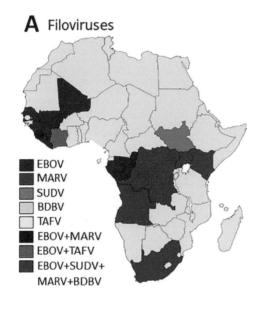

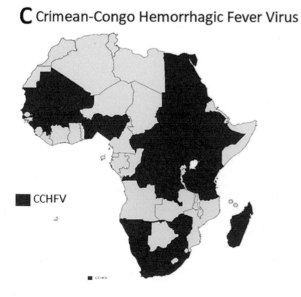

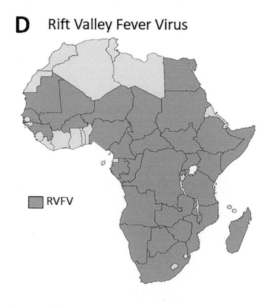

FIGURE 50.1 Emerging hemorrhagic fever viruses in Africa. Filoviruses (Panel A) are colored as follows: red (EBOV), blue (MARV), green (SUDV), purple (EBOV and MARV), orange (EBOV and TAFV), and dark gray (SUDV, EBOV, MARV, and BDBV). Mammarenaviruses (Panel B) are colored as follows: light green (LASV) and pink (LUJV). Crimean-Congo hemorrhagic fever virus distribution (Panel C) is shown in dark green, and Rift Valley fever virus distribution (Panel D) is shown in light blue.

FIGURE 50.2 Emerging hemorrhagic fever viruses in the Americas. Viral distributions are shown by color coding as follows: red (JUNV), blue (MACV), yellow (YFV), green (MACV and YFV), and orange (JUNV and YFV).

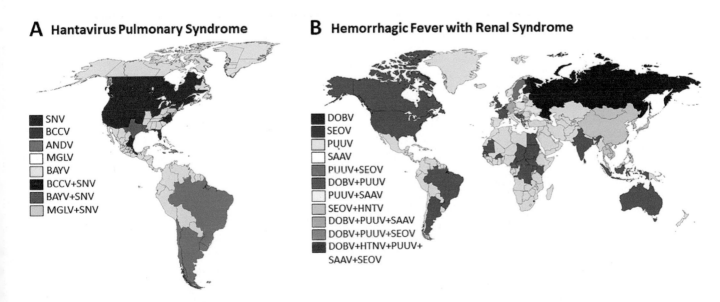

FIGURE 50.3 Global distribution of hantaviruses. Viruses causing hantavirus pulmonary syndrome (Panel A) are colored as follows: red (SNV), green (ANDV), orange (BAYV and SNV), purple (BCCV and SNV), and pink (MGLV and SNV). Viruses causing hemorrhagic fever with renal syndrome (Panel B) are colored as follows: blue (SEOV), red (DOBV), yellow (PUUV), white (SAAV), orange (DOBV and PUUV), pink (DOV and SAAV), light yellow (PUUV and SIAAV), green (PUUV and SEOV), light blue (PUUV and SAAV), light orange (DOBV, PUUV, and SAAV), light gray (DOBV, PUUV, and SEOV), and dark gray (DOBV, HTNV, PUUV, SAAV, and SEOV).

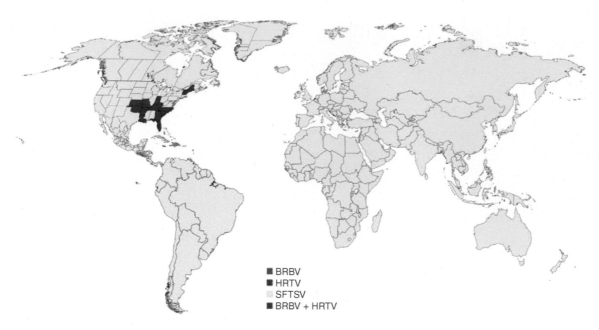

FIGURE 50.4 Emerging viral causes of febrile thrombocytopenia. Viral distributions are shown by color coding as follows: red (HRTV), yellow (SFTSV), and purple (HRTV and BRBV).

FIGURE 50.5 Crimean-Congo hemorrhagic fever virus activity and vector distribution. The distribution of CCHFV and *Hyalomma* tick vectors are shown in red, and areas with *Hyalomma* ticks where autochthonous transmission of CCHFV has not been documented are shown in green. Green areas should be considered potential sites of future CCHFV activity.

TABLE 50.6

Notable Re-Emerging Viruses

Virus	Abbreviation	Disease	Drivers of Re-Emergence
Rubeola virus	MV	Measles	Vaccine coverage; international travel
Rubula virus	MuV	Mumps	Vaccine escape; vaccine coverage
Yellow fever virus	YFV	Yellow fever	Primate ecology; vector control; vaccine coverage
Rubella virus	RuV	Rubella/CRS (?)	Vaccine coverage; international travel
Dengue virus	DENV	DF/DHF	Population growth; urbanization; vector control

viruses is clear. Additional diversity can be rapidly achieved by genome reassortment for segmented viruses, many of which fall into the category of emerging viruses (Table 50.7). Both point mutations and genome segment reassortment can produce viral variants with novel abilities to interact with different tissues, new hosts, or new vectors, allow viruses to escape immune responses raised against parallel strains, and, arguably, to escape clinical interventions that curb transmission such as vaccines or diagnostic tests [2, 5, 7].

Spectrum of Disease Presentation, Diagnostic Approaches, Treatment, and Prevention Strategies

Many emerging and re-emerging human-pathogenic viruses cause acute febrile disease processes that exhibit overlapping clinical signs and symptoms. In general, arthropod-borne viruses (arboviruses) cause disseminated diseases that are characterized either by a nonspecific prodrome accompanied by rash, arthralgia, myalgia, malaise, and/or gastrointestinal symptoms. Meningitis and/or encephalitis are not infrequent complications of infection with some arboviruses, including members of the genera *Alphavirus*, *Orthobunyavirus*, and *Phlebovirus*. Infection with the same or other arboviruses can cause severe organ disease (*e.g.*, hepatitis) and systemic infections characterized by coagulopathy that can lead to bleeding diatheses; this syndrome is referred to as viral hemorrhagic fever. Arboviruses associated with systemic diseases accompanied by coagulopathy or hemorrhagic fever include members of the genera *Banyangvirus*, *Flavivirus*, *Orthonairovirus*, *Phlebovirus*, and *Thogotovirus*. Specific viruses and their associated disease presentations are given in Tables 50.1 and 50.2 and arthropod vectors are listed in Table 50.4.

Many emerging and re-emerging nonarthropod-borne zoonotic viruses cause diseases similar to those caused by arboviruses.

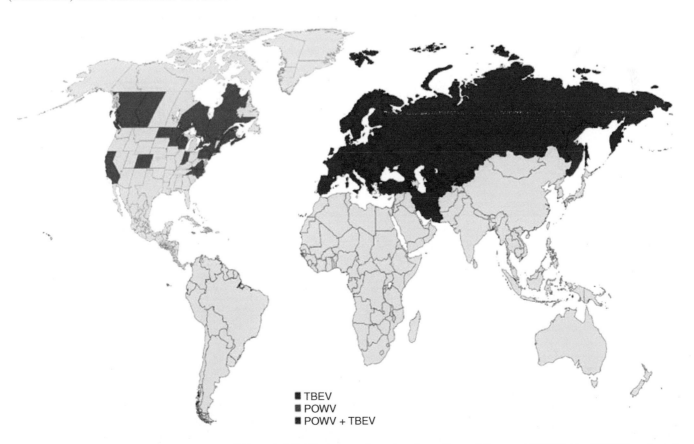

FIGURE 50.6 Emerging tick-borne viral encephalitides. Viral distributions are shown by color coding as follows: red (TBEV), blue (POWV), and purple (TBEV and POWV).

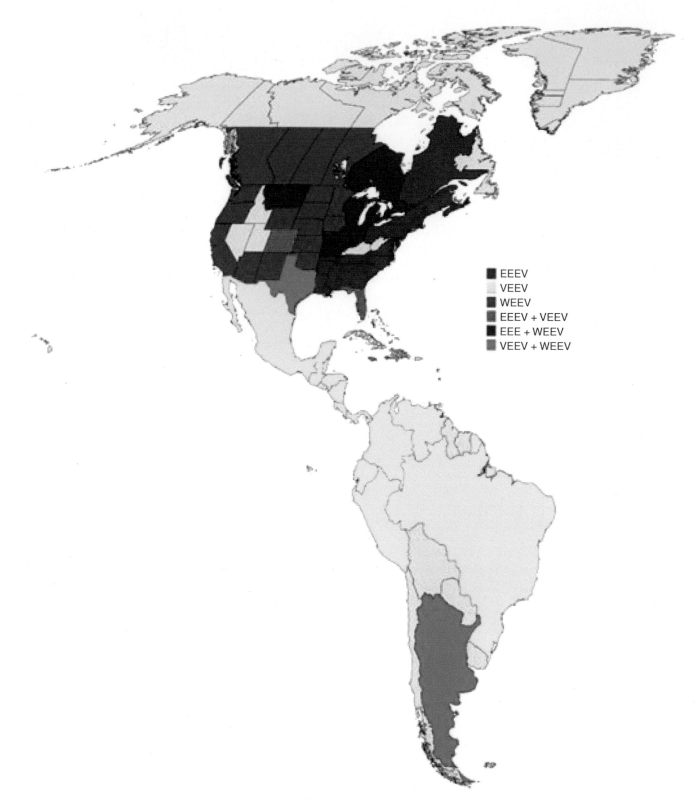

FIGURE 50.7 Equine encephalitis viruses. Viral distributions are shown by color coding as follows: red (EEEV), blue (WEEV), yellow (VEEV), green (VEEV and WEEV), orange (EEEV and VEEV), and purple (EEEV and WEEV).

In addition, some of these viruses cause severe respiratory disease that, depending upon the virus, may be followed by invasive and often fatal central nervous system disease. Rodent-borne viruses, primarily members of the genera *Mammarenavirus* and *Orthohantavirus*, are frequently implicated in outbreaks of hemorrhagic fever, hemorrhagic fever with renal syndrome, or cardiopulmonary disease. Viruses reservoired by bats that are either directly transmitted from bats to humans or are indirectly transmitted from bats to humans through one or more other animal hosts are among the deadliest viruses known.

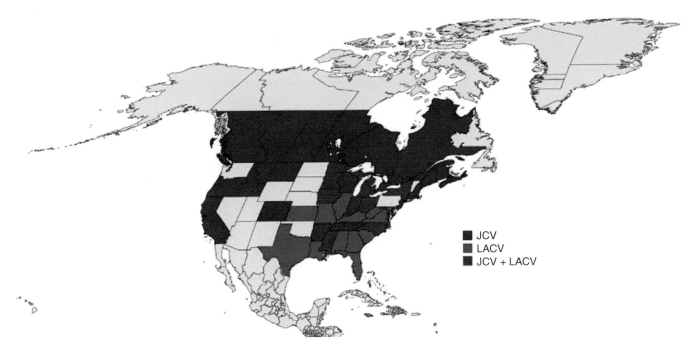

FIGURE 50.8 Distribution of California serogroup viruses. Viral distributions are shown by color coding as follows: red (JCV), blue (LACV), and purple (JCV and LACV).

Four members of the genus *Ebolavirus*, family *Filoviridae* (e.g., Bundibugyo virus, Ebola virus, Sudan virus, and Tai Forest virus), which are believed to be carried by Old World fruit bats, cause severe human systemic infections that can progress to hemorrhagic fever and often death. These viruses are also highly contagious, so large outbreaks associated with single-index cases are not uncommon [8]. Infection caused by the related filovirus Marburg virus (genus *Marburgvirus*) shares disease symptomatology, ecology, epidemiology, and transmission dynamics with ebolaviruses, and case fatality rates can also be quite high (up to ~90% in some outbreaks) [9].

Two species comprising the genus *Henipavirus* (Hendra virus and Nipah virus) are the causative agents of severe respiratory and central nervous system diseases (encephalitis); Nipah virus is frequently associated with outbreaks in Bangladesh and India [10]. Case fatality rates with Nipah and Hendra viruses range from 40–90% [11]. These two viruses are carried by large fruit bats belonging to the genus *Pteropus*, which are commonly

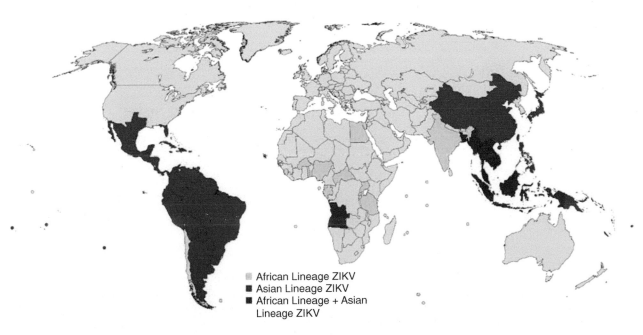

FIGURE 50.9 Zika virus distribution by lineage. The distributions of African lineage, Asian lineage, and both African and Asian lineage ZIKV strains are shown in pink, orange, and red, respectively.

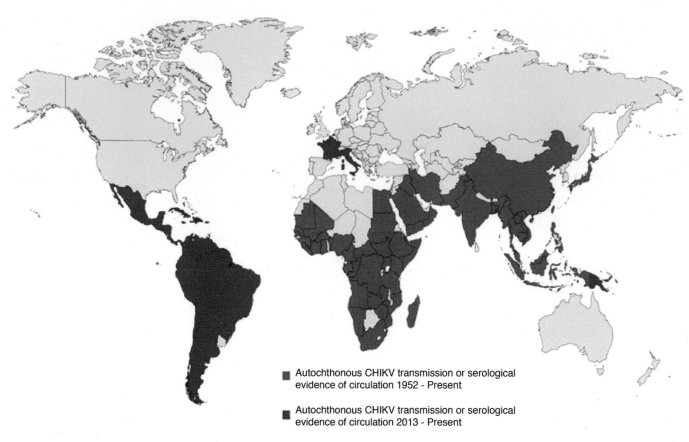

FIGURE 50.10 Global chikungunya distribution. Areas with autochthonous transmission of CHIKV as early as 1952 are shown in blue, whereas areas with local transmission beginning in 2013 are shown in orange. All shaded areas continue to have CHIKV activity at this time.

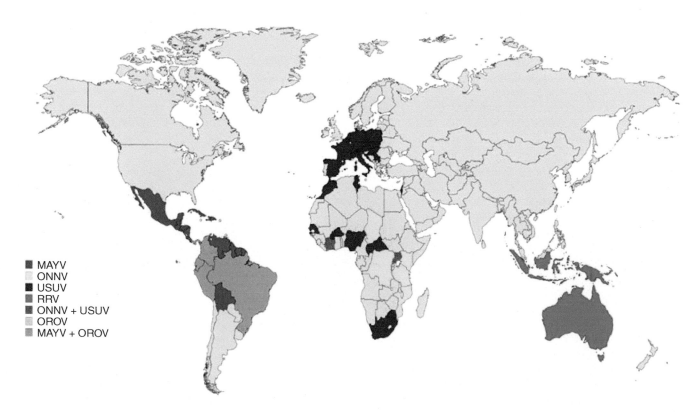

FIGURE 50.11 Emerging viruses causing febrile arthralgia. Viral distributions are shown by color coding as follows: red (USUV), blue (MAYV), yellow (ONNV), green (RRV), orange (ONNV and USUV), pink (OROV), and purple (MAYV and OROV).

FIGURE 50.12 Distribution of emerging henipaviruses causing disease in humans. Areas with NiV activity are shown in red, and areas with HeV activity are shown in blue.

called flying foxes [12]. Outbreaks of both viruses usually involve transmission of the virus from bats to horses to humans (Hendra virus) or bats to pigs to humans or bats to humans through bat-urine/excrement-contaminated foodstuffs (Nipah virus); however, some Nipah virus outbreaks have undetermined sources [12, 13]. The human-pathogenic filoviruses and henipaviruses are geographically spread across Africa, Asia, and Australia; to date, these viruses have not been found to naturally occur in the Western Hemisphere. However, cases of filoviral disease have been exported to the United States and Europe.

The betacoronaviruses associated with severe acute respiratory syndrome (SARS; SARS-CoV) and Middle East respiratory syndrome (MERS; MERS-CoV) have emerged in the 21st century and have proven to be formidable public health threats. SARS-CoV appeared in 2003 in China and resulted in a pandemic that caused more than 8,000 human cases of lower respiratory tract infection and 774 deaths, globally [14]. This pandemic is believed to have originated from one or more spillover events

from infected palm civets to humans in Guangdong Province, China during 2002; however, the exact source has not been identified [15]. The discovery of related viruses from Chinese bats (e.g., Chinese horseshoe bats, genus *Rhinolophus*) suggests a bat reservoir of SARS-CoV [16]. MERS-CoV, which was first recognized in the human population in 2012, has since caused over 2,400 cases and nearly 860 deaths [17]. All cases have been traced back to exposures that took place on the Arabian Peninsula; however, cases have been exported from this region to Europe, Asia, and North America. Most severe and fatal cases have occurred in individuals with underlying health conditions, including structural lung disease, chronic heart disease, and diabetes [18]. Like with SARS-CoV, investigations of bats indicate a likely chiropteran source for MERS-CoV [19]. Dromedary camels appear to serve as the main source of human infections; although, human-to-human transmission via inhalation of infectious aerosols acquired at close proximity to patients is documented [20]. In 2019 a novel coronavirus emerged in Wuhan,

FIGURE 50.13 Distribution of emerging coronaviruses causing respiratory distress syndrome in humans. Areas with MERS-CoV activity are shown in blue, and areas with presumed local reservoirs of SARS-CoV are shown in orange. At this time, SARS-CoV-2 has been detected worldwide and is therefore not depicted.

China (SARS-CoV-2) and has spread around the world. As of this writing (November, 2020), it has affected more than sixty million people worldwide, and caused more than 1.5 million deaths [21]. Cases outside of China were originally among travelers from Wuhan.

Because the early clinical signs and symptoms of emerging and re-emerging viral diseases are often vague and overlap with each other and with other infectious diseases (e.g., malaria, leptospirosis, etc.), their diagnosis requires a high index of clinical suspicion. In areas outside of endemic regions, clinicians should obtain a thorough travel history, animal and vector exposure history, and sick-contact exposure history from patients to assess their risk of infection with specific emerging agents. Care providers must also be aware of ongoing or recent outbreaks of specific pathogens and must be cognizant of dangers posed to them by treating patients with diseases caused by high-consequence pathogens. In response to the 2014–2016 West African Ebola virus disease (EVD) epidemic, healthcare facilities throughout the

world implemented questionnaires to gauge a patient's risk for having EVD prior to clinical assessment. In addition, dedicated patient containment rooms or units were constructed throughout many hospitals in the United States and abroad to permit the isolation and care of patients with EVD and other high-consequence pathogens. Barrier precautions, including the use of full-body personal protective equipment, became standard for the care of persons under investigation of EVD or other viral hemorrhagic fevers.

Clinical laboratory testing of patients should include general health screening (e.g., complete blood count, biochemical markers of renal and hepatic function, and markers of hemostasis [D-dimer, prothrombin time/international normalized ration]) and agent-specific testing. Because of the general nature of the prodromal signs and symptoms of most emerging viral diseases, testing for common pathogens (e.g., influenza, malaria, etc.) should also be performed to rule out the involvement of such pathogens. If high-consequence pathogens are suspected, clinical specimens should be handled according to recommendations

TABLE 50.7

Emerging Viruses Featuring Segmented Genomes

Virus	Segments	Names	Encoded Proteins
ANDV	3	S, M, L	S: nucleocapsid; M: glycoproteins; L: polymerase
BAV	12	1 - 12	1: VP1 polymerase; 2–8: VP1–VP8; 9: VP9, VP9′; 10–12: VP10–VP12
BAYV	3	S, M, L	S: nucleocapsid; M: glycoproteins; L: polymerase
BCCV	3	S, M, L	S: nucleocapsid; M: glycoproteins; L: polymerase
BRBV	6	1 - 6	1: PB2 pol; 2: PB1 pol; 3: PA pol; 4: envelope; 5: nucleoprotein; 6: M protein
CCHFV	3	S, M, L	S: nucleocapsid; M: glycoproteins; L: polymerase
CTFV	12	1 - 12	1: VP1 polymerase; 2-8: VP1 - VP8; 9: VP9, VP9′; 10-12: VP10 – VP12
DOBV	3	S, M, L	S: nucleocapsid; M: glycoproteins; L: polymerase
HTNV	3	S, M, L	S: nucleocapsid; M: glycoproteins; L: polymerase
HRTV	3	S, M, L	S: nucleocapsid; M: glycoproteins; L: polymerase
JCV	3	S, M, L	S: nucleocapsid; M: glycoproteins; L: polymerase
JUNV	2	S, L	S: nucleocapsid, glycoproteins; L: polymerase
LACV	3	S, M, L	S: nucleocapsid; M: glycoproteins; L: polymerase
LASV	2	S, L	S: nucleocapsid, glycoproteins; L: polymerase
LUJV	2	S, L	S: nucleocapsid, glycoproteins; L: polymerase
MACV	2	S, L	S: nucleocapsid, glycoproteins; L: polymerase
MGLV	3	S, M, L	S: nucleocapsid; M: glycoproteins; L: polymerase
OROV	3	S, M, L	S: nucleocapsid; M: glycoproteins; L: polymerase
PUUV	3	S, M, L	S: nucleocapsid; M: glycoproteins; L: polymerase
RVFV	3	S, M, L	S: nucleocapsid; M: glycoproteins; L: polymerase
SAAV	3	S, M, L	S: nucleocapsid; M: glycoproteins; L: polymerase
SEOV	3	S, M, L	S: nucleocapsid; M: glycoproteins; L: polymerase
SFTSV	3	S, M, L	S: nucleocapsid; M: glycoproteins; L: polymerase
SNV	3	S, M, L	S: nucleocapsid; M: glycoproteins; L: polymerase

published by various professional societies (e.g., American Society for Microbiology) and public health agencies (e.g., CDC). If the healthcare institution supports biocontainment patient care rooms or units, testing should be performed in the containment suite, if space permits.

For the detection of specific agents in human clinical specimens, one or more serological, cultivation-based, molecular, and/or histopathological methods can be used; specific methods for individual viruses are described elsewhere. Serological tests include assays designed to detect virus-specific antibodies (IgM and IgG) and antigens and are usually enzyme immunoassays, such as enzyme-linked immunosorbent assays (ELISA). Because IgM is produced soon after infection with a pathogen, IgM-based assays are among the first to be used to attempt diagnosis. Several IgM-based assays have been developed to assist with the diagnosis of an array of viral pathogens, and many are available from commercial suppliers; however, assays for most high-consequence and esoteric viruses are limited to public health and research laboratories. Testing for pathogen-specific IgG is often diagnostic: a fourfold increase in IgG titer between paired acute and convalescent serum specimens indicates infection [22]. Immunochromatographic tests designed to detect either viral antigens or antiviral antibodies have also been developed to aid in the diagnosis of some emerging or re-emerging viral diseases (e.g., EVD) and offer a rapid answer when patients initially present for treatment; however, the performance characteristics of many of these tests have not been well-established because of the relatively low prevalence of many of the diseases they have been designed to diagnose.

Major shortcomings of serological tests include potential cross-reactivity of assays and the length of time often required to obtain a definitive diagnosis. Because closely related viruses share antigenic similarity, antibodies generated against common antigens can cross-react with those of one or more related viruses. Assays designed to detect pathogen-specific antibodies usually cannot readily distinguish the exact specificity of the antibodies present in clinical specimens (e.g., serum), so arbitration is required to determine their specificity. The most common arbiter is plaque reduction neutralization testing, which is described in-depth elsewhere. The length of time required to obtain a definitive diagnosis using serological tests is also an issue, as serum obtained at both the acute phase and convalescent phase (if the patient survives the disease) must be tested together to detect a fourfold rise in IgG titer. Although serum specimens are readily obtainable at the time of initial patient presentation to a healthcare facility, which generally occurs during the acute phase of the illness, obtaining a serum specimen 10–14 days or more later, during the convalescent phase, is sometimes problematic. In addition, this attribute makes some serological tests impractical for clinical use, as the diagnosis is only available long after acute infection, a therapeutically relevant timeframe, and/or death [22].

Cultivation-based detection of viral pathogens has traditionally been regarded as the gold standard for the diagnosis of infections caused by culturable viruses. Viral culture requires procurement of clinical specimens that are the most likely to contain viable viruses: selection of specimens should be guided by published recommendations for specific agents. Although viral culture is a very sensitive method for diagnosing many viral infections, it is slow, laborious, requires highly skilled laboratory scientists to perform and interpret, and can pose a significant biohazard to laboratory personnel if exposures occur. Because many emerging and re-emerging viruses are classified as either risk group-3 or risk group-4 pathogens, which require biosafety level (BSL)-3 and BSL-4, respectively, containment and work practices, cultivation attempts should not be made unless appropriate containment facilities are available. To that end, cultured isolates are needed to thoroughly characterize novel viruses and variants of known viruses.

Molecular methods, including nucleic acid amplification tests such as the polymerase chain reaction (PCR) and its variations (e.g., real-time PCR), have supplanted most traditional diagnostic testing approaches such as viral culture. PCR and other molecular methods usually require smaller clinical specimen volumes, offer superior analytical sensitivities and specificities, and significantly reduce the result turnaround time, often from days or weeks to hours [23]. In addition, some molecular methods such as next-generation nucleotide sequencing permit rapid characterization of viruses and other pathogens present in clinical specimens. Among other benefits, the information gleaned from nucleotide sequencing permits discovery of new viruses, detection of novel viral variants, and provides insight into the evolution and phylogeny of viruses. Although these methods have become commonplace in diagnostic laboratories throughout the world, most often, molecular diagnostics for emerging viruses are not. Instead, hospitals and other healthcare facilities must rely upon specialized laboratories, including commercial reference laboratories or public health laboratories, to meet testing needs.

The analysis of human tissues by light microscopic and immunohistochemical methods is uncommon for the diagnosis of acute emerging viral infections, except for diseases caused by poxviruses [24]. Instead, histopathological analyses are most commonly performed on tissues collected at autopsy. Microscopic examination of 5-μm-thick, hematoxylin and eosin-stained sections of formalin-fixed, paraffin-embedded tissue is essential for the description of the microscopic effects of emerging viral diseases. Features such as inflammation, alteration of tissue architecture, necrosis, hemorrhage, and others become apparent. Localization of viral antigens and nucleic acids within tissue specimens is possible using virus-specific immunohistochemical and *in situ* hybridization methods, respectively. Observation of positive results permits pathologists to determine the identity of the virus present within the tissue. In cases of death associated with a novel virus, histopathological examination permits characterization of the pathology of the disease process caused by the virus. Electron microscopy is also applicable to viral detection and characterization, as the high magnification afforded by transmission electron microscopy permits visualization of viruses.

Due to a paucity of effective antiviral therapeutics, treatment of emerging viral diseases consists largely of supportive care measures. Supportive treatments for most mild emerging arboviral diseases (e.g., non-neuroinvasive La Crosse virus infections) include antipyretics, analgesics, and antinausea medications. If patients are dehydrated due to diarrhea and/or vomiting, oral hydration, if tolerated, should be emphasized. If oral hydration is not tolerated, intravenous fluid administration is required. If mild infections affect underlying health conditions (e.g., diabetes), additional disease-specific treatments (e.g., insulin) may be needed. For more serious arboviral infections, such as encephalitis and hemorrhagic fever, hospitalization is needed to closely monitor signs and symptoms and to provide intensive care and laboratory-guided assessment of patient status. Additionally, if patients are infected with high-consequence arboviral pathogens (e.g., Crimean-Congo hemorrhagic fever virus), isolation of the sick is paramount to limiting further spread. Similar isolation and supportive treatment measures are used for the care of patients afflicted with most other emerging viral infections. For those viruses that cause respiratory diseases, care may also include mechanical ventilation or extracorporeal membrane oxygenation, if available.

Some antiviral medications such as ribavirin have proven useful for the treatment of arenaviral hemorrhagic fevers, including Lassa fever and Venezuelan hemorrhagic fever [25]. Experimental antiviral drugs such as favipiravir offer promise that better treatments for at least some emerging viral diseases are within reach [26]. However, what is known regarding the antiviral efficacy of this and other experimental drugs is mostly derived from *in vitro* and animal model studies. Data concerning the efficacy of these drugs in treating human cases is severely lacking. Recently, the novel antiviral drug remdesivir (GS-5734) was shown to protect African green monkeys from experimental infection with Nipah virus, showing promise that another effective drug may be added to the antiviral armamentarium used to combat emerging infectious diseases [27]. The EVD drug ZMapp and other biopharmaceuticals offer potential alternatives to conventional antiviral compounds, as these drugs comprise monoclonal antibodies or other biological molecules that effect virus neutralization or inhibit viral infectious cycles [28]. Significant research is still needed to prove the utility of these drugs for combating human infections, but what little data has been generated is promising.

Prevention of emerging viral diseases using vaccines and other prophylactics has already proven useful for abating outbreaks of some diseases, including Japanese encephalitis, yellow fever, Kyasanur Forest disease, and, most recently, EVD [29–32]. The use of genetic engineering has permitted researchers to create recombinant rhabdoviruses (e.g., vesicular stomatitis virus [VSV]) that express immunodominant antigens (e.g., viral peplomers) of viruses that can stimulate production of a humoral antiviral response. These recombinant viruses can be used to vaccinate susceptible hosts. One example is rVSV-ZEBOV, a vaccine licensed by Merck and approved by the United States Food and Drug Administration for the prevention of EVD. This vaccine is highly effective and has been in wide use throughout the 2018–2020 EVD epidemics in the Democratic Republic of the Congo [33]. Additional vaccines using recombinant VSV are expected to be produced in the years to come. Another area of vaccine research is that involving the development of veterinary vaccines against emerging viruses. Because animals serve as sources of many human infectious diseases, the idea of vaccinating animals is to prevent infection in them, thereby eliminating the risk of animal-to-human transmission. Such a vaccine is already available for the prevention of Hendra virus disease in horses [34].

Other prevention strategies include vector and reservoir avoidance and elimination, proper animal husbandry and food handling practices, and education of at-risk individuals about ways to prevent infection and curb the transmission of emerging viral diseases. Vector and reservoir abatement measures include use of chemical pest repellants, use of mosquito netting, elimination of vector and reservoir breeding sites, pesticide use, and avoidance of vector- and reservoir-inhabited areas. During the recent Zika virus epidemic in the Western Hemisphere, genetically altered male mosquitoes were introduced into regions where stable autochthonous transmission of Zika virus occurred in an effort to decrease the *Aedes aegypti* vector population. Data from earlier studies performed in Panama, where sterile male *A. aegypti* were released, showed a massive reduction in numbers, suggesting a possible means to curb the spread of Zika in Brazil [35]. The aim of improving animal husbandry and food handling practices is to prevent zoonotic- and food-borne transmission of viruses from food and companion animals to human. Animal biosecurity measures, including animal health surveillance and screening of sentinel animals, provides valuable information regarding the presence or absence of potentially pathogenic viruses in agriculturally relevant, companion, and wild animals [36]. Education of at-risk populations about the modes of disease transmission (e.g., sexual, airborne, etc.), methods of disease prevention (e.g., safe sex practices, safe animal and food handling practices, vector avoidance measures, etc.), symptom recognition, and treatment is also key to the prevention of infectious disease transmission.

Key to the prevention of emerging viral disease outbreaks is vigilant surveillance for known and novel agents among vector and potential reservoir species. Surveillance is useful for the detection and characterization of known viruses in endemic and nonendemic regions, and it offers insight into predicting where and when viruses might next emerge. It also permits determination of whether currently available diagnostics and prophylactics are able to detect viruses and prevent disease they cause, respectively, and it reveals information about the evolution and adaptability of these viruses. Surveillance also enables detection of novel agents that could pose health risks to human and/or animal populations. Together, prevention strategies are a front-line defense against threats posed by emerging viruses.

Atypical Presentations, Permanent Sequelae, and Complications of Emerging Viruses

While primarily known to cause acute febrile illnesses or hemorrhagic fevers, an increasing number of emerging viruses have been associated with congenital syndromes or perinatal complications. The emergence of ZIKV in the Western Hemisphere, first recognized in 2015, was accompanied by a dramatic spike in the incidence of stillbirth and babies born with microcephaly [37]. Postmortem examination of stillborn fetuses and miscarriage remains showed systemic presence of ZIKV [38], demonstrating its involvement in what is now termed congenital Zika

syndrome. VEEV also causes a congenital syndrome featuring microcephaly, but the rate of occurrence and neuronal targets are poorly understood [39]. A recent report indicated poor neurocognitive development in children whose mothers were infected with CHIKV during pregnancy, suggesting that this rapidly emerging virus also potentially causes a congenital syndrome [40, 41].

Pregnancy is also a significant risk factor for enhanced mortality of both women and fetuses during Lassa fever. The pathophysiological mechanism leading to more severe disease is somewhat unclear, but the increase in blood volume in pregnancy causing enhanced viremia and LASV affinity for placental vasculature have been proposed [42, 43]. The intervention leading to substantially improved mortality in pregnant women is expulsion of the fetus by delivery, spontaneous abortion, or pregnancy termination [44]. The effect of ribavirin on maternal and fetal outcomes during pregnancy is unclear, likely because the potential teratogenic effects of ribavirin make it unsuitable for use during pregnancy [45]. The impact of infection with other arenaviruses (e.g., MACV, LUJV) on mortality during pregnancy is a question of clinical interest but has not yet been explored.

Multiple emerging viruses have been associated with permanent sequelae or ongoing, postinfectious symptoms. Permanent loss of hearing has been reported following recovery from Lassa fever [46], and loss of eyesight can result from Rift Valley fever [47]. Some patients experience chronic and debilitating joint pain following infection with RRV, CHIKV, MAYV, and OROV [48–51]. A significant proportion of patients who survive central nervous system (CNS) infection with NiV, POWV, EEEV, and numerous other viruses are left with severely debilitating neurological and cognitive damage. It is notable that the majority of infections caused by these and related viruses (e.g., SLEV, WEEV, WNV, POWV) result in subclinical or mild illnesses in the majority of exposed patients, but those experiencing CNS infection can have dramatic or fatal outcomes. In addition to decline in cognitive function, a proportion of Nipah virus-associated encephalitis patients also develop severe and permanent psychiatric symptoms [52, 53].

Factors Introducing Epidemiological Complexities

Outbreaks stemming from emerging viruses often evoke emergency response protocols based on prior observations of transmission dynamics, incubation period, and persistence of pathogen shedding. Risk assessment models for the scope and duration of outbreaks or epidemics are also calculated based on current understandings of these factors. As repeated outbreaks occur or animal models are developed for individual emerging viruses, more nuanced understandings can dramatically alter methods of prediction.

An unexpected finding of the West African Ebola virus epidemic of 2014 was the capacity of viable EBOV to persist in tears and semen long after patient recovery [54–57]. A subsequent study of Marburg virus disease using a macaque model demonstrated persistence of MARV in testicular tissue, indicating that the clinical observation reflects a previously unrecognized capacity of multiple filoviruses [58]. It is not yet clear if SUDV, BDBV, or TAFV are similarly capable of crossing the blood-testicle barrier and maintaining a reservoir in infected patients. Relapse of encephalitis has been reported in survivors of Nipah virus disease months to years after acute disease and apparent recovery, suggesting that a reservoir of infectious NiV persists within the CNS [55, 59]. This observation is bolstered by a nonhuman primate model showing viable NiV in the brain of infected grivets surviving experimental exposure [60]. The unexpected persistence of these viruses in surviving patients raises the question of whether follow-up outbreaks can result from convalescent or relapsing patients, as human-to-human transmission has been clearly documented for EBOV, MARV, and NiV [61–63].

Disease risk assessment for ZIKV focused initially on the ranges of the competent transmission vectors *Aedes aegypti* and *Aedes albopictus* during the 2015 Western Hemisphere epidemic. A single report from 2011 had documented sexual transmission of ZIKV [64], necessitating the evaluation of this second modality of viral spread. Sexual transmission of ZIKV was thereafter confirmed to occur in the 2015 epidemic [65], and epidemiological models were then adapted to reflect these additional dynamics [66]. Viable filoviruses in semen historically suggested that sexual transmission is plausible for these viruses as well, and this was definitively confirmed for EBOV during the 2014 epidemic in West Africa [54]. Though sexual transmission has not been described, recent case reports document the detection of YFV, CHIKV, RVFV, LASV, ANDV, WNV, NiV, and SFTSV in the semen of convalescent patients [67–73]. Additionally, sexual transmission of EEEV and JEV following experimental infection or artificial insemination of livestock have been reported for avian and porcine systems [74–75], though neither virus has been reported in human semen. It is notable that the detection of these viruses in human semen has all been described within 3 years of this writing, indicating that persistence of emerging viruses in semen may be more common than previously appreciated. Evaluating the extent of their sexual transmission therefore become a matter of urgency.

Conclusions

The emergence of novel viral pathogens, previously unreported pathologies associated with known viral agents, and the re-emergence of previously well-controlled viruses, is a constant factor in human health. Greater recognition of this phenomenon and highly sensitive molecular diagnostic methods lead to broader and faster recognition of novel viral agents than in the past. This improved capacity is offset by greater density in the human population with its increased interconnectedness and rapid movement across the globe, against the backdrop of ecological flux due to climate change. Understanding the factors that drive disease emergence and pathogenic mechanisms of emerging viruses can help to predict or rapidly recognize these viruses and those that have yet to emerge.

REFERENCES

1. Bhatt, S., et al., *The global distribution and burden of dengue. Nature*, 2013. **496**(7446): p. 504–507.
2. Romero-Tejeda, A. and I. Capua, *Virus-specific factors associated with zoonotic and pandemic potential. Influenza Other Respir Viruses*, 2013. **7** (Suppl 2): p. 4–14.
3. Sartin, J.S., *Contagious horror: infectious themes in fiction and film. Clin Med Res*, 2019. **17**(1–2): p. 41–46.
4. Wolfe, N.D., C.P. Dunavan, and J. Diamond, *Origins of major human infectious diseases. Nature*, 2007. **447**(7142): p. 279–283.
5. Taylor, L.H., S.M. Latham, and M.E. Woolhouse, *Risk factors for human disease emergence. Philos Trans R Soc Lond B Biol Sci*, 2001. **356**(1411): p. 983–989.
6. Parvez, M.K. and S. Parveen, *Evolution and emergence of Pathogenic viruses: past, present, and future. Intervirology*, 2017. **60**(1–2): p. 1–7.
7. Scagnolari, C., et al., *Consolidation of molecular testing in clinical virology. Expert Rev Anti Infect Ther*, 2017. **15**(4): p. 387–400.
8. Feldmann, H. and T.W. Geisbert, *Ebola haemorrhagic fever. Lancet*, 2011. **377**(9768): p. 849–862.
9. Jeffs, B., et al., *The Medecins sans Frontieres intervention in the Marburg hemorrhagic fever epidemic, Uige, Angola, 2005. I. Lessons learned in the hospital. J Infect Dis*, 2007. **196** (Suppl 2): p. S154–161.
10. Ang, B.S.P., T.C.C. Lim, and L. Wang, *Nipah virus infection. J Clin Microbiol*, 2018. 56(6).
11. Safronetz, D., H. Feldmann, and E. de Wit, *Birth and pathogenesis of rogue respiratory viruses. Annu Rev Pathol*, 2015. **10**: p. 449–471.
12. Weatherman, S., H. Feldmann, and E. de Wit, *Transmission of henipaviruses. Curr Opin Virol*, 2018. **28**: p. 7–11.
13. Paul, L., *Nipah virus in Kerala: a deadly zoonosis. Clin Microbiol Infect*, 2018. **24**(10): p. 1113–1114.
14. Centers for Disease Control and Prevention. *Severe Acute Respiratory Syndrome (SARS)*, 2017.
15. World Health Organization. *Severe Acute Respiratory Syndrome (SARS)*. 2004.
16. Cui, J., F. Li, and Z.L. Shi, *Origin and evolution of pathogenic coronaviruses. Nat Rev Microbiol*, 2019. **17**(3): p. 181–192.
17. Organization, W.H. *Middle Eastern Respiratory Syndrome Coronavirus (MERS-CoV)*. 2019.
18. Centers for Disease Control and Prevention. *Middle Eastern Respiratory Syndrome (MERS)*, 2019.
19. Ramadan, N. and H. Shaib, *Middle east respiratory syndrome coronavirus (MERS-CoV): a review. Germs*, 2019. **9**(1): p. 35–42.
20. Rubbenstroth, D., et al., *Human Bornavirus research: back on track! PLOS Pathog*, 2019. **15**(8): p. e1007873.
21. WorldOMeter Info. Coronavirus Cases, 2020.
22. Theel, E.S., *Immunoassays for diagnosis of infectious diseases*, in *Manual of Clinical Microbiology*, Carroll C.K., et al., Editors. 2019, ASM Press: Washington, DC, p. 124–138.
23. Relich, R.F. and A.N. Abbott, *Syndromic and point-of-care molecular testing. Adv Mol Pathol*, 2018. **1**(1): p. 97–113.
24. Goldsmith, C.S. and S.E. Miller, *Modern uses of electron microscopy for detection of viruses. Clin Microbiol Rev*, 2009. **22**(4): p. 552–563.
25. Eberhardt, K.A., et al., *Ribavirin for the treatment of Lassa fever: a systematic review and meta-analysis. Int J Infect Dis*, 2019. **87**: p. 15–20.
26. De Clercq, E., *New nucleoside analogues for the treatment of Hemorrhagic fever virus infections. Chem Asian J*, 2019. **14**(22): p. 3962–3968.

27. Lo, M.K., et al., *Remdesivir (GS-5734) protects African green monkeys from Nipah virus challenge. Sci Transl Med*, 2019. **11**(494).

28. Espeland, E.M., et al., *Safeguarding against Ebola: vaccines and therapeutics to be stockpiled for future outbreaks. PLOS Negl Trop Dis*, 2018. **12**(4): p. e0006275.

29. Chen, L.H. and M.E. Wilson, *Yellow fever control: current epidemiology and vaccination strategies. Trop Dis Travel Med Vaccines*, 2020. **6**: p. 1.

30. de La Vega, M.A. and G.P. Kobinger, *Safety and immunogenicity of Vesicular Stomatitis virus-based vaccines for Ebola virus disease. Lancet Infect Dis*, 2020. **20**(4): p. 388–389.

31. Lee, P.I., et al., *Recommendations for the use of Japanese encephalitis vaccines. Pediatr Neonatol*, 2019.

32. Rajaiah, P., *Kyasanur forest disease in India: innovative options for intervention. Hum Vaccin Immunother*, 2019. **15**(10): p. 2243–2248.

33. Merler, S., et al., *Containing Ebola at the source with ring vaccination. PLOS Negl Trop Dis*, 2016. **10**(11): p. e0005093.

34. Peel, A.J., et al., *The equine Hendra virus vaccine remains a highly effective preventative measure against infection in horses and humans: 'the imperative to develop a human vaccine for the Hendra virus in Australia'. Infect Ecol Epidemiol*, 2016. **6**: p. 31658.

35. Gorman, K., et al., *Short-term suppression of Aedes aegypti using genetic control does not facilitate aedes albopictus. Pest Manag Sci*, 2016. **72**(3): p. 618–628.

36. Muellner, P., et al., *SurF: an innovative framework in biosecurity and animal health surveillance evaluation. Transbound Emerg Dis*, 2018. **65**(6): p. 1545–1552.

37. Schuler-Faccini, L., et al., *Possible association between Zika virus infection and Microcephaly - Brazil, 2015. MMWR Morb Mortal Wkly Rep*, 2016. **65**(3): p. 59–62.

38. Sarno, M., et al., *Zika virus infection and stillbirths: a case of hydrops fetalis, hydranencephaly and fetal demise. PLOS Negl Trop Dis*, 2016. **10**(2): p. e0004517.

39. García-Tamayo, J., *[Teratogenic effect of the Venezuelan equine Encephalitis virus: a review of the problem]. Invest Clin*, 1992. **33**(2): p. 81–86.

40. Del Carpio-Orantes, L. and M.C. González-Clemente, *Microcephaly and arbovirus. Rev Med Inst Mex Seguro Soc*, 2018. **56**(2): p. 186–188.

41. Gérardin, P., et al., *Neurocognitive outcome of children exposed to perinatal mother-to-child Chikungunya virus infection: the CHIMERE cohort study on Reunion Island. PLOS Negl Trop Dis*, 2014. **8**(7): p. e2996.

42. Okogbenin, S., et al., *Retrospective cohort study of Lassa fever in pregnancy, Southern Nigeria. Emerg Infect Dis*, 2019. **25**(8).

43. Walker, D.H., et al., *Pathologic and virologic study of fatal Lassa fever in man. Am J Pathol*, 1982. **107**(3): p. 349–356.

44. Price, M.E., et al., *A prospective study of maternal and fetal outcome in acute Lassa fever infection during pregnancy. BMJ*, 1988. **297**(6648): p. 584–587.

45. Sinclair, S.M., et al., *The ribavirin pregnancy registry: an interim analysis of potential teratogenicity at the mid-point of enrollment. Drug Saf*, 2017. **40**(12): p. 1205–1218.

46. Mateer, E.J., et al., *Lassa fever-induced sensorineural hearing loss: a neglected public health and social burden. PLOS Negl Trop Dis*, 2018. **12**(2): p. e0006187.

47. Al-Hazmi, A., et al., *Ocular complications of rift valley fever outbreak in Saudi Arabia. Ophthalmology*, 2005. **112**(2): p. 313–318.

48. Abdelnabi, R., et al., *Antiviral drug discovery against arthritogenic Alphaviruses: tools and molecular targets. Biochem Pharmacol*, 2019. **174**: p. 113777.

49. Acosta-Ampudia, Y., et al., *Mayaro: an emerging viral threat? Emerg Microbes Infect*, 2018. **7**(1): p. 163.

50. Chang, A.Y., et al., *Frequency of chronic joint pain following Chikungunya virus infection: a Colombian cohort study. Arthritis Rheumatol*, 2018. **70**(4): p. 578–584.

51. Sakkas, H., et al., *Oropouche fever: a review. Viruses*, 2018. **10**(4).

52. Ng, B.Y., et al., *Neuropsychiatric sequelae of Nipah virus encephalitis. J Neuropsychiatry Clin Neurosci*, 2004. **16**(4): p. 500–504.

53. Sejvar, J.J., et al., *Long-term neurological and functional outcome in Nipah virus infection. Ann Neurol*, 2007. **62**(3): p. 235–242.

54. Schindell, B.G., A.L. Webb, and J. Kindrachuk, *Persistence and sexual Transmission of filoviruses. Viruses*, 2018. **10**(12).

55. Sissoko, D., et al., *Persistence and clearance of Ebola virus RNA from seminal fluid of Ebola virus disease survivors: a longitudinal analysis and modelling study. Lancet Glob Health*, 2017. **5**(1): p. e80–e88.

56. Uyeki, T.M., et al., *Ebola virus persistence in semen of male survivors. Clin Infect Dis*, 2016. **62**(12): p. 1552–1555.

57. Varkey, J.B., et al., *Persistence of Ebola virus in ocular fluid during convalescence. N Engl J Med*, 2015. **372**(25): p. 2423–2427.

58. Coffin, K.M., et al., *Persistent Marburg virus infection in the testes of nonhuman primate survivors. Cell Host Microbe*, 2018. **24**(3): p. 405–416.e3.

59. Aditi and M. Shariff, *Nipah virus infection: a review. Epidemiol Infect*, 2019. **147**: p. e95.

60. Liu, J., et al., *Nipah virus persists in the brains of nonhuman primate survivors. JCI Insight*, 2019. **4**(14).

61. Gurley, E.S., et al., *Person-to-person transmission of Nipah virus in a Bangladeshi community. Emerg Infect Dis*, 2007. **13**(7): p. 1031–1037.

62. Luby, S.P., E.S. Gurley, and M.J. Hossain, *Transmission of human infection with Nipah virus. Clin Infect Dis*, 2009. **49**(11): p. 1743–1748.

63. Murray, M.J., *Ebola virus disease: a review of its past and present. Anesth Analg*, 2015. **121**(3): p. 798–809.

64. Foy, B.D., et al., *Probable non-vector-borne transmission of Zika virus, Colorado, USA. Emerg Infect Dis*, 2011. **17**(5): p. 880–882.

65. McCarthy, M., *Zika virus was transmitted by sexual contact in Texas, health officials report. BMJ*, 2016. **352**: p. i720.

66. Allard, A., et al., *Asymmetric percolation drives a double transition in sexual contact networks. Proc Natl Acad Sci U S A*, 2017. **114**(34): p. 8969–8973.

67. Arunkumar, G., et al., *Persistence of Nipah virus RNA in semen of survivor. Clin Infect Dis*, 2019. **69**(2): p. 377–378.

68. Bandeira, A.C., et al., *Prolonged shedding of Chikungunya virus in semen and urine: a new perspective for diagnosis and implications for transmission. IDCases*, 2016. **6**: p. 100–103.

69. Barbosa, C.M., et al., *Yellow fever virus RNA in urine and semen of convalescent patient, Brazil. Emerg Infect Dis*, 2018. **24**(1).

70. Gorchakov, R., et al., *Optimizing PCR detection of West Nile virus from body fluid specimens to delineate natural history in an infected human cohort. Int J Mol Sci*, 2019. **20**(8).

71. Haneche, F., et al., *Rift valley fever in kidney transplant recipient returning from mali with viral RNA detected in semen up to four months from symptom onset, France, autumn 2015. Euro Surveill*, 2016. **21**(18).

72. Koga, S., et al., *Severe fever with thrombocytopenia Syndrome virus RNA in semen, Japan. Emerg Infect Dis*, 2019. **25**(11): p. 2127–2128.

73. McElroy, A.K., et al., *A case of human Lassa virus infection with robust acute T-cell activation and long-term virus-specific T-cell responses. J Infect Dis*, 2017. **215**(12): p. 1862–1872.

74. Guérin, B. and N. Pozzi, *Viruses in boar semen: detection and clinical as well as epidemiological consequences regarding disease transmission by artificial insemination. Theriogenology*, 2005. **63**(2): p. 556–572.

75. Guy, J.S., et al., *Experimental transmission of eastern equine Encephalitis virus and highlands J virus via semen of infected tom Turkeys. Avian Dis*, 1995. **39**(2): p. 337–342.

Part III

Applied Practical Microbiology

51

Mechanisms of Action of Antibacterial Agents

Ammara Mushtaq, Joseph Adrian L. Buensalido, Carmen E. DeMarco, Rimsha Sohail, and Stephen A. Lerner

CONTENTS

DOI: 10.1201/9781003099277-54

Introduction

This chapter will attempt to summarize current information about, and working models for, the mechanisms of action of antimicrobial agents, which are in clinical use or in advanced stages of investigation, or which illustrate important points or are used in laboratory research. We shall provide references to more extensive information.

The antimicrobial agents presented here will be grouped into two major classes according to the target microbes, antibacterials and antimycobacterials, and then, where applicable, by mechanisms of action into five major categories:

- Inhibition of cell wall synthesis
- Interference with membrane integrity
- Inhibition of nucleic acid synthesis
- Inhibition of protein synthesis
- inhibition of synthesis of essential small molecules

A brief discussion of monoclonal antibacterial antitoxins appear at the end of this chapter.

Antibacterial Agents

Inhibition of Cell Wall Synthesis

The bacterial cell wall is a unique structure responsible for multiple functions: maintenance of cell shape, protection against osmotic pressure, or serving as a platform for bacterial appendages, such as flagella, fimbriae, and pili.

The main component of the bacterial cell wall is the peptidoglycan layer. In Gram-positive bacteria, cell walls are much thicker than those of Gram-negative bacteria (Figure 51.1). The Gram-negative cell wall is surrounded by an outer membrane, consisting of a phospholipid bilayer with incorporated protein

molecules. Some of these proteins, the so-called porins, form aqueous channels through which hydrophilic molecules, including most antimicrobial agents, diffuse across the outer membrane to enter the periplasmic space, which lies between the outer membrane and inner (or cytoplasmic) membrane. The peptidoglycan lies in the periplasmic space, but it does not generally pose a physical barrier to penetration by most antimicrobials, since such small molecules easily pass through the cell wall. Interference with cell wall biosynthesis generally has bactericidal consequences, since cell wall lysis generally ensues. In some cases, the interference with cell wall synthesis leads to liberation of lipoteichoic acids, which normally regulate the activity of hydrolytic enzymes that play roles in the dynamic changes of the cell wall in cell growth and division. When such regulation is lifted by the loss of the lipoteichoic acids, these enzymes may actively hydrolyze the cell wall, leading to its disintegration.

Drugs That Inhibit Intracellular Cell Wall Biosynthetic Enzymes

Phosphonomycin (Fosfomycin)

Phosphonomycin (now more commonly known as fosfomycin) acts as an analog of phosphoenolpyruvate and binds covalently to a cysteinyl residue of MurA (the phosphoenolpyruvate, UDP-GlcNAc-3-enolpyruvyl transferase), inhibiting the enzyme and interfering with the formation of UDP-*N*-acetyl muramic acid (UDP-MurNAc), early in the synthesis of the bacterial cell wall (Figure 51.2).[1-4] Fosfomycin is rapidly bactericidal against a wide bacterial spectrum, including MRSA, vancomycin-resistant enterococci, and many Gram-negative bacteria, including *Pseudomonas aeruginosa* and ESBL-producing and other multidrug-resistant *Enterobacterales*. The activity of fosfomycin is dependent on two different transport systems for its uptake into the bacterial cell: the 1-α-glycero-phosphate transport system and the hexose monophosphate system, responsible for uptake of glucose-6-phosphate and other hexose monophosphates.[3]

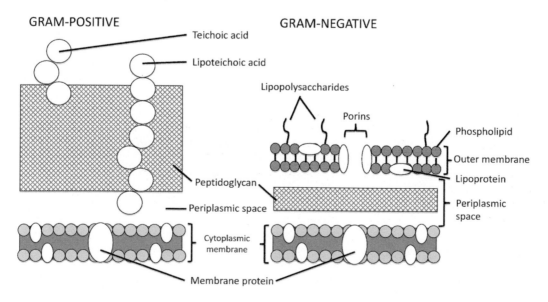

FIGURE 51.1   Cell envelopes of Gram-positive and Gram-negative bacteria.

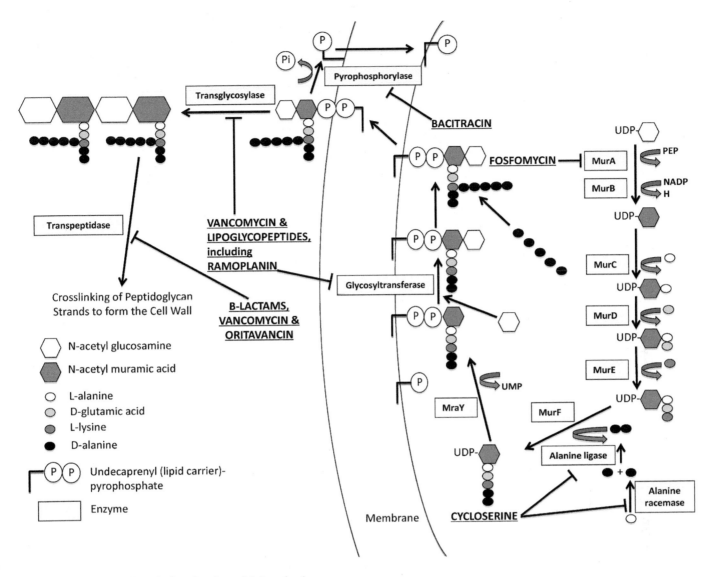

FIGURE 51.2 Cell Wall synthesis and antibacterial sites of action.

Cycloserine

Cycloserine is an inhibitor of cell wall peptidoglycan synthesis and is active against many Gram-positive and Gram-negative bacteria. Its intracellular targets are alanine racemase, which converts L-alanine to D-alanine, and D-alanine-D-alanine ligase, which links two D-alanine molecules to form the dipeptide (Figure 51.2). This dipeptide unit is then added to UDP-MurNAc-tripeptide to form the pentapeptide, which is essential for the eventual cross-linking to form the intact cell wall.[5] The principal use of cycloserine is as an antimycobacterial agent, so it is also discussed later in this chapter.

Drugs That Bind to Carrier Molecules

Bacitracin

Bacitracin is a mixture of at least 22 structurally related dodeca-peptides produced by some strains of *Bacillus licheniformis* and *Bacillus subtilis*. Its basic structure consists of a cyclic peptide of seven amino acids plus a linear peptide side chain with five amino acids. At the N-terminus of the peptide, there is a thiazoline ring

that is formed either by L-cysteine and L-isoleucine or L-cysteine and L-valine, as a result of substitutions of hydrophobic amino acids in the peptide sequence and from oxidative transformation of the thiazoline ring.[6]

More than four decades ago, it was observed that in the presence of divalent cations, bacitracin complexes with the membrane-bound pyrophosphate form of the undecaprenyl (C_{55}-isoprenyl) lipid carrier molecule that remains after the disaccharide-peptide unit (UDP-GlcNAC-[pentapeptide-MurNAC]) is transferred to the nascent peptidoglycan chain. This binding of bacitracin inhibits the enzymatic dephosphorylation of this lipid carrier to its monophosphate form, which is required for another round of transfer of the disaccharide-peptide unit to the growing peptido-glycan chain. Similar complex formation with intermediates of sterol biosynthesis in animal tissues (e.g., farnesyl pyrophosphate and mevalonic acid) may account in part for the toxicity of baci-tracin in animal cells.[7–9]

There has also been discussion of alternative activity of bac-itracin. It was reported to be a PDI inhibitor, but *in vitro* evi-dence for specificity of bacitracin for inhibiting PDI was scarce.[6] In fact, in 2010, Karala and Ruddock presented *in vitro* data showing that bacitracin was not effective in PDI inhibition, failing to introduce or isomerize disulfide bonds in a folding protein, and unable to act effectively as a chaperone. What they did show was that bacitracin inhibited rhodanese aggregation in the non-catalyzed refolding of rhodanese. It inhibited chaperone activity of bacitracin-induced protein, and at the same time inhibited the reductive activity of PDI by competing with substrate binding.[10] But testing by Dickerhof et al. suggested that bacitracin, indeed, interacts with PDI in a robust, covalent manner. Their study reconciled with the work of Karala et al. by explaining that bacitracin targets not the catalytic domain, but the substrate-binding domain, of PDI. The thiazoline ring of bacitracin opens and interacts with free cysteines (Cys314 and 345) on PDI to form disulfide bonds, although it is non-specific to these amino acids.[6] This research reaffirms that bacitracin is a PDI inhibitor, but it is unclear whether or how this activity produces the antibacterial effect.

Drugs That Bind to Cell Wall Precursors

Glycopeptide: Vancomycin

Vancomycin is a glycopeptide antibiotic that interferes with bacterial cell wall synthesis. It is bactericidal and may even kill cell wall-deficient bacteria (L forms), suggesting secondary effects not limited to cell wall synthesis.

Vancomycin is a 1450-dalton tricyclic glycopeptide, in which are linked two chlorinated tyrosines, three substituted phenylglycines, glucose, vancosamine (a unique amino-sugar), N-methyl leucine, and aspartic acid amide. This structure can undergo hydrogen bonding to the D-alanyl-D-alanine terminus of the pentapeptide in various peptidoglycan precursors (e.g., UDP-MurNAc-pentapeptide, MurNAc-pentapeptide pyrophospholipid, and GlcNAc-MurNAc-pentapeptide pyrophospholipid). Binding of vancomycin occurs through a set of hydrogen bonds between the D-Ala-D-Ala dipeptide and the amides that line a cleft formed by the cross-linked heptapeptide of the vancomycin. The bound glycopeptide acts as a steric impediment, preventing the attack of the transglycosylase that links N-acetyl glucosamine with N-acetyl muramic acid-pentapeptide to form the repeating subunits of the peptidoglycan, and the subsequent attack of the transpeptidase that normally completes the peptidoglycan by cross-linking two peptidoglycan strands. The latter cross-linking normally occurs with binding of the transpeptidase to the penultimate D-Ala to the N-terminal amino acid of the species-specific cross-linking peptide (glycine in *Staphylococcus aureus*). Thus, the impairment of cross-linking by the transglycosylase of the disaccharide units to elongate the glycan strand and the impairment of transpeptidation that normally cross-links the strands via the peptides, lead to incomplete cell wall with poor integrity.[11–18]

The bactericidal activity of vancomycin is generally concentration-independent, although it carries out some (minimal) concentration-dependent killing, with a long postantibiotic effect that has been demonstrated *in vitro*.[16,19]

Because of the frequent use of vancomycin, certain drug-resistant Gram-positive bacterial pathogens have emerged, such as vancomycin-resistant *enterococci* (VRE), which continue to assemble peptidoglycan precursors because they are able to synthesize and place D-Lactate in the place of the terminal D-Alanine in the peptide chain of the nascent peptidoglycan. This drastically reduces the binding affinity of vancomycin, making it ineffective. Such resistance to vancomycin has stimulated efforts to examine the structure and activity of vancomycin derivatives in search of more effective drugs.[20,21]

Lipoglycopeptides: Telavancin, Teicoplanin, Ramoplanin, Dalbavancin, and Oritavancin

Telavancin is a semisynthetic derivative of vancomycin, featuring the same glycopeptide core as vancomycin, but with the addition of an extended lipophilic (decylaminoethyl) side chain to the vancosamine sugar and a hydrophilic side chain represented by the negatively charged phosphonomethyl aminomethyl moiety to the resorcinol-like 4'-position of amino acid 7. In contrast to vancomycin, which binds to the cell wall, telavancin binds predominantly to the cell membrane. Telavancin thus has a dual mode of action. First, the lipophilic moiety promotes interaction with the cell membrane and anchors the molecule to the cell surface, providing improved binding affinity of the glycopeptide core for D-alanyl-D-alanine-containing peptidoglycan, inhibiting peptidoglycan biosynthesis. Second, telavancin disrupts the barrier function of the cell membrane, producing a rapid bactericidal effect leading to both concentration- and time-dependent loss of membrane potential and increased membrane permeability, although most reports speak more of its concentration-dependent killing in both intra- and extracellular *S. aureus*.[22–24] This is a result of telavancin binding to lipid II, which places it in a good position to insert its lipophilic side chain into the membrane, thus disrupting the lipid bilayer.[25] Meanwhile, the hydrophilic substituent of the drug increases its distribution in the body and promotes rapid clearance, thus reducing the potential for nephrotoxicity. Global gene expression studies, which showed an increase in expression of the genes *vraS* and *vraR* (principal controllers of the cell wall stress regulon), and upregulation of cell-envelope genes *lytM*, *murZ*, *pbp2*, SAS1987, *murI*, *fmt*, *bacA*, and *llm/tarO*, among others, were supportive of inhibition of peptidoglycan synthesis as one mechanism of action. Furthermore, telavancin also causes upregulation of two important genes, *lrgA* and *lrgB*, which have been shown in the past to be induced by other known membrane depolarizers, daptomycin, and chitosan, supporting membrane depolarization as another mode of action. Lastly, this laboratory method also showed that telavancin administration led to a strong induction of *vraD* and *vraE* genes, which have been associated with detoxification and resistance (as in previous studies with bacitracin and nisin), and are, in turn, responses to membrane depolarization.[26–30]

Teicoplanin is a lipoglycopeptide antibiotic approved for human use outside the United States, and so its use has been mostly reported in European studies.[31–33] The activity profile of teicoplanin is similar to that of vancomycin.[16] Like vancomycin, teicoplanin acts by interacting and forming a complex with the substrate peptidoglycan units that have the pentapeptidyl tails terminating in D-Ala$_4$-D-Ala$_5$. This substrate sequestration effectively shuts down transpeptidation by making the D-Ala-D-Ala acceptor unavailable to the transpeptidases.[34,35] Teicoplanin has a long elimination half-life because of its high serum protein binding, and combined with good renal tubular resorption, this drug is able to reach high and prolonged serum concentrations.[36]

Ramoplanin is a novel lipoglycopeptide with a unique mechanism of action, causing alteration of cell wall peptidoglycan linkages and membrane permeability of Gram-positive bacteria, thereby disrupting the cell wall. Ramoplanin acts by inhibiting the late-stage assembly of the peptidoglycan monomer and its polymerization into mature peptidoglycan, by blocking two sequential enzymatic steps (Figure 51.2).[37] First, it inhibits the glycosyltransferase that links the N-acetyl-glucosamine (GlcNAC) unit with the N-acetyl-muramyl (MurNAC)-pentapeptide to form the recurring disaccharide-pentapeptide unit of the peptidoglycan chain.[28,38,39] Next, ramoplanin inhibits the transglycosylation step in peptidoglycan synthesis, joining the disaccharide peptide subunits into a peptidoglycan chain.[37]

Dalbavancin is a semisynthetic lipoglycopeptide structurally related to vancomycin and teicoplanin, with an unusually prolonged half-life of six to ten days. Dalbavancin interrupts Gram-positive bacterial cell wall biosynthesis by binding, like vancomycin, to the terminal D-alanyl-D-alanine of the pentapeptide-glycosyl cell wall intermediate, thus inhibiting transpeptidation and leading to cell death. A unique property of dalbavancin is the ability to dimerize and anchor itself into bacterial membranes, improving its interaction with, and binding affinity to, D-alanyl-D-alanine residues of nascent peptidoglycan precursors and producing more rapid and potent bactericidal activity against MRSA and coagulase-negative staphylococci, as compared with vancomycin and teicoplanin.[16,28,40,41]

Oritavancin is a novel semisynthetic lipoglycopeptide antibiotic with concentration-dependent bactericidal activity against Gram-positive cocci, including glycopeptide-resistant enterococci and *S. aureus* with intermediate susceptibility to glycopeptides. In comparison with vancomycin, it also displays a more prolonged post-antibiotic effect and elevated serum protein binding. The mechanism of action of oritavancin is similar to that of vancomycin (Figure 51.2), with some important differences that explain its activity against vancomycin-resistant pathogens. The mechanism of action of oritavancin has three steps: (1) inhibits the transglycosylation step of cell wall biosynthesis like vancomycin, (2) inhibits transpeptidation by binding to the peptidic cross-linking part (pentaglycine bridge in *S. aureus*, and Aspartate/Asparagine bridge in *E. faecium*), and (3) has certain cell membrane effects, particularly increased cell membrane anchoring and dimer formation (due to the presence of a hydrophobic side chain 4'-chlorobiphenylmethyl), which lead to concentration-dependent membrane depolarization, increased permeability, membrane disruption, and then cell death. The latter two actions allow oritavancin to remain active against vancomycin-resistant *S. aureus* and enterococci. Furthermore, it has been reported that oritavancin is able to sterilize biofilms of *S. aureus*, whether MSSA, MRSA, or VRSA.[28,42–45]

Drugs That Inhibit Enzymatic Polymerization and Attachment of New Peptidoglycan Units to the Nascent Cell Wall (β-Lactams)

All β-lactam antibiotics contain the highly reactive β-lactam ring that binds to, and inactivates, various penicillin-binding proteins (PBPs), a family of enzymes (transpeptidases and carboxypeptidases) in the cytoplasmic membrane that catalyze various steps in the biosynthesis and reshaping of the cell wall peptidoglycan.

Notably, these target proteins are in the cytoplasmic membrane and accessible on the outside, which means that these antibiotics need not cross the membrane to act on them.[46]

Penicillins and Cephalosporins

Penicillins and cephalosporins interfere with the terminal step in the cross-linking of peptidoglycan chains (i.e., transpeptidation of the N-terminus of the amino acid or peptide bridge onto the penultimate D-alanine of the pentapeptide chain, with release of the terminal D-alanine). It has been suggested that penicillin is a structural analog of a transition state of the D-alanyl-D-alanine terminus of the peptidoglycan strand which is involved in transpeptidation. It thereby competes with the D-Ala-D-Ala bond in binding to the transpeptidase enzyme, compromising the formation of an acyl-enzyme intermediate with the penultimate D-alanine. The highly reactive CO-N bond of the β-lactam ring acylates the transpeptidase irreversibly, thereby inactivating it, so peptide cross-linking of peptidoglycan chains does not occur.[47] Bacteria display in their cytoplasmic membrane a variety of PBPs, which include other penicillin-sensitive enzymes besides the transpeptidases and D-alanine carboxypeptidases. The concept of multiple target sites for the actions of penicillins and cephalosporins is supported by the observations that various β-lactam antibiotics produce differential effects on the processes of cell division, elongation, and lysis. Specific β-lactam antibiotics bind different PBPs preferentially, and the differences in physiological and morphological effects of various β-lactams correlate with the affinity of their binding to different PBPs. The lytic and killing actions of penicillins and other β-lactams may not result directly from their inhibition of peptidoglycan synthesis, but may also involve the action of bacterial murein hydrolases. It has been observed that lipoteichoic acids, which inhibit these autolytic enzymes or autolysins, are released from some bacteria upon treatment with penicillin *in vitro*. It has thus been postulated that the release of these endogenous inhibitors of murein hydrolase enzymes interferes with the regulation of these enzymes in the cell, so the unregulated hydrolases destroy the integrity of the cell wall and contribute to cell lysis and death. Modifications of the side chains of some penicillins and cephalosporins have permitted improved penetration through the porins in cell envelope structures, resulting in an enhanced antibacterial spectrum.[48–52]

The β-lactam ring, which is the key to the success of this class of antibiotics, is hydrolyzed by β-lactamase enzymes that confer resistance to β-lactams and have evolved from penicillin-binding proteins by mutation and selective pressure. Fortunately, continued research has come up with ways to combat this resistance mechanism, such as combining β-lactamase inhibitors with a penicillin.[46] These inhibitors (such as clavulanate, and the penicillin sulfones, sulbactam, and tazobactam) inactivate serine-dependent β-lactamases, and when combined with a penicillin, the resulting drug combination has extended therapeutic effectiveness by inactivating many serine ß-lactamases and thereby protecting the associated penicillin from inactivation.[53]

The resistance conferred by ß-lactamases can also be overcome by modifying the side chain of penicillin, which may reduce the binding of the resulting penicillin to the ß-lactamase. Furthermore, cephalosporins and carbapenems (other classes of ß-lactams) may also bind poorly to existing ß-lactamases, and thereby retain activity against some ß-lactamase-containing bacteria.

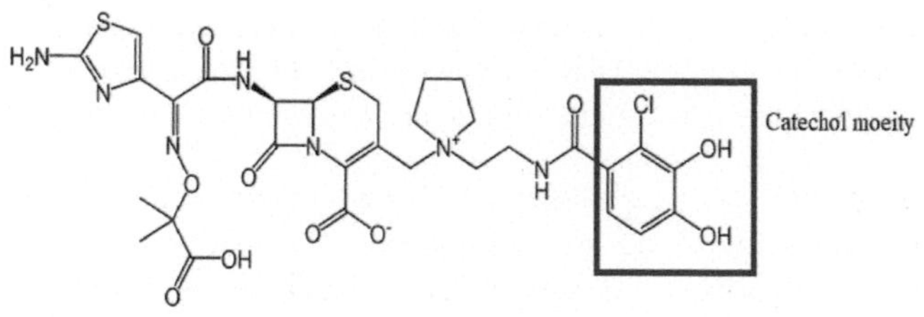

FIGURE 51.3   Chemical structure of cefiderocol.

New cephalosporins have been made by pairing the N-(α-oxyimino)acyl side chain of third-generation cephalosporins with structurally complex heterocycles at the C-3 position. Ceftaroline and ceftolozane have very good activity against methicillin-resistant *S. aureus* and against *P. aeruginosa*, respectively. The ability of ceftaroline to access the open cleft that leads to the PBP active site, particularly PBP 2a and PBP 2x in MRSA and *Streptococcus pneumoniae*, respectively, explains its improved activity against these organisms.[53–55]

Cefiderocol is a novel catechol-substituted siderophore cephalosporin that has potent activity against Gram-negative bacilli including multidrug-resistant Pseudomonas, multidrug-resistant *Acinetobacter*, and carbapenem-resistant *Enterobacterales* (Figure 51.3).[56] The presence of a halogenated catechol group at the C-3 position of cefiderocol enhances the *in vitro* potency against IMP-1, KPC-2, and OXA-producing strains of *P. aeruginosa*. Cefiderocol utilizes the bacterial iron transport systems and penetrates the outer membrane of the bacteria along with ferric iron.[56] It forms a chelating complex with free iron which is actively transported into the periplasm via iron transporters.[57] The iron transport systems as well as small molecules called siderophores are upregulated to sequester iron in an iron-depleted environment. It inhibits bacterial cell wall synthesis by binding mainly to PBP-3 of Gram-negative bacteria. There was increased intracellular accumulation of cefiderocol in *P. aeruginosa* cells incubated under iron-depleted conditions compared with those incubated under iron-sufficient conditions.[57] This transport of extracellular ferric iron was not enhanced when one of the hydroxyl groups of the catechol moiety of cefiderocol was replaced with a methoxy group.[57] It is unclear whether the molecule is stable by itself or only when bound to iron. The molecule has high stability against a range of ESBLs and carbapenemases.[57]

Ceftobiprole is a parenteral cephalosporin that is available in its prodrug form, ceftobiprole medocaril.[58] Like other β-lactams, it inhibits transpeptidation by binding to penicillin-binding proteins. It is an extended-spectrum antibiotic that is effective against both Gram-positive and Gram-negative bacteria, including MRSA, β-hemolytic streptococci, methicillin-resistant coagulase-negative staphylococci, *S. pneumoniae*, and *P. aeruginosa*.[58] Its activity against *P. aeruginosa* is an important distinguishing feature from ceftaroline. It is also active against AmpC-producing strains of *P. aeruginosa*, as ceftobiprole is a poor inducer and is hydrolyzed slowly by AmpC. Its activity for *P. aeruginosa* is attributed to its enhanced binding to PBP3. It has remarkably strong affinity of binding to PBP 2A, which is responsible for imparting resistance to β-lactam antibiotics in MRSA and methicillin-resistant CoNS.[58] In addition, ceftobiprole can also saturate PBPs 1, 3, and 4 in low concentrations,

a property not shared by other β-lactams. It also has potent activity against penicillin- and ceftriaxone-resistant strains of *S. pneumoniae*.[58] It has activity against *E. faecalis*, but not against *E. faecium*. It has limited activity against *Acinetobacter* sp. and *Stenotrophomonas maltophilia*. Ceftobiprole is now approved for treatment in Europe of adults with hospital-acquired and community-acquired pneumonia.

Carbapenems: Imipenem, Meropenem, Ertapenem, Doripenem, Faropenem, Tebipenem, and Tomopenem

Carbapenems are β-lactams with the widest antibacterial spectrum, and the most potent agents versus Gram-positive and Gram-negative bacteria, although they are not active against MRSA. They have a 4:5 fused ring lactam, a double bond between C-2 and C-3, a sulfur (instead of a carbon at C-1), and a hydroxyethyl side chain, which is the key to their broad activity[59] and provides the mechanism of resistance to most β-lactamases.[60]

Carbapenems diffuse more readily through the outer membrane proteins (porins) of most Gram-negative bacteria, in part because of their zwitterionic characteristics. Furthermore, in *P. aeruginosa*, carbapenems enter more rapidly through pores formed by a specific porin, outer membrane protein D2, whose mutational loss or alteration produces carbapenem resistance by reducing entry into the cell. Once inside the periplasmic space, these drugs acylate multiple PBPs, inhibiting peptide crosslinking and peptidase action. This leads to peptidoglycan weakening and disruption of formation until the cell bursts from osmotic pressure.[59] Carbapenems are also generally less readily inactivated by bacterial β-lactamases than other β-lactams.[51,61,62]

The first carbapenem was thienamycin, which was unstable in aqueous solution, leading to the development of the more stable imipenem, which possesses a high affinity for PBPs.[63]

Imipenem inhibits bacterial cell wall synthesis by binding to and inactivating specific PBPs. In *Escherichia coli* imipenem inhibits the transpeptidase activities of PBPs-1A, -1B, and -2, and the D-alanine carboxypeptidase activities of PBP-4 and PBP-5. It also inhibits the transglycosylase activity of PBP-1A, and it is a weak inhibitor of the transpeptidase activity of PBP-3, in contrast to all other β-lactams and other carbapenems that preferentially bind to PBP-1 and –3. Imipenem induces sphere formation with subsequent cell rupture, and it does not inhibit septum formation by the cells, reducing the amount of lipopolysaccharide liberated during bacteriolysis. For human use, imipenem is administered in combination with cilastatin, an inhibitor of dehydropeptidase (DHP) I, an enzyme produced by the renal epithelial cells which hydrolyzes imipenem.[62,64] The inhibition of the renal dehydropeptidase preserves the activity of imipenem in urine, thereby permitting its use in treating urinary-tract infections.

Meropenem contains a C-methyl substituent at the 1-β position, and thus is not susceptible to hydrolysis by the renal DHP-I. Its bactericidal action is similar to that of imipenem, although various bacterial mechanisms of resistance may affect susceptibility to these two carbapenems differentially.[50,59]

Ertapenem binds most strongly to PBP-2 of *E. coli*, followed by PBP-3, and it also has good affinity for PBP-1A and -1B, achieving rapid bactericidal action without the prior filamentation that occurs with third-generation cephalosporins that bind primarily to PBP-3.[65,66] Although it has broad activity against *Enterobacterales*, it does not have activity against *P. aeruginosa*.

Doripenem, a 1b-methyl-carbapenem, is similar to meropenem in its stability to many β-lactamases and its resistance to inactivation by renal dehydropeptidases.[67]

Faropenem is an orally administered penem antibiotic which displays both bactericidal and β-lactamase inhibitor effects and is less susceptible to the renal dipeptidase enzyme DHP-1 as compared with imipenem. It has preferential affinity for PBP-2 but also affinity for PBP-1A, -1B, and -3. Faropenem exposure at concentrations below the minimum inhibitory concentration (MIC) for *S. aureus* affected septum formation with a decrease in the number of viable cells at increasing antibiotic concentrations; exposure to concentrations equivalent to the MIC and above produces cell lysis. A similar effect is observed in *E. coli*, with changes in cell shape at concentrations below the MIC and cell lysis at or above the MIC.[66,68]

Faropenem was approved for use in 2009 in Japan, where it has been used for dermatologic conditions, such as acne vulgaris (alone orally, or in combination with topical agents),[69,70] and even in clarithromycin-resistant cutaneous *Mycobacterium chelonae* infections.[71]

Tebipenem is another oral carbapenem that is being developed in Japan, but it lacks activity versus *P. aeruginosa* and MRSA.[59]

Tomopenem is another broader spectrum intravenous carbapenem with activity against both *P. aeruginosa* and MRSA. The presence of a basic group at position 6 is responsible for the activity against the latter.[67,72]

As with cephalosporins, new carbapenems continue to be developed by incorporating heterocycles at the C-2 position, resulting in even broader Gram-negative activity, such as razupenem and ME-1036.[53] Due its increased adverse effects, the development of razupenem in North America was terminated.[215]

Penicillins combined with β-Lactamase Inhibitors: Amoxicillin-clavulanate, Ampicillin-sulbactam, and Piperacillin-tazobactam

Amoxicillin and Clavulanic Acid. Amoxicillin is an orally administered penicillin that inhibits cell wall synthesis and has bactericidal activity. It works by binding to penicillin binding protein and thus inhibiting the transpeptidation reaction, which ultimately impairs cell wall synthesis and causes lysis of the cell wall. The major cause of bacterial resistance to amoxicillin is the presence of β-lactamase enzymes that hydrolyze the β-lactam structure of the antibiotic rendering it ineffective. An effective way to combat resistance conferred by β-lactamases is to use agents such as β-lactams, which bind the active site of the β-lactamase and inactivate it.[73] This could be achieved by either developing substrates that with high affinity to bind the enzyme (reversibly and/or irreversibly)form unfavorable steric interactions or by creating irreversible "suicide inhibitors."[73] Clavulanic acid belongs to the second category.[73] It functions by deactivating the β-lactamase enzyme permanently through secondary chemical reactions in the active site of the enzyme.[73] The reaction includes the development of a noncovalent complex between the enzyme and the inhibitor, which then leads to the formation of the acyl-enzyme intermediate by the nucleophilic attack by a serine residue of the enzyme.[74] This then develops into a catalytically inactive inhibitor-enzyme complex.[74] One problem that might arise by the use of these suicide inhibitors is that they themselves have β-lactam structure and are, therefore, liable to hydrolysis by

β-lactamase enzyme, which could then lead to the regeneration of the enzyme.[74] Therefore, greater formation of permanently inactivated enzyme-inhibitor complex is required for the greater efficiency of suicide inhibitors.[74] Amoxicillin-clavulanate is a widely used parenteral β-lactam-β-lactamase-inhibitor combination. Clavulanic acid is the β-lactamase inhibitor and protects amoxicillin from hydrolysis by β-lactamase enzymes and, therefore, amplifies its antibacterial spectrum.

Ampicillin and Sulbactam. Ampicillin is a penicillin with an antibacterial spectrum of activity similar to that of amoxicillin, but it is generally administered intravenously. Ampicillin and sulbactam is the β-lactam-β-lactamase-inhibitor combination. Sulbactam, like clavulanic acid is a suicide inhibitor. Addition of sulbactam to ampicillin enhances antibacterial spectrum of ampicillin by protecting it from hydrolysis by certain β-lactamases. The spectrum of activity of ampicillin-sulbactam is similar to that of the orally administered amoxicillin-clavulanate.

Piperacillin and Tazobactam. Piperacillin-tazobactam is another β-lactam-β-lactamase-inhibitor combination. Piperacillin is a cell wall synthesis inhibitor and bactericidal in nature while tazobactam protects its partner from hydrolysis by certain β-lactamase enzymes. Piperacillin has a broader spectrum of activity against Gram-negative bacteria, including *P. aeruginosa*. Since sulbactam and tazobactam inhibit β-lactams in many gut anaerobes, these β-lactamase inactivators endow their respective combinations with penicillins with activity also against many gut anaerobes.

Newer β-Lactams combined with β-Lactamase Inhibitors: Ceftazidime-avibactam, Meropenem-vaborbactam, Ceftolozane-tazobactam, Imipenem-relebactam, and Cefepime-zidebactam

Resistance to the older inhibitors (clavulanic acid, sulbactam, and tazobactam) is now quite prevalent for several reasons. First, they only target the class A serine β-lactamases. Secondly, there are variants of class A β-lactamases that have acquired single amino acid substitutions, resulting in older inhibitors failing to inactivate these enzymes. Lastly, newer class A β-lactamases like KPC-2 were identified that can hydrolyze older β-lactamases. This led to the need to develop new β-lactam-β-lactamase inhibitor combinations discussed later.

Ceftazidime is a third-generation cephalosporin that is active against *P. aeruginosa* and other Gram-negative bacilli. The most common mechanism of resistance to β-lactams in Gram-negative bacteria is by production of β-lactamases that hydrolyze the β-lactam ring. Avibactam is a novel non-β-lactam β-lactamase inhibitor that is active against almost all members of class A and class C β-lactamase enzymes and some class D β-lactamase enzymes (Figure 51.4).[75] Inhibition of class C β-lactamase enzymes is the most important attribute of avibactam. Unlike other β-lactamase inhibitors, avibactam does not contain a β-lactam core in its structure. Furthermore, the mechanism of action through which avibactam resists hydrolysis and thus protects its associated ceftazidime differs from other β-lactamase inhibitors that are themselves β-lactams.[75]

The covalent inhibition of the catalytic serine occurs in a similar fashion via the opening of the avibactam ring, but the

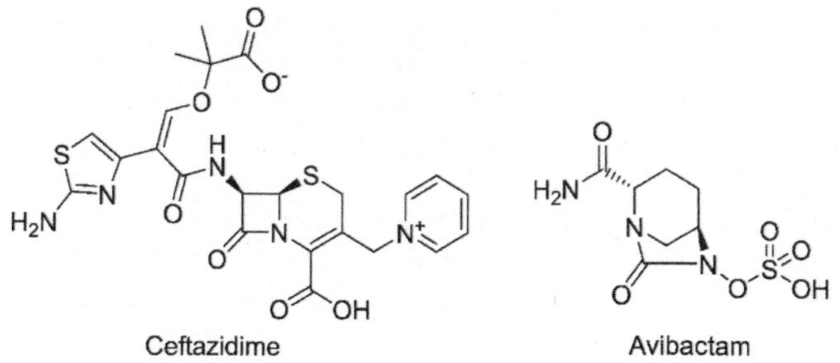

FIGURE 51.4 Chemical structure of ceftazidime and avibactam.

reaction is reversible. The deacylation of the opened avibactam ring leads to the regeneration of the original compound and therefore hydrolysis does not occur.[75] Therefore, reversible binding of avibactam to β-lactamases allowing for recyclization and inhibition of additional β-lactamase molecules is a unique feature of avibactam, compared to earlier generation inhibitors. The protection of ceftazidime by inactivation of a broad range of chromosomal and plasmidic β-lactamases including ESBLs, AmpCs, class D carbapenemases, and KPC-2 enzymes yields a broad spectrum of the combination of β-lactam-resistant Gram-negative bacterial pathogens.[76] *The key limitation of avibactam is that it is not active against class B metallo-β-lactamases.* Compared to meropenem-vaborbactam (discussed later), ceftazidime-avibactam is active against pan-β-lactam- and carbapenem- resistant *P. aeruginosa* and has a role for the same. Avibactam is active against OXA-48 enzymes, but vaborbactam and relebactam are not. In addition, avibactam has some intrinsic antibacterial activity because of its ability to bind to some PBPs like PBP2 of *E. coli* and *H. influenzae*, PBPs 2 and 3 of *P. aeruginosa* and *S. aureus,* and PBP3 of *S. pneumoniae*.[76] The combination can be used for treatment of infections caused by MDR *P. aeruginosa* and *Enterobacterales*. The combination has proven to be effective for the treatment of intra-abdominal and complicated urinary tract infection, and was shown to be noninferior to meropenem for the treatment of intra-abdominal infections.[76] Although its activity against anaerobic bacteria is restricted, studies in animal models suggest that avibactam is potent against BlaMab, which is a broad-spectrum β-lactamase produced by *Mycobacterium abscessus*. Resistance mechanisms against ceftazidime-avibactam involve efflux pumps and reduction of porins, resulting in reduced intracellular concentrations of ceftazidime-avibactam.[76]

Vaborbactam, a new β-lactamase inhibitor, is based on a cyclic boronic acid pharmacophore with high-affinity to serine β-lactamases.[77,78] The boron in vaborbactam is electrophilic, and mimics the carbonyl carbon of the β-lactam ring and forms a reversible covalent bond with the catalytic serine of some β-lactamases.[78] It is a potent broad-spectrum inhibitor of class A (e.g., KPC, CTX-M, SHV, TEM) and class C (e.g., P99, MIR, FOX) β-lactamases.[77] It even inhibits the newly discovered class A carbapenemases BKC-1 and FRI-1.[78] Vaborbactam achieves greatest reduction in MICs against KPC-producing strains when combined with carbapenems, compared to other β-lactams.[77] Vaborbactam, however, does not reduce the meropenem MIC of *A. baumannii, P. aeruginosa*, or *S. maltophilia*.[78] Vaborbactam also does not restore the activity of meropenem in the presence

of OX-48 or MBLs. Inactivation of the porin protein OmpK36 in *K. pneumoniae* confers resistance to meropenem-vaborbactam (as well as imipenem-relebactam and ceftazidime-avibactam).[78] Pathogens that are resistant to meropenem-vaborbactam include *S. maltophilia, Elizabethkingia meningoseptica,* Aeromonas spp, *A. baumannii,* and class D β-lactamase-producing *K. pneumoniae.*[78] Furthermore, pathogens that are intrinsically resistant to carbapenems due to mechanisms other than β-lactamase production, are resistant to meropenem-vaborbactam, e.g., MRSA and VRE.[78] Meropenem-vaborbactam has a more potent *in vitro* activity against KPC producers and is less likely to acquire resistance, which may make it a preferred agent for KPC-producing CRE, compared to ceftazidime-avibactam.[79] Ceftazidime-avibactam, on the other hand, is the preferred agent for treatment of OXA-48 producing CRE.

Ceftolozane is a fifth-generation cephalosporin and it is available in combination with tazobactam (Figure 51.5).[80] Ceftolozane is structurally similar to late-generation cephalosporins, but has a bulkier side chain at the 3-position of the dihydrothiazine ring (R1 in Figure 51.5). This allows ceftolozane to be the most potent inhibitor of PBP1b and PBP3, which are among the PBPs essential for pseudomonal cell wall synthesis.[81,82] The ceftolozane molecule has a modified side chain at the 3-position of the cephem nucleus. This modification is responsible for its potent antipseudomonal activity, *which is the key feature of ceftolozane.* Ceftolozane is more stable against the Amp-C β-lactamase. Ceftolozane therefore then acts on PBPs, TEMs, SHVs, and AmpC and tazobactam targets class A serine β-lactamases and ESBLs. It is active against carbapenem-, piperacillin/tazobactam-, and ceftazidime-resistant isolates of *P. aeruginosa.* It can evade resistance mechanisms of efflux pumps, reduced uptake through porins, and modification of PBPs employed by Pseudomonas.[38] Ceftolozane has activity against Gram-negative bacilli including those that harbor classical β-lactamase inhibitors, e.g. TEM-1 and SHV-1. Ceftolozane is not active against strains with ESBLs and carbapenemases. Tazobactam, the

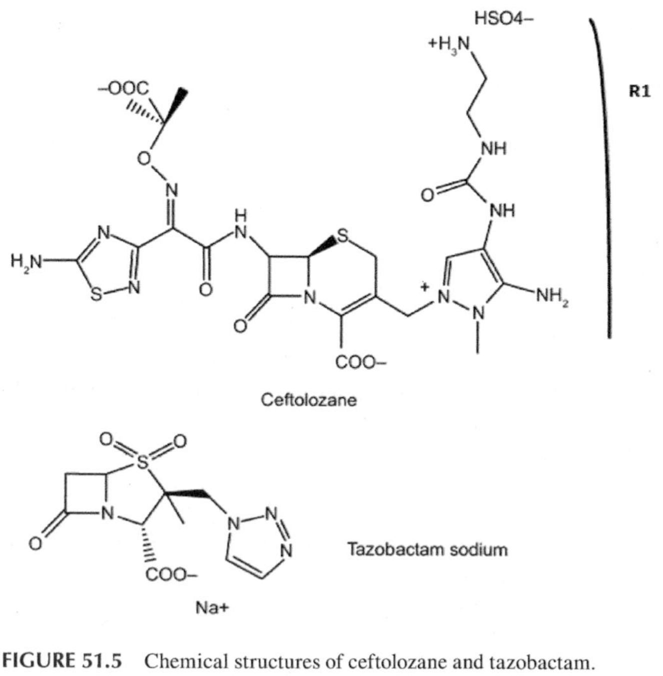

FIGURE 51.5 Chemical structures of ceftolozane and tazobactam.

β-lactamase inhibitor, extends the activity of ceftolozane to most ESBL producers and some anaerobes. Tazobactam inactivates most class A β-lactamases (including many ESBLs) and some class C β-lactamases (cephalosporinases), and ceftolozane is also more resistant to hydrolysis by Pseudomonas-derived cephalosporinase 1 (PDC-1) (a class C β-lactamase) as compared to the previous penicillin that was used in combination with tazobactam (piperacillin).[80] *P. aeruginosa* produces PDC-3, which robustly hydrolyzes early-generation cephalosporins, making it resistant to many β-lactam antibiotics.[80] The combination of ceftolozane and tazobactam is active against extended-spectrum β-lactamase (ESBL)-producing *Enterobacterales*, but it lacks useful activity against carbapenem-resistant strains. [81,82] A study showed that 94.4% of *Enterobacterales* were susceptible to ceftolozane-tazobactam.[83] Ceftolozane-tazobactam was more active than cefepime (90.2% susceptible), ceftazidime (87.3% susceptible), and piperacillin-tazobactam (91.7% susceptible) but less active than meropenem (98.0% susceptible) and amikacin (99.0% susceptible).[83] This study included 9.5% ESBL organisms that were non-CRE phenotype and 87.5% of those were susceptible to ceftolozane-tazobactam.[83] The combination is available for injection and has been approved for the treatment of complicated intra-abdominal and urinary tract infections. Resistance to this combination in *P. aeruginosa* is due to the mutations leading to expression of the AmpC gene and due to horizontally acquired β-lactamases.[82]

Relebactam is a non-β-lactam β-lactamase inhibitor.[84] Structurally, it differs from avibactam in that it contains a piperidine ring in the R1 side chain.[85] The presence of a highly strained bicyclic urea core and electron-withdrawing aminooxy sulfate moiety in its structure makes relebactam more reactive, which in turn makes it an effective inhibitor of β-lactamases.[84] Relebactam is effective against class A and C β-lactamases but not metallo-β-lactamases of class B (VIM, IMP, etc.) and β-lactamases of class D (e.g., OXA). Therefore, the organisms that produce class B (e.g., *S. maltophilia, E. meningoseptica, Aeromonas* spp.) and class D β-lactamases (e.g., *A. baumannii*) are resistant to the imipenem-cilastatin plus relebactam combination.[84]

Additionally, organisms in which resistance is not due to the presence of β-lactamases but of some other factor, for example, the presence of efflux pumps such as the MexA-MexB-OprM efflux system, or a mutated or altered PBP to which the binding of β-lactams is reduced, as observed in MRSA, remain unaffected by the combination.[84]

Cefepime is a fourth-generation cephalosporin and is effective against both Gram-positive (but not MRSA) and Gram-negative organisms like *E. coli, K. pneumoniae,* and *P. aeruginosa.* Zidebactam is a non-β-lactam bicyclo-acyl hydrazide β-lactamase inhibitor that also has some intrinsic antibacterial activity because of its ability to bind to PBP 2.[86] It inhibits class A, C, and D β-lactamases and also metallo-β-lactamases.[87]

The combination, cefepime, and zidebactam, also known as WCK 5222, is being developed for the treatment of serious infections caused by multidrug-resistant (MDR) and extensively drug-resistant (XDR) Gram-negative bacteria like *Enterobacterales,* and *P. aeruginosa.*

Cefepime-zidebactam was shown to be very active against *Enterobacterales,* producing CTX-M-15 (MIC90, 1 mg/L), SHV (MIC90, 0.25 mg/L), other ESBLs (MIC90, 1 mg/L), plasmidic AmpC (MIC90, ≤0.06 mg/L), derepressed AmpC (MIC90, 0.5 mg/L), KPC (MIC90, 1 mg/L) and metallo-β-lactamases (MIC90, 8 mg/L) in *in vitro* studies.[87] It also showed activity against *P. aeruginosa* with overexpression of AmpC (MIC90, 8 mg/L) and MBLs [MIC90, 8 mg/L].[87] Cefepime-zidebactam showed moderate activity against *A. baumannii* with OXA-23, -24, and -58 with MIC90 of 32 mg/L.[87] It has also been shown to have *in vitro* activity against strains of colistin-nonsusceptible mcr-1- and MBL-producing Gram-negative pathogens.[86]

Monobactam: Aztreonam

The binding of aztreonam to PBP-3 of Gram-negative aerobic organisms results in the formation of nonviable filamentous structures, resulting from the inhibition of cell wall synthesis, and ultimately, bacterial cell death. Its bactericidal activity is limited to aerobic Gram-negative bacteria, because it does not bind to PBPs in Gram-positive or anaerobic bacteria.[88–92] Aztreonam displays an intrinsic stability to metallo-β-lactamase-catalyzed hydrolysis.[93] Aztreonam is currently under development in combination with avibactam. Aztreonam, however, can be inactivated by ESBLs, which can frequently co-occur with MBLs. A current strategy to treat such strains is aztreonam with ceftazidime/avibactam.[79]

Interference with Cytoplasmic Membrane Integrity

Drugs That Disorganize the Cytoplasmic Membrane

Polymyxins: Polymyxin B, Colistin (Polymyxin E)

Polymyxins are cationic antimicrobial peptides, not synthesized by ribosomes, that bind to Lipid A found in the outer membrane of Gram-negative bacteria. Polymyxin is extremely bactericidal, which is related to its high lipopolysaccharide affinity. Positively charged amino acid residues in the drug form a ring and interact with the negatively charged residues in Lipid A, interacting electrostatically and competitively displacing covalent cations (calcium and magnesium) that normally stabilize the lipopolysaccharide molecules in the outer leaflet of the bacterial outer membrane. As a result of studies with liposome systems, it has been postulated that an ionic interaction occurs between the phospholipid phosphate and the ammonium group of polymyxin. Simultaneous proton transfers between these two groups and also between an ammonium group on the phospholipid and a γ-amino group on polymyxin result in neutralization of charge on the polar head of the phospholipid, which can alter the electrostatic and hydrophobic stabilizing forces of the membrane. Liposomes derived from methylated phospholipids are not sensitive to polymyxin; possible explanations are prevention of proper alignment of the phospholipid and antibiotic by the methyl groups, or an increase in the basicity of the phospholipid amino group, which would make proton transfer less favored. A unique and very beneficial property of colistin is its antiendotoxin activity, resulting from its ability to neutralize bacterial lipopolysaccharides (LPS).[94–99] In any case, the interaction also stimulates LPS to aggregate in a concentration-dependent fashion. In fact, it has been shown that at lower concentrations, polymyxins bind to membranes superficially, while at higher concentrations, they insert themselves deeper, as a result of the increase in the surface charge of LPS and the membrane. The drug and its hydrophobic

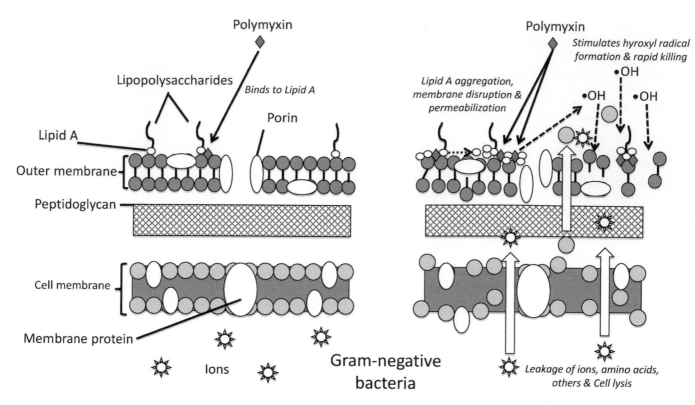

FIGURE 51.6 Mechanism of action of polymyxins.

amino acids then cause membrane permeabilization and disruptions, which result in the release of lipopolysaccharides, and the leakage of cytoplasmic components, including amino acids, uracil, and potassium. All this leads to death of the resulting bacterial cell (Figure 51.6).[100–103]

There is also evidence that polymyxins stimulate hydroxyl radical production which mediates rapid killing (Figure 51.6), holding true even in multidrug-resistant clinical isolates of *A. baumannii*, but not in colistin-resistant bacteria. Interestingly, the presence of significant intracellular iron increases the induction of hydroxyl radicals, whereas low levels of iron are associated with more resistance to oxidative killing.[101,102]

Polymyxin B is regarded as a model compound of polymyxins, which are polypeptides; colistin (polymyxin E) is believed to have an identical mechanism of action. There is evidence though that colistin may also have an added ability to inhibit cell division.[104]

Outer Membrane Protein-Targeting Antibiotic: Murepavadin

Murepavadin is a targeted drug for *P. aeruginosa* and has a novel and nonlytic mechanism of action (Figure 51.7).[105] Murepavadin is a first-in-class outer membrane protein-targeting antibiotic (OMPTA) currently under development for the treatment of hospital-acquired bacterial pneumonia (HABP) and ventilator-associated bacterial pneumonia (VABP).[105,106] The outer membrane (OM) of Gram-negatives is composed of lipopolysaccharides (LPS) on the outer surface and phospholipids on the inner surface.[19] The core-lipid A portion of LPS in *P. aeruginosa* is synthesized on the cytoplasmic side of the inner membrane (IM). It is then flipped to the outer surface of the IM by the ABC

transporter MsbA, and more oligosaccharides are added there. Proteins LptD and LptE form a complex to transport LPS from periplasm to the OM. Murepavadin is bactericidal by binding to lipoprotein D (LptD).[105,107] Therefore, it interferes with the transport processes of LptD, leading to lethal modifications in lipopolysaccharides of the OM and cell death.[105] It is highly specific, and disturbances of the patient's natural bacterial flora are not expected.[105]

Drugs That Produce Pores in Membranes

Lipopeptides: Daptomycin

Lipopeptides are a class of antibiotics that are active against multidrug-resistant Gram-positive bacteria. Lipopeptides consist of a linear or cyclic peptide sequence, with either net positive or negative charge, to which a fatty acid moiety is covalently attached at its N-terminus.

Daptomycin is a cyclic lipopeptide with rapid bactericidal action that results from its binding to the cytoplasmic membrane in a calcium-dependent manner. The acyl chain of the daptomycin molecule partly inserts into the membrane. In the presence of Ca^{2+}, which binds to daptomycin, a conformational change occurs leading to membrane binding and oligomerization. This happens by bridging between the anionic daptomycin and the anionic headgroups of bacterial outer leaflet lipids, before the daptomycin molecule is anchored deeper into the membrane and forms aggregates. Daptomycin actually generates random patches on the cell membranes, and the drug-membrane "aggregates" that are formed affect cell division proteins that lead to severe alteration of the cell shapes and the cell membrane itself. These changes occur in multiple sites, producing local membrane

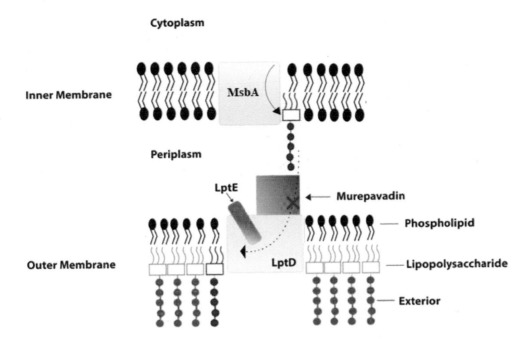

FIGURE 51.7 LPS transport in Pseudomonas aeruginosa. The arrow depicts LPS assembly in the cytoplasm and on the inner membrane and its subsequent transport to outer membrane.

activities, overwhelming the cells, which are unable to respond to the changes. These produce membrane discontinuities that cause leakage of ions and loss of membrane potential, with local dysregulation of cell division or cell wall synthesis. There is also activation of autolysis. Subsequently the bacterial membrane is depolarized and perforated, and cell death ensues (Figure 51.8). Daptomycin is rapidly bactericidal in a concentration-dependent fashion.[16,108–111]

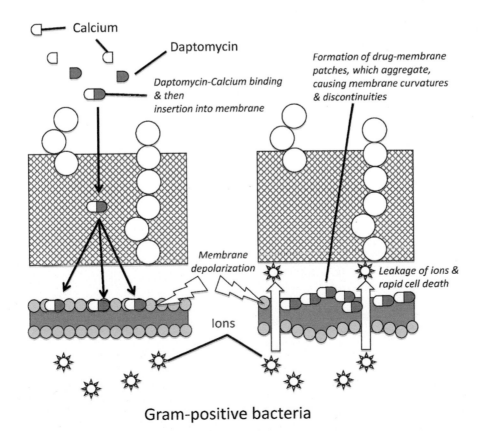

FIGURE 51.8 Mechanism of action of daptomycin.

Peptide Antibiotics: Gramicidin A, Gramicidin S

Gramicidin A is a linear pentadecapeptide with an alternating D- and L-amino acid structural sequence, which is active in a dimer form. In membranes, gramicidin A forms transient channels specific for monovalent cations.[112–116]

Gramicidin S is a cyclic decapeptide that interacts electrostatically with the membrane phospholipids, without incorporating into the lipid region of the membrane. It achieves its bactericidal action by promoting leakage of cytoplasmic solutes by disrupting the barrier properties of cell membranes. It may also cause uncoupling of oxidative phosphorylation as a secondary effect. The interaction of this peptide with the phospholipid bilayer is nonspecific, and in addition to the antimicrobial activity, it exhibits a highly hemolytic activity by interacting with the lipid bilayer of erythrocytes, limiting its use to topical preparations.[117–119]

These peptides kill bacteria by making the cell membranes permeable after interacting with the lipid bilayers. To date, the mechanism by which they do it is either through forming a peptide "carpet" on the membrane surface or by making a transmembrane pore. However, a different study suggests that gramicidin S does not form membrane pores.[216] Studies on the structural properties of Gramicidin S 14 gave some insight into the mechanism by which these peptides disrupt the membrane. Molecular models showed that since Gramicidin has a much shorter hydrophobic length compared to the hydrophobic lipid layers, each peptide displaced six to eight lipid molecules. Therefore, multiple peptides together, would form a "carpet" that would displace even more lipids, producing curvature strain at first, then disruption of the membrane. One of the latest structural studies also suggests that Gramicidin S self-assembles an oligomeric β-barrel pore that is stabilized by hydrogen bonding.[119,120]

Inhibition of Nucleic Acid Synthesis

Inhibitors of DNA Replication

Quinolones/Fluoroquinolones

The principal targets for the antibacterial activity of quinolones are DNA gyrase and topoisomerase IV, which are both bacterial type II topoisomerases.[121]

DNA gyrase is a tetramer enzyme with two GyrA and GyrB subunits (encoded by the *gyrA* and *gyrB* genes) that catalyzes the introduction of negative superhelical turns and elimination of positive supercoils at the replication fork into the double-stranded covalently closed circular bacterial DNA. DNA gyrase acts before the replicating fork, preventing replication-induced structural changes.[122]

Topoisomerase IV is also an enzyme composed of pairs of two subunits, named ParC and ParE (homologous to GyrA and B, but encoded by genes *parC/parE*) in *E. coli* and with GrlA and GrlB (encoded by genes *grlA/grlB*) in *S. aureus*. A *parF* locus has been identified in *Salmonella typhimurium* and is suggested to be involved in topoisomerase IV activity. Topoisomerase IV acts behind the replicating fork and catalyzes removal of supercoils and decatenation (unlinking) of interlocked daughter DNA molecules.[123]

Both DNA gyrase and topoisomerase IV appear to be targeted by all quinolones, regardless of bacterial species. However, the preferred drug target depends on the organisms and on the drug. For example, the preferred site of action for levofloxacin is topoisomerase IV, while the main target for moxifloxacin and gatifloxacin is DNA gyrase. In Gram-negative bacteria, DNA gyrase is generally bound more readily by quinolones. In Gram-positive bacteria, topoisomerase IV is generally preferred.[124,125]

To summarize (Figure 51.9), the first central step is the formation of the fluoroquinolone-enzyme-DNA complex. Next is

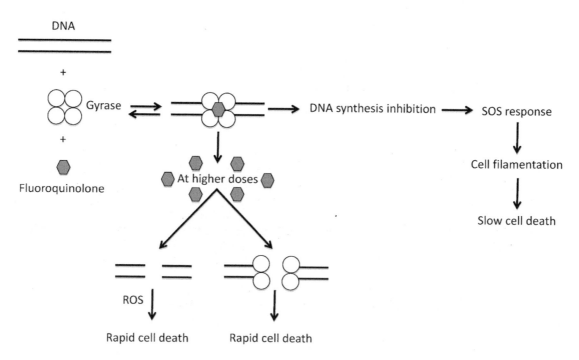

FIGURE 51.9 Mechanism of action of fluoroquinolones.

the bacteriostatic step, where there is formation of cleaved complexes. The quinolones turn these type II enzymes into toxins that keep the cleavage complexes in an open form by creating breaks in the chromosome.[121] This blocks bacterial DNA replication, stimulates the SOS response, and causes cell filamentation, and presumably slow cell death. With higher concentrations, DNA breaks occur via two pathways: the protein synthesis-dependent pathway, where the drug-gyrase complex is removed from the DNA and there is liberation of double-stranded DNA breaks; and the protein synthesis independent pathway, where the gyrase subunits dissociate and the DNA ends attached to them are released. These pathways result in stimulation of reactive oxygen species to act and cause irreversible oxidative DNA damage.[126]

Nalidixic acid, the first quinolone discovered and used in clinical practice, has a narrow spectrum of activity (active against Gram-negative bacteria).[126–128] But chemical structure changes at positions 1, 6, 7, and 8 resulted in more potent antibacterials.[129] The second-generation quinolones are "fluorinated" quinolones (hence the term "fluoroquinolones") and were discovered in the mid-1980s. They (e.g., norfloxacin, ciprofloxacin, enoxacin, ofloxacin, and pefloxacin) all have a fluorine atom at position 6, plus a piperidine at position 7, resulting in a much broader spectrum of activity that includes *P. aeruginosa* and some Gram-positive bacteria, such as *S. aureus*, and improved bioavailability. The third-generation fluoroquinolones (e.g., sparfloxacin, grepafloxacin, temafloxacin, and levofloxacin) have even better Gram-positive coverage, especially against pneumococci, and good potency versus anaerobes because of substitutions at positions 7 and 8. The fourth-generation fluoroquinolones (garenoxacin, gemifloxacin, gatifloxacin, moxifloxacin, and trovafloxacin) have even greater activity against pneumococci and anaerobes. Several fluoroquinolones, such as ciprofloxacin, levofloxacin, and moxifloxacin, have extensive usage in clinical practice for treatment of Gram-negative and some Gram-positive bacterial pathogens.[122,124–128,130–132]

Ozenoxacin is a novel topical, nonfluorinated quinolone antimicrobial.[133] It is used topically for the treatment of impetigo.[134] It binds with greater affinity than other established quinolones to both DNA gyrase and topoisomerase IV. It is active against Gram-positive organisms including MSSA, MRSA, methicillin-resistant Staphylococcus epidermidis, and Streptococcus pyogenes in vitro, and in clinical trials.[10,11] It also has activity against aerobic Gram-negative bacteria, but it is inferior to the activities of older quinolones. Resistance mechanisms are similar to those of other quinolones, including loss of binding affinity to its targets, efflux pumps, and modified bacterial porin channels.[11]

Delafloxacin is an anionic synthetic fluoroquinolone with broad-spectrum bactericidal activity against Gram-positive, Gram-negative, atypical (Legionella, Chlamydia, and Mycoplasma), and anaerobic bacteria in a concentration-dependent manner.[135,136] Other quinolones have binding affinity primarily for either DNA gyrase or topoisomerase IV, whereas delafloxacin is a dual-targeting fluoroquinolone that binds to both enzymes, leading to broader activity against Gram-positive and Gram-negative bacteria.[137] It forms cleavable complexes with topoisomerase IV or DNA gyrase.[27] It has similar affinity for both types of enzymes as other quinolones, however, due to mechanisms not completely understood, it has greater potency for Gram-positives.[27] It has been hypothesized that this improved

affinity is due to structural features of the delafloxacin molecule, namely the presence of a heteroaromatic substituent at position 1, a weak polarity associated with the presence of a chlorine in position 8, and a lack of basic group in position 7 next to the fluorinated ring, which contributes to its weakly acidic properties.[135] This anionic structure of delafloxacin helps to enhance its potency in the acidic infectious or inflammatory environment, which is not seen with other quinolones.[137] It has activity against levofloxacin nonsusceptible (i.e., intermediate and resistant) strains of MRSA and MSSA, methicillin-susceptible and resistant coagulase-negative staphylococci, *S. pneumoniae*, β-hemolytic streptococci, viridans group streptococci, *P. aeruginosa*, and against isolates with documented mutations in the quinolone resistance-determining region.[136,137] It is active against *E. faecalis* but not against *E. faecium*.[137] *S. aureus* was eradicated in 98.4% of patients with acute bacterial skin and skin structure infections receiving delafloxacin, and the eradication rates were similar against levofloxacin-nonsusceptible and methicillin-resistant strains of *S. aureus*.[136] Mutations in both the gyrA and gyrB were required for resistance to delafloxacin in a mutant selection study.[17]

Triazaacenaphthylene: Gepotidacin

Gepotidacin is the first member of the class triazaacenaphthylene antibacterial and is a topoisomerase II inhibitor.[138] It has a unique mechanism of action that is currently not known for any other antibacterial agent. It arrests bacterial DNA replication, and consequently cell division, by acting on GyrA and ParC subunits of bacterial DNA gyrase and topoisomerase IV, respectively.[139] It is bactericidal against many drug-resistant Gram-positives, including MRSA, and Gram-negatives, including ESBL-producing strains of *E. coli*.[138] Its novel mode of action that is different from that of fluoroquinolones allows activity against fluoroquinolone-resistant strains, including fluoroquinolone-resistant *Neisseria gonorrhoeae*.[138] The novelty is that it has a binding site that is close to but distinct from that of quinolones and it does not form "cleavage complex". Gepotidacin binds to a distinct site and stabilizes the precleavage type II topoisomerase enzyme-DNA complex prior to DNA cleavage, generating single-strand breaks, instead of stabilizing DNA double-strand breaks as fluoroquinolones do. A study suggests that it is safe and effective for the treatment of Gram-positive acute bacterial skin and skin structure infections by both oral and intravenous routes.[139] Phase 2 trials of the use of gepotidacin as a single, oral dose for the treatment of uncomplicated urogenital gonorrhea due to *N. gonorrhoeae* was shown to be almost 95% effective for bacterial eradication.[140]

Spiropyrimidinetrione: Zoliflodacin

Zoliflodacin is an experimental spiropyrimidinetrione antibiotic that has completed phase II clinical trials for the treatment of gonorrhea. Its mode of action is to inhibit bacterial DNA gyrase/topoisomerase. It works by blocking re-ligation of the double-stranded cleaved DNA to form fused circular DNA.[141] It has shown *in vitro* activity against *Chlamydia trachomatis* and *Chlamydia pneumoniae*, and *Mycoplasma* and *Ureaplasma* species.[142,143] It also has potent *in vitro* activity against a number of Gram-positives (*S. aureus*, *S. epidermidis*, *S. pneumoniae*, *S. pyogenes*, and *Streptococcus agalactiae*), fastidious Gram-negatives (*Haemophilus influenzae* and

N. gonorrhoeae), atypical (*Legionella pneumophila*), and anaerobic (*Clostridioides difficile*) organisms, including those resistant to fluoroquinolones.[141]

Nitroimidazoles: Metronidazole and Secnidazole

Nitroimidazoles—with the class representative metronidazole—are cytotoxic to obligate anaerobes and facultative anaerobic bacteria through a four-step process: (1) entry into the bacterial cell by diffusion across the cell membrane; (2) reduction of the nitro group by the intracellular electron transport proteins; the nitro group of metronidazole acts as an electron recipient, capturing electrons that are usually transferred to hydrogen ions in the oxidative decarboxylation of pyruvate cycle; the concentration gradient created by this process promotes further uptake of metronidazole into the cell and formation of intermediate compounds and free radicals toxic to the cell; (3) interaction of cytotoxic byproducts with host cell DNA, resulting in DNA strand breakage, inhibited repair, and ultimately disrupted transcription and cell death; (4) the toxic intermediate compounds decay into inactive end-products.[144–147]

Aerobic bacteria are intrinsically resistant to metronidazole because these microbes lack electron-transport proteins with negative redox potential, making the drug inactive. It has activity versus some microaerophiles (e.g., *Helicobacter pylori*), although reoxidation can occur with exposure to oxygen, which will return it to its inactivated state.[148]

Secnidazole (a,2-Dimethyl-5-nitroimidazole-1-ethanol) is a 5-nitroimidazole and has a longer half-life than metronidazole. It is approved for the treatment of bacterial vaginosis (BV).[149] BV is characterized by a decrease in lactobacilli and an increase in other bacteria like *Gardnerella vaginalis*, *Atopobium vaginae*, and other anaerobes.[150] Secnidazole has activity against most bacteria implicated in BV and has limited activity against lactobacilli, the microbes that are important in maintaining vaginal health.[149] The MIC$_{90}$ for secnidazole was similar to metronidazole and tinidazole for *Atopobium vaginae*, *Bacteroides* species, *Finegoldia magna*, *G. vaginalis*, *Mageeibacillus indolicus*, Megasphaera-like bacteria, *Mobiluncus curtisii* and *mulieris*, *Peptoniphilus lacrimalis* and *hare*, *Porphyromonas* species, *Prevotella bivia*, *Prevotella amnii*, and *Prevotella timonensis*.[151] The MIC$_{90}$ of all lactobacilli tested was >128 mg/mL for secnidazole and metronidazole.[151] While not currently approved for such treatment in the United States, it is used for trichomoniasis, amebiasis, and giardiasis in other parts of the world.[151]

Inhibitors of RNA Polymerase: Rifamycins, Rifampin, and Rifaximin

Rifamycins, with rifampin as a class representative, specifically inhibit bacterial DNA-dependent RNA polymerases by blocking the RNA chain initiation step of bacterial DNA transcription to messenger RNA that is needed to synthesize bacterial proteins.[152] Rifampin is used as part of a combination regimen in the treatment of mycobacterial infections and some Gram-positive infections (e.g., *S. aureus*, *S. epidermidis*). Rifampin is bactericidal against Gram-positive bacteria, and in addition, diffuses well into biofilm where it continues to be bactericidal. It has the ability to disrupt biofilm, but one of the downsides of its use is that it may select fairly readily rifampin-resistant mutants. This is why

rifampin should always be used in combination with other antibiotics. *In vitro* studies have shown that there is rapid synergy when rifampin is given with a β-lactam, and slower synergy with vancomycin.[152–154]

Rifamycins bind very tightly to the β-subunit of the RNA polymerase in a molar ratio of 1:1. They do not bind well to β-subunits of RNA polymerase isolated from rifamycin-resistant mutants, and mutations are found in the β-subunits of RNA polymerase in such mutants that reduce such binding of rifampin to ß-subunits of RNA polymerase. Mammalian enzymes are not affected by rifampin. Rifamycins also modify bacterial pathogenicity and may alter attachment and tissue toxicity.[155–157]

Rifaximin is a novel semisynthetic derivative of rifamycin, used for treatment of enteric infections because it is virtually nonabsorbable from the gastrointestinal tract and is active against enteric infections. Like other rifamycins, the mechanism of action involves binding of rifaximin to the β-subunit of the bacterial DNA-dependent RNA polymerase, inhibiting initiation of chain formation in RNA synthesis. In contrast to rifampin, rifaximin does not diffuse readily, is poorly absorbed, and so acts topically in the lumen of the gut. Another distinction from rifampin is that it is much less prone to select microbial resistance.[157,158]

Inhibition of Ribosome Function and Protein Synthesis

Inhibitors of 30S Ribosomal Subunits

Aminoglycosides

The aminoglycoside antibiotics (also known as aminocyclitols) are hydrophilic sugars with multiple amino groups, protonated at physiological pH to function as polycations. Two main categories of aminoglycosides are exemplified by the streptidine-containing class (primarily, streptomycin) and the 2-deoxystreptamine-containing aminoglycosides that include neomycin, kanamycin (with its derivative amikacin), gentamicin, and tobramycin.

Aminoglycosides are bactericidal antibiotics that target accessible regions of polyanionic 16S rRNA on the 30S ribosomal subunit, decreasing the accuracy of the translation of mRNA into protein by contributing to the misreading of codons in the mRNA. As a result of the tight binding of 2-deoxystreptamine aminoglycosides in the major groove of helix H44 of 16S rRNA, the A site in the 30S ribosomal subunit undergoes a conformational change. Incorporation of noncognate (or near-cognate) aminoacyl-tRNAs is allowed and the error rate of the ribosome is increased. Elongation is also retarded, but less drastically. As for the 2-deoxystreptamine aminoglycosides, the binding of streptomycin also allows the binding of near-cognate tRNAs to the mRNA, so mistranslation ensues. Studies with ribosomes from streptomycin-resistant mutants have shown that protein S12 of the 30S ribosomal subunit is altered; therefore, it plays a role in the binding of streptomycin to normal ribosomes and the resulting effects. Studies have shown that aminoglycosides also have a role in the inhibition of assembly of the 30S ribosomal subunit.[159–163]

The uptake of aminoglycosides into Gram-negative bacteria is energy-dependent, but is inhibited by divalent cations, hyperosmolarity, low pH, and anaerobiosis, explaining why this drug class is ineffective versus anaerobic bacteria, and is poor against

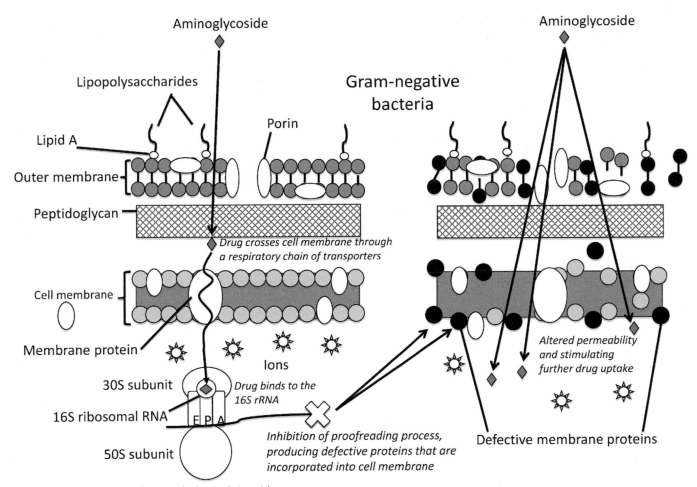

FIGURE 51.10 Mechanism of action of aminoglycosides.

localized infections with low pH and low pO$_2$, such as abscesses. The two phases of aminoglycoside uptake are: when small amounts of the drug cross the cell membrane by using a membrane-bound respiratory chain of electron transporters and when the drug binds to the 16S rRNA component of the 30S subunit of the bacterial ribosome. All this leads to the inhibition of the proofreading process in the rRNA, producing defective proteins that are incorporated in the cell membrane, causing increased permeability and stimulating further aminoglycoside uptake. This process will lead to bacterial cell death (Figure 51.10).[163,164]

Some aminoglycosides that are available in other countries include: Isepamicin, a derivative of gentamicin, is not degraded by the aminoglycoside-modifying enzyme (AME) called type I 6'-N-acetyltrasferase (AAC(6')-I), which acetylates, and this inactivates amikacin; and Arbekacin, a derivative of dibekacin, is another aminoglycoside that is also resistant to many AMEs, and it has been shown to have good activity against MRSA in *in vitro* studies. Arbekacin is licensed in Japan for systemic use while in United States, it is under developmental stages as an inhalation solution.[217] In addition, plazomicin, an aminoglycoside, was approved by the U.S. Food and Drug Administration (FDA) for treatment of cUTIs.[218] The drug is also resistant to inactivation by most AMEs and so may be a promising new agent for MDR bacteria.[164–166]

New laboratory research on aminoglycosides has come up with two new classes, pyranmycin, which is structurally related to neomycins, but with a 4,5-disubstituted 2-deoxystreptomycin core and a pyranose as the ring component; and amphiphilic neomycin (NEOF004). The former possesses the traditional mode of action already described, but with an additional antifungal activity, while the latter, a structurally modified neomycin, was shown to disrupt the bacterial cell wall. Another novel aminoglycoside, FG08, which is similar to kanamycin (except for a C-8 alkyl chain at the 4''-O position at ring III), acts as a broad-spectrum antifungal, but surprisingly has no antibacterial activity.[164,167]

Plazomicin is a novel semisynthetic aminoglycoside that is derived from sisomicin.[168] As with other aminoglycosides, it binds to the 30S ribosomal subunit of bacteria, thereby inhibiting protein synthesis. It is active against aerobic Gram-positive and Gram-negative bacteria, including organisms producing ESBLs, carbapenemases, and/or AMEs, and also organisms with site mutations conferring fluoroquinolone resistance.[168–170] Plazomicin was active against 99.3% of Enterobacterales (*E. coli*, *Klebsiella* spp., *Proteus* spp., and *Enterobacter* spp) isolates carrying AMEs, while amikacin, gentamicin, and tobramycin were active against 90.1%, 20.9%, and 18.3%, respectively.[25] CREs can be resistant to plazomicin due to production of 16S rRNA methyltransferases, which are often associated with New Delhi metallo-ß-lactamase-1 (NDM-1), which is so far not yet common in the United States.[171] It has activity against colistin-resistant *Enterobacterales* as well, which are commonly resistant to older aminoglycosides.[168] A study showed that plazomicin inhibited 93.7% of colistin-resistant *Enterobacterales*.[168]

The mechanism of aminoglycoside resistance is generally by production of AMEs, which are encoded by genes on plasmids and can be present in various combinations in a given isolate.[171] Plazomicin is an aminoglycoside that is not inactivated by the vast majority of AMEs, such as acetyl-, phosphoryl- and nucleotidyltransferases.[172]

A study tested 110 CRE isolates for susceptibility to plazomicin and reference aminoglycosides and carbapenems. Of these, 97.3% and 100% were nonsusceptible to meropenem and imipenem, respectively. Similarly, many isolates were nonsusceptible to amikacin (76.4%), gentamicin (18.2%), kanamycin (91.8%), and tobramycin (96.4%). On the other hand, MIC50/ MIC90 values for plazomicin were 0.5/1 mg/L for all 110 carbapenemase-producing isolates. Only one isolate, which was an NDM-1-producing *K. pneumoniae*, was resistant to plazomicin. Across the literature, plazomicin MICs against all tested isolates have been ≤4 µg/mL, except for a set of CRE isolates producing NDM-1 MBL because of associated 16S rRNA methyltransferases.[7] A large phase III clinical trial showed plazomicin to be noninferior to meropenem in the treatment of cUTI and it is FDA-approved for this indication.[26]

Tetracycline, Doxycycline, Minocycline, Omadacycline, Sarecycline, and Eravacycline

Tetracycline and the more recent members of the class, doxycycline and minocycline, exert their antibacterial effect by initially traversing the outer membrane of Gram-negative bacteria through the OmpF and OmpC porin channels as positively charged cation-tetracycline coordination complexes. The metal ion-tetracycline dissociates in the periplasm and tetracycline diffuses through the inner (cytoplasmic) membrane and binds to the acceptor A site on the 30S ribosomal subunit, subsequently preventing the binding of the incoming aminoacyl-tRNA (Figure 51.11). Since this interference is reversible, tetracyclines are bacteriostatic. A specific major binding site and a lower affinity binding site of tetracycline have been identified by determining the crystal structure of the 30S ribosomal subunit with the bound drug.[173–175]

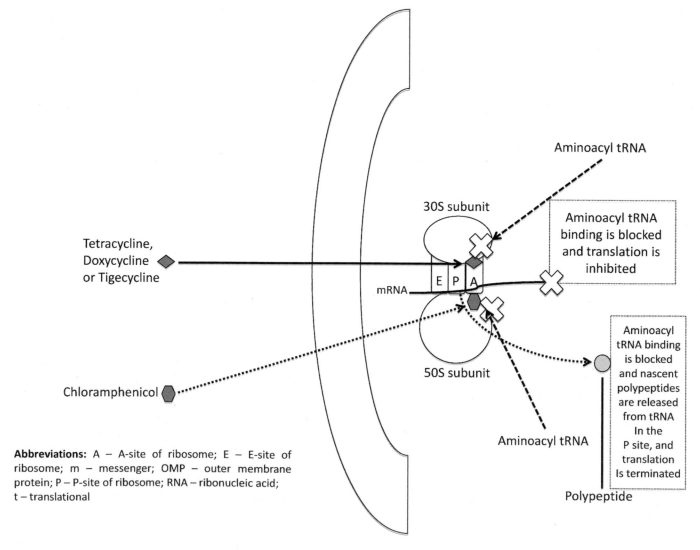

Abbreviations: A – A-site of ribosome; E – E-site of ribosome; m – messenger; OMP – outer membrane protein; P – P-site of ribosome; RNA – ribonucleic acid; t – translational

FIGURE 51.11 Mechanism of action of tetracycline, doxycycline, minocycline, tigecycline, and chloramphenicol.

Minocycline is more potent versus Staphylococcus (especially, CA-MRSA strains) than tetracycline or doxycycline, and *in vitro* data have shown that it may be active versus multidrug-resistant *Acinetobacter baumannii*.[176]

A novel semi-synthetic aminomethylcycline that is first in its class is omadacycline, a noteworthy approach to tetracycline resistance. It is derived from minocycline, and structural modifications include the presence of an aminoethyl group at the C9 position of the basic tetracycline structure.[182] As a result of this modification, omadacycline can evade bacterial resistance mechanisms, such as tetracycline efflux and ribosomal protection.[182] Omadacycline has potent activity versus skin and pneumonia pathogens, including CA-MRSA, β-hemolytic streptococci, penicillin-resistant *S. pneumoniae*, *H. influenzae,* and *Legionella*. It was approved by the FDA in 2018 for acute bacterial skin and skin structure infections (ABSSI) and community-acquired bacterial pneumonia (CABP).[219] Interestingly, it is not affected by the presence of either efflux (tet[K]) or ribosome protection (tet[M]) mechanisms that confer resistance to tetracycline. Omadacycline is active with or without Tet(O), unlike tetracycline, which is inhibited when the ribosomal protection protein, Tet(O), is present.[177]

In the tetracycline class, sarecycline is the first narrow-spectrum antibiotic for acne treatment. It has a stable modification at position C7-7-[methoxy(methyl)amino]methyl. This modification can potentially overcome a tetracycline resistance mechanism.[178] Sarecycline was also shown to have anti-inflammatory effects *in vitro*.[179] Unlike other tetracyclines, it has targeted activity against *Cutibacterium acnes* and is less active against aerobic Gram-negative bacilli and anaerobic bacteria that constitute normal gut microbiota.[178] Due to its targeted activity, it thus has fewer complications of diarrhea and fungal overgrowth.[178] This is especially important in patients undergoing treatment of acne, which requires longer duration of treatment. Sarecycline also has a low level of acquired resistance in *C. acnes* strains.[178]

Eravacycline is a fully synthetic fluorocycline in the tetracycline class that has activity against most Gram-negative and Gram-positive aerobic and facultative bacteria. Like other tetracyclines, it is a potent inhibitor of the bacterial ribosome. There are modifications at both the C-7 (fluorine) and C-9 [2-(pyrrolidin-1-yl) ethanamido] positions on the tetracyclic core.[180] It has activity against strains that are resistant to cephalosporins, β-lactam-β-lactamase inhibitors, carbapenems, and fluoroquinolones.[181] It is not active against *P. aeruginosa* and *Burkholderia cenocepacia*, but is active against *Acinetobacter baumannii*.[180] There are two main acquired tetracycline-specific resistance mechanisms that do not affect eravacycline activity: drug efflux and ribosomal protection. The activity of eravacycline was minimally affected by these resistance mechanisms.[180] Extended-spectrum β-lactamases are a common problem in healthcare settings, and eravacycline has *in vitro* activity against ESBL-producing *Enterobacterales*. It is also active against methicillin-resistant *S. aureus*, carbapenem-resistant *Enterobacterales*, and vancomycin-resistant enterococci.[180,181] Eravacycline was shown to be noninferior to ertapenem in complicated intra-abdominal infections requiring surgery or drainage and is FDA-approved for this indication.[181,182]

Tetracyclines have been found to have antimalarial activity by specifically blocking the expression of the apicoplast gene, leading to the distribution of nonfunctioning apicoplasts into merozoites, which is responsible for the slow but potent antimalarial activity.[183]

Glycylcycline: Tigecycline

Tigecycline is the 9-*t*-butylglycylamido derivative of minocycline. The central 4-ring carbocyclic skeleton and the addition of the large *N*-alkyl-glycylamido group provide the drug with a broad spectrum of activity (Gram-positive bacteria broadly, including MRSA, and Gram-negative bacteria [primarily *Enterobacterales*, not *P. aeruginosa*], anaerobes, and even many multidrug-resistant pathogens) and allow evasion from two major mechanisms of tetracycline resistance: tetracycline-specific efflux pump acquisition and ribosomal protection by production of a protein that sterically blocks the binding of tetracycline but not of tigecycline. Tigecycline exerts its effects by binding to the H34 helical region on the 30S ribosomal subunit of bacteria and blocking entry of amino-acyl transfer RNA into the A site of the ribosome. Studies have shown that protein S7 and 16S rRNA bases are essential to binding action, and that the ribosome-binding actually causes structural changes in the 16S rRNA. Compared with tetracyclines, tigecycline binds five times more efficiently to this ribosomal target site and subsequently amino acid residues are prevented from incorporation into elongating peptide chains and protein synthesis is inhibited. All these explain the bacteriostatic nature of tigecycline, which is the end-result of protein synthesis inhibition.[184–188]

Inhibitors of 50S Ribosomal Subunits

Chloramphenicol

Chloramphenicol inhibits peptide bond formation in bacterial ribosomes by binding to the 50S ribosomal subunit at low- and high- affinity sites. This inhibition is explained by the slow binding of chloramphenicol to the peptidyl transferase center of the 50S ribosomal subunit in the A-site, inhibiting that enzyme's activity, and subsequently blocking the interaction of the aminoacyl-tRNA with the A-site. Specifically, the entry of the 3'-terminus of the aminoacyl-tRNA is blocked from entering the ribosomal catalytic cavity. This results in translation termination and translation inaccuracies, producing nascent polypeptides that are released prematurely from the tRNA in the P site (Figure 51.7).[189–192]

Lincomycin and Clindamycin

Lincomycin and its chlorinated derivative, clindamycin, inhibit the peptidyl transferase function of the 50S ribosomal subunit, resulting in alteration of the P-site to produce an enhanced binding capacity for the fMet-tRNA. At lower concentrations, these antibiotics also interact with the process of peptide chain initiation in cell free systems. Both agents are protein synthesis inhibitors and are bacteriostatic. Clindamycin is more active than lincomycin in treating bacterial infections, especially those caused by staphylococci, streptococci, and anaerobes. An additional advantage of clindamycin is its ability to inhibit exotoxin production in toxic shock syndromes. Furthermore, it has been demonstrated that both drugs can act on peripheral ribosomes, causing selective inhibition of hemolysin and streptolysin S synthesis.[60,193–198]

Macrolides: Erythromycin, Azithromycin, Clarithromycin, and Fidaxomicin

Macrolides bind to the 50S ribosomal subunit of susceptible bacteria, inhibiting RNA-dependent protein synthesis. In addition, they appear to have anti-inflammatory activity as suggested by clinical trials in patients with chronic inflammatory lung diseases, in whom macrolides effect significant improvements in lung function, quality of life, and frequency of exacerbations.[60,199]

Erythromycin inhibits prokaryotic, but not eukaryotic, protein synthesis as a result of its binding to the 50S ribosomal subunit. Ribosomal protein L4 may be involved in this interaction since some erythromycin-resistant bacterial mutants have exhibited an altered L4 protein. It has been postulated that erythromycin inhibits translocation by preventing the positioning of peptidyl-tRNA in the donor site. Other studies have indicated that the peptidyl transferase reaction may also be involved.[195,200–203]

Azithromycin and clarithromycin bind to the same receptor on the bacterial 50S ribosomal subunit and inhibit RNA-dependent protein synthesis by the same mechanism as erythromycin. Azithromycin appears to penetrate better through the outer envelope of Gram-negative bacteria, such as *H. influenzae* and *Mycoplasma catarrhalis*.[203–205]

A new important macrolide antibiotic is fidaxomicin, which is the first medication for the treatment of *Clostridioides difficile*-associated diarrhea that has been approved (on May 27, 2011) by the FDA in over 20 years.[206] It is considered novel, and the first in a new class of macrocyclic antibiotics that is distinctly narrow-spectrum.[207] At present, fidaxomicin is the only other antibiotic (aside from vancomycin) that is approved by the FDA for treatment of *C. difficile*-associated diarrhea. Its active metabolite is OP-118, which is formed from fidaxomicin after hydrolysis of the isobutyl ester at the 4' position. Fidaxomicin and OP-1118 stop bacterial protein synthesis by inhibiting DNA transcription, specifically, inhibiting RNA polymerase at the σ subunit through inactivation of holoenzyme gene transcription.[206] Its absorption via the gut is minimal and it is considered specific in targeting *C. difficile*.[207] When compared with vancomycin, fidaxomicin use was shown to preserve the intestinal microbiome and, at the same time, it resulted in significantly less reappearance of toxin in stools, which is consistent with a decrease in recurrence of *C. difficile* infection.[208] Similarly, fidaxomicin and OP-1118 were found to be very effective in inhibiting *C. difficile* spore formation, even in strains with high sporulation efficiency. These compounds act even in the early stationary phase of growth, and their action is by potent inhibition of transcription of some, or even all, of the sporulation genes. All these contribute to fidaxomicin's clinical efficacy and ability to reduce recurrences and reduce pathogen shedding and transmission.[209,210]

Ketolides: Telithromycin, Solithromycin, Cethromycin, and Nafithromycin

Telithromycin was the first ketolide antibacterial to receive FDA approval. It is a derivative of erythromycin, and so has structural similarities to it and the other macrolides, except that ketolides have a keto group instead of an α-L-cladinose at the 3 position. This structural change has kept the drug active versus MLS$_B$ (macrolide-lincomycin-streptogramin B)-resistant strains.[211]

The mechanism of action of telithromycin is essentially similar to that of erythromycin and other macrolides, by inhibiting bacterial protein synthesis via a close interaction with the peptidyl transferase site of the 50S ribosomal subunit, specifically binding to the 23S ribosomal RNA. Telithromycin has a higher binding affinity and mechanisms that may overcome the methylation of binding sites of the peptidyl transferase loop. At higher concentrations, telithromycin is also able to inhibit the formation of the 30S ribosomal subunit in addition to the inhibition of formation of the 50S ribosomal subunit. It is a bactericidal agent with a postantibiotic effect, able to suppress bacterial growth even after serum concentrations have fallen below the MIC.[203,211]

Solithromycin (CEM-101) is a "fluoroketolide," having a fluorine at the C-2 position of the 14-member macrocyclic lactone and an alkyl-aryl side chain, responsible for tighter ribosome-binding and enhanced antimicrobial activity, and it is less prone to resistance via post-translational methylation of the 23S rRNA in the 50S ribosomal subunit. It binds to the 50S subunit of the ribosome and inhibits protein biosynthesis, impairing ribosomal subunit function.[212,213] Solithromycin is active against *S. pneumoniae*, *H. influenzae*, *M. catarrhalis*, and *S. aureus*. It also has activity against macrolide-resistant strains of the above bacteria due to interaction with three ribosomal sites.[214] It is active against some resistant strains of *N. gonorrhoeae* and *Mycoplasma genitalium*.[214] *In vitro* studies also suggest antimalarial activity associated with solithromycin.[214] Like nafithromycin, it is under development for the treatment of CABP.

A new ketolide called cethromycin has the same mechanism of action as the other ketolides, but with greater potency than telithromycin, plus a superior safety profile. These differences may arise from its unique structural differences from the other ketolides, including (1) the macrolactone ring linkage via the 6(O)-position, and (2) the O-propyl allyl linkage instead of the aminopropyl linkage of telithromycin. These changes make the drug bind much more tightly to the ribosomes, preventing its efflux from the bacteria and allowing it to act on its target more efficiently.[215–218]

Nafithromycin is a lactone ketolide which works as a protein synthesis inhibitor. As described under telithromycin, ketolides are derivatives of the erythromycin A class that lack the L-cladinose sugar at position 3 and have a ketone in its place in the erythronolide ring (Figure 51.12). Ketolides interact at multiple positions on the ribosome, leading to activity against macrolide-resistant strains. *In vitro* studies suggest activity

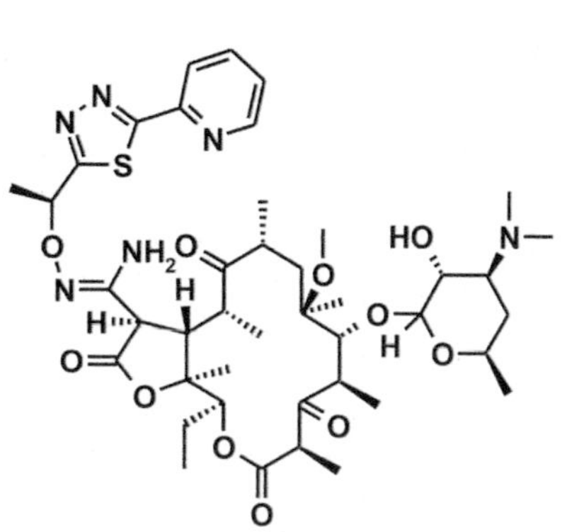

FIGURE 51.12 Chemical structure of nafithromycin.

against *S. aureus, H. influenzae, Moraxella catarrhalis*, atypicals (*Legionella, Mycoplasma, Chlamydophila*), and also macrolide-resistant and telithromycin-nonsusceptible strains of *S. pneumoniae*.[210] It is presently under development for the treatment of community-acquired bacterial pneumonia (CABP), which may be caused by multidrug-resistant Gram-positive bacteria.[210]

Pleuromutilin: Lefamulin

Pleuromutilins are natural antimicrobial substances obtained from *Pleurotus mutilus* (*Clitophilus scyphoides*), an edible mushroom.[219] Pleuromutilins are antibacterial agents that inhibit protein synthesis by binding to the 50S subunits of bacterial ribosomes. Lefamulin is a novel semisynthetic pleuromutilin antibiotic that works by inhibiting protein synthesis by binding to the 50S ribosomal subunit at the peptidyl transferase center (PTC), which leads to incorrect positioning of the CCA ends of tRNAs for peptide transfer in the A- and P-sites.[219] Lefamulin has a tricyclene mutilin core that interferes with 23S rRNA with high affinity and specificity.[220] A unique induced-fit mechanism is employed by lefamulin to close the binding pocket within the ribosome, leading to tight binding of lefamulin to its target.[220] This inhibits bacterial peptide chain elongation, leading to bacteriostatic activity of lefamulin. It is active against a number of aerobic Gram-positives (including MRSA, VISA, and vancomycin-resistant *Enterococcus faecium*), some anaerobic Gram-positives (e.g., *Cutibacterium acnes* and *Peptostreptococcus* sp.), and organisms related to community-acquired bacterial pneumonia (*S. pneumoniae, Chlamydia pneumoniae, Mycoplasma pneumoniae, L. pneumophila,* and *H. influenzae*).[220,221] It is not active against *E. faecalis. In vitro* studies show favorable results for the treatment of sexually transmitted infections caused by *Chlamydia trachomatis, Mycoplasma genitalium,* and *N. gonorrhoeae.*[222]

Streptogramins

Streptogramins are very potent against many highly resistant bacteria, and so are considered as last-resort treatment.[223] These drugs have a bacteriostatic effect individually, by binding to the 50S ribosomal subunit, which blocks translation. In combination, different streptogramins may demonstrate synergy of up to a hundred times more than as single agents, and in the process they become bactericidal. Their synergy arises after direct hydrophobic interaction between the streptogramins (e.g., quinupristin-dalfopristin) and a single 23S rRNA nucleotide, resulting in a stable ternary drug-ribosome complex that narrows the ribosomal exit tunnel, preventing nearly synthesized nascent peptide chains from extrusion from the peptide transferase site. These chains accumulate there, possibly blocking peptidyl-tRNA hydrolase activity, leading to a decrease in free tRNAs, and subsequently, bacterial death.[16,223,224] The conformational changes in the 50S subunit have been reported to be induced by the attachment of type A compounds.[225–227]

Type A streptogramins, such as dalfopristin and virginiamycin M, bind to a tight pocket in the peptidyl-transferase catalytic center, preventing attachment of the tRNA to both the acceptor (A) site and donor (D) site. This leads to inhibition of the early stages of protein synthesis, particularly peptide bond formation, and the inhibition of the elongation phase. Ribosomal initiation complexes are assembled in an apparently normal fashion but are functionally inactive.[225–227]

Type B streptogramins, such as quinupristin, inhibit late stages of protein synthesis by inhibiting the first two of the three elongation steps (AA-tRNA binding to the A site and peptidyl transfer from site P) and do not affect the third step (translocation from the A site to the P site). They act by binding to the 23S ribosomal RNA, thereby inhibiting peptide elongation after a few cycles of peptide bond formation. The nascent protein chain is prevented from extending, and the peptidyl tRNA is released from the ribosomes as incomplete peptides.[223,228]

Type A and B streptogramins exert a synergistic inhibitory activity due to the conformational changes in the 50S subunit induced by the attachment of type A compounds,[225–227] which enhances the binding of type B compounds.

Fusidic Acid

Fusidic acid is a steroidal antibiotic that inhibits protein synthesis in prokaryotic and eukaryotic subcellular systems. The antibiotic binds to the ternary complex of prokaryotic EF-G (or EF-2 in eukaryotes) and GDP with the ribosome, preventing release of the elongation factor for further rounds of translocation. This sequestration of EF-G may not explain the inhibitory action of fusidic acid *in vivo*, since growing cells appear to contain an excess of elongation factors; alternative explanations include interference with subsequent complex formation between the ribosome and aminoacyl-tRNA-EF-Tu-GTP and possible effects of fusidic acid on both ribosomal A and P sites.[229,230]

Oxazolidinones: Linezolid, Tedizolid, and Ranbezolid

Linezolid, the first oxazolidinone antibiotic approved for human use, inhibits bacterial growth by interfering with protein synthesis. The drug acts early in protein synthesis, binding to the 23S rRNA of the 50S ribosomal subunit, but not to the 30S ribosomal subunit.[231] Several ribosomal activities (such as formation of the initiation complex, synthesis of the first peptide bond, elongation factor G-dependent translocation, and termination of translation) may be inhibited by oxazolidinones. Linezolid acts early in protein synthesis, preventing binding to the 30S ribosome and formation of the 70S initiation complex. Some of these activities depend on the functions of the ribosomal peptidyl transferase center. Interaction of oxazolidinones with the mitochondrial ribosomes provides a structural basis for the inhibition of mitochondrial protein synthesis, which correlates with clinical side effects associated with oxazolidinone therapy. Molecular modeling has shown that oxazolidinones go "squarely" into the A site of the bacterial ribosome, specifically at the center of the phospholipase transferase center, in the site the aminoacyl-tRNA normally occupies, thereby preventing the latter from binding there.[232] This unique mechanism of action of linezolid explains the absence of cross-resistance with other antibiotics. Linezolid also exhibits antimycobacterial activity.[16,227,232,233]

Tedizolid is a novel oxazolidinone that inhibits protein synthesis. Tedizolid phosphate is a prodrug that is converted by serum phosphatases to the active form, tedizolid.[234] It binds to the 50S ribosomal subunit, thereby blocking the binding between the aminoacyl t-RNA and the A-site of the peptidyl transferase center. Its binding site is very similar to that of linezolid, although tedizolid has an additional D-ring that also engages onto other ribosomal sites, and may explain its improved antimicrobial potency and the development of fewer target mutations and thus less resistance, when compared to linezolid.[235] Due to its increased affinity for the target site, tedizolid has more anti-Gram-positive activity than linezolid.[234] It is active against MRSA, *S. pneumoniae*, VRE, and some anaerobes, including some linezolid-resistant strains of these bacteria.[234]

Ranbezolid is a new investigational oxazolidinone that inhibits protein synthesis as well, by acting on the bacterial ribosome, at the same place where linezolid acts. In addition, it has been shown to inhibit both cell wall and lipid synthesis in a dose-dependent manner in *S. epidermidis*, where it has a bactericidal effect.[236]

Inhibition of Synthesis of Essential Small Molecules

Inhibitors of Dihydropteroate Synthetase

Sulfonamides

The sulfonamides are structural analogs of para-aminobenzoic acid (PABA) and compete with this natural substrate for dihydropteroate synthetases in the bacterial biosynthesis of dihydrofolate (Figure 51.13). In bacterial and mammalian cells, dihydrofolate is reduced by distinct reductases to tetrahydrofolate, which serves as a cofactor for one-carbon transfer in the biosynthesis of essential metabolites such as purines, thymidylate, glycine, and methionine. The susceptibility of bacterial dihydrofolate reductases to inhibition by trimethoprim thus permits sequential blockade of the biosynthesis of tetrahydrofolate in bacteria by the combination of a sulfonamide plus trimethoprim (Figure 51.13). Mammalian cells, which cannot synthesize dihydrofolate from PABA, derive preformed folates (folate, dihydrofolate, and also tetrahydrofolate) from the medium (or diet), whereas bacteria are unable to take up these compounds and are, thus, dependent on the endogenous synthesis of tetrahydrofolate. Sulfonamides are generally used clinically in combination with trimethoprim.[237,238] Like mammalian cells, enterococci are able to take up exogenous folates, and are likewise not inhibited by blockade of the synthesis of dihydro- and tetrahydrofolate.

Inhibitors of Dihydrofolate Reductase

Trimethoprim

Trimethoprim is a potent inhibitor of bacterial dihydrofolate reductases (Figure 51.13), which catalyze the reduction of dihydrofolate to the active tetrahydrofolate. Cofactors of tetrahydrofolate function as donors of one-carbon fragments in the synthesis of thymidylate and other essential compounds such as purines, methionine, panthotenate, glycine, other folate-dependent metabolites, and *N*-formyl methionyl-tRNA. The synthesis of thymidylate from uridylate involves not only C transfer, but also the oxidation of tetrahydrofolate to dihydrofolate (Figure 51.13), which must be reactivated to tetrahydrofolate by dihydrofolate reductase. The toxic effect of the limitation of tetrahydrofolate is primarily due to the deprivation of thymine, which results in

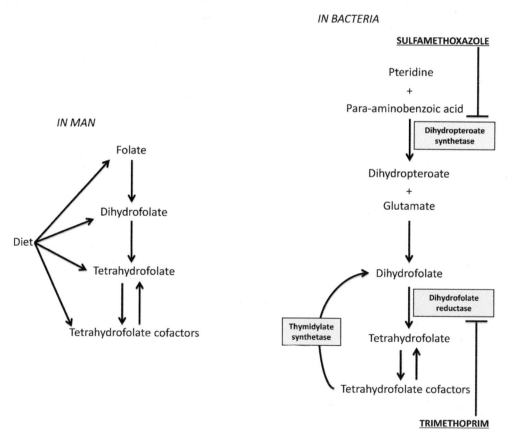

FIGURE 51.13   Folate metabolism in humans and in bacteria, and the sites of action of trimethoprim and sulfa drugs (sulfamethoxazole).

thymineless death because of the resulting "near complete cessation of DNA, RNA and protein synthesis." There is also a depletion of S-adenosyl methionine (SAM), which is an important methyl group donor, and this limits the synthesis of other essential cofactors, fatty acids, and polyamines, contributing to its action. It has been reported that treatment conditions, particularly the availability of folate-dependent metabolites, affect whether trimethoprim will be bacteriostatic or bactericidal. In general, mammalian cells are protected by the extraordinarily low affinity of their dihydrofolate reductase for trimethoprim. The inhibition of dihydrofolate reduction to tetrahydrofolate in bacteria and the resulting DNA damage areessential for the bactericidal effect of trimethoprim. This occurs at the start of the lag phase of treatment and the duration of which is dependent on mutations in the deoxyribose nucleotide salvage pathway and by the ability of cells to buffer themselves from the thymine starvation. It is the distribution of drug susceptibilities in the bacterial population that determines the rate of killing.[238–240]

Iclaprim

Dihydrofolate reductase (DHFR) is an enzyme required for the synthesis of tetrahydrofolate. It participates in the conversion of dihydrofolate to tetrahydrofolate. Like trimethoprim, iclaprim inhibits tetrahydrofolate synthesis by arresting the terminal step in this process, the conversion of dihydrofolateto tetrahydrofolateby acting on DHFR enzyme.[241] Iclaprim is active against many Gram-positive organisms, including methicillin-, vancomycin- and macrolide-resistant strains of *S. aureus*.[242,243] It is also active against some Gram-negatives, including *H. influenzae, M. catarrhalis, E. coli,* and *P. vulgaris*.[241] Iclaprim was modestly active against *K. pneumoniae* and *A. baumannii*, with MICs of 8 mg/L. It does not have activity against *P. aeruginosa*.

Iclaprim is one of the only two members of its class, the other one is trimethoprim.[241] Although both have the same mechanism of action, iclaprim is active against some trimethoprim-resistant strains because of increased hydrophobic interactions between iclaprim and the target enzyme.[241,242]

Iclaprim was synergistic when used with the sulfonamides, sulfamethoxazole, and sulfadiazine, and the combination has been studied as an alternative to TMP-SMX for treatment of *Pneumocystis jirovecii*.[244] Synergy between iclaprim and sulfamethoxazole was also observed for MSSA, MRSA, penicillin-susceptible and penicillin-resistant *S. pneumoniae, H. influenzae,* and *M. catarrhalis*. Synergy was not observed for *K. pneumoniae*.[243]

Antimycobacterial Agents

Inhibitors of Cell Wall Synthesis

Isoniazid (Isonicotinic Acid Hydrazide, INH)

Isoniazid (INH) is a bactericidal agent most active against *Mycobacterium tuberculosis, Mycobacterium bovis,* and *Mycobacterium kansasii*. INH enters the mycobacterial cell by passive diffusion. INH is not toxic to the bacterial cell, but acts as a prodrug. The primary target of action for INH is InhA.[245]

The *inhA* gene of *M. tuberculosis* encodes an NADH-specific enoyl-acyl carrier protein reductase, which was identified in 1994 as a putative target for both INH and the related drug, ethionamide.[246]

INH is activated by the mycobacterial enzyme KatG, a multifunctional catalase-peroxidase, leading to generation of reactive species that form adducts with NAD+ and NADP+, which in turn leads to inhibition of critical enzymes in the biosynthesis of cell wall lipid and nucleic acid.[247] These compounds are potent inhibitors of critical enzymes in the biosynthesis of cell wall lipid and nucleic acid. Some INH-derived reactive species, such as nitric oxide, have a direct role in inhibiting mycobacterial metabolic enzymes.[248]

Cycloserine

D-Cycloserine, a structural analog of D-alanine, is a broadspectrum antibiotic with activity against mycobacteria including *M. tuberculosis* and strains of other organisms, such as *E. coli, S. aureus, Nocardia,* and *Chlamydia* species. At concentrations of 100–200 µg/mL, D-cycloserine inhibits the synthesis of bacterial cell wall peptidoglycan by competitively inhibiting L-alanine racemase and D-alanyl-D-alanine synthetase. This prevents the formation of D-alanine from L-alanine and the synthesis of the D-alanyl-D-alanine dipeptide prior to its addition to the UDP-MurNAc tripeptide. The rigidly planar D-cycloserine has greater affinity for alanine racemase than does either D- or L-alanine.[93,106] (Please see "Cycloserine" section and Figure 51.2 discussed previously in this chapter.)

Ethambutol (EMB)

The mechanism of action of ethambutol (EMB) is not completely understood. It appears to inhibit arabinosyl transferase enzymes that are involved in the biosynthesis of arabinoglycan and lipoarabinomannan, which are essential elements within the mycobacterial cell wall.[249]

EMB differentially affects a regulon (GC82) containing genes within the FAS-II (fatty acid synthase) pathway as well as a regulon (GC17) that contains genes implicated in fatty acid modification. This mechanistic divergence was confirmed by the observation that *M. tuberculosis* cells treated with EMB rapidly lost acid-fastness, while cells treated with the analogs did not.[250]

Pretomanid

Pretomanid is active against both rapidly replicating as well as hypoxic *M. tuberculosis*.[251] Activation of pretomanid is dependent on deazaflavin-dependent nitroreductase.[251] Pretomanid exhibits its antibacterial effects by the production of nitric oxide that results in nonspecific damage of intracellular macromolecules, including proteins and cell wall lipids.[251] This might explain its activity in anaerobic environment.[252] Its activity on replicating bacteria is explained in various studies that suggest that this drug affects mycolic acid synthesis through depletion and accumulation of ketomycolates and hydroxymycolates, respectively.[252]

It is now approved for treatment of multidrug-resistant and extensively drug-resistant tuberculosis in combination with linezolid and bedaquiline.

Interference with Cytoplasmic Membrane Function

Pyrazinamide (PZA)

Pyrazinamide (PZA), an analog of nicotinamide, is preferentially active against nonreplicating persistent mycobacteria with low metabolism at acid pH *in vitro*[253] or *in vivo* during active inflammation.[254] PZA is a prodrug that enters *M. tuberculosis* by passive diffusion, where it is converted into the active form, pyrazinoic acid, by bacterial nicotinamidase/pyrazinamidase.

The target of pyrazinoic acid is the ribosomal protein S1 (RpsA), a vital protein involved in protein translation and the ribosome-sparing process of trans-translation experienced by the bacteria under stress conditions. Inhibition of trans-translation may explain the ability of PZA to eradicate persisting organisms.[255] The remarkable *in vitro* sterilizing effect of PZA is also due to disruption of membrane energetics and inhibited membrane function in *M. tuberculosis*.[256]

Inhibition of Nucleic Acid Synthesis

Rifamycins: Rifampin, Rifabutin, and Rifapentine

The mechanism of action of rifamycins involves inhibition of DNA-dependent RNA synthesis caused by strong binding to the β-subunit of the DNA-dependent RNA polymerase of prokaryotes, with a binding constant in the range of 10^{-8} M. Eukaryotic enzymes are at least 10^2 to 10^4 times less sensitive to inhibition by rifampin. Rifabutin and rifapentine are semisynthetic derivatives of rifampin. Rifabutin is active against several rifampin-resistant clinical pathogens, and rifapentine has a long half-life of approximately 12 hours, allowing its administration in once-weekly regimens.[257-259] (Please see "Inhibitors of RNA Polymerase" section discussed previously in this chapter.)

Fluoroquinolones as Antimycobacterial Agents

Fluorinated quinolones have been studied and used as second-line antituberculous drugs. Some fluoroquinolones, such as levofloxacin and moxifloxacin, are bactericidal against *M. tuberculosis*, presumably by inhibition of its DNA gyrase at concentrations within achievable serum levels.[260-264]

Inhibitors of Ribosome Function and Protein Synthesis

Streptomycin and Amikacin

For streptomycin and amikacin, see previous section entitled "Aminoglycosides" in this chapter.

Capreomycin

Capreomycin is a polypeptide antibiotic with a poorly understood mechanism of action, even active on nonreplicating Mycobacteria. It is thought to inhibit protein synthesis by binding to the 50S ribosomal subunit of the 70S ribosome, leading to abnormal protein synthesis that is ultimately fatal to the bacteria. Microarray studies have suggested that capreomycin also affects biochemical and metabolic pathways.[265]

Inhibition of Mycobacterial ATP Synthase

Bedaquiline

Bedaquiline (formerly known as R207910 and TMC207) is a diarylquinoline compound with a unique spectrum specific to mycobacteria, including atypical Mycobacteria that are important in humans such as *Mycobacterium avium*, *Mycobacterium intracellulare complex*, *Mycobacterium kansasii*, and the fast growers, *Mycobacterium fortuitum* and *Mycobacterium abscessus*.

Due to its unique mechanism of action, bedaquiline is active not only on drug-sensitive *M. tuberculosis*, but also against isolates resistant to other TB agents: Isoniazid, rifampin, streptomycin, ethambutol, pyrazinamide, and moxifloxacin.[266,267] *In vitro*, it is also bactericidal against dormant, nonreplicating tubercle bacilli[266,268,269] and it has been shown to have synergistic activity against murine tuberculosis when combined with pyrazinamide.[270]

The target of bedaquiline in mycobacteria is the *atpE* gene product, the oligomeric subunit c of ATP synthase.[266-267]

Inhibition of the proton pump of ATP synthase may lead to ATP depletion and imbalance in ATP homeostasis, both contributing to decreased survival.[266,271,272]

Monoclonal Antibodies

Anthrax Antitoxin

Obiltoxaximab

Obiltoxaximab is a chimeric immunoglobulin G1 κ monoclonal antibody used for the prevention and treatment of inhalational anthrax due to *Bacillus anthracis* in combination with appropriate antibacterials.[273] The protective antigen (PA) of *B. anthracis* binds to host cell surface receptors and facilitates the transport of exotoxins, lethal factor (LF), and edema factor (EF) into the host cell.[273] The PA therefore plays a key role in *B. anthracis* pathogenesis.[273] Obiltoxaximab neutralizes the free PA and thereby prevents the cell entry of LF and EF, the two exotoxins that mediate the pathogenesis of anthrax.[273] It carries the risk of hypersensitivity and anaphylaxis, and therefore patients are premedicated with diphenhydramine.[273] It does not have direct antibacterial activity and does not cross the blood-brain barrier.[273]

Bezlotoxumab

Bezlotoxumab is a monoclonal antibody that binds to two separate sites in the N-terminal half of the combined repetitive oligopeptide (CROP) domain of toxin B produced by *Clostridioides difficile*.[274] Once it binds to toxin B, it prevents interaction of toxin B with the host receptor CSPG-4 (chondroitin sulfate proteoglycan 4) and resulting intoxication of the mammalian cells.[274] It can neutralize toxins from several strains of *C. difficile*, including the BI/NAP1/027 and BK/NAP7/078 strains.[274] A post-hoc analysis of pooled data revealed that patients with age >65, history of CDI, immunocompromise, or severe CDI had reduced risk of recurrent CDI, and 30-day readmission for CDI, and avoided fecal transplant if they received adjunctive treatment with a single-dose infusion of bezlotoxumab.[274] The risk of first post-treatment relapse is nearly 40% lower with bezlotoxumab than that associated with standard treatment alone with oral vancomycin, metronidazole, or fidaxomicin.[275]

REFERENCES

1. Kahan, F. M., Kahan, J. S., Cassidy, P. J. & Kropp, H. The mechanism of action of fosfomycin (phosphonomycin). *Ann N Y Acad Sci* **235**, 364–386 (1974).
2. Wu, H. C. & Venkateswaran, P. S. Fosfomycin-resistant mutant of *Escherichia coli*. *Ann N Y Acad Sci* **235**, 587–592 (1974).
3. Raz, R. Fosfomycin: an old–new antibiotic. *Clin Microbiol Infect* **18**, 4–7 (2012).
4. Silver, L. L. Does the cell wall of bacteria remain a viable source of targets for novel antibiotics? *Biochem Pharmacol* **71**, 996–1005 (2006).
5. Kim, M. G., Strych, U., Krause, K., Benedik, M. & Kohn, H. N(2)-substituted D, L-cycloserine derivatives: synthesis and evaluation as alanine racemase inhibitors. *J Antibiot (Tokyo)* **56**, 160–168 (2003).
6. Dickerhof, N., Kleffmann, T., Jack, R. & McCormick, S. Bacitracin inhibits the reductive activity of protein disulfide isomerase by disulfide bond formation with free cysteines in the substrate-binding domain. *FEBS J* **278**, 2034–2043 (2011).
7. Siewert, G. & Strominger, J. L. Bacitracin: an inhibitor of the dephosphorylation of lipid pyrophosphate, an intermediate in the biosynthesis of the peptidoglycan of bacterial cell walls. *Proc Natl Acad Sci U S A* **57**, 767–773 (1967).
8. Stone, K. J. & Strominger, J. L. Mechanism of action of bacitracin: complexation with metal ion and C 55 -isoprenyl pyrophosphate. *Proc Natl Acad Sci U S A* **68**, 3223–3227 (1971).
9. Stone, K. J. & Strominger, J. L. Inhibition of sterol biosynthesis by bacitracin. *Proc Natl Acad Sci U S A* **69**, 1287–1289 (1972).
10. Karala, A. R. & Ruddock, L. W. Bacitracin is not a specific inhibitor of protein disulfide isomerase. *FEBS J* **277**, 2454–2462 (2010).
11. Hammes, W. P. & Neuhaus, F. C. On the mechanism of action of vancomycin: inhibition of peptidoglycan synthesis in Gaffkya homari. *Antimicrob Agents Chemother* **6**, 722–728 (1974).
12. Perkins, H. R. & Nieto, M. The chemical basis for the action of the vancomycin group of antibiotics. *Ann N Y Acad Sci* **235**, 348–363 (1974).
13. Williams, D. H. & Kalman, J. Structural and mode of action studies on the antibiotic vancomycin. Evidence from 270-MHz proton magnetic resonance. *J Am Chem Soc* **99**, 2768–2774 (1977).
14. Sheldrick, G. M., Jones, P. G., Kennard, O., Williams, D. H. & Smith, G. A. Structure of vancomycin and its complex with acetyl-D-alanyl-D-alanine. *Nature* **271**, 223–225 (1978).
15. Kahne, D., Leimkuhler, C., Lu, W. & Walsh, C. Glycopeptide and lipoglycopeptide antibiotics. *Chem Rev* **105**, 425–448 (2005).
16. Nailor, M. D. & Sobel, J. D. Antibiotics for gram-positive bacterial infections: vancomycin, teicoplanin, quinupristin/ dalfopristin, oxazolidinones, daptomycin, dalbavancin, and telavancin. *Infect Dis Clin North Am* **23**, 965–982, ix (2009).
17. Jordan, D. C. & Inniss, W. E. Selective inhibition of ribonucleic acid synthesis in Staphylococcus aureus by vancomycin. *Nature* **184**(Suppl 24), 1894–1895 (1959).
18. Jordan, D. C. & Mallory, H. D. Site of action of vancomycin on Staphylococcus Aureus. *Antimicrob Agents Chemother* **10**, 489–494 (1964).
19. Cunha, B. A., Quintiliani, R., Deglin, J. M., Izard, M. W. & Nightingale, C. H. Pharmacokinetics of vancomycin in anuria. *Rev Infect Dis* **3**(suppl), S269–272 (1981).
20. Xing, B. *et al.* Molecular interactions between glycopeptide vancomycin and bacterial cell wall peptide analogues. *Chemistry* **17**, 14170–14177 (2011).
21. Koteva, K. *et al.* A vancomycin photoprobe identifies the histidine kinase VanSsc as a vancomycin receptor. *Nat Chem Biol* **6**, 327–329 (2010).
22. Saravolatz, L. D., Stein, G. E. & Johnson, L. B. Telavancin: a novel lipoglycopeptide. *Clinical Infect Dis* **49**, 1908–1914 (2009).
23. Pace, J. L. *et al.* In vitro activity of TD-6424 against Staphylococcus aureus. *Antimicrob Agents Chemother* **47**, 3602–3604 (2003).
24. Barcia-Macay, M., Lemaire, S., Mingcot-Leclercq, M. P., Tulkens, P. M. & Van Bambeke, F. Evaluation of the extracellular and intracellular activities (human THP-1 macrophages) of telavancin versus vancomycin against methicillin-susceptible, methicillin-resistant, vancomycin-intermediate and vancomycin-resistant Staphylococcus aureus. *J Antimicrob Chemother* **58**, 1177–1184, doi:10.1093/jac/dkl424 (2006).
25. Lunde, C. S. *et al.* Telavancin disrupts the functional integrity of the bacterial membrane through targeted interaction with the cell wall precursor lipid II. *Antimicrob Agents Chemother* **53**, 3375–3383 (2009).
26. Pace, J. L. & Judice, J. K. Telavancin (Theravance). *Curr Opin Investig Drugs* **6**, 216–225 (2005).
27. Higgins, D. L. *et al.* Telavancin, a multifunctional lipoglycopeptide, disrupts both cell wall synthesis and cell membrane integrity in methicillin-resistant Staphylococcus aureus. *Antimicrob Agents Chemother* **49**, 1127–1134 (2005).
28. Van Bambeke, F. Glycopeptides and glycodepsipeptides in clinical development: a comparative review of their antibacterial spectrum, pharmacokinetics and clinical efficacy. *Curr Opin Investig Drugs* **7**, 740–749 (2006).
29. Laohavaleeson, S., Kuti, J. L. & Nicolau, D. P. Telavancin: a novel lipoglycopeptide for serious gram-positive infections. *Expert Opin Investig Drugs* **16**, 347–357 (2007).
30. Song, Y., Lunde, C. S., Benton, B. M. & Wilkinson, B. J. Further insights into the mode of action of the lipoglycopeptide telavancin through global gene expression studies. *Antimicrob Agents Chemother* **56**, 3157–3164 (2012).
31. Alfandari, S. *et al.* Evaluation of glycopeptide use in nine French hospitals. *Med Mal Infect* **40**, 232–237 (2010).
32. Meyer, E., Schwab, F., Schroeren-Boersch, B. & Gastmeier, P. Increasing consumption of MRSA-active drugs without increasing MRSA in German ICUs. *Intensive Care Med* **37**, 1628–1632, (2011).
33. Seetulsingh, P. & Collier, S. An audit of teicoplanin use in a district general hospital. *J Hosp Infect* **64**, 194–196 (2006).
34. Borghi, A. *et al.* Teichomycins, new antibiotics from Actinoplanes teichomyceticus nov. sp. IV. Separation and characterization of the components of teichomycin (teicoplanin). *J Antibiot (Tokyo)* **37**, 615–620 (1984).
35. Van Bambeke, F., Van Laethem, Y., Courvalin, P. & Tulkens, P. M. Glycopeptide antibiotics: from conventional molecules to new derivatives. *Drugs* **64**, 913–936 (2004).

36. Lode, H. *et al.* Comparative pharmacokinetics of glycopeptide antibiotics, and the influence of teicoplanin on granulocyte function. *Scand J Infect Dis Suppl* **72**, 9–13 (1990).

37. Fang, X. *et al.* The mechanism of action of ramoplanin and enduracidin. *Mol Biosyst* **2**, 69–76, doi:10.1039/b515328j (2006).

38. McCafferty, D. G. *et al.* Chemistry and biology of the ramoplanin family of peptide antibiotics. *Biopolymers* **66**, 261–284 (2002).

39. Fulco, P. & Wenzel, R. P. Ramoplanin: a topical lipoglycodepsipeptide antibacterial agent. *Expert Rev Anti Infect Ther* **4**, 939–945 (2006).

40. Malabarba, A. & Goldstein, B. P. Origin, structure, and activity in vitro and in vivo of dalbavancin. *J Antimicrob Chemother* **55**(Suppl 2), ii15–20 (2005).

41. Lin, S.-W., Carver, P. L. & DePestel, D. D. Dalbavancin: a new option for the treatment of gram-positive infections. *Ann Pharmacother* **40**, 449–460 (2006).

42. Boylan, C. J. *et al.* Pharmacodynamics of oritavancin (LY333328) in a neutropenic-mouse thigh model of Staphylococcus aureus infection. *Antimicrob Agents Chemother* **47**, 1700–1706 (2003).

43. Malabarba, A. & Ciabatti, R. Glycopeptide derivatives. *Curr Med Chem* **8**, 1759–1773 (2001).

44. Allen, N. E. & Nicas, T. I. Mechanism of action of oritavancin and related glycopeptide antibiotics. *FEMS Microbiol Rev* **26**, 511–532 (2003).

45. Zhanel, G. G., Schweizer, F. & Karlowsky, J. A. Oritavancin: mechanism of action. *Clin Infect Dis* **54**(Suppl 3), S214–219 (2012).

46. Schneider, T. & Sahl, H. G. An oldie but a goodie - cell wall biosynthesis as antibiotic target pathway. *Int J Med Microbiol* **300**, 161–169 (2010).

47. Kong, K. F., Schneper, L. & Mathee, K. Beta-lactam antibiotics: from antibiosis to resistance and bacteriology. *APMIS* **118**, 1–36 (2010).

48. Strominger, J. L., Willoughby, E., Kamiryo, T., Blumberg, P. M. & Yocum, R. R. Penicillin-sensitive enzymes and penicillin-binding components in bacterial cells. *Ann N Y Acad Sci* **235**, 210–224 (1974).

49. Blumberg, P. M. & Strominger, J. L. Interaction of penicillin with the bacterial cell: penicillin-binding proteins and penicillin-sensitive enzymes. *Bacteriol Rev* **38**, 291–335 (1974).

50. Walsh, C. *Antibiotics. Actions, Origins, Resistance* (Washington, DC: ASM Press, 2003).

51. Asbel, L. E. & Levison, M. E. Cephalosporins, carbapenems, and monobactams. *Infect Dis Clin North Am* **14**, 435–447, ix (2000).

52. Kohanski, M. A., Dwyer, D. J. & Collins, J. J. How antibiotics kill bacteria: from targets to networks. *Nat Rev Microbiol* **8**, 423–435 (2010).

53. Llarrull, L. I., Testero, S. A., Fisher, J. F. & Mobashery, S. The future of the beta-lactams. *Curr Opin Microbiol* **13**, 551–557 (2010).

54. Villegas-Estrada, A., Lee, M., Hesek, D., Vakulenko, S. B. & Mobashery, S. Co-opting the cell wall in fighting methicillin-resistant Staphylococcus aureus: potent inhibition of PBP 2a by two anti-MRSA beta-lactam antibiotics. *J Am Chem Soc* **130**, 9212–9213 (2008).

55. Moya, B. *et al.* Activity of a new cephalosporin, CXA-101 (FR264205), against beta-lactam-resistant *Pseudomonas aeruginosa* mutants selected in vitro and after antipseudomonal treatment of intensive care unit patients. *Antimicrob Agents Chemother* **54**, 1213–1217 (2010).

56. Matsumoto, S. *et al.* Efficacy of cefiderocol against carbapenem-resistant gram-negative bacilli in immunocompetent-rat respiratory tract infection models recreating human plasma pharmacokinetics. *Antimicrob Agents Chemother* **61**, e00700–00717 (2017).

57. Ito, A. *et al.* Siderophore cephalosporin cefiderocol utilizes ferric iron transporter systems for antibacterial activity against pseudomonas aeruginosa. *Antimicrob Agents Chemother* **60**, 7396–7401 (2016).

58. Farrell, D. J., Flamm, R. K., Sader, H. S. & Jones, R. N. Ceftobiprole activity against over 60,000 clinical bacterial pathogens isolated in Europe, Turkey, and Israel from 2005 to 2010. *Antimicrob Agents Chemother* **58**, 3882–3888 (2014).

59. Papp-Wallace, K. M., Endimiani, A., Taracila, M. A. & Bonomo, R. A. Carbapenems: past, present, and future. *Antimicrob Agents Chemother* **55**, 4943–4960 (2011).

60. Chavez-Bueno, S. & Stull, T. L. Antibacterial agents in pediatrics. *Infect Dis Clin North Am* **23**, 865–880, viii (2009).

61. Trias, J. & Nikaido, H. Outer membrane protein D2 catalyzes facilitated diffusion of carbapenems and penems through the outer membrane of Pseudomonas aeruginosa. *Antimicrob Agents Chemother* **34**, 52–57 (1990).

62. Zhanel, G. G. *et al.* Comparative review of the carbapenems. *Drugs* **67**, 1027–1052 (2007).

63. Hashizume, T., Ishino, F., Nakagawa, J., Tamaki, S. & Matsuhashi, M. Studies on the mechanism of action of imipenem (N-formimidoylthienamycin) in vitro: binding to the penicillin-binding proteins (PBPs) in Escherichia coli and Pseudomonas aeruginosa, and inhibition of enzyme activities due to the PBPs in E. coli. *J Antibiot (Tokyo)* **37**, 394–400 (1984).

64. Rodloff, A. C., Goldstein, E. J. & Torres, A. Two decades of imipenem therapy. *J Antimicrob Chemother* **58**, 916–929 (2006).

65. Livermore, D. M., Sefton, A. M. & Scott, G. M. Properties and potential of ertapenem. *J Antimicrob Chemother* **52**, 331–344 (2003).

66. Dalhoff, A., Nasu, T. & Okamoto, K. Target affinities of faropenem to and its impact on the morphology of gram-positive and gram-negative bacteria. *Chemotherapy* **49**, 172–183 (2003).

67. Bassetti, M., Righi, E. & Viscoli, C. Novel beta-lactam antibiotics and inhibitor combinations. *Expert Opin Investig Drugs* **17**, 285–296 (2008).

68. Hamilton-Miller, J. M. Chemical and microbiologic aspects of penems, a distinct class of beta-lactams: focus on faropenem. *Pharmacotherapy* **23**, 1497–1507 (2003).

69. Hayashi, N. & Kawashima, M. Multicenter randomized controlled trial on combination therapy with 0.1% adapalene gel and oral antibiotics for acne vulgaris: comparison of the efficacy of adapalene gel alone and in combination with oral faropenem. *J Dermatol* **39**, 511–515 (2012).

70. Hayashi, N. & Kawashima, M. Efficacy of oral antibiotics on acne vulgaris and their effects on quality of life: a multicenter randomized controlled trial using minocycline, roxithromycin and faropenem. *J Dermatol* **38**, 111–119 (2011).

71. Morihara, K., Takenaka, H., Morihara, T. & Katoh, N. Cutaneous Mycobacterium chelonae infection successfully treated with faropenem. *J Dermatol* **38**, 211–213 (2011).

72. Maruyama, T. *et al.* CP5484, a novel quaternary carbapenem with potent anti-MRSA activity and reduced toxicity. *Bioorg Med Chem* **15**, 6379–6387 (2007).

73. Drawz, S. M. & Bonomo, R. A. Three decades of beta-lactamase inhibitors. *Clin Microbiol Rev* **23**, 160–201 (2010).

74. Crass, R. L. & Pai, M. P. Pharmacokinetics and pharmacodynamics of beta-lactamase inhibitors. *Pharmacotherapy* **39**, 182–195 (2019).

75. Lahiri, S. D. *et al.* Avibactam and class C beta-lactamases: mechanism of inhibition, conservation of the binding pocket, and implications for resistance. *Antimicrob Agents Chemother* **58**, 5704–5713 (2014).

76. Tuon, F. F., Rocha, J. L. & Formigoni-Pinto, M. R. Pharmacological aspects and spectrum of action of ceftazidime-avibactam: a systematic review. *Infection* **46**, 165–181 (2018).

77. Lomovskaya, O. *et al.* Vaborbactam: Spectrum of beta-lactamase inhibition and impact of resistance mechanisms on activity in Enterobacteriaceae. *Antimicrob Agents Chemother* **61**, e01443–01417 (2017).

78. Zhanel, G. G. *et al.* Imipenem–relebactam and meropenem–vaborbactam: Two novel carbapenem-β-lactamase inhibitor combinations. *Drugs* **78**, 65–98 (2018).

79. Pogue, J. M., Bonomo, R. A. & Kaye, K. S. Ceftazidime/Avibactam, Meropenem/Vaborbactam, or Both? Clinical and formulary considerations. *Clin Infect Dis* **68**, 519–524 (2019).

80. Barnes, M. D. *et al.* Deciphering the evolution of cephalosporin resistance to ceftolozane-tazobactam in pseudomonas aeruginosa. *mBio* **9** (2018).

81. Giacobbe, D. R. *et al.* Ceftolozane/tazobactam: place in therapy. *Expert Rev Anti Infect Ther* **16**, 307–320 (2018).

82. Fraile-Ribot, P. A. *et al.* Mechanisms leading to in vivo ceftolozane/tazobactam resistance development during the treatment of infections caused by MDR Pseudomonas aeruginosa. *J Antimicrob Chemother* **73**, 658–663 (2018).

83. Shortridge, D., Pfaller, M. A., Castanheira, M. & Flamm, R. K. Antimicrobial activity of ceftolozane-tazobactam tested against enterobacteriaceae and pseudomonas aeruginosa with various resistance patterns isolated in U.S. Hospitals (2013-2016) as part of the surveillance program: Program to assess ceftolozane-tazobactam susceptibility. *Microb Drug Resist* **24**, 563–577 (2018).

84. Zhanel, G. G. *et al.* Imipenem-relebactam and meropenem-vaborbactam: Two novel carbapenem-beta-lactamase inhibitor combinations. *Drugs* **78**, 65–98 (2018).

85. Papp-Wallace, K. M. *et al.* Relebactam is a potent inhibitor of the KPC-2 beta-lactamase and restores imipenem susceptibility in KPC-producing enterobacteriaceae. *Antimicrob Agents Chemother* **62** (2018).

86. Rodvold, K. A. *et al.* Plasma and intrapulmonary concentrations of cefepime and zidebactam following intravenous administration of WCK 5222 to healthy adult subjects. *Antimicrob Agents Chemother* **62** (2018).

87. Sader, H. S., Rhomberg, P. R., Flamm, R. K., Jones, R. N. & Castanheira, M. WCK 5222 (cefepime/zidebactam) antimicrobial activity tested against gram-negative organisms producing clinically relevant beta-lactamases. *J Antimicrob Chemother* **72**, 1696–1703 (2017).

88. Sykes, R. B. & Bonner, D. P. Aztreonam: the first monobactam. *Am J Med* **78**, 2–10 (1985).

89. Sykes, R. B., Bonner, D. P., Bush, K. & Georgopapadakou, N. H. Azthreonam (SQ 26,776), a synthetic monobactam specifically active against aerobic gram-negative bacteria. *Antimicrob Agents Chemother* **21**, 85–92 (1982).

90. Hutchinson, D., Barclay, M., Prescott, W. A. & Brown, J. Inhaled aztreonam lysine: an evidence-based review. *Expert Opin Pharmacother* **14**, 2115–2124 (2013).

91. Georgopapadakou, N. H., Smith, S. A. & Sykes, R. B. Mode of action of azthreonam. *Antimicrob Agents Chemother* **21**, 950–956 (1982).

92. Gibson, R. L. *et al.* Microbiology, safety, and pharmacokinetics of aztreonam lysinate for inhalation in patients with cystic fibrosis. *Pediatr Pulmonol* **41**, 656–665 (2006).

93. Queenan, A. M. & Bush, K. Carbapenemases: the versatile beta-lactamases. *Clin Microbiol Rev* **20**, 440–458 (2007).

94. HsuChen, C. C. & Feingold, D. S. The mechanism of polymyxin B action and selectivity toward biologic membranes. *Biochemistry* **12**, 2105–2111 (1973).

95. Feingold, D. S., HsuChen, C. C. & Sud, I. J. Basis for the selectivity of action of the polymyxin antibiotics on cell membranes. *Ann N Y Acad Sci* **235**, 480–492 (1974).

96. Lounatmaa, K., Makela, P. H. & Sarvas, M. Effect of polymyxin on the ultrastructure of the outer membrane of wild-type and polymyxin-resistant strain of Salmonella. *J Bacteriol* **127**, 1400–1407 (1976).

97. Li, J., Nation, R. L., Milne, R. W., Turnidge, J. D. & Coulthard, K. Evaluation of colistin as an agent against multi-resistant gram-negative bacteria. *Int J Antimicrob Agents* **25**, 11–25 (2005).

98. Evans, M. E., Feola, D. J. & Rapp, R. P. Polymyxin B sulfate and colistin: old antibiotics for emerging multiresistant gram-negative bacteria. *Ann Pharmacother* **33**, 960–967 (1999).

99. Clausell, A. *et al.* Gram-negative outer and inner membrane models: insertion of cyclic cationic lipopeptides. *J Phys Chem B* **111**, 551–563 (2007).

100. Domingues, M. M. *et al.* Biophysical characterization of polymyxin B interaction with LPS aggregates and membrane model systems. *Biopolymers* **98**, 338–344 (2012).

101. Sampson, T. R. *et al.* Rapid killing of *Acinetobacter baumannii* by polymyxins is mediated by a hydroxyl radical death pathway. *Antimicrob Agents Chemother* **56**, 5642–5649 (2012).

102. Dixon, R. A. & Chopra, I. Polymyxin B and polymyxin B nonapeptide alter cytoplasmic membrane permeability in Escherichia coli. *J Antimicrob Chemother* **18**, 557–563 (1986).

103. Clausell, A., Pujol, M., Alsina, M. A. & Cajal, Y. Influence of polymyxins on the structural dynamics of Escherichia coli lipid membranes. *Talanta* **60**, 225–234 (2003).

104. Mortensen, N. P. *et al.* Effects of colistin on surface ultrastructure and nanomechanics of Pseudomonas aeruginosa cells. *Langmuir* **25**, 3728–3733 (2009).

105. Wach, A., Dembowsky, K. & Dale, G. E. Pharmacokinetics and safety of intravenous murepavadin infusion in healthy adult subjects administered single and multiple ascending doses. *Antimicrob Agents Chemother* **62** (2018).

106. Martin-Loeches, I., Dale, G. E. & Torres, A. Murepavadin: a new antibiotic class in the pipeline. *Expert Rev Anti Infect Ther* **16**, 259–268 (2018).

107. Werneburg, M. *et al.* Inhibition of lipopolysaccharide transport to the outer membrane in Pseudomonas aeruginosa by peptidomimetic antibiotics. *Chembiochem* **13**, 1767–1775 (2012).

108. Straus, S. K. & Hancock, R. E. Mode of action of the new antibiotic for gram-positive pathogens daptomycin: comparison with cationic antimicrobial peptides and lipopeptides. *Biochim Biophys Acta* **1758**, 1215–1223 (2006).

109. Muraih, J. K., Pearson, A., Silverman, J. & Palmer, M. Oligomerization of daptomycin on membranes. *Biochim Biophys Acta* **1808**, 1154–1160 (2011).

110. Tally, F. P. & DeBruin, M. F. Development of daptomycin for gram-positive infections. *J Antimicrob Chemother* **46**, 523–526 (2000).

111. Pogliano, J., Pogliano, N. & Silverman, J. A. Daptomycin-mediated reorganization of membrane architecture causes mislocalization of essential cell division proteins. *J Bacteriol* **194**, 4494–4504 (2012).

112. Duax, W. L. *et al.* Molecular structure and mechanisms of action of cyclic and linear ion transport antibiotics. *Biopolymers* **40**, 141–155 (1996).

113. Wallace, B. A. Recent advances in the high resolution structures of bacterial channels: Gramicidin A. *J Struct Biol* **121**, 123–141 (1998).

114. Urban, B. W., Hladky, S. B. & Haydon, D. A. Ion movements in gramicidin pores. An example of single-file transport. *Biochim Biophys Acta* **602**, 331–354 (1980).

115. Urry, D. W. A molecular theory of ion-conductng channels: a field-dependent transition between conducting and non-conducting conformations. *Proc Natl Acad Sci U S A* **69**, 1610–1614 (1972).

116. Urry, D. W., Goodall, M. C., Glickson, J. D. & Mayers, D. F. The gramicidin A transmembrane channel: characteristics of head-to-head dimerized (L,D) helices. *Proc Natl Acad Sci U S A* **68**, 1907–1911 (1971).

117. Pache, W., Chapman, D. & Hillaby, R. Interaction of antibiotics with membranes: Polymyxin B and gramicidin S. *Biochim Biophys Acta* **255**, 358–364 (1972).

118. Prenner, E. J., Lewis, R. N. & McElhaney, R. N. The interaction of the antimicrobial peptide gramicidin S with lipid bilayer model and biological membranes. *Biochim Biophys Acta* **1462**, 201–221 (1999).

119. Prenner, E. J. *et al.* Structure-activity relationships of diastereomeric lysine ring size analogs of the antimicrobial peptide gramicidin S: mechanism of action and discrimination between bacterial and animal cell membranes. *J Biol Chem* **280**, 2002–2011 (2005).

120. Afonin, S., Durr, U. H., Wadhwani, P., Salgado, J. & Ulrich, A. S. Solid state NMR structure analysis of the antimicrobial peptide gramicidin S in lipid membranes: Concentration-dependent re-alignment and self-assembly as a beta-barrel. *Top Curr Chem* **273**, 139–154 (2008).

121. Pommier, Y., Leo, E., Zhang, H. & Marchand, C. DNA topoisomerases and their poisoning by anticancer and antibacterial drugs. *Chem Biol* **17**, 421–433 (2010).

122. Drlica, K. Mechanism of fluoroquinolone action. *Curr Opin Microbiol* **2**, 504–508 (1999).

123. Levine, C., Hiasa, H. & Marians, K. J. DNA gyrase and topoisomerase IV: biochemical activities, physiological roles during chromosome replication, and drug sensitivities. *Biochim Biophys Acta* **1400**, 29–43 (1998).

124. Mitscher, L. A. Bacterial topoisomerase inhibitors: quinolone and pyridone antibacterial agents. *Chem Rev* **105**, 559–592 (2005).

125. Khodursky, A. B. & Cozzarelli, N. R. The mechanism of inhibition of topoisomerase IV by quinolone antibacterials. *J Biol Chem* **273**, 27668–27677 (1998).

126. Cheng, G., Hao, H., Dai, M., Liu, Z. & Yuan, Z. Antibacterial action of quinolones: from target to network. *Eur J Med Chem* **66**, 555–562 (2013).

127. Bourguignon, G. J., Levitt, M. & Sternglanz, R. Studies on the mechanism of action of nalidixic acid. *Antimicrob Agents Chemother* **4**, 479–486 (1973).

128. Sugino, A., Peebles, C. L., Kreuzer, K. N. & Cozzarelli, N. R. Mechanism of action of nalidixic acid: purification of Escherichia coli nalA gene product and its relationship to DNA gyrase and a novel nicking-closing enzyme. *Proc Natl Acad Sci U S A* **74**, 4767–4771 (1977).

129. Fabrega, A., Madurga, S., Giralt, E. & Vila, J. Mechanism of action of and resistance to quinolones. *Microb Biotechnol* **2**, 40–61 (2009).

130. Emmerson, A. M. & Jones, A. M. The quinolones: decades of development and use. *J Antimicrob Chemother* **51**(Suppl 1), 13–20 (2003).

131. Stein, G. E. & Goldstein, E. J. Fluoroquinolones and anaerobes. *Clin Infect Dis* **42**, 1598–1607 (2006).

132. Buensalido, J. A. L., Schindler, B. D. & Kaatz, G. W. Fluoroquinolone Resistance in Bacteria, in *Antimicrobial Drug Resistance* (Springer, 2017).

133. López, Y. *et al.* In vitro activity of Ozenoxacin against quinolone-susceptible and quinolone-resistant gram-positive bacteria. *Antimicrob Agents Chemother* **57**, 6389–6392 (2013).

134. Wren, C., Bell, E. & Eiland, L. S. Ozenoxacin: A novel topical quinolone for impetigo. *Ann Pharmacother* **52**, 1233–1237 (2018).

135. Cho, J. C., Crotty, M. P., White, B. P. & Worley, M. V. What is old is new again: Delafloxacin, a modern fluoroquinolone. *Pharmacotherapy* **38**, 108–121 (2018).

136. McCurdy, S. *et al.* In Vitro activity of Delafloxacin and microbiological response against fluoroquinolone-susceptible and nonsusceptible staphylococcus aureus isolates from two phase 3 studies of acute bacterial skin and skin structure infections. *Antimicrob Agents Chemother* **61**, e00772–00717 (2017).

137. Pfaller, M. A., Sader, H. S., Rhomberg, P. R. & Flamm, R. K. In vitro activity of delafloxacin against contemporary bacterial pathogens from the United States and Europe, 2014. *Antimicrob Agents Chemother* **61**, e02609–02616 (2017).

138. Flamm, R. K., Farrell, D. J., Rhomberg, P. R., Scangarella-Oman, N. E. & Sader, H. S. Gepotidacin (GSK2140944) in vitro activity against gram-positive and gram-negative bacteria. *Antimicrob Agents Chemother* **61** (2017).

139. O'Riordan, W. *et al.* Efficacy, safety, and tolerability of gepotidacin (GSK2140944) in the treatment of patients with suspected or confirmed gram-positive acute bacterial skin and skin structure infections. *Antimicrob Agents Chemother* **61** (2017).

140. Taylor, S. N. *et al.* Gepotidacin for the treatment of uncomplicated urogenital gonorrhea: A phase 2, randomized, dose-ranging, single-oral dose evaluation. *Clin Infect Dis* **67**, 504–512 (2018).

141. Huband, M. D. *et al.* In vitro antibacterial activity of AZD0914, a new spiropyrimidinetrione DNA gyrase/topoisomerase inhibitor with potent activity against gram-positive, fastidious gram-negative, and atypical bacteria. *Antimicrob Agents Chemother* **59**, 467–474 (2015).

142. Kohlhoff, S. A., Huband, M. D. & Hammerschlag, M. R. In Vitro Activity of AZD0914, a novel DNA gyrase inhibitor, against Chlamydia trachomatis and *Chlamydia pneumoniae*. *Antimicrob Agents Chemother* **58**, 7595–7596 (2014).

143. Waites, K. B., Crabb, D. M., Duffy, L. B. & Huband, M. D. In vitro antibacterial activity of AZD0914 against human Mycoplasmas and Ureaplasmas. *Antimicrob Agents Chemother* **59**, 3627–3629 (2015).

144. Lamp, K. C., Freeman, C. D., Klutman, N. E. & Lacy, M. K. Pharmacokinetics and pharmacodynamics of the nitroimidazole antimicrobials. *Clin Pharmacokinet* **36**, 353–373 (1999).

145. Tocher, J. H. & Edwards, D. I. Evidence for the direct interaction of reduced metronidazole derivatives with DNA bases. *Biochem Pharmacol* **48**, 1089–1094 (1994).

146. Edwards, D. I. Nitroimidazole drugs–action and resistance mechanisms. I. Mechanisms of action. *J Antimicrob Chemother* **31**, 9–20 (1993).

147. Tocher, J. H. & Edwards, D. I. The interaction of nitroaromatic drugs with aminothiols. *Biochem Pharmacol* **50**, 1367–1371 (1995).

148. Lofmark, S., Edlund, C. & Nord, C. E. Metronidazole is still the drug of choice for treatment of anaerobic infections. *Clin Infect Dis* **50**(Suppl 1), S16–23 (2010).

149. Hillier, S. L. *et al.* Secnidazole treatment of bacterial vaginosis: A randomized controlled trial. *Obstet Gynecol* **130**, 379–386 (2017).

150. Petrina, M. A. B., Cosentino, L. A., Rabe, L. K. & Hillier, S. L. Susceptibility of bacterial vaginosis (BV)-associated bacteria to secnidazole compared to metronidazole, tinidazole and clindamycin. *Anaerobe* **47**, 115–119 (2017).

151. Ghosh, A., Aycock, C. & Schwebke, J. Susceptibility of trichomonas vaginalis clinical isolates to secnidazole and metronidazole. *Am J Obstet Gynecol* **217**, 714–715 (2017).

152. Coiffier, G., Albert, J. D., Arvieux, C. & Guggenbuhl, P. Optimizing combination rifampin therapy for staphylococcal osteoarticular infections. *Joint Bone Spine* **80**, 11–17 (2013).

153. Richards, G. K., Gagnon, R. F. & Prentis, J. Comparative rates of antibiotic action against Staphylococcus epidermidis biofilms. *ASAIO Trans* **37**, M160–162 (1991).

154. Gagnon, R. F., Richards, G. K. & Kostiner, G. B. Time-kill efficacy of antibiotics in combination with rifampin against Staphylococcus epidermidis biofilms. *Adv Perit Dial* **10**, 189–192 (1994).

155. Riva, S., Fietta, A., Berti, M., Silvestri, L. G. & Romero, E. Relationships between curing of the F episome by rifampin and by acridine orange in Escherichia coli. *Antimicrob Agents Chemother* **3**, 456–462 (1973).

156. Wehrli, W. Rifampin: mechanisms of action and resistance. *Rev Infect Dis* **5**(Suppl 3), S407–411 (1983).

157. Adachi, J. A. & DuPont, H. L. Rifaximin: a novel nonabsorbed rifamycin for gastrointestinal disorders. *Clin Infect Dis* **42**, 541–547 (2006).

158. Ojetti, V. *et al.* Rifaximin pharmacology and clinical implications. *Expert Opin Drug Metab Toxicol* **5**, 675–682 (2009).

159. Carter, A. P. *et al.* Functional insights from the structure of the 30S ribosomal subunit and its interactions with antibiotics. *Nature* **407**, 340–348 (2000).

160. Vakulenko, S. B. & Mobashery, S. Versatility of aminoglycosides and prospects for their future. *Clin Microbiol Rev* **16**, 430–450 (2003).

161. Mehta, R. & Champney, W. S. 30S ribosomal subunit assembly is a target for inhibition by aminoglycosides in Escherichia coli. *Antimicrob Agents Chemother* **46**, 1546–1549 (2002).

162. Magnet, S. & Blanchard, J. S. Molecular insights into aminoglycoside action and resistance. *Chem Rev* **105**, 477–498 (2005).

163. Mingeot-Leclercq, M. P., Glupczynski, Y. & Tulkens, P. M. Aminoglycosides: activity and resistance. *Antimicrob Agents Chemother* **43**, 727–737 (1999).

164. Poulikakos, P. & Falagas, M. E. Aminoglycoside therapy in infectious diseases. *Expert Opin Pharmacother* **14**, 1585–1597 (2013).

165. Hamilton-Miller, J. M & Shah, S. Activity of the semisynthetic kanamycin B derivative, arbekacin against methicillin-resistant Staphylococcus aureus. *J Antimicrob Chemother* **35**, 865–868 (1995).

166. Hwang, J. H. *et al.* The usefulness of arbekacin compared to vancomycin. *Eur J Clin Microbiol Infect Dis* **31**, 1663–1666 (2012).

167. Shrestha, S., Grilley, M., Fosso, M. Y., Chang, C. W. & Takemoto, J. Y. Membrane lipid-modulated mechanism of action and non-cytotoxicity of novel fungicide aminoglycoside FG08. *PLOS One* **8**, e73843 (2013).

168. Denervaud-Tendon, V., Poirel, L., Connolly, L. E., Krause, K. M. & Nordmann, P. Plazomicin activity against polymyxin-resistant Enterobacteriaceae, including MCR-1-producing isolates. *J Antimicrob Chemother* **72**, 2787–2791 (2017).

169. Shaeer, K. M., Zmarlicka, M. T., Chahine, E. B., Piccicacco, N. & Cho, J. C. Plazomicin, a next-generation aminoglycoside. *Pharmacotherapy* **39**, 77–93 (2019).

170. Zhang, Y., Kashikar, A. & Bush, K. In vitro activity of plazomicin against β-lactamase-producing carbapenem-resistant Enterobacteriaceae (CRE). *J Antimicrob Chemother* **72**, 2792–2795 (2017).

171. Haidar, G. *et al.* Association between the presence of aminoglycoside-modifying enzymes and in vitro activity of gentamicin, tobramycin, amikacin, and plazomicin against klebsiella pneumoniae carbapenemase- and extended-spectrum-β-lactamase-producing enterobacter species. *Antimicrob Agents Chemother* **60**, 5208–5214 (2016).

172. López-Diaz, M. D. C. *et al.* Plazomicin activity against 346 extended-spectrum β Lactamase/AmpC-producing *Escherichia coli* urinary isolates in relation to aminoglycoside-modifying enzymes. *Antimicrob Agents Chemother* **61**, e02454–02416 (2017).

173. Chopra, I. & Roberts, M. Tetracycline antibiotics: mode of action, applications, molecular biology, and epidemiology of bacterial resistance. *Microbiol Mol Biol Rev* **65**, 232–260 (2001).

174. Brodersen, D. E. *et al.* The structural basis for the action of the antibiotics tetracycline, pactamycin, and hygromycin B on the 30S ribosomal subunit. *Cell* **103**, 1143–1154 (2000).

175. Pioletti, M. *et al.* Crystal structures of complexes of the small ribosomal subunit with tetracycline, edeine and IF3. *EMBO J* **20**, 1829–1839 (2001).

176. Bishburg, E. & Bishburg, K. Minocycline–an old drug for a new century: emphasis on methicillin-resistant Staphylococcus aureus (MRSA) and Acinetobacter baumannii. *Int J Antimicrob Agents* **34**, 395–401 (2009).

177. Draper, M. P. *et al.* The mechanism of action of the novel aminomethylcycline antibiotic omadacycline. *Antimicrob Agents Chemother* **58**,1279–1283 (2014).

178. Zhanel, G., Critchley, I., Lin, L.-Y. & Alvandi, N. Microbiological profile of sarecycline: A novel targeted spectrum tetracycline for the treatment of acne vulgaris. *Antimicrob Agents Chemother* **63** (2018).

179. Deeks, E. D. Sarecycline: First global approval. *Drugs* **79**, 325–329 (2019).

180. Sutcliffe, J. A., O'Brien, W., Fyfe, C. & Grossman, T. H. Antibacterial activity of eravacycline (TP-434), a novel fluorocycline, against hospital and community pathogens. *Antimicrob Agents Chemother* **57**, 5548–5558 (2013).

181. Solomkin, J., Evans, D., Slepavicius, A. et al. Assessing the efficacy and safety of eravacycline vs ertapenem in complicated intra-abdominal infections in the investigating gram-negative infections treated with eravacycline (ignite 1) trial: A randomized clinical trial. *JAMA Surgery* **152**, 224–232 (2017).

182. Barber, K. E., Bell, A. M., Wingler, M. J. B., Wagner, J. L. & Stover, K. R. Omadacycline enters the ring: A new antimicrobial contender. *Pharmacotherapy* **38**, 1194–1204 (2018).

183. Dahl, E. L. *et al.* Tetracyclines specifically target the apicoplast of the malaria parasite Plasmodium falciparum. *Antimicrob Agents Chemother* **50**, 3124–3131 (2006).

184. Noskin, G. A. Tigecycline: a new glycylcycline for treatment of serious infections. *Clin Infect Dis* **41**(Suppl **5**), S303–314 (2005).

185. Bergeron, J. *et al.* Glycylcyclines bind to the high-affinity tetracycline ribosomal binding site and evade Tet(M)- and Tet(O)-mediated ribosomal protection. *Antimicrob Agents Chemother* **40**, 2226–2228 (1996).

186. da Silva, L. M. & Nunes Salgado, H. R. Tigecycline: a review of properties, applications, and analytical methods. *Ther Drug Monit* **32**, 282–288 (2010).

187. Schnappinger, D. & Hillen, W. Tetracyclines: antibiotic action, uptake, and resistance mechanisms. *Arch Microbiol* **165**, 359–369 (1996).

188. Oehler, R., Polacek, N., Steiner, G. & Barta, A. Interaction of tetracycline with RNA: photoincorporation into ribosomal RNA of Escherichia coli. *Nucleic Acids Res* **25**, 1219–1224 (1997).

189. Jardetzky, O. Studies on the mechanism of action of chloramphenicol. I. The conformation of chloramphenicol in solution. *J Biol Chem* **238**, 2498–2508 (1963).

190. Schlunzen, F. *et al.* Structural basis for the interaction of antibiotics with the peptidyl transferase centre in eubacteria. *Nature* **413**, 814–821 (2001).

191. Xaplanteri, M. A., Andreou, A., Dinos, G. P. & Kalpaxis, D. L. Effect of polyamines on the inhibition of peptidyltransferase by antibiotics: revisiting the mechanism of chloramphenicol action. *Nucleic Acids Res* **31**, 5074–5083 (2003).

192. Thompson, J., O'Connor, M., Mills, J. A. & Dahlberg, A. E. The protein synthesis inhibitors, oxazolidinones and chloramphenicol, cause extensive translational inaccuracy in vivo. *J Mol Biol* **322**, 273–279 (2002).

193. Pestka, S. Inhibitors of ribosome functions. *Annu Rev Microbiol* **25**, 487–562 (1971).

194. Reusser, F. Effect of lincomycin and clindamycin on peptide chain initiation. *Antimicrob Agents Chemother* **7**, 32–37 (1975).

195. Champney, W. S. & Tober, C. L. Specific inhibition of 50S ribosomal subunit formation in Staphylococcus aureus cells by 16-membered macrolide, lincosamide, and streptogramin B antibiotics. *Curr Microbiol* **41**, 126–135 (2000).

196. Spizek, J. & Rezanka, T. Lincomycin, clindamycin and their applications. *Appl Microbiol Biotechnol* **64**, 455–464 (2004).

197. Shibl, A. M. & Al-Sowaygh, I. A. Differential inhibition of bacterial growth and hemolysin production by lincosamide antibiotics. *J Bacteriol* **137**, 1022–1023 (1979).

198. Spizek, J., Novotna, J. & Rezanka, T. Lincosamides: chemical structure, biosynthesis, mechanism of action, resistance, and applications. *Adv Appl Microbiol* **56**, 121–154 (2004).

199. Equi, A., Balfour-Lynn, I. M., Bush, A. & Rosenthal, M. Long term azithromycin in children with cystic fibrosis: a randomised, placebo-controlled crossover trial. *Lancet* **360**, 978–984 (2002).

200. Cannon, M. & Burns, K. Modes of action of erythromycin and thiostrepton as inhibitors of protein synthesis. *FEBS Lett* **18**, 1–5 (1971).

201. Tanaka, S., Otaka, T. & Kaji, A. Further studies on the mechanism of erythromycin action. *Biochim Biophys Acta* **331**, 128–140 (1973).

202. Poehlsgaard, J. & Douthwaite, S. The bacterial ribosome as a target for antibiotics. *Nat Rev Microbiol* **3**, 870–881 (2005).

203. Katz, L. & Ashley, G. W. Translation and protein synthesis: macrolides. *Chem Rev* **105**, 499–528 (2005).

204. Piscitelli, S. C., Danziger, L. H. & Rodvold, K. A. Clarithromycin and azithromycin: new macrolide antibiotics. *Clin Pharm* **11**, 137–152 (1992).

205. Neu, H. C. Clinical microbiology of azithromycin. *Am J Med* **91**, 12S–18S (1991).

206. Lancaster, J. W. & Matthews, S. J. Fidaxomicin: the newest addition to the armamentarium against Clostridium difficile infections. *Clin Ther* **34**, 1–13 (2012).

207. Louie, T. J. *et al.* Fidaxomicin versus vancomycin for Clostridium difficile infection. *N Engl J Med* **364**, 422–431 (2011).

208. Louie, T. J. *et al.* Fidaxomicin preserves the intestinal microbiome during and after treatment of Clostridium difficile infection (CDI) and reduces both toxin reexpression and recurrence of CDI. *Clin Infect Dis* **55**(Suppl **2**), S132–142 (2012).

209. Babakhani, F. *et al.* Fidaxomicin inhibits spore production in Clostridium difficile. *Clin Infect Dis* **55**(Suppl **2**), S162–169 (2012).

210. Rodvold, K. A. *et al.* Comparison of plasma and intrapulmonary concentrations of nafithromycin (WCK 4873) in healthy adult subjects. *Antimicrob Agents Chemother* **61** (2017).

211. Nguyen, M. & Chung, E. P. Telithromycin: the first ketolide antimicrobial. *Clin Ther* **27**, 1144–1163 (2005).

212. Rodgers, W., Frazier, A. D. & Champney, W. S. Solithromycin inhibition of protein synthesis and ribosome biogenesis in Staphylococcus aureus, Streptococcus pneumoniae, and *Haemophilus influenzae*. *Antimicrob Agents Chemother* **57**, 1632–1637 (2013).

213. Llano-Sotelo, B. *et al.* Binding and action of CEM-101, a new fluoroketolide antibiotic that inhibits protein synthesis. *Antimicrob Agents Chemother* **54**, 4961–4970 (2010).

214. Buege, M. J., Brown, J. E. & Aitken, S. L. Solithromycin: A novel ketolide antibiotic. *Am J Health Syst Pharm* **74**, 875–887 (2017).

215. Mansour, H., Chahine, E. B., Karaoui, L. R. & El-Lababidi, R. M. Cethromycin: a new ketolide antibiotic. *Ann Pharmacother* **47**, 368–379 (2013).

216. Rafie, S., MacDougall, C. & James, C. L. Cethromycin: a promising new ketolide antibiotic for respiratory infections. *Pharmacotherapy* **30**, 290–303 (2010).

217. Hammerschlag, M. R. & Sharma, R. Use of cethromycin, a new ketolide, for treatment of community-acquired respiratory infections. *Expert Opin Investig Drugs* **17**, 387–400 (2008).

218. Cethromycin: A-195773, A-195773-0, A-1957730, Abbott-195773, ABT 773. *Drugs R D* **8**, 95–102 (2007).

219. Paukner, S. & Riedl, R. Pleuromutilins: Potent drugs for resistant bugs-mode of action and resistance. *Cold Spring Harb Perspect Med* **7** (2017).

220. Eyal, Z. *et al.* A novel pleuromutilin antibacterial compound, its binding mode and selectivity mechanism. *Sci Rep* **6**, 39004 (2016).

221. Veve, M. P. & Wagner, J. L. Lefamulin: Review of a promising novel pleuromutilin antibiotic. *Pharmacotherapy* **38**, 935–946 (2018).

222. Paukner, S., Gruss, A. & Jensen, J. S. In vitro activity of lefamulin against sexually transmitted bacterial pathogens. *Antimicrob Agents Chemother* **62** (2018).

223. Mast, Y. & Wohlleben, W. Streptogramins - two are better than one! *Int J Med Microbiol* **304**, 44–50 (2014).

224. Di Giambattista, M., Chinali, G. & Cocito, C. The molecular basis of the inhibitory activities of type A and type B synergimycins and related antibiotics on ribosomes. *J Antimicrob Chemother* **24**, 485–507 (1989).

225. Cocito, C., Di Giambattista, M., Nyssen, E. & Vannuffel, P. Inhibition of protein synthesis by streptogramins and related antibiotics. *J Antimicrob Chemother* **39**(Suppl A), 7–13 (1997).

226. Livermore, D. M. Quinupristin/dalfopristin and linezolid: where, when, which and whether to use? *J Antimicrob Chemother* **46**, 347–350 (2000).

227. Mukhtar, T. A. & Wright, G. D. Streptogramins, oxazolidinones, and other inhibitors of bacterial protein synthesis. *Chem Rev* **105**, 529–542 (2005).

228. Harms, J. M., Schlunzen, F., Fucini, P., Bartels, H. & Yonath, A. Alterations at the peptidyl transferase centre of the ribosome induced by the synergistic action of the streptogramins dalfopristin and quinupristin. *BMC Biol* **2**, 4 (2004).

229. Cundliffe, E. & Burns, D. J. Long term effects of fusidic acid on bacterial protein synthesis in vivo. *Biochem Biophys Res Commun* **49**, 766–774 (1972).

230. Burns, K., Cannon, M. & Cundliffe, E. A resolution of conflicting reports concerning the mode of action of fusidic acid. *FEBS Lett* **40**, 219–223 (1974).

231. Metaxas, E. I. & Falagas, M. E. Update on the safety of linezolid. *Expert Opin Drug Saf* **8**, 485–491 (2009).

232. Leach, K. L. *et al.* The site of action of oxazolidinone antibiotics in living bacteria and in human mitochondria. *Mol Cell* **26**, 393–402 (2007).

233. Nagiec, E. E. *et al.* Oxazolidinones inhibit cellular proliferation via inhibition of mitochondrial protein synthesis. *Antimicrob Agents Chemother* **49**, 3896–3902 (2005).

234. Kisgen, J. J., Mansour, H., Unger, N. R. & Childs, L. M. Tedizolid: a new oxazolidinone antimicrobial. *Am J Health Syst Pharm* **71**, 621–633 (2014).

235. Locke, J. B., Zurenko, G. E., Shaw, K. J. & Bartizal, K. Tedizolid for the management of human infections: in vitro characteristics. *Clin Infect Dis* **58**(Suppl 1), S35–42 (2014).

236. Kalia, V. *et al.* Mode of action of Ranbezolid against staphylococci and structural modeling studies of its interaction with ribosomes. *Antimicrob Agents Chemother* **53**, 1427–1433 (2009).

237. Kompis, I. M., Islam, K. & Then, R. L. DNA and RNA synthesis: antifolates. *Chem Rev* **105**, 593–620 (2005).

238. Hitchings, G. H. Mechanism of action of trimethoprim-sulfamethoxazole. I. *J Infect Dis* **128**(Suppl 3), S433–436 (1973).

239. Quinlivan, E. P., McPartlin, J., Weir, D. G. & Scott, J. Mechanism of the antimicrobial drug trimethoprim revisited. *FASEB J* **14**, 2519–2524 (2000).

240. Sangurdekar, D. P., Zhang, Z. & Khodursky, A. B. The association of DNA damage response and nucleotide level modulation with the antibacterial mechanism of the anti-folate drug trimethoprim. *BMC Genomics* **12**, 583 (2011).

241. Huang, D. B. *et al.* An updated review of iclaprim: A potent and rapidly bactericidal antibiotic for the treatment of skin and skin structure infections and nosocomial pneumonia caused by gram-positive including multidrug-resistant bacteria. *Open Forum Infect Dis* **5**, ofy003 (2018).

242. Oefner, C. *et al.* Increased hydrophobic interactions of iclaprim with Staphylococcus aureus dihydrofolate reductase are responsible for the increase in affinity and antibacterial activity. *J Antimicrob Chemother* **63**, 687–698 (2009).

243. Kohlhoff, S. A. & Sharma, R. Iclaprim. *Expert Opin Investig Drugs* **16**, 1441–1448 (2007).

244. Laue, H. *et al.* In vitro activity of the novel diaminopyrimidine, iclaprim, in combination with folate inhibitors and other antimicrobials with different mechanisms of action. *J Antimicrob Chemother* **60**, 1391–1394 (2007).

245. Vilcheze, C. *et al.* Transfer of a point mutation in Mycobacterium tuberculosis inhA resolves the target of isoniazid. *Nat Med* **12**, 1027–1029 (2006).

246. Banerjee, A. *et al.* inhA, a gene encoding a target for isoniazid and ethionamide in *Mycobacterium tuberculosis*. *Science* **263**, 227–230 (1994).

247. Wilming, M. & Johnsson, K. Spontaneous formation of the bioactive form of the tuberculosis drug isoniazid. *Angew Chem Int Ed Engl* **38**, 2588–2590 (1999).

248. Timmins, G. S. & Deretic, V. Mechanisms of action of isoniazid. *Mol Microbiol* **62**, 1220–1227 (2006).

249. Belanger, A. E. *et al.* The embAB genes of Mycobacterium avium encode an arabinosyl transferase involved in cell wall arabinan biosynthesis that is the target for the antimycobacterial drug ethambutol. *Proc Natl Acad Sci U S A* **93**, 11919–11924 (1996).

250. Boshoff, H. I. *et al.* The transcriptional responses of *Mycobacterium tuberculosis* to inhibitors of metabolism: novel insights into drug mechanisms of action. *J Biol Chem* **279**, 40174–40184 (2004).

251. Rakesh, Bruhn, D. F., Scherman, M. S., Singh, A. P., Yang, L., Liu, J., Lenaerts, A. J., Lee, R. E. Synthesis and evaluation of pretomanid (PA-824) oxazolidinone hybrids. *Bioorg Med Chem Lett* **26**, 388–391 (2016).

252. Baptista, R., Fazakerley, D. M., Beckmann, M., Baillie, L. & Mur, L. A. J. Untargeted metabolomics reveals a new mode of action of pretomanid (PA-824). *Sci Rep* **8**, 5084 (2018).

253. Mc, D. W. & Tompsett, R. Activation of pyrazinamide and nicotinamide in acidic environments in vitro. *Am Rev Tuberc* **70**, 748–754 (1954).

254. McCune, R. M., Feldmann, F. M., Lambert, H. P. & McDermott, W. Microbial persistence. I. The capacity of tubercle bacilli to survive sterilization in mouse tissues. *J Exp Med* **123**, 445–468 (1966).

255. Shi, W. *et al.* Pyrazinamide inhibits trans-translation in *Mycobacterium tuberculosis*. *Science* **333**, 1630–1632 (2011).

256. Zhang, Y., Wade, M. M., Scorpio, A., Zhang, H. & Sun, Z. Mode of action of pyrazinamide: disruption of *Mycobacterium tuberculosis* membrane transport and energetics by pyrazinoic acid. *J Antimicrob Chemother* **52**, 790–795 (2003).

257. Nahid, P., Pai, M. & Hopewell, P. C. Advances in the diagnosis and treatment of tuberculosis. *Proc Am Thorac Soc* **3**, 103–110 (2006).

258. Floss, H. G. & Yu, T. W. Rifamycin-mode of action, resistance, and biosynthesis. *Chem Rev* **105**, 621–632 (2005).

259. Munsiff, S. S., Kambili, C. & Ahuja, S. D. Rifapentine for the treatment of pulmonary tuberculosis. *Clin Infect Dis* **43**, 1468–1475 (2006).

260. Kubendiran, G., Paramasivan, C. N., Sulochana, S. & Mitchison, D. A. Moxifloxacin and gatifloxacin in an acid model of persistent *Mycobacterium tuberculosis*. *J Chemother* **18**, 617–623 (2006).

261. Shandil, R. K. *et al.* Moxifloxacin, ofloxacin, sparfloxacin, and ciprofloxacin against *Mycobacterium tuberculosis*: evaluation of in vitro and pharmacodynamic indices that best predict in vivo efficacy. *Antimicrob Agents Chemother* **51**, 576–582 (2007).

262. Sriram, D., Bal, T. R., Yogeeswari, P., Radha, D. R. & Nagaraja, V. Evaluation of antimycobacterial and DNA gyrase inhibition of fluoroquinolone derivatives. *J Gen Appl Microbiol* **52**, 195–200 (2006).

263. De Souza, M. V., Vasconcelos, T. R., de Almeida, M. V. & Cardoso, S. H. Fluoroquinolones: an important class of antibiotics against tuberculosis. *Curr Med Chem* **13**, 455–463 (2006).

264. Leysen, D. C., Haemers, A. & Pattyn, S. R. Mycobacteria and the new quinolones. *Antimicrob Agents Chemother* **33**, 1–5 (1989).

265. Fu, L. M. & Shinnick, T. M. Genome-wide exploration of the drug action of capreomycin on *Mycobacterium tuberculosis* using Affymetrix oligonucleotide GeneChips. *J Infect* **54**, 277–284 (2007).

266. Andries, K. *et al.* A diarylquinoline drug active on the ATP synthase of *Mycobacterium tuberculosis*. *Science* **307**, 223–227 (2005).

267. Koul, A. *et al.* Diarylquinolines target subunit c of mycobacterial ATP synthase. *Nat Chem Biol* **3**, 323–324 (2007).

268. Huitric, E., Verhasselt, P., Andries, K. & Hoffner, S. E. In vitro antimycobacterial spectrum of a diarylquinoline ATP synthase inhibitor. *Antimicrob Agents Chemother* **51**, 4202–4204 (2007).

269. Koul, A. *et al.* Diarylquinolines are bactericidal for dormant mycobacteria as a result of disturbed ATP homeostasis. *J Biol Chem* **283**, 25273–25280 (2008).

270. Ibrahim, M. *et al.* Synergistic activity of R207910 combined with pyrazinamide against murine tuberculosis. *Antimicrob Agents Chemother* **51**, 1011–1015 (2007).

271. Rao, M., Streur, T. L., Aldwell, F. E. & Cook, G. M. Intracellular pH regulation by Mycobacterium smegmatis and Mycobacterium bovis BCG. *Microbiology* **147**, 1017–1024 (2001).

272. Deckers-Hebestreit, G. & Altendorf, K. The F0F1-type ATP synthases of bacteria: structure and function of the F0 complex. *Annu Rev Microbiol* **50**, 791–824 (1996).

273. Greig, S. L. Obiltoxaximab: First global approval. *Drugs* **76**, 823–830 (2016).

274. Johnson, S. & Gerding, D. N. Bezlotoxumab. *Clin Infect Dis*, **68**, 699–704 (2019).

275. Wilcox, M. H. *et al.* Bezlotoxumab for prevention of recurrent clostridium difficile infection. *N Engl J Med* **376**, 305–317 (2017).

52

Mechanisms of Action of Antifungal Agents

Jeffrey M. Rybak and P. David Rogers

CONTENTS

Introduction

There are currently only four structurally distinct antifungal classes available for the treatment of invasive fungal infections. As both fungal cells and mammalian cells are eukaryotic, anti-infective drug targets that are fungal specific are few. The most widely exploited strategy for antifungal therapy makes use of differences in sterol biosynthesis and targets either the ergosterol biosynthesis pathway or ergosterol itself, which is incorporated into the fungal cell membrane. Other strategies include inhibiting RNA and DNA synthesis by compounds that are selectively transported into the fungal cell and inhibition of the fungal cell wall. Mortality rates in neutropenic patients for yeast and mold infections are still unacceptably high and time to implementation of the optimal antifungal has a significant impact on patient outcomes (Garey, Rege et al. 2006).

Polyenes: Amphotericin B and Nystatin *Amphotericin B*

Amphotericin B

Amphotericin B is a polyene macrolide antifungal that was first isolated from *Streptomyces nodosus* in the 1950s. Over the past five decades, amphotericin B deoxycholate (Fungizone) has been the gold standard for the treatment of life-threatening fungal infections and generally remains refractory to resistance. Its primary fungicidal mechanism of action is the direct binding to ergosterol; the primary sterol found in the fungal cell membranes. Its dual complementary action of aggregating to form membrane-permeabilizing ion channels increases its potency but is not required for fungicidal activity (Gray, Palacios et al. 2012). Pharmacodynamic studies have demonstrated that fungicidal activity of amphotericin B can be enhanced by optimizing

DOI: 10.1201/9781003099277-55

the concentration-to-MIC ratio (also called the area under the concentration curve to MIC ratio [AUC:MIC]) (Klepser, Wolfe et al. 1997). Unfortunately, the clinical utility of amphotericin B deoxycholate is often times limited by its associated adverse effects such as nephrotoxicity and infusion-related reactions (Deray 2002). Because of these treatment-limiting adverse effects, newer lipid-associated formulations of amphotericin B (Ambisome, Abelcet, and Amphotec) have replaced conventional amphotericin B deoxycholate as therapy for treatment of life-threatening fungal infections, specifically if the causative organism is suspected to be non-*Candida* or a resistant organism.

Spectrum of Activity

Amphotericin B exhibits strong activity against most medically relevant fungal species as well as certain protozoa, including *Leishmania spp.* Species such as *Candida auris, C. lusitaniae, C. guilliermondii, Trichosporon spp., Geotrichum spp., Scedosporium apiospermum*, and the causative organisms of dermatomycoses are less susceptible to amphotericin B.

Adverse Effects

Although more specific for fungal ergosterol, amphotericin B can also bind mammalian cells causing toxic effects in patients. The most prominent treatment-limiting adverse effect is nephrotoxicity. Amphotericin B-induced nephrotoxicity is dose dependent and is primarily characterized by a decrease in glomerular filtration rate (GFR) by vasoconstriction, sodium, potassium, and magnesium wasting, as well as renal tubular acidosis (Sawaya, Briggs et al. 1995). The mechanism of this effect has been studied but has not been completely elucidated. There is evidence to show that the reduction in GFR may be mediated by the tubuloglomerular feedback system (TGF). Under normal conditions, the TGF system is triggered by the perceived loss of NaCl from the proximal tubule by macula densa cells located in the early distal tubule (Ito and Abe 1996). The TGF response triggers afferent arteriolar vasoconstriction to minimize excessive NaCl loss in the urine. Amphotericin B increases the permeability of the macula densa cells and inappropriately triggers the TGF response resulting in afferent arteriolar vasoconstriction and results in a decrease of GFR. Due to this proposed mechanism, the infusion of intravenous NaCl, known as "salt loading" has been utilized to prevent the effects of amphotericin B on GFR as volume expansion decreases the sensitivity of the TGF system (Gardner, Godley et al. 1990). In a human model, salt loading was shown to decrease the effects of amphotericin B on serum creatinine (SCr) level when compared to amphotericin B delivered in dextrose; however, the saline group required statistically significant K+ supplementation and was observed to have decreased ability to acidify the urine (Llanos, Cieza et al. 1991).

Another common side effect of amphotericin B is infusion-related reactions such as nausea, vomiting, chills, rigors, and phlebitis (Grasela, Goodwin et al. 1990). Although the exact mechanism of infusion toxicity is not completely known, amphotericin B has been shown to stimulate the immune system via toll-like receptors resulting in an increased inflammatory cytokine production (Sau, Mambula et al. 2003). A wide variety of pretreatment medications, which commonly include a drug cocktail of diphenhydramine, acetaminophen and corticosteroids, are used to decrease these symptoms (Grasela, Goodwin et al. 1990).

Lipid Formulations

To decrease the nephrotoxic effects of amphotericin B deoxycholate, newer lipid formulations have been developed as a drug delivery system. Currently, there are three lipid-based amphotericin B products available, which have largely replaced the amphotericin B deoxycholate formulation as first-line therapy of life-threatening fungal infections. In 1995 and 1996, amphotericin B lipid complex, or ABLC (Abelcet; The Liposome Co.), and amphotericin B colloidal dispersion, or ABCD (Amphotec; Sequus Pharmaceuticals) became available, respectively. In 1997, the U.S. Food and Drug Administration (FDA) approved liposomal amphotericin B, or AmBisome (NeXstar Pharmaceuticals/ Fujisawa). Although there are distinct differences between the formulations in regard to the lipid composition, particle size, and pharmacokinetics, the clinical significance of these variances is not yet known (Dupont 2002). All lipid formulations have been shown to be less nephrotoxic than amphotericin B deoxycholate formulation but vary in regard to infusion-related toxicity (Bowden, Cays et al. 1996; Leenders, Reiss et al. 1997; Prentice, Hann et al. 1997; Wingard, White et al. 2000).

Amphotericin B Resistance

Amphotericin B-resistant clinical fungal isolates are rare and resistance mechanisms are poorly understood. Mutations in genes involved in the ergosterol biosynthesis pathway, specifically *ERG3* and *ERG6*, have been identified in amphotericin B-resistant *C. albicans* isolates leading to decreased levels of ergosterol in the cell membrane (Kelly, Lamb et al. 1997). Decreased ergosterol content in the fungal cell membrane decreases the amphotericin B-ergosterol interaction and results in decreased polyene activity (Klepser 2001). Polyene resistant isolates of *Candida* and *Cryptococcus* have lower ergosterol content than in susceptible isolates (Espinel-Ingroff 2008). Isolates of *C. albicans*, the most common etiological agent of IFIs, have been documented to develop resistance to amphotericin B but this has only been observed in combination with resistance to azole antifungals (Kelly, Lamb et al. 1997; Nolte, Parkinson et al. 1997; Martel, Parker et al. 2010). Additionally, a large proportion of *C. auris* clinical isolates are found to have increased amphotericin B resistance at the time of isolation, and *C. lusitaniae* and *C. guilliermondii* can develop resistance secondarily during amphotericin B treatment.

Nystatin

Nystatin, also a polyene antifungal with the same mechanism of action of amphotericin B, is not formulated as an intravenous product. However, as it is not absorbed through mucocutaneous membranes, this agent can be applied either topically or orally to treat infections such as oropharyngeal candidiasis (OPC) with minimal risk.

Azole Antifungals

Members of the azole class are the most widely used of all antifungals. Their utility is mainly due to their activity against a broad spectrum of fungal pathogens and a mild spectrum of side effects, specifically lacking the treatment limiting nephrotoxicity experienced by patients undergoing treatment with amphotericin B. There are five azoles currently used for the treatment of invasive fungal infections. Although these agents are all of the same class, they vary individually with regard to spectrum of activity, pharmacokinetic and toxicity profiles. Appropriate utilization of these agents is based upon these unique characteristics.

The primary target of all azole antifungals is 14-α sterol-demethylase, which is the product of the *ERG11* gene (Yoshida 1988). Sterol demethylase is a member of the cytochrome P450 (CYP450) class of enzymes and is a critical enzyme in the biosynthesis of ergosterol. As ergosterol is not found in the mammalian cell membrane, inhibition of this pathway is a fungal specific mechanism of action for this class. Treatment with azoles results in decreased cellular ergosterol and leads to the accumulation of toxic methylated sterols (Ghannoum and Rice 1999). Azoles are considered fungistatic for *Candida* and other yeast species but some are fungicidal to *Aspergillus* species (Manavathu, Cutright et al. 1998).

Fluconazole (Diflucan, Pfizer)

Fluconazole is available in both an oral and intravenous formulations and has been widely studied in treating invasive fungal disease. When compared to amphotericin B, fluconazole is defined by a limited toxicity profile, but its spectrum of activity is limited to yeast. Species such as *C. krusei* display inherent resistance to this agent and *C. glabrata* has been observed to display inherent resistance and acquire resistance secondary to treatment (Orozco, Higginbotham et al. 1998; Pfaller 2012). Additionally, an approximate 90% of clinical isolates of the emerging fungal pathogen, *C. auris*, are found to have increased resistance to fluconazole (Chowdhary, Prakash et al. 2018). Fluconazole has excellent bioavailability independent of pH or concomitant meals and exhibits linear kinetics (Zimmermann, Yeates et al. 1994; Zimmermann, Yeates et al. 1994; Groll, Gea-Banacloche et al. 2003). This agent also exhibits limited protein binding (~10%), a high volume of distribution and achieves appreciable concentrations in the cerebrospinal fluid (CSF) (Brammer, Farrow et al. 1990; Thaler, Bernard et al. 1995). Fluconazole is primarily eliminated unchanged in the urine and requires dosage adjustment in renal insufficiency (Brammer, Coakley et al. 1991). As fluconazole is a moderate inhibitor of CYP450 enzymes 3A4, CYP2C9 and CYP2C19, it potentially interacts with the metabolism of a wide variety of drugs (Niwa, Shiraga et al. 2005). In addition, fluconazole also inhibits uridine diphosphate (UDP)-glucuronosyltransferase (UGTs), which mediates drug metabolism by glucuronidation (Uchaipichat, Winner et al. 2006). Like all azole antifungals, AUC:MIC is the best pharmacodynamic parameter that predicts azole antifungal activity (Klepser, Wolfe et al. 1997).

Itraconazole (Sporanox, Janssen Pharmaceuticals)

Itraconazole, voriconazole, posaconazole, and isavuconazole are all considered extended-spectrum azoles as they exhibit greater potency than fluconazole against most fungal species. Itraconazole is effective against most yeast however, *C. krusei* and *C. glabrata* may show cross resistance with fluconazole (Pfaller 2012). When compared to fluconazole, itraconazole gains activity against certain molds such as *Aspergillus* and dimorphic endemic fungi. Itraconazole is only available orally and is formulated as either a capsule or a solution. Oral bioavailability of itraconazole is significantly less than fluconazole (Zimmermann, Yeates et al. 1994). The oral solution exhibits better bioavailability than the capsule, and this improves under acidic fasting conditions (Zimmermann, Yeates et al. 1994). Itraconazole is highly protein bound and is metabolized in the liver by CYP3A4, which results in the formation of the pharmacologically active metabolite hydroxyitraconazole (Poirier and Cheymol 1998; Templeton, Thummel et al. 2008). Itraconazole is a strong inhibitor of CYP3A4 and has the potential for numerous drug-drug interactions (Isoherranen, Kunze et al. 2004).

Voriconazole (Vfend, Pfizer)

Voriconazole is also an extended spectrum azole that is structurally related to fluconazole. Voriconazole displays excellent activity against *Candida* spp. including *C. krusei* and is the drug of choice for treatment of invasive aspergillosis (Herbrecht, Denning et al. 2002). Voriconazole exhibits activity against endemic fungi but notably lacks activity against Zygomycetes. Voriconazole is formulated orally and intravenously with a cyclodextrin carrier. Like fluconazole, voriconazole is well-absorbed and displays high bioavailability upon oral administration (Johnson and Kauffman 2003). Voriconazole displays nonlinear kinetics and absorption is independent of pH and food administration. Voriconazole is moderately protein bound (~58%) and demonstrates CSF penetration (Lutsar, Roffey et al. 2003). Unlike itraconazole, voriconazole is primarily metabolized by CYP2C19. The argument has been made for therapeutic drug monitoring (TDM) of voriconazole levels due to inter-patient genetic differences in CYP2C19 activity (Desta, Zhao et al. 2002; Weiss, Ten Hoevel et al. 2009). Renal function should be monitored in patients receiving the intravenous formulation due to the sulfobutylether-beta-cyclodextrin carrier (Oude Lashof, Sobel et al. 2012).

Posaconazole (Noxafil, Schering Corporation)

Posaconazole is structurally related to itraconazole, and global surveillance studies have noted that posaconazole is more active than itraconazole and fluconazole, and as effective as voriconazole against a diverse population of clinically relevant *Candida* spp. (Pfaller, Messer et al. 2001; Pfaller, Messer et al. 2004). Posaconazole is now available in multiple formulations, including both those for oral and intravenous administration. Due to the lipophilicity of posaconazole, the original oral suspension formulation requires administered with a full meal or a nutritional supplement with high fat content (Courtney, Wexler et al. 2004). Absorption of posaconazole suspension is saturable so the administration of this formulation in divided doses has been shown to increase the relative bioavailability when compared to a single dose (Ezzet, Wexler et al. 2005). Additionally, gastric pH also plays a role in posaconazole absorption so administration with an acidic beverage is recommended while coadministration with acid suppressing drugs may decrease absorption. (Courtney, Radwanski et al. 2004). The more recently introduced delayed release oral tablet formulation of posaconazole allows for once-daily administration following the initial loading dose, and is more minimally impacted by co-administration with high-fat-content food or agents altering gastric pH. Additionally, the intravenous formulation of posaconazole allows for administration to patients who are unable to take oral formulations (Cornely, Duarte et al. 2016; Cornely, Robertson et al. 2017). Posaconazole is highly protein bound and distributes widely to the tissues. Metabolism is largely mediated by UGT (glucuronosyltransferase), which allelic polymorphisms have been shown to alter pharmacokinetics (Ghosal, Hapangama et al. 2004). Excretion of posaconazole has also been shown to be mediated in part by P-glycoprotein and inducers of P-glycoprotein have been shown to increase clearance (Courtney 2004). Posaconazole is not metabolized significantly by CYP450 enzymes but remains an inhibitor of CYP isoenzyme 3A4 which can contribute to drug interactions (Krieter, Flannery et al. 2004).

Isavuconazole (Cresemba, Astellas)

Isavuconazole is the newest addition to the azole class and is indicated for the treatment of both invasive aspergillosis and invasive mucormycosis. Structurally similar to voriconazole, isavuconazole possesses a compact triazole pharmacophore as well as a lipophilic thiazolylbenzonitrile side chain. The spectrum of activity of isavuconazole is also most similar to that of voriconazole, but with improved activity against rare mold including both Mucoralean and endemic fungi (Rybak, Marx et al. 2015). Isavuconazole is however less active against *Fusarium* spp. than is voriconazole, and while demonstrating excellent *in vitro* activity against *Candida* spp., isavuconazole failed to meet the definition of noninferiority as compared to caspofungin in a Phase 3 clinical trial studying the treatment of invasive candidiasis, and multiple case of breakthrough *Candida* infections among patients receiving isavuconazole have been reported (Rausch, DiPippo et al. 2018; Kullberg, Viscoli et al. 2019).

Isavuconazole is available in both intravenous and oral capsule formulations, formulated as isavuconazonium sulfate, a prodrug that significantly improves water solubility and negates the need for the inclusion of cyclodextrin in the intravenous product. Upon intravenous administration, isavuconazonium sulfate is rapidly converted to the active isavuconazole by serum esterases. When administered orally, isavuconazonium sulfate is believed to be converted to the active isavuconazole by gut microbiota and only the active isavuconazole is detected in serum. Bioavailability of the oral formulation is excellent, approximating 98%, and absorption is not significantly affected by coadministration with food. Isavuconazole is extensively protein bound (~99%), and predominantly cleared via hepatic metabolism by CYP3A4 and CYP3A5 enzymes. Isavuconazole is generally well tolerated, exhibiting significantly less hepatobiliary toxicity and fewer drug metabolism interactions as compared with voriconazole (Rybak, Marx et al. 2015; Maertens, Raad et al. 2016).

Adverse Effects

Azoles are generally well tolerated, with the most common adverse effect being gastrointestinal symptoms such as nausea, vomiting, and diarrhea. Elevations in hepatic transaminases are associated with the use of all azoles, but cases of fulminant hepatotoxicity are rare. Cases of QT prolongation and torsades de pointes have also been associated with azoles; however, reports are usually in patients with comorbidities and on concomitant medications also associated with QT prolongation (Alkan, Haefeli et al. 2004; Philips, Marty et al. 2007). Voriconazole has been linked with a number of unique side effects, such as vision and neurologic changes, which are correlated with high voriconazole serum concentrations (Zonios, Gea-Banacloche et al. 2008).

Azole Resistance

Azole resistance in *Candida* spp. has been extensively studied and is commonly multifactorial in an azole resistant clinical isolate. The efficacy of azoles is decreased in clinical isolates of *Candida* spp. by the interplay of several mechanisms of resistance (Lopez-Ribot, McAtee et al. 1998; Perea, Lopez-Ribot et al. 2001; Morschhauser 2002). The mechanism of azole uptake into the fungal cell is unclear but recent kinetic data supports the hypothesis that its uptake is facilitated in *C. albicans* (Mansfield, Oltean et al. 2010). The constitutive overexpression of 14-α sterol demethylase requires higher concentrations of intracellular fluconazole for sufficient inhibition (Sanglard, Kuchler et al. 1995; Perea, Lopez-Ribot et al. 2001). Increased expression of *ERG11* from azole-resistant isolates can result from two documented methods. Gain-of-function (GOF) mutations in transcription factor *UPC2* results in increased expression of the gene encoding 14-α sterol demethylase, *ERG11*, and increased MICs to fluconazole (Flowers, Barker et al. 2012). Alternately, as *ERG11* resides on Chromosome 5, instances of aneuploidy, or specifically, increased chromosome 5 copy number can result in the increased expression of *ERG11* (Selmecki, Forche et al. 2006). Nonsynonymous mutations in 14-α sterol demethylase have also been shown to contribute to azole resistance (Sanglard, Ischer et al. 1998; Franz, Ruhnke et al. 1999). To date, over 30 amino acid substitutions have been described in 14-α sterol demethylase that have associated with fluconazole resistance, and the individual contribution of many of these mutations on azole susceptibility in *C. albicans* has been well described (Wang, Kong et al. 2009; Flowers, Colon et al. 2015).

Importantly, overexpression of efflux pumps is a common mechanism of resistance to drugs toxic to the fungal cell. In *C. albicans*, *CDR1*, *CDR2*, and *MDR1* are usually expressed in a low or nondetectable level, but expression is stimulated in response to azoles and other chemicals. In fluconazole-resistant strains of *C. albicans*, it is common to observe the constitutive overexpression of the *CDR1*, *CDR2*, and *MDR1* genes. In *C. glabrata*, efflux transporters *PDH1*, *CDR1*, and *SNQ2* mediate efflux of azoles (Cannon, Lamping et al. 2009; Whaley, Zhang et al. 2018). In *C. dubliniensis*, the Mdr1p transporter is the primary mediator of fluconazole efflux however, CdCdr1 and CdCdr2 can contribute to azole resistance in some isolates (Pinjon, Jackson et al. 2005; Schubert, Rogers et al. 2008). In *C. auris*, increased expression of both *CDR1* and *MDR1* has been observed among fluconazole-resistant clinical isolates, and *CDR1* has been demonstrated to significantly contribute to clinical fluconazole resistance (Rybak, Doorley et al. 2019). These established mechanisms of resistance are all able to individually contribute to azole resistance but high-level resistance is mainly due to multiple mechanisms of resistance that accumulate over time.

In the opportunistic fungal pathogen *Cryptococcus neoformans*, azole resistance is acquired from similar mechanisms as *Candida* spp. Mutations in 14-α sterol demethylase in addition to overexpression of the *ERG11* gene by chromosomal aneuploidy can contribute to an azole-resistant phenotype (Shapiro, Robbins et al. 2011). Additionally, overexpression of drug efflux transporter *AFR1* contributes to azole resistance (Shapiro, Robbins et al. 2011).

While previously uncommon, azole resistance in *Aspergillus* species such as *A. fumigatus* is increasingly reported worldwide, and its emergence has been associated with increasing use of agricultural demethylase inhibitor antifungals (Ashu, Hagen et al. 2017; Rybak, Fortwendel et al. 2019; Sewell, Zhu et al. 2019). To date, this azole resistance has primarily been linked to mutations in the *cyp51A* gene, which encodes one of the two *A. fumigatus* 14-α sterol demethylase genes (Balashov, Gardiner et al. 2005). In fact, the vast majority of azole-resistant *A. fumigatus* isolates of both environmental and clinical origin are found to possess mutations in *cyp51A*. The two most encountered mutations, TR_{34}/L98H and TR_{46}/Y121F/T289A, encode both amino acid substitutions and promoter region alterations. The former of which decreases azole binding affinity for the mutant enzyme, and the latter of which has been shown to increase the expression *cyp51A* significantly (Snelders, Karawajczyk et al. 2011; Warrilow, Parker et al. 2015). More recently, however, mutations in the sterol sensing domain of the HMG-CoA reductase encoding gene, *hmg1*, have been found among a large number of azole-resistant clinical isolates and these mutations have been shown to directly contribute to greatly increased resistance to the entire azole class (Rybak, Ge et al. 2019). Additionally, studies have shown that overexpression of efflux transporters such as *abcC* and *mdr1* may also contribute to clinical azole resistance in *A. fumigatus* (Mellado, Garcia-Effron et al. 2007; Shapiro, Robbins et al. 2011; Fraczek, Bromley et al. 2013).

Allylamines: Terbinafine

Terbinafine is an allylamine antifungal that inhibits the first enzyme in the ergosterol biosynthesis pathway, squalene epoxidase, resulting in a decrease of cellular ergosterol (Ryder 1992). Unlike azoles, which target an alternate enzyme in the same pathway, terbinafine is fungicidal and is the drug of choice for dermatophytes. Terbinafine exhibits less activity against nondermatophyte molds, which precludes its use in the treatment of invasive fungal disease (Trovato, Rapisarda et al. 2009). Oral terbinafine (Lamisil; Novartis) is primarily eliminated by the kidneys but is an inhibitor of CYP2D6. The toxicity profile of terbinafine is minimal with the most common adverse effects

being nausea, diarrhea, and dermatologic events, such as rash and pruritus (Hall, Monka et al. 1997).

Resistance

Squalene epoxidase is encoded by ergosterol biosynthesis gene *ERG1*. Studies in *Saccharomyces cerevisiae* show specific mutations in *ERG1* result in amino acid substitutions in squalene epoxidase that result in terbinafine resistance (Leber, Fuchsbichler et al. 2003).

Echinocandins

Echinocandins are the newest class of antifungals to be introduced to the market. Current approved agents by the FDA are caspofungin (Cancidas, Merck and Co.), micafungin (Mycamine, Astellas Pharmaceuticals) and anidulafungin (Eraxis, Pfizer Pharmaceuticals). Echinocandins are now recommended as frontline therapy for treatment of invasive candidiasis based on results of three defining clinical trials in which caspofungin, micafungin, and anidulafungin were shown noninferior to amphotericin B, liposomal amphotericin B and fluconazole, respectively (Mora-Duarte, Betts et al. 2002; Kuse, Chetchotisakd et al. 2007; Reboli, Rotstein et al. 2007; Pappas, Kauffman et al. 2016). These agents are characterized by their broad spectrum of activity, low incidence of drug-drug interactions and a relative lack of adverse effects (Zaas and Alexander 2005). Echinocandins are lipopeptides that bind to 1,3-β-D glucan synthase, inhibiting production of β-1,3 glucan which, along with chitin, is responsible for integrity and shape of the fungal cell wall. 1,3-β-D glucan synthase is a multiunit enzyme that is made of a catalytic subunit Fksp, the target of echinocandins, and the regulatory unit Rho1p. All three echinocandin agents are fungicidal in a dose dependent fashion against the *Candida* species including those isolates resistant to azole antifungals (Ostrosky-Zeichner, Rex et al. 2003; Pfaller, Diekema et al. 2003; Pfaller, Boyken et al. 2005). Unlike the azole antifungal class, echinocandins are noted to retain activity against fungal biofilm and biofilm formation (Bachmann, VandeWalle et al. 2002; Soustre, Rodier et al. 2004).

Formulations

All agents in the echinocandin class are available only as a parenteral formulation. Caspofungin requires no dosage adjustment in patients with renal failure however this agent requires a dosage reduction in patients with moderate hepatic impairment (Childs-Pugh score of 7–9). Micafungin and anidulafungin require no dosage adjustment for hepatic or renal impairment. Echinocandins are generally well tolerated however, cases of hepatotoxicity and infusion related reactions are associated with this class (Sucher, Chahine et al. 2009).

Spectrum of Activity

Echinocandins are fungicidal in a dose dependent manner against *Candida* species and are clinically important because they retain activity against azole-resistant isolates. Although echinocandins are fungistatic against *Aspergillus* species, they still retain excellent activity against *A. fumigatus, A. flavus, A. niger,* and *A. terreus.* Zygomycetes lack the 1,3-β-D glucan synthase component of the fungal cell wall, therefore, echinocandins show little activity for these species. Although the cell wall of *Cryptococcus neoformans* contains 1,3-β-D glucan synthase, these agents display little activity against this species.

The activity of the echinocandins against *Candida* species is divided into two broad groups. Echinocandins are particularly active against *C. albicans, C. glabrata, C. dubliniensis, C. tropicalis, C. auris,* and *C. krusei* (Pfaller, Boyken et al. 2008; Pfaller, Diekema et al. 2008; Berkow and Lockhart 2018; Chowdhary, Prakash et al. 2018). Though, as a class, echinocandins have reduced activity against *C. parapsilosis* and *C. guilliermondii* (Pfaller, Boyken et al. 2008; Pfaller, Diekema et al. 2008). However, while echinocandin MIC among clinical isolates of *C. parapsilosis* and *C. guilliermondii* are generally higher, 90–100% of isolates remain classified as echinocandin susceptible based on the MIC breakpoint of ≤ 2 mg/L (Pfaller, Boyken et al. 2008). Additionally, there is a lack of a clear relationship regarding echinocandin MIC to overall treatment response as even species with reduced susceptibilities can still be treated effectively with current echinocandin therapy (Mora-Duarte, Betts et al. 2002; Bennett 2006; Kartsonis, Killar et al. 2005). Thus, the clinical implications of the reduced susceptibility of certain *Candida* species to echinocandins remain unclear.

Data from the SENTRY Antimicrobial Surveillance Program (2008–2009) were recently published and compared the antifungal resistance patterns and species distributions from patients with community onset and nosocomial bloodstream infections (BSI) in 79 medical centers. For the 1,354 isolates examined, resistance to echinocandins was uncommon but was most prevalent among nosocomial bloodstream infection isolates of *C. glabrata* (Pfaller, Castanheira et al. 2010; Lewis, Wiederhold et al. 2013). Furthermore, previous prospective sentinel surveillance for the emergence of *in vitro* resistance to echinocandins comprised of 5,346 clinical *Candida* isolates from 91 medical centers from 2001 to 2006 showed no evidence of emerging echinocandin resistance among invasive isolates of *Candida* species (Pfaller, Boyken et al. 2008). However, reports of increasing echinocandin resistance among some institutions have been described. A recent review of echinocandin susceptibility among bloodstream isolates of *C. glabrata* over a 10-year period at a single academic medical center observed echinocandin resistance to rise from 5% to 12%. Notably, fluconazole resistance was also observed to rise from 18% to 30% over this same study period (Alexander, Johnson et al. 2013).

Echinocandin Resistance

Documentation of individual resistant *Candida* isolates is rare but as these agents are used more frequently, incidence of echinocandin resistance will assuredly increase. Unlike the azole antifungals, drug efflux transporters have not been documented to contribute to echinocandin resistance (Bachmann, Patterson et al. 2002; Niimi, Maki et al. 2006). Reduced susceptibility to echinocandins has been attributed to assorted alterations in 'hot spot' sequences within *FKS* genes; the target of the echinocandins and whose gene product is part of the β-1,3 glucan synthase complex (Park, Kelly et al. 2005; Laverdiere, Lalonde

et al. 2006). The mechanism for echinocandin resistance was first documented in *C. albicans* by identifying mutations within the *FKS1* gene in isolates that were less susceptible to the drug (Park, Kelly et al. 2005). Mutations in *FKS1* generally map onto two locations and occur between amino acid positions 641-649 and 1345-1365 of Fks1p (Park, Kelly et al. 2005). As a consequence of these mutations, the Fks1p amino acid protein sequence is altered resulting in an enzyme with reduced catalytic activity (Garcia-Effron, Lee et al. 2009). For species such as *C. guilliermondii* and *C. glabrata*, mutations in the *CaFKS1* ortholog, *FKS2*, has also contributed to elevated echinocandin MICs and potential treatment failure (Katiyar, Pfaller et al. 2006, Garcia-Effron; Lee et al. 2009). Additionally, clinical isolates of *C. auris* with reduced susceptibility to echinocandins both possessing or lacking mutations in the homolog of *CaFKS1* have been described, and clinical isolates have been documented to acquire resistance while patients are receiving echinocandin therapy (Sharma, Kumar et al. 2016; Biagi, Wiederhold et al. 2019). Additionally, clinical isolates of *C. parapsilosis* and *C. guilliermondii* are commonly found to exhibit elevated echinocandin MIC attributable to naturally occurring polymorphisms within the *FKS1* gene. However, the clinical significance of these elevated MIC remains unclear (Katiyar, Pfaller et al. 2006; Garcia-Effron, Katiyar et al. 2008).

Some clinical isolates of *Candida* have shown a reproducible pattern of having reduced susceptibility to caspofungin at higher concentrations, allowing growth, while lower concentrations remaining effective at inhibiting growth (Stevens, Espiritu et al. 2004). This observation has been referred to as a paradoxical or "Eagle" effect, which is similar to that previously observed for other cell-wall active antimicrobial agents (Shah 1982; Stevens, Gibbons et al. 1988). Although the mechanisms responsible for this phenomenon have not been completely elucidated, investigations have shown that paradoxical growth is not associated with overexpression or mutations in the *FKS* gene. Some isolates capable of surviving in high concentrations of caspofungin have a documented increase in their cell wall chitin content. (Drakulovski, Dunyach et al. 2011; Stevens, Ichinomiya et al. 2006). Paradoxical attenuation of caspofungin activity may also be linked to calcineurin and PKC stress response pathways (Wiederhold, Kontoyiannis et al. 2005). Chamilos et al. showed that paradoxical growth was more frequent in caspofungin than with micafungin and anidulafungin and was completely absent in *C. glabrata* isolates (Chamilos, Lewis et al. 2007). Currently, the *in vivo* documentation of the paradoxical effect in *Candida* spp. is not significant and may not be relevant to patient treatment at this time.

Flucytosine

Flucytosine (5FC) is a pyrimidine analog that inhibits DNA replication and protein synthesis. Flucytosine is selectively transported into the fungal cell by cytosine permease and is then deaminated by cytosine deaminase to 5-fluorouracil. It is then converted to its active form, 5-fluorodeoxyuridylic acid to inhibit DNA synthesis (Vermes, Guchelaar et al. 2000). Due to the rapid development of resistance, 5FC is rarely used as monotherapy. In a murine model of disseminated candidiasis, time over MIC was the best pharmacodynamic parameter correlated with efficacy (Andes and van Ogtrop 2000). Although the Clinical Practice Guidelines for the Management of Candidiasis recommend 5FC in combination with amphotericin B or fluconazole for a variety of infection types in the setting of refractory or resistant disease, concerns regarding toxicity such as bone marrow suppression and hepatic transaminase elevation often times prevent its use (Pappas, Kauffman et al. 2009). Its current utility is primarily limited to treatment of cryptococcal meningitis (Saag, Graybill et al. 2000).

Flucytosine Resistance

Intrinsic resistance to flucytosine is a result of mutations in purine-cytosine permease (Fcy2), which impairs cellular uptake of the drug (Kanafani and Perfect 2008). Secondary resistance often correlates with mutations in the enzyme uracil phosphoribosyltransferase (Fur1) or in cytosine deaminase (Fca1) thereby preventing conversion of 5-fluorouracil to its active form (Kanafani and Perfect 2008).

REFERENCES

Alexander, B. D., M. D. Johnson, C. D. Pfeiffer, C. Jimenez-Ortigosa, J. Catania, R. Booker, M. Castanheira, S. A. Messer, D. S. Perlin and M. A. Pfaller (2013). "Increasing echinocandin resistance in Candida glabrata: clinical failure correlates with presence of FKS mutations and elevated minimum inhibitory concentrations." *Clin Infect Dis* **56**(12): 1724–1732.

Alkan, Y., W. E. Haefeli, J. Burhenne, J. Stein, I. Yaniv and I. Shalit (2004). "Voriconazole-induced QT interval prolongation and ventricular tachycardia: a non-concentration-dependent adverse effect." *Clin Infect Dis* **39**(6): e49–52.

Andes, D. and M. van Ogtrop (2000). "In vivo characterization of the pharmacodynamics of flucytosine in a neutropenic murine disseminated candidiasis model." *Antimicrob Agents Chemother* **44**(4): 938–942.

Ashu, E. E., F. Hagen, A. Chowdhary, J. F. Meis and J. Xu (2017). "Global population genetic analysis of aspergillus fumigatus." *mSphere* **2**(1).

Bachmann, S. P., T. F. Patterson and J. L. Lopez-Ribot (2002). "In vitro activity of caspofungin (MK-0991) against candida albicans clinical isolates displaying different mechanisms of azole resistance." *J Clin Microbiol* **40**(6): 2228–2230.

Bachmann, S. P., K. VandeWalle, G. Ramage, T. F. Patterson, B. L. Wickes, J. R. Graybill and J. L. Lopez-Ribot (2002). "In vitro activity of caspofungin against candida albicans biofilms." *Antimicrob Agents Chemother* **46**(11): 3591–3596.

Balashov, S. V., R. Gardiner, S. Park and D. S. Perlin (2005). "Rapid, high-throughput, multiplex, real-time PCR for identification of mutations in the cyp51A gene of aspergillus fumigatus that confer resistance to itraconazole." *J Clin Microbiol* **43**(1): 214–222.

Bennett, J. E. (2006). "Echinocandins for candidemia in adults without neutropenia." *N Engl J Med* **355**(11): 1154–1159.

Berkow, E. L. and S. R. Lockhart (2018). "Activity of CD101, a long-acting echinocandin, against clinical isolates of candida auris." *Diagn Microbiol Infect Dis* **90**(3): 196–197.

Biagi, M. J., N. P. Wiederhold, C. Gibas, B. L. Wickes, V. Lozano, S. C. Bleasdale and L. Danziger (2019). "Development of high-level echinocandin resistance in a patient with recurrent candida auris candidemia secondary to chronic candiduria." *Open Forum Infect Dis* **6**(7): ofz262.

Bowden, R. A., M. Cays, T. Gooley, R. D. Mamelok and J. A. van Burik (1996). "Phase I study of amphotericin B colloidal dispersion for the treatment of invasive fungal infections after marrow transplant." *J Infect Dis* **173**(5): 1208–1215.

Brammer, K. W., A. J. Coakley, S. G. Jezequel and M. H. Tarbit (1991). "The disposition and metabolism of [14C]fluconazole in humans." *Drug Metab Dispos* **19**(4): 764–767.

Brammer, K. W., P. R. Farrow and J. K. Faulkner (1990). "Pharmacokinetics and tissue penetration of fluconazole in humans." *Rev Infect Dis* **12**(Suppl 3): S318–326.

Cannon, R. D., E. Lamping, A. R. Holmes, K. Niimi, P. V. Baret, M. V. Keniya, K. Tanabe, M. Niimi, A. Goffeau and B. C. Monk (2009). "Efflux-mediated antifungal drug resistance." *Clin Microbiol Rev* **22**(2): 291–321 (Table of Contents).

Chamilos, G., R. E. Lewis, N. Albert and D. P. Kontoyiannis (2007). "Paradoxical effect of Echinocandins across candida species in vitro: evidence for echinocandin-specific and candida species-related differences." *Antimicrob Agents Chemother* **51**(6): 2257–2259.

Chowdhary, A., A. Prakash, C. Sharma, M. Kordalewska, A. Kumar, S. Sarma, B. Tarai, A. Singh, G. Upadhyaya, S. Upadhyay, P. Yadav, P. K. Singh, V. Khillan, N. Sachdeva, D. S. Perlin and J. F. Meis (2018). "A multicentre study of antifungal susceptibility patterns among 350 Candida auris isolates (2009-17) in India: role of the ERG11 and FKS1 genes in azole and echinocandin resistance." *J Antimicrob Chemother* **73**(4): 891–899.

Cornely, O. A., R. F. Duarte, S. Haider, P. Chandrasekar, D. Helfgott, J. L. Jimenez, A. Candoni, I. Raad, M. Laverdiere, A. Langston, N. Kartsonis, M. Van Iersel, N. Connelly and H. Waskin (2016). "Phase 3 pharmacokinetics and safety study of a posaconazole tablet formulation in patients at risk for invasive fungal disease." *J Antimicrob Chemother* **71**(3): 718–726.

Cornely, O. A., M. N. Robertson, S. Haider, A. Grigg, M. Geddes, M. Aoun, W. J. Heinz, I. Raad, U. Schanz, R. G. Meyer, S. P. Hammond, K. M. Mullane, H. Ostermann, A. J. Ullmann, S. Zimmerli, M. Van Iersel, D. A. Hepler, H. Waskin, N. A. Kartsonis and J. Maertens (2017). "Pharmacokinetics and safety results from the Phase 3 randomized, open-label, study of intravenous posaconazole in patients at risk of invasive fungal disease." *J Antimicrob Chemother* **72**(12): 3406–3413.

Courtney, R., E. Radwanski, J. Lim and M. Laughlin (2004). "Pharmacokinetics of posaconazole coadministered with antacid in fasting or nonfasting healthy men." *Antimicrob Agents Chemother* **48**(3): 804–808.

Courtney, R., A. Sansone-Parsons, D. Devlin, P. Soni, M. Laughlin and J. Simon (2004). "P-Glycoprotein (P-gp) expression and gentype: Exploratory analysis of posaconazole (POS) in healthy volunteers." *Intrsci Conf Antimicrob Agents Chemother* (Abstract A-40).

Courtney, R., D. Wexler, E. Radwanski, J. Lim and M. Laughlin (2004). "Effect of food on the relative bioavailability of two oral formulations of posaconazole in healthy adults." *Br J Clin Pharmacol* **57**(2): 218–222.

Deray, G. (2002). "Amphotericin B nephrotoxicity." *J Antimicrob Chemother* **49**(Suppl 1): 37–41.

Desta, Z., X. Zhao, J. G. Shin and D. A. Flockhart (2002). "Clinical significance of the cytochrome P450 2C19 genetic polymorphism." *Clin Pharmacokinet* **41**(12): 913–958.

Drakulovski, P., C. Dunyach, S. Bertout, J. Reynes and M. Mallie (2011). "A candida albicans strain with high MIC for caspofungin and no FKS1 mutations exhibits a high chitin content and mutations in two chitinase genes." *Med Mycol* **49**(5): 467–474.

Dupont, B. (2002). "Overview of the lipid formulations of amphotericin B." *J Antimicrob Chemother* **49**(Suppl 1): 31–36.

Espinel-Ingroff, A. (2008). "Mechanisms of resistance to antifungal agents: yeasts and filamentous fungi." *Rev Iberoam Micol* **25**(2): 101–106.

Ezzet, F., D. Wexler, R. Courtney, G. Krishna, J. Lim and M. Laughlin (2005). "Oral bioavailability of posaconazole in fasted healthy subjects: comparison between three regimens and basis for clinical dosage recommendations." *Clin Pharmacokinet* **44**(2): 211–220.

Flowers, S. A., K. S. Barker, E. L. Berkow, G. Toner, S. G. Chadwick, S. E. Gygax, J. Morschhauser and P. D. Rogers (2012). "Gain-of-function mutations in UPC2 are a frequent cause of ERG11 upregulation in azole-resistant clinical isolates of candida albicans." *Eukaryot Cell* **11**(10): 1289–1299.

Flowers, S. A., B. Colon, S. G. Whaley, M. A. Schuler and P. D. Rogers (2015). "Contribution of clinically derived mutations in ERG11 to azole resistance in candida albicans." *Antimicrob Agents Chemother* **59**(1): 450–460.

Fraczek, M. G., M. Bromley, A. Buied, C. B. Moore, R. Rajendran, R. Rautemaa, G. Ramage, D. W. Denning and P. Bowyer (2013). "The cdr1B efflux transporter is associated with non-cyp51a-mediated itraconazole resistance in aspergillus fumigatus." *J Antimicrob Chemother* **68**(7): 1486–1496.

Franz, R., M. Ruhnke and J. Morschhauser (1999). "Molecular aspects of fluconazole resistance development in candida albicans." *Mycoses* **42**(7-8): 453–458.

Garcia-Effron, G., S. K. Katiyar, S. Park, T. D. Edlind and D. S. Perlin (2008). "A naturally occurring proline-to-alanine amino acid change in Fks1p in candida parapsilosis, candida orthopsilosis, and candida metapsilosis accounts for reduced echinocandin susceptibility." *Antimicrob Agents Chemother* **52**(7): 2305–2312.

Garcia-Effron, G., S. Lee, S. Park, J. D. Cleary and D. S. Perlin (2009). "Effect of candida glabrata FKS1 and FKS2 mutations on echinocandin sensitivity and kinetics of 1,3-beta-D-glucan synthase: implication for the existing susceptibility breakpoint." *Antimicrob Agents Chemother* **53**(9): 3690–3699.

Gardner, M. L., P. J. Godley and S. M. Wasan (1990). "Sodium loading treatment for amphotericin B-induced nephrotoxicity." *DICP* **24**(10): 940–946.

Garey, K. W., M. Rege, M. P. Pai, D. E. Mingo, K. J. Suda, R. S. Turpin and D. T. Bearden (2006). "Time to initiation of fluconazole therapy impacts mortality in patients with candidemia: a multi-institutional study." *Clin Infect Dis* **43**(1): 25–31.

Ghannoum, M. A. and L. B. Rice (1999). "Antifungal agents: mode of action, mechanisms of resistance, and correlation of these mechanisms with bacterial resistance." *Clin Microbiol Rev* **12**(4): 501–517.

Ghosal, A., N. Hapangama, Y. Yuan, J. Achanfuo-Yeboah, R. Iannucci, S. Chowdhury, K. Alton, J. E. Patrick and S. Zbaida (2004). "Identification of human UDP-glucuronosyltransferase enzyme(s) responsible for the glucuronidation of posaconazole (Noxafil)." *Drug Metab Dispos* **32**(2): 267–271.

Grasela, T. H., Jr., S. D. Goodwin, M. K. Walawander, R. L. Cramer, D. W. Fuhs and V. P. Moriarty (1990). "Prospective surveillance of intravenous amphotericin B use patterns." *Pharmacotherapy* **10**(5): 341–348.

Gray, K. C., D. S. Palacios, I. Dailey, M. M. Endo, B. E. Uno, B. C. Wilcock and M. D. Burke (2012). "Amphotericin primarily kills yeast by simply binding ergosterol." *Proc Natl Acad Sci U S A* **109**(7): 2234–2239.

Groll, A. H., J. C. Gea-Banacloche, A. Glasmacher, G. Just-Nuebling, G. Maschmeyer and T. J. Walsh (2003). "Clinical pharmacology of antifungal compounds." *Infect Dis Clin North Am* **17**(1): 159–191.

Hall, M., C. Monka, P. Krupp and D. O'Sullivan (1997). "Safety of oral terbinafine: results of a postmarketing surveillance study in 25,884 patients." *Arch Dermatol* **133**(10): 1213–1219.

Herbrecht, R., D. W. Denning, T. F. Patterson, J. E. Bennett, R. E. Greene, J. W. Oestmann, W. V. Kern, K. A. Marr, P. Ribaud, O. Lortholary, R. Sylvester, R. H. Rubin, J. R. Wingard, P. Stark, C. Durand, D. Caillot, E. Thiel, P. H. Chandrasekar, M. R. Hodges, H. T. Schlamm, P. F. Troke and B. de Pauw (2002). "Voriconazole versus amphotericin B for primary therapy of invasive aspergillosis." *N Engl J Med* **347**(6): 408–415.

Isoherranen, N., K. L. Kunze, K. E. Allen, W. L. Nelson and K. E. Thummel (2004). "Role of itraconazole metabolites in CYP3A4 inhibition." *Drug Metab Dispos* **32**(10): 1121–1131.

Ito, S. and K. Abe (1996). "Tubuloglomerular feedback." *Jpn Heart J* **37**(2): 153–163.

Johnson, L. B. and C. A. Kauffman (2003). "Voriconazole: a new triazole antifungal agent." *Clin Infect Dis* **36**(5): 630–637.

Kanafani, Z. A. and J. R. Perfect (2008). "Antimicrobial resistance: resistance to antifungal agents: mechanisms and clinical impact." *Clin Infect Dis* **46**(1): 120–128.

Kartsonis, N., J. Killar, L. Mixson, C. M. Hoe, C. Sable, K. Bartizal and M. Motyl (2005). "Caspofungin susceptibility testing of isolates from patients with esophageal candidiasis or invasive candidiasis: relationship of MIC to treatment outcome." *Antimicrob Agents Chemother* **49**(9): 3616–3623.

Katiyar, S., M. Pfaller and T. Edlind (2006). "Candida albicans and candida glabrata clinical isolates exhibiting reduced echinocandin susceptibility." *Antimicrob Agents Chemother* **50**(8): 2892–2894.

Kelly, S. L., D. C. Lamb, D. E. Kelly, N. J. Manning, J. Loeffler, H. Hebart, U. Schumacher and H. Einsele (1997). "Resistance to fluconazole and cross-resistance to amphotericin B in candida albicans from AIDS patients caused by defective sterol delta5,6-desaturation." *FEBS Lett* **400**(1): 80–82.

Klepser, M. E. (2001). "Antifungal resistance among candida species." *Pharmacotherapy* **21**(8 Pt 2): 124S–132S.

Klepser, M. E., E. J. Wolfe, R. N. Jones, C. H. Nightingale and M. A. Pfaller (1997). "Antifungal pharmacodynamic characteristics of fluconazole and amphotericin B tested against candida albicans." *Antimicrob Agents Chemother* **41**(6): 1392–1395.

Krieter, P., B. Flannery, T. Musick, M. Gohdes, M. Martinho and R. Courtney (2004). "Disposition of posaconazole following single-dose oral administration in healthy subjects." *Antimicrob Agents Chemother* **48**(9): 3543–3551.

Kullberg, B. J., C. Viscoli, P. G. Pappas, J. Vazquez, L. Ostrosky-Zeichner, C. Rotstein, J. D. Sobel, R. Herbrecht, G. Rahav, S. Jaruratanasirikul, P. Chetchotisakd, E. Van Wijngaerden, J. De Waele, C. Lademacher, M. Engelhardt, L. Kovanda, R. Croos-Dabrera, C. Fredericks and G. R. Thompson (2019).

"Isavuconazole versus caspofungin in the treatment of candidemia and other invasive candida infections: the ACTIVE trial." *Clin Infect Dis* **68**(12): 1981–1989.

Kuse, E. R., P. Chetchotisakd, C. A. da Cunha, M. Ruhnke, C. Barrios, D. Raghunadharao, J. S. Sekhon, A. Freire, V. Ramasubramanian, I. Demeyer, M. Nucci, A. Leelarasamee, F. Jacobs, J. Decruyenaere, D. Pittet, A. J. Ullmann, L. Ostrosky-Zeichner, O. Lortholary, S. Koblinger, H. Diekmann-Berndt, O. A. Cornely and G. Micafungin Invasive Candidiasis Working (2007). "Micafungin versus liposomal amphotericin B for candidaemia and invasive candidosis: a phase III randomised double-blind trial." *Lancet* **369**(9572): 1519–1527.

Laverdiere, M., R. G. Lalonde, J. G. Baril, D. C. Sheppard, S. Park and D. S. Perlin (2006). "Progressive loss of echinocandin activity following prolonged use for treatment of candida albicans oesophagitis." *J Antimicrob Chemother* **57**(4): 705–708.

Leber, R., S. Fuchsbichler, V. Klobucnikova, N. Schweighofer, E. Pitters, K. Wohlfarter, M. Lederer, K. Landl, C. Ruckenstuhl, I. Hapala and F. Turnowsky (2003). "Molecular mechanism of terbinafine resistance in saccharomyces cerevisiae." *Antimicrob Agents Chemother* **47**(12): 3890–3900.

Leenders, A. C., P. Reiss, P. Portegies, K. Clezy, W. C. Hop, J. Hoy, J. C. Borleffs, T. Allworth, R. H. Kauffmann, P. Jones, F. P. Kroon, H. A. Verbrugh and S. de Marie (1997). "Liposomal amphotericin B (AmBisome) compared with amphotericin B both followed by oral fluconazole in the treatment of AIDS-associated cryptococcal meningitis." *AIDS* **11**(12): 1463–1471.

Lewis, J. S., 2nd, N. P. Wiederhold, B. L. Wickes, T. F. Patterson and J. H. Jorgensen (2013). "Rapid emergence of echinocandin resistance in candida glabrata resulting in clinical and microbiologic failure." *Antimicrob Agents Chemother.*

Llanos, A., J. Cieza, J. Bernardo, J. Echevarria, I. Biaggioni, R. Sabra and R. A. Branch (1991). "Effect of salt supplementation on amphotericin B nephrotoxicity." *Kidney Int* **40**(2): 302–308.

Lopez-Ribot, J. L., R. K. McAtee, L. N. Lee, W. R. Kirkpatrick, T. C. White, D. Sanglard and T. F. Patterson (1998). "Distinct patterns of gene expression associated with development of fluconazole resistance in serial candida albicans isolates from human immunodeficiency virus-infected patients with oropharyngeal candidiasis." *Antimicrob Agents Chemother* **42**(11): 2932–2937.

Lutsar, I., S. Roffey and P. Troke (2003). "Voriconazole concentrations in the cerebrospinal fluid and brain tissue of guinea pigs and immunocompromised patients." *Clin Infect Dis* **37**(5): 728–732.

Maertens, J. A., Raad, II, K. A. Marr, T. F. Patterson, D. P. Kontoyiannis, O. A. Cornely, E. J. Bow, G. Rahav, D. Neofytos, M. Aoun, J. W. Baddley, M. Giladi, W. J. Heinz, R. Herbrecht, W. Hope, M. Karthaus, D. G. Lee, O. Lortholary, V. A. Morrison, I. Oren, D. Selleslag, S. Shoham, G. R. Thompson, 3rd, M. Lee, R. M. Maher, A. H. Schmitt-Hoffmann, B. Zeiher and A. J. Ullmann (2016). "Isavuconazole versus voriconazole for primary treatment of invasive mould disease caused by aspergillus and other filamentous fungi (SECURE): a phase 3, randomised-controlled, non-inferiority trial." *Lancet* **387**(10020): 760–769.

Manavathu, E. K., J. L. Cutright and P. H. Chandrasekar (1998). "Organism-dependent fungicidal activities of azoles." *Antimicrob Agents Chemother* **42**(11): 3018–3021.

Mansfield, B. E., H. N. Oltean, B. G. Oliver, S. J. Hoot, S. E. Leyde, L. Hedstrom and T. C. White (2010). "Azole drugs are imported by facilitated diffusion in candida albicans and other pathogenic fungi." *PLOS Pathog* **6**(9): e1001126.

Martel, C. M., J. E. Parker, O. Bader, M. Weig, U. Gross, A. G. Warrilow, D. E. Kelly and S. L. Kelly (2010). "A clinical isolate of candida albicans with mutations in ERG11 (encoding sterol 14alpha-demethylase) and ERG5 (encoding C22 desaturase) is cross resistant to azoles and amphotericin B." *Antimicrob Agents Chemother* **54**(9): 3578–3583.

Mellado, E., G. Garcia-Effron, L. Alcazar-Fuoli, W. J. Melchers, P. E. Verweij, M. Cuenca-Estrella and J. L. Rodriguez-Tudela (2007). "A new aspergillus fumigatus resistance mechanism conferring in vitro cross-resistance to azole antifungals involves a combination of cyp51A alterations." *Antimicrob Agents Chemother* **51**(6): 1897–1904.

Mora-Duarte, J., R. Betts, C. Rotstein, A. L. Colombo, L. Thompson-Moya, J. Smietana, R. Lupinacci, C. Sable, N. Kartsonis and J. Perfect (2002). "Comparison of caspofungin and amphotericin B for invasive candidiasis." *N Engl J Med* **347**(25): 2020–2029.

Morschhauser, J. (2002). "The genetic basis of fluconazole resistance development in candida albicans." *Biochim Biophys Acta* **1587**(2-3): 240–248.

Niimi, K., K. Maki, F. Ikeda, A. R. Holmes, E. Lamping, M. Niimi, B. C. Monk and R. D. Cannon (2006). "Overexpression of candida albicans CDR1, CDR2, or MDR1 does not produce significant changes in echinocandin susceptibility." *Antimicrob Agents Chemother* **50**(4): 1148–1155.

Niwa, T., T. Shiraga and A. Takagi (2005). "Effect of antifungal drugs on cytochrome P450 (CYP) 2C9, CYP2C19, and CYP3A4 activities in human liver microsomes." *Biol Pharm Bull* **28**(9): 1805–1808.

Nolte, F. S., T. Parkinson, D. J. Falconer, S. Dix, J. Williams, C. Gilmore, R. Geller and J. R. Wingard (1997). "Isolation and characterization of fluconazole- and amphotericin B-resistant candida albicans from blood of two patients with leukemia." *Antimicrob Agents Chemother* **41**(1): 196–199.

Orozco, A. S., L. M. Higginbotham, C. A. Hitchcock, T. Parkinson, D. Falconer, A. S. Ibrahim, M. A. Ghannoum and S. G. Filler (1998). "Mechanism of fluconazole resistance in *Candida krusei*." *Antimicrob Agents Chemother* **42**(10): 2645–2649.

Ostrosky-Zeichner, L., J. H. Rex, P. G. Pappas, R. J. Hamill, R. A. Larsen, H. W. Horowitz, W. G. Powderly, N. Hyslop, C. A. Kauffman, J. Cleary, J. E. Mangino and J. Lee (2003). "Antifungal susceptibility survey of 2,000 bloodstream candida isolates in the United States." *Antimicrob Agents Chemother* **47**(10): 3149–3154.

Oude Lashof, A. M., J. D. Sobel, M. Ruhnke, P. G. Pappas, C. Viscoli, H. T. Schlamm, J. H. Rex and B. J. Kullberg (2012). "Safety and tolerability of voriconazole in patients with baseline renal insufficiency and candidemia." *Antimicrob Agents Chemother* **56**(6): 3133–3137.

Pappas, P. G., C. A. Kauffman, D. Andes, D. K. Benjamin, Jr., T. F. Calandra, J. E. Edwards, Jr., S. G. Filler, J. F. Fisher, B. J. Kullberg, L. Ostrosky-Zeichner, A. C. Reboli, J. H. Rex, T. J. Walsh and J. D. Sobel (2009). "Clinical practice guidelines for the management of candidiasis: 2009 update by the infectious diseases society of America." *Clin Infect Dis* **48**(5): 503–535.

Pappas, P. G., C. A. Kauffman, D. R. Andes, C. J. Clancy, K. A. Marr, L. Ostrosky-Zeichner, A. C. Reboli, M. G. Schuster, J. A. Vazquez, T. J. Walsh, T. E. Zaoutis and J. D. Sobel (2016). "Clinical Practice guideline for the management of candidiasis: 2016 update by the infectious diseases society of America." *Clin Infect Dis* **62**(4): e1–50.

Park, S., R. Kelly, J. N. Kahn, J. Robles, M. J. Hsu, E. Register, W. Li, V. Vyas, H. Fan, G. Abruzzo, A. Flattery, C. Gill, G. Chrebet, S. A. Parent, M. Kurtz, H. Teppler, C. M. Douglas and D. S. Perlin (2005). "Specific substitutions in the echinocandin target Fks1p account for reduced susceptibility of rare laboratory and clinical candida sp. isolates." *Antimicrob Agents Chemother* **49**(8): 3264–3273.

Perea, S., J. L. Lopez-Ribot, W. R. Kirkpatrick, R. K. McAtee, R. A. Santillan, M. Martinez, D. Calabrese, D. Sanglard and T. F. Patterson (2001). "Prevalence of molecular mechanisms of resistance to azole antifungal agents in candida albicans strains displaying high-level fluconazole resistance isolated from human immunodeficiency virus-infected patients." *Antimicrob Agents Chemother* **45**(10): 2676–2684.

Pfaller, M. A. (2012). "Antifungal drug resistance: mechanisms, epidemiology, and consequences for treatment." *Am J Med* **125**(1 Suppl): S3–13.

Pfaller, M. A., L. Boyken, R. J. Hollis, J. Kroeger, S. A. Messer, S. Tendolkar and D. J. Diekema (2008). "In vitro susceptibility of invasive isolates of candida spp. to anidulafungin, caspofungin, and micafungin: six years of global surveillance." *J Clin Microbiol* **46**(1): 150–156.

Pfaller, M. A., L. Boyken, R. J. Hollis, S. A. Messer, S. Tendolkar and D. J. Diekema (2005). "In vitro activities of anidulafungin against more than 2,500 clinical isolates of candida spp., including 315 isolates resistant to fluconazole." *J Clin Microbiol* **43**(11): 5425–5427.

Pfaller, M. A., M. Castanheira, S. A. Messer, G. J. Moet and R. N. Jones (2010). "Variation in candida spp. distribution and antifungal resistance rates among bloodstream infection isolates by patient age: report from the SENTRY Antimicrobial Surveillance Program (2008-2009)." *Diagn Microbiol Infect Dis* **68**(3): 278–283.

Pfaller, M. A., D. J. Diekema, S. A. Messer, R. J. Hollis and R. N. Jones (2003). "In vitro activities of caspofungin compared with those of fluconazole and itraconazole against 3,959 clinical isolates of candida spp., including 157 fluconazole-resistant isolates." *Antimicrob Agents Chemother* **47**(3): 1068–1071.

Pfaller, M. A., D. J. Diekema, L. Ostrosky-Zeichner, J. H. Rex, B. D. Alexander, D. Andes, S. D. Brown, V. Chaturvedi, M. A. Ghannoum, C. C. Knapp, D. J. Sheehan and T. J. Walsh (2008). "Correlation of MIC with outcome for candida species tested against caspofungin, anidulafungin, and micafungin: analysis and proposal for interpretive MIC breakpoints." *J Clin Microbiol* **46**(8): 2620–2629.

Pfaller, M. A., S. A. Messer, L. Boyken, R. J. Hollis, C. Rice, S. Tendolkar and D. J. Diekema (2004). "In vitro activities of voriconazole, posaconazole, and fluconazole against 4,169 clinical isolates of Candida spp. and cryptococcus neoformans collected during 2001 and 2002 in the ARTEMIS global antifungal surveillance program." *Diagn Microbiol Infect Dis* **48**(3): 201–205.

Pfaller, M. A., S. A. Messer, R. J. Hollis and R. N. Jones (2001). "In vitro activities of posaconazole (Sch 56592) compared with those of itraconazole and fluconazole against 3,685 clinical isolates of Candida spp. and cryptococcus neoformans." *Antimicrob Agents Chemother* **45**(10): 2862–2864.

Philips, J. A., F. M. Marty, R. M. Stone, B. A. Koplan, J. T. Katz and L. R. Baden (2007). "Torsades de pointes associated with voriconazole use." *Transpl Infect Dis* **9**(1): 33–36.

Pinjon, E., C. J. Jackson, S. L. Kelly, D. Sanglard, G. Moran, D. C. Coleman and D. J. Sullivan (2005). "Reduced azole susceptibility in genotype 3 candida dubliniensis isolates associated with increased CdCDR1 and CdCDR2 expression." *Antimicrob Agents Chemother* **49**(4): 1312–1318.

Poirier, J. M. and G. Cheymol (1998). "Optimisation of itraconazole therapy using target drug concentrations." *Clin Pharmacokinet* **35**(6): 461–473.

Prentice, H. G., I. M. Hann, R. Herbrecht, M. Aoun, S. Kvaloy, D. Catovsky, C. R. Pinkerton, S. A. Schey, F. Jacobs, A. Oakhill, R. F. Stevens, P. J. Darbyshire and B. E. Gibson (1997). "A randomized comparison of liposomal versus conventional amphotericin B for the treatment of pyrexia of unknown origin in neutropenic patients." *Br J Haematol* **98**(3): 711–718.

Rausch, C. R., A. J. DiPippo, P. Bose and D. P. Kontoyiannis (2018). "Breakthrough fungal infections in patients with leukemia receiving isavuconazole." *Clin Infect Dis* **67**(10): 1610–1613.

Reboli, A. C., C. Rotstein, P. G. Pappas, S. W. Chapman, D. H. Kett, D. Kumar, R. Betts, M. Wible, B. P. Goldstein, J. Schranz, D. S. Krause and T. J. Walsh (2007). "Anidulafungin versus fluconazole for invasive candidiasis." *N Engl J Med* **356**(24): 2472–2482.

Rybak, J. M., L. A. Doorley, A. T. Nishimoto, K. S. Barker, G. E. Palmer and P. D. Rogers (2019). "Abrogation of triazole resistance upon deletion of CDR1 in a clinical isolate of candida auris." *Antimicrob Agents Chemother* **63**(4).

Rybak, J. M., J. R. Fortwendel and P. D. Rogers (2019). "Emerging threat of triazole-resistant Aspergillus fumigatus." *J Antimicrob Chemother* **74**(4): 835–842.

Rybak, J. M., W. Ge, N. P. Wiederhold, J. E. Parker, S. L. Kelly, P. D. Rogers and J. R. Fortwendel (2019). "Mutations in hmg1, challenging the paradigm of clinical triazole resistance in aspergillus fumigatus." *mBio* **10**(2).

Rybak, J. M., K. R. Marx, A. T. Nishimoto and P. D. Rogers (2015). "Isavuconazole: pharmacology, pharmacodynamics, and current clinical experience with a new triazole antifungal agent." *Pharmacotherapy* **35**(11): 1037–1051.

Ryder, N. S. (1992). "Terbinafine: mode of action and properties of the squalene epoxidase inhibition." *Br J Dermatol* **126**(Suppl 39): 2–7.

Saag, M. S., R. J. Graybill, R. A. Larsen, P. G. Pappas, J. R. Perfect, W. G. Powderly, J. D. Sobel and W. E. Dismukes (2000). "Practice guidelines for the management of cryptococcal disease. infectious diseases society of America." *Clin Infect Dis* **30**(4): 710–718.

Sanglard, D., F. Ischer, L. Koymans and J. Bille (1998). "Amino acid substitutions in the cytochrome P-450 lanosterol 14alpha-demethylase (CYP51A1) from azole-resistant candida albicans clinical isolates contribute to resistance to azole antifungal agents." *Antimicrob Agents Chemother* **42**(2): 241–253.

Sanglard, D., K. Kuchler, F. Ischer, J. L. Pagani, M. Monod and J. Bille (1995). "Mechanisms of resistance to azole antifungal agents in candida albicans isolates from AIDS patients involve specific multidrug transporters." *Antimicrob Agents Chemother* **39**(11): 2378–2386.

Sau, K., S. S. Mambula, E. Latz, P. Henneke, D. T. Golenbock and S. M. Levitz (2003). "The antifungal drug amphotericin B promotes inflammatory cytokine release by a Toll-like receptor- and CD14-dependent mechanism." *J Biol Chem* **278**(39): 37561–37568.

Sawaya, B. P., J. P. Briggs and J. Schnermann (1995). "Amphotericin B nephrotoxicity: the adverse consequences of altered membrane properties." *J Am Soc Nephrol* **6**(2): 154–164.

Schubert, S., P. D. Rogers and J. Morschhauser (2008). "Gain-of-function mutations in the transcription factor MRR1 are responsible for overexpression of the MDR1 efflux pump in fluconazole-resistant candida dubliniensis strains." *Antimicrob Agents Chemother* **52**(12): 4274–4280.

Selmecki, A., A. Forche and J. Berman (2006). "Aneuploidy and isochromosome formation in drug-resistant candida albicans." *Science* **313**(5785): 367–370.

Sewell, T. R., J. Zhu, J. Rhodes, F. Hagen, J. F. Meis, M. C. Fisher and T. Jombart (2019). "Nonrandom distribution of azole resistance across the global population of aspergillus fumigatus." *MBio* **10**(3).

Shah, P. M. (1982). "Paradoxical effect of antibiotics. I. The 'Eagle effect.'" *J Antimicrob Chemother* **10**(4): 259–260.

Shapiro, R. S., N. Robbins and L. E. Cowen (2011). "Regulatory circuitry governing fungal development, drug resistance, and disease." *Microbiol Mol Biol Rev* **75**(2): 213–267.

Sharma, C., N. Kumar, R. Pandey, J. F. Meis and A. Chowdhary (2016). "Whole genome sequencing of emerging multidrug resistant candida auris isolates in India demonstrates low genetic variation." *New Microbes New Infect* **13**: 77–82.

Snelders, E., A. Karawajczyk, R. J. Verhoeven, H. Venselaar, G. Schaftenaar, P. E. Verweij and W. J. Melchers (2011). "The structure-function relationship of the aspergillus fumigatuscyp51A L98H conversion by site-directed mutagenesis: the mechanism of L98H azole resistance." *Fungal Genet Biol* **48**(11): 1062–1070.

Soustre, J., M. H. Rodier, S. Imbert-Bouyer, G. Daniault and C. Imbert (2004). "Caspofungin modulates in vitro adherence of candida albicans to plastic coated with extracellular matrix proteins." *J Antimicrob Chemother* **53**(3): 522–525.

Stevens, D. A., M. Espiritu and R. Parmar (2004). "Paradoxical effect of caspofungin: reduced activity against Candida albicans at high drug concentrations." *Antimicrob Agents Chemother* **48**(9): 3407–3411.

Stevens, D. A., M. Ichinomiya, Y. Koshi and H. Horiuchi (2006). "Escape of candida from caspofungin inhibition at concentrations above the MIC (paradoxical effect) accomplished by increased cell wall chitin; evidence for beta-1,6-glucan synthesis inhibition by caspofungin." *Antimicrob Agents Chemother* **50**(9): 3160–3161.

Stevens, D. L., A. E. Gibbons, R. Bergstrom and V. Winn (1988). "The Eagle effect revisited: efficacy of clindamycin, erythromycin, and penicillin in the treatment of streptococcal myositis." *J Infect Dis* **158**(1): 23–28.

Sucher, A. J., E. B. Chahine and H. E. Balcer (2009). "Echinocandins: the newest class of antifungals." *Ann Pharmacother* **43**(10): 1647–1657.

Templeton, I. E., K. E. Thummel, E. D. Kharasch, K. L. Kunze, C. Hoffer, W. L. Nelson and N. Isoherranen (2008). "Contribution of itraconazole metabolites to inhibition of CYP3A4 in vivo." *Clin Pharmacol Ther* **83**(1): 77–85.

Thaler, F., B. Bernard, M. Tod, C. P. Jedynak, P. Derome and P. Loirat (1995). "Fluconazole penetration in cerebral parenchyma in humans at steady state." *Antimicrob Agents Chemother* **39**(5): 1154–1156.

Trovato, L., M. F. Rapisarda, A. M. Greco, F. Galata and S. Oliveri (2009). "In vitro susceptibility of nondermatophyte molds isolated from onychomycosis to antifungal drugs." *J Chemother* **21**(4): 403–407.

Uchaipichat, V., L. K. Winner, P. I. Mackenzie, D. J. Elliot, J. A. Williams and J. O. Miners (2006). "Quantitative prediction of in vivo inhibitory interactions involving glucuronidated drugs from in vitro data: the effect of fluconazole on zidovudine glucuronidation." *Br J Clin Pharmacol* **61**(4): 427–439.

Vermes, A., H. J. Guchelaar and J. Dankert (2000). "Flucytosine: a review of its pharmacology, clinical indications, pharmacokinetics, toxicity and drug interactions." *J Antimicrob Chemother* **46**(2): 171–179.

Wang, H., F. Kong, T. C. Sorrell, B. Wang, P. McNicholas, N. Pantarat, D. Ellis, M. Xiao, F. Widmer and S. C. Chen (2009). "Rapid detection of ERG11 gene mutations in clinical candida albicans isolates with reduced susceptibility to fluconazole by rolling circle amplification and DNA sequencing." *BMC Microbiol* **9**: 167.

Warrilow, A. G., J. E. Parker, C. L. Price, W. D. Nes, S. L. Kelly and D. E. Kelly (2015). "In vitro biochemical study of CYP51-mediated azole resistance in aspergillus fumigatus." *Antimicrob Agents Chemother* **59**(12): 7771–7778.

Weiss, J., M. M. Ten Hoevel, J. Burhenne, I. Walter-Sack, M. M. Hoffmann, J. Rengelshausen, W. E. Haefeli and G. Mikus (2009). "CYP2C19 genotype is a major factor contributing to the highly variable pharmacokinetics of voriconazole." *J Clin Pharmacol* **49**(2): 196–204.

Whaley, S. G., Q. Zhang, K. E. Caudle and P. D. Rogers (2018). "Relative contribution of the ABC transporters Cdr1, Pdh1, and Snq2 to azole resistance in candida glabrata." *Antimicrob Agents Chemother* **62**(10).

Wiederhold, N. P., D. P. Kontoyiannis, R. A. Prince and R. E. Lewis (2005). "Attenuation of the activity of caspofungin at high concentrations against candida albicans: possible role of cell wall integrity and calcineurin pathways." *Antimicrob Agents Chemother* **49**(12): 5146–5148.

Wingard, J. R., M. H. White, E. Anaissie, J. Raffalli, J. Goodman and A. Arrieta (2000). "A randomized, double-blind comparative trial evaluating the safety of liposomal amphotericin B versus amphotericin B lipid complex in the empirical treatment of febrile neutropenia. L Amph/ABLC collaborative study group." *Clin Infect Dis* **31**(5): 1155–1163.

Yoshida, Y. (1988). "Cytochrome P450 of fungi: primary target for azole antifungal agents." *Curr Top Med Mycol* **2**: 388–418.

Zaas, A. K. and B. D. Alexander (2005). "Echinocandins: role in antifungal therapy, 2005." *Expert Opin Pharmacother* **6**(10): 1657–1668.

Zimmermann, T., R. A. Yeates, M. Albrecht, H. Laufen and A. Wildfeuer (1994). "Influence of concomitant food intake on the gastrointestinal absorption of fluconazole and itraconazole in Japanese subjects." *Int J Clin Pharmacol Res* **14**(3): 87–93.

Zimmermann, T., R. A. Yeates, H. Laufen, G. Pfaff and A. Wildfeuer (1994). "Influence of concomitant food intake on the oral absorption of two triazole antifungal agents, itraconazole and fluconazole." *Eur J Clin Pharmacol* **46**(2): 147–150.

Zimmermann, T., R. A. Yeates, K. D. Riedel, P. Lach and H. Laufen (1994). "The influence of gastric pH on the pharmacokinetics of fluconazole: the effect of omeprazole." *Int J Clin Pharmacol Ther* **32**(9): 491–496.

Zonios, D. I., J. Gea-Banacloche, R. Childs and J. E. Bennett (2008). "Hallucinations during voriconazole therapy." *Clin Infect Dis* **47**(1): e7–e10.

53

Mechanisms of Action of Antiviral Agents

Guido Antonelli, Francesca Falasca, and Ombretta Turriziani

CONTENTS

Introduction

Today, more than 50 compounds have been formally approved for the treatment of viral infections. They are formally licensed for clinical use in the treatment of several viral infections caused by several human viruses such as human immunodeficiency virus (HIV), herpes simplex viruses, varicella–zoster virus, respiratory syncytial virus, cytomegalovirus, hepatitis B virus, hepatitis C virus, or influenza A virus. The chapter first describes the mechanism of action of the main approved antiviral drugs, summarized in Table 53.1, and then discusses issues stemming from the broad use of the antiviral drugs that we have seen in the last decade.

Mechanism of Action of Antiviral Agents

A description of the mechanism of action of the various antiviral agents focused on the target they affect on the virus life cycle is presented. Figure 53.1 summarizes the steps of the viral life cycle that can be affected by the antiviral drugs. We will handle only drugs made by chemical synthesis approved for clinical use in western countries omitting the many antiviral drugs that are in the process of being approved. The following report neglects also an in-depth description of the mechanism of action of interferons that would deserve an extensive discussion, being an endogenous and a very potent mechanism of defense against viral diseases and have been the topic of several recent publications but are beyond the scope of this review (Scagnolari and Antonelli 2013; Antonelli et al. 2015).

Antiviral Agents Affecting Absorption and Entry (Entry Inhibitors)

The first stage of viral infection involves attachment of the virus to host cellular membranes. This phase is mediated by the interaction between viral surface molecules and cellular receptors. Following adsorption, virus entry into host cell occurs by different routes such as fusion or endocytosis.

Inhibitors that specifically target this first stage of the viral cycle have been approved for HIV infection treatment (Esté and Telenti 2007; Soriano et al. 2009; Tilton and Doms 2010) and in the prevention of RSV infection (Hu and Robinson 2010; Lanari et al. 2010; Welliver 2010).

HIV-1 entry is a multistep process. Binding of the HIV-1 envelope protein gp120 to the CD4 receptor on host cells is followed by engagement of specific chemokine receptors, either CCR5 or CXCR4, leading to rearrangements in the envelope transmembrane subunit that result in membrane fusion. An important step in the fusion process is a conformational change in the viral transmembrane glycoprotein gp41. HIV-1 entry inhibitors comprise three drugs: enfuvirtide, a fusion inhibitor, maraviroc, a CCR5 antagonist (Matthews et al. 2004; Perros 2007) and Ibalizumab, a genetically engineered IgG4 MAb which binds to CD4 T cell receptors (Burkly et al. 1992; Reimann et al. 1997; Iacob and Iacob 2017).

Enfuvirtide binds to a region of gp41 and disrupts the conformational changes associated with membrane fusion, thereby blocking virus entry. Maraviroc is an allosteric inhibitor of HIV-1 entry, binding to and altering the conformation of CCR5 such that it is no longer recognized by gp120 and membrane fusion

TABLE 53.1

Main Approved Antiviral Drugs and Their Mechanism of Action**

Anti-HIV	Mechanism of Action (Main)	*Pyrophosphate Analog*	
Entry Inhibitors		Foscarnet	It selectively inhibits the pyrophosphate binding site on viral DNA polymerases.
Enfuvirtide	It interferes with glycoprotein 41-dependent membrane fusion blocking virus entry.	***DNA Maturation and Packaging Inhibitor***	
Ibalizumab	It binds CD4 extracellular domain, preventing conformational changes in the CD4–gp120 complex essential for viral entry.	Letermovir	It inhibits CMV replication by binding to components of the terminase complex.
Maraviroc	Negative allosteric modulator of the CCR5 receptor that is an essential coreceptor for the entry process of HIV.	**Anti HBV**	
		Reverse Transcriptase Inhibitors	
		Nucleoside Analogs	
Reverse Transcriptase Inhibitors		Entecavir	Active as triphosphate derivates, which prematurely terminate viral DNA synthesis.
Nucleoside Analogs	Active as triphosphate derivates, which prematurely terminate DNA synthesis.	Telbivudine	
Abacavir		Lamivudine	
Emtricitabine		*Nucleotide Analog*	
Lamivudine		Tenofovir	
Stavudine		**Anti-HCV**	
Zidovudine		***NS5B RNA Polymerase Inhibitor***	
Nucleotide analog		*Non-Nucleoside Inhibitor*	
Tenofovir		Dasabuvir	It binds HCV NS5B polymerase and blocks viral RNA synthesis and replication.
Non-nucleoside inhibitors	They bind to a hydrophobic pocket of HIV-1 reverse transcriptase, blocking polymerization of viral DNA.	*Nucleotide Analog*	
Doravirine		Sofosbuvir	Potent inhibitor of the NS5B polymerase. It blocks viral RNA synthesis and replication.
Efavirenz			
Etravirine			
Nevirapine		*Nucleoside Analog*	
Rilpivirine			
Protease inhibitors		Ribavirin	Active as triphosphate derivates, which prematurely terminate viral nucleic acid synthesis.[c]
Atazanavir	They bind to the active site of viral protease.		
Fosamprenavir			
Lopinavir		*NS5A Protein Inhibitor*	
Darunavir	They bind strongly and selectively to the HIV-1 protease.	Elbasvir	They bind to the replication complex protein NS5A, disrupting HCV RNA replication and virion assembly.
Tipranavir		Ledipasvir	
Pharmacodynamic enhancer		Ombitasvir	
Cobicistat	These drugs inhibit CYP3A4 leading to higher drug exposure.	Pibrentasvir	
Ritonavir		Velpatasvir	
Integrase inhibitor		*NS3/4A Protease Inhibitor*	
Dolutegravir	They bind viral integrase, inhibiting the strand transfer step of HIV-1 integration.	Glecaprevir	They prevent viral replication by inhibiting the NS3/4A serine protease of HCV.
Bictegravir[a]		Grazoprevir	
Elvitegravi[a]		Paritaprevir	
Raltegravir		Voxilaprevir	
Anti-Herpes viruses		**Anti-influenza Virus**	
DNA Polymerase Inhibitors		*Uncoating Inhibitor*	
Nucleoside Analogs		Amantadine	Inhibitors of the viral M2 channel. The activity of channel is required for uncoating.
Aciclovir	Active as triphosphate derivates, which prematurely terminate DNA synthesis.	Rimantadine	
Brivudine[b]		*Neuraminidase Inhibitors*	
Famciclovir		Oseltamivir	Block influenza neuraminidase and prevent the cleavage of sialic acid residues, thus interfering with progeny virus release.
Ganciclovir		Peramivir	
Valaciclovir		Zanamivir	
Valganciclovir		*Endonuclease Inhibitor*	
Nucleotide Analog		Baloxavir marboxil	It inhibits cap-dependent endonuclease, a key enzyme involved in the initiation of influenza virus mRNA synthesis.
Cidofovir			

** See text for details.

[a] Used in fixed-dose combination formulations.

[b] Brivudine is approved in some European countries and in Central America.

TABLE 53.1 *(Continued)*

Main Approved Antiviral Drugs and Their Mechanism of Action**

Anti-RSV		**Anti-SARS-CoV-2**	
Entry Inhibitors		*RNA Polymerase inhibitor*	
Palivizumab[c]	It Binds to RSV F protein, which plays a role in virus attachment and fusion.	Redmesivir[d]	It is a prodrug of a nucleotide analogue that is intracellularly metabolized to an analogue of adenosine triphosphate that inhibits a number of viral RNA polymerases
RNA Inhibitor			
Ribavirin	It interferes with viral RNA synthesis.[e]		
Anti-variola virus			
Maturation Inhibitor			
Tecovirimat[d]	It binds and inhibits P37 envelope protein, preventing the formation of egress-competent enveloped virions.		

[c] Palivizumab is indicated for the prevention of serious lower respiratory-tract disease caused by respiratory syncytial virus (RSV) in pediatric patients at high risk of RSV disease.

[d] Currently approved only by FDA-USA and Japan's authorities..

[e] Ribavirin has been reported also to inhibit inosine monophosphate dehydrogenase (IMPDH), thus reducing intracellular pools of GTP, and to induce mutagenesis, thus leading to lethal mutations. Furthermore, ribavirin has been used in the treatment of chronic hepatitis C in combination with IFN alpha (in this framework, some authors reported that it has immunomodulatory activity) and with the early DAAs before the introduction of the very effective new ones.

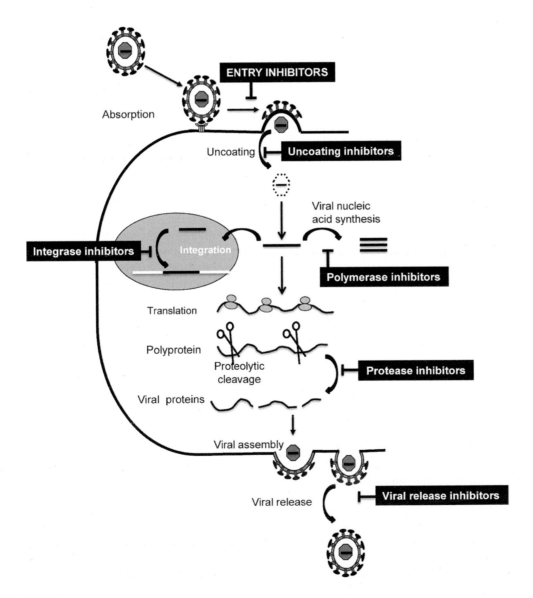

FIGURE 53.1 Entry inhibitors.

cannot proceed. Ibalizumab specifically leads to conformational changes of the CD4 T cell receptor–gp120 complex thus preventing HIV fusion and entry. Ibalizumab is the first monoclonal antibody for the treatment of multidrug resistant HIV-1 infection in combination with other forms of antiretroviral therapy in heavily treatment-experienced adults who are failing their current antiretroviral regimen (Beccari et al. 2019).

Fusion inhibitory activity is also displayed by palivizumab, a humanized monoclonal antibody directed to an epitope in the A antigenic site of the RSV fusion (F) protein (Scott and Lamb 1999). Palivizumab is a composite of primarily human antibody sequences (95%) and murine antibody sequences (5%) providing both neutralizing and fusion inhibitory activity against RSV, resulting in inhibition of RSV replication (see footnote to Table 53.1 for further details).

Antiviral Agents Affecting Uncoating (Uncoating Inhibitors)

Viral uncoating, an essential step in the viral replication occurring after penetration, leads to the removal of capsid from the viral structure. In most cases, it is a process of disaggregation that sometimes occurs with the aid of cellular enzymes. At the end of this process, only the nucleic acid or a nucleic acid-protein complex remains into the cytoplasm. Basically, the process enables viral genes to become available for transcription and translation.

Inhibitors targeting this phase have been identified against influenza virus (Davies et al. 1964). Specifically, following endocytosis and before hemagglutinin-mediated fusion between viral and endosomal membranes, the influenza virus M2 channels are activated by the low pH of the endosome and conduct protons to acidify the viral interior (Pinto et al. 1992). The acidification has been suggested to weaken the electrostatic interaction between matrix proteins and ribonucleoprotein complexes such that subsequent membrane fusion can release the uncoated ribonucleoproteins into the cytosol for transport into the nucleus (Martin and Helenius 1991). The tricyclic amines amantadine hydrochloride and rimantadine hydrochloride block replication of influenza A by inhibiting the function of its M2 protein. The adamantine derivatives are effective only against influenza A because influenza B does not have an M2 protein but has an NB protein that is not affected by amantadine (Hay et al. 1986).

Antiviral Agents Affecting Viral Nucleic Acid Synthesis (Polymerase Complex Inhibitor)

After uncoating, a series of complex key events occur that finally leads to the synthesis of viral proteins and genomes by the host synthesizing machinery. Irrespective of the structure, the size, the genome organization, and the replicative strategies, all the viruses must proceed to a synthesis of specific viral nucleic acids that are in some cases synthesized by a viral enzyme such as viral DNA polymerase, RNA polymerase, or reverse transcriptase. These polymerases are crucial in the viral life cycle and, as such, were and still are attractive drug targets for the treatment of many viral infections (Tsai et al. 2006).

Nucleoside analogs (NAs) represent the first polymerase inhibitors that showed clinical efficacy (Douglas et al. 1984; Straus et al. 1984; Fischl et al. 1987) and are nowadays broadly used to treat viral infections such as HBV, HSV-1, HSV-2, and HIV-1 infection (Coen and Schaffer 2003; Dienstag 2009; von Kleist et al. 2012). NAs are typically formulated as prodrugs, requiring intracellular phosphorylation to form an analog of (deoxy-)nucleoside triphosphate, which can be incorporated into nascent viral DNA by the viral polymerase. Most NAs are phosphorylated by cellular kinases, except the compounds used against herpetic infections in which the initial phosphorylation step is carried out by viral thymidine kinase and the two subsequent steps by cellular enzymes (Miller and Miller 1982; Elion 1993). After incorporation, NAs bring the polymerization machinery to a halt as they lack the 3-hydroxyl group on the deoxyribose moiety required to attach the next incoming nucleotide (Coen and Schaffer 2003; Painter et al. 2004). NAs such as adefovir and tenofovir show the same mode of action, but they need only two phosphorylation steps because they are already phosphorylated (De Clercq and Hol 2005). With regard to HIV infection, viral polymerase inhibitors are classified into three groups (Table 53.1): nucleoside reverse transcriptase inhibitors (NRTIs), such as zidovudine, lamivudine, and abacavir; nucleotide reverse transcriptase inhibitors (NtRTIs), of which tenofovir is the only one approved; and non-nucleoside reverse transcriptase inhibitors (NNRTIs) such as nevirapine and efavirenz (De Clercq 2009).

Since zidovudine was approved in 1987, another six NRTIs have been approved (see Table 53.1). Some of these, such as didanosine and zalcitabine are no longer used due the introduction in clinical practice of new drugs with a better pharmacological/toxicity profile.

NNRTIs have a completely different mode of action from NRTIs and NtRTIs. They are allosteric inhibitors of the HIV-1 reverse transcriptase. They bind to an allosteric pocket in the p66 subunit of the reverse transcriptase and inhibit DNA synthesis reactions by a noncompetitive mechanism of action (Kohlstaedt et al. 1992; de Béthune 2010). Five NNRTIs have currently been approved for the clinical treatment of acquired immune deficiency syndrome (AIDS) disease: nevirapine, doravirine, efavirenz, etravirine, and rilpivirine (Table 53.1). The rapid emergence of drug resistance mutations for the first-generation NNRTIs (nevirapine and efavirenz) dramatically reduced their use (de Béthune 2010). Alternatively, the second-generation NNRTIs, such as etravirine, rilpivirine, and the recently approved doravirine, possess a high genetic barrier and, thanks to their structurally intrinsic flexibility, are able to adapt their binding orientation and overcome common NNRTI resistance-associated mutations (Schiller and Youssef-Bessler 2009; Adams et al. 2010; Feng et al. 2016).

Ribavirin is a guanosine analog that has inhibitory activity against a broad spectrum of RNA viruses and is usually included among the different drugs able to affect viral nucleic acid synthesis. Although ribavirin is a NA, various mechanisms of action have been suggested to account for its antiviral activity (Hong and Cameron 2002), including (1) depletion of intracellular GTP pools by inhibition of the cellular inosine-5-monophosphate dehydrogenase by the 5-monophosphate metabolite of ribavirin, (2) inhibition of viral polymerase activity by the 5-triphosphate metabolite of ribavirin, (3) inhibition of viral capping by inhibition of (viral or cellular) guanylyltransferase activity by ribavirin 5-triphosphate, and (4) induction of error catastrophe as a result of accumulation of mutations (some of them lethal) in the viral genome. In addition to its effect on viral replication, ribavirin is also thought to have immunomodulatory properties, including the regulation of macrophage- and T-helper cell-produced cytokines, modulation of the Th1/Th2 subset balance and enhancement of interferon stimulated gene expression (Stevenson et al. 2011).

The detailed characterization of the viral life cycle HCV of the different genotype of allowed the development of directly acting antiviral drugs (DAAs). Currently, NS5A phosphoprotein inhibitors and NS5B polymerase inhibitors are available for clinical use. Most of these drugs are administered as fixed-dose combinations of different DAAs (Bourlière and Pietri 2019). The combination of sofosbuvir and daclatasvir was the first pangenotypic combination achieving SVR rates over 90–95% in patients infected by genotypes 1–4 (Sulkowski et al. 2014; Alavian and Rezaee-Zavareh 2016). Since 2017, another two pangenotypic fixed-dose combinations, sofosbuvir/velpatasvir (Fcld et al. 2015) and glecaprevir/pibrentasvir (Forns et al. 2017), were approved (see also the section entitled "Antiviral Agents Affecting Viral Proteases" later in this chapter).

In 2018, the first polymerase complex inhibitor, baloxavir marboxil, was approved for the treatment of uncomplicated influenza in the United States and in Japan. The polymerase complex of influenza viruses is a heterotrimer composed of three subunits: polymerase basic protein 1 (PB1), polymerase basic protein 2 (PB2), and polymerase acidic protein (PA). The PA protein of influenza A and B viruses possesses endonuclease activity that is necessary to cleave host mRNA caps in order to initiate viral transcription baloxavir marboxil is a prodrug that is metabolized into baloxavir marboxil acid, which directly inhibits the cap-dependent endonuclease activity of the PA protein of influenza, thereby inhibiting the cleavage of mRNA during the "cap-snatching" process (Mifsud et al. 2019).

At the time of writing, there are only few antiviral therapies of proven effectiveness in patients with coronavirus disease 2019 (COVID-19) caused by Severe Acute Respiratory Syndrome Coronavirus 2 (SARS-CoV-2). In vitro testing has shown that Remdesivir, an inhibitor which has broad-spectrum activity against members of several virus families, including filoviruses (e.g., Ebola) and coronaviruses (e.g., SARS-CoV and Middle East respiratory syndrome coronavirus [MERS-CoV]), has also activity against SARS-CoV-2 (Li G, De Clercq E, 2020) and has been authorized for potential COVID-19 treatment (US Food & Drug Administration, 2020).

Antiviral Agents Affecting Integration of Viral Genomes (Integrase Inhibitors)

Many viruses are able to insert their genetic material into host chromosomes. This process is called integration. For most of the members of *Retroviridae* family, such as HIV, the integration process is mandatory and results in a permanent insertion of the viral genome into the host chromosomal DNA. At this stage, the virus is referred as a provirus. It can remain latent and then be passively replicated along with the host genome and passed on to the cell's offspring. Changes of host's environmental condition, such as lymphocyte activation in the case of HIV, can

reactivate the virus leading to a viral transcription, translation, and yield of new infectious viruses.

Inhibitors targeting this phase have been identified against HIV whose integrase is one of three essential enzymes encoded by the viral pol gene (Krishnan and Engelman 2012; Métifiot et al. 2013). Integrase is responsible for an essential step in HIV-1 replication involving integration of the viral genome into the host cell DNA. This viral enzyme catalyzes two reactions: 3' end processing of the viral DNA during which the terminal GpT dinucleotide is cleaved from the 3' end of each long terminal repeats, and strand transfer, which joins the viral genome to the host chromosomal DNA forming a functional integrated proviral DNA. Since integrase is indispensable for HIV replication, it has become a validated target for developing anti-HIV agents (Hazuda et al. 2000; Pommier et al. 2005) and inhibitors that specifically target integration have been approved for the treatment of HIV infection. One integrase inhibitor, raltegravir, was approved in late 2007 for the treatment of HIV-1 infection in patients with prior antiretroviral treatment experience and it was recently also approved for first-line therapy (Grinsztejn et al. 2007). Three other effective integrase inhibitors (elvitegravir, dolutegravir, and the recently approved bictegravir) are today available for the HIV infection treatment (Shimura and Kodama 2009; Fantauzzi et al. 2013; Stellbrink et al. 2019).

Antiviral Agents Affecting Viral Proteases (Protease Inhibitors)

Once viral genomes and proteins have been newly synthesized, they must be assembled into new virions. For many viruses, this phase occurs through a maturation of viral precursors that include proteolytic cleavage of capsid or envelope polyproteins/precursors. Specific viral proteases catalyze this process and their catalytic activity is required to produce new infectious virions.

For this reason, viral proteases represented an attractive target for the development of novel antiviral agents, and indeed, compounds that block these enzymes have been approved for HIV and HCV infection. The clinical relevance of viral protease inhibitors (PIs) was demonstrated with the successful application of HIV PIs (Eron 2000). Currently, 6 PIs, including ritonavir that is used as a pharmacokinetic booster, are available in clinical practice (Table 53.1). Indeed, over the past years other PIs were approved for HIV infection treatment, such as Nelfinavir, Amprenavir, and Indinavir, but their use was overcome with the introduction of more effective and safer drugs. Most HIV-1 PIs were designed as peptidomimetics of the viral peptide substrates with not cleavable structures in the scissile bond (Eron 2000). These compounds bind to the site where protein cutting occurs, thereby preventing the enzyme from releasing the individual core proteins so the new viral particles are unable to mature or become infectious. Two agents from a new chemical class of PI, namely tipranavir and darunavir, were approved by the U.S. Food and Drug Administration (FDA) in 2006. These two PIs are unique in their ability to remain active against the increasing number and combinations of PI mutations that render other PIs partially inactive (Luna and Townsend 2007; McCoy 2007).

HCV nonstructural (NS)3/NS4A protease inhibitors have been the target of intensive research for the past 20 years.

First generation of HCV PIs, telaprevir and boceprevir, had to be combined with interferons. These compounds were mainly specific for HCV genotype 1, showed a low genetic barrier to viral resistance and were associated with a plethora of side effects. Therefore, these drugs have been abandoned meanwhile. The scenario has been revolutionized by the introduction of new antiviral compounds, exhibiting excellent efficacy (80–100% sustained response) and improved safety profiles. In fact, new generation PIs, such as paritaprevir, grazoprevir, glecaprevir, and voxilaprevir, have increased intrinsic antiviral potency, wider coverage of viral genotypes, a high genetic barrier, and better pharmacokinetics (Bartenschlager et al. 2013). Specifically, third-generation pangenotypic NS3/4A protease inhibitors (mainly glecaprevir and voxilaprevir) show high antiviral activity and a genetic resistance barrier with cure rates of more than 95% when combined with an NS5A inhibitor, irrespective of baseline resistance associated variants being present. It is common opinion that, due to these new options, HCV will not represent a significant health issue in the future (Pawlotsky 2014). The World Health Organization (WHO) has already aimed for the elimination of hepatitis C as a public health threat by 2030 (defined as a 90% reduction in new infections and a 65% reduction in mortality) (WHO 2016).

Antiviral Agents Affecting Virus Maturation

In some viruses, a key step of maturation is the packaging of unit-length genomes into capsids. Taking human cytomegalovirus (HCMV), it is known that such a step occurs through the action of specific viral enzymes, so-called terminases. HCMV-terminase complex is a heterotrimer consisting of the ATPase pUL56, the nuclease pUL89 and a third component, pUL51 (Borst et al. 2013). Letermovir is a 3,4-dihydro-quinazoline-4-yl-acetic acid derivative that inhibits CMV replication by binding to components of the terminase complex (UL51, UL56, or both) (Goldner et al. 2011; Chou 2017). To date, Letermovir is registered and used to prevent illness caused by HCMV in adults having an allogeneic hematopoietic stem-cell transplant to replace their bone marrow.

Recently, the FDA has approved tecovirimat, the first drug with an indication for treatment of variola (smallpox) virus. Though the WHO declared smallpox, a contagious and sometimes fatal infectious disease, eradicated in 1980, there have been longstanding concerns that smallpox could be used as a bioweapon (Chan-Tack et al. 2019). Tecovirimat [4-trifluoromethyl-N-(3,3a,4,4a,5,5a,6,6a-octahydro-1,3-dioxo-4,6-ethenocycloprop[f] isoindol-2(1H)-yl)-benzamide] (TPOXX) is a small molecule that has been shown to impede the envelopment of virions and the egress of enveloped viruses from the infected cell. Specifically, this compound interferes with F13L, that is one of the pivotal proteins required for the production of enveloped virions, blocks its interaction with cellular Rab9 GTPase and TIP47, which prevents the formation of egress-competent enveloped virions necessary for cell-to-cell and long-range dissemination of virus Smith et al. 2009; Duraffour et al. 2015). The efficacy of TPOXX for the treatment of smallpox is based on studies in monkeys (cynomolgus macaques) and New Zealand White (NZW) rabbits. The primary efficacy endpoint was proportion

of infected animals that survived to the prespecified end-of-study. It was not possible to conduct controlled clinical trials in humans with smallpox because of the ethical issue (Chan-Tack et al. 2019).

Antiviral Agents Affecting Viral Release (Release Inhibitors)

The release of the newly synthesized viral particles is the final stage of the virus' life cycle. In this stage, the naked viruses are released rapidly after maturation, in most cases when the cells undergo lysis while the enveloped virus pushes itself out of the host cell, taking with it part of the membrane of the cell. The newly synthesized viruses contain all the structures necessary to bind to a new cell and receptors and begin the process again.

In some cases, the released virions become infectious only after the intervention of a viral enzyme. Some drugs are able to inhibit this phase, acting on the viral protein involved in such a process, and have been approved for the treatment of influenza virus infection (Hayden et al. 1997; Gubareva et al. 2000). The hemagglutinin of influenza A and B viruses binds to neuraminic acid-containing receptors (Gubareva et al. 2000). The neuraminidase enzymatic activity promotes influenza virus disbursement within mucosal secretions of the respiratory tract and prevents progeny virions from self-aggregating by removing neuraminic (sialic) acids from oligosaccharide chains of receptors (Colman 1994). Neuraminidase inhibitors (NAIs) are designed to limit influenza virus infection by blocking the enzyme active site. Two NAIs, inhaled zanamivir and oral oseltamivir (prodrug), have been licensed for influenza treatment in several countries.

Interferons

As stated, interferon would deserve a special attention being an important mechanism of defense of the organism against viral infections.

Interferon (IFN) was discovered in 1957 (Isaacs and Lindenmann 1957) as a substance able to interfere with virus replication. It is now known that IFN was not only a protein but a group of proteins with parallel, but not analogous or identical, biochemical, and biological properties. It is now recognized that there are more than 20 nonallelic genes that codify proteins possessing properties comparable to those of IFN (i.e., direct antiviral activity, species specificity, relatively low molecular weights, and pleiotropic activity). All these genes can be divided into three main groups: IFN type I, which includes many different IFNs among which many IFN-alpha and IFN-beta; IFN type II, which includes IFN-gamma; and IFN type III, which includes IFN-lambda. In the last decades, interferons (IFNs) have been widely studied from the chemical and biochemical points of view. The mechanisms through which IFNs interact with the cell have been elucidated and many of their biological activities have been characterized. It is now known that IFNs, other than a direct antiviral effect elicited through the (over)expression of several cellular antiviral proteins, play a pivotal role in shaping innate and adaptive immune response to viral infections (Scagnolari and Antonelli 2013). Altogether, these acquisitions led to the clinical use of different types of IFNs with an appreciable success in several viral diseases alone or in combination with other antiviral drugs (such as chronic hepatitis B and C, genital warts), marking a major advance in modern medicine.

Originally the commercial preparations of IFN-alpha currently available for clinical use were either of recombinant IFN (rIFN) alpha 2 obtained from transfected bacteria or were mixtures of many subtypes of IFN-alpha obtained from transformed cell cultures (lymphoblastoid IFN-alpha), or primary human blood leukocytes (leukocyte IFN-alpha), both stimulated by Sendai virus (Scagnolari and Antonelli 2013). The composition of the mixtures depended on the methods used for their production and purification. The short half-life of IFN-alpha has led to the development of IFN-alpha preparations that have a longer half-life. This was achieved by the attachment of a polyethylene glycol (PEG) molecule to the IFN-alpha molecule. This allows a once-weekly administration.

Recent, but consolidated, findings, prompted us to reconsider the application of the different type of IFNs (especially the most used type I IFNs) in therapy of infectious diseases (Antonelli et al. 2015).

As stated, it is well established that all IFNs have a strong and direct antiviral effect and that all type of IFNs are fundamental for host defense against virus infections. However, a better understanding of the IFN response is crucial to understand the natural course of viral infections. Indeed, it has been established that during viral infections (especially during chronic viral infections), definite levels of specific type of IFNs may be required at an early stage, to initiate cell-mediated immune responses but continuous production and high concentrations of the same type or other types of IFNs may lead to the induction of immunosuppressive molecules and then to immunosuppression and tissue damage. Then, IFN (especially type I)-induced negative regulatory pathways are emerging as pivotal determinants of chronic inflammation and progression of diseases in chronic viral infections, such as HIV or HCV (Snell et al. 2017; Scagnolari and Antonelli, 2018). More recently, type I IFNs have been shown to cause tissue damage in some acute viral infections, such as influenza or SARS virus infection (Galani et al. 2017).

It can be concluded that a precise and definite, but still unknown, balanced network of IFNs does exist during viral infection/disease to provide protection with minimum damage to the host and recovery from viral infections. It is our opinion that understanding in detail such a network, and its complex regulation, is essential for a future design of interferon-based therapies.

New Strategies in the Antiviral Chemotherapy Field

Although the primary goal of the chapter is to give basic knowledge and practical information on the mechanism action of the drugs currently used to treat viral infections, we believe it is worth to mention that the scenario of antiviral drugs is rapidly changing. Indeed, new classes of antiviral drugs directed not only to viral targets but also to host factors involved in virus replication, the immune response, or virus-cell interactions are starting to be considered. Some of the advances in this background will be briefly discussed later.

It is now clear that the life cycle of all viruses strictly depends on the participation of many cellular proteins that are largely unknown. However, some advances have been made in exactly identifying those cellular proteins participating to the replication of some viruses, such as HIV. Obviously, such proteins may provide an additional target for antiviral drugs that might, for definition, represent a potential target for broad-spectrum antiviral activity as many viruses may share a dependency on that host function.

The first example of the application of such a strategy is the CCR5 inhibitor (see earlier text), which can have a dramatic impact on HIV replication *in vivo* without considerably affecting essential cellular functions with the important exception of the immune response to West Nile virus and tick-borne encephalitis virus infections (Kindberg et al. 2008; Lim and Murphy 2011).

The entry inhibitor Myrcludex-B is a first-in-class virus entry inhibitor for patients with chronic hepatitis B or B/D infections. This compound inhibits the entry of HDV into hepatocytes by blocking its binding to the sodium taurocholate cotransporting polypeptide (NTCP), thereby depriving HDV of key functions provided by HBV (Bogomolov et al. 2016; Rizzetto 2018).

Another example of new target is provided by the observation indicating that two antifungal compounds, namely, ciclopirox and deferiprone, are able to *in vitro* cure HIV-1 infected cells through eradication of viral infection. Specifically, it appears that these drugs, inhibiting hydroxylation of cellular proteins essential for apoptosis, activate apoptosis, thus leading to the selective death of HIV-1-infected cells (Hanauske-Abel et al. 2013; Saxena et al. 2016). However other studies reported that ciclopirox inhibits HIV-1 replication through control of HIV-1 transcription. Recently it has also been demonstrated that ciclopirox may act as an anti-HBV agent by inhibiting normal HBV capsid assembly. However, minor effects of ciclopirox on HBV replication cannot be entirely excluded because it affects the regulation of several intracellular processes that account for most of the anticancer actions of ciclopirox.

Another example is Gleevec®, a drug approved for the treatment of cancer that may block the replication of some poxviruses and the Ebola virus because of its ability to inhibit the cellular tyrosine kinase (Reeves et al. 2005; Uebelhoer et al. 2014).

Host kinase inhibitors represent a category of compounds with a great potential to combat viral infections. Several kinase pathways are essential for virus cycle and thus represent attractive targets for broad-spectrum therapy. Recently, the repurposing of approved kinase inhibitors has been considered by some authors as a new cost- and time-effective strategy for the development of broad-spectrum antivirals (Schor and Einav, 2018).

Another known approach to block host factors involved in virus replication is represented by zinc-finger nucleases (ZFNs), transcription activator like effector nucleases (TALEN), and clustered regularly interspaced short palindromic repeats (CRISPR)-Cas9. Notably, first generation genome-editing technologies (i.e., ZFNs and TALENs) have been replaced by CRISPR/Cas9, which work with a short guide RNA (gRNA) to hybridize to a target DNA site and recruit the Cas9 endonuclease. All these technologies have been utilized to successfully disrupt the HIV-1 co-receptors CCR5 or CXCR4, thereby restricting HIV-1 infection (Tebas et al. 2014; Liu et al. 2017, Shi et al. 2017).

In recent years, lipidomics has gained importance in virology. Lipids play important roles in cellular signaling and regulatory processes in both uninfected and infected cells. The study of modifications in lipid biosynthesis pathways and transport in infected cells is a promising research area because many viruses manipulate host membranes and lipid flows (Toledo and Benach 2015; Altan-Bonnet 2017).

Viruses generally unsettle lipid flows by three different mechanisms: the translocation and/or regulation of lipid biosynthesis enzymes, the interference with lipid-mediated signaling through the regulation of phosphatidylinositol (PI) kinases, and the regulation of lipid trafficking at membrane contact sites (MCSs) between adjacent organelles.

It has been observed that Dengue virus, Usutu virus and West Nile virus, are sensitive to the inhibitory effect of cerulenin a specific inhibitors of fatty acid synthase, the key multienzyme complex for fatty acid synthesis (Fernández-Oliva et al. 2019). Furthermore, it is known that viruses take advantage of cholesterol properties. Cholesterol is a critical membrane component that determines membrane fluidity, regulates the formation and function of membrane-bound complexes of proteins and lipids. Viruses exploit these properties to make specialized membrane platforms for RNA replication and virus assembly.

Lipid rafts have a key role in the life cycle of viruses, such as entry, RNA translation, virus progeny assembly, and egress (Kim et al. 2017; Verma et al. 2018; Chen et al. 2019). It is known that statins deplete cholesterol and disrupt lipid rafts. West Nile virus and Dengue virus are inhibited by lovastatin (Mackenzie et al. 2007; Soto–Acosta et al. 2017), and DENV and HIV-1 are inhibited by simvastatin (Bryan-Marrugo et al. 2016; DeLucia et al. 2018).

These findings confirm that host factors may positively modulate the replication of viruses thus indicating that their inhibition may eventually block viral replication. However, in some well-documented cases, host factors (i.e., APOBEC, TRIM, Tetherin, SAMDH1, etc.) have been developed by the cell machinery to affect viral replication (Huthoff and Towers 2008). Hence, it is tempting to speculate that another approach to achieve an effective antiviral strategy is to improve the action of these natural defense mechanisms, but only preliminary results are available.

Altogether these findings prompt speculation that the traditional paradigm "antiviral therapy must be directed only to the viral target" is being challenged. However, while this approach may represent an interesting and useful advance for acute infections, people are worried about their potential use in chronic infections due to the possibility of their interference with vital cellular functions.

A special and borderline situation is represented by the antiviral approaches through nucleic acids. Indeed, a number of nucleic acid-based antiviral therapies have been targeted to viral genes, including ribozymes (small, catalytically active RNA molecules, the most well studied being the hammerhead and the hairpin ribozymes), aptamers (nucleic acid sequences that bind to a specific target molecule and are synthesized by an amplification technique), antisense oligonucleotides and RNA interference (RNAi)-based therapeutic strategy (which utilizes short RNAs to mediate degradation of viral mRNA in a sequence-specific manner) (Ivacik et al. 2011; Zeller et al. 2011; Asha et al. 2018). Specifically, small interfering RNAs (siRNAs), the main effectors of RNAi, are now routinely used to assess gene function, both *in vitro* and *in vivo*, and many innovative

screens have been reported on the use of RNAi to identify potential drug targets. Some of these approaches have demonstrated that siRNAs can effectively inhibit the replication of viruses despite different mechanisms they evolved to resist the pressure imposed by immune system and antiviral drugs (Levanova and Poranen 2018). However, the possibility of appearance of escape mutants, recently discovered inhibitors of RNAi in human viruses (Fabozzi et al. 2011; Qiu et al. 2018), and complex interactions between RNAi and interferon pathways (Seo et al. 2013; Maillard et al. 2016) must be taken into consideration when designing new antiviral siRNA molecules. Despite several technical advances, many challenges remain in determining the ideal design of nucleic acid-based antiviral therapy products, including pharmacokinetic and pharmacodynamic issues and competition with endogenous microRNAs. Although several clinical trials are ongoing (specifically against HIV, HCV, HBV, Ebola virus, Marburg virus, and influenza virus), translation of nucleic acid technology into the clinic depends on resolving these challenges.

Apart from targeting the host factors directly involved in virus replication, the drug market is also directed toward finding molecules able to modulate the host's inherent immune response. Here, we have not described any of these attempts because the discussion would deserve a description also of determinants of the immune response to viral infections, which is out of the scope of this chapter.

Forthcoming Issues on Antiviral Chemotherapy

Over the last two decades, beside the unceasing identification and development of new antiviral drugs above all against HIV and HCV, we have witnessed a concomitant and tremendous progress in understanding all the aspects of the modern medical virology: molecular steps of viral life cycle, pathogenesis, immune response, epidemiology, and diagnosis. From such concurrence, new and often unexpected issues in the antiviral chemotherapy field stemmed during the last few years. A detailed and exhaustive description of all these aspects is outside the scope of this chapter, but it is worth at least to discuss the main and most interesting one.

For instance, it became clear that one drug may be inadequate to control viral infections while combination therapy with several antiviral drugs may be therapeutically effective in controlling viral infections such as HIV, HCV, and HBV. Thus, the general strategy of antiviral therapy is changed drastically considering always as occurred in other diseases the use of a combination of different antiviral drugs targeting different steps of the viral life cycle.

At the same time, we have also learned that some viruses possess an extraordinary propensity to undergo continuous diversification of the viral population and this, other than making difficult to carry out the prevention strategies, raises issues related to the antiviral field because of the possibility to develop drug-resistant viral variants under antiviral drug pressure.

It is well known that administration of standard doses of most of the current antiviral drugs results in plasma drug concentrations that may differ significantly among different individuals. Obviously, this may influence the antiviral activity, the emergence of resistant variants and the incidence of drug-related toxicities. In some viral infection, treatment compliance became one of the main aspects that may affect drug concentration and, consequently, the success of therapy. Optimal adherence to antiretroviral therapy is central to achieving suppression of the HIV viral load in plasma, slowing disease progression, and decreasing HIV transmission rates. In addition, low rates of adherence have also reduced the effectiveness of orally administered pre-exposure prophylaxis (PrEP). It's important to remember that PrEP is a prevention strategy of giving antiretroviral drugs to an uninfected individual at risk for HIV infection, which reduces the risk of acquiring infection by over 85% as indicated by some clinical trials (McCormack et al. 2016). Long-acting (LA) injectable antiretroviral formulations, that are under investigation, could improve success rates while also helping to prevent transmission (Stellbrink and Hoffmann 2018; Gulick and Flexner 2019). Currently, cabotegravir, an HIV integrase inhibitor, and rilpivirine are the compounds available as LA and under investigation as coformulation (Margolis et al. 2017; Murray et al. 2019).

Furthermore, it is now clear that antiviral therapy may have a role in preventing infection at individual and community levels by decreasing virus transmission. Indeed, in the case of HIV infection, it has been demonstrated that virus transmission in serodiscordant couples is reduced in those who achieved lower viral loads through an effective antiviral therapy, or that HIV suppression prevents virus transmission from infected women to their newborn babies (Grant et al. 2010; Anglemyer et al. 2013). More recently, "partner" studies demonstrated that the risk of HIV transmission in serodifferent heterosexual or gay couples through condomless sex is effectively zero when the HIV-positive partner was on virally suppressive ART (Rodger et al. 2016; Rodger et al. 2019).

Although treatment adherence is crucial to maintain the benefit of therapy, the high interindividual variability in drug concentrations are multifactorial and include differences in concomitant medications, underlying diseases, gender-related metabolism, and genetic factors (Turriziani and Antonelli 2004; Mahungu et al. 2009).

Another issue is represented by the genetic determinants, which include polymorphisms of genes encoding for drug-transporting proteins, receptors, and/or drug targets. To date, pharmacogenetic studies conducted with antiviral drugs have focused on metabolism-involved enzymes and transporter proteins on cell membranes. A few antiviral drugs used for the treatment of HIV infection can cause drug hypersensitivity reactions, which vary in severity, clinical manifestations, and frequency. In most cases, the pathogenesis of these manifestations is unknown, but there is increasing evidence that many are mediated by a combination of immunological and genetic factors through the major histocompatibility complex (MHC) (Chaponda and Pirmohamed 2011). One of the best-known cases is abacavir, which represents the gold standard example of the clinical utility of pharmacogenetic screening in HIV therapy. Indeed, there is a strong association between abacavir hypersensitivity reaction and HLA-B*5701 in racially diverse populations. Genotyping for HLA-B*5701 before prescribing an abacavir-containing regimen has been introduced into routine clinical practice as the standard of care for all patients (Chaponda and Pirmohamed 2011). The safety assessment of antiviral drugs is mandatory, especially in the case of life-long therapy. However, these considerations indicate the need for

further research in this area to increase our understanding of the mechanisms underlying therapy-induced side effects and to try to develop predictive genetic biomarkers leading to better management of antiviral agents. Since genetic variations between human beings also account for the variability in the antiviral action of the drug, the same considerations apply to the pharmacokinetic and pharmacodynamic data of individual drugs. We now know that administration of drugs with a narrow therapeutic index may readily be associated with changes in pharmacokinetics and ultimately in variations in antiviral efficacy.

One of the possibilities to tackle the pharmacokinetics issue is therapeutic drug monitoring (TDM), which is widely used at least for those drugs whose plasma concentration is known to be directly correlated to the antiviral action (Wertheimer et al. 2006; Kredo et al. 2009). This approach has been extrapolated from well-known examples in other areas of medicine and again requires close collaboration between virologist, pharmacologist, and clinician for prompt decision making to improve individual patient management. Again, the possibility of TDM is available only for some antiviral drugs and this represents another issue that deserves attention and development in the future.

Finally, there are other challenges that affect the importance of the antiviral chemotherapy and lower, in some way, the extraordinary success obtained in the last decades on this topic: the low rate of viral diagnosis and the unavailability of effective drugs in developing countries; the difficulty in establishing the balance between benefits of treatment and keeping people chronically in care; the delay in development of safe and effective vaccine against some viral diseases. These challenges require additional research but, above all, policy and community discussions.

REFERENCES

Adams, J., Patel, N., Mankaryous, N., Tadros, M., Miller, C.D. 2010. Nonnucleoside reverse transcriptase inhibitor resistance and the role of the second-generation agents. *Ann Pharmacother* 44: 157–165.

Alavian, S.M., Rezaee-Zavareh, M.S. 2016. Daclatasvir-based treatment regimens for hepatitis C virus infection: a systematic review and meta-analysis. *Hepat Mon* 16: e41077.

Altan-Bonnet, N. 2017. Lipid tales of viral replication and transmission. *Trends Cell Biol* 27: 201–213.

Anglemyer, A., Rutherford, G.W., Horvath, T., Baggaley, R.C., Egger, M., Siegfried, N. 2013. Antiretroviral therapy for prevention of HIV transmission in HIV-discordant couples. *Cochrane Database Syst Rev* 30(4): CD009153.

Antonelli, G., Scagnolari, C., Moschella, F., Proietti, E. 2015. Twenty-five years of type I interferon-based treatment: a critical analysis of its therapeutic use. *Cytokine Growth Factor Rev* 26: 121–131.

Asha, K., Kumar, P., Sanicas, M., Meseko, C.A., Khanna, M., Kumar, B. 2018. Advancements in nucleic acid based therapeutics against respiratory viral infections. *J Clin Med* 8(1): E6.

Bartenschlager, R., Lohmann, V., Penin, F. 2013. The molecular and structural basis of advanced antiviral therapy for hepatitis C virus infection. *Nat Rev Microbiol* 11(7): 482–496.

Beccari, M.V., Mogle, B.T., Sidman, E.F., Mastro, K.A., Asiago-Reddy, E., Kufel, W.D. 2019. Ibalizumab, a novel monoclonal antibody for the management of multidrug-resistant HIV-1 infection. *Antimicrob Agents Chemother* 63(6): e00110–19.

Bogomolov, P., Alexandrov, A., Voronkova, N., et al. 2016. Treatment of chronic hepatitis D with the entry inhibitor my rcludex B: First results of a phase Ib/IIa study. *J Hepatol* 65(3): 490–498.

Borst EM, Kleine-Albers J, Gabaev I., et al. 2013. The human cytomegalovirus UL51 protein is essential for viral genome cleavage-packaging and interacts with the terminase subunits pUL56 and pUL89. *J Virol* 87(3): 1720–1732.

Bourlière, M., Pietri, O. 2019. Hepatitis C virus therapy: No one will be left behind. *Int J Antimicrob Agents* 53: 755–760.

Bryan-Marrugo, O. L., Arellanos-Soto, D., Rojas-Martinez, et al. 2016. The antidengue virus properties of statins may be associated with alterations in the cellular antiviral profile expression. *Mol Med Rep* 14: 2155–2163.

Burkly, L.C., Olson, D., Shapiro, R., et al. 1992. Inhibition of HIV infection by a novel CD4 domain 2-specific monoclonal antibody: dissecting the basis for its inhibitory effect on HIV-induced cell fusion. *J Immunol* 149: 1779–1787.

Chan-Tack, K.M., Harrington, P.R., Choi, S.Y., et al. 2019. Assessing a drug for an eradicated human disease: US Food and Drug Administration review of tecovirimat for the treatment of smallpox. *Lancet Infect Dis* 19: e221–e224.

Chaponda, M., Pirmohamed, M. 2011. Hypersensitivity reactions to HIV therapy. *Br J Clin Pharmacol* 71: 659–671.

Chen, S., He, H., Yang, H., et al. 2019. The role of lipid rafts in cell entry of human Metapneumovirus. *J Med Virol* 91(6): 949–957.

Chou S. 2017. A third component of the human cytomegalovirus terminase complex is involved in letermovir resistance. *Antiviral Res* 148:1–4.

Coen, D.M., Schaffer, P.A. 2003. Antiherpesvirus drugs: A promising spectrum of new drugs and drug targets. *Nat Rev Drug Discov* 2: 278–288.

Colman, P.M. 1994. Influenza virus neuraminidase: Structure, antibodies, and inhibitors. *Protein Sci* 3: 1687–1696.

Davies, W.L., Grunert, R.R., Haff, F. et al. 1964. Antiviral activity of 1-adamantanamine amantadine. *Science* 144: 862–863.

de Béthune, M.P. 2010. Non-nucleoside reverse transcriptase inhibitors NNRTIs, their discovery, development, and use in the treatment of HIV-1 infection: A review of the last 20 years 1989–2009. *Antiviral Res* 85: 75–90.

De Clercq, E. 2009. The history of antiretrovirals: Key discoveries over the past 25 years. *Rev Med Virol* 19: 287–299.

De Clercq, E., Hol, A. 2005. Acyclic nucleoside phosphonates: A key class of antiviral drugs. *Nat Rev Drug Discov* 4: 928–940.

DeLucia, D. C., Rinaldo, C. R., Rappocciolo, G. 2018. Inefficient HIV-1 trans infection of CD4+ T cells by macrophages from HIV-1 non-progressors is associated with altered membrane cholesterol and DC-SIGN. *J Virol* 92 pii: e00092–18.

Dienstag, J.L. 2009. Benefits and risks of nucleoside analog therapy for hepatitis B. *Hepatology* 495: S112–S121.

Douglas, J.M., Critchlow, C., Benedetti, J. et al. 1984. A double-blind study of oral acyclovir for suppression of recurrences of genital herpes simplex 3 virus infection. *N Engl J Med* 310: 1551–1556.

Duraffour, S., Lorenzo, M.M., Zöller, G., et al. 2015. ST-246 is a key antiviral to inhibit the viral F13L phospholipase, one of the essential proteins for orthopoxvirus wrapping. *J Antimicrob Chemother* 70: 1367–1380.

Elion, G.B. 1993. Acyclovir: Discovery, mechanism of action and selectivity. *J Med Virol* 41(Suppl. 1): 2–6.

Eron, Jr., J.J. 2000. HIV-1 protease inhibitors. *Clin Infect Dis* 30: S160–S170.

Esté, J.A., Telenti, A. 2007. HIV entry inhibitors. *Lancet* 370: 81–88.

Fabozzi, G., Nabel, C.S., Dolan, M.A., Sullivan, N.J. 2011. Ebolavirus proteins suppress the effects of small interfering RNA by direct interaction with the mammalian RNA interference pathway. *J Virol* 85(6): 2512–2523.

Fantauzzi, A., Turriziani, O., Mezzaroma, I. 2013. Potential benefit of dolutegravir once daily: Efficacy and safety. *HIV AIDS (Auckl.)* 5: 29–40.

Feld J.J., Jacobson I.M., Hezode C., et al. 2015. Sofosbuvir and velpatasvir for HCV genotype 1, 2, 4, 5, and 6 infection. *N Engl J Med* 373; 2599–2607.

Feng, M., Sachs, N.A., Xu, M., et al. 2016. Doravirine suppresses common nonnucleoside reverse transcriptase inhibitor-associated mutants at clinically relevant concentrations. *Antimicrob Agents Chemother* 60: 2241–2247.

Fernández-Oliva, A., Ortega-González, P., Risco C. 2019. Targeting host lipid flows: Exploring new antiviral and antibiotic strategies. *Cell Microbiol* 21: e12996.

Fischl, M.A., Richman, D.D., Grieco, M.H. et al. 1987. The efficacy of azidothymidine (AZT) in the treatment of patients with AIDS and AIDS-related complex. A double-blind, placebo-controlled trial. *N Engl J Med* 317: 185–191.

Forns X., Lee S.S., Valdes J., et al. 2017. Glecaprevir plus pibrentasvir for chronic hepatitis C virus genotype 1, 2, 4, 5, or 6 infection in adults with compensated cirrhosis (EXPEDITION-1): a single-arm, open-label, multicentre phase 3 trial. *Lancet Infect Dis* 17: 1062–1068.

Galani, I.E., Triantafyllia, V., Eleminiadou, et al. 2017. Interferon-λ mediates non-redundant front-line antiviral protection against influenza virus infection without compromising host fitness. *Immunity* 46: 875–890.

Goldner, T., Hewlett, G., Ettischer, N., Ruebsamen-Schaeff, H., Zimmermann, H., Lischka P. 2011. The novel anticytomegalovirus compound AIC246 (Letermovir) inhibits human cytomegalovirus replication through a specific antiviral mechanism that involves the viral terminase. *J Virol* 85: 10884–10893.

Grant, R.M., Lama, J.R., Anderson, P.L. et al. 2010. Preexposure chemoprophylaxis for HIV prevention in men who have sex with men. *N Engl J Med* 363: 2587–2599.

Grinsztejn, B., Nguyen, B.Y., Katlama, C. et al. 2007. Safety and efficacy of the HIV-1 integrase inhibitor raltegravir (MK-0518) in treatment-experienced patients with multidrug-resistant virus: A phase II randomized controlled trial. *Lancet* 369: 1261–1269.

Gubareva, L.V., Kaiser, L., Hayden, F.G. 2000. Influenza virus neuraminidase inhibitors. *Lancet* 355: 827–835.

Gulick, R.M., Flexner, C. 2019. Long-acting HIV drugs for treatment and prevention. *Annu Rev Med* 70: 137–150.

Hanauske-Abel, H.M., Saxena, D., Palumbo, P.E. et al. 2013. Drug-induced reactivation of apoptosis abrogates HIV-1 infection. *PLOS ONE* 8(9): e74414.

Hay, A.J., Zambon, M.C., Wolstenholme, A.J., Skehel, J.J., Smith, M.H. 1986. Molecular basis of resistance of influenza A viruses to amantadine. *J Antimicrob Chemother* 18(Suppl. B): 19–29.

Hayden, F.G., Osterhaus, A.D.M.E., Treanor, J.J. et al. 1997. Efficacy and safety of the neuraminidase inhibitor zanamivir in the treatment of influenza virus infections. GG167 Influenza Study Group. *N Engl J Med* 337: 874–880.

Hazuda, D.J., Felock, P., Witmer, M. et al. 2000. Inhibitors of strand transfer that prevent integration and inhibit HIV-1 replication in cells. *Science* 287: 646–650.

Hong, Z., Cameron, C.E. 2002. Pleiotropic mechanisms of ribavirin antiviral activities. *Prog Drug Res* 59: 41–69.

Hu, J., Robinson, J.L. 2010. Treatment of respiratory syncytial virus with palivizumab: A systematic review. *World J Pediatr* 6: 296–300.

Huthoff, H., Towers, G.J. 2008. Restriction of retroviral replication by APOBEC3G/F and TRIM5. *Trends Microbiol* 16: 612–619.

Iacob, S.A., Iacob, D.G. 2017. Ibalizumab targeting CD4 receptors, An emerging molecule in HIV therapy. *Front Microbiol* 8: 2323.

Isaacs, A., Lindenmann, J. 1957. Virus interference. I. The interferon. *Proc R Soc Lond B Biol Sci* 147: 258–267.

Ivacik, D., Ely, A., Arbuthnot, P. 2011. Countering hepatitis B virus infection using RNAi: How far are we from the clinic? *Rev Med Virol* 21: 383–396.

Kim, J.Y., Wang, L., Lee, J., Ou, J.J. 2017. Hepatitis C virus induces the localization of lipid rafts to autophagosomes for its RNA replication. *J Virol* 91(20): e00541–17.

Kindberg, E., Mickiene, A., Ax, C. et al. 2008. A deletion in the chemokine receptor 5 (CCR5) gene is associated with tickborne encephalitis. *J Infect Dis* 197: 266–269.

Kohlstaedt, L.A., Wang, J., Steitz, T.A. 1992. Crystal structure at 3.5°A resolution of HIV-1 reverse transcriptase complexed with an inhibitor. *Science* 256: 1783–1790.

Kredo, T., Van der Walt, J.S., Siegfried, N., Cohen, K. 2009. Therapeutic drug monitoring of antiretrovirals for people with HIV. *Cochrane Database Syst Rev* (3): CD007268.

Krishnan, L., Engelman, A. 2012. Retroviral integrase proteins and HIV-1 DNA integration. *J Biol Chem* 287: 40858–40866.

Lanari, M., Silvestri, M., Rossi, G.A. 2010. Palivizumab prophylaxis in 'late preterm' newborns. *J Matern Fetal Neonatal Med* 23(Suppl. 3): 53–55.

Levanova, A., Poranen, M.M. 2018. RNA interference as a prospective tool for the control of human viral infections. *Front Microbiol* 9: 2151.

Li G, De Clercq E. 2020. Therapeutic options for the 2019 novel coronavirus (2019-nCoV). *Nat Rev Drug Discov.* 19:149.

Lim, J.K., Murphy, P.M. 2011. Chemokine control of west Nile virus infection. *Exp Cell Res* 317: 569–574.

Liu, Z., Chen, S., Jin, X., et al. 2017. Genome editing of the HIV co-receptors CCR5 and CXCR4 by CRISPR-Cas9 protects CD4(+) T cells from HIV-1 infection. *Cell Biosci* 7: 47.

Luna, B., Townsend, M.U. 2007. Tipranavir: The first nonpeptidic protease inhibitor for the treatment of protease resistance. *Clin Ther* 29: 2309–2318.

Mackenzie, J. M., Khromykh, A. A., Parton, R. G. (2007). Cholesterol manipulation by West Nile virus perturbs the cellular immune response. *Cell Host Microbe* 2: 229–239.

Mahungu, T.W., Johnson, M.A., Owen, A., Back, D.J. 2009. The impact of pharmacogenetics on HIV therapy. *Int J STD AIDS* 20: 145–151.

Maillard, P.V., Van der Veen, A.G., Deddouche-Grass, S., Rogers, N.C., Merits, A., Reis e Sousa, C. 2016. Inactivation of the type I interferon pathway reveals long double-stranded RNA-mediated RNA interference in mammalian cells. *EMBO J* 35: 2505–2518.

Margolis, D.A., Gonzalez-Garcia, J., Stellbrink, H.J. et al. 2017. Long-acting intramuscular cabotegravir and rilpivirine in adults with HIV-1 infection (LATTE-2): 96-week results of a randomised, open-label, phase 2b, non-inferiority trial. *Lancet* 390(10101): 1499–1510.

Martin, K., Helenius, A. 1991. Nuclear transport of influenza virus ribonucleoproteins: The viral matrix protein (M1) promotes export and inhibits import. *Cell* 67: 117–130.

Matthews, T., Salgo, M., Greenberg, M., Chung, J., DeMasi, R., Bolognesi, D. 2004. Enfuvirtide: The first therapy to inhibit the entry of HIV-1 into host CD4 lymphocytes. *Nat Rev Drug Discov* 3: 215–225.

McCormack, S., Dunn, D.T., Desai, M., et al. 2016. Pre-exposure prophylaxis to prevent the acquisition of HIV-1 infection (PROUD): effectiveness results from the pilot phase of a pragmatic open-label randomised trial. *Lancet* 387: 53–60.

McCoy, C. 2007. Darunavir: A nonpeptidic antiretroviral protease inhibitor. *Clin Ther* 29: 1559–1576.

Métifiot, M., Marchand, C., Pommier, Y. 2013. HIV integrase inhibitors: 20-year landmark and challenges. *Adv Pharmacol* 67: 75–105.

Mifsud, E.J., Hayden, F.G., Hurt, A.C. 2019. Antivirals targeting the polymerase complex of influenza viruses. *Antiviral Res* 169: 104545.

Miller, W.H., Miller, R.L. 1982. Phosphorylation of acyclovir diphosphate by cellular enzymes. *Biochem Pharmacol* 31: 3879–3884.

Murray, M., Pulido, F., Mills, A, et al. 2019. Patient-reported tolerability and acceptability of cabotegravir+rilpivirine long-acting injections for the treatment of HIV-1 infection: 96-week results from the randomized LATTE-2 study. *HIV Res Clin Pract* 20: 111–122.

Painter, G.R., Almond, M.R., Mao, S., Liotta, D.C. 2004. Biochemical and mechanistic basis for the activity of nucleoside analogue inhibitors of HIV reverse transcriptase. *Curr Top Med Chem* 4: 1035–1044.

Pawlotsky, J.M. 2014. New hepatitis C therapies: The toolbox, strategies, and challenges. *Gastroenterology* 146: 1176–1192.

Perros, M. 2007. CCR5 antagonists for the treatment of HIV infection and AIDS. *Adv Antivir Drug Des* 5: 185–212.

Pinto, L.H., Holsinger, L.J., Lamb, R.A. 1992. Influenza virus M2 protein has ion channel activity. *Cell* 69: 517–528.

Pommier, Y., Johnson, A.A., Marchand, C. 2005. Integrase inhibitors to treat HIV/AIDS. *Nat Rev Drug Discov* 4: 236–248.

Qiu, Y., Xu, Y., Zhang, Y., Zhou, H. et al. 2018. Human virus-derived small RNAs can confer antiviral immunity in mammals. *Immunity* 49(4): 780–781.

Reeves, P.M., Bommarius, B., Lebeis, S. et al. 2005. Disabling poxvirus pathogenesis by inhibition of Abl-family tyrosine kinases. *Nat Med* 11: 731–739.

Reimann, K.A., Lin, W., Bixler, S., et al. 1997. A humanized form of a CD4-specific monoclonal antibody exhibits decreased antigenicity and prolonged plasma half-life in rhesus monkeys while retaining its unique biological and antiviral properties. *AIDS Res Hum Retroviruses* 13(11): 933–943.

Rizzetto, M. 2018. Targeting Hepatitis D. *Semin Liver Dis* 38(1): 66–72.

Rodger, A.J., Cambiano, V., Bruun, T., et al. 2016. Sexual activity without condoms and risk of HIV transmission in serodifferent couples when the HIV-positive partner is using suppressive antiretroviral therapy. *JAMA* 316(2): 171–81.

Rodger, A.J., Cambiano, V., Bruun, T., et al. 2019. Risk of HIV transmission through condomless sex in serodifferent gay couples with the HIV-positive partner taking suppressive antiretroviral therapy (PARTNER): final results of a multicentre, prospective, observational study. *Lancet* 393(10189): 2428–2438.

Saxena, D., Spino, M., Tricta, F., et a al. 2016. Drug-based lead discovery: The novel ablative antiretroviral profile of deferiprone in HIV-1-Infected cells and in HIV-Infected treatment-naive subjects of a Double-Blind, Placebo-Controlled, Randomized Exploratory Trial. *PLOS One* 11: e0154842.

Scagnolari, C., Antonelli, G. 2018. Type I interferon and HIV: Subtle balance between antiviral activity, immunopathogenesis and the microbiome. *Cytokine Growth Factor Rev* 40: 19–31.

Scagnolari, C., Antonelli, G. 2013. Antiviral activity of the interferon α family: Biological and pharmacological aspects of the treatment of chronic hepatitis C. *Exp Opin Biol Ther* 13: 693–711.

Schiller, D.S., Youssef-Bessler, M. 2009. Etravirine: A second-generation nonnucleoside reverse transcriptase inhibitor (NNRTI) active against NNRTI-resistant strains of HIV. *Clin Ther* 31: 692–704.

Schor, S., Einav, S. 2018. Repurposing of Kinase inhibitors as broad-spectrum antiviral drugs. *DNA Cell Biol* 37: 63–69.

Scott, L.J., Lamb, H.M. 1999. Palivizumab. *Drugs* 58: 305–311.

Seo, G. J., Kincaid, R.P., Phanaksri, T. et al. 2013. Reciprocal inhibition between intracellular antiviral signaling and the RNAi machinery in mammalian cells. *Cell Host Microbe* 14: 435–445.

Shi, B., Li, J., Shi, X., Jia, W., et al. 2017. TALEN-mediated knockout of CCR5 confers protection against infection of human immunodeficiency virus. *J Acquir Immune Defic Syndr* 74: 229–241.

Shimura, K., Kodama, E.N. 2009. Elvitegravir: A new HIV integrase inhibitor. *Antivir Chem Chemother* 20: 79–85.

Smith, S.K., Olson, V.A., Karem, K.L., Jordan, R., Hruby, D.E., Damon, I.K. 2009. In vitro efficacy of ST246 against smallpox and monkeypox. *Antimicrob Agents Chemother* 53: 1007–1012.

Snell, M., McGaha, T.L., Brooks, D.G. 2017. Type I interferon in chronic virus infection and Cancer. *Trends Immunol* 38: 542–557.

Soriano, V., Perno, C.F., Kaiser, R. et al. 2009. When and how to use maraviroc in HIV-infected patients. *AIDS* 23: 2377–2385.

Soto-Acosta, R., Bautista-Carbajal, P., Cervantes-Salazar, M., Angel-Ambrocio, A. H., Del Angel, R. M. 2017. DENV up-regulates the HMG-CoA reductase activity through the impairment of AMPK phosphorylation: A potential antiviral target. *PLOS Pathogens* 13: e1006257.

Stellbrink, H.J., Arribas, J.R., Stephens, J.L., et al. 2019. Co-formulated bictegravir, emtricitabine, and tenofovir alafenamide versus dolutegravir with emtricitabine and tenofovir alafenamide for initial treatment of HIV-1 infection: week 96 results from a randomised, double-blind, multicentre, phase 3, non-inferiority trial. *Lancet HIV* 6(6), e364–e372.

Stellbrink, H.J., Hoffmann, C. 2018. Cabotegravir: its potential for antiretroviral therapy and preexposure prophylaxis. *Curr Opin HIV AIDS* 13: 334–340.

Stevenson, N.J., Murphy, A.G., Bourke, N.M., Keogh, C.A., Hegarty, J.E., O'Farrelly, C. 2011. Ribavirin enhances IFN-signalling and MxA expression: A novel immune modulation mechanism during treatment of HCV. *PLOS ONE* 6: e27866.

Straus, S.E., Takiff, H.E., Seidlin, M. et al. 1984. Suppression of frequently recurring genital herpes. A placebo-controlled double-blind trial of oral acyclovir. *N Engl J Med* 310: 1545–1550.

Sulkowski, M.S., Gardiner, D.F., Rodriguez-Torres, M., et al. 2014. Daclatasvir plus sofosbuvir for previously treated or untreated chronic HCV infection. *N Engl J Med* 370: 211–221.

Tebas, P., Stein, D., Tang, W.W., et al. 2014. Gene editing of CCR5 in autologous CD4 T cells of persons infected with HIV. *N Engl J Med* 370(10): 901–910.

Tilton, J.C., Doms, R.W. 2010. Entry inhibitors in the treatment of HIV-1 infection. *Antiviral Res* 85: 91–100.

Toledo, A., Benach, J. L. 2015. Hijacking and use of host lipids by intracellular pathogens. *Microbiol Spectr* 3(6).

Tsai, C.H., Lee, P.Y., Stollar, V. et al. 2006. Antiviral therapy targeting viral polymerase. *Curr Pharm Des* 12: 1339–1355.

Turriziani, O., Antonelli, G. 2004. Host factors and efficacy of antiretroviral treatment. *New Microbiol* 27(2 Suppl. 1): 63–69.

Uebelhoer, L.S., Albariño, C.G., McMullan, L.K., et al. 2014. High-throughput, luciferase-based reverse genetics systems for identifying inhibitors of Marburg and Ebola viruses. *Antiviral Res* 106: 86–94.

US Food & Drug Administration. Coronavirus (COVID-19) Update: FDA Issues Emergency Use Authorization for Potential COVID-19 Treatment. https://www.fda.gov/news-events/press-announcements/coronavirus-covid-19-update-fda-issues-emergency-use-authorization-potential-covid-19-treatment.Date: May 1, 2020

Verma, D.K., Gupta, D., Lal, S.K. 2018. Host lipid rafts play a major role in binding and Endocytosis of Influenza A Virus. *Viruses* 10(11): 650.

von Kleist, M., Metzner, P., Marquet, R., Schütte, C. 2012. HIV-1 polymerase inhibition by nucleoside analogs: Cellular- and kinetic parameters of efficacy, susceptibility and resistance selection. *PLOS Comput Biol* 8: e1002359.

Welliver, R.C. 2010. Pharmacotherapy of respiratory syncytial virus infection. *Curr Opin Pharmacol* 10: 289–293.

Wertheimer, B.Z., Freedberg, K.A., Walensky, R.P., Yazdanapah, Y., Losina, E. 2006. Therapeutic drug monitoring in HIV treatment: A literature review. *HIV Clin Trials* 7: 59–69.

World Health Organization. 2016. Combating hepatitis B and C to reach elimination by 2030: Advocacy brief. Geneva: World Health Organization; Available at: https://www.who.int/hepatitis/publications/hep-elimination-by-2030-brief/en/.

Zeller, S.J., Kumar, P., Yale, J. 2011. RNA-based gene therapy for the treatment and prevention of HIV: From bench to bedside. *Biol Med* 84: 301–309.

54

Phage Therapy: Bacteriophages as Natural, Self-Replicating Antimicrobials

Naomi Hoyle and Elizabeth Kutter

CONTENTS

Introduction

Bacteriophages—specific kinds of viruses that can only replicate in bacteria—have been discussed in much detail in Chapter 48 (Introduction to Bacteriophages) of this book. The art of using these phages to kill pathogenic microorganisms was first developed early in the last century, but since chemical antibiotics became available in the 1940s, phage therapy has been little used in the West. Today, however, the growing incidence of bacteria that are resistant to most or all available antibiotics is leading to widespread renewed interest in the possibilities of phage therapy.[1–16]

In 2014, *Nature* published a news story entitled "Phage Therapy Gets Revitalized."[17] Particular emphasis there was placed on Phagoburn—"the first large, multicenter clinical trial of phage therapy for human infections, largely funded by the European Commission," which invested €3.8 million in the study. The phage

DOI: 10.1201/9781003099277-57

cocktails being used in this project had been developed by the French company Phagoburn, but the project also involved the French military, the patient advocacy group Phagospoir, the production company Clean Cells, and three major burn-care units in France, three in Belgium, and one in Switzerland. We will later discuss the outcome of this study. Collaborative approaches of this sort may well be required to rapidly harness phage to help fight the ever-growing serious antibiotic-resistance crisis. Progress there has been very slow, since phage therapy does not fit well into our standard corporate pharmaceutical model.

This overview is designed to put phage therapy into historical context and to bring the reader up to date with the current status of this approach. We will briefly explore some of the most interesting and extensive applications carried out in Eastern Europe, conducted primarily in a clinical rather than controlled-research mode, and we will look at the original decline and current renaissance of phage therapy work in the West. Ironically, this renaissance takes full advantage of the advances in molecular biology that were themselves made possible by the shift from therapeutic to molecular work with phages in the 1940s under the leadership of Max Delbrück. The rapidly increasing emergence of multiantibiotic-resistant pathogenic bacteria has rekindled the interest of the Western scientific community, industry, and the public in this century-old approach. This is highlighted by several new phage therapy-related publications in the Western peer-reviewed and popular literature and the formation of new biotechnology firms commercializing phage-based technology. Recent work also suggests potential for using phage lysins in accessible places such as nasal passages and in some systemic infections and biofilms as well.[18-21] Detailed discussions of phage ecology, engineering and interesting phage characteristics for human, agricultural and biocontrol applications can be found in the books by Kutter & Sulakvelidze[7] and by Borysowski et al.[21] However, in this review we will focus only on the use of natural, unmodified phages targeting human pathogens.

Biomedical technology today is very different from what it was in the early days of phage therapy research, and our understanding of the biological properties of phages and the basic mechanisms of phage-bacterial host interaction has improved dramatically. These advances can have a profound impact on the development of safe therapeutic phage preparations with optimal efficacy against their specific bacterial hosts and on designing science-based strategies for integrating phage therapy into our arsenal of tools for preventing and treating bacterial infections.

A major benchmark was reached in 2006 with the approval by the U.S. Food and Drug Administration (FDA) of an Intralytix phage cocktail targeting *Listeria monocytogenes* in packaged meats and other food applications. The FDA later also gave GRAS (generally regarded as safe) status to a similar product developed by the European branch of Exponential Biotherapies, EBI Food Safety (now named Micreos Food Safety). Intralytix also supplied the fully sequenced phages that were used in the first FDA-approved clinical trials in the United States, a physician-initiated study adding a cocktail containing two phages against *Staphylococcus aureus*, three against *Pseudomonas aeruginosa*, and one *Escherichia coli* phage to the standard treatment of leg ulcers. Randy Wolcott carried out that study using patients presenting for treatment in his Wound Care Center in Lubbock, Texas in 2006–2007. All safety criteria were met, but the population

was too small and varied to establish efficacy.[22] In England, 2006 saw the advent of a clinical trial of phage against *P. aeruginosa* in human ear infections, led by James Soothill and building on a small successful trial in dog-ear infections.[23] In Poland, a phage therapy clinic approved under European Union experimental-therapies guidelines was set up by Andrzej Gorski, director of the Institute of Immunology and Experimental Therapy in Wroclaw, which had long been a center of phage therapy research, and it has been very successful in exploring a variety of phage applications,[24] as discussed in some detail later in this chapter.

Early History

In 1915 and 1917, respectively, Edward Twort and Felix d'Herelle independently reported isolating these filterable entities that could lyse bacterial cultures; they are jointly given credit for the discovery. It was d'Herelle, a Canadian working at the Paris Pasteur Institute, who gave them the name *bacteriophages*—using the suffix *phage* "not in its strict sense of *to eat*, but in that of *developing at the expense of*."[25] From the time he first discovered phages, d'Herelle was excited about their relationship to disease and their potential as therapeutic agents. Their discovery came while he was doing clinical work at the Pasteur Institute exploring why enteric bacteria are only sometimes pathogenic. He was called to investigate an outbreak of bacillary dysentery in a group of French soldiers and examined their filtered stool samples for signs of invisible viruses that might alter the pathogenicity of the bacteria from the dysentery patients. He observed clear, round spots in the confluent bacterial cultures covering his agar slants and determined that the responsible agent, which he called bacteriophage, multiplied indefinitely as long as appropriate living cells were present, and that cell lysis was required for multiplication. Importantly, the presence of the phage correlated with the recovery of sick soldiers.

Shortly after publishing his first paper describing bacteriophages, d'Herelle used specially selected phages to successfully treat avian typhoid in chickens and then human bacterial dysentery. The latter studies were conducted in 1919 with Victor Hutinel, chief of pediatrics at the Hôpital des Enfants-Malades in Paris.[6,7] A dose 100-fold higher than the chosen therapeutic dose was first ingested by d'Herelle, Hutinel, and several interns, and when none of the volunteers showed any side effects a day later, Hutinel approved the first known human therapeutic use of phage. The first patient, a 12-year-old boy with severe dysentery (10–12 bloody stools per day), was given 2 mL of phage orally after taking samples for microbiological analysis. The patient's condition rapidly improved after phage ingestion: he passed three more bloody stools that afternoon and one nonbloody stool during the night, and all symptoms disappeared by the next morning. Three brothers, aged 3, 7, and 10, were then treated the following month; they had been admitted in grave condition after their sister died of dysentery at home and were each given one dose of phage. All three started to recover within 24 hours.

Before attempting further clinical phage trials, d'Herelle turned to basic research, working out the details of phage infection of different bacterial hosts under a variety of environmental conditions. His 300-page 1922 book *The Bacteriophage*[25] includes excellent descriptions of plaque formation and composition, infective centers, the lysis process, host specificity of adsorption and multiplication, the dependence of phage production on the precise state of the host, isolation of phages from sources of infectious bacteria, and the factors controlling stability of the free phage. He quickly became fascinated with the apparent role of phages in the natural control of microbial infections. He noted, for example, the frequent specificities of the phages isolated from recuperating patients for disease organisms infecting them and the rather rapid variations over time of the phage populations. However, others did not wait for a better understanding of phage biology before applying them clinically. The first known published report of successful phage therapy came from Bruynoghe & Maisin[26] who used phage to treat staphylococcal skin infections.

Throughout his life, d'Herelle worked to develop the therapeutic potential of properly selected phages against the most devastating health problems of the day. After travel to study epidemics in Latin America and a year at the Pasteur branch in Saigon, d'Herelle left the Pasteur Institute in 1922. He worked in Holland and then in Alexandria, Egypt as a health officer for the League of Nations, applying phage therapy and sanitation measures to deal with major outbreaks of infectious disease throughout the Middle East and India. In 1928, he was invited to Stanford to give the prestigious Lane lectures, which were published as the monograph, *The Bacteriophage and its Clinical Applications*.[27] He had accepted a regular faculty position at Yale, carrying on his phage work and helping inspire others to get involved in working with phage. However, he continued to spend summers in Paris working to strengthen the successful company he established there to produce high-quality therapeutic phage cocktails—a company which was run by his son-in-law. In 1933, he returned permanently to France, with excursions to Tbilisi, Georgia to help further establish phage therapy work there.

George Eliava, director of the Georgian Institute of Microbiology, had seen the bactericidal action of the water of the Koura river in Tbilisi, which he could not explain until he became familiar with d'Herelle's work while spending 1920–1921 at the Pasteur Institute.[7,25] The two developed the dream of founding an Institute of Bacteriophage Research in Tbilisi—to be a world center of phage therapy for infectious disease, including scientific and industrial facilities and supplied with its own experimental clinics. The dream quickly became a reality due to the support of Sergo Orjonikidze, the People's Commissar of Heavy Industry, despite KGB opposition to this "foreign project." A large campus on the river Mtkvari was allotted for the project in 1926 and d'Herelle sent supplies, equipment, and library materials.

In 1934 and 1935, d'Herelle visited Tbilisi for 6 months and wrote a book on *The Bacteriophage and the Phenomenon of Recovery*.[28] He intended to move to Georgia; a cottage built for his use still stands on the Eliava Institute's grounds. However, in 1937, Eliava was arrested as a "people's enemy" by Beria, then head of the KGB in Georgia and soon to direct the Soviet KGB. Eliava was executed, sharing the tragic fate of many Georgian and Russian progressive intellectuals of the time, and d'Herelle, disillusioned, never returned from France to Georgia. However, their Institute in Tbilisi survived, led by the people they had trained so well, and it is still functioning at its original site on the Mtkvari. In 1951, it was formally transferred to the Soviet All-Union

Ministry of Health set of Institutes of Vaccine and Sera, taking on the leadership role in providing bacteriophages for therapy and bacterial typing throughout the Soviet Union. Under orders from the Ministry of Health, hundreds of thousands of samples of pathogenic bacteria were sent to the institute over the years from throughout the Soviet Union to isolate more effective phage strains and to better characterize their efficacy. The Institute's industrial branch eventually made up to two tons of phage products twice a week, 80% of it for the Soviet Army.[7,12,29] The challenges they faced following the break-up of the Soviet Union, including loss of their primary markets, forced privatization of the industrial arm, the Institute's more recent evolution into the Eliava Institute of Bacteriophage, Microbiology, and Virology, and their current activities are further discussed later in this chapter.

Early World-Wide Attempts at Commercialization

Since the 1930s, a major medically important commercial use of phages has been for so-called *phage typing*, using patterns of sensitivity to a specific battery of phages with particularly narrow host ranges to precisely characterize bacterial strains. This technique takes advantage of the fine specificity of many phages for their hosts and is still in common use around the world for epidemiological and diagnostic purposes, using various approved carefully defined sets of commercially prepared phages (see Chapter 8 in this book).

Therapeutic administration of phages was explored extensively in many parts of the world, with successes being reported for a variety of diseases, including dysentery, typhoid and paratyphoid fevers, pyogenic and urinary-tract infections, and cholera. Phages have been given orally, through colon infusion, as aerosols, and poured directly in lesions. They have also been given as injections: intradermal, intravascular, intramuscular, intraduodenal, intraperitoneal, and even (very rarely) into the lung, carotid artery, and pericardium. The early strong interest in phage therapy is reflected in some 800 papers published on the topic between 1917 and 1956, many of which have been reviewed by Ackermann & DuBow,[30] Kutter & Sulakvelidze,[7] Kutter et al.,[12] and Abedon et al.[14] The reported results were quite variable. Many of the early physicians and entrepreneurs who became so excited by the potential clinical implications jumped into applications with very little understanding of phages, microbiology, or basic scientific process. Thus, many of the studies were anecdotal and/or poorly controlled; many of the failures were predictable, and some of the reported successes did not make scientific sense.

Much of the understanding gained by d'Herelle was ignored in this early work, and inappropriate methods of preparation, supposed "preservatives" and storage procedures were often used. On one occasion, d'Herelle reported testing 20 preparations from various companies and finding that none of them contained active phages! On another occasion, a preparation was advertised as containing a number of different phages, but the technician responsible had decided it was easiest to grow them up in one large batch and one phage had out-competed all the others, so this was not, in fact, a polyvalent preparation. In general, there was little or no quality control except in a few research centers. It is not surprising that virtually all the published successful work in the United States in the 1920s–1940s involved phage preparations made in academic labs that were collaborating with clinical researchers, not the phage produced by corporations. Large clinical studies were rare, and the results of those that were carried out were largely inaccessible outside of Eastern Europe until 1990, when Dr. Kutter spent 4 months in the Soviet Union on a U.S.-USSR Academies of Science exchange, including two visits to the Eliava Institute, establishing a collaborative partnership. Eliava scientists soon began participating in her biennial Evergreen International Phage Meetings. Accessibility to the classical Georgian phage therapy work further improved in 2012, when Nino Chanishvili published an extensive review[31] of the material in the library of the Eliava Institute, supported by the UK Global Partnership Program. Her book provides badly needed access to the extensive studies and clinical applications carried out in the early years in Georgia and, to some degree, in Russia.

Specific Problems of Early Phage Therapy Work

Many still believe (erroneously) that phage therapy was proven not to work during the explorations that began in the 1920s at the Pasteur Institute and went on until the 1940s in Europe and the United States; however, it simply was never adequately researched, having ended before most tools of microbiology and molecular biology became available, and the work that was done well is not widely enough known. It is thus important to carefully consider the reasons for the early problems and the questioning of efficacy. These included:

- Paucity of understanding of the heterogeneity and ecology of either the phages or the bacteria involved

- Lack of availability or reliability of bacterial laboratories for carefully identifying the pathogens involved (necessitated by the relative specificity of phage therapy)

- Frequent failure to select phages of high virulence against the target bacteria and to test them *in vitro* before using them in patients

- Use of single phages in infections, which involved mixtures of different bacterial species and strains

- Emergence of resistant bacterial strains, (especially problematic if only one phage strain is used against a particular bacterium); this can also happen by lysogenization if one uses temperate phages, discussed in the previous chapter.

- Failure to neutralize gastric pH prior to oral phage administration

- Inactivation of phages by both specific and nonspecific factors in bodily fluids, especially if phages were used intravenously

Key technical developments that helped to clarify the general nature and properties of bacteriophages included:

- The concentration and purification of some large phages by means of very high-speed centrifugation, developed in the early 1940s, and the demonstration that they were made up of approximately equal amounts of DNA and protein[32]

- Visualization of phages by means of the newly developed electron microscope (EM)[33,34]

A much better understanding of the interactions between lytic phages and bacteria began with the one-step growth curve experiments of Ellis and Delbrück[35] and the studies of Doermann.[36] These experiments demonstrated an eclipse period during which the DNA began replicating and there were no free phages in the cell; a period of accumulation of intracellular phages; and a lysis process that released the phage to go in search of new hosts, as discussed in Chapter 48 (Introduction to Bacteriophages). While this basic pattern had been described by d'Herelle,[25] few people were aware of the details of his work.

In 1943, an event happened that was to have a major impact on the orientation of phage research in the United States and much of western Europe, strongly shifting the emphasis from practical applications to basic science. Physicist-turned-phage biologist Max Delbrück met with Alfred Hershey and Salvador Luria and formed the "Phage Group," which rapidly expanded through the influence of the annual summer "Phage Course" and phage meetings at Cold Spring Harbor, Long Island that began in 1945. Their influence on the origins of molecular biology has been well documented.[37,38]

A major element of the successes of phage as model systems for working out fundamental biological principles was that Delbrück convinced most phage biologists in the United States to just focus on one bacterial host (*E. coli* B) and seven of its most lytic phages, which he renamed T(type)1–T7. As it turned out, T2, T4, and T6, originally isolated for potential therapeutic applications, were quite similar to each other, defining a family now called the "T-even phages." These phages were key in demonstrating that DNA is the genetic material, that viruses can encode enzymes, that gene expression is mediated through special copies in the form of "messenger RNA," that the genetic code is triplet in nature, and many other fundamental concepts. The negative side of this strong focus on a few phages growing under rich laboratory conditions, however, was that there was very little study or awareness of the ranges, roles, and properties of bacteriophages in the natural environment, or of potential therapeutic applications.

Early Studies of Phage Behavior in the Animal Body

A number of fairly early experiments involving the injection of phages into animals led to the widespread impression that phage therapy could not in fact succeed because the phage were too rapidly cleared by the immune system; the remarks in this regard by Gunter Stent[39] had a particularly strong impact on the phage community. For example, two early experiments involving rabbits showed rapid disappearance of the particular phages used from the blood and organs, but long-term survival in the spleen.[40,41] Subsequent experiments in rats and mice also showed rapid loss from the circulation. When Nungester & Watrous in 1934 injected 10^9 PFU of a staph phage intravenously into albino rats, a blood concentration of only 10^5 PFU/mL was seen after 5 minutes, and this dropped to 40 PFU/mL by 2 hours.[42,43]

The pessimistic conclusions broadly reached based on this research, however, had two serious flaws. First, the experiments were done in the absence of host bacteria in which the phage could multiply and find protection. Furthermore, they were carried out by the very unnatural mode of intravenous injection, subjecting the phage almost immediately to the reticuloendothelial system. Many more recent results have made it clear that phage are often seen in the mammalian circulatory system; however, this generally occurs under conditions where they are entering the circulatory system from some sort of reservoir in other tissues and where the mammal is dealing with an infection by a bacterium that the phage can infect—precisely the sort of situation seen in phage therapy as currently practiced in Eastern Europe.

One of the best early sets of experiments was published in 1943 by the noted Harvard bacteriologist, René Dubos.[44] He injected white mice intracerebrally with a dose of a smooth *Shigella dysenteriae* strain that was sufficient to kill >95% of the mice in 2–5 days and then treated them by intraperitoneal injection of a phage mixture isolated from New York City sewage, grown in the same bacteria and purified only by sterile filtration. With no treatment, when treated with filtrates of uninfected bacterial cultures, or with heat-killed phage, only three of 84 (3.6%) survived, whereas 46 of 64 (72%) of those given 10^7–10^9 active phage survived. Dubos also carried out pharmacokinetic studies. When phage were given to uninfected mice, they appeared in the blood stream almost immediately, but the levels started to drop within hours and very few were seen in the brain. In contrast, brain levels in the infected animals quickly greatly exceeded blood levels; around 10^7–10^9 phage per gram were often seen between 8 and 110 hours, starting to drop anywhere between 75 and 140 hours. After the first 18 hours, the levels in the blood were far lower than in the brain, but still present at 10^4–10^5 PFU per mL in those cases where the brain levels were still over 10^9 per gram. Their data are shown graphically in the review by Kutter et al.[12] Dubos' studies clearly established that the phage themselves were responsible, not something in the lysate that just stimulated normal immune mechanisms; that phage could rapidly find and multiply in pockets of infection anywhere in the body; and that phage could be maintained in the circulation as long as there was a reservoir of infection where phage were continually being produced. Without providing data or pharmacokinetics, they mention that the mice were also rescued by subcutaneous or intravenous administration of the phage, but not by stomach tube or in drinking water.

Carefully controlled experiments carried out in 1943–1945 by Henry Morton's group at the University of Pennsylvania[45] supported those of Dubos. They further showed the lack of any protection when lysates of phage with inappropriate host specificities were used. A final review authorized by the Council on Pharmacy and Chemistry discussed the major advantages of phage, such as the ability to enter problem areas and multiple there if appropriate bacteria were present, thus treating localized infections that are relatively inaccessible via the circulatory system. Also, it emphasized that the high specificity of phage greatly aided in reducing later resistance problems and that almost all of the earlier research had been so poorly conceived and/or carried out that it offered no proof either for or against the promise of phage as antibiotics.

U.S. work with dysentery phages largely ended in 1944 when penicillin was made available to the general public. The military secrecy, the end of the war emergency funding, the rapid rise in antibiotic availability and their broad spectra, and Max Delbrück's success in convincing the phage community to shift its focus to basic mechanistic research involving a few model systems all contributed to the fact that there was little U.S. follow-up to these interesting and successful results, and few even knew about them—or about two successful subsequent human applications.

Penicillin only worked against some kinds of bacterial infections. Typhoid, for example, was not treatable, and some excellent phage work was carried out in the interim. It was known that the strains of *Salmonella typhi* that created the main pathogenicity problems were those carrying one particular antigen, named Vi (for "virulence"). In 1936, a pair of Canadians had identified phages specific against cells bearing the Vi antigen. In the early 1940s, Walter Ward, of the Los Angeles County Hospital, was trying to deal with repeated serious outbreaks of typhoid fever that were killing one in five of those afflicted.[46] He tested the Vi-specific phages against mouse typhoid, and found that the death rate fell to 6%, versus 93% in the controls. Some of his colleagues then used these phages to treat typhoid patients; only 3 of 56 of their treated patients died, versus the 20% mortality they were seeing with the other treatments available at the time.[47] Most impressively, the rest of the patients rapidly went from being largely comatose to full of vigor, with renewed appetite, in 24–48 hours. In 1948–1949, near Montreal, Desranleau treated nearly 100 dysentery patients with a cocktail of six Vi-specific phages, and the deaths dropped from 20% to 2%.[48] However, by 1947, chloramphenicol had been shown to work well against typhoid and it was much easier for pharmaceutical companies to deal with, so that seems to have been the (hopefully temporary) end of phage clinical trials in the Western Hemisphere.

Clinical Application: Phage Therapy Work in the Age of Antibiotics

The strong understanding of phage biology gained in recent years has the potential to facilitate far more rational thinking about the therapeutic process and the selection of therapeutic phages. During the evolution of molecular biology, there was little interaction between those who were so effectively developing the field using phage as tools to understand molecular mechanisms and structures and those working on phage ecology and therapeutic applications. As discussed extensively in the books by Kutter & Sulakvelidze,[7] Häusler,[11] McGrath & van Sinderen,[49] and Borysowski et al.,[21] the latter fields have grown greatly in recent years, spurred on by an increasing awareness of phage variety and roles in maintaining microbial balance in every ecosystem, and by concerns about the increasing incidence of nosocomial infections and of bacteria resistant against most or all known antibiotics, as well as by the fact that phages are far more effective than antibiotics in areas where the circulation is bad, and they do not disrupt normal flora. This strong sense of the potential importance of phage was seen particularly early on in Poland, France, Switzerland, and the former Soviet Union, where use of therapeutic phages never fully died out and where there has been some ongoing research and clinical experience. In France, Dr. Jean-François Vieu led the therapeutic phage efforts until his retirement in 1980. He worked in the *Service des Entirobactiries* of the Pasteur Institute in Paris and, for example, prepared *Pseudomonas* phages on a case-by-case basis for patients. The experience there is discussed in Vieu[50] and Vieu et al.[51] In Vevey, Switzerland, the small pharmaceutical firm *Saphal* made "Coliphagine," "Intestiphagine," "Pyophagine," and "Staphagine" in drinkable and injectable forms, salves, and sprays into the 1960s.[11] The preparations were officially approved and were paid for by insurance there. However, in the Western world they soon were totally superseded by the explosive development and commercialization of modern antibiotics, with their broad host ranges.

Phage therapy continued to be extensively used in the Soviet Union. Much of the research and strong early therapeutic phage preparation and application came through the Eliava Institute in Tbilisi, Georgia, strongly stimulated and supported by Felix d'Herelle, as described in some detail later in this chapter. The Eliava Institute still plays a leading role in phage therapy education, exploration and application and its reintroduction to the rest of the world. By the 1980s, they had helped develop production centers in the Russian towns Ufa, Alma Ata, and Nizhniy Novgorod.[7] In recent years, these formed the kernel of Russian companies that are now again developing and marketing complex therapeutic phage preparations, led by the pharmaceutical giant Microgen and available in pharmacies throughout Russia as well as on line. They also held very interesting phage therapy conferences in 2016 in Moscow and 2018 in Nizhniy Novgorod, with significant international participation in addition to speakers actively involved in the field from all over Russia; Dr. Kutter, coauthor of this chapter, was keynote speaker at the latter.

Serious phage therapy research work in Wroclaw, Poland expanded markedly in the 1970s, and they hosted the 2018 Viruses of Microbes meeting, which drew more than 400 participants. In both Tbilisi and Wroclaw, the close interactions between research scientists and physicians played an important role in the high degree of success obtained—just as appears to have been the case for d'Herelle's early work and the many cases of successful phage therapy applications reported in the United States in the 1930s and 1940s.

Institute of Immunology and Experimental Medicine, Polish Academy of Sciences

The most detailed publications documenting successful early phage therapy have come from the group of Stefan Slopek, director for many years of the Institute of Immunology and Experimental Medicine, Polish Academy of Sciences, Wroclaw. They published a series of extensive papers describing work carried out from 1981 to 1986 with 550 patients.[52–54] This set of studies involved 10 Polish medical centers, including the Wroclaw Medical Academy Institute of Surgery Cardiosurgery Clinic, Children's Surgery Clinic and Orthopedic Clinic, the Institute of Internal Diseases Nephrology Clinic, and the Clinic of Pulmonary Diseases. The patients ranged in age from 1 week to 86 years. In 518 of the cases, phage use followed unsuccessful treatment with all available other antimicrobials. The major categories of infections treated were:

- long-persisting suppurative fistulas
- septicemia
- abscesses
- respiratory tract suppurative infections and bronchopneumonia
- purulent peritonitis
- furunculosis

In a final summary paper, the authors carefully analyzed the results with regard to such factors as nature and severity of the infection and monoinfection versus infection with multiple bacteria.[54] Rates of success ranged from 75% to 100% (92% overall), as measured by marked improvement, wound healing, and disappearance of titratable bacteria; 84% demonstrated full elimination of the suppurative process and healing of local wounds. Infants and children did particularly well. Not surprisingly, the poorest results came with the elderly and those in the final stages of extended serious illnesses, two groups with weakened immune systems and generally poor resistance.

The phages all came from the extensive collection of the Bacteriophage Laboratory of the Institute of Immunology and Experimental Therapy, Polish Academy of Sciences, Wroclaw. In the later studies, some of the specific phages they employed were identified. All were virulent, capable of completely lysing the bacteria being treated. In the first study alone, 259 different phages were tested (116 for *Staphylococcus*, 42 for *Klebsiella*, 11 for *Proteus*, 39 for *Escherichia*, 30 for *Shigella*, 20 for *Pseudomonas*, and one for *Salmonella*); 40% of them were selected to be used directly for therapy. All the treatments were carried out in a research mode, with the phage prepared at the institute by standard methods and tested for sterility. Treatment generally involved 10 mL of sterile phage lysate administered orally half an hour before each meal, with gastric juices neutralized by taking (basic) Vichy water, baking soda, or gelatum. In addition, phage-soaked compresses were often applied three times a day where dictated by localized infection. Treatment ran for 1.5–14 weeks, with an average of 5.3 weeks. For gastrointestinal problems, short treatment usually sufficed, while long-term use was necessary for such problems as pneumonia with pleural fistula and pyogenic arthritis. Bacterial levels and phage sensitivity were continually monitored, and the phage(s) were changed if the bacteria being targeted developed resistance to the particular phages, which was very common for gram-negative bacteria. Therapy was generally continued for 2 weeks beyond the last positive test for the bacteria.

Few side effects were observed; those that were seen seemed to be directly associated with the therapeutic process. On about days 3–5, pain in the liver area was often reported, lasting several hours. The authors suggested that this might be related to the extensive liberation of endotoxins as the phage are destroying the bacteria most effectively. In severe cases with sepsis, patients often ran a fever for 24 hours on about days 7 to 8.[52]

Various methods of administration were successfully used, including oral, aerosol, and infusion rectally or in surgical wounds. Intravenous administration was not recommended for fear of possible toxic shock from bacterial debris in the lysates.[52] However, it was clear that the phages readily entered the body from the digestive tract and multiplied internally wherever appropriate bacteria were present, as measured by the level of their occurrence in blood and urine as well as by therapeutic effects.[55] This interesting and rather unexpected finding has been replicated in many studies and systems,[56–59] while control studies in the absence of bacterial infection seldom show any phage in the urine.

Detailed notes were kept throughout on each patient. The final evaluating therapist also filled out a special inquiry form that was sent to the Polish Academy of Science research team along with the notes. The Computer Center at Wroclaw Technical University carried out extensive analyses of the data. They used the categories established in the WHO (1977) International Classification of Diseases in assessing results. They also looked at the effects of age, severity of initial condition, type(s) of bacteria involved, length of treatment, and other concomitant treatments. The papers included many specific details on individual patients, which helped to give some insight into the ways phage therapy was used as well as an in-depth analysis of difficult cases.

After Slopek's retirement, Dr. Beatta Weber-Dabrowska carried on with the phage preparation and treatment work. She published a very interesting and useful summary in English of the results with the next 1,600 patients[60] and continues to play a very major role in the Institute's ongoing phage therapy work and the relevant training of excellent, dedicated young scientists. In 1998, immunologist A. Górski took over as Institute Director and helped her build up a very strong focus on their phage therapy work, with special emphasis on the immunological consequences of phage treatment.[10,15,21,61] He now has passed on the directorship of the Institute and focuses all of his energy on the Laboratory of Bacteriophages, which he still heads (http://www.iitd.pan.wroc.pl/en/Phages/). Under the experimental therapeutics rules of the European Union, to which Poland now belongs, they have opened their own small clinic at the Institute and are treating both Polish and occasional international patients there while collecting data in a carefully controlled fashion.[15,21,24,62]

Recent publications by this Polish group have focused on phage interactions with the immune system, ranging from studies on phage applications in allergy challenges and graft versus[63–77] host disease, to phage therapy in autoimmune liver disease and phage as an immune modulator. These important works give insight into the additional significant approaches and therapeutic properties phage can offer, including for example down regulation of inflammatory cascades, among others. They also have long worked closely with Dr. M. Lobocka's strong phage molecular biology group in Warsaw to sequence and further characterize their key therapeutic phages and explore a broad range of new candidates and targets, very effectively supporting their therapeutic work.[68]

Eliava Institute of Bacteriophage, Microbiology, and Virology, Tbilisi, Georgia

The most extensive work on phage therapy is still being carried out under the auspices of the Bacteriophage Institute in Tbilisi, in the former Soviet republic of Georgia, as discussed previously in this chapter; much more detail can be found in several reviews.[7,12–15] In Georgia, phage therapy was long part of the general standard of care, especially in pediatric, burn, and surgical hospital settings. Phage preparation was carried out on an industrial scale, employing 700 people in the factory and several hundred more in the research arm of the Institute just before the 1990 break-up of the Soviet Union, and many tons of various products were regularly shipped throughout the former Soviet Union, mainly for use by the military. They were also available both over the counter and through physicians. The largest use was in hospitals, to treat both primary and nosocomial infections, alone or in conjunction with other antibiotics and particularly when antibiotic-resistant organisms were found. The Georgian military was among the strongest supporters of phage therapy research and development because phage have proven so useful for wound and burn infections as well as for preventing debilitating gastrointestinal epidemics among the troops. The International Science and Technology Centers (ISTC) program, set up jointly by the United States, Europe, and Japan to give constructive opportunities to scientists formerly working with Soviet military projects, also became a strong supporter of basic and applied research in this area in Tbilisi in the 1990s, as did the Civilian Research and Defense Fund (CRDF) and the Science & Technology Center in Ukraine (STCU), with similar goals. National versions of the two very complex major therapeutic phage cocktails, Pyophage (for wounds) and Intestiphage, are still very widely sold in pharmacies in both Georgia and Russia.

From quite early on, the industrial arm of the Eliava was run on a self-supporting basis, making multiple tons weekly of the major phage cocktails, while its scientific branch was government-supported. The latter included the electron microscope facility, permanent strain collections, laboratories studying a variety of phages of the enterobacteria, staphylococci, and pseudomonads, constantly updating the commercial phage cocktails and formulating new ones. In addition, it had groups involved in immunology, vaccine production, *Lactobacillus* work, and other therapeutic approaches. It also carried out the very extensive studies of each new strain, addition to the therapeutic cocktail, host strain if any and means of delivery needed for approval by the Ministry of Health in Moscow. This careful study of the host ranges, lytic spectra, cross-resistance, and other fundamental properties of the phages being used was a major factor in the reported successes of the phage therapy work carried out through the institute, as was their ability to initially select particularly useful, highly virulent phages from among the myriads potentially available against any given host. All the phages used there for therapy are obligatorily lytic, avoiding the problems engendered by lysogeny. The problems of bacterial resistance were largely solved using well-chosen mixtures of phages with different receptor specificities against each target type of bacterium as well as of phages against the various bacteria likely to be causing the problem in multiple infections. The situation was further improved whenever the clinicians typed the pathogenic bacteria and monitored their phage sensitivity. Where necessary, they added additional phages to which the given bacteria were sensitive. Not infrequently, using a phage in conjunction with carefully chosen antibiotics was found to give better results than either the phage or the antibiotic alone.

A great deal of work went into developing and providing the documentation for Ministry of Health approval of specialized new cocktails and delivery systems, such as a spray for use in respiratory tract infections, in treating the incision area before surgery, and in sanitation of hospital problem areas such as operating rooms. An enteric-coated pill was also developed, using phage strains that could survive the drying process, and accounted for the bulk of the shipments to other parts of the former Soviet Union.

The depth and extent of the work involved are very impressive. For example, in 1983–85 alone, the Institute's Laboratory of Morphology and Biology of Bacteriophages carried out studies of growth, biochemical features, and phage sensitivity on 2,038 strains of *Staphylococcus*, 1,128 of *Streptococcus*, 328 of *Proteus*,

373 of *P. aeruginosa*, and 622 of *Clostridium* received from clinics and hospitals in towns across the former Soviet Union. New broader-acting phage strains were isolated using these and other Institute cultures and were included in a reformulation of their extensively used Pyophage and Intestiphage preparations. In the years since, both formulations have continued to be extensively improved based on further studies, and phages against *Klebsiella* and *Acinetobacter* have also been isolated and developed into therapeutic cocktails. The other major product, *Intestiphage*, was used very extensively by the military, the pediatric centers, and in regions with extensive diarrheal problems; it includes phages active against a range of enteric bacteria, which were often also prepared in tablet form.

Much of the focus in the last 50 years has been on combating nosocomial infections, where multidrug-resistant organisms have become a particularly lethal problem and where it is also easier to carry out proper long-term research. Special mixtures were developed for dealing with strains causing nosocomial infections in various hospitals, and they were very effectively used in sanitizing operating rooms and equipment, water taps and other sources of spread of the infections (most of them involving predominantly *Staphylococcus*). The number of sites testing positive for the problem bacteria decreased by orders of magnitude over the several months of the trial at each site. Clinical and prophylactic studies in collaboration with institutions such as the Leningrad (now St. Petersburg) Intensive Burn Therapy Center, the Academy of Military Medicine in Leningrad, the Karan Trauma Center, and the Kemerovo Maternity Hospital as well as in Tbilisi at the Pediatric Hospital, the Burn Center, the Center for Sepsis and the Institute for Surgery, were used to further develop treatment protocols and phage cocktails. During the 1991–1992 invasion by Russian troops, every soldier carried a special version of Pyophage formulated and constantly updated through a collaboration between the chief military surgeon Dr. Guram Gvasalia and the Eliava's Dr. Zemphira Alavidze to best target the major pathogens being encountered by the soldiers in the field. The soldiers could immediately use this in the field to treat any wounds received; this approach was highly successful. However, Western-style double-blind clinical trials are still needed before Intestiphage, Pyophage, and their other narrower-spectrum cocktails can be marketed in the rest of the world.

An exciting new product completed the Georgian approval process and was licensed in 2000 by the Georgian Ministry of Health. PhagoBioDerm™, developed by Dr. Alavidze in collaboration with chemist Ramaz Katsarava of Tbilisi's Technical University, is a biodegradable, nontoxic polymer composite that is impregnated with the Pyophage cocktail of phages, along with other antimicrobial agents.[69] Markoishvili et al.[70] reported the results of a study of PhagoBioDerm™ involving 107 patients with ulcers that had failed to respond to conventional therapy, including systemic antibiotics, antibiotic-containing ointments, and various phlebotonic and vascular-protecting agents. The patients were treated with PhagoBioDerm™ alone or in combination with other interventions during 1999 and 2000. The wounds or ulcers healed completely in 70% of the 96 patients for whom there was follow-up data. In the 22 cases for which complete microbiologic analyses were available, healing was associated with the concomitant elimination or very marked reduction of the pathogenic bacteria in the ulcers. Other versions of the product are now available in pharmacies or in the final developmental stages. Further studies investigating the immobilization of bacteriophage in nanostructure wound dressings are ongoing by both Portuguese and Georgian researchers.[71]

In 2005, Dr. Gvasalia, by then head surgeon at the central Georgian state medical school, started a Surgical Infections and Phage Therapy training program with support from the PhageBiotics Research Foundation to help prepare young Georgian surgeons for treating major wound infections and for carrying out research on phage therapy. Much of Dr. Gvasalia's expertise has been transferred to several of his former resident surgeons, but the main practical legacy is in the two phage-therapy clinics in Tbilisi. The Eliava Phage Therapy Center is a part of the Eliava Foundation suite of companies, treating both Georgian and international patients.

The older Tbilisi Phage Therapy Center was started in 2003 by the members of Dr. Alavidze's lab and the primary Bioderm co-developer, chemist Ramaz Katsarava, working with Dr. Gvasalia. In 2005, it became part of a new California-based American company, Phage International. In 2009, the Eliava Foundation was founded by the Eliava Institute group of senior scientists as a nongovernmental body to promote the development and marketing of Eliava products and services for human, animal, plant, and environmental applications (eliava-institute.org/eliava-foundation). It has spun off multiple small related companies, including under its umbrella the already established Diagnostics Center and Eliava BioPreparations, the phage cocktail production company. It has also formed new subsidiary enterprises such as an Eliava Media Production Center, the Eliava Authorized (Compounding) Pharmacy, the Eliava Phage Therapy Center, and the Eliava Management group. The purpose of the Foundation is to promote the development and marketing of Eliava products and services for human, animal, plant, and environmental applications, independently from the Institute's status under the Georgian Academy of Sciences (eliava-institute.org/eliava-foundation). A very important newly approved Eliava product soon joined their Intestiphage and Pyophage in most Georgian pharmacies.[72] This is a highly purified solution of their sequenced phage Sb1, which targets 99% of *S. aureus* strains.

Of particular importance, in 2010 the Eliava Foundation opened its own outpatient clinic on its grounds to treat both Georgians and foreign patients, and its own compounding pharmacy, whose services include isolating and preparing specific phages for patients when the commercial cocktails aren't efficacious.

Many Georgian pharmacies also market a variety of licensed phage cocktails produced by Biochimpharm. This is a privately-owned phage therapy company that was developed out of the commercial section of the old Eliava Institute, which was privatized in 1994 along with many other former Soviet companies. Biochimpharm also markets its phage cocktails online.

The extensive clinical observations in Georgia and Poland confirm that phages have many potential advantages:

- They are self-replicating but also self-limiting, since they multiply only as long as sensitive bacteria are present.

- They can be targeted far more specifically than most antibiotics to the problem bacteria, causing much less damage to the normal microbial balance in the gut. Particular resultant problems of antibiotic treatment include *Pseudomonads*, which are especially difficult to treat, and *Clostridium difficile*, the cause of serious diarrhea and pseudomembranous colitis.[73]

- Phages can often be targeted to receptors on the bacterial surface that are involved in pathogenesis, so any resistant mutants are attenuated in virulence.

- Few side-effects have been observed, and none of them were serious.

- Phage therapy is particularly useful for people with allergies to antibiotics.

- Appropriately selected phages can easily be used prophylactically to help prevent bacterial disease at times of exposure or to sanitize hospitals and to help protect against hospital-acquired (nosocomial) infections.

- Especially for external applications, phages can be prepared fairly inexpensively and locally, facilitating their potential applications to underserved populations.

- Phages can be used either independently or in conjunction with antibiotics used intravenously and help reduce the development of bacterial resistance.

The Western World Rediscovers Phage Therapy: The 1980s and 1990s

Levin & Bull,[1] Barrow & Soothill,[3] and Kutter & Sulakvelidze[7] have extensively reviewed the animal research carried out in Britain and the United States after interest in the possibilities of phage therapy began resurfacing in the early 1980s. The results, in general, are in very good agreement with the clinical work described above in terms of efficacy, safety and importance of appropriate attention to the biology of the host–phage interactions, reinforcing trust in the reported extensive eastern European results.

In Britain, Smith & Huggins[56–59] carried out a series of excellent, well-controlled studies on the use of phages in systemic *E. coli* infections in mice and then in diarrheic disease in young calves and pigs. For example, they found that injecting 10^6 colony-forming units (CFU) of a particular pathogenic *E. coli* strain intramuscularly killed 10 out of 10 mice, but none died if they simultaneously injected 10^4 plaque-forming units of a phage selected against the K1 capsule antigen of that bacterial strain. A single treatment with this phage was far more effective than using such antibiotics as tetracycline, streptomycin, ampicillin, or trimethoprim/sulfafurazole, even when many doses of the antibiotics were used. Furthermore, the resistant bacteria that emerged had lost their capsule and were far less virulent. In calves, they also found very high and specific levels of protection. They had to isolate different phages for each of their pathogenic bacterial strains, since they were focusing on high specificity against the pathogens and did not succeed in isolating phages specific for more general pathogenicity-related surface receptors such as the K88 or K99 adhesive fimbriae, which play key roles in attachment to the small intestine. Still, the phage treatment was able to reduce the number of bacteria bound there by many orders of magnitude and to virtually stop fluid loss. The results were particularly effective if the phage were present before or at the time of bacterial infection, and if multiple phages with different attachment specificities were used. Furthermore, the phage could be transferred from animal to animal, supporting the possibility of prophylactic use in a herd. If the phage were given only after the development of diarrhea, the severity of the infection was still substantially reduced, and none of the animals died.[59]

Levin & Bull[1] conducted a detailed analysis of the population dynamics and tissue phage distribution of the 1982 Smith & Huggins[56] study, which can be helpful in assessing the parameters involved in successful phage therapy and its apparent superiority to antibiotics. They have gone on to do interesting animal studies of their own and conclude that phage therapy is at least well worth further study.[1]

Barrow & Soothill[3] carried out a series of studies preparatory to using phages for treating infections of burn patients. Using guinea pigs, they showed that skin graft rejection could be prevented by prior treatment with phages against *P. aeruginosa*. They also saw excellent protection of mice against systemic infections with both *Pseudomonas* and *Acinetobacter* when appropriate phages were used.[3] In the latter case, they reported that as few as 100 phages protected against infection with 10^8 bacteria—several times the LD_{50}.

Merrill et al.[74] carried out a series of experiments designed to better understand the interactions of phages with the human immune system and helped MD Richard Carlton start a company called Exponential Biotherapies to explore the possibilities of phage therapy. Their initial work was with lytic derivatives of the temperate phages λ and P22—very poor choices for therapeutic applications, as discussed in this chapter—but they gathered some very interesting data about factors affecting interactions between these phages and the innate immune system and patented a process for isolating longer-circulating mutants of phage lambda. They published successful animal studies with phages against Vancomycin-resistant *Enterococcus* and carried out a successful (but unpublished) small stage-one clinical safety trial with these phages, but focused more on work in the area of food safety, which required much less venture capital. In the spring of 2006, a spin-off Dutch branch, EBI food safety (now called Micreos food safety), became the first company to receive governmental approval of phage as a food additive. Their phage preparation to destroy *Listeria monocytogenes* on cheeses was also granted GRAS status by the United States Food and Drug Administration (FDA) that fall. Earlier in 2006, after taking 4 years for a very thorough review, the FDA had also approved a Listeria-phage cocktail made by Baltimore-based Intralytix for use on packaged meats.

Phage and Bacterial Pathogenicity

Most bacteria are not pathogenic; in fact, they play crucial roles in the ecological balance in the digestive system, mucous membranes, and all body surfaces. They often actually help to protect against pathogens. This is one reason why broad-spectrum antibiotics have such a large range of side-effects and why more narrowly targeted bactericidal agents would be highly advantageous. Interestingly, most of the serious pathogens are close relatives of nonpathogenic strains; there, antibiotics can't differentiate between pathogenic and nonpathogenic strains, but phages can be isolated that are appropriately selective.

Studies clarifying the mechanisms of pathogenesis at the molecular level have progressed remarkably in recent years, crowned by determination of the complete sequence of many bacteria and extensive cloning and sequencing of pathogenicity determinants. Generally, several genes are involved, and many of these are clustered in so-called "pathogenicity islands," or PAIS, which may be 50,000–200,000 base pairs long. They generally have some unique properties indicating that the bacterium itself probably acquired them as a sort of "infectious disease" at some time in the past, and then kept them because they helped the bacterium to infect new ecological niches where there was less competition. Many of these PAIS are carried on small extrachromosomal circles of DNA called *plasmids*, which can also be carriers of drug-resistance genes. Others reside in the chromosome where they are often found embedded in defective temperate prophages which have lost some key genes in the process and cannot be induced to form phage particles. However, they can sometimes recombine with related infecting phages. Therefore, it makes sense to avoid using temperate phages or their lytic derivatives for phage therapy to avoid any chance of picking up such pathogenicity islands.

For bacteria in the human gut, pathogenicity involves two main factors:

- The production of toxin molecules, such as Shiga toxin (from *Shigella* and some pathogenic *E. coli*) or cholera toxin; these toxins modify proteins in the target host cells and thereby cause the problem.

- The acquisition of cell-surface adhesins, which allow the bacterium to bind to specific receptor sites in the small intestine, rather than just moving on through to the colon.

Pathogenic bacteria also all contain the components of so-called type III secretion machinery, related to structures involved in the assembly of flagella (for motility) and of filamentous phages, and instrumental in many plant pathogens. For all of the pathogenic enteric bacteria, the infection process triggers changes in the neighboring intestinal cells. These include degeneration of the microvilli, formation of individual "pedestals" cupping each bacterium above the cell surface, and, in the case of *Salmonella* and *Shigella*, induction of cell-signaling molecules that trigger engulfment of the bacterium and its subsequent growth inside the cell.

For several years, *E. coli* O157:H7 has been the subject of much concern, with contamination of such products as hamburgers and unpasteurized fruit juices leading to very serious problems. Particularly in young children and the elderly, there is a high probability of death when O157 infections evolve into hemorrhagic colitis (bloody diarrhea) and hemolytic-uremic syndrome, where the kidneys are affected. Antibiotic therapy has shown no benefit; it is actually generally contraindicated because it leads to increased toxin release.[75] We have isolated phages that look promising for reducing the O157 load in livestock from sheep naturally *resistant* to inoculation with *E. coli* O157:H7, working with collaborators at the U.S. Department of Agriculture in College Station, Texas.[76–78] A major job there is to eliminate this human pathogen that is found in the normal flora of one-third of the cattle brought to slaughter in the United States, contaminating water sources and creating massive recalls for industries from meat packers to juices to packaged spinach. The first phage so isolated turns out to also be T4-related, and to infect virtually all tested O157 strains.[76] Studies are suggesting that this and additional phage we have isolated from sheep are good candidates for controlling O157 in the GI tracts of ruminants[78]; one of them, CBA120, is very highly specific for *E.coli* O157 and is closely related to *Salmonella typhi* phage ViI, used to treat typhoid fever in Los Angeles and Montreal in the 1940s.[47,79] Intralytix and Omnilytix have now both started marketing products that use anti-O157 phages for cleaning the hides of cattle prior to slaughter, and several commercial labs are making use of our CBA120 to explore further applications, such as detection of coli O157.

Phage Therapy with Well-Characterized, Professionally Lytic Phages

A substantial fraction of the phages in current various therapeutic mixes for gram-negative bacteria turn out to also be relatives of bacteriophage T4,[80] which has played such a key role in the development of molecular biology. This family is often called the T-even phages, an historical accident reflecting the fact that T2, T4 and T6 from the original collection of Delbrück's Phage Group all turned out to be related. T4-like phages are found infecting all of the enteric bacteria and their relatives.[81] Large sets have been isolated from all over the world: Long Island sewage treatment plants, animals in the Denver zoo, African lagoons and dysentery patients in eastern Europe (the latter using *Shigella* as host); pediatric diarrhea patients in Bangladesh.[80,82,83]

Most of the T4-like phages targeting enteric bacteria use 5-hydroxymethylcytosine (5-HMdC) instead of cytosine in their DNA, which protects them against most of the restriction enzymes that bacteria make to protect themselves against invaders; this gives the phage a much more effective host range. T4's DNA sequence was the first completed for a large phage, aided by its very extensive system of mutants and the general knowledge of its genome.[80,84,85] We know a great deal about its infection process in standard laboratory conditions and about the methods it uses to target bacteria so effectively.[85,86] We can potentially use this knowledge to develop more targeted approaches to phage therapy, particularly as more is learned about the similarities and differences in its extended family.[80,87] We now know the sequences of many T4-related phages and that certain ones of them use different outer membrane proteins and/or oligosaccharides as their receptors, and we understand the tail-fiber structures involved well enough to potentially predict which phages will work on given bacteria and to engineer phages with new specificities.[88,89]

There have still been far too few studies of T4 ecology and its behavior under conditions more closely approaching the natural environment and the circumstances it will encounter in phage therapy—often anaerobic and/or with frequent periods of starvation. The limited available information in that regard was summarized by Kutter et al.[86] A variety of studies are shedding light on the ability of these highly virulent phages infecting *E. coli* or *Pseudomonas* to coexist in balance with their hosts in nature.[90] For example, they can reproduce in the absence of oxygen as long as their *Pseudomonas* or *E. coli* bacterial host has been growing anaerobically for several generations (Kutter and Brabban lab students, unpublished data, 2010-2015), or infect *E. coli K803* or *O157* several days into stationary phase. Most of the T4-related bacteriophages share a unique ability that contributes significantly to their widespread occurrence in nature and to their competitive advantage. They are able to control the timing of lysis in response to the relative availability of bacterial hosts in their environment. When *E. coli* are singly infected with T4, they lyse after 25–30 minutes at body temperature in rich media, releasing about 200–300 phage/cell. However, when additional T-even phages attack the cell more than 4 minutes after the initial infection, the cell does not lyse at the normal time. Instead, it continues to make phage for as long as 6 hours.[91,92] We have found that T4 can also survive for days in a hibernation-like state inside starved cells, allowing its host to readapt when appropriate nutrients are again supplied, and then very rapidly produce up to 40 phage per cell.[93] The precise patterns depend on the genetics of both the phage and the host, including on whether or not the host has a functional stationary-phase sigma factor and on the multiplicity of infection. This is particularly interesting and important since bacteria undergo many drastic changes to survive periods of starvation, which increase their resistance to a variety of environmental insults.[94]

Thus, for many reasons, the T4-like phages make excellent candidates for therapeutic and prophylactic use against enteric and other gram-negative bacteria, and widespread studies of their ecology and infection properties are now being carried out world-wide with these goals in mind. The most extensive clinical trial to date—the Nestlé study of phage targeting infant diarrhea in Bangladesh—provides the strongest safety studies available; in this case, the broad T4 family of phages. This trial grew out of Nestlé's efforts to address the serious developing world problem of infant diarrhea, 27% of which is due to coliform bacteria. Immunological-based treatments had worked for Nestlé against diarrheal viruses but were unsuccessful against these pathogenic *E. coli*, so Harald Brüssow decided to try phage. The Evergreen phage lab students had recently tested nearly 100 T4-related phages from around the world on the 72-member ECOR collection of *E. coli*, selected to represent as wide a variety as possible, and sent Brüssow those phages with the broadest host ranges. He found some that infected his Bangladesh strains, but not very efficiently, so he tried isolating phages from the stools of infant diarrheal patients at the pediatric diarrhea center in Dhaka. Initially, he used just two *E. coli* hosts for this isolation: lab strain K803 and a very widespread and well-studied enteropathogenic (EPEC) strain.

To their surprise, the phages Brüssow's lab isolated using the EPEC strain were all siphoviruses and none of them were able to infect more than a few of the Dhaka strains. All the useful broad-spectrum phages they obtained from the Dhaka patient samples were isolated using lab strain K12, and all of them were related to T4. Their broader host ranges presumably reflect the fact that the lab strains are missing the complex O-antigens that protect clinical coliform bacteria against the mammalian immune system; over 150 of these O-antigens have been identified for *E. coli*, and at most one of them is produced by each bacterial strain. Phages selected against lab strains that lack these O-antigens must either use one of the outer-membrane proteins or the highly conserved inner stretches of the lipopolysaccharide (LPS) that are still present in K12 and most other lab strains as their receptors. The versatile adhesin that T4 uses to bind to these primary receptors is located near the tip of its long tail fibers, which can penetrate the O-antigen forest on the clinical strains to reach those more conserved internal sites. During initial adsorption, these fibers bind the phage to the cell, bending in the process to bring the baseplate down close to the surface, at which point the short tail fibers are deployed to bind irreversibly to the secondary receptor, which is another very well-conserved internal component of the LPS. Eventually, Brüssow isolated and characterized nearly 100 of these T4-related phages, carrying out extensive studies of their genomics and their host ranges on broad sets of pathogenic enteric bacteria. He also explored the properties of several in a mouse-gut model.

After more than a decade of laying such groundwork, Brüssow was ready to initiate clinical trials in Dhaka. He prepared a cocktail using 15 of these phages that had a very broad spectrum on the set of Dhaka diarrheal strains. Metagenomic sequencing both of this phage cocktail and of a commercial Russian *E. coli/Proteus* phage cocktail that they used as an extra control established that there were no genes that could confer virulence to the host.[95,96] They complemented their molecular analyses with years of extensive safety studies of the phages in adults and older children in Bangladesh, following a range of physiological parameters. They then simply added either their phage cocktail, the Russian cocktail or a placebo to the routine rehydration fluid that was the standard treatment for these infants, with the amount each patient received thus depending on the amount of rehydration fluid they needed.

The Nestle study in Bangladesh was a long-awaited and carefully planned controlled clinical trial to look at the effects of *E.coli* T4-related phages on infant diarrhea.[97] The study results were inconclusive as to the effect of bacteriophage for several possible reasons. Initial research toward this study stretched over more than a 10-year period[98] and during this time development of techniques in microbiology and genetics significantly advanced.[96,98] Full genome sequencing after the trial provided a new perspective on the gut microbiota in acute infant diarrhea, which had not previously studied to any extent beyond determining the presence of *E. coli*. Sarker and Brüssow explain that while *E.coli* was isolated from stool cultures of all 120 children enrolled in the study, there were not high titers, especially compared to those of two *Streptococcal* bacteria, which were very high.[99] The ratio of the *Streptococcus* to *Bifidobacterium* titers was associated with disease progression and resolution. Symptoms of diarrhea were correlated with increased titers of *Streptococcus* with low *Bifidobacterium*, while improvement correlates more with an increase in *Bifidobacterium*. Sarker et al. noted, "Oral coliphages showed a safe gut transit in children but failed to achieve intestinal amplification and to improve diarrhea outcome, possibly due to insufficient phage coverage and too low *E. coli* pathogen titers requiring higher oral phage doses".[97] The authors noted the need to deepen our general understanding of the etiology of diarrhea and take into consideration the interaction within the fecal microbial community the best biomarker for their data group was *Bifidobacterium*.[97] A recent study from our Evergreen lab suggested another possible contributing factor. Stationary phase studies carried out on T4 indicated that phage require nutrient rich media to maintain efficient infection ability,[93] the phage in the Nestle study were mixed with ORS, which could be another contributing factor to the lack of efficacy seen in the study. This very expensive trial has been extremely important as a safety study and for working through various sorts of procedures, regardless of the lack of demonstration of significant efficacy, many lessons were learned through this experience.

For gram-positive bacteria, the family of phages related to the genetically very well-characterized *B. subtilis* phage SBO1 appears to have similarly broad advantages to those of T4. The Listeria phages recently approved for food-safety use belong to this genus, as do virtually all of the staph phages therapeutically used in Poland and Georgia.[68,72] The best studied include staph phage K, isolated in Berlin in the 1930s and sequenced and extensively studied in Cork a decade ago,[100] the sequenced Eliava phage Sb1 used for their highly purified commercial staph phage preparation,[101] and the set of Polish staph phages sequenced and analyzed by Lobocka et al.[68] All of these staph phages that have been looked at in detail have far broader host ranges than any of the phages infecting gram-negative bacteria and give very good clinical results, with little development of resistance, making them excellent candidates for the modern clinical trials needed for FDA approval.

Other Modern Clinical Studies

For a variety of regulatory with financial and attitudinal reasons, disappointingly little progress has been made to date in moving forward with formal clinical trials of phage therapy.

In 2009, British company Biocontrol reported significant clinical efficacy of a phage treatment for intractable *Pseudomonas* ear infections, with no reported safety issues in a small phase I/IIa trial.[23] Biocontrol later joined with the American company Targeted Genetics (which had been involved in human gene-therapy work) and the Australian company Special Phage Services (SPS) to form AmpliPhi Biosciences. In 2019, they merged with the company C3J Therapeutics to form Armata Pharmaceuticals, which has both natural and synthetic phage preparations in its pipeline and purportedly plans to eventually proceed with clinical trials in both areas. SPS had focused mainly on aquaculture, agriculture and race-horse applications, but was also working toward human phage therapy. Their scientists had carried out a number of compassionate-use case studies in a local Sydney hospital using phage from Tbilisi and published one of them.[102]

A physician-initiated phase I trial of a phage cocktail designed to treat leg ulcer infections at the Wound Care Center in Lubbock, Texas received FDA approval. The formal trial built on earlier encouraging work at the Wound Care Center there, which had employed the commercial Pyophage cocktail from the Eliava Institute on a compassionate-use basis, but the FDA required that a set of fully sequenced phages be used for this formal study. Thus, they arranged with Intralytix to supply them with a cocktail of eight sequenced phages (five targeting *P. aeruginosa*, two targeting *S. aureus,* and one specific for *E. coli*). No safety concerns were observed,[103] but the conditions used were not such as to establish efficacy and no funding was available to proceed with a phase II trial. Randy Wolcott, the surgeon involved, has now primarily turned to other ways of treating the biofilms that make the wounds he treats so challenging.

Phagoburn Clinical Trial

The United States, Belgian, and French militaries have all shown an active interest in phage therapy. The Belgians have carried out a small clinical safety trial of phage therapy for the treatment of burns in their major military hospital using three phages—one targeting *S. aureus* and two against *P. aeruginosa*—that had been isolated in Georgia and Russia and were produced as a highly purified and fully defined cocktail in Belgium.[104] The trial used a second, similar burn site in the same patient, treated identically except for the phage, as the control in each case. No problems were reported.[12,105] This work helped lay the groundwork for the multisite clinical trial Phagoburn, mentioned at the start of this chapter. The trial used phage cocktails developed by Pherecydes that target either *P. aeruginosa* or *E. coli*. It was largely funded by grants from the European Union and the French military, with impressive levels of cooperation between public, private, military, and other governmental entities as part of the FP7 program of projects targeting solutions to fight antibiotic resistance. The needed GMP phages were produced, gradually the various logistic regulatory details worked out with all three governments, and the protocol registered at http://clinicaltrials.gov. Jault et al. in 2018 reported enrollment of 27 patients with a confirmed burn wound in the study, this group was randomly divided into two groups.[106] One group received topical application of a cocktail of 12 lytic bacteriophages against *P. aeruginosa* (PP1131, 1×10^6 plaque-forming units [PFU] per mL) while the other, the current standard of care 1% sulfadiazine silver emulsion cream once daily for 7 days with two weeks of follow up. The groups were divided as follows: 13 were designated in the phage treatment group and 14 were standard of care. In the standard of care group, one patient was not exposed to treatment, in the phage treatment group one patient was found to not have an infection, leaving the phage treated group $n = 12$ and the standard treatment group $n = 13$. However, very unusually and unexpectedly, the titer of the phage preparation decreased markedly during storage after manufacturing, so the patients' actual dosing was only 10^2 PFU/mL per daily dose).[106] The study was not continued because of the lack of efficacy at that concentration. This was a high-interest study, to proceed with such a low titer preparation predestined its failure. Pirnay notes that despite the study outcome, this study gave a good indication to regulators as to what kind of documentation requirements are necessary to guarantee a timely supply of phage product.[107]

Weastmead Clinical Trial: Safety of Bacteriophage in Severe Staphylococcal Infections

Petrovic Fabijan et al. in 2020 published work on safety of intravenous phage in severe Staphylococcal infection.[108] In a single-arm noncomparative trial, 13 patients were administered GMP-quality phage cocktail against *S. aureus* in patients with septic shock or infective prosthetic valve endocarditis as an adjunctive therapy to standard antibiotic treatment.[108] Patients were administered 50–100ml 10^9 phage every 12 hours for 14 days. Patient selection was determined based on clinical criteria (seriously ill patients were included but not those where imminent death (<48 hours) was expected. Some of the patient samples *in vitro* sensitivity to the phage preparation was determined prior to administration while others were started on treatment empirically in parallel to testing—one patients' treatment was discontinued after showing resistance on initial testing. The primary endpoint of the study was safety and tolerability. Secondary outcomes were efficacy—specifically clinical improvement, inflammatory response, and mortality.[108] This study followed pharmacokinetics of the administered phage using qPCR. The levels of both phage and bacterial DNA were measured throughout treatment in addition to studying inflammatory markers overtime in patient samples. Inflammatory markers decreased in 11 of 13 patients soon after commencement of phage therapy.[108] The development of phage resistance was not noted during treatment. No adverse reactions including, fever, tachycardia, hypotension, diarrhea, or abdominal pain were observed. The authors concluded this preparation was safe for use in severe infections.[108] This study is an important stepping stone for a larger scale controlled clinical study.

Clinical Trial in Urology

A landmark clinical trial in urology took place at the Tsulukidze National Center of Urology, Tbilisi, Georgia in collaboration between the Eliava Institute and several Swiss institutions. This study was a double-blind placebo-controlled trial to treat urinary tract infections with bacteriophage in patients who were undergoing transurethral proctectomy.[109] The study had three arms, one group was to be treated with commercially available Pyophage adapted to the treated population, the second group was a placebo control, and the third received antibiotic therapy. A total of 81 patients were included in the study. While the results of the study have not yet been published, indications to positive results have been eluded to. Dr. Leitner, one of the authors of the study and physician at Neuro-Urology, Spinal Cord Injury Center & Research, University of Zürich, Balgrist University Hospital, Zürich, Switzerland recently published a review article discussing the potential of phage therapy to address the unmet need for chronic urinary tract infections in neuro-urologic conditions.[110]

Safety Issues

From a clinical standpoint, phages appear to be very safe, as discussed extensively in Chapter 14 of Kutter & Sulakvelidze[7] and chapters in Borysowski et al.[21] and by Kutter et al.,[12,15] Abedon et al.,[14] Chanishvili[31] and most other recent reviews. This is not surprising, given that humans are exposed to phages from birth. Bergh et al. reported in 1989 that nonpolluted water contains about 2×10^8 phages/mL.[111] They are normally found in the gastrointestinal tract, urine, mouth, and on the skin.[112–114] Phages also were shown to be unintentional contaminants of sera and thence of commercially available vaccines[115–118] that were allowed to be sold despite this discovery, due to the general consensus that phage are safe for humans.

Extensive preclinical animal testing was required for approving new phage formulations in the former Soviet Union, including Georgia, but few of these studies were published. Bogovazova et al.[119,120] evaluated the safety and efficacy of *Klebsiella* phages produced by the Russian company Immunopreparat. Pharmacokinetic and toxicological studies were carried out in mice and guinea pigs using intramuscular, intraperitoneal, or intravenous administration of phages. They found no signs of toxicity or gross histological changes, even using a dose/g 3,500-fold higher than the projected human dose. They then evaluated the safety and efficacy of the phages in treating 109 patients infected with *Klebsiella*. The phage preparation was reported to be nontoxic for humans and effective in treating *Klebsiella* infections, as manifested by marked clinical improvements and bacterial clearance in the phage-treated patients. Chanishvili[31] also provides very extensive data relevant to safety issues.

Side effects such as occasional liver pain and fever reported in the early days of Western phage therapy may well have been due to bacterial byproducts contaminating phage preparations used intravenously.[121–123] The Polish phage therapy group never administers their phage intravenously because of this concern. The same is true for almost all therapeutic work carried out in Tbilisi, including all that has been performed in the last 30 years, and helps explain their virtually total lack of significant problems.

Phage administration has never been reported to adversely affect the efficacy or safety of other drugs in the long history of work in Eastern Europe. No systematic studies have been carried out in this regard, but phages are so specific in their activity that it is hard to predict where such interactions might occur. On the other hand, at least some local antibiotics tend to interfere with phage treatment of infections in areas with poor circulation by killing off the most-accessible of the bacteria in which the phage need to multiply as they work their way deeper into the lesion. This would be a particular problem in cases where the phage can still attach and infect but cannot complete their replication cycle. (Many of the Georgian physicians feel that antibiotics should never be used topically for wounds with deep-seated infections, since the decrease in antibiotic concentration below the surface provides a strong selection for antibiotic resistance, while this problem does not occur with phages.)

Phage Interactions with Mammalian Systems: Modern Studies

Phage can never reproduce in mammalian cells due to different biosynthetic mechanisms between prokaryotic and eukaryotic cells. However, that does not mean that they cannot interact with or be taken up by mammalian cells. One of the most interesting and important areas of study is the exploration of direct phage interactions with the mammalian immune system, in ways that go beyond the production of antiphage antibodies.

The major studies relevant to this important question have been carried out at the Hirszfeld Institute in Poland, which has supported so much clinical phage work. These studies often take advantage of the extensive literature on bacteriophage T4 as well as the strong immunological focus of the whole Institute. Since phage can only replicate in their specific host bacteria, much of the rest of the world long discounted the work Gorski and his colleagues were doing on this topic.[124] However, the Polish studies and those of others now make it clear that some phages, including T4, have direct effects on mammalian cells. As Krystyna Dabrowska of the Hirszfeld Institute writes:

> Mammals have become 'an environment' for enterobacterial phage life cycles. Therefore, it could be expected that bacteriophage adapt to them. (T4's)

gpHoc, not found in most related phages, seems to have significance neither for phage particle structure nor for phage antibacterial activity. ...But the rules of evolution make it improbable that gpHoc really has no function ... More interesting is the apparent eukaryotic origin of gpHoc: a strong resemblance to immunoglobulin-like proteins appears likely to reflect their evolutionary relation. Substantial differences in biological activity between T4 and a mutant that lacks gpHoc were observed in a mammalian system. Hoc protein seems to be one of the molecules predicted to interact with mammalian organisms and/or modulate these interactions.[125]

Further supporting this perspective is the fact that the Hoc protein binds in the center of each cluster of capsid proteins of all of the T4-like phages that infect *E. coli* and related animal pathogens, but not on other T4-related phages such as those infecting marine cyanobacteria.

Barr and colleagues[126] have identified a very interesting additional and potentially related role for this "nonessential" T4 Hoc protein. They showed that Hoc mediates interaction of the phage head with the mucus layer on various animal tissues, an interaction which leaves the phage tails sticking out, ready to interact with incoming bacteria. It appears that the phage help protect the underlying tissues from bacterial attack while benefiting from the rich source of bacterial hosts to be found in mucus layers.

A Possible Path Forward for Phage Therapy

Phage therapy can be very effective in certain conditions and has some unique advantages over antibiotics. With the increasing incidence of antibiotic-resistant bacteria and a deficit in the development of new classes of antibiotics to counteract them, there is a need to investigate the use of phage in a range of infections.[3]

Clearly, the time has come to re-evaluate the potential of phage therapy, both by supporting new research and by evaluating the research already available.[5] Earlier research that deserves much more attention includes the extensive use of phage to very successfully treat typhoid fever over a number of years at a major hospital in Los Angeles in the 1940s[47] (which only came to light this century), as well as one in Montreal and in earlier extensive French and Polish work, and applications in the former Soviet Union.[6,7,12,14]

Many aspects of working therapeutically with phage are extensively discussed by both the Polish and Georgian groups and a number of others in the book *Phage Therapy: Current Research and Applications,* edited by J. Borysowski, R Miedzybrodzki, and A. Gorski.[127] Much more is being learned about various methods of phage delivery, interactions of phage in animal husbandry applications, and even applications involving intracellular pathogens.[128] The special complexities of the pharmacology of phage therapy are being more thoroughly explored.[129] Also, exciting new work is happening engineering phages for a variety of diagnostic and therapeutic applications.[130]

Phages are quite specific as to the bacteria they infect, and the stipulations of Ackermann & DuBow[30] are important here. The specificity of phages means that:

Phages have to be tested against the patient's bacteria just as antibiotics [should be], and the indications have to be right, but this holds everywhere in medicine. However, phage therapy requires the creation of phage banks and a close collaboration between the clinician and the laboratory. Phages have at least one advantage.... While the concentration of antibiotics decreases from the moment of application, phage numbers increase whenever their targeted bacteria are present. Another advantage is that phages are able to spread and thus prevent disease. Nonetheless, much research remains to be done... on the stability of therapeutic preparations; clearance of phages from blood and tissues; their multiplication in the human body; inactivation by antibodies, serum or pus; and the release of bacterial endotoxins by lysis.[30]

Furthermore, compassionate use of appropriate phages seems warranted in cases where bacteria resistant against all available antibiotics cause life-threatening illness.[131] They are especially useful for treating recalcitrant nosocomial infections, where large numbers of vulnerable people are exposed to the same strains of bacteria in a closed hospital setting. In this case, the environment as well as the patients can be treated with phage. Multiple case studies published in recent years illustrate the widespread interest of physicians and researchers in using phage in a clinical setting, especially for patients who have failed antibiotic therapy.

One of the first publications on intravenous phage therapy for compassionate use is described by Belgian physicians at the Queen Astrid Military Hospital in Brussels. Jennes et al. reported the first contemporary intravenous use of bacteriophage monotherapy against multidrug resistant *P. aeruginosa* in septicemia.[132] This was a case of a hospitalized Belgian patient who, after a three-month complicated hospital course was transferred to the Queen Astrid Military Hospital to receive wound care for his infected pressure sores. During his stay, he developed a *P. aeruginosa* infection sensitive only to Colistin. Intravenous Colistin was started, but because of a rapid decline in kidney function, it was stopped after 10 days. The *P. aeruginosa* septicemia soon returned, and the patient went into a coma. At that time, the decision was made to use a purified bacteriophage preparation[93] to which the *Pseudomonas* strain had *in vitro* sensitivity. Daily six-hour IV infusions were performed over a 10-day period, and the same preparation was also applied to his wounds. During that time, the patient's body temperature and C-reactive protein levels normalized, and blood cultures came back sterile. The patient's kidney function was restored within a few days. Bacteriophage treatment was thus seen to be safe and effective in this case.[132] However, it should be noted that the infected wounds persisted significantly even after phage treatment in this case and there were several episodes of sepsis involving other bacteria, which were all treated with empirical antibiotic therapy since no appropriate phage were available. Unfortunately, four months later, the patient developed *Klebsiella* sepsis and died due to sudden cardiac arrest.[132]

Intravenous Bacteriophage for Prosthetic Valve Endocarditis

Gilbey and Ho et. al describe a case of a 65-year-old male patient with a staphylococcal prosthetic valve endocarditis.[133] The patient was given a 14-day course of intravenous phage therapy as an adjunct to antibiotic therapy. Shortly after commencement of phage therapy, the patient's blood cultures were negative. Within 24 hours, the patient's white blood cell count, body temperature and CRP were significantly reduced. No adverse reactions from the treatment were noted.[133]

Phage Against MDR Carbapenemase-Producing *Klebsiella Pneumoniae*

Physician-initiated advocacy for patients with antibiotic resistant infections is becoming more common. The Eliava Clinic was contacted regarding a 58-year-old woman with frequent hospitalization due to MDR carbapenemase-producing *Klebsiella pneumoniae* (Kp) by her physician, infectious disease specialist Dr. Mario Corbellino. The patient, a 58-year-old woman with Crohn's disease diagnosed at the age of 20, had a history of seven major surgical interventions, with clinical remission after Adalimumab therapy (2009–2015). From 2011 to 2017, she had 30 hospital admissions for recurrent bilateral renal calculosis, obstructive nephropathy and urinary tract infections complicated by massive fibrosis of the bladder (residual volume: 25 mL), with the creation of a left cutaneous ureterostomy and positioning of a ureteral stent. She also had multisite colonization of the gastrointestinal tract and urine and a ureteral stent of 1-month duration with an MDR *Kp* strain.[134]

The infecting strain was *Kp* Sequence Type ST307. It was resistant to all ß-lactams, including carbapenems and produced the KPC-3 carbapenemase, remaining department susceptible only to ceftazidime-avibactam (CZA), amikacin, fosfomycin, and tigecycline. The patient was admitted for sepsis; despite last resort antibiotic therapy, the proximal tip of the ureteral stent remained positive for Kp culture. Reappearance of the microorganism was observed in cultures of the urine, the ureteral stent, and the rectal swabs of the patient a few days after interruption of the antibiotic therapy. Patient samples had been sent to the Eliava Institute in the meantime and a custom phage preparation against Klebsiella was prepared. The phage is a 161,728 bp Myoviridae, which was sequenced in collaboration with the Medical and Molecular Microbiology, Faculty of Science and Medicine group at the University of Fribourg. They found no virulance or antibioitic resistance or integrase sequences in the genome of the phage using *in-silico* analysis.[134] A 10 mL dose of phage preparation *bid* (twice a day) was administered on an empty stomach, preceded by alkalinization of the gastric pH with the use of carbonated water. Intrarectal administration *via* suppository for the first 14 days was also carried out. The phage therapy did not result in any untoward effects. After the stent substitution and 14 days of phage treatment, both stent and rectal surveillance swabs were negative. Molecular screening for Carbapenemase genes since then were also negative (Cepheid Gene Xpert, carba R).[134] The patient has not had another case of Klebsiella infection in two years and was hospitalized only twice since for UTI involving other organisms, compared to more than 30 times prior to phage therapy (Hoyle, Aug. 2019, personal communication).

Phage Therapy in a Renal Transplant Patient

Dutch physicians reported a case study of a renal transplant patient with ESBL-*Klebsiella* UTI and epididymitis that was eradicated using a custom Eliava phage preparation of a lytic bacteriophage to which the patient's isolate was sensitive. Kuipers et al. reported that the patient underwent more than seven treatment courses of meropenem, fosfomycin, and intravesical amikacin without success. A combination of meropenem and 8 weeks of custom oral and intravesical application of anti-*Klebsiella* phage therapy successfully and permanently eradicated the infection.[135]

Refractory *Pseudomonas* UTI

Khawaldeh et al. discuss a successful case of treatment of a *P. aeruginosa* urinary tract infection, using combined phage-antibiotic therapy.[102] A 67-year-old woman developed a refractory UTI following a bilateral ureteric stent placement that was resistant to multiple courses of gentamicin, ceftazidime, ciprofloxacin, and meropenem. Several isolates of bacterial strains were screened against the bacteriophage library at the Eliava Institute in Tbilisi and six lytic bacteriophages were then combined into a filter-sterilized phage product. Two antibiotics—colistin and meropenem—were then used in combination with bacteriophage therapy. The therapy resulted in symptomatic relief as well as microbiological cure.[102] Importantly, a year posttreatment, the urine cultures were still negative.

Phage Therapy for Netherton's Syndrome

A very interesting and rare case was treated at the Eliava Phage Therapy Center and has been under continued periodic observation led by Dr. Pikria Zhvania since 2016.[136] Netherton's syndrome is a genetic disease that has a hallmark triad of ichthyosis, allergy, and chronic bacterial infections. The patient presented as a 16-year-old boy from France with a significant history of strong allergy to all groups of antibiotics and frequent hospitalizations due to *Staphylococcal* sepsis.[136] After initial phage treatment, the patient's skin began to restore, and most notably, the patient did not have any recurrences of sepsis for several years. He continues to use phage prophylactically both orally and topically, and periodically visits the clinic in Tbilisi for evaluation. As the patient has been taking phage steadily for several years, his serum was studied for antiphage antibodies. Neutralizing antibodies were found, but do not seem to interfere with the patient's continued clinical improvement, indicating phage efficacy continues despite antibody development (Hoyle, unpublished data). This phenomenon has also been confirmed by Lusiak-Szelachowska et al., who demonstrated that the level of antiphage antibodies does not correlate with a negative clinical outcome.[137]

Cystic Fibrosis and Bacteriophage Therapy

Cystic fibrosis has been an area of particular interest and continued research for phage therapy since the late 1990s, likely due to the combination of frequent respiratory infections, need for constant chronic antibiotic use and impaired mucous clearance.

Numerous international as well as Georgian cystic fibrosis patients have been treated at the Eliava Institute and its Phage Therapy Center since it opened in 2011.[15,138] Some of the first patients treated there were from the National CF center, where eight patients ranging from infancy to adulthood were treated using inhaled phage in combination with antibiotics and other standard CF care.[13] The bacterial titers in sputum samples of these patients reduced significantly. This work was part of a collaboration with the Evergreen State College phage lab, which acquired a collection of 200 CF bacterial strains from the Seattle Children's Hospital in 1995. Through the years, undergraduate students have participated in phage research learning to isolate phage from sewage samples and further characterize these phages in the lab. The student-generated Pseudomonas phage collection against these isolates has grown to over 100 phages, which have interesting host range patterns against the CF collection. A set of six of these phages, representing all of the lytic Pseudomonas lytic phage types that have been described, were gradually tested against all 200 of the Children's Hospital strains, giving many different infectivity patterns, and have since been shared with a number of labs in various parts of the world. The most widely used are PEV2 and PEV40, which have for a number of years (and papers) been the most successful in a series of Australian pharmacology studies focused on optimizing phage type, dosage, and delivery system details for the treatment of cystic fibrosis and lung infections.[139]

Phage are also useful for a variety of lung infections. One particularly interesting case was of a 17-year-old girl with a multiantibiotic resistant *Achromobacter xylosoxidans* infection.[140] On presentation, the patient had dyspnea, cough, and generalized weakness. Well-characterized phage against *Achromobacter* were obtained from the Liebniz Institute DSMZ in Germany under a collaboration with the Eliava Institute and were prepared as a therapeutic preparation for the patient. The patient carried out phage therapy courses of treatment four times in one year, showing gradual improvement in both symptoms and lung function. Sputum cultures showed a two-fold reduction in bacterial titers over time.[140]

So far, most Western use of phage for treating CF and other lung patients has been as a rather last-minute endeavor when all else has failed. However, a group at the Yale School of Medicine have started treating CF patients with Pseudomonas infections with phage inhalation therapy. They have treated 13 patients with phage as a personalized medicine under the emergent indication mechanism at the FDA. Their largely successful research has been much discussed in a range of venues but has not been published in a scientific journal to date.[141]

Intravenous Phage Therapy in CF

Law et al. argue about the importance of bacteriophage therapy for multidrug-resistant *P. aeruginosa* infection in cystic fibrosis patients.[142] The researchers have successfully treated *P. aeruginosa* caused pneumonia in a 26-year-old patient awaiting lung transplantation surgery. Importantly, prior to receiving phage therapy, the patient had developed colistin-induced renal failure, further narrowing the possible therapeutic options. Intravenous administration of phage therapy was successful in resolving the bacterial infection and the patient has not had a recurrence of pseudomonal pneumonia and CF exacerbation within 100 days following the end of the bacteriophage therapy.

Novel Synthetic Bacteriophage Treatment Against *Mycobacteria abscessus*

We have refrained from including information about genetically engineered bacteriophages, as the topic area is significantly different from the use of natural bacteriophages traditionally used for therapeutic purposes. However, we must make an exception to note the groundbreaking case of a cystic fibrosis patient treated with phage against intracellular *Mycobacteria abscessus*, a nontuberculosis mycobacteria (NTM). *Mycobacteria* has been known to be a significant cause of morbidity and mortality.[143] A 15-year-old patient was the first patient to be treated with bacteriophage against a mycobacterial infection, and reflected the first use of genetically engineered phages, as described by Dedrick and colleagues.[144] The patient had a double lung transplant and was on immunosuppressive therapy. She had taken anti-NTM treatment for 8 years prior to transplant. One week after discontinuing postoperative antibiotic therapy, signs of the returning Mycobacterial infection were evident, including multiple skin lesions. The patient was sent home for palliative care after 7 months of treatment, with a diagnosis of disseminated mycobacterial infection. A three-phage cocktail was developed and was applied both topically and intravenously. The patient responded well to the therapy and had gradual improvement of both lung function and the skin lesions. No *Mycobacteria* were detected in serum or sputum after initiation of phage therapy.[144]

German CF Trial Expected

A clinical trial on the safety of phage preparations against *P. aeruginosa* in chronically infected CF and non-CF bronchiectasis patients by inhalation is to be conducted at the Charité University Hospital in Berlin (http://phage4cure.de/en/). The trial will also focus on the establishment of a GMP purification process for German phage production in the future.[145]

Many groups are interested in the idea of phage therapy for cystic fibrosis-associated infections as reflected by many review articles on the topic.[146–150] Further advancement in this area would greatly benefit the patient population.

Pharmacology and Delivery Technology

Much work and advancement in therapeutic phage pharmacology has occurred in recent years. Four major areas of focus are reflected in the literature: (1) high quality intravenous phage preparations, both phage cocktails and monophage; (2) Inhalation phage, both powder and liquid forms, spray; and (3) Encapsulation delivery. These modes of delivery arc in addition to already well-used applications, such as oral, topical, and rectal, including wound care. During the development of a new pharmaceutical, attention would be paid to things like development of a delivery vehicle and optimization of delivery. As phage are a long-standing medicine, many studies are using the traditional formulas to conduct clinical studies. As we saw in the

Phagoburn study, adequate manufacturing process is critical in receiving a preparation that will sufficiently represent its therapeutic potential.[106] Malik et al. review the current work for phage as a pharmaceutical, highlighting encapsulation of phage into liposomes, which gives the phage preparation added stability, and allows for extended release and intracellular distribution.[151]

Preparatory work in terms of optimizing dosage and delivery has been extensively studied by Dr. Kim Chan's group at the University of Sydney. They have investigated stability and efficacy of spray dried inhalable phage powders using Kutter-lab Pseudomonas phages, particularly PEV2, a podophage.[139,152] Leung et al. also investigated structural effects of different Kutter-lab lytic-tailed phages after jet nebulization, noting that titer loss after nebulization was associated with phage-tail length.[153] Encapsulation methods for inhalation phage are also being investigated.[154]

The complexities of phage selection and designing appropriate phage cocktails are becoming better understood.[155] With the increasing availability and decreasing costs of genomic analysis, it is now generally expected that phage for therapeutic cocktails be sequenced, enabling the exclusion of temperate phages and/or phages likely to carry or acquire genes related to lysogeny, pathogenicity, or toxin production. This is now standard for therapeutic phages being developed in the West. The application of metagenomic analyses to longstanding complex therapeutic cocktails was illustrated above for the Russian phage cocktail being used as a control in the Nestlé study.[96] This awareness also is being applied to some of the Georgian phage cocktails through collaborations with scientists in a variety of countries as well as access to a new sequencing center at the Georgian Center for Disease Control. Metagenomic analyses of several of the key broad Georgian and Russian cocktails have found them to contain no potentially undesirable/harmful genes.[156–158] This is an important step in considering their importation for oral, topical, and surgical use in the Western world, taking advantage of the very extensive clinical experience with these phage preparations. The availability of a cocktail of eight sequenced Intralytix phages clearly played a role in the FDA's approval of the physician-initiated clinical trial of phage targeting bacteria in leg ulcers.[22] This was an important beginning in terms of phage being approved for safety studies, but this narrow cocktail was reportedly much less successful therapeutically than the complex cocktails obtained earlier from Tbilisi for extensive compassionate use applications there at the Wound Care Center in Lubbock, Texas.

Bacteriophage Therapy in Gastrointestinal Health

Many studies have been conducted in recent years on the microbiome. As we learned earlier from Brüssow's work discussed earlier, there is much to discover in terms of etiology and pathogenicity for various gastrointestinal disorders. Much of the progress that has been made in the study of the microbiome has come through metagenomics, which has provided much insight into the extent of diversity in the human gut. Phage scientists have started looking seriously at the virome of the distal GI tract, which, like all other phage ecology, reflects the diversity of the microbiome. Shkoporov and Hill review recent advances in the study of gut microbiota and note the importance of the phageome. There are

estimated to be more than 10^{12} phage in the human gut.[159] Much of the viral taxonomy is unknown and the term "viral dark matter" has been coined to illustrate how much is left to be studied in this area.[160] Phages have been used for intestinal disorders virtually since their discovery, although understanding of some of the mechanisms for their efficacy is only now coming to light. Research in the areas of immunity and the role of the bacteria, viruses, and fungi is becoming better understood.[161–164] Fecal transplant quickly became an area of interest for physicians and phage enthusiasts, who were delighted at the work of Ott et al. demonstrating that sterile filtrated fecal transfer was effective in treating *Clostridia difficile* infection.[165] From this, it was determined that bacteriophages present in the filtrate were responsible for the inhibition of pathobionts.

Initial isolation of the needed phages to infect this *anaerobic* host was particularly challenging; virtually all the phage therapy-related work through the years has been carried out aerobically. The group at the University of Leicester lead by Dr. Martha Clokie has managed to isolate some potentially useful phages, largely from the environment, and has developed efficient *in vitro*[166] and *in vivo*[167] models for studying these infections in conditions similar to those in human infection. Movement toward a clinical trial in this area is much awaited, as it should have a significant impact on morbidity and mortality for those infected with *C. difficile*.

American Case Studies

A few American physicians and researchers are already employing phage therapy on a compassionate use basis. A very well-publicized case is that of an American UC San Diego professor who acquired a resistant *Acinetobacter* infection while on holiday in Egypt. The spell-binding book about their experiences by Strathdee and Patterson, called The Perfect Predator,[168] is on the 2019 best-seller lists—and Dr. Schooley et al.[169] describe how this 68-year-old patient of his presented over months with necrotizing pancreatitis and MDR *Acinetobacter* pseudocyst. Despite multiple courses of antibiotic therapy and drainage, his condition continued to worsen, and he was near death in a coma for several months before Strathdee came across a paper describing the strange idea of using bacteriophage to treat *Acinetobacter*. This reminded her of explorations with coliphage in a lab back in college, and she dove intensely into exploring this possible option. Virtually all of the phage therapy papers dealt with aerobic applications, but she fairly quickly found a few mentioning work with *Acinetobacter* and reached out to the researchers. Mobilizing all her many contacts, *Acinetobacter* phage isolates to which his strain was found to be sensitive were identified from two separate laboratories, some from a U.S. Navy project that had been precipitated by extensive *Acinetobacter* infections seen in soldiers fighting in Iraq, the other in Texas but originating from Intralytix in Australia. With Herculean efforts from many people, these phages were highly purified within days and administered both IV and percutaneously into the abscess cavity. Phage pharmacokinetics and antiphage bacterial mutants were studied in detail as the patient's condition quickly improved. The success of this case gave impetus to introduction of finding more common phages over the next 2 years to treat five organ-transplant candidates who were fighting infections when an appropriate organ had been found. Its success led them to establish the first U.S. phage therapy research center there at UC San Diego—a very exciting step forward.

Phage Therapy of Aortic Graft

Yale University has also been involved in work to make phage therapy accessible for emergency cases in the United States, including both a variety of unusual surgical cases and treatment of CF patients presenting for lung transplant. For example, Chan et al. have published a case study of a 76-year old man who was treated for a major *P. aeruginosa* infection following aortic arch replacement surgery with a Dacron graft for an aortic aneurysm.[170] At presentation, the patient complained of increased serosanguinous drainage from his mediastinal fistula, which despite multiple antibiotic treatments, did not resolve. Initially, the team attempted to administer a combination of phage OMKO1 and ceftazidime by needle puncture into the perigraft collection. However, due to heavy scarring in this area, the combination was delivered into the mediastinal fistula, which was in continuity with the perigraft collection.[170] The patient has not had any evidence of recurrent infections, despite being off of antibiotic therapy. This study highlights the efficacy of the synergic use of antibiotics and bacteriophages, as well as the capability of phage therapy in resolving biofilm-associated infections.

Craniotomy Complicated by an MDR *Acinetobacter baumannii* Infection

LaVergne et al.[171] present a case study of a 77-year-old man who underwent craniotomy complicated by postoperative infection with cerebritis and a subdural and epidural empyema. Intraoperative cultures grew MDR *Acinetobacter baumannii*. Despite a combination therapy of Colistin, Azithromycin, and Rifampin, the patient's clinical condition did not improve. The phage cocktail was selected using a library with 104 *A. baumannii* bacteriophages, screened for activity against his *A. baumannii* strain cultured intraoperatively. The patient received 8 days of phage therapy, but due to the severity of his condition, the patient's family decided to withdraw care.[171] The researchers were not allowed to deliver phages via an intrathecal route, which would allow phages to enter the CSF. It can be argued that the phage titers were not high enough to have reached an effective therapeutic dose through IV delivery.

Scientists such as Brüssow[172] and Knoll Mylonakis[173] are very thoughtfully exploring what steps are most urgently needed for human phage therapy to become a reality in Western medicine. They argue persuasively that industry alone cannot carry out the studies necessary for approval under our current regulatory system, which is designed for nonbiologics, and suggest that a much broader partnership is required to evaluate and implement the promise of phage therapy. We strongly agree with Brüssow:[172]

> Organizations like the US or European CDC should establish phage collections for antibiotic-resistant pathogens which soon are likely to become untreatable in a sizable number of patients. The CDC should characterize, amplify and purify the relevant phages

> according to safety standards agreed with the FDA .... Until the value of PT is confirmed, it is also desirable that (such) international organizations supervise the controlled clinical trials ... the first trials should target medical priorities. The lower economic prospects in treating these infections will necessitate that the first phase is paid by public money ... It is only once the feasibility, regulatory, and legal parameters have been set that one might expect private companies to take the lead ... for a variety of bacterial infections ... we cannot simply wait another 10 years until non-coordinated PT approaches yield reliable answers...Even when it should turn out that PT does not work for a particular application, knowing this is better than cultivating the illusion that we have a secret weapon in our tool box against antibiotic-resistant infections ...

Diabetic Foot Ulcers as an Early Target

Given the body of existing documentation, broad phage host range, and the small number of phage types involved, phage targeting *S. aureus* to treat diabetic ulcers and MRSA are among the most obvious candidates for this approach. (It should help that this would also be among the hardest applications to patent, with so much use of staph phage in the literature.)

Podiatrist Randy Fish[174] documents the use of the Eliava commercial staph phage cocktail to treat this series of eight patients with diabetic toe ulcers where all else had failed, including extensive IV antibiotics as well as local antibiotics, and the only other option was amputation. There had been no plan to publish this data—just to save these patients from amputation, which at least half the time leads to death within a few years. Dr. Fish had worked in the phage lab while an Evergreen student in 1975, tracked us down a few years earlier when he heard that the phage lab was still open and we had become interested in therapeutic applications of phage, and became a member of the Board of our nonprofit Phagebiotics Research Foundation, complementing well the other members of the board who included an medical doctor, veterinarian, and the head microbiologist of the Northwest Indian Fisheries Commission. A poster on the work was selected for top honors at both the 2014 Academy of Physicians and Wound Healing Scientific Conference in Philadelphia and the Desert Foot conference that year in Phoenix, Arizona, drawing much interest. This work is particularly significant in that it clearly recorded the regeneration of bone in the weeks after phage therapy and has long-term follow up showing no recurrence in treated patients. Further successful follow-up work was reported by Fish et al.[175,176] This is clearly an area that should be explored expeditiously as there is a large unmet need and could significantly reduce morbidity and mortality for diabetic patients.

It is clear that collaboration among governmental entities/ health agencies, doctors and patients, and academia and industry is crucial to expeditiously bring phage therapy to fruition in the West. One major effort in that regard was the Phagoburn project (discussed previously in this chapter) that targeted *Pseudomonas* and *E. coli* in burn wounds. This effort has involved two companies, several other institutions, and burn units in seven hospitals in Belgium, France, and Switzerland. The French military, a very active phage-oriented patient support group, and €3.8 million in funding from the EU also play important roles.

A Belgian/French/German nonprofit group called P.H.A.G.E. (Phages for Human Application Group Europe) has put in years of crucial educational and political advocacy work related to the regulatory challenges[177] and the experience of a related small-scale phase I clinical trial at the Queen Astrid Military Hospital in Brussels—one of the participants—all have helped the cause, but progress is still slow. Another key area is the education of doctors and potential patients about the possibilities of phage therapy, as started to happen more than two decades ago with BBC and CBS documentaries, but such exposure has bred frustration and dashed hopes when the only option to obtain phage therapy is medical tourism in Georgia or Poland. This has just further emphasized the importance of establishing clearer and faster paths forward for at least one or two clinical phage applications in the West.

Important gains were made in Belgium by the official approval of a magisterial product designation for bacteriophages, which allows phage to be prescribed by physicians after testing samples from the patient against an appropriate library of very well-characterized phages that is being, and then prepared for administration on an individual basis in a compounding pharmacy." In addition, nonauthorized ingredients may also be used in magistral preparations, providing that they are accompanied by a certificate of analysis issued by a Belgian Approved Laboratory. The so-called "Belgian Approved Laboratories" are quality control laboratories that are granted an accreditation by the Belgian regulatory authorities. This status allows them to perform the batch release testing of medicinal products.[178] The hurdle for this approach is to obtain an adequate phage bank that provides the active pharmaceutical ingredient (API) necessary to make the phage preparation.

To move the field of phage therapy forward as a viable international component of the war against antibiotic resistance, there is a pressing need for an unequivocal success leading to at least one approval by the FDA and/or the parallel EU regulatory body. For various reasons, phages targeting *S. aureus* in nosocomial hospital MRSA infections, wounds and/or diabetic ulcers offer the most clear-cut route to reaching this goal in timely fashion. This is most likely to be accomplished through collaborative public/private efforts leading to a product for external applications that will be widely available as a public good. First, phage therapy of staphylococcal infections is the application for which the most data is available based on work in various countries over the last 70 years. This also makes it one of the most difficult for which to get patent protection. At the same time, it is the application requiring the smallest number of individual phages due to the very broad host ranges of these particular phages and well-characterized, satisfactory phages are already available commercially or in the public domain. Some compassionate-care case studies are already being carried out in the United States using the DNA-sequenced, highly purified staph phage that is approved and commercially available in Tbilisi. It is simply added to standard treatments of, for example, diabetic foot ulcers that so often lead to amputation, but accomplishing that is challenging when the only way to access phage therapy is to travel to Poland or Georgia. Furthermore, staphylococcal phage therapy is straightforward enough that the requisite supportive infrastructure could easily be built. Learning to work effectively

with these phages could help educate both the community and the medical community as to the potential benefits of phage therapy. Expanding this approach would help to deal with a major conundrum: much work is needed to raise the awareness of phage therapy among physicians and prospective patients and to determine the best conditions for definitive clinical trials. It could also lay the groundwork for developing the infrastructure for characterizing and treating more complex infections. Beyond the well-known targets, there is a wide range of possible new applications that would be more patentable. One already hears about phage targeting the etiological agents of periodontal disease, urinary tract infections, and prostatitis, the *Propionibacteria* that cause acne, and *C. difficile*, among others.

Support for phage therapy is evidenced by the active work from all corners of the world to utilize this treatment in patients via compassionate use, and in the multiple clinical trials being carried out that are necessary for broader access. Numerous reviews call for a closer look and highlight much of what we already know.[179–182] All in all, we need to somehow find our way through the regulatory labyrinth and the financial hurdles and further explore the possibilities of having phage help rescue us from the growing antibiotic resistance crisis and the health challenges resulting from excessive antibiotic use.

REFERENCES

1. Levin B and Bull JJ (1996). Phage therapy revisited: the population biology of a bacterial infection and its treatment with bacteriophage and antibiotics. *Am Nat* 147: 881–898.
2. Radetsky P (1996). The good virus. *Discover* 17: 50–58.
3. Barrow PA and Soothill JS. (1997). Bacteriophage therapy and prophylaxis: rediscovery and renewed assessment of the potential. *Trends Microbiol* 5: 268–271.
4. Alisky J, Iczkonski K., Rapoport A and Troitsky N. (1998). Bacteriophage shows promise as antimicrobial agents. *J Infection* 36: 5–13.
5. Sulakvelidze A., Alavidze Z and Morris J. (2001). Bacteriophage therapy. *Antimicrobial Agents Chemother* 45: 649–659.
6. Summers WC. (1999) *Felix d'Herelle and the Origins of Molecular Biology.* Yale University Press, New Haven, CT.
7. Kutter E and Sulakvelidze A. (2005) *Bacteriophages: Biology and Applications.* CRC Press, Boca Raton, FL.
8. Merril CR, Scholl D and Adhya SL. (2003). The prospect for bacteriophage therapy in Western medicine. *Nat Rev Drug Discov* 2: 489–497.
9. Bruessow H. (2007) Phage Therapy: The Western Perspective. In: *Bacteriophage: Genetics and Molecular Biology.* S McGrath and D van Sinderen, editors. Caister Academic Press, Norfolk, UK, pp. 159–192.
10. Gorski A, Borysowski J, Miedzybrodzki R. and Weber-Dabrowska B. (2007). Bacteriophages in Medicine. In: *Bacteriophage: Genetics and Molecular Biology.* S McGrath and D van Sinderen, editors, Caister Academic Press, Norfolk, UK, pp. 125–158.
11. Hausler T. (2006). *Viruses vs. Superbugs: A Solution to the Antibiotics Crisis?* Macmillan, New York, NY.
12. Kutter E, De Vos D, Gvasalia G, Alavidze Z, Gogokhia L, Kuhl S and Abedon ST. (2010). Phage therapy in clinical practice: treatment of human infections. *Curr Pharm Biotechnol* 11: 69–86.

13. Kutateladze M and Adamia R. (2010). Bacteriophages as potential new therapeutics to replace or supplement antibiotics. *Trends Biotechnol* 28, 591–595.

14. Abedon S, Kuhl S, Blasdel B and Kutter E. (2011). Phage treatment of human infections. *Bacteriophage* 1: 66–85.

15. Kutter E, Borysowski J, Międzybrodzki R, Górski A, Weber-Dabrowska B, Kutateladze M, Alavidze Z and Goderdzishvili M. (2014). *Phage Therapy: Current Research and Applications.* J. Borysowski and R. Miedzybrodzki, editors. Caister Academic Press, Norfolk, UK.

16. Kuchment A. (2012). *The Forgotten Cure: The Past and Future of Phage Therapy.* Copernicus Books, New York.

17. Reardon S. (2014). Phage therapy gets revitalized. *Nature* 510: 15–16.

18. O'Flaherty S, Ross RP and Coffey A. (2009). Bacteria and their lysins for elimination of infectious bacteria. *FEMS Microbiol Rev* 33: 801–819.

19. Entenza JM, Loeffler JM, Grandgirard D., Fischetti VA and Moreillon P. (2005). Therapeutic effects of bacteriophage CPL-1 lysin against streptococcus pneumoniae endocarditis in rats. *Antimicrob Agents Chemother* 49: 4789–4792.

20. Nelson D, Loomis L and Fischetti VA. (2001). Prevention and elimination of upper respiratory colonization of mice by group A streptococci by using a bacteriophage lytic enzyme. *Proc Natl Acad Sci USA* 98: 4107–4112.

21. Borysowski J, Międzybrodzki R and Gorski A. (2014). *Phage Therapy: Current Research and Applications.* J. Borysowski and R. Miedzybrodzki, editors. Caister Academic Press, Norfolk, UK.

22. Rhoads DD, Wolcott RD, Kuskowski MA, Wolcott BM, Ward LS and Sulakvelidze A. (2009). Bacterophage therapy of venous leg ulcers in humans: results of a phase I safety trial. *Journal of Wound Care* 18: 237–243.

23. Wright A., Hawkins CH, Anggård EE and Harper DR. (2009). A controlled clinical trial of a therapeutic bacteriophage preparation in chronic otitis due to antibiotic-resistant *Pseudomonas aeruginosa*; a preliminary report of efficacy. *Clin. Otolaryngol.* 34: 349–57.

24. Międzybrodzki R, Borysowski J, Weber-Dąbrowska B, Fortuna W, Letkiewicz S, Szufnarowski K, Pawełczyk Z, Rogóż P, Kłak M, Wojtasik E and Górski A. (2012). Clinical aspects of phage therapy. *Adv Virus Res* 83: 73–121.

25. D'Herelle F. (Smith GH, translation). (1922) *The Bacteriophage: Its Role in Immunity.* Williams and Wilkins Company, Baltimore, MD.

26. Bruynoghe R and Maisin J. (1921). Essais de therapeutique au moyen du bacteriophage du staphylocoque. *C R Soc Biol* 85: 1120–1121.

27. D'Herelle F. (1930) *The Bacteriophage and its Clinical Applications.* CC Thomas, Springfield MA.

28. D'Herelle F. (1935). *The Bacteriophage and the Phenomenon of Recovery.* Tiflis University Press, Tiblisi, Georgia.

29. Saunders ME. (1994). Bacteriophages in Industrial Fermentations. In: *Encyclopedia of Virology.* RG Webster, and A Granoff, editors, Academic Press, San Diego, CA, pp. 116–121.

30. Ackermann HW and DuBow MS. (1987). Viruses of Prokaryotes I: General Properties of Bacteriophages, Ch. 7. *Practical Applications of Bacteriophages.* CRC Press, Boca Raton, FL.

31. Chanishvili N. (2012). *A Literature Review of the Practical Application of Bacteriophage Research.* Nova Science Publishers, New York, NY.

32. Schlesinger M. (1933). Reindarstellung eines Bakteriophagen in mit freiem Auge sichtbaren Mengen. *Biochem Z.* 264: 6.

33. Ruska H. (1940). Die Sichtbarmachung der Bakteriophagen Lyse im Übermikroskop. *Naturwissenschaften* 28: 45–46.

34. Pfankuch E and Kausche G. (1940). Isolierung u. Üebermikroskopische abbildung eines Bakteriophagen. *Naturwissenschaften* 28: 46.

35. Ellis EL and Delbrück M. (1939). The growth of bacteriophage. *J Gen Physiol* 22: 365–384.

36. Doermann AD. (1952). The intracellular growth of bacteriophages. I. Liberation of intracellular bacteriophage T4 by premature lysis with another phage or with cyanide. *J Gen Physiol* 35: 645–656.

37. Cairns J, Stent GS and Watson JD. (1966). *Phage and the Origins of Molecular Biology.* Cold Spring Harbor Laboratory Press, Cold Spring Harbor, NY.

38. Fischer E and Lipson C. (1988). *Thinking About Science: Max Delbrück and the Origins of Molecular Biology.* Norton, New York, NY.

39. Stent G. (1963). *Molecular Biology of Bacterial Viruses.* W. H. Freeman, San Francisco, CA, p. 8.

40. Appelmans R (1921). Le bacteriophage dans l'organisme. *C R Seances Soc Biol Fil* 85: 722–724.

41. Evans AC. (1933). Inactivation of Antistreptococcus Bacteriophage by Animal Fluids. *Pub Health Rep* 48: 411–446.

42. Nungester WJ and Watrous R.M. (1934). Accumulation of bacteriophage in spleen and liver following its intravenous inoculation. *Proc Soc Exp Biol Med* 31: 901–905.

43. Geier MR, Trigg ME and Merril CR. (1973). Fate of bacteriophage lambda in non-immune germ-free mice. *Nature* 246: 221–223.

44. Dubos R, Straus JH and Pierce C. (1943). The multiplication of bacteriophage in vivo and its protective effects against an experimental infection with *Shigella dysenteriae. J Exp Med* 78: 161–168.

45. Morton HE and Engely FB. (1945). Dysentery bacteriophage: review of the literature on its prophylactic and therapeutic uses in man and in experimental infections in animals. *J Am Med Assoc* 127: 584–591.

46. Ward WE. (1943). Protective action of VI bacteriophage in eberthella typhi infections in mice. *J Infect Dis* 72: 172–176.

47. Knouf EG, Ward WE, Reichle PA, Bower AG and Hamilton PM. (1946). Treatment of typhoid fever with type specific bacteriophage. *J Am Med Assoc* 132: 134–138.

48. Desranleu, J-M. (1949). Progress in the treatment of typhoid fever with Vi phages. *Can J Public Health*, pp. 473–478.

49. McGrath S. and van Sinderen D., editors. (2007). *Bacteriophage Genetics and Molecular Biology.* Caister Academic Press, Norfolk, UK.

50. Vieu JF. (1975). Les bacteriophages. In: Fabre J, ed. *Traite de Therapeutique, Vol. Serums et vaccins.* Flammarion, Paris, pp. 337–340b.

51. Vieu J.F., Guillermet F, Minck R and Nicolle P. (1979). Données actuelles sur les applications therapeutiques des bacteriophages. *Bull Acad Natl Med* 163: 61–66.

52. Slopek S, Durlakova I, Weber-Dabrowska B, Kucharewicz-Krukowska A, Dabrowski M and Bisikiewicz R. (1983). Results of bacteriophage treatment of suppurative bacterial infections I. General evaluation of the results. *Arch Immunol Ther Exp* 31: 267–291.

53. Slopek S., Kucharewica-Krukowska A., Weber-Dabrowska B. and Dabrowski M. (1985). Results of bacteriophage treatment of suppurative bacterial infections VI. Analysis of treatment of suppurative staphylococcal infections. *Arch Immunol Ther Exp* 33: 261–273.

54. Slopek S, Weber-Dabrowska B, Dabrowski M and Kucharewica-Krukowska A. (1987). Results of bacteriophage treatment of suppurative bacterial infections in the years 1981–1986. *Arch Immunol Ther Exp* 35: 569–583.

55. Weber-Dabrowska B, Dabrowski M. and Slopek S. (1987). Studies on bacteriophage penetration in patients subjected to phage therapy. *Arch Immunol Ther Exp* 35: 363–368.

56. Smith H.W. and Huggins M.B. (1982). Successful treatment of experimental *E. coli* infections in mice using phage: its general superiority over antibiotics. *J Gen Microbiology* 128: 307–318.

57. Smith HW and Huggins MB. (1983). Effectiveness of phages in treating experimental *E. coli* diarrhoea in calves, piglets and lambs. *J Gen Microbiology* 129: 2659–2675.

58. Smith HW and Huggins MB. (1987). The control of experimental *E. coli* diarrhea in calves by means of bacteriophage. *J Gen Microbiology* 133: 1111–1126.

59. Smith HW, Huggins MB and Shaw KM. (1987). Factors influencing the survival and multiplication of bacteriophages in calves and in their environment. *J Gen Microbiology* 133: 1127–1135.

60. Weber-Dabrowska B, Mulczyk M and Gorski A. (2000). Bacteriophage therapy of bacterial infections: an update of our institute's experience. *Arch Immunol Ther Exp* 48: 547–551.

61. Weber-Dabrowska B, Zimecki M., Mulczyk M and Gorski A. (2002). Effect of phage therapy on the turnover and function of peripheral neutrophils. *FEMS Immunol Med Microbiol* 34: 135–138.

62. Międzybrodzki, Fortuna W, Weber-Dąbrowska B and Górski A. (2007). Phage therapy of staphylococcal infections (including MRSA) may be less expensive than antibiotic treatment. *Postepy Hig Med Dosw. (online)*, 61: 461–465, <www.phmd.pl> e-ISSN 1732-2693.

63. Górski A, Jończyk-Matysiak E, Łusiak-Szelachowska M, Międzybrodzki R, Weber-Dąbrowska B and Borysowski J. (2018). Phage therapy in allergic disorders?. *Exp Biol Med (Maywood)*; 243(6): 534–537. doi:10.1177/1535370218755658.

64. Górski A, Jończyk-Matysiak E, Międzybrodzki R, Weber-Dąbrowska B and Borysowski J. (2018). Phage transplantation in allotransplantation: possible treatment in graft-versus-host disease? *Front Immunol*; 9: 941.

65. Górski A, Jończyk-Matysiak E, Łusiak-Szelachowska M, Weber-Dąbrowska B, Międzybrodzki R and Borysowski J. (2018). Therapeutic potential of phages in autoimmune liver diseases. *Clin Exp Immunol* 192(1): 1–6.

66. Górski A, Dąbrowska K, Międzybrodzki R, et al. (2017). Phages and immunomodulation. *Future Microbiol* 12: 905–914.

67. Górski A, Międzybrodzki R, Łobocka M, et al. (2018). Phage therapy: what have we learned? *Viruses* 10(6):288.

68. Lobocka M, Hejnowicz MS, Dabrowski K, Gozdek A, Kosakowski J, Witkowska M, Ulatowska MI, Weber-Dabrowska B, Kwiatek M, Parasion S, Gawor J, Kosowska H and Glowacka A. (2012). Genomics of staphylococcal Twort-like phages—potential therapeutics of the post-antibiotic era. *Adv Virus Res* 83: 143–216.

69. Katsarava R., Beridze V., Arabuli N., Kharadze D., Chu C.C. and Won C.Y. (1999). Amino acid-based bioanalogous polymers. Synthesis, and study of regular poly(ester amide)s based on bis(α-amino acid) α,θ-alkylene diesters, and aliphatic dicarboxylic acids. *J Polymer Sci* 37: 391–407.

70. Markoishvili K., Tsitlanadze G, Katsarava R, Morris JG Jr. and Sulakvelidze A. (2002). A novel sustained-release matrix based on biodegradable poly(ester amide)s and impregnated with bacteriophages and an antibiotic shows promise in management of infected venous stasis ulcers and other poorly healing wounds. *Int J Dermatol* 41: 453–458.

71. Nogueira F, Karumidze N, Kusradze I, Goderdzishvili M, Teixeira P and Gouveia IC. (2017). Immobilization of bacteriophage in wound-dressing nanostructure. *Nanomedicine* 13(8):2475–2484.

72. Kvachadze L, Balarjishvili N, Meskhi T, Tevdoradze E, Skhirtladze N, Pataridze T, Adamia R, Topuria T, Kutter E, Rohde C and Kutateladze M. (2011). Evaluation of lytic activity of staphylococcal bacteriophage Sb-1 against freshly isolated clinical pathogens. *Microb Biotechnol* 4: 643–650.

73. Fekety R. (1995). Antibiotic-associated diarrhea and colitis. *Cur Opin Infect Dis* 8: 391–397.

74. Merrill C, Biswis B, Carlton R, Jensen NC, Creed GJ, Zullo S. and Adhya S. (1996). Long-circulating bacteriophages as antibacterial agents. *Proc Natl Acad Sci USA* 93: 3188–3192.

75. Greenwald D and Brandt L. (1997). Recognizing *E. coli* O157:H7 infection. *Hosp Pract (1995)* 32: 123–140.

76. Raya RR, Varey P, Oot RA, Dyen MR, Callaway T, Edrington T.S, Kutter E.M. and Brabban A. D. (2006). Isolation and characterization of a new t-even bacteriophage, CEV1, and determination of its potential to reduce *Escherichia coli* O157:H7 levels in sheep. *App Environ Microbiol* 72: 6405–6410.

77. Brabban AD, Nelson DA, Kutter E, Edrington T.S. and Callaway TR. (2004). Approaches to controlling *Escherichia coli* O157:H7, a food-borne pathogen and an emerging environmental hazard. *Environ Pract* 6: 208–229.

78. Raya RR, Oot RA, Moore-Maley B, Wieland S, Callaway TR, Kutter EM and Brabban AD. (2011). Naturally resident and exogenously applied T4-like and T5-like bacteriophages can reduce *Escherichia coli* O157:H7 Levels in Sheep Guts. *Bacteriophage* 1: 15–24.

79. Kutter E, Skutt-Kakaria K, Blasdel B, El-Shibiny A, Castano A, Bryan D, Kropinski A, Villegas A, Ackermann H-D, Toribio AL, Pickard D, Anany H, T Callaway and Brabban AD. (2011). Characterization of a ViI-like Phage Specific to *E. coli* O157. *Virology J* 8(430): 1–14.

80. Kutter E, Ketevan K, Gachechiladze K, Poglazov A, Marusich E, Shneider M, Aronsson P, Napuli A, Porter D and Mesyanzhinov V. (1995). Evolution of T4-related phages. *Virus Genes* 11: 285–297.

81. Ackermann H and Krisch H. (1997). A catalogue of T4-type bacteriophages. *Archi Virol* 142: 2329–2345.

82. Chibani-Chennoufi S, Sidoti J, Dillmann M-L, Bruttin A, Kutter E, Krisch H, Sarker S and Brüssow H. (2004). Isolation of T4-like Bacteriophages from the stool of pediatric diarrhea patients in Bangladesh. *J Bacteriol* 186: 8287–8294.

83. Chibani-Chennoufi S, Sidoti J, Bruttin A, Kutter E, Sarker S and Brüssow H. (2004). Comparison of in vitro and in vivo bacteriolytic activities of *Escherichia coli* phages: implications for phage therapy. *Antimicrob Agents Chemother* 48: 2558–2569.

84. Kutter E, Stidham T, Guttman B, Kutter E, Batts D, Peterson S, Djavachishvili T, Arisaka F, Mesyanzhinov V, Rüger W and Mosig G. (1994). Genomic Map of Bacteriophage T4. In: *Molecular Biology of Bacteriophage T4*. J.D. Karam, editor. American Society for Microbiology, Washington, DC, pp. 491–519.

85. Miller E, Kutter E, Mosig G, Arisaka F, Kunisawa T and Rueger W. (2003). Bacteriophage T4 Genome. *Microbiol Mol Biol Rev.* 67: 86–156.

86. Kutter E, Kellenberger E, Carlson K., Eddy S, Neitzel J, Messinger L, North J and Guttman B. (1994). Effects of Bacterial Growth Conditions and Physiology on T4 Infection. In: *Molecular Biology of Bacteriophage T4*. J.D. Karam, editor. American Society for Microbiology, Washington, DC, pp. 406–420.

87. Nolan JM, Petrov V, Bertrand C, Krisch HM and Karam JD. (2006). Genetic diversity among five T4-like bacteriophages. *Virol J.* 3: 1–31.

88. Henning U and Hashemolhosseini S. (1994). Receptor Recognition by T-Even-Type Coliphages. In: *Molecular Biology of Bacteriophage T4*. J.D. Karam, editor. American Society for Microbiology, Washington, DC, pp. 291–298.

89. Tetart F, Desplats C and Krisch HM. (1998). Genome plasticity in the distal tail fiber locus of the T-even bacteriophage: recombination between conserved motifs swaps adhesin specificity. *J Mol Biol* 282: 543–556.

90. Kutter E, Bryan D, Ray G, Brewster E, Blasdel B and Guttman B. (2018). From host to phage metabolism: hot tales of phage T4's takeover of *E. coli*. *Viruses* 10(7): 387.

91. Doermann AH. (1948). Lysis and lysis inhibition with *E. coli* bacteriophage. *J Bacteriol* 55: 257–275

92. Abedon S. (1994). Lysis and the Interaction Between Free Phages and Infected Cells. In: *Molecular Biology of Bacteriophage T4*. J.D. Karam, editor. American Society for Microbiology, Washington, DC, pp. 397–405.

93. Bryan D, El-Shibiny A, Hobbs Z, Porter J, Kutter EM. (2016). Bacteriophage T4 Infection of Stationary Phase E. coli: Life after Log from a Phage Perspective. *Front Microbiol* 7: 1391.

94. Kolter R (1992). Life and death in stationary phase. *ASM News* 58: 75–79.

95. Sarker SA, McCallin S, Barretto C, Berger B, Pittet AC, Sultana S, Krause L, Huq S, Bibiloni R, Bruttin A, Reuteler G and Brüssow H. (2012). Oral T4-like phage cocktail application to healthy adult volunteers from Bangladesh. *Virology* 434: 222–232.

96. McCallin S, Alam Sarker S, Barretto C, Sultana S, Berger B, Huq S, Krause L, Bibiloni R, Schmitt B, Reuteler G and Brüssow H. (2013). Safety analysis of a Russian phage cocktail: from metagenomic analysis to oral application in healthy human subjects. *Virology* 443: 187–96.

97. Sarker SA, Sultana S, Reuteler G, et al. (2016). Oral phage therapy of acute bacterial diarrhea with two coliphage preparations: a randomized trial in children from Bangladesh. *EBioMedicine* 4: 124–137.

98. Brüssow H. (2005). Phage therapy: the *Escherichia coli* experience. *Microbiology* 151(Pt 7): 2133–2140.

99. Sarker SA and Brüssow H. (2016). From bench to bed and back again: phage therapy of childhood *Escherichia coli* diarrhea. *Ann N Y Acad Sci* 1372(1): 42–52.

100 O'Flaherty S., Coffey A., Edwards R., Meaney W., Fitzgerald G.F. and Ross R.P. (2004). Genome of staphylococcal phage K: a new lineage of Myoviridae infecting gram-positive bacteria with a low G+C content. *J Bacteriol* 186: 2862–2871.

101. Kvachadze L, Balarjishvili N, Meskhi T, Tevdoradze E, Skhirtladze N, Pataridze T, Adamia R, Topuria T, Kutter E, Rohde C and Kutateladze M. (2011). Evaluation of lytic activity of staphylococcal bacteriophage Sb-1 against freshly isolated clinical pathogens. *Microb Biotechnol* 4: 643–650.

102. Khawaldeh A., Morales S., Dillon B., Alavidze Z., Ginn A.N., Thomas L., Chapman S.J., Dublanchet A., Smithyman A. and Iredell JR. (2011). Bacteriophage therapy for refractory Pseudomonas aeruginosa urinary tract infection. *J Med Microbiol* 60(Pt 11): 1697–1700.

103. Rhoads DD, Wolcott RD, Kuskowski MA, Wolcott BM, Ward LS and Sulakvelidze A. (2009). Bacterophage therapy of venous leg ulcers in humans: results of a phase I safety trial. *J Wound Care* 18(6): 237–243.

104. Merabishvili M, Pirnay J-P, Verbeken G, Chanishvili N, Tediashvili M, et al. (2009). Quality-controlled small-scale production of a well-defined bacteriophage cocktail for use in human clinical trials. *PLOS One* 4(3): e4944.

105. Pirnay J-P, Verbeken G, Rose T, Jennes S, Zizi M, Huys I, Lavigne R, Merabishvill M, Vaneechoute M, Buckling A and De Vos D. (2012). Introducing yesterday's phage therapy in today's medicine. *Future Virol* 7: 379–390.

106. Jault P, Leclerc T, Jennes S, et al. (2019). Efficacy and tolerability of a cocktail of bacteriophages to treat burn wounds infected by Pseudomonas aeruginosa (PhagoBurn): a randomised, controlled, double-blind phase 1/2 trial. *Lancet Infect Dis* 19(1): 35–45.

107. Pirnay JP, Verbeken G, Ceyssens PJ, et al. (2018). The magistral phage. *Viruses* 10(2):64.

108. Petrovic Fabijan A, Lin RCY, Ho J, et al. (2020). Safety of bacteriophage therapy in severe *Staphylococcus aureus* infection. *Nat Microbiol* 5(3): 465–472. (Correction published April 2020, Nat Microbiol 5(4): 652.)

109. Leitner L, Sybesma W, Chanishvili N, et al. (2017). Bacteriophages for treating urinary tract infections in patients undergoing transurethral resection of the prostate: a randomized, placebo-controlled, double-blind clinical trial. *BMC Urol* 17(1):90.

110. Leitner L, Kessler TM and Klumpp J. (2020). Bacteriophages: a panacea in neuro-urology? *Eur Urol Focus* 6(3): 518–521.

111. Bergh O, Borsheim KY, Bratbak G. et al. (1989). High abundance of viruses found in aquatic environments. *Nature* 340: 467–468.

112. Caldwell, JA. (1928). Bacteriologic and bacteriophagic study of infected urines. *J Infect Dis* 43: 353–362.

113. Yeung MK. and Kozelsky CS. (1997). Transfection of Actinomyces spp. by genomic DNA of bacteriophages from human dental plaque. *Plasmid* 37: 141–153.

114. Bachrach G, Leizerovici-Zigmond, M, Zlotkin A, et al. (2003). Bacteriophage isolation from human saliva. *Lett Appl Microbiol* 36: 50–53.

114. Merril CR, Friedman TB, Attallah AF, et al. (1972). Isolation of bacteriophages from commercial sera. *In Vitro* 8: 91–93.

116. Geier MR, Attallah AF and Merril CR (1975). Characterization of *Escherichia coli* bacterial viruses in commercial sera. *In Vitro* 11: 55–58.

117. Milch H and Fornosi F. (1975). Bacteriophage contamination in live poliovirus vaccine. *J Biol Stand* 3: 307–310.

118. Moody EE, Trousdale MD, Jorgensen JH, et al. (1975). Bacteriophages and endotoxin in licensed live-virus vaccines. *J Infect Dis* 131: 588–591.

119. Bogovazova GG, Voroshilova NN and Bondarenko VM. (1991). The efficacy of Klebsiella pneumoniae bacteriophage in the therapy of experimental Klebsiella infection. *Zh Mikrobiol Epidemiol Immunobiol* pp. 5–8.

120. Bogovazova GG, Voroshilova NN, Bondarenko VM, et al. (1992). Immunobiological properties and therapeutic effectiveness of preparations from Klebsiella bacteriophages. *Zh Mikrobiol Epidemiol Immunobiol*, 30–33.

121. Larkum NW. (1929). Bacteriophage as a substitute for typhoid vaccine. *J Bacteriol* 17: 42.

122. Larkum NW. (1929). Bacteriophage from public health standpoint. *Am J Pub Health* 19: 31–36.

123. King WE, Boyd DA and Conlin JH. (1934). The cause of local reactions following the administration of Staphylococcus bacteriophage. *Am J Clin Pathol* 4: 336–345.

124. Górski A. Międzybrodzki R, Borysowski J, Dąbrowska K., Wierzbicki P, Ohams M, Korczak-Kowalska G, Olszowska-Zaremba N, Łusiak-Szelachowska M, Kłak M, Jończyk E, Kaniuga E, Gołaś A., Purchla S, Weber-Dąbrowska B, Letkiewicz S, Fortuna W, Szufnarowsk K., Pawełczyk Z, Rogóź P and Kłosowska D. (2012). Phage as a modulator of immune responses: practical implications for phage therapy. *Adv Virus Res* 83, 41–71.

125. Dàbrowska K, Switala-Jeleñ K, Opolski A and Górski A. (2006). Possible association between phages, Hoc protein, and the immune system. *Arch Virol* 151: 209–215.

126. Barr JJ, Auro, R Furlan M, Whiteson KL, Erb ML, Pogliano J, Stotland A, Wolkowicz R, Cutting AS and Doran KS. (2013). Bacteriophage adhering to mucus provide a non–host-derived immunity. *Proc Natl Acad Sci USA* 110: 10771–10776.

127. Borysowski J, Miedzybrodzki R and Gorski A (editors). (2014). *Phage Therapy. Current Research and Applications.* Caister Academic Press, Norfolk, UK.

128. Tsonos J, Vandenheuvel D, Briers Y, H De Greve, Hernalsteens J-P and Lavigne R. (2014). Hurdles in bacteriophate therapy: deconstructing the parameters. *Vet. Microbiol* 171: 460–469.

129. Abedon S. (2014). Bacteriophages as Drugs: The Pharmacology of Phage Therapy. In: *Phage Therapy: Current Research and Applications.* J. Borysowski and R Miedzybrodzki, editor. Caister Academic Press, Norfolk, UK, pp. 69–100.

130. Lu TK and Koeris MS. (2011). The next generation of bacteriophage therapy. *Curr Opin Microbiol* 14(5): 524–531.

131. McCallin S, Sacher JC, Zheng J and Chan BK. (2019). Current state of compassionate phage therapy. *Viruses* 11(4): 343.

132. Jennes S, Merabishvili M, Soentjens P, et al. (2017). Use of bacteriophages in the treatment of colistin-only-sensitive *Pseudomonas aeruginosa* septicaemia in a patient with acute kidney injury-a case report. *Crit Care* 21(1): 129.

133. Gilbey T, Ho J, Cooley LA, Petrovic Fabijan A and Iredell JR. (2019). Adjunctive bacteriophage therapy for prosthetic valve endocarditis due to *Staphylococcus aureus*. *Med J Aust* 211(3): 142–143.e1.

134. Corbellino M, Kieffer N, Kutateladze M, et al. (2020). Eradication of a multi-drug resistant, carbapenemase-producing Klebsiella pneumoniae isolate following oral and intra-rectal therapy with a custom-made, lytic bacteriophage preparation. *Clin Infect Dis* 70(9): 1998–2001.

135. Kuipers S, Ruth MM, Mientjes M, de Sévaux RGL and van Ingen J. (2019). A Dutch case report of successful treatment of chronic relapsing urinary tract infection with bacteriophages in a renal transplant patient. *Antimicrob Agents Chemother* 64(1): e01281–19.

136. Zhvania P, Hoyle NS, Nadareishvili L, Nizharadze D and Kutateladze M. (2017). Phage therapy in a 16-year-old boy with netherton syndrome. *Front Med (Lausanne)* 4: 94.

137. Lusiak-Szelachowska M, Zaczek M, Weber-Dabrowska B, Miedzybrodzki R, Letkiewicz S, Fortuna W, Rogoz P, Szufnarowski K, Jonczyk-Matysiak E, Olchawa E, Walaszek KM and Gorski A. (2017). Antiphage activity of sera during phage therapy in relation to its outcome, *Future Microbiol* 12:109–117.

138. Międzybrodzki R, Hoyle N, Zhvaniya F, et al. (2018). Current Updates from the Long-Standing Phage Research Centers in Georgia, Poland and Russia. In: *Bacteriophages*, D Harper, S Abedon, B Burrowes, et al, editors. Springer, Cham, Switzerland, pp. 1–31.

139. Lin Y, Chang RYK, Britton WJ, Morales S, Kutter E and Chan HK. (2018). Synergy of nebulized phage PEV20 and ciprofloxacin combination against *Pseudomonas aeruginosa*. *Int J Pharm* 551(1–2): 158–165.

140. Hoyle N, Zhvaniya P, Balarjishvili N, et al. (2018). Phage therapy against *Achromobacter xylosoxidans* lung infection in a patient with cystic fibrosis: a case report. *Res Microbiol* 169(9): 540–542.

141. Yale Medicine website. How phage therapy kills superbugs: Weaponizing viruses to fight infections. Available at: https://www.yalemedicine.org/stories/phage-therapy/

142. Law N, Logan C, Yung G, et al. (2019). Successful adjunctive use of bacteriophage therapy for treatment of multidrug-resistant *Pseudomonas aeruginosa* infection in a cystic fibrosis patient. *Infection* 47(4): 665–668.

143. Adjemian J, Olivier KN and Prevots DR. (2018). Epidemiology of pulmonary nontuberculous mycobacterial sputum positivity in patients with cystic fibrosis in the United States, 2010-2014. *Ann Am Thorac Soc* 15(7): 817–826. (Correction published September 2018, Ann Am Thorac Soc. 15(9):1114–1115.)

144. Dedrick RM, Guerrero-Bustamante CA, Garlena RA, et al. (2019). Engineered bacteriophages for treatment of a patient with a disseminated drug-resistant Mycobacterium abscessus. *Nat Med* 25(5): 730–733.

145. Huber I, Potapova K, Kuhn A, et al. (2018). 1st German phage symposium-conference report. *Viruses* 10(4): 158.

146. Trend S, Fonceca AM, Ditcham WG, Kicic A and Arest CF. (2017). The potential of phage therapy in cystic fibrosis: Essential human-bacterial-phage interactions and delivery considerations for use in Pseudomonas aeruginosa-infected airways. *J Cyst Fibros*;16(6): 663–670.

147. Hraiech S, Brégeon F and Rolain JM. (2015). Bacteriophage-based therapy in cystic fibrosis-associated Pseudomonas aeruginosa infections: rationale and current status. *Drug Des Devel Ther* 9: 3653–3663.

148. Roach DR, Leung CY, Henry M, et al. (2017). Synergy between the host immune system and bacteriophage is essential for successful phage therapy against an acute respiratory pathogen. *Cell Host Microbe* 22(1): 38–47.e4.

149. Rossitto M, Fiscarelli EV and Rosati P. (2018). Challenges and promises for planning future clinical research into bacteriophage therapy against *Pseudomonas aeruginosa* in cystic fibrosis. an argumentative review. *Front Microbiol* 9:775.

150. Waters EM, Neill DR, Kaman B, et al. (2017). Phage therapy is highly effective against chronic lung infections with Pseudomonas aeruginosa. *Thorax* 72(7): 666–667.

151. Malik DJ, Sokolov IJ, Vinner GK, et al. (2017). Formulation, stabilisation and encapsulation of bacteriophage for phage therapy. *Adv Colloid Interface Sci* 249: 100–133.

152. Chang RYK, Das T, Manos J, Kutter E, Morales S and Chan HK. (2019). Bacteriophage PEV20 and ciprofloxacin combination treatment enhances removal of pseudomonas aeruginosa biofilm isolated from cystic fibrosis and wound patients. *AAPS J* 21(3): 49.

153. Leung SSY, Carrigy NB, Vehring R, et al. (2019). Jet nebulization of bacteriophages with different tail morphologies—Structural effects. *Int J Pharm* 554: 322–326.

154. Leung SSY, Morales S, Britton W, Kutter E and Chan HK. (2018). Microfluidic-assisted bacteriophage encapsulation into liposomes. *Int J Pharm* 545(1–2): 176–182.

155. Chan BK, Abedon ST and Loc-Carrillo C. (2013). Phage cocktails and the future of phage therapy. *Future Microbiol* 8: 769–783.

156. McCallin S, Sarker SA, Sultana S, Oechslin F and Brüssow H. (2018). Metagenome analysis of Russian and Georgian Pyophage cocktails and a placebo-controlled safety trial of single phage versus phage cocktail in healthy *Staphylococcus aureus* carriers. *Environ Microbiol* 20(9): 3278–3293.

157. Villarroel J, Larsen MV, Kilstrup M and Nielsen M. (2017). Metagenomic analysis of therapeutic PYO phage cocktails from 1997 to 2014. *Viruses* 9(11): 328.

158. Zschach H, Joensen KG, Lindhard B, et al. (2015). What can we learn from a metagenomic analysis of a Georgian bacteriophage cocktail? *Viruses*;7(12): 6570–6589.

159. Shkoporov AN and Hill C. (2019). Bacteriophages of the human gut: the "known unknown" of the microbiome. *Cell Host Microbe*;25(2): 195–209.

160. Weiss GA and Hennet T. (2017). Mechanisms and consequences of intestinal dysbiosis. *Cell Mol Life Sci*;74(16): 2959–2977.

161. Van Belleghem JD, Dąbrowska K, Vaneechoutte M, Barr JJ and Bollyky PL. (2018). Interactions between bacteriophage, bacteria, and the mammalian immune system. *Viruses*;11(1): 10.

162. Coyne MJ, Comstock LE. (2019). Type VI secretion systems and the gut microbiota. *Microbiol Spectr* 7(2):10.1128/microbiolspec.PSIB-0009-2018.

163. Sutton TDS and Hill C. (2019). Gut bacteriophage: current understanding and challenges. *Front Endocrinol (Lausanne)* 10: 784

164. Shkoporov AN, Clooney AG, Sutton TDS, et al. (2019). The human gut virome is highly diverse, stable, and individual specific. *Cell Host Microbe* 26(4): 527–541.e5.

165. Ott SJ, Waetzig GH, Rehman A, et al. (2017). Efficacy of sterile fecal filtrate transfer for treating patients with clostridium difficile infection. *Gastroenterology* 152(4): 799–811.e7.

166. Nale JY, Redgwell TA, Millard A and Clokie MRJ. (2018). Efficacy of an optimised bacteriophage cocktail to clear clostridium difficile in a batch fermentation model. *Antibiotics (Basel)* 7(1): 13.

167. Nale JY, Chutia M, Carr P, Hickenbotham PT and Clokie MR. (2016). "Get in Early"; biofilm and wax moth (galleria mellonela) models reveal new insights into the therapeutic potential of clostridium difficile bacteriophages. *Front Microbiol* 7: 1383.

168. Strathdee S and Patterson T. (2019). *The Perfect Predator—A Scientist's Race to Save Her Husband from a Deadly Superbug*. Hachette Books, New York, NY.

169. Schooley RT, Biswas B, Gill JJ, et al. (2017). Development and use of personalized bacteriophage-based therapeutic cocktails to treat a patient with a disseminated resistant acinetobacter baumannii infection *Antimicrob Agents Chemother* 61(10):e00954–17. (Correction published November 2018, Antimicrob Agents chemother 62[12]:e02221-18.)

170. Chan BK, Turner PE, Kim S, Mojibian HR, Elefteriades JA and Narayan D. (2018). Phage treatment of an aortic graft infected with *Pseudomonas aeruginosa*. *Evol Med Public Health* 2018(1): 60–66.

171. LaVergne S, Hamilton T, Biswas B, Kumaraswamy M, Schooley RT and Wooten D. (2018). Phage therapy for a multidrug-resistant *Acinetobacter baumannii* craniectomy site infection. *Open Forum Infect Dis* 5(4):ofy064.

172. Brüssow H. (2012). What is needed for phage therapy to become a reality in Western medicine? *Virology* 434: 138–142.

173. Knoll B and Mylonakis E. (2014). Antibacterial bioagents based on principles of bacteriophage biology: an overview. *Clin Infect Dis* 58: 528–534.

174. Fish R, Kutter E, Bryan D, Wheat G and Kuhl S. (2018). Resolving digital staphylococcal osteomyelitis using bacteriophage-a case report. *Antibiotics (Basel)* 7(4): 87.

175. Fish R, Kutter E, Wheat G, Blasdel B, Kutateladze M and Kuhl S. (2016). Bacteriophage treatment of intransigent diabetic toe ulcers: a case series. *J Wound Care* 25(Suppl 7) S27–S33.

176. Fish R, Kutter E, Wheat G, Blasdel B, Kutateladze M and Kuhl S. (2018). Compassionate use of bacteriophage therapy for foot ulcer treatment as an effective step for moving toward clinical trials. *Methods Mol Biol* 1693:159–170.

177. Pirnay J-P, Verbeken G, Rose T, Jennes S, Zizi M, Huys I, Lavigne R, Merabishvill M, Vaneechoute M, Buckling A and De Vos D. (2012). Introducing yesterday's phage therapy in today's medicine. *Future Virology* 7: 379–390.

178. Pirnay JP, Verbeken G and Ceyssens PJ, et al (2018). The magistral phage. *Viruses* 10(2):64.

179. Górski A, Międzybrodzki R, Łobocka M, et al. (2018). Phage therapy: what have we learned? *Viruses* 10(6):288.

180. Kortright KE, Chan BK, Koff JL and Turner PE. (2019). Phage therapy: a renewed approach to combat antibiotic-resistant bacteria. *Cell Host Microbe* 25(2): 219–232.

181. Rohde C, Resch G, Pirnay JP, et al. (2018). Expert opinion on three phage therapy related topics: bacterial phage resistance, phage training and prophages in bacterial production strains. *Viruses* 10(4):178.

182. Taati Moghadam M, Amirmozafari N, Shariati A, et al. (2020). How phages overcome the challenges of drug resistant bacteria in clinical infections. *Infect Drug Resist* 13: 45–61.

55

Emergence of Antimicrobial Resistance in Hospitals

Paramita Basu, Joshua Garcia, and Priyank Kumar

CONTENTS

Introduction

Since the discovery of penicillin by Alexander Fleming in 1928, the world has been revolutionized by the further discovery and synthesis of antibiotics. This increased life expectancy, and reduced morbidity and mortality, which allowed for continued growth of the normal human lifespan well past its original bounds. Unfortunately, despite the rapid implementation of antibiotics into the healthcare field, there has been a continuous growth in the emergence of resistance in pathogenic organisms.[1] In Darwin-like fashion, continued use of suppressive agents has created a selective pressure for pathogens with the capabilities to survive most antibiotics, allowing them to continue to infect patients, which over a few years spread across different countries. The discovery of each new generation of antibiotic quickly followed the same trend (Figure 55.1). Several studies have demonstrated the association of antibiotic use with the emergence of resistance. In the past century, the majority of newly developed antibiotics have seen resistance within a few years of its initial introduction into the healthcare field.[2]

Antibiotic resistance and use have always been a consideration in clinical medicine, but it was not until 2013 that the impact of antibiotic resistance in patient health was thrust into the spotlight. In 2013, the Centers for Disease Control and Prevention (CDC) famously released a report detailing the escalating resistance threats in our healthcare system. By their estimates there were 2,049,442 illnesses and 23,000 deaths attributable to antibiotic resistance in that year.[3] Such a staggering number put the U.S. healthcare system on notice prompting a number of antimicrobial related initiatives. Some of these initiatives include Executive Order 13676, which established the Interagency Task Force for Combating Antibiotic-Resistant Bacteria and the Presidential Advisory Council on Combating Antibiotic-Resistant Bacteria (PACCARB).[4] These advocacy groups allowed for the creation of government incentives for developing new antibiotics and pushed for the increased utilization of formalized Antimicrobial Stewardship Programs (ASP) in hospitals.

Despite the massive amounts of media coverage that antibiotic stewardship is receiving now compared to the past, antimicrobial resistance remains a clinically relevant consideration today. According to the most recently available data from the CDC in conjuncture with the National Healthcare Safety Network (NHSN), national resistance rates to healthcare associated organisms remains high across the board. For example of common hospital pathogens methicillin-resistant *Staphylococcus aureus* (MRSA) and multidrug resistant (MDR) *Pseudomonas aeruginosa,* the most recently reported prevalence numbers were 45.1% and 14.2%, respectively.[5] When taking a deeper look at the prevalence of more resistant isolates, the percentage of carbapenem nonsusceptible *P. aeruginosa* was 19.3%, carbapenem nonsusceptible *Acinetobacter* spp. was 52.7%, and vancomycin-resistant *Enterococcus faecium* was 77.3%, respectively.[5] The high prevalence of resistance is a demonstration of continued antibiotic exposure and could very likely be associated with fewer available antibiotics to use in patients in individual patient cases and worse patient outcomes.[3,6,7]

In addition to negative clinical implications, there are several additional costs associated with antibiotic resistance. Antibiotic resistance can lead to increased morbidity, prolonged antibiotic treatment durations, and increased length of hospital stays, all of which are associated with additional costs to the healthcare

DOI: 10.1201/9781003099277-58

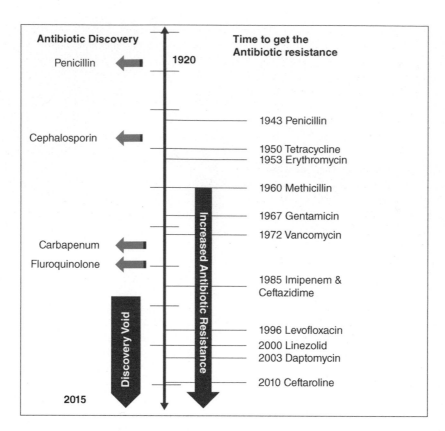

FIGURE 55.1 Graphical representation of onset of antibiotic resistance versus time to get antibiotic resistance. (Reproduced from Review on Antibiotic Resistance: Alarm Bells are Ringing, authored by Zaman et al., 2017).[1] History of antibiotic discovery (green arrow) and time of first reported year of antibiotic resistance (right side). The red arrow (lower direction) indicates the discovery void and increased antibiotic resistance. The blue line indicates the time flow.

system and patient.[7,8] In terms of dollars, the CDC report from 2013 estimated that the overall economic burden of antibiotic resistance to the U.S. healthcare system ranges as high as $20 billion in excess medical costs with additional costs to society of up to $35 billion in lost productivity costs.[3] Resistance of individual organisms have varying cost burdens as well. For example, the additional excess cost of treatment of MRSA as compared to Methicillin-sensitive *Staphylococcus aureus* (MSSA) can range up to $29,030 annually, whereas the cost range for sensitive compared to resistant *P. aeruginosa* can be up to $45,256 annually.[8] Overall, the majority of cost associated with antibiotic resistance is more heavily weighted toward the hospital setting.

Managing antibiotic resistance in the healthcare system remains challenging, particularly in the hospital setting. We will continue to explore this concept throughout the rest of the chapter.

History of Antimicrobial Resistance in the Hospitals

One of the major threats to public health is the constant rise in the progression of resistance to antimicrobial drugs, which is a leading cause of development and relapse of infectious disease. The advent of penicillin has led to the reduction in the numbers of susceptible strains and the rise and propagation of more resistant strains.[9,10] The first signs of antibiotic resistance

became apparent soon after the discovery of penicillin. In 1940, Abraham and Chain reported that an *E. coli* strain was able to inactivate penicillin by producing penicillinase. The spread of penicillin resistance was already documented by 1942, when four *S. aureus* strains were found to resist the action of penicillin in hospitalized patients and was found to have developed an ability to produce β-lactamase, which inactivates both natural penicillin and aminopenicillins making *S. aureus* resistant to antimicrobials.[11] During the next few years, the proportion of infections caused by penicillin-resistant *S. aureus* rapidly rose, spreading quickly from hospitals to communities. By the late 1960s, more than 80% of both community and hospital-acquired strains of *S. aureus* were penicillin-resistant.[22]. Initial sporadic appearance of penicillin resistance was observed in hospitals, however, with time it spread to the community. Emergence of penicillin resistant *S. aureus* fueled the development of second generation, β-lactamase-resistant, antimicrobial agents such as methicillin in 1960.[12] Soon after, the first case of MRSA appeared in the late 1960s, which spread to become endemic in hospitals during the 1980s, reaching 29% of hospitalized *S. aureus*-infected patients.[13–15] Earlier reports on parameters affecting this trend of emergence of resistance had indicated that the rate of increase in MRSA was proportional to the size of the hospital[14]; however, current trends show that rate of rise in methicillin resistance is similar irrespective of the hospital size.[15] Another such example is the emergence of resistance to vancomycin, which is the drug of choice to treat MRSA. In 1997, initial reports of reduced vancomycin susceptibility in clinical isolates of *S. aureus* from Japan generated significant concern in the medical community, though, according to a report in 2004, the number of cases reporting reduced vancomycin susceptibility is low overall and hemodialysis patients are more prone to antibiotic resistance from vancomycin-intermediate *S. aureus* (VISA) and vancomycin-resistant *S. aureus* (VRSA).[16] Various mutations in VISA result in strains with impenetrable cell walls that make it difficult for vancomycin to enter the bacterial cell.[16,17] In 2003 a report by Fridkin et al. evaluated risk factors for appearance of Infections caused by *S. aureus* with reduced vancomycin susceptibility, and demonstrated that these include antecedent vancomycin use and prior oxacillin-resistant *S. aureus* infection 2–3 months before the current infection.[18] By 2015, only 14 reports of VRSA were reported in the United States, though a CDC report on antibiotic resistance threats in the United States published in 2013 considered VRSA to be a concerning threat.[19,20] Along with VRSA, other pathogenic microbes also started demonstrating vancomycin resistance, for example, vancomycin resistance in enterococci (VRE) emerged in 1988 in Europe and around the same time in United States.[21] By 1998, the prevalence of VRE increased by 20-fold in the intensive care setting in the United States. Flokas et al. reported an increased prevalence of VRE colonization among hospitalized children in hematology/oncology units. Additionally, the colonized children were reported to be nine times more likely to develop consequent VRE infections.[22] However, Kampmeier et al. published a study in 2018 that reported disassociation of surgical Intensive care units (ICU)from VRE colonization.[23]

In past few years, resistance related to Gram-negative bacteria has significantly increased, which has become a serious challenge for the healthcare community.[24] For example, a nine-year surveillance study in 2004, reported the appearance of multidrug-resistant (MDR) *P. aeruginosa* with a frequency of 1–16%, along with the appearance of MDR *Klebsiella* species with a frequency of 0.5–17%.[25] As per the NNIS database, *P. aeruginosa* samples from ICU patients in 2003 were found to have a rate of resistance of 20% to carbapenems and about 30% to third-generation cephalosporins as well as quinolones.[16] Another Gram-negative bacterium, *Acinetobacter baumannii*, is known for causing pneumonia in hospitalized patients. A rising number of cases related to MDR *A. Baumannii* have been reported in military field hospitals, tertiary care centers for treating military personnel in the United States and abroad, as well as U.S. Navy hospital ships.[26] It would be worthwhile to mention that *Clostridium difficile*-associated disease (CDAD) and antimicrobial-associated disease are amongst the most frequently occurring side effects due to antibiotic therapy. In the 1990s, a steady increase in antibiotic associated-CDAD was reported in NNIS hospitals.[27] Additionally, data from National Center for Health in 2001 suggested an alarming increase in the rate of *C. difficile* associated infections in U.S. hospitals.[28] An 11-year study conducted by Dallal et al. demonstrated that the incidence of *C. difficile* colitis accompanied by life threatening symptoms increased significantly in the hospitalized patients.[29] According to a report by the NHSN in 2009–2010, the infections associated with MRSA showed a slight dip, however, the same report showed an increase in MDR Gram-negative bacteria.[30] Furthermore, a report by CDC in 2013 stated that the fatalities related to antibiotic resistance are more prevalent in the healthcare settings as compared to community.[31]

Causative Mechanisms of Antimicrobial Resistance in the Hospitals

The cellular mechanism of acquisition of antibiotic resistance occurs through the interaction of antibiotics and bacteria in the environments. Certain resistance genes ward off destruction by giving rise to enzymes that degrade antibiotics or that chemically modify, and so inactivate, the drugs. Alternatively, some resistance genes cause bacteria to alter or replace molecules that are normally bound by an antibiotic, that is, changes that essentially eliminate the drug's targets in bacterial cells. Bacteria might also eliminate entry ports for the drugs by modifying the target receptor or their genes, or they may activate the synthesis protein pumps that export antibiotics before the medicines have a chance to find their intracellular targets.[32,33] Antibiotic resistant bacteria owe their drug insensitivity to resistance genes. For example, such genes might code for "efflux" pumps that eject antibiotics from cells, or the genes might give rise to enzymes that degrade the antibiotics or chemically alter and inactivate the drugs.

Bacteria can acquire resistance genes through a few routes. In some cases, they are obtained through vertical gene transmission. Other times, genetic mutations spontaneously occur and produce a new resistance trait or strengthen an existing one. Frequently, bacteria may also become resistant through horizontal gene transfer from other microorganisms in the environment. In these cases, genes located on plasmids or small pieces of chromosomal DNA from dead microorganisms must be integrated into the recipient bacteria. Resistance genes can also be transferred by viruses through transduction. Such integration occurs frequently, though, because resistance genes are often embedded in small units of DNA, called transposons that readily hop into other DNA molecules. Many bacteria play host to specialized transposons, termed integrons, that can consist of several different resistance genes that are passed to other bacteria as whole regiments of antibiotic-defying arsenals. Many bacteria possessed resistance genes even before commercial antibiotics came into use. Later, these protective genes found their way into other species, some of which were pathogenic. Bacteria acquire resistance genes from other bacterial cells in three main ways.[32] Often they receive whole plasmids bearing one or more such genes from a donor cell. Other times, a virus will pick up a resistance gene from one bacterium and inject it into a different bacterial cell. Alternatively, bacteria sometimes scavenge gene-bearing snippets of DNA from dead cells in their vicinity. Genes obtained through viruses or from dead cells persist in their new owner if they become incorporated stably into the recipient's chromosome or into a plasmid. See Table 55.1.

Regardless of how bacteria acquire resistance genes today, excessive and improper usage of commercial antibiotics can promote the survival and propagation of antibiotic-resistant strains. Excessive and unnecessary use of antibiotics favors rapid selection of antibiotic-resistant organisms in the microbial population. Exchange of genetic materials between different antibiotic resistant populations has also enabled emergence of multidrug-resistant strains. When antibiotics are given as a treatment for

TABLE 55.1

Summary of the Mechanism of Drug Resistance of Common Antibiotics

Antibiotic Class	Example(s)	Mode(s) of Resistance
P-lactams	Penicillins, cephalosporins, penems, monobactams	Hydrolysis, efflux, altered target
Aminoglycosides	Gentamicin, streptomycin, spectinomycin	Phosphorylation, acetylation, nucleotidylation, efflux, altered target
Glycopeptides	Vancomycin, teicoplanin	Reprogramming peptidoglycan biosynthesis
Tetracyclines	Minocycline, tigecycline	Monooxygenation, efflux, altered target
Macrolides	Erythromycin, azithromycin	Hydrolysis, glycosylation, phosphorylation, efflux, altered target
Lincosamides	Clindamycin	Nucleotidylation, efflux, altered target
Streptogramins	Synercid	Carbon-oxygen lyase, acetylation, efflux, altered target
Oxazolidinones	Linezolid	Efflux, altered target
Phenicols	Chloramphenicol	Acetylation, efflux, altered target
Quinolones	Ciprofloxacin	Acetylation, efflux, altered target
Pyrimidines	Trimethoprim	Efflux, altered target
Sulfonamides	Sulfamethoxazole	Efflux, altered target
Rifamycins	Rifampin	ADP-ribosylation, efflux, altered target
Lipopeptides	Daptomycin	Altered target
Cationic peptides	Colistin	Altered target, efflux

Source: Reproduced from Review on Antibiotic Resistance: Alarm Bells are Ringing, authored by Zaman et al. (2017).[1]

disease-causing bacteria, they also affect nonpathogenic bacteria in their vicinity. They eliminate these drug-susceptible bystanders that might otherwise limit the expansion of pathogens, and they simultaneously encourage the growth of resistant bystanders. Propagation of these resistant, nonpathogenic bacteria increases the reservoir of resistance traits in the bacterial population as a whole and raises the odds that such traits will spread to pathogens. There are two main biological routes of acquiring antibiotic resistance.[34] Microbes reproduce by dividing every few hours, allowing them to evolve rapidly and adapt quickly to new environmental conditions. With each replication, mutations arise, and some of these mutations may help an individual microbe survive exposure to an antimicrobial. Microbes may also acquire genes from each other through spontaneous genetic transfer, including genes that make the microbe drug resistant.

In addition, sometimes the growing populations of bystanders themselves become agents of disease. A significant example of this phenomenon is how the widespread use of cephalosporin antibiotics has promoted the proliferation of the once nonpathogenic intestinal bacterium *Enterococcus faecalis*, which is naturally resistant to those drugs. Subsequently, regular exposure to vancomycin persistently over the years has turned *E. faecalis* into a reservoir of vancomycin-resistance traits due to constant

selection pressure. Since some strains of the pathogen *S. aureus* are multidrug-resistant, and are responsive only to vancomycin, public-health experts fear that it will soon deliver strong vancomycin resistance to those *S. aureus* strains, making them incurable.[35]

Human treatment accounts for roughly half the antibiotics consumed every year in the United States. Perhaps only half of that use is appropriate and meant to cure bacterial infections and administered correctly in ways that do not strongly encourage resistance. Reports from the CDC and other researchers have estimated that some 50 million of the 150 million outpatient prescriptions for antibiotics every year are unneeded.[36-38]

In the industrial world, most antibiotics are available only by prescription, but this restriction does not ensure proper use. People often fail to finish the full course of treatment. Patients then often store the leftover doses and medicate themselves, or their family and friends, in less than therapeutic amounts. In both circumstances, the improper dosing will fail to eliminate the disease agent completely and will, furthermore, encourage growth of the most resistant strains, which may later produce hard-to-treat disorders. Many studies have reported presence of resistance markers and resistance bacteria in the environment, which includes the presence of *vanA*, *mecA*, *AmpC*, and gentamicin resistance containing genetic material in hospital effluents and waste water, as well as presence of antibiotic resistant mutants of *P. aeruginosa*, *Acinetobacter* spp., *E. coli*, *Campylobacter* spp., *Pseudomonas* spp., *Enterococci* spp., etc.[39] In the developing world, antibiotic use is even less controlled. Many of the same drugs marketed in the industrial nations are available over the counter. Unfortunately, when resistance becomes a clinical problem, these countries, which often do not have access to expensive drugs, may have no substitutes available.

Antibiotics are used extensively to prevent or to treat microbial infections in human and veterinary medicine. Over-prescribing antibiotics for viral-caused conditions like the flu or common cold, against which antibiotics are useless, contributes to antibiotic resistance. Most of the compounds used in medicine are only partially metabolized by patients and are then discharged into the hospital sewage system or directly into municipal wastewater if used at home. Along with excreta, they flow with municipal wastewater to the sewage treatment plant. They may pass through the sewage system and end up in the environment, mainly in the water compartment. Antibacterial substances used for livestock enter the environment when manure is applied to fields. These antibiotics may either end up in soil or sediment or in ground water.[40-43] In addition, resistant bacteria themselves are excreted by humans and animals[44-46] and are emitted into sewage or manure and other environmental compartments.

Antibiotics affect the mixed population of resistant and non-resistant bacteria both in the individual being treated, and in that individual's environment. This is because the rise of resistant bacteria in treated individuals spread readily to the surrounding populations and to new hosts. Also, regular use of antimicrobials in such settings as hospitals, daycare centers, and even in the household (cleaning products, disinfectants, etc.), increases the levels of resistant bacteria in people and others who are not being treated, including in individuals who live near those epicenters of high consumption or who pass through the centers. So, these drugs also affect the kinds of bacteria in the environment

TABLE 55.2

Bacteria and Fungi Listed in the 2013 Antibiotic Resistant Threats Report

Urgent Threats	Serious Threats	Concerning Threats	Watch List
• Carbapenem-resistant *Acinetobacter* • *Candida auris* • *Clostridioides difficile* • Carbapenem-resistant *Enterobacteriaceae* • Drug-resistant *Neisseria gonorrhoeae*	• Drug-resistant *Campylobacter* • Drug-resistant *Candida* • ESBL-producing *Enterobacteriaceae* • Vancomycin-resistant *Enterococci* (VRE) • Multidrug-resistant *Pseudomonas aeruginosa* • Drug-resistant nontyphoidal *Salmonella* • Drug-resistant *Salmonella* serotype Typhi • Drug-resistant *Shigella* • Methicillin-resistant *Staphylococcus aureus* (MRSA) • Drug-resistant *Streptococcus pneumoniae* • Drug-resistant Tuberculosis	• Erythromycin-resistant Group A *Streptococcus* • Clindamycin-resistant Group B *Streptococcus*	• Azole-resistant *Aspergillus fumigatus* • Drug-resistant *Mycoplasma genitalium*

Source: Centers for Disease Control and Prevention Website.[3]

and people around the individual being treated, resulting in a widespread societal effect.[47] On a larger scale, antibiotic resistance that emerges in one place can often spread far and wide. The ever increasing volume of international travel has hastened transfer to the United States of multidrug-resistant tuberculosis from other countries.[37] And investigators have documented the migration of a strain of multidrug-resistant *Streptococcus pneumoniae* from Spain to the United Kingdom, the United States, South Africa, and elsewhere. This bacterium, also known as the pneumococcus, is a cause of pneumonia, meningitis, and other diseases. Antibiotic and other drug resistance is already a major issue in medicine, with significant health and economic impacts, as shown by the resurgence of tuberculosis and malaria, and multidrug resistance is easily developed.[48] What has not been appreciated is the potential of drugs as environmental pollutants to create new pools of resistance which, by genetic transfers across species, can accelerate the development of resistance in many disease organisms. If new forms of resistance start to come not only from human treatments and hospital settings, but by interspecies transfers from the environment, the growing ineffectiveness of present forms of treatment and a revival of major epidemics is a distinct possibility. Since the emergence of resistance renders that particular antibiotic unavailable to treat infections for a long time it means that the product's life cycle may be short. The result of this is that the costs of antibiotic research and development must be recouped quickly, raising the price of the drug.[49] If the current patterns of high levels of antibiotic utilization persist (while there remains a paucity of new antibiotics), then the availability of effective antibiotics will decline. It is unlikely that the rapidly growing problem of antibiotic resistance will be solved by new antibiotics in the near future, so in order to sustain the therapeutic functions of antibiotics, it is necessary to preserve the efficacy of those antibiotics that are currently available, at least for the next few decades.[50–54]

Antibiotic-Resistant Organisms in U.S. Hospitals

Out of the billions of microorganisms on this planet, only a smaller subset is known to be pathogenic to humans, and an even smaller subset still is known to be problematic in terms of hospital resistance. The CDC outlined a number of potentially problematic organisms in their 2013 report and organized them by level of threat level severity to the U.S. population.[3] The classification process describes three different levels: Urgent, serious, and concerning.[3] Of these organisms, we will focus on a select few based upon their prevalence in the hospital system. These organisms include: Carbapenem-resistant *Enterobacteriaceae* (CRE), *Clostridioides difficile*, multidrug-resistant *Acinetobacter* spp., multidrug-resistant *P. aeruginosa*, extended spectrum beta-lactamase (ESBL)-producing *Enterobacteriaceae*, and *S. aureus*. Information regarding other strains of resistant bacteria that have been identified as threats by the CDC can be found in the Drug Resistance Threat Report[3] published by CDC accessible in the CDC website. See Table 55.2.

Carbapenem-resistant *Enterobacteriaceae* (CRE)

Carbapenem-resistance *Enterobacteriaceae* are classified as an "Urgent" threat by the CDC 2013 report, meaning that it represents an immediate public-health threat that requires aggressive action.[3] They have estimated that of the 140,000 healthcare-associated *Enterobacteriaceae* infections in the United States in the last year, roughly 9,300 were due to CRE organisms.[3] Although CRE organisms have characteristically been identified in other countries around the world, increases in the use of broad-spectrum antibiotics and international travel have contributed to increased isolations of this organism. CRE organisms are Gram-negative aerobic bacilli (such as *E. coli, Klebsiella* spp., etc.) and are appropriately named for their resistance to the carbapenem medication class (such as meropenem, ertapenem, imipenem-cilastatin, and doripenem), one of the broadest and most effective class of antibiotics. In addition, CRE organisms often express additional mechanisms of resistance that confer nonsusceptibility to other nonbeta-lactam medication classes, such as sulfamethoxazole/trimethoprim, aminoglycosides, and fluoroquinolones.[55,56] This often leads to near-pan resistance among CRE organisms. This lack of available medication options has led to a high mortality (nearly 50%) of CRE bloodstream infections.[3,57] These organisms obtain resistance to the carbapenem class via production of carbapenemases, a subtype of beta-lactam hydrolyzing enzyme, which cleaves the structural integrity of the beta-lactam ring. These carbapenemases provide the organism with resistance to not only, carbapenems but also the majority of other beta-lactams such as penicillins, cephalosporins, and monobactams.[56] Although the nomenclature and categorization have changed throughout the years, carbapenemases have been associated with both chromosomal and plasmid-mediated genes.[56] Genes associated with the production of carbapenemases include the molecular class A, B, and D classes, which include, but are not limited to, KPC, IMP, VIM, and OXA genes.[56,58,59]

The lack effectiveness of current agents has prompted the use of older or more last line antibiotics, such as colistin, polymyxin b, and tigecycline.[60] Colistin is a surface active agent that exerts its mechanism of action by disrupting the cell membrane of Gram-negative bacteria.[61] Although colistin retains activity against a majority of Gram-negative agents ($MIC_{90} \leq 0.5–1$ mg/L and resistance rates <0.1–1.5%), complicated pharmacokinetic dosing and renal toxicity severely limits its use in treatment.[61–63] Polymyxin B is in the same class as colistin (colistin is also known as polymyxin E) and works in the same fashion but does not exhibit the same type of complicated renal toxicity.[62–64] Tigecycline is another agent that also demonstrates activity against CRE. Unlike, polymyxin B and colistin, tigecycline is a glycylcycline that inhibits protein synthesis.[65] Although tigecycline was developed later, it is not used as frequently today due to postmarketing studies revealing an increased risk of mortality as compared to the standard of therapy.[65] Today, tigecycline is reserved for CRE infections.[63] Combination therapy has shown favorable outcomes when compared to monotherapy, often involving a combination of tigecycline, polymyxin b, colistin, and/or carbapenems if the MIC ≤8mcg/mL.[66–69]

Unfortunately, the trials involving treatment of CREs are scarce, and the data reported may not often be reliable, due to the small sample size, often performed retrospectively or observationally. The dearth of available antibiotics to treat CREs combined with the looming resistance growth in hospitals has prompted several initiatives for the development of new antibiotics. Many newly approved and pipeline antibiotics are being developed with the hope to treat CREs and will be discussed later in the chapter.

Antibiotic-Resistant *Clostridioides difficile* Infections

Clostridioides (formerly clostridium) difficile is classified as an "Urgent" threat by the CDC 2013 report, meaning that it represents an immediate public health threat that requires aggressive action not necessarily due to innate resistance and inability to treat the disease, but rather due to frequent recurrence of disease.[3] *C. difficile* is a Gram-positive, anaerobic, spore-forming, bacilli that most commonly manifests with clinical symptoms of diarrhea, abdominal pain, and in more serious cases, toxic megacolon, pseudomembranous colitis, septic shock, and death.[70] *C. difficile* can be both a pathogenic organism as well as a normal colonizer of the gastrointestinal tract. Infection is associated with overgrowth of *C. difficile* due to elimination of commensal gut flora from antibiotic exposure.[70,71] Although other risk factors have been associated with *C. difficile*-associated diarrhea (CDAD), antibiotic exposure (particularly exposure to erythromycin, penicillin, clindamycin, higher generation cephalosporins, and fluoroquinolones) remains the primary factor causing CDI to spread more widely in and out of the hospital and increase the resistance of the epidemic strains of *C. difficile* to those

drugs.[72,73] Current treatments available for CDAD include oral vancomycin (intravenous vancomycin does not achieve adequate concentrations in the gastrointestinal lumen), fidaxomicin, and metronidazole (in rare cases).[73] Strain 027 is the most frequently observed drug-resistant *C. difficile* strain and shows a statistically high incidence in Europe and the United States. Several countries have reported the resistance of strain 027 to moxifloxacin and, more recently, to metronidazole and vancomycin.[70,74]

In recent years, an increase in CDI has been reported in different countries. China, Sweden, and several other countries as well as the United States have reported a CDI incidence rate of ~17.1 cases/10,000 admitted to a hospital.[70] Initial cure rates with fidaxomicin and oral vancomycin are relatively high, around 87% for both.[74] However, recurrence rates have been observed to be anywhere between 10–40%.[73,74] The high recurrence rate is due to the spores that are produced from the *C. difficile* organism. Spores are not eliminated by antibacterial therapy and can remain in the gut for a substantial amount of time before germinating and reinfecting the patient. With every concurrent recurrence, clinicians cannot use a therapy that has been used in previous episodes, effectively narrowing the available treatments.[73] Although other modalities of management such as fecal microbiota transplantation and bezlotoxumab have been developed, a large part of *C. difficile* management involves discontinuing insulting antibiotics and isolation precautions such as gowns, gloves, and handwashing.[73]

Multidrug-resistant *Acinetobacter* spp.

Multidrug-resistant *Acinetobacter* spp. is classified as a "Serious" threat by the CDC 2013 report, meaning that it represents a concern that requires prompt and sustained action to ensure that the problem does not grow.[3] A recent report by the CDC stated that carbapenem-resistant *Acinetobacter* spp. were responsible for an estimated 8,500 hospitalizations and 700 estimated deaths in 2017.[75] Like *P. aeruginosa*, *Acinetobacter* spp. (most commonly *Acinetobacter baumannii*) are Gram-negative, glucose nonfermenting bacteria. Although wild strains of *Acinetobacter* spp. have been associated with fresh water and soil environments, multidrug-resistant strains have been more widely associated with hospital infections such as ventilator-associated pneumonia (VAP), urinary tract infections, and bacteremia.[75] *Acinetobacter baumannii* has accumulated a number of different resistance mechanisms such as beta-lactamases, aminoglycoside-modifying enzymes, efflux pumps, porin channel degradation, and modification of antibiotic target sites. In terms of beta-lactamases, *Acinetobacter* spp. strains have exhibited all four molecular subgroups included prominent ESBL and CRE beta-lactamases.[76] Combined with the aforementioned resistance mechanisms, *Acinetobacter baumannii* represents a formidable problem in U.S. healthcare today.

Treatment of susceptible strains of Acinetobacter is consistent with treatment of susceptible *P. aeruginosa* strains. Antibiotics that tend to retain effectiveness against wild-type *Acinetobacter baumannii* strains include piperacillin/tazobactam, ceftazidime, cefepime, aztreonam, carbapenems (except ertapenem), aminoglycosides, and fluoroquinolones. There are a few medications that also have well-documented activity to *A. baumannii* but not *P. aeruginosa*. These antibiotics include ampicillin/sulbactam (just the sulbactam component), minocycline, and tigecycline. Although sulbactam is primarily used in combination with ampicillin as a beta-lactamase inhibitor to protect the active antibiotic from Ambler class A beta-lactams, sulbactam has shown *in vitro* and *in vivo* activity against *A. baumannii* alone.[77] Since sulbactam is only available as a co-formulated product in the United States, ampicillin/sulbactam is a viable option whereas in Europe sulbactam is available as a single product. In addition, there are no evidence-based guideline recommendations for optimal therapy against multidrug resistant strains of *A. baumannii*. However, current practice generally involves the use of combination therapy of sulbactam, tigecycline, and colistin/polymyxin B as compared to monotherapy with any singular agent.[78] Of the new antibiotic agents, few have substantial evidence showing them to be effective against multidrug-resistant Acinetobacter, but eravacycline and omadacycline have shown promising *in vitro* activity and may be potential future options for treatment as resistance of *Acinetobacter* spp. grows.[79,80]

Multidrug-Resistant *Pseudomonas aeruginosa*

Multidrug-resistant *P. aeruginosa* is classified as a "Serious" threat by the CDC 2013 report, meaning that it represents a concern that requires prompt and sustained action to ensure that the problem does not grow.[3] It is estimated to be responsible for up to 6,700 infections per year (51,000 infections of non-MDR *P. aeruginosa*) and 440 deaths per year.[3] *P. aeruginosa* is an aerobic, glucose nonfermenting, Gram-negative bacillus that is commonly found in hospital fluid reservoirs.[81] *P. aeruginosa* represents one of the most common nosocomial pathogens in the U.S. healthcare systems today and display a high level of resistance (roughly 19.3% were found to have resistant or intermediate susceptibility to at least one carbapenem and approximately 14% of isolates are classified as multidrug-resistant).[5,82] These pathogens contain multiple resistance mechanisms combinations (such as beta-lactamase production, efflux pump overexpression, porin channel degradation, and target site alteration), which confer high levels of resistance to most commercially available antibiotics.[83-87] The high frequency of multiple resistance mechanisms expressed by *P. aeruginosa* infections make the organism unique compared to other gram negative pathogens.

Medications that have potential efficacy against *P. aeruginosa* include piperacillin/tazobactam, ceftazidime, cefepime, imipenem-cilastatin, meropenem, doripenem, aztreonam, ciprofloxacin, levofloxacin, the aminoglycosides, polymyxin b, and colistin.[88] Although these medications have the potential to treat *P. aeruginosa* infections, individual strains may be resistant to one, multiple, or all of these agents, which creates a clinically difficult management plan.[3] In combination with high virulence and mortality rates with insufficient treatment, some infectious disease guideline updates (such as the Hospital Acquired/Ventilator Associated Pneumonia Infectious Disease Society of America Guidelines) have suggested treatment regimens using two different antipseudomonal agents of different medication classes as a first-line option to ensure adequate coverage of the organism in the acute phase.[89] MDR *P. aeruginosa* infections are also an area of interest in the development of new Gram-negative antibiotics. Of the new agents mentioned above for CRE

infections, almost all of them have *in vitro* and clinically relevant activity against wild-type *P. aeruginosa*, but only some have shown promising activity in MDR strains. Notably, meropenem/vaborbactam, plazomicin, and aztreonam/avibactam (currently in development) did not show additional activity against MDR *P. aeruginosa* isolates.[88] Of the newer agents, ceftazidime/avibactam, ceftolozane/tazobactam, imipenem-cilastatin/relebactam, and cefiderocol show promise against MDR *P. aeruginosa* isolates. Ceftazidime/avibactam shows clinical good *in vitro* activity against *P. aeruginosa* in general with susceptibility rates close to 90% in the United States.[90-92] The addition of avibactam to ceftazidime allows for additional stability against beta-lactamase produced by *P. aeruginosa* with roughly 80% of ceftazidime/avibactam susceptible isolates exhibiting resistance to ceftazidime alone.[91,92] Ceftolozane/tazobactam utilizes a novel cephalosporin with a historical beta-lactamase inhibitor. Ceftolozane is structurally similar to ceftazidime but is more stable against AmpC, allowing for better activity against AmpC derepressed strains of *P. aeruginosa*, and has a greater affinity for *P. aeruginosa* penicillin binding proteitns.[93,94] *In vitro* activity of ceftolozane/tazobactam to *P. aeruginosa* is high with susceptibility to 96.1% of isolates in general and 77% to ceftazidime.[95] Imipenem-cilastatin/relebactam is currently undergoing phase 3 trials, but utilizes a novel beta-lactamase inhibitor, relebactam, for additional Gram-negative coverage. *In vitro*, the addition of relebactam to imipenem-cilastatin has demonstrated increased susceptibility of *P. aeruginosa* from 70% to 94%.[96]

Extended Spectrum Beta-lactamase (ESBL)-Producing *Enterobaceriaceae*

Extended spectrum Beta-lactamase (ESBL)-producing *Enterobaceriaceae* are classified as "Serious" threats by the CDC 2013 report, meaning that it represents a concern that requires prompt and sustained action to ensure that the problem does not grow.[3] ESBLs were initially defined as beta-lactamases that could hydrolyze penicillins and third-generation cephalosporins, the latest generation of that time. Now ESBLs can be better defined as multiple plasma-mediated ambler class A beta-lactamases (most commonly mutations of the SHV, TEM, and CTX enzymes) that hydrolyze almost all beta-lactams except carbapenems, but remain susceptible to select beta-lactamase inhibitors.[97-99] In addition, the ESBL plasmids have also been found to contain other genes that code for resistance mechanisms of nonbeta lactam antibiotics, nullifying other empiric therapy options. Because ESBLs are plasma-mediated, the ability for resistance to transfer between pathogens (even different organisms entirely) creating infection prevention concerns as well.

Traditionally, carbapenems have been the treatment of choice for ESBL infections due to their resilience to hydrolysis and the aforementioned ineffectiveness of other beta-lactam and nonbeta-lactam antibiotics. This of course, only applies to empiric therapy, when susceptibilities of the particular infection strain are unknown. Once susceptibilities are available, other nonbeta-lactam antibiotic, such as sulfamethoxazole/trimethoprim and fluoroquinolones, may be used.[100] For a time, piperacillin/tazobactam was thought to be a potential carbapenem-sparing regimen due to the continued activity of beta-lactamase inhibitors against ESBLs.[99] This practice was further supported by small clinical trials showing potential benefit in the using of carbapenem-sparring piperacillin/tazobactam, most notably in ESBL urinary tract infections.[101] However, these trials were generally observational studies or had low enrollment, lacking the ability to demonstrate effectiveness comparisons to a carbapenem-based regimen. A recent randomized controlled trail compared the efficacy of meropenem and piperacillin/tazobactam in ESBL bacteremia. The results of the trial showed that piperacillin/tazobactam failed to demonstrate noninferiority to meropenem in ESBL bacteremia.[102] Therefore, carbapenems remain the standard of therapy for ESBL infections. Luckily, the majority of newly approved antibiotics have activity against ESBLs.

Staphylococcus aureus

Methicillin-resistant *Staphylococcus aureus* (MRSA) is classified as a "Serious" threat by the CDC 2013 report, meaning that it represents a concern that requires prompt and sustained action to ensure that the problem does not grow.[3] A 2019 report by the CDC estimated that there were 323,700 cases and 10,600 deaths due to MRSA in the United States.[75] MRSA is the Gram-positive pathogen of greatest concern in hospital-related infections, often necessitating the addition of vancomycin to most empiric regimens. In addition to vancomycin, several other antibiotics have MRSA activity, such as ceftaroline, dalbavancin, oritavancin, doxycycline, tigecycline, sulfamethoxazole/trimethoprim, clindamycin, lefamulin, delafloxacin, linezolid, and tedizolid. In addition to hospital-acquired infections, community-acquired MRSA has been associated with intravenous drug users, confined living spaces, residence in nursing homes and assisted living facilities, hemodialysis, previous antimicrobial therapy, and previous hospitalization. However, more recent guidelines have begun to better define the risks for MRSA, elevating its previous isolation (via previous infection or colonization) as primary risk factors for MRSA infection.[89,103]

Although all the agents listed above may be used in MRSA infections, vancomycin remains the standard of therapy for the majority of hospitals in the United States. This is primarily due to the low cost, relatively high efficacy, and large amounts of experience and clinical data associated with vancomycin use for MRSA. However, vancomycin's adverse effect profile and a slowly increasing level of MRSA vancomycin resistance have prompted the use of alternative agents. Vancomycin is associated largely with nephrotoxicity and infusion related red man syndrome (a combination of symptoms including torso erythema, itching, and fever).[104] While the latter can be mitigated with antihistamine premedication and decreasing the rate of infusion, nephrotoxicity is dose related and can be difficult to avoid even with appropriate therapeutic drug monitoring. Recently a phenomenon, known as the "MIC creep," has been observed in MRSA isolates in which the MIC of MRSA isolates to vancomycin has started to rise, causing clinical failure despite MICs that remain below the CLSI breakpoint.[105,106] The decision to continue to treat with vancomycin or switch to an alternative agent is currently divisive amongst clinicians and requires more clinical data before a consensus can be made regarding optimal therapy.

The Clinician's Role in Combating Antimicrobial Resistance

Antimicrobial Stewardship Programs

The data from all the studies listed above, on incidence, causative factors, and mechanisms of action of antimicrobial resistance warrants a multifaceted approach to curb the effects of MDR that will include educating patients as well as physicians about the appropriate antimicrobial use. What this means is that the clinician's current antibiotic armamentarium is all they can expect in the foreseeable future. It also means that special care needs to be taken to optimally use currently available agents to ensure continued activity against the pathogens encountered in the hospital (and community) setting, now and in the future. Maximizing clinical outcomes, while minimizing the emergence and spread of antimicrobial resistance (and other adverse effects associated with suboptimal antimicrobial drug use), falls under the purview of antimicrobial stewardship.[107]

Early in the onset of many infections, the data needed to make a rational, informed decision about specific antibiotic therapy is usually unavailable. For many infections, therapy cannot be delayed waiting for microbiology or other findings, and broad-spectrum empiric therapy is begun based on educated guesses made from the patient's presentation and characteristics, and local or hospital antibiograms. In addition, for many serious infections, delay in antimicrobial therapy will increase patient morbidity and mortality.[108] Examining antibiotic usage at the hospital level, one notices that approximately 60% of adult patients admitted to U.S. hospitals receive at least one dose of an antibiotic agent during their stay (range: 44–74% for individual hospitals).[109,110] Similarly, at Wake Forest University Baptist Medical Center, approximately 75% of inpatients receive antimicrobials at some point during their hospitalization.[119] One recent example by Hecker and colleagues conducted in a 650-bed, university-affiliated U.S. hospital reported 30% of the total days of antibiotic therapy received by adult non-ICU inpatients was unnecessary.[111] The most common reasons for unnecessary use were administration for longer than recommended durations, administration for a noninfectious or nonbacterial syndrome, and treatment of colonizing or contaminating microorganisms. Unwanted consequences of antimicrobial therapy include increased morbidity and mortality, adverse drug reactions, increased length of hospital stay and hospitalization costs, predisposition to secondary infections, and emergence and selection of drug-resistant organisms.[112,113] Selection or induction of antimicrobial resistance and promotion of secondary infection with *C. difficile*, particularly with new, more toxigenic strains, are of particular concern in the current hospital environment.[113] These untoward consequences can be seen as a calculated risk of antibiotic therapy for any single-treated patient, or as an undesired outcome measure for excessive use at the level of the healthcare institution.

Antimicrobial stewardship programs in hospitals seek to optimize antimicrobial prescribing to improve individual patient care as well as reduce hospital costs and slow the spread of antimicrobial resistance. The various strategies used as part of these programs also help to reduce unwanted consequences of antimicrobial overuse or misuse, including lowered efficacy, development of secondary infections, adverse drug reactions, increased length of hospital stay, and additional healthcare costs.[114,115] The role of an effective control of infections in achievement of optimized clinical outcomes and decreased potential pharmacotherapy-related damage is also very fundamental. Infection control is clearly necessary and often enough to reduce hospital-acquired infections (HAIs). However, a comprehensive infection-control program, combined with an effective antimicrobial stewardship agenda, synergistically limit the emergence and spread of antimicrobial-resistant bacteria, reduce HAIs, control resistance, and improve overall inpatient care.[114,116] With antimicrobial resistance on the rise worldwide and few new agents in development, antimicrobial stewardship programs are more important than ever in ensuring the continued efficacy of available antimicrobials.

The design of antimicrobial management programs should be based on the best current understanding of the relationship between antimicrobial use and resistance. Such programs should be administered by multidisciplinary teams composed of infectious diseases physicians, clinical pharmacists, clinical microbiologists, and infection control practitioners and should be actively supported by hospital administrators. Strategies for changing antimicrobial prescribing behavior include education of prescribers regarding proper antimicrobial usage, creation of an antimicrobial formulary with restricted prescribing of targeted agents, and review of antimicrobial prescribing with feedback to prescribers.[117] Clinical computer systems can aid in the implementation of each of these strategies, especially as expert systems able to provide patient-specific data and suggestions at the point of care. Antibiotic rotation strategies control the prescribing process by scheduled changes of antimicrobial classes used for empirical therapy. When instituting an antimicrobial stewardship program, a hospital should tailor its choice of strategies to its needs and available resources.

Recent guidelines established by the Infectious Diseases Society of America and Society for Healthcare Epidemiology of America (IDSA/SHEA) make specific recommendations for the development of institutional programs to enhance antimicrobial stewardship.[116,118] Optimally, such programs should be comprehensive, multidisciplinary, supported by hospital and medical staff leadership, and should employ evidence-based strategies that best fit local needs and resources.[114,115] Principal proactive strategies—with evidence supporting their consideration—include prospective audits, with intervention and feedback, formulary restriction, and preauthorization.[107,117] In addition, other strategies include persistent one-on-one and large group education, guidelines adapted to local needs; computer based informatics to support clinical decision making, using the antimicrobial stewardship program to adapt national guidelines to local antimicrobial use; microbiology, and resistance patterns[118,119]; and the use of rapid molecular diagnostic testing. All of these have also been shown to improve antimicrobial utilization at hospitals. A summarized list of potential strategies or elements of an antimicrobial stewardship program[107,117] is listed in Table 55.3.

The size and nature of the institution can make a big difference in determining what program to set up and what elements it should entail. The program components and effectiveness of each will differ for community hospitals compared to academic medical centers. A comprehensive program includes active monitoring, fostering of appropriate antimicrobial use, and collaboration with an effective infection control program as well as other hospital entities.[114] The role of a multidisciplinary team, with administrative support, is particularly underscored in the guidelines. According to the guidelines, core members of the multidisciplinary team should include an infectious diseases physician and a clinical pharmacist with infectious diseases training. It should also ideally include a clinical microbiologist, information system specialist, infection control professional, and hospital epidemiologist,[115,117] and all members understanding their responsibility for the successful implementation of the program and oversight of its implementation and daily functions.

Most community hospitals, if sufficiently resourced, should be able to implement a successful antimicrobial stewardship program. Evidence suggests that good antimicrobial stewardship can lead to less overall and inappropriate antimicrobial use, lower drug-related costs, reductions in antimicrobial use, improve patient outcomes, and have the potential to prevent or control the emergence of antimicrobial-resistant organisms.[120] For these programs to be successful, they must address the specific needs of individual institutions and be built on available resources, the limitations and advantages of each institution, and the available staffing and technological infrastructure.

Summary

The emergence of antibiotic-resistant organisms is a major public health concern that continues to increase and is acknowledged to be a major threat to the treatment of infectious diseases, particularly among patients in hospitals and other healthcare settings. Antibiotic-resistant organisms appear to be biologically fit and can cause serious, life-threatening infections that are difficult to manage because of limited treatment options. This increase in the prevalence of drug-resistant pathogens is occurring at a time when the discovery and development of new anti-infective agents is slowing down dramatically. Consequently, there is concern that in the not-too-distant future, we may be faced with a growing number of potentially untreatable infections. Over the years, with the development of antimicrobial agents, several antibiotic-resistant pathogens have been identified as causes of serious infections among patients in hospital. These organisms are typically resistant to multiple classes of antimicrobial agents and are, therefore, called multidrug-resistant organisms and may also be considered a nosocomial antibiotic resistant organism because of their prevalence in hospital settings. Infections caused by antimicrobial-resistant organisms are almost always associated with increased attributable mortality, prolonged hospital stays and excess costs. Rates of emergence of antibiotic-resistant organisms will continue to increase unless aggressive control measures are implemented. These interventions must include enhanced surveillance of antibiotic resistance, attention to hand hygiene and other standard infection prevention and control measures, and antibiotic stewardship to ensure appropriate use of antimicrobial agents.

Overuse or misuse of antibiotics and other antimicrobials for hospital inpatients is relatively common and can be associated with several unintended negative consequences. Improving medical

TABLE 55.3

Summary of Strategies used in Antimicrobial Stewardship Programs[107,117]

Program Element	Strategy Procedure	Personnel	Advantages	Disadvantages
Guidelines and pathways	• Creation of guidelines for antimicrobial use • Best if local data and conditions are used to adapt guidelines to a specific institution	Antimicrobial committee to create guidelines	• May alter behavior patterns • Limits variation in therapy of infectious diseases • Best evidence-based • Assists in adherence with regulatory and third-party-payer stipulations	• Passive education likely ineffective • Often not utilized unless combined with other stewardship strategies or elements
Education and promotion of awareness in large group	• Group or individual education of clinicians by educators • Grand rounds or clinical staff meeting venues • Provides information to prescribers and thought-leader clinicians on justification for stewardship • Feedback antimicrobial susceptibility and use data to clinicians	Educators (physicians, pharmacists)	• Avoids loss of prescriber autonomy • Can reach many prescribers in a short period of time • Effective for communicating the need and rationale for subsequent stewardship interventions	• Not particularly effective in changing prescribing behavior without other interventions • Rapid extinction of gained knowledge
Formulary restriction or preauthorization	• Restrict dispensing of targeted antimicrobials to approved indications • "Front-end approach" • Formulary restriction or contact a stewardship team member to obtain authorization to prescribe a selected antimicrobial • Each intervention is a "mini-consult"	Antimicrobial committee to create guidelines	• Most direct control over antimicrobial use • Proven in clinical studies to reduce and modify antimicrobial consumption, improve selected clinical outcomes, and decrease antimicrobial expenditures • Together with infection control, effective in controlling outbreaks of resistant or secondary pathogens (such as *C. difficile*)	• Perceived loss of autonomy for prescribers • Less appealing to clinicians • Potential need for after-hours service • Time intensive • Potential for delay in antimicrobial administration
		Approval personnel (physician, infectious diseases fellow, clinical pharmacist)	• Individual educational opportunities	• Need for all-hours consultant availability
Prospective audit and feedback	• Daily review of targeted antimicrobials for appropriateness • "Back-end approach" • Identify and intervene on patients already started on antimicrobials • Interventions include changing, streamlining, de-escalation, pharmacodynamic/dose optimization, IV to PO switch, and limitation of duration of therapy	Antimicrobial committee to create guidelines	• Avoids loss of autonomy for prescribers	• Compliance with recommendations and adherence to stewardship interventions by the clinician is voluntary • Resource intensive • Requires a greater amount of team-member training and experience in anti-infective therapy
	• Contact prescribers with recommendations for alternative therapy	Review personnel (usually clinical pharmacist)	• Individual educational opportunities	
Computer assistance in physician order Entry and clinical Decision support	• Use of information technology to implement previous strategies • Often entails modification of existing or purchasing of additional informatics	Antimicrobial committee to create rules for computer systems	• Provides patient-specific data where most likely to impact (point of care) • Shown in limited clinical studies to reduce and modify antimicrobial consumption, improve selected clinical outcomes, and decrease antimicrobial expenditures • Once established, can greatly assist with implementation of guidelines and best-evidence therapy. • Reduces adverse events related to antimicrobials	• Significant time and resource investment to design and implement sophisticated systems • Expensive • Not readily available

TABLE 55.3 *(Continued)*

Summary of Strategies used in Antimicrobial Stewardship Programs[107,117]

Program Element	Strategy Procedure	Personnel	Advantages	Disadvantages
	• Expert systems provide patient-specific recommendations at point of care (order entry)	Personnel for approval or review (physicians, pharmacists) Computer programmers	• Facilitates other strategies	
Antimicrobial cycling	• Scheduled rotation of antimicrobials used in hospital or unit (e.g., intensive care unit) • Changing antimicrobial protocols periodically in an attempt to reduce selection pressure for resistance	Antimicrobial committee to create cycling protocol	• Potential to decrease antimicrobial resistance for an institution or geographic unit by changing selective pressure	• Difficult and labor-intensive to ensure adherence to cycling protocol • Not consistently shown in clinical trials to improve clinical outcomes or decrease resistance • Often increases antimicrobial consumption
		Personnel to oversee adherence (pharmacist, physicians)		• Theoretical concerns about effectiveness
Microbiology interventions	• Includes cascade reporting to "hide" antimicrobial susceptibilities that might promote suboptimal therapy (e.g., fluoroquinolone susceptibility for invasive *S. aureus*) • Assistance with choices of automated susceptibility profile, communication of new or changes in testing protocols • Preauthorization of susceptibility testing for unconventional antibiotics	Clinical microbiologist	• Potential to improve antimicrobial use and anti-infective therapy for the individual patient	• Not well studied
Rapid diagnostics	• Includes PCR and antigen testing of clinical specimens or early culture growth with rapid turnaround of test results	Clinical microbiology and diagnostics laboratory personnel	• Provides opportunity for early targeted therapy • Assists with de-escalation • Shown in very limited studies to decrease antimicrobial consumption and improve clinical outcomes	• Not readily available • Expensive

care necessarily includes better use of antimicrobials to optimize outcomes and preserve the effectiveness of currently available agents. Further, an important additional consequence of effective antimicrobial stewardship and improved patient care is typically a lowering of overall healthcare costs. The recent 2007 IDSA/SHEA guidelines provide recommendations for developing an institutional program to enhance antimicrobial stewardship. However, individual institutions need to look closely at their own systems and patients to develop an antimicrobial stewardship program that best serves the needs of their hospital and the people it serves.

REFERENCES

1. Zaman SB, Hussain MA, Nye R, Mehta V, Mamun KT, Hossain N. A review on antibiotic resistance: Alarm bells are ringing. *Cureus*. 2017;9(6):e1403.
2. Clatworthy AE, Pierson E, Hung DT. Targeting virulence: a new paradigm for antimicrobial therapy. *Nat Chem Biol*. 2007;3(9):541–548.
3. Centers of Disease Control. Antibiotic Resistance Threats in the United States, 2013. https://www.cdc.gov/drugresistance/pdf/ar-threats-2013-508.pdf. Accessed September 7, 2016.
4. Executive Order – Combating Antibiotic-Resistant Bacteria. whitehouse.gov. https://obamawhitehouse.archives.gov/the-press-office/2014/09/18/executive-order-combating-antibiotic-resistant-bacteria. Published September 18, 2014. Accessed October 15, 2019.
5. Patient Safety Atlas (PSA). https://gis-cdc-gov.eproxy.ketchum.edu/grasp/PSA/MapView.html. Accessed October 17, 2019.
6. MacGowan AP, on behalf of the BSAC Working Parties on Resistance Surveillance. Clinical implications of antimicrobial resistance for therapy. *J Antimicrob Chemother*. 2008;62(Supl 2):ii105–ii114.
7. Friedman ND, Temkin E, Carmeli Y. The negative impact of antibiotic resistance. *Clin Microbiol Infect*. 2016;22(5):416–422.
8. Gandra S, Barter DM, Laxminarayan R. Economic burden of antibiotic resistance: How much do we really know? *Clin Microbiol Infect*. 2014;20(10):973–980.

9. Heymann D. Emerging Infections. In: Schaechter M, Ed. *The Desk Encyclopedia of Microbiology.* Amsterdam: Elsevier Academic Press; 2004.

10. Levy S. Antibiotic resistance: An ecological imbalance. In: Chadwick DJ, Goode J, Eds. *Antibiotic Resistance: Origins, Evolution, Selection and Spread.* Chichester, England: John Wiley and Sons; 1997.

11. Kirby WMM. Extraction of a highly potent penicillin inactivator from penicillin resistant *Staphylococci. Science.* 1944;99(2579):452–453.

12. Chambers HF. The changing epidemiology of *Staphylococcus aureus? Emerg Infect Dis.* 2001;7(2):178–182.

13. Barber M. Methicillin-resistant staphylococci. *J Clin Pathol.* 1961;14:385–393.

14. Panlilio AL, Culver DH, Gaynes RP, et al. Methicillin-resistant *Staphylococcus aureus* in U.S. hospitals, 1975-1991. *Infect Control Hosp Epidemiol.* 1992;13(10):582–586.

15. Klevins M, Edwards J, Tokars J. The National Nosocomial Infections Surveillance System. Changes in the epidemiology of methicillin/oxacillin-resistant *Staphylococcus aureus* (ORSA) in U.S. intensive care units (ICUs): 1992–2002. Abstract 74. 2004; Proceedings of the International Conference on Emerging Infectious Diseases (Atlanta).

16. McDonald LC, Hageman JC. Vancomycin intermediate and resistant *Staphylococcus aureus.* What the nephrologist needs to know. *Nephrol News Issues.* 2004;18(11):63–64, 66–67, 71–72 passim.

17. McGuinness WA, Malachowa N, DeLeo FR. Vancomycin resistance in *Staphylococcus aureus. Yale J Biol Med.* 2017;90(2):269–281.

18. Fridkin SK, Hageman J, McDougal LK, et al. Epidemiological and microbiological characterization of infections caused by *Staphylococcus aureus* with reduced susceptibility to vancomycin, United States, 1997-2001. *Clin Infect Dis.* 2003;36(4):429–439.

19. McDonald LC. Trends in antimicrobial resistance in health care–associated pathogens and effect on treatment. *Clinical Infectious Diseases.* 2006;42(Supplement_2):S65–S71.

20. Centers for Disease Control and Prevention (CDC). Notes from the Field: Vancomycin-Resistant Staphylococcus aureus — Delaware, 2015. MMWR Morb Mortal Wkly Rep. September 25, 2015. Available at: https://www.cdc.gov/mmwr/preview/mmwrhtml/mm6437a6.htm.

21. Leclercq R, Derlot E, Duval J, Courvalin P. Plasmid-mediated resistance to vancomycin and teicoplanin in *Enterococcus faecium. N Engl J Med.* 1988;319(3):157–161.

22. Flokas ME, Karageorgos SA, Detsis M, Alevizakos M, Mylonakis E. Vancomycin-resistant enterococci colonisation, risk factors and risk for infection among hospitalised paediatric patients: A systematic review and meta-analysis. *Int J Antimicrob Agents.* 2017;49(5):565–572.

23. Kampmeier S, Kossow A, Clausen LM, et al. Hospital acquired vancomycin resistant enterococci in surgical intensive care patients – a prospective longitudinal study. *Antimicrob Resist Infect Control.* 2018;7(1):103.

24. Exner M, Bhattacharya S, Christiansen B, et al. Antibiotic resistance: What is so special about multidrug-resistant Gram-negative bacteria? *GMS Hyg Infect Control.* 2017;12:Doc05.

25. D'Agata EMC. Rapidly rising prevalence of nosocomial multidrug-resistant, Gram-negative bacilli: A 9-year surveillance study. *Infect Control Hosp Epidemiol.* 2004;25(10):842–846.

26. CDC. *Acinetobacter baumannii* infections among patients at military medical facilities treating injured U.S. service members, 2002-2004. *MMWR Morb Mortal Wkly Rep.* 2004;53(45):1063–1066.

27. Archibald LK, Banerjee SN, Jarvis WR. Secular trends in hospital-acquired *Clostridium difficile* disease in the United States, 1987-2001. *J Infect Dis.* 2004;189(9):1585–1589.

28. McDonald LC, Owings M, Jernigan DB. *Clostridium difficile* infection in patients discharged from US Short-stay hospitals, 1996–20031. *Emerg Infect Dis.* 2006;12(3):409–415.

29. Layton B, McDonald L, Gerding D, Liedtke L, Strausbaugh L. Perceived increases in the incidence and severity of *Clostridium difficile* disease: an emerging threat that continues to unfold [abstract 66], Proceedings of the 14th Annual Scientific Meeting of the Society for Healthcare Epidemiology of America (Los Angeles). Alexandria, VA. 2005; Society for Healthcare Epidemiology of America.

30. Sievert DM, Ricks P, Edwards JR, et al. Antimicrobial-resistant pathogens associated with healthcare-associated infections: summary of data reported to the national healthcare safety network at the centers for disease control and prevention, 2009-2010. *Infect Control Hosp Epidemiol.* 2013;34(1):1–14.

31. Vital Signs: Carbapenem-Resistant Enterobacteriaceae. http://www.cdc.gov/mmwr/preview/mmwrhtml/mm6209a3.htm. Accessed January 27, 2020.

32. Levy SB. The challenge of antibiotic resistance. *Sci Am.* 1998;278(3):46–53.

33. Harrison J, Svec T (1998). The beginning of the end of the antibiotic era? Part I. The problem: Abuse of the "miracle drugs." *Quintessence Int.* 29(3):151–162.

34. Cleveland Clinic. 2011. Health Information. Diseases & Conditions: Antibiotic Resistance. Available at: https://my.clevelandclinic.org/health/diseases/16142-antimicrobial-resistance.

35. Nemecek S. Beating bacteria. New ways to fend off antibiotic-resistant pathogens. *Sci Am.* 1997;276(2):38–39.

36. WHO. 1998. World Health Organization. Fact Sheet; 1998.

37. CDC. 2011. Centers for Disease Control and Prevention. Get Smart for Healthcare Topics. Fast facts. Nov 2011.

38. Harrison JW, Svec TA. The beginning of the end of the antibiotic era? Part II. Proposed solutions to antibiotic abuse. *Quintessence Int.* 1998;29(4):223–229.

39. Kummerer K. Resistance in the environment. *J Antimicrob Chemother.* 2004;54(2):311–320.

40. Kummerer K. Significance of antibiotics in the environment. *J Antimicrob Chemother.* 2003;52(1):5–7.

41. Chapin A, Rule A, Gibson K, Buckley T, Schwab K. Airborne multidrug-resistant bacteria isolated from a concentrated swine feeding operation. *Environ Health Perspect.* 2005;113(2):137–142.

42. APUA, 1999. Alliance for the Prudent Use of Antibiotics. Questions and Answers about Antibiotics and Resistance. About Resistance: A Societal Problem. June 1999.

43. Horrigan L, Lawrence RS, Walker P. How sustainable agriculture can address the environmental and human health harms of industrial agriculture. *Environ Health Perspect.* 2002;110(5):445–456.

44. Zuccato E, Calamari D, Natangelo M, Fanelli R. Presence of therapeutic drugs in the environment. *Lancet.* 2000; 355(9217):1789–1790.

45. Golet EM, Alder AC, Hartmann A, Ternes TA, Giger W. Trace determination of fluoroquinolone antibacterial agents in urban wastewater by solid-phase extraction and liquid chromatography with fluorescence detection. *Anal Chem.* 2001;73(15):3632–3638.

46. Richardson ML, Bowron JM. The fate of pharmaceutical chemicals in the aquatic environment. *J Pharm Pharmacol.* 1985;37(1):1–12.

47. Basu P. Antibiotic resistance: The impact on the environment and sustainability. *Int J Environment Sustain.* 2013;8(3):117–135.

48. Atkinson BA, Abu-Al-Jaibat A, LeBlanc DJ. Antibiotic resistance among enterococci isolated from clinical specimens between 1953 and 1954. *Antimicrob Agents Chemother.* 1997;41(7):1598–1600.

49. Morel CM, Mossialos E. Stoking the antibiotic pipeline. *BMJ.* 2010;340(May 18 2):c2115–c2115.

50. So AD, Gupta N, Cars O. Tackling antibiotic resistance. *BMJ.* 2010;340:c2071.

51. Edgar T, Boyd SD, Palame MJ. Sustainability for behaviour change in the fight against antibiotic resistance: A social marketing framework. *J Antimicrob Chemother.* 2008;63(2):230–237.

52. CBS News. 2009. "Superbug" Deaths in U.S. may Surpass AIDS. Associated Press. February 11, 2007. Available at: https://www.cbsnews.com/news/superbug-deaths-in-us-may-surpass-aids/.

53. Institute of Medicine (US) Forum on Emerging Infections. *The Resistance Phenomenon in Microbes and Infectious Disease Vectors: Implications for Human Health and Strategies for Containment: Workshop Summary.* Knobler SL, Lemon SM, Najafi M, Burroughs T, Eds. Washington, DC: National Academies Press; 2003. http://www.ncbi.nlm.nih.gov/books/NBK97138/. Accessed January 27, 2020.

54. Shea K, Florini K, Barlam T. When Wonder Drugs Don't Work: How Antibiotic resistance Threatens Children, Seniors, and the Medically Vulnerable, 48. Available at: https://www.edf.org/sites/default/files/162_abrreport.pdf

55. van Duin D, Perez F, Rudin SD, et al. Surveillance of carbapenem-resistant Klebsiella pneumoniae: tracking molecular epidemiology and outcomes through a regional network. *Antimicrob Agents Chemother.* 2014;58(7):4035–4041.

56. Queenan AM, Bush K. Carbapenemases: The versatile beta-lactamases. *Clin Microbiol Rev.* 2007;20(3):440–458.

57. Logan LK, Weinstein RA. The epidemiology of carbapenem-resistant Enterobacteriaceae: The impact and evolution of a global menace. *J Infect Dis.* 2017;215(suppl_1):S28–S36.

58. Bush K. The ABCD's of β-lactamase nomenclature. *J Infect Chemother.* 2013;19(4):549–559.

59. Bush K. Past and present perspectives on β Lactamases. *Antimicrob Agents Chemother.* 2018;62(10).

60. Perez F, El Chakhtoura NG, Papp-Wallace KM, Wilson BM, Bonomo RA. Treatment options for infections caused by carbapenem-resistant Enterobacteriaceae: Can we apply "precision medicine" to antimicrobial chemotherapy?. *Expert Opin Pharmacother.* 2016;17(6):761–781.

61. Monarch Pharmaceuticals. Colistin Package Insert. 2002.

62. Gales AC, Jones RN, Sader HS. Contemporary activity of colistin and polymyxin B against a worldwide collection of Gram-negative pathogens: Results from the SENTRY antimicrobial surveillance program (2006-09). *J Antimicrob Chemother.* 2011;66(9):2070–2074.

63. Ozbek B, Mataracı-Kara E, Er S, Ozdamar M, Yilmaz M. In vitro activities of colistin, tigecycline and tobramycin, alone or in combination, against carbapenem-resistant Enterobacteriaceae strains. *J Global Antimicrob Res.* 2015;3(4):278–282.

64. SteriMax Inc. Polymyxin B Package Insert. 2016.

65. Wyeth Pharmaceuticals Inc. Tigecycline Package Insert. 2010.

66. Paul M, Carmeli Y, Durante-Mangoni E, et al. Combination therapy for carbapenem-resistant Gram-negative bacteria. *J Antimicrob Chemother.* 2014;69(9):2305–2309.

67. Daikos GL, Tsaousi S, Tzouvelekis LS, et al. Carbapenemase-producing Klebsiella pneumoniae bloodstream infections: lowering mortality by antibiotic combination schemes and the role of carbapenems. *Antimicrob Agents Chemother.* 2014;58(4):2322–2328.

68. Qureshi ZA, Paterson DL, Potoski BA, et al. Treatment outcome of bacteremia due to KPC-producing Klebsiella pneumoniae: superiority of combination antimicrobial regimens. *Antimicrob Agents Chemother.* 2012;56(4):2108–2113.

69. Tumbarello M, Viale P, Viscoli C, et al. Predictors of mortality in bloodstream infections caused by Klebsiella pneumoniae carbapenemase-producing K. pneumoniae: Importance of combination therapy. *Clin Infect Dis.* 2012;55(7):943–950.

70. Gerding DN, Johnson S, Peterson LR, Mulligan ME, Silva J. *Clostridium difficile*-associated diarrhea and colitis. *Infect Control Hosp Epidemiol.* 1995;16(8):459–477.

71. Manabe YC, Vinetz JM, Moore RD, Merz C, Charache P, Bartlett JG. *Clostridium difficile* colitis: An efficient clinical approach to diagnosis. *Ann Intern Med.* 1995;123(11):835–840.

72. Brown KA, Khanafer N, Daneman N, Fisman DN. Meta-analysis of antibiotics and the risk of community-associated *Clostridium difficile* infection. *Antimicrob Agents Chemother.* 2013;57(5):2326–2332.

73. McDonald LC, Gerding DN, Johnson S, et al. Clinical practice guidelines for *Clostridium difficile* infection in adults and children: 2017 Update by the infectious diseases society of America (IDSA) and society for healthcare epidemiology of America (SHEA). *Clin Infect Dis.* 2018;66(7): e1–e48.

74. Louie TJ, Miller MA, Mullane KM, et al. Fidaxomicin versus vancomycin for *Clostridium difficile* infection. *N Engl J Med.* 2011;364(5):422–431.

75. CDC. The biggest antibiotic-resistant threats in the U.S. Centers for Disease Control and Prevention. https://www.cdc.gov/drugresistance/biggest-threats.html. Published November 14, 2019. Accessed November 29, 2019.

76. Lee C-R, Lee JH, Park M, et al. Biology of *Acinetobacter baumannii*: Pathogenesis, antibiotic resistance mechanisms, and prospective treatment options. *Front Cell Infect Microbiol.* 2017;7:55.

77. Noguchi JK, Gill MA. Sulbactam: A beta-lactamase inhibitor. *Clin Pharm.* 1988;7(1):37–51.

78. Kengkla K, Kongpakwattana K, Saokaew S, Apisarnthanarak A, Chaiyakunapruk N. Comparative efficacy and safety of treatment options for MDR and XDR *Acinetobacter baumannii* infections: A systematic review and network meta-analysis. *J Antimicrob Chemother.* 2018;73(1):22–32.

79. TetraPhase Pharmaceuticals. Xerava Package Insert.

80. Paratek Pharmaceuticals. Nuzyra Package Insert.

81. Jefferies JMC, Cooper T, Yam T, Clarke SC. *Pseudomonas aeruginosa* outbreaks in the neonatal intensive care unit - a systematic review of risk factors and environmental sources. *J Med Microbiol.* 2012;61(Pt_8):1052–1061.

82. Weiner LM, Webb AK, Limbago B, et al. Antimicrobial-resistant pathogens associated with healthcare-associated infections: Summary of data reported to the national health-care safety network at the centers for disease control and prevention, 2011-2014. *Infect Control Hosp Epidemiol.* 2016;37(11):1288–1301.

83. Wolter DJ, Lister PD. Mechanisms of β-lactam resistance among *Pseudomonas aeruginosa*. *Curr Pharm Des.* 2013;19(2):209–222.

84. Juan C, Moyá B, Pérez JL, Oliver A. Stepwise upregulation of the *Pseudomonas aeruginosa* chromosomal cephalosporinase conferring high-level beta-lactam resistance involves three AmpD homologues. *Antimicrob Agents Chemother.* 2006;50(5):1780–1787.

85. Hancock REW, Brinkman FSL. Function of pseudomonas porins in uptake and efflux. *Annu Rev Microbiol.* 2002;56:17–38.

86. Livermore DM. Of Pseudomonas, porins, pumps and carbapenems. *Journal of Antimicrobial Chemotherapy.* 2001; 47(3):247–250.

87. Doi Y, Arakawa Y. 16S ribosomal RNA methylation: Emerging resistance mechanism against aminoglycosides. *Clin Infect Dis.* 2007;45(1):88–94.

88. Nguyen L, Garcia J, Gruenberg K, MacDougall C. Multidrug-resistant pseudomonas infections: Hard to treat, but hope on the horizon?. *Curr Infect Dis Rep.* 2018;20(8):23.

89. Kalil AC, Metersky ML, Klompas M, et al. Management of adults with hospital-acquired and ventilator-associated pneumonia: 2016 clinical practice guidelines by the infectious diseases society of America and the American thoracic society. *Clin Infect Dis.* 2016;63(5):e61–e111.

90. Huband MD, Castanheira M, Flamm RK, Farrell DJ, Jones RN, Sader HS. In vitro activity of ceftazidime-avibactam against contemporary *Pseudomonas aeruginosa* isolates from U.S. Medical centers by census region, 2014. *Antimicrob Agents Chemother.* 2016;60(4):2537–2541.

91. Sader HS, Castanheira M, Flamm RK, Farrell DJ, Jones RN. Antimicrobial activity of ceftazidime-avibactam against Gram-negative organisms collected from U.S. medical centers in 2012. *Antimicrob Agents Chemother.* 2014;58(3):1684–1692.

92. Sader HS, Castanheira M, Mendes RE, Flamm RK, Farrell DJ, Jones RN. Ceftazidime-avibactam activity against multidrug-resistant *Pseudomonas aeruginosa* isolated in U.S. medical centers in 2012 and 2013. *Antimicrob Agents Chemother.* 2015;59(6):3656–3659.

93. van Duin D, Bonomo RA. Ceftazidime/avibactam and ceftolozane/tazobactam: Second-generation β-Lactam/β-Lactamase inhibitor combinations. *Clin Infect Dis.* 2016;63(2):234–241.

94. Moyá B, Zamorano L, Juan C, Ge Y, Oliver A. Affinity of the new cephalosporin CXA-101 to penicillin-binding proteins of *Pseudomonas aeruginosa*. *Antimicrob Agents Chemother.* 2010;54(9):3933–3937.

95. Farrell DJ, Flamm RK, Sader HS, Jones RN. Antimicrobial activity of ceftolozane-tazobactam tested against enterobacteriaceae and *Pseudomonas aeruginosa* with various resistance patterns isolated in U.S. Hospitals (2011-2012). *Antimicrob Agents Chemother.* 2013;57(12):6305–6310.

96. Lob SH, Hackel MA, Kazmierczak KM, et al. In vitro activity of imipenem-relebactam against gram-negative ESKAPE pathogens isolated by clinical laboratories in the united states in 2015 (Results from the SMART global surveillance program). *Antimicrob Agents Chemother.* 2017;61(6):e02209-16,/aac/61/6/e02209-16.atom.

97. Bush K, Jacoby GA. Updated functional classification of -Lactamases. *Antimicrob Agents Chemother.* 2010;54(3): 969–976.

98. Paterson DL, Bonomo RA. Extended-spectrum beta-lactamases: A clinical update. *Clin Microbiol Rev.* 2005;18(4): 657–686.

99. Livermore DM, Hope R, Mushtaq S, Warner M. Orthodox and unorthodox clavulanate combinations against extended-spectrum beta-lactamase producers. *Clin Microbiol Infect.* 2008;14(Suppl 1):189–193.

100. Meije Y, Pigrau C, Fernández-Hidalgo N, et al. Non-intravenous carbapenem-sparing antibiotics for definitive treatment of Bacteraemia due to Enterobacteriaceae producing extended-spectrum β-lactamase (ESBL) or AmpC β-lactamase: A propensity score study. *Int J Antimicrob Agents.* 2019;54(2):189–196.

101. Retamar P, López-Cerero L, Muniain MA, Pascual Á, Rodríguez-Baño J. ESBL-REIPI/GEIH Group. Impact of the MIC of piperacillin-tazobactam on the outcome of patients with bacteremia due to extended-spectrum-β-lactamase-producing *Escherichia coli*. *Antimicrob Agents Chemother.* 2013;57(7):3402–3404.

102. Harris PNA, Tambyah PA, Lye DC, et al. Effect of piper-acillin-tazobactam vs meropenem on 30-day mortality for patients with *E coli* or klebsiella pneumoniae bloodstream infection and ceftriaxone resistance: A randomized clinical trial. *JAMA.* 2018;320(10):984.

103. Metlay JP, Waterer GW, Long AC, et al. Diagnosis and treatment of adults with community-acquired pneumonia. An official clinical practice guideline of the American Thoracic Society and Infectious Diseases Society of America. *Am J Respir Crit Care Med.* 2019;200(7):e45–e67.

104. Hospira Inc. Vancomycin Package Insert. June 2012.

105. Sakoulas G, Moise-Broder PA, Schentag J, Forrest A, Moellering RC, Eliopoulos GM. Relationship of MIC and bactericidal activity to efficacy of vancomycin for treatment of methicillin-resistant *Staphylococcus aureus* bacteremia. *J Clin Microbiol.* 2004;42(6):2398–2402.

106. Lodise TP, Graves J, Evans A, et al. Relationship between Vancomycin MIC and failure among patients with methicillin-resistant *Staphylococcus aureus* bacteremia treated with vancomycin. *Antimicrobial Agents and Chemotherapy.* 2008;52(9):3315–3320.

107. MacDougall C, Polk RE. Antimicrobial stewardship programs in health care systems. *Clin Microbiol Rev.* 2005;18(4):638–656.

108. Kim JH, Gallis HA. Observations on spiraling empiricism: Its causes, allure, and perils, with particular reference to antibiotic therapy. *Am J Med.* 1989;87:201–206.

109. MacDougall C, Polk RE. Variability in rates of use of antibacterials among 130 US hospitals and risk-adjustment models for interhospital comparison. *Infect Control Hosp Epidemiol.* 2008;29:203–211.

110. Pakyz AL, MacDougall C, Oinonen M, Polk RE. Trends in antibacterial use in US academic health centers: 2002 to 2006. *Arch Intern Med.* 2008; 168:2254–2260.

111. Hecker MT, Aron DC, Patel NP, Lehmann MK, Donskey CJ. Unnecessary use of antimicrobials in hospitalized patients: Current patterns of misuse with an emphasis on the antianaerobic spectrum of activity. *Arch Intern Med.* 2003;163:972–978.

112. Polk RE, Fishman NO. Antimicrobial Stewardship. In: Mandell GL, Bennett JE, Dolin R, Eds. *Mandell, Douglas, and Bennett's Principles and Practice of Infectious Diseases, No. 1.* 7th ed. Philadelphia, PA: Churchill Livingstone Elsevier; 2010.

113. Weber DJ. Collateral damage and what the future might hold. The need to balance prudent antibiotic utilization and stewardship with effective patient management. *Int J Infect Dis.* 2006;10:S17–S24.

114. Loo VG, Poirier L, Miller MA, et al. A predominantly clonal multi-institutional outbreak of *Clostridium difficile*-associated diarrhea with high morbidity and mortality. *N Engl J Med.* 2005;353:2442–2449.

115. Dellit TH, Owens RC, McGowan JE, Jr et al. Infectious diseases society of America and the society for healthcare epidemiology of America guidelines for developing an institutional program to enhance antimicrobial stewardship. *Clin Infect Dis.* 2007;44:159–177.

116. Ohl CA. Antimicrobial stewardship. *Semin Infect Contr.* 2001;1:210–221.

117. Ohl CA, Luther VP, Antimicrobial stewardship for inpatient facilities. *J. Hosp. Med.* 2011;1;S4–S15.

118. Ibrahim EH, Ward S, Sherman G, Schaiff R, Fraser VJ, Kollef MH. Experience with a clinical guideline for the treatment of ventilator-associated pneumonia. *Crit Care Med.* 2001;29:1109–1115.

119. Singh N, Rogers P, Atwood CW, Wagener MM, Yu VL. Short-course empiric antibiotic therapy for patients with pulmonary infiltrates in the intensive care unit: A proposed solution for indiscriminate antibiotic prescription. *Am J Respir Crit Care Med.* 2000;162:505–511.

120. Ohl CA, Dodds Ashley ES. Antimicrobial stewardship programs in community hospitals: The evidence base and case studies, *Clin Infect Dis.* 2011;53(Suppl 1):S23–S28.

56

Emerging Antimicrobial-Resistant Microorganisms in the Community

Negin Alizadeh Shaygh, Divya Sarvaiya, and Paramita Basu

CONTENTS

Introduction

The discovery of penicillin by Sir Alexander Fleming in 1928 marked the beginning of the modern era of antibiotics.[1] Following this discovery, the first penicillin resistance was reported in 1950.[2] Despite the availability of numerous antimicrobial agents available to treat microbial infections, sometimes these agents become ineffective against microorganisms such as bacteria, viruses, fungi, and protozoa. As a result infections persisted in the body increasing the risk of spread to others. Resistant microorganisms are emerging globally and are threatening our ability to fight common infections resulting in increased morbidity and mortality. Over time, changes occur naturally in the genetic composition of microbes due to overuse and misuse of antimicrobial agents which is one of the primary reasons for the emergence of antimicrobial resistance (AMR). Nearly two million Americans per year develop healthcare-associated infections (HAIs), resulting in 99,000 deaths. Most of the mortalities are due to antimicrobial-resistant pathogens. In 2006, two common HAIs, sepsis and pneumonia, were responsible for the deaths of nearly 50,000 Americans, costing the U.S. healthcare systems more than $8 billion. Estimates regarding the medical cost per patient with an antibiotic-resistant infection range from $18,588 to $29,069. The total economic burden placed on the U.S. economy is estimated to be as high as $20 billion in healthcare and $35 billion in lost productivity because of AMR.[2] AMR has emerged in Europe and North America as well due to inappropriate use of antibiotics in humans and animals. Resistant pathogens such as methicillin-resistant *Staphylococcus aureus* (MRSA), penicillin nonsusceptible *Streptococcus pneumoniae* (PNSSP), vancomycin-resistant enterococci (VRE), and extended-spectrum β-lactamase (ESBL)-producing *Enterobacteriaceae* have emerged and spread into communities and hospitals.[3] In a recent study conducted at Children's hospital in China, it was found that there are five main pathogens responsible for multidrug-resistant infections. Extended spectrum β-lactamase (ESBL) was found in 1.9% cases making it the most common pathogen.[4] The second most common was MRSA, accounting for 1.51%.[4] The third, fourth, and fifth most common pathogens were Carbapenem-resistant *Enterobacteriaceae* (CRE) in 0.17% of the cases, Carbapenem-resistant *Acinetobacter baumannii* (CRAB) in 0.10% of the cases, and multidrug-resistant/pandrug-resistant *Pseudomonas aeruginosa* (MDR/PDR-PA) in 0.03% of the cases, respectively.[4]

AMR Mechanisms

Medical practice was transformed after discovery of antibiotics in early 20th century; however, antibiotic lethality is a complex system level process that is sensitive to the external environment.[5] Antibiotics inhibit cellular processes in microorganisms, retarding their growth and causing cell death. Microorganisms can develop resistance via resistance conferring mutation when they are exposed to concentrations of drugs below their minimum bactericidal concentration (MBC). Genetic mutations in these microorganisms arise from plasmids. These plasmids code for mutations or resistant genes in microbial chromosomes itself.[6]

Resistance Prevalence and Their Significance

The Infectious Diseases Society of America (IDSA) Emerging Infectious Network conducted a national survey in 2011 and found that more than 60% of participants had seen a pan-resistant, untreatable bacterial infection within a period of 1 year. In 2013, the Centers for Disease Control and Prevention (CDC) declared that the human race is now in the "post antibiotic era." CDC assesses antibiotic resistance based on the following factors: Clinical impact, economic impact, incidence, 10-year projection of incidence, transmissibility, availability of effective antibiotics, and barriers to prevention. Based on the impact, as assessed by CDC, Table 56.1 classifies the threat level of each bacteria as Urgent, Serious, Concerning, or to be put on the watch list.

In Europe, the combined resistance to fluoroquinolones, third-generation cephalosporins, and aminoglycosides increased between 2011 and 2014 in isolates of Gram-negative bacteria, including *Klebsiella pneumoniae* (from 16.7% to 19.6%) and *Escherichia coli* (from 3.8% to 4.8%).[7] Also, according to data published by CDC in 2002, foodborne diseases account for approximately 76 million illnesses, 325,000 hospitalizations, and 5,000 deaths each year in the United States alone.[52] Antimicrobial-resistant zoonotic foodborne bacterial pathogens are an undesired consequence of use of antimicrobials in animals, and subsequent transmission to humans as food causes foodborne diseases. In this regard, five pathogens account for more than 90% of estimated food-related deaths: Salmonella (31%), listeria (28%), toxoplasma (21%), Norwalk-like viruses (7%), *Campylobacter* (5%), and *E. coli* (3%).[29]

Causative Factors

Factors that contribute toward AMR are as following: Over use and misuse of antimicrobial agents, inappropriate prescribing methods, extensive agricultural use, availability of few new antibiotics, and regulatory barriers.[2] Level of education, low income, housing conditions, water, and sanitation are some of the other factors that are positively associated with AMR. In low- and middle-income countries, issues like housing, clean water and sanitation, lack of infrastructure, and access to vaccines leads to recurring infections and the use of antimicrobial agents, which also contributes to the emergence of resistance.[9] For example, before 1995 in countries like Taiwan, drugs were available in drug stores without prescriptions, which greatly contributed to AMR.[3] It is difficult and expensive to treat infections and harder to control epidemics when the effectiveness of antibiotics is declining. Of all the infections, bacterial infections contribute the most to human and animal diseases in developing countries and hence the emergence of antibiotic resistance. Also, unfortunately, not much has been accomplished in developing countries due to poverty and the lack of adequate resources when compared with developed countries.[12]

TABLE 56.1

Bacteria and Fungi Listed in the 2019 Antibiotic Resistance Threats Report

Urgent Threats	Serious Threats	Concerning Threats	Watch List
• Carbapenem-resistant *Acinetobacter* • *Candida auris* • *Clostridioides difficile* • Carbapenem-resistant *Enterobacteriaceae* • Drug-resistant *Neisseria gonorrhoeae*	• Drug-resistant *Campylobacter* • Drug-resistant *Candida* • ESBL-producing *Enterobacteriaceae* • Vancomycin-resistant *Enterococci* (VRE) • Multidrug-resistant *Pseudomonas aeruginosa* • Drug-resistant nontyphoidal *Salmonella* • Drug-resistant *Salmonella* serotype Typhi • Drug-resistant *Shigella* • Methicillin-resistant *Staphylococcus aureus* (MRSA) • Drug-resistant *Streptococcus pneumoniae* • Drug-resistant tuberculosis	• Erythromycin-resistant Group A *Streptococcus* • Clindamycin-resistant Group B *Streptococcus*	• Azole-resistant *Aspergillus fumigatus* • Drug-resistant *Mycoplasma genitalium*

Source: Centers for Disease Control and Prevention website, www.cdc.gov/drugresistance/pdf/threats-report/2019-ar-threats-report-508.pdf.

Immune System Response and Duration of Antimicrobial Therapy

The correct length and timing of antimicrobial therapy are two important factors in effectively treating bacterial infections with minimal risk of the development of resistance. The immune system is effective in killing bacteria when bacterial load is below the minimal infecting dose and no infection develops in patient. Antimicrobial therapy of nine days reduces the bacterial load below that threshold and prevents the progression of infection. However, when the duration of therapy is less, for example six days, then bacterial load is not reduced below the threshold and bacterial infection persists. Such short therapies and early interruption of antibiotics are advantageous for resistant strains causing infections by these resistant bacteria to progress. Thus, it is important to develop optimal antibiotic regimens which can prevent the progression of infection with antibiotic-resistant strains. Delayed start of therapy is also one of the key factors that promotes a rapid rise in resistant strains.[16]

The quantity of antimicrobial drugs prescribed is overall highest in the community healthcare setting. Although the link between human antimicrobial use and resistance seems clear cut, this association is complex. The confounding factors influencing antimicrobial resistance in the community include pathogen-drug interactions, pathogen-host interactions, mutation rates of the pathogen, emergence of successful antimicrobial-resistant clones, the transmission rates of pathogens between human beings, animals, and the environment, cross-resistance, and selection of co-resistance to unrelated drugs. It is also proven by various studies that emergence in AMR can be the result of selective pressure exerted by antimicrobial use outside of human medicine. Use of antimicrobial agents in veterinary medicine, food-animal and fish production, and agriculture contributes towards AMR.[27]

Role of Environment

Antibacterial agents can enter the environment via several routes, namely sewage from communities or hospitals, as well as through manure and water bodies. The accumulation of antimicrobial agents in the environment makes it a gigantic reservoir for resistant microorganisms. A hotspot for horizontal gene transfer and coselection of genetic determinants is wastewater treatment plants, which provide organisms that are resistant to antibiotics, pollutants, heavy metals, biocides, disinfectants, or detergents. To address the risks of environmental exposure strategies should aim at improving industrial systems for sanitation and decontamination of hospital sewage water.[44]

Antibiotic Resistance and Epidemiological Trends

MRSA

S. aureus is the most commonly isolated human bacterial pathogen and an important cause of skin and soft-tissue infections (SSTIs), endovascular infections, pneumonia, septic arthritis, endocarditis, osteomyelitis, foreign-body infections, and sepsis.[94] MRSA isolates are resistant to all available penicillin and other β-lactam antimicrobial drugs. They were once confined largely to hospitals, other healthcare environments, and patients visiting these facilities. Since the mid-1990s, however, there has been an explosion in the number of MRSA infections reported for populations lacking risk factors for exposure to the healthcare system.[17] This increase has been associated with the recognition of new MRSA strains, often called community-associated MRSA (CA-MRSA) strains that have been responsible for a large proportion of the increased disease burden observed in the last decade. These CA-MRSA strains appear to have rapidly disseminated among the general population in most areas of the United States and affect patients with and without exposure to the healthcare environment.

The first case of MRSA was reported in the United Kingdom in 1962 and in the United States in 1968.[2] Since MRSA is resistant to penicillin and a number of similar β-lactam antibiotics, a number of other antibiotics including glycopeptides (e.g., vancomycin and teicoplanin), linezolid, tigecycline, daptomycin, and even some new β-lactams, such as ceftaroline and ceftobiprole are still active against MRSA, which shows tremendous versatility at emerging and spreading in different epidemiological settings over time (in hospitals, the community, and, more recently, in animals). Resistance to anti-MRSA agents usually occurs through bacterial mutation, but there have been reports of the horizontal transfer of resistance to linezolid and glycopeptide antibiotics, eliciting major concern. An earlier study conducted in Taiwan in 2002 reported that the prevalence of MRSA increased from 26.7% to 75% from 1980 to 1998 and reached to 84% in 2000. Also, 40% of out-patients reported incidence of MRSA in community settings.[3] But aggressive preventive hygiene measures in the community and hospitals have led to a decline in the incidence of MRSA infections between 2005 and 2011, where overall rates of invasive MRSA dropped 31%, while the largest declines (around 54%) were observed in healthcare–associated infections.[2] Data reported in a systematic review by Nellums et al. showed that prevalence of MRSA is 0.3–60% in Europe, 12–89% in Africa, and 10–53% in eastern Mediterranean countries.[7]

Resistance to first-line drugs used to treat infections caused by *S. aureus*, a common cause of severe infections in health facilities and the community, is widespread. People with MRSA are estimated to be 64% more likely to die than people with a nonresistant form of the infection per reports from WHO in 2018.[1] A recent shift in the epidemiological profile of MRSA has resulted not only in healthcare–associated infections, but now, it has recently been appreciated as a cause of community-associated infections. Some reports have also shown decreased susceptibility, and even resistance of these pathogens to vancomycin. The increasing prevalence of MRSA in hospitals has led to the increased use of vancomycin for treatment, leading to the emergence of decreased susceptibility and even resistance to vancomycin. There are now 14 isolates of vancomycin-intermediate *S. aureus* in the United States.[37] Vancomycin resistant *S. aureus* are susceptible to daptomycin, which is an antibacterial agent approved by the Food and Drug Administration (FDA) since 2006. MRSA infection in the community setting has been documented in patients without known risk factors for MRSA infection (e.g., recent hospitalization, recent surgery, residence in a long-term care facility, or injection drug use). The prevalence of community-associated MRSA (as a proportion of all MRSA infections) varies according to the geographic region of the United States (from 9% in

Maryland to 20% in Georgia) and according to race within the geographic regions (43% and 17% among African Americans and whites, respectively, in Georgia, and 12% and 5%, respectively, in Maryland.)[53,93] According to data collected during the U.S. influenza season from 2003 to 2004, 17 cases of community-acquired staphylococcal pneumonia were reported to the CDC, in which 15 of the cases were caused by MRSA.[95]

The emergence of MRSA in the community has an obvious effect on the choices of empirical treatment for skin and soft-tissue infections and community-acquired pneumonia. In regions where MRSA is prevalent in the community, clindamycin or trimethoprim-sulfamethoxazole has been proposed for the treatment of skin and soft-tissue infections. Sometimes this includes incision and drainage of purulent lesions where possible.[36] Cosgrove et al. evaluated two cohort studies that reported that methicillin resistance is also associated with significant increases in the median length of hospital stay after acquisition of infection and hospital charges after *S. aureus* bacteremia.[14]

Strains of *S. aureus* belonging to a few phage types that possessed an inducible penicillinase gradually spread throughout the world's hospitals in the 1950s and throughout communities everywhere in the 1960s.[40] In addition, 6 months after methicillin was marketed in October, 1960, three methicillin-resistant isolates were reported after the screening of 5,000 clinical isolates.[23,29] However, since methicillin resistance remained rare and was expressed only by the resistant organisms in conditions that seemed to be very different from those prevailing at the site of infections (such as at low temperatures and high salt concentrations), methicillin and its congeners were deemed successful as antistaphylococcal agents with long-term effectiveness. But, by 1967, the situation started to change and multidrug-resistant MRSA was reported from Switzerland, France, Denmark, England, Australia, and India. About 15% of all these *S. aureus* isolates were methicillin resistant with combined resistance to penicillin, streptomycin, tetracycline, and occasionally to erythromycin. Most of these isolates belonged to the same phage complex.[31] After the rapid dissemination of this clone, MRSA began to fall in Europe in the 1970s and early 1980s. There were renewed concerns in the early 1980s, when a rise in the frequency of gentamicin-resistant MRSA was reported from several countries, including the United States, Ireland, and the United Kingdom.[10,19,23,32]

Community-Acquired MRSA

The majority of documented MRSA infections are acquired nosocomially, with community acquired MRSA restricted to patients with frequent contact with health facilities, such as residents of long-term care facilities and intravenous drug users. In 1993, novel MRSA strains were reported from Western Australia, which seemed to be community acquired, since they had been isolated from Indigenous Australian patients who had not been previously exposed to the healthcare system. These strains were characteristically different from hospital-acquired MRSA in many ways. Firstly, they were more susceptible to antibiotic classes other than β-lactam antibiotics; secondly, their genotypes were not the same as isolates from local hospitals; thirdly, they harbored different methicillin-resistance cassettes; and finally, community isolates were more likely to encode a putative virulence factor called Panton-Valentine leukocidin.[20]

Community-acquired MRSA has been reported most often from indigenous populations,[22] homeless people,[23] men who have sex with men,[45] jailed inmates (Control and Prevention 2003), military recruits,[53] children in daycare centers,[47] and competitive athletes (Control and Prevention 2003). Common to all these groups is high intensity physical contact which might help with transmission. A publication from Texas, USA showed that more than 70% of community-acquired *S. aureus* infections in clinical settings were MRSA.[30] In 2002, MRSA with vancomycin resistance (VRSA) was isolated from two independent patients who were coinfected with vancomycin-resistant enterococci.[92] Although few publications have addressed this issue, heterogeneously expressed vancomycin resistance has proven to be associated with treatment failure, which is defined as persistent bacteremia and fever for longer than seven days. However, successful treatment with alternative agents (e.g., linezolid) has been documented after vancomycin treatment had failed.[28] MRSA is at present, the most commonly identified antibiotic-resistant pathogen in many parts of the world, including Europe, the Americas, north Africa, the middle east, and east Asia. Moreover, MRSA rates have been swiftly increasing worldwide over the past decades, as shown by data from continuing surveillance initiatives such as the National Nosocomial Infection Surveillance System and European Antimicrobial Resistance Surveillance Systems.[21,50,51] MRSA originates from the introduction of a large mobile genetic element called staphylococcal cassette chromosome mec (SCCmec) into a methicillin-susceptible *S. aureus* strain. Up to now, five types of SCCmec (SCCmec I to V) and several variants have been described according to the combination of the mec and ccr gene complexes they contain.[23]

Carbapenem-Resistant Enterobacteriaceae (CRE)

Out of 140,000 healthcare-associated *Enterobacteriaceae* infections occurring in the United States each year, 9,300 are caused by CRE, resulting in an average of 600 deaths. The two most common types of CRE causing these fatalities are carbapenem-resistant *Klebsiella* species and carbapenem resistant *E. coli*.[2]

Vancomycin-Resistant Enterococci (VRE)

The progression of vancomycin resistance amongst enterococci is increasing; however, vancomycin resistance in *Enterococcus faecium* is already at high levels. The first case of vancomycin resistance was reported in Europe in 1988. VRE on its own is not only an important reason of blood stream, wound, and urinary tract infection in immunocompromised and severely unwell patients, but also can serve as a supply of genetic determinants of vancomycin resistance genes, which can be transferred *in vivo* to MRSA in similar patients.[36] Enterococci mainly causes blood stream, surgical site infections, and urinary tract infections in hospital as well as other healthcare settings, making it a pathogen that poses a major therapeutic challenge. Every year, it is estimated that 66,000 HAI enterococci infections occur in the United States, of which approximately 20,000 (30%) are resistant to vancomycin. This leads to more than 1,300 deaths. The agents currently used to fight VRE are linezolid and quinupristin/dalfopristin since the roles of daptomycin and tigecycline are not very well-defined in the treatment of VRE infections. Novel drugs like oritavancin might have bactericidal activity against VRE, however there is tremendous interest in developing others.[2] Vancomycin was being used for more than two decades before resistance was seen by a species of bacteria that were supposed to be susceptible. Not only was resistance-found on plasmids and transposons, but the resistance genes may have assembled an intricate, elaborate genetic construct.[40]

Multidrug-Resistant (MDR) Streptococcus pneumoniae

One of the major causes of bacterial pneumonia and meningitis, bloodstream, ear, and sinus infections is *S. pneumoniae*. The majority of morbidity and mortality occurs in adult patients above the age of 50, with the highest susceptibility in people 65 and above. Bacteria are fully resistant to one or more antibiotics in almost 30% of severe *S. pneumoniae* cases. Hence, in 2010, a new version of pneumococcal conjugate vaccine PCV13 was prepared against the most resistant pneumococcal strains. Thus, there is a decline in rates of resistant *S. pneumoniae* infections. PCV7, an earlier pneumococcal conjugate vaccine protected against only seven strains compared to PCV13, which protects against 13 strains. This vaccine not only protects against pneumococci, but also it reduces the spread of antibiotic resistance by blocking the transmission of resistant *S. pneumoniae* strains.[2] In Taiwan in 1999, the prevalence of MDR *S. pneumoniae* was 60%, which increased to 80% in 2000. Of these, 20–30% were penicillin-intermediate-resistant and 40–50% were penicillin-resistant strains. Of these penicillin-nonsusceptible *S. pneumoniae* (PSSNP), 60% also showed nonsusceptibility to extended spectrum cephalosporins and carbapenems, as well showing resistance to multiple antibiotics.[3] The problem of the emergence of AMR worsens due the ease of transmission and asymptomatic colonization in the upper respiratory tract of children. The most important determinant in the emergence of AMR in *S. pneumoniae* is selective pressure of the antimicrobial agents. The more the potent a drug, the less likely will be its selection for resistance.[18]

Mycobacterium Tuberculosis (TB) Drug Resistance

In 2012, 170,000 mortalities were reported from drug-resistant TB infections by WHO.[1] In 2011, 10,528 TB cases were reported in the United States, of which 1,042 (9.9%) cases were identified as antibiotic resistant. Incomplete, incorrect, or unavailability of treatment and lack of new drugs were identified as major causes.[2] WHO reported 480,000 cases of multidrug-resistant tuberculosis (MDR-TB) worldwide in 2014, of which only half of the MDR-TB cases were treated successfully. A form of TB that is resistant to at least four of the core anti-TB drugs is said to be extensively drug-resistant TB (XDR-TB). Prevalence of XDR-TB has been identified in 105 countries. It is also estimated that 9.7% people with MDR-TB have XDR-TB.[1]

Salmonella Drug Resistance

Human salmonellosis is one of the main causes of hospitalization and community outbreaks in developing countries. Despite an incidence of global morbidity, mortality is restricted to developing countries. A systematic review of the global morbidity and mortality of typhoid fever and paratyphoid fever found 13.5 million cases of typhoidal salmonella (TS) serotypes. Typhi, para typhi A and B have been reported especially in low- and middle-income countries in 2012.[11] However, nontyphoidal salmonella (NTS) infections are most notorious and represent 80% of infectious cases.[46] NTS is generally limited to gastroenteritis, but sometimes can get into the bloodstream and cause life-threatening bacteremia both in children and adults. Serotypes Typhimurium and Enteritidis are the most common agents of NTS outbreaks globally. Salmonellosis treatment is usually not required for straightforward gastroenteritis, but it is necessary for an invasive nontyphoidal salmonella (iNTS) infection. Drugs of choice for many years have been chloramphenicol, ampicillin, and Bactrim, but worldwide microbial resistance has been seen for these agents. Fluoroquinolones (such as ciprofloxacin and ofloxacin) and extended-spectrum cephalosporins (such as ceftriaxone and cefotaxime) have been preferred by the clinicians in the most recent years, although most iNTS have acquired resistance to these both antibiotics too.[41,49]

The situation has further deteriorated with the emergence of multidrug-resistant strains of salmonella, which are associated with long-term hospitalizations and increased risk of more severe outcomes (death). AMR to salmonella is a serious threat to public health in the United States. Enteritidis accounts for about half the incidence of ciprofloxacin nonsusceptible infections; Newport, Typhimurium, and Heidelberg for three-fourths of the incidence of infections with resistance to both ceftriaxone and ampicillin; and Typhimurium for more than half the incidence of infections with resistance to ampicillin only.[39,96] The main source of infection has been foods, such as beef, poultry, eggs, dairy products, etc. Most nontyphoidal salmonella infections that are AMR are caused by four out of five serotypes that are commonly isolated: Typhimurium, Enteritidis, Newport, and Heidelberg. They persist in food animals, are transmitted through food supplies, and cause illness in humans.[15,35,39,96] According to data reported by the National Antibiotic Resistance Monitoring System (NARMS) during the periods of 2004–2012, the percentage of Ceftriaxone-resistant infections increased from 9% to 22%. From 2004 through 2012, the 48 contiguous states reported 369,254 culture-confirmed salmonella infections to LEDS. During this period, NARMS tested 19,410 salmonella isolates from the 48 states for resistance. Overall resistance was detected in 2,320 (12%) isolates. Ampicillin-only resistance was the most common pattern, detected in 1,254 (6.5%) isolates, of which 60% were Typhimurium. Figure 56.1 summarizes the data of resistance of all the serotypes reported.[38,96]

Antimicrobial drug use in food-producing animals is a major driver of—although not the only contributor to—resistant salmonella infections. An example is the contribution of third-generation cephalosporin use in poultry to ceftriaxone resistance among Heidelberg infections of humans. The FDA is taking strict actions to prevent the spread of antimicrobial resistance and prolong the use of antimicrobial agents, which includes strategies for controlled use of agents under veterinary supervision.[38] In a retrospective case control study of sporadic NewportMDRAmpC infections in residents of New England by Gupta et al., it has been demonstrated that infections are domestically acquired and are associated with direct exposure to a dairy farm.[24] They found

Serotype

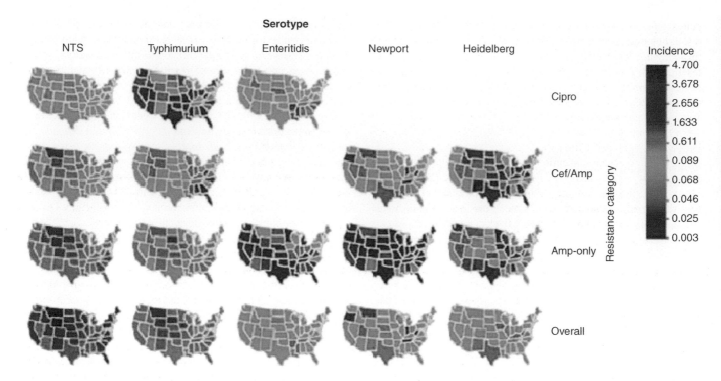

FIGURE 56.1 Estimated incidence of infection with all NTS and major serotypes with clinically important resistance (no. infections per 100,000 person-years), by state and resistance category, United States, 2004–2012. Estimates were derived by using Bayesian hierarchical models. All NTS includes the four major and other serotypes. Isolates in each category may have resistance to other agents. Data on Cipro among Newport (8 isolates), Cipro among Heidelberg (7), and Cef/Amp among Enteritidis (2) were too sparse to use in the Bayesian hierarchical models. Overall resistance was defined as Cipro, Cef/Amp, or Amp-only. Amp-only, resistant to ampicillin (MIC >32 µg/mL) but susceptible to ceftriaxone and ciprofloxacin; Cef/Amp, resistant to ceftriaxone (MIC >4 µg/mL) and ampicillin; Cipro, nonsusceptible to ciprofloxacin (MIC >0.12 µg /mL) but susceptible to ceftriaxone; NTS, nontyphoidal Salmonella. (Reproduced from Medalla et al. [2016], accessed from the Centers for Disease Control and Prevention website).[38]

that yogurt consumption is protective, which is intriguing, since some studies suggested that bacteria found in yogurt may prevent salmonellosis. Multistate outbreaks of MDR salmonella Newport caused by other, less resistant strains during the 1970s and 1980s were associated with the consumption of ground beef, particularly from dairy cattle. Subsequent to their investigation of sporadic cases of infection, multistate outbreaks of NewportMDRAmpC infection have been associated with foods, including cheese made from unpasteurized milk and ground beef produced at a large processing plant in the Mid-Atlantic that slaughters New England dairy cattle.

Prevention and control of Newport-MDRAmpC infection in humans and cattle requires an understanding of how this strain is introduced onto farms and disseminated among cattle. The management practices that promote its spread are not known but could include the use of therapeutic or prophylactic antimicrobial agents on farms, as has been suggested elsewhere. Ceftiofur, an expanded-spectrum cephalosporin, is commonly used therapeutically in dairy cattle, whereas tetracycline and neomycin are used prophylactically in calf milk replacers. Approximately 55% of dairy farms surveyed in 21 major dairy states during 2002 reported using calf milk replacer that contained medication; oxytetracycline and neomycin were the most commonly used. On the farms visited by the authors, they documented the therapeutic use of ceftiofur in ill cattle and the use of tetracycline- and neomycin-supplemented calf-milk replacers. Use of these or other antimicrobial agents on farms, especially if used frequently, create a

selective pressure that is likely to promote transmission and dissemination of Newport-MDRAmpC. This study confirmed that Newport-MDRAmpC isolates contains genes of the blaCMY family, which produce AmpC-type β-lactamases.[24]

Clinically important resistance is strongly linked to specific serotypes. A study conducted in Brazil in 2018 described the serotype distribution and AMR of NTS isolated from enteric and nonenteric human samples and also investigated the AMR patterns.[42] The results of the study showed that salmonellosis frequency was higher in adults above the age of 45. Also, serotypes Typhimurium and Enteritidis were dominant in the study samples collected. This study showed that most sporadic cases of salmonella isolates were from hospitalized patients and that Typhimurium was the most frequently isolated serotypes. Resistance to more than three antibiotics was showed by most Typhimurium isolates whereas Enteritidis showed resistance to two antibiotics. All the other remaining serotypes were susceptible to antibiotics.[42] Earlier, another study conducted by Chen, Zhao et al. in 2004 characterized multiple-antimicrobial-resistant salmonella serovars isolates from retail meats in United States and the People's Republic of China.[13] Molecular genetic techniques were used to characterize antimicrobial-resistant salmonellae, especially *Salmonella enterica* serovar Typhimurium DT104. A total of 133 salmonella isolates were recovered from retail meats purchased in the United States and the People's Republic of China. These were assayed for antimicrobial susceptibility, the presence of integrons and antimicrobial resistance genes, and horizontal transfer of characterized antimicrobial resistance determinants via conjugation. Their findings showed that 73 (82%) of the total salmonella isolates were resistant to at least one antimicrobial agent. Resistance to the following antibiotics was common among the U.S. isolates: Tetracycline (68% of the isolates were resistant), streptomycin (61%), sulfamethoxazole (42%), and ampicillin (29%). Eight salmonella isolates (6%) were resistant to ceftriaxone. Fourteen isolates (11%) from the People's Republic of China were resistant to nalidixic acid and displayed decreased susceptibility to ciprofloxacin. A total of 19 different antimicrobial resistance genes were identified in 30 multidrug-resistant salmonella isolates. The study concluded that salmonella isolates in retail meats are commonly resistant to multiple antimicrobials, including tetracycline, sulfamethoxazole, and streptomycin. Resistance to ceftriaxone is a concern because of the importance of this agent for treatment of salmonellosis in children.[13]

In the United States, antimicrobial-resistant Salmonella strains have emerged in the last two decades, causing increased morbidity and hospitalization especially in children. A multidrug-resistant (MDR) strain of Salmonella serotype Typhimurium definitive type 104(DT104) (resistant to ampicillin, chloramphenicol, streptomycin, sulfamethoxazole, and tetracycline) emerged across the United States during the 1990s. Alarmingly, between 1998 and 2001 alone, Newport-MDRAmpC (resistant) strains of Salmonella have rapidly emerged throughout the United States increasing from 1–25% in prevalence. These strains are resistant to nine antimicrobials, have either decreased susceptibility to or resistance to extended-spectrum cephalosporins, such as ceftriaxone, and are sometimes resistant to trimethoprim-sulfamethoxazole.[24]

E. coli Drug Resistance

E. coli is the most prominent enteropathogenic bacterial cause of infectious diseases that span from the gastrointestinal tract to extraintestinal sites such as urinary-tract infection, septicemia, and neonatal meningitis.[56,57] The emergence and spread of antibiotic resistance in *E. coli* is an increasing public-health concern across the world. An important example of antibiotic resistance is multidrug-resistant (MDR) and ESBL-producing *E. coli*, which can cause life-threatening infections. In 1982, *E. coli* was suspected in two epidemics of hemorrhagic colitis that were associated with consumption of undercooked ground beef. There have been more than 200 *E. coli* serotypes that have been identified and isolated from animals, food, and other resources since then. Sixty of these serotypes are associated with human diseases. In 2002, it was reported that 73,000 illnesses were caused by *E. coli* and 61 deaths were reported in the United States.[52] In the recent estimates of global antibiotic resistance published by the WHO in 2014, *E. coli* was named as one of the biggest concerns associated with hospital- and community-acquired infections.[54,55] In Europe, antimicrobial resistance in Gram-negative bacteria, particularly in *E. coli* continues to increase, which constitutes a majority of invasive Gram-negative isolates in European countries.[58] The emergence and diffusion of multidrug-resistant strains of *E. coli* is complicating the treatment of several serious infections. *Enterobacteriaceae*, particularly *E. coli*, are the most frequent cause of hospital- and community-acquired

infections.[60–62] A recent annual surveillance report of the European Centre for Disease Prevention and Control showed the presence of isolates resistant to third-generation cephalosporins, fluoroquinolones, and aminoglycosides as well as isolates with resistance to all three antimicrobial classes. This was observed in all EU member states and Norway, Iceland, and Liechtenstein, but only 0.04% of 59,326 isolates of *E. coli* studied were found to be resistant to carbapenems ([63]).

Multidrug-resistant *E. coli* strains are also commonly isolated from animals and food.[64–70] The use of antibiotics in animals contributed to the emergence and spread of the number of antibiotic-resistant strains, including *E. coli*, which can also infect humans through either direct contact with animals or through consumption of contaminated food. *E. coli* is able to survive and adapt in various extra-intestinal habitats and to spread resistances between humans, animals, their products and the environment through several transmission pathways.[59]

The environment plays a key role in the spread of antimicrobial resistance serving as an unlimited reservoir of antimicrobial resistance genes.[71] Therefore, *E. coli* may acquire other drug resistance traits from environmental bacteria, and conversely, can spread its resistance genes to potential pathogens in different habitats.[72] A source of multidrug-resistant *E. coli* could be hospital effluents. Several studies have reported the presence of multidrug-resistant strains in hospital effluents contributing to the spread of their antibiotic resistance genes in the municipal sewage systems and in the environment.[73]

A very recent meta-analysis assessed the prevalence of antibiotic resistance in *E. coli* strains simultaneously isolated from human, animal, food, and environmental samples from 2000 to 2018 worldwide, and provided a comprehensive systematic review on the prevalence of antimicrobial resistance in *E. coli* from different sources.[74] The data reported on the comparison of MDR *E. coli* strains isolated from different sources. It showed higher prevalence in animal and environmental sources than humans, prevalence of drug resistance in different sources, and documented increases in *E. coli* drug resistance. Overall, the results showed that antibiotic resistance in American and European countries is lower than other regions of the world. Subgroup analysis from 2000 to 2018 also indicated a significant increase in ciprofloxacin resistance. Moreover, the results also showed that the prevalence of ESBL antibiotic resistance and MDR *E. coli* strains in animal isolates was higher than in human isolates. According to their findings, systematic surveillance of hospital-associated infections, proper monitoring of disposal processes in hospitals, monitoring the use of antibiotics in animals, monitoring and evaluation of antibiotic-sensitivity patterns, and preparation of reliable antibiotic strategies may ease more corrective actions for the inhibition and control of *E. coli* infections in different parts of the world.[74]

Neisseria gonorrhea Drug Resistance

Annually, more than 800,000 cases of gonorrhea are estimated by CDC, which makes it the second most frequently reported infectious disease in the United States. It is projected over 10 years that an additional 75,000 cases of pelvic inflammatory disease, 15,000 cases of epididymitis and 222 additional human immunodeficiency virus infections will be caused by drug resistant

N. gonorrhoeae, making it more widespread.[2] The WHO has recently updated the treatment guidelines to address the issue of emerging resistance in *N. gonorrhea*. As per the new guideline, quinolones are no longer recommended for the treatment of gonorrhea due to the high levels of resistance.[1]

Malaria Drug Resistance

Resistant strains of malaria can spread from their origins in Cambodia, Myanmar, Thailand, and Vietnam to other parts of the world and become a major challenge to public health, threatening the recent gains in malaria control.[1] In 1955, the WHO launched the "Global Malaria Eradication Program" and chloroquine chemotherapy was implemented to complement the initial vector control measures leading to eradication of malaria from 27 countries.[97]

Unfortunately, elimination of malaria could not be achieved in most underdeveloped countries, resulting in the current predominant distribution of malaria to subtropical and tropical regions.[1] Among the reasons for the eventual halt to the eradication effort were widespread resistance to available insecticides and the emergence of chloroquine-resistant malaria in southeast Asia and South America around 1960.[98] The subsequent spread of chloroquine-resistant *P. falciparum* to Africa and the lack of an effective and affordable alternative ultimately led to a two-to-three-fold increase in malaria-related deaths in the 1980s.[99] The only viable alternative to chloroquine at that time was sulfadoxine-pyrimethamine, however, it also encountered drug-resistant parasites about a year after implementation.[100] Several other antimalarial drugs have since been deployed to combat parasites resistant to chloroquine and sulfadoxine-pyrimethamine, including mefloquine, amodiaquine, and quinine. The historic usage of these replacement drugs in monotherapy has now similarly resulted in the selection of resistant parasites, at least in some parts of the world.

Despite causing an estimated 72–80 million malaria infections yearly, *P. vivax* has not received as much public and scientific attention as *P. falciparum*.[101] One of the main reasons for this is that *P. vivax* usually produces less severe symptoms. Nevertheless, *P. vivax* leads to a disabling disease that can be fatal, and exacts a similar economic burden as falciparum malaria. Additionally, it has been recently noted that the severity of the disease caused by *P. vivax* is increasing in some parts of the world and the development of drug resistance could result in an expansion of this debilitating and sometimes deadly infection.[102] To assess drug resistance in *P. vivax* in the field, results from clinical studies and a limited number of laboratory studies using modified drug susceptibility assays, which are recommended by the WHO for *P. falciparum*, have been used.[103]

Human Immunodeficiency Virus (HIV) Drug Resistance

HIV drug resistance is reported in 7% of people by WHO, even before they start their treatment. Similar figures are reported to be 15–20% in developing countries. Levels above 15% have been reported by some countries amongst those starting HIV treatments and up to 40% in those restarting the treatment.[1]

Other Drug-Resistant Infections

ESBL-Producing Enterobacteriaceae

26,000 HAIs and 1,700 deaths per year are caused by ESBL-producing *Enterobacteriaceae*.[2] A study conducted by Sun, Liu et al. in 2019 at Children's Hospital in China showed that 52.21% of all the MDR infections are caused by ESBL.[4] When pathogens are isolated from all types of clinical specimens, species of the family *Enterobacteriaceae* are very commonly identified. *E.coli* was the third most prevalent bacterial species out of the 15 found in intensive care unit (ICU) patients of 25 European hospitals in 1997–1998. Among the bloodstream isolates, *E. coli* ranked third, *Enterobacter* species ranked sixth, *Klebsiella pneumonia* ranked eighth with *Proteus mirabilis* ranking the tenth most frequently isolated pathogens. The highest rates of carbapenem resistance were found in *K. pneumonia*. Greece has the highest proportion of this phenotype with 46% of tested isolates in 2007 that were nonsusceptible to carbapenems among the 33 European countries.[48] ESBLs have been identified in 70–90% of *Enterobacteriaceae* in India in recent surveys. These statistics suggest a serious problem, making it necessary to use reserved antibiotics like carbapenems. *Enterobacteriaceae* with NDM-1 carbapenemases show high resistance to a lot of antibiotics and potentially indicate that treatment with β-lactams, fluoroquinolones, and aminoglycosides, which are the main classes of antibiotics used for treatment of Gram-negative infections, may soon end. Owing to the loss of resistant genes, a few isolates remain sensitive to individual aminoglycosides and aztreonam. Susceptibility to colistin and tigecycline is still present in most isolates. Since there are only few new anti-Gram-negative antibiotics in the pipeline and none are active against NDM-1 producers, treatment options in the future are of utmost concern.

On the advice of the Health Protection Agency, the Department of Health in the United Kingdom had prompted the release of a National Resistance Alert 3 notice due to introduction of NDM-1 in the United Kingdom. It is not surprising that the United Kingdom is the first Western country to register the widespread presence of NDM-1 positive bacteria due to its historical link to India.[33] After birth, neonates are rapidly colonized by *Enterobacteriaceae* irrespective of whether they are breast fed. In a cohort study on breast-fed babies in India, it was noted that 14.3% 1-day olds harbored *Enterobacteriaceae* that contained enzymes that could inactivate β-lactam drugs (ESBL). By day 60, this increased to 41.5% of babies. To establish a normal healthy gut microflora environment, drinking water and food are the most important means. In an ecosystem free from external antimicrobial selection pressure, antimicrobial-resistant and nonresistant species coexist in a stable balance. Human microbiota is exposed to unprecedented high concentrations of antimicrobials due to their use in clinical settings. Within the human body, the *in vivo* development of *de novo* resistance to antimicrobial agents has been recorded during the course of treatment, which also includes carbapenems.

MDR *Pseudomonas aeruginosa*

Each year in the United States, 51,000 healthcare-associated *P. aeruginosa* infections are reported, of which more than

6,000 (13%) are MDR, causing roughly 400 deaths.[2] They are one of the causative pathogens of nosocomial infections in the immunocompromised and the immunocompetent host associated with invasive devices, mechanical ventilation, burn wounds, or surgery is *P. aeruginosa*. Between 1997 and 2002, the prevalence of *P. aeruginosa* in bloodstream infections increased slightly from 5.5–6.8% in Europe according to the SENTRY Antimicrobial Surveillance Program. Thirteen medical centers from 15 European countries participated in the survey. *P. aeruginosa* resistance to carbapenems appeared to be higher all over Europe in 2007. As per that date, carbapenem resistance above 25% was seen in Croatia, Turkey, Germany, Italy, Czech Republic, and Greece, whereas carbapenem resistance was below 10% in Denmark, the Netherlands, Switzerland, Sweden, and Finland.[48] The emergence of MDR *P. aeruginosa* was seen in a tertiary care center in the United States between 1994 and 2002 with a prevalence of 1–16%. This was seen along with the emergence of MDR *Klebsiella* species with prevalence of 0.5–17%.

MDR Actinobacteria

Each year in the United States, approximately 12,000 healthcare-acquired *Acinetobacter* infections occur. Of those, 7,300 (63%) are resistant to at least three different classes of antibiotics, causing 500 deaths.[2]

Campylobacter

Another indistinguishable form of acute diarrheal disease is *Campylobacterial* gastroenteritis. It is similar to those caused by *Salmonella* spp. and *Shigella* spp. More than 95% of *Campylobacter enteritis* associated with diarrheal illness is caused by *C. Jejuni* and *C. coli*, which are clinically indistinguishable. Fecal *C. jejuni* or *C. coli* are regarded as a commensal organism in cattle, poultry, and swine. It is routinely recovered from retail chicken and poultry products. As per a study in 1986 by Washington State, 57% of poultry processing-plant samples and 23% of retail chicken carried *C. jejuni*. However, no reports of ciprofloxacin-resistant human *Campylobacter* strains are reported in isolates prior to 1992. Countries in which use of antimicrobials is not restricted have fluoroquinolone-resistant *Campylobacter* as a threat. In Spain, 88% of *Campylobacter* isolates displayed fluoroquinolone resistance. Due to high levels of fluoroquinolone-resistance in *Campylobacter*, the FDA and the Center for Veterinary Medicine (CVM) in 2000 had proposed to withdraw the approval of the new animal drug application for use of the fluoroquinolones in poultry. There was a temporal relation that was cited between the approval of these drugs for use in poultry in the United States and the increase in resistance to fluoroquinolones in *Campylobacter* species associated with human infections.[8]

Listeria monocytogenes

Several large foodborne epidemics of listeriosis have been reported in several nations, which include England, Germany, Sweden, New Zealand, Switzerland, Australia, France, and the United States. It is predicted that there are 2,500 cases of listeriosis in the United States every 12 months, resulting in 500 deaths. Most manifestations of listeriosis are treated with ampicillin,

rifampin, or penicillin plus gentamycin. Co-trimoxazole is considered as a second-choice therapy. If a pregnant woman is diagnosed with listeriosis or if bacteremia is present in a patient, then Vancomycin and erythromycin are used for treatment. The first strains of *L. monocytogenes* that were antibiotic resistant were reported in 1988. Since then, resistance of *L. monocytogenes* to antibiotics has emerged. Studies have shown that this pathogen acquires its antibiotic resistance genes from foreign sources through mobile genetic elements, such as plasmids and transposons. Enterococci and streptococci have appeared to be a common source of resistant genes for *L. monocytogenes*.[8]

Yersinia

Out of the 11 species of *Yersinia* that are recognized presently, only three species are medically important. *Y. pestis* is a pathogen that causes bubonic and pneumonic plague whereas *Y. enterocolitica* and *Y. pseudotuberculosis* both can induce severe gastroenteritis. The susceptibility pattern of *Y. enterocolitica* to antibiotics is biogroup dependent. Strains of biogroup 1B are generally susceptible to ampicillin but resistant to carbenicillin, ticarcillin, and cephalothin. Biogroup 1A and biogroup 3 strains are resistant to amoxicillin/clavulanic acid and cefoxitin. However, biogroup 3 strains are reported to be susceptible to carbenicillin and ticarcillin. Nonetheless, *Yersinia* remain susceptible to tetracycline and quinolones in contrast to the above susceptibility and resistant patterns.[8]

Clinical and Economic Burden of Antimicrobial Effectiveness

A systematic review by Nellums, Thompson et al. in Europe, showed that there is an increased risk of antibiotic resistance in migrant groups.[7] This could be because of poor sanitation, overcrowded living conditions, barriers to accessing health services including antibiotics or vaccinations, disruptions in treatment during migration and on arrival in host countries, high prevalence rates of AMR in the countries of origin or transit. The risk of bacterial resistance increases with insufficient health infrastructure, antibiotic stewardship, infection prevention and control, vaccine coverage, or surveillance. Return travel to the countries of origin and contact with healthcare services in those settings can also pose a threat. The prevalence of AMR carriage and infection is elevated in refugees and asylum seekers and in high-migrant community settings. AMR organisms are acquired by immigrants during migration or in the host country from local populations or between migrants.[7]

AMR is not only a medical problem, but also an economic problem. There is enormous potential in finding new antimicrobials in virgin tropical forests with plant resources in developing countries. These may provide solutions to a lot of current resistance problems. However, their exploitation requires a great financial investment. The risk factors for antimicrobial resistance are poverty and inadequate resources, natural calamities, human population growth, healthcare-provider-related factors, health service provision centers, user related factors, policy and regulatory issues, and use of antimicrobial drugs in food animal production.[12]

The steady increase in resistance is seen to quinolones, carbapenems, and third-generation cephalosporins in community settings. Identification, assessment, and communication of the threats of current and emerging antimicrobial resistance are a crucial task for CDC. The updating of local prescribing guidelines, active reporting on antibiotic prescribing and consumption, or enforcement of local surveillance programs on antibiotic resistance are some measures that can be implemented. However, substantial legislative action and increased funding is needed to implement these measures. This requires a strong commitment from policy makers at the national and international levels. Appropriate use of antibiotics is important in low-income countries since there are additional factors that contribute to the emergence of resistance. These include, (a) less potent activity of some antibacterial agents (including counterfeit drugs), (b) over-the-counter availability, with insufficient dosages, (c) lack of diagnostic laboratories, and (d) poor level of sanitation. This leads to familial and community spread of resistance.

Global food commercialization and international travel also needs special attention.[44] A review of the available data on the relationship between antimicrobial resistance and patient outcomes by Cosgrove et al. describes the most sensitive patient outcomes and factors affected by AMR. These include decreased functional status, loss of work and fewer antimicrobial agent options available for therapy for the patient, and societal issues such as increase of total healthcare costs, as well as loss of antimicrobial classes usable for treatment.[4]

Resolution of AMR

In spite of the challenges faced by developing countries, it is imperative that they use antimicrobial agents responsibly. AMR is greatly influenced by poverty-associated factors, particularly because there are more important problems like providing basic healthcare or sanitation in developing countries. Hence, developing countries have to depend upon partnerships with developed nations and global organizations to formulate and implement mechanisms that will allow their healthcare providers, policy makers, and users to understand the problems and the consequences of lack of control in AMR[12] (Table 56.2).

Uncertainty in diagnosis and laboratory detection of microbial infections is one of the key drivers for drug misuse and overuse. To reduce the antimicrobial selection pressure, diagnosis of microbial infections can be improved by incorporation of new biological markers such as procalcitonin, soluble triggering receptors expressed on myeloid cells, use of newer molecular diagnostic techniques, amplification technology with DNA microarray, etc. These improvements in diagnostic techniques not only improve accuracy in results but increase the speed of testing. Diagnostic tools that can rapidly distinguish between bacterial and viral infections should be made available especially in hospitals. Table 56.3 gives a list of countries, many of which are industrialized, where either the number of antimicrobial agent prescriptions or the volume of antimicrobial use has decreased over the last 20 years, especially in the ambulatory setting[26]. Greater antimicrobial drug use and the spread of resistant microbes increase due to an increase in immunocompromised patients, the growing life-expectancy, and the susceptibility of

TABLE 56.2

Potential Determinants Influencing Future Dissemination and Control of Antimicrobial Resistance

Dimension	Determinant	Potential Control Measures and Interventions
Pathogen and microbial ecology	Evolution Survival fitness Virulence Commensal flora Laboratory detection and identification	Evolutionary engineering Inhibition of microbial gene expression Antibodies, antipathogenic drugs, biologic response modifiers Probiotics Improved rapid diagnostic tests
Physician's prescribing practice	Antimicrobial drug usage pattern Diversity of antimicrobial drug prescribing Training and knowledge	Multimodal interventions Decision support tools Academic detailing and educational campaigns
Population characteristics	Migration, travel, and globalization Case mix and host susceptibility to infections Antimicrobial demand and health beliefs Transmission and infection rates	Screening and improved surveillance Immunization; better control of chronic diseases Public information campaigns Hand hygiene and barrier precautions
Politics and healthcare policy	Healthcare policy Promotional activities by industry Technologic development	Change in reimbursement patterns Regulation New prevention and treatment approaches

Source: Ref. [26].

older persons to infections. Increased use of medical devices, gene therapy, and better management of conditions like diabetes and cancer are key trends in clinical care and greatly influence AMR. Prescribing of antimicrobial drugs can be affected by global threats such as the influenza pandemic or the current coronavirus outbreak, by reversing the decreasing trend of

TABLE 56.3

Countries with Decreased Number of Antimicrobial Drug Prescriptions or Total Volume of Outpatient Antimicrobial Drugs Used Within the Last 20 Years[26]

Continent	Country
Europe	France Belgium Spain Germany United Kingdom Sweden
Asian-Pacific region	South Korea Taiwan Australia
Americas	Canada United States Chile

Source: Ref. [26].

antimicrobial drug consumption. A decrease in transmission rates and the impact of AMR can be improved by developing new vaccines in near future.[19]

It is not only important to optimize antimicrobial drug use, but it is also important to understand how to minimize the selection of antimicrobial resistance. Defining optimum duration and dosage of a therapy needs detailed analysis especially in specific patient populations (infants, pregnant women, undernourished, obese, and coinfected patients). Secondly, rapid diagnosis of infections has to be enabled to improve targeting. Due to technical and financial barriers and issues around the adoption of innovative techniques, implementation of the methods is slow despite them being widely advocated. Thirdly, research needs to be focused on improving the quality of current antimicrobials along with the discovery of new drugs.[27]

Nanomedical Approach

Recently, nanoparticles (NPs) have emerged as new tools that can be used to combat deadly bacterial infections. Nanoparticle-based strategies can overcome the barriers faced by traditional antimicrobials, including antibiotic resistance. The various strategies by which nanoparticles can overcome this resistance, include the following: Nitric oxide-releasing nanoparticles (NO NPs), chitosan-containing nanoparticles (chitosan NPs), and metal-containing nanoparticles—all of which use multiple mechanisms simultaneously to combat microbes. This makes the development of resistance to these nanoparticles unlikely. These strategies have been studied in an oral/dental setting in ongoing clinical trials (NCT03588351),[75] which compare *in vivo* susceptibility of root canal bacteria to chitosan, chitosan nanoparticles, and chlorhexidine gluconate as intracanal medicaments in necrotic primary molars. Other approaches include ongoing clinical trials (NCT03666195)[76] studying the antimicrobial effect of titanium dioxide nanoparticles in complete dentures made for edentulous patients to control oral biofilm formation. Another area where antimicrobial properties of nanomaterials and their use in treatment of antibiotic resistance are tested is chronic wound/ulcer infections. One such study reports that a Carbopol® 934-based nanogold formulation prevents microbial adhesion, reducing fungal infections and accelerates the wound healing potential of the WfAuNPs.[77] Antibacterial activities of carbon nanotubes coated with different concentrations of iron oxide NPs on cultures of Gram-negative and Gram-positive bacteria are being examined in animal models by Khashan et al. and show good healing properties.[78] It is possible to package multiple antimicrobial agents within the same.[79,85] Development of resistance to the multiple antimicrobial agents within these nanoparticles is, again, unlikely,[86] possibly because it would require multiple simultaneous gene mutations in the same microbial cell. Nanoparticles can also overcome drug-resistance mechanisms of microbes, including decreased uptake and increased efflux of drug from the microbial cell,[85,83,87] biofilm formation[83,88] and intracellular bacteria.[81,83,87] Finally, nanoparticles have been used to target antimicrobial agents to the site of infection, so that higher doses of drug can be given at the infected site, thereby overcoming resistance with fewer adverse effects upon the patient.[89]

The use of multiple simultaneous mechanisms of antimicrobial action makes the development of resistance to these nanoparticles unlikely, because multiple simultaneous gene mutations in the same microbial cell would be required for that resistance to develop.[79–84]

Needs

Due to global nature of antimicrobial resistance and the failure to control the emergence of resistant organisms, a global surveillance program involving both developed and developing countries had to be implemented. To address this problem, CDC initiated the International Network for the Study and Prevention of Emerging Antimicrobial Resistance (INSPEAR). INSPEAR serves as an early warning system for emergence of drug-resistant pathogens, as outlined in Table 56.4 and serves to facilitate rapid distribution of information about emerging multidrug-resistant pathogens to hospitals and public health authorities worldwide. It serves as a model for the development and implementation of infection control interventions.[43]

Resolution

There is a need for better ways of dosing and sequencing antibiotics to prevent resistance, since there are a limited number of

TABLE 56.4

Important Types of Antimicrobial Resistance of Public Health Importance

Microorganism	Resistance
Early warning system: Local and regional sentinel events	
Staphylococcus aureus	Methicillin, intermediate susceptibility to glycopeptides
Enterococcus spp.	Vancomycin
Enterobacteriaceae	Extended-spectrum β-lactamase-mediated resistance, carbapenems, fluoroquinolones
Acinetobacter baumannii	Carbapenem
Any bacteria	All antimicrobials available at the regional and local settings
Early warning system: Global sentinel events	
Streptococcus spp.	Penicillinase, gentamicin, glycopeptides, fluoroquinolones
S. pneumoniae	Vancomycin, third-generation cephalosporins, new fluoroquinolones (gemifloxacin, grepafloxacin, levofloxacin, trovafloxacin)
Staphylococcus spp.	Glycopeptides (high level)
Enterobacteriaceae	Carbapenemase
Neisseria meningitidis	Penicillinase, chloramphenicol, cephalosporins, fluoroquinolones
Acinetobacter baumannii	Carbapenemase
Salmonella typhi	Third-generation cephalosporins, fluoroquinolones
Haemophilus influenzae	Cephalosporins
Brucella spp.	Tetracycline, rifampin, streptomycin
Clostridium difficile	Glycopeptides
Clostridium perfringens	Penicillinase

Source: Ref. [43].

antibiotics available. Alternating between antimicrobial agents is one way to improve dosing. Maximizing collateral sensitivity and minimizing cross resistance is the key goal of alternating drug therapy. Sometimes the efficacy of synergistic antibiotics is more than sum of individual efficacies. However, if a single-drug regimen is used, then proper dosing has to be followed. Immune system clearance of antimicrobial resistant infections also plays an important role.[6] Conflicting recommendations are often being made to reduce antimicrobial resistance. For instance, to reduce the burden of resistance in hospitals, strategies incorporate reduced the use of all antimicrobial classes, increased use of prophylactic antimicrobials, rotation of different antibiotic classes in temporal sequence, and simultaneous use of different antibiotics for different patients. There is conflicting evidence about the relationship between antibiotic treatment and resistance patterns. It is difficult to show that antibiotic treatment cuts a patient's risk of acquiring resistant organisms. However, the cumulative effect of using antibiotics has shown to increase the prevalence of AMR in the population as a whole.[34] Some of the ways in which MRSA can be controlled in hospital settings are screening of patients, screening of staff, isolation and barrier nursing, hand hygiene, environmental cleaning, ionization therapy, and multi-faceted MRSA control programs.[23]

Summary

Higher awareness of the prevalence of antimicrobial resistance is needed among the medical community and the general public. With the emergence of drug resistance, it is of critical importance to implement antidrug-resistance policies to contain, and hopefully curtail, the emergence and spread of resistance. Failure to do so would lead to a tragic setback in the current efforts to eliminate the diseases that are being spread due to drug resistance and the achieved reductions in disease-related morbidity and mortality. Coordinated action is needed by all countries to minimize the emergence and spread of antimicrobial resistance. Knowledge of the effect of antimicrobial resistance on patient outcomes can have several benefits. Healthcare providers can initiate and support infection control programs and antimicrobial-agent management programs if patient outcomes are a priority. Influencing healthcare providers to follow guidelines about isolation and making rational decisions with regards to use of antimicrobial agents can stimulate their interest in developing new antimicrobial agents and therapies. Knowledge of the implications of resistance on patient outcomes, which include decreased functional status, loss of work, fewer antimicrobial agent options available for therapy for the patient, and societal issues (such as increase of total healthcare costs)—as well as loss of antimicrobial classes usable for treatment—is needed to overcome these effects.

REFERENCES

1. World Health Organization (2018). "Antimicrobial resistance." Retrieved July 2019, from. https://www.who.int/news-room/fact-sheets/detail/antimicrobial-resistance.

2. Ventola, C. L., (2015). "The antibiotic resistance crisis: part 1: causes and threats." *P & T: A Peer-Reviewed Journal for Formulary Management* 40(4): 277–283.

3. Hsueh, P.-R., et al. (2002). "Current status of antimicrobial resistance in Taiwan." *Emerging Infectious Diseases* 8(2): 132–137.

4. Sun, L., et al. (2019). "Analysis of risk factors for multiantibiotic-resistant infections among surgical patients at a children's hospital." *Microbial Drug Resistance* 25(2): 297–303.

5. Yang, Bening, et al. (2017). "Antibiotic-induced changes to the host metabolic environment inhibit drug efficacy and alter immune function." *Cell Host & Microbe* 22(6): 757–765.e3.

6. Richardson, L. A., (2017). "Understanding and overcoming antibiotic resistance." *PLOS Biology* 15(8): e2003775–e2003775.

7. Nellums, L. B., et al. (2018). "Antimicrobial resistance among migrants in Europe: a systematic review and meta-analysis." *The Lancet Infectious Diseases* 18(7): 796–811.

8. White, Zhao, et al. (2004). "Characterization of multiple-antimicrobial resistant salmonella serovars isolated from retail meats." *Applied and Environmental Microbiology* 70(1): 1–7.

9. Alividza, V., et al. (2018). "Investigating the impact of poverty on colonization and infection with drug-resistant organisms in humans: a systematic review." *Infectious Diseases of Poverty* 7(1).

10. Boyce, J. M., et al. (1983). "Impact of methicillin-resistant *Staphylococcus aureus* on the incidence of nosocomial staphylococcal infections." *The Journal of Infectious Diseases* 148(4): 763.

11. Buckle, G. C., et al. (2012). "Typhoid fever and paratyphoid fever: Systematic review to estimate global morbidity and mortality for 2010." *Journal of Global Health* 2(1): 010401–010401.

12. Byarugaba, D. K., (2004). "Antimicrobial resistance in developing countries and responsible risk factors." *International Journal of Antimicrobial Agents* 24(2): 105–110.

13. Chen, S., et al. (2004). "Characterization of multiple-antimicrobial-resistant salmonella serovars isolated from retail meats." *Applied and Environmental Microbiology* 70(1): 1–7.

14. Cosgrove, S. E., (2006). "The relationship between antimicrobial resistance and patient outcomes: mortality, length of hospital stay, and health care costs." *Clinical Infectious Diseases* 42(Supplement_2): S82–S89.

15. Crim, S. M., et al. (2015). "Preliminary incidence and trends of infection with pathogens transmitted commonly through food - Foodborne Diseases Active Surveillance Network, 10 U.S. sites, 2006-2014." *MMWR. Morbidity and Mortality Weekly Report* 64(18): 495–499.

16. D'Agata, E. M. C., et al. (2008). "The impact of different antibiotic regimens on the emergence of antimicrobial-resistant bacteria." *PLOS ONE* 3(12): e4036.

17. David MZ, Daum RS., (2010) "Community-associated methicillin-resistant *Staphylococcus aureus*: epidemiology and clinical consequences of an emerging epidemic." *Clinical Microbiology Reviews*. 2010;23(3): 616–687.

18. Doern, Gary V., (2001). "Antimicrobial use and the emergence of antimicrobial resistance with *Streptococcus pneumoniae* in the United States." *Clinical Infectious Diseases* 33(s3): S187–S192.

19. Duckworth, G. J., et al. (1988). "Methicillin-resistant *Staphylococcus aureus*: report of an outbreak in a London teaching hospital." *Journal of Hospital Infection* 11(1): 1–15.

20. Dufour, P., et al. (2002). "Community-acquired methicillin-resistant *Staphylococcus aureus* infections in France: emergence of a single clone that produces Panton-Valentine leukocidin." *Clinical Infectious Diseases* 35(7): 819–824.

21. Fridkin, S. K., et al. (2002). "Temporal changes in prevalence of antimicrobial resistance in 23 US hospitals." *Emerging Infectious Diseases* 8(7): 697–701.

22. Groom, A. V., et al. (2001). "Community-acquired methicillin-resistant *Staphylococcus aureus* in a rural American Indian community." *JAMA* 286(10): 1201–1205.

23. Grundmann, H., et al. (2006). "Emergence and resurgence of meticillin-resistant *Staphylococcus aureus* as a public-health threat." *The Lancet* 368(9538): 874–885.

24. Gupta, A., et al. (2003). "Emergence of multidrug-resistant Salmonella enterica serotype newport infections resistant to expanded-spectrum cephalosporins in the United States." *The Journal of Infectious Diseases* 188(11): 1707–1716.

25. Hageman JC, Francis J, Uyeki TM, et al. (2004). Emergence of methicillin-resistant *Staphylococcus aureus* as a cause of community-acquired pneumonia during the influenza season, 2003–4 [abstract LB-8], Proceedings of the 42nd Annual Meeting of the Infectious Diseases Society of America Alexandria, VA Infectious Diseases Society of America 2004: 60.

26. Harbarth, S. and M. H., Samore, (2005). "Antimicrobial resistance determinants and future control." *Emerging Infectious Diseases* 11(6): 794–801.

27. Holmes, A. H., et al. (2016). "Understanding the mechanisms and drivers of antimicrobial resistance." *The Lancet* 387(10014): 176–187.

28. Howden, B. P., et al. (2004). "Treatment outcomes for serious infections caused by methicillin-resistant *Staphylococcus aureus* with reduced vancomycin susceptibility." *Clinical Infectious Diseases:* 38(4): 521–528.

29. Jevons, M. P., (1961). " 'Celbenin' - resistant Staphylococci." *British Medical Journal* 1(5219): 124–125.

30. Kaplan, S. L., et al. (2005). "Three-year surveillance of community-acquired *Staphylococcus aureus* infections in children." *Clinical Infectious Diseases* 40(12): 1785–1791.

31. Kayser, F. H., (1975). "Methicillin-resistant staphylococci 1965-75." *The Lancet* 306(7936): 650–653.

32. Keane, C. T. and M. T., Cafferkey, (1984). "Re-emergence of methicillin-resistant *staphylococcus aureus* causing severe infection." *Journal of Infection* 9(1): 6–16.

33. Kumarasamy, K. K., et al. (2010). "Emergence of a new antibiotic resistance mechanism in India, Pakistan, and the UK: a molecular, biological, and epidemiological study." *The Lancet. Infectious Diseases* 10(9): 597–602.

34. Lipsitch, M. and M. H., Samore, (2002). "Antimicrobial use and antimicrobial resistance: a population perspective." *Emerging Infectious Diseases* 8(4): 347–354.

35. McDermott, P. F., et al. (2018). "Antimicrobial resistance in nontyphoidal salmonella." *Microbiology Spectrum* 6(4).

36. McDonald, L. C., (2006). "Trends in antimicrobial resistance in health care–associated pathogens and effect on treatment." *Clinical Infectious Diseases* 42(Supplement 2): S65–S71.

37. McDonald, L. C. and J. C., Hageman, (2004). "Vancomycin intermediate and resistant *Staphylococcus aureus*. What the nephrologist needs to know." *Nephrology News & Issues* 18(11): 63–71.

38. Medalla, F., et al. (2016). "Estimated incidence of antimicrobial drug-resistant nontyphoidal salmonella infections, United States, 2004-2012." *Emerging Infectious Diseases* 23(1): 29–37.

39. Medalla, F., et al. (2013). "Increase in resistance to ceftriaxone and nonsusceptibility to ciprofloxacin and decrease in multidrug resistance among Salmonella strains, United States, 1996-2009." *Foodborne Pathogens and Disease* 10(4): 302–309.

40. O'Brien, Thomas F., (2002). "Emergence, spread, and environmental effect of antimicrobial resistance: How use of an antimicrobial anywhere can increase resistance to any antimicrobial anywhere else." *Clinical Infectious Diseases* 34(s3): S78–S84.

41. Pribul, B. R., et al. (2016). "Characterization of quinolone resistance in Salmonella spp. isolates from food products and human samples in Brazil." *Brazilian Journal of Microbiology*: [publication of the Brazilian Society for Microbiology] 47(1): 196–201.

42. Reis, R. O. D., et al. (2018). "Increasing prevalence and dissemination of invasive nontyphoidal Salmonella serotype Typhimurium with multidrug resistance in hospitalized patients from southern Brazil." *The Brazilian Journal of Infectious Diseases* 22(5): 424–432.

43. Richet, H. M., et al. (2001). "Building communication networks: International network for the study and prevention of emerging antimicrobial resistance." *Emerging Infectious Diseases* 7(2): 319–322.

44. Roca, I., et al. (2015). "The global threat of antimicrobial resistance: science for intervention." *New Microbes and New Infections* 6: 22–29.

45. Romano, R., et al. (2006). "Outbreak of community-acquired methicillin-resistant *Staphylococcus aureus* skin infections among a collegiate football team." *Journal of Athletic Training* 41(2): 141.

46. Scallan, E., et al. (2011). "Foodborne illness acquired in the United States–major pathogens." *Emerging Infectious Diseases* 17(1): 7–15.

47. Shahin, R., et al. (1999). "Methicillin-resistant *Staphylococcus aureus* carriage in a child care center following a case of disease." *Archives of Pediatrics & Adolescent Medicine* 153(8): 864–868.

48. Souli, M., et al. (2008). "Emergence of extensively drug-resistant and pandrug-resistant Gram-negative bacilli in Europe." *Eurosurveillance* 13(47): 19045.

49. Threlfall, E. J., (2002). "Antimicrobial drug resistance in Salmonella: Problems and perspectives in food- and waterborne infections." *FEMS Microbiology Reviews* 26(2): 141–148.

50. Tiemersma, E. W., et al. (2004). "Methicillin-resistant *Staphylococcus aureus* in Europe, 1999-2002." *Emerging Infectious Diseases* 10(9): 1627–1634.

51. Turnidge, J. D., and J. M., Bell, (2000). "Methicillin-resistant *Staphylococcal aureus* evolution in Australia over 35 years." *Microbial Drug Resistance* 6(3): 223–229.

52. White, D. G., et al. (2004). "Antimicrobial resistance among gram-negative foodborne bacterial pathogens associated with foods of animal origin." *Foodborne Pathogens and Disease* 1(3): 137–152.

53. Zinderman, C. E., et al. (2004). "Community-acquired methicillin-resistant *Staphylococcus aureus* among military recruits." *Emerging Infectious Diseases* 10(5): 941–944.

54. WHO (1983). "Antimicrobial resistance." *Bull World Health Organ* 61: 383–94.

55. Alizade, H., (2018). "*Escherichia coli* in Iran: An overview of antibiotic resistance." *Iranian Journal of Public Health* 47(1): 1–12.

56. Piatti, G, Mannini, A, Balistreri, M, Schito, AM, (2008). "Virulence factors in urinary *Escherichia coli* strains: Phylogenetic background and quinolone and fluoroquinolone resistance." *Journal of Clinical Microbiology* 46: 480–7.

57. Alizade, H, Ghanbarpour, R, Aflatoonian, MR, (2014). "Molecular study on diarrheagenic *Escherichia coli* pathotypes isolated from under 5 years old children in southeast of Iran." *Asian Pacific Journal of Tropical Diseases* 4(Suppl 2): S813–S817.

58. Allocati, N, MAsulli, M, Alexeyev, M, and Di Ilio, C, (2013). "*Escherichia coli* in Europe: An overview." *International Journal of Environmental Research and Public Health* 2013, 10, 6235–6254.

59. Ewers, C., Bethe, A., Semmler, T., Guenther, S., Wieler, L.H., (2012). "Extended-spectrum beta-lactamase-producing and AmpC-producing *Escherichia coli* from livestock and companion animals, and their putative impact on public health: A global perspective." *Clinical Microbiology and Infection* 18, 646–655.

60. Pitout, J.D., (2008). "Multiresistant Enterobacteriaceae: New threat of an old problem." *Expert Review of Anti-infective Therapy* 6, 657–669.

61. Pitout, J.D., (2012). "Extraintestinal pathogenic *Escherichia coli*: An update on antimicrobial resistance, laboratory diagnosis and treatment." *Expert Review of Anti-infective Therapy* 10, 1165–1176.

62. Paterson, D.L., (2006). "Resistance in gram-negative bacteria: Enterobacteriaceae." *The American Journal of Medicine* 119, S20–S28.

63. European Centre for Disease Prevention and Control. Antimicrobial Resistance Surveillance in Europe 2011. Annual Report of the European Antimicrobial Resistance Surveillance Network; Stockholm, Sweden: 2012.

64. Kirchner, M., Wearing, H., Teale, C., (2011). "Plasmid-mediated quinolone resistance gene detected in *Escherichia coli* from cattle." *Veterinary Microbiology* 148, 434–435.

65. Snow, L.C., Warner, R.G., Cheney, T., Wearing, H., Stokes, M., Harris, K., Teale, C.J., Coldham, N.G., (2012). "Risk factors associated with extended spectrum beta-lactamase *Escherichia coli* (CTX-M) on dairy farms in North West England and North Wales." *Preventative Veterinary Medicine* 106, 225–234.

66. Platell, J.L., Johnson, J.R., Cobbold, R.N., Trott, D.J., (2011). "Multidrug-resistant extraintestinal pathogenic *Escherichia coli* of sequence type ST131 in animals and foods." *Veterinary Microbiology* 153, 99–108.

67. Hendriksen, R.S., Mevius, D.J., Schroeter, A., Teale, C., Jouy, E., Butaye, P., Franco, A., Utinane, A., Amado, A., Moreno, M., et al. (2008). "Occurrence of antimicrobial resistance among bacterial pathogens and indicator bacteria in pigs in different European countries from year 2002–2004: The ARBAO-II study." *Acta Veterinaria Scandinavica* 50, 19.

68. Giufrè, M., Graziani, C., Accogli, M., Luzzi, I., Busani, L., Cerquetti, M., (2012). "*Escherichia coli* Study Group. *Escherichia coli* of human and avian origin: Detection of clonal groups associated with fluoroquinolone and multidrug resistance in Italy." *Journal of Antimicrobial Chemotherapy* 67, 860–867.

69. Cengiz, M., Buyukcangaz, E., Arslan, E., Mat, B., Sahinturk, P., Sonal, S., Gocmen, H., Sen, A., (2012). "Molecular characterisation of quinolone resistance in *Escherichia coli* from animals in Turkey." *Veterinary Record* 171, 151–154.

70. Aleisa, A.M., Ashgan, M.H., Alnasserallah, A.A., Mahmoud, M.H., Moussa, I.M., (2013). "Molecular detection of β-lactamases and aminoglycoside resistance genes among *Escherichia coli* isolates recovered from medicinal plant." *African Journal of Microbiology Research* 7, 2305–2310.

71. Gonzalez-Zorn, B., Escudero, J.A., (2012). "Ecology of antimicrobial resistance: Humans, animals, food and environment." *International Microbiology* 15, 101–109.

72. Da Costa, P.M., Loureiro, L., Matos, A.J., (2013). "Transfer of multidrug-resistant bacteria between intermingled ecological niches: The interface between humans, animals and the environment." *International Journal of Environmental Research and Public Health* 10, 278–294.

73. Korzeniewska, E., Korzeniewska, A., Harnisz, M., (2013). "Antibiotic resistant *Escherichia coli* in hospital and municipal sewage and their emission to the environment." *Ecotoxicology and Environmental Safety* 91, 96–102.

74. Pormohammad, A., Nasiri, M., and Azimi, T., (2019). *Infection and Drug Resistance* 2019:12, 1181–1197.

75. ClinicalTrials.gov Identifier: NCT03588351. Chitosan, Chitosan Nanoparticles, and Chlorhexidine Gluconate, as Intra Canal Medicaments in Primary Teeth.

76. ClinicalTrials.gov Identifier: NCT03666195. The Antimicrobial Effect of Titanium Dioxide Nano Particles in Complete Dentures Made for Edentulous Patients.

77. Raghuwanshi, N., et al. (2017). "Synergistic effects of Woodfordia fruticosa gold nanoparticles in preventing microbial adhesion and accelerating wound healing in Wistar albino rats in vivo." *Materials Science & Engineering. C, Materials for Biological Applications* 80: 252–262.

78. Khashan, K. S., et al. (2017). "Preparation of iron oxide nanoparticles-decorated carbon nanotube using laser ablation in liquid and their antimicrobial activity." *Artificial Cells, Nanomedicine, and Biotechnology* 45(8): 1699–1709.

79. Schairer, D., et al. (2012). "Nitric oxide nanoparticles: preclinical utility as a therapeutic for intramuscular abscesses." *Virulence* 3(1): 62–67.

80. Schairer, D. O., et al. (2012). "The potential of nitric oxide releasing therapies as antimicrobial agents." *Virulence* 3(3): 271–279.

81. Lecher, K., et al. (2012). "Nitric oxide-releasing nanoparticles accelerate wound healing in NOD-SCID mice." *Nanomedicine: Nanotechnology, Biology and Medicine* 8(8): 1364–1371.

82. Hindi, K. M., et al. (2009). "The antimicrobial efficacy of sustained release silver-carbene complex-loaded L-tyrosine polyphosphate nanoparticles: characterization, in vitro and in vivo studies." *Biomaterials* 30(22): 3771–3779.

83. Huh, A. J. and Y. J., Kwon, (2011). "'Nanoantibiotics': A new paradigm for treating infectious diseases using nanomaterials in the antibiotics resistant era." *Journal of Controlled Release: Official Journal of the Controlled Release Society* 156(2): 128–145.

84. Knetsch, M. and L. H., Koole, (2011). "New strategies in the development of antimicrobial coatings: the example of increasing usage of silver and silver nanoparticles." *Polymers* 3.

85. Zhang, L., et al. (2010). "Development of nanoparticles for antimicrobial drug delivery." *Current Medicinal Chemistry* 17(6): 585–594.

86. Friedman, A. D., et al. (2013). "The smart targeting of nanoparticles." *Current Pharmaceutical Design* 19(35): 6315–6329.

87. Huang, S.-H. and R.-S., Juang, (2011). "Biochemical and biomedical applications of multifunctional magnetic nanoparticles: a review." *Journal of Nanoparticle Research* 13(10): 4411.

88. Hajipour, M. J., et al. (2012). "Antibacterial properties of nanoparticles." *Trends in Biotechnology* 30(10): 499–511.

89. Leid, J. G., et al. (2011). "In vitro antimicrobial studies of silver carbene complexes: activity of free and nanoparticle carbene formulations against clinical isolates of pathogenic bacteria." *Journal of Antimicrobial Chemotherapy* 67(1): 138–148.

90. U.S. Centers for Disease Control and Prevention. (2003). "Methicillin-resistant *staphylococcus aureus* infections among competitive sports participants–Colorado, Indiana, Pennsylvania, and Los Angeles County, 2000-2003." *MMWR. Morbidity and Mortality Weekly Report* 52(33): 793–5.

91. U.S. Centers for Disease Control and Prevention. (2003). "Methicillin-resistant *Staphylococcus aureus* infections in correctional facilities—Georgia, California, and Texas, 2001-2003." *MMWR. Morbidity and Mortality Weekly Report* 52(41): 992–6.

92. U.S. Centers for Disease Control and Prevention. (2002) "Vancomycin- resistant *Staphylococcus aureus*: Pennsylvania 2002." *MMWR. Morbidity and Mortality Weekly Report* 51: 902.

93. U.S. Centers for Disease Control and Prevention. (2002) "*Staphylococcus aureus* resistant to vancomycin: United States 2002. *MMWR. Morbidity and Mortality Weekly Report* 51: 565–67.

94. Lowy, F. D., (1998). *Staphylococcus aureus* infections. *New England Journal of Medicine*. 339:520–532.

95. Stewart, J., et al. (2003). Impact of Community-Acquired MRSA Colonization on MRSA Infection in Hospitalized Patients. Abstract 257. Proceedings of the IDSA Annual Meeting 2003.

96. US Centers for Disease Control and Prevention. (2018). National Antimicrobial Resistance Monitoring System for Enteric Bacteria (NARMS): Human Isolates Surveillance Report for 2015 (Final Report). Retrieved July 2019 from https://www.cdc.gov/narms/pdf/2015-NARMS-Annual-Report-cleared_508.pdf?.

97. Hay, S. I., et al. (2004). "The global distribution and population at risk of malaria: past, present, and future." *The Lancet. Infectious Diseases* 4(6): 327–336.

98. Nájera, J. A., et al. (2011). "Some lessons for the future from the Global Malaria Eradication Programme (1955-1969)." *PLOS Medicine* 8(1): e1000412–e1000412.

99 Trape, J. F., (2001). "The public health impact of chloroquine resistance in Africa." *The American Journal of Tropical Medicine and Hygiene* 64(1–2 Suppl): 12–17.

100. Clyde, D. F. and G. T., Shute, (1957). "Resistance of *Plasmodium falciparum* in Tanganyika to pyrimethamine administered at weekly intervals." *Transactions of the Royal Society of Tropical Medicine and Hygiene* 51(6): 505–513.

101. Price, R. N., et al. (2007). "Vivax malaria: neglected and not benign." *The American Journal of Tropical Medicine and Hygiene* 77(6 Suppl): 79–87.

102. Price, R. N., et al. (2009). "New developments in Plasmodium vivax malaria: severe disease and the rise of chloroquine resistance." *Current Opinion in Infectious Diseases* 22(5): 430–435.

103. Suwanarusk, R., et al. (2007). "Chloroquine resistant *Plasmodium vivax*: in vitro characterisation and association with molecular polymorphisms." *Plos One* 2(10): e1089–e1089.

57

Overview of Biofilms and Some Key Methods for Their Study

Paramita Basu, Michael Boadu, and Irvin N. Hirshfield

CONTENTS

DOI: 10.1201/9781003099277-60

Introduction

Perhaps we can say that microbiology began in the late 17th century through the curiosity of Antony van Leeuwenhoek, who was the first to describe accurately the creatures of the "invisible world" we now recognize as bacteria and eukaryotic microbes. In the earlier era of the study of microbes, the growth of bacteria and their study were predicated on the concept that bacteria are free-living, planktonic cells. This view dominated the study of bacteria for decades [1–3]. But by the 1970s, the concept of biofilms, which have multiple definitions, became known. They generally refer to communities of microbes (prokaryotic and eukaryotic) that adhere to a surface to form a slimy coat and are enclosed in an extracellular polysaccharide matrix [3–5]. A structured community of bacterial cells that attach to surfaces aggregates in a hydrated polymeric matrix of their own synthesis to form biofilms. Biofilms constitute a protected mode of growth that allows survival in a hostile environment leading to their inherent resistance to antimicrobial agents and host immune responses, which are at the root of many persistent and chronic bacterial infections [6]. In both the healthcare and the food industry, the number of bacterial infections and noninfectious complications are rising due to the bacterial biofilms formation and the subsequent failure of many medical devices [7].

Microscopy, especially light microscopy [8, 9], and transmission and scanning electron microscopy [8, 10, 11] played a pivotal role in the realization of the biofilm concept. Through the application of transmission electron microscopy, morphological evidence was found for an attachment of *Escherichia coli* to the intestines of newborn calves. Subsequently, it was shown that biofilms form in aquatic ecosystems under high-shear conditions [12]. Application of scanning confocal laser microscopy to the study of biofilms showed that morphologically a biofilm is an open system of cells, exopolymeric material, and extracellular spaces. The spatial arrangement of the cells, the extracellular polymeric substances (EPS), and the spaces may be referred to as the architecture of the biofilm [13, 14]. Major advantages of this technology are that the samples are living and hydrated. Consequently, it was discovered that there is a complex organization to biofilms with mushroom-like and tower-like structures enclosed in a matrix with intervening water channels [15, 16]. Biofilm is also composed of rigid bacteria (analogous to colloids), soft viscoelastic extracellular matrix, and small molecules that may provide signaling or generate forces for spreading and survival have been observed [17, 18].

The process of biofilm formation starts with colonization by microorganisms. They do so by adhering to living or inert surfaces, followed by multiplication and forming a self-produced polymeric matrix, which may trap both organic and inorganic debris and numerous microbial species [19]. Biofilms, therefore, may be made up of monocultures, multiple species, or combinations of phenotypes of different species [20]. Attached bacteria grow as a multicellular community, forming microcolonies in which they proliferate and mature. This microbial organization results in the development of a mature biofilm and eventually serves as a bacterial reservoir, from which bacterial cells are transmitted back to the environment through biofilm dispersal and then colonize new surfaces [17].

It has been established that biofilm formation and survival is not restricted to specific species of bacteria, rather, under adequate environmental conditions, many microbes will develop biofilms [21]. Bacteria form biofilm when the environment becomes stressful. These stressful environmental factors include a change in pH or temperature, and adhesive properties, nutrient availability, composition of the microbial community, hydrodynamics, quorum sensing, and cellular transport [22]. Biofilm formation in response to stress is much faster in pace when compared to planktonic cells, which occurs to overcome the adverse effects of the stress agent to avoid the complete extinction of the microbial community.

The arrangement of bacterial biofilm can be extremely variable and complex. For instance, intracellular bacterial communities (IBCs) including uropathogenic *E. coli* (UPEC) and *Klebsiella pneumoniae* bacteria can attach directly to their host or form bare or immersed biofilms on either biotic or abiotic surfaces and under static or shear-flow conditions [23]. Biofilm communities may also demonstrate differences in response to incoming signals such as nutritional availability and may play a role in bacterial biofilm formation [10]. For example, it has been demonstrated that D-amino acids act individually or synergistically to elicit dispersion of biofilms and inhibit Bacillus subtilis and other species pellicle formation [24]. Similarly, cyclic diguanylate monophosphate (c-di-GMP) concentration affects matrix and structural component production, motility, cell attachment, and eventually biofilm formation in a number of bacterial species [25, 26].

Numerous investigations have shown that biofilms have properties, probably derived from their organization, that are of both biological and clinical importance. It has been found consistently that biofilms are far more resistant to static and cidal agents than planktonic cells [27–29]. This applies to a broad spectrum of pathogens, including *Legionella*, *Listeria*, *Mycobacteria*, *Pseudomonas*, and *Vibrio*. There can be as much as a 100–1000 times greater resistance of the biofilms compared to their planktonic counterparts [27, 30, 31]. Treating biofilms of *P. aeruginosa* is of medical importance because of its pathogenesis in cystic fibrosis patients, and this has proved a difficult task [29–32]. The increased resistance of biofilms to antimicrobials is not the result of mutation or horizontal gene transfer because, upon removal of the cells from biofilms, and their subsequent growth as planktonic cells, the sensitivity to the agents re-emerges [3, 27, 32].

The resistance phenomenon also applies to fungal biofilms and has been documented with *Candida albicans* [33–35]. *Candida* is cited as being the fourth most-common cause of bloodstream infections [36], and is also a major clinical problem because of its ability to form biofilms on indwelling medical devices [33].

Multiple explanations have been advanced for the high resistance of biofilms to antimicrobials [27]. These include (1) depletion of the antimicrobial agent through its interaction and neutralization by the biofilm; (2) slow penetration of the antimicrobial into the biofilm [29]; (3) the presence of slow-growing and/or nongrowing cells (e.g., due to oxygen, pH, or nutrient gradients in the biofilm) [15, 37]; (4) stress responses, induced or constitutive; and more controversially, (5) persister cells. Bacterial populations produce persisters—cells that neither grow nor die in the presence of bactericidal agents—and, thus, exhibit multidrug tolerance. Research done by Spoering and Lewis showed that persisters are largely responsible for the recalcitrance of infections caused by bacterial biofilms. The persister cells, if they exist, constitute only a small percentage of the population, but represent a cell state that is tolerant to antimicrobial assault [38].

Due to the ability of microbial cells to adhere to surfaces, both biotic and abiotic, a major problem has developed in medicine with the colonization of indwelling medical devices and nosocomial infections. As the technology of medicine has advanced, and concomitantly the population of the aged is increasing, there has been increased utilization of indwelling medical devices. One major reason for this is that there has been a reduction in the length of hospital stays and a subsequent reliance on ambulatory care. For people of all ages, there are catheters to deliver medications or fluids. In addition, there are devices for those with heart defects and prostheses for joint replacements, particularly hips and knees [31, 39].

A negative consequence of these advances is that biofilms, whether bacterial or fungal, can form on these indwelling medical devices [40, 41]. These biofilms are difficult to treat and can lead to infection, which leads to a serious economic impact. For example, millions of intravascular catheters are commonly used in today's medicine in the United States. However, there are a huge number of bloodstream infections that arise from this use, which result in raising costs of treatment billions of dollars per annum [40, 42]. Often, the only recourse is to remove the device and replace it.

An understanding of the above-mentioned features of bacterial biofilms is utterly imperative in the discovery of novel means of controlling and preventing their resistance. This brief introduction gives a taste of the importance of biofilms, and why methods are needed to assess their formation as well as to test their response to antimicrobials.

Methods to Assess Biofilm Formation in Lab Conditions

There are various strategies used to study biofilm formation. Many of these generally employ stationary phase cells. These modes of study used static systems, which are better for the study of early events in biofilm formation, and continuous flow or chemostat systems, which are preferable for the examination of mature biofilms [43].

Microtiter Plate Biofilm Assay

A popular static method that has the advantage of high throughput is the microtiter plate biofilm assay. This method is based on its development and modification by Christensen et al. [44], Mack et al. [45], and O'Toole and Kolter [46].

Procedure

1. The bacterial strain is grown in batch culture overnight at the desired temperature to stationary phase in the desired medium, such as minimal medium or Luria-Bertani medium. A 10 mL culture or less will suffice.
2. Dilute the cells 1:100 with the desired growth medium into the wells of a 96-well polystyrene microtiter plate. It is recommended that each strain be assayed in

quadruplicate. Some wells should be filled with sterile water, and the plate should be covered with a lid and/or parafilm to prevent dehydration.
3. Incubate the cells in the plate at the appropriate temperature, 25–37°C is typically used [43], and the length of the incubation should be between 2 and 48 hours, depending on the strain, the medium, and the growth temperature.
4. After the incubation, the planktonic bacteria should be separated from those cells that have formed the biofilm. First, invert the plate and shake firmly over a wastewater tray. Alternatively, the cells can be washed using a water bottle that gently delivers the liquid. The wells should then be washed three more times, either by immersing them into water washes or gently applying water. Finally, invert the plates and tap them to remove any remaining water.
5. Air-dry the wells for a few minutes at room temperature, and then add 0.1% crystal violet solution, 20–200 uL, to each well. Allow the staining to proceed for 10–20 minutes at room temperature.
6. Wash the wells with water to remove crystal violet that has not adhered to the cells. Repeat three more times and air-dry the microtiter plate at room temperature.
7. Add 200 uL of 95% ethanol to each well, and incubate for 10–15 minutes at room temperature to solubilize the crystal violet. Mix with a pipette. Other solvents, such as dimethyl sulfoxide, have been used instead of ethanol [47].
8. Transfer 100–150 uL of crystal violet solution to corresponding wells of an optically clear, flat-bottom 96-well plate, and measure the absorbance at 500–600 nm with a plate reader.
9. Alternatively, the crystal violet stain can be quantified by another means [46]:
   a. Add 200 uL of 95% ethanol twice to each microtiter plate well and mix.
   b. Transfer the ethanol to a 1.5-mL microcentrifuge tube.
   c. Bring the volume to 1 mL with deionized water.
   d. Transfer to a cuvette and read spectrophotometrically at 540 nm.

An *In Vitro* Cell Culture Method for Assessing Biofilm Formation [48]

Batch culture is commonly used in studying biofilms because it is easy to grow the cells in a variety of media and vessels, and the cost is low [8]. In this method, the biofilm is grown on a coverslip that has been inserted into the culture medium.

Procedure

1. Cells are grown overnight in the appropriate medium. The bacteria can be grown aerobically or anaerobically, depending on the organism and the objective of the study.

2. Add 2.5 mL medium to individual sterile culture dishes.

3. To each dish, add a sterile 18 mm-diameter glass microscope coverslip and cover the dish.

4. Add the requisite amount of overnight culture to the dish and incubate overnight.

5. Remove the coverslips and rinse briefly with water. These can be viewed by phase-contrast microscopy.

6. For fluorescence microscopy, use the following approach:

   a. Remove the coverslip from the culture and transfer to a dry dish.

   b. Add 20 uL monoclonal antibody solution and incubate 30 minutes at room temperature [49].

   c. Add 5 uL fluorescein-conjugated goat secondary antibody and incubate at room temperature for 10 minutes.

   d. Briefly rinse the coverslips with water to remove unbound antibodies and cells, and then use immediately for microscopy to demonstrate biofilm development.

A Static Method for Assessment of *Staphylococcus* Biofilms

It has been well established that indwelling medical devices provide excellent surfaces for biofilm formation [33]. *Staphylococcus* spp.—in particular *Staphylococcus aureus* and *Staphylococcus epidermidis* as well as other coagulase-negative staphylococci—can produce a mucoid exopolysaccharide (referred to as slime) that allows them to form biofilms on the surfaces of these devices [50–52]. It is these staphylococci that are frequently involved in troublesome infections because of the difficulty in treating biofilms with antibiotics [27, 30, 31]. Biofilm formation can be assessed by a microtiter plate method [53] similar to that described above [44, 46]. Also, it was found that adding 1% glucose to trypticase soy broth or 2% sucrose to brain-heart infusion broth markedly increased the number of species that formed biofilms [53].

As an alternative, a method that requires simpler equipment and detects the slime production is the tube method [52, 53]. This is considered a qualitative method, but Mathur et al. [53] have shown that it is equivalent to the plate method, at least for strong biofilm formers.

Tube Method

Procedure

1. Grow staphylococcus on culture plate overnight.

2. Inoculate 10 mL TSB containing 1% glucose with a loopful of cells from the culture plate and incubate statically for 18–24 hours at 37°C in a glass or polystyrene test tube.

3. Decant the tube and wash with phosphate buffered saline (pH 7.3).

4. Air dry up to 10 minutes.

5. Stain with 0.1% crystal violet for 5–10 minutes at room temperature.

6. Wash with deionized water to remove the excess stain.

7. Invert the tube and allow it to air dry.

8. A positive result is a visible film that lines the wall and the bottom of the tube [52, 53]. In assessing the results, the formation of a ring at the liquid interface is not considered a positive result.

Flow Cell Method

Microscopic studies have been an important tool in the study of biofilm development. An effective protocol for the microscopic examination of developing biofilms is via the flow cell technique. This method was developed to allow direct, online observation of biofilm structure by light microscopy and scanning confocal laser microscopy [8, 54, 55]. Initially, light microscopy can be used; but as the biofilm thickens, scanning confocal laser microscopy (SCLM) is required to obtain interpretable images.

The scanning consists of five units. A container of medium is connected to a flow pump that is connected to a bubble trap; the medium then flows into a flow cell (chamber), and the effluent then goes to a waste bottle [54, 55]. The flow cell has four parallel flow channels (for details of its construction, see Wolfaardt et al. [54] or Christensen et al. [55]). The bubble chamber is present to prevent air bubbles from reaching the biofilm and disrupting its structure. The tubing between the bubble chamber and the flow cell must be of sufficient length to allow the flow cell to be mounted on a microscope. A critical component of the apparatus is a microscope glass coverslip that serves as a substratum for biofilm development as well as for microscopic observation. According to Christensen et al. [55], the coverslip is turned upward to prevent the sedimentation of flocs detached from other parts of the biofilm.

Procedure [55]

1. Sterilize flow cell with 0.5% (vol/vol) hypochlorite overnight, and then rinse with distilled water.

2. Using a syringe, 250 uL of an overnight culture diluted to an A_{450} of 0.05 is injected into the flow cell while the pump is turned off.

3. The flow cell is turned upside down.

4. The influent line is clamped to prevent back-growth.

5. After 1 hour, the flow chamber is righted and medium is pumped through the system at a rate of 0.2 mm/second.

6. The development of the biofilm can be followed by microscopy.

The Modified Robbins Device

A motivation for the development of the Robbins device derived from the problem of biofouling in industry [56–58]. This instrument, to a large degree, has been supplanted by the modified Robbins device [8, 30]. With the modified device, materials such as catheter surfaces can be employed to monitor the growth of biofilms [30]. Consequently, depending on the material chosen, the device can be used for industrial purposes, or for clinical or laboratory investigations [8, 30, 59]. It has also been used for the determination of the cidal effects of antibiotics on biofilms [30]. A key aspect in the design of this instrument is that the sampling ports have disks of the test surface attached at the bottom and in contact with the medium, but the flow of the medium does not disturb the biofilm. Because this is a multiport system, samples can be removed at various times or in multiples. The biofilms that grow can be examined by microscopy, or cells can be scraped off the surface of the disk material for enumeration [30]. The modified Robbins device can be used with continuous cultures systems as well as with batch culture [60].

Procedure

The experimental design is that of Nickel et al. [30], which is the seminal publication with respect to the modified Robbins device. In this work, a batch culture was used. The disks attached to the sampling ports were derived from a urinary catheter, and the system was organized with a flask that contained the medium fitted with a line containing a valve going to a peristaltic pump. From the pump, medium flowed through tubing into the modified Robbins device and the effluent was collected in an appropriate vessel.

1. The entire experimental apparatus was sterilized using ethylene oxide.
2. Cells *(Pseudomonas aeruginosa)* were added to the medium so that they would be in log phase for the duration of the experiment.
3. The pump rate was 60 mL per hour.
4. During the course of the experiment, up to 8 hours, the disks were aseptically removed and examined by scanning electron microscopy, or cells were scraped off the disks into medium for enumeration.

Microscopy

Microscopy has been an invaluable tool in studying the development of biofilms [3, 8–11]. Two major modes of microscopy used are scanning electron microscopy (SEM) and scanning confocal laser microscopy (SCLM). The major advantage of SEM is its high resolution; but because of fixation and dehydration of the samples in the preparative phases, it gives a distorted image of biofilm structure. In contrast, SCLM yields a lower resolution; but because the samples are more representative of their natural state, the image obtained is more accurate and has allowed for a much-improved understanding of biofilm structure [3].

Scanning Electron Microscopy (SEM)

Preparation of Samples for SEM [61–65]

1. Place the sample in 2.5 to 5% glutaraldehyde in 0.05 to 0.1 M cacodylate buffer (pH 6.2–7.0) for 1–24 hours at 20°C. The mixture often contains 0.15% ruthenium red to stain the exopolysaccharide matrix.
2. If desired, the sample can be washed two to three times with the buffer for 1 hour each.
3. Treat the sample with 1% osmium tetroxide [66, 67].

4. Dehydrate in an ethanol series from 25% to 100%. Each ethanol wash is for 10 minutes at room temperature.
5. The samples are critical-point dried.
6. The samples are coated with gold or platinum-palladium.

A variation on this procedure, which was used for the study of *Acinetobacter baumannii*, was published more recently by Tomaras et al. [65]. In this study, the bacteria were grown in Petri dishes without shaking.

1. After the appropriate time period, the dishes were flooded with 2.5% glutaraldehyde in 0.05 M cacodylate buffer.
2. This was incubated at room temperature for 2 hours.
3. The glutaraldehyde was removed and the cells were washed with distilled water.
4. Samples of 1 cm × 1 cm were cut from the Petri dish and rinsed gently with distilled water.
5. The preparation was treated with an ethanol series of washes as above.
6. The samples were then CO_2-critical-point dried for 60–120 minutes.
7. The samples were then gold-coated.

Scanning Confocal Laser Microscopy (SCLM)

Sample Preparation and Analysis for SCLM [68–72]

With SCLM it is possible to view live samples that have not been distorted by fixation. This has allowed for a much better understanding of the three-dimensional organization of biofilms. In addition, SCLM has been an effective experimental tool in conjunction with *in situ* hybridization [73, 74].

Treatment of the sample in order to view it using SCLM depends on the type of instrument used:

1. Upright microscope: With this type of microscope, the samples are mounted with a material such as acid-free silicone glue.
   a. Grow biofilm overnight in the appropriate medium on a 35 mm, collagen-coated, glass coverslip that is sealed to the bottom of a culture dish [72]. An alternative is to use a poly-L-lysine-coated coverslip [71] glued to an appropriate vessel.
   b. Remove the medium and wash once with water or 0.85% saline.
   c. Mount the coverslip.
2. Inverted microscope:
   a. Grow biofilm on an appropriate substratum. Flat surfaces are better because they are easily mounted and stained. Flow cells and the disks of the modified Robbins device are well suited for this.
   b. Wash the biofilm in 0.85% saline or the medium of growth.
   c. The use of an inverted microscope dictates that the growth chamber is mounted upside down using a

coverslip (sealed around with nail polish) at the bottom [58]. This way, the laser light penetrates the coverslip from the bottom of the biofilm to the top, producing an inverted image [72].

3. Staining:

   a. There are a variety of fluors for nucleic acids. Positive stains include SYTO dyes, acridine orange, and the DAPI stain and propidium iodide [68]. Fluorescein can be used for either negative or positive staining, depending on the pH [54]. Depending on the experimental system, the amount of fluor added must be empirically determined.

   b. When more than one fluor stain is used, factors such as relative intensity emission characteristics, photostability, interference, additive effects, and quenching effects must be considered [68].

4. The mountant: The primary purposes served by the mountant are to preserve the sample and prevent photobleaching. Dimethyl sulfoxide (DMSO) is often used with bacterial samples and mixed with the fluor before its addition to the sample.

   a. The mountant should have a pH that is compatible with the optimum pH for the fluor.

   b. The mountant must have an additive such as DMSO to prevent photobleaching.

   c. The refractive index of the mountant should be as close to that of the sample as possible. For bacteria, this is 1.517 [69].

5. Magnification: This is much less than with SEM and is 40× or 100× with an oil immersion lens. Another option is a 63× water immersion objective lens [72]. The sample (with the fluor) is then excited with light of appropriate wavelength, causing it to emit light (fluorescence).

6. Imaging:

   a. After locating suitable fields (usually with phase contrast or epifluorescence), the appropriate excitation and emission filters are set before scanning the sample.

   b. Either a single thin section (x-y plane) is scanned or a series of x-y optically thin sections through the biofilm are scanned. This is called a z series or section. It is also referred to as sagittal imaging [54]. The distance between the z sections depends on the thickness of the biofilm. For example, if the biofilm is approximately 30–40 u thick, then z slices are obtained every 2 u [72]. Another option is to collect a single or a series of x-z vertical sections through the specimen [68].

   c. These two-dimensional (2D) cross-sectional images are stacked using the SCLM software (provided by the manufacturer) to get 3D information. With the quantitative extraction of information contained in each image, a three-dimensional reconstruction of the principal biological events can be achieved [70].

7. Image processing and analysis:

   a. The SCLM images may require basic processing and enhancement before analysis. These include histogram analysis, gray-level transformation, normalization, contrast enhancement, application of median or lowpass filters or image addition, or multiplication. These are done using specialized digital image processing and analysis software. The processing applied also depends on the information desired. For example, when two different fluors are used (e.g., green and red), images of the biofilms can be converted to Adobe Photoshop files [70]. A region near the center of the biofilm is selected, a histogram of the region obtained, and the mean intensity value of the green pixels is divided by the mean intensity of the red pixels to obtain a semiquantitative measure of the accumulation of the fluorophore in the biofilm. Ratios of green/red intensity values obtained for the same region at different time points can be used to calculate the rate of fluorophore transport.

   b. Alternatively, images can be generated interactively by calling appropriate command-line scripts using the dialog and menu facilities of the SCLM software (provided by the manufacturer). Then digital image processing and analysis are performed with analytical packages such as the QUANTIMET 570 (Leica) computer system, which enhances images, discards unwanted details, and accelerates gray-scale processing [68].

   c. Computerized image analysis comprises a sequence of operations that depend on the specimen being studied. Before the automated image processing is activated, pixel calibration is done through the QUANTIMET program. The size and position of the area of interest for measurement are specified, and the computer is then programmed to perform a sequence of automated operations, such as image acquisition, gray image analysis, detection, measurements of signal intensity, volume integration, and data recording and display [68, 70].

   d. Deconvolution can also be applied to SCLM images to sharpen the image by mathematical removal of "out-of-focus" information. All these functions help in smoothing the image, reducing the noise, and sharpening the image, thus defining the objects more accurately.

8. Image output: After obtaining the images, they are stored as TIFF, GIFF, PICT, or JPEG, depending on the degree of image fidelity and the degree of image compression required.

Atomic Force Microscopy

An atomic force microscope has a nanoprobe or tip fixed on a sensitive cantilever, which can handle the micrometer scale. Atomic force microscopy (AFM) is an advanced technology for observing the force interactions between the tip or probe and the surface of the sample under observation [75, 76]. When the probe is very close to the sample, the top surface of the probe on the cantilever will bend from the original location. A photo detector receives the tiny signal from a laser beam deflected from the cantilever, and transfers/converts it into images viewed in a computer. In AFM, the force between the sample and the tip is used to sense the proximity of the tip to the sample. AFM offers considerable advantages in many scientific areas, including mapping of biological samples such as bacterial biofilms. Chemical modification of the AFM tip by using functional groups such as hydrophilic, hydrophobic, alkyl, or carboxyl groups can detect microbial surface functional groups, such as surface hydrophobicity or surface charge [77–79].

Recently, electrochemical experimental techniques have been used to monitor the growth of bacteria in the process of biofilm growth. Factors such as biofilm age, cell density, and pH as well as macromolecules in the EPS, such as polysaccharides, protein, humic substances and nucleic acids, as discussed later in this chapter) were studied. Biofilms on the electrode surface have been studied using cyclic voltammetry (CV) [80], impedance spectroscopy [81], and infrared reflection spectroscopy [82]. Electrochemistry has been used to control biofilm growth, such as stimulation on positively charged solid surfaces and inhibition on negatively charged surfaces [83, 84]. Some studies have shown that the negatively charged bacterial biofilms are more likely to grow faster on positively charged electrode surfaces; whereas, negatively charged surfaces inhibit bacterial growth [85]. Other research has shown that static electric fields play a major role in the adhesion of bacterial biofilms to solid surfaces [76, 86, 87].

The combination of AFM and electrochemistry is a new approach to track and observe the growth of live bacterial biofilms in their natural complex environment with high resolution. Advanced techniques such as AFM and electrochemical methods have been used to carry out investigations of the adhesion mechanisms of biofilms on various substrates, e.g., mica, glass, and pure and self-assembled monolayer (SAM) gold AU (111) surfaces (parentheses indicate the most stable isotope of gold). A SAM, an organized molecular assembly of amphiphilic molecules, is formed by adsorption onto a solid surface from a homogeneous solution [88]. This organization is achieved due to the affinity of the head group for the surface, and from the slow two-dimensional orientations of the hydrophobic tail groups.

Preparation of the Sample and the AFM Tip

Immobilization of Cells

AFM requires the sample to be attached to a solid substrate, which can be a glass slides, mica, or metal surface. Glass is composed of silicon dioxide and is not atomically flat; whereas mica can provide atomically flat surfaces that are ideal as a supporting substrate for AFM studies. Biosamples can be immobilized on a solid surface by physical adsorption or chemical adsorption. For physical adsorption, a drop of sample solution is put on a surface and dried in air. For chemical adsorption, a functional chemical is used to immobilize the target sample on a surface through formation of chemical bonds. L-lysine, fibronectin, and agglutinin are examples of linkers for chemical adsorption or electrostatic adsorption. Along this line, a surface modified by a chemical can

change its original physical or chemical properties and increase the adhesion of cells. Denaturation of biological samples is an essential problem for any immobilization method. Much effort has been put into seeking an approach with high AFM resolution and little denaturation. Mechanical fixation of individual cells in a polymer membrane is among the most successful developments. Some approaches with modified surfaces, like SAMs on AU (111) or the change of molecular charge on a surface, offer a promise to improve bacterial adhesion [89]. Some modes of attachment on a solid surface use extracellular polymeric substance or agar gel or a porous membrane to immobilize the cells mechanically. In the latter, spherical cells are trapped in a polycarbonate membrane with a pore size similar to that of the cells, to obtain high-resolution images without cell attachment or cell damage [90]. This approach has two advantages: (1), it is simple and does not involve drying, coating or chemical fixation; and (2), it can be used for AFM observations *in situ*, which is close to the natural environment during bacterial growth.

Chemically Modified Tips

A chemically modified tip is a prerequisite for a reliable, quantitative surface force measurement. The surfaces of commercial tips have defined surface coatings, such as silicon nitride, or gold-coated films, which are easily contaminated. Functional tips with SAMs of alkanethiols give high-resolution images due to their chemical sensitivity. The procedure is described in the experimental section (subsection describing preparation of hydrophobic tips and surfaces under the procedures section). Alkanethiols with different terminal functional groups: -OH, -CH3, -COOH, -NH2, can be used to detect the physical and chemical properties (hydrophobicity, charge) of biological surfaces [79].

Biologically Modified Tips

Some biomolecules are used to modify tips with the purpose of measuring intermolecular forces between individual ligand-receptor contacts, the adhesion of the biofilms, or the receptor protein location on single cell surfaces [91]. The binding of the biomolecules to the tips should be much stronger than the intermolecular force.

Cell-Modified Probes

A method has been developed to attach cells directly onto an AFM cantilever. This approach can probe cell-solid and cell-cell interactions. This method preserves the native ultrastructure of the cells during measuring. First, cells were immobilized by using glutaraldehyde treatment to create covalent crosslinking between the bacterial layer and the polyethylene glycol, polyethyleneimine (PEG-PEI)-coated tip, or by attaching a single yeast cell with a small amount of glue [92]. Lower and coworkers [93] adsorbed bacteria onto poly-L-lysine-coated glass beads, and subsequently attached a cell-coated bead to a cantilever by using a small amount of epoxy resin. Another approach is to attach individual cells to an AFM cantilever via lectins, such as wheat germ agglutinin [94]. However, these treatments are likely to affect the structure and properties of the cell surface.

General Procedure [94–96]

Cleaning

All glassware and the cells for both electrochemistry and AFM are boiled in 15% nitric acid, and washed thoroughly with ultrapure deionized water. The glassware and the cells are then placed in an ultrasound bath for at least 20 minutes, and washed again to ensure that all impurities have been removed. Both the electrochemical cell and the AFM cell were kept in a beaker filled with ultrapure deionized water, and sealed with Parafilm after cleaning.

Procedure for Cleaning Gold Surfaces

All gold electrodes were annealed in an oven at 860°C overnight. Prior to use, the surfaces were flame-annealed in a hydrogen flame, and quenched in ultrapure deionized water saturated with hydrogen gas. Gold surfaces that had been in contact with metal ions were electropolished in 3M sulfuric acid before the annealing procedures. The electro-oxidation produces a thin crust of gold oxide containing possible surface impurities, which are subsequently removed by dissolution in concentrated (36–38%) hydrochloric acid.

Self-Assembled Monolayers (SAM)

Self-assembled monolayers on Au (111) surfaces were prepared by immersion of the freshly annealed metal surfaces into a 1–2 mM solution of the monolayer-forming compound for 24 hours. The solvent was either ultrapure deionized water or ethanol as required. The surfaces were thoroughly rinsed in both solvent and ultrapure deionized water before further use.

Procedure

1. Biofilm growth:
   a. Mica is used for growing biofilms. 2.0×1.0 cm mica sheets were cut and freshly cleaved by tape on both sides. Prior to use, the sheets were sterilized by immersion into sodium hypochlorite solution (0.5%) for 2 hours.
   b. The mica slices were then washed three times with sterilized ultrapure deionized water, autoclaved at 121°C and 15 psi for 30 minutes, and then immersed into a Petri dish containing 10 mL of nutrient medium. Typically, bacteria are cultivated on mica in Trypticase Soy Broth (TSB) for 4 or 24 hours.
   c. Overnight cultures grown in appropriate medium are diluted to $OD_{600} = 0.05$ into fresh TSB medium, added over the mica, sealed by Parafilm, and incubated for given lengths of time at 37°C. The mica was then washed three times with ultrapure deionized water, and put into a new Petri dish to dehydrate naturally for one day.
2. Preparation of hydrophobic tips and surfaces. Gold-coated surfaces are obtained as follow:
   a. Silicon wafers are coated with electron beam thermal evaporation with a 5 nm thick chromium layer followed by a 30 nm thick gold [Au (111)] layer.
   b. The Au (111) substrate is coated with a SAM of octadecanethiol. The gold-coated surfaces and gold-coated AFM tips are cleaned for 15 minutes by UV and ozone treatment, rinsed with ethanol, dried with a gentle nitrogen flow, and immersed overnight in a 1 mM solution of octadecanethiol in ethanol, and then rinsed in ethanol.
3. AFM measurements and image acquisition:
   a. AFM measurements in ultrapure deionized water are performed using a Nanoscope 111a AFM and oxide-sharpened microfabricated silicon nitride cantilevers. To probe the microbes in their native state by AFM, the biofilm bacteria are immobilized by mechanical trapping into the mica surface. The mounted sample is transferred into the AFM liquid cell while avoiding dewetting, which can occur at a solid-liquid or liquid-liquid interface. Generally, dewetting describes the rupture of a thin liquid film on the substrate (either a liquid itself or a solid) and the formation of droplets. The imaging force is kept as low as possible (=250 pN) to minimize sample damage.
   b. To study adhesion properties of biofilms, both topographical and deflection images were recorded as they yield complementary information. Topographical images provide quantitative information on sample surface topography (surface roughness and height measurements). Deflection images often exhibit a higher contrast of the morphological details, and are especially useful for rough surfaces. Real time images were collected at time intervals, for example every 15 minutes.
   c. To study biofilm structure and dynamics in a natural aqueous medium, and other aspects of the chemical as well as cell surface properties important in biofilm formation and growth, a dualscope microscope equipped with a C-21 controller can be used. The tapping mode is used to record all images. A second instrument that can be employed is a PicoScan 5500 with a 100 um scanner and cantilevers (NP-S). This instrument is used in the contact mode.

AFM-Electrochemistry

The AFM-electrochemical approach is used to study electrostatic interactions between the bacteria and both the electrode surface and electroactive probe molecules [95, 96].

1. Cultivation of biofilms on solid surfaces:
   a. Biofilms were cultivated on either bare or SAM-modified Au (111) in Brain-heart infusion (BHI) medium for 24 hours. Thereafter, the cells were diluted to $OD_{600} = 0.05$ and regrown in fresh BHI medium. The development of the biofilm was sampled by AFM after suitable time periods (four hours and 24 hours).
   b. The clean gold sample is added, sealed with Parafilm, and incubated for given lengths of time

at 37°C. The Au (111) sample is then washed three times with Millipore-grade water.

c. The gold (biofilm) sample was then placed in a clean Petri dish to dehydrate naturally for 1 day before being imaged by AFM, but used directly without drying for cyclic voltammetry and linear sweep voltammetry. Bare Au (111) or SAM-modified Au (111) surfaces without bacteria are used as a reference.

2. AFM sample preparation and image recording: An atomic force microscope with a 100 um scanner and cantilevers (NP-S) in the contact mode is used. Twelve mm diameter Au (111) discs are used for AFM observations.

3. Electrochemistry:

a. Cyclic voltammetry is carried out using an Autolab system. The electrochemical cell is kept in a Faraday cage. The hanging meniscus method is used.

b. A freshly prepared reversible hydrogen electrode is used as a reference electrode and checked against a saturated calomel electrode (SCE) after each measurement. A clean, coiled, platinum wire serves as a counter electrode. Purified argon is used to deoxygenate all solutions and an argon stream is maintained over the solutions.

c. Phosphate buffered saline, pH 7.4, is the supporting electrolyte solution. Potassium ferrocyanide or potassium hexachloroiridate (negatively charged molecular probes), or ruthenium ammonium chloride or cobalt terpy perchlorate (terpy = 2, 2', 2''-terpyridine), which are positively charged molecular probes, are added independently (as needed) to the buffer.

Methods for Examining the Effect of Antibiotics or Biocides on Biofilms

It is now accepted that bacterial growth in natural settings is most commonly in the form of biofilms. From a clinical perspective, this presents a problem because biofilms grow on indwelling medical devices, particularly central venous catheters [40], and may be involved in the persistence of chronic infections [31]. A common experience is that the biofilms are considerably more resistant to antibiotics and biocides than planktonic cells [30, 97, 98]. Consequently, methods have been developed to examine the inhibition or killing of biofilms by antimicrobials, which have led to the term minimal biofilm eradication concentrations (MBECs) [97, 98]. These include the use of the modified Robbins device, the Calgary Biofilm Device, and simpler methods.

The Modified Robbins Device [30, 99]

As detailed previously, the modified Robbins device was designed primarily to measure biofilm formation. However, it can be used to determine the ability of antibiotics to inhibit the growth or kill cells in a biofilm [30, 99].

Procedure

Grow cells in the modified Robbins device to form a biofilm as described previously.

1. After the biofilm has formed, open the valve of a parallel reservoir containing antibiotic in the same medium.

2. After the appropriate period of antibiotic exposure, aseptically scrape the cells on a disk (urinary catheter) from a specimen plug into a tube containing sterile phosphate-buffered saline.

3. Subject the cells along with the disk to low-output sonication to disperse the cells.

4. Dilute the cells as needed and plate on nutrient agar for plate counting.

The Calgary Biofilm Device [8, 97–100]

Although the modified Robbins device could be used to determine the susceptibility of biofilms to antibiotics, it was deemed less than optimal for rapid antibiotic susceptibility testing for clinical purposes [97]. The Calgary biofilm device is relatively simple in that it consists of two parts that are composed of plastic. One is the bottom, which contains 96 troughs (channels) for bacterial growth, and the other is a lid into which pegs are placed to fit into the troughs. The biofilms grow on the pegs. The pegs will also fit into standard 96-well polystyrene microtiter plates [98]. Ceri et al. [97, 98] demonstrated that the device can produce 96 equivalent biofilms and is highly reproducible. Advantages of this system are that it requires no pumps or tubing, thereby reducing problems of contamination; allows for multiple testing; has a platform that uses standard 96-well plates; and is considerably faster than the modified Robbins device [99, 100]. This system is also referred to as the MBEC Assay System, and can be obtained from MBEC Biofilms Technologies Limited, Calgary, Alberta, Canada.

Procedure

1. Establish biofilm conditions:

a. Add a standardized inoculum to the channels in the bottom plate using TSB as the growth medium.

b. Attach the lid and place on a rocking table. The rocking action creates shear forces that promote biofilm formation. For aerobic cultures, the temperature is usually 35°C with 95% relative humidity.

c. At selected times, remove duplicate pegs, place in 200 uL sterile buffer, and sonicate for five minutes using a sonic cleaning system.

d. Plate on an appropriate medium to determine viable counts.

2. Susceptibility testing:

a. After generating a standardized biofilm, then its susceptibility to antibiotics or biocides can be tested.

b. Prepare working solutions of antibiotics in cation-adjusted Mueller-Hinton broth (CAMHB) to a concentration of 1024 ug mL^{-1}.

c. Serially dilute the antibiotics in CAMHB and add to the wells of a 96-well plate.

d. Wash the pegs (in the lid) on which biofilms have grown in a wash tray containing sterile buffer.

e. Place the lid (with the pegs) over the wells to expose the pegs with the biofilms to the antibiotics. The assembly is usually incubated overnight.

f. Determine the MBEC value (the minimal concentration of antibiotic that prevents growth of the biofilm).

3. Take the lid with the pegs and wash as previously discussed. Then place the lid over a 96-well plate in which the wells contain recovery medium.

4. Sonicate as above to disperse the cells into the recovery medium.

5. Determine the viable cell number by plating onto the medium of choice.

6. Alternatively, measure the turbidity in the wells using a 96-well plate reader.

Other Methods for Assessing the Effects of Antibiotics or Biocides on Biofilm Viability [32, 101–103]

Method 1

Stone et al. [101] examined the effect of tetracycline on biofilms of uropathogenic *E. coli* grown in Plexiglas flow cells.

1. For enumeration, first aseptically scrape the biofilm into 400 uL of 0.1M phosphate buffer at pH 7.0.

2. Vigorously vortex the cells.

3. Serially dilute the suspensions and plate on LB agar for enumeration.

Method 2

Anderl et al. [102] examined the effect of ampicillin and ciprofloxacin on *Klebsiella pneumoniae*. In this study, the biofilms were propagated on polycarbonate filters.

1. To enumerate the biofilm, first immerse the membrane in 9 mL phosphate-buffered water.

2. Vigorously vortex the membrane to suspend the cells.

3. Plate serial dilutions on R2A agar by the drop plating method [103].

Method 3

Elkins et al. [32] studied the killing of planktonic cells and biofilms of *Pseudomonas aeruginosa* by hydrogen peroxide. The biofilms were generated using a drip-flow reactor with stainless steel slides as the substratum.

1. To enumerate the biofilms, remove the slides and scrape aseptically into 50 mL phosphate-buffered saline (pH 7.2) containing 0.2% sodium thiosulfate to neutralize the hydrogen peroxide.

2. To suspend the cells, homogenize the above mixture in a PT 10/35 Brinkman homogenizer for 15 seconds at setting 4.

3. Appropriately dilute the homogenate and plate on R2A agar medium.

Molecular Genetic Methods

Polymerase Chain Reaction (PCR)

Based on PCR, various analytical techniques can be used to identify biofilm organisms without prior cultivation [104–107] in order to study the composition of the biofilm.

1. DNA fragments have to be extracted from the sample. It may be suitable to disrupt the organisms within the biofilm, (for example, by ultrasonic treatment), then they are concentrated by membrane filtration and lysed afterward again, e.g., by using organic solvents like chloroform, methanol, or detergents like Triton etc. (biofilms require mechanical lysis followed by chemical lysis to disrupt the membranes for complete extraction of DNA). It is also possible to lyse the cells directly in the sample, and obtain the DNA by high-speed centrifugation. The use of polyvinylpolypyrrolidone is recommended to remove humic substances if these are present in the sample, as they can interfere with the PCR reaction [104, 105].

2. After amplification of the 16S rRNA genes from the extracted genomic DNA by PCR, a mixture of the amplified products is obtained, which has to be separated by molecular biological techniques (like gel electrophoresis) and subsequently cloned [106] into vectors to get the corresponding rDNA.

3. Restriction analysis of the cloned rDNA fragments is performed, and the pattern of the fragment analysis can be compared to that of the culturable organisms of the population. In this way, it is possible to obtain genetic fingerprints of bacteria, protozoa, and fungi possibly present in the sample. An efficient analysis of the microorganisms by means of PCR fingerprinting can be supported by using DNA sequencing, capillary electrophoresis, and databases with reference patterns for the identification of unknown isolates [107]. Facultative bacteria can also be identified by polymerase chain reaction amplification and sequencing of the 16S rRNA gene [115].

Real-Time Quantitative Reverse-transcription-PCR (qRT-PCR)

In biofilms, the use of this method allows the investigator to identify efficiently the presence of specific genetic sequences related to individual bacterial species. Traditional PCR as such, is not suitable for quantitative studies of biofilms. Another difficulty is that PCR amplifies the DNA of both viable and dead cells. Thus PCR cannot be used accurately for the enumeration of living cells in a biofilm. To overcome these problems, qRT-PCR has been adopted. While in traditional PCR analysis, results are collected at the end of the reaction, during qRT-PCR, the fluorescent signal is measured in real time at each amplification cycle, and is directly proportional to the number of amplicons generated. Moreover, PCR amplifies all DNA present in the sample, whereas qRT-PCR detects the bacterial mRNA, with its short half-life. Detection of the mRNA is a promising indicator of cell viability [108]; therefore qRT-PCR has been applied not only to detect but also to quantify a specific microorganism in a biofilm [109–111].

Fluorescence *In Situ* Hybridization (FISH)

Fluorescence *in situ* hybridization (FISH) technique is a genetic approach using oligonucleotide probes labelled with fluorescent dyes. With the use of rDNA directed and fluorescence labeled oligonucleotide probes, it is possible to detect and localize microorganisms in a biofilm that may not be culturable and identify them phylogenetically. More than one oligonucleotide probe can be applied in order to label more than one kind of organism. Combining the FISH technique with SCLM facilitates the identification and topographical visualization of different species in a multispecies biofilm. This method provides not only information about the species present in the undisturbed sample, but also reveals their position in relation to each other, using the potential of SCLM to demonstrate three dimensional structures. This approach makes it particularly suitable to investigate the functional architecture of a biofilm, and the intensity of the signal depends on the ribosome content and thus reflects the physiological status [112–114].

Though the gene probe (FISH) strategy has brought significant progress to biofilm research, a drawback is that, in order to apply the probes, the cells must be permeabilized. The procedure required for this purpose may kill them, and therefore the possibility of affecting the architecture of the biofilm cannot be completely excluded.

Immunological Methods

Detection and identification of biofilm population members can also be performed by immunological coupling reactions between fluorescent labeled antibodies that are specific to certain organisms. This method requires cultivation of the target organisms, and the availability of antibodies that can detect suitable epitopes in a thin biofilm. In multilayered biofilms, the EPS matrix acts as a diffusion barrier for large molecules and the antibodies will be impeded in reaching the antigenic sites of the target organisms [114, 116]. The advantage in the use of mono- or polyclonal antibodies is that the cells are not killed by this method, and thus cellular processes in the biofilm still can be observed. Also, the combination with other nondestructive fluorescence staining methods is possible, e.g., using the tetrazolium salt 5-cyano-2,3-ditolyltetrazolium chloride [117].

Methods to Study Biofilms in Clinical Conditions

There is a great deal of literature dealing with how to cultivate and quantify biofilms, however, there is no real standard method. This is partly due to the differences in bacterial response to internal and environmental triggers before and during biofilm production and genetic variability in the community [118].

Vortexing-sonication-vortexing Method

The approach detailed in this section is a combination of conventional quantitative methods to optimize biofilm growth and quantification, particularly on curved surfaces, (catheter) to mimic the lumen of the urethra. The methods involve vortexing, sonication, and vortexing. It has been established that when biofilm cells are vortexed before sonication it aids in dislodging any loosely attached layers of biofilm and boosts the effect of sonication on deep-seated layers that might be sturdily attached to catheter surfaces. Furthermore, when the cells are vortexed after limited sonication, it helps in breaking down the bacterial communities into individual cells, which improves bacterial isolation. This facilitates a more accurate and uniform quantification without excessively destroying bacterial cells [119].

Although this method has advantages over others, it is not without limitations. First, the extent of sonication may affect the cell structure and/or the metabolism of bacteria. It has been reported that sonication more than 5 minutes can be detrimental to cell viability [120]. In addition, destructive techniques can have negative effects on biofilm-forming bacteria by producing viable but nonculturable bacteria. It is worthy to note that Gram-negative bacteria are more sensitive to sonication than are Gram-positive bacteria [121].

Another factor that needs to be considered is the intensity of the sonicator. For example, it is important to note that the intensity of cavitation in a water bath sonicator is low and the effect can be uneven [122] when compared to a probe-based ultrasonic processor. This difference in output can considerably affect the outcome.

Imaging is a prominent nondestructive method used to study biofilms. Thus, the use of microscopy with phase contrast or bacteria-specific staining is commonly used. One of the advantages of microscopy is the high-resolution images that can be directly used in publications or even quantified using imaging software. Another major advantage of microscopy is that it is able to be used to quantitatively analyze biofilms without the harvesting or disrupting and resuspension the biofilms. This permits the natural structures to be preserved [123].

The methods listed have been modified to cover other bacteria that can cause urinary-tract infections.

1. Growing bacterial biofilm on a urinary catheter segment *in vitro*:
   a. Obtain a single bacterial colony and inoculate in 10 mL of LB broth.
   b. Incubate the cells overnight at 37°C in static condition.
   c. Dilute the overnight culture to ~5 × 10⁵ CFU/mL and subculture again overnight.

   Two static serial passages have been shown to be essential for ideal type 1-pilus expression, which plays a critical role in attachment of cells to catheter surface [124–125].

d. Immerse Foley catheter segments (1 cm long) in a tube with 5 mL culture and provide sufficient aeration.

e. Allow the biofilm to develop on samples by incubating at 37°C (nonshaking) for 7 days using the fed batch culture method, by replacing the growth medium with fresh medium every 24 hours [126].

2. Biofilm growth *in vivo:*

a. Collect catheter samples from mice with associated urinary-tract infections (CAUTI) after 10–14 days of bacterial inoculation as described previously by Kadurugamuwa et al. [127].

b. Collect full-length indwelling catheters from female micropigs after 25 days of catheter placement, as described [119, 128].

c. Collect full-length indwelling urinary catheters from human patients and treat them as (b) above.

3. Extraction of biofilms:

a. Wash the loosely attached planktonic bacterial cells from catheter segments.

   i. Cut full length catheters from micropigs and humans into segments of 0.5 cm and 1 cm, respectively, for ease of extraction and calculation.

   ii. Gently dip samples and wash once in 5 mL sterile 1x PBS.

   iii. Gently tap the Foley catheter on a sterile absorbent to remove remaining liquid from the lumen.

b. Extract the biofilm using vortexing-sonication-vortexing method.

   i. Transfer samples to 10 mL 1x PBS in a 50 mL tube.

   ii. Execute a continuous vortex for 1 minute at full speed.

   iii. Using Ultrasonic Cell Disruptor XL-2000 probe P-1 (with a tip diameter of 3.2 mm), perform sonication at 10 W (RMS) for 50–60 seconds.

       – Enhance the strength and timing for different sonicators.

       – Keep the PBS with segments on ice to prevent heating due to sonication.

       – Clean the probe with 70% ethanol and ddH$_2$O between each sample.

   iv. Conduct another round of continuous vortex for one minute at full.

c. Verify nearly complete removal of biofilm.

   i. Dip-wash samples once in sterile PBS and then transfer to 5 mL sterile LB broth.

   ii. Incubate the samples in LB broth at 37°C with shaking at 200 rpm for the log phase period determined.

       – Incubation period depends on the doubling time for the bacteria.

   iii. Serial dilute the culture and plate on nutrient agar.

4. Plating:

a. Concentrate the sonicate by centrifuging at 3,100 g for 10 minutes.

   i. Carefully discard 9 mL of supernatant

   ii. Resuspend the pellet in the remaining 1 mL PBS.

b. Prepare ten-fold serial dilutions of the sonication supernatant.

   i. Plate serial dilutions including undiluted sonicate on LB Agar plate by 20 μL drop (Miles and Misra) method in triplicate [129]

   ii. Use CLED agar for all porcine samples to prevent swarming of *P. mirabilis*.

   iii. Streak and plate 20 μL sonicate of human samples on CHROMagarTM to identify other organisms.

   iv. At least triplicate samples all serial dilutions and plate.

c. Invert the agar plates and incubate at 37°C (nonshaking) for 14–18 hours.

5. Quantification of bacteria:

a. Count the number of colonies per 20 μL drop and note the dilution for each drop.

b. Calculate the average number of colonies.

   i. Count between 5 and 50 colonies per drop.

c. CFU per cm of a segment can be calculated as final quantification using the formula: CFU/cm = Average number of colonies for a dilution × 50 × dilution factor.

   i. The formula is adjusted per the length of the catheter segment.

Scanning Electron Microscopy (SEM) of Clinical Biofilm Samples

To qualitatively assess the biofilms from clinical samples, SEM should be performed as published elsewhere [130] with some modifications.

1. Fix the samples in 2.5% glutaraldehyde in PBS at 4°C overnight.

2. Wash the samples with PBS three times, then fix the samples once again with 1% OsO$_4$ reagent for 1–2 hours.

3. Dehydrate the samples in a stepwise manner in increasing percentages of ethanol from 25% up to 100%.

4. Dry the dehydrated samples with CO$_2$ in a critical point drier.

5. Mount the dried samples on stubs and sputter coat with a thin layer of gold.

6. Visualize the coated samples using a field emission scanning electron microscope.

REFERENCES

1. Prescott, L.M., Harley, J.P. and Klein, D.A. *Microbiology*, 6th ed., McGraw-Hill, New York, 2005, chap. 1.
2. Enge, R.S. Early developments in medical microbiology. *SCIEH Weekly Report*, 34, 7, 2000.
3. Costerton, J.W. A short history of the development of the biofilm concept, in *Microbial Biofilms*, Ghannoum, M. and O'Toole, G.A., Eds., ASM Press, Washington, DC, 2004, chap. 1.
4. Costerton, J.W., Cheng, K.J., Geesy, G.G., Ladd, T.I., Nickel, J.C., Dasgupta, M. and Marrie, T.J. Bacterial biofilms in nature and disease. *Annu Rev Microbiol.*, 41, 435, 1987.
5. Costerton, J.W., Lewandowski, Z., De Beer, D., Caldwell, D., Korber, D. and James, G. Biofilms, the customized microniche. *J Bacteriol.*, 176, 2137, 1994.
6. Costerton, J.W., Stewart, P.S. and Greenberg, E.P. Bacterial biofilms: a common cause of persistent infections. *Science*, 284, 1318, 1999.
7. Donelli, G. *Biofilm-Based Healthcare-Associated Infections*, vol. I, Springer, Switzerland, 2014.
8. McLean, R.J.C., Bates, C.L., Barnes, M.B., McGowin, C.L. and Aron, G.M. Methods of studying biofilms, in *Microbial Biofilms*, Ghannoum, M. and O'Toole, G.A., Eds., ASM Press, Washington, DC, 2004, chap. 20.
9. Loeb, G.I. Measurement of microbial marine fouling films by light section microscopy. *Mar Technol Soc J*, 14, 14, 1980.
10. Marrie, T., Nelligan, J. and Costerton, J.W. A scanning and transmission electron study of an infected endocardial pacemaker lead. *Circulation*, 66, 1339, 1982.
11. Kinner, N.E., Balkwell, D.L. and Bishop, P.L. Light and electron microscopic studies of microorganisms growing in rotating biological contactor biofilms. *Appl Environ Microbiol.*, 45, 1659, 1983.
12. Geesy, G.G., Richardson, W.T., Yeomans, H.G., Irvin, R.T. and Costerton, J.W. Microscopic examination of natural sessile bacterial populations from an alpine stream. *Can J Microbiol.*, 23, 1733, 1977.
13. Lawrence, J.R., Korber, D.R., Hoyle, B.D., Costerton, J.W. and Caldwell, D.E. Optical sectioning of microbial biofilms. *J Bacteriol.*, 173, 6558, 1991.
14. Andersen, J.B., Sternberg, C., Poulson, L.K., Bjorn, S.P., Givskov, M. and Molin, S. New unstable variants of green fluorescent protein for studies of transient gene expression in bacteria. *Appl Environ Microbiol.*, 64, 2420, 1998.
15. de Beer, D., Stoodley, P. and Lewandowski, Z., Effects of biofilm structure on oxygen distribution and mass transport, *Biotech Bioeng.*, 44, 636, 1994.
16. Potera, C., Studying slime, *Environ Health Perspect.*, 106, A604, 1998.
17. Magana, M., Sereti, C., Ioannidis, A. et al. Options and limitations in clinical investigation of bacterial biofilms. *Clin Microbiol Rev.*, 31(3), 2018.
18. Seminara, A., Angelini, T.E., Wilking, J.N., Vlamakis, H., Ebrahim, S., Kolter, R., Weitz, D.A. and Brenner, M.P. Osmotic spreading of Bacillus subtilis biofilms driven by an extracellular matrix. *Proc Natl Acad Sci USA*, 109:1116–1121, 2011
19. Satpathy, S., Sen, S.K., Pattanaik, S. and Raut, S. Review on bacterial biofilm: an universal cause of contamination. *Biocatalysis Agricultur Biotechnol.*, 7:56–66, 2016.
20. González-Rivas, F., Ripolles-Avila, C., Fontecha-Umaña, F., Ríos-Castillo, A.G. and Rodríguez-Jerez, J.J. Biofilms in the spotlight: detection, quantification, and removal methods. *Comp Rev Food Sci Food Safety*, 17(5): 1261–1276, 2018.
21. Kumar, C.G. and Anand, S.K. Significance of microbial biofilms in food industry: a review. *Int J Food Microbiol.*, 42(1–2): 9–27, 1998.
22. Böhmler, J., Haidara, H., Ponche, A. and Ploux, L. Impact of chemical heterogeneities of surfaces on colonization by bacteria. *ACS Biomat Sci Eng.*, 1(8): 693–704, 2015.
23. Justice, S.S., Hung, C., Theriot, J.A., Fletcher, D.A., Anderson, G.G., Footer, M.J. and Hultgren, S.J. Differentiation and developmental pathways of uropathogenic *Escherichia coli* in urinary tract pathogenesis. *Proc Natl Acad Sci USA*, 101: 1333–1338, 2004.
24. Leiman, S.A., May, J.M., Lebar, M.D., Kahne, D., Kolter, R. and Losick, R. D-Amino acids indirectly inhibit biofilm formation in Bacillus subtilis by interfering with protein synthesis. *J Bacteriol.*, 195: 5391–5395, 2013.
25. Simm, R., Morr, M., Kader, A., Nimtz, M. and Romling, U. GGDEF and EAL domains inversely regulate cyclic di-GMP levels and transition from sessility to motility. *Mol Microbiol.*, 53: 1123–1134, 2004.
26. Karaolis, D.K., Rashid, M.H., Chythanya, R., Luo W., Hyodo, M. and Hayakawa, Y. c-di-GMP (3'-5'-cyclic diguanylic acid) inhibits *Staphylococcus aureus* cell-cell interactions and biofilm formation. *Antimicrob Agents Chemother.*, 49: 1029–1038, 2005.
27. Stewart, P.S., Mukherjee, P.K. and Ghannoum, M.A. Biofilm antimicrobial resistance, in *Microbial Biofilms*, Ghannoum, M. and O'Toole, G.A., Eds., ASM Press, Washington, DC, 2004.
28. Zheng, Z. and Stewart, P.S. Penetration of rifampin through *Staphylococcus epidermis* biofilms. *Antimicrob Agents Chemother.*, 46, 900, 2002.
29. Walters, M.C., Roe, F., Bugnicourt, A., Franklin, M.J. and Stewart, P.S. Contributions of antibiotic penetration, oxygen limitation, and low metabolic activity to the tolerance of *Pseudomonas aeruginosa* biofilms to ciprofloxacin and tobramycin. *Antimicrob Agents Chemother.*, 47, 317, 2003.
30. Nickel, J.C., Ruseska, I., Wright, J.B. and Costerton, J.W. Tobramycin resistance of cells of *Pseudomonas aeruginosa* growing as a biofilm on urinary catheter material. *Antimicrob Agents Chemother.*, 27, 619, 1985.
31. Donlan, R.M. and Costerton, J.W. Biofilms: survival mechanisms of clinically relevant microorganisms. *Clin Microbiol Rev.*, 15, 167, 2002.
32. Elkins, J.G., Hassett, D.J., Stewart, P.S., Schweizer, H.P. and McDermott, T.R. Protective role of catalase in *Pseudomonas aeruginosa* biofilm resistance to hydrogen peroxide. *Appl Environ Microbiol.*, 65, 4594, 1999.
33. Hawser, S.P. and Douglas, L.J. Resistance of *Candida albicans* biofilms to antifungal agents *in vitro*. *Antimicrob Agents Chemother.*, 39, 2128, 1995.
34. Chandra, J.P., Mukherjee, P.K., Leidich, S.D., Faddoul, F.F., Hoyer, L.L., Douglas, L.J. and Ghannoum, M.A. Antifungal resistance of *Candida* biofilms formed on denture acrylic *in vitro*. *J Dent Res.*, 80, 903, 2001.
35. Kuhn, D.M., George, T., Chandra, J., Mukherjee, P.K. and Ghannoum, M.A. Antifungal susceptibility of *Candida* biofilms: unique efficacy of amphotericin B lipid formulations and echinocandins. *Antimicrob Agents Chemother.*, 46, 1773, 2002.

36. Banerjee, S.N. et al. Secular trends in nosocomial primary bloodstream infections in the United States 1980-1989: National Nosocomial Infections Surveillance Systems. *Am J Med.*, 91, 86S, 1991.

37. Zhang, T.C., Fu, Y.-C. and Bishop, P.L. Competition for substrate and space in biofilms. *Water Environ Res.*, 67, 992, 1995.

38. Spoering, A.L. and Lewis, K. Biofilms and planktonic cells of *Pseudomonas aeruginosa* have similar resistance to killing by antimicrobials. *J Bacteriol.*, 183, 6746, 2001.

39. Thomas, J.G., Ramage, G. and Lopez-Ribot, J.L. Biofilms and implant infections, in *Microbial Biofilms*, Ghannoum, M. and O'Toole, G.A., Eds., ASM Press, Washington, DC, 2004, chap. 15.

40. Maki, D.G. Infections caused by intravascular devices used for infusion therapy: pathogenesis, prevention, and management, in *Infections Associated with Indwelling Medical Devices*, 2nd ed., Bisno, A.L. and Waldovogel, F.A., Eds., ASM Press, Washington, DC, 1994, p. 155.

41. Nguyen, M.H. et al. Therapeutic approaches in patients with candidemia: evaluation in a multicenter prospective observational study. *Arch Intern Med.*, 155, 2429, 1995.

42. Raad, I.I. Intravascular-catheter-related infections. *Lancet*, 351, 893, 1998.

43. Merritt, J.H., Kadori, D.E. and O'Toole, G.A. Growing and analyzing static biofilms, in *Current Protocols in Microbiology*, John Wiley & Sons, Inc., New York, 205, pp. IB-1.1, 2005.

44. Christensen, G.D., Simpson, W.A., Younger, J.J., Badour, L.M., Barrett, F.F., Melton, D.M., and Beachy, E.H. Adherence of coagulase negative staphylococci to plastic tissue culture plates: a quantitative model for the adherence of staphylococci to medical devices. *J Clin Microbiol.*, 22, 996, 1985.

45. Mack, D., Nedelmann, M., Krokotsch, A., Schwarzkopf, A., Heesemann, J. and Laufs, R. Characterization of transposon mutants of biofilm-producing *Staphylococcus epidermidis* impaired in the accumulative phase of biofilm production: Genetic identification of a hexosamine-containing polysaccharide intracellular adhesin. *Infect Immun.*, 62, 3244, 1994.

46. O'Toole, G.A. and Kolter, R. Initiation of biofilm formation in *Pseudomonas fluorescens* WCS365 proceeds via multiple, convergent signaling pathways: a genetic analysis. *Mol. Microbiol.*, 28, 449, 1998.

47. Danhorn, T., Hentzer, M., Givskov, M., Parsek, M.R. and Fuqua, C. Phosphorous limitation enhances biofilm formation of the plant pathogen *Agrobacterium tumefaciens* through the PhoR-PhoB regulatory system. *J Bacteriol.*, 186, 4492, 2004.

48. Merritt, J., Qi, F., Goodman, S.D., Anderson, M.H. and Shi, W. Mutation of *luxS* affects biofilm formation in *Streptococcus mutans*. *Infect Immun.*, 71, 1972, 2003.

49. Shi, W., Jewitt, A. and Hume, W.R. Rapid and quantitative detection of *Streptococcus mutans* with species specific monoclonal antibodies. *Hybridoma*, 17, 365, 1998.

50. Raad, I. Management of intravascular catheter-related infection. *J Antimicrob Chemother.*, 45, 267, 2000.

51. Donlan, R.M. Biofilms and device associated infection. *Emerg Infect Dis.*, 7, 277, 2001.

52. Christensen, G.D., Simpson, W.A., Bisno, A.L. and Beachey, E.H. Adherence of slime-producing strains of *Staphylococcus epidermidis* to smooth surfaces. *Infect Immun.*, 37, 318, 1982.

53. Mathur, T., Singhal, S., Khan, S., Upadhyay, D.J., Fatma, T. and Ratta, A. Detection of biofilm formation among the clinical isolates of staphylococci: an evaluation of three different screening methods. *Indian J Med Microbiol.*, 24, 25, 2006.

54. Wolfaardt, G.M., Lawrence, J.R., Robarts, R.D., Caldwell, S.J. and Caldwell, D.E., Multicellular organization in a degradative biofilm community. *Appl Environ Microbiol.*, 60, 434, 1994.

55. Christensen, B.B. et al. Molecular tools for study of biofilm physiology. *Methods Enzymol.*, 310, 20, 1999.

56. McCoy, W.F., Bryers, J.D., Robbins, J. and Costerton, D.W. Observations of fouling biofilm formations. *Can J Microbiol.*, 27, 910, 1981.

57. McCoy, W.F. and Costerton, J.W. Fouling biofilm development in tubular flow systems. *Dev Indust Microbiol.*, 23, 551, 1982.

58. Costerton, J.W. and Lashen, E.S. Influence of biofilm on efficacy of biocides on corrosion-causing bacteria. *Mater Perform.*, 23, 34, 1984.

59. Domingue, G., Ellis, B., Dasgupta, M. and Costerton, J.W. Testing antimicrobial susceptibilities of adherent bacteria by a method that incorporates guidelines in the national committee for clinical laboratory standards. *J Clin Microbiol.*, 32, 2564, 1994.

60. Millar, M.R., Linton, C.J. and Sherriff, A. Use of a continuous culture system linked to a modified Robbins device or flow cells to study attachment of bacteria to surfaces. *Methods Enzymol.*, 337, 43, 2001.

61. Marrie, T.J. and Costerton, J.W. Scanning electron microscopic study of uropathogen adherence to a plastic surface. *Appl Environ Microbiol.*, 45, 1018, 1983.

62. Marrie, T.J. and Costerton, J.W. Scanning and transmission electron microscopy of *in situ* bacterial colonization of intravenous and intraarterial catheters. *J Clin Microbiol.*, 19, 687, 1984.

63. Ladd, T.I., Schmiel, D., Nickel, J.C. and Costerton, J.W. Rapid method for detection of adherent bacteria on Foley urinary catheters. *J Clin Microbiol.*, 21, 1004, 1985.

64. Kumo, H., Ono, N., Iida, M. and Nickel, J.C. Combination effect of fosfomycin and ofloxacin against *Pseudomonas aeruginosa* growing in a biofilm. *Antimicrob. Agents Chemother.*, 39, 1038, 1995.

65. Tomaras, A.P., Dorsey, C.W., Edelmann, R.E. and Actis, L.A. Attachment to and biofilm formation on abiotic surfaces by *Acinetobacter baumannii*: involvement of a novel chaperone-usher pili assembly system. *Microbiology*, 149, 3473, 2003.

66. Malik, L.E. and Wilson, B.W. Modified thiocarbohydrazide procedure for scanning electron microscopy: routine use for normal, pathological or experimental tissues. *Stain Tech.*, 50, 265, 1975.

67. Kumon, H., Kaneshige, T. and Omari, H. T_4-labeling technique and its application with particular reference to blood group antigens in bladder tumors. *Scanning Electron Microsc.*, II, 939, 1983.

68. Lawrence, J.R. and Neu, T.R. Confocal laser scanning microscopy for analysis of microbial biofilms. *Methods Enzymol.*, 310, 131, 1999.

69. W.M. Keck Microscopy Center. Confocal Microscopes. http://depts.washington.edu/keck/confocal-microscopes/. Accessed on 3/26/2020.

70. Kuehn, M., Hausner, M., Bungartz, H.J., Wagner, M., Wilderer, P.A. and Wuertz, S., Automated confocal laser scanning microscopy and semiautomated image processing for the analysis of biofilms. *Appl. Environ. Microbiol.*, 64, 4115, 1998.

71. Cowan, E.S., Gilbert, E., Khlebnikov, A. and Keasling, J.D. Dual labeling with green fluorescent proteins for confocal microscopy. *Appl. Environ. Microbiol.*, 66, 4113, 2000.

72. Jefferson, K.K., Goldmann, D.A. and Pier, G.B. Use of confocal microscopy to analyze the rate of vancomycin penetration through *Staphylococcus aureus* biofilms. *Antimicrob. Agents Chemother.*, 49, 2467, 2005.

73. Amann, R., Snaidr, J., Wagner, M., Ludwig, W. and Schleifer, K.H. *In situ* visualization of high genetic diversity in a natural microbial community. *J Bacteriol.*, 178, 3496, 1996.

74. Møller, S., Pedersen, A.R., Poulsen, L.K., Arvin, E., Molin, S. Activity and three-dimensional distribution of toluene-degrading *Pseudomonas putida*, in a multispecies biofilm assessed by quantitative *in situ* hybridization and scanning confocal laser microscopy. *Appl. Environ. Microbiol.*, 62, 4632, 1996.

75. Binnig, G., Quate, C.F. and Gerber, C. Atomic force microscopy. *Phys Rev Lett.*, 56: 930–933, 1986.

76. Kasemo, B. Biological surface science. *Surf Sci.*, 500: 656–677, 2002.

77. Wright, C.J., Shah, M.K., Powell, L.C. and Armstrong, I. Application of AFM from microbial cell to biofilm. *Scanning*, 32(3): 134–49, 2010.

78. Alsteens, D., Dague, E., Rouxhet, P.G., Baulard, A.R. and Dufrêne, Y.F. Direct measurement of hydrophobic forces on cell surfaces using AFM. *Langmuir*, 23: 11977–11979, 2007.

79. Ahimou, F., Denis, F.A., Touhami A. and Dufrene, Y.F. Probing microbial cell surface charges by atomic force microscopy. *Langmuir*, 18: 9937–9941, 2002.

80. Giao, M.S., Montenegro, M.I. and Vieira, M.J. Monitoring biofilm formation by using cyclic voltammetry-effect of the experimental conditions on biofilm removal and activity. *Water Sci Tech.*, 47: 51–56, 2003.

81. Hu, Z., Jin, J., Abruna, H.D., Houston, P.L., Hey, A.G., Ghiorse, W.C., Shuler, M.L., Hidalgo, G. and Lion, L.W. Spatial distributions of copper in microbial biofilms by scanning electrochemical microscopy. *Environment Sci Tech.*, 41: 936–941, 2007.

82. Vostiar, I., Ferapontova, E.E. and Gorton, L. Electrical "wiring" of viable Gluconobacter oxydans cells with a flexible osmium-redox polyelectrolyte. *Electrochem Comm.*, 6: 621–626, 2004.

83. Van der Mei., H.C. and Busscher, H.J. Electrophoretic mobility distributions of single-strain microbial populations. *App Environment Microbiol.*, 67(2): 491–494, 2001.

84. Bos, R., Van der Mei, H.C. and Busscher, H.J. Physicochemistry of initial microbial adhesive interactions—its mechanisms and methods for study. *FEMS Microbiol Rev.*, 23: 179–229, 1999.

85. Lower, S.K., Tadanier, C.J. and Hochella, M.F. Measuring interfacial and adhesion forces between bacteria and mineral surfaces with biological force microscopy. *Geochimica et Cosmochimica Acta*, 64: 3133–3139, 2000.

86. Timur, S., Haghigghi, B., Tkac, J., Pazarholu, N., Telefoncu, A. and Gorton, L. Electrical wring of *Pseudomonas putida* and *Pseudomonas flluorescens* with osmium redox polymers. *Bioelectrochemistry*, 71: 38–45, 2007.

87. Marsili, E., Rollefson, J.B., Baron, D.B., Hozalski, R.M. and Bond, D.R. Microbial biofilm voltamemetry: direct electrochemical characterization of catalytic electrode-attached biofilms. *App Environment Microbiol.*, 74(23): 7329–733, 2008.

88. Busalmen,J.P.,Berna,A.andFeliu,J.M.Spectroelectrochemical examination of the interaction between bacterial cells and gold electrodes. *Langmuir*, 23: 6459–6466, 2007.

89. Hu, Y., Zhang, J. and Ulstrup, J. Interfacial electrochemical electron transfer processes in bacterial biofilm environments on Au(111). *Langmuir*, 26(11): 9094–10103, 2010.

90. Gaboriaud, F. and Dufrene, Y.F. Atomic force microscopy of microbial cells: application to nanomechanical properties, surface forces and molecular recognition forces. *Colloids Surf B Biointerfaces.*, 54: 10–19, 2007.

91. Benoit, M. Cell adhesion measured by force spectroscopy on living cells. *Methods Cell Biol.*, 68: 91–114, 2002.

92. Razatos, A., Ong, Y.L., Sharma, M.M. and Georgiou, G. Molecular determinants of bacterial adhesion monitored by atomic force microscopy. *App Biol Sci.*, 95(19): 11059–11064, 1998.

93. Lower, S.K., Tadanier, C.J. and Hochella, M.F. Measuring interfacial and adhesion forces between bacteria and mineral surfaces with biological force microscopy. *Geochimica et Cosmochimica Acta,* 64: 3133–3139, 2000.

94. Oh, Y.J., Lee, N.R., Jo, W., Jung, W.K. and Lim, J.S. Effects of substrates on biofilm formation observed by atomic force microscopy. *Ultramicroscopy*, 109(8): 874–880, 2009.

95. Beech, I.B., Smith, J.R., Steele, A.A., Penegar, I. and Campbel, S.A. The use of atomic force microscopy for studying interactions of bacterial biofilms with surfaces. *Colloids Surf B Biointerfaces*, 23 (2–3): 231–247, 2002.

96. Oh, Y.J., Jo, W., Yang, Y. and Park, S. Influence of culture conditions on *Escherichia coli* O157:H7 biofilm formation by atomic force microscopy. *Ultramicroscopy*, 107(10–11): 869–874, 2007.

97. Ceri, H., Olson, M.E., Stremick, C., Read, R.R., Morck, D. and Buret, A. The Calgary biofilm device: new technology for rapid determination of antibiotic susceptibilities of bacterial biofilms. *J. Clin. Microbiol.*, 37, 1771, 1999.

98. Ceri, H. et al. The MBEC assay system: multiple equivalent biofilms for antibiotic and biocide susceptibility testing. *Methods Enzymol.*, 337, 377, 2001.

99. Raad, I., Darouiche, R., Hachem, R., Sacilowski, M., and Bodey, G.P., Antibiotics and prevention of microbial colonization of catheters, *Antimicrob. Agents Chemother.*, 39, 2397, 1995.

100. Bardouniotis, E., Huddleston, W., Ceri, H. and Olson, M.E. Characterization of biofilm growth and biocide susceptibility testing of *Mycobacterium phlei* using the MBEC® assay system. *FEMS Microbiol Lett.*, 203, 263, 2001.

101. Stone, G., Wood, P., Dixon, L., Keyhan, M. and Matin, A. Tetracycline rapidly reaches all the constituent cells of uropathogenic *Escherichia coli* biofilms. *Antimicrob Agents Chemother.*, 46, 2458, 2002.

102. Anderl, J.N., Franklin, M.J. and Stewart, P.S. Role of antibiotic penetration limitation in *Klebsiella pneumoniae* biofilm resistance to ampicillin and ciprofloxacin. *Antimicrob. Agents Chemother.*, 44, 1818, 2000.

103. Hoben, H.J. and Somasegaran, P., Comparison of the pour, spread, and drop plate methods for enumeration of *Rhizobium* spp. in inoculants made from presterilized peat, *Appl Environ Microbiol.*, 44, 1246, 1982.

104. Ogram, A.V., Sayler, G.S. and Barkay, T. The extraction and purification of microbial DNA from sediments. *J Microbiol Methods*, 7: 57–66, 1987.

105. Holben, W.E. and Tiedje, J.M. Applications of nucleic acid hybridization in microbial ecology. *Ecology*, 561–568, 1988.

106. Ogram, A. and Sayler, G.S. The use of gene probes in the rapid analysis of natural microbial communities. *J Indust Microbiol.*, 3: 281–292, 1988.

107. Tichy, H-V., Wiesner, P. and Simon, R. PCR-Fingerprint-Yerfahren zur Analyse von. Mikroorganismen-Populationen, in *Ökologie der Abwasserorganismen*. H. Lemmer, T. Ariebe and H.C. Flemming, Eds., Springer-Verlag, Berlin, Germany, 1996, pp. 111–122.

108. Sheridan, G., Masters, C., Shallcross, J. and MacKey, B. Detection of mRNA by reverse transcription-PCR as an indicator of viability in *Escherichia coli* cells. *Appl Environ Microbiol.*, 64: 1313–1318, 1998.

109. Warwick, S., Wilks, M., Hennessy, E. et al. Use of quantitative 16S ribosomal DNA detection for diagnosis of central vascular catheter-associated bacterial infection. *J Clin Microbiol.*, 42: 1402–1408, 2004.

110. Xie, Z., Thompson, A., Kashleva, H. and Dongari-Bagtzoglou, A. A quantitative real-time RT-PCR assay for mature *C. albicans* biofilms. *BMC Microbiol.*, 11: 93, 2011.

111. Guilbaud, M., de Coppet, P., Bourion, F., Rachman, C., Prévost, H. and Dousset, X. Quantitative detection of *Listeria monocytogenes* in biofilms by real time PCR. *Appl Environ Microbiol.*, 71: 2190–2194, 2005.

112. Thurnheer, T., Gmür, R. and Guggenheim, B. Multiplex FISH analysis of a six-species bacterial biofilm. *J Microbiol Methods*, 56: 37–47, 2004.

113. Fazli, M., Bjarnsholt, T., Kirketerp-Møller, K. et al. Quantitative analysis of the cellular inflammatory response against biofilm bacteria in chronic wounds. *Wound Repair Regen.*, 19: 387–391, 2011.

114. Machado, F., Cesar, D., Assis, A., Diniz, C. and Ribeiro, R. Detection and enumeration of periodontopathogenic bacteria in subgingival biofilm of pregnant women. *Braz Oral Res.*, 26: 443–449, 2012.

115. Gaylarde, C. and Cook, P. New rapid methods for the identification of sulphate-reducing bacteria. *Int Biodeterior.*, 26(5): 337–345, 1990.

116. Antloga, K.M. and Griffin, W.M. Characterization of sulphate-reducing bacteria isolated from oilfield waters. *Dev Ind Microbiol.*, 26: 597–610, 1985.

117. Schaule, G., Flemming, H. and Ridgway, H. The use of CTC (5-cyano-2,3-ditolyl tetrazolium chloride) in the quantification of respiratory active bacteria in biofilms. *Appl Environ Microbiol.*, 59: 3850–3857, 1993.

118. Allegrucci, M. and Sauer, K. Characterization of colony morphology variants isolated from Streptococcus pneumoniae biofilms. *J Bacteriol.*, 189: 2030–2038, 2007.

119. Mandakhalikar, K.D., Rahmat, J.N., Chiong, E., Neoh, K.G., Shen, L. and Tambyah, P.A. Extraction and quantification of biofilm bacteria: method optimized for urinary catheters. *Sci Rep.*, 8(1), 2018.

120. Kobayashi, N., Bauer, T.W., Tuohy, M.J., Fujishiro, T. and Procop, G.W. Brief ultrasonication improves detection of biofilm-formative bacteria around a metal implant. *Clin Orthop Relat Res.*, 457: 210–213, 2007.

121. Monsen, T., Lovgren, E., Widerstrom, M. and Wallinder, L. In vitro effect of ultrasound on bacteria and suggested protocol for sonication and diagnosis of prosthetic infections. *J Clin Microbiol.*, 47: 2496–2501, 2009.

122. Nascentesa, C.C., Mauro, K., Sousa, C.S. and Arruda, MarcoA. Z. Use of ultrasonic baths for analytical applications: A new approach for optimisation conditions. *J Braz Chem Soc.*, 12: 57–63, 2001.

123. Bjarnsholt, T, et al. Applying insights from biofilm biology to drug development — can a new approach be developed? *Drug Discov Nat Rev.*, 12: 791–808, 2013.

124. Hung, C.S., Dodson, K.W. and Hultgren, S.J. A murine model of urinary tract infection. *Nat Protoc.*, 4: 1230–1243, 2009.

125. Pratt, L.A. and Kolter, R. Genetic analysis of *Escherichia coli* biofilm formation: roles of flagella, motility, chemotaxis and type I pili. *Mol Microbiol.*, 30: 285–293, 1998.

126. Cerca, N., Pier, G.B., Vilanova, M., Oliveira, R. and Azeredo, J. Influence of batch or fed-batch growth on Staphylococcus epidermidis biofilm formation. *Lett Appl Microbiol.*, 39: 420–424, 2004.

127. Kadurugamuwa, J.L. et al. Noninvasive biophotonic imaging for monitoring of catheter-associated urinary tract infections and therapy in mice. *Infect Immun.*, 73: 3878–3887, 2005.

128. Umscheid, C.A. et al. Estimating the proportion of healthcare-associated infections that are reasonably preventable and the related mortality and costs. *Infect Control Hosp Epidemiol.*, 32: 101–114, 2011.

129. Miles, A.A., Misra, S.S. and Irwin, J.O. The estimation of the bactericidal power of the blood. *J Hyg (Lond)*, 38: 732–749, 1938.

130. Mukherjee, D., Zou, H., Liu, S., Beuerman, R. and Dick, T. Membrane-targeting AM-0016 kills mycobacterial persisters and shows low propensity for resistance development. *Future Microbiol.*, 11: 643–650, 2016.

58

Biofilms in Healthcare

Rebecca K. Kavanagh, Arindam Mitra, and Paramita Basu

CONTENTS

Introduction

Antibiotics have revolutionized the treatment of infections, saving millions of lives in the process. Antibiotic resistance occurs when microorganisms acquire the ability to resist the effects of an antimicrobial agent. As bacteria develop new ways to resist antibiotic effects, researchers have been driven to develop more sophisticated antibiotics. There are many mechanisms by which microorganisms become resistant to antibiotics, one of which being formation of a biofilm. These three-dimensional formations of bacteria can form on as varied surfaces as human teeth, lungs, intravenous catheters, and surgical implants.

Biofilms are colonies of bacteria, or more rarely fungi, existing within a hydrated extracellular matrix of polymeric substances (EPS) produced by the organisms.[1,2] The microorganisms within the biofilm adhere strongly to each other, as well as any surrounding surfaces. Biofilms form on organic or artificial surfaces within the human body. This process becomes problematic for healthcare providers, as formation of biofilms renders bacteria more resistant to antimicrobials, less susceptible to host immune responses, and allows the bacteria to survive harsh conditions. Once bacteria have formed a biofilm, the microbes are exceedingly difficult to eradicate. Biofilms provide microorganisms protection from environmental toxins and antibiotics via two general mechanisms: genetically inherited antibiotic resistance and phenotypes that confer drug resistance.[3–7] These biofilms can seed further bacterial expansion into other protected areas of the human body and potentiate relapse of infectious processes. Biofilm-related disease is associated with significant morbidity and mortality.[2] Understanding the nature of biofilm-related disease, the locations commonly implicated in biofilm formation, typical causative organisms, and how to treat and eradicate biofilms once they form is critical to reducing the risk of infection relapse and improving patient outcomes.

Pathophysiology

Bacterial biofilms are usually pathogenic in nature and can cause persistent and chronic bacterial infections as well as nosocomial infections. The National Institutes of Health (NIH) revealed that among all microbial and chronic infections; 65% and 80%, respectively, are associated with biofilm formation, which includes both device-related and nondevice-related infections. These numbers become cause for greater concern since most bacterial biofilms show resistance against the human immune system, as well as against antibiotics, increasing the potential of biofilms to cause infections and reduce the ability to treat them.[1] During biofilm formation, an organized aggregate of microorganisms adhere to one another and to a static surface (living or nonliving) by producing a matrix composed extracellular polymeric substances (EPS), which provides strength to the interaction of the microorganisms in the biofilm.[8] This attachment leads to formation of a micro-colony, giving rise to three-dimensional structures and ending up, after maturation, with detachment caused by producing saccharolytic enzymes that help to release the surface of the microbes into a new area for colonization. During the formation of a biofilm, bacteria communicate within and between species, using quorum sensing through small secreted autoinducer signal molecules.[9,10] This activates certain genes responsible to produce virulence factors. To some extent, microbe structure within a biofilm will appear different from the planktonic forms of the same microbial organisms.[11] Since bacteria that adhere to implanted medical devices or damaged tissue are able to encase themselves in a hydrated matrix of polysaccharide and protein, they are highly resistant to antimicrobial agents due to delay in penetration. This antibiotic resistance of bacteria in the biofilm mode of growth contributes to the chronicity of infections. The mechanism of resistance depends on multicellular strategies.

Biofilm-associated microorganisms have been shown to be associated with several human diseases, such as native valve endocarditis and cystic fibrosis; and to colonize a wide variety of medical devices. Though epidemiologic evidence points to biofilms as a source of several infectious diseases, the exact mechanisms by which biofilm-associated microorganisms elicit disease are poorly understood. Detachment of cells or cell aggregates, production of endotoxin, increased resistance to the host immune system, and provision of a niche for the generation of resistant organisms are all biofilm processes which can initiate the disease process.[12] Mechanisms responsible for resistance may be one or more of the following: (i) delayed penetration of the antimicrobial agent through the biofilm matrix, (ii) altered growth rate of biofilm organisms, and (iii) other physiological changes due to the biofilm mode of growth. Several pathogens associated with chronic infections include *Pseudomonas aeruginosa* in cystic fibrosis pneumonia, *Haemophilus influenzae* and *Streptococcus pneumoniae* in chronic otitis media, *Staphylococcus aureus* in chronic rhinosinusitis, and enteropathogenic *Escherichia coli* in recurrent urinary-tract infections, are linked to biofilm formation.[13]

Biofilm infections are important clinically because bacteria in biofilms exhibit recalcitrance to antimicrobial compounds and persistence despite sustained host defenses. Nutrient-depleted zones can result in a stationary phase-like dormancy within the biofilm, which may be responsible for the general resistance of biofilms to antibiotics.[14,15] Therefore, limited penetration of nutrients, rather than restricted antibiotic diffusion, may contribute to a generalized resistance or tolerance to antibiotics.[16] Although the matrix may not inhibit the penetration of antibiotics into the biofilm altogether, it may retard the rate of penetration long enough to induce the expression of genes within the biofilm that mediate resistance.

Typical Locations

In the medical setting, bacterial biofilms lead to high levels of antibiotic tolerance on medical devices. Biofilms are also responsible for the pathophysiology of the chronic infections many healthcare-associated infections (see Table 58.1).

Device-Related Biofilm Infections

Biofilms usually occur on or within indwelling medical devices, such as contact lenses, central venous catheters, mechanical heart valves, peritoneal dialysis catheters, prosthetic joints, pacemakers, urinary catheters, and voice prostheses.[19,20] Biofilms may be composed of only a single species or may also include several species. This will depend on the devices and their duration of action.[9]

Contact lenses are categorized as soft and hard, and microorganisms can adhere to both. Their classification is based on construction materials, frequency of disposal, wear schedule, and design. The type of microorganisms attached to contact lenses are mainly *E. coli*, *P. aeruginosa*, *Staphylococcus aureus*, *Staphylococcus epidermidis*, species of *Candida*, *Serratia* and *Proteus*. The most important factor in infection, the degree of adherence to the lenses, depends on the water content, substrate nature, electrolyte concentration, type of microbial strain

TABLE 58.1

Infections Caused by Biofilms

Device-Related Infections	Tissue-Related Infections
Ventricular derivations	Chronic otitis media, chronic sinusitis
Contact lenses	Chronic tonsillitis, dental plaque, chronic laryngitis
Voice prosthesis	Periodontitis
Endotracheal tubes	Endocarditis
Central vascular catheters	Lung infection in cystic fibrosis
Prosthetic cardiac valves, pacemakers, and vascular grafts	Chronic obstructive pulmonary disease (COPD) and diffuse panbronchiolitis (DPB)
Tissue fillers, breast implants	Intestinal and gallbladder Infections
Peripheral vascular catheters	Kidney stones
Urinary catheters	Biliary tract infections
Orthopedic implants and prosthetic joints	Urinary tract infections
Peritoneal dialysis catheters	Osteomyelitis
Ventilator-associated pneumonia	Chronic wounds
Surgical meshes	Chronic bacterial prostatitis
Intrauterine devices	Vaginosis

Source: Adapted from selected articles.[1,12,13,17,18]

involved, and lastly the composition of the polymer. Under scanning electron microscopy biofilms have been observed on contact lenses of a patient diagnosed with keratitis, produced by *P. aeruginosa*. Biofilms can also form more frequently on contact lenses that are usually kept in lens storage cases. The lens storage cases, therefore, have been declared as a source of lens contamination.[12]

Formation of biofilm is universal on central venous catheters, but the location and extent of biofilm formation depends upon the duration of catheterization. For example, short-term (<10 days) catheters have more biofilm formation on the external surface, whereas long-term catheters (30 days) have greater biofilm formation in the catheter lumen. The growth of microbes may be affected by the nature of the fluid which is administered through the central venous catheter. For example, Gram-positive bacteria, such as *S. epidermidis* and *S. aureus*, do not grow well in intravenous fluids, whereas Gram-negative aquatic bacteria, such as *P. aeruginosa*, *Enterobacter* species and *Klebsiella* species sustain growth in such fluids.[21,22]

Microbial cells attach and produce biofilms on mechanical heart valves and surrounding tissues, a condition known as prosthetic valve endocarditis. The types of bacteria responsible for this unpleasant condition are *Streptococcus* species, *S. aureus*, *S. epidermidis*, Gram-negative *Bacillus*, *Enterococcus*, and *Candida* spp. The origin of these microorganisms may be from the skin or from other indwelling devices like central venous catheters or dental work.[23] At the time of surgical implantation of prosthetic heart valves, tissue damage may occur as a result of the accumulation of platelets and fibrin at the location of suture and on the devices. Microbial cells have better ability to colonize at these locations.[12]

Urinary catheters are usually made of silicon or latex and are normally used during surgical operations to measure the urine generation and excretion. Urinary catheters are administered through the urethra into the urinary bladder. They may be closed or open systems. In an open catheter system, urine is drained into an open collector. This type of system increases the chances of contamination and this may lead to urinary tract infections (UTI) within days. In closed catheter systems, urine accumulates in a plastic bag, and this type of system results in a smaller incidence of UTIs.[24] Microbial cells that commonly contaminate and form biofilms on these devices are *E. coli, Enterococcus faecalis, S. epidermidis, P. aeruginosa, Proteus mirabilis, Klebsiella pneumoniae*, and other Gram-negative bacteria.[25]

Nondevice-Related Biofilm Infections

Periodontitis is an infection of the gums. In this infection damaging of soft tissues, as well as that of bones supporting the teeth occurs. Normally, it is caused by poor oral hygiene. Tooth loss is also possible.[26] *P. aerobicus* and *Fusobacterium nucleatum* are among the causative agents of periodontitis. These microbes also have the ability to form biofilms on a variety of surfaces, including mucosal surfaces in the oral cavity.[27] Microbial colonization of teeth surfaces may allow microorganisms to invade mucosal cells and alter the flow of calcium in the epithelial cells. The microorganisms can also release toxins. A plaque can then develop within 2–3 weeks. The plaque may mineralize with calcium and phosphate ions, forming the so-called tartar or calculus.[28]

Osteomyelitis is a disease of bones, which may be caused by bacterial cells or fungi. Bacteria enter the bones through the bloodstream, trauma, or through previous infections.[29] When microbes enter through the bloodstream and infect the metaphysis of the bone, this leads to the recruitment of white blood cells (WBCs) at the site. These WBCs attempt to phagocytose or lyse the pathogens by secreting enzymes, which may lyse the bone. This can result in the formation of pus, and spread infection through the bone blood vessels, thus stopping the proper flow of blood and causing tissue damage and deterioration of the function of the affected bone areas.[30]

Typical Organisms

A diverse population of microbes colonizes wounds, which can cause a delay in wound healing. However, microbial composition and diversity will vary depending upon the type of wound and may vary from wound to wound.[31] Major pathogens frequently isolated from acute and chronic wounds that can be grown in biofilms under lab conditions are *Staphylococcus aureus* and *Pseudomonas aeruginosa*.[32–35] One study described how *Staphylococcus* can be found on the surface whereas *Pseudomonas* can be detected in deeper tissues. Most chronic wounds contain at least one or more bacterial species and may cause synergistic virulence activity between the species. Other organisms such as *Escherichia coli, Enterococcus faecalis* and various species of *Klebsiella, Peptoniphilus, Enterobacter, Stenotrophomonas, Finegoldia, Serratia, Proteus, Moraxella,* and *Staphylococci* can be also be detected in clinical wounds using molecular techniques.[36,37] Polymicrobial species consisting of bacteria, fungus, and anaerobic bacteria can be identified

in chronic nonhealing wounds. Some chronic wounds may be devoid of an oxygen supply and this may permit anaerobic bacteria to form microcolonies. A diverse microbial population makes treatment challenging in chronic wounds and particularly when there are antibiotic resistant strains within biofilms. Few studies have indicated that biofilm forming potential of microbes isolated from wounds are greater than that found in skin.

Etiology

Breaks in the skin expose the subcutaneous tissue and provide microbes an opportunity to colonize open wounds due to suitable pH, moisture, nutrient, temperature, and other growth conditions. The subsequent formation of a biofilm requires multiple virulence factors such as flagella, pili, self-synthesized exopolysaccharides (such as alginate), lipopolysaccharides, specific membrane proteins, receptors, and toxins among others.[38] Initially pili, flagella, and other surface appendages assist microbes in making reversible attachments to each other or to a surface. This is followed by EPS-mediated irreversible attachment. The third step is cell proliferation leading to the formation of an aggregate or microcolony. The fourth step is growth and differentiation leading to establishment of a mature three-dimensional biofilm with structural features such as towers and channels. Quorum sensing is also required for biofilm formation with many wound pathogens such as *Pseudomonas aeruginosa* and *Staphylococcus aureus*.[39–41] Biofilm formation in wounds requires quorum sensing mediated diffusible signals termed autoinducers (AI) secreted in response to population density. Formation of a microcolony or aggregate of bacteria is dependent upon the presence and concentration of AI. In case of *Pseudomonas*, secretion of alginate protects the organism from immune defenses, which leads to the persistence of bacteria in wounds.

Treatment

Biofilm-associated infections can cause chronic illness, despite antibiotic therapy and innate and adaptive immune responses of the infected individual.[42] One of the reasons that bacteria protected by biofilms are so difficult to eradicate is the varied metabolic activity of cells inside the matrix. Biofilm clusters contain persister cells, which are neither growing nor dying.[43] These unique cells, which are metabolically inactive, are resistant to antibiotics that may rely on cellular activity to exert their bactericidal effects. Exposure of these persister cells to antibiotics leads to rapid development of antibiotic resistance.[44] Biofilms have been implicated in many different infectious processes, and treatments vary based on the causative pathogen and the location of the suspected biofilm. Whenever possible, surgical or non-surgical removal of the biofilm-infected object (implant, prosthetic, catheter etc.) should be performed to make the underlying bacterial load more susceptible to targeted antimicrobial therapy.[45]

It is important that clinicians consider the goals of therapy when treating patients with known or suspected biofilms.[46] Typically, when using antibiotics for patients with infections, clinicians are aiming for a cure and eradication of the pathogen.

This may be different for patients with biofilm formation. For example, for patients with cystic fibrosis and colonization with *Pseudomonas aeruginosa*, the goal of therapy may be first to eradicate the active infection. If the same patient develops a *Pseudomonas* biofilm in their lung tissue, the goal of antibiotic therapy may instead to target suppression of bacterial growth and maintenance of lung function. Wound-associated biofilms are particularly problematic because they not only permit antibiotic tolerance, but they also may sequester the source of infection from host immune responses. An exciting development in treatment of wound-associated biofilms is the use of biofilm-disrupting topical therapies.[47] A study by Miller et al. found that surfactants dissolved a *Pseudomonas aeruginosa* biofilm and associated bacteria in as little as one day after application of the surfactant. Another study by Wolcott et al. demonstrated that a biofilm-disrupting gel showed a robust wound-volume reduction within four weeks.[48] This wound-volume reduction was even more pronounced when the biofilm-disrupting gel was combined with topical antimicrobial agents. Authors noted that wound-associated biofilms should be treated with debridement in addition to these topical antibiofilm and antimicrobial agents. They also note that wound volume reduction is a surrogate marker of efficacy and does not take into account the full picture of wound improvement. A panel of wound management experts convened in 2017, echoed these recommendations. They also noted that there is positive data on use of high-osmolarity surfactant solution to assist in the disruption of the EPS and the prevention of biofilm formation after debridement.[49]

A combination of antimicrobials that target the causative pathogen with biofilm controllers can be used to treat and disperse biofilms. Most biofilm-controlling agents do not eradicate bacteria, so this combination could be promising for future research and development.[50] Some antimicrobials that can penetrate a biofilm matrix include tetracyclines, rifampin, daptomycin, amikacin, and ciprofloxacin. Although rifampin is commonly used in practice for biofilm-related disease, there is little clinical data to support this. Most rifampin studies are limited to *in vitro* and *in vivo* studies or retrospective observations.[51]

Biofilms remain some of the most difficult infectious processes for clinicians to treat effectively. This continues to be an area for ongoing *in vitro* and *in vivo* research. Clinicians should consider the source of biofilm formation, and seriously consider if this source can be safely removed from the patient. If not, they might consider a combination of antibiotics that are shown to have efficacy in penetrating biofilms. Clinicians may not be able to realistically achieve bacterial eradication, so their goal of therapy may need to be bacterial suppression and maintenance of biologic function.

REFERENCES

1. Jamal M., Ahmad W., Andleeb S., et al. Bacterial biofilm and associated infections. *J Chin Med Assoc*. 2018;81(1):7–11.
2. Del Pozo J.L. Biofilm-related disease. *Expert Rev Anti Infect Ther*. 2018;16(1):51–65.
3. Jefferson K.K. What drives bacteria to produce a biofilm? *FEMS Microbiol Lett*. 2004;236(2):163–73.
4. Olsen I. Biofilm-specific antibiotic tolerance and resistance. *Eur J Clin Microbiol Infect Dis*. 2015;34:877–86.
5. Parsek M.R., Singh P.K. Bacterial biofilms: an emerging link to disease pathogenesis. *Ann. Rev. Microbiol*. 2003;57:677–701.
6. Stewart P.S., Costerton J.W. Antibiotic resistance of bacteria in biofilms. *Lancet*. 2001;358:135.
7. Stewart P.S. Antimicrobial tolerance in biofilms. *Microbiol Spectr*. 2015;3(3) Doi: 10.1128/microbiolspec.MB-0010-2014.
8. Branda S.S., Vik Å., Friedman L., Kolter R. Biofilms: The matrix revisited. *Trends Microbiol*. 2005;13(1):20–6.
9. Donlan R.M. Biofilms: microbial life on surfaces. *Emerg Infect Dis*. 2002;8(9):881–90.
10. Federle M.J., Bassler B.L. Interspecies communication in bacteria. *J Clin Invest*. 2003;112(9):1291–9.
11. Høiby N., Ciofu O., Johansen H.K., Song Z-J., Moser C., Jensen P.Ø., Molin S., Givskov M., Tolker-Nielsen T., Bjarnsholt T. The clinical impact of bacterial biofilms. *Int J Oral Sci*. 2011;3(2):55–65.
12. Donlan R.M., Costerton J.W. Biofilms: Survival mechanisms of clinically relevant microorganisms. *Clin Microbiol Rev*. 2002;15(2):167.
13. Hall-Stoodley L., Costerton J.W., Stoodley P. Bacterial biofilms: From the natural environment to infectious diseases. *Nat Rev Microbiol*. 2004;2(2):95–108.
14. Walters M.C., Roe F., Bugnicourt A., Franklin M.J., Stewart P.S. Contributions of antibiotic penetration, oxygen limitation, and low metabolic activity to tolerance of pseudomonas aeruginosa biofilms to ciprofloxacin and tobramycin. *Antimicrob Agents Chemother*. 2003;47(1):317–23.
15. Fux C.A., Wilson S., Stoodley P. Detachment characteristics and oxacillin resistance of *Staphylococcus aureus* biofilm emboli in an in vitro catheter infection model. *J Bacteriol*. 2004;186(14):4486–91.
16. Shah D., Zhang Z., Khodursky A., Kaldalu N., Kurg K., Lewis K. Persisters: A distinct physiological state of *E. coli*. *BMC Microbiol*. 2006;6:53.
17. Lebeaux D., Chauhan A., Rendueles O., Beloin C. From in vitro to in vivo models of bacterial biofilm-related infections. *Pathogens*. 2013;2(2):288–356.
18. Lebeaux D., Ghigo J.M., Beloin C. Biofilm-related infections: Bridging the gap between clinical management and fundamental aspects of recalcitrance toward antibiotics. *Microbiol Mol Bio Rev*. 2014;78(3):510–43.
19. Donlan R.M. Biofilms and device-associated infections. *Emerg Infect Dis*. 2001;7:277–81.
20. Percival S.L., Kite P. Intravascular catheters and biofilm control. *J Vasc Access*. 2008;2:69–80.
21. Maki D.G., Mermel L.A. Infections due to infusion therapy. In: J.V. Bennett, P.S. Brachman (Eds.), *Hospital Infections*, 4th ed., Lippincott-Raven, Philadelphia, PA (1998), pp. 689–724.
22. Donlan R., Murga R., Carson L. Growing biofilms in intravenous fluids. In: J. Wimpenny, P. Gilbert, J. Walker, M. Brading, R. Bayston (Eds.), *Biofilms, the Good, the Bad, and the Ugly*. Presented at the fourth meeting of the Biofilm Club; Powys, UK (1999), pp. 23–29.
23. Braunwald E. Valvular heart disease, 5th ed. In: E. Braunwald (Ed.), *Heart Disease*, vol. 2, W.B. Saunders Co., Philadelphia, PA (1997), pp. 1007–1066.
24. Kokare C.R., Chakraborty S., Khopade A.N., Mahadik K.R. Biofilm: Importance and applications. *Indian J Biotechnol*. 2009;8:159–68.
25. Stickler D.J. Bacterial biofilms and the encrustation of urethral catheters. *Biofouling*. 1996;9:293–305.

26. Kokare C.R., Kadam S.S., Mahadik K.R., Chopade B.A. Studies on bioemulsifier production from marine Streptomyces sp. S1. *Indian J Biotechnol.* 2007;6:78–84.

27. Lamont R.J., Jenkinson H.F. Life below gum line: pathogenetic mechanisms of *Porphyromonas gingivalis*. *Microbiol Mol Biol Rev.* 1998;62:1244–63.

28. Overman P.R. Biofilm: A new view of plaque. *J Contemp Dent Pract.* 2000;1:18–29.

29. Ziran B.H. Osteomyelitis. *J Trauma.* 2007;62:559–60.

30. Kumar V., Abbas A.K., Fausto N., Mitchell R. *Robbins Basic Pathology*, 8th ed., Elsevier, Philadelphia (2007), pp. 810–811.

31. Percival S.L., Emanuel C., Cutting K.F., Williams D.W. Microbiology of the skin and the role of biofilms in infection. *Int Wound J.* 2012;9:14–32.

32. Anderson M.J., Schaaf E., Breshears L.M., Wallis H.W., Johnson J.R., Tkaczyk C., Sellman B.R., Sun J., Peterson M.L. Alpha-toxin contributes to biofilm formation among *Staphylococcus aureus* wound isolates. *Toxins (Basel).* 2018;10(4):157.

33. Fazli M., Bjarnsholt T., Kirketerp-Moller K., Jorgensen B., Andersen A.S., Krogfelt K.A., Givskov M., Tolker-Nielsen T. Nonrandom distribution of *Pseudomonas aeruginosa* and *Staphylococcus aureus* in chronic wounds. *J Clin Microbiol.* 2009;47:4084–9.

34. Harrison-Balestra C., Cazzaniga A.L., Davis S.C., Mertz P.M. A wound-isolated *Pseudomonas aeruginosa* grows a biofilm in vitro within 10 hours and is visualized by light microscopy. *Dermatol Surg.* 2003;29:631–5.

35. Pastar I., Nusbaum A.G., Gil J., Patel S.B., Chen J., Valdes J., Stojadinovic O., Plano L.R., Tomic-Canic M., Davis S.C. Interactions of methicillin resistant *Staphylococcus aureus* USA300 and pseudomonas aeruginosa in polymicrobial wound infection. *PLOS One.* 2013;8:e56846.

36. Dowd S.E., Sun Y., Secor P.R., Rhoads D.D., Wolcott B.M., James G.A., Wolcott R.D. Survey of bacterial diversity in chronic wounds using pyrosequencing, DGGE, and full ribosome shotgun sequencing. *BMC Microbiol.* 2008;8:43.

37. Gjodsbol K., Christensen J.J., Karlsmark T., Jorgensen B., Klein B.M., Krogfelt K.A. Multiple bacterial species reside in chronic wounds: a longitudinal study. *Int Wound J.* 2006;3:225–31.

38. Lindsay D., von Holy A. Bacterial biofilms within the clinical setting: what healthcare professionals should know. *J Hosp Infect.* 2006;64:313–25.

39. Davies D.G., Parsek M.R., Pearson J.P., Iglewski B.H., Costerton J.W., Greenberg E.P. The involvement of cell-to-cell signals in the development of a bacterial biofilm. *Science.* 1998;280:295–8.

40. Hammer B.K., Bassler B.L. Quorum sensing controls biofilm formation in *Vibrio cholerae*. *Mol Microbiol.* 2003;50:101–4.

41. Rumbaugh K.P., Griswold J.A., Iglewski B.H., Hamood A.N. Contribution of quorum sensing to the virulence of *Pseudomonas aeruginosa* in burn wound infections. *Infect Immun.* 1999;67:5854–62.

42. Høiby N., Bjarnsholt T., Givskov M., Molin S., Ciofu O. Antibiotic resistance of bacterial biofilms. *Int. J. Antimicrob. Agents.* 2010;35:322–32.

43. Miyaue S., Suzuki E., Komiyama Y., Kondo Y., Morikawa M. Bacterial memory of persisters: Bacterial persister cells can retain their phenotype for days or weeks after withdrawal from colony – biofilm culture. *Front. Microbiol.* 2018;9:1–6.

44. Lewis K. Persister cells, dormancy and infectious disease. *Nat. Rev. Microbiol.* 2007;5:48–56.

45. Kim D., Namen Ii W., Moore J., et al. Clinical assessment of a biofilm-disrupting agent for the management of chronic wounds compared with standard of care: A therapeutic approach. *Wounds.* 2018;30(5):120–30.

46. Ciofu O., Rojo-Molinero E., Macià MD., Oliver A. Antibiotic treatment of biofilm infections. *APMIS.* 2017;125(4):304–19.

47. Miller K.G., Tran P.L., Haley C.L., et al. Next science wound gel technology, a novel agent that inhibits biofilm development by gram-positive and gram-negative wound pathogens. *Antimicrob Agents Chemother.* 2014;58(6):3060–72.

48. Wolcott R. Disrupting the biofilm matrix improves wound healing outcomes. *J Wound Care.* 2015;24(8):366–71.

49. Snyder R.J., Bohn G., Hanft J., et al. Wound biofilm: Current perspectives and strategies on biofilm disruption and treatments. *Wounds.* 2017;29(6):S1–S17.

50. Gebreyohannes G., Nyerere A., Bii C., Sbhatu D.B. Challenges of intervention, treatment, and antibiotic resistance of biofilm-forming microorganisms. *Heliyon.* 2019;5(8):e02192.

51. Lee C.Y., Huang C.H., Lu P.L., Ko W.C., Chen Y.H., Hsueh P.R. Role of rifampin for the treatment of bacterial infections other than mycobacteriosis. *J Infect.* 2017;75(5):395–408.

59

The Business of Microbiology

Michael C. Nugent and Lorrence H. Green

CONTENTS

Introduction

This chapter was introduced in the previous edition of this book to help assist businesspeople requiring an understanding of the microbiology and underlying technologies associated with the companies they were considering supporting. This was important because microorganisms and infectious disease can oftentimes form a main portion of the intellectual property (IP) or the strategic basis of many start-up companies in the healthcare industry. This updated chapter is meant to provide businesspeople with a few basic insights into the industry regarding the various commercialization opportunities. A secondary goal of this chapter is to again introduce academic scientists to some of the factors that businesspeople consider when deciding on making an investment in a company, a technology, or a product. Academic scientists that are considering commercializing their research should find this information useful.

Scope of the Industry

With the advent of the cloning techniques developed in the mid-1970s that allowed scientists to insert the genes coding for proteins into the plasmids of bacteria, the scope of microbiology rapidly shifted from one of identifying and eliminating the causes of disease, to one in which microorganisms could be used as reagents and raw materials in the synthesis of medicines. Microbiology-based companies expanded from the healthcare industry, into industries ranging from bioremediation to agriculture to the energy industry. While this book will provide useful information to businesspeople interested in those areas, this chapter will confine itself to the industry of diagnosing the causes of infectious disease.

There are myriad opportunities for commercialization in the infectious disease diagnostics space from traditional culture, microscopy, and biochemical detection to novel technologies such as molecular detection methods, mass spectrometry, and genetic sequencing. The opportunities continue to grow as the need for fast, accurate diagnostic information will only become more important in the era of emerging diseases and new pandemic outbreaks. This challenge has been quite evident during the ongoing COVID-19 pandemic in which the scarcity of robust diagnostic tests and immunological assays led the FDA to quickly clear several new tests under their emergency use authorization (EUA).[1] Companies with open architecture on their molecular diagnostics platforms and those that had early prototype designs were able to move fast to meet the critical issue, but this experience highlights the need for robust pipelines and ongoing development to try to preempt the detection of emerging infections.

Genetic sequencing has become increasingly effective for research applications, but the development of useful bioinformatics tools has allowed this technology to penetrate deeper into clinical diagnostics microbiology.[2] Bioinformatics will ultimately help clinicians translate the wealth of data that is gained from sequencing applications into targeted useful information that can affect therapeutic decision making. Molecular detection methods are becoming more widely used and are replacing traditional methodologies for those clinical conditions that require either a rapid result (e.g. hours vs days) or more accurate ones (infections associated with healthcare workers, sexually transmitted diseases, etc.). In addition, molecular detection methods are ideal when the clinician is interested in a screening for a specific target or gene. The newest molecular detection methods consist of highly multiplexed assays and arrays that can detect many different analytes in a single sample. These multiplex assays are gaining rapid clinical acceptance in U.S. hospitals. One example is in screening for the cause of complex respiratory pathogens.[3] This "syndromic" approach to molecular diagnostics is particularly important when patients have conditions with similar clinical presentation. For instance, there are many pathogens that are the etiology of "flu like" illnesses. With the advent of these highly multiplexed molecular detection methods, clinicians can order a test for upwards of 20+ pathogens, collecting a single sample and providing a broad range of pathogen information based on patient symptoms.[4]

Although the development of innovative technologies in microbiology continues at a rapid pace, there are still many opportunities to commercialize traditional methodologies. Organism cultivation, isolation, and phenotypical differentiation for antibiotic resistance characteristics, for example, are still commonly used and in some instances are a critical requirement in making therapeutic decisions or tracking organisms from an epidemiological perspective. If a new method can produce an accurate result, is reasonably fast (most growth-based methods require at least 24-hour incubation) and is inexpensive to use, many laboratories will implement it for the cost effectiveness alone. Most newer technologies are not inexpensive, however, and in instances when a result is not urgent, laboratorians may be willing to accept a longer turnaround time as a tradeoff to a more cost-effective approach. Ultimately, the cost benefit analysis would be conducted by the laboratory leadership in conjunction with medical staff to determine the best clinical approach that is also cost effective and implementable in the facility. In many instances, hospitals will utilize several technologies and take a front-line approach, then cascade to other methods if the most likely pathogen is determined to be negative. When commercializing novel technologies, it is important that the marketing on these devices encompasses a clear positioning around the quantifiable economic and clinical outcome benefits that will be gained to offset the increased cost of the device. These benefits could be reduction in readmissions, more efficient bed utilization, reduction in healthcare-associated infections or better, more cost-effective antibiotic usage. There will always be a need to bring in a diagnostic test that offers the same accuracy, but may be able to provide a result better, faster, or cheaper. In recent years, declining reimbursement rates for diagnostic tests have

put a downward pressure on laboratory operational costs. Even in an era in which new technologies are entering the market at rapid pace, traditional methodologies, such as culture and microscopy, are widely considered the "gold standard" for reference methods in clinical trial and FDA 510K submissions for diagnostic tests.

History of Microbiology Technologies

Perhaps the most important aspect of any microbiology company's IP is its technology. In the 135 years since Koch first developed his postulates to identify the microorganisms that caused cholera and tuberculosis, the technologies used to identify microorganisms has been evolving. This evolution initially began at a slow pace, but much like other mainstream innovations, it has accelerated rapidly in the past 60 years. Unlike biological evolution, however, none of the ancestral forms of technology used to identify microorganisms, have gone extinct. The reason for this is that while technologies have evolved to meet ever changing marketing demands, the expanse of the field still provides plenty of niches that allow for the survival of older technologies. It is important that an investor in a company have a clear view of the market conditions, target opportunities, and the benefits of the technology to the ultimate user. In many cases, new, better technologies have not replaced the traditional methods. Clinical microbiology has historically been somewhat more subjective with microbiologists relying on observation of morphology through microscopic and growth-based methodologies to categorize the cause of an infection. Because of the qualitative and subjective nature of clinical microbiology, there will continue to be a place for traditional methods in the market for the foreseeable future. Despite rapid innovation, clinical microbiologists can oftentimes be resistant to change, at times preferring the tried and true techniques of growth-based observation for diagnosing pathogens. Despite this tendency, molecular methodologies have become widely available and prevalent in even the most ardent traditionalist microbiology laboratories. The key element in adoption of these molecular methods in some of the more traditional culture-oriented microbiology laboratories is full sample to result integration and ease of use.

As shown in Figure 59.1 at the outset of the study of microbiology, the only technologies available were microscopy. It was almost 200 years before the techniques of culture and biochemical testing were added to identify microorganisms. The combination of these three technologies remained at the cutting edge of microbial identification for decades. Due to their low cost and relative ease to perform, they still account for a large share of the microbiology testing that is done today. Even the widespread introduction of antibiotics in the mid-to-late 1940s did not really alter the use of microscopy, culture, and biochemical testing in identifying microorganisms or in ascertaining which antibiotics would be effective in treating a patient infection.

In the 1970s, the API strip introduced an age of miniaturization of biochemical testing (United States Patent 3,936,356 Janin February 3, 1976).[5] The API strip transformed biochemical testing from a cumbersome procedure that required the inoculation of 20 or more test tubes and the use of large incubators, to one in which all of the biochemical tests could be performed on a small plastic strip. While this was a great improvement for the

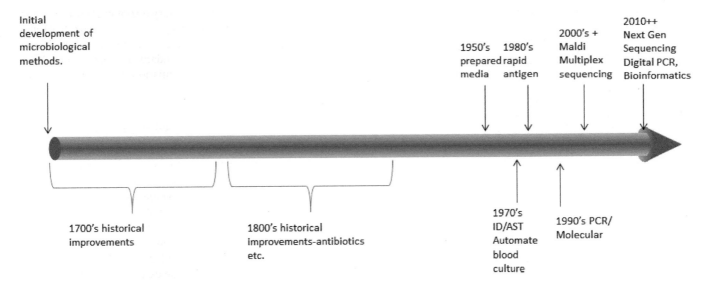

FIGURE 59.1 Timeline of innovations in the microbiology field.

identification of microorganisms, it still used the same microscopy, culturing, and biochemical technologies.

The first major shift in technology in identifying microorganisms came as this field was nearing the end of its first century, in the early 1960s, with the advent of antibody-based tests. These tests were based on the use of antibody molecules that were specific for certain organisms. First, was the introduction of the radioimmunoassays, which was followed by enzyme immunoassays in the 1970s.[6,7]

The ability to bind antibodies to inert particles such as latex, charcoal, or fixed red blood cells, led to the widespread use of easy to perform agglutination tests. The problem with these antibody-based technologies was that while they were fast and easy to perform, they were limited to only identifying a single pathogen. Thus, if the patient did not have the suspected infection, no additional useful information was provided. Immunoassays were further improved in the 1980s and 1990s with the development of lateral flow and membrane filtration immunoassays that would capture organism antigen on a filter and bind antigens directly on the membrane for detection of a specific microorganism.

The next major change came in the late 1970s and early 1980s and was driven by the revolution in isolating specific nucleic acid sequences that could be used to identify small numbers of microorganisms in samples, as detailed in Chapter 10 (Diagnostic Medical Microbiology), since this time many variations of these techniques have been developed. These include simple DNA probe tests and amplification methods such as polymerase chain reaction. While these Nucleic Acid based tests were originally developed to identify microorganisms, recently they have also been applied to determining antibiotic susceptibility patterns.[8] Today's molecular methods with multiplexing capability will not only identify pathogen strains with a single sample input, but also identify novel or otherwise well-understood resistance markers to provide additional clinical benefit to the diagnostic tool.

Reagents and test methods were not the only new technologies to enter the field at this time. By the late 1980s, many companies in the diagnostics industry had introduced completely automated testing units, capable of processing dozens of samples at one time. These instruments combined technologies such as spectrophotometry, video image processing, infrared detection, fluorescent detection, and computer software to develop instruments that required nothing more than that a technician prepare a sample and put it in the machine.

Marketing Forces in Diagnostic Microbiology

Several marketing forces in this industry have been at work at different times during the past 160 years. In most instances, the marketing forces have driven the technology rather than the other way around.

Microorganism Identification

Since the dawn of the human race, people have been getting sick and dying of infectious diseases. There has been a basic concomitant human need to determine the cause of these diseases and thus save the lives of loved ones. Long before the discovery that microorganisms were the causative agent of disease, people ascribed the causes to many things. Malaria, still a major worldwide killer, derives its name from the contraction of the Italian words *mala* (bad) and *aria* (air), since this was believed to be its cause.[9] Influenza derives its name from the Italian *influenza* (influence) since it was thought that the disease was caused by the influence of the stars.[10]

In the 1850s while attempting to determine why grapes were producing vinegar instead of the fine wines of France, Pasteur demonstrated that microbes, which had been known to exist since the 1660s, could affect their environment. It was soon realized that they were also capable of causing disease. Thus, the impetus of medical microbiology since that time has been to identify which microorganisms caused which diseases, and this became a primary marketing force. Today, the emergence of many new, primarily viral pathogens, and bacteria that are resistant to antibiotics, are creating new challenges to develop infectious disease diagnostics.

Population Centers

The late 18th and early 19th centuries witnessed the start of the Industrial Revolution. Over the next several decades, this led to more and more people leaving the farms and moving to population centers where the factories were located. Population centers continued to grow and expand, which eventually led to the large metropolitan areas that exist today.[11] This also led to an increase in the number of people that had to be treated in a small area, which then gave rise to the large hospitals. For example, today, Northwell Health System in New York has more than 1,000 beds and performs hundreds of tests each day.[12] Many hospitals are coming together to form major health systems and many smaller community hospitals are either disappearing or being absorbed into these systems, which will certainly continue to have an impact on diagnostic testing requirements and where this testing occurs.

Population centers have been responsible for several marketing forces. As populations centers grew and the number of laboratory tests grew, the amount of space required for multiple racks of tubes in biochemical identification procedures increased as well. The need to streamline these identification procedures became necessary due to the sheer volume of chambers required to incubate standard tests. When API strips were developed, they were not only easy to perform and interpret, but they also offered the benefit of providing a drastic reduction in the amount of incubator space needed to perform a microbial identification. This is probably one of the reasons why the test was, and after more than half a century, still is so popular. In addition to space constraints, as volume in testing grew, the need to develop tests that could be batched (e.g. tests in which many patient samples are tested at the same time) grew as well. This led to more automation of some of the highest volume samples in the microbiology lab, such as blood cultures.

As laboratory volume grew, there was also a need to develop systems that were easier to use, more efficient, and automated. This led to the development of automated identification and susceptibility systems, such as the Microscan Walkaway®, the BD Phoenix™ and Vitek® systems. Many of these systems could now perform biochemical testing to identify bacterial isolates in hours or overnight. In addition, these systems now provided the ability to conduct susceptibility testing more quickly and accurately using the minimum inhibitory concentration (MIC) determinations versus the less accurate antibiotic disc diffusion testing that was subject to overnight incubation, space constraints, and manual interpretation by the microbiologist. New developments continue to emerge in MIC testing as microfluidics now allows for growth-based detection of a bacterial response to an antibiotic more quickly.[13] There are several new technologies entering the market that promise "rapid" MIC detection. These technologies may be combined with resistant marker detection using molecular methodologies that will allow for demonstration of a resistance factor. In some instances, the gene for a resistance factor may be present, but not active. MIC detection in a growth-based system allows for laboratorians to determine if there is gene expression with the particular pathogen.

As a result of having large treatment facilities, large quantities of patient data were generated. These fully automated systems eventually included sophisticated database systems that allowed databases that could not only collect and store data, but could also be used to help interpret this data, thus, minimizing error

through manual interpretation. The use of these databases also offered a more efficient electronic method of updating changing MIC values and endpoints for antibiotic resistance routinely issued by the Clinical Laboratory Standards Institute (CLSI). Electronic updates also ensured that the personnel in the microbiology lab were using the latest guidelines.

Hospital-based testing remains somewhat less urgent than point-of-care or emergency room-based testing and has not resulted in significant pressure to develop rapid tests. Oftentimes, patients with chronic or longer-term conditions may not need a result very quickly or the therapeutic intervention based on the result may be limited. This is evident in cases of infection monitoring such as transplant infections and viral load assessments. When a patient is in a more acute condition (i.e. emergency room) and a therapeutic intervention can be made, there is a greater need for a rapid, even syndromic approach to testing. It is important to note that while there is a general desire to have the results of tests on patient samples provided quickly, when weighing the speed of obtaining a result against the costs of obtaining that speed; tradeoffs may result in a hospital adopting a slower test, particularly for a less urgent sample or a less acute patient condition. In addition, hospitals currently deal with several delays that are built into their systems. For instance, samples are collected and brought down to a central lab for testing, which oftentimes can take hours. Since tests must fit into a hospital routine, rapid test (less than 30 minutes) are actually something that is not desired. Generally, tests that can be completed in less than 30 minutes will ideally be CLIA waivable (discussed later) and can be performed in point-of-care settings such as doctors' offices and urgent care settings. Again, these diagnostics should also allow the ability to change a therapy or add a therapy while the patient is still onsite.

Epidemics and Bioterrorism

Today's population is much more mobile than the previous generations. As a result, diseases that may have once taken days or months to spread can now be spread in a matter of hours across the world. Additionally, diseases are now expanding beyond their traditional endemic areas. Globalization and a more mobile population have now forced laboratories, primarily those in large cities, to be prepared for a wider range of pathogens and drug-resistance factors. In the current pandemic, the CDC has issued guidance documents for airline employees on how to deal with possible COVID-19 infections.[14] In addition to epidemics, as was discovered in 2001 with the anthrax attacks in the United States, large segments of the population may be vulnerable to microbial terrorist attacks. Both factors have contributed to a marketing force that is driving the development of very rapid, accurate, and specific tests. This same force is also being seen in the doctor's office, where a test that can provide a result in a short enough amount of time to allow the doctor to begin treatment has great value.

Fewer and Less-Skilled Technicians

In recent years, there has been a shortage in the clinical laboratory workforce. This has been due to retirement of the aging workforce, an increase in demand for laboratory services, changes in the field, and vacancy rates that exceed the number of graduates.[15] This is partially due to the fact that students who might have originally considered this profession with a bachelor's degree are now opting to spend an extra year or two in school to become a Physician's Assistant (PA), which has a significantly higher starting salary. In addition, the implementation of more stringent educational requirements by some states has also led to a migration to PA programs.

With fewer and younger people entering this profession, laboratory tests need to become automated and simplified. This is particularly important in the microbiology lab because it has traditionally been the most labor intensive of the diagnostic laboratory areas. These issues, coupled with the ever-increasing cost of healthcare, have resulted in laboratories trying to replace more expensive staff with less expensive staff. As a result, this has led to a force that is driving to the development of tests that do not require a lot of expertise to run or that are automated.

In 1988, the Clinical Laboratory Improvement Amendments (CLIA) set the standards that apply to all clinical laboratory testing performed on humans in the United States. Tests were assigned to one of three categories, based on the simplicity of performing the test and of the amount of harm that might be caused to a patient if a test produced an erroneous result.

Category 1, CLIA Waived: These tests are the simplest to perform and interpret. They generally require a limited number of steps and if they produce an erroneous result, it will result in very little harm to the patient. CLIA Waived tests can be performed in a doctor's office by staff with less technical training. These include pregnancy tests, dipstick urine analysis tests, and some tests for influenza or strep throat diagnosis.

Category 2, Moderate Complexity: These tests include automated tests for *S. aureus*, MRSA, *C. difficile*, or various other infectious agents. They require a greater degree of technical intervention, but interpretation is relatively straightforward.

Category 3, High Complexity: These tests include automated tests for antibiotic susceptibility testing that require the greatest amount of technical interpretation or intervention and processing. To run testing of this designation, laboratory directors must have doctoral degrees in chemical, physical, biological, or clinical laboratory science from an accredited institution and be certified, and continue to be certified, by a board-approved by the U.S. Department of Health and Human Services.[16]

Since manufacturing a test that falls into the Waived Category means that a laboratory running it has fewer constraints and can hire lesser trained technicians to perform it, the impetus on the companies has been to develop tests that are much less complex to run. Waived tests also find use in the less-developed areas of the world, where fully trained laboratory technicians are scarce.

42 CFR 493 describing CLIA regulations can be accessed at: http://www.gpo.gov/fdsys/pkg/CFR-2012-title42-vol5/pdf/CFR-2012-title42-vol5-part493.pdf (Accessed October 23, 2020).

FDA guidance documents can be accessed at: http://www.fda.gov/medicaldevices/deviceregulationandguidance/guidancedocuments/ucm079632.htm#1 (Accessed October 23, 2020)

Hospital Cost Cutting

While this section has focused on marketing forces that are confined to the microbiology laboratory, this is not the only place where an effective test can save the institution money. There is great pressure to have people released from the hospital as soon as possible. Managing admissions and readmissions are important metrics that are being considered by today's hospital administrators. This means that even though a specific test might cost more than its immediate competitor in the microbiology laboratory, if using it results in a patient getting out of the hospital a day sooner or reduces the likelihood that that patient will return with the same condition, hospitals will be interested in considering that test. The overall savings to hospitals can be upward of hundreds of thousands of dollars in reducing readmissions alone. Companies will need to invest in outcomes studies and healthcare economics analysis to demonstrate and quantify the value of the specific diagnostic.

Developing Countries

As described above, many of the newer technologies are being used in the developed world. Billions of people, however, live in undeveloped countries and this has led to the survival of the older technologies as well as the development of easy-to-use technologies that do not require a great deal of technical expertise to run. Unlike the 1800s in which treatment options were limited, today, identifying a pathogen means that a physician can prescribe an antimicrobial. Developing markets need inexpensive accurate diagnostic tests to influence treatment and patient management. These tests need to be rapid since oftentimes patients may spend half a day walking to a clinic for testing and treatment. In addition, many of these clinicians need portable technology that they can take with them since they are often taking diagnostics and treatment directly to the patient in remote areas.

Today, antigen testing provides a good opportunity to pursue these markets, but as molecular diagnostics are developed, that dynamic is changing fast. There are more molecular diagnostics companies in the market than ever before, and they are beginning to offer technologies to developing markets to pursue this vast potential. As more of these molecular-based companies commercialize and develop their technologies, costs will continue to come down quickly for molecular diagnostics. As these costs decrease, these markets are becoming increasingly realistic growth opportunities. To fully unlock the potential in the developing markets, a fast, accurate diagnostic that can be purchased for a few dollars will be required. Portable molecular diagnostic systems for diseases such as HIV, TB, malaria and dengue fever are great opportunities to establish growth in developing markets.[17,18]

Future Marketing Forces

Government-Mandated Health Insurance

Healthcare in the United States is changing rapidly. There are many aspects of these changes that create both challenges and opportunities for people considering investments in companies that focus on the diagnosis of infectious diseases. Some challenges will arise as pools of money shrink and the number of insured patients increases. These two factors will create significant price pressures on diagnostic manufacturers to lower costs, maintain profit margins, and allow for the price reductions that hospital, hospital systems, and purchasing coalitions are requiring. With the additional patients being covered under mandatory insurance programs, it is possible that specific test markets, once deemed too small for investors to consider, may now have enough consumers to provide lucrative opportunities.

Government-required cost reductions may also provide an opportunity to tap into the downstream cost savings that hospitals and integrated delivery networks are going to need to stay viable in the new healthcare environment. Diagnostic tests that can show the economic benefits of a reduced length of hospital stay, a reduction in readmissions, and a reduction in costly drug therapies will have an opportunity to differentiate themselves from their competition. This data will need to be supported by peer-reviewed white papers and clinical evidence that demonstrates these quantifiable benefits. Producing this material can be costly for a small, startup company, but will be a critical component of any market strategy in the new environment.

Consolidation

Hospital consolidation is happening on a global level. In the United States as more hospitals lose funding, it becomes more and more challenging to stay in business. Larger integrated delivery networks, some with global locations, are actively pursuing doctors' practices, urgent-care centers, and community hospitals to gain access to markets they haven't been supplying previously. This could eventually lead to organizations with the ability to purchase tests not only for local networks, but for global ones as well.[19]

Reimbursement

Changing reimbursement in the United States will have an impact on diagnostic tests in the future. Congress passed the Protecting Access to Medicare Act (PAMA) in 2014 to help safeguard Medicare beneficiaries' access to needed health services, including laboratory tests. According to the American Clinical Laboratory Association (ACLA) laboratory testing cuts in 2018 amounted to close to $670 million in reduction of reimbursement for diagnostic testing in which most laboratory test reimbursement was reduced by close to 10%.[20] This will create additional price pressures on novel diagnostic technologies and the ability to bring new solutions to market. Selling tests using these technologies will require the additional associated costs of new product development, scale up to manufacturing-sized lots, and market introduction and acceptance. In the case of novel technologies where the consumer has multiple options and where there is limited differentiation from competitors, vendors will be forced to drastically lower pricing.[21]

Product Research and Development

Historically, companies that developed diagnostic test kits did this using research generated internally. Most diagnostic companies continue to have large research and development structures within their organizations, but starting in the late 1980s, a trend began in which these larger companies acquired technologies and

product pipelines from smaller, independent companies. This led to entrepreneurs and innovators financing beginning diagnostic start-up companies. Unfortunately, most of these small companies were at a very early stage of development, where the risk of being successful was small. Companies of this size could only take a diagnostic tool so far given expansive resource required for development, clinical evaluation, and regulatory clearance. As the number of these start-up companies grew, so too did the competition to get a corporate buyer or sponsor. As a result, big companies now demand that technologies are much further along in development. Today, while there are still plenty of opportunities for entrepreneurs to invest in diagnostic startups, the investment is much larger, and the chance of actually being able to sell the technology is much smaller. There remain some pathways that include licensure, third-party development, or pursuing several rounds of investment—all of which are capital and time intensive. A clear exit strategy needs to be considered for these endeavors and the assumption that the "big guys" will come along and scoop up a good idea or novel technology may not be realistic.

Factors That Investors Need to Consider

When large, established diagnostic manufacturers begin a product development process, the strategic marketing team in conjunction with business development will assess opportunities and propose these technologies for development or acquisition. The first step in this market assessment requires a great deal of research into the market need, the clinical disease considerations, and significant competition analysis. Many companies will assemble a target product profile (TPP). This is a master document outlining the proposed development. This document will answer many questions concerning whether to undertake a product development project for a specific diagnostic test. Critical for the success of a TPP or any similar methodology that may be employed, is ensuring that all key details are included in the opportunity assessment. Factors such as the total available market and the potential available market are important. The total available market is the total amount of revenue that is available to a product assuming that there are no competitors. Potential available market is that portion of the revenue that the product might possibly capture. Many times, startup companies will mistakenly assume that both the total and the potential available market for a product are the same. The market numbers should also consider the annual expected rate of growth of the market.

Market size is a very important factor that investors need to consider when assessing an opportunity, but it is only the first step. While the company with the technology must also have conducted research, it is always important for the investor to independently verify the opportunity. A potential investor should also perform an assessment of the attributes of the product that will provide a point of differentiation that hospitals and laboratories will embrace. Oftentimes it is this step that requires some more significant scientific and medical consultation to determine that key point of differentiation. It should be reviewed by key opinion leaders (KOLs) with significant expertise in the specific disease area. In many cases, laboratory directors or KOLs will have broad knowledge and experience across a multitude of disciplines that will allow for some efficiency in looking at technologies.

Current and potential competitors in the field are also another area that must be considered. What are the advantages that a product will offer that are not offered by the competitor? How will competitors react when the new product is introduced? Are there future improvements in the design of the product being considered that may help give it an advantage over competitors? Are there other future competitors on the horizon?

If one looks at a list of the current pathogens in the news today, you soon realize that at least some of them were almost unknown as few as 10 years ago and some are new strains of otherwise known pathogens. Emerging pathogens are a reality in this field, and investors should also consider how adaptable a product or a technology may be. This is particularly important when pursuing diagnostic test kits in respiratory disease and healthcare-associated infections. Pathogens in these spaces change and emerge quickly. Bird Flu (H7N9), SARS, MERS, and COVID-19 are very good current examples of how a pathogen can emerge and result in the almost immediate need for a diagnostic test.

Emerging antibiotic resistance is another area that evolves very quickly. While methicillin-resistant *Staphylococcus aureus* (MRSA) was known as a leading cause of nosocomial infection, as early as the 1980s, it wasn't until more than 10 years later that it was realized that this organism was now out in the community.[22] Rapidly identifying MRSA remains a huge opportunity to commercialize a variety of methods and procedures even beyond simply diagnosing the infection. It is important to continually scan the horizon for the next emerging antibiotic resistances since tracking and identifying these dangerous pathogens early is an area of great concern.

Technology

What innovation is being considered for commercialization? Is it a completely new technology that may present unique challenges during its development, or is it an adaptation of an existing technology? New technology can be a very exciting commercial opportunity, but truly novel technology often has challenges in adoption in the clinical lab. This is due partly to the regulatory nature of the medical-device industry and to the lack of innovators and early adopters in the microbiology lab. There are also cost considerations. New technologies tend to be very expensive and unless the differentiation and value are clear and quantifiable, market uptake can be slow in a price-focused hospital environment. An example of this is MALDI-TOF MS (matrix-assisted laser desorption/ionization time of flight mass spectrometry) technology for the rapid identification of microbiological organisms. This novel technology has gained significant traction in Europe with fewer regulatory restrictions, but has been slower to develop in the United States.

There are still many opportunities to commercialize upon existing technologies. An example of this is in the area of chromogenic growth media. Several manufacturers have now entered the market with a number of new products. Chromogenic media are being used to detect antibiotic resistance for organisms such as MRSA, VRE, and Gram-negative antibiotic resistance.[23] Hospitals are very cost conscious and if they can get an accurate result and time is not critical, these methodologies have proven to be very useful to the lab and hospital.

Estimated Life Cycle for Product

As is true of all products, diagnostic tests have a defined life cycle. Due to several factors mentioned above, however, these cycles can last from just a few years to more than four decades in this industry for a product such as the API 20E.

Government Regulations

The diagnostics industry comes under the control of the FDA using the regulations described under the Code of Federal Regulations Title 21 (CFR 21, available at http://www.accessdata.fda.gov/scripts/cdrh/cfdocs/cfcfr/cfrsearch.cfm. Accessed October 23, 2020). Links from this website will also direct an investor to databases concerning device classification, product life cycle, and tests approved by the 510(k) route.

Time to Get to Market: Window of Opportunity

These are three critical components when considering an investment opportunity. Questions that need to be considered are:

1. The research and development plan. Is it realistic? What obstacles need to be overcome?
2. Product manufacturing. Is this a product that can be produced on a laboratory scale, but may have limitations when produced in manufacturing sized lots?
3. The time required for all clinical trials and government approvals.

These items are critical in today's environment because new opportunities appeal to a great number of small entrepreneurial start-up companies; and a project that arrives late may find a marketplace already crowded with competitors.

How Will the Product Penetrate the Market?

Consumers of diagnostic tests are supplied either by a company's direct sales supply chain or through a distributor. Hospitals and buying groups negotiate contracts with most of the major distributors and oftentimes choose a single distributor for their facility or system. Partnering with a distributor may allow a company quicker market penetration for its product, but there are tradeoffs in using this supply channel. There are distributors ranging in size from multibillion-dollar organizations, such as Cardinal Health and Fisher Scientific with large national contracts and group purchasing agreements, to smaller, local and regional partners who may have contracts with regional medical centers and laboratories.

In either case, some of the key negotiating points in distributing a product, regardless of the supply chain used, revolve around a company's revenue and gross profit goals. Distributors will require a margin to carry and distribute the product and this must be considered when assessing the gross profit of the product. If the product is highly complex or the positioning and value proposition is highly technical, the distribution sales teams may not have the correct skills to adequately inform the clinical staff, doctors, or laboratory personnel. On the other hand, distributors of almost all sizes will give significantly more market access than going through a direct channel with a direct sales force. In a distributor model, the company providing the technology to the distributor will need field-based associates who can train their distribution partners on the important aspects of the technical and economic value proposition so they can articulate it to the end user. In addition, if a company decides to use a distributor to sell its products, it should place restrictions on the distributor to prevent or limit it from also selling other products that directly compete.

Even in today's challenging healthcare economics environment, hospitals will continue to purchase products directly from a company. This is particularly true if the product is a very specific, valued niche technology. If this is the case, a direct sales channel may be a good direction to take, particularly in the early stages of market adoption. While there is more control over profitability and value proposition in the direct model, hospital invoicing and collections can be challenging. A strong gross-profit position loses significant value when payments are delayed beyond 120 days. Another factor to consider in looking at the direct-chain path is that most hospitals and laboratories buy tests from a single supplier since vendor consolidation is a benefit for them. If a company has a great test for identifying *Streptococcus pyogenes*, but the hospital or even doctor's office is buying all of their other tests from someone else, even if a test is better, the laboratories may not buy it directly unless the company is partnered with an incumbent and established distributor.

In short, there is a great deal of variables to consider when approaching an investment in the business of microbiology. There is far more detail than can be communicated through a single chapter. The goal of this chapter was to provide some overview of key areas to investigate and be aware of when pursuing opportunities in microbiology. It is important to note that this chapter only scratches the surfaces of the opportunities; there continues to be significant growth in industrial, food, brewing, and a significant number of other microbiology disciplines outside of the infectious disease diagnostics.

REFERENCES

1. FDA. (2020, April 18). Coronavirus (COVID-19) Update: Serological Test Validation and Education Efforts [Statement]. Retrieved from https://www.fda.gov/news-events/press-announcements/coronavirus-covid-19-update-serological-test-validation-and-education-efforts

2. Bauer, D.C., Tay, A.P., Wilson L.O.W., Reti, D., Hosking, C., McAuley, A.J., Pharo, E., Todd, S., Stevens, V., Neave, M.J., Tachedjian, M., Drew, T.W., and Vasan, S.S. Supporting pandemic response using genomics and bioinformatics: a case study on the emergent SARS-CoV-2 outbreak. *Transbound Emerg Dis.* 67(4), 453–1462, 2020.

3. Popowitch, E.B., O'Neill, S.S and Miller, M.B. Comparison of four multiplex assays for the detection of respiratory viruses: Biofire FilmArray RP, Genmark eSensor RVP, Luminex xTAG RVPv1 and Luminex xTAG RVP FAST. *J. Clin. Microbiol.* 51(5), 1528, 2013.

4. May, L., Tatro, G., Poltavskiy, E., Mooso, B., Hon, S., Bang, H. and Polage, C. Rapid multiplex testing for upper respiratory pathogens in the emergency department: a randomized controlled trial. *Open Forum Infect Dis.* 6, ofz481, 2019.

5. http://patft.uspto.gov/netacgi/nph-Parser?Sect1=PTO2&Sect2=HITOFF&p=1&u=%2Fnetahtml%2FPTO%2Fsearch-bool.html&r=1&f=G&l=50&col=AND&d=PTXT&s1=3,936,356.PN.&OS=PN/3,936,356&RS=PN/3,936,356 (Accessed October 23, 2020)

6. Yalow, R.S. and Berson, S.A. Immunoassay of endogenous plasma insulin in man. *J Clin Invest.* 39, 1157, 1960.

7. Van Weemen B.K. and Schuurs A.H. Immunoassay using antigen-enzyme conjugates. *FEBS Lett.* 15, 232, 1971.

8. Kaplan, S., Marlowe, E.M., Hogan, J.J., Doymaz, M., Bruckner, D.A. and Simor, A.E. Sensitivity and specificity of a rapid rRNA gene probe assay for simultaneous identification of *Staphylococcus aureus* and detection of *mecA. J. Clin. Microbiol.* 43, 3438, 2005.

9. Hempelmann, E. and Krafts, K. Bad air, amulets and mosquitoes: 2,000 years of changing perspectives on malaria. *Malaria J.* 12, 232, 2013.

10. Quinion, M., 1998. World Wide Words. Influenza. http://www.worldwidewords.org/topicalwords/tw-inf1.htm (Accessed October 23, 2020.)

11. Stearns, P.N. *The Industrial Revolution in World History*, 4th ed., Westview Press, Boulder CO, 2013.

12. http://www.ahd.com (Accessed October 23, 2020.)

13. Wistrand-Yuen, P, Malmberg, C, FatsisKavalopoulos, N, Lübke, M, Tängdén, T, Kreuger, J. A multiplex fluidic chip for rapid phenotypic antibiotic susceptibility testing. *mBio.* 11, e03109–19, 2020.

14. https://www.cdc.gov/quarantine/air/managing-sick-travelers/ncov-airlines.html (Accessed October 23, 2020.)

15. ALCLS (2018, August 2) Addressing the Clinical Workforce Shortage. [Position paper] Retrieved from https://www.ascls.org/position-papers/321-laboratory-workforce/440-addressing-the-clinical-laboratory-workforce-shortage

16. https://www.cms.gov/Regulations-and-Guidance/Legislation/CLIA/Certification_Boards_Laboratory_Directors (Accessed October 23, 2020.)

17. Ahou Tayoun, A.N., Burchard, P.R., Malik, I., Scherer, A. and Gregory Tsongalis, G.J. Democratizing molecular diagnostics for the developing world. *Am J Clin Pathol.* 141, 17, 2014.

18. http://www.marketsandmarkets.com/Market-Reports/point-of-care-diagnostic-market-106829185.html?gclid=CKvU54SLu70CFTIV7AodN20Asg (Accessed October 23, 2020.)

19. http://www.healthcareitnews.com/news/global-model-care-coordination (Accessed October 23, 2020.)

20. ACLA (2018, July 13) Protecting Access to Medicare Act [Statement]. Retrieved from https://www.acla.com/acla-statement-on-proposed-2019-medicare-payment-changes

21. https://www.pathologyphotostore.com/attack-by-centers-for-medicare-services-cms-on-diagnostic-immunohistochemistry-it-is-here-now/ (Accessed October 23, 2020)

22. Shopsin, B., Mathema, B., Martinez, J., Ha, E., Campo, M. L., Fierman, A., Krasinski, K., Kornblum, J., Alcabes, P., Waddington, M., Riehman, M. and Kreiswirth, B. N. Prevalence of methicillin-resistant and methicillin-susceptible *Staphylococcus aureus* in the community. *J Infect Dis.* 182, 359, 2000.

23. Malhotra-Kumar, S., Abrahantes, J.C., Sabiiti, W., Lammens, C., Vercauteren, G., Ieven, M., Molenberghs, G., Aerts, M., and Goossens, H. and the MOSAR WP2 Study Team. Evaluation of chromogenic media for detection of methicillin-resistant *Staphylococcus aureus. J Clin Microbiol.* 48, 1040, 2010.

60

Launching a Microbiology-Based Company

Leonard Osser

In preparation for writing this chapter, I read and then reread the chapter on the Business of Microbiology written by Lorrence H. Green and Michael C. Nugent in the *Practical Handbook of Microbiology*, third edition.[1] While it was a scholarly chapter written with a strong focus on microbiology, many of the points made would apply to any industry. Having a background in many diverse industries only some of which involve microbiology, this chapter will focus on the aspects of what I have learned when starting out on the quest to begin a business.

My first observation: If you believe the adage, "if you build a better mouse trap, the world will beat a path to your door," you are mistaken. There are many examples of companies that have had better technologies, but have not succeeded. We invested in such a technology more than 20 years ago that dealt with a common problem: fear of going to the dentist. Studies indicated that one in seven Americans are dental phobic.[2] One of the main reasons for this is the fear of the hypodermic syringe.[3] Our technology not only delivered a drug without pain, but avoided the problem of causing collateral numbness. If this product were marketed directly to the patient, one would have predicted that it would have been readily and happily accepted.

The cost of the small computer was under $3,000, and the disposable cost was approximately $3. Studies indicate that there are over 300 million dental injections in the United States per year, with a potential market of approaching $1 billion.[4] As such, one would have expected sales of this instrument to have been significant. Why then is it only a $9 million business? The reason quite possibly is that while it was the patient that most benefited from the product, it was the dentist that was actually the customer. In the case of dental injections, the current standard of use is the hypodermic, which was introduced in 1885.[5] The barrier is not technology, funding, regulation, or the cost of a product that does not deliver; it is overcoming market inertia. To be successful, you must make certain that the market needs, wants, and must have your product.

In pursuing a career as someone who both starts and develops companies, I have not only read business books, but I have read the works of Marcus Aurelius. While getting an MBA will help, getting actual experience will be invaluable. In addition, I have found these four books to be almost required reading for any business person: *The Art of Worldly Wisdom,* by Baltasar Gracian; *The Prince,* by Nicolo Machiavelli; *The Art of War,* by Sun Tzu; and *Why Smart People Do Dumb Things,* by John Tarrant and Mortimer Feinberg. These books have taught me that this career path is one in which you will make mistakes, many

mistakes. When this happens, you must fix them immediately. Do not wait. Sometimes you may find yourself making mistakes until you get it right. In the companies that I have been involved with, I have found that the biggest mistake that CEOs made was keeping the wrong people for too long. Another thing these books taught me is to always tell the truth—even when you find that no one believes you. If nothing else, it easier to remember. Also, spend time just thinking. Finally, if people think that you are wasting time, then you are probably succeeding.

When starting a new business or investigating an opportunity, whether it be a general company or one that specifically involves microbiology, one should work backward. What is the need? What is the need that is perceived by the market? What is the value of this need to the buyer? Is the buyer the one who can write the check? In the case of the dental instrument, is it the buyer that will actually be driving the market?

Another observation is that while the people reading this chapter might think that the microbiology industry is not like any other, that is not true. All businesses are the same. I have been involved in several health-related companies, and I am by no means a microbiologist. This means that I must hire the right experts, people who can explain the science to me and to other investors in a way that we will understand. I have found that the best people for this are the creative people in large companies. While working in stable positions, they usually tire of large companies due to the wasted time dealing with bureaucracy and politics. Get the best possible people, then watch them and learn from them. Let them fulfill your agreed-upon goals. Enable them, and don't get in their way.

Before starting your company, figure out how much time and money will be required to make the company self-sustaining. Once you have figured out what that number is, then double it. Many great and useful ideas do not come to fruition because the entrepreneur simply runs out of capital. Always try to get capital from people that have it, not from people that must raise it. This is less complicated and almost always less costly in the long run, to say nothing of much less risk and time spent. Do not raise money from relatives or friends.

In the arena of funding options, we have rather large groups of sources. All the following funding options come with their unique risks: Venture capitalists, vulture capitalists, venture funds, family funds, tertiary market, bankers, crowd funding, angel investors, etc. These people fully understand your financial situation. They also know that you will not perform on time. They know that you will need more money than you have planned for.

DOI: 10.1201/9781003099277-63

Do not, however, mistake their helicopter for your umbrella. If you manage to attract a true angel long-term investor who will be supportive and ethical, you have achieved nirvana. There are such people, but not many. To varying degrees, the other investors will set up arrangements that anticipate that you will not reach your goals on time. Depending on their level of mendacity, they will acquire more and more of the company each time you do not meet a goal. Their models anticipate this happening. If you have a public offering too soon for too little net money, the same will occur to a lesser degree.

If you engage someone to help you find funding (called a finder), interview all of their clients going back at least five years. Pay on performance only—do not pay any upfront money when raising capital. If they want upfront money, that is their profit center. This means that even if the company does not succeed, they will.

Do not be impressed by opulent offices or firm names. Many excellent investors work from offices no bigger than closets. It is the person that you work with who is key, not their firms. This is also true for attorneys, advisers, and bankers. When interviewing, get the names of all their past deals, not just the specific references they supply. While hearing success stories is important, most companies do not succeed. You will also learn about your investors by finding out about their failures. While you're raising money, you're not running your business—this means that you need to raise sufficient money before you need it. When people want to give you too much money, take it, you'll probably need it.

Form a board of people who are willing to invest. Pay them in stock, not money. Get people who know more than you do. A top name in the field always helps. Investigate them. A public underwriting will be impossible should anyone on your team have regulatory issues, such as a bankruptcy. Board members are not there to agree with you, they are there to support the company by expressing their views. Once a decision is made, the board is there to support it. Again, they do not have to be your friends, they must, however, be able to function well together.

When you plan the amount of time it will take until the company is self-sufficient, triple it. Nothing goes as planned you will require much more time and money than you envision. You have to be prepared for numerous setbacks prior to reaching your goal, and you must have enough time and money to deal with them. If you then do not have access to either of the multiples of time and money, do not start the business. This can be difficult given other obligations that you may have in life, but I have found maintaining a positive mental attitude to be very helpful.

While the science and technology are important to any company, what is more important are investors. You must develop a plan that shows success along the way so investors can feel positive about what they are doing. Investors must see that goals are being achieved. This will enhance the opportunity to raise more capital. Raise much more money before you need it, even if you are certain you will not need it. If you don't, it will cost you much more of the company when you do need it. Make sure that your backers have significantly greater resources than they are investing.

I have found especially in trying to bring new healthcare products to the market that the biggest enemy I encounter is obstinance. To get the market to accept a new product, I have to deal with potential users who, for a good part of their lives, have trained and used the products that you want to replace. They are experts on the old technology, and if you challenge them, you will not be well received. Even if what they have been doing is problematic, even if your solution is obviously superior, they will find reasons not to change, including worries of your product being too slow, too expensive, too difficult to learn, etc. However, after you address all these issues, they probably will still not buy your product because they are being challenged. I have found that the solution is to make certain that you show them that you are enhancing their abilities and making them more valuable, rather than challenging their abilities.

To get new products or technologies accepted you will probably have to partner with or at least convince those opinion leaders that can force the issue. These may include hospitals, insurance companies, regulatory agencies, etc. Have successful experts evaluate your projects. Meet with the decision makers, end users, buyers, salespeople, and owners of distribution companies. Do they all want your product? What do they envision the barriers will be? Will they buy it for the price that will make your venture successful?

Of course, with any new product you must always have top patent attorneys first review your ideas. Without proper patent protection, it will be far more difficult to raise money. This is not an inexpensive proposition. You will also need excellent attorneys to help you with fundraising. Before selecting one, interview many. Focus on those that have securities and fundraising backgrounds. Remember in your search that the best attorney and the best surgeon are not necessarily the one that you like the best. Their ability, not their bedside manner, is what you're looking for. Their fees, although they may tell you are carved in stone, in fact they are negotiable like everything else. Estimates are exactly that; get a firm commitment on the high end. Interview others for whom they have raised money.

Everything that you do will depend on people—investors, staff, attorneys, accountants, auditors, regulators, etc.—so you must hire the best. To do otherwise will not serve you well. Be 100% loyal to your partners and staff. If not, you will never have their loyalty. Also, remember that no one, including you, is indispensable. Your shareholders want you in the business for the money, not ego gratification. The investors want a return. If you are reading this, then you are probably thinking of being an entrepreneur for the first time. All those that you encounter will, in their area of expertise, have more knowledge and experience than you. You may think that you're smarter, and in fact you may be, but the odds are on their side. The larger and more established organizations are almost always favored.

Something that may not be obvious is that while having the best test kit in a market might be important, it is not necessarily sufficient. Sometimes, the finances might cause a distributor, for instance, to accept a product for sale that isn't the best in the marketplace because it offers them the best financial return. Sometimes they may accept an inferior product and even put up a very small amount of money for it, if they can be assured that they will get the product as soon as it is available. There are many examples of companies that have had inferior technologies but have succeeded for other reasons.

I have found that a very good indicator of how well-accepted a product may ultimately be is seeing if a distributor will invest any of their own money at the beginning or give you an irrevocable order. While many will be kind and encouraging, an investment no matter how small, will tell you what they really think of your product.

Having the product completed, functioning perfectly, approved by regulators, and patented is not the end of the process, it is just the beginning. It has to sell at a profit and it must continue to sell. Once you are in this position you then must start thinking about what the next generation of your product is going to look like. Competitors are always examining each other's products in great detail to see what they can learn. In the case of our painless injection device mentioned at the beginning of this chapter, I once had a meeting with the marketing people from one of the potential competitors in which we were shown *their* new needle. They were certain it would significantly reduce pain because it was sharper than the needle of their competitors. However, needle sharpness is almost irrelevant regarding pain. Pain is controlled by the pressure and flow rate of the drug entering the body. In order to show how much better their sharper needle could penetrate tissue, they produced a small device to be used as a demonstration for potential customers. They were able to show that their needle penetrated more easily. Unfortunately, it did not reduce pain when used. They were thinking in a small, irrelevant box. You need to develop the next generation before one of your competitors does.

While this might appear counterintuitive, to be successful in developing new companies or promoting new technologies—whether they are microbiologically based or not—the most successful people are not going to be the smartest, they are going to be the ones who can escape the confines or their own subjectivity and do things others do not think are possible. Perhaps Einstein said it best:

> Imagination is more important than knowledge. For knowledge is limited, whereas imagination embraces the entire world, stimulating progress, giving birth to evolution.[6]

REFERENCES

1. Nugent, M.C., and Green, L.H. 2015. Business of microbiology. In *Practical Handbook of Microbiology*, 3rd ed., E. Goldman, and L.H. Green (eds.), pp. 293–303, Boca Raton, FL: Taylor & Francis.
2. Armfield, J.M., and Heaton, L.J. 2013. Management of fear and anxiety in the dental clinic: a review. *Aust Dent J.* 58:390.
3. Armfield, J.M., and Milgrom, P. 2011. A clinician guide to patients afraid of dental injections and numbness. *SAAD Dig.* 27:33.
4. Hass, D.A. 2002. An update on local anesthetics in dentistry. *Can Dent Assoc.* 68:546.
5. Lopez-Valverde, A., De Vincente, J., and Cutando, A. 2011. The surgeons Halsted and Hall, cocaine and the discovery of dental anaesthesia by nerve blocking. *Br Dent J.* 211:485.
6. Viereck, G.S. 1929. What life means to Einstein. *Saturday Evening Post.* 17.

61

Microbiology for Dental Hygienists

Victoria Benvenuto and Donna L Catapano

CONTENTS

History and Role of the Dental Hygienist

The first dental hygienist on record was in 1906 and the first dental hygiene courses began in 1913 in Bridgeport, Connecticut. In those early days, full academic courses ran for a duration of one year with dental hygiene students only permitted to remove "lime deposits, accretions, and stain" on teeth above the gumline.[1] Fast forward to today, and the roles of the dental hygienist have vastly improved. Graduates can now earn associate, baccalaureate, and master's degrees in dental hygiene with emphasis not only on removing said accretions from the teeth above and below the gumline, but to now having a full understanding of how the body reacts to such accretions and the necessary measures for disease prevention for both the patient and the clinician.[1]

Currently, the dental hygienist's role is one that is collaborative amongst dentists, physicians, chiropractors, nutritionists, nurses, and other healthcare professionals. This intra and interprofessional collaboration is partly what makes their role so valuable in maintaining optimal health. For example, by collecting all necessary medical history data from a patient, the dental hygienist may need to confer with pacemaker companies, physicians, such as cardiologists, gastroenterologists, chiropractors, and other healthcare professionals to ascertain if modified treatment is required. Dental hygienists provide educational, clinical, and consultative services to individuals and populations of all ages in a variety of settings and capacities, including private dental practices, community hospitals, university dental clinics, prison facilities, long-term care facilities, and schools.[2]

The following are many responsibilities of current hygienists.[2,3]

- Patient screening procedures, such as assessment of oral health conditions, review of the health history, oral cancer screening, head and neck inspection, skin inspection, dental charting and taking blood pressure and pulse

DOI: 10.1201/9781003099277-64

- Taking and developing dental radiographs (x-rays)
- Assessing, diagnosing, planning, implementing, evaluating, and documenting treatment for prevention, intervention, and control of oral diseases
- Removing calculus and oral biofilm/dental plaque (hard and soft deposits) from all surfaces above and below the teeth
- Applying preventive materials to the teeth (e.g., sealants and fluorides)
- Teaching patients appropriate oral hygiene strategies to maintain oral health; (e.g., tooth brushing, flossing, and nutritional counseling)
- Counseling patients about good nutrition and its impact on oral health
- Making impressions of patient teeth for study casts (models of teeth used by dentists to evaluate patient treatment needs)
- Performing documentation and office management activities
- Practicing proper disinfection and sterilization techniques

Dental Hygienists' Educational Role in Disease Prevention

Studies consistently support the correlation between poor oral health and several systemic diseases, such as periodontal disease, cardiovascular and cerebrovascular disease, respiratory disease, certain cancers, diabetes, low birth-weight babies, and possibly dementia.[4,5] Dental hygienists use evidence-based research to inform patients of such connections. The dental professional's role in prevention is through thorough oral education and assisting patients in devising specific home-care regimens to reduce and prevent oral inflammation in hopes of preventing many systemic diseases.[4]

Oral-Pathogens and Dental Hygiene Exposure

The oral cavity plays a central role in the process of infection.[6] More than 700 bacterial species inhabit the oral cavity, as well as species of several other types of microorganisms. Many of these can cause both local and systemic disease. Specific microorganisms and diseases that dental healthcare professionals and patients are at risk for include but are not limited to staphylococci, streptococci, *Mycobacterium tuberculosis*, hepatitis B and hepatitis C viruses, herpes simplex virus type 1 and type 2, human immunodeficiency virus, mumps, influenza, rubella, pneumonia, tuberculosis, influenza, Legionnaire's disease and severe acute respiratory syndrome (SARS).[9]

The majority of dental procedures allow such microorganisms to become airborne in the form of aerosols and stay suspended contaminating the air within treatment areas.[7] According to Aas et al., "Aerosols can originate from a variety of patient sources including plaque, calculus, dental materials, blood, saliva, nasopharynx, and inadequately disinfected dental unit waterlines".[7]

These aerosols place dental healthcare professionals and patients at risk for illness due to the exposure of many infectious agents, including bacteria, viruses, and fungi.[8] The teeth also serve as a reservoir for respiratory pathogen colonization and subsequent nosocomial pneumonia.

Influenza and Pneumonia

The influenza virus is the cause of the majority of acute respiratory illnesses annually.[10] The virus is transmitted from person to person through close contact, which is something that dental professionals cannot avoid. Transmission can occur as a result of droplet exposure from coughing or sneezing, contact with a contaminated surface and then self-inoculation onto mucosal areas, such as the mouth, nose or eyes, or through aerosols. This can be spread from one clinician to another, from clinician to patient, from patient to patient, or patient to clinician.[10] As soon as a person opens their mouth, aerosols are spread into the air. Additional aerosols are spread via use of ultrasonic prophylactic devices and dental instruments, which can cause aerosols to remain in the air for up to 2 hours.[9] Dental professionals can take the following precautions to limit the transmission of the influenza virus: Educating staff members about symptoms and prevention, following appropriate infection control protocols, such as wearing a (NIOSH)-certified particulate-filter respirator (e.g., N95, N99, or N100), and donning other proper personal protective equipment (PPE), practicing proper hand hygiene and sterilization methods, avoiding persons if they have been diagnosed with the virus, actively screening patients who enter the waiting area before treatment is performed, and employing the use of air purifiers.[8–11]

The same precautions can be taken by dental professionals to prevent the spread of pneumonia within a dental setting amongst each other and patients. Since dental hygienists are also permitted to perform prophylactic procedures in long term-care facilities, it is vital that aspiration pneumonia/ventilator-associated pneumonia be prevented. By performing thorough and routine oral care to remove oral biofilm, the number of oral bacteria can be reduced, possibly inhibiting the development and mortality rates of pneumonia.[12]

Tuberculosis

Mycobacterium tuberculosis (*M. tuberculosis*) is the pathogen responsible for causing tuberculosis (TB). The infection is spread through inhalation when an individual with active pulmonary TB coughs, sneezes, or even sings, causing droplet nuclei.[13,14] A clinician or patient can transmit the disease in less than an hour in some cases. The risk of developing TB is higher in those born outside the United States, residents and employees who work in high-risk facilities such as correctional facilities, long-term care facilities and homeless shelters, and in healthcare professionals who care for those who are at high risk. Although the risk of transmission is low amongst dental professionals, contraction is still possible since they can legally care for patients in correctional institutions and long-term care facilities. There are even documented cases of transmission within the dental setting.[11] The Centers for Disease Control and Prevention (CDC) updated their guidelines in 2005 for preventing transmission of

M. tuberculosis in healthcare settings.[13] All dental settings need to follow a TB infection control program based on three levels of controls. The most important component of this program is the use of administrative measures to reduce the risk of exposure to potentially infectious persons. Environmental controls reduce the spread and concentration of infectious droplet nuclei in the air. And the use respiratory protection and respiratory hygiene to reduce the risk of exposure to infectious droplet nuclei that may be expelled into the air.[14]

Although the CDC does not recommend routine immunization of United States healthcare workers against TB, it does recommend all persons in the dental office who have the potential for exposure to *M. tuberculosis* through air space shared with persons with infectious tuberculosis disease (which essentially means all personnel) be tested for infection. This can be done by a two-step baseline tuberculin skin test at the beginning of employment. The CDC recommends annual TB education for all healthcare workers, which is to include information about TB exposure risk.[14]

Hepatitis B Virus (HBV)

The hepatitis virus causes liver disease, which left untreated may develop into chronic diseases, such as cirrhosis, liver cancer, liver failure, and even death.[15] In 2016, 3,218 cases of acute hepatitis B (HBV) infection were reported in the United States. And between 847,000 to 2.2 million people in the United States are estimated to be infected with chronic HBV infection. The HBV can be transmitted through direct and indirect contact with the virus. Direct contact includes a percutaneous injury (such as a needlestick or cut with a sharp object), contact with mucous membranes, nonintact skin with HBV-positive blood, or other bodily fluids. Indirect contact includes operatory equipment or contact with airborne contaminants (blood and saliva) present in aerosols and respiratory fluid.[15,16] The use of ultrasonic scalers increases the amount of aerosols produced, allowing the virus to become more easily airborne.[16] Although blood is the most efficient mode of transmission, the virus has been found in other body fluids, such as bile, nasopharyngeal secretions, saliva, and sweat. The risk of infection after a needlestick with HBV-positive blood ranges from 23% to 62% depending on the individual's immunity. Although percutaneous injuries are the most common mode of transmission, the majority of HBV infections among dental practitioners occur from infected blood or body fluids coming in contact with mucosa or existing breaks in the surface of the skin.[15]

Dental practitioners need to follow proper infection control protocols to avoid direct contact with blood and saliva during dental procedures, as well as receive the HBV vaccine. In the past, dental practitioners had a three- to four-fold higher risk of HBV infection than the general population, but vaccines and precautionary methods have contributed to a decrease in risk.[15]

Hepatitis C Virus (HCV)

Although hepatitis C (HCV) is a short-term illness, it is still deadly. In 2016, 18,153 people died from the hepatitis C virus and the death toll continues to rise annually.[15] The virus can be transmitted through a percutaneous injury (such as a needlestick or cut with a sharp object) or contact between mucous membranes or nonintact skin with blood, tissue, or other bodily fluids. After a needlestick exposure to HCV-positive blood, the risk of HCV infection is approximately 1.8%. HCV has been detected in saliva, but as of yet, no undisputed case of HCV salivary transmission has been documented. However, this may change, therefore, dental professionals should take the same infection control precautions as all other infectious diseases. Unlike HBV infection, there is no effective vaccine or postexposure prophylaxis available for HCV infection.[15]

Cytomegalovirus (CMV)

Cytomegalovirus (CMV) is a virus that belongs to the *Herpesviridae* family.[17] It is frequently excreted in saliva, particularly in children and individuals who suffer from significant immune suppression. However, 11–33.5% of the healthy population can also excrete this virus. Therefore, dental practitioners may be at higher risk of developing the CMV infection due to frequent contact with saliva of infected persons.[17]

Epstein-Barr Virus (EBV)

The Epstein-Barr virus (EBV) also belongs to the *Herpesviridae* family.[18] More than 90% of the adult population is infected with EBV and it is associated with the nasopharyngeal carcinoma, oral hairy leukoplakia, and lymphoma.[19–21] Transmission occurs mainly by exchange of saliva.[20] The infection is asymptomatic or mild in children, but it causes infectious mononucleosis in adolescents and young adults. Once established, the virus often remains dormant and individuals become lifelong carriers. The virus persists in epithelial cells of the oropharynx and possibly also in salivary glands. The primary site of EBV infection is the oropharynx. It has been noted that at any point in time, as many as 20–30% of healthy adults who are previously infected with EBV shed the virus in low concentrations in their oral secretions.[18]

The EBV has also been implicated in the pathogenesis of advanced types of periodontal disease.[20] EBV DNA is detected in 60–80% of aggressive periodontitis lesions and in 15–20% of gingivitis lesions or normal periodontal sites. EBV and cytomegalovirus often co-exist in marginal and apical periodontitis.[20] Dental practitioners should look for signs of periodontal disease and oral cancer during routine examinations.

Herpetic Whitlow

Herpetic whitlow is an acute viral infection of the hand caused by the herpes simplex virus (herpes simplex virus 1 or 2). Its characteristic findings are significant pain and localized erythema followed by development of small nonpurulent vesicles.[22] This infection can be found in any age group. However, it is most common in children who suck their thumbs and in healthcare providers (medical or dental) who are exposed to patients' oral mucosa while not wearing gloves. The infection is most common in dental hygienists and respiratory therapists. Its incidence has been reported at 2.4 cases per 100,000 people per year.[23] Prevention includes donning proper PPE and practicing proper hand hygiene.

Legionnaires' Disease

Legionnaires' disease is a type of pneumonia.[24] Patients and dental healthcare workers may contract a *Legionella* infection through inhalation of potentially contaminated aerosols produced by ultrasonic dental instruments and through exposure from contaminated dental unit water lines. Those at greatest risk for developing the infection are individuals who are current or previous smokers, are older than 50, and/or are immune-compromised.[24] Ensuring that dental unit water lines are properly cared for can help prevent contraction of this infection in the dental office.

Human Papillomavirus (HPV)

The human papillomavirus (HPV) is the most common sexually transmitted disease in America and is known to cause very serious health issues, including oropharyngeal cancer. Human papilloma cancers are associated with 72% of cancers in the back of the throat, base of tongue and tonsils in men alone. Reports show that 79 million individuals are infected with some strain on HPV.[25] While dental practitioners already screen for oral cancer, their prevention efforts can expand to education and recommendations regarding additional prevention efforts, such as the HPV vaccines.

Human Immunodeficiency Virus (HIV)

The human immunodeficiency virus (HIV) continues to be a worldwide health issue affecting 36.7 million people globally.[26] HIV infection may result in destruction in the oral mucosal tissues. It is possible that immunological changes following HIV infection compromise the integrity of physical barrier in the oral cavity and contributes to chronic inflammation.[26]

Blood has the greatest amount of infectious viral particles but all bodily fluids, secretions, and excretions other than sweat may contain transmissible infectious agents. Avoiding exposure to blood and bodily fluids is the primary key to prevent transmission of the virus in dental care settings.[27] During dental procedures, the risk of HIV transmission is increased because saliva tends to become contaminated with blood. Standard precautions should be followed with all patients, whether or not they have been diagnosed with HIV and dental personnel should don PPE and utilize appropriate barriers (e.g., gloves, masks, protective eyewear, and barrier tape and covers) whenever there is potential for contact with body fluids, nonintact skin, or mucous membranes. Though the occupational source of greatest risk of HIV transmission is percutaneous injuries, needlestick exposures with HIV-infected blood has a low transmission rate of approximately 0.3% in dental settings.[27]

Periodontal Organisms Associated with Systemic Disease

Porphyromonas gingivalis

Human subgingival plaque harbors more than 700 bacterial species and considerable research has shown that *Porphyromonas gingivalis (P. gingivalis)*, a Gram-negative anaerobic bacterium that resides subgingivally, is the major contributing etiologic agent causing chronic periodontitis, as well as several systemic diseases.[4,28] *P. gingivalis* causes direct and indirect destruction of the periodontium by evoking an inflammation response by producing a myriad of virulence factors.[28]

P. gingivalis has also been found in the brain plaques of Alzheimer's patients and has been shown to impair cognitive functions.[29] Recent animal models demonstrated that inflammation that results from *P. gingivalis* causes alterations in neurovascular functions, resulting in an increase in the blood-brain barrier permeability, reduction of nutrient supplements, and aggregation of toxins. Elevated levels of proinflammatory mediators in the blood can lead to their direct or indirect transport to the brain, which could accelerate brain impairment.[30]

P. gingivalis has also been implicated as the main pathogen in several cancers, including pancreatic cancer, oral squamous cell carcinoma, head and neck cancer, and precancerous gastric and colon lesions.[31]

Aggregatibacter actinomycetemcomitans

Epidemiological studies have a proven association between the periodontal pathogens *Aggregatibacter actinomycetemcomitans* (formerly *Actinobacillus actinomycetemcomitans*) and *P. gingivalis*, and coronary artery disease (atherosclerosis) and cerebrovascular disease (stroke).[32] Both species have been shown to increase proinflammatory cytokines causing local and systemic inflammation and contributing to atherosclerotic plaques in the coronary arteries and cerebrovascular arteries. Endarterectomy studies show that more than 40% of atheromatous plaques in the carotid arteries presented with antigens from periodontal pathogens.[4,32] Individuals with active periodontal disease are twice as likely to suffer from heart attacks and at least three times as likely to suffer a stroke.[33,34]

Several oral bacterial species have been investigated for their role in inflammation and the link between chronic periodontitis and diabetes. *A. actinomycetemcomitans* and *P. gingivalis* have both been implicated in such a link through their ability to interfere with the immune response.[35]

Methods of Disease Prevention

Not only do dental hygienists/dental professionals need to take precautions to prevent oral disease for their patient's through education and devising effective home care regimens and treatments, but they must also take necessary precautions to prevent infectious disease from spreading from patient to patient, clinician to patient, patient to clinician, clinician to clinician, and clinician to other workers. Dental professionals should follow the CDC current recommendations that are outlined in the *Guidelines for Infection Control in Dental Health-Care Settings* and *The Summary of Infection Prevention Practices in Dental Settings: Basic Expectations for Safe Care*.[9,36] These guidelines highlight the existing CDC recommendations, state basic infection prevention principles and recommendations for dental professionals, and reiterate standard precautions as the foundation for preventing transmission of infectious agents.

Prevention of Aerosols

Dental patients and professionals are constantly exposed to tens of thousands of aerosols per cubic meter and the potential to breathe contaminated air during procedures is almost unavoidable.[37] Aerosols are extremely small particles (>50 μm) that can stay suspended in air for up to two hours after completion of dental procedures and inhalation can be deadly.[7] As previously noted, aerosols consist of water, saliva, blood, microorganisms (bacteria, fungi, viruses, protozoa), and other toxins and can originate from several patient sources, such as oral biofilm, calculus, and improperly disinfected dental unit water lines. The use of a dental turbine handpieces, air-polishers, air/water syringes, hand instrumentation, laser or electrosurgery units, prophy angles, and ultrasonic scaling devices have been reported to increase bioaerosol levels dramatically.[6,7] Inhalation of aerosols by patients and clinicians produced during dental procedures have the potential to cause adverse respiratory health effects.[37] Dental professionals should to be cognizant of the dangers of aerosols within treatment areas and need to assume the responsibility of preventing the risk of transmission before, during, and after patient care.

Many procedures that dental professionals can incorporate into daily practice to reduce aerosol production include utilizing high-velocity air evacuation, preprocedural antimicrobial mouth rinses, flushing water lines at the beginning of the workday and between each patient, donning PPE, and employing air purifications systems.[7,36]

High-Volume Evacuation

The use of a high-volume evacuator/high-velocity air evacuation device is commonly used in dental settings to help reduce the transmission of aerosols. Its suction capability and wide opening draw a large volume of air into it, diminishing aerosol emission by 100 cubic feet per minute and reducing the amount of bioaerosols created during patient treatment by as much as 90%.[38]

Preprocedural Rinse

Providing patients with preprocedural antimicrobial rinses for 60 seconds, such as chlorhexidine gluconate, essential oils, or povidone-iodine, has shown to decrease the level of oral microorganisms in aerosols generated during routine dental procedures, as well as reduce the amount of microorganisms that may be introduced into the patient's bloodstream during certain invasive dental procedures.[9]

Air Purification

Control of aerosols in confined, poorly ventilated spaces presents a challenge and can spread microorganisms from patient to patient.[39] The aerosols in dental treatment areas can be purified through the use of air cleaning systems, i.e., high efficiency particulate air (HEPA) filters, gas filter cartridges, and electrostatic filters. HEPA filters continuously trap aerosols and retain particles as small as 0.3 μm while electrostatically charged post-filters help to purify air by minimizing dust, particulates, and vapors.[6]

Dental Unit Water Quality

Dental hygienists use both air and water devices, such as dental handpieces, air/water syringes, and ultrasonic/sonic scalers that are attached to the dental unit for use during routine treatments. Several species of bacteria, fungi, and protozoa have been cultivated from these dental unit waterlines and thus are a potential source of infection.[9,40] The specific bacteria known to cause adverse health conditions include species of *Staphylococcus, Streptococcus, Legionella, Pseudomonas, Propionibacterium, Stenotrophomonas,* and *Mycobacterium tuberculosis.*[41,42] To ensure the delivery of quality water, the dental hygienist should follow the manufacturer instructions for proper disinfection to avoid cross-infection during treatment. The CDC recommends air and water be discharged for a minimum of 20–30 seconds after every patient. This should be completed for all devices that connect to a water line and enters the patient's mouth. Some manufacturers require purging water lines at the beginning of the workday as well as between patients for two minutes. Monitoring water quality can be completed through laboratory or in-office testing kits. Independent reservoirs, chemical treatment, filtration, sterile water delivery system, or combinations of technologies are modalities of improving/maintaining water quality.[9]

Disposable Items

Single-use devices, otherwise known as disposable devices, are meant for use on one patient during a single procedure. Such items are not meant to be cleaned, disinfected, or sterilized and used on additional patients. Disposable devices/items eliminate the risk of patient-to-patient contamination because the item is disposed of and not reused, therefore, such items should be used when applicable.[9] Examples of common single-use items/devices employed by dental hygienists include but are not limited to: High- and low-speed volume evacuation devices, such as saliva ejectors and high-speed evacuation tips, air/water syringe tips, barrier wrap, masks, gloves, paper drapes, paper bibs and paper bib holders, bracket table paper, single-use needles, prophylaxis cups and brushes, prophylaxis angles, plastic impression trays, retraction devices, cotton rolls, gauze, and irrigating syringes.[43]

Personal Protective Equipment

As noted by the CDC, personal protective equipment is a means of protecting skin and mucous membranes (eyes, nose, and mouth) from transmission of pathogens found in blood and other potentially infectious material. Donning PPE, such as proper fitting gloves and surgical masks, protective eyewear with solid side shields or face shield, and protective clothing/disposable arm length gowns are necessary measures.[9] Intact skin is the primary defense against disease; lotions and petroleum based lubricants can breakdown the integrity of gloves, and therefore, are only recommended for use at the end of the day and not during patient care. The dental professional should don PPE whenever there is a potential to encounter splatter during patient care and also during disinfecting treatment areas because bioaerosols stay suspended in the atmosphere after commencement of patient care. Providing protective eye wear for patients is a necessary prophylactic measure meant to reduce the risk of infection if splatter is generated during procedures. Masks and gloves should be changed between every patient. Additionally, PPE should be changed if torn, wet, or visibly soiled.[9] The CDC and the National Institute for Occupational Safety and Health (NIOSH) recommends that healthcare workers wear a NIOSH-certified N95 or improved respirator (N99 or N100) if providing care for patients with known infectious diseases, such as influenza and other infectious diseases.[44] To reduce disease transmission, it is imperative to remove all PPE and wash hands prior to exiting the treatment area.[9]

Sterilization Methods and Testing

Cleaning, disinfecting, and sterilization methods are necessary precautions for the prevention of disease in a dental setting and dental hygienists must be trained to properly follow these precautions.[36] According to the CDC's sterilizing practices guide[45]:

> Ensuring consistency of sterilization practices requires a comprehensive program that ensures operator competence and proper methods of cleaning and wrapping instruments, loading the sterilizer, operating the sterilizer, and monitoring of the entire process. Such procedures should take place in a central processing area to minimize disease transmission.

Sterilization monitoring is done through biological, mechanical, and chemical indicators. Biological indicators, aka spore tests, are the most accepted means of monitoring sterilization because they assess the sterilization process directly by killing highly resistant microorganisms. Spore tests are only performed weekly, therefore, mechanical and chemical monitoring should also be employed.[45,46]

Conclusion

Any healthcare profession that involves contact with mucosa, blood, or blood contaminated with body fluids should be compliant and follow standard precautions and other methods to minimize infection risks amongst themselves and patients, such as training on bloodborne pathogens as outlined by the Occupational Safety and Health Administration (OSHA).[16,47] According to the CDC's recommended infection control practices for dentistry:[48]

> Dental personnel may be exposed to a wide variety of microorganisms in the blood and saliva of patients they treat in the dental operatory. Infections may be transmitted in dental practice by blood or saliva through direct contact, droplets, or aerosols. Patients and dental health-care workers have the potential of transmitting infections to each other.

To minimize transmission potential, precautions include following proper hand-hygiene protocol (washing hands before and after wearing gloves and prior to touching potentially contaminated surfaces), changing gloves when torn or soiled, utilizing proper PPE and barrier techniques during treatment, properly disinfecting and sterilizing dental instruments and operatory

surfaces, using disposable items when possible, instituting the use of high-volume evacuation systems, maintaining dental unit water quality, providing patients with preprocedural mouth rinses, and employing air purification systems to help reduce aerosols.[6,9,27,36–40,45] By practicing a combination of these protective measures, the risk of infection from patient to patient and dental professional to patient can be greatly reduced.

REFERENCES

1. Fones AC. The origin and history of the dental hygienists. 1926. *J. Dent Hyg.* 2013;87(Suppl 1):58–62.
2. American Dental Hygienists' Association. Professional roles of the dental hygienist. Retrieved from https://www.adha.org/resources-docs/714112_DHiCW_Roles_Dental_Hygienist.pdf (Accessed December 18, 2019.)
3. American Dental Association. Dental hygienist. Retrieved from https://www.ada.org/en/education-careers/careers-in-dentistry/dental-team-careers/dental-hygienist (Accessed December 18, 2019.)
4. Gurelian, J. Inflammation: The relationship between oral and systemic disease. *Dent Assist.* 2009;78(2):8–10.
5. Pazos, P, Leira, Y, Dominguez, C, Pias-Peleteiro, JM, Blanco, J, Aldrey, JM. Association between periodontal disease and dementia: A literature review. *Neurolgia.* 2016;33(9):602–613.
6. Scannapieco F. Role of oral bacteria in respiratory infection. *J Periodontol.* 1999;70(7):793–801.
7. Aas JA, Paster BJ, Stokes LN, Olsen I, Dewhirst FE. Defining the normal bacterial flora of the oral cavity. *J Clin Microbiol.* 2005;43(11):5721–5732.
8. Hallier C, Williams DW, Potts AJC, Lewis MAO. A pilot study of bioaerosol reduction using an air cleaning system during dental procedures. *Br Dent J.* 2010; 209(8):E14.
9. Harrel SK, Molinari J. Aerosols and splatter in dentistry: A brief review of the literature and infection control implications. *J Am Dent Assoc.* 2004;135(4):429–437.
10. Fotedar S, Sharma KR, Bhardwaj V, Fotedar V. Precaution in dentistry against swine flu. *SRM J Res Dent Sci.* 2013;4:161–3.
11. Centers for Disease Control and Prevention. Guidelines for infection control in dental health-care settings-2003. *MMWR.* 2003;52(No. RR-17).
12. Kikutani K, Tamura F, Takahashi Y, Konishi K, Hamada R. A novel rapid oral bacteria detection apparatus for effective oral care to prevent pneumonia. *Gerodontology.* 2011;29(2):e560–565.
13. Merte J, Kroll C, Collins A, Melnick A. An epidemiologic investigation of occupational transmission of mycobacterium tuberculosis infection to dental health care personnel: Infection prevention and control implications. *J Am Dent Assoc.* 2014; 145(5):464–471.
14. American Dental Association. Tuberculosis. 2019. Retrieved from https://www.ada.org/en/member-center/oral-health-topics/tuberculosis-overview-and-dental-treatment-conside (Accessed January 3, 2020.)
15. American Dental Association. Hepatitis viruses. 2019. Retrieved from https://www.ada.org/en/member-center/oral-health-topics/hepatitis-viruses (Accessed January 3, 2020.)
16. Dahiya P, Kamal R, Sharma V, Kaur S. "Hepatitis" - Prevention and management in dental practice. *J Educ Health Promot.* 2015;4:33.

17. De la Tejera-Hernandez C, Noyola D, Sanchez-Vargas L, Nava-Zarate N, De le Cruz-Mendoza E, Gomez-Hernandez A, Aranda-Romo S. Analysis of risk factors associated to cytomegalovirus infection in dentistry students. *J Oral Res*. 2015. 4(3):197–204.

18. Centers for Disease Control and Prevention. About Epstein-Barr virus and Infectious Mononucleosis. 2018. Retrieved from https://www.cdc.gov/epstein-barr/about-ebv.html (Accessed January 3, 2020.)

19. Gao, Z, Lv, J, Wany, M. Epstein-Barr virus is associated with periodontal disease. *Medicine*. 2017. 96(6):e5980.

20. Slots, J, Saygun, I, Sabeti, M, Kubar, A. Epstein-Barr virus in oral diseases. *J of Perio Res*. 2006;41(4):325–244.

21. Prabhu, S, Wilson F. Evidence of Epstein–Barr virus association with head and neck cancers: a review. *Can Dent Assoc*. 2016;82:g2.

22. Brkljac M, Bitar S, Naqui Z. A case report of herpetic whitlow with positive Kanavel's cardinal signs: a diagnostic and treatment difficulty. *Case Rep Orthop*. 2014;2014:906487.

23. Betz D and Fane, K. Herpetic Whitlow. StatPearls, 2019. Retrieved from https://www.ncbi.nlm.nih.gov/books/NBK482379/. (Accessed December 18, 2019.)

24. Petti S, Vitali, M. Occupational risk for legionella infection among dental healthcare workers: meta-analysis in occupational epidemiology. *BMJ Open*. 2017;7(7):e015374.

25. Han, J, Beltran, T, Song, J. et al. Prevalence of genital human papillomavirus infection and human papillomavirus vaccination rates among US adult men. National health and nutrition examination survey (NHANES) 2013-2014. *JAMA*. 2017;3(6):810–816.

26. Heron SE, Elahi S. HIV infection and compromised mucosal immunity: Oral manifestations and systemic inflammation. *Front Immunol*. 2017;8:241.

27. American Dental Association. Human immunodeficiency virus. 2019. Retrieved from https://www.ada.org/en/member-center/oral-health-topics/hiv (Accessed January 3, 2020.)

28. How KY, Song KP, Chan KG. Porphyromonas gingivalis: An overview of periodontopathic pathogen below the gum line. *Front Microbiol*. 2016;7:53.

29. Dominy SS, Lynch C, Ermini F. et al. Porphyromonas gingivalis in Alzheimer's disease brains: Evidence for disease causation and treatment with small-molecule inhibitors. *Sci Adv*. 2019;5(1):eaau3333.

30. Ding Y, Ren J, Yu H, Yu W, Zhou Y. Porphyromonas gingivalis, a periodontitis causing bacterium, induces memory impairment and age-dependent neuroinflammation in mice. *Immun Ageing*. 2018;15:6.

31. Olsen I, Yilmaz Ö. Possible role of porphyromonas gingivalis in orodigestive cancers. *J Oral Microbiol*. 2019;11(1):1563410.

32. Mougeot JC, Stevens CB, Paster BJ, Brennan MT, Lockhart PB, Mougeot FK. Porphyromonas gingivalis is the most abundant species detected in coronary and femoral arteries. *J Oral Microbiol*. 2017;9(1):1281562.

33. Harvard Medical School. Gum disease and heart disease: The common thread. 2018. Retrieved from https://www.health.harvard.edu/heart-health/gum-disease-and-heart-disease-the-common-thread. (Accessed January 3, 2020.)

34. Grau, A. *et al*. Periodontal disease as a risk factor for ischemic stroke. *Stroke*. 2004;35(2):496–501.

35. Preshaw, PM, Alba, AL, Herrera, D. et al. Periodontitis and diabetes: A two-way relationship. *Diabetologia*. 2012;55(1):21–31.

36. Centers for Disease Control and Prevention. Summary of infection prevention practices in dental settings: basic expectations for safe care. Atlanta, GA: US Dept of Health and Human Services; 2016.

37. Gomes-Filho IS, Passos JS, Seixas da Cruz S. Respiratory disease and the role of oral bacteria. *J Oral Microbiol*. 2010;2(1):5811.

38. Emmons, L, Wu, C, Shutter, T. High-volume evacuation: Aerosols—it's what you can't see that can hurt you. *RDH*. 2017. Retrieved from https://www.rdhmag.com/patient-care/article/16409779/highvolume-evacuation-aerosolsits-what-you-cant-see-that-can-hurt-you (Accessed December 3, 2019.)

39. Grinshpun SA, Adhikari A, Honda T. et al. Control of aerosol contaminants in indoor air: Combining the particle concentration reduction with microbial inactivation. *Environ Sci Technol*. 2007; 41(2):606–612.

40. Leoni, E, Dallolio, L, Stagni, F, Sanna, T, D'Alessandro, G, Piana, G. Impact of a risk management plan on legionella contamination of dental unit water. *Int. J. Environ. Res. Public Health*. 2015;12(3):2344–2358.

41. Costa, D, Mercier A, Gravouli, K, Lesobre, J, Delafont, V, Bousseau, A, et al. Pyrosequencing analysis of bacterial diversity in dental unit water lines. *Water Research*. 2015;81(15):223–231.

42. Zhang, Y, Ping, Y, Zhou, R, Wang, J, Zhang, G. High throughput sequencing-based analysis of microbial diversity in dental unit waterlines supports the importance of providing safe water for clinical use. *J Infect Pub Health*. 2018;11(3):357–363.

43. Kelsch, N. Single-use (disposable) devices. *RDH*. 2012. Retrieved from https://www.rdhmag.com/infection-control/disinfection/article/16405788/singleuse-disposable-devices (Accessed January 3, 2020.)

44. Centers for Disease Control and Prevention. N95 respirators and surgical masks. 2009. Retrieved from https://blogs.cdc.gov/niosh-science-blog/2009/10/14/n95/ (Accessed January 3, 2020.)

45. Centers for Disease Control and Prevention. Sterilizing practices. 2016. Retrieved from https://www.cdc.gov/infectioncontrol/guidelines/disinfection/sterilization/sterilizing-practices.html (Accessed February 18, 2019.)

46. Centers for Disease Control and Prevention. Sterilization: Monitoring. Retrieved from https://www.cdc.gov/oralhealth/infectioncontrol/faqs/monitoring.html. (Accessed January 3, 2020.)

47. ECRI Institute. Preventing transmission of infectious disease in the dental clinic. 2013. Retrieved from https://www.ecri.org/components/HRSA/Pages/GetSafe_052213.aspx (Accessed January 3, 2020.)

48. Centers for Disease Control and Prevention. Recommended infection-control practices for dentistry. *MMWR*. 1986;35(15): 237–242.

62

Microbiology for Pre-College Teachers

Madge Nanney and Scott Sowell

CONTENTS

Our Middle School Program

Introduction

> Many of tomorrow's jobs... don't exist today. These jobs will most certainly require a workforce of highly educated workers, utilizing skills that have not yet been identified in fields and operations that, at the moment, are only being discussed in theory.
>
> Hearing on Careers for the 21st century, 2004[1]

While meeting the national/state standards is foremost in our minds, educating students is so much more than how they perform on high-stakes tests at the end of the year. Embedded in each assignment we give them are the intangible skills that students learn that aren't measured on a test. Some of these skills are:

- Working in a group. Students must learn how to function as a group. This also teaches them that if work does not get done, the entire group fails.
- Communication. Students must learn how to get the rest of the group to listen to them, and how to listen to others.
- Materials and methods. How are the parts put together to work? The directions are very specific (on purpose).
- Creativity. What if we try this? Are there alternate but viable solutions to a testable question that I haven't even thought of yet?

Attracting Students to Our Program

As classroom teachers, we all know the dreaded email from the principal that starts with "Thank you so much for volunteering…." For us, this year's volunteering opportunity was to develop an engaging hands-on science experience for our Open House Meeting whose purpose was to recruit students to our medical magnet middle/high school. Open House at our school is a family affair. Parents, siblings, and extended family all attend. We had to design an activity that would appeal to all. We chose to do a project involving microbiology. What's more engaging that viewing drops of water teeming with microorganisms? We recruited students that we taught last year to be lab assistants. With a sample of pond water, our students interacted with the student visitors to help them make and view their slides in our classroom microscopes. Parents were impressed as their children looked up from their microscopes and exclaimed, "Mom you have got to see this!" There was minimal set up and clean up. Our former students were excited to assist potential students. We overheard more than one visitor state, "This is the school I want to attend."

First Day of School

In secondary schools across the nation, the first day of school is where students are sitting class after class listening to classroom/school rules and procedures. This year, we switched the lesson plan to include learning stations. Around the classroom we designed short self-directed lessons that a group of two to four students finished in 15 minutes. They rotated through each station in a clockwise rotation. Of course, there was a station that included classroom expectations, as well as one that included an ice breaker to meet your new classmates.

We wanted stations that would give students a hint of what concepts we would study this year. Microscopes were set up with prepared plant and animal slides, and students were asked to look into the microscopes and decide whether what they were observing on the slides were plants or animals. This exercise also served as a preassessment for later exercises that meet one of our elementary benchmarks: having students identify structures in plant versus animal cells. (Today's public-school educators are given a curriculum that is written to meet the science benchmarks that the State Board of Education has determined that will be taught. National benchmarks are described in Appendix 1.)

The Microbiology Unit

Many of the microbiology investigations that are assigned in our middle school classrooms are common to middle schools across the nation. The excitement for us as educators is when a student looks up from their microscope declaring, "Is there a job that includes microscopes?" Although not a specific benchmark, students need to learn how to operate a microscope to have a successful experience. Learning the specific parts and how to focus a microscope is accomplished through the "d lab." Using a wet mount technique, students make their own slides using a typed letter d and drops of water. Students learn that the microscope reverses and inverts the image when they observe that the "letter d" becomes the "letter p" when viewed under the microscope. To address the benchmarks that compare animal and plant cells, students will prepare their own slides using onion cells. This "tried and true" investigation allows students to make their own slides learning proper technique for making wet mounts.

Making animal-cell slides is a little more difficult, because for several years, our district has banned the traditional cheek cell lab due to possible pathogen transmittal. Looking for a creative way to examine epithelial cells, we have students rub clear tape on their inner arms. Following this, the tape is then pressed onto a clean slide. Using methylene blue, the transferred epithelial cells are stained and then observed.

One of our benchmarks is learning the levels of taxonomy from species up to domains. Knowing the different kingdoms is essential to mastering these benchmarks. Middle school students are familiar with Kingdom Plantae and Kingdom Animalia, but Kingdom Protista, Kingdom Fungi and Kingdom Monera are a mystery.

Students are thrilled to look at prepared slides of paramecium, euglena, and volvox. But what is even more exciting is investigating life in a drop of pond water. Aside from freshwater that is available near school, a filter from a classroom fish tank is an interesting source for Protista.

Last Days of School

The end of the school year is always hectic. Our students must turn in their textbooks, clean out their lockers, and, of course, complete the exams that will be important in determining their final grades. Here again, microbiology is an important part of

the process. Every school has a collection of prepared microscope slides that can be organized into stations. As students rotate through the stations, there are task cards at each station that direct the students to observe a specific part of the slide. Students are engaged and learning new content.

Our High School Program

Introduction

One of our goals as teachers of high school environmental science is to instill in our students a sense of stewardship for our planet's natural resources. It is not enough that the students leave the school year knowing more about the soil, the groundwater, or the atmosphere. Rather, we want them to be able to act on that knowledge as environmentally aware citizens, either in a science-related profession or in everyday decisions. Therefore, the content must be meaningful. It must have purposeful and meaningful utility in their daily lives. That is especially true when it comes to the intersection of environmental science and microbiology.

To explicitly work on this goal, we try and make microbiology a thread within our course. While we do not have a particular unit that covers microbiology itself, we do have a clear reoccurrence of microorganisms in various topics. In the past, we did not connect these micro-moments; rather, we would leave them as isolated comments within lessons and labs. However, drawing a curricular thread among these lessons can be very useful. That way, students walk away with a clearer understanding of the roles the microorganisms play in the environment.

Microbial Misconception

The main misconception that our students have regarding microbiology is that they think that there is no biodiversity within microbes. The understanding typically brought to the table is that all microbes are a homogenous group. While they may have a tentative grasp of photosynthetic versus heterotrophic microbes, there typically is no further discernment within these two groups. To many of our students, microorganisms are simply little living things in the soil that do everything and nothing at the same time. Students may be familiar with the different shapes, sizes, colors, and structures of microorganisms from looking at collections of prepared slides, but they see those differences more as varieties within a species rather than distinct species themselves. While they can clearly discuss the different ecological niches of a hummingbird and a pelican, they lack that discernment when it comes to microbes. It is as if the hummingbird might at any point dive into the ocean and catch a fish with its beak.

This misconception is counterproductive to strong learning within environmental science, as it limits the students' understanding of many Earth systems and the anthropogenic effects on them (e.g., high-input agriculture's effects on the nitrogen cycle).

To address important misconceptions, we often use the conceptual change model.[2] This pedagogical strategy provides a framework for constructing learning experiences within the course. In a nutshell, the conceptual change model has four distinct stages that work to replace a student's misconception with more scientifically correct concepts.

- *Dissatisfaction:* In this first stage, the student sees the limitations of their current understanding. This is where teachers can use discrepant events or inquiry to explicitly point out why the current understanding is incorrect. The goal here is to disrupt with cognitive dissonance.
- *Intelligibility:* The students can then understand and explain the new (and more scientifically correct) understanding. In our classes, this is achieved via more traditional, teacher-centered explanations. The goal here is to clearly explain the new idea.
- *Plausibility:* The student then needs to see that this new understanding fits within their current ideas about the world. The goal here is for students to see the new concept as true.
- *Fruitfulness:* In the final stage, the student needs to be able to use this new understanding in a novel context. The goal is for the teacher to provide experiences in which the student can see that their new understanding is meaningful.

We use these cognitive stages as simple reminders about how to orchestrate learning experiences with our students. It shapes the lectures that are given, the classroom discussions that we have, and the laboratory investigations we work through. If this model is used successfully, the students will abandon their initial misconception about a homogeneous group of microorganisms, and ideally replace that misconception with the idea that there are myriad different species of microorganisms; each with a unique niche, and a fruitful role in the discipline of environmental science.

Here is an example of a lesson sequence that uses the conceptual change model:

Discuss the resource Nutrient Cycling in the Serengeti
- Dissatisfaction: Point out that different things need to happen in order for nutrients to cycle (i.e., different microbes need to do different things); because if they were all the same, the cycle would not work.
- Intelligibility: Define/explain the different microorganisms and their niches.
- Plausibility: Do the inquiry activity where students work through the nutrient cycling in the Serengeti, seeing that microbes play a central role in allowing plants to have access to limiting nutrients.
- Fruitfulness: Use the new-tank-syndrome as the new context; allow students to work through a novel situation, using the different roles microorganisms play in the nitrogen cycle; OR discuss the issue of over-irrigation and denitrification during agriculture.

Exploring Fruitfulness

This last phase, fruitfulness, is the one we would like to expand upon a bit more. If a student has a robust understanding of the vast biodiversity of microbes on the planet, then they are more likely to appreciate the role certain microorganisms play in environmental issues. It is a powerful moment when a student realizes that the ecological niche of *Rhizobium* is vastly different from that of *Alcanivorax*. *Rhizobium*'s nitrogen fixation in the soil is dramatically distinct from *Alcanivorax*'s oil-degrading in the ocean. Two very different organisms with very little niche overlap. To mitigate the effects of high-input, industrialized agriculture or the effects of an oil spill, we need to understand the roles these microorganisms play. In fact, once we understand their roles in nature, we can use those roles to improve our lives and the environment. Thus, students in environmental science find meaningfulness and fruitfulness with their microbe knowledge.

Looking across our environmental science courses, there are many moments in which microbes play meaningful roles. For example, take the simple idea of decomposition, both terrestrial and aquatic. If students understand the roles microorganisms play in the decomposition of organic matter, they are more likely to be able to understand things such as wastewater treatment plants, biogas digesters, or sanitary landfills.

Here in Florida, algal blooms are an environmental menace. It is important for our students to see how nutrient pollution and microorganisms all play in an algal bloom event. Connecting fertilizer runoff, failing septic tanks, and decomposing microorganisms gives students more agency in addressing this environmental issue.

Constructing Systems Models

Once a student's misconceptions about microorganisms are challenged and replaced and the new ideas are seen as meaningful and fruitful, they can strengthen the fruitfulness of their ideas using systems models.

Students are now ready to construct, revise, and use systems models to understand how anthropogenic factors can affect natural processes that will result in measurable changes. An example is: Understanding Global Change (https://cleanet.org/clean/literacy/tools/UGC/index.html Accessed April 19, 2020).

Other useful resources are listed in Appendix 2.

REFERENCES

1. Hearing on Careers for the 21st Century: The Importance of Education and Worker Training for Small Business. Committee on Small Business. 108th Congressional Hearings. Government Publishing Office, Washington DC, (2004).
2. Posner, G. J., Strike, K. A., Hewson, P. W., & Gertzog, W. A. Accommodation of a scientific conception: towards a theory of conceptual change. *Science Education*, 66, 211,1982.

APPENDIX 1: NATIONAL BENCHMARKS

MS-LS1-1. Conduct an investigation to provide evidence that living things are made of cells; either one cell or many different numbers and types of cells. (Clarification Statement: Emphasis is on developing evidence that living things are made of cells, distinguishing between living and nonliving things, and understanding that living things may be made of one cell or many and varied cells.)

MS-LS1-2. Develop and use a model to describe the function of a cell as a whole and ways the parts of cells contribute to the function. (Clarification Statement: Emphasis is on the cell functioning as a whole system and the primary role of identified parts of the cell, specifically the nucleus, chloroplasts, mitochondria, cell membrane, and cell wall.) (Assessment Boundary: Assessment of organelle structure/function relationships is limited to the cell wall and cell membrane. Assessment of the function of the other organelles is limited to their relationship to the whole cell. Assessment does not include the biochemical function of cells or cell parts.)

MS-LS1-3. Use argument supported by evidence for how the body is a system of interacting subsystems composed of groups of cells. (Clarification Statement: Emphasis is on the conceptual understanding that cells form tissues and tissues form organs specialized for particular body functions. Examples could include the interaction of subsystems within a system and the normal functioning of those systems.) (Assessment Boundary: Assessment does not include the mechanism of one body system independent of others. Assessment is limited to the circulatory, excretory, digestive, respiratory, muscular, and nervous systems.)

MS-LS1-4. Use argument based on empirical evidence and scientific reasoning to support an explanation for how characteristic animal behaviors and specialized plant structures affect the probability of successful reproduction of animals and plants respectively. (Clarification Statement: Examples of behaviors that affect the probability of animal reproduction could include nest building to protect young from cold, herding of animals to protect young from predators, and vocalization of animals and colorful plumage to attract mates for breeding. Examples of animal behaviors that affect the probability of plant reproduction could include transferring pollen or seeds and creating conditions for seed germination and growth. Examples of plant structures could include bright flowers attracting butterflies that transfer pollen, flower nectar and odors that attract insects that transfer pollen, and hard shells on nuts that squirrels bury.)

MS-LS1-5. Construct a scientific explanation based on evidence for how environmental and genetic factors influence the growth of organisms. (Clarification Statement: Examples of local environmental conditions could include availability of food, light, space, and water. Examples of genetic factors could include large breed cattle and species of grass affecting growth of organisms. Examples of evidence could include drought decreasing plant growth, fertilizer increasing plant growth, different varieties of plant seeds growing at different rates in different conditions, and fish growing larger in large ponds than they do in small ponds.) (Assessment Boundary: Assessment does not include genetic mechanisms, gene regulation, or biochemical processes.)

Disciplinary Core Ideas

LS1.A: Structure and Function

- All living things are made up of cells, which is the smallest unit that can be said to be alive. An organism may consist of one single cell (unicellular) or many different numbers and types of cells (multicellular). (MS-LS1-1).

- Within cells, special structures are responsible for particular functions, and the cell membrane forms the boundary that controls what enters and leaves the cell. (MS-LS1-2)

- In multicellular organisms, the body is a system of multiple interacting subsystems. These subsystems are groups of cells that work together to form tissues and organs that are specialized for particular body functions (MS-LS1-3)

LS1.C: Organization for Matter and Energy Flow in Organisms

- Plants, algae (including phytoplankton), and many microorganisms use the energy from light to make sugars (food) from carbon dioxide from the atmosphere and water through the process of photosynthesis, which also releases oxygen. These sugars can be used immediately or stored for growth or later use. (MS-LS1-6)

- Within individual organisms, food moves through a series of chemical reactions in which it is broken down and rearranged to form new molecules, to support growth, or to release energy. (MS-LS1-7)

High School

HS-LS1: From Molecules to Organisms: Structures and Processes. Students who demonstrate understanding can:

- *HS-LS1-1.* Construct an explanation based on evidence for how the structure of DNA determines the structure of proteins which carry out the essential functions of life through systems of specialized cells. (Assessment Boundary: Assessment does not include identification of specific cell or tissue ty pes, whole body systems, specific protein structures and functions, or the biochemistry of protein synthesis.)

- *HS-LS1-2.* Develop and use a model to illustrate the hierarchical organization of interacting systems that provide specific functions within multicellular organisms. (Clarification Statement: Emphasis is on functions at the organism system level, such as nutrient uptake, water delivery, and organism movement in response to neural stimuli. An example of an interacting system could be an artery depending on the proper function of elastic tissue and smooth muscle to regulate and deliver the proper amount of blood within the circulatory system.) (Assessment Boundary: Assessment does not include interactions and functions at the molecular or chemical reaction level.)

- *HS-LS1-3.* Plan and conduct an investigation to provide evidence that feedback mechanisms maintain homeostasis. (Clarification Statement: Examples of investigations could include heart rate response to exercise, stomate response to moisture and temperature, and root development in response to water levels.) (Assessment Boundary: Assessment does not include the cellular processes involved in the feedback mechanism.)

- *HS-LS1-4.* Use a model to illustrate the role of cellular division (mitosis) and differentiation in producing and maintaining complex organisms. (Assessment Boundary: Assessment does not include specific gene control mechanisms or rote memorization of the steps of mitosis.)

- *HS-LS1-5.* Use a model to illustrate how photosynthesis transforms light energy into stored chemical energy. (Clarification Statement: Emphasis is on illustrating inputs and outputs of matter and the transfer and transformation of energy in photosynthesis by plants and other photosynthesizing organisms. Examples of models could include diagrams, chemical equations, and conceptual models. Assessment Boundary: Assessment does not include specific biochemical steps.)

- *HS-LS1-6.* Construct and revise an explanation based on evidence for how carbon, hydrogen, and oxygen from sugar molecules may combine with other elements to form amino acids and/or other large carbon-based molecules. (Clarification Statement: Emphasis is on using evidence from models and simulations to support explanations.) (Assessment Boundary: Assessment does not include the details of the specific chemical reactions or identification of macromolecules.)

- *HS-LS1-7.* Use a model to illustrate that cellular respiration is a chemical process whereby the bonds of food molecules and oxygen molecules are broken and the bonds in new compounds are formed resulting in a net transfer of energy. (Clarification Statement: Emphasis is on the conceptual understanding of the inputs and outputs of the process of cellular respiration.) (Assessment Boundary: Assessment should not include identification of the steps or specific processes involved in cellular respiration.)

APPENDIX 2: SECONDARY SCIENCE PROFESSIONAL ORGANIZATIONS

- **(National Science Teachers Association www.nsta.org).** From peer-reviewed journals to national conferences, this educational organization has something for every level and content area of secondary science. Although membership is needed for full access, podcasts, resources, and conference materials are free.

- **(National Biology Teachers Association www.nabt.org).** Established in 1938, this association offers invaluable material for biology teachers in microbiology. Links to a wide assortment of instructional materials that could be adapted to any classroom.

- **(National Education Association www.nea.org).** The Microbe Passport link has virtual slides that allow students to view microbes online. Free lesson plans are also available on this site.

- **(American Society for Microbiology www.asm.org).** The secondary science teacher can find free lesson plans that are ready for classroom use.

- **(CPALMS www.cpalms.org).** This website contains lesson plans, student video tutorials, and articles that are classroom-ready and fully editable. There 63 lessons for microbiology in middle school. For high school biology, there are 265 lessons that align to cellular biology.

63

Microbiology for Home Inspectors

William E. Herrmann

CONTENTS

As a home inspector, I have dealt with many different types of microorganisms: Bacteria parasites, nematodes, protozoa, fungi, algae, and viruses. As it turned out, it was not that safe. I retired recently after chest pains led to a 7-hour heart procedure. The doctors also found asbestos in my lung, which contributed to breathing problems developed over the past few years. One of my strengths is putting complex subjects into simple practical terms, and I'm happy for the opportunity to relate microbiology to real-world application in my career as a Home Inspector.

When we walk into an unfamiliar home, we never know what is inside that may harm us. Our job is to evaluate almost everything inside and outside of the home. While we are worried about the health, safety, and welfare of our clients, we are exposed to many things that we need to worry about for ourselves. By industry standards and most state regulations, Home Inspectors evaluate heat, plumbing, insulation, foundation, electric, roof, windows, doors, etc. (As an example, New York State regulations may be found at https://www.dos.ny.gov/licensing/homeinspect/hinspector.html. Accessed April 1, 2020). These items are primarily mechanical things. Water, septic, radon levels, and air quality testing are ancillary services that are usually not covered within the scope of a formal Home Inspection. High school and college science courses have greatly helped my understanding of the challenges that I face—the research to understand, explain to the client, and recommend or oversee solutions becomes easier because of a firm science-based approach.

I have performed many initial investigations into "sick homes." This can be tricky because sometimes the diagnosis is straight forward—mold, back-drafting fuel burning appliances (exhaust byproducts including carbon monoxide), sewer gases, etc.—and sometimes it isn't. I have vacuumed carpets and looked through the debris for anything of interest to the challenge at hand. Insect fragments, unidentifiable strange dusts, dander, and mouse or insect droppings have all helped contribute to causing sick homes. Sometimes an industrial engineer is needed. Knowing my limits when I have reached them, I unapologetically recommend the next level of expertise.

Inspecting the Attic

The importance of wearing a dust mask cannot be overemphasized, because of what can be encountered when testing an attic. Bat, bird, squirrel, and raccoon droppings can introduce not only ectoparasites (organisms that live on the skin, such as fleas and lice), but also microorganisms that can cause diseases. These include leptospirosis and histoplasmosis, which are respiratory diseases; candidiasis, which are yeast infections spread by pigeons; cryptococcosis, which is a yeast found in the intestinal tract of pigeons; St. Louis encephalitis (inflammation of the nervous system); salmonellosis, caused by the salmonella bacteria; *Escherichia. coli*, also a bacteria; rabies, bubonic plague, West Nile encephalitis, and Legionnaires' disease.

Diseases passed from animals to humans are called zoonoses and we encounter many of them. Among these are bubonic plague, which is spread by rodents; Q fever, which is spread by barnyard animals; *Campylobacter* infection, which is spread by animal feces; hantavirus, which is spread by mice; lymphocytic choriomeningitis, which is spread by house mice; tularemia, which is spread by rabbits; raccoon roundworm; and Giardia.

Infections Caused by Pets

Cats allowed to roam outdoors and then return home, often pick up a parasite known as *Toxoplasma gondii*. Toxoplasmosis is of particular danger to small children, people with damaged immune systems, people undergoing cancer treatment, and pregnant women. Cats are also a vector for *Bartonella henselae*, which causes cat scratch fever. Dogs can come into contact with rats, mice, cows, pigs, or other dogs and then spread infection when they greet you. They can expose the Home Inspector to leptospirosis, which can lead to liver and kidney damage and even death. In addition, rabies is another occupational hazard that can be encountered.

DOI: 10.1201/9781003099277-66

Water Testing

Water testing includes testing for both organic and inorganic compounds. One must specify what is to be tested as there are specific lab procedures for each test agent. Almost all wells are routinely tested for total coliform and *E. coli* bacteria. Total coliforms are a group of closely related, mostly harmless bacteria that live in soil and water as well as the gut of animals (including humans). Total coliforms are currently controlled by drinking water regulations (i.e., Total Coliform Rule which can be found at https://www.epa. gov/dwreginfo/revised-total-coliform-rule-and-total-coliform-rule. Accessed April 1, 2020). Their presence above the standard indicates problems in water treatment or in the distribution system. If total coliforms are found, then the water system sample is further analyzed to determine if specific types of coliforms (i.e., fecal coliforms or *E. coli*) are present. *E. coli* are fecal coliforms commonly found in the intestines of animals and humans. The presence of E. coli in water is a strong indication of recent sewage or animal waste contamination of the drinking water. It is also an indicator for the presence of other disease-causing organisms.

If we turn the water on and smell "rotten eggs," it means that hydrogen sulfide (H_2S) is present. It comes from the sulfur bacteria that occur naturally from the decay and chemical reactions within soil and rocks. Although this is normally a nuisance and not a real concern, in extremely high concentrations it can be both poisonous and flammable.

It is important to understand the proper operation of septic systems. Aerobic bacteria require oxygen to live while anaerobic bacteria do not. A very important function of the septic tank is to allow bacteria to convert organic material into energy that the bacteria then use for their life functions. This conversion is what causes the breakdown of organic solids in the wastewater into simpler solids and gas, making the wastewater leaving the septic tank comparatively cleaner than what enters. However, since most septic tanks do not contain oxygen, an anaerobic environment, the aerobic bacteria quickly die off, leaving the job of breaking down waste to the anaerobic bacteria. There are alternative systems to the traditional, which may cost an additional $20,000–$30,000, that have pumps in the tank to encourage the aerobic activity. (I have one of these in my home in the Georgia mountains. They are required by the local codes because of the poor soil types. In addition, instead of discharging through traditional 4-inch perforated PVC, it is distributed below the soil at a shallower depth through small diameter drip irrigation tubes.) When performing a rudimentary septic dye test, a nontoxic fluorescent dye is put into the tank and water is introduced for a while to put pressure on the leach field. If the color (yellow, blue, green, red) comes to the surface or if standing water by the leach field or the river stream lake close by turns color, you have a problem. Dye in the toilet ending up on the surface of the lawn means an inadequate septic system.

Water may also spread Legionnaires' disease. This form of pneumonia caused by a bacterium, spreads mostly as a mist leaking from air conditioning, but can come from a shower, hot tubs, etc. It causes disease through aspiration.

Mold

Mold is common and in fact, in my years of mold testing, I have discovered it is uncommon to find a zero level of mold in a home.

It is everywhere and there are many types. Some are more toxic and dangerous than others. It is not the mold itself that is toxic, but the toxins the mold produces that are the concern. All molds produce toxins (called mycotoxins). They are the organism's natural defense system. Some are worse than others. They can cause toxicity. These mycotoxins, depending on exposure levels, create various symptoms: brain fog, fatigue, asthmas, muscle cramps, weakness, cognitive problems, headaches, numbness, sinus problems, shortness of breath, eye problems, infections, toxic reactions, flu-like symptoms, and skin infections. Aflatoxins, produced by *Aspergillus*, are among the most poisonous; ochratoxin produced by *Aspergillus* and *Penicillium*; patulin produced from *Aspergillus*, *Penicillium*, and *Byssochlamys* are other toxins that are encountered.

The mold spores are an irritant to the respiratory system. We look at air-quality test results in comparative terms. The most common molds we find (in alphabetical not in order of pathogenicity) are Alternaria (respiratory response), *Aspergillus* (lung), *Cladosporium* (asthma), *Penicillium* (allergy and asthma), and *Stachybotrys*. *Stachybotrys*, aka black mold, is probably the most serious as it causes serious health problems including bleeding of the lungs.

Off-Gassing

Finally, we are concerned with the off-gassing of construction materials. Off-gassing is the release of chemicals, such as volatile organic chemicals (VOCs) or airborne particles. It is correlated with congestion, coughing, lymphomas, cognitive issues, skin irritation, asthma, leukemia, and fatigue. It can be caused by carpeting (five plus years of off-gassing), cabinetry, furniture (plywood and particleboard off-gas for two years), paint, nail polish remover, dryer sheets, household cleaners, and other household goods. Gasses can include benzene, ammonia, ozone (electronics), toluene, and the carcinogen formaldehyde. Many of these can induce the same symptoms, which makes identifying a specific compound difficult. It is estimated that there have been about 80,000 chemicals introduced into our environment in the past 50 years and the current building science instructs us to make houses "tight" to avoid heat loss. The problem is that this limits air exchange with the outside unless this was addressed in the design, which is often doubtful since it requires expensive systems. This is why some people believe the old drafty houses were healthier.

Every day, while processing the many thousands of mechanical elements it takes to build a home, I am always extremely mindful of the health-related issues, most of which are microbiologically based. These elements have mostly been created or introduced after the initial home construction. Higher temperatures and humidity make most things off-gas more quickly. These can be slowed by using a dehumidifier to keep the humidity levels at about 45%. Since mold will grow above 60%, the dehumidifier will prevent both off-gassing and as well as mold growth. Solid wood furniture is better than particleboard. If you remove wall to wall carpeting to get rid of VOCs you will also have the added benefit of getting rid of mold and dust mites.

Writing this chapter has made me consider all the things that I have come across in the past 29 years, and all the mental gymnastics I executed daily. It has been an interesting, challenging, and rewarding career, in which I have helped a lot of people make what might be the largest purchase in their life. My experience has taught me that this can be a dangerous career and the Home Inspectors need to make sure that they adequately protect themselves.

Survey of Selected Clinical, Commercial, and Research-Model Eubacterial Species

Emanuel Goldman

Organism	Growth	Staining	Habitat	Pathogenesis (in Humans)	Comments
Acetobacter spp.	aerobic	gram⁻ rod	plants and soil	nonpathogenic	converts ethanol to acetic acid
Acidithiobacillus ferrooxidans	aerobic	gram⁻ rod	soil, water	nonpathogenic	thermo- & acidophilic; mine waste recovery
Acinetobacter spp.	aerobic	gram⁻ rod	soil and other environments	lung, blood, wound infections	nosocomial in debilitated patients
Actinomyces israelii	anaerobic[1]	gram⁺ rod	normal mouth/ intestine flora	lung, abdomen, cervicofacial abscess	opportunistic pathogen, sulfur-like granules
Actinomyces naeslundii	facultative anaerobe	gram⁺ rod	normal mouth flora	periodontal disease, dental caries	networked filaments, sulfur-like granules
Aeromonas spp.	facultative	gram⁻ rod	fresh & slightly salted water	gastroenteritis, wound infections	systemic disease in immunocompromised
Agrobacterium tumefaciens	facultative	gram⁻ rod	plants	rarely pathogenic (catheters)	plant tumors; plant genetic engineering
Alcaligenes spp.	aerobic	gram⁻ rod	soil and water	lung infections in cystic fibrosis	used to produce non-standard amino acids
Anabaena spp.	facultative aerobe	gram⁻ oval	water, soil, plankton	nonpathogenic but some make toxin	heterocysts; photosynthetic & N₂-fixing
Anaplasma spp.	intracellular	gram⁻ rod	ticks, animals	pathogenic primarily in animals	transmission by insect vectors
Arthrobacter spp.	aerobic	gram⁺ variable[2]	soil	nonpathogenic	cleans some environmental pollutants
Bacillus spp.	facultative and aerobes	gram⁺ rod	soil	nonpathogenic	forms spores (all *Bacillus*)
Bacillus anthracis	facultative aerobe	gram⁺ rod	soil	anthrax	found on animal skins
Bacillus cereus	facultative aerobe	gram⁺ rod	soil	gastroenteritis	food-poisoning both by intoxication & growth
Bacillus megaterium	facultative aerobe	gram⁺ rod	soil	nonpathogenic	large organism; soil inoculant & biotech uses
Bacillus subtilis	facultative aerobe	gram⁺ rod	soil	nonpathogenic	model organism for research
Bacillus thuringiensis	facultative aerobe	gram⁺ rod	soil	nonpathogenic	produces insecticide, agricultural use
Bacteroides fragilis	anaerobic	gram⁻ rod	normal intestinal flora	peritonitis, abdominal & other sites	causes disease in inappropriate sites
Bartonella spp.	intracellular	gram⁻ rod	ticks, insects, animals	cat scratch disease, trench fever, carditis	transmission to humans by insect vectors
Bdellovibrio bacteriovorus	aerobic	gram⁻ rod	fresh water and soil	nonpathogenic	kills other gram neg bacteria, in periplasm
Beggiatoa spp.	facultative anaerobe	gram⁻ rod	fresh water	nonpathogenic	filaments; sulfur granules; pollution indicator
Beijerinckia spp.	aerobic	gram⁻ rod	soil	nonpathogenic	nitrogen-fixing
Bifidobacterium spp.	anaerobic	gram⁺ rod	normal intestinal flora	nonpathogenic	symbiotic to humans; used as a probiotic
Bordetella pertussis	aerobic	gram⁻ coccobacilli	humans	whooping cough	effective vaccine has reduced incidence

(Continued)

Organism	Growth	Staining	Habitat	Pathogenesis (in Humans)	Comments
Borrelia burgdorferi	microaerophile	gram⁻ spirochete	animals, insects	Lyme disease	transmission to humans by ticks
Brucella melitensis	facultative aerobe	gram⁻ rod	animals	brucellosis	zoonotic disease in humans
Burkholderia spp.	aerobic	gram⁻ rod	soil, water, animals	melioidosis; lung in cystic fibrosis	degrades polychlorinated biphenyls
Campylobacter jejuni	microaerophile	gram⁻ rod	animals, chickens	gastroenteritis	widespread cause of food-poisoning
Caulobacter spp.	aerobic	gram⁻ rod	fresh water	nonpathogenic	asymmetric division; stalks; research model
Chlamydia pneumoniae	intracellular	gram⁻ rod	animals, insects, protozoa	pneumonia	elementary and reticulate body forms
Chlamydia trachomatis	intracellular	gram⁻ rod	humans	chlamydia	STD³ that can cause infertility
Chlorobium spp.	facultative anaerobe	gram⁻ coccobacilli	warm water	nonpathogenic	photosynthetic
Citrobacter spp.	facultative	gram⁻ rod	normal intestinal flora, soil	rarely pathogenic; some UTIs⁴	uses citrate as sole carbon source
Clostridium botulinum	anaerobic	gram⁺ rod	soil	botulism (food-poisoning)	forms spores; toxin used cosmetically
Clostridium difficile	anaerobic	gram⁺ rod	endogenous intestinal flora	pseudomembranous colitis	forms spores; antibiotic-induced disease
Clostridium perfringens	anaerobic	gram⁺ rod	soil and other environments	gas gangrene; food-poisoning	forms spores; also normal intestinal flora
Clostridium tetani	anaerobic	gram⁺ rod	soil	tetanus	spores; wound infection; effective vaccine
Corynebacterium diphtheriae	facultative anaerobe	gram⁺ rod	mammals	diphtheria (upper respiratory disease)	pathogenicity due to phage; effective vaccine
Corynebacterium glutamicum	facultative anaerobe	gram+ rod	soil	non-pathogenic	industrial production of amino acids
Corynebacterium spp.	facultative anaerobe	gram⁺ rod	normal skin flora	most are non-pathogenic	can cause disease in debilitated patients
Coxiella burnetii	intracellular	gram⁻ rod-like	animals	Q-fever	forms spores; transmitted by inhalation
Cyanobacteria genera & spp.	facultative	gram⁻ mixed	water and soil	a few species produce toxins	"blue-green algae"; some in food; "bio-concrete"
Cytophaga spp.	aerobic	gram⁻ rod	water and soil	nonpathogenic	gliders; degrade cellulose; make antibiotics
Desulfovibrio spp.	anaerobic	gram⁻ rod	water and soil	nonpathogenic	sulfate-reducing; pollution control
Ehrlichia spp.	intracellular	gram⁻ coccus	ticks, animals	ehrlichiosis; blood disease	transmission by insect vectors; dog pathogen
Enterobacter spp.	facultative	gram⁻ rod	normal intestinal flora	skin and tissue	opportunistic; nosocomial; debilitated patient
Enterococcus faecalis	facultative	gram⁺ coccus	normal intestinal flora	UTIs⁴, endocarditis	causes disease in inappropriate sites
Enterococcus faecium	facultative	gram⁺ coccus	normal intestinal flora	UTIs⁴, endocarditis	causes disease in inappropriate sites
Erwinia spp.	facultative	gram⁻ rod	soil, water, insects, plants	nonpathogenic	plant pathogen affects fruits & vegetables
Erysipelothrix rhusiopathiae	facultative anaerobe	gram⁺ rod	animals, soil	skin infections	primarily an animal pathogen
Escherichia coli	facultative	gram⁻ rod	normal intestinal flora	gastroenteritis, UTI⁴, septicemia	fast grower; diseases at inappropriate sites
Escherichia coli O157:H7	facultative	gram⁻ rod	cattle	HUS⁵, hemorrhagic colitis	food-poisoning: beef, manure runoff in water
Eubacterium spp.	anaerobic	gram⁺ rod	endogenous mouth flora	periodontal disease	historically genus of convenience; diverse
Flavobacterium spp.	aerobic	gram⁻ rod	fresh water, soil	nonpathogenic	serious pathogen of fish
Francisella tularensis	aerobic	gram⁻ coccobacilli	animals; rodents, rabbits	tularemia	transmitted by insects, animals, food, air

Organism	Growth	Staining	Habitat	Pathogenesis (in Humans)	Comments
Frankia spp.	microaerophile	gram⁺ rod	soil, plant roots	nonpathogenic	N_2-fixing filamentous plant symbiont; spores
Fusobacterium spp.	anaerobic	gram⁻ rod	endogenous mouth flora	periodontal disease, part of ANUG[6]	biofilms; can cause disease at other sites
Gallionella ferruginea	aerobic	no peptidoglycan	fresh and salt water	nonpathogenic	chemolithotrophic; iron-fixing; biofilms
Gardnerella vaginalis	facultative anaerobe	gram variable rod	endogenous vaginal flora	vaginosis	not exclusively an STD[3]
Haemophilus ducreyi	facultative aerobe	gram⁻ coccobacilli	humans	chancroid (genital lesions)	STD[3] facilitates HIV transmission
Haemophilus influenzae	facultative aerobe	gram⁻ coccobacilli	normal nasopharynx flora	lung, ear, eye, meningitis in children	effective vaccine for children
Haemophilus parainfluenzae	facultative aerobe	gram⁻ coccobacilli	normal oral flora	endocarditis, meningitis, lung	infrequent disease, secondary to trauma
Helicobacter pylori	microaerophile	gram⁻ helical	stomach and duodenum	gastric ulcers	survives in high acidity; carcinogen
Klebsiella granulomatis[7]	aerobic	gram⁻ rod	humans, tropical/subtropical	granuloma inguinale	STD[3] forms intracellular Donovan bodies
Klebsiella pneumoniae	facultative	gram⁻ rod	normal gut flora; water, soil	pneumonia, UTIs[4], septicemia	often nosocomial; model research organism
Lactobacillus acidophilus	facultative anaerobe	gram⁺ rod	normal flora gut, vagina etc.	nonpathogenic	probiotic, control of yeast infections
Lactobacillus spp.	facultative anaerobe	gram⁺ rod	flora of humans & animals	dental caries	used in production of various foods
Lactococcus lactis	facultative	gram⁺ coccus	plants, human gut flora	nonpathogenic	probiotic used in dairy food production
Legionella pneumophila	facultative aerobe	gram⁻ rod	fresh water; in amoeba	Legionnaire's disease (pneumonia)	transmitted by inhalation of aerosols
Leptospirillum ferrooxidans	aerobic	gram⁻ spiral	fresh water, acidic	nonpathogenic	chemolithotrophic; iron-fixing; industrial use
Leptospira spp.	aerobic	gram⁻ spirochete	animals, water	leptospirosis	transmitted by contaminated water or food
Listeria monocytogenes	facultative aerobe	gram⁺ rod	soil, animals	listeriosis; food-poisoning, meningitis	danger to pregnancy; grows in cold temperature
Magnetospirillum magneticum	microaerophilic	gram⁻ rod	fresh water, sediments	nonpathogenic	magnetic organelles have commercial potential
Methylococcus capsulatus	aerobic	gram⁻ coccus	soil, animals	nonpathogenic	methanotroph may help control pollution
Methylomonas spp.	aerobic	gram⁻ rod	soil, water	nonpathogenic	methane primary carbon and energy source
Micrococcus spp.	aerobic	gram⁺ coccus	normal skin flora, soil, water	rarely pathogenic	pathogen in immunocompromised hosts
Moraxella catarrhalis	aerobic	gram⁻ coccus	endogenous to nasopharynx	respiratory tract infections	other *Moraxella* spp. are gram⁻ rods
Morganella morganii	facultative	gram⁻ rod	human gut, animals, water	UTIs[4] and other infections	nosocomial, opportunist
Mycobacterium avium	aerobic	gm⁺ acid fast rod	soil, water, animals, birds	disseminated disease (wasting, etc.)	immunocompromised patients at risk
Mycobacterium bovis	aerobic	gm⁺ acid fast rod	cattle	tuberculosis (infrequent)	source of the BCG vaccination; TB in cattle
Mycobacterium intracellulare	aerobic	gm⁺ acid fast rod	soil, water, animals, birds	lung & disseminated disease	in the *Mycobacterium avium* complex, MAC
Mycobacterium leprae	aerobic	gm⁺ acid fast rod	soil, humans, animals	leprosy (Hansen's disease)	intracellular infection of many tissues
Mycobacterium marinum	aerobic	gm⁺ acid fast rod	fresh and salt water	wound infections	swimming/fish tank granuloma; fish pathogen

(Continued)

Organism	Growth	Staining	Habitat	Pathogenesis (in Humans)	Comments
Mycobacterium smegmatis	aerobic	gm⁺ acid fast rod	water, soil, plants, biofilms	nonpathogenic	fast growth enables research model for genus
Mycobacterium tuberculosis	aerobic	gm⁺ acid fast rod	humans	tuberculosis ("TB")	slow growth; drug-resistance a problem
Mycoplasma capricolum	aerobic	no cell wall, round	goats	nonpathogenic	goat pathogen; needs cholesterol for growth
Mycoplasma pneumoniae	aerobic	no cell wall, rod	humans	atypical pneumonia	small; common cell culture contaminant
Myxococcus xanthus	aerobic	gram⁻ rod[8]	soil, plants, fresh water	nonpathogenic	developmental cycle, fruiting bodies, spores
Neisseria gonorrhoeae	aerobic	gram⁻ coccus	humans	gonorrhea	STD[3]; intracellular; antigenic variation
Neisseria meningitides	aerobic	gram⁻ coccus	humans	meningitis	entry to blood via nasopharyngeal mucosa
Nitrobacter spp.	facultative	gram⁻ rod	soil, fresh water	nonpathogenic	nitrite-oxidizing; wastewater treatment
Nitrosomonas spp.	aerobic	gram⁻ rod	soil, water	nonpathogenic	oxidizes ammonia to nitrite; treat wastewater
Nitrosomonas eutropha	facultative	gram⁻ coccobacilli	soil, water	nonpathogenic	used as skin cleanser in place of soap
Nocardia asteroids	aerobic	gm⁺ acid fast rod	soil, water	lung infection	infection by inhalation, not person-to-person
Pantoea spp.	facultative	gram⁻ rod	plants, water, animal guts	skin, blood, endocarditis	plant pathogen, opportunistic in humans
Pasteurella multocida	facultative anaerobe	gram⁻ coccobacilli	animals	inflammation, possible systemic	zoonotic disease
Peptostreptococcus spp.	anaerobic	gram⁺ coccus	endogenous mouth, gut flora	multiple sites: skin, mouth, blood etc.	also contributes to caries
Photobacterium spp.	facultative	gram⁻ rod	salt water, marine organisms	mostly nonpathogenic	luminescent fish symbiont, some pathogenic
Porphyromonas gingivalis	anaerobic	gram⁻ rod	endogenous mouth flora	periodontal disease, gingivitis	biofilms; black pigment from iron
Prevotella spp.	anaerobic	gram⁻ rod	mouth flora, cattle	periodontal disease, caries	used to treat acidosis in cattle
Propionibacterium acnes	anaerobic	gram⁺ rod	skin, gut	acne and other infections	as the name implies, produces propionic acid
Proteus mirabilis	facultative	gram⁻ rod	soil, water, humans	kidney stones, UTIs[4]	swarming motility, biofilms; urease
Proteus vulgaris	facultative	gram⁻ rod	normal intestinal flora	UTIs[4], wound infections	swarming motility, putrefies meat
Providencia stuartii	facultative	gram⁻ rod	Soil, water, sewage	UTIs[4] from catheters, burn wounds	source for restriction endonuclease PstI
Providencia spp.	facultative	gram⁻ rod	humans, animals, water	UTIs[4] from catheters, burn wounds	opportunist can also cause gastroenteritis
Pseudomonas aeruginosa	facultative[9] aerobe	gram⁻ rod	soil, water, humans, plants	lung and many other sites	opportunist harmful for cystic fibrosis, others
Pseudomonas syringae	facultative aerobe	gram⁻ rod	plants, water	nonpathogenic	crop spoilage; promotes ice nucleation
Rhizobium spp.	facultative aerobe	gram⁻ rod	soil, plant nodules	nonpathogenic	major terrestrial N_2-fixing plant symbiont
Rhodopseudomonas spp.	facultative aerobe	gram⁻ rod	water, soil	nonpathogenic	photosynthetic; clean environmental waste
Rhodospirillum spp.	facultative	gram⁻ spiral	marine environments, soil	non-pathogenic	photosynthetic; N_2-fixing
Rickettsia prowazekii	intracellular	gram⁻ rod	lice, humans	epidemic typhus	Brill-Zinsser disease a relapse long afterward
Rickettsia rickettsii	intracellular	gram⁻ rod	ticks, humans	rocky mountain spotted fever	zoonotic disease (applies to all *Rickettsia*)
Rickettsia typhi	intracellular	gram⁻ rod	fleas, rodents, cats	murine (endemic) typhus	obligate intracellular parasite (all *Rickettsia*)

Organism	Growth	Staining	Habitat	Pathogenesis (in Humans)	Comments
Roseburia spp.	anaerobic	gram⁺ rod	normal intestinal flora	nonpathogenic	breakdown polysaccharides, produce butyrate
Salmonella enterica	facultative	gram⁻ rod	animals, birds, water	mostly gastrointestinal diseases	>2,000 subspecies including 5 serovars below
Salmonella choleraesuis[10]	facultative	gram⁻ rod	swine	Septicemia	mycotic aneurysm a dangerous complication
Salmonella enteritidis	facultative	gram⁻ rod	animals, birds, water	gastroenteritis	ovarian transmission from chicken to eggs
Salmonella paratyphi	facultative	gram⁻ rod	animals, birds, water	paratyphoid enteric fever	fecal-oral transmission (all *Salmonella*)
Salmonella typhi	facultative	gram⁻ rod	animals, birds, water	typhoid fever	asymptomatic carriers in gall bladder
Salmonella typhimurium	facultative	gram⁻ rod	animals, birds, water	gastroenteritis	grows fast (all *Salmonella*); research-model
Sarcina spp.	anaerobic	gram⁺ coccus	soil, water, humans	dental caries, periodontal disease	ferment carbohydrates; one sp. in stomach
Serratia marcescens	facultative	gram⁻ rod	soil, water, humans	UTIs[4], wound infections, other sites	nosocomial; red pigment
Shigella boydii	facultative	gram⁻ rod	humans, primates, water	dysentery, shigellosis	fecal-oral transmission (all *Shigella*)
Shigella dysenteriae	facultative	gram⁻ rod	humans, primates, water	dysentery, shigellosis, HUS[5]	makes Shiga toxin, inhibits protein synthesis
Shigella flexneri	facultative	gram⁻ rod	humans, primates, water	dysentery, shigellosis	Reiter's syndrome sequel to primary disease
Shigella sonnei	facultative	gram⁻ rod	humans, primates, water	dysentery, shigellosis	10 cells enough to cause disease (all *Shigella*)
Sphaerotilus natans	aerobic	gram⁻ rod	flowing and waste water	nonpathogenic	filamentous; sheathed; pipe-clogging
Spirillum volutans	microaerophile	gram⁻ helical	stagnant water	nonpathogenic	use in motility assay due to active flagella
Spiroplasma spp.	intracellular	no cell wall, helical	insects, plants	nonpathogenic[11]	insect and plant pathogen
Spirulina spp.	facultative aerobe	gram⁻ spiral	water, soil, plankton	nonpathogenic	food supplement with reputed health benefits
Staphylococcus aureus	facultative	gram⁺ coccus	endogenous human flora	skin infection, food-poisoning, TSS[12]	+ other diseases; MRSA[13] a major problem
Staphylococcus epidermidis	facultative	gram⁺ coccus	normal skin flora	septicemia, endocarditis	nosocomial from catheters & implants
Staphylococcus saprophyticus	facultative	gram⁺ coccus	animals, human urinary tract	UTIs[4]	primarily affects females
Streptobacillus moniliformis	facultative	gram⁻ rod	rodents	rat-bite fever, Haverhill fever	also transmitted by food; develops L-forms[14]
Streptococcus agalactiae	facultative	gram⁺ coccus	gut & urogenital flora	neonatal meningitis	CAMP[15] hemolysis test for identification
Streptococcus bovis	facultative	gram⁺ coccus	human gut, cows, sheep	endocarditis	bacteremia associated with colon cancer
Streptococcus gordonii	facultative	gram⁺ coccus	human mouth flora	dental plaque; endocarditis	member of the *S. viridans* group of spp.
Streptococcus mutans	facultative	gram⁺ coccus	human mouth flora	dental caries	member of the *S. viridans* group of spp.
Streptococcus pneumoniae	facultative	gram⁺ coccus	human nasopharynx flora	pneumonia, meningitis, others	phage lysin has potential to eradicate species
Streptococcus pyogenes	facultative	gram⁺ coccus	endogenous human flora	strep throat; toxic shock; flesh-eating	rheumatic fever & glomerulonephritis sequelae
Streptococcus thermophilus[16]	facultative	gram⁺ coccus	dairy, cow mammary glands	nonpathogenic	manufacture of yogurt and some cheeses
Streptococcus viridans (spp.)	facultative	gram⁺ coccus	normal human flora	generally nonpathogenic	large group mostly α-hemolytic *Streptococci*
Streptomyces spp.	aerobic	gram⁺ rod	soil, water	nonpathogenic (a few exceptions)	fungi-like filaments, mycelium, spores

(Continued)

Organism	Growth	Staining	Habitat	Pathogenesis (in Humans)	Comments
Streptomyces avermitilis	aerobic	gram+ rod	Soil	nonpathogenic	producer of anti-parasitic agent avermectin
Streptomyces coelicolor	aerobic	gram+ rod	soil, water	nonpathogenic	major producer of antibiotics (all *Streptomyces*)
Streptomyces somaliensis	aerobic	gram+ rod	soil	bacterial mycetoma (Madura foot)	other bacterial & fungal causes of this disease
Streptomyces verticillus	aerobic	gram+ rod	soil	nonpathogenic	produces anticancer drug bleomycin
Thiobacillus denitrificans	facultative anaerobe	gram− rod	water, soil	nonpathogenic	autotrophic denitrification; sulfur oxidation
Thiothrix spp.	aerobic	gram− rod	water	nonpathogenic	filaments; oxidize sulfur; wastewater problem
Treponema denticola	anaerobic	gram− spirochete	endogenous mouth flora	periodontal disease, part of ANUG[6]	forms biofilms with other organisms
Treponema pallidum	microaerophile	gram− spirochete	humans	syphilis	slow growth; obligate intracellular parasite
Ureaplasma spp.	facultative	no cell wall, oval	human urogenital tract	urethritis, neonatal illness, others	hydrolyze urea; opportunist; debilitated hosts
Veillonella spp.	anaerobic	gram− coccus	mouth, gut, vagina flora	oral, bone & other infections	dental plaque biofilm; infrequent pathogen
Vibrio cholera	facultative	gram− rod	salt water	cholera	transmission by seafood & oral-fecal
Vibrio fischeri	facultative	gram− rod	salt water, marine animals	nonpathogenic	bioluminescent; symbiotic with fish
Vibrio harveyi	facultative	gram− rod	salt water	nonpathogenic	bioluminescent; research-model organism
Vibrio parahaemolyticus	facultative	gram− rod	salt water	gastroenteritis	from undercooked seafood, shellfish
Vibrio vulnificus	facultative	gram− rod	salt water	gastroenteritis, wounds, septicemia	contaminated seafood; debilitated patients
Xanthomonas spp.	facultative aerobe	gram− rod	plants	nonpathogenic	plant pathogen; agricultural crop spoilage
Yersinia enterocolitica	facultative anaerobe	gram− coccobacilli	rodents, animals, water, soil	gastroenteritis	poisoning from contaminated animal foods
Yersinia pestis	facultative	gram− coccobacilli	rodents, animals, insects	bubonic/pneumonic plague	transmitted from rodents via fleas
Yersinia pseudotuberculosis	facultative	gram− coccobacilli	rodents, animals	disseminated disease from the gut	fecal-oral transmission; intracellular in host
Zoogloea spp.	aerobic	gram− rod	wastewater	nonpathogenic	form flocs (gels); sewage & water treatment
Zymomonas mobilis	facultative anaerobe	gram− rod	plant and tree saps	nonpathogen; protects against sepsis	industrial use for ethanol production

[1] Some research papers describe *Actinomyces israelii* as a facultative anaerobe

[2] Rods in log phase, cocci in stationary phase

[3] STD, sexually transmitted disease

[4] UTI, urinary-tract infection

[5] HUS, hemolytic uremic syndrome

[6] ANUG, acute necrotizing ulcerative gingivitis

[7] Formerly named *Calymmatobacterium granulomatis*

[8] The rod shape of *Myxococcus xanthus* refers to the vegetative form. The "coccus" in the species name refers to the shape of the myxospore.

[9] Usually described as an obligate aerobe, *Pseudomonas aeruginosa* can also grow anaerobically, but does not have the ability to conduct fermentation.

[10] *S. choleraesuis* is the former species name now replaced with *S. enterica*. However, there also remains a subspecies serovar named *S. choleraesuis*.

[11] Some controversy as there is evidence from some workers that *Spiroplasma* spp. can cause transmissible spongiform encephalopathies

[12] TSS, toxic shock syndrome

[13] MRSA, methicillin-resistant *Staphylococcus aureus*

[14] L-forms are cell-wall deficient derivatives that render the bacteria resistant to penicillin.

[15] CAMP, acronym for Christie, Atkins and Munch-Petersen, who discovered the basis for this test in 1944.

[16] Despite its name, *S. thermophilus* does not appear to be particularly thermophilic, with optimal growth from 35°C to 42°C.

Index

Note: Locators in *italics* represent figures and **bold** indicate tables in the text.

A